A Brief Guide to Get the Most from This Book

1 Read the Book

Feature	Description	Benefit	Page
Section-Opening Scenarios	Every section opens with a scenario presenting a unique application of mathematics in your life outside the classroom.	Realizing that mathematics is everywhere will help motivate your learning.	354
Learning Objectives	Every section begins with a list of objectives that specify what you are supposed to learn. Each objective is restated in the margin where the objective is covered.	The objectives focus your reading by emphasizing what is most important and where to find it.	708
Detailed Worked-Out Examples	Examples are clearly written and provide step-by-step solutions. No steps are omitted, and each step is thoroughly explained to the right of the mathematics.	The blue annotations will help you understand the solutions by providing the reason why every mathematics step is true.	372
Applications Using Real-World Data	Interesting applications from nearly every discipline, supported by up-to-date real-world data, are included in every section.	Ever wondered how you'll use mathematics? This feature will show you how mathematics can solve real problems.	350
Great Question!	Answers to students' questions offer suggestions for problem solving, point out common errors to avoid, and provide informal hints and suggestions.	By seeing common mistakes, you'll be able to avoid them. This feature should help you not to feel anxious or threatened when asking questions in class.	357
Explanatory Voice Balloons	Voice balloons help to demystify mathematics. They translate mathematical language into plain English, clarify problem-solving procedures, and present alternative ways of understanding.	Does math ever look foreign to you? This feature often translates math into everyday English.	346

2 Work the Problems

Feature	Description	Benefit	Page
Check Point Examples	Each example is followed by a similar matched problem, called a Check Point, that offers you the opportunity to work a similar exercise. The answers to the Check Points are provided in the answer section.	You learn best by doing. You'll solidify your understanding of worked examples if you try a similar problem right away to be sure you understand what you've just read.	516
Concept and Vocabulary Checks	These short-answer questions, mainly fill-in-the-blank and true/false items, assess your understanding of the definitions and concepts presented in each section.	It is difficult to learn mathematics without knowing its special language. These exercises test your understanding of the vocabulary and concepts.	855
Extensive and Varied Exercise Sets	An abundant collection of exercises is included in an Exercise Set at the end of each section. Exercises are organized within several categories. Your instructor will usually provide guidance on which exercises to work. The exercises in the first category, Practice Exercises, follow the same order as the section's worked examples.	The parallel order of the Practice Exercises lets you refer to the worked examples and use them as models for solving these problems.	174
Practice Plus Problems	This category of exercises contains more challenging problems that often require you to combine several skills or concepts.	It is important to dig in and develop your problem-solving skills. Practice Plus Exercises provide you with ample opportunity to do so.	138

Review for Quizzes and Tests

Feature	Description	Benefit	Page
Chapter Review Grids	Each chapter contains a review chart that summarizes the definitions and concepts in every section of the chapter. Examples that illustrate these key concepts are also referenced in the chart.	Review this chart and you'll know the most important material in the chapter!	110
Chapter Review Exercises	A comprehensive collection of review exercises for each of the chapter's sections follows the review grid.	Practice makes perfect. These exercises contain the most significant problems for each of the chapter's sections.	575
Chapter Tests	Each chapter contains a practice test with approximately 25 problems that cover the important concepts in the chapter.	You can use the chapter test to determine whether you have mastered the material covered in the chapter.	896
Chapter Test Prep Videos	These videos contain worked-out solutions to every exercise in each chapter test and can be found in MyMathLab and on YouTube.	The videos let you review any exercises you miss on the chapter test.	114

Thinking Mathematically

Seventh Edition

Robert Blitzer
Miami Dade College

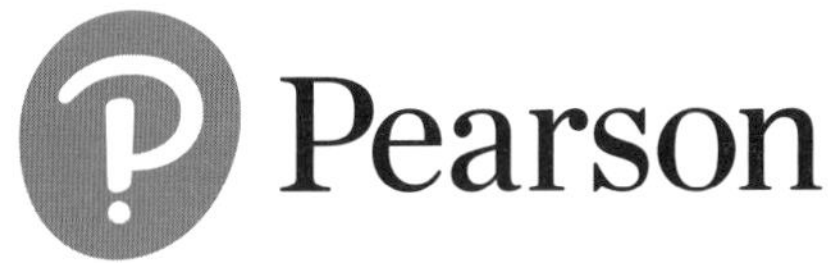

Director, Portfolio Management	Anne Kelly
Courseware Portfolio Managers	Marnie Greenhut and Dawn Murrin
Courseware Portfolio Management Assistant	Stacey Miller
Content Producer	Kathleen A. Manley
Managing Producer	Karen Wernholm
Producer	Nick Sweeny
Manager, Courseware QA	Mary Durnwald
Product Marketing Manager	Kyle DiGiannantonio
Field Marketing Manager	Andrew Noble
Marketing Assistant	Brooke Imbornone
Senior Author Support/Technology Specialist	Joe Vetere
Manager, Rights and Permissions	Gina Cheselka
Manufacturing Buyer	Carol Melville, LSC Communications
Text and Cover Design	Studio Montage
Production Coordination and Composition	codeMantra
Illustrations	Scientific Illustrations
Cover Images	Catherine Ledner/Iconica/Getty Images (cow) and Hunter Bliss/Shutterstock (frame)

Library of Congress Cataloging-in-Publication Data
Names: Blitzer, Robert, author.
Title: Thinking mathematically / Robert F. Blitzer.
Description: Seventh edition. | Boston : Pearson, [2019]
Identifiers: LCCN 2017046337 | ISBN 9780134683713 (alk. paper) | ISBN 0134683714 (alk. paper)
Subjects: LCSH: Mathematics–Textbooks.
Classification: LCC QA39.3 .B59 2019 | DDC 510–dc23
LC record available at https://lccn.loc.gov/2017046337

2 18

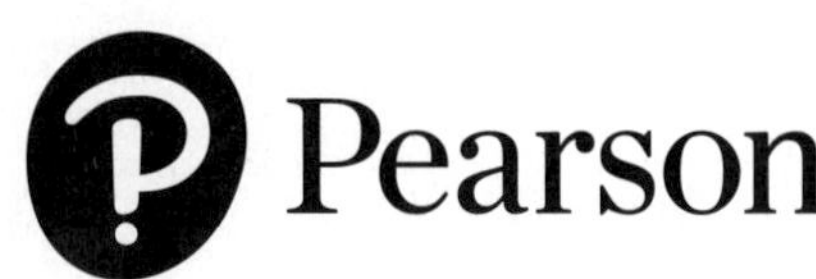

ISBN-13: 978-0-13-468371-3
ISBN-10: 0-13-468371-4

Contents

About the Author

Bob Blitzer is a native of Manhattan and received a Bachelor of Arts degree with dual majors in mathematics and psychology (minor: English literature) from the City College of New York. His unusual combination of academic interests led him toward a Master of Arts in mathematics from the University of Miami and a doctorate in behavioral sciences from Nova University. Bob's love for teaching mathematics was nourished for nearly 30 years at Miami Dade College, where he received numerous teaching awards, including Innovator of the Year from the League for Innovations in the Community College and an endowed chair based on excellence in the classroom. In addition to *Thinking Mathematically*, Bob has written textbooks covering introductory algebra, intermediate algebra, college algebra, algebra and trigonometry, precalculus, trigonometry, and liberal arts mathematics for high school students, all published by Pearson. When not secluded in his Northern California writer's cabin, Bob can be found hiking the beaches and trails of Point Reyes National Seashore, and tending to the chores required by his beloved entourage of horses, chickens, and irritable roosters.

Preface

Thinking Mathematically, Seventh Edition provides a general survey of mathematical topics that are useful in our contemporary world. My primary purpose in writing the book was to show students how mathematics can be applied to their lives in interesting, enjoyable, and meaningful ways. The book's variety of topics and flexibility of sequence make it appropriate for a one- or two-term course in liberal arts mathematics, quantitative reasoning, finite mathematics, as well as for courses specifically designed to meet state-mandated requirements in mathematics.

I wrote the book to help diverse students, with different backgrounds and career plans, to succeed. **Thinking Mathematically, Seventh Edition,** has four major goals:

1. To help students acquire knowledge of fundamental mathematics.
2. To show students how mathematics can solve authentic problems that apply to their lives.
3. To enable students to understand and reason with quantitative issues and mathematical ideas they are likely to encounter in college, career, and life.
4. To enable students to develop problem-solving skills, while fostering critical thinking, within an interesting setting.

One major obstacle in the way of achieving these goals is the fact that very few students actually read their textbook. This has been a regular source of frustration for me and my colleagues in the classroom. Anecdotal evidence gathered over years highlights two basic reasons why students do not take advantage of their textbook:

"I'll never use this information."
"I can't follow the explanations."

I've written every page of the Seventh Edition with the intent of eliminating these two objections. The ideas and tools I've used to do so are described for the student in "A Brief Guide to Getting the Most from This Book," which appears inside the front cover.

What's New in the Seventh Edition?

- **New and Updated Applications and Real-World Data.** I'm on a constant search for real-world data that can be used to illustrate unique mathematical applications. I researched hundreds of books, magazines, newspapers, almanacs, and online sites to prepare the Seventh Edition. This edition contains 110 worked-out examples and exercises based on new data sets and 104 examples and exercises based on updated data. New applications include student-loan debt (Exercise Set 1.2), movie rental options (Exercise Set 1.3), impediments to academic performance (Section 2.1), measuring racial prejudice, by age (Exercise Set 2.1), generational support for legalized adult marijuana use (Exercise Set 2.3), different cultural values among nations (Exercise Set 2.5), episodes from the television series *The Twilight Zone* (Section 3.6) and the film *Midnight Express* (Exercise Set 3.7), excuses by college students for not meeting assignment deadlines (Exercise Set 5.3), fraction of jobs requiring various levels of education by 2020 (Exercise Set 5.3), average earnings by college major (Exercise Set 6.5), the pay gap (Exercise Set 7.2), inmates in federal prisons for drug offenses and all other crimes (Exercise Set 7.3), time breakdown for an average 90-minute NFL broadcast (Section 11.6), Scrabble tiles (Exercise Set 11.5), and are inventors born or made? (Section 12.2).
- **New Blitzer Bonuses.** The Seventh Edition contains a variety of new but optional enrichment essays. There are more new Blitzer Bonuses in this edition than in any previous revision of *Thinking Mathematically*. These include "Surprising Friends with Induction" (Section 1.1), "Predicting Your Own Life Expectancy" (Section 1.2), "Is College Worthwhile?" (Section 1.2), "Yogi-isms" (Section 3.4), "Quantum Computers" (Section 4.3), "Slope and Applauding Together" (Section 7.2), "A Brief History of U.S. Income Tax" (Section 8.2) "Three Decades of Mortgages" (Section 8.7), "Up to Our Ears in Debt" (Section 8.8), "The Best Financial Advice for College Graduates" (Section 8.8), "Three Weird Units of Measure" (Section 9.1), "Screen Math" (Section 10.2), "Senate Voting Power" (Section 13.3), "Hamilton Mania" (Section 13.3), "Dirty Presidential Elections" (Section 13.3), "Campaign Posters as Art" (Section 13.4), and "The 2016 Presidential Election" (Section 13.4).
- **New Graphing Calculator Screens.** All screens have been updated using the TI-84 Plus C.
- **Updated Tax Tables.** Section 8.2 (Income Tax) contains the most current federal marginal tax tables and FICA tax rates available for the Seventh Edition.
- **New MyLab™ Math.** In addition to the new functionalities within an updated MyLab Math, the new items specific to *Thinking Mathematically, Seventh Edition* MyLab Math include
  - All new objective-level videos with assessment
  - Interactive concept videos with assessment
  - Animations with assessment
  - StatCrunch integration.

What Familiar Features Have Been Retained in the Seventh Edition?

- **Chapter-Opening and Section-Opening Scenarios.** Every chapter and every section open with a scenario presenting a unique application of mathematics in students' lives outside the classroom. These scenarios are revisited in the course of the chapter or section in an example, discussion, or exercise. The often humorous tone of these openers is intended to help fearful and reluctant students overcome their negative perceptions about math. A feature called "Here's Where You'll Find These Applications" is included with each chapter opener.
- **Section Objectives (What Am I Supposed to Learn?).** Learning objectives are clearly stated at the beginning of each section. These objectives help students recognize and focus on the section's most important ideas. The objectives are restated in the margin at their point of use.
- **Detailed Worked-Out Examples.** Each example is titled, making the purpose of the example clear. Examples are clearly written and provide students with detailed step-by-step solutions. No steps are omitted and each step is thoroughly explained to the right of the mathematics.
- **Explanatory Voice Balloons.** Voice balloons are used in a variety of ways to demystify mathematics. They translate mathematical language into everyday English, help clarify problem-solving procedures, present alternative ways of understanding concepts, and connect problem solving to concepts students have already learned.
- **Check Point Examples.** Each example is followed by a similar matched problem, called a Check Point, offering students the opportunity to test for conceptual understanding by working a similar exercise. The answers to the Check Points are provided in the answer section in the back of the book. Worked-out video solutions for many Check Points are in the MyLab Math course.
- **Great Question!** This feature presents study tips in the context of students' questions. Answers to the questions offer suggestions for problem solving, point out common errors to avoid, and provide informal hints and suggestions. As a secondary benefit, this feature should help students not to feel anxious or threatened when asking questions in class.
- **Brief Reviews.** The book's Brief Review boxes summarize mathematical skills that students should have learned previously, but which many students still need to review. This feature appears whenever a particular skill is first needed and eliminates the need to reteach that skill.
- **Concept and Vocabulary Checks.** The Seventh Edition contains 653 short-answer exercises, mainly fill-in-the blank and true/false items, that assess students' understanding of the definitions and concepts presented in each section. The Concept and Vocabulary Checks appear as separate features preceding the Exercise Sets. These are assignable in the MyLab Math course.
- **Extensive and Varied Exercise Sets.** An abundant collection of exercises is included in an Exercise Set at the end of each section. Exercises are organized within seven category types: Practice Exercises, Practice Plus Exercises, Application Exercises, Explaining the Concepts, Critical Thinking Exercises, Technology Exercises, and Group Exercises.
- **Practice Plus Problems.** This category of exercises contains practice problems that often require students to combine several skills or concepts, providing instructors the option of creating assignments that take Practice Exercises to a more challenging level.
- **Chapter Summaries.** Each chapter contains a review chart that summarizes the definitions and concepts in every section of the chapter. Examples that illustrate these key concepts are also referenced in the chart.
- **End-of-Chapter Materials.** A comprehensive collection of review exercises for each of the chapter's sections follows the Summary. This is followed by a Chapter Test that enables students to test their understanding of the material covered in the chapter. Worked-out video solutions are available for every Chapter Test Prep problem in the MyLab Math course or on YouTube.
- **Learning Guide.** This study aid is organized by objective and provides support for note-taking, practice, and video review. The Learning Guide is available as PDFs in MyLab Math. It can also be packaged with the textbook and MyLab Math access code.

I hope that my love for learning, as well as my respect for the diversity of students I have taught and learned from over the years, is apparent throughout this new edition. By connecting mathematics to the whole spectrum of learning, it is my intent to show students that their world is profoundly mathematical, and indeed, π is in the sky.

Robert Blitzer

Resources for Success

Pearson MyLab

MyLab^{TM} Math Online Course for *Thinking Mathematically, Seventh Edition* by Robert Blitzer (access code required)

MyLab Math is available to accompany Pearson's market leading text offerings. To give students a consistent tone, voice, and teaching method each text's flavor and approach are tightly integrated throughout the accompanying MyLab Math course, making learning the material as seamless as possible.

NEW! Video Program

All new objective-level videos provide a new level of coverage throughout the text. Videos at the objective level allow students to get support just where they need it. Instructors can assign these as media assignments or use the provided assessment questions for each video.

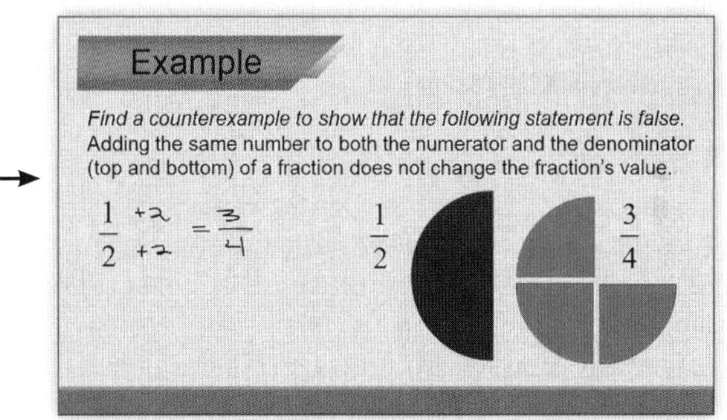

NEW! Interactive Concept Videos

New Interactive Concept Videos are also available in MyLab Math. After a brief explanation, the video pauses to ask students to try a problem on their own. Incorrect answers are followed by further explanation, taking into consideration what may have led to the student selecting that particular wrong answer. Incorrect answer 'A' goes down one path while incorrect answer 'B' provides a different explanation based on why the student may have selected that option.

NEW! Animations

New animations let students interact with the math in a visual, tangible way. These animations allow students to explore and manipulate the mathematical concepts, leading to more durable understanding. Corresponding exercises in MyLab Math make these truly assignable.

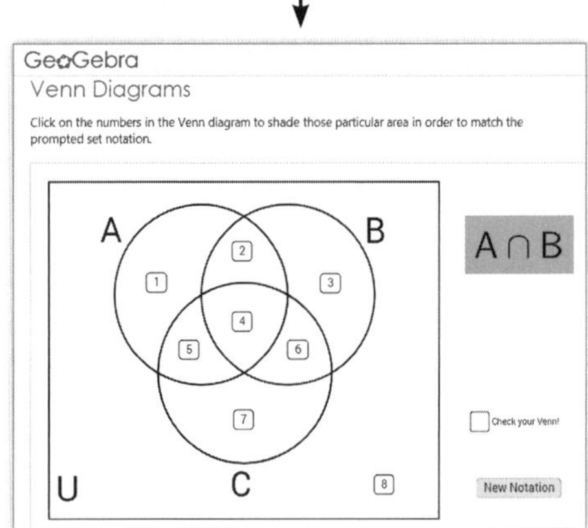

StatCrunch

Newly integrated StatCrunch allows students to harness technology to perform complex analyses on data.

pearson.com/mylab/math

Resources for Success

Instructor Resources

Annotated Instructor's Edition (AIE)

ISBN-10: 0-13-468454-0
ISBN-13: 978-0-13-468454-3

The AIE includes answers to all exercises presented in the book, most on the page with the exercise and the remainder in the back of the book.

The following resources can be downloaded from MyLab Math or the Instructor's Resource Center on **www.pearsonhighered.com.**

MyLab Math with Integrated Review

Provides a full suite of supporting resources for the collegiate course content plus additional assignments and study aids for students who will benefit from remediation. Assignments for the integrated review content are preassigned in MyLab™ Math, making it easier than ever to create your course.

Instructor's Solutions Manual

This manual contains detailed, worked-out solutions to all the exercises in the text.

PowerPoint Lecture Presentation

These editable slides present key concepts and definitions from the text. Instructors can add art from the text located in the Image Resource Library in MyLab Math or slides that they create on their own. PointPoint slides are fully accessible.

Image Resource Library

This resource in MyLab Math contains all art from the text, for instructors to use in their own presentations and handouts.

Instructor's Testing Manual

The Testing Manual includes two alternative tests per chapter. These items may be used as actual tests or as references for creating actual tests.

TestGen

TestGen® (www.pearsoned.com/testgen) enables instructors to build, edit, print, and administer tests using a computerized bank of questions developed to cover all the objectives of the text. TestGen is algorithmically based, allowing instructors to create multiple but equivalent versions of the same question or test with the click of a button. Instructors can also modify test bank questions or add new questions. The software are available for download from Pearson's Instructor Resource Center.

Student Resources

Learning Guide with Integrated Review Worksheets
ISBN 10: 0-13-470508-4
ISBN 13: 978-0-13470508-8
Bonnie Rosenblatt, Reading Area Community College

This workbook is organized by objective and provides support for note-taking, practice, and video review and includes the Integrated Review worksheets from the Integrated Review version of the MyLab Math course. The Learning Guide is also available as PDFs in MyLab Math. It can also be packaged with the textbook and MyLab Math access code.

Student's Solutions Manual

ISBN 10: 0-13-468650-0
ISBN 13: 978-0-13-468650-9
Daniel Miller, Niagara County Community College

This manual provides detailed, worked-out solutions to odd-numbered exercises, as well as solutions to all Check Points, Concept and Vocabulary Checks, Chapter Reviews, and Chapter Tests.

pearson.com/mylab/math

To the Student

The bar graph shows some of the qualities that students say make a great teacher. It was my goal to incorporate each of these qualities throughout the pages of this book to help you gain control over the part of your life that involves numbers and mathematical ideas.

Explains Things Clearly

I understand that your primary purpose in reading *Thinking Mathematically* is to acquire a solid understanding of the required topics in your liberal arts math course. In order to achieve this goal, I've carefully explained each topic. Important definitions and procedures are set off in boxes, and worked-out examples that present solutions in a step-by-step manner appear in every section. Each example is followed by a similar matched problem, called a Check Point, for you to try so that you can actively participate in the learning process as you read the book. (Answers to all Check Points appear in the back of the book and video solutions are in MyLab Math.)

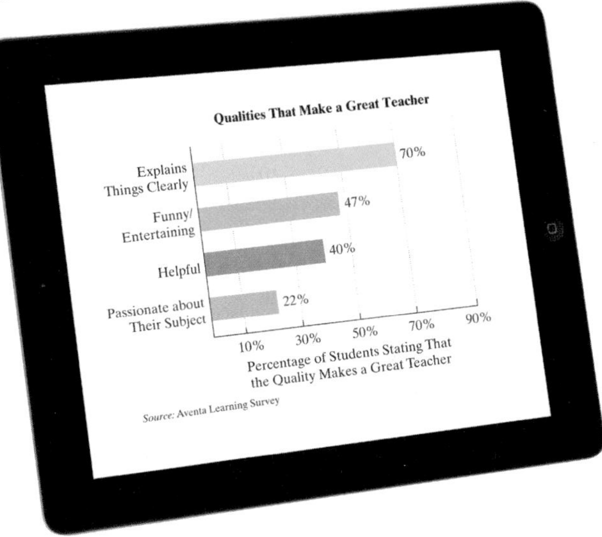

Funny & Entertaining

Who says that a math textbook can't be entertaining? From our engaging cover to the photos in the chapter and section openers, prepare to expect the unexpected. I hope some of the book's enrichment essays, called Blitzer Bonuses, will put a smile on your face from time to time.

Helpful

I designed the book's features to help you acquire knowledge of fundamental mathematics, as well as to show you how math can solve authentic problems that apply to your life. These helpful features include

- **Explanatory Voice Balloons:** Voice balloons are used in a variety of ways to make math less intimidating. They translate mathematical language into everyday English, help clarify problem-solving procedures, present alternative ways of understanding concepts, and connect new concepts to concepts you have already learned.
- **Great Question!:** The book's Great Question! boxes are based on questions students ask in class. The answers to these questions give suggestions for problem solving, point out common errors to avoid, and provide informal hints and suggestions.
- **Chapter Summaries:** Each chapter contains a review chart that summarizes the definitions and concepts in every section of the chapter. Examples from the chapter that illustrate these key concepts are also referenced in the chart. Review these summaries and you'll know the most important material in the chapter!

Passionate about the Subject

I passionately believe that no other discipline comes close to math in offering a more extensive set of tools for application and development of your mind. I wrote the book in Point Reyes National Seashore, 40 miles north of San Francisco. The park consists of 75,000 acres with miles of pristine surf-washed beaches, forested ridges, and bays bordered by white cliffs. It was my hope to convey the beauty and excitement of mathematics using nature's unspoiled beauty as a source of inspiration and creativity. Enjoy the pages that follow as you empower yourself with the mathematics needed to succeed in college, your career, and in your life.

Regards,

Bob

Robert Blitzer

Acknowledgments

An enormous benefit of authoring a successful textbook is the broad-based feedback I receive from students, dedicated users, and reviewers. Every change to this edition is the result of their thoughtful comments and suggestions. I would like to express my appreciation to all the reviewers, whose collective insights form the backbone of this revision. In particular, I would like to thank the following people for reviewing *Thinking Mathematically* for this Seventh Edition.

Deana Alexander, *Indiana University—Purdue University*
Nina Bohrod, *Anoka-Ramsey Community College*
Kim Caldwell, *Volunteer State Community College*
Kevin Charlwood, *Washburn University*
Elizabeth T. Dameron, *Tallahassee Community College*
Darlene O. Diaz, *Santiago Canyon College*
Cornell Grant, *Georgia Piedmont Technical College*
Theresa Jones, *Texas State University*
Elizabeth Kiedaisch, *College of DuPage*
Lauren Kieschnick, *Mineral Area College*
Alina Klein, *University of Dubuque*
Susan Knights, *College of Western Idaho*
Isabelle Kumar, *Miami Dade College*
Dennine LaRue, *Farmont State University*
David Miller, *William Paterson University*
Carla A. Monticelli, *Camden County College*
Tonny Sangutei, *North Carolina Central University*
Cindy Vanderlaan, *Indiana Purdue University—Fort Wayne*
Alexandra Verkhovtseva, *Anoka-Ramsey Community College*

Each reviewer from every edition has contributed to the success of this book and I would like to also continue to offer my thanks to them.

David Allen, *Iona College*; Carl P. Anthony, *Holy Family University*; Laurel Berry, *Bryant and Stratton College*; Kris Bowers, *Florida State University*; Gerard Buskes, *University of Mississippi*; Fred Butler, *West Virginia University*; Jimmy Chang, *St. Petersburg College*; Jerry Chen, *Suffolk County Community College*; Ivette Chuca, *El Paso Community College*; David Cochener, *Austin Peay State University*; Stephanie Costa, *Rhode Island College*; Tristen Denley, *University of Mississippi*; Suzanne Feldberg, *Nassau Community College*; Margaret Finster, *Erie Community College*; Maryanne Frabotta, *Community Campus of Beaver County*; Lyn Geisler III, *Randolph-Macon College*; Patricia G. Granfield, *George Mason University*; Dale Grussing, *Miami Dade College*; Cindy Gubitose, *Southern Connecticut State University*; Virginia Harder, *College at Oneonta*; Joseph Lloyd Harris, *Gulf Coast Community College*; Julia Hassett, *Oakton Community College*; Sonja Hensler, *St. Petersburg College*; James Henson, *Edinboro University of Pennsylvania*; Larry Hoehn, *Austin Peay State University*; Diane R. Hollister, *Reading Area Community College*; Kalynda Holton, *Tallahassee Community College*; Alec Ingraham, *New Hampshire College*; Linda Kuroski, *Erie Community College—City Campus*; Jamie Langille, *University of Nevada, Las Vegas*; Veronique Lanuqueitte, *St. Petersburg College*; Julia Ledet, *Louisiana State University*; Mitzi Logan, *Pitt Community College*; Dmitri Logvnenko, *Phoenix College*; Linda Lohman, *Jefferson Community College*; Richard J. Marchand, *Slippery Rock University*; Mike Marcozzi, *University of Nevada, Las Vegas*; Diana Martelly, *Miami Dade College*; Jim Matovina, *Community College of Southern Nevada*; Erik Matsuoka, *Leeward Community College*; Marcel Maupin, *Oklahoma State University*; Carrie McCammon, *Ivy Tech Community College*; Diana McCammon, *Delgado Community College*; Mcx McKinley, *Florida Keys Community College*; Taranna Amani Miller, *Indian River State College*; Paul Mosbos, *State University of New York—Cortland*; Tammy Muhs, *University of Central Florida*; Cornelius Nelan, *Quinnipiac University*; Lawrence S. Orilia, *Nassau Community College*; Richard F. Patterson, *University of North Florida*; Frank Pecchioni, *Jefferson Community College*; Stan Perrine, *Charleston Southern University*; Anthony Pettofrezzo, *University of Central Florida*; Val Pinciu, *Southern Connecticut State University*; Evelyn Pupplo-Cody, *Marshall University*; Virginia S. Powell, *University of Louisiana at Monroe*; Kim Query, *Lindenwood College*; Anne Quinn, *Edinboro University of Pennsylvania*; Bill Quinn, *Frederick Community College*; Sharonda Ragland, *ECPI College of Technology*; Shawn Robinson, *Valencia Community College*; Gary Russell, *Brevard Community College*; Mary Lee Seitz, *Erie Community College*; Laurie A. Stahl, *State University of New York—Fredonia*; Abolhassan Taghavy, *Richard J. Daley College & Chicago State University*; Diane Tandy, *New Hampshire Technical Institute*; Ann Thrower, *Kilgore College*; Mike Tomme, *Community College of Southern Nevada*; Sherry Tornwall, *University of Florida*; Linda Tully, *University of Pittsburgh at Johnstown*; Christopher Scott Vaughen, *Miami Dade College*; Bill Vaughters, *Valencia Community College*; Karen Villareal, *University of New Orleans*; Don Warren, *Edison Community College*; Shirley Wilson, *North Central College*; James Wooland, *Florida State University*; Clifton E. Webb, *Virginia Union University*; Cindy Zarske, *Fullerton College*; Marilyn Zopp, *McHenry County College*

Additional acknowledgments are extended to Brad Davis, for preparing the answer section and annotated answers and serving as accuracy checker; Bonnie Rosenblatt for writing the Learning Guide;

Dan Miller and Kelly Barber, for preparing the solutions manuals; the codeMantra formatting team for the book's brilliant paging; Brian Morris and Kevin Morris at Scientific Illustrators, for superbly illustrating the book; and Francesca Monaco, project manager, and Kathleen Manley, production editor, whose collective talents kept every aspect of this complex project moving through its many stages.

I would like to thank my editors at Pearson, Dawn Murrin and Marnie Greenhut, and editorial assistant, Stacey Miller, who guided and coordinated the book from manuscript through production. Finally, thanks to marketing manager Kyle DiGiannantonio and marketing assistant Brooke Imbornone for your innovative marketing efforts, and to the entire Pearson sales force, for your confidence and enthusiasm about the book.

Robert Blitzer

Index of Applications

W

Y

Z

Problem Solving and Critical Thinking

1

HOW WOULD YOUR LIFESTYLE CHANGE IF A GALLON OF GAS COST $9.15? OR IF THE PRICE OF A STAPLE SUCH AS MILK WAS $15? THAT'S HOW much those products would cost if their prices had increased at the same rate college tuition has increased since 1980.

TUITION AND FEES AT FOUR-YEAR COLLEGES

	School Year Ending 2000	School Year Ending 2016
Public	$3349	$9410
Private	$15,518	$33,480

Source: The College Board

If these trends continue, what can we expect in the 2020s and beyond? We can answer this question by using estimation techniques that allow us to represent the data mathematically. With such representations, called *mathematical models*, we can gain insights and predict what might occur in the future on a variety of issues, ranging from college costs to global warming.

Here's where you'll find these applications:

Mathematical models involving college costs are developed in Example 8 and Check Point 8 of Section 1.2. In Exercises 51 and 52 in Exercise Set 1.2, you will approach our climate crisis mathematically by developing models for data related to global warming.

1.1

Inductive and Deductive Reasoning

WHAT AM I SUPPOSED TO LEARN?

After studying this section, you should be able to:

1 Understand and use inductive reasoning.

2 Understand and use deductive reasoning.

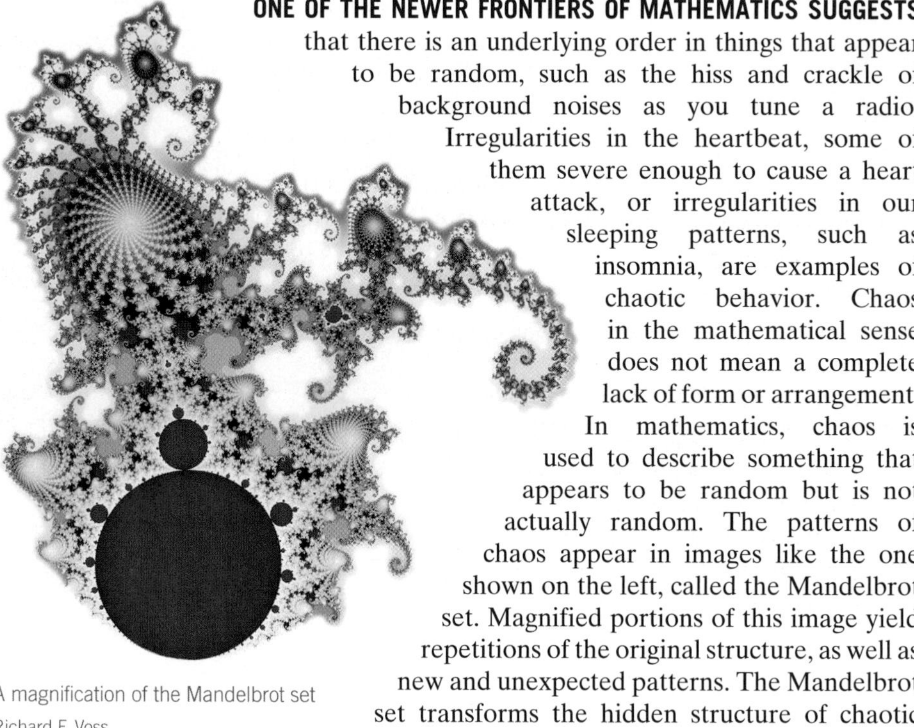

A magnification of the Mandelbrot set

Richard F. Voss

ONE OF THE NEWER FRONTIERS OF MATHEMATICS SUGGESTS that there is an underlying order in things that appear to be random, such as the hiss and crackle of background noises as you tune a radio. Irregularities in the heartbeat, some of them severe enough to cause a heart attack, or irregularities in our sleeping patterns, such as insomnia, are examples of chaotic behavior. Chaos in the mathematical sense does not mean a complete lack of form or arrangement. In mathematics, chaos is used to describe something that appears to be random but is not actually random. The patterns of chaos appear in images like the one shown on the left, called the Mandelbrot set. Magnified portions of this image yield repetitions of the original structure, as well as new and unexpected patterns. The Mandelbrot set transforms the hidden structure of chaotic events into a source of wonder and inspiration.

Many people associate mathematics with tedious computation, meaningless algebraic procedures, and intimidating sets of equations. The truth is that mathematics is the most powerful means we have of exploring our world and describing how it works. The word *mathematics* comes from the Greek word *mathematikos*, which means "inclined to learn." To be mathematical literally means to be inquisitive, open-minded, and interested in a lifetime of pursuing knowledge!

Mathematics and Your Life

A major goal of this book is to show you how mathematics can be applied to your life in interesting, enjoyable, and meaningful ways. The ability to think mathematically and reason with quantitative issues will help you so that you can:

- order and arrange your world by using sets to sort and classify information (Chapter 2, Set Theory);
- use logic to evaluate the arguments of others and become a more effective advocate for your own beliefs (Chapter 3, Logic);
- understand the relationship between cutting-edge technology and ancient systems of number representation (Chapter 4, Number Representation and Calculation);
- put the numbers you encounter in the news, from contemplating the national debt to grasping just how colossal \$1 trillion actually is, into perspective (Chapter 5, Number Theory and the Real Number System);
- use mathematical models to gain insights into a variety of issues, including the positive benefits that humor and laughter can have on your life (Chapter 6, Algebra: Equations and Inequalities);
- use basic ideas about savings, loans, and investments to achieve your financial goals (Chapter 8, Personal Finance);
- use geometry to study the shape of your world, enhancing your appreciation of nature's patterns and beauty (Chapter 10, Geometry);
- develop an understanding of the fundamentals of statistics and how these numbers are used to make decisions (Chapter 12, Statistics);

- understand the mathematical paradoxes of voting in a democracy, increasing your ability to function as a more fully aware citizen (Chapter 13, Voting and Apportionment);
- use graph theory to examine how mathematics is used to solve problems in the business world (Chapter 14, Graph Theory).

Mathematics and Your Career

"It is better to take what may seem to be too much math rather than too little. Career plans change, and one of the biggest roadblocks in undertaking new educational or training goals is poor preparation in mathematics. Furthermore, not only do people qualify for more jobs with more math, they are also better able to perform their jobs."
—Occupational Outlook Quarterly

Generally speaking, the income of an occupation is related to the amount of education required. This, in turn, is usually related to the skill level required in language and mathematics. With our increasing reliance on technology, the more mathematics you know, the more career choices you will have.

Mathematics and Your World

Mathematics is a science that helps us recognize, classify, and explore the hidden patterns of our universe. Focusing on areas as different as planetary motion, animal markings, shapes of viruses, aerodynamics of figure skaters, and the very origin of the universe, mathematics is the most powerful tool available for revealing the underlying structure of our world. Within the last 40 years, mathematicians have even found order in chaotic events such as the uncontrolled storm of noise in the nerve cells of the brain during an epileptic seizure.

Inductive Reasoning

1 *Understand and use inductive reasoning.*

Mathematics involves the study of patterns. In everyday life, we frequently rely on patterns and routines to draw conclusions. Here is an example:

> The last six times I went to the beach, the traffic was light on Wednesdays and heavy on Sundays. My conclusion is that weekdays have lighter traffic than weekends.

This type of reasoning process is referred to as *inductive reasoning*, or *induction*.

INDUCTIVE REASONING

Inductive reasoning is the process of arriving at a general conclusion based on observations of specific examples.

Although inductive reasoning is a powerful method of drawing conclusions, we can never be absolutely certain that these conclusions are true. For this reason, the conclusions are called **conjectures**, **hypotheses**, or educated guesses. A strong inductive argument does not guarantee the truth of the conclusion, but rather provides strong support for the conclusion. If there is just one case for which the conjecture does not hold, then the conjecture is false. Such a case is called a **counterexample**.

EXAMPLE 1 *Finding a Counterexample*

The ten symbols that we use to write numbers, namely 0, 1, 2, 3, 4, 5, 6, 7, 8, and 9, are called **digits**. In each example shown below, the sum of two two-digit numbers is a three-digit number.

$$\begin{array}{r} 47 \\ +73 \\ \hline 120 \end{array} \qquad \begin{array}{r} 56 \\ +46 \\ \hline 102 \end{array}$$

Two-digit numbers

Three-digit sums

Is the sum of two two-digit numbers always a three-digit number? Find a counterexample to show that the statement

> The sum of two two-digit numbers is a three-digit number

is false.

SOLUTION

There are many counterexamples, but we need to find only one. Here is an example that makes the statement false:

$$\begin{array}{r} 56 \\ +43 \\ \hline 99 \end{array}$$

This is a two-digit sum, not a three-digit sum.

This example is a counterexample that shows the statement

The sum of two two-digit numbers is a three-digit number

is false.

GREAT QUESTION!

Why is it so important to work each of the book's Check Points?

You learn best by doing. Do not simply look at the worked examples and conclude that you know how to solve them. To be sure you understand the worked examples, try each Check Point. Check your answer in the answer section before continuing your reading. Expect to read this book with pencil and paper handy to work the Check Points.

CHECK POINT 1 Find a counterexample to show that the statement

The product of two two-digit numbers is a three-digit number

is false.

Here are two examples of inductive reasoning:

- **Strong Inductive Argument** In a random sample of 380,000 freshmen at 722 four-year colleges, 25% said they frequently came to class without completing readings or assignments (*Source*: National Survey of Student Engagement). We can conclude that there is a 95% probability that between 24.84% and 25.15% of all college freshmen frequently come to class unprepared.

In Chapter 12, you will learn how observations from a randomly selected group, one in which each member of the population has an equal chance of being selected, can provide probabilities of what is true about an entire population.

- **Weak Inductive Argument** Neither my dad nor my boyfriend has ever cried in front of me. Therefore, men have difficulty expressing their feelings.

When generalizing from observations about your own circumstances and experiences, avoid jumping to hasty conclusions based on a few observations. Psychologists theorize that we do this—that is, place everyone in a neat category—to feel more secure about ourselves and our relationships to others.

Inductive reasoning is extremely important to mathematicians. Discovery in mathematics often begins with an examination of individual cases to reveal patterns about numbers.

EXAMPLE 2 *Using Inductive Reasoning*

Identify a pattern in each list of numbers. Then use this pattern to find the next number.

a. 3, 12, 21, 30, 39, _____ **b.** 3, 12, 48, 192, 768, _____

c. 3, 4, 6, 9, 13, 18, _____ **d.** 3, 6, 18, 36, 108, 216, _____

SOLUTION

a. Because 3, 12, 21, 30, 39, _____ is increasing relatively slowly, let's use addition as the basis for our individual observations.

3, 12, 21, 30, 39, _____

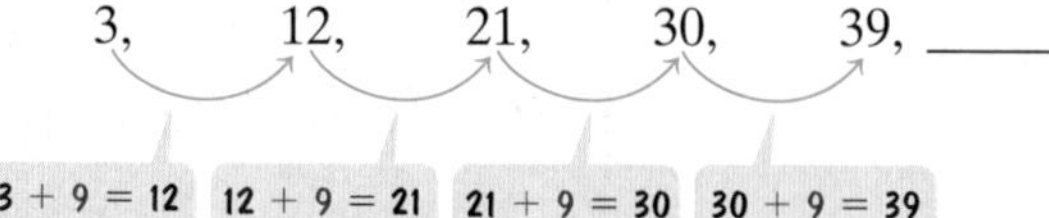

"For thousands of years, people have loved numbers and found patterns and structures among them. The allure of numbers is not limited to or driven by a desire to change the world in a practical way. When we observe how numbers are connected to one another, we are seeing the inner workings of a fundamental concept."
—Edward B. Burger and Michael Starbird, *Coincidences, Chaos, and All That Math Jazz*, W. W. Norton and Company, 2005

Generalizing from these observations, we conclude that each number after the first is obtained by adding 9 to the previous number. Using this pattern, the next number is 39 + 9, or 48.

b. Because 3, 12, 48, 192, 768, _____ is increasing relatively rapidly, let's use multiplication as the basis for our individual observations.

3, 12, 48, 192, 768, _____

$3 \times 4 = 12$ $\quad 12 \times 4 = 48$ $\quad 48 \times 4 = 192$ $\quad 192 \times 4 = 768$

Generalizing from these observations, we conclude that each number after the first is obtained by multiplying the previous number by 4. Using this pattern, the next number is 768×4, or 3072.

c. Because 3, 4, 6, 9, 13, 18, _____ is increasing relatively slowly, let's use addition as the basis for our individual observations.

3, 4, 6, 9, 13, 18, _____

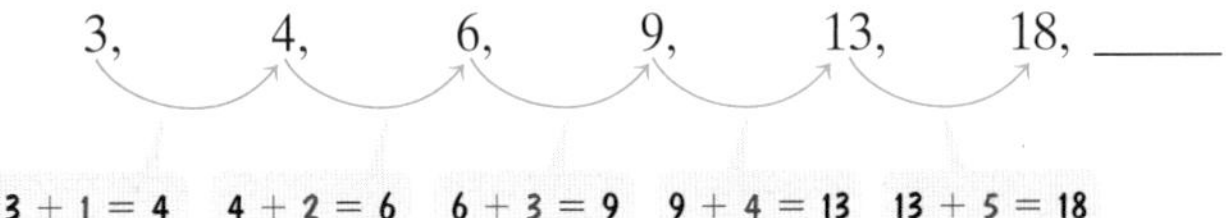

Generalizing from these observations, we conclude that each number after the first is obtained by adding a counting number to the previous number. The additions begin with 1 and continue through each successive counting number. Using this pattern, the next number is 18 + 6, or 24.

d. Because 3, 6, 18, 36, 108, 216, _____ is increasing relatively rapidly, let's use multiplication as the basis for our individual observations.

3, 6, 18, 36, 108, 216, _____

$3 \times 2 = 6$ $\quad 6 \times 3 = 18$ $\quad 18 \times 2 = 36$ $\quad 36 \times 3 = 108$ $\quad 108 \times 2 = 216$

Generalizing from these observations, we conclude that each number after the first is obtained by multiplying the previous number by 2 or by 3. The multiplications begin with 2 and then alternate, multiplying by 2, then 3, then 2, then 3, and so on. Using this pattern, the next number is 216×3, or 648.

CHECK POINT 2 Identify a pattern in each list of numbers. Then use this pattern to find the next number.

a. 3, 9, 15, 21, 27, _____

b. 2, 10, 50, 250, _____

c. 3, 6, 18, 72, 144, 432, 1728, _____

d. 1, 9, 17, 3, 11, 19, 5, 13, 21, _____

In our next example, the patterns are a bit more complex than the additions and multiplications we encountered in Example 2.

EXAMPLE 3 *Using Inductive Reasoning*

Identify a pattern in each list of numbers. Then use this pattern to find the next number.

a. 1, 1, 2, 3, 5, 8, 13, 21, _____ **b.** 23, 54, 95, 146, 117, 98, _____

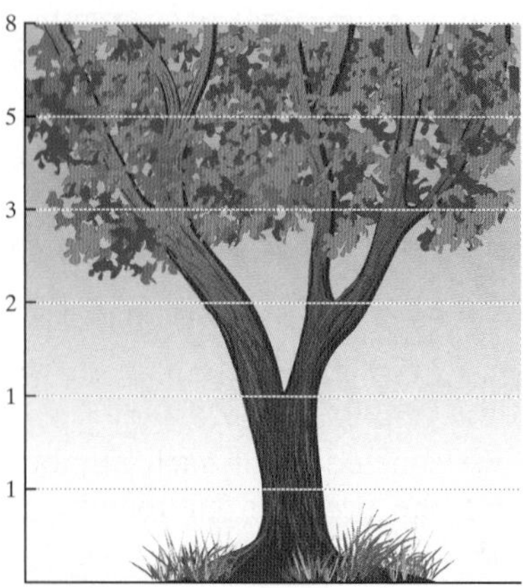

As this tree branches, the number of branches forms the Fibonacci sequence.

SOLUTION

a. We begin with 1, 1, 2, 3, 5, 8, 13, 21. Starting with the third number in the list, let's form our observations by comparing each number with the two numbers that immediately precede it.

1, 1, 2, 3, 5, 8, 13, 21, _____

preceded by 1 and 1: $1 + 1 = 2$ | preceded by 1 and 2: $1 + 2 = 3$ | preceded by 2 and 3: $2 + 3 = 5$ | preceded by 3 and 5: $3 + 5 = 8$ | preceded by 5 and 8: $5 + 8 = 13$ | preceded by 8 and 13: $8 + 13 = 21$

The first two numbers are 1. Generalizing from these observations, we conclude that each number thereafter is the sum of the two preceding numbers. Using this pattern, the next number is $13 + 21$, or 34. (The numbers 1, 1, 2, 3, 5, 8, 13, 21, and 34 are the first nine terms of the *Fibonacci sequence*, discussed in Chapter 5, Section 5.7.)

b. Now, we consider 23, 54, 95, 146, 117, 98. Let's use the digits that form each number as the basis for our individual observations. Focus on the sum of the digits, as well as the final digit increased by 1.

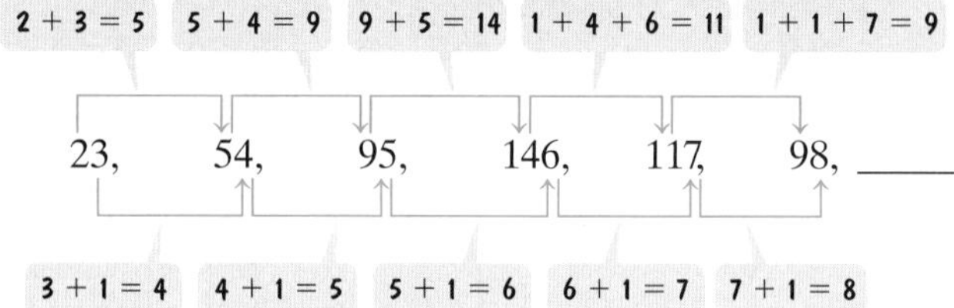

Generalizing from these observations, we conclude that for each number after the first, we obtain the first digit or the first two digits by adding the digits of the previous number. We obtain the last digit by adding 1 to the final digit of the preceding number. Applying this pattern to find the number that follows 98, the first two digits are $9 + 8$, or 17. The last digit is $8 + 1$, or 9. Thus, the next number in the list is 179.

GREAT QUESTION!

Can a list of numbers have more than one pattern?

Yes. Consider the illusion in **Figure 1.1**. This ambiguous figure contains two patterns, where it is not clear which pattern should predominate. Do you see a wine goblet or two faces looking at each other? Like this ambiguous figure, some lists of numbers can display more than one pattern, particularly if only a few numbers are given. Inductive reasoning can result in more than one probable next number in a list.

Example: 1, 2, 4, _________

Pattern: Each number after the first is obtained by multiplying the previous number by 2. The missing number is 4×2, or 8.

Pattern: Each number after the first is obtained by adding successive counting numbers, starting with 1, to the previous number. The second number is $1 + 1$, or 2. The third number is $2 + 2$, or 4. The missing number is $4 + 3$, or 7.

FIGURE 1.1

Inductive reasoning can also result in different patterns that produce the same probable next number in a list.

Example: 1, 4, 9, 16, 25, _________

Pattern: Start by adding 3 to the first number. Then add successive odd numbers, 5, 7, 9, and so on. The missing number is $25 + 11$, or 36.

Pattern: Each number is obtained by squaring its position in the list: The first number is $1^2 = 1 \times 1 = 1$, the second number is $2^2 = 2 \times 2 = 4$, the third number is $3^2 = 3 \times 3 = 9$, and so on. The missing sixth number is $6^2 = 6 \times 6$, or 36.

The numbers that we found in Examples 2 and 3 are probable numbers. Perhaps you found patterns other than the ones we pointed out that might have resulted in different answers.

 **CHECK POINT 3** Identify a pattern in each list of numbers. Then use this pattern to find the next number.

a. 1, 3, 4, 7, 11, 18, 29, 47, ______

b. 2, 3, 5, 9, 17, 33, 65, 129, ______

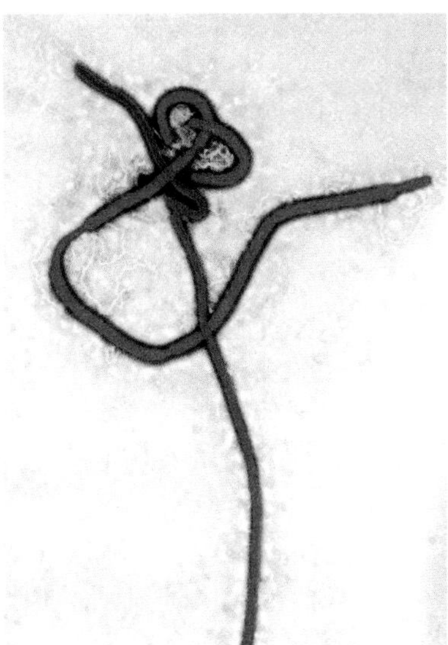

This electron microscope photograph shows the knotty shape of the Ebola virus.

Mathematics is more than recognizing number patterns. It is about the patterns that arise in the world around us. For example, by describing patterns formed by various kinds of knots, mathematicians are helping scientists investigate the knotty shapes and patterns of viruses. One of the weapons used against viruses is based on recognizing visual patterns in the possible ways that knots can be tied.

Our next example deals with recognizing visual patterns.

EXAMPLE 4 *Finding the Next Figure in a Visual Sequence*

Describe two patterns in this sequence of figures. Use the patterns to draw the next figure in the sequence.

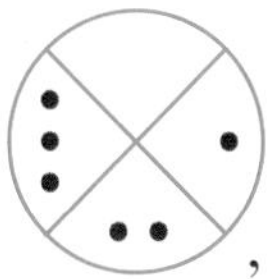, 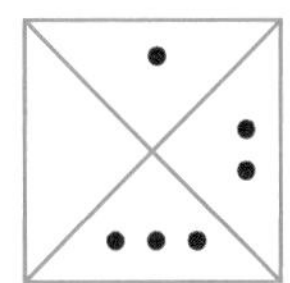, 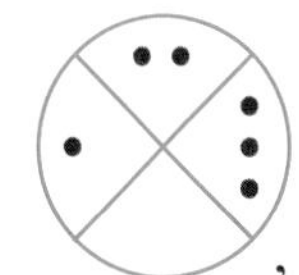, 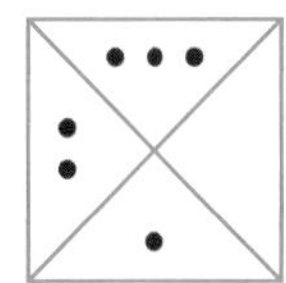, ______

SOLUTION

The more obvious pattern is that the figures alternate between circles and squares. We conclude that the next figure will be a circle. We can identify the second pattern in the four regions containing no dots, one dot, two dots, and three dots. The dots are placed in order (no dots, one dot, two dots, three dots) in a clockwise direction. However, the entire pattern of the dots rotates counterclockwise as we follow the figures from left to right. This means that the next figure should be a circle with a single dot in the right-hand region, two dots in the bottom region, three dots in the left-hand region, and no dots in the top region.

The missing figure in the visual sequence, a circle with a single dot in the right-hand region, two dots in the bottom region, three dots in the left-hand region, and no dots in the top region, is drawn in **Figure 1.2**.

FIGURE 1.2

 **CHECK POINT 4** Describe two patterns in this sequence of figures. Use the patterns to draw the next figure in the sequence.

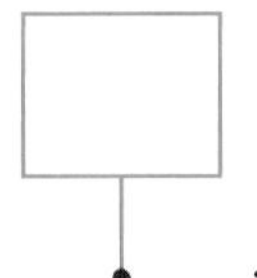, , , , ______

Blitzer Bonus

Are You Smart Enough to Work at Google?

In *Are You Smart Enough to Work at Google?* (Little, Brown, and Company, 2012), author William Poundstone guides readers through the surprising solutions to challenging job-interview questions. The book covers the importance of creative thinking in inductive reasoning, estimation, and problem solving. Best of all, Poundstone explains the answers.

Whether you're preparing for a job interview or simply want to increase your critical thinking skills, we highly recommend tackling the puzzles in *Are You Smart Enough to Work at Google?* Here is a sample of two of the book's problems that involve inductive reasoning. We've provided hints to help you recognize the pattern in each sequence. The answers appear in the answer section.

1. Determine the next entry in the sequence.
   SSS, SCC, C, SC, ___?___

   Hint: Think of the capital letters in the English alphabet. A is made up of three straight lines. B consists of one straight line and two curved lines. C is made up of one curved line.
2. Determine the next line in this sequence of digits.

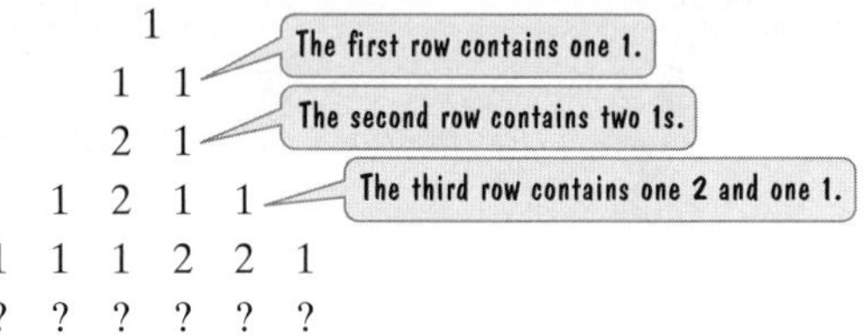

2 *Understand and use deductive reasoning.*

Deductive Reasoning

We use inductive reasoning in everyday life. Many of the conjectures that come from this kind of thinking seem highly likely, although we can never be absolutely certain that they are true. Another method of reasoning, called *deductive reasoning*, or *deduction*, can be used to prove that some conjectures are true.

> **DEDUCTIVE REASONING**
>
> **Deductive reasoning** is the process of proving a specific conclusion from one or more general statements. A conclusion that is proved to be true by deductive reasoning is called a **theorem**.

Deductive reasoning allows us to draw a specific conclusion from one or more general statements. Two examples of deductive reasoning are shown below. Notice that in both everyday situations, the general statement from which the conclusion is drawn is implied rather than directly stated.

Everyday Situation	Deductive Reasoning
One player to another in a Scrabble game: "You have to remove those five letters. You can't use TEXAS as a word."	• All proper names are prohibited in Scrabble. (general statement) TEXAS is a proper name. Therefore, TEXAS is prohibited in Scrabble. (conclusion)
Advice to college freshmen on choosing classes: "Never sign up for a 7 A.M. class. Yes, you did it in high school, but Mom was always there to keep waking you up, and if by some miracle you do make it to an early class, you will sleep through the lecture when you get there." (*Source: How to Survive Your Freshman Year*, Hundreds of Heads Books, 2004)	• All people need to sleep at 7 A.M. (general statement) You sign up for a class at 7 A.M. Therefore, you'll sleep through the lecture or not even make it to class. (conclusion) In Chapter 3, you'll learn how to prove this conclusion from the general statement in the first line. But is the general statement really true? Can we make assumptions about the sleeping patterns of all people, or are we using deductive reasoning to reinforce an untrue reality assumption?

Our next example illustrates the difference between inductive and deductive reasoning. The first part of the example involves reasoning that moves from specific examples to a general statement, illustrating inductive reasoning. The second part of the example begins with the general case rather than specific examples and illustrates deductive reasoning. To begin the general case, we use a letter to represent any one of various numbers. A letter used to represent any number in a collection of numbers is called a **variable**. Variables and other mathematical symbols allow us to work with the general case in a very concise manner.

A BRIEF REVIEW

In case you have forgotten some basic terms of arithmetic, the following list should be helpful.

Sum:	the result of addition
Difference:	the result of subtraction
Product:	the result of multiplication
Quotient:	the result of division

EXAMPLE 5 *Using Inductive and Deductive Reasoning*

Consider the following procedure:

Select a number. Multiply the number by 6. Add 8 to the product. Divide this sum by 2. Subtract 4 from the quotient.

a. Repeat this procedure for at least four different numbers. Write a conjecture that relates the result of this process to the original number selected.

b. Use the variable n to represent the original number and use deductive reasoning to prove the conjecture in part (a).

SOLUTION

a. First, let us pick our starting numbers. We will use 4, 7, 11, and 100, but we could pick any four numbers. Next we will apply the procedure given in this example to 4, 7, 11, and 100, four individual cases, in **Table 1.1**.

TABLE 1.1 Applying a Procedure to Four Individual Cases

Select a number.	4	7	11	100
Multiply the number by 6.	$4 \times 6 = 24$	$7 \times 6 = 42$	$11 \times 6 = 66$	$100 \times 6 = 600$
Add 8 to the product.	$24 + 8 = 32$	$42 + 8 = 50$	$66 + 8 = 74$	$600 + 8 = 608$
Divide this sum by 2.	$\frac{32}{2} = 16$	$\frac{50}{2} = 25$	$\frac{74}{2} = 37$	$\frac{608}{2} = 304$
Subtract 4 from the quotient.	$16 - 4 = 12$	$25 - 4 = 21$	$37 - 4 = 33$	$304 - 4 = 300$

Because we are asked to write a conjecture that relates the result of this process to the original number selected, let us focus on the result of each case.

Original number selected	4	7	11	100
Result of the process	12	21	33	300

Do you see a pattern? Our conjecture is that the result of the process is three times the original number selected. We have used inductive reasoning.

b. Now we begin with the general case rather than specific examples. We use the variable n to represent any number.

Select a number.	n
Multiply the number by 6.	$6n$ (This means 6 times n.)
Add 8 to the product.	$6n + 8$
Divide this sum by 2.	$\frac{6n + 8}{2} = \frac{6n}{2} + \frac{8}{2} = 3n + 4$
Subtract 4 from the quotient.	$3n + 4 - 4 = 3n$

Using the variable n to represent any number, the result is $3n$, or three times the number n. This proves that the result of the procedure is three times the original number selected for any number. We have used deductive reasoning. Observe how algebraic notation allows us to work with the general case quite efficiently through the use of a variable.

CHECK POINT 5 Consider the following procedure:

Select a number. Multiply the number by 4. Add 6 to the product. Divide this sum by 2. Subtract 3 from the quotient.

a. Repeat this procedure for at least four different numbers. Write a conjecture that relates the result of this process to the original number selected.

b. Use the variable n to represent the original number and use deductive reasoning to prove the conjecture in part (a).

Blitzer Bonus

Surprising Friends with Induction

Ask a few friends to follow this procedure:

Write down a whole number from 2 to 10. Multiply the number by 9. Add the digits. Subtract 3. Assign a letter to this result using $A = 1, B = 2, C = 3$, and so on. Write down the name of a state that begins with this letter. Select the name of an insect that begins with the last letter of the state. Name a fruit or vegetable that begins with the last letter of the insect.

After following this procedure, surprise your friend by asking, "Are you thinking of an ant in Florida eating a tomato?" (Try using inductive reasoning to determine how you came up with this "astounding" question. Are other less-probable "astounding" questions possible using inductive reasoning?)

Concept and Vocabulary Check

GREAT QUESTION!

What am I supposed to do with the exercises in the Concept and Vocabulary Check?

An important component of thinking mathematically involves knowing the special language and notation used in mathematics. The exercises in the Concept and Vocabulary Check, mainly fill-in-the-blank and true/false items, test your understanding of the definitions and concepts presented in each section. **Work all of the exercises in the Concept and Vocabulary Check** regardless of which exercises your professor assigns in the Exercise Set that follows.

Fill in each blank so that the resulting statement is true.

1. The statement $3 + 3 = 6$ serves as a/an ____________ to the conjecture that the sum of two odd numbers is an odd number.
2. Arriving at a specific conclusion from one or more general statements is called __________ reasoning.
3. Arriving at a general conclusion based on observations of specific examples is called __________ reasoning.
4. True or False: A theorem cannot have counterexamples. ______

Exercise Set 1.1

GREAT QUESTION!

Any way that I can perk up my brain before working the book's Exercise Sets?

Researchers say the mind can be strengthened, just like your muscles, with regular training and rigorous practice. Think of the book's Exercise Sets as brain calisthenics. If you're feeling a bit sluggish before any of your mental workouts, try this warmup:

In the list below, say the color the word is printed in, not the word itself. Once you can do this in 15 seconds without an error, the warmup is over and it's time to move on to the assigned exercises.

Blue Yellow Red Green Yellow Green Blue Red Yellow Red

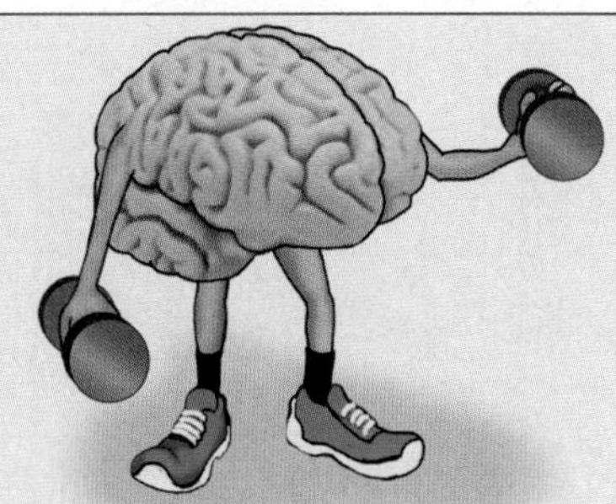

Practice Exercises

In Exercises 1–8, find a counterexample to show that each of the statements is false.

1. No U.S. president has been younger than 65 at the time of his inauguration.
2. No singers appear in movies.
3. If a number is multiplied by itself, the result is even.
4. The sum of two three-digit numbers is a four-digit number.
5. Adding the same number to both the numerator and the denominator (top and bottom) of a fraction does not change the fraction's value.
6. If the difference between two numbers is odd, then the two numbers are both odd.
7. If a number is added to itself, the sum is greater than the original number.
8. If 1 is divided by a number, the quotient is less than that number.

In Exercises 9–38, identify a pattern in each list of numbers. Then use this pattern to find the next number. (More than one pattern might exist, so it is possible that there is more than one correct answer.)

9. 8, 12, 16, 20, 24, ______
10. 19, 24, 29, 34, 39, ______
11. 37, 32, 27, 22, 17, ______
12. 33, 29, 25, 21, 17, ______
13. 3, 9, 27, 81, 243, ______
14. 2, 8, 32, 128, 512, ______
15. 1, 2, 4, 8, 16, ______
16. 1, 5, 25, 125, ______
17. 1, 4, 1, 8, 1, 16, 1, ______
18. 1, 4, 1, 7, 1, 10, 1, ______
19. 4, 2, 0, −2, −4, ______
20. 6, 3, 0, −3, −6, ______
21. $\frac{1}{2}, \frac{1}{6}, \frac{1}{10}, \frac{1}{14}, \frac{1}{18}$, ______
22. $1, \frac{1}{2}, \frac{1}{3}, \frac{1}{4}, \frac{1}{5}$, ______
23. $1, \frac{1}{3}, \frac{1}{9}, \frac{1}{27}$, ______
24. $1, \frac{1}{2}, \frac{1}{4}, \frac{1}{8}$, ______
25. 3, 7, 12, 18, 25, 33, ______
26. 2, 5, 9, 14, 20, 27, ______
27. 3, 6, 11, 18, 27, 38, ______
28. 2, 5, 10, 17, 26, 37, ______
29. 3, 7, 10, 17, 27, 44, ______
30. 2, 5, 7, 12, 19, 31, ______
31. 2, 7, 12, 5, 10, 15, 8, 13, ______
32. 3, 9, 15, 5, 11, 17, 7, 13, ______
33. 3, 6, 5, 10, 9, 18, 17, 34, ______
34. 2, 6, 5, 15, 14, 42, 41, 123, ______
35. 64, −16, 4, −1, ______
36. 125, −25, 5, −1, ______
37. $(6, 2), (0, -4), \left(7\frac{1}{2}, 3\frac{1}{2}\right), (2, -2), (3, \_\_\_\_)$
38. $\left(\frac{2}{3}, \frac{4}{9}\right), \left(\frac{1}{5}, \frac{1}{25}\right), (7, 49), \left(-\frac{5}{6}, \frac{25}{36}\right), \left(-\frac{4}{7}, \_\_\_\_\right)$

In Exercises 39–42, identify a pattern in each sequence of figures. Then use the pattern to find the next figure in the sequence.

39.

40.

41.

42.

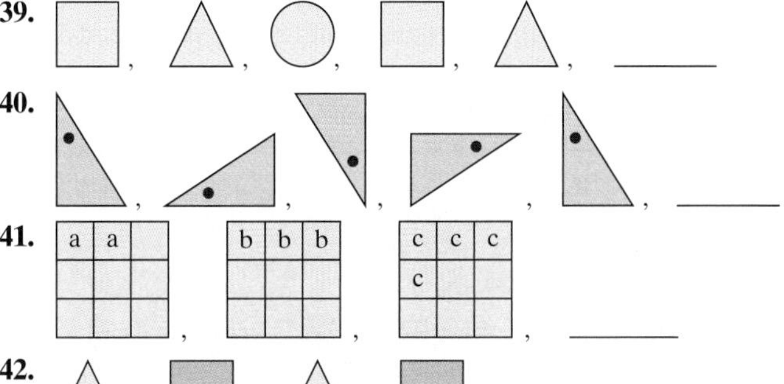

Exercises 43–46 describe procedures that are to be applied to numbers. In each exercise,

a. *Repeat the procedure for four numbers of your choice. Write a conjecture that relates the result of the process to the original number selected.*

b. *Use the variable n to represent the original number and use deductive reasoning to prove the conjecture in part (a).*

43. Select a number. Multiply the number by 4. Add 8 to the product. Divide this sum by 2. Subtract 4 from the quotient.
44. Select a number. Multiply the number by 3. Add 6 to the product. Divide this sum by 3. Subtract the original selected number from the quotient.
45. Select a number. Add 5. Double the result. Subtract 4. Divide by 2. Subtract the original selected number.
46. Select a number. Add 3. Double the result. Add 4. Divide by 2. Subtract the original selected number.

In Exercises 47–52, use inductive reasoning to predict the next line in each sequence of computations. Then use a calculator or perform the arithmetic by hand to determine whether your conjecture is correct.

47.

$$1 + 2 = \frac{2 \times 3}{2}$$
$$1 + 2 + 3 = \frac{3 \times 4}{2}$$
$$1 + 2 + 3 + 4 = \frac{4 \times 5}{2}$$
$$1 + 2 + 3 + 4 + 5 = \frac{5 \times 6}{2}$$

48.

$$3 + 6 = \frac{6 \times 3}{2}$$
$$3 + 6 + 9 = \frac{9 \times 4}{2}$$
$$3 + 6 + 9 + 12 = \frac{12 \times 5}{2}$$
$$3 + 6 + 9 + 12 + 15 = \frac{15 \times 6}{2}$$

49.

$$1 + 3 = 2 \times 2$$
$$1 + 3 + 5 = 3 \times 3$$
$$1 + 3 + 5 + 7 = 4 \times 4$$
$$1 + 3 + 5 + 7 + 9 = 5 \times 5$$

50.

$$1 \times 9 + (1 + 9) = 19$$
$$2 \times 9 + (2 + 9) = 29$$
$$3 \times 9 + (3 + 9) = 39$$
$$4 \times 9 + (4 + 9) = 49$$

51.

$$9 \times 9 + 7 = 88$$
$$98 \times 9 + 6 = 888$$
$$987 \times 9 + 5 = 8888$$
$$9876 \times 9 + 4 = 88{,}888$$

52.

$$1 \times 9 - 1 = 8$$
$$21 \times 9 - 1 = 188$$
$$321 \times 9 - 1 = 2888$$
$$4321 \times 9 - 1 = 38{,}888$$

Practice Plus

In Exercises 53–54, use inductive reasoning to predict the next line in each sequence of computations. Then use a calculator or perform the arithmetic by hand to determine whether your conjecture is correct.

53. $33 \times 3367 = 111{,}111$
$66 \times 3367 = 222{,}222$
$99 \times 3367 = 333{,}333$
$132 \times 3367 = 444{,}444$

54. $1 \times 8 + 1 = 9$
$12 \times 8 + 2 = 98$
$123 \times 8 + 3 = 987$
$1234 \times 8 + 4 = 9876$
$12{,}345 \times 8 + 5 = 98{,}765$

55. Study the pattern in these examples:

$$a^2 \# a^4 = a^{10} \quad a^3 \# a^2 = a^7 \quad a^5 \# a^3 = a^{11}.$$

Select the equation that describes the pattern.

a. $a^x \# a^y = a^{2x+y}$ **b.** $a^x \# a^y = a^{x+2y}$
c. $a^x \# a^y = a^{x+y+4}$ **d.** $a^x \# a^y = a^{xy+2}$

56. Study the pattern in these examples:

$$a^5 * a^3 * a^2 = a^5 \quad a^3 * a^7 * a^2 = a^6 \quad a^2 * a^4 * a^8 = a^7.$$

Select the equation that describes the pattern.

a. $a^x * a^y * a^z = a^{x+y+z}$ **b.** $a^x * a^y * a^z = a^{\frac{xyz}{2}}$
c. $a^x * a^y * a^z = a^{\frac{x+y+z}{2}}$ **d.** $a^x * a^y * a^z = a^{\frac{xy}{2}+z}$

Application Exercises

In Exercises 57–60, identify the reasoning process, induction or deduction, in each example. Explain your answer.

57. It can be shown that

$$1 + 2 + 3 + \cdots + n = \frac{n(n+1)}{2}.$$

I can use this formula to conclude that the sum of the first one hundred counting numbers, $1 + 2 + 3 + \cdots + 100$, is

$$\frac{100(100+1)}{2} = \frac{100(101)}{2} = 50(101), \text{ or } 5050.$$

58. An HMO does a follow-up study on 200 randomly selected patients given a flu shot. None of these people became seriously ill with the flu. The study concludes that all HMO patients be urged to get a flu shot in order to prevent a serious case of the flu.

59. The data in the graph are from a random sample of 1200 full-time four-year undergraduate college students on 100 U.S. campuses.

The Greatest Problems on Campus

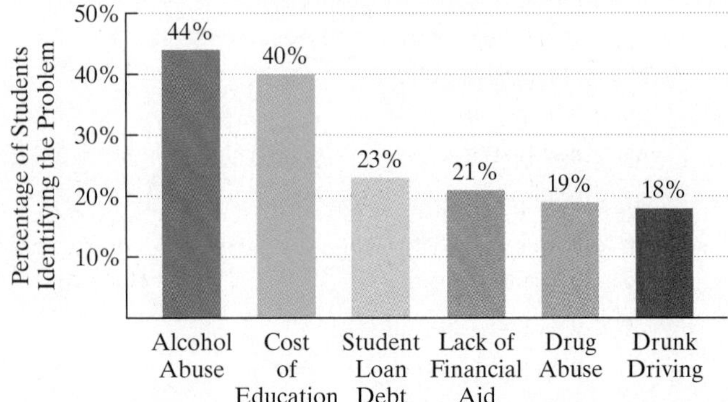

Source: Student Monitor LLC

Using the graph at the bottom of the previous column, we can conclude that there is a high probability that approximately 44% of all full-time four-year college students in the United States believe that alcohol abuse is the greatest problem on campus.

60. The course policy states that work turned in late will be marked down a grade. I turned in my report a day late, so it was marked down from B to C.

61. The ancient Greeks studied **figurate numbers**, so named because of their representations as geometric arrangements of points.

Triangular Numbers

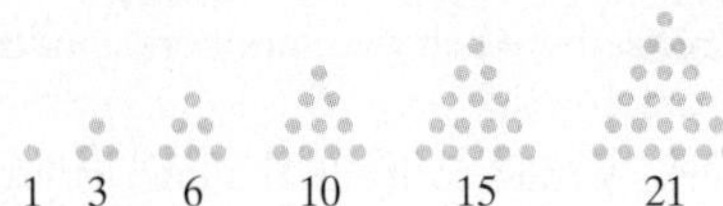

Square Numbers

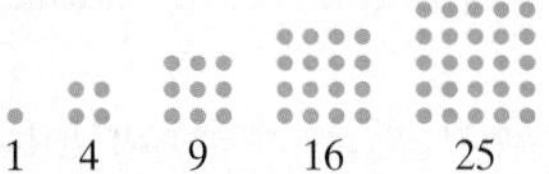

Pentagonal Numbers

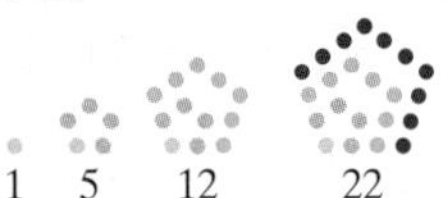

a. Use inductive reasoning to write the five triangular numbers that follow 21.

b. Use inductive reasoning to write the five square numbers that follow 25.

c. Use inductive reasoning to write the five pentagonal numbers that follow 22.

d. Use inductive reasoning to complete this statement: If a triangular number is multiplied by 8 and then 1 is added to the product, a _______ number is obtained.

62. The triangular arrangement of numbers shown below is known as **Pascal's triangle**, credited to French mathematician Blaise Pascal (1623–1662). Use inductive reasoning to find the six numbers designated by question marks.

1
1 1
1 2 1
1 3 3 1
1 4 6 4 1
? ? ? ? ? ?

Explaining the Concepts

An effective way to understand something is to explain it to someone else. *You can do this by using the Explaining the Concepts exercises that ask you to respond with verbal or written explanations. Speaking about a new concept uses a different part of your brain than thinking about the concept. Explaining new ideas verbally will quickly reveal any gaps in your understanding. It will also help you to remember new concepts for longer periods of time.*

63. The word *induce* comes from a Latin term meaning *to lead*. Explain what leading has to do with inductive reasoning.

64. Describe what is meant by deductive reasoning. Give an example.

65. Give an example of a decision that you made recently in which the method of reasoning you used to reach the decision was induction. Describe your reasoning process.

Critical Thinking Exercises

Make Sense? *In Exercises 66–69, determine whether each statement makes sense or does not make sense, and explain your reasoning.*

66. I use deductive reasoning to draw conclusions that are not certain, but likely.

67. Additional information may strengthen or weaken the probability of my inductive arguments.

68. I used the data shown in the bar graph, which summarizes a random sample of 752 college seniors, to conclude with certainty that 51% of all graduating college females expect to earn $30,000 or less after graduation.

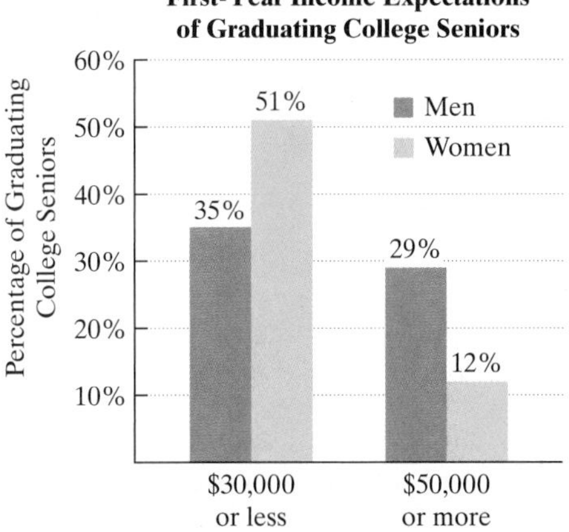

Source: Duquesne University Seniors' Economic Expectation Research Survey

69. I used the data shown in the bar graph for Exercise 68, which summarizes a random sample of 752 college seniors, to conclude inductively that a greater percentage of male graduates expect higher first-year income than female graduates.

70. If $(6 - 2)^2 = 36 - 24 + 4$ and $(8 - 5)^2 = 64 - 80 + 25$, use inductive reasoning to write a compatible expression for $(11 - 7)^2$.

71. The rectangle shows an array of nine numbers represented by combinations of the variables a, b, and c.

$a + b$	$a - b - c$	$a + c$
$a - b + c$	a	$a + b - c$
$a - c$	$a + b + c$	$a - b$

a. Determine the nine numbers in the array for $a = 10$, $b = 6$, and $c = 1$. What do you observe about the sum of the numbers in all rows, all columns, and the two diagonals?

b. Repeat part (a) for $a = 12$, $b = 5$, and $c = 2$.

c. Repeat part (a) for values of a, b, and c of your choice.

d. Use the results of parts (a) through (c) to make an inductive conjecture about the rectangular array of nine numbers represented by a, b, and c.

e. Use deductive reasoning to prove your conjecture in part (d).

72. Write a list of numbers that has two patterns so that the next number in the list can be 15 or 20.

73. a. Repeat the following procedure with at least five people. Write a conjecture that relates the result of the procedure to each person's birthday.

> Take the number of the month of your birthday (January = 1, February = 2, . . . , December = 12), multiply by 5, add 6, multiply this sum by 4, add 9, multiply this new sum by 5, and add the number of the day on which you were born. Finally, subtract 165.

b. Let M represent the month number and let D represent the day number of any person's birthday. Use deductive reasoning to prove your conjecture in part (a).

Technology Exercises

74. a. Use a calculator to find 6×6, 66×66, 666×666, and 6666×6666.

b. Describe a pattern in the numbers being multiplied and the resulting products.

c. Use the pattern to write the next two multiplications and their products. Then use your calculator to verify these results.

d. Is this process an example of inductive or deductive reasoning? Explain your answer.

75. a. Use a calculator to find 3367×3, 3367×6, 3367×9, and 3367×12.

b. Describe a pattern in the numbers being multiplied and the resulting products.

c. Use the pattern to write the next two multiplications and their products. Then use your calculator to verify these results.

d. Is this process an example of inductive or deductive reasoning? Explain your answer.

Group Exercise

76. Stereotyping refers to classifying people, places, or things according to common traits. Prejudices and stereotypes can function as assumptions in our thinking, appearing in inductive and deductive reasoning. For example, it is not difficult to find inductive reasoning that results in generalizations such as these, as well as deductive reasoning in which these stereotypes serve as assumptions:

> School has nothing to do with life.
>
> Intellectuals are nerds.
>
> People on welfare are lazy.

Each group member should find one example of inductive reasoning and one example of deductive reasoning in which stereotyping occurs. Upon returning to the group, present each example and then describe how the stereotyping results in faulty conjectures or prejudging situations and people.

1.2 Estimation, Graphs, and Mathematical Models

WHAT AM I SUPPOSED TO LEARN?

After studying this section, you should be able to:

1. Use estimation techniques to arrive at an approximate answer to a problem.
2. Apply estimation techniques to information given by graphs.
3. Develop mathematical models that estimate relationships between variables.

IF PRESENT TRENDS CONTINUE, IS IT POSSIBLE THAT OUR DESCENDANTS COULD LIVE to be 200 years of age? To answer this question, we need to examine data for life expectancy and develop estimation techniques for representing the data mathematically. In this section, you will learn estimation methods that will enable you to obtain mathematical representations of data displayed by graphs, using these representations to predict what might occur in the future.

1 *Use estimation techniques to arrive at an approximate answer to a problem.*

Estimation

Estimation is the process of arriving at an approximate answer to a question. For example, companies estimate the amount of their products consumers are likely to use, and economists estimate financial trends. If you are about to cross a street, you may estimate the speed of oncoming cars so that you know whether or not to wait before crossing. Rounding numbers is also an estimation method. You might round a number without even being aware that you are doing so. You may say that you are 20 years old, rather than 20 years 5 months, or that you will be home in about a half-hour, rather than 25 minutes.

You will find estimation to be equally valuable in your work for this class. Making mistakes with a calculator or a computer is easy. Estimation can tell us whether the answer displayed for a computation makes sense.

In this section, we demonstrate several estimation methods. In the second part of the section, we apply these techniques to information given by graphs.

Rounding Whole Numbers

The numbers that we use for counting, 1, 2, 3, 4, 5, 6, 7, and so on, are called **natural numbers**. When we combine 0 with the natural numbers, we obtain the whole numbers.

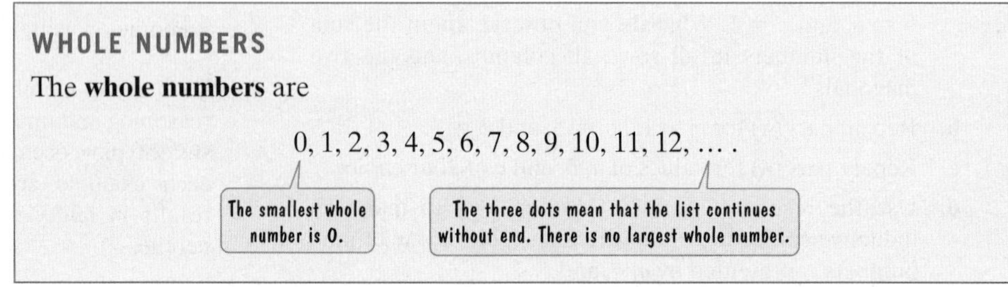

WHOLE NUMBERS

The **whole numbers** are

$$0, 1, 2, 3, 4, 5, 6, 7, 8, 9, 10, 11, 12, \ldots .$$

The smallest whole number is 0.

The three dots mean that the list continues without end. There is no largest whole number.

The numbers 0, 1, 2, 3, 4, 5, 6, 7, 8, and 9 are called **digits**, from the Latin word for fingers. Digits are used to write whole numbers.

> **GREAT QUESTION!**
>
> **When do I need to use hyphens to write the names of numbers?**
>
> Hyphenate the names for the numbers 21 (twenty-one) through 99 (ninety-nine), except 30, 40, 50, 60, 70, 80, and 90.

The position of each digit in a whole number tells us the value of that digit. Here is an example using world population at 7:35 A.M. Eastern Time on January 9, 2017.

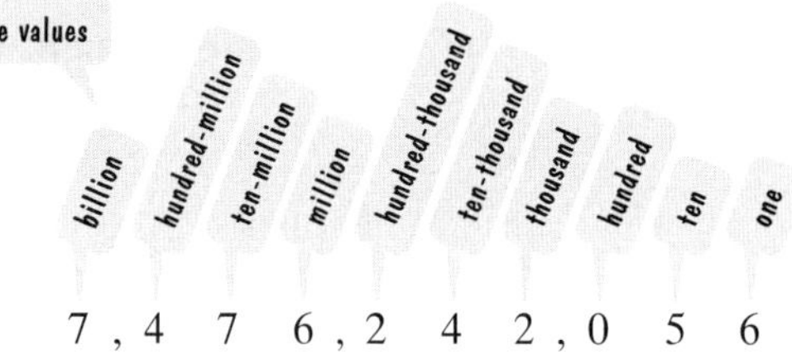

This number is read "seven billion, four hundred seventy-six million, two hundred forty-two thousand, fifty-six."

ROUNDING WHOLE NUMBERS

1. Look at the digit to the right of the digit where rounding is to occur.
2. **a.** If the digit to the right is 5 or greater, add 1 to the digit to be rounded. Replace all digits to the right with zeros.

   b. If the digit to the right is less than 5, do not change the digit to be rounded. Replace all digits to the right with zeros.

The symbol $\approx$ means *is approximately equal to*. We will use this symbol when rounding numbers.

EXAMPLE 1 *Rounding a Whole Number*

Round world population (7,476,242,056) as follows:

a. to the nearest hundred-million
b. to the nearest million
c. to the nearest hundred-thousand.

SOLUTION

a. $7{,}476{,}242{,}056 \approx 7{,}500{,}000{,}000$

Hundred-millions digit, where rounding is to occur | Digit to the right is greater than 5. | Add 1 to the digit to be rounded. | Replace all digits to the right with zeros.

World population to the nearest hundred-million is seven billion, five hundred-million.

b. $7{,}476{,}242{,}056 \approx 7{,}476{,}000{,}000$

Millions digit, where rounding is to occur | Digit to the right is less than 5. | Do not change the digit to be rounded. | Replace all digits to the right with zeros.

World population to the nearest million is seven billion, four hundred seventy-six million.

c. $7{,}476{,}242{,}056 \approx 7{,}476{,}200{,}000$

Hundred-thousands digit, where rounding is to occur | Digit to the right is less than 5. | Do not change the digit to be rounded. | Replace all digits to the right with zeros.

World population to the nearest hundred-thousand is seven billion, four hundred seventy-six million, two hundred thousand.

 **CHECK POINT 1** Round world population (7,476,242,056) as follows:

a. to the nearest billion

b. to the nearest ten-million.

Rounding can also be applied to decimal notation, used to denote a part of a whole. Once again, the place that a digit occupies tells us its value. Here's an example using the first seven digits of the number π (pi). (We'll have more to say about π, whose digits extend endlessly with no repeating pattern, in Chapter 5.)

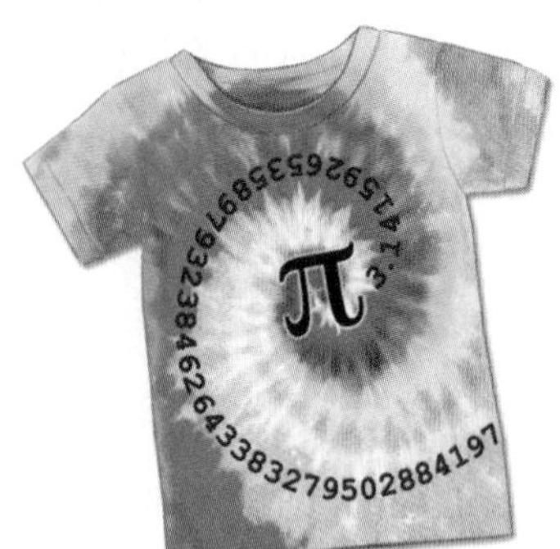

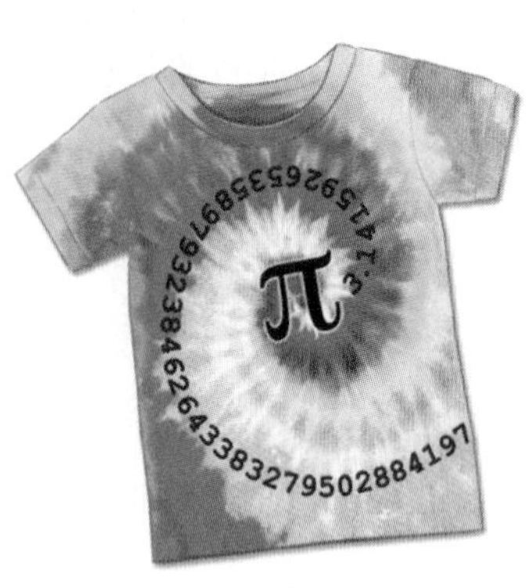

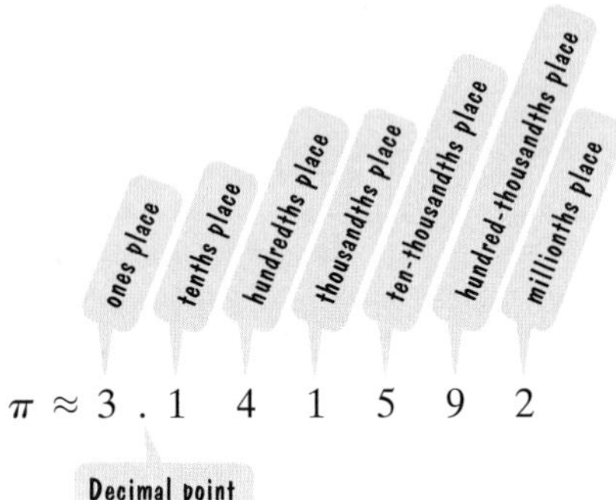

$\pi \approx 3\,.\,1\;\;4\;\;1\;\;5\;\;9\;\;2$

Decimal point

We round the decimal part of a decimal number in nearly the same way that we round whole numbers. The only difference is that we drop the digits to the right of the rounding place rather than replacing these digits with zeros.

EXAMPLE 2 Rounding the Decimal Part of a Number

Round 3.141592, the first seven digits of π, as follows:

a. to the nearest hundredth

b. to the nearest thousandth.

SOLUTION

a. $3.141592 \approx 3.14$

Hundredths digit, where rounding is to occur | Digit to the right is less than 5. | Do not change the digit to be rounded. | Drop all digits to the right.

The number π to the nearest hundredth is three and fourteen hundredths.

b. $3.141592 \approx 3.142$

Thousandths digit, where rounding is to occur | Digit to the right is 5. | Add 1 to the digit to be rounded. | Drop all digits to the right.

The number π to the nearest thousandth is three and one hundred forty-two thousandths.

GREAT QUESTION!

Could you please explain how the decimal numbers in Example 2 are read?

Of course! The whole-number part to the left of the decimal point is read like any whole number, which is *three* in both parts of Example 2. The decimal point is read as *and.* The decimal part to the right of the decimal point is read like a whole number followed by the place value of the rightmost digit. In 3.14, the 4 is in the hundredths place, so there are *fourteen hundredths*. In 3.142, the 2 is in the thousandths place, so there are *one hundred forty-two thousandths.*

 **CHECK POINT 2** Round 3.141592, the first seven digits of π, as follows:

a. to the nearest tenth

b. to the nearest ten-thousandth.

Blitzer Bonus

Estimating Support for a Cause

Police often need to estimate the size of a crowd at a political demonstration. One way to do this is to select a reasonably sized rectangle within the crowd and estimate (or count) the number of people within the rectangle. Police then estimate the number of such rectangles it would take to completely fill the area occupied by the crowd. The police estimate is obtained by multiplying the number of such rectangles by the number of demonstrators in the representative rectangle. The organizers of the demonstration might give a larger estimate than the police to emphasize the strength of their support.

EXAMPLE 3 *Estimation by Rounding*

You purchased bread for \$2.59, detergent for \$5.17, a sandwich for \$3.65, an apple for \$0.47, and coffee for \$8.79. The total bill was given as \$24.67. Is this amount reasonable?

SOLUTION

If you are in the habit of carrying a calculator to the store, you can answer the question by finding the exact cost of the purchase. However, estimation can be used to determine if the bill is reasonable even if you do not have a calculator. We will round the cost of each item to the nearest dollar.

Round to the nearest dollar. | Use digits in the tenths place to do the rounding.

Bread	\$2.59 ≈	\$3.00
Detergent	\$5.17 ≈	\$5.00
Sandwich	\$3.65 ≈	\$4.00
Apple	\$0.47 ≈	\$0.00
Coffee	\$8.79 ≈	\$9.00
		\$21.00

The total bill that you were given, \$24.67, seems a bit high compared to the \$21.00 estimate. You should check the bill before paying it. Adding the prices of all five items gives the true total bill of \$20.67.

CHECK POINT 3 You and a friend ate lunch at Ye Olde Cafe. The check for the meal showed soup for \$3.40, tomato juice for \$2.25, a roast beef sandwich for \$5.60, a chicken salad sandwich for \$5.40, two coffees totaling \$3.40, apple pie for \$2.85, and chocolate cake for \$3.95.

a. Round the cost of each item to the nearest dollar and obtain an estimate for the food bill.

b. The total bill before tax was given as \$29.85. Is this amount reasonable?

EXAMPLE 4 *Estimation by Rounding*

A carpenter who works full time earns \$28 per hour.

a. Estimate the carpenter's weekly salary.

b. Estimate the carpenter's annual salary.

SOLUTION

a. In order to simplify the calculation, we can round the hourly rate of \$28 to \$30. Be sure to write out the units for each number in the calculation. The work week is 40 hours per week, and the rounded salary is \$30 per hour. We express this as

$$\frac{40 \text{ hours}}{\text{week}} \quad \text{and} \quad \frac{\$30}{\text{hour}}.$$

The word *per* is represented by the division bar. We multiply these two numbers to estimate the carpenter's weekly salary. We cancel out units that are identical if they are above and below the division bar.

$$\frac{40\ \cancel{\text{hours}}}{\text{week}} \times \frac{\$30}{\cancel{\text{hour}}} = \frac{\$1200}{\text{week}}$$

Thus, the carpenter earns approximately \$1200 per week, written $\approx \$1200$.

b. For the estimate of annual salary, we may round 52 weeks to 50 weeks. The annual salary is approximately the product of \$1200 per week and 50 weeks per year:

$$\frac{\$1200}{\cancel{\text{week}}} \times \frac{50\ \cancel{\text{weeks}}}{\text{year}} = \frac{\$60{,}000}{\text{year}}.$$

Thus, the carpenter earns approximately \$60,000 per year, or \$60,000 annually, written $\approx \$60{,}000$.

GREAT QUESTION!

Is it OK to cancel identical units if one unit is singular and the other is plural?

Yes. It does not matter whether a unit is singular, such as *week*, or plural, such as *weeks*. *Week* and *weeks* are identical units and can be canceled out, as shown on the right.

CHECK POINT 4 A landscape architect who works full time earns \$52 per hour.

a. Estimate the landscape architect's weekly salary.

b. Estimate the landscape architect's annual salary.

2 *Apply estimation techniques to information given by graphs.*

Estimation with Graphs

Magazines, newspapers, and websites often display information using circle, bar, and line graphs. The following examples illustrate how rounding and other estimation techniques can be applied to data displayed in each of these types of graphs.

Circle graphs, also called **pie charts**, show how a whole quantity is divided into parts. Circle graphs are divided into pieces, called **sectors**. **Figure 1.3** shows a circle graph that indicates how Americans disagree as to when "old age" begins.

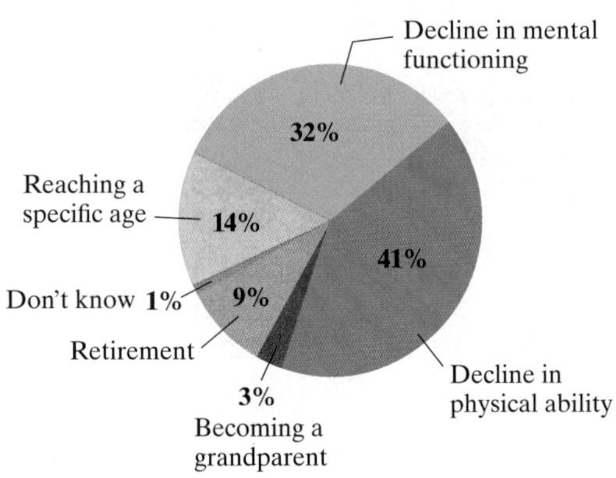

FIGURE 1.3
Source: *American Demographics*

A BRIEF REVIEW *Percents*

- **Percents** are the result of expressing numbers as part of 100. The word *percent* means *per hundred*. For example, the circle graph in **Figure 1.3** shows that 41% of Americans define old age by a decline in physical ability. Thus, 41 out of every 100 Americans define old age in this manner: $41\% = \frac{41}{100}$.
- To convert a number from percent form to decimal form, move the decimal point two places to the left and drop the percent sign. Example:

$$41\% = 41.\% = 0.41\cancel{\%}$$

Thus, $41\% = 0.41$.

- Many applications involving percent are based on the following formula:

A is *P* percent of *B*.

$$A = P \cdot B.$$

Note that the word *of* implies multiplication.

In our next example, we will use the information in the circle graph on page 18 to estimate a quantity. Although different rounding results in different estimates, the whole idea behind the rounding process is to make calculations simple.

EXAMPLE 5 *Applying Estimation Techniques to a Circle Graph*

According to the U.S. Census Bureau, in 2016, there were 219,345,624 Americans 25 years and older. Assuming the circle graph in **Figure 1.3** is representative of this age group,

a. Use the appropriate information displayed by the graph to determine a calculation that shows the number of Americans 25 years and older who define old age by a decline in physical ability.

b. Use rounding to find a reasonable estimate for this calculation.

SOLUTION

a. The circle graph in **Figure 1.3** indicates that 41% of Americans define old age by a decline in physical ability. Among the 219,345,624 Americans 25 years and older, the number who define old age in this manner is determined by finding 41% of 219,345,624.

The number of Americans 25 and older who define old age by a decline in physical ability	is	41%	of	the number of Americans 25 and older.
	=	0.41	×	219,345,624

b. We can use rounding to obtain a reasonable estimate of $0.41 \times 219,345,624$.

Round to the nearest ten million.

$$0.41 \times 219,345,624 \approx 0.4 \times 220,000,000 = 88,000,000$$

Round to the nearest tenth.

$$\begin{array}{r} 220,000,000 \\ \times \quad 0.4 \\ \hline 88,000,000.0 \end{array}$$

Our answer indicates that approximately 88,000,000 (88 million) Americans 25 years and older define old age by a decline in physical ability.

The Home Energy Pie

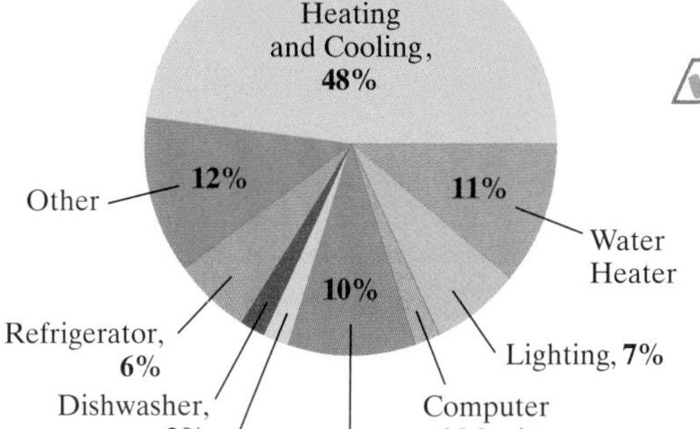

FIGURE 1.4
Source: Natural Home and Garden

CHECK POINT 5 Being aware of which appliances and activities in your home use the most energy can help you make sound decisions that allow you to decrease energy consumption and increase savings. The circle graph in **Figure 1.4** shows how energy consumption is distributed throughout a typical home.

Suppose that last year your family spent \$2148.72 on natural gas and electricity. Assuming the circle graph in **Figure 1.4** is representative of your family's energy consumption,

a. Use the appropriate information displayed by the graph to determine a calculation that shows the amount your family spent on heating and cooling for the year.

b. Use rounding to find a reasonable estimate for this calculation.

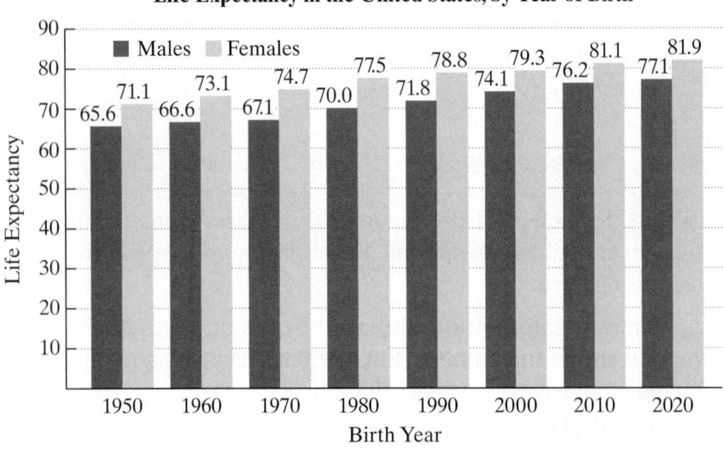

FIGURE 1.5
Source: National Center for Health Statistics

Bar graphs are convenient for comparing some measurable attribute of various items. The bars may be either horizontal or vertical, and their heights or lengths are used to show the amounts of different items. **Figure 1.5** is an example of a typical bar graph. The graph shows life expectancy for American men and American women born in various years from 1950 through 2020.

EXAMPLE 6 Applying Estimation and Inductive Reasoning to Data in a Bar Graph

Use the data for men in **Figure 1.5** to estimate each of the following:

a. a man's increased life expectancy, rounded to the nearest hundredth of a year, for each subsequent birth year

b. the life expectancy of a man born in 2030.

SOLUTION

a. One way to estimate increased life expectancy for each subsequent birth year is to generalize from the information given for 1950 (male life expectancy: 65.6 years) and for 2020 (male life expectancy: 77.1 years). The average yearly increase in life expectancy is the change in life expectancy from 1950 to 2020 divided by the change in time from 1950 to 2020.

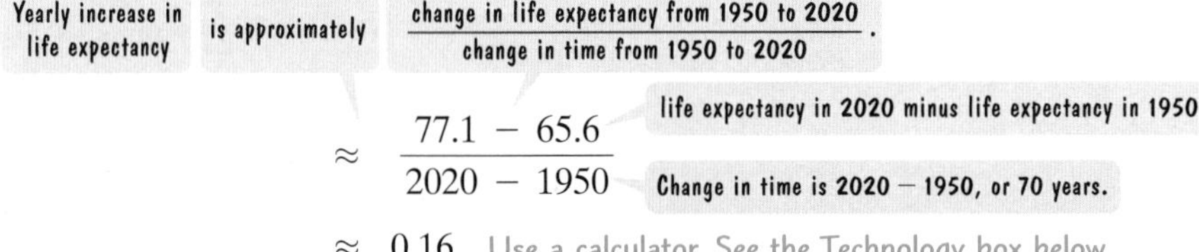

$$\approx \frac{77.1 - 65.6}{2020 - 1950}$$

$$\approx 0.16$$

For each subsequent birth year, a man's life expectancy is increasing by approximately 0.16 year.

TECHNOLOGY

Here is the calculator keystroke sequence needed to perform the computation in Example 6(a).

(77.1 − 65.6) ÷ (2020 − 1950)

Press = on a scientific calculator or ENTER on a graphing calculator to display the answer. As specified, we round to the nearest hundredth.

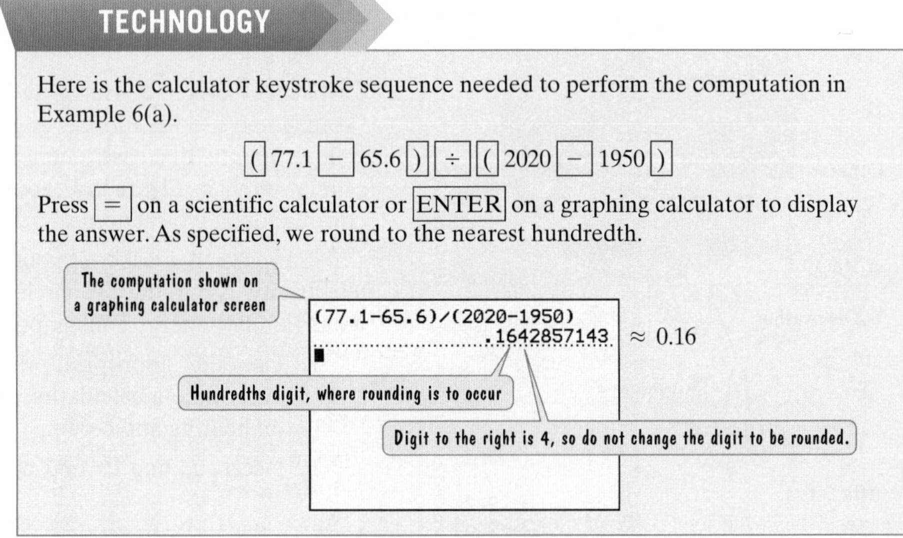

b. We can use our computation in part (a) to estimate the life expectancy of an American man born in 2030. The bar graph indicates that men born in 1950 had a life expectancy of 65.6 years. The year 2030 is 80 years after 1950, and life expectancy is increasing by approximately 0.16 year for each subsequent birth year.

Life expectancy for a man born in 2030	is approximately	life expectancy for a man born in 1950	plus	yearly increase in life expectancy	times	the number of years from 1950 to 2030.
	$\approx$	65.6	+	0.16	$\times$	80

$$\approx 65.6 + 0.16 \times 80$$
$$= 65.6 + 12.8 = 78.4$$

An American man born in 2030 will have a life expectancy of approximately 78.4 years.

GREAT QUESTION!

In the calculation at the right, you multiplied before adding. Would it be ok if I performed the operations from left to right and added before multiplying?

No. Arithmetic operations should be performed in a specific order. When there are no grouping symbols, such as parentheses, multiplication is always done before addition. We will have more to say about the order of operations in Chapter 5.

CHECK POINT 6 Use the data for women in **Figure 1.5** to estimate each of the following:

a. a woman's increased life expectancy, rounded to the nearest hundredth of a year, for each subsequent birth year

b. the life expectancy, to the nearest tenth of a year, of a woman born in 2050.

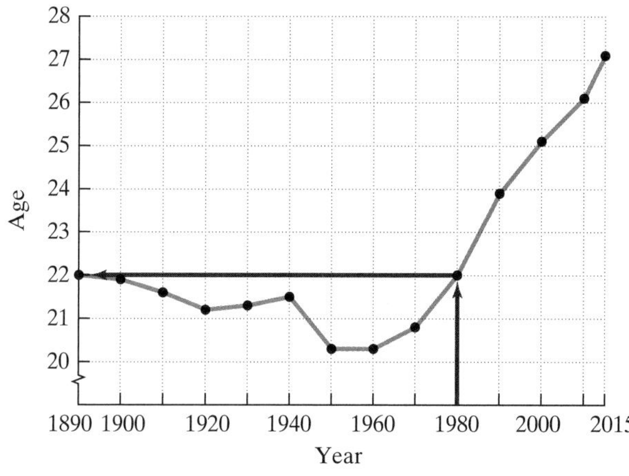

FIGURE 1.6
Source: U.S. Census Bureau

Line graphs are often used to illustrate trends over time. Some measure of time, such as months or years, frequently appears on the horizontal axis. Amounts are generally listed on the vertical axis. Points are drawn to represent the given information. The graph is formed by connecting the points with line segments.

Figure 1.6 is an example of a typical line graph. The graph shows the average age at which women in the United States married for the first time from 1890 through 2015. The years are listed on the horizontal axis, and the ages are listed on the vertical axis. The symbol ⌇ on the vertical axis shows that there is a break in values between 0 and 20. Thus, the first tick mark on the vertical axis represents an average age of 20.

Figure 1.6 shows how to find the average age at which women married for the first time in 1980.

Step 1 Locate 1980 on the horizontal axis.

Step 2 Locate the point on the line graph above 1980.

Step 3 Read across to the corresponding age on the vertical axis.

The age is 22. Thus, in 1980, women in the United States married for the first time at an average age of 22.

EXAMPLE 7 *Using a Line Graph*

The line graph in **Figure 1.7** shows the percentage of U.S. college students who smoked cigarettes from 1982 through 2014.

a. Find an estimate for the percentage of college students who smoked cigarettes in 2010.

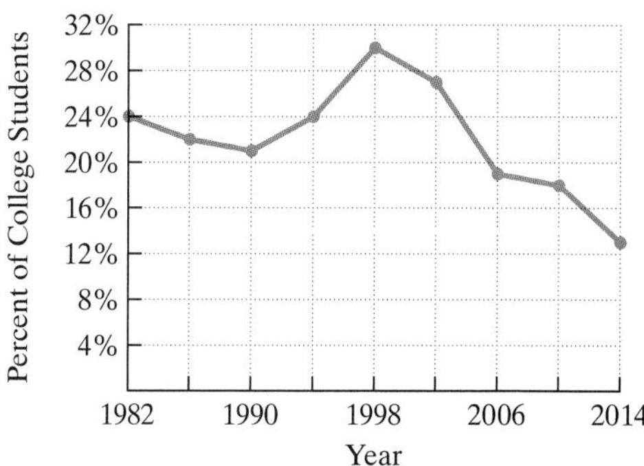

FIGURE 1.7
Source: Rebecca Donatelle, *Health The Basics*, 10th Edition, Pearson; *Monitoring the Future Study*, University of Michigan.

b. In which four-year period did the percentage of college students who smoked cigarettes decrease at the greatest rate?

c. In which year did 30% of college students smoke cigarettes?

SOLUTION

a. Estimating the Percentage Smoking Cigarettes in 2010

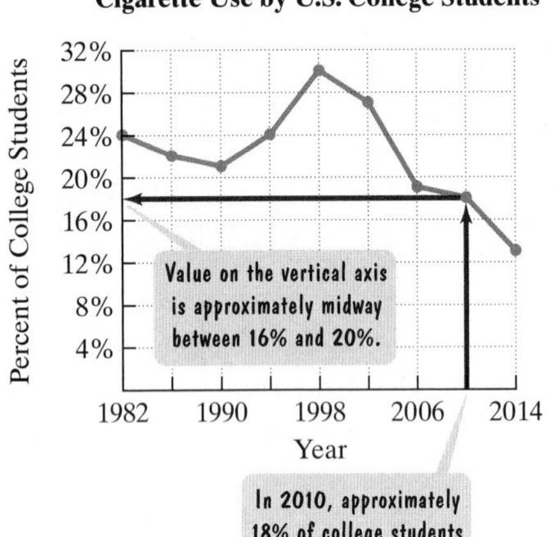

b. Identifying the Period of the Greatest Rate of Decreasing Cigarette Smoking

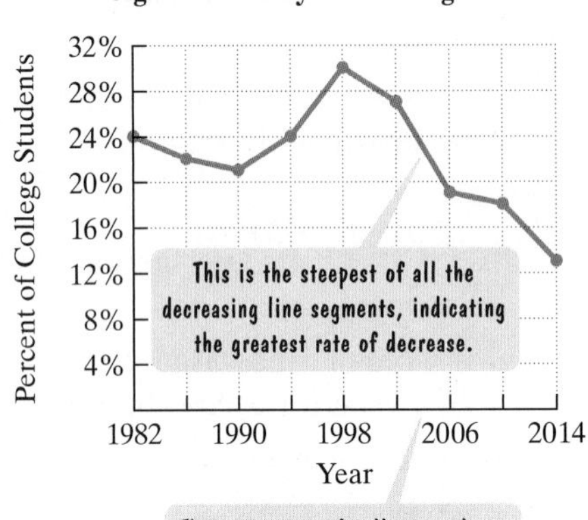

c. Identifying the Year when 30% of College Students Smoked Cigarettes

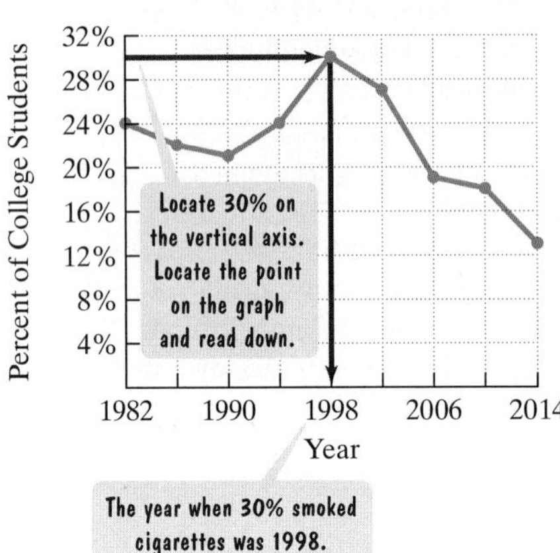

CHECK POINT 7 Use the line graph in **Figure 1.7** at the bottom of the previous page to solve this exercise.

a. Find an estimate for the percentage of college students who smoked cigarettes in 1986.

b. In which four-year period did the percentage of college students who smoked cigarettes increase at the greatest rate?

c. In which years corresponding to a tick mark on the horizontal axis did 24% of college students smoke cigarettes?

d. In which year did the least percentage of college students smoke cigarettes? What percentage of students smoked in that year?

3 *Develop mathematical models that estimate relationships between variables.*

Mathematical Models

We have seen that American men born in 1950 have a life expectancy of 65.6 years, increasing by approximately 0.16 year for each subsequent birth year. We can use variables to express the life expectancy, E, for American men born x years after 1950.

Life expectancy for American men	is	life expectancy for a man born in 1950	plus	yearly increase in life expectancy	times the number of birth years after 1950.
E	$=$	65.6	$+$	$0.16x$	

A **formula** is a statement of equality that uses letters to express a relationship between two or more variables. Thus, $E = 65.6 + 0.16x$ is a formula describing life expectancy, E, for American men born x years after 1950. Be aware that this formula provides *estimates* of life expectancy, as shown in **Table 1.2**.

Predicting Your Own Life Expectancy

The formula in **Table 1.2** does not take into account your current health, lifestyle, and family history, all of which could increase or decrease your life expectancy. Thomas Perls at Boston University Medical School, who studies centenarians, developed a much more detailed formula for life expectancy at livingto100.com. The model takes into account everything from your stress level to your sleep habits and gives you the exact age it predicts you will live to.

TABLE 1.2 Comparing Given Data with Estimates Determined by a Formula

Birth Year	Life Expectancy: Given Data	Life Expectancy: Formula Estimate $E = 65.6 + 0.16x$
1950	65.6	$E = 65.6 + 0.16(0) = 65.6 + 0 = 65.6$
1960	66.6	$E = 65.6 + 0.16(10) = 65.6 + 1.6 = 67.2$
1970	67.1	$E = 65.6 + 0.16(20) = 65.6 + 3.2 = 68.8$
1980	70.0	$E = 65.6 + 0.16(30) = 65.6 + 4.8 = 70.4$
1990	71.8	$E = 65.6 + 0.16(40) = 65.6 + 6.4 = 72.0$
2000	74.1	$E = 65.6 + 0.16(50) = 65.6 + 8.0 = 73.6$
2010	76.2	$E = 65.6 + 0.16(60) = 65.6 + 9.6 = 75.2$
2020	77.1	$E = 65.6 + 0.16(70) = 65.6 + 11.2 = 76.8$

In each row, we substitute the number of years after 1950 for x. The better estimates occur in 1950, 1990, and 2020.

The process of finding formulas to describe real-world phenomena is called **mathematical modeling**. Such formulas, together with the meaning assigned to the variables, are called **mathematical models**. We often say that these formulas model, or describe, the relationships among the variables.

EXAMPLE 8 *Modeling the Cost of Attending a Public College*

The bar graph in **Figure 1.8** shows the average cost of tuition and fees for public four-year colleges, adjusted for inflation.

a. Estimate the yearly increase in tuition and fees. Round to the nearest dollar.

b. Write a mathematical model that estimates the average cost of tuition and fees, T, at public four-year colleges for the school year ending x years after 2000.

c. Use the mathematical model from part (b) to project the average cost of tuition and fees at public four-year colleges for the school year ending in 2020.

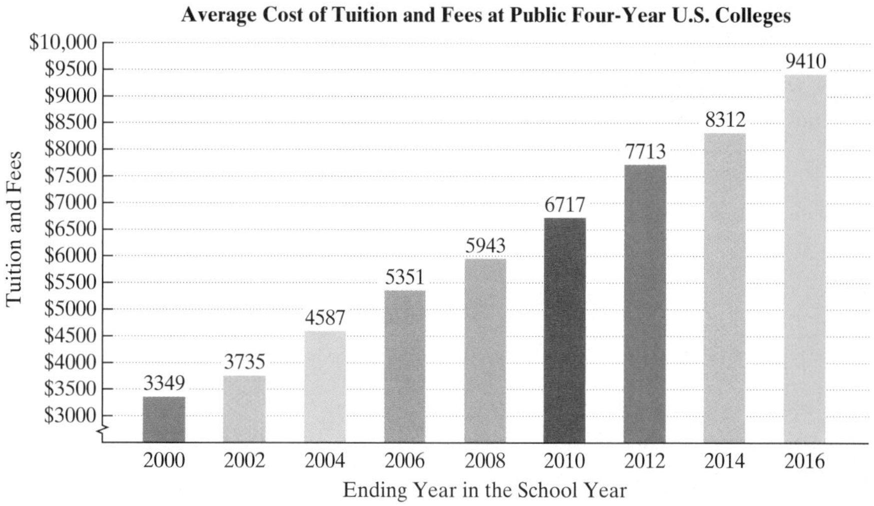

Average Cost of Tuition and Fees at Public Four-Year U.S. Colleges

Ending Year in the School Year	Tuition and Fees
2000	3349
2002	3735
2004	4587
2006	5351
2008	5943
2010	6717
2012	7713
2014	8312
2016	9410

FIGURE 1.8
Source: U.S. Department of Education

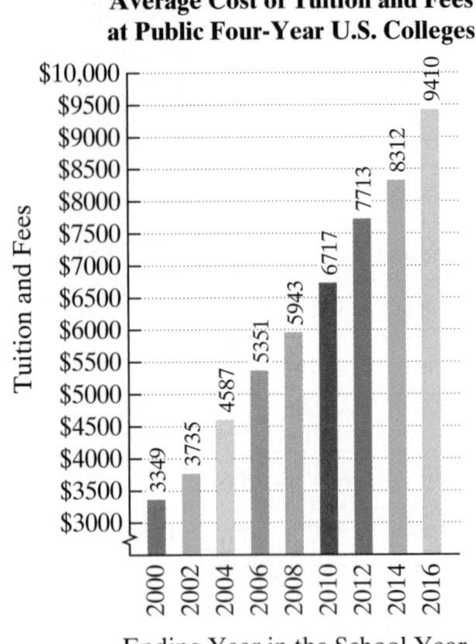

FIGURE 1.8 (repeated)

SOLUTION

a. We can use the data in **Figure 1.8** from 2000 and 2016 to estimate the yearly increase in tuition and fees.

Yearly increase in tuition and fees | is approximately | $\frac{\text{change in tuition and fees from 2000 to 2016}}{\text{change in time from 2000 to 2016}}$.

$$\approx \frac{9410 - 3349}{2016 - 2000}$$

$$= \frac{6061}{16} = 378.8125 \approx 379$$

Each year the average cost of tuition and fees for public four-year colleges is increasing by approximately $379.

b. Now we can use variables to obtain a mathematical model that estimates the average cost of tuition and fees, T, for the school year ending x years after 2000.

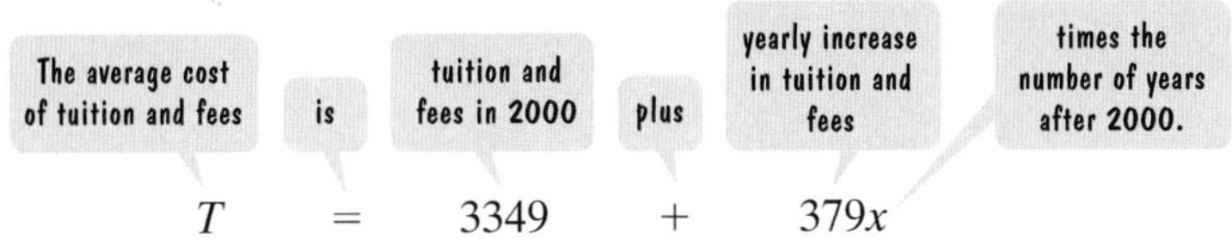

$$T \quad = \quad 3349 \quad + \quad 379x$$

The mathematical model $T = 3349 + 379x$ estimates the average cost of tuition and fees, T, at public four-year colleges for the school year ending x years after 2000.

c. Now let's use the mathematical model to project the average cost of tuition and fees for the school year ending in 2020. Because 2020 is 20 years after 2000, we substitute 20 for x.

$T = 3349 + 379x$ This is the mathematical model from part (b).

$T = 3349 + 379(20)$ Substitute 20 for x.

$= 3349 + 7580$ Multiply: $379(20) = 7580$.

$= 10{,}929$ Add. On a calculator, enter 3349 [+] 379 [×] 20 and press [=] or [ENTER].

Our model projects that the average cost of tuition and fees at public four-year colleges for the school year ending in 2020 will be $10,929.

CHECK POINT 8 The bar graph in **Figure 1.9** on the next page shows the average cost of tuition and fees for private four-year colleges, adjusted for inflation.

a. Estimate the yearly increase in tuition and fees. Round to the nearest dollar.

b. Write a mathematical model that estimates the average cost of tuition and fees, T, at private four-year colleges for the school year ending x years after 2000.

c. Use the mathematical model from part (b) to project the average cost of tuition and fees at private four-year colleges for the school year ending in 2020.

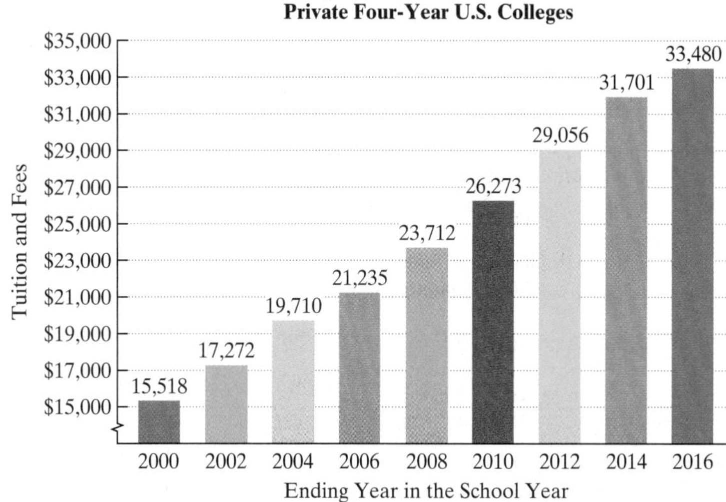

FIGURE 1.9
Source: U.S. Department of Education

Blitzer Bonus

Is College Worthwhile?

"Questions have intensified about whether going to college is worthwhile," says *Education Pays*, released by the College Board Advocacy & Policy Center. "For the typical student, the investment pays off very well over the course of a lifetime, even considering the expense."

Among the findings in *Education Pays*:

- Mean (average) full-time earnings with a bachelor's degree are approximately $63,000, which is $28,000 more than high school graduates.
- Compared with a high school graduate, a four-year college graduate who enrolled in a public university at age 18 will break even by age 33. The college graduate will have earned enough by then to compensate for being out of the labor force for four years and for borrowing enough to pay tuition and fees, shown in **Figure 1.8**.

Sometimes a mathematical model gives an estimate that is not a good approximation or is extended to include values of the variable that do not make sense. In these cases, we say that **model breakdown** has occurred. Models that accurately describe data for the past 10 years might not serve as reliable predictions for what can reasonably be expected to occur in the future. Model breakdown can occur when formulas are extended too far into the future.

Concept and Vocabulary Check

Fill in each blank so that the resulting statement is true.

1. The process of arriving at an approximate answer to a computation such as 0.79×403 is called __________.
2. A graph that shows how a whole quantity is divided into parts is called a/an ___________.
3. A formula that approximates real-world phenomena is called a/an _________________.
4. True or False: Decimal numbers are rounded by using the digit to the right of the digit where rounding is to occur. ______
5. True or False: Line graphs are often used to illustrate trends over time. ______
6. True or False: Mathematical modeling results in formulas that give exact values of real-world phenomena over time. ______

Exercise Set 1.2

Practice Exercises

The bar graph gives the populations of the ten most populous states in the United States. Use the appropriate information displayed by the graph to solve Exercises 1–2.

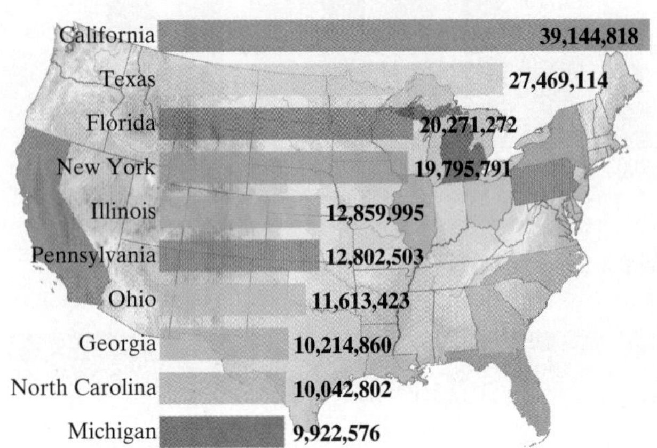

Source: U.S. Census Bureau

1. Round the population of California to the nearest **a.** hundred, **b.** thousand, **c.** ten-thousand, **d.** hundred-thousand, **e.** million, **f.** ten-million.

2. Select any state other than California. For the state selected, round the population to the nearest **a.** hundred, **b.** thousand, **c.** ten-thousand, **d.** hundred-thousand, **e.** million, **f.** ten million.

Pi goes on and on and on ...
And e is just as cursed.
I wonder: Which is larger
When their digits are reversed?
Martin Gardner

Although most people are familiar with π, the number e is more significant in mathematics, showing up in problems involving population growth and compound interest, and at the heart of the statistical bell curve. One way to think of e is the dollar amount you would have in a savings account at the end of the year if you invested $1 at the beginning of the year and the bank paid an annual interest rate of 100% compounded continuously (compounding interest every trillionth of a second, every quadrillionth of a second, etc.). Although continuous compounding sounds terrific, at the end of the year your $1 would have grown to a mere $e, or $2.72, rounded to the nearest cent. Here is a better approximation for e.

$$e \approx 2.718281828459045$$

In Exercises 3–8, use this approximation to round e as specified.

3. to the nearest thousandth
4. to the nearest ten-thousandth
5. to the nearest hundred-thousandth
6. to the nearest millionth
7. to nine decimal places
8. to ten decimal places

In Exercises 9–34, because different rounding results in different estimates, there is not one single, correct answer to each exercise.

In Exercises 9–22, obtain an estimate for each computation by rounding the numbers so that the resulting arithmetic can easily be performed by hand or in your head. Then use a calculator to perform the computation. How reasonable is your estimate when compared to the actual answer?

9. $359 + 596$
10. $248 + 797$
11. $8.93 + 1.04 + 19.26$
12. $7.92 + 3.06 + 24.36$
13. $32.15 - 11.239$
14. $46.13 - 15.237$
15. 39.67×5.5
16. 78.92×6.5
17. 0.79×414
18. 0.67×211
19. $47.83 \div 2.9$
20. $54.63 \div 4.7$
21. 32% of 187,253
22. 42% of 291,506

In Exercises 23–34, determine each estimate without using a calculator. Then use a calculator to perform the computation necessary to obtain an exact answer. How reasonable is your estimate when compared to the actual answer?

23. Estimate the total cost of six grocery items if their prices are $3.47, $5.89, $19.98, $2.03, $11.85, and $0.23.
24. Estimate the total cost of six grocery items if their prices are $4.23, $7.79, $28.97, $4.06, $13.43, and $0.74.
25. A full-time employee who works 40 hours per week earns $19.50 per hour. Estimate that person's annual income.
26. A full-time employee who works 40 hours per week earns $29.85 per hour. Estimate that person's annual income.
27. You lease a car at $605 per month for 3 years. Estimate the total cost of the lease.
28. You lease a car at $415 per month for 4 years. Estimate the total cost of the lease.
29. A raise of $310,000 is evenly distributed among 294 professors. Estimate the amount each professor receives.
30. A raise of $310,000 is evenly distributed among 196 professors. Estimate the amount each professor receives.
31. If a person who works 40 hours per week earns $61,500 per year, estimate that person's hourly wage.
32. If a person who works 40 hours per week earns $38,950 per year, estimate that person's hourly wage.
33. The average life expectancy in Canada is 80.1 years. Estimate the country's life expectancy in hours.
34. The average life expectancy in Mozambique is 40.3 years. Estimate the country's life expectancy in hours.

Practice Plus

In Exercises 35–36, obtain an estimate for each computation without using a calculator. Then use a calculator to perform the computation. How reasonable is your estimate when compared to the actual answer?

35. $\dfrac{0.19996 \times 107}{0.509}$
36. $\dfrac{0.47996 \times 88}{0.249}$

37. Ten people ordered calculators. The least expensive was \$19.95 and the most expensive was \$39.95. Half ordered a \$29.95 calculator. Select the best estimate of the amount spent on calculators.

a. \$240 **b.** \$310 **c.** \$345 **d.** \$355

38. Ten people ordered calculators. The least expensive was \$4.95 and the most expensive was \$12.95. Half ordered a \$6.95 calculator. Select the best estimate of the amount spent on calculators.

a. \$160 **b.** \$105 **c.** \$75 **d.** \$55

39. Traveling at an average rate of between 60 and 70 miles per hour for 3 to 4 hours, select the best estimate for the distance traveled.

a. 90 miles **b.** 190 miles **c.** 225 miles **d.** 275 miles

40. Traveling at an average rate of between 40 and 50 miles per hour for 3 to 4 hours, select the best estimate for the distance traveled.

a. 120 miles **b.** 160 miles **c.** 195 miles **d.** 210 miles

41. Imagine that you counted 60 numbers per minute and continued to count nonstop until you reached 10,000. Determine a reasonable estimate of the number of hours it would take you to complete the counting.

42. Imagine that you counted 60 numbers per minute and continued to count nonstop until you reached one million. Determine a reasonable estimate of the number of days it would take you to complete the counting.

Application Exercises

The circle graph shows the most important problems for the 16,503,611 high school teenagers in the United States. Use this information to solve Exercises 43–44.

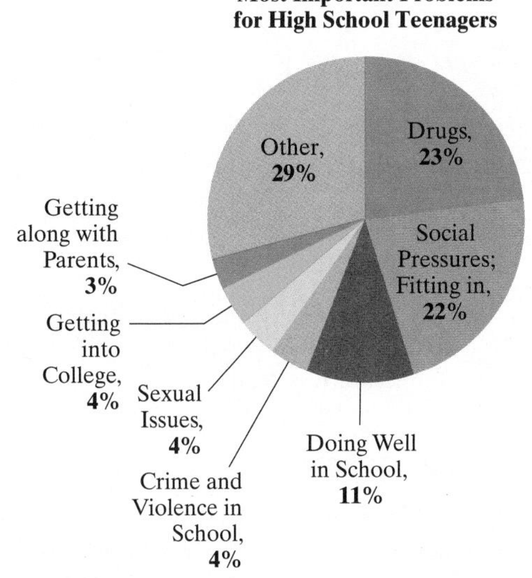

Source: Columbia University

43. Without using a calculator, estimate the number of high school teenagers for whom doing well in school is the most important problem.

44. Without using a calculator, estimate the number of high school teenagers for whom social pressures and fitting in is the most important problem.

An online test of English spelling looked at how well people spelled difficult words. The bar graph shows how many people per 100 spelled each word correctly. Use this information to solve Exercises 45–46.

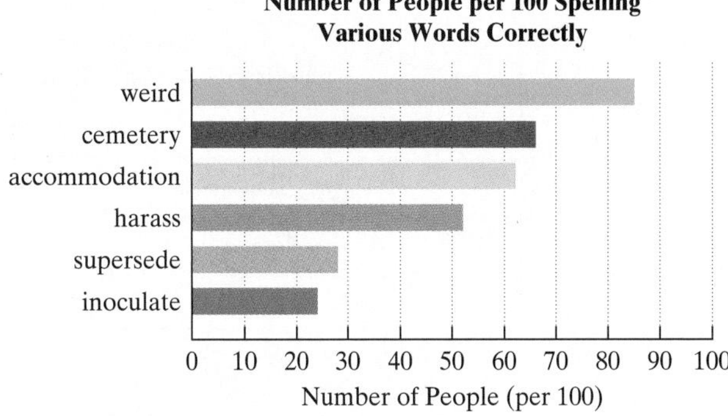

Source: Vivian Cook, *Accomodating Brocolli in the Cemetary or Why Can't Anybody Spell?*, Simon and Schuster, 2004

45. a. Estimate the number of people per 100 who spelled *weird* correctly.

b. In a group consisting of 8729 randomly selected people, estimate how many more people can correctly spell *weird* than *inoculate*.

46. a. Estimate the number of people per 100 who spelled *cemetery* correctly.

b. In a group consisting of 7219 randomly selected people, estimate how many more people can correctly spell *cemetery* than *supersede*.

The percentage of U.S. college freshmen claiming no religious affiliation has risen in recent decades. The bar graph shows the percentage of first-year college students claiming no religious affiliation for four selected years from 1980 through 2012. Use this information to solve Exercises 47–48.

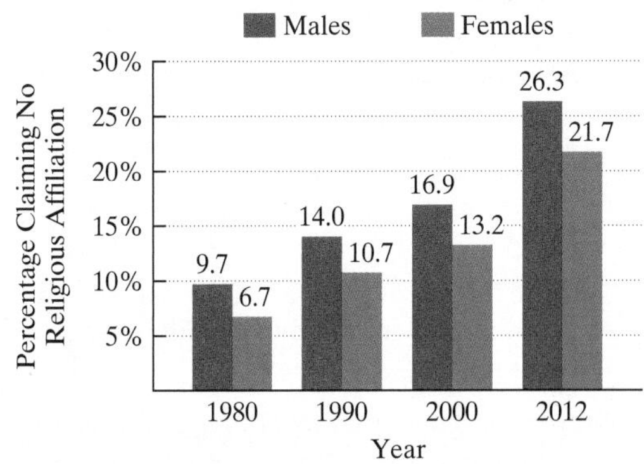

Source: John Macionis, *Sociology*, 15th Edition, Pearson, 2014.

47. a. Estimate the average yearly increase in the percentage of first-year college males claiming no religious affiliation. Round the percentage to the nearest tenth.

b. Estimate the percentage of first-year college males who will claim no religious affiliation in 2020.

48. a. Estimate the average yearly increase in the percentage of first-year college females claiming no religious affiliation. Round the percentage to the nearest tenth.

b. Estimate the percentage of first-year college females who will claim no religious affiliation in 2020.

With aging, body fat increases and muscle mass declines. The line graphs show the percent body fat in adult women and men as they age from 25 to 75 years. Use the graphs to solve Exercises 49–50.

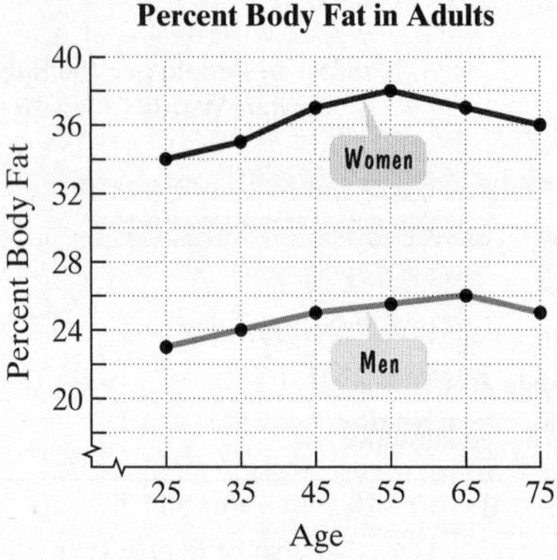

Source: Thompson et al., *The Science of Nutrition*, Benjamin Cummings, 2008.

49. a. Find an estimate for the percent body fat in 45-year-old women.

b. At what age does the percent body fat in women reach a maximum? What is the percent body fat for that age?

c. At what age do women have 34% body fat?

50. a. Find an estimate for the percent body fat in 25-year-old men.

b. At what age does the percent body fat in men reach a maximum? What is the percent body fat for that age?

c. At what age do men have 24% body fat?

There is a strong scientific consensus that human activities are changing the Earth's climate. Scientists now believe that there is a striking correlation between atmospheric carbon dioxide concentration and global temperature. As both of these variables increase at significant rates, there are warnings of a planetary emergency that threatens to condemn coming generations to a catastrophically diminished future. The bar graphs give the average atmospheric concentration of carbon dioxide and the average global temperature for eight selected years. Use this information to solve Exercises 51–52.

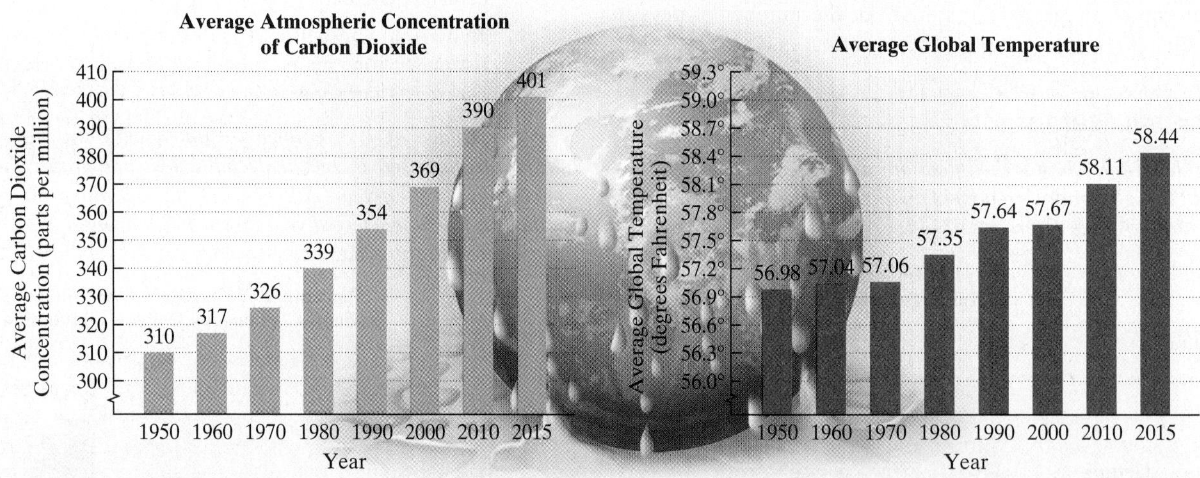

Source: National Oceanic and Atmospheric Administration

51. a. Estimate the yearly increase in the average atmospheric concentration of carbon dioxide. Express the answer in parts per million.

b. Write a mathematical model that estimates the average atmospheric concentration of carbon dioxide, C, in parts per million, x years after 1950.

c. If the trend shown by the data continues, use your mathematical model from part (b) to project the average atmospheric concentration of carbon dioxide in 2050.

52. a. Estimate the yearly increase in the average global temperature, rounded to the nearest hundredth of a degree.

b. Write a mathematical model that estimates the average global temperature, T, in degrees Fahrenheit, x years after 1950.

c. If the trend shown by the data continues, use your mathematical model from part (b) to project the average global temperature in 2050.

Explaining the Concepts

53. What is estimation? When is it helpful to use estimation?

54. Explain how to round 218,543 to the nearest thousand and to the nearest hundred-thousand.

55. Explain how to round 14.26841 to the nearest hundredth and to the nearest thousandth.

56. What does the $\approx$ symbol mean?

57. In this era of calculators and computers, why is there a need to develop estimation skills?

58. Describe a circle graph.

59. Describe a bar graph.

60. Describe a line graph.

61. What does it mean when we say that a formula models real-world phenomena?

62. College students are graduating with the highest debt burden in history. The bar graph shows the mean, or average, student-loan debt in the United States for six selected graduating years from 2001 through 2016.

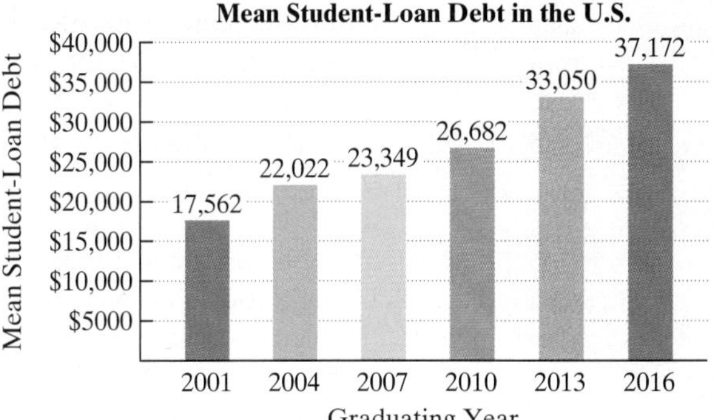

Source: Pew Research Center

Describe how to use the data for 2001 and 2016 to estimate the yearly increase in mean student-loan debt.

63. Explain how to use the estimate from Exercise 62 to write a mathematical model that estimates mean student-loan debt, D, in dollars, x years after 2001. How can this model be used to project mean student-loan debt in 2020?

64. Describe one way in which you use estimation in a nonacademic area of your life.

65. A forecaster at the National Hurricane Center needs to estimate the time until a hurricane with high probability of striking South Florida will hit Miami. Is it better to overestimate or underestimate? Explain your answer.

Critical Thinking Exercises

Make Sense? *In Exercises 66–69, determine whether each statement makes sense or does not make sense, and explain your reasoning.*

66. When buying several items at the market, I use estimation before going to the cashier to be sure I have enough money to pay for the purchase.

67. It's not necessary to use estimation skills when using my calculator.

68. Being able to compute an exact answer requires a different ability than estimating the reasonableness of the answer.

69. My mathematical model estimates the data for the past 10 years extremely well, so it will serve as an accurate prediction for what will occur in 2050.

70. Take a moment to read the verse preceding Exercises 3–8 that mentions the numbers π and e, whose decimal representations continue infinitely with no repeating patterns. The verse was written by the American mathematician (and accomplished amateur magician!) Martin Gardner (1914–2010), author of more than 60 books and best known for his "Mathematical Games" column, which ran in *Scientific American* for 25 years. Explain the humor in Gardner's question.

In Exercises 71–74, match the story with the correct graph. The graphs are labeled (a), (b), (c), and (d).

71. As the blizzard got worse, the snow fell harder and harder.

72. The snow fell more and more softly.

73. It snowed hard, but then it stopped. After a short time, the snow started falling softly.

74. It snowed softly, and then it stopped. After a short time, the snow started falling hard.

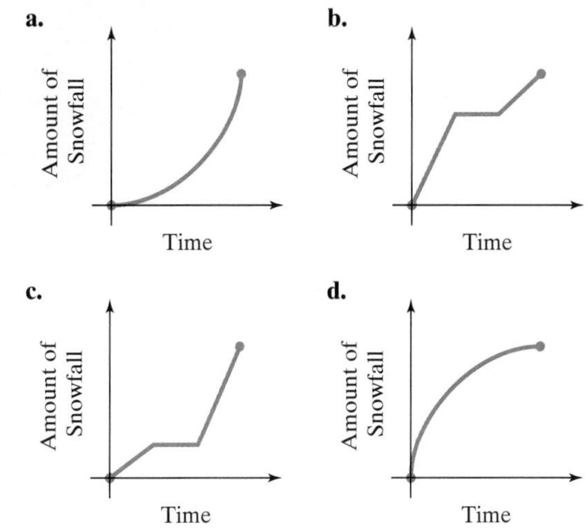

75. American children ages 2 to 17 spend 19 hours 40 minutes per week watching television. (*Source*: TV-Turnoff Network) From ages 2 through 17, inclusive, estimate the number of days an American child spends watching television. How many years, to the nearest tenth of a year, is that?

76. If you spend $1000 each day, estimate how long it will take to spend a billion dollars.

Group Exercises

77. Group members should devise an estimation process that can be used to answer each of the following questions. Use input from all group members to describe the best estimation process possible.

a. Is it possible to walk from San Francisco to New York in a year?

b. How much money is spent on ice cream in the United States each year?

78. Group members should begin by consulting an almanac, newspaper, magazine, or the Internet to find two graphs that show "intriguing" data changing from year to year. In one graph, the data values should be increasing relatively steadily. In the second graph, the data values should be decreasing relatively steadily. For each graph selected, write a mathematical model that estimates the changing variable x years after the graph's starting date. Then use each mathematical model to make predictions about what might occur in the future. Are there circumstances that might affect the accuracy of the prediction? List some of these circumstances.

1.3 Problem Solving

WHAT AM I SUPPOSED TO LEARN?

After studying this section, you should be able to:

1 Solve problems using the organization of the four-step problem-solving process.

CRITICAL THINKING AND problem solving are essential skills in both school and work. A model for problem solving was established by the charismatic teacher and mathematician George Polya (1887–1985) in *How to Solve It* (Princeton University Press, Princeton, NJ, 1957). This book, first published in 1945, has sold more than one million copies and is available in 17 languages. Using a four-step procedure for problem solving, Polya's book demonstrates how to think clearly in any field.

1 *Solve problems using the organization of the four-step problem-solving process.*

"If you don't know where you're going, you'll probably end up some place else."
—Yogi Berra

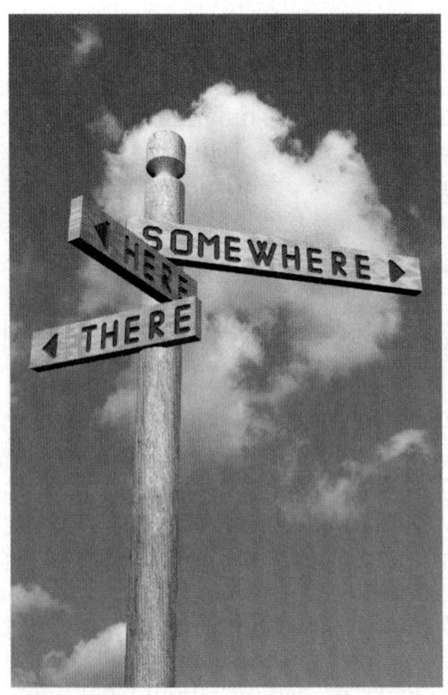

POLYA'S FOUR STEPS IN PROBLEM SOLVING

Step 1 Understand the problem. Read the problem several times. The first reading can serve as an overview. In the second reading, write down what information is given and determine exactly what it is that the problem requires you to find.

Step 2 Devise a plan. The plan for solving the problem might involve one or more of these suggested problem-solving strategies:

- Use inductive reasoning to look for a pattern.
- Make a systematic list or a table.
- Use estimation to make an educated guess at the solution. Check the guess against the problem's conditions and work backward to eventually determine the solution.
- Try expressing the problem more simply and solve a similar simpler problem.
- Use trial and error.
- List the given information in a chart or table.
- Try making a sketch or a diagram to illustrate the problem.
- Relate the problem to a similar problem that you have seen before. Try applying the procedures used to solve the similar problem to the new one.
- Look for a "catch" if the answer seems too obvious. Perhaps the problem involves some sort of trick question deliberately intended to lead the problem solver in the wrong direction.
- Use the given information to eliminate possibilities.
- Use common sense.

Step 3 Carry out the plan and solve the problem.

Step 4 Look back and check the answer. The answer should satisfy the conditions of the problem. The answer should make sense and be reasonable. If this is not the case, recheck the method and any calculations. Perhaps there is an alternate way to arrive at a correct solution.

GREAT QUESTION!

Should I memorize Polya's four steps in problem solving?

Not necessarily. Think of Polya's four steps as guidelines that will help you organize the process of problem solving, rather than a list of rigid rules that need to be memorized. You may be able to solve certain problems without thinking about or using every step in the four-step process.

The very first step in problem solving involves evaluating the given information in a deliberate manner. Is there enough given to solve the problem? Is the information relevant to the problem's solution, or are some facts not necessary to arrive at a solution?

EXAMPLE 1 *Finding What Is Missing*

Which necessary piece of information is missing and prevents you from solving the following problem?

> A man purchased five shirts, each at the same discount price. How much did he pay for them?

SOLUTION

Step 1 Understand the problem. Here's what is given:

> Number of shirts purchased: 5.

We must find how much the man paid for the five shirts.

Step 2 Devise a plan. The amount that the man paid for the five shirts is the number of shirts, 5, times the cost of each shirt. The discount price of each shirt is not given. This missing piece of information makes it impossible to solve the problem.

 **CHECK POINT 1** Which necessary piece of information is missing and prevents you from solving the following problem?

> The bill for your meal totaled $20.36, including the tax. How much change should you receive from the cashier?

EXAMPLE 2 *Finding What Is Unnecessary*

In the following problem, one more piece of information is given than is necessary for solving the problem. Identify this unnecessary piece of information. Then solve the problem.

> A roll of E-Z Wipe paper towels contains 100 sheets and costs $1.38. A comparable brand, Kwik-Clean, contains five dozen sheets per roll and costs $1.23. If you need three rolls of paper towels, which brand is the better value?

SOLUTION

Step 1 Understand the problem. Here's what is given:

> E-Z Wipe: 100 sheets per roll; $1.38
> Kwik-Clean: 5 dozen sheets per roll; $1.23
> Needed: 3 rolls.

We must determine which brand offers the better value.

Unit Prices and Sneaky Pricejacks

In *200% of Nothing* (John Wiley & Sons, 1993), author A. K. Dewdney writes, "It must be something of a corporate dream come true when a company charges more for a product and no one notices." He gives two examples of "sneaky pricejacks," both easily detected using unit prices. The manufacturers of Mennen Speed Stick deodorant increased the size of the package that held the stick, left the price the same, and reduced the amount of actual deodorant in the stick from 2.5 ounces to 2.25. Fabergé's Brut left the price and size of its cologne jar the same, but reduced its contents from 5 ounces to 4. Surprisingly, the new jar read, "Now, more Brut!" *Consumer Reports* contacted Fabergé to see how this could be possible. Their response: The new jar contained "more fragrance." *Consumer Reports* moaned, "Et tu Brut?"

Step 2 Devise a plan. The brand with the better value is the one that has the lower price per sheet. Thus, we can compare the two brands by finding the cost for one sheet of E-Z Wipe and one sheet of Kwik-Clean. The price per sheet, or the *unit price*, is the price of a roll divided by the number of sheets in the roll. The fact that three rolls are required is not relevant to the problem. This unnecessary piece of information is not needed to find which brand is the better value.

Step 3 Carry out the plan and solve the problem.

E-Z Wipe:

$$\text{price per sheet} = \frac{\text{price of a roll}}{\text{number of sheets per roll}} = \frac{\$1.38}{100\text{ sheets}} = \$0.0138 \approx \$0.01$$

Kwik-Clean:

$$\text{price per sheet} = \frac{\text{price of a roll}}{\text{number of sheets per roll}} = \frac{\$1.23}{60\text{ sheets}} = \$0.0205 \approx \$0.02$$

5 dozen = 5×12, or 60 sheets

By comparing unit prices, we see that E-Z Wipe, at approximately \$0.01 per sheet, is the better value.

Step 4 Look back and check the answer. We can double-check the arithmetic in each of our unit-price computations. We can also see if these unit prices satisfy the problem's conditions. The product of each brand's price per sheet and the number of sheets per roll should result in the given price for a roll.

E-Z Wipe: Check \$0.0138 Kwik-Clean: Check \$0.0205

$\$0.0138 \times 100 = \$1.38$ $\$0.0205 \times 60 = \$1.23$

These are the given prices for a roll of each respective brand.

The unit prices satisfy the problem's conditions.

A generalization of our work in Example 2 allows you to compare different brands and make a choice among various products of different sizes. When shopping at the supermarket, a useful number to keep in mind is a product's *unit price*. The **unit price** is the total price divided by the total units. Among comparable brands, the best value is the product with the lowest unit price, assuming that the units are kept uniform.

The word *per* is used to state unit prices. For example, if a 12-ounce box of cereal sells for \$3.00, its unit price is determined as follows:

$$\text{Unit price} = \frac{\text{total price}}{\text{total units}} = \frac{\$3.00}{12\text{ ounces}} = \$0.25 \text{ per ounce.}$$

CHECK POINT 2 Solve the following problem. If the problem contains information that is not relevant to its solution, identify this unnecessary piece of information.

A manufacturer packages its apple juice in bottles and boxes. A 128-ounce bottle costs \$5.39, and a 9-pack of 6.75-ounce boxes costs \$3.15. Which packaging option is the better value?

EXAMPLE 3 Applying the Four-Step Procedure

By paying \$100 cash up front and the balance at \$20 a week, how long will it take to pay for a bicycle costing \$680?

SOLUTION

Step 1 Understand the problem. Here's what is given:

Cost of the bicycle: \$680
Amount paid in cash: \$100
Weekly payments: \$20.

If necessary, consult a dictionary to look up any unfamiliar words. The word *balance* means the amount still to be paid. We must find the balance to determine the number of weeks required to pay off the bicycle.

Step 2 Devise a plan. Subtract the amount paid in cash from the cost of the bicycle. This results in the amount still to be paid. Because weekly payments are \$20, divide the amount still to be paid by 20. This will give the number of weeks required to pay for the bicycle.

Step 3 Carry out the plan and solve the problem. Begin by finding the balance, the amount still to be paid for the bicycle.

$$\begin{array}{rl} \$680 & \text{cost of the bicycle} \\ -\$100 & \text{amount paid in cash} \\ \hline \$580 & \text{amount still to be paid} \end{array}$$

Now divide the \$580 balance by \$20, the payment per week. The result of the division is the number of weeks needed to pay off the bicycle.

$$\frac{\$580}{\frac{\$20}{\text{week}}} = \$580 \times \frac{\text{week}}{\$20} = \frac{580 \text{ weeks}}{20} = 29 \text{ weeks}$$

It will take 29 weeks to pay for the bicycle.

Step 4 Look back and check the answer. We can certainly double-check the arithmetic either by hand or with a calculator. We can also see if the answer, 29 weeks to pay for the bicycle, satisfies the condition that the bicycle costs \$680.

This is the answer we are checking.

$$\begin{array}{rl} \$20 & \text{weekly payment} \\ \times\ 29 & \text{number of weeks} \\ \hline \$580 & \text{total of weekly payments} \end{array} \qquad \begin{array}{rl} \$580 & \text{total of weekly payments} \\ +\$100 & \text{amount paid in cash} \\ \hline \$680 & \text{cost of bicycle} \end{array}$$

The answer of 29 weeks satisfies the condition that the cost of the bicycle is \$680.

> **GREAT QUESTION!**
>
> **Is there a strategy I can use to determine whether I understand a problem?**
>
> An effective way to see if you understand a problem is to restate the problem in your own words.
>
> "A problem well stated is a problem half solved."
>
> —*Charles Franklin Kettering*

CHECK POINT 3 By paying \$350 cash up front and the balance at \$45 per month, how long will it take to pay for a computer costing \$980?

Making lists is a useful strategy in problem solving.

EXAMPLE 4 Solving a Problem by Making a List

Suppose you are an engineer programming the automatic gate for a 50-cent toll. The gate should accept exact change only. It should not accept pennies. How many coin combinations must you program the gate to accept?

SOLUTION

Step 1 Understand the problem. The total change must always be 50 cents. One possible coin combination is two quarters. Another is five dimes. We need to count all such combinations.

Step 2 Devise a plan. Make a list of all possible coin combinations. Begin with the coins of larger value and work toward the coins of smaller value.

Step 3 Carry out the plan and solve the problem. First we must find all of the coins that are not pennies but can combine to form 50 cents. This includes half-dollars, quarters, dimes, and nickels. Now we can set up a table. We will use these coins as table headings.

Half-Dollars	Quarters	Dimes	Nickels

Each row in the table will represent one possible combination for exact change. We start with the largest coin, the half-dollar. Only one half-dollar is needed to make exact change. No other coins are needed. Thus, we put a 1 in the half-dollars column and 0s in the other columns to represent the first possible combination.

Half-Dollars	Quarters	Dimes	Nickels
1	0	0	0

Likewise, two quarters are also exact change for 50 cents. We put a 0 in the half-dollars column, a 2 in the quarters column, and 0s in the columns for dimes and nickels.

Half-Dollars	Quarters	Dimes	Nickels
1	0	0	0
0	2	0	0

In this manner, we can find all possible combinations for exact change for the 50-cent toll. These combinations are shown in **Table 1.3**.

TABLE 1.3 Exact Change for 50 Cents: No Pennies

Half-Dollars	Quarters	Dimes	Nickels
1	0	0	0
0	2	0	0
0	1	2	1
0	1	1	3
0	1	0	5
0	0	5	0
0	0	4	2
0	0	3	4
0	0	2	6
0	0	1	8
0	0	0	10

Blitzer Bonus

Trick Questions

Think about the following questions carefully before answering because each contains some sort of trick or catch.

Sample: Do they have a fourth of July in England?

Answer: Of course they do. However, there is no national holiday on that date!

See if you can answer the questions that follow without developing mental whiplash. The answers appear in the answer section.

1. A farmer had 17 sheep. All but 12 died. How many sheep does the farmer have left?
2. Some months have 30 days. Some have 31. How many months have 28 days?
3. A doctor had a brother, but this brother had no brothers. What was the relationship between doctor and brother?
4. If you had only one match and entered a log cabin in which there was a candle, a fireplace, and a woodburning stove, which should you light first?

Count the coin combinations shown in **Table 1.3**. How many coin combinations must the gate accept? You must program the gate to accept 11 coin combinations.

Step 4 Look back and check the answer. Double-check **Table 1.3** to make sure that no possible combinations have been omitted and that the total in each row is 50 cents. Double-check your count of the number of combinations.

CHECK POINT 4 Suppose you are an engineer programming the automatic gate for a 30-cent toll. The gate should accept exact change only. It should not accept pennies. How many coin combinations must you program the gate to accept?

Sketches and diagrams are sometimes useful in problem solving.

EXAMPLE 5 *Solving a Problem by Using a Diagram*

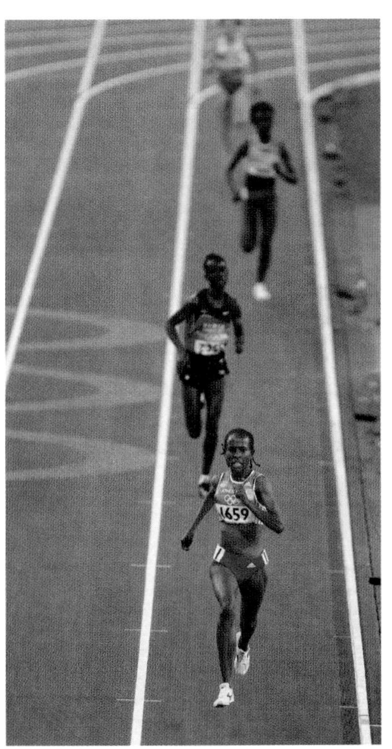

Four runners are in a one-mile race: Maria, Aretha, Thelma, and Debbie. Points are awarded only to the women finishing first or second. The first-place winner gets more points than the second-place winner. How many different arrangements of first- and second-place winners are possible?

SOLUTION

Step 1 Understand the problem. Three possibilities for first and second position are

Maria-Aretha
Maria-Thelma
Aretha-Maria.

Notice that Maria finishing first and Aretha finishing second is a different outcome than Aretha finishing first and Maria finishing second. Order makes a difference because the first-place winner gets more points than the second-place winner. We must count all possibilities for first and second position.

Step 2 Devise a plan. If Maria finishes first, then each of the other three runners could finish second:

First place	Second place	Possibilities for first and second place
	Aretha	Maria-Aretha
Maria	Thelma	Maria-Thelma
	Debbie	Maria-Debbie

Similarly, we can list each woman as the possible first-place runner. Then we will list the other three women as possible second-place runners. Next we will determine the possibilities for first and second place. This diagram will show how the runners can finish first or second.

Step 3 Carry out the plan and solve the problem. Now we complete the diagram started in step 2. The diagram is shown in **Figure 1.10**.

First place	Second place	Possibilities for first and second place
Maria	Aretha	Maria-Aretha
	Thelma	Maria-Thelma
	Debbie	Maria-Debbie
Aretha	Maria	Aretha-Maria
	Thelma	Aretha-Thelma
	Debbie	Aretha-Debbie
Thelma	Maria	Thelma-Maria
	Aretha	Thelma-Aretha
	Debbie	Thelma-Debbie
Debbie	Maria	Debbie-Maria
	Aretha	Debbie-Aretha
	Thelma	Debbie-Thelma

Because of the way **Figure 1.10** branches from first to second place, it is called a **tree diagram.** We will be using tree diagrams in Chapter 11 as a problem-solving tool in the study of uncertainty and probability.

FIGURE 1.10 Possible ways for four runners to finish first and second

Count the number of possibilities shown under the third column, "Possibilities for first and second place." Can you see that there are 12 possibilities? Therefore, 12 different arrangements of first- and second-place winners are possible.

Step 4 Look back and check the answer. Check the diagram in **Figure 1.10** to make sure that no possible first- and second-place outcomes have been left out. Double-check your count for the winning pairs of runners.

CHECK POINT 5 Your "lecture wardrobe" is rather limited—just two pairs of jeans to choose from (one blue, one black) and three T-shirts to choose from (one beige, one yellow, and one blue). How many different outfits can you form?

In Chapter 14, we will be studying diagrams, called *graphs*, that provide structures for describing relationships. In Example 6, we use such a diagram to illustrate the relationship between cities and one-way airfares between them.

EXAMPLE 6 *Using a Reasonable Option to Solve a Problem with More Than One Solution*

A sales director who lives in city A is required to fly to regional offices in cities B, C, D, and E. Other than starting and ending the trip in city A, there are no restrictions as to the order in which the other four cities are visited.

The one-way fares between each of the cities are given in **Table 1.4**. A diagram that illustrates this information is shown in **Figure 1.11**.

TABLE 1.4 One-Way Airfares

	A	*B*	*C*	*D*	*E*
A	*	$180	$114	$147	$128
B	$180	*	$116	$145	$195
C	$114	$116	*	$169	$115
D	$147	$145	$169	*	$194
E	$128	$195	$115	$194	*

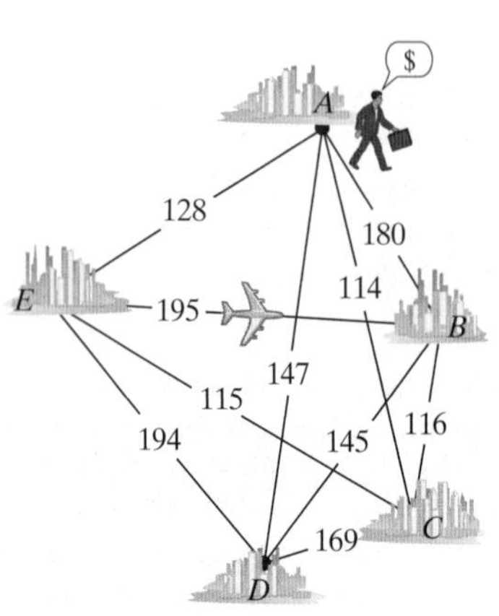

FIGURE 1.11

Give the sales director an order for visiting cities B, C, D, and E once, returning home to city A, for less than $750.

SOLUTION

Step 1 Understand the problem. There are many ways to visit cities B, C, D, and E once, and return home to A. One route is

$$A, E, D, C, B, A.$$

Fly from A to E to D to C to B and then back to A.

The cost of this trip involves the sum of five costs, shown in both **Table 1.4** and **Figure 1.11**:

$$\$128 + \$194 + \$169 + \$116 + \$180 = \$787.$$

We must find a route that costs less than \$750.

Step 2 Devise a plan. The sales director starts at city A. From there, fly to the city to which the airfare is cheapest. Then from there fly to the next city to which the airfare is cheapest, and so on. From the last of the cities, fly home to city A. Compute the cost of this trip to see if it is less than \$750. If it is not, use trial and error to find other possible routes and select an order (if there is one) whose cost is less than \$750.

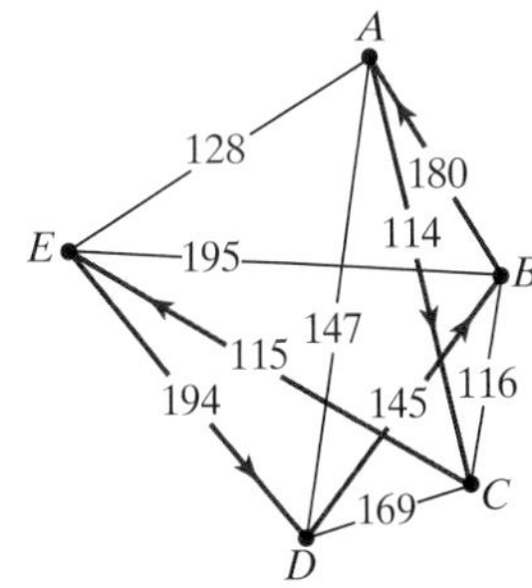

FIGURE 1.12

Step 3 Carry out the plan and solve the problem. See **Figure 1.12**. The route is indicated using red lines with arrows.

- Start at A.
- Choose the line segment with the smallest number: 114. Fly from A to C. (cost: \$114)
- From C, choose the line segment with the smallest number that does not lead to A: 115. Fly from C to E. (cost: \$115)
- From E, choose the line segment with the smallest number that does not lead to a city already visited: 194. Fly from E to D. (cost: \$194)
- From D, there is little choice but to fly to B, the only city not yet visited. (cost: \$145)
- From B, return home to A. (cost: \$180)

The route that we are considering is

$$A, C, E, D, B, A.$$

Let's see if the cost is less than \$750. The cost is

$$\$114 + \$115 + \$194 + \$145 + \$180 = \$748.$$

Because the cost is less than \$750, the sales director can follow the order A, C, E, D, B, A.

Step 4 Look back and check the answer. Use **Table 1.4** on the previous page or **Figure 1.12** to verify that the five numbers used in the sum shown above are correct. Use estimation to verify that \$748 is a reasonable cost for the trip.

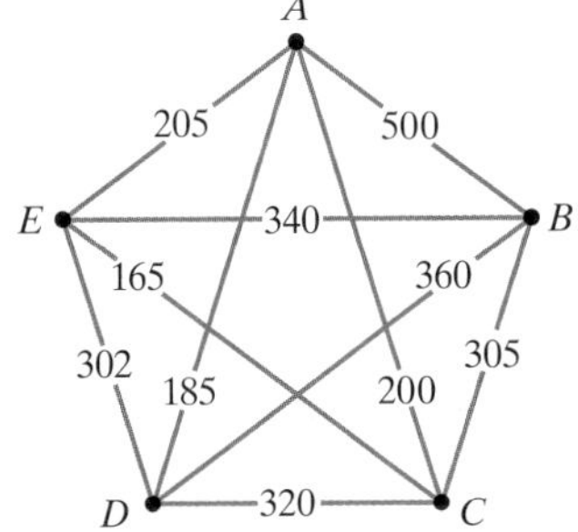

FIGURE 1.13

CHECK POINT 6 As in Example 6, a sales director who lives in city A is required to fly to regional offices in cities B, C, D, and E. The diagram in **Figure 1.13** shows the one-way airfares between any two cities. Give the sales director an order for visiting cities B, C, D, and E once, returning home to city A, for less than \$1460.

Concept and Vocabulary Check

Fill in each blank so that the resulting statement is true.

1. The first step in problem solving is to read the problem several times in order to __________ the problem.
2. The second step in problem solving is to __________ for solving the problem.
3. True or False: Polya's four steps in problem solving make it possible to obtain answers to problems even if necessary pieces of information are missing. ______
4. True or False: When making a choice between various sizes of a product, the best value is the size with the lowest price. ______

Exercise Set 1.3

Everyone can become a better, more confident problem solver. As in learning any other skill, learning problem solving requires hard work and patience. Work as many problems as possible in this Exercise Set. You may feel confused once in a while, but do not be discouraged. Thinking about a particular problem and trying different methods can eventually lead to new insights. Be sure to check over each answer carefully!

Practice and Application Exercises

In Exercises 1–4, what necessary piece of information is missing that prevents solving the problem?

1. If a student saves $35 per week, how long will it take to save enough money to buy a computer?
2. If a steak sells for $8.15, what is the cost per pound?
3. If it takes you 4 minutes to read a page in a book, how many words can you read in one minute?
4. By paying $1500 cash and the balance in equal monthly payments, how many months would it take to pay for a car costing $12,495?

In Exercises 5–8, one more piece of information is given than is necessary for solving the problem. Identify this unnecessary piece of information. Then solve the problem.

5. A salesperson receives a weekly salary of $350. In addition, $15 is paid for every item sold in excess of 200 items. How much extra is received from the sale of 212 items?
6. You have $250 to spend and you need to purchase four new tires. If each tire weighs 21 pounds and costs $42 plus $2.50 tax, how much money will you have left after buying the tires?
7. A parking garage charges $2.50 for the first hour and $0.50 for each additional hour. If a customer gave the parking attendant $20.00 for parking from 10 A.M. to 3 P.M., how much did the garage charge?
8. An architect is designing a house. The scale on the plan is 1 inch = 6 feet. If the house is to have a length of 90 feet and a width of 30 feet, how long will the line representing the house's length be on the blueprint?

Use Polya's four-step method in problem solving to solve Exercises 9–44.

9. **a.** Which is the better value: a 15.3-ounce box of cereal for $3.37 or a 24-ounce box of cereal for $4.59?

   b. The supermarket displays the unit price for the 15.3-ounce box in terms of cost per ounce, but displays the unit price for the 24-ounce box in terms of cost per pound. What are the unit prices, to the nearest cent, given by the supermarket?

   c. Based on your work in parts (a) and (b), does the better value always have the lower displayed unit price? Explain your answer.

10. **a.** Which is the better value: a 12-ounce jar of honey for $2.25 or an 18-ounce jar of honey for $3.24?

    b. The supermarket displays the unit price for the 12-ounce jar in terms of cost per ounce, but displays the unit price for the 18-ounce jar in terms of cost per quart. Assuming 32 ounces in a quart, what are the unit prices, to the nearest cent, given by the supermarket?

    c. Based on your work in parts (a) and (b), does the better value always have the lower displayed unit price? Explain your answer.

11. One person earns $48,000 per year. Another earns $3750 per month. How much more does the first person earn in a year than the second?
12. At the beginning of a year, the odometer on a car read 25,124 miles. At the end of the year, it read 37,364 miles. If the car averaged 24 miles per gallon, how many gallons of gasoline did it use during the year?

Use the following movie-rental options to solve Exercises 13–14.

Redbox

- *Rent DVDs from vending machines: $1.00 per DVD per night*

iTunes

- *New films (watching online): $3.99/24 hours*
- *Other films (watching online): $2.99/24 hours*

Netflix

- *Unlimited streaming (watching online): $7.99/month*
- *One DVD at a time by mail: $7.99/month*

13. In one month, you rent seven DVDs from a Redbox machine. You return four of the movies after one night, but keep the other three for two nights. Would you have spent more or less on Netflix's unlimited streaming option? How much more or less?

14. Suppose that you have the Netflix unlimited streaming plan. Because iTunes has two new films that are not available on Netflix, you download the movies on iTunes, each for 24 hours. What is your total movie-rental cost for the month?

Acetaminophen is in many non-prescription medications, making it easy to get more than the 4000 milligrams per day linked to liver damage and the recommended 3250-milligram daily maximum. Tylenol Extra Strength contains 500 milligrams of acetaminophen per pill. NyQuil Cold and Flu contains 325 milligrams of acetaminophen per pill. Use this information to solve Exercises 15–16.

15. **a.** What is the maximum number of Tylenol Extra Strength pills that should be taken in 24 hours?

b. If you take one Tylenol Extra Strength pill per hour for three hours, what is the maximum number of NyQuil Cold and Flu pills that should be taken for the remainder of 24 hours?

16. **a.** What is the maximum number of NyQuil Cold and Flu pills that should be taken should be taken in 24 hours?

b. If you take one Tylenol Extra Strength pill per hour for four hours, what is the maximum number of NyQuil Cold and Flu pills that should be taken for the remainder of 24 hours?

17. A television sells for $750. Instead of paying the total amount at the time of the purchase, the same television can be bought by paying $100 down and $50 a month for 14 months. How much is saved by paying the total amount at the time of the purchase?

18. In a basketball game, the Bulldogs scored 34 field goals, each counting 2 points, and 13 foul goals, each counting 1 point. The Panthers scored 38 field goals and 8 foul goals. Which team won? By how many points did it win?

19. Calculators were purchased at $65 per dozen and sold at $20 for three calculators. Find the profit on six dozen calculators.

20. Pens are bought at $0.95 per dozen and sold in groups of four for $2.25. Find the profit on 15 dozen pens.

21. Each day a small business owner sells 200 pizza slices at $1.50 per slice and 85 sandwiches at $2.50 each. If business expenses come to $60 per day, what is the owner's profit for a 10-day period?

22. A college tutoring center pays math tutors $8.15 per hour. Tutors earn an additional $2.20 per hour for each hour over 40 hours per week. A math tutor worked 42 hours one week and 45 hours the second week. How much did the tutor earn in this two-week period?

23. A car rents for $220 per week plus $0.25 per mile. Find the rental cost for a two-week trip of 500 miles for a group of three people.

24. A college graduate receives a salary of $2750 a month for her first job. During the year she plans to spend $4800 for rent, $8200 for food, $3750 for clothing, $4250 for household expenses, and $3000 for other expenses. With the money that is left, she expects to buy as many shares of stock at $375 per share as possible. How many shares will she be able to buy?

25. Charlene decided to ride her bike from her home to visit her friend Danny. Three miles away from home, her bike got a flat tire and she had to walk the remaining two miles to Danny's home. She could not repair the tire and had to walk all the way back home. How many more miles did Charlene walk than she rode?

26. A store received 200 containers of juice to be sold by April 1. Each container cost the store $0.75 and sold for $1.25. The store signed a contract with the manufacturer in which the manufacturer agreed to a $0.50 refund for every container not sold by April 1. If 150 containers were sold by April 1, how much profit did the store make?

27. A storeowner ordered 25 calculators that cost $30 each. The storeowner can sell each calculator for $35. The storeowner sold 22 calculators to customers. He had to return 3 calculators and pay a $2 charge for each returned calculator. Find the storeowner's profit.

28. New York City and Washington, D.C. are about 240 miles apart. A car leaves New York City at noon traveling directly south toward Washington, D.C. at 55 miles per hour. At the same time and along the same route, a second car leaves Washington, D.C. bound for New York City traveling directly north at 45 miles per hour. How far has each car traveled when the drivers meet for lunch at 2:24 P.M.?

29. An automobile purchased for $23,000 is worth $2700 after 7 years. Assuming that the car's value depreciated steadily from year to year, what was it worth at the end of the third year?

30. An automobile purchased for $34,800 is worth $8550 after 7 years. Assuming that the car's value depreciated steadily from year to year, what was it worth at the end of the third year?

31. A vending machine accepts nickels, dimes, and quarters. Exact change is needed to make a purchase. How many ways can a person with five nickels, three dimes, and two quarters make a 45-cent purchase from the machine?

32. How many ways can you make change for a quarter using only pennies, nickels, and dimes?

33. The members of the Student Activity Council on your campus are meeting to select two speakers for a month-long event celebrating artists and entertainers. The choices are Emma Watson, George Clooney, Leonardo DiCaprio, and Jennifer Lawrence. How many different ways can the two speakers be selected?

34. The members of the Student Activity Council on your campus are meeting to select two speakers for a month-long event exploring why some people are most likely to succeed. The choices are Bill Gates, Oprah Winfrey, Mark Zuckerberg, Hillary Clinton, and Steph Curry. How many different ways can the two speakers be selected?

35. If you spend \$4.79, in how many ways can you receive change from a five-dollar bill?

36. If you spend \$9.74, in how many ways can you receive change from a ten-dollar bill?

37. You throw three darts at the board shown. Each dart hits the board and scores a 1, 5, or 10. How many different total scores can you make?

38. Suppose that you throw four darts at the board shown. With these four darts, there are 16 ways to hit four different numbers whose sum is 100. Describe one way you can hit four different numbers on the board that total 100.

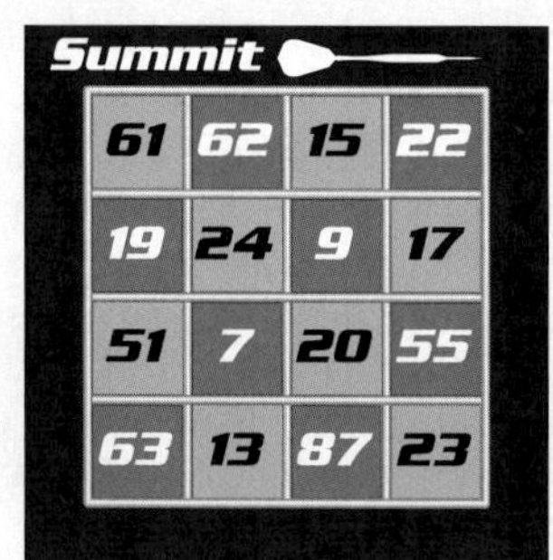

39. Five housemates (A, B, C, D, and E) agreed to share the expenses of a party equally. If A spent \$42, B spent \$10, C spent \$26, D spent \$32, and E spent \$30, who owes money after the party and how much do they owe? To whom is money owed, and how much should they receive? In order to resolve these discrepancies, who should pay how much to whom?

40. Six houses are spaced equally around a circular road. If it takes 10 minutes to walk from the first house to the third house, how long would it take to walk all the way around the road?

41. If a test has four true/false questions, in how many ways can there be three answers that are false and one answer that is true?

42. There are five people in a room. Each person shakes the hand of every other person exactly once. How many handshakes are exchanged?

43. Five runners, Andy, Beth, Caleb, Darnell, and Ella, are in a one-mile race. Andy finished the race 7 seconds before Caleb. Caleb finished the race 2 seconds before Beth. Beth finished the race 6 seconds after Darnell. Ella finished the race 8 seconds after Darnell. In which order did the runners finish the race?

44. Eight teams are competing in a volleyball tournament. Any team that loses a game is eliminated from the tournament. How many games must be played to determine the tournament winner?

In Exercises 45–46, you have three errands to run around town, although in no particular order. You plan to start and end at home. You must go to the bank, the post office, and the dry cleaners. Distances, in miles, between any two of these locations are given in the diagram.

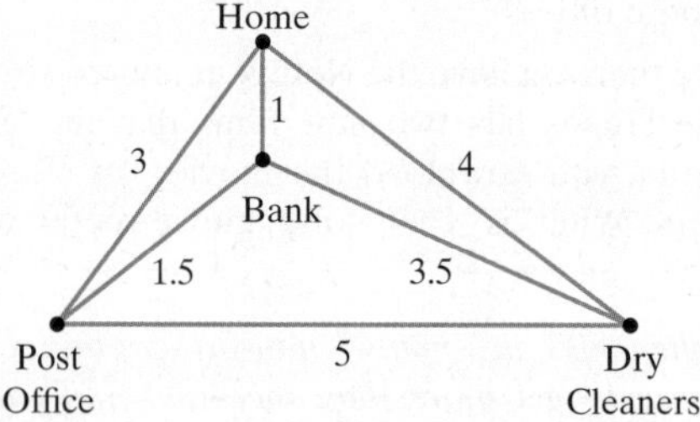

45. Determine a route whose distance is less than 12 miles for running the errands and returning home.

46. Determine a route whose distance exceeds 12 miles for running the errands and returning home.

47. The map shows five western states. Trace a route on the map that crosses each common state border exactly once.

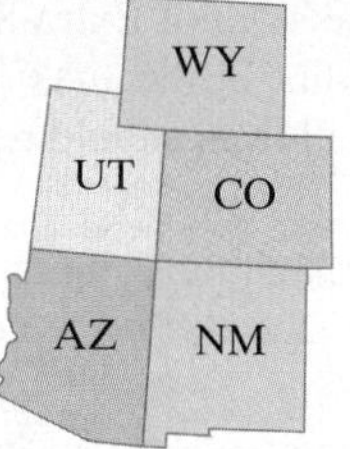

48. The layout of a city with land masses and bridges is shown. Trace a route that shows people how to walk through the city so as to cross each bridge exactly once.

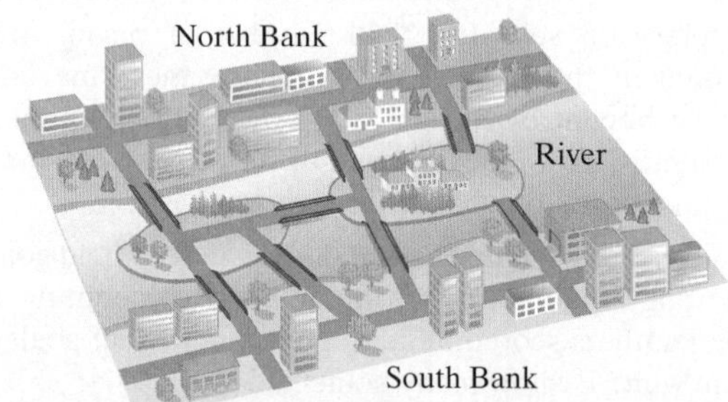

49. Jose, Bob, and Tony are college students living in adjacent dorm rooms. Bob lives in the middle dorm room. Their majors are business, psychology, and biology, although not necessarily in that order. The business major frequently uses the new computer in Bob's dorm room when Bob is in class. The psychology major and Jose both have 8 A.M. classes, and the psychology major knocks on Jose's wall to make sure he is awake. Determine Bob's major.

50. The figure represents a map of 13 countries. If countries that share a common border cannot be the same color, what is the minimum number of colors needed to color the map?

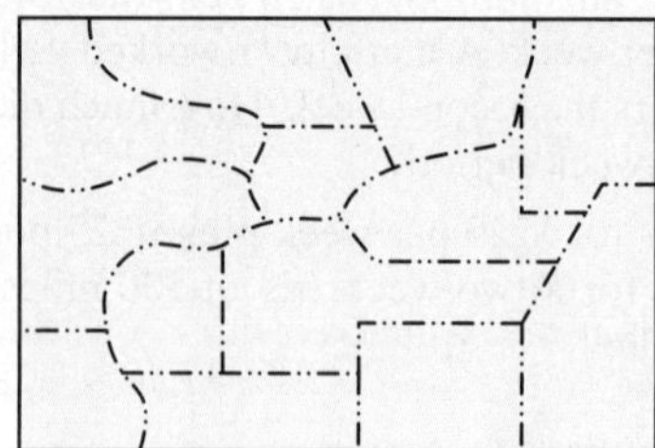

The sudoku (pronounced: sue-DOE-koo) craze, a number puzzle popular in Japan, hit the United States in 2005. A sudoku ("single number") puzzle consists of a 9-by-9 grid of 81 boxes subdivided into nine 3-by-3 squares. Some of the square boxes contain numbers. Here is an example:

2						6	7	
4	5							1
	9		7					
			3					
		4				2	8	
	1			2	4			
	8	3	9		7			
6	2			5		4	1	
							9	

The objective is to fill in the remaining squares so that every row, every column, and every 3-by-3 square contains each of the digits from 1 through 9 exactly once. (You can work this puzzle in Exercise 70, perhaps consulting one of the dozens of sudoku books in which the numerals 1 through 9 have created a cottage industry for publishers. There's even a Sudoku for Dummies.)

Trying to slot numbers into small checkerboard grids is not unique to sudoku. In Exercises 51–54, we explore some of the intricate patterns in other arrays of numbers, including magic squares. A ***magic square*** *is a square array of numbers arranged so that the numbers in all rows, all columns, and the two diagonals have the same sum. Here is an example of a magic square in which the sum of the numbers in each row, each column, and each diagonal is 15:*

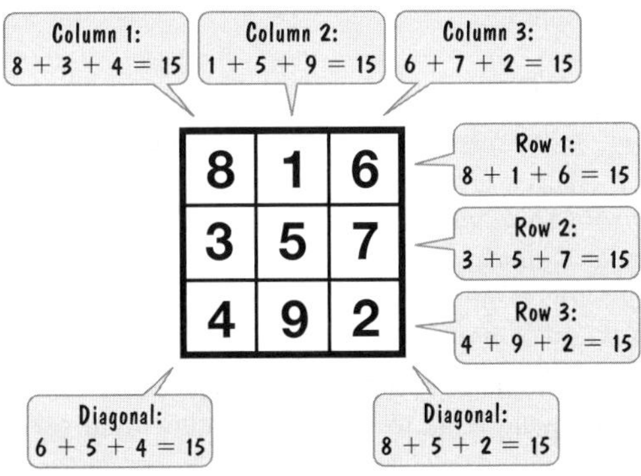

Exercises 51–52 are based on magic squares. (Be sure you have read the preceding discussion.)

51. a. Use the properties of a magic square to fill in the missing numbers.

5		18
	15	
		25

b. Show that the number of letters in the word for each number in the square in part (a) generates another magic square.

52. a. Use the properties of a magic square to fill in the missing numbers.

96		37	45
	43		
		25	57
23		82	78

b. Show that if you reverse the digits for each number in the square in part (a), another magic square is generated. (*Source* for the *alphamagic square* in Exercise 51 and the *mirrormagic square* in Exercise 52: Clifford A. Pickover, *A Passion for Mathematics*, John Wiley & Sons, Inc., 2005)

53. As in sudoku, fill in the missing numbers in the 3-by-3 square so that it contains each of the digits from 1 through 9 exactly once. Furthermore, in this *antimagic square*, the rows, the columns, and the two diagonals must have *different sums*.

9		7
	1	
3		5

54. The missing numbers in the 4-by-4 array are one-digit numbers. The sums for each row, each column, and one diagonal are listed in the voice balloons outside the array. Find the missing numbers.

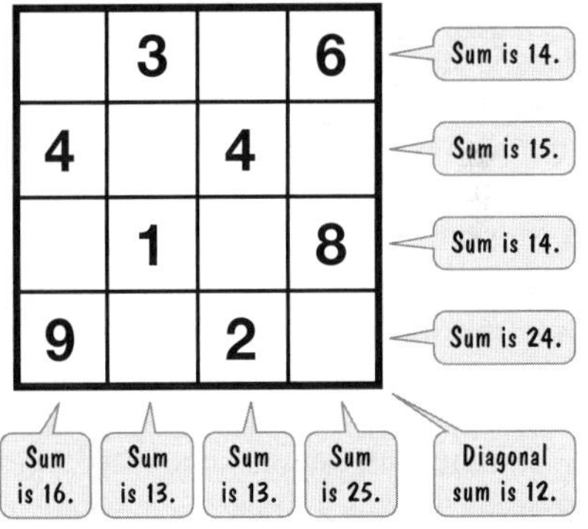

55. Some numbers in the printing of a division problem have become illegible. They are designated below by *. Fill in the blanks.

```
     1**
**)4***
   28
   *56
   ***
    ***
    ***
      0
```

Explaining the Concepts

In Exercises 56–58, explain the plan needed to solve the problem.

56. If you know how much was paid for several pounds of steak, find the cost of one pound.

57. If you know a person's age, find the year in which that person was born.

58. If you know how much you earn each hour, find your yearly income.

59. Write your own problem that can be solved using the four-step procedure. Then use the four steps to solve the problem.

Critical Thinking Exercises

Make Sense? *In Exercises 60–63, determine whether each statement makes sense or does not make sense, and explain your reasoning.*

60. Polya's four steps in problem solving make it possible for me to solve any mathematical problem easily and quickly.

61. I used Polya's four steps in problem solving to deal with a personal problem in need of a creative solution.

62. I find it helpful to begin the problem-solving process by restating the problem in my own words.

63. When I get bogged down with a problem, there's no limit to the amount of time I should spend trying to solve it.

64. Gym lockers are to be numbered from 1 through 99 using metal numbers to be nailed onto each locker. How many 7s are needed?

65. You are on vacation in an isolated town. Everyone in the town was born there and has never left. You develop a toothache and check out the two dentists in town. One dentist has gorgeous teeth and one has teeth that show the effects of poor dental work. Which dentist should you choose and why?

66. India Jones is standing on a large rock in the middle of a square pool filled with hungry, man-eating piranhas. The edge of the pool is 20 feet away from the rock. India's mom wants to rescue her son, but she is standing on the edge of the pool with only two planks, each $19\frac{1}{2}$ feet long. How can India be rescued using the two planks?

67. One person tells the truth on Monday, Tuesday, Wednesday, and Thursday, but lies on all other days. A second person lies on Tuesday, Wednesday, and Thursday, but tells the truth on all other days. If both people state "I lied yesterday," then what day of the week is it today?

68. (This logic problem dates back to the eighth century.) A farmer needs to take his goat, wolf, and cabbage across a stream. His boat can hold him and one other passenger (the goat, wolf, or cabbage). If he takes the wolf with him, the goat will eat the cabbage. If he takes the cabbage, the wolf will eat the goat. Only when the farmer is present are the cabbage and goat safe from their respective predators. How does the farmer get everything across the stream?

69. As in sudoku, fill in the missing numbers along the sides of the triangle so that it contains each of the digits from 1 through 9 exactly once. Furthermore, each side of the triangle should contain four digits whose sum is 17.

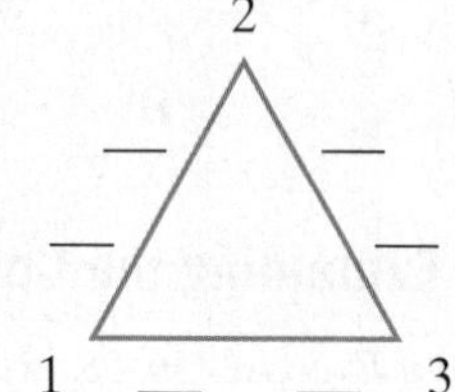

70. Solve the sudoku puzzle in the top of the left column on page 41.

71. A version of this problem, called the *missing dollar problem*, first appeared in 1933. Three people eat at a restaurant and receive a total bill for \$30. They divide the amount equally and pay \$10 each. The waiter gives the bill and the \$30 to the manager, who realizes there is an error: The correct charge should be only \$25. The manager gives the waiter five \$1 bills to return to the customers, with the restaurant's apologies. However, the waiter is dishonest, keeping \$2 and giving back only \$3 to the customers. In conclusion, each of the three customers has paid \$9 and the waiter has stolen \$2, giving a total of \$29. However, the original bill was \$30. Where has the missing dollar gone?

72. A firefighter spraying water on a fire stood on the middle rung of a ladder. When the smoke became less thick, the firefighter moved up 4 rungs. However it got too hot, so the firefighter backed down 6 rungs. Later, the firefighter went up 7 rungs and stayed until the fire was out. Then, the firefighter climbed the remaining 4 rungs and entered the building. How many rungs does the ladder have?

73. The Republic of Margaritaville is composed of four states: A, B, C, and D. According to the country's constitution, the congress will have 30 seats, divided among the four states according to their respective populations. The table shows each state's population.

POPULATION OF MARGARITAVILLE BY STATE

State	A	B	C	D	Total
Population (in thousands)	275	383	465	767	1890

Allocate the 30 congressional seats among the four states in a fair manner.

Group Exercises

Exercises 74–78 describe problems that have many plans for finding an answer. Group members should describe how the four steps in problem solving can be applied to find a solution. It is not necessary to actually solve each problem. Your professor will let the group know if the four steps should be described verbally by a group spokesperson or in essay form.

74. How much will it cost to install bicycle racks on campus to encourage students to use bikes, rather than cars, to get to campus?

75. How many new counselors are needed on campus to prevent students from waiting in long lines for academic advising?

76. By how much would taxes in your state have to be increased to cut tuition at community colleges and state universities in half?

77. Is your local electric company overcharging its customers?

78. Should solar heating be required for all new construction in your community?

79. Group members should describe a problem in need of a solution. Then, as in Exercises 74–78, describe how the four steps in problem solving can be applied to find a solution.

Chapter Summary, Review, and Test

SUMMARY – DEFINITIONS AND CONCEPTS	EXAMPLES
1.1 Inductive and Deductive Reasoning	
a. Inductive reasoning is the process of arriving at a general conclusion based on observations of specific examples. The conclusion is called a conjecture or a hypothesis. A case for which a conjecture is false is called a counterexample.	Ex. 1, p. 3; Ex. 2, p. 4; Ex. 3, p. 5; Ex. 4, p. 7
b. Deductive reasoning is the process of proving a specific conclusion from one or more general statements. The statement that is proved is called a theorem.	Ex. 5, p. 9
1.2 Estimation, Graphs, and Mathematical Models	
a. The procedure for rounding whole numbers is given in the box on page 15. The symbol ≈ means *is approximately equal to*.	Ex. 1, p. 15
b. Decimal parts of numbers are rounded in nearly the same way as whole numbers. However, digits to the right of the rounding place are dropped.	Ex. 2, p. 16
c. Estimation is the process of arriving at an approximate answer to a question. Computations can be estimated by using rounding that results in simplified arithmetic.	Ex. 3, p. 17; Ex. 4, p. 17
d. Estimation is useful when interpreting information given by circle, bar, or line graphs.	Ex. 5, p. 19; Ex. 6, p. 20; Ex. 7, p. 21
e. The process of finding formulas to describe real-world phenomena is called mathematical modeling. Such formulas, together with the meaning assigned to the variables, are called mathematical models.	Ex. 8, p. 23
1.3 Problem Solving	
Polya's Four Steps in Problem Solving **1.** Understand the problem. **2.** Devise a plan. **3.** Carry out the plan and solve the problem. **4.** Look back and check the answer.	Ex. 1, p. 31; Ex. 2, p. 31; Ex. 3, p. 33; Ex. 4, p. 34; Ex. 5, p. 35; Ex. 6, p. 36

Review Exercises

1.1

1. Which reasoning process is shown in the following example? Explain your answer.

All books by Stephen King have made the best-seller list. *Carrie* is a novel by Stephen King. Therefore, *Carrie* was on the best-seller list.

2. Which reasoning process is shown in the following example? Explain your answer.

All books by Stephen King have made the best-seller list. Therefore, it is highly probable that the novel King is currently working on will make the best-seller list.

In Exercises 3–10, identify a pattern in each list of numbers. Then use this pattern to find the next number.

3. 4, 9, 14, 19, ____

4. 7, 14, 28, 56, ____

5. 1, 3, 6, 10, 15, ____

6. $\frac{3}{4}, \frac{3}{5}, \frac{1}{2}, \frac{3}{7},$ ____

7. 40, −20, 10, −5, ____

8. 40, −20, −80, −140, ____

9. 2, 2, 4, 6, 10, 16, 26, ____

10. 2, 6, 12, 36, 72, 216, ____

11. Identify a pattern in the following sequence of figures. Then use the pattern to find the next figure in the sequence.

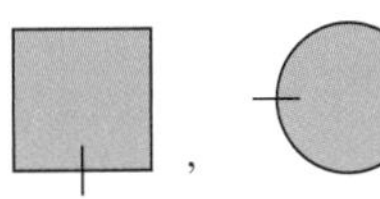 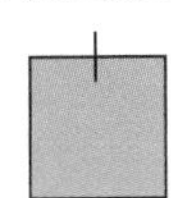 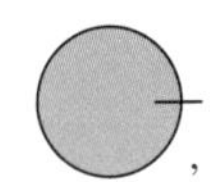 ____

In Exercises 12–13, use inductive reasoning to predict the next line in each sequence of computations. Then perform the arithmetic to determine whether your conjecture is correct.

12.
$$2 = 4 - 2$$
$$2 + 4 = 8 - 2$$
$$2 + 4 + 8 = 16 - 2$$
$$2 + 4 + 8 + 16 = 32 - 2$$

13.
$$111 \div 3 = 37$$
$$222 \div 6 = 37$$
$$333 \div 9 = 37$$

14. Consider the following procedure:

Select a number. Double the number. Add 4 to the product. Divide the sum by 2. Subtract 2 from the quotient.

a. Repeat the procedure for four numbers of your choice. Write a conjecture that relates the result of the process to the original number selected.

b. Represent the original number by the variable n and use deductive reasoning to prove the conjecture in part (a).

1.2

15. The number 923,187,456 is called a *pandigital square* because it uses all the digits from 1 to 9 once each and is the square of a number:

$$30{,}384^2 = 30{,}384 \times 30{,}384 = 923{,}187{,}456.$$

(*Source*: David Wells, *The Penguin Dictionary of Curious and Interesting Numbers*)

Round the pandigital square 923,187,456 to the nearest

a. hundred.

b. thousand.

c. hundred-thousand.

d. million.

e. hundred-million.

16. A magnified view of the boundary of this black "buglike" shape, called the Mandelbrot set, was illustrated in the Section 1.1 opener on page 2.

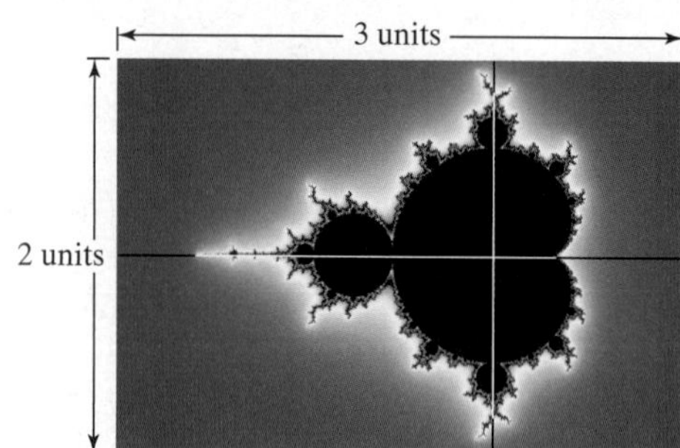

The area of the blue rectangular region is the product of its length, 3 units, and its width, 2 units, or 6 square units. It is conjectured that the area of the black buglike region representing the Mandelbrot set is

$$\sqrt{6\pi - 1} - e \approx 1.5065916514855 \text{ square units.}$$

(*Source:* Robert P. Munafo, *Mandelbrot Set Glossary and Encyclopedia*)

Round the area of the Mandelbrot set to

a. the nearest tenth.

b. the nearest hundredth.

c. the nearest thousandth.

d. seven decimal places.

In Exercises 17–20, obtain an estimate for each computation by rounding the numbers so that the resulting arithmetic can easily be performed by hand or in your head. Then use a calculator to perform the computation. How reasonable is your estimate when compared to the actual answer?

17. $1.57 + 4.36 + 9.78$

18. 8.83×49

19. $19.894 \div 4.179$

20. 62.3% of 3847.6

In Exercises 21–24, determine each estimate without using a calculator. Then use a calculator to perform the computation necessary to obtain an exact answer. How reasonable is your estimate when compared to the actual answer?

21. Estimate the total cost of six grocery items if their prices are \$8.47, \$0.89, \$2.79, \$0.14, \$1.19, and \$4.76.

22. Estimate the salary of a worker who works for 78 hours at \$9.95 per hour.

23. At a yard sale, a person bought 21 books at \$0.85 each, two chairs for \$11.95 each, and a ceramic plate for \$14.65. Estimate the total amount spent.

24. The circle graph shows how the 20,207,375 students enrolled in U.S. colleges and universities in 2015 funded college costs. Estimate the number of students who covered these costs through grants and scholarships.

How Students Cover College Costs

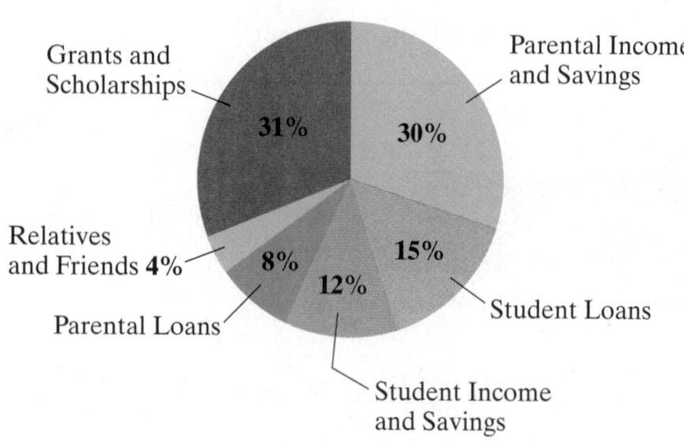

Source: The College Board

25. A small private school employs 10 teachers with salaries ranging from \$817 to \$992 per week. Which of the following is the best estimate of the monthly payroll for the teachers?

a. \$30,000 **b.** \$36,000

c. \$42,000 **d.** \$50,000

26. Select the best estimate for the number of seconds in a day.

a. 1500 **b.** 15,000

c. 86,000 **d.** 100,000

27. Imagine the entire global population as a village of precisely 200 people. The bar graph shows some numeric observations based on this scenario.

Earth's Population as a Village of 200 People

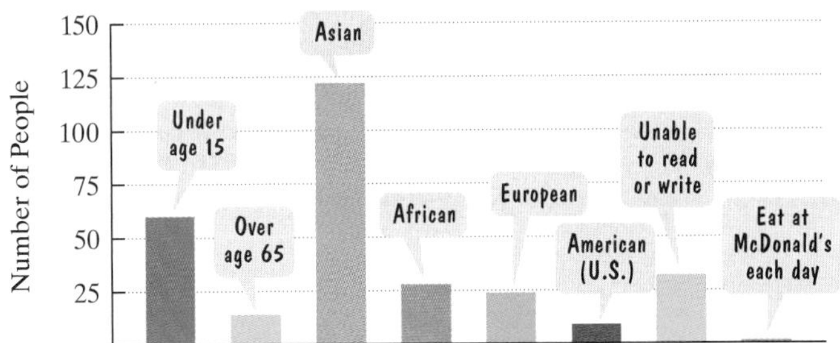

Source: Gary Rimmer, *Number Freaking*, The Disinformation Company Ltd.

a. Which group in the village has a population that exceeds 100? Estimate this group's population.

b. World population is approximately 33 million times the population of the village of 200 people. Use this observation to estimate the number of people in the world, in millions, unable to read or write.

28. The bar graph shows the percentage of people 25 years of age and older who were college graduates in the United States for eight selected years.

Percentage of College Graduates, Among People Ages 25 and Older, in the United States

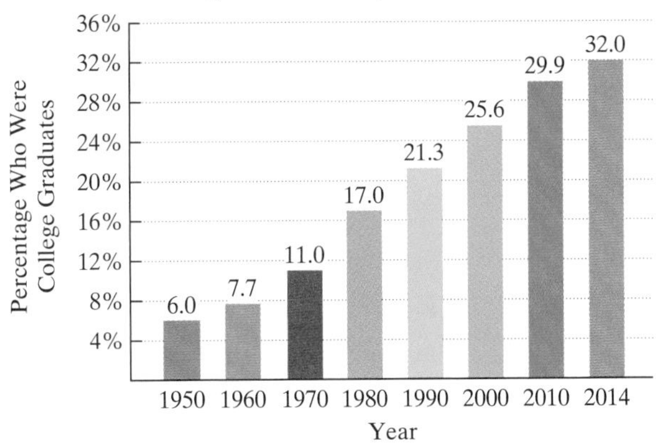

Source: U.S. Census Bureau

a. Estimate the average yearly increase in the percentage of college graduates. Round to the nearest tenth of a percent.

b. If the trend shown by the graph continues, estimate the percentage of people 25 years of age and older who will be college graduates in 2020.

29. During a diagnostic evaluation, a 33-year-old woman experienced a panic attack a few minutes after she had been asked to relax her whole body. The graph at the top of the next column shows the rapid increase in heart rate during the panic attack.

Heart Rate before and during a Panic Attack

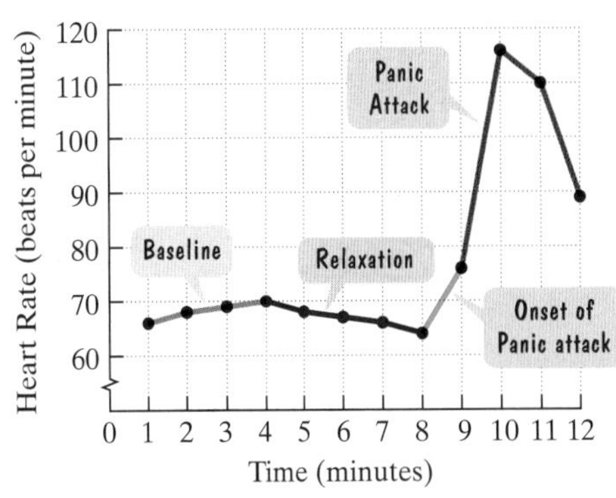

Source: Davis and Palladino, *Psychology*, Fifth Edition, Prentice Hall, 2007.

a. Use the graph to estimate the woman's maximum heart rate during the first 12 minutes of the diagnostic evaluation. After how many minutes did this occur?

b. Use the graph to estimate the woman's minimum heart rate during the first 12 minutes of the diagnostic evaluation. After how many minutes did this occur?

c. During which time period did the woman's heart rate increase at the greatest rate?

d. After how many minutes was the woman's heart rate approximately 75 beats per minute?

30. The bar graph shows the population of the United States, in millions, for five selected years.

Population of the United States

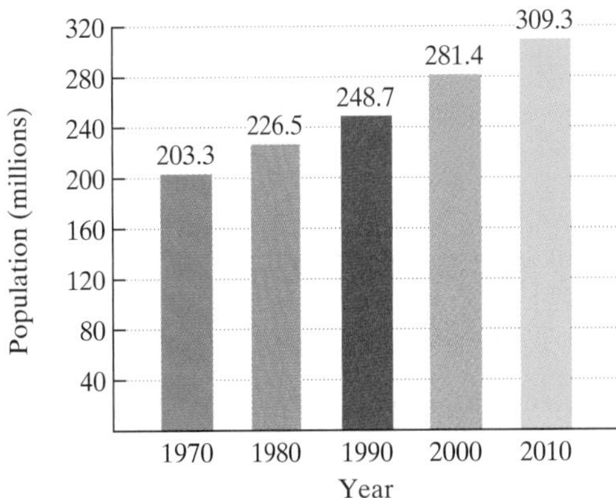

Source: U.S. Census Bureau

a. Estimate the yearly increase in the U.S. population. Express the answer in millions and do not round.

b. Write a mathematical model that estimates the U.S. population, p, in millions, x years after 1970.

c. Use the mathematical model from part (b) to project the U.S. population, in millions, in 2020.

1.3

31. What necessary piece of information is missing that prevents solving the following problem?

If 3 milligrams of a medicine is given for every 20 pounds of body weight, how many milligrams should be given to a 6-year-old child?

32. In the following problem, there is one more piece of information given than is necessary for solving the problem. Identify this unnecessary piece of information. Then solve the problem.

A taxicab charges \$3.00 for the first mile and \$0.50 for each additional half-mile. After a 6-mile trip, a customer handed the taxi driver a \$20 bill. Find the cost of the trip.

Use the four-step method in problem solving to solve Exercises 33–39.

33. A company offers the following text message monthly price plans.

Pay-per-Text

\$0.20 per regular text

\$0.30 per photo or video text

Packages (include photo and video texts)

200 messages: \$5.00 per month

1500 messages: \$15.00 per month

Unlimited messages: \$20.00 per month

Suppose that you send 40 regular texts and 35 photo texts in a month. With which plan (pay-per-text or a package) will you pay less money? How much will you save over the other plan?

34. If there are seven frankfurters in one pound, how many pounds would you buy for a picnic to supply 28 people with two frankfurters each?

35. A car rents for \$175 per week plus \$0.30 per mile. Find the rental cost for a three-week trip of 1200 miles.

36. The costs for two different kinds of heating systems for a two-bedroom home are given in the following table.

System	Cost to install	Operating cost per year
Solar	\$29,700	\$200
Electric	\$5500	\$1800

After 12 years, which system will have the greater total costs (installation cost plus operating cost)? How much greater will the total costs be?

37. Miami is on Eastern Standard Time and San Francisco is on Pacific Standard Time, three hours earlier than Eastern Standard Time. A flight leaves Miami at 10 A.M. Eastern Standard Time, stops for 45 minutes in Houston, Texas, and arrives in San Francisco at 1:30 P.M. Pacific time. What is the actual flying time from Miami to San Francisco?

38. An automobile purchased for \$37,000 is worth \$2600 after eight years. Assuming that the value decreased steadily each year, what was the car worth at the end of the fifth year?

39. Suppose you are an engineer programming the automatic gate for a 35-cent toll. The gate is programmed for exact change only and will not accept pennies. How many coin combinations must you program the gate to accept?

Chapter 1 Test

1. Which reasoning process is shown in the following example?

The course policy states that if you turn in at least 80% of the homework, your lowest exam grade will be dropped. I turned in 90% of the homework, so my lowest grade will be dropped.

2. Which reasoning process is shown in the following example?

We examine the fingerprints of 1000 people. No two individuals in this group of people have identical fingerprints. We conclude that for all people, no two people have identical fingerprints.

In Exercises 3–6, find the next number, computation, or figure, as appropriate.

3. 0, 5, 10, 15, ______

4. $\frac{1}{6}, \frac{1}{12}, \frac{1}{24}, \frac{1}{48}$, ______

5.
$3367 \times 3 = 10{,}101$
$3367 \times 6 = 20{,}202$
$3367 \times 9 = 30{,}303$
$3367 \times 12 = 40{,}404$
______

6. , 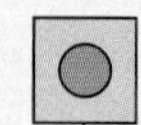, 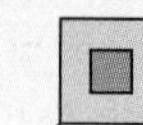, 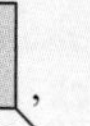, 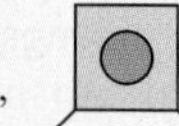, ______

7. Consider the following procedure:

Select a number. Multiply the number by 4. Add 8 to the product. Divide the sum by 2. Subtract 4 from the quotient.

a. Repeat this procedure for three numbers of your choice. Write a conjecture that relates the result of the process to the original number selected.

b. Represent the original number by the variable n and use deductive reasoning to prove the conjecture in part (a).

8. Round 3,279,425 to the nearest hundred-thousand.

9. Round 706.3849 to the nearest hundredth.

In Exercises 10–13, determine each estimate without using a calculator. Different rounding results in different estimates, so there is not one single correct answer to each exercise. Use rounding to make the resulting calculations simple.

10. For a spring break vacation, a student needs to spend \$47.00 for gas, \$311.00 for food, and \$405.00 for a hotel room. If the student takes \$681.79 from savings, estimate how much more money is needed for the vacation.

11. The cost for opening a restaurant is \$485,000. If 19 people decide to share equally in the business, estimate the amount each must contribute.

12. Find an estimate of 0.48992×121.976.

13. The graph shows the composition of a typical American community's trash.

Types of Trash in an American Community by Percentage of Total Weight

Source: U.S. Environmental Protection Agency

Across the United States, people generate approximately 512 billion pounds of trash per year. Estimate the number of pounds of trash in the form of plastic.

14. If the odometer of a car reads 71,911.5 miles and it averaged 28.9 miles per gallon, select the best estimate for the number of gallons of gasoline used.

a. 2400 **b.** 3200 **c.** 4000 **d.** 4800 **e.** 5600

15. The stated intent of the 1994 "don't ask, don't tell" policy was to reduce the number of discharges of gay men and lesbians from the military. Nearly 14,000 active-duty gay servicemembers were dismissed under the policy, which officially ended in 2011, after 18 years. The line graph at the top of the next column shows the number of discharges under "don't ask, don't tell" from 1994 through 2010.

Number of Active-Duty Gay Servicemembers Discharged from the Military for Homosexuality

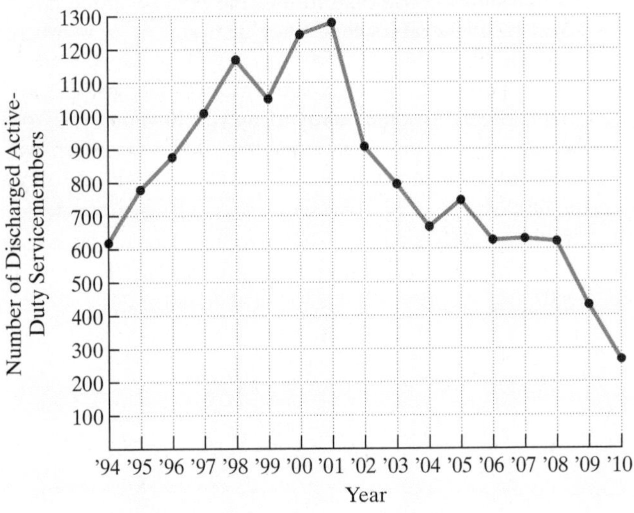

Source: General Accountability Office

a. For the period shown, in which year did the number of discharges reach a maximum? Find a reasonable estimate of the number of discharges for that year.

b. For the period shown, in which year did the number of discharges reach a minimum? Find a reasonable estimate of the number of discharges for that year.

c. In which one-year period did the number of discharges decrease at the greatest rate?

d. In which year were approximately 1000 gay servicemembers discharged under the "don't ask, don't tell" policy?

16. Grade Inflation. The bar graph shows the percentage of U.S. college freshmen with an average grade of A in high school.

Percentage of U.S. College Freshmen with an Average Grade of A (A− to A+) in High School

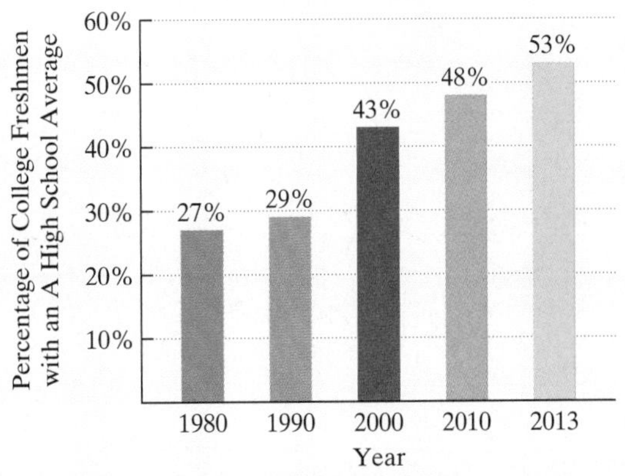

Source: Higher Education Research Institute

a. Estimate the average yearly increase in the percentage of high school grades of A. Round to the nearest tenth of a percent.

b. Write a mathematical model that estimates the percentage of high school grades of A, p, x years after 1980.

c. If the trend shown by the graph continues, use your mathematical model from part (b) to project the percentage of high school grades of A in 2020.

17. The cost of renting a boat from Estes Rental is $9 per 15 minutes. The cost from Ship and Shore Rental is $20 per half-hour. If you plan to rent the boat for three hours, which business offers the better deal and by how much?

18. A bus operates between Miami International Airport and Miami Beach, 10 miles away. It makes 20 round trips per day carrying 32 passengers per trip. If the fare each way is $11.00, how much money is taken in from one day's operation?

19. By paying $50 cash up front and the balance at $35 a week, how long will it take to pay for a computer costing $960?

20. In 2000, the population of Greece was 10,600,000, with projections of a population decrease of 28,000 people per year. In the same year, the population of Belgium was 10,200,000, with projections of a population decrease of 12,000 people per year. (*Source:* United Nations)

According to these projections, which country will have the greater population in 2035 and by how many more people?

Set Theory

2

OUR BODIES ARE FRAGILE AND COMPLEX, VULNERABLE TO DISEASE AND EASILY DAMAGED. THE SEQUENCING OF THE HUMAN GENOME IN 2003—ALL 140,000 GENES— should lead to rapid advances in treating heart disease, cancer, depression, Alzheimer's, and AIDS. Neural stem cell research could make it possible to repair brain damage and even re-create whole parts of the brain. There appears to be no limit to the parts of our bodies that can be replaced. By contrast, at the start of the twentieth century, we lacked even a basic understanding of the different types of human blood. The discovery of blood types, organized into collections called *sets* and illustrated by a special set diagram, rescued surgery patients from random, often lethal, transfusions. In this sense, the set diagram for blood types that you will encounter in this chapter reinforces our optimism that life does improve and that we are better off today than we were one hundred years ago.

Here's where you'll find this application:

Organizing and visually representing sets of human blood types is presented in the Blitzer Bonus on page 94. The vital role that this representation plays in blood transfusions is developed in Exercises 113–117 of Exercise Set 2.4.

2.1

WHAT AM I SUPPOSED TO LEARN?

After studying this section, you should be able to:

1 Use three methods to represent sets.
2 Define and recognize the empty set.
3 Use the symbols $\in$ and $\notin$.
4 Apply set notation to sets of natural numbers.
5 Determine a set's cardinal number.
6 Recognize equivalent sets.
7 Distinguish between finite and infinite sets.
8 Recognize equal sets.

Basic Set Concepts

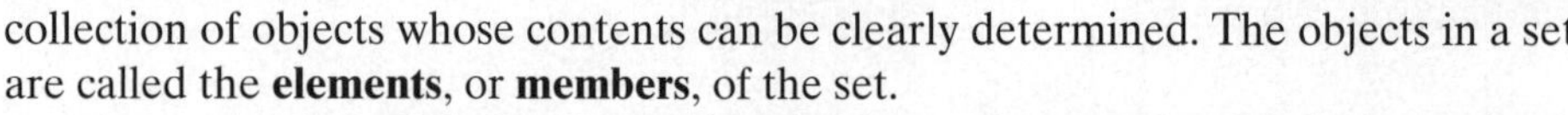

WE TEND TO PLACE THINGS IN categories, which allows us to order and structure the world. For example, to which populations do you belong? Do you categorize yourself as a college student? What about your gender? What about your academic major or your ethnic background? Our minds cannot find order and meaning without creating collections. Mathematicians call such collections *sets*. A **set** is a collection of objects whose contents can be clearly determined. The objects in a set are called the **elements**, or **members**, of the set.

A set must be **well defined**, meaning that its contents can be clearly determined. Using this criterion, the collection of actors who have won Academy Awards is a set. We can always determine whether or not a particular actor is an element of this collection. By contrast, consider the collection of great actors. Whether or not a person belongs to this collection is a matter of how we interpret the word *great*. In this text, we will only consider collections that form well-defined sets.

1 *Use three methods to represent sets.*

Methods for Representing Sets

An example of a set is the set of the days of the week, whose elements are Monday, Tuesday, Wednesday, Thursday, Friday, Saturday, and Sunday.

Capital letters are generally used to name sets. Let's use W to represent the set of the days of the week.

Three methods are commonly used to designate a set. One method is a **word description**. We can describe set W as the set of the days of the week. A second method is the **roster method**. This involves listing the elements of a set inside a pair of braces, { }. The braces at the beginning and end indicate that we are representing a set. The roster form uses commas to separate the elements of the set. Thus, we can designate the set W by listing its elements:

$W = \{$Monday, Tuesday, Wednesday, Thursday, Friday, Saturday, Sunday$\}$.

Grouping symbols such as parentheses, ( ), and square brackets, [ ], are not used to represent sets. Only commas are used to separate the elements of a set. Separators such as colons or semicolons are not used. Finally, the order in which the elements are listed in a set is not important. Thus, another way of expressing the set of the days of the week is

$W = \{$Saturday, Sunday, Monday, Tuesday, Wednesday, Thursday, Friday$\}$.

EXAMPLE 1 Representing a Set Using a Description

Write a word description of the set

$P = \{$Washington, Adams, Jefferson, Madison, Monroe$\}$.

SOLUTION

Set P is the set of the first five presidents of the United States.

CHECK POINT 1 Write a word description of the set

$L = \{\text{a, b, c, d, e, f}\}$.

EXAMPLE 2 Representing a Set Using the Roster Method

Set C is the set of U.S. coins with a value of less than a dollar. Express this set using the roster method.

SOLUTION

$$C = \{\text{penny, nickel, dime, quarter, half-dollar}\}$$

 **CHECK POINT 2** Set M is the set of months beginning with the letter A. Express this set using the roster method.

> **GREAT QUESTION!**
>
> **Do I have to use x to represent the variable in set-builder notation?**
>
> No. Any letter can be used to represent the variable. Thus, $\{x \mid x \text{ is a day of the week}\}$, $\{y \mid y \text{ is a day of the week}\}$, and $\{z \mid z \text{ is a day of the week}\}$ all represent the same set.

The third method for representing a set is with **set-builder notation**. Using this method, the set of the days of the week can be expressed as

$$W = \{x \mid x \text{ is a day of the week}\}.$$

Set W | is | the set of | all elements x | such that

We read this notation as "Set W is the set of all elements x such that x is a day of the week." Before the vertical line is the variable x, which represents an element in general. After the vertical line is the condition x must meet in order to be an element of the set.

Table 2.1 contains two examples of sets, each represented with a word description, the roster method, and set-builder notation.

The Beatles climbed to the top of the British music charts in 1963, conquering the United States a year later.

TABLE 2.1 Sets Using Three Designations

Word Description	Roster Method	Set-Builder Notation
B is the set of members of the Beatles in 1963.	B = {George Harrison, John Lennon, Paul McCartney, Ringo Starr}	$B = \{x \mid x$ was a member of the Beatles in 1963$\}$
S is the set of states whose names begin with the letter A.	S = {Alabama, Alaska, Arizona, Arkansas}	$S = \{x \mid x$ is a U.S. state whose name begins with the letter A$\}$

EXAMPLE 3 Converting from Set-Builder to Roster Notation

Express set

$$A = \{x \mid x \text{ is a month that begins with the letter M}\}$$

using the roster method.

SOLUTION

Set A is the set of all elements x such that x is a month beginning with the letter M. There are two such months, namely March and May. Thus,

$$A = \{\text{March, May}\}.$$

 **CHECK POINT 3** Express the set

$$O = \{x \mid x \text{ is a positive odd number less than } 10\}$$

using the roster method.

The representation of some sets by the roster method can be rather long, or even impossible, if we attempt to list every element. For example, consider the set of all lowercase letters of the English alphabet. If L is chosen as a name for this set, we can use set-builder notation to represent L as follows:

$$L = \{x \mid x \text{ is a lowercase letter of the English alphabet}\}.$$

A complete listing using the roster method is rather tedious:

$$L = \{\text{a, b, c, d, e, f, g, h, i, j, k, l, m, n, o, p, q, r, s, t, u, v, w, x, y, z}\}.$$

We can shorten the listing in set L by writing

$$L = \{\text{a, b, c, d}, \ldots, \text{z}\}.$$

The three dots after the element d, called an *ellipsis*, indicate that the elements in the set continue in the same manner up to and including the last element z.

Blitzer Bonus

The Loss of Sets

Have you ever considered what would happen if we suddenly lost our ability to recall categories and the names that identify them? This is precisely what happened to Alice, the heroine of Lewis Carroll's *Through the Looking Glass*, as she walked with a fawn in "the woods with no names."

> So they walked on together through the woods, Alice with her arms clasped lovingly round the soft neck of the Fawn, till they came out into another open field, and here the Fawn gave a sudden bound into the air, and shook itself free from Alice's arm. "I'm a Fawn!" it cried out in a voice of delight. "And, dear me! you're a human child!" A sudden look of alarm came into its beautiful brown eyes, and in another moment it had darted away at full speed.

By realizing that Alice is a member of the set of human beings, which in turn is part of the set of dangerous things, the fawn is overcome by fear. Thus, the fawn's experience is determined by the way it structures the world into sets with various characteristics.

2 *Define and recognize the empty set.*

The Empty Set

Consider the following sets:

$$\{x \mid x \text{ is a fawn that speaks}\}$$
$$\{x \mid x \text{ is a number greater than 10 and less than 4}\}.$$

Can you see what these sets have in common? They both contain no elements. There are no fawns that speak. There are no numbers that are both greater than 10 and also less than 4. Sets such as these that contain no elements are called the *empty set*, or the *null set*.

THE EMPTY SET

The **empty set**, also called the **null set**, is the set that contains no elements. The empty set is represented by { } or $\varnothing$.

Notice that **{ } and $\varnothing$ have the same meaning**. However, **the empty set is not represented by {$\varnothing$}**. This notation represents a set containing the element $\varnothing$.

Blitzer Bonus

The Musical Sounds of the Empty Set

John Cage (1912–1992), the American avant-garde composer, translated the empty set into the quietest piece of music ever written. His piano composition *4′33″* requires the musician to sit frozen in silence at a piano stool for 4 minutes, 33 seconds, or 273 seconds. (The significance of 273 is that at approximately −273°C, all molecular motion stops.) The set

$\{x \mid x$ is a musical sound from *4′33″*$\}$

is the empty set. There are no musical sounds in the composition. Mathematician Martin Gardner wrote, "I have not heard *4′33″* performed, but friends who have tell me it is Cage's finest composition."

EXAMPLE 4 ***Recognizing the Empty Set***

Which one of the following is the empty set?

a. $\{0\}$ **b.** 0

c. $\{x \mid x$ is a number less than 4 or greater than 10$\}$

d. $\{x \mid x$ is a square with exactly three sides$\}$

SOLUTION

a. $\{0\}$ is a set containing one element, 0. Because this set contains an element, it is not the empty set.

b. 0 is a number, not a set, so it cannot possibly be the empty set. It does, however, represent the number of members of the empty set.

c. $\{x \mid x$ is a number less than 4 or greater than 10$\}$ contains all numbers that are either less than 4, such as 3, or greater than 10, such as 11. Because some elements belong to this set, it cannot be the empty set.

d. $\{x \mid x$ is a square with exactly three sides$\}$ contains no elements. There are no squares with exactly three sides. This set is the empty set.

CHECK POINT 4 Which one of the following is the empty set?

a. $\{x \mid x$ is a number less than 3 or greater than 5$\}$

b. $\{x \mid x$ is a number less than 3 and greater than 5$\}$

c. nothing

d. $\{\varnothing\}$

3 *Use the symbols $\in$ and $\notin$.*

Notations for Set Membership

We now consider two special notations that indicate whether or not a given object belongs to a set.

THE NOTATIONS $\in$ AND $\notin$

The symbol $\in$ is used to indicate that an object is an element of a set. The symbol $\in$ is used to replace the words "is an element of."

The symbol $\notin$ is used to indicate that an object is *not* an element of a set. The symbol $\notin$ is used to replace the words "is not an element of."

EXAMPLE 5 ***Using the Symbols $\in$ and $\notin$***

Determine whether each statement is true or false:

a. $\text{r} \in \{\text{a, b, c}, \dots, \text{z}\}$ **b.** $7 \notin \{1, 2, 3, 4, 5\}$ **c.** $\{\text{a}\} \in \{\text{a, b}\}$.

SOLUTION

a. Because r is an element of the set $\{\text{a, b, c}, \dots, \text{z}\}$, the statement

$$\text{r} \in \{\text{a, b, c}, \dots, \text{z}\}$$

is true.

Observe that an element can belong to a set in roster notation when three dots appear even though the element is not listed.

GREAT QUESTION!

Can a set ever belong to another set—sort of a set within a set?

Yes. A set can be an element of another set. For example, $\{\{a, b\}, c\}$ is a set with two elements. One element is the set $\{a, b\}$ and the other element is the letter c. Thus, $\{a, b\} \in \{\{a, b\}, c\}$ and $c \in \{\{a, b\}, c\}$.

b. Because 7 is not an element of the set $\{1, 2, 3, 4, 5\}$, the statement

$$7 \notin \{1, 2, 3, 4, 5\}$$

is true.

c. Because $\{\text{a}\}$ is a set and the set $\{\text{a}\}$ is not an element of the set $\{\text{a, b}\}$, the statement

$$\{\text{a}\} \in \{\text{a, b}\}$$

is false.

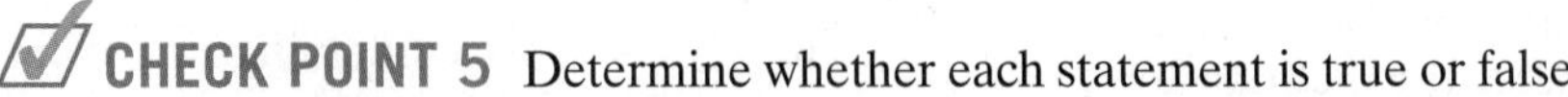

CHECK POINT 5 Determine whether each statement is true or false:

a. $8 \in \{1, 2, 3, \ldots, 10\}$

b. $\text{r} \notin \{\text{a, b, c, z}\}$

c. $\{\text{Monday}\} \in \{x \mid x \text{ is a day of the week}\}$.

4 *Apply set notation to sets of natural numbers.*

Sets of Natural Numbers

For much of the remainder of this section, we will focus on the set of numbers used for counting:

$$\{1, 2, 3, 4, 5, 6, 7, 8, 9, 10, 11, \ldots\}.$$

The set of counting numbers is also called the set of **natural numbers**. We represent this set by the bold face letter **N**.

THE SET OF NATURAL NUMBERS

$$\mathbf{N} = \{1, 2, 3, 4, 5, \ldots\}$$

The three dots, or ellipsis, after the 5 indicate that there is no final element and that the listing goes on forever.

EXAMPLE 6 *Representing Sets of Natural Numbers*

Express each of the following sets using the roster method:

a. Set A is the set of natural numbers less than 5.

b. Set B is the set of natural numbers greater than or equal to 25.

c. $E = \{x \mid x \in \mathbf{N} \text{ and } x \text{ is even}\}$.

SOLUTION

a. The natural numbers less than 5 are 1, 2, 3, and 4. Thus, set A can be expressed using the roster method as

$$A = \{1, 2, 3, 4\}.$$

b. The natural numbers greater than or equal to 25 are 25, 26, 27, 28, and so on. Set B in roster form is

$$B = \{25, 26, 27, 28, \ldots\}.$$

The three dots show that the listing goes on forever.

c. The set-builder notation

$$E = \{x \mid x \in \mathbf{N} \text{ and } x \text{ is even}\}$$

indicates that we want to list the set of all x such that x is an element of the set of natural numbers and x is even. The set of numbers that meets both conditions is the set of even natural numbers. The set in roster form is

$$E = \{2, 4, 6, 8, \ldots\}.$$

CHECK POINT 6 Express each of the following sets using the roster method:

a. Set A is the set of natural numbers less than or equal to 3.

b. Set B is the set of natural numbers greater than 14.

c. $O = \{x \mid x \in \mathbf{N} \text{ and } x \text{ is odd}\}$.

A BRIEF REVIEW *Inequality Notation*

Inequality symbols are frequently used to describe sets of natural numbers. **Table 2.2** reviews basic inequality notation.

TABLE 2.2 Inequality Notation and Sets

Inequality Symbol and Meaning	Example: Set-Builder Notation	Example: Roster Method
$x < a$ — x is less than a.	$\{x \mid x \in \mathbf{N} \text{ and } x < 4\}$ — x is a natural number less than 4.	$\{1, 2, 3\}$
$x \leq a$ — x is less than or equal to a.	$\{x \mid x \in \mathbf{N} \text{ and } x \leq 4\}$ — x is a natural number less than or equal to 4.	$\{1, 2, 3, 4\}$
$x > a$ — x is greater than a.	$\{x \mid x \in \mathbf{N} \text{ and } x > 4\}$ — x is a natural number greater than 4.	$\{5, 6, 7, 8, \ldots\}$
$x \geq a$ — x is greater than or equal to a.	$\{x \mid x \in \mathbf{N} \text{ and } x \geq 4\}$ — x is a natural number greater than or equal to 4.	$\{4, 5, 6, 7, \ldots\}$
$a < x < b$ — x is greater than a and less than b.	$\{x \mid x \in \mathbf{N} \text{ and } 4 < x < 8\}$ — x is a natural number greater than 4 and less than 8.	$\{5, 6, 7\}$
$a \leq x \leq b$ — x is greater than or equal to a and less than or equal to b.	$\{x \mid x \in \mathbf{N} \text{ and } 4 \leq x \leq 8\}$ — x is a natural number greater than or equal to 4 and less than or equal to 8.	$\{4, 5, 6, 7, 8\}$
$a \leq x < b$ — x is greater than or equal to a and less than b.	$\{x \mid x \in \mathbf{N} \text{ and } 4 \leq x < 8\}$ — x is a natural number greater than or equal to 4 and less than 8.	$\{4, 5, 6, 7\}$
$a < x \leq b$ — x is greater than a and less than or equal to b.	$\{x \mid x \in \mathbf{N} \text{ and } 4 < x \leq 8\}$ — x is a natural number greater than 4 and less than or equal to 8.	$\{5, 6, 7, 8\}$

EXAMPLE 7 Representing Sets of Natural Numbers

Express each of the following sets using the roster method:

a. $\{x \mid x \in \mathbf{N} \text{ and } x \leq 100\}$

b. $\{x \mid x \in \mathbf{N} \text{ and } 70 \leq x < 100\}$.

SOLUTION

a. $\{x \mid x \in \mathbf{N} \text{ and } x \leq 100\}$ represents the set of natural numbers less than or equal to 100. This set can be expressed using the roster method as

$$\{1, 2, 3, 4, \ldots, 100\}.$$

b. $\{x \mid x \in \mathbf{N} \text{ and } 70 \leq x < 100\}$ represents the set of natural numbers greater than or equal to 70 and less than 100. This set in roster form is $\{70, 71, 72, 73, \ldots, 99\}$.

CHECK POINT 7 Express each of the following sets using the roster method:

a. $\{x \mid x \in \mathbf{N} \text{ and } x < 200\}$

b. $\{x \mid x \in \mathbf{N} \text{ and } 50 < x \leq 200\}$.

5 *Determine a set's cardinal number.*

Cardinality and Equivalent Sets

The number of elements in a set is called the **cardinal number**, or **cardinality**, of the set. For example, the set {a, e, i, o, u} contains five elements and therefore has the cardinal number 5. We can also say that the set has a cardinality of 5.

DEFINITION OF A SET'S CARDINAL NUMBER

The **cardinal number** of set A, represented by $n(A)$, is the number of distinct elements in set A. The symbol $n(A)$ is read "n of A."

Notice that the cardinal number of a set refers to the number of *distinct*, or different, elements in the set. **Repeating elements in a set neither adds new elements to the set nor changes its cardinality.** For example, $A = \{3, 5, 7\}$ and $B = \{3, 5, 5, 7, 7, 7\}$ represent the same set with three distinct elements, 3, 5, and 7. Thus, $n(A) = 3$ and $n(B) = 3$.

EXAMPLE 8 Determining a Set's Cardinal Number

Find the cardinal number of each of the following sets:

a. $A = \{7, 9, 11, 13\}$ **b.** $B = \{0\}$

c. $C = \{13, 14, 15, \ldots, 22, 23\}$ **d.** $\varnothing$.

SOLUTION

The cardinal number for each set is found by determining the number of elements in the set.

a. $A = \{7, 9, 11, 13\}$ contains four distinct elements. Thus, the cardinal number of set A is 4. We also say that set A has a cardinality of 4, or $n(A) = 4$.

b. $B = \{0\}$ contains one element, namely, 0. The cardinal number of set B is 1. Therefore, $n(B) = 1$.

c. Set $C = \{13, 14, 15, \ldots, 22, 23\}$ lists only five elements. However, the three dots indicate that the natural numbers from 16 through 21 are also in the set. Counting the elements in the set, we find that there are 11 natural numbers in set C. The cardinality of set C is 11, and $n(C) = 11$.

d. The empty set, $\varnothing$, contains no elements. Thus, $n(\varnothing) = 0$.

CHECK POINT 8 Find the cardinal number of each of the following sets:

a. $A = \{6, 10, 14, 15, 16\}$ **b.** $B = \{872\}$

c. $C = \{9, 10, 11, \ldots, 15, 16\}$ **d.** $D = \{\ \}$.

---

6 *Recognize equivalent sets.*

Sets that contain the same number of elements are said to be *equivalent.*

DEFINITION OF EQUIVALENT SETS

Set A is **equivalent** to set B means that set A and set B contain the same number of elements. For equivalent sets, $n(A) = n(B)$.

Here is an example of two equivalent sets:

$n(A) = n(B) = 5$

$$A = \{x \mid x \text{ is a vowel}\} = \{\text{a, e, i, o, u}\}$$

$$\updownarrow\ \updownarrow\ \updownarrow\ \updownarrow\ \updownarrow$$

$$B = \{x \mid x \in \mathbf{N} \text{ and } 3 \le x \le 7\} = \{3, 4, 5, 6, 7\}.$$

It is not necessary to count elements and arrive at 5 to determine that these sets are equivalent. The lines with arrowheads, $\updownarrow$, indicate that each element of set A can be paired with exactly one element of set B and each element of set B can be paired with exactly one element of set A. We say that the sets can be placed in a **one-to-one correspondence**.

ONE-TO-ONE CORRESPONDENCES AND EQUIVALENT SETS

1. If set A and set B can be placed in a one-to-one correspondence, then A is equivalent to B: $n(A) = n(B)$.
2. If set A and set B cannot be placed in a one-to-one correspondence, then A is not equivalent to B: $n(A) \ne n(B)$.

EXAMPLE 9 *Determining If Sets Are Equivalent*

Figure 2.1 shows the top five impediments to academic performance for U.S. college students.

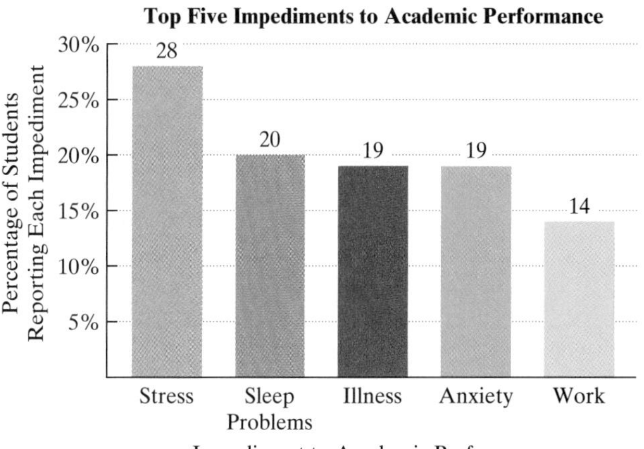

FIGURE 2.1
Source: American College Health Association

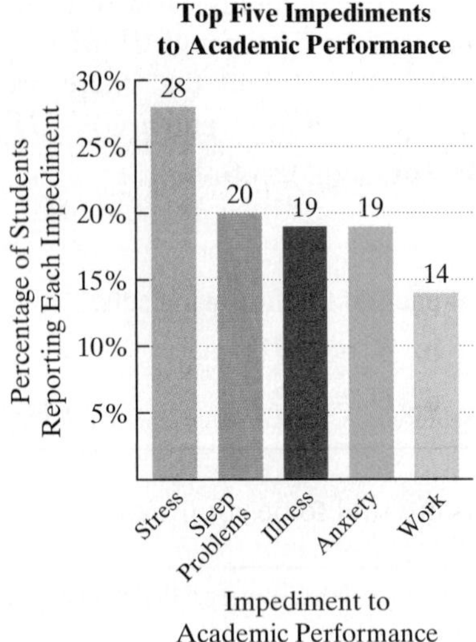

FIGURE 2.1 (repeated)

Let

A = the set of five impediments shown in **Figure 2.1**

B = the set of the percentage of college students reporting each impediment.

Are these sets equivalent? Explain.

SOLUTION

Let's begin by expressing each set in roster form.

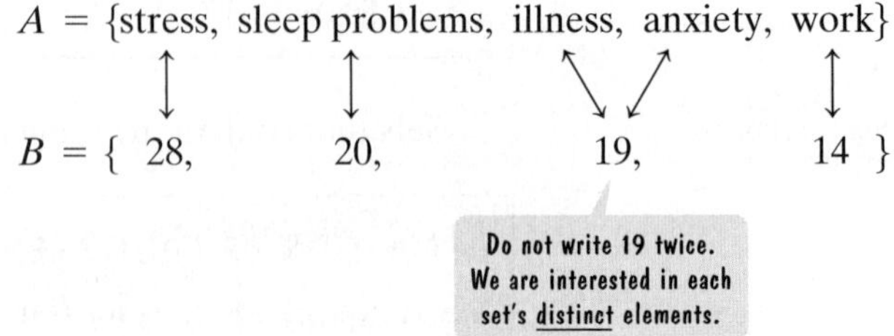

There are two ways to determine that these sets are not equivalent.

Method 1. Trying to Set Up a One-to-One Correspondence

The lines with arrowheads between the sets in roster form indicate that the correspondence between the sets is not one-to-one. The elements illness and anxiety from set A are both paired with the element 19 from set B. These sets are not equivalent.

Method 2. Counting Elements

Set A contains five distinct elements: $n(A) = 5$. Set B contains four distinct elements: $n(B) = 4$. Because the sets do not contain the same number of elements, they are not equivalent.

CHECK POINT 9 **Figure 2.2** shows the percentage of Americans optimistic about the future for each region of the country. Let

A = the set of the four regions shown in **Figure 2.2**

B = the set of the percentage of Americans in each region optimistic about the future.

Are these sets equivalent? Explain.

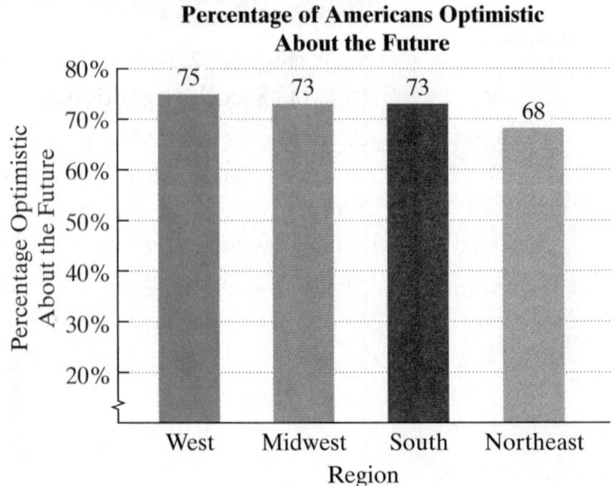

FIGURE 2.2
Source: The Harris Poll (2016 data)

7 *Distinguish between finite and infinite sets.*

Finite and Infinite Sets

Example 9 illustrated that to compare the cardinalities of two sets, pair off their elements. If there is not a one-to-one correspondence, the sets have different cardinalities and are not equivalent. Although this idea is obvious in the case of *finite sets*, some unusual conclusions emerge when dealing with *infinite sets*.

> **FINITE SETS AND INFINITE SETS**
>
> Set A is a **finite set** if $n(A) = 0$ (that is, A is the empty set) or $n(A)$ is a natural number. A set whose cardinality is not 0 or a natural number is called an **infinite set**.

An example of an infinite set is the set of natural numbers, $\mathbf{N} = \{1, 2, 3, 4, 5, 6, \ldots\}$, where the ellipsis indicates that there is no last, or final, element. Does this set have a cardinality? The answer is yes, albeit one of the strangest numbers you've ever seen. The set of natural numbers is assigned the infinite cardinal number $\aleph_0$ (read: "aleph-null," aleph being the first letter of the Hebrew alphabet). What follows is a succession of mind-boggling results, including a hierarchy of different infinite numbers in which $\aleph_0$ is the smallest infinity:

$$\aleph_0 < \aleph_1 < \aleph_2 < \aleph_3 < \aleph_4 < \aleph_5 \ldots.$$

These ideas, which are impossible for our imaginations to grasp, are developed in Section 2.2 and the Blitzer Bonus at the end of that section.

8 *Recognize equal sets.*

Equal Sets

We conclude this section with another important concept of set theory, equality of sets.

> **DEFINITION OF EQUALITY OF SETS**
>
> Set A is **equal** to set B means that set A and set B contain exactly the same elements, regardless of order or possible repetition of elements. We symbolize the equality of sets A and B using the statement $A = B$.

GREAT QUESTION!

Can you clarify the difference between equal sets and equivalent sets?

In English, the words *equal* and *equivalent* often mean the same thing. This is not the case in set theory. **Equal sets** contain the **same elements. Equivalent sets** contain the **same number of elements**. If two sets are equal, then they must be equivalent. However, if two sets are equivalent, they are not necessarily equal.

For example, if $A = \{w, x, y, z\}$ and $B = \{z, y, w, x\}$, then $A = B$ because the two sets contain exactly the same elements.

Because equal sets contain the same elements, they also have the same cardinal number. For example, the equal sets $A = \{w, x, y, z\}$ and $B = \{z, y, w, x\}$ have four elements each. Thus, both sets have the same cardinal number: 4. Notice that a possible one-to-one correspondence between the equal sets A and B can be obtained by pairing each element with itself:

$$A = \{w, x, y, z\}$$

$$B = \{z, y, w, x\}$$

This illustrates an important point: **If two sets are equal, then they must be equivalent.**

EXAMPLE 10 Determining Whether Sets Are Equal

Determine whether each statement is true or false:

a. $\{4, 8, 9\} = \{8, 9, 4\}$

b. $\{1, 3, 5\} = \{0, 1, 3, 5\}$.

SOLUTION

a. The sets $\{4, 8, 9\}$ and $\{8, 9, 4\}$ contain exactly the same elements. Therefore, the statement

$$\{4, 8, 9\} = \{8, 9, 4\}$$

is true.

b. As we look at the given sets, $\{1, 3, 5\}$ and $\{0, 1, 3, 5\}$, we see that 0 is an element of the second set, but not the first. The sets do not contain exactly the same elements. Therefore, the sets are not equal. This means that the statement

$$\{1, 3, 5\} = \{0, 1, 3, 5\}$$

is false.

 **CHECK POINT 10** Determine whether each statement is true or false:

a. $\{O, L, D\} = \{D, O, L\}$

b. $\{4, 5\} = \{5, 4, \varnothing\}$.

Concept and Vocabulary Check

Fill in each blank so that the resulting statement is true.

1. The set {California, Colorado, Connecticut} is expressed using the _______ method. The set $\{x|x$ is a U.S. state whose name begins with the letter C$\}$ is expressed using __________ notation.
2. A set that contains no elements is called the null set or the _______ set. This set is represented by { } or _____.
3. The symbol $\in$ is used to indicate that an object ___________ of a set.
4. The set $\mathbf{N} = \{1, 2, 3, 4, 5, \ldots\}$ is called the set of ______________.
5. The number of distinct elements in a set is called the _________ number of the set. If A represents the set, this number is represented by _______.
6. Two sets that contain the same number of elements are called ____________ sets.
7. Two sets that contain the same elements are called _______ sets.

Exercise Set 2.1

Practice Exercises

In Exercises 1–6, determine which collections are not well defined and therefore not sets.

1. The collection of U.S. presidents
2. The collection of part-time and full-time students currently enrolled at your college
3. The collection of the five worst U.S. presidents
4. The collection of elderly full-time students currently enrolled at your college
5. The collection of natural numbers greater than one million
6. The collection of even natural numbers greater than 100

In Exercises 7–14, write a word description of each set. (More than one correct description may be possible.)

7. {Mercury, Venus, Earth, Mars, Jupiter, Saturn, Uranus, Neptune}

8. {Saturday, Sunday}

9. {January, June, July}

10. {April, August}

11. $\{6, 7, 8, 9, \ldots\}$

12. $\{9, 10, 11, 12, \ldots\}$

13. $\{6, 7, 8, 9, \ldots, 20\}$

14. $\{9, 10, 11, 12, \ldots, 25\}$

In Exercises 15–32, express each set using the roster method.

15. The set of the four seasons in a year

16. The set of months of the year that have exactly 30 days

17. $\{x \mid x$ is a month that ends with the letters b-e-r$\}$

18. $\{x \mid x$ is a lowercase letter of the alphabet that follows d and comes before j$\}$

19. The set of natural numbers less than 4

20. The set of natural numbers less than or equal to 6

21. The set of odd natural numbers less than 13

22. The set of even natural numbers less than 10

23. $\{x \mid x \in \mathbf{N} \text{ and } x \leq 5\}$

24. $\{x \mid x \in \mathbf{N} \text{ and } x \leq 4\}$

25. $\{x \mid x \in \mathbf{N} \text{ and } x > 5\}$

26. $\{x \mid x \in \mathbf{N} \text{ and } x > 4\}$

27. $\{x \mid x \in \mathbf{N} \text{ and } 6 < x \leq 10\}$

28. $\{x \mid x \in \mathbf{N} \text{ and } 7 < x \leq 11\}$

29. $\{x \mid x \in \mathbf{N} \text{ and } 10 \leq x < 80\}$

30. $\{x \mid x \in \mathbf{N} \text{ and } 15 \leq x < 60\}$

31. $\{x \mid x + 5 = 7\}$

32. $\{x \mid x + 3 = 9\}$

In Exercises 33–46, determine which sets are the empty set.

33. $\{\varnothing, 0\}$

34. $\{0, \varnothing\}$

35. $\{x \mid x$ is a woman who served as U.S. president before 2016$\}$

36. $\{x \mid x$ is a living U.S. president born before 1200$\}$

37. $\{x \mid x$ is the number of women who served as U.S. president before 2016$\}$

38. $\{x \mid x$ is the number of living U.S. presidents born before 1200$\}$

39. $\{x \mid x$ is a U.S. state whose name begins with the letter X$\}$

40. $\{x \mid x$ is a month of the year whose name begins with the letter X$\}$

41. $\{x \mid x < 2 \text{ and } x > 5\}$

42. $\{x \mid x < 3 \text{ and } x > 7\}$

43. $\{x \mid x \in \mathbf{N} \text{ and } 2 < x < 5\}$

44. $\{x \mid x \in \mathbf{N} \text{ and } 3 < x < 7\}$

45. $\{x \mid x$ is a number less than 2 or greater than 5$\}$

46. $\{x \mid x$ is a number less than 3 or greater than 7$\}$

In Exercises 47–66, determine whether each statement is true or false.

47. $3 \in \{1, 3, 5, 7\}$

48. $6 \in \{2, 4, 6, 8, 10\}$

49. $12 \in \{1, 2, 3, \ldots, 14\}$

50. $10 \in \{1, 2, 3, \ldots, 16\}$

51. $5 \in \{2, 4, 6, \ldots, 20\}$

52. $8 \in \{1, 3, 5, \ldots 19\}$

53. $11 \notin \{1, 2, 3, \ldots, 9\}$

54. $17 \notin \{1, 2, 3, \ldots, 16\}$

55. $37 \notin \{1, 2, 3, \ldots, 40\}$

56. $26 \notin \{1, 2, 3, \ldots, 50\}$

57. $4 \notin \{x \mid x \in \mathbf{N}$ and x is even$\}$

58. $2 \in \{x \mid x \in \mathbf{N}$ and x is odd$\}$

59. $13 \notin \{x \mid x \in \mathbf{N} \text{ and } x < 13\}$

60. $20 \notin \{x \mid x \in \mathbf{N} \text{ and } x < 20\}$

61. $16 \notin \{x \mid x \in \mathbf{N} \text{ and } 15 \leq x < 20\}$

62. $19 \notin \{x \mid x \in \mathbf{N} \text{ and } 16 \leq x < 21\}$

63. $\{3\} \in \{3, 4\}$

64. $\{7\} \in \{7, 8\}$

65. $-1 \notin \mathbf{N}$

66. $-2 \notin \mathbf{N}$

In Exercises 67–80, find the cardinal number for each set.

67. $A = \{17, 19, 21, 23, 25\}$

68. $A = \{16, 18, 20, 22, 24, 26\}$

69. $B = \{2, 4, 6, \ldots, 30\}$

70. $B = \{1, 3, 5, \ldots, 21\}$

71. $C = \{x \mid x$ is a day of the week that begins with the letter A$\}$

72. $C = \{x \mid x$ is a month of the year that begins with the letter W$\}$

73. $D = \{\text{five}\}$

74. $D = \{\text{six}\}$

75. $A = \{x \mid x$ is a letter in the word *five*$\}$

76. $A = \{x \mid x$ is a letter in the word *six*$\}$

77. $B = \{x \mid x \in \mathbf{N} \text{ and } 2 \leq x < 7\}$

78. $B = \{x \mid x \in \mathbf{N} \text{ and } 3 \leq x < 10\}$

79. $C = \{x \mid x < 4 \text{ and } x \geq 12\}$

80. $C = \{x \mid x < 5 \text{ and } x \geq 15\}$

In Exercises 81–90,

a. *Are the sets equivalent? Explain.*

b. *Are the sets equal? Explain.*

81. A is the set of students at your college. B is the set of students majoring in business at your college.

82. A is the set of states in the United States. B is the set of people who are now governors of the states in the United States.

83. $A = \{1, 2, 3, 4, 5\}$
$B = \{0, 1, 2, 3, 4\}$

84. $A = \{1, 3, 5, 7, 9\}$
$B = \{2, 4, 6, 8, 10\}$

85. $A = \{1, 1, 1, 2, 2, 3, 4\}$
$B = \{4, 3, 2, 1\}$

86. $A = \{0, 1, 1, 2, 2, 2, 3, 3, 3, 3\}$
$B = \{3, 2, 1, 0\}$

87. $A = \{x \mid x \in \mathbf{N} \text{ and } 6 \le x < 10\}$
$B = \{x \mid x \in \mathbf{N} \text{ and } 9 < x \le 13\}$

88. $A = \{x \mid x \in \mathbf{N} \text{ and } 12 < x \le 17\}$
$B = \{x \mid x \in \mathbf{N} \text{ and } 20 \le x < 25\}$

89. $A = \{x \mid x \in \mathbf{N} \text{ and } 100 \le x \le 105\}$
$B = \{x \mid x \in \mathbf{N} \text{ and } 99 < x < 106\}$

90. $A = \{x \mid x \in \mathbf{N} \text{ and } 200 \le x \le 206\}$
$B = \{x \mid x \in \mathbf{N} \text{ and } 199 < x < 207\}$

In Exercises 91–96, determine whether each set is finite or infinite.

91. $\{x \mid x \in \mathbf{N} \text{ and } x \ge 100\}$

92. $\{x \mid x \in \mathbf{N} \text{ and } x \ge 50\}$

93. $\{x \mid x \in \mathbf{N} \text{ and } x \le 1{,}000{,}000\}$

94. $\{x \mid x \in \mathbf{N} \text{ and } x \le 2{,}000{,}000\}$

95. The set of natural numbers less than 1

96. The set of natural numbers less than 0

Practice Plus

In Exercises 97–100, express each set using set-builder notation. Use inequality notation to express the condition x must meet in order to be a member of the set. (More than one correct inequality may be possible.)

97. $\{61, 62, 63, 64, \ldots\}$

98. $\{36, 37, 38, 39, \ldots\}$

99. $\{61, 62, 63, 64, \ldots, 89\}$

100. $\{36, 37, 38, 39, \ldots, 59\}$

In Exercises 101–104, give examples of two sets that meet the given conditions. If the conditions are impossible to satisfy, explain why.

101. The two sets are equivalent but not equal.

102. The two sets are equivalent and equal.

103. The two sets are equal but not equivalent.

104. The two sets are neither equivalent nor equal.

Application Exercises

Although you want to choose a career that fits your interests and abilities, it is good to have an idea of what jobs pay when looking at career options. The bar graph shows the average yearly earnings of full-time employed college graduates with only a bachelor's degree based on their college major.

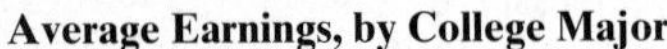

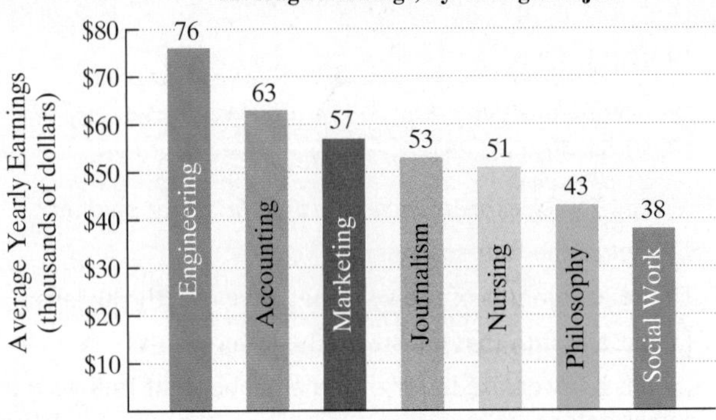

Source: Arthur J. Keown, *Personal Finance*, Pearson

In Exercises 105–108, use the information given by the graph to represent each set by the roster method.

105. The set of college majors with average yearly earnings that exceed $57,000

106. The set of college majors with average yearly earnings that exceed $63,000

107. $\{x \mid x$ is a major with
$38,000 < average yearly earnings ≤ $53,000$\}$

108. $\{x \mid x$ is a major with
$38,000 ≤ average yearly earnings < $53,000$\}$

The bar graph shows the differences among age groups on the Implicit Association Test that measures levels of racial prejudice. Higher scores indicate stronger bias.

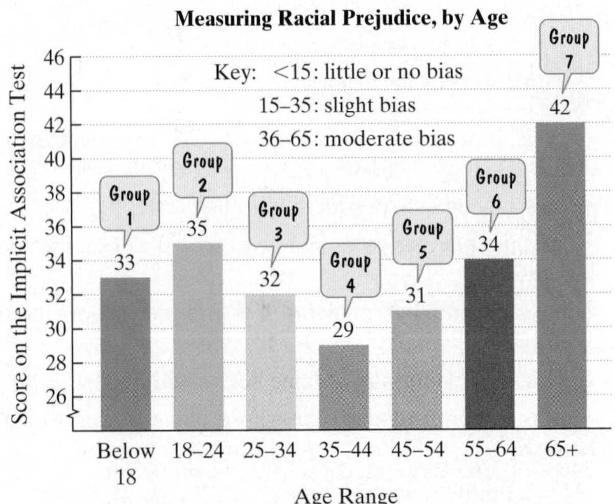

Source: The Race Implicit Association Test on the Project Implicit Demonstration Website

In Exercises 109–112, use the information given by the graph at the bottom of the previous page to represent each set by the roster method, or use the appropriate notation to indicate that the set is the empty set.

109. $\{x \mid x$ is a group whose score indicates little or no bias$\}$

110. $\{x \mid x$ is a group whose score indicates slight bias$\}$

111. $\{x \mid x$ is a group whose score indicates moderate bias$\}$

112. $\{x \mid x$ is a group whose score is at least 30 and at most 40$\}$

A study of 900 working women in Texas showed that their feelings changed throughout the day. The following line graph shows 15 different times in a day and the average level of happiness for the women at each time. Based on the information given by the graph, represent each of the sets in Exercises 113–116 using the roster method.

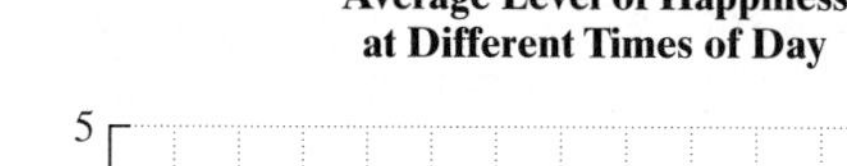

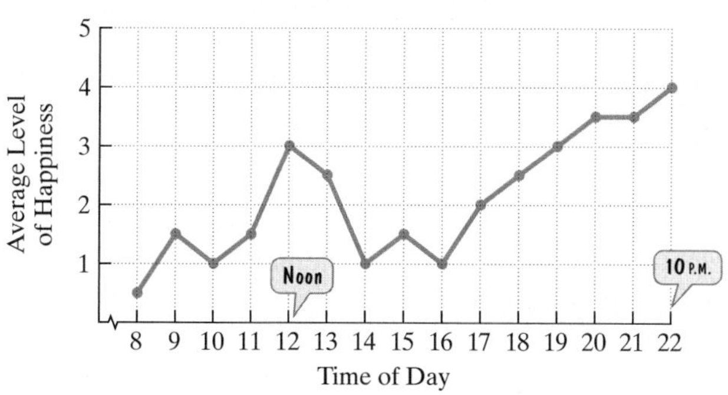

Source: D. Kahneman et al. "A Survey Method for Characterizing Daily Life Experience," *Science*

113. $\{x \mid x$ is a time of the day when the average level of happiness was $3\}$

114. $\{x \mid x$ is a time of the day when the average level of happiness was $1\}$

115. $\{x \mid x$ is a time of the day when

$3 <$ average level of happiness $< 4\}$

116. $\{x \mid x$ is a time of the day when

$3 <$ average level of happiness $\leq 4\}$

117. Do the results of Exercise 113 or 114 indicate a one-to-one correspondence between the set representing the time of day and the set representing average level of happiness? Are these sets equivalent?

Explaining the Concepts

118. What is a set?

119. Describe the three methods used to represent a set. Give an example of a set represented by each method.

120. What is the empty set?

121. Explain what is meant by *equivalent sets*.

122. Explain what is meant by *equal sets*.

123. Use cardinality to describe the difference between a finite set and an infinite set.

Critical Thinking Exercises

Make Sense? *In Exercises 124–127, determine whether each statement makes sense or does not make sense, and explain your reasoning.*

124. I used the roster method to express the set of countries that I have visited.

125. I used the roster method and natural numbers to express the set of average daily Fahrenheit temperatures throughout the month of July in Vostok Station, Antarctica, the coldest month in one of the coldest locations in the world.

126. Using this bar graph that shows the average number of hours that Americans sleep per day, I can see that there is a one-to-one correspondence between the set of six ages on the horizontal axis and the set of the average number of hours that men sleep per day.

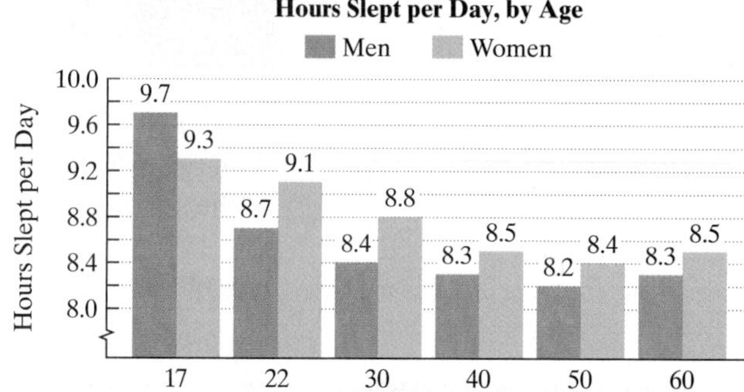

Source: ATUS, Bureau of Labor Statistics

127. Using the bar graph in Exercise 126, I can see that there is a one-to-one correspondence between the set of the average number of hours that men sleep per day and the set of the average number of hours that women sleep per day.

In Exercises 128–135, determine whether each statement is true or false. If the statement is false, make the necessary change(s) to produce a true statement.

128. Two sets can be equal but not equivalent.

129. Any set in roster notation that contains three dots must be an infinite set.

130. $n(\varnothing) = 1$

131. Some sets that can be written in set-builder notation cannot be written in roster form.

132. The set of fractions between 0 and 1 is an infinite set.

133. The set of multiples of 4 between 0 and 4,000,000,000 is an infinite set.

134. If the elements in a set cannot be counted in a trillion years, the set is an infinite set.

135. Because 0 is not a natural number, it can be deleted from any set without changing the set's cardinality.

136. In a certain town, a barber shaves all those men and only those men who do not shave themselves. Consider each of the following sets:

$A = \{x \mid x$ is a man of the town who shaves himself$\}$

$B = \{x \mid x$ is a man of the town who does not shave himself$\}$.

The one and only barber in the town is Sweeney Todd. If s represents Sweeney Todd,

a. is $s \in A$? **b.** is $s \in B$?

2.2 Subsets

WHAT AM I SUPPOSED TO LEARN?

After studying this section, you should be able to:

1. Recognize subsets and use the notation $\subseteq$.
2. Recognize proper subsets and use the notation $\subset$.
3. Determine the number of subsets of a set.
4. Apply concepts of subsets and equivalent sets to infinite sets.

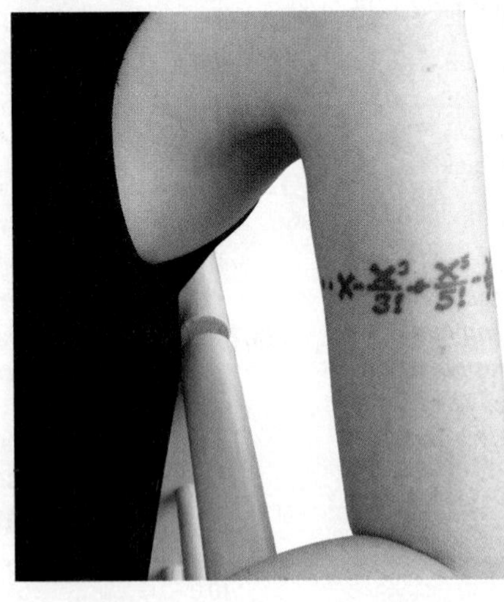

MATH TATTOOS. WHO KNEW? EMERGING from their often unsavory reputation of the recent past, tattoos have gained increasing prominence as a form of body art and self-expression. A recent Harris poll estimated that 45 million Americans, or 21% of the adult population, have at least one tattoo.

Table 2.3 shows the percentage of Americans, by age group, with tattoos. The categories in the table divide the set of tattooed Americans into smaller sets, called *subsets*, based on age. The age subsets can be broken into still-smaller subsets. For example, tattooed Americans ages 25–29 can be categorized by gender, political party affiliation, race/ethnicity, or any other area of interest. This suggests numerous possible subsets of the set of Americans with tattoos. Every American in each of these subsets is also a member of the set of tattooed Americans.

1 *Recognize subsets and use the notation $\subseteq$.*

TABLE 2.3 Percentage of Tattooed Americans, by Age Group

Age Group	Percent Tattooed
18–24	22%
25–29	30%
30–39	38%
40–49	27%
50–64	11%
65+	5%

Source: Harris Interactive

Subsets

Situations in which all the elements of one set are also elements of another set are described by the following definition:

DEFINITION OF A SUBSET OF A SET

Set A is a **subset** of set B, expressed as

$$A \subseteq B,$$

if every element in set A is also an element in set B.

Let's apply this definition to the set of people ages 25–29 in **Table 2.3**.

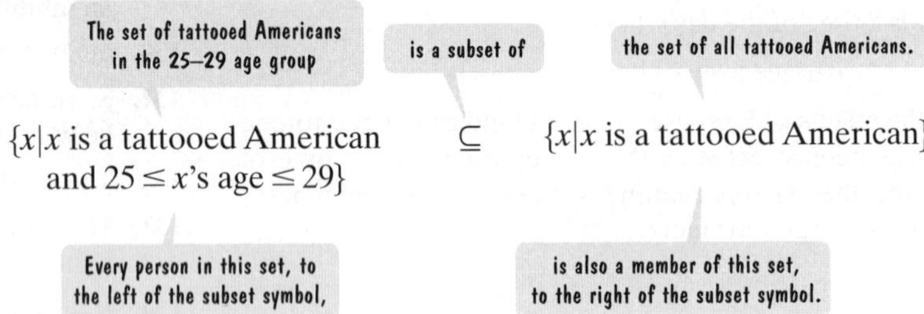

Observe that a subset is itself a set.

The notation $A \not\subseteq B$ means that A **is not a subset** of B. Set A is not a subset of set B if there is at least one element of set A that is not an element of set B. For example, consider the following sets:

$$A = \{1, 2, 3\} \quad \text{and} \quad B = \{1, 2\}.$$

Can you see that 3 is an element of set A that is not in set B? Thus, set A is not a subset of set B: $A \not\subseteq B$.

We can show that $A \subseteq B$ by showing that every element of set A also occurs as an element of set B. We can show that $A \not\subseteq B$ by finding one element of set A that is not in set B.

EXAMPLE 1 Using the Symbols $\subseteq$ and $\nsubseteq$

Write $\subseteq$ or $\nsubseteq$ in each blank to form a true statement:

a. $A = \{1, 3, 5, 7\}$
$B = \{1, 3, 5, 7, 9, 11\}$
A ____ B

b. $A = \{x \mid x$ is a letter in the word *proof*$\}$
$B = \{y \mid y$ is a letter in the word *roof*$\}$
A ____ B

c. $A = \{x \mid x$ is a planet of Earth's solar system$\}$
$B = \{$Mercury, Venus, Earth, Mars, Jupiter, Saturn, Uranus, Neptune$\}$
A ____ B

SOLUTION

a. All the elements of $A = \{1, 3, 5, 7\}$ are also contained in $B = \{1, 3, 5, 7, 9, 11\}$. Therefore, set A is a subset of set B:

$$A \subseteq B.$$

b. Let's write the set of letters in the word *proof* and the set of letters in the word *roof* in roster form. In each case, we consider only the distinct elements, so there is no need to repeat the o.

$$A = \{\text{p, r, o, f}\} \qquad B = \{\text{r, o, f}\}$$

The element p is in set A but not in set B.

Because there is an element in set A that is not in set B, set A is not a subset of set B:

$$A \nsubseteq B.$$

c. All the elements of

$$A = \{x \mid x \text{ is a planet of Earth's solar system}\}$$

are contained in

$$B = \{\text{Mercury, Venus, Earth, Mars, Jupiter, Saturn, Uranus, Neptune}\}.$$

Because all elements in set A are also in set B, set A is a subset of set B:

$$A \subseteq B.$$

Furthermore, the sets are equal ($A = B$).

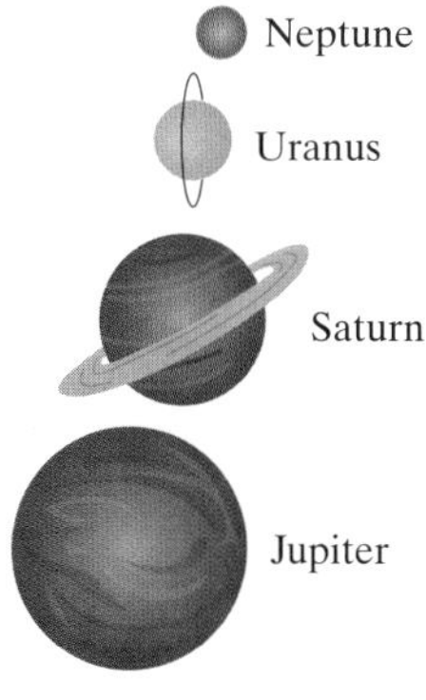

The eight planets in Earth's solar system

No, we did not forget Pluto. In 2006, based on the requirement that a planet must dominate its own orbit (Pluto is slave to Neptune's orbit), the International Astronomical Union removed Pluto from the list of planets and decreed that it belongs to a new category of heavenly body, a "dwarf planet."

CHECK POINT 1 Write $\subseteq$ or $\nsubseteq$ in each blank to form a true statement:

a. $A = \{1, 3, 5, 6, 9, 11\}$
$B = \{1, 3, 5, 7\}$
A ____ B

b. $A = \{x \mid x$ is a letter in the word *roof*$\}$
$B = \{y \mid y$ is a letter in the word *proof*$\}$
A ____ B

c. $A = \{x \mid x$ is a day of the week$\}$
$B = \{$Monday, Tuesday, Wednesday, Thursday, Friday, Saturday, Sunday$\}$
A ____ B

2 *Recognize proper subsets and use the notation ⊂.*

Proper Subsets

In Example 1(c) and Check Point 1(c), the given sets are equal and illustrate that **every set is a subset of itself.** If A is any set, then $A \subseteq A$ because it is obvious that each element of A is a member of A.

If we know that set A is a subset of set B and we exclude the possibility of the sets being equal, then set A is called a *proper subset* of set B, written $A \subset B$.

DEFINITION OF A PROPER SUBSET OF A SET

Set A is a **proper subset** of set B, expressed as $A \subset B$, if set A is a subset of set B and sets A and B are not equal ($A \neq B$).

Try not to confuse the symbols for subset, ⊆, and proper subset, ⊂. In some subset examples, both symbols can be placed between sets:

Set A Set B Set A Set B

$$\{1, 3\} \subseteq \{1, 3, 5\} \quad \text{and} \quad \{1, 3\} \subset \{1, 3, 5\}.$$

A is a subset of B. Every element in A is also an element in B.

A is a proper subset of B because A and B are not equal sets.

By contrast, there are subset examples where only the symbol ⊆ can be placed between sets:

Set A Set B

$$\{1, 3, 5\} \subseteq \{1, 3, 5\}.$$

A is a subset of B. Every element in A is also an element in B. A is not a proper subset of B because $A = B$. The symbol ⊂ should not be placed between the sets.

Because the lower part of the subset symbol in $A \subseteq B$ suggests an equal sign, it is *possible* that sets A and B are equal, although they do not have to be. By contrast, the missing lower line for the proper subset symbol in $A \subset B$ indicates that sets A and B *cannot* be equal.

GREAT QUESTION!

Is there a relationship between the symbols ⊆ and ⊂ and the inequality symbols ≤ and <?

Great observation!

- The notation for "is a subset of," ⊆, is similar to the notation for "is less than or equal to," ≤. Because the notations share similar ideas, $A \subseteq B$ applies to finite sets only if the cardinal number of set A is less than or equal to the cardinal number of set B.
- The notation for "is a proper subset of," ⊂, is similar to the notation for "is less than," <. Because the notations share similar ideas, $A \subset B$ applies to finite sets only if the cardinal number of set A is less than the cardinal number of set B.

EXAMPLE 2 *Using the Symbols ⊆ and ⊂*

Write ⊆, ⊂, or both, in each blank to form a true statement:

a. $A = \{x \mid x \text{ is a person and } x \text{ lives in San Francisco}\}$
$B = \{x \mid x \text{ is a person and } x \text{ lives in California}\}$
A ____ B

b. $A = \{2, 4, 6, 8\}$
$B = \{2, 8, 4, 6\}$
A ____ B.

SOLUTION

a. We begin with $A = \{x \mid x \text{ is a person and } x \text{ lives in San Francisco}\}$ and $B = \{x \mid x \text{ is a person and } x \text{ lives in California}\}$. Every person living in San Francisco is also a person living in California. Because each person in set A is contained in set B, set A is a subset of set B:

$$A \subseteq B.$$

Can you see that the two sets, $A = \{x|x$ is a person and x lives in San Francisco$\}$ and $B = \{x|x$ is a person and x lives in California$\}$, do not contain exactly the same elements and, consequently, are not equal? A person living in California outside San Francisco is in set B, but not in set A. Because there is at least one such person, the sets are not equal and set A is a proper subset of set B:

$$A \subset B.$$

The symbols $\subseteq$ and $\subset$ can both be placed in the blank to form a true statement.

b. Every number in $A = \{2, 4, 6, 8\}$ is contained in $B = \{2, 8, 4, 6\}$, so set A is a subset of set B:

$$A \subseteq B.$$

Because the sets contain exactly the same elements and are equal, set A is *not* a proper subset of set B. The symbol $\subset$ cannot be placed in the blank if we want to form a true statement. (Because set A is not a proper subset of set B, it is correct to write $A \not\subset B$.)

CHECK POINT 2 Write $\subseteq$, $\subset$, or both in each blank to form a true statement:

a. $A = \{2, 4, 6, 8\}$
$B = \{2, 8, 4, 6, 10\}$
A ____ B

b. $A = \{x|x$ is a person and x lives in Atlanta$\}$
$B = \{x|x$ is a person and x lives in Georgia$\}$
A ____ B

GREAT QUESTION!

All the symbols used in set theory make me feel that I'm an element of the set of the notationally confused! For example, what's the difference between the symbols $\in$ and $\subseteq$?

The symbol $\in$ means "is an element of" and the symbol $\subseteq$ means "is a subset of." Notice the differences among the following true statements:

$4 \in \{4, 8\}$ — 4 is an element of the set {4, 8}.

$\{4\} \subseteq \{4, 8\}$ — The set containing 4 is a subset of the set {4, 8}.

$\{4\} \notin \{4, 8\}$. — The set containing 4 is not an element of {4, 8}, although $\{4\} \in \{\{4\}, \{8\}\}$.

Blitzer Bonus

Science and Math Tattoos

We opened the section by considering subsets of the set of tattooed Americans, based on age. We'll continue dividing the set of tattooed Americans into subsets using party affiliation and gender in the Exercise Set at the end of this section (see Exercises 83–92).

In *Science Ink* (Sterling, 2011), science writer Carl Zimmer presents more than 300 thought-provoking science and math tattoos, explaining the significance of the body art. Many of the tattooed images in Zimmer's book relate to topics you'll encounter in *Thinking Mathematically*, including the empty set, numerals in base two (Section 4.2), the golden ratio (Section 6.5), and Σ, a symbol of summation, that appears in many statistical formulas (Section 12.2). Check out *Science Ink* and prepare to be dazzled by the images and the stories behind them.

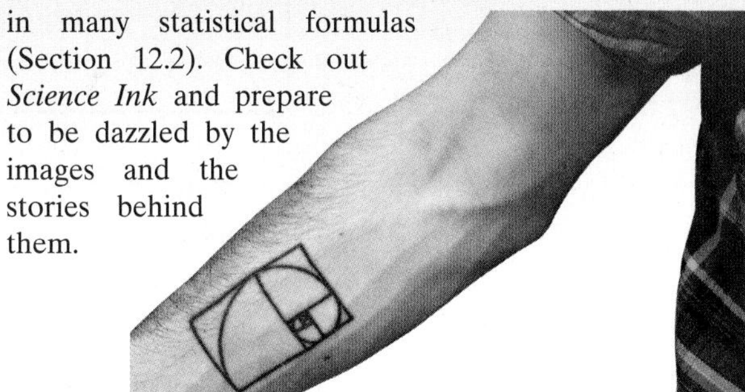

Subsets and the Empty Set

The meaning of $A \subseteq B$ leads to some interesting properties of the empty set.

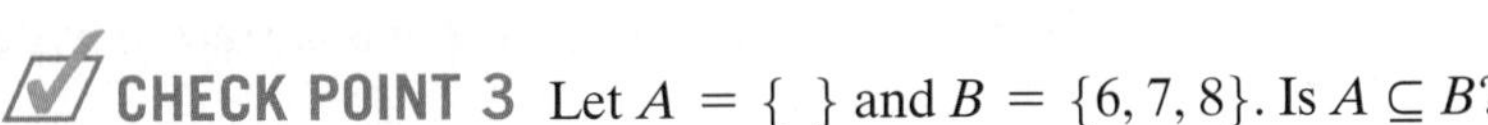

EXAMPLE 3 *The Empty Set as a Subset*

Let $A = \{\ \}$ and $B = \{1, 2, 3, 4, 5\}$. Is $A \subseteq B$?

SOLUTION

A is not a subset of B ($A \not\subseteq B$) if there is at least one element of set A that is not an element of set B. Because A represents the empty set, there are no elements in set A, period, much less elements in A that do not belong to B. Because we cannot find an element in $A = \{\ \}$ that is not contained in $B = \{1, 2, 3, 4, 5\}$, this means that $A \subseteq B$. Equivalently, $\varnothing \subseteq B$.

CHECK POINT 3 Let $A = \{\ \}$ and $B = \{6, 7, 8\}$. Is $A \subseteq B$?

Example 3 illustrates the principle that **the empty set is a subset of every set.** Furthermore, the empty set is a proper subset of every set except itself.

THE EMPTY SET AS A SUBSET

1. For any set B, $\varnothing \subseteq B$.
2. For any set B other than the empty set, $\varnothing \subset B$.

3 *Determine the number of subsets of a set.*

The Number of Subsets of a Given Set

If a set contains n elements, how many distinct subsets can be formed? Let's observe some special cases, namely sets with 0, 1, 2, and 3 elements. We can use inductive reasoning to arrive at a general conclusion. We begin by listing subsets and counting the number of subsets in our list. This is shown in **Table 2.4**.

TABLE 2.4 The Number of Subsets: Some Special Cases

Set	Number of Elements	List of All Distinct Subsets	Number of Subsets
$\{\ \}$	0	$\{\ \}$	1
$\{a\}$	1	$\{a\}, \{\ \}$	2
$\{a, b\}$	2	$\{a, b\}, \{a\}, \{b\}, \{\ \}$	4
$\{a, b, c\}$	3	$\{a, b, c\}$, $\{a, b\}, \{a, c\}, \{b, c\}$, $\{a\}, \{b\}, \{c\}, \{\ \}$	8

Table 2.4 suggests that when we increase the number of elements in the set by one, the number of subsets doubles. The number of subsets appears to be a power of 2.

Number of elements	0	1	2	3
Number of subsets	$1 = 2^0$	$2 = 2^1$	$4 = 2 \times 2 = 2^2$	$8 = 2 \times 2 \times 2 = 2^3$

A BRIEF REVIEW
Powers of 2

If powers of 2 have you in an exponentially increasing state of confusion, here's a list of values that should be helpful. Observe how rapidly these values are increasing.

Powers of 2

$2^0 = 1$
$2^1 = 2$
$2^2 = 2 \times 2 = 4$
$2^3 = 2 \times 2 \times 2 = 8$
$2^4 = 2 \times 2 \times 2 \times 2 = 16$
$2^5 = 2 \times 2 \times 2 \times 2 \times 2 = 32$
$2^6 = 64$
$2^7 = 128$
$2^8 = 256$
$2^9 = 512$
$2^{10} = 1024$
$2^{11} = 2048$
$2^{12} = 4096$
$2^{15} = 32{,}768$
$2^{20} = 1{,}048{,}576$
$2^{25} = 33{,}554{,}432$
$2^{30} = 1{,}073{,}741{,}824$

The power of 2 is the same as the number of elements in the set. Using inductive reasoning, if the set contains n elements, then the number of subsets that can be formed is 2^n.

NUMBER OF SUBSETS

The number of distinct subsets of a set with n elements is 2^n.

For a given set, we know that every subset except the set itself is a proper subset. In **Table 2.4**, we included the set itself when counting the number of subsets. If we want to find the number of proper subsets, we must exclude counting the given set, thereby decreasing the number by 1.

NUMBER OF PROPER SUBSETS

The number of distinct proper subsets of a set with n elements is $2^n - 1$.

EXAMPLE 4 *Finding the Number of Subsets and Proper Subsets*

Find the number of distinct subsets and the number of distinct proper subsets for each set:

a. $\{a, b, c, d, e\}$

b. $\{x \mid x \in \mathbf{N} \text{ and } 9 \le x \le 15\}$.

SOLUTION

a. A set with n elements has 2^n subsets. Because the set $\{a, b, c, d, e\}$ contains 5 elements, there are $2^5 = 2 \times 2 \times 2 \times 2 \times 2 = 32$ subsets. Of these, we must exclude counting the given set as a proper subset, so there are $2^5 - 1 = 32 - 1 = 31$ proper subsets.

b. We can write $\{x \mid x \in \mathbf{N} \text{ and } 9 \le x \le 15\}$ in roster form as $\{9, 10, 11, 12, 13, 14, 15\}$. Because this set contains 7 elements, there are $2^7 = 2 \times 2 \times 2 \times 2 \times 2 \times 2 \times 2 = 128$ subsets. Of these, there are $2^7 - 1 = 128 - 1 = 127$ proper subsets.

CHECK POINT 4 Find the number of distinct subsets and the number of distinct proper subsets for each set:

a. $\{a, b, c, d\}$

b. $\{x \mid x \in \mathbf{N} \text{ and } 3 \le x \le 8\}$.

4 *Apply concepts of subsets and equivalent sets to infinite sets.*

The Number of Subsets of Infinite Sets

In Section 2.1, we mentioned that the infinite set of natural numbers, $\{1, 2, 3, 4, 5, 6, \ldots\}$, is assigned the cardinal number $\aleph_0$ (read "aleph-null"), called a *transfinite* cardinal number. Equivalently, there are $\aleph_0$ natural numbers.

Once we accept the cardinality of sets with infinitely many elements, a surreal world emerges in which there is no end to an ascending hierarchy of infinities. Because the set of natural numbers contains $\aleph_0$ elements, it has $2^{\aleph_0}$ subsets, where $2^{\aleph_0} > \aleph_0$. Denoting $2^{\aleph_0}$ by $\aleph_1$, we have $\aleph_1 > \aleph_0$. Because the set of subsets of the natural numbers contains $\aleph_1$ elements, it has $2^{\aleph_1}$ subsets, where $2^{\aleph_1} > \aleph_1$. Denoting $2^{\aleph_1}$ by $\aleph_2$, we now have $\aleph_2 > \aleph_1 > \aleph_0$. Continuing in this manner, $\aleph_0$ is the "smallest" transfinite cardinal number in an infinite hierarchy of different infinities!

"Infinity is where things happen that don't."
—W. W. Sawyer, *Prelude to Mathematics*, Penguin Books, 1960

Blitzer Bonus

Cardinal Numbers of Infinite Sets

Time and Time Again (1981), P.J. Crook/Bridgeman Art Library

The mirrors in the painting *Time and Time Again* have the effect of repeating the image infinitely many times, creating an endless tunnel of mirror images. There is something quite fascinating about the idea of endless infinity. Did you know that for thousands of years religious leaders warned that human beings should not examine the nature of the infinite? Religious teaching often equated infinity with the concept of a Supreme Being. One of the last victims of the Inquisition, Giordano Bruno, was burned at the stake for his explorations into the characteristics of infinity. It was not until the 1870s that the German mathematician Georg Cantor (1845–1918) began a careful analysis of the mathematics of infinity.

It was Cantor who assigned the transfinite cardinal number $\aleph_0$ to the set of natural numbers. He used one-to-one correspondences to establish some surprising equivalences between the set of natural numbers and its proper subsets. Here are two examples:

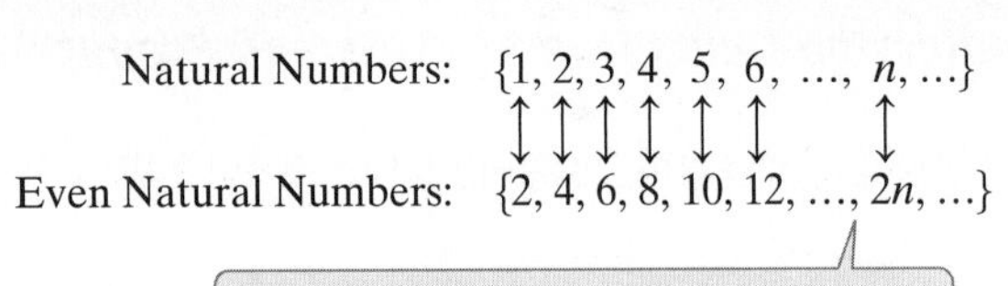

Each natural number, n, is paired with its double, $2n$, in the set of even natural numbers.

Natural Numbers: $\{1, 2, 3, 4, 5, 6, \ldots, n, \ldots\}$

Odd Natural Numbers: $\{1, 3, 5, 7, 9, 11, \ldots, 2n - 1, \ldots\}$

Each natural number, n, is paired with 1 less than its double, $2n - 1$, in the set of odd natural numbers.

These one-to-one correspondences indicate that the set of even natural numbers and the set of odd natural numbers are equivalent to the set of all natural numbers. In fact, an infinite set, such as the natural numbers, can be *defined* as any set that can be placed in a one-to-one correspondence with a proper subset of itself. This definition boggles the mind because it implies that part of a set has the same number of objects as the entire set. There are $\aleph_0$ even natural numbers, $\aleph_0$ odd natural numbers, and $\aleph_0$ natural numbers. Because the even and odd natural numbers combined make up the entire set of natural numbers, we are confronted with an unusual statement of transfinite arithmetic:

$$\aleph_0 + \aleph_0 = \aleph_0.$$

As Cantor continued studying infinite sets, his observations grew stranger and stranger. It was Cantor who showed that some infinite sets contain more elements than others. This was too much for his colleagues, who considered this work ridiculous. Cantor's mentor, Leopold Kronecker, told him, "Look at the crazy ideas that are now surfacing with your work with infinite sets. How can one infinity be greater than another? Best to ignore such inconsistencies. By considering these monsters and infinite numbers mathematics, I will make sure that you never gain a faculty position at the University of Berlin." Although Cantor was not burned at the stake, universal condemnation of his work resulted in numerous nervous breakdowns. His final days, sadly, were spent in a psychiatric hospital. However, Cantor's work later regained the respect of mathematicians. Today, he is seen as a great mathematician who demystified infinity.

Concept and Vocabulary Check

Fill in each blank so that the resulting statement is true.

1. Set A is a subset of set B, expressed as ________, means that ______________________________.
2. Set A is a proper subset of set B, expressed as ________, means that set A is a subset of set B and ______________________.
3. The statement $\varnothing \subseteq B$ tells us that __________ set is a ________ of every set.
4. The number of distinct subsets of a set with n elements is _____.
5. The number of distinct proper subsets of a set with n elements is ________.

Exercise Set 2.2

Practice Exercises

In Exercises 1–18, write $\subseteq$ or $\nsubseteq$ in each blank so that the resulting statement is true.

1. {1, 2, 5} ____ {1, 2, 3, 4, 5, 6, 7}
2. {2, 3, 7} ____ {1, 2, 3, 4, 5, 6, 7}
3. {−3, 0, 3} ____ {−3, −1, 1, 3}
4. {−4, 0, 4} ____ {−4, −3, −1, 1, 3, 4}
5. {Monday, Friday} ____ {Saturday, Sunday, Monday, Tuesday, Wednesday}
6. {Mercury, Venus, Earth} ____ {Venus, Earth, Mars, Jupiter}
7. $\{x \mid x$ is a cat$\}$ ____ $\{x \mid x$ is a black cat$\}$
8. $\{x \mid x$ is a dog$\}$ ____ $\{x \mid x$ is a pure-bred dog$\}$
9. {c, o, n, v, e, r, s, a, t, i, o, n} ____ {v, o, i, c, e, s, r, a, n, t, o, n}
10. {r, e, v, o, l, u, t, i, o, n} ____ {t, o, l, o, v, e, r, u, i, n}
11. $\{\frac{4}{7}, \frac{9}{13}\}$ ____ $\{\frac{7}{4}, \frac{13}{9}\}$
12. $\{\frac{1}{2}, \frac{1}{3}\}$ ____ {2, 3, 5}
13. $\varnothing$ ____ {2, 4, 6}
14. $\varnothing$ ____ {1, 3, 5}
15. {2, 4, 6} ____ $\varnothing$
16. {1, 3, 5} ____ $\varnothing$
17. { } ____ $\varnothing$
18. $\varnothing$ ____ { }

In Exercises 19–40, determine whether $\subseteq$, $\subset$, both, or neither can be placed in each blank to form a true statement.

19. {V, C, R} ____ {V, C, R, S}
20. {F, I, N} ____ {F, I, N, K}
21. {0, 2, 4, 6, 8} ____ {8, 0, 6, 2, 4}
22. {9, 1, 7, 3, 4} ____ {1, 3, 4, 7, 9}
23. $\{x \mid x$ is a man$\}$ ____ $\{x \mid x$ is a woman$\}$
24. $\{x \mid x$ is a woman$\}$ ____ $\{x \mid x$ is a man$\}$
25. $\{x \mid x$ is a man$\}$ ____ $\{x \mid x$ is a person$\}$
26. $\{x \mid x$ is a woman$\}$ ____ $\{x \mid x$ is a person$\}$
27. $\{x \mid x$ is a man or a woman$\}$ ____ $\{x \mid x$ is a person$\}$
28. $\{x \mid x$ is a woman or a man$\}$ ____ $\{x \mid x$ is a person$\}$
29. $A = \{x \mid x \in \mathbf{N}$ and $5 < x < 12\}$
    $B =$ the set of natural numbers between 5 and 12
    A ____ B
30. $A = \{x \mid x \in \mathbf{N}$ and $3 < x < 10\}$
    $B =$ the set of natural numbers between 3 and 10
    A ____ B
31. $A = \{x \mid x \in \mathbf{N}$ and $5 < x < 12\}$
    $B =$ the set of natural numbers between 3 and 17
    A ____ B
32. $A = \{x \mid x \in \mathbf{N}$ and $3 < x < 10\}$
    $B =$ the set of natural numbers between 2 and 16
    A ____ B
33. $A = \{x \mid x \in \mathbf{N}$ and $5 < x < 12\}$
    $B = \{x \mid x \in \mathbf{N}$ and $2 \le x \le 11\}$
    A ____ B
34. $A = \{x \mid x \in \mathbf{N}$ and $3 < x < 10\}$
    $B = \{x \mid x \in \mathbf{N}$ and $2 \le x \le 8\}$
    A ____ B
35. $\varnothing$ ____ {7, 8, 9, . . . , 100}
36. $\varnothing$ ____ {101, 102, 103, . . . , 200}
37. {7, 8, 9, . . . } ____ $\varnothing$
38. {101, 102, 103, . . . } ____ $\varnothing$
39. $\varnothing$ ____ { }
40. { } ____ $\varnothing$

In Exercises 41–54, determine whether each statement is true or false. If the statement is false, explain why.

41. Ralph $\in$ {Ralph, Alice, Trixie, Norton}
42. Canada $\in$ {Mexico, United States, Canada}
43. Ralph $\subseteq$ {Ralph, Alice, Trixie, Norton}
44. Canada $\subseteq$ {Mexico, United States, Canada}
45. {Ralph} $\subseteq$ {Ralph, Alice, Trixie, Norton}
46. {Canada} $\subseteq$ {Mexico, United States, Canada}
47. $\varnothing \in$ {Archie, Edith, Mike, Gloria}
48. $\varnothing \subseteq$ {Charlie Chaplin, Groucho Marx, Woody Allen}
49. $\{5\} \in \{\{5\}, \{9\}\}$
50. $\{1\} \in \{\{1\}, \{3\}\}$
51. $\{1, 4\} \nsubseteq \{4, 1\}$
52. $\{1, 4\} \not\subset \{4, 1\}$
53. $0 \notin \varnothing$
54. $\{0\} \nsubseteq \varnothing$

In Exercises 55–60, list all the subsets of the given set.

55. {border collie, poodle}
56. {Romeo, Juliet}
57. {t, a, b}
58. {I, II, III}
59. {0}
60. $\varnothing$

In Exercises 61–68, calculate the number of distinct subsets and the number of distinct proper subsets for each set.

61. {2, 4, 6, 8}
62. $\{\frac{1}{2}, \frac{1}{3}, \frac{1}{4}, \frac{1}{5}\}$
63. {2, 4, 6, 8, 10, 12}
64. {a, b, c, d, e, f}
65. $\{x \mid x$ is a day of the week$\}$
66. $\{x \mid x$ is a U.S. coin worth less than a dollar$\}$
67. $\{x \mid x \in \mathbf{N}$ and $2 < x < 6\}$
68. $\{x \mid x \in \mathbf{N}$ and $2 \le x \le 6\}$

Practice Plus

In Exercises 69–82, determine whether each statement is true or false. If the statement is false, make the necessary change(s) to produce a true statement.

69. The set {1, 2, 3, . . . , 1000} has 2^{1000} proper subsets.
70. The set {1, 2, 3, . . . , 10,000} has $2^{10,000}$ proper subsets.
71. $\{x \mid x \in \mathbf{N}$ and $30 < x < 50\} \subseteq \{x \mid x \in \mathbf{N}$ and $30 \le x \le 50\}$
72. $\{x \mid x \in \mathbf{N}$ and $20 \le x \le 60\} \nsubseteq \{x \mid x \in \mathbf{N}$ and $20 < x < 60\}$
73. $\varnothing \nsubseteq \{\varnothing, \{\varnothing\}\}$
74. $\{\varnothing\} \nsubseteq \{\varnothing, \{\varnothing\}\}$
75. $\varnothing \in \{\varnothing, \{\varnothing\}\}$

76. $\{\emptyset\} \in \{\emptyset, \{\emptyset\}\}$

77. If $A \subseteq B$ and $d \in A$, then $d \in B$.

78. If $A \subseteq B$ and $B \subseteq C$, then $A \subseteq C$.

79. If set A is equivalent to the set of natural numbers, then $n(A) = \aleph_0$.

80. If set A is equivalent to the set of even natural numbers, then $n(A) = \aleph_0$.

81. The set of subsets of {a, e, i, o, u} contains 64 elements.

82. The set of subsets of {a, b, c, d, e, f} contains 128 elements.

Application Exercises

We opened this section citing a Harris poll that estimated 45 million Americans have at least one tattoo. The bar graph on the left shows the percentage of tattooed Americans, by party affiliation and gender.

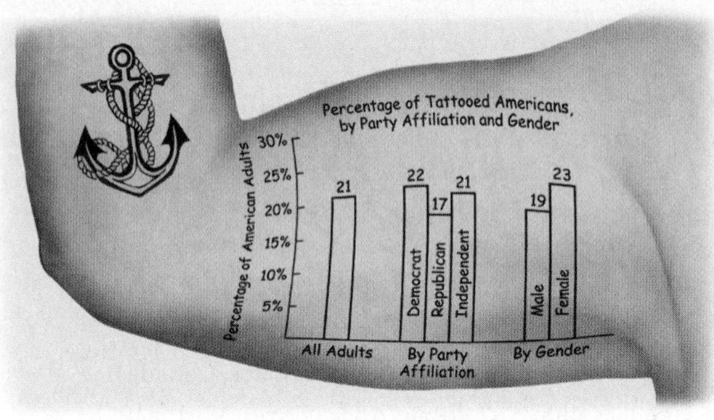

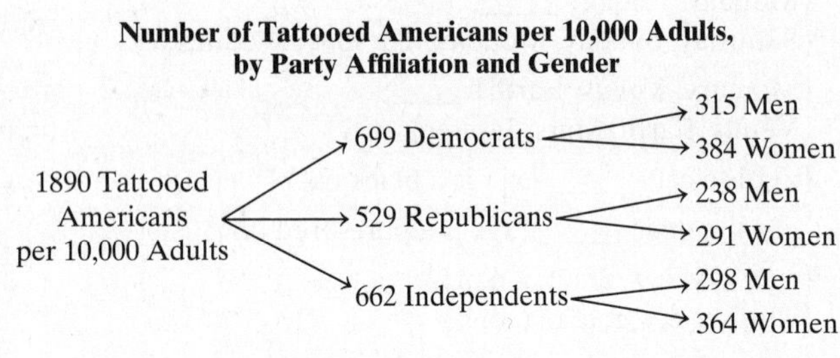

Source: Harris Interactive

Sets and subsets allow us to order and structure the data. On the right, the set of tattooed Americans is divided into subsets categorized by party affiliation. These subsets are further broken down into subsets categorized by gender. All numbers in the branching tree diagram are based on the number of people per 10,000 American adults. Based on the tree diagram, let

T = the set of tattooed Americans

R = the set of tattooed Republicans

D = the set of tattooed Democrats

M = the set of tattooed Democratic men

W = the set of tattooed Democratic women.

In Exercises 83–92, determine whether each statement is true or false. If the statement is false, make the necessary change(s) to produce a true statement.

83. $D \in T$

84. $R \in T$

85. $M \subset T$

86. $W \subset T$

87. If $x \in D$, then $x \in W$.

88. If $x \in D$, then $x \in M$.

89. If $x \in R$, then $x \notin D$.

90. If $x \in D$, then $x \notin R$.

91. The set of elements in M and W combined is equal to set D.

92. The set of elements in M and W combined is equivalent to set D.

93. Houses in Euclid Estates are all identical. However, a person can purchase a new house with some, all, or none of a set of options. This set includes {pool, screened-in balcony, lake view, alarm system, upgraded landscaping}. How many options are there for purchasing a house in this community?

94. A cheese pizza can be ordered with some, all, or none of the following set of toppings: {beef, ham, mushrooms, sausage, peppers, pepperoni, olives, prosciutto, onion}. How many different variations are available for ordering a pizza?

95. Based on more than 1500 ballots sent to film notables, the American Film Institute rated the top U.S. movies. The Institute selected *Citizen Kane* (1941), *The Godfather* (1972), *Casablanca* (1942), *Raging Bull* (1980), *Singin' in the Rain* (1952), and *Gone with the Wind* (1939) as the top six films. Suppose that you have all six films on DVD and decide to view some, all, or none of these films. How many viewing options do you have?

96. A small town has four police cars. If a radio dispatcher receives a call, depending on the nature of the situation, no cars, one car, two cars, three cars, or all four cars can be sent. How many options does the dispatcher have for sending the police cars to the scene of the caller?

97. According to the U.S. Census Bureau, the most ethnically diverse U.S. cities are New York City, Los Angeles, Miami, Chicago, Washington, D.C., Houston, San Diego, and Seattle. If you decide to visit some, all, or none of these cities, how many travel options do you have?

98. Film documentaries with the highest box office grosses include

Fahrenheit 9/11 ($222 million), *March of the Penguins* ($127 million), *Earth* ($109 million), *Justin Bieber: Never Say Never* ($99 million), *Oceans* ($83 million), *One Direction: This Is Us* ($69 million), and *Bowling for Columbine* ($58 million).
(*Source: Top 10 of Everything 2017*, Portable Press)

Suppose that you have all seven documentaries on DVD and decide, over the course of a week, to view some, all, or none of these films. How many viewing options do you have?

Explaining the Concepts

99. Explain what is meant by a subset.

100. What is the difference between a subset and a proper subset?

101. Explain why the empty set is a subset of every set.

102. Describe the difference between the symbols $\in$ and $\subseteq$. Explain how each symbol is used.

103. Describe the formula for finding the number of distinct subsets for a given set. Give an example.

104. Describe how to find the number of distinct proper subsets for a given set. Give an example.

Critical Thinking Exercises

Make Sense? *In Exercises 105–108, determine whether each statement makes sense or does not make sense, and explain your reasoning.*

105. The set of my six rent payments from January through June is a subset of the set of my 12 cable television payments from January through December.

106. Every time I increase the number of elements in a set by one, I double the number of distinct subsets.

107. Because Exercises 93–98 involve different situations, I cannot solve them by the same method.

108. I recently purchased a set of books and am deciding which books, if any, to take on vacation. The number of subsets of my set of books gives me the number of different combinations of the books that I can take.

In Exercises 109–112, determine whether each statement is true or false. If the statement is false, make the necessary change(s) to produce a true statement.

109. The set $\{3\}$ has 2^3, or eight, subsets.

110. All sets have subsets.

111. Every set has a proper subset.

112. The set $\{3, \{1, 4\}\}$ has eight subsets.

113. Suppose that a nickel, a dime, and a quarter are on a table. You may select some, all, or none of the coins. Specify all of the different amounts of money that can be selected.

114. If a set has 127 proper subsets, how many elements are there in the set?

Group Exercises

115. This activity is a group research project and should result in a presentation made by group members to the entire class. Georg Cantor was certainly not the only genius in history who faced criticism during his lifetime, only to have his work acclaimed as a masterpiece after his death. Describe the life and work of three other people, including at least one mathematician, who faced similar circumstances.

116. Research useful websites and present a report on infinite sets and their cardinalities. Explain why the sets of whole numbers, integers, and rational numbers each have cardinal number $\aleph_0$. Be sure to define these sets and show the one-to-one correspondences between each set and the set of natural numbers. Then explain why the set of real numbers does not have cardinal number $\aleph_0$ by describing how a real number can always be left out in a pairing with the natural numbers. Spice up the more technical aspects of your report with ideas you discovered about infinity that you find particularly intriguing.

2.3 Venn Diagrams and Set Operations

WHAT AM I SUPPOSED TO LEARN?

After studying this section, you should be able to:

1. Understand the meaning of a universal set.
2. Understand the basic ideas of a Venn diagram.
3. Use Venn diagrams to visualize relationships between two sets.
4. Find the complement of a set.
5. Find the intersection of two sets.
6. Find the union of two sets.
7. Perform operations with sets.
8. Determine sets involving set operations from a Venn diagram.
9. Understand the meaning of *and* and *or*.
10. Use the formula for $n(A \cup B)$.

LATINOS MAKE UP APPROXIMATELY 17% OF THE U.S. population and pump an estimated \$1.3 trillion into the economy each year, equal to the GDP of Mexico, the Dominican Republic, Guatemala, and El Salvador combined. (*Source*: *Time*) As Latino spending power steadily rises, corporate America has discovered that Hispanic Americans, particularly young spenders between the ages of 14 and 34, want to be spoken to in English, even as they stay true to their Latino identity.

What is the primary language spoken at home by U.S. Hispanics? In this section, we use sets to analyze the answer to this question. By doing so, you will see how sets and their visual representations provide precise ways of organizing, classifying, and describing a wide variety of data.

Universal Sets and Venn Diagrams

The circle graph in **Figure 2.3** on the next page categorizes America's 55 million Hispanics by the primary language spoken at home. The graph's sectors define four sets:

- the set of U.S. Hispanics who speak Spanish at home.
- the set of U.S. Hispanics who speak English at home.

1 *Understand the meaning of a universal set.*

- the set of U.S. Hispanics who speak both Spanish and English at home.
- the set of U.S. Hispanics who speak neither Spanish nor English at home.

In discussing sets, it is convenient to refer to a general set that contains all elements under discussion. This general set is called the *universal set*. A **universal set**, symbolized by U, is a set that contains all the elements being considered in a given discussion or problem. Thus, a convenient universal set for the sets described above is

$$U = \text{the set of U.S. Hispanics.}$$

Notice how this universal set restricts our attention so that we can divide it into the four subsets shown by the circle graph in **Figure 2.3**.

Languages Spoken at Home by U.S. Hispanics

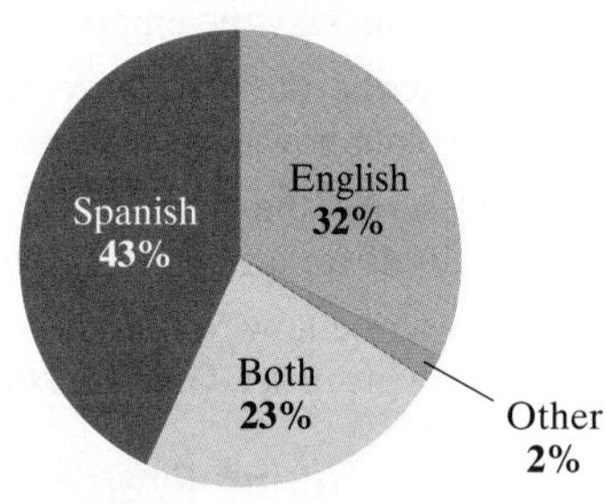

FIGURE 2.3
Source: Time

2 *Understand the basic ideas of a Venn diagram.*

We can obtain a more thorough understanding of sets and their relationship to a universal set by considering diagrams that allow visual analysis. **Venn diagrams**, named for the British logician John Venn (1834–1923), are used to show the visual relationship among sets.

Figure 2.4 is a Venn diagram. The universal set is represented by a region inside a rectangle. Subsets within the universal set are depicted by circles, or sometimes by ovals or other shapes. In this Venn diagram, set A is represented by the light blue region inside the circle.

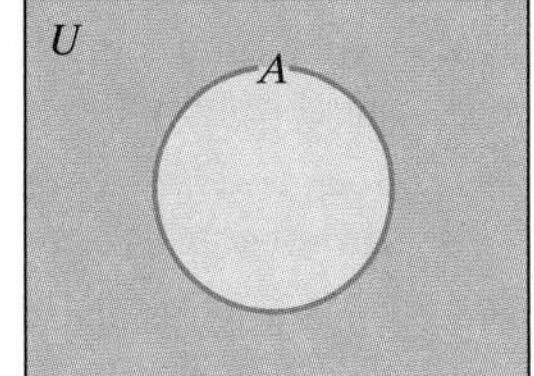

FIGURE 2.4

The dark blue region in **Figure 2.4** represents the set of elements in the universal set U that are not in set A. By combining the regions shown by the light blue shading and the dark blue shading, we obtain the universal set, U.

GREAT QUESTION!

Is the size of the circle in a Venn diagram important?

No. The size of the circle representing set A in a Venn diagram has nothing to do with the number of elements in set A.

EXAMPLE 1 Determining Sets from a Venn Diagram

Use the Venn diagram in **Figure 2.5** to determine each of the following sets:

a. U **b.** A **c.** the set of elements in U that are not in A.

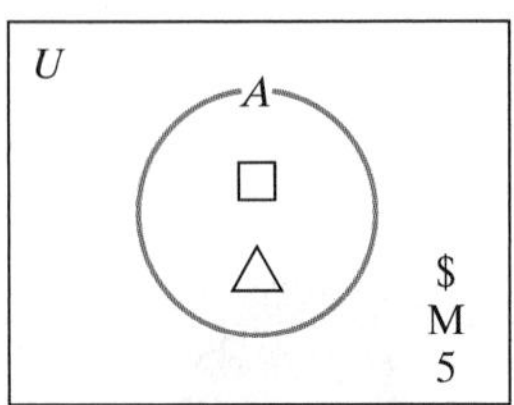

FIGURE 2.5

SOLUTION

a. Set U, the universal set, consists of all the elements within the rectangle. Thus, $U = \{\square, \triangle, \$, \text{M}, 5\}$.

b. Set A consists of all the elements within the circle. Thus, $A = \{\square, \triangle\}$.

c. The set of elements in U that are not in A, shown by the set of all the elements outside the circle, is $\{\$, \text{M}, 5\}$.

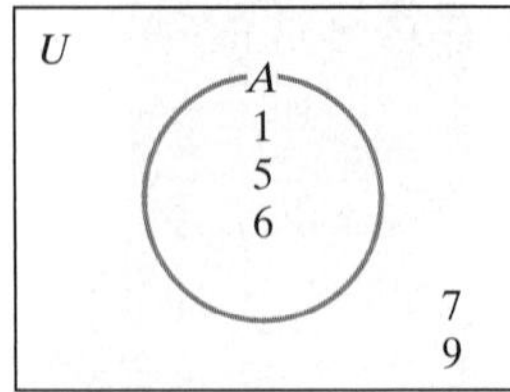

FIGURE 2.6

CHECK POINT 1 Use the Venn diagram in **Figure 2.6** to determine each of the following sets:

a. U **b.** A

c. the set of elements in U that are not in A.

3 *Use Venn diagrams to visualize relationships between two sets.*

Representing Two Sets in a Venn Diagram

There are a number of different ways to represent two subsets of a universal set in a Venn diagram. To help understand these representations, consider the following scenario:

You need to determine whether there is sufficient support on campus to have a blood drive. You take a survey to obtain information, asking students

Would you be willing to donate blood?

Would you be willing to help serve a free breakfast to blood donors?

Set A represents the set of students willing to donate blood. Set B represents the set of students willing to help serve breakfast to donors. Possible survey results include the following:

- No students willing to donate blood are willing to serve breakfast, and vice versa.
- All students willing to donate blood are willing to serve breakfast.
- The same students who are willing to donate blood are willing to serve breakfast.
- Some of the students willing to donate blood are willing to serve breakfast.

We begin by using Venn diagrams to visualize these results. To do so, we consider four basic relationships and their visualizations.

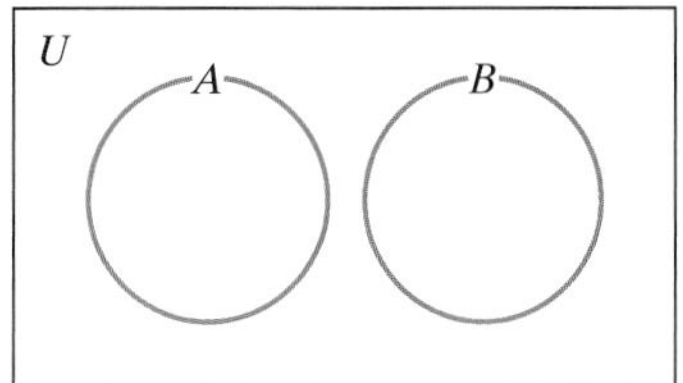

FIGURE 2.7

Relationship 1: Disjoint Sets Two sets that have no elements in common are called **disjoint sets**. Two disjoint sets, A and B, are shown in the Venn diagram in **Figure 2.7**. Disjoint sets are represented as circles that do not overlap. No elements of set A are elements of set B, and vice versa.

Since set A represents the set of students willing to donate blood and set B represents the set of students willing to serve breakfast to donors, the set diagram illustrates

No students willing to donate blood are willing to serve breakfast, and vice versa.

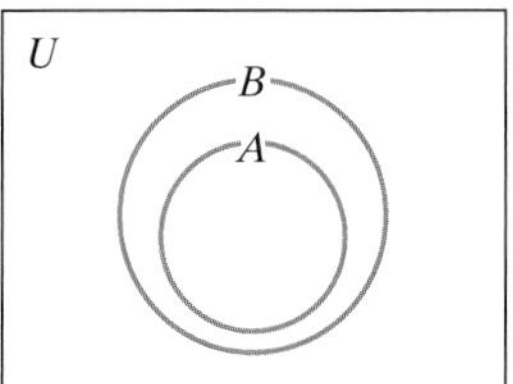

FIGURE 2.8

Relationship 2: Proper Subsets If set A is a proper subset of set B ($A \subset B$), the relationship is shown in the Venn diagram in **Figure 2.8**. All elements of set A are elements of set B. If an x representing an element is placed inside circle A, it automatically falls inside circle B.

Since set A represents the set of students willing to donate blood and set B represents the set of students willing to serve breakfast to donors, the set diagram illustrates

All students willing to donate blood are willing to serve breakfast.

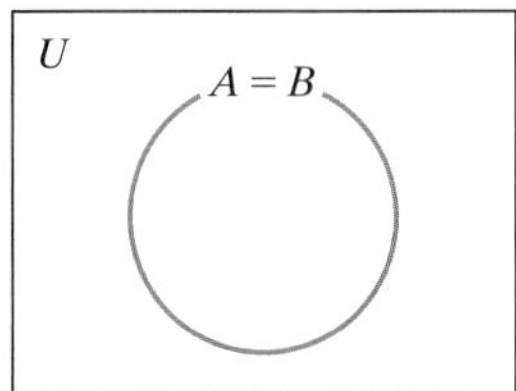

FIGURE 2.9

Relationship 3: Equal Sets If $A = B$, then set A contains exactly the same elements as set B. This relationship is shown in the Venn diagram in **Figure 2.9**. Because all elements in set A are in set B, and vice versa, this diagram illustrates that when $A = B$, then $A \subseteq B$ and $B \subseteq A$.

Since set A represents the set of students willing to donate blood and set B represents the set of students willing to serve breakfast to donors, the set diagram illustrates

The same students who are willing to donate blood are willing to serve breakfast.

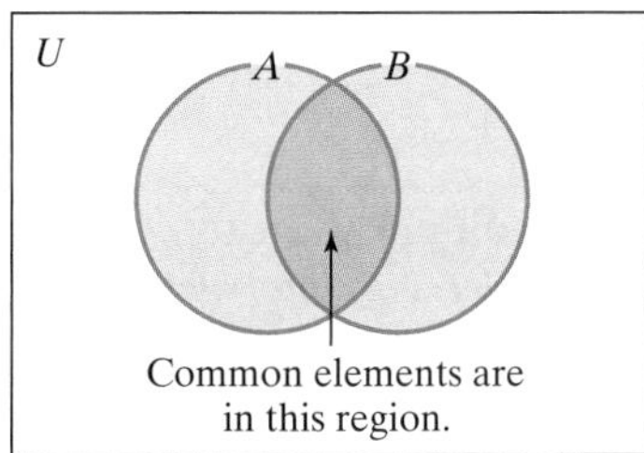

FIGURE 2.10

Relationship 4: Sets with Some Common Elements In mathematics, the word *some* means *there exists at least one*. If set A and set B have at least one element in common, then the circles representing the sets must overlap. This is illustrated in the Venn diagram in **Figure 2.10**.

Since set A represents the set of students willing to donate blood and set B represents the set of students willing to serve breakfast to donors, the presence of at least one student in the dark blue region in **Figure 2.10** illustrates

Some students willing to donate blood are willing to serve breakfast.

In **Figure 2.11** at the top of the next page, we've numbered each of the regions in the Venn diagram in **Figure 2.10**. Let's make sure we understand what these regions represent in terms of the campus blood drive scenario. Remember that A is the set of blood donors and B is the set of breakfast servers.

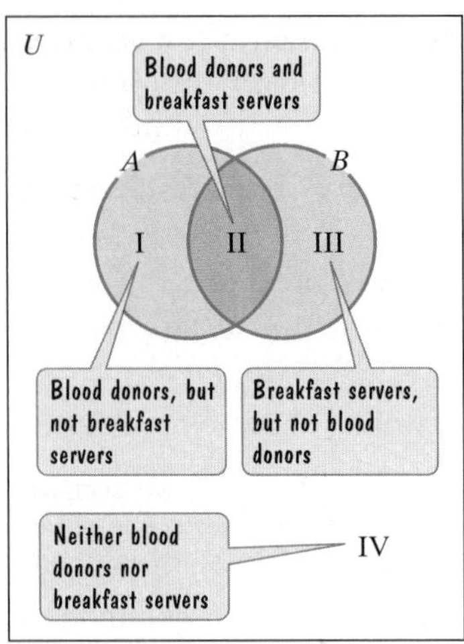

A: Set of blood donors
B: Set of breakfast servers

FIGURE 2.11

In **Figure 2.11**, we'll start with the innermost region, region II, and work outward to region IV.

Region II	This region represents the set of students willing to donate blood and serve breakfast. The elements that belong to both set A and set B are in this region.
Region I	This region represents the set of students willing to donate blood but not serve breakfast. The elements that belong to set A but not to set B are in this region.
Region III	This region represents the set of students willing to serve breakfast but not donate blood. The elements that belong to set B but not to set A are in this region.
Region IV	This region represents the set of students surveyed who are not willing to donate blood and are not willing to serve breakfast. The elements that belong to the universal set U that are not in sets A or B are in this region.

EXAMPLE 2 *Determining Sets from a Venn Diagram*

Use the Venn diagram in **Figure 2.12** to determine each of the following sets:

a. U **b.** B

c. the set of elements in A but not B

d. the set of elements in U that are not in B

e. the set of elements in both A and B.

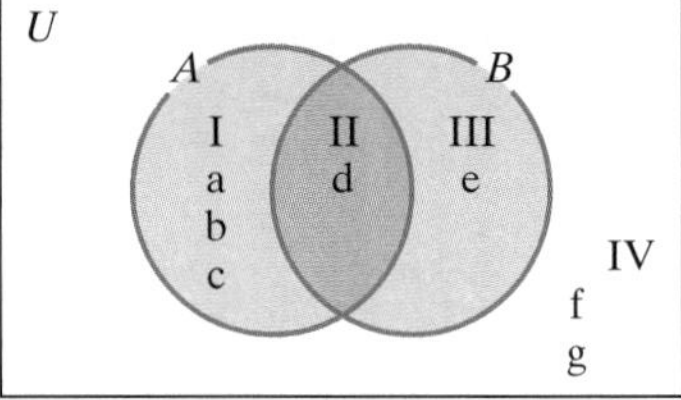

FIGURE 2.12

SOLUTION

a. Set U, the universal set, consists of all elements within the rectangle. Taking the elements in regions I, II, III, and IV, we obtain $U = \{a, b, c, d, e, f, g\}$.

b. Set B consists of the elements in regions II and III. Thus, $B = \{d, e\}$.

c. The set of elements in A but not B, found in region I, is $\{a, b, c\}$.

d. The set of elements in U that are not in B, found in regions I and IV, is $\{a, b, c, f, g\}$.

e. The set of elements in both A and B, found in region II, is $\{d\}$.

CHECK POINT 2 Use the Venn diagram in **Figure 2.12** to determine each of the following sets:

a. A

b. the set of elements in B but not A

c. the set of elements in U that are not in A

d. the set of elements in U that are not in A or B.

4 *Find the complement of a set.*

The Complement of a Set

In arithmetic, we use operations such as addition and multiplication to combine numbers. We now turn to three set operations, called *complement, intersection*, and *union*. We begin by defining a set's complement.

DEFINITION OF THE COMPLEMENT OF A SET

The **complement** of set A, symbolized by A', is the set of all elements in the universal set that are *not* in A. This idea can be expressed in set-builder notation as follows:

$$A' = \{x \mid x \in U \quad \text{and} \quad x \notin A\}.$$

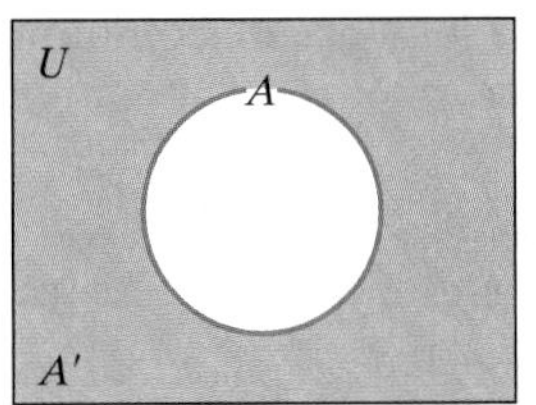

FIGURE 2.13

The shaded region in **Figure 2.13** represents the complement of set A, or A'. This region lies outside circle A, but within the rectangular universal set.

In order to find A', a universal set U must be given. A fast way to find A' is to cross out the elements in U that are given to be in set A. A' is the set that remains.

EXAMPLE 3 Finding a Set's Complement

Let $U = \{1, 2, 3, 4, 5, 6, 7, 8, 9\}$ and $A = \{1, 3, 4, 7\}$. Find A'.

SOLUTION

Set A' contains all the elements of set U that are not in set A. Because set A contains the elements 1, 3, 4, and 7, these elements cannot be members of set A':

$$\{\not{1}, 2, \not{3}, \not{4}, 5, 6, \not{7}, 8, 9\}.$$

Thus, set A' contains 2, 5, 6, 8, and 9:

$$A' = \{2, 5, 6, 8, 9\}.$$

A Venn diagram illustrating A and A' is shown in **Figure 2.14**.

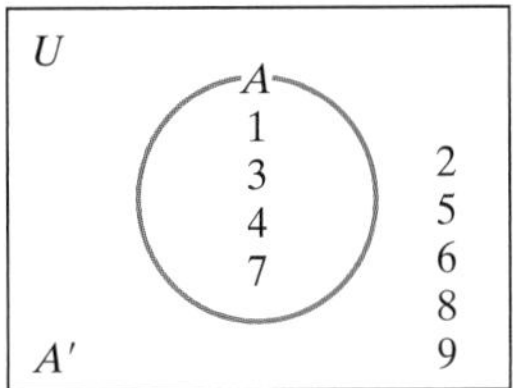

FIGURE 2.14

CHECK POINT 3 Let $U = \{a, b, c, d, e\}$ and $A = \{a, d\}$. Find A'.

5 *Find the intersection of two sets.*

The Intersection of Sets

If A and B are sets, we can form a new set consisting of all elements that are in both A and B. This set is called the *intersection* of the two sets.

DEFINITION OF THE INTERSECTION OF SETS

The **intersection** of sets A and B, written $A \cap B$, is the set of elements common to both set A and set B. This definition can be expressed in set-builder notation as follows:

$$A \cap B = \{x \mid x \in A \quad \text{and} \quad x \in B\}.$$

In Example 4, we are asked to find the intersection of two sets. This is done by listing the common elements of both sets. Because the intersection of two sets is also a set, we enclose these elements with braces.

EXAMPLE 4 Finding the Intersection of Two Sets

Find each of the following intersections:

a. $\{7, 8, 9, 10, 11\} \cap \{6, 8, 10, 12\}$

b. $\{1, 3, 5, 7, 9\} \cap \{2, 4, 6, 8\}$

c. $\{1, 3, 5, 7, 9\} \cap \varnothing$.

SOLUTION

a. The elements common to $\{7, 8, 9, 10, 11\}$ and $\{6, 8, 10, 12\}$ are 8 and 10. Thus,

$$\{7, 8, 9, 10, 11\} \cap \{6, 8, 10, 12\} = \{8, 10\}.$$

The Venn diagram in **Figure 2.15** illustrates this situation.

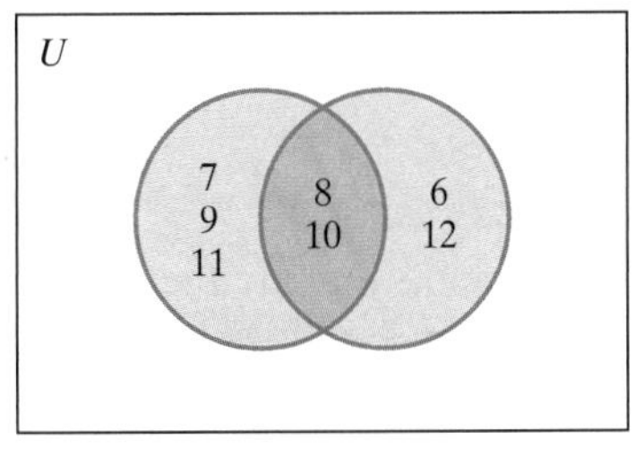

FIGURE 2.15 The numbers 8 and 10 belong to both sets.

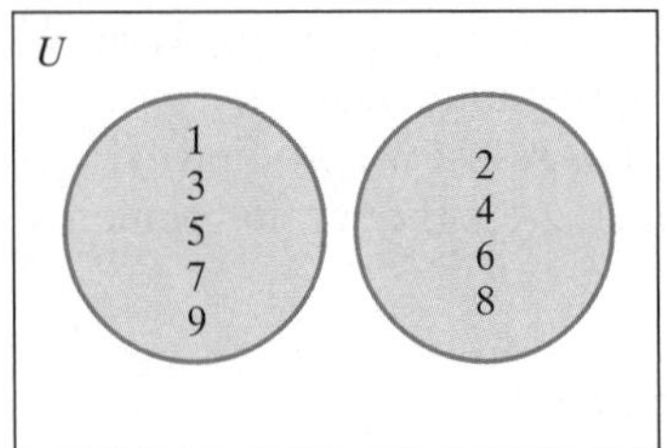

FIGURE 2.16 These disjoint sets have no common elements.

b. The sets $\{1, 3, 5, 7, 9\}$ and $\{2, 4, 6, 8\}$ have no elements in common. Thus,

$$\{1, 3, 5, 7, 9\} \cap \{2, 4, 6, 8\} = \varnothing.$$

The Venn diagram in **Figure 2.16** illustrates this situation. The sets are disjoint.

c. There are no elements in $\varnothing$, the empty set. This means that there can be no elements belonging to both $\{1, 3, 5, 7, 9\}$ and $\varnothing$. Therefore,

$$\{1, 3, 5, 7, 9\} \cap \varnothing = \varnothing.$$

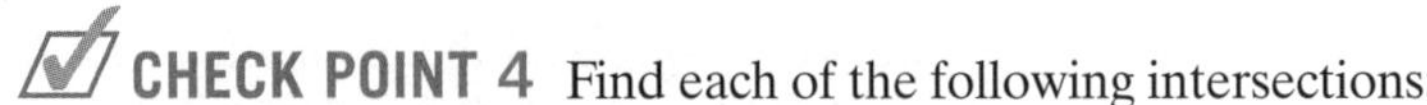

CHECK POINT 4 Find each of the following intersections:

a. $\{1, 3, 5, 7, 10\} \cap \{6, 7, 10, 11\}$

b. $\{1, 2, 3\} \cap \{4, 5, 6, 7\}$

c. $\{1, 2, 3\} \cap \varnothing$.

GREAT QUESTION!

Set theory seems so abstract. For instance, how do I come across the intersection of two sets in my daily life?

Here's an example: TV celebrities earning more than \$80 million. This is the intersection of the set of TV celebrities and the set of people earning more than \$80 million. It's easy not to notice set theory, but if you look at the media and listen closely to conversations, it's all over the place.

TV Celebrities Earning More Than \$80 Million between June 2013 and June 2014

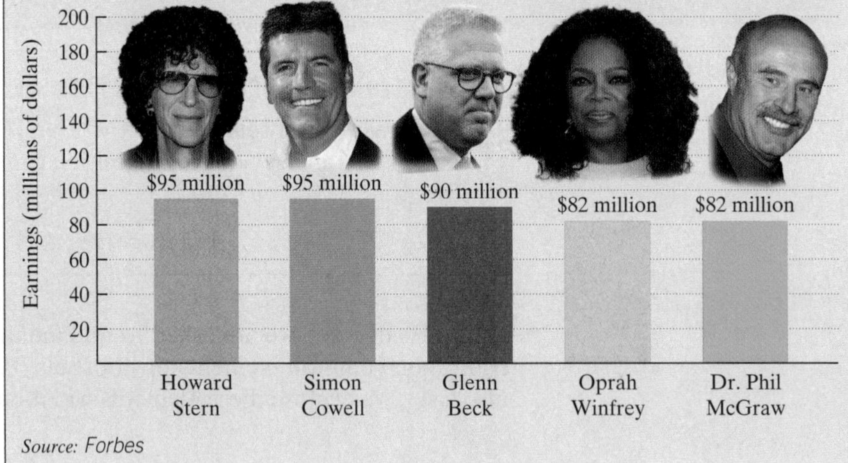

Source: Forbes

6 *Find the union of two sets.*

The Union of Sets

Another set that we can form from sets A and B consists of elements that are in A or B or in both sets. This set is called the *union* of the two sets.

DEFINITION OF THE UNION OF SETS

The **union** of sets A and B, written $A \cup B$, is the set of elements that are members of set A or of set B or of both sets. This definition can be expressed in set-builder notation as follows:

$$A \cup B = \{x \mid x \in A \quad \text{or} \quad x \in B\}.$$

We can find the union of set A and set B by listing the elements of set A. Then, we include any elements of set B that have not already been listed. Enclose all elements that are listed with braces. This shows that the union of two sets is also a set.

EXAMPLE 5 *Finding the Union of Two Sets*

Find each of the following unions:

a. $\{7, 8, 9, 10, 11\} \cup \{6, 8, 10, 12\}$

b. $\{1, 3, 5, 7, 9\} \cup \{2, 4, 6, 8\}$

c. $\{1, 3, 5, 7, 9\} \cup \varnothing$.

SOLUTION

This example uses the same sets as in Example 4. However, this time we are finding the unions of the sets, rather than their intersections.

a. To find $\{7, 8, 9, 10, 11\} \cup \{6, 8, 10, 12\}$, start by listing all the elements from the first set, namely 7, 8, 9, 10, and 11. Now list all the elements from the second set that are not in the first set, namely 6 and 12. The union is the set consisting of all these elements. Thus,

$$\{7, 8, 9, 10, 11\} \cup \{6, 8, 10, 12\} = \{6, 7, 8, 9, 10, 11, 12\}.$$

b. To find $\{1, 3, 5, 7, 9\} \cup \{2, 4, 6, 8\}$, list the elements from the first set, namely 1, 3, 5, 7, and 9. Now add to the list the elements in the second set that are not in the first set. This includes every element in the second set, namely 2, 4, 6, and 8. The union is the set consisting of all these elements, so

$$\{1, 3, 5, 7, 9\} \cup \{2, 4, 6, 8\} = \{1, 2, 3, 4, 5, 6, 7, 8, 9\}.$$

c. To find $\{1, 3, 5, 7, 9\} \cup \varnothing$, list the elements from the first set, namely 1, 3, 5, 7, and 9. Because there are no elements in $\varnothing$, the empty set, there are no additional elements to add to the list. Thus,

$$\{1, 3, 5, 7, 9\} \cup \varnothing = \{1, 3, 5, 7, 9\}.$$

GREAT QUESTION!

When finding the union of two sets, what should I do if some elements appear in both sets?

List these common elements only once, *not twice*, in the union of the sets.

Examples 4 and 5 illustrate the role that the empty set plays in intersection and union.

THE EMPTY SET IN INTERSECTION AND UNION

For any set A,

1. $A \cap \varnothing = \varnothing$

2. $A \cup \varnothing = A$.

 **CHECK POINT 5** Find each of the following unions:

a. $\{1, 3, 5, 7, 10\} \cup \{6, 7, 10, 11\}$

b. $\{1, 2, 3\} \cup \{4, 5, 6, 7\}$

c. $\{1, 2, 3\} \cup \varnothing$.

7 *Perform operations with sets.*

GREAT QUESTION!

How can I use the words *union* and *intersection* to help me distinguish between these two operations?

Union, as in a marriage union, suggests joining things, or uniting them. Intersection, as in the intersection of two crossing streets, brings to mind the area common to both, suggesting things that overlap.

Performing Set Operations

Some problems involve more than one set operation. The set notation specifies the order in which we perform these operations. **Always begin by performing any operations inside parentheses.** Here are two examples involving sets we will find in Example 6.

- Finding $(A \cup B)'$

Step 1. Parentheses indicate to first find the union of A and B.

Step 2. Find the complement of $A \cup B$.

- Finding $A' \cap B'$

Step 1. Find the complement of A.

Step 2. Find the complement of B.

Step 3. Find the intersection of A' and B'.

EXAMPLE 6 *Performing Set Operations*

Given

$$U = \{1, 2, 3, 4, 5, 6, 7, 8, 9, 10\}$$
$$A = \{1, 3, 7, 9\}$$
$$B = \{3, 7, 8, 10\},$$

find each of the following sets:

a. $(A \cup B)'$ **b.** $A' \cap B'$.

SOLUTION

a. To find $(A \cup B)'$, we will first work inside the parentheses and determine $A \cup B$. Then we'll find the complement of $A \cup B$, namely $(A \cup B)'$.

$A \cup B = \{1, 3, 7, 9\} \cup \{3, 7, 8, 10\}$ These are the given sets.

$= \{1, 3, 7, 8, 9, 10\}$ Join (unite) the elements, listing the common elements (3 and 7) only once.

Now find $(A \cup B)'$, the complement of $A \cup B$.

$(A \cup B)' = \{1, 3, 7, 8, 9, 10\}'$

$= \{2, 4, 5, 6\}$ List the elements in the universal set that are not listed in $\{1, 3, 7, 8, 9, 10\}$: $\{\not{1}, 2, \not{3}, 4, 5, 6, \not{7}, \not{8}, \not{9}, \not{10}\}$.

b. To find $A' \cap B'$, we must first identify the elements in A' and B'. Set A' is the set of elements of U that are not in set A:

$A' = \{2, 4, 5, 6, 8, 10\}$. List the elements in the universal set that are not listed in $A = \{1, 3, 7, 9\}$: $\{\not{1}, 2, \not{3}, 4, 5, 6, \not{7}, 8, \not{9}, 10\}$.

Set B' is the set of elements of U that are not in set B:

$B' = \{1, 2, 4, 5, 6, 9\}$. List the elements in the universal set that are not listed in $B = \{3, 7, 8, 10\}$: $\{1, 2, \not{3}, 4, 5, 6, \not{7}, \not{8}, 9, \not{10}\}$.

Now we can find $A' \cap B'$, the set of elements belonging to both A' and to B':

$A' \cap B' = \{2, 4, 5, 6, 8, 10\} \cap \{1, 2, 4, 5, 6, 9\}$

$= \{2, 4, 5, 6\}$. The numbers 2, 4, 5, and 6 are common to both sets.

CHECK POINT 6 Given $U = \{a, b, c, d, e\}$, $A = \{b, c\}$, and $B = \{b, c, e\}$, find each of the following sets:

a. $(A \cup B)'$ **b.** $A' \cap B'$.

8 *Determine sets involving set operations from a Venn diagram.*

EXAMPLE 7 Determining Sets from a Venn Diagram

The Venn diagram in **Figure 2.17** percolates with interesting numbers. Use the diagram to determine each of the following sets:

a. $A \cup B$ **b.** $(A \cup B)'$ **c.** $A \cap B$

d. $(A \cap B)'$ **e.** $A' \cap B$ **f.** $A \cup B'$.

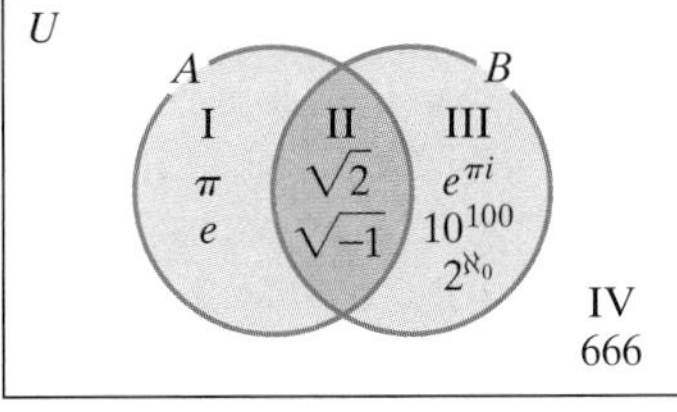

FIGURE 2.17

SOLUTION

Refer to **Figure 2.17.**

Set to Determine	Description of Set	Regions in Venn Diagram in Figure 2.17	Set in Roster Form
a. $A \cup B$	set of elements in A or B or both	I, II, III	$\{\pi, e, \sqrt{2}, \sqrt{-1}, e^{\pi i}, 10^{100}, 2^{\aleph_0}\}$
b. $(A \cup B)'$	set of elements in U that are not in $A \cup B$	IV	$\{666\}$
c. $A \cap B$	set of elements in both A and B	II	$\{\sqrt{2}, \sqrt{-1}\}$
d. $(A \cap B)'$	set of elements in U that are not in $A \cap B$	I, III, IV	$\{\pi, e, e^{\pi i}, 10^{100}, 2^{\aleph_0}, 666\}$
e. $A' \cap B$	set of elements that are not in A and are in B	III	$\{e^{\pi i}, 10^{100}, 2^{\aleph_0}\}$
f. $A \cup B'$	set of elements that are in A or not in B or both	I, II, IV	$\{\pi, e, \sqrt{2}, \sqrt{-1}, 666\}$

CHECK POINT 7 Use the Venn diagram in **Figure 2.18** to determine each of the following sets:

a. $A \cap B$

b. $(A \cap B)'$

c. $A \cup B$

d. $(A \cup B)'$

e. $A' \cup B$

f. $A \cap B'$.

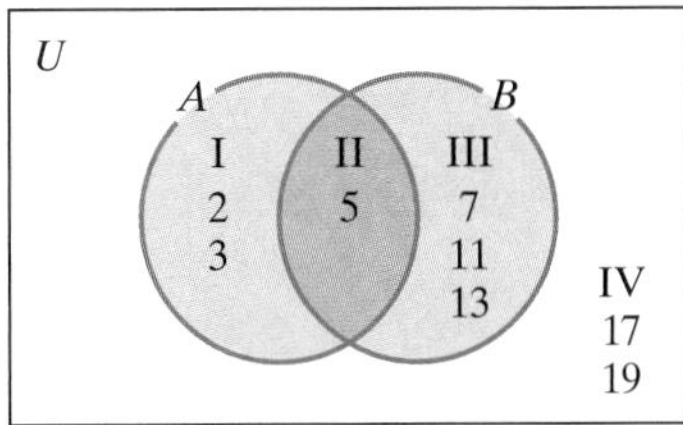

FIGURE 2.18

9 *Understand the meaning of* and *and* or.

Sets and Precise Use of Everyday English

Set operations and Venn diagrams provide precise ways of organizing, classifying, and describing the vast array of sets and subsets we encounter every day. Let's see how this applies to the sets from the beginning of this section:

U = the set of U.S. Hispanics

S = the set of U.S. Hispanics who speak Spanish at home

E = the set of U.S. Hispanics who speak English at home.

When describing collections in everyday English, the word **or** refers to the **union** of sets. Thus, U.S. Hispanics who speak Spanish or English at home means those who speak Spanish or English or both. The word **and** refers to the **intersection** of sets. Thus, U.S. Hispanics who speak Spanish and English at home means those who speak both languages.

In **Figure 2.19**, we revisit the circle graph showing languages spoken at home by U.S. Hispanics. To the right of the circle graph, we've organized the data using a Venn diagram. The voice balloons indicate how the Venn diagram provides a more accurate understanding of the subsets and their data.

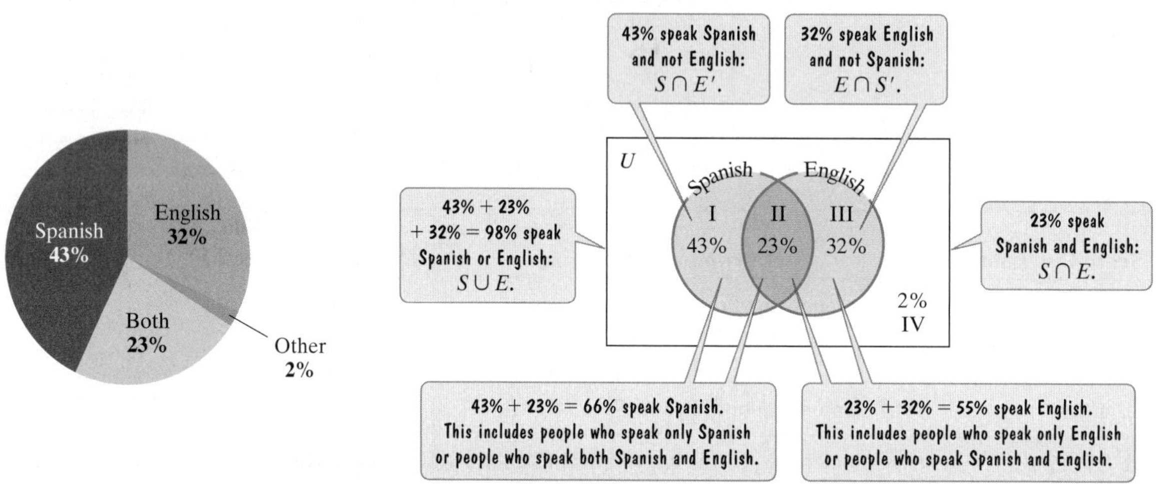

FIGURE 2.19 Comparing a circle graph and a Venn diagram
Source: Time

10 *Use the formula for $n(A \cup B)$.*

The Cardinal Number of the Union of Two Finite Sets

Can the number of elements in A or B, $n(A \cup B)$, be determined by adding the number of elements in A and the number of elements in B, $n(A) + n(B)$? The answer is no. **Figure 2.20** illustrates that by doing this, we are counting elements in both sets, $A \cap B$, or region II, twice.

To find the number of elements in the union of finite sets A and B, add the number of elements in A and the number of elements in B. Then subtract the number of elements common to both sets. We perform this subtraction so that we do not count the number of elements in the intersection twice, once for $n(A)$, and again for $n(B)$.

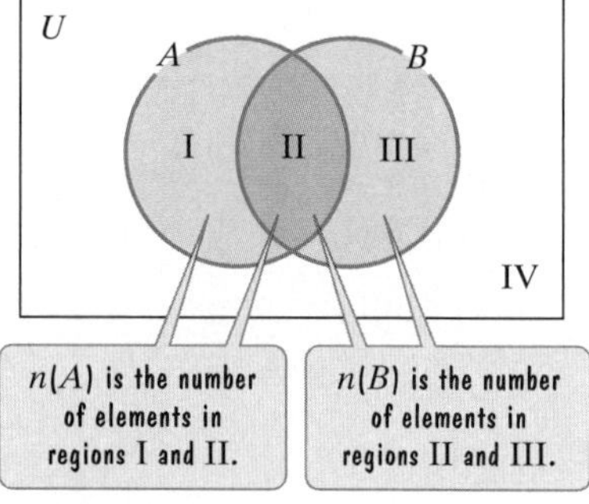

FIGURE 2.20

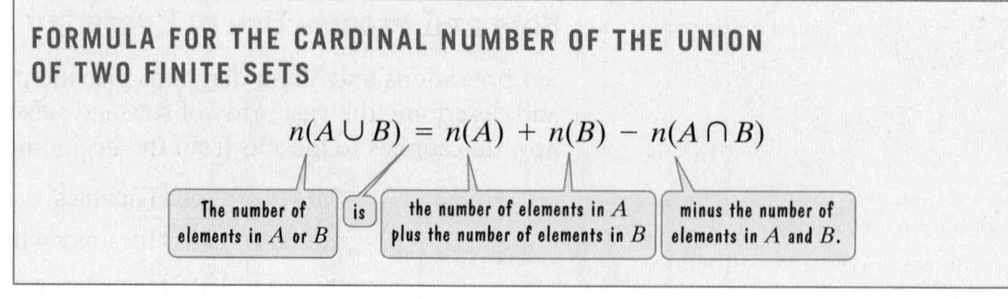

EXAMPLE 8 Using the Formula for $n(A \cup B)$

Some of the results of the campus blood drive survey indicated that 490 students were willing to donate blood, 340 students were willing to help serve a free breakfast to blood donors, and 120 students were willing to donate blood and serve breakfast. How many students were willing to donate blood or serve breakfast?

SOLUTION

Let A = the set of students willing to donate blood and B = the set of students willing to serve breakfast. We are interested in how many students were willing to donate blood or serve breakfast. Thus, we need to determine $n(A \cup B)$.

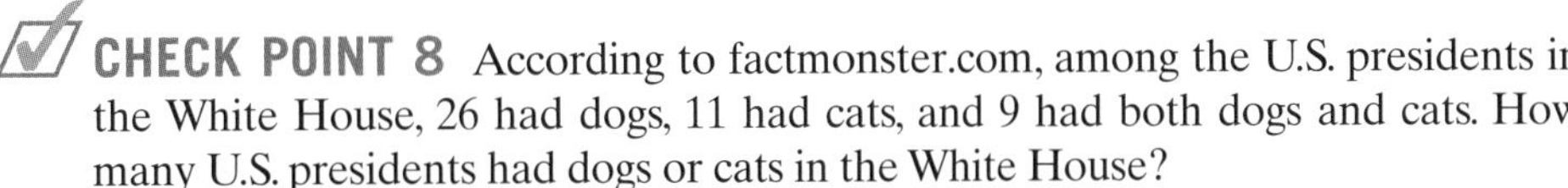

$$
\begin{aligned}
n(A \cup B) &= n(A) + n(B) - n(A \cap B) \\
&= 490 + 340 - 120 \\
&= 830 - 120 \\
&= 710
\end{aligned}
$$

We see that 710 students were willing to donate blood or serve a free breakfast.

CHECK POINT 8 According to factmonster.com, among the U.S. presidents in the White House, 26 had dogs, 11 had cats, and 9 had both dogs and cats. How many U.S. presidents had dogs or cats in the White House?

Concept and Vocabulary Check

Fill in each blank so that the resulting statement is true.

1. Visual relationships among sets are shown by ______________.
2. The set of all elements in the universal set that are not in set A is called the ____________ of set A, and is symbolized by _____.
3. The set of elements common to both set A and set B is called the ____________ of sets A and B, and is symbolized by ________.
4. The set of elements that are members of set A or set B or of both sets is called the _______ of sets A and B, and is symbolized by ________.
5. The formula for the cardinal number of elements in set A or set B is $n(A \cup B)$ = ________________________.
6. True or False: Disjoint sets are represented by circles that do not overlap. _______
7. True or False: If set A is a proper subset of set B, the sets are represented by two circles where circle A is drawn outside of circle B. ______
8. True or False: Equal sets are represented by the same circle. ______
9. True or False: As the number of elements in a set increases, larger circles are needed to represent the set. ______

Exercise Set 2.3

Practice Exercises

In Exercises 1–4, describe a universal set U that includes all elements in the given sets. Answers may vary.

1. $A = \{\text{Bach, Mozart, Beethoven}\}$
$B = \{\text{Brahms, Schubert}\}$

2. $A = \{\text{William Shakespeare, Charles Dickens}\}$
$B = \{\text{Mark Twain, Robert Louis Stevenson}\}$

3. $A = \{\text{Pepsi, Sprite}\}$
$B = \{\text{Coca-Cola, Seven-Up}\}$

4. $A = \{\text{Acura RDX, Toyota Camry, Mitsubishi Lancer}\}$
$B = \{\text{Dodge Ram, Chevrolet Impala}\}$

In Exercises 5–8, let $U = \{a, b, c, d, e, f, g\}$, $A = \{a, b, f, g\}$, $B = \{c, d, e\}$, $C = \{a, g\}$, *and* $D = \{a, b, c, d, e, f\}$. *Use the roster method to write each of the following sets.*

5. A' **6.** B' **7.** C' **8.** D'

In Exercises 9–12, let $U = \{1, 2, 3, 4, \ldots, 20\}$, $A = \{1, 2, 3, 4, 5\}$, $B = \{6, 7, 8, 9\}$, $C = \{1, 3, 5, 7, \ldots, 19\}$, *and* $D = \{2, 4, 6, 8, \ldots, 20\}$. *Use the roster method to write each of the following sets.*

9. A' **10.** B' **11.** C' **12.** D'

In Exercises 13–16, let $U = \{1, 2, 3, 4, \ldots\}$, $A = \{1, 2, 3, 4, \ldots, 20\}$, $B = \{1, 2, 3, 4, \ldots, 50\}$, $C = \{2, 4, 6, 8, \ldots\}$, *and* $D = \{1, 3, 5, 7, \ldots\}$. *Use the roster method to write each of the following sets.*

13. A' **14.** B' **15.** C' **16.** D'

In Exercises 17–40, let

$$U = \{1, 2, 3, 4, 5, 6, 7\}$$
$$A = \{1, 3, 5, 7\}$$
$$B = \{1, 2, 3\}$$
$$C = \{2, 3, 4, 5, 6\}.$$

Find each of the following sets.

17. $A \cap B$ **18.** $B \cap C$ **19.** $A \cup B$
20. $B \cup C$ **21.** A' **22.** B'
23. $A' \cap B'$ **24.** $B' \cap C$ **25.** $A \cup C'$
26. $B \cup C'$ **27.** $(A \cap C)'$ **28.** $(A \cap B)'$
29. $A' \cup C'$ **30.** $A' \cup B'$ **31.** $(A \cup B)'$
32. $(A \cup C)'$ **33.** $A \cup \varnothing$ **34.** $C \cup \varnothing$
35. $A \cap \varnothing$ **36.** $C \cap \varnothing$ **37.** $A \cup U$
38. $B \cup U$ **39.** $A \cap U$ **40.** $B \cap U$

In Exercises 41–66, let

$$U = \{a, b, c, d, e, f, g, h\}$$
$$A = \{a, g, h\}$$
$$B = \{b, g, h\}$$
$$C = \{b, c, d, e, f\}.$$

Find each of the following sets.

41. $A \cap B$ **42.** $B \cap C$ **43.** $A \cup B$
44. $B \cup C$ **45.** A' **46.** B'
47. $A' \cap B'$ **48.** $B' \cap C$ **49.** $A \cup C'$
50. $B \cup C'$ **51.** $(A \cap C)'$ **52.** $(A \cap B)'$
53. $A' \cup C'$ **54.** $A' \cup B'$ **55.** $(A \cup B)'$
56. $(A \cup C)'$ **57.** $A \cup \varnothing$ **58.** $C \cup \varnothing$
59. $A \cap \varnothing$ **60.** $C \cap \varnothing$ **61.** $A \cup U$
62. $B \cup U$ **63.** $A \cap U$ **64.** $B \cap U$
65. $(A \cap B) \cup B'$ **66.** $(A \cup B) \cap B'$

In Exercises 67–78, use the Venn diagram to represent each set in roster form.

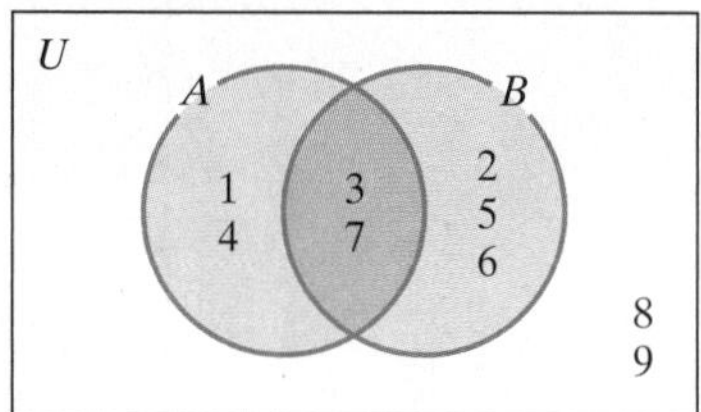

67. A **68.** B
69. U **70.** $A \cup B$
71. $A \cap B$ **72.** A'
73. B' **74.** $(A \cap B)'$
75. $(A \cup B)'$ **76.** $A' \cap B$
77. $A \cap B'$ **78.** $A \cup B'$

In Exercises 79–92, use the Venn diagram to determine each set or cardinality.

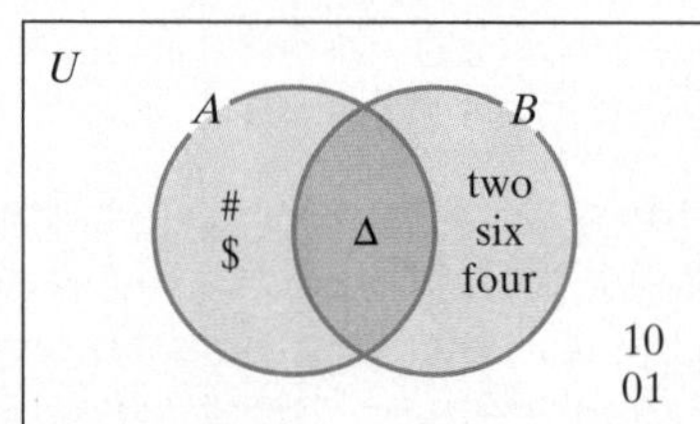

79. B **80.** A **81.** $A \cup B$
82. $A \cap B$ **83.** $n(A \cup B)$ **84.** $n(A \cap B)$
85. $n(A')$ **86.** $n(B')$ **87.** $(A \cap B)'$
88. $(A \cup B)'$ **89.** $A' \cap B$ **90.** $A \cap B'$
91. $n(U) - n(B)$ **92.** $n(U) - n(A)$

Use the formula for the cardinal number of the union of two sets to solve Exercises 93–96.

93. Set A contains 17 elements, set B contains 20 elements, and 6 elements are common to sets A and B. How many elements are in $A \cup B$?

94. Set A contains 30 elements, set B contains 18 elements, and 5 elements are common to sets A and B. How many elements are in $A \cup B$?

95. Set A contains 8 letters and 9 numbers. Set B contains 7 letters and 10 numbers. Four letters and 3 numbers are common to both sets A and B. Find the number of elements in set A or set B.

96. Set A contains 12 numbers and 18 letters. Set B contains 14 numbers and 10 letters. One number and 6 letters are common to both sets A and B. Find the number of elements in set A or set B.

Practice Plus

In Exercises 97–104, let

$U = \{x | x \in \mathbf{N} \text{ and } x < 9\}$

$A = \{x | x \text{ is an odd natural number and } x < 9\}$

$B = \{x | x \text{ is an even natural number and } x < 9\}$

$C = \{x | x \in \mathbf{N} \text{ and } 1 < x < 6\}$.

Find each of the following sets.

97. $A \cup B$ **98.** $B \cup C$

99. $A \cap U$ **100.** $A \cup U$

101. $A \cap C'$ **102.** $A \cap B'$

103. $(B \cap C)'$ **104.** $(A \cap C)'$

In Exercises 105–108, use the Venn diagram to determine each set or cardinality.

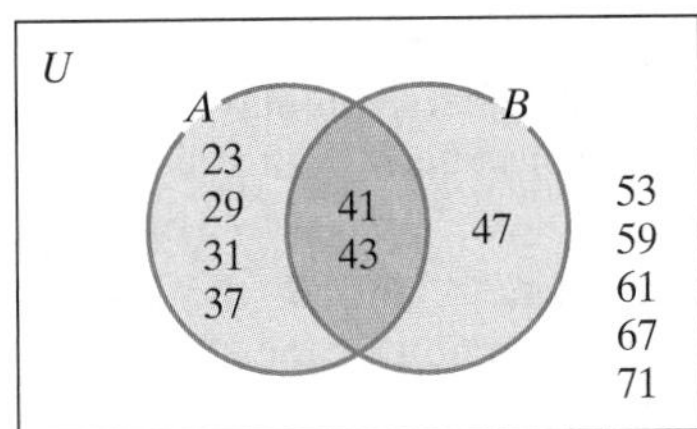

105. $A \cup (A \cup B)'$

106. $(A' \cap B) \cup (A \cap B)$

107. $n(U)[n(A \cup B) - n(A \cap B)]$

108. $n(A \cap B)[n(A \cup B) - n(A')]$

Application Exercises

A math tutor working with a small group of students asked each student when he or she had studied for class the previous weekend. Their responses are shown in the Venn diagram.

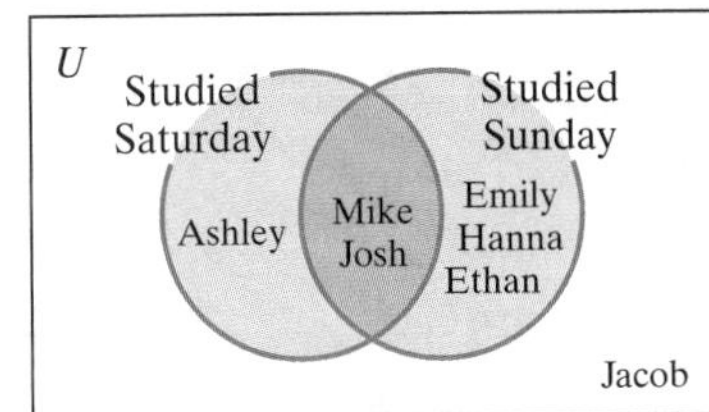

In Exercises 109–116, use the Venn diagram to list the elements of each set in roster form.

109. The set of students who studied Saturday

110. The set of students who studied Sunday

111. The set of students who studied Saturday or Sunday

112. The set of students who studied Saturday and Sunday

113. The set of students who studied Saturday and not Sunday

114. The set of students who studied Sunday and not Saturday

115. The set of students who studied neither Saturday nor Sunday

116. The set of students surveyed by the math tutor

The bar graph shows the percentage of Americans with gender preferences for various jobs.

Gender and Jobs: Percentage of Americans Who Prefer Men or Women in Various Jobs

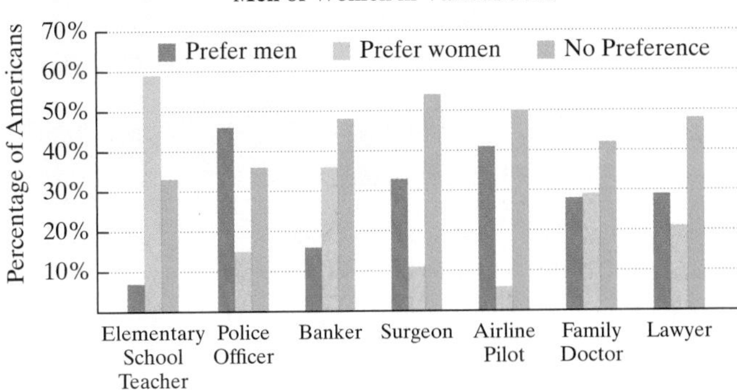

Source: Pew Research Center

In Exercises 117–122, use the information in the graph to place the indicated job in the correct region of the following Venn diagram.

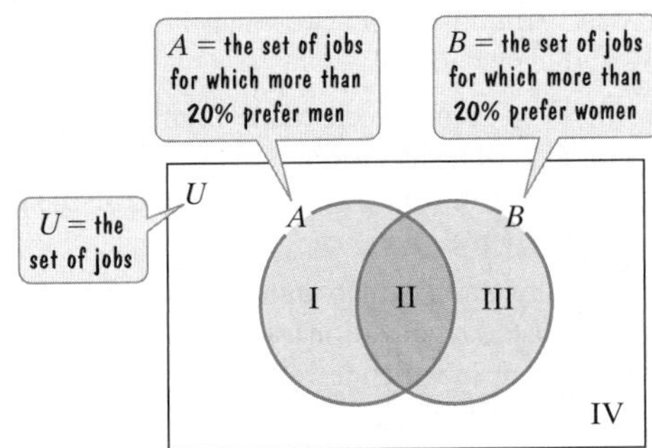

117. elementary school teacher **118.** police officer

119. surgeon **120.** banker

121. family doctor **122.** lawyer

A **palindromic number** *is a natural number whose value does not change if its digits are reversed. Examples of palindromic numbers are 11, 454, and 261,162. In Exercises 123–132, use this definition to place the indicated natural number in the correct region of the following Venn diagram.*

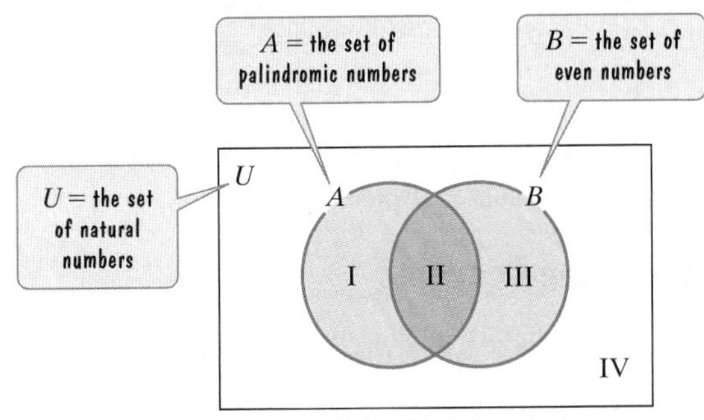

123. 11 **124.** 22 **125.** 15 **126.** 17

127. 454 **128.** 101 **129.** 9558 **130.** 9778

131. 9559 **132.** 9779

The bar graph shows the percentage of Americans, by age group, supporting legalized marijuana for four selected years from 1969 through 2015. Use the information in the graph to write each set in Exercises 133–138 in roster form or express the set as ∅.

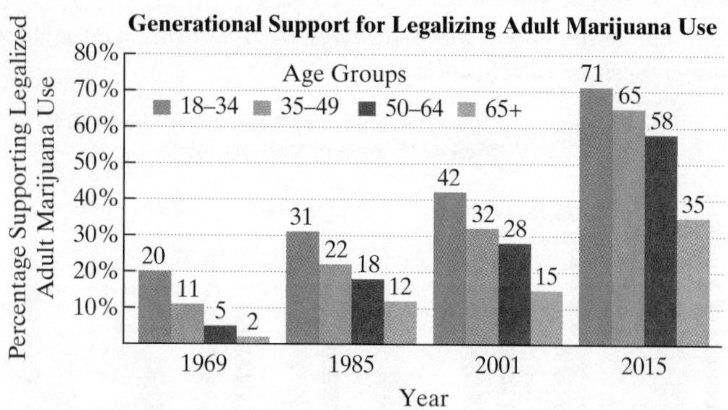

Source: USA TODAY

133. $\{x \mid x$ was a year in which more than 40% of age group 18–34 supported legalization$\} \cap$
$\{x \mid x$ was a year in which fewer than 20% of age group 65+ supported legalization$\}$

134. $\{x \mid x$ was a year in which more than 30% of age group 18–34 supported legalization$\} \cap$
$\{x \mid x$ was a year in which fewer than 14% of age group 65+ supported legalization$\}$

135. $\{x \mid x$ was a year in which more than 40% of age group 18–34 supported legalization$\} \cup$
$\{x \mid x$ was a year in which fewer than 20% of age group 65+ supported legalization$\}$

136. $\{x \mid x$ was a year in which more than 30% of age group 18–34 supported legalization$\} \cup$
$\{x \mid x$ was a year in which fewer than 14% of age group 65+ supported legalization$\}$

137. The set of years in which more than 50% of age group 18–34 supported legalization and fewer than 35% of age group 35–49 supported legalization

138. The set of years in which more than 50% of age group 18–34 supported legalization or fewer than 35% of age group 35–49 supported legalization

139. A winter resort took a poll of its 350 visitors to see which winter activities people enjoyed. The results were as follows: 178 people liked to ski, 154 people liked to snowboard, and 49 people liked to ski and snowboard. How many people in the poll liked to ski or snowboard?

140. A pet store surveyed 200 pet owners and obtained the following results: 96 people owned cats, 97 people owned dogs, and 29 people owned cats and dogs. How many people in the survey owned cats or dogs?

Explaining the Concepts

141. Describe what is meant by a universal set. Provide an example.

142. What is a Venn diagram and how is it used?

143. Describe the Venn diagram for two disjoint sets. How does this diagram illustrate that the sets have no common elements?

144. Describe the Venn diagram for proper subsets. How does this diagram illustrate that the elements of one set are also in the second set?

145. Describe the Venn diagram for two equal sets. How does this diagram illustrate that the sets are equal?

146. Describe the Venn diagram for two sets with common elements. How does the diagram illustrate this relationship?

147. Describe what is meant by the complement of a set.

148. Is it possible to find a set's complement if a universal set is not given? Explain your answer.

149. Describe what is meant by the intersection of two sets. Give an example.

150. Describe what is meant by the union of two sets. Give an example.

151. Describe how to find the cardinal number of the union of two finite sets.

Critical Thinking Exercises

Make Sense? *In Exercises 152–155, determine whether each statement makes sense or does not make sense, and explain your reasoning.*

152. Set A and set B share only one element, so I don't need to use overlapping circles to visualize their relationship.

153. Even if I'm not sure how mathematicians define irrational and complex numbers, telling me how these sets are related, I can construct a Venn diagram illustrating their relationship.

154. If I am given sets A and B, the set $(A \cup B)'$ indicates I should take the union of the complement of A and the complement of B.

155. I suspect that at least 90% of college students have no preference whether their professor is a man or a woman, so I should place college professors in region IV of the Venn diagram that precedes Exercises 117–122.

In Exercises 156–163, determine whether each statement is true or false. If the statement is false, make the necessary change(s) to produce a true statement.

156. $n(A \cup B) = n(A) + n(B)$

157. $A \cap A' = \varnothing$

158. $(A \cup B) \subseteq A$

159. If $A \subseteq B$, then $A \cap B = B$.

160. $A \cap U = U$

161. $A \cup \varnothing = \varnothing$

162. If $A \subseteq B$, then $A \cap B = \varnothing$.

163. If $B \subseteq A$, then $A \cap B = B$.

In Exercises 164–167, assume $A \neq B$. Draw a Venn diagram that correctly illustrates the relationship between the sets.

164. $A \cap B = A$

165. $A \cap B = B$

166. $A \cup B = A$

167. $A \cup B = B$

2.4 Set Operations and Venn Diagrams with Three Sets

WHAT AM I SUPPOSED TO LEARN?

After studying this section, you should be able to:

1 Perform set operations with three sets.

2 Use Venn diagrams with three sets.

3 Use Venn diagrams to prove equality of sets.

SHOULD YOUR BLOOD TYPE determine what you eat? The blood-type diet, developed by naturopathic physician Peter D'Adamo, is based on the theory that people with different blood types require different diets for optimal health. D'Adamo gives very detailed recommendations for what people with each type should and shouldn't eat. For example, he says shitake mushrooms are great for type B's, but bad for type O's. Type B? Type O? In this section, we present a Venn diagram with three sets that will give you a unique perspective on the different types of human blood. Despite this perspective, we'll have nothing to say about shitakes, avoiding the question as to whether or not the blood-type diet really works.

1 *Perform set operations with three sets.*

Set Operations with Three Sets

We now know how to find the union and intersection of two sets. We also know how to find a set's complement. In Example 1, we apply set operations to situations containing three sets.

EXAMPLE 1 ***Set Operations with Three Sets***

Given

$$U = \{1, 2, 3, 4, 5, 6, 7, 8, 9\}$$
$$A = \{1, 2, 3, 4, 5\}$$
$$B = \{1, 2, 3, 6, 8\}$$
$$C = \{2, 3, 4, 6, 7\},$$

find each of the following sets:

a. $A \cup (B \cap C)$ **b.** $(A \cup B) \cap (A \cup C)$ **c.** $A \cap (B \cup C')$.

SOLUTION

Before determining each set, let's be sure we perform the operations in the correct order. Remember that we begin by performing any set operations inside parentheses.

- Finding $A \cup (B \cap C)$

  Step 1. Find the intersection of B and C.

  Step 2. Find the union of A and $(B \cap C)$.

- Finding $(A \cup B) \cap (A \cup C)$

  Step 1. Find the union of A and B.

  Step 2. Find the union of A and C.

  Step 3. Find the intersection of $(A \cup B)$ and $(A \cup C)$.

- Finding $A \cap (B \cup C')$

Step 1. Find the complement of C.

Step 2. Find the union of B and C'.

Step 3. Find the intersection of A and $(B \cup C')$.

$U = \{1, 2, 3, 4, 5, 6, 7, 8, 9\}$
$A = \{1, 2, 3, 4, 5\}$
$B = \{1, 2, 3, 6, 8\}$
$C = \{2, 3, 4, 6, 7\}$,

The given sets (repeated)

a. To find $A \cup (B \cap C)$, first find the set within the parentheses, $B \cap C$:

$$B \cap C = \{1, 2, 3, 6, 8\} \cap \{2, 3, 4, 6, 7\} = \{2, 3, 6\}.$$

Common elements are 2, 3, and 6.

Now finish the problem by finding $A \cup (B \cap C)$:

$$A \cup (B \cap C) = \{1, 2, 3, 4, 5\} \cup \{2, 3, 6\} = \{1, 2, 3, 4, 5, 6\}.$$

List all elements in A and then add the only unlisted element in $B \cap C$, namely 6.

b. To find $(A \cup B) \cap (A \cup C)$, first find the sets within parentheses. Start with $A \cup B$:

$$A \cup B = \{1, 2, 3, 4, 5\} \cup \{1, 2, 3, 6, 8\} = \{1, 2, 3, 4, 5, 6, 8\}.$$

List all elements in A and then add the unlisted elements in B, namely 6 and 8.

Now find $A \cup C$:

$$A \cup C = \{1, 2, 3, 4, 5\} \cup \{2, 3, 4, 6, 7\} = \{1, 2, 3, 4, 5, 6, 7\}.$$

List all elements in A and then add the unlisted elements in C, namely 6 and 7.

Now finish the problem by finding $(A \cup B) \cap (A \cup C)$:

$$(A \cup B) \cap (A \cup C) = \{1, 2, 3, 4, 5, 6, 8\} \cap \{1, 2, 3, 4, 5, 6, 7\} = \{1, 2, 3, 4, 5, 6\}.$$

Common elements are 1, 2, 3, 4, 5, and 6.

c. As in parts (a) and (b), to find $A \cap (B \cup C')$, begin with the set in parentheses. First we must find C', the set of elements in U that are not in C:

$$C' = \{1, 5, 8, 9\}.$$

List the elements in U that are not in $C = \{2, 3, 4, 6, 7\}$: $\{1, \not{2}, \not{3}, \not{4}, 5, \not{6}, \not{7}, 8, 9\}$.

Now we can identify elements of $B \cup C'$:

$$B \cup C' = \{1, 2, 3, 6, 8\} \cup \{1, 5, 8, 9\} = \{1, 2, 3, 5, 6, 8, 9\}.$$

List all elements in B and then add the unlisted elements in C', namely 5 and 9.

Now finish the problem by finding $A \cap (B \cup C')$:

$$A \cap (B \cup C') = \{1, 2, 3, 4, 5\} \cap \{1, 2, 3, 5, 6, 8, 9\} = \{1, 2, 3, 5\}.$$

Common elements are 1, 2, 3, and 5.

CHECK POINT 1 Given $U = \{a, b, c, d, e, f\}$, $A = \{a, b, c, d\}$, $B = \{a, b, d, f\}$, and $C = \{b, c, f\}$, find each of the following sets:

a. $A \cup (B \cap C)$

b. $(A \cup B) \cap (A \cup C)$

c. $A \cap (B \cup C')$.

2 *Use Venn diagrams with three sets.*

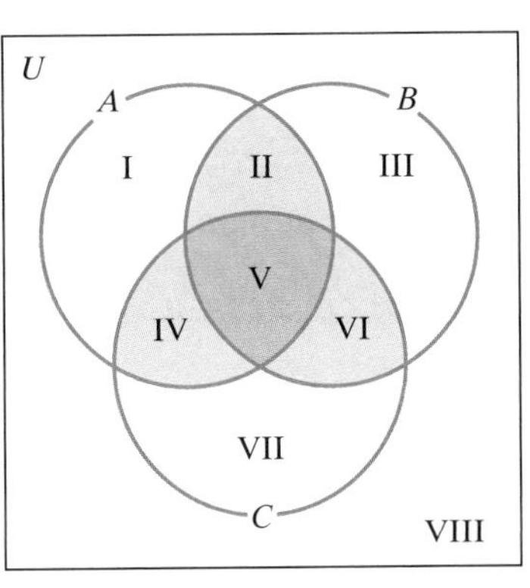

FIGURE 2.21 Three intersecting sets separate the universal set into eight regions.

Venn Diagrams with Three Sets

Venn diagrams can contain three or more sets, such as the diagram in **Figure 2.21**. The three sets in the figure separate the universal set, U, into eight regions. The numbering of these regions is arbitrary—that is, we can number any region as I, any region as II, and so on. Here is a description of each region, starting with the innermost region, region V, and working outward to region VIII.

The Region Shown in Dark Blue

Region V — This region represents elements that are common to sets A, B, and C: $A \cap B \cap C$.

The Regions Shown in Light Blue

Region II — This region represents elements in both sets A and B that are not in set C: $(A \cap B) \cap C'$.

Region IV — This region represents elements in both sets A and C that are not in set B: $(A \cap C) \cap B'$.

Region VI — This region represents elements in both sets B and C that are not in set A: $(B \cap C) \cap A'$.

The Regions Shown in White

Region I — This region represents elements in set A that are in neither sets B nor C: $A \cap (B' \cap C')$.

Region III — This region represents elements in set B that are in neither sets A nor C: $B \cap (A' \cap C')$.

Region VII — This region represents elements in set C that are in neither sets A nor B: $C \cap (A' \cap B')$.

Region VIII — This region represents elements in the universal set U that are not in sets A, B, or C: $A' \cap B' \cap C'$.

EXAMPLE 2 Determining Sets from a Venn Diagram with Three Intersecting Sets

Use the Venn diagram in **Figure 2.22** to determine each of the following sets:

a. A **b.** $A \cup B$ **c.** $B \cap C$ **d.** C' **e.** $A \cap B \cap C$.

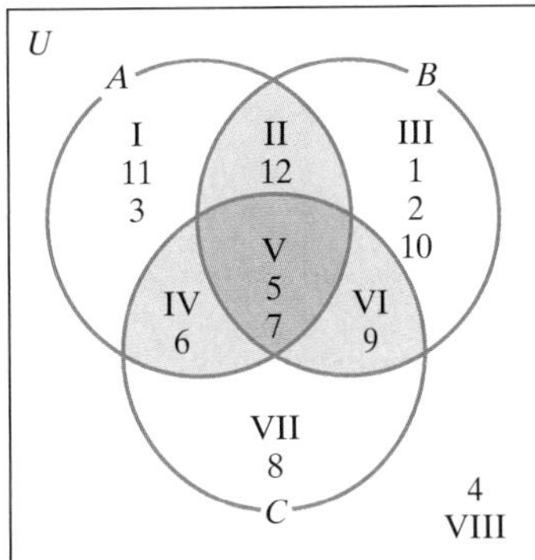

FIGURE 2.22

SOLUTION

Set to Determine	Description of Set	Regions in Venn Diagram	Set in Roster Form
a. A	set of elements in A	I, II, IV, V	{11, 3, 12, 6, 5, 7}
b. $A \cup B$	set of elements in A or B or both	I, II, III, IV, V, VI	{11, 3, 12, 1, 2, 10, 6, 5, 7, 9}
c. $B \cap C$	set of elements in both B and C	V, VI	{5, 7, 9}
d. C'	set of elements in U that are not in C	I, II, III, VIII	{11, 3, 12, 1, 2, 10, 4}
e. $A \cap B \cap C$	set of elements in A and B and C	V	{5, 7}

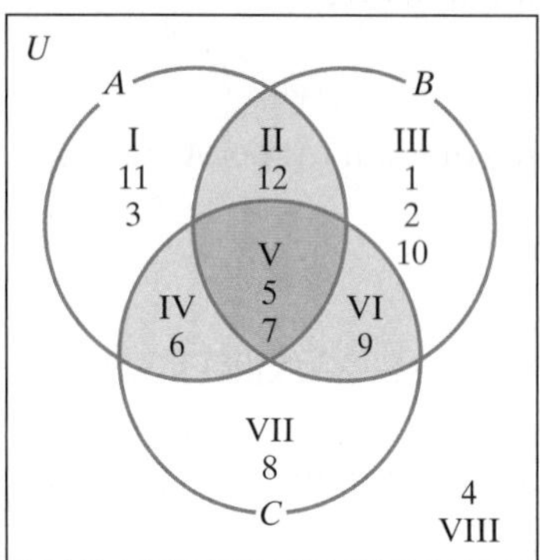

FIGURE 2.22 (repeated)

CHECK POINT 2 Use the Venn diagram in **Figure 2.22** to determine each of the following sets:

a. C

b. $B \cup C$

c. $A \cap C$

d. B'

e. $A \cup B \cup C$.

In Example 2, we used a Venn diagram showing elements in the regions to determine various sets. Now we are going to reverse directions. We'll use sets A, B, C, and U to determine the elements in each region of a Venn diagram.

To construct a Venn diagram illustrating the elements in A, B, C, and U, **start by placing elements into the innermost region and work outward**. Because the four inner regions represent various intersections, find $A \cap B$, $A \cap C$, $B \cap C$, and $A \cap B \cap C$. Then use these intersections and the given sets to place the various elements into regions. This procedure is illustrated in Example 3.

EXAMPLE 3 *Determining a Venn Diagram from Sets*

Construct a Venn diagram illustrating the following sets:

$$A = \{a, d, e, g, h, i, j\}$$
$$B = \{b, e, g, h, l\}$$
$$C = \{a, c, e, h\}$$
$$U = \{a, b, c, d, e, f, g, h, i, j, k, l\}.$$

SOLUTION

We begin by finding four intersections. In each case, common elements are shown in red.

- $A \cap B = \{a, d, e, g, h, i, j\} \cap \{b, e, g, h, l\} = \{e, g, h\}$
- $A \cap C = \{a, d, e, g, h, i, j\} \cap \{a, c, e, h\} = \{a, e, h\}$
- $B \cap C = \{b, e, g, h, l\} \cap \{a, c, e, h\} = \{e, h\}$
- $A \cap B \cap C = \{e, g, h\} \cap \{a, c, e, h\} = \{e, h\}$

This is $A \cap B$ from above.

Now we can place elements into regions, starting with the innermost region, region V, and working outward.

Before placing elements into regions, let's repeat the four intersections that we found:

$$A \cap B \cap C = \{e, h\}, A \cap B = \{e, g, h\}, A \cap C = \{a, e, h\}, B \cap C = \{e, h\}.$$

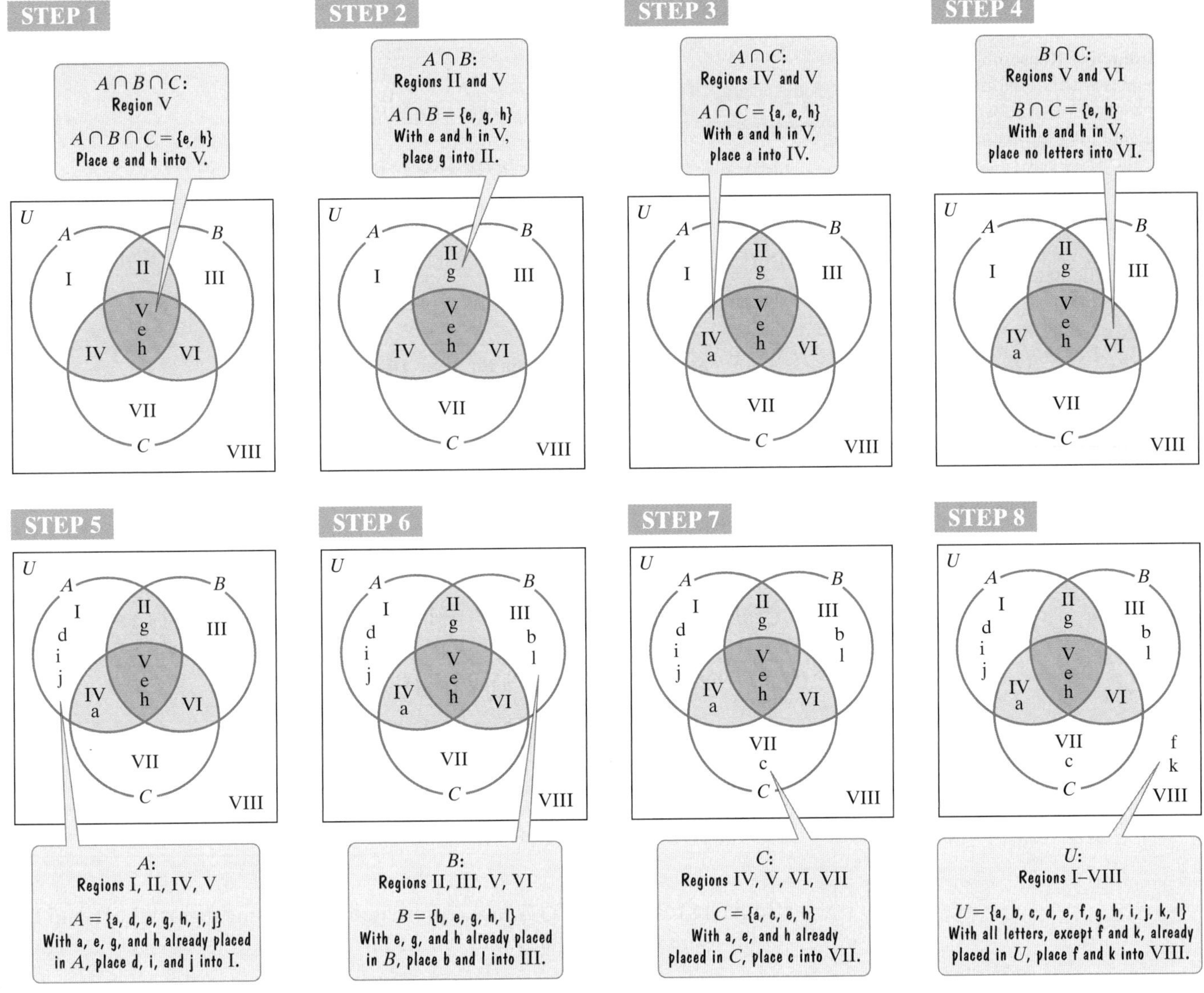

The completed Venn diagram in step 8 illustrates the given sets.

CHECK POINT 3 Construct a Venn diagram illustrating the following sets:

$$A = \{1, 3, 6, 10\}$$
$$B = \{4, 7, 9, 10\}$$
$$C = \{3, 4, 5, 8, 9, 10\}$$
$$U = \{1, 2, 3, 4, 5, 6, 7, 8, 9, 10\}.$$

3 *Use Venn diagrams to prove equality of sets.*

Proving the Equality of Sets

Throughout Section 2.3, you were given two sets A and B and their universal set U and asked to find $(A \cap B)'$ and $A' \cup B'$. In each example, $(A \cap B)'$ and $A' \cup B'$ resulted in the same set. This occurs regardless of which sets we choose for A and B in a universal set U. Examining these individual cases and applying inductive reasoning, a conjecture (or educated guess) is that $(A \cap B)' = A' \cup B'$.

A BRIEF REVIEW

In summary, here are the two forms of reasoning discussed in Chapter 1.

- **Inductive Reasoning:** Starts with individual observations and works to a general conjecture (or educated guess)
- **Deductive Reasoning:** Starts with general cases and works to the proof of a specific statement (or theorem)

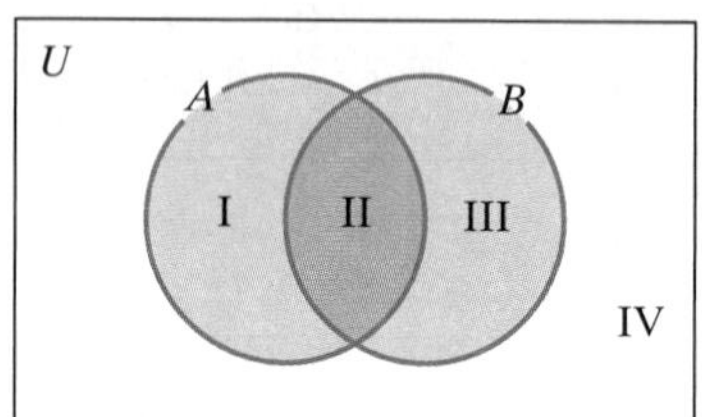

FIGURE 2.23

We can apply deductive reasoning to *prove* the statement $(A \cap B)' = A' \cup B'$ for *all* sets A and B in any universal set U. To prove that $(A \cap B)'$ and $A' \cup B'$ are equal, we use a Venn diagram. If both sets are represented by the same regions in this general diagram, then this proves that they are equal. Example 4 shows how this is done.

EXAMPLE 4 Proving the Equality of Sets

Use the Venn diagram in **Figure 2.23** to prove that

$$(A \cap B)' = A' \cup B'.$$

SOLUTION

Begin by identifying the regions representing $(A \cap B)'$.

Set	Regions in the Venn Diagram
A	I, II
B	II, III
$A \cap B$	II (This is the region common to A and B.)
$(A \cap B)'$	I, III, IV (These are the regions in U that are not in $A \cap B$.)

Next, find the regions in **Figure 2.23** representing $A' \cup B'$.

Set	Regions in the Venn Diagram
A'	III, IV (These are the regions not in A.)
B'	I, IV (These are the regions not in B.)
$A' \cup B'$	I, III, IV (These are the regions obtained by uniting the regions representing A' and B'.)

Both $(A \cap B)'$ and $A' \cup B'$ are represented by the same regions, I, III, and IV, of the Venn diagram. This result proves that

$$(A \cap B)' = A' \cup B'$$

for all sets A and B in any universal set U.

Can you see how we applied deductive reasoning in Example 4? We started with the two general sets in the Venn diagram in **Figure 2.23** and worked to the specific conclusion that $(A \cap B)'$ and $A' \cup B'$ represent the same regions in the diagram. Thus, the statement $(A \cap B)' = A' \cup B'$ is a theorem.

CHECK POINT 4 Use the Venn diagram in **Figure 2.23** to solve this exercise.

a. Which region represents $(A \cup B)'$?

b. Which region represents $A' \cap B'$?

c. Based on parts (a) and (b), what can you conclude?

GREAT QUESTION!

Why did you put De Morgan's laws in a box? Do I need to memorize them?

Learn the pattern of De Morgan's laws. They are extremely important in logic and will be discussed in more detail in Chapter 3, Logic.

The statements proved in Example 4 and Check Point 4 are known as *De Morgan's laws*, named for the British logician Augustus De Morgan (1806–1871).

DE MORGAN'S LAWS

$(A \cap B)' = A' \cup B'$:	The complement of the intersection of two sets is the union of the complements of those sets.
$(A \cup B)' = A' \cap B'$:	The complement of the union of two sets is the intersection of the complements of those sets.

EXAMPLE 5 *Proving the Equality of Sets*

Use a Venn diagram to prove that

$$A \cup (B \cap C) = (A \cup B) \cap (A \cup C).$$

SOLUTION

Use a Venn diagram with three sets A, B, and C, as shown in **Figure 2.24**. Begin by identifying the regions representing $A \cup (B \cap C)$.

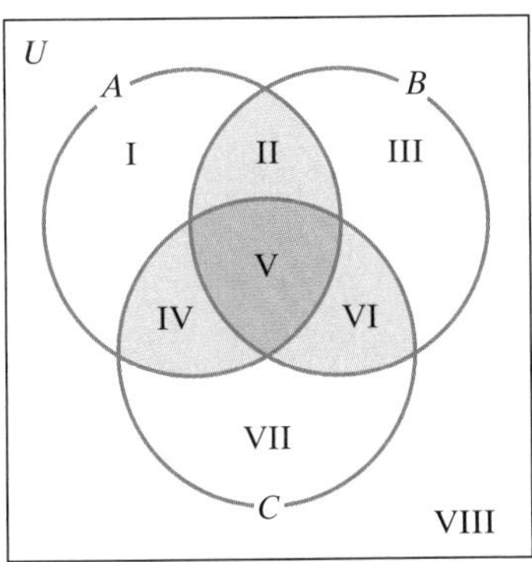

FIGURE 2.24

Set	Regions in the Venn Diagram
A	I, II, IV, V
$B \cap C$	V, VI (These are the regions common to B and C.)
$A \cup (B \cap C)$	I, II, IV, V, VI (These are the regions obtained by uniting the regions representing A and $B \cap C$.)

Next, find the regions representing $(A \cup B) \cap (A \cup C)$.

Set	Regions in the Venn Diagram
A	I, II, IV, V
B	II, III, V, VI
C	IV, V, VI, VII
$A \cup B$	I, II, III, IV, V, VI (Unite the regions representing A and B.)
$A \cup C$	I, II, IV, V, VI, VII (Unite the regions representing A and C.)
$(A \cup B) \cap (A \cup C)$	I, II, IV, V, VI (These are the regions common to $A \cup B$ and $A \cup C$.)

Both $A \cup (B \cap C)$ and $(A \cup B) \cap (A \cup C)$ are represented by the same regions, I, II, IV, V, and VI, of the Venn diagram. This result proves that

$$A \cup (B \cap C) = (A \cup B) \cap (A \cup C)$$

for all sets A, B, and C in any universal set U. Thus, the statement is a theorem.

CHECK POINT 5 Use the Venn diagram in **Figure 2.24** to solve this exercise.

a. Which regions represent $A \cap (B \cup C)$?

b. Which regions represent $(A \cap B) \cup (A \cap C)$?

c. Based on parts (a) and (b), what can you conclude?

Blitzer Bonus

Blood Types and Venn Diagrams

In the early 1900s, the Austrian immunologist Karl Landsteiner discovered that all blood is not the same. Blood serum drawn from one person often clumped when mixed with the blood cells of another. The clumping was caused by different antigens, proteins, and carbohydrates that trigger antibodies and fight infection. Landsteiner classified blood types based on the presence or absence of the antigens A, B, and Rh in red blood cells. The Venn diagram in **Figure 2.25** contains eight regions representing the eight common blood groups.

In the Venn diagram, blood with the Rh antigen is labeled positive and blood lacking the Rh antigen is labeled negative. The region where the three circles intersect represents type AB^+, indicating that a person with this blood type has the antigens A, B, and Rh. Observe that type O blood (both positive and negative) lacks A and B antigens. Type O^- lacks all three antigens, A, B, and Rh.

In blood transfusions, the recipient must have all or more of the antigens present in the donor's blood. This discovery rescued surgery patients from random, often lethal, transfusions. This knowledge made the massive blood drives during World War I possible. Eventually, it made the modern blood bank possible as well.

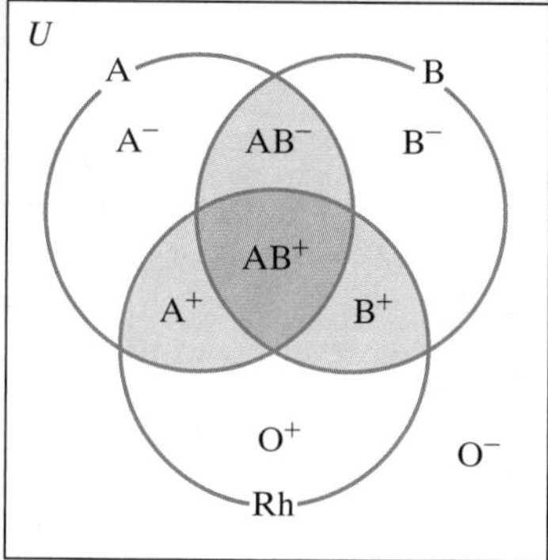

FIGURE 2.25 Human blood types

Concept and Vocabulary Check

Fill in each blank so that the resulting statement is true.

1. In order to perform set operations such as $(A \cup B) \cap (A \cup C)$, begin by performing any set operations ______________.
2. The three sets in the Venn diagrams that appeared throughout this section separate the universal set into ______ regions.
3. True or False: Inductive reasoning is used to prove the equality of sets. ______
4. True or False: In this section we proved that $(A \cap B)' = A' \cup B'$ is a theorem, so this means that the complement of the intersection of two sets is the union of the complement of those sets. ______

Exercise Set 2.4

Practice Exercises

In Exercises 1–12, let

$$U = \{1, 2, 3, 4, 5, 6, 7\}$$
$$A = \{1, 3, 5, 7\}$$
$$B = \{1, 2, 3\}$$
$$C = \{2, 3, 4, 5, 6\}.$$

Find each of the following sets.

1. $A \cup (B \cap C)$
2. $A \cap (B \cup C)$
3. $(A \cup B) \cap (A \cup C)$
4. $(A \cap B) \cup (A \cap C)$
5. $A' \cap (B \cup C')$
6. $C' \cap (A \cup B')$
7. $(A' \cap B) \cup (A' \cap C')$
8. $(C' \cap A) \cup (C' \cap B')$
9. $(A \cup B \cup C)'$
10. $(A \cap B \cap C)'$
11. $(A \cup B)' \cap C$
12. $(B \cup C)' \cap A$

In Exercises 13–24, let

$$U = \{a, b, c, d, e, f, g, h\}$$
$$A = \{a, g, h\}$$
$$B = \{b, g, h\}$$
$$C = \{b, c, d, e, f\}.$$

Find each of the following sets.

13. $A \cup (B \cap C)$
14. $A \cap (B \cup C)$
15. $(A \cup B) \cap (A \cup C)$
16. $(A \cap B) \cup (A \cap C)$
17. $A' \cap (B \cup C')$
18. $C' \cap (A \cup B')$
19. $(A' \cap B) \cup (A' \cap C')$
20. $(C' \cap A) \cup (C' \cap B')$
21. $(A \cup B \cup C)'$
22. $(A \cap B \cap C)'$
23. $(A \cup B)' \cap C$
24. $(B \cup C)' \cap A$

In Exercises 25–32, use the Venn diagram shown to answer each question.

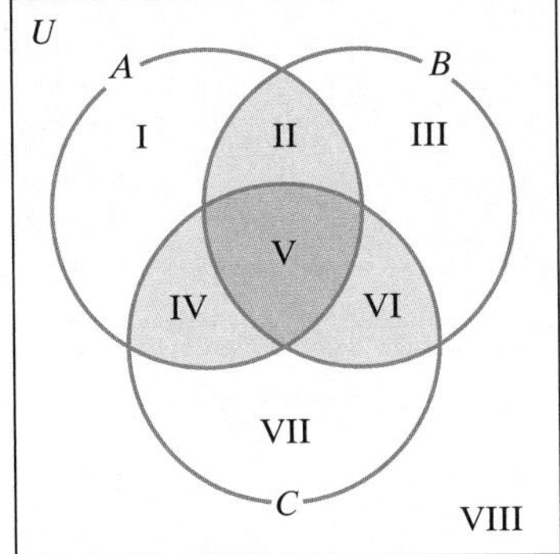

25. Which regions represent set B?
26. Which regions represent set C?
27. Which regions represent $A \cup C$?
28. Which regions represent $B \cup C$?
29. Which regions represent $A \cap B$?
30. Which regions represent $A \cap C$?
31. Which regions represent B'?
32. Which regions represent C'?

In Exercises 33–44, use the Venn diagram to represent each set in roster form.

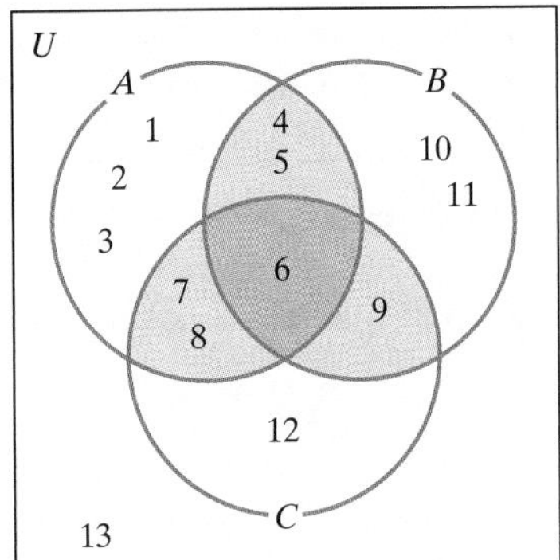

33. A	34. B
35. $A \cup B$	36. $B \cup C$
37. $(A \cup B)'$	38. $(B \cup C)'$
39. $A \cap B$	40. $A \cap C$
41. $A \cap B \cap C$	42. $A \cup B \cup C$
43. $(A \cap B \cap C)'$	44. $(A \cup B \cup C)'$

In Exercises 45–48, construct a Venn diagram illustrating the given sets.

45. $A = \{4, 5, 6, 8\}, B = \{1, 2, 4, 5, 6, 7\},$
$C = \{3, 4, 7\}, U = \{1, 2, 3, 4, 5, 6, 7, 8, 9\}$

46. $A = \{a, e, h, i\}, B = \{b, c, e, f, h, i\},$
$C = \{e, f, g\}, U = \{a, b, c, d, e, f, g, h, i\}$

47. $A = \{+, -, \times, \div, \rightarrow, \leftrightarrow\}$
$B = \{\times, \div, \rightarrow\}$
$C = \{\wedge, \vee, \rightarrow, \leftrightarrow\}$
$U = \{+, -, \times, \div, \wedge, \vee, \rightarrow, \leftrightarrow, \sim\}$

48. $A = \{x_3, x_9\}$
$B = \{x_1, x_2, x_3, x_5, x_6\}$
$C = \{x_3, x_4, x_5, x_6, x_9\}$
$U = \{x_1, x_2, x_3, x_4, x_5, x_6, x_7, x_8, x_9\}$

Use the Venn diagram shown to solve Exercises 49–52.

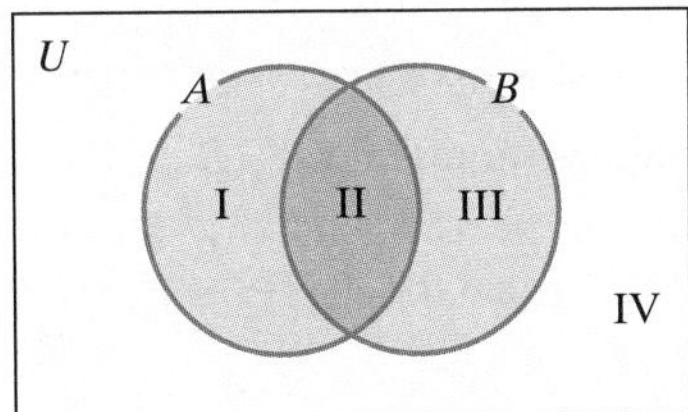

49. **a.** Which region represents $A \cap B$?
b. Which region represents $B \cap A$?
c. Based on parts (a) and (b), what can you conclude?

50. **a.** Which regions represent $A \cup B$?
b. Which regions represent $B \cup A$?
c. Based on parts (a) and (b), what can you conclude?

51. **a.** Which region(s) represent(s) $(A \cap B)'$?
b. Which region(s) represent(s) $A' \cap B'$?
c. Based on parts (a) and (b), are $(A \cap B)'$ and $A' \cap B'$ equal for all sets A and B? Explain your answer.

52. **a.** Which region(s) represent(s) $(A \cup B)'$?
b. Which region(s) represent(s) $A' \cup B'$?
c. Based on parts (a) and (b), are $(A \cup B)'$ and $A' \cup B'$ equal for all sets A and B? Explain your answer.

In Exercises 53–58, use the Venn diagram for Exercises 49–52 to determine whether the given sets are equal for all sets A and B.

53. $A' \cup B, \quad A \cap B'$	54. $A' \cap B, \quad A \cup B'$
55. $(A \cup B)', \quad (A \cap B)'$	56. $(A \cup B)', \quad A' \cap B$
57. $(A' \cap B)', \quad A \cup B'$	58. $(A \cup B')', \quad A' \cap B$

Use the Venn diagram shown to solve Exercises 59–62.

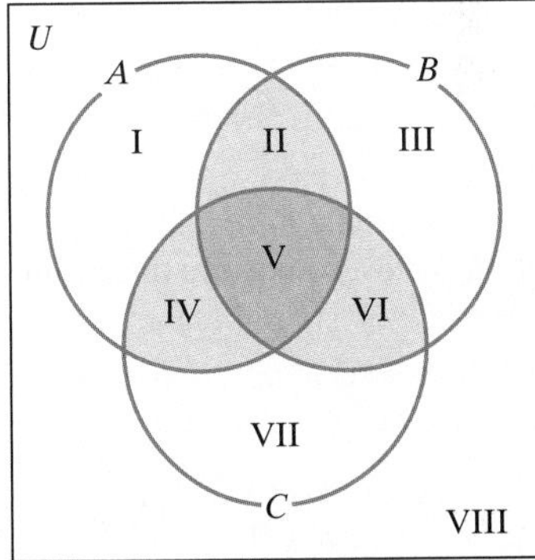

59. **a.** Which regions represent $(A \cap B) \cup C$?
b. Which regions represent $(A \cup C) \cap (B \cup C)$?
c. Based on parts (a) and (b), what can you conclude?

60. **a.** Which regions represent $(A \cup B) \cap C$?
b. Which regions represent $(A \cap C) \cup (B \cap C)$?
c. Based on parts (a) and (b), what can you conclude?

61. **a.** Which regions represent $A \cap (B \cup C)$?
b. Which regions represent $A \cup (B \cap C)$?
c. Based on parts (a) and (b), are $A \cap (B \cup C)$ and $A \cup (B \cap C)$ equal for all sets $A, B,$ and C? Explain your answer.

Continue to refer to the Venn diagram at the bottom of the previous page.

62. a. Which regions represent $C \cup (B \cap A)$?

b. Which regions represent $C \cap (B \cup A)$?

c. Based on parts (a) and (b), are $C \cup (B \cap A)$ and $C \cap (B \cup A)$ equal for all sets A, B, and C? Explain your answer.

In Exercises 63–68, use the Venn diagram shown at the bottom of the previous page to determine which statements are true for all sets A, B, and C, and, consequently, are theorems.

63. $A \cap (B \cup C) = (A \cap B) \cup C$

64. $A \cup (B \cap C) = (A \cup B) \cap C$

65. $B \cup (A \cap C) = (A \cup B) \cap (B \cup C)$

66. $B \cap (A \cup C) = (A \cap B) \cup (B \cap C)$

67. $A \cap (B \cup C)' = A \cap (B' \cap C')$

68. $A \cup (B \cap C)' = A \cup (B' \cup C')$

Practice Plus

69. a. Let $A = \{c\}$, $B = \{a, b\}$, $C = \{b, d\}$, and $U = \{a, b, c, d, e, f\}$. Find $A \cup (B' \cap C')$ and $(A \cup B') \cap (A \cup C')$.

b. Let $A = \{1, 3, 7, 8\}$, $B = \{2, 3, 6, 7\}$, $C = \{4, 6, 7, 8\}$, and $U = \{1, 2, 3, \ldots, 8\}$. Find $A \cup (B' \cap C')$ and $(A \cup B') \cap (A \cup C')$.

c. Based on your results in parts (a) and (b), use inductive reasoning to write a conjecture that relates $A \cup (B' \cap C')$ and $(A \cup B') \cap (A \cup C')$.

d. Use deductive reasoning to determine whether your conjecture in part (c) is a theorem.

70. a. Let $A = \{3\}$, $B = \{1, 2\}$, $C = \{2, 4\}$, and $U = \{1, 2, 3, 4, 5, 6\}$. Find $(A \cup B)' \cap C$ and $A' \cap (B' \cap C)$.

b. Let $A = \{d, f, g, h\}$, $B = \{a, c, f, h\}$, $C = \{c, e, g, h\}$, and $U = \{a, b, c, \ldots, h\}$. Find $(A \cup B)' \cap C$ and $A' \cap (B' \cap C)$.

c. Based on your results in parts (a) and (b), use inductive reasoning to write a conjecture that relates $(A \cup B)' \cap C$ and $A' \cap (B' \cap C)$.

d. Use deductive reasoning to determine whether your conjecture in part (c) is a theorem.

In Exercises 71–78, use the symbols A, B, C, ∩, ∪, and ′, as necessary, to describe each shaded region. More than one correct symbolic description may be possible.

71.

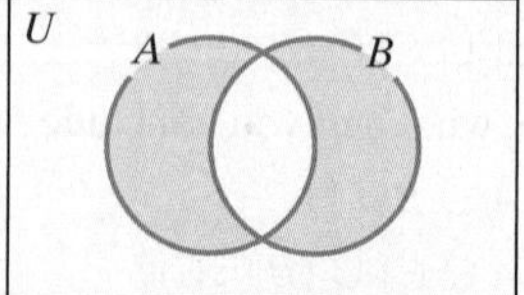

72.

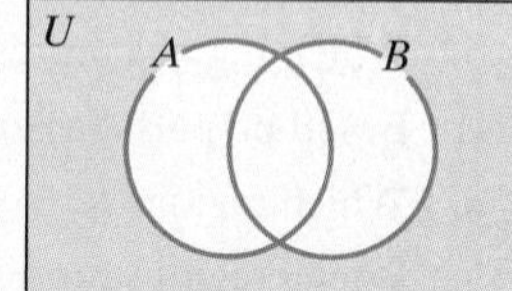

73.

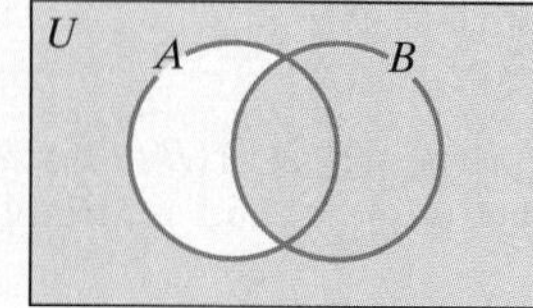

74.

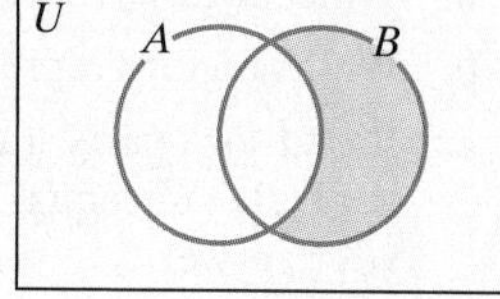

75.

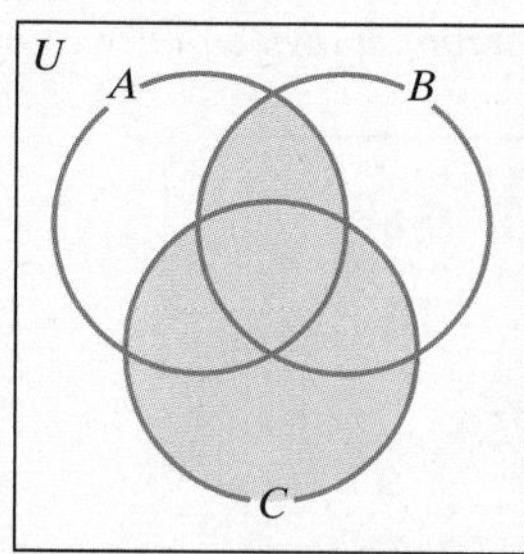

76.

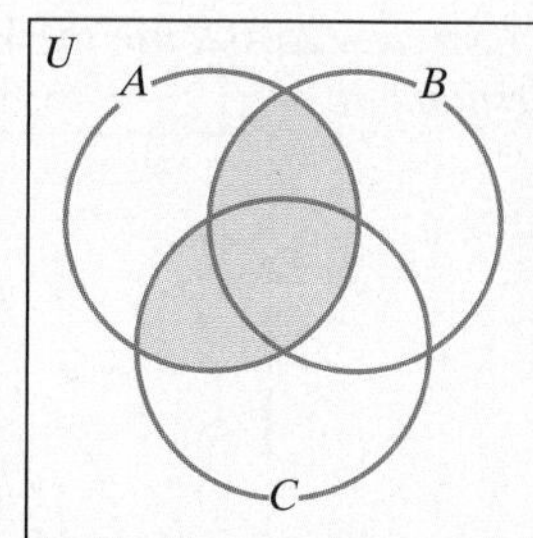

77.

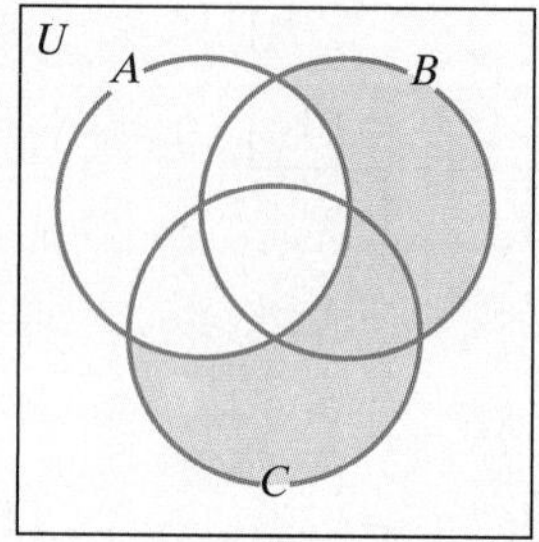

78.

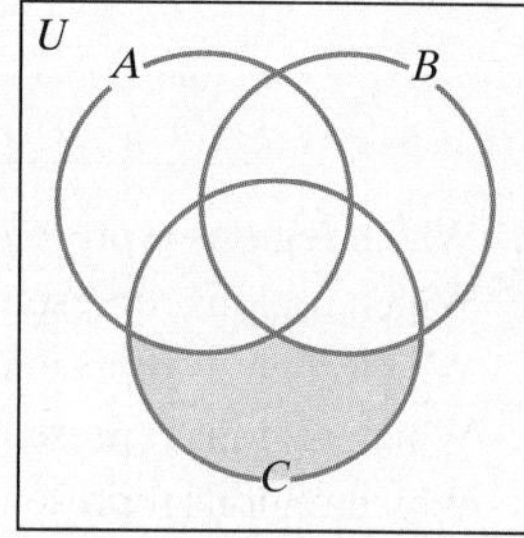

Application Exercises

A math tutor working with a small study group has classified students in the group by whether or not they scored 90% or above on each of three tests. The results are shown in the Venn diagram.

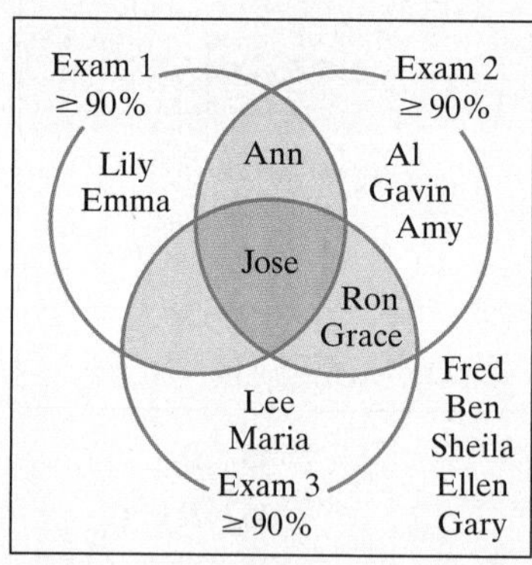

In Exercises 79–90, use the Venn diagram to represent each set in roster form.

79. The set of students who scored 90% or above on exam 2

80. The set of students who scored 90% or above on exam 3

81. The set of students who scored 90% or above on exam 1 and exam 3

82. The set of students who scored 90% or above on exam 1 and exam 2

83. The set of students who scored 90% or above on exam 1 and not on exam 2

84. The set of students who scored 90% or above on exam 3 and not on exam 1

85. The set of students who scored 90% or above on exam 1 or not on exam 2

86. The set of students who scored 90% or above on exam 3 or not on exam 1

87. The set of students who scored 90% or above on *exactly one* test

(In Exercises 88–90, be sure to refer to the Venn diagram at the bottom of the previous page in order to represent each set in roster form.)

88. The set of students who scored 90% or above on *at least two* tests

89. The set of students who scored 90% or above on exam 2 and not on exam 1 and exam 3

90. The set of students who scored 90% or above on exam 1 and not on exam 2 and exam 3

91. Use the Venn diagram shown at the bottom of the previous page to describe a set of students that is the empty set.

92. Use the Venn diagram shown at the bottom of the previous page to describe the set {Fred, Ben, Sheila, Ellen, Gary}.

The chart shows the highest-rated prime time television programs for the 2013–2014, 2014–2015, and 2015–2016 seasons.

HIGHEST-RATED PRIME TIME TV SHOWS

2013–2014	2014–2015	2015–2016
1. *The Big Bang Theory*	1. *The Big Bang Theory*	1. *NCIS*
2. *NCIS*	2. *NCIS*	2. *The Big Bang Theory*
3. *NCIS: Los Angeles*	3. *NCIS: New Orleans*	3. *Empire*
4. *Dancing with the Stars*	4. *Empire*	4. *NCIS: New Orleans*
5. *The Blacklist*	5. *Dancing with the Stars*	5. *Dancing with the Stars*
6. *The OT*	6. *Madam Secretary*	6. *Blue Bloods*

In Exercises 93–98, use the Venn diagram to indicate in which region, I through VIII, each television show should be placed.

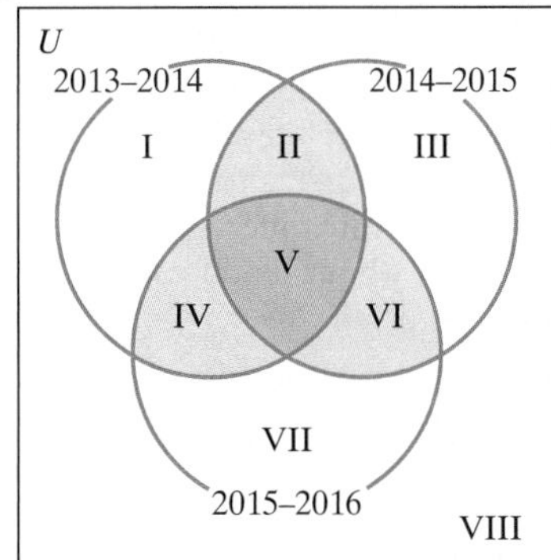

93. *The Blacklist*

94. *Madam Secretary*

95. *The Big Bang Theory*

96. *NCIS: New Orleans*

97. *Empire*

98. *Two and a Half Men*

The chart shows some of the top single recordings of all time.

TOP SINGLE RECORDINGS

Title	Artist or Group	Sales	Year Released
"White Christmas"	Bing Crosby	50 million	1942
"Candle in the Wind"	Elton John	37 million	1997
"Rock Around the Clock"	Bill Haley and His Comets	25 million	1954
"Little Drummer Boy"	The Harry Simeone Chorale	25 million	1958
"It's Now or Never"	Elvis Presley	20 million	1960
"We Are the World"	USA for Africa	20 million	1985
"Yes Sir, I Can Boogie"	Baccara	18 million	1977
"Wind of Change"	The Scorpions	14 million	1991
"Sukiyaki"	Kyu Sakamoto	13 million	1963
"I Want to Hold Your Hand"	The Beatles	12 million	1963
"I Will Always Love You"	Whitney Houston	12 million	1992

Source: Music Information Database

In Exercises 99–104, use the Venn diagram to indicate in which region, I through VIII, each recording should be placed.

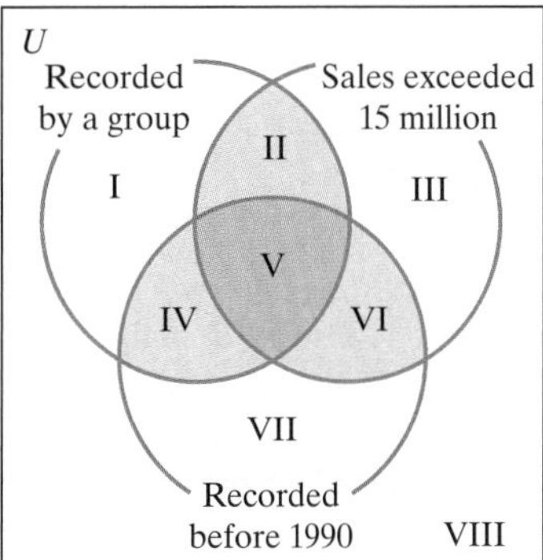

99. "Candle in the Wind"

100. "White Christmas"

101. "I Want to Hold Your Hand"

102. "I Will Always Love You"

103. "It's Now or Never"

104. "Wind of Change"

105. The chart shows three health indicators for seven countries or regions.

WORLDWIDE HEALTH INDICATORS

Country/Region	Male Life Expectancy	Female Life Expectancy	Persons per Doctor
United States	77.5	82.1	360
Italy	79.6	85.0	180
Russia	65.0	76.8	240
East Africa	46.9	48.2	13,620
Japan	81.7	88.5	530
England	78.5	83.0	720
Iran	69.8	73.1	1200

Source: Time Almanac 2013; *World Almanac* 2017

Let U = the set of countries/regions shown in the chart, A = the set of countries/regions with male life expectancy that exceeds 75 years, B = the set of countries/regions with female life expectancy that exceeds 80 years, and C = the set of countries/regions with fewer than 400 persons per doctor. Use the information in the chart to construct a Venn diagram that illustrates these sets.

106. The chart shows the 12 films nominated for the most Oscars.

FILMS WITH THE MOST OSCAR NOMINATIONS

Film	Nominations	Awards	Year
All About Eve	14	6	1950
Titanic	14	11	1997
La La Land	14	6	2017
Gone with the Wind	13	8	1939
From Here to Eternity	13	8	1953
Shakespeare in Love	13	7	1998
Mary Poppins	13	5	1964
Who's Afraid of Virginia Woolf?	13	5	1966
Forrest Gump	13	6	1994
The Lord of the Rings: The Fellowship of the Ring	13	4	2001
Chicago	13	6	2004
The Curious Case of Benjamin Button	13	3	2008

Source: Academy of Motion Picture Arts and Sciences

Using abbreviated film titles, let $U = \{Eve, Titanic, La, Wind, Eternity, Love, Poppins, Woolf, Gump, Ring, Chicago, Curious\}$, $A =$ the set of films nominated for 14 Oscars, $B =$ the set of films that won at least 7 Oscars, and $C =$ the set of films that won Oscars after 1965. Use the information in the chart to construct a Venn diagram that illustrates these sets.

Explaining the Concepts

107. If you are given four sets, A, B, C, and U, describe what is involved in determining $(A \cup B)' \cap C$. Be as specific as possible in your description.

108. Describe how a Venn diagram can be used to prove that $(A \cup B)'$ and $A' \cap B'$ are equal sets.

Critical Thinking Exercises

Make Sense? *In Exercises 109–112, determine whether each statement makes sense or does not make sense, and explain your reasoning.*

109. I constructed a Venn diagram for three sets by placing elements into the outermost region and working inward.

110. This Venn diagram showing color combinations from red, green, and blue illustrates that white is a combination of all three colors and black uses none of the colors.

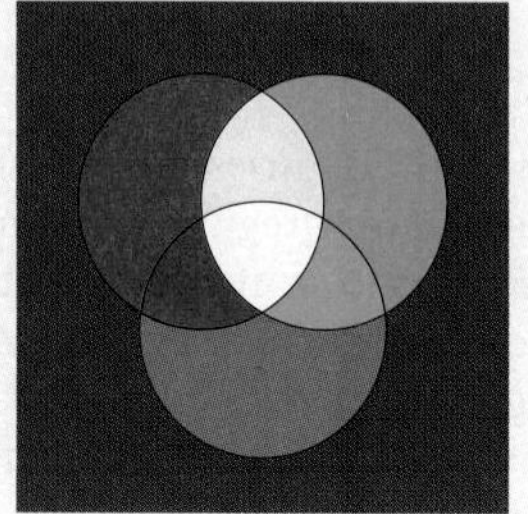

111. I used a Venn diagram to prove that $(A \cup B)'$ and $A' \cup B'$ are not equal.

112. I found 50 examples of two sets, A and B, for which $(A \cup B)'$ and $A' \cap B'$ resulted in the same set, so this proves that $(A \cup B)' = A' \cap B'$.

The eight blood types discussed in the Blitzer Bonus on page 94 are shown once again in the Venn diagram. In blood transfusions, the set of antigens in a donor's blood must be a subset of the set of antigens in a recipient's blood. Thus, the recipient must have all or more of the antigens present in the donor's blood. Use this information to solve Exercises 113–116.

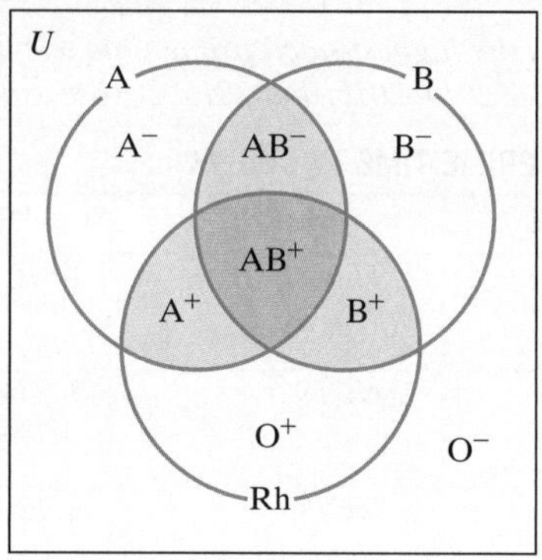

Human blood types

113. What is the blood type of a universal recipient?

114. What is the blood type of a universal donor?

115. Can an A^+ person donate blood to an A^- person?

116. Can an A^- person donate blood to an A^+ person?

Group Exercises

117. Each group member should find out his or her blood type. (If you cannot obtain this information, select a blood type that you find appealing!) Read the introduction to Exercises 113–116. Referring to the Venn diagram for these exercises, each group member should determine all other group members to whom blood can be donated and from whom it can be received.

118. The group should define three sets, each of which categorizes U, the set of students in the group, in different ways. Examples include the set of students with blonde hair, the set of students no more than 23 years old, and the set of students whose major is undecided. Once you have defined the sets, construct a Venn diagram with three intersecting sets and eight regions. Each student should determine to which region he or she belongs. Illustrate the sets by writing each first name in the appropriate region.

2.5 Survey Problems

WHAT AM I SUPPOSED TO LEARN?

After studying this section, you should be able to:

1 Use Venn diagrams to visualize a survey's results.

2 Use survey results to complete Venn diagrams and answer questions about the survey.

THERE ARE NOTABLE DIFFERENCES among people of different nations on some basic cultural attitudes. For example, **Figure 2.26** illustrates differences among five selected countries on assessing the causes of poverty. The graph shows that Mexicans tend to see societal injustice, rather than personal laziness, as the primary cause of poverty.

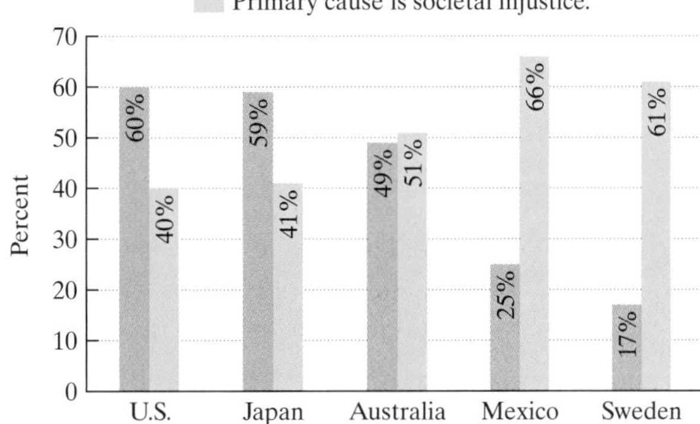

FIGURE 2.26 Percentages for each country may not total 100% because less frequently identified primary causes of poverty were omitted from the graph.
Source: Ronald Inglehart et al., *World Values Surveys and European Values Surveys*

Suppose a survey is taken that asks randomly selected adults in the United States and Mexico the following question:

Do you agree or disagree that the primary cause of poverty is societal injustice?

In this section, you will see how sets and Venn diagrams are used to tabulate information collected in such a survey. In survey problems, it is helpful to remember that **and** means **intersection**, **or** means **union**, and **not** means **complement**. Furthermore, *but* means the same thing as *and*. Thus, **but** means **intersection**.

1 *Use Venn diagrams to visualize a survey's results.*

Visualizing the Results of a Survey

In Section 2.1, we defined the cardinal number of set A, denoted by $n(A)$, as the number of elements in set A. Venn diagrams are helpful in determining a set's cardinality.

EXAMPLE 1 Using a Venn Diagram to Visualize the Results of a Survey

We return to the campus survey in which students were asked two questions:

Would you be willing to donate blood?

Would you be willing to help serve a free breakfast to blood donors?

Set A represents the set of students willing to donate blood. Set B represents the set of students willing to help serve breakfast to donors. The survey results are summarized in **Figure 2.27** at the top of the next page.

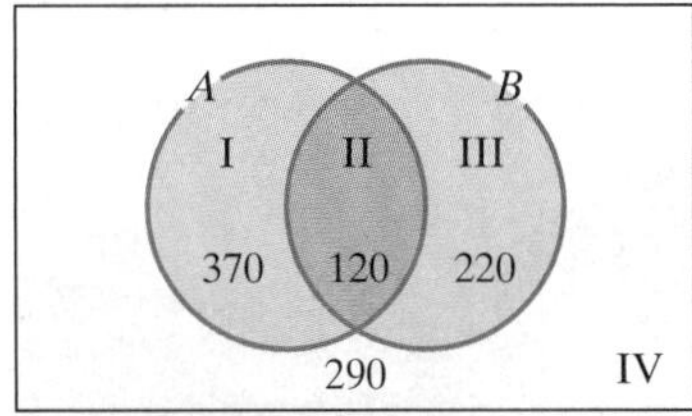

A: Set of students willing to donate blood
B: Set of students willing to serve breakfast to donors

FIGURE 2.27 Results of a survey

GREAT QUESTION!

Hold on! In part (d), we added three numbers to find $n(A \cup B)$. In a previous section, didn't we add two numbers and subtract a third to find $n(A \cup B)$?

You are correct! We could also use the formula

$$n(A \cup B) = n(A) + n(B) - n(A \cap B)$$

and our results from parts (a)–(c) to find $n(A \cup B)$.

$$n(A \cup B) = 490 + 340 - 120 = 710$$

Notice that we get the same answer.
Pay attention to what information is given and what information is needed to use the formula. While the formula gives the correct answer, it is not the most direct method to calculate $n(A \cup B)$ from the numbers given in the Venn diagram.

Use the diagram to answer the following questions:

a. How many students are willing to donate blood?

b. How many students are willing to help serve a free breakfast to blood donors?

c. How many students are willing to donate blood and serve breakfast?

d. How many students are willing to donate blood or serve breakfast?

e. How many students are willing to donate blood but not serve breakfast?

f. How many students are willing to serve breakfast but not donate blood?

g. How many students are neither willing to donate blood nor serve breakfast?

h. How many students were surveyed?

SOLUTION

a. The number of students willing to donate blood can be determined by adding the numbers in regions I and II. Thus, $n(A) = 370 + 120 = 490$. There are 490 students willing to donate blood.

b. The number of students willing to help serve a free breakfast to blood donors can be determined by adding the numbers in regions II and III. Thus, $n(B) = 120 + 220 = 340$. There are 340 students willing to help serve breakfast.

c. The number of students willing to donate blood and serve breakfast appears in region II, the region representing the intersection of the two sets. Thus, $n(A \cap B) = 120$. There are 120 students willing to donate blood and serve breakfast.

d. The number of students willing to donate blood or serve breakfast is found by adding the numbers in regions I, II, and III, representing the union of the two sets. We see that $n(A \cup B) = 370 + 120 + 220 = 710$. Therefore, 710 students in the survey are willing to donate blood or serve breakfast.

e. The region representing students who are willing to donate blood but not serve breakfast, $A \cap B'$, is region I. We see that 370 of the students surveyed are willing to donate blood but not serve breakfast.

f. Region III represents students willing to serve breakfast but not donate blood: $B \cap A'$. We see that 220 students surveyed are willing to help serve breakfast but not donate blood.

g. Students who are neither willing to donate blood nor serve breakfast, $A' \cap B'$, fall within the universal set, but outside circles A and B. These students fall in region IV, where the Venn diagram indicates that there are 290 elements. There are 290 students in the survey who are neither willing to donate blood nor serve breakfast.

h. We can find the number of students surveyed by adding the numbers in regions I, II, III, and IV. Thus, $n(U) = 370 + 120 + 220 + 290 = 1000$. There were 1000 students surveyed.

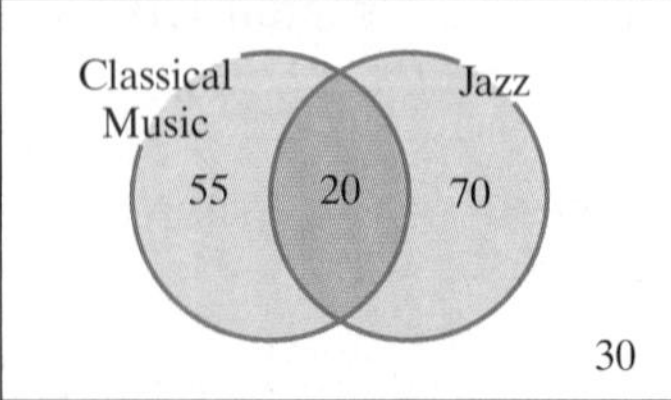

FIGURE 2.28

CHECK POINT 1 In a survey on musical tastes, respondents were asked: Do you listen to classical music? Do you listen to jazz? The survey results are summarized in **Figure 2.28**. Use the diagram to answer the following questions.

a. How many respondents listened to classical music?

b. How many respondents listened to jazz?

c. How many respondents listened to both classical music and jazz?

d. How many respondents listened to classical music or jazz?

e. How many respondents listened to classical music but not jazz?

f. How many respondents listened to jazz but not classical music?

g. How many respondents listened to neither classical music nor jazz?

h. How many people were surveyed?

2 *Use survey results to complete Venn diagrams and answer questions about the survey.*

Solving Survey Problems

Venn diagrams are used to solve problems involving surveys. Here are the steps needed to solve survey problems:

SOLVING SURVEY PROBLEMS

1. Use the survey's description to define sets and draw a Venn diagram.
2. Use the survey's results to determine the cardinality for each region in the Venn diagram. **Start with the intersection of the sets, the innermost region, and work outward.**
3. Use the completed Venn diagram to answer the problem's questions.

EXAMPLE 2 *Surveying People's Attitudes*

A survey is taken that asks 2000 randomly selected U.S. and Mexican adults the following question:

> Do you agree or disagree that the primary cause of poverty is societal injustice?

The results of the survey showed that

> 1060 people agreed with the statement.
> 400 Americans agreed with the statement.
> *Source: World Values Surveys*

If half the adults surveyed were Americans,

a. How many Mexicans agreed with the statement?

b. How many Mexicans disagreed with the statement?

SOLUTION

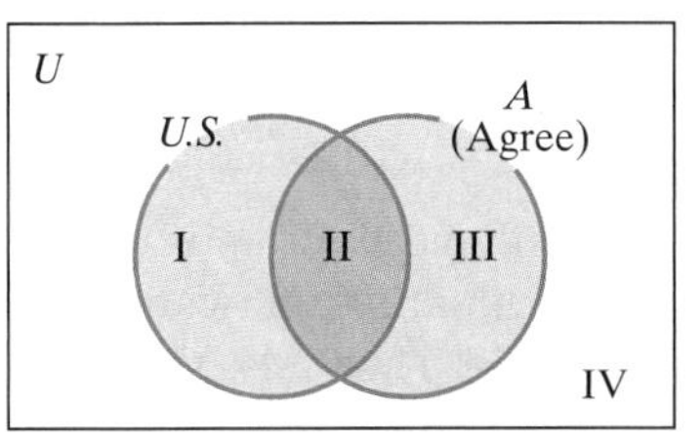

FIGURE 2.29

Step 1 Define sets and draw a Venn diagram. The Venn diagram in **Figure 2.29** shows two sets. Set *U.S.* is the set of Americans surveyed. Set *A* (labeled "Agree") is the set of people surveyed who agreed with the statement. By representing the Americans surveyed with circle *U.S.*, we do not need a separate circle for the Mexicans. The group of people outside circle *U.S.* must be the set of Mexicans. Similarly, by visualizing the set of people who agreed with the statement as circle *A*, we do not need a separate circle for those who disagreed. The group of people outside circle *A* (Agree) must be the set of people disagreeing with the statement.

Step 2 Determine the cardinality for each region in the Venn diagram, starting with the innermost region and working outward. We are given the following cardinalities:

> There were 2000 people surveyed: $n(U) = 2000$.
>
> Half the people surveyed were Americans: $n(U.S.) = 1000$.
>
> The number of people who agreed with the statement was 1060: $n(A) = 1060$.
>
> There were 400 Americans who agreed with the statement: $n(U.S. \cap A) = 400$.

Now let's use these numbers to determine the cardinality of each region, starting with region II, moving outward to regions I and III, and ending with region IV.

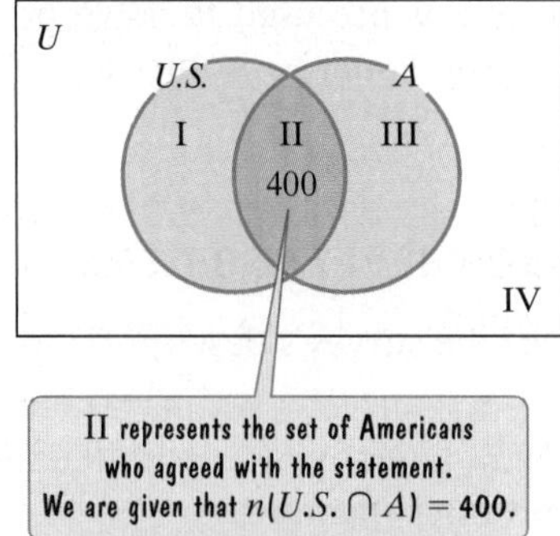

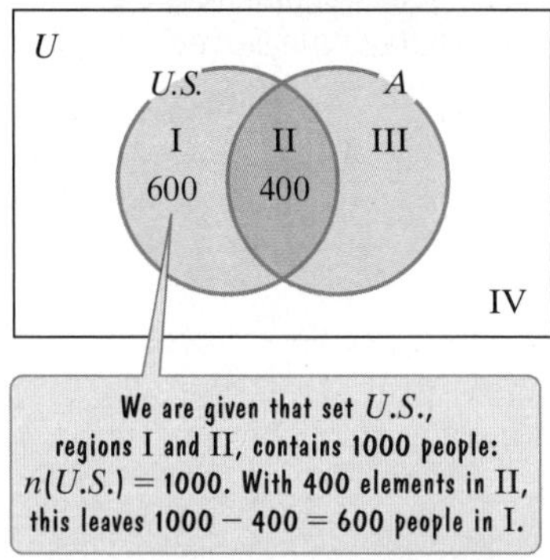

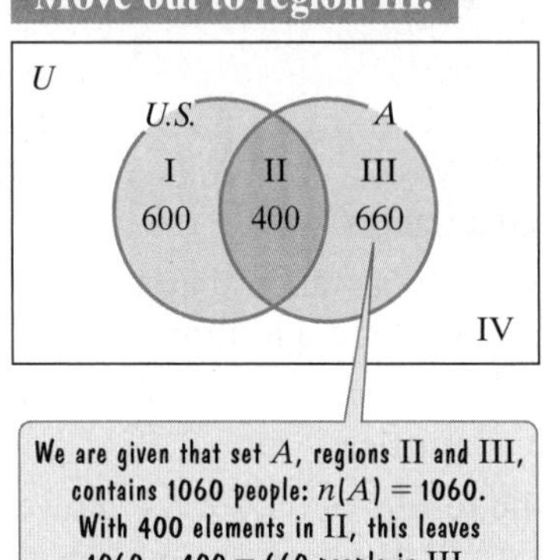

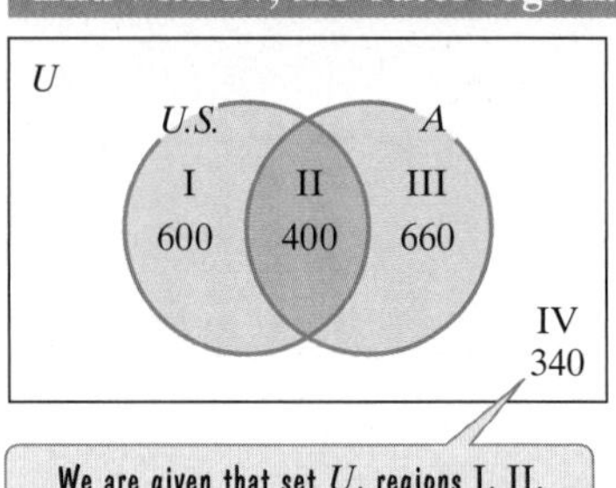

Is the Primary Cause of Poverty Societal Injustice?

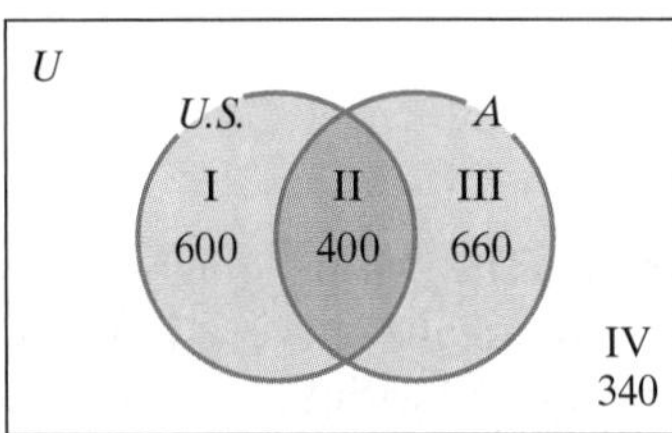

FIGURE 2.30

Step 3 Use the completed Venn diagram to answer the problem's questions. The completed Venn diagram that illustrates the survey's results is shown in **Figure 2.30**.

a. The Mexicans who agreed with the statement are those members of the set of people who agreed who are not Americans, shown in region III. This means that 660 Mexicans agreed that societal injustice is the primary cause of poverty.

b. The Mexicans who disagreed with the statement can be found outside the circles of people who agreed and people who are Americans. This corresponds to region IV, whose cardinality is 340. Thus, 340 Mexicans disagreed that societal injustice is the primary cause of poverty.

CHECK POINT 2 A survey was taken that asked 1700 randomly selected U.S. and French adults to agree or disagree with the following statement:

It is the responsibility of the government to reduce income differences.

The results of the survey showed that

1040 people agreed with the statement.
290 Americans agreed with the statement.
Source: Based on data from *The Sociology Project 2.0*, Pearson

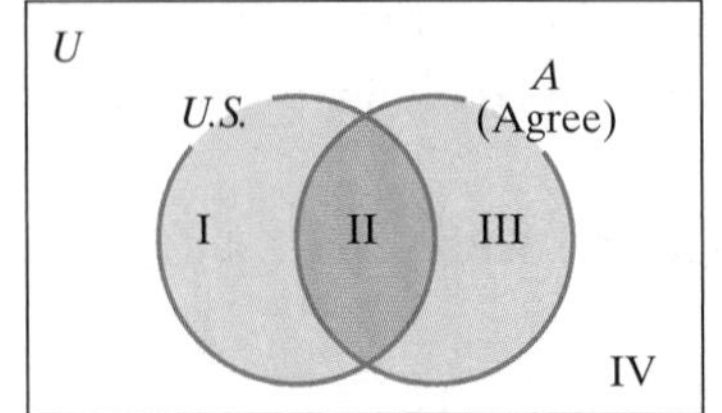

If 810 adults surveyed were Americans,

a. How many French adults agreed with the statement?

b. How many French adults disagreed with the statement?

When tabulating survey results, more than two circles within a Venn diagram are often needed. For example, consider a *Time*/CNN poll that sought to determine how Americans felt about reserving a certain number of college scholarships exclusively for minorities and women. Respondents were asked the following question:

> Do you agree or disagree with the following statement: Colleges should reserve a certain number of scholarships exclusively for minorities and women? *Source: Time Almanac*

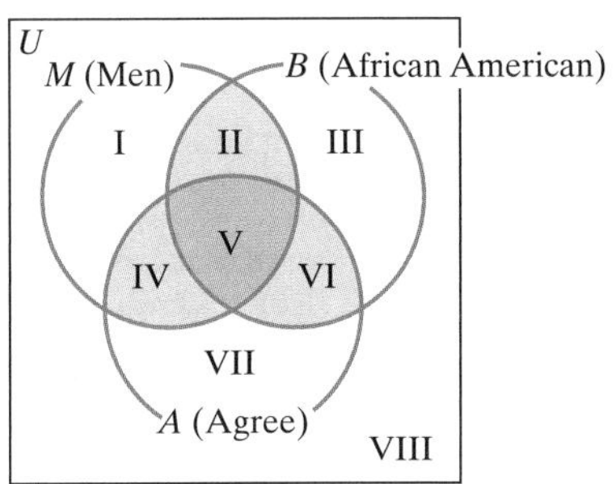

FIGURE 2.31

Suppose that we want the respondents to the poll to be identified by gender (man or woman), ethnicity (African American or other), and whether or not they agreed with the statement. A Venn diagram into which the results of the survey can be tabulated is shown in **Figure 2.31**.

Based on our work in Example 2, we only used one circle in the Venn diagram to indicate the gender of the respondent. We used M for men, so the set of women respondents, M', consists of the regions outside circle M. Similarly, we used B for the set of African American respondents, so the regions outside circle B account for all other ethnicities. Finally, we used A for the set of respondents who agreed with the statement. Those who disagreed lie outside circle A.

In the next example, we create a Venn diagram with three intersecting sets to illustrate a survey's results. In our final example, we use this Venn diagram to answer questions about the survey.

EXAMPLE 3 *Constructing a Venn Diagram for a Survey*

Sixty people were contacted and responded to a movie survey. The following information was obtained:

a. 6 people liked comedies, dramas, and science fiction.
b. 13 people liked comedies and dramas.
c. 10 people liked comedies and science fiction.
d. 11 people liked dramas and science fiction.
e. 26 people liked comedies.
f. 21 people liked dramas.
g. 25 people liked science fiction.

Use a Venn diagram to illustrate the survey's results.

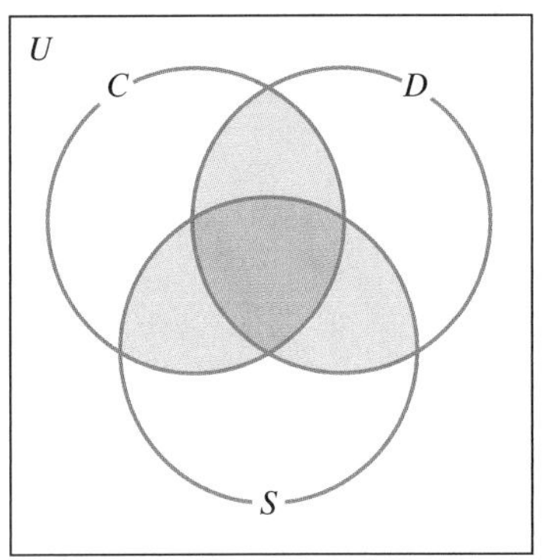

FIGURE 2.32

SOLUTION

The set of people surveyed is a universal set with 60 elements containing three subsets:

$$C = \text{the set of those who like comedies}$$
$$D = \text{the set of those who like dramas}$$
$$S = \text{the set of those who like science fiction.}$$

We draw these sets in **Figure 2.32**. Now let's use the numbers in (a) through (g), as well as the fact that 60 people were surveyed, which we call condition (h), to determine the cardinality of each region in the Venn diagram.

GREAT QUESTION!

Can you briefly summarize the procedure used to construct a Venn diagram for a survey?

After entering the cardinality of the innermost region, work outward and use subtraction to obtain subsequent cardinalities. The phrase

SURVEY SUBTRACT

is a helpful reminder of these repeated subtractions.

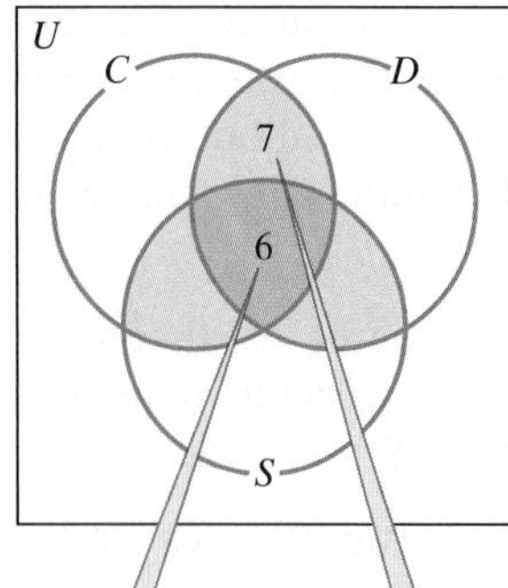

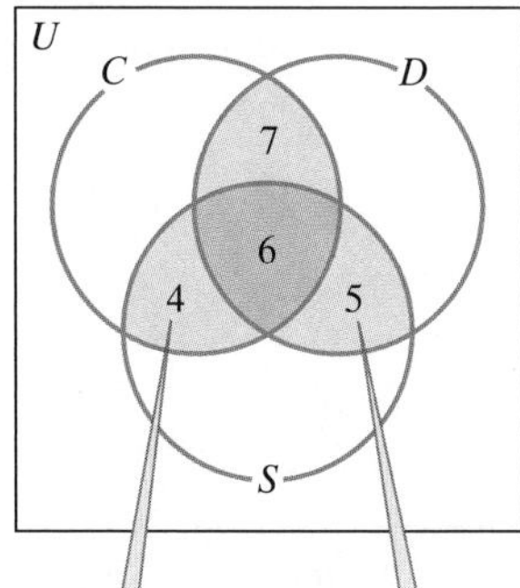

(a) 6 people liked comedies, drama, and science fiction: $n(C \cap D \cap S) = 6$.

(b) 13 people liked comedies and drama: $n(C \cap D) = 13$. With 6 counted, there are $13 - 6 = 7$ people in this region.

(c) 10 people liked comedies and science fiction: $n(C \cap S) = 10$. With 6 counted, there are $10 - 6 = 4$ people in this region.

(d) 11 people liked drama and science fiction: $n(D \cap S) = 11$. With 6 counted, there are $11 - 6 = 5$ people in this region.

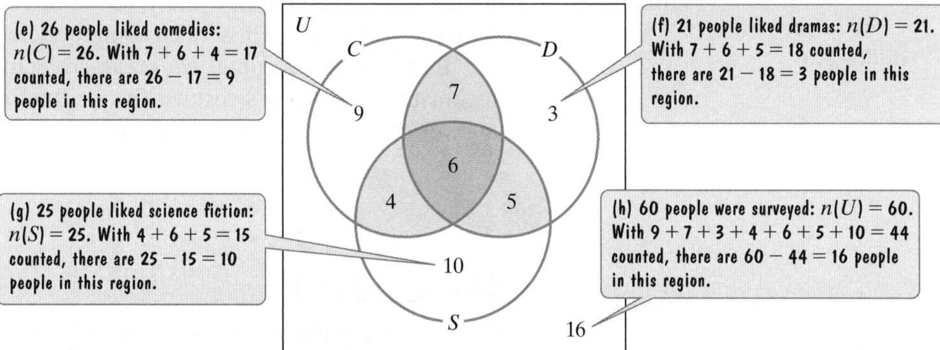

With a cardinality in each region, we have completed the Venn diagram that illustrates the survey's results.

CHECK POINT 3 A survey of 250 memorabilia collectors showed the following results: 108 collected baseball cards. 92 collected comic books. 62 collected stamps. 29 collected baseball cards and comic books. 5 collected baseball cards and stamps. 2 collected comic books and stamps. 2 collected all three types of memorabilia. Use a Venn diagram to illustrate the survey's results.

EXAMPLE 4 *Using a Survey's Venn Diagram*

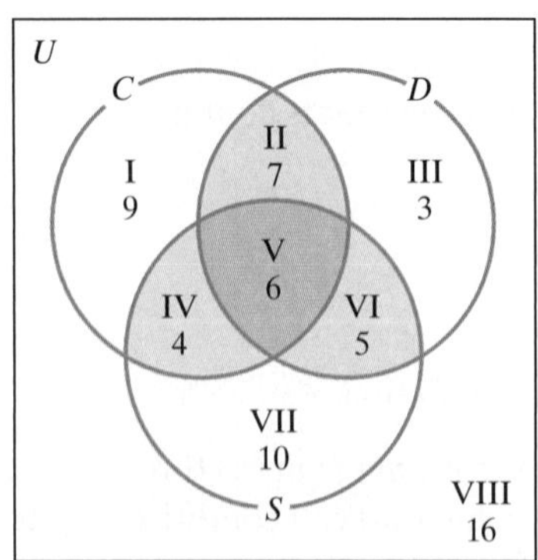

FIGURE 2.33

The Venn diagram in **Figure 2.33** shows the results of the movie survey in Example 3. How many of those surveyed liked

a. comedies, but neither dramas nor science fiction?

b. dramas and science fiction, but not comedies?

c. dramas or science fiction, but not comedies?

d. exactly one movie style?

e. at least two movie styles?

f. none of the movie styles?

SOLUTION

$C \cap (D' \cap S')$

a. Those surveyed who liked comedies, but neither dramas nor science fiction, are represented in region I. There are 9 people in this category.

$(D \cap S) \cap C'$

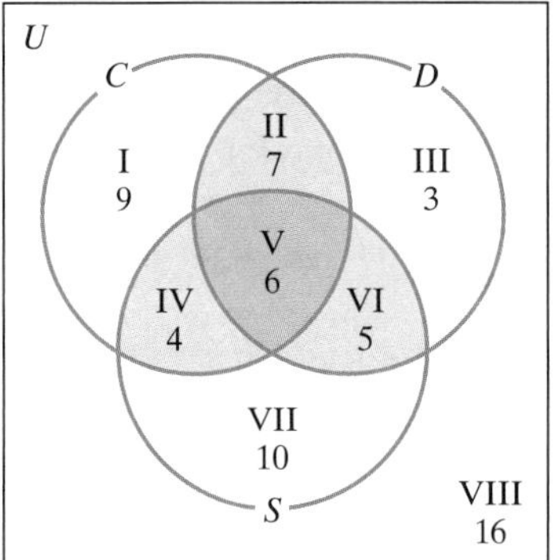

FIGURE 2.33 (repeated)

b. Those surveyed who liked dramas and science fiction, but not comedies, are represented in region VI. There are 5 people in this category.

c. We are interested in those surveyed who liked dramas or science fiction, but not comedies:

Dramas or science fiction, but not comedies

$$(D \cup S) \cap C'.$$

Regions II, III, IV, V, VI, VII Regions III, VI, VII, VIII

The intersection of the regions in the voice balloons consists of the common regions shown in red, III, VI, and VII. There are $3 + 5 + 10 = 18$ elements in these regions. There are 18 people who liked dramas or science fiction, but not comedies.

d. Those surveyed who liked exactly one movie style are represented in regions I, III, and VII. There are $9 + 3 + 10 = 22$ elements in these regions. Thus, 22 people liked exactly one movie style.

e. Those surveyed who liked at least two movie styles are people who liked two or more types of movies. People who liked two movie styles are represented in regions II, IV, and VI. Those who liked three movie styles are represented in region V. Thus, we add the number of elements in regions II, IV, V, and VI: $7 + 4 + 6 + 5 = 22$. Thus, 22 people liked at least two movie styles.

f. Those surveyed who liked none of the movie styles are represented in region VIII. There are 16 people in this category.

CHECK POINT 4 Use the Venn diagram you constructed in Check Point 3 to determine how many of those surveyed collected

a. comic books, but neither baseball cards nor stamps.

b. baseball cards and stamps, but not comic books.

c. baseball cards or stamps, but not comic books.

d. exactly two types of memorabilia.

e. at least one type of memorabilia.

f. none of the types of memorabilia.

Concept and Vocabulary Check

Fill in each blank so that the resulting statement is true.

1. In survey problems, the word __________ means "intersection."
2. In survey problems, the word _____ means "union."
3. In survey problems, the word ______ means "complement."
4. In order to construct a Venn diagram for a survey, first enter the cardinality of the __________ region. Then use ___________ to obtain subsequent cardinalities.

The Venn diagram on the right shows the results of a GfK Roper Public Affairs and Media survey of 1967 American adults taken on October 9, 2012. Use the Venn diagram to determine whether each statement in Exercises 5–8 is true or false. If the statement is false, make the necessary change(s) to produce a true statement.

How Americans Feel about Pets

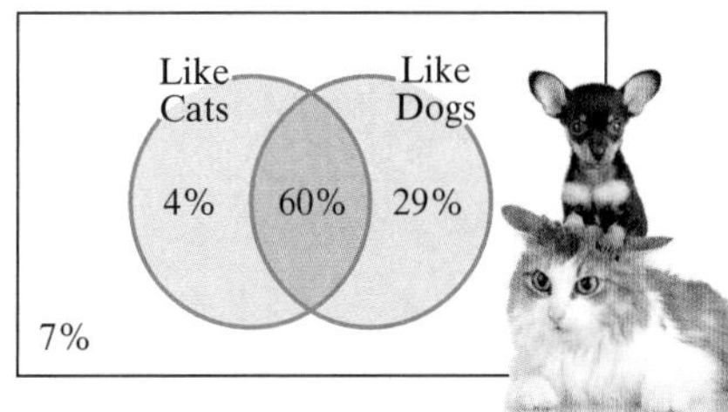

Source: GfK Roper Public Affairs and Media Survey

5. 64% of those surveyed liked cats. ______
6. 29% of those surveyed liked dogs but not cats. ______
7. 60% of those surveyed liked cats or dogs. ______
8. 7% of those surveyed liked neither cats nor dogs. ______

Exercise Set 2.5

Practice Exercises

Use the accompanying Venn diagram, which shows the number of elements in regions I through IV, to answer the questions in Exercises 1–8.

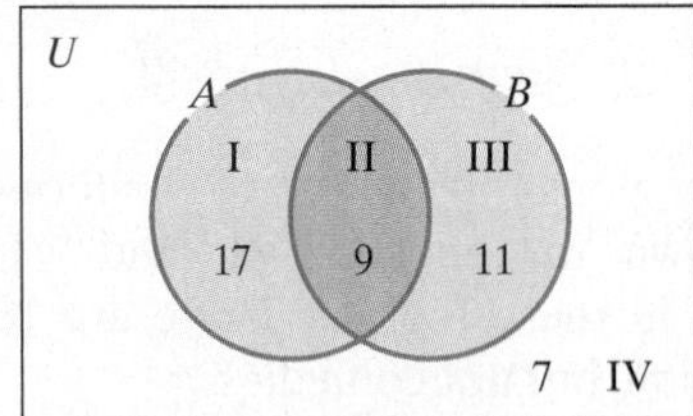

1. How many elements belong to set A?
2. How many elements belong to set B?
3. How many elements belong to set A but not set B?
4. How many elements belong to set B but not set A?
5. How many elements belong to set A or set B?
6. How many elements belong to set A and set B?
7. How many elements belong to neither set A nor set B?
8. How many elements are there in the universal set?

Use the accompanying Venn diagram, which shows the number of elements in region II, to answer Exercises 9–10.

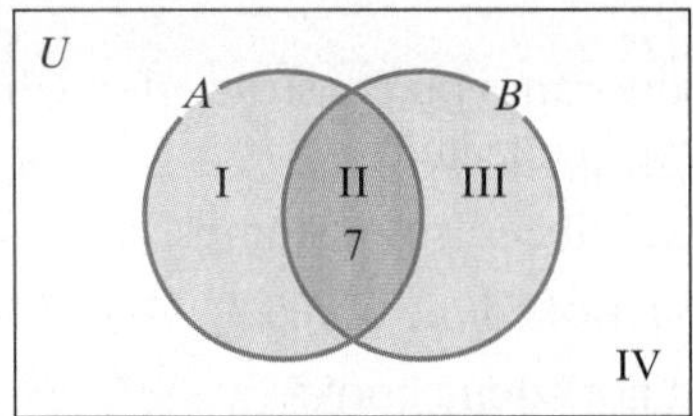

9. If $n(A) = 21$, $n(B) = 29$, and $n(U) = 48$, find the number of elements in each of regions I, III, and IV.
10. If $n(A) = 23$, $n(B) = 27$, and $n(U) = 53$, find the number of elements in each of regions I, III, and IV.

Use the accompanying Venn diagram, which shows the cardinality of each region, to answer Exercises 11–26.

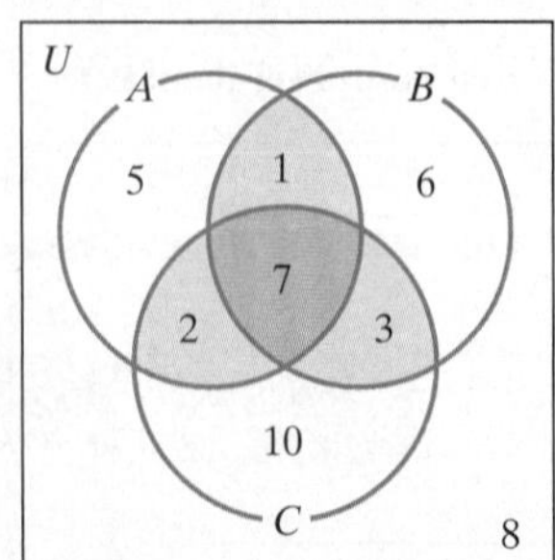

11. How many elements belong to set B?
12. How many elements belong to set A?
13. How many elements belong to set A but not set C?
14. How many elements belong to set B but not set A?
15. How many elements belong to set A or set C?
16. How many elements belong to set A or set B?
17. How many elements belong to set A and set C?
18. How many elements belong to set A and set B?
19. How many elements belong to set B and set C, but not to set A?
20. How many elements belong to set A and set C, but not to set B?
21. How many elements belong to set B or set C, but not to set A?
22. How many elements belong to set A or set C, but not to set B?
23. Considering sets A, B, and C, how many elements belong to exactly one of these sets?
24. Considering sets A, B, and C, how many elements belong to exactly two of these sets?
25. Considering sets A, B, and C, how many elements belong to at least one of these sets?
26. Considering sets A, B, and C, how many elements belong to at least two of these sets?

The accompanying Venn diagram shows the number of elements in region V. In Exercises 27–28, use the given cardinalities to determine the number of elements in each of the other seven regions.

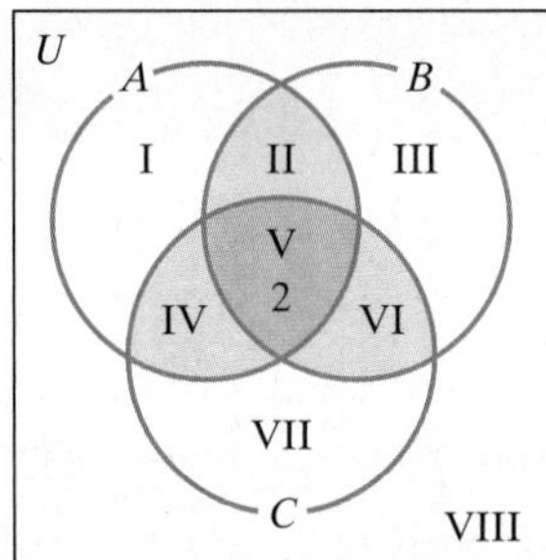

27. $n(U) = 30, n(A) = 11, n(B) = 8, n(C) = 14,$
$n(A \cap B) = 3, n(A \cap C) = 5, n(B \cap C) = 3$
28. $n(U) = 32, n(A) = 21, n(B) = 15, n(C) = 14,$
$n(A \cap B) = 6, n(A \cap C) = 7, n(B \cap C) = 8$

Practice Plus

In Exercises 29–32, use the Venn diagram and the given conditions to determine the number of elements in each region, or explain why the conditions are impossible to meet.

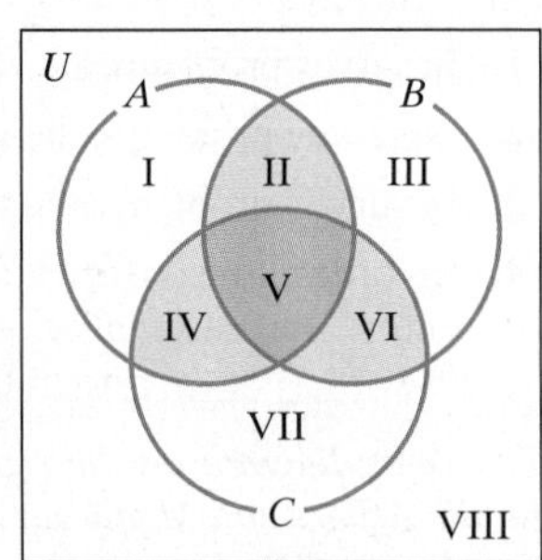

29. $n(U) = 38, n(A) = 26, n(B) = 21, n(C) = 18,$
$n(A \cap B) = 17, n(A \cap C) = 11, n(B \cap C) = 8,$
$n(A \cap B \cap C) = 7$

(In Exercises 30–32, continue to refer to the Venn diagram at the bottom of the previous page.)

30. $n(U) = 42, n(A) = 26, n(B) = 22, n(C) = 25,$
$n(A \cap B) = 17, n(A \cap C) = 11, n(B \cap C) = 9,$
$n(A \cap B \cap C) = 5$

31. $n(U) = 40, n(A) = 10, n(B) = 11, n(C) = 12,$
$n(A \cap B) = 6, n(A \cap C) = 9, n(B \cap C) = 7,$
$n(A \cap B \cap C) = 2$

32. $n(U) = 25, n(A) = 8, n(B) = 9, n(C) = 10,$
$n(A \cap B) = 6, n(A \cap C) = 9, n(B \cap C) = 8,$
$n(A \cap B \cap C) = 5$

Application Exercises

As discussed in the text on page 103, a poll asked respondents if they agreed with the statement

Colleges should reserve a certain number of scholarships exclusively for minorities and women.

Hypothetical results of the poll are tabulated in the Venn diagram. Use these cardinalities to solve Exercises 33–38.

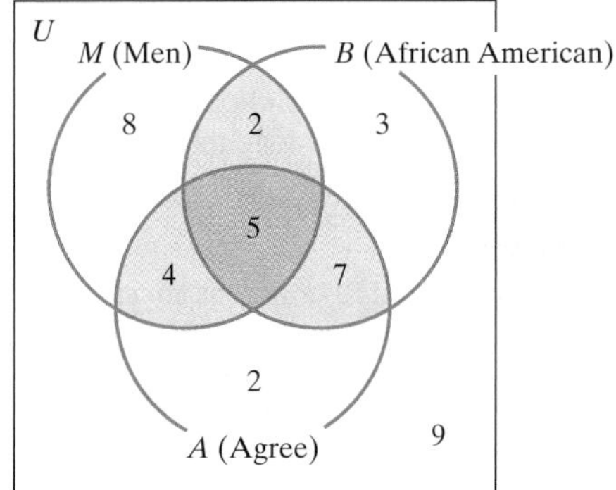

33. How many respondents agreed with the statement?

34. How many respondents disagreed with the statement?

35. How many women agreed with the statement?

36. How many people who are not African American agreed with the statement?

37. How many women who are not African American disagreed with the statement?

38. How many men who are not African American disagreed with the statement?

Solve the survey problems in Exercises 39–40 using the following Venn diagram.

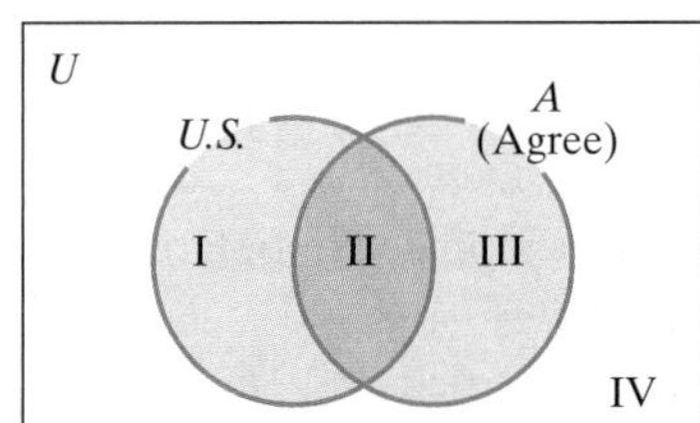

39. A survey was taken that asked 1620 randomly selected U.S. and British adults to agree or disagree with the following statement:

Government should provide only limited healthcare services.

The results of the survey showed that

370 people agreed with the statement.
270 Americans agreed with the statement.
Source: Based on data from *The Sociology Project 2.0*, Pearson

If 780 adults surveyed were Americans,

a. How many British adults agreed with the statement?

b. How many British adults disagreed with the statement?

40. A survey was taken that asked 1510 randomly selected U.S. and German adults to agree or disagree with the following statement:

It is the husband's job to earn money and the wife's job is the family.

The results of the survey showed that

440 people agreed with the statement.
260 Americans agreed with the statement.
Source: Based on data from *The Sociology Project 2.0*, Pearson

If 770 adults surveyed were Americans,

a. How many German adults agreed with the statement?

b. How many German adults disagreed with the statement?

41. A pollster conducting a telephone poll of a city's residents asked two questions:

1. Do you currently smoke cigarettes?
2. Regardless of your answer to question 1, would you support a ban on smoking in all city parks?

a. Construct a Venn diagram that allows the respondents to the poll to be identified by whether or not they smoke cigarettes and whether or not they support the ban.

b. Write the letter b in every region of the diagram that represents smokers polled who support the ban.

c. Write the letter c in every region of the diagram that represents nonsmokers polled who support the ban.

d. Write the letter d in every region of the diagram that represents nonsmokers polled who do not support the ban.

42. A pollster conducting a telephone poll at a college campus asked students two questions:

1. Do you binge drink three or more times per month?
2. Regardless of your answer to question 1, are you frequently behind in your school work?

a. Construct a Venn diagram that allows the respondents to the poll to be identified by whether or not they binge drink and whether or not they frequently fall behind in school work.

b. Write the letter b in every region of the diagram that represents binge drinkers who are frequently behind in school work.

c. Write the letter c in every region of the diagram that represents students polled who do not binge drink but who are frequently behind in school work.

d. Write the letter d in every region of the diagram that represents students polled who do not binge drink and who do not frequently fall behind in their school work.

43. A pollster conducting a telephone poll asked three questions:

1. Are you religious?
2. Have you spent time with a person during his or her last days of a terminal illness?
3. Should assisted suicide be an option for terminally ill people?

a. Construct a Venn diagram with three circles that can assist the pollster in tabulating the responses to the three questions.

b. Write the letter b in every region of the diagram that represents all religious persons polled who are not in favor of assisted suicide for the terminally ill.

c. Write the letter c in every region of the diagram that represents the people polled who do not consider themselves religious, who have not spent time with a terminally ill person during his or her last days, and who are in favor of assisted suicide for the terminally ill.

d. Write the letter d in every region of the diagram that represents the people polled who consider themselves religious, who have not spent time with a terminally ill person during his or her last days, and who are not in favor of assisted suicide for the terminally ill.

e. Write the letter e in a region of the Venn diagram other than those in parts (b)–(d) and then describe who in the poll is represented by this region.

44. A poll asks respondents the following question:

Do you agree or disagree with this statement: In order to address the trend in diminishing male enrollment, colleges should begin special efforts to recruit men?

a. Construct a Venn diagram with three circles that allows the respondents to be identified by gender (man or woman), education level (college or no college), and whether or not they agreed with the statement.

b. Write the letter b in every region of the diagram that represents men with a college education who agreed with the statement.

c. Write the letter c in every region of the diagram that represents women who disagreed with the statement.

d. Write the letter d in every region of the diagram that represents women without a college education who agreed with the statement.

e. Write the letter e in a region of the Venn diagram other than those in parts (b)–(d) and then describe who in the poll is represented by this region.

In Exercises 45–50, construct a Venn diagram and determine the cardinality for each region. Use the completed Venn diagram to answer the questions.

45. A survey of 75 college students was taken to determine where they got the news about what's going on in the world. Of those surveyed, 29 students got the news from newspapers, 43 from television, and 7 from both newspapers and television.

Of those surveyed,

a. How many got the news from only newspapers?

b. How many got the news from only television?

c. How many got the news from newspapers or television?

d. How many did not get the news from either newspapers or television?

46. A survey of 120 college students was taken at registration. Of those surveyed, 75 students registered for a math course, 65 for an English course, and 40 for both math and English.

Of those surveyed,

a. How many registered only for a math course?

b. How many registered only for an English course?

c. How many registered for a math course or an English course?

d. How many did not register for either a math course or an English course?

47. A survey of 80 college students was taken to determine the musical styles they listened to. Forty-two students listened to rock, 34 to classical, and 27 to jazz. Twelve students listened to rock and jazz, 14 to rock and classical, and 10 to classical and jazz. Seven students listened to all three musical styles.

Of those surveyed,

a. How many listened to only rock music?

b. How many listened to classical and jazz, but not rock?

c. How many listened to classical or jazz, but not rock?

d. How many listened to music in exactly one of the musical styles?

e. How many listened to music in at least two of the musical styles?

f. How many did not listen to any of the musical styles?

48. A survey of 180 college men was taken to determine participation in various campus activities. Forty-three students were in fraternities, 52 participated in campus sports, and 35 participated in various campus tutorial programs. Thirteen students participated in fraternities and sports, 14 in sports and tutorial programs, and 12 in fraternities and tutorial programs. Five students participated in all three activities.

Of those surveyed,

a. How many participated in only campus sports?

b. How many participated in fraternities and sports, but not tutorial programs?

c. How many participated in fraternities or sports, but not tutorial programs?

d. How many participated in exactly one of these activities?

e. How many participated in at least two of these activities?

f. How many did not participate in any of the three activities?

49. An anonymous survey of college students was taken to determine behaviors regarding alcohol, cigarettes, and illegal drugs. The results were as follows: 894 drank alcohol regularly, 665 smoked cigarettes, 192 used illegal drugs, 424 drank alcohol regularly and smoked cigarettes, 114 drank alcohol regularly and used illegal drugs, 119 smoked cigarettes and used illegal drugs, 97 engaged in all three behaviors, and 309 engaged in none of these behaviors.
Source: Jamie Langille, University of Nevada Las Vegas

a. How many students were surveyed?

Of those surveyed,

b. How many drank alcohol regularly or smoked cigarettes?

c. How many used illegal drugs only?

d. How many drank alcohol regularly and smoked cigarettes, but did not use illegal drugs?

e. How many drank alcohol regularly or used illegal drugs, but did not smoke cigarettes?

f. How many engaged in exactly two of these behaviors?

g. How many engaged in at least one of these behaviors?

50. In the August 2005 issue of *Consumer Reports*, readers suffering from depression reported that alternative treatments were less effective than prescription drugs. Suppose that 550 readers felt better taking prescription drugs, 220 felt better through meditation, and 45 felt better taking St. John's wort. Furthermore, 95 felt better using prescription drugs and meditation, 17 felt better using prescription drugs and St. John's wort, 35 felt better using meditation and St. John's wort, 15 improved using all three treatments, and 150 improved using none of these treatments. (Hypothetical results are partly based on percentages given in *Consumer Reports*.)

a. How many readers suffering from depression were included in the report?

Of those included in the report,

b. How many felt better using prescription drugs or meditation?

c. How many felt better using St. John's wort only?

d. How many improved using prescription drugs and meditation, but not St. John's wort?

e. How many improved using prescription drugs or St. John's wort, but not meditation?

f. How many improved using exactly two of these treatments?

g. How many improved using at least one of these treatments?

Explaining the Concepts

51. Suppose that you are drawing a Venn diagram to sort and tabulate the results of a survey. If results are being tabulated along gender lines, explain why only a circle representing women is needed, rather than two separate circles representing the women surveyed and the men surveyed.

52. Suppose that you decide to use two sets, M and W, to sort and tabulate the responses for men and women in a survey. Describe the set of people represented by regions II and IV in the Venn diagram shown. What conclusion can you draw?

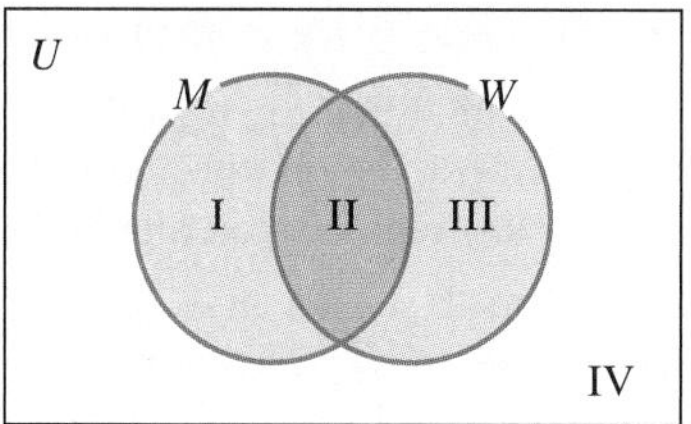

Critical Thinking Exercises

Make Sense? *In Exercises 53–56, determine whether each statement makes sense or does not make sense, and explain your reasoning.*

53. A survey problem must present the information in exactly the same order in which I determine cardinalities from innermost to outermost region.

Exercises 54–56 are based on the graph that shows the percentage of smokers and nonsmokers suffering from various ailments. Use the graph to determine whether each statement makes sense.

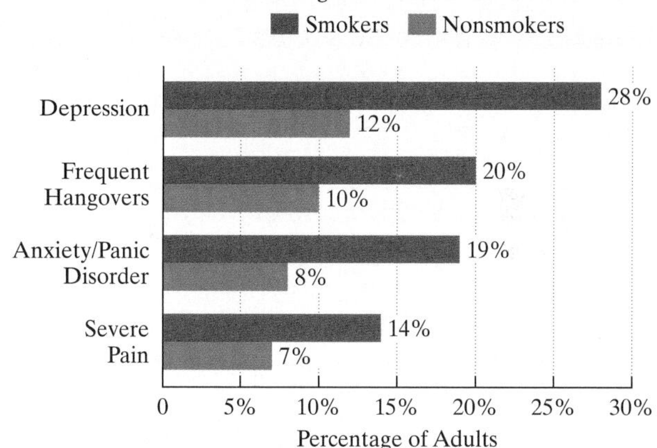

Source: MARS OTC/DTC

54. I represented the data for depression using the following Venn diagram:

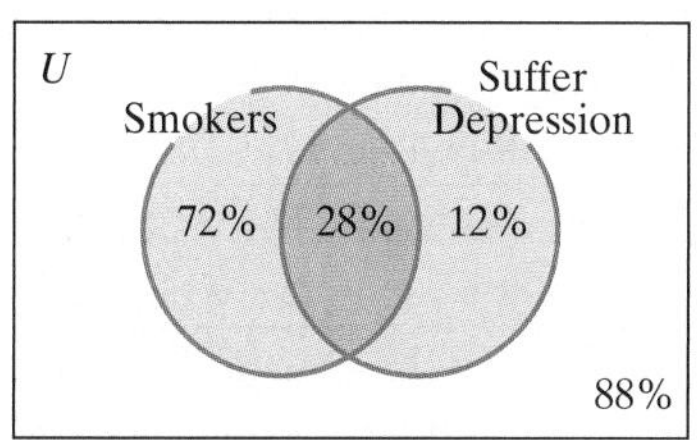

55. I improved the Venn diagram in Exercise 54 by adding a third circle for nonsmokers.

56. I used a single Venn diagram to represent all the data displayed by the bar graph.

In Exercises 57–60, determine whether each statement is true or false. If the statement is false, make the necessary change(s) to produce a true statement.

57. In a survey, 110 students were taking mathematics, 90 were taking psychology, and 20 were taking neither. Thus, 220 students were surveyed.

58. If $A \cap B = \varnothing$, then $n(A \cup B) = n(A) + n(B)$.

59. When filling in cardinalities for regions in a two-set Venn diagram, the innermost region, the intersection of the two sets, should be the last region to be filled in.

60. For a finite universal set U, $n(A')$ can be obtained by subtracting $n(A)$ from $n(U)$.

61. In a survey of 150 students, 90 were taking mathematics and 30 were taking psychology.

a. What is the least number of students who could have been taking both courses?

b. What is the greatest number of students who could have been taking both courses?

c. What is the greatest number of students who could have been taking neither course?

62. A person applying for the position of college registrar submitted the following report to the college president on 90 students: 31 take math; 28 take chemistry; 42 take psychology; 9 take math and chemistry; 10 take chemistry and psychology; 6 take math and psychology; 4 take all three subjects; and 20 take none of these courses. The applicant was not hired. Explain why.

Group Exercise

63. This group activity is intended to provide practice in the use of Venn diagrams to sort responses to a survey. The group will determine the topic of the survey. Although you will not actually conduct the survey, it might be helpful to imagine carrying out the survey using the students on your campus.

a. In your group, decide on a topic for the survey.

b. Devise three questions that the pollster will ask to the people who are interviewed.

c. Construct a Venn diagram that will assist the pollster in sorting the answers to the three questions. The Venn diagram should contain three intersecting circles within a universal set and eight regions.

d. Describe what each of the regions in the Venn diagram represents in terms of the questions in your poll.

Chapter Summary, Review, and Test

SUMMARY – DEFINITIONS AND CONCEPTS | EXAMPLES

2.1 Basic Set Concepts

a. A set is a collection of objects whose contents can be clearly determined. The objects in a set are called the elements, or members, of the set.

b. Sets can be designated by word descriptions, the roster method (a listing within braces, separating elements with commas), or set-builder notation: Ex. 1, p. 50; Ex. 2, p. 51; Ex. 3, p. 51

$$\{ \quad x \quad | \quad \text{condition (s)} \}.$$

The set of | all elements x | such that | x meets these conditions

c. The empty set, or the null set, represented by { } or $\varnothing$, is a set that contains no elements. Ex. 4, p. 53

d. The symbol $\in$ means that an object is an element of a set. The symbol $\notin$ means that an object is not an element of a set. Ex. 5, p. 53

e. The set of natural numbers is $\mathbf{N} = \{1, 2, 3, 4, 5, \ldots\}$. Inequality symbols, summarized in **Table 2.2** on page 55, are frequently used to describe sets of natural numbers. Ex. 6, p. 54; Ex. 7, p. 56

f. The cardinal number of a set A, $n(A)$, is the number of distinct elements in set A. Repeating elements in a set neither adds new elements to the set nor changes its cardinality. Ex. 8, p. 56

g. Equivalent sets have the same number of elements, or the same cardinality. A one-to-one correspondence between sets A and B means that each element in A can be paired with exactly one element in B, and vice versa. If two sets can be placed in a one-to-one correspondence, then they are equivalent. Ex. 9, p. 57

h. Set A is a finite set if $n(A) = 0$ or if $n(A)$ is a natural number. A set that is not finite is an infinite set.

i. Equal sets have exactly the same elements, regardless of order or possible repetition of elements. If two sets are equal, then they must be equivalent. Ex. 10, p. 60

2.2 Subsets

a. Set A is a subset of set B, expressed as $A \subseteq B$, if every element in set A is also in set B. The notation $A \nsubseteq B$ means that set A is not a subset of set B, so there is at least one element of set A that is not an element of set B. Ex. 1, p. 65

b. Set A is a proper subset of set B, expressed as $A \subset B$, if A is a subset of B and $A \neq B$. Ex. 2, p. 66

c. The empty set is a subset of every set. Ex. 3, p. 68

d. A set with n elements has 2^n distinct subsets and $2^n - 1$ distinct proper subsets. Ex. 4, p. 69

2.3 Venn Diagrams and Set Operations

a. A universal set, symbolized by U, is a set that contains all the elements being considered in a given discussion or problem. Ex. 1, p. 74

b. Venn Diagrams: Representing Two Subsets of a Universal Set Ex. 2, p. 76

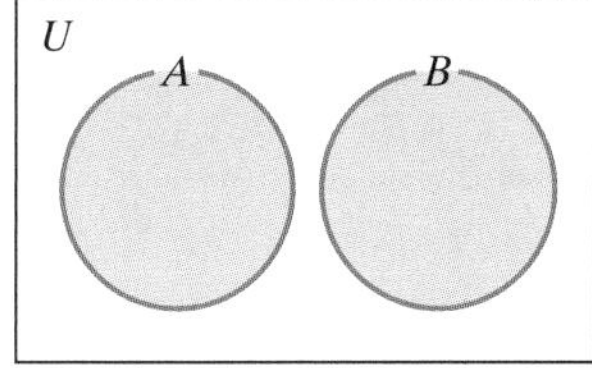

No A are B.
A and B are disjoint.

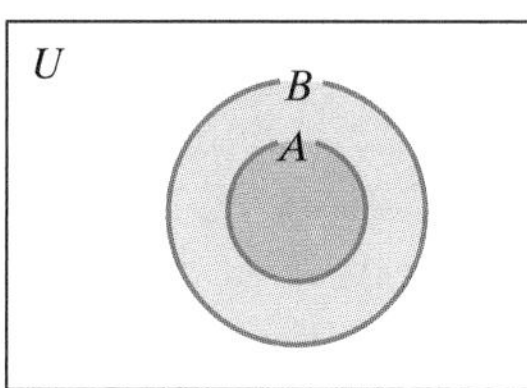

All A are B.
$A \subset B$

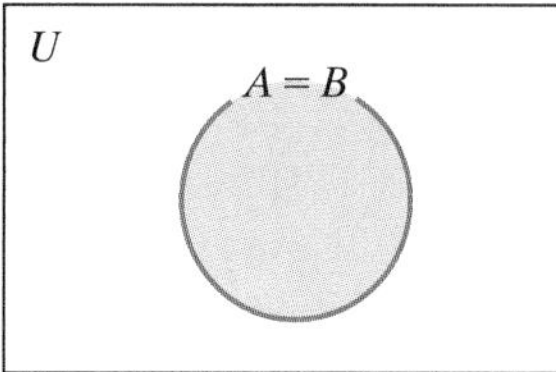

A and B are equal sets.

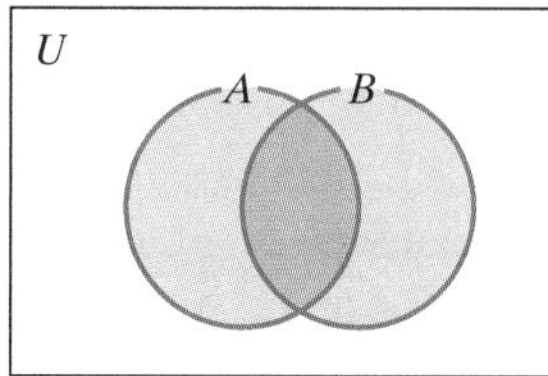

Some (at least one) A are B.

c. A' (the complement of set A), which can be read A prime or **not** A, is the set of all elements in the universal set that are not in A. Ex. 3, p. 77

d. $A \cap B$ (A intersection B), which can be read set A **and** set B, is the set of elements common to both set A and set B. Ex. 4, p. 77

e. $A \cup B$ (A union B), which can be read set A **or** set B, is the set of elements that are members of set A or of set B or of both sets. Ex. 5, p. 79

f. Some problems involve more than one set operation. Begin by performing any operations inside parentheses. Ex. 6, p. 80

g. Elements of sets involving a variety of set operations can be determined from Venn diagrams. Ex. 7, p. 81

h. Cardinal Number of the Union of Two Finite Sets Ex. 8, p. 83

$$n(A \cup B) = n(A) + n(B) - n(A \cap B)$$

2.4 Set Operations and Venn Diagrams with Three Sets

a. When using set operations involving three sets, begin by performing operations within parentheses. Ex. 1, p. 87

b. The figure below shows a Venn diagram with three intersecting sets that separate the universal set, U, into eight regions. Elements of sets involving a variety of set operations can be determined from this Venn diagram. Ex. 2, p. 89

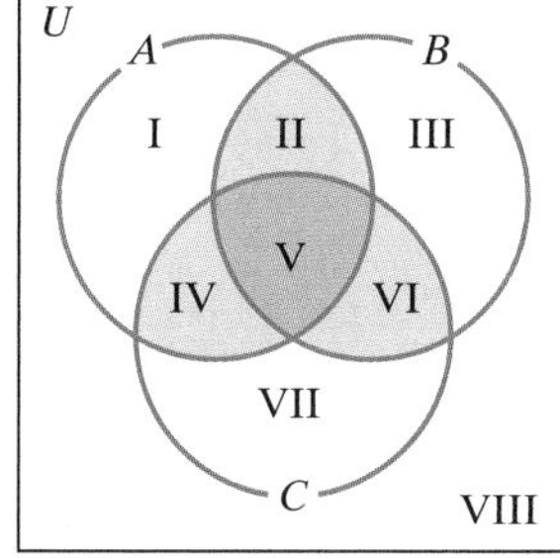

c. To construct a Venn diagram illustrating the elements in A, B, C, and U, first find $A \cap B$, $A \cap C$, $B \cap C$, and $A \cap B \cap C$. Then place elements into the eight regions shown above, starting with the innermost region, region V, and working outward to region VIII. Ex. 3, p. 90

d. If two specific sets represent the same regions of a general Venn diagram, then this deductively proves that the two sets are equal.	Ex. 4, p. 92; Ex. 5, p. 93

2.5 Survey Problems

a. Venn diagrams can be used to organize information collected in surveys. When interpreting cardinalities in such diagrams, *and* and *but* mean intersection, *or* means union, and *not* means complement.	Ex. 1, p. 99
b. To solve a survey problem, **1.** Define sets and draw a Venn diagram. **2.** Fill in the cardinality of each region, starting with the innermost region and working outward. **3.** Use the completed diagram to answer the problem's questions.	Ex. 2, p. 101; Ex. 3, p. 103; Ex. 4, p. 104

Review Exercises

2.1

In Exercises 1–2, write a word description of each set. (More than one correct description may be possible.)

1. {Tuesday, Thursday}

2. $\{1, 2, 3, \ldots, 10\}$

In Exercises 3–5, express each set using the roster method.

3. $\{x \mid x$ is a letter in the word *miss*$\}$

4. $\{x \mid x \in \mathbf{N}$ and $8 \le x < 13\}$

5. $\{x \mid x \in \mathbf{N}$ and $x \le 30\}$

In Exercises 6–7, determine which sets are the empty set.

6. $\{\varnothing\}$

7. $\{x \mid x < 4$ and $x \ge 6\}$

In Exercises 8–9, fill in the blank with either $\in$ or $\notin$ to make each statement true.

8. 93 ______ $\{1, 2, 3, 4, \ldots, 99, 100\}$

9. {d} ______ {a, b, c, d, e}

In Exercises 10–11, find the cardinal number for each set.

10. $A = \{x \mid x$ is a month of the year$\}$

11. $B = \{18, 19, 20, \ldots, 31, 32\}$

In Exercises 12–13, fill in the blank with either $=$ or $\ne$ to make each statement true.

12. $\{0, 2, 4, 6, 8\}$ ______ $\{8, 2, 6, 4\}$

13. $\{x \mid x \in \mathbf{N}$ and $x > 7\}$ ______ $\{8, 9, 10, \ldots, 100\}$

In Exercises 14–15, determine if the pairs of sets are equivalent, equal, both, or neither.

14. $A = \{x \mid x$ is a lowercase letter that comes before f in the English alphabet$\}$
$B = \{2, 4, 6, 8, 10\}$

15. $A = \{x \mid x \in \mathbf{N}$ and $3 < x < 7\}$
$B = \{4, 5, 6\}$

In Exercises 16–17, determine whether each set is finite or infinite.

16. $\{x \mid x \in \mathbf{N}$ and $x < 50{,}000\}$

17. $\{x \mid x \in \mathbf{N}$ and x is even$\}$

2.2

In Exercises 18–20, write $\subseteq$ or $\not\subseteq$ in each blank so that the resulting statement is true.

18. {penny, nickel, dime}
______ {half-dollar, quarter, dime, nickel, penny}

19. $\{-1, 0, 1\}$ ______ $\{-3, -2, -1, 1, 2, 3\}$

20. $\varnothing$ ______ $\{x \mid x$ is an odd natural number$\}$

In Exercises 21–22, determine whether $\subseteq$, $\subset$, both, or neither can be placed in each blank to form a true statement.

21. $\{1, 2\}$ ______ $\{1, 1, 2, 2\}$

22. $\{x \mid x$ is a person living in the United States$\}$
______ $\{y \mid y$ is a person living on planet Earth$\}$

In Exercises 23–29, determine whether each statement is true or false. If the statement is false, explain why.

23. Texas $\in$ {Oklahoma, Louisiana, Georgia, South Carolina}

24. $4 \subseteq \{2, 4, 6, 8, 10, 12\}$

25. $\{e, f, g\} \subset \{d, e, f, g, h, i\}$

26. $\{\ominus, \varnothing\} \subset \{\varnothing, \ominus\}$

27. $\{3, 7, 9\} \subseteq \{9, 7, 3, 1\}$

28. {six} has 2^6 subsets.

29. $\varnothing \subseteq \{\ \}$

30. List all subsets for the set $\{1, 5\}$. Which one of these subsets is not a proper subset?

In Exercises 31–32, find the number of distinct subsets and the number of distinct proper subsets for each set.

31. $\{2, 4, 6, 8, 10\}$

32. $\{x | x$ is a month that begins with the letter J$\}$

2.3

In Exercises 33–37, let $U = \{1, 2, 3, 4, 5, 6, 7, 8\}$, $A = \{1, 2, 3, 4\}$, *and* $B = \{1, 2, 4, 5\}$. *Find each of the following sets.*

33. $A \cap B$

34. $A \cup B'$

35. $A' \cap B$

36. $(A \cup B)'$

37. $A' \cap B'$

In Exercises 38–45, use the Venn diagram to represent each set in roster form.

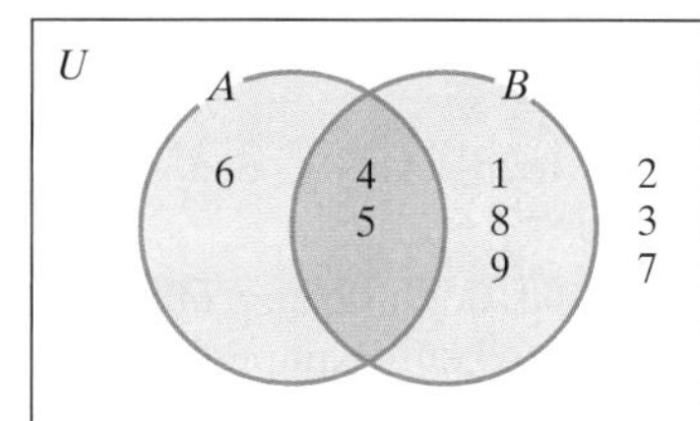

38. A

39. B'

40. $A \cup B$

41. $A \cap B$

42. $(A \cap B)'$

43. $(A \cup B)'$

44. $A \cap B'$

45. U

46. Set A contains 25 elements, set B contains 17 elements, and 9 elements are common to sets A and B. How many elements are in $A \cup B$?

2.4

In Exercises 47–48, let

$$U = \{1, 2, 3, 4, 5, 6, 7, 8\}$$
$$A = \{1, 2, 3, 4\}$$
$$B = \{1, 2, 4, 5\}$$
$$C = \{1, 5\}.$$

Find each of the following sets.

47. $A \cup (B \cap C)$

48. $(A \cap C)' \cup B$

In Exercises 49–54, use the Venn diagram to represent each set in roster form.

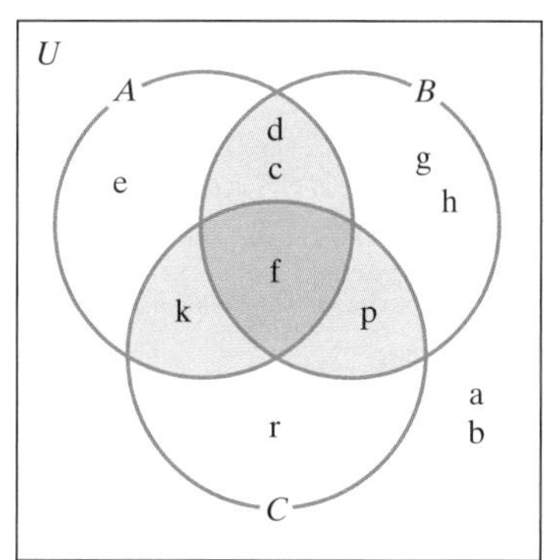

49. $A \cup C$

50. $B \cap C$

51. $(A \cap B) \cup C$

52. $A \cap C'$

53. $(A \cap C)'$

54. $A \cap B \cap C$

55. Construct a Venn diagram illustrating the following sets: $A = \{q, r, s, t, u\}$, $B = \{p, q, r\}$, $C = \{r, u, w, y\}$, and $U = \{p, q, r, s, t, u, v, w, x, y\}$.

56. Use a Venn diagram with two intersecting circles to prove that

$$(A \cup B)' = A' \cap B'.$$

57. Use a Venn diagram with three intersecting circles to determine whether the following statement is a theorem:

$$A \cap (B \cup C) = A \cup (B \cap C).$$

58. *The Penguin Atlas of Women in the World* uses maps and graphics to present data on how women live across continents and cultures. The table is based on data from the atlas.

WOMEN IN THE WORLD: SCHOOL, WORK, AND LITERACY

	Percentage of College Students Who Are Women	Percentage of Women Working for Pay	Percentage of Women Who Are Illiterate
United States	57%	59%	1%
Italy	57%	37%	2%
Turkey	43%	28%	20%
Norway	60%	63%	1%
Pakistan	46%	33%	65%
Iceland	65%	71%	1%
Mexico	51%	40%	10%

Source: The Penguin Atlas of Women in the World

The data can be organized in the following Venn diagram:

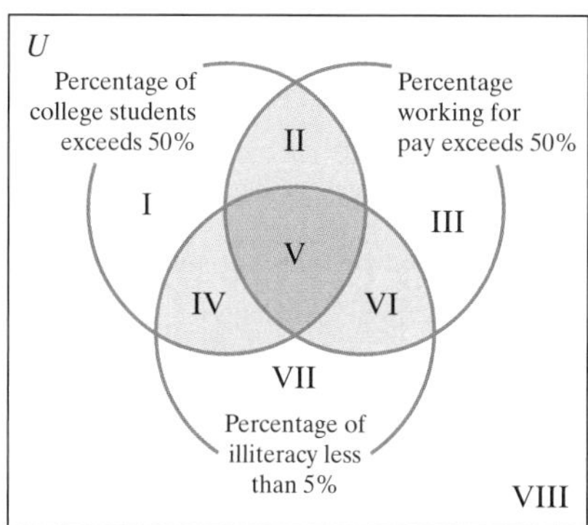

Use the data to determine the region in the Venn diagram into which each of the seven countries should be placed.

2.5

59. In a Gallup poll, 2000 U.S. adults were selected at random and asked to agree or disagree with the following statement:

> Job opportunities for women are not equal to those for men.

The results of the survey showed that

> 1190 people agreed with the statement.
> 700 women agreed with the statement.
> *Source: The People's Almanac*

If half the people surveyed were women,

a. How many men agreed with the statement?

b. How many men disagreed with the statement?

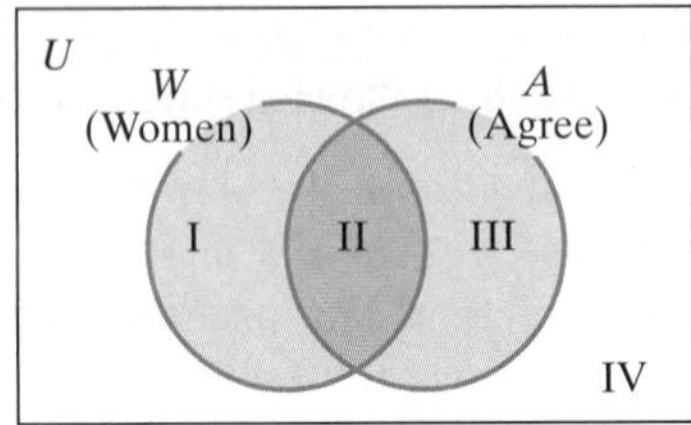

In Exercises 60–61, construct a Venn diagram and determine the cardinality for each region. Use the completed Venn diagram to answer the questions.

60. A survey of 1000 American adults was taken to analyze their investments. Of those surveyed, 650 had invested in stocks, 550 in bonds, and 400 in both stocks and bonds. Of those surveyed,

a. How many invested in only stocks?

b. How many invested in stocks or bonds?

c. How many did not invest in either stocks or bonds?

61. A survey of 200 students at a nonresidential college was taken to determine how they got to campus during the fall term. Of those surveyed, 118 used cars, 102 used public transportation, and 70 used bikes. Forty-eight students used cars and public transportation, 38 used cars and bikes, and 26 used public transportation and bikes. Twenty-two students used all three modes of transportation.
Of those surveyed,

a. How many used only public transportation?

b. How many used cars and public transportation, but not bikes?

c. How many used cars or public transportation, but not bikes?

d. How many used exactly two of these modes of transportation?

e. How many did not use any of the three modes of transportation to get to campus?

Chapter 2 Test

1. Express the following set using the roster method:
$\{x \mid x \in \mathbf{N} \text{ and } 17 < x \leq 24\}$.

In Exercises 2–9, determine whether each statement is true or false. If the statement is false, explain why.

2. $\{6\} \in \{1, 2, 3, 4, 5, 6, 7\}$

3. If $A = \{x \mid x \text{ is a day of the week}\}$ and $B = \{2, 4, 6, \ldots, 14\}$, then sets A and B are equivalent.

4. $\{2, 4, 6, 8\} = \{8, 8, 6, 6, 4, 4, 2\}$

5. $\{\text{d, e, f, g}\} \subseteq \{\text{a, b, c, d, e, f}\}$

6. $\{3, 4, 5\} \subset \{x \mid x \in \mathbf{N} \text{ and } x < 6\}$

7. $14 \notin \{1, 2, 3, 4, \ldots, 39, 40\}$

8. $\{\text{a, b, c, d, e}\}$ has 25 subsets.

9. The empty set is a proper subset of any set, including itself.

10. List all subsets for the set $\{6, 9\}$. Which of these subsets is not a proper subset?

In Exercises 11–15, let

$$U = \{\text{a, b, c, d, e, f, g}\}$$
$$A = \{\text{a, b, c, d}\}$$
$$B = \{\text{c, d, e, f}\}$$
$$C = \{\text{a, e, g}\}.$$

Find each of the following sets or cardinalities.

11. $A \cup B$

12. $(B \cap C)'$

13. $A \cap C'$

14. $(A \cup B) \cap C$

15. $n(A \cup B')$

In Exercises 16–18, use the Venn diagram to represent each set in roster form.

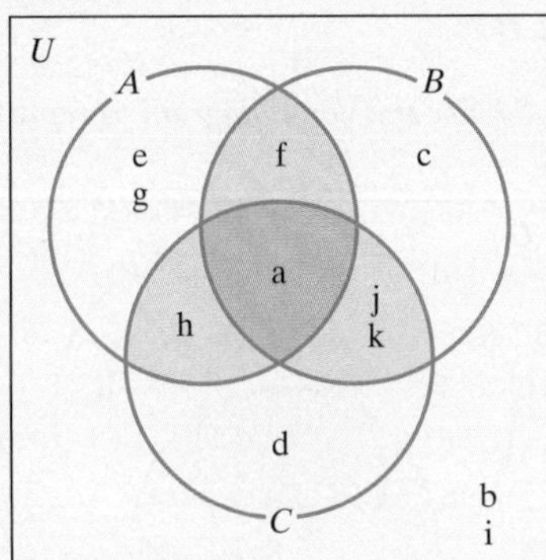

16. A'

17. $A \cap B \cap C$

18. $(A \cap B) \cup (A \cap C)$

19. Construct a Venn diagram illustrating the following sets: $A = \{1, 4, 5\}$, $B = \{1, 5, 6, 7\}$, and $U = \{1, 2, 3, 4, 5, 6, 7\}$.

20. Use the Venn diagram shown to determine whether the following statement is a theorem:

$$A' \cap (B \cup C) = (A' \cap B) \cup (A' \cap C).$$

Show work clearly as you develop the regions representing each side of the statement.

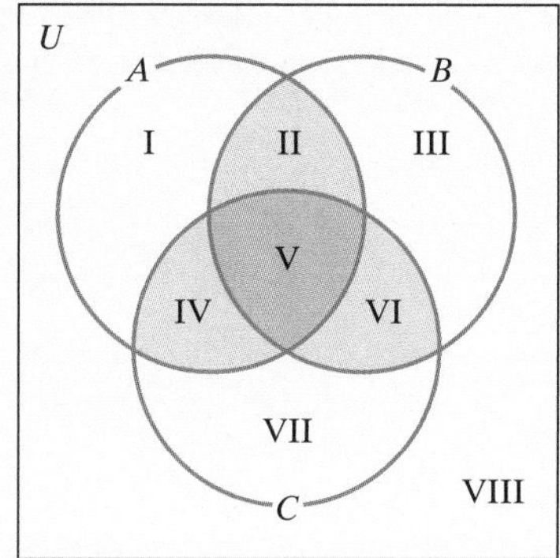

21. Here is a list of some famous people on whom the FBI kept files:

Famous Person	Length of FBI File
Bud Abbott (entertainer)	14 pages
Charlie Chaplin (entertainer)	2063 pages
Albert Einstein (scientist)	1800 pages
Martin Luther King, Jr. (civil rights leader)	17,000 pages
Elvis Presley (entertainer)	663 pages
Jackie Robinson (athlete)	131 pages
Eleanor Roosevelt (first lady; U.N. representative)	3000 pages
Frank Sinatra (entertainer)	1275 pages

Source: Paul Grobman, *Vital Statistics*, Plume, 2005.

The data can be organized in the following Venn diagram:

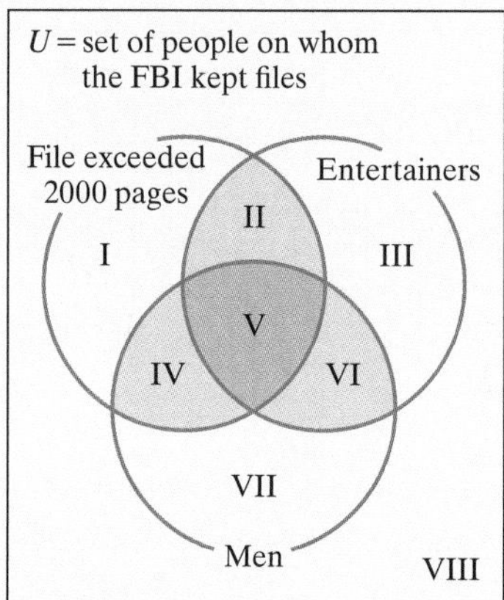

Use the data to determine the region in the Venn diagram into which each of the following people should be placed.

a. Chaplin

b. Einstein

c. King

d. Roosevelt

e. Sinatra

22. A winter resort took a poll of its 350 visitors to see which winter activities people enjoyed. The results were as follows: 178 liked to ski, 154 liked to snowboard, 57 liked to ice skate, 49 liked to ski and snowboard, 15 liked to ski and ice skate, 2 liked to snowboard and ice skate, and 2 liked all three activities.

a. Use a Venn diagram to illustrate the survey's results.

Use the Venn diagram to determine how many of those surveyed enjoyed

b. exactly one of these activities.

c. none of these activities.

d. at least two of these activities.

e. snowboarding and ice skating, but not skiing.

f. snowboarding or ice skating, but not skiing.

g. only skiing.

Logic

3

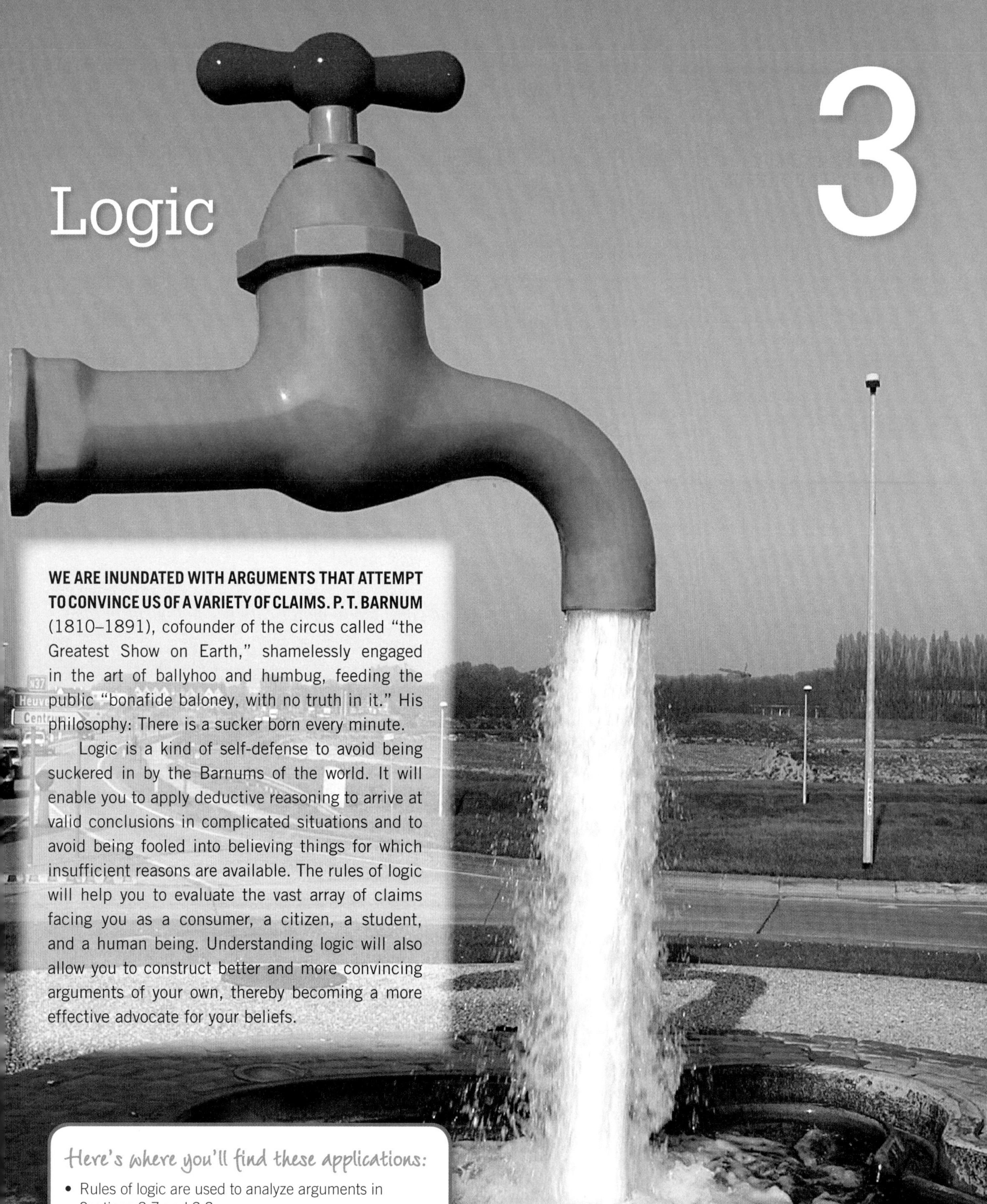

WE ARE INUNDATED WITH ARGUMENTS THAT ATTEMPT TO CONVINCE US OF A VARIETY OF CLAIMS. P. T. BARNUM (1810–1891), cofounder of the circus called "the Greatest Show on Earth," shamelessly engaged in the art of ballyhoo and humbug, feeding the public "bonafide baloney, with no truth in it." His philosophy: There is a sucker born every minute.

Logic is a kind of self-defense to avoid being suckered in by the Barnums of the world. It will enable you to apply deductive reasoning to arrive at valid conclusions in complicated situations and to avoid being fooled into believing things for which insufficient reasons are available. The rules of logic will help you to evaluate the vast array of claims facing you as a consumer, a citizen, a student, and a human being. Understanding logic will also allow you to construct better and more convincing arguments of your own, thereby becoming a more effective advocate for your beliefs.

Here's where you'll find these applications:

- Rules of logic are used to analyze arguments in Sections 3.7 and 3.8.
- In Exercise 81 of Exercise Set 3.7, you'll use logical tools to construct valid arguments of your own.

3.1 Statements, Negations, and Quantified Statements

WHAT AM I SUPPOSED TO LEARN?

After studying this section, you should be able to:

1. Identify English sentences that are statements.
2. Express statements using symbols.
3. Form the negation of a statement.
4. Express negations using symbols.
5. Translate a negation represented by symbols into English.
6. Express quantified statements in two ways.
7. Write negations of quantified statements.

HISTORY IS FILLED WITH BAD PREDICTIONS. HERE ARE EXAMPLES OF STATEMENTS that turned out to be notoriously false:

"Television won't be able to hold onto any market. People will soon get tired of staring at a plywood box every night."
—DARRYL F. ZANUCK, 1949

"The actual building of roads devoted to motor cars will not occur in the future."
—*Harper's Weekly*, August 2, 1902

"Everything that can be invented has been invented."
—CHARLES H. DUELL, Commissioner, U.S. Office of Patents, 1899

"The abdomen, the chest, and the brain will forever be shut from the intrusion of the wise and humane surgeon."
—JOHN ERICKSEN, Queen Victoria's surgeon, 1873

"We don't like their sound and guitar music is on the way out."
—Decca Recording Company, rejecting the Beatles, 1962

"When the President does it, that means that it is not illegal."
—RICHARD M. NIXON, TV interview with David Frost, May 20, 1977

Understanding that these statements are false enables us to negate each statement mentally and, with the assistance of historical perspective, obtain a true statement. We begin our study of logic by looking at statements and their negations.

Statements and Using Symbols to Represent Statements

1 *Identify English sentences that are statements.*

In everyday English, we use many different kinds of sentences. Some of these sentences are clearly true or false. Others are opinions, questions, and exclamations such as *Help*! or *Fire*! However, in logic we are concerned solely with statements, and not all English sentences are statements.

DEFINITION OF A STATEMENT

A **statement** is a sentence that is either true or false, but not both simultaneously.

Here are two examples of statements:

1. London is the capital of England.
2. William Shakespeare wrote the television series *Modern Family*.

Statement 1 is true and statement 2 is false. Shakespeare had nothing to do with *Modern Family* (perhaps writer's block after *Macbeth*).

As long as a sentence is either true or false, *even if we do not know which it is*, then that sentence is a statement. For example, the sentence

The United States has the world's highest divorce rate

is a statement. It's clearly either true or false, and it's not necessary to know which it is.

Some sentences, such as commands, questions, and opinions, are not statements because they are not either true or false. The following sentences are not statements:

1. Think before you speak. (This is an order or command.)
2. Does "just sayin'" annoy you in everyday conversation? (This is a question.)
3. *Titanic* is the greatest movie of all time. (This is an opinion.)

2 Express statements using symbols.

In symbolic logic, we use lowercase letters such as p, q, r, and s to represent statements. Here are two examples:

p: London is the capital of England.

q: William Shakespeare wrote the television series *Modern Family*.

The letter p represents the first statement.
The letter q represents the second statement.

3 Form the negation of a statement.

Negating Statements

The sentence "London is the capital of England" is a true statement. The *negation* of this statement, "London is not the capital of England," is a false statement. The **negation** of a true statement is a false statement and the negation of a false statement is a true statement.

EXAMPLE 1 Forming Negations

Form the negation of each statement:

a. Shakespeare wrote the television series *Modern Family*.

b. Today is not Monday.

SOLUTION

a. The most common way to negate "Shakespeare wrote the television series *Modern Family*" is to introduce *not* into the sentence. The negation is

Shakespeare did not write the television series *Modern Family*.

The English language provides many ways of expressing a statement's meaning. Here is another way to express the negation:

It is not true that Shakespeare wrote the television series *Modern Family*.

b. The negation of "Today is not Monday" is

It is not true that today is not Monday.

The negation is more naturally expressed in English as

Today is Monday.

 **CHECK POINT 1** Form the negation of each statement:

a. Paris is the capital of Spain.

b. July is not a month.

---

4 *Express negations using symbols.*

The negation of statement p is expressed by writing $\sim p$. We read this as "not p" or "It is not true that p."

EXAMPLE 2 Expressing Negations Symbolically

Let p and q represent the following statements:

p: Shakespeare wrote the television series *Modern Family*.

q: Today is not Monday.

Express each of the following statements symbolically:

a. Shakespeare did not write the television series *Modern Family*.

b. Today is Monday.

SOLUTION

a. Shakespeare did not write the television series *Modern Family* is the negation of statement p. Therefore, it is expressed symbolically as $\sim p$.

b. Today is Monday is the negation of statement q. Therefore, it is expressed symbolically as $\sim q$.

GREAT QUESTION!

Can letters such as *p*, *q*, or *r* represent any statement, including English statements containing the word *not*?

Yes. When choosing letters to represent statements, you may prefer to use the symbol ~ with negated English statements. However, it is not wrong to let p, q, or r represent such statements.

CHECK POINT 2 Let p and q represent the following statements:

p: Paris is the capital of Spain.

q: July is not a month.

Express each of the following statements symbolically:

a. Paris is not the capital of Spain.

b. July is a month.

---

5 *Translate a negation represented by symbols into English.*

In Example 2, we translated English statements into symbolic statements. In Example 3, we reverse the direction of our translation.

EXAMPLE 3 Translating a Symbolic Statement into Words

Let p represent the following statement:

p: The United States has the world's highest divorce rate.

Express the symbolic statement $\sim p$ in words.

SOLUTION

The symbol ~ is translated as "not." Therefore, $\sim p$ represents

The United States does not have the world's highest divorce rate.

This can also be expressed as

It is not true that the United States has the world's highest divorce rate.

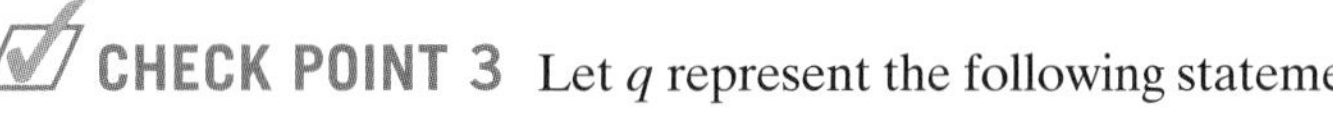

CHECK POINT 3 Let q represent the following statement:

q: Chicago O'Hare is the world's busiest airport.

Express the symbolic statement $\sim q$ in words.

6 *Express quantified statements in two ways.*

Quantified Statements

In English, we frequently encounter statements containing the words **all, some**, and **no** (or **none**). These words are called **quantifiers**. A statement that contains one of these words is a **quantified statement**. Here are some examples:

All poets are writers.
Some people are bigots.
No common colds are fatal.
Some students do not work hard.

Using our knowledge of the English language, we can express each of these quantified statements in two equivalent ways, that is, in two ways that have exactly the same meaning. These equivalent statements are shown in **Table 3.1**.

TABLE 3.1 Equivalent Ways of Expressing Quantified Statements

Statement	An Equivalent Way to Express the Statement	Example (Two Equivalent Quantified Statements)
All A are B.	There are no A that are not B.	All poets are writers. There are no poets that are not writers.
Some A are B.	There exists at least one A that is a B.	Some people are bigots. At least one person is a bigot.
No A are B.	All A are not B.	No common colds are fatal. All common colds are not fatal.
Some A are not B.	Not all A are B.	Some students do not work hard. Not all students work hard.

7 *Write negations of quantified statements.*

GREAT QUESTION!

It seems to me that the negation of "All writers are poets" should be "No writers are poets." What's wrong with my thinking?

The negation of "All writers are poets" cannot be "No writers are poets" because both statements are false. The negation of a false statement must be a true statement. In general, the negation of "All A are B" is *not* "No A are B."

Forming the negation of a quantified statement can be a bit tricky. Suppose we want to negate the statement "All writers are poets." Because this statement is false, its negation must be true. The negation is "Not all writers are poets." This means the same thing as "Some writers are not poets." Notice that the negation is a true statement.

Statement **Negation**

All writers are poets. Some writers are not poets.

Poets Writers Poets Writers

In general, the negation of "All A are B" is "Some A are not B." Likewise, the negation of "Some A are not B" is "All A are B."

Now let's investigate how to negate a statement with the word *some*. Consider the statement "Some canaries weigh 50 pounds." Because *some* means "there exists at least one," the negation is "It is not true that there is at least one canary that weighs 50 pounds." Because it is not true that there is even one such critter, we can express the negation as "No canary weighs 50 pounds."

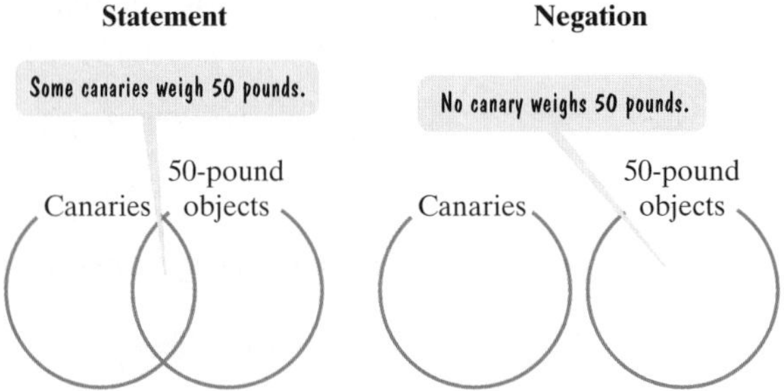

In general, the negation of "Some *A* are *B*" is "No *A* are *B*." Likewise, the negation of "No *A* are *B*" is "Some *A* are *B*."

Negations of quantified statements are summarized in **Table 3.2**.

TABLE 3.2 Negations of Quantified Statements

Statement	Negation	Example (A Quantified Statement and Its Negation)
All *A* are *B*.	Some *A* are not *B*.	All people take exams honestly. Negation: Some people do not take exams honestly.
Some *A* are *B*.	No *A* are *B*.	Some roads are open. Negation: No roads are open.

(The negations of the statements in the second column are the statements in the first column.)

GREAT QUESTION!

Is there a way to help me remember the negations of quantified statements?

This diagram should help. The statements diagonally opposite each other are negations.

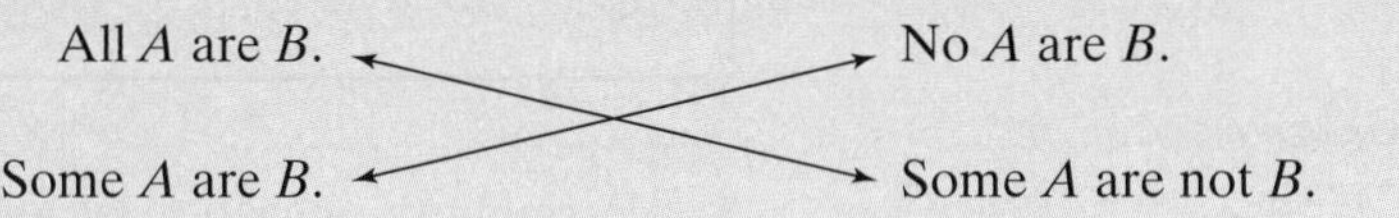

Table 3.3 contains examples of negations for each of the four kinds of quantified statements.

TABLE 3.3 Examples of Negations of Quantified Statements

Statement	Negation
All humans are mortal.	Some humans are not mortal.
Some students do not come to class prepared.	All students come to class prepared.
Some psychotherapists are in therapy.	No psychotherapists are in therapy.
No well-behaved dogs shred couches.	Some well-behaved dogs shred couches.

Blitzer Bonus

Logic and Star Trek

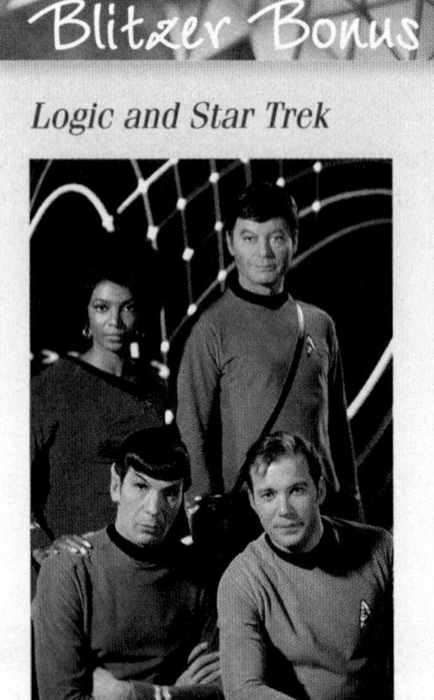

Four Enterprise *officers (clockwise from lower left): Spock, Uhura, McCoy, Kirk*

In the television series *Star Trek*, the crew of the *Enterprise* is held captive by an evil computer. The crew escapes after one of them tells the computer, "I am lying to you." If he says that he is lying and he *is* lying, then he isn't lying. But consider this: If he says that he is lying and he *isn't* lying, then he is lying. The sentence "I am lying to you" is not a statement because it is true and false simultaneously. Thinking about this sentence destroys the computer.

EXAMPLE 4 Negating a Quantified Statement

The mechanic told me, "All piston rings were replaced." I later learned that the mechanic never tells the truth. What can I conclude?

SOLUTION

Let's begin with the mechanic's statement:

All piston rings were replaced.

Because the mechanic never tells the truth, I can conclude that the truth is the negation of what I was told. The negation of "All *A* are *B*" is "Some *A* are not *B*." Thus, I can conclude that

Some piston rings were not replaced.

Because *some* means *at least one*, I can also correctly conclude that

At least one piston ring was not replaced.

CHECK POINT 4 The board of supervisors told us, "All new tax dollars will be used to improve education." I later learned that the board of supervisors never tells the truth. What can I conclude? Express the conclusion in two equivalent ways.

Concept and Vocabulary Check

Fill in each blank so that the resulting statement is true.

1. A statement is a sentence that is either ______ or ______, but not both simultaneously.
2. The negation of a true statement is a/an ______ statement, and the negation of a false statement is a/an ______ statement.
3. The negation of statement *p* is expressed by writing ______. We read this as ______.
4. Statements that contain the words *all*, *some*, and *no* are called __________ statements.
5. The statement "All *A* are *B*" can be expressed equivalently as ______________________.
6. The statement "Some *A* are *B*" can be expressed equivalently as ______________________.
7. The statement "No *A* are *B*" can be expressed equivalently as ______________.
8. The statement "Some *A* are not *B*" can be expressed equivalently as ______________.
9. The negation of "All *A* are *B*" is ______________.
10. The negation of "Some *A* are *B*" is ___________.

Exercise Set 3.1

Practice Exercises

In Exercises 1–14, determine whether or not each sentence is a statement.

1. René Descartes came up with the theory of analytic geometry by watching a fly walk across a ceiling.
2. The number of U.S. patients killed annually by medical errors is equivalent to four jumbo jets crashing each week.
3. On January 20, 2017, Hillary Clinton became America's 45th president.
4. On January 20, 2017, Donald Trump became America's first Hispanic president.
5. Take the most interesting classes you can find.
6. Don't try to study on a Friday night in the dorms.
7. The average human brain contains 100 billion neurons.
8. There are 2,500,000 rivets in the Eiffel Tower.

9. Is the unexamined life worth living?

10. Is this the best of all possible worlds?

11. All U.S. presidents with beards have been Republicans.

12. No U.S. president was an only child.

13. The shortest sentence in the English language is "Go!"

14. Go!

In Exercises 15–20, form the negation of each statement.

15. It is raining.

16. It is snowing.

17. "Facts do not cease to exist because they are ignored."
—Aldous Huxley

18. "I'm not anti-social. I'm just not user friendly."
—T-Shirt
(In this case, form the negation of both statements.)

19. It is not true that chocolate in moderation is good for the heart.

20. It is not true that Albert Einstein was offered the presidency of Israel.

In Exercises 21–24, let p, q, r, and s represent the following statements:

p: One works hard.
q: One succeeds.
r: The temperature outside is not freezing.
s: It is not true that the heater is working.

Express each of the following statements symbolically.

21. One does not work hard.

22. One does not succeed.

23. The temperature outside is freezing.

24. The heater is working.

According to Condensed Knowledge: A Deliciously Irreverent Guide to Feeling Smart Again *(Harper Collins, 2004), each statement listed below is false.*

p: Listening to classical music makes infants smarter.
q: Subliminal advertising makes you buy things.
r: Sigmund Freud's father was not 20 years older than his mother.
s: Humans and bananas do not share approximately 60% of the same DNA structure.

In Exercises 25–28, use these representations to express each symbolic statement in words. What can you conclude about the resulting verbal statement?

25. $\sim p$ **26.** $\sim q$ **27.** $\sim r$ **28.** $\sim s$

In Exercises 29–42,

a. *Express the quantified statement in an equivalent way, that is, in a way that has exactly the same meaning.*

b. *Write the negation of the quantified statement. (The negation should begin with "all," "some," or "no.")*

29. All whales are mammals.

30. All journalists are writers.

31. Some students are business majors.

32. Some movies are comedies.

33. Some thieves are not criminals.

34. Some pianists are not keyboard players.

35. No Democratic presidents have been impeached.

36. No women have served as Supreme Court justices.

37. There are no seniors who did not graduate.

38. There are no applicants who were not hired.

39. Not all parrots are pets.

40. Not all dogs are playful.

41. All atheists are not churchgoers.

42. All burnt muffins are not edible.

Here's another list of false statements from Condensed Knowledge.

p: No Africans have Jewish ancestry.
q: No religious traditions recognize sexuality as central to their understanding of the sacred.
r: All rap is hip-hop.
s: Some hip-hop is not rap.

In Exercises 43–46, use the representations shown to express each symbolic statement in words. Verbal statements should begin with "all," "some," or "no." What can you conclude about each resulting verbal statement?

43. $\sim p$ **44.** $\sim q$ **45.** $\sim r$ **46.** $\sim s$

Practice Plus

Exercises 47–50 contain diagrams that show relationships between two sets. (These diagrams are just like the Venn diagrams studied in Chapter 2. However, the circles are not enclosed in a rectangle representing a universal set.)

a. *Use each diagram to write a statement beginning with the word "all," "some," or "no" that illustrates the relationship between the sets.*

b. *Determine if the statement in part (a) is true or false. If it is false, write its negation.*

47. Parrots
Birds

48.

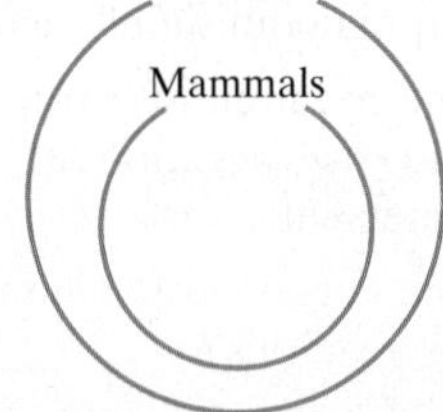

49. College Students Business Majors

50.

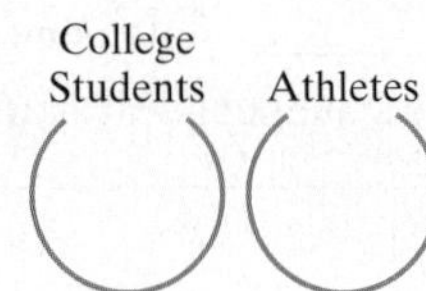

In Exercises 51–56,

a. *Express each statement in an equivalent way that begins with "all," "some," or "no."*

b. *Write the negation of the statement in part (a).*

51. Nobody doesn't like Sara Lee.

52. A problem well stated is a problem half solved.

53. Nothing is both safe and exciting.

54. Many a person has lived to regret a misspent youth.

55. Not every great actor is a Tom Hanks.

56. Not every generous philanthropist is a Bill Gates.

Application Exercises

In Exercises 57 and 58, choose the correct statement.

57. The City Council of a large northern metropolis promised its citizens that in the event of snow, all major roads connecting the city to its airport would remain open. The City Council did not keep its promise during the first blizzard of the season. Therefore, during the first blizzard:

a. No major roads connecting the city to the airport were open.

b. At least one major road connecting the city to the airport was not open.

c. At least one major road connecting the city to the airport was open.

d. The airport was forced to close.

58. During the Watergate scandal in 1974, President Richard Nixon assured the American people that "In all my years of public service, I have never obstructed justice." Later, events indicated that the president was not telling the truth. Therefore, in his years of public service:

a. Nixon always obstructed justice.

b. Nixon sometimes did not obstruct justice.

c. Nixon sometimes obstructed justice.

d. Nixon never obstructed justice.

Our culture celebrates romantic love—affection and sexual passion for another person—as the basis for marriage. However, the bar graph illustrates that in some countries, romantic love plays a less important role in marriage.

Percentage of College Students Willing to Marry without Romantic Love

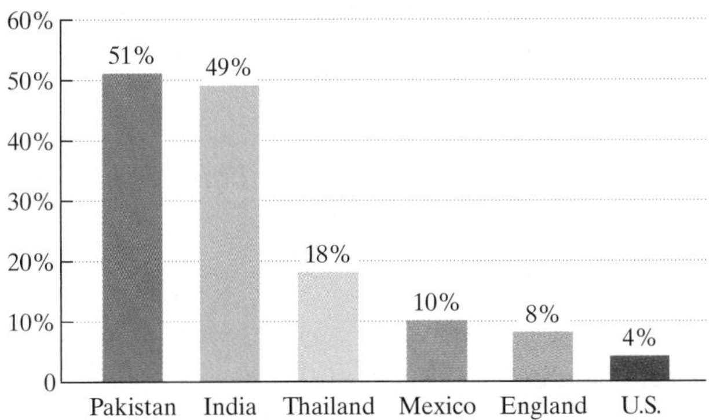

Source: Robert Levine, "Is Love a Luxury?" *American Demographics*

In Exercises 59–66, use the graph to determine whether each statement is true or false. If the statement is false, write its negation.

59. A majority of college students in Pakistan are willing to marry without romantic love.

60. Nearly half of India's college students are willing to marry without romantic love.

61. No college students in the United States are willing to marry without romantic love.

62. All college students in Pakistan are willing to marry without romantic love.

63. Not all college students in Mexico are willing to marry without romantic love.

64. Not all college students in England are willing to marry without romantic love.

65. The sentence "5% of college students in Australia are willing to marry without romantic love" is not a statement.

66. The sentence "12% of college students in the Philippines are willing to marry without romantic love" is not a statement.

Explaining the Concepts

67. What is a statement? Explain why commands, questions, and opinions are not statements.

68. Explain how to form the negation of a given English statement. Give an example.

69. Describe how the negation of statement p is expressed using symbols.

70. List the words identified as quantifiers. Give an example of a statement that uses each of these quantifiers.

71. Explain how to write the negation of a quantified statement in the form "All A are B." Give an example.

72. Explain how to write the negation of a quantified statement in the form "Some A are B." Give an example.

73. If the ancient Greek god Zeus could do anything, could he create a rock so huge that he could not move it? Explain your answer.

Critical Thinking Exercises

Make Sense? *In Exercises 74–77, determine whether each statement makes sense or does not make sense, and explain your reasoning.*

74. I have no idea if a particular sentence is true or false, so I cannot determine whether or not that sentence is a statement.

75. "All beagles are dogs" is true and "no beagles are dogs" is false, so the second statement must be the negation of the first statement.

76. Little Richard's "A-wop-bop-a-lula-a-wop-bam-boom!" is an exclamation and not a statement.

77. Researchers at Cambridge University made the following comments on how we read:

> It deson't mtater waht oerdr the ltteres in a wrod are, so lnog as the frist and lsat ltteer are in the crorcet pclae. Tihs is bcuseae we dno't raed ervey lteter but the wrod as a wlohe.

Because of the incorrect spellings in these sentences, neither sentence is a statement.

78. Give an example of a sentence that is not a statement because it is true and false simultaneously.

79. Give an example in which the statement "Some A are not B" is true, but the statement "Some B are not A" is false.

80. The statement

> She isn't dating him because he is muscular

is confusing because it can mean two different things. Describe the two different meanings that make this statement ambiguous.

3.2 Compound Statements and Connectives

WHAT AM I SUPPOSED TO LEARN?

After studying this section, you should be able to:

1 Express compound statements in symbolic form.
2 Express symbolic statements with parentheses in English.
3 Use the dominance of connectives.

WHAT CONDITIONS ENABLE US TO FLOURISH and our hearts to sing? Researchers in the new science of happiness have learned some surprising things about what it doesn't take to ring our inner chimes. Neither wealth nor a good education is sufficient for happiness. Put in another way, we should not rely on the following statement:

If you're wealthy or well educated, then you'll be happy.

We can break this statement down into three basic sentences:

You're wealthy. You're well educated. You'll be happy.

These sentences are called **simple statements** because each one conveys one idea with no connecting words. Statements formed by combining two or more simple statements are called **compound statements**. Words called **connectives** are used to join simple statements to form a compound statement. Connectives include words such as **and, or, if . . . then**, and **if and only if**.

Compound statements appear throughout written and spoken language. We need to be able to understand the logic of such statements to analyze information objectively. In this section, we analyze four kinds of compound statements.

1 *Express compound statements in symbolic form.*

And Statements

If p and q represent two simple statements, then **the compound statement "p and q" is symbolized by $p \wedge q$.** The compound statement formed by connecting statements with the word **and** is called a **conjunction**. The symbol for *and* is $\wedge$.

GREAT QUESTION!

How can I remember that $\wedge$ is the symbol for *and*?

Think of $\wedge$ as an A (for And) with the horizontal bar missing.

EXAMPLE 1 *Translating from English to Symbolic Form*

Let p and q represent the following simple statements:

p: It is after 5 P.M.
q: They are working.

Write each compound statement below in symbolic form:

a. It is after 5 P.M. and they are working.
b. It is after 5 P.M. and they are not working.

SOLUTION

a. It is after 5 P.M. | and | they are working.

p $\wedge$ q

The symbolic form is $p \wedge q$.

b. It is after 5 P.M. | and | they are not working.

p $\wedge$ $\sim q$

The symbolic form is $p \wedge \sim q$.

CHECK POINT 1 Use the representations in Example 1 to write each compound statement below in symbolic form:

a. They are working and it is after 5 P.M.

b. It is not after 5 P.M. and they are working.

The English language has a variety of ways to express the connectives that appear in compound statements. **Table 3.4** shows a number of ways to translate $p \wedge q$ into English.

TABLE 3.4 Common English Expressions for $p \wedge q$

Symbolic Statement	**English Statement**	**Example** ***p*: It is after 5 P.M.** ***q*: They are working.**
$p \wedge q$	p and q.	It is after 5 P.M. and they are working.
$p \wedge q$	p but q.	It is after 5 P.M., but they are working.
$p \wedge q$	p yet q.	It is after 5 P.M., yet they are working.
$p \wedge q$	p nevertheless q.	It is after 5 P.M.; nevertheless, they are working.

GREAT QUESTION!

When the word *and* appears in English, does that mean the statement is a conjunction?

Not necessarily. Some English statements with the word *and* are not conjunctions.

- Not a conjunction:

  "Nonviolence and truth are inseparable."

  —GANDHI

  This statement cannot be broken down into two simple statements. It is itself a simple statement.

- Conjunction:

  Pizza and beer are not recommended for people with ulcers.

  Can be broken down as follows: Pizza is not recommended for people with ulcers and beer is not recommended for people with ulcers.

Or Statements

The connective *or* can mean two different things. For example, consider this statement:

I visited London or Paris.

The statement can mean

I visited London or Paris, but not both.

This is an example of the **exclusive or**, which means "one or the other, but not both." By contrast, the statement can mean

I visited London or Paris or both.

This is an example of the **inclusive or**, which means "either or both."

In mathematics, when the connective *or* appears, it means the *inclusive or*. If p and q represent two simple statements, then the compound statement "p or q" means p or q or both. The compound statement formed by connecting statements with the word *or* is called a **disjunction**. **The symbol for *or* is $\vee$. Thus, we can symbolize the compound statement "p or q or both" by $p \vee q$.**

EXAMPLE 2 Translating from English to Symbolic Form

Let p and q represent the following simple statements:

p: The bill receives majority approval.

q: The bill becomes a law.

Write each compound statement below in symbolic form:

a. The bill receives majority approval or the bill becomes a law.

b. The bill receives majority approval or the bill does not become a law.

SOLUTION

a.

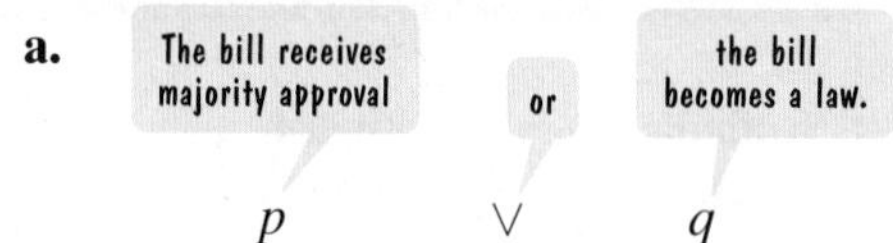

$p \quad \vee \quad q$

The symbolic form is $p \vee q$.

b.

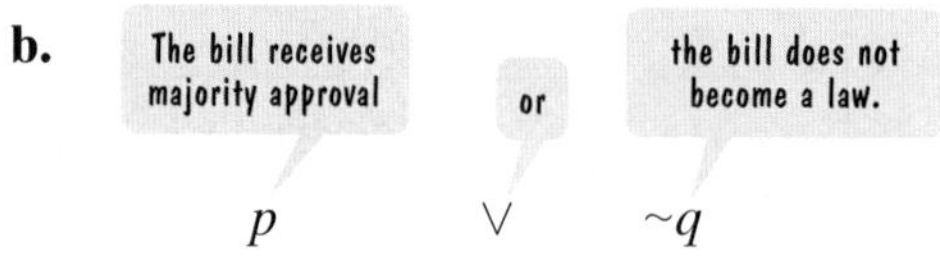

$p \quad \vee \quad {\sim}q$

The symbolic form is $p \vee {\sim}q$.

CHECK POINT 2 Let p and q represent the following simple statements:

p: You graduate.

q: You satisfy the math requirement.

Write each compound statement below in symbolic form:

a. You graduate or you satisfy the math requirement.

b. You satisfy the math requirement or you do not graduate.

If–Then Statements

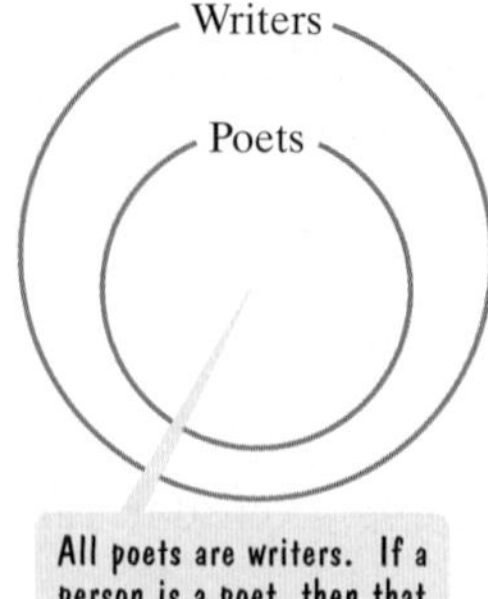

FIGURE 3.1

The diagram in **Figure 3.1** shows that

All poets are writers. *The set of poets is a subset of the set of writers.*

In Section 3.1, we saw that this can be expressed as

There are no poets that are not writers.

Another way of expressing this statement is

If a person is a poet, then that person is a writer.

The form of this statement is "If p, then q." **The compound statement "If p, then q" is symbolized by $p \rightarrow q$.** The compound statement formed by connecting statements with "if–then" is called a **conditional statement**. The symbol for "if–then" is $\rightarrow$.

In a conditional statement, the statement before the $\rightarrow$ connective is called the **antecedent**. The statement after the $\rightarrow$ connective is called the **consequent**:

antecedent $\rightarrow$ consequent.

EXAMPLE 3 Translating from English to Symbolic Form

Let p and q represent the following simple statements:

p: A person is a father.

q: A person is a male.

Write each compound statement below in symbolic form:

a. If a person is a father, then that person is a male.

b. If a person is a male, then that person is a father.

c. If a person is not a male, then that person is not a father.

SOLUTION

We use p: A person is a father and q: A person is a male.

a. 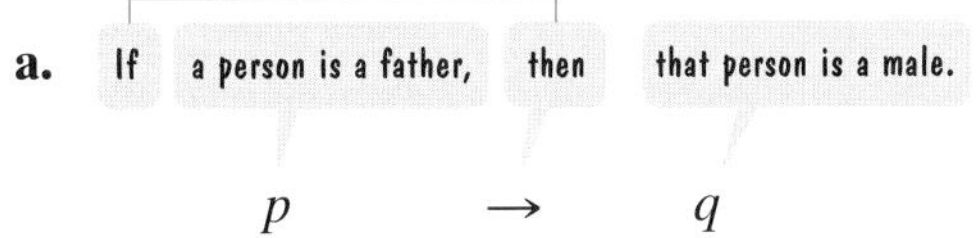

$p \quad \rightarrow \quad q$

The symbolic form is $p \rightarrow q$.

b. 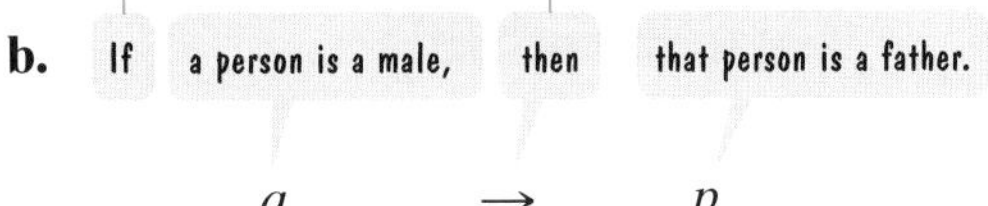

$q \quad \rightarrow \quad p$

The symbolic form is $q \rightarrow p$.

c. 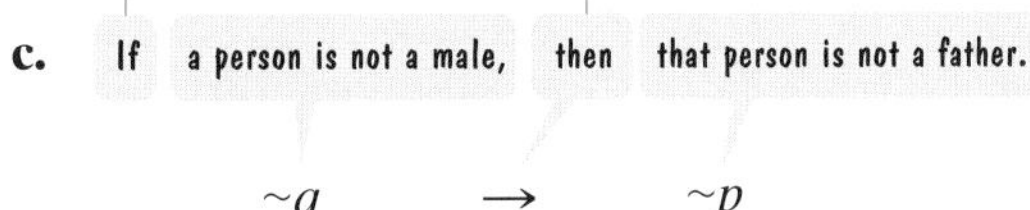

$\sim q \quad \rightarrow \quad \sim p$

The symbolic form is $\sim q \rightarrow \sim p$.

 **CHECK POINT 3** Use the representations in Example 3 to write each compound statement below in symbolic form:

a. If a person is not a father, then that person is not a male.

b. If a person is a male, then that person is not a father.

Conditional statements in English often omit the word *then* and simply use a comma. When *then* is included, the comma can be included or omitted. Here are some examples:

If a person is a father, then that person is a male.
If a person is a father then that person is a male.
If a person is a father, that person is a male.

Table 3.5 shows some of the common ways to translate $p \rightarrow q$ into English.

TABLE 3.5 Common English Expressions for $p \rightarrow q$

Symbolic Statement	English Statement	Example p: A person is a father. q: A person is a male.
$p \rightarrow q$	If p then q.	If a person is a father, then that person is a male.
$p \rightarrow q$	q if p.	A person is a male if that person is a father.
$p \rightarrow q$	p is sufficient for q.	Being a father is sufficient for being a male.
$p \rightarrow q$	q is necessary for p.	Being a male is necessary for being a father.
$p \rightarrow q$	p only if q.	A person is a father only if that person is a male.
$p \rightarrow q$	Only if q, p.	Only if a person is a male is that person a father.

GREAT QUESTION!

In a conditional statement, how do I tell the sufficient part from the necessary part?

The sufficient condition of a conditional statement is the part that precedes →, or the antecedent. The necessary condition of a conditional statement is the part that follows →, or the consequent.

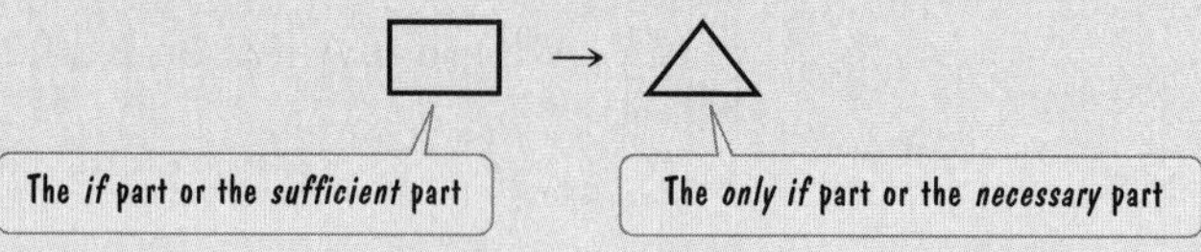

EXAMPLE 4 *Translating from English to Symbolic Form*

Let p and q represent the following simple statements:

p: We suffer huge budget deficits.

q: We control military spending.

Write the following compound statement in symbolic form:

Controlling military spending is necessary for not suffering huge budget deficits.

SOLUTION

The necessary part of a conditional statement follows the *if–then* connective. Because "controlling military spending" is the necessary part, we can rewrite the compound statement as follows:

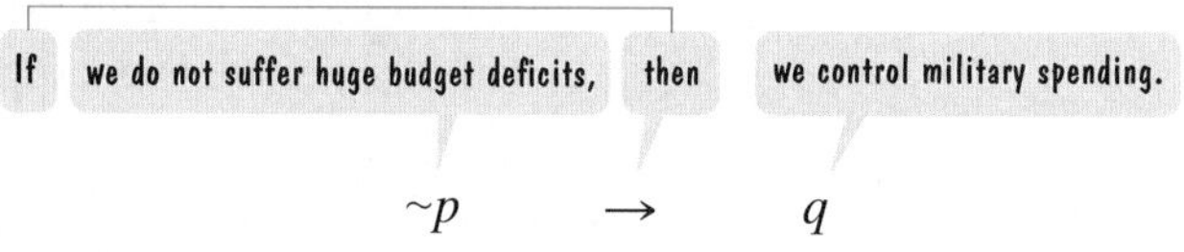

The symbolic form is $\sim p \rightarrow q$.

CHECK POINT 4 Let p and q represent the following simple statements:

p: You exercise regularly.

q: You increase your chances of a heart attack.

Write the following compound statement in symbolic form:

Not exercising regularly is sufficient for increasing your chances of a heart attack.

If-and-Only-If Statements

If a conditional statement is true, reversing the antecedent and consequent may result in a statement that is not necessarily true:

- If a person is a father, then that person is a male.
  true
- If a person is a male, then that person is a father.
  not necessarily true

However, some true conditional statements are still true when the antecedent and consequent are reversed:

- If a person is an unmarried male, then that person is a bachelor. (true)
- If a person is a bachelor, then that person is an unmarried male. (also true)

Rather than deal with two separate conditionals, we can combine them into one *biconditional statement:*

A person is an unmarried male if and only if that person is a bachelor.

If p and q represent two simple statements, then **the compound statement "p if and only if q" is symbolized by $p \leftrightarrow q$.** The compound statement formed by connecting statements with *if and only if* is called a **biconditional**. The symbol for *if and only if* is $\leftrightarrow$. The phrase *if and only if* can be abbreviated as *iff*.

EXAMPLE 5 Translating from English to Symbolic Form

TABLE 3.6 Words with the Most Meanings in the *Oxford English Dictionary*

Word	Meanings
Set	464
Run	396
Go	368
Take	343
Stand	334

Table 3.6 shows that the word *set* has 464 meanings, making it the word with the most meanings in the English language. Let p and q represent the following simple statements:

p: The word is *set*.

q: The word has 464 meanings.

Write each of the compound statements below in its symbolic form:

a. The word is *set* if and only if the word has 464 meanings.

b. The word does not have 464 meanings if and only if the word is not *set*.

SOLUTION

a. The word is *set* | if and only if | the word has 464 meanings.

$p \qquad \leftrightarrow \qquad q$

The symbolic form is $p \leftrightarrow q$. Observe that each of the following statements is true:

If the word is *set*, then it has 464 meanings.

If the word has 464 meanings, then it is *set*.

b.

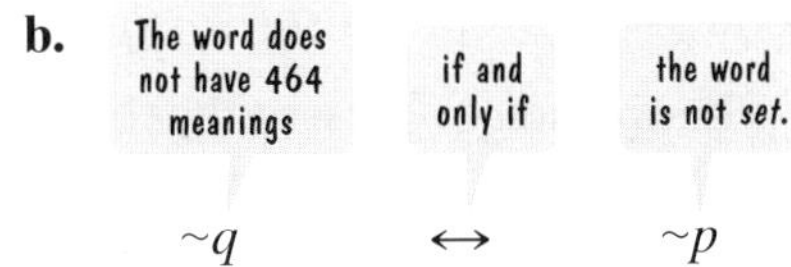

$\sim q \qquad \leftrightarrow \qquad \sim p$

The symbolic form is $\sim q \leftrightarrow \sim p$.

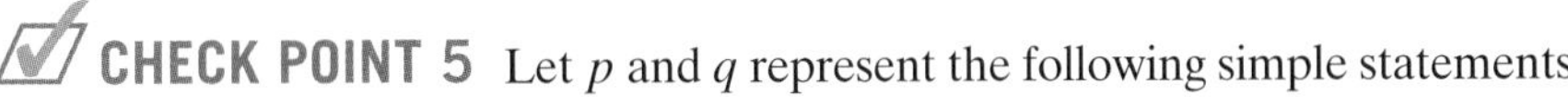

CHECK POINT 5 Let p and q represent the following simple statements:

p: The word is *run*.

q: The word has 396 meanings.

Write each of the compound statements below in its symbolic form:

a. The word has 396 meanings if and only if the word is *run*.

b. The word is not *run* if and only if the word does not have 396 meanings.

Table 3.7 shows some of the common ways to translate $p \leftrightarrow q$ into English.

TABLE 3.7 Common English Expressions for $p \leftrightarrow q$

Symbolic Statement	English Statement	Example *p*: A person is an unmarried male. *q*: A person is a bachelor.
$p \leftrightarrow q$	p if and only if q.	A person is an unmarried male if and only if that person is a bachelor.
$p \leftrightarrow q$	q if and only if p.	A person is a bachelor if and only if that person is an unmarried male.
$p \leftrightarrow q$	If p then q, and if q then p.	If a person is an unmarried male then that person is a bachelor, and if a person is a bachelor then that person is an unmarried male.
$p \leftrightarrow q$	p is necessary and sufficient for q.	Being an unmarried male is necessary and sufficient for being a bachelor.
$p \leftrightarrow q$	q is necessary and sufficient for p.	Being a bachelor is necessary and sufficient for being an unmarried male.

GREAT QUESTION!

In arithmetic, I know that $a + b$ means the same thing as $b + a$. Can I switch p and q in each of the four compound statements in Table 3.8 without changing the statement's meaning?

It depends on the connective between p and q.

- $p \wedge q$ means the same thing as $q \wedge p$.
- $p \vee q$ means the same thing as $q \vee p$.
- $p \leftrightarrow q$ means the same thing as $q \leftrightarrow p$.

However,

- $p \rightarrow q$ does not necessarily mean the same thing as $q \rightarrow p$.

We'll have more to say about these observations throughout the chapter.

Table 3.8 summarizes the statements discussed in the first two sections of this chapter.

TABLE 3.8 Statements of Symbolic Logic

Name	Symbolic Form	Common English Translations
Negation	$\sim p$	Not p. It is not true that p.
Conjunction	$p \wedge q$	p and q. p but q.
Disjunction	$p \vee q$	p or q.
Conditional	$p \rightarrow q$	If p, then q. p is sufficient for q. q is necessary for p.
Biconditional	$p \leftrightarrow q$	p if and only if q. p is necessary and sufficient for q.

Symbolic Statements with Parentheses

Parentheses in symbolic statements indicate which statements are to be grouped together. For example, $\sim(p \wedge q)$ means the negation of the entire statement $p \wedge q$. By contrast, $\sim p \wedge q$ means that only statement p is negated. We read $\sim(p \wedge q)$ as "it is not true that p and q." We read $\sim p \wedge q$ as "not p and q." Unless parentheses appear in a symbolic statement, the symbol $\sim$ negates only the statement that immediately follows it.

2 Express symbolic statements with parentheses in English.

EXAMPLE 6 Expressing Symbolic Statements with and without Parentheses in English

Let p and q represent the following simple statements:

p: She is wealthy.

q: She is happy.

Write each of the following symbolic statements in words:

a. $\sim(p \wedge q)$

b. $\sim p \wedge q$

c. $\sim(p \vee q)$.

SOLUTION

The voice balloons illustrate the differences among the three statements.

- $\sim(p \wedge q)$
- $\sim p \wedge q$
- $\sim(p \vee q)$

GREAT QUESTION!

Do I always have to use p and q to represent two simple statements?

No. You can select letters that help you remember what the statements represent. For example, in Example 6, you may prefer using w and h rather than p and q:

w: She is wealthy.

h: She is happy.

a. The symbolic statement $\sim(p \wedge q)$ means the negation of the entire statement $p \wedge q$. A translation of $\sim(p \wedge q)$ is

It is not true that she is wealthy and happy.

We can also express this statement as

It is not true that she is both wealthy and happy.

b. A translation of $\sim p \wedge q$ is

She is not wealthy and she is happy.

c. The symbolic statement $\sim(p \vee q)$ means the negation of the entire statement $p \vee q$. A translation of $\sim(p \vee q)$ is

It is not true that she is wealthy or happy.

We can express this statement as

She is neither wealthy nor happy.

CHECK POINT 6 Let p and q represent the following simple statements:

p: He earns $105,000 yearly.

q: He is often happy.

Write each of the following symbolic statements in words:

a. $\sim(p \wedge q)$ **b.** $\sim q \wedge p$ **c.** $\sim(q \rightarrow p)$.

Many compound statements contain more than one connective. When expressed symbolically, parentheses are used to indicate which simple statements are grouped together. When expressed in English, commas are used to indicate the groupings. Here is a table that illustrates groupings using parentheses in symbolic statements and commas in English statements:

Symbolic Statement	Statements to Group Together	English Translation
$(q \wedge \sim p) \rightarrow \sim r$	$q \wedge \sim p$	If q and not p, then not r.
$q \wedge (\sim p \rightarrow \sim r)$	$\sim p \rightarrow \sim r$	q, and if not p then not r.

The statement in the first row is an *if–then* conditional statement. Notice that the symbol $\rightarrow$ is outside the parentheses. By contrast, the statement in the second row is an *and* conjunction. In this case, the symbol $\wedge$ is outside the parentheses. Notice that when we translate the symbolic statement into English, **the simple statements in parentheses appear on the same side of the comma**.

EXAMPLE 7 *Expressing Symbolic Statements with Parentheses in English*

Let p, q, and r represent the following simple statements:

p: A student misses lecture.

q: A student studies.

r: A student fails.

Write each of the symbolic statements below in words:

a. $(q \wedge \sim p) \rightarrow \sim r$

b. $q \wedge (\sim p \rightarrow \sim r)$.

p: A student misses lecture.
q: A student studies.
r: A student fails.
The given simple statements (repeated)

SOLUTION

a. $(q \wedge \sim p) \rightarrow \sim r$

If [A student studies] and [A student does not miss lecture], then [A student does not fail.]

One possible English translation for the symbolic statement is

If a student studies and does not miss lecture, then the student does not fail.

Observe how the symbolic statements in parentheses appear on the same side of the comma in the English translation.

b. $q \wedge (\sim p \rightarrow \sim r)$

[A student studies], and if [A student does not miss lecture] then [A student does not fail.]

One possible English translation for the symbolic statement is

A student studies, and if a student does not miss lecture then the student does not fail.

Once again, the symbolic statements in parentheses appear on the same side of the comma in the English statement.

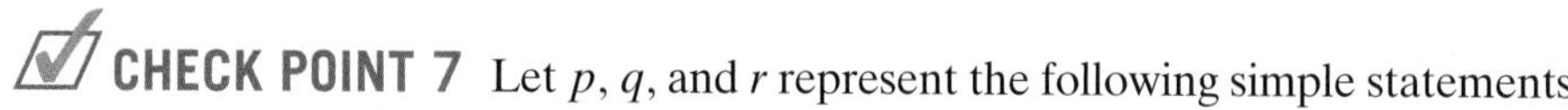

CHECK POINT 7 Let p, q, and r represent the following simple statements:

p: The plant is fertilized.
q: The plant is not watered.
r: The plant wilts.

Write each of the symbolic statements in words:

a. $(p \wedge \sim q) \rightarrow \sim r$ **b.** $p \wedge (\sim q \rightarrow \sim r)$.

3 *Use the dominance of connectives.*

Dominance of Connectives

In Example 7, the statements $(q \wedge \sim p) \rightarrow \sim r$ and $q \wedge (\sim p \rightarrow \sim r)$ had different meanings. If we are given $q \wedge \sim p \rightarrow \sim r$ without parentheses, how do we know which statements to group together?

If a symbolic statement appears without parentheses, statements before and after the most *dominant connective* should be grouped. Symbolic connectives are categorized from the least dominant, negation, to the most dominant, the biconditional.

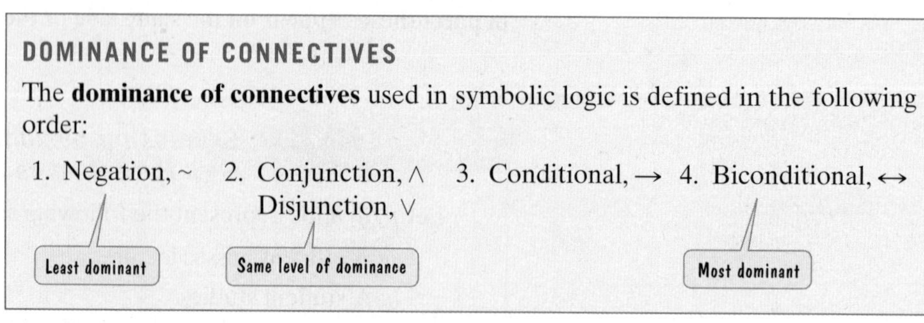

DOMINANCE OF CONNECTIVES

The **dominance of connectives** used in symbolic logic is defined in the following order:

1. Negation, ~
2. Conjunction, $\wedge$
   Disjunction, $\vee$
3. Conditional, $\rightarrow$
4. Biconditional, $\leftrightarrow$

Table 3.9 at the top of the next page shows a number of symbolic statements without parentheses. The meaning of each statement is then clarified by placing grouping symbols (parentheses), as needed, before and after the most dominant connective used.

TABLE 3.9 Using the Dominance of Connectives

Statement	Most Dominant Connective Highlighted in Red	Statement's Meaning Clarified with Grouping Symbols	Type of Statement
$p \to q \wedge \sim r$	$p \to q \wedge \sim r$	$p \to (q \wedge \sim r)$	Conditional
$p \wedge q \to \sim r$	$p \wedge q \to \sim r$	$(p \wedge q) \to \sim r$	Conditional
$p \leftrightarrow q \to r$	$p \leftrightarrow q \to r$	$p \leftrightarrow (q \to r)$	Biconditional
$p \to q \leftrightarrow r$	$p \to q \leftrightarrow r$	$(p \to q) \leftrightarrow r$	Biconditional
$p \wedge \sim q \to r \vee s$	$p \wedge \sim q \to r \vee s$	$(p \wedge \sim q) \to (r \vee s)$	Conditional
$p \wedge q \vee r$	$\wedge$ and $\vee$ have the same level of dominance.	The meaning is ambiguous.	?

Grouping symbols must be given with this statement to determine whether it means $(p \wedge q) \vee r$, a disjunction, or $p \wedge (q \vee r)$, a conjunction.

EXAMPLE 8 Using the Dominance of Connectives

Write each compound statement below in symbolic form:

a. I do not fail the course if and only if I study hard and I pass the final.

b. I do not fail the course if and only if I study hard, and I pass the final.

SOLUTION

We begin by assigning letters to the simple statements. Let each letter represent an English statement that is not negated. We can then represent any negated simple statement with the negation symbol, ~. Use the following representations:

p: I fail the course.
q: I study hard.
r: I pass the final.

a. I do not fail the course | iff | I study hard | and | I pass the final.

$\sim p \quad \leftrightarrow \quad q \quad \wedge \quad r$

Because the most dominant connective that appears is $\leftrightarrow$, the symbolic form with parentheses is $\sim p \leftrightarrow (q \wedge r)$.

b. I do not fail the course | iff | I study hard | , | and | I pass the final.

$(\sim p \quad \leftrightarrow \quad q) \quad \wedge \quad r$

In this statement, the comma indicates the grouping, so it is not necessary to apply the dominance of connectives. The symbolic form of the statement is $(\sim p \leftrightarrow q) \wedge r$.

GREAT QUESTION!

When am I supposed to use the dominance of connectives?

Only apply the dominance of connectives if grouping symbols (parentheses) are not given in compound symbolic statements or commas do not appear in compound English statements.

CHECK POINT 8 Write each compound statement below in symbolic form:

a. If there is too much homework or a teacher is boring then I do not take that class.

b. There is too much homework, or if a teacher is boring then I do not take that class.

Concept and Vocabulary Check

Fill in each blank so that the resulting statement is true.

1. The compound statement "p and q" is symbolized by _______ and is called a/an ____________.
2. The compound statement "p or q" is symbolized by _______ and is called a/an ___________.
3. The compound statement "If p, then q" is symbolized by _______ and is called a/an ___________ statement.
4. The compound statement "p if and only if q" is symbolized by ______ and is called a/an ___________.

In Exercises 5–13, determine whether each statement is true or false. If the statement is false, make the necessary change(s) to produce a true statement.

5. $p \wedge q$ can be translated as "p but q." ______
6. $p \vee q$ means p or q, but not both. ______
7. $p \rightarrow q$ can be translated as "p is sufficient for q." ______
8. $p \rightarrow q$ can be translated as "p is necessary for q." ______
9. The consequent is the necessary condition in a conditional statement. ______
10. $p \leftrightarrow q$ can be translated as "If p then q, and if q then p." ______
11. $p \leftrightarrow q$ can be translated as "p is necessary and sufficient for q." ______
12. When symbolic statements are translated into English, the simple statements in parentheses appear on the same side of the comma. ______
13. Using the dominance of connectives, $p \rightarrow q \wedge r$ means $(p \rightarrow q) \wedge r$. ______

Exercise Set 3.2

Practice Exercises

In Exercises 1–6, let p and q represent the following simple statements:

p: I'm leaving.

q: You're staying.

Write each compound statement in symbolic form.

1. I'm leaving and you're staying.
2. You're staying and I'm leaving.
3. You're staying and I'm not leaving.
4. I'm leaving and you're not staying.
5. You're not staying, but I'm leaving.
6. I'm not leaving, but you're staying.

In Exercises 7–10, let p and q represent the following simple statements:

p: I study.

q: I pass the course.

Write each compound statement in symbolic form.

7. I study or I pass the course.
8. I pass the course or I study.
9. I study or I do not pass the course.
10. I do not study or I do not pass the course.

In Exercises 11–18, let p and q represent the following simple statements:

p: This is an alligator.

q: This is a reptile.

Write each compound statement in symbolic form.

11. If this is an alligator, then this is a reptile.
12. If this is a reptile, then this is an alligator.
13. If this is not an alligator, then this is not a reptile.
14. If this is not a reptile, then this is not an alligator.
15. This is not an alligator if it's not a reptile.
16. This is a reptile if it's an alligator.
17. Being a reptile is necessary for being an alligator.
18. Being an alligator is sufficient for being a reptile.

In Exercises 19–26, let p and q represent the following simple statements:

p: You are human.

q: You have feathers.

Write each compound statement in symbolic form.

19. You do not have feathers if you are human.
20. You are not human if you have feathers.
21. Not being human is necessary for having feathers.
22. Not having feathers is necessary for being human.
23. Being human is sufficient for not having feathers.
24. Having feathers is sufficient for not being human.
25. You have feathers only if you're not human.
26. You're human only if you do not have feathers.

In Exercises 27–32, let p and q represent the following simple statements:

p: The campus is closed.

q: It is Sunday.

Write each compound statement in symbolic form.

27. The campus is closed if and only if it is Sunday.
28. It is Sunday if and only if the campus is closed.
29. It is not Sunday if and only if the campus is not closed.

30. The campus is not closed if and only if it is not Sunday.

31. Being Sunday is necessary and sufficient for the campus being closed.

32. The campus being closed is necessary and sufficient for being Sunday.

In Exercises 33–40, let p and q represent the following simple statements:

p: The heater is working.

q: The house is cold.

Write each symbolic statement in words.

33. $\sim p \wedge q$ **34.** $p \wedge \sim q$ **35.** $p \vee \sim q$

36. $\sim p \vee q$ **37.** $p \rightarrow \sim q$ **38.** $q \rightarrow \sim p$

39. $p \leftrightarrow \sim q$ **40.** $\sim p \leftrightarrow q$

In Exercises 41–48, let q and r represent the following simple statements:

q: It is July 4th.

r: We are having a barbecue.

Write each symbolic statement in words.

41. $q \wedge \sim r$ **42.** $\sim q \wedge r$ **43.** $\sim q \vee r$

44. $q \vee \sim r$ **45.** $r \rightarrow \sim q$ **46.** $q \rightarrow \sim r$

47. $\sim q \leftrightarrow r$ **48.** $q \leftrightarrow \sim r$

In Exercises 49–58, let p and q represent the following simple statements:

p: Romeo loves Juliet.

q: Juliet loves Romeo.

Write each symbolic statement in words.

49. $\sim(p \wedge q)$ **50.** $\sim(q \wedge p)$ **51.** $\sim p \wedge q$

52. $\sim q \wedge p$ **53.** $\sim(q \vee p)$ **54.** $\sim(p \vee q)$

55. $\sim q \vee p$ **56.** $\sim p \vee q$ **57.** $\sim p \wedge \sim q$

58. $\sim q \wedge \sim p$

In Exercises 59–66, let p, q, and r represent the following simple statements:

p: The temperature outside is freezing.

q: The heater is working.

r: The house is cold.

Write each compound statement in symbolic form.

59. The temperature outside is freezing and the heater is working, or the house is cold.

60. If the temperature outside is freezing, then the heater is working or the house is not cold.

61. If the temperature outside is freezing or the heater is not working, then the house is cold.

62. It is not the case that if the house is cold then the heater is not working.

63. The house is cold, if and only if the temperature outside is freezing and the heater isn't working.

64. If the heater is working, then the temperature outside is freezing if and only if the house is cold.

65. Sufficient conditions for the house being cold are freezing outside temperatures and a heater not working.

66. A freezing outside temperature is both necessary and sufficient for a cold house if the heater is not working.

In Exercises 67–80, let p, q, and r represent the following simple statements:

p: The temperature is above 85°.

q: We finished studying.

r: We go to the beach.

Write each symbolic statement in words. If a symbolic statement is given without parentheses, place them, as needed, before and after the most dominant connective and then translate into English.

67. $(p \wedge q) \rightarrow r$ **68.** $(q \wedge r) \rightarrow p$

69. $p \wedge (q \rightarrow r)$ **70.** $p \wedge (r \rightarrow q)$

71. $\sim r \rightarrow \sim p \vee \sim q$ **72.** $\sim p \rightarrow q \vee r$

73. $(\sim r \rightarrow \sim q) \vee p$ **74.** $(\sim p \rightarrow \sim r) \vee q$

75. $r \leftrightarrow p \wedge q$ **76.** $r \leftrightarrow q \wedge p$

77. $(p \leftrightarrow q) \wedge r$ **78.** $q \rightarrow (r \leftrightarrow p)$

79. $\sim r \rightarrow \sim(p \wedge q)$ **80.** $\sim(p \wedge q) \rightarrow \sim r$

In Exercises 81–90, write each compound statement in symbolic form. Let letters assigned to the simple statements represent English sentences that are not negated. If commas do not appear in compound English statements, use the dominance of connectives to show grouping symbols (parentheses) in symbolic statements.

81. If I like the teacher or the course is interesting then I do not miss class.

82. If the lines go down or the transformer blows then we do not have power.

83. I like the teacher, or if the course is interesting then I do not miss class.

84. The lines go down, or if the transformer blows then we do not have power.

85. I miss class if and only if it's not true that both I like the teacher and the course is interesting.

86. We have power if and only if it's not true that both the lines go down and the transformer blows.

87. If I like the teacher I do not miss class if and only if the course is interesting.

88. If the lines go down we do not have power if and only if the transformer blows.

89. If I do not like the teacher and I miss class then the course is not interesting or I spend extra time reading the textbook.

90. If the lines do not go down and we have power then the transformer does not blow or there is an increase in the cost of electricity.

Practice Plus

In Exercises 91–96, write each compound statement in symbolic form. Assign letters to simple statements that are not negated and show grouping symbols in symbolic statements.

91. If it's not true that being French is necessary for being a Parisian then it's not true that being German is necessary for being a Berliner.

92. If it's not true that being English is necessary for being a Londoner then it's not true that being American is necessary for being a New Yorker.

93. Filing an income tax report and a complete statement of earnings is necessary for each taxpayer or an authorized tax preparer.

94. Falling in love with someone in your class or picking someone to hate are sufficient conditions for showing up to vent your emotions and not skipping.
(*Source:* Paraphrased from a student at the University of Georgia)

95. It is not true that being wealthy is a sufficient condition for being happy and living contentedly.

96. It is not true that being happy and living contentedly are necessary conditions for being wealthy.

In Exercises 97–100, use grouping symbols to clarify the meaning of each symbolic statement.

97. $p \rightarrow q \vee r \leftrightarrow p \wedge r$

98. $p \wedge q \rightarrow r \leftrightarrow p \vee r$

99. $p \rightarrow p \leftrightarrow p \wedge p \rightarrow \sim p$

100. $p \rightarrow p \leftrightarrow p \vee p \rightarrow \sim p$

Application Exercises

Exercises 101–106 contain statements made by well-known people. Use letters to represent each non-negated simple statement and rewrite the given compound statement in symbolic form.

101. "If you cannot get rid of the family skeleton, you may as well make it dance." (George Bernard Shaw)

102. "If my doctor told me I had only six minutes to live, I wouldn't brood and I'd type a little faster." (Isaac Asimov)

103. "If you know what you believe then it makes it a lot easier to answer questions, and I can't answer your question." (George W. Bush)

104. "If you don't like what you're doing, you can always pick up your needle and move to another groove." (Timothy Leary)

105. "If I were an intellectual, I would be pessimistic about America, but since I'm not an intellectual, I am optimistic about America." (General Lewis B. Hershey, Director of the Selective Service during the Vietnam war) (For simplicity, regard "optimistic" as "not pessimistic.")

106. "You cannot be both a good socializer and a good writer." (Erskine Caldwell)

Explaining the Concepts

107. Describe what is meant by a compound statement.

108. What is a conjunction? Describe the symbol that forms a conjunction.

109. What is a disjunction? Describe the symbol that forms a disjunction.

110. What is a conditional statement? Describe the symbol that forms a conditional statement.

111. What is a biconditional statement? Describe the symbol that forms a biconditional statement.

112. Discuss the difference between the symbolic statements $\sim(p \wedge q)$ and $\sim p \wedge q$.

113. If a symbolic statement does not contain parentheses, how are the grouping symbols determined?

114. Suppose that a friend tells you, "This summer I plan to visit Paris or London." Under what condition can you conclude that if your friend visits Paris, London will not be visited? Under what condition can you conclude that if your friend visits Paris, London might be visited? Assuming your friend has told you the truth, what can you conclude if you know that Paris will not be visited? Explain each of your answers.

Critical Thinking Exercises

Make Sense? *In Exercises 115–118, determine whether each statement makes sense or does not make sense, and explain your reasoning.*

115. When the waiter asked if I would like soup or salad, he used the exclusive *or*. However, when he asked if I would like coffee or dessert, he used the inclusive *or*.

116. When you wrote me that you planned to enroll in English and math or chemistry, I knew that if you didn't enroll in math, you'd be taking chemistry.

117. In China, the bride wears red, so wearing red is sufficient for being a Chinese bride.

118. Earth is the only planet not named after a god, so not being named after a god is both necessary and sufficient for a planet being Earth.

119. Use letters to represent each simple statement in the compound statement that follows. Then express the compound statement in symbolic form.

Shooting unarmed civilians is morally justifiable if and only if bombing them is morally justifiable, and as the former is not morally justifiable, neither is the latter.

120. Using a topic on which you have strong opinions, write a compound statement that contains at least two different connectives. Then express the statement in symbolic form.

Group Exercise

121. Each group member should find a legal document that contains at least six connectives in one paragraph. The connectives should include at least three different kinds of connectives, such as *and*, *or*, *if–then*, and *if and only if*. Share your example with other members of the group and see if the group can explain what some of the more complicated statements actually mean.

3.3 Truth Tables for Negation, Conjunction, and Disjunction

WHAT AM I SUPPOSED TO LEARN?

After studying this section, you should be able to:

1. Use the definitions of negation, conjunction, and disjunction.
2. Construct truth tables.
3. Determine the truth value of a compound statement for a specific case.

HERE'S LOOKING AT YOU. ACCORDING TO University of Texas economist Daniel Hamermesh (*Beauty Pays: Why Attractive People Are More Successful*), strikingly attractive and good-looking men and women can expect to earn an average of \$230,000 more in a lifetime than a person who is homely or plain. (Your author feels the need to start affirmative action for the beauty-bereft, consoled by the reality that looks are only one of many things that matter.)

In this section, you will work with a bar graph that shows the distribution of looks for American men and women, ranging from homely to strikingly attractive. By determining when statements involving negation, $\sim$ (not), conjunction, $\wedge$ (and), and disjunction, $\vee$ (or), are true and when they are false, you will be able to draw conclusions from the data. Classifying a statement as true or false is called **assigning a truth value to the statement**.

1 *Use the definitions of negation, conjunction, and disjunction.*

Negation, ~

The negation of a true statement is a false statement. We can express this in a table in which T represents true and F represents false.

p	$\sim p$
T	F

The negation of a false statement is a true statement. This, too, can be shown in table form.

p	$\sim p$
F	T

Combining the two tables results in **Table 3.10**, called the **truth table for negation**. This truth table expresses the idea that $\sim p$ has the opposite truth value from p.

TABLE 3.10 Negation

p	$\sim p$
T	F
F	T

$\sim p$ has the opposite truth value from p.

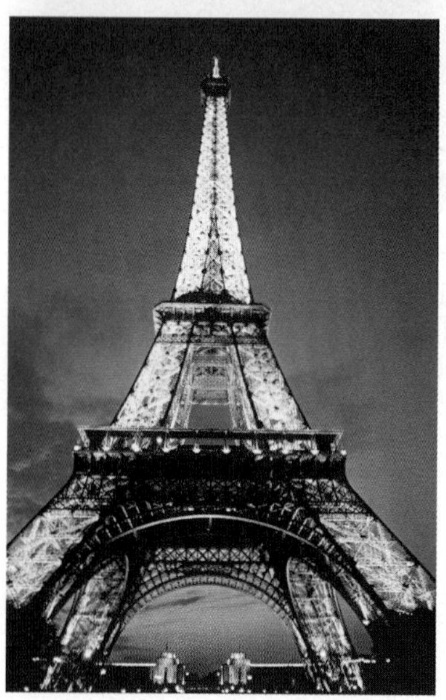

I visited London and I visited Paris.

Conjunction, $\wedge$

A friend tells you, "I visited London and I visited Paris." In order to understand the truth values for this statement, let's break it down into its two simple statements:

p: I visited London.
q: I visited Paris.

There are four possible cases to consider.

Case 1. Your friend actually visited both cities, so p is true and q is true. The conjunction "I visited London and I visited Paris" is true because your friend did both things. If both p and q are true, the conjunction $p \wedge q$ is true. We can show this in truth table form:

p	q	$p \wedge q$
T	T	T

Case 2. Your friend actually visited London, but did not tell the truth about visiting Paris. In this case, p is true and q is false. Your friend didn't do what was stated, namely visit both cities, so $p \wedge q$ is false. If p is true and q is false, the conjunction $p \wedge q$ is false.

p	q	$p \wedge q$
T	F	F

Case 3. This time, London was not visited, but Paris was. This makes p false and q true. As in case 2, your friend didn't do what was stated, namely visit both cities, so $p \wedge q$ is false. If p is false and q is true, the conjunction $p \wedge q$ is false.

p	q	$p \wedge q$
F	T	F

Case 4. This time your friend visited neither city, so p is false and q is false. The statement that both were visited, $p \wedge q$, is false.

p	q	$p \wedge q$
F	F	F

Let's use a truth table to summarize all four cases. Only in the case that your friend visited London and visited Paris is the conjunction true. Each of the four cases appears in **Table 3.11**, the truth table for conjunction, $\wedge$. The definition of conjunction is given in words to the right of the table.

THE DEFINITION OF CONJUNCTION

TABLE 3.11 Conjunction

p	q	$p \wedge q$
T	T	T
T	F	F
F	T	F
F	F	F

A conjunction is true only when both simple statements are true.

Table 3.12 contains an example of each of the four cases in the conjunction truth table.

GREAT QUESTION!

When is a conjunction false?

As soon as you find one simple statement in a conjunction that is false, then the conjunction is false.

TABLE 3.12 Statements of Conjunction and Their Truth Values

Statement	Truth Value	Reason
$3 + 2 = 5$ and London is in England.	T	Both simple statements are true.
$3 + 2 = 5$ and London is in France.	F	The second simple statement is false.
$3 + 2 = 6$ and London is in England.	F	The first simple statement is false.
$3 + 2 = 6$ and London is in France.	F	Both simple statements are false.

The statements that come before and after the main connective in a compound statement do not have to be simple statements. Consider, for example, the compound statement

$$(\sim p \vee q) \wedge \sim q.$$

The statements that make up this conjunction are $\sim p \vee q$ and $\sim q$. The conjunction is true only when both $\sim p \vee q$ and $\sim q$ are true. Notice that $\sim p \vee q$ is not a simple statement. We call $\sim p \vee q$ and $\sim q$ the *component statements* of the conjunction. The statements making up a compound statement are called **component statements**.

Disjunction, ∨

Now your friend states, "I will visit London or I will visit Paris." Because we assume that this is the inclusive "or," if your friend visits either or both of these cities, the truth has been told. The disjunction is false only in the event that neither city is visited. An *or* statement is true in every case, except when both component statements are false.

The truth table for disjunction, ∨, is shown in **Table 3.13**. The definition of disjunction is given in words to the right of the table.

THE DEFINITION OF DISJUNCTION

TABLE 3.13 Disjunction

p	q	$p \vee q$
T	T	T
T	F	T
F	T	T
F	F	F

A disjunction is false only when both component statements are false.

Table 3.14 contains an example of each of the four cases in the disjunction truth table.

TABLE 3.14 Statements of Disjunction and Their Truth Values

Statement	Truth Value	Reason
$3 + 2 = 5$ or London is in England.	T	Both component statements are true.
$3 + 2 = 5$ or London is in France.	T	The first component statement is true.
$3 + 2 = 6$ or London is in England.	T	The second component statement is true.
$3 + 2 = 6$ or London is in France.	F	Both component statements are false.

GREAT QUESTION!

When is a disjunction true?

As soon as you find one component statement in a disjunction that is true, then the disjunction is true.

EXAMPLE 1 ***Using the Definitions of Negation, Conjunction, and Disjunction***

Let p and q represent the following statements:

p: $10 > 4$

q: $3 < 5$.

Determine the truth value for each statement:

a. $p \wedge q$ **b.** $\sim p \wedge q$ **c.** $p \vee \sim q$ **d.** $\sim p \vee \sim q$.

SOLUTION

a. $p \wedge q$ translates as

$10 > 4$ and $3 < 5$.

10 is greater than 4 is true. 3 is less than 5 is true.

By definition, a conjunction, $\wedge$, is true only when both component statements are true. Thus, $p \wedge q$ is a true statement.

b. $\sim p \wedge q$ translates as

$10 \not> 4$ and $3 < 5$.

10 is not greater than 4 is false. 3 is less than 5 is true.

By definition, a conjunction, $\wedge$, is true only when both component statements are true. In this conjunction, only one of the two component statements is true. Thus, $\sim p \wedge q$ is a false statement.

c. $p \vee \sim q$ translates as

$10 > 4$ or $3 \not< 5$.

10 is greater than 4 is true. 3 is not less than 5 is false.

By definition, a disjunction, $\vee$, is false only when both component statements are false. In this disjunction, only one of the two component statements is false. Thus, $p \vee \sim q$ is a true statement.

d. $\sim p \vee \sim q$ translates as

$10 \not> 4$ or $3 \not< 5$.

10 is not greater than 4 is false. 3 is not less than 5 is false.

By definition, a disjunction, $\vee$, is false only when both component statements are false. Thus, $\sim p \vee \sim q$ is a false statement.

 **CHECK POINT 1** Let p and q represent the following statements:

p: $3 + 5 = 8$

q: $2 \times 7 = 20$.

Determine the truth value for each statement:

a. $p \wedge q$ **b.** $p \wedge \sim q$

c. $\sim p \vee q$ **d.** $\sim p \vee \sim q$.

2 *Construct truth tables.*

Constructing Truth Tables

Truth tables can be used to gain a better understanding of English statements. The truth tables in this section are based on the definitions of negation, ~, conjunction, $\wedge$, and disjunction, $\vee$. It is helpful to remember these definitions in words.

DEFINITIONS OF NEGATION, CONJUNCTION, AND DISJUNCTION

1. Negation ~: not
   The negation of a statement has the opposite truth value from the statement.
2. Conjunction $\wedge$: and
   The only case in which a conjunction is true is when both component statements are true.
3. Disjunction $\vee$: or
   The only case in which a disjunction is false is when both component statements are false.

Breaking compound statements into component statements and applying these definitions will enable you to construct truth tables.

GREAT QUESTION!

Why should I list the four possible combinations for *p* and *q* in the order shown on the right?

Always using this order makes it easier to follow another student's work and to check your truth tables with those in the answer section.

CONSTRUCTING TRUTH TABLES FOR COMPOUND STATEMENTS CONTAINING ONLY THE SIMPLE STATEMENTS *p* AND *q*

- List the four possible combinations of truth values for p and q.

p	q	
T	T	
T	F	
F	T	
F	F	

We will always list the combinations in this order. Although any order can be used, this standard order makes for a consistent presentation.

- Determine each column heading by reconstructing the given compound statement one component statement at a time. The final column heading should be the given compound statement.
- Use each column heading to fill in the four truth values.
  → If a column heading involves negation, ~ (not), fill in the column by looking back at the column that contains the statement that must be negated. Take the opposite of the truth values in this column.
  → If a column heading involves the symbol for conjunction, $\wedge$ (and), fill in the truth values in the column by looking back at two columns—the column for the statement before the $\wedge$ connective and the column for the statement after the $\wedge$ connective. Fill in the column by applying the definition of conjunction, writing T only when both component statements are true.
  → If a column heading involves the symbol for disjunction, $\vee$ (or), fill in the truth values in the column by looking back at two columns—the column for the statement before the $\vee$ connective and the column for the statement after the $\vee$ connective. Fill in the column by applying the definition of disjunction, writing F only when both component statements are false.

EXAMPLE 2 *Constructing a Truth Table*

Construct a truth table for

$$\sim(p \wedge q)$$

to determine when the statement is true and when the statement is false.

Lewis Carroll in Logical Wonderland

One of the world's most popular logicians was Lewis Carroll (1832–1898), author of *Alice's Adventures in Wonderland* and *Through the Looking Glass*. Carroll's passion for word games and logic is evident throughout the Alice stories.

> "Speak when you're spoken to!" the Queen sharply interrupted her.
> "But if everybody obeyed that rule," said Alice, who was always ready for a little argument, "and if you only spoke when you were spoken to, and the other person always waited for *you* to begin, you see nobody would ever say anything, so that—"

> "Take some more tea," the March Hare said to Alice, very earnestly.
> "I've had nothing yet," Alice replied in an offended tone, "so I can't take more."
> "You mean you can't take *less*," said the Hatter. "It's very easy to take *more* than nothing."

SOLUTION

The parentheses in $\sim(p \wedge q)$ indicate that we must first determine the truth values for the conjunction $p \wedge q$. After this, we determine the truth values for the negation $\sim(p \wedge q)$ by taking the opposite of the truth values for $p \wedge q$.

Step 1 As with all truth tables, first list the simple statements on top. Then show all the possible truth values for these statements. In this case there are two simple statements and four possible combinations, or cases.

p	q	
T	T	
T	F	
F	T	
F	F	

Step 2 Make a column for $p \wedge q$, the statement within the parentheses in $\sim(p \wedge q)$. Use $p \wedge q$ as the heading for the column, and then fill in the truth values for the conjunction by looking back at the p and q columns. A conjunction is true only when both component statements are true.

p	q	$p \wedge q$
T	T	T
T	F	F
F	T	F
F	F	F

p and q are true, so $p \wedge q$ is true.

Step 3 Construct one more column for $\sim(p \wedge q)$. Fill in this column by negating the values in the $p \wedge q$ column. Using the negation definition, take the opposite of the truth values in the third column.

p	q	$p \wedge q$	$\sim(p \wedge q)$
T	T	T	F
T	F	F	T
F	T	F	T
F	F	F	T

Opposite truth values because we are negating column 3

This completes the truth table for $\sim(p \wedge q)$.

The final column in the truth table for $\sim(p \wedge q)$ tells us that the statement is false only when both p and q are true. For example, using

p: Harvard is a college (true)

q: Yale is a college (true),

the statement $\sim(p \wedge q)$ translates as

It is not true that Harvard and Yale are colleges.

This compound statement is false. It *is* true that Harvard and Yale are colleges.

CHECK POINT 2 Construct a truth table for $\sim(p \vee q)$ to determine when the statement is true and when the statement is false.

EXAMPLE 3 *Constructing a Truth Table*

Construct a truth table for

$$\sim p \vee \sim q$$

to determine when the statement is true and when the statement is false.

SOLUTION

Without parentheses, the negation symbol, ~, negates only the statement that immediately follows it. Therefore, we first determine the truth values for $\sim p$ and for $\sim q$. Then we determine the truth values for the *or* disjunction, $\sim p \vee \sim q$.

Step 1 List the simple statements on top and show the four possible cases for the truth values.

p	q	
T	T	
T	F	
F	T	
F	F	

Step 2 Make columns for $\sim p$ and for $\sim q$. Fill in the $\sim p$ column by looking back at the p column, the first column, and taking the opposite of the truth values in that column. Fill in the $\sim q$ column by taking the opposite of the truth values in the second column, the q column.

Opposite truth values

p	q	$\sim p$	$\sim q$
T	T	F	F
T	F	F	T
F	T	T	F
F	F	T	T

Opposite truth values

Step 3 Construct one more column for $\sim p \vee \sim q$. To determine the truth values of $\sim p \vee \sim q$, look back at the $\sim p$ column, column 3, and the $\sim q$ column, column 4. Now use the disjunction definition on the entries in columns 3 and 4. Disjunction definition: An *or* statement is false only when both component statements are false. This occurs only in the first row.

p	q	$\sim p$	$\sim q$	$\sim p \vee \sim q$
T	T	F	F	F
T	F	F	T	T
F	T	T	F	T
F	F	T	T	T

$\sim p$ is false and $\sim q$ is false, so $\sim p \vee \sim q$ is false.

column 3 $\vee$ column 4

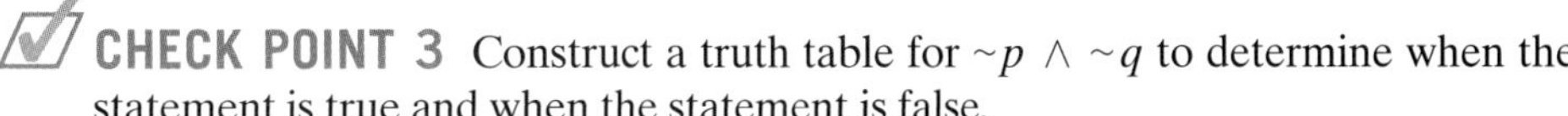

CHECK POINT 3 Construct a truth table for $\sim p \wedge \sim q$ to determine when the statement is true and when the statement is false.

EXAMPLE 4 Constructing a Truth Table

Construct a truth table for

$$(\sim p \vee q) \wedge \sim q$$

to determine when the statement is true and when the statement is false.

SOLUTION

The statement $(\sim p \vee q) \wedge \sim q$ is an *and* conjunction because the conjunction symbol, $\wedge$, is outside the parentheses. We cannot determine the truth values for the statement until we first determine the truth values for $\sim p \vee q$ and for $\sim q$, the component statements before and after the $\wedge$ connective:

$$\boxed{(\sim p \vee q)} \wedge \boxed{\sim q}.$$

We'll need a column with truth values for this component statement.

We'll need a column with truth values for this component statement.

$\boxed{(\sim p \vee q)} \wedge \boxed{\sim q}$

Column needed

Column needed

Step 1 The compound statement involves two simple statements and four possible cases.

p	q	
T	T	
T	F	
F	T	
F	F	

Step 2 Because we need a column with truth values for $\sim p \vee q$, begin with $\sim p$. Use $\sim p$ as the heading. Fill in the column by looking back at the p column, column 1, and take the opposite of the truth values in that column.

p	q	$\sim p$
T	T	F
T	F	F
F	T	T
F	F	T

Opposite truth values

Step 3 Now add a $\sim p \vee q$ column. To determine the truth values of $\sim p \vee q$, look back at the $\sim p$ column, column 3, and the q column, column 2. Now use the disjunction definition on the entries in columns 3 and 2. Disjunction definition: An *or* statement is false only when both component statements are false. This occurs only in the second row.

p	q	$\sim p$	$\sim p \vee q$
T	T	F	T
T	F	F	F
F	T	T	T
F	F	T	T

$\sim p$ is false and q is false, so $\sim p \vee q$ is false.

column 3 $\vee$ column 2

Step 4 The statement following the $\wedge$ connective in $(\sim p \vee q) \wedge \sim q$ is $\sim q$, so add a $\sim q$ column. Fill in the column by looking back at the q column, column 2, and take the opposite of the truth values in that column.

p	q	$\sim p$	$\sim p \vee q$	$\sim q$
T	T	F	T	F
T	F	F	F	T
F	T	T	T	F
F	F	T	T	T

Opposite truth values

Step 5 The final column heading is

$$(\sim p \vee q) \wedge \sim q,$$

which is our given statement. To determine its truth values, look back at the $\sim p \vee q$ column, column 4, and the $\sim q$ column, column 5. Now use the conjunction definition on the entries in columns 4 and 5. Conjunction definition: An *and* statement is true only when both component statements are true. This occurs only in the last row.

p	q	$\sim p$	$\sim p \vee q$	$\sim q$	$(\sim p \vee q) \wedge \sim q$
T	T	F	T	F	F
T	F	F	F	T	F
F	T	T	T	F	F
F	F	T	T	T	T

$\sim p \vee q$ is true and $\sim q$ is true, so $(\sim p \vee q) \wedge \sim q$ is true.

column 4 $\wedge$ column 5

The truth table is now complete. By looking at the truth values in the last column, we can see that the compound statement

$$(\sim p \vee q) \wedge \sim q$$

is true only in the fourth row, that is, when p is false and q is false.

 **CHECK POINT 4** Construct a truth table for $(p \wedge \sim q) \vee \sim p$ to determine when the statement is true and when the statement is false.

Some compound statements, such as $p \vee \sim p$, consist of only one simple statement. In cases like this, there are only two true–false possibilities: p can be true or p can be false.

EXAMPLE 5 *Constructing a Truth Table*

Construct a truth table for

$$p \vee \sim p$$

to determine when the statement is true and when the statement is false.

SOLUTION

In order to construct a truth table for $p \vee \sim p$, we first determine the truth values for $\sim p$. Then we determine the truth values for the *or* disjunction, $p \vee \sim p$.

Step 1 The compound statement involves one simple statement and two possible cases.

p	
T	
F	

Step 2 Add a column for $\sim p$.

p	$\sim p$
T	F
F	T

Take the opposite of the truth values in the first column.

Step 3 Construct one more column for $p \vee \sim p$.

p	$\sim p$	$p \vee \sim p$
T	F	T
F	T	T

Look back at columns 1 and 2 and apply the disjunction definition: An *or* statement is false only when both component statements are false. This does not occur in either row.

The truth table is now complete. By looking at the truth values in the last column, we can see that the compound statement $p \vee \sim p$ is true in all cases.

A compound statement that is always true is called a **tautology**. Example 5 proves that $p \vee \sim p$ is a tautology.

Blitzer Bonus

Tautologies and Commercials

Think about the logic in the following statement from a commercial for a high-fiber cereal:

> Some studies suggest a high-fiber, low-fat diet may reduce the risk of some kinds of cancer.

From a strictly logical point of view, "may reduce" means "It may reduce the risk of cancer or it may not." The symbolic translation of this sentence, $p \vee \sim p$, is a tautology. The claim made in the cereal commercial can be made about any substance that is not known to actually cause cancer and, like $p \vee \sim p$, will always be true.

CHECK POINT 5 Construct a truth table for

$$p \wedge \sim p$$

to determine when the statement is true and when the statement is false.

Some compound statements involve three simple statements, usually represented by p, q, and r. In this situation, there are eight different true–false combinations, shown in **Table 3.15**. The first column has four Ts followed by four Fs. The second column has two Ts, two Fs, two Ts, and two Fs. Under the third statement, r, T alternates with F. It is not necessary to list the eight cases in this order, but this systematic method ensures that no case is repeated and that all cases are included.

TABLE 3.15

	p	q	r
Case 1	T	T	T
Case 2	T	T	F
Case 3	T	F	T
Case 4	T	F	F
Case 5	F	T	T
Case 6	F	T	F
Case 7	F	F	T
Case 8	F	F	F

There are eight different true–false combinations for compound statements consisting of three simple statements.

EXAMPLE 6 ***Constructing a Truth Table with Eight Cases***

a. Construct a truth table for the following statement:
I study hard and ace the final, or I fail the course.

b. Suppose that you study hard, you do not ace the final, and you fail the course. Under these conditions, is the compound statement in part (a) true or false?

SOLUTION

a. We begin by assigning letters to the simple statements. Use the following representations:

p: I study hard.
q: I ace the final.
r: I fail the course.

Now we can write the given statement in symbolic form.

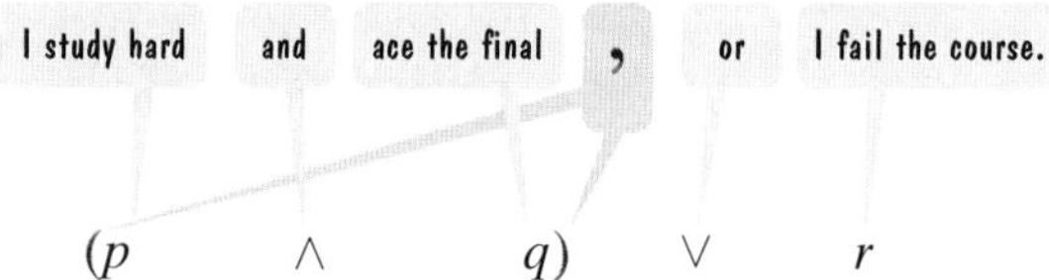

The statement $(p \wedge q) \vee r$ is a disjunction because the *or* symbol, $\vee$, is outside the parentheses. We cannot determine the truth values for this disjunction until we have determined the truth values for $p \wedge q$ and for r, the statements before and after the $\vee$ connective. The completed truth table appears as follows:

The conjunction is true only when p, the first column, is true and q, the second column, is true.

The disjunction is false only when $p \wedge q$, column 4, is false and r, column 3, is false.

Show eight possible cases.

These are the conditions in part (b).

p	q	r	$p \wedge q$	$(p \wedge q) \vee r$
T	T	T	T	T
T	T	F	T	T
T	F	T	F	T
T	F	F	F	F
F	T	T	F	T
F	T	F	F	F
F	F	T	F	T
F	F	F	F	F

b. We are given the following:

p: I study hard. — This is true. We are told you study hard.

q: I ace the final. — This is false. We are told you do not ace the final.

r: I fail the course. — This is true. We are told you fail the course.

The given conditions, T F T, correspond to case 3 of the truth table, indicated by the voice balloon on the far left. Under these conditions, the original compound statement is true, shown by the red T in the truth table.

CHECK POINT 6

a. Construct a truth table for the following statement:

I study hard, and I ace the final or fail the course.

b. Suppose that you do not study hard, you ace the final, and you fail the course. Under these conditions, is the compound statement in part (a) true or false?

3 *Determine the truth value of a compound statement for a specific case.*

Determining Truth Values for Specific Cases

A truth table shows the truth values of a compound statement for every possible case. In our next example, we will determine the truth value of a compound statement for a specific case in which the truth values of the simple statements are known. This does not require constructing an entire truth table. By substituting the truth values of the simple statements into the symbolic form of the compound statement and using the appropriate definitions, we can determine the truth value of the compound statement.

EXAMPLE 7 *Determining the Truth Value of a Compound Statement*

The bar graph in **Figure 3.2** shows the distribution of looks for American men and women, ranging from homely to strikingly attractive.

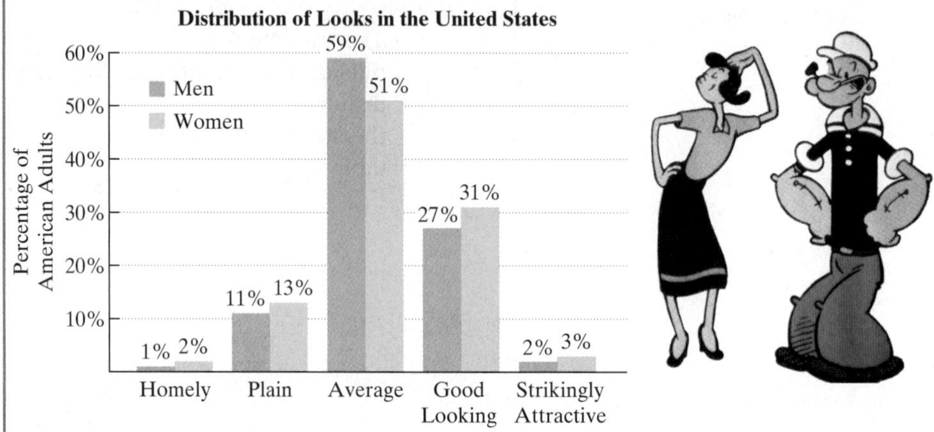

FIGURE 3.2
Source: Time

Use the information in the bar graph to determine the truth value of the following statement:

It is not true that 1% of American men are homely and more than half are average, or it is not true that 5% of American women are strikingly attractive.

SOLUTION

We begin by assigning letters to the simple statements, using the graph to determine whether each simple statement is true or false. As always, we let these letters represent statements that are not negated.

p: One percent of American men are homely. — This statement is true.

q: More than half of American men are average looking. — This statement is true. 59% of American men are average, which is more than half: $\frac{1}{2} = 50\%$.

r: Five percent of American women are strikingly attractive. — This statement is false. 3% of American women are strikingly attractive.

Using these representations, the given compound statement can be expressed in symbolic form as

$$\sim(p \quad \wedge \quad q) \quad \vee \quad \sim r.$$

It is not true that | 1% of American men are homely | and | more than half are average, | or | it is not true that 5% of American women are strikingly attractive.

Now we substitute the truth values for p, q, and r that we obtained from the bar graph to determine the truth value for the given compound statement.

$\sim(p \wedge q) \vee \sim r$	This is the given compound statement in symbolic form.
$\sim(\text{T} \wedge \text{T}) \vee \sim\text{F}$	Substitute the truth values obtained from the graph.
$\sim\text{T} \vee \sim\text{F}$	Replace T ∧ T with T. Conjunction is true when both parts are true.
$\text{F} \vee \text{T}$	Replace ~T with F and ~F with T. Negation gives the opposite truth value.
T	Replace F ∨ T with T. Disjunction is true when at least one part is true.

We conclude that the given statement is true.

 **CHECK POINT 7** Use the information in the bar graph in **Figure 3.2** to determine the truth value for the following statement:

Two percent of American women are homely or more than half are good looking, and it is not true that 5% of American men are strikingly attractive.

Concept and Vocabulary Check

Fill in each blank so that the resulting statement is true.

1. $\sim p$ has the ________ truth value from p.
2. A conjunction, $p \wedge q$, is true only when ________________.
3. A disjunction, $p \vee q$, is false only when ________________.

In Exercises 4–8, determine whether each statement is true or false. If the statement is false, make the necessary change(s) to produce a true statement.

4. If one component statement in a conjunction is false, the conjunction is false. _______
5. If one component statement in a disjunction is true, the disjunction is true. _______
6. A truth table for $p \vee \sim q$ requires four possible combinations of truth values. _______
7. A truth table for $p \vee \sim p$ requires four possible combinations of truth values. _______
8. A truth table for $(p \vee \sim q) \wedge r$ requires eight possible combinations of truth values. _______

Exercise Set 3.3

Practice Exercises

In Exercises 1–16, let p and q represent the following statements:

p: $4 + 6 = 10$
q: $5 \times 8 = 80$

Determine the truth value for each statement.

1. $\sim q$
2. $\sim p$
3. $p \wedge q$
4. $q \wedge p$
5. $\sim p \wedge q$
6. $p \wedge \sim q$
7. $\sim p \wedge \sim q$
8. $q \wedge \sim q$
9. $q \vee p$
10. $p \vee q$
11. $p \vee \sim q$
12. $\sim p \vee q$
13. $p \vee \sim p$
14. $q \vee \sim q$
15. $\sim p \vee \sim q$
16. $\sim q \vee \sim p$

In Exercises 17–24, complete the truth table for the given statement by filling in the required columns.

17. $\sim p \wedge p$

p	$\sim p$	$\sim p \wedge p$
T		
F		

18. $\sim(\sim p)$

p	$\sim p$	$\sim(\sim p)$
T		
F		

19. $\sim p \wedge q$

p	q	$\sim p$	$\sim p \wedge q$
T	T		
T	F		
F	T		
F	F		

20. $\sim p \vee q$

p	q	$\sim p$	$\sim p \vee q$
T	T		
T	F		
F	T		
F	F		

21. $\sim(p \vee q)$

p	q	$p \vee q$	$\sim(p \vee q)$
T	T		
T	F		
F	T		
F	F		

22. $\sim(p \vee \sim q)$

p	q	$\sim q$	$p \vee \sim q$	$\sim(p \vee \sim q)$
T	T			
T	F			
F	T			
F	F			

23. $\sim p \wedge \sim q$

p	q	$\sim p$	$\sim q$	$\sim p \wedge \sim q$
T	T			
T	F			
F	T			
F	F			

24. $p \wedge \sim q$

p	q	$\sim q$	$p \wedge \sim q$
T	T		
T	F		
F	T		
F	F		

In Exercises 25–42, construct a truth table for the given statement.

25. $p \vee \sim q$

26. $\sim q \wedge p$

27. $\sim(\sim p \vee q)$

28. $\sim(p \wedge \sim q)$

29. $(p \vee q) \wedge \sim p$

30. $(p \wedge q) \vee \sim p$

31. $\sim p \vee (p \wedge \sim q)$

32. $\sim p \wedge (p \vee \sim q)$

33. $(p \vee q) \wedge (\sim p \vee \sim q)$

34. $(p \wedge \sim q) \vee (\sim p \wedge q)$

35. $(p \wedge \sim q) \vee (p \wedge q)$

36. $(p \vee \sim q) \wedge (p \vee q)$

37. $p \wedge (\sim q \vee r)$

38. $p \vee (\sim q \wedge r)$

39. $(r \wedge \sim p) \vee \sim q$

40. $(r \vee \sim p) \wedge \sim q$

41. $\sim(p \vee q) \wedge \sim r$

42. $\sim(p \wedge q) \vee \sim r$

In Exercises 43–52,

a. *Write each statement in symbolic form. Assign letters to simple statements that are not negated.*

b. *Construct a truth table for the symbolic statement in part (a).*

c. *Use the truth table to indicate one set of conditions that makes the compound statement true, or state that no such conditions exist.*

43. You did not do the dishes and you left the room a mess.

44. You did not do the dishes, but you did not leave the room a mess.

45. It is not true that I bought a meal ticket and did not use it.

46. It is not true that I ordered pizza while watching late-night TV and did not gain weight.

47. The student is intelligent or an overachiever, and not an overachiever.

48. You're blushing or sunburned, and you're not sunburned.

49. Married people are healthier than single people and more economically stable than single people, and children of married people do better on a variety of indicators.

50. You walk or jog, or engage in something physical.

51. I go to office hours and ask questions, or my professor does not remember me.

52. You marry the person you love, but you do not always love that person or do not always have a successful marriage.

In Exercises 53–62, determine the truth value for each statement when p is false, q is true, and r is false.

53. $p \wedge (q \vee r)$

54. $p \vee (q \wedge r)$

55. $\sim p \vee (q \wedge \sim r)$

56. $\sim p \wedge (\sim q \wedge r)$

57. $\sim(p \wedge q) \vee r$

58. $\sim(p \vee q) \wedge r$

59. $\sim(p \vee q) \wedge \sim(p \wedge r)$

60. $\sim(p \wedge q) \vee \sim(p \vee r)$

61. $(\sim p \wedge q) \vee (\sim r \wedge p)$

62. $(\sim p \vee q) \wedge (\sim r \vee p)$

Practice Plus

In Exercises 63–66, construct a truth table for each statement.

63. $\sim[\sim(p \wedge \sim q) \vee \sim(\sim p \vee q)]$

64. $\sim[\sim(p \vee \sim q) \wedge \sim(\sim p \wedge q)]$

65. $[(p \wedge \sim r) \vee (q \wedge \sim r)] \wedge \sim(\sim p \vee r)$

66. $[(p \vee \sim r) \wedge (q \vee \sim r)] \vee \sim(\sim p \vee r)$

In Exercises 67–70, write each statement in symbolic form and construct a truth table. Then indicate under what conditions, if any, the compound statement is true.

67. You notice this notice or you do not, and you notice this notice is not worth noticing.

68. You notice this notice and you notice this notice is not worth noticing, or you do not notice this notice.

69. It is not true that $x \leq 3$ or $x \geq 7$, but $x > 3$ and $x < 7$.

70. It is not true that $x < 5$ or $x > 8$, but $x \geq 5$ and $x \leq 8$.

Application Exercises

The line graph shows the average U.S. gasoline prices from 1990 through 2015.

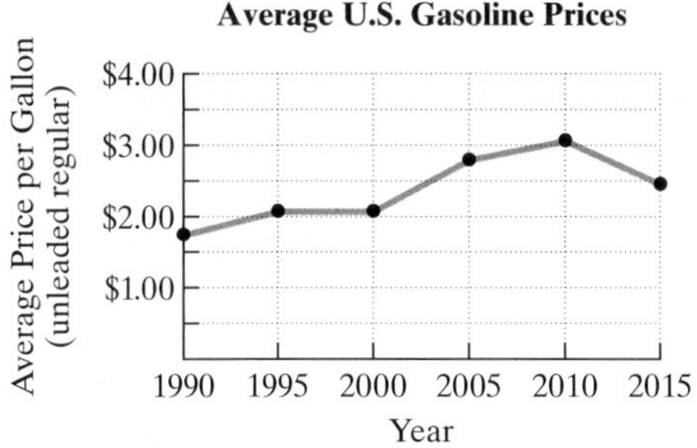

Source: U.S. Department of Energy

In Exercises 71–80, let p, q, and r represent the following simple statements:

p: Prices peaked in 2010.

q: Prices remained steady from 1995 through 2000.

r: Prices were more than $2.00 per gallon in 1990.

Write each symbolic statement in words. Then use the information given by the graph to determine the truth value of the statement.

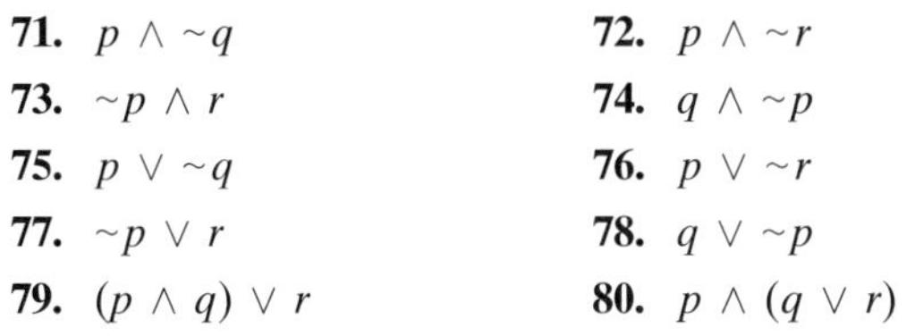

71. $p \wedge \sim q$ **72.** $p \wedge \sim r$

73. $\sim p \wedge r$ **74.** $q \wedge \sim p$

75. $p \vee \sim q$ **76.** $p \vee \sim r$

77. $\sim p \vee r$ **78.** $q \vee \sim p$

79. $(p \wedge q) \vee r$ **80.** $p \wedge (q \vee r)$

The bar graph shows the careers named as probable by U.S. college freshmen in 2012.

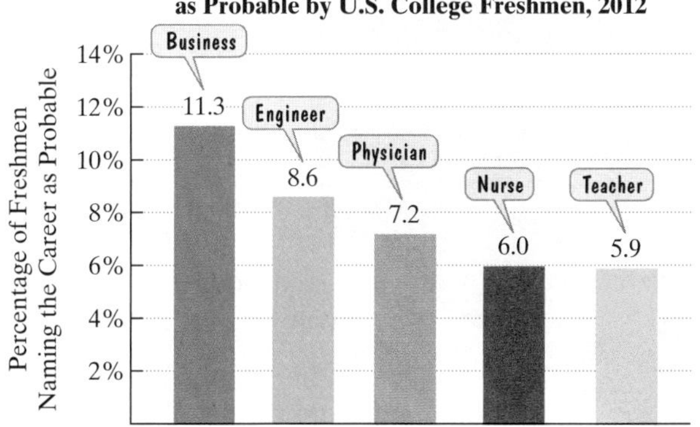

Source: The American Freshman: National Norms

In Exercises 81–84, write each statement in symbolic form. Then use the information in the graph to determine the truth value of the compound statement.

81. More than 12% named business or it is not true that 9% named engineering.

82. More than 12% named business and it is not true that 9% named engineering.

83. 5.9% named teaching or 8.9% named nursing, and it is not true that 12% named business.

84. 5.9% named teaching, but 8.9% named nursing or it is not true that 12% named business.

85. To qualify for the position of music professor, an applicant must have a master's degree in music, and be able to play at least three musical instruments or have at least five years' experience playing with a symphony orchestra.

There are three applicants for the position:

- *Bolero Mozart* has a master's degree in journalism and plays the piano, tuba, and violin.
- *Cha-Cha Bach* has a master's degree in music, plays the piano and the harp, and has two years' experience with a symphony orchestra.
- *Hora Gershwin* has a master's degree in music, plays 14 instruments, and has two years' experience in a symphony orchestra.

a. Which of the applicants qualifies for the position?

b. Explain why each of the other applicants does not qualify for the position.

86. To qualify for the position of art professor, an applicant must have a master's degree in art, and a body of work judged as excellent by at least two working artists or at least two works on public display in the United States.

There are three applicants for the position:

- *Adagio Picasso* needs two more courses to complete a master's degree in art, has a body of work judged as excellent by ten working artists, and has over 50 works on public display in a number of different American cities.
- *Rondo Seurat* has a master's degree in art, a body of work judged as excellent by a well-known working artist, and two works on public display in New York City.
- *Yodel Van Gogh* has a master's degree in art, is about to complete a doctorate in art history, has 20 works on public display in Paris, France, and has a body of work judged as excellent by a working artist.

a. Which of the applicants qualifies for the position?

b. Explain why each of the other applicants does not qualify for the position.

Explaining the Concepts

87. Under which conditions is a conjunction true?

88. Under which conditions is a conjunction false?

89. Under which conditions is a disjunction true?

90. Under which conditions is a disjunction false?

91. Describe how to construct a truth table for a compound statement.

92. Describe the information given by the truth values in the final column of a truth table.

93. Describe how to set up the eight different true–false combinations for a compound statement consisting of three simple statements.

94. Television commercials often flash disclaimers (in tiny print, of course) that read "Individual results will vary" and "Individual results may vary." What is the difference between these two statements? Which statement is a tautology and which negates the advertised product's effectiveness?

Critical Thinking Exercises

Make Sense? *In Exercises 95–98, determine whether each statement makes sense or does not make sense, and explain your reasoning.*

95. I'm filling in the truth values for a column in a truth table that requires me to look back at three columns.

96. If I know that p is true, q is false, and r is false, the most efficient way to determine the truth value of $(p \wedge \sim q) \vee r$ is to construct a truth table.

97. My truth table for $\sim(\sim p)$ has four possible combinations of truth values.

98. Using inductive reasoning, I conjecture that a truth table for a compound statement consisting of n simple statements has 2^n true–false combinations.

99. Use the bar graph for Exercises 81–84 to write a true compound statement with each of the following characteristics. Do not use any of the simple statements that appear in Exercises 81–84.

a. The statement contains three different simple statements.

b. The statement contains two different connectives.

c. The statement contains one simple statement with the word *not*.

100. If $\sim(p \vee q)$ is true, determine the truth values for p and q.

101. The truth table that defines $\vee$, the *inclusive or*, indicates that the compound statement is true if one or both of its component statements are true. The symbol for the *exclusive or* is $\underline{\vee}$. The *exclusive or* means *either p or q*, but *not both*. Use this meaning to construct the truth table that defines $p \underline{\vee} q$.

Group Exercise

102. Each member of the group should find a graph that is of particular interest to that person. Share the graphs. The group should select the three graphs that it finds most intriguing. For the graphs selected, group members should write four compound statements. Two of the statements should be true and two should be false. One of the statements should contain three different simple statements and two different connectives.

3.4 Truth Tables for the Conditional and the Biconditional

WHAT AM I SUPPOSED TO LEARN?

After studying this section, you should be able to:

1 Understand the logic behind the definition of the conditional.
2 Construct truth tables for conditional statements.
3 Understand the definition of the biconditional.
4 Construct truth tables for biconditional statements.
5 Determine the truth value of a compound statement for a specific case.

YOUR AUTHOR RECEIVED junk mail with this claim:

If your Super Million Dollar Prize Entry Number matches the winning preselected number and you return the number before the deadline stated below, you will win $1,000,000.00.

Should he obediently return the number before the deadline or trash the whole thing?

In this section, we will use logic to analyze the claim in the junk mail. By understanding when statements involving the conditional, $\rightarrow$ (if–then), and the biconditional, $\leftrightarrow$ (if and only if), are true and when they are false, you will be able to determine the truth value of the claim.

1 *Understand the logic behind the definition of the conditional.*

Conditional Statements, $\rightarrow$

We begin by looking at the truth table for conditional statements. Suppose that your professor promises you the following:

If you pass the final, then you pass the course.

Break the statement down into its two component statements:

p: You pass the final.

q: You pass the course.

Translated into symbolic form, your professor's statement is $p \rightarrow q$. We now look at the four cases shown in **Table 3.16**, the truth table for the conditional.

TABLE 3.16 Conditional

	p	q	$p \rightarrow q$
Case 1	T	T	T
Case 2	T	F	F
Case 3	F	T	T
Case 4	F	F	T

Case 1 (T, T) You do pass the final and you do pass the course. Your professor did what was promised, so the conditional statement is true.

Case 2 (T, F) You pass the final, but you do not pass the course. Your professor did not do what was promised, so the conditional statement is false.

Case 3 (F, T) You do not pass the final, but you do pass the course. Your professor's original statement talks about only what would happen if you passed the final. It says nothing about what would happen if you did not pass the final. Your professor did not break the promise of the original statement, so the conditional statement is true.

Case 4 (F, F) You do not pass the final and you do not pass the course. As with case 3, your professor's original statement talks about only what would happen if you passed the final. The promise of the original statement has not been broken. Therefore, the conditional statement is true.

Table 3.16 illustrates that a conditional statement is false only when the antecedent, the statement before the $\rightarrow$ connective, is true and the consequent, the statement after the $\rightarrow$ connective, is false. A conditional statement is true in all other cases.

THE DEFINITION OF THE CONDITIONAL

p	q	$p \rightarrow q$
T	T	T
T	F	F
F	T	T
F	F	T

A conditional is false only when the antecedent is true and the consequent is false.

2 *Construct truth tables for conditional statements.*

Constructing Truth Tables

Our first example shows how truth tables can be used to gain a better understanding of conditional statements.

EXAMPLE 1 *Constructing a Truth Table*

Construct a truth table for

$$\sim q \rightarrow \sim p$$

to determine when the statement is true and when the statement is false.

SOLUTION

Remember that without parentheses, the symbol ~ negates only the statement that immediately follows it. Therefore, we cannot determine the truth values for this conditional statement until we first determine the truth values for $\sim q$ and for $\sim p$, the statements before and after the $\rightarrow$ connective.

Step 1 List the simple statements on top and show the four possible cases for the truth values.

p	q	
T	T	
T	F	
F	T	
F	F	

Step 2 In order to construct a truth table for $\sim q \rightarrow \sim p$, we need to make columns for $\sim q$ and for $\sim p$. Fill in the $\sim q$ column by looking back at the q column, the second column, and taking the opposite of the truth values in this column. Fill in the $\sim p$ column by taking the opposite of the truth values in the first column, the p column.

Opposite truth values

p	q	$\sim q$	$\sim p$
T	T	F	F
T	F	T	F
F	T	F	T
F	F	T	T

Opposite truth values

Step 3 Construct one more column for $\sim q \rightarrow \sim p$. Look back at the $\sim q$ column, column 3, and the $\sim p$ column, column 4. Now use the conditional definition to determine the truth values for $\sim q \rightarrow \sim p$ based on columns 3 and 4. Conditional definition: An *if–then* statement is false only when the antecedent is true and the consequent is false. This occurs only in the second row.

p	q	$\sim q$	$\sim p$	$\sim q \rightarrow \sim p$
T	T	F	F	T
T	F	T	F	F
F	T	F	T	T
F	F	T	T	T

$\sim q$ is true and $\sim p$ is false, so $\sim q \rightarrow \sim p$ is false.

column 3 $\rightarrow$ column 4

TABLE 3.17

p	q	$p \rightarrow q$	$\sim q \rightarrow \sim p$
T	T	T	T
T	F	F	F
F	T	T	T
F	F	T	T

$p \rightarrow q$ and $\sim q \rightarrow \sim p$ have the same truth values.

CHECK POINT 1 Construct a truth table for $\sim p \rightarrow \sim q$ to determine when the statement is true and when the statement is false.

The truth values for $p \rightarrow q$, as well as those for $\sim q \rightarrow \sim p$ from Example 1, are shown in **Table 3.17**. Notice that $p \rightarrow q$ and $\sim q \rightarrow \sim p$ have the same truth value in each of the four cases. What does this mean? **Every time you hear or utter a**

conditional statement, you can reverse and negate the antecedent and consequent, and the statement's truth value will not change. Here's an example from a student providing advice on campus fashion:

- If you're cool, you won't wear clothing with your school name on it.
- If you wear clothing with your school name on it, you're not cool.

If the fashion tip above is true then so is this, and if it's false then this is false as well.

We'll have lots more to say about this (that is, variations of conditional statements, not tips on dressing up and down around campus) in the next section.

EXAMPLE 2 Constructing a Truth Table

Construct a truth table for

$$[(p \vee q) \wedge \sim p] \rightarrow q$$

to determine when the statement is true and when the statement is false.

SOLUTION

The statement is a conditional statement because the *if–then* symbol, $\rightarrow$, is outside the grouping symbols. We cannot determine the truth values for this conditional until we first determine the truth values for the statements before and after the $\rightarrow$ connective.

$$\boxed{[(p \vee q) \wedge \sim p]} \rightarrow \boxed{q}$$

We'll need a column with truth values for this statement. Prior to this column, we'll need columns for $p \vee q$ and for $\sim p$.

We'll need a column with truth values for this statement. This will be the second column of the truth table.

The disjunction, $\vee$, is false only when both component statements are false.

The truth value of $\sim p$ is opposite that of p.

The conjunction, $\wedge$, is true only when both $p \vee q$ and $\sim p$ are true.

The conditional, $\rightarrow$, is false only when $(p \vee q) \wedge \sim p$ is true and q is false.

Show four possible cases.

p	q	$p \vee q$	$\sim p$	$(p \vee q) \wedge \sim p$	$[(p \vee q) \wedge \sim p] \rightarrow q$
T	T	T	F	F	T
T	F	T	F	F	T
F	T	T	T	T	T
F	F	F	T	F	T

The completed truth table shows that the conditional statement in the last column, $[(p \vee q) \wedge \sim p] \rightarrow q$, is true in all cases.

In Section 3.3, we defined a **tautology** as a compound statement that is always true. Example 2 proves that the conditional statement $[(p \vee q) \wedge \sim p] \rightarrow q$ is a tautology.

Conditional statements that are tautologies are called **implications**. For the conditional statement

$$[(p \vee q) \wedge \sim p] \rightarrow q$$

we can say that

$$(p \vee q) \wedge \sim p \textit{ implies } q.$$

Using p: I am visiting London and q: I am visiting Paris, we can say that

> I am visiting London or Paris, and I am not visiting London, implies that I am visiting Paris.

 **CHECK POINT 2** Construct a truth table for $[(p \rightarrow q) \wedge \sim q] \rightarrow \sim p$ and show that the compound statement is a tautology.

Blitzer Bonus

Yogi-isms

Baseball legend Yogi Berra, who died in 2015 at age 90, will forever be remembered for the accidental comedy of his statements. Here's a sampling of wisdom credited to Berra over the years. Many of these yogi-isms can be expressed as compound statements that are self-contradictions or tautologies.

- "Baseball is 90% mental and the other half is physical."
- "I usually take a two-hour nap from 1 to 4."
- "The future ain't what it used to be."
- "It gets late early out here."
- "You can observe a lot by watching."
- "When you come to a fork in the road, take it."
- "It's déjà vu all over again."
- "It ain't over till it's over."
- "We made too many wrong mistakes."
- "I really didn't say everything I said."

Some compound statements are false in all possible cases. Such statements are called **self-contradictions**. An example of a self-contradiction is the statement $p \wedge \sim p$:

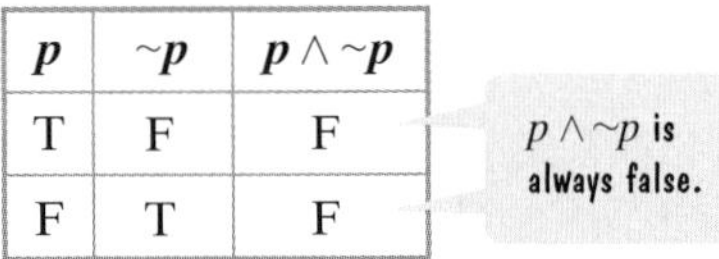

p	$\sim p$	$p \wedge \sim p$
T	F	F
F	T	F

If p represents "I am going," then $p \wedge \sim p$ translates as "I am going and I am not going." Such a translation sounds like a contradiction.

EXAMPLE 3 *Constructing a Truth Table with Eight Cases*

The following is from an editorial that appeared in *The New York Times*:

> *Our entire tax system depends upon the vast majority of taxpayers who attempt to pay the taxes they owe having confidence that they're being treated fairly and that their competitors and neighbors are also paying what is due. If the public concludes that the IRS cannot meet these basic expectations, the risk to the tax system will become very high and the effects very difficult to reverse.*
>
> —*The New York Times*, February 13, 2000

a. Construct a truth table for the underlined statement.

b. Suppose that the public concludes that the IRS cannot meet basic expectations, the risk to the tax system becomes very high, but the effects are not very difficult to reverse. Under these conditions, is the underlined statement true or false?

SOLUTION

a. We begin by assigning letters to the simple statements. Use the following representations:

p: The public concludes that the IRS *can* meet basic expectations (of fair treatment and others paying what is due).

q: The risk to the tax system will become very high.

r: The effects will be very difficult to reverse.

The underlined statement (If the public concludes that the IRS cannot meet these basic expectations, the risk to the tax system will become very high and the effects very difficult to reverse) in symbolic form is

$$\sim p \rightarrow (q \wedge r).$$

... cannot meet basic expectations | ... high risk | ... difficult to reverse effects

The statement $\sim p \rightarrow (q \wedge r)$ is a conditional statement because the *if–then* symbol, $\rightarrow$, is outside the parentheses. We cannot determine the truth values for this conditional until we have determined the truth values for $\sim p$ and for $q \wedge r$, the statements before and after the $\rightarrow$ connective. Because the compound statement consists of three simple statements, represented by p, q, and r, the truth table must contain eight cases. The completed truth table appears as follows:

The conjunction is true only when q, column 2, is true and r, column 3, is true.

Take the opposite of the truth values in column 1.

The conditional is false only when the $\sim p$ column is true and the $q \wedge r$ column is false.

Show eight possible cases.

These are the conditions in part (b).

p	q	r	$\sim p$	$q \wedge r$	$\sim p \rightarrow (q \wedge r)$
T	T	T	F	T	T
T	T	F	F	F	T
T	F	T	F	F	T
T	F	F	F	F	T
F	T	T	T	T	T
F	T	F	T	F	F
F	F	T	T	F	F
F	F	F	T	F	F

b. We are given that p (... can meet basic expectations) is false, q (... high risk) is true, and r (... difficult to reverse effects) is false. The given conditions, F T F, correspond to case 6 of the truth table, indicated by the voice balloon on the far left. Under these conditions, the original compound statement is false, shown by the red F in the truth table.

CHECK POINT 3 An advertisement makes the following claim:

If you use Hair Grow and apply it daily, then you will not go bald.

a. Construct a truth table for the claim.

b. Suppose you use Hair Grow, forget to apply it every day, and you go bald. Under these conditions, is the claim in the advertisement false?

3 *Understand the definition of the biconditional.*

Biconditional Statements

In Section 3.2, we introduced the biconditional connective, $\leftrightarrow$, translated as "if and only if." The biconditional statement $p \leftrightarrow q$ means that $p \rightarrow q$ and $q \rightarrow p$. We write this symbolically as

$$(p \rightarrow q) \wedge (q \rightarrow p).$$

To create the truth table for $p \leftrightarrow q$, we will first make a truth table for the conjunction of the two conditionals $p \rightarrow q$ and $q \rightarrow p$. The truth table for $(p \rightarrow q) \wedge (q \rightarrow p)$ is shown as follows:

The conditional is false only when p is true and q is false.

The conditional is false only when q is true and p is false.

The conjunction is true only when both $p \rightarrow q$ and $q \rightarrow p$ are true.

Show four possible cases.

p	q	$p \rightarrow q$	$q \rightarrow p$	$(p \rightarrow q) \wedge (q \rightarrow p)$
T	T	T	T	T
T	F	F	T	F
F	T	T	F	F
F	F	T	T	T

col. 1 → col. 2 | col. 2 → col. 1 | col. 3 ∧ col. 4

The truth values in the column for $(p \rightarrow q) \wedge (q \rightarrow p)$ show the truth values for the biconditional statement $p \leftrightarrow q$.

THE DEFINITION OF THE BICONDITIONAL

p	q	$p \leftrightarrow q$
T	T	T
T	F	F
F	T	F
F	F	T

A biconditional is true only when the component statements have the same truth value.

Before we continue our work with truth tables, let's take a moment to summarize the basic definitions of symbolic logic.

THE DEFINITIONS OF SYMBOLIC LOGIC

1. Negation ~: not
   The negation of a statement has the opposite meaning, as well as the opposite truth value, from the statement.
2. Conjunction ∧: and
   The only case in which a conjunction is true is when both component statements are true.
3. Disjunction ∨: or
   The only case in which a disjunction is false is when both component statements are false.
4. Conditional →: if–then
   The only case in which a conditional is false is when the first component statement, the antecedent, is true and the second component statement, the consequent, is false.
5. Biconditional ↔: if and only if
   A biconditional is true only when the component statements have the same truth value.

4 *Construct truth tables for biconditional statements.*

EXAMPLE 4 *Constructing a Truth Table*

Construct a truth table for

$$(p \vee q) \leftrightarrow (\sim q \rightarrow p)$$

to determine whether the statement is a tautology.

SOLUTION

The statement is a biconditional because the biconditional symbol, $\leftrightarrow$, is outside the parentheses. We cannot determine the truth values for this biconditional until we determine the truth values for the statements in parentheses.

$$\boxed{(p \vee q)} \leftrightarrow \boxed{(\sim q \rightarrow p)}$$

We need a column with truth values for this statement.

We need a column with truth values for this statement.

The completed truth table for $(p \vee q) \leftrightarrow (\sim q \rightarrow p)$ appears as follows:

p	q	$p \vee q$	$\sim q$	$\sim q \rightarrow p$	$(p \vee q) \leftrightarrow (\sim q \rightarrow p)$
T	T	T	F	T	T
T	F	T	T	T	T
F	T	T	F	T	T
F	F	F	T	F	T
		col. 1 $\vee$ col. 2	$\sim$ col. 2	col. 4 $\rightarrow$ col. 1	col. 3 $\leftrightarrow$ col. 5

We applied the definition of the biconditional to fill in the last column. In each case, the truth values of $p \vee q$ and $\sim q \rightarrow p$ are the same. Therefore, the biconditional $(p \vee q) \leftrightarrow (\sim q \rightarrow p)$ is true in each case. Because all cases are true, the biconditional is a tautology.

CHECK POINT 4 Construct a truth table for $(p \vee q) \leftrightarrow (\sim p \rightarrow q)$ to determine whether the statement is a tautology.

5 *Determine the truth value of a compound statement for a specific case.*

EXAMPLE 5 *Determining the Truth Value of a Compound Statement*

Your author recently received a letter from his credit card company that began as follows:

> Dear Mr. Bob Blitzer,
> I am pleased to inform you that a personal Super Million Dollar Prize Entry Number—665567010—has been assigned in your name as indicated above. <u>If your Super Million Dollar Prize Entry Number matches the winning preselected number and you return the number before the deadline stated below, you will win $1,000,000.00</u>. It's as simple as that.

Consider the claim in the underlined conditional statement: If your Super Million Dollar Prize Entry Number matches the winning preselected number and you return the number before the deadline stated below, you will win \$1,000,000.00. Suppose that your Super Million Dollar Prize Entry Number does not match the winning preselected number (those dirty rotten scoundrels!), you obediently return the number before the deadline, and you win only a free issue of a magazine (with the remaining 11 issues billed to your credit card). Under these conditions, can you sue the credit card company for making a false claim?

SOLUTION

Let's begin by assigning letters to the simple statements in the claim. We'll also indicate the truth value of each simple statement.

p: Your Super Million Dollar Prize Entry Number matches the winning preselected number. — **false**

q: You return the number before the stated deadline. — **true**

r: You win \$1,000,000.00. — **false; to make matters worse, you were duped into buying a magazine subscription.**

Now we can write the underlined claim in the letter to Bob in symbolic form.

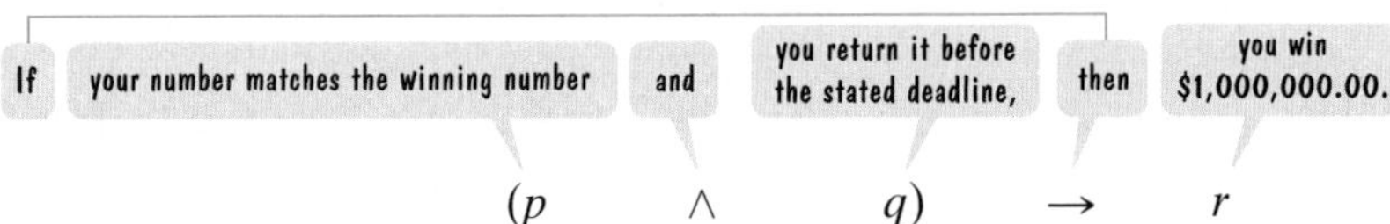

We substitute the truth values for p, q, and r to determine the truth value for the credit card company's claim.

$(p \wedge q) \rightarrow r$	This is the claim in symbolic form.
$(\text{F} \wedge \text{T}) \rightarrow \text{F}$	Substitute the truth values for the simple statements.
$\text{F} \rightarrow \text{F}$	Replace F ∧ T with F. Conjunction is false when one part is false.
T	Replace F → F with T. The conditional is true with a false antecedent and a false consequent.

Our truth-value analysis indicates that you cannot sue the credit card company for making a false claim. Call it conditional trickery, but the company's claim is true.

CHECK POINT 5 Consider the underlined claim in the letter in Example 5: If your Super Million Dollar Prize Entry Number matches the winning preselected number and you return the number before the deadline stated below, you will win \$1,000,000.00. Suppose that your number actually matches the winning preselected number, you do not return the number, and you win nothing. Under these conditions, determine the claim's truth value.

Blitzer Bonus

Conditional Wishful Thinking

ASSIGNED EXCLUSIVELY TO MR. BOB BLITZER

SUPER MILLION DOLLAR PRIZE ENTRY NUMBER 665567010

IF I WIN,

I PREFER MY $1,000,000 PRIZE:

☐ in installments of $33,333.34 annually until the entire $1,000,000.00 is paid out
☐ lump-sum settlement of $546,783.00

Bob's credit card company is too kind. It even offers options as to how he wants to receive his million-dollar winnings. With this lure, people who do not think carefully might interpret the conditional claim in the letter to read as follows:

> If you return your winning number and do so before the stated deadline, you win $1,000,000.00.

This misreading is wishful thinking. There is no winning number to return. What there is, of course, is a deceptive attempt to sell magazines.

Concept and Vocabulary Check

Fill in each blank so that the resulting statement is true.

1. A conditional statement, $p \rightarrow q$, is false only when ________.
2. A compound statement that is always true is called a/an ________. Conditional statements that are always true are called ________. A compound statement that is always false is called a/an ________.
3. A biconditional statement, $p \leftrightarrow q$, is true only when ________.

In Exercises 4–7, determine whether each statement is true or false. If the statement is false, make the necessary change(s) to produce a true statement.

4. A conditional statement is false only when the consequent is true and the antecedent is false. ______
5. Some implications are not tautologies. ______
6. An equivalent form for a conditional statement is obtained by reversing and negating the antecedent and consequent. ______
7. A compound statement consisting of two simple statements that are both false can be true. ______

Exercise Set 3.4

Practice Exercises

In Exercises 1–16, construct a truth table for the given statement.

1. $p \rightarrow \sim q$
2. $\sim p \rightarrow q$
3. $\sim(q \rightarrow p)$
4. $\sim(p \rightarrow q)$
5. $(p \wedge q) \rightarrow (p \vee q)$
6. $(p \vee q) \rightarrow (p \wedge q)$
7. $(p \rightarrow q) \wedge \sim q$
8. $(p \rightarrow q) \wedge \sim p$
9. $(p \vee q) \rightarrow r$
10. $p \rightarrow (q \vee r)$
11. $r \rightarrow (p \wedge q)$
12. $r \rightarrow (p \vee q)$
13. $\sim r \wedge (\sim q \rightarrow p)$
14. $\sim r \wedge (q \rightarrow \sim p)$
15. $\sim(p \wedge r) \rightarrow (\sim q \vee r)$
16. $\sim(p \vee r) \rightarrow (\sim q \wedge r)$

In Exercises 17–32, construct a truth table for the given statement.

17. $p \leftrightarrow \sim q$
18. $\sim p \leftrightarrow q$
19. $\sim(p \leftrightarrow q)$
20. $\sim(q \leftrightarrow p)$
21. $(p \leftrightarrow q) \rightarrow p$
22. $(p \leftrightarrow q) \rightarrow q$
23. $(\sim p \leftrightarrow q) \rightarrow (\sim p \rightarrow q)$
24. $(p \leftrightarrow \sim q) \rightarrow (q \rightarrow \sim p)$
25. $[(p \wedge q) \wedge (q \rightarrow p)] \leftrightarrow (p \wedge q)$
26. $[(p \rightarrow q) \vee (p \wedge \sim p)] \leftrightarrow (\sim q \rightarrow \sim p)$
27. $(p \leftrightarrow q) \rightarrow \sim r$
28. $(p \rightarrow q) \leftrightarrow \sim r$
29. $(p \wedge r) \leftrightarrow \sim(q \vee r)$
30. $(p \vee r) \leftrightarrow \sim(q \wedge r)$
31. $[r \vee (\sim q \wedge p)] \leftrightarrow \sim p$
32. $[r \wedge (q \vee \sim p)] \leftrightarrow \sim q$

In Exercises 33–56, use a truth table to determine whether each statement is a tautology, a self-contradiction, or neither.

33. $[(p \rightarrow q) \wedge q] \rightarrow p$
34. $[(p \rightarrow q) \wedge p] \rightarrow q$
35. $[(p \rightarrow q) \wedge \sim q] \rightarrow \sim p$
36. $[(p \rightarrow q) \wedge \sim p] \rightarrow \sim q$
37. $[(p \vee q) \wedge p] \rightarrow \sim q$

38. $[(p \vee q) \wedge \sim q] \rightarrow p$

39. $(p \rightarrow q) \rightarrow (\sim p \vee q)$

40. $(q \rightarrow p) \rightarrow (p \vee \sim q)$

41. $(p \wedge q) \wedge (\sim p \vee \sim q)$

42. $(p \vee q) \wedge (\sim p \wedge \sim q)$

43. $\sim(p \wedge q) \leftrightarrow (\sim p \wedge \sim q)$

44. $\sim(p \vee q) \leftrightarrow (\sim p \wedge \sim q)$

45. $(p \rightarrow q) \leftrightarrow (q \rightarrow p)$

46. $(p \rightarrow q) \leftrightarrow (\sim p \rightarrow \sim q)$

47. $(p \rightarrow q) \leftrightarrow (\sim p \vee q)$

48. $(p \rightarrow q) \leftrightarrow (p \vee \sim q)$

49. $(p \leftrightarrow q) \leftrightarrow [(q \rightarrow p) \wedge (p \rightarrow q)]$

50. $(q \leftrightarrow p) \leftrightarrow [(p \rightarrow q) \wedge (q \rightarrow p)]$

51. $(p \wedge q) \leftrightarrow (\sim p \vee r)$

52. $(p \wedge q) \rightarrow (\sim q \vee r)$

53. $[(p \rightarrow q) \wedge (q \rightarrow r)] \rightarrow (p \rightarrow r)$

54. $[(p \rightarrow q) \wedge (q \rightarrow r)] \rightarrow (\sim r \rightarrow \sim p)$

55. $[(q \rightarrow r) \wedge (r \rightarrow \sim p)] \leftrightarrow (q \wedge p)$

56. $[(q \rightarrow \sim r) \wedge (\sim r \rightarrow p)] \leftrightarrow (q \wedge \sim p)$

In Exercises 57–64,

a. *Write each statement in symbolic form. Assign letters to simple statements that are not negated.*

b. *Construct a truth table for the symbolic statement in part (a).*

c. *Use the truth table to indicate one set of conditions that makes the compound statement false, or state that no such conditions exist.*

57. If you do homework right after class then you will not fall behind, and if you do not do homework right after class then you will.

58. If you do a little bit each day then you'll get by, and if you do not do a little bit each day then you won't.

59. If you "cut-and-paste" from the Internet and do not cite the source, then you will be charged with plagiarism.

60. If you take more than one class with a lot of reading, then you will not have free time and you'll be in the library until 1 A.M.

61. You'll be comfortable in your room if and only if you're honest with your roommate, or you won't enjoy the college experience.

62. I fail the course if and only if I rely on a used book with highlightings by an idiot, or I do not buy a used book.

63. I enjoy the course if and only if I choose the class based on the professor and not the course description.

64. I do not miss class if and only if they take attendance or there are pop quizzes.

In Exercises 65–74, determine the truth value for each statement when p is false, q is true, and r is false.

65. $\sim(p \rightarrow q)$

66. $\sim(p \leftrightarrow q)$

67. $\sim p \leftrightarrow q$

68. $\sim p \rightarrow q$

69. $q \rightarrow (p \wedge r)$

70. $(p \wedge r) \rightarrow q$

71. $(\sim p \wedge q) \leftrightarrow \sim r$

72. $\sim p \leftrightarrow (\sim q \wedge r)$

73. $\sim[(p \rightarrow \sim r) \leftrightarrow (r \wedge \sim p)]$

74. $\sim[(\sim p \rightarrow r) \leftrightarrow (p \vee \sim q)]$

Practice Plus

In Exercises 75–78, use grouping symbols to clarify the meaning of each statement. Then construct a truth table for the statement.

75. $p \rightarrow q \leftrightarrow p \wedge q \rightarrow \sim p$

76. $q \rightarrow p \leftrightarrow p \vee q \rightarrow \sim p$

77. $p \rightarrow \sim q \vee r \leftrightarrow p \wedge r$

78. $\sim p \rightarrow q \wedge r \leftrightarrow p \vee r$

In Exercises 79–82, construct a truth table for each statement. Then use the table to indicate one set of conditions that makes the compound statement false, or state that no such conditions exist.

79. Loving a person is necessary for marrying that person, but not loving someone is sufficient for not marrying that person.

80. Studying hard is necessary for getting an A, but not studying hard is sufficient for not getting an A.

81. It is not true that being happy and living contentedly are necessary conditions for being wealthy.

82. It is not true that being wealthy is a sufficient condition for being happy and living contentedly.

Application Exercises

The bar graph shows the percentage of American adults who believed in God, Heaven, the devil, and Hell in 2013 compared with 2003.

Percentage of American Adults Believing in God, Heaven, the Devil, and Hell

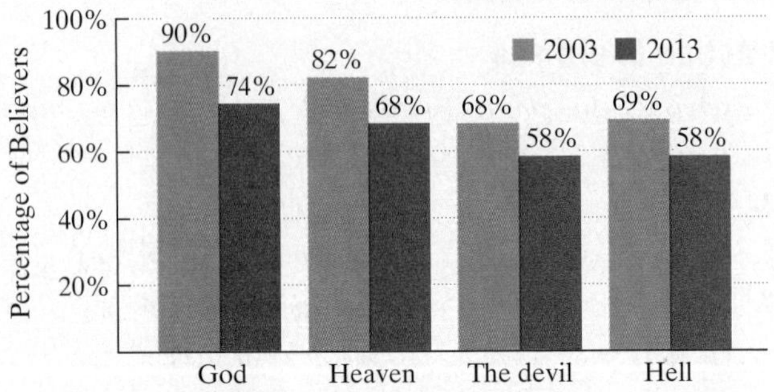

Source: Harris Interactive

In Exercises 83–86, write each statement in symbolic form. (Increases or decreases in each simple statement refer to 2013 compared with 2003.) Then use the information in the graph to determine the truth value of each compound statement.

83. If there was an increase in the percentage who believed in God and a decrease in the percentage who believed in Heaven, then there was an increase in the percentage who believed in the devil.

84. If there was a decrease in the percentage who believed in God, then it is not the case that there was an increase in the percentage who believed in the devil or in Hell.

85. There was a decrease in the percentage who believed in God if and only if there was an increase in the percentage who believed in Heaven, or the percentage believing in the devil decreased.

86. There was an increase in the percentage who believed in God if and only if there was an increase in the percentage who believed in Heaven, and the percentage believing in Hell decreased.

Sociologists Joseph Kahl and Dennis Gilbert developed a six-tier model to portray the class structure of the United States. The bar graph gives the percentage of Americans who are members of each of the six social classes.

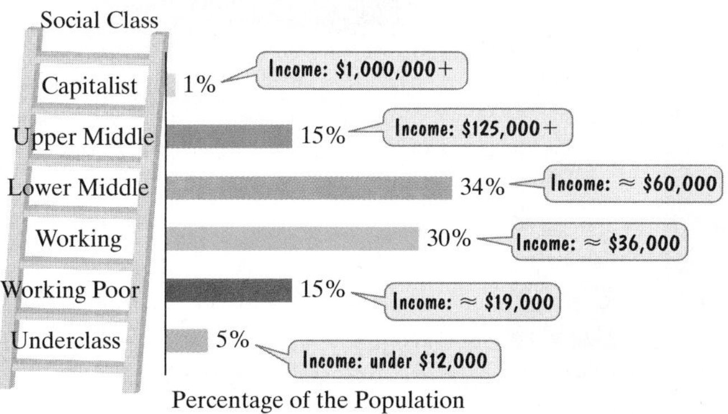

Source: James Henslin, *Sociology*, Twelfth Edition, Pearson, 2014.

In Exercises 87–90, write each statement in symbolic form. Then use the information given by the graph to determine the truth value of each compound statement.

87. Fifteen percent are capitalists or it is not true that 34% are members of the upper-middle class, if and only if the number of working poor exceeds the number belonging to the working class.

88. Fifteen percent are capitalists and it is not true that 34% are members of the upper-middle class, if and only if the number of working poor exceeds the number belonging to the working class.

89. If there are more people in the lower-middle class than in the capitalist and upper-middle classes combined, then 1% are capitalists and 34% belong to the upper-middle class.

90. If there are more people in the lower-middle class than in the capitalist and upper-middle classes combined, then 1% are capitalists or 34% belong to the upper-middle class.

Explaining the Concepts

91. Explain when conditional statements are true and when they are false.

92. Explain when biconditional statements are true and when they are false.

93. What is the difference between a tautology and a self-contradiction?

94. Based on the meaning of the inclusive *or*, explain why it is reasonable that if $p \vee q$ is true, then $\sim p \rightarrow q$ must also be true.

95. Based on the meaning of the inclusive *or*, explain why if $p \vee q$ is true, then $p \rightarrow \sim q$ is not necessarily true.

Critical Thinking Exercises

Make Sense? *In Exercises 96–99, determine whether each statement makes sense or does not make sense, and explain your reasoning.*

96. The statement "If 2 + 2 = 5, then the moon is made of green cheese" is true in logic, but does not make much sense in everyday speech.

97. I'm working with a true conditional statement, but when I reverse the antecedent and the consequent, my new conditional statement is no longer true.

98. When asked the question "What is time?", the fourth-century Christian philosopher St. Augustine replied,

> "If you don't ask me, I know, but if you ask me, I don't know."

I constructed a truth table for St. Augustine's statement and discovered it is a tautology.

99. In "Computing Machines and Intelligence," the English mathematician Alan Turing (1912–1954) wrote,

> "If each man had a definite set of rules of conduct by which he regulated his life, he would be a machine, but there are no such rules, so men cannot be machines."

I constructed a truth table for Turing's statement and discovered it is a tautology.

100. Consider the statement "If you get an A in the course, I'll take you out to eat." If you complete the course and I do take you out to eat, can you conclude that you got an A? Explain your answer.

In Exercises 101–102, the headings for the columns in the truth tables are missing. Fill in the statements to replace the missing headings. (More than one correct statement may be possible.)

101.

Do not repeat the statement from the third column.

p	q				
T	T	T	F	T	T
T	F	F	F	F	T
F	T	T	T	T	T
F	F	T	T	T	T

102.

Do not repeat the previous statement.

p	q				
T	T	T	T	F	F
T	F	T	F	T	T
F	T	T	F	T	T
F	F	F	F	T	T

3.5 Equivalent Statements and Variations of Conditional Statements

WHAT AM I SUPPOSED TO LEARN?

After studying this section, you should be able to:

1 Use a truth table to show that statements are equivalent.
2 Write the contrapositive for a conditional statement.
3 Write the converse and inverse of a conditional statement.

FIGURE 3.3 INDICATES THAT JACK NICHOLSON, Laurence Olivier, Paul Newman, and Spencer Tracy are the four male actors with the most Academy Award (Oscar) nominations.

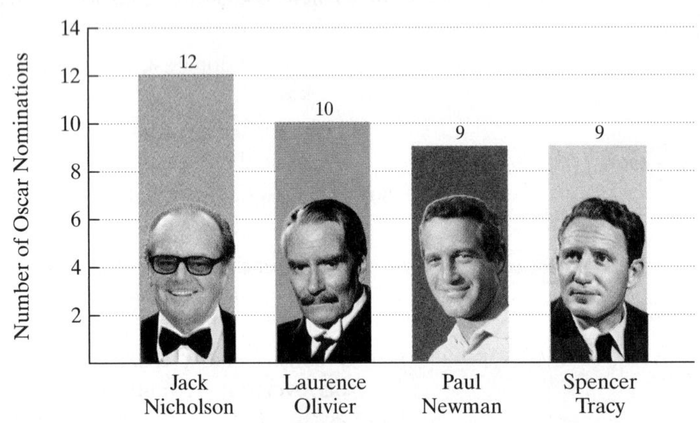

FIGURE 3.3
Source: Russell Ash, *The Top Ten of Everything*, 2013

The bar graph shows that Paul Newman received nine Oscar nominations, so the following statement is true:

If the actor is Paul Newman, he received nine Oscar nominations. — true

Now consider three variations of this conditional statement:

If he received nine Oscar nominations, the actor is Paul Newman. — not necessarily true: Spencer Tracy also received nine nominations.

If the actor is not Paul Newman, he did not receive nine Oscar nominations. — not necessarily true: Spencer Tracy isn't Paul Newman, but he received nine nominations.

If he did not receive nine Oscar nominations, the actor is not Paul Newman. — true: Jack Nicholson and Laurence Olivier did not receive nine nominations, and they are not Paul Newman.

In this section, we will use truth tables and logic to unravel this verbal morass of conditional statements.

1 *Use a truth table to show that statements are equivalent.*

Equivalent Statements

Equivalent compound statements are made up of the same simple statements and have the same corresponding truth values for all true–false combinations of these simple statements. If a compound statement is true, then its equivalent statement must also be true. Likewise, if a compound statement is false, its equivalent statement must also be false.

Truth tables are used to show that two statements are equivalent. When translated into English, equivalencies can be used to gain a better understanding of English statements.

EXAMPLE 1 Showing That Statements Are Equivalent

a. Show that $p \vee \sim q$ and $\sim p \rightarrow \sim q$ are equivalent.

b. Use the result from part (a) to write a statement that is equivalent to

The bill receives majority approval or the bill does not become law.

SOLUTION

a. Construct a truth table that shows the truth values for $p \vee \sim q$ and $\sim p \rightarrow \sim q$. The truth values for each statement are shown below.

p	q	$\sim q$	$p \vee \sim q$	$\sim p$	$\sim p \rightarrow \sim q$
T	T	F	T	F	T
T	F	T	T	F	T
F	T	F	F	T	F
F	F	T	T	T	T

Corresponding truth values are the same.

The table shows that the truth values for $p \vee \sim q$ and $\sim p \rightarrow \sim q$ are the same. Therefore, the statements are equivalent.

b. The statement

The bill receives majority approval or the bill does not become law

can be expressed in symbolic form using the following representations:

p: The bill receives majority approval.

q: The bill becomes law.

In symbolic form, the statement is $p \vee \sim q$. Based on the truth table in part (a), we know that an equivalent statement is $\sim p \rightarrow \sim q$. The equivalent statement can be expressed in words as

If the bill does not receive majority approval, then the bill does not become law.

Notice that the given statement and its equivalent are both true.

CHECK POINT 1

a. Show that $p \vee q$ and $\sim q \rightarrow p$ are equivalent.

b. Use the result from part (a) to write a statement that is equivalent to

I attend classes or I lose my scholarship.

A special symbol, $\equiv$, is used to show that two statements are equivalent. Because $p \vee \sim q$ and $\sim p \rightarrow \sim q$ are equivalent, we can write

$$p \vee \sim q \equiv \sim p \rightarrow \sim q \quad \text{or} \quad \sim p \rightarrow \sim q \equiv p \vee \sim q.$$

EXAMPLE 2 Showing That Statements Are Equivalent

Show that $\sim(\sim p) \equiv p$.

SOLUTION

p	$\sim p$	$\sim(\sim p)$
T	F	T
F	T	F

Corresponding truth values are the same.

Determine the truth values for $\sim(\sim p)$ and p. These are shown in the truth table at the left.

The truth values for $\sim(\sim p)$ were obtained by taking the opposite of each truth value for $\sim p$. The table shows that the truth values for $\sim(\sim p)$ and p are the same. Therefore, the statements are equivalent:

$$\sim(\sim p) \equiv p.$$

The equivalence in Example 2, $\sim(\sim p) \equiv p$, illustrates that **the double negation of a statement is equivalent to the statement**. For example, the statement "It is not true that Ernest Hemingway was not a writer" means the same thing as "Ernest Hemingway was a writer."

CHECK POINT 2 Show that $\sim[\sim(\sim p)] \equiv \sim p$.

GREAT QUESTION!

Can you give me a realistic example of when I might come across $\sim(\sim p) \equiv p$? And while we're on the topic, why does this equivalence look so familiar?

Here's an example from former U.S. President Barack Obama referring to a congressional debate over the debt ceiling:

We can't not pay bills that we've already incurred.

Equivalently,

We can (and must) pay bills that we've already incurred.

The reason that $\sim(\sim p) \equiv p$ looks familiar is because you encountered a similar statement in algebra:

$-(-a) = a$, where a represents a number.

For example, $-(-4) = 4$.

EXAMPLE 3 Equivalencies and Truth Tables

Select the statement that is not equivalent to

Miguel is blushing or sunburned.

a. If Miguel is blushing, then he is not sunburned.
b. Miguel is sunburned or blushing.
c. If Miguel is not blushing, then he is sunburned.
d. If Miguel is not sunburned, then he is blushing.

SOLUTION

To determine which of the choices is not equivalent to the given statement, begin by writing the given statement and the choices in symbolic form. Then construct a truth table and compare each statement's truth values to those of the given statement. The nonequivalent statement is the one that does not have exactly the same truth values as the given statement.

The simple statements that make up "Miguel is blushing or sunburned" can be represented as follows:

p: Miguel is blushing.
q: Miguel is sunburned.

Here are the symbolic representations for the given statement and the four choices:

Miguel is blushing or sunburned: $p \vee q$.

a. If Miguel is blushing, then he is not sunburned: $p \rightarrow \sim q$.
b. Miguel is sunburned or blushing: $q \vee p$.
c. If Miguel is not blushing, then he is sunburned: $\sim p \rightarrow q$.
d. If Miguel is not sunburned, then he is blushing: $\sim q \rightarrow p$.

Next, construct a truth table that contains the truth values for the given statement, $p \vee q$, as well as those for the four options. The truth table is shown as follows:

Equivalent (same corresponding truth values)

		Given		(a)	(b)		(c)	(d)
p	q	$p \vee q$	$\sim q$	$p \to \sim q$	$q \vee p$	$\sim p$	$\sim p \to q$	$\sim q \to p$
T	T	T	F	F	T	F	T	T
T	F	T	T	T	T	F	T	T
F	T	T	F	T	T	T	T	T
F	F	F	T	T	F	T	F	F

Not equivalent

The statement in option (a) does not have the same corresponding truth values as those for $p \vee q$. Therefore, this statement is not equivalent to the given statement.

In Example 3, we used a truth table to show that $p \vee q$ and $p \to \sim q$ are not equivalent. We can use our understanding of the inclusive *or* to see why the following English translations for these symbolic statements are not equivalent:

Miguel is blushing or sunburned.

If Miguel is blushing, then he is not sunburned.

Let us assume that the first statement is true. The inclusive *or* tells us that Miguel might be both blushing and sunburned. This means that the second statement might not be true. The fact that Miguel is blushing does not indicate that he is not sunburned; he might be both.

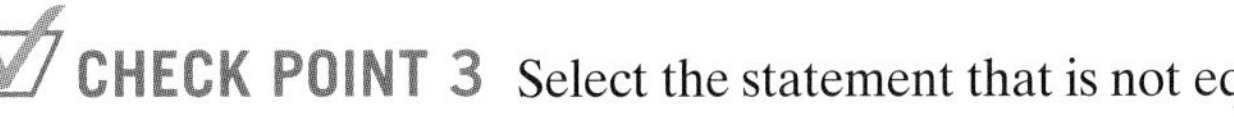

CHECK POINT 3 Select the statement that is not equivalent to

If it's raining, then I need a jacket.

a. It's not raining or I need a jacket.

b. I need a jacket or it's not raining.

c. If I need a jacket, then it's raining.

d. If I do not need a jacket, then it's not raining.

2 *Write the contrapositive for a conditional statement.*

Variations of the Conditional Statement $p \to q$

In Section 3.4, we learned that $p \to q$ is equivalent to $\sim q \to \sim p$. The truth value of a conditional statement does not change if the antecedent and the consequent are reversed and then both of them are negated. The **contrapositive** of a conditional statement is a statement obtained by reversing and negating the antecedent and the consequent.

p	q	$p \to q$	$\sim q \to \sim p$
T	T	T	T
T	F	F	F
F	T	T	T
F	F	T	T

$p \to q$ and $\sim q \to \sim p$ are equivalent.

A CONDITIONAL STATEMENT AND ITS EQUIVALENT CONTRAPOSITIVE

$$p \to q \equiv \sim q \to \sim p$$

The truth value of a conditional statement does not change if the antecedent and consequent are reversed and both are negated. The statement $\sim q \to \sim p$ is called the **contrapositive** of the conditional $p \to q$.

EXAMPLE 4 Writing Equivalent Contrapositives

Write the contrapositive for each of the following statements:

a. If you live in Los Angeles, then you live in California.

b. If the patient is not breathing, then the patient is dead.

c. If all people obey the law, then prisons are not needed.

d. $\sim(p \wedge q) \rightarrow r$

SOLUTION

In parts (a)–(c), we write each statement in symbolic form. Then we form the contrapositive by reversing and negating the antecedent and the consequent. Finally, we translate the symbolic form of the contrapositive back into English.

a. Use the following representations:

p: You live in Los Angeles.

q: You live in California.

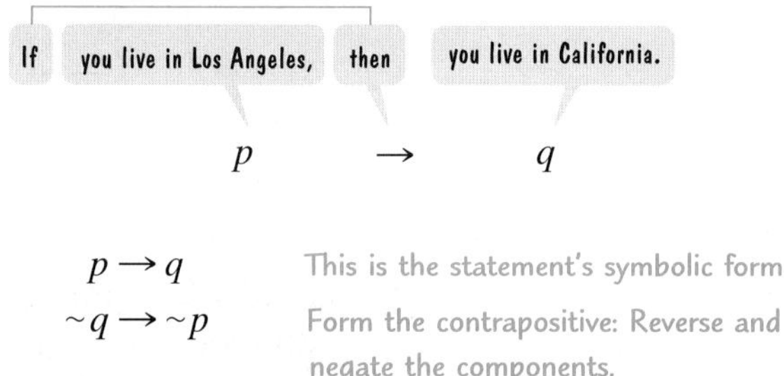

$p \rightarrow q$	This is the statement's symbolic form.
$\sim q \rightarrow \sim p$	Form the contrapositive: Reverse and negate the components.

Translating $\sim q \rightarrow \sim p$ into English, the contrapositive is

If you do not live in California, then you do not live in Los Angeles.

Notice that the given conditional statement and its contrapositive are both true.

b. Use the following representations:

p: The patient is breathing.

q: The patient is dead.

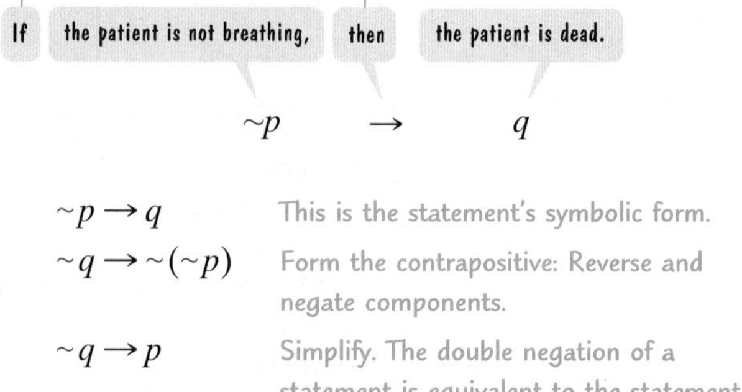

$\sim p \rightarrow q$	This is the statement's symbolic form.
$\sim q \rightarrow \sim(\sim p)$	Form the contrapositive: Reverse and negate components.
$\sim q \rightarrow p$	Simplify. The double negation of a statement is equivalent to the statement.

Translating $\sim q \rightarrow p$ into English, the contrapositive is

If the patient is not dead, then the patient is breathing.

c. Use the following representations:

p: All people obey the law.

q: Prisons are needed.

If all people obey the law, then prisons are not needed.

$$p \quad \rightarrow \quad {\sim}q$$

$p \rightarrow {\sim}q$ This is the statement's symbolic form.

${\sim}({\sim}q) \rightarrow {\sim}p$ Form the contrapositive: Reverse and negate components.

$q \rightarrow {\sim}p$ Simplify: ${\sim}({\sim}q) \equiv q$.

Negations of Quantified Statements

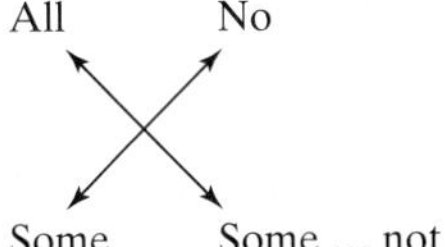

Recall, as shown in the margin, that the negation of *all* is *some* ... *not*. Using this negation and translating $q \rightarrow {\sim}p$ into English, the contrapositive is

If prisons are needed, then some people do not obey the law.

d. ${\sim}(p \wedge q) \rightarrow r$ This is the given symbolic statement.

${\sim}r \rightarrow {\sim}[{\sim}(p \wedge q)]$ Form the contrapositive: Reverse and negate the components.

${\sim}r \rightarrow (p \wedge q)$ Simplify: ${\sim}[{\sim}(p \wedge q)] \equiv p \wedge q$.

The contrapositive of ${\sim}(p \wedge q) \rightarrow r$ is ${\sim}r \rightarrow (p \wedge q)$. Using the dominance of connectives, the contrapositive can be expressed as ${\sim}r \rightarrow p \wedge q$.

CHECK POINT 4 Write the contrapositive for each of the following statements:

a. If you can read this, then you're driving too closely.

b. If you do not have clean underwear, it's time to do the laundry.

c. If all students are honest, then supervision during exams is not required.

d. ${\sim}(p \vee r) \rightarrow {\sim}q$

---

3 *Write the converse and inverse of a conditional statement.*

The truth value of a conditional statement does not change if the antecedent and the consequent are reversed and then both of them are negated. But what happens to the conditional's truth value if just one, but not both, of these changes is made? If the antecedent and the consequent are reversed but not negated, the resulting statement is called the **converse** of the conditional statement. By negating both the antecedent and the consequent but not reversing them, we obtain the **inverse** of the conditional statement.

VARIATIONS OF THE CONDITIONAL STATEMENT

Name	Symbolic Form	English Translation
Conditional	$p \rightarrow q$	If p, then q.
Converse	$q \rightarrow p$	If q, then p.
Inverse	${\sim}p \rightarrow {\sim}q$	If not p, then not q.
Contrapositive	${\sim}q \rightarrow {\sim}p$	If not q, then not p.

Let's see what happens to the truth value of a true conditional statement when we form its converse and its inverse.

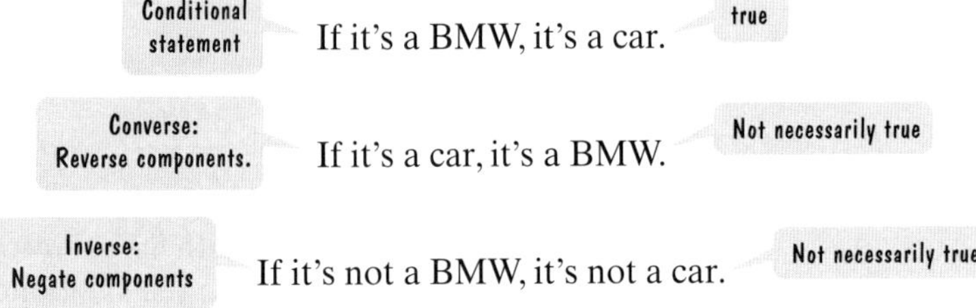

These statements illustrate that if a conditional statement is true, its converse and inverse are not necessarily true. Because the equivalent of a true statement must be true, we see that a conditional statement is not equivalent to its converse or its inverse.

The relationships among the truth values for a conditional statement, its converse, its inverse, and its contrapositive are shown in the truth table that follows:

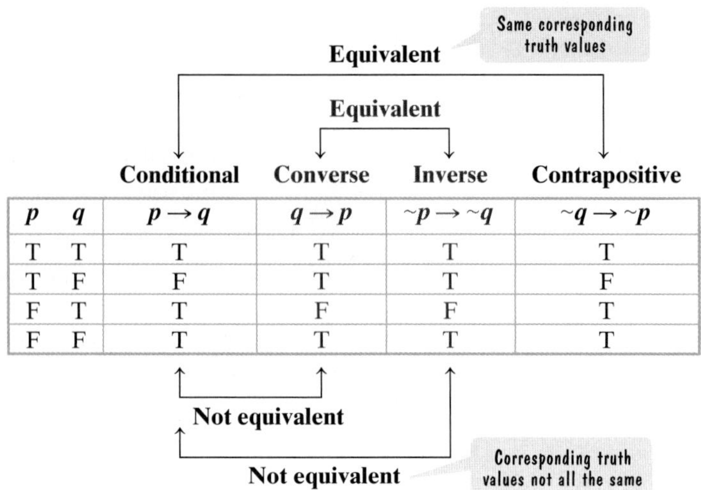

p	q	$p \to q$	$q \to p$	$\sim p \to \sim q$	$\sim q \to \sim p$
T	T	T	T	T	T
T	F	F	T	T	F
F	T	T	F	F	T
F	F	T	T	T	T

The above truth table confirms that a conditional statement is equivalent to its contrapositive. The table also shows that a conditional statement is not equivalent to its converse; in some cases they have the same truth value, but in other cases they have opposite truth values. Also, a conditional statement is not equivalent to its inverse. By contrast, the converse and the inverse are equivalent to each other.

EXAMPLE 5 *Writing Variations of a Conditional Statement*

The following conditional statement regarding state and national elections in the United States is true:

If you are 17, then you are not eligible to vote.

(In 1971, the voting age was lowered from 21 to 18 with the ratification of the 26th Amendment.) Write the statement's converse, inverse, and contrapositive.

SOLUTION

Use the following representations for "If you are 17, then you are not eligible to vote."

p: You are 17.

q: You are eligible to vote.

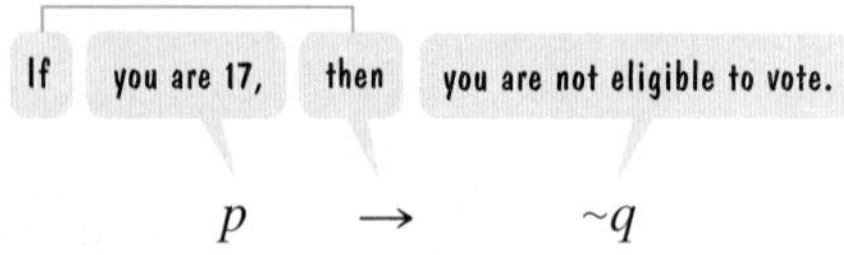

We now work with $p \to \sim q$ to form the converse, inverse, and contrapositive. We then translate the symbolic form of each statement back into English.

	Symbolic Statement	English Translation
Given Conditional Statement	$p \rightarrow \sim q$	If you are 17, then you are not eligible to vote. **true**
Converse: Reverse the components of $p \rightarrow \sim q$.	$\sim q \rightarrow p$	If you are not eligible to vote, then you are 17. **not necessarily true**
Inverse: Negate the components of $p \rightarrow \sim q$.	$\sim p \rightarrow \sim(\sim q)$ simplifies to $\sim p \rightarrow q$	If you are not 17, then you are eligible to vote. **not necessarily true**
Contrapositive: Reverse and negate the components of $p \rightarrow \sim q$.	$\sim(\sim q) \rightarrow \sim p$ simplifies to $q \rightarrow \sim p$	If you are eligible to vote, then you are not 17. **true**

CHECK POINT 5 Write the converse, inverse, and contrapositive of the following statement:

If you are in Iran, then you don't see a Club Med.

Blitzer Bonus

Converses in Alice's Adventures in Wonderland

Alice has a problem with logic: She believes that a conditional and its converse mean the same thing. In the passage on the right, she states that

If I say it, I mean it

is the same as

If I mean it, I say it.

She is corrected, told that

If I eat it, I see it

is not the same as

If I see it, I eat it.

"Come, we shall have some fun now," thought Alice. "I'm glad they've begun asking riddles—I believe I can guess that," she added aloud.

"Do you mean that you think you can find out the answer to it?" said the March Hare.

"Exactly so," said Alice.

"Then you should say what you mean," the March Hare went on.

"I do," Alice hastily replied; "at least—at least I mean what I say—that's the same thing, you know."

"Not the same thing a bit!" said the Hatter. "Why, you might just as well say that 'I see what I eat' is the same thing as 'I eat what I see'!"

"You might just as well say," added the March Hare, "that 'I like what I get' is the same thing as 'I get what I like'!"

"You might just as well say," added the Dormouse, which seemed to be talking in its sleep, "that 'I breathe when I sleep' is the same thing as 'I sleep when I breathe'!"

Concept and Vocabulary Check

Fill in each blank so that the resulting statement is true.

1. Compound statements that are made up of the same simple statements and have the same corresponding truth values for all true–false combinations of these simple statements are said to be __________, connected by the symbol _____.
2. The contrapositive of $p \rightarrow q$ is __________.
3. The converse of $p \rightarrow q$ is _______.
4. The inverse of $p \rightarrow q$ is __________.
5. A conditional statement is ___________ to its contrapositive, but not to its __________ or its ________.
6. True or False: The double negation of a statement is equivalent to the statement's negation. ______
7. True or False: If a conditional statement is true, its inverse must be false. ______

Exercise Set 3.5

Practice Exercises

1. a. Use a truth table to show that $\sim p \rightarrow q$ and $p \vee q$ are equivalent.
   b. Use the result from part (a) to write a statement that is equivalent to

      If the United States does not energetically support the development of solar-powered cars, then it will suffer increasing atmospheric pollution.

2. a. Use a truth table to show that $p \rightarrow q$ and $\sim p \vee q$ are equivalent.
   b. Use the result from part (a) to write a statement that is equivalent to

      If a number is even, then it is divisible by 2.

In Exercises 3–14, use a truth table to determine whether the two statements are equivalent.

3. $\sim p \rightarrow q, q \rightarrow \sim p$
4. $\sim p \rightarrow q, p \rightarrow \sim q$
5. $(p \rightarrow \sim q) \wedge (\sim q \rightarrow p), p \leftrightarrow \sim q$
6. $(\sim p \rightarrow q) \wedge (q \rightarrow \sim p), \sim p \leftrightarrow q$
7. $(p \wedge q) \wedge r, p \wedge (q \wedge r)$
8. $(p \vee q) \vee r, p \vee (q \vee r)$
9. $(p \wedge q) \vee r, p \wedge (q \vee r)$
10. $(p \vee q) \wedge r, p \vee (q \wedge r)$
11. $(p \vee r) \rightarrow \sim q, (\sim p \wedge \sim r) \rightarrow q$
12. $(p \wedge \sim r) \rightarrow q, (\sim p \vee r) \rightarrow \sim q$
13. $\sim p \rightarrow (q \vee \sim r), (r \wedge \sim q) \rightarrow p$
14. $\sim p \rightarrow (\sim q \wedge r), (\sim r \vee q) \rightarrow p$
15. Select the statement that is equivalent to

    I saw the original *King Kong* or the 2005 version.

    a. If I did not see the original *King Kong*, I saw the 2005 version.
    b. I saw both the original *King Kong* and the 2005 version.
    c. If I saw the original *King Kong*, I did not see the 2005 version.
    d. If I saw the 2005 version, I did not see the original *King Kong*.
16. Select the statement that is equivalent to

    Citizen Kane or *Howard the Duck* appears in a list of greatest U.S. movies.

    a. If *Citizen Kane* appears in the list of greatest U.S. movies, *Howard the Duck* does not.
    b. If *Howard the Duck* does not appear in the list of greatest U.S. movies, then *Citizen Kane* does.
    c. Both *Citizen Kane* and *Howard the Duck* appear in a list of greatest U.S. movies.
    d. If *Howard the Duck* appears in the list of greatest U.S. movies, *Citizen Kane* does not.
17. Select the statement that is *not* equivalent to

    It is not true that Sondheim and Picasso are both musicians.

    a. Sondheim is not a musician or Picasso is not a musician.
    b. If Sondheim is a musician, then Picasso is not a musician.
    c. Sondheim is not a musician and Picasso is not a musician.
    d. If Picasso is a musician, then Sondheim is not a musician.
18. Select the statement that is *not* equivalent to

    It is not true that England and Africa are both countries.

    a. If England is a country, then Africa is not a country.
    b. England is not a country and Africa is not a country.
    c. England is not a country or Africa is not a country.
    d. If Africa is a country, then England is not a country.

In Exercises 19–30, write the converse, inverse, and contrapositive of each statement.

19. If I am in Chicago, then I am in Illinois.
20. If I am in Birmingham, then I am in the South.
21. If the stereo is playing, then I cannot hear you.
22. If it is blue, then it is not an apple.
23. "If you don't laugh, you die." (humorist Alan King)
24. "If it doesn't fit, you must acquit." (lawyer Johnnie Cochran)
25. If the president is telling the truth, then all troops were withdrawn.
26. If the review session is successful, then no students fail the test.
27. If all institutions place profit above human need, then some people suffer.
28. If all hard workers are successful, then some people are not hard workers.
29. $\sim q \rightarrow \sim r$
30. $\sim p \rightarrow r$

Practice Plus

In Exercises 31–38, express each statement in "if... then" form. (More than one correct wording in "if... then" form may be possible.) Then write the statement's converse, inverse, and contrapositive.

31. All scientists know some math.
32. All senators are politicians.
33. There are no atheists in foxholes.
34. All people who are not fearful are crazy.

35. Passing the bar exam is a necessary condition for being an attorney.

36. Being a citizen is a necessary condition for voting.

37. Being a pacifist is sufficient for not being a warmonger.

38. Being a writer is sufficient for not being illiterate.

Application Exercises

The Corruption Perceptions Index uses perceptions of the general public, business people, and risk analysts to rate countries by how likely they are to accept bribes. The ratings are on a scale from 0 to 10, where higher scores represent less corruption. The graph shows the corruption ratings for the world's least corrupt and most corrupt countries. (The rating for the United States is 7.6.) Use the graph to solve Exercises 39–40.

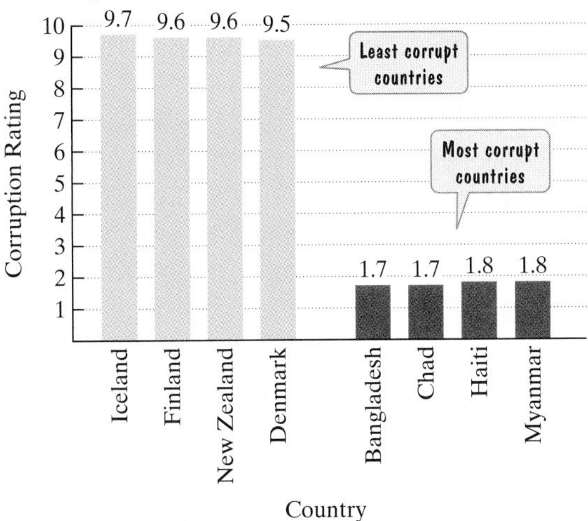

Source: Transparency International, *Corruption Perceptions Index*

39. a. Consider the statement

If the country is Finland, then the corruption rating is 9.6.

Use the information given by the graph to determine the truth value of this conditional statement.

b. Write the converse, inverse, and contrapositive of the statement in part (a). Then use the information given by the graph to determine whether each statement is true or not necessarily true.

40. a. Consider the statement

If the country is Haiti, then the corruption rating is 1.8.

Use the information given by the graph to determine the truth value of this conditional statement.

b. Write the converse, inverse, and contrapositive of the statement in part (a). Then use the information given by the graph to determine whether each statement is true or not necessarily true.

Explaining the Concepts

41. What are equivalent statements?

42. Describe how to determine if two statements are equivalent.

43. Describe how to obtain the contrapositive of a conditional statement.

44. Describe how to obtain the converse and the inverse of a conditional statement.

45. Give an example of a conditional statement that is true, but whose converse and inverse are not necessarily true. Try to make the statement somewhat different from the conditional statements that you have encountered throughout this section. Explain why the converse and the inverse that you wrote are not necessarily true.

46. Read the Blitzer Bonus on page 173. The Dormouse's last statement is the setup for a joke. The punchline, delivered by the Hatter to the Dormouse, is, "For you, it's the same thing." Explain the joke. What does this punchline have to do with the difference between a conditional and a biconditional statement?

Critical Thinking Exercises

Make Sense? *In Exercises 47–50, determine whether each statement makes sense or does not make sense, and explain your reasoning.*

47. A conditional statement can sometimes be true if its contrapositive is false.

48. A conditional statement can never be false if its converse is true.

49. The inverse of a statement's converse is the statement's contrapositive.

50. Groucho Marx stated, "I cannot say that I do not disagree with you," which is equivalent to asserting that I disagree with you.

Group Exercise

51. Can you think of an advertisement in which the person using a product is extremely attractive or famous? It is true that if you are this attractive or famous person, then you use the product. (Or at least pretend, for monetary gain, that you use the product!) In order to get you to buy the product, here is what the advertisers would *like* you to believe: If I use this product, then I will be just like this attractive or famous person. This, the converse, is not necessarily true and, for most of us, is unfortunately false. Each group member should find an example of this kind of deceptive advertising to share with the other group members.

3.6 Negations of Conditional Statements and De Morgan's Laws

WHAT AM I SUPPOSED TO LEARN?

After studying this section, you should be able to:

1 Write the negation of a conditional statement.
2 Use De Morgan's laws.

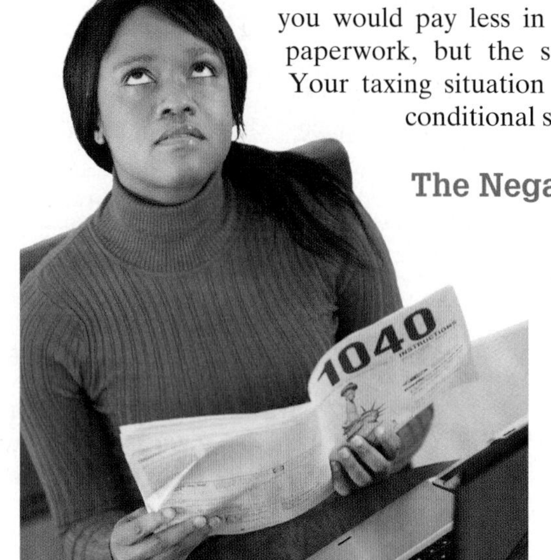

IT WAS SUGGESTED THAT BY ITEMIZING DEDUCTIONS, you would pay less in taxes. Not only did you drown in paperwork, but the suggestion turned out to be false. Your taxing situation can be summarized by negating a conditional statement.

1 *Write the negation of a conditional statement.*

The Negation of the Conditional Statement $p \rightarrow q$

Suppose that your accountant makes the following statement:

If you itemize deductions, then you pay less in taxes.

When will your accountant have told you a lie? The only case in which you have been lied to is when you itemize deductions and you do *not* pay less in taxes. We can analyze this situation symbolically with the following representations:

p: You itemize deductions.

q: You pay less in taxes.

We represent each compound statement in symbolic form.

$p \rightarrow q$: If you itemize deductions, then you pay less in taxes.

$p \wedge \sim q$: You itemize deductions and you do not pay less in taxes.

The truth table that follows shows that the negation of $p \rightarrow q$ is $p \wedge \sim q$.

p	q	$p \rightarrow q$	$\sim q$	$p \wedge \sim q$
T	T	T	F	F
T	F	F	T	T
F	T	T	F	F
F	F	T	T	F

These columns have opposite truth values, so $p \wedge \sim q$ negates $p \rightarrow q$.

THE NEGATION OF A CONDITIONAL STATEMENT

The negation of $p \rightarrow q$ is $p \wedge \sim q$. This can be expressed as

$$\sim(p \rightarrow q) \equiv p \wedge \sim q.$$

To form the negation of a conditional statement, leave the antecedent (the first part) unchanged, change the *if–then* connective to *and*, and negate the consequent (the second part).

EXAMPLE 1 Writing the Negation of a Conditional Statement

Write the negation of

If too much homework is given, a class should not be taken.

SOLUTION

Use the following representations:

p: Too much homework is given.

q: A class should be taken.

The symbolic form of the conditional statement is $p \rightarrow \sim q$.

$p \rightarrow \sim q$	This is the given statement in symbolic form.
$p \wedge \sim(\sim q)$	Form the negation: Copy the antecedent, change $\rightarrow$ to $\wedge$, and negate the consequent.
$p \wedge q$	Simplify: $\sim(\sim q) \equiv q$.

Translating $p \wedge q$ into English, the negation of the given statement is

Too much homework is given and the class should be taken.

CHECK POINT 1 Write the negation of

If you do not have a fever, you do not have the flu.

GREAT QUESTION!

What's the difference between the inverse of $p \rightarrow q$ and the negation of $p \rightarrow q$?

They're easy to confuse. You obtain the inverse, $\sim p \rightarrow \sim q$, which is an *if–then* statement, by negating the two component statements. However, this process does not make the inverse the negation. The negation of $p \rightarrow q$ is $p \wedge \sim q$, which is an *and* statement.

The box that follows summarizes what we have learned about conditional statements:

THE CONDITIONAL STATEMENT $p \rightarrow q$

Contrapositive

$p \rightarrow q$ is equivalent to $\sim q \rightarrow \sim p$ (the contrapositive).

Converse and Inverse

1. $p \rightarrow q$ is not equivalent to $q \rightarrow p$ (the converse).
2. $p \rightarrow q$ is not equivalent to $\sim p \rightarrow \sim q$ (the inverse).

Negation

The negation of $p \rightarrow q$ is $p \wedge \sim q$.

2 *Use De Morgan's laws.*

De Morgan's Laws

De Morgan's laws, named after the English mathematician Augustus De Morgan (1806–1871), were introduced in Chapter 2, where they applied to sets:

$$(A \cap B)' = A' \cup B'$$
$$(A \cup B)' = A' \cap B'.$$

Similar relationships apply to the statements of symbolic logic:

$$\sim(p \wedge q) \equiv \sim p \vee \sim q$$
$$\sim(p \vee q) \equiv \sim p \wedge \sim q.$$

Here is a truth table that serves as a deductive proof for $\sim(p \wedge q) \equiv \sim p \vee \sim q$, the first of De Morgan's two equivalences:

p	q	$p \wedge q$	$\sim(p \wedge q)$	$\sim p$	$\sim q$	$\sim p \vee \sim q$
T	T	T	F	F	F	F
T	F	F	T	F	T	T
F	T	F	T	T	F	T
F	F	F	T	T	T	T

Corresponding truth values are the same, proving that $\sim(p \wedge q) \equiv \sim p \vee \sim q$.

We can prove that $\sim(p \vee q) \equiv \sim p \wedge \sim q$ in a similar manner. Do this now by constructing a truth table and showing that $\sim(p \vee q)$ and $\sim p \wedge \sim q$ have the same corresponding truth values.

DE MORGAN'S LAWS

1. $\sim(p \wedge q) \equiv \sim p \vee \sim q$
2. $\sim(p \vee q) \equiv \sim p \wedge \sim q$

EXAMPLE 2 *Using a De Morgan Law*

Write a statement that is equivalent to

It is not true that Atlanta and California are cities.

SOLUTION

Let p and q represent the following statements:

p: Atlanta is a city.

q: California is a city.

The statement is of the form $\sim(p \wedge q)$. An equivalent statement is $\sim p \vee \sim q$. We can translate this as

Atlanta is not a city or California is not a city.

CHECK POINT 2 Write a statement that is equivalent to

It is not true that Bart Simpson and Tony Soprano are cartoon characters.

EXAMPLE 3 *Using a De Morgan Law*

"The Monsters Are Due on Maple Street" was crowned the all-time best episode of the television series *The Twilight Zone* by *Time* magazine. The episode hinges on an alien invasion that employs human beings' own prejudices and suspicion to turn an otherwise ordinary neighborhood against itself and do the invaders' job for them. The underlined portion of the following passage is an equivalent paraphrase of writer Rod Serling's wrap-up to the episode:

"There are weapons that are simply thoughts. For the record, it is not the case that prejudices cannot kill or suspicion cannot destroy."

Write a statement that is equivalent to the underlined passage.

SOLUTION

Let p and q represent the following statements:

p: Prejudices can kill.

q: Suspicion can destroy.

It is not the case that | prejudices cannot kill | or | suspicion cannot destroy.

$\sim(\sim p \quad \vee \quad \sim q)$

$\sim(\sim p \vee \sim q)$ This is the underlined passage in symbolic form.

$\sim(\sim p) \wedge \sim(\sim q)$ Use a De Morgan law to write an equivalent statement. Negate each component statement and change *or* to *and*.

$p \wedge q$ Simplify: $\sim(\sim p) \equiv p$ and $\sim(\sim q) \equiv q$.

Translating $p \wedge q$ into English, a statement that is equivalent to the underlined passage is

Prejudices can kill and suspicion can destroy.

 **CHECK POINT 3** Write a statement that is equivalent to

It is not true that you leave by 5 P.M. or you do not arrive home on time.

De Morgan's laws can be used to write the negation of a compound statement that is a conjunction ($\wedge$, *and*) or a disjunction ($\vee$, *or*).

GREAT QUESTION!

The box shows the procedures for negating conjunctions and disjunctions. What about negating conditionals?

Remember that we also have a rule for negating conditionals: $\sim(p \rightarrow q) \equiv p \wedge \sim q$.

DE MORGAN'S LAWS AND NEGATIONS

1. $\sim(p \wedge q) \equiv \sim p \vee \sim q$
   The negation of $p \wedge q$ is $\sim p \vee \sim q$. To negate a conjunction, negate each component statement and change *and* to *or*.
2. $\sim(p \vee q) \equiv \sim p \wedge \sim q$
   The negation of $p \vee q$ is $\sim p \wedge \sim q$. To negate a disjunction, negate each component statement and change *or* to *and*.

We can apply these rules to English conjunctions or disjunctions and immediately obtain their negations without having to introduce symbolic representations.

EXAMPLE 4 *Negating Conjunctions and Disjunctions*

Write the negation for each of the following statements:

a. All students do laundry on weekends and I do not.

b. Some college professors are entertaining lecturers or I'm bored.

Negations of Quantified Statements

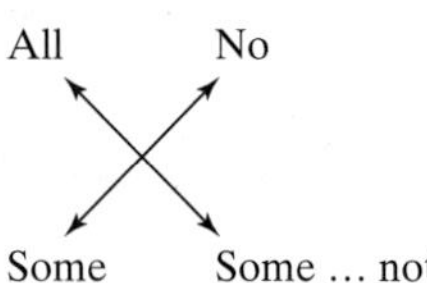

SOLUTION

To negate some of the simple statements, we use the negations of quantified statements shown in the margin.

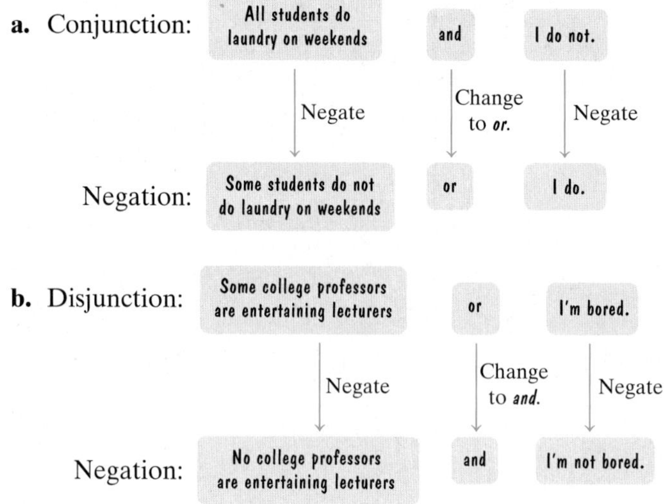

CHECK POINT 4 Write the negation for each of the following statements:

a. All horror movies are scary and some are funny.

b. Your workouts are strenuous or you do not get stronger.

EXAMPLE 5 *Using a De Morgan Law to Formulate a Contrapositive*

Write a statement that is equivalent to

If it rains, I do not go outdoors and I study.

SOLUTION

We begin by writing the conditional statement in symbolic form. Let p, q, and r represent the following simple statements:

p: It rains.
q: I go outdoors.
r: I study.

Using these representations, the given conditional statement can be expressed in symbolic form as

$$p \rightarrow (\sim q \wedge r).$$ If it rains, I do not go outdoors and I study.

An equivalent statement is the contrapositive.

$$\sim(\sim q \wedge r) \rightarrow \sim p$$ Form the contrapositive: Reverse and negate the components.

$$[\sim(\sim q) \vee \sim r] \rightarrow \sim p$$ Use a De Morgan law to negate the conjunction: Negate each component and change $\wedge$ to $\vee$.

$$(q \vee \sim r) \rightarrow \sim p$$ Simplify: $\sim(\sim q) \equiv q$.

Thus,

$$p \rightarrow (\sim q \wedge r) \equiv (q \vee \sim r) \rightarrow \sim p.$$

Using the representations for p, q, and r, a statement that is equivalent to "If it rains, I do not go outdoors and I study" is

If I go outdoors or I do not study, it is not raining.

CHECK POINT 5 Write a statement that is equivalent to:
If it is not windy, we can swim and we cannot sail.

Blitzer Bonus

A Gödelian Universe

At the age of 10, Czech mathematician Kurt Gödel (1906–1978) was studying mathematics, religion, and several languages. By age 25 he had produced what many mathematicians consider the most important result of twentieth-century mathematics: Gödel proved that all deductive systems eventually give rise to statements that cannot be proved to be either true or false within that system. Take someone who says "I am lying." If he is then he isn't, and if he isn't then he is. There is no way to determine whether the statement is true or false. There are similar undecidable statements in every branch of mathematics, from number theory to algebra.

Gödel's Theorem suggests infinitely many layers, none of which are capable of capturing all truth in one logical system. Gödel showed that statements arise in a system that cannot be proved or disproved within that system. To prove them, one must ascend to a "richer" system in which the previous undecidable statement can now be proved, but this richer system will in turn lead to new statements that cannot be proved, and so on. The process goes on forever.

Is the universe Gödelian in the sense there is no end to the discovery of its mathematical laws and in which the ultimate reality is always out of reach? The situation is echoed in the painting *The Two Mysteries* by René Magritte. A small picture of a pipe is shown with a caption that asserts (to translate from the French) "This is not a pipe." Above the fake pipe is a presumably genuine larger pipe, but it too is painted on the canvas. In Magritte's Gödelian universe, reality is infinitely layered, and it is impossible to say what reality really is.

Concept and Vocabulary Check

Fill in each blank so that the resulting statement is true.

1. The negation of $p \rightarrow q$ is ________. To form the negation of a conditional statement, leave the ___________ unchanged, change the *if–then* connective to _____, and negate the ___________.
2. De Morgan's laws state that $\sim(p \wedge q) \equiv$ __________ and $\sim(p \vee q) \equiv$ __________.
3. To negate a conjunction, negate each of the component statements and change *and* to ____.
4. To negate a disjunction, negate each of the component statements and change *or* to _____.
5. True or False: The negation for $p \rightarrow q$ is $\sim p \rightarrow \sim q$. ______

Exercise Set 3.6

Practice Exercises

In Exercises 1–10, write the negation of each conditional statement.

1. If I am in Los Angeles, then I am in California.
2. If I am in Houston, then I am in Texas.
3. If it is purple, then it is not a carrot.
4. If the TV is playing, then I cannot concentrate.
5. If he doesn't, I will.
6. If she says yes, he says no.
7. If there is a blizzard, then all schools are closed.
8. If there is a tax cut, then all people have extra spending money.
9. $\sim q \rightarrow \sim r$
10. $\sim p \rightarrow r$

In Exercises 11–26, use De Morgan's laws to write a statement that is equivalent to the given statement.

11. It is not true that Australia and China are both islands.
12. It is not true that Florida and California are both peninsulas.
13. It is not the case that my high school encouraged creativity and diversity.
14. It is not the case that the course covers logic and dream analysis.
15. It is not the case that Jewish scripture gives a clear indication of a heaven or an afterlife.
16. It is not true that Martin Luther King, Jr. supported violent protest or the Vietnam War.
17. It is not the case that the United States has eradicated poverty or racism.
18. It is not the case that the movie is interesting or entertaining.
19. $\sim(\sim p \wedge q)$
20. $\sim(p \vee \sim q)$
21. If you attend lecture and study, you succeed.
22. If you suffer from synesthesia, you can literally taste music and smell colors.
23. If he does not cook, his wife or child does.
24. If it is Saturday or Sunday, I do not work.
25. $p \rightarrow (q \vee \sim r)$
26. $p \rightarrow (\sim q \wedge \sim r)$

In Exercises 27–38, write the negation of each statement.

27. I'm going to Seattle or San Francisco.
28. This course covers logic or statistics.
29. I study or I do not pass.
30. I give up tobacco or I am not healthy.
31. I am not going and he is going.
32. I do not apply myself and I succeed.
33. A bill becomes law and it does not receive majority approval.
34. They see the show and they do not have tickets.
35. $p \vee \sim q$
36. $\sim p \vee q$
37. $p \wedge (q \vee r)$
38. $p \vee (q \wedge r)$

In Exercises 39–46, determine which, if any, of the three given statements are equivalent. You may use information about a conditional statement's converse, inverse, or contrapositive, De Morgan's laws, or truth tables.

39. **a.** If he is guilty, then he does not take a lie-detector test.
    b. He is not guilty or he takes a lie-detector test.
    c. If he is not guilty, then he takes a lie-detector test.
40. **a.** If the train is late, then I am not in class on time.
    b. The train is late or I am in class on time.
    c. If I am in class on time, then the train is not late.
41. **a.** It is not true that I have a ticket and cannot go.
    b. I do not have a ticket and can go.
    c. I have a ticket or I cannot go.
42. **a.** I work hard or I do not succeed.
    b. It is not true that I do not work hard and succeed.
    c. I do not work hard and I do succeed.
43. **a.** If the grass turns yellow, you did not use fertilizer or water.
    b. If you use fertilizer and water, the grass will not turn yellow.
    c. If the grass does not turn yellow, you used fertilizer and water.
44. **a.** If you do not file or provide fraudulent information, you will be prosecuted.
    b. If you file and do not provide fraudulent information, you will not be prosecuted.
    c. If you are not prosecuted, you filed or did not provide fraudulent information.
45. **a.** I'm leaving, and Tom is relieved or Sue is relieved.
    b. I'm leaving, and it is false that Tom and Sue are not relieved.
    c. If I'm leaving, then Tom is relieved or Sue is relieved.
46. **a.** You play at least three instruments, and if you have a master's degree in music then you are eligible.
    b. You are eligible, if and only if you have a master's degree in music and play at least three instruments.
    c. You play at least three instruments, and if you are not eligible then you do not have a master's degree in music.

Practice Plus

In Exercises 47–50, express each statement in "if . . . then" form. (More than one correct wording in "if . . . then" form is possible.) Then write the statement's converse, inverse, contrapositive, and negation.

47. No pain is sufficient for no gain.
48. Not observing the speed limit is necessary for getting a speeding ticket.

49. Being neither hedonistic nor ascetic is necessary for following Buddha's "Middle Way."

50. Going into heat and not finding a mate are sufficient for a female ferret's death.

In Exercises 51–54, write the negation of each statement. Express each negation in a form such that the symbol ~ negates only simple statements.

51. $p \rightarrow (r \wedge \sim s)$

52. $p \rightarrow (\sim r \vee s)$

53. $p \wedge (r \rightarrow \sim s)$

54. $p \vee (\sim r \rightarrow s)$

Application Exercises

The bar graph shows ten leading causes of death in the United States, along with the average number of days of life lost for each hazard.

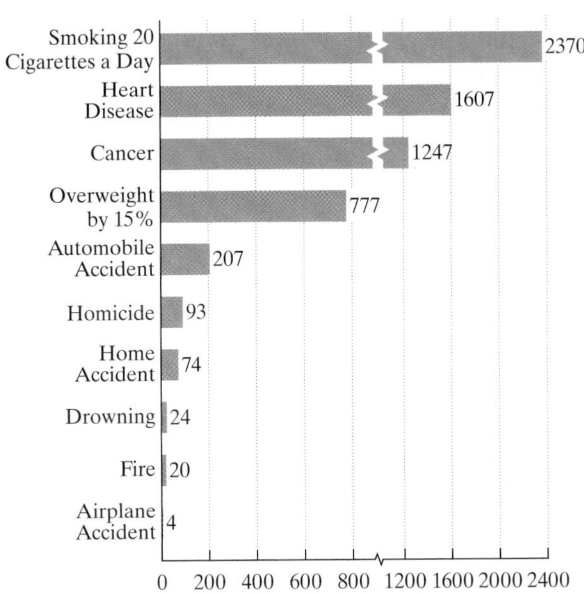

Source: Withgott and Brennan, *Essential Environment*, Third Edition, Pearson, 2009.

In Exercises 55–60,

a. *Use the information given by the graph to determine the truth value of the compound statement.*

b. *Write the compound statement's negation.*

c. *Use the information given by the graph to determine the truth value of the negation in part (b).*

55. Smoking reduces life expectancy by 2370 days, and heart disease reduces life expectancy by 1247 days.

56. Cancer reduces life expectancy by 1607 days, and being overweight reduces life expectancy by 777 days.

57. Homicide reduces life expectancy by 74 days or fire does not reduce life expectancy by 25 days.

58. Automobile accidents reduce life expectancy by 500 days or drowning does not reduce life expectancy by 30 days.

59. If drowning reduces life expectancy by 10 times the number of days as airplane accidents, then drowning does not reduce life expectancy by 24 days.

60. If fire reduces life expectancy by 10 times the number of days as airplane accidents, then fire does not reduce life expectancy by 20 days.

Explaining the Concepts

61. Explain how to write the negation of a conditional statement.

62. Explain how to write the negation of a conjunction.

63. Give an example of a disjunction that is true, even though one of its component statements is false. Then write the negation of the disjunction and explain why the negation is false.

Critical Thinking Exercises

Make Sense? *In Exercises 64–67, determine whether each statement makes sense or does not make sense, and explain your reasoning.*

64. Too much time was spent explaining how to negate $p \rightarrow q$, $p \wedge q$, and $p \vee q$, when all I have to do is to negate p and negate q.

65. If I know that a conditional statement is false, I can obtain a true statement by taking the conjunction of its antecedent and negated consequent.

66. I took the contrapositive of $\sim q \rightarrow (p \wedge r)$ and obtained $\sim(p \vee r) \rightarrow q$.

67. The Chinese Taoist philosopher Lao Tzu (*The Way of Life*) wrote, "If one man leads, another must follow. How silly that is and how false!" Based on my understanding of the conditional and its negation, I concluded that Lao Tzu is saying that one man leads and another need not follow.

68. Write the negation for the following conjunction:

We will neither replace nor repair the roof, and we will sell the house.

69. Write the contrapositive and the negation for the following statement:

Some people eating turkey is necessary for it to be Thanksgiving.

3.7 Arguments and Truth Tables

WHAT AM I SUPPOSED TO LEARN?

After studying this section, you should be able to:

1 Use truth tables to determine validity.

2 Recognize and use forms of valid and invalid arguments.

A NOTED CRIMINAL CASE IN 1995 involved Lyle and Erik Menendez, who shot and killed their parents. Although everyone agreed that the brothers committed the crime, it took two trials before they were convicted. The *arguments* in the trial centered around the boys' motivation: Was the killing a premeditated act by two children hoping to receive an inheritance, or was it an act motivated by years of abuse and a desperate sense of helplessness and rage?

1 *Use truth tables to determine validity.*

An **argument** consists of two parts: the given statements, called the **premises**, and a **conclusion**. Here's the prosecutor's argument from the Menendez brothers' criminal case:

Premise 1: If children murder their parents in cold blood, they deserve to be punished to the full extent of the law.

Premise 2: These children murdered their parents in cold blood.

Conclusion: Therefore, these children deserve to be punished to the full extent of the law.

(*Source:* Sherry Diestler, *Becoming a Critical Thinker*, Fourth Edition, Prentice Hall, 2005.)

It appears that if the premises are true, then the jurors must decide to punish the brothers to the full extent of the law. The true premises force the conclusion to be true, making this an example of a *valid argument*.

DEFINITION OF A VALID ARGUMENT

An argument is **valid** if the conclusion is true whenever the premises are assumed to be true. An argument that is not valid is said to be an **invalid argument**, also called a **fallacy**.

Truth tables can be used to test validity. We begin by writing the argument in symbolic form. Let's do this for the prosecutor's argument in the Menendez case. Represent each simple statement with a letter:

p: Children murder their parents in cold blood.

q: They deserve to be punished to the full extent of the law.

Now we write the two premises and the conclusion in symbolic form:

Premise 1:	$p \rightarrow q$	If children murder their parents in cold blood, they deserve to be punished to the full extent of the law.
Premise 2:	p	These children murdered their parents in cold blood.
Conclusion:	$\therefore q$	Therefore, these children need to be punished to the full extent of the law.

(The three-dot triangle, $\therefore$, is read "therefore.")

To decide whether this argument is valid, we rewrite it as a conditional statement that has the following form:

$$[(p \rightarrow q) \wedge p] \rightarrow q.$$

If premise 1 and premise 2, then conclusion.

At this point, we can determine whether the conjunction of the premises implies that the conclusion is true for all possible truth values for p and q. We construct a truth table for the statement

$$[(p \rightarrow q) \wedge p] \rightarrow q.$$

If the final column in the truth table for $[(p \rightarrow q) \wedge p] \rightarrow q$ is true in every case, then the statement is a tautology and the argument is valid. If the conditional statement in the last column is false in at least one case, then the statement is not a tautology, and the argument is invalid. The truth table is shown below.

p	q	$p \rightarrow q$	$(p \rightarrow q) \wedge p$	$[(p \rightarrow q) \wedge p] \rightarrow q$
T	T	T	T	T
T	F	F	F	T
F	T	T	F	T
F	F	T	F	T

The final column in the table is true in every case. The conditional statement is a tautology. This means that the premises imply the conclusion. The conclusion necessarily follows from the premises. Therefore, the argument is valid.

The form of the prosecutor's argument in the Menendez case

$$\begin{array}{l} p \rightarrow q \\ \underline{p \quad} \\ \therefore q \end{array}$$

is called **direct reasoning**. All arguments that have the direct reasoning form are valid regardless of the English statements that p and q represent.

Here's a step-by-step procedure to test the validity of an argument using truth tables:

TESTING THE VALIDITY OF AN ARGUMENT WITH A TRUTH TABLE

1. Use a letter to represent each simple statement in the argument.
2. Express the premises and the conclusion symbolically.
3. Write a symbolic conditional statement of the form

   $$[(\text{premise 1}) \wedge (\text{premise 2}) \wedge \cdots \wedge (\text{premise } n)] \rightarrow \text{conclusion},$$

   where n is the number of premises.
4. Construct a truth table for the conditional statement in step 3.
5. If the final column of the truth table has all trues, the conditional statement is a tautology and the argument is valid. If the final column does not have all trues, the conditional statement is not a tautology and the argument is invalid.

EXAMPLE 1 Did the Pickiest Logician in the Galaxy Foul Up?

In an episode of the television series *Star Trek*, the starship *Enterprise* is hit by an ion storm, causing the power to go out. Captain Kirk wonders if Mr. Scott, the engineer, is aware of the problem. Mr. Spock, the paragon of extraterrestrial intelligence, replies, "If Mr. Scott is still with us, the power should be on momentarily." Moments later, the ship's power comes on and Spock arches his Vulcan brow: "Ah, Mr. Scott is still with us."

Spock's logic can be expressed in the form of an argument:

If Mr. Scott is still with us, then the power will come on.
The power comes on.
____
Therefore, Mr. Scott is still with us.

Determine whether this argument is valid or invalid.

SOLUTION

Step 1 Use a letter to represent each simple statement in the argument. We introduce the following representations:

p: Mr. Scott is still with us.

q: The power will come on.

Step 2 Express the premises and the conclusion symbolically.

$p \rightarrow q$ If Mr. Scott is still with us, then the power will come on.
q The power comes on.
____
$\therefore p$ Mr. Scott is still with us.

Step 3 Write a symbolic statement of the form

$$[(\text{premise 1}) \wedge (\text{premise 2})] \rightarrow \text{conclusion.}$$

The symbolic statement is

$$[(p \rightarrow q) \wedge q] \rightarrow p.$$

Step 4 Construct a truth table for the conditional statement in step 3.

p	q	$p \rightarrow q$	$(p \rightarrow q) \wedge q$	$[(p \rightarrow q) \wedge q] \rightarrow p$
T	T	T	T	T
T	F	F	F	T
F	T	T	T	F
F	F	T	F	T

Step 5 Use the truth values in the final column to determine if the argument is valid or invalid. The entries in the final column of the truth table are not all true, so the conditional statement is not a tautology. Spock's argument is invalid, or a fallacy.

GREAT QUESTION!

Is Example 1 for real? Did the writers of the television series *Star Trek* actually construct an invalid argument for the super-intelligent Mr. Spock?

Indeed they did. And if Spock's fallacious reasoning represents television writers' best efforts at being logical, do not expect much from their commercials.

The form of the argument in Spock's logical foul-up

$$\begin{array}{l} p \rightarrow q \\ \underline{q \quad\quad} \\ \therefore p \end{array}$$

is called the **fallacy of the converse**. It should remind you that a conditional statement is not equivalent to its converse. All arguments that have this form are invalid regardless of the English statements that p and q represent.

You may also recall that a conditional statement is not equivalent to its inverse. Another common invalid form of argument is called the **fallacy of the inverse**:

$$\begin{array}{l} p \rightarrow q \\ \underline{\sim p \quad} \\ \therefore \sim q. \end{array}$$

An example of the fallacy of the inverse is "If I study, I pass. I do not study. Therefore, I do not pass." For most students, the conclusion is true, but it does not have to be. If an argument is invalid, then the conclusion is not necessarily true. This, however, does not mean that the conclusion must be false.

 **CHECK POINT 1** Use a truth table to determine whether the following argument is valid or invalid:

The United States must energetically support the development of solar-powered cars or suffer increasing atmospheric pollution.

The United States must not suffer increasing atmospheric pollution.

Therefore, the United States must energetically support the development of solar-powered cars.

EXAMPLE 2 Determining Validity with a Truth Table

Determine whether the following argument is valid or invalid:

"I can't have anything more to do with the operation. If I did, I'd have to lie to the Ambassador. And I can't do that."

—Henry Bromell, "I Know Your Heart, Marco Polo," *The New Yorker*

SOLUTION

We can express the argument as follows:

If I had anything more to do with the operation, I'd have to lie to the Ambassador.

I can't lie to the Ambassador.

Therefore, I can't have anything more to do with the operation.

Step 1 Use a letter to represent each statement in the argument. We introduce the following representations:

p: I have more to do with the operation.

q: I have to lie to the Ambassador.

Step 2 Express the premises and the conclusion symbolically.

$p \rightarrow q$	If I had anything more to do with the operation, I'd have to lie to the Ambassador.
$\sim q$	I can't lie to the Ambassador.
$\therefore \sim p$	Therefore, I can't have anything more to do with the operation.

Step 3 Write a symbolic statement of the form

[(premise 1) $\wedge$ (premise 2)] $\rightarrow$ conclusion.

The symbolic statement is

$$[(p \rightarrow q) \wedge \sim q] \rightarrow \sim p.$$

Step 4 Construct a truth table for the conditional statement in step 3. We need to construct a truth table for $[(p \to q) \wedge {\sim}q] \to {\sim}p$.

p	q	$p \to q$	${\sim}q$	$(p \to q) \wedge {\sim}q$	${\sim}p$	$[(p \to q) \wedge {\sim}q] \to {\sim}p$
T	T	T	F	F	F	T
T	F	F	T	F	F	T
F	T	T	F	F	T	T
F	F	T	T	T	T	T

Step 5 Use the truth values in the final column to determine if the argument is valid or invalid. The entries in the final column of the truth table are all true, so the conditional statement is a tautology. The given argument is valid.

The form of the argument in Example 2

$$\begin{array}{l} p \to q \\ \underline{{\sim}q} \\ \therefore {\sim}p \end{array}$$

should remind you that a conditional statement is equivalent to its contrapositive:

$$p \to q \equiv {\sim}q \to {\sim}p.$$

The form of this argument is called **contrapositive reasoning**.

CHECK POINT 2 Use a truth table to determine whether the following argument is valid or invalid:

I study for 5 hours or I fail.

I did not study for 5 hours.

Therefore, I failed.

EXAMPLE 3 *The Defense Attorney's Argument at the Menendez Trial*

The defense attorney at the Menendez trial admitted that the brothers murdered their parents. However, she presented the following argument that resulted in a different conclusion about sentencing:

If children murder parents because they fear abuse, there are mitigating circumstances to the murder.

If there are mitigating circumstances, then children deserve a lighter sentence.

Therefore, if children murder parents because they fear abuse, they deserve a lighter sentence.

(*Source:* Sherry Diestler, *Becoming a Critical Thinker*, Fourth Edition, Prentice Hall, 2005.)

Determine whether this argument is valid or invalid.

SOLUTION

Step 1 Use a letter to represent each statement in the argument. We introduce the following representations:

p: Children murder parents because they fear abuse.

q: There are mitigating circumstances to the murder.

r: Children deserve a lighter sentence.

Step 2 Express the premises and the conclusion symbolically.

$p \to q$ If children murder parents because they fear abuse, there are mitigating circumstances to the murder.

$q \to r$ If there are mitigating circumstances, then children deserve a lighter sentence.

$\therefore p \to r$ Therefore, if children murder parents because they fear abuse, they deserve a lighter sentence.

Step 3 Write a symbolic statement of the form

[(premise 1) $\wedge$ (premise 2)] $\to$ conclusion.

The symbolic statement is

$$[(p \to q) \wedge (q \to r)] \to (p \to r).$$

Step 4 Construct a truth table for the conditional statement in step 3.

p	q	r	$p \to q$	$q \to r$	$p \to r$	$(p \to q) \wedge (q \to r)$	$[(p \to q) \wedge (q \to r)] \to (p \to r)$
T	T	T	T	T	T	T	T
T	T	F	T	F	F	F	T
T	F	T	F	T	T	F	T
T	F	F	F	T	F	F	T
F	T	T	T	T	T	T	T
F	T	F	T	F	T	F	T
F	F	T	T	T	T	T	T
F	F	F	T	T	T	T	T

Step 5 Use the truth values in the final column to determine if the argument is valid or invalid. The entry in each of the eight rows in the final column of the truth table is true, so the conditional statement is a tautology. The defense attorney's argument is valid.

The form of the defense attorney's argument

$$\begin{array}{l} p \to q \\ \underline{q \to r} \\ \therefore p \to r \end{array}$$

is called **transitive reasoning**. If p implies q and q implies r, then p must imply r. Because $p \to r$ is a valid conclusion, the contrapositive, $\sim r \to \sim p$, is also a valid conclusion. Not necessarily true are the converse, $r \to p$, and the inverse, $\sim p \to \sim r$.

CHECK POINT 3 Use a truth table to determine whether the following argument is valid or invalid:

If you lower the fat in your diet, you lower your cholesterol.

If you lower your cholesterol, you reduce the risk of heart disease.

Therefore, if you do not lower the fat in your diet, you do not reduce the risk of heart disease.

We have seen two valid arguments that resulted in very different conclusions. The prosecutor in the Menendez case concluded that the brothers needed to be punished to the full extent of the law. The defense attorney concluded that they deserved a lighter sentence. This illustrates that the conclusion of a valid argument is true *relative to the premises*. The conclusion may follow from the premises, although one or more of the premises may not be true.

A valid argument with true premises is called a **sound argument**. The conclusion of a sound argument is true relative to the premises, but it is also true as a separate statement removed from the premises. When an argument is sound, its conclusion represents perfect certainty. Knowing how to assess the validity and soundness of arguments is a very important skill that will enable you to avoid being fooled into thinking that something is proven with certainty when it is not.

2 *Recognize and use forms of valid and invalid arguments.*

Table 3.18 contains the standard forms of commonly used valid and invalid arguments. If an English argument translates into one of these forms, you can immediately determine whether or not it is valid without using a truth table.

GREAT QUESTION!

Should I memorize the forms of the valid and invalid arguments in Table 3.18?

Yes. If only the writers of *Star Trek* had done so! (See Example 1.)

TABLE 3.18 Standard Forms of Arguments

Valid Arguments			
Direct Reasoning	**Contrapositive Reasoning**	**Disjunctive Reasoning**	**Transitive Reasoning**
$p \to q$ p $\therefore q$	$p \to q$ $\sim q$ $\therefore \sim p$	$p \vee q$ $p \vee q$ $\sim p$ $\sim q$ $\therefore q$ $\therefore p$	$p \to q$ $q \to r$ $\therefore p \to r$ $\therefore \sim r \to \sim p$
Invalid Arguments			
Fallacy of the Converse	**Fallacy of the Inverse**	**Misuse of Disjunctive Reasoning**	**Misuse of Transitive Reasoning**
$p \to q$ q $\therefore p$	$p \to q$ $\sim p$ $\therefore \sim q$	$p \vee q$ $p \vee q$ p q $\therefore \sim q$ $\therefore \sim p$	$p \to q$ $q \to r$ $\therefore r \to p$ $\therefore \sim p \to \sim r$

EXAMPLE 4 *Determining Validity without Truth Tables*

Determine whether each argument is valid or invalid. Identify any sound arguments.

a. There is no need for surgery. I know this because if there is a tumor then there is need for surgery, but there is no tumor.

b. The emergence of democracy is a cause for hope or environmental problems will overshadow any promise of a bright future. Because environmental problems will overshadow any promise of a bright future, it follows that the emergence of democracy is not a cause for hope.

c. If evidence of the defendant's DNA is found at the crime scene, we can connect him with the crime. If we can connect him with the crime, we can have him stand trial. Therefore, if the defendant's DNA is found at the crime scene, we can have him stand trial.

SOLUTION

a. We introduce the following representations:

p: There is a tumor.

q: There is need for surgery.

We express the premises and conclusion symbolically.

If there is a tumor then there is need for surgery.	$p \to q$
There is no tumor.	$\sim p$
Therefore, there is no need for surgery.	$\therefore \sim q$

The argument is in the form of the fallacy of the inverse. Therefore, the argument is invalid.

b. We introduce the following representations:

p: The emergence of democracy is a cause for hope.

q: Environmental problems will overshadow any promise of a bright future.

We express the premises and conclusion symbolically.

The emergence of democracy is a cause for hope or environmental problems will overshadow any promise of a bright future.	$p \vee q$
Environmental problems will overshadow any promise of a bright future.	q
Therefore, the emergence of democracy is not a cause for hope.	$\therefore \sim p$

The argument is in a form that represents a misuse of disjunctive reasoning. Therefore, the argument is invalid.

c. We introduce the following representations:

p: Evidence of the defendant's DNA is found at the crime scene.

q: We can connect him with the crime.

r: We can have him stand trial.

The argument can now be expressed symbolically.

If evidence of the defendant's DNA is found at the crime scene, we can connect him with the crime.	$p \rightarrow q$
If we can connect him with the crime, we can have him stand trial.	$q \rightarrow r$
Therefore, if the defendant's DNA is found at the crime scene, we can have him stand trial.	$\therefore p \rightarrow r$

The argument is in the form of transitive reasoning. Therefore, the argument is valid. Furthermore, the premises appear to be true statements, so this is a sound argument.

CHECK POINT 4 Determine whether each argument is valid or invalid.

a. The emergence of democracy is a cause for hope or environmental problems will overshadow any promise of a bright future. Environmental problems will not overshadow any promise of a bright future. Therefore, the emergence of democracy is a cause for hope.

b. If the defendant's DNA is found at the crime scene, then we can have him stand trial. He is standing trial. Consequently, we found evidence of his DNA at the crime scene.

c. If you mess up, your self-esteem goes down. If your self-esteem goes down, everything else falls apart. So, if you mess up, everything else falls apart.

Richard Nixon's resignation on August 8, 1974, was the sixth anniversary of the day he had triumphantly accepted his party's nomination for his first term as president.

EXAMPLE 5 Nixon's Resignation

"The decision of the Supreme Court in U.S. v. Nixon *(1974), handed down the first day of the Judiciary Committee's final debate, was critical. If the President defied the order, he would be impeached. If he obeyed the order, it was increasingly apparent he would be impeached on the evidence."*

—Victoria Schuck, "Watergate," *The Key Reporter*

Based on the above paragraph, we can formulate the following argument:

If Nixon did not obey the Supreme Court order, he would be impeached.
If Nixon obeyed the Supreme Court order, he would be impeached.
______
Therefore, Nixon's impeachment was certain.

Determine whether this argument is valid or invalid.

SOLUTION

Step 1 Use a letter to represent each simple statement in the argument. We introduce the following representations:

p: Nixon obeys the Supreme Court order.

q: Nixon is impeached.

Step 2 Express the premises and the conclusion symbolically.

$\sim p \rightarrow q$ If Nixon did not obey the Supreme Court order, he would be impeached.

$p \rightarrow q$ If Nixon obeyed the Supreme Court order, he would be impeached.

______

$\therefore q$ Therefore, Nixon's impeachment was certain.

Because this argument is not in the form of a recognizable valid or invalid argument, we will use a truth table to determine validity.

Step 3 Write a symbolic statement of the form

$$[(\text{premise 1}) \wedge (\text{premise 2})] \rightarrow \text{conclusion}.$$

The symbolic statement is

$$[(\sim p \rightarrow q) \wedge (p \rightarrow q)] \rightarrow q.$$

Step 4 Construct a truth table for the conditional statement in step 3.

p	q	$\sim p$	$\sim p \rightarrow q$	$p \rightarrow q$	$(\sim p \rightarrow q) \wedge (p \rightarrow q)$	$[(\sim p \rightarrow q) \wedge (p \rightarrow q)] \rightarrow q$
T	T	F	T	T	T	T
T	F	F	T	F	F	T
F	T	T	T	T	T	T
F	F	T	F	T	F	T

Step 5 Use the truth values in the final column to determine if the argument is valid or invalid. The entries in the final column of the truth table are all true, so the conditional statement is a tautology. Thus, the given argument is valid. Because the premises are true statements, this is a sound argument, with impeachment a certainty. In a 16-minute broadcast on August 8, 1974, Richard Nixon yielded to the inevitability of the argument's conclusion and, staring sadly into the cameras, announced his resignation.

CHECK POINT 5 Determine whether the following argument is valid or invalid:

If people are good, laws are not needed to prevent wrongdoing.
If people are not good, laws will not succeed in preventing wrongdoing.
___
Therefore, laws are not needed to prevent wrongdoing or laws will not succeed in preventing wrongdoing.

A **logical** or **valid conclusion** is one that forms a valid argument when it follows a given set of premises. Suppose that the premises of an English argument translate into any one of the symbolic forms of premises for the valid arguments in **Table 3.18** on page 190. The symbolic conclusion can be used to find a valid English conclusion. Example 6 shows how this is done.

EXAMPLE 6 Drawing a Logical Conclusion

Draw a valid conclusion from the following premises:

If all students get requirements out of the way early, then no students take required courses in their last semester. Some students take required courses in their last semester.

SOLUTION

Let p be: All students get requirements out of the way early.

Let q be: No students take required courses in their last semester.

The form of the premises is

$p \rightarrow q$	If all students get requirements out of the way early, then no students take required courses in their last semester.
$\sim q$	Some students take required courses in their last semester. (Recall that the negation of *no* is *some*.)
$\therefore$?	

The conclusion $\sim p$ is valid because it forms the contrapositive reasoning of a valid argument when it follows the given premises. The conclusion $\sim p$ translates as

Not all students get requirements out of the way early.

Because the negation of *all* is *some . . . not*, we can equivalently conclude that

Some students do not get requirements out of the way early.

CHECK POINT 6 Draw a valid conclusion from the following premises:

If all people lead, then no people follow. Some people follow.

Concept and Vocabulary Check

Fill in each blank so that the resulting statement is true.

1. An argument is ______ if the conclusion is true whenever the premises are assumed to be true.

2. The argument

$$\begin{array}{l} p \rightarrow q \\ \underline{p \quad\quad} \\ \therefore \text{\_\_\_} \end{array}$$

is called direct reasoning and is ______ because ________________ is a tautology.

3. The argument

$$\begin{array}{l} p \rightarrow q \\ \underline{\sim q \quad\quad} \\ \therefore \text{\_\_\_} \end{array}$$

is called contrapositive reasoning and is ______ because ________________ is a tautology.

4. The argument

$$\begin{array}{l} p \rightarrow q \\ \underline{q \rightarrow r \quad\quad} \\ \therefore \text{\_\_\_} \end{array}$$

is called transitive reasoning and is ______ because ________________ is a tautology.

5. The argument

$$\begin{array}{l} p \vee q \\ \underline{\sim p \quad\quad} \\ \therefore \text{\_\_\_} \end{array}$$

is called disjunctive reasoning and is ______ because ________________ is a tautology.

6. The fallacy of the converse has the form

$$\begin{array}{l} p \rightarrow q \\ \underline{q \quad\quad} \\ \therefore \text{\_\_\_}. \end{array}$$

7. The fallacy of the inverse has the form

$$\begin{array}{l} p \rightarrow q \\ \underline{\sim p \quad\quad} \\ \therefore \text{\_\_\_}. \end{array}$$

8. True or False: Any argument with true premises is valid. ______

9. True or False: The conclusion of a sound argument is true relative to the premises, but it is also true as a separate statement removed from the premises.______

10. True or False: Any argument whose premises are $p \rightarrow q$ and $q \rightarrow r$ is valid regardless of the conclusion.______

Exercise Set 3.7

Practice Exercises

In Exercises 1–14, use a truth table to determine whether the symbolic form of the argument is valid or invalid.

1. $\begin{array}{l} p \rightarrow q \\ \underline{\sim p \quad} \\ \therefore \sim q \end{array}$

2. $\begin{array}{l} p \rightarrow q \\ \underline{\sim p \quad} \\ \therefore q \end{array}$

3. $\begin{array}{l} p \rightarrow \sim q \\ \underline{q \quad\quad} \\ \therefore \sim p \end{array}$

4. $\begin{array}{l} \sim p \rightarrow q \\ \underline{\sim q \quad\quad} \\ \therefore p \end{array}$

5. $\begin{array}{l} p \wedge \sim q \\ \underline{p \quad\quad} \\ \therefore \sim q \end{array}$

6. $\begin{array}{l} \sim p \vee q \\ \underline{p \quad\quad} \\ \therefore q \end{array}$

7. $\begin{array}{l} p \rightarrow q \\ \underline{q \rightarrow p \quad\quad} \\ \therefore p \wedge q \end{array}$

8. $\begin{array}{l} (p \rightarrow q) \wedge (q \rightarrow p) \\ \underline{p \quad\quad\quad\quad} \\ \therefore p \vee q \end{array}$

9. $\begin{array}{l} p \rightarrow q \\ \underline{q \rightarrow r \quad\quad} \\ \therefore r \rightarrow p \end{array}$

10. $\begin{array}{l} p \rightarrow q \\ \underline{q \rightarrow r \quad\quad} \\ \therefore \sim p \rightarrow \sim r \end{array}$

11. $\begin{array}{l} p \rightarrow q \\ \underline{q \wedge r \quad\quad} \\ \therefore p \vee r \end{array}$

12. $\begin{array}{l} \sim p \wedge q \\ \underline{p \leftrightarrow r \quad\quad} \\ \therefore p \wedge r \end{array}$

13. $\begin{array}{l} p \leftrightarrow q \\ \underline{q \rightarrow r \quad\quad} \\ \therefore \sim r \rightarrow \sim p \end{array}$

14. $\begin{array}{l} q \rightarrow \sim p \\ \underline{q \wedge r \quad\quad} \\ \therefore r \rightarrow p \end{array}$

In Exercises 15–42, translate each argument into symbolic form. Then determine whether the argument is valid or invalid. You may use a truth table or, if applicable, compare the argument's symbolic form to a standard valid or invalid form. (You can ignore differences in past, present, and future tense.)

15. If it is cold, my motorcycle will not start.
My motorcycle started.
∴ It is not cold.

16. If a metrorail system is not in operation, there are traffic delays.
Over the past year there have been no traffic delays.
∴ Over the past year a metrorail system has been in operation.

17. There must be a dam or there is flooding.
This year there is flooding.
∴ This year there is no dam.

18. You must eat well or you will not be healthy.
I eat well.
∴ I am healthy.

19. If we close the door, then there is less noise.
There is less noise.
∴ We closed the door.

20. If an argument is in the form of the fallacy of the inverse, then it is invalid.
This argument is invalid.
∴ This argument is in the form of the fallacy of the inverse.

21. If he was disloyal, his dismissal was justified.
If he was loyal, his dismissal was justified.
∴ His dismissal was justified.

22. If I tell you I cheated, I'm miserable.
If I don't tell you I cheated, I'm miserable.
∴ I'm miserable.

23. We criminalize drugs or we damage the future of young people.
We will not damage the future of young people.
∴ We criminalize drugs.

24. He is intelligent or an overachiever.
He is not intelligent.
∴ He is an overachiever.

25. If all people obey the law, then no jails are needed.
Some people do not obey the law.
∴ Some jails are needed.

26. If all people obey the law, then no jails are needed.
Some jails are needed.
∴ Some people do not obey the law.

27. If I'm tired, I'm edgy.
If I'm edgy, I'm nasty.
∴ If I'm tired, I'm nasty.

28. If I am at the beach, then I swim in the ocean.
If I swim in the ocean, then I feel refreshed.
∴ If I am at the beach, then I feel refreshed.

29. If I'm tired, I'm edgy.
If I'm edgy, I'm nasty.
∴ If I'm nasty, I'm tired.

30. If I'm at the beach, then I swim in the ocean.
If I swim in the ocean, then I feel refreshed.
∴ If I'm not at the beach, then I don't feel refreshed.

31. If Tim and Janet play, then the team wins.
Tim played and the team did not win.
∴ Janet did not play.

32. If *The Graduate* and *Midnight Cowboy* are shown, then the performance is sold out.
Midnight Cowboy was shown and the performance was not sold out.
∴ *The Graduate* was not shown.

33. If it rains or snows, then I read.
I am not reading.
∴ It is neither raining nor snowing.

34. If I am tired or hungry, I cannot concentrate.
I can concentrate.
∴ I am neither tired nor hungry.

35. If it rains or snows, then I read.
I am reading.
∴ It is raining or snowing.

36. If I am tired or hungry, I cannot concentrate.
I cannot concentrate.
∴ I am tired or hungry.

37. If it is hot and humid, I complain.
It is not hot or it is not humid.
∴ I am not complaining.

38. If I watch *Schindler's List* and *Milk*, I am aware of the destructive nature of intolerance.
Today I did not watch *Schindler's List* or I did not watch *Milk*.
∴ Today I am not aware of the destructive nature of intolerance.

39. If you tell me what I already understand, you do not enlarge my understanding.
If you tell me something that I do not understand, then your remarks are unintelligible to me.
∴ Whatever you tell me does not enlarge my understanding or is unintelligible to me.

40. If we are to have peace, we must not encourage the competitive spirit.
If we are to make progress, we must encourage the competitive spirit.
∴ We do not have peace and we do not make progress.

41. If some journalists learn about the invasion, the newspapers will print the news.
If the newspapers print the news, the invasion will not be a secret.
The invasion was a secret.
∴ No journalists learned about the invasion.

42. If some journalists learn about the invasion, the newspapers will print the news.
If the newspapers print the news, the invasion will not be a secret.
No journalists learned about the invasion.
∴ The invasion was a secret.

In Exercises 43–50, use the standard forms of valid arguments to draw a valid conclusion from the given premises.

43. If a person is a chemist, then that person has a college degree.
My best friend does not have a college degree.
Therefore, . . .

44. If the Westway Expressway is not in operation, automobile traffic makes the East Side Highway look like a parking lot.
On June 2, the Westway Expressway was completely shut down because of an overturned truck.
Therefore, . . .

45. The writers of *My Mother the Car* were told by the network to improve their scripts or be dropped from prime time.

The writers of *My Mother the Car* did not improve their scripts.

Therefore, . . .

46. You exercise or you do not feel energized.

I do not exercise.

Therefore, . . .

47. If all electricity is off, then no lights work.

Some lights work.

Therefore, . . .

48. If all houses meet the hurricane code, then none of them are destroyed by a category 4 hurricane.

Some houses were destroyed by Andrew, a category 4 hurricane.

Therefore, . . .

49. If I vacation in Paris, I eat French pastries.

If I eat French pastries, I gain weight.

Therefore, . . .

50. If I am a full-time student, I cannot work.

If I cannot work, I cannot afford a rental apartment costing more than $500 per month.

Therefore, . . .

Practice Plus

In Exercises 51–58, translate each argument into symbolic form. Then determine whether the argument is valid or invalid.

51. If it was any of your business, I would have invited you. It is not, and so I did not.

52. If it was any of your business, I would have invited you. I did, and so it is.

53. It is the case that $x < 5$ or $x > 8$, but $x \geq 5$, so $x > 8$.

54. It is the case that $x < 3$ or $x > 10$, but $x \leq 10$, so $x < 3$.

55. Having a college degree is necessary for obtaining a teaching position. You have a college degree, so you have a teaching position.

56. Having a college degree is necessary for obtaining a teaching position. You do not obtain a teaching position, so you do not have a college degree.

57. "I do know that this pencil exists; but I could not know this if Hume's principles were true. Therefore, Hume's principles, one or both of them, are false."

—G. E. Moore, *Some Main Problems of Philosophy*

58. (In this exercise, determine if the argument is sound, valid but not sound, or invalid.)

If an argument is invalid, it does not produce truth, whereas a valid unsound argument also does not produce truth. Arguments are invalid or they are valid but unsound. Therefore, no arguments produce truth.

Application Exercises

59. From *Alice in Wonderland*:

"This time she found a little bottle and tied around the neck of the bottle was a paper label, with the words DRINK ME beautifully printed on it in large letters.

It was all very well to say DRINK ME, but the wise little Alice was not going to do that in a hurry. 'No, I'll look first,' she said, 'and see whether it's marked poison or not,' for she had never forgotten that if you drink much from a bottle marked poison, it is almost certain to disagree with you, sooner or later.

However, this bottle was not marked poison, so Alice ventured to taste it."

Alice's argument:

If the bottle is marked poison, I should not drink from it.

This bottle is not marked poison.

∴ I should drink from it.

Translate this argument into symbolic form and determine whether it is valid or invalid.

60. From *Alice in Wonderland*:

"Alice noticed, with some surprise, that the pebbles were all turning into little cakes as they lay on the floor, and a bright idea came into her head. 'If I eat one of these cakes,' she thought, 'it's sure to make some change in my size; and as it can't possibly make me larger, it must make me smaller, I suppose.' "

Alice's argument:

If I eat the cake, it will make me larger or smaller.

It can't make me larger.

∴ If I eat the cake, it will make me smaller.

Translate this argument into symbolic form and determine whether it is valid or invalid.

61. Billy Hayes, author of *Midnight Express*, told a college audience of his decision to escape from the Turkish prison in which he had been confined for five years:

"My thoughts were that if I made it, I would be free. If they shot and killed me, I would also be free."

(*Source:* Rodes and Pospesel, *Premises and Conclusions*, Pearson, 1997)

Hayes's dilemma can be expressed in the form of an argument:

If I escape, I will be free.

If they kill me, I will be free.

I escape or they kill me.

∴ I will be free.

Translate this argument into symbolic form and determine whether it is valid or invalid.

62. Conservative commentator Rush Limbaugh directed this passage at liberals and the way they think about crime.

> Of course, liberals will argue that these actions [contemporary youth crime] can be laid at the foot of socioeconomic inequities, or poverty. However, the Great Depression caused a level of poverty unknown to exist in America today, and yet I have been unable to find any accounts of crime waves sweeping our large cities. Let the liberals chew on that.
>
> (*See, I Told You So*, p. 83)

Limbaugh's passage can be expressed in the form of an argument:

> If poverty causes crime, then crime waves would have swept American cities during the Great Depression.
>
> Crime waves did not sweep American cities during the Great Depression.
>
> ---
>
> ∴ Poverty does not cause crime. (Liberals are wrong.)

Translate this argument into symbolic form and determine whether it is valid or invalid.

In addition to the forms of invalid arguments, fallacious reasoning occurs in everyday logic. Some people use the fallacies described below to intentionally deceive. Others use fallacies innocently; they are not even aware they are using them. Match each description below with the example from Exercises 63–74 that illustrates the fallacy. The matching is one-to-one.

Common Fallacies in Everyday Reasoning

a. *The* **fallacy of emotion** *consists of appealing to emotion (pity, force, etc.) in an argument.*

b. *The* **fallacy of inappropriate authority** *consists of claiming that a statement is true because a person cited as an authority says it's true or because most people believe it's true.*

c. *The* **fallacy of hasty generalization** *occurs when an inductive generalization is made on the basis of a few observations or an unrepresentative sample.*

d. *The* **fallacy of questionable cause** *consists of claiming that A caused B when it is known that A occurred before B.*

e. *The* **fallacy of ambiguity** *occurs when the conclusion of an argument is based on a word or phrase that is used in two different ways in the premises.*

f. *The* **fallacy of ignorance** *consists of claiming that a statement is true simply because it has not been proven false, or vice versa.*

g. *The* **mischaracterization fallacy** *consists of misrepresenting an opponent's position or attacking the opponent rather than that person's ideas in order to refute his or her argument.*

h. *The* **slippery slope fallacy** *occurs when an argument reasons without justification that an event will set off a series of events leading to an undesirable consequence.*

i. *The* **either/or fallacy** *mistakenly presents only two solutions to a problem, negates one of these either/or alternatives, and concludes that the other must be true.*

j. *The* **fallacy of begging the question** *assumes that the conclusion is true within the premises.*

k. *The* **fallacy of composition** *occurs when an argument moves from premises about the parts of a group to a conclusion about the whole group. It also occurs when characteristics of an entire group are mistakenly applied to parts of the group.*

l. *The* **fallacy of the complex question** *consists of drawing a conclusion from a self-incriminating question.*

63. If we allow physician-assisted suicide for those who are terminally ill and request it, it won't be long before society begins pressuring the old and infirm to get out of the way and make room for the young. Before long the government will be deciding who should live and who should die.

64. Of course there are extraterrestrials. Haven't you read that article in the *National Enquirer* about those UFOs spotted in Texas last month?

65. Either you go to college and make something of yourself, or you'll end up as an unhappy street person. You cannot be an unhappy street person, so you should go to college.

66. Scientists have not proved that AIDS cannot be transmitted through casual contact. Therefore, we should avoid casual contact with suspected AIDS carriers.

67. Each of my three uncles smoked two packs of cigarettes every day and they all lived into their 90s. Smoking can't be that bad for your health.

68. You once cheated on tests. I know this because when I asked you if you had stopped cheating on tests, you said yes.

69. My paper is late, but I know you'll accept it because I've been sick and my parents will kill me if I flunk this course.

70. We've all heard Professor Jones tell us about how economic systems should place human need above profit. But I'm not surprised that he neglected to tell you that he's a communist who has visited Cuba twice. How can he possibly speak the truth about economic systems?

71. It's easy to see that suicide is wrong. After all, no one is ever justified in taking his or her own life.

72. The reason I hurt your arm is because you hurt me just as much by telling Dad.

73. Statistics show that nearly every heroin user started out by using marijuana. It's reasonable to conclude that smoking marijuana leads to harder drugs.

74. I know, without even looking, that question #17 on this test is difficult. This is the case because the test was made up by Professor Flunkem and Flunkem's exams are always difficult.

Explaining the Concepts

75. Describe what is meant by a valid argument.

76. If you are given an argument in words that contains two premises and a conclusion, describe how to determine if the argument is valid or invalid.

77. Write an original argument in words for the direct reasoning form.

78. Write an original argument in words for the contrapositive reasoning form.

79. Write an original argument in words for the transitive reasoning form.

80. What is a valid conclusion?

81. Write a valid argument on one of the following questions. If you can, write valid arguments on both sides.

a. Should the death penalty be abolished?

b. Should *Roe v. Wade* be overturned?

c. Are online classes a good idea?

d. Should recreational marijuana be legalized?

e. Should grades be abolished?

f. Should the Electoral College be replaced with a popular vote?

Critical Thinking Exercises

Make Sense? *Exercises 82–85 are based on the following argument by conservative radio talk show host Rush Limbaugh and directed at former vice president Al Gore.*

> You would think that if Al Gore and company believe so passionately in their environmental crusading that [sic] they would first put these ideas to work in their own lives, right? . . . Al Gore thinks the automobile is one of the greatest threats to the planet, but he sure as heck still travels in one of them—a gas guzzler too.
>
> (*See, I Told You So*, p. 168)

Limbaugh's passage can be expressed in the form of an argument:

If Gore really believed that the automobile were a threat to the planet, he would not travel in a gas guzzler.

Gore does travel in a gas guzzler.

Therefore, Gore does not really believe that the automobile is a threat to the planet.

In Exercises 82–85, use Limbaugh's argument to determine whether each statement makes sense or does not make sense, and explain your reasoning.

82. I know for a fact that Al Gore does not travel in a gas guzzler, so Limbaugh's argument is invalid.

83. I think Limbaugh is a fanatic and all his arguments are invalid.

84. In order to avoid a long truth table and instead use a standard form of an argument, I tested the validity of Limbaugh's argument using the following representations:

p: Gore really believes that the automobile is a threat to the planet.

q: He does not travel in a gas guzzler.

85. Using my representations in Exercise 84, I determined that Limbaugh's argument is invalid.

86. Write an original argument in words that has a true conclusion, yet is invalid.

87. Draw a valid conclusion from the given premises. Then use a truth table to verify your answer.

If you only spoke when spoken to and I only spoke when spoken to, then nobody would ever say anything. Some people do say things. Therefore, . . .

88. Translate the argument below into symbolic form. Then use a truth table to determine if the argument is valid or invalid.

It's wrong to smoke in public if secondary cigarette smoke is a health threat. If secondary cigarette smoke were not a health threat, the American Lung Association would not say that it is. The American Lung Association says that secondary cigarette smoke is a health threat. Therefore, it's wrong to smoke in public.

89. Draw what you believe is a valid conclusion in the form of a disjunction for the following argument. Then verify that the argument is valid for your conclusion.

> *"Inevitably, the use of the placebo involved built-in contradictions. A good patient–doctor relationship is essential to the process, but what happens to that relationship when one of the partners conceals important information from the other? If the doctor tells the truth, he destroys the base on which the placebo rests. If he doesn't tell the truth, he jeopardizes a relationship built on trust."*
>
> —Norman Cousins, *Anatomy of an Illness*

Group Exercise

90. In this section, we used a variety of examples, including arguments from the Menendez trial, the inevitability of Nixon's impeachment, and Spock's (fallacious) logic on *Star Trek*, to illustrate symbolic arguments.

a. From any source that is of particular interest to you (these can be the words of someone you truly admire or a person who really gets under your skin), select a paragraph or two in which the writer argues a particular point. (An intriguing source is *What Is Your Dangerous Idea?*, edited by John Brockman, published by Harper Perennial, 2007.) Rewrite the reasoning in the form of an argument using words. Then translate the argument into symbolic form and use a truth table to determine if it is valid or invalid.

b. Each group member should share the selected passage with other people in the group. Explain how it was expressed in argument form. Then tell why the argument is valid or invalid.

3.8 Arguments and Euler Diagrams

WHAT AM I SUPPOSED TO LEARN?

After studying this section, you should be able to:

1 Use Euler diagrams to determine validity.

William Shakespeare

Leonhard Euler

HE IS THE SHAKESPEARE OF MATHEMATICS, YET HE IS UNKNOWN BY THE GENERAL public. Most people cannot even correctly pronounce his name. The Swiss mathematician Leonhard Euler (1707–1783), whose last name rhymes with *boiler*, not *ruler*, is the most prolific mathematician in history. His collected books and papers fill some 80 volumes; Euler published an average of 800 pages of new mathematics per year over a career that spanned six decades. Euler was also an astronomer, botanist, chemist, physicist, and linguist. His productivity was not at all slowed down by the total blindness he experienced the last 17 years of his life. An equation discovered by Euler, $e^{\pi i} + 1 = 0$, connected five of the most important numbers in mathematics in a totally unexpected way.

1 *Use Euler diagrams to determine validity.*

Euler invented an elegant way to determine the validity of arguments whose premises contain the words *all, some*, and *no*. The technique for doing this uses geometric ideas and involves four basic diagrams, known as **Euler diagrams**. **Figure 3.4** illustrates how Euler diagrams represent four quantified statements.

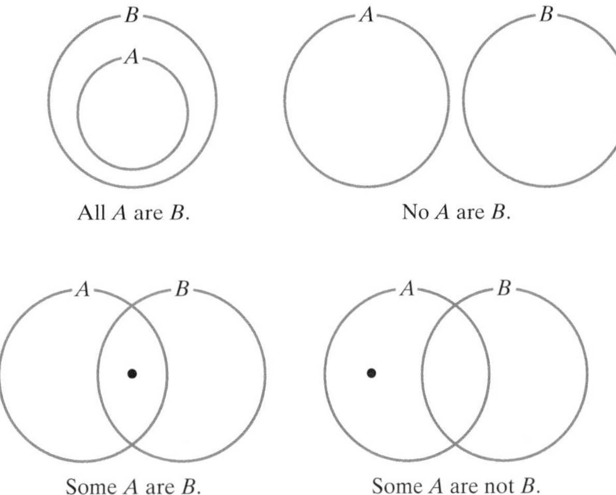

FIGURE 3.4 Euler diagrams for quantified statements

The Euler diagrams in **Figure 3.4** are just like the Venn diagrams that we used in studying sets. However, there is no need to enclose the circles inside a rectangle representing a universal set. In these diagrams, circles are used to indicate relationships of premises to conclusions.

Here's a step-by-step procedure for using Euler diagrams to determine whether or not an argument is valid:

EULER DIAGRAMS AND ARGUMENTS

1. Make an Euler diagram for the first premise.
2. Make an Euler diagram for the second premise on top of the one for the first premise.
3. The argument is valid if and only if every possible diagram illustrates the conclusion of the argument. If there is even *one* possible diagram that contradicts the conclusion, this indicates that the conclusion is not true in every case, so the argument is invalid.

The goal of this procedure is to produce, if possible, *a diagram that does* **not** *illustrate the argument's conclusion*. The method of Euler diagrams boils down to determining whether such a diagram is possible. If it is, this serves as a counterexample to the argument's conclusion, and the argument is immediately declared invalid. By contrast, if no such counterexample can be drawn, the argument is valid.

The technique of using Euler diagrams is illustrated in Examples 1–6.

EXAMPLE 1 *Arguments and Euler Diagrams*

Use Euler diagrams to determine whether the following argument is valid or invalid:

All people who arrive late cannot perform.

All people who cannot perform are ineligible for scholarships.

Therefore, all people who arrive late are ineligible for scholarships.

GREAT QUESTION!

In step 1, does it matter how large I draw the "arrive late" circle as long as it's inside the "cannot perform" circle?

It does not matter. When making Euler diagrams, remember that the size of a circle is not relevant. It is the circle's location that counts.

SOLUTION

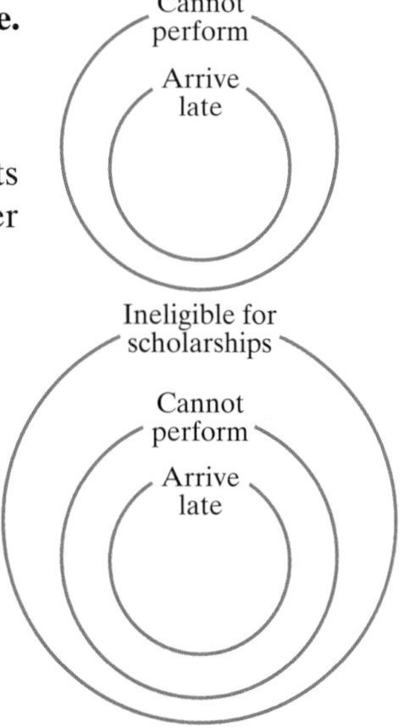

Step 1 Make an Euler diagram for the first premise. We begin by diagramming the premise

All people who arrive late cannot perform.

The region inside the smaller circle represents people who arrive late. The region inside the larger circle represents people who cannot perform.

Step 2 Make an Euler diagram for the second premise on top of the one for the first premise. We add to our previous figure the diagram for the second premise:

All people who cannot perform are ineligible for scholarships.

A third, larger, circle representing people who are ineligible for scholarships is drawn surrounding the circle representing people who cannot perform.

Step 3 The argument is valid if and only if every possible diagram illustrates the argument's conclusion. There is only one possible diagram. Let's see if this diagram illustrates the argument's conclusion, namely

All people who arrive late are ineligible for scholarships.

This is indeed the case because the Euler diagram shows the circle representing the people who arrive late contained within the circle of people who are ineligible for scholarships. The Euler diagram supports the conclusion, and the given argument is valid.

CHECK POINT 1 Use Euler diagrams to determine whether the following argument is valid or invalid:

All U.S. voters must register.

All people who register must be U.S. citizens.

Therefore, all U.S. voters are U.S. citizens.

EXAMPLE 2 Arguments and Euler Diagrams

Use Euler diagrams to determine whether the following argument is valid or invalid:

All poets appreciate language.

All writers appreciate language.

Therefore, all poets are writers.

SOLUTION

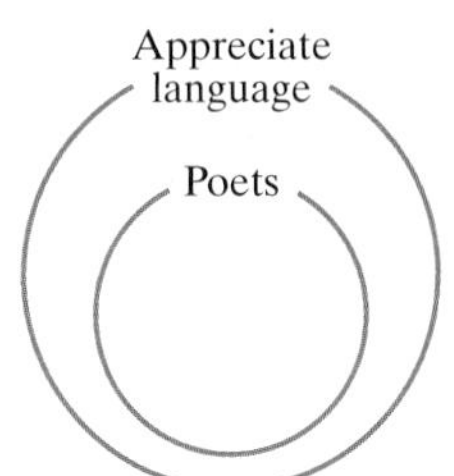

Step 1 Make an Euler diagram for the first premise. We begin by diagramming the premise

All poets appreciate language.

Up to this point, our work is similar to what we did in Example 1.

Step 2 Make an Euler diagram for the second premise on top of the one for the first premise. We add to our previous figure the diagram for the second premise:

All writers appreciate language.

A third circle representing writers must be drawn inside the circle representing people who appreciate language. There are four ways to do this.

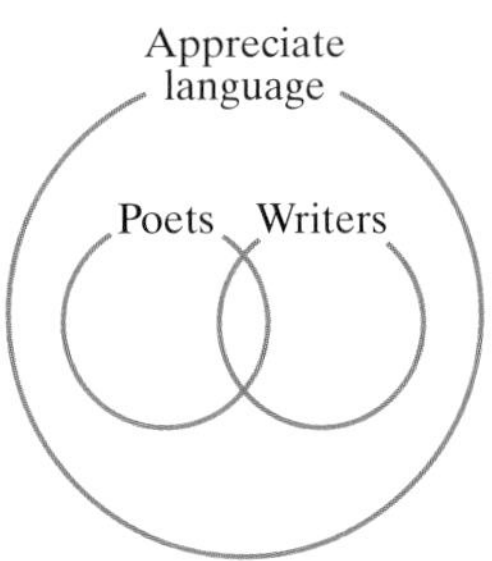

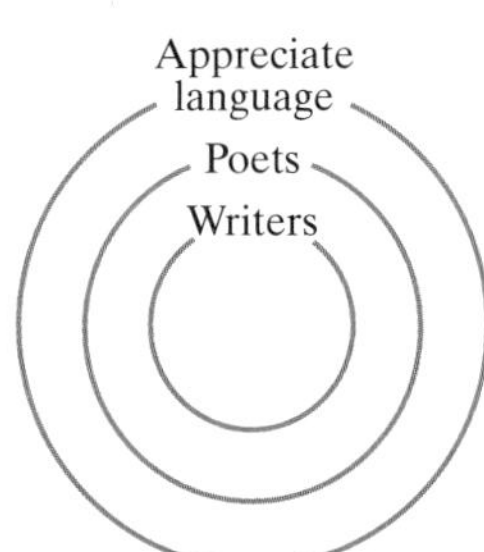

Step 3 The argument is valid if and only if every possible diagram illustrates the argument's conclusion. The argument's conclusion is

All poets are writers.

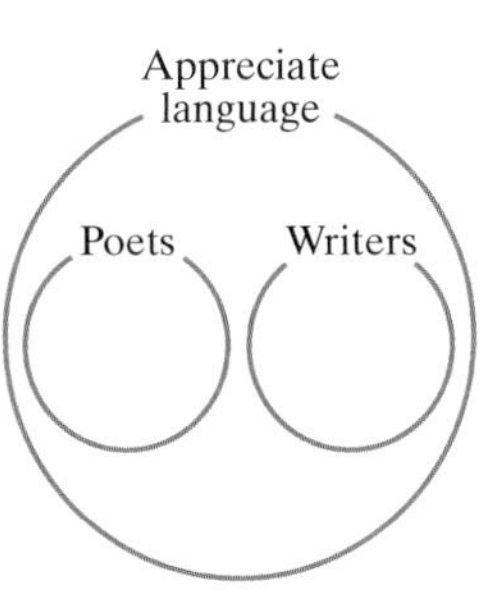

This conclusion is not illustrated by every possible diagram shown above. One of these diagrams is repeated on the right. This diagram shows "no poets are writers." There is no need to examine the other three diagrams.

The diagram on the right above serves as a counterexample to the argument's conclusion. This means that the given argument is invalid. It would have sufficed to draw only the counterexample to determine that the argument is invalid.

CHECK POINT 2 Use Euler diagrams to determine whether the following argument is valid or invalid:

All baseball players are athletes.
All ballet dancers are athletes.
____
Therefore, no baseball players are ballet dancers.

EXAMPLE 3 Arguments and Euler Diagrams

Use Euler diagrams to determine whether the following argument is valid or invalid:

All freshmen live on campus.
No people who live on campus can own cars.
____
Therefore, no freshmen can own cars.

SOLUTION

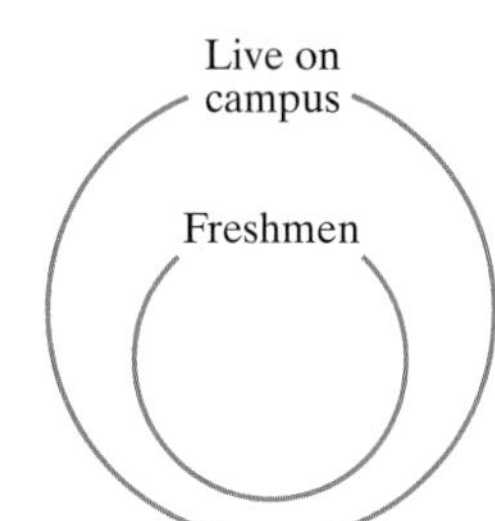

Step 1 Make an Euler diagram for the first premise. The diagram for

All freshmen live on campus

is shown on the right. The region inside the smaller circle represents freshmen. The region inside the larger circle represents people who live on campus.

Step 2 Make an Euler diagram for the second premise on top of the one for the first premise. We add to our previous figure the diagram for the second premise:

No people who live on campus can own cars.

A third circle representing people who own cars is drawn outside the circle representing people who live on campus.

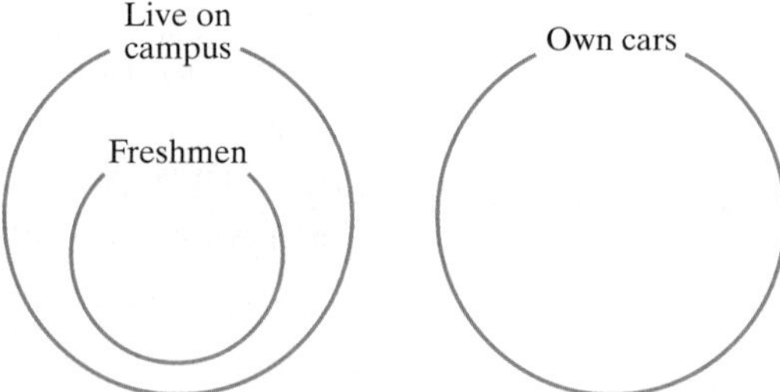

Step 3 The argument is valid if and only if every possible diagram illustrates the argument's conclusion. There is only one possible diagram. The argument's conclusion is

No freshmen can own cars.

This is supported by the diagram shown above because it shows the circle representing freshmen drawn outside the circle representing people who own cars. The Euler diagram supports the conclusion, and it is impossible to find a counterexample that does not. The given argument is valid.

CHECK POINT 3 Use Euler diagrams to determine whether the following argument is valid or invalid:

All mathematicians are logical.
No poets are logical.
____
Therefore, no poets are mathematicians.

Let's see what happens to the validity if we reverse the second premise and the conclusion of the argument in Example 3.

EXAMPLE 4 *Euler Diagrams and Validity*

Use Euler diagrams to determine whether the following argument is valid or invalid:

All freshmen live on campus.
No freshmen can own cars.

Therefore, no people who live on campus can own cars.

SOLUTION

Live on campus
Freshmen

Step 1 Make an Euler diagram for the first premise. We once again begin with the diagram for

All freshmen live on campus.

So far, our work is exactly the same as in the previous example.

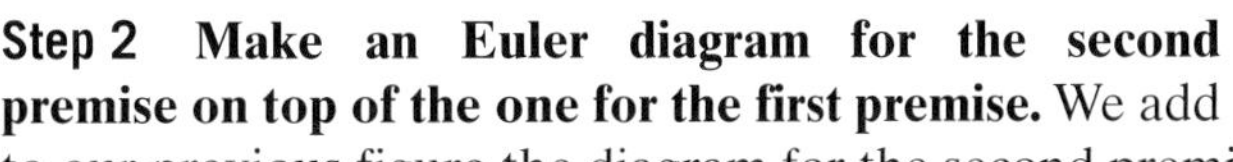

Step 2 Make an Euler diagram for the second premise on top of the one for the first premise. We add to our previous figure the diagram for the second premise:

No freshmen can own cars.

The circle representing people who own cars is drawn outside the freshmen circle. At least two Euler diagrams are possible.

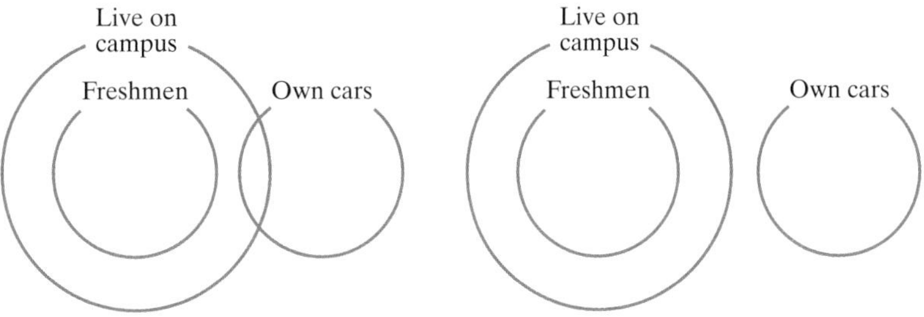

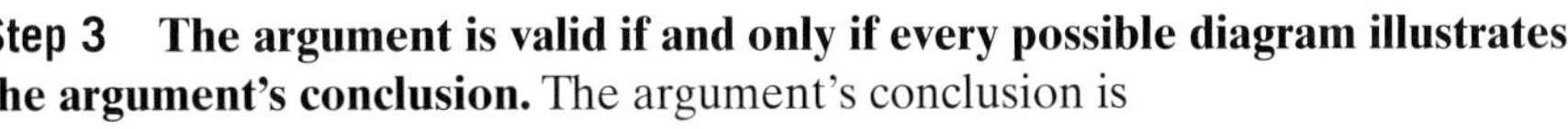

Step 3 The argument is valid if and only if every possible diagram illustrates the argument's conclusion. The argument's conclusion is

No people who live on campus can own cars.

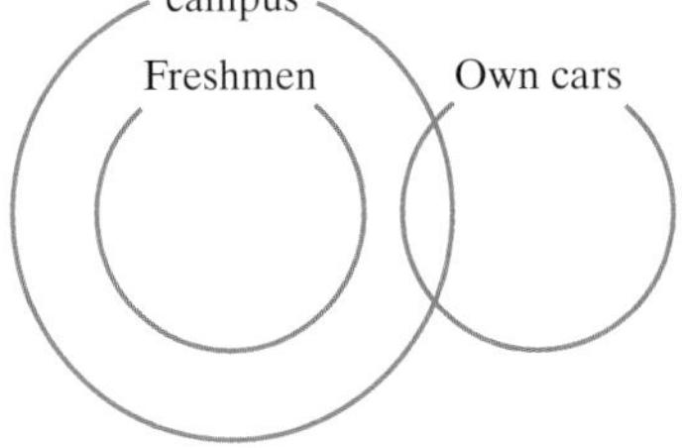

This conclusion is not supported by both diagrams shown above. The diagram that does not support the conclusion is repeated in the margin. Notice that the "live on campus" circle and the "own cars" circle intersect. This diagram serves as a counterexample to the argument's conclusion. This means that the argument is invalid. Once again, only the counterexample on the left is needed to conclude that the argument is invalid.

CHECK POINT 4 Use Euler diagrams to determine whether the following argument is valid or invalid:

All mathematicians are logical.
No poets are mathematicians.

Therefore, no poets are logical.

So far, the arguments that we have looked at have contained "all" or "no" in the premises and conclusions. The quantifier "some" is a bit trickier to work with.

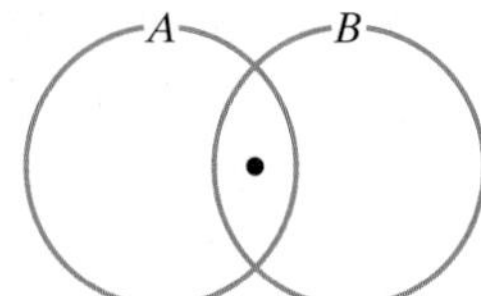

FIGURE 3.5 Some *A* are *B*

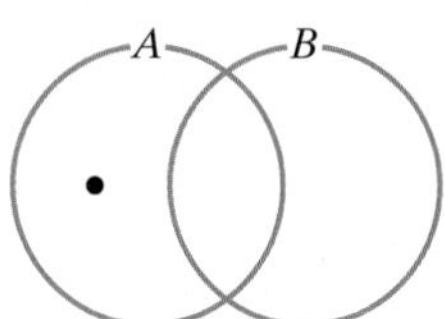

FIGURE 3.6 Illustrated by the dot is some *A* are not *B*. We cannot validly conclude that some *B* are not *A*.

Because the statement "Some *A* are *B*" means there exists at least one *A* that is a *B*, we diagram this existence by showing a dot in the region where *A* and *B* intersect, illustrated in **Figure 3.5**.

Suppose that it is true that "Some *A* are not *B*," illustrated by the dot in **Figure 3.6**. This Euler diagram does not let us conclude that "Some *B* are not *A*" because there is not a dot in the part of the *B* circle that is not in the *A* circle. Conclusions with the word "some" must be shown by existence of at least one element represented by a dot in an Euler diagram.

Here is an example that shows the premise "Some *A* are not *B*" does not enable us to logically conclude that "Some *B* are not *A*."

Some U.S. citizens are not U.S. senators. (true)

∴ Some U.S. senators are not U.S. citizens. (false)

EXAMPLE 5 *Euler Diagrams and the Quantifier "Some"*

Use Euler diagrams to determine whether the following argument is valid or invalid:

All people are mortal.

Some mortals are students.

Therefore, some people are students.

SOLUTION

Step 1 Make an Euler diagram for the first premise. Begin with the premise

All people are mortal.

The Euler diagram is shown on the right.

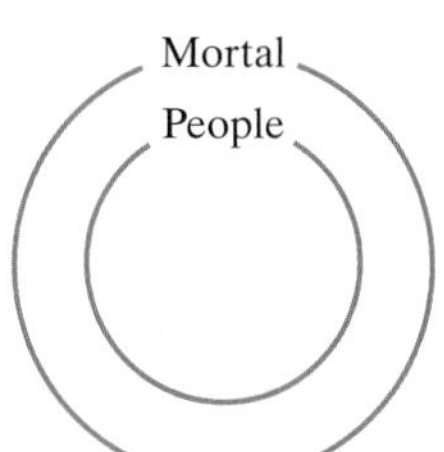

Step 2 Make an Euler diagram for the second premise on top of the one for the first premise. We add to our previous figure the diagram for the second premise:

Some mortals are students.

The circle representing students intersects the circle representing mortals. The dot in the region of intersection shows that at least one mortal is a student. Another diagram is possible, but if this serves as a counterexample then it is all we need. Let's check if it is a counterexample.

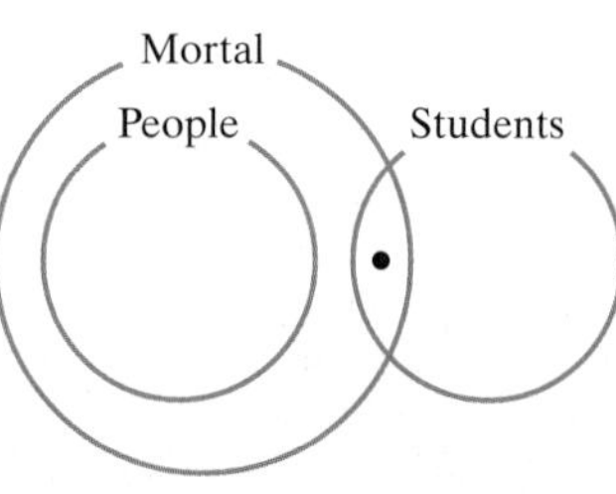

Step 3 The argument is valid if and only if every possible diagram illustrates the conclusion of the argument. The argument's conclusion is

Some people are students.

This conclusion is not supported by the Euler diagram. The diagram does not show the "people" circle and the "students" circle intersecting with a dot in the region of intersection. Although this conclusion is true in the real world, the Euler diagram serves as a counterexample that shows it does not follow from the premises. Therefore, the argument is invalid.

 **CHECK POINT 5** Use Euler diagrams to determine whether the following argument is valid or invalid:

All mathematicians are logical.

Some poets are logical.

Therefore, some poets are mathematicians.

Some arguments show existence without using the word "some." Instead, a particular person or thing is mentioned in one of the premises. This particular person or thing is represented by a dot. Here is an example:

All men are mortal.

Aristotle is a man.

Therefore, Aristotle is mortal.

The two premises can be represented by the following Euler diagrams:

The Euler diagram on the right uses a dot labeled A (for Aristotle). The diagram shows Aristotle (•) winding up in the "mortal" circle. The diagram supports the conclusion that Aristotle is mortal. This argument is valid.

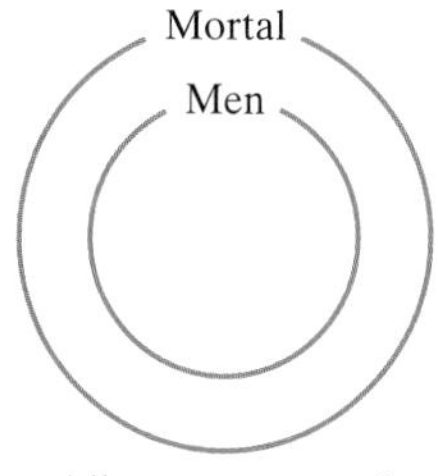

All men are mortal.

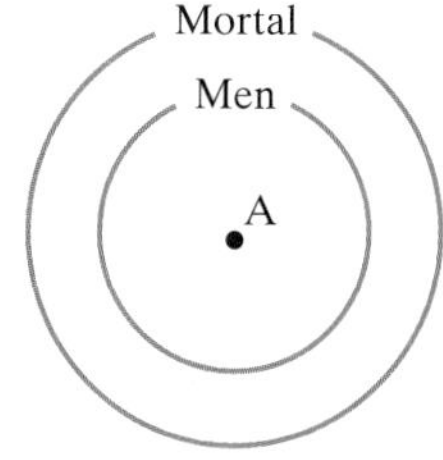

Aristotle (•) is a man.

EXAMPLE 6 An Argument Mentioning One Person

Use Euler diagrams to determine whether the following argument is valid or invalid:

All children love to swim.

Michael Phelps loves to swim.

Therefore, Michael Phelps is a child.

SOLUTION

Step 1 Make an Euler diagram for the first premise. Begin with the premise

All children love to swim.

The Euler diagram is shown on the right.

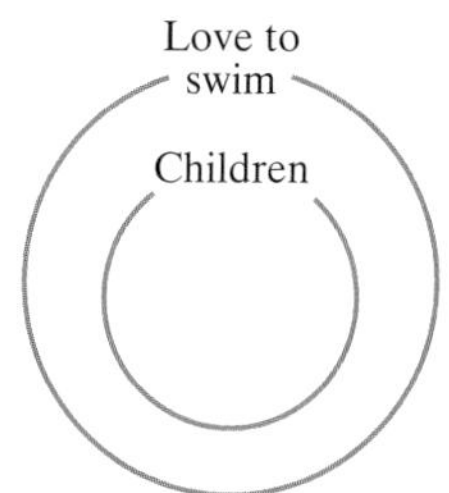

Step 2 Make an Euler diagram for the second premise on top of the one for the first premise. We add to our previous figure the diagram for the second premise:

Michael Phelps loves to swim.

Michael Phelps is represented by a dot labeled M. The dot must be placed in the "love to swim" circle. At least two Euler diagrams are possible.

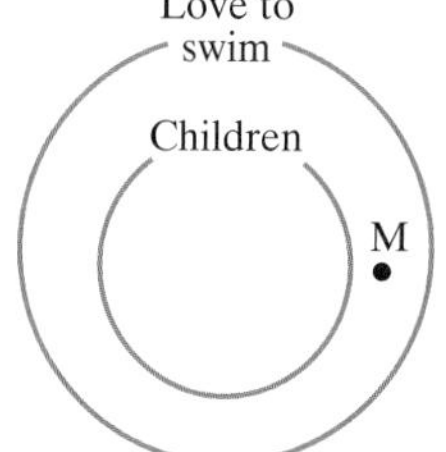

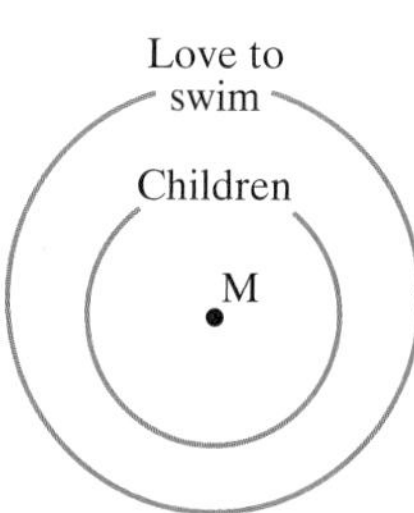

Step 3 The argument is valid if and only if every possible diagram illustrates the conclusion of the argument. The argument's conclusion is

Michael Phelps is a child.

This conclusion is not supported by the Euler diagram shown above on the left. The dot representing Michael Phelps is outside the "children" circle. Michael Phelps might not be a child. This diagram serves as a counterexample to the argument's conclusion. The argument is invalid.

 **CHECK POINT 6** Use Euler diagrams to determine whether the following argument is valid or invalid:

All mathematicians are logical.

Euclid was logical.

Therefore, Euclid was a mathematician.

Blitzer Bonus

Aristotle 384–322 B.C.

The first systematic attempt to describe the logical rules that may be used to arrive at a valid conclusion was made by the ancient Greeks, in particular Aristotle. Aristotelian forms of valid arguments are built into the ways that Westerners think and view the world. In this detail of Raphael's painting *The School of Athens*, Aristotle (on the left) is debating with his teacher and mentor, Plato.

School of Athens, (Detail) (1510), Raphael. Stanza della Segnatura, Stanze di Raffaello, Vatican Palace. Scala/Art Resource, New York.

GREAT QUESTION!

We've now devoted two sections to arguments. What's the bottom line on how to determine whether an argument is valid or invalid?

- Use Euler diagrams when an argument's premises contain quantified statements. (All A are B. No A are B. Some A are B. Some A are not B.)
- Use (memorized) standard forms of arguments if an English argument translates into one of the forms in **Table 3.18** on page 190.
- Use truth tables when an argument's premises are not quantified statements and the argument is not in one of the standard valid or invalid forms.

Concept and Vocabulary Check

Fill in each blank so that the resulting statement is true. Refer to parts (a) through (d) in the following figure.

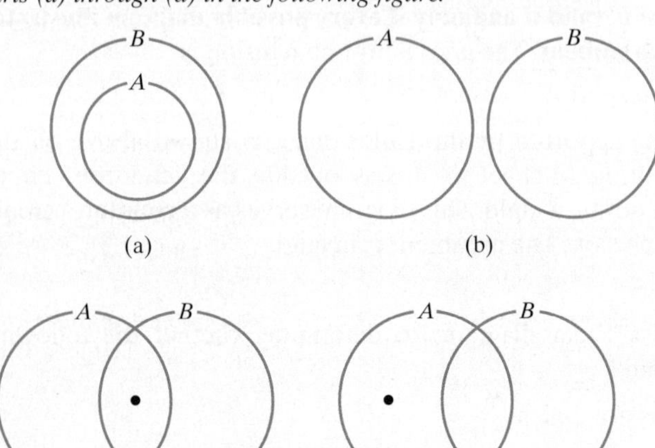

1. The figure in part (a) illustrates the quantified statement __________.
2. The figure in part (b) illustrates the quantified statement __________.
3. The figure in part (c) illustrates the quantified statement __________.
4. The figure in part (d) illustrates the quantified statement __________.
5. True or False: Truth tables are used to represent quantified statements. ______
6. True or False: The most important part in a quantified statement's representation is the size of each circle. ______

Exercise Set 3.8

Practice Exercises

In Exercises 1–24, use Euler diagrams to determine whether each argument is valid or invalid.

1. All writers appreciate language.
   All poets are writers.
   Therefore, all poets appreciate language.
2. All physicists are scientists.
   All scientists attended college.
   Therefore, all physicists attended college.
3. All clocks keep time accurately.
   All time-measuring devices keep time accurately.
   Therefore, all clocks are time-measuring devices.
4. All cowboys live on ranches.
   All cowherders live on ranches.
   Therefore, all cowboys are cowherders.
5. All insects have six legs.
   No spiders have six legs.
   Therefore, no spiders are insects.
6. All humans are warm-blooded.
   No reptiles are warm-blooded.
   Therefore, no reptiles are human.
7. All insects have six legs.
   No spiders are insects.
   Therefore, no spiders have six legs.
8. All humans are warm-blooded.
   No reptiles are human.
   Therefore, no reptiles are warm-blooded.
9. All professors are wise people.
   Some wise people are actors.
   Therefore, some professors are actors.
10. All comedians are funny people.
    Some funny people are professors.
    Therefore, some comedians are professors.
11. All professors are wise people.
    Some professors are actors.
    Therefore, some wise people are actors.
12. All comedians are funny people.
    Some comedians are professors.
    Therefore, some funny people are professors.
13. All dancers are athletes.
    Savion Glover is a dancer.
    Therefore, Savion Glover is an athlete.
14. All actors are artists.
    Sean Penn is an actor.
    Therefore, Sean Penn is an artist.
15. All dancers are athletes.
    Savion Glover is an athlete.
    Therefore, Savion Glover is a dancer.
16. All actors are artists.
    Sean Penn is an artist.
    Therefore, Sean Penn is an actor.
17. Some people enjoy reading.
    Some people enjoy TV.
    Therefore, some people who enjoy reading enjoy TV.
18. All thefts are immoral acts.
    Some thefts are justifiable.
    Therefore, some immoral acts are justifiable.
19. All dogs have fleas.
    Some dogs have rabies.
    Therefore, all dogs with rabies have fleas.
20. All logic problems make sense.
    Some jokes make sense.
    Therefore, some logic problems are jokes.
21. No blank disks contain data.
    Some blank disks are formatted.
    Therefore, some formatted disks do not contain data.
22. Some houses have two stories.
    Some houses have air conditioning.
    Therefore, some houses with air conditioning have two stories.
23. All multiples of 6 are multiples of 3.
    Eight is not a multiple of 3.
    Therefore, 8 is not a multiple of 6.
24. All multiples of 6 are multiples of 3.
    Eight is not a multiple of 6.
    Therefore, 8 is not a multiple of 3.

Practice Plus

In Exercises 25–36, determine whether each argument is valid or invalid.

25. All natural numbers are whole numbers, all whole numbers are integers, and −4006 is not a whole number. Thus, −4006 is not an integer.
26. Some natural numbers are even, all natural numbers are whole numbers, and all whole numbers are integers. Thus, some integers are even.

27. All natural numbers are real numbers, all real numbers are complex numbers, but some complex numbers are not real numbers. The number $19 + 0i$ is a complex number, so it is not a natural number.

28. All rational numbers are real numbers, all real numbers are complex numbers, but some complex numbers are not real numbers. The number $\frac{1}{2} + 0i$ is a complex number, so it is not a rational number.

29. All *A* are *B*, all *B* are *C*, and all *C* are *D*. Thus, all *A* are *D*.

30. All *A* are *B*, no *C* are *B*, and all *D* are *C*. Thus, no *A* are *D*.

31. No *A* are *B*, some *A* are *C*, and all *C* are *D*. Thus, some *D* are *B*.

32. No *A* are *B*, some *A* are *C*, and all *C* are *D*. Thus, some *D* are *C*.

33. No *A* are *B*, no *B* are *C*, and no *C* are *D*. Thus, no *A* are *D*.

34. Some *A* are *B*, some *B* are *C*, and some *C* are *D*. Thus, some *A* are *D*.

35. All *A* are *B*, all *A* are *C*, and some *B* are *D*. Thus, some *A* are *D*.

36. Some *A* are *B*, all *B* are *C*, and some *C* are *D*. Thus, some *A* are *D*.

Application Exercises

37. This is an excerpt from a 1967 speech in the U.S. House of Representatives by Representative Adam Clayton Powell:

> He who is without sin should cast the first stone. There is no one here who does not have a skeleton in his closet. I know, and I know them by name.

Powell's argument can be expressed as follows:

No sinner is one who should cast the first stone.
All people here are sinners.
____
Therefore, no person here is one who should cast the first stone.

Use an Euler diagram to determine whether the argument is valid or invalid.

38. In the *Sixth Meditation*, Descartes writes

> I first take notice here that there is a great difference between the mind and the body, in that the body, from its nature, is always divisible and the mind is completely indivisible.

Descartes's argument can be expressed as follows:

All bodies are divisible.
No minds are divisible.
____
Therefore, no minds are bodies.

Use an Euler diagram to determine whether the argument is valid or invalid.

39. In *Symbolic Logic*, Lewis Carroll presents the following argument:

Babies are illogical. (All babies are illogical persons.)

Illogical persons are despised. (All illogical persons are despised persons.)

Nobody is despised who can manage a crocodile. (No persons who can manage crocodiles are despised persons.)
____
Therefore, babies cannot manage crocodiles.

Use an Euler diagram to determine whether the argument is valid or invalid.

Explaining the Concepts

40. Explain how to use Euler diagrams to determine whether or not an argument is valid.

41. Under what circumstances should Euler diagrams rather than truth tables be used to determine whether or not an argument is valid?

Critical Thinking Exercises

Make Sense? *In Exercises 42–45, determine whether each statement makes sense or does not make sense, and explain your reasoning.*

42. I made Euler diagrams for the premises of an argument and one of my possible diagrams illustrated the conclusion, so the argument is valid.

43. I made Euler diagrams for the premises of an argument and one of my possible diagrams did not illustrate the conclusion, so the argument is invalid.

44. I used Euler diagrams to determine that an argument is valid, but when I reverse one of the premises and the conclusion, this new argument is invalid.

45. I can't use Euler diagrams to determine the validity of an argument if one of the premises is false.

46. Write an example of an argument with two quantified premises that is invalid but that has a true conclusion.

47. No animals that eat meat are vegetarians.

No cat is a vegetarian.

Felix is a cat.

Therefore, . . .

a. Felix is a vegetarian.

b. Felix is not a vegetarian.

c. Felix eats meat.

d. All animals that do not eat meat are vegetarians.

48. Supply the missing first premise that will make this argument valid.

Some opera singers are terrible actors.
____
Therefore, some people who take voice lessons are terrible actors.

49. Supply the missing first premise that will make this argument valid.

All amusing people are entertaining.
____
Therefore, some teachers are entertaining.

Chapter Summary, Review, and Test

SUMMARY – DEFINITIONS AND CONCEPTS — EXAMPLES

3.1 Statements, Negations, and Quantified Statements

3.2 Compound Statements and Connectives

a. A statement is a sentence that is either true or false, but not both simultaneously.

b. Negations and equivalences of quantified statements are given in the following diagram. Each quantified statement's equivalent is written in parentheses below the statement. The statements diagonally opposite each other are negations. Table 3.2, p. 122; Ex. 4, p. 123

All A are B. (There are no A that are not B.)	No A are B. (All A are not B.)
Some A are B. (There exists at least one A that is a B.)	Some A are not B. (Not all A are B.)

c. The statements of symbolic logic and their translations are given as follows: Ex. 1, p. 126; Ex. 2, p. 128; Ex. 3, p. 128; Ex. 4, p. 130; Ex. 5, p. 131

- Negation
  $\sim p$: Not p. It is not true that p.
- Conjunction
  $p \wedge q$: p and q. p but q. p yet q. p nevertheless q.
- Disjunction
  $p \vee q$: p or q.
- Conditional
  $p \rightarrow q$: If p, then q. q if p. p is sufficient for q. q is necessary for p. p only if q. Only if q, p.
- Biconditional
  $p \leftrightarrow q$: p if and only if q. q if and only if p. If p then q, and if q then p. p is necessary and sufficient for q. q is necessary and sufficient for p.

d. Groupings in symbolic statements are determined as follows: Ex. 6, p. 132; Ex. 7, p. 133; Ex. 8, p. 135

- Unless parentheses follow the negation symbol, $\sim$, only the statement that immediately follows it is negated.
- When translating symbolic statements into English, the simple statements in parentheses appear on the same side of the comma.
- If a symbolic statement appears without parentheses, group statements before and after the most dominant connective, where dominance is defined as follows:

1. Negation 2. Conjunction Disjunction 3. Conditional 4. Biconditional.

Least dominant (Negation) — Most dominant (Biconditional)

3.3 Truth Tables for Negation, Conjunction, and Disjunction

3.4 Truth Tables for the Conditional and the Biconditional

a. The definitions of symbolic logic are given by the truth values in the following table: Table 3.12, p. 141; Table 3.14, p. 141; Ex. 1, p. 142

p	q	Negation $\sim p$	Conjunction $p \wedge q$	Disjunction $p \vee q$	Conditional $p \rightarrow q$	Biconditional $p \leftrightarrow q$
T	T	F	T	T	T	T
T	F	F	F	T	F	F
F	T	T	F	T	T	F
F	F	T	F	F	T	T
		Opposite truth values from p	True only when both component statements are true	False only when both component statements are false	False only when the antecedent is true and the consequent is false	True only when the component statements have the same truth value

b. A truth table for a compound statement shows when the statement is true and when it is false. The first few columns show the simple statements that comprise the compound statement and their possible truth values. The final column heading is the given compound statement. The truth values in each column are determined by looking back at appropriate columns and using one of the five definitions of symbolic logic. If a compound statement is always true, it is called a tautology.

Ex. 2, p. 143; Ex. 3, p. 145; Ex. 4, p. 146; Ex. 5, p. 147; Ex. 6, p. 148; Ex. 1, p. 155; Ex. 2, p. 157; Ex. 3, p. 158; Ex. 4, p. 161

c. To determine the truth value of a compound statement for a specific case, substitute the truth values of the simple statements into the symbolic form of the compound statement and then use the appropriate definitions.

Ex. 7, p. 150; Ex. 5, p. 161

3.5 Equivalent Statements and Variations of Conditional Statements

3.6 Negations of Conditional Statements and De Morgan's Laws

a. Two statements are equivalent, symbolized by $\equiv$, if they have the same truth value in every possible case.

Ex. 1, p. 167; Ex. 2, p. 167; Ex. 3, p. 168

b. Variations of the Conditional Statement $p \rightarrow q$

- $p \rightarrow q$ is equivalent to $\sim q \rightarrow \sim p$, the contrapositive: $p \rightarrow q \equiv \sim q \rightarrow \sim p$.
- $p \rightarrow q$ is not equivalent to $q \rightarrow p$, the converse.
- $p \rightarrow q$ is not equivalent to $\sim p \rightarrow \sim q$, the inverse.
- The negation of $p \rightarrow q$ is $p \wedge \sim q$: $\sim(p \rightarrow q) \equiv p \wedge \sim q$.

Ex. 4, p. 170; Ex. 5, p. 172; Ex. 1, p. 177

c. De Morgan's Laws

- $\sim(p \wedge q) \equiv \sim p \vee \sim q$: The negation of $p \wedge q$ is $\sim p \vee \sim q$.
- $\sim(p \vee q) \equiv \sim p \wedge \sim q$: The negation of $p \vee q$ is $\sim p \wedge \sim q$.

Ex. 2, p. 178; Ex. 3, p. 178; Ex. 4, p. 179; Ex. 5, p. 180

3.7 Arguments and Truth Tables

a. An argument consists of two parts: the given statements, called the premises, and a conclusion. An argument is valid if the conclusion is true whenever the premises are assumed to be true. An argument that is not valid is called an invalid argument or a fallacy. A valid argument with true premises is called a sound argument.

b. A procedure to test the validity of an argument using a truth table is described in the box on page 185. If the argument contains n premises, write a conditional statement of the form

$$[(\text{premise } 1) \wedge (\text{premise } 2) \wedge \cdots \wedge (\text{premise } n)] \rightarrow \text{conclusion}$$

and construct a truth table. If the conditional statement is a tautology, the argument is valid; if not, the argument is invalid.

Ex. 1, p. 186; Ex. 2, p. 187; Ex. 3, p. 188; Ex. 5, p. 192

c. Table 3.18 on page 190 contains the standard forms of commonly used valid and invalid arguments.

Ex. 4, p. 190; Ex. 6, p. 193

3.8 Arguments and Euler Diagrams

a. Euler diagrams for quantified statements are given as follows:

All *A* are *B*. No *A* are *B*. Some *A* are *B*. Some *A* are not *B*.

b. To test the validity of an argument with an Euler diagram,

1. Make an Euler diagram for the first premise.
2. Make an Euler diagram for the second premise on top of the one for the first premise.
3. The argument is valid if and only if every possible diagram illustrates the conclusion of the argument.

Ex. 1, p. 200; Ex. 2, p. 201; Ex. 3, p. 202; Ex. 4, p. 203; Ex. 5, p. 204; Ex. 6, p. 205

Review Exercises

3.1 and 3.2

In Exercises 1–6, let p, q, and r represent the following simple statements:

p: The temperature is below 32°.

q: We finished studying.

r: We go to the movies.

Express each symbolic compound statement in English. If a symbolic statement is given without parentheses, place them, as needed, before and after the most dominant connective and then translate into English.

1. $p \wedge q \rightarrow r$

2. $\sim r \rightarrow \sim p \vee \sim q$

3. $p \wedge (q \rightarrow r)$

4. $r \leftrightarrow (p \wedge q)$

5. $\sim(p \wedge q)$

6. $\sim r \leftrightarrow (\sim p \vee \sim q)$

In Exercises 7–12, let p, q, and r represent the following simple statements:

p: The outside temperature is at least 80°.

q: The air conditioner is working.

r: The house is hot.

Express each English statement in symbolic form.

7. The outside temperature is at least 80° and the air conditioner is working, or the house is hot.

8. If the outside temperature is at least 80° or the air conditioner is not working, then the house is hot.

9. If the air conditioner is working, then the outside temperature is at least 80° if and only if the house is hot.

10. The house is hot, if and only if the outside temperature is at least 80° and the air conditioner is not working.

11. Having an outside temperature of at least 80° is sufficient for having a hot house.

12. Not having a hot house is necessary for the air conditioner to be working.

In Exercises 13–16, write the negation of each statement.

13. All houses are made with wood.

14. No students major in business.

15. Some crimes are motivated by passion.

16. Some Democrats are not registered voters.

17. The speaker stated that, "All new taxes are for education." We later learned that the speaker was not telling the truth. What can we conclude about new taxes and education?

3.3 and 3.4

In Exercises 18–25, construct a truth table for each statement. Then indicate whether the statement is a tautology, a self-contradiction, or neither.

18. $p \vee (\sim p \wedge q)$

19. $\sim p \vee \sim q$

20. $p \rightarrow (\sim p \vee q)$

21. $p \leftrightarrow \sim q$

22. $\sim(p \vee q) \rightarrow (\sim p \wedge \sim q)$

23. $(p \vee q) \rightarrow \sim r$

24. $(p \wedge q) \leftrightarrow (p \wedge r)$

25. $p \wedge [q \vee (r \rightarrow p)]$

In Exercises 26–27,

a. *Write each statement in symbolic form. Assign letters to simple statements that are not negated.*

b. *Construct a truth table for the symbolic statement in part (a).*

c. *Use the truth table to indicate one set of conditions that makes the compound statement true, or state that no such conditions exist.*

26. I'm in class or I'm studying, and I'm not in class.

27. If you spit from a truck then it's legal, but if you spit from a car then it's not. (This law is still on the books in Georgia!)

In Exercises 28–31, determine the truth value for each statement when p is true, q is false, and r is false.

28. $\sim(q \leftrightarrow r)$

29. $(p \wedge q) \rightarrow (p \vee r)$

30. $(\sim q \rightarrow p) \vee (r \wedge \sim p)$

31. $\sim[(\sim p \vee r) \rightarrow (q \wedge r)]$

The diversity index, from 0 (no diversity) to 100, measures the chance that two randomly selected people are a different race or ethnicity. The diversity index in the United States varies widely from region to region, from as high as 81 in Hawaii to as low as 11 in Vermont. The bar graph shows the national diversity index for the United States for four years in the period from 1980 through 2010.

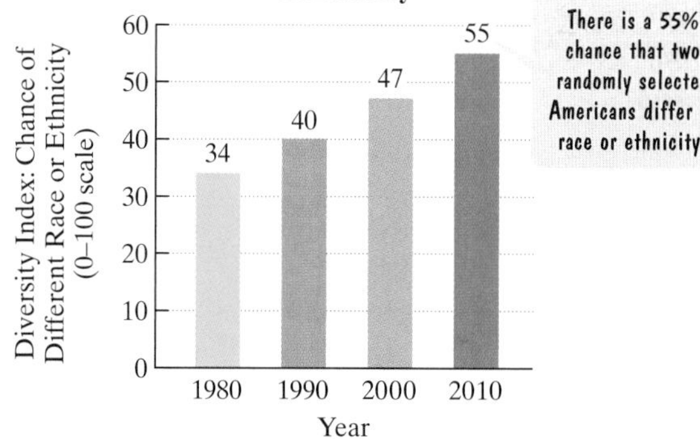

Source: *USA TODAY*

In Exercises 32–34, write each statement in symbolic form. Then use the information displayed by the graph to determine the truth value of the compound statement.

32. The 2000 diversity index was 47, and it is not true that the index increased from 2000 to 2010.

33. If the diversity index decreased from 1980 through 2010, then the index was 55 in 1980 and 34 in 2010.

34. The diversity index increased by 6 from 1980 to 1990 if and only if it increased by 7 from 1990 to 2000, or it is not true that the index was at a maximum in 2010.

3.5 and 3.6

35. **a.** Use a truth table to show that $\sim p \vee q$ and $p \rightarrow q$ are equivalent.

b. Use the result from part (a) to write a statement that is equivalent to

The triangle is not isosceles or it has two equal sides.

36. Select the statement that is equivalent to

Joe grows mangos or oranges.

a. If Joe grows mangos, he does not grow oranges.

b. If Joe grows oranges, he does not grow mangos.

c. If Joe does not grow mangos, he grows oranges.

d. Joe grows both mangos and oranges.

In Exercises 37–38, use a truth table to determine whether the two statements are equivalent.

37. $\sim(p \leftrightarrow q), \sim p \vee \sim q$

38. $\sim p \wedge (q \vee r), (\sim p \wedge q) \vee (\sim p \wedge r)$

In Exercises 39–42, write the converse, inverse, and contrapositive of each statement.

39. If I am in Atlanta, then I am in the South.

40. If I am in class, then today is not a holiday.

41. If I work hard, then I pass all courses.

42. $\sim p \rightarrow \sim q$

In Exercises 43–45, write the negation of each conditional statement.

43. If an argument is sound, then it is valid.

44. If I do not work hard, then I do not succeed.

45. $\sim r \rightarrow p$

In Exercises 46–48, use De Morgan's laws to write a statement that is equivalent to each statement.

46. It is not true that both Chicago and Maine are cities.

47. It is not true that Ernest Hemingway was a musician or an actor.

48. If a number is not positive and not negative, the number is 0.

In Exercises 49–51, use De Morgan's laws to write the negation of each statement.

49. I work hard or I do not succeed.

50. She is not using her car and she is taking a bus.

51. $\sim p \vee q$

In Exercises 52–55, determine which, if any, of the three given statements are equivalent.

52. **a.** If it is hot, then I use the air conditioner.

b. If it is not hot, then I do not use the air conditioner.

c. It is not hot or I use the air conditioner.

53. **a.** If she did not play, then we lost.

b. If we did not lose, then she played.

c. She did not play and we did not lose.

54. **a.** He is here or I'm not.

b. If I'm not here, he is.

c. It is not true that he isn't here and I am.

55. **a.** If the class interests me and I like the teacher, then I enjoy studying.

b. If the class interests me, then I like the teacher and I enjoy studying.

c. The class interests me, or I like the teacher and I enjoy studying.

3.7

In Exercises 56–57, use a truth table to determine whether the symbolic form of the argument is valid or invalid.

56. $p \rightarrow q$
$\underline{\sim q}$
$\therefore p$

57. $p \wedge q$
$\underline{q \rightarrow r}$
$\therefore p \rightarrow r$

In Exercises 58–63, translate each argument into symbolic form. Then determine whether the argument is valid or invalid. You may use a truth table or, if applicable, compare the argument's symbolic form to a standard valid or invalid form.

58. If Tony plays, the team wins.
The team won.
∴ Tony played.

59. My plant is fertilized or it turns yellow.
My plant is turning yellow.
∴ My plant is not fertilized.

60. A majority of legislators vote for a bill or that bill does not become law.
A majority of legislators did not vote for bill x.
∴ Bill x did not become law.

61. Having good eye–hand coordination is necessary for being a good baseball player.
Todd does not have good eye–hand coordination.
∴ Todd is not a good baseball player.

62. If you love the person you marry, you can fall out of love with that person.
If you do not love the person you marry, you can fall in love with that person.
∴ You love the person you marry if and only if you can fall out of love with that person.

63. If I purchase season tickets to the football games, then I do not attend all lectures.
If I do well in school, then I attend all lectures.
∴ If I do not do well in school, then I purchased season tickets to the football games.

3.8

In Exercises 64–69, use Euler diagrams to determine whether each argument is valid or invalid.

64. All birds have feathers.
All parrots have feathers.
∴ All parrots are birds.

65. All botanists are scientists.
All scientists have college degrees.
∴ All botanists have college degrees.

66. All native desert plants can withstand severe drought.
No tree ferns can withstand severe drought.
∴ No tree ferns are native desert plants.

67. All native desert plants can withstand severe drought.
No tree ferns are native desert plants.
∴ No tree ferns can withstand severe drought.

68. All poets are writers.
Some writers are wealthy.
∴ Some poets are wealthy.

69. Some people enjoy reading.
All people who enjoy reading appreciate language.
∴ Some people appreciate language.

Chapter 3 Test

Use the following representations in Exercises 1–6:

p: I'm registered.
q: I'm a citizen.
r: I vote.

Express each compound statement in English.

1. $(p \wedge q) \rightarrow r$

2. $\sim r \leftrightarrow (\sim p \vee \sim q)$

3. $\sim(p \vee q)$

Express each English statement in symbolic form.

4. I am registered and a citizen, or I do not vote.

5. If I am not registered or not a citizen, then I do not vote.

6. Being a citizen is necessary for voting.

In Exercises 7–8, write the negation of the statement.

7. All numbers are divisible by 5.

8. Some people wear glasses.

In Exercises 9–11, construct a truth table for the statement.

9. $p \wedge (\sim p \vee q)$

10. $\sim(p \wedge q) \leftrightarrow (\sim p \vee \sim q)$

11. $p \leftrightarrow q \vee r$

12. Write the following statement in symbolic form and construct a truth table. Then indicate one set of conditions that makes the compound statement false.

If you break the law and change the law, then you have not broken the law.

In Exercises 13–14, determine the truth value for each statement when p is false, q is true, and r is false.

13. $\sim(q \rightarrow r)$

14. $(p \vee r) \leftrightarrow (\sim r \wedge p)$

15. The bar graph shows that as costs changed over the decades, Americans devoted less of their budget to groceries and more to health care.

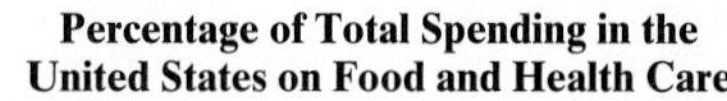

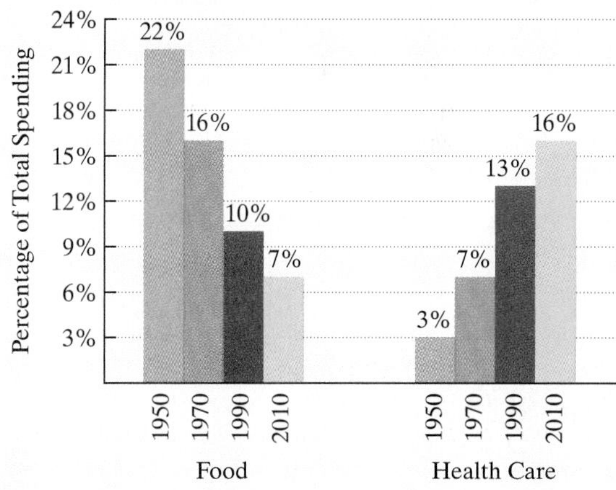

Source: *Time*, October 10, 2011

Write the following statement in symbolic form. Then use the information displayed by the graph to determine the truth value of the compound statement.

There was no increase in the percentage of their budget that Americans spent on food, or there was an increase in the percentage spent on health care and by 2010 the percentage spent on health care was more than triple the percentage spent on food.

16. Select the statement below that is equivalent to

Gene is an actor or a musician.

a. If Gene is an actor, then he is not a musician.

b. If Gene is not an actor, then he is a musician.

c. It is false that Gene is not an actor or not a musician.

d. If Gene is an actor, then he is a musician.

17. Write the contrapositive of
If it is August, it does not snow.

18. Write the converse and the inverse of the following statement:
If the radio is playing, then I cannot concentrate.

19. Write the negation of the following statement:
If it is cold, we do not use the pool.

20. Write a statement that is equivalent to
It is not true that the test is today or the party is tonight.

21. Write the negation of the following statement:
The banana is green and it is not ready to eat.

In Exercises 22–23, determine which, if any, of the three given statements are equivalent.

22. a. If I'm not feeling well, I'm grouchy.

b. I'm feeling well or I'm grouchy.

c. If I'm feeling well, I'm not grouchy.

23. a. It is not true that today is a holiday or tomorrow is a holiday.

b. If today is not a holiday, then tomorrow is not a holiday.

c. Today is not a holiday and tomorrow is not a holiday.

Determine whether each argument in Exercises 24–29 is valid or invalid.

24. If a parrot talks, it is intelligent.
This parrot is intelligent.
∴ This parrot talks.

25. I am sick or I am tired.
I am not tired.
∴ I am sick.

26. I am going if and only if you are not.
You are going.
∴ I'm going.

27. All mammals are warm-blooded.
All dogs are warm-blooded.
∴ All dogs are mammals.

28. All conservationists are advocates of solar-powered cars.
No oil company executives are advocates of solar-powered cars.
∴ No conservationists are oil company executives.

29. All rabbis are Jewish.
Some Jews observe kosher dietary traditions.
∴ Some rabbis observe kosher dietary traditions.

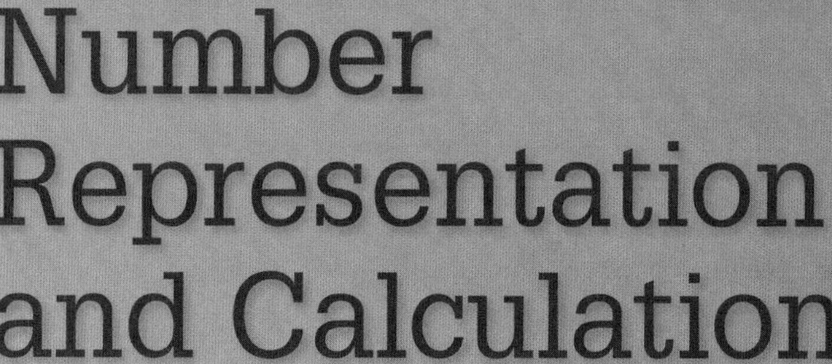

Number Representation and Calculation

4

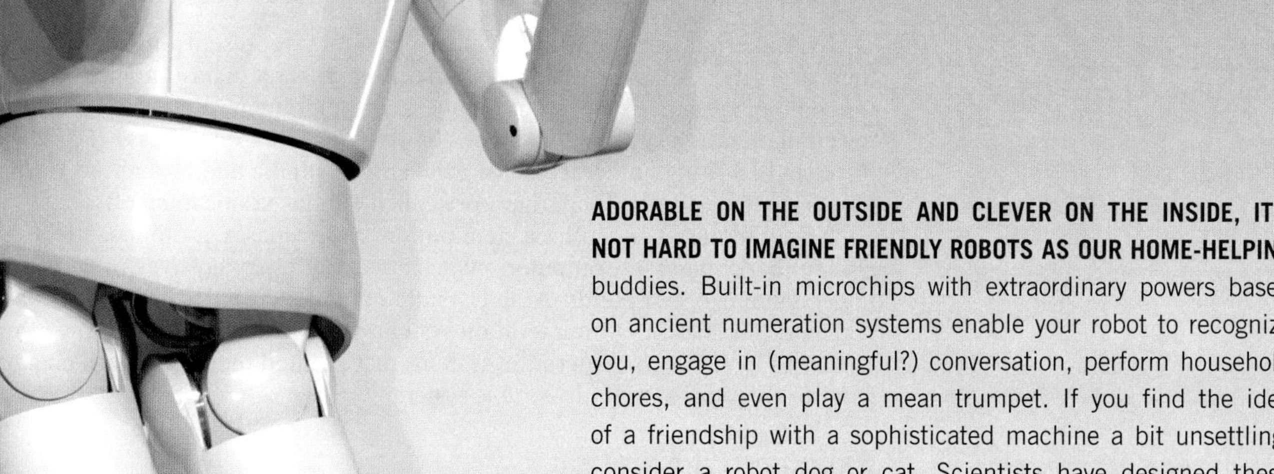

ADORABLE ON THE OUTSIDE AND CLEVER ON THE INSIDE, IT'S NOT HARD TO IMAGINE FRIENDLY ROBOTS AS OUR HOME-HELPING buddies. Built-in microchips with extraordinary powers based on ancient numeration systems enable your robot to recognize you, engage in (meaningful?) conversation, perform household chores, and even play a mean trumpet. If you find the idea of a friendship with a sophisticated machine a bit unsettling, consider a robot dog or cat. Scientists have designed these critters to blend computer technology with the cuddly appeal of animals. They move, play, and sleep like real pets, and can even be programmed to sing and dance. Without an understanding of how we represent numbers, none of this technology could exist.

Here's where you'll find these applications:

Connections between binary numeration systems and computer technology are discussed in "Letters and Words in Base Two" on page 226, "Music in Base Two" on page 228, and "Base Two, Logic, and Computers" on page 237.

4.1 Our Hindu-Arabic System and Early Positional Systems

WHAT AM I SUPPOSED TO LEARN?

After studying this section, you should be able to:

1. Evaluate an exponential expression.
2. Write a Hindu-Arabic numeral in expanded form.
3. Express a number's expanded form as a Hindu-Arabic numeral.
4. Understand and use the Babylonian numeration system.
5. Understand and use the Mayan numeration system.

ALL OF US HAVE AN INTUITIVE understanding of *more* and *less*. As humanity evolved, this sense of more and less was used to develop a system of counting. A tribe needed to know how many sheep it had and whether the flock was increasing or decreasing in number. The earliest way of keeping count probably involved some tally method, using one vertical mark on a cave wall for each sheep. Later, a variety of vocal sounds developed as a tally for the number of things in a group. Finally, written symbols, or numerals, were used to represent numbers.

A **number** is an abstract idea that addresses the question, "How many?" A **numeral** is a symbol used to represent a number. For example, the answer to "How many dots: • • • • •?" is a number, but as soon as we use a word or symbol to describe that number we are using a numeral.

Different symbols may be used to represent the same number. Numerals used to represent how many buffalo are shown in **Figure 4.1** include

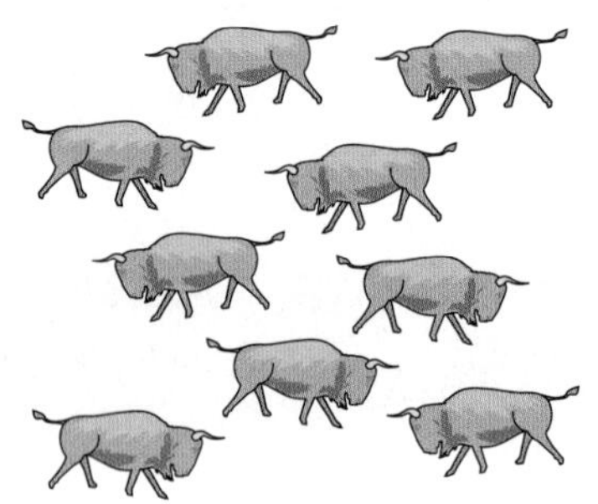

FIGURE 4.1

𝍸 ||||	IX	9
Tally method	Roman numeral	Hindu-Arabic numeral

We take numerals and the numbers that they represent for granted and use them every day. A **system of numeration** consists of a set of basic numerals and rules for combining them to represent numbers. It took humanity thousands of years to invent numeration systems that made computation a reasonable task. Today we use a system of writing numerals that was invented in India and brought to Europe by the Arabs. Our numerals are therefore called **Hindu-Arabic numerals**.

Like literature or music, a numeration system has a profound effect on the culture that created it. Computers, which affect our everyday lives, are based on an understanding of our Hindu-Arabic system of numeration. In this section, we study the characteristics of our numeration system. We also take a brief journey through history to look at two numeration systems that pointed the way toward an amazing cultural creation, our Hindu-Arabic system.

1 *Evaluate an exponential expression.*

Exponential Notation

An understanding of *exponents* is important in understanding the characteristics of our numeration system.

A BRIEF REVIEW *Exponents*

- If n is a natural number,

Exponent or Power

$$b^n = \underbrace{b \cdot b \cdot b \cdot \cdots \cdot b}_{b \text{ appears as a factor } n \text{ times.}}$$

Base

- b^n is read "the nth power of b" or "b to the nth power." Thus, the nth power of b is defined as the product of n factors of b. The expression b^n is called an **exponential expression**. Furthermore, $b^1 = b$.

Exponential Expression	Read	Evaluation
8^1	8 to the first power	$8^1 = 8$
5^2	5 to the second power or 5 squared	$5^2 = 5 \cdot 5 = 25$
6^3	6 to the third power or 6 cubed	$6^3 = 6 \cdot 6 \cdot 6 = 216$
10^4	10 to the fourth power	$10^4 = 10 \cdot 10 \cdot 10 \cdot 10 = 10{,}000$
2^5	2 to the fifth power	$2^5 = 2 \cdot 2 \cdot 2 \cdot 2 \cdot 2 = 32$

- Powers of 10 play an important role in our system of numeration.

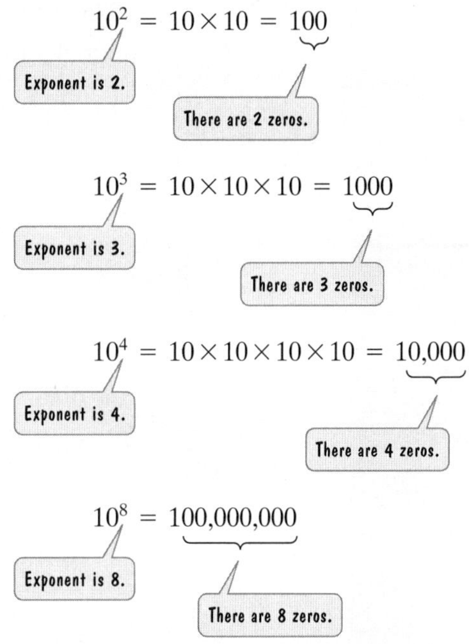

In general, the number of zeros appearing to the right of the 1 in any numeral that represents a power of 10 is the same as the exponent on that power of 10.

2 *Write a Hindu-Arabic numeral in expanded form.*

Our Hindu-Arabic Numeration System

An important characteristic of our Hindu-Arabic system is that we can write the numeral for any number, large or small, using only ten symbols. The ten symbols that we use are

$$0, 1, 2, 3, 4, 5, 6, 7, 8, \text{and } 9.$$

These symbols are called **digits**, from the Latin word for fingers.

With the use of exponents, Hindu-Arabic numerals can be written in **expanded form** in which the value of the digit in each position is made clear. In a Hindu-Arabic numeral, the place value of the first digit on the right is 1. The place value of the second digit from the right is 10. The place value of the third digit from the right is 100, or 10^2. For example, we can write 663 in expanded form by thinking of 663 as six 100s plus six 10s plus three 1s. This means that 663 in expanded form is

$$\begin{aligned} 663 &= (6 \times 100) + (6 \times 10) + (3 \times 1) \\ &= (6 \times 10^2) + (6 \times 10^1) + (3 \times 1). \end{aligned}$$

Because the value of a digit varies according to the position it occupies in a numeral, the Hindu-Arabic numeration system is called a **positional-value**, or **place-value**, system. The positional values in the system are based on powers of 10 and are

$$\ldots, 10^5, 10^4, 10^3, 10^2, 10^1, 1.$$

EXAMPLE 1 Writing Hindu-Arabic Numerals in Expanded Form

Write each of the following in expanded form:

a. 3407 **b.** 53,525.

SOLUTION

a. $3407 = (3 \times 10^3) + (4 \times 10^2) + (0 \times 10^1) + (7 \times 1)$

or $= (3 \times 1000) + (4 \times 100) + (0 \times 10) + (7 \times 1)$

Because $0 \times 10^1 = 0$, this term could be left out, but the expanded form is clearer when it is included.

b. $53{,}525 = (5 \times 10^4) + (3 \times 10^3) + (5 \times 10^2) + (2 \times 10^1) + (5 \times 1)$

or $= (5 \times 10{,}000) + (3 \times 1000) + (5 \times 100) + (2 \times 10) + (5 \times 1)$

CHECK POINT 1 Write each of the following in expanded form:

a. 4026

b. 24,232.

3 *Express a number's expanded form as a Hindu-Arabic numeral.*

EXAMPLE 2 Expressing a Number's Expanded Form as a Hindu-Arabic Numeral

Express each expanded form as a Hindu-Arabic numeral:

a. $(7 \times 10^3) + (5 \times 10^1) + (4 \times 1)$

b. $(6 \times 10^5) + (8 \times 10^1)$.

SOLUTION

For clarification, we begin by showing all powers of 10, starting with the highest exponent given. Any power of 10 that is left out is expressed as 0 times that power of 10.

a. $(7 \times 10^3) + (5 \times 10^1) + (4 \times 1)$

$= (7 \times 10^3) + (0 \times 10^2) + (5 \times 10^1) + (4 \times 1)$

$= 7054$

b. $(6 \times 10^5) + (8 \times 10^1)$

$= (6 \times 10^5) + (0 \times 10^4) + (0 \times 10^3) + (0 \times 10^2) + (8 \times 10^1) + (0 \times 1)$

$= 600{,}080$

CHECK POINT 2 Express each expanded form as a Hindu-Arabic numeral:

a. $(6 \times 10^3) + (7 \times 10^1) + (3 \times 1)$

b. $(8 \times 10^4) + (9 \times 10^2)$.

Examples 1 and 2 show how there would be no Hindu-Arabic system without an understanding of zero and the invention of a symbol to represent nothingness. The system must have a symbol for zero to serve as a placeholder in case one or more powers of 10 are not needed. The concept of zero was a new and radical invention, one that changed our ability to think about the world.

Tally Sticks

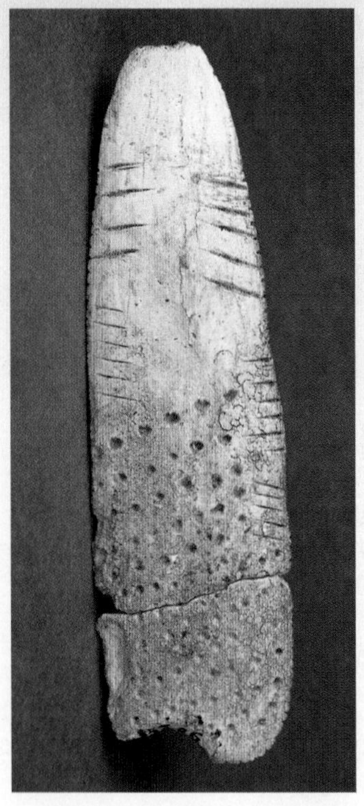

This notched reindeer antler dates from 15,000 B.C. Humans learned how to keep track of numbers by tallying notches on bones with the same intelligence that led them to preserve and use fire, and at around the same time. Using tally sticks, early people grasped the idea that nine buffalo and nine sheep had something in common: the abstract idea of *nine*. As the human mind conceived of numbers separately from the things they represented, systems of numeration developed.

Early Positional Systems

"It took men about five thousand years, counting from the beginning of number symbols, to think of a symbol for nothing."
—Isaac Asimov, *Asimov on Numbers*

Our Hindu-Arabic system developed over many centuries. Its digits can be found carved on ancient Hindu pillars over 2200 years old. In 1202, the Italian mathematician Leonardo Fibonacci (1170–1250) introduced the system to Europe, writing of its special characteristic: "With the nine Hindu digits and the Arab symbol 0, any number can be written." The Hindu-Arabic system came into widespread use only when printing was invented in the fifteenth century.

The Hindu-Arabic system uses powers of 10. However, positional systems can use powers of any number, not just 10. Think about our system of time, based on powers of 60:

$$1 \text{ minute} = 60 \text{ seconds}$$

$$1 \text{ hour} = 60 \text{ minutes} = 60 \times 60 \text{ seconds} = 60^2 \text{ seconds}.$$

What is significant in a positional system is position and the powers that positions convey. The first early positional system that we will discuss uses powers of 60, just like those used for units of time.

4 *Understand and use the Babylonian numeration system.*

The Babylonian Numeration System

TABLE 4.1 Babylonian Numerals		
Babylonian numerals	∨	<
Hindu-Arabic numerals	1	10

The city of Babylon, 55 miles south of present-day Baghdad, was the center of Babylonian civilization that lasted for about 1400 years between 2000 B.C. and 600 B.C. The Babylonians used wet clay as a writing surface. Their clay tablets were heated and dried to give a permanent record of their work, which we are able to decipher and read today. **Table 4.1** gives the numerals of this civilization's numeration system. Notice that the system uses only two symbols, ∨ for 1 and < for 10.

The place values in the Babylonian system use powers of 60. The place values are

$$\ldots, \quad 60^3, \quad 60^2, \quad 60^1, \quad 1.$$

$60^3 = 60 \times 60 \times 60 = 216{,}000$ $\qquad$ $60^2 = 60 \times 60 = 3600$

The Babylonians left a space to distinguish the various place values in a numeral from one another. For example,

∨ < ∨∨

means

$$= (1 \times 60^2) + (10 \times 60^1) + (1 + 1) \times 1$$
$$= (1 \times 3600) + (10 \times 60) + (2 \times 1)$$
$$= 3600 + 600 + 2 = 4202.$$

EXAMPLE 3 ***Converting from Babylonian Numerals to Hindu-Arabic Numerals***

Write each Babylonian numeral as a Hindu-Arabic numeral:

a. ∨∨ <∨ <<∨∨

b. << ∨ ∨∨ <<∨.

Blitzer Bonus

Numbers and Bird Brains

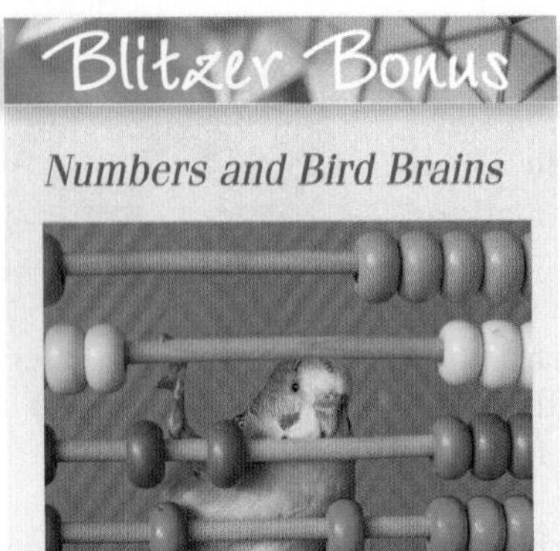

Birds have large, well-developed brains and are more intelligent than is suggested by the slur "bird brain." Parakeets can learn to count to seven. They have been taught to identify a box of food by counting the number of small objects in front of the box. Some species of birds can tell the difference between two and three. If a nest contains four eggs and one is taken, the bird will stay in the nest to protect the remaining three eggs. However, if two of the four eggs are taken, the bird recognizes that only two remain and will desert the nest, leaving the remaining two eggs unprotected.

Birds easily master complex counting problems. The sense of *more* and *less* that led to the development of numeration systems is not limited to the human species.

SOLUTION

Represent the numeral in each place as a familiar Hindu-Arabic numeral using 1 for ∨ and 10 for <. Multiply each Hindu-Arabic numeral by its respective place value. Then find the sum of these products.

a. ∨∨   <∨   <<∨∨

(Place value: 60^2; Place value: 60; Place value: 1. Symbol for 1; Symbol for 10)

$$= (1 + 1) \times 60^2 + (10 + 1) \times 60^1 + (10 + 10 + 1 + 1) \times 1$$
$$= (2 \times 60^2) + (11 \times 60^1) + (22 \times 1)$$
$$= (2 \times 3600) + (11 \times 60) + (22 \times 1)$$
$$= 7200 + 660 + 22 = 7882$$

This sum indicates that the given Babylonian numeral is 7882 when written as a Hindu-Arabic numeral.

b. <<   ∨   ∨∨   <<∨

(Place value: 60^3; Place value: 60^2; Place value: 60; Place value: 1. Symbol for 10; Symbol for 1)

$$= (10 + 10) \times 60^3 + 1 \times 60^2 + (1 + 1) \times 60 + (10 + 10 + 1) \times 1$$
$$= (20 \times 60^3) + (1 \times 60^2) + (2 \times 60) + (21 \times 1)$$
$$= (20 \times 216{,}000) + (1 \times 3600) + (2 \times 60) + (21 \times 1)$$
$$= 4{,}320{,}000 + 3600 + 120 + 21 = 4{,}323{,}741$$

This sum indicates that the given Babylonian numeral is 4,323,741 when written as a Hindu-Arabic numeral.

A major disadvantage of the Babylonian system is that it did not contain a symbol for zero. Some Babylonian tablets have a larger gap between the numerals or the insertion of the symbol ⪡ to indicate a missing place value, but this led to some ambiguity and confusion.

CHECK POINT 3 Write each Babylonian numeral as a Hindu-Arabic numeral:

a. ∨∨∨   <<   <<<∨   **b.** ∨∨   <   <∨∨   ∨.

The Mayan Numeration System

The Maya, a tribe of Central American Indians, lived on the Yucatan Peninsula. At its peak, between A.D. 300 and 1000, their civilization covered an area including parts of Mexico, all of Belize and Guatemala, and part of Honduras. They were famous for their magnificent architecture, their astronomical and mathematical knowledge, and their excellence in the arts. Their numeration system was the first to have a symbol for zero. **Table 4.2** gives the Mayan numerals.

TABLE 4.2 Mayan Numerals

0	1	2	3	4	5	6	7	8	9
⬭	•	••	•••	••••	—	• —	•• —	••• —	•••• —
10	**11**	**12**	**13**	**14**	**15**	**16**	**17**	**18**	**19**
— —	• — —	•• — —	••• — —	•••• — —	— — —	• — — —	•• — — —	••• — — —	•••• — — —

The place values in the Mayan system are

$$\ldots,\quad 18\times 20^3,\quad 18\times 20^2,\quad 18\times 20,\quad 20,\quad 1.$$

$18 \times 20 \times 20 \times 20 = 144{,}000$ $18 \times 20 \times 20 = 7200$ $18 \times 20 = 360$

Notice that instead of giving the third position a place value of 20^2, the Mayans used 18×20. This was probably done so that their calendar year of 360 days would be a basic part of the numeration system.

Numerals in the Mayan system are expressed vertically. The place value at the bottom of the column is 1.

5 *Understand and use the Mayan numeration system.*

EXAMPLE 4 Using the Mayan Numeration System

Write each Mayan numeral as a Hindu-Arabic numeral:

a.  ••••
═
⊖
••
—
••
═

b.  ═
≡
••
⊖
—

SOLUTION

Represent the numeral in each row as a familiar Hindu-Arabic numeral using **Table 4.2**. Multiply each Hindu-Arabic numeral by its respective place value. Then find the sum of these products.

a.

Mayan numeral		Hindu-Arabic numeral		Place value				
•••• ═	=	14	×	18×20^2	=	14×7200	=	100,800
⊖	=	0	×	18×20	=	0×360	=	0
•• —	=	7	×	20	=	7×20	=	140
•• ═	=	12	×	1	=	12×1	=	12
								100,952

The sum on the right indicates that the given Mayan numeral is 100,952 when written as a Hindu-Arabic numeral.

b.

Mayan numeral		Hindu-Arabic numeral		Place value				
═	=	10	×	18×20^3	=	$10\times 144{,}000$	=	1,440,000
≡	=	15	×	18×20^2	=	15×7200	=	108,000
••	=	2	×	18×20	=	2×360	=	720
⊖	=	0	×	20	=	0×20	=	0
—	=	5	×	1	=	5×1	=	5
								1,548,725

The sum on the right indicates that the given Mayan numeral is 1,548,725 when written as a Hindu-Arabic numeral.

CHECK POINT 4 Write each Mayan numeral as a Hindu-Arabic numeral:

a.  •
═
•••
⊖
•••
═

b.  ••
⊖
•
—
•
≡
═

Concept and Vocabulary Check

Fill in each blank so that the resulting statement is true.

1. A number addresses the question "How many?" A symbol used to represent a number is called a ________.
2. Our system of numeration is called the __________ system.
3. 10^7 = 1 followed by ____ zeros = __________.
4. When we write 547 as $(5 \times 10^2) + (4 \times 10^1) + (7 \times 1)$, we are using an _________ form in which the value of the digit in each position is made clear. Consequently, ours is a place-value or __________-value system.
5. Using the form described in Exercise 4,
   $74{,}716 = (7 \times \_\_\_\_) + (4 \times \_\_\_\_) + (7 \times \_\_\_\_) + (1 \times \_\_\_\_) + (6 \times \_\_\_\_)$
6. Our numeration system uses powers of ____, whereas the Babylonian numeration system uses powers of ____.
7. Using v for 1 and < for 10, <v vv <<vv
   $= (\_ + \_) \times 60^2 + (\_ + \_) \times 60^1 + (\_ + \_ + \_ + \_) \times 1$
8. Using v for 1 and < for 10, < v vv <v
   $= (10 \times \_) + (1 \times \_) + (2 \times \_) + (\_ \times \_)$
9. The place values in the Mayan numeration system are
   $\ldots$, ______, 18×20^3, 18×20^2, 18×20, 20, 1.
10.

				Place value		
••	=	2	×	______	=	______
•••	=	3	×	______	=	______
•••• (with bar)	=	9	×	______	=	______

The sum of the three numbers on the right is _____, so the given Mayan numeral is _____ in our numeration system.

Exercise Set 4.1

Practice Exercises

In Exercises 1–8, evaluate the expression.

1. 5^2
2. 6^2
3. 2^3
4. 4^3
5. 3^4
6. 2^4
7. 10^5
8. 10^6

In Exercises 9–22, write each Hindu-Arabic numeral in expanded form.

9. 36
10. 65
11. 249
12. 698
13. 703
14. 902
15. 4856
16. 5749
17. 3070
18. 9007
19. 34,569
20. 67,943
21. 230,007,004
22. 909,006,070

In Exercises 23–32, express each expanded form as a Hindu-Arabic numeral.

23. $(7 \times 10^1) + (3 \times 1)$
24. $(9 \times 10^1) + (4 \times 1)$
25. $(3 \times 10^2) + (8 \times 10^1) + (5 \times 1)$
26. $(7 \times 10^2) + (5 \times 10^1) + (3 \times 1)$
27. $(5 \times 10^5) + (2 \times 10^4) + (8 \times 10^3) + (7 \times 10^2) + (4 \times 10^1) + (3 \times 1)$
28. $(7 \times 10^6) + (4 \times 10^5) + (2 \times 10^4) + (3 \times 10^3) + (1 \times 10^2) + (9 \times 10^1) + (6 \times 1)$
29. $(7 \times 10^3) + (0 \times 10^2) + (0 \times 10^1) + (2 \times 1)$
30. $(9 \times 10^4) + (0 \times 10^3) + (0 \times 10^2) + (4 \times 10^1) + (5 \times 1)$
31. $(6 \times 10^8) + (2 \times 10^3) + (7 \times 1)$
32. $(3 \times 10^8) + (5 \times 10^4) + (4 \times 1)$

*In Exercises 34–46, use **Table 4.1** on page 219 to write each Babylonian numeral as a Hindu-Arabic numeral.*

33. <<vvv
34. <<<vv
35. <<v vv
36. << <vv
37. <<< <<<vvv
38. <<v <<vvvv
39. vvv <vv vvv
40. vv <v <<vv
41. <<v vvvv <v
42. <vv <vvvv <<
43. <v <v <v <v
44. << << <vv <vv
45. vvv vv v v
46. v v vv vvv

*In Exercises 47–60, use **Table 4.2** on page 220 to write each Mayan numeral as a Hindu-Arabic numeral.*

47.
48.
49.
50.
51.
52.
53.
54.
55.
56.
57.
58.
59.
60.

Practice Plus

In Exercises 61–64, express the result of each addition as a Hindu-Arabic numeral in expanded form.

61. ∨ ‹‹ ‹‹∨ + ‹∨ ‹‹‹ ∨∨∨∨

62. ‹∨ ‹ ‹∨∨∨ + ∨∨∨ ‹‹ ∨∨

63. 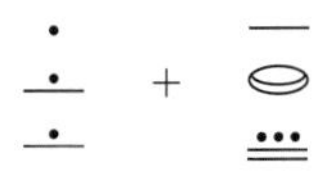 •
•̲
•̲ + —
⊖
•••̳

64.  ••
—
••̲ + •̲
═
⊖

If n is a natural number, then $10^{-n} = \frac{1}{10^n}$. *Negative powers of 10 can be used to write the decimal part of Hindu-Arabic numerals in expanded form. For example,*

$$\begin{aligned} 0.8302 &= (8 \times 10^{-1}) + (3 \times 10^{-2}) + (0 \times 10^{-3}) + (2 \times 10^{-4}) \\ &= \left(8 \times \frac{1}{10^1}\right) + \left(3 \times \frac{1}{10^2}\right) + \left(0 \times \frac{1}{10^3}\right) + \left(2 \times \frac{1}{10^4}\right) \\ &= \left(8 \times \frac{1}{10}\right) + \left(3 \times \frac{1}{100}\right) + \left(0 \times \frac{1}{1000}\right) + \left(2 \times \frac{1}{10{,}000}\right). \end{aligned}$$

In Exercises 65–72, express each expanded form as a Hindu-Arabic numeral.

65. $(4 \times 10^{-1}) + (7 \times 10^{-2}) + (5 \times 10^{-3}) + (9 \times 10^{-4})$

66. $(6 \times 10^{-1}) + (8 \times 10^{-2}) + (1 \times 10^{-3}) + (2 \times 10^{-4})$

67. $(7 \times 10^{-1}) + (2 \times 10^{-4}) + (3 \times 10^{-6})$

68. $(8 \times 10^{-1}) + (3 \times 10^{-4}) + (7 \times 10^{-6})$

69. $(5 \times 10^{3}) + (3 \times 10^{-2})$

70. $(7 \times 10^{4}) + (5 \times 10^{-3})$

71. $(3 \times 10^{4}) + (7 \times 10^{2}) + (5 \times 10^{-2}) + (8 \times 10^{-3}) + (9 \times 10^{-5})$

72. $(7 \times 10^{5}) + (3 \times 10^{2}) + (2 \times 10^{-1}) + (2 \times 10^{-3}) + (1 \times 10^{-5})$

Application Exercises

The Chinese "rod system" of numeration is a base ten positional system. The digits for 1 through 9 are shown as follows:

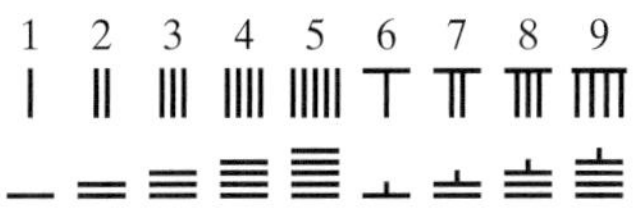

1	2	3	4	5	6	7	8	9
𝍩	𝍪	𝍫	𝍬	𝍭	𝍮	𝍯	𝍰	𝍱
𝍠	𝍡	𝍢	𝍣	𝍤	𝍥	𝍦	𝍧	𝍨

The vertical digits in the second row are used for place values of 1, 10^2, 10^4, *and all even powers of 10. The horizontal digits in the third row are used for place values of* 10^1, 10^3, 10^5, 10^7, *and all odd powers of 10. A blank space is used for the digit zero. In Exercises 73–76, write each Chinese "rod system" numeral as a Hindu-Arabic numeral.*

73. 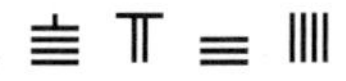 𝍧 𝍯 𝍢 𝍬

74. 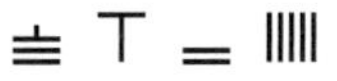 𝍧 𝍮 𝍡 𝍭

75. 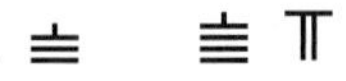 𝍧   𝍨 𝍯

76.  𝍦 𝍰   𝍪

77. Humans have debated for decades about what messages should be sent to the stars to grab the attention of extraterrestrials and demonstrate our mathematical prowess. In the 1970s, Soviet scientists suggested we send the exponential message

$$10^2 + 11^2 + 12^2 = 13^2 + 14^2.$$

The Soviets called this equation "mind-catching." Evaluate the exponential expressions and verify that the sums on the two sides are equal. What is the significance of this sum?

Explaining the Concepts

78. Describe the difference between a number and a numeral.

79. Explain how to evaluate 7^3.

80. What is the base in our Hindu-Arabic numeration system? What are the digits in the system?

81. Why is a symbol for zero needed in a positional system?

82. Explain how to write a Hindu-Arabic numeral in expanded form.

83. Describe one way that the Babylonian system is similar to the Hindu-Arabic system and one way that it is different from the Hindu-Arabic system.

84. Describe one way that the Mayan system is similar to the Hindu-Arabic system and one way that it is different from the Hindu-Arabic system.

85. **Research activity**. Write a report on the history of the Hindu-Arabic system of numeration. Useful references include history of mathematics books, encyclopedias, and the Internet.

Critical Thinking Exercises

Make Sense? *In Exercises 86–89, determine whether each statement makes sense or does not make sense, and explain your reasoning.*

86. I read that a certain star is 10^4 light-years from Earth, which means 100,000 light-years.

87. When expressing $(4 \times 10^6) + (3 \times 10^2)$ as a Hindu-Arabic numeral, only two digits, 4 and 3, are needed.

88. I write Babylonian numerals horizontally, using spaces to distinguish place values.

89. When I write a Mayan numeral as a Hindu-Arabic numeral, if ⊖ appears in any row, I ignore the place value of that row and immediately write 0 for the product.

90. Write ∨ ‹∨∨ ‹∨ as a Mayan numeral.

91. Write ••̲
••̲
••̲ as a Babylonian numeral.

92. Use Babylonian numerals to write the numeral that precedes and follows the numeral

‹∨ ‹‹‹‹‹∨∨∨∨∨∨∨∨∨.

Group Exercise

93. Your group task is to create an original positional numeration system that is different from the three systems discussed in this section.

a. Construct a table showing your numerals and the corresponding Hindu-Arabic numerals.

b. Explain how to represent numbers in your system, and express a three-digit and a four-digit Hindu-Arabic numeral in your system.

4.2 Number Bases in Positional Systems

WHAT AM I SUPPOSED TO LEARN?

After studying this section, you should be able to:

1. Change numerals in bases other than ten to base ten.
2. Change base ten numerals to numerals in other bases.

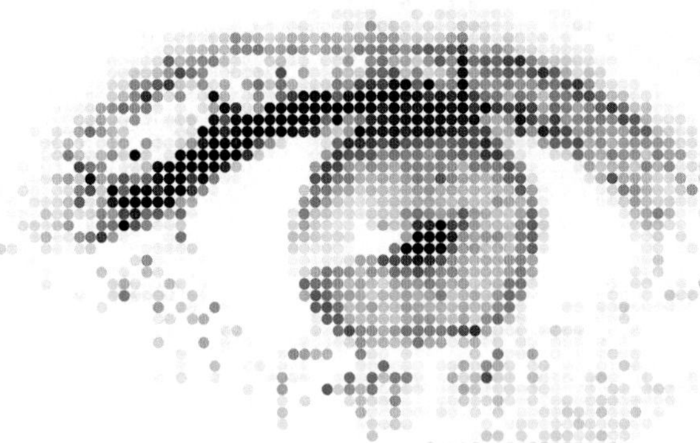

YOU ARE BEING DRAWN DEEPER into cyberspace, spending more time online each week. With constantly improving high-resolution images, cyberspace is reshaping your life by nourishing shared enthusiasms. The people who built your computer talk of bandwidth that will give you the visual experience, in high-definition 3-D format, of being in the same room with a person who is actually in another city.

Because of our ten fingers and ten toes, the base ten Hindu-Arabic system seems to be an obvious choice. However, it is not base ten that computers use to process information and communicate with one another. Your experiences in cyberspace are sustained with a binary, or base two, system. In this section, we study numeration systems with bases other than ten. An understanding of such systems will help you to appreciate the nature of a positional system. You will also attain a better understanding of the computations you have used all of your life. You will even get to see how the world looks from a computer's point of view.

1 *Change numerals in bases other than ten to base ten.*

Changing Numerals in Bases Other Than Ten to Base Ten

The base of a positional numeration system refers to the number of individual digit symbols that can be used in that system as well as to the number whose powers define the place values. For example, the digit symbols in a base two system are 0 and 1. The place values in a base two system are powers of 2:

$$\ldots, 2^4, 2^3, 2^2, 2^1, 1$$
$$\text{or } \ldots, 2 \times 2 \times 2 \times 2, 2 \times 2 \times 2, 2 \times 2, 2, 1$$
$$\text{or } \ldots, 16, 8, 4, 2, 1.$$

When a numeral appears without a subscript, it is assumed that the base is ten. Bases other than ten are indicated with a spelled-out subscript, as in the numeral

$$1001_{\text{two}}.$$

This numeral is read "one zero zero one base two." Do not read it as "one thousand one" because that terminology implies a base ten numeral, naming 1001 in base ten.

We can convert 1001_{two} to a base ten numeral by following the same procedure used in Section 4.1 to change the Babylonian and Mayan numerals to base ten Hindu-Arabic numerals. In the case of 1001_{two}, the numeral has four places. From left to right, the place values are 2^3, 2^2, 2^1, and 1. Multiply each digit in the numeral by its respective place value. Then add these products.

$$\begin{aligned} 1001_{\text{two}} &= (1 \times 2^3) + (0 \times 2^2) + (0 \times 2^1) + (1 \times 1) \\ &= (1 \times 8) + (0 \times 4) + (0 \times 2) + (1 \times 1) \\ &= 8 + 0 + 0 + 1 \\ &= 9 \end{aligned}$$

Thus,

$$1001_{\text{two}} = 9.$$

In base two, we do not need a digit symbol for 2 because

$$10_{\text{two}} = (1 \times 2^1) + (0 \times 1) = 2.$$

Likewise, the base ten numeral 3 is represented as 11_{two}, the base ten numeral 4 as 100_{two}, and so on. **Table 4.3** shows base ten numerals from 0 through 20 and their base two equivalents.

TABLE 4.3

Base Ten	Base Two
0	0
1	1
2	10
3	11
4	100
5	101
6	110
7	111
8	1000
9	1001
10	1010
11	1011
12	1100
13	1101
14	1110
15	1111
16	10000
17	10001
18	10010
19	10011
20	10100

In any base, the digit symbols begin at 0 and go up to one less than the base. In base b, the digit symbols begin at 0 and go up to $b - 1$. The place values in a base b system are powers of b:

$$\ldots, b^4, b^3, b^2, b, 1.$$

Table 4.4 shows the digit symbols and place values in various bases.

TABLE 4.4 Digit Symbols and Place Values in Various Bases

Base	Digit Symbols	Place Values
two	0, 1	$\ldots, 2^4, 2^3, 2^2, 2^1, 1$
three	0, 1, 2	$\ldots, 3^4, 3^3, 3^2, 3^1, 1$
four	0, 1, 2, 3	$\ldots, 4^4, 4^3, 4^2, 4^1, 1$
five	0, 1, 2, 3, 4	$\ldots, 5^4, 5^3, 5^2, 5^1, 1$
six	0, 1, 2, 3, 4, 5	$\ldots, 6^4, 6^3, 6^2, 6^1, 1$
seven	0, 1, 2, 3, 4, 5, 6	$\ldots, 7^4, 7^3, 7^2, 7^1, 1$
eight	0, 1, 2, 3, 4, 5, 6, 7	$\ldots, 8^4, 8^3, 8^2, 8^1, 1$
nine	0, 1, 2, 3, 4, 5, 6, 7, 8	$\ldots, 9^4, 9^3, 9^2, 9^1, 1$
ten	0, 1, 2, 3, 4, 5, 6, 7, 8, 9	$\ldots, 10^4, 10^3, 10^2, 10^1, 1$

We have seen that in base two, 10_{two} represents one group of 2 and no groups of 1. Thus, $10_{\text{two}} = 2$. Similarly, in base six, 10_{six} represents one group of 6 and no groups of 1. Thus, $10_{\text{six}} = 6$. In general $10_{\text{base } b}$ represents one group of b and no groups of 1. This means that $10_{\text{base } b} = b$.

Here is the procedure for changing a numeral in a base other than ten to base ten:

CHANGING TO BASE TEN

To change a numeral in a base other than ten to a base ten numeral,

1. Find the place value for each digit in the numeral.
2. Multiply each digit in the numeral by its respective place value.
3. Find the sum of the products in step 2.

TECHNOLOGY

You can use a calculator to convert to base ten. For example, to convert 4726_{eight} to base ten as in Example 1, press the following keys:

Many Scientific Calculators

4 [×] 8 [y^x] 3 [+] 7 [×] 8 [y^x] 2 [+] 2 [×] 8 [+] 6 [=].

Many Graphing Calculators

4 [×] 8 [∧] 3 [+] 7 [×] 8 [∧] 2 [+] 2 [×] 8 [+] 6 [ENTER].

Some graphing calculators require that you press the right arrow key to exit the exponent after entering the exponent.

EXAMPLE 1 Converting to Base Ten

Convert 4726_{eight} to base ten.

SOLUTION

The given base eight numeral has four places. From left to right, the place values are

$$8^3, 8^2, 8^1, \text{ and } 1.$$

Multiply each digit in the numeral by its respective place value. Then find the sum of these products.

Place value: 8^3	Place value: 8^2	Place value: 8^1	Place value: 1
4	7	2	6_{eight}

$$\begin{aligned} 4726_{\text{eight}} &= (4 \times 8^3) + (7 \times 8^2) + (2 \times 8^1) + (6 \times 1) \\ &= (4 \times 8 \times 8 \times 8) + (7 \times 8 \times 8) + (2 \times 8) + (6 \times 1) \\ &= 2048 + 448 + 16 + 6 \\ &= 2518 \end{aligned}$$

CHECK POINT 1 Convert 3422_{five} to base ten.

EXAMPLE 2 Converting to Base Ten

Convert 100101_{two} to base ten.

SOLUTION

Multiply each digit in the numeral by its respective place value. Then find the sum of these products.

Place value: 2^5	Place value: 2^4	Place value: 2^3	Place value: 2^2	Place value: 2^1	Place value: 1
1	0	0	1	0	1_{two}

$$\begin{aligned}100101_{\text{two}} &= (1 \times 2^5) + (0 \times 2^4) + (0 \times 2^3) + (1 \times 2^2) + (0 \times 2^1) + (1 \times 1)\\ &= (1 \times 32) + (0 \times 16) + (0 \times 8) + (1 \times 4) + (0 \times 2) + (1 \times 1)\\ &= 32 + 0 + 0 + 4 + 0 + 1\\ &= 37\end{aligned}$$

CHECK POINT 2 Convert 110011_{two} to base ten.

Blitzer Bonus

Letters and Words in Base Two

Letters are converted into base two numbers for computer processing. The capital letters A through Z are assigned 65 through 90, with each number expressed in base two. Thus, the binary code for A(65) is 1000001. Similarly, the lowercase letters a through z are assigned 97 through 122 in base two.

The German mathematician Wilhelm Leibniz was the first modern thinker to promote the base two system. He never imagined that one day the base two system would enable computers to process information and communicate with one another.

Wilhelm Leibniz (1646–1716)

The word *digital* in computer technology refers to a method of encoding numbers, letters, visual images, and sounds using a **binary**, or base two, **system** of 0s and 1s. Because computers use electrical signals that are groups of on–off pulses of electricity, the digits in base two are convenient. In binary code, 1 indicates the passage of an electrical pulse ("on") and 0 indicates its interruption ("off"). For example, the number 37 (100101_{two}) becomes the binary code on–off–off–on–off–on. Microchips in a computer store and process these binary signals.

In addition to base two, computer applications often involve base eight, called an **octal system**, and base sixteen, called a **hexadecimal system**. Base sixteen presents a problem because digit symbols are needed from 0 up to one less than the base. This means that we need more digit symbols than the ten (0, 1, 2, 3, 4, 5, 6, 7, 8, and 9) used in our base ten system. Computer programmers use the letters A, B, C, D, E, and F as base sixteen digit symbols for the numbers ten through fifteen, respectively.

Additional digit symbols in base sixteen:

A = 10	B = 11
C = 12	D = 13
E = 14	F = 15

EXAMPLE 3 Converting to Base Ten

Convert $EC7_{\text{sixteen}}$ to base ten.

SOLUTION

From left to right, the place values are

$$16^2, 16^1, \text{ and } 1.$$

The digit symbol E represents 14 and the digit symbol C represents 12. Although this numeral looks a bit strange, follow the usual procedure: Multiply each digit in the numeral by its respective place value. Then find the sum of these products.

Place value: 16^2 | Place value: 16^1 | Place value: 1

E C 7_{sixteen}

E = 14 C = 12

$$\begin{aligned} EC7_{\text{sixteen}} &= (14 \times 16^2) + (12 \times 16^1) + (7 \times 1) \\ &= (14 \times 16 \times 16) + (12 \times 16) + (7 \times 1) \\ &= 3584 + 192 + 7 \\ &= 3783 \end{aligned}$$

CHECK POINT 3 Convert $AD4_{\text{sixteen}}$ to base ten.

GREAT QUESTION!

I understand why the on–off pulses of electricity result in computers using a binary, base two system. But what's the deal with octal, base eight, and hexadecimal, base sixteen? Why are these systems used by computer programmers?

Octal and hexadecimal systems provide a compact way of representing binary numerals. With fewer digit symbols to read and fewer operations to perform, the computer's operating speed is increased and space in its memory is saved. In particular:

- Every three-digit binary numeral can be replaced by a one-digit octal numeral. ($2^3 = 8$)

Binary Numeral	Octal Equivalent
000	0
001	1
010	2
011	3
100	4
101	5
110	6
111	7

Computer programmers use this table to go back and forth between binary and octal.

Example

$\underbrace{110}_{6}\ \underbrace{111}_{7}{}_{\text{two}} = 67_{\text{eight}}$

Example

$\underbrace{2}_{010}\ \underbrace{3}_{011}{}_{\text{eight}} = 010011_{\text{two}}$

- Every four-digit binary numeral can be replaced by a one-digit hexadecimal numeral. ($2^4 = 16$)

Binary Numeral	Hex Equivalent
0000	0
0001	1
0010	2
0011	3
0100	4
0101	5
0110	6
0111	7

Binary Numeral	Hex Equivalent
1000	8
1001	9
1010	A
1011	B
1100	C
1101	D
1110	E
1111	F

Computer programmers use these tables to convert between binary and hexadecimal.

Starting on the right, group digits into groups of four, adding zeros in front as needed.

Example

$1111001101_{\text{two}} = \underbrace{0011}_{3}\ \underbrace{1100}_{C}\ \underbrace{1101}_{D} = 3CD_{\text{sixteen}}$

Example

$\underbrace{6}_{0110}\ \underbrace{F}_{1111}\ \underbrace{A}_{1010}{}_{\text{sixteen}} = 011011111010_{\text{two}}$

2 *Change base ten numerals to numerals in other bases.*

Changing Base Ten Numerals to Numerals in Other Bases

To convert a base ten numeral to a numeral in a base other than ten, we need to find how many groups of each place value are contained in the base ten numeral. When the base ten numeral consists of one or two digits, we can do this mentally. For example, suppose that we want to convert the base ten numeral 6 to a base four numeral. The place values in base four are

$$\ldots, 4^3, 4^2, 4, 1.$$

The place values that are less than 6 are 4 and 1. We can express 6 as one group of four and two ones:

$$6_{\text{ten}} = (1 \times 4) + (2 \times 1) = 12_{\text{four}}.$$

EXAMPLE 4 *A Mental Conversion from Base Ten to Base Five*

Convert the base ten numeral 8 to a base five numeral.

SOLUTION

The place values in base five are

$$\ldots, 5^3, 5^2, 5, 1.$$

The place values that are less than 8 are 5 and 1. We can express 8 as one group of five and three ones:

$$8_{\text{ten}} = (1 \times 5) + (3 \times 1) = 13_{\text{five}}.$$

CHECK POINT 4 Convert the base ten numeral 6 to a base five numeral.

If a conversion cannot be performed mentally, you can use divisions to determine how many groups of each place value are contained in a base ten numeral.

EXAMPLE 5 *Using Divisions to Convert from Base Ten to Base Eight*

Convert the base ten numeral 299 to a base eight numeral.

SOLUTION

The place values in base eight are

$$\ldots, 8^3, 8^2, 8^1, 1, \quad \text{or} \quad \ldots, 512, 64, 8, 1.$$

The place values that are less than 299 are 64, 8, and 1. We can use divisions to show how many groups of each of these place values are contained in 299. Divide 299 by 64. Divide the remainder by 8.

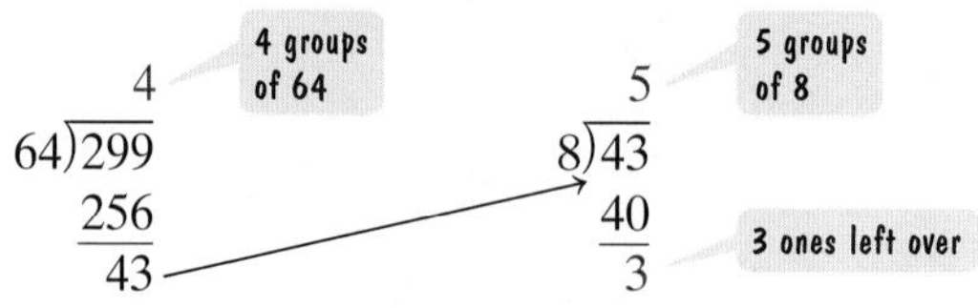

Blitzer Bonus

Music in Base Two

Digital technology can represent musical sounds as a series of numbers in binary code. The music is recorded by a microphone that converts it into an electric signal in the form of a sound wave. The computer measures the height of the sound wave thousands of times per second, converting each discrete measurement into a base two number. The resulting series of binary numbers is used to encode the music. Compact discs store these digital sequences using millions of bumps (1) and spaces (0) on a tiny spiral that is just one thousand-millionth of a meter wide and almost 3 miles (5 kilometers) long. A CD player converts the digital data back into sound.

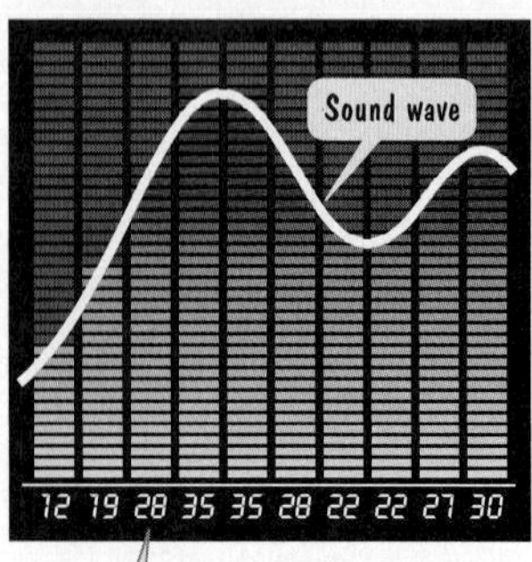

These divisions show that 299 can be expressed as four groups of 64, five groups of 8, and three ones:

$$299 = (4 \times 64) + (5 \times 8) + (3 \times 1)$$
$$= (4 \times 8^2) + (5 \times 8^1) + (3 \times 1)$$
$$= 453_{\text{eight}}.$$

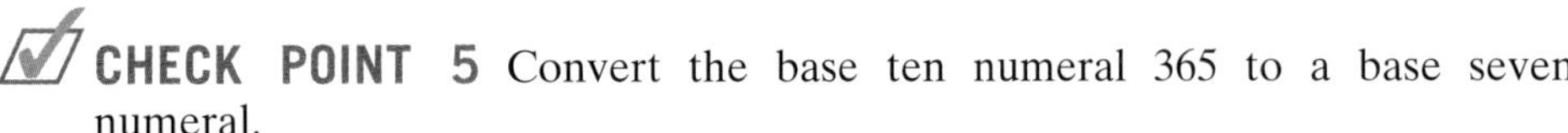

CHECK POINT 5 Convert the base ten numeral 365 to a base seven numeral.

EXAMPLE 6 *Using Divisions to Convert from Base Ten to Base Two*

Convert the base ten numeral 26 to a base two numeral.

SOLUTION

The place values in base two are

$$\ldots, 2^5, 2^4, 2^3, 2^2, 2^1, 1 \quad \text{or} \quad \ldots, 32, 16, 8, 4, 2, 1.$$

We use the powers of 2 that are less than 26 and perform successive divisions by these powers.

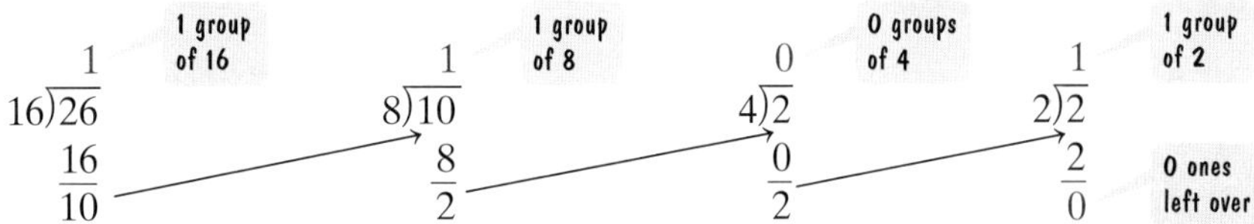

Using these four quotients and the final remainder, we can immediately write the answer.

$$26 = 11010_{\text{two}}$$

CHECK POINT 6 Convert the base ten numeral 51 to a base two numeral.

EXAMPLE 7 *Using Divisions to Convert from Base Ten to Base Six*

Convert the base ten numeral 3444 to a base six numeral.

SOLUTION

The place values in base six are

$$\ldots, 6^5, 6^4, 6^3, 6^2, 6^1, 1, \quad \text{or} \quad \ldots, 7776, 1296, 216, 36, 6, 1.$$

We use the powers of 6 that are less than 3444 and perform successive divisions by these powers.

- $1296\overline{)3444}$ = 2; 2592; remainder 852 — 2 groups of 1296
- $216\overline{)852}$ = 3; 648; remainder 204 — 3 groups of 216
- $36\overline{)204}$ = 5; 180; remainder 24 — 5 groups of 36
- $6\overline{)24}$ = 4; 24; remainder 0 — 4 groups of 6; 0 ones left over

Using these four quotients and the final remainder, we can immediately write the answer.

$$3444 = 23540_{\text{six}}$$

CHECK POINT 7 Convert the base ten numeral 2763 to a base five numeral.

Concept and Vocabulary Check

Fill in each blank so that the resulting statement is true.

1. In the numeral 324_{five}, the base is ____. In this base, the digit symbols are ________.
2. $324_{\text{five}} = (3 \times \_\_) + (2 \times \_\_) + (4 \times \_\_)$
3. In the numeral 1101_{two}, the base is ____. In this base, the digit symbols are _____.
4. $1101_{\text{two}} = (1 \times \_\_) + (1 \times \_\_) + (0 \times \_\_) + (1 \times \_\_)$
5. To mentally convert 9 from base ten to base six, we begin with the place values in base six:. . . , $6^2, 6^1$, 1. We then express 9 as one group of six and three ones:

$$9_{\text{ten}} = (1 \times 6) + (3 \times 1).$$

Thus, $9_{\text{ten}} = \_\__{\text{six}}$.

6. To convert 473 from base ten to base eight, we begin with the place values in base eight:. . . 8^2 (or 64), 8, 1. We then perform two divisions.

$$\begin{array}{r} 7 \\ 64\overline{)473} \\ \underline{448} \\ 25 \end{array} \qquad \begin{array}{r} 3 \\ 8\overline{)25} \\ \underline{24} \\ 1 \end{array}$$

Thus, $473_{\text{ten}} = \_\__{\text{eight}}$.

7. Computers use three bases to perform operations. They are the binary system, or base _____, the octal system, or base _____, and the hexadecimal system, or base _______.

Exercise Set 4.2

Practice Exercises

In Exercises 1–18, convert the numeral to a numeral in base ten.

1. 43_{five} **2.** 34_{five} **3.** 52_{eight}
4. 67_{eight} **5.** 132_{four} **6.** 321_{four}
7. 1011_{two} **8.** 1101_{two} **9.** 2035_{six}
10. 2073_{nine} **11.** 70355_{eight}
12. 41502_{six} **13.** 2096_{sixteen}
14. 3104_{fifteen} **15.** 110101_{two}
16. 101101_{two} **17.** $\text{ACE5}_{\text{sixteen}}$
18. $\text{EDF7}_{\text{sixteen}}$

In Exercises 19–32, mentally convert each base ten numeral to a numeral in the given base.

19. 7 to base five **20.** 9 to base five
21. 11 to base seven **22.** 12 to base seven
23. 2 to base two **24.** 3 to base two
25. 5 to base two **26.** 6 to base two
27. 8 to base two **28.** 9 to base two
29. 13 to base four **30.** 19 to base four
31. 37 to base six **32.** 25 to base six

In Exercises 33–48, convert each base ten numeral to a numeral in the given base.

33. 87 to base five **34.** 85 to base seven
35. 108 to base four **36.** 199 to base four
37. 19 to base two **38.** 23 to base two
39. 57 to base two **40.** 63 to base two
41. 90 to base two **42.** 87 to base two
43. 138 to base three **44.** 129 to base three
45. 386 to base six **46.** 428 to base nine
47. 1599 to base seven
48. 1346 to base eight

Practice Plus

*In Exercises 49–52, use **Table 4.1** on page 219 to write each Hindu-Arabic numeral as a Babylonian numeral.*

49. 3052 **50.** 6704
51. 23,546 **52.** 41,265

*In Exercises 53–56, use **Table 4.2** on page 220 to write each Hindu-Arabic numeral as a Mayan numeral.*

53. 9307
54. 8703
55. 28,704
56. 34,847
57. Convert 34_{five} to base seven.
58. Convert 46_{eight} to base five.
59. Convert 110010011_{two} to base eight.
60. Convert 101110001_{two} to base eight.

Application Exercises

Read the Blitzer Bonus on page 226. Then use the information in the essay to solve Exercises 61–68.

In Exercises 61–64, write the binary representation for each letter.

61. F **62.** Y
63. m **64.** p

In Exercises 65–66, break each binary sequence into groups of seven digits and write the word represented by the sequence.

65. 101000010000011001100
66. 1001100101010110000111001011

In Exercises 67–68, write a sequence of binary digits that represents each word.

67. Mom
68. Dad

Explaining the Concepts

69. Explain how to determine the place values for a four-digit numeral in base six.
70. Describe how to change a numeral in a base other than ten to a base ten numeral.
71. Describe how to change a base ten numeral to a numeral in another base.
72. The illustration in the Great Question! feature on page 227 includes the following sentence:

    There are 10 kinds of people in the world—those who understand binary and those who don't.

    Explain the joke.

Critical Thinking Exercises

Make Sense? *In Exercises 73–76, determine whether each statement makes sense or does not make sense, and explain your reasoning.*

73. Base b contains $b - 1$ digit symbols.
74. Bases greater than ten are not possible because we are limited to ten digit symbols.
75. Because the binary system has only two available digit symbols, representing numbers in binary form requires more digits than in any other base.
76. I converted 28 to base two by performing successive divisions by powers of 2, starting with 2^5.

In Exercises 77–78, write, in the indicated base, the counting numbers that precede and follow the number expressed by the given numeral.

77. 888_{nine}
78. $\text{EC5}_{\text{sixteen}}$
79. Arrange from smallest to largest:

    $11111011_{\text{two}}, 3\text{A}6_{\text{twelve}}, 673_{\text{eight}}$.

Group Exercises

The following topics are appropriate for either individual or group research projects. A report should be given to the class on the researched topic. Useful references include history of mathematics books, books whose purpose is to excite the reader about mathematics, encyclopedias, and the Internet.

80. Societies That Use Numeration Systems with Bases Other Than Ten
81. The Use of Fingers to Represent Numbers
82. Applications of Bases Other Than Ten
83. Binary, Octal, Hexadecimal Bases and Computers
84. Babylonian and Mayan Civilizations and Their Contributions

4.3 Computation in Positional Systems

WHAT AM I SUPPOSED TO LEARN?

After studying this section, you should be able to:

1. Add in bases other than ten.
2. Subtract in bases other than ten.
3. Multiply in bases other than ten.
4. Divide in bases other than ten.

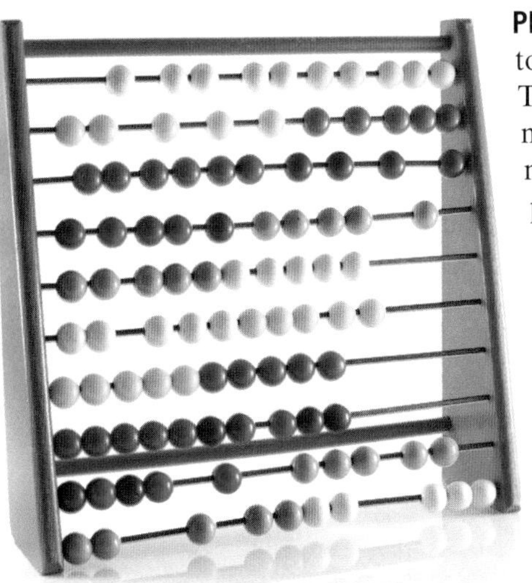

PEOPLE HAVE ALWAYS LOOKED FOR WAYS to make calculations faster and easier. The Hindu-Arabic system of numeration made computation simpler and less mysterious. More people were able to perform computation with ease, leading to the widespread use of the system.

All computations in bases other than ten are performed exactly like those in base ten. However, when a computation is equal to or exceeds the given base, use the mental conversions discussed in the previous section to convert from the base ten numeral to a numeral in the desired base.

1 *Add in bases other than ten.*

Addition

EXAMPLE 1 *Addition in Base Four*

Add:

$$\begin{array}{r} 33_{\text{four}} \\ +\ 13_{\text{four}} \\ \hline \end{array}.$$

The 4^1 or fours' column

The ones' column

$$\begin{array}{r} 33_{four} \\ +13_{four} \\ \hline \end{array}$$

SOLUTION

We will begin by adding the numbers in the right-hand column. In base four, the digit symbols are 0, 1, 2, and 3. If a sum in this, or any, column exceeds 3, we will have to convert this base ten number to base four. We begin by adding the numbers in the right-hand, or ones', column:

$$3_{four} + 3_{four} = 6.$$

6 is not a digit symbol in base four. However, we can express 6 as one group of four and two ones left over:

$$3_{four} + 3_{four} = 6_{ten} = (1 \times 4) + (2 \times 1) = 12_{four}.$$

Now we record the sum of the right-hand column, 12_{four}:

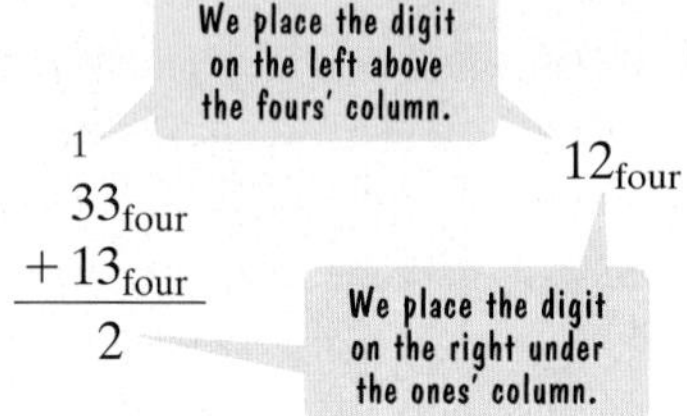

$$\begin{array}{r} 1 \\ 33_{four} \\ +13_{four} \\ \hline 2\phantom{_{four}} \end{array} \qquad 12_{four}$$

Next, we add the three digits in the fours' column:

$$1_{four} + 3_{four} + 1_{four} = 5.$$

5 is not a digit symbol in base four. However, we can express 5 as one group of four and one left over:

$$1_{four} + 3_{four} + 1_{four} = 5_{ten} = (1 \times 4) + (1 \times 1) = 11_{four}.$$

Record the 11_{four}.

$$\begin{array}{r} 1 \\ 33_{four} \\ +13_{four} \\ \hline 112_{four} \end{array}$$

This is the desired sum.

You can check the sum by converting 33_{four}, 13_{four}, and 112_{four} to base ten: $33_{four} = 15$, $13_{four} = 7$, and $112_{four} = 22$. Because $15 + 7 = 22$, our work is correct.

CHECK POINT 1 Add:

$$\begin{array}{r} 32_{five} \\ +44_{five}. \\ \hline \end{array}$$

EXAMPLE 2 *Addition in Base Two*

Add:

$$\begin{array}{r} 111_{two} \\ +101_{two}. \\ \hline \end{array}$$

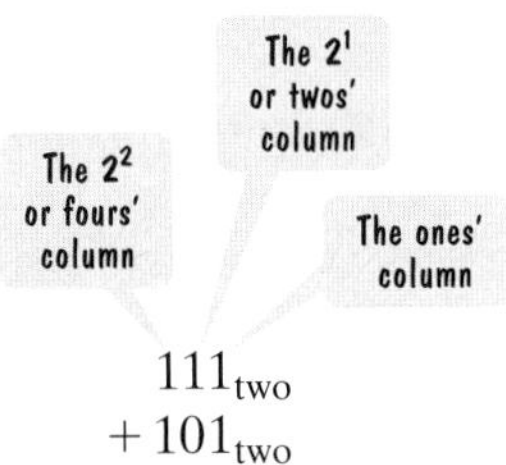

SOLUTION

We begin by adding the numbers in the right-hand, or ones', column:

$$1_{two} + 1_{two} = 2.$$

2 is not a digit symbol in base two. We can express 2 as one group of 2 and zero ones left over:

$$1_{two} + 1_{two} = 2_{ten} = (1 \times 2) + (0 \times 1) = 10_{two}.$$

Now we record the sum of the right-hand column, 10_{two}:

$$\begin{array}{r} {\scriptstyle 1} \\ 111_{two} \\ +\,101_{two} \\ \hline 0\phantom{_{two}} \end{array} \qquad 10_{two}$$

We place the digit on the left above the twos' column.

We place the digit on the right under the ones' column.

Next, we add the three digits in the twos' column:

$$1_{two} + 1_{two} + 0_{two} = 2_{ten} = (1 \times 2) + (0 \times 1) = 10_{two}.$$

Now we record the sum of the middle column, 10_{two}:

$$\begin{array}{r} {\scriptstyle 1\,1} \\ 111_{two} \\ +\,101_{two} \\ \hline 00\phantom{_{two}} \end{array} \qquad 10_{two}$$

We place the digit on the left above the fours' column.

We place the digit on the right under the twos' column.

Finally, we add the three digits in the fours' column:

$$1_{two} + 1_{two} + 1_{two} = 3.$$

3 is not a digit symbol in base two. We can express 3 as one group of 2 and one 1 left over:

$$1_{two} + 1_{two} + 1_{two} = 3_{ten} = (1 \times 2) + (1 \times 1) = 11_{two}.$$

Record the 11_{two}.

$$\begin{array}{r} {\scriptstyle 1\,1} \\ 111_{two} \\ +\,101_{two} \\ \hline 1100_{two} \end{array}$$

This is the desired sum.

You can check the sum by converting to base ten: $111_{two} = 7$, $101_{two} = 5$, and $1100_{two} = 12$. Because $7 + 5 = 12$, our work is correct.

CHECK POINT 2 Add:

$$\begin{array}{r} 111_{two} \\ +\,111_{two}. \\ \hline \end{array}$$

2 *Subtract in bases other than ten.*

Subtraction

To subtract in bases other than ten, we line up the digits with the same place values and subtract column by column, beginning with the column on the right. If "borrowing" is necessary to perform the subtraction, borrow the amount of the base. For example, when we borrow in base ten subtraction, we borrow 10s. Likewise, we borrow 2s in base two, 3s in base three, 4s in base four, and so on.

EXAMPLE 3 Subtraction in Base Four

Subtract:

$$\begin{array}{r} 31_{\text{four}} \\ -\ 12_{\text{four}}. \\ \hline \end{array}$$

SOLUTION

We start by performing subtraction in the right column, $1_{\text{four}} - 2_{\text{four}}$. Because 2_{four} is greater than 1_{four}, we need to borrow from the preceding column. We are working in base four, so we borrow one group of 4. This gives a sum of $4 + 1$, or 5, in base ten. Now we subtract 2 from 5, obtaining a difference of 3:

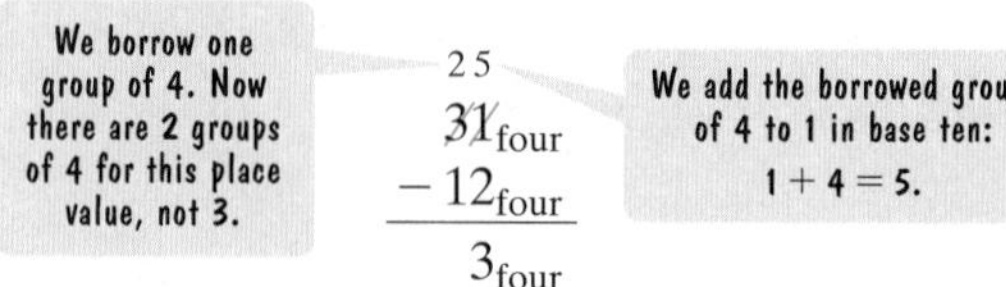

Now we perform the subtraction in the second column from the right.

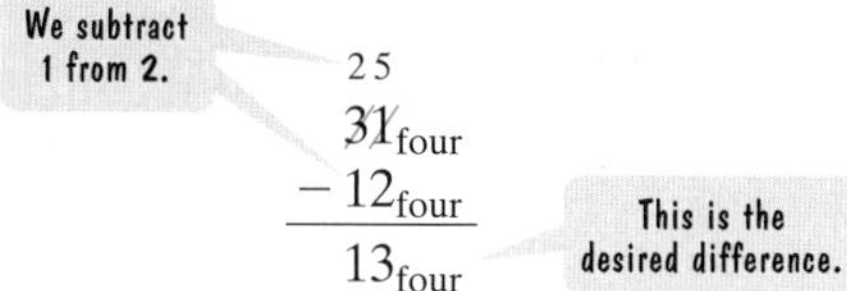

You can check the difference by converting to base ten: $31_{\text{four}} = 13$, $12_{\text{four}} = 6$, and $13_{\text{four}} = 7$. Because $13 - 6 = 7$, our work is correct.

CHECK POINT 3 Subtract:

$$\begin{array}{r} 41_{\text{five}} \\ -\ 23_{\text{five}}. \\ \hline \end{array}$$

EXAMPLE 4 Subtraction in Base Five

Subtract:

$$3431_{\text{five}} - 1242_{\text{five}}.$$

SOLUTION

Step 1. Borrow a group of 5 from the preceding column. This gives a sum of $5 + 1$, or 6, in base ten.

$$\begin{array}{r} {\scriptstyle 2\,6} \\ 343\not{1}_{\text{five}} \\ -\ 1242_{\text{five}} \\ \hline 4_{\text{five}} \end{array}$$

Step 2. $6 - 2 = 4$

Step 3. Borrow a group of 5 from the preceding column. This gives a sum of $5 + 2$, or 7, in base ten.

$$\begin{array}{r} {\scriptstyle 7} \\ {\scriptstyle 3\,2\,6} \\ 3\not{4}\not{3}\not{1}_{\text{five}} \\ -\ 1242_{\text{five}} \\ \hline 34_{\text{five}} \end{array}$$

Step 4. $7 - 4 = 3$

Step 5. No borrowing is needed for these two columns.

$$\begin{array}{r} {\scriptstyle 7} \\ {\scriptstyle 3\,2\,6} \\ 3\not{4}\not{3}\not{1}_{\text{five}} \\ -\ 1242_{\text{five}} \\ \hline 2134_{\text{five}} \end{array}$$

Step 6. $3 - 2 = 1$

Step 7. $3 - 1 = 2$

Thus, $3431_{\text{five}} - 1242_{\text{five}} = 2134_{\text{five}}$.

CHECK POINT 4 Subtract: $5144_{\text{seven}} - 3236_{\text{seven}}$.

A Revolution at the Supermarket

Computerized scanning registers "read" the universal product code on packaged goods and convert it to a base two numeral that is sent to the scanner's computer. The computer calls up the appropriate price and subtracts the sale from the supermarket's inventory.

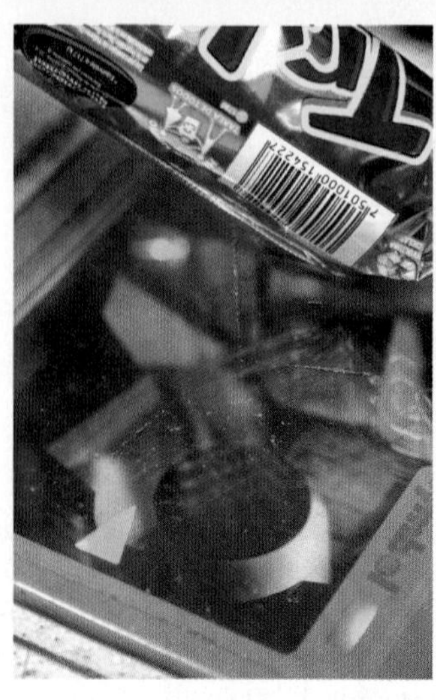

3 *Multiply in bases other than ten.*

Multiplication

EXAMPLE 5 Multiplication in Base Six

Multiply:

$$\begin{array}{r} 34_{\text{six}} \\ \times\ 2_{\text{six}} \\ \hline \end{array}.$$

SOLUTION

We multiply just as we do in base ten. That is, first we will multiply the digit 2 by the digit 4 directly above it. Then we will multiply the digit 2 by the digit 3 in the left column. Keep in mind that only the digit symbols 0, 1, 2, 3, 4, and 5 are permitted in base six. We begin with

$$2_{\text{six}} \times 4_{\text{six}} = 8_{\text{ten}} = (1 \times 6) + (2 \times 1) = 12_{\text{six}}.$$

Record the 2 and carry the 1:

$$\begin{array}{r} \scriptstyle 1\ \ \\ 34_{\text{six}} \\ \times\ 2_{\text{six}} \\ \hline 2_{\text{six}} \end{array}.$$

Our next computation involves both multiplication and addition:

$$(2_{\text{six}} \times 3_{\text{six}}) + 1_{\text{six}} = 6 + 1 = 7_{\text{ten}} = (1 \times 6) + (1 \times 1) = 11_{\text{six}}.$$

Record the 11_{six}.

$$\begin{array}{r} 34_{\text{six}} \\ \times\ 2_{\text{six}} \\ \hline 112_{\text{six}} \end{array}$$

This is the desired product.

Let's check the product by converting to base ten: $34_{\text{six}} = 22$, $2_{\text{six}} = 2$, and $112_{\text{six}} = 44$. Because $22 \times 2 = 44$, our work is correct.

CHECK POINT 5 Multiply:

$$\begin{array}{r} 45_{\text{seven}} \\ \times\ 3_{\text{seven}} \\ \hline \end{array}.$$

4 *Divide in bases other than ten.*

Division

The answer in a division problem is called a **quotient**. A multiplication table showing products in the same base as the division problem is helpful.

EXAMPLE 6 Division in Base Four

Use **Table 4.5**, showing products in base four, to perform the following division:

$$3_{\text{four}}\overline{)222_{\text{four}}}.$$

TABLE 4.5 Multiplication: Base Four

×	0	1	2	3
0	0	0	0	0
1	0	1	2	3
2	0	2	10	12
3	0	3	12	21

SOLUTION

We can use the same method to divide in base four that we use in base ten. Begin by dividing 22_{four} by 3_{four}. Use **Table 4.5** to find, in the vertical column headed by 3, the largest product that is less than or equal to 22_{four}. This product is 21_{four}. Because $3_{\text{four}} \times 3_{\text{four}} = 21_{\text{four}}$, the first number in the quotient is 3_{four}.

Quantum Computers

Rapid calculating without thought and skill was the motivating factor in the history of mechanical computing. Classical computers process data in binary bits that can be 1 or 0. Quantum computers rely on quantum bits, called qubits (pronounced *cubits*) that can be 1 or 0 *at the same time.* A quantum computer with a chip that has n qubits can perform 2^n calculations and operations simultaneously.

The quantum computer D-Wave Two (costing approximately \$10 million a pop!) has a niobium chip with 512 qubits and can perform 2^{512} calculations simultaneously. That's more operations than there are atoms in the universe. Corporations and government agencies believe that quantum computers will change how we cure disease, explore the heavens, and do business here on Earth. "The kind of physical effects that our machine has access to are simply not available to supercomputers, no matter how big you make them," says Colin Williams, D-Wave's director of development, who once worked as Stephen Hawking's research assistant. "We're tapping into the fabric of reality in a fundamentally new way, to make a kind of computer that the world has never seen."

Divisor: 3_{four}; First digit in the quotient: 3; Dividend: 222_{four}

$$3_{\text{four}}\overline{)222_{\text{four}}} \quad \text{quotient } 3$$

Now multiply $3_{\text{four}} \times 3_{\text{four}}$ and write the product, 21_{four}, under the first two digits of the dividend.

$$\begin{array}{r} 3\phantom{_{\text{four}}} \\ 3_{\text{four}}\overline{)222_{\text{four}}} \\ \underline{21}\phantom{2_{\text{four}}} \end{array}$$

Subtract: $22_{\text{four}} - 21_{\text{four}} = 1_{\text{four}}$.

$$\begin{array}{r} 3\phantom{_{\text{four}}} \\ 3_{\text{four}}\overline{)222_{\text{four}}} \\ \underline{21}\phantom{2_{\text{four}}} \\ 1\phantom{2_{\text{four}}} \end{array}$$

Bring down the next digit in the dividend, 2_{four}.

$$\begin{array}{r} 3\phantom{_{\text{four}}} \\ 3_{\text{four}}\overline{)222_{\text{four}}} \\ \underline{21}\phantom{2_{\text{four}}} \\ 12\phantom{_{\text{four}}} \end{array}$$

TABLE 4.5 Multiplication: Base Four (repeated)

×	0	1	2	3
0	0	0	0	0
1	0	1	2	3
2	0	2	10	12
3	0	3	12	21

We now return to **Table 4.5**. Find, in the vertical column headed by 3, the largest product that is less than or equal to 12_{four}. Because $3_{\text{four}} \times 2_{\text{four}} = 12_{\text{four}}$, the next numeral in the quotient is 2_{four}. We use this information to finish the division.

This is the desired quotient.

$$\begin{array}{r} 32_{\text{four}} \\ 3_{\text{four}}\overline{)222_{\text{four}}} \\ \underline{21}\phantom{2_{\text{four}}} \\ 12\phantom{_{\text{four}}} \\ \underline{12}\phantom{_{\text{four}}} \\ 0\phantom{_{\text{four}}} \end{array}$$

Let's check the quotient by converting to base ten: $3_{\text{four}} = 3$, $222_{\text{four}} = 42$, and $32_{\text{four}} = 14$. Because $3\overline{)42}$ with quotient 14, our work is correct.

CHECK POINT 6 Use **Table 4.5**, showing products in base four, to perform the following division:

$$2_{\text{four}}\overline{)112_{\text{four}}}.$$

Blitzer Bonus

Base Two, Logic, and Computers

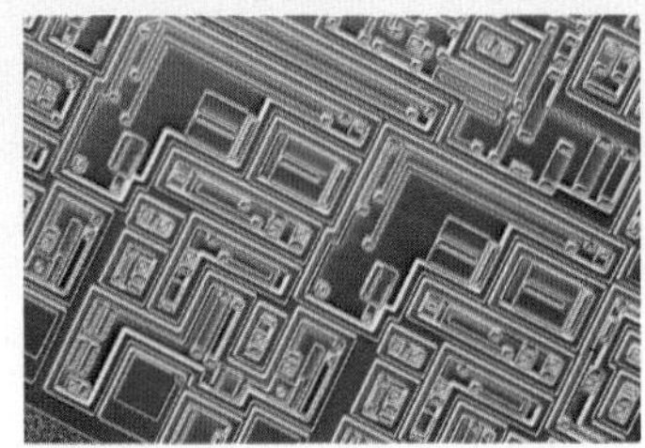
FIGURE 4.2

Smaller than a fingernail, a computer's microchip operates like a tiny electronic brain. The microchip in **Figure 4.2** is magnified almost 1200 times, revealing transistors with connecting tracks positioned above them. These tiny transistors switch on and off to control electronic signals, processing thousands of pieces of information per second. Since 1971, the number of transistors that can fit onto a single chip has increased from over 2000 to a staggering 2 billion in 2010.

We have seen that communication inside a computer takes the form of sequences of on–off electric pulses that digitally represent numbers, words, sounds, and visual images. These binary streams are manipulated when they pass through the microchip's gates, shown in **Figure 4.3**. The **not gate** takes a digital sequence and changes all the 0s to 1s and all the 1s to 0s.

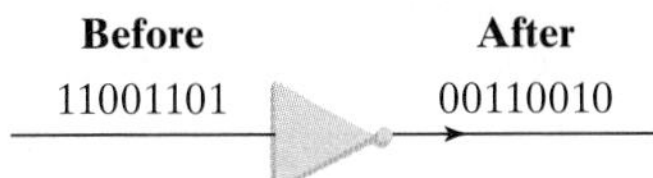

The *and* and *or gates* take two input sequences and produce one output sequence. The **and gate** outputs a 1 if both sequences have a 1; otherwise, it outputs a 0.

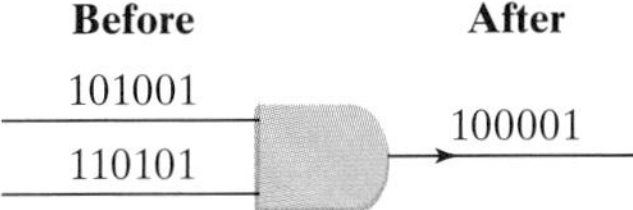

The **or gate** outputs a 1 if either sequence has a 1; otherwise, it outputs a 0.

Before **After**

101001
110101 111101

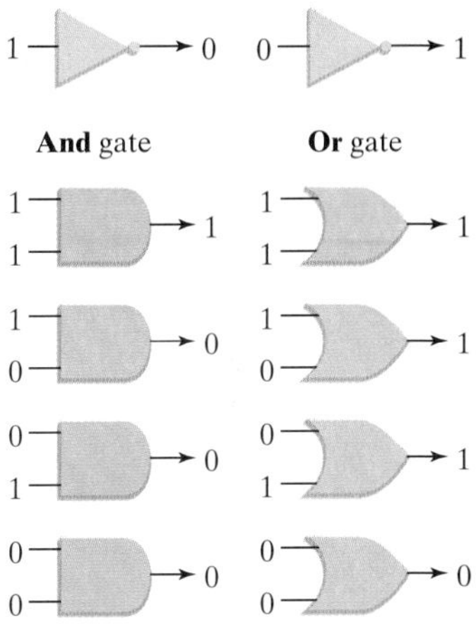

FIGURE 4.3

These gates are at the computational heart of a computer. They should remind you of negation, conjunction, and disjunction in logic, except that T is now 1 and F is now 0. Without the merging of base two and logic, computers as we know them would not exist.

Concept and Vocabulary Check

Fill in each blank so that the resulting statement is true.

1. $4_{\text{five}} + 2_{\text{five}} = 6_{\text{ten}} = (\_ \times 5) + (\_ \times 1) = \_\,_{\text{five}}$
2. $1_{\text{two}} + 1_{\text{two}} + 1_{\text{two}} = 3_{\text{ten}} = (\_ \times 2) + (\_ \times 1) = \_\,_{\text{two}}$
3. Consider the following addition in base eight:

$$\begin{array}{r} 57_{\text{eight}} \\ +\ 26_{\text{eight}} \\ \hline \end{array}$$

Step 1. $7_{\text{eight}} + 6_{\text{eight}} = 13_{\text{ten}} = (\_ \times 8) + (\_ \times 1) = \_\,_{\text{eight}}$

Step 2. Place the ____ above the eights' column and place the ____ under the ones' column.

4. Consider the following subtraction in base seven:

$$\begin{array}{r} 43_{\text{seven}} \\ -\ 25_{\text{seven}} \\ \hline \end{array}$$

We begin with the right column and borrow one group of ____ from the preceding column. This gives a sum of ____ + 3, or ____, in base ten. Then we subtract ____ from ____, obtaining a difference of ____. Now we perform the subtraction in the second column from the right. We subtract 2 from ____ and obtain ____. The answer to this subtraction problem is ____$_{\text{seven}}$.

5. Consider the following multiplication in base four:

$$\begin{array}{r} 23_{\text{four}} \\ \times\ \ 3_{\text{four}} \\ \hline \end{array}$$

We begin with $3_{\text{four}} \times 3_{\text{four}} = 9_{\text{ten}} = (\_ \times 4) + (\_ \times 1) = \_\,_{\text{four}}$. We record the ____ and carry the ____. Our next computation involves both multiplication and addition.

$(3_{\text{four}} \times 2_{\text{four}}) + \_\,_{\text{four}} = 8_{\text{ten}} = (\_ \times 4) + (\_ \times 1) = \_\,_{\text{four}}$

Recording this last computation, the desired product is ____$_{\text{four}}$.

6. We can use products in base three to perform the following division:

$$2_{\text{three}}\overline{)110_{\text{three}}}.$$

Multiplication: Base Three

×	0	1	2
0	0	0	0
1	0	1	2
2	0	2	11

Using the multiplication table, the first number in the quotient is ____$_{\text{three}}$. Completing the division, the quotient is ____$_{\text{three}}$.

7. True or False: Computation in bases other than ten is similar to the base ten arithmetic I learned as a child. ______

Exercise Set 4.3

Practice Exercises

In Exercises 1–12, add in the indicated base.

1. $23_{\text{four}} + 13_{\text{four}}$
2. $31_{\text{four}} + 22_{\text{four}}$
3. $11_{\text{two}} + 11_{\text{two}}$
4. $101_{\text{two}} + 11_{\text{two}}$
5. $342_{\text{five}} + 413_{\text{five}}$
6. $323_{\text{five}} + 421_{\text{five}}$
7. $645_{\text{seven}} + 324_{\text{seven}}$
8. $632_{\text{seven}} + 564_{\text{seven}}$
9. $6784_{\text{nine}} + 7865_{\text{nine}}$
10. $1021_{\text{three}} + 2011_{\text{three}}$
11. $14632_{\text{seven}} + 5604_{\text{seven}}$
12. $53B_{\text{sixteen}} + 694_{\text{sixteen}}$

In Exercises 13–24, subtract in the indicated base.

13. $32_{\text{four}} - 13_{\text{four}}$
14. $21_{\text{four}} - 12_{\text{four}}$
15. $23_{\text{five}} - 14_{\text{five}}$
16. $32_{\text{seven}} - 16_{\text{seven}}$
17. $475_{\text{eight}} - 267_{\text{eight}}$
18. $712_{\text{nine}} - 483_{\text{nine}}$
19. $563_{\text{seven}} - 164_{\text{seven}}$
20. $462_{\text{eight}} - 177_{\text{eight}}$
21. $1001_{\text{two}} - 111_{\text{two}}$
22. $1000_{\text{two}} - 101_{\text{two}}$
23. $1200_{\text{three}} - 1012_{\text{three}}$
24. $4C6_{\text{sixteen}} - 198_{\text{sixteen}}$

In Exercises 25–34, multiply in the indicated base.

25. $25_{\text{six}} \times 4_{\text{six}}$
26. $34_{\text{five}} \times 3_{\text{five}}$
27. $11_{\text{two}} \times 1_{\text{two}}$
28. $21_{\text{four}} \times 3_{\text{four}}$
29. $543_{\text{seven}} \times 5_{\text{seven}}$
30. $243_{\text{nine}} \times 6_{\text{nine}}$
31. $623_{\text{eight}} \times 4_{\text{eight}}$
32. $543_{\text{six}} \times 5_{\text{six}}$
33. $21_{\text{four}} \times 12_{\text{four}}$
34. $32_{\text{four}} \times 23_{\text{four}}$

In Exercises 35–38, use the multiplication tables shown below to divide in the indicated base.

MULTIPLICATION: BASE FOUR

×	0	1	2	3
0	0	0	0	0
1	0	1	2	3
2	0	2	10	12
3	0	3	12	21

MULTIPLICATION: BASE FIVE

×	0	1	2	3	4
0	0	0	0	0	0
1	0	1	2	3	4
2	0	2	4	11	13
3	0	3	11	14	22
4	0	4	13	22	31

35. $2_{\text{four}}\overline{)100_{\text{four}}}$
36. $2_{\text{four}}\overline{)321_{\text{four}}}$
37. $3_{\text{five}}\overline{)224_{\text{five}}}$
38. $4_{\text{five}}\overline{)134_{\text{five}}}$

Practice Plus

In Exercises 39–46, perform the indicated operations.

39. $10110_{\text{two}} + 10100_{\text{two}} + 11100_{\text{two}}$
40. $11100_{\text{two}} + 11111_{\text{two}} + 10111_{\text{two}}$
41. $11111_{\text{two}} + 10110_{\text{two}} - 101_{\text{two}}$
42. $10111_{\text{two}} + 11110_{\text{two}} - 111_{\text{two}}$
43. $1011_{\text{two}} \times 101_{\text{two}}$
44. $1101_{\text{two}} \times 110_{\text{two}}$
45. $D3_{\text{sixteen}} \times 8A_{\text{sixteen}}$
46. $B5_{\text{sixteen}} \times 2C_{\text{sixteen}}$

Application Exercises

Read the Blitzer Bonus on page 237. Then use the information in the essay to solve Exercises 47–52. Each exercise shows the binary sequences 10011 and 11001 about to be manipulated by passing through a microchip's series of gates. Provide the result(s) of these computer manipulations, designated by ? *in each diagram.*

47.

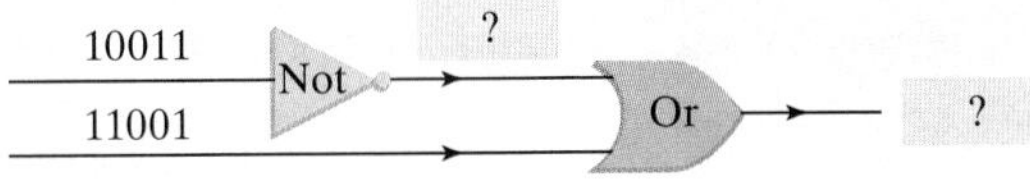

48.

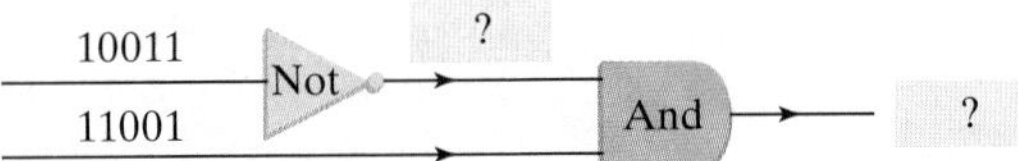

49.

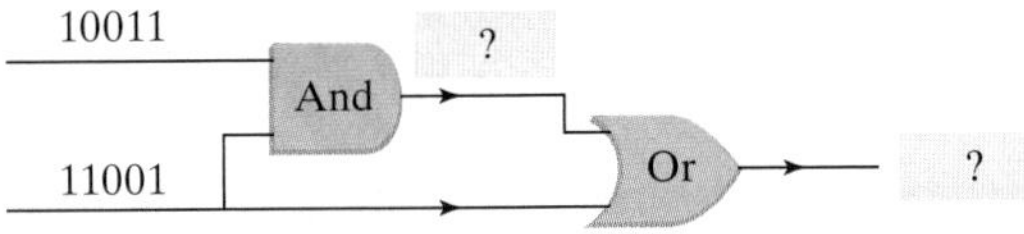

50.

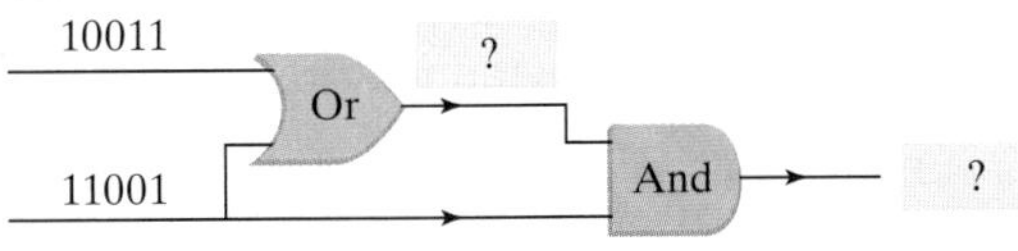

51.

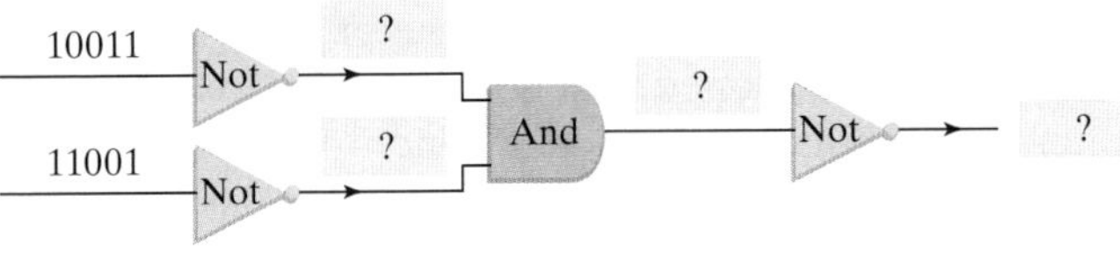

52.

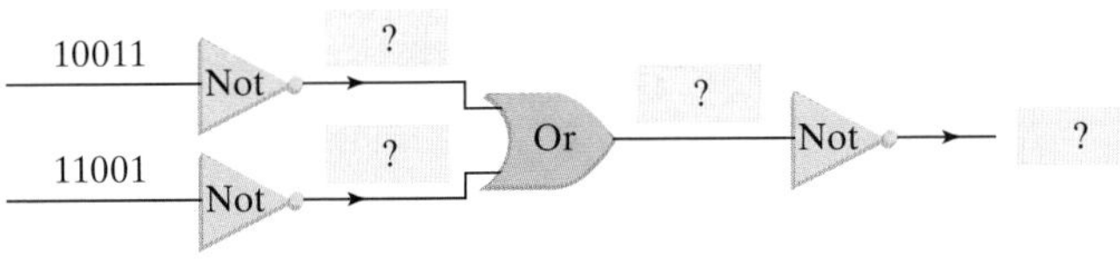

53. Use the equivalence $p \rightarrow q \equiv \sim p \vee q$ to select the circuit in Exercises 47–52 that illustrates a conditional gate.

Explaining the Concepts

54. Describe how to add two numbers in a base other than ten. How do you express and record the sum of numbers in a column if that sum exceeds the base?

55. Describe how to subtract two numbers in a base other than ten. How do you subtract a larger number from a smaller number in the same column?

56. Describe two difficulties that youngsters encounter when learning to add, subtract, multiply, and divide using Hindu-Arabic numerals. Base your answer on difficulties that are encountered when performing these computations in bases other than ten.

Critical Thinking Exercises

Make Sense? *In Exercises 57–60, determine whether each statement makes sense or does not make sense, and explain your reasoning.*

57. Arithmetic in bases other than ten works just like arithmetic in base ten.

58. When I perform subtraction problems that require borrowing, I always borrow the amount of the base given in the problem.

59. Performing the following addition problem reminds me of adding in base sixty.

$$\begin{array}{r} \text{4 hours, 26 minutes, 57 seconds} \\ +\ \text{3 hours, 46 minutes, 39 seconds} \\ \hline \end{array}$$

60. Performing the following subtraction problem reminds me of subtracting in base sixty.

$$\begin{array}{r} \text{8 hours, 45 minutes, 28 seconds} \\ -\ \text{2 hours, 47 minutes, 53 seconds} \\ \hline \end{array}$$

61. Perform the addition problem in Exercise 59. Do not leave more than 59 seconds or 59 minutes in the sum.

62. Perform the subtraction problem in Exercise 60.

63. Divide: $31_{\text{seven}}\overline{)2426_{\text{seven}}}$.

64. Use the Mayan numerals in **Table 4.2** on page 220 to solve this exercise. Add and without converting to Hindu-Arabic numerals.

Group Exercises

65. Group members should research various methods that societies have used to perform computations. Include finger multiplication, the galley method (sometimes called the Gelosia method), Egyptian duplation, subtraction by complements, Napier's bones, and other methods of interest in your presentation to the entire class.

66. Organize a debate. One side represents people who favor performing computations by hand, using the methods and procedures discussed in this section, but applied to base ten numerals. The other side represents people who favor the use of calculators for performing all computations. Include the merits of each approach in the debate.

4.4 Looking Back at Early Numeration Systems

WHAT AM I SUPPOSED TO LEARN?

After studying this section, you should be able to:

1. Understand and use the Egyptian system.
2. Understand and use the Roman system.
3. Understand and use the traditional Chinese system.
4. Understand and use the Ionic Greek system.

SUPER BOWL XXV, PLAYED ON January 27, 1991, resulted in the closest score of all time: NY Giants: 20; Buffalo: 19. If you are intrigued by sports facts and figures, you are probably aware that major sports events, such as the Super Bowl, are named using Roman numerals. Perhaps you have seen the use of Roman numerals in dating movies and television shows, or on clocks and watches.

In this section, we embark on a brief journey through time and numbers. Our Hindu-Arabic numeration system, the focus of this chapter, is successful because it expresses numbers with just ten symbols and makes computation with these numbers relatively easy. By these standards, the early numeration systems discussed in this section, such as Roman numerals, are unsuccessful. By looking briefly at these systems, you will see that our system is outstanding when compared with other historical systems.

1 *Understand and use the Egyptian system.*

The Egyptian Numeration System

Like most great civilizations, ancient Egypt had several numeration systems. The oldest is hieroglyphic notation, which developed around 3400 B.C. **Table 4.6** lists the Egyptian hieroglyphic numerals with the equivalent Hindu-Arabic numerals. Notice that the numerals are powers of ten. Their numeral for 1,000,000, or 10^6, looks like someone who just won the lottery!

GREAT QUESTION!

Do I have to memorize the symbols for the four numeration systems discussed in this section?

No. Focus your attention on understanding the idea behind each system and how these ideas have been incorporated into our Hindu-Arabic system.

TABLE 4.6 Egyptian Hieroglyphic Numerals

Hindu-Arabic Numeral	Egyptian Numeral	Description
1	𓏺	Staff
10	𓎆	Heel bone
100	𓍢	Spiral
1000	𓆼	Lotus blossom
10,000	𓂭	Pointing finger
100,000	𓆐	Tadpole
1,000,000	𓁨	Astonished person

It takes far more space to represent most numbers in the Egyptian system than in our system. This is because a number is expressed by repeating each numeral the required number of times. However, no numeral, except perhaps the astonished person, should be repeated more than nine times. If we were to use the Egyptian system to represent 764, we would need to write

100 100 100 100 100 100 100 10 10 10 10 10 10 1 1 1 1

and then represent each of these symbols with the appropriate hieroglyphic numeral from **Table 4.6**. Thus, 764 as an Egyptian numeral is

𓍢𓍢𓍢𓍢𓍢𓍢𓍢𓎆𓎆𓎆𓎆𓎆𓎆𓏺𓏺𓏺𓏺.

Blitzer Bonus

Hieroglyphic Numerals on Ancient Egyptian Tombs

Egyptian tombs from as early as 2600 B.C. contained hieroglyphic numerals. The funeral rites of ancient Egypt provided the dead with food and drink. The numerals showed how many of each item were included in the offering. Thus, the deceased had nourishment in symbolic form even when the offerings of the rite itself were gone.

The ancient Egyptian system is an example of an **additive system**, one in which the number represented is the sum of the values of the numerals.

EXAMPLE 1 Using the Egyptian Numeration System

Write the following numeral as a Hindu-Arabic numeral:

𓁨𓂭𓂭𓎆𓎆𓎆𓏺𓏺𓏺𓏺.

SOLUTION

Using **Table 4.6**, find the value of each of the Egyptian numerals. Then add them.

$$1{,}000{,}000 + 10{,}000 + 10{,}000 + 10 + 10 + 10 + 1 + 1 + 1 + 1 = 1{,}020{,}034$$

CHECK POINT 1 Write the following numeral as a Hindu-Arabic numeral:

𓆐𓆐𓆐𓍢𓍢𓎆𓎆𓏺𓏺.

EXAMPLE 2 Using the Egyptian Numeration System

Write 1752 as an Egyptian numeral.

SOLUTION

First break down the Hindu-Arabic numeral into quantities that match the Egyptian numerals:

$$\begin{aligned} 1752 &= 1000 + 700 + 50 + 2 \\ &= 1000 + 100 + 100 + 100 + 100 + 100 + 100 + 100 \\ &\quad + 10 + 10 + 10 + 10 + 10 + 1 + 1. \end{aligned}$$

Now, use **Table 4.6** to find the Egyptian symbol that matches each quantity. For example, the lotus blossom, 𓆼, matches 1000. Write each of these symbols and leave out the addition signs. Thus, the number 1752 can be expressed as

𓆼𓍢𓍢𓍢𓍢𓍢𓍢𓍢𓎆𓎆𓎆𓎆𓎆𓏺𓏺.

CHECK POINT 2 Write 2563 as an Egyptian numeral.

2 *Understand and use the Roman system.*

The Roman Numeration System

The Roman numeration system was developed between 500 B.C. and 100 A.D. It evolved as a result of tax collecting and commerce in the vast Roman Empire. The Roman numerals shown in **Table 4.7** were used throughout Europe until the eighteenth century. They are still commonly used in outlining, on clocks, for certain copyright dates, and in numbering some pages in books. Roman numerals are selected letters from the Roman alphabet.

TABLE 4.7 Roman Numerals

Roman numeral	I	V	X	L	C	D	M
Hindu-Arabic numeral	1	5	10	50	100	500	1000

GREAT QUESTION!

When I encounter Roman numerals, I often forget what the letters represent. Can you help me out?

Here's a sentence to help remember the letters used for Roman numerals in increasing order:

If **V**al's **X**-ray **L**ooks **C**lear, **D**on't **M**edicate.

TABLE 4.7 Roman Numerals (repeated)

Roman numeral	I	V	X	L	C	D	M
Hindu-Arabic numeral	1	5	10	50	100	500	1000

If the symbols in **Table 4.7** decrease in value from left to right, then add their values to obtain the value of the Roman numeral as a whole. For example, CX = 100 + 10 = 110. On the other hand, if symbols increase in value from left to right, then subtract the value of the symbol on the left from the symbol on the right. For example, IV means 5 − 1 = 4 and IX means 10 − 1 = 9.

Only the Roman numerals representing 1, 10, 100, 1000, . . . , can be subtracted. Furthermore, they can be subtracted only from their next two greater Roman numerals.

Roman numeral (values that can be subtracted are shown in red)	I	V	X	L	C	D	M
Hindu-Arabic numeral	1	5	10	50	100	500	1000

I can be subtracted only from V and X.

X can be subtracted only from L and C.

C can be subtracted only from D and M.

Blitzer Bonus

Do Not Offend Jupiter

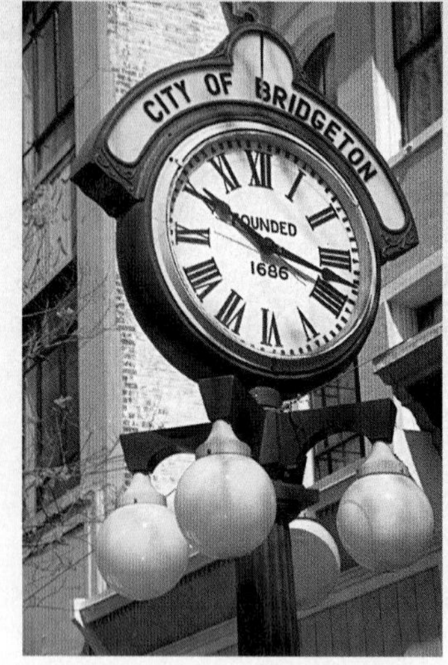

Have you ever noticed that clock faces with Roman numerals frequently show the number 4 as IIII instead of IV? One possible reason is that IIII provides aesthetic balance when visually paired with VIII on the other side. A more intriguing reason (although not necessarily true) is that the Romans did not want to offend the god Jupiter (spelled IVPITER) by daring to place the first two letters of his name on the face of a clock.

EXAMPLE 3 *Using Roman Numerals*

Write CLXVII as a Hindu-Arabic numeral.

SOLUTION

Because the numerals decrease in value from left to right, we add their values to find the value of the Roman numeral as a whole.

$$\text{CLXVII} = 100 + 50 + 10 + 5 + 1 + 1 = 167$$

CHECK POINT 3 Write MCCCLXI as a Hindu-Arabic numeral.

EXAMPLE 4 *Using Roman Numerals*

Write MCMXCVI as a Hindu-Arabic numeral.

SOLUTION

$$\begin{array}{l} \text{M} \quad \text{CM} \quad \text{XC} \quad \text{V} \quad \text{I} \\ = 1000 + (1000 - 100) + (100 - 10) + 5 + 1 \\ = 1000 + 900 + 90 + 5 + 1 = 1996 \end{array}$$

CHECK POINT 4 Write MCDXLVII as a Hindu-Arabic numeral.

Because Roman numerals involve subtraction as well as addition, it takes far less space to represent most numbers than in the Egyptian system. It is never necessary to repeat any symbol more than three consecutive times. For example, we write 46 as a Roman numeral using

XLVI rather than XXXXVI.

XL = 50 − 10 = 40

EXAMPLE 5 Using Roman Numerals

Write 249 as a Roman numeral.

SOLUTION

$$\begin{aligned} 249 &= 200 + 40 + 9 \\ &= 100 + 100 + (50 - 10) + (10 - 1) \\ &= \text{C} \quad \text{C} \quad \text{XL} \quad \text{IX} \end{aligned}$$

Thus, 249 = CCXLIX.

CHECK POINT 5 Write 399 as a Roman numeral.

The Roman numeration system uses bars above numerals or groups of numerals to show that the numbers are to be multiplied by 1000. For example,

$$\overline{\text{L}} = 50 \times 1000 = 50{,}000 \quad \text{and} \quad \overline{\text{CM}} = 900 \times 1000 = 900{,}000.$$

Placing bars over Roman numerals reduces the number of symbols needed to represent large numbers.

3 *Understand and use the traditional Chinese system.*

The Traditional Chinese Numeration System

The numerals used in the traditional Chinese numeration system are given in **Table 4.8**. At least two things are missing—a symbol for zero and a surprised lottery winner!

3
1000
2
100
6
10
4

Representing 3264 vertically is the first step in expressing it as a Chinese numeral.

TABLE 4.8 Traditional Chinese Numerals

Traditional Chinese numerals	一	二	三	四	五	六	七	八	九	十	百	千
Hindu-Arabic numerals	1	2	3	4	5	6	7	8	9	10	100	1000

So, how are numbers represented with this set of symbols? Chinese numerals are written vertically. Using our digits, the number 3264 is expressed as shown in the margin.

The next step is to replace each of these seven symbols with a traditional Chinese numeral from **Table 4.8**. Our next example illustrates this procedure.

3000: { 3 三, 1000 千 }
200: { 2 二, 100 百 }
60: { 6 六, 10 十 }
4: 4 四

Writing 3264 as a Chinese numeral

EXAMPLE 6 Using the Traditional Chinese Numeration System

Write 3264 as a Chinese numeral.

SOLUTION

First, break down the Hindu-Arabic numeral into quantities that match the Chinese numerals. Represent each quantity vertically. Then, use **Table 4.8** to find the Chinese symbol that matches each quantity. This procedure, with the resulting Chinese numeral, is shown in the margin.

The Chinese system does not need a numeral for zero because it is not positional. For example, we write 8006, using zeros as placeholders, to indicate that two powers of ten, namely 10^2, or 100, and 10^1, or 10, are not needed. The Chinese leave this out, writing

8		八
1000		千
6	or	六.

CHECK POINT 6 Write 2693 as a Chinese numeral.

4 Understand and use the Ionic Greek system.

The Ionic Greek Numeration System

The ancient Greeks, masters of art, architecture, theater, literature, philosophy, geometry, and logic, were not masters when it came to representing numbers. The Ionic Greek numeration system, which can be traced back as far as 450 B.C., used letters of their alphabet for numerals. **Table 4.9** shows the many symbols (too many symbols!) used to represent numbers.

TABLE 4.9 Ionic Greek Numerals

1	α	alpha	10	ι	iota	100	ρ	rho
2	β	beta	20	κ	kappa	200	σ	sigma
3	γ	gamma	30	λ	lambda	300	τ	tau
4	δ	delta	40	μ	mu	400	υ	upsilon
5	ε	epsilon	50	ν	nu	500	ϕ	phi
6	ɩ	vau	60	ξ	xi	600	χ	chi
7	ζ	zeta	70	o	omicron	700	ψ	psi
8	η	eta	80	π	pi	800	ω	omega
9	θ	theta	90	Q	koph	900	╥	sampi

To represent a number from 1 to 999, the appropriate symbols are written next to one another. For example, the number $21 = 20 + 1$. When 21 is expressed as a Greek numeral, the plus sign is left out:

$$21 = \kappa\alpha.$$

Similarly, the number 823 written as a Greek numeral is $\omega\kappa\gamma$.

EXAMPLE 7 ***Using the Ionic Greek Numeration System***

Write $\psi\lambda\delta$ as a Hindu-Arabic numeral.

SOLUTION

$\psi = 700$, $\lambda = 30$, and $\delta = 4$. Adding these numbers gives 734.

CHECK POINT 7 Write $\omega\pi\varepsilon$ as a Hindu-Arabic numeral.

One of the many unsuccessful features of the Greek numeration system is that new symbols have to be added to represent higher numbers. It is like an alphabet that gets bigger each time a new word is used and has to be written.

Concept and Vocabulary Check

Fill in each blank so that the resulting statement is true.

In Exercises 1–2, consider a system that represents numbers exactly like the Egyptian numeration system, but with different symbols:

a = 1, b = 10, c = 100, and d = 1000.

1. dccbaa = ___ + ___ + ___ + __ + __ + __ = ___

2. 1423 = 1000 + 100 + 100 + 100 + 100 + __ + __ + __ + __ + __ = ________

3. True or False: Like the system in Exercises 1–2, the Egyptian hieroglyphic system represents numbers as the sum of the values of the numerals. ______

Exercises 4–7 involve Roman numerals.

Roman numeral	I	V	X	L	C	D	M
Hindu-Arabic numeral	1	5	10	50	100	500	1000

4. If the symbols in the Roman numeral system decrease in value from left to right, then _____ their values to obtain the value of the Roman numeral. For example, CL = 100 __ 10 = ___.

5. If the symbols in the Roman numeral system increase in value from left to right, then ________ the value of the symbol on the left from the symbol on the right. For example, XL = 50 __ 10 = __.

6. A bar above a Roman numeral means to multiply that numeral by ______. For example, $\overline{\text{L}}$ = __ × ___ = ______.

7. True or False: When writing Roman numerals, it is never necessary to repeat any symbol more than three consecutive times. ______

In Exercises 8–9, assume a system that represents numbers exactly like the traditional Chinese system, but with different symbols. The symbols are shown as follows:

Numerals in the System	A	B	C	D	E	F	G	H	I	X	Y	Z
Hindu-Arabic Numerals	1	2	3	4	5	6	7	8	9	10	100	1000

8. 846 = 8 =
100
4
10
6 ______

9. True or False: Like the system in Exercise 8, Chinese numerals are written vertically. ______

In Exercises 10–11 assume a system that represents numbers exactly like the Greek Ionic system, but with different symbols. The symbols are shown as follows:

Decimal	1	2	3	4	5	6	7	8	9
Ones	A	B	C	D	E	F	G	H	I
Tens	J	K	L	M	N	O	P	Q	R
Hundreds	S	T	U	V	W	X	Y	Z	a
Thousands	b	c	d	e	f	g	h	i	j
Ten thousands	k	l	m	n	o	p	q	r	s

10. 5473 = f____

11. mgWLE = 30,000 + ___ + ___ + __ + __ = ______

12. Like the system in Exercises 10–11, the Greek Ionic system requires that new symbols be added to represent higher numbers. ______

Exercise Set 4.4

Practice Exercises

*Use **Table 4.6** on page 240 to solve Exercises 1–12.*

In Exercises 1–6, write each Egyptian numeral as a Hindu-Arabic numeral.

1.  𓍢𓍢𓍢𓎆𓎆𓏺𓏺

2.  𓁨𓁨𓁨𓂭𓂭𓂭𓂭𓍢𓍢𓎆𓏺𓏺𓏺𓏺

3.  𓆐𓆐𓆐𓍢𓍢𓍢𓍢𓎆𓎆𓏺𓏺𓏺

4.  𓆐𓆐𓍢𓍢𓎆𓏺𓏺𓏺

5.  𓍢𓎆𓎆𓎆𓏺𓏺

6.  𓆐𓂭𓂭𓆼𓍢𓍢𓍢𓏺𓏺

In Exercises 7–12, write each Hindu-Arabic numeral as an Egyptian numeral.

7. 423 **8.** 825 **9.** 1846

10. 1425 **11.** 23,547 **12.** 2,346,031

*Use **Table 4.7** on page 241 to solve Exercises 13–36.*

In Exercises 13–28, write each Roman numeral as a Hindu-Arabic numeral.

13. XI **14.** CL **15.** XVI

16. LVII **17.** XL **18.** CM

19. LIX **20.** XLIV **21.** CXLVI

22. CLXI **23.** MDCXXI

24. MMCDXLV **25.** MMDCLXXVII

26. MDCXXVI **27.** $\overline{\text{IX}}$CDLXVI

28. $\overline{\text{V}}$MCCXI

In Exercises 29–36, write each Hindu-Arabic numeral as a Roman numeral.

29. 43 **30.** 96

31. 129 **32.** 469

33. 1896 **34.** 4578

35. 6892 **36.** 5847

*Use **Table 4.8** on page 243 to solve Exercises 37–48.*

In Exercises 37–42, write each traditional Chinese numeral as a Hindu-Arabic numeral.

37. 八十八 **38.** 七百五 **39.** 五百二十七

40. 三千八十一 **41.** 二千七百七十六 **42.** 八千二百三十六

In Exercises 43–48, write each Hindu-Arabic numeral as a traditional Chinese numeral.

43. 43 **44.** 269 **45.** 583

46. 2965 **47.** 4870 **48.** 7605

*Use **Table 4.9** on page 244 to solve Exercises 49–56.*

In Exercises 49–52, write each Ionic Greek numeral as a Hindu-Arabic numeral.

49. $\iota\beta$ **50.** $\phi\varepsilon$ **51.** $\sigma\lambda\delta$ **52.** $\psi o\theta$

In Exercises 53–56, write each Hindu-Arabic numeral as an Ionic Greek numeral.

53. 43 **54.** 257 **55.** 483 **56.** 895

Practice Plus

57. Write 𓆼𓆼𓍢𓍢𓍢𓎆𓎆|||| as a Roman numeral and as a traditional Chinese numeral.

58. Write 𓆼𓆼𓆼𓍢𓍢𓍢𓍢𓎆||| as a Roman numeral and as a traditional Chinese numeral.

59. Write MDCCXLI as an Egyptian numeral and as a traditional Chinese numeral.

60. Write MMCCXLV as an Egyptian numeral and as a traditional Chinese numeral.

In Exercises 61–64, write each numeral as a numeral in base five.

61. 𓍢𓍢𓍢𓍢||||

62. 𓍢𓍢𓍢𓎆𓎆𓎆||

63. CXCII

64. CMLXXIV

In Exercises 65–66, perform each subtraction without converting to Hindu-Arabic numerals.

65.
```
   𓍢𓍢 𓎆𓎆𓎆 |
 −   𓍢 𓎆𓎆𓎆 |||
```

66.
```
   𓍢𓍢 𓎆𓎆 ||
 −   𓍢 𓎆𓎆 |||
```

Application Exercises

67. Look at the back of a U.S. one dollar bill. What date is written in Roman numerals along the base of the pyramid with an eye? What is this date's significance?

68. A construction crew demolishing a very old building was surprised to find the numeral MCMLXXXIX inscribed on the cornerstone. Explain why they were surprised.

The Braille numeration system is a base ten positional system that uses raised dots in 2-by-3 cells as digit symbols.

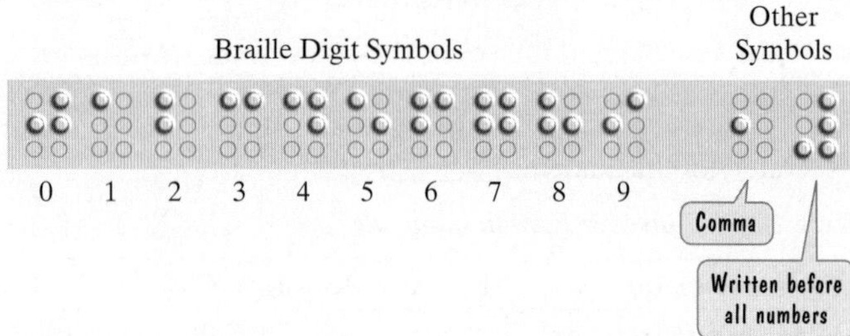

In Exercises 69–70, use the digit symbols to express each Braille numeral as a Hindu-Arabic numeral.

69.

70.

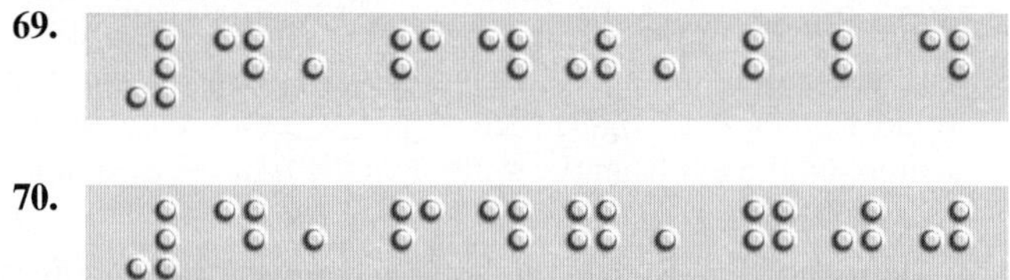

Explaining the Concepts

71. Describe how a number is represented in the Egyptian numeration system.

72. If you are interpreting a Roman numeral, when do you add values and when do you subtract them? Give an example to illustrate each case.

73. Describe how a number is represented in the traditional Chinese numeration system.

74. Describe one disadvantage of the Ionic Greek numeration system.

75. If you could use only one system of numeration described in this section, which would you prefer? Discuss the reasons for your choice.

Critical Thinking Exercises

Make Sense? *In Exercises 76–79, determine whether each statement makes sense or does not make sense, and explain your reasoning.*

76. In order to understand the early numeration systems presented in this section, it's important that I take the time to memorize the various symbols.

77. Because 𓍢 represents 100 and 𓎆 represents 10 in the Egyptian numeration system,

𓍢𓎆 represents 100 + 10, or 110, and

𓎆𓍢 represents 100 − 10, or 90.

78. It takes far more space to represent numbers in the Roman numeration system than in the Egyptian numeration system.

79. In terms of the systems discussed in this chapter, the Babylonian numeration system uses the least number of symbols and the Ionic Greek numeration system uses the most.

80. Arrange these three numerals from smallest to largest.

81. Use Egyptian numerals to write the numeral that precedes and the numeral that follows

82. After reading this section, a student had a numeration nightmare about selling flowers in a time-warped international market. She started out with 200 flowers, selling XLVI of them to a Roman,

to an Egyptian,

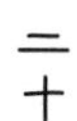

to a traditional Chinese family, and the remainder to a Greek. How many flowers were sold to the Greek? Express the answer in the Ionic Greek numeration system.

Group Exercises

Take a moment to read the introduction to the group exercises on page 231. Exercises 83–87 list some additional topics for individual or group research projects.

83. A Time Line Showing Significant Developments in Numeration Systems
84. Animals and Number Sense
85. The Hebrew Numeration System (or any system not discussed in this chapter)
86. The Rhind Papyrus and What We Learned from It
87. Computation in an Early Numeration System

Chapter Summary, Review, and Test

SUMMARY – DEFINITIONS AND CONCEPTS	EXAMPLES
4.1 Our Hindu-Arabic System and Early Positional Systems	
a. In a positional-value, or place-value, numeration system, the value of each symbol, called a digit, varies according to the position it occupies in the number.	
b. The Hindu-Arabic numeration system is a base ten system with the digits 0, 1, 2, 3, 4, 5, 6, 7, 8, and 9. The place values in the system are $\ldots, 10^5, 10^4, 10^3, 10^2, 10^1, 1.$	Ex. 1, p. 218; Ex. 2, p. 218
c. The Babylonian numeration system is a base sixty system, with place values given by $\ldots, 60^3, 60^2, 60^1, 1.$ (60^3 or 216,000; 60^2 or 3600; 60^1 or 60) Babylonian numerals are given in Table 4.1 on page 219.	Ex. 3, p. 219
d. The Mayan numeration system has place values given by $\ldots, 18 \times 20^3, 18 \times 20^2, 18 \times 20, 20, 1.$ (18×20^3 or 144,000; 18×20^2 or 7200; 18×20 or 360) Mayan numerals are given in Table 4.2 on page 220.	Ex. 4, p. 221
4.2 Number Bases in Positional Systems	
a. The base of a positional numeration system refers to the number of individual digit symbols used in the system as well as to the powers of the numbers used in place values. In base b, there are b digit symbols (from 0 through $b - 1$, inclusive) with place values given by $\ldots, b^4, b^3, b^2, b^1, 1.$	

b. To change a numeral in a base other than ten to a base ten numeral, **1.** Multiply each digit in the numeral by its respective place value. **2.** Find the sum of the products in step 1.	Ex. 1, p. 225; Ex. 2, p. 226; Ex. 3, p. 226
c. To change a base ten numeral to a base b numeral, use mental conversions or repeated divisions by powers of b to find how many groups of each place value are contained in the base ten numeral.	Ex. 4, p. 228; Ex. 5, p. 228; Ex. 6, p. 229; Ex. 7, p. 229

4.3 Computation in Positional Systems

a. Computations in bases other than ten are performed using the same procedures as in ordinary base ten arithmetic. When a computation is equal to or exceeds the given base, use mental conversions to convert from the base ten numeral to a numeral in the desired base.	Ex. 1, p. 231; Ex. 2, p. 232; Ex. 3, p. 234; Ex. 4, p. 234; Ex. 5, p. 235
b. To divide in bases other than ten, it is convenient to use a multiplication table for products in the required base.	Ex. 6, p. 235

4.4 Looking Back at Early Numeration Systems

a. A successful numeration system expresses numbers with relatively few symbols and makes computation with these numbers fairly easy.	
b. By the standard in (a), the Egyptian system (Table 4.6 on page 240), the Roman system (Table 4.7 on page 241), the Chinese system (Table 4.8 on page 243), and the Greek system (Table 4.9 on page 244) are all unsuccessful. Unlike our Hindu-Arabic system, these systems are not positional and contain no symbol for zero.	Ex. 1, p. 241; Ex. 2, p. 241; Ex. 3, p. 242; Ex. 4, p. 242; Ex. 5, p. 243; Ex. 6, p. 243; Ex. 7, p. 244

Review Exercises

4.1

In Exercises 1–2, evaluate the expression.

1. 11^2 **2.** 7^3

In Exercises 3–5, write each Hindu-Arabic numeral in expanded form.

3. 472 **4.** 8076 **5.** 70,329

In Exercises 6–7, express each expanded form as a Hindu-Arabic numeral.

6. $(7 \times 10^5) + (0 \times 10^4) + (6 \times 10^3) + (9 \times 10^2) + (5 \times 10^1) + (3 \times 1)$

7. $(7 \times 10^8) + (4 \times 10^7) + (3 \times 10^2) + (6 \times 1)$

*Use **Table 4.1** on page 219 to write each Babylonian numeral in Exercises 8–9 as a Hindu-Arabic numeral.*

8. < ∨   < ∨ ∨ ∨

9. ∨ ∨   < <   < < <

*Use **Table 4.2** on page 220 to write each Mayan numeral in Exercises 10–11 as a Hindu-Arabic numeral.*

10.
•
—
•••
—
•
═

11.
••••
—
••
⬭
•
≡

12. Describe how a positional system is used to represent a number.

4.2

In Exercises 13–18, convert the numeral to a numeral in base ten.

13. 34_{five} **14.** 110_{two}

15. 643_{seven} **16.** 1084_{nine}

17. $FD3_{\text{sixteen}}$ **18.** 202202_{three}

In Exercises 19–24, convert each base ten numeral to a numeral in the given base.

19. 89 to base five

20. 21 to base two

21. 473 to base three

22. 7093 to base seven

23. 9348 to base six

24. 554 to base twelve

4.3

In Exercises 25–28, add in the indicated base.

25. $\begin{array}{r} 46_{\text{seven}} \\ \underline{+53_{\text{seven}}} \end{array}$

26. $\begin{array}{r} 574_{\text{eight}} \\ \underline{+605_{\text{eight}}} \end{array}$

27. $\begin{array}{r} 11011_{\text{two}} \\ \underline{+10101_{\text{two}}} \end{array}$

28. $\begin{array}{r} 43C_{\text{sixteen}} \\ \underline{+694_{\text{sixteen}}} \end{array}$

In Exercises 29–32, subtract in the indicated base.

29. 34_{six}
-25_{six}

30. 624_{seven}
-246_{seven}

31. 1001_{two}
$-\ 110_{\text{two}}$

32. 4121_{five}
-1312_{five}

In Exercises 33–35, multiply in the indicated base.

33. 32_{four}
$\times\ 3_{\text{four}}$

34. 43_{seven}
$\times\ 6_{\text{seven}}$

35. 123_{five}
$\times\ \ 4_{\text{five}}$

In Exercises 36–37, divide in the indicated base. Use the multiplication tables on page 238.

36. $2_{\text{four}}\overline{)332_{\text{four}}}$

37. $4_{\text{five}}\overline{)103_{\text{five}}}$

4.4

*Use **Table 4.6** on page 240 to solve Exercises 38–41.*

In Exercises 38–39, write each Egyptian numeral as a Hindu-Arabic numeral.

38. 𓆼𓍢𓍢𓎆𓎆𓎆𓎆𓏺𓏺𓏺𓏺𓏺𓏺

39. 𓂭𓆼𓆼𓍢𓍢𓍢𓍢𓎆𓎆𓎆𓏺𓏺

In Exercises 40–41, write each Hindu-Arabic numeral as an Egyptian numeral.

40. 2486

41. 34,573

In Exercises 42–43, assume a system that represents numbers exactly like the Egyptian system, but with different symbols. In particular, A = 1, B = 10, C = 100, *and* D = 1000.

42. Write DDCCCBAAAA as a Hindu-Arabic numeral.

43. Write 5492 as a numeral in terms of A, B, C, and D.

44. Describe how the Egyptian system or the system in Exercises 42–43 is used to represent a number. Discuss one disadvantage of such a system when compared to our Hindu-Arabic system.

*Use **Table 4.7** on page 241 to solve Exercises 45–49.*

In Exercises 45–47, write each Roman numeral as a Hindu-Arabic numeral.

45. CLXIII

46. MXXXIV

47. MCMXC

In Exercises 48–49, write each Hindu-Arabic numeral as a Roman numeral.

48. 49

49. 2965

50. Explain when to subtract the value of symbols when interpreting a Roman numeral. Give an example.

*Use **Table 4.8** on page 243 to solve Exercises 51–54.*

In Exercises 51–52, write each traditional Chinese numeral as a Hindu-Arabic numeral.

51.
五
百
五
十
四

52.
八
千
二
百
五
十
三

In Exercises 53–54, write each Hindu-Arabic numeral as a traditional Chinese numeral.

53. 274

54. 3587

In Exercises 55–58, assume a system that represents numbers exactly like the traditional Chinese system, but with different symbols. The symbols are shown as follows:

Numerals in the System	A	B	C	D	E	F	G	H	I	X	Y	Z
Hindu-Arabic Numerals	1	2	3	4	5	6	7	8	9	10	100	1000

Express each numeral in Exercises 55–56 as a Hindu-Arabic numeral.

55.
C
Y
F
X
E

56.
D
Z
E
Y
B
X

Express each Hindu-Arabic numeral in Exercises 57–58 as a numeral in the system used for Exercises 55–56.

57. 793

58. 6854

59. Describe how the Chinese system or the system in Exercises 55–58 is used to represent a number. Discuss one disadvantage of such a system when compared to our Hindu-Arabic system.

*Use **Table 4.9** on page 244 to solve Exercises 60–63.*

In Exercises 60–61, write each Ionic Greek numeral as a Hindu-Arabic numeral.

60. $\chi\nu\gamma$

61. $\chi o\eta$

In Exercises 62–63, write each Hindu-Arabic numeral as an Ionic Greek numeral.

62. 453

63. 902

In Exercises 64–68, assume a system that represents numbers exactly like the Greek Ionic system, but with different symbols. The symbols are shown as follows:

Decimal	**1**	**2**	**3**	**4**	**5**	**6**	**7**	**8**	**9**
Ones	A	B	C	D	E	F	G	H	I
Tens	J	K	L	M	N	O	P	Q	R
Hundreds	S	T	U	V	W	X	Y	Z	a
Thousands	b	c	d	e	f	g	h	i	j
Ten thousands	k	l	m	n	o	p	q	r	s

(In Exercises 64–68, be sure to refer to the table at the bottom of the previous page.)

In Exercises 64–66, express each numeral as a Hindu-Arabic numeral.

64. UNG

65. mhZRD

66. rXJH

In Exercises 67–68, express each Hindu-Arabic numeral as a numeral in the system used for Exercises 64–66.

67. 597

68. 25,483

69. Discuss one disadvantage of the Greek Ionic system or the system described in Exercises 64–68 when compared to our Hindu-Arabic system.

Chapter 4 Test

1. Evaluate 9^3.

2. Write 567 in expanded form.

3. Write 63,028 in expanded form.

4. Express as a Hindu-Arabic numeral:
$(7 \times 10^3) + (4 \times 10^2) + (9 \times 10^1) + (3 \times 1)$.

5. Express as a Hindu-Arabic numeral:
$(4 \times 10^5) + (2 \times 10^2) + (6 \times 1)$.

6. What is the difference between a number and a numeral?

7. Explain why a symbol for zero is needed in a positional system.

8. Place values in the Babylonian system are $\ldots, 60^3, 60^2, 60^1, 1$.

Use the numerals shown to write the following Babylonian numeral as a Hindu-Arabic numeral:

<< <∨∨ <∨.

Babylonian	∨	<
Hindu-Arabic	1	10

9. Place values in the Mayan system are $\ldots, 18 \times 20^3, 18 \times 20^2, 18 \times 20, 20, 1$.

Use the numerals shown to write the following Mayan numeral as a Hindu-Arabic numeral:

••••
•
⊖.

Mayan	⊖	•	••	•••	••••	—	• —
Hindu-Arabic	0	1	2	3	4	5	6

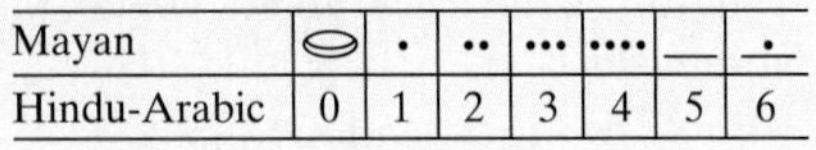

In Exercises 10–12, convert the numeral to a numeral in base ten.

10. 423_{five} **11.** 267_{nine} **12.** 110101_{two}

In Exercises 13–15, convert each base ten numeral to a numeral in the given base.

13. 77 to base three **14.** 56 to base two

15. 1844 to base five

In Exercises 16–18, perform the indicated operation.

16. $\begin{array}{r} 234_{\text{five}} \\ +\ 423_{\text{five}} \\ \hline \end{array}$

17. $\begin{array}{r} 562_{\text{seven}} \\ -\ 145_{\text{seven}} \\ \hline \end{array}$

18. $\begin{array}{r} 54_{\text{six}} \\ \times\ 3_{\text{six}} \\ \hline \end{array}$

19. Use the multiplication table shown to perform this division: $3_{\text{five}}\overline{)1213_{\text{five}}}$.

A MULTIPLICATION TABLE FOR BASE FIVE

×	**0**	**1**	**2**	**3**	**4**
0	0	0	0	0	0
1	0	1	2	3	4
2	0	2	4	11	13
3	0	3	11	14	22
4	0	4	13	22	31

Use the symbols in the tables shown below to solve Exercises 20–23.

Hindu-Arabic Numeral	**Egyptian Numeral**
1	𓏺
10	𓎆
100	𓍢
1000	𓆼
10,000	𓂭
100,000	𓆐
1,000,000	𓁨

Hindu-Arabic Numeral	Roman Numeral
1	I
5	V
10	X
50	L
100	C
500	D
1000	M

20. Write the following numeral as a Hindu-Arabic numeral:

𓂭𓂭𓍢𓍢𓏺𓏺𓏺.

21. Write 32,634 as an Egyptian numeral.

22. Write the Roman numeral MCMXCIV as a Hindu-Arabic numeral.

23. Express 459 as a Roman numeral.

24. Describe one difference between how a number is represented in the Egyptian system and the Roman system.

5

Number Theory and the Real Number System

SURFING THE WEB, YOU HEAR POLITICIANS DISCUSSING THE PROBLEM OF THE NATIONAL debt that exceeds $18 trillion. They state that the interest on the debt equals government spending on veterans, homeland security, education, and transportation combined. They make it seem like the national debt is a real problem, but later you realize that you don't really know what a number like 18 trillion means. If the national debt were evenly divided among all citizens of the country, how much would every man, woman, and child have to pay? Is economic doomsday about to arrive?

Here's where you'll find this application:

Literacy with numbers, called numeracy, is a prerequisite for functioning in a meaningful way personally, professionally, and as a citizen. In this chapter, our focus is on understanding numbers, their properties, and their applications.

- The problem of placing a national debt that exceeds $18 trillion in perspective appears as Example 9 in Section 5.6.
- Confronting a national debt in excess of $18 trillion starts with grasping just how colossal $1 trillion actually is. The Blitzer Bonus on page 323 should help provide insight into this mind-boggling number.

5.1 Number Theory: Prime and Composite Numbers

WHAT AM I SUPPOSED TO LEARN?

After studying this section, you should be able to:

1 Determine divisibility.
2 Write the prime factorization of a composite number.
3 Find the greatest common divisor of two numbers.
4 Solve problems using the greatest common divisor.
5 Find the least common multiple of two numbers.
6 Solve problems using the least common multiple.

Number Theory and Divisibility

YOU ARE ORGANIZING AN intramural league at your school. You need to divide 40 men and 24 women into all-male and all-female teams so that each team has the same number of people. The men's teams should have the same number of players as the women's teams. What is the largest number of people that can be placed on a team?

This problem can be solved using a branch of mathematics called **number theory**. Number theory is primarily concerned with the properties of numbers used for counting, namely 1, 2, 3, 4, 5, and so on. The set of counting numbers is also called the set of **natural numbers**. As we saw in Chapter 2, we represent this set by the letter **N**.

THE SET OF NATURAL NUMBERS

$$\mathbf{N} = \{1, 2, 3, 4, 5, 6, 7, 8, 9, 10, 11, \dots\}$$

We can solve the intramural league problem. However, to do so we must understand the concept of divisibility. For example, there are a number of different ways to divide the 24 women into teams, including

1 team with all 24 women:	$1 \times 24 = 24$
2 teams with 12 women per team:	$2 \times 12 = 24$
3 teams with 8 women per team:	$3 \times 8 = 24$
4 teams with 6 women per team:	$4 \times 6 = 24$
6 teams with 4 women per team:	$6 \times 4 = 24$
8 teams with 3 women per team:	$8 \times 3 = 24$
12 teams with 2 women per team:	$12 \times 2 = 24$
24 teams with 1 woman per team:	$24 \times 1 = 24.$

The natural numbers that are multiplied together resulting in a product of 24 are called *factors* of 24. Any natural number can be expressed as a product of two or more natural numbers. The natural numbers that are multiplied are called the **factors** of the product. Notice that a natural number may have many factors.

$$2 \times 12 = 24 \qquad 3 \times 8 = 24 \qquad 6 \times 4 = 24$$

Factors of 24   Factors of 24   Factors of 24

The numbers 1, 2, 3, 4, 6, 8, 12, and 24 are all factors of 24. Each of these numbers divides 24 without a remainder.

In general, let a and b represent natural numbers. We say that a is **divisible** by b if the operation of dividing a by b leaves a remainder of 0.

A natural number is divisible by all of its factors. Thus, 24 is divisible by 1, 2, 3, 4, 6, 8, 12, and 24. Using the factor 8, we can express this divisibility in a number of ways:

24 is **divisible** by 8.

8 is a **divisor** of 24.

8 **divides** 24.

Mathematicians use a special notation to indicate divisibility.

GREAT QUESTION!

What's the difference between a factor and a divisor?

There is no difference. The words *factor* and *divisor* mean the same thing. Thus, 8 is a factor and a divisor of 24.

DIVISIBILITY

If a and b are natural numbers, a is **divisible** by b if the operation of dividing a by b leaves a remainder of 0. This is the same as saying that b is a **divisor** of a, or b **divides** a. All three statements are symbolized by writing

$$b \mid a.$$

GREAT QUESTION!

What's the difference between $b \mid a$ and b/a?

It's easy to confuse these notations. The symbol $b \mid a$ means b divides a. The symbol b/a means b divided by a (that is, $b \div a$, the quotient of b and a). For example, $5 \mid 35$ means 5 divides 35, whereas $5/35$ means 5 divided by 35, which is equivalent to the fraction $\frac{1}{7}$.

Using this new notation, we can write

$$12 \mid 24.$$

Twelve divides 24 because 24 divided by 12 leaves a remainder of 0. By contrast, 13 does not divide 24 because 24 divided by 13 does not leave a remainder of 0. The notation

$$13 \nmid 24$$

means that 13 does not divide 24.

1 *Determine divisibility.*

Table 5.1 shows some common rules for divisibility. Divisibility rules for 7 and 11 are difficult to remember and are not included in the table.

TABLE 5.1 Rules of Divisibility

Divisible By	Test	Example
2	The last digit is 0, 2, 4, 6, or 8.	5,892,796 is divisible by 2 because the last digit is 6.
3	The sum of the digits is divisible by 3.	52,341 is divisible by 3 because the sum of the digits is $5 + 2 + 3 + 4 + 1 = 15$, and 15 is divisible by 3.
4	The last two digits form a number divisible by 4.	3,947,136 is divisible by 4 because 36 is divisible by 4.
5	The number ends in 0 or 5.	28,160 and 72,805 end in 0 and 5, respectively. Both are divisible by 5.
6	The number is divisible by both 2 and 3. (In other words, the number is even and the sum of its digits is divisible by 3.)	954 is divisible by 2 because it ends in 4. 954 is also divisible by 3 because the digit sum is 18, which is divisible by 3. Because 954 is divisible by both 2 and 3, it is divisible by 6.
8	The last three digits form a number that is divisible by 8.	593,777,832 is divisible by 8 because 832 is divisible by 8.
9	The sum of the digits is divisible by 9.	5346 is divisible by 9 because the sum of the digits, 18, is divisible by 9.
10	The last digit is 0.	998,746,250 is divisible by 10 because the number ends in 0.
12	The number is divisible by both 3 and 4. (In other words, the sum of the digits is divisible by 3 and the last two digits form a number divisible by 4.)	614,608,176 is divisible by 3 because the digit sum is 39, which is divisible by 3. It is also divisible by 4 because the last two digits form 76, which is divisible by 4. Because 614,608,176 is divisible by both 3 and 4, it is divisible by 12.

EXAMPLE 1 Using the Rules of Divisibility

Which one of the following statements is true?

a. $4|3{,}754{,}086$ **b.** $9\nmid 4{,}119{,}706{,}413$ **c.** $8|677{,}840$

SOLUTION

a. $4|3{,}754{,}086$ states that 4 divides 3,754,086. **Table 5.1**, on the previous page, indicates that for 4 to divide a number, the last two digits must form a number that is divisible by 4. Because 86 is not divisible by 4, the given statement is false.

b. $9\nmid 4{,}119{,}706{,}413$ states that 9 does *not* divide 4,119,706,413. Based on **Table 5.1**, if the sum of the digits is divisible by 9, then 9 does indeed divide this number. The sum of the digits is $4 + 1 + 1 + 9 + 7 + 0 + 6 + 4 + 1 + 3 = 36$, which is divisible by 9. Because 4,119,706,413 is divisible by 9, the given statement is false.

c. $8|677{,}840$ states that 8 divides 677,840. **Table 5.1** indicates that for 8 to divide a number, the last three digits must form a number that is divisible by 8. Because 840 is divisible by 8, then 8 divides 677,840, and the given statement is true.

The statement given in part (c) is the only true statement.

TECHNOLOGY

Calculators and Divisibility

You can use a calculator to verify each result in Example 1. Consider part (a):

$4|3{,}754{,}086.$

Divide 3,754,086 by 4 using the following keystrokes:

3754086 [÷] 4.

Press [=] or [ENTER].

The number displayed is 938521.5. This is not a natural number. The 0.5 shows that the division leaves a nonzero remainder. Thus, 4 does not divide 3,754,086. The given statement is false.

Now consider part (b):

$9\nmid 4{,}119{,}706{,}413.$

Use your calculator to divide the number on the right by 9:

4119706413 [÷] 9.

Press [=] or [ENTER].

The display is 457745157. This is a natural number. The remainder of the division is 0, so 9 does divide 4,119,706,413. The given statement is false.

CHECK POINT 1 Which one of the following statements is true?

a. $8|48{,}324$ **b.** $6|48{,}324$ **c.** $4\nmid 48{,}324$

Prime Factorization

By developing some other ideas of number theory, we will be able to solve the intramural league problem. We begin with the definition of a prime number.

PRIME NUMBERS

A **prime number** is a natural number greater than 1 that has only itself and 1 as factors.

GREAT QUESTION!

Can a prime number be even?

The number 2 is the only even prime number. Every other even number has at least three factors: 1, 2, and the number itself.

Using this definition, we see that the number 7 is a prime number because it has only 1 and 7 as factors. Said in another way, 7 is prime because it is divisible by only 1 and 7. The first ten prime numbers are 2, 3, 5, 7, 11, 13, 17, 19, 23, and 29. Each number in this list has exactly two divisors, itself and 1. By contrast, 9 is not a prime number; in addition to being divisible by 1 and 9, it is also divisible by 3. The number 9 is an example of a *composite number*.

COMPOSITE NUMBERS

A **composite number** is a natural number greater than 1 that is divisible by a number other than itself and 1.

Using this definition, the first ten composite numbers are 4, 6, 8, 9, 10, 12, 14, 15, 16, and 18. Each number in this list has at least three distinct divisors.

By the definitions above, both prime numbers and composite numbers must be natural numbers *greater than* 1, so **the natural number 1 is neither prime nor composite.**

2 *Write the prime factorization of a composite number.*

Every composite number can be expressed as the product of prime numbers. For example, the composite number 45 can be expressed as

$$45 = 3 \times 3 \times 5.$$

Note that 3 and 5 are prime numbers. Expressing a composite number as the product of prime numbers is called **prime factorization**. The prime factorization of 45 is $3 \times 3 \times 5$. The order in which we write these factors does not matter. This means that

$$\begin{aligned} 45 &= 3 \times 3 \times 5 \\ \text{or } 45 &= 5 \times 3 \times 3 \\ \text{or } 45 &= 3 \times 5 \times 3. \end{aligned}$$

In Chapter 1, we defined a **theorem** as a statement that can be proved using deductive reasoning. The ancient Greeks proved that if the order of the factors is disregarded, there is only one prime factorization possible for any given composite number. This statement is called the **Fundamental Theorem of Arithmetic**.

THE FUNDAMENTAL THEOREM OF ARITHMETIC

Every composite number can be expressed as a product of prime numbers in one and only one way (if the order of the factors is disregarded).

One method used to find the prime factorization of a composite number is called a **factor tree**. To use this method, begin by selecting any two numbers, other than 1, whose product is the number to be factored. One or both of the factors may not be prime numbers. Continue to factor composite numbers. Stop when all numbers are prime.

GREAT QUESTION!

In Example 2, do I have to start the factor tree for 700 with 7 · 100?

No. It does not matter how you begin a factor tree. For example, in Example 2 you can factor 700 by starting with 5 and 140. $(5 \times 140 = 700)$

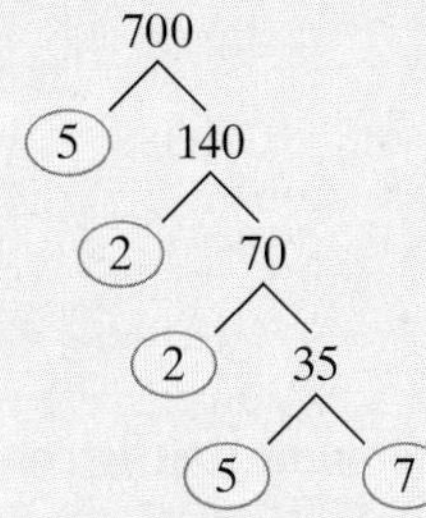

The prime factorization of 700 is

$$\begin{aligned} 700 &= 5 \times 2 \times 2 \times 5 \times 7 \\ &= 2^2 \times 5^2 \times 7. \end{aligned}$$

This is the same prime factorization we obtained in Example 2.

EXAMPLE 2 Prime Factorization Using a Factor Tree

Find the prime factorization of 700.

SOLUTION

Start with any two numbers, other than 1, whose product is 700, such as 7 and 100. This forms the first branch of the tree. Continue factoring the composite number or numbers that result (in this case 100), branching until each branch ends with a prime number.

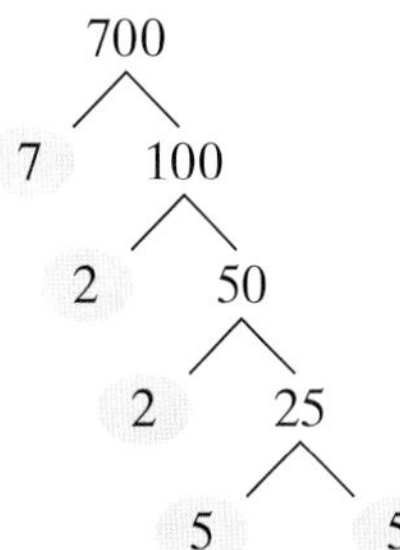

The prime factors are shown on light blue ovals. Thus, the prime factorization of 700 is

$$700 = 7 \times 2 \times 2 \times 5 \times 5.$$

We can use exponents to show the repeated prime factors:

$$700 = 7 \times 2^2 \times 5^2.$$

Using a dot to indicate multiplication and arranging the factors from least to greatest, we can write

$$700 = 2^2 \cdot 5^2 \cdot 7.$$

CHECK POINT 2 Find the prime factorization of 120.

3 *Find the greatest common divisor of two numbers.*

Greatest Common Divisor

The greatest common divisor of two or more natural numbers is the largest number that is a divisor (or factor) of all the numbers. For example, 8 is the greatest common divisor of 32 and 40 because it is the largest natural number that divides both 32 and 40. Some pairs of numbers have 1 as their greatest common divisor. Such number pairs are said to be **relatively prime**. For example, the greatest common divisor of 5 and 26 is 1. Thus, 5 and 26 are relatively prime.

The greatest common divisor can be found using prime factorizations.

FINDING THE GREATEST COMMON DIVISOR USING PRIME FACTORIZATIONS

To find the greatest common divisor of two or more numbers,

1. Write the prime factorization of each number.
2. Select each prime factor with the smallest exponent that is common to each of the prime factorizations.
3. Form the product of the numbers from step 2. The greatest common divisor is the product of these factors.

Blitzer Bonus

Simple Questions with No Answers

In number theory, a good problem is one that can be stated quite simply, but whose solution turns out to be particularly difficult, if not impossible.

In 1742, the mathematician Christian Goldbach (1690–1764) wrote a letter to Leonhard Euler (1707–1783) in which he proposed, without a proof, that every even number greater than 2 is the sum of two primes. For example,

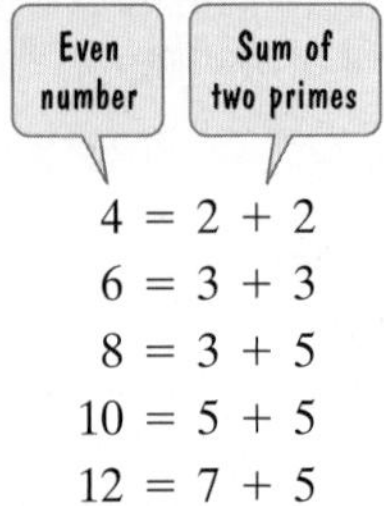

and so on.

Two and a half centuries later, it is still not known if this conjecture is true or false. Inductively, it appears to be true; computer searches have written even numbers as large as 400 trillion as the sum of two primes. Deductively, no mathematician has been able to prove that the conjecture is true. Even a reward of one million dollars for a proof offered by the publishing house Farber and Farber in 2000 to help publicize the novel *Uncle Petros and Goldbach's Conjecture* went unclaimed.

EXAMPLE 3 *Finding the Greatest Common Divisor*

Find the greatest common divisor of 216 and 234.

SOLUTION

Step 1 Write the prime factorization of each number. Begin by writing the prime factorizations of 216 and 234.

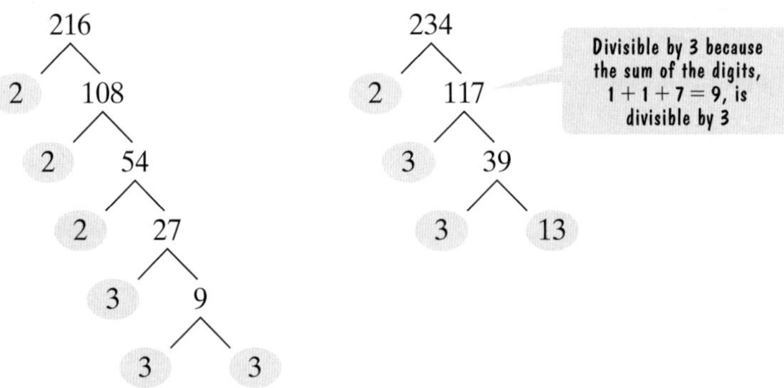

The factor tree on the left indicates that

$$216 = 2^3 \times 3^3.$$

The factor tree on the right indicates that

$$234 = 2 \times 3^2 \times 13.$$

Step 2 Select each prime factor with the smaller exponent that is common to each of the prime factorizations. Look at the factorizations of 216 and 234 from step 1. Can you see that 2 is a prime number common to the factorizations of 216 and 234? Likewise, 3 is also a prime number common to the two factorizations. By contrast, 13 is a prime number that is not common to both factorizations.

$$216 = 2^3 \times 3^3$$
$$234 = 2 \times 3^2 \times 13$$

2 is a prime number common to both factorizations.

3 is a prime number common to both factorizations.

Now we need to use these prime factorizations to determine which exponent is appropriate for 2 and which exponent is appropriate for 3. The appropriate exponent is the smaller exponent associated with the prime number in the factorizations. The exponents associated with 2 in the factorizations are 1 and 3, so we select 1. Therefore, one factor for the greatest common divisor is 2^1, or 2. The exponents associated with 3 in the factorizations are 2 and 3, so we select 2. Therefore, another factor for the greatest common divisor is 3^2.

$$216 = 2^3 \times 3^3$$

The smaller exponent on 2 is 1. The smaller exponent on 3 is 2.

$$234 = 2^1 \times 3^2 \times 13$$

Step 3 Form the product of the numbers from step 2. The greatest common divisor is the product of these factors.

$$\text{Greatest common divisor} = 2 \times 3^2 = 2 \times 9 = 18$$

The greatest common divisor of 216 and 234 is 18.

CHECK POINT 3 Find the greatest common divisor of 225 and 825.

4 *Solve problems using the greatest common divisor.*

GIMPS

A prime number of the form $2^p - 1$, where p is prime, is called a **Mersenne prime**, named for the seventeenth-century monk Marin Mersenne (1588–1648), who stated that $2^p - 1$ is prime for $p = 2, 3, 5, 7, 13, 17, 19, 31, 67, 127$, and 257. Without calculators and computers, it is not known how Mersenne arrived at these assertions, although it is now known that $2^p - 1$ is *not* prime for $p = 67$ and 257; it *is* prime for $p = 61, 89$, and 107.

In 1995, the American computer scientist George Woltman began the *Great Internet Mersenne Prime Search (GIMPS)*. By pooling the combined efforts of thousands of people interested in finding new Mersenne primes, GIMPS's participants have yielded several important results, including the record prime $2^{74,207,281} - 1$, a number with 22,338,618 digits that was discovered in 2016.

EXAMPLE 4 *Solving a Problem Using the Greatest Common Divisor*

For an intramural league, you need to divide 40 men and 24 women into all-male and all-female teams so that each team has the same number of people. What is the largest number of people that can be placed on a team?

SOLUTION

Because 40 men are to be divided into teams, the number of men on each team must be a divisor of 40. Because 24 women are to be divided into teams, the number of women placed on a team must be a divisor of 24. Although the teams are all-male and all-female, the same number of people must be placed on each team. The largest number of people that can be placed on a team is the largest number that will divide into 40 and 24 without a remainder. This is the greatest common divisor of 40 and 24.

To find the greatest common divisor of 40 and 24, begin with their prime factorizations.

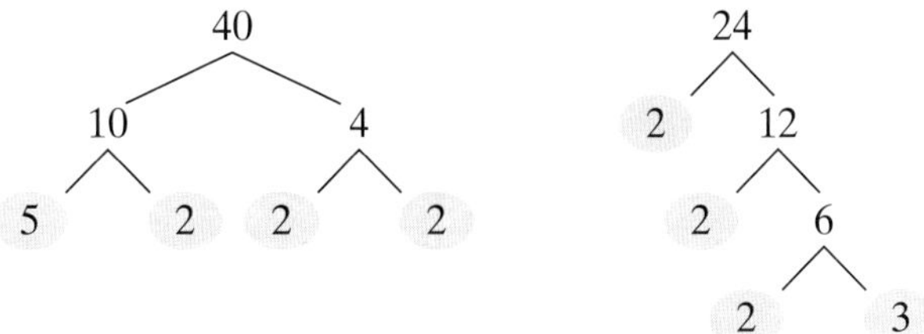

The factor trees indicate that

$$40 = 2^3 \times 5 \quad \text{and} \quad 24 = 2^3 \times 3.$$

We see that 2 is a prime number common to both factorizations. The exponents associated with 2 in the factorizations are 3 and 3, so we select 3.

$$\text{Greatest common divisor} = 2^3 = 2 \times 2 \times 2 = 8$$

The largest number of people that can be placed on a team is 8. Thus, the 40 men can form five teams with 8 men per team. The 24 women can form three teams with 8 women per team.

CHECK POINT 4 A choral director needs to divide 192 men and 288 women into all-male and all-female singing groups so that each group has the same number of people. What is the largest number of people that can be placed in each singing group?

5 *Find the least common multiple of two numbers.*

Least Common Multiple

The **least common multiple** of two or more natural numbers is the smallest natural number that is divisible by all of the numbers. One way to find the least common multiple is to make a list of the numbers that are divisible by each number. This list represents the **multiples** of each number. For example, if we wish to find the least common multiple of 15 and 20, we can list the sets of multiples of 15 and multiples of 20.

Numbers Divisible by 15: Multiples of 15:	{15, 30, 45, 60, 75, 90, 105, 120, . . .}
Numbers Divisible by 20: Multiples of 20:	{20, 40, 60, 80, 100, 120, 140, 160, . . .}

Two common multiples of 15 and 20 are 60 and 120. The least common multiple is 60. Equivalently, 60 is the smallest number that is divisible by both 15 and 20.

Sometimes a partial list of the multiples for each of two numbers does not reveal the smallest number that is divisible by both given numbers. A more efficient method for finding the least common multiple is to use prime factorizations.

FINDING THE LEAST COMMON MULTIPLE USING PRIME FACTORIZATIONS

To find the least common multiple of two or more numbers,

1. Write the prime factorization of each number.
2. Select every prime factor that occurs, raised to the greatest power to which it occurs, in these factorizations.
3. Form the product of the numbers from step 2. The least common multiple is the product of these factors.

EXAMPLE 5 *Finding the Least Common Multiple*

Find the least common multiple of 144 and 300.

SOLUTION

Step 1 Write the prime factorization of each number. Write the prime factorizations of 144 and 300.

$$144 = 2^4 \times 3^2$$
$$300 = 2^2 \times 3 \times 5^2$$

Step 2 Select every prime factor that occurs, raised to the greater power to which it occurs, in these factorizations. The prime factors that occur are 2, 3, and 5. The greater exponent that appears on 2 is 4, so we select 2^4. The greater exponent that appears on 3 is 2, so we select 3^2. The only exponent that occurs on 5 is 2, so we select 5^2. Thus, we have selected 2^4, 3^2, and 5^2.

Step 3 Form the product of the numbers from step 2. The least common multiple is the product of these factors.

$$\text{Least common multiple} = 2^4 \times 3^2 \times 5^2 = 16 \times 9 \times 25 = 3600$$

The least common multiple of 144 and 300 is 3600. The smallest natural number divisible by both 144 and 300 is 3600.

CHECK POINT 5 Find the least common multiple of 18 and 30.

Blitzer Bonus

Palindromic Primes

May a moody baby doom a yam? Leaving aside the answer to this question, what makes the sentence interesting is that it reads the same from left to right and from right to left! Such a sentence is called a **palindrome**. Some prime numbers are also palindromic. For example, the prime number 11 reads the same forward and backward, although a more provocative example containing all ten digits is

1,023,456,987,896,543,201.

In the following pyramid of palindromic primes, each number is obtained by adding two digits to the beginning and end of the previous prime.

2
30203
133020331
1713302033171
12171330203317121
151217133020331712151
1815121713302033171215181
16181512171330203317121518161

A huge palindromic prime was discovered in 2003 by David Broadhurst, a retired electrical engineer. The number contains 30,803 digits (and, yes, 30,803 is also a palindromic prime!).

Source: Clifford A. Pickover, *A Passion for Mathematics*, John Wiley and Sons, Inc., 2005.

6 *Solve problems using the least common multiple.*

EXAMPLE 6 *Solving a Problem Using the Least Common Multiple*

A movie theater runs its films continuously. One movie runs for 80 minutes and a second runs for 120 minutes. Both movies begin at 4:00 P.M. When will the movies begin again at the same time?

SOLUTION

The shorter movie lasts 80 minutes, or 1 hour, 20 minutes. It begins at 4:00, so it will be shown again at 5:20. The longer movie lasts 120 minutes, or 2 hours. It begins at 4:00, so it will be shown again at 6:00. We are asked to find when the movies will begin again at the same time. Therefore, we are looking for the least common multiple of 80 and 120. Find the least common multiple and then add this number of minutes to 4:00 P.M.

Begin with the prime factorizations of 80 and 120:

$$80 = 2^4 \times 5$$
$$120 = 2^3 \times 3 \times 5.$$

Now select each prime factor, with the greater exponent from each factorization.

$$\text{Least common multiple} = 2^4 \times 3 \times 5 = 16 \times 3 \times 5 = 240$$

Therefore, it will take 240 minutes, or 4 hours, for the movies to begin again at the same time. By adding 4 hours to 4:00 P.M., they will start together again at 8:00 P.M.

GREAT QUESTION!

Can I solve Example 6 by making a partial list of starting times for each movie?

Yes. Here's how it's done:

Shorter Movie (Runs 1 hour, 20 minutes):

4:00, 5:20, 6:40, 8:00, . . .

Longer Movie (Runs 2 hours):

4:00, 6:00, 8:00, . . .

The list reveals that both movies start together again at 8:00 P.M.

CHECK POINT 6 A movie theater runs two documentary films continuously. One documentary runs for 40 minutes and a second documentary runs for 60 minutes. Both movies begin at 3:00 P.M. When will the movies begin again at the same time?

Concept and Vocabulary Check

Fill in each blank so that the resulting statement is true.

1. A natural number greater than 1 that has only itself and 1 as factors is called a/an _______ number.
2. A natural number greater than 1 that is divisible by a number other than itself and 1 is called a/an __________ number.
3. The largest number that is a factor of two or more natural numbers is called their ____________________.
4. The smallest number that is divisible by two or more natural numbers is called their ____________________.

In Exercises 5–8, determine whether each statement is true or false. If the statement is false, make the necessary change(s) to produce a true statement.

5. The notation $b|a$ means that b is divisible by a. ______
6. $b \nmid a$ means that b does not divide a. ______
7. The words *factor* and *divisor* have opposite meanings. ______
8. A number can only be divisible by exactly one number. ______

Exercise Set 5.1

Practice Exercises

Use rules of divisibility to determine whether each number given in Exercises 1–10 is divisible by

a. 2 **b.** 3 **c.** 4 **d.** 5 **e.** 6
f. 8 **g.** 9 **h.** 10 **i.** 12.

1. 6944 **2.** 7245 **3.** 21,408 **4.** 25,025
5. 26,428 **6.** 89,001 **7.** 374,832 **8.** 347,712
9. 6,126,120 **10.** 5,941,221

In Exercises 11–24, use a calculator to determine whether each statement is true or false. If the statement is true, explain why this is so using one of the rules of divisibility in ***Table 5.1*** *on page 253.*

11. $3|5958$ **12.** $3|8142$ **13.** $4|10{,}612$
14. $4|15{,}984$ **15.** $5|38{,}814$ **16.** $5|48{,}659$
17. $6|104{,}538$ **18.** $6|163{,}944$ **19.** $8|20{,}104$
20. $8|28{,}096$ **21.** $9|11{,}378$ **22.** $9|23{,}772$
23. $12|517{,}872$ **24.** $12|785{,}172$

In Exercises 25–44, find the prime factorization of each composite number.

25. 75 **26.** 45 **27.** 56
28. 48 **29.** 105 **30.** 180
31. 500 **32.** 360 **33.** 663
34. 510 **35.** 885 **36.** 999
37. 1440 **38.** 1280 **39.** 1996
40. 1575 **41.** 3675 **42.** 8316
43. 85,800 **44.** 30,600

In Exercises 45–56, find the greatest common divisor of the numbers.

45. 42 and 56 **46.** 25 and 70 **47.** 16 and 42
48. 66 and 90 **49.** 60 and 108 **50.** 96 and 212
51. 72 and 120 **52.** 220 and 400 **53.** 342 and 380
54. 224 and 430 **55.** 240 and 285 **56.** 150 and 480

In Exercises 57–68, find the least common multiple of the numbers.

57. 42 and 56 **58.** 25 and 70 **59.** 16 and 42
60. 66 and 90 **61.** 60 and 108 **62.** 96 and 212
63. 72 and 120 **64.** 220 and 400 **65.** 342 and 380
66. 224 and 430 **67.** 240 and 285 **68.** 150 and 480

Practice Plus

In Exercises 69–74, determine all values of d that make each statement true.

69. $9|12{,}34d$ **70.** $9|23{,}42d$ **71.** $8|76{,}523{,}45d$
72. $8|88{,}888{,}82d$ **73.** $4|963{,}23d$ **74.** $4|752{,}67d$

A ***perfect number*** *is a natural number that is equal to the sum of its factors, excluding the number itself. In Exercises 75–78, determine whether or not each number is perfect.*

75. 28 **76.** 6 **77.** 20 **78.** 50

A prime number is an ***emirp*** *("prime" spelled backward) if it becomes a different prime number when its digits are reversed. In Exercises 79–82, determine whether or not each prime number is an emirp.*

79. 41 **80.** 43 **81.** 107 **82.** 113

A prime number p such that 2p + 1 is also a prime number is called a ***Germain prime****, named after the German mathematician Sophie Germain (1776–1831), who made major contributions to number theory. In Exercises 83–86, determine whether or not each prime number is a Germain prime.*

83. 13 **84.** 11 **85.** 241 **86.** 97

87. Find the product of the greatest common divisor of 24 and 27 and the least common multiple of 24 and 27. Compare this result to the product of 24 and 27. Write a conjecture based on your observation.

88. Find the product of the greatest common divisor of 48 and 72 and the least common multiple of 48 and 72. Compare this result to the product of 48 and 72. Write a conjecture based on your observation.

Application Exercises

89. In Carl Sagan's novel *Contact*, Ellie Arroway, the book's heroine, has been working at SETI, the Search for Extraterrestrial Intelligence, listening to the crackle of the cosmos. One night, as the radio telescopes are turned toward Vega, they suddenly pick up strange pulses through the background noise. Two pulses are followed by a pause, then three pulses, five, seven,

$$11, \; 13, \; 17, \; 19, \; 23, \; 29, \; 31, \ldots$$

2 3 5 7 11 13 17 19 23 29 31 37

continuing through 97. Then it starts all over again. Ellie is convinced that only intelligent life could generate the structure in the sequence of pulses. "It's hard to imagine some radiating plasma sending out a regular set of mathematical signals like this." What is it about the structure of the pulses that the book's heroine recognizes as the sign of intelligent life? Asked in another way, what is significant about the numbers of pulses?

90. There are two species of insects, *Magicicada septendecim* and *Magicicada tredecim*, that live in the same environment. They have a life cycle of exactly 17 and 13 years, respectively. For all but their last year, they remain in the ground feeding on the sap of tree roots. Then, in their last year, they emerge en masse from the ground as fully formed cricketlike insects, taking over the forest in a single night. They chirp loudly, mate, eat, lay eggs, then die six weeks later.

(*Source:* Marcus du Sautoy, *The Music of the Primes*, HarperCollins, 2003)

a. Suppose that the two species have life cycles that are not prime, say 18 and 12 years, respectively. List the set of multiples of 18 that are less than or equal to 216. List the set of multiples of 12 that are less than or equal to 216. Over a 216-year period, how many times will the two species emerge in the same year and compete to share the forest?

b. Recall that both species have evolved prime-number life cycles, 17 and 13 years, respectively. Find the least common multiple of 17 and 13. How often will the two species have to share the forest?

c. Compare your answers to parts (a) and (b). What explanation can you offer for each species having a prime number of years as the length of its life cycle?

91. A relief worker needs to divide 300 bottles of water and 144 cans of food into groups that each contain the same number of items. Also, each group must have the same type of item (bottled water or canned food). What is the largest number of relief supplies that can be put in each group?

92. A choral director needs to divide 180 men and 144 women into all-male and all-female singing groups so that each group has the same number of people. What is the largest number of people that can be placed in each singing group?

93. You have in front of you 310 five-dollar bills and 460 ten-dollar bills. Your problem: Place the five-dollar bills and the ten-dollar bills in stacks so that each stack has the same number of bills, and each stack contains only one kind of bill (five-dollar or ten-dollar). What is the largest number of bills that you can place in each stack?

94. Harley collects sports cards. He has 360 football cards and 432 baseball cards. Harley plans to arrange his cards in stacks so that each stack has the same number of cards. Also, each stack must have the same type of card (football or baseball). Every card in Harley's collection is to be placed in one of the stacks. What is the largest number of cards that can be placed in each stack?

95. You and your brother both work the 4:00 P.M. to midnight shift. You have every sixth night off. Your brother has every tenth night off. Both of you were off on June 1. Your brother would like to see a movie with you. When will the two of you have the same night off again?

96. A movie theater runs its films continuously. One movie is a short documentary that runs for 40 minutes. The other movie is a full-length feature that runs for 100 minutes. Each film is shown in a separate theater. Both movies begin at noon. When will the movies begin again at the same time?

97. Two people are jogging around a circular track in the same direction. One person can run completely around the track in 15 minutes. The second person takes 18 minutes. If they both start running in the same place at the same time, how long will it take them to be together at this place if they continue to run?

98. Two people are in a bicycle race around a circular track. One rider can race completely around the track in 40 seconds. The other rider takes 45 seconds. If they both begin the race at a designated starting point, how long will it take them to be together at this starting point again if they continue to race around the track?

Explaining the Concepts

99. If a is a factor of c, what does this mean?

100. How do you know that 45 is divisible by 5?

101. What does "a is divisible by b" mean?

102. Describe the difference between a prime number and a composite number.

103. What does the Fundamental Theorem of Arithmetic state?

104. What is the greatest common divisor of two or more natural numbers?

105. Describe how to find the greatest common divisor of two numbers.

106. What is the least common multiple of two or more natural numbers?

107. Describe how to find the least common multiple of two natural numbers.

108. The process of finding the greatest common divisor of two natural numbers is similar to finding the least common multiple of the numbers. Describe how the two processes differ.

109. What does the Blitzer Bonus on page 256 have to do with Gödel's discovery about mathematics and logic, described on page 181?

Critical Thinking Exercises

Make Sense? *In Exercises 110–113, determine whether each statement makes sense or does not make sense, and explain your reasoning.*

110. I'm working with a prime number that intrigues me because it has three natural number factors.

111. When I find the greatest common factor, I select common prime factors with the greatest exponent and when I find the least common multiple, I select common prime factors with the smallest exponent.

112. I need to separate 70 men and 175 women into all-male or all-female teams with the same number of people on each team. By finding the least common multiple of 70 and 175, I can determine the largest number of people that can be placed on a team.

113. (If you have not yet done so, read the Blitzer Bonus "GIMPS" on page 257.) I can find a prime number larger than the record prime $2^{74,207,281} - 1$ by simply writing $2^{74,207,282} - 1$.

114. Write a four-digit natural number that is divisible by 4 and not by 8.

115. Find the greatest common divisor and the least common multiple of $2^{17} \cdot 3^{25} \cdot 5^{31}$ and $2^{14} \cdot 3^{37} \cdot 5^{30}$. Express answers in the same form as the numbers given.

116. A middle-aged man observed that his present age was a prime number. He also noticed that the number of years in which his age would again be prime was equal to the number of years ago in which his age was prime. How old is the man?

117. A movie theater runs its films continuously. One movie runs for 85 minutes and a second runs for 100 minutes. The theater has a 15-minute intermission after each movie, at which point the movie is shown again. If both movies start at noon, when will the two movies start again at the same time?

118. The difference between consecutive prime numbers is always an even number, except for two particular prime numbers. What are those numbers?

119. A question from *Who Wants to Be a Millionaire?* How many total years during one's teen years is a person an age that's a prime number?

a. 1 **b.** 2 **c.** 3 **d.** 4

Technology Exercises

Use the divisibility rules listed in **Table 5.1** *on page 253 to answer the questions in Exercises 120–122. Then, using a calculator, perform the actual division to determine whether your answer is correct.*

120. Is 67,234,096 divisible by 4?

121. Is 12,541,750 divisible by 3?

122. Is 48,201,651 divisible by 9?

Group Exercises

The following topics from number theory are appropriate for either individual or group research projects. A report should be given to the class on the researched topic. Useful references include liberal arts mathematics textbooks, books about numbers and number theory, books whose purpose is to excite the reader about mathematics, history of mathematics books, encyclopedias, and the Internet.

123. Euclid and Number Theory

124. An Unsolved Problem from Number Theory

125. Perfect Numbers

126. Deficient and Abundant Numbers

127. Formulas That Yield Primes

128. The Sieve of Eratosthenes

5.2 The Integers; Order of Operations

WHAT AM I SUPPOSED TO LEARN?

After studying this section, you should be able to:

1. Define the integers.
2. Graph integers on a number line.
3. Use the symbols < and >.
4. Find the absolute value of an integer.
5. Perform operations with integers.
6. Use the order of operations agreement.

CAN YOU CHEAT DEATH? LIFE expectancy for the average American man is 77.1 years; for a woman, it's 81.9. But what's in your hands if you want to eke out a few more birthday candles? In this section, we use operations on a set of numbers called the *integers* to indicate factors within your control that can stretch your probable life span. Start by flossing. (See Example 5 on page 268.)

George Tooker (1920–2011) "*Mirror II*" Addison Gallery of American Art, Phillips Academy, Andover, MA/Art Resource, NY; © Estate of George Tooker. Courtesy of DC Moore Gallery, New York

Defining the Integers

In Section 5.1, we applied some ideas of number theory to the set of natural, or counting, numbers:

$$\text{Natural numbers} = \{1, 2, 3, 4, 5, \ldots\}.$$

1 *Define the integers.*

When we combine the number 0 with the natural numbers, we obtain the set of **whole numbers**:

$$\text{Whole numbers} = \{0, 1, 2, 3, 4, 5, \dots\}.$$

The whole numbers do not allow us to describe certain everyday situations. For example, if the balance in your checking account is \$30 and you write a check for \$35, your checking account is overdrawn by \$5. We can write this as -5, read *negative* 5. The set consisting of the natural numbers, 0, and the negatives of the natural numbers is called the set of **integers**.

$$\text{Integers} = \{\underbrace{\dots, -4, -3, -2, -1}_{\text{Negative integers}}, 0, \underbrace{1, 2, 3, 4, \dots}_{\text{Positive integers}}\}.$$

Notice that the term *positive integers* is another name for the natural numbers. The positive integers can be written in two ways:

1. Use a "+" sign. For example, +4 is "positive four."
2. Do not write any sign. For example, 4 is assumed to be "positive four."

The Number Line; The Symbols < and >

The **number line** is a graph we use to visualize the set of integers, as well as sets of other numbers. The number line is shown in **Figure 5.1**.

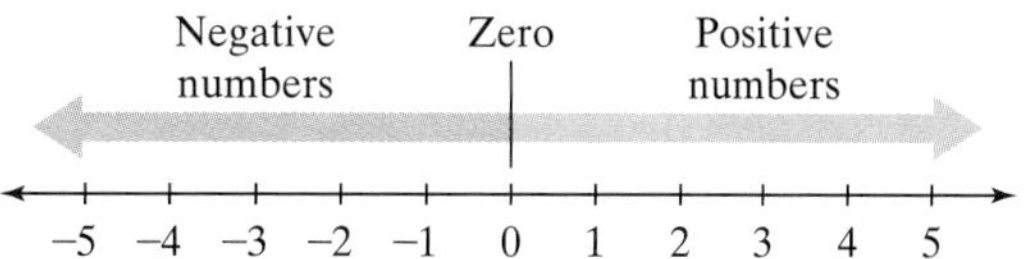

FIGURE 5.1 The number line

The number line extends indefinitely in both directions, shown by the arrows on the left and the right. Zero separates the positive numbers from the negative numbers on the number line. The positive integers are located to the right of 0 and the negative integers are located to the left of 0. **Zero is neither positive nor negative.** For every positive integer on a number line, there is a corresponding negative integer on the opposite side of 0.

2 *Graph integers on a number line.*

Integers are graphed on a number line by placing a dot at the correct location for each number.

EXAMPLE 1 Graphing Integers on a Number Line

Graph: **a.** -3 **b.** 4 **c.** 0.

SOLUTION

Place a dot at the correct location for each integer.

(a) (c) (b)

−5 −4 −3 −2 −1 0 1 2 3 4 5

CHECK POINT 1 Graph:

a. -4 **b.** 0 **c.** 3.

3 *Use the symbols < and >.*

We will use the following symbols for comparing two integers:

< means "is less than."
> means "is greater than."

On the number line, the integers increase from left to right. The *lesser* of two integers is the one farther to the *left* on a number line. The *greater* of two integers is the one farther to the *right* on a number line.

Look at the number line in **Figure 5.2**. The integers -4 and -1 are graphed.

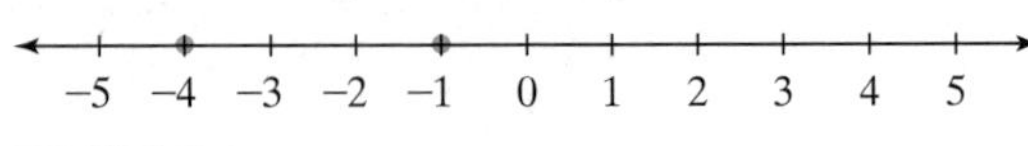

FIGURE 5.2

Observe that -4 is to the left of -1 on the number line. This means that -4 is less than -1.

$$-4 < -1$$
-4 is less than -1 because -4 is to the left of -1 on the number line.

In **Figure 5.2**, we can also observe that -1 is to the right of -4 on the number line. This means that -1 is greater than -4.

$$-1 > -4$$
-1 is greater than -4 because -1 is to the right of -4 on the number line.

The symbols < and > are called **inequality symbols**. These symbols always point to the lesser of the two integers when the inequality statement is true.

-4 is less than -1. $\quad -4 < -1 \quad$ The symbol points to -4, the lesser number.

-1 is greater than -4. $\quad -1 > -4 \quad$ The symbol still points to -4, the lesser number.

EXAMPLE 2 Using the Symbols < and >

Insert either < or > in the shaded area between the integers to make each statement true:

a. -4 ▒ 3 **b.** -1 ▒ -5 **c.** -5 ▒ -2 **d.** 0 ▒ -3.

SOLUTION

The solution is illustrated by the number line in **Figure 5.3**.

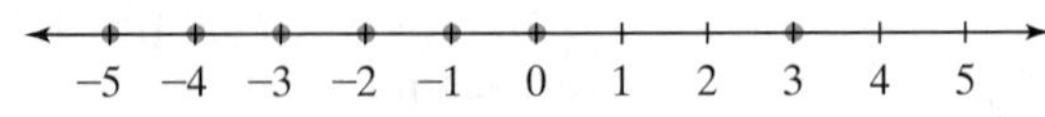

FIGURE 5.3

a. $-4 < 3$ (negative 4 is less than 3) because -4 is to the left of 3 on the number line.

b. $-1 > -5$ (negative 1 is greater than negative 5) because -1 is to the right of -5 on the number line.

c. $-5 < -2$ (negative 5 is less than negative 2) because -5 is to the left of -2 on the number line.

d. $0 > -3$ (zero is greater than negative 3) because 0 is to the right of -3 on the number line.

GREAT QUESTION!

Other than using a number line, is there another way to remember that -1 is greater than -5?

Yes. Think of negative integers as amounts of money that you *owe*. It's better to owe less, so

$$-1 > -5.$$

CHECK POINT 2 Insert either < or > in the shaded area between the integers to make each statement true:

a. 6 ▒ -7 **b.** -8 ▒ -1 **c.** -25 ▒ -2 **d.** -14 ▒ 0.

The symbols $<$ and $>$ may be combined with an equal sign, as shown in the following table:

This inequality is true if either the $<$ part or the $=$ part is true.

This inequality is true if either the $>$ part or the $=$ part is true.

Symbols	Meaning	Examples	Explanation
$a \le b$	*a* is less than or equal to *b*.	$2 \le 9$ $9 \le 9$	Because $2 < 9$ Because $9 = 9$
$b \ge a$	*b* is greater than or equal to *a*.	$9 \ge 2$ $2 \ge 2$	Because $9 > 2$ Because $2 = 2$

4 *Find the absolute value of an integer.*

Absolute Value

Absolute value describes distance from 0 on a number line. If a represents an integer, the symbol $|a|$ represents its absolute value, read "the absolute value of a." For example,

$$|-5| = 5.$$

The absolute value of -5 is 5 because -5 is 5 units from 0 on a number line.

ABSOLUTE VALUE

The **absolute value** of an integer a, denoted by $|a|$, is the distance from 0 to a on the number line. Because absolute value describes a distance, it is never negative.

EXAMPLE 3 *Finding Absolute Value*

Find the absolute value:

a. $|-3|$ **b.** $|5|$ **c.** $|0|$.

SOLUTION

The solution is illustrated in **Figure 5.4**.

a. $|-3| = 3$ The absolute value of -3 is 3 because -3 is 3 units from 0.

b. $|5| = 5$ 5 is 5 units from 0.

c. $|0| = 0$ 0 is 0 units from itself.

$|-3| = 3$ $|5| = 5$

$-5\ -4\ -3\ -2\ -1\ 0\ 1\ 2\ 3\ 4\ 5$

$|0| = 0$

FIGURE 5.4 Absolute value describes distance from 0 on a number line.

Example 3 illustrates that the absolute value of a positive integer or 0 is the number itself. The absolute value of a negative integer, such as -3, is the number without the negative sign. Zero is the only real number whose absolute value is 0: $|0| = 0$. **The absolute value of any integer other than 0 is always positive.**

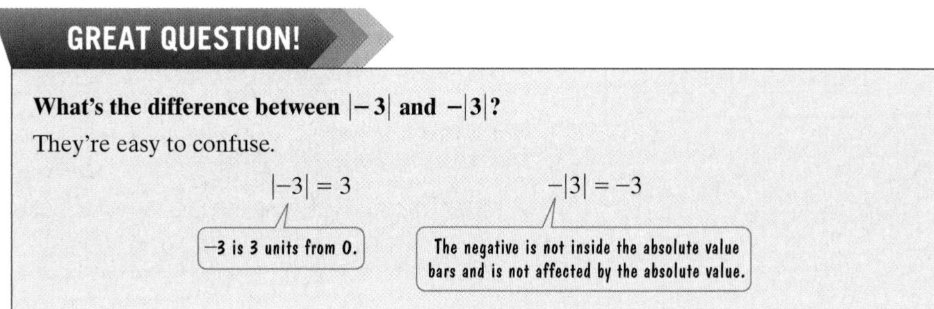
GREAT QUESTION!

What's the difference between $|-3|$ and $-|3|$?

They're easy to confuse.

$|-3| = 3$ $-|3| = -3$

-3 is 3 units from 0.

The negative is not inside the absolute value bars and is not affected by the absolute value.

CHECK POINT 3 Find the absolute value:

a. $|-8|$ **b.** $|6|$ **c.** $-|8|$.

5 *Perform operations with integers.*

Addition of Integers

It has not been a good day! First, you lost your wallet with \$50 in it. Then, you borrowed \$10 to get through the day, which you somehow misplaced. Your loss of \$50 followed by a loss of \$10 is an overall loss of \$60. This can be written

$$-50 + (-10) = -60.$$

The result of adding two or more numbers is called the **sum** of the numbers. The sum of -50 and -10 is -60.

You can think of gains and losses of money to find sums. For example, to find $17 + (-13)$, think of a gain of \$17 followed by a loss of \$13. There is an overall gain of \$4. Thus, $17 + (-13) = 4$. In the same way, to find $-17 + 13$, think of a loss of \$17 followed by a gain of \$13. There is an overall loss of \$4, so $-17 + 13 = -4$.

Using gains and losses, we can develop the following rules for adding integers:

RULES FOR ADDITION OF INTEGERS

Rule	Examples
If the integers have the same sign, **1.** Add their absolute values. **2.** The sign of the sum is the same as the sign of the two numbers.	$-11 + (-15) = -26$ Add absolute values: $11 + 15 = 26$. Use the common sign.
If the integers have different signs, **1.** Subtract the smaller absolute value from the larger absolute value. **2.** The sign of the sum is the same as the sign of the number with the larger absolute value.	$-13 + 4 = -9$ Subtract absolute values: $13 - 4 = 9$. Use the sign of the number with the greater absolute value. $13 + (-6) = 7$ Subtract absolute values: $13 - 6 = 7$. Use the sign of the number with the greater absolute value.

TECHNOLOGY

Calculators and Adding Integers

You can use a calculator to add integers. Here are the keystrokes for finding $-11 + (-15)$:

Scientific Calculator

11 [+/−] [+] 15 [+/−] [=]

Graphing Calculator

[(−)] 11 [+] [(−)] 15 [ENTER]

Here are the keystrokes for finding $-13 + 4$:

Scientific Calculator

13 [+/−] [+] 4 [=]

Graphing Calculator

[(−)] 13 [+] 4 [ENTER]

GREAT QUESTION!

Other than gains and losses of money, is there another good analogy for adding integers?

Yes. Think of temperatures above and below zero on a thermometer. Picture the thermometer as a number line standing straight up. For example,

$-11 + (-15) = -26$ — If it's 11 below zero and the temperature falls 15 degrees, it will then be 26 below zero.

$-13 + 4 = -9$ — If it's 13 below zero and the temperature rises 4 degrees, the new temperature will be 9 below zero.

$13 + (-6) = 7.$ — If it's 13 above zero and the temperature falls 6 degrees, it will then be 7 above zero.

Using the analogies of gains and losses of money or temperatures can make the formal rules for addition of integers easy to use.

Can you guess what number is displayed if you use a calculator to find a sum such as $18 + (-18)$? If you gain 18 and then lose 18, there is neither an overall gain nor loss. Thus,

$$18 + (-18) = 0.$$

We call 18 and −18 **additive inverses**. Additive inverses have the same absolute value, but lie on opposite sides of zero on the number line. Thus, −7 is the additive inverse of 7, and 5 is the additive inverse of −5. In general, the sum of any integer and its additive inverse is 0:

$$a + (-a) = 0.$$

Subtraction of Integers

Suppose that a computer that normally sells for $1500 has a price reduction of $600. The computer's reduced price, $900, can be expressed in two ways:

$$1500 - 600 = 900 \quad \text{or} \quad 1500 + (-600) = 900.$$

This means that

$$1500 - 600 = 1500 + (-600).$$

To subtract 600 from 1500, we add 1500 and the additive inverse of 600. Generalizing from this situation, we define subtraction as follows:

DEFINITION OF SUBTRACTION

For all integers a and b,

$$a - b = a + (-b).$$

In words, to subtract b from a, add the additive inverse of b to a. The result of subtraction is called the **difference**.

TECHNOLOGY

Calculators and Subtracting Integers

You can use a calculator to subtract integers. Here are the keystrokes for finding 17 − (−11):

Scientific Calculator

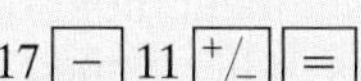

17 [−] 11 [+/−] [=]

Graphing Calculator

17 [−] [(−)] 11 [ENTER]

Here are the keystrokes for finding −18 − (−5):

Scientific Calculator

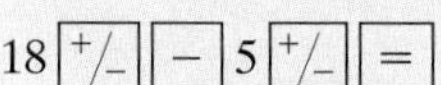

18 [+/−] [−] 5 [+/−] [=]

Graphing Calculator

[(−)] 18 [−] [(−)] 5 [ENTER]

Don't confuse the subtraction key on a graphing calculator, [−], with the sign change or additive inverse key, [(−)]. What happens if you do?

EXAMPLE 4 Subtracting Integers

Subtract:

a. 17 − (−11) **b.** −18 − (−5) **c.** −18 − 5.

SOLUTION

a. $17 - (-11) = 17 + 11 = 28$

Change the subtraction to addition. Replace −11 with its additive inverse.

b. $-18 - (-5) = -18 + 5 = -13$

Change the subtraction to addition. Replace −5 with its additive inverse.

c. $-18 - 5 = -18 + (-5) = -23$

Change the subtraction to addition. Replace 5 with its additive inverse.

CHECK POINT 4 Subtract:

a. 30 − (−7) **b.** −14 − (−10) **c.** −14 − 10.

GREAT QUESTION!

Is there a practical way to think about what it means to subtract a negative integer?

Yes. Think of taking away a debt. Let's apply this analogy to 17 − (−11). Your checking account balance is $17 after an erroneous $11 charge was made against your account. When you bring this error to the bank's attention, they will take away the $11 debit and your balance will go up to $28:

$$17 - (-11) = 28.$$

Subtraction is used to solve problems in which the word *difference* appears. The difference between integers a and b is expressed as $a - b$.

EXAMPLE 5 An Application of Subtraction Using the Word Difference

Life expectancy for the average American man is 77.1 years; for a woman, it's 81.9 years. The number line in **Figure 5.5**, with points representing eight integers, indicates factors, many within our control, that can stretch or shrink one's probable life span.

Stretching or Shrinking One's Life Span

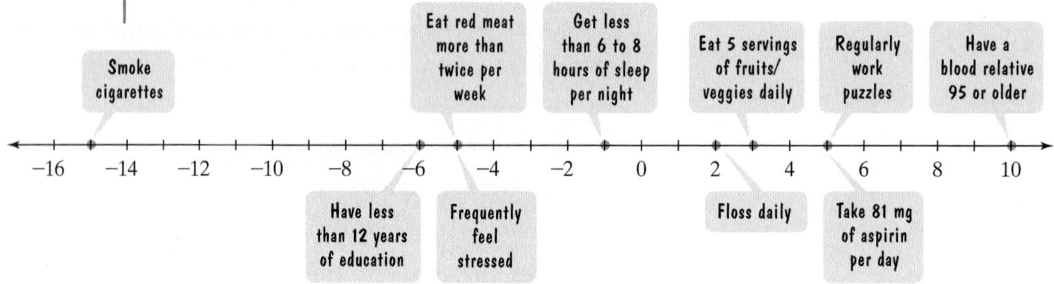

FIGURE 5.5
Source: Newsweek

a. What is the difference in the life span between a person who regularly works puzzles and a person who eats red meat more than twice per week?

b. What is the difference in the life span between a person with less than 12 years of education and a person who smokes cigarettes?

SOLUTION

a. We begin with the difference in the life span between a person who regularly works puzzles and a person who eats red meat more than twice per week. Refer to **Figure 5.5** to determine years of life gained or lost.

$$= 5 - (-5)$$
$$= 5 + 5 = 10$$

The difference in the life span is 10 years.

b. Now we consider the difference in the life span between a person with less than 12 years of education and a person who smokes cigarettes.

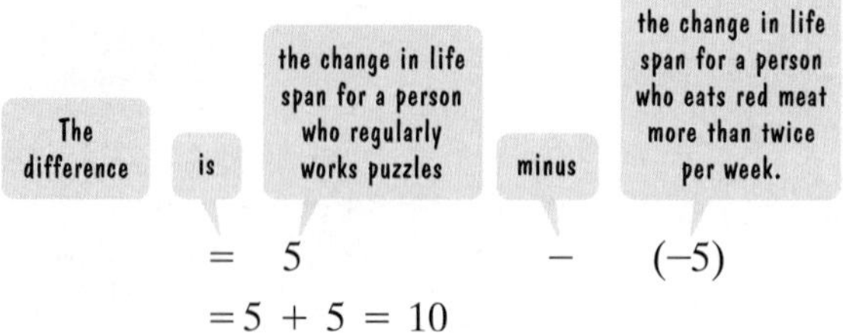

$$= -6 - (-15)$$
$$= -6 + 15 = 9$$

The difference in the life span is 9 years.

CHECK POINT 5 Use the number line in **Figure 5.5** to answer the following questions:

a. What is the difference in the life span between a person who eats five servings of fruits/veggies daily and a person who frequently feels stressed?

b. What is the difference in the life span between a person who gets less than 6 to 8 hours of sleep per night and a person who smokes cigarettes?

Multiplication of Integers

The result of multiplying two or more numbers is called the **product** of the numbers. You can think of multiplication as repeated addition or subtraction that starts at 0. For example,

$$3(-4) = 0 + (-4) + (-4) + (-4) = -12$$

The numbers have different signs and the product is negative.

and

$$(-3)(-4) = 0 - (-4) - (-4) - (-4) = 0 + 4 + 4 + 4 = 12.$$

The numbers have the same sign and the product is positive.

These observations give us the following rules for multiplying integers:

RULES FOR MULTIPLYING INTEGERS

Rule	Examples
1. The product of two integers with different signs is found by multiplying their absolute values. The product is negative.	• $7(-5) = -35$
2. The product of two integers with the same sign is found by multiplying their absolute values. The product is positive.	• $(-6)(-11) = 66$
3. The product of 0 and any integer is 0: $a \cdot 0 = 0$ and $0 \cdot a = 0$.	• $-17(0) = 0$
4. If no number is 0, a product with an odd number of negative factors is found by multiplying absolute values. The product is negative.	• $-2(-3)(-5) = -30$ Three (odd) negative factors
5. If no number is 0, a product with an even number of negative factors is found by multiplying absolute values. The product is positive.	• $-2(3)(-5) = 30$ Two (even) negative factors

Exponential Notation

Because exponents indicate repeated multiplication, rules for multiplying integers can be used to evaluate exponential expressions.

EXAMPLE 6 *Evaluating Exponential Expressions*

Evaluate:

a. $(-6)^2$ **b.** -6^2 **c.** $(-5)^3$ **d.** $(-2)^4$.

SOLUTION

a. $(-6)^2 = (-6)(-6) = 36$

Base is −6. Same signs give positive product.

b. $-6^2 = -(6 \cdot 6) = -36$

Base is 6. The negative is not inside parentheses and is not taken to the second power.

c. $(-5)^3 = (-5)(-5)(-5) = -125$

An odd number of negative factors gives a negative product.

d. $(-2)^4 = (-2)(-2)(-2)(-2) = 16$

An even number of negative factors gives a positive product.

CHECK POINT 6 Evaluate:

a. $(-5)^2$ **b.** -5^2 **c.** $(-4)^3$ **d.** $(-3)^4$.

Blitzer Bonus

Integers, Karma, and Exponents

On Friday the 13th, are you a bit more careful crossing the street even if you don't consider yourself superstitious? Numerology, the belief that certain integers have greater significance and can be lucky or unlucky, is widespread in many cultures.

Integer	Connotation	Culture	Origin	Example
4	Negative	Chinese	The word for the number 4 sounds like the word for death.	Many buildings in China have floor-numbering systems that skip 40–49.
7	Positive	United States	In dice games, this prime number is the most frequently rolled number with two dice.	There was a spike in the number of couples getting married on 7/7/07.
8	Positive	Chinese	It's considered a sign of prosperity.	The Beijing Olympics began at 8 P.M. on 8/8/08.
13	Negative	Various	Various reasons, including the number of people at the Last Supper	Many buildings around the world do not label any floor "13."
18	Positive	Jewish	The Hebrew letters spelling *chai*, or living, are the 8th and 10th in the alphabet, adding up to 18	Monetary gifts for celebrations are often given in multiples of 18.
666	Negative	Christian	The New Testament's Book of Revelation identifies 666 as the "number of the beast," which some say refers to Satan.	In 2008, Reeves, Louisiana, eliminated 666 as the prefix of its phone numbers.

Source: The New York Times

Although your author is not a numerologist, he is intrigued by curious exponential representations for 666:

$$666 = 6 + 6 + 6 + 6^3 + 6^3 + 6^3$$

$$666 = 1^3 + 2^3 + 3^3 + 4^3 + 5^3 + 6^3 + 5^3 + 4^3 + 3^3 + 2^3 + 1^3$$

$$666 = 2^2 + 3^2 + 5^2 + 7^2 + 11^2 + 13^2 + 17^2$$

Sum of the squares of the first seven prime numbers

$$666 = 1^6 - 2^6 + 3^6$$

The number 666 is even interesting in Roman numerals:

666 = DCLXVI.

Contains all Roman numerals from D (500) to I (1) in decending order

Division of Integers

The result of dividing the integer a by the nonzero integer b is called the **quotient** of the numbers. We can write this quotient as $a \div b$ or $\frac{a}{b}$.

A relationship exists between multiplication and division. For example,

$$\frac{-12}{4} = -3 \text{ means that } 4(-3) = -12.$$

$$\frac{-12}{-4} = 3 \text{ means that } -4(3) = -12.$$

Because of the relationship between multiplication and division, the rules for obtaining the sign of a quotient are the same as those for obtaining the sign of a product.

TECHNOLOGY

Multiplying and Dividing on a Calculator

Example: $(-173)(-256)$

Scientific Calculator

173 [+/−] [×] 256 [+/−] [=]

Graphing Calculator

[(−)] 173 [×] [(−)] 256 [ENTER]

The number 44288 should be displayed.

Division is performed in the same manner, using [÷] instead of [×]. What happens when you divide by 0? Try entering

8 [÷] 0

and pressing [=] or [ENTER].

RULES FOR DIVIDING INTEGERS

Rule	Examples
1. The quotient of two integers with different signs is found by dividing their absolute values. The quotient is negative.	• $\frac{80}{-4} = -20$ • $\frac{-15}{5} = -3$
2. The quotient of two integers with the same sign is found by dividing their absolute values. The quotient is positive.	• $\frac{27}{9} = 3$ • $\frac{-45}{-3} = 15$
3. Zero divided by any nonzero integer is zero.	• $\frac{0}{-5} = 0$ (because $-5(0) = 0$)
4. Division by 0 is undefined.	• $\frac{-8}{0}$ is undefined (because 0 cannot be multiplied by an integer to obtain -8).

6 *Use the order of operations agreement.*

Order of Operations

Suppose that you want to find the value of $3 + 7 \cdot 5$. Which procedure shown below is correct?

$$3 + 7 \cdot 5 = 3 + 35 = 38 \quad \text{or} \quad 3 + 7 \cdot 5 = 10 \cdot 5 = 50$$

If you know the answer, you probably know certain rules, called the **order of operations**, that make sure there is only one correct answer. One of these rules states that if a problem contains no parentheses, perform multiplication before addition. Thus, the procedure on the left is correct because the multiplication of 7 and 5 is done first. Then the addition is performed. The correct answer is 38.

Here are the rules for determining the order in which operations should be performed:

GREAT QUESTION!

How can I remember the order of operations?

This sentence may help: Please excuse my dear Aunt Sally.

Please	Parentheses
Excuse	Exponents
{My Dear	{Multiplication Division
{Aunt Sally	{Addition Subtraction

ORDER OF OPERATIONS

1. Perform all operations within grouping symbols.
2. Evaluate all exponential expressions.
3. Do all multiplications and divisions in the order in which they occur, working from left to right.
4. Finally, do all additions and subtractions in the order in which they occur, working from left to right.

In the third step in the order of operations, be sure to do all multiplications and divisions *as they occur* from left to right. For example,

$$8 \div 4 \cdot 2 = 2 \cdot 2 = 4$$ Do the division first because it occurs first.

$$8 \cdot 4 \div 2 = 32 \div 2 = 16.$$ Do the multiplication first because it occurs first.

EXAMPLE 7 Using the Order of Operations

Simplify: $6^2 - 24 \div 2^2 \cdot 3 + 1$.

SOLUTION

There are no grouping symbols. Thus, we begin by evaluating exponential expressions. Then we multiply or divide. Finally, we add or subtract.

$6^2 - 24 \div 2^2 \cdot 3 + 1$

$= 36 - 24 \div 4 \cdot 3 + 1$ Evaluate exponential expressions: $6^2 = 6 \cdot 6 = 36$ and $2^2 = 2 \cdot 2 = 4$.

$= 36 - 6 \cdot 3 + 1$ Perform the multiplications and divisions from left to right. Start with $24 \div 4 = 6$.

$= 36 - 18 + 1$ Now do the multiplication: $6 \cdot 3 = 18$.

$= 18 + 1$ Finally, perform the additions and subtractions from left to right. Subtract: $36 - 18 = 18$.

$= 19$ Add: $18 + 1 = 19$.

CHECK POINT 7 Simplify: $7^2 - 48 \div 4^2 \cdot 5 + 2$.

EXAMPLE 8 Using the Order of Operations

Simplify: $(-6)^2 - (5 - 7)^2(-3)$.

SOLUTION

Because grouping symbols appear, we perform the operation within parentheses first.

$(-6)^2 - (5 - 7)^2(-3)$

$= (-6)^2 - (-2)^2(-3)$ Work inside parentheses first: $5 - 7 = 5 + (-7) = -2$.

$= 36 - 4(-3)$ Evaluate exponential expressions: $(-6)^2 = (-6)(-6) = 36$ and $(-2)^2 = (-2)(-2) = 4$.

$= 36 - (-12)$ Multiply: $4(-3) = -12$.

$= 48$ Subtract: $36 - (-12) = 36 + 12 = 48$.

CHECK POINT 8 Simplify: $(-8)^2 - (10 - 13)^2(-2)$.

Concept and Vocabulary Check

Fill in each blank so that the resulting statement is true.

1. The integers are defined by the set ____________________.
2. If $a < b$, then a is located to the _____ of b on a number line.
3. On a number line, the absolute value of a, denoted $|a|$, represents __________________.
4. Two integers that have the same absolute value, but lie on opposite sides of zero on a number line, are called ______________.

In Exercises 5–8, determine whether each statement is true or false. If the statement is false, make the necessary change(s) to produce a true statement.

5. The sum of a positive integer and a negative integer is always a positive integer. ______
6. The difference between 0 and a negative integer is always a positive integer. ______
7. The product of a positive integer and a negative integer is never a positive integer. ______
8. The quotient of 0 and a negative integer is undefined. ______

Exercise Set 5.2

Practice Exercises

In Exercises 1–4, start by drawing a number line that shows integers from −5 to 5. Then graph each of the following integers on your number line.

1. 3 **2.** 5 **3.** −4 **4.** −2

In Exercises 5–12, insert either < or > in the shaded area between the integers to make the statement true.

5. $-2 \,\blacksquare\, 7$ **6.** $-1 \,\blacksquare\, 13$ **7.** $-13 \,\blacksquare\, -2$

8. $-1 \,\blacksquare\, -13$ **9.** $8 \,\blacksquare\, -50$ **10.** $7 \,\blacksquare\, -9$

11. $-100 \,\blacksquare\, 0$ **12.** $0 \,\blacksquare\, -300$

In Exercises 13–18, find the absolute value.

13. $|-14|$ **14.** $|-16|$ **15.** $|14|$

16. $|16|$ **17.** $|-300{,}000|$ **18.** $|-1{,}000{,}000|$

In Exercises 19–30, find each sum.

19. $-7 + (-5)$ **20.** $-3 + (-4)$

21. $12 + (-8)$ **22.** $13 + (-5)$

23. $6 + (-9)$ **24.** $3 + (-11)$

25. $-9 + (+4)$ **26.** $-7 + (+3)$

27. $-9 + (-9)$ **28.** $-13 + (-13)$

29. $9 + (-9)$ **30.** $13 + (-13)$

In Exercises 31–42, perform the indicated subtraction.

31. $13 - 8$ **32.** $14 - 3$

33. $8 - 15$ **34.** $9 - 20$

35. $4 - (-10)$ **36.** $3 - (-17)$

37. $-6 - (-17)$ **38.** $-4 - (-19)$

39. $-12 - (-3)$ **40.** $-19 - (-2)$

41. $-11 - 17$ **42.** $-19 - 21$

In Exercises 43–52, find each product.

43. $6(-9)$ **44.** $5(-7)$

45. $(-7)(-3)$ **46.** $(-8)(-5)$

47. $(-2)(6)$ **48.** $(-3)(10)$

49. $(-13)(-1)$ **50.** $(-17)(-1)$

51. $0(-5)$ **52.** $0(-8)$

In Exercises 53–66, evaluate each exponential expression.

53. 5^2 **54.** 6^2

55. $(-5)^2$ **56.** $(-6)^2$

57. 4^3 **58.** 2^3

59. $(-5)^3$ **60.** $(-4)^3$

61. $(-5)^4$ **62.** $(-4)^4$

63. -3^4 **64.** -1^4

65. $(-3)^4$ **66.** $(-1)^4$

In Exercises 67–80, find each quotient, or, if applicable, state that the expression is undefined.

67. $\frac{-12}{4}$ **68.** $\frac{-40}{5}$

69. $\frac{21}{-3}$ **70.** $\frac{60}{-6}$

71. $\frac{-90}{-3}$ **72.** $\frac{-66}{-6}$

73. $\frac{0}{-7}$ **74.** $\frac{0}{-8}$

75. $\frac{-7}{0}$ **76.** $\frac{0}{0}$

77. $(-480) \div 24$ **78.** $(-300) \div 12$

79. $(465) \div (-15)$ **80.** $(-594) \div (-18)$

In Exercises 81–100, use the order of operations to find the value of each expression.

81. $7 + 6 \cdot 3$ **82.** $-5 + (-3) \cdot 8$

83. $(-5) - 6(-3)$ **84.** $-8(-3) - 5(-6)$

85. $6 - 4(-3) - 5$ **86.** $3 - 7(-1) - 6$

87. $3 - 5(-4 - 2)$ **88.** $3 - 9(-1 - 6)$

89. $(2 - 6)(-3 - 5)$ **90.** $9 - 5(6 - 4) - 10$

91. $3(-2)^2 - 4(-3)^2$ **92.** $5(-3)^2 - 2(-2)^3$

93. $(2 - 6)^2 - (3 - 7)^2$

94. $(4 - 6)^2 - (5 - 9)^3$

95. $6(3 - 5)^3 - 2(1 - 3)^3$

96. $-3(-6 + 8)^3 - 5(-3 + 5)^3$

97. $8^2 - 16 \div 2^2 \cdot 4 - 3$

98. $10^2 - 100 \div 5^2 \cdot 2 - (-3)$

99. $24 \div [3^2 \div (8 - 5)] - (-6)$

100. $30 \div [5^2 \div (7 - 12)] - (-9)$

Practice Plus

In Exercises 101–110, use the order of operations to find the value of each expression.

101. $8 - 3[-2(2 - 5) - 4(8 - 6)]$

102. $8 - 3[-2(5 - 7) - 5(4 - 2)]$

103. $-2^2 + 4[16 \div (3 - 5)]$

104. $-3^2 + 2[20 \div (7 - 11)]$

105. $4|10 - (8 - 20)|$

106. $-5|7 - (20 - 8)|$

107. $[-5^2 + (6 - 8)^3 - (-4)] - [|-2|^3 + 1 - 3^2]$

108. $[-4^2 + (7 - 10)^3 - (-27)] - [|-2|^5 + 1 - 5^2]$

109. $\dfrac{12 \div 3 \cdot 5|2^2 + 3^2|}{7 + 3 - 6^2}$

110. $\dfrac{-3 \cdot 5^2 + 89}{(5 - 6)^2 - 2|3 - 7|}$

In Exercises 111–114, express each sentence as a single numerical expression. Then use the order of operations to simplify the expression.

111. Cube −2. Subtract this exponential expression from −10.

112. Cube −5. Subtract this exponential expression from −100.

113. Subtract 10 from 7. Multiply this difference by 2. Square this product.

114. Subtract 11 from 9. Multiply this difference by 2. Raise this product to the fourth power.

Application Exercises

115. The peak of Mount McKinley, the highest point in the United States, is 20,320 feet above sea level. Death Valley, the lowest point in the United States, is 282 feet below sea level. What is the difference in elevation between the peak of Mount McKinley and Death Valley?

116. The peak of Mount Kilimanjaro, the highest point in Africa, is 19,321 feet above sea level. Qattara Depression, Egypt, the lowest point in Africa, is 436 feet below sea level. What is the difference in elevation between the peak of Mount Kilimanjaro and the Qattara Depression?

In Exercises 117–126, we return to the number line that shows factors that can stretch or shrink one's probable life span.

Stretching or Shrinking One's Life Span

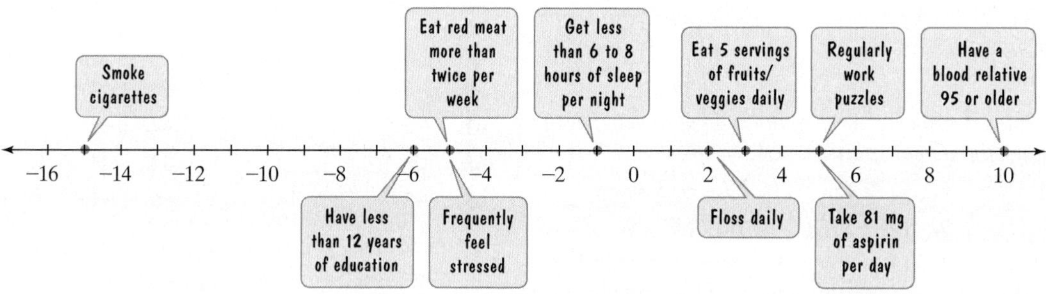

Years of Life Gained or Lost

Source: Newsweek

117. If you have a blood relative 95 or older and you smoke cigarettes, do you stretch or shrink your life span? By how many years?

118. If you floss daily and eat red meat more than twice per week, do you stretch or shrink your life span? By how many years?

119. If you frequently feel stressed and have less than 12 years of education, do you stretch or shrink your life span? By how many years?

120. If you get less than 6 to 8 hours of sleep per night and smoke cigarettes, do you stretch or shrink your life span? By how many years?

121. What happens to the life span for a person who takes 81 mg of aspirin per day and eats red meat more than twice per week?

122. What happens to the life span for a person who regularly works puzzles and a person who frequently feels stressed?

123. What is the difference in the life span between a person who has a blood relative 95 or older and a person who smokes cigarettes?

124. What is the difference in the life span between a person who has a blood relative 95 or older and a person who has less than 12 years of education?

125. What is the difference in the life span between a person who frequently feels stressed and a person who has less than 12 years of education?

126. What is the difference in the life span between a person who gets less than 6 to 8 hours of sleep per night and a person who frequently feels stressed?

The accompanying bar graph shows the amount of money, in billions of dollars, collected and spent by the U.S. government in selected years from 2001 through 2016. Use the information from the graph to solve Exercises 127–130. Express answers in billions of dollars.

Money Collected and Spent by the United States Government

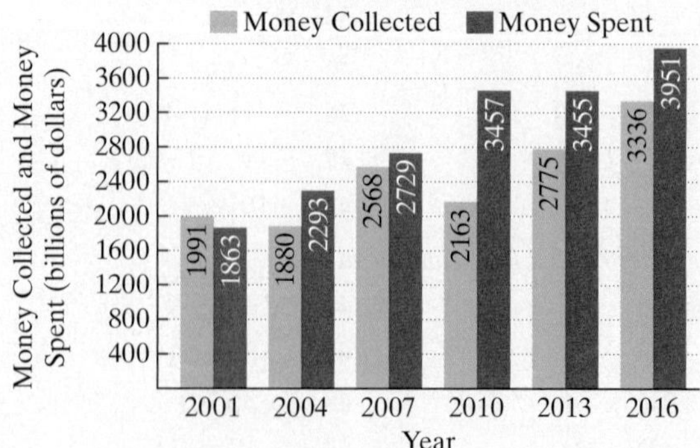

Source: Budget of the U.S. Government

127. **a.** In 2001, what was the difference between the amount of money collected and the amount spent? Was there a budget surplus or deficit in 2001?

b. In 2016, what was the difference between the amount of money collected and the amount spent? Was there a budget surplus or deficit in 2016?

c. What is the difference between the 2001 surplus and the 2016 deficit?

128. **a.** In 2001, what was the difference between the amount of money collected and the amount spent? Was there a budget surplus or deficit in 2001?

b. In 2013, what was the difference between the amount of money collected and the amount spent? Was there a budget surplus or deficit in 2013?

c. What is the difference between the 2001 surplus and the 2013 deficit?

129. What is the difference between the 2016 deficit and the 2013 deficit?

130. What is the difference between the 2013 deficit and the 2010 deficit?

The way that we perceive the temperature on a cold day depends on both air temperature and wind speed. The windchill is what the air temperature would have to be with no wind to achieve the same chilling effect on the skin. In 2002, the National Weather Service issued new windchill temperatures, shown in the table below. Use the information from the table to solve Exercises 131–134.

New Windchill Temperature Index

Wind Speed (miles per hour)	Air Temperature (F)											
	30	25	20	15	10	5	0	−5	−10	−15	−20	−25
5	25	19	13	7	1	−5	−11	−16	−22	−28	−34	−40
10	21	15	9	3	−4	−10	−16	−22	−28	−35	−41	−47
15	19	13	6	0	−7	−13	−19	−26	−32	−39	−45	−51
20	17	11	4	−2	−9	−15	−22	−29	−35	−42	−48	−55
25	16	9	3	−4	−11	−17	−24	−31	−37	−44	−51	−58
30	15	8	1	−5	−12	−19	−26	−33	−39	−46	−53	−60
35	14	7	0	−7	−14	−21	−27	−34	−41	−48	−55	−62
40	13	6	−1	−8	−15	−22	−29	−36	−43	−50	−57	−64
45	12	5	−2	−9	−16	−23	−30	−37	−44	−51	−58	−65
50	12	4	−3	−10	−17	−24	−31	−38	−45	−52	−60	−67
55	11	4	−3	−11	−18	−25	−32	−39	−46	−54	−61	−68
60	10	3	−4	−11	−19	−26	−33	−40	−48	−55	−62	−69

Frostbite occurs in 15 minutes or less. (shaded cells)

Source: National Weather Service

131. What is the difference between how cold the temperature feels with winds at 10 miles per hour and 25 miles per hour when the air temperature is 15°F?

132. What is the difference between how cold the temperature feels with winds at 5 miles per hour and 30 miles per hour when the air temperature is 10°F?

133. What is the difference in the windchill temperature between an air temperature of 5°F with winds at 50 miles per hour and an air temperature of −10°F with winds at 5 miles per hour?

134. What is the difference in the windchill temperature between an air temperature of 5°F with winds at 55 miles per hour and an air temperature of −5°F with winds at 10 miles per hour?

Explaining the Concepts

135. How does the set of integers differ from the set of whole numbers?

136. Explain how to graph an integer on a number line.

137. If you are given two integers, explain how to determine which one is smaller.

138. Explain how to add integers.

139. Explain how to subtract integers.

140. Explain how to multiply integers.

141. Explain how to divide integers.

142. Describe what it means to raise a number to a power. In your description, include a discussion of the difference between -5^2 and $(-5)^2$.

143. Why is $\frac{0}{4}$ equal to 0, but $\frac{4}{0}$ undefined?

Critical Thinking Exercises

Make Sense? *In Exercises 144–147, determine whether each statement makes sense or does not make sense, and explain your reasoning.*

144. Without adding integers, I can see that the sum of −227 and 319 is greater than the sum of 227 and −319.

145. I found the variation in U.S. temperature by subtracting the record low temperature, a negative integer, from the record high temperature, a positive integer.

146. I've noticed that the sign rules for dividing integers are slightly different than the sign rules for multiplying integers.

147. The rules for the order of operations avoid the confusion of obtaining different results when I simplify the same expression.

In Exercises 148–149, insert one pair of parentheses to make each calculation correct.

148. $8 - 2 \cdot 3 - 4 = 10$

149. $8 - 2 \cdot 3 - 4 = 14$

Technology Exercises

Scientific calculators that have parentheses keys allow for the entry and computation of relatively complicated expressions in a single step. For example, the expression $15 + (10 - 7)^2$ *can be evaluated by entering the following keystrokes:*

15 [+] [(] 10 [−] 7 [)] [y^x] 2 [=].

Find the value of each expression in Exercises 150–152 in a single step on your scientific calculator.

150. $8 - 2 \cdot 3 - 9$

151. $(8 - 2) \cdot (3 - 9)$

152. $5^3 + 4 \cdot 9 - (8 + 9 \div 3)$

5.3 The Rational Numbers

WHAT AM I SUPPOSED TO LEARN?

After studying this section, you should be able to:

1. Define the rational numbers.
2. Reduce rational numbers.
3. Convert between mixed numbers and improper fractions.
4. Express rational numbers as decimals.
5. Express decimals in the form $\frac{a}{b}$.
6. Multiply and divide rational numbers.
7. Add and subtract rational numbers.
8. Use the order of operations agreement with rational numbers.
9. Apply the density property of rational numbers.
10. Solve problems involving rational numbers.

YOU ARE MAKING EIGHT DOZEN CHOCOLATE chip cookies for a large neighborhood block party. The recipe lists the ingredients needed to prepare five dozen cookies, such as $\frac{3}{4}$ cup sugar. How do you adjust the amount of sugar, as well as the amounts of each of the other ingredients, given in the recipe?

Adapting a recipe to suit a different number of portions usually involves working with numbers that are not integers. For example, the number describing the amount of sugar, $\frac{3}{4}$ (cup), is not an integer, although it consists of the quotient of two integers, 3 and 4. Before returning to the problem of changing the size of a recipe, we study a new set of numbers consisting of the quotients of integers.

1 *Define the rational numbers.*

Defining the Rational Numbers

If two integers are added, subtracted, or multiplied, the result is always another integer. This, however, is not always the case with division. For example, 10 divided by 5 is the integer 2. By contrast, 5 divided by 10 is $\frac{1}{2}$, and $\frac{1}{2}$ is not an integer. To permit divisions such as $\frac{5}{10}$, we enlarge the set of integers, calling the new collection the *rational numbers*. The set of **rational numbers** consists of all the numbers that can be expressed as a quotient of two integers, with the denominator not 0.

THE RATIONAL NUMBERS

The set of **rational numbers** is the set of all numbers which can be expressed in the form $\frac{a}{b}$, where a and b are integers and b is not equal to 0. The integer a is called the **numerator**, and the integer b is called the **denominator**.

GREAT QUESTION!

Is the rational number $\frac{-3}{4}$ the same as $-\frac{3}{4}$?

We know that the quotient of two numbers with different signs is a negative number. Thus,

$$\frac{-3}{4} = -\frac{3}{4} \quad \text{and} \quad \frac{3}{-4} = -\frac{3}{4}.$$

The following numbers are examples of rational numbers:

$$\frac{1}{2}, \frac{-3}{4}, 5, 0.$$

The integer 5 is a rational number because it can be expressed as the quotient of integers: $5 = \frac{5}{1}$. Similarly, 0 can be written as $\frac{0}{1}$.

In general, every integer a is a rational number because it can be expressed in the form $\frac{a}{1}$.

2 *Reduce rational numbers.*

Reducing Rational Numbers

A rational number is **reduced to its lowest terms**, or **simplified**, when the numerator and denominator have no common divisors other than 1. Reducing rational numbers to lowest terms is done using the **Fundamental Principle of Rational Numbers**.

THE FUNDAMENTAL PRINCIPLE OF RATIONAL NUMBERS

If $\frac{a}{b}$ is a rational number and c is any number other than 0,

$$\frac{a \cdot c}{b \cdot c} = \frac{a}{b}.$$

The rational numbers $\frac{a}{b}$ and $\frac{a \cdot c}{b \cdot c}$ are called **equivalent fractions**.

When using the Fundamental Principle to reduce a rational number, the simplification can be done in one step by finding the greatest common divisor of the numerator and the denominator, and using it for c. Thus, **to reduce a rational number to its lowest terms, divide both the numerator and the denominator by their greatest common divisor.**

For example, consider the rational number $\frac{12}{100}$. The greatest common divisor of 12 and 100 is 4. We reduce to lowest terms as follows:

$$\frac{12}{100} = \frac{3 \cdot \cancel{4}}{25 \cdot \cancel{4}} = \frac{3}{25} \quad \text{or} \quad \frac{12}{100} = \frac{12 \div 4}{100 \div 4} = \frac{3}{25}.$$

EXAMPLE 1 *Reducing a Rational Number*

Reduce $\frac{130}{455}$ to lowest terms.

SOLUTION

Begin by finding the greatest common divisor of 130 and 455.

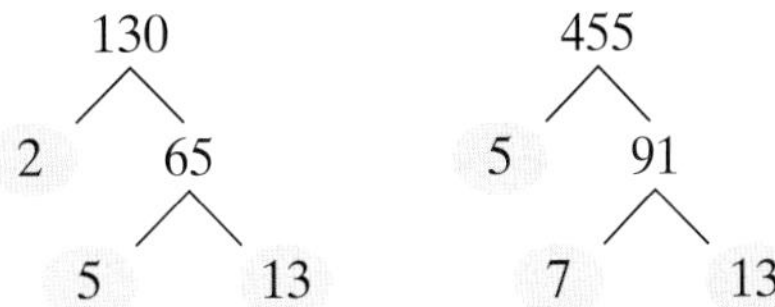

Thus, $130 = 2 \cdot 5 \cdot 13$ and $455 = 5 \cdot 7 \cdot 13$. The greatest common divisor is $5 \cdot 13$, or 65. Divide the numerator and the denominator of the given rational number by $5 \cdot 13$ or by 65.

$$\frac{130}{455} = \frac{2 \cdot \cancel{5} \cdot \cancel{13}}{\cancel{5} \cdot 7 \cdot \cancel{13}} = \frac{2}{7} \quad \text{or} \quad \frac{130}{455} = \frac{130 \div 65}{455 \div 65} = \frac{2}{7}$$

There are no common divisors of 2 and 7 other than 1. Thus, the rational number $\frac{2}{7}$ is in its lowest terms.

CHECK POINT 1 Reduce $\frac{72}{90}$ to lowest terms.

3 *Convert between mixed numbers and improper fractions.*

Mixed Numbers and Improper Fractions

A **mixed number** consists of the sum of an integer and a rational number, expressed without the use of an addition sign. Here is an example of a mixed number:

$$3\frac{4}{5}.$$

The integer is 3 and the rational number is $\frac{4}{5}$. $3\frac{4}{5}$ means $3 + \frac{4}{5}$.

GREAT QUESTION!

How do I read the mixed number $3\frac{4}{5}$?

It's read "three and four-fifths."

An **improper fraction** is a rational number whose numerator is greater than its denominator. An example of an improper fraction is $\frac{19}{5}$.

The mixed number $3\frac{4}{5}$ can be converted to the improper fraction $\frac{19}{5}$ using the following procedure:

CONVERTING A POSITIVE MIXED NUMBER TO AN IMPROPER FRACTION

1. Multiply the denominator of the rational number by the integer and add the numerator to this product.
2. Place the sum in step 1 over the denominator in the mixed number.

EXAMPLE 2 Converting from a Mixed Number to an Improper Fraction

Convert $3\frac{4}{5}$ to an improper fraction.

SOLUTION

$$3\frac{4}{5} = \frac{5 \cdot 3 + 4}{5}$$

Multiply the denominator by the integer and add the numerator.

Place the sum over the mixed number's denominator.

$$= \frac{15 + 4}{5} = \frac{19}{5}$$

CHECK POINT 2 Convert $2\frac{5}{8}$ to an improper fraction.

GREAT QUESTION!

Does $-2\frac{3}{4}$ mean that I need to add $\frac{3}{4}$ to -2?

No. $-2\frac{3}{4}$ means

$$-(2\tfrac{3}{4}) \quad \text{or} \quad -(2 + \tfrac{3}{4}).$$

$-2\frac{3}{4}$ does not mean

$$-2 + \tfrac{3}{4}.$$

When converting a negative mixed number to an improper fraction, copy the negative sign and then follow the previous procedure. For example,

$$-2\frac{3}{4} = -\frac{4 \cdot 2 + 3}{4} = -\frac{8 + 3}{4} = -\frac{11}{4}.$$

Copy the negative sign from step to step and convert $2\frac{3}{4}$ to an improper fraction.

A positive improper fraction can be converted to a mixed number using the following procedure:

CONVERTING A POSITIVE IMPROPER FRACTION TO A MIXED NUMBER

1. Divide the denominator into the numerator. Record the quotient and the remainder.
2. Write the mixed number using the following form:

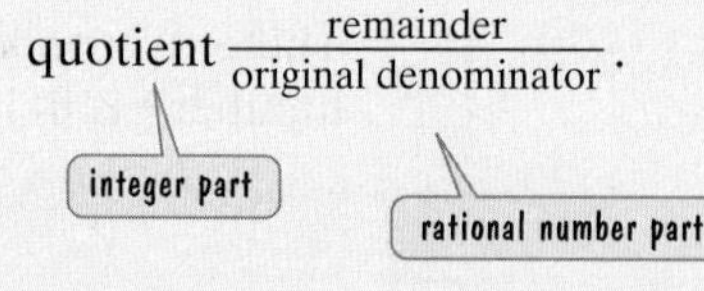

EXAMPLE 3 Converting from an Improper Fraction to a Mixed Number

Convert $\frac{42}{5}$ to a mixed number.

SOLUTION

Step 1 Divide the denominator into the numerator.

$$\begin{array}{r} 8 \\ 5\overline{)42} \\ \underline{40} \\ 2 \end{array}$$

quotient

remainder

Step 2 Write the mixed number using quotient $\frac{\textbf{remainder}}{\textbf{original denominator}}$. Thus,

$$\frac{42}{5} = 8\frac{2}{5}.$$

remainder

original denominator

quotient

GREAT QUESTION!

When should I use mixed numbers and when are improper fractions preferable?

In applied problems, answers are usually expressed as mixed numbers, which many people find more meaningful than improper fractions. However, improper fractions are often easier to work with when performing operations with fractions.

CHECK POINT 3 Convert $\frac{5}{3}$ to a mixed number.

When converting a negative improper fraction to a mixed number, copy the negative sign and then follow the previous procedure. For example,

$$-\frac{29}{8} = -3\frac{5}{8}.$$

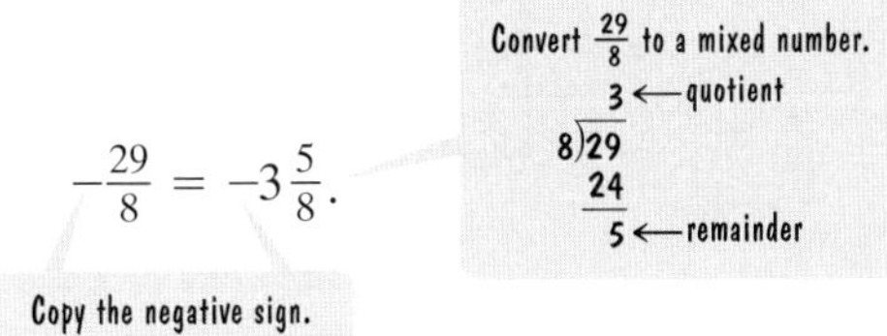

4 *Express rational numbers as decimals.*

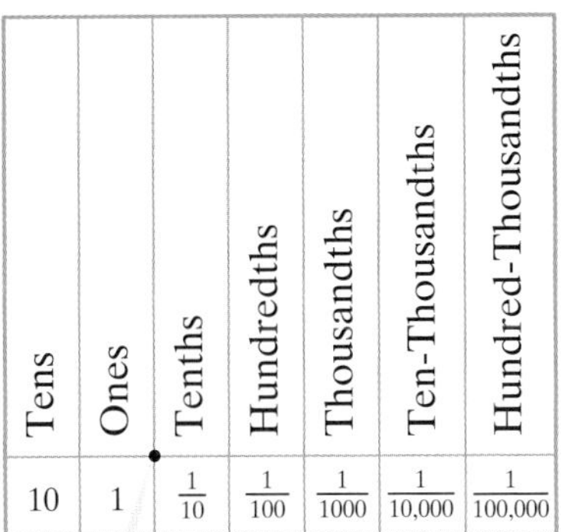

Tens	Ones	Tenths	Hundredths	Thousandths	Ten-Thousandths	Hundred-Thousandths
10	1	$\frac{1}{10}$	$\frac{1}{100}$	$\frac{1}{1000}$	$\frac{1}{10,000}$	$\frac{1}{100,000}$

Rational Numbers and Decimals

We have seen that a rational number is the quotient of integers. Rational numbers can also be expressed as decimals. As shown in the place-value chart in the margin, it is convenient to represent rational numbers with denominators of 10, 100, 1000, and so on as decimals. For example,

$$\frac{7}{10} = 0.7, \quad \frac{3}{100} = 0.03, \quad \text{and} \quad \frac{8}{1000} = 0.008.$$

Any rational number $\frac{a}{b}$ can be expressed as a decimal by dividing the denominator, b, into the numerator, a.

EXAMPLE 4 Expressing Rational Numbers as Decimals

Express each rational number as a decimal:

a. $\frac{5}{8}$ **b.** $\frac{7}{11}$.

SOLUTION

In each case, divide the denominator into the numerator.

a. $\frac{5}{8} = 0.625$

```
  0.625
8)5.000
  4 8
    20
    16
     40
     40
      0
```

b. $\frac{7}{11} = 0.6363\ldots$

```
   0.6363...
11)7.0000...
   6 6
     40
     33
      70
      66
       40
       33
        70
         ⋮
```

In Example 4, the decimal for $\frac{5}{8}$, namely 0.625, stops and is called a **terminating decimal**. Other examples of terminating decimals are

$$\frac{1}{4} = 0.25, \quad \frac{2}{5} = 0.4, \quad \text{and} \quad \frac{7}{8} = 0.875.$$

By contrast, the division process for $\frac{7}{11}$ results in 0.6363..., with the digits 63 repeating over and over indefinitely. To indicate this, write a bar over the digits that repeat. Thus,

$$\frac{7}{11} = 0.\overline{63}.$$

The decimal for $\frac{7}{11}$, $0.\overline{63}$, is called a **repeating decimal**. Other examples of repeating decimals are

$$\frac{1}{3} = 0.333\ldots = 0.\overline{3} \quad \text{and} \quad \frac{2}{3} = 0.666\ldots = 0.\overline{6}.$$

RATIONAL NUMBERS AND DECIMALS

Any rational number can be expressed as a decimal. The resulting decimal will either terminate (stop), or it will have a digit that repeats or a block of digits that repeats.

CHECK POINT 4 Express each rational number as a decimal:

a. $\frac{3}{8}$ **b.** $\frac{5}{11}$.

5 *Express decimals in the form $\frac{a}{b}$.*

Reversing Directions: Expressing Decimals as Quotients of Two Integers

Terminating decimals can be expressed with denominators of 10, 100, 1000, 10,000, and so on. Use the place-value chart shown in the margin. The digits to the right of the decimal point are the numerator of the rational number. To find the denominator, observe the last digit to the right of the decimal point. The place-value of this digit will indicate the denominator.

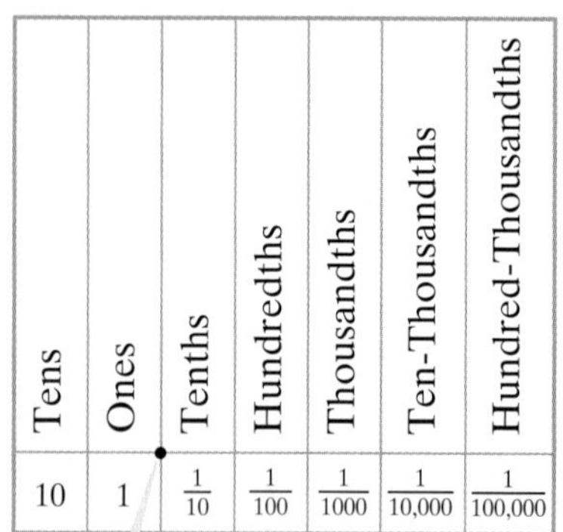

EXAMPLE 5 Expressing Terminating Decimals in $\frac{a}{b}$ Form

Express each terminating decimal as a quotient of integers:

a. 0.7 **b.** 0.49 **c.** 0.048.

SOLUTION

a. $0.7 = \frac{7}{10}$ because the 7 is in the tenths position.

b. $0.49 = \frac{49}{100}$ because the last digit on the right, 9, is in the hundredths position.

c. $0.048 = \frac{48}{1000}$ because the digit on the right, 8, is in the thousandths position. Reducing to lowest terms, $\frac{48}{1000} = \frac{48 \div 8}{1000 \div 8} = \frac{6}{125}$.

CHECK POINT 5 Express each terminating decimal as a quotient of integers, reduced to lowest terms:

a. 0.9 **b.** 0.86 **c.** 0.053.

A BRIEF REVIEW ***Solving One-Step Equations***

- Solving an equation involves determining all values that result in a true statement when substituted into the equation. Such values are solutions of the equation.

  Example

  The solution of $x - 4 = 10$ is 14 because $14 - 4 = 10$ is a true statement.

- Two basic rules can be used to solve equations:
  1. We can add or subtract the same quantity on both sides of an equation.
  2. We can multiply or divide both sides of an equation by the same quantity, as long as we do not multiply or divide by zero.

Examples of Equations That Can Be Solved in One Step

Equation	How to Solve	Solving the Equation	The Equation's Solution
$x - 4 = 10$	Add 4 to both sides.	$x - 4 + 4 = 10 + 4$ $x = 14$	14
$y + 12 = 17$	Subtract 12 from both sides.	$y + 12 - 12 = 17 - 12$ $y = 5$	5
$99n = 53$	Divide both sides by 99.	$\frac{99n}{99} = \frac{53}{99}$ $n = \frac{53}{99}$	$\frac{53}{99}$
$\frac{z}{5} = 9$	Multiply both sides by 5.	$5 \cdot \frac{z}{5} = 5 \cdot 9$ $z = 45$	45

Equations whose solutions require more than one step are discussed in Chapter 6.

Why have we provided this brief review of equations that can be solved in one step? If you are given a rational number as a repeating decimal, there is a technique for expressing the number as a quotient of integers that requires solving a one-step equation. We begin by illustrating the technique with an example. Then we will summarize the steps in the procedure and apply them to another example.

EXAMPLE 6 Expressing a Repeating Decimal in $\frac{a}{b}$ Form

Express $0.\overline{6}$ as a quotient of integers.

SOLUTION

Step 1 Let n equal the repeating decimal. Let $n = 0.\overline{6}$, so that $n = 0.66666\ldots$.

Step 2 If there is one repeating digit, multiply both sides of the equation in step 1 by 10.

$$n = 0.66666\ldots \quad \text{This is the equation from step 1.}$$
$$10n = 10(0.66666\ldots) \quad \text{Multiply both sides by 10.}$$
$$10n = 6.66666\ldots \quad \text{Multiplying by 10 moves the decimal point one place to the right.}$$

Step 3 Subtract the equation in step 1 from the equation in step 2. Be sure to line up the decimal points before subtracting.

Remember from algebra that n means $1n$. Thus, $10n - 1n = 9n$.

$$\begin{aligned} 10n &= 6.66666\ldots && \text{This is the equation from step 2.} \\ -\ n &= 0.66666\ldots && \text{This is the equation from step 1.} \\ \hline 9n &= 6 \end{aligned}$$

Step 4. Divide both sides of the equation in step 3 by the number in front of n and solve for n. We solve $9n = 6$ for n by dividing both sides by 9.

$$9n = 6 \quad \text{This is the equation from step 3.}$$
$$\frac{9n}{9} = \frac{6}{9} \quad \text{Divide both sides by 9.}$$
$$n = \frac{6}{9} = \frac{2}{3} \quad \text{Reduce } \tfrac{6}{9} \text{ to lowest terms: } \frac{6}{9} = \frac{2 \cdot \cancel{3}}{3 \cdot \cancel{3}} = \frac{2}{3}.$$

We began the solution process with $n = 0.\overline{6}$, and now we have $n = \frac{2}{3}$. Therefore,

$$0.\overline{6} = \frac{2}{3}.$$

Here are the steps for expressing a repeating decimal as a quotient of integers. Assume that the repeating digit or digits begin directly to the right of the decimal point.

EXPRESSING A REPEATING DECIMAL AS A QUOTIENT OF INTEGERS

Step 1 Let n equal the repeating decimal.

Step 2 Multiply both sides of the equation in step 1 by 10 if one digit repeats, by 100 if two digits repeat, by 1000 if three digits repeat, and so on.

Step 3 Subtract the equation in step 1 from the equation in step 2.

Step 4 Divide both sides of the equation in step 3 by the number in front of n and solve for n.

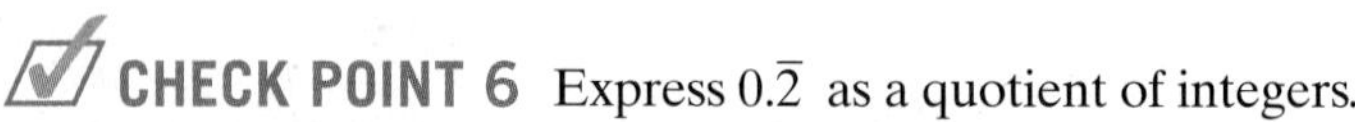

CHECK POINT 6 Express $0.\overline{2}$ as a quotient of integers.

EXAMPLE 7 Expressing a Repeating Decimal in $\frac{a}{b}$ Form

Express $0.\overline{53}$ as a quotient of integers.

SOLUTION

Step 1 Let n equal the repeating decimal. Let $n = 0.\overline{53}$, so that $n = 0.535353\ldots$.

Step 2 If there are two repeating digits, multiply both sides of the equation in step 1 by 100.

$$n = 0.535353\ldots \quad \text{This is the equation from step 1.}$$
$$100n = 100(0.535353\ldots) \quad \text{Multiply both sides by 100.}$$
$$100n = 53.535353\ldots \quad \text{Multiplying by 100 moves the decimal point two places to the right.}$$

Step 3 Subtract the equation in step 1 from the equation in step 2.

$$\begin{aligned} 100n &= 53.535353\ldots && \text{This is the equation from step 2.} \\ -\quad n &= 0.535353\ldots && \text{This is the equation from step 1.} \\ \hline 99n &= 53 \end{aligned}$$

Step 4 Divide both sides of the equation in step 3 by the number in front of n and solve for n. We solve $99n = 53$ for n by dividing both sides by 99.

$$99n = 53 \quad \text{This is the equation from step 3.}$$
$$\frac{99n}{99} = \frac{53}{99} \quad \text{Divide both sides by 99.}$$
$$n = \frac{53}{99}$$

Because n equals $0.\overline{53}$ and n equals $\frac{53}{99}$,

$$0.\overline{53} = \frac{53}{99}.$$

CHECK POINT 7 Express $0.\overline{79}$ as a quotient of integers.

6 *Multiply and divide rational numbers.*

Multiplying and Dividing Rational Numbers

The product of two rational numbers is found as follows:

MULTIPLYING RATIONAL NUMBERS

The product of two rational numbers is the product of their numerators divided by the product of their denominators.

If $\frac{a}{b}$ and $\frac{c}{d}$ are rational numbers, then $\frac{a}{b} \cdot \frac{c}{d} = \frac{a \cdot c}{b \cdot d}$.

GREAT QUESTION!

Is it OK if I divide by common factors before I multiply?

Yes. You can divide numerators and denominators by common factors *before* performing multiplication. Then multiply the remaining factors in the numerators and multiply the remaining factors in the denominators. For example,

$$\frac{7}{15} \cdot \frac{20}{21} = \frac{\overset{1}{\cancel{7}}}{\underset{3}{\cancel{15}}} \cdot \frac{\overset{4}{\cancel{20}}}{\underset{3}{\cancel{21}}} = \frac{1 \cdot 4}{3 \cdot 3} = \frac{4}{9}.$$

EXAMPLE 8 Multiplying Rational Numbers

Multiply. If possible, reduce the product to its lowest terms:

a. $\frac{3}{8} \cdot \frac{5}{11}$ **b.** $\left(-\frac{2}{3}\right)\left(-\frac{9}{4}\right)$ **c.** $\left(3\frac{2}{3}\right)\left(1\frac{1}{4}\right)$.

SOLUTION

a. $\frac{3}{8} \cdot \frac{5}{11} = \frac{3 \cdot 5}{8 \cdot 11} = \frac{15}{88}$

b. $\left(-\frac{2}{3}\right)\left(-\frac{9}{4}\right) = \frac{(-2)(-9)}{3 \cdot 4} = \frac{18}{12} = \frac{3 \cdot \cancel{6}}{2 \cdot \cancel{6}} = \frac{3}{2}$ or $1\frac{1}{2}$

c. $\left(3\frac{2}{3}\right)\left(1\frac{1}{4}\right) = \frac{11}{3} \cdot \frac{5}{4} = \frac{11 \cdot 5}{3 \cdot 4} = \frac{55}{12}$ or $4\frac{7}{12}$

CHECK POINT 8 Multiply. If possible, reduce the product to its lowest terms:

a. $\frac{4}{11} \cdot \frac{2}{3}$ **b.** $\left(-\frac{3}{7}\right)\left(-\frac{14}{4}\right)$ **c.** $\left(3\frac{2}{5}\right)\left(1\frac{1}{2}\right)$.

Two numbers whose product is 1 are called **reciprocals**, or **multiplicative inverses**, of each other. Thus, the reciprocal of 2 is $\frac{1}{2}$ and the reciprocal of $\frac{1}{2}$ is 2 because $2 \cdot \frac{1}{2} = 1$. In general, if $\frac{c}{d}$ is a nonzero rational number, its reciprocal is $\frac{d}{c}$ because $\frac{c}{d} \cdot \frac{d}{c} = 1$.

Reciprocals are used to find the quotient of two rational numbers.

DIVIDING RATIONAL NUMBERS

The quotient of two rational numbers is the product of the first number and the reciprocal of the second number.

If $\frac{a}{b}$ and $\frac{c}{d}$ are rational numbers and $\frac{c}{d}$ is not 0, then $\frac{a}{b} \div \frac{c}{d} = \frac{a}{b} \cdot \frac{d}{c} = \frac{a \cdot d}{b \cdot c}$.

EXAMPLE 9 Dividing Rational Numbers

Divide. If possible, reduce the quotient to its lowest terms:

a. $\frac{4}{5} \div \frac{1}{10}$ **b.** $-\frac{3}{5} \div \frac{7}{11}$ **c.** $4\frac{3}{4} \div 1\frac{1}{2}$.

SOLUTION

a. $\frac{4}{5} \div \frac{1}{10} = \frac{4}{5} \cdot \frac{10}{1} = \frac{4 \cdot 10}{5 \cdot 1} = \frac{40}{5} = 8$

b. $-\frac{3}{5} \div \frac{7}{11} = -\frac{3}{5} \cdot \frac{11}{7} = \frac{-3(11)}{5 \cdot 7} = -\frac{33}{35}$

c. $4\frac{3}{4} \div 1\frac{1}{2} = \frac{19}{4} \div \frac{3}{2} = \frac{19}{4} \cdot \frac{2}{3} = \frac{19 \cdot 2}{4 \cdot 3} = \frac{38}{12} = \frac{19 \cdot \cancel{2}}{6 \cdot \cancel{2}} = \frac{19}{6}$ or $3\frac{1}{6}$

CHECK POINT 9 Divide. If possible, reduce the quotient to its lowest terms:

a. $\frac{9}{11} \div \frac{5}{4}$ **b.** $-\frac{8}{15} \div \frac{2}{5}$ **c.** $3\frac{3}{8} \div 2\frac{1}{4}$.

7 *Add and subtract rational numbers.*

Adding and Subtracting Rational Numbers

Rational numbers with identical denominators are added and subtracted using the following rules:

ADDING AND SUBTRACTING RATIONAL NUMBERS WITH IDENTICAL DENOMINATORS

The sum or difference of two rational numbers with identical denominators is the sum or difference of their numerators over the common denominator.

If $\frac{a}{b}$ and $\frac{c}{b}$ are rational numbers, then $\frac{a}{b} + \frac{c}{b} = \frac{a+c}{b}$ and $\frac{a}{b} - \frac{c}{b} = \frac{a-c}{b}$.

EXAMPLE 10 *Adding and Subtracting Rational Numbers with Identical Denominators*

Perform the indicated operations:

a. $\frac{3}{7} + \frac{2}{7}$ **b.** $\frac{11}{12} - \frac{5}{12}$ **c.** $-5\frac{1}{4} - \left(-2\frac{3}{4}\right)$.

SOLUTION

a. $\frac{3}{7} + \frac{2}{7} = \frac{3+2}{7} = \frac{5}{7}$

b. $\frac{11}{12} - \frac{5}{12} = \frac{11-5}{12} = \frac{6}{12} = \frac{1 \cdot 6}{2 \cdot 6} = \frac{1}{2}$

c. $-5\frac{1}{4} - \left(-2\frac{3}{4}\right) = -\frac{21}{4} - \left(-\frac{11}{4}\right) = -\frac{21}{4} + \frac{11}{4} = \frac{-21+11}{4} = \frac{-10}{4} = -\frac{5}{2}$ or $-2\frac{1}{2}$

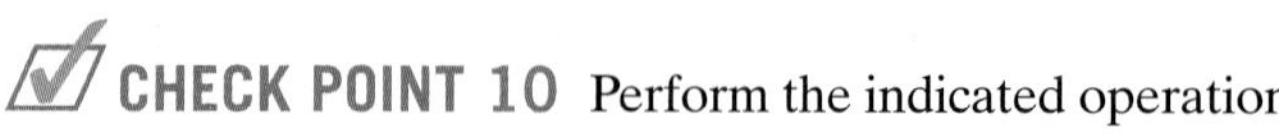

CHECK POINT 10 Perform the indicated operations:

a. $\frac{5}{12} + \frac{3}{12}$ **b.** $\frac{7}{4} - \frac{1}{4}$ **c.** $-3\frac{3}{8} - \left(-1\frac{1}{8}\right)$.

If the rational numbers to be added or subtracted have different denominators, we use the least common multiple of their denominators to rewrite the rational numbers. The least common multiple of the denominators is called the **least common denominator.**

Rewriting rational numbers with a least common denominator is done using the Fundamental Principle of Rational Numbers, discussed at the beginning of this section. Recall that if $\frac{a}{b}$ is a rational number and c is a nonzero number, then

$$\frac{a}{b} = \frac{a}{b} \cdot \frac{c}{c} = \frac{a \cdot c}{b \cdot c}.$$

Multiplying the numerator and the denominator of a rational number by the same nonzero number is equivalent to multiplying by 1, resulting in an equivalent fraction.

EXAMPLE 11 *Adding Rational Numbers with Unlike Denominators*

Find the sum: $\frac{3}{4} + \frac{1}{6}$.

SOLUTION

The smallest number divisible by both 4 and 6 is 12. Therefore, 12 is the least common multiple of 4 and 6, and will serve as the least common denominator. To obtain a denominator of 12, multiply the denominator and the numerator of the first rational number, $\frac{3}{4}$, by 3. To obtain a denominator of 12, multiply the denominator and the numerator of the second rational number, $\frac{1}{6}$, by 2.

$$\frac{3}{4}+\frac{1}{6}=\frac{3}{4}\cdot\frac{3}{3}+\frac{1}{6}\cdot\frac{2}{2}$$

Rewrite each rational number as an equivalent fraction with a denominator of 12; $\frac{3}{3}=1$ and $\frac{2}{2}=1$, and multiplying by 1 does not change a number's value.

$$=\frac{9}{12}+\frac{2}{12}$$

Multiply.

$$=\frac{11}{12}$$

Add numerators and put this sum over the least common denominator.

CHECK POINT 11 Find the sum: $\frac{1}{5}+\frac{3}{4}$.

If the least common denominator cannot be found by inspection, use prime factorizations of the denominators and the method for finding their least common multiple, discussed in Section 5.1.

EXAMPLE 12 *Subtracting Rational Numbers with Unlike Denominators*

Perform the indicated operation: $\frac{1}{15}-\frac{7}{24}$.

SOLUTION

We need to first find the least common denominator, which is the least common multiple of 15 and 24. What is the smallest number divisible by both 15 and 24? The answer is not obvious, so we begin with the prime factorization of each number.

$$15 = 5\cdot 3$$
$$24 = 8\cdot 3 = 2^3\cdot 3$$

The different factors are 5, 3, and 2. Using the greater number of times each factor appears in either factorization, we find that the least common multiple is $5\cdot 3\cdot 2^3 = 5\cdot 3\cdot 8 = 120$. We will now express each rational number with a denominator of 120, which is the least common denominator. For the first rational number, $\frac{1}{15}$, 120 divided by 15 is 8. Thus, we will multiply the numerator and the denominator by 8. For the second rational number, $\frac{7}{24}$, 120 divided by 24 is 5. Thus, we will multiply the numerator and the denominator by 5.

$$\frac{1}{15}-\frac{7}{24}=\frac{1}{15}\cdot\frac{8}{8}-\frac{7}{24}\cdot\frac{5}{5}$$

Rewrite each rational number as an equivalent fraction with a denominator of 120.

$$=\frac{8}{120}-\frac{35}{120}$$

Multiply.

$$=\frac{8-35}{120}$$

Subtract the numerators and put this difference over the least common denominator.

$$=\frac{-27}{120}$$

Perform the subtraction.

$$=\frac{-9\cdot \cancel{3}}{40\cdot \cancel{3}}$$

Reduce to lowest terms.

$$=-\frac{9}{40}$$

TECHNOLOGY

Here is a possible sequence on a graphing calculator for the subtraction problem in Example 12:

1 [÷] 15 [−] 7 [÷] 24 [▶ Frac] [ENTER].

```
1/15-7/24▸Frac
                -9/40
```

The calculator display reads $-\frac{9}{40}$, serving as a check for our answer in Example 12.

CHECK POINT 12 Perform the indicated operation: $\frac{3}{10}-\frac{7}{12}$.

8 *Use the order of operations agreement with rational numbers.*

Order of Operations with Rational Numbers

In the previous section, we presented rules for determining the order in which operations should be performed: operations in grouping symbols; exponential expressions; multiplication/division (left to right); addition/subtraction (left to right). In our next example, we apply the order of operations to an expression with rational numbers.

EXAMPLE 13 *Using the Order of Operations*

Simplify: $\left(\frac{1}{2}\right)^3 - \left(\frac{1}{2} - \frac{3}{4}\right)^2(-4)$.

SOLUTION

Because grouping symbols appear, we perform the operation within parentheses first.

$$\left(\frac{1}{2}\right)^3 - \left(\frac{1}{2} - \frac{3}{4}\right)^2(-4)$$

$$= \left(\frac{1}{2}\right)^3 - \left(-\frac{1}{4}\right)^2(-4)$$ Work inside parentheses first: $\frac{1}{2} - \frac{3}{4} = \frac{2}{4} - \frac{3}{4} = \frac{2}{4} + \left(-\frac{3}{4}\right) = -\frac{1}{4}$.

$$= \frac{1}{8} - \frac{1}{16}(-4)$$ Evaluate exponential expressions: $\left(\frac{1}{2}\right)^3 = \frac{1}{2}\cdot\frac{1}{2}\cdot\frac{1}{2} = \frac{1}{8}$ and $\left(-\frac{1}{4}\right)^2 = \left(-\frac{1}{4}\right)\left(-\frac{1}{4}\right) = \frac{1}{16}$.

$$= \frac{1}{8} - \left(-\frac{1}{4}\right)$$ Multiply: $\frac{1}{16}\cdot\left(\frac{-4}{1}\right) = -\frac{4}{16} = -\frac{1}{4}$.

$$= \frac{3}{8}$$ Subtract: $\frac{1}{8} - \left(-\frac{1}{4}\right) = \frac{1}{8} + \frac{1}{4} = \frac{1}{8} + \frac{2}{8} = \frac{3}{8}$.

CHECK POINT 13 Simplify: $\left(-\frac{1}{2}\right)^2 - \left(\frac{7}{10} - \frac{8}{15}\right)^2(-18)$.

9 *Apply the density property of rational numbers.*

Density of Rational Numbers

It is always possible to find a rational number between any two distinct rational numbers. Mathematicians express this idea by saying that the set of rational numbers is **dense**.

> **DENSITY OF THE RATIONAL NUMBERS**
>
> If r and t represent rational numbers, with $r < t$, then there is a rational number s such that s is between r and t:
>
> $$r < s < t.$$

One way to find a rational number between two given rational numbers is to find the rational number halfway between them. Add the given rational numbers and divide their sum by 2, thereby finding the average of the numbers.

EXAMPLE 14 *Illustrating the Density Property*

Find the rational number halfway between $\frac{1}{2}$ and $\frac{3}{4}$.

SOLUTION

First, add $\frac{1}{2}$ and $\frac{3}{4}$.

$$\frac{1}{2}+\frac{3}{4}=\frac{2}{4}+\frac{3}{4}=\frac{5}{4}$$

Next, divide this sum by 2.

$$\frac{5}{4}\div\frac{2}{1}=\frac{5}{4}\cdot\frac{1}{2}=\frac{5}{8}$$

The number $\frac{5}{8}$ is halfway between $\frac{1}{2}$ and $\frac{3}{4}$. Thus,

$$\frac{1}{2}<\frac{5}{8}<\frac{3}{4}.$$

GREAT QUESTION!

Why does $\frac{1}{2}<\frac{5}{8}<\frac{3}{4}$ seem meaningless to me?

The inequality $\frac{1}{2}<\frac{5}{8}<\frac{3}{4}$ is more obvious if all denominators are changed to 8:

$$\frac{4}{8}<\frac{5}{8}<\frac{6}{8}.$$

We can repeat the procedure of Example 14 and find the rational number halfway between $\frac{1}{2}$ and $\frac{5}{8}$. Repeated application of this procedure implies the following surprising result:

Between any two distinct rational numbers are *infinitely many* rational numbers.

CHECK POINT 14 Find the rational number halfway between $\frac{1}{3}$ and $\frac{1}{2}$.

10 *Solve problems involving rational numbers.*

Problem Solving with Rational Numbers

A common application of rational numbers involves preparing food for a different number of servings than what the recipe gives. The amount of each ingredient can be found as follows:

$$\text{Amount of ingredient needed} = \frac{\text{desired serving size}}{\text{recipe serving size}} \times \text{ingredient amount in the recipe.}$$

EXAMPLE 15 *Adjusting the Size of a Recipe*

A chocolate-chip cookie recipe for five dozen cookies requires $\frac{3}{4}$ cup sugar. If you want to make eight dozen cookies, how much sugar is needed?

SOLUTION

$$\text{Amount of sugar needed} = \frac{\text{desired serving size}}{\text{recipe serving size}} \times \text{sugar amount in recipe}$$

$$= \frac{8 \text{ dozen}}{5 \text{ dozen}} \times \frac{3}{4}\text{ cup}$$

The amount of sugar needed, in cups, is determined by multiplying the rational numbers:

$$\frac{8}{5}\times\frac{3}{4}=\frac{8\cdot 3}{5\cdot 4}=\frac{24}{20}=\frac{6\cdot \cancel{4}}{5\cdot \cancel{4}}=1\frac{1}{5}.$$

Thus, $1\frac{1}{5}$ cups of sugar is needed. (Depending on the measuring cup you are using, you may need to round the sugar amount to $1\frac{1}{4}$ cups.)

CHECK POINT 15 A chocolate-chip cookie recipe for five dozen cookies requires two eggs. If you want to make seven dozen cookies, exactly how many eggs are needed? Now round your answer to a realistic number that does not involve a fractional part of an egg.

Blitzer Bonus

NUMB3RS: Solving Crime with Mathematics

NUMB3RS was a prime-time TV crime series. The show's hero, Charlie Eppes, is a brilliant mathematician who uses his powerful skills to help the FBI identify and catch criminals. The episodes are entertaining and the basic premise shows how math is a powerful weapon in the never-ending fight against crime. *NUMB3RS* is significant because it was the first popular weekly drama that revolved around mathematics. A team of mathematician advisors ensured that the equations seen in the scripts were real and relevant to the episodes. The mathematical content of the show included many topics from this book, ranging from prime numbers, probability theory, and basic geometry.

Episodes of *NUMB3RS* begin with a spoken tribute about the importance of mathematics:

"We all use math everywhere. To tell time, to predict the weather, to handle money ... Math is more than formulas and equations. Math is more than numbers. It is logic. It is rationality. It is using your mind to solve the biggest mysteries we know."

Concept and Vocabulary Check

Fill in each blank so that the resulting statement is true.

1. The set of ________ numbers is the set of all numbers which can be expressed in the form $\frac{a}{b}$, where a and b are ________ and b is not equal to ______.
2. The number $\frac{17}{5}$ is an example of a/an ________ fraction because ________________________.
3. Numbers in the form $\frac{a}{b}$ (see Exercise 1) can be expressed as decimals. The decimals either ____________ or ____________.
4. The quotient of two fractions is the product of the first number and the ________________ of the second number.

In Exercises 5–8, determine whether each statement is true or false. If the statement is false, make the necessary change(s) to produce a true statement.

5. $\frac{1}{2} + \frac{1}{5} = \frac{2}{7}$ ______
6. $\frac{1}{2} \div 4 = 2$ ______
7. Every fraction has infinitely many equivalent fractions. ______
8. $\frac{3+7}{30} = \frac{\overset{1}{\cancel{3}}+7}{\underset{10}{\cancel{30}}} = \frac{8}{10} = \frac{4}{5}$ ______

Exercise Set 5.3

Practice Exercises

In Exercises 1–12, reduce each rational number to its lowest terms.

1. $\frac{10}{15}$
2. $\frac{18}{45}$
3. $\frac{15}{18}$
4. $\frac{16}{64}$
5. $\frac{24}{42}$
6. $\frac{32}{80}$
7. $\frac{60}{108}$
8. $\frac{112}{128}$
9. $\frac{342}{380}$
10. $\frac{210}{252}$
11. $\frac{308}{418}$
12. $\frac{144}{300}$

In Exercises 13–18, convert each mixed number to an improper fraction.

13. $2\frac{3}{8}$
14. $2\frac{7}{9}$
15. $-7\frac{3}{5}$
16. $-6\frac{2}{5}$
17. $12\frac{7}{16}$
18. $11\frac{5}{16}$

In Exercises 19–24, convert each improper fraction to a mixed number.

19. $\frac{23}{5}$
20. $\frac{47}{8}$
21. $-\frac{76}{9}$
22. $-\frac{59}{9}$
23. $\frac{711}{20}$
24. $\frac{788}{25}$

In Exercises 25–36, express each rational number as a decimal.

25. $\frac{3}{4}$
26. $\frac{3}{5}$
27. $\frac{7}{20}$
28. $\frac{3}{20}$
29. $\frac{7}{8}$
30. $\frac{5}{16}$
31. $\frac{9}{11}$
32. $\frac{3}{11}$
33. $\frac{22}{7}$
34. $\frac{20}{3}$
35. $\frac{2}{7}$
36. $\frac{5}{7}$

In Exercises 37–48, express each terminating decimal as a quotient of integers. If possible, reduce to lowest terms.

37. 0.3
38. 0.9
39. 0.4
40. 0.6
41. 0.39
42. 0.59
43. 0.82
44. 0.64
45. 0.725
46. 0.625
47. 0.5399
48. 0.7006

In Exercises 49–56, express each repeating decimal as a quotient of integers. If possible, reduce to lowest terms.

49. $0.\overline{7}$ **50.** $0.\overline{1}$ **51.** $0.\overline{9}$ **52.** $0.\overline{3}$

53. $0.\overline{36}$ **54.** $0.\overline{81}$ **55.** $0.\overline{257}$ **56.** $0.\overline{529}$

In Exercises 57–104, perform the indicated operations. If possible, reduce the answer to its lowest terms.

57. $\frac{3}{8}\cdot\frac{7}{11}$ **58.** $\frac{5}{8}\cdot\frac{3}{11}$ **59.** $\left(-\frac{1}{10}\right)\left(\frac{7}{12}\right)$

60. $\left(-\frac{1}{8}\right)\left(\frac{5}{9}\right)$ **61.** $\left(-\frac{2}{3}\right)\left(-\frac{9}{4}\right)$ **62.** $\left(-\frac{5}{4}\right)\left(-\frac{6}{7}\right)$

63. $\left(3\frac{3}{4}\right)\left(1\frac{3}{5}\right)$ **64.** $\left(2\frac{4}{5}\right)\left(1\frac{1}{4}\right)$ **65.** $\frac{5}{4}\div\frac{3}{8}$

66. $\frac{5}{8}\div\frac{4}{3}$ **67.** $-\frac{7}{8}\div\frac{15}{16}$ **68.** $-\frac{13}{20}\div\frac{4}{5}$

69. $6\frac{3}{5}\div 1\frac{1}{10}$ **70.** $1\frac{3}{4}\div 2\frac{5}{8}$ **71.** $\frac{2}{11}+\frac{3}{11}$

72. $\frac{5}{13}+\frac{2}{13}$ **73.** $\frac{5}{6}-\frac{1}{6}$ **74.** $\frac{7}{12}-\frac{5}{12}$

75. $\frac{7}{12}-\left(-\frac{1}{12}\right)$ **76.** $\frac{5}{16}-\left(-\frac{5}{16}\right)$ **77.** $\frac{1}{2}+\frac{1}{5}$

78. $\frac{1}{3}+\frac{1}{5}$ **79.** $\frac{3}{4}+\frac{3}{20}$ **80.** $\frac{2}{5}+\frac{2}{15}$

81. $\frac{5}{24}+\frac{7}{30}$ **82.** $\frac{7}{108}+\frac{55}{144}$ **83.** $\frac{13}{18}-\frac{2}{9}$

84. $\frac{13}{15}-\frac{2}{45}$ **85.** $\frac{4}{3}-\frac{3}{4}$ **86.** $\frac{3}{2}-\frac{2}{3}$

87. $\frac{1}{15}-\frac{27}{50}$ **88.** $\frac{4}{15}-\frac{1}{6}$ **89.** $2\frac{2}{3}+1\frac{3}{4}$

90. $2\frac{1}{8}+3\frac{3}{4}$ **91.** $3\frac{2}{3}-2\frac{1}{2}$

92. $3\frac{3}{4}-2\frac{1}{3}$ **93.** $-5\frac{2}{3}+3\frac{1}{6}$

94. $-2\frac{1}{2}+1\frac{3}{4}$ **95.** $-1\frac{4}{7}-\left(-2\frac{5}{14}\right)$

96. $-1\frac{4}{9}-\left(-2\frac{5}{18}\right)$ **97.** $\left(\frac{1}{2}-\frac{1}{3}\right)\div\frac{5}{8}$

98. $\left(\frac{1}{2}+\frac{1}{4}\right)\div\left(\frac{1}{2}+\frac{1}{3}\right)$ **99.** $-\frac{9}{4}\left(\frac{1}{2}\right)+\frac{3}{4}\div\frac{5}{6}$

100. $\left[-\frac{4}{7}-\left(-\frac{2}{5}\right)\right]\left[-\frac{3}{8}+\left(-\frac{1}{9}\right)\right]$

101. $\dfrac{\frac{7}{9}-3}{\frac{5}{6}}\div\dfrac{3}{2}+\dfrac{3}{4}$ **102.** $\dfrac{\frac{17}{25}}{\frac{3}{5}-4}\div\dfrac{1}{5}+\dfrac{1}{2}$

103. $\frac{1}{4}-6(2+8)\div\left(-\frac{1}{3}\right)\left(-\frac{1}{9}\right)$

104. $\frac{3}{4}-4(2+7)\div\left(-\frac{1}{2}\right)\left(-\frac{1}{6}\right)$

In Exercises 105–110, find the rational number halfway between the two numbers in each pair.

105. $\frac{1}{4}$ and $\frac{1}{3}$ **106.** $\frac{2}{3}$ and $\frac{5}{6}$ **107.** $\frac{1}{2}$ and $\frac{2}{3}$

108. $\frac{3}{5}$ and $\frac{2}{3}$ **109.** $-\frac{2}{3}$ and $-\frac{5}{6}$ **110.** -4 and $-\frac{7}{2}$

Different operations with the same rational numbers usually result in different answers. Exercises 111–112 illustrate some curious exceptions.

111. Show that $\frac{13}{4}+\frac{13}{9}$ and $\frac{13}{4}\times\frac{13}{9}$ give the same answer.

112. Show that $\frac{169}{30}+\frac{13}{15}$ and $\frac{169}{30}\div\frac{13}{15}$ give the same answer.

Practice Plus

In Exercises 113–116, perform the indicated operations. Leave denominators in prime factorization form.

113. $\dfrac{5}{2^2\cdot 3^2}-\dfrac{1}{2\cdot 3^2}$ **114.** $\dfrac{7}{3^2\cdot 5^2}-\dfrac{1}{3\cdot 5^3}$

115. $\dfrac{1}{2^4\cdot 5^3\cdot 7}+\dfrac{1}{2\cdot 5^4}-\dfrac{1}{2^3\cdot 5^2}$

116. $\dfrac{1}{2^3\cdot 17^8}+\dfrac{1}{2\cdot 17^9}-\dfrac{1}{2^2\cdot 3\cdot 17^8}$

In Exercises 117–120, express each rational number as a decimal. Then insert either < or > in the shaded area between the rational numbers to make the statement true.

117. $\frac{6}{11}$ ■ $\frac{7}{12}$ **118.** $\frac{29}{36}$ ■ $\frac{28}{35}$

119. $-\frac{5}{6}$ ■ $-\frac{8}{9}$ **120.** $-\frac{1}{125}$ ■ $-\frac{3}{500}$

Application Exercises

The Dog Ate My Calendar. *The bar graph shows seven common excuses by college students for not meeting assignment deadlines and the number of excuses for every 500 excuses that fall into each of these categories. Use the information displayed by the graph to solve Exercises 121–122. Reduce fractions to lowest terms.*

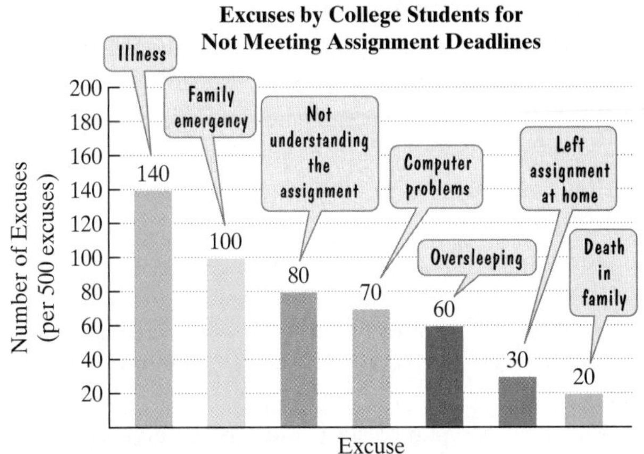

Source: Roig and Caso, "Lying and Cheating: Fraudulent Excuse Making, Cheating, and Plagiarism," *Journal of Psychology*

121. What fraction of the excuses involve not understanding the assignment?

122. What fraction of excuses involve illness?

To meet the demands of an economy that values computer and technical skills, the United States will continue to need more workers with a postsecondary education (education beyond high school). The circle graph shows the fraction of jobs in the United States requiring various levels of education by 2020. Use the information displayed by the graph to solve Exercises 123–124.

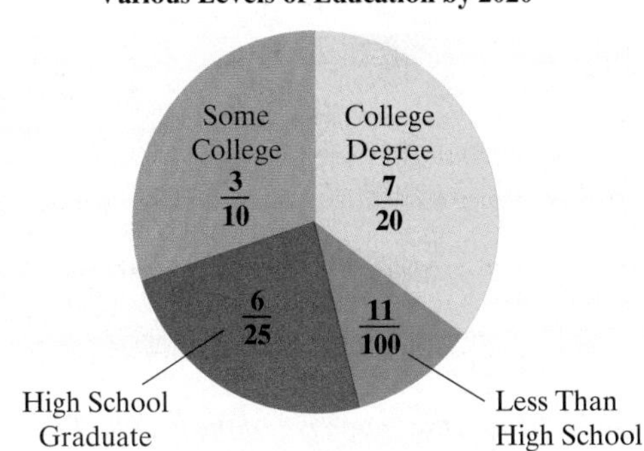

Source: U.S. Bureau of Labor Statistics

(In Exercises 123–124, refer to the circle graph at the bottom of the previous page.)

123. a. What fraction of jobs will require postsecondary education by 2020?

b. How much greater is the fraction of jobs that will require a college degree than a high school diploma by 2020?

124. a. What fraction of jobs will not require any college by 2020?

b. How much greater is the fraction of jobs that will require a college degree than some college by 2020?

Use the following list of ingredients for chocolate brownies to solve Exercises 125–130.

Ingredients for 16 Brownies

$\frac{2}{3}$ cup butter, 5 ounces unsweetened chocolate, $1\frac{1}{2}$ cups sugar, 2 teaspoons vanilla, 2 eggs, 1 cup flour

125. How much of each ingredient is needed to make 8 brownies?

126. How much of each ingredient is needed to make 12 brownies?

127. How much of each ingredient is needed to make 20 brownies?

128. How much of each ingredient is needed to make 24 brownies?

129. With only one cup of butter, what is the greatest number of brownies that you can make? (Ignore part of a brownie.)

130. With only one cup of sugar, what is the greatest number of brownies that you can make? (Ignore part of a brownie.)

A mix for eight servings of instant potatoes requires $2\frac{2}{3}$ cups of water. Use this information to solve Exercises 131–132.

131. If you want to make 11 servings, how much water is needed?

132. If you want to make six servings, how much water is needed?

The sounds created by plucked or bowed strings of equal diameter and tension produce various notes depending on the lengths of the strings. If a string is half as long as another, its note will be an octave higher than the longer string. Using a length of 1 unit to represent middle C, the diagram shows different fractions of the length of this unit string needed to produce the notes D, E, F, G, A, B, and c one octave higher than middle C.

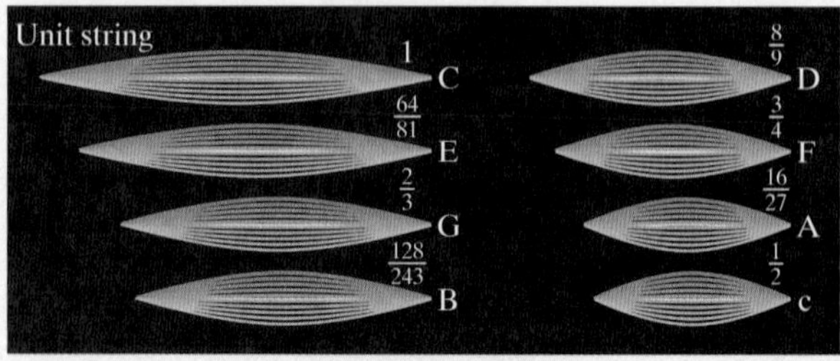

For many of the strings, the length is $\frac{8}{9}$ of the length of the previous string. For example, the A string is $\frac{8}{9}$ of the length of the G string: $\frac{8}{9} \cdot \frac{2}{3} = \frac{16}{27}$. Use this information to solve Exercises 133–134.

133. a. Which strings from D through c are $\frac{8}{9}$ of the length of the preceding string?

b. How is your answer to part (a) shown on this one-octave span of the piano keyboard?

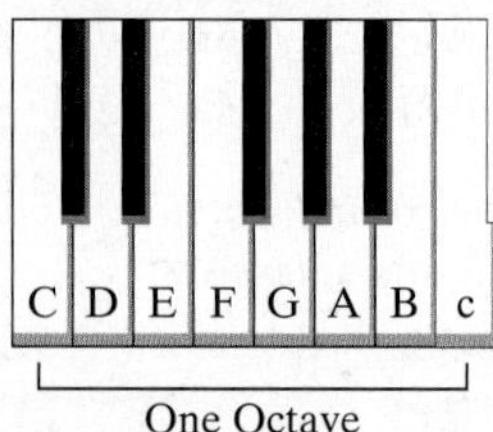

134. a. Which strings from D through c are not $\frac{8}{9}$ of the length of the preceding string?

b. How is your answer to part (a) shown on the one-octave span on the piano keyboard in Exercise 133(b)?

135. A board $7\frac{1}{2}$ inches long is cut from a board that is 2 feet long. If the width of the saw cut is $\frac{1}{16}$ inch, what is the length of the remaining piece?

136. A board that is $7\frac{1}{4}$ inches long is cut from a board that is 3 feet long. If the width of the saw cut is $\frac{1}{16}$ inch, what is the length of the remaining piece?

137. A franchise is owned by three people. The first owns $\frac{5}{12}$ of the business and the second owns $\frac{1}{4}$ of the business. What fractional part of the business is owned by the third person?

138. At a workshop on enhancing creativity, $\frac{1}{4}$ of the participants are musicians, $\frac{2}{5}$ are artists, $\frac{1}{10}$ are actors, and the remaining participants are writers. What fraction of the people attending the workshop are writers?

139. If you walk $\frac{3}{4}$ mile and then jog $\frac{2}{5}$ mile, what is the total distance covered? How much farther did you walk than jog?

140. Some companies pay people extra when they work more than a regular 40-hour work week. The overtime pay is often $1\frac{1}{2}$ times the regular hourly rate. This is called time and a half. A summer job for students pays $12 an hour and offers time and a half for the hours worked over 40. If a student works 46 hours during one week, what is the student's total pay before taxes?

141. A will states that $\frac{3}{5}$ of the estate is to be divided among relatives. Of the remaining estate, $\frac{1}{4}$ goes to charity. What fraction of the estate goes to charity?

142. The legend of a map indicates that 1 inch = 16 miles. If the distance on the map between two cities is $2\frac{3}{8}$ inches, how far apart are the cities?

Explaining the Concepts

143. What is a rational number?

144. Explain how to reduce a rational number to its lowest terms.

145. Explain how to convert from a mixed number to an improper fraction. Use $7\frac{2}{3}$ as an example.

146. Explain how to convert from an improper fraction to a mixed number. Use $\frac{47}{5}$ as an example.

147. Explain how to write a rational number as a decimal.

148. Explain how to write 0.083 as a quotient of integers.

149. Explain how to write $0.\overline{9}$ as a quotient of integers.

150. Explain how to multiply rational numbers. Use $\frac{5}{6} \cdot \frac{1}{2}$ as an example.

151. Explain how to divide rational numbers. Use $\frac{5}{6} \div \frac{1}{2}$ as an example.

152. Explain how to add rational numbers with different denominators. Use $\frac{5}{6} + \frac{1}{2}$ as an example.

153. What does it mean when we say that the set of rational numbers is dense?

Critical Thinking Exercises

Make Sense? *In Exercises 154–157, determine whether each statement makes sense or does not make sense, and explain your reasoning.*

154. I saved money by buying a computer for $\frac{3}{2}$ of its original price.

155. I find it easier to multiply $\frac{1}{5}$ and $\frac{3}{4}$ than to add them.

156. My calculator shows the decimal form for the rational number $\frac{3}{11}$ as 0.2727273, so $\frac{3}{11} = 0.2727273$.

157. The value of $\frac{|3 - 7| - 2^3}{(-2)(-3)}$ is the rational number that results when $\frac{1}{3}$ is subtracted from $-\frac{1}{3}$.

158. Shown below is a short excerpt from "The Star-Spangled Banner." The time is $\frac{3}{4}$, which means that each measure must contain notes that add up to $\frac{3}{4}$. The values of the different notes tell musicians how long to hold each note.

$$\text{𝅝} = 1 \quad \text{𝅗𝅥} = \frac{1}{2} \quad \text{♩} = \frac{1}{4} \quad \text{♪} = \frac{1}{8}$$

Use vertical lines to divide this line of "The Star-Spangled Banner" into measures.

159. Use inductive reasoning to predict the addition problem and the sum that will appear in the fourth row. Then perform the arithmetic to verify your conjecture.

$$\frac{1}{1\cdot 2} + \frac{1}{2\cdot 3} = \frac{2}{3}$$
$$\frac{1}{1\cdot 2} + \frac{1}{2\cdot 3} + \frac{1}{3\cdot 4} = \frac{3}{4}$$
$$\frac{1}{1\cdot 2} + \frac{1}{2\cdot 3} + \frac{1}{3\cdot 4} + \frac{1}{4\cdot 5} = \frac{4}{5}$$

Technology Exercises

160. Use a calculator to express the following rational numbers as decimals.

a. $\frac{197}{800}$ **b.** $\frac{4539}{3125}$ **c.** $\frac{7}{6250}$

161. Some calculators have a fraction feature. This feature allows you to perform operations with fractions and displays the answer as a fraction reduced to its lowest terms. If your calculator has this feature, use it to verify any five of the answers that you obtained in Exercises 57–104.

Group Exercise

162. Each member of the group should present an application of rational numbers. The application can be based on research or on how the group member uses rational numbers in his or her life. If you are not sure where to begin, ask yourself how your life would be different if fractions and decimals were concepts unknown to our civilization.

5.4 The Irrational Numbers

WHAT AM I SUPPOSED TO LEARN?

After studying this section, you should be able to:

1. Define the irrational numbers.
2. Simplify square roots.
3. Perform operations with square roots.
4. Rationalize denominators.

Shown here is Renaissance artist Raphael Sanzio's (1483–1520) image of Pythagoras from *The School of Athens* mural. Detail of left side.

FOR THE FOLLOWERS OF THE GREEK MATHEMATICIAN PYTHAGORAS IN THE SIXTH century B.C., numbers took on a life-and-death importance. The "Pythagorean Brotherhood" was a secret group whose members were convinced that properties of whole numbers were the key to understanding the universe. Members of the Brotherhood (which admitted women) thought that all numbers that were not whole numbers could be represented as the ratio of whole numbers. A crisis occurred for the Pythagoreans when they discovered the existence of a number that was not rational. Because the Pythagoreans viewed numbers with reverence and awe, the punishment for speaking about this number was death. However, a member of the Brotherhood revealed the secret of the number's existence. When he later died in a shipwreck, his death was viewed as punishment from the gods.

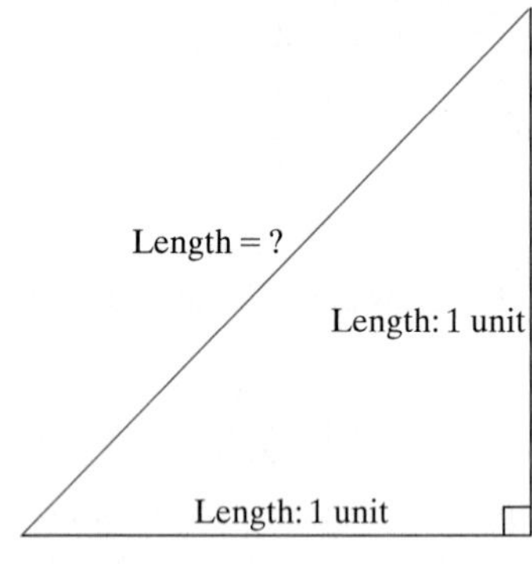

FIGURE 5.6

The triangle in **Figure 5.6** led the Pythagoreans to the discovery of a number that could not be expressed as the quotient of integers. Based on their understanding of the relationship among the sides of this triangle, they knew that the length of the side shown in red had to be a number that, when squared, is equal to 2. The Pythagoreans discovered that this number seemed to be close to the rational numbers

$$\frac{14}{10}, \frac{141}{100}, \frac{1414}{1000}, \frac{14,142}{10,000}, \text{ and so on.}$$

However, they were shocked to find that there is no quotient of integers whose square is equal to 2.

The positive number whose square is equal to 2 is written $\sqrt{2}$. We read this "the square root of 2," or "radical 2." The symbol $\sqrt{\ }$ is called the **radical sign**. The number under the radical sign, in this case 2, is called the **radicand**. The entire symbol $\sqrt{2}$ is called a **radical**.

Using deductive reasoning, mathematicians have proved that $\sqrt{2}$ cannot be represented as a quotient of integers. This means that there is no terminating or repeating decimal that can be multiplied by itself to give 2. We can, however, give a decimal approximation for $\sqrt{2}$. We use the symbol $\approx$, which means "is approximately equal to." Thus,

$$\sqrt{2} \approx 1.414214.$$

We can verify that this is only an approximation by multiplying 1.414214 by itself. The product is not exactly 2:

$$1.414214 \times 1.414214 = 2.000001237796.$$

1 *Define the irrational numbers.*

A number like $\sqrt{2}$, whose decimal representation does not come to an end and does not have a block of repeating digits, is an example of an **irrational number**.

THE IRRATIONAL NUMBERS

The set of **irrational numbers** is the set of numbers whose decimal representations are neither terminating nor repeating.

Perhaps the best known of all the irrational numbers is π (pi). This irrational number represents the distance around a circle (its circumference) divided by the diameter of the circle. In the *Star Trek* episode "Wolf in the Fold," Spock foils an evil computer by telling it to "compute the last digit in the value of π." Because π is an irrational number, there is no last digit in its decimal representation:

$$\pi = 3.14159265358979323846264338327 95....$$

The nature of the irrational number π has fascinated mathematicians for centuries. Amateur and professional mathematicians have taken up the challenge of calculating π to more and more decimal places. Although such an exercise may seem pointless, it serves as the ultimate stress test for new high-speed computers and also as a test for the long-standing, but still unproven, conjecture that the distribution of digits in π is completely random.

Blitzer Bonus

The Best and Worst of π

In 2014, mathematicians calculated π to more than thirteen trillion decimal places. The calculations used 208 days of computer time.

The most inaccurate version of π came from the 1897 General Assembly of Indiana. Bill No. 246 stated that "π was by law 4."

TECHNOLOGY

You can obtain decimal approximations for irrational numbers using a calculator. For example, to approximate $\sqrt{2}$, use the following keystrokes:

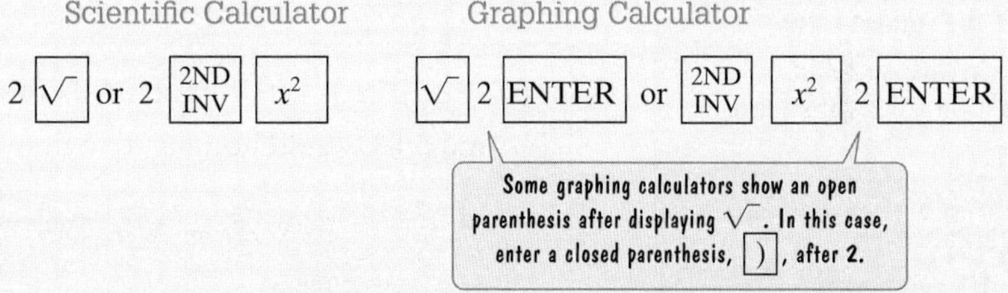

The display may read 1.41421356237, although your calculator may show more or fewer digits. Between which two integers would you graph $\sqrt{2}$ on a number line?

Square Roots

The United Nations Building in New York was designed to represent its mission of promoting world harmony. Viewed from the front, the building looks like three rectangles stacked upon each other. In each rectangle, the width divided by the height is $\sqrt{5} + 1$ to 2, approximately 1.618 to 1. The ancient Greeks believed that such a rectangle, called a **golden rectangle**, was the most pleasing of all rectangles. The comparison 1.618 to 1 is approximate because $\sqrt{5}$ is an irrational number.

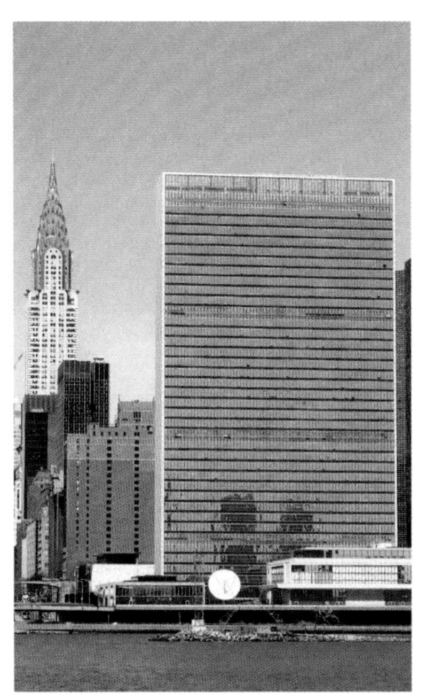

The U.N. building is designed with three golden rectangles.

The **principal square root** of a nonnegative number n, written $\sqrt{n}$, is the nonnegative number that when multiplied by itself gives n. Thus,

$$\sqrt{36} = 6 \text{ because } 6 \cdot 6 = 36$$

and

$$\sqrt{81} = 9 \text{ because } 9 \cdot 9 = 81.$$

Notice that both $\sqrt{36}$ and $\sqrt{81}$ are rational numbers because 6 and 9 are terminating decimals. Thus, **not all square roots are irrational.**

Numbers such as 36 and 81 are called *perfect squares*. A **perfect square** is a number that is the square of a whole number. The first few perfect squares are as follows.

$\mathbf{0} = 0^2$ $\quad \mathbf{16} = 4^2$ $\quad \mathbf{64} = 8^2$ $\quad \mathbf{144} = 12^2$

$\mathbf{1} = 1^2$ $\quad \mathbf{25} = 5^2$ $\quad \mathbf{81} = 9^2$ $\quad \mathbf{169} = 13^2$

$\mathbf{4} = 2^2$ $\quad \mathbf{36} = 6^2$ $\quad \mathbf{100} = 10^2$ $\quad \mathbf{196} = 14^2$

$\mathbf{9} = 3^2$ $\quad \mathbf{49} = 7^2$ $\quad \mathbf{121} = 11^2$ $\quad \mathbf{225} = 15^2$

The principal square root of a perfect square is a whole number. For example,

$$\sqrt{0} = 0, \sqrt{1} = 1, \sqrt{4} = 2, \sqrt{9} = 3, \sqrt{16} = 4, \sqrt{25} = 5, \sqrt{36} = 6,$$

and so on.

2 *Simplify square roots.*

Simplifying Square Roots

A rule for simplifying square roots can be generalized by comparing $\sqrt{25 \cdot 4}$ and $\sqrt{25} \cdot \sqrt{4}$. Notice that

$$\sqrt{25 \cdot 4} = \sqrt{100} = 10 \quad \text{and} \quad \sqrt{25} \cdot \sqrt{4} = 5 \cdot 2 = 10.$$

Because we obtain 10 in both situations, the original radicals must be equal. That is,

$$\sqrt{25 \cdot 4} = \sqrt{25} \cdot \sqrt{4}.$$

This result is a particular case of the **product rule for square roots** that can be generalized as follows:

THE PRODUCT RULE FOR SQUARE ROOTS

If a and b represent nonnegative numbers, then

$$\sqrt{ab} = \sqrt{a} \cdot \sqrt{b} \quad \text{and} \quad \sqrt{a} \cdot \sqrt{b} = \sqrt{ab}.$$

The square root of a product is the product of the square roots.

GREAT QUESTION!

Is the square root of a sum the sum of the square roots?

No. There are no addition or subtraction rules for square roots:

$$\sqrt{a+b} \neq \sqrt{a} + \sqrt{b}$$
$$\sqrt{a-b} \neq \sqrt{a} - \sqrt{b}.$$

For example, if $a = 9$ and $b = 16$,

$$\sqrt{9+16} = \sqrt{25} = 5$$

and

$$\sqrt{9} + \sqrt{16} = 3 + 4 = 7.$$

Thus,

$$\sqrt{9+16} \neq \sqrt{9} + \sqrt{16}.$$

Example 1 shows how the product rule is used to remove from the square root any perfect squares that occur as factors.

EXAMPLE 1 *Simplifying Square Roots*

Simplify, if possible:

a. $\sqrt{75}$ **b.** $\sqrt{500}$ **c.** $\sqrt{17}$.

SOLUTION

a. $\sqrt{75} = \sqrt{25 \cdot 3}$ — 25 is the greatest perfect square that is a factor of 75.

$= \sqrt{25} \cdot \sqrt{3}$ — $\sqrt{ab} = \sqrt{a} \cdot \sqrt{b}$

$= 5\sqrt{3}$ — Write $\sqrt{25}$ as 5.

b. $\sqrt{500} = \sqrt{100 \cdot 5}$ — 100 is the greatest perfect square factor of 500.

$= \sqrt{100} \cdot \sqrt{5}$ — $\sqrt{ab} = \sqrt{a} \cdot \sqrt{b}$

$= 10\sqrt{5}$ — Write $\sqrt{100}$ as 10.

c. Because 17 has no perfect square factors (other than 1), $\sqrt{17}$ cannot be simplified.

CHECK POINT 1 Simplify, if possible:

a. $\sqrt{12}$ **b.** $\sqrt{60}$ **c.** $\sqrt{55}$.

3 *Perform operations with square roots.*

Multiplying Square Roots

If a and b are nonnegative, then we can use the product rule

$$\sqrt{a} \cdot \sqrt{b} = \sqrt{a \cdot b}$$

to multiply square roots. The product of the square roots is the square root of the product. Once the square roots are multiplied, simplify the square root of the product when possible.

EXAMPLE 2 *Multiplying Square Roots*

Multiply:

a. $\sqrt{2} \cdot \sqrt{5}$ **b.** $\sqrt{7} \cdot \sqrt{7}$ **c.** $\sqrt{6} \cdot \sqrt{12}$.

SOLUTION

a. $\sqrt{2} \cdot \sqrt{5} = \sqrt{2 \cdot 5} = \sqrt{10}$

b. $\sqrt{7} \cdot \sqrt{7} = \sqrt{7 \cdot 7} = \sqrt{49} = 7$

It is possible to multiply irrational numbers and obtain a rational number for the product.

c. $\sqrt{6} \cdot \sqrt{12} = \sqrt{6 \cdot 12} = \sqrt{72} = \sqrt{36 \cdot 2} = \sqrt{36} \cdot \sqrt{2} = 6\sqrt{2}$

CHECK POINT 2 Multiply:

a. $\sqrt{3} \cdot \sqrt{10}$ **b.** $\sqrt{10} \cdot \sqrt{10}$ **c.** $\sqrt{6} \cdot \sqrt{2}$.

Dividing Square Roots

Another property for square roots involves division.

THE QUOTIENT RULE FOR SQUARE ROOTS

If a and b represent nonnegative numbers and $b \neq 0$, then

$$\frac{\sqrt{a}}{\sqrt{b}} = \sqrt{\frac{a}{b}} \quad \text{and} \quad \sqrt{\frac{a}{b}} = \frac{\sqrt{a}}{\sqrt{b}}.$$

The quotient of two square roots is the square root of the quotient.

Once the square roots are divided, simplify the square root of the quotient when possible.

EXAMPLE 3 *Dividing Square Roots*

Find the quotient:

a. $\frac{\sqrt{75}}{\sqrt{3}}$ **b.** $\frac{\sqrt{90}}{\sqrt{2}}$.

SOLUTION

a. $\frac{\sqrt{75}}{\sqrt{3}} = \sqrt{\frac{75}{3}} = \sqrt{25} = 5$

b. $\frac{\sqrt{90}}{\sqrt{2}} = \sqrt{\frac{90}{2}} = \sqrt{45} = \sqrt{9 \cdot 5} = \sqrt{9} \cdot \sqrt{5} = 3\sqrt{5}$

CHECK POINT 3 Find the quotient:

a. $\dfrac{\sqrt{80}}{\sqrt{5}}$ **b.** $\dfrac{\sqrt{48}}{\sqrt{6}}$.

Adding and Subtracting Square Roots

The number that multiplies a square root is called the square root's **coefficient**. For example, in $3\sqrt{5}$, 3 is the coefficient of the square root.

Square roots with the same radicand can be added or subtracted by adding or subtracting their coefficients:

$$a\sqrt{c} + b\sqrt{c} = (a + b)\sqrt{c} \qquad a\sqrt{c} - b\sqrt{c} = (a - b)\sqrt{c}.$$

Sum of coefficients times the common square root

Difference of coefficients times the common square root

EXAMPLE 4 *Adding and Subtracting Square Roots*

Add or subtract as indicated:

a. $7\sqrt{2} + 5\sqrt{2}$ **b.** $2\sqrt{5} - 6\sqrt{5}$ **c.** $3\sqrt{7} + 9\sqrt{7} - \sqrt{7}$.

SOLUTION

a. $7\sqrt{2} + 5\sqrt{2} = (7 + 5)\sqrt{2}$
$= 12\sqrt{2}$

b. $2\sqrt{5} - 6\sqrt{5} = (2 - 6)\sqrt{5}$
$= -4\sqrt{5}$

c. $3\sqrt{7} + 9\sqrt{7} - \sqrt{7} = 3\sqrt{7} + 9\sqrt{7} - 1\sqrt{7}$ Write $\sqrt{7}$ as $1\sqrt{7}$.
$= (3 + 9 - 1)\sqrt{7}$
$= 11\sqrt{7}$

CHECK POINT 4 Add or subtract as indicated:

a. $8\sqrt{3} + 10\sqrt{3}$

b. $4\sqrt{13} - 9\sqrt{13}$

c. $7\sqrt{10} + 2\sqrt{10} - \sqrt{10}$.

In some situations, it is possible to add and subtract square roots that do not contain a common square root by first simplifying.

EXAMPLE 5 *Adding and Subtracting Square Roots by First Simplifying*

Add or subtract as indicated:

a. $\sqrt{2} + \sqrt{8}$ **b.** $4\sqrt{50} - 6\sqrt{32}$.

GREAT QUESTION!

Can I combine $\sqrt{2} + \sqrt{7}$?

No. Sums or differences of square roots that cannot be simplified and that do not contain a common radicand cannot be combined into one term by adding or subtracting coefficients. Some examples:

- $5\sqrt{3} + 3\sqrt{5}$ cannot be combined by adding coefficients. The square roots, $\sqrt{3}$ and $\sqrt{5}$, are different.
- $28 + 7\sqrt{3}$, or $28\sqrt{1} + 7\sqrt{3}$, cannot be combined by adding coefficients. The square roots, $\sqrt{1}$ and $\sqrt{3}$, are different.

SOLUTION

a. $\sqrt{2} + \sqrt{8}$

$= \sqrt{2} + \sqrt{4 \cdot 2}$ Split 8 into two factors such that one factor is a perfect square.

$= 1\sqrt{2} + 2\sqrt{2}$ $\sqrt{4 \cdot 2} = \sqrt{4} \cdot \sqrt{2} = 2\sqrt{2}$

$= (1 + 2)\sqrt{2}$ Add coefficients and retain the common square root.

$= 3\sqrt{2}$ Simplify.

b. $4\sqrt{50} - 6\sqrt{32}$

$= 4\sqrt{25 \cdot 2} - 6\sqrt{16 \cdot 2}$ 25 is the greatest perfect square factor of 50 and 16 is the greatest perfect square factor of 32.

$= 4 \cdot 5\sqrt{2} - 6 \cdot 4\sqrt{2}$ $\sqrt{25 \cdot 2} = \sqrt{25}\sqrt{2} = 5\sqrt{2}$ and $\sqrt{16 \cdot 2} = \sqrt{16}\sqrt{2} = 4\sqrt{2}$

$= 20\sqrt{2} - 24\sqrt{2}$ Multiply.

$= (20 - 24)\sqrt{2}$ Subtract coefficients and retain the common square root.

$= -4\sqrt{2}$ Simplify.

CHECK POINT 5 Add or subtract as indicated:

a. $\sqrt{3} + \sqrt{12}$ **b.** $4\sqrt{8} - 7\sqrt{18}$.

4 *Rationalize denominators.*

Rationalizing Denominators

The calculator screen in **Figure 5.7** shows approximate values for $\frac{1}{\sqrt{3}}$ and $\frac{\sqrt{3}}{3}$. The two approximations are the same. This is not a coincidence:

$$\frac{1}{\sqrt{3}} = \frac{1}{\sqrt{3}} \cdot \boxed{\frac{\sqrt{3}}{\sqrt{3}}} = \frac{\sqrt{3}}{\sqrt{9}} = \frac{\sqrt{3}}{3}$$

Any number divided by itself is 1. Multiplication by 1 does not change the value of $\frac{1}{\sqrt{3}}$.

```
1/√3
                .5773502692
√3/3
                .5773502692
```

FIGURE 5.7 The calculator screen shows approximate values for $\frac{1}{\sqrt{3}}$ and $\frac{\sqrt{3}}{3}$.

This process involves rewriting a radical expression as an equivalent expression in which the denominator no longer contains any radicals. The process is called **rationalizing the denominator**. If the denominator contains the square root of a natural number that is not a perfect square, **multiply the numerator and the denominator by the smallest number that produces the square root of a perfect square in the denominator**.

EXAMPLE 6 *Rationalizing Denominators*

Rationalize the denominator:

a. $\frac{15}{\sqrt{6}}$ **b.** $\sqrt{\frac{3}{5}}$ **c.** $\frac{12}{\sqrt{8}}$.

GREAT QUESTION!

What exactly does rationalizing a denominator do to an irrational number in the denominator?

Rationalizing a numerical denominator makes that denominator a rational number.

SOLUTION

a. If we multiply the numerator and the denominator of $\frac{15}{\sqrt{6}}$ by $\sqrt{6}$, the denominator becomes $\sqrt{6}\cdot\sqrt{6} = \sqrt{36} = 6$. Therefore, we multiply by 1, choosing $\frac{\sqrt{6}}{\sqrt{6}}$ for 1.

$$\frac{15}{\sqrt{6}} = \frac{15}{\sqrt{6}}\cdot\frac{\sqrt{6}}{\sqrt{6}} = \frac{15\sqrt{6}}{\sqrt{36}} = \frac{15\sqrt{6}}{6} = \frac{5\sqrt{6}}{2}$$

Multiply by 1.

Simplify: $\frac{15}{6} = \frac{5\cdot\cancel{3}}{2\cdot\cancel{3}} = \frac{5}{2}$.

b. $$\sqrt{\frac{3}{5}} = \frac{\sqrt{3}}{\sqrt{5}} = \frac{\sqrt{3}}{\sqrt{5}}\cdot\frac{\sqrt{5}}{\sqrt{5}} = \frac{\sqrt{15}}{\sqrt{25}} = \frac{\sqrt{15}}{5}$$

Multiply by 1.

GREAT QUESTION!

Can I rationalize the denominator of $\frac{12}{\sqrt{8}}$ by multiplying by $\frac{\sqrt{8}}{\sqrt{8}}$?

Yes. However, it takes more work to simplify the result.

c. The *smallest* number that will produce a perfect square in the denominator of $\frac{12}{\sqrt{8}}$ is $\sqrt{2}$, because $\sqrt{8}\cdot\sqrt{2} = \sqrt{16} = 4$. We multiply by 1, choosing $\frac{\sqrt{2}}{\sqrt{2}}$ for 1.

$$\frac{12}{\sqrt{8}} = \frac{12}{\sqrt{8}}\cdot\frac{\sqrt{2}}{\sqrt{2}} = \frac{12\sqrt{2}}{\sqrt{16}} = \frac{12\sqrt{2}}{4} = 3\sqrt{2}$$

CHECK POINT 6 Rationalize the denominator:

a. $\frac{25}{\sqrt{10}}$ **b.** $\sqrt{\frac{2}{7}}$ **c.** $\frac{5}{\sqrt{18}}$.

Blitzer Bonus

Golden Rectangles

The early Greeks believed that the most pleasing of all rectangles were **golden rectangles**, whose ratio of width to height is

$$\frac{w}{h} = \frac{\sqrt{5}+1}{2}.$$

The Parthenon at Athens fits into a golden rectangle once the triangular pediment is reconstructed.

Irrational Numbers and Other Kinds of Roots

Irrational numbers appear in the form of roots other than square roots. The symbol $\sqrt[3]{\ }$ represents the **cube root** of a number. For example,

$$\sqrt[3]{8} = 2 \text{ because } 2\cdot2\cdot2 = 8 \quad \text{and} \quad \sqrt[3]{64} = 4 \text{ because } 4\cdot4\cdot4 = 64.$$

Although these cube roots are rational numbers, most cube roots are not. For example,

$$\sqrt[3]{217} \approx 6.0092 \text{ because } (6.0092)^3 \approx 216.995, \text{ not exactly } 217.$$

There is no end to the kinds of roots for numbers. For example, $\sqrt[4]{\ }$ represents the **fourth root** of a number. Thus, $\sqrt[4]{81} = 3$ because $3\cdot3\cdot3\cdot3 = 81$. Although the fourth root of 81 is rational, most fourth roots, fifth roots, and so on tend to be irrational.

Blitzer Bonus

A Radical Idea: Time Is Relative

Digital Image © The Museum of Modern Art/ Licensed by Scala/Art Resource, NY; © 2017 Salvador Dalí, Fundació Gala-Salvador Dalí, Artists Rights Society

What does travel in space have to do with square roots? Imagine that in the future we will be able to travel at velocities approaching the speed of light (approximately 186,000 miles per second). According to Einstein's theory of special relativity, time would pass more quickly on Earth than it would in the moving spaceship. The special-relativity equation

$$R_a = R_f\sqrt{1 - \left(\frac{v}{c}\right)^2}$$

gives the aging rate of an astronaut, R_a, relative to the aging rate of a friend, R_f, on Earth. In this formula, v is the astronaut's speed and c is the speed of light. As the astronaut's speed approaches the speed of light, we can substitute c for v.

$$R_a = R_f\sqrt{1 - \left(\frac{v}{c}\right)^2}$$ Einstein's equation gives the aging rate of an astronaut, R_a, relative to the aging rate of a friend, R_f, on Earth.

$$R_a = R_f\sqrt{1 - \left(\frac{c}{c}\right)^2}$$ The velocity, v, is approaching the speed of light, c, so let $v = c$.

$$= R_f\sqrt{1 - 1}$$ $\left(\frac{c}{c}\right)^2 = 1^2 = 1\cdot1 = 1$

$$= R_f\sqrt{0}$$ Simplify the radicand: $1 - 1 = 0$.

$$= R_f\cdot 0$$ $\sqrt{0} = 0$

$$= 0$$ Multiply: $R_f\cdot 0 = 0$.

Close to the speed of light, the astronaut's aging rate, R_a, relative to that of a friend, R_f, on Earth is nearly 0. What does this mean? As we age here on Earth, the space traveler would barely get older. The space traveler would return to an unknown futuristic world in which friends and loved ones would be long gone.

Concept and Vocabulary Check

Fill in each blank so that the resulting statement is true.

1. The set of irrational numbers is the set of numbers whose decimal representations are neither __________ nor __________.
2. The irrational number ____ represents the circumference of a circle divided by the diameter of the circle.
3. The square root of n, represented by ______, is the nonnegative number that when multiplied by itself gives ____.
4. $\sqrt{49 \cdot 6} = \sqrt{\_} \cdot \sqrt{\_} = \_\_\_$
5. The number that multiplies a square root is called the square root's __________.
6. $8\sqrt{3} + 10\sqrt{3} = (\_ + \_)\sqrt{3} = \_\_\_$
7. $\sqrt{50} + \sqrt{32} = \sqrt{25 \cdot 2} + \sqrt{16 \cdot 2}$ $= \sqrt{25} \cdot \sqrt{2} + \sqrt{16} \cdot \sqrt{2} = \_\sqrt{2} + \_\sqrt{2} = \_\_\_$
8. The process of rewriting a radical expression as an equivalent expression in which the denominator no longer contains any radicals is called __________________.
9. The number $\sqrt{\frac{2}{7}}$ can be rewritten without a radical in the denominator by multiplying the numerator and denominator by ______.
10. The number $\frac{5}{\sqrt{12}}$ can be rewritten without a radical in the denominator by multiplying the numerator and denominator by ______, which is the smallest number that will produce a perfect square in the denominator.

Exercise Set 5.4

Practice Exercises

Evaluate each expression in Exercises 1–10.

1. $\sqrt{9}$
2. $\sqrt{16}$
3. $\sqrt{25}$
4. $\sqrt{49}$
5. $\sqrt{64}$
6. $\sqrt{100}$
7. $\sqrt{121}$
8. $\sqrt{144}$
9. $\sqrt{169}$
10. $\sqrt{225}$

In Exercises 11–16, use a calculator with a square root key to find a decimal approximation for each square root. Round the number displayed to the nearest **a.** *tenth,* **b.** *hundredth,* **c.** *thousandth.*

11. $\sqrt{173}$
12. $\sqrt{3176}$
13. $\sqrt{17{,}761}$
14. $\sqrt{779{,}264}$
15. $\sqrt{\pi}$
16. $\sqrt{2\pi}$

In Exercises 17–24, simplify the square root.

17. $\sqrt{20}$
18. $\sqrt{50}$
19. $\sqrt{80}$
20. $\sqrt{12}$
21. $\sqrt{250}$
22. $\sqrt{192}$
23. $7\sqrt{28}$
24. $3\sqrt{52}$

In Exercises 25–56, perform the indicated operation. Simplify the answer when possible.

25. $\sqrt{7} \cdot \sqrt{6}$
26. $\sqrt{19} \cdot \sqrt{3}$
27. $\sqrt{6} \cdot \sqrt{6}$
28. $\sqrt{5} \cdot \sqrt{5}$
29. $\sqrt{3} \cdot \sqrt{6}$
30. $\sqrt{12} \cdot \sqrt{2}$
31. $\sqrt{2} \cdot \sqrt{26}$
32. $\sqrt{5} \cdot \sqrt{50}$
33. $\frac{\sqrt{54}}{\sqrt{6}}$
34. $\frac{\sqrt{75}}{\sqrt{3}}$
35. $\frac{\sqrt{90}}{\sqrt{2}}$
36. $\frac{\sqrt{60}}{\sqrt{3}}$
37. $\frac{-\sqrt{96}}{\sqrt{2}}$
38. $\frac{-\sqrt{150}}{\sqrt{3}}$
39. $7\sqrt{3} + 6\sqrt{3}$
40. $8\sqrt{5} + 11\sqrt{5}$
41. $4\sqrt{13} - 6\sqrt{13}$
42. $6\sqrt{17} - 8\sqrt{17}$
43. $\sqrt{5} + \sqrt{5}$
44. $\sqrt{3} + \sqrt{3}$
45. $4\sqrt{2} - 5\sqrt{2} + 8\sqrt{2}$
46. $6\sqrt{3} + 8\sqrt{3} - 16\sqrt{3}$
47. $\sqrt{5} + \sqrt{20}$
48. $\sqrt{3} + \sqrt{27}$
49. $\sqrt{50} - \sqrt{18}$
50. $\sqrt{63} - \sqrt{28}$
51. $3\sqrt{18} + 5\sqrt{50}$
52. $4\sqrt{12} + 2\sqrt{75}$
53. $\frac{1}{4}\sqrt{12} - \frac{1}{2}\sqrt{48}$
54. $\frac{1}{5}\sqrt{300} - \frac{2}{3}\sqrt{27}$
55. $3\sqrt{75} + 2\sqrt{12} - 2\sqrt{48}$
56. $2\sqrt{72} + 3\sqrt{50} - \sqrt{128}$

In Exercises 57–66, rationalize the denominator.

57. $\frac{5}{\sqrt{3}}$ **58.** $\frac{12}{\sqrt{5}}$ **59.** $\frac{21}{\sqrt{7}}$

60. $\frac{30}{\sqrt{5}}$ **61.** $\frac{12}{\sqrt{30}}$ **62.** $\frac{15}{\sqrt{50}}$

63. $\frac{15}{\sqrt{12}}$ **64.** $\frac{13}{\sqrt{40}}$ **65.** $\sqrt{\frac{2}{5}}$

66. $\sqrt{\frac{5}{7}}$

Practice Plus

In Exercises 67–74, perform the indicated operations. Simplify the answer when possible.

67. $3\sqrt{8} - \sqrt{32} + 3\sqrt{72} - \sqrt{75}$

68. $3\sqrt{54} - 2\sqrt{24} - \sqrt{96} + 4\sqrt{63}$

69. $3\sqrt{7} - 5\sqrt{14} \cdot \sqrt{2}$

70. $4\sqrt{2} - 8\sqrt{10} \cdot \sqrt{5}$

71. $\frac{\sqrt{32}}{5} + \frac{\sqrt{18}}{7}$

72. $\frac{\sqrt{27}}{2} + \frac{\sqrt{75}}{7}$

73. $\frac{\sqrt{2}}{\sqrt{3}} + \frac{\sqrt{3}}{\sqrt{2}}$

74. $\frac{\sqrt{2}}{\sqrt{7}} + \frac{\sqrt{7}}{\sqrt{2}}$

Application Exercises

The formula

$$d = \sqrt{\frac{3h}{2}}$$

models the distance, d, in miles, that a person h feet high can see to the horizon. Use this formula to solve Exercises 75–76.

75. The pool deck on a cruise ship is 72 feet above the water. How far can passengers on the pool deck see? Write the answer in simplified radical form. Then use the simplified radical form and a calculator to express the answer to the nearest tenth of a mile.

76. The captain of a cruise ship is on the star deck, which is 120 feet above the water. How far can the captain see? Write the answer in simplified radical form. Then use the simplified radical form and a calculator to express the answer to the nearest tenth of a mile.

Police use the formula $v = 2\sqrt{5L}$ to estimate the speed of a car, v, in miles per hour, based on the length, L, in feet, of its skid marks upon sudden braking on a dry asphalt road. Use the formula to solve Exercises 77–78.

77. A motorist is involved in an accident. A police officer measures the car's skid marks to be 245 feet long. Estimate the speed at which the motorist was traveling before braking. If the posted speed limit is 50 miles per hour and the motorist tells the officer he was not speeding, should the officer believe him? Explain.

78. A motorist is involved in an accident. A police officer measures the car's skid marks to be 45 feet long. Estimate the speed at which the motorist was traveling before braking. If the posted speed limit is 35 miles per hour and the motorist tells the officer she was not speeding, should the officer believe her? Explain.

79. The graph shows the median heights for boys of various ages in the United States from birth through 60 months, or five years old.

Boys' Heights

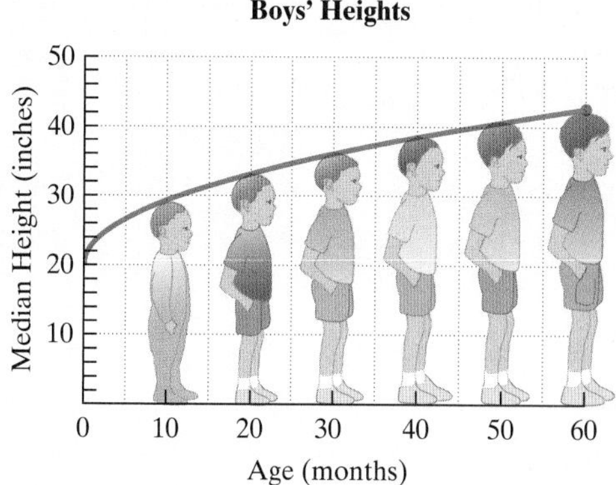

Source: Laura Walther Nathanson, *The Portable Pediatrician for Parents*

a. Use the graph to estimate the median height, to the nearest inch, of boys who are 50 months old.

b. The formula $h = 2.9\sqrt{x} + 20.1$ models the median height, h, in inches, of boys who are x months of age. According to the formula, what is the median height of boys who are 50 months old? Use a calculator and round to the nearest tenth of an inch. How well does your estimate from part (a) describe the median height obtained from the formula?

80. The graph shows the median heights for girls of various ages in the United States from birth through 60 months, or five years old.

Girls' Heights

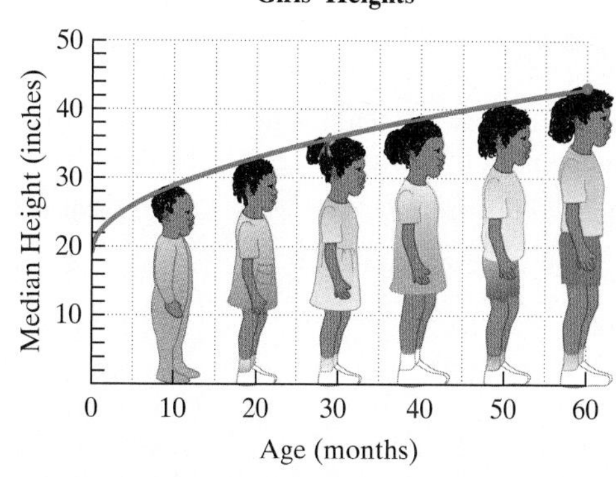

Source: Laura Walther Nathanson, *The Portable Pediatrician for Parents*

a. Use the graph to estimate the median height, to the nearest inch, of girls who are 50 months old.

b. The formula $h = 3.1\sqrt{x} + 19$ models the median height, h, in inches, of girls who are x months of age. According to the formula, what is the median height of girls who are 50 months old? Use a calculator and round to the nearest tenth of an inch. How well does your estimate from part (a) describe the median height obtained from the formula?

81. America is getting older. The graph shows the projected elderly U.S. population for ages 65–84 and for ages 85 and older.

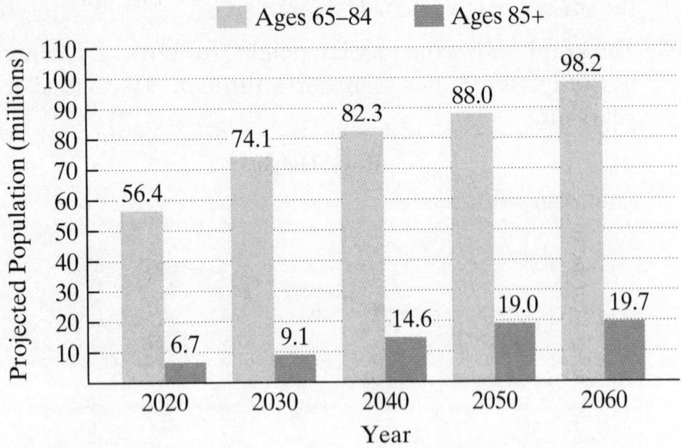

Source: U.S. Census Bureau

The formula $E = 5.8\sqrt{x} + 56.4$ models the projected number of elderly Americans ages 65–84, E, in millions, x years after 2020.

a. Use the formula to find the projected increase in the number of Americans ages 65–84, in millions, from 2030 to 2060. Express this difference in simplified radical form.

b. Use a calculator and write your answer in part (a) to the nearest tenth. Does this rounded decimal overestimate or underestimate the difference in the projected data shown by the bar graph? By how much?

82. Read the Blitzer Bonus on page 298.

The early Greeks believed that the most pleasing of all rectangles were **golden rectangles**, whose ratio of width to height is

$$\frac{w}{h} = \frac{2}{\sqrt{5} - 1}.$$

Use a calculator and find the ratio of width to height, correct to the nearest hundredth, in golden rectangles.

The Blitzer Bonus on page 299 gives Einstein's special-relativity equation

$$R_a = R_f\sqrt{1 - \left(\frac{v}{c}\right)^2}$$

for the aging rate of an astronaut, R_a, relative to the aging rate of a friend on Earth, R_f, where v is the astronaut's speed and c is the speed of light. Take a few minutes to read the essay and then solve Exercises 83–84.

83. You are moving at 80% of the speed of light. Substitute $0.8c$ in the equation shown above. What is your aging rate relative to a friend on Earth? If 100 weeks have passed for your friend, how long were you gone?

84. You are moving at 90% of the speed of light. Substitute $0.9c$ in the equation shown at the bottom of the previous column. What is your aging rate, correct to two decimal places, relative to a friend on Earth? If 100 weeks have passed for your friend, how long, to the nearest week, were you gone?

Explaining the Concepts

85. Describe the difference between a rational number and an irrational number.

86. Describe what is wrong with this statement: $\pi = \frac{22}{7}$.

87. Using $\sqrt{50}$, explain how to simplify a square root.

88. Describe how to multiply square roots.

89. Explain how to add square roots with the same radicand.

90. Explain how to add $\sqrt{3} + \sqrt{12}$.

91. Describe what it means to rationalize a denominator. Use $\frac{2}{\sqrt{5}}$ in your explanation.

92. Read the Blitzer Bonus on page 299. The future is now: You have the opportunity to explore the cosmos in a starship traveling near the speed of light. The experience will enable you to understand the mysteries of the universe in deeply personal ways, transporting you to unimagined levels of knowing and being. The down side: You return from your two-year journey to a futuristic world in which friends and loved ones are long gone. Do you explore space or stay here on Earth? What are the reasons for your choice?

Critical Thinking Exercises

Make Sense? *In Exercises 93–96, determine whether each statement makes sense or does not make sense, and explain your reasoning.*

93. The irrational number π is equal to $\frac{22}{7}$.

94. I rationalized a numerical denominator and the simplified denominator still contained an irrational number.

95. I simplified $\sqrt{20}$ and $\sqrt{75}$, and then I was able to perform the addition $2\sqrt{20} + 4\sqrt{75}$ by combining the sum into one square root.

96. Using my calculator, I determined that $6^7 = 279{,}936$, so 6 must be a seventh root of 279,936.

In Exercises 97–100, determine whether each statement is true or false. If the statement is false, make the necessary change(s) to produce a true statement.

97. The product of any two irrational numbers is always an irrational number.

98. $\sqrt{9} + \sqrt{16} = \sqrt{25}$

99. $\sqrt{\sqrt{16}} = 2$

100. $\frac{\sqrt{64}}{2} = \sqrt{32}$

In Exercises 101–103, insert either < or > in the shaded area between the numbers to make each statement true.

101. $\sqrt{2} \blacksquare 1.5$

102. $-\pi \blacksquare -3.5$

103. $-\frac{3.14}{2} \blacksquare -\frac{\pi}{2}$

104. How does doubling a number affect its square root?

105. Between which two consecutive integers is $-\sqrt{47}$?

106. Simplify: $\sqrt{2} + \sqrt{\frac{1}{2}}$.

107. Create a counterexample to show that the following statement is false: The difference between two irrational numbers is always an irrational number.

Group Exercises

The following topics related to irrational numbers are appropriate for either individual or group research projects. A report should be given to the class on the researched topic.

108. A History of How Irrational Numbers Developed

109. Pi: Its History, Applications, and Curiosities

110. Proving That $\sqrt{2}$ Is Irrational

111. Imaginary Numbers: Their History, Applications, and Curiosities

112. The Golden Rectangle in Art and Architecture

5.5 Real Numbers and Their Properties; Clock Addition

WHAT AM I SUPPOSED TO LEARN?

After studying this section, you should be able to:

1. Recognize subsets of the real numbers.
2. Recognize properties of real numbers.
3. Apply properties of real numbers to clock addition.

The Set of Real Numbers

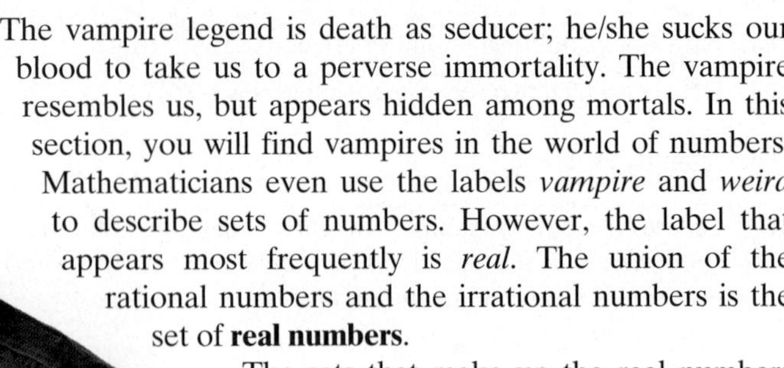

The vampire legend is death as seducer; he/she sucks our blood to take us to a perverse immortality. The vampire resembles us, but appears hidden among mortals. In this section, you will find vampires in the world of numbers. Mathematicians even use the labels *vampire* and *weird* to describe sets of numbers. However, the label that appears most frequently is *real*. The union of the rational numbers and the irrational numbers is the set of **real numbers**.

1 *Recognize subsets of the real numbers.*

The sets that make up the real numbers are summarized in **Table 5.2**. We refer to these sets as **subsets** of the real numbers, meaning that all elements in each subset are also elements in the set of real numbers.

Real numbers

Rational numbers | Irrational numbers

Integers

Whole numbers

Natural numbers

This diagram shows that every real number is rational or irrational.

TABLE 5.2 Important Subsets of the Real Numbers

Name	Description	Examples
Natural numbers	$\{1, 2, 3, 4, 5, \ldots\}$ These are the numbers that we use for counting.	2, 3, 5, 17
Whole numbers	$\{0, 1, 2, 3, 4, 5, \ldots\}$ The set of whole numbers includes 0 and the natural numbers.	0, 2, 3, 5, 17
Integers	$\{\ldots, -5, -4, -3, -2, -1, 0, 1, 2, 3, 4, 5, \ldots\}$ The set of integers includes the whole numbers and the negatives of the natural numbers.	$-17, -5, -3, -2, 0, 2, 3, 5, 17$
Rational numbers	$\left\{\frac{a}{b} \middle\vert a \text{ and } b \text{ are integers and } b \neq 0\right\}$ The set of rational numbers is the set of all numbers that can be expressed as a quotient of two integers, with the denominator not 0. Rational numbers can be expressed as terminating or repeating decimals.	$-17 = \frac{-17}{1}, -5 = \frac{-5}{1}, -3,$ $-2, 0, 2, 3, 5, 17,$ $\frac{2}{5} = 0.4,$ $\frac{-2}{3} = -0.6666\ldots = -0.\overline{6}$
Irrational numbers	The set of irrational numbers is the set of all numbers whose decimal representations are neither terminating nor repeating. Irrational numbers cannot be expressed as a quotient of integers.	$\sqrt{2} \approx 1.414214$ $-\sqrt{3} \approx -1.73205$ $\pi \approx 3.142$ $-\frac{\pi}{2} \approx -1.571$

EXAMPLE 1 *Classifying Real Numbers*

Consider the following set of numbers:

$$\left\{-7, -\frac{3}{4}, 0, 0.\overline{6}, \sqrt{5}, \pi, 7.3, \sqrt{81}\right\}.$$

List the numbers in the set that are

a. natural numbers. **b.** whole numbers. **c.** integers.
d. rational numbers. **e.** irrational numbers. **f.** real numbers.

Weird Numbers

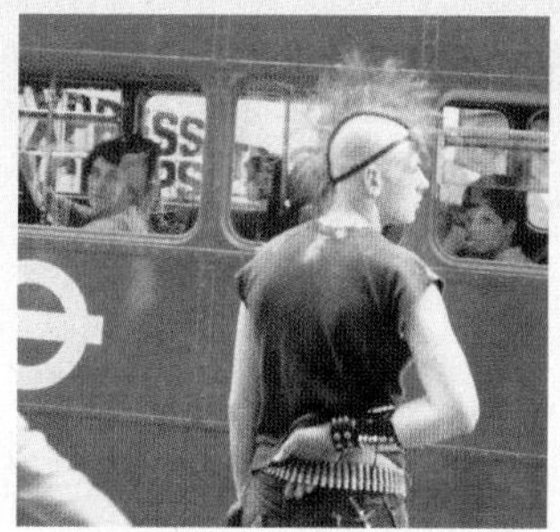

Mathematicians use the label **weird** to describe a number if

1. The sum of its factors, excluding the number itself, is greater than the number.
2. No partial collection of the factors adds up to the number.

The number 70 is weird. Its factors other than itself are 1, 2, 5, 7, 10, 14, and 35. The sum of these factors is 74, which is greater than 70. Two or more numbers in the list of factors cannot be added to obtain 70.

Weird numbers are rare. Below 10,000, the weird numbers are 70, 836, 4030, 5830, 7192, 7912, and 9272. It is not known whether an odd weird number exists.

SOLUTION

a. Natural numbers: The natural numbers are the numbers used for counting. The only natural number in the set is $\sqrt{81}$ because $\sqrt{81} = 9$. (9 multiplied by itself, or 9^2, is 81.)

b. Whole numbers: The whole numbers consist of the natural numbers and 0. The elements of the set that are whole numbers are 0 and $\sqrt{81}$.

c. Integers: The integers consist of the natural numbers, 0, and the negatives of the natural numbers. The elements of the set that are integers are $\sqrt{81}$, 0, and -7.

d. Rational numbers: All numbers in the set that can be expressed as the quotient of integers are rational numbers. These include $-7\left(-7 = \frac{-7}{1}\right), -\frac{3}{4}, 0\left(0 = \frac{0}{1}\right)$, and $\sqrt{81}\left(\sqrt{81} = \frac{9}{1}\right)$. Furthermore, all numbers in the set that are terminating or repeating decimals are also rational numbers. These include $0.\overline{6}$ and 7.3.

e. Irrational numbers: The irrational numbers in the set are $\sqrt{5}(\sqrt{5} \approx 2.236)$ and $\pi(\pi \approx 3.14)$. Both $\sqrt{5}$ and π are only approximately equal to 2.236 and 3.14, respectively. In decimal form, $\sqrt{5}$ and π neither terminate nor have blocks of repeating digits.

f. Real numbers: All the numbers in the given set are real numbers.

CHECK POINT 1 Consider the following set of numbers:

$$\left\{-9, -1.3, 0, 0.\overline{3}, \frac{\pi}{2}, \sqrt{9}, \sqrt{10}\right\}.$$

List the numbers in the set that are

a. natural numbers.
b. whole numbers.
c. integers.
d. rational numbers.
e. irrational numbers.
f. real numbers.

Blitzer Bonus

Vampire Numbers

Like legendary vampires that lie concealed among humans, vampire numbers lie hidden within the set of real numbers, mostly undetected. By definition, vampire numbers have an even number of digits. Furthermore, they are the product of two numbers whose digits all survive, in scrambled form, in the vampire. For example, 1260, 1435, and 2187 are vampire numbers.

$21 \times 60 = 1260$ — The digits 2, 1, 6, and 0 lie scrambled in the vampire number.

$35 \times 41 = 1435$ — The digits 3, 5, 4, and 1 lurk within the vampire number.

$27 \times 81 = 2187$ — The digits 2, 7, 8, and 1 survive in the vampire number.

As the real numbers grow increasingly larger, is it necessary to pull out a wooden stake with greater frequency? How often can you expect to find vampires hidden among the giants? And is it possible to find a weird vampire?

On the right of the equal sign is a 40-digit vampire number that was discovered using a Pascal program on a personal computer:

98,765,432,198,765,432,198 × 98,765,432,198,830,604,534 = 9,754,610,597,415,368,368,844,499,268,390,128,385,732.

Source: Clifford Pickover, *Wonders of Numbers*, Oxford University Press, 2001.

2 *Recognize properties of real numbers.*

Properties of the Real Numbers

When you use your calculator to add two real numbers, you can enter them in either order. The fact that two real numbers can be added in either order is called the **commutative property of addition**. You probably use this property, as well as other properties of the real numbers listed in **Table 5.3**, without giving it much thought. The properties of the real numbers are especially useful in algebra, as we shall see in Chapter 6.

Blitzer Bonus

The Associative Property and the English Language

In the English language, phrases can take on different meanings depending on the way the words are associated with commas.

Here are three examples.

- Woman, without her man, is nothing.

  Woman, without her, man is nothing.
- In the parade will be several hundred students carrying flags, and many teachers.

  In the parade will be several hundred students, carrying flags and many teachers.
- Population of Amsterdam broken down by age and sex

  Population of Amsterdam, broken down by age and sex

TABLE 5.3 Properties of the Real Numbers

Name	Meaning	Examples
Closure Property of Addition	The sum of any two real numbers is a real number.	$4\sqrt{2}$ is a real number and $5\sqrt{2}$ is a real number, so $4\sqrt{2}+5\sqrt{2}$, or $9\sqrt{2}$, is a real number.
Closure Property of Multiplication	The product of any two real numbers is a real number.	10 is a real number and $\frac{1}{2}$ is a real number, so $10\cdot\frac{1}{2}$, or 5, is a real number.
Commutative Property of Addition	Changing order when adding does not affect the sum. $a+b=b+a$	• $13+7=7+13$ • $\sqrt{2}+\sqrt{5}=\sqrt{5}+\sqrt{2}$
Commutative Property of Multiplication	Changing order when multiplying does not affect the product. $ab=ba$	• $13\cdot 7=7\cdot 13$ • $\sqrt{2}\cdot\sqrt{5}=\sqrt{5}\cdot\sqrt{2}$
Associative Property of Addition	Changing grouping when adding does not affect the sum. $(a+b)+c=a+(b+c)$	$(7+2)+5=7+(2+5)$ $9+5=7+7$ $14=14$
Associative Property of Multiplication	Changing grouping when multiplying does not affect the product. $(ab)c=a(bc)$	$(7\cdot 2)\cdot 5=7\cdot(2\cdot 5)$ $14\cdot 5=7\cdot 10$ $70=70$
Distributive Property of Multiplication over Addition	Multiplication distributes over addition. $a\cdot(b+c)=a\cdot b+a\cdot c$	$7(4+\sqrt{3})=7\cdot 4+7\cdot\sqrt{3}$ $=28+7\sqrt{3}$
Identity Property of Addition	Zero can be deleted from a sum. $a+0=a$ $0+a=a$	• $\sqrt{3}+0=\sqrt{3}$ • $0+\pi=\pi$
Identity Property of Multiplication	One can be deleted from a product. $a\cdot 1=a$ $1\cdot a=a$	• $\sqrt{3}\cdot 1=\sqrt{3}$ • $1\cdot\pi=\pi$
Inverse Property of Addition	The sum of a real number and its additive inverse gives 0, the additive identity. $a+(-a)=0$ $(-a)+a=0$	• $\sqrt{3}+(-\sqrt{3})=0$ • $-\pi+\pi=0$
Inverse Property of Multiplication	The product of a nonzero real number and its multiplicative inverse gives 1, the multiplicative identity. $a\cdot\frac{1}{a}=1, a\neq 0$ $\frac{1}{a}\cdot a=1, a\neq 0$	• $\sqrt{3}\cdot\frac{1}{\sqrt{3}}=1$ • $\frac{1}{\pi}\cdot\pi=1$

GREAT QUESTION!

Is there an easy way to distinguish between the commutative and associative properties?

Commutative: Changes *order.*
Associative: Changes *grouping.*

EXAMPLE 2 *Identifying Properties of the Real Numbers*

Name the property illustrated:

a. $\sqrt{3} \cdot 7 = 7 \cdot \sqrt{3}$

b. $(4 + 7) + 6 = 4 + (7 + 6)$

c. $2(3 + \sqrt{5}) = 6 + 2\sqrt{5}$

d. $\sqrt{2} + (\sqrt{3} + \sqrt{7}) = \sqrt{2} + (\sqrt{7} + \sqrt{3})$

e. $17 + (-17) = 0$

f. $\sqrt{2} \cdot 1 = \sqrt{2}.$

SOLUTION

a. $\sqrt{3} \cdot 7 = 7 \cdot \sqrt{3}$ Commutative property of multiplication

b. $(4 + 7) + 6 = 4 + (7 + 6)$ Associative property of addition

c. $2(3 + \sqrt{5}) = 6 + 2\sqrt{5}$ Distributive property of multiplication over addition

d. $\sqrt{2} + (\sqrt{3} + \sqrt{7}) = \sqrt{2} + (\sqrt{7} + \sqrt{3})$ The only change between the left and the right sides is in the order that $\sqrt{3}$ and $\sqrt{7}$ are added. The order is changed from $\sqrt{3} + \sqrt{7}$ to $\sqrt{7} + \sqrt{3}$ using the commutative property of addition.

e. $17 + (-17) = 0$ Inverse property of addition

f. $\sqrt{2} \cdot 1 = \sqrt{2}$ Identity property of multiplication

CHECK POINT 2 Name the property illustrated:

a. $(4 \cdot 7) \cdot 3 = 4 \cdot (7 \cdot 3)$

b. $3(\sqrt{5} + 4) = 3(4 + \sqrt{5})$

c. $3(\sqrt{5} + 4) = 3\sqrt{5} + 12$

d. $2(\sqrt{3} + \sqrt{7}) = (\sqrt{3} + \sqrt{7})2$

e. $1 + 0 = 1$

f. $-4\left(-\frac{1}{4}\right) = 1.$

Although the entire set of real numbers is closed with respect to addition and multiplication, some of the subsets of the real numbers do not satisfy the closure property for a given operation. If an operation on a set results in just one number that is not in that set, then the set is not closed for that operation.

EXAMPLE 3 *Verifying Closure*

a. Are the integers closed with respect to multiplication?

b. Are the irrational numbers closed with respect to multiplication?

c. Are the natural numbers closed with respect to division?

SOLUTION

a. Consider some examples of the multiplication of integers:

$$3 \cdot 2 = 6 \qquad 3(-2) = -6 \qquad -3(-2) = 6 \qquad -3 \cdot 0 = 0.$$

The product of any two integers is always a positive integer, a negative integer, or zero, which is an integer. Thus, the integers are closed under the operation of multiplication.

b. If we multiply two irrational numbers, must the product always be an irrational number? The answer is no. Here is an example:

$$\sqrt{7} \cdot \sqrt{7} = \sqrt{49} = 7.$$

Both irrational — Not an irrational number

This means that the irrational numbers are not closed under the operation of multiplication.

c. If we divide any two natural numbers, must the quotient always be a natural number? The answer is no. Here is an example:

$$4 \div 8 = \frac{1}{2}.$$

Both natural numbers — Not a natural number

Thus, the natural numbers are not closed under the operation of division.

CHECK POINT 3

a. Are the natural numbers closed with respect to multiplication?

b. Are the integers closed with respect to division?

The commutative property involves a change in order with no change in the final result. However, changing the order in which we subtract and divide real numbers can produce different answers. For example,

$$7 - 4 \neq 4 - 7 \quad \text{and} \quad 6 \div 2 \neq 2 \div 6.$$

Because the real numbers are not commutative with respect to subtraction and division, it is important that you enter numbers in the correct order when using a calculator to perform these operations.

The associative property does not hold for the operations of subtraction and division. The examples below show that if we change groupings when subtracting or dividing three numbers, the answer may change.

$$\begin{aligned}(6-1)-3 &\neq 6-(1-3)\\ 5-3 &\neq 6-(-2)\\ 2 &\neq 8\end{aligned} \qquad \begin{aligned}(8 \div 4) \div 2 &\neq 8 \div (4 \div 2)\\ 2 \div 2 &\neq 8 \div 2\\ 1 &\neq 4\end{aligned}$$

Blitzer Bonus

Beyond the Real Numbers

Only real numbers greater than or equal to zero have real number square roots. The square root of -1, $\sqrt{-1}$, is not a real number. This is because there is no real number that can be multiplied by itself that results in -1. Multiplying any real number by itself can never give a negative product. In the sixteenth century, mathematician Girolamo Cardano (1501–1576) wrote that square roots of negative numbers would cause "mental tortures." In spite of these "tortures," mathematicians invented a new number, called i, to represent $\sqrt{-1}$. The number i is not a real number; it is called an **imaginary number**. Thus, $\sqrt{9} = 3$, $-\sqrt{9} = -3$, but $\sqrt{-9}$ is not a real number. However, $\sqrt{-9}$ is an imaginary number, represented by $3i$. The adjective *real* as a way of describing what we now call the real numbers was first used by the French mathematician and philosopher René Descartes (1596–1650) in response to the concept of imaginary numbers.

3 *Apply properties of real numbers to clock addition.*

Properties of the Real Numbers and Clock Arithmetic

Mathematics is about the patterns that arise in the world about us. Mathematicians look at a snowflake in terms of its underlying structure. Notice that if the snowflake in **Figure 5.8** is rotated by any multiple of 60° ($\frac{1}{6}$ of a rotation), it will always look about the same. A **symmetry** of an object is a motion that moves the object back onto itself. In a symmetry, you cannot tell, at the end of the motion, that the object has been moved.

The snowflake in **Figure 5.8** has **sixfold rotational symmetry**. After six 60° turns, the snowflake is back to its original position. If it takes m equal turns to restore an object to its original position and each of these turns is a figure that is identical to the original figure, the object has ***m*-fold rotational symmetry**.

FIGURE 5.8

The sixfold rotational symmetry of the snowflake in **Figure 5.8** can be studied using the set $\{0, 1, 2, 3, 4, 5\}$ and an operation called *clock addition*. The "6-hour clock" in **Figure 5.9** exhibits the sixfold rotational symmetry of the snowflake.

FIGURE 5.9 Sixfold rotational symmetry in the face of a clock and a flower

Using **Figure 5.9**, we can define **clock addition** as follows: Add by moving the hour hand in a clockwise direction. The symbol $\oplus$ is used to designated clock addition. **Figure 5.10** illustrates that

$$2 \oplus 3 = 5, \quad 4 \oplus 5 = 3, \quad \text{and} \quad 3 \oplus 4 = 1.$$

FIGURE 5.10 Addition in a 6-hour clock system

TABLE 5.4 6-Hour Clock Addition

$\oplus$	**0**	**1**	**2**	**3**	**4**	**5**
0	0	1	2	3	4	5
1	1	2	3	4	5	0
2	2	3	4	5	0	1
3	3	4	5	0	1	2
4	4	5	0	1	2	3
5	5	0	1	2	3	4

Table 5.4 is the addition table for clock addition in a 6-hour clock system.

TABLE 5.4 (repeated)

⊕	0	1	2	3	4	5
0	0	1	2	3	4	5
1	1	2	3	4	5	0
2	2	3	4	5	0	1
3	3	4	5	0	1	2
4	4	5	0	1	2	3
5	5	0	1	2	3	4

EXAMPLE 4 *Properties of the Real Numbers Applied to the 6-Hour Clock System*

Table 5.4, the table for clock addition in a 6-hour clock system, is repeated in the margin.

a. How can you tell that the set $\{0, 1, 2, 3, 4, 5\}$ is closed under the operation of clock addition?

b. Verify one case of the associative property:

$$(2 \oplus 3) \oplus 4 = 2 \oplus (3 \oplus 4).$$

c. What is the identity element in the 6-hour clock system?

d. Find the inverse of each element in the 6-hour clock system.

e. Verify two cases of the commutative property:

$$4 \oplus 3 = 3 \oplus 4 \quad \text{and} \quad 5 \oplus 4 = 4 \oplus 5.$$

SOLUTION

a. The Closure Property. The set $\{0, 1, 2, 3, 4, 5\}$ is closed under the operation of clock addition because the entries in the body of **Table 5.4** are all elements of the set.

b. The Associative Property. We were asked to verify one case of the associative property.

Locate 2 on the left and 3 on the top of **Table 5.4.** Intersecting lines show $2 \oplus 3 = 5$.

$$(2 \oplus 3) \oplus 4 = 2 \oplus (3 \oplus 4)$$
$$5 \oplus 4 = 2 \oplus 1$$
$$3 = 3$$

Locate 3 on the left and 4 on the top of **Table 5.4.** Intersecting lines show $3 \oplus 4 = 1$.

c. The Identity Property. Look for the element in **Table 5.4** that does not change anything when used in clock addition. **Table 5.4** shows that the column under 0 is the same as the column with boldface numbers on the left. Thus, $0 \oplus 0 = 0$, $1 \oplus 0 = 1$, $2 \oplus 0 = 2$, $3 \oplus 0 = 3$, $4 \oplus 0 = 4$, and $5 \oplus 0 = 5$. The table also shows that the row next to 0 is the same as the row with boldface numbers on top. Thus, $0 \oplus 0 = 0$, $0 \oplus 1 = 1$, $0 \oplus 2 = 2$, up through $0 \oplus 5 = 5$. Each element of the set does not change when we perform clock addition with 0. Thus, 0 is the identity element. The identity property is satisfied because 0 is contained in the given set.

d. The Inverse Property. When an element is added to its inverse, the result is the identity element. Because the identity element is 0, we can find the inverse of each element in $\{0, 1, 2, 3, 4, 5\}$ by answering the question: What must be added to each element to obtain 0?

$$\text{element} + ? = 0$$

Figure 5.11 illustrates how we answer the question. If each element in the set has an inverse, then a 0 will appear in every row and column of the table. This is, indeed, the case. Use the 0 in each row. Because each element in $\{0, 1, 2, 3, 4, 5\}$ has an inverse within the set, the inverse property is satisfied.

⊕	0	1	2	3	4	5	
0	0	1	2	3	4	5	$0 \oplus 0 = 0$: The inverse of 0 is 0.
1	1	2	3	4	5	0	$1 \oplus 5 = 0$: The inverse of 1 is 5.
2	2	3	4	5	0	1	$2 \oplus 4 = 0$: The inverse of 2 is 4.
3	3	4	5	0	1	2	$3 \oplus 3 = 0$: The inverse of 3 is 3.
4	4	5	0	1	2	3	$4 \oplus 2 = 0$: The inverse of 4 is 2.
5	5	0	1	2	3	4	$5 \oplus 1 = 0$: The inverse of 5 is 1.

FIGURE 5.11

e. The Commutative Property. We were asked to verify two cases of the commutative property.

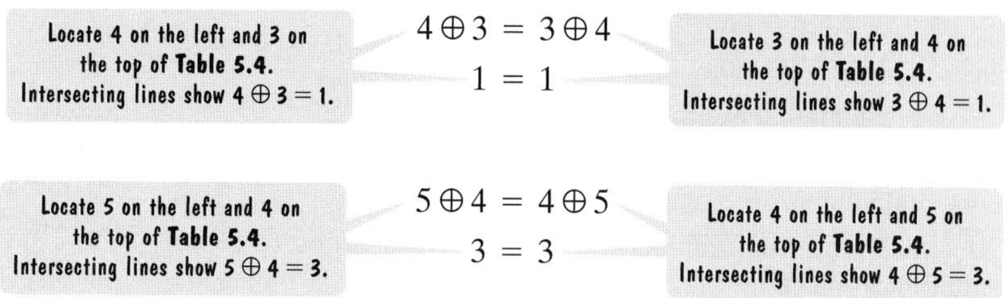

Figure 5.12 illustrates four types of rotational symmetry.

Fourfold rotational symmetry

Fivefold rotational symmetry

Eightfold rotational symmetry

18–fold rotational symmetry

FIGURE 5.12 Types of rotational symmetry

The fourfold rotational symmetry shown on the left in **Figure 5.12** can be explored using the 4-hour clock in **Figure 5.13** and **Table 5.5**, the table for clock addition in the 4-hour clock system.

FIGURE 5.13 A 4-hour clock

TABLE 5.5 4-Hour Clock Addition

$\oplus$	**0**	**1**	**2**	**3**
0	0	1	2	3
1	1	2	3	0
2	2	3	0	1
3	3	0	1	2

CHECK POINT 4 Use **Table 5.5** which shows clock addition in the 4-hour clock system to solve this exercise.

a. How can you tell that the set $\{0, 1, 2, 3\}$ is closed under the operation of clock addition?

b. Verify one case of the associative property: $(2 \oplus 2) \oplus 3 = 2 \oplus (2 \oplus 3)$.

c. What is the identity element in the 4-hour clock system?

d. Find the inverse of each element in the 4-hour clock system.

e. Verify two cases of the commutative property: $1 \oplus 3 = 3 \oplus 1$ and $3 \oplus 2 = 2 \oplus 3$.

Concept and Vocabulary Check

Fill in each blank so that the resulting statement is true.

1. Every real number is either ________ or ________.
2. The ________ property of addition states that the sum of any two real numbers is a real number.
3. If a and b are real numbers, the commutative property of multiplication states that ________.
4. If a, b, and c are real numbers, the associative property of addition states that ________________.
5. If a, b, and c are real numbers, the distributive property states that ________________.
6. The ________ property of addition states that zero can be deleted from a sum.
7. The ________ property of multiplication states that ___ can be deleted from a product.
8. The product of a nonzero real number and its ________ ________ gives 1, the ________.
9. Shown in the figure is a 5-hour clock. Clock addition is performed by moving the hour hand in a clockwise direction. Thus,

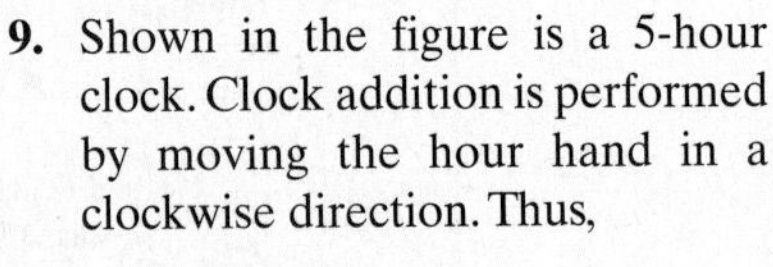

$1 \oplus 4 =$ ___

$3 \oplus 3 =$ ___

and $4 \oplus 2 =$ ___.

10. True or False: The 5-hour clock in Exercise 9 could be used to describe the rotational symmetry of this flower. ______

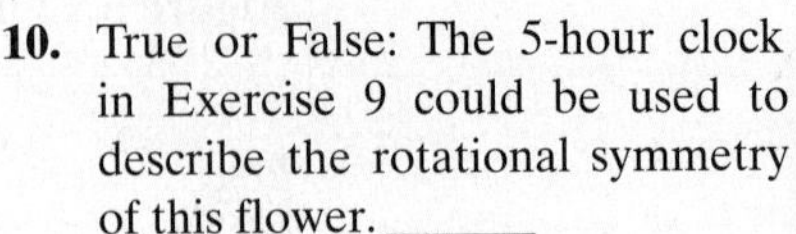

Exercise Set 5.5

Practice Exercises

In Exercises 1–4, list all numbers from the given set that are

a. *natural numbers.* **b.** *whole numbers.*
c. *integers.* **d.** *rational numbers.*
e. *irrational numbers.* **f.** *real numbers.*

1. $\{-9, -\frac{4}{5}, 0, 0.25, \sqrt{3}, 9.2, \sqrt{100}\}$
2. $\{-7, -0.\overline{6}, 0, \sqrt{49}, \sqrt{50}\}$
3. $\{-11, -\frac{5}{6}, 0, 0.75, \sqrt{5}, \pi, \sqrt{64}\}$
4. $\{-5, -0.\overline{3}, 0, \sqrt{2}, \sqrt{4}\}$
5. Give an example of a whole number that is not a natural number.
6. Give an example of an integer that is not a whole number.
7. Give an example of a rational number that is not an integer.
8. Give an example of a rational number that is not a natural number.
9. Give an example of a number that is an integer, a whole number, and a natural number.
10. Give an example of a number that is a rational number, an integer, and a real number.
11. Give an example of a number that is an irrational number and a real number.
12. Give an example of a number that is a real number, but not an irrational number.

Complete each statement in Exercises 13–15 to illustrate the commutative property.

13. $3 + (4 + 5) = 3 + (5 +$ ___$)$
14. $\sqrt{5} \cdot 4 = 4 \cdot$ ___
15. $9 \cdot (6 + 2) = 9 \cdot (2 +$ ___$)$

Complete each statement in Exercises 16–17 to illustrate the associative property.

16. $(3 + 7) + 9 =$ ___ $+ (7 +$ ___$)$
17. $(4 \cdot 5) \cdot 3 =$ ___ $\cdot (5 \cdot$ ___$)$

Complete each statement in Exercises 18–20 to illustrate the distributive property.

18. $3 \cdot (6 + 4) = 3 \cdot 6 + 3 \cdot$ ___
19. ___ $\cdot (4 + 5) = 7 \cdot 4 + 7 \cdot 5$
20. $2 \cdot ($___ $+ 3) = 2 \cdot 7 + 2 \cdot 3$

Use the distributive property to simplify the radical expressions in Exercises 21–28.

21. $5(6 + \sqrt{2})$
22. $4(3 + \sqrt{5})$
23. $\sqrt{7}(3 + \sqrt{2})$
24. $\sqrt{6}(7 + \sqrt{5})$
25. $\sqrt{3}(5 + \sqrt{3})$
26. $\sqrt{7}(9 + \sqrt{7})$
27. $\sqrt{6}(\sqrt{2} + \sqrt{6})$
28. $\sqrt{10}(\sqrt{2} + \sqrt{10})$

In Exercises 29–44, state the name of the property illustrated.

29. $6 + (-4) = (-4) + 6$
30. $11 \cdot (7 + 4) = 11 \cdot 7 + 11 \cdot 4$
31. $6 + (2 + 7) = (6 + 2) + 7$
32. $6 \cdot (2 \cdot 3) = 6 \cdot (3 \cdot 2)$
33. $(2 + 3) + (4 + 5) = (4 + 5) + (2 + 3)$
34. $7 \cdot (11 \cdot 8) = (11 \cdot 8) \cdot 7$
35. $2(-8 + 6) = -16 + 12$

36. $-8(3 + 11) = -24 + (-88)$

37. $(2\sqrt{3}) \cdot \sqrt{5} = 2(\sqrt{3} \cdot \sqrt{5})$

38. $\sqrt{2\pi} = \pi\sqrt{2}$

39. $\sqrt{17} \cdot 1 = \sqrt{17}$

40. $\sqrt{17} + 0 = \sqrt{17}$

41. $\sqrt{17} + (-\sqrt{17}) = 0$

42. $\sqrt{17} \cdot \dfrac{1}{\sqrt{17}} = 1$

43. $\dfrac{1}{\sqrt{2} + \sqrt{7}}(\sqrt{2} + \sqrt{7}) = 1$

44. $(\sqrt{2} + \sqrt{7}) + -(\sqrt{2} + \sqrt{7}) = 0$

In Exercises 45–49, use two numbers to show that

45. the natural numbers are not closed with respect to subtraction.

46. the natural numbers are not closed with respect to division.

47. the integers are not closed with respect to division.

48. the irrational numbers are not closed with respect to subtraction.

49. the irrational numbers are not closed with respect to multiplication.

50. Shown in the figure is a 7-hour clock and the table for clock addition in the 7-hour clock system.

⊕	0	1	2	3	4	5	6
0	0	1	2	3	4	5	6
1	1	2	3	4	5	6	0
2	2	3	4	5	6	0	1
3	3	4	5	6	0	1	2
4	4	5	6	0	1	2	3
5	5	6	0	1	2	3	4
6	6	0	1	2	3	4	5

a. How can you tell that the set $\{0, 1, 2, 3, 4, 5, 6\}$ is closed under the operation of clock addition?

b. Verify one case of the associative property: $(3 \oplus 5) \oplus 6 = 3 \oplus (5 \oplus 6)$.

c. What is the identity element in the 7-hour clock system?

d. Find the inverse of each element in the 7-hour clock system.

e. Verify two cases of the commutative property: $4 \oplus 5 = 5 \oplus 4$ and $6 \oplus 1 = 1 \oplus 6$.

51. Shown in the figure is an 8-hour clock and the table for clock addition in the 8-hour clock system.

⊕	0	1	2	3	4	5	6	7
0	0	1	2	3	4	5	6	7
1	1	2	3	4	5	6	7	0
2	2	3	4	5	6	7	0	1
3	3	4	5	6	7	0	1	2
4	4	5	6	7	0	1	2	3
5	5	6	7	0	1	2	3	4
6	6	7	0	1	2	3	4	5
7	7	0	1	2	3	4	5	6

a. How can you tell that the set $\{0, 1, 2, 3, 4, 5, 6, 7\}$ is closed under the operation of clock addition?

b. Verify one case of the associative property: $(4 \oplus 6) \oplus 7 = 4 \oplus (6 \oplus 7)$.

c. What is the identity element in the 8-hour clock system?

d. Find the inverse of each element in the 8-hour clock system.

e. Verify two cases of the commutative property: $5 \oplus 6 = 6 \oplus 5$ and $4 \oplus 7 = 7 \oplus 4$.

Practice Plus

In Exercises 52–55, determine whether each statement is true or false. Do not use a calculator.

52. $468(787 + 289) = 787 + 289(468)$

53. $468(787 + 289) = 787(468) + 289(468)$

54. $58 \cdot 9 + 32 \cdot 9 = (58 + 32) \cdot 9$

55. $58 \cdot 9 \cdot 32 \cdot 9 = (58 \cdot 32) \cdot 9$

In Exercises 56–57, name the property used to go from step to step each time that (why?) occurs.

56. $7 + 2(x + 9)$
$= 7 + (2x + 18)$ (why?)
$= 7 + (18 + 2x)$ (why?)
$= (7 + 18) + 2x$ (why?)
$= 25 + 2x$
$= 2x + 25$ (why?)

57. $5(x + 4) + 3x$
$= (5x + 20) + 3x$ (why?)
$= (20 + 5x) + 3x$ (why?)
$= 20 + (5x + 3x)$ (why?)
$= 20 + (5 + 3)x$ (why?)
$= 20 + 8x$
$= 8x + 20$ (why?)

The tables show the operations □ and △ on the set {a, b, c, d, e}. Use these tables to solve Exercises 58–65.

□	*a*	*b*	*c*	*d*	*e*
a	a	b	c	d	e
b	b	c	d	e	a
c	c	d	e	a	b
d	d	e	a	b	c
e	e	a	b	c	d

△	*a*	*b*	*c*	*d*	*e*
a	a	a	a	a	a
b	a	b	c	d	e
c	a	c	e	b	d
d	a	d	b	e	c
e	a	e	d	c	b

58. **a.** Show that $e \triangle (c \square d) = (e \triangle c) \square (e \triangle d)$.

b. What property of the real numbers is illustrated in part (a)?

59. **a.** Show that $c \triangle (d \square e) = (c \triangle d) \square (c \triangle e)$.

b. What property of the real numbers is illustrated in part (a)?

60. Find $c \triangle [c \square (c \triangle c)]$.

61. Find $d \triangle [d \square (d \triangle d)]$.

In Exercises 62–65, replace x with a, b, c, d, or e to form a true statement.

62. $x \square d = e$

63. $x \square d = a$

64. $x \triangle (e \square c) = d$

65. $x \triangle (e \square d) = b$

66. If $\begin{bmatrix} a & b \\ c & d \end{bmatrix} \times \begin{bmatrix} e & f \\ g & h \end{bmatrix} = \begin{bmatrix} ae + bg & af + bh \\ ce + dg & cf + dh \end{bmatrix}$, find

a. $\begin{bmatrix} 2 & 3 \\ 4 & 7 \end{bmatrix} \times \begin{bmatrix} 0 & 1 \\ 5 & 6 \end{bmatrix}$

b. $\begin{bmatrix} 0 & 1 \\ 5 & 6 \end{bmatrix} \times \begin{bmatrix} 2 & 3 \\ 4 & 7 \end{bmatrix}$

c. Draw a conclusion about one of the properties discussed in this section in terms of these arrays of numbers under multiplication.

Application Exercises

In Exercises 67–70, use the definition of vampire numbers from the Blitzer Bonus on page 305 to determine which products are vampires.

67. $15 \times 93 = 1395$

68. $80 \times 86 = 6880$

69. $20 \times 51 = 1020$

70. $146 \times 938 = 136{,}948$

A **narcissistic number** *is an n-digit number equal to the sum of each of its digits raised to the nth power. Here's an example:*

$$153 = 1^3 + 5^3 + 3^3.$$

Three digits, so exponents are 3

In Exercises 71–74, determine which real numbers are narcissistic.

71. 370

72. 371

73. 372

74. 9474

75. The algebraic expressions

$$\frac{D(A + 1)}{24} \quad \text{and} \quad \frac{DA + D}{24}$$

describe the drug dosage for children between the ages of 2 and 13. In each algebraic expression, D stands for an adult dose and A represents the child's age.

a. Name the property that explains why these expressions are equal for all values of D and A.

b. If an adult dose of ibuprofen is 200 milligrams, what is the proper dose for a 12-year-old child? Use both forms of the algebraic expressions to answer the question. Which form is easier to use?

76. Closure illustrates that a characteristic of a set is not necessarily a characteristic of all of its subsets. The real numbers are closed with respect to multiplication, but the irrational numbers, a subset of the real numbers, are not. Give an example of a set that is not mathematical that has a particular characteristic, but which has a subset without this characteristic.

Name the kind of rotational symmetry shown in Exercises 77–78.

77.

Native American design

78.

Mercedes Benz symbol

Explaining the Concepts

79. What does it mean when we say that the rational numbers are a subset of the real numbers?

80. What does it mean if we say that a set is closed under a given operation?

81. State the commutative property of addition and give an example.

82. State the commutative property of multiplication and give an example.

83. State the associative property of addition and give an example.

84. State the associative property of multiplication and give an example.

85. State the distributive property of multiplication over addition and give an example.

86. Does $7 \cdot (4 \cdot 3) = 7 \cdot (3 \cdot 4)$ illustrate the commutative property or the associative property? Explain your answer.

87. Explain how to use the 8-hour clock shown in Exercise 51 to find $6 \oplus 5$.

Critical Thinking Exercises

Make Sense? *In Exercises 88–91, determine whether each statement makes sense or does not make sense, and explain your reasoning.*

88. The humor in this cartoon is based on the fact that "rational" and "real" have different meanings in mathematics and in everyday speech.

89. The number of pages in this book is a real number.

90. The book that I'm reading on the history of π appropriately contains an irrational number of pages.

91. Although the integers are closed under the operation of addition, I was able to find a subset that is not closed under this operation.

In Exercises 92–99, determine whether each statement is true or false. If the statement is false, make the necessary change(s) to produce a true statement.

92. Every rational number is an integer.

93. Some whole numbers are not integers.

94. Some rational numbers are not positive.

95. Irrational numbers cannot be negative.

96. Subtraction is a commutative operation.

97. $(24 \div 6) \div 2 = 24 \div (6 \div 2)$

98. $7 \cdot a + 3 \cdot a = a \cdot (7 + 3)$

99. $2 \cdot a + 5 = 5 \cdot a + 2$

5.6 Exponents and Scientific Notation

WHAT AM I SUPPOSED TO LEARN?

After studying this section, you should be able to:

1. Use properties of exponents.
2. Convert from scientific notation to decimal notation.
3. Convert from decimal notation to scientific notation.
4. Perform computations using scientific notation.
5. Solve applied problems using scientific notation.

Bigger than the biggest thing ever and then some. Much bigger than that in fact, really amazingly immense, a totally stunning size, real 'wow, that's big', time...Gigantic multiplied by colossal multiplied by staggeringly huge is the sort of concept we're trying to get across here.

—Douglas Adams, *The Restaurant at the End of the Universe*

Although Adams's description may not quite apply to this \$18.9 trillion national debt, exponents can be used to explore the meaning of this "staggeringly huge" number. In this section, you will learn to use exponents to provide a way of putting large and small numbers in perspective.

Properties of Exponents

1 *Use properties of exponents.*

We have seen that exponents are used to indicate repeated multiplication. Now consider the multiplication of two exponential expressions, such as $b^4 \cdot b^3$. We are multiplying 4 factors of b and 3 factors of b. We have a total of 7 factors of b:

4 factors of b | 3 factors of b

$$b^4 \cdot b^3 = (b \cdot b \cdot b \cdot b)(b \cdot b \cdot b) = b^7.$$

Total: 7 factors of b

The product is exactly the same if we add the exponents:

$$b^4 \cdot b^3 = b^{4+3} = b^7.$$

Properties of exponents allow us to perform operations with exponential expressions without having to write out long strings of factors. Three such properties are given in **Table 5.6**.

TABLE 5.6 Properties of Exponents

Property	Meaning	Examples
The Product Rule $b^m \cdot b^n = b^{m+n}$	When multiplying exponential expressions with the same base, add the exponents. Use this sum as the exponent of the common base.	$9^6 \cdot 9^{12} = 9^{6+12} = 9^{18}$
The Power Rule $(b^m)^n = b^{m \cdot n}$	When an exponential expression is raised to a power, multiply the exponents. Place the product of the exponents on the base and remove the parentheses.	$(3^4)^5 = 3^{4 \cdot 5} = 3^{20}$ $(5^3)^8 = 5^{3 \cdot 8} = 5^{24}$
The Quotient Rule $\frac{b^m}{b^n} = b^{m-n}$	When dividing exponential expressions with the same base, subtract the exponent in the denominator from the exponent in the numerator. Use this difference as the exponent of the common base.	$\frac{5^{12}}{5^4} = 5^{12-4} = 5^8$ $\frac{9^{40}}{9^5} = 9^{40-5} = 9^{35}$

The third property in **Table 5.6**, $\frac{b^m}{b^n} = b^{m-n}$, called the quotient rule, can lead to a zero exponent when subtracting exponents. Here is an example:

$$\frac{4^3}{4^3} = 4^{3-3} = 4^0.$$

We can see what this zero exponent means by evaluating 4^3 in the numerator and the denominator:

$$\frac{4^3}{4^3} = \frac{4 \cdot 4 \cdot 4}{4 \cdot 4 \cdot 4} = \frac{64}{64} = 1.$$

This means that 4^0 must equal 1. This example illustrates the zero exponent rule.

THE ZERO EXPONENT RULE

If b is any real number other than 0,

$$b^0 = 1.$$

EXAMPLE 1 *Using the Zero Exponent Rule*

Use the zero exponent rule to simplify:

a. 7^0 **b.** π^0 **c.** $(-5)^0$ **d.** -5^0.

SOLUTION

a. $7^0 = 1$ **b.** $\pi^0 = 1$ **c.** $(-5)^0 = 1$ **d.** $-5^0 = -1$

Only 5 is raised to the 0 power.

CHECK POINT 1 Use the zero exponent rule to simplify:

a. 19^0 **b.** $(3\pi)^0$ **c.** $(-14)^0$ **d.** -14^0.

The quotient rule can result in a negative exponent. Consider, for example, $4^3 \div 4^5$:

$$\frac{4^3}{4^5} = 4^{3-5} = 4^{-2}.$$

We can see what this negative exponent means by evaluating the numerator and the denominator:

$$\frac{4^3}{4^5} = \frac{\cancel{4} \cdot \cancel{4} \cdot \cancel{4}}{\cancel{4} \cdot \cancel{4} \cdot \cancel{4} \cdot 4 \cdot 4} = \frac{1}{4^2}.$$

Notice that $\frac{4^3}{4^5}$ equals both 4^{-2} and $\frac{1}{4^2}$. This means that 4^{-2} must equal $\frac{1}{4^2}$. This example is a particular case of the negative exponent rule.

GREAT QUESTION!

What's the difference between $\frac{4^3}{4^5}$ and $\frac{4^5}{4^3}$?

$\frac{4^3}{4^5}$ and $\frac{4^5}{4^3}$ represent different numbers:

$$\frac{4^3}{4^5} = 4^{3-5} = 4^{-2} = \frac{1}{4^2} = \frac{1}{16}$$

$$\frac{4^5}{4^3} = 4^{5-3} = 4^2 = 16.$$

THE NEGATIVE EXPONENT RULE

If b is any real number other than 0 and m is a natural number,

$$b^{-m} = \frac{1}{b^m}.$$

EXAMPLE 2 Using the Negative Exponent Rule

Use the negative exponent rule to simplify:

a. 8^{-2} **b.** 5^{-3} **c.** 7^{-1}.

SOLUTION

a. $8^{-2} = \frac{1}{8^2} = \frac{1}{8 \cdot 8} = \frac{1}{64}$

b. $5^{-3} = \frac{1}{5^3} = \frac{1}{5 \cdot 5 \cdot 5} = \frac{1}{125}$

c. $7^{-1} = \frac{1}{7^1} = \frac{1}{7}$

CHECK POINT 2 Use the negative exponent rule to simplify:

a. 9^{-2} **b.** 6^{-3} **c.** 12^{-1}.

Powers of Ten

Exponents and their properties allow us to represent and compute with numbers that are large or small. For example, one billion, or 1,000,000,000 can be written as 10^9. In terms of exponents, 10^9 might not look very large, but consider this: If you can count to 200 in one minute and decide to count for 12 hours a day at this rate, it would take you in the region of 19 years, 9 days, 5 hours, and 20 minutes to count to 10^9!

Powers of ten follow two basic rules:

1. **A positive exponent tells how many 0s follow the 1.** For example, 10^9 (one billion) is a 1 followed by nine zeros: 1,000,000,000. A googol, 10^{100}, is a 1 followed by one hundred zeros. (A googol far exceeds the number of protons, neutrons, and electrons in the universe.) A googol is a veritable pipsqueak compared to the googolplex, 10 raised to the googol power, or $10^{10^{100}}$; that's a 1 followed by a googol zeros. (If each zero in a googolplex were no larger than a grain of sand, there would not be enough room in the universe to represent the number.)
2. **A negative exponent tells how many places there are to the right of the decimal point.** For example, 10^{-9} (one billionth) has nine places to the right of the decimal point. The nine places include eight 0s and the 1.

$$10^{-9} = 0.\underbrace{000000001}_{\text{nine places}}$$

Blitzer Bonus

Earthquakes and Powers of Ten

The earthquake that ripped through northern California on October 17, 1989, measured 7.1 on the Richter scale, killed more than 60 people, and injured more than 2400. Shown here is San Francisco's Marina district, where shock waves tossed houses off their foundations and into the street.

The Richter scale is misleading because it is not actually a 1 to 8, but rather a 1 to 10 million scale. Each level indicates a tenfold increase in magnitude from the previous level, making a 7.0 earthquake a million times greater than a 1.0 quake.

The following is a translation of the Richter scale:

Richter Number (R)	**Magnitude** (10^{R-1})
1	$10^{1-1} = 10^0 = 1$
2	$10^{2-1} = 10^1 = 10$
3	$10^{3-1} = 10^2 = 100$
4	$10^{4-1} = 10^3 = 1000$
5	$10^{5-1} = 10^4 = 10,000$
6	$10^{6-1} = 10^5 = 100,000$
7	$10^{7-1} = 10^6 = 1,000,000$
8	$10^{8-1} = 10^7 = 10,000,000$

TABLE 5.7 Names of Large Numbers

10^2	hundred
10^3	thousand
10^6	million
10^9	billion
10^{12}	trillion
10^{15}	quadrillion
10^{18}	quintillion
10^{21}	sextillion
10^{24}	septillion
10^{27}	octillion
10^{30}	nonillion
10^{100}	googol
10^{googol}	googolplex

Scientific Notation

Earth is a 4.5-billion-year-old ball of rock orbiting the Sun. Because a billion is 10^9 (see **Table 5.7**), the age of our world can be expressed as

$$4.5 \times 10^9.$$

The number 4.5×10^9 is written in a form called *scientific notation*.

> **SCIENTIFIC NOTATION**
>
> A positive number is written in **scientific notation** when it is expressed in the form
>
> $$a \times 10^n,$$
>
> where a is a number greater than or equal to 1 and less than 10 ($1 \le a < 10$) and n is an integer.

It is customary to use the multiplication symbol, $\times$, rather than a dot, when writing a number in scientific notation.

Here are three examples of numbers in scientific notation:

- The universe is 1.375×10^{10} years old.
- In 2010, humankind generated 1.2 zettabytes, or 1.2×10^{21} bytes, of digital information. (A byte consists of eight binary digits, or bits, 0 or 1.)
- The length of the AIDS virus is 1.1×10^{-4} millimeter.

2 Convert from scientific notation to decimal notation.

We can use n, the exponent on the 10 in $a \times 10^n$, to change a number in scientific notation to decimal notation. If n is **positive**, move the decimal point in a to the **right** n places. If n is **negative**, move the decimal point in a to the **left** $|n|$ places.

EXAMPLE 3 Converting from Scientific to Decimal Notation

Write each number in decimal notation:

a. 1.375×10^{10} **b.** 1.1×10^{-4}.

SOLUTION

In each case, we use the exponent on the 10 to move the decimal point. In part (a), the exponent is positive, so we move the decimal point to the right. In part (b), the exponent is negative, so we move the decimal point to the left.

a. $1.375 \times 10^{10} = 13{,}750{,}000{,}000$

$n = 10$

Move the decimal point 10 places to the right.

b. $1.1 \times 10^{-4} = 0.00011$

$n = -4$

Move the decimal point $|-4|$ places, or 4 places, to the left.

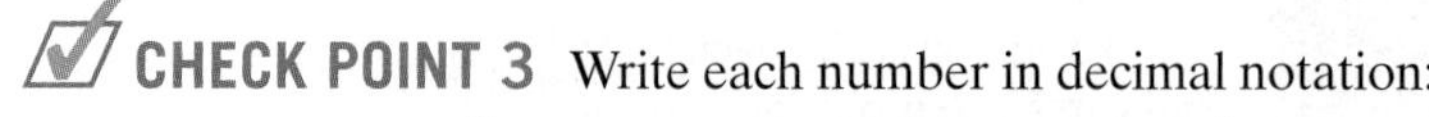

CHECK POINT 3 Write each number in decimal notation:

a. 7.4×10^9 **b.** 3.017×10^{-6}.

3 *Convert from decimal notation to scientific notation.*

To convert a positive number from decimal notation to scientific notation, we reverse the procedure of Example 3.

CONVERTING FROM DECIMAL TO SCIENTIFIC NOTATION

Write the number in the form $a \times 10^n$.

- Determine a, the numerical factor. Move the decimal point in the given number to obtain a number greater than or equal to 1 and less than 10.
- Determine n, the exponent on 10^n. The absolute value of n is the number of places the decimal point was moved. The exponent n is positive if the given number is greater than 10 and negative if the given number is between 0 and 1.

TECHNOLOGY

You can use your calculator's EE (enter exponent) or EXP key to convert from decimal to scientific notation. Here is how it's done for 0.000023:

Many Scientific Calculators

Keystrokes	Display
.000023 EE =	2.3 − 05

Many Graphing Calculators

Use the mode setting for scientific notation.

Keystrokes	Display
.000023 ENTER	2.3 E −5

EXAMPLE 4 Converting from Decimal Notation to Scientific Notation

Write each number in scientific notation:

a. 4,600,000 **b.** 0.000023.

SOLUTION

a. $4{,}600{,}000 = 4.6 \times 10^6$

This number is greater than 10, so n is positive in $a \times 10^n$.

Move the decimal point in 4,600,000 to get $1 \le a < 10$.

The decimal point moved 6 places from 4,600,000 to 4.6.

b. $0.000023 = 2.3 \times 10^{-5}$

This number is less than 1, so n is negative in $a \times 10^n$.

Move the decimal point in 0.000023 to get $1 \le a < 10$.

The decimal point moved 5 places from 0.000023 to 2.3.

CHECK POINT 4 Write each number in scientific notation:

a. 7,410,000,000 **b.** 0.000000092.

EXAMPLE 5 Expressing the U.S. Population in Scientific Notation

As of January 2016, the population of the United States was approximately 322 million. Express the population in scientific notation.

SOLUTION

Because a million is 10^6, the 2016 population can be expressed as

$$322 \times 10^6.$$

This factor is not between 1 and 10, so the number is not in scientific notation.

The voice balloon indicates that we need to convert 322 to scientific notation.

$$322 \times 10^6 = (3.22 \times 10^2) \times 10^6 = 3.22 \times 10^{2+6} = 3.22 \times 10^8$$

$322 = 3.22 \times 10^2$

In scientific notation, the population is 3.22×10^8.

GREAT QUESTION!

I read that the U.S. population exceeds $\frac{3}{10}$ of a billion. Yet you described it as 322 million. Which description is correct?

Both descriptions are correct. We can use exponential properties to express 322 million in billions.

$$322 \text{ million} = 322 \times 10^6 = (0.322 \times 10^3) \times 10^6 = 0.322 \times 10^{3+6} = 0.322 \times 10^9$$

Because 10^9 is a billion, U.S. population exceeds $\frac{3}{10}$ of a billion.

CHECK POINT 5 Express 410×10^7 in scientitic notation.

4 *Perform computations using scientific notation.*

Computations with Scientific Notation

We use the product rule for exponents to multiply numbers in scientific notation:

$$(a \times 10^n) \times (b \times 10^m) = (a \times b) \times 10^{n+m}.$$

Add the exponents on 10 and multiply the other parts of the numbers separately.

TECHNOLOGY

$(3.4 \times 10^9)(2 \times 10^{-5})$ on a Calculator:

Many Scientific Calculators

3.4 [EE] 9 [×] 2 [EE] 5 [+/−] [=]

Display: 6.8 04

Many Graphing Calculators

3.4 [EE] 9 [×] 2 [EE] [(−)] 5 [ENTER]

Display: 6.8 E 4

EXAMPLE 6 *Multiplying Numbers in Scientific Notation*

Multiply: $(3.4 \times 10^9)(2 \times 10^{-5})$. Write the product in decimal notation.

SOLUTION

$(3.4 \times 10^9)(2 \times 10^{-5}) = (3.4 \times 2) \times (10^9 \times 10^{-5})$ Regroup factors.

$= 6.8 \times 10^{9+(-5)}$ Add the exponents on 10 and multiply the other parts.

$= 6.8 \times 10^4$ Simplify.

$= 68{,}000$ Write the product in decimal notation.

CHECK POINT 6 Multiply: $(1.3 \times 10^7)(4 \times 10^{-2})$. Write the product in decimal notation.

We use the quotient rule for exponents to divide numbers in scientific notation:

$$\frac{a \times 10^n}{b \times 10^m} = \left(\frac{a}{b}\right) \times 10^{n-m}.$$

Subtract the exponents on 10 and divide the other parts of the numbers separately.

EXAMPLE 7 Dividing Numbers in Scientific Notation

Divide: $\dfrac{8.4 \times 10^{-7}}{4 \times 10^{-4}}$. Write the quotient in decimal notation.

SOLUTION

$$\frac{8.4 \times 10^{-7}}{4 \times 10^{-4}} = \left(\frac{8.4}{4}\right) \times \left(\frac{10^{-7}}{10^{-4}}\right)$$ Regroup factors.

$$= 2.1 \times 10^{-7-(-4)}$$ Subtract the exponents on 10 and divide the other parts.

$$= 2.1 \times 10^{-3}$$ Simplify: $-7 - (-4) = -7 + 4 = -3$.

$$= 0.0021$$ Write the quotient in decimal notation.

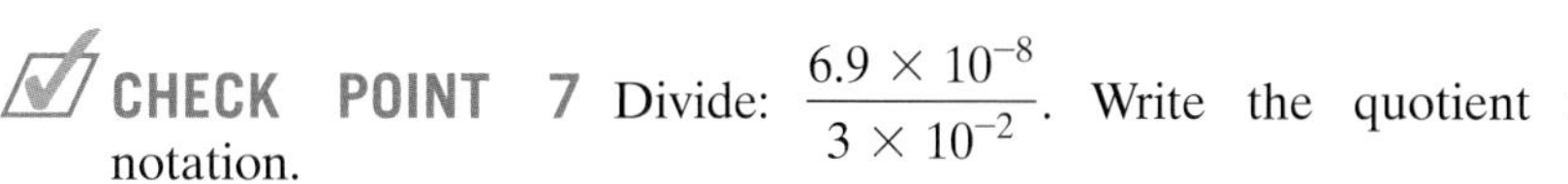

CHECK POINT 7 Divide: $\dfrac{6.9 \times 10^{-8}}{3 \times 10^{-2}}$. Write the quotient in decimal notation.

Multiplication and division involving very large or very small numbers can be performed by first converting each number to scientific notation.

EXAMPLE 8 Using Scientific Notation to Multiply

Multiply: $0.00064 \times 9{,}400{,}000{,}000$. Express the product in **a.** scientific notation and **b.** decimal notation.

SOLUTION

a. $0.00064 \times 9{,}400{,}000{,}000$

$$= 6.4 \times 10^{-4} \times 9.4 \times 10^{9}$$ Write each number in scientific notation.

$$= (6.4 \times 9.4) \times (10^{-4} \times 10^{9})$$ Regroup factors.

$$= 60.16 \times 10^{-4+9}$$ Add the exponents on 10 and multiply the other parts.

$$= 60.16 \times 10^{5}$$ Simplify.

$$= (6.016 \times 10) \times 10^{5}$$ Express 60.16 in scientific notation.

$$= 6.016 \times 10^{6}$$ Add exponents on 10: $10^1 \times 10^5 = 10^{1+5} = 10^6$.

b. The answer in decimal notation is obtained by moving the decimal point in 6.016 six places to the right. The product is 6,016,000.

CHECK POINT 8 Multiply: $0.0036 \times 5{,}200{,}000$. Express the product in **a.** scientific notation and **b.** decimal notation.

5 *Solve applied problems using scientific notation.*

Applications: Putting Numbers in Perspective

Due to tax cuts and spending increases, the United States began accumulating large deficits in the 1980s. To finance the deficit, the government had borrowed \$18.9 trillion as of January 2016. The graph in **Figure 5.14** on the next page shows the national debt increasing over time.

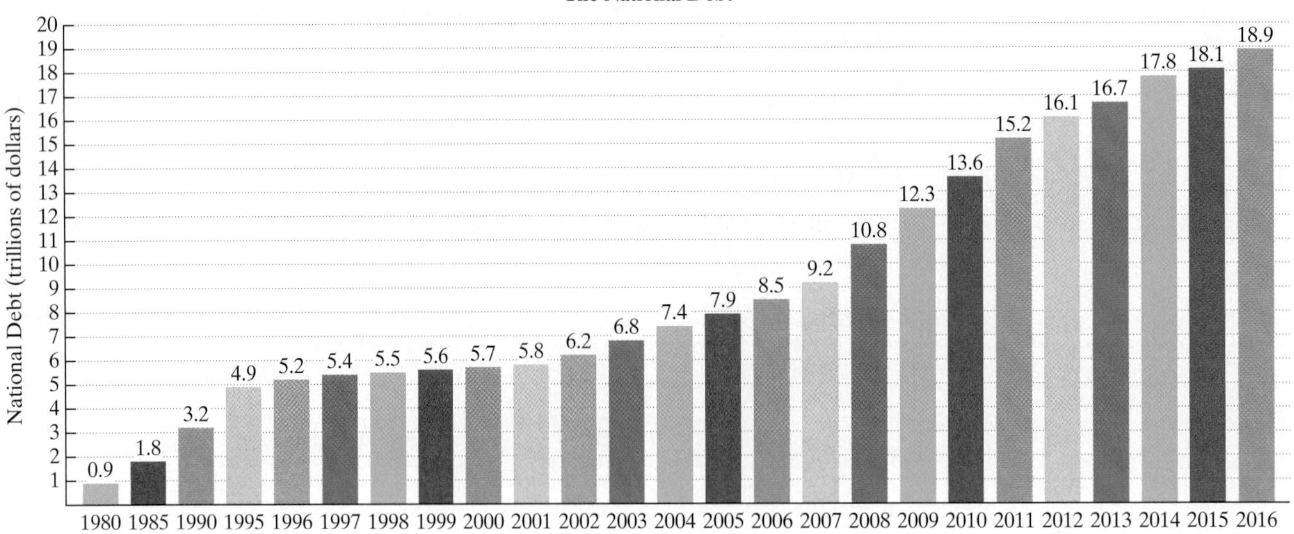

FIGURE 5.14
Source: Office of Management and Budget

Example 9 shows how we can use scientific notation to comprehend the meaning of a number such as 18.9 trillion.

EXAMPLE 9 *The National Debt*

As of January 2016, the national debt was \$18.9 trillion, or 18.9×10^{12} dollars. At that time, the U.S. population was approximately 322,000,000 (322 million), or 3.22×10^8. If the national debt was evenly divided among every individual in the United States, how much would each citizen have to pay?

SOLUTION

The amount each citizen must pay is the total debt, 18.9×10^{12} dollars, divided by the number of citizens, 3.22×10^8.

$$\begin{aligned}\frac{18.9 \times 10^{12}}{3.22 \times 10^8} &= \left(\frac{18.9}{3.22}\right) \times \left(\frac{10^{12}}{10^8}\right) \\ &\approx 5.87 \times 10^{12-8} \\ &= 5.87 \times 10^4 \\ &= 58{,}700\end{aligned}$$

Every U.S. citizen would have to pay approximately \$58,700 to the federal government to pay off the national debt.

> **TECHNOLOGY**
>
> Here is the keystroke sequence for solving Example 9 using a calculator:
>
> 18.9 [EE] 12 [÷] 3.22 [EE] 8.
>
> The quotient is displayed by pressing [=] on a scientific calculator or [ENTER] on a graphing calculator. The answer can be displayed in scientific or decimal notation. Consult your manual.

If a number is written in scientific notation, $a \times 10^n$, the digits in a are called **significant digits**.

National Debt: 18.9×10^{12} (Three significant digits)   U.S. Population: 3.22×10^8 (Three significant digits)

Because these were the given numbers in Example 9, we rounded the answer, 5.87×10^4, to three significant digits. When multiplying or dividing in scientific notation where rounding is necessary and rounding instructions are not given, **round the scientific notation answer to the least number of significant digits found in any of the given numbers.**

CHECK POINT 9 In 2015, there were 680,000 police officers in the United States with yearly wages totaling $\$4.08 \times 10^{10}$. If these wages were evenly divided among all police officers, find the mean, or average, salary of a U.S. police officer. (*Source:* Bureau of Justice Statistics)

Blitzer Bonus

Seven Ways to Spend $1 Trillion

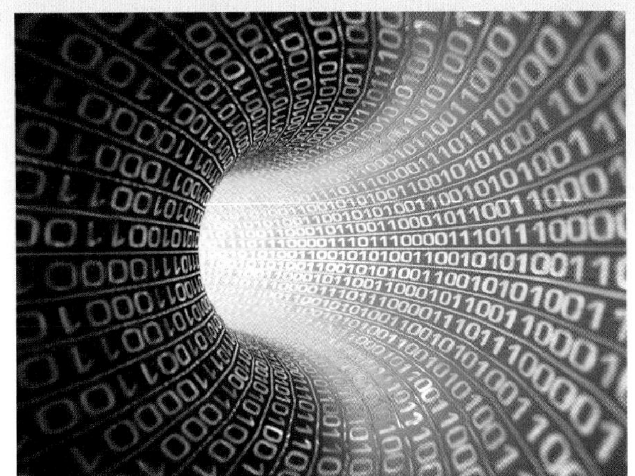

Confronting a national debt of $18.9 trillion starts with grasping just how colossal $1 trillion (1×10^{12}) actually is. To help you wrap your head around this mind-boggling number, and to put the national debt in further perspective, consider what $1 trillion will buy:

- 40,816,326 new cars based on an average sticker price of $24,500 each
- 5,574,136 homes based on the national median price of $179,400 for existing single-family homes
- one year's salary for 14.7 million teachers based on the average teacher salary of $68,000 in California
- the annual salaries of all 535 members of Congress for the next 10,742 years based on current salaries of $174,000 per year
- the salary of basketball superstar LeBron James for 50,000 years based on an annual salary of $20 million
- annual base pay for 59.5 million U.S. privates (that's 100 times the total number of active-duty soldiers in the Army) based on basic pay of $16,794 per year
- salaries to hire all 2.8 million residents of the state of Kansas in full-time minimum-wage jobs for the next 23 years based on the federal minimum wage of $7.25 per hour

Source: Kiplinger.com

Concept and Vocabulary Check

Fill in each blank so that the resulting statement is true.

1. When multiplying exponential expressions with the same base, ______ the exponents.
2. When an exponential expression is raised to a power, _________ the exponents.
3. When dividing exponential expressions with the same base, _________ the exponents.
4. Any nonzero real number raised to the zero power is equal to ______.
5. A positive number is written in scientific notation when the first factor is ______________________ and the second factor is _______________.

In Exercises 6–10, determine whether each statement is true or false. If the statement is false, make the necessary change(s) to produce a true statement.

6. $2^3 \cdot 2^5 = 4^8$ ______
7. $\dfrac{10^8}{10^4} = 10^2$ ______
8. $5^{-2} = -5^2$ ______
9. A trillion is one followed by 12 zeros. ______
10. According to *Mother Jones* magazine, sending all U.S. high school graduates to private colleges would cost $347 billion. Because a billion is 10^9, the cost in scientific notation is 347×10^9 dollars. ______

Exercise Set 5.6

Practice Exercises

In Exercises 1–12, use properties of exponents to simplify each expression. First express the answer in exponential form. Then evaluate the expression.

1. $2^2 \cdot 2^3$
2. $3^3 \cdot 3^2$
3. $4 \cdot 4^2$
4. $5 \cdot 5^2$
5. $(2^2)^3$
6. $(3^3)^2$
7. $(1^4)^5$
8. $(1^3)^7$
9. $\dfrac{4^7}{4^5}$
10. $\dfrac{6^7}{6^5}$
11. $\dfrac{2^8}{2^4}$
12. $\dfrac{3^8}{3^4}$

In Exercises 13–24, use the zero and negative exponent rules to simplify each expression.

13. 3^0 **14.** 9^0 **15.** $(-3)^0$ **16.** $(-9)^0$

17. -3^0 **18.** -9^0 **19.** 2^{-2} **20.** 3^{-2}

21. 4^{-3} **22.** 2^{-3} **23.** 2^{-5} **24.** 2^{-6}

In Exercises 25–30, use properties of exponents to simplify each expression. First express the answer in exponential form. Then evaluate the expression.

25. $3^4 \cdot 3^{-2}$ **26.** $2^5 \cdot 2^{-2}$ **27.** $3^{-3} \cdot 3$

28. $2^{-3} \cdot 2$ **29.** $\frac{2^3}{2^7}$ **30.** $\frac{3^4}{3^7}$

In Exercises 31–42, use properties of exponents to simplify each expression. Express answers in exponential form with positive exponents only. Assume that any variables in denominators are not equal to zero.

31. $(x^5 \cdot x^3)^{-2}$ **32.** $(x^2 \cdot x^4)^{-3}$ **33.** $\frac{(x^3)^4}{(x^2)^7}$

34. $\frac{(x^2)^5}{(x^3)^4}$ **35.** $\left(\frac{x^5}{x^2}\right)^{-4}$ **36.** $\left(\frac{x^7}{x^2}\right)^{-3}$

37. $\frac{2x^5 \cdot 3x}{15x^6}$ **38.** $\frac{4x^7 \cdot 5x}{10x^8}$

39. $(-2x^3y^{-4})(3x^{-1}y)$ **40.** $(-5x^4y^{-3})(4x^{-1}y)$

41. $\frac{30x^2y^5}{-6x^8y^{-3}}$ **42.** $\frac{24x^2y^{13}}{-8x^5y^{-2}}$

In Exercises 43–58, express each number in decimal notation.

43. 2.7×10^2 **44.** 4.7×10^3

45. 9.12×10^5 **46.** 8.14×10^4

47. 8×10^7 **48.** 7×10^6

49. 1×10^5 **50.** 1×10^8

51. 7.9×10^{-1} **52.** 8.6×10^{-1}

53. 2.15×10^{-2} **54.** 3.14×10^{-2}

55. 7.86×10^{-4} **56.** 4.63×10^{-5}

57. 3.18×10^{-6} **58.** 5.84×10^{-7}

In Exercises 59–78, express each number in scientific notation.

59. 370 **60.** 530

61. 3600 **62.** 2700

63. 32,000 **64.** 64,000

65. 220,000,000 **66.** 370,000,000,000

67. 0.027 **68.** 0.014

69. 0.0037 **70.** 0.00083

71. 0.00000293 **72.** 0.000000647

73. 820×10^5 **74.** 630×10^8

75. 0.41×10^6 **76.** 0.57×10^9

77. 2100×10^{-9} **78.** $97{,}000 \times 10^{-11}$

In Exercises 79–92, perform the indicated operation and express each answer in decimal notation.

79. $(2 \times 10^3)(3 \times 10^2)$ **80.** $(5 \times 10^2)(4 \times 10^4)$

81. $(2 \times 10^9)(3 \times 10^{-5})$ **82.** $(4 \times 10^8)(2 \times 10^{-4})$

83. $(4.1 \times 10^2)(3 \times 10^{-4})$ **84.** $(1.2 \times 10^3)(2 \times 10^{-5})$

85. $\frac{12 \times 10^6}{4 \times 10^2}$ **86.** $\frac{20 \times 10^{20}}{10 \times 10^{15}}$

87. $\frac{15 \times 10^4}{5 \times 10^{-2}}$ **88.** $\frac{18 \times 10^2}{9 \times 10^{-3}}$

89. $\frac{6 \times 10^3}{2 \times 10^5}$ **90.** $\frac{8 \times 10^4}{2 \times 10^7}$

91. $\frac{6.3 \times 10^{-6}}{3 \times 10^{-3}}$ **92.** $\frac{9.6 \times 10^{-7}}{3 \times 10^{-3}}$

In Exercises 93–102, perform the indicated operation by first expressing each number in scientific notation. Write the answer in scientific notation.

93. (82,000,000)(3,000,000,000)

94. (94,000,000)(6,000,000,000)

95. (0.0005)(6,000,000) **96.** (0.000015)(0.004)

97. $\frac{9{,}500{,}000}{500}$ **98.** $\frac{30{,}000}{0.0005}$

99. $\frac{0.00008}{200}$ **100.** $\frac{0.0018}{0.0000006}$

101. $\frac{480{,}000{,}000{,}000}{0.00012}$ **102.** $\frac{0.000000096}{16{,}000}$

Practice Plus

In Exercises 103–106, perform the indicated operations. Express each answer as a fraction reduced to its lowest terms.

103. $\frac{2^4}{2^5} + \frac{3^3}{3^5}$ **104.** $\frac{3^5}{3^6} + \frac{2^3}{2^6}$

105. $\frac{2^6}{2^4} - \frac{5^4}{5^6}$ **106.** $\frac{5^6}{5^4} - \frac{2^4}{2^6}$

In Exercises 107–110, perform the indicated computations. Express answers in scientific notation.

107. $(5 \times 10^3)(1.2 \times 10^{-4}) \div (2.4 \times 10^2)$

108. $(2 \times 10^2)(2.6 \times 10^{-3}) \div (4 \times 10^3)$

109. $\frac{(1.6 \times 10^4)(7.2 \times 10^{-3})}{(3.6 \times 10^8)(4 \times 10^{-3})}$

110. $\frac{(1.2 \times 10^6)(8.7 \times 10^{-2})}{(2.9 \times 10^6)(3 \times 10^{-3})}$

Application Exercises

The bar graph shows the total amount Americans paid in federal taxes, in trillions of dollars, and the U.S. population, in millions, from 2012 through 2015. Exercises 111–112 are based on the numbers displayed by the graph.

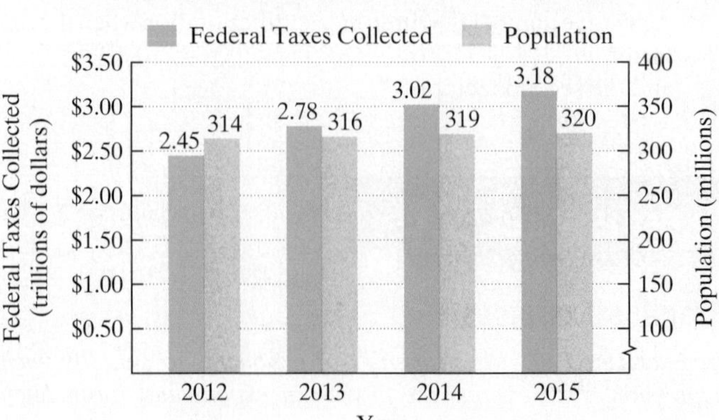

Sources: Internal Revenue Service and U.S. Census Bureau

111. **a.** In 2015, the United States government collected \$3.18 trillion in taxes. Express this number in scientific notation.

b. In 2015, the population of the United States was approximately 320 million. Express this number in scientific notation.

c. Use your scientific notation answers from parts (a) and (b) to answer this question: If the total 2015 tax collections were evenly divided among all Americans, how much would each citizen pay? Express the answer in decimal notation, rounded to the nearest dollar.

112. **a.** In 2014, the United States government collected \$3.02 trillion in taxes. Express this number in scientific notation.

b. In 2014, the population of the United States was approximately 319 million. Express this number in scientific notation.

c. Use your scientific notation answers from parts (a) and (b) to answer this question: If the total 2014 tax collections were evenly divided among all Americans, how much would each citizen pay? Express the answer in decimal notation, rounded to the nearest dollar.

We have seen that the 2016 U.S. national debt was \$18.9 trillion. In Exercises 113–114, you will use scientific notation to put a number like 18.9 trillion in perspective.

113. **a.** Express 18.9 trillion in scientific notation.

b. Four years of tuition, fees, and room and board at a public U.S. college cost approximately \$60,000. Express this number in scientific notation.

c. Use your answers from parts (a) and (b) to determine how many Americans could receive a free college education for \$18.9 trillion.

114. **a.** Express 18.9 trillion in scientific notation.

b. Each year, Americans spend \$254 billion on summer vacations. Express this number in scientific notation.

c. Use your answers from parts (a) and (b) to determine how many years Americans can have free summer vacations for \$18.9 trillion.

115. The mass of one oxygen molecule is 5.3×10^{-23} gram. Find the mass of 20,000 molecules of oxygen. Express the answer in scientific notation.

116. The mass of one hydrogen atom is 1.67×10^{-24} gram. Find the mass of 80,000 hydrogen atoms. Express the answer in scientific notation.

117. There are approximately 3.2×10^7 seconds in a year. According to the United States Department of Agriculture, Americans consume 127 chickens per second. How many chickens are eaten per year in the United States? Express the answer in scientific notation.

118. Convert 365 days (one year) to hours, to minutes, and, finally, to seconds, to determine how many seconds there are in a year. Express the answer in scientific notation.

Explaining the Concepts

119. Explain the product rule for exponents. Use $2^3 \cdot 2^5$ in your explanation.

120. Explain the power rule for exponents. Use $(3^2)^4$ in your explanation.

121. Explain the quotient rule for exponents. Use $\frac{5^8}{5^2}$ in your explanation.

122. Explain the zero exponent rule and give an example.

123. Explain the negative exponent rule and give an example.

124. How do you know if a number is written in scientific notation?

125. Explain how to convert from scientific to decimal notation and give an example.

126. Explain how to convert from decimal to scientific notation and give an example.

127. Suppose you are looking at a number in scientific notation. Describe the size of the number you are looking at if the exponent on ten is **a.** positive, **b.** negative, **c.** zero.

128. Describe one advantage of expressing a number in scientific notation over decimal notation.

Critical Thinking Exercises

Make Sense? *In Exercises 129–132, determine whether each statement makes sense or does not make sense, and explain your reasoning.*

129. If 5^{-2} is raised to the third power, the result is a number between 0 and 1.

130. The expression $\frac{a^n}{b^0}$ is undefined because division by 0 is undefined.

131. For a recent year, total tax collections in the United States were $\$2.02 \times 10^7$.

132. I just finished reading a book that contained approximately 1.04×10^5 words.

In Exercises 133–140, determine whether each statement is true or false. If the statement is false, make the necessary change(s) to produce a true statement.

133. $4^{-2} < 4^{-3}$

134. $5^{-2} > 2^{-5}$

135. $(-2)^4 = 2^{-4}$

136. $5^2 \cdot 5^{-2} > 2^5 \cdot 2^{-5}$

137. $534.7 = 5.347 \times 10^3$

138. $\frac{8 \times 10^{30}}{4 \times 10^{-5}} = 2 \times 10^{25}$

139. $(7 \times 10^5) + (2 \times 10^{-3}) = 9 \times 10^2$

140. $(4 \times 10^3) + (3 \times 10^2) = 43 \times 10^2$

141. Give an example of a number for which there is no advantage to using scientific notation instead of decimal notation. Explain why this is the case.

142. The mad Dr. Frankenstein has gathered enough bits and pieces (so to speak) for $2^{-1} + 2^{-2}$ of his creature-to-be. Write a fraction that represents the amount of his creature that must still be obtained.

Technology Exercises

143. Use a calculator in a fraction mode to check your answers in Exercises 19–24.

144. Use a calculator to check any three of your answers in Exercises 43–58.

145. Use a calculator to check any three of your answers in Exercises 59–78.

146. Use a calculator with an EE or EXP key to check any four of your computations in Exercises 79–102. Display the result of the computation in scientific notation and in decimal notation.

Group Exercises

147. **Putting Numbers into Perspective.** A large number can be put into perspective by comparing it with another number. For example, we put the \$18.9 trillion national debt (Example 9) and the \$3.18 trillion the government collected in taxes (Exercise 111) into perspective by comparing these numbers to the number of U.S. citizens.

For this project, each group member should consult an almanac, a newspaper, or the Internet to find a number greater than one million. Explain to other members of the group the context in which the large number is used. Express the number in scientific notation. Then put the number into perspective by comparing it with another number.

148. Refer to the Blitzer Bonus on page 323. Group members should use scientific notation to verify any three of the bulleted items on ways to spend \$1 trillion.

5.7 Arithmetic and Geometric Sequences

WHAT AM I SUPPOSED TO LEARN?

After studying this section, you should be able to:

1. Write terms of an arithmetic sequence.
2. Use the formula for the general term of an arithmetic sequence.
3. Write terms of a geometric sequence.
4. Use the formula for the general term of a geometric sequence.

Blitzer Bonus

Fibonacci Numbers on the Piano Keyboard

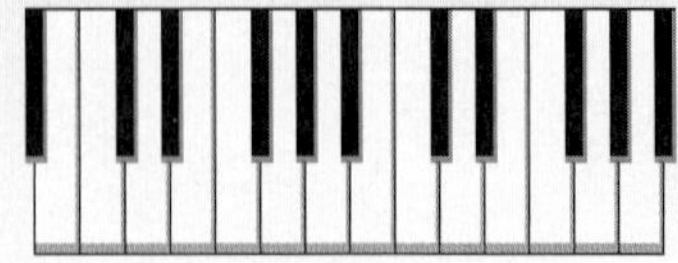

One Octave

Numbers in the Fibonacci sequence can be found in an octave on the piano keyboard. The octave contains 2 black keys in one cluster, 3 black keys in another cluster, a total of 5 black keys, 8 white keys, and a total of 13 keys altogether. The numbers 2, 3, 5, 8, and 13 are the third through seventh terms of the Fibonacci sequence.

Sequences

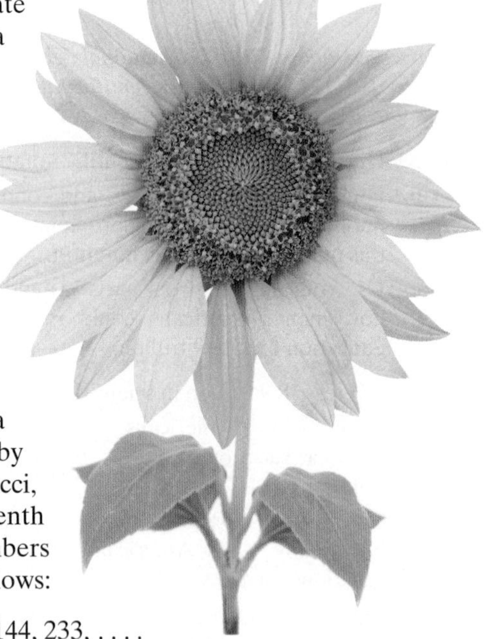

Many creations in nature involve intricate mathematical designs, including a variety of spirals. For example, the arrangement of the individual florets in the head of a sunflower forms spirals. In some species, there are 21 spirals in the clockwise direction and 34 in the counterclockwise direction. The precise numbers depend on the species of sunflower: 21 and 34, or 34 and 55, or 55 and 89, or even 89 and 144.

This observation becomes even more interesting when we consider a sequence of numbers investigated by Leonardo of Pisa, also known as Fibonacci, an Italian mathematician of the thirteenth century. The **Fibonacci sequence** of numbers is an infinite sequence that begins as follows:

$$1, 1, 2, 3, 5, 8, 13, 21, 34, 55, 89, 144, 233, \ldots.$$

The first two terms are 1. Every term thereafter is the sum of the two preceding terms. For example, the third term, 2, is the sum of the first and second terms: $1 + 1 = 2$. The fourth term, 3, is the sum of the second and third terms: $1 + 2 = 3$, and so on. Did you know that the number of spirals in a daisy or a sunflower, 21 and 34, are two Fibonacci numbers? The number of spirals in a pinecone, 8 and 13, and a pineapple, 8 and 13, are also Fibonacci numbers.

We can think of a **sequence** as a list of numbers that are related to each other by a rule. The numbers in a sequence are called its **terms**. The letter a with a subscript is used to represent the terms of a sequence. Thus, a_1 represents the first term of the sequence, a_2 represents the second term, a_3 the third term, and so on. This notation is shown for the first six terms of the Fibonacci sequence:

$$1, \quad 1, \quad 2, \quad 3, \quad 5, \quad 8.$$

$a_1 = 1$ $a_2 = 1$ $a_3 = 2$ $a_4 = 3$ $a_5 = 5$ $a_6 = 8$

Arithmetic Sequences

The bar graph in **Figure 5.15** is based on a mathematical model that shows how much Americans spent on their pets, to the nearest billion dollars, each year from 2001 through 2012.

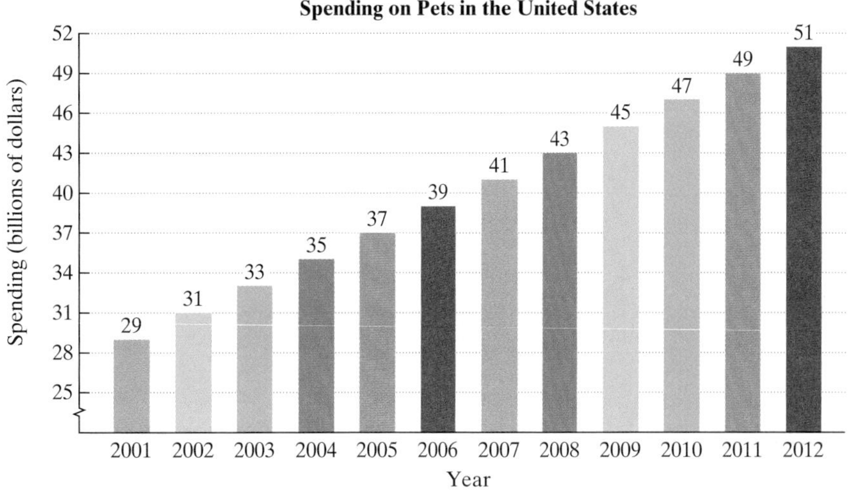

FIGURE 5.15
Source: American Pet Products Manufacturers Association

The graph illustrates that each year spending increased by \$2 billion. The sequence of annual spending

$$29, 31, 33, 35, 37, 39, 41, \ldots$$

shows that each term after the first, 29, differs from the preceding term by a constant amount, namely 2. This sequence is an example of an *arithmetic sequence*.

DEFINITION OF AN ARITHMETIC SEQUENCE

An **arithmetic sequence** is a sequence in which each term after the first differs from the preceding term by a constant amount. The difference between consecutive terms is called the **common difference** of the sequence.

The common difference, d, is found by subtracting any term from the term that directly follows it. In the following examples, the common difference is found by subtracting the first term from the second term: $a_2 - a_1$.

Arithmetic Sequence	Common Difference
$29, 31, 33, 35, 37, \ldots$	$d = 31 - 29 = 2$
$-5, -2, 1, 4, 7, \ldots$	$d = -2 - (-5) = -2 + 5 = 3$
$8, 3, -2, -7, -12, \ldots$	$d = 3 - 8 = -5$

If the first term of an arithmetic sequence is a_1, each term after the first is obtained by adding d, the common difference, to the previous term.

1 *Write terms of an arithmetic sequence.*

EXAMPLE 1 *Writing the Terms of an Arithmetic Sequence*

Write the first six terms of the arithmetic sequence with first term 6 and common difference 4.

SOLUTION

The first term is 6. The second term is $6 + 4$, or 10. The third term is $10 + 4$, or 14, and so on. The first six terms are

$$6, 10, 14, 18, 22, \text{ and } 26.$$

CHECK POINT 1 Write the first six terms of the arithmetic sequence with first term 100 and common difference 20.

EXAMPLE 2 Writing the Terms of an Arithmetic Sequence

Write the first six terms of the arithmetic sequence with $a_1 = 5$ and $d = -2$.

SOLUTION

The first term, a_1, is 5. The common difference, d, is -2. To find the second term, we add -2 to 5, giving 3. For the next term, we add -2 to 3, and so on. The first six terms are

$$5, 3, 1, -1, -3, \text{ and } -5.$$

CHECK POINT 2 Write the first six terms of the arithmetic sequence with $a_1 = 8$ and $d = -3$.

2 *Use the formula for the general term of an arithmetic sequence.*

The General Term of an Arithmetic Sequence

Consider an arithmetic sequence whose first term is a_1 and whose common difference is d. We are looking for a formula for the general term, a_n. Let's begin by writing the first six terms. The first term is a_1. The second term is $a_1 + d$. The third term is $a_1 + d + d$, or $a_1 + 2d$. Thus, we start with a_1 and add d to each successive term. The first six terms are

$$a_1, \quad a_1 + d, \quad a_1 + 2d, \quad a_1 + 3d, \quad a_1 + 4d, \quad a_1 + 5d.$$

a_1, first term	a_2, second term	a_3, third term	a_4, fourth term	a_5, fifth term	a_6, sixth term

Applying inductive reasoning to the pattern of the terms results in the following formula for the general term, or the nth term, of an arithmetic sequence:

GENERAL TERM OF AN ARITHMETIC SEQUENCE

The nth term (the general term) of an arithmetic sequence with first term a_1 and common difference d is

$$a_n = a_1 + (n - 1)d.$$

EXAMPLE 3 Using the Formula for the General Term of an Arithmetic Sequence

Find the eighth term of the arithmetic sequence whose first term is 4 and whose common difference is -7.

SOLUTION

To find the eighth term, a_8, we replace n in the formula with 8, a_1 with 4, and d with -7.

$$a_n = a_1 + (n - 1)d$$
$$a_8 = 4 + (8 - 1)(-7) = 4 + 7(-7) = 4 + (-49) = -45$$

The eighth term is -45. We can check this result by writing the first eight terms of the sequence:

$$4, -3, -10, -17, -24, -31, -38, -45.$$

CHECK POINT 3 Find the ninth term of the arithmetic sequence whose first term is 6 and whose common difference is -5.

In Chapter 1, we saw that the process of finding formulas to describe real-world phenomena is called mathematical modeling. Such formulas, together with the meaning assigned to the variables, are called mathematical models. Example 4 illustrates how the formula for the general term of an arithmetic sequence can be used to develop a mathematical model.

EXAMPLE 4 *Using an Arithmetic Sequence to Model Changes in the U.S. Population*

The graph in **Figure 5.16** shows the percentage of the U.S. population by race/ethnicity for 2010, with projections by the U.S. Census Bureau for 2050.

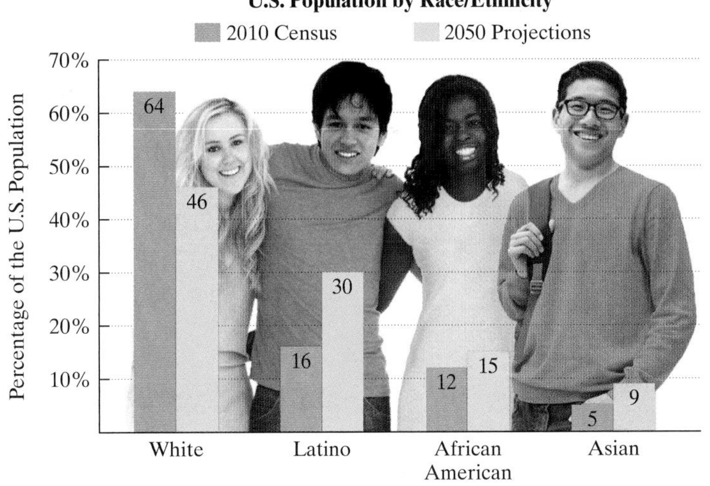

FIGURE 5.16
Source: U.S. Census Bureau

The data show that in 2010, 64% of the U.S. population was white. On average, this is projected to decrease by approximately 0.45% per year.

a. Write a formula for the nth term of the arithmetic sequence that describes the percentage of the U.S. population that will be white n years after 2009.

b. What percentage of the U.S. population is projected to be white in 2030?

SOLUTION

a. With a yearly decrease of 0.45%, we can express the percentage of the white population by the following arithmetic sequence:

$$64, \quad 64 - 0.45 = 63.55, \quad 63.55 - 0.45 = 63.10, \ \ldots .$$

a_1: percentage of whites in the population in 2010, 1 year after 2009

a_2: percentage of whites in the population in 2011, 2 years after 2009

a_3: percentage of whites in the population in 2012, 3 years after 2009

In this sequence, 64, 63.55, 63.10, . . . , the first term, a_1, represents the percentage of the population that was white in 2010. Each subsequent year this amount decreases by 0.45%, so $d = -0.45$. We use the formula for the general term of an arithmetic sequence to write the nth term of the sequence that describes the percentage of whites in the population n years after 2009.

$a_n = a_1 + (n - 1)d$ — This is the formula for the general term of an arithmetic sequence.

$a_n = 64 + (n - 1)(-0.45)$ — $a_1 = 64$ and $d = -0.45$.

$a_n = 64 - 0.45n + 0.45$ — Distribute -0.45 to each term in parentheses.

$a_n = -0.45n + 64.45$ — Simplify.

Thus, the percentage of the U.S. population that will be white n years after 2009 can be described by

$$a_n = -0.45n + 64.45.$$

b. Now we need to project the percentage of the population that will be white in 2030. The year 2030 is 21 years after 2009. Thus, $n = 21$. We substitute 21 for n in $a_n = -0.45n + 64.45$.

$$a_{21} = -0.45(21) + 64.45 = 55$$

The 21st term of the sequence is 55. Thus, 55% of the U.S. population is projected to be white in 2030.

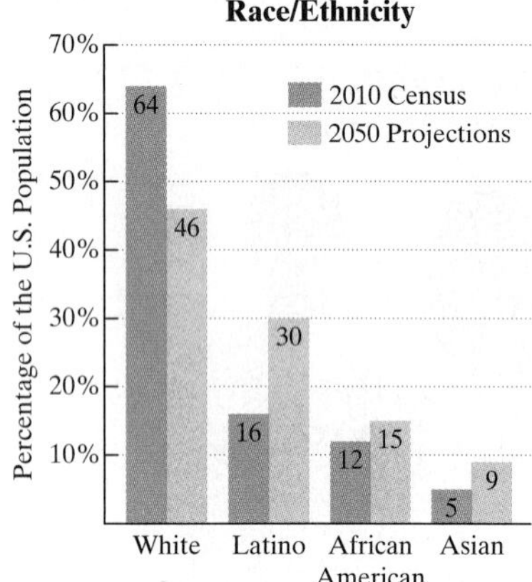

FIGURE 5.16 (repeated)

CHECK POINT 4 The data in **Figure 5.16**, repeated in the margin, show that in 2010, 16% of the U.S. population was Latino. On average, this is projected to increase by approximately 0.35% per year.

a. Write a formula for the nth term of the arithmetic sequence that describes the percentage of the U.S. population that will be Latino n years after 2009.

b. What percentage of the U.S. population is projected to be Latino in 2030?

Geometric Sequences

Figure 5.17 shows a sequence in which the number of squares is increasing. From left to right, the number of squares is 1, 5, 25, 125, and 625. In this sequence, each term after the first, 1, is obtained by multiplying the preceding term by a constant amount, namely 5. This sequence of increasing numbers of squares is an example of a *geometric sequence*.

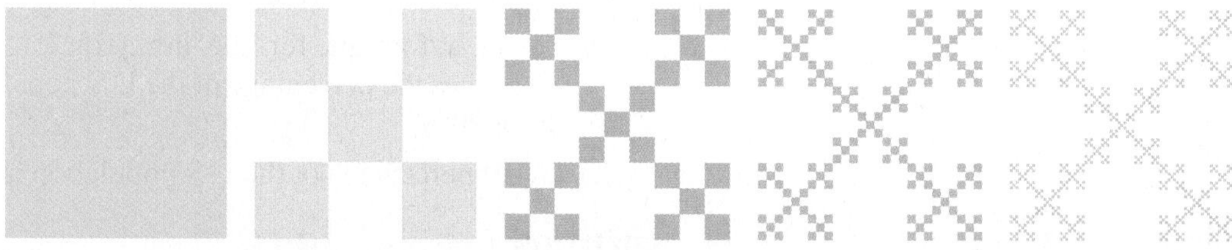

FIGURE 5.17 A geometric sequence of squares

DEFINITION OF A GEOMETRIC SEQUENCE

A **geometric sequence** is a sequence in which each term after the first is obtained by multiplying the preceding term by a fixed nonzero constant. The amount by which we multiply each time is called the **common ratio** of the sequence.

The common ratio, r, is found by dividing any term after the first term by the term that directly precedes it. In the examples below, the common ratio is found by dividing the second term by the first term: $\frac{a_2}{a_1}$.

GREAT QUESTION!

What happens to the terms of a geometric sequence when the common ratio is negative?

When the common ratio of a geometric sequence is negative, the signs of the terms alternate.

Geometric sequence	Common ratio
$1, 5, 25, 125, 625, \ldots$	$r = \frac{5}{1} = 5$
$4, 8, 16, 32, 64, \ldots$	$r = \frac{8}{4} = 2$
$6, -12, 24, -48, 96, \ldots$	$r = \frac{-12}{6} = -2$
$9, -3, 1, -\frac{1}{3}, \frac{1}{9}, \ldots$	$r = \frac{-3}{9} = -\frac{1}{3}$

3 *Write terms of a geometric sequence.*

How do we write out the terms of a geometric sequence when the first term and the common ratio are known? We multiply the first term by the common ratio to get the second term, multiply the second term by the common ratio to get the third term, and so on.

EXAMPLE 5 *Writing the Terms of a Geometric Sequence*

Write the first six terms of the geometric sequence with first term 6 and common ratio $\frac{1}{3}$.

SOLUTION

The first term is 6. The second term is $6 \cdot \frac{1}{3}$, or 2. The third term is $2 \cdot \frac{1}{3}$, or $\frac{2}{3}$. The fourth term is $\frac{2}{3} \cdot \frac{1}{3}$, or $\frac{2}{9}$, and so on. The first six terms are

$$6, 2, \tfrac{2}{3}, \tfrac{2}{9}, \tfrac{2}{27}, \text{ and } \tfrac{2}{81}.$$

CHECK POINT 5 Write the first six terms of the geometric sequence with first term 12 and common ratio $-\frac{1}{2}$.

4 *Use the formula for the general term of a geometric sequence.*

The General Term of a Geometric Sequence

Consider a geometric sequence whose first term is a_1 and whose common ratio is r. We are looking for a formula for the general term, a_n. Let's begin by writing the first six terms. The first term is a_1. The second term is $a_1 r$. The third term is $a_1 r \cdot r$, or $a_1 r^2$. The fourth term is $a_1 r^2 \cdot r$, or $a_1 r^3$, and so on. Starting with a_1 and multiplying each successive term by r, the first six terms are

$$a_1, \quad a_1 r, \quad a_1 r^2, \quad a_1 r^3, \quad a_1 r^4, \quad a_1 r^5.$$

a_1, first term; a_2, second term; a_3, third term; a_4, fourth term; a_5, fifth term; a_6, sixth term

Applying inductive reasoning to the pattern of the terms results in the following formula for the general term, or the nth term, of a geometric sequence:

GENERAL TERM OF A GEOMETRIC SEQUENCE

The nth term (the general term) of a geometric sequence with first term a_1 and common ratio r is

$$a_n = a_1 r^{n-1}.$$

EXAMPLE 6 *Using the Formula for the General Term of a Geometric Sequence*

Find the eighth term of the geometric sequence whose first term is -4 and whose common ratio is -2.

SOLUTION

To find the eighth term, a_8, we replace n in the formula with 8, a_1 with -4, and r with -2.

$$a_n = a_1 r^{n-1}$$
$$a_8 = -4(-2)^{8-1} = -4(-2)^7 = -4(-128) = 512$$

The eighth term is 512. We can check this result by writing the first eight terms of the sequence: $-4, 8, -16, 32, -64, 128, -256, 512$.

GREAT QUESTION!

When using $a_1 r^{n-1}$ to find the nth term of a geometric sequence, what should I do first?

Be careful with the order of operations when evaluating

$$a_1 r^{n-1}.$$

First, subtract 1 in the exponent and then raise r to that power. Finally, multiply the result by a_1.

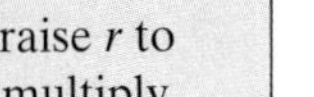

CHECK POINT 6 Find the seventh term of the geometric sequence whose first term is 5 and whose common ratio is -3.

EXAMPLE 7 Geometric Population Growth

The table shows the population of the United States in 2000 and 2010, with estimates given by the Census Bureau for 2001 through 2009.

Year	2000	2001	2002	2003	2004	2005	2006	2007	2008	2009	2010
Population (millions)	281.4	284.0	286.6	289.3	292.0	294.7	297.4	300.2	303.0	305.8	308.7

a. Show that the population is increasing geometrically.

b. Write the general term for the geometric sequence modeling the population of the United States, in millions, n years after 1999.

c. Project the U.S. population, in millions, for the year 2020.

SOLUTION

a. First, we use the sequence of population growth, 281.4, 284.0, 286.6, 289.3, and so on, to divide the population for each year by the population in the preceding year.

$$\frac{284.0}{281.4} \approx 1.009, \quad \frac{286.6}{284.0} \approx 1.009, \quad \frac{289.3}{286.6} \approx 1.009$$

Continuing in this manner, we will keep getting approximately 1.009. This means that the population is increasing geometrically with $r \approx 1.009$. The population of the United States in any year shown in the sequence is approximately 1.009 times the population the year before.

b. The sequence of the U.S. population growth is

$$281.4, 284.0, 286.6, 289.3, 292.0, 294.7, \ldots.$$

Because the population is increasing geometrically, we can find the general term of this sequence using

$$a_n = a_1 r^{n-1}.$$

In this sequence, $a_1 = 281.4$ and [from part (a)] $r \approx 1.009$. We substitute these values into the formula for the general term. This gives the general term for the geometric sequence modeling the U.S. population, in millions, n years after 1999.

$$a_n = 281.4(1.009)^{n-1}$$

c. We can use the formula for the general term, a_n, in part (b) to project the U.S. population for the year 2020. The year 2020 is 21 years after 1999—that is, $2020 - 1999 = 21$. Thus, $n = 21$. We substitute 21 for n in $a_n = 281.4(1.009)^{n-1}$.

$$a_{21} = 281.4(1.009)^{21-1} = 281.4(1.009)^{20} \approx 336.6$$

The model projects that the United States will have a population of approximately 336.6 million in the year 2020.

CHECK POINT 7 Write the general term for the geometric sequence

$$3, 6, 12, 24, 48, \ldots.$$

Then use the formula for the general term to find the eighth term.

Geometric Population Growth

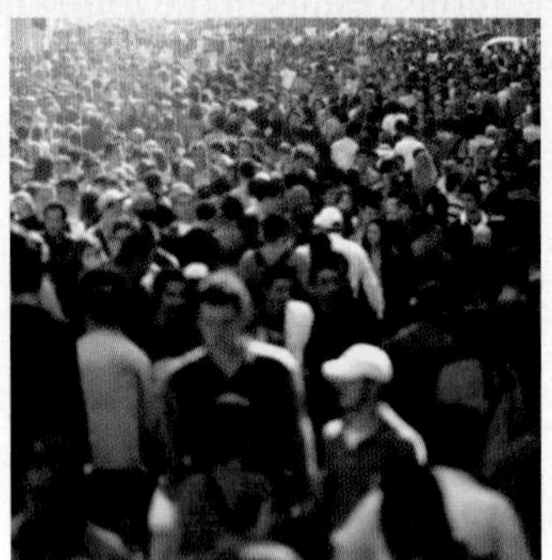

Economist Thomas Malthus (1766–1834) predicted that population growth would increase as a geometric sequence and food production would increase as an arithmetic sequence. He concluded that eventually population would exceed food production. If two sequences, one geometric and one arithmetic, are increasing, the geometric sequence will eventually overtake the arithmetic sequence, regardless of any head start that the arithmetic sequence might initially have.

Concept and Vocabulary Check

Fill in each blank so that the resulting statement is true.

1. A sequence in which each term after the first differs from the preceding term by a constant amount is called a/an __________ sequence. The difference between consecutive terms is called the __________ of the sequence.
2. The nth term of the sequence described in Exercise 1 is given by the formula __________, where a_1 is __________ and d is __________.
3. A sequence in which each term after the first is obtained by multiplying the preceding term by a fixed nonzero number is called a/an __________ sequence. The amount by which we multiply each time is called the __________ of the sequence.
4. The nth term of the sequence described in Exercise 3 is given by the formula __________, where a_1 is __________ and r is __________.

Exercise Set 5.7

Practice Exercises

In Exercises 1–20, write the first six terms of the arithmetic sequence with the first term, a_1, and common difference, d.

1. $a_1 = 8, d = 2$
2. $a_1 = 5, d = 3$
3. $a_1 = 200, d = 20$
4. $a_1 = 300, d = 50$
5. $a_1 = -7, d = 4$
6. $a_1 = -8, d = 5$
7. $a_1 = -400, d = 300$
8. $a_1 = -500, d = 400$
9. $a_1 = 7, d = -3$
10. $a_1 = 9, d = -5$
11. $a_1 = 200, d = -60$
12. $a_1 = 300, d = -90$
13. $a_1 = \frac{5}{2}, d = \frac{1}{2}$
14. $a_1 = \frac{3}{4}, d = \frac{1}{4}$
15. $a_1 = \frac{3}{2}, d = \frac{1}{4}$
16. $a_1 = \frac{3}{2}, d = -\frac{1}{4}$
17. $a_1 = 4.25, d = 0.3$
18. $a_1 = 6.3, d = 0.25$
19. $a_1 = 4.5, d = -0.75$
20. $a_1 = 3.5, d = -1.75$

In Exercises 21–40, find the indicated term for the arithmetic sequence with first term, a_1, and common difference, d.

21. Find a_6, when $a_1 = 13, d = 4$.
22. Find a_{16}, when $a_1 = 9, d = 2$.
23. Find a_{50}, when $a_1 = 7, d = 5$.
24. Find a_{60}, when $a_1 = 8, d = 6$.
25. Find a_9, when $a_1 = -5, d = 9$.
26. Find a_{10}, when $a_1 = -8, d = 10$.
27. Find a_{200}, when $a_1 = -40, d = 5$.
28. Find a_{150}, when $a_1 = -60, d = 5$.
29. Find a_{10}, when $a_1 = 8, d = -10$.
30. Find a_{11}, when $a_1 = 10, d = -6$.
31. Find a_{60}, when $a_1 = 35, d = -3$.
32. Find a_{70}, when $a_1 = -32, d = 4$.
33. Find a_{12}, when $a_1 = 12, d = -5$.
34. Find a_{20}, when $a_1 = -20, d = -4$.
35. Find a_{90}, when $a_1 = -70, d = -2$.
36. Find a_{80}, when $a_1 = 106, d = -12$.
37. Find a_{12}, when $a_1 = 6, d = \frac{1}{2}$.
38. Find a_{14}, when $a_1 = 8, d = \frac{1}{4}$.
39. Find a_{50}, when $a_1 = 14, d = -0.25$.
40. Find a_{110}, when $a_1 = -12, d = -0.5$.

In Exercises 41–48, write a formula for the general term (the nth term) of each arithmetic sequence. Then use the formula for a_n to find a_{20}, the 20th term of the sequence.

41. $1, 5, 9, 13, \ldots$
42. $2, 7, 12, 17, \ldots$
43. $7, 3, -1, -5, \ldots$
44. $6, 1, -4, -9, \ldots$
45. $a_1 = 9, d = 2$
46. $a_1 = 6, d = 3$
47. $a_1 = -20, d = -4$
48. $a_1 = -70, d = -5$

In Exercises 49–70, write the first six terms of the geometric sequence with the first term, a_1, and common ratio, r.

49. $a_1 = 4, r = 2$
50. $a_1 = 2, r = 3$
51. $a_1 = 1000, r = 1$
52. $a_1 = 5000, r = 1$
53. $a_1 = 3, r = -2$
54. $a_1 = 2, r = -3$
55. $a_1 = 10, r = -4$
56. $a_1 = 20, r = -4$
57. $a_1 = 2000, r = -1$
58. $a_1 = 3000, r = -1$
59. $a_1 = -2, r = -3$
60. $a_1 = -4, r = -2$
61. $a_1 = -6, r = -5$
62. $a_1 = -8, r = -5$
63. $a_1 = \frac{1}{4}, r = 2$
64. $a_1 = \frac{1}{2}, r = 2$
65. $a_1 = \frac{1}{4}, r = \frac{1}{2}$
66. $a_1 = \frac{1}{5}, r = \frac{1}{2}$
67. $a_1 = -\frac{1}{16}, r = -4$
68. $a_1 = -\frac{1}{8}, r = -2$
69. $a_1 = 2, r = 0.1$
70. $a_1 = -1000, r = 0.1$

In Exercises 71–90, find the indicated term for the geometric sequence with first term, a_1, and common ratio, r.

71. Find a_7, when $a_1 = 4, r = 2$.
72. Find a_5, when $a_1 = 4, r = 3$.
73. Find a_{20}, when $a_1 = 2, r = 3$.
74. Find a_{20}, when $a_1 = 2, r = 2$.
75. Find a_{100}, when $a_1 = 50, r = 1$.
76. Find a_{200}, when $a_1 = 60, r = 1$.
77. Find a_7, when $a_1 = 5, r = -2$.
78. Find a_4, when $a_1 = 4, r = -3$.
79. Find a_{30}, when $a_1 = 2, r = -1$.
80. Find a_{40}, when $a_1 = 6, r = -1$.
81. Find a_6, when $a_1 = -2, r = -3$.
82. Find a_5, when $a_1 = -5, r = -2$.
83. Find a_8, when $a_1 = 6, r = \frac{1}{2}$.
84. Find a_8, when $a_1 = 12, r = \frac{1}{2}$.

85. Find a_6, when $a_1 = 18, r = -\frac{1}{3}$.

86. Find a_4, when $a_1 = 9, r = -\frac{1}{3}$.

87. Find a_{40}, when $a_1 = 1000, r = -\frac{1}{2}$.

88. Find a_{30}, when $a_1 = 8000, r = -\frac{1}{2}$.

89. Find a_8, when $a_1 = 1{,}000{,}000, r = 0.1$.

90. Find a_8, when $a_1 = 40{,}000, r = 0.1$.

In Exercises 91–98, write a formula for the general term (the nth term) of each geometric sequence. Then use the formula for a_n to find a_7, the seventh term of the sequence.

91. $3, 12, 48, 192, \ldots$

92. $3, 15, 75, 375, \ldots$

93. $18, 6, 2, \frac{2}{3}, \ldots$

94. $12, 6, 3, \frac{3}{2}, \ldots$

95. $1.5, -3, 6, -12, \ldots$

96. $5, -1, \frac{1}{5}, -\frac{1}{25}, \ldots$

97. $0.0004, -0.004, 0.04, -0.4, \ldots$

98. $0.0007, -0.007, 0.07, -0.7, \ldots$

Determine whether each sequence in Exercises 99–114 is arithmetic or geometric. Then find the next two terms.

99. $2, 6, 10, 14, \ldots$

100. $3, 8, 13, 18, \ldots$

101. $5, 15, 45, 135, \ldots$

102. $15, 30, 60, 120, \ldots$

103. $-7, -2, 3, 8, \ldots$

104. $-9, -5, -1, 3, \ldots$

105. $3, \frac{3}{2}, \frac{3}{4}, \frac{3}{8}, \ldots$

106. $6, 3, \frac{3}{2}, \frac{3}{4}, \ldots$

107. $\frac{1}{2}, 1, \frac{3}{2}, 2, \ldots$

108. $\frac{2}{3}, 1, \frac{4}{3}, \frac{5}{3}, \ldots$

109. $7, -7, 7, -7, \ldots$

110. $6, -6, 6, -6, \ldots$

111. $7, -7, -21, -35, \ldots$

112. $6, -6, -18, -30, \ldots$

113. $\sqrt{5}, 5, 5\sqrt{5}, 25, \ldots$

114. $\sqrt{3}, 3, 3\sqrt{3}, 9, \ldots$

Practice Plus

The sum, S_n, of the first n terms of an arithmetic sequence is given by

$$S_n = \frac{n}{2}(a_1 + a_n),$$

in which a_1 is the first term and a_n is the nth term. The sum, S_n, of the first n terms of a geometric sequence is given by

$$S_n = \frac{a_1(1 - r^n)}{1 - r},$$

in which a_1 is the first term and r is the common ratio ($r \neq 1$). In Exercises 115–122, determine whether each sequence is arithmetic or geometric. Then use the appropriate formula to find S_{10}, the sum of the first ten terms.

115. $4, 10, 16, 22, \ldots$

116. $7, 19, 31, 43, \ldots$

117. $2, 6, 18, 54, \ldots$

118. $3, 6, 12, 24, \ldots$

119. $3, -6, 12, -24, \ldots$

120. $4, -12, 36, -108, \ldots$

121. $-10, -6, -2, 2, \ldots$

122. $-15, -9, -3, 3, \ldots$

123. Use the appropriate formula shown above to find $1 + 2 + 3 + 4 + \cdots + 100$, the sum of the first 100 natural numbers.

124. Use the appropriate formula shown above to find $2 + 4 + 6 + 8 + \cdots + 200$, the sum of the first 100 positive even integers.

Application Exercises

The bar graph shows changes in the percentage of college graduates for Americans ages 25 and older from 1990 to 2015. Exercises 125–126 involve developing arithmetic sequences that model the data.

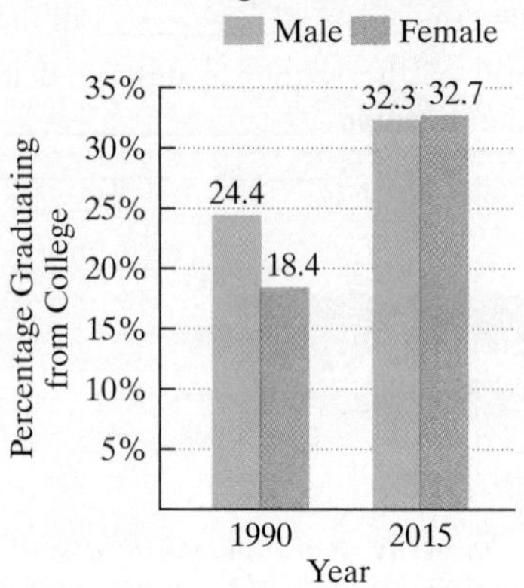

Source: U.S. Census Bureau

125. In 1990, 18.4% of American women ages 25 and older had graduated from college. On average, this percentage has increased by approximately 0.6 each year.

a. Write a formula for the nth term of the arithmetic sequence that models the percentage of American women ages 25 and older who had graduated from college n years after 1989.

b. Use the model from part (a) to project the percentage of American women ages 25 and older who will be college graduates by 2029.

126. In 1990, 24.4% of American men ages 25 and older had graduated from college. On average, this percentage has increased by approximately 0.3 each year.

a. Write a formula for the nth term of the arithmetic sequence that models the percentage of American men ages 25 and older who had graduated from college n years after 1989.

b. Use the model from part (a) to project the percentage of American men ages 25 and older who will be college graduates by 2029.

127. Company A pays \$44,000 yearly with raises of \$1600 per year. Company B pays \$48,000 yearly with raises of \$1000 per year. Which company will pay more in year 10? How much more?

128. Company A pays \$53,000 yearly with raises of \$1600 per year. Company B pays \$56,000 yearly with raises of \$1200 per year. Which company will pay more in year 10? How much more?

In Exercises 129–130, suppose you save \$1 the first day of a month, \$2 the second day, \$4 the third day, and so on. That is, each day you save twice as much as you did the day before.

129. What will you put aside for savings on the fifteenth day of the month?

130. What will you put aside for savings on the thirtieth day of the month?

131. A professional baseball player signs a contract with a beginning salary of \$3,000,000 for the first year with an annual increase of 4% per year beginning in the second year. That is, beginning in year 2, the athlete's salary will be 1.04 times what it was in the previous year. What is the athlete's salary for year 7 of the contract? Round to the nearest dollar.

132. You are offered a job that pays \$50,000 for the first year with an annual increase of 3% per year beginning in the second year. That is, beginning in year 2, your salary will be 1.03 times what it was in the previous year. What can you expect to earn in your sixth year on the job? Round to the nearest dollar.

In Exercises 133–134, you will develop geometric sequences that model the population growth for California and Texas, the two most populated U.S. states.

133. The table shows the population of California for 2000 and 2010, with estimates given by the U.S. Census Bureau for 2001 through 2009.

Year	2000	2001	2002	2003	2004
Population in millions	33.87	34.21	34.55	34.90	35.25

Year	2005	2006	2007	2008	2009	2010
Population in millions	35.60	36.00	36.36	36.72	37.09	37.25

a. Divide the population for each year by the population in the preceding year. Round to two decimal places and show that California has a population increase that is approximately geometric.

b. Write the general term of the geometric sequence modeling California's population, in millions, n years after 1999.

c. Use your model from part (b) to project California's population, in millions, for the year 2020. Round to two decimal places.

134. The table shows the population of Texas for 2000 and 2010, with estimates given by the U.S. Census Bureau for 2001 through 2009.

Year	2000	2001	2002	2003	2004	2005
Population in millions	20.85	21.27	21.70	22.13	22.57	23.02

Year	2006	2007	2008	2009	2010
Population in millions	23.48	23.95	24.43	24.92	25.15

a. Divide the population for each year by the population in the preceding year. Round to two decimal places and show that Texas has a population increase that is approximately geometric.

b. Write the general term of the geometric sequence modeling Texas's population, in millions, n years after 1999.

c. Use your model from part (b) to project Texas's population, in millions, for the year 2020. Round to two decimal places.

Explaining the Concepts

135. What is a sequence? Give an example with your description.

136. What is an arithmetic sequence? Give an example with your description.

137. What is the common difference in an arithmetic sequence?

138. What is a geometric sequence? Give an example with your description.

139. What is the common ratio in a geometric sequence?

140. If you are given a sequence that is arithmetic or geometric, how can you determine which type of sequence it is?

141. For the first 30 days of a flu outbreak, the number of students on your campus who become ill is increasing. Which is worse: The number of students with the flu is increasing arithmetically or is increasing geometrically? Explain your answer.

Critical Thinking Exercises

Make Sense? *In Exercises 142–145, determine whether each statement makes sense or does not make sense, and explain your reasoning.*

142. Now that I've studied sequences, I realize that the joke in the accompanying cartoon is based on the fact that you can't have a negative number of sheep.

When math teachers can't sleep.

143. The sequence for the number of seats per row in our movie theater as the rows move toward the back is arithmetic with $d = 1$ so people don't block the view of those in the row behind them.

144. There's no end to the number of geometric sequences that I can generate whose first term is 5 if I pick nonzero numbers r and multiply 5 by each value of r repeatedly.

145. I've noticed that the big difference between arithmetic and geometric sequences is that arithmetic sequences are based on addition and geometric sequences are based on multiplication.

In Exercises 146–153, determine whether each statement is true or false. If the statement is false, make the necessary change(s) to produce a true statement.

146. The common difference for the arithmetic sequence given by $1, -1, -3, -5, \ldots$ is 2.

147. The sequence $1, 4, 8, 13, 19, 26, \ldots$ is an arithmetic sequence.

148. The nth term of an arithmetic sequence whose first term is a_1 and whose common difference is d is $a_n = a_1 + nd$.

149. If the first term of an arithmetic sequence is 5 and the third term is -3, then the fourth term is -7.

150. The sequence $2, 6, 24, 120, \ldots$ is an example of a geometric sequence.

151. Adjacent terms in a geometric sequence have a common difference.

152. A sequence that is not arithmetic must be geometric.

153. If a sequence is geometric, we can write as many terms as we want by repeatedly multiplying by the common ratio.

154. A person is investigating two employment opportunities. They both have a beginning salary of \$20,000 per year. Company A offers an increase of \$1000 per year. Company B offers 5% more than during the preceding year. Which company will pay more in the sixth year?

Group Exercise

155. Enough curiosities involving the Fibonacci sequence exist to warrant a flourishing Fibonacci Association. It publishes a quarterly journal. Do some research on the Fibonacci sequence by consulting the research department of your library or the Internet, and find one property that interests you. After doing this research, get together with your group to share these intriguing properties.

Chapter Summary, Review, and Test

SUMMARY – DEFINITIONS AND CONCEPTS	EXAMPLES
5.1 Number Theory: Prime and Composite Numbers	
a. The set of natural numbers is $\{1, 2, 3, 4, 5, \ldots\}$. $b \mid a$ (b divides a: a is divisible by b) for natural numbers a and b if the operation of dividing a by b leaves a remainder of 0. Rules of divisibility are given in Table 5.1 on page 253.	Ex. 1, p. 254
b. A prime number is a natural number greater than 1 that has only itself and 1 as factors. A composite number is a natural number greater than 1 that is divisible by a number other than itself and 1. The Fundamental Theorem of Arithmetic: Every composite number can be expressed as a product of prime numbers in one and only one way (if the order of the factors is disregarded).	Ex. 2, p. 255
c. The greatest common divisor of two or more natural numbers is the largest number that is a divisor (or factor) of all the numbers. The procedure for finding the greatest common divisor is given in the box on page 256.	Ex. 3, p. 256; Ex. 4, p. 257
d. The least common multiple of two or more natural numbers is the smallest natural number that is divisible by all of the numbers. The procedure for finding the least common multiple is given in the box on page 258.	Ex. 5, p. 258; Ex. 6, p. 259
5.2 The Integers; Order of Operations	
a. The set of whole numbers is $\{0, 1, 2, 3, 4, 5, \ldots\}$. The set of integers is $\{\ldots, -3, -2, -1, 0, 1, 2, 3, \ldots\}$. Integers are graphed on a number line by placing a dot at the correct location for each number.	Ex. 1, p. 263
b. $a < b$ (a is less than b) means a is to the left of b on a number line. $a > b$ (a is greater than b) means a is to the right of b on a number line.	Ex. 2, p. 264
c. $\lvert a \rvert$, the absolute value of a, is the distance of a from 0 on a number line. The absolute value of a positive number is the number itself. The absolute value of 0 is 0: $\lvert 0 \rvert = 0$. The absolute value of a negative number is the number without the negative sign. For example, $\lvert -8 \rvert = 8$.	Ex. 3, p. 265
d. Rules for performing operations with integers are given in the boxes on pages 266, 269, and 271.	Ex. 4, p. 267; Ex. 5, p. 268; Ex. 6, p. 269
e. Order of Operations **1.** Perform all operations within grouping symbols. **2.** Evaluate all exponential expressions. **3.** Do all multiplications and divisions from left to right. **4.** Do all additions and subtractions from left to right.	Ex. 7, p. 272; Ex. 8, p. 272

5.3 The Rational Numbers

a. The set of rational numbers is the set of all numbers which can be expressed in the form $\frac{a}{b}$, where a and b are integers and b is not equal to 0.	
b. A rational number is reduced to its lowest terms, or simplified, by dividing both the numerator and the denominator by their greatest common divisor.	Ex. 1, p. 277
c. A mixed number consists of the sum of an integer and a rational number, expressed without the use of an addition sign. An improper fraction is a rational number whose numerator is greater than its denominator. Procedures for converting between these forms are given in the boxes on pages 277 and 278.	Ex. 2, p. 278; Ex. 3, p. 278
d. Any rational number can be expressed as a decimal. The resulting decimal will either terminate (stop), or it will have a digit that repeats or a block of digits that repeats. The rational number $\frac{a}{b}$ is expressed as a decimal by dividing b into a.	Ex. 4, p. 279
e. To express a terminating decimal as a quotient of integers, the digits to the right of the decimal point are the numerator. The place-value of the last digit to the right of the decimal point determines the denominator.	Ex. 5, p. 280
f. To express a repeating decimal as a quotient of integers, use the boxed procedure on page 282.	Ex. 6, p. 281; Ex. 7, p. 282
g. The product of two rational numbers is the product of their numerators divided by the product of their denominators.	Ex. 8, p. 283
h. Two numbers whose product is 1 are called reciprocals, or multiplicative inverses, of each other. The quotient of two rational numbers is the product of the first number and the reciprocal of the second number.	Ex. 9, p. 283
i. The sum or difference of two rational numbers with identical denominators is the sum or difference of their numerators over the common denominator.	Ex. 10, p. 284
j. Add or subtract rational numbers with unlike denominators by first expressing each rational number with the least common denominator and then following item (i) above.	Ex. 11, p. 284; Ex. 12, p. 285
k. The order of operations can be applied to an expression with rational numbers.	Ex. 13, p. 286
l. Density of the Rational Numbers Given any two distinct rational numbers, there is always a rational number between them. To find the rational number halfway between two rational numbers, add the rational numbers and divide their sum by 2.	Ex. 14, p. 286

5.4 The Irrational Numbers

a. The set of irrational numbers is the set of numbers whose decimal representations are neither terminating nor repeating. Examples of irrational numbers are $\sqrt{2} \approx 1.414$ and $\pi \approx 3.142$.	
b. Simplifying square roots: Use the product rule, $\sqrt{ab} = \sqrt{a} \cdot \sqrt{b}$, to remove from the square root any perfect squares that occur as factors.	Ex. 1, p. 294
c. Multiplying square roots: $\sqrt{a} \cdot \sqrt{b} = \sqrt{ab}$. The product of square roots is the square root of the product.	Ex. 2, p. 295
d. Dividing square roots: $\frac{\sqrt{a}}{\sqrt{b}} = \sqrt{\frac{a}{b}}$. The quotient of square roots is the square root of the quotient.	Ex. 3, p. 295
e. Adding and subtracting square roots: If the radicals have the same radicand, add or subtract their coefficients. The answer is the sum or difference of the coefficients times the common square root. Addition or subtraction is sometimes possible by first simplifying the square roots.	Ex. 4, p. 296; Ex. 5, p. 296
f. Rationalizing denominators: Multiply the numerator and the denominator by the smallest number that produces a perfect square radicand in the denominator.	Ex. 6, p. 297

5.5 Real Numbers and Their Properties; Clock Addition

a. The set of real numbers is obtained by combining the rational numbers with the irrational numbers. The important subsets of the real numbers are summarized in Table 5.2 on page 304. A diagram representing the relationships among the subsets of the real numbers is given to the left of Table 5.2. Ex. 1, p. 304

b. Properties of real numbers, including closure properties ($a + b$ and ab are real numbers), commutative properties ($a + b = b + a; ab = ba$), associative properties $[(a + b) + c = a + (b + c);\ (ab)c = a(bc)]$, the distributive property $[a(b + c) = ab + ac]$, identity properties ($a + 0 = a; 0 + a = a;\ a \cdot 1 = a; 1 \cdot a = a$), and inverse properties $[a + (-a) = 0; (-a) + a = 0; a \cdot \frac{1}{a} = 1, a \neq 0;\ \frac{1}{a} \cdot a = 1, a \neq 0]$ are summarized in Table 5.3 on page 306. Ex. 2, p. 307; Ex. 3, p. 307

c. Clock addition is defined by moving a clock's hour hand in a clockwise direction. Tables for clock addition show that the operation satisfies closure, associative, identity, inverse, and commutative properties. Clock addition can be used to explore various kinds of rotational symmetry. Ex. 4, p. 310

5.6 Exponents and Scientific Notation

a. Properties of Exponents Table 5.6, p. 315; Ex. 1, p. 316; Ex. 2, p. 317

- Product rule: $b^m \cdot b^n = b^{m+n}$
- Zero exponent rule: $b^0 = 1, b \neq 0$
- Power rule: $(b^m)^n = b^{m \cdot n}$
- Negative exponent rule: $b^{-m} = \frac{1}{b^m}, b \neq 0$
- Quotient rule: $\frac{b^m}{b^n} = b^{m-n}, b \neq 0$

b. A positive number in scientific notation is expressed as $a \times 10^n$, where $1 \leq a < 10$ and n is an integer.

c. Changing from Scientific to Decimal Notation: If n is positive, move the decimal point in a to the right n places. If n is negative, move the decimal point in a to the left $|n|$ places. Ex. 3, p. 318

d. Changing from Decimal to Scientific Notation: Move the decimal point in the given number to obtain a, where $1 \leq a < 10$. The number of places the decimal point moves gives the absolute value of n in $a \times 10^n$; n is positive if the number is greater than 10 and negative if the number is less than 1. Ex. 4, p. 319; Ex. 5, p. 319

e. The product and quotient rules for exponents are used to multiply and divide numbers in scientific notation. If a number is written in scientific notation, $a \times 10^n$, the digits in a are called significant digits. If rounding is necessary, round the scientific notation answer to the least number of significant digits found in any of the given numbers. Ex. 6, p. 320; Ex. 7, p. 321; Ex. 8, p. 321; Ex. 9, p. 322

5.7 Arithmetic and Geometric Sequences

a. In an arithmetic sequence, each term after the first differs from the preceding term by a constant, the common difference. Subtract any term from the term that directly follows it to find the common difference. Ex. 1, p. 327; Ex. 2, p. 328

b. The general term, or the nth term, of an arithmetic sequence is Ex. 3, p. 328; Ex. 4, p. 329

$$a_n = a_1 + (n - 1)d,$$

where a_1 is the first term and d is the common difference.

c. In a geometric sequence, each term after the first is obtained by multiplying the preceding term by a nonzero constant, the common ratio. Divide any term after the first by the term that directly precedes it to find the common ratio. Ex. 5, p. 331

d. The general term, or the nth term, of a geometric sequence is Ex. 6, p. 331; Ex. 7, p. 332

$$a_n = a_1 r^{n-1},$$

where a_1 is the first term and r is the common ratio.

Review Exercises

5.1

In Exercises 1 and 2, determine whether the number is divisible by each of the following numbers: 2, 3, 4, 5, 6, 8, 9, 10, and 12. If you are using a calculator, explain the divisibility shown by your calculator using one of the rules of divisibility.

1. 238,632 **2.** 421,153,470

In Exercises 3–5, find the prime factorization of each composite number.

3. 705 **4.** 960 **5.** 6825

In Exercises 6–8, find the greatest common divisor and the least common multiple of the numbers.

6. 30 and 48

7. 36 and 150

8. 216 and 254

9. For an intramural league, you need to divide 24 men and 60 women into all-male and all-female teams so that each team has the same number of people. What is the largest number of people that can be placed on a team?

10. The media center at a college runs videotapes of two lectures continuously. One videotape runs for 42 minutes and a second runs for 56 minutes. Both videotapes begin at 9:00 A.M. When will the videos of the two lectures begin again at the same time?

5.2

In Exercises 11–12, insert either < or > in the shaded area between the integers to make the statement true.

11. $-93 \blacksquare 17$ 12. $-2 \blacksquare -200$

In Exercises 13–15, find the absolute value.

13. $|-860|$ 14. $|53|$ 15. $|0|$

Perform the indicated operations in Exercises 16–28.

16. $8 + (-11)$ 17. $-6 + (-5)$
18. $-7 - 8$ 19. $-7 - (-8)$
20. $(-9)(-11)$ 21. $5(-3)$
22. $\dfrac{-36}{-4}$ 23. $\dfrac{20}{-5}$
24. $-40 \div 5 \cdot 2$ 25. $-6 + (-2) \cdot 5$
26. $6 - 4(-3 + 2)$ 27. $28 \div (2 - 4^2)$
28. $36 - 24 \div 4 \cdot 3 - 1$

29. For the year 2015, the Congressional Budget Office projected a budget deficit of −$57 billion. For the same year, the Brookings Institution forecast a budget deficit of −$715 billion. What is the difference between the CBO projection and the Brookings projection?

5.3

In Exercises 30–32, reduce each rational number to its lowest terms.

30. $\frac{40}{75}$ 31. $\frac{36}{150}$ 32. $\frac{165}{180}$

In Exercises 33–34, convert each mixed number to an improper fraction.

33. $5\frac{9}{11}$ 34. $-3\frac{2}{7}$

In Exercises 35–36, convert each improper fraction to a mixed number.

35. $\frac{27}{5}$ 36. $-\frac{17}{9}$

In Exercises 37–40, express each rational number as a decimal.

37. $\frac{4}{5}$ 38. $\frac{3}{7}$ 39. $\frac{5}{8}$ 40. $\frac{9}{16}$

In Exercises 41–44, express each terminating decimal as a quotient of integers in lowest terms.

41. 0.6 42. 0.68 43. 0.588 44. 0.0084

In Exercises 45–47, express each repeating decimal as a quotient of integers in lowest terms.

45. $0.\overline{5}$ 46. $0.\overline{34}$ 47. $0.\overline{113}$

In Exercises 48–58, perform the indicated operations. Where possible, reduce the answer to lowest terms.

48. $\frac{3}{5} \cdot \frac{7}{10}$ 49. $\left(3\frac{1}{3}\right)\left(1\frac{3}{4}\right)$ 50. $\frac{4}{5} \div \frac{3}{10}$
51. $-1\frac{2}{3} \div 6\frac{2}{3}$ 52. $\frac{2}{9} + \frac{4}{9}$ 53. $\frac{7}{9} + \frac{5}{12}$
54. $\frac{3}{4} - \frac{2}{15}$ 55. $\frac{1}{3} + \frac{1}{2} \cdot \frac{4}{5}$ 56. $\frac{3}{8}\left(\frac{1}{2} + \frac{1}{3}\right)$
57. $\frac{1}{2} - \frac{2}{3} \div \frac{5}{9} + \frac{3}{10}$
58. $\left(\frac{1}{2} + \frac{1}{3}\right) \div \left(\frac{1}{4} - \frac{3}{8}\right)$

In Exercises 59–60, find the rational number halfway between the two numbers in each pair.

59. $\frac{1}{7}$ and $\frac{1}{8}$ 60. $\frac{3}{4}$ and $\frac{3}{5}$

61. A recipe for coq au vin is meant for six people and requires $4\frac{1}{2}$ pounds of chicken. If you want to serve 15 people, how much chicken is needed?

62. The gas tank of a car is filled to its capacity. The first day, $\frac{1}{4}$ of the tank's gas is used for travel. The second day, $\frac{1}{3}$ of the tank's original amount of gas is used for travel. What fraction of the tank is filled with gas at the end of the second day?

5.4

In Exercises 63–66, simplify the square root.

63. $\sqrt{28}$ 64. $\sqrt{72}$
65. $\sqrt{150}$ 66. $\sqrt{300}$

In Exercises 67–75, perform the indicated operation. Simplify the answer when possible.

67. $\sqrt{6} \cdot \sqrt{8}$ 68. $\sqrt{10} \cdot \sqrt{5}$
69. $\dfrac{\sqrt{24}}{\sqrt{2}}$ 70. $\dfrac{\sqrt{27}}{\sqrt{3}}$
71. $\sqrt{5} + 4\sqrt{5}$ 72. $7\sqrt{11} - 13\sqrt{11}$
73. $\sqrt{50} + \sqrt{8}$ 74. $\sqrt{3} - 6\sqrt{27}$
75. $2\sqrt{18} + 3\sqrt{8}$

In Exercises 76–77, rationalize the denominator.

76. $\dfrac{30}{\sqrt{5}}$ 77. $\sqrt{\dfrac{2}{3}}$

78. Paleontologists use the mathematical model $W = 4\sqrt{2x}$ to estimate the walking speed of a dinosaur, W, in feet per second, where x is the length, in feet, of the dinosaur's leg. What is the walking speed of a dinosaur whose leg length is 6 feet? Express the answer in simplified radical form. Then use your calculator to estimate the walking speed to the nearest tenth of a foot per second.

5.5

79. Consider the set

$$\left\{-17, -\tfrac{9}{13}, 0, 0.75, \sqrt{2}, \pi, \sqrt{81}\right\}.$$

List all numbers from the set that are **a.** natural numbers, **b.** whole numbers, **c.** integers, **d.** rational numbers, **e.** irrational numbers, **f.** real numbers.

80. Give an example of an integer that is not a natural number.

81. Give an example of a rational number that is not an integer.

82. Give an example of a real number that is not a rational number.

In Exercises 83–90, state the name of the property illustrated.

83. $3 + 17 = 17 + 3$

84. $(6 \cdot 3) \cdot 9 = 6 \cdot (3 \cdot 9)$

85. $\sqrt{3}(\sqrt{5} + \sqrt{3}) = \sqrt{15} + 3$

86. $(6 \cdot 9) \cdot 2 = 2 \cdot (6 \cdot 9)$

87. $\sqrt{3}(\sqrt{5} + \sqrt{3}) = (\sqrt{5} + \sqrt{3})\sqrt{3}$

88. $(3 \cdot 7) + (4 \cdot 7) = (4 \cdot 7) + (3 \cdot 7)$

89. $-3\left(-\frac{1}{3}\right) = 1$

90. $\sqrt{7} \cdot 1 = \sqrt{7}$

In Exercises 91–92, give an example to show that

91. The natural numbers are not closed with respect to division.

92. The whole numbers are not closed with respect to subtraction.

93. Shown in the figure is a 5-hour clock and the table for clock addition in the 5-hour system.

$\oplus$	0	1	2	3	4
0	0	1	2	3	4
1	1	2	3	4	0
2	2	3	4	0	1
3	3	4	0	1	2
4	4	0	1	2	3

a. How can you tell that the set {0, 1, 2, 3, 4} is closed under the operation of clock addition?

b. Verify the associative property:

$$(4 \oplus 2) \oplus 3 = 4 \oplus (2 \oplus 3)$$

c. What is the identity element in the 5-hour clock system?

d. Find the inverse of each element in the 5-hour clock system.

e. Verify two cases of the commutative property:

$$3 \oplus 4 = 4 \oplus 3 \quad \text{and} \quad 3 \oplus 2 = 2 \oplus 3$$

5.6

In Exercises 94–104, evaluate each expression.

94. $6 \cdot 6^2$ **95.** $2^3 \cdot 2^3$ **96.** $(2^2)^2$

97. $(3^3)^2$ **98.** $\frac{5^6}{5^4}$ **99.** 7^0

100. $(-7)^0$ **101.** 6^{-3} **102.** 2^{-4}

103. $\frac{7^4}{7^6}$ **104.** $3^5 \cdot 3^{-2}$

In Exercises 105–108, express each number in decimal notation.

105. 4.6×10^2 **106.** 3.74×10^4

107. 2.55×10^{-3} **108.** 7.45×10^{-5}

In Exercises 109–114, express each number in scientific notation.

109. 7520 **110.** 3,590,000

111. 0.00725 **112.** 0.000000409

113. 420×10^{11} **114.** 0.97×10^{-4}

In Exercises 115–118, perform the indicated operation and express each answer in decimal notation.

115. $(3 \times 10^7)(1.3 \times 10^{-5})$ **116.** $(5 \times 10^3)(2.3 \times 10^2)$

117. $\frac{6.9 \times 10^3}{3 \times 10^5}$ **118.** $\frac{2.4 \times 10^{-4}}{6 \times 10^{-6}}$

In Exercises 119–122, perform the indicated operation by first expressing each number in scientific notation. Write the answer in scientific notation.

119. (60,000)(540,000)

120. (91,000)(0.0004)

121. $\frac{8{,}400{,}000}{4000}$ **122.** $\frac{0.000003}{0.00000006}$

In 2011, the United States government spent more than it had collected in taxes, resulting in a budget deficit of \$1.3 trillion. In Exercises 123–125, you will use scientific notation to put a number like 1.3 trillion in perspective. Use 10^{12} for 1 trillion.

123. Express 1.3 trillion in scientific notation.

124. There are approximately 32,000,000 seconds in a year. Express this number in scientific notation.

125. Use your scientific notation answers from Exercises 123 and 124 to answer this question: How many years is 1.3 trillion seconds? (*Note*: 1.3 trillion seconds would take us back in time to a period when Neanderthals were using stones to make tools.)

126. The human body contains approximately 3.2×10^4 microliters of blood for every pound of body weight. Each microliter of blood contains approximately 5×10^6 red blood cells. Express in scientific notation the approximate number of red blood cells in the body of a 180-pound person.

5.7

In Exercises 127–129, write the first six terms of the arithmetic sequence with the first term, a_1, and common difference, d.

127. $a_1 = 7, d = 4$

128. $a_1 = -4, d = -5$

129. $a_1 = \frac{3}{2}, d = -\frac{1}{2}$

In Exercises 130–132, find the indicated term for the arithmetic sequence with first term, a_1, and common difference, d.

130. Find a_6, when $a_1 = 5, d = 3$.

131. Find a_{12}, when $a_1 = -8, d = -2$.

132. Find a_{14}, when $a_1 = 14, d = -4$.

In Exercises 133–134, write a formula for the general term (the nth term) of each arithmetic sequence. Then use the formula for a_n to find a_{20}, the 20th term of the sequence.

133. $-7, -3, 1, 5, \ldots$

134. $a_1 = 200, d = -20$

In Exercises 135–137, write the first six terms of the geometric sequence with the first term, a_1, and common ratio, r.

135. $a_1 = 3, r = 2$

136. $a_1 = \frac{1}{2}, r = \frac{1}{2}$

137. $a_1 = 16, r = -\frac{1}{2}$

In Exercises 138–140, find the indicated term for the geometric sequence with first term, a_1, and common ratio, r.

138. Find a_4, when $a_1 = 2, r = 3$.

139. Find a_6, when $a_1 = 16, r = \frac{1}{2}$.

140. Find a_5, when $a_1 = -3, r = 2$.

In Exercises 141–142, write a formula for the general term (the nth term) of each geometric sequence. Then use the formula for a_n to find a_8, the eighth term of the sequence.

141. $1, 2, 4, 8, \ldots$

142. $100, 10, 1, \frac{1}{10}, \ldots$

Determine whether each sequence in Exercises 143–146 is arithmetic or geometric. Then find the next two terms.

143. $4, 9, 14, 19, \ldots$

144. $2, 6, 18, 54, \ldots$

145. $1, \frac{1}{4}, \frac{1}{16}, \frac{1}{64}, \ldots$

146. $0, -7, -14, -21, \ldots$

147. In 2014, the average ticket price for top rock concerts, adjusted for inflation, had increased by 77% since 1995. This was greater than the percent increase in the cost of tuition at private four-year colleges during the same time period. The bar graph shows the average ticket price, in 2015 dollars, for rock concerts in 1995 and 2014.

Average Ticket Price for Rock Concerts

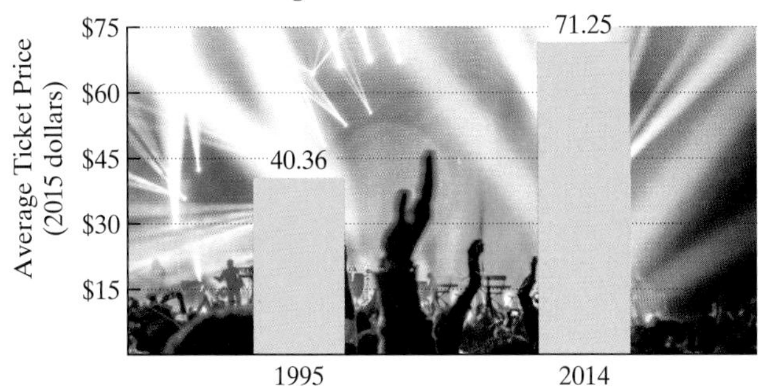

Source: *Rolling Stone*

In 1995, the average ticket price, in 2015 dollars, for a rock concert was \$40.36. On average, this has increased by approximately \$1.63 per year since then.

a. Write a formula for the nth term of the arithmetic sequence that describes the average ticket price, in 2015 dollars, for rock concerts n years after 1994.

b. Use the model to project the average ticket price, in 2015 dollars, for rock concerts in 2020.

148. The table shows the population of Florida for 2000 and 2010, with estimates given by the U.S. Census Bureau for 2001 through 2009.

Year	2000	2001	2002	2003	2004	2005
Population in millions	15.98	16.24	16.50	16.76	17.03	17.30

Year	2006	2007	2008	2009	2010
Population in millions	17.58	17.86	18.15	18.44	18.80

a. Divide the population for each year by the population in the preceding year. Round to two decimal places and show that Florida has a population increase that is approximately geometric.

b. Write the general term of the geometric sequence modeling Florida's population, in millions, n years after 1999.

c. Use your the model from part (b) to project Florida's population, in millions, for the year 2030. Round to two decimal places.

Chapter 5 Test

1. Which of the numbers 2, 3, 4, 5, 6, 8, 9, 10, and 12 divide 391,248?

2. Find the prime factorization of 252.

3. Find the greatest common divisor and the least common multiple of 48 and 72.

Perform the indicated operations in Exercises 4–6.

4. $-6 - (5 - 12)$

5. $(-3)(-4) \div (7 - 10)$

6. $(6 - 8)^2(5 - 7)^3$

7. Express $\frac{7}{12}$ as a decimal.

8. Express $0.\overline{64}$ as a quotient of integers in lowest terms.

In Exercises 9–11, perform the indicated operations. Where possible, reduce the answer to its lowest terms.

9. $\left(-\frac{3}{7}\right) \div \left(-2\frac{1}{7}\right)$

10. $\frac{19}{24} - \frac{7}{40}$

11. $\frac{1}{2} - 8\left(\frac{1}{4} + 1\right)$

12. Find the rational number halfway between $\frac{1}{2}$ and $\frac{2}{3}$.

13. Multiply and simplify: $\sqrt{10} \cdot \sqrt{5}$.

14. Add: $\sqrt{50} + \sqrt{32}$.

15. Rationalize the denominator: $\frac{6}{\sqrt{2}}$.

16. List all the rational numbers in this set:

$\left\{-7, -\frac{4}{5}, 0, 0.25, \sqrt{3}, \sqrt{4}, \frac{22}{7}, \pi\right\}$.

In Exercises 17–18, state the name of the property illustrated.

17. $3(2 + 5) = 3(5 + 2)$

18. $6(7 + 4) = 6 \cdot 7 + 6 \cdot 4$

In Exercises 19–21, evaluate each expression.

19. $3^3 \cdot 3^2$

20. $\frac{4^6}{4^3}$

21. 8^{-2}

22. Multiply and express the answer in decimal notation.

$$(3 \times 10^8)(2.5 \times 10^{-5})$$

23. Divide by first expressing each number in scientific notation. Write the answer in scientific notation.

$$\frac{49{,}000}{0.007}$$

In Exercises 24–26 use 10^6 for one million and 10^9 for one billion to rewrite the number in each statement in scientific notation.

24. The 2009 economic stimulus package allocated \$53.6 billion for grants to states for education.

25. The population of the United States at the time the economic stimulus package was voted into law was approximately 307 million.

26. Use your scientific notation answers from Exercises 24 and 25 to answer this question:

If the cost for grants to states for education was evenly divided among every individual in the United States, how much would each citizen have to pay?

27. Write the first six terms of the arithmetic sequence with first term, a_1, and common difference, d.

$$a_1 = 1, d = -5$$

28. Find a_9, the ninth term of the arithmetic sequence, with the first term, a_1, and common difference, d.

$$a_1 = -2, d = 3$$

29. Write the first six terms of the geometric sequence with first term, a_1, and common ratio, r.

$$a_1 = 16, r = \frac{1}{2}$$

30. Find a_7, the seventh term of the geometric sequence, with the first term, a_1, and common ratio, r.

$$a_1 = 5, r = 2$$

Algebra: Equations and Inequalities

6

THE BELIEF THAT HUMOR AND LAUGHTER CAN HAVE POSITIVE EFFECTS ON OUR LIVES IS NOT NEW. THE BIBLE TELLS US, "A MERRY HEART DOETH GOOD LIKE A MEDICINE, BUT A BROKEN spirit drieth the bones" (Proverbs 17:22).

Some random humor factoids: • The average adult laughs 15 times each day (Newhouse News Service). • Forty-six percent of people who are telling a joke laugh more than the people they are telling it to (*U.S. News and World Report*). • Eighty percent of adult laughter does not occur in response to jokes or funny situations (*Independent*). • Algebra can be used to model the influence that humor plays in our responses to negative life events (Bob Blitzer, *Thinking Mathematically*).

That last tidbit that your author threw into the list is true. Based on our sense of humor, there is actually a formula that predicts how we will respond to difficult life events. Formulas can be used to explain what is happening in the present and to make predictions about what might occur in the future. In this chapter, you will learn to use formulas and mathematical models in new ways that will help you to recognize patterns, logic, and order in a world that can appear chaotic to the untrained eye.

Here's where you'll find this application:

Humor opens Section 6.2, and the advantage of having a sense of humor becomes laughingly evident in the models in Example 6 on page 360.

6.1 Algebraic Expressions and Formulas

WHAT AM I SUPPOSED TO LEARN?

After studying this section, you should be able to:

1. Evaluate algebraic expressions.
2. Use mathematical models.
3. Understand the vocabulary of algebraic expressions.
4. Simplify algebraic expressions.

YOU ARE THINKING ABOUT BUYING A high-definition television. How much distance should you allow between you and the TV for pixels to be undetectable and the image to appear smooth?

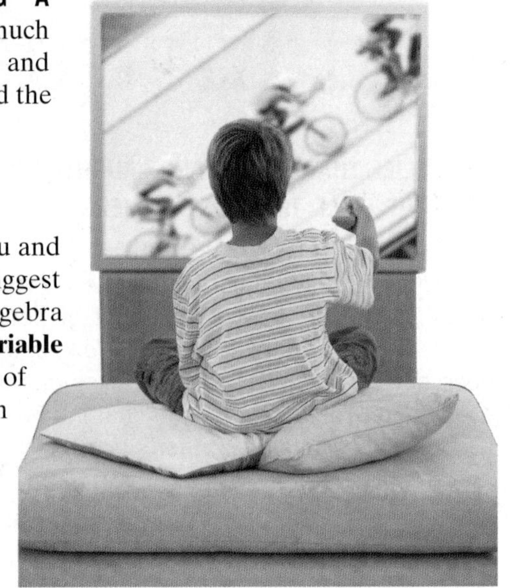

Algebraic Expressions

Let's see what the distance between you and your TV has to do with algebra. The biggest difference between arithmetic and algebra is the use of *variables* in algebra. A **variable** is a letter that represents a variety of different numbers. For example, we can let x represent the diagonal length, in inches, of a high-definition television. The industry rule for most of the current HDTVs on the market is to multiply this diagonal length by 2.5 to get the distance, in inches, at which a person with perfect vision can see a smooth image. This can be written $2.5 \cdot x$, but it is usually expressed as $2.5x$. Placing a number and a letter next to one another indicates multiplication.

Notice that $2.5x$ combines the number 2.5 and the variable x using the operation of multiplication. A combination of variables and numbers using the operations of addition, subtraction, multiplication, or division, as well as powers or roots, is called an **algebraic expression.** Here are some examples of algebraic expressions:

$x + 2.5$	$x - 2.5$	$2.5x$	$\frac{x}{2.5}$	$3x + 5$	$\sqrt{x} + 7.$
The variable x increased by 2.5	The variable x decreased by 2.5	2.5 times the variable x	The variable x divided by 2.5	5 more than 3 times the variable x	7 more than the square root of the variable x

1 Evaluate algebraic expressions.

Evaluating Algebraic Expressions

Evaluating an algebraic expression means finding the value of the expression for a given value of the variable. For example, we can evaluate $2.5x$ (the ideal distance between you and your x-inch TV) for $x = 50$. We substitute 50 for x. We obtain $2.5 \cdot 50$, or 125. This means that if the diagonal length of your TV is 50 inches, your distance from the screen should be 125 inches. Because 12 inches = 1 foot, this distance is $\frac{125}{12}$ feet, or approximately 10.4 feet.

Many algebraic expressions contain more than one operation. Evaluating an algebraic expression correctly involves carefully applying the order of operations agreement that we studied in Chapter 5.

THE ORDER OF OPERATIONS AGREEMENT

1. Perform operations within the innermost parentheses and work outward. If the algebraic expression involves a fraction, treat the numerator and the denominator as if they were each enclosed in parentheses.
2. Evaluate all exponential expressions.
3. Perform multiplications and divisions as they occur, working from left to right.
4. Perform additions and subtractions as they occur, working from left to right.

EXAMPLE 1 Evaluating an Algebraic Expression

Evaluate $7 + 5(x - 4)^3$ for $x = 6$.

SOLUTION

$$\begin{aligned} 7 + 5(x-4)^3 &= 7 + 5(6-4)^3 && \text{Replace } x \text{ with 6.} \\ &= 7 + 5(2)^3 && \text{First work inside parentheses: } 6 - 4 = 2. \\ &= 7 + 5(8) && \text{Evaluate the exponential expression: } 2^3 = 2 \cdot 2 \cdot 2 = 8. \\ &= 7 + 40 && \text{Multiply: } 5(8) = 40. \\ &= 47 && \text{Add: } 7 + 40 = 47. \end{aligned}$$

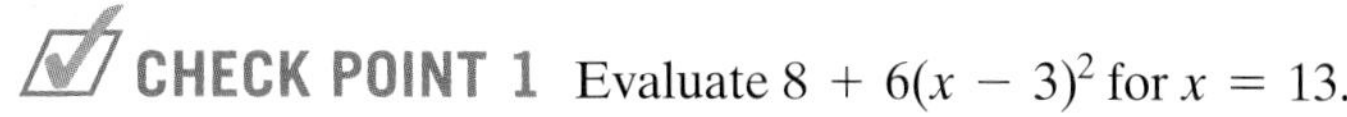

CHECK POINT 1 Evaluate $8 + 6(x - 3)^2$ for $x = 13$.

EXAMPLE 2 Evaluating an Algebraic Expression

Evaluate $x^2 + 5x - 3$ for $x = -6$.

SOLUTION

We substitute -6 for each of the two occurrences of x. Then we use the order of operations to evaluate the algebraic expression.

$$\begin{aligned} & x^2 + 5x - 3 && \text{This is the given algebraic expression.} \\ &= (-6)^2 + 5(-6) - 3 && \text{Substitute } -6 \text{ for each } x. \\ &= 36 + 5(-6) - 3 && \text{Evaluate the exponential expression: } (-6)^2 = (-6)(-6) = 36. \\ &= 36 + (-30) - 3 && \text{Multiply: } 5(-6) = -30. \\ &= 6 - 3 && \text{Add and subtract from left to right. First add: } 36 + (-30) = 6. \\ &= 3 && \text{Subtract: } 6 - 3 = 3. \end{aligned}$$

CHECK POINT 2 Evaluate $x^2 + 4x - 7$ for $x = -5$.

GREAT QUESTION!

Is there a difference between evaluating x^2 for $x = -6$ and evaluating $-x^2$ for $x = 6$?

Yes. Notice the difference between these evaluations:

- x^2 for $x = -6$
  $x^2 = (-6)^2 = (-6)(-6) = 36$
- $-x^2$ for $x = 6$
  $-x^2 = -6^2 = -6 \cdot 6 = -36$

The negative is not inside parentheses and is not taken to the second power.

Work carefully when evaluating algebraic expressions with exponents and negatives.

EXAMPLE 3 Evaluating an Algebraic Expression

Evaluate $-2x^2 + 5xy - y^3$ for $x = 4$ and $y = -2$.

SOLUTION

We substitute 4 for each x and -2 for each y. Then we use the order of operations to evaluate the algebraic expression.

$$\begin{aligned} & -2x^2 + 5xy - y^3 && \text{This is the given algebraic expression.} \\ &= -2 \cdot 4^2 + 5 \cdot 4(-2) - (-2)^3 && \text{Substitute 4 for } x \text{ and } -2 \text{ for } y. \\ &= -2 \cdot 16 + 5 \cdot 4(-2) - (-8) && \text{Evaluate the exponential expressions: } 4^2 = 4 \cdot 4 = 16 \text{ and } (-2)^3 = (-2)(-2)(-2) = -8. \\ &= -32 + (-40) - (-8) && \text{Multiply: } -2 \cdot 16 = -32 \text{ and } 5(4)(-2) = 20(-2) = -40. \\ &= -72 - (-8) && \text{Add and subtract from left to right. First add: } -32 + (-40) = -72. \\ &= -64 && \text{Subtract: } -72 - (-8) = -72 + 8 = -64. \end{aligned}$$

CHECK POINT 3 Evaluate $-3x^2 + 4xy - y^3$ for $x = 5$ and $y = -1$.

2 *Use mathematical models.*

Formulas and Mathematical Models

An **equation** is formed when an equal sign is placed between two algebraic expressions. One aim of algebra is to provide a compact, symbolic description of the world. These descriptions involve the use of *formulas*. A **formula** is an equation that uses variables to express a relationship between two or more quantities.

Here are two examples of formulas related to heart rate and exercise.

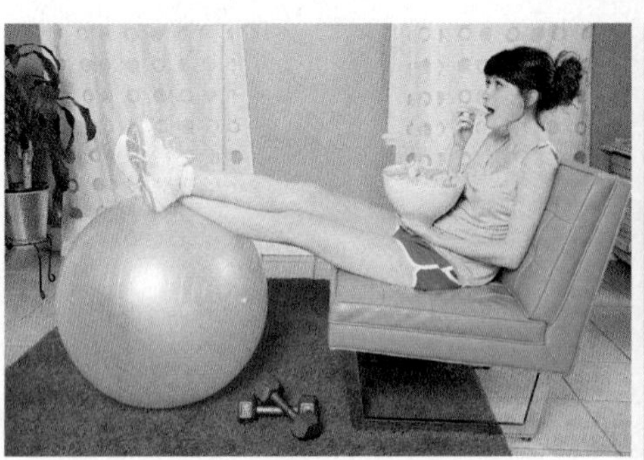

Couch-Potato Exercise

$$H = \frac{1}{5}(220 - a)$$

Heart rate, in beats per minute, is $\frac{1}{5}$ of the difference between 220 and your age.

Working It

$$H = \frac{9}{10}(220 - a)$$

Heart rate, in beats per minute, is $\frac{9}{10}$ of the difference between 220 and your age.

These important definitions are repeated from earlier chapters in case your course did not cover this material.

The process of finding formulas to describe real-world phenomena is called **mathematical modeling**. Such formulas, together with the meaning assigned to the variables, are called **mathematical models**. We often say that these formulas model, or describe, the relationships among the variables.

EXAMPLE 4 *Modeling Caloric Needs*

The bar graph in **Figure 6.1** shows the estimated number of calories per day needed to maintain energy balance for various gender and age groups for moderately active lifestyles. (Moderately active means a lifestyle that includes physical activity equivalent to walking 1.5 to 3 miles per day at 3 to 4 miles per hour, in addition to the light physical activity associated with typical day-to-day life.)

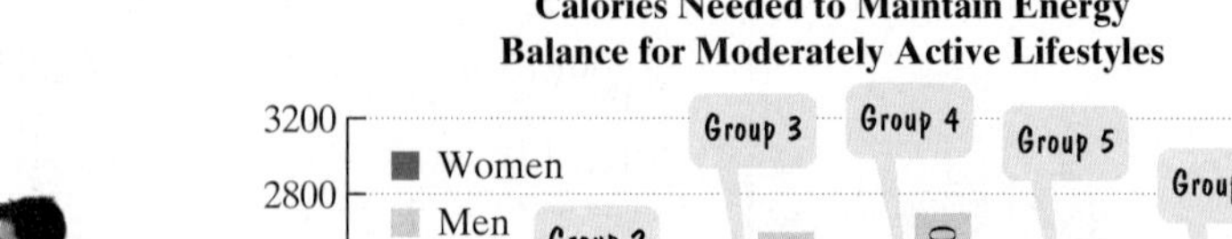

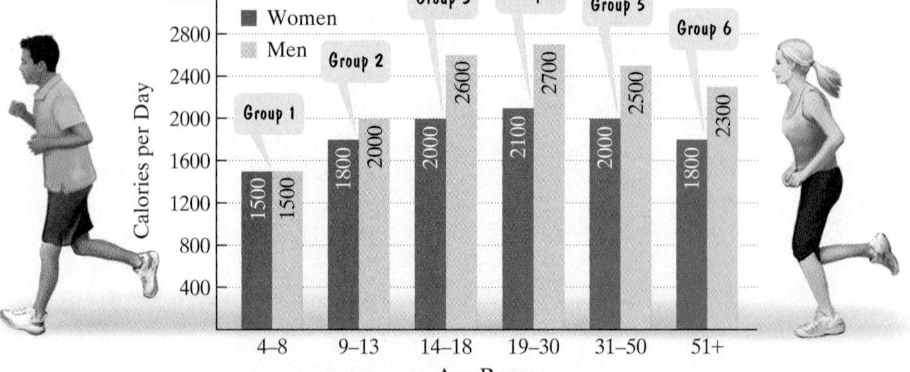

FIGURE 6.1
Source: USDA

The mathematical model

$$W = -66x^2 + 526x + 1030$$

describes the number of calories needed per day, W, by women in age group x with moderately active lifestyles. According to the model, how many calories per day are needed by women between the ages of 19 and 30, inclusive, with this lifestyle? Does this underestimate or overestimate the number shown by the graph in **Figure 6.1**? By how much?

SOLUTION

Because the 19–30 age range is designated as group 4, we substitute 4 for x in the given model. Then we use the order of operations to find W, the number of calories needed per day by women between the ages of 19 and 30.

$W = -66x^2 + 526x + 1030$ This is the given mathematical model.

$W = -66 \cdot 4^2 + 526 \cdot 4 + 1030$ Replace each occurrence of x with 4.

$W = -66 \cdot 16 + 526 \cdot 4 + 1030$ Evaluate the exponential expression: $4^2 = 4 \cdot 4 = 16$.

$W = -1056 + 2104 + 1030$ Multiply from left to right: $-66 \cdot 16 = -1056$ and $526 \cdot 4 = 2104$.

$W = 2078$ Add.

The formula indicates that women in the 19–30 age range with moderately active lifestyles need 2078 calories per day. **Figure 6.1** indicates that 2100 calories are needed. Thus, the mathematical model underestimates caloric needs by $2100 - 2078$ calories, or by 22 calories per day.

CHECK POINT 4 The mathematical model

$$M = -120x^2 + 998x + 590$$

describes the number of calories needed per day, M, by men in age group x with moderately active lifestyles. According to the model, how many calories per day are needed by men between the ages of 19 and 30, inclusive, with this lifestyle? Does this underestimate or overestimate the number shown by the graph in **Figure 6.1**? By how much?

3 *Understand the vocabulary of algebraic expressions.*

The Vocabulary of Algebraic Expressions

We have seen that an algebraic expression combines numbers and variables. Here is another example of an algebraic expression:

$$7x - 9y - 3.$$

The **terms** of an algebraic expression are those parts that are separated by addition. For example, we can rewrite $7x - 9y - 3$ as

$$7x + (-9y) + (-3).$$

This expression contains three terms, namely $7x$, $-9y$, and -3.

The numerical part of a term is called its **coefficient**. In the term $7x$, the 7 is the coefficient. In the term $-9y$, the -9 is the coefficient.

Coefficients of 1 and -1 are not written. Thus, the coefficient of x, meaning $1x$, is 1. Similarly, the coefficient of $-y$, meaning $-1y$, is -1.

A term that consists of just a number is called a **numerical term** or a **constant**. The numerical term of $7x - 9y - 3$ is -3.

The parts of each term that are multiplied are called the **factors** of the term. The factors of the term $7x$ are 7 and x.

Like terms are terms that have the same variable factors. For example, $3x$ and $7x$ are like terms.

4 *Simplify algebraic expressions.*

Simplifying Algebraic Expressions

The properties of real numbers that we discussed in Chapter 5 can be applied to algebraic expressions.

PROPERTIES OF REAL NUMBERS

Property	Example
Commutative Property of Addition $a + b = b + a$	$13x^2 + 7x = 7x + 13x^2$
Commutative Property of Multiplication $ab = ba$	$x \cdot 6 = 6x$
Associative Property of Addition $(a + b) + c = a + (b + c)$	$3 + (8 + x) = (3 + 8) + x = 11 + x$
Associative Property of Multiplication $(ab)c = a(bc)$	$-2(3x) = (-2 \cdot 3)x = -6x$
Distributive Property $a(b + c) = ab + ac$	$5(3x + 7) = 5 \cdot 3x + 5 \cdot 7 = 15x + 35$
$a(b - c) = ab - ac$	$4(2x - 5) = 4 \cdot 2x - 4 \cdot 5 = 8x - 20$

GREAT QUESTION!

Do I have to use the distributive property to combine like terms? Can't I just do it in my head?

Yes, you can combine like terms mentally. Add or subtract the coefficients of the terms. Use this result as the coefficient of the terms' variable factor(s).

The distributive property in the form

$$ba + ca = (b + c)a$$

enables us to add or subtract like terms. For example,

$$3x + 7x = (3 + 7)x = 10x$$
$$7y^2 - y^2 = 7y^2 - 1y^2 = (7 - 1)y^2 = 6y^2.$$

This process is called **combining like terms**.

An algebraic expression is **simplified** when parentheses have been removed and like terms have been combined.

EXAMPLE 5 *Simplifying an Algebraic Expression*

Simplify: $5(3x - 7) - 6x$.

SOLUTION

$5(3x - 7) - 6x$

$= 5 \cdot 3x - 5 \cdot 7 - 6x$ Use the distributive property to remove the parentheses.

$= 15x - 35 - 6x$ Multiply.

$= (15x - 6x) - 35$ Group like terms.

$= 9x - 35$ Combine like terms: $15x - 6x = (15 - 6)x = 9x$.

CHECK POINT 5 Simplify: $7(2x - 3) - 11x$.

EXAMPLE 6 Simplifying an Algebraic Expression

Simplify: $6(2x^2 + 4x) + 10(4x^2 + 3x)$.

SOLUTION

$6(2x^2 + 4x) + 10(4x^2 + 3x)$

$= 6 \cdot 2x^2 + 6 \cdot 4x + 10 \cdot 4x^2 + 10 \cdot 3x$ — Use the distributive property to remove the parentheses.

$= 12x^2 + 24x + 40x^2 + 30x$ — Multiply.

$= (12x^2 + 40x^2) + (24x + 30x)$ — Group like terms.

$= 52x^2 + 54x$ — Combine like terms: $12x^2 + 40x^2 = (12 + 40)x^2 = 52x^2$ and $24x + 30x = (24 + 30)x = 54x$.

$52x^2$ and $54x$ are not like terms. They contain different variable factors, x^2 and x, and cannot be combined.

CHECK POINT 6 Simplify: $7(4x^2 + 3x) + 2(5x^2 + x)$.

It is not uncommon to see algebraic expressions with parentheses preceded by a negative sign or subtraction. An expression of the form $-(a + b)$ can be simplified as follows:

$$-(a + b) = -1(a + b) = (-1)a + (-1)b = -a + (-b) = -a - b.$$

Do you see a fast way to obtain the simplified expression on the right? **If a negative sign or a subtraction symbol appears outside parentheses, drop the parentheses and change the sign of every term within the parentheses.** For example,

$$-(3x^2 - 7x - 4) = -3x^2 + 7x + 4.$$

EXAMPLE 7 Simplifying an Algebraic Expression

Simplify: $8x + 2[5 - (x - 3)]$.

SOLUTION

$8x + 2[5 - (x - 3)]$

$= 8x + 2[5 - x + 3]$ — Drop parentheses and change the sign of each term in parentheses: $-(x - 3) = -x + 3$.

$= 8x + 2[8 - x]$ — Simplify inside brackets: $5 + 3 = 8$.

$= 8x + 16 - 2x$ — Apply the distributive property: $2[8 - x] = 2 \cdot 8 - 2x = 16 - 2x$.

$= (8x - 2x) + 16$ — Group like terms.

$= 6x + 16$ — Combine like terms: $8x - 2x = (8 - 2)x = 6x$.

CHECK POINT 7 Simplify: $6x + 4[7 - (x - 2)]$.

Blitzer Bonus

Using Algebra to Measure Blood-Alcohol Concentration

The amount of alcohol in a person's blood is known as blood-alcohol concentration (BAC), measured in grams of alcohol per deciliter of blood. A BAC of 0.08, meaning 0.08%, indicates that a person has 8 parts alcohol per 10,000 parts blood. In every state in the United States, it is illegal to drive with a BAC of 0.08 or higher.

How Do I Measure My Blood-Alcohol Concentration?
Here's a formula that models BAC for a person who weighs w pounds and who has n drinks* per hour.

$$\text{BAC} = \frac{600n}{w(0.6n + 169)}$$

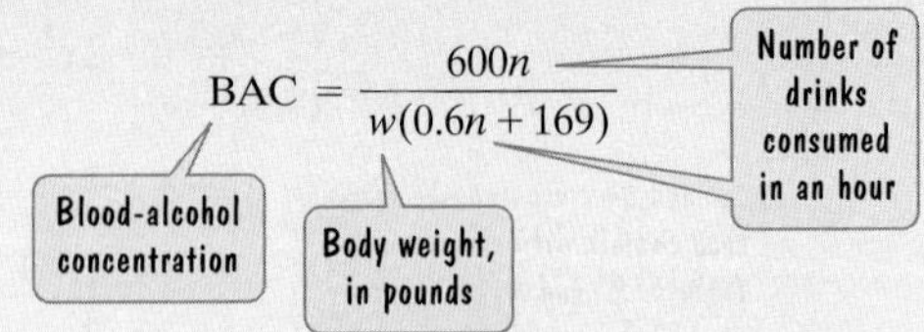

* A drink can be a 12-ounce can of beer, a 5-ounce glass of wine, or a 1.5-ounce shot of liquor. Each contains approximately 14 grams, or $\frac{1}{2}$ ounce, of alcohol.

Blood-alcohol concentration can be used to quantify the meaning of "tipsy."

BAC	Effects on Behavior
0.05	Feeling of well-being; mild release of inhibitions; absence of observable effects
0.08	Feeling of relaxation; mild sedation; exaggeration of emotions and behavior; slight impairment of motor skills; increase in reaction time
0.12	Muscle control and speech impaired; difficulty performing motor skills; uncoordinated behavior
0.15	Euphoria; major impairment of physical and mental functions; irresponsible behavior; some difficulty standing, walking, and talking
0.35	Surgical anesthesia; lethal dosage for a small percentage of people
0.40	Lethal dosage for 50% of people; severe circulatory and respiratory depression; alcohol poisoning/overdose

Source: National Clearinghouse for Alcohol and Drug Information

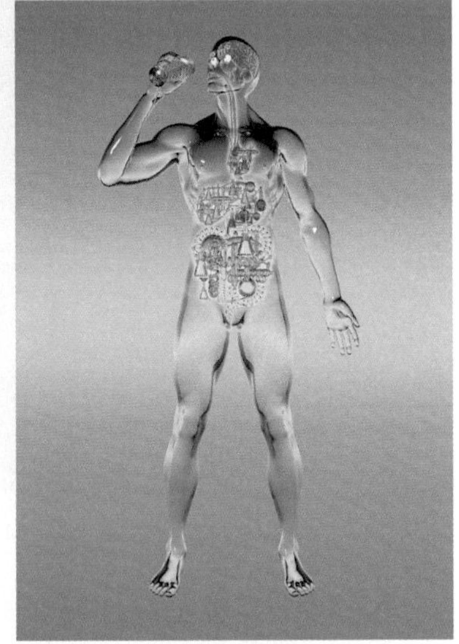

Keeping in mind the meaning of "tipsy," we can use our model to compare blood-alcohol concentrations of a 120-pound person and a 200-pound person for various numbers of drinks. We determined each BAC using a calculator, rounding to three decimal places.

Blood-Alcohol Concentrations of a 120-Pound Person

$$\text{BAC} = \frac{600n}{120(0.6n + 169)}$$

n **(number of drinks per hour)**	1	2	3	4	5	6	7	8	9	10
BAC (blood-alcohol concentration)	0.029	0.059	0.088	0.117	0.145	0.174	0.202	0.230	0.258	0.286

Illegal to drive (n = 3 through 10)

Blood-Alcohol Concentrations of a 200-Pound Person

$$\text{BAC} = \frac{600n}{200(0.6n + 169)}$$

n **(number of drinks per hour)**	1	2	3	4	5	6	7	8	9	10
BAC (blood-alcohol concentration)	0.018	0.035	0.053	0.070	0.087	0.104	0.121	0.138	0.155	0.171

Illegal to drive (n = 5 through 10)

Like all mathematical models, the formula for BAC gives approximate rather than exact values. There are other variables that influence blood-alcohol concentration that are not contained in the model. These include the rate at which an individual's body processes alcohol, how quickly one drinks, sex, age, physical condition, and the amount of food eaten prior to drinking.

Concept and Vocabulary Check

Fill in each blank so that the resulting statement is true.

1. Finding the value of an algebraic expression for a given value of the variable is called __________ the expression.
2. When an equal sign is placed between two algebraic expressions, an __________ is formed.
3. The parts of an algebraic expression that are separated by addition are called the _______ of the expression.
4. In the algebraic expression $7x$, 7 is called the __________ because it is the numerical part.
5. In the algebraic expression $7x$, 7 and x are called ________ because they are multiplied together.
6. The algebraic expressions $3x$ and $7x$ are called __________ because they contain the same variable to the same power.

Exercise Set 6.1

Practice Exercises

In Exercises 1–34, evaluate the algebraic expression for the given value or values of the variables.

1. $5x + 7;\ x = 4$
2. $9x + 6;\ x = 5$
3. $-7x - 5;\ x = -4$
4. $-6x - 13;\ x = -3$
5. $x^2 + 4;\ x = 5$
6. $x^2 + 9;\ x = 3$
7. $x^2 - 6;\ x = -2$
8. $x^2 - 11;\ x = -3$
9. $-x^2 + 4;\ x = 5$
10. $-x^2 + 9;\ x = 3$
11. $-x^2 - 6;\ x = -2$
12. $-x^2 - 11;\ x = -3$
13. $x^2 + 4x;\ x = 10$
14. $x^2 + 6x;\ x = 9$
15. $8x^2 + 17;\ x = 5$
16. $7x^2 + 25;\ x = 3$
17. $x^2 - 5x;\ x = -11$
18. $x^2 - 8x;\ x = -5$
19. $x^2 + 5x - 6;\ x = 4$
20. $x^2 + 7x - 4;\ x = 6$
21. $4 + 5(x - 7)^3;\ x = 9$
22. $6 + 5(x - 6)^3;\ x = 8$
23. $x^2 - 3(x - y);\ x = 2, y = 8$
24. $x^2 - 4(x - y);\ x = 3, y = 8$
25. $2x^2 - 5x - 6;\ x = -3$
26. $3x^2 - 4x - 9;\ x = -5$
27. $-5x^2 - 4x - 11;\ x = -1$
28. $-6x^2 - 11x - 17;\ x = -2$
29. $3x^2 + 2xy + 5y^2;\ x = 2, y = 3$
30. $4x^2 + 3xy + 2y^2;\ x = 3, y = 2$
31. $-x^2 - 4xy + 3y^3;\ x = -1, y = -2$
32. $-x^2 - 3xy + 4y^3;\ x = -3, y = -1$
33. $\dfrac{2x + 3y}{x + 1};\ x = -2, y = 4$
34. $\dfrac{2x + y}{xy - 2x};\ x = -2, y = 4$

The formula

$$C = \frac{5}{9}(F - 32)$$

expresses the relationship between Fahrenheit temperature, F, and Celsius temperature, C. In Exercises 35–36, use the formula to convert the given Fahrenheit temperature to its equivalent temperature on the Celsius scale.

35. 50°F
36. 86°F

A football was kicked vertically upward from a height of 4 feet with an initial speed of 60 feet per second. The formula

$$h = 4 + 60t - 16t^2$$

describes the ball's height above the ground, h, in feet, t seconds after it was kicked. Use this formula to solve Exercises 37–38.

37. What was the ball's height 2 seconds after it was kicked?
38. What was the ball's height 3 seconds after it was kicked?

In Exercises 39–40, name the property used to go from step to step each time that "(why?)" occurs.

39. $7 + 2(x + 9)$
$= 7 + (2x + 18)$ (why?)
$= 7 + (18 + 2x)$ (why?)
$= (7 + 18) + 2x$ (why?)
$= 25 + 2x$
$= 2x + 25$ (why?)

40. $5(x + 4) + 3x$
$= (5x + 20) + 3x$ (why?)
$= (20 + 5x) + 3x$ (why?)
$= 20 + (5x + 3x)$ (why?)
$= 20 + (5 + 3)x$ (why?)
$= 20 + 8x$
$= 8x + 20$ (why?)

In Exercises 41–62, simplify each algebraic expression.

41. $7x + 10x$
42. $5x + 13x$
43. $5x^2 - 8x^2$
44. $7x^2 - 10x^2$
45. $3(x + 5)$
46. $4(x + 6)$
47. $4(2x - 3)$
48. $3(4x - 5)$
49. $5(3x + 4) - 4$
50. $2(5x + 4) - 3$
51. $5(3x - 2) + 12x$
52. $2(5x - 1) + 14x$
53. $7(3y - 5) + 2(4y + 3)$
54. $4(2y - 6) + 3(5y + 10)$
55. $5(3y - 2) - (7y + 2)$
56. $4(5y - 3) - (6y + 3)$
57. $3(-4x^2 + 5x) - (5x - 4x^2)$
58. $2(-5x^2 + 3x) - (3x - 5x^2)$
59. $7 - 4[3 - (4y - 5)]$

60. $6 - 5[8 - (2y - 4)]$

61. $8x - 3[5 - (7 - 6x)]$

62. $7x - 4[6 - (8 - 5x)]$

Practice Plus

In Exercises 63–66, simplify each algebraic expression.

63. $18x^2 + 4 - [6(x^2 - 2) + 5]$

64. $14x^2 + 5 - [7(x^2 - 2) + 4]$

65. $2(3x^2 - 5) - [4(2x^2 - 1) + 3]$

66. $4(6x^2 - 3) - [2(5x^2 - 1) + 1]$

Application Exercises

The maximum heart rate, in beats per minute, that you should achieve during exercise is 220 minus your age:

$$220 - a.$$

This algebraic expression gives maximum heart rate in terms of age, a.

The bar graph shows the target heart rate ranges for four types of exercise goals. The lower and upper limits of these ranges are fractions of the maximum heart rate, 220 – a. Exercises 67–68 are based on the information in the graph.

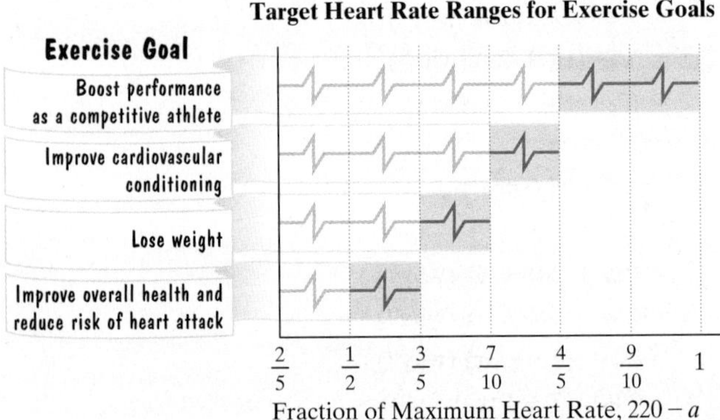

67. If your exercise goal is to improve cardiovascular conditioning, the graph shows the following range for target heart rate, H, in beats per minute:

Lower limit of range: $H = \frac{7}{10}(220 - a)$

Upper limit of range: $H = \frac{4}{5}(220 - a).$

a. What is the lower limit of the heart rate range, in beats per minute, for a 20-year-old with this exercise goal?

b. What is the upper limit of the heart rate range, in beats per minute, for a 20-year-old with this exercise goal?

68. If your exercise goal is to improve overall health, the graph in the previous column shows the following range for target heart rate, H, in beats per minute:

Lower limit of range: $H = \frac{1}{2}(220 - a)$

Upper limit of range: $H = \frac{3}{5}(220 - a).$

a. What is the lower limit of the heart rate range, in beats per minute, for a 30-year-old with this exercise goal?

b. What is the upper limit of the heart rate range, in beats per minute, for a 30-year-old with this exercise goal?

The bar graph shows the estimated number of calories per day needed to maintain energy balance for various gender and age groups for sedentary lifestyles. (Sedentary means a lifestyle that includes only the light physical activity associated with typical day-to-day life.)

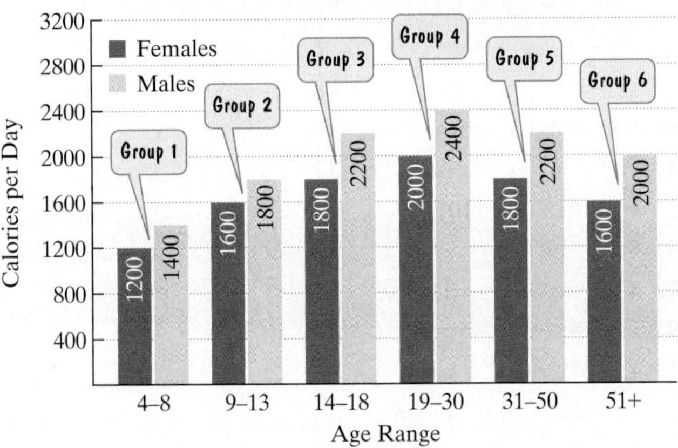

Source: USDA

Use the appropriate information displayed by the graph to solve Exercises 69–70.

69. The mathematical model

$$F = -82x^2 + 654x + 620$$

describes the number of calories needed per day, F, by females in age group x with sedentary lifestyles. According to the model, how many calories per day are needed by females between the ages of 19 and 30, inclusive, with this lifestyle? Does this underestimate or overestimate the number shown by the graph? By how much?

70. The mathematical model

$$M = -96x^2 + 802x + 660$$

describes the number of calories needed per day, M, by males in age group x with sedentary lifestyles. According to the model, how many calories per day are needed by males between the ages of 19 and 30, inclusive, with this lifestyle? Does this underestimate or overestimate the number shown by the graph? By how much?

Salary after College. *In 2010, MonsterCollege surveyed 1250 U.S. college students expecting to graduate in the next several years. Respondents were asked the following question:*

> *What do you think your starting salary will be at your first job after college?*

The line graph shows the percentage of college students who anticipated various starting salaries.

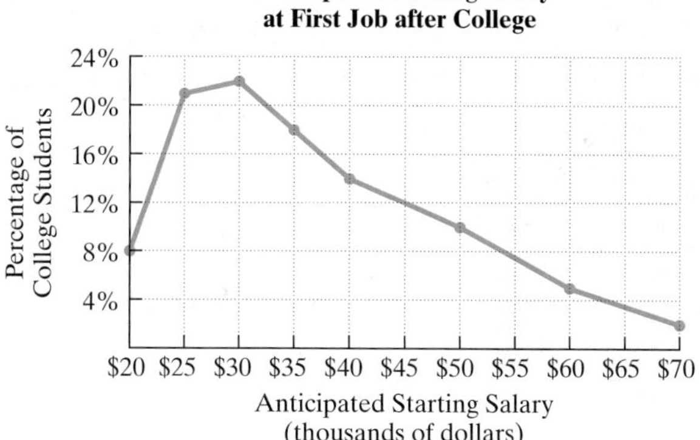

Source: MonsterCollege™

The mathematical model

$$p = -0.01s^2 + 0.8s + 3.7$$

describes the percentage of college students, p, who anticipated a starting salary, s, in thousands of dollars. Use this information to solve Exercises 71–72.

71. a. Use the line graph to estimate the percentage of students who anticipated a starting salary of $30 thousand.

b. Use the formula to find the percentage of students who anticipated a starting salary of $30 thousand. How does this compare with your estimate in part (a)?

72. a. Use the line graph to estimate the percentage of students who anticipated a starting salary of $40 thousand.

b. Use the formula to find the percentage of students who anticipated a starting salary of $40 thousand. How does this compare with your estimate in part (a)?

73. Read the Blitzer Bonus on page 350. Use the formula

$$\text{BAC} = \frac{600n}{w(0.6n + 169)}$$

and replace w with your body weight. Using this formula and a calculator, compute your BAC for integers from $n = 1$ to $n = 10$. Round to three decimal places. According to this model, how many drinks can you consume in an hour without exceeding the legal measure of drunk driving?

Explaining the Concepts

74. What is an algebraic expression? Provide an example with your description.

75. What does it mean to evaluate an algebraic expression? Provide an example with your description.

76. What is a term? Provide an example with your description.

77. What are like terms? Provide an example with your description.

78. Explain how to add like terms. Give an example.

79. What does it mean to simplify an algebraic expression?

80. An algebra student incorrectly used the distributive property and wrote $3(5x + 7) = 15x + 7$. If you were that student's teacher, what would you say to help the student avoid this kind of error?

Critical Thinking Exercises

Make Sense? *In Exercises 81–84, determine whether each statement makes sense or does not make sense, and explain your reasoning.*

81. I did not use the distributive property to simplify $3(2x + 5x)$.

82. The terms $13x^2$ and $10x$ both contain the variable x, so I can combine them to obtain $23x^3$.

83. Regardless of what real numbers I substitute for x and y, I will always obtain zero when evaluating $2x^2y - 2yx^2$.

84. A model that describes the number of lobbyists x years after 2000 cannot be used to estimate the number in 2000.

In Exercises 85–92, determine whether each statement is true or false. If the statement is false, make the necessary change(s) to produce a true statement.

85. The term x has no coefficient.

86. $5 + 3(x - 4) = 8(x - 4) = 8x - 32$

87. $-x - x = -x + (-x) = 0$

88. $x - 0.02(x + 200) = 0.98x - 4$

89. $3 + 7x = 10x$

90. $b \cdot b = 2b$

91. $(3y - 4) - (8y - 1) = -5y - 3$

92. $-4y + 4 = -4(y + 4)$

93. A business that manufactures small alarm clocks has weekly fixed costs of $5000. The average cost per clock for the business to manufacture x clocks is described by

$$\frac{0.5x + 5000}{x}.$$

a. Find the average cost when $x = 100$, 1000, and 10,000.

b. Like all other businesses, the alarm clock manufacturer must make a profit. To do this, each clock must be sold for at least 50¢ more than what it costs to manufacture. Due to competition from a larger company, the clocks can be sold for $1.50 each and no more. Our small manufacturer can only produce 2000 clocks weekly. Does this business have much of a future? Explain.

6.2 Linear Equations in One Variable and Proportions

WHAT AM I SUPPOSED TO LEARN?

After studying this section, you should be able to:

1. Solve linear equations.
2. Solve linear equations containing fractions.
3. Solve proportions.
4. Solve problems using proportions.
5. Identify equations with no solution or infinitely many solutions.

FIGURE 6.2
Source: Steven Davis and Joseph Palladino, *Psychology*, 5th Edition. Prentice Hall, 2007.

The belief that humor and laughter can have positive benefits on our lives is not new. The graphs in **Figure 6.2** indicate that persons with a low sense of humor have higher levels of depression in response to negative life events than those with a high sense of humor. These graphs can be modeled by the following formulas:

Low-Humor Group: $$D = \frac{10}{9}x + \frac{53}{9}$$

High-Humor Group: $$D = \frac{1}{9}x + \frac{26}{9}.$$

In each formula, x represents the intensity of a negative life event (from 1, low, to 10, high) and D is the level of depression in response to that event.

Suppose that the low-humor group averages a level of depression of 10 in response to a negative life event. We can determine the intensity of that event by substituting 10 for D in the low-humor model, $D = \frac{10}{9}x + \frac{53}{9}$:

$$10 = \frac{10}{9}x + \frac{53}{9}.$$

The two sides of an equation can be reversed. So, we can also express this equation as

$$\frac{10}{9}x + \frac{53}{9} = 10.$$

Notice that the highest exponent on the variable is 1. Such an equation is called a *linear equation in one variable*. In this section, we will study how to solve such equations. We return to the models for sense of humor and depression later in the section.

1 *Solve linear equations.*

Solving Linear Equations in One Variable

We begin with the general definition of a linear equation in one variable.

DEFINITION OF A LINEAR EQUATION

A **linear equation in one variable** x is an equation that can be written in the form

$$ax + b = 0,$$

where a and b are real numbers, and $a \neq 0$.

An example of a linear equation in one variable is

$$4x + 12 = 0.$$

Solving an equation in x involves determining all values of x that result in a true statement when substituted into the equation. Such values are **solutions**, or **roots**, of the equation. For example, substitute -3 for x in $4x + 12 = 0$. We obtain

$$4(-3) + 12 = 0, \quad \text{or} \quad -12 + 12 = 0.$$

This simplifies to the true statement $0 = 0$. Thus, -3 is a solution of the equation $4x + 12 = 0$. We also say that -3 **satisfies** the equation $4x + 12 = 0$, because when we substitute -3 for x, a true statement results. The set of all such solutions is called the equation's **solution set**. For example, the solution set of the equation $4x + 12 = 0$ is $\{-3\}$.

Two or more equations that have the same solution set are called **equivalent equations**. For example, the equations

$$4x + 12 = 0 \quad \text{and} \quad 4x = -12 \quad \text{and} \quad x = -3$$

are equivalent equations because the solution set for each is $\{-3\}$. To solve a linear equation in x, we transform the equation into an equivalent equation one or more times. Our final equivalent equation should be of the form

$$x = \text{a number}.$$

The solution set of this equation is the set consisting of the number.

To generate equivalent equations, we will use the following properties:

THE ADDITION AND MULTIPLICATION PROPERTIES OF EQUALITY

The Addition Property of Equality

The same real number or algebraic expression may be added to both sides of an equation without changing the equation's solution set.

$a = b$ and $a + c = b + c$ are equivalent equations.

The Multiplication Property of Equality

The same nonzero real number may multiply both sides of an equation without changing the equation's solution set.

$a = b$ and $ac = bc$ are equivalent equations as long as $c \neq 0$.

Because subtraction is defined in terms of addition, the addition property also lets us subtract the same number from both sides of an equation without changing the equation's solution set. Similarly, because division is defined in terms of multiplication, the multiplication property of equality can be used to divide both sides of an equation by the same nonzero number to obtain an equivalent equation.

Table 6.1 illustrates how these properties are used to isolate x to obtain an equation of the form $x =$ a number.

These equations are solved using the Addition Property of Equality.

These equations are solved using the Multiplication Property of Equality.

TABLE 6.1 Using Properties of Equality to Solve Equations

Equation	How to Isolate x	Solving the Equation	The Equation's Solution Set
$x - 3 = 8$	Add 3 to both sides.	$x - 3 + 3 = 8 + 3$ $x = 11$	$\{11\}$
$x + 7 = -15$	Subtract 7 from both sides.	$x + 7 - 7 = -15 - 7$ $x = -22$	$\{-22\}$
$6x = 30$	Divide both sides by 6 (or multiply both sides by $\frac{1}{6}$).	$\frac{6x}{6} = \frac{30}{6}$ $x = 5$	$\{5\}$
$\frac{x}{5} = 9$	Multiply both sides by 5.	$5 \cdot \frac{x}{5} = 5 \cdot 9$ $x = 45$	$\{45\}$

EXAMPLE 1 Using Properties of Equality to Solve an Equation

Solve and check: $2x + 3 = 17$.

SOLUTION

Our goal is to obtain an equivalent equation with x isolated on one side and a number on the other side.

$2x + 3 = 17$	This is the given equation.
$2x + 3 - 3 = 17 - 3$	Subtract 3 from both sides.
$2x = 14$	Simplify.
$\frac{2x}{2} = \frac{14}{2}$	Divide both sides by 2.
$x = 7$	Simplify: $\frac{2x}{2} = 1x = x$ and $\frac{14}{2} = 7$.

Now we check the proposed solution, 7, by replacing x with 7 in the original equation.

$2x + 3 = 17$	This is the original equation.
$2 \cdot 7 + 3 \stackrel{?}{=} 17$	Substitute 7 for x. The question mark indicates that we do not yet know if the two sides are equal.
$14 + 3 \stackrel{?}{=} 17$	Multiply: $2 \cdot 7 = 14$.
This statement is true. $17 = 17$	Add: $14 + 3 = 17$.

Because the check results in a true statement, we conclude that the solution set of the given equation is $\{7\}$.

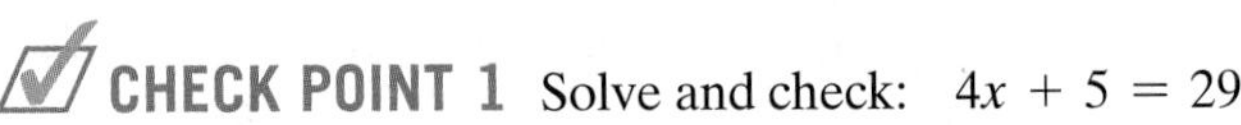

CHECK POINT 1 Solve and check: $4x + 5 = 29$.

Here is a step-by-step procedure for solving a linear equation in one variable. Not all of these steps are necessary to solve every equation.

SOLVING A LINEAR EQUATION

1. Simplify the algebraic expression on each side by removing grouping symbols and combining like terms.
2. Collect all the variable terms on one side and all the constants, or numerical terms, on the other side.
3. Isolate the variable and solve.
4. Check the proposed solution in the original equation.

EXAMPLE 2 Solving a Linear Equation

Solve and check: $2(x - 4) - 5x = -5$.

SOLUTION

Step 1 Simplify the algebraic expression on each side.

$2(x - 4) - 5x = -5$	This is the given equation.
$2x - 8 - 5x = -5$	Use the distributive property.
$-3x - 8 = -5$	Combine like terms: $2x - 5x = -3x$.

Step 2 Collect variable terms on one side and constants on the other side. The only variable term in $-3x - 8 = -5$ is $-3x$, and $-3x$ is already on the left side. We will collect constants on the right side by adding 8 to both sides.

$$-3x - 8 + 8 = -5 + 8 \quad \text{Add 8 to both sides.}$$
$$-3x = 3 \quad \text{Simplify.}$$

Step 3 Isolate the variable and solve. We isolate the variable, x, by dividing both sides of $-3x = 3$ by -3.

$$\frac{-3x}{-3} = \frac{3}{-3} \quad \text{Divide both sides by } -3.$$
$$x = -1 \quad \text{Simplify: } \frac{-3x}{-3} = 1x = x \text{ and } \frac{3}{-3} = -1.$$

Step 4 Check the proposed solution in the original equation. Substitute -1 for x in the original equation.

$$2(x - 4) - 5x = -5 \quad \text{This is the original equation.}$$
$$2(-1 - 4) - 5(-1) \stackrel{?}{=} -5 \quad \text{Substitute } -1 \text{ for } x.$$
$$2(-5) - 5(-1) \stackrel{?}{=} -5 \quad \text{Simplify inside parentheses: } -1 - 4 = -1 + (-4) = -5.$$
$$-10 - (-5) \stackrel{?}{=} -5 \quad \text{Multiply: } 2(-5) = -10 \text{ and } 5(-1) = -5.$$
This statement is true. $$-5 = -5 \quad -10 - (-5) = -10 + 5 = -5$$

Because the check results in a true statement, we conclude that the solution set of the given equation is $\{-1\}$.

CHECK POINT 2 Solve and check: $6(x - 3) - 10x = -10$.

GREAT QUESTION!

What are the differences between what I'm supposed to do with algebraic expressions and algebraic equations?

We simplify algebraic expressions. We solve algebraic equations. Although basic rules of algebra are used in both procedures, notice the differences between the procedures:

Simplifying an Algebraic Expression

Simplify: $3(x - 7) - (5x - 11)$.

This is not an equation. There is no equal sign.

Solution
$$\begin{aligned} &3(x - 7) - (5x - 11) \\ &= 3x - 21 - 5x + 11 \\ &= (3x - 5x) + (-21 + 11) \\ &= -2x + (-10) \\ &= -2x - 10 \end{aligned}$$

Stop! Further simplification is not possible. Avoid the common error of setting $-2x - 10$ equal to 0.

Solving an Algebraic Equation

Solve: $3(x - 7) - (5x - 11) = 14$.

This is an equation. There is an equal sign.

Solution
$$\begin{aligned} 3(x - 7) - (5x - 11) &= 14 \\ 3x - 21 - 5x + 11 &= 14 \\ -2x - 10 &= 14 \\ -2x - 10 + 10 &= 14 + 10 \quad \text{Add 10 to both sides.} \\ -2x &= 24 \\ \frac{-2x}{-2} &= \frac{24}{-2} \quad \text{Divide both sides by } -2. \\ x &= -12 \end{aligned}$$

The solution set is $\{-12\}$.

GREAT QUESTION!

Do I have to solve $5x - 12 = 8x + 24$ by collecting variable terms on the left and numbers on the right?

No. If you prefer, you can solve the equation by collecting variable terms on the right and numbers on the left. To collect variable terms on the right, subtract $5x$ from both sides:

$$5x - 12 - 5x = 8x + 24 - 5x$$
$$-12 = 3x + 24.$$

To collect numbers on the left, subtract 24 from both sides:

$$-12 - 24 = 3x + 24 - 24$$
$$-36 = 3x.$$

Now isolate x by dividing both sides by 3:

$$\frac{-36}{3} = \frac{3x}{3}$$
$$-12 = x.$$

This is the same solution that we obtained in Example 3.

EXAMPLE 3 *Solving a Linear Equation*

Solve and check: $5x - 12 = 8x + 24$.

SOLUTION

Step 1 Simplify the algebraic expression on each side. Neither side contains grouping symbols or like terms that can be combined. Therefore, we can skip this step.

Step 2 Collect variable terms on one side and constants on the other side. One way to do this is to collect variable terms on the left and constants on the right. This is accomplished by subtracting $8x$ from both sides and adding 12 to both sides.

$$5x - 12 = 8x + 24$$ This is the given equation.

$$5x - 12 - 8x = 8x + 24 - 8x$$ Subtract $8x$ from both sides.

$$-3x - 12 = 24$$ Simplify: $5x - 8x = -3x$.

$$-3x - 12 + 12 = 24 + 12$$ Add 12 to both sides and collect constants on the right side.

$$-3x = 36$$ Simplify.

Step 3 Isolate the variable and solve. We isolate the variable, x, by dividing both sides of $-3x = 36$ by -3.

$$\frac{-3x}{-3} = \frac{36}{-3}$$ Divide both sides by -3.

$$x = -12$$ Simplify.

Step 4 Check the proposed solution in the original equation. Substitute -12 for x in the original equation.

$$5x - 12 = 8x + 24$$ This is the original equation.

$$5(-12) - 12 \stackrel{?}{=} 8(-12) + 24$$ Substitute -12 for x.

$$-60 - 12 \stackrel{?}{=} -96 + 24$$ Multiply: $5(-12) = -60$ and $8(-12) = -96$.

This statement is true. $$-72 = -72$$ Add: $-60 + (-12) = -72$ and $-96 + 24 = -72$.

Because the check results in a true statement, we conclude that the solution set of the given equation is $\{-12\}$.

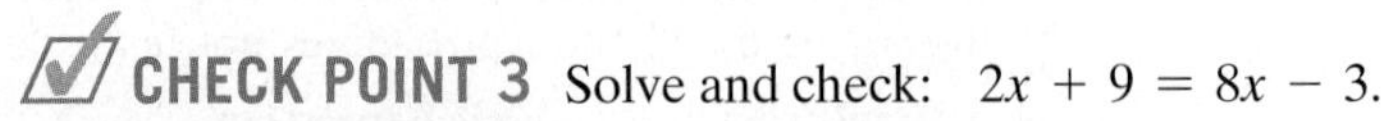

CHECK POINT 3 Solve and check: $2x + 9 = 8x - 3$.

EXAMPLE 4 *Solving a Linear Equation*

Solve and check: $2(x - 3) - 17 = 13 - 3(x + 2)$.

SOLUTION

Step 1 Simplify the algebraic expression on each side.

Do not begin with $13 - 3$. Multiplication (the distributive property) is applied before subtraction.

$$2(x - 3) - 17 = 13 - 3(x + 2)$$ This is the given equation.

$$2x - 6 - 17 = 13 - 3x - 6$$ Use the distributive property.

$$2x - 23 = -3x + 7$$ Combine like terms.

Step 2 Collect variable terms on one side and constants on the other side. We will collect variable terms of $2x - 23 = -3x + 7$ on the left by adding $3x$ to both sides. We will collect the numbers on the right by adding 23 to both sides.

$$2x - 23 + 3x = -3x + 7 + 3x \quad \text{Add } 3x \text{ to both sides.}$$
$$5x - 23 = 7 \quad \text{Simplify: } 2x + 3x = 5x.$$
$$5x - 23 + 23 = 7 + 23 \quad \text{Add 23 to both sides.}$$
$$5x = 30 \quad \text{Simplify.}$$

Step 3 Isolate the variable and solve. We isolate the variable, x, by dividing both sides of $5x = 30$ by 5.

$$\frac{5x}{5} = \frac{30}{5} \quad \text{Divide both sides by 5.}$$
$$x = 6 \quad \text{Simplify.}$$

Step 4 Check the proposed solution in the original equation. Substitute 6 for x in the original equation.

$$2(x - 3) - 17 = 13 - 3(x + 2) \quad \text{This is the original equation.}$$
$$2(6 - 3) - 17 \stackrel{?}{=} 13 - 3(6 + 2) \quad \text{Substitute 6 for } x.$$
$$2(3) - 17 \stackrel{?}{=} 13 - 3(8) \quad \text{Simplify inside parentheses.}$$
$$6 - 17 \stackrel{?}{=} 13 - 24 \quad \text{Multiply.}$$
$$-11 = -11 \quad \text{Subtract.}$$

The true statement $-11 = -11$ verifies that the solution set is $\{6\}$.

CHECK POINT 4 Solve and check: $4(2x + 1) = 29 + 3(2x - 5)$.

2 Solve linear equations containing fractions.

Linear Equations with Fractions

Equations are easier to solve when they do not contain fractions. How do we remove fractions from an equation? We begin by multiplying both sides of the equation by the least common denominator of any fractions in the equation. The least common denominator is the smallest number that all denominators will divide into. Multiplying every term on both sides of the equation by the least common denominator will eliminate the fractions in the equation. Example 5 shows how we "clear an equation of fractions."

EXAMPLE 5 *Solving a Linear Equation Involving Fractions*

Solve and check: $\frac{3x}{2} = \frac{8x}{5} - 4$.

SOLUTION

The denominators are 2 and 5. The smallest number that is divisible by 2 and 5 is 10. We begin by multiplying both sides of the equation by 10, the least common denominator.

$$\frac{3x}{2} = \frac{8x}{5} - 4 \quad \text{This is the given equation.}$$
$$10 \cdot \frac{3x}{2} = 10\left(\frac{8x}{5} - 4\right) \quad \text{Multiply both sides by 10.}$$
$$10 \cdot \frac{3x}{2} = 10 \cdot \frac{8x}{5} - 10 \cdot 4 \quad \text{Use the distributive property. Be sure to multiply all terms by 10.}$$

$$\overset{5}{\cancel{10}} \cdot \frac{3x}{\underset{1}{\cancel{2}}} = \overset{2}{\cancel{10}} \cdot \frac{8x}{\underset{1}{\cancel{5}}} - 40$$ Divide out common factors in the multiplications.

$$15x = 16x - 40$$ Complete the multiplications. The fractions are now cleared.

At this point, we have an equation similar to those we have previously solved. Collect the variable terms on one side and the constants on the other side.

$$15x - 16x = 16x - 40 - 16x$$ Subtract $16x$ from both sides to get the variable terms on the left.

$$-x = -40$$ Simplify.

We're not finished. A negative sign should not precede x.

Isolate x by multiplying or dividing both sides of this equation by -1.

$$\frac{-x}{-1} = \frac{-40}{-1}$$ Divide both sides by -1.

$$x = 40$$ Simplify.

Check the proposed solution. Substitute 40 for x in the original equation. You should obtain $60 = 60$. This true statement verifies that the solution set is $\{40\}$.

CHECK POINT 5 Solve and check: $\frac{2x}{3} = 7 - \frac{x}{2}$.

EXAMPLE 6 *An Application: Responding to Negative Life Events*

In the section opener, we introduced line graphs, repeated in **Figure 6.2**, indicating that persons with a low sense of humor have higher levels of depression in response to negative life events than those with a high sense of humor. These graphs can be modeled by the following formulas:

Low-Humor Group

$$D = \frac{10}{9}x + \frac{53}{9}$$

High-Humor Group

$$D = \frac{1}{9}x + \frac{26}{9}.$$

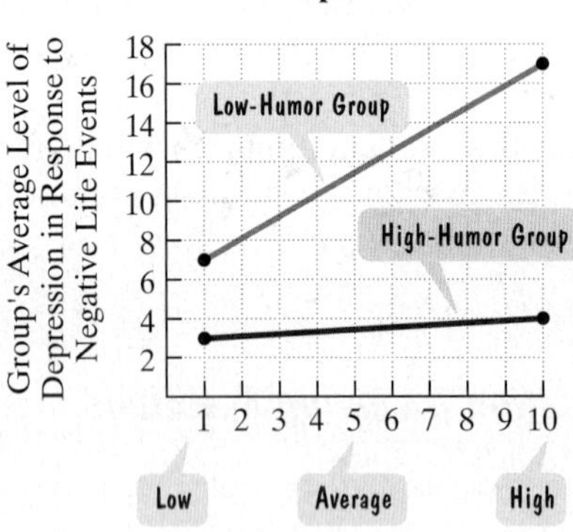

FIGURE 6.2 (repeated)

In each formula, x represents the intensity of a negative life event (from 1, low, to 10, high) and D is the average level of depression in response to that event. If the high-humor group averages a level of depression of 3.5, or $\frac{7}{2}$, in response to a negative life event, what is the intensity of that event? How is the solution shown on the red line graph in **Figure 6.2**?

SOLUTION

We are interested in the intensity of a negative life event with an average level of depression of $\frac{7}{2}$ for the high-humor group. We substitute $\frac{7}{2}$ for D in the high-humor model and solve for x, the intensity of the negative life event.

$$D = \frac{1}{9}x + \frac{26}{9}$$ This is the given formula for the high-humor group.

$$\frac{7}{2} = \frac{1}{9}x + \frac{26}{9}$$ Replace D with $\frac{7}{2}$.

$$18 \cdot \frac{7}{2} = 18\left(\frac{1}{9}x + \frac{26}{9}\right)$$ Multiply both sides by 18, the least common denominator.

$$18 \cdot \frac{7}{2} = 18 \cdot \frac{1}{9}x + 18 \cdot \frac{26}{9}$$ Use the distributive property.

$$\overset{9}{\cancel{18}} \cdot \frac{7}{\underset{1}{\cancel{2}}} = \overset{2}{\cancel{18}} \cdot \frac{1}{\underset{1}{\cancel{9}}}x + \overset{2}{\cancel{18}} \cdot \frac{26}{\underset{1}{\cancel{9}}}$$ Divide out common factors in the multiplications.

$$63 = 2x + 52$$ Complete the multiplications. The fractions are now cleared.

$$63 - 52 = 2x + 52 - 52$$ Subtract 52 from both sides to get constants on the left.

$$11 = 2x$$ Simplify.

$$\frac{11}{2} = \frac{2x}{2}$$ Divide both sides by 2.

$$\frac{11}{2} = x$$ Simplify.

The formula indicates that if the high-humor group averages a level of depression of 3.5 in response to a negative life event, the intensity of that event is $\frac{11}{2}$, or 5.5. This is illustrated on the line graph for the high-humor group in **Figure 6.3**.

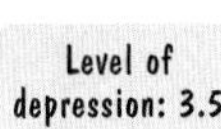
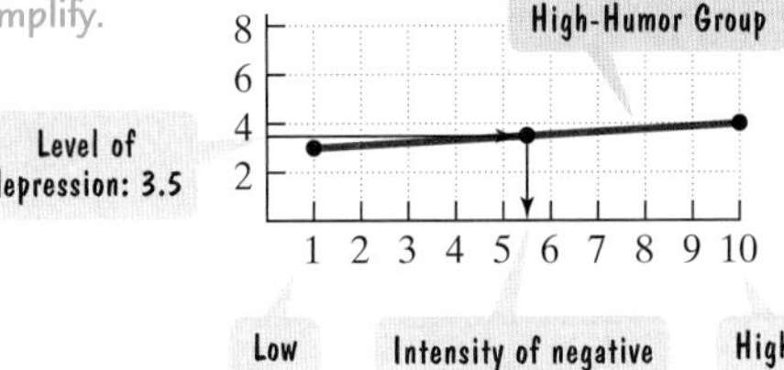

FIGURE 6.3

CHECK POINT 6 Use the model for the low-humor group given in Example 6 on the previous page to solve this problem. If the low-humor group averages a level of depression of 10 in response to a negative life event, what is the intensity of that event? How is the solution shown on the blue line graph in **Figure 6.2**?

3 *Solve proportions.*

Proportions

A **ratio** compares quantities by division. For example, a group contains 60 women and 30 men. The ratio of women to men is $\frac{60}{30}$. We can express this ratio as a fraction reduced to lowest terms:

$$\frac{60}{30} = \frac{2 \cdot \cancel{30}}{1 \cdot \cancel{30}} = \frac{2}{1}.$$

This ratio can be expressed as 2:1, or 2 to 1.

A **proportion** is a statement that says that two ratios are equal. If the ratios are $\frac{a}{b}$ and $\frac{c}{d}$, then the proportion is

$$\frac{a}{b} = \frac{c}{d}.$$

We can clear this equation of fractions by multiplying both sides by bd, the least common denominator:

$$\frac{a}{b} = \frac{c}{d}$$ This is the given proportion.

$$bd \cdot \frac{a}{b} = bd \cdot \frac{c}{d}$$ Multiply both sides by bd ($b \neq 0$ and $d \neq 0$). Then simplify.

$$ad = bc.$$ On the left, $\frac{\cancel{b}d}{1} \cdot \frac{a}{\cancel{b}} = da = ad$. On the right, $\frac{b\cancel{d}}{1} \cdot \frac{c}{\cancel{d}} = bc$.

We see that the following principle is true for any proportion:

bc

$\frac{a}{b} = \frac{c}{d}$

ad

The cross-products principle: $ad = bc$

THE CROSS-PRODUCTS PRINCIPLE FOR PROPORTIONS

If $\frac{a}{b} = \frac{c}{d}$, then $ad = bc$. ($b \neq 0$ and $d \neq 0$)

The cross products ad and bc are equal.

For example, since $\frac{2}{3} = \frac{6}{9}$, we see that $2 \cdot 9 = 3 \cdot 6$, or $18 = 18$. We can also use $\frac{2}{3} = \frac{6}{9}$ and conclude that $3 \cdot 6 = 2 \cdot 9$. When using the cross-products principle, it does not matter on which side of the equation each product is placed.

If three of the numbers in a proportion are known, the value of the missing quantity can be found by using the cross-products principle. This idea is illustrated in Example 7(a).

EXAMPLE 7 *Solving Proportions*

Solve each proportion and check:

a. $\dfrac{63}{x} = \dfrac{7}{5}$ **b.** $\dfrac{20}{x - 10} = \dfrac{30}{x}$.

SOLUTION

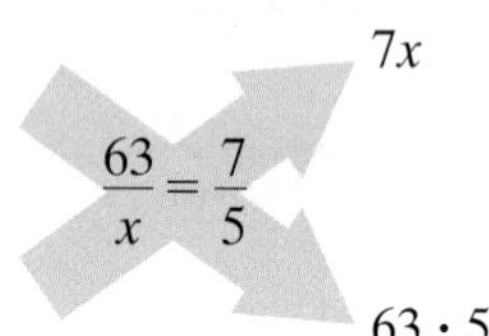

Cross products

a.

$$\frac{63}{x} = \frac{7}{5} \quad \text{This is the given proportion.}$$

$$63 \cdot 5 = 7x \quad \text{Apply the cross-products principle.}$$

$$315 = 7x \quad \text{Simplify.}$$

$$\frac{315}{7} = \frac{7x}{7} \quad \text{Divide both sides by 7.}$$

$$45 = x \quad \text{Simplify.}$$

The solution set is $\{45\}$.

Check

$$\frac{63}{45} \stackrel{?}{=} \frac{7}{5} \quad \text{Substitute 45 for } x \text{ in } \frac{63}{x} = \frac{7}{5}.$$

$$\frac{7 \cdot \cancel{9}}{5 \cdot \cancel{9}} \stackrel{?}{=} \frac{7}{5} \quad \text{Reduce } \frac{63}{45} \text{ to lowest terms.}$$

$$\frac{7}{5} = \frac{7}{5} \quad \text{This true statement verifies that the solution set is } \{45\}.$$

b.

$$\frac{20}{x - 10} = \frac{30}{x} \quad \text{This is the given proportion.}$$

$$20x = 30(x - 10) \quad \text{Apply the cross-products principle.}$$

$$20x = 30x - 30 \cdot 10 \quad \text{Use the distributive property.}$$

$$20x = 30x - 300 \quad \text{Simplify.}$$

$$20x - 30x = 30x - 300 - 30x \quad \text{Subtract } 30x \text{ from both sides.}$$

$$-10x = -300 \quad \text{Simplify.}$$

$$\frac{-10x}{-10} = \frac{-300}{-10} \quad \text{Divide both sides by } -10.$$

$$x = 30 \quad \text{Simplify.}$$

The solution set is $\{30\}$.

Check

$$\frac{20}{30 - 10} \stackrel{?}{=} \frac{30}{30} \quad \text{Substitute 30 for } x \text{ in } \frac{20}{x - 10} = \frac{30}{x}.$$

$$\frac{20}{20} \stackrel{?}{=} \frac{30}{30} \quad \text{Subtract: } 30 - 10 = 20.$$

$$1 = 1 \quad \text{This true statement verifies that the solution is 30.}$$

CHECK POINT 7 Solve each proportion and check:

a. $\frac{10}{x} = \frac{2}{3}$ **b.** $\frac{22}{60 - x} = \frac{2}{x}$.

4 *Solve problems using proportions.*

Applications of Proportions

We now turn to practical application problems that can be solved using proportions. Here is a procedure for solving these problems:

SOLVING APPLIED PROBLEMS USING PROPORTIONS

1. Read the problem and represent the unknown quantity by x (or any letter).
2. Set up a proportion by listing the given ratio on one side and the ratio with the unknown quantity on the other side. Each respective quantity should occupy the same corresponding position on each side of the proportion.
3. Drop units and apply the cross-products principle.
4. Solve for x and answer the question.

EXAMPLE 8 *Applying Proportions: Calculating Taxes*

The property tax on a house with an assessed value of \$480,000 is \$5760. Determine the property tax on a house with an assessed value of \$600,000, assuming the same tax rate.

SOLUTION

Step 1 Represent the unknown by x. Let x = the tax on the \$600,000 house.

Step 2 Set up a proportion. We will set up a proportion comparing taxes to assessed value.

$$\frac{\text{Tax on \$480,000 house}}{\text{Assessed value (\$480,000)}} \quad \text{equals} \quad \frac{\text{Tax on \$600,000 house}}{\text{Assessed value (\$600,000)}}$$

$$\text{Given ratio} \left\{ \frac{\$5760}{\$480,000} \right. = \frac{\$x}{\$600,000}$$

($\$x$ ← Unknown; \$600,000 ← Given quantity)

Step 3 Drop the units and apply the cross-products principle. We drop the dollar signs and begin to solve for x.

$$\frac{5760}{480,000} = \frac{x}{600,000}$$ This is the proportion that models the problem's conditions.

$$480,000x = (5760)(600,000)$$ Apply the cross-products principle.

$$480,000x = 3,456,000,000$$ Multiply.

Step 4 Solve for x and answer the question.

$$\frac{480,000x}{480,000} = \frac{3,456,000,000}{480,000}$$ Divide both sides by 480,000.

$$x = 7200$$ Simplify.

The property tax on the \$600,000 house is \$7200.

GREAT QUESTION!

Are there other proportions that I can use in step 2 to model the problem's conditions?

Yes. Here are three other correct proportions you can use:

- $\frac{\$480,000 \text{ value}}{\$5760 \text{ tax}} = \frac{\$600,000 \text{ value}}{\$x \text{ tax}}$
- $\frac{\$480,000 \text{ value}}{\$600,000 \text{ value}} = \frac{\$5760 \text{ tax}}{\$x \text{ tax}}$
- $\frac{\$600,000 \text{ value}}{\$480,000 \text{ value}} = \frac{\$x \text{ tax}}{\$5760 \text{ tax}}$

Each proportion gives the same cross product obtained in step 3.

CHECK POINT 8 The property tax on a house with an assessed value of \$250,000 is \$3500. Determine the property tax on a house with an assessed value of \$420,000, assuming the same tax rate.

EXAMPLE 9 *Applying Proportions: Estimating Wildlife Population*

Wildlife biologists catch, tag, and then release 135 deer back into a wildlife refuge. Two weeks later they observe a sample of 140 deer, 30 of which are tagged. Assuming the ratio of tagged deer in the sample holds for all deer in the refuge, approximately how many deer are in the refuge?

SOLUTION

Step 1 Represent the unknown by x. Let x = the total number of deer in the refuge.

Step 2 Set up a proportion.

Unknown → (Original number of tagged deer) / (Total number of deer) equals (Number of tagged deer in the observed sample) / (Total number of deer in the observed sample) } Known ratio

$$\frac{135}{x} = \frac{30}{140}$$

Steps 3 and 4 Apply the cross-products principle, solve, and answer the question.

$$\frac{135}{x} = \frac{30}{140} \quad \text{This is the proportion that models the problem's conditions.}$$

$$(135)(140) = 30x \quad \text{Apply the cross-products principle.}$$

$$18{,}900 = 30x \quad \text{Multiply.}$$

$$\frac{18{,}900}{30} = \frac{30x}{30} \quad \text{Divide both sides by 30.}$$

$$630 = x \quad \text{Simplify.}$$

There are approximately 630 deer in the refuge.

 **CHECK POINT 9** Wildlife biologists catch, tag, and then release 120 deer back into a wildlife refuge. Two weeks later they observe a sample of 150 deer, 25 of which are tagged. Assuming the ratio of tagged deer in the sample holds for all deer in the refuge, approximately how many deer are in the refuge?

5 *Identify equations with no solution or infinitely many solutions.*

Equations with No Solution or Infinitely Many Solutions

Thus far, each equation or proportion that we have solved has had a single solution. However, some equations are not true for even one real number. By contrast, other equations are true for all real numbers.

If you attempt to solve an equation with no solution, you will eliminate the variable and obtain a false statement, such as $2 = 5$. If you attempt to solve an equation that is true for every real number, you will eliminate the variable and obtain a true statement, such as $4 = 4$.

EXAMPLE 10 *Attempting to Solve an Equation with No Solution*

Solve: $2x + 6 = 2(x + 4)$.

SOLUTION

$$2x + 6 = 2(x + 4) \quad \text{This is the given equation.}$$

$$2x + 6 = 2x + 8 \quad \text{Use the distributive property.}$$

$$2x + 6 - 2x = 2x + 8 - 2x \quad \text{Subtract } 2x \text{ from both sides.}$$

$$6 = 8 \quad \text{Simplify.}$$

Keep reading. $6 = 8$ is not the solution.

The original equation, $2x + 6 = 2(x + 4)$, is equivalent to the statement $6 = 8$, which is false for every value of x. The equation has no solution. The solution set is $\varnothing$, the empty set.

CHECK POINT 10 Solve: $3x + 7 = 3(x + 1)$.

EXAMPLE 11 ***Solving an Equation for Which Every Real Number Is a Solution***

Solve: $4x + 6 = 6(x + 1) - 2x$.

SOLUTION

$$4x + 6 = 6(x + 1) - 2x$$ This is the given equation.

$$4x + 6 = 6x + 6 - 2x$$ Apply the distributive property on the right side.

$$4x + 6 = 4x + 6$$ Combine like terms on the right side: $6x - 2x = 4x$.

Can you see that the equation $4x + 6 = 4x + 6$ is true for every value of x? Let's continue solving the equation by subtracting $4x$ from both sides.

$$4x + 6 - 4x = 4x + 6 - 4x$$

$$6 = 6$$

Keep reading. $6 = 6$ is not the solution.

The original equation is equivalent to the statement $6 = 6$, which is true for every value of x. Thus, the solution set consists of the set of all real numbers, expressed in set-builder notation as $\{x \mid x \text{ is a real number}\}$. Try substituting any real number of your choice for x in the original equation. You will obtain a true statement.

GREAT QUESTION!

Do I have to use sets to write the solution of an equation?

Because of the fundamental role that sets play in mathematics, it's a good idea to use set notation to express an equation's solution. If an equation has no solution, its solution set is $\varnothing$, the empty set. If an equation with variable x is true for every real number, its solution set is $\{x \mid x \text{ is a real number}\}$.

CHECK POINT 11 Solve: $7x + 9 = 9(x + 1) - 2x$.

Concept and Vocabulary Check

Fill in each blank so that the resulting statement is true.

1. An equation in the form $ax + b = 0$, $a \neq 0$, such as $3x + 17 = 0$, is called a/an _______ equation in one variable.
2. Two or more equations that have the same solution set are called __________ equations.
3. The addition property of equality states that if $a = b$, then $a + c =$ _______.
4. The multiplication property of equality states that if $a = b$ and $c \neq 0$, then $ac =$ ____.
5. The first step in solving $7 + 3(x - 2) = 2x + 10$ is to ________________________.
6. The algebraic expression $7(x - 4) + 2x$ can be __________, whereas the algebraic equation $7(x - 4) + 2x = 35$ can be _______.
7. The equation
$$\frac{x}{4} = 2 + \frac{x}{3}$$
can be cleared of fractions by multiplying both sides by the ____________________ of $\frac{x}{4}$ and $\frac{x}{3}$, which is ____.
8. A statement that two ratios are equal is called a/an _________.
9. The cross-products principle states that if $\frac{a}{b} = \frac{c}{d}$ ($b \neq 0$ and $d \neq 0$), then ________.

10. In solving an equation, if you eliminate the variable and obtain a statement such as $2 = 3$, the equation has _____ solution. The solution set can be expressed using the symbol _____.

11. In solving an equation with variable x, if you eliminate the variable and obtain a statement such as $6 = 6$, the equation is ______ for every value of x. The solution set can be expressed in set-builder notation as __________________.

In Exercises 12–15, determine whether each statement is true or false. If the statement is false, make the necessary change(s) to produce a true statement.

12. The equation $2x + 5 = 0$ is equivalent to $2x = 5$. ______

13. The equation $x + \frac{1}{3} = \frac{1}{2}$ is equivalent to $x + 2 = 3$. ______

14. The equation $3x = 2x$ has no solution. ______

15. The equation $3(x + 4) = 3(4 + x)$ has precisely one solution. ______

Exercise Set 6.2

Practice Exercises

In Exercises 1–58, solve and check each equation.

1. $x - 7 = 3$

2. $x - 3 = -17$

3. $x + 5 = -12$

4. $x + 12 = -14$

5. $\frac{x}{3} = 4$

6. $\frac{x}{5} = 3$

7. $5x = 45$

8. $6x = 18$

9. $8x = -24$

10. $5x = -25$

11. $-8x = 2$

12. $-6x = 3$

13. $5x + 3 = 18$

14. $3x + 8 = 50$

15. $6x - 3 = 63$

16. $5x - 8 = 72$

17. $4x - 14 = -82$

18. $9x - 14 = -77$

19. $14 - 5x = -41$

20. $25 - 6x = -83$

21. $9(5x - 2) = 45$

22. $10(3x + 2) = 70$

23. $5x - (2x - 10) = 35$

24. $11x - (6x - 5) = 40$

25. $3x + 5 = 2x + 13$

26. $2x - 7 = 6 + x$

27. $8x - 2 = 7x - 5$

28. $13x + 14 = -5 + 12x$

29. $7x + 4 = x + 16$

30. $8x + 1 = x + 43$

31. $8y - 3 = 11y + 9$

32. $5y - 2 = 9y + 2$

33. $2(4 - 3x) = 2(2x + 5)$

34. $3(5 - x) = 4(2x + 1)$

35. $8(y + 2) = 2(3y + 4)$

36. $3(3y - 1) = 4(3 + 3y)$

37. $3(x + 1) = 7(x - 2) - 3$

38. $5x - 4(x + 9) = 2x - 3$

39. $5(2x - 8) - 2 = 5(x - 3) + 3$

40. $7(3x - 2) + 5 = 6(2x - 1) + 24$

41. $6 = -4(1 - x) + 3(x + 1)$

42. $100 = -(x - 1) + 4(x - 6)$

43. $10(z + 4) - 4(z - 2) = 3(z - 1) + 2(z - 3)$

44. $-2(z - 4) - (3z - 2) = -2 - (6z - 2)$

45. $\frac{2x}{3} - 5 = 7$

46. $\frac{3x}{4} - 9 = -6$

47. $\frac{x}{3} + \frac{x}{2} = \frac{5}{6}$

48. $\frac{x}{4} - \frac{x}{5} = 1$

49. $20 - \frac{z}{3} = \frac{z}{2}$

50. $\frac{z}{5} - \frac{1}{2} = \frac{z}{6}$

51. $\frac{y}{3} + \frac{2}{5} = \frac{y}{5} - \frac{2}{5}$

52. $\frac{y}{12} + \frac{1}{6} = \frac{y}{2} - \frac{1}{4}$

53. $\frac{3x}{4} - 3 = \frac{x}{2} + 2$

54. $\frac{3x}{5} - \frac{2}{5} = \frac{x}{3} + \frac{2}{5}$

55. $\frac{3x}{5} - x = \frac{x}{10} - \frac{5}{2}$

56. $2x - \frac{2x}{7} = \frac{x}{2} + \frac{17}{2}$

57. $\frac{x - 3}{5} - 1 = \frac{x - 5}{4}$

58. $\frac{x - 2}{3} - 4 = \frac{x + 1}{4}$

In Exercises 59–72, solve each proportion and check.

59. $\frac{24}{x} = \frac{12}{7}$

60. $\frac{56}{x} = \frac{8}{7}$

61. $\frac{x}{6} = \frac{18}{4}$

62. $\frac{x}{32} = \frac{3}{24}$

63. $\frac{-3}{8} = \frac{x}{40}$

64. $\frac{-3}{8} = \frac{6}{x}$

65. $\frac{x}{12} = -\frac{3}{4}$

66. $\frac{x}{64} = -\frac{9}{16}$

67. $\frac{x - 2}{12} = \frac{8}{3}$

68. $\frac{x - 4}{10} = \frac{3}{5}$

69. $\frac{x}{7} = \frac{x + 14}{5}$

70. $\frac{x}{5} = \frac{x - 3}{2}$

71. $\frac{y + 10}{10} = \frac{y - 2}{4}$

72. $\frac{2}{y - 5} = \frac{3}{y + 6}$

In Exercises 73–92, solve each equation. Use set notation to express solution sets for equations with no solution or equations that are true for all real numbers.

73. $3x - 7 = 3(x + 1)$

74. $2(x - 5) = 2x + 10$

75. $2(x + 4) = 4x + 5 - 2x + 3$

76. $3(x - 1) = 8x + 6 - 5x - 9$

77. $7 + 2(3x - 5) = 8 - 3(2x + 1)$

78. $2 + 3(2x - 7) = 9 - 4(3x + 1)$

79. $4x + 1 - 5x = 5 - (x + 4)$

80. $5x - 5 = 3x - 7 + 2(x + 1)$

81. $4(x + 2) + 1 = 7x - 3(x - 2)$

82. $5x - 3(x + 1) = 2(x + 3) - 5$

83. $3 - x = 2x + 3$

84. $5 - x = 4x + 5$

85. $\frac{x}{3} + 2 = \frac{x}{3}$

86. $\frac{x}{4} + 3 = \frac{x}{4}$

87. $\frac{x}{3} = \frac{x}{2}$

88. $\frac{x}{4} = \frac{x}{3}$

89. $\frac{x-2}{5} = \frac{3}{10}$

90. $\frac{x+4}{8} = \frac{3}{16}$

91. $\frac{x}{2} - \frac{x}{4} + 4 = x + 4$

92. $\frac{x}{2} + \frac{2x}{3} + 3 = x + 3$

Practice Plus

93. Evaluate $x^2 - x$ for the value of x satisfying $4(x - 2) + 2 = 4x - 2(2 - x)$.

94. Evaluate $x^2 - x$ for the value of x satisfying $2(x - 6) = 3x + 2(2x - 1)$.

95. Evaluate $x^2 - (xy - y)$ for x satisfying $\frac{x}{5} - 2 = \frac{x}{3}$ and y satisfying $-2y - 10 = 5y + 18$.

96. Evaluate $x^2 - (xy - y)$ for x satisfying $\frac{3x}{2} + \frac{3x}{4} = \frac{x}{4} - 4$ and y satisfying $5 - y = 7(y + 4) + 1$.

In Exercises 97–104, solve each equation.

97. $[(3 + 6)^2 \div 3] \cdot 4 = -54x$

98. $2^3 - [4(5 - 3)^3] = -8x$

99. $5 - 12x = 8 - 7x - [6 \div 3(2 + 5^3) + 5x]$

100. $2(5x + 58) = 10x + 4(21 \div 3.5 - 11)$

101. $0.7x + 0.4(20) = 0.5(x + 20)$

102. $0.5(x + 2) = 0.1 + 3(0.1x + 0.3)$

103. $4x + 13 - \{2x - [4(x - 3) - 5]\} = 2(x - 6)$

104. $-2\{7 - [4 - 2(1 - x) + 3]\} = 10 - [4x - 2(x - 3)]$

Application Exercises

The latest guidelines, which apply to both men and women, give healthy weight ranges, rather than specific weights, for your height. The further you are above the upper limit of your range, the greater are the risks of developing weight-related health problems. The bar graph shows these ranges for various heights for people between the ages of 19 and 34, inclusive.

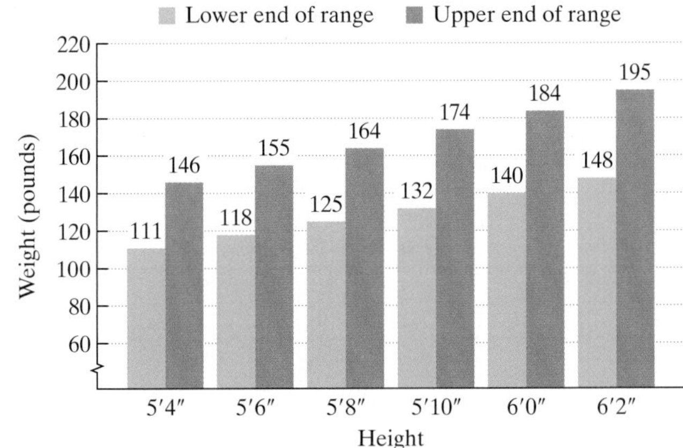

Source: U.S. Department of Health and Human Services

The mathematical model

$$\frac{W}{2} - 3H = 53$$

describes a weight, W, in pounds, that lies within the healthy weight range for a person whose height is H inches over 5 feet. Use this information to solve Exercises 105–106.

105. Use the formula to find a healthy weight for a person whose height is 5′6″. (*Hint:* $H = 6$ because this person's height is 6 inches over 5 feet.) How many pounds is this healthy weight below the upper end of the range shown by the bar graph at the bottom of the previous column?

106. Use the formula to find a healthy weight for a person whose height is 6′0″. (*Hint:* $H = 12$ because this person's height is 12 inches over 5 feet.) How many pounds is this healthy weight below the upper end of the range shown by the bar graph at the bottom of the previous column?

Grade Inflation. *The bar graph shows the percentage of U.S. college freshmen with an average grade of A in high school.*

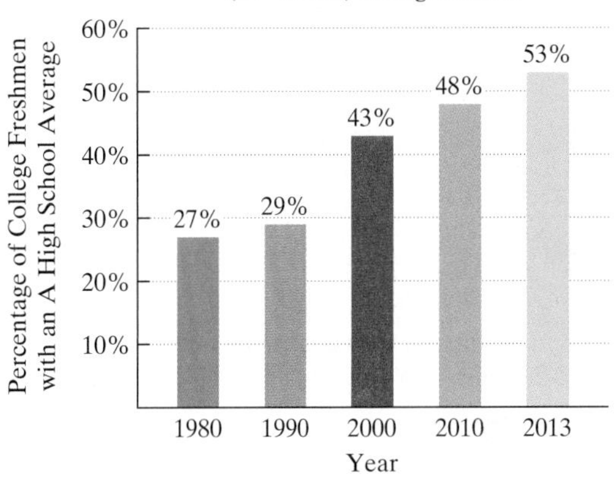

Source: Higher Education Research Institute

The data displayed by the bar graph can be described by the mathematical model

$$p = \frac{4x}{5} + 25,$$

where x is the number of years after 1980 and p is the percentage of U.S. college freshmen who had an average grade of A in high school. Use this information to solve Exercises 107–108.

107. **a.** According to the formula, in 2010, what percentage of U.S. college freshmen had an average grade of A in high school? Does this underestimate or overestimate the percent displayed by the bar graph? By how much?

b. If trends shown by the formula continue, project when 57% of U.S. college freshmen will have had an average grade of A in high school.

108. **a.** According to the formula, in 2000, what percentage of U.S. college freshmen had an average grade of A in high school? Does this underestimate or overestimate the percent displayed by the bar graph? By how much?

b. If trends shown by the formula continue, project when 65% of U.S. college freshmen will have had an average grade of A in high school.

109. The volume of blood in a person's body is proportional to body weight. A person who weighs 160 pounds has approximately 5 quarts of blood. Estimate the number of quarts of blood in a person who weighs 200 pounds.

110. The number of gallons of water used when taking a shower is proportional to the time in the shower. A shower lasting 5 minutes uses 30 gallons of water. How much water is used in a shower lasting 11 minutes?

111. An alligator's tail length is proportional to its body length. An alligator with a body length of 4 feet has a tail length of 3.6 feet. What is the tail length of an alligator whose body length is 6 feet?

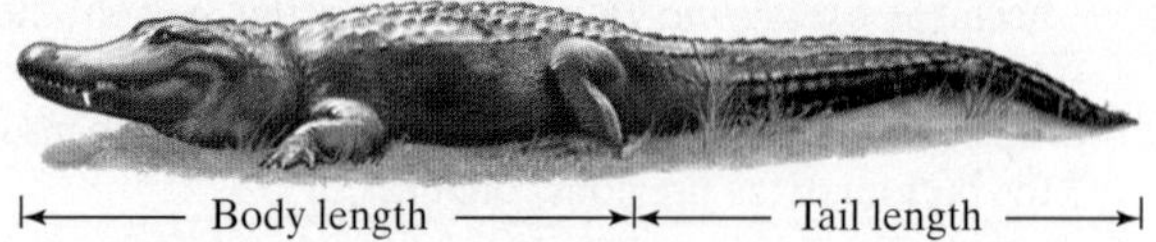

112. An object's weight on the Moon is proportional to its weight on Earth. Neil Armstrong, the first person to step on the Moon on July 20, 1969, weighed 360 pounds on Earth (with all of his equipment on) and 60 pounds on the Moon. What is the Moon weight of a person who weighs 186 pounds on Earth?

113. St. Paul Island in Alaska has 12 fur seal rookeries (breeding places). In 1961, to estimate the fur seal pup population in the Gorbath rookery, 4963 fur seal pups were tagged in early August. In late August, a sample of 900 pups was observed and 218 of these were found to have been previously tagged. Estimate the total number of fur seal pups in this rookery.

114. To estimate the number of bass in a lake, wildlife biologists tagged 50 bass and released them in the lake. Later they netted 108 bass and found that 27 of them were tagged. Approximately how many bass are in the lake?

Explaining the Concepts

115. What is the solution set of an equation?

116. State the addition property of equality and give an example.

117. State the multiplication property of equality and give an example.

118. What is a proportion? Give an example with your description.

119. Explain how to solve a proportion. Illustrate your explanation with an example.

120. How do you know whether a linear equation has one solution, no solution, or infinitely many solutions?

121. What is the difference between solving an equation such as $2(x - 4) + 5x = 34$ and simplifying an algebraic expression such as $2(x - 4) + 5x$? If there is a difference, which topic should be taught first? Why?

122. Suppose that you solve $\frac{x}{5} - \frac{x}{2} = 1$ by multiplying both sides by 20, rather than the least common denominator of $\frac{x}{5}$ and $\frac{x}{2}$ (namely, 10). Describe what happens. If you get the correct solution, why do you think we clear the equation of fractions by multiplying by the *least* common denominator?

123. Suppose you are an algebra teacher grading the following solution on an examination:

$$\begin{aligned} \text{Solve: } -3(x - 6) &= 2 - x. \\ \text{Solution: } -3x - 18 &= 2 - x \\ -2x - 18 &= 2 \\ -2x &= -16 \\ x &= 8. \end{aligned}$$

You should note that 8 checks, and the solution set is $\{8\}$. The student who worked the problem therefore wants full credit. Can you find any errors in the solution? If full credit is 10 points, how many points should you give the student? Justify your position.

124. Although the formulas in Example 6 on page 360 are correct, some people object to representing the variables with numbers, such as a 1-to-10 scale for the intensity of a negative life event. What might be their objection to quantifying the variables in this situation?

Critical Thinking Exercises

Make Sense? *In Exercises 125–128, determine whether each statement makes sense or does not make sense, and explain your reasoning.*

125. Although I can solve $3x + \frac{1}{5} = \frac{1}{4}$ by first subtracting $\frac{1}{5}$ from both sides, I find it easier to begin by multiplying both sides by 20, the least common denominator.

126. Because I know how to clear an equation of fractions, I decided to clear the equation $0.5x + 8.3 = 12.4$ of decimals by multiplying both sides by 10.

127. The number 3 satisfies the equation $7x + 9 = 9(x + 1) - 2x$, so $\{3\}$ is the equation's solution set.

128. I can solve $\frac{x}{9} = \frac{4}{6}$ by using the cross-products principle or by multiplying both sides by 18, the least common denominator.

129. Write three equations whose solution set is $\{5\}$.

130. If x represents a number, write an English sentence about the number that results in an equation with no solution.

131. A woman's height, h, is related to the length of the femur, f (the bone from the knee to the hip socket), by the formula $f = 0.432h - 10.44$. Both h and f are measured in inches. A partial skeleton is found of a woman in which the femur is 16 inches long. Police find the skeleton in an area where a woman slightly over 5 feet tall has been missing for over a year. Can the partial skeleton be that of the missing woman? Explain.

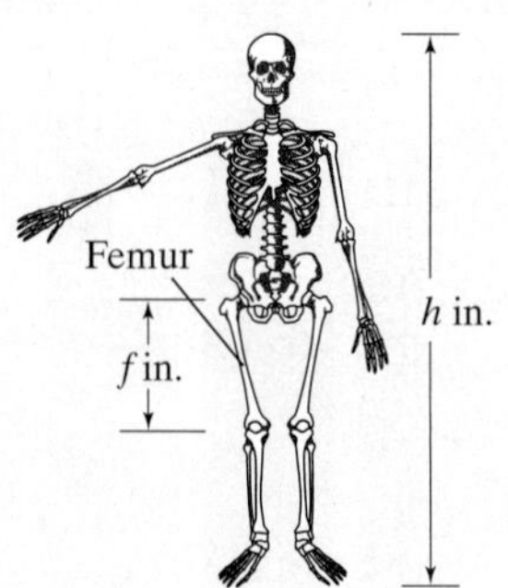

6.3 Applications of Linear Equations

WHAT AM I SUPPOSED TO LEARN?

After studying this section, you should be able to:

1 Use linear equations to solve problems.
2 Solve a formula for a variable.

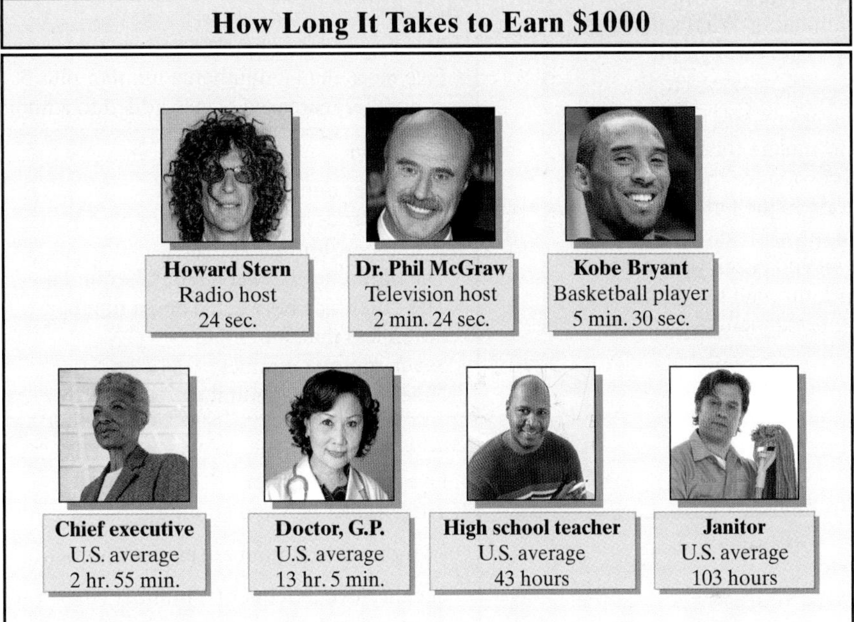

Source: *Time*

In this section, you'll see examples and exercises focused on how much money Americans earn. These situations illustrate a step-by-step strategy for solving problems. As you become familiar with this strategy, you will learn to solve a wide variety of problems.

1 *Use linear equations to solve problems.*

Problem Solving with Linear Equations

We have seen that a model is a mathematical representation of a real-world situation. In this section, we will be solving problems that are presented in English. This means that we must obtain models by translating from the ordinary language of English into the language of algebraic equations. To translate, however, we must understand the English prose and be familiar with the forms of algebraic language. Following are some general steps we will follow in solving word problems.

GREAT QUESTION!

Why are word problems important?

There is great value in reasoning through the steps for solving a word problem. This value comes from the problem-solving skills that you will attain and is often more important than the specific problem or its solution.

STRATEGY FOR SOLVING WORD PROBLEMS

Step 1 Read the problem carefully several times until you can state in your own words what is given and what the problem is looking for. Let x (or any variable) represent one of the unknown quantities in the problem.

Step 2 If necessary, write expressions for any other unknown quantities in the problem in terms of x.

Step 3 Write an equation in x that models the verbal conditions of the problem.

Step 4 Solve the equation and answer the problem's question.

Step 5 Check the solution *in the original wording* of the problem, not in the equation obtained from the words.

The most difficult step in this process is step 3 because it involves translating verbal conditions into an algebraic equation. Translations of some commonly used English phrases are listed in **Table 6.2** on the next page. We choose to use x to represent the variable, but we can use any letter.

GREAT QUESTION!

Table 6.2 looks long and intimidating. What's the best way to get through the table?

Cover the right column with a sheet of paper and attempt to formulate the algebraic expression for the English phrase in the left column on your own. Then slide the paper down and check your answer. Work through the entire table in this manner.

TABLE 6.2 Algebraic Translations of English Phrases

English Phrase	Algebraic Expression
Addition	
The sum of a number and 7	$x + 7$
Five more than a number; a number plus 5	$x + 5$
A number increased by 6; 6 added to a number	$x + 6$
Subtraction	
A number minus 4	$x - 4$
A number decreased by 5	$x - 5$
A number subtracted from 8	$8 - x$
The difference between a number and 6	$x - 6$
The difference between 6 and a number	$6 - x$
Seven less than a number	$x - 7$
Seven minus a number	$7 - x$
Nine fewer than a number	$x - 9$
Multiplication	
Five times a number	$5x$
The product of 3 and a number	$3x$
Two-thirds of a number (used with fractions)	$\frac{2}{3}x$
Seventy-five percent of a number (used with decimals)	$0.75x$
Thirteen multiplied by a number	$13x$
A number multiplied by 13	$13x$
Twice a number	$2x$
Division	
A number divided by 3	$\frac{x}{3}$
The quotient of 7 and a number	$\frac{7}{x}$
The quotient of a number and 7	$\frac{x}{7}$
The reciprocal of a number	$\frac{1}{x}$
More than one operation	
The sum of twice a number and 7	$2x + 7$
Twice the sum of a number and 7	$2(x + 7)$
Three times the sum of 1 and twice a number	$3(1 + 2x)$
Nine subtracted from 8 times a number	$8x - 9$
Twenty-five percent of the sum of 3 times a number and 14	$0.25(3x + 14)$
Seven times a number, increased by 24	$7x + 24$
Seven times the sum of a number and 24	$7(x + 24)$

EXAMPLE 1 *Education Pays Off*

The bar graph in **Figure 6.4** shows average yearly earnings in the United States by highest educational attainment.

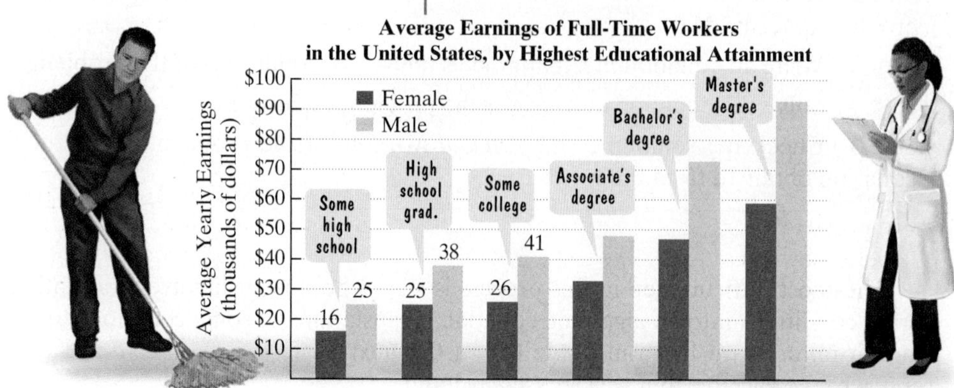

FIGURE 6.4
Source: U.S. Census Bureau

The average yearly salary of a man with a bachelor's degree exceeds that of a man with an associate's degree by \$25 thousand.The average yearly salary of a man with a master's degree exceeds that of a man with an associate's degree by \$45 thousand. Combined, three men with each of these degrees earn \$214 thousand. Find the average yearly salary of men with each of these levels of education.

SOLUTION

Step 1 Let x represent one of the unknown quantities. We know something about salaries of men with bachelor's degrees and master's degrees: They exceed the salary of a man with an associate's degree by \$25 thousand and \$45 thousand, respectively. We will let

x = the average yearly salary of a man with an associate's degree (in thousands of dollars).

Step 2 Represent other unknown quantities in terms of x. Because a man with a bachelor's degree earns \$25 thousand more than a man with an associate's degree, let

$x + 25$ = the average yearly salary of a man with a bachelor's degree.

Because a man with a master's degree earns \$45 thousand more than a man with an associate's degree, let

$x + 45$ = the average yearly salary of a man with a master's degree.

Step 3 Write an equation in x that models the conditions. Combined, three men with each of these degrees earn \$214 thousand.

Salary: associate's degree	plus	salary: bachelor's degree	plus	salary: master's degree	equals	\$214 thousand.
x	$+$	$(x + 25)$	$+$	$(x + 45)$	$=$	214

Step 4 Solve the equation and answer the question.

$$x + (x + 25) + (x + 45) = 214$$ This is the equation that models the problem's conditions.

$$3x + 70 = 214$$ Remove parentheses, regroup, and combine like terms.

$$3x = 144$$ Subtract 70 from both sides.

$$x = 48$$ Divide both sides by 3.

Because we isolated the variable in the model and obtained $x = 48$,

average salary with an associate's degree $= x = 48$

average salary with a bachelor's degree $= x + 25 = 48 + 25 = 73$

average salary with a master's degree $= x + 45 = 48 + 45 = 93$.

Men with associate's degrees average \$48 thousand per year, men with bachelor's degrees average \$73 thousand per year, and men with master's degrees average \$93 thousand per year.

Step 5 Check the proposed solution in the original wording of the problem. The problem states that combined, three men with each of these educational attainments earn \$214 thousand. Using the salaries we determined in step 4, the sum is

\$48 thousand + \$73 thousand + \$93 thousand, or \$214 thousand,

which satisfies the problem's conditions.

CHECK POINT 1 The average yearly salary of a woman with a bachelor's degree exceeds that of a woman with an associate's degree by \$14 thousand. The average yearly salary of a woman with a master's degree exceeds that of a woman with an associate's degree by \$26 thousand. Combined, three women with each of these educational attainments earn \$139 thousand. Find the average yearly salary of women with each of these levels of education. (These salaries are illustrated by the bar graph on page 370.)

GREAT QUESTION!

Example 1 involves using the word *exceeds* to represent two of the unknown quantities. Can you help me to write algebraic expressions for quantities described using *exceeds*?

Modeling with the word *exceeds* can be a bit tricky. It's helpful to identify the smaller quantity. Then add to this quantity to represent the larger quantity. For example, suppose that Tim's height exceeds Tom's height by a inches. Tom is the shorter person. If Tom's height is represented by x, then Tim's height is represented by $x + a$.

Your author teaching math in 1969

EXAMPLE 2 *Modeling Attitudes of College Freshmen*

Researchers have surveyed college freshmen every year since 1969. **Figure 6.5** shows that attitudes about some life goals have changed dramatically over the years. In particular, the freshman class of 2013 was more interested in making money than the freshmen of 1969 had been. In 1969, 42% of first-year college students considered "being well-off financially" essential or very important. For the period from 1969 through 2013, this percentage increased by approximately 0.9 each year. If this trend continues, by which year will all college freshmen consider "being well-off financially" essential or very important?

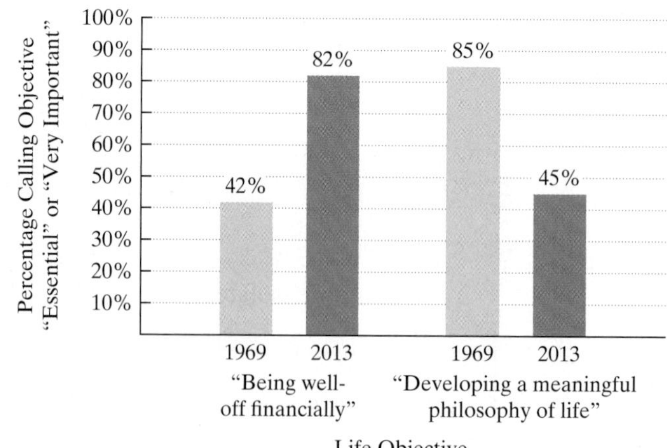

FIGURE 6.5
Source: Higher Education Research Institute

SOLUTION

Step 1 Let x represent one of the unknown quantities. We are interested in the year when all college freshmen, or 100% of the freshmen, will consider this life objective essential or very important. Let

x = the number of years after 1969 when all freshmen will consider "being well-off financially" essential or very important.

Step 2 Represent other unknown quantities in terms of x. There are no other unknown quantities to find, so we can skip this step.

Step 3 Write an equation in x that models the conditions. The problem states that the 1969 percentage increased by approximately 0.9 each year.

The 1969 percentage	increased by	0.9 each year for x years	equals	100% of the freshmen.
42	+	$0.9x$	=	100

Step 4 Solve the equation and answer the question.

$$42 + 0.9x = 100$$ This is the equation that models the problem's conditions.

$$42 - 42 + 0.9x = 100 - 42$$ Subtract 42 from both sides.

$$0.9x = 58$$ Simplify.

$$\frac{0.9x}{0.9} = \frac{58}{0.9}$$ Divide both sides by 0.9.

$$x = 64.\overline{4} \approx 64$$ Simplify and round to the nearest whole number.

Using current trends, by approximately 64 years after 1969, or in 2033, all freshmen will consider "being well-off financially" essential or very important.

Step 5 Check the proposed solution in the original wording of the problem. The problem states that all freshmen (100%, represented by 100 using the model) will consider the objective essential or very important. Does this approximately occur if we increase the 1969 percentage, 42, by 0.9 each year for 64 years, our proposed solution?

$$42 + 0.9(64) = 42 + 57.6 = 99.6 \approx 100$$

This verifies that using trends shown in **Figure 6.5**, all first-year college students will consider the objective of being well-off financially essential or very important approximately 64 years after 1969.

GREAT QUESTION!

Why should I check the proposed solution in the original wording of the problem and not in the equation?

If you made a mistake and your equation does not correctly model the problem's conditions, you'll be checking your proposed solution in an incorrect model. Using the original wording allows you to catch any mistakes you may have made in writing the equation as well as solving it.

A BRIEF REVIEW *Clearing an Equation of Decimals*

- You can clear an equation of decimals by multiplying each side by a power of 10. The exponent on 10 will be equal to the greatest number of digits to the right of any decimal point in the equation.
- Multiplying a decimal number by 10^n has the effect of moving the decimal point n places to the right.

Example

$$42 + 0.9x = 100$$

The greatest number of digits to the right of any decimal point in the equation is 1. Multiply each side by 10^1, or 10.

$$10(42 + 0.9x) = 10(100)$$
$$10(42) + 10(0.9x) = 10(100)$$
$$420 + 9x = 1000$$
$$420 - 420 + 9x = 1000 - 420$$
$$9x = 580$$
$$\frac{9x}{9} = \frac{580}{9}$$
$$x = 64.\overline{4} \approx 64$$

It is not a requirement to clear decimals before solving an equation. Compare this solution to the one in step 4 of Example 2. Which method do you prefer?

CHECK POINT 2 **Figure 6.5** on page 372 shows that the freshman class of 2013 was less interested in developing a philosophy of life than the freshmen of 1969 had been. In 1969, 85% of the freshmen considered this objective essential or very important. Since then, this percentage has decreased by approximately 0.9 each year. If this trend continues, by which year will only 25% of college freshmen consider "developing a meaningful philosophy of life" essential or very important?

EXAMPLE 3 *Modeling Options for a Toll*

The toll to a bridge costs \$7. Commuters who use the bridge frequently have the option of purchasing a monthly discount pass for \$30. With the discount pass, the toll is reduced to \$4. For how many bridge crossings per month will the total monthly cost without the discount pass be the same as the total monthly cost with the discount pass?

SOLUTION

Step 1 **Let x represent one of the unknown quantities.** Let

$$x = \text{the number of bridge crossings per month.}$$

Step 2 **Represent other unknown quantities in terms of x.** There are no other unknown quantities, so we can skip this step.

Step 3 **Write an equation in x that models the conditions.** The monthly cost without the discount pass is the toll, \$7, times the number of bridge crossings per month, x. The monthly cost with the discount pass is the cost of the pass, \$30, plus the toll, \$4, times the number of bridge crossings per month, x.

The monthly cost without the discount pass | must equal | the monthly cost with the discount pass.

$$7x = 30 + 4x$$

Step 4 **Solve the equation and answer the question.**

$7x = 30 + 4x$ This is the equation that models the problem's conditions.

$3x = 30$ Subtract $4x$ from both sides.

$x = 10$ Divide both sides by 3.

Because x represents the number of bridge crossings per month, the total monthly cost without the discount pass will be the same as the total monthly cost with the discount pass for 10 bridge crossings per month.

Step 5 **Check the proposed solution in the original wording of the problem.** The problem states that the monthly cost without the discount pass should be the same as the monthly cost with the discount pass. Let's see if they are the same with 10 bridge crossings per month.

Cost without the discount pass = \$7(10) = \$70

Cost of the pass | Toll

Cost with the discount pass = \$30 + \$4(10) = \$30 + \$40 = \$70

With 10 bridge crossings per month, both options cost \$70 for the month. Thus the proposed solution, 10 bridge crossings, satisfies the problem's conditions.

CHECK POINT 3 The toll to a bridge costs \$5. Commuters who use the bridge frequently have the option of purchasing a monthly discount pass for \$40. With the discount pass, the toll is reduced to \$3. For how many bridge crossings per month will the total monthly cost without the discount pass be the same as the total monthly cost with the discount pass?

EXAMPLE 4 A Price Reduction on a Digital Camera

Your local computer store is having a terrific sale on digital cameras. After a 40% price reduction, you purchase a digital camera for \$276. What was the camera's price before the reduction?

SOLUTION

Step 1 Let x represent one of the unknown quantities. We will let

$x =$ the original price of the digital camera prior to the reduction.

Step 2 Represent other unknown quantities in terms of x. There are no other unknown quantities to find, so we can skip this step.

Step 3 Write an equation in x that models the conditions. The camera's original price minus the 40% reduction is the reduced price, \$276.

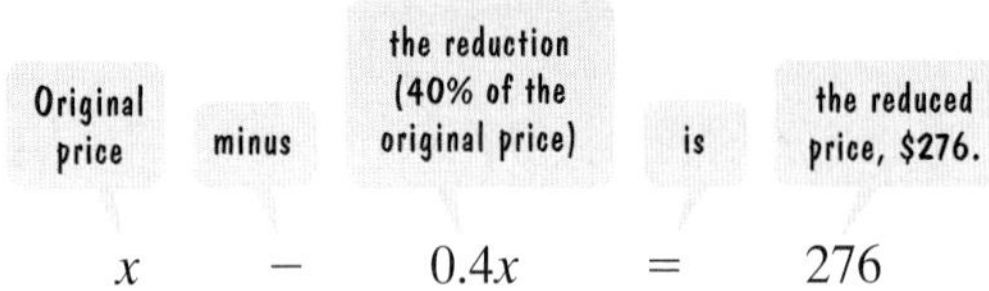

$$x - 0.4x = 276$$

Step 4 Solve the equation and answer the question.

$$x - 0.4x = 276 \quad \text{This is the equation that models the problem's conditions.}$$

$$0.6x = 276 \quad \text{Combine like terms: } x - 0.4x = 1x - 0.4x = 0.6x.$$

$$\frac{0.6x}{0.6} = \frac{276}{0.6} \quad \text{Divide both sides by 0.6.}$$

$$x = 460 \quad \text{Simplify: } 0.6\overline{)276.0} = 460.$$

The digital camera's price before the reduction was \$460.

Step 5 Check the proposed solution in the original wording of the problem. The price before the reduction, \$460, minus the 40% reduction should equal the reduced price given in the original wording, \$276:

$$460 - 40\% \text{ of } 460 = 460 - 0.4(460) = 460 - 184 = 276.$$

This verifies that the digital camera's price before the reduction was \$460.

> **GREAT QUESTION!**
>
> **Why is the 40% reduction written as $0.4x$ in Example 4?**
>
> - 40% is written 0.40 or 0.4.
> - "Of" represents multiplication, so 40% of the original price is $0.4x$.
>
> Notice that the original price, x, reduced by 40% is $x - 0.4x$ and *not* $x - 0.4$.

CHECK POINT 4 After a 30% price reduction, you purchase a new computer for \$840. What was the computer's price before the reduction?

2 *Solve a formula for a variable.*

Solving a Formula for One of Its Variables

We know that solving an equation is the process of finding the number (or numbers) that make the equation a true statement. All of the equations we have solved contained only one letter, x.

By contrast, formulas contain two or more letters, representing two or more variables. An example is the formula for the perimeter of a rectangle:

$$P = 2l + 2w.$$

A rectangle's perimeter is the sum of twice its length and twice its width.

We say that this formula is solved for the variable P because P is alone on one side of the equation and the other side does not contain a P.

Solving a formula for a variable means rewriting the formula so that the variable is isolated on one side of the equation. It does not mean obtaining a numerical value for that variable.

To solve a formula for one of its variables, treat that variable as if it were the only variable in the equation. Think of the other variables as if they were numbers. Isolate all terms with the specified variable on one side of the equation and all terms without the specified variable on the other side. Then divide both sides by the same nonzero quantity to get the specified variable alone. The next two examples show how to do this.

EXAMPLE 5 Solving a Formula for a Variable

Solve the formula $P = 2l + 2w$ for l.

SOLUTION

First, isolate $2l$ on the right by subtracting $2w$ from both sides. Then solve for l by dividing both sides by 2.

We need to isolate l.

$P = 2l + 2w$	This is the given formula.
$P - 2w = 2l + 2w - 2w$	Isolate $2l$ by subtracting $2w$ from both sides.
$P - 2w = 2l$	Simplify.
$\frac{P - 2w}{2} = \frac{2l}{2}$	Solve for l by dividing both sides by 2.
$\frac{P - 2w}{2} = l$	Simplify.

Equivalently, $l = \frac{P - 2w}{2}$.

CHECK POINT 5 Solve the formula $P = 2l + 2w$ for w.

EXAMPLE 6 Solving a Formula for a Variable

The total price of an article purchased on a monthly deferred payment plan is described by the following formula:

$$T = D + pm.$$

In this formula, T is the total price, D is the down payment, p is the monthly payment, and m is the number of months one pays. Solve the formula for p.

SOLUTION

First, isolate pm on the right by subtracting D from both sides. Then, isolate p from pm by dividing both sides of the formula by m.

We need to isolate p.

$T = D + pm$	This is the given formula. We want p alone.
$T - D = D - D + pm$	Isolate pm by subtracting D from both sides.
$T - D = pm$	Simplify.
$\frac{T - D}{m} = \frac{pm}{m}$	Now isolate p by dividing both sides by m.
$\frac{T - D}{m} = p$	Simplify: $\frac{pm}{m} = \frac{p\cancel{m}}{\cancel{m}} = \frac{p}{1} = p$.

CHECK POINT 6 Solve the formula $T = D + pm$ for m.

Concept and Vocabulary Check

Fill in each blank so that the resulting statement is true.

1. According to the U.S. Office of Management and Budget, the 2011 budget for defense exceeded the budget for education by \$658.6 billion. If x represents the budget for education, in billions of dollars, the budget for defense can be represented by __________.
2. In 2000, 31% of U.S. adults viewed a college education as essential for success. For the period from 2000 through 2010, this percentage increased by approximately 2.4 each year. The percentage of U.S. adults who viewed a college education as essential for success x years after 2000 can be represented by __________.
3. A text message plan costs \$4.00 per month plus \$0.15 per text. The monthly cost for x text messages can be represented by __________.
4. I purchased a computer after a 15% price reduction. If x represents the computer's original price, the reduced price can be represented by __________.
5. Solving a formula for a variable means rewriting the formula so that the variable is __________.
6. In order to solve $y = mx + b$ for x, we first __________ and then __________.

Exercise Set 6.3

Practice Exercises

Use the five-step strategy for solving word problems to find the number or numbers described in Exercises 1–10.

1. When five times a number is decreased by 4, the result is 26. What is the number?
2. When two times a number is decreased by 3, the result is 11. What is the number?
3. When a number is decreased by 20% of itself, the result is 20. What is the number?
4. When a number is decreased by 30% of itself, the result is 28. What is the number?
5. When 60% of a number is added to the number, the result is 192. What is the number?
6. When 80% of a number is added to the number, the result is 252. What is the number?
7. 70% of what number is 224?
8. 70% of what number is 252?
9. One number exceeds another by 26. The sum of the numbers is 64. What are the numbers?
10. One number exceeds another by 24. The sum of the numbers is 58. What are the numbers?

Practice Plus

In Exercises 11–18, write each English phrase as an algebraic expression. Then simplify the expression. Let x represent the number.

11. A number decreased by the sum of the number and four
12. A number decreased by the difference between eight and the number
13. Six times the product of negative five and a number
14. Ten times the product of negative four and a number
15. The difference between the product of five and a number and twice the number
16. The difference between the product of six and a number and negative two times the number
17. The difference between eight times a number and six more than three times the number
18. Eight decreased by three times the sum of a number and six

Application Exercises

How will you spend your average life expectancy of 78 years? The bar graph shows the average number of years you will devote to each of your most time-consuming activities. Exercises 19–20 are based on the data displayed by the graph.

How You Will Spend Your Average Life Expectancy of 78 Years

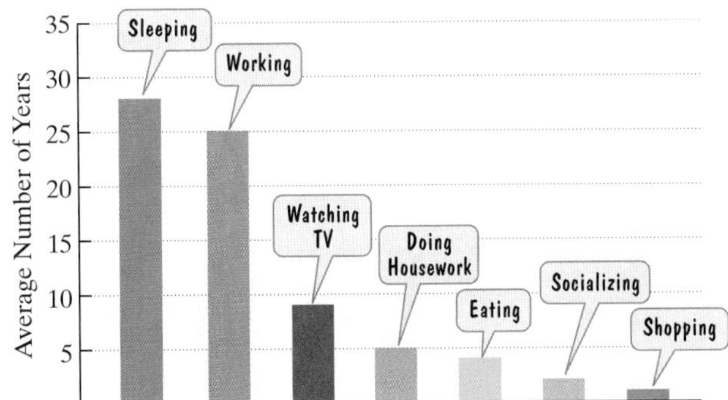

Source: U.S. Bureau of Labor Statistics

19. According to the U.S. Bureau of Labor Statistics, you will devote 37 years to sleeping and watching TV. The number of years sleeping will exceed the number of years watching TV by 19. Over your lifetime, how many years will you spend on each of these activities?
20. According to the U.S. Bureau of Labor Statistics, you will devote 32 years to sleeping and eating. The number of years sleeping will exceed the number of years eating by 24. Over your lifetime, how many years will you spend on each of these activities?

The bar graph shows average yearly earnings in the United States for four selected occupations. Exercises 21–22 are based on the data displayed by the graph.

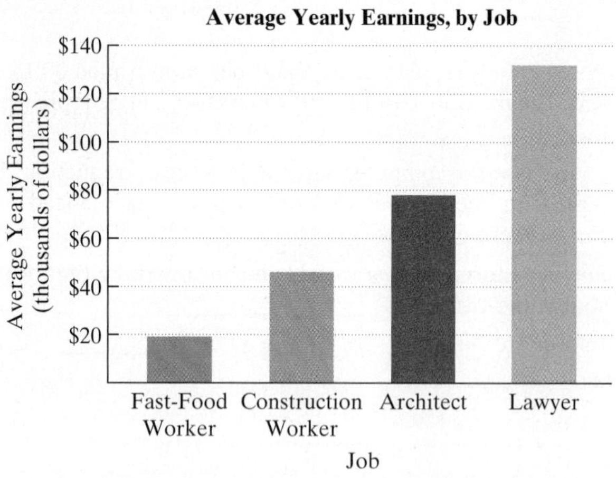

Source: U.S. Department of Labor

21. The average yearly salary of a lawyer is $24 thousand less than twice that of an architect. Combined, an architect and a lawyer earn $210 thousand. Find the average yearly salary of an architect and a lawyer.

22. The average yearly salary of a construction worker is $11 thousand less than three times that of a fast-food worker. Combined, a fast-food worker and a construction worker earn $65 thousand. Find the average yearly salary of a fast-food worker and a construction worker.

Despite booming new car sales with their cha-ching sounds, the average age of vehicles on U.S. roads is not going down. The bar graph shows the average price of new cars in the United States and the average age of cars on U.S. roads for two selected years. Exercises 23–24 are based on the information displayed by the graph.

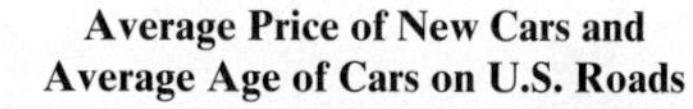

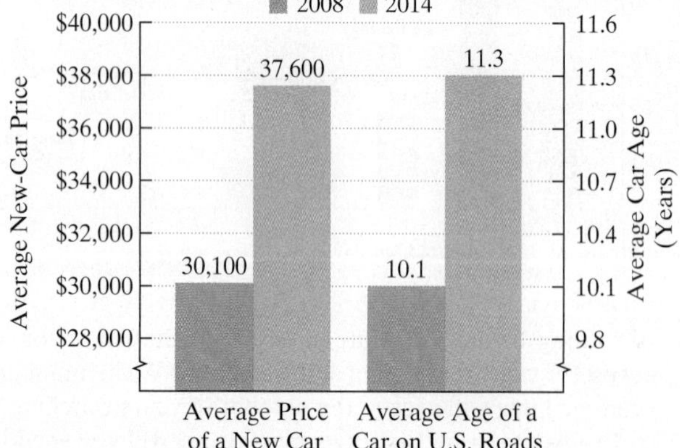

Source: Kelley Blue Book, IHS Automotive/Polk

23. In 2014, the average price of a new car was $37,600. For the period shown, new-car prices increased by approximately $1250 per year. If this trend continues, how many years after 2014 will the price of a new car average $46,350? In which year will this occur?

24. In 2014, the average age of cars on U.S. roads was 11.3 years. For the period shown, this average age increased by approximately 0.2 year per year. If this trend continues, how many years after 2014 will the average age of vehicles on U.S. roads be 12.3 years? In which year will this occur?

On average, every minute of every day, 158 babies are born. The bar graph represents the results of a single day of births, deaths, and population increase worldwide. Exercises 25–26 are based on the information displayed by the graph.

Daily Growth of World Population

Births

Deaths

Population Increase

0 50 100 150 200 250 300 350 400

Number of People (thousands)

Source: James Henslin, *Sociology*, Eleventh Edition, Pearson, 2012.

25. Each day, the number of births in the world is 84 thousand less than three times the number of deaths.

a. If the population increase in a single day is 228 thousand, determine the number of births and deaths per day.

b. If the population increase in a single day is 228 thousand, by how many millions of people does the worldwide population increase each year? Round to the nearest million.

c. Based on your answer to part (b), approximately how many years does it take for the population of the world to increase by an amount greater than the entire U.S. population (315 million)?

26. Each day, the number of births in the world exceeds twice the number of deaths by 72 thousand.

a. If the population increase in a single day is 228 thousand, determine the number of births and deaths per day.

b. If the population increase in a single day is 228 thousand, by how many millions of people does the worldwide population increase each year? Round to the nearest million.

c. Based on your answer to part (b), approximately how many years does it take for the population of the world to increase by an amount greater than the entire U.S. population (315 million)?

27. A new car worth $24,000 is depreciating in value by $3000 per year. After how many years will the car's value be $9000?

28. A new car worth $45,000 is depreciating in value by $5000 per year. After how many years will the car's value be $10,000?

29. You are choosing between two health clubs. Club A offers membership for a fee of $40 plus a monthly fee of $25. Club B offers membership for a fee of $15 plus a monthly fee of $30. After how many months will the total cost at each health club be the same? What will be the total cost for each club?

30. You need to rent a rug cleaner. Company A will rent the machine you need for \$22 plus \$6 per hour. Company B will rent the same machine for \$28 plus \$4 per hour. After how many hours of use will the total amount spent at each company be the same? What will be the total amount spent at each company?

31. The bus fare in a city is \$1.25. People who use the bus have the option of purchasing a monthly discount pass for \$15.00. With the discount pass, the fare is reduced to \$0.75. Determine the number of times in a month the bus must be used so that the total monthly cost without the discount pass is the same as the total monthly cost with the discount pass.

32. A discount pass for a bridge costs \$30.00 per month. The toll for the bridge is normally \$5.00, but it is reduced to \$3.50 for people who have purchased the discount pass. Determine the number of times in a month the bridge must be crossed so that the total monthly cost without the discount pass is the same as the total monthly cost with the discount pass.

33. You are choosing between two plans at a discount warehouse. Plan A offers an annual membership fee of \$100 and you pay 80% of the manufacturer's recommended list price. Plan B offers an annual membership fee of \$40 and you pay 90% of the manufacturer's recommended list price. How many dollars of merchandise would you have to purchase in a year to pay the same amount under both plans? What will be the cost for each plan?

34. You are choosing between two plans at a discount warehouse. Plan A offers an annual membership fee of \$300 and you pay 70% of the manufacturer's recommended list price. Plan B offers an annual membership fee of \$40 and you pay 90% of the manufacturer's recommended list price. How many dollars of merchandise would you have to purchase in a year to pay the same amount under both plans? What will be the cost for each plan?

35. In 2010, there were 13,300 students at college A, with a projected enrollment increase of 1000 students per year. In the same year, there were 26,800 students at college B, with a projected enrollment decline of 500 students per year. According to these projections, when will the colleges have the same enrollment? What will be the enrollment in each college at that time?

36. In 2000, the population of Greece was 10,600,000, with projections of a population decrease of 28,000 people per year. In the same year, the population of Belgium was 10,200,000, with projections of a population decrease of 12,000 people per year. (*Source:* United Nations) According to these projections, when will the two countries have the same population? What will be the population at that time?

37. After a 20% reduction, you purchase a television for \$336. What was the television's price before the reduction?

38. After a 30% reduction, you purchase a dictionary for \$30.80. What was the dictionary's price before the reduction?

39. Including 8% sales tax, an inn charges \$162 per night. Find the inn's nightly cost before the tax is added.

40. Including 5% sales tax, an inn charges \$252 per night. Find the inn's nightly cost before the tax is added.

Exercises 41–42 involve markup, the amount added to the dealer's cost of an item to arrive at the selling price of that item.

41. The selling price of a refrigerator is \$584. If the markup is 25% of the dealer's cost, what is the dealer's cost of the refrigerator?

42. The selling price of a scientific calculator is \$15. If the markup is 25% of the dealer's cost, what is the dealer's cost of the calculator?

In Exercises 43–60, solve each formula for the specified variable. Do you recognize the formula? If so, what does it describe?

43. $A = LW$ for L

44. $D = RT$ for R

45. $A = \frac{1}{2}bh$ for b

46. $V = \frac{1}{3}Bh$ for B

47. $I = Prt$ for P

48. $C = 2\pi r$ for r

49. $E = mc^2$ for m

50. $V = \pi r^2 h$ for h

51. $y = mx + b$ for m

52. $P = C + MC$ for M

53. $A = \frac{1}{2}h(a + b)$ for a

54. $A = \frac{1}{2}h(a + b)$ for b

55. $S = P + Prt$ for r

56. $S = P + Prt$ for t

57. $Ax + By = C$ for x

58. $Ax + By = C$ for y

59. $a_n = a_1 + (n - 1)d$ for n

60. $a_n = a_1 + (n - 1)d$ for d

Explaining the Concepts

61. In your own words, describe a step-by-step approach for solving algebraic word problems.

62. Write an original word problem that can be solved using a linear equation. Then solve the problem.

63. Explain what it means to solve a formula for a variable.

64. Did you have difficulties solving some of the problems that were assigned in this Exercise Set? Discuss what you did if this happened to you. Did your course of action enhance your ability to solve algebraic word problems?

Critical Thinking Exercises

Make Sense? *In Exercises 65–68, determine whether each statement makes sense or does not make sense, and explain your reasoning.*

65. By modeling attitudes of college freshmen from 1969 through 2010, I can make precise predictions about the attitudes of the freshman class of 2020.

66. I find the hardest part in solving a word problem is writing the equation that models the verbal conditions.

67. I solved a formula for one of its variables, so now I have a numerical value for that variable.

68. After a 35% reduction, a computer's price is \$780, so I determined the original price, x, by solving $x - 0.35 = 780$.

69. The price of a dress is reduced by 40%. When the dress still does not sell, it is reduced by 40% of the reduced price. If the price of the dress after both reductions is \$72, what was the original price?

70. In a film, the actor Charles Coburn plays an elderly "uncle" character criticized for marrying a woman when he is 3 times her age. He wittily replies, "Ah, but in 20 years time I shall only be twice her age." How old is the "uncle" and the woman?

71. Suppose that we agree to pay you 8¢ for every problem in this chapter that you solve correctly and fine you 5¢ for every problem done incorrectly. If at the end of 26 problems we do not owe each other any money, how many problems did you solve correctly?

72. It was wartime when the Ricardos found out Mrs. Ricardo was pregnant. Ricky Ricardo was drafted and made out a will, deciding that \$14,000 in a savings account was to be divided between his wife and his child-to-be. Rather strangely, and certainly with gender bias, Ricky stipulated that if the child were a boy, he would get twice the amount of the mother's portion. If it were a girl, the mother would get twice the amount the girl was to receive. We'll never know what Ricky was thinking of, for (as fate would have it) he did not return from the war. Mrs. Ricardo gave birth to twins—a boy and a girl. How was the money divided?

73. A thief steals a number of rare plants from a nursery. On the way out, the thief meets three security guards, one after another. To each security guard, the thief is forced to give one-half the plants that he still has, plus 2 more. Finally, the thief leaves the nursery with 1 lone palm. How many plants were originally stolen?

In Exercises 74–75, solve each proportion for x.

74. $\dfrac{x+a}{a} = \dfrac{b+c}{c}$

75. $\dfrac{ax-b}{b} = \dfrac{c-d}{d}$

Group Exercise

76. One of the best ways to learn how to *solve* a word problem in algebra is to *design* word problems of your own. Creating a word problem makes you very aware of precisely how much information is needed to solve the problem. You must also focus on the best way to present information to a reader and on how much information to give. As you write your problem, you gain skills that will help you solve problems created by others.

The group should design five different word problems that can be solved using linear equations. All of the problems should be on different topics. For example, the group should not have more than one problem on price reduction. The group should turn in both the problems and their algebraic solutions.

6.4 Linear Inequalities in One Variable

WHAT AM I SUPPOSED TO LEARN?

After studying this section, you should be able to:

1 Graph subsets of real numbers on a number line.
2 Solve linear inequalities.
3 Solve applied problems using linear inequalities.

RENT-A-HEAP, A CAR RENTAL company, charges \$125 per week plus \$0.20 per mile to rent one of their cars. Suppose you are limited by how much money you can spend for the week: You can spend at most \$335. If we let x represent the number of miles you drive the heap in a week, we can write an inequality that models the given conditions:

The weekly charge of \$125	plus	the charge of \$0.20 per mile for x miles	must be less than or equal to	\$335.
125	+	$0.20x$	$\leq$	335.

Notice that the highest exponent on the variable in $125 + 0.20x \leq 335$ is 1. Such an inequality is called a *linear inequality in one variable*. The symbol between the two sides of an inequality can be $\leq$ (is less than or equal to), $<$ (is less than), $\geq$ (is greater than or equal to), or $>$ (is greater than).

In this section, we will study how to solve linear inequalities such as $125 + 0.20x \leq 335$. **Solving an inequality** is the process of finding the set of numbers that makes the inequality a true statement. These numbers are called the **solutions** of the inequality and we say that they **satisfy** the inequality. The set of all solutions is called the **solution set** of the inequality. We begin by discussing how to represent these solution sets, which are subsets of real numbers, on a number line.

1 *Graph subsets of real numbers on a number line.*

Graphing Subsets of Real Numbers on a Number Line

Table 6.3 shows how to represent various subsets of real numbers on a number line. Open dots (∘) indicate that a number is not included in a set. Closed dots (•) indicate that a number is included in a set.

TABLE 6.3 Graphs of Subsets of Real Numbers

Let a and b be real numbers such that $a < b$.

Set-Builder Notation		Graph
$\{x \mid x < a\}$	x is a real number less than a.	a b
$\{x \mid x \leq a\}$	x is a real number less than or equal to a.	a b
$\{x \mid x > b\}$	x is a real number greater than b.	a b
$\{x \mid x \geq b\}$	x is a real number greater than or equal to b.	a b
$\{x \mid a < x < b\}$	x is a real number greater than a and less than b.	a b
$\{x \mid a \leq x \leq b\}$	x is a real number greater than or equal to a and less than or equal to b.	a b
$\{x \mid a \leq x < b\}$	x is a real number greater than or equal to a and less than b.	a b
$\{x \mid a < x \leq b\}$	x is a real number greater than a and less than or equal to b.	a b

EXAMPLE 1 Graphing Subsets of Real Numbers

Graph each set:

a. $\{x|x < 3\}$ **b.** $\{x|x \geq -1\}$ **c.** $\{x|-1 < x \leq 3\}$.

SOLUTION

a. $\{x \mid x < 3\}$ — x is a real number less than 3.

−4 −3 −2 −1 0 1 2 3 4

b. $\{x \mid x \geq -1\}$ — x is a real number greater than or equal to −1.

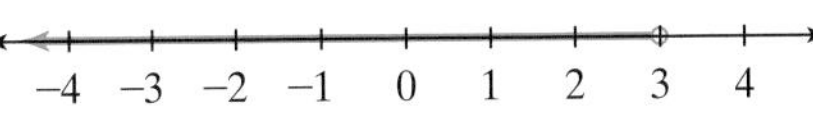

c. $\{x \mid -1 < x \leq 3\}$ — x is a real number greater than −1 and less than or equal to 3.

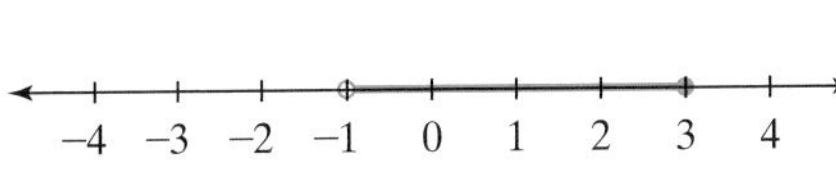

CHECK POINT 1 Graph each set:

a. $\{x|x < 4\}$ **b.** $\{x|x \geq -2\}$ **c.** $\{x|-4 \leq x < 1\}$.

2 *Solve linear inequalities.*

Solving Linear Inequalities in One Variable

We know that a linear equation in x can be expressed as $ax + b = 0$. A **linear inequality in x** can be written in one of the following forms:

$$ax + b < 0, \quad ax + b \leq 0, \quad ax + b > 0, \quad ax + b \geq 0.$$

In each form, $a \neq 0$.

Back to our question that opened this section: How many miles can you drive your Rent-a-Heap car if you can spend at most \$335? We answer the question by solving

$$0.20x + 125 \leq 335$$

for x. The solution procedure is nearly identical to that for solving

$$0.20x + 125 = 335.$$

Our goal is to get x by itself on the left side. We do this by subtracting 125 from both sides to isolate $0.20x$:

$$0.20x + 125 \leq 335 \quad \text{This is the given inequality.}$$
$$0.20x + 125 - 125 \leq 335 - 125 \quad \text{Subtract 125 from both sides.}$$
$$0.20x \leq 210. \quad \text{Simplify.}$$

Finally, we isolate x from $0.20x$ by dividing both sides of the inequality by 0.20:

$$\frac{0.20x}{0.20} \leq \frac{210}{0.20} \quad \text{Divide both sides by 0.20.}$$
$$x \leq 1050. \quad \text{Simplify.}$$

With at most \$335 per week to spend, you can travel at most 1050 miles.

We started with the inequality $0.20x + 125 \leq 335$ and obtained the inequality $x \leq 1050$ in the final step. Both of these inequalities have the same solution set, namely $\{x|x \leq 1050\}$. Inequalities such as these, with the same solution set, are said to be **equivalent**.

We isolated x from $0.20x$ by dividing both sides of $0.20x \leq 210$ by 0.20, a positive number. Let's see what happens if we divide both sides of an inequality by a negative number. Consider the inequality $10 < 14$. Divide 10 and 14 by -2:

$$\frac{10}{-2} = -5 \quad \text{and} \quad \frac{14}{-2} = -7.$$

Because -5 lies to the right of -7 on the number line, -5 is greater than -7:

$$-5 > -7.$$

Notice that the direction of the inequality symbol is reversed:

$$10 < 14$$
$$\updownarrow$$
$$-5 > -7.$$

Dividing by -2 changes the direction of the inequality symbol.

In general, **when we multiply or divide both sides of an inequality by a negative number, the direction of the inequality symbol is reversed**. When we reverse the direction of the inequality symbol, we say that we change the *sense* of the inequality.

GREAT QUESTION!

What are some common English phrases and sentences that I can model with linear inequalities?

English phrases such as "at least" and "at most" can be represented by inequalities.

English Sentence	Inequality
x is at least 5.	$x \geq 5$
x is at most 5.	$x \leq 5$
x is no more than 5.	$x \leq 5$
x is no less than 5.	$x \geq 5$

We can summarize our discussion with the following statement:

SOLVING LINEAR INEQUALITIES

The procedure for solving linear inequalities is the same as the procedure for solving linear equations, with one important exception: When multiplying or dividing both sides of the inequality by a negative number, reverse the direction of the inequality symbol, changing the sense of the inequality.

EXAMPLE 2 *Solving a Linear Inequality*

Solve and graph the solution set: $4x - 7 \geq 5$.

SOLUTION

Our goal is to get x by itself on the left side. We do this by first getting $4x$ by itself, adding 7 to both sides.

$$4x - 7 \geq 5 \quad \text{This is the given inequality.}$$
$$4x - 7 + 7 \geq 5 + 7 \quad \text{Add 7 to both sides.}$$
$$4x \geq 12 \quad \text{Simplify.}$$

Next, we isolate x from $4x$ by dividing both sides by 4. The inequality symbol stays the same because we are dividing by a positive number.

$$\frac{4x}{4} \geq \frac{12}{4} \quad \text{Divide both sides by 4.}$$
$$x \geq 3 \quad \text{Simplify.}$$

The solution set consists of all real numbers that are greater than or equal to 3, expressed in set-builder notation as $\{x | x \geq 3\}$. The graph of the solution set is shown as follows:

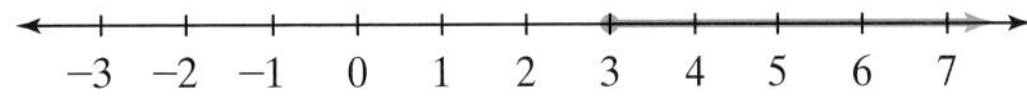

We cannot check all members of an inequality's solution set, but we can take a few values to get an indication of whether or not it is correct. In Example 2, we found that the solution set of $4x - 7 \geq 5$ is $\{x | x \geq 3\}$. Show that 3 and 4 satisfy the inequality, whereas 2 does not.

 **CHECK POINT 2** Solve and graph the solution set: $5x - 3 \leq 17$.

EXAMPLE 3 *Solving Linear Inequalities*

Solve and graph the solution set:

a. $\frac{1}{3}x < 5$ **b.** $-3x < 21$.

SOLUTION

In each case, our goal is to isolate x. In the first inequality, this is accomplished by multiplying both sides by 3. In the second inequality, we can do this by dividing both sides by -3.

a. $\frac{1}{3}x < 5$ This is the given inequality.

$3 \cdot \frac{1}{3}x < 3 \cdot 5$ Isolate x by multiplying by 3 on both sides. The symbol $<$ stays the same because we are multiplying both sides by a positive number.

$x < 15$ Simplify.

The solution set is $\{x | x < 15\}$. The graph of the solution set is shown as follows:

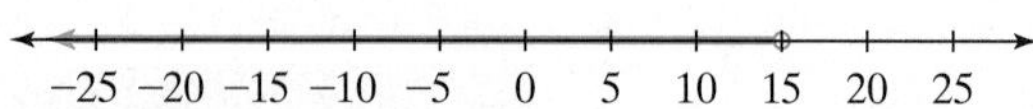

b. $-3x < 21$ This is the given inequality.

$\frac{-3x}{-3} > \frac{21}{-3}$ Isolate x by dividing by -3 on both sides. The symbol $<$ must be reversed because we are dividing both sides by a negative number.

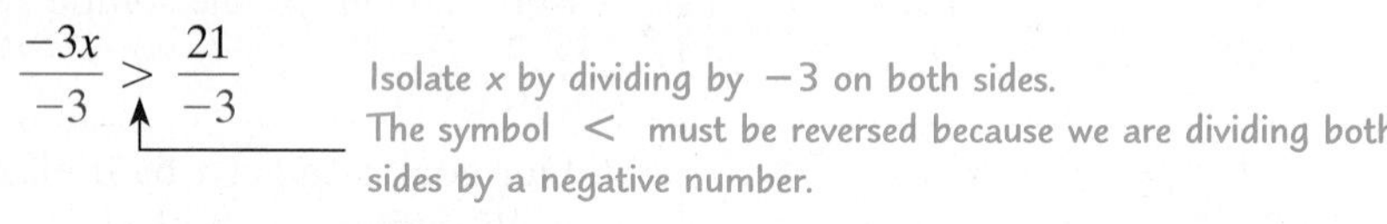

$x > -7$ Simplify.

The solution set is $\{x | x > -7\}$. The graph of the solution set is shown as follows:

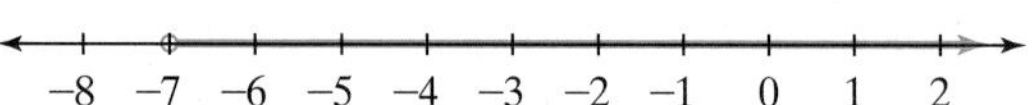

CHECK POINT 3 Solve and graph the solution set:

a. $\frac{1}{4}x < 2$ **b.** $-6x < 18$.

EXAMPLE 4 *Solving a Linear Inequality*

Solve and graph the solution set: $6x - 12 > 8x + 2$.

SOLUTION

We will get x by itself on the left side. We begin by subtracting $8x$ from both sides so that the variable term appears on the left.

$6x - 12 > 8x + 2$ This is the given inequality.

$6x - 8x - 12 > 8x - 8x + 2$ Subtract $8x$ on both sides with the goal of isolating x on the left.

$-2x - 12 > 2$ Simplify.

Next, we get $-2x$ by itself, adding 12 to both sides.

$-2x - 12 + 12 > 2 + 12$ Add 12 to both sides.

$-2x > 14$ Simplify.

In order to solve $-2x > 14$, we isolate x from $-2x$ by dividing both sides by -2. The direction of the inequality symbol must be reversed because we are dividing by a negative number.

$\frac{-2x}{-2} < \frac{14}{-2}$ Divide both sides by -2 and change the sense of the inequality.

$x < -7$ Simplify.

The solution set is $\{x | x < -7\}$. The graph of the solution set is shown as follows:

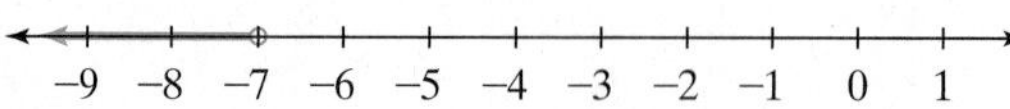

CHECK POINT 4 Solve and graph the solution set: $7x - 3 > 13x + 33$.

EXAMPLE 5 *Solving a Linear Inequality*

Solve and graph the solution set:

$$2(x - 3) + 5x \leq 8(x - 1).$$

SOLUTION

Begin by simplifying the algebraic expression on each side.

$2(x - 3) + 5x \leq 8(x - 1)$	This is the given inequality.
$2x - 6 + 5x \leq 8x - 8$	Use the distributive property.
$7x - 6 \leq 8x - 8$	Add like terms on the left: $2x + 5x = 7x$.

We will get x by itself on the left side. Subtract $8x$ from both sides.

$$7x - 8x - 6 \leq 8x - 8x - 8$$
$$-x - 6 \leq -8$$

Next, we get $-x$ by itself, adding 6 to both sides.

$$-x - 6 + 6 \leq -8 + 6$$
$$-x \leq -2$$

To isolate x, we must eliminate the negative sign in front of the x. Because $-x$ means $-1x$, we can do this by dividing both sides of the inequality by -1. This reverses the direction of the inequality symbol.

$\frac{-x}{-1} \geq \frac{-2}{-1}$	Divide both sides by -1 and change the sense of the inequality.
$x \geq 2$	Simplify.

The solution set is $\{x \mid x \geq 2\}$. The graph of the solution set is shown as follows:

−5 −4 −3 −2 −1 0 1 2 3 4 5

GREAT QUESTION!

Do I have to solve $7x - 6 \leq 8x - 8$ by isolating the variable on the left?

No. You can solve

$$7x - 6 \leq 8x - 8$$

by isolating x on the right side. Subtract $7x$ from both sides and add 8 to both sides:

$$7x - 6 - 7x \leq 8x - 8 - 7x$$
$$-6 \leq x - 8$$
$$-6 + 8 \leq x - 8 + 8$$
$$2 \leq x.$$

This last inequality means the same thing as

$$x \geq 2.$$

Solution sets, in this case $\{x \mid x \geq 2\}$, are expressed with the variable on the left and the constant on the right.

CHECK POINT 5 Solve and graph the solution set:

$$2(x - 3) - 1 \leq 3(x + 2) - 14.$$

In our next example, the inequality has three parts:

$$-3 < 2x + 1 \leq 3.$$

$2x + 1$ is greater than -3 and less than or equal to 3.

By performing the same operation on all three parts of the inequality, our goal is to **isolate x in the middle**.

EXAMPLE 6 *Solving a Three-Part Inequality*

Solve and graph the solution set:

$$-3 < 2x + 1 \leq 3.$$

SOLUTION

We would like to isolate x in the middle. We can do this by first subtracting 1 from all three parts of the inequality. Then we isolate x from $2x$ by dividing all three parts of the inequality by 2.

$-3 < 2x + 1 \leq 3$	This is the given inequality.
$-3 - 1 < 2x + 1 - 1 \leq 3 - 1$	Subtract 1 from all three parts.
$-4 < 2x \leq 2$	Simplify.
$\frac{-4}{2} < \frac{2x}{2} \leq \frac{2}{2}$	Divide each part by 2.
$-2 < x \leq 1$	Simplify.

The solution set consists of all real numbers greater than -2 and less than or equal to 1, represented by $\{x \mid -2 < x \leq 1\}$. The graph is shown as follows:

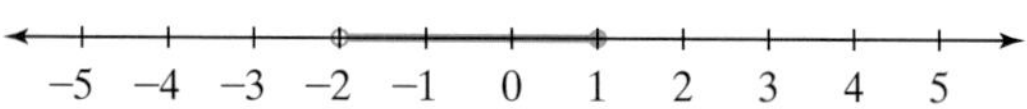

CHECK POINT 6 Solve and graph the solution set on a number line: $1 \leq 2x + 3 < 11$.

As you know, different professors may use different grading systems to determine your final course grade. Some professors require a final examination; others do not. In our next example, a final exam is required *and* it counts as two grades.

3 *Solve applied problems using linear inequalities.*

EXAMPLE 7 *An Application: Final Course Grade*

To earn an A in a course, you must have a final average of at least 90%. On the first four examinations, you have grades of 86%, 88%, 92%, and 84%. If the final examination counts as two grades, what must you get on the final to earn an A in the course?

SOLUTION

We will use our five-step strategy for solving algebraic word problems.

Steps 1 and 2 Represent unknown quantities in terms of x. Let

$$x = \text{your grade on the final examination.}$$

Step 3 Write an inequality in x that models the conditions. The average of the six grades is found by adding the grades and dividing the sum by 6.

$$\text{Average} = \frac{86 + 88 + 92 + 84 + x + x}{6}$$

Because the final counts as two grades, the x (your grade on the final examination) is added twice. This is also why the sum is divided by 6.

To get an A, your average must be at least 90. This means that your average must be greater than or equal to 90.

$$\frac{86 + 88 + 92 + 84 + x + x}{6} \geq 90$$

Step 4 Solve the inequality and answer the problem's question.

$$\frac{86 + 88 + 92 + 84 + x + x}{6} \geq 90$$ This is the inequality that models the given conditions.

$$\frac{350 + 2x}{6} \geq 90$$ Combine like terms in the numerator.

$$6\left(\frac{350 + 2x}{6}\right) \geq 6\,(90)$$ Multiply both sides by 6, clearing the fraction.

$$350 + 2x \geq 540$$ Multiply.

$$350 + 2x - 350 \geq 540 - 350$$ Subtract 350 from both sides.

$$2x \geq 190$$ Simplify.

$$\frac{2x}{2} \geq \frac{190}{2}$$ Divide both sides by 2.

$$x \geq 95$$ Simplify.

You must get at least 95% on the final examination to earn an A in the course.

Step 5 Check. We can perform a partial check by computing the average with any grade that is at least 95. We will use 96. If you get 96% on the final examination, your average is

$$\frac{86 + 88 + 92 + 84 + 96 + 96}{6} = \frac{542}{6} = 90\frac{1}{3}.$$

Because $90\frac{1}{3} > 90$, you earn an A in the course.

 **CHECK POINT 7** To earn a B in a course, you must have a final average of at least 80%. On the first three examinations, you have grades of 82%, 74%, and 78%. If the final examination counts as two grades, what must you get on the final to earn a B in the course?

Concept and Vocabulary Check

Fill in each blank so that the resulting statement is true.

1. On a number line, an open dot indicates that a number ______________ in a solution set, and a closed dot indicates that a number ______________ in a solution set.
2. If an inequality's solution set consists of all real numbers, x, that are less than a, the solution set is represented in set-builder notation as ___________.
3. If an inequality's solution set consists of all real numbers, x, that are greater than a and less than or equal to b, the solution set is represented in set-builder notation as ______________.
4. When both sides of an inequality are multiplied or divided by a/an _________ number, the direction of the inequality symbol is reversed.

In Exercises 5–8, determine whether each statement is true or false. If the statement is false, make the necessary change(s) to produce a true statement.

5. The inequality $x - 3 > 0$ is equivalent to $x < 3$. ________
6. The statement "x is at most 5" is written $x < 5$. ________
7. The inequality $-4x < -20$ is equivalent to $x > -5$. ______
8. The statement "the sum of x and 6% of x is at least 80" is modeled by $x + 0.06x \geq 80$. ______

Exercise Set 6.4

Practice Exercises

In Exercises 1–12, graph each set of real numbers on a number line.

1. $\{x|x > 6\}$
2. $\{x|x > -2\}$
3. $\{x|x < -4\}$
4. $\{x|x < 0\}$
5. $\{x|x \geq -3\}$
6. $\{x|x \geq -5\}$
7. $\{x|x \leq 4\}$
8. $\{x|x \leq 7\}$
9. $\{x|-2 < x \leq 5\}$
10. $\{x|-3 \leq x < 7\}$
11. $\{x|-1 < x < 4\}$
12. $\{x|-7 \leq x \leq 0\}$

In Exercises 13–66, solve each inequality and graph the solution set on a number line.

13. $x - 3 > 2$
14. $x + 1 < 5$
15. $x + 4 \leq 9$
16. $x - 5 \geq 1$
17. $x - 3 < 0$
18. $x + 4 \geq 0$
19. $4x < 20$
20. $6x \geq 18$
21. $3x \geq -15$
22. $7x < -21$
23. $2x - 3 > 7$
24. $3x + 2 \leq 14$
25. $3x + 3 < 18$
26. $8x - 4 > 12$
27. $\frac{1}{2}x < 4$
28. $\frac{1}{2}x > 3$
29. $\frac{x}{3} > -2$
30. $\frac{x}{4} < -1$
31. $-3x < 15$
32. $-7x > 21$
33. $-3x \geq -15$
34. $-7x \leq -21$
35. $3x + 4 \leq 2x + 7$
36. $2x + 9 \leq x + 2$
37. $5x - 9 < 4x + 7$
38. $3x - 8 < 2x + 11$
39. $-2x - 3 < 3$
40. $14 - 3x > 5$
41. $3 - 7x \leq 17$
42. $5 - 3x \geq 20$
43. $-x < 4$
44. $-x > -3$
45. $5 - x \leq 1$
46. $3 - x \geq -3$
47. $2x - 5 > -x + 6$
48. $6x - 2 \geq 4x + 6$
49. $2x - 5 < 5x - 11$
50. $4x - 7 > 9x - 2$
51. $3(x + 1) - 5 < 2x + 1$
52. $4(x + 1) + 2 \geq 3x + 6$
53. $8x + 3 > 3(2x + 1) - x + 5$
54. $7 - 2(x - 4) < 5(1 - 2x)$
55. $\frac{x}{4} - \frac{3}{2} \leq \frac{x}{2} + 1$
56. $\frac{3x}{10} + 1 \geq \frac{1}{5} - \frac{x}{10}$
57. $1 - \frac{x}{2} > 4$
58. $7 - \frac{4}{5}x < \frac{3}{5}$
59. $6 < x + 3 < 8$
60. $7 < x + 5 < 11$
61. $-3 \leq x - 2 < 1$
62. $-6 < x - 4 \leq 1$
63. $-11 < 2x - 1 \leq -5$
64. $3 \leq 4x - 3 < 19$
65. $-3 \leq \frac{2}{3}x - 5 < -1$
66. $-6 \leq \frac{1}{2}x - 4 < -3$

Practice Plus

In Exercises 67–70, write an inequality with x isolated on the left side that is equivalent to the given inequality.

67. $Ax + By > C$; Assume $A > 0$.
68. $Ax + By \leq C$; Assume $A > 0$.
69. $Ax + By > C$; Assume $A < 0$.
70. $Ax + By \leq C$; Assume $A < 0$.

In Exercises 71–76, use set-builder notation to describe all real numbers satisfying the given conditions.

71. A number increased by 5 is at least two times the number.
72. A number increased by 12 is at least four times the number.
73. Twice the sum of four and a number is at most 36.
74. Three times the sum of five and a number is at most 48.
75. If the quotient of three times a number and five is increased by four, the result is no more than 34.
76. If the quotient of three times a number and four is decreased by three, the result is no less than 9.

Application Exercises

The graphs show that the three components of love, namely passion, intimacy, and commitment, progress differently over time. Passion peaks early in a relationship and then declines. By contrast, intimacy and commitment build gradually. Use the graphs to solve Exercises 77–84. Assume that x represents years in a relationship.

The Course of Love Over Time

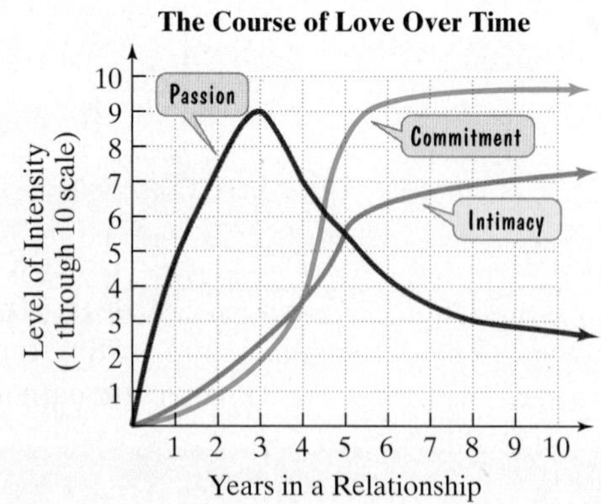

Source: R. J. Sternberg, A Triangular Theory of Love, *Psychological Review*, 93, 119–135.

77. Use set-builder notation to write an inequality that expresses for which years in a relationship intimacy is greater than commitment.
78. Use set-builder notation to write an inequality that expresses for which years in a relationship passion is greater than or equal to intimacy.
79. What is the relationship between passion and intimacy for $\{x|5 \leq x < 7\}$?
80. What is the relationship between intimacy and commitment for $\{x|4 \leq x < 7\}$?

81. What is the relationship between passion and commitment for $\{x|6 < x < 8\}$?

82. What is the relationship between passion and commitment for $\{x|7 < x < 9\}$?

83. What is the maximum level of intensity for passion? After how many years in a relationship does this occur?

84. After approximately how many years do levels of intensity for commitment exceed the maximum level of intensity for passion?

In more U.S. marriages, spouses have different faiths. The bar graph shows the percentage of households with an interfaith marriage in 1988 and 2012. Also shown is the percentage of households in which a person of faith is married to someone with no religion.

Percentage of U.S. Households in Which Married Couples Do Not Share the Same Faith

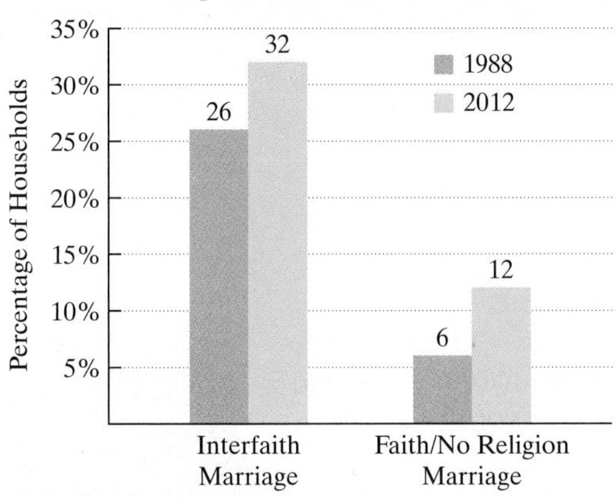

Source: General Social Survey, University of Chicago

The formula

$$I = \frac{1}{4}x + 26$$

models the percentage of U.S. households with an interfaith marriage, I, x years after 1988. The formula

$$N = \frac{1}{4}x + 6$$

models the percentage of U.S. households in which a person of faith is married to someone with no religion, N, x years after 1988. Use these models to solve Exercises 85–86.

85. a. In which years will more than 33% of U.S. households have an interfaith marriage?

b. In which years will more than 14% of U.S. households have a person of faith married to someone with no religion?

c. Based on your answers to parts (a) and (b), in which years will more than 33% of households have an interfaith marriage and more than 14% have a faith/no religion marriage?

86. a. In which years will more than 34% of U.S. households have an interfaith marriage?

b. In which years will more than 15% of U.S. households have a person of faith married to someone with no religion?

c. Based on your answers to parts (a) and (b), in which years will more than 34% of households have an interfaith marriage and more than 15% have a faith/no religion marriage?

87. On two examinations, you have grades of 86 and 88. There is an optional final examination, which counts as one grade. You decide to take the final in order to get a course grade of A, meaning a final average of at least 90.

a. What must you get on the final to earn an A in the course?

b. By taking the final, if you do poorly, you might risk the B that you have in the course based on the first two exam grades. If your final average is less than 80, you will lose your B in the course. Describe the grades on the final that will cause this to happen.

88. On three examinations, you have grades of 88, 78, and 86. There is still a final examination, which counts as one grade.

a. In order to get an A, your average must be at least 90. If you get 100 on the final, compute your average and determine if an A in the course is possible.

b. To earn a B in the course, you must have a final average of at least 80. What must you get on the final to earn a B in the course?

89. A car can be rented from Continental Rental for \$80 per week plus 25 cents for each mile driven. How many miles can you travel if you can spend at most \$400 for the week?

90. A car can be rented from Basic Rental for \$60 per week plus 50 cents for each mile driven. How many miles can you travel if you can spend at most \$600 for the week?

91. An elevator at a construction site has a maximum capacity of 3000 pounds. If the elevator operator weighs 245 pounds and each cement bag weighs 95 pounds, up to how many bags of cement can be safely lifted on the elevator in one trip?

92. An elevator at a construction site has a maximum capacity of 2800 pounds. If the elevator operator weighs 265 pounds and each cement bag weighs 65 pounds, up to how many bags of cement can be safely lifted on the elevator in one trip?

93. A phone plan costs \$20 per month for 60 calling minutes. Additional time costs \$0.40 per minute. The formula

$$C = 20 + 0.40(x - 60)$$

gives the monthly cost for this plan, C, for x calling minutes, where $x > 60$. How many calling minutes are possible for a monthly cost of at least \$28 and at most \$40?

94. The formula for converting Fahrenheit temperature, F, to Celsius temperature, C, is

$$C = \frac{5}{9}(F - 32).$$

If Celsius temperature ranges from 15° to 35°, inclusive, what is the range for the Fahrenheit temperature?

Explaining the Concepts

95. When graphing the solutions of an inequality, what is the difference between an open dot and a closed dot?

96. When solving an inequality, when is it necessary to change the direction of the inequality symbol? Give an example.

97. Describe ways in which solving a linear inequality is similar to solving a linear equation.

98. Describe ways in which solving a linear inequality is different than solving a linear equation.

Critical Thinking Exercises

Make Sense? *In Exercises 99–102, determine whether each statement makes sense or does not make sense, and explain your reasoning.*

99. I can check inequalities by substituting 0 for the variable: When 0 belongs to the solution set, I should obtain a true statement, and when 0 does not belong to the solution set, I should obtain a false statement.

100. In an inequality such as $5x + 4 < 8x - 5$, I can avoid division by a negative number depending on which side I collect the variable terms and on which side I collect the constant terms.

101. I solved $-2x + 5 \geq 13$ and concluded that -4 is the greatest integer in the solution set.

102. I began the solution of $5 - 3(x + 2) > 10x$ by simplifying the left side, obtaining $2x + 4 > 10x$.

103. A car can be rented from Basic Rental for \$260 per week with no extra charge for mileage. Continental charges \$80 per week plus 25 cents for each mile driven to rent the same car. How many miles must be driven in a week to make the rental cost for Basic Rental a better deal than Continental's?

104. A company manufactures and sells personalized stationery. The weekly fixed cost is \$3000 and it cost \$3.00 to produce each package of stationery. The selling price is \$5.50 per package. How many packages of stationery must be produced and sold each week for the company to generate a profit?

6.5 Quadratic Equations

WHAT AM I SUPPOSED TO LEARN?

After studying this section, you should be able to:

1. Multiply binomials using the FOIL method.
2. Factor trinomials.
3. Solve quadratic equations by factoring.
4. Solve quadratic equations using the quadratic formula.
5. Solve problems modeled by quadratic equations.

I'm very well acquainted, too, with matters mathematical,
I understand equations, both simple and quadratical.
About binomial theorem I'm teeming with a lot of news,
With many cheerful facts about the square of the hypotenuse.

—Gilbert and Sullivan, *The Pirates of Penzance*

EQUATIONS QUADRATICAL? CHEERFUL NEWS ABOUT THE SQUARE OF THE HYPOTENUSE? You've come to the right place. In this section, we study two methods for solving *quadratic equations*, equations in which the highest exponent on the variable is 2. (Yes, it's *quadratic* and not *quadratical*, despite the latter's rhyme with mathematical.) In Chapter 10 (Section 10.2), we look at an application of quadratic equations, introducing (cheerfully, of course) the Pythagorean Theorem and the square of the hypotenuse.

1 *Multiply binomials using the FOIL method.*

Multiplying Two Binomials Using the FOIL Method

Before we learn about the first method for solving quadratic equations, factoring, we need to consider the FOIL method for multiplying two binomials. A **binomial** is a simplified algebraic expression that contains two terms in which each exponent that appears on a variable is a whole number.

Examples of Binomials

$$x + 3, \quad x + 4, \quad 3x + 4, \quad 5x - 3$$

Two binomials can be quickly multiplied by using the FOIL method, in which **F** represents the product of the **first** terms in each binomial, **O** represents the product of the **outside** terms, **I** represents the product of the two **inside** terms, and **L** represents the product of the **last,** or second, terms in each binomial.

USING THE FOIL METHOD TO MULTIPLY BINOMIALS

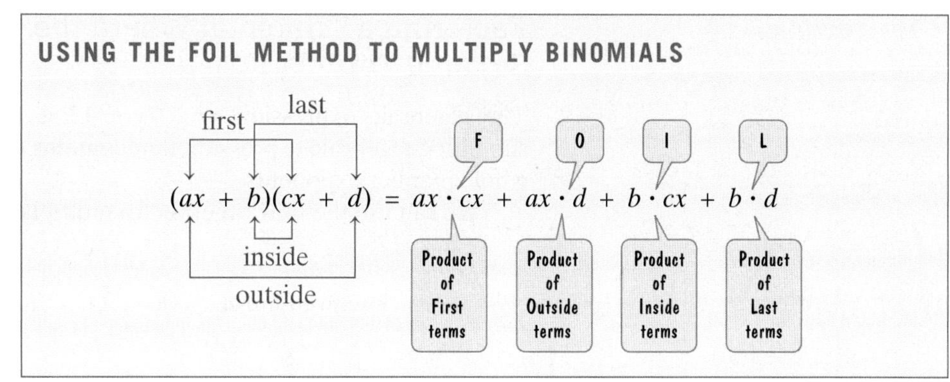

Once you have multiplied first, outside, inside, and last terms, combine all like terms.

EXAMPLE 1 Using the FOIL Method

Multiply: $(x + 3)(x + 4)$.

SOLUTION

F : First terms $= x \cdot x = x^2$ $(x + 3)(x + 4)$

O: Outside terms $= x \cdot 4 = 4x$ $(x + 3)(x + 4)$

I : Inside terms $= 3 \cdot x = 3x$ $(x + 3)(x + 4)$

L : Last terms $= 3 \cdot 4 = 12$ $(x + 3)(x + 4)$

$$
\begin{aligned}
(x + 3)(x + 4) &= x \cdot x + x \cdot 4 + 3 \cdot x + 3 \cdot 4 \\
&= x^2 + 4x + 3x + 12 \\
&= x^2 + 7x + 12 \quad \text{Combine like terms.}
\end{aligned}
$$

CHECK POINT 1 Multiply: $(x + 5)(x + 6)$.

EXAMPLE 2 Using the FOIL Method

Multiply: $(3x + 4)(5x - 3)$.

SOLUTION

$$
\begin{aligned}
(3x + 4)(5x - 3) &= 3x \cdot 5x + 3x(-3) + 4 \cdot 5x + 4(-3) \\
&= 15x^2 - 9x + 20x - 12 \\
&= 15x^2 + 11x - 12 \quad \text{Combine like terms.}
\end{aligned}
$$

CHECK POINT 2 Multiply: $(7x + 5)(4x - 3)$.

2 *Factor trinomials.*

Factoring a Trinomial Where the Coefficient of the Squared Term Is 1

The algebraic expression $x^2 + 7x + 12$ is called a trinomial. A **trinomial** is a simplified algebraic expression that contains three terms in which all variables have whole number exponents.

We can use the FOIL method to multiply two binomials to obtain the trinomial $x^2 + 7x + 12$:

Factored Form		F	O	I	L		Trinomial Form
$(x+3)(x+4)$	$=$	x^2	$+ 4x$	$+ 3x$	$+ 12$	$=$	$x^2 + 7x + 12$

Because the product of $x + 3$ and $x + 4$ is $x^2 + 7x + 12$, we call $x + 3$ and $x + 4$ the **factors** of $x^2 + 7x + 12$. **Factoring** an algebraic expression containing the sum or difference of terms means finding an equivalent expression that is a product. Thus, to factor $x^2 + 7x + 12$, we write

$$x^2 + 7x + 12 = (x + 3)(x + 4).$$

We can make several important observations about the factors on the right side.

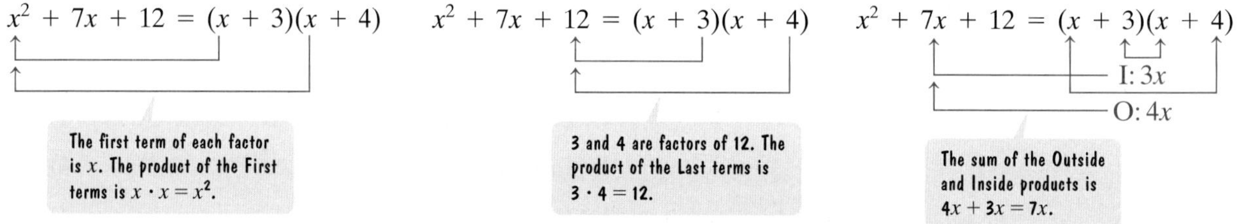

These observations provide us with a procedure for factoring $x^2 + bx + c$.

A STRATEGY FOR FACTORING $x^2 + bx + c$

1. Enter x as the first term of each factor.

$$(x \quad)(x \quad) = x^2 + bx + c$$

2. List pairs of factors of the constant c.
3. Try various combinations of these factors as the second term in each set of parentheses. Select the combination in which the sum of the Outside and Inside products is equal to bx.

$$(x + \square)(x + \square) = x^2 + bx + c$$

I

O

Sum of O + I

4. Check your work by multiplying the factors using the FOIL method. You should obtain the original trinomial.

If none of the possible combinations yield an Outside product and an Inside product whose sum is equal to bx, the trinomial cannot be factored using integers and is called **prime**.

EXAMPLE 3 Factoring a Trinomial in $x^2 + bx + c$ Form

Factor: $x^2 + 6x + 8$.

SOLUTION

Step 1 Enter x as the first term of each factor.

$$x^2 + 6x + 8 = (x \quad)(x \quad)$$

To find the second term of each factor, we must find two integers whose product is 8 and whose sum is 6.

Step 2 List all pairs of factors of the constant, 8.

Factors of 8	8, 1	4, 2	−8, −1	−4, −2

Step 3 Try various combinations of these factors. The correct factorization of $x^2 + 6x + 8$ is the one in which the sum of the Outside and Inside products is equal to $6x$. Here is a list of the possible factorizations:

Possible Factorizations of $x^2 + 6x + 8$	Sum of Outside and Inside Products (Should Equal $6x$)
$(x + 8)(x + 1)$	$x + 8x = 9x$
$(x + 4)(x + 2)$	$2x + 4x = 6x$
$(x - 8)(x - 1)$	$-x - 8x = -9x$
$(x - 4)(x - 2)$	$-2x - 4x = -6x$

This is the required middle term.

Thus, $x^2 + 6x + 8 = (x + 4)(x + 2)$.

Step 4 Check this result by multiplying the right side using the FOIL method. You should obtain the original trinomial. Because of the commutative property, the factorization can also be expressed as

$$x^2 + 6x + 8 = (x + 2)(x + 4).$$

GREAT QUESTION!

Is there a way to eliminate some of the combinations of factors for $x^2 + bx + c$ when c is positive?

Yes. To factor $x^2 + bx + c$ when c is positive, find two numbers with the same sign as the middle term.

$$x^2 + 6x + 8 = (x + 2)(x + 4)$$

Same signs

$$x^2 - 5x + 6 = (x - 3)(x - 2)$$

Same signs

Using this observation, it is not necessary to list the last two factorizations in step 3 on the right.

CHECK POINT 3 Factor: $x^2 + 5x + 6$.

EXAMPLE 4 Factoring a Trinomial in $x^2 + bx + c$ Form

Factor: $x^2 + 2x - 35$.

SOLUTION

Step 1 Enter x as the first term of each factor.

$$x^2 + 2x - 35 = (x \quad)(x \quad)$$

To find the second term of each factor, we must find two integers whose product is −35 and whose sum is 2.

Step 2 List pairs of factors of the constant, −35.

Factors of −35	35, −1	−35, 1	−7, 5	7, −5

Step 3 Try various combinations of these factors. The correct factorization of $x^2 + 2x - 35$ is the one in which the sum of the Outside and Inside products is equal to $2x$. Here is a list of the possible factorizations:

Possible Factorizations of $x^2 + 2x - 35$	Sum of Outside and Inside Products (Should Equal $2x$)
$(x - 1)(x + 35)$	$35x - x = 34x$
$(x + 1)(x - 35)$	$-35x + x = -34x$
$(x - 7)(x + 5)$	$5x - 7x = -2x$
$(x + 7)(x - 5)$	$-5x + 7x = 2x$

This is the required middle term.

Thus, $x^2 + 2x - 35 = (x + 7)(x - 5)$ or $(x - 5)(x + 7)$.

Step 4 Verify the factorization using the FOIL method.

F O I L

$$(x + 7)(x - 5) = x^2 - 5x + 7x - 35 = x^2 + 2x - 35$$

Because the product of the factors is the original trinomial, the factorization is correct.

GREAT QUESTION!

Is there a way to eliminate some of the combinations of factors for $x^2 + bx + c$ when c is negative?

Yes. To factor $x^2 + bx + c$ when c is negative, find two numbers with opposite signs whose sum is the coefficient of the middle term.

$$x^2 + 2x - 35 = (x + 7)(x - 5)$$

Negative — Opposite signs

CHECK POINT 4 Factor: $x^2 + 3x - 10$.

Factoring a Trinomial Where the Coefficient of the Squared Term Is Not 1

How do we factor a trinomial such as $3x^2 - 20x + 28$? Notice that the coefficient of the squared term is 3. We must find two binomials whose product is $3x^2 - 20x + 28$. The product of the First terms must be $3x^2$:

$$(3x \quad)(x \quad).$$

From this point on, the factoring strategy is exactly the same as the one we use to factor trinomials for which the coefficient of the squared term is 1.

EXAMPLE 5 *Factoring a Trinomial*

Factor: $3x^2 - 20x + 28$.

SOLUTION

Step 1 Find two First terms whose product is $3x^2$.

$$3x^2 - 20x + 28 = (3x \quad)(x \quad)$$

Step 2 List all pairs of factors of the constant, 28. The number 28 has pairs of factors that are either both positive or both negative. Because the middle term, $-20x$, is negative, both factors must be negative. The negative factorizations of 28 are $-1(-28)$, $-2(-14)$, and $-4(-7)$.

GREAT QUESTION!

When factoring trinomials, must I list every possible factorization before getting the correct one?

With practice, you will find that it is not necessary to list every possible factorization of the trinomial. As you practice factoring, you will be able to narrow down the list of possible factors to just a few. When it comes to factoring, practice makes perfect.

Step 3 Try various combinations of these factors. The correct factorization of $3x^2 - 20x + 28$ is the one in which the sum of the Outside and Inside products is equal to $-20x$. Here is a list of the possible factorizations:

Possible Factorizations of $3x^2 - 20x + 28$	Sum of Outside and Inside Products (Should Equal $-20x$)
$(3x - 1)(x - 28)$	$-84x - x = -85x$
$(3x - 28)(x - 1)$	$-3x - 28x = -31x$
$(3x - 2)(x - 14)$	$-42x - 2x = -44x$
$(3x - 14)(x - 2)$	$-6x - 14x = -20x$
$(3x - 4)(x - 7)$	$-21x - 4x = -25x$
$(3x - 7)(x - 4)$	$-12x - 7x = -19x$

This is the required middle term.

Thus,

$$3x^2 - 20x + 28 = (3x - 14)(x - 2) \quad \text{or} \quad (x - 2)(3x - 14).$$

Step 4 Verify the factorization using the FOIL method.

F O I L

$$\begin{aligned}(3x - 14)(x - 2) &= 3x \cdot x + 3x(-2) + (-14) \cdot x + (-14)(-2)\\ &= 3x^2 - 6x - 14x + 28\\ &= 3x^2 - 20x + 28\end{aligned}$$

Because this is the trinomial we started with, the factorization is correct.

CHECK POINT 5 Factor: $5x^2 - 14x + 8$.

EXAMPLE 6 *Factoring a Trinomial*

Factor: $8y^2 - 10y - 3$.

SOLUTION

Step 1 Find two first terms whose product is $8y^2$.

$$8y^2 - 10y - 3 \stackrel{?}{=} (8y \quad)(y \quad)$$
$$8y^2 - 10y - 3 \stackrel{?}{=} (4y \quad)(2y \quad)$$

Step 2 List all pairs of factors of the constant, -3. The possible factorizations are $1(-3)$ and $-1(3)$.

Step 3 Try various combinations of these factors. The correct factorization of $8y^2 - 10y - 3$ is the one in which the sum of the Outside and Inside products is equal to $-10y$. Here is a list of the possible factorizations:

These four factorizations are $(8y \quad)(y \quad)$ with $1(-3)$ and $-1(3)$ as factorizations of -3.

These four factorizations are $(4y \quad)(2y \quad)$ with $1(-3)$ and $-1(3)$ as factorizations of -3.

Possible Factorizations of $8y^2 - 10y - 3$	Sum of Outside and Inside Products (Should Equal $-10y$)
$(8y + 1)(y - 3)$	$-24y + y = -23y$
$(8y - 3)(y + 1)$	$8y - 3y = 5y$
$(8y - 1)(y + 3)$	$24y - y = 23y$
$(8y + 3)(y - 1)$	$-8y + 3y = -5y$
$(4y + 1)(2y - 3)$	$-12y + 2y = -10y$
$(4y - 3)(2y + 1)$	$4y - 6y = -2y$
$(4y - 1)(2y + 3)$	$12y - 2y = 10y$
$(4y + 3)(2y - 1)$	$-4y + 6y = 2y$

This is the required middle term.

Thus, $8y^2 - 10y - 3 = (4y + 1)(2y - 3)$.

By the commutative property,

$$8y^2 - 10y - 3 = (4y + 1)(2y - 3) \quad \text{or} \quad (2y - 3)(4y + 1).$$

Show that either of these factorizations is correct by multiplying the factors using the FOIL method. You should obtain the original trinomial.

CHECK POINT 6 Factor: $6y^2 + 19y - 7$.

3 *Solve quadratic equations by factoring.*

Solving Quadratic Equations by Factoring

We have seen that in a linear equation, the highest exponent on the variable is 1. We now define a quadratic equation, in which the greatest exponent on the variable is 2.

DEFINITION OF A QUADRATIC EQUATION

A **quadratic equation** in x is an equation that can be written in the form

$$ax^2 + bx + c = 0,$$

where a, b, and c are real numbers, with $a \neq 0$.

Here is an example of a quadratic equation:

$$x^2 - 7x + 10 = 0.$$

$a = 1$ $b = -7$ $c = 10$

Notice that we can factor the left side of this equation.

$$x^2 - 7x + 10 = 0$$
$$(x - 5)(x - 2) = 0$$

If a quadratic equation has zero on one side and a factored trinomial on the other side, it can be solved using the **zero-product principle**:

THE ZERO-PRODUCT PRINCIPLE

If the product of two factors is zero, then one (or both) of the factors must have a value of zero.

If $AB = 0$, then $A = 0$ or $B = 0$.

EXAMPLE 7 ***Solving a Quadratic Equation Using the Zero-Product Principle***

Solve: $(x - 5)(x - 2) = 0$.

SOLUTION

The product $(x - 5)(x - 2)$ is equal to zero. By the zero-product principle, the only way that this product can be zero is if at least one of the factors is zero. We set each individual factor equal to zero and solve each resulting equation for x.

$$(x - 5)(x - 2) = 0$$

$$x - 5 = 0 \quad \text{or} \quad x - 2 = 0$$

$$x = 5 \qquad\qquad x = 2$$

Check the proposed solutions by substituting each one separately for x in the original equation.

Check 5:		**Check 2:**	
$(x - 5)(x - 2) = 0$		$(x - 5)(x - 2) = 0$	
$(5 - 5)(5 - 2) \stackrel{?}{=} 0$		$(2 - 5)(2 - 2) \stackrel{?}{=} 0$	
$0(3) \stackrel{?}{=} 0$		$-3(0) \stackrel{?}{=} 0$	
$0 = 0,$	true	$0 = 0,$	true

The resulting true statements indicate that the solutions are 5 and 2. The solution set is $\{2, 5\}$.

CHECK POINT 7 Solve: $(x + 6)(x - 3) = 0$.

SOLVING A QUADRATIC EQUATION BY FACTORING

1. If necessary, rewrite the equation in the form $ax^2 + bx + c = 0$, moving all terms to one side, thereby obtaining zero on the other side.
2. Factor.
3. Apply the zero-product principle, setting each factor equal to zero.
4. Solve the equations in step 3.
5. Check the solutions in the original equation.

GREAT QUESTION!

Can all quadratic equations be solved by factoring?

No. The method on the right does not apply if $ax^2 + bx + c$ is not factorable, or prime.

EXAMPLE 8 Solving a Quadratic Equation by Factoring

Solve: $x^2 - 2x = 35$.

SOLUTION

Step 1 Move all terms to one side and obtain zero on the other side. Subtract 35 from both sides and write the equation in $ax^2 + bx + c = 0$ form.

$$x^2 - 2x = 35$$
$$x^2 - 2x - 35 = 35 - 35$$
$$x^2 - 2x - 35 = 0$$

Step 2 Factor.

$$(x - 7)(x + 5) = 0$$

Steps 3 and 4 Set each factor equal to zero and solve each resulting equation.

$$x - 7 = 0 \quad \text{or} \quad x + 5 = 0$$
$$x = 7 \qquad\qquad x = -5$$

Step 5 Check the solutions in the original equation.

Check 7:		**Check −5:**	
$x^2 - 2x = 35$		$x^2 - 2x = 35$	
$7^2 - 2 \cdot 7 \stackrel{?}{=} 35$		$(-5)^2 - 2(-5) \stackrel{?}{=} 35$	
$49 - 14 \stackrel{?}{=} 35$		$25 + 10 \stackrel{?}{=} 35$	
$35 = 35,$	true	$35 = 35,$	true

The resulting true statements indicate that the solutions are 7 and −5. The solution set is $\{-5, 7\}$.

CHECK POINT 8 Solve: $x^2 - 6x = 16$.

EXAMPLE 9 Solving a Quadratic Equation by Factoring

Solve: $5x^2 - 33x + 40 = 0$.

SOLUTION

All terms are already on the left and zero is on the other side. Thus, we can factor the trinomial on the left side. $5x^2 - 33x + 40$ factors as $(5x - 8)(x - 5)$.

$$5x^2 - 33x + 40 = 0 \quad \text{This is the given quadratic equation.}$$

$$(5x - 8)(x - 5) = 0 \quad \text{Factor.}$$

$$5x - 8 = 0 \quad \text{or} \quad x - 5 = 0 \quad \text{Set each factor equal to zero.}$$

$$5x = 8 \qquad x = 5 \quad \text{Solve the resulting equations.}$$

$$x = \frac{8}{5}$$

Check these values in the original equation to confirm that the solution set is $\{\frac{8}{5}, 5\}$.

CHECK POINT 9 Solve: $2x^2 + 7x - 4 = 0$.

4 *Solve quadratic equations using the quadratic formula.*

Solving Quadratic Equations Using the Quadratic Formula

The solutions of a quadratic equation cannot always be found by factoring. Some trinomials are difficult to factor, and others cannot be factored (that is, they are prime). However, there is a formula that can be used to solve all quadratic equations, whether or not they contain factorable trinomials. The formula is called the *quadratic formula.*

GREAT QUESTION!

Is it ok if I write

$$x = -b \pm \frac{\sqrt{b^2 - 4ac}}{2a}?$$

No. The entire numerator of the quadratic formula must be divided by $2a$. Always write the fraction bar all the way across the numerator.

$$x = \frac{-b \pm \sqrt{b^2 - 4ac}}{2a}$$

THE QUADRATIC FORMULA

The solutions of a quadratic equation in the form $ax^2 + bx + c = 0$, with $a \neq 0$, are given by the **quadratic formula**

$$x = \frac{-b \pm \sqrt{b^2 - 4ac}}{2a}.$$

x equals negative b plus or minus the square root of $b^2 - 4ac$, all divided by $2a$.

To use the quadratic formula, be sure that the quadratic equation is expressed with all terms on one side and zero on the other side. It may be necessary to begin by rewriting the equation in this form. Then determine the numerical values for a (the coefficient of the x^2-term), b (the coefficient of the x-term), and c (the constant term). Substitute the values of a, b, and c into the quadratic formula and evaluate the expression. The $\pm$ sign indicates that there are two solutions of the equation.

EXAMPLE 10 Solving a Quadratic Equation Using the Quadratic Formula

Solve using the quadratic formula: $2x^2 + 9x - 5 = 0$.

SOLUTION

The given equation is in the desired form, with all terms on one side and zero on the other side. Begin by identifying the values for a, b, and c.

$$2x^2 + 9x - 5 = 0$$

$a = 2$ $\quad$ $b = 9$ $\quad$ $c = -5$

Substituting these values into the quadratic formula and simplifying gives the equation's solutions.

$$x = \frac{-b \pm \sqrt{b^2 - 4ac}}{2a}$$ Use the quadratic formula.

$$x = \frac{-9 \pm \sqrt{9^2 - 4(2)(-5)}}{2(2)}$$ Substitute the values for a, b, and c: $a = 2$, $b = 9$, and $c = -5$.

$$= \frac{-9 \pm \sqrt{81 + 40}}{4}$$ $9^2 - 4(2)(-5) = 81 - (-40) = 81 + 40$

$$= \frac{-9 \pm \sqrt{121}}{4}$$ Add under the radical sign.

$$= \frac{-9 \pm 11}{4}$$ $\sqrt{121} = 11$

Now we will evaluate this expression in two different ways to obtain the two solutions. On the left, we will *add* 11 to -9. On the right, we will *subtract* 11 from -9.

$$x = \frac{-9 + 11}{4} = \frac{2}{4} = \frac{1}{2} \quad \text{or} \quad x = \frac{-9 - 11}{4} = \frac{-20}{4} = -5$$

The solution set is $\left\{-5, \frac{1}{2}\right\}$.

CHECK POINT 10 Solve using the quadratic formula: $8x^2 + 2x - 1 = 0$.

GREAT QUESTION!

What's the bottom line on whether I can use factoring to solve a quadratic equation?

Compute $b^2 - 4ac$, which appears under the radical sign in the quadratic formula. If $b^2 - 4ac$ is a perfect square, such as 4, 25, or 121, then the equation can be solved by factoring.

The quadratic equation in Example 10 has rational solutions, namely -5 and $\frac{1}{2}$. The equation can also be solved by factoring. Take a few minutes to do this now and convince yourself that you will arrive at the same two solutions.

Any quadratic equation that has rational solutions can be solved by factoring or using the quadratic formula. However, quadratic equations with irrational solutions cannot be solved by factoring. These equations can be readily solved using the quadratic formula.

EXAMPLE 11 *Solving a Quadratic Equation Using the Quadratic Formula*

Solve using the quadratic formula: $2x^2 = 4x + 1$.

SOLUTION

The quadratic equation must have zero on one side to identify the values for a, b, and c. To move all terms to one side and obtain zero on the right, we subtract $4x + 1$ from both sides. Then we can identify the values for a, b, and c.

$$2x^2 = 4x + 1$$ This is the given equation.

$$2x^2 - 4x - 1 = 0$$ Subtract $4x + 1$ from both sides.

$a = 2$ $b = -4$ $c = -1$

Substituting these values, $a = 2, b = -4$, and $c = -1$, into the quadratic formula and simplifying gives the equation's solutions.

$$x = \frac{-b \pm \sqrt{b^2 - 4ac}}{2a}$$ Use the quadratic formula.

$$x = \frac{-(-4) \pm \sqrt{(-4)^2 - 4(2)(-1)}}{2(2)}$$ Substitute the values for a, b, and c: $a = 2$, $b = -4$, and $c = -1$.

$$= \frac{4 \pm \sqrt{16 - (-8)}}{4}$$ $-(-4) = 4$, $(-4)^2 = (-4)(-4) = 16$, and $4(2)(-1) = -8$

$$= \frac{4 \pm \sqrt{24}}{4}$$ $16 - (-8) = 16 + 8 = 24$

The solutions are $\frac{4 + \sqrt{24}}{4}$ and $\frac{4 - \sqrt{24}}{4}$. These solutions are irrational numbers. You can use a calculator to obtain a decimal approximation for each solution. However, in situations such as this that do not involve applications, it is better to leave the irrational solutions in radical form as exact answers. In some cases, we can simplify this radical form. Using methods for simplifying square roots discussed in Section 5.4, we can simplify $\sqrt{24}$:

$$\sqrt{24} = \sqrt{4 \cdot 6} = \sqrt{4}\sqrt{6} = 2\sqrt{6}.$$

Now we can use this result to simplify the two solutions. First, use the distributive property to factor out 2 from both terms in the numerator. Then, divide the numerator and the denominator by 2.

$$x = \frac{4 \pm \sqrt{24}}{4} = \frac{4 \pm 2\sqrt{6}}{4} = \frac{\overset{1}{\cancel{2}}(2 \pm \sqrt{6})}{\underset{2}{\cancel{4}}} = \frac{2 \pm \sqrt{6}}{2}$$

In simplified radical form, the equation's solution set is

$$\left\{\frac{2 + \sqrt{6}}{2}, \frac{2 - \sqrt{6}}{2}\right\}.$$

TECHNOLOGY

Using a Calculator to Approximate $\frac{4 + \sqrt{24}}{4}$:

Many Scientific Calculators

[(] 4 [+] 24 [√] [)] [÷] 4 [=]

Many Graphing Calculators

[(] 4 [+] [√] 24 [)] [÷] 4

[ENTER]

If your calculator displays an open parenthesis after √, you'll need to enter another closed parenthesis here. Some calculators require that you press the right arrow key to exit the radical after entering the radicand.

Correct to the nearest tenth,

$$\frac{4 + \sqrt{24}}{4} \approx 2.2.$$

GREAT QUESTION!

The simplification of the irrational solutions in Example 11 was kind of tricky. Any suggestions to guide the process?

Many students use the quadratic formula correctly until the last step, where they make an error in simplifying the solutions. Be sure to factor the numerator before dividing the numerator and the denominator by the greatest common factor.

$$\frac{4 \pm 2\sqrt{6}}{4} = \frac{2(2 \pm \sqrt{6})}{4} = \frac{\overset{1}{\cancel{2}}(2 \pm \sqrt{6})}{\underset{2}{\cancel{4}}} = \frac{2 \pm \sqrt{6}}{2}$$

You cannot divide just one term in the numerator and the denominator by their greatest common factor.

Incorrect!

$$\frac{\overset{1}{\cancel{4}} \pm 2\sqrt{6}}{\underset{1}{\cancel{4}}} = 1 \pm 2\sqrt{6} \qquad \frac{4 \pm \overset{1}{\cancel{2}}\sqrt{6}}{\underset{2}{\cancel{4}}} = \frac{4 \pm \sqrt{6}}{2}$$

Examples 10 and 11 illustrate that the solutions of quadratic equations can be rational or irrational numbers. In Example 10, the expression under the square root was 121, a perfect square ($\sqrt{121} = 11$), and we obtained rational solutions. In Example 11, this expression was 24, which is not a perfect square (although we simplified $\sqrt{24}$ to $2\sqrt{6}$), and we obtained irrational solutions. If the expression under the square root simplifies to a negative number, then the quadratic equation has **no real solution**. The solution set consists of *imaginary numbers*, discussed in the Blitzer Bonus on page 308.

CHECK POINT 11 Solve using the quadratic formula: $2x^2 = 6x - 1$.

5 *Solve problems modeled by quadratic equations.*

Applications

EXAMPLE 12 Blood Pressure and Age

The graphs in **Figure 6.6** illustrate that a person's normal systolic blood pressure, measured in millimeters of mercury (mm Hg), depends on his or her age. The formula

$$P = 0.006A^2 - 0.02A + 120$$

models a man's normal systolic pressure, P, at age A.

a. Find the age, to the nearest year, of a man whose normal systolic blood pressure is 125 mm Hg.

b. Use the graphs in **Figure 6.6** to describe the differences between the normal systolic blood pressures of men and women as they age.

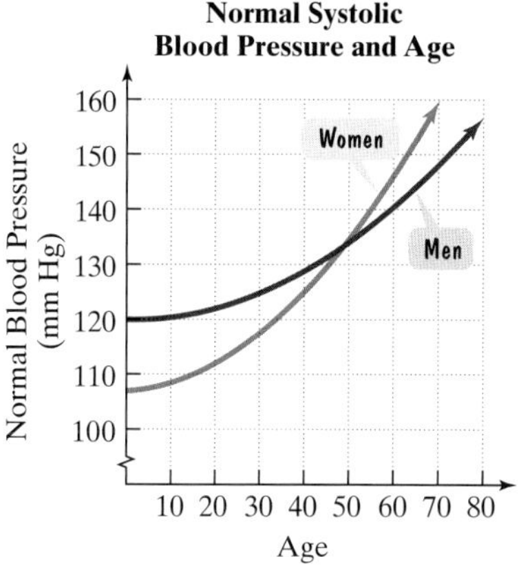

FIGURE 6.6

SOLUTION

a. We are interested in the age of a man with a normal systolic blood pressure of 125 millimeters of mercury. Thus, we substitute 125 for P in the given formula for men. Then we solve for A, the man's age.

$$P = 0.006A^2 - 0.02A + 120$$ This is the given formula for men.

$$125 = 0.006A^2 - 0.02A + 120$$ Substitute 125 for P.

$$0 = 0.006A^2 - 0.02A - 5$$ Subtract 125 from both sides and obtain zero on one side.

$a = \mathbf{0.006}$ $b = \mathbf{-0.02}$ $c = \mathbf{-5}$

Because the trinomial on the right side of the equation is prime, we solve using the quadratic formula.

Notice that the variable is A, rather than the usual x.

$$A = \frac{-b \pm \sqrt{b^2 - 4ac}}{2a}$$ Use the quadratic formula.

$$= \frac{-(-0.02) \pm \sqrt{(-0.02)^2 - 4(0.006)(-5)}}{2(0.006)}$$ Substitute the values for a, b, and c: $a = 0.006$, $b = -0.02$, and $c = -5$.

$$= \frac{0.02 \pm \sqrt{0.1204}}{0.012}$$ Use a calculator to simplify the expression under the square root.

$$\approx \frac{0.02 \pm 0.347}{0.012}$$ Use a calculator: $\sqrt{0.1204} \approx 0.347$.

$$A \approx \frac{0.02 + 0.347}{0.012} \quad \text{or} \quad A \approx \frac{0.02 - 0.347}{0.012}$$

$$A \approx 31 \qquad\qquad A \approx -27$$ Use a calculator and round to the nearest integer.

Reject this solution. Age cannot be negative.

TECHNOLOGY

On most calculators, here is how to approximate

$$\frac{0.02 + \sqrt{0.1204}}{0.012}.$$

Many Scientific Calculators

[(] .02 [+] .1204 [√] [)]

[÷] .012 [=]

Many Graphing Calculators

[(] .02 [+] [√] .1204 [)]

[÷] .012 [ENTER]

If your calculator displays an open parenthesis after √, you'll need to enter another closed parenthesis here. Some calculators require that you press the right arrow key to exit the radical after entering the radicand.

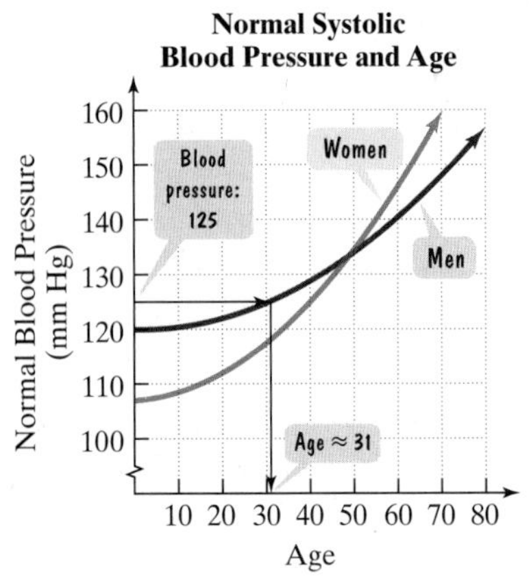

FIGURE 6.7

The positive solution, $A \approx 31$, indicates that 31 is the approximate age of a man whose normal systolic blood pressure is 125 mm Hg. This is illustrated by the black lines with the arrows on the red graph representing men in **Figure 6.7**.

b. Take a second look at the graphs in **Figure 6.6** or **Figure 6.7**. Before approximately age 50, the blue graph representing women's normal systolic blood pressure lies below the red graph representing men's normal systolic blood pressure. Thus, up to age 50, women's normal systolic blood pressure is lower than men's, although it is increasing at a faster rate. After age 50, women's normal systolic blood pressure is higher than men's.

CHECK POINT 12 The formula $P = 0.01A^2 + 0.05A + 107$ models a woman's normal systolic blood pressure, P, at age A. Use this formula to find the age, to the nearest year, of a woman whose normal systolic blood pressure is 115 mm Hg. Use the blue graph in **Figure 6.6** on the previous page to verify your solution.

Blitzer Bonus

Art, Nature, and Quadratic Equations

A **golden rectangle** can be a rectangle of any size, but its long side must be Φ times as long as its short side, where $\Phi \approx 1.6$. Artists often use golden rectangles in their work because they are considered to be more visually pleasing than other rectangles.

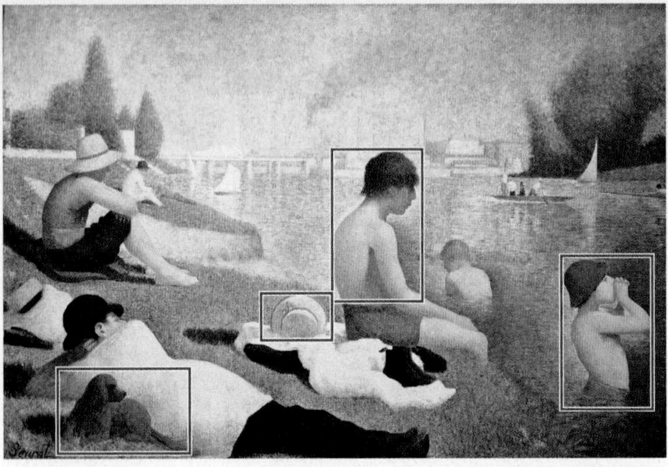

In *The Bathers at Asnières*, by the French impressionist Georges Seurat (1859–1891), the artist positions parts of the painting as though they were inside golden rectangles.

If a golden rectangle is divided into a square and a rectangle, as in **Figure 6.8(a)**, the smaller rectangle is a golden rectangle. If the smaller golden rectangle is divided again, the same is true of the yet smaller rectangle, and so on. The process of repeatedly dividing each golden rectangle in this manner is illustrated in **Figure 6.8(b)**. We've also created a spiral by connecting the opposite corners of all the squares with a smooth curve. This spiral matches the spiral shape of the chambered nautilus shell shown in **Figure 6.8(c)**. The shell spirals out at an ever-increasing rate that is governed by this geometry.

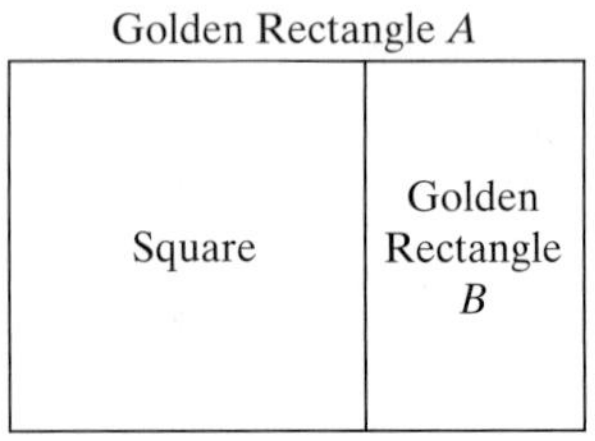

FIGURE 6.8(a)

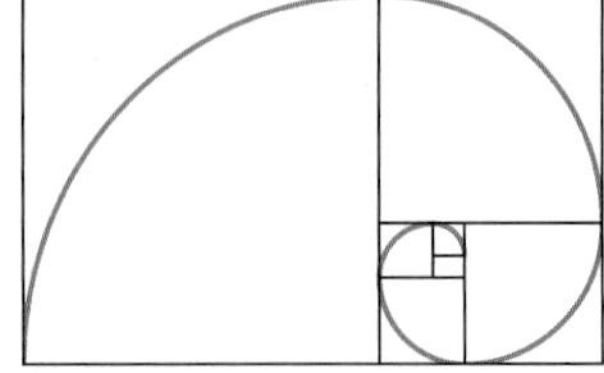

FIGURE 6.8(b)

FIGURE 6.8(c)

In the Exercise Set that follows, you will use the golden rectangles in **Figure 6.8(a)** to obtain an exact value for Φ, the ratio of the long side to the short side in a golden rectangle of any size. Your model will involve a quadratic equation that can be solved by the quadratic formula. (See Exercise 87.)

Concept and Vocabulary Check

Fill in each blank so that the resulting statement is true.

1. For $(x+5)(2x+3)$, the product of the first terms is _____, the product of the outside terms is _____, the product of the inside terms is _____, and the product of the last terms is _____.
2. $x^2 + 13x + 30 = (x+3)(x$ ____$)$
3. $x^2 - 9x + 18 = (x-3)(x$ ____$)$
4. $x^2 - x - 30 = (x-6)(x$ ____$)$
5. $x^2 - 5x - 14 = (x+2)(x$ ____$)$
6. $8x^2 - 10x - 3 = (4x+1)(2x$ ____$)$
7. $12x^2 - x - 20 = (4x+5)(3x$ ____$)$
8. $2x^2 - 5x + 3 = (x-1)($______$)$
9. $6x^2 + 17x + 12 = (2x+3)($______$)$
10. An equation that can be written in the form $ax^2 + bx + c = 0$, $a \neq 0$, is called a/an _________ equation.
11. The zero-product principle states that if $AB = 0$, then _____________.
12. The equation $5x^2 + x = 18$ can be written in the form $ax^2 + bx + c = 0$ by _____________ on both sides.
13. The solutions of $ax^2 + bx + c = 0$, $a \neq 0$, are given by

    ____________________, called the ________________.

In Exercises 14–18, determine whether each statement is true or false. If the statement is false, make the necessary change(s) to produce a true statement.

14. One factor of $x^2 + x + 20$ is $x + 5$. ______
15. If $(x+3)(x-4) = 2$, then $(x+3) = 0$ or $(x-4) = 0$. ______
16. In using the quadratic formula to solve the quadratic equation $5x^2 = 2x - 7$, we have $a = 5$, $b = 2$, and $c = -7$. ______
17. The quadratic formula can be expressed as
$$x = -b \pm \frac{\sqrt{b^2 - 4ac}}{2a}.$$ ______
18. The solutions $\frac{4 \pm \sqrt{3}}{2}$ can be simplified to $2 \pm \sqrt{3}$. ______

Exercise Set 6.5

Practice Exercises

Use FOIL to find the products in Exercises 1–8.

1. $(x+3)(x+5)$
2. $(x+7)(x+2)$
3. $(x-5)(x+3)$
4. $(x-1)(x+2)$
5. $(2x-1)(x+2)$
6. $(2x-5)(x+3)$
7. $(3x-7)(4x-5)$
8. $(2x-9)(7x-4)$

Factor the trinomials in Exercises 9–32, or state that the trinomial is prime. Check your factorization using FOIL multiplication.

9. $x^2 + 5x + 6$
10. $x^2 + 8x + 15$
11. $x^2 - 2x - 15$
12. $x^2 - 4x - 5$
13. $x^2 - 8x + 15$
14. $x^2 - 14x + 45$
15. $x^2 - 9x - 36$
16. $x^2 - x - 90$
17. $x^2 - 8x + 32$
18. $x^2 - 9x + 81$
19. $x^2 + 17x + 16$
20. $x^2 - 7x - 44$
21. $2x^2 + 7x + 3$
22. $3x^2 + 7x + 2$
23. $2x^2 - 17x + 30$
24. $5x^2 - 13x + 6$
25. $3x^2 - x - 2$
26. $2x^2 + 5x - 3$
27. $3x^2 - 25x - 28$
28. $3x^2 - 2x - 5$
29. $6x^2 - 11x + 4$
30. $6x^2 - 17x + 12$
31. $4x^2 + 16x + 15$
32. $8x^2 + 33x + 4$

In Exercises 33–36, solve each equation using the zero-product principle.

33. $(x-8)(x+3) = 0$
34. $(x+11)(x-5) = 0$
35. $(4x+5)(x-2) = 0$
36. $(x+9)(3x-1) = 0$

Solve the quadratic equations in Exercises 37–52 by factoring.

37. $x^2 + 8x + 15 = 0$
38. $x^2 + 5x + 6 = 0$
39. $x^2 - 2x - 15 = 0$
40. $x^2 + x - 42 = 0$
41. $x^2 - 4x = 21$
42. $x^2 + 7x = 18$
43. $x^2 + 9x = -8$
44. $x^2 - 11x = -10$

45. $x^2 - 12x = -36$

46. $x^2 - 14x = -49$

47. $2x^2 = 7x + 4$

48. $3x^2 = x + 4$

49. $5x^2 + x = 18$

50. $3x^2 - 4x = 15$

51. $x(6x + 23) + 7 = 0$

52. $x(6x + 13) + 6 = 0$

Solve the equations in Exercises 53–72 using the quadratic formula.

53. $x^2 + 8x + 15 = 0$

54. $x^2 + 8x + 12 = 0$

55. $x^2 + 5x + 3 = 0$

56. $x^2 + 5x + 2 = 0$

57. $x^2 + 4x = 6$

58. $x^2 + 2x = 4$

59. $x^2 + 4x - 7 = 0$

60. $x^2 + 4x + 1 = 0$

61. $x^2 - 3x = 18$

62. $x^2 - 3x = 10$

63. $6x^2 - 5x - 6 = 0$

64. $9x^2 - 12x - 5 = 0$

65. $x^2 - 2x - 10 = 0$

66. $x^2 + 6x - 10 = 0$

67. $x^2 - x = 14$

68. $x^2 - 5x = 10$

69. $6x^2 + 6x + 1 = 0$

70. $3x^2 = 5x - 1$

71. $4x^2 = 12x - 9$

72. $9x^2 + 6x + 1 = 0$

Practice Plus

In Exercises 73–80, solve each equation by the method of your choice.

73. $\frac{3x^2}{4} - \frac{5x}{2} - 2 = 0$

74. $\frac{x^2}{3} - \frac{x}{2} - \frac{3}{2} = 0$

75. $(x - 1)(3x + 2) = -7(x - 1)$

76. $x(x + 1) = 4 - (x + 2)(x + 2)$

77. $(2x - 6)(x + 2) = 5(x - 1) - 12$

78. $7x(x - 2) = 3 - 2(x + 4)$

79. $2x^2 - 9x - 3 = 9 - 9x$

80. $3x^2 - 6x - 3 = 12 - 6x$

81. When the sum of 6 and twice a positive number is subtracted from the square of the number, 0 results. Find the number.

82. When the sum of 1 and twice a negative number is subtracted from twice the square of the number, 0 results. Find the number.

Application Exercises

The formula

$$N = \frac{t^2 - t}{2}$$

describes the number of football games, N, that must be played in a league with t teams if each team is to play every other team once. Use this information to solve Exercises 83–84.

83. If a league has 36 games scheduled, how many teams belong to the league, assuming that each team plays every other team once?

84. If a league has 45 games scheduled, how many teams belong to the league, assuming that each team plays every other team once?

A substantial percentage of the United States population is foreign-born. The bar graph shows the percentage of foreign-born Americans for selected years from 1920 through 2014.

Percentage of the United States Population That Was Foreign-Born, 1920–2014

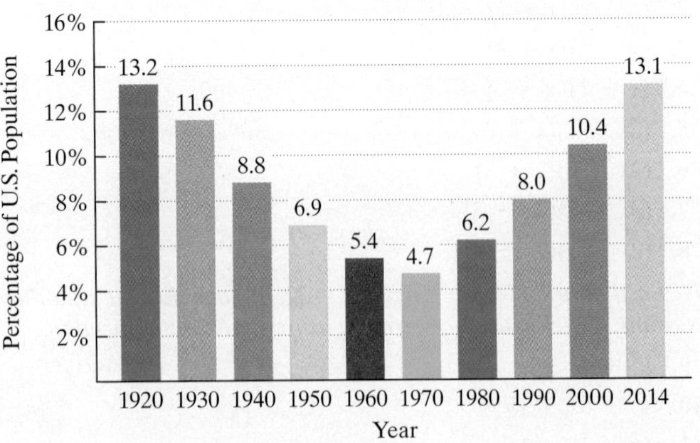

Source: U.S. Census Bureau

The percentage, p, of the United States population that was foreign-born x years after 1920 can be modeled by the formula

$$p = 0.004x^2 - 0.35x + 13.9.$$

Use the formula to solve Exercises 85–86.

85. **a.** According to the model, what percentage of the U.S. population was foreign-born in 2000? Does the model underestimate or overestimate the actual number displayed by the bar graph? By how much?

b. If trends shown by the model continue, in which year will 18.9% of the U.S. population be foreign-born?

86. a. According to the model, what percentage of the U.S. population was foreign-born in 1990? Does the model underestimate or overestimate the actual number displayed by the bar graph on the previous page? By how much?

b. If trends shown by the model continue, in which year will 23.8% of the U.S. population be foreign-born?

87. If you have not yet done so, read the Blitzer Bonus on page 402. In this exercise, you will use the golden rectangles shown to obtain an exact value for Φ, the ratio of the long side to the short side in a golden rectangle of any size.

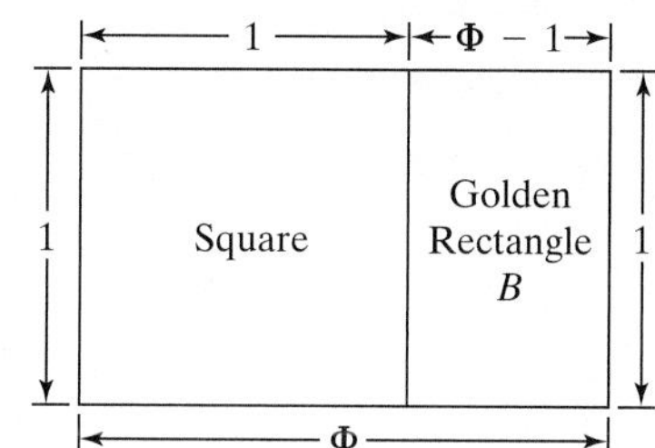

a. The golden ratio in rectangle A, or the ratio of the long side to the short side, can be modeled by $\frac{\Phi}{1}$. Write a fractional expression that models the golden ratio in rectangle B.

b. Set the expression for the golden ratio in rectangle A equal to the expression for the golden ratio in rectangle B. Solve the resulting proportion using the quadratic formula. Express Φ as an exact value in simplified radical form.

c. Use your solution from part (b) to complete this statement: The ratio of the long side to the short side in a golden rectangle of any size is _________ to 1.

Explaining the Concepts

88. Explain how to multiply two binomials using the FOIL method. Give an example with your explanation.

89. Explain how to factor $x^2 - 5x + 6$.

90. Explain how to solve a quadratic equation by factoring. Use the equation $x^2 + 6x + 8 = 0$ in your explanation.

91. Explain how to solve a quadratic equation using the quadratic formula. Use the equation $x^2 + 6x + 8 = 0$ in your explanation.

92. Describe the trend shown by the data for the percentage of foreign-born Americans in the graph for Exercises 85–86. Do you believe that this trend is likely to continue or might something occur that would make it impossible to extend the model into the future? Explain your answer.

Critical Thinking Exercises

Make Sense? *In Exercises 93–96, determine whether each statement makes sense or does not make sense, and explain your reasoning.*

93. I began factoring $x^2 - 17x + 72$ by finding all number pairs with a sum of -17.

94. It's easy to factor $x^2 + x + 1$ because of the relatively small numbers for the constant term and the coefficient of x.

95. The fastest way for me to solve $x^2 - x - 2 = 0$ is to use the quadratic formula.

96. I simplified $\frac{3 + 2\sqrt{3}}{2}$ to $3 + \sqrt{3}$ because 2 is a factor of $2\sqrt{3}$.

97. The radicand of the quadratic formula, $b^2 - 4ac$, can be used to determine whether $ax^2 + bx + c = 0$ has solutions that are rational, irrational, or not real numbers. Explain how this works. Is it possible to determine the kinds of answers that one will obtain to a quadratic equation without actually solving the equation? Explain.

In Exercises 98–99, find all positive integers b so that the trinomial can be factored.

98. $x^2 + bx + 15$

99. $x^2 + 4x + b$

100. Factor: $x^{2n} + 20x^n + 99$.

101. Solve: $x^2 + 2\sqrt{3}x - 9 = 0$.

Chapter Summary, Review, and Test

SUMMARY – DEFINITIONS AND CONCEPTS	EXAMPLES
6.1 Algebraic Expressions and Formulas	
a. An algebraic expression combines variables and numbers using addition, subtraction, multiplication, division, powers, or roots.	
b. Evaluating an algebraic expression means finding its value for a given value of the variable or for given values of the variables. Once these values are substituted, follow the order of operations agreement in the box on page 344.	Ex. 1, p. 345; Ex. 2, p. 345; Ex. 3, p. 345
c. An equation is a statement that two expressions are equal. Formulas are equations that express relationships among two or more variables. Mathematical modeling is the process of finding formulas to describe real-world phenomena. Such formulas, together with the meaning assigned to the variables, are called mathematical models. The formulas are said to model, or describe, the relationships among the variables.	Ex. 4, p. 346

d. Terms of an algebraic expression are separated by addition. Like terms have the same variables with the same exponents on the variables. To add or subtract like terms, add or subtract the coefficients and copy the common variable.	
e. An algebraic expression is simplified when parentheses have been removed (using the distributive property) and like terms have been combined.	Ex. 5, p. 348; Ex. 6, p. 349; Ex. 7, p. 349

6.2 Linear Equations in One Variable and Proportions

a. A linear equation in one variable can be written in the form $ax + b = 0$, where a and b are real numbers, and $a \neq 0$.	
b. Solving a linear equation is the process of finding the set of numbers that makes the equation a true statement. These numbers are the solutions. The set of all such solutions is the solution set.	
c. Equivalent equations have the same solution set. Properties for generating equivalent equations are given in the box on page 355.	Ex. 1, p. 356
d. A step-by-step procedure for solving a linear equation is given in the box on page 356.	Ex. 2, p. 356; Ex. 3, p. 358; Ex. 4, p. 358
e. If an equation contains fractions, begin by multiplying both sides of the equation by the least common denominator of the fractions in the equation, thereby clearing fractions.	Ex. 5, p. 359; Ex. 6, p. 360
f. The ratio of a to b is written $\frac{a}{b}$, or $a : b$.	
g. A proportion is a statement in the form $\frac{a}{b} = \frac{c}{d}$.	
h. The cross-products principle states that if $\frac{a}{b} = \frac{c}{d}$, then $ad = bc$.	Ex. 7, p. 362
i. A step-by-step procedure for solving applied problems using proportions is given in the box on page 363.	Ex. 8, p. 363; Ex. 9, p. 364
j. If a false statement (such as $-6 = 7$) is obtained in solving an equation, the equation has no solution. The solution set is $\varnothing$, the empty set.	Ex. 10, p. 364
k. If a true statement (such as $-6 = -6$) is obtained in solving an equation, the equation has infinitely many solutions. The solution set is the set of all real numbers, written $\{x \mid x \text{ is a real number}\}$.	Ex. 11, p. 365

6.3 Applications of Linear Equations

a. Algebraic translations of English phrases are given in Table 6.2 on page 370.	
b. A step-by-step strategy for solving word problems using linear equations is given in the box on page 369.	Ex. 1, p. 370; Ex. 2, p. 372; Ex. 3, p. 374; Ex. 4, p. 375
c. Solving a formula for a variable means rewriting the formula so that the variable is isolated on one side of the equation.	Ex. 5, p. 376; Ex. 6, p. 376

6.4 Linear Inequalities in One Variable

A procedure for solving a linear inequality is given in the box on page 383. Remember to reverse the direction of the inequality symbol when multiplying or dividing both sides of an inequality by a negative number, thereby changing the sense of the inequality.	Ex. 2, p. 383; Ex. 3, p. 383; Ex. 4, p. 384; Ex. 5, p. 385; Ex. 6, p. 386

6.5 Quadratic Equations

a. A quadratic equation can be written in the form $ax^2 + bx + c = 0, a \neq 0$.	
b. Some quadratic equations can be solved using factoring and the zero-product principle. A step-by-step procedure is given in the box on page 397.	Ex. 8, p. 397; Ex. 9, p. 398
c. All quadratic equations in the form $ax^2 + bx + c = 0$ can be solved using the quadratic formula: $x = \frac{-b \pm \sqrt{b^2 - 4ac}}{2a}$.	Ex. 10, p. 398; Ex. 11, p. 399; Ex. 12, p. 401

Review Exercises

6.1

In Exercises 1–3, evaluate the algebraic expression for the given value of the variable.

1. $6x + 9; x = 4$

2. $7x^2 + 4x - 5; x = -2$

3. $6 + 2(x - 8)^3; x = 5$

4. The diversity index, from 0 (no diversity) to 100, measures the chance that two randomly selected people are a different race or ethnicity. The diversity index in the United States varies widely from region to region, from as high as 81 in Hawaii to as low as 11 in Vermont. The bar graph shows the national diversity index for the United States for four years in the period from 1980 through 2010.

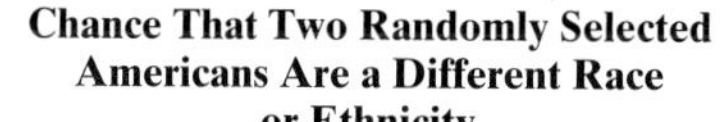

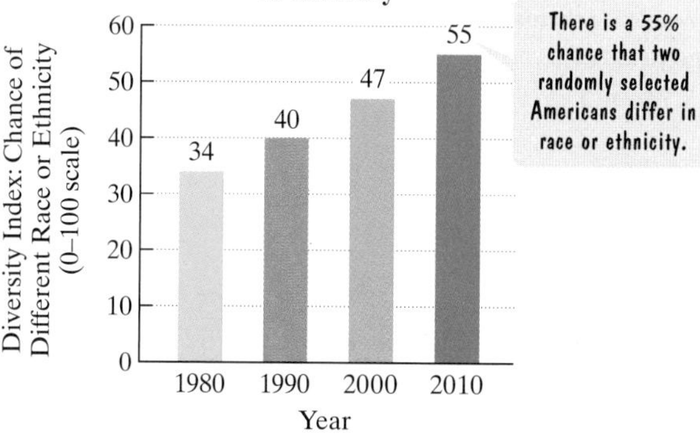

Source: USA Today

The data in the graph can be modeled by the formula

$$D = 0.005x^2 + 0.55x + 34,$$

where D is the national diversity index in the United States x years after 1980. According to the formula, what was the U.S. diversity index in 2010? How does this compare with the index displayed by the bar graph?

In Exercises 5–7, simplify each algebraic expression.

5. $5(2x - 3) + 7x$

6. $3(4y - 5) - (7y - 2)$

7. $2(x^2 + 5x) + 3(4x^2 - 3x)$

6.2

In Exercises 8–14, solve each equation.

8. $4x + 9 = 33$

9. $5x - 3 = x + 5$

10. $3(x + 4) = 5x - 12$

11. $2(x - 2) + 3(x + 5) = 2x - 2$

12. $\frac{2x}{3} = \frac{x}{6} + 1$

13. $7x + 5 = 5(x + 3) + 2x$

14. $7x + 13 = 2(2x - 5) + 3x + 23$

In Exercises 15–18, solve each proportion.

15. $\frac{3}{x} = \frac{15}{25}$

16. $\frac{-7}{5} = \frac{91}{x}$

17. $\frac{x + 2}{3} = \frac{4}{5}$

18. $\frac{5}{x + 7} = \frac{3}{x + 3}$

19. If a school board determines that there should be 3 teachers for every 50 students, how many teachers are needed for an enrollment of 5400 students?

20. To determine the number of trout in a lake, a conservationist catches 112 trout, tags them, and returns them to the lake. Later, 82 trout are caught, and 32 of them are found to be tagged. How many trout are in the lake?

21. The line graph shows the cost of inflation. What cost \$10,000 in 1984 would cost the amount shown by the graph in subsequent years.

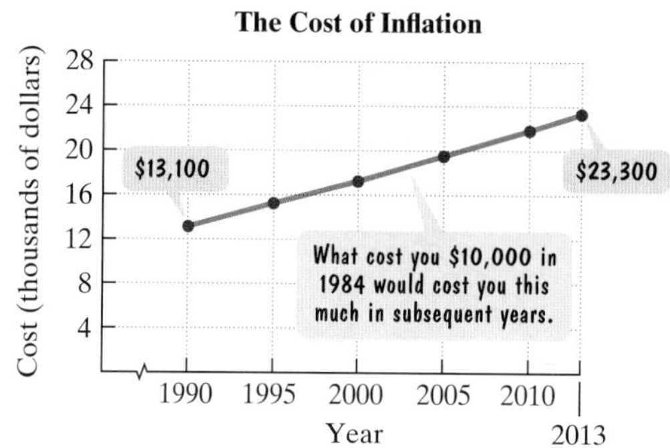

Source: U.S. Bureau of Labor Statistics

Here are two mathematical models for the data shown by the graph. In each formula, C represents the cost x years after 1990 of what cost \$10,000 in 1984.

Model 1 $C = 442x + 12{,}969$

Model 2 $C = 2x^2 + 390x + 13{,}126$

a. Use the graph to estimate the cost in 2010, to the nearest thousand dollars, of what cost \$10,000 in 1984.

b. Use model 1 to determine the cost in 2010. How well does this describe your estimate from part (a)?

c. Use model 2 to determine the cost in 2010. How well does this describe your estimate from part (a)?

d. Use model 1 to determine in which year the cost will be \$26,229 for what cost \$10,000 in 1984.

6.3

22. Although you want to choose a career that fits your interests and abilities, it is good to have an idea of what jobs pay when looking at career options. The bar graph shows the average yearly earnings of full-time employed college graduates with only a bachelor's degree based on their college major.

Average Earnings, by College Major

Average Yearly Earnings (thousands of dollars)
$80
$70
$60
$50
$40
$30
$20
$10
Engineering
Accounting
Marketing
Journalism 53
Nursing 51
Philosophy 43
Social Work 38

Source: Arthur J. Keown, *Personal Finance*, Pearson.

The average yearly earnings of engineering majors exceed the earnings of marketing majors by $19 thousand. The average yearly earnings of accounting majors exceed the earnings of marketing majors by $6 thousand. Combined, the average yearly earnings for these three college majors are $196 thousand. Determine the average yearly earnings, in thousands of dollars, for each of these three college majors.

23. The bar graph shows the average price of a movie ticket for selected years from 1980 through 2013. The graph indicates that in 1980, the average movie ticket price was $2.69. For the period from 1980 through 2013, the price increased by approximately $0.17 per year. If this trend continues, by which year will the average price of a movie ticket be $9.49?

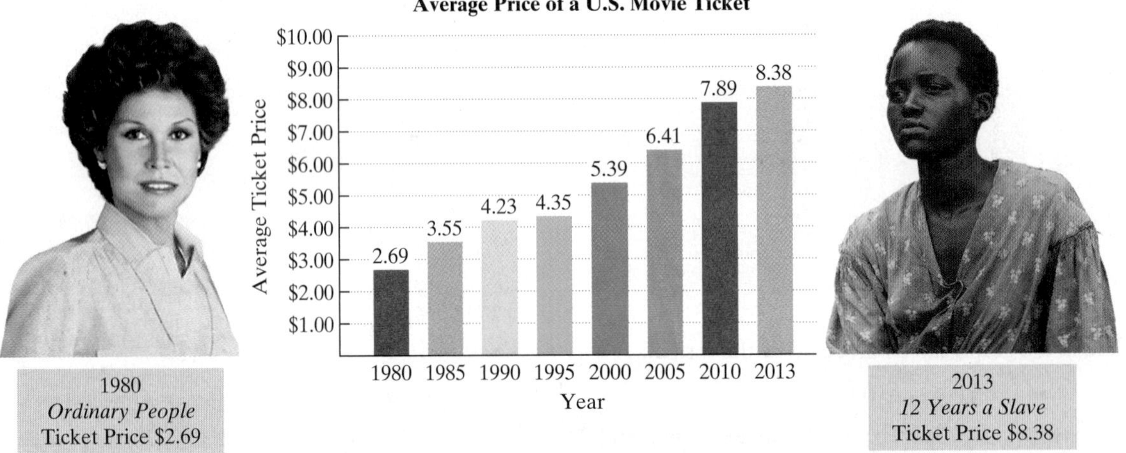

Sources: Motion Picture Association of America, National Association of Theater Owners (NATO), and Bureau of Labor Statistics (BLS)

24. You are choosing between two cellphone plans. Data Plan A has a monthly fee of $52 with a charge of $18 per gigabyte (GB). Data Plan B has a monthly fee of $32 with a charge of $22 per GB. For how many GB of data will the costs for the two data plans be the same?

25. After a 20% price reduction, a cordless phone sold for $48. What was the phone's price before the reduction?

26. A salesperson earns $300 per week plus 5% commission on sales. How much must be sold to earn $800 in a week?

In Exercises 27–30, solve each formula for the specified variable.

27. $Ax - By = C$ for x

28. $A = \frac{1}{2}bh$ for h

29. $A = \dfrac{B + C}{2}$ for B

30. $vt + gt^2 = s$ for g

6.4

In Exercises 31–37, solve each inequality and graph the solution set on a number line.

31. $2x - 5 < 3$

32. $\frac{x}{2} > -4$

33. $3 - 5x \le 18$

34. $4x + 6 < 5x$

35. $6x - 10 \ge 2(x + 3)$

36. $4x + 3(2x - 7) \le x - 3$

37. $-1 < 4x + 2 \le 6$

38. To pass a course, a student must have an average on three examinations of at least 60. If a student scores 42 and 74 on the first two tests, what must be earned on the third test to pass the course?

6.5

Use FOIL to find the products in Exercises 39–40.

39. $(x + 9)(x - 5)$

40. $(4x - 7)(3x + 2)$

Factor the trinomials in Exercises 41–46, or state that the trinomial is prime.

41. $x^2 - x - 12$

42. $x^2 - 8x + 15$

43. $x^2 + 2x + 3$

44. $3x^2 - 17x + 10$

45. $6x^2 - 11x - 10$

46. $3x^2 - 6x - 5$

Solve the quadratic equations in Exercises 47–50 by factoring.

47. $x^2 + 5x - 14 = 0$

48. $x^2 - 4x = 32$

49. $2x^2 + 15x - 8 = 0$

50. $3x^2 = -21x - 30$

Solve the quadratic equations in Exercises 51–54 using the quadratic formula.

51. $x^2 - 4x + 3 = 0$

52. $x^2 - 5x = 4$

53. $2x^2 + 5x - 3 = 0$

54. $3x^2 - 6x = 5$

55. As gas prices surge, more Americans are cycling as a way to save money, stay fit, or both. In 2010, Boston installed 20 miles of bike lanes and New York City added more than 50 miles. The bar graph shows the number of bicycle-friendly U.S. communities, as designated by the League of American Bicyclists, for selected years from 2003 through 2015.

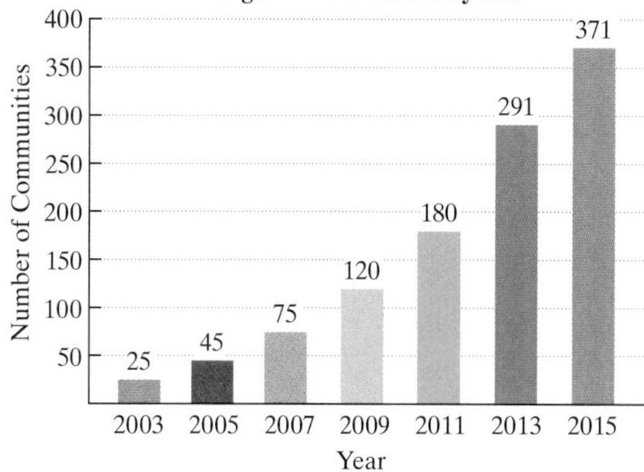

Source: League of American Bicyclists

The formula

$$B = 2.2x^2 + 3x + 27$$

models the number of bicycle-friendly communities, B, x years after 2003.

a. Use the formula to find the number of bicycle-friendly communities in 2013. Does this value underestimate or overestimate the number shown by the graph? By how much?

b. Use the formula to determine the year in which 967 U.S. communities will be bicycle friendly.

Chapter 6 Test

1. Evaluate $x^3 - 4(x - 1)^2$ when $x = -2$.

2. Simplify: $5(3x - 2) - (x - 6)$.

In Exercises 3–6, solve each equation.

3. $12x + 4 = 7x - 21$

4. $3(2x - 4) = 9 - 3(x + 1)$

5. $3(x - 4) + x = 2(6 + 2x)$

6. $\frac{x}{5} - 2 = \frac{x}{3}$

7. Solve for y: $By - Ax = A$.

8. The bar graph in the next column shows the percentage of American adults reporting personal gun ownership for selected years from 1980 through 2010.

Here are two mathematical models for the data shown by the graph. In each formula, p represents the percentage of American adults who reported personal gun ownership x years after 1980.

Model 1: $p = -0.3x + 30$

Model 2: $p = -0.003x^2 - 0.22x + 30$

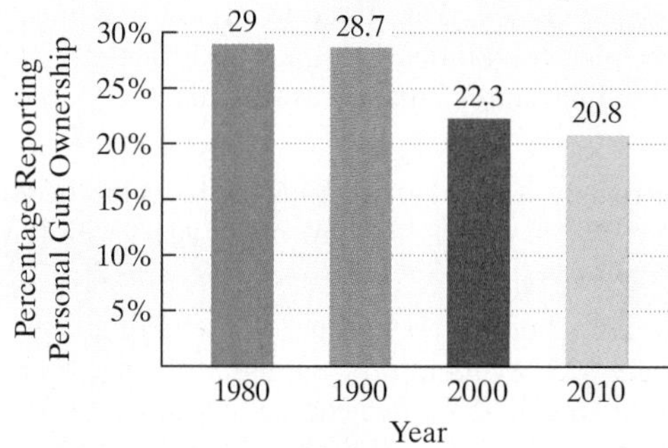

Source: General Social Survey

a. According to model 1, what percentage of American adults reported personal gun ownership in 2010? Does this underestimate or overestimate the percentage shown by the graph? By how much?

b. According to model 2, what percentage of American adults reported personal gun ownership in 2010? Does this underestimate or overestimate the percentage shown by the graph? By how much?

c. If trends shown by the data continue, use model 1 to determine in which year 17.7% of American adults will report personal gun ownership.

In Exercises 9–10, solve each proportion.

9. $\frac{5}{8} = \frac{x}{12}$

10. $\frac{x+5}{8} = \frac{x+2}{5}$

11. Park rangers catch, tag, and release 200 tule elk back into a wildlife refuge. Two weeks later they observe a sample of 150 elk, of which 5 are tagged. Assuming that the ratio of tagged elk in the sample holds for all elk in the refuge, how many elk are there in the park?

12. What's the last word in capital punishment? An analysis of the final statements of all men and women Texas has executed since the Supreme Court reinstated the death penalty in 1976 revealed that "love" is by far the most frequently uttered word. The bar graph shows the number of times various words were used in final statements by Texas death-row inmates.

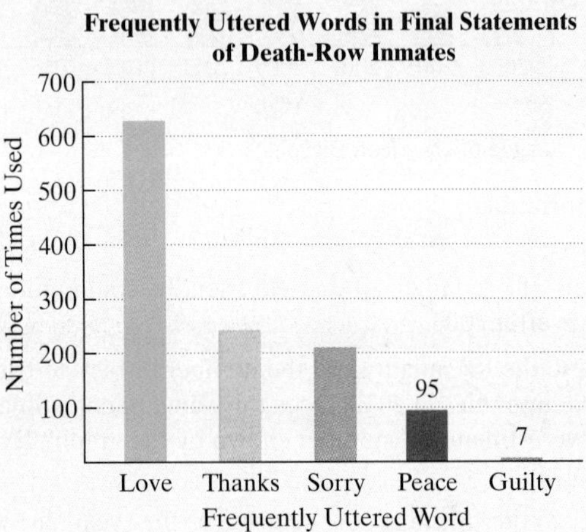

Source: Texas Department of Criminal Justice

The number of times "love" was used exceeded the number of times "sorry" was used by 419. The number of utterances of "thanks" exceeded the number of utterances of "sorry" by 32. Combined, these three words were used 1084 times. Determine the number of times each of these words was used in final statements by Texas inmates.

13. You bought a new car for $50,750. Its value is decreasing by $5500 per year. After how many years will its value be $12,250?

14. You are choosing between two texting plans. Plan A charges $25 per month for unlimited texting. Plan B has a monthly fee of $13 with a charge of $0.06 per text. For how many text messages, will the costs for the two plans be the same?

15. After a 60% reduction, a jacket sold for $20. What was the jacket's price before the reduction?

In Exercises 16–18, solve each inequality and graph the solution set on a number line.

16. $6 - 9x \geq 33$

17. $4x - 2 > 2(x + 6)$

18. $-3 \leq 2x + 1 < 6$

19. A student has grades on three examinations of 76, 80, and 72. What must the student earn on a fourth examination in order to have an average of at least 80?

20. Use FOIL to find this product: $(2x - 5)(3x + 4)$.

21. Factor: $2x^2 - 9x + 10$.

22. Solve by factoring: $x^2 + 5x = 36$.

23. Solve using the quadratic formula: $2x^2 + 4x = -1$.

The graphs show the amount being paid in Social Security benefits and the amount going into the system. All data are expressed in billions of dollars. Amounts from 2016 through 2024 are projections.

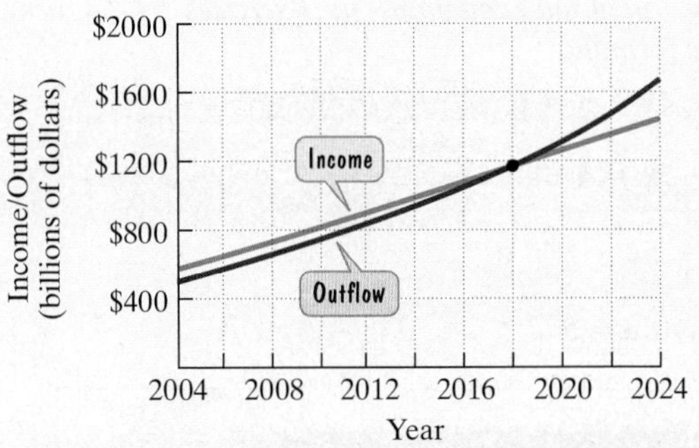

Source: Social Security Trustees Report

Exercises 24–26 are based on the data shown by the graphs.

24. In 2004, the system's income was $575 billion, projected to increase at an average rate of $43 billion per year. In which year will the system's income be $1177 billion?

25. The data for the system's outflow can be modeled by the formula

$$B = 0.07x^2 + 47.4x + 500,$$

where B represents the amount paid in benefits, in billions of dollars, x years after 2004. According to this model, when will the amount paid in benefits be $1177 billion? Round to the nearest year.

26. How well do your answers to Exercises 24 and 25 model the data shown by the graphs?

Algebra: Graphs, Functions, and Linear Systems

7

TELEVISION, MOVIES, AND MAGAZINES PLACE GREAT EMPHASIS ON PHYSICAL BEAUTY. OUR CULTURE emphasizes physical appearance to such an extent that it is a central factor in the perception and judgment of others. The modern emphasis on thinness as the ideal body shape has been suggested as a major cause of eating disorders among adolescent women.

Cultural values of physical attractiveness change over time. During the 1950s, actress Jayne Mansfield embodied the postwar ideal: curvy, buxom, and big-hipped. Men, too, have been caught up in changes of how they "ought" to look. The 1960s' ideal was the soft and scrawny hippie. Today's ideal man is tough and muscular.

Given the importance of culture in setting standards of attractiveness, how can you establish a healthy weight range for your age and height? In this chapter, we will use systems of inequalities to explore these skin-deep issues.

Here's where you'll find these applications:

You'll find a weight that fits you using the models (mathematical, not fashion) in Example 4 of Section 7.4 and Exercises 45–48 in Exercise Set 7.4. Exercises 51–52 use graphs and a formula for body-mass index to indicate whether you are obese, overweight, borderline overweight, normal weight, or underweight.

7.1 Graphing and Functions

WHAT AM I SUPPOSED TO LEARN?

After studying this section, you should be able to:

1 Plot points in the rectangular coordinate system.
2 Graph equations in the rectangular coordinate system.
3 Use function notation.
4 Graph functions.
5 Use the vertical line test.
6 Obtain information about a function from its graph.

THE BEGINNING OF THE SEVENTEENTH CENTURY WAS a time of innovative ideas and enormous intellectual progress in Europe. English theatergoers enjoyed a succession of exciting new plays by Shakespeare. William Harvey proposed the radical notion that the heart was a pump for blood rather than the center of emotion. Galileo, with his new-fangled invention called the telescope, supported the theory of Polish astronomer Copernicus that the Sun, not the Earth, was the center of the solar system. Monteverdi was writing the world's first grand operas. French mathematicians Pascal and Fermat invented a new field of mathematics called probability theory.

Into this arena of intellectual electricity stepped French aristocrat René Descartes (1596–1650). Descartes (pronounced "day cart"), propelled by the creativity surrounding him, developed a new branch of mathematics that brought together algebra and geometry in a unified way—a way that visualized numbers as points on a graph, equations as geometric figures, and geometric figures as equations. This new branch of mathematics, called *analytic geometry*, established Descartes as one of the founders of modern thought and among the most original mathematicians and philosophers of any age. We begin this section by looking at Descartes's deceptively simple idea, called the **rectangular coordinate system** or (in his honor) the **Cartesian coordinate system**.

1 *Plot points in the rectangular coordinate system.*

Points and Ordered Pairs

Descartes used two number lines that intersect at right angles at their zero points, as shown in **Figure 7.1**. The horizontal number line is the ***x*-axis**. The vertical number line is the ***y*-axis**. The point of intersection of these axes is their zero points, called the **origin**. Positive numbers are shown to the right and above the origin. Negative numbers are shown to the left and below the origin. The axes divide the plane into four quarters, called **quadrants**. The points located on the axes are not in any quadrant.

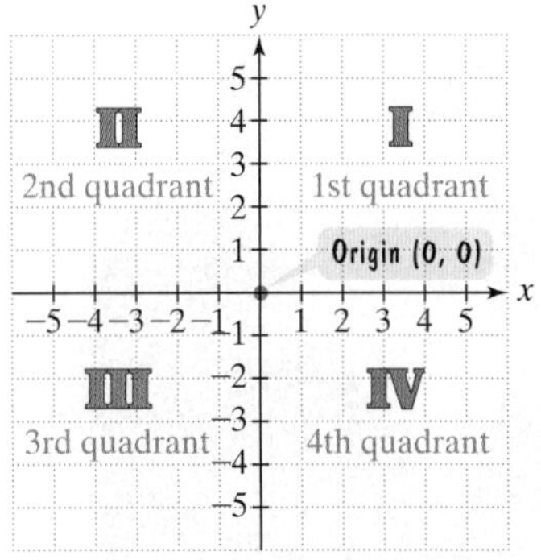

FIGURE 7.1 The rectangular coordinate system.

Each point in the rectangular coordinate system corresponds to an **ordered pair** of real numbers, (x, y). Examples of such pairs are $(-5, 3)$ and $(3, -5)$. The first number in each pair, called the ***x*-coordinate**, denotes the distance and direction from the origin along the x-axis. The second number in each pair, called the ***y*-coordinate**, denotes vertical distance and direction along a line parallel to the y-axis or along the y-axis itself.

Figure 7.2 shows how we **plot**, or locate, the points corresponding to the ordered pairs $(-5, 3)$ and $(3, -5)$. We plot $(-5, 3)$ by going 5 units from 0 to the left along the x-axis. Then we go 3 units up parallel to the y-axis. We plot $(3, -5)$ by going 3 units from 0 to the right along the x-axis and 5 units down parallel to the y-axis. The phrase "the points corresponding to the ordered pairs $(-5, 3)$ and $(3, -5)$" is often abbreviated as "the points $(-5, 3)$ and $(3, -5)$."

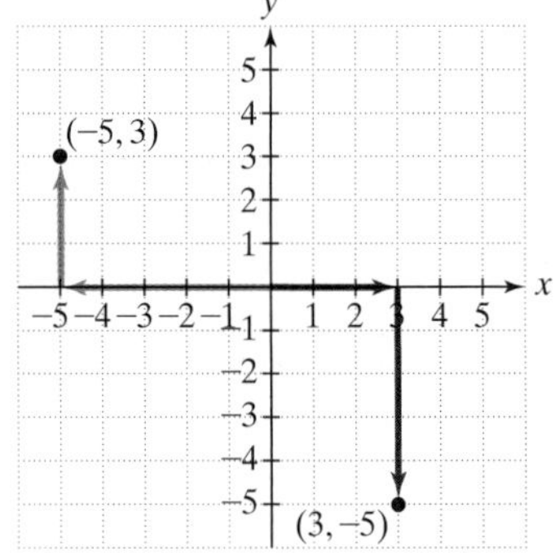

FIGURE 7.2 Plotting $(-5, 3)$ and $(3, -5)$.

GREAT QUESTION!

What's the significance of the word *ordered* when describing a pair of real numbers?

The phrase *ordered pair* is used because order is important. The order in which coordinates appear makes a difference in a point's location. This is illustrated in **Figure 7.2**.

EXAMPLE 1 ***Plotting Points in the Rectangular Coordinate System***

Plot the points: $A(-3, 5), B(2, -4), C(5, 0), D(-5, -3), E(0, 4)$, and $F(0, 0)$.

SOLUTION

See **Figure 7.3**. We move from the origin and plot the points in the following way:

$A(-3, 5)$: 3 units left, 5 units up

$B(2, -4)$: 2 units right, 4 units down

$C(5, 0)$: 5 units right, 0 units up or down

$D(-5, -3)$: 5 units left, 3 units down

$E(0, 4)$: 0 units right or left, 4 units up

$F(0, 0)$: 0 units right or left, 0 units up or down

Notice that the origin is represented by (0, 0).

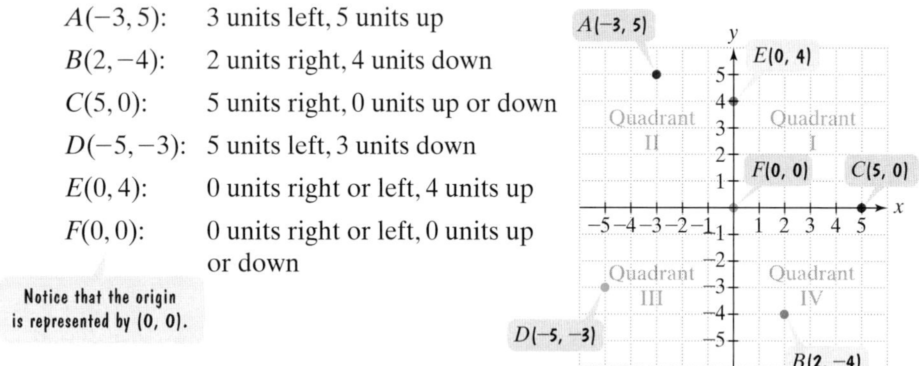

FIGURE 7.3 Plotting points

CHECK POINT 1 Plot the points: $A(-2, 4), B(4, -2), C(-3, 0)$, and $D(0, -3)$.

2 *Graph equations in the rectangular coordinate system.*

Graphs of Equations

A relationship between two quantities can sometimes be expressed as an **equation in two variables**, such as

$$y = 4 - x^2.$$

A **solution of an equation in two variables**, x and y, is an ordered pair of real numbers with the following property: When the x-coordinate is substituted for x and the y-coordinate is substituted for y in the equation, we obtain a true statement. For example, consider the equation $y = 4 - x^2$ and the ordered pair $(3, -5)$. When 3 is substituted for x and -5 is substituted for y, we obtain the statement $-5 = 4 - 3^2$, or $-5 = 4 - 9$, or $-5 = -5$. Because this statement is true, the ordered pair $(3, -5)$ is a solution of the equation $y = 4 - x^2$. We also say that $(3, -5)$ **satisfies** the equation.

We can generate as many ordered-pair solutions as desired to $y = 4 - x^2$ by substituting numbers for x and then finding the corresponding values for y. For example, suppose we let $x = 3$:

Start with x. Compute y. Form the ordered pair (x, y).

x	$y = 4 - x^2$	**Ordered Pair (x, y)**
3	$y = 4 - 3^2 = 4 - 9 = -5$	$(3, -5)$

Let $x = 3$. $(3, -5)$ is a solution of $y = 4 - x^2$.

The **graph of an equation in two variables** is the set of all points whose coordinates satisfy the equation. One method for graphing such equations is the **point-plotting method**. First, we find several ordered pairs that are solutions of the equation. Next, we plot these ordered pairs as points in the rectangular coordinate system. Finally, we connect the points with a smooth curve or line. This often gives us a picture of all ordered pairs that satisfy the equation.

EXAMPLE 2 Graphing an Equation Using the Point-Plotting Method

Graph $y = 4 - x^2$. Select integers for x, starting with -3 and ending with 3.

SOLUTION

For each value of x, we find the corresponding value for y.

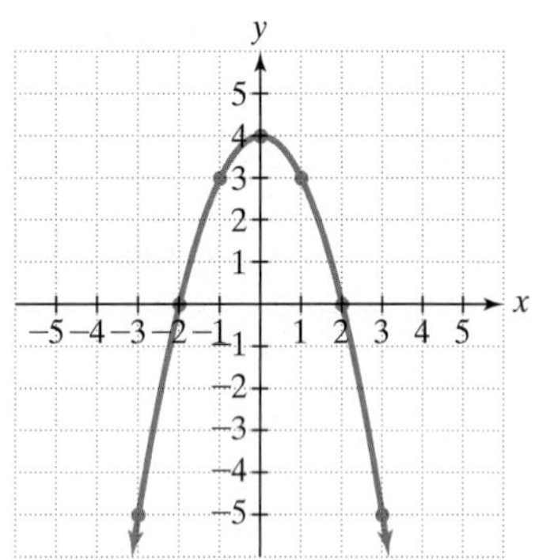

FIGURE 7.4 The graph of $y = 4 - x^2$

Start with x. Compute y. Form the ordered pair (x, y).

We selected integers from -3 to 3, inclusive, to include three negative numbers, 0, and three positive numbers. We also wanted to keep the resulting computations for y relatively simple.

x	$y = 4 - x^2$	Ordered Pair (x, y)
-3	$y = 4 - (-3)^2 = 4 - 9 = -5$	$(-3, -5)$
-2	$y = 4 - (-2)^2 = 4 - 4 = 0$	$(-2, 0)$
-1	$y = 4 - (-1)^2 = 4 - 1 = 3$	$(-1, 3)$
0	$y = 4 - 0^2 = 4 - 0 = 4$	$(0, 4)$
1	$y = 4 - 1^2 = 4 - 1 = 3$	$(1, 3)$
2	$y = 4 - 2^2 = 4 - 4 = 0$	$(2, 0)$
3	$y = 4 - 3^2 = 4 - 9 = -5$	$(3, -5)$

Now we plot the seven points and join them with a smooth curve, as shown in **Figure 7.4**. The graph of $y = 4 - x^2$ is a curve where the part of the graph to the right of the y-axis is a reflection of the part to the left of it and vice versa. The arrows on the left and the right of the curve indicate that it extends indefinitely in both directions.

CHECK POINT 2 Graph $y = 4 - x$. Select integers for x, starting with -3 and ending with 3.

Part of the beauty of the rectangular coordinate system is that it allows us to "see" formulas and visualize the solution to a problem. This idea is demonstrated in Example 3.

EXAMPLE 3 An Application Using Graphs of Equations

The toll to a bridge costs \$2.50. Commuters who use the bridge frequently have the option of purchasing a monthly discount pass for \$21.00. With the discount pass, the toll is reduced to \$1.00. The monthly cost, y, of using the bridge x times can be described by the following formulas:

Without the discount pass:

$y = 2.5x$ — The monthly cost, y, is \$2.50 times the number of times, x, that the bridge is used.

With the discount pass:

$y = 21 + 1 \cdot x$

$y = 21 + x.$

The monthly cost, y, is \$21 for the discount pass plus \$1 times the number of times, x, that the bridge is used.

a. Let $x = 0, 2, 4, 10, 12, 14,$ and 16. Make a table of values for each equation showing seven solutions for the equation.

b. Graph the equations in the same rectangular coordinate system.

c. What are the coordinates of the intersection point for the two graphs? Interpret the coordinates in practical terms.

SOLUTION

a. Tables of values showing seven solutions for each equation follow.

WITHOUT THE DISCOUNT PASS

x	$y = 2.5x$	(x, y)
0	$y = 2.5(0) = 0$	$(0, 0)$
2	$y = 2.5(2) = 5$	$(2, 5)$
4	$y = 2.5(4) = 10$	$(4, 10)$
10	$y = 2.5(10) = 25$	$(10, 25)$
12	$y = 2.5(12) = 30$	$(12, 30)$
14	$y = 2.5(14) = 35$	$(14, 35)$
16	$y = 2.5(16) = 40$	$(16, 40)$

WITH THE DISCOUNT PASS

x	$y = 21 + x$	(x, y)
0	$y = 21 + 0 = 21$	$(0, 21)$
2	$y = 21 + 2 = 23$	$(2, 23)$
4	$y = 21 + 4 = 25$	$(4, 25)$
10	$y = 21 + 10 = 31$	$(10, 31)$
12	$y = 21 + 12 = 33$	$(12, 33)$
14	$y = 21 + 14 = 35$	$(14, 35)$
16	$y = 21 + 16 = 37$	$(16, 37)$

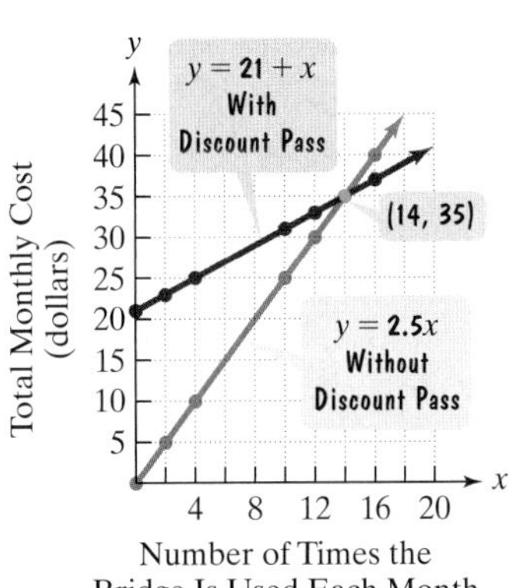

FIGURE 7.5 Options for a toll

b. Now we are ready to graph the two equations. Because the x- and y-coordinates are nonnegative, it is only necessary to use the origin, the positive portions of the x- and y-axes, and the first quadrant of the rectangular coordinate system. The x-coordinates begin at 0 and end at 16. We will let each tick mark on the x-axis represent two units. However, the y-coordinates begin at 0 and get as large as 40 in the formula that describes the monthly cost without the coupon book. So that our y-axis does not get too long, we will let each tick mark on the y-axis represent five units. Using this setup and the two tables of values, we construct the graphs of $y = 2.5x$ and $y = 21 + x$, shown in **Figure 7.5**.

c. The graphs intersect at $(14, 35)$. This means that if the bridge is used 14 times in a month, the total monthly cost without the discount pass is the same as the total monthly cost with the discount pass, namely $35.

In **Figure 7.5**, look at the two graphs to the right of the intersection point $(14, 35)$. The red graph of $y = 21 + x$ lies below the blue graph of $y = 2.5x$. This means that if the bridge is used more than 14 times in a month $(x > 14)$, the (red) monthly cost, y, with the discount pass is less than the (blue) monthly cost, y, without the discount pass.

 **CHECK POINT 3** The toll to a bridge costs $2. If you use the bridge x times in a month, the monthly cost, y, is $y = 2x$. With a $10 discount pass, the toll is reduced to $1. The monthly cost, y, of using the bridge x times in a month with the discount pass is $y = 10 + x$.

a. Let $x = 0, 2, 4, 6, 8, 10,$ and 12. Make tables of values showing seven solutions of $y = 2x$ and seven solutions of $y = 10 + x$.

b. Graph the equations in the same rectangular coordinate system.

c. What are the coordinates of the intersection point for the two graphs? Interpret the coordinates in practical terms.

3 *Use function notation.*

Functions

Reconsider one of the equations from Example 3, $y = 2.5x$. Recall that this equation describes the monthly cost, y, of using the bridge x times, with a toll cost of $2.50 each time the bridge is used. The monthly cost, y, depends on the number of times the bridge is used, x. For each value of x, there is one and only one value of y. If an equation in two variables (x and y) yields precisely one value of y for each value of x, we say that y is a **function** of x.

The notation $y = f(x)$ indicates that the variable y is a function of x. The notation $f(x)$ is read "f of x."

For example, the formula for the cost of the bridge

$$y = 2.5x$$

can be expressed in function notation as

$$f(x) = 2.5x.$$

We read this as "f of x is equal to $2.5x$." If, say, x equals 10 (meaning that the bridge is used 10 times), we can find the corresponding value of y (monthly cost) using the equation $f(x) = 2.5x$.

$$f(x) = 2.5x$$

$$f(10) = 2.5(10)$$ To find $f(10)$, read "f of 10," replace x with 10.

$$= 25$$

Because $f(10) = 25$ (f of 10 equals 25), this means that if the bridge is used 10 times in a month, the total monthly cost is \$25.

Table 7.1 compares our previous notation with the new notation of functions.

TABLE 7.1 Function Notation

"y Equals" Notation	"$f(x)$ Equals" Notation
$y = 2.5x$	$f(x) = 2.5x$
If $x = 10$,	$f(10) = 2.5(10) = 25$
$y = 2.5(10) = 25.$	f of 10 equals 25.

GREAT QUESTION!

Doesn't $f(x)$ indicate that I need to multiply f and x?

The notation $f(x)$ does *not* mean "f times x." The notation describes the "output" for the function f when the "input" is x. Think of $f(x)$ as another name for y.

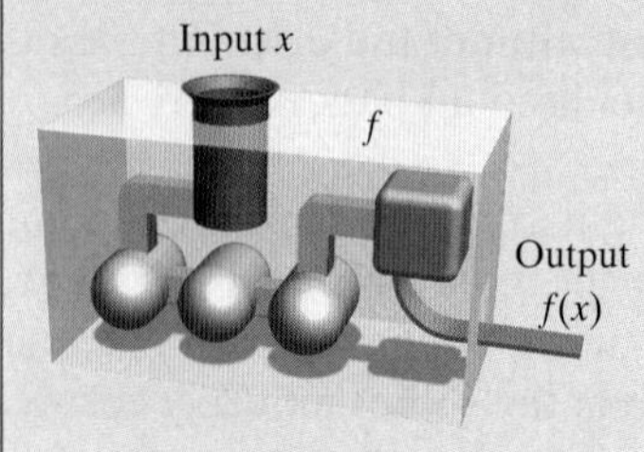

In our next example, we will apply function notation to three different functions. It would be awkward to call all three functions f. We will call the first function f, the second function g, and the third function h. These are the letters most frequently used to name functions.

EXAMPLE 4 *Using Function Notation*

Find each of the following:

a. $f(4)$ for $f(x) = 2x + 3$

b. $g(-2)$ for $g(x) = 2x^2 - 1$

c. $h(-5)$ for $h(r) = r^3 - 2r^2 + 5$.

SOLUTION

a.

$f(x) = 2x + 3$ — This is the given function.

$f(4) = 2 \cdot 4 + 3$ — To find f of 4, replace x with 4.

$= 8 + 3$ — Multiply: $2 \cdot 4 = 8$.

$f(4) = 11$ (f of 4 is 11.) — Add.

b.

$g(x) = 2x^2 - 1$ — This is the given function.

$g(-2) = 2(-2)^2 - 1$ — To find g of -2, replace x with -2.

$= 2(4) - 1$ — Evaluate the exponential expression: $(-2)^2 = 4$.

$= 8 - 1$ — Multiply: $2(4) = 8$.

$g(-2) = 7$ (g of -2 is 7.) — Subtract.

c.

$h(r) = r^3 - 2r^2 + 5$ — The function's name is h and r represents the function's input.

$h(-5) = (-5)^3 - 2(-5)^2 + 5$ — To find h of -5, replace each occurrence of r with -5.

$= -125 - 2(25) + 5$ — Evaluate exponential expressions: $(-5)^3 = -125$ and $(-5)^2 = 25$.

$= -125 - 50 + 5$ — Multiply: $2(25) = 50$.

$h(-5) = -170$ (h of -5 is -170.) — $-125 - 50 = -175$ and $-175 + 5 = -170$

CHECK POINT 4 Find each of the following:

a. $f(6)$ for $f(x) = 4x + 5$

b. $g(-5)$ for $g(x) = 3x^2 - 10$

c. $h(-4)$ for $h(r) = r^2 - 7r + 2$.

EXAMPLE 5 An Application Involving Function Notation

Tailgaters beware: If your car is going 35 miles per hour on dry pavement, your required stopping distance is 160 feet, or the width of a football field. At 65 miles per hour, the distance required is 410 feet, or approximately the length of one and one-tenth football fields. **Figure 7.6** shows stopping distances for cars at various speeds on dry roads and on wet roads. **Figure 7.7** uses a line graph to represent stopping distances at various speeds on dry roads.

© Warren Miller/The New Yorker Collection/The Cartoon Bank

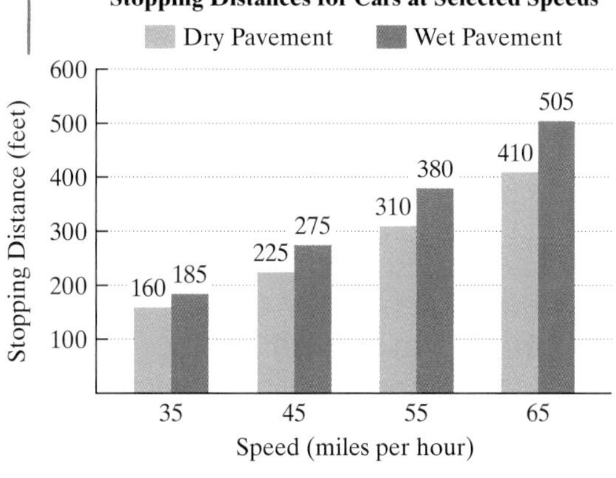

FIGURE 7.6
Source: National Highway Traffic Safety Administration

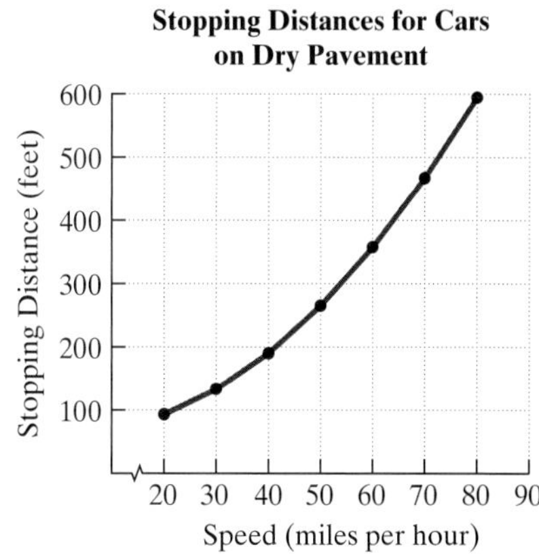

FIGURE 7.7

a. Use the line graph in **Figure 7.7** to estimate a car's required stopping distance at 60 miles per hour on dry pavement. Round to the nearest 10 feet.

b. The function

$$f(x) = 0.0875x^2 - 0.4x + 66.6$$

models a car's required stopping distance, $f(x)$, in feet, on dry pavement traveling at x miles per hour. Use this function to find the required stopping distance at 60 miles per hour. Round to the nearest foot.

SOLUTION

a. The required stopping distance at 60 miles per hour is estimated using the point shown in **Figure 7.8**. The second coordinate of this point extends slightly more than midway between 300 and 400 on the vertical axis. Thus, 360 is a reasonable estimate. We conclude that at 60 miles per hour on dry pavement, the required stopping distance is approximately 360 feet.

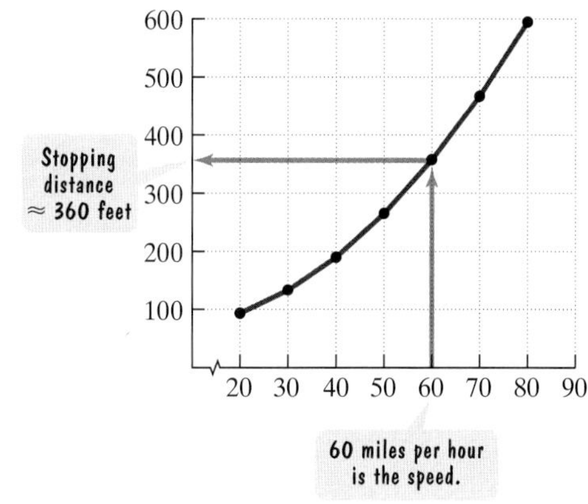

FIGURE 7.8

TECHNOLOGY

On most calculators, here is how to find

$0.0875(60)^2 - 0.4(60) + 66.6.$

Many Scientific Calculators

.0875 [×] 60 [x^2] [−]

.4 [×] 60 [+] 66.6 [=]

Many Graphing Calculators

.0875 [×] 60 [∧] 2 [−]

.4 [×] 60 [+] 66.6 [ENTER]

Some calculators require that you press the right arrow key after entering the exponent 2.

b. Now we use the given function to determine the required stopping distance at 60 miles per hour. We need to find $f(60)$. The arithmetic gets somewhat "messy," so it is probably a good idea to use a calculator.

$f(x) = 0.0875x^2 - 0.4x + 66.6$ — This function models stopping distance, $f(x)$, at x miles per hour.

$f(60) = 0.0875(60)^2 - 0.4(60) + 66.6$ — Replace each x with 60.

$= 0.0875(3600) - 0.4(60) + 66.6$ — Use the order of operations, first evaluating the exponential expression.

$= 315 - 24 + 66.6$ — Perform the multiplications.

$= 357.6$ — Subtract and add as indicated.

≈ 358 — As directed, we've rounded to the nearest foot.

We see that $f(60) \approx 358$—that is, f of 60 is approximately 358. The model indicates that the required stopping distance on dry pavement at 60 miles per hour is approximately 358 feet.

CHECK POINT 5 **a.** Use the line graph in **Figure 7.7** on the previous page to estimate a car's required stopping distance at 40 miles per hour on dry pavement. Round to the nearest ten feet.

b. Use the function in Example 5(b), $f(x) = 0.0875x^2 - 0.4x + 66.6$, to find the required stopping distance at 40 miles per hour. Round to the nearest foot.

4 *Graph functions.*

Graphing Functions

The **graph of a function** is the graph of its ordered pairs. In our next example, we will graph two functions.

EXAMPLE 6 ***Graphing Functions***

Graph the functions $f(x) = 2x$ and $g(x) = 2x + 4$ in the same rectangular coordinate system. Select integers for x from −2 to 2, inclusive.

SOLUTION

For each function, we use the suggested values for x to create a table of some of the coordinates. These tables are shown below. Then, we plot the five points in each table and connect them, as shown in **Figure 7.9**. The graph of each function is a straight line. Do you see a relationship between the two graphs? The graph of g is the graph of f shifted vertically up 4 units.

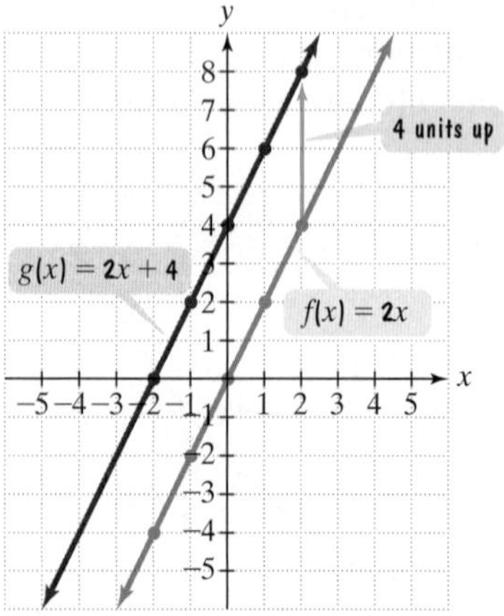

x	$f(x) = 2x$	(x, y) or $(x, f(x))$
−2	$f(-2) = 2(-2) = -4$	$(-2, -4)$
−1	$f(-1) = 2(-1) = -2$	$(-1, -2)$
0	$f(0) = 2 \cdot 0 = 0$	$(0, 0)$
1	$f(1) = 2 \cdot 1 = 2$	$(1, 2)$
2	$f(2) = 2 \cdot 2 = 4$	$(2, 4)$

Choose x. Compute $f(x)$ by evaluating f at x. Form the ordered pair.

x	$g(x) = 2x + 4$	(x, y) or $(x, g(x))$
−2	$g(-2) = 2(-2) + 4 = 0$	$(-2, 0)$
−1	$g(-1) = 2(-1) + 4 = 2$	$(-1, 2)$
0	$g(0) = 2 \cdot 0 + 4 = 4$	$(0, 4)$
1	$g(1) = 2 \cdot 1 + 4 = 6$	$(1, 6)$
2	$g(2) = 2 \cdot 2 + 4 = 8$	$(2, 8)$

Choose x. Compute $g(x)$ by evaluating g at x. Form the ordered pair.

4 units up

$g(x) = 2x + 4$

$f(x) = 2x$

FIGURE 7.9

CHECK POINT 6 Graph the functions $f(x) = 2x$ and $g(x) = 2x - 3$ in the same rectangular coordinate system. Select integers for x from −2 to 2, inclusive. How is the graph of g related to the graph of f?

5 *Use the vertical line test.*

TECHNOLOGY

A graphing calculator is a powerful tool that quickly generates the graph of an equation in two variables. Here is the graph of $y = 4 - x^2$ that we drew by hand in **Figure 7.4** on page 414.

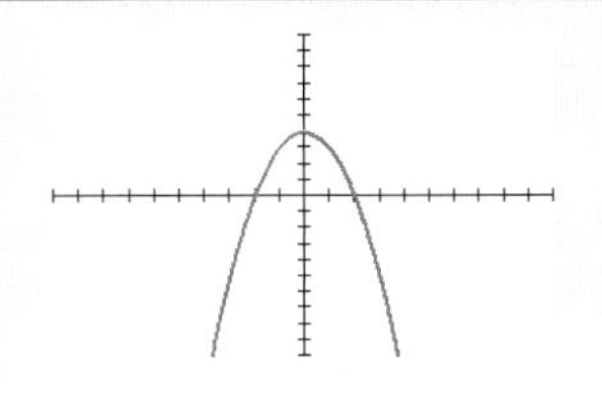

What differences do you notice between this graph and the graph we drew by hand? This graph seems a bit "jittery." Arrows do not appear on the left and right ends of the graph. Furthermore, numbers are not given along the axes. For the graph shown above, the x-axis extends from -10 to 10 and the y-axis also extends from -10 to 10. The distance represented by each consecutive tick mark is one unit. We say that the **viewing window** is $[-10, 10, 1]$ by $[-10, 10, 1]$.

To graph an equation in x and y using a graphing calculator, enter the equation, which must be solved for y, and specify the size of the viewing window. The size of the viewing window sets minimum and maximum values for both the x- and y-axes. Enter these values, as well as the values between consecutive tick marks, on the respective axes. The $[-10, 10, 1]$ by $[-10, 10, 1]$ viewing window used above is called the **standard viewing window**.

The Vertical Line Test

Not every graph in the rectangular coordinate system is the graph of a function. The definition of a function specifies that no value of x can be paired with two or more different values of y. Consequently, if a graph contains two or more different points with the same first coordinate, the graph cannot represent a function. This is illustrated in **Figure 7.10**. Observe that points sharing a common first coordinate are vertically above or below each other.

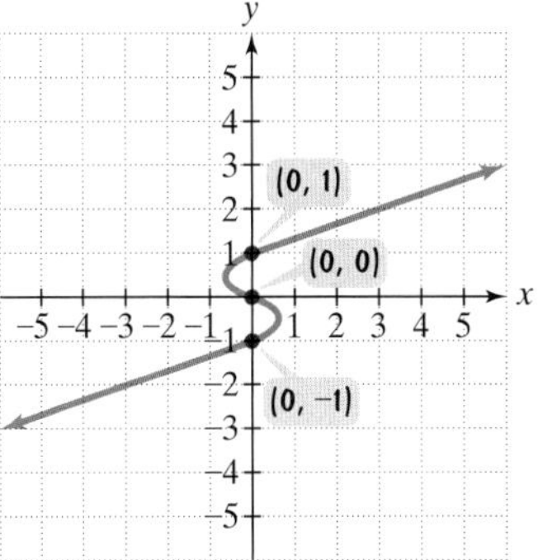

FIGURE 7.10 y is not a function of x because 0 is paired with three values of y, namely, 1, 0, and -1.

This observation is the basis of a useful test for determining whether a graph defines y as a function of x. The test is called the **vertical line test**.

THE VERTICAL LINE TEST FOR FUNCTIONS

If any vertical line intersects a graph in more than one point, the graph does not define y as a function of x.

EXAMPLE 7 *Using the Vertical Line Test*

Use the vertical line test to identify graphs in which y is a function of x.

a. **b.** 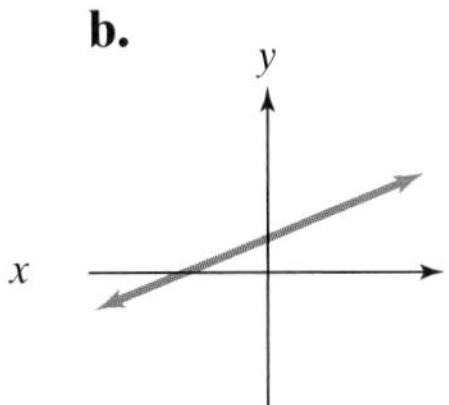 **c.** 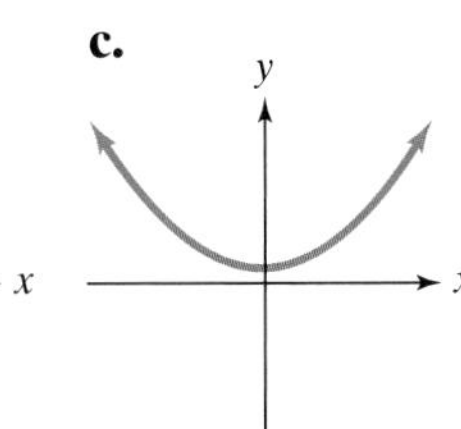 **d.** 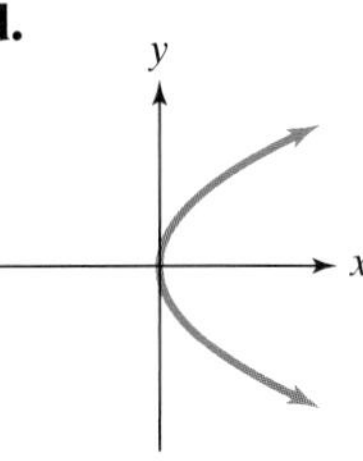

SOLUTION

y is a function of x for the graphs in (b) and (c).

a. 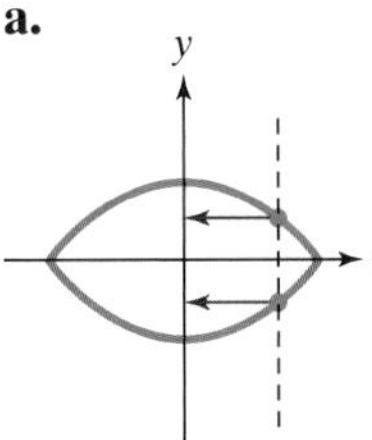

y **is not a function** of x. Two values of y correspond to one x-value.

b. y **is a function** of x.

c. 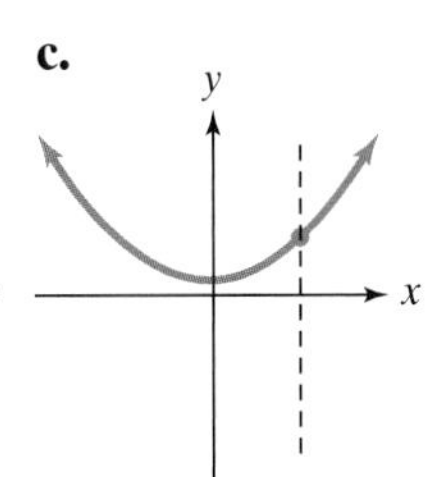

y **is a function** of x.

d. y **is not a function** of x. Two values of y correspond to one x-value.

CHECK POINT 7 Use the vertical line test to identify graphs in which y is a function of x.

a. 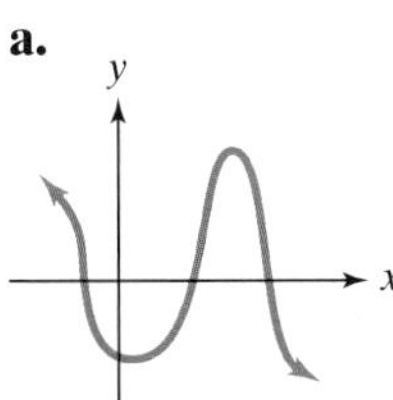 **b.** 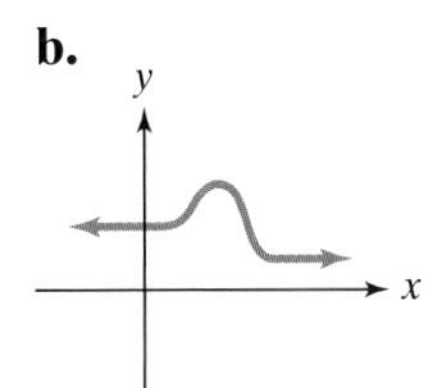 **c.** 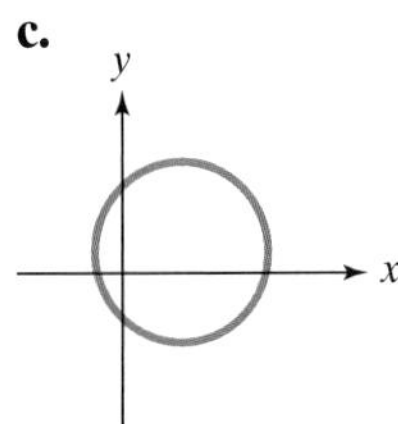

6 *Obtain information about a function from its graph.*

Obtaining Information from Graphs

Example 8 illustrates how to obtain information about a function from its graph.

FIGURE 7.11 Body temperature from 8 A.M. through 3 P.M.

EXAMPLE 8 *Analyzing the Graph of a Function*

Too late for that flu shot now! It's only 8 A.M. and you're feeling lousy. Fascinated by the way that algebra models the world (your author is projecting a bit here), you construct a graph showing your body temperature from 8 A.M. through 3 P.M. You decide to let x represent the number of hours after 8 A.M. and y represent your body temperature at time x. The graph is shown in **Figure 7.11**. The symbol on the y-axis shows that there is a break in values between 0 and 98. Thus, the first tick mark on the y-axis represents a temperature of 98°F.

a. What is your temperature at 8 A.M.?

b. During which period of time is your temperature decreasing?

c. Estimate your minimum temperature during the time period shown. How many hours after 8 A.M. does this occur? At what time does this occur?

d. During which period of time is your temperature increasing?

e. Part of the graph is shown as a horizontal line segment. What does this mean about your temperature and when does this occur?

f. Explain why the graph defines y as a function of x.

SOLUTION

a. Because x is the number of hours after 8 A.M., your temperature at 8 A.M. corresponds to $x = 0$. Locate 0 on the horizontal axis and look at the point on the graph above 0. **Figure 7.12** shows that your temperature at 8 A.M. is 101°F.

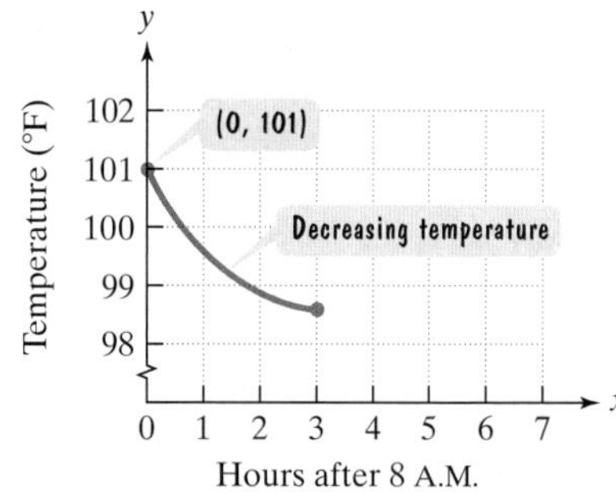

FIGURE 7.12

b. Your temperature is decreasing when the graph falls from left to right. This occurs between $x = 0$ and $x = 3$, also shown in **Figure 7.12**. Because x represents the number of hours after 8 A.M., your temperature is decreasing between 8 A.M. and 11 A.M.

c. Your minimum temperature can be found by locating the lowest point on the graph. This point lies above 3 on the horizontal axis, shown in **Figure 7.13**. The y-coordinate of this point falls more than midway between 98 and 99, at approximately 98.6. The lowest point on the graph, (3, 98.6), shows that your minimum temperature, 98.6°F, occurs 3 hours after 8 A.M., at 11 A.M.

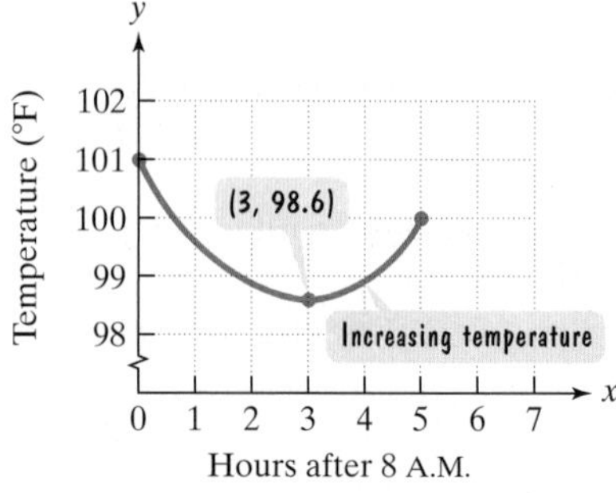

FIGURE 7.13

d. Your temperature is increasing when the graph rises from left to right. This occurs between $x = 3$ and $x = 5$, shown in **Figure 7.13**. Because x represents the number of hours after 8 A.M., your temperature is increasing between 11 A.M. and 1 P.M.

e. The horizontal line segment shown in **Figure 7.14** indicates that your temperature is neither increasing nor decreasing. Your temperature remains the same, 100°F, between $x = 5$ and $x = 7$. Thus, your temperature is at a constant 100°F between 1 P.M. and 3 P.M.

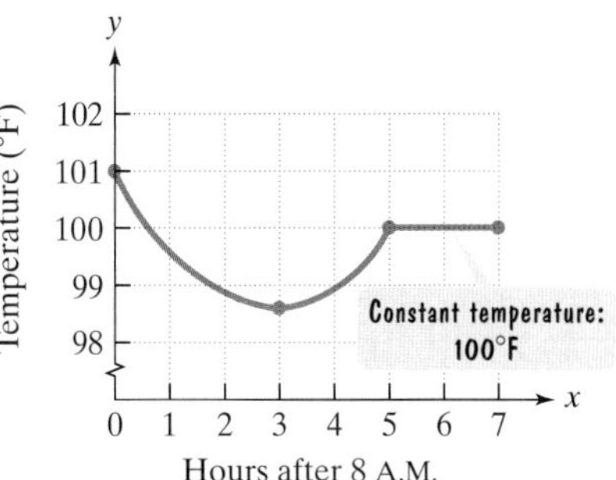

FIGURE 7.14

f. The complete graph of your body temperature from 8 A.M. through 3 P.M. is shown in **Figure 7.14**. No vertical line can be drawn that intersects this blue graph more than once. By the vertical line test, the graph defines y as a function of x. In practical terms, this means that your body temperature is a function of time. Each hour (or fraction of an hour) after 8 A.M., represented by x, yields precisely one body temperature, represented by y.

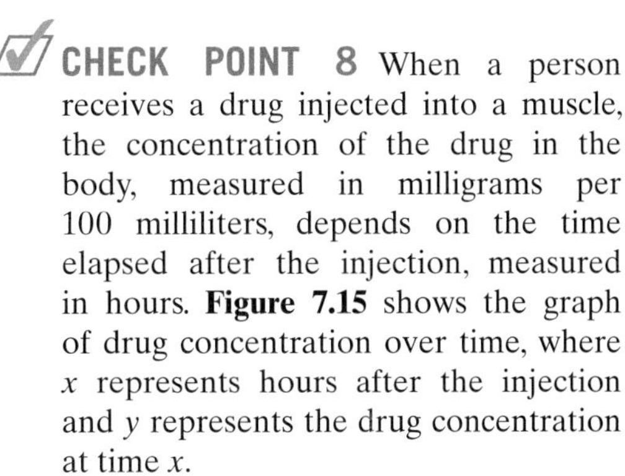

CHECK POINT 8 When a person receives a drug injected into a muscle, the concentration of the drug in the body, measured in milligrams per 100 milliliters, depends on the time elapsed after the injection, measured in hours. **Figure 7.15** shows the graph of drug concentration over time, where x represents hours after the injection and y represents the drug concentration at time x.

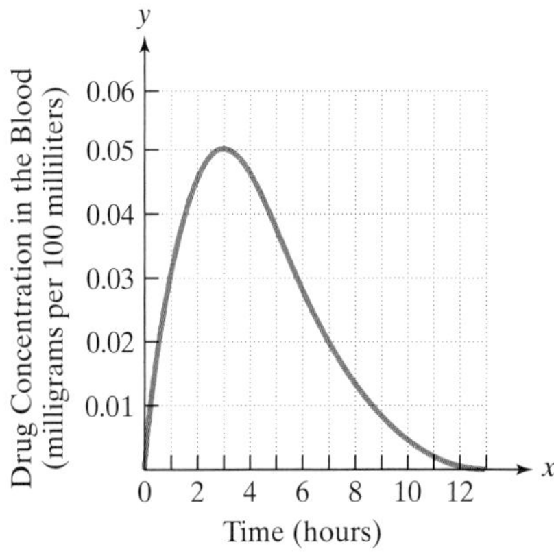

FIGURE 7.15

a. During which period of time is the drug concentration increasing?

b. During which period of time is the drug concentration decreasing?

c. What is the drug's maximum concentration and when does this occur?

d. What happens by the end of 13 hours?

e. Explain why the graph defines y as a function of x.

Concept and Vocabulary Check

Fill in each blank so that the resulting statement is true.

1. In the rectangular coordinate system, the horizontal number line is called the _______.
2. In the rectangular coordinate system, the vertical number line is called the _______.
3. In the rectangular coordinate system, the point of intersection of the horizontal axis and the vertical axis is called the_______.
4. The axes of the rectangular coordinate system divide the plane into regions, called __________. There are ______ of these regions.
5. The first number in an ordered pair such as (3, 8) is called the ___________. The second number in such an ordered pair is called the ___________.
6. The ordered pair (1, 3) is a/an ________ of the equation $y = 5x - 2$ because when 1 is substituted for x and 3 is substituted for y, we obtain a true statement. We also say that (1, 3) ________ the equation.
7. If an equation in two variables (x and y) yields precisely one value of ____ for each value of ____, we say that y is a/an _________ of x.
8. If $f(x) = 3x + 5$, we can find $f(6)$ by replacing ___ with ___.
9. If any vertical line intersects a graph ____________, the graph does not define y as a/an ________ of x.

Exercise Set 7.1

Practice Exercises

In Exercises 1–20, plot the given point in a rectangular coordinate system.

1. $(1, 4)$ **2.** $(2, 5)$
3. $(-2, 3)$ **4.** $(-1, 4)$
5. $(-3, -5)$ **6.** $(-4, -2)$
7. $(4, -1)$ **8.** $(3, -2)$
9. $(-4, 0)$ **10.** $(-5, 0)$
11. $(0, -3)$ **12.** $(0, -4)$
13. $(0, 0)$ **14.** $\left(-3, -1\frac{1}{2}\right)$
15. $\left(-2, -3\frac{1}{2}\right)$ **16.** $(-5, -2.5)$
17. $(3.5, 4.5)$ **18.** $(2.5, 3.5)$
19. $(1.25, -3.25)$ **20.** $(2.25, -4.25)$

Graph each equation in Exercises 21–32. Select integers for x from −3 to 3, inclusive.

21. $y = x^2 - 2$ **22.** $y = x^2 + 2$
23. $y = x - 2$ **24.** $y = x + 2$
25. $y = 2x + 1$ **26.** $y = 2x - 4$
27. $y = -\frac{1}{2}x$ **28.** $y = -\frac{1}{2}x + 2$
29. $y = x^3$ **30.** $y = x^3 - 1$
31. $y = |x| + 1$ **32.** $y = |x| - 1$

In Exercises 33–46, evaluate each function at the given value of the variable.

33. $f(x) = x - 4$ **a.** $f(8)$ **b.** $f(1)$
34. $f(x) = x - 6$ **a.** $f(9)$ **b.** $f(2)$
35. $f(x) = 3x - 2$ **a.** $f(7)$ **b.** $f(0)$
36. $f(x) = 4x - 3$ **a.** $f(7)$ **b.** $f(0)$
37. $g(x) = x^2 + 1$ **a.** $g(2)$ **b.** $g(-2)$
38. $g(x) = x^2 + 4$ **a.** $g(3)$ **b.** $g(-3)$
39. $g(x) = -x^2 + 2$ **a.** $g(4)$ **b.** $g(-3)$
40. $g(x) = -x^2 + 1$ **a.** $g(5)$ **b.** $g(-4)$
41. $h(r) = 3r^2 + 5$ **a.** $h(4)$ **b.** $h(-1)$
42. $h(r) = 2r^2 - 4$ **a.** $h(5)$ **b.** $h(-1)$
43. $f(x) = 2x^2 + 3x - 1$ **a.** $f(3)$ **b.** $f(-4)$
44. $f(x) = 3x^2 + 4x - 2$ **a.** $f(2)$ **b.** $f(-1)$
45. $f(x) = \dfrac{x}{|x|}$ **a.** $f(6)$ **b.** $f(-6)$
46. $f(x) = \dfrac{|x|}{x}$ **a.** $f(5)$ **b.** $f(-5)$

In Exercises 47–54, evaluate $f(x)$ for the given values of x. Then use the ordered pairs $(x, f(x))$ from your table to graph the function.

47. $f(x) = x^2 - 1$

x	$f(x) = x^2 - 1$
−2	
−1	
0	
1	
2	

48. $f(x) = x^2 + 1$

x	$f(x) = x^2 + 1$
−2	
−1	
0	
1	
2	

49. $f(x) = x - 1$

x	$f(x) = x - 1$
−2	
−1	
0	
1	
2	

50. $f(x) = x + 1$

x	$f(x) = x + 1$
−2	
−1	
0	
1	
2	

51. $f(x) = (x - 2)^2$

x	$f(x) = (x - 2)^2$
0	
1	
2	
3	
4	

52. $f(x) = (x + 1)^2$

x	$f(x) = (x + 1)^2$
−3	
−2	
−1	
0	
1	

53. $f(x) = x^3 + 1$

x	$f(x) = x^3 + 1$
−3	
−2	
−1	
0	
1	

54. $f(x) = (x + 1)^3$

x	$f(x) = (x + 1)^3$
−3	
−2	
−1	
0	
1	

For Exercises 55–62, use the vertical line test to identify graphs in which y is a function of x.

55.

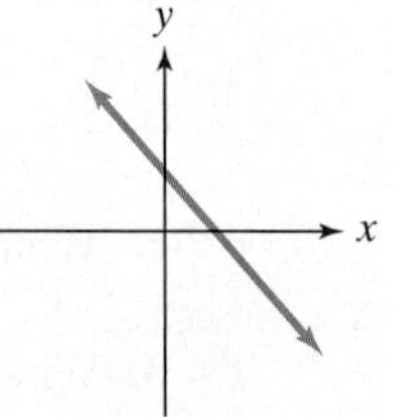

56.

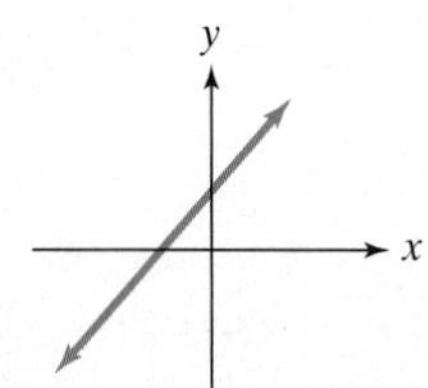

57.

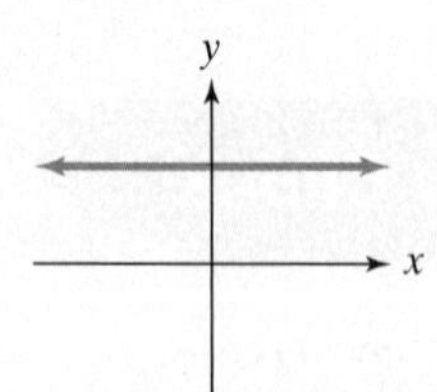

58.

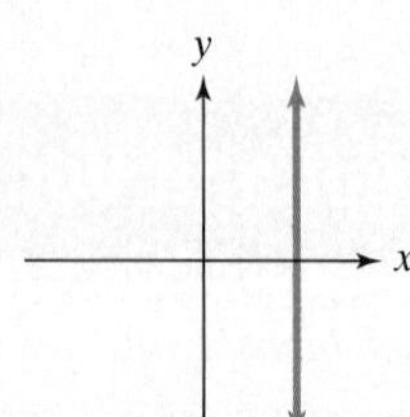

59.

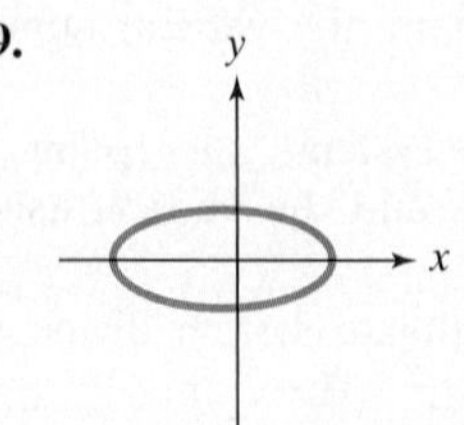

60.

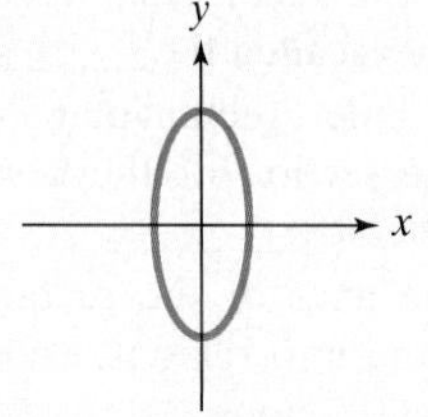

61. **62.**

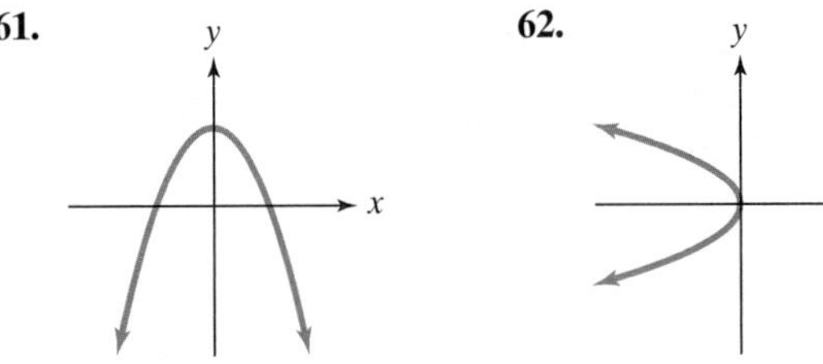

Practice Plus

In Exercises 63–64, let $f(x) = x^2 - x + 4$ *and* $g(x) = 3x - 5$.

63. Find $g(1)$ and $f(g(1))$.

64. Find $g(-1)$ and $f(g(-1))$.

In Exercises 65–66, let f and g be defined by the following table:

x	$f(x)$	$g(x)$
−2	6	0
−1	3	4
0	−1	1
1	−4	−3
2	0	−6

65. Find $\sqrt{f(-1) - f(0)} - [g(2)]^2 + f(-2) \div g(2) \cdot g(-1)$.

66. Find $|f(1) - f(0)| - [g(1)]^2 + g(1) \div f(-1) \cdot g(2)$.

In Exercises 67–70, write each English sentence as an equation in two variables. Then graph the equation.

67. The y-value is four more than twice the x-value.

68. The y-value is the difference between four and twice the x-value.

69. The y-value is three decreased by the square of the x-value.

70. The y-value is two more than the square of the x-value.

Application Exercises

A football is thrown by a quarterback to a receiver. The points in the figure show the height of the football, in feet, above the ground in terms of its distance, in yards, from the quarterback. Use this information to solve Exercises 71–76.

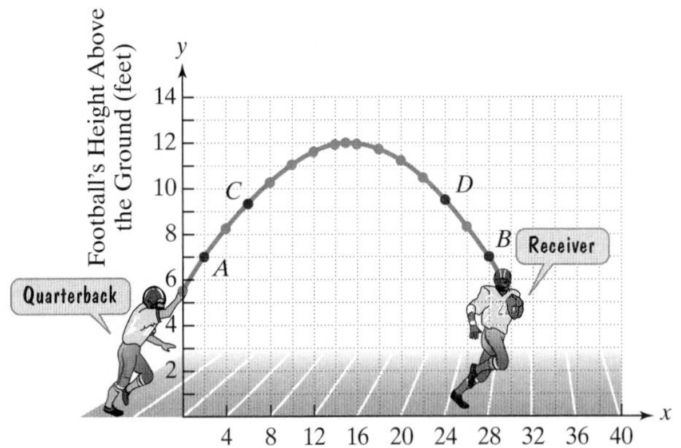

71. Find the coordinates of point A. Then interpret the coordinates in terms of the information given.

72. Find the coordinates of point B. Then interpret the coordinates in terms of the information given.

73. Estimate the coordinates of point C.

74. Estimate the coordinates of point D.

75. What is the football's maximum height? What is its distance from the quarterback when it reaches its maximum height?

76. What is the football's height when it is caught by the receiver? What is the receiver's distance from the quarterback when he catches the football?

The wage gap is used to compare the status of women's earnings relative to men's. The wage gap is expressed as a percent and is calculated by dividing the median, or middlemost, annual earnings for women by the median annual earnings for men. The bar graph shows the wage gap for selected years from 1980 through 2010.

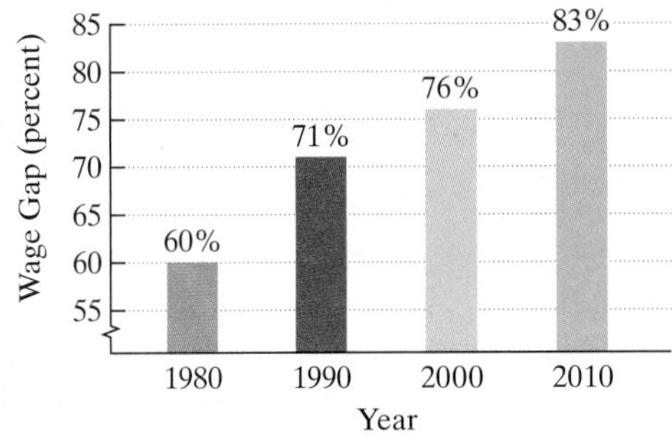

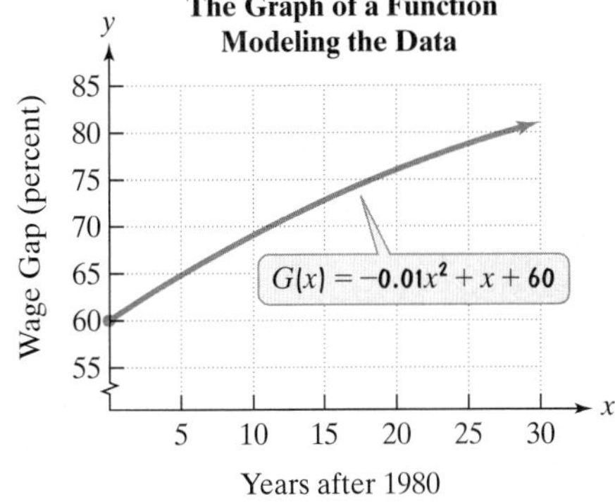

Source: U.S. Bureau of Labor Statistics

The function $G(x) = -0.01x^2 + x + 60$ *models the wage gap, as a percent, x years after 1980. The graph of function G is shown to the right of the actual data. Use this information to solve Exercises 77–78.*

77. a. Find and interpret $G(30)$. Identify this information as a point on the graph of the function.

b. Does $G(30)$ overestimate or underestimate the actual data shown by the bar graph? By how much?

78. a. Find and interpret $G(10)$. Identify this information as a point on the graph of the function.

b. Does $G(10)$ overestimate or underestimate the actual data shown by the bar graph? By how much?

The function $f(x) = 0.4x^2 - 36x + 1000$ models the number of accidents, $f(x)$, per 50 million miles driven as a function of a driver's age, x, in years, where x includes drivers from ages 16 through 74, inclusive. The graph of f is shown. Use the equation for f to solve Exercises 79–82.

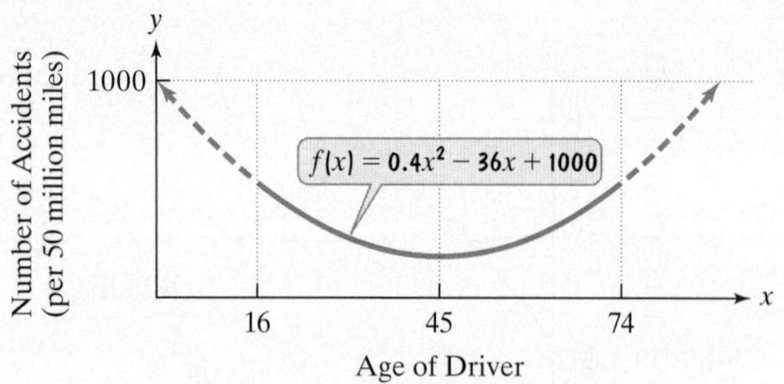

79. Find and interpret $f(20)$. Identify this information as a point on the graph of f.

80. Find and interpret $f(50)$. Identify this information as a point on the graph of f.

81. For what value of x does the graph reach its lowest point? Use the equation for f to find the minimum value of y. Describe the practical significance of this minimum value.

82. Use the graph to identify two different ages for which drivers have the same number of accidents. Use the equation for f to find the number of accidents for drivers at each of these ages.

Explaining the Concepts

83. What is the rectangular coordinate system?

84. Explain how to plot a point in the rectangular coordinate system. Give an example with your explanation.

85. Explain why $(5, -2)$ and $(-2, 5)$ do not represent the same ordered pair.

86. Explain how to graph an equation in the rectangular coordinate system.

87. What is a function?

88. Explain how the vertical line test is used to determine whether a graph represents a function.

Critical Thinking Exercises

Make Sense? *In Exercises 89–92, determine whether each statement makes sense or does not make sense, and explain your reasoning.*

89. My body temperature is a function of the time of day.

90. Using $f(x) = 3x + 2$, I found $f(50)$ by applying the distributive property to $(3x + 2)50$.

91. I knew how to use point plotting to graph the equation $y = x^2 - 1$, so there was really nothing new to learn when I used the same technique to graph the function $f(x) = x^2 - 1$.

92. The graph of my function revealed aspects of its behavior that were not obvious by just looking at its equation.

In Exercises 93–96, use the graphs of f and g to find each number.

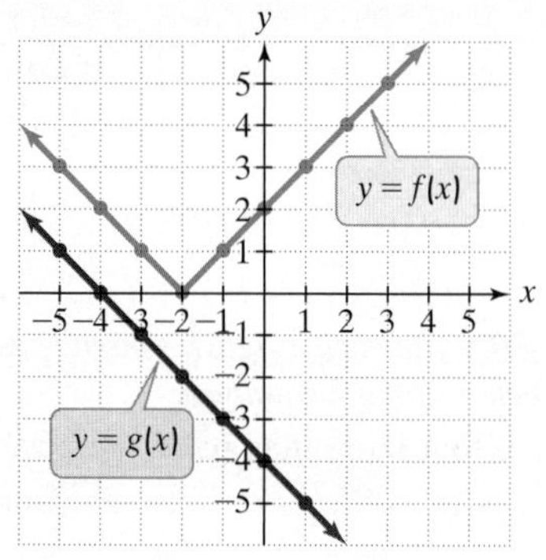

93. $f(-1) + g(-1)$

94. $f(1) + g(1)$

95. $f(g(-1))$

96. $f(g(1))$

Technology Exercise

97. Use a graphing calculator to verify the graphs that you drew by hand in Exercises 47–54.

7.2 Linear Functions and Their Graphs

WHAT AM I SUPPOSED TO LEARN?

After studying this section, you should be able to:

1. Use intercepts to graph a linear equation.
2. Calculate slope.
3. Use the slope and y-intercept to graph a line.
4. Graph horizontal or vertical lines.
5. Interpret slope as rate of change.
6. Use slope and y-intercept to model data.

IT'S HARD TO BELIEVE THAT THIS gas-guzzler, with its huge fins and overstated design, was available in 1957 for approximately \$1800. Sadly, its elegance quickly faded, depreciating by \$300 per year, often sold for scrap just six years after its glorious emergence from the dealer's showroom.

From these casual observations, we can obtain a mathematical model and its graph. The model is

$$y = -300x + 1800.$$

The car is depreciating by \$300 per year for x years.

The new car is worth \$1800.

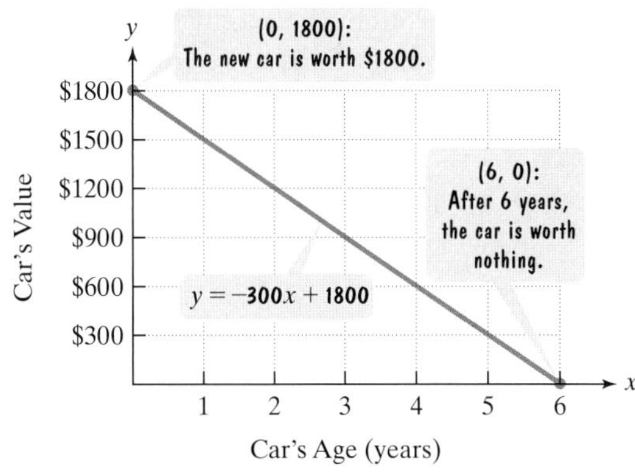

FIGURE 7.16

In this model, y is the car's value after x years. **Figure 7.16** shows the equation's graph. Using function notation, we can rewrite the equation as

$$f(x) = -300x + 1800.$$

A function such as this, whose graph is a straight line, is called a **linear function**. In this section, we will study linear functions and their graphs.

Graphing Using Intercepts

There is another way that we can write the equation

$$y = -300x + 1800.$$

We will collect the x- and y-terms on the left side. This is done by adding $300x$ to both sides:

$$300x + y = 1800.$$

1 *Use intercepts to graph a linear equation.*

All equations of the form $Ax + By = C$ are straight lines when graphed, as long as A and B are not both zero. Such equations are called **linear equations in two variables**. We can quickly obtain the graph for equations in this form when none of A, B, or C is zero by finding the points where the graph intersects the x-axis and the y-axis. The x-coordinate of the point where the graph intersects the x-axis is called the ***x*-intercept**. The y-coordinate of the point where the graph intersects the y-axis is called the ***y*-intercept**.

The graph of $300x + y = 1800$ in **Figure 7.16** intersects the x-axis at $(6, 0)$, so the x-intercept is 6. The graph intersects the y-axis at $(0, 1800)$, so the y-intercept is 1800.

LOCATING INTERCEPTS

To locate the x-intercept, set $y = 0$ and solve the equation for x.

To locate the y-intercept, set $x = 0$ and solve the equation for y.

An equation of the form $Ax + By = C$ as described above can be graphed by finding the x- and y-intercepts, plotting the intercepts, and drawing a straight line through these points. When graphing using intercepts, it is a good idea to use a third point, a checkpoint, before drawing the line. A checkpoint can be obtained by selecting a value for x, other than 0 or the x-intercept, and finding the corresponding value for y. The checkpoint should lie on the same line as the x- and y-intercepts. If it does not, recheck your work and find the error.

EXAMPLE 1 *Using Intercepts to Graph a Linear Equation*

Graph: $3x + 2y = 6$.

SOLUTION

Note that $3x + 2y = 6$ is of the form $Ax + By = C$.

$$3x + 2y = 6$$

$A = 3$ $B = 2$ $C = 6$

In this case, none of A, B, or C is zero.

Find the x-intercept by letting $y = 0$ and solving for x.	Find the y-intercept by letting $x = 0$ and solving for y.
$3x + 2y = 6$	$3x + 2y = 6$
$3x + 2 \cdot 0 = 6$	$3 \cdot 0 + 2y = 6$
$3x = 6$	$2y = 6$
$x = 2$	$y = 3$

The x-intercept is 2, so the line passes through the point $(2, 0)$. The y-intercept is 3, so the line passes through the point $(0, 3)$.

For our checkpoint, we choose a value for x other than 0 or the x-intercept, 2. We will let $x = 1$ and find the corresponding value for y.

$3x + 2y = 6$	This is the given equation.
$3 \cdot 1 + 2y = 6$	Substitute 1 for x.
$3 + 2y = 6$	Simplify.
$2y = 3$	Subtract 3 from both sides.
$y = \frac{3}{2}$	Divide both sides by 2.

FIGURE 7.17 The graph of $3x + 2y = 6$

The checkpoint is the ordered pair $\left(1, \frac{3}{2}\right)$, or $(1, 1.5)$.

The three points in **Figure 7.17** lie along the same line. Drawing a line through the three points results in the graph of $3x + 2y = 6$. The arrowheads at the ends of the line show that the line continues indefinitely in both directions.

CHECK POINT 1 Graph: $2x + 3y = 6$.

2 *Calculate slope.*

Slope

Mathematicians have developed a useful measure of the steepness of a line, called the *slope* of the line. Slope compares the vertical change (the **rise**) to the horizontal change (the **run**) when moving from one fixed point to another along the line. To calculate the slope of a line, we use a ratio that compares the change in y (the rise) to the change in x (the run).

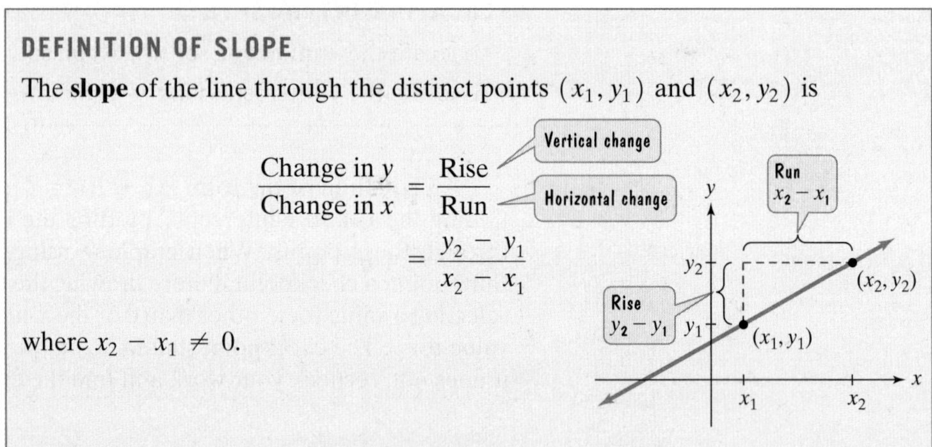

DEFINITION OF SLOPE

The **slope** of the line through the distinct points (x_1, y_1) and (x_2, y_2) is

$$\frac{\text{Change in } y}{\text{Change in } x} = \frac{\text{Rise}}{\text{Run}} = \frac{y_2 - y_1}{x_2 - x_1}$$

where $x_2 - x_1 \neq 0$.

It is common notation to let the letter m represent the slope of a line. The letter m is used because it is the first letter of the French verb *monter*, meaning "to rise," or "to ascend."

EXAMPLE 2 *Using the Definition of Slope*

Find the slope of the line passing through each pair of points:

a. $(-3, -1)$ and $(-2, 4)$ **b.** $(-3, 4)$ and $(2, -2)$.

SOLUTION

a. Let $(x_1, y_1) = (-3, -1)$ and $(x_2, y_2) = (-2, 4)$. We obtain the slope as follows:

$$m = \frac{\text{Change in } y}{\text{Change in } x} = \frac{y_2 - y_1}{x_2 - x_1} = \frac{4 - (-1)}{-2 - (-3)} = \frac{5}{1} = 5.$$

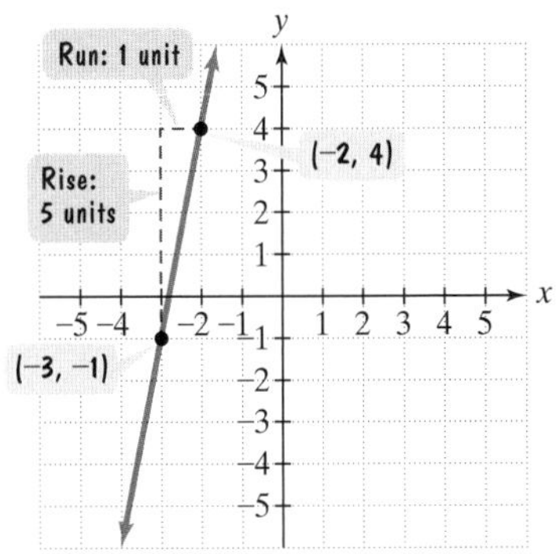

FIGURE 7.18 Visualizing a slope of 5

The situation is illustrated in **Figure 7.18**. The slope of the line is 5, indicating that there is a vertical change, a rise, of 5 units for each horizontal change, a run, of 1 unit. The slope is positive and the line rises from left to right.

GREAT QUESTION!

When using the definition of slope, how do I know which point to call (x_1, y_1) and which point to call (x_2, y_2)?

When computing slope, it makes no difference which point you call (x_1, y_1) and which point you call (x_2, y_2). If we let $(x_1, y_1) = (-2, 4)$ and $(x_2, y_2) = (-3, -1)$, the slope is still 5:

$$m = \frac{\text{Change in } y}{\text{Change in } x} = \frac{y_2 - y_1}{x_2 - x_1} = \frac{-1 - 4}{-3 - (-2)} = \frac{-5}{-1} = 5.$$

However, you should not subtract in one order in the numerator $(y_2 - y_1)$ and then in the opposite order in the denominator $(x_1 - x_2)$. The slope is *not* −5:

$$\frac{-1-4}{-2-(-3)} = \frac{-5}{1} = -5. \quad \text{incorrect}$$

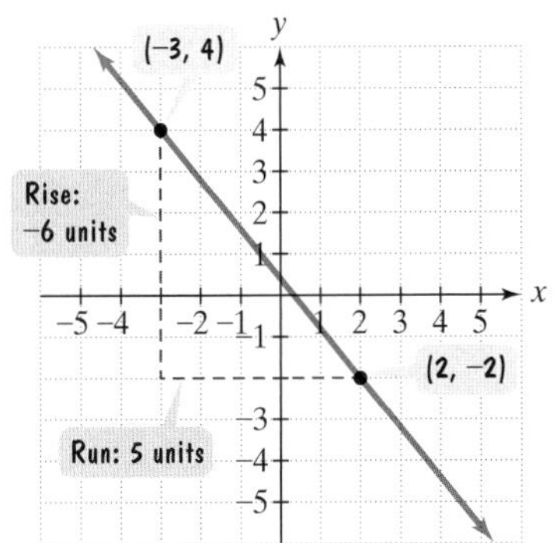

FIGURE 7.19 Visualizing a slope of $-\frac{6}{5}$

b. We can let $(x_1, y_1) = (-3, 4)$ and $(x_2, y_2) = (2, -2)$. The slope of the line shown in **Figure 7.19** is computed as follows:

$$m = \frac{\text{Change in } y}{\text{Change in } x} = \frac{y_2 - y_1}{x_2 - x_1} = \frac{-2 - 4}{2 - (-3)} = \frac{-6}{5} = -\frac{6}{5}.$$

The slope of the line is $-\frac{6}{5}$. For every vertical change of −6 units (6 units down), there is a corresponding horizontal change of 5 units. The slope is negative and the line falls from left to right.

CHECK POINT 2 Find the slope of the line passing through each pair of points:

a. $(-3, 4)$ and $(-4, -2)$

b. $(4, -2)$ and $(-1, 5)$.

Example 2 illustrates that a line with a positive slope is rising from left to right and a line with a negative slope is falling from left to right. By contrast, a horizontal line neither rises nor falls and has a slope of zero. A vertical line has no horizontal change, so $x_2 - x_1 = 0$ in the formula for slope. Because we cannot divide by zero, the slope of a vertical line is undefined. This discussion is summarized in **Table 7.2**.

TABLE 7.2 Possibilities for a Line's Slope

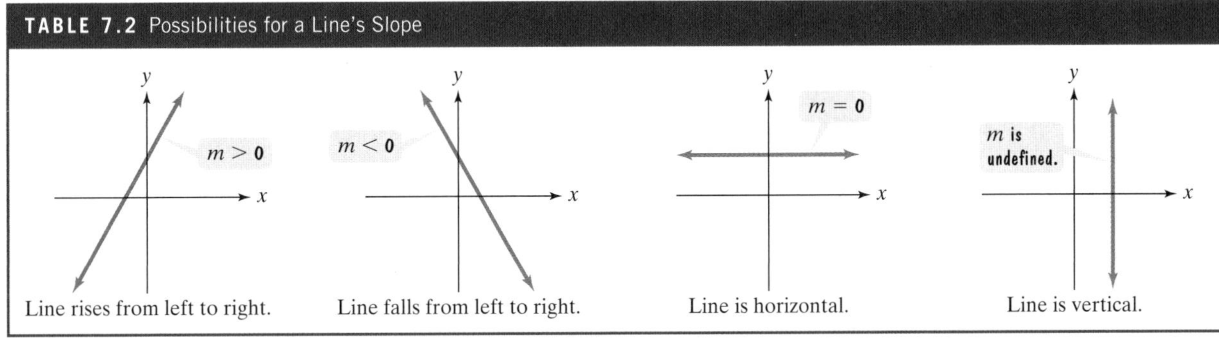

3 *Use the slope and y-intercept to graph a line.*

The Slope-Intercept Form of the Equation of a Line

We can use the definition of slope to write the equation of any nonvertical line with slope m and y-intercept b. Because the y-intercept is b, the point $(0, b)$ lies on the line. Now, let (x, y) represent any other point on the line, shown in **Figure 7.20**. Keep in mind that the point (x, y) is arbitrary and is not in one fixed position. By contrast, the point $(0, b)$ is fixed.

Regardless of where the point (x, y) is located, the steepness of the line in **Figure 7.20** remains the same. Thus, the ratio for slope stays a constant m. This means that for all points along the line

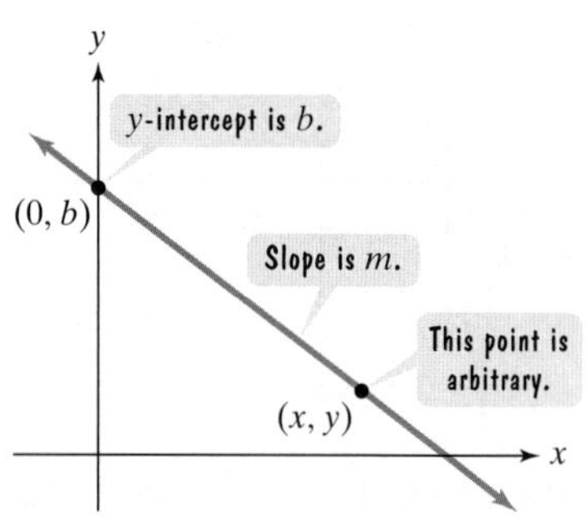

FIGURE 7.20 A line with slope m and y-intercept b.

$$m = \frac{\text{Change in } y}{\text{Change in } x} = \frac{y - b}{x - 0} = \frac{y - b}{x}.$$

We can clear the fraction by multiplying both sides by x, the denominator. Note that x is not zero since (x, y) is distinct from $(0, b)$, the only point on the line with first coordinate 0.

$$m = \frac{y - b}{x} \quad \text{This is the slope of the line in Figure 7.20.}$$

$$mx = \frac{y - b}{x} \cdot x \quad \text{Multiply both sides by } x.$$

$$mx = y - b \quad \text{Simplify: } \frac{y - b}{x} \cdot x = y - b.$$

$$mx + b = y - b + b \quad \text{Add } b \text{ to both sides and solve for } y.$$

$$mx + b = y \quad \text{Simplify.}$$

Now, if we reverse the two sides, we obtain the *slope-intercept form* of the equation of a line.

SLOPE-INTERCEPT FORM OF THE EQUATION OF A LINE

The **slope-intercept form of the equation** of a nonvertical line with slope m and y-intercept b is

$$y = mx + b.$$

The slope-intercept form of a line's equation, $y = mx + b$, can be expressed in function notation by replacing y with $f(x)$:

$$f(x) = mx + b.$$

We have seen that functions in the form $f(x) = mx + b$ are called **linear functions**. Thus, in the equation of a linear function, the x-coefficient is the line's slope and the constant term is the y-intercept. Here are two examples:

$$y = 2x - 4 \qquad\qquad f(x) = \frac{1}{2}x + 2.$$

The slope is 2. The y-intercept is −4. The slope is $\frac{1}{2}$. The y-intercept is 2.

If a linear function's equation is in slope-intercept form, we can use the y-intercept and the slope to obtain its graph.

GREAT QUESTION!

If the slope is an integer, such as 2, why should I express it as $\frac{2}{1}$ for graphing purposes?

Writing the slope, m, as a fraction allows you to identify the rise (the fraction's numerator) and the run (the fraction's denominator).

GRAPHING $y = mx + b$ USING THE SLOPE AND y-INTERCEPT

1. Plot the point containing the y-intercept on the y-axis. This is the point $(0, b)$.
2. Obtain a second point using the slope, m. Write m as a fraction, and use rise over run, starting at the point containing the y-intercept, to plot the second point.
3. Use a straightedge to draw a line through the two points. Draw arrowheads at the ends of the line to show that the line continues indefinitely in both directions.

EXAMPLE 3 Graphing by Using the Slope and y-Intercept

Graph the linear function $y = \frac{2}{3}x + 2$ by using the slope and y-intercept.

SOLUTION

The equation of the linear function is in the form $y = mx + b$. We can find the slope, m, by identifying the coefficient of x. We can find the y-intercept, b, by identifying the constant term.

$$y = \frac{2}{3}x + 2$$

The slope is $\frac{2}{3}$. The y-intercept is 2.

Now that we have identified the slope and the y-intercept, we use the three-step procedure to graph the equation.

Step 1 Plot the point containing the y-intercept on the y-axis. The y-intercept is 2. We plot $(0, 2)$, shown in **Figure 7.21**.

Step 2 Obtain a second point using the slope, m. Write m as a fraction, and use rise over run, starting at the point containing the y-intercept, to plot the second point. The slope, $\frac{2}{3}$, is already written as a fraction:

$$m = \frac{2}{3} = \frac{\text{Rise}}{\text{Run}}.$$

We plot the second point on the line by starting at $(0, 2)$, the first point. Based on the slope, we move 2 units *up* (the rise) and 3 units to the *right* (the run). This puts us at a second point on the line, $(3, 4)$, shown in **Figure 7.21**.

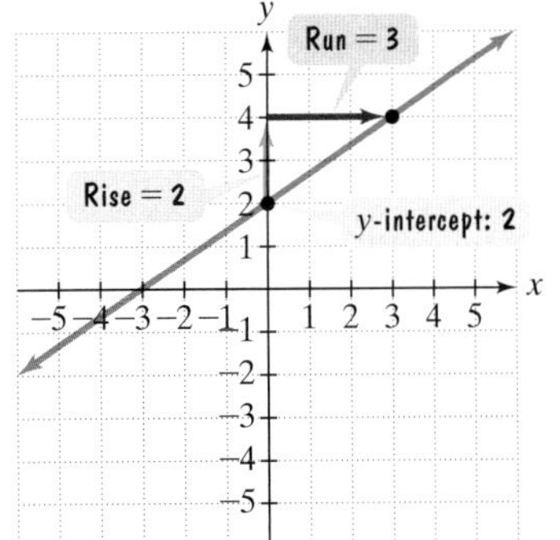

FIGURE 7.21 The graph of $y = \frac{2}{3}x + 2$

Step 3 Use a straightedge to draw a line through the two points. The graph of $y = \frac{2}{3}x + 2$ is shown in **Figure 7.21**.

CHECK POINT 3 Graph the linear function $y = \frac{3}{5}x + 1$ by using the slope and y-intercept.

Earlier in this section, we considered linear functions of the form $Ax + By = C$. We used x- and y-intercepts, as well as checkpoints, to graph these functions. It is also possible to obtain the graphs by using the slope and y-intercept. To do this, begin by solving $Ax + By = C$ for y. This will put the equation in slope-intercept form. Then use the three-step procedure to graph the equation. This is illustrated in Example 4.

EXAMPLE 4 Graphing by Using the Slope and y-Intercept

Graph the linear function $2x + 5y = 0$ by using the slope and y-intercept.

SOLUTION

We put the equation in slope-intercept form by solving for y.

$$2x + 5y = 0 \quad \text{This is the given equation.}$$

$$2x - 2x + 5y = 0 - 2x \quad \text{Subtract } 2x \text{ from both sides.}$$

$$5y = -2x + 0 \quad \text{Simplify.}$$

$$\frac{5y}{5} = \frac{-2x + 0}{5} \quad \text{Divide both sides by 5.}$$

$$y = \frac{-2x}{5} + \frac{0}{5} \quad \text{Divide each term in the numerator by 5.}$$

$$y = -\frac{2}{5}x + 0 \quad \text{Simplify. Equivalently, } f(x) = -\tfrac{2}{5}x + 0.$$

Now that the equation is in slope-intercept form, we can use the slope and y-intercept to obtain its graph. Examine the slope-intercept form:

$$y = -\frac{2}{5}x + 0.$$

slope: $-\frac{2}{5}$ y-intercept: 0

Note that the slope is $-\frac{2}{5}$ and the y-intercept is 0. Use the y-intercept to plot $(0, 0)$ on the y-axis. Then locate a second point by using the slope.

$$m = -\frac{2}{5} = \frac{-2}{5} = \frac{\text{Rise}}{\text{Run}}$$

Because the rise is -2 and the run is 5, move *down* 2 units and to the *right* 5 units, starting at the point $(0, 0)$. This puts us at a second point on the line, $(5, -2)$. The graph of $2x + 5y = 0$ is the line drawn through these points, shown in **Figure 7.22**.

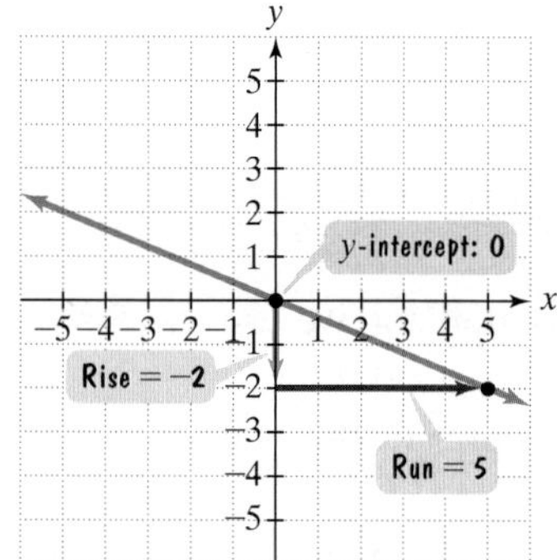

FIGURE 7.22 The graph of $2x + 5y = 0$, or $y = -\frac{2}{5}x$.

The equation $2x + 5y = 0$ in Example 4 is of the form $Ax + By = C$ with $C = 0$. If you try graphing $2x + 5y = 0$ by using intercepts, you will find that the x-intercept is 0 and the y-intercept is 0. This means that the graph passes through the origin. A second point must be found to graph the line. In Example 4, the line's slope gave us the second point.

CHECK POINT 4 Graph the linear function $3x + 4y = 0$ by using the slope and y-intercept.

4 *Graph horizontal or vertical lines.*

Equations of Horizontal and Vertical Lines

If a line is horizontal, its slope is zero: $m = 0$. Thus, the equation $y = mx + b$ becomes $y = b$, where b is the y-intercept. All horizontal lines have equations of the form $y = b$.

EXAMPLE 5 Graphing a Horizontal Line

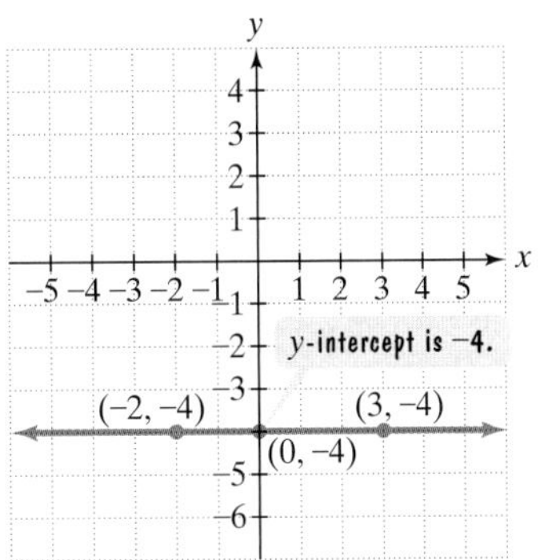

FIGURE 7.23 The graph of $y = -4$ or $f(x) = -4$.

Graph $y = -4$ in the rectangular coordinate system.

SOLUTION

All ordered pairs that are solutions of $y = -4$ have a value of y that is always -4. Any value can be used for x. (Think of $y = -4$ as $0x + 1y = -4$.) In the table at the right, we have selected three of the possible values for x: -2, 0, and 3. The table shows that three ordered pairs that are solutions of $y = -4$ are $(-2, -4)$, $(0, -4)$, and $(3, -4)$. Drawing a line that passes through the three points gives the horizontal line shown in **Figure 7.23**.

x	$y = -4$	(x, y)
−2	−4	(−2, −4)
0	−4	(0, −4)
3	−4	(3, −4)

For all choices of x, y is a constant -4.

CHECK POINT 5 Graph $y = 3$ in the rectangular coordinate system.

Next, let's see what we can discover about the graph of an equation of the form $x = a$ by looking at an example.

EXAMPLE 6 Graphing a Vertical Line

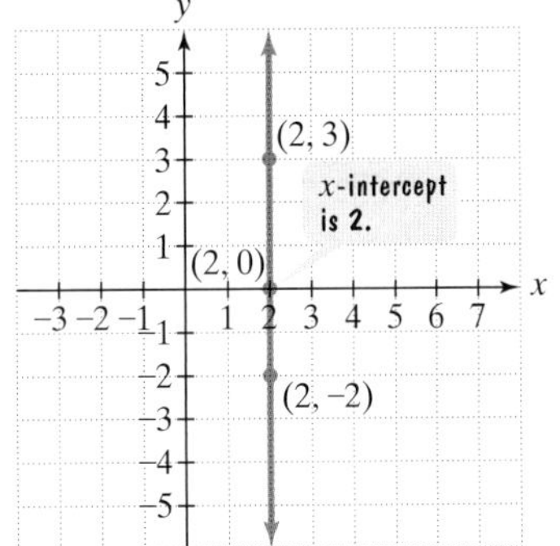

FIGURE 7.24 The graph of $x = 2$

Graph $x = 2$ in the rectangular coordinate system.

SOLUTION

All ordered pairs that are solutions of $x = 2$ have a value of x that is always 2. Any value can be used for y. (Think of $x = 2$ as $1x + 0y = 2$.) In the table at the right, we have selected three of the possible values for y: -2, 0, and 3. The table shows that three ordered pairs that are solutions of $x = 2$ are $(2, -2)$, $(2, 0)$, and $(2, 3)$. Drawing a line that passes through the three points gives the vertical line shown in **Figure 7.24**.

For all choices of y, x is always 2.

$x = 2$	y	(x, y)
2	−2	(2, −2)
2	0	(2, 0)
2	3	(2, 3)

Does a vertical line represent the graph of a linear function? No. Look at the graph of $x = 2$ in **Figure 7.24**. A vertical line drawn through $(2, 0)$ intersects the graph infinitely many times. This shows that infinitely many outputs are associated with the input 2. **No vertical line represents a linear function.** All other lines are graphs of functions.

HORIZONTAL AND VERTICAL LINES

The graph of $y = b$ or $f(x) = b$ is a horizontal line. The y-intercept is b.

The graph of $x = a$ is a vertical line. The x-intercept is a.

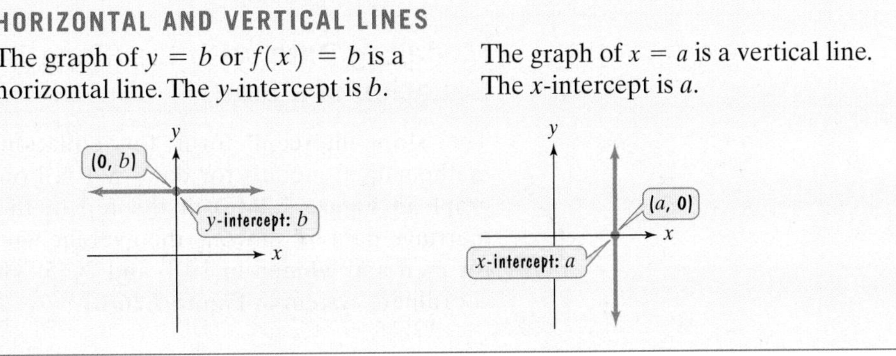

CHECK POINT 6 Graph $x = -2$ in the rectangular coordinate system.

5 *Interpret slope as rate of change.*

Slope as Rate of Change

Slope is defined as the ratio of a change in y to a corresponding change in x. Our next example shows how slope can be interpreted as a **rate of change** in an applied situation.

EXAMPLE 7 *Slope as a Rate of Change*

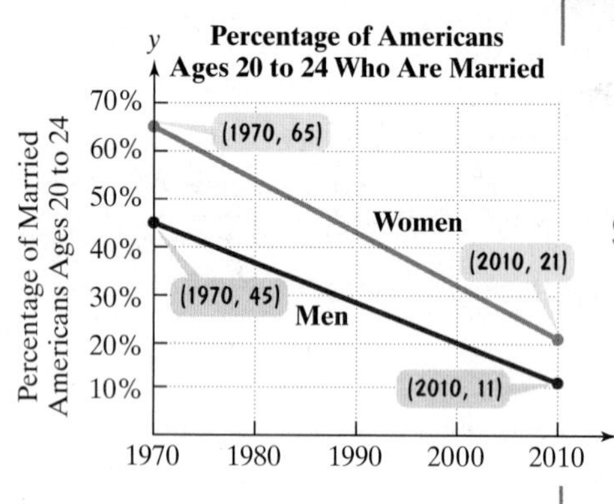

FIGURE 7.25
Source: U.S. Census Bureau

The line graphs in **Figure 7.25** show the percentage of American men and women ages 20 to 24 who were married from 1970 through 2010. Find the slope of the line segment representing women. Describe what the slope represents.

SOLUTION

We let x represent a year and y the percentage of married women ages 20–24 in that year. The two points shown on the line segment for women have the following coordinates:

$$(1970, 65) \quad \text{and} \quad (2010, 21).$$

In 1970, 65% of American women ages 20 to 24 were married.

In 2010, 21% of American women ages 20 to 24 were married.

Now we compute the slope.

$$m = \frac{\text{Change in } y}{\text{Change in } x} = \frac{21 - 65}{2010 - 1970} = \frac{-44}{40} = -1.1$$

The unit in the numerator is the *percentage of married women ages 20 to 24.*

The unit in the denominator is *year.*

The slope indicates that for the period from 1970 through 2010, the percentage of married women ages 20 to 24 decreased by 1.1 per year. The rate of change is −1.1% per year.

CHECK POINT 7 Find the slope of the line segment representing men in **Figure 7.25**. Use your answer to complete this statement:

For the period from 1970 through 2010, the percentage of married men ages 20 to 24 decreased by ______ per year. The rate of change is _________ per _______.

6 *Use slope and y-intercept to model data.*

Modeling Data with the Slope-Intercept Form of the Equation of a Line

The slope-intercept form for equations of lines is useful for obtaining mathematical models for data that fall on or near a line. For example, the bar graph in **Figure 7.26(a)** at the top of the next page continues our work with marriage data by showing the average age of first marriage in the United States for men and women in 1970 and 2015. The data are displayed in a rectangular coordinate system in **Figure 7.26(b)**.

Average Age of First Marriage in the United States

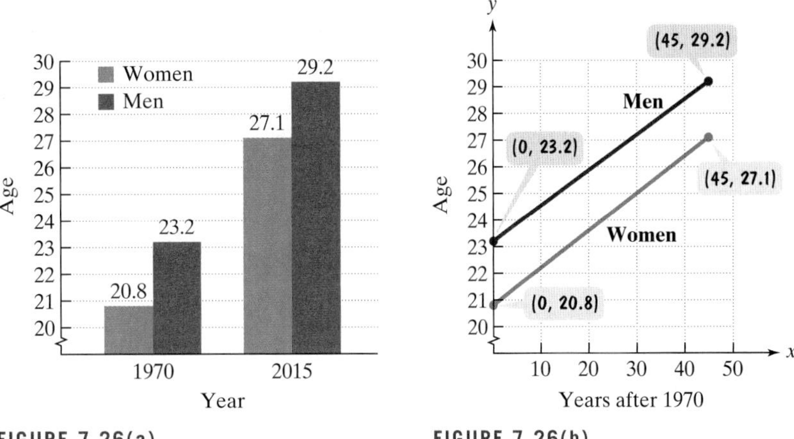

FIGURE 7.26(a)
Source: U.S. Census Bureau

FIGURE 7.26(b)

Example 8 illustrates how we can use the equation $y = mx + b$ to obtain a model for the data and make predictions about what might occur in the future.

EXAMPLE 8 *Modeling with the Slope-Intercept Form of the Equation*

a. Use the two points for women in **Figure 7.26(b)** to find a function in the form $W(x) = mx + b$ that models the average age of first marriage for U.S. women, $W(x)$, x years after 1970.

b. Use the model to project the average age of first marriage for U.S. women in 2030.

SOLUTION

a. We will use the line segment for women with the points (0, 20.8) and (45, 27.1) to obtain a model. We need to find values for m, the slope, and b, the y-intercept.

$$y = mx + b$$

$m = \dfrac{\text{Change in } y}{\text{Change in } x} = \dfrac{27.1 - 20.8}{45 - 0} = 0.14$

The point (0, 20.8) lies on the line segment for women, so the y-intercept is 20.8: $b = 20.8$.

The average age of first marriage for U.S. women, $W(x)$, x years after 1970 can be modeled by the linear function

$$W(x) = 0.14x + 20.8.$$

The slope, 0.14, indicates an increase in age of first marriage of 0.14 per year from 1970 through 2015.

b. Now let's use this model to project the average age of first marriage for U.S. women in 2030. Because 2030 is 60 years after 1970, substitute 60 for x in $W(x) = 0.14x + 20.8$ and evaluate the function at 60.

$$W(60) = 0.14(60) + 20.8 = 8.4 + 20.8 = 29.2$$

Our model projects that the average age of first marriage for U.S. women will be 29.2 in 2030.

CHECK POINT 8 **a.** Use the two points for men in **Figure 7.26(b)** on the previous page to find a function in the form $M(x) = mx + b$ that models the average age of first marriage for U.S. men, $M(x)$, x years after 1970. Round the value of m to two decimal places.

b. Use the model to project the average age of first marriage for U.S. men in 2030.

Blitzer Bonus

Slope and Applauding Together

Using a decibel meter, sociologist Max Atkinson found a function that models the intensity of a group's applause, $d(t)$, in decibels, over time, t, in seconds.

- Applause starts very fast, reaching a full crescendo of 30 decibels in one second. $(m = 30)$
- Applause remains level at 30 decibels for 5.5 seconds. $(m = 0)$
- Applause trails off by 15 decibels per second for two seconds. $(m = -15)$

These verbal conditions can be modeled by a function with three equations.

$$d(t) = \begin{cases} 30t & 0 \le t \le 1 \\ 30 & 1 < t \le 6.5 \\ -15t + 127.5 & 6.5 < t \le 8.5 \end{cases}$$

Intensity of applause, $d(t)$, in decibels, as a function of time, t, in seconds

The graph of this function is shown below. The function suggests how strongly people work to coordinate their applause with one another, careful to start clapping at the "right" time and careful to stop when it seems others are stopping as well.

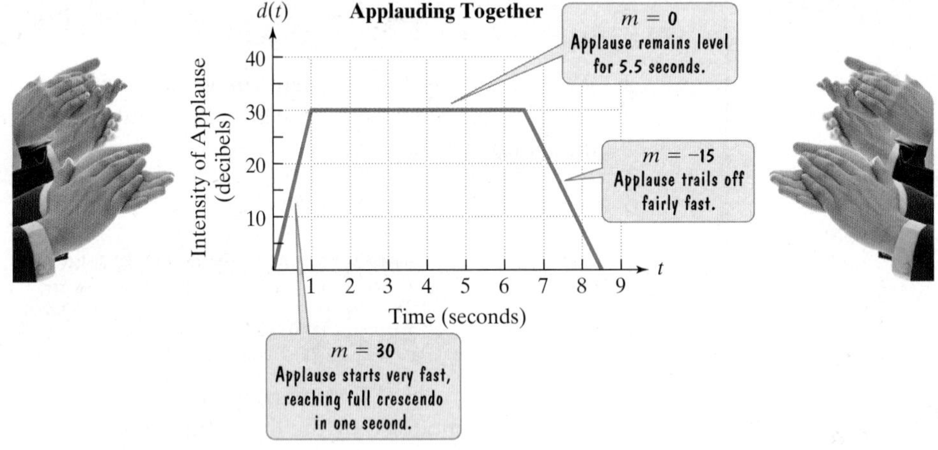

Source: The Sociology Project 2.0, Pearson, 2016.

Concept and Vocabulary Check

Fill in each blank so that the resulting statement is true.

1. The x-coordinate of a point where a graph crosses the x-axis is called a/an ________.
2. The point (0, 3) lies along a line, so 3 is a/an ________ of that line.
3. The slope of the line through the distinct points (x_1, y_1) and (x_2, y_2) is ________.
4. The slope-intercept form of the equation of a line is ________, where m represents the ________ and b represents the ________.
5. The slope of the linear function whose equation is $f(x) = -4x + 3$ is ____ and the y-intercept of its graph is ____.

6. In order to graph the line whose equation is $y = \frac{2}{5}x + 3$, begin by plotting the point ________. From this point, we move ____ units up (the rise) and ____ units to the right (the run).

7. The graph of the equation $y = 3$ is a/an __________ line.
8. The graph of the equation $x = -2$ is a/an ________ line.

Exercise Set 7.2

Practice Exercises

In Exercises 1–8, use the x- and y-intercepts to graph each linear equation.

1. $x - y = 3$
2. $x + y = 4$
3. $3x - 4y = 12$
4. $2x - 5y = 10$
5. $2x + y = 6$
6. $x + 3y = 6$
7. $5x = 3y - 15$
8. $3x = 2y + 6$

In Exercises 9–20, calculate the slope of the line passing through the given points. If the slope is undefined, so state. Then indicate whether the line rises, falls, is horizontal, or is vertical.

9. $(2, 6)$ and $(3, 5)$
10. $(4, 2)$ and $(3, 4)$
11. $(-2, 1)$ and $(2, 2)$
12. $(-1, 3)$ and $(2, 4)$
13. $(-2, 4)$ and $(-1, -1)$
14. $(6, -4)$ and $(4, -2)$
15. $(5, 3)$ and $(5, -2)$
16. $(3, -4)$ and $(3, 5)$
17. $(2, 0)$ and $(0, 8)$
18. $(3, 0)$ and $(0, -9)$
19. $(5, 1)$ and $(-2, 1)$
20. $(-2, 3)$ and $(1, 3)$

In Exercises 21–32, graph each linear function using the slope and y-intercept.

21. $y = 2x + 3$
22. $y = 2x + 1$
23. $y = -2x + 4$
24. $y = -2x + 3$
25. $y = \frac{1}{2}x + 3$
26. $y = \frac{1}{2}x + 2$
27. $f(x) = \frac{2}{3}x - 4$
28. $f(x) = \frac{3}{4}x - 5$
29. $y = -\frac{3}{4}x + 4$
30. $y = -\frac{2}{3}x + 5$
31. $f(x) = -\frac{5}{3}x$
32. $f(x) = -\frac{4}{3}x$

In Exercises 33–40,

a. *Put the equation in slope-intercept form by solving for y.*
b. *Identify the slope and the y-intercept.*
c. *Use the slope and y-intercept to graph the line.*

33. $3x + y = 0$
34. $2x + y = 0$
35. $3y = 4x$
36. $4y = 5x$
37. $2x + y = 3$
38. $3x + y = 4$
39. $7x + 2y = 14$
40. $5x + 3y = 15$

In Exercises 41–48, graph each horizontal or vertical line.

41. $y = 4$
42. $y = 2$
43. $y = -2$
44. $y = -3$
45. $x = 2$
46. $x = 4$
47. $x + 1 = 0$
48. $x + 5 = 0$

Practice Plus

In Exercises 49–52, find the slope of the line passing through each pair of points or state that the slope is undefined. Assume that all variables represent positive real numbers. Then indicate whether the line through the points rises, falls, is horizontal, or is vertical.

49. $(0, a)$ and $(b, 0)$
50. $(-a, 0)$ and $(0, -b)$
51. (a, b) and $(a, b + c)$
52. $(a - b, c)$ and $(a, a + c)$

In Exercises 53–54, find the slope and y-intercept of each line whose equation is given. Assume that $B \neq 0$.

53. $Ax + By = C$
54. $Ax = By - C$

In Exercises 55–56, find the value of y if the line through the two given points is to have the indicated slope.

55. $(3, y)$ and $(1, 4)$, $m = -3$
56. $(-2, y)$ and $(4, -4)$, $m = \frac{1}{3}$

Use the figure to make the lists in Exercises 57–58.

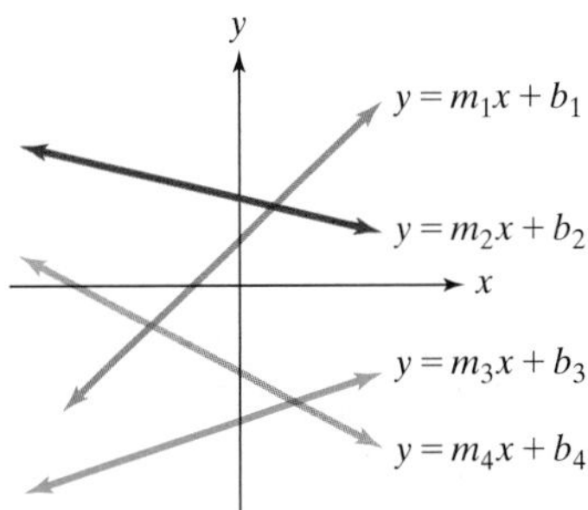

57. List the slopes m_1, m_2, m_3, and m_4 in order of decreasing size.
58. List the y-intercepts b_1, b_2, b_3, and b_4 in order of decreasing size.

Application Exercises

59. Older, Calmer. As we age, daily stress and worry decrease and happiness increases, according to an analysis of 340,847 U.S. adults, ages 18–85, in the journal *Proceedings of the National Academy of Sciences*. The graphs show a portion of the research.

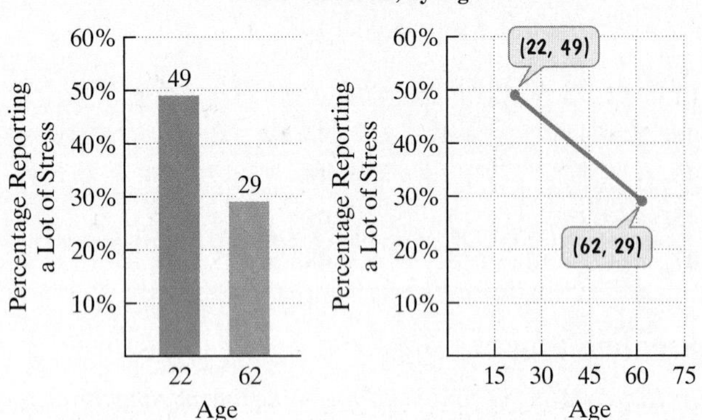

Source: National Academy of Sciences

a. Find the slope of the line passing through the two points shown by the voice balloons. Express the slope as a decimal.

b. Use your answer from part (a) to complete the statement:

For each year of aging, the percentage of Americans reporting "a lot" of stress decreases by _____. The rate of change is _____% per ____________.

60. Exercise is useful not only in preventing depression, but also as a treatment. The graphs show the percentage of patients with depression in remission when exercise (brisk walking) was used as a treatment. (The control group that engaged in no exercise had 11% of the patients in remission.)

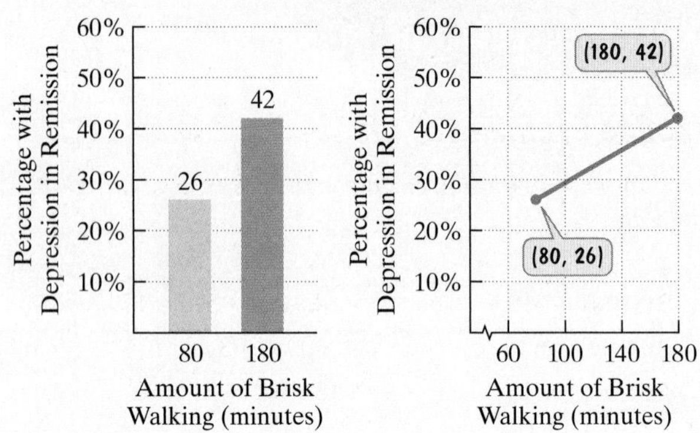

Source: *Newsweek*, March 26, 2007

a. Find the slope of the line passing through the two points shown by the voice balloons. Express the slope as a decimal.

b. Use your answer from part (a) to complete this statement: For each minute of brisk walking, the percentage of patients with depression in remission increased by _______. The rate of change is _______% per ________________________.

The bar graph shows that as costs changed over the decades, Americans devoted less of their budget to groceries and more to health care.

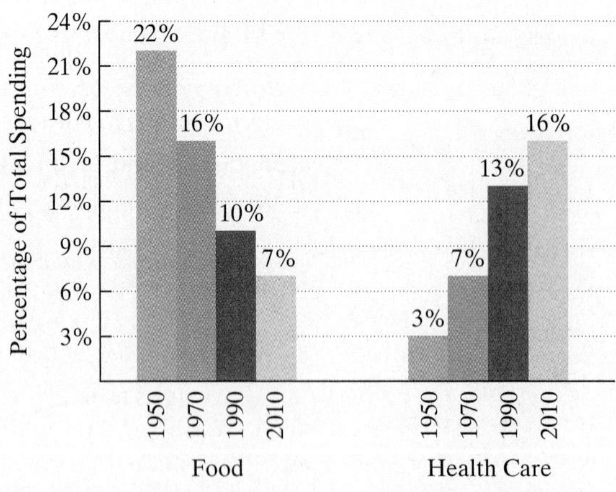

Source: *Time*, October 10, 2011

In Exercises 61–62, find a linear function in slope-intercept form that models the given description. Each function should model the percentage of total spending, $p(x)$, by Americans x years after 1950.

61. In 1950, Americans spent 22% of their budget on food. This has decreased at an average rate of approximately 0.25% per year since then.

62. In 1950, Americans spent 3% of their budget on health care. This has increased at an average rate of approximately 0.22% per year since then.

The Pay Gap. *How wide is the chasm between what men and women earn in the workplace? According to a 2015 analysis from the National Women's Law Center, women lose \$435,049 over the course of a career because of the pay gap. The bar graph shows the average earnings in the United States for men and women at ages 22 and 52.*

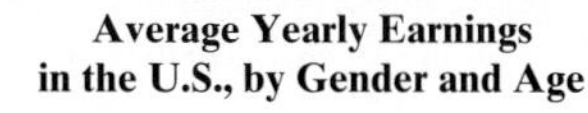

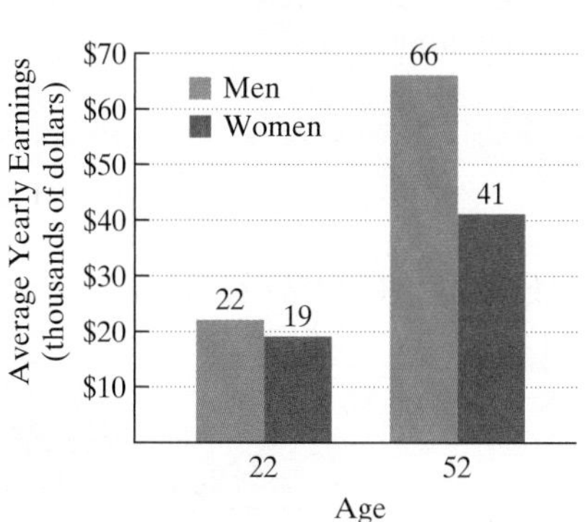

Source: *Time*, March 14, 2016

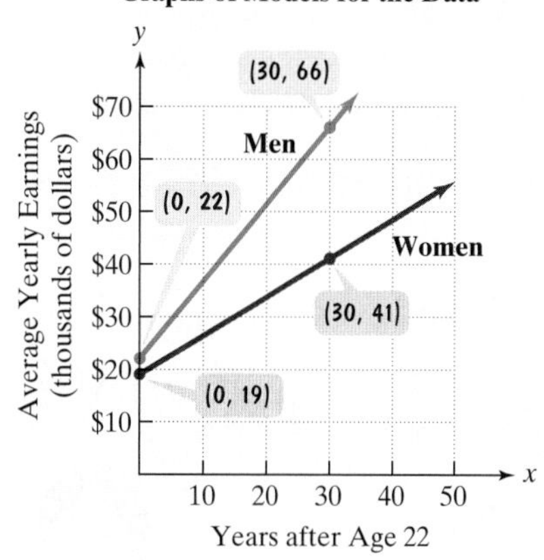

Exercises 63–64 involve the graphs of models for the data shown in the rectangular coordinate system.

63. a. Use the two points for men shown by the blue voice balloons to find a function in the form $M(x) = mx + b$ that models average yearly earnings for men x years after age 22. Round the value for m to two decimal places.

b. Use the two points for women shown by the red voice balloons to find a function in the form $W(x) = mx + b$ that models average yearly earnings for women x years after age 22. Round the value for m to two decimal places.

c. Use the models in parts (a) and (b) to find the average yearly earnings for men and women at age 32. How are these values shown on the graphs of the models for the data? What is the difference between earnings for men and women at that age?

64. a. Use the two points for men shown by the blue voice balloons to find a function in the form $M(x) = mx + b$ that models average yearly earnings for men x years after age 22. Round the value for m to two decimal places.

b. Use the two points for women shown by the red voice balloons to find a function in the form $W(x) = mx + b$ that models average yearly earnings for women x years after age 22. Round the value for m to two decimal places.

c. Use the models in parts (a) and (b) to find the average yearly earnings for men and women at age 42. How are these values shown on the graphs of the models for the data? What is the difference between earnings for men and women at that age?

Explaining the Concepts

65. Describe how to find the x-intercept of a linear equation.

66. Describe how to find the y-intercept of a linear equation.

67. What is the slope of a line?

68. Describe how to calculate the slope of a line passing through two points.

69. Describe how to graph a line using the slope and y-intercept. Provide an original example with your description.

70. What does it mean if the slope of a line is 0?

71. What does it mean if the slope of a line is undefined?

72. What is the least number of points needed to graph a line? How many should actually be used? Explain.

73. Explain why the y-values can be any number for the equation $x = 5$. How is this shown in the graph of the equation?

Critical Thinking Exercises

Make Sense? *In Exercises 74–77, determine whether each statement makes sense or does not make sense, and explain your reasoning.*

74. When finding the slope of the line passing through $(-1, 5)$ and $(2, -3)$, I must let (x_1, y_1) be $(-1, 5)$ and (x_2, y_2) be $(2, -3)$.

75. A linear function that models tuition and fees at public four-year colleges from 2000 through 2010 has negative slope.

76. Because the variable m does not appear in $Ax + By = C$, equations in this form make it impossible to determine the line's slope.

77. If I drive m miles in a year, the function $c(x) = 0.61m + 3500$ models the annual cost, $c(x)$, in dollars, of operating my car, so the function shows that with no driving at all, the cost is \$3500, and the rate of increase in this cost is \$0.61 for each mile that I drive.

In Exercises 78–81, determine whether each statement is true or false. If the statement is false, make the necessary change(s) to produce a true statement.

78. The equation $y = mx + b$ shows that no line can have a y-intercept that is numerically equal to its slope.

79. Every line in the rectangular coordinate system has an equation that can be expressed in slope-intercept form.

80. The line $3x + 2y = 5$ has slope $-\frac{3}{2}$.

81. The line $2y = 3x + 7$ has a y-intercept of 7.

82. The relationship between Celsius temperature, C, and Fahrenheit temperature, F, can be described by a linear equation in the form $F = mC + b$. The graph of this equation contains the point $(0, 32)$: Water freezes at 0°C or at 32°F. The line also contains the point $(100, 212)$: Water boils at 100°C or at 212°F. Write the linear equation expressing Fahrenheit temperature in terms of Celsius temperature.

Technology Exercises

83. Use a graphing utility to verify any three of your hand-drawn graphs in Exercises 21–32.

84. Use a graphing utility to verify any three of your hand-drawn graphs in Exercises 33–40. Solve the equation for y before entering it.

7.3 Systems of Linear Equations in Two Variables

WHAT AM I SUPPOSED TO LEARN?

After studying this section, you should be able to:

1 Determine whether an ordered pair is a solution of a linear system.
2 Solve linear systems by graphing.
3 Solve linear systems by substitution.
4 Solve linear systems by addition.
5 Identify systems that do not have exactly one ordered-pair solution.
6 Solve problems using systems of linear equations.

RESEARCHERS IDENTIFIED college students who generally were procrastinators or nonprocrastinators. The students were asked to report throughout the semester how many symptoms of physical illness they had experienced. **Figure 7.27** shows that by late in the semester, all students experienced increases in symptoms. Early in the semester, procrastinators reported fewer symptoms, but late in the semester, as work came due, they reported more symptoms than their nonprocrastinating peers.

Symptoms of Physical Illness among College Students

Average Number of Symptoms (y: 1–9); Week in a 16-Week Semester (x: 2, 4, 6, 8, 10, 12, 14, 16); Procrastinators; Nonprocrastinators

FIGURE 7.27
Source: Richard Gerrig, *Psychology and Life*, 20th Edition, Pearson, 2013.

The data in **Figure 7.27** can be analyzed using a pair of linear models in two variables. The figure shows that by week 6, both groups reported the same number of symptoms of illness, an average of approximately 3.5 symptoms per group. In this section, you will learn two algebraic methods, called *substitution* and *addition*, that will reinforce this graphic observation, verifying $(6, 3.5)$ as the point of intersection.

1 Determine whether an ordered pair is a solution of a linear system.

Systems of Linear Equations and Their Solutions

We have seen that all equations in the form $Ax + By = C$, A and B not both zero, are straight lines when graphed. Two such equations are called a **system of linear equations** or a **linear system**. A **solution to a system of linear equations in two variables** is an ordered pair that satisfies both equations in the system. For example, $(3, 4)$ satisfies the system

$$\begin{cases} x + y = 7 & (3 + 4 \text{ is, indeed, } 7.) \\ x - y = -1. & (3 - 4 \text{ is, indeed, } -1.) \end{cases}$$

Thus, $(3, 4)$ satisfies both equations and is a solution of the system. The solution can be described by saying that $x = 3$ and $y = 4$. The solution can also be described using set notation. The solution set of the system is $\{(3, 4)\}$—that is, the set consisting of the ordered pair $(3, 4)$.

A system of linear equations can have exactly one solution, no solution, or infinitely many solutions. We begin with systems having exactly one solution.

EXAMPLE 1 Determining Whether an Ordered Pair Is a Solution of a System

Determine whether $(1, 2)$ is a solution of the system:

$$\begin{cases} 2x - 3y = -4 \\ 2x + y = 4. \end{cases}$$

SOLUTION

Because 1 is the x-coordinate and 2 is the y-coordinate of $(1, 2)$, we replace x with 1 and y with 2.

$2x - 3y = -4$	$2x + y = 4$
$2(1) - 3(2) \stackrel{?}{=} -4$	$2(1) + 2 \stackrel{?}{=} 4$
$2 - 6 \stackrel{?}{=} -4$	$2 + 2 \stackrel{?}{=} 4$
$-4 = -4$, true	$4 = 4$, true

The pair $(1, 2)$ satisfies both equations: It makes each equation true. Thus, the pair is a solution of the system.

CHECK POINT 1 Determine whether $(-4, 3)$ is a solution of the system:

$$\begin{cases} x + 2y = 2 \\ x - 2y = 6. \end{cases}$$

2 *Solve linear systems by graphing.*

Solving Linear Systems by Graphing

The solution to a system of linear equations can be found by graphing both of the equations in the same rectangular coordinate system. For a system with one solution, **the coordinates of the point of intersection of the lines is the system's solution.**

EXAMPLE 2 Solving a Linear System by Graphing

Solve by graphing:

$$\begin{cases} x + 2y = 2 \\ x - 2y = 6. \end{cases}$$

SOLUTION

We find the solution by graphing both $x + 2y = 2$ and $x - 2y = 6$ in the same rectangular coordinate system. We will use intercepts to graph each equation.

$$x + 2y = 2$$

x-intercept: Set $y = 0$.	**y-intercept: Set $x = 0$.**
$x + 2 \cdot 0 = 2$	$0 + 2y = 2$
$x = 2$	$2y = 2$
	$y = 1$
The line passes through $(2, 0)$.	The line passes through $(0, 1)$.

We graph $x + 2y = 2$ as a blue line in **Figure 7.28**.

$$x - 2y = 6$$

x-intercept: Set $y = 0$.	**y-intercept: Set $x = 0$.**
$x - 2 \cdot 0 = 6$	$0 - 2y = 6$
$x = 6$	$-2y = 6$
	$y = -3$
The line passes through $(6, 0)$.	The line passes through $(0, -3)$.

We graph $x - 2y = 6$ as a red line in **Figure 7.28**.

The system is graphed in **Figure 7.28**. To ensure that the graph is accurate, check the coordinates of the intersection point, $(4, -1)$, in both equations.

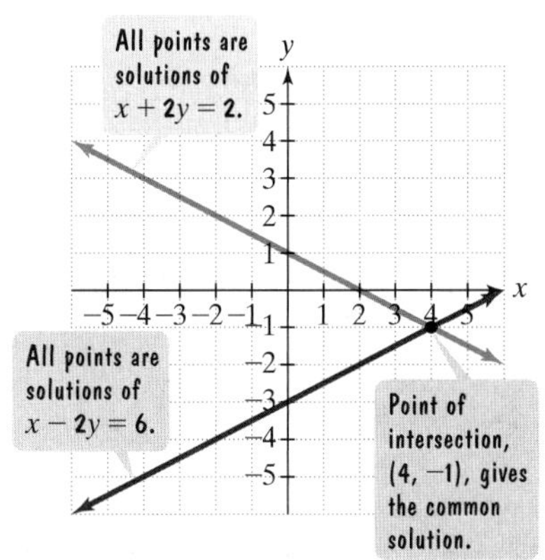

FIGURE 7.28 Visualizing a system's solution

We replace x with 4 and y with -1.

$$x + 2y = 2 \qquad\qquad x - 2y = 6$$
$$4 + 2(-1) \stackrel{?}{=} 2 \qquad\qquad 4 - 2(-1) \stackrel{?}{=} 6$$
$$4 + (-2) \stackrel{?}{=} 2 \qquad\qquad 4 - (-2) \stackrel{?}{=} 6$$
$$2 = 2, \text{ true} \qquad\qquad 4 + 2 \stackrel{?}{=} 6$$
$$6 = 6, \text{ true}$$

The pair $(4, -1)$ satisfies both equations—that is, it makes each equation true. This verifies that the system's solution set is $\{(4, -1)\}$.

GREAT QUESTION!

Can I use a rough sketch on scratch paper to solve a linear system by graphing?

No. When solving linear systems by graphing, neatly drawn graphs are essential for determining points of intersection.

- Use rectangular coordinate graph paper.
- Use a ruler or straightedge.
- Use a pencil with a sharp point.

CHECK POINT 2 Solve by graphing:

$$\begin{cases} 2x + 3y = 6 \\ 2x + y = -2. \end{cases}$$

3 *Solve linear systems by substitution.*

Solving Linear Systems by the Substitution Method

Finding the solution to a linear system by graphing equations may not be easy to do. For example, a solution of $\left(-\frac{2}{3}, \frac{157}{29}\right)$ would be difficult to "see" as an intersection point on a graph.

Let's consider a method that does not depend on finding a system's solution visually: the substitution method. This method involves converting the system to one equation in one variable by an appropriate substitution.

GREAT QUESTION!

In the first step of the substitution method, how do I know which variable to isolate and in which equation?

You can choose both the variable and the equation. If possible, solve for a variable whose coefficient is 1 or −1 to avoid working with fractions.

SOLVING LINEAR SYSTEMS BY SUBSTITUTION

1. Solve either of the equations for one variable in terms of the other. (If one of the equations is already in this form, you can skip this step.)
2. Substitute the expression found in step 1 into the *other* equation. This will result in an equation in one variable.
3. Solve the equation containing one variable.
4. Back-substitute the value found in step 3 into the equation from step 1. Simplify and find the value of the remaining variable.
5. Check the proposed solution in both of the system's given equations.

EXAMPLE 3 *Solving a System by Substitution*

Solve by the substitution method:

$$\begin{cases} y = -x - 1 \\ 4x - 3y = 24. \end{cases}$$

SOLUTION

Step 1 Solve either of the equations for one variable in terms of the other. This step has already been done for us. The first equation, $y = -x - 1$, is solved for y in terms of x.

Step 2 Substitute the expression from step 1 into the other equation. We substitute the expression $-x - 1$ for y into the other equation:

$$y = \boxed{-x - 1} \qquad 4x - 3\boxed{y} = 24 \qquad \text{Substitute } -x - 1 \text{ for } y.$$

This gives us an equation in one variable, namely

$$4x - 3(-x - 1) = 24.$$

The variable y has been eliminated.

Step 3 Solve the resulting equation containing one variable.

$$4x - 3(-x - 1) = 24$$ This is the equation containing one variable.

$$4x + 3x + 3 = 24$$ Apply the distributive property.

$$7x + 3 = 24$$ Combine like terms.

$$7x = 21$$ Subtract 3 from both sides.

$$x = 3$$ Divide both sides by 7.

Step 4 Back-substitute the obtained value into the equation from step 1. We now know that the x-coordinate of the solution is 3. To find the y-coordinate, we back-substitute the x-value into the equation from step 1.

$$y = -x - 1$$ This is the equation from step 1.

Substitute 3 for x.

$$y = -3 - 1$$

$$y = -4$$ Simplify.

With $x = 3$ and $y = -4$, the proposed solution is $(3, -4)$.

Step 5 Check. Check the proposed solution, $(3, -4)$, in both of the system's given equations. Replace x with 3 and y with -4.

$$y = -x - 1 \qquad\qquad 4x - 3y = 24$$

$$-4 \stackrel{?}{=} -3 - 1 \qquad\qquad 4(3) - 3(-4) \stackrel{?}{=} 24$$

$$-4 = -4, \text{ true} \qquad\qquad 12 + 12 \stackrel{?}{=} 24$$

$$24 = 24, \text{ true}$$

The pair $(3, -4)$ satisfies both equations. The system's solution set is $\{(3, -4)\}$.

TECHNOLOGY

A graphing calculator can be used to solve the system in Example 3. Graph each equation and use the intersection feature. The calculator displays the solution $(3, -4)$, as

$$x = 3, y = -4.$$

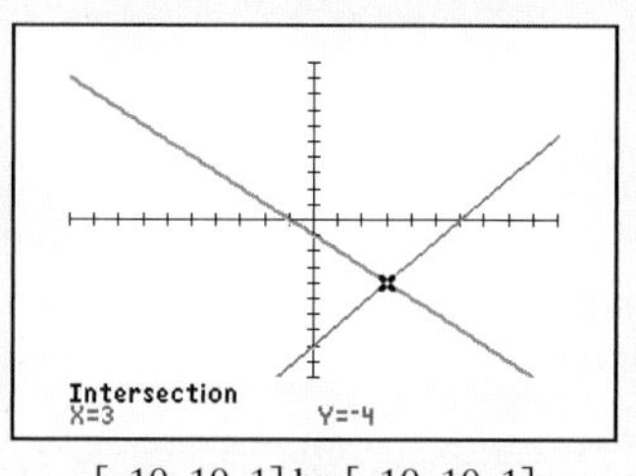

$[-10, 10, 1]$ by $[-10, 10, 1]$

CHECK POINT 3 Solve by the substitution method:

$$\begin{cases} y = 3x - 7 \\ 5x - 2y = 8. \end{cases}$$

EXAMPLE 4 *Solving a System by Substitution*

Solve by the substitution method:

$$\begin{cases} 5x - 4y = 9 \\ x - 2y = -3. \end{cases}$$

SOLUTION

Step 1 Solve either of the equations for one variable in terms of the other. We begin by isolating one of the variables in either of the equations. By solving for x in the second equation, which has a coefficient of 1, we can avoid fractions.

$$x - 2y = -3$$ This is the second equation in the given system.

$$x = 2y - 3$$ Solve for x by adding $2y$ to both sides.

Step 2 Substitute the expression from step 1 into the other equation. We substitute $2y - 3$ for x in the first equation.

$$x = \boxed{2y - 3} \qquad 5\boxed{x} - 4y = 9$$

Substituting $2y - 3$ for x in $5x - 4y = 9$ gives us an equation in one variable, namely

$$5(2y - 3) - 4y = 9.$$

The variable x has been eliminated.

Step 3 Solve the resulting equation containing one variable.

$5(2y - 3) - 4y = 9$	This is the equation containing one variable.
$10y - 15 - 4y = 9$	Apply the distributive property.
$6y - 15 = 9$	Combine like terms: $10y - 4y = 6y$.
$6y = 24$	Add 15 to both sides.
$y = 4$	Divide both sides by 6.

Step 4 Back-substitute the obtained value into the equation from step 1. Now that we have the y-coordinate of the solution, we back-substitute 4 for y in the equation $x = 2y - 3$.

$x = 2y - 3$	Use the equation obtained in step 1.
$x = 2(4) - 3$	Substitute 4 for y.
$x = 8 - 3$	Multiply.
$x = 5$	Subtract.

With $x = 5$ and $y = 4$, the proposed solution is $(5, 4)$.

Step 5 Check. Take a moment to show that (5, 4) satisfies both given equations, $5x - 4y = 9$ and $x - 2y = -3$. The solution set is $\{(5, 4)\}$.

GREAT QUESTION!

If my solution satisfies one of the equations in the system, do I have to check the solution in the other equation?

Yes. Get into the habit of checking ordered-pair solutions in both equations of the system.

CHECK POINT 4 Solve by the substitution method:

$$\begin{cases} 3x + 2y = -1 \\ x - y = 3. \end{cases}$$

4 *Solve linear systems by addition.*

Solving Linear Systems by the Addition Method

The substitution method is most useful if one of the given equations has an isolated variable. A third, and frequently the easiest, method for solving a linear system is the addition method. Like the substitution method, the addition method involves eliminating a variable and ultimately solving an equation containing only one variable. However, this time we eliminate a variable by adding the equations.

For example, consider the following system of linear equations:

$$\begin{cases} 3x - 4y = 11 \\ -3x + 2y = -7. \end{cases}$$

When we add these two equations, the x-terms are eliminated. This occurs because the coefficients of the x-terms, 3 and -3, are opposites (additive inverses) of each other:

$$\begin{cases} 3x - 4y = 11 \\ -3x + 2y = -7 \end{cases}$$

Add: $-2y = 4$ — The sum is an equation in one variable.

$y = -2$ — Divide both sides by -2 and solve for y.

Now we can back-substitute -2 for y into one of the original equations to find x. It does not matter which equation we use: We will obtain the same value for x in either case. If we use either equation, we can show that $x = 1$ and the solution $(1, -2)$ satisfies both equations in the system.

When we use the addition method, we want to obtain two equations whose sum is an equation containing only one variable. The key step is to **obtain, for one of the variables, coefficients that differ only in sign**. To do this, we may need to multiply one or both equations by some nonzero number so that the coefficients of one of the variables, x or y, become opposites. Then when the two equations are added, this variable is eliminated.

GREAT QUESTION!

Isn't the addition method also called the elimination method?

Although the addition method is also known as the elimination method, variables are eliminated when using both the substitution and addition methods. The name *addition method* specifically tells us that the elimination of a variable is accomplished by adding two equations.

SOLVING LINEAR SYSTEMS BY ADDITION

1. If necessary, rewrite both equations in the form $Ax + By = C$.
2. If necessary, multiply either equation or both equations by appropriate nonzero numbers so that the sum of the x-coefficients or the sum of the y-coefficients is 0.
3. Add the equations in step 2. The sum is an equation in one variable.
4. Solve the equation in one variable.
5. Back-substitute the value obtained in step 4 into either of the given equations and solve for the other variable.
6. Check the solution in both of the original equations.

EXAMPLE 5 Solving a System by the Addition Method

Solve by the addition method:

$$\begin{cases} 3x + 2y = 48 \\ 9x - 8y = -24. \end{cases}$$

SOLUTION

Step 1 Rewrite both equations in the form $Ax + By = C$. Both equations are already in this form. Variable terms appear on the left and constants appear on the right.

Step 2 If necessary, multiply either equation or both equations by appropriate numbers so that the sum of the x-coefficients or the sum of the y-coefficients is 0. We can eliminate x or y. Let's eliminate x. Consider the terms in x in each equation, that is, $3x$ and $9x$. To eliminate x, we can multiply each term of the first equation by -3 and then add the equations.

$$\begin{cases} 3x + 2y = 48 \\ 9x - 8y = -24 \end{cases} \xrightarrow[\text{No change}]{\text{Multiply by } -3.} \begin{cases} -9x - 6y = -144 \\ 9x - 8y = -24 \end{cases}$$

Step 3 Add the equations. Add: $-14y = -168$

Step 4 Solve the equation in one variable. We solve $-14y = -168$ by dividing both sides by -14.

$$\frac{-14y}{-14} = \frac{-168}{-14} \quad \text{Divide both sides by } -14.$$

$$y = 12 \quad \text{Simplify.}$$

Step 5 Back-substitute and find the value for the other variable. We can back-substitute 12 for y into either one of the given equations. We'll use the first one.

$$3x + 2y = 48 \quad \text{This is the first equation in the given system.}$$
$$3x + 2(12) = 48 \quad \text{Substitute 12 for } y.$$
$$3x + 24 = 48 \quad \text{Multiply.}$$
$$3x = 24 \quad \text{Subtract 24 from both sides.}$$
$$x = 8 \quad \text{Divide both sides by 3.}$$

We found that $y = 12$ and $x = 8$. The proposed solution is $(8, 12)$.

Step 6 Check. Take a few minutes to show that $(8, 12)$ satisfies both of the original equations in the system. The solution set is $\{(8, 12)\}$.

CHECK POINT 5 Solve by the addition method:

$$\begin{cases} 4x + 5y = 3 \\ 2x - 3y = 7. \end{cases}$$

EXAMPLE 6 *Solving a System by the Addition Method*

Solve by the addition method:

$$\begin{cases} 7x = 5 - 2y \\ 3y = 16 - 2x. \end{cases}$$

SOLUTION

Step 1 Rewrite both equations in the form $Ax + By = C$. We first arrange the system so that variable terms appear on the left and constants appear on the right. We obtain

$$\begin{cases} 7x + 2y = 5 & \text{Add } 2y \text{ to both sides of the first equation.} \\ 2x + 3y = 16. & \text{Add } 2x \text{ to both sides of the second equation.} \end{cases}$$

Step 2 If necessary, multiply either equation or both equations by appropriate numbers so that the sum of the x-coefficients or the sum of the y-coefficients is 0. We can eliminate x or y. Let's eliminate y by multiplying the first equation by 3 and the second equation by -2.

$$\begin{cases} 7x + 2y = 5 \xrightarrow{\text{Multiply by 3.}} \\ 2x + 3y = 16 \xrightarrow{\text{Multiply by } -2.} \end{cases} \begin{cases} 21x + 6y = 15 \\ -4x - 6y = -32 \end{cases}$$

Step 3 Add the equations.

$$\text{Add: } 17x + 0y = -17$$
$$17x = -17$$

Step 4 Solve the equation in one variable. We solve $17x = -17$ by dividing both sides by 17.

$$\frac{17x}{17} = \frac{-17}{17} \quad \text{Divide both sides by 17.}$$
$$x = -1 \quad \text{Simplify.}$$

Step 5 Back-substitute and find the value for the other variable. We can back-substitute -1 for x into either one of the given equations. We'll use the second one.

$$3y = 16 - 2x \quad \text{This is the second equation in the given system.}$$
$$3y = 16 - 2(-1) \quad \text{Substitute } -1 \text{ for } x.$$
$$3y = 16 + 2 \quad \text{Multiply.}$$
$$3y = 18 \quad \text{Add.}$$
$$y = 6 \quad \text{Divide both sides by 3.}$$

With $x = -1$ and $y = 6$, the proposed solution is $(-1, 6)$.

Step 6 Check. Take a moment to show that $(-1, 6)$ satisfies both given equations. The solution is $(-1, 6)$ and the solution set is $\{(-1, 6)\}$.

CHECK POINT 6 Solve by the addition method:

$$\begin{cases} 3x = 2 - 4y \\ 5y = -1 - 2x. \end{cases}$$

5 *Identify systems that do not have exactly one ordered-pair solution.*

Linear Systems Having No Solution or Infinitely Many Solutions

We have seen that a system of linear equations in two variables represents a pair of lines. The lines either intersect at one point, are parallel, or are identical. Thus, there are three possibilities for the number of solutions to a system of two linear equations.

THE NUMBER OF SOLUTIONS TO A SYSTEM OF TWO LINEAR EQUATIONS

The number of solutions to a system of two linear equations in two variables is given by one of the following. (See **Figure 7.29**.)

Number of Solutions	What This Means Graphically
Exactly one ordered-pair solution	The two lines intersect at one point.
No solution	The two lines are parallel.
Infinitely many solutions	The two lines are identical.

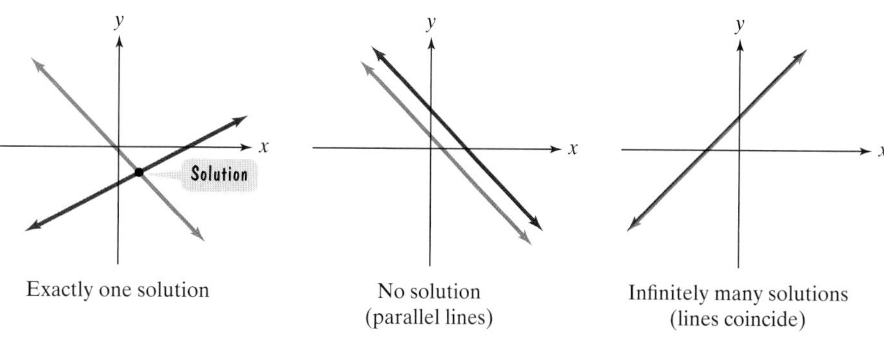

FIGURE 7.29 Possible graphs for a system of two linear equations in two variables

EXAMPLE 7 *A System with No Solution*

Solve the system:

$$\begin{cases} 4x + 6y = 12 \\ 6x + 9y = 12. \end{cases}$$

SOLUTION

Because no variable is isolated, we will use the addition method. To obtain coefficients of x that differ only in sign, we multiply the first equation by 3 and the second equation by -2.

$$\begin{cases} 4x + 6y = 12 \\ 6x + 9y = 12 \end{cases} \xrightarrow[\text{Multiply by } -2.]{\text{Multiply by 3.}} \begin{cases} 12x + 18y = 36 \\ -12x - 18y = -24 \end{cases}$$

$$\text{Add: } \quad 0 = 12$$

There are no values of x and y for which $0 = 12$. No values of x and y satisfy $0x + 0y = 12$.

The false statement $0 = 12$ indicates that the system has no solution. The solution set is the empty set, $\varnothing$.

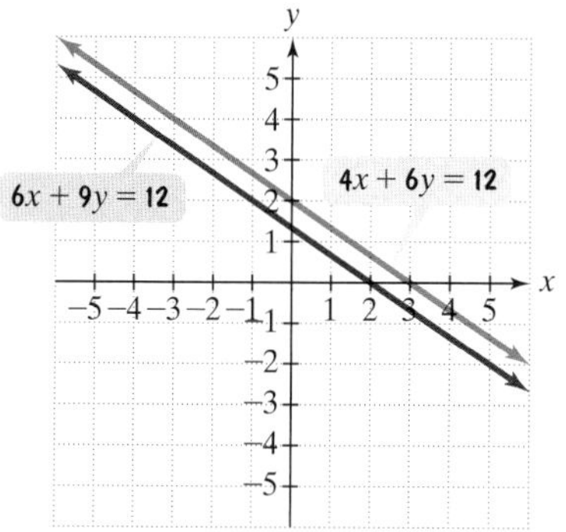

FIGURE 7.30 The graph of a system with no solution

The lines corresponding to the two equations in Example 7 are shown in **Figure 7.30**. The lines are parallel and have no point of intersection.

CHECK POINT 7 Solve the system:

$$\begin{cases} x + 2y = 4 \\ 3x + 6y = 13. \end{cases}$$

EXAMPLE 8 A System with Infinitely Many Solutions

Solve the system:

$$\begin{cases} y = 3x - 2 \\ 15x - 5y = 10. \end{cases}$$

SOLUTION

Because the variable y is isolated in $y = 3x - 2$, the first equation, we can use the substitution method. We substitute the expression for y into the second equation.

$y = \boxed{3x - 2}$

$15x - 5\boxed{y} = 10$	Substitute $3x - 2$ for y.
$15x - 5(3x - 2) = 10$	The substitution results in an equation in one variable.
$15x - 15x + 10 = 10$	Apply the distributive property.
$10 = 10$	Simplify.

This statement is true for all values of x and y.

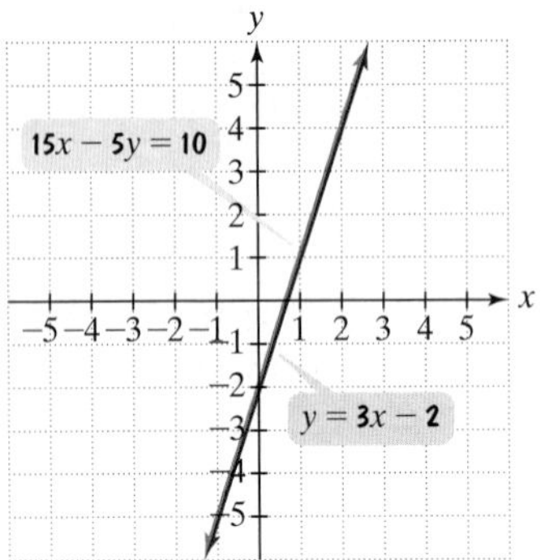

FIGURE 7.31 The graph of a system with infinitely many solutions

In our final step, both variables have been eliminated and the resulting statement, $10 = 10$, is true. This true statement indicates that the system has infinitely many solutions. The solution set consists of all points (x, y) lying on either of the coinciding lines, $y = 3x - 2$ or $15x - 5y = 10$, as shown in **Figure 7.31**.

We express the solution set for the system in one of two equivalent ways:

$$\{(x, y) \mid y = 3x - 2\} \quad \text{or} \quad \{(x, y) \mid 15x - 5y = 10\}.$$

The set of all ordered pairs (x, y) such that $y = 3x - 2$

The set of all ordered pairs (x, y) such that $15x - 5y = 10$

GREAT QUESTION!

The system in Example 8 has infinitely many solutions. Does that mean that any ordered pair of numbers is a solution?

No. Although the system in Example 8 has infinitely many solutions, this does not mean that any ordered pair of numbers you can form will be a solution. The ordered pair (x, y) must satisfy one of the system's equations, $y = 3x - 2$ or $15x - 5y = 10$, and there are infinitely many such ordered pairs. Because the graphs are coinciding lines, the ordered pairs that are solutions of one of the equations are also solutions of the other equation.

CHECK POINT 8 Solve the system:

$$\begin{cases} y = 4x - 4 \\ 8x - 2y = 8. \end{cases}$$

LINEAR SYSTEMS HAVING NO SOLUTION OR INFINITELY MANY SOLUTIONS

If both variables are eliminated when solving a system of linear equations by substitution or addition, one of the following applies:

1. There is no solution if the resulting statement is false.
2. There are infinitely many solutions if the resulting statement is true.

6 *Solve problems using systems of linear equations.*

Modeling with Systems of Equations: Making Money (and Losing It)

What does every entrepreneur, from a kid selling lemonade to Mark Zuckerberg, want to do? Generate profit, of course. The profit made is the money taken in, or the revenue, minus the money spent, or the cost.

> **REVENUE AND COST FUNCTIONS**
>
> A company produces and sells x units of a product. Its **revenue** is the money generated by selling x units of the product. Its **cost** is the cost of producing x units of the product.
>
> Revenue Function
>
> $$R(x) = (\text{price per unit sold})x$$
>
> Cost Function
>
> $$C(x) = \text{fixed cost} + (\text{cost per unit produced})x$$

The point of intersection of the graphs of the revenue and cost functions is called the **break-even point**. The x-coordinate of the point reveals the number of units that a company must produce and sell so that money coming in, the revenue, is equal to money going out, the cost. The y-coordinate of the break-even point gives the amount of money coming in and going out. Example 9 illustrates the use of the substitution method in determining a company's break-even point.

EXAMPLE 9 *Finding a Break-Even Point*

Technology is now promising to bring light, fast, and beautiful wheelchairs to millions of disabled people. A company is planning to manufacture these radically different wheelchairs. Fixed cost will be \$500,000 and it will cost \$400 to produce each wheelchair. Each wheelchair will be sold for \$600.

a. Write the cost function, C, of producing x wheelchairs.

b. Write the revenue function, R, from the sale of x wheelchairs.

c. Determine the break-even point. Describe what this means.

SOLUTION

a. The cost function is the sum of the fixed cost and variable cost.

Fixed cost of \$500,000 | plus | Variable cost: \$400 for each chair produced

$$C(x) = 500{,}000 + 400x$$

b. The revenue function is the money generated from the sale of x wheelchairs. We are given that each wheelchair will be sold for \$600.

Revenue per chair, \$600, times | the number of chairs sold

$$R(x) = 600x$$

c. The break-even point occurs where the graphs of C and R intersect. Thus, we find this point by solving the system

$$\begin{cases} C(x) = 500{,}000 + 400x \\ R(x) = 600x \end{cases} \quad \text{or} \quad \begin{cases} y = 500{,}000 + 400x \\ y = 600x. \end{cases}$$

Using substitution, we can substitute $600x$ for y in the first equation:

$600x = 500{,}000 + 400x$ Substitute $600x$ for y in $y = 500{,}000 + 400x$.

$200x = 500{,}000$ Subtract $400x$ from both sides.

$x = 2500$ Divide both sides by 200.

Back-substituting 2500 for x in either of the system's equations (or functions), $C(x) = 500{,}000 + 400x$ or $R(x) = 600x$, we obtain

$$R(2500) = 600(2500) = 1{,}500{,}000.$$

We used $R(x) = 600x$.

The break-even point is $(2500, 1{,}500{,}000)$. This means that the company will break even if it produces and sells 2500 wheelchairs. At this level, the money coming in is equal to the money going out: $1,500,000.

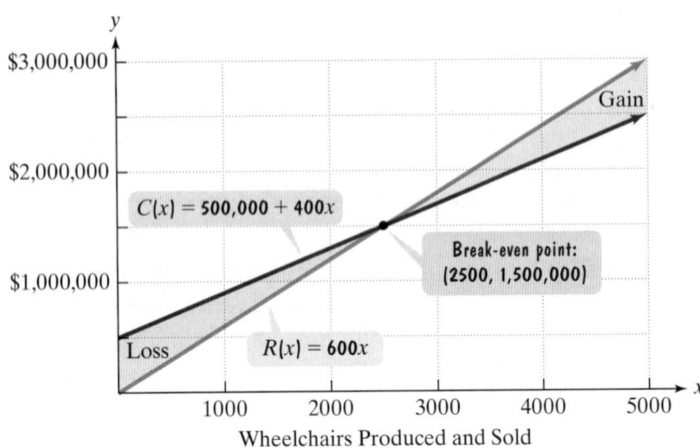

FIGURE 7.32

Figure 7.32 shows the graphs of the revenue and cost functions for the wheelchair business. Similar graphs and models apply no matter how small or large a business venture may be.

The intersection point confirms that the company breaks even by producing and selling 2500 wheelchairs. Can you see what happens for $x < 2500$? The red cost graph lies above the blue revenue graph. The cost is greater than the revenue and the business is losing money. Thus, if they sell fewer than 2500 wheelchairs, the result is a *loss*. By contrast, look at what happens for $x > 2500$. The blue revenue graph lies above the red cost graph. The revenue is greater than the cost and the business is making money. Thus, if they sell more than 2500 wheelchairs, the result is a *gain*.

CHECK POINT 9 A company that manufactures running shoes has a fixed cost of $300,000. Additionally, it costs $30 to produce each pair of shoes. They are sold at $80 per pair.

a. Write the cost function, C, of producing x pairs of running shoes.

b. Write the revenue function, R, from the sale of x pairs of running shoes.

c. Determine the break-even point. Describe what this means.

The profit generated by a business is the money taken in (its revenue) minus the money spent (its cost). Thus, once a business has modeled its revenue and cost with a system of equations, it can determine its *profit function*, $P(x)$.

THE PROFIT FUNCTION

The profit, $P(x)$, generated after producing and selling x units of a product is given by the **profit function**

$$P(x) = R(x) - C(x),$$

where R and C are the revenue and cost functions, respectively.

The profit function for the wheelchair business in Example 9 is

$$\begin{aligned} P(x) &= R(x) - C(x) \\ &= 600x - (500{,}000 + 400x) \\ &= 200x - 500{,}000. \end{aligned}$$

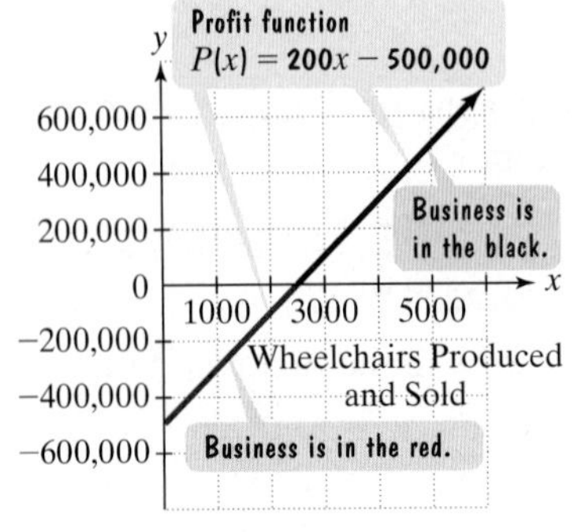

FIGURE 7.33

The graph of this profit function is shown in **Figure 7.33**. The red portion lies below the x-axis and shows a loss when fewer than 2500 wheelchairs are sold. The business is "in the red." The black portion lies above the x-axis and shows a gain when more than 2500 wheelchairs are sold. The wheelchair business is "in the black."

Concept and Vocabulary Check

Fill in each blank so that the resulting statement is true.

1. A solution to a system of linear equations in two variables is an ordered pair that ________.
2. When solving a system of linear equations by graphing, the system's solution is determined by locating ________.
3. When solving
$$\begin{cases} 3x - 2y = 5 \\ y = 3x - 3 \end{cases}$$
by the substitution method, we obtain $x = \frac{1}{3}$, so the solution set is ________.
4. When solving
$$\begin{cases} 2x + 10y = 9 \\ 8x + 5y = 7 \end{cases}$$
by the addition method, we can eliminate y by multiplying the second equation by ______ and then adding the equations.
5. When solving
$$\begin{cases} 4x - 3y = 15 \\ 3x - 2y = 10 \end{cases}$$
by the addition method, we can eliminate y by multiplying the first equation by 2 and the second equation by ______ and then adding the equations.
6. When solving
$$\begin{cases} 12x - 21y = 24 \\ 4x - 7y = 7 \end{cases}$$
by the addition method, we obtain $0 = 3$, so the solution set is ____. If you attempt to solve such a system by graphing, you will obtain two lines that are ________.
7. When solving
$$\begin{cases} x = 3y + 2 \\ 5x - 15y = 10 \end{cases}$$
by the substitution method, we obtain $10 = 10$, so the solution set is ________. If you attempt to solve such a system by graphing, you will obtain two lines that ________.
8. A company's ________ function is the money generated by selling x units of its product. The difference between this function and the company's cost function is called its ________ function.
9. A company has a graph that shows the money it generates by selling x units of its product. It also has a graph that shows its cost of producing x units of its product. The point of intersection of these graphs is called the company's ________.

Exercise Set 7.3

Practice Exercises

In Exercises 1–4, determine whether the given ordered pair is a solution of the system.

1. $(2, 3)$
$$\begin{cases} x + 3y = 11 \\ x - 5y = -13 \end{cases}$$

2. $(-3, 5)$
$$\begin{cases} 9x + 7y = 8 \\ 8x - 9y = -69 \end{cases}$$

3. $(2, 5)$
$$\begin{cases} 2x + 3y = 17 \\ x + 4y = 16 \end{cases}$$

4. $(8, 5)$
$$\begin{cases} 5x - 4y = 20 \\ 3y = 2x + 1 \end{cases}$$

In Exercises 5–12, solve each system by graphing. Check the coordinates of the intersection point in both equations.

5. $\begin{cases} x + y = 6 \\ x - y = 2 \end{cases}$

6. $\begin{cases} x + y = 2 \\ x - y = 4 \end{cases}$

7. $\begin{cases} 2x - 3y = 6 \\ 4x + 3y = 12 \end{cases}$

8. $\begin{cases} 4x + y = 4 \\ 3x - y = 3 \end{cases}$

9. $\begin{cases} y = x + 5 \\ y = -x + 3 \end{cases}$

10. $\begin{cases} y = x + 1 \\ y = 3x - 1 \end{cases}$

11. $\begin{cases} y = -x - 1 \\ 4x - 3y = 24 \end{cases}$

12. $\begin{cases} y = 3x - 4 \\ 2x + y = 1 \end{cases}$

In Exercises 13–24, solve each system by the substitution method. Be sure to check all proposed solutions.

13. $\begin{cases} x + y = 4 \\ y = 3x \end{cases}$

14. $\begin{cases} x + y = 6 \\ y = 2x \end{cases}$

15. $\begin{cases} x + 3y = 8 \\ y = 2x - 9 \end{cases}$

16. $\begin{cases} 2x - 3y = -13 \\ y = 2x + 7 \end{cases}$

17. $\begin{cases} x + 3y = 5 \\ 4x + 5y = 13 \end{cases}$

18. $\begin{cases} y = 3x - 17 \\ 2x - y = 11 \end{cases}$

19. $\begin{cases} 2x - y = -5 \\ x + 5y = 14 \end{cases}$

20. $\begin{cases} 2x + 3y = 11 \\ x - 4y = 0 \end{cases}$

21. $\begin{cases} 2x - y = 3 \\ 5x - 2y = 10 \end{cases}$

22. $\begin{cases} -x + 3y = 10 \\ 2x + 8y = -6 \end{cases}$

23. $\begin{cases} x + 8y = 6 \\ 2x + 4y = -3 \end{cases}$

24. $\begin{cases} -4x + y = -11 \\ 2x - 3y = 5 \end{cases}$

In Exercises 25–36, solve each system by the addition method. Be sure to check all proposed solutions.

25. $\begin{cases} x + y = 1 \\ x - y = 3 \end{cases}$

26. $\begin{cases} x + y = 6 \\ x - y = -2 \end{cases}$

27. $\begin{cases} 2x + 3y = 6 \\ 2x - 3y = 6 \end{cases}$

28. $\begin{cases} 3x + 2y = 14 \\ 3x - 2y = 10 \end{cases}$

29. $\begin{cases} x + 2y = 2 \\ -4x + 3y = 25 \end{cases}$

30. $\begin{cases} 2x - 7y = 2 \\ 3x + y = -20 \end{cases}$

31. $\begin{cases} 4x + 3y = 15 \\ 2x - 5y = 1 \end{cases}$

32. $\begin{cases} 3x - 7y = 13 \\ 6x + 5y = 7 \end{cases}$

33. $\begin{cases} 3x - 4y = 11 \\ 2x + 3y = -4 \end{cases}$

34. $\begin{cases} 2x + 3y = -16 \\ 5x - 10y = 30 \end{cases}$

35. $\begin{cases} 2x = 3y - 4 \\ -6x + 12y = 6 \end{cases}$

36. $\begin{cases} 5x = 4y - 8 \\ 3x + 7y = 14 \end{cases}$

In Exercises 37–44, solve by the method of your choice. Identify systems with no solution and systems with infinitely many solutions, using set notation to express their solution sets.

37. $\begin{cases} x = 9 - 2y \\ x + 2y = 13 \end{cases}$

38. $\begin{cases} 6x + 2y = 7 \\ y = 2 - 3x \end{cases}$

39. $\begin{cases} y = 3x - 5 \\ 21x - 35 = 7y \end{cases}$

40. $\begin{cases} 9x - 3y = 12 \\ y = 3x - 4 \end{cases}$

41. $\begin{cases} 3x - 2y = -5 \\ 4x + y = 8 \end{cases}$

42. $\begin{cases} 2x + 5y = -4 \\ 3x - y = 11 \end{cases}$

43. $\begin{cases} x + 3y = 2 \\ 3x + 9y = 6 \end{cases}$

44. $\begin{cases} 4x - 2y = 2 \\ 2x - y = 1 \end{cases}$

Practice Plus

Use the graphs of the linear functions to solve Exercises 45–46.

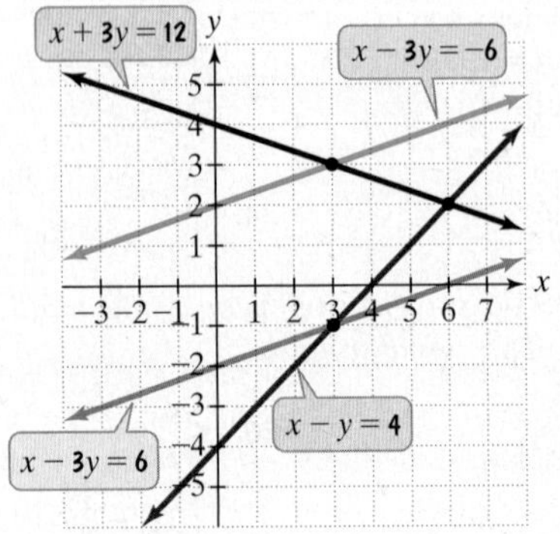

45. Write the linear system whose solution set is $\{(6, 2)\}$. Express each equation in the system in slope-intercept form.

46. Write the linear system whose solution set is $\varnothing$. Express each equation in the system in slope-intercept form.

In Exercises 47–48, solve each system for x and y, expressing either value in terms of a or b, if necessary. Assume that $a \neq 0$ and $b \neq 0$.

47. $\begin{cases} 5ax + 4y = 17 \\ ax + 7y = 22 \end{cases}$

48. $\begin{cases} 4ax + by = 3 \\ 6ax + 5by = 8 \end{cases}$

49. For the linear function $f(x) = mx + b$, $f(-2) = 11$ and $f(3) = -9$. Find m and b.

50. For the linear function $f(x) = mx + b$, $f(-3) = 23$ and $f(2) = -7$. Find m and b.

Application Exercises

The figure shows the graphs of the cost and revenue functions for a company that manufactures and sells small radios. Use the information in the figure to solve Exercises 51–56.

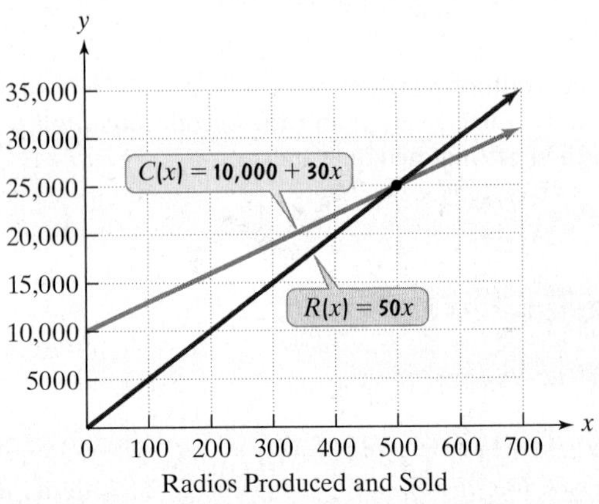

51. How many radios must be produced and sold for the company to break even?

52. More than how many radios must be produced and sold for the company to have a profit?

53. Use the formulas shown in the voice balloons to find $R(200) - C(200)$. Describe what this means for the company.

54. Use the formulas shown in the voice balloons to find $R(300) - C(300)$. Describe what this means for the company.

55. **a.** Use the formulas shown in the voice balloons to write the company's profit function, P, from producing and selling x radios.

    b. Find the company's profit if 10,000 radios are produced and sold.

56. **a.** Use the formulas shown in the voice balloons to write the company's profit function, P, from producing and selling x radios.

    b. Find the company's profit if 20,000 radios are produced and sold.

Exercises 57–60 describe a number of business ventures. For each exercise,

a. *Write the cost function, C.*

b. *Write the revenue function, R.*

c. *Determine the break-even point. Describe what this means.*

57. A company that manufactures small canoes has a fixed cost of \$18,000. It costs \$20 to produce each canoe. The selling price is \$80 per canoe. (In solving this exercise, let x represent the number of canoes produced and sold.)

58. A company that manufactures bicycles has a fixed cost of \$100,000. It costs \$100 to produce each bicycle. The selling price is \$300 per bike. (In solving this exercise, let x represent the number of bicycles produced and sold.)

59. You invest in a new play. The cost includes an overhead of \$30,000, plus production costs of \$2500 per performance. A sold-out performance brings in \$3125. (In solving this exercise, let x represent the number of sold-out performances.)

60. You invested \$30,000 and started a business writing greeting cards. Supplies cost 2 cents per card and you are selling each card for 50 cents. (In solving this exercise, let x represent the number of cards produced and sold.)

An important application of systems of equations arises in connection with supply and demand. As the price of a product increases, the demand for that product decreases. However, at higher prices, suppliers are willing to produce greater quantities of the product. The price at which supply and demand are equal is called the **equilibrium price**. *The quantity supplied and demanded at that price is called the* **equilibrium quantity**. *Exercises 61–62 involve supply and demand.*

61. The table shows the price of a gallon of unleaded premium gasoline. For each price, the table lists the number of gallons per day that a gas station sells and the number of gallons per day that can be supplied.

SUPPLY AND DEMAND FOR UNLEADED PREMIUM GASOLINE

Price per Gallon	Gallons Demanded per Day	Gallons Supplied per Day
\$3.20	1400	200
\$3.60	1200	600
\$4.40	800	1400
\$4.80	600	1800

The data in the table are described by the following demand and supply models:

Demand Model

$p = -0.002x + 6$

Price per gallon — Number of gallons demanded per day

Supply Model

$p = 0.001x + 3.$

Price per gallon — Number of gallons supplied per day

a. Solve the system and find the equilibrium quantity and the equilibrium price for a gallon of unleaded premium gasoline.

b. Use your answer from part (a) to complete this statement: If unleaded premium gasoline is sold for _____ per gallon, there will be a demand for _______ gallons per day and _______ gallons will be supplied per day.

62. The table shows the price of a package of cookies. For each price, the table lists the number of packages that consumers are willing to buy and the number of packages that bakers are willing to supply.

SUPPLY AND DEMAND FOR PACKAGES OF COOKIES

Price of a Package of Cookies	Quantity Demanded (millions of packages) per Week	Quantity Supplied (millions of packages) per Week
30¢	150	70
40¢	130	90
60¢	90	130
70¢	70	150

The data in the table can be described by the following demand and supply models:

Demand Model

$p = -0.5x + 105$

Price per package (cents) — Number of packages demanded per week (millions)

Supply Model

$p = 0.5x - 5.$

Price per package (cents) — Number of packages supplied per week (millions)

a. Solve the system and find the equilibrium quantity and the equilibrium price for a package of cookies.

b. Use your answer from part (a) to complete this statement: If cookies are sold for ______ per package, there will be a demand for ______ million packages per week and bakers will supply ______ million packages per week.

63. We opened this section with a study showing that late in the semester, procrastinating students reported more symptoms of physical illness than their nonprocrastinating peers.

a. At the beginning of the semester, procrastinators reported an average of 0.8 symptoms, increasing at a rate of 0.45 symptoms per week. Write a function that models the average number of symptoms, y, after x weeks.

b. At the beginning of the semester, nonprocrastinators reported an average of 2.6 symptoms, increasing at a rate of 0.15 symptoms per week. Write a function that models the average number of symptoms, y, after x weeks.

c. By which week in the semester did both groups report the same number of symptoms of physical illness? For that week, how many symptoms were reported by each group? How is this shown in **Figure 7.27** on page 438?

64. Harsh, mandatory minimum sentences for drug offenses account for more than half the population in U.S. federal prisons. The bar graph shows the number of inmates in federal prisons, in thousands, for drug offenses and all other crimes in 1998 and 2010. (Other crimes include murder, robbery, fraud, burglary, weapons offenses, immigration offenses, racketeering, and perjury.)

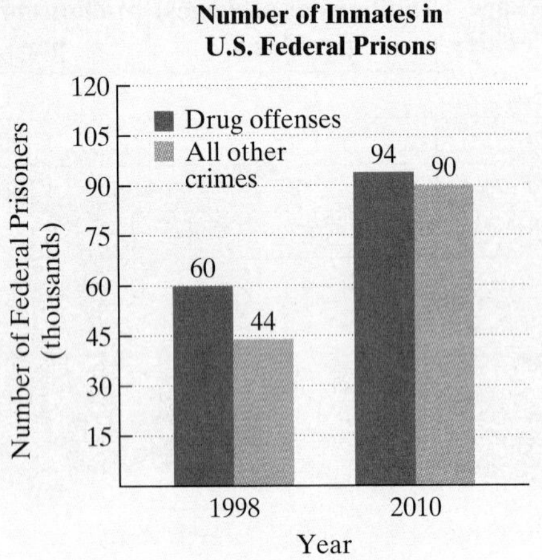

Source: Bureau of Justice Statistics

a. In 1998, there were 60 thousand inmates in federal prisons for drug offenses. For the period shown by the graph, this number increased by approximately 2.8 thousand inmates per year. Write a function that models the number of inmates, y, in thousands, for drug offenses x years after 1998.

b. In 1998, there were 44 thousand inmates in federal prisons for all crimes other than drug offenses. For the period shown by the graph, this number increased by approximately 3.8 thousand inmates per year. Write a function that models the number of inmates, y, in thousands, for all crimes other than drug offenses x years after 1998.

c. Use the models from parts (a) and (b) to determine in which year the number of federal inmates for drug offenses was the same as the number of federal inmates for all other crimes. How many inmates were there for drug offenses and for all other crimes in that year?

Explaining the Concepts

65. What is a system of linear equations? Provide an example with your description.

66. What is the solution to a system of linear equations?

67. Explain how to solve a system of equations using graphing.

68. Explain how to solve a system of equations using the substitution method. Use $y = 3 - 3x$ and $3x + 4y = 6$ to illustrate your explanation.

69. Explain how to solve a system of equations using the addition method. Use $3x + 5y = -2$ and $2x + 3y = 0$ to illustrate your explanation.

70. What is the disadvantage to solving a system of equations using the graphing method?

71. When is it easier to use the addition method rather than the substitution method to solve a system of equations?

72. When using the addition or substitution method, how can you tell whether a system of linear equations has infinitely many solutions? What is the relationship between the graphs of the two equations?

73. When using the addition or substitution method, how can you tell whether a system of linear equations has no solution? What is the relationship between the graphs of the two equations?

74. Describe the break-even point for a business.

Critical Thinking Exercises

Make Sense? *In Exercises 75–78, determine whether each statement makes sense or does not make sense, and explain your reasoning.*

75. Even if a linear system has a solution set involving fractions, such as $\{(\frac{8}{11}, \frac{43}{11})\}$, I can use graphs to determine if the solution set is reasonable.

76. Each equation in a system of linear equations has infinitely many ordered-pair solutions.

77. Every system of linear equations has infinitely many ordered-pair solutions.

78. I find it easiest to use the addition method when one of the equations has a variable on one side by itself.

79. Write a system of equations having $\{(-2, 7)\}$ as a solution set. (More than one system is possible.)

80. One apartment is directly above a second apartment. The resident living downstairs calls his neighbor living above him and states, "If one of you is willing to come downstairs, we'll have the same number of people in both apartments." The upstairs resident responds, "We're all too tired to move. Why don't one of you come up here? Then we will have twice as many people up here as you've got down there." How many people are in each apartment?

81. A set of identical twins can only be distinguished by the characteristic that one always tells the truth and the other always lies. One twin tells you of a lucky number pair: "When I multiply my first lucky number by 3 and my second lucky number by 6, the addition of the resulting numbers produces a sum of 12. When I add my first lucky number and twice my second lucky number, the sum is 5." Which twin is talking?

7.4

WHAT AM I SUPPOSED TO LEARN?

After studying this section, you should be able to:

1. Graph a linear inequality in two variables.
2. Use mathematical models involving linear inequalities.
3. Graph a system of linear inequalities.

Linear Inequalities in Two Variables

WE OPENED THE CHAPTER NOTING THAT THE modern emphasis on thinness as the ideal body shape has been suggested as a major cause of eating disorders. In this section (Example 4), as well as in the Exercise Set (Exercises 45–48), we use systems of linear inequalities in two variables that will enable you to establish a healthy weight range for your height and age.

Linear Inequalities in Two Variables and Their Solutions

We have seen that equations in the form $Ax + By = C$, where A and B are not both zero, are straight lines when graphed. If we change the symbol $=$ to $>$, $<$, $\geq$, or $\leq$, we obtain a **linear inequality in two variables**. Some examples of linear inequalities in two variables are $x + y > 2$, $3x - 5y \leq 15$, and $2x - y < 4$.

A **solution of an inequality in two variables**, x and y, is an ordered pair of real numbers with the following property: When the x-coordinate is substituted for x and the y-coordinate is substituted for y in the inequality, we obtain a true statement. For example, $(3, 2)$ is a solution of the inequality $x + y > 1$. When 3 is substituted for x and 2 is substituted for y, we obtain the true statement $3 + 2 > 1$, or $5 > 1$. Because there are infinitely many pairs of numbers that have a sum greater than 1, the inequality $x + y > 1$ has infinitely many solutions. Each ordered-pair solution is said to **satisfy** the inequality. Thus, $(3, 2)$ satisfies the inequality $x + y > 1$.

1 *Graph a linear inequality in two variables.*

The Graph of a Linear Inequality in Two Variables

We know that the graph of an equation in two variables is the set of all points whose coordinates satisfy the equation. Similarly, the **graph of an inequality in two variables** is the set of all points whose coordinates satisfy the inequality.

Let's use **Figure 7.34** to get an idea of what the graph of a linear inequality in two variables looks like. Part of the figure shows the graph of the linear equation $x + y = 2$. The line divides the points in the rectangular coordinate system into three sets. First, there is the set of points along the line satisfying $x + y = 2$. Next, there is the set of points in the green region above the line. Points in the green region satisfy the linear inequality $x + y > 2$. Finally, there is the set of points in the purple region below the line. Points in the purple region satisfy the linear inequality $x + y < 2$.

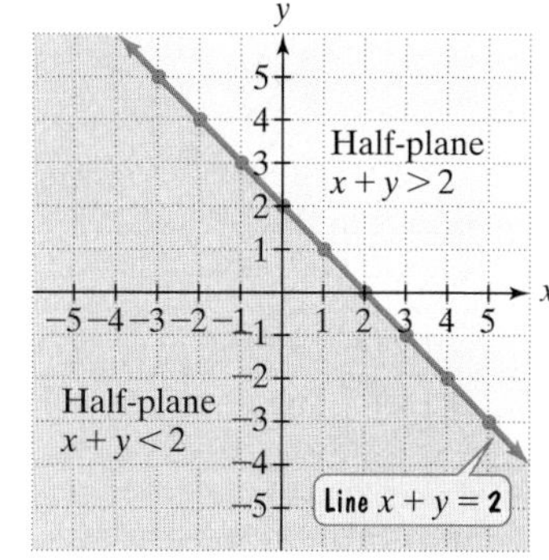

FIGURE 7.34

A **half-plane** is the set of all the points on one side of a line. In **Figure 7.34**, the green region is a half-plane. The purple region is also a half-plane. A half-plane is the graph of a linear inequality that involves $>$ or $<$. The graph of an inequality that involves $\geq$ or $\leq$ is a half-plane and a line. A solid line is used to show that a line is part of a graph. A dashed line is used to show that a line is not part of a graph.

GRAPHING A LINEAR INEQUALITY IN TWO VARIABLES

1. Replace the inequality symbol with an equal sign and graph the corresponding linear equation. Draw a solid line if the original inequality contains a $\leq$ or $\geq$ symbol. Draw a dashed line if the original inequality contains a $<$ or $>$ symbol.
2. Choose a test point from one of the half-planes. (Do not choose a point on the line.) Substitute the coordinates of the test point into the inequality.
3. If a true statement results, shade the half-plane containing this test point. If a false statement results, shade the half-plane not containing this test point.

EXAMPLE 1 Graphing a Linear Inequality in Two Variables

Graph: $3x - 5y \geq 15$.

SOLUTION

Step 1 Replace the inequality symbol by = and graph the linear equation. We need to graph $3x - 5y = 15$. We can use intercepts to graph this line.

We set $y = 0$ to find the x-intercept.	**We set $x = 0$ to find the y-intercept.**
$3x - 5y = 15$	$3x - 5y = 15$
$3x - 5 \cdot 0 = 15$	$3 \cdot 0 - 5y = 15$
$3x = 15$	$-5y = 15$
$x = 5$	$y = -3$

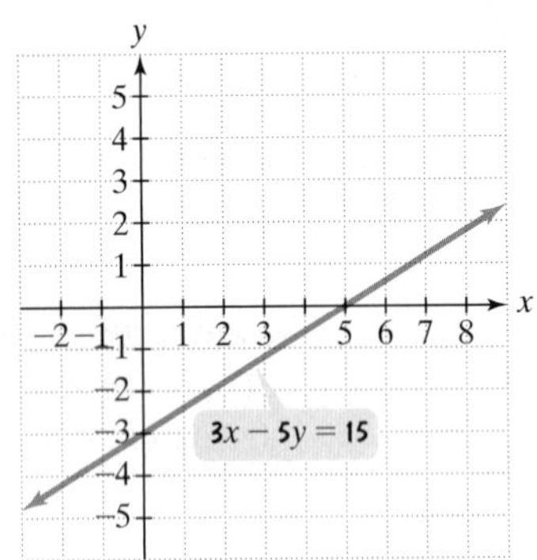

FIGURE 7.35 Preparing to graph $3x - 5y \geq 15$

The x-intercept is 5, so the line passes through $(5, 0)$. The y-intercept is -3, so the line passes through $(0, -3)$. Using the intercepts, the line is shown in **Figure 7.35** as a solid line. The line is solid because the inequality $3x - 5y \geq 15$ contains a $\geq$ symbol, in which equality is included.

Step 2 Choose a test point from one of the half-planes and not from the line. Substitute its coordinates into the inequality. The line $3x - 5y = 15$ divides the plane into three parts—the line itself and two half-planes. The points in one half-plane satisfy $3x - 5y > 15$. The points in the other half-plane satisfy $3x - 5y < 15$. We need to find which half-plane belongs to the solution of $3x - 5y \geq 15$. To do so, we test a point from either half-plane. The origin, $(0, 0)$, is the easiest point to test.

$3x - 5y \geq 15$ This is the given inequality.

$3 \cdot 0 - 5 \cdot 0 \overset{?}{\geq} 15$ Test (0, 0) by substituting 0 for x and 0 for y.

$0 - 0 \overset{?}{\geq} 15$ Multiply.

$0 \geq 15$ This statement is false.

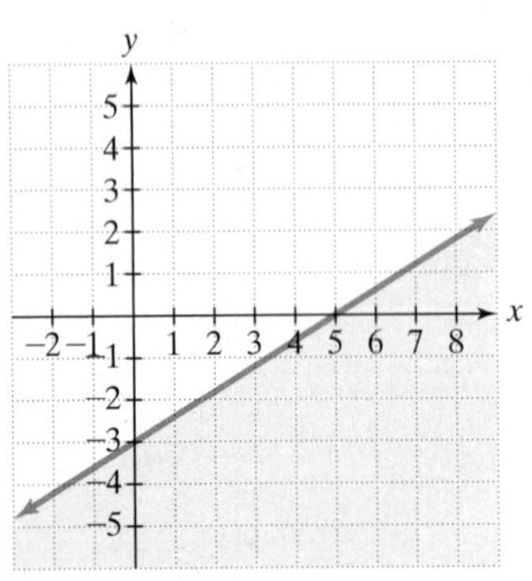

FIGURE 7.36 The graph of $3x - 5y \geq 15$

Step 3 If a false statement results, shade the half-plane not containing the test point. Because 0 is not greater than or equal to 15, the test point, (0, 0), is not part of the solution set. Thus, the half-plane below the solid line $3x - 5y = 15$ is part of the solution set. The solution set is the line and the half-plane that does not contain the point (0, 0), indicated by shading this half-plane. The graph is shown using green shading and a blue line in **Figure 7.36**.

CHECK POINT 1 Graph: $2x - 4y \geq 8$.

When graphing a linear inequality, test a point that lies in one of the half-planes and *not on the line separating the half-planes*. The test point $(0, 0)$ is convenient because it is easy to calculate when 0 is substituted for each variable. However, if $(0, 0)$ lies on the dividing line and not in a half-plane, a different test point must be selected.

EXAMPLE 2 Graphing a Linear Inequality in Two Variables

Graph: $y > -\frac{2}{3}x$.

SOLUTION

Step 1 Replace the inequality symbol by = and graph the linear equation. Because we are interested in graphing $y > -\frac{2}{3}x$, we begin by graphing $y = -\frac{2}{3}x$. We can use the slope and the y-intercept to graph this linear function.

$$y = -\frac{2}{3}x + 0$$

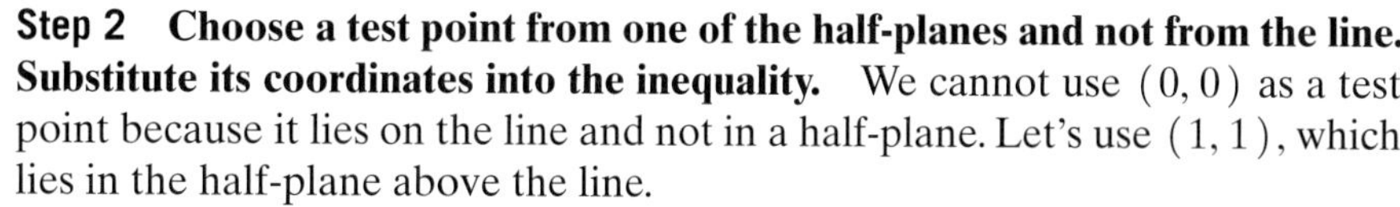

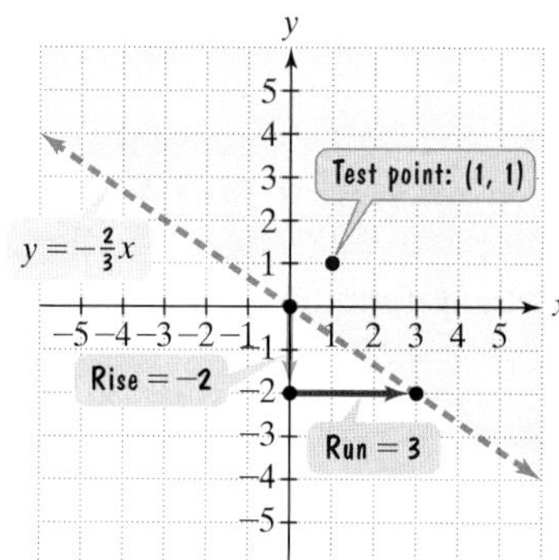

FIGURE 7.37 The graph of $y > -\frac{2}{3}x$

The y-intercept is 0, so the line passes through (0, 0). Using the y-intercept and the slope, the line is shown in **Figure 7.37** as a dashed line. The line is dashed because the inequality $y > -\frac{2}{3}x$ contains a $>$ symbol, in which equality is not included.

Step 2 Choose a test point from one of the half-planes and not from the line. Substitute its coordinates into the inequality. We cannot use $(0, 0)$ as a test point because it lies on the line and not in a half-plane. Let's use $(1, 1)$, which lies in the half-plane above the line.

$$y > -\frac{2}{3}x \quad \text{This is the given inequality.}$$

$$1 \overset{?}{>} -\frac{2}{3} \cdot 1 \quad \text{Test } (1, 1) \text{ by substituting 1 for } x \text{ and 1 for } y.$$

$$1 > -\frac{2}{3} \quad \text{This statement is true.}$$

Step 3 If a true statement results, shade the half-plane containing the test point. Because 1 is greater than $-\frac{2}{3}$, the test point, $(1, 1)$, is part of the solution set. All the points on the same side of the line $y = -\frac{2}{3}x$ as the point $(1, 1)$ are members of the solution set. The solution set is the half-plane that contains the point $(1, 1)$, indicated by shading this half-plane. The graph is shown using green shading and a dashed blue line in **Figure 7.37**.

CHECK POINT 2 Graph: $y > -\frac{3}{4}x$.

Graphing Linear Inequalities without Using Test Points

You can graph inequalities in the form $y > mx + b$ or $y < mx + b$ without using test points. The inequality symbol indicates which half-plane to shade.

- If $y > mx + b$, shade the half-plane above the line $y = mx + b$.
- If $y < mx + b$, shade the half-plane below the line $y = mx + b$.

Observe how this is illustrated in **Figure 7.37**. The graph of $y > -\frac{2}{3}x$ is the half-plane above the line $y = -\frac{2}{3}x$.

It is also not necessary to use test points when graphing inequalities involving half-planes on one side of a vertical or a horizontal line.

GREAT QUESTION!

When is it important to use test points to graph linear inequalities?

Continue using test points to graph inequalities in the form $Ax + By > C$ or $Ax + By < C$. The graph of $Ax + By > C$ can lie above or below the line given by $Ax + By = C$, depending on the values of A and B. The same comment applies to the graph of $Ax + By < C$.

For the Vertical Line $x = a$:

- If $x > a$, shade the half-plane to the right of $x = a$.
- If $x < a$, shade the half-plane to the left of $x = a$.

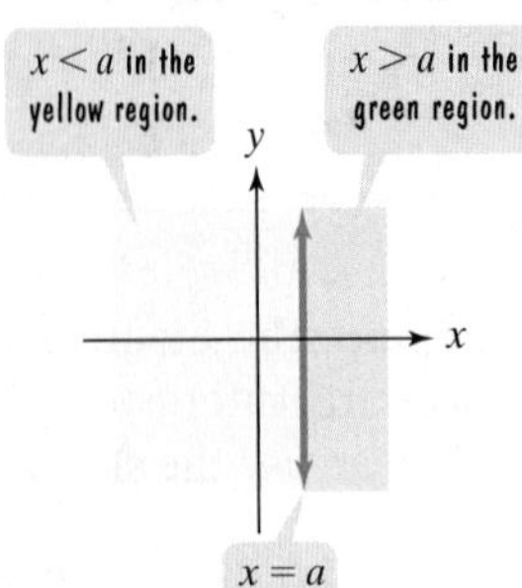

For the Horizontal Line $y = b$:

- If $y > b$, shade the half-plane above $y = b$.
- If $y < b$, shade the half-plane below $y = b$.

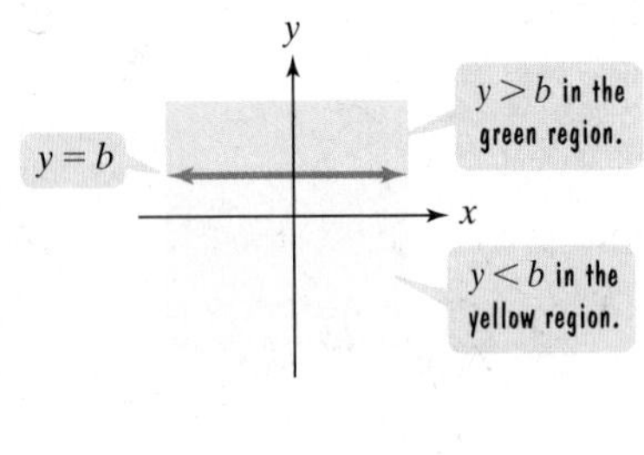

EXAMPLE 3 Graphing Inequalities without Using Test Points

Graph each inequality in a rectangular coordinate system:

a. $y \le -3$ **b.** $x > 2$.

SOLUTION

a. $y \le -3$

Graph $y = -3$, a horizontal line with y-intercept -3. The line is solid because equality is included in $y \le -3$. Because of the less than part of $\le$, shade the half-plane below the horizontal line.

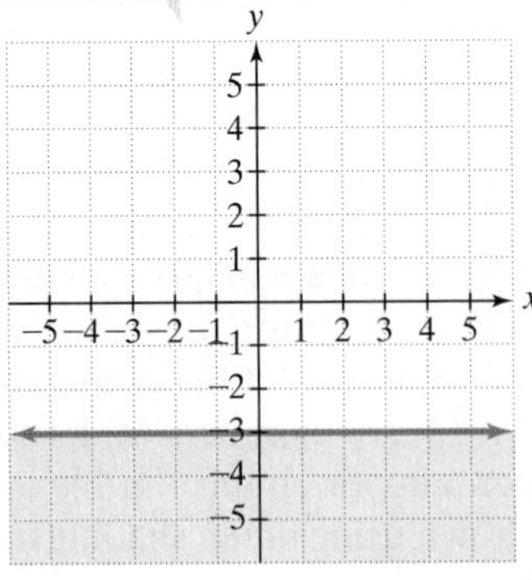

b. $x > 2$

Graph $x = 2$, a vertical line with x-intercept 2. The line is dashed because equality is not included in $x > 2$. Because of $>$, the greater than symbol, shade the half-plane to the right of the vertical line.

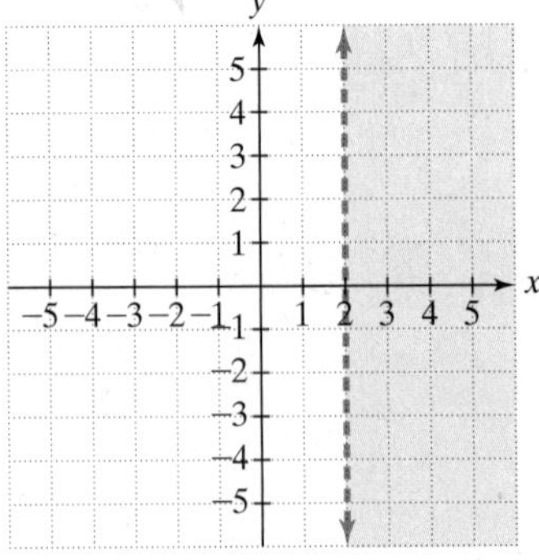

CHECK POINT 3 Graph each inequality in a rectangular coordinate system:

a. $y > 1$ **b.** $x \le -2$.

2 *Use mathematical models involving linear inequalities.*

Modeling with Systems of Linear Inequalities

Just as two or more linear equations make up a system of linear equations, two or more linear inequalities make up a **system of linear inequalities**. A **solution of a system of linear inequalities** in two variables is an ordered pair that satisfies each inequality in the system.

EXAMPLE 4 Does Your Weight Fit You?

The latest guidelines, which apply to both men and women, give healthy weight ranges, rather than specific weights, for your height. **Figure 7.38** shows the healthy weight region for various heights for people between the ages of 19 and 34, inclusive.

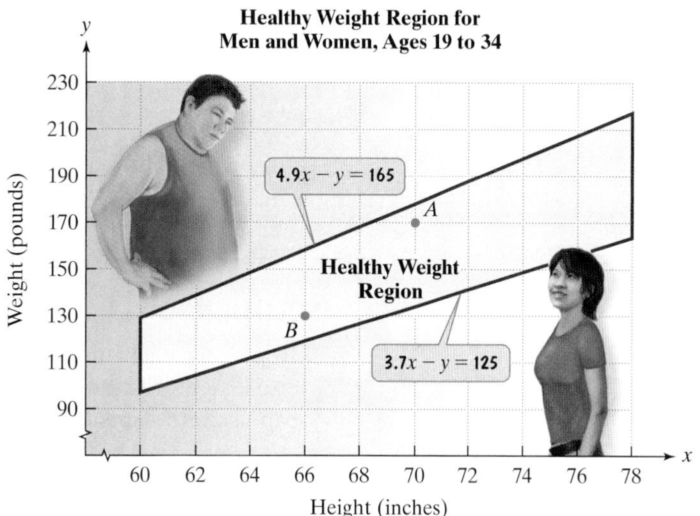

FIGURE 7.38
Source: U.S. Department of Health and Human Services

If x represents height, in inches, and y represents weight, in pounds, the healthy weight region in **Figure 7.38** can be modeled by the following system of linear inequalities:

$$\begin{cases} 4.9x - y \geq 165 \\ 3.7x - y \leq 125. \end{cases}$$

Show that point A in **Figure 7.38** is a solution of the system of inequalities that describes healthy weight.

SOLUTION

Point A has coordinates $(70, 170)$. This means that if a person is 70 inches tall, or 5 feet 10 inches, and weighs 170 pounds, then that person's weight is within the healthy weight region. We can show that $(70, 170)$ satisfies the system of inequalities by substituting 70 for x and 170 for y in each inequality in the system.

$$\begin{aligned} 4.9x - y &\geq 165 \\ 4.9(70) - 170 &\geq 165 \\ 343 - 170 &\geq 165 \\ 173 &\geq 165, \quad \text{true} \end{aligned} \qquad \begin{aligned} 3.7x - y &\leq 125 \\ 3.7(70) - 170 &\leq 125 \\ 259 - 170 &\leq 125 \\ 89 &\leq 125, \quad \text{true} \end{aligned}$$

The coordinates $(70, 170)$ make each inequality true. Thus, $(70, 170)$ satisfies the system for the healthy weight region and is a solution of the system.

CHECK POINT 4 Show that point B in **Figure 7.38** is a solution of the system of inequalities that describes healthy weight.

3 *Graph a system of linear inequalities.*

Graphing Systems of Linear Inequalities

The **solution set of a system of linear inequalities in two variables** is the set of all ordered pairs that satisfy each inequality in the system. Thus, to graph a system of inequalities in two variables, begin by graphing each individual inequality in the same rectangular coordinate system. Then find the region, if there is one, that is common to every graph in the system. This region of intersection gives a picture of the system's solution set.

EXAMPLE 5 Graphing a System of Linear Inequalities

Graph the solution set of the system:

$$\begin{cases} x - y < 1 \\ 2x + 3y \geq 12. \end{cases}$$

SOLUTION

Replacing each inequality symbol in $x - y < 1$ and $2x + 3y \geq 12$ with an equal sign indicates that we need to graph $x - y = 1$ and $2x + 3y = 12$. We can use intercepts to graph these lines.

$x - y = 1$		$2x + 3y = 12$
x-intercept: $x - 0 = 1$ $x = 1$ The line passes through (1, 0).	Set $y = 0$ in each equation.	x-intercept: $2x + 3 \cdot 0 = 12$ $2x = 12$ $x = 6$ The line passes through (6, 0).
y-intercept: $0 - y = 1$ $-y = 1$ $y = -1$ The line passes through (0, −1).	Set $x = 0$ in each equation.	y-intercept: $2 \cdot 0 + 3y = 12$ $3y = 12$ $y = 4$ The line passes through (0, 4).

Now we are ready to graph the solution set of the system of linear inequalities.

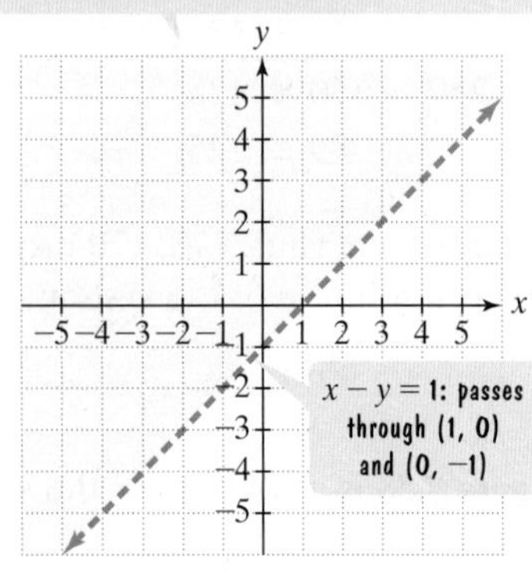

The graph of $x - y < 1$

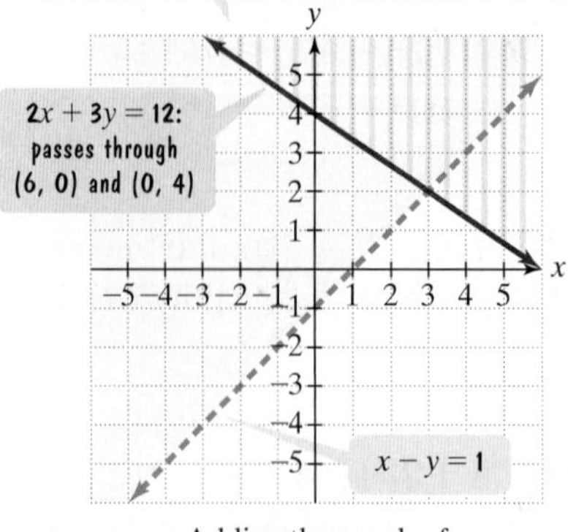

Adding the graph of $2x + 3y \geq 12$

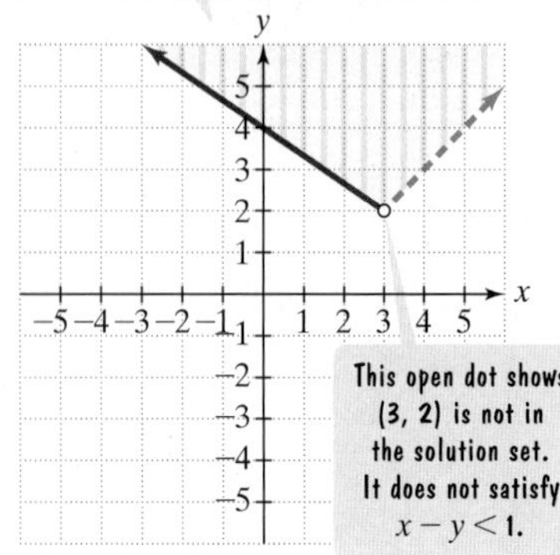

The graph of $x - y < 1$ and $2x + 3y \geq 12$

CHECK POINT 5 Graph the solution set of the system:

$$\begin{cases} x + 2y > 4 \\ 2x - 3y \le -6. \end{cases}$$

EXAMPLE 6 *Graphing a System of Linear Inequalities*

Graph the solution set of the system:

$$\begin{cases} x \le 4 \\ y > -2. \end{cases}$$

SOLUTION

Graph $x \le 4$. The blue vertical line, $x = 4$, is solid. Graph $x < 4$, the half-plane to the left of the blue line, using yellow shading.

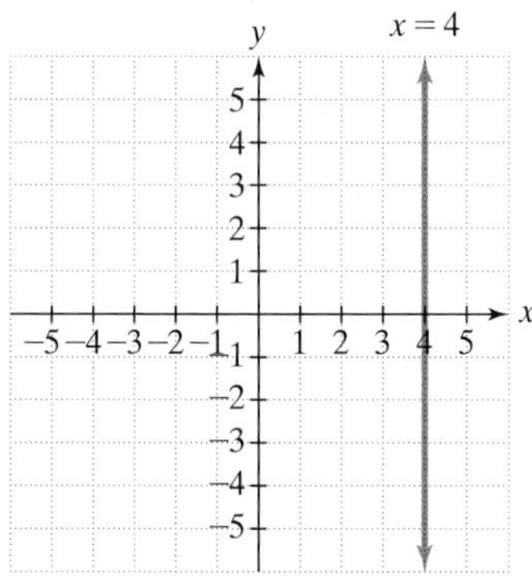

The graph of $x \le 4$

Add the graph of $y > -2$. The red horizontal line, $y = -2$, is dashed. Graph $y > -2$, the half-plane above the dashed red line, using green vertical shading.

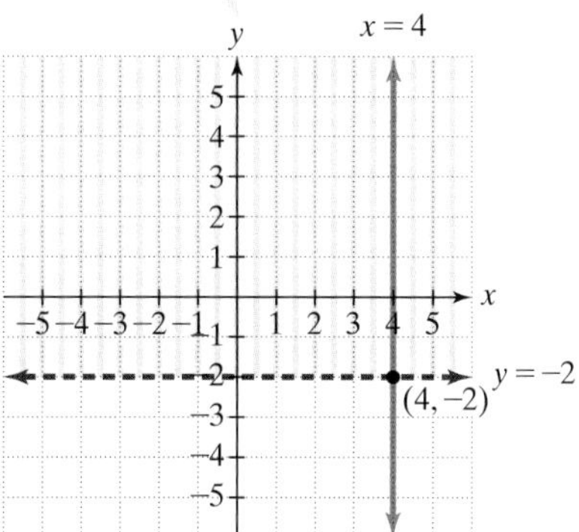

Adding the graph of $y > -2$

The solution set of the system is graphed as the intersection (the overlap) of the two half-planes. This is the region in which the yellow shading and the green vertical shading overlap.

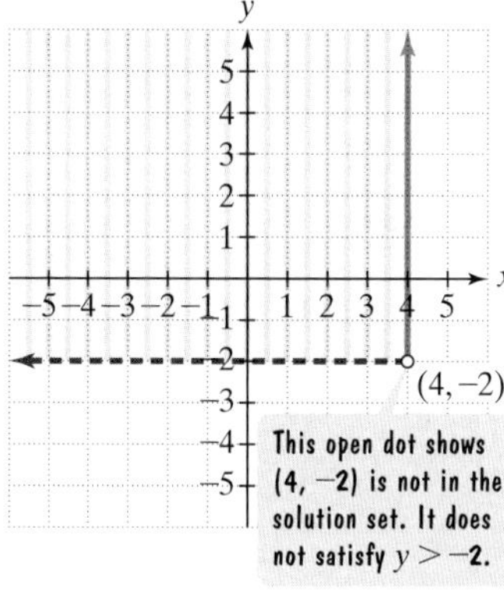

The graph of $x \le 4$ and $y > -2$

CHECK POINT 6 Graph the solution set of the system:

$$\begin{cases} x < 3 \\ y \ge -1. \end{cases}$$

Concept and Vocabulary Check

Fill in each blank so that the resulting statement is true.

1. The ordered pair (3, 2) is a/an ________ of the inequality $x + y > 1$ because when 3 is substituted for ________ and 2 is substituted for ________, the true statement ________ is obtained.
2. The set of all points that satisfy a linear inequality in two variables is called the ________ of the inequality.
3. The set of all points on one side of a line is called a/an ________.
4. True or False: The graph of $2x - 3y > 6$ includes the line $2x - 3y = 6$. ________
5. True or False: The graph of the linear equation $2x - 3y = 6$ is used to graph the linear inequality $2x - 3y > 6$. ________
6. True or False: When graphing $4x - 2y \ge 8$, to determine which side of the line to shade, choose a test point on $4x - 2y = 8$. ________
7. The solution set of the system
$$\begin{cases} x - y < 1 \\ 2x + 3y \ge 12 \end{cases}$$
is the set of ordered pairs that satisfy ________ and ________.
8. True or False: The graph of the solution set of the system
$$\begin{cases} x - 3y < 6 \\ 2x + 3y \ge -6 \end{cases}$$
includes the intersection point of $x - 3y = 6$ and $2x + 3y = -6$. ________

Exercise Set 7.4

Practice Exercises

In Exercises 1–22, graph each linear inequality.

1. $x + y \geq 2$ **2.** $x - y \leq 1$

3. $3x - y \geq 6$ **4.** $3x + y \leq 3$

5. $2x + 3y > 12$ **6.** $2x - 5y < 10$

7. $5x + 3y \leq -15$ **8.** $3x + 4y \leq -12$

9. $2y - 3x > 6$ **10.** $2y - x > 4$

11. $y > \frac{1}{3}x$ **12.** $y > \frac{1}{4}x$

13. $y \leq 3x + 2$ **14.** $y \leq 2x - 1$

15. $y < -\frac{1}{4}x$ **16.** $y < -\frac{1}{3}x$

17. $x \leq 2$ **18.** $x \leq -4$

19. $y > -4$ **20.** $y > -2$

21. $y \geq 0$ **22.** $x \geq 0$

In Exercises 23–38, graph the solution set of each system of inequalities.

23. $\begin{cases} 3x + 6y \leq 6 \\ 2x + y \leq 8 \end{cases}$ **24.** $\begin{cases} x - y \geq 4 \\ x + y \leq 6 \end{cases}$

25. $\begin{cases} 2x + y < 3 \\ x - y > 2 \end{cases}$ **26.** $\begin{cases} x + y < 4 \\ 4x - 2y < 6 \end{cases}$

27. $\begin{cases} 2x + y < 4 \\ x - y > 4 \end{cases}$ **28.** $\begin{cases} 2x - y < 3 \\ x + y < 6 \end{cases}$

29. $\begin{cases} x \geq 2 \\ y \leq 3 \end{cases}$ **30.** $\begin{cases} x \geq 4 \\ y \leq 2 \end{cases}$

31. $\begin{cases} x \leq 5 \\ y > -3 \end{cases}$ **32.** $\begin{cases} x \leq 3 \\ y > -1 \end{cases}$

33. $\begin{cases} x - y \leq 1 \\ x \geq 2 \end{cases}$ **34.** $\begin{cases} 4x - 5y \geq -20 \\ x \geq -3 \end{cases}$

35. $\begin{cases} y > 2x - 3 \\ y < -x + 6 \end{cases}$ **36.** $\begin{cases} y < -2x + 4 \\ y < x - 4 \end{cases}$

37. $\begin{cases} x + 2y \leq 4 \\ y \geq x - 3 \end{cases}$ **38.** $\begin{cases} x + y \leq 4 \\ y \geq 2x - 4 \end{cases}$

Practice Plus

In Exercises 39–40, write each sentence as an inequality in two variables. Then graph the inequality.

39. The y-variable is at least 4 more than the product of -2 and the x-variable.

40. The y-variable is at least 2 more than the product of -3 and the x-variable.

In Exercises 41–42, write the given sentences as a system of inequalities in two variables. Then graph the system.

41. The sum of the x-variable and the y-variable is at most 4. The y-variable added to the product of 3 and the x-variable does not exceed 6.

42. The sum of the x-variable and the y-variable is at most 3. The y-variable added to the product of 4 and the x-variable does not exceed 6.

The graphs of solution sets of systems of inequalities involve finding the intersection of the solution sets of two or more inequalities. By contrast, in Exercises 43–44, you will be graphing the union of the solution sets of two inequalities.

43. Graph the union of $y > \frac{3}{2}x - 2$ and $y < 4$.

44. Graph the union of $x - y \geq -1$ and $5x - 2y \leq 10$.

Application Exercises

The figure shows the healthy weight region for various heights for people ages 35 and older.

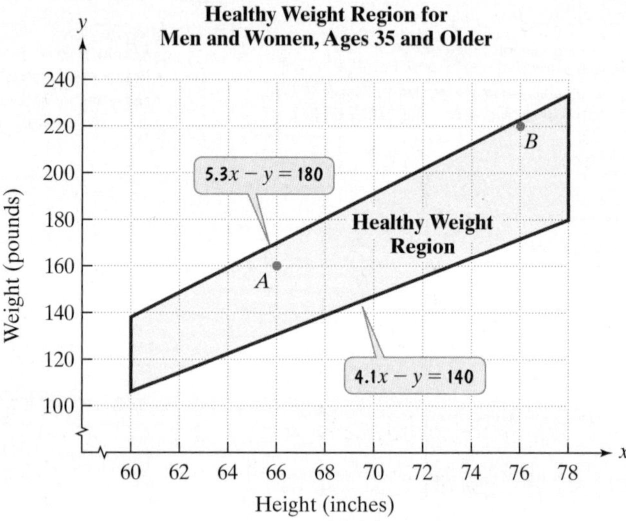

Source: U.S. Department of Health and Human Services

If x represents height, in inches, and y represents weight, in pounds, the healthy weight region can be modeled by the following system of linear inequalities:

$$\begin{cases} 5.3x - y \geq 180 \\ 4.1x - y \leq 140. \end{cases}$$

Use this information to solve Exercises 45–48.

45. Show that point A is a solution of the system of inequalities that describes healthy weight for this age group.

46. Show that point B is a solution of the system of inequalities that describes healthy weight for this age group.

47. Is a person in this age group who is 6 feet tall weighing 205 pounds within the healthy weight region?

48. Is a person in this age group who is 5 feet 8 inches tall weighing 135 pounds within the healthy weight region?

49. Many elevators have a capacity of 2000 pounds.

a. If a child averages 50 pounds and an adult 150 pounds, write an inequality that describes when x children and y adults will cause the elevator to be overloaded.

b. Graph the inequality. Because x and y must be positive, limit the graph to quadrant I only.

c. Select an ordered pair satisfying the inequality. What are its coordinates and what do they represent in this situation?

50. A patient is not allowed to have more than 330 milligrams of cholesterol per day from a diet of eggs and meat. Each egg provides 165 milligrams of cholesterol. Each ounce of meat provides 110 milligrams.

a. Write an inequality that describes the patient's dietary restrictions for x eggs and y ounces of meat.

b. Graph the inequality. Because x and y must be positive, limit the graph to quadrant I only.

c. Select an ordered pair satisfying the inequality. What are its coordinates and what do they represent in this situation?

The graph of an inequality in two variables is a region in the rectangular coordinate system. Regions in coordinate systems have numerous applications. For example, the regions in the following two graphs indicate whether a person is obese, overweight, borderline overweight, normal weight, or underweight.

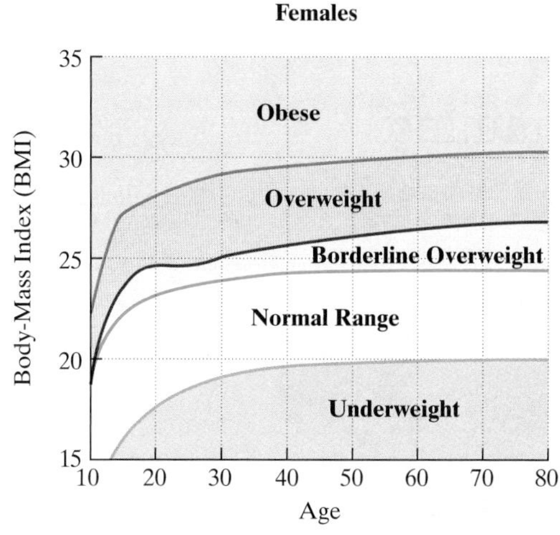

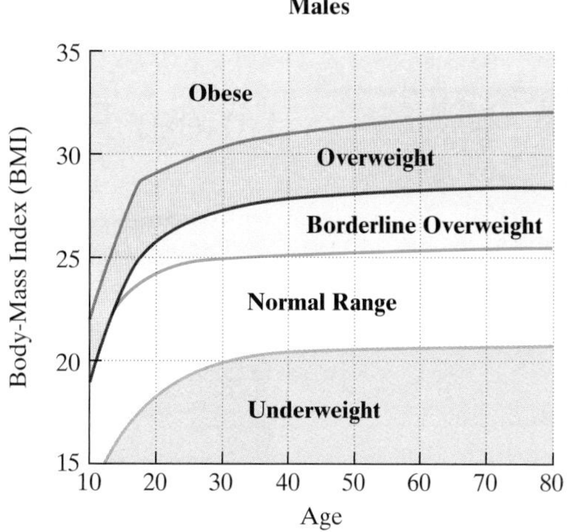

Source: Centers for Disease Control and Prevention

In these graphs, each horizontal axis shows a person's age. Each vertical axis shows that person's body-mass index (BMI), computed using the following formula:

$$\text{BMI} = \frac{703W}{H^2}.$$

The variable W represents weight, in pounds. The variable H represents height, in inches. Use this information and the graphs shown above to solve Exercises 51–52.

51. A man is 20 years old, 72 inches (6 feet) tall, and weighs 200 pounds.

a. Compute the man's BMI. Round to the nearest tenth.

b. Use the man's age and his BMI to locate this information as a point in the coordinate system for males. Is this person obese, overweight, borderline overweight, normal weight, or underweight?

52. A woman is 25 years old, 66 inches (5 feet, 6 inches) tall, and weighs 105 pounds.

a. Compute the woman's BMI. Round to the nearest tenth.

b. Use the woman's age and her BMI to locate this information as a point in the coordinate system for females. Is this person obese, overweight, borderline overweight, normal weight, or underweight?

Explaining the Concepts

53. What is a half-plane?

54. What does a dashed line mean in the graph of an inequality?

55. Explain how to graph $2x - 3y < 6$.

56. Compare the graphs of $3x - 2y > 6$ and $3x - 2y \le 6$. Discuss similarities and differences between the graphs.

57. Describe how to solve a system of linear inequalities.

Critical Thinking Exercises

Make Sense? *In Exercises 58–61, determine whether each statement makes sense or does not make sense, and explain your reasoning.*

58. When graphing a linear inequality, I should always use (0, 0) as a test point because it's easy to perform the calculations when 0 is substituted for each variable.

59. When graphing $3x - 4y < 12$, it's not necessary for me to graph the linear equation $3x - 4y = 12$ because the inequality contains a $<$ symbol, in which equality is not included.

60. Systems of linear inequalities are appropriate for modeling healthy weight because guidelines give healthy weight ranges, rather than specific weights, for various heights.

61. I graphed the solution set of $y \ge x + 2$ and $x \ge 1$ without using test points.

In Exercises 62–63, write a system of inequalities for each graph.

62. **63.**

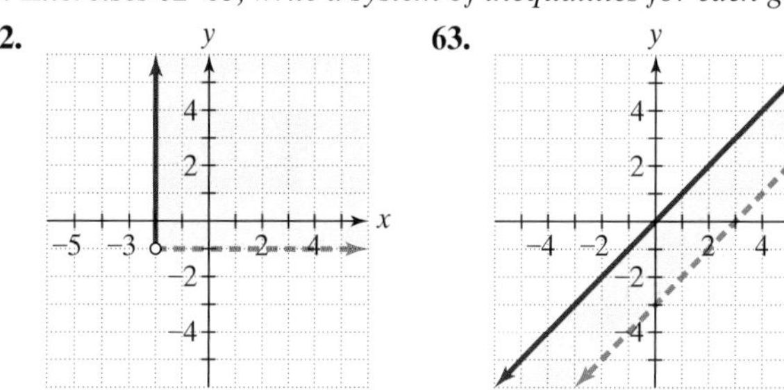

Without graphing, in Exercises 64–67, determine if each system has no solution or infinitely many solutions.

64. $\begin{cases} 3x + y < 9 \\ 3x + y > 9 \end{cases}$

65. $\begin{cases} 6x - y \le 24 \\ 6x - y > 24 \end{cases}$

66. $\begin{cases} 3x + y \le 9 \\ 3x + y \ge 9 \end{cases}$

67. $\begin{cases} 6x - y \le 24 \\ 6x - y \ge 24 \end{cases}$

7.5 Linear Programming

WHAT AM I SUPPOSED TO LEARN?

After studying this section, you should be able to:

1 Write an objective function describing a quantity that must be maximized or minimized.

2 Use inequalities to describe limitations in a situation.

3 Use linear programming to solve problems.

West Berlin children at Tempelhof airport watch fleets of U.S. airplanes bringing in supplies to circumvent the Soviet blockade. The airlift began June 28, 1948, and continued for 15 months.

THE BERLIN AIRLIFT (1948–1949) was an operation by the United States and Great Britain in response to military action by the former Soviet Union: Soviet troops closed all roads and rail lines between West Germany and Berlin, cutting off supply routes to the city. The Allies used a mathematical technique developed during World War II to maximize the amount of supplies transported. During the 15-month airlift, 278,228 flights provided basic necessities to blockaded Berlin, saving one of the world's great cities.

In this section, we will look at an important application of systems of linear inequalities. Such systems arise in **linear programming**, a method for solving problems in which a particular quantity that must be maximized or minimized is limited by other factors. Linear programming is one of the most widely used tools in management science. It helps businesses allocate resources to manufacture products in a way that will maximize profit. Linear programming accounts for more than 50% and perhaps as much as 90% of all computing time used for management decisions in business. The Allies used linear programming to save Berlin.

1 *Write an objective function describing a quantity that must be maximized or minimized.*

Objective Functions in Linear Programming

Many problems involve quantities that must be maximized or minimized. Businesses are interested in maximizing profit. An operation in which bottled water and medical kits are shipped to earthquake survivors needs to maximize the number of survivors helped by this shipment. An **objective function** is an algebraic expression in two or more variables describing a quantity that must be maximized or minimized.

EXAMPLE 1 *Writing an Objective Function*

Bottled water and medical supplies are to be shipped to survivors of an earthquake by plane. Each container of bottled water will serve ten people and each medical kit will aid six people. If x represents the number of bottles of water to be shipped and y represents the number of medical kits, write the objective function that describes the number of people who can be helped.

SOLUTION

Because each bottle of water serves ten people and each medical kit aids six people, we have

The number of people helped	is	10 times the number of bottles of water	plus	6 times the number of medical kits.
	$=$	$10x$	$+$	$6y$.

Using z to represent the number of people helped, the objective function is

$$z = 10x + 6y.$$

Unlike the functions that we have seen so far, the objective function is an equation in three variables. For a value of x and a value of y, there is one and only one value of z. Thus, z is a function of x and y.

CHECK POINT 1 A company manufactures bookshelves and desks for computers. Let x represent the number of bookshelves manufactured daily and y the number of desks manufactured daily. The company's profits are \$25 per bookshelf and \$55 per desk. Write the objective function that describes the company's total daily profit, z, from x bookshelves and y desks. (Check Points 2 through 4 are also related to this situation, so keep track of your answers.)

2 *Use inequalities to describe limitations in a situation.*

Constraints in Linear Programming

Ideally, the number of earthquake survivors helped in Example 1 should increase without restriction so that every survivor receives water and medical kits. However, the planes that ship these supplies are subject to weight and volume restrictions. In linear programming problems, such restrictions are called **constraints**. Each constraint is expressed as a linear inequality. The list of constraints forms a system of linear inequalities.

EXAMPLE 2 *Writing a Constraint*

Each plane can carry no more than 80,000 pounds. The bottled water weighs 20 pounds per container and each medical kit weighs 10 pounds. Let x represent the number of bottles of water to be shipped and y the number of medical kits. Write an inequality that describes this constraint.

SOLUTION

Because each plane can carry no more than 80,000 pounds, we have

The total weight of the water bottles	plus	the total weight of the medical kits	must be less than or equal to	80,000 pounds.
$20x$	$+$	$10y$	$\leq$	80,000.
↑ Each bottle weighs 20 pounds.		↑ Each kit weighs 10 pounds.		

The plane's weight constraint is described by the inequality

$$20x + 10y \leq 80{,}000.$$

CHECK POINT 2 To maintain high quality, the company in Check Point 1 should not manufacture more than a total of 80 bookshelves and desks per day. Write an inequality that describes this constraint.

In addition to a weight constraint on its cargo, each plane has a limited amount of space in which to carry supplies. Example 3 demonstrates how to express this constraint.

EXAMPLE 3 *Writing a Constraint*

Each plane can carry a total volume of supplies that does not exceed 6000 cubic feet. Each water bottle is 1 cubic foot and each medical kit also has a volume of 1 cubic foot. With x still representing the number of water bottles and y the number of medical kits, write an inequality that describes this second constraint.

SOLUTION

Because each plane can carry a volume of supplies that does not exceed 6000 cubic feet, we have

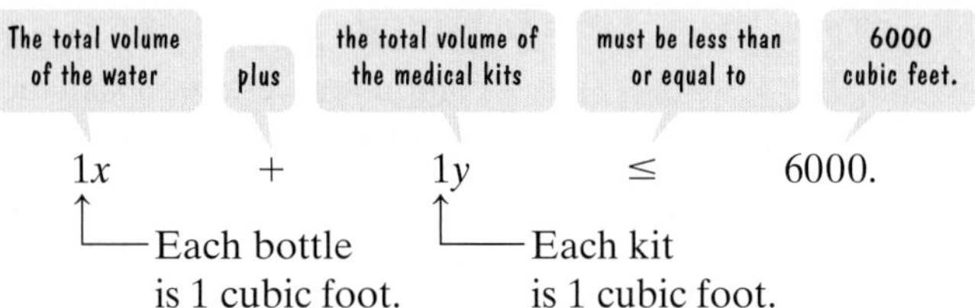

$$1x \quad + \quad 1y \quad \leq \quad 6000.$$

Each bottle is 1 cubic foot. Each kit is 1 cubic foot.

The plane's volume constraint is described by the inequality $x + y \leq 6000$.

In summary, here's what we have described so far in this aid-to-earthquake-survivors situation:

$$z = 10x + 6y$$ This is the objective function describing the number of people helped with x bottles of water and y medical kits.

$$\begin{cases} 20x + 10y \leq 80{,}000 \\ x + y \leq 6000. \end{cases}$$ These are the constraints based on each plane's weight and volume limitations.

CHECK POINT 3 To meet customer demand, the company in Check Point 1 must manufacture between 30 and 80 bookshelves per day, inclusive. Furthermore, the company must manufacture at least 10 and no more than 30 desks per day. Write an inequality that describes each of these sentences. Then summarize what you have described about this company by writing the objective function for its profits and the three constraints.

3 *Use linear programming to solve problems.*

Solving Problems with Linear Programming

The problem in the earthquake situation described previously is to maximize the number of survivors who can be helped, subject to each plane's weight and volume constraints. The process of solving this problem is called *linear programming*, based on a theorem that was proven during World War II.

SOLVING A LINEAR PROGRAMMING PROBLEM

Let $z = ax + by$ be an objective function that depends on x and y. Furthermore, z is subject to a number of constraints on x and y. If a maximum or minimum value of z exists, it can be determined as follows:

1. Graph the system of inequalities representing the constraints.
2. Find the value of the objective function at each corner, or **vertex**, of the graphed region. The maximum and minimum of the objective function occur at one or more of the corner points.

EXAMPLE 4 Solving a Linear Programming Problem

Determine how many bottles of water and how many medical kits should be sent on each plane to maximize the number of earthquake survivors who can be helped.

SOLUTION

We must maximize $z = 10x + 6y$ subject to the following constraints:

$$\begin{cases} 20x + 10y \le 80{,}000 \\ x + y \le 6000. \end{cases}$$

Step 1 Graph the system of inequalities representing the constraints. Because x (the number of bottles of water per plane) and y (the number of medical kits per plane) must be nonnegative, we need to graph the system of inequalities in quadrant I and its boundary only.

To graph the inequality $20x + 10y \le 80{,}000$, we graph the equation $20x + 10y = 80{,}000$ as a solid blue line (**Figure 7.39**). Setting $y = 0$, the x-intercept is 4000 and setting $x = 0$, the y-intercept is 8000. Using $(0, 0)$ as a test point, the inequality is satisfied, so we shade below the blue line, as shown in yellow in **Figure 7.39**.

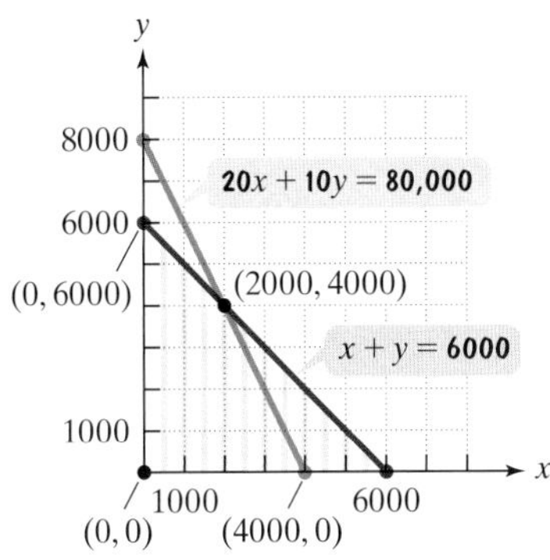

FIGURE 7.39 The region in quadrant I representing the constraints $20x + 10y \le 80{,}000$ and $x + y \le 6000$

Now we graph $x + y \le 6000$ by first graphing $x + y = 6000$ as a solid red line. Setting $y = 0$, the x-intercept is 6000. Setting $x = 0$, the y-intercept is 6000. Using $(0, 0)$ as a test point, the inequality is satisfied, so we shade below the red line, as shown using green vertical shading in **Figure 7.39**.

We use the addition method to find where the lines $20x + 10y = 80{,}000$ and $x + y = 6000$ intersect.

$$\begin{cases} 20x + 10y = 80{,}000 \\ x + y = 6000 \end{cases} \xrightarrow[\text{Multiply by } -10.]{\text{No change}} \begin{cases} 20x + 10y = 80{,}000 \\ -10x - 10y = -60{,}000 \end{cases}$$

$$\text{Add: } 10x = 20{,}000$$

$$x = 2000$$

Back-substituting 2000 for x in $x + y = 6000$, we find $y = 4000$, so the intersection point is $(2000, 4000)$.

The system of inequalities representing the constraints is shown by the region in which the yellow shading and the green vertical shading overlap in **Figure 7.39**. The graph of the system of inequalities is shown again in **Figure 7.40**. The red and blue line segments are included in the graph.

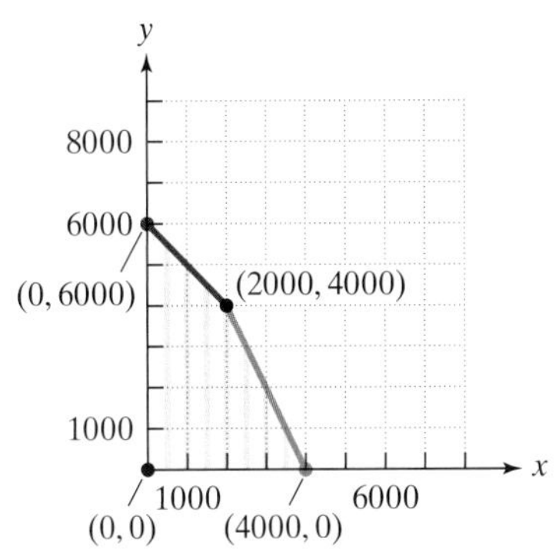

FIGURE 7.40

Step 2 Find the value of the objective function at each corner of the graphed region. The maximum and minimum of the objective function occur at one or more of the corner points. We must evaluate the objective function, $z = 10x + 6y$, at the four corners, or vertices, of the region in **Figure 7.40**.

Corner (x, y)	Objective Function $z = 10x + 6y$	
$(0, 0)$	$z = 10(0) + 6(0) = 0$	
$(4000, 0)$	$z = 10(4000) + 6(0) = 40{,}000$	
$(2000, 4000)$	$z = 10(2000) + 6(4000) = 44{,}000$	← maximum
$(0, 6000)$	$z = 10(0) + 6(6000) = 36{,}000$	

Thus, the maximum value of z is 44,000 and this occurs when $x = 2000$ and $y = 4000$. In practical terms, this means that the maximum number of earthquake survivors who can be helped with each plane shipment is 44,000. This can be accomplished by sending 2000 water bottles and 4000 medical kits per plane.

CHECK POINT 4 For the company in Check Points 1–3, how many bookshelves and how many desks should be manufactured per day to obtain a maximum profit? What is the maximum daily profit?

Concept and Vocabulary Check

Fill in each blank so that the resulting statement is true.

1. A method for finding the maximum or minimum value of a quantity that is subject to various limitations is called ________.
2. An algebraic expression in two or more variables describing a quantity that must be maximized or minimized is called a/an ________ function.
3. A system of linear inequalities is used to represent restrictions, or ________, on a function that must be maximized or minimized. Using the graph of such a system of inequalities, the maximum and minimum values of the function occur at one or more of the ________ points.

Exercise Set 7.5

Practice Exercises

In Exercises 1–4, find the value of the objective function at each corner of the graphed region. What is the maximum value of the objective function? What is the minimum value of the objective function?

1. Objective Function
   $z = 5x + 6y$

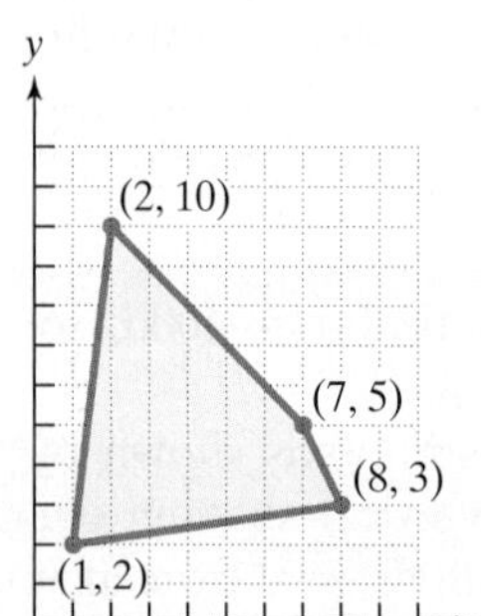

2. Objective Function
   $z = 3x + 2y$

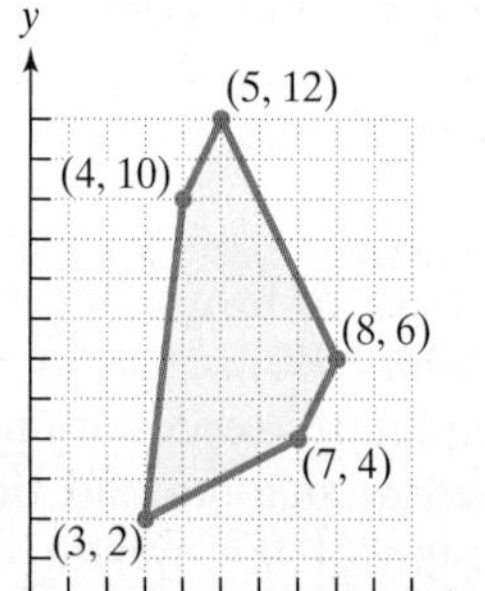

3. Objective Function
   $z = 40x + 50y$

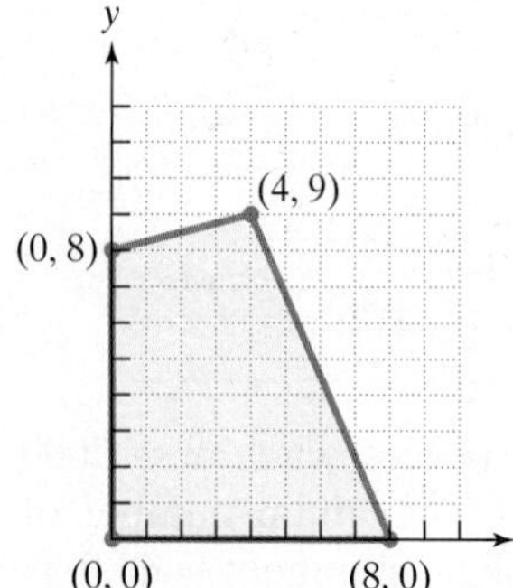

4. Objective Function
   $z = 30x + 45y$

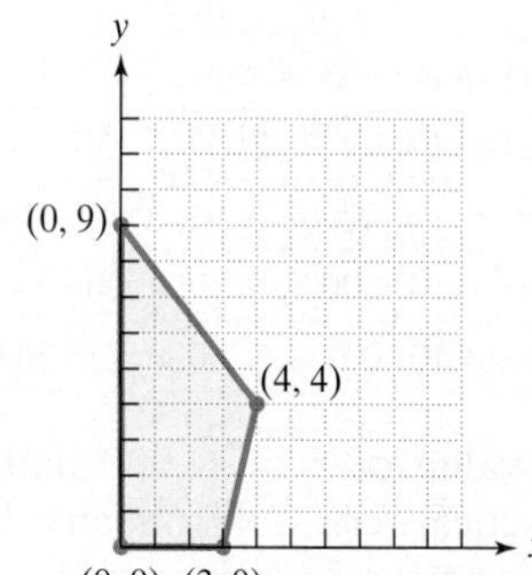

In Exercises 5–8, an objective function and a system of linear inequalities representing constraints are given.

a. *Graph the system of inequalities representing the constraints.*

b. *Find the value of the objective function at each corner of the graphed region.*

c. *Use the values in part (b) to determine the maximum value of the objective function and the values of x and y for which the maximum occurs.*

5. Objective Function
   $z = x + y$
   Constraints
   $$\begin{cases} x \le 6 \\ y \ge 1 \\ 2x - y \ge -1 \end{cases}$$

6. Objective Function
   $z = 3x - 2y$
   Constraints
   $$\begin{cases} x \ge 1 \\ x \le 5 \\ y \ge 2 \\ x - y \ge -3 \end{cases}$$

7. Objective Function
   $z = 6x + 10y$
   Constraints
   $$\begin{cases} x + y \le 12 \\ x + 2y \le 20 \\ \left.\begin{matrix} x \ge 0 \\ y \ge 0 \end{matrix}\right\} \text{Quadrant I and its boundary} \end{cases}$$

8. Objective Function
   $z = x + 3y$
   Constraints
   $$\begin{cases} x + y \ge 2 \\ x \le 6 \\ y \le 5 \\ \left.\begin{matrix} x \ge 0 \\ y \ge 0 \end{matrix}\right\} \text{Quadrant I and its boundary} \end{cases}$$

Practice Plus

Use the directions for Exercises 5–8 to solve Exercises 9–12.

9. Objective Function
   $z = 5x - 2y$
   Constraints
   $$\begin{cases} 0 \le x \le 5 \\ 0 \le y \le 3 \\ x + y \ge 2 \end{cases}$$

10. Objective Function
    $z = 3x - 2y$
    Constraints
    $$\begin{cases} 1 \le x \le 5 \\ y \ge 2 \\ x - y \ge -3 \end{cases}$$

11. Objective Function

$z = 10x + 12y$

Constraints

$$\begin{cases} x \geq 0, y \geq 0 \\ x + y \leq 7 \\ 2x + y \leq 10 \\ 2x + 3y \leq 18 \end{cases}$$

12. Objective Function

$z = 5x + 6y$

Constraints

$$\begin{cases} x \geq 0, y \geq 0 \\ 2x + y \geq 10 \\ x + 2y \geq 10 \\ x + y \leq 10 \end{cases}$$

Application Exercises

13. a. A student earns \$15 per hour for tutoring and \$10 per hour as a teacher's aide. Let x = the number of hours each week spent tutoring and y = the number of hours each week spent as a teacher's aide. Write the objective function that describes total weekly earnings.

b. The student is bound by the following constraints:

- To have enough time for studies, the student can work no more than 20 hours per week.
- The tutoring center requires that each tutor spend at least three hours per week tutoring.
- The tutoring center requires that each tutor spend no more than eight hours per week tutoring.

Write a system of three inequalities that describes these constraints.

c. Graph the system of inequalities in part (b). Use only the first quadrant and its boundary, because x and y are nonnegative.

d. Evaluate the objective function for total weekly earnings at each of the four vertices of the graphed region. [The vertices should occur at $(3, 0)$, $(8, 0)$, $(3, 17)$, and $(8, 12)$.]

e. Complete the missing portions of this statement: The student can earn the maximum amount per week by tutoring for ________ hours per week and working as a teacher's aide for ________ hours per week. The maximum amount that the student can earn each week is \$ ________.

14. A television manufacturer makes rear-projection and plasma televisions. The profit per unit is \$125 for the rear-projection televisions and \$200 for the plasma televisions.

a. Let x = the number of rear-projection televisions manufactured in a month and y = the number of plasma televisions manufactured in a month. Write the objective function that describes the total monthly profit.

b. The manufacturer is bound by the following constraints:

- Equipment in the factory allows for making at most 450 rear-projection televisions in one month.
- Equipment in the factory allows for making at most 200 plasma televisions in one month.
- The cost to the manufacturer per unit is \$600 for the rear-projection televisions and \$900 for the plasma televisions. Total monthly costs cannot exceed \$360,000.

Write a system of three inequalities that describes these constraints.

c. Graph the system of inequalities in part (b). Use only the first quadrant and its boundary, because x and y must both be nonnegative.

d. Evaluate the objective function for total monthly profit at each of the five vertices of the graphed region. [The vertices should occur at $(0, 0)$, $(0, 200)$, $(300, 200)$, $(450, 100)$, and $(450, 0)$.]

e. Complete the missing portions of this statement: The television manufacturer will make the greatest profit by manufacturing ________ rear-projection televisions each month and ________ plasma televisions each month. The maximum monthly profit is \$ ________.

15. Food and clothing are shipped to survivors of a natural disaster. Each carton of food will feed 12 people, while each carton of clothing will help 5 people. Each 20-cubic-foot box of food weighs 50 pounds and each 10-cubic-foot box of clothing weighs 20 pounds. The commercial carriers transporting food and clothing are bound by the following constraints:

- The total weight per carrier cannot exceed 19,000 pounds.
- The total volume must be no more than 8000 cubic feet.

How many cartons of food and clothing should be sent with each plane shipment to maximize the number of people who can be helped?

16. You are about to take a test that contains computation problems worth 6 points each and word problems worth 10 points each. You can do a computation problem in 2 minutes and a word problem in 4 minutes. You have 40 minutes to take the test and may answer no more than 12 problems. Assuming you answer all the problems attempted correctly, how many of each type of problem must you answer to maximize your score? What is the maximum score?

17. A theater is presenting a program on drinking and driving for students and their parents. The proceeds will be donated to a local alcohol information center. Admission is \$2 for parents and \$1 for students. However, the situation has two constraints: The theater can hold no more than 150 people and every two parents must bring at least one student. How many parents and students should attend to raise the maximum amount of money?

18. On June 24, 1948, the former Soviet Union blocked all land and water routes through East Germany to Berlin. A gigantic airlift was organized using American and British planes to bring food, clothing, and other supplies to the more than 2 million people in West Berlin. The cargo capacity was 30,000 cubic feet for an American plane and 20,000 cubic feet for a British plane. To break the Soviet blockade, the Western Allies had to maximize cargo capacity, but were subject to the following restrictions:

- No more than 44 planes could be used.
- The larger American planes required 16 personnel per flight, double that of the requirement for the British planes. The total number of personnel available could not exceed 512.
- The cost of an American flight was \$9000 and the cost of a British flight was \$5000. Total weekly costs could not exceed \$300,000.

Find the number of American and British planes that were used to maximize cargo capacity.

Explaining the Concepts

19. What kinds of problems are solved using the linear programming method?
20. What is an objective function in a linear programming problem?
21. What is a constraint in a linear programming problem? How is a constraint represented?
22. In your own words, describe how to solve a linear programming problem.
23. Describe a situation in your life in which you would like to maximize something, but you are limited by at least two constraints. Can linear programming be used in this situation? Explain your answer.

Critical Thinking Exercises

Make Sense? *In Exercises 24–27, determine whether each statement makes sense or does not make sense, and explain your reasoning.*

24. In order to solve a linear programming problem, I use the graph representing the constraints and the graph of the objective function.
25. I use the coordinates of each vertex from my graph representing the constraints to find the values that maximize or minimize an objective function.
26. I need to be able to graph systems of linear inequalities in order to solve linear programming problems.
27. An important application of linear programming for businesses involves maximizing profit.
28. Suppose that you inherit \$10,000. The will states how you must invest the money. Some (or all) of the money must be invested in stocks and bonds. The requirements are that at least \$3000 be invested in bonds, with expected returns of \$0.08 per dollar, and at least \$2000 be invested in stocks, with expected returns of \$0.12 per dollar. Because the stocks are medium risk, the final stipulation requires that the investment in bonds should never be less than the investment in stocks. How should the money be invested so as to maximize your expected returns?

Group Exercises

29. Group members should choose a particular field of interest. Research how linear programming is used to solve problems in that field. If possible, investigate the solution of a specific practical problem. Present a report on your findings, including the contributions of George Dantzig, Narendra Karmarkar, and L. G. Khachion to linear programming.
30. Members of the group should interview a business executive who is in charge of deciding the product mix for a business. How are production policy decisions made? Are other methods used in conjunction with linear programming? What are these methods? What sort of academic background, particularly in mathematics, does this executive have? Present a group report addressing these questions, emphasizing the role of linear programming for the business.

7.6 Modeling Data: Exponential, Logarithmic, and Quadratic Functions

WHAT AM I SUPPOSED TO LEARN?

After studying this section, you should be able to:

1 Graph exponential functions.
2 Use exponential models.
3 Graph logarithmic functions.
4 Use logarithmic models.
5 Graph quadratic functions.
6 Use quadratic models.
7 Determine an appropriate function for modeling data.

IS THERE A RELATIONSHIP BETWEEN LITERACY AND CHILD mortality? As the percentage of adult females who are literate increases, does the mortality of children under five decrease? **Figure 7.41**, based on data from the United Nations, indicates that this is, indeed, the case. Each point in the figure represents one country.

Data presented in a visual form as a set of points are called a **scatter plot**. Also shown in **Figure 7.41** is a line that passes through or near the points. The line that best fits the data points in a scatter plot is called a **regression line**. We can use the line's slope and y-intercept to obtain a linear model for under-five mortality, y, per thousand, as a function of the percentage of literate adult females, x. The model is given at the top of the next page.

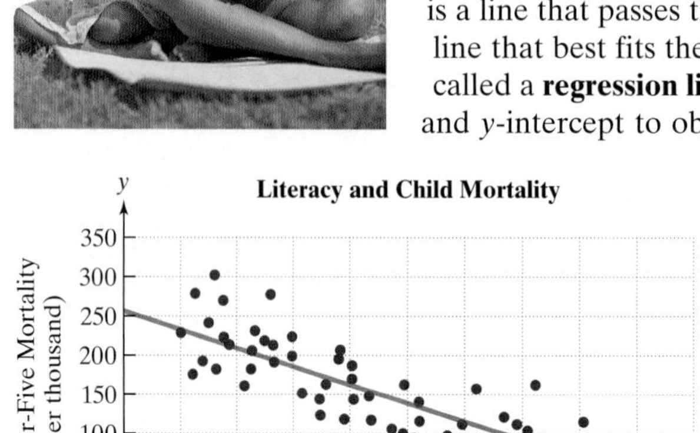

FIGURE 7.41
Source: United Nations

$$y = -2.3x + 255$$

For each percent increase in adult female literacy, under-five mortality decreases by 2.3 per thousand.

Using this model, we can make predictions about child mortality based on the percentage of literate adult females in a country.

In **Figure 7.41**, the data fall on or near a line. However, scatter plots are often curved in a way that indicates that the data do not fall near a line. In this section, we will use functions that are not linear functions to model such data and make predictions.

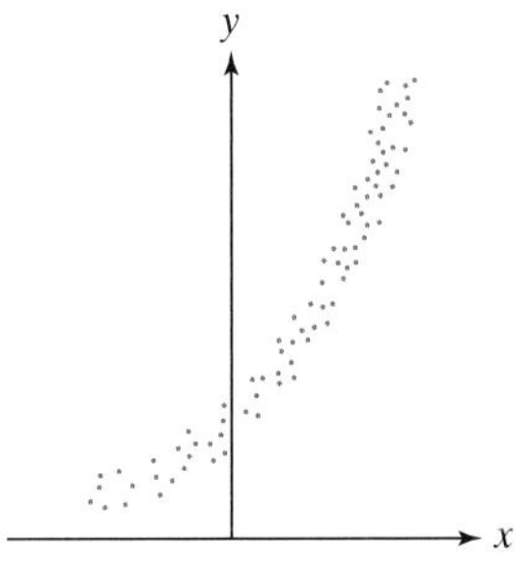

FIGURE 7.42

Modeling with Exponential Functions

The scatter plot in **Figure 7.42** has a shape that indicates the data are increasing more and more rapidly. *Exponential functions* can be used to model this explosive growth, typically associated with populations, epidemics, and interest-bearing bank accounts.

DEFINITION OF THE EXPONENTIAL FUNCTION

The **exponential function with base b** is defined by

$$y = b^x \quad \text{or} \quad f(x) = b^x,$$

where b is a positive constant other than 1 ($b > 0$ and $b \neq 1$) and x is any real number.

1 *Graph exponential functions.*

EXAMPLE 1 *Graphing an Exponential Function*

Graph: $f(x) = 2^x$.

SOLUTION

We start by selecting numbers for x and finding the corresponding values for $f(x)$.

We selected integers from −3 to 3, inclusive, to include three negative numbers, 0, and three positive numbers. We also wanted to keep the resulting computations for y relatively simple.

x	$f(x) = 2^x$	(x, y)
−3	$f(-3) = 2^{-3} = \frac{1}{8}$	$(-3, \frac{1}{8})$
−2	$f(-2) = 2^{-2} = \frac{1}{4}$	$(-2, \frac{1}{4})$
−1	$f(-1) = 2^{-1} = \frac{1}{2}$	$(-1, \frac{1}{2})$
0	$f(0) = 2^0 = 1$	$(0, 1)$
1	$f(1) = 2^1 = 2$	$(1, 2)$
2	$f(2) = 2^2 = 4$	$(2, 4)$
3	$f(3) = 2^3 = 8$	$(3, 8)$

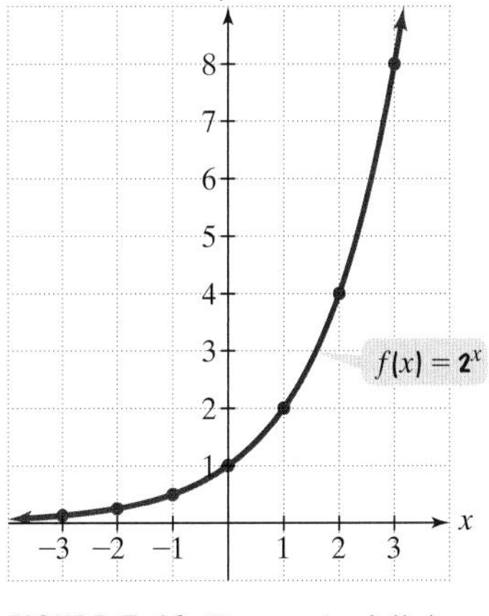

FIGURE 7.43 The graph of $f(x) = 2^x$

We plot these points, connecting them with a smooth curve. **Figure 7.43** shows the graph of $f(x) = 2^x$.

All exponential functions of the form $y = b^x$, or $f(x) = b^x$, where b is a number greater than 1, have the shape of the graph shown in **Figure 7.43**. The graph approaches, but never touches, the negative portion of the x-axis.

CHECK POINT 1 Graph: $f(x) = 3^x$.

Blitzer Bonus

Exponential Growth: The Year Humans Become Immortal

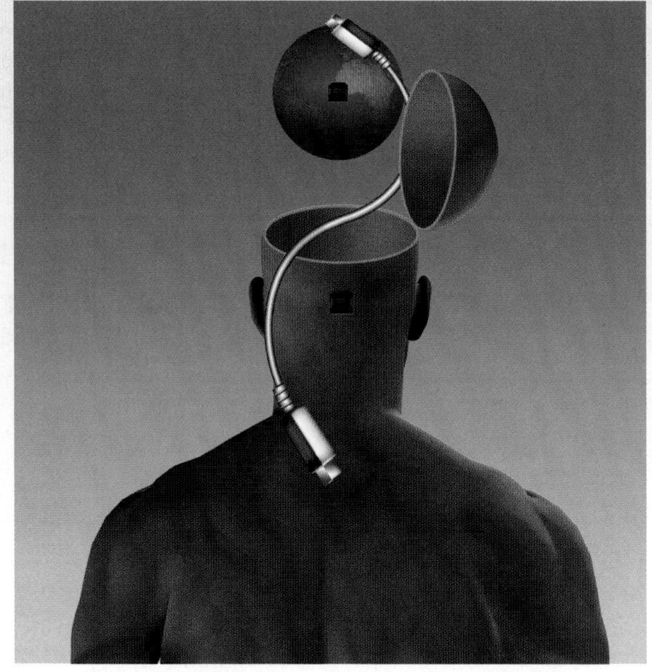

In 2011, *Jeopardy!* aired a three-night match between a personable computer named Watson and the show's two most successful players. The winner: Watson. In the time it took each human contestant to respond to one trivia question, Watson was able to scan the content of one million books. It was also trained to understand the puns and twists of phrases unique to *Jeopardy!* clues.

Watson's remarkable accomplishments can be thought of as a single data point on an exponential curve that models growth in computing power. According to inventor, author, and computer scientist Ray Kurzweil (1948–), computer technology is progressing exponentially, doubling in power each year. What does this mean in terms of the accelerating pace of the graph of $y = 2^x$ that starts slowly and then rockets skyward toward infinity? According to Kurzweil, by 2023, a supercomputer will surpass the brainpower of a human. As progress accelerates exponentially and every hour brings a century's worth of scientific breakthroughs, by 2045, computers will surpass the brainpower equivalent to that of all human brains combined. Here's where it gets exponentially weird: In that year (says Kurzweil), we will be able to scan our consciousness into computers and enter a virtual existence, or swap our bodies for immortal robots. Indefinite life extension will become a reality and people will die only if they choose to.

2 *Use exponential models.*

Figure 7.44(a) shows world population, in billions, for seven selected years from 1950 through 2010. A scatter plot of the data is shown in **Figure 7.44(b)**.

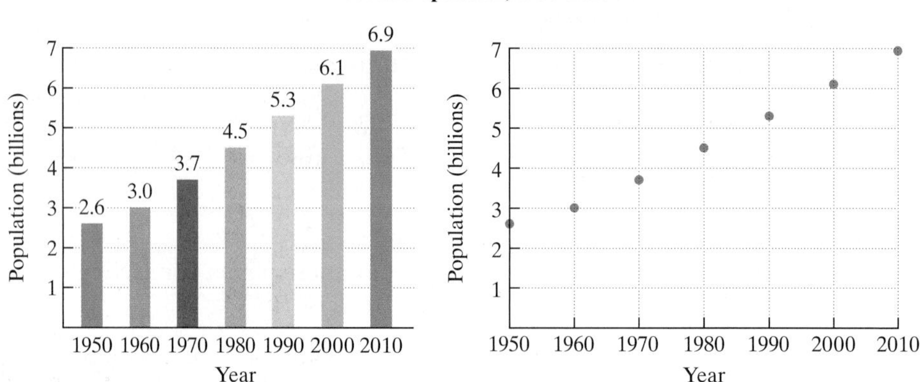

FIGURE 7.44(a) **FIGURE 7.44(b)**

Source: U.S. Census Bureau, International Database

Because the data in the scatter plot appear to increase more and more rapidly, the shape suggests that an exponential function might be a good choice for modeling the data. Furthermore, we can probably draw a line that passes through or near the seven points. Thus, a linear function would also be a good choice for a model.

EXAMPLE 2 Comparing Linear and Exponential Models

The data for world population are shown in **Table 7.3**. Using a graphing utility's linear regression feature and exponential regression feature, we enter the data and obtain the models shown in **Figure 7.45**.

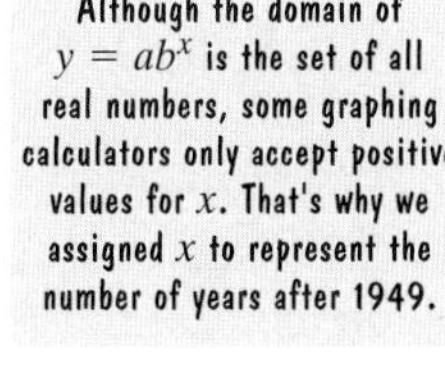
Although the domain of $y = ab^x$ is the set of all real numbers, some graphing calculators only accept positive values for x. That's why we assigned x to represent the number of years after 1949.

TABLE 7.3

x, Number of Years after 1949	y, World Population (billions)
1 (1950)	2.6
11 (1960)	3.0
21 (1970)	3.7
31 (1980)	4.5
41 (1990)	5.3
51 (2000)	6.1
61 (2010)	6.9

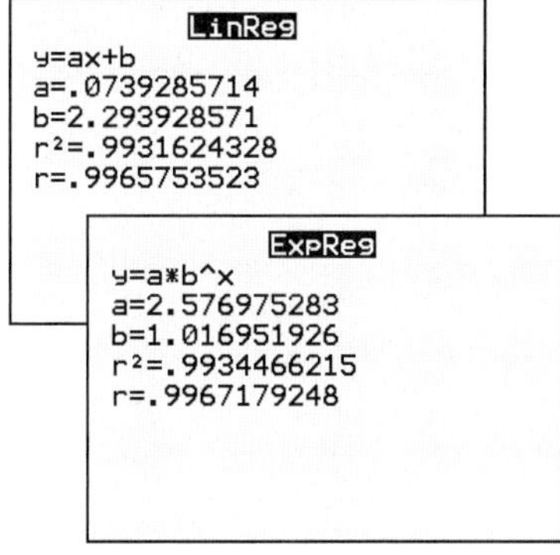

FIGURE 7.45 A linear model and an exponential model for the data in Table 7.3

a. Use **Figure 7.45** to express each model in function notation, with numbers rounded to three decimal places.

b. How well do the functions model world population in 2000?

c. By one projection, world population is expected to reach 8 billion in the year 2026. Which function serves as a better model for this prediction?

SOLUTION

a. Using **Figure 7.45** and rounding to three decimal places, the functions

$$f(x) = 0.074x + 2.294 \quad \text{and} \quad g(x) = 2.577(1.017)^x$$

model world population, in billions, x years after 1949. We named the linear function f and the exponential function g, although any letters can be used.

b. **Table 7.3** shows that world population in 2000 was 6.1 billion. The year 2000 is 51 years after 1949. Thus, we substitute 51 for x in each function's equation and then evaluate the resulting expressions with a calculator to see how well the functions describe world population in 2000.

$f(x) = 0.074x + 2.294$ This is the linear model.

$f(51) = 0.074(51) + 2.294$ Substitute 51 for x.

≈ 6.1 Use a calculator.

$g(x) = 2.577(1.017)^x$ This is the exponential model.

$g(51) = 2.577(1.017)^{51}$ Substitute 51 for x.

≈ 6.1 Use a calculator: 2.577 [×] 1.017 [y^x] (or [∧]) 51 [=].

Because 6.1 billion was the actual world population in 2000, both functions model world population in 2000 extremely well.

Global Population Increase

Exponential functions of the form $y = ab^x, b > 1$, model growth in which quantities increase at a rate proportional to their size. Populations that are growing exponentially grow extremely rapidly as they get larger because there are more adults to have offspring. Here's a way to put this idea into perspective:

By the time you finish reading Example 2 and working Check Point 2, more than 1000 people will have been added to our planet. By this time tomorrow, world population will have increased by more than 220,000.

c. Let's see which model comes closer to projecting a world population of 8 billion in the year 2026. Because 2026 is 77 years after 1949 $(2026 - 1949 = 77)$, we substitute 77 for x in each function's equation.

$$f(x) = 0.074x + 2.294 \quad \text{This is the linear model.}$$
$$f(77) = 0.074(77) + 2.294 \quad \text{Substitute 77 for } x.$$
$$\approx 8.0 \quad \text{Use a calculator.}$$
$$g(x) = 2.577(1.017)^x \quad \text{This is the exponential model.}$$
$$g(77) = 2.577(1.017)^{77} \quad \text{Substitute 77 for } x.$$
$$\approx 9.4 \quad \text{Use a calculator:}$$

2.577 [×] 1.017 [y^x] (or [∧]) 77 [=].

The linear function $f(x) = 0.074x + 2.294$ serves as a better model for a projected world population of 8 billion in 2026.

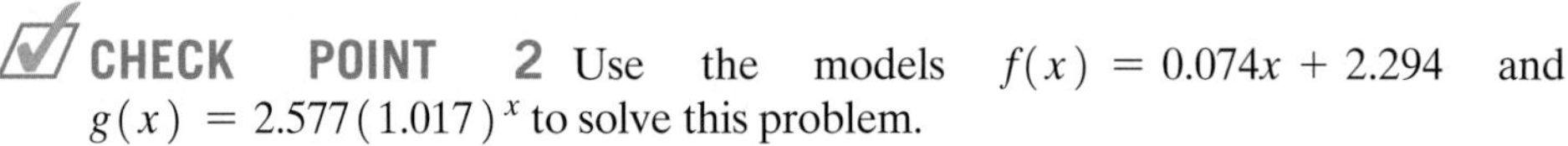

CHECK POINT 2 Use the models $f(x) = 0.074x + 2.294$ and $g(x) = 2.577(1.017)^x$ to solve this problem.

a. World population in 1970 was 3.7 billion. Which function serves as a better model for this year?

b. By one projection, world population is expected to reach 9.3 billion by 2050. Which function serves as a better model for this projection?

The Role of *e* in Applied Exponential Functions

An irrational number, symbolized by the letter e, appears as the base in many applied exponential functions. This irrational number is approximately equal to 2.72. More accurately,

$$e \approx 2.71828\ldots.$$

The number e is called the **natural base**. The function $f(x) = e^x$ is called the **natural exponential function.**

Use a scientific or graphing calculator with an [e^x] key to evaluate e to various powers. For example, to find e^2, press the following keys on most calculators:

Scientific calculator: 2 [e^x]

Graphing calculator: [e^x] 2 [ENTER].

The calculator display for e^2 should be approximately 7.389.

$$e^2 \approx 7.389$$

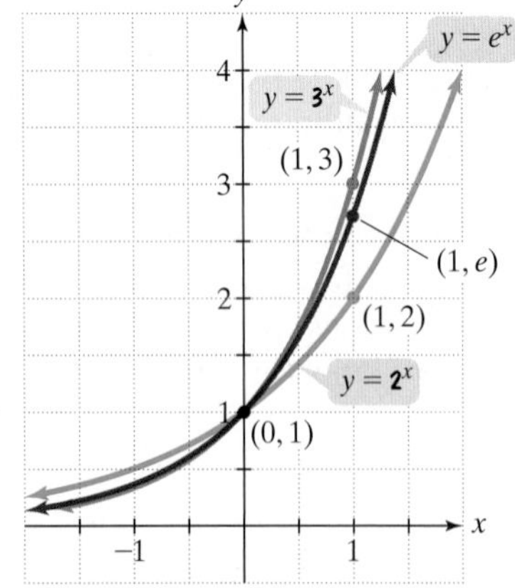

FIGURE 7.46 Graphs of three exponential functions

The number e lies between 2 and 3. Because $2^2 = 4$ and $3^2 = 9$, it makes sense that e^2, approximately 7.389, lies between 4 and 9.

Because $2 < e < 3$, the graph of $y = e^x$ lies between the graphs of $y = 2^x$ and $y = 3^x$, shown in **Figure 7.46.**

EXAMPLE 3 ***Alcohol and Risk of a Car Accident***

Medical research indicates that the risk of having a car accident increases exponentially as the concentration of alcohol in the blood increases. The risk is modeled by

$$R = 6e^{12.77x},$$

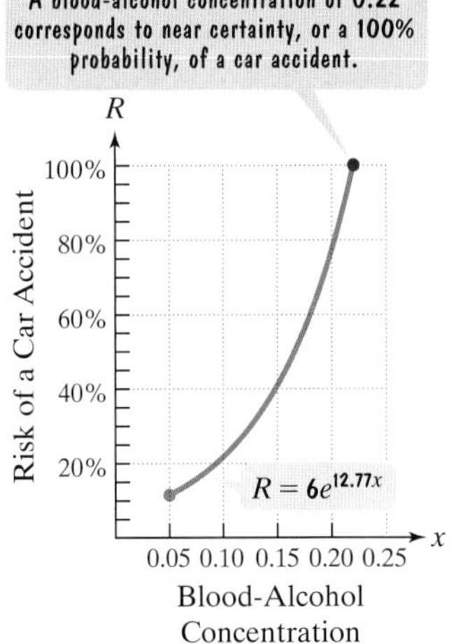

FIGURE 7.47

where x is the blood-alcohol concentration and R, given as a percent, is the risk of having a car accident. In every state, it is illegal to drive with a blood-alcohol concentration of 0.08 or greater. What is the risk of a car accident with a blood-alcohol concentration of 0.08? How is this shown on the graph of R in **Figure 7.47**?

SOLUTION

For a blood-alcohol concentration of 0.08, we substitute 0.08 for x in the exponential model's equation. Then we use a calculator to evaluate the resulting expression.

$$R = 6e^{12.77x} \quad \text{This is the given exponential model.}$$

$$R = 6e^{12.77(0.08)} \quad \text{Substitute 0.08 for } x.$$

Perform this computation on your calculator.

Scientific calculator: 6 [×] [(] 12.77 [×] .08 [)] [e^x] [=]

Graphing calculator: 6 [×] [e^x] [(] 12.77 [×] .08 [)] [ENTER]

The display should be approximately 16.665813. Rounding to one decimal place, the risk of a car accident is approximately 16.7% with a blood-alcohol concentration of 0.08. This can be visualized as the point $(0.08, 16.7)$ on the graph of R in **Figure 7.47**. Take a moment to locate this point on the curve.

CHECK POINT 3 Use the model in Example 3 to solve this problem. In many states, it is illegal for drivers under 21 years old to drive with a blood-alcohol concentration of 0.01 or greater. What is the risk of a car accident with a blood-alcohol concentration of 0.01? Round to one decimal place.

Modeling with Logarithmic Functions

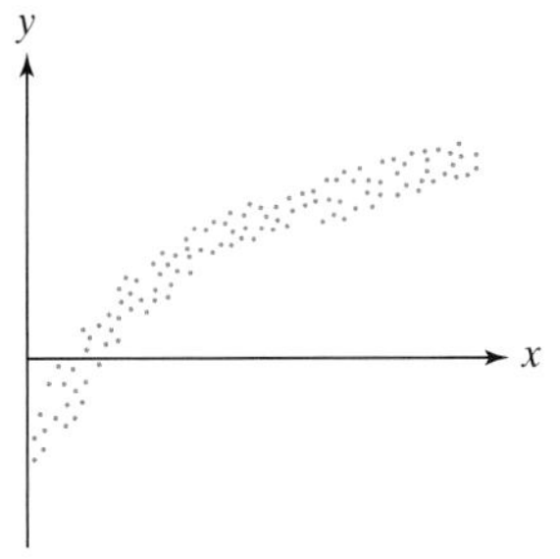

FIGURE 7.48

The scatter plot in **Figure 7.48** starts with rapid growth and then the growth begins to level off. This type of behavior can be modeled by *logarithmic functions*.

DEFINITION OF THE LOGARITHMIC FUNCTION

For $x > 0$ and $b > 0, b \neq 1$,

$$y = \log_b x \text{ is equivalent to } b^y = x.$$

The function $f(x) = \log_b x$ is the **logarithmic function with base *b*.**

The equations

$$y = \log_b x \quad \text{and} \quad b^y = x$$

are different ways of expressing the same thing. The first equation is in **logarithmic form**, and the second equivalent equation is in **exponential form**.

Notice that a **logarithm**, y, is an **exponent**. You should learn the location of the base and exponent in each form.

LOCATION OF BASE AND EXPONENT IN EXPONENTIAL AND LOGARITHMIC FORMS

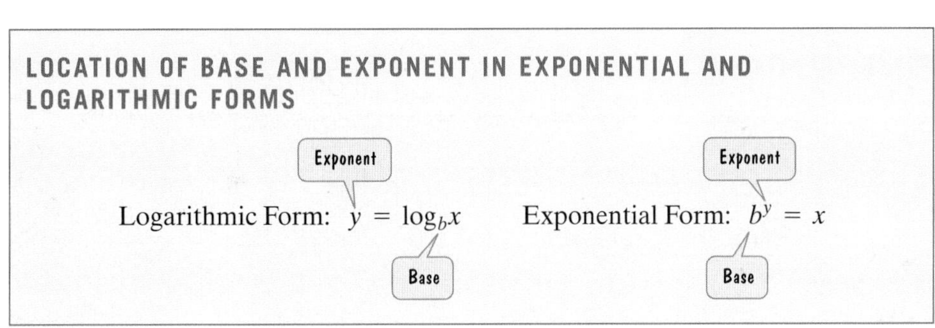

3 *Graph logarithmic functions.*

EXAMPLE 4 Graphing a Logarithmic Function

Graph: $y = \log_2 x$.

SOLUTION

Because $y = \log_2 x$ means $2^y = x$, we will use the exponential form of the equation to obtain the function's graph. Using $2^y = x$, we start by selecting numbers for y and then we find the corresponding values for x.

GREAT QUESTION!

I know that $y = \log_2 x$ means $2^y = x$. But what's the relationship between $y = \log_2 x$ and $y = 2^x$?

The coordinates for the logarithmic function with base 2 are the reverse of those for the exponential function with base 2. In general, $y = \log_b x$ reverses the x- and y-coordinates of $y = b^x$.

Start with values for y.

$x = 2^y$	y	(x, y)
$2^{-2} = \frac{1}{4}$	-2	$\left(\frac{1}{4}, -2\right)$
$2^{-1} = \frac{1}{2}$	-1	$\left(\frac{1}{2}, -1\right)$
$2^0 = 1$	0	(1, 0)
$2^1 = 2$	1	(2, 1)
$2^2 = 4$	2	(4, 2)
$2^3 = 8$	3	(8, 3)

Compute x using $x = 2^y$.

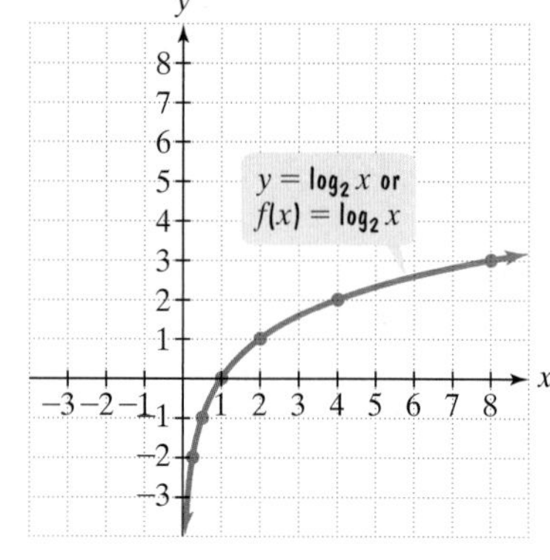

FIGURE 7.49 The graph of the logarithmic function with base 2

We plot the six ordered pairs in the table, connecting the points with a smooth curve. **Figure 7.49** shows the graph of $y = \log_2 x$.

All logarithmic functions of the form $y = \log_b x$, or $f(x) = \log_b x$, where $b > 1$, have the shape of the graph shown in **Figure 7.49**. The graph approaches, but never touches, the negative portion of the y-axis. Observe that the graph is increasing from left to right. However, the rate of increase is slowing down as the graph moves to the right. This is why logarithmic functions are often used to model growing phenomena with growth that is leveling off.

CHECK POINT 4 Rewrite $y = \log_3 x$ in exponential form. Then use the exponential form of the equation to obtain the function's graph. Select integers from -2 to 2, inclusive, for y.

Scientific and graphing calculators contain keys that can be used to evaluate the logarithmic function with base 10 and the logarithmic function with base e.

Key	Function the Key Is Used to Evaluate
LOG	$y = \log_{10} x$ — This is called the **common logarithmic function**, usually expressed as $y = \log x$.
LN	$y = \log_e x$ — This is called the **natural logarithmic function**, usually expressed as $y = \ln x$.

4 *Use logarithmic models.*

EXAMPLE 5 Dangerous Heat: Temperature in an Enclosed Vehicle

When the outside air temperature is anywhere from 72° to 96° Fahrenheit, the temperature in an enclosed vehicle climbs by 43° in the first hour. The bar graph in **Figure 7.50(a)** shows the temperature increase throughout the hour. A scatter plot of the data is shown in **Figure 7.50(b)**.

Temperature Increase in an Enclosed Vehicle

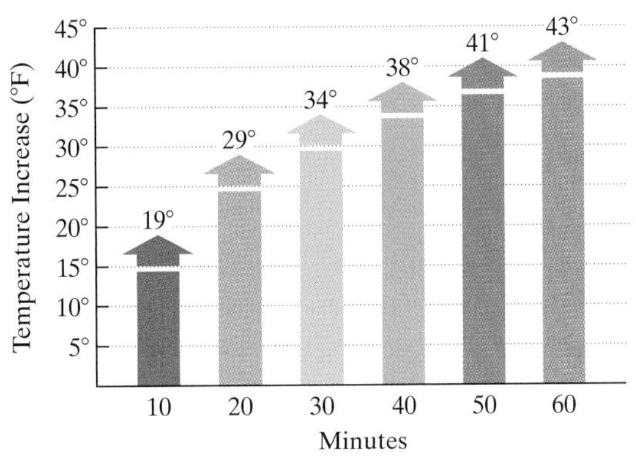

FIGURE 7.50(a)
Source: Professor Jan Null, San Francisco State University

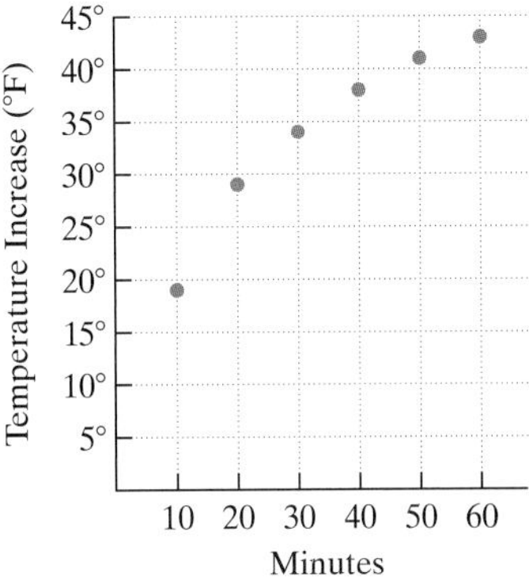

FIGURE 7.50(b)

Because the data in the scatter plot increase rapidly at first and then begin to level off a bit, the shape suggests that a logarithmic function is a good choice for a model. After entering the data, a graphing calculator displays the logarithmic model, $y = a + b \ln x$, shown in **Figure 7.51**.

```
LnReg
y=a+blnx
a=-11.62862899
b=13.42363186
```

FIGURE 7.51 Data (10, 19), (20, 29), (30, 34),(40, 38), (50, 41). (60, 43)

a. Express the model in function notation, with numbers rounded to one decimal place.

b. Use the function to find the temperature increase, to the nearest degree, after 50 minutes. How well does the function model the actual increase shown in **Figure 7.50(a)**?

SOLUTION

a. Using **Figure 7.51** and rounding to one decimal place, the function

$$f(x) = -11.6 + 13.4 \ln x$$

models the temperature increase, $f(x)$, in degrees Fahrenheit, after x minutes.

b. We find the temperature increase after 50 minutes by substituting 50 for x and evaluating the function at 50.

$f(x) = -11.6 + 13.4 \ln x$ This is the logarithmic model from part (a).

$f(50) = -11.6 + 13.4 \ln 50$ Substitute 50 for x.

Perform this computation on your calculator.

Scientific calculator: 11.6 [+/−] [+] 13.4 [×] 50 [LN] [=]

Graphing calculator: [(−)] 11.6 [+] 13.4 [LN] 50 [ENTER]

The display should be approximately 40.821108. Rounding to the nearest degree, the logarithmic model indicates that the temperature will have increased by approximately 41° after 50 minutes. Because the increase shown in **Figure 7.50(a)** is 41°, the function models the actual increase extremely well.

GREAT QUESTION!

How can I use a graphing calculator to see how well my models describe the data?

Once you have obtained one or more models for the data, you can use a graphing calculator's [TABLE] feature to numerically see how well each model describes the data. Enter the models as y_1, y_2, and so on. Create a table, scroll through the table, and compare the table values given by the models to the actual data.

CHECK POINT 5 Use the model obtained in Example 5(a) on the previous page to find the temperature increase, to the nearest degree, after 30 minutes. How well does the function model the actual increase shown in **Figure 7.50(a)**?

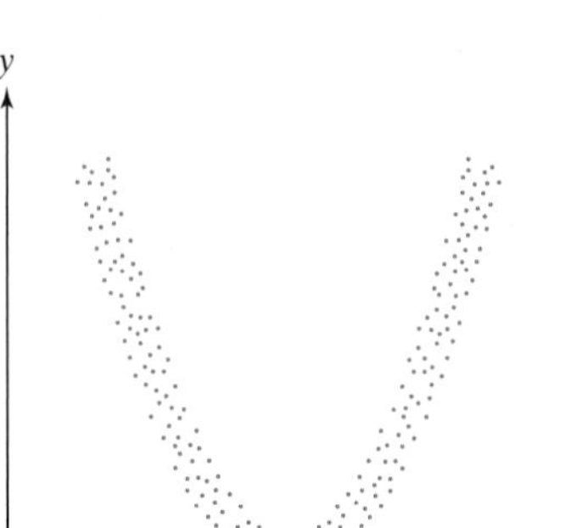

FIGURE 7.52

Modeling with Quadratic Functions

The scatter plot in **Figure 7.52** has a shape that indicates the data are first decreasing and then increasing. This type of behavior can be modeled by a *quadratic function*.

> **DEFINITION OF THE QUADRATIC FUNCTION**
>
> A **quadratic function** is any function of the form
>
> $$y = ax^2 + bx + c \quad \text{or} \quad f(x) = ax^2 + bx + c,$$
>
> where a, b, and c are real numbers, with $a \neq 0$.

The graph of any quadratic function is called a **parabola**. Parabolas are shaped like bowls or inverted bowls, as shown in **Figure 7.53**. If the coefficient of x^2 (the value of a in $ax^2 + bx + c$) is positive, the parabola opens upward. If the coefficient of x^2 is negative, the graph opens downward. The **vertex** (or turning point) of the parabola is the lowest point on the graph when it opens upward and the highest point on the graph when it opens downward.

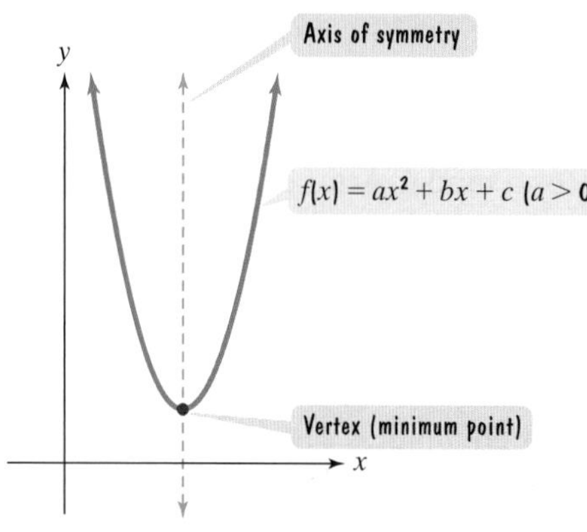

$a > 0$: Parabola opens upward.

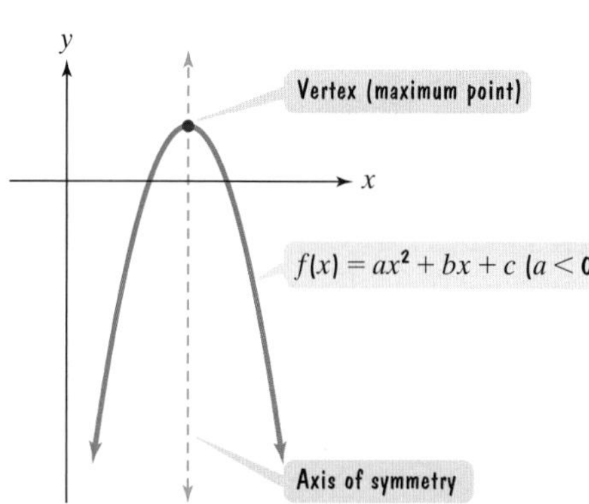

$a < 0$: Parabola opens downward.

FIGURE 7.53 Characteristics of graphs of quadratic functions

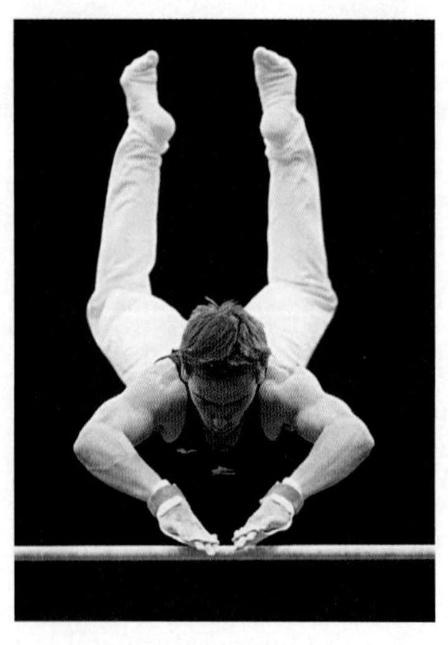

Look at the unusual image of the word *mirror* shown below. The artist, Scott Kim, has created the image so that the two halves of the whole are mirror images of each other. A parabola shares this kind of symmetry, in which a line through the vertex divides the figure in half. Parabolas are symmetric with respect to this line, called the **axis of symmetry**. If a parabola is folded along its axis of symmetry, the two halves match exactly.

© 2011 Scott Kim, scottkim.com

When graphing quadratic functions or using them as models, it is frequently helpful to determine where the vertex, or turning point, occurs.

> **THE VERTEX OF A PARABOLA**
>
> The vertex of a parabola whose equation is $y = ax^2 + bx + c$ occurs where
>
> $$x = \frac{-b}{2a}.$$

5 *Graph quadratic functions.*

Several points are helpful when graphing a quadratic function. These points are the x-intercepts (although not every parabola has x-intercepts), the y-intercept, and the vertex.

GRAPHING QUADRATIC FUNCTIONS

The graph of $y = ax^2 + bx + c$ or $f(x) = ax^2 + bx + c$, called a parabola, can be graphed using the following steps:

1. Determine whether the parabola opens upward or downward. If $a > 0$, it opens upward. If $a < 0$, it opens downward.
2. Determine the vertex of the parabola. The x-coordinate is $\frac{-b}{2a}$. The y-coordinate is found by substituting the x-coordinate into the parabola's equation and evaluating.
3. Find any x-intercepts by replacing y or $f(x)$ with 0. Solve the resulting quadratic equation for x.
4. Find the y-intercept by replacing x with 0. Because $f(0) = c$ (the constant term in the function's equation), the y-intercept is c and the parabola passes through $(0, c)$.
5. Plot the intercepts and the vertex.
6. Connect these points with a smooth curve.

EXAMPLE 6 *Graphing a Parabola*

Graph the quadratic function: $y = x^2 - 2x - 3$.

SOLUTION

We can graph this function by following the steps in the box.

Step 1 Determine how the parabola opens. Note that a, the coefficient of x^2, is 1. Thus, $a > 0$; this positive value tells us that the parabola opens upward.

Step 2 Find the vertex. We know that the x-coordinate of the vertex is $\frac{-b}{2a}$. Let's identify the numbers a, b, and c in the given equation, which is in the form $y = ax^2 + bx + c$.

$$y = x^2 - 2x - 3$$

$a = 1$ $b = -2$ $c = -3$

Now we substitute the values of a and b into the expression for the x-coordinate:

$$x\text{-coordinate of vertex} = \frac{-b}{2a} = \frac{-(-2)}{2(1)} = \frac{2}{2} = 1.$$

The x-coordinate of the vertex is 1. We substitute 1 for x in the equation $y = x^2 - 2x - 3$ to find the y-coordinate:

$$y\text{-coordinate of vertex} = 1^2 - 2 \cdot 1 - 3 = 1 - 2 - 3 = -4.$$

The vertex is $(1, -4)$, shown in **Figure 7.54**.

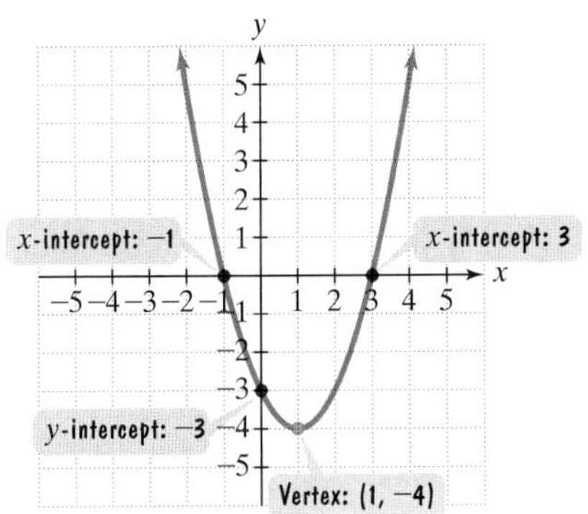

FIGURE 7.54 The graph of $y = x^2 - 2x - 3$

Step 3 Find the x-intercepts. Replace y with 0 in $y = x^2 - 2x - 3$. We obtain $0 = x^2 - 2x - 3$ or $x^2 - 2x - 3 = 0$. We can solve this equation by factoring.

$$x^2 - 2x - 3 = 0$$
$$(x - 3)(x + 1) = 0$$
$$x - 3 = 0 \quad \text{or} \quad x + 1 = 0$$
$$x = 3 \qquad\qquad x = -1$$

The x-intercepts are 3 and -1. The parabola passes through $(3, 0)$ and $(-1, 0)$, shown in **Figure 7.54**.

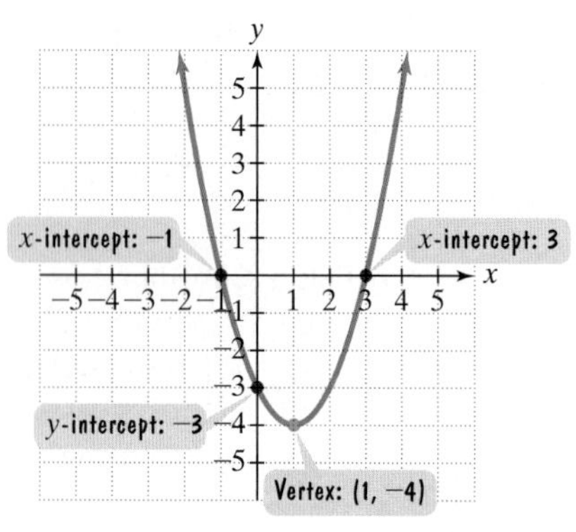

FIGURE 7.54 (repeated) The graph of $y = x^2 - 2x - 3$

Step 4 Find the *y*-intercept. Replace x with 0 in $y = x^2 - 2x - 3$:

$$y = 0^2 - 2 \cdot 0 - 3 = 0 - 0 - 3 = -3.$$

The y-intercept is -3. The parabola passes through $(0, -3)$, shown in **Figure 7.54**.

Steps 5 and 6 Plot the intercepts and the vertex. Connect these points with a smooth curve. The intercepts and the vertex are shown as the four labeled points in **Figure 7.54**. Also shown is the graph of the quadratic function, obtained by connecting the points with a smooth curve.

CHECK POINT 6 Graph the quadratic function: $y = x^2 + 6x + 5$.

6 *Use quadratic models.*

EXAMPLE 7 *Modeling the Parabolic Path of a Punted Football*

Figure 7.55 shows that when a football was kicked, the nearest defensive player was 6 feet from the point of impact with the kicker's foot. **Table 7.4** shows five measurements indicating the football's height at various horizontal distances from its point of impact. A scatter plot of the data is shown in **Figure 7.56**.

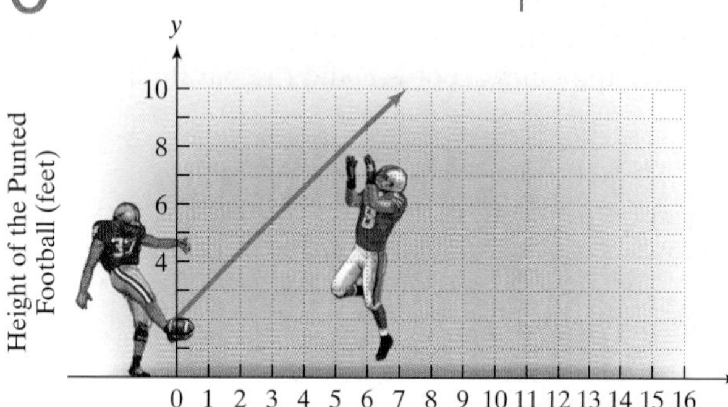

FIGURE 7.55

TABLE 7.4

x, Football's Horizontal Distance (feet)	*y*, Football's Height (feet)
0	2
30	28.4
60	36.8
90	27.2
110	10.8

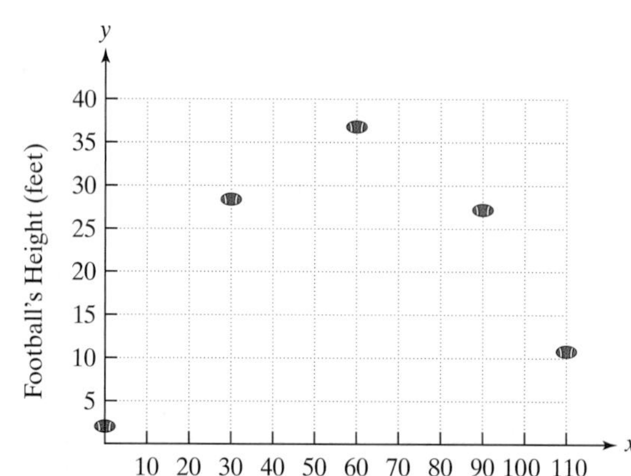

FIGURE 7.56 A scatter plot for the data in Table 7.4

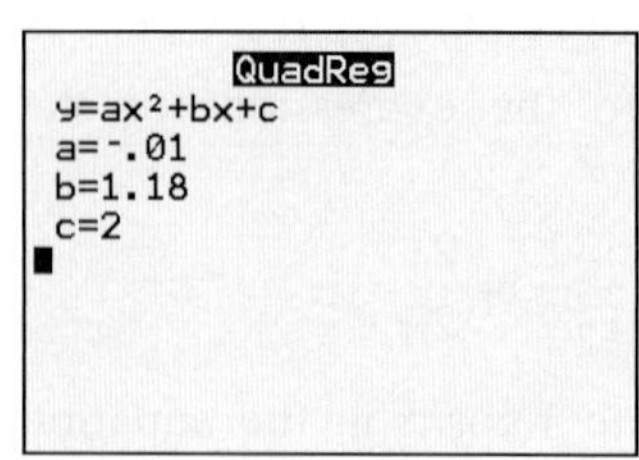

FIGURE 7.57 Data (0, 2), (30, 28.4), (60, 36.8), (90, 27.2), (110, 10.8)

Because the data in the scatter plot first increase and then decrease, the shape suggests that a quadratic function is a good choice for a model. Using the data in **Table 7.4**, a graphing calculator displays the quadratic function, $y = ax^2 + bx + c$, shown in **Figure 7.57**.

a. Express the model in function notation.

b. How far would the nearest defensive player, who was 6 feet from the kicker's point of impact, have needed to reach to block the punt?

SOLUTION

a. Using **Figure 7.57**, the function

$$f(x) = -0.01x^2 + 1.18x + 2$$

models the football's height, $f(x)$, in feet, in terms of its horizontal distance, x, in feet.

b. **Figure 7.55** shows that the defensive player was 6 feet from the point of impact. To block the punt, he needed to touch the football along its parabolic path. This means that we must find the height of the ball 6 feet from the kicker. Replace x with 6 in the function, $f(x) = -0.01x^2 + 1.18x + 2$.

$$f(6) = -0.01(6)^2 + 1.18(6) + 2 = -0.36 + 7.08 + 2 = 8.72$$

The defensive player would have needed to reach 8.72 feet above the ground to block the punt.

Assuming that the football was not blocked by the defensive player, the graph of the function that models the football's parabolic path is shown in **Figure 7.58**. The graph is shown only for $x \geq 0$, indicating horizontal distances that begin at the football's impact with the kicker's foot and end with the ball hitting the ground. Notice how the graph provides a visual story of the punted football's parabolic path.

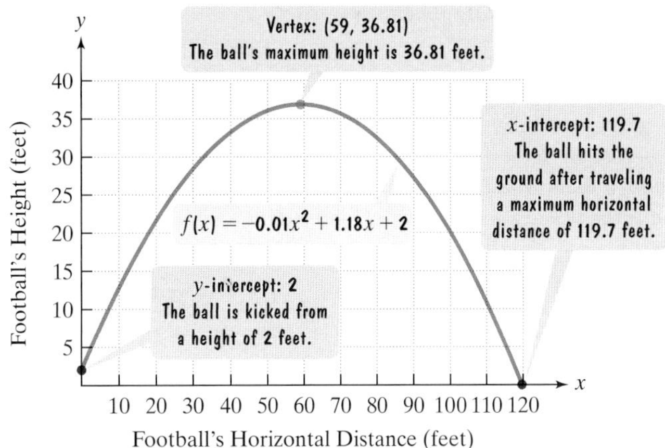

FIGURE 7.58 The parabolic path of a punted football

CHECK POINT 7 Use the model obtained in Example 7(a) to answer this question: If the defensive player had been 8 feet from the kicker's point of impact, how far would he have needed to reach to block the punt? Does this seem realistic? Identify the solution as a point on the graph in **Figure 7.58**.

7 *Determine an appropriate function for modeling data.*

Table 7.5 contains a description of the scatter plots we have encountered in this section, as well as the type of function that can serve as an appropriate model for each description.

TABLE 7.5 Modeling Data

Description of Data Points in a Scatter Plot	Model
Lie on or near a line	Linear Function: $y = mx + b$ or $f(x) = mx + b$
Increasing more and more rapidly	Exponential Function: $y = b^x$ or $f(x) = b^x, b > 1$
Increasing, although rate of increase is slowing down	Logarithmic Function: $y = \log_b x$ or $f(x) = \log_b x, b > 1$ ($y = \log_b x$ means $b^y = x$.)
Decreasing and then increasing	Quadratic Function: $y = ax^2 + bx + c$ or $f(x) = ax^2 + bx + c, a > 0$ The vertex, $\left(\frac{-b}{2a}, f\left(\frac{-b}{2a}\right)\right)$, is a minimum point on the parabola.
Increasing and then decreasing	Quadratic Function: $y = ax^2 + bx + c$ or $f(x) = ax^2 + bx + c, a < 0$ The vertex, $\left(\frac{-b}{2a}, f\left(\frac{-b}{2a}\right)\right)$, is a maximum point on the parabola.

Once the type of model has been determined, the data can be entered into a graphing calculator. The calculator's regression feature will display the specific function of the type requested that best fits the data. That, in short, is how your author obtained the algebraic models you have encountered throughout this book. In this era of technology, the process of determining models that approximate real-world situations is based on a knowledge of functions and their graphs, and has nothing to do with long and tedious computations.

Concept and Vocabulary Check

Fill in each blank so that the resulting statement is true.

1. Data presented in a visual form as a set of points are called a/an ________. The line that best fits the data points is called a/an ________ line.

For each set of points in Exercises 2–5, determine whether an exponential function, a logarithmic function, a linear function, or a quadratic function is the best choice for modeling the data.

2.

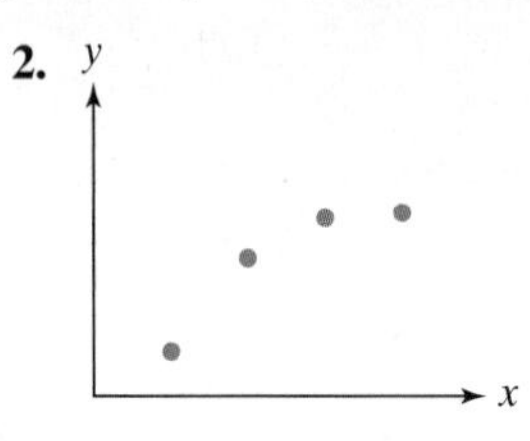

____

3.

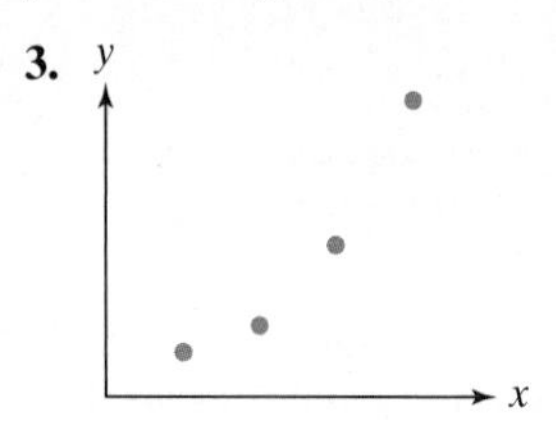

____

4.

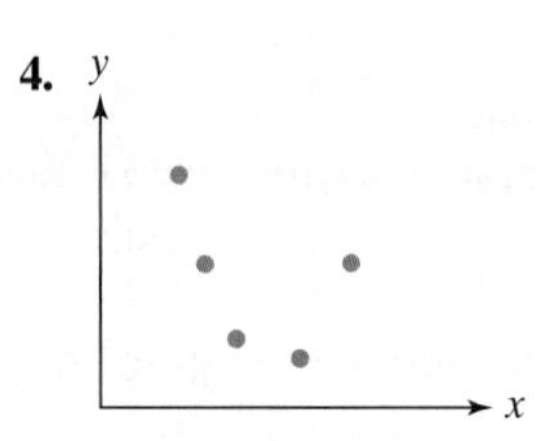

____

5.

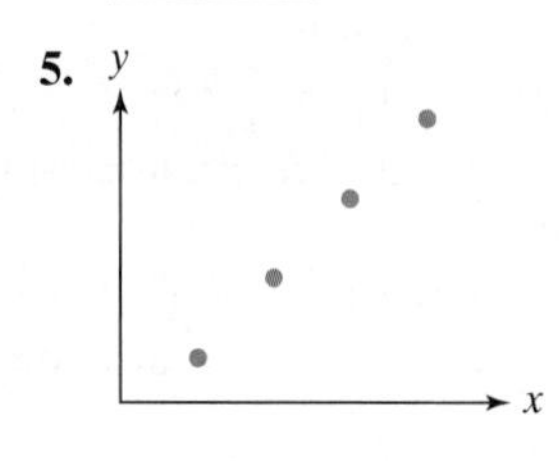

____

6. The exponential function with base b is defined by $y =$ ________, $b > 0$ and $b \neq 1$.

7. The irrational number e is approximately equal to ________. The function $y = e^x$ or $f(x) = e^x$ is called the ________ exponential function.

8. $y = \log_b x$ is equivalent to the exponential form ________, $x > 0, b > 0, b \neq 1$.

9. $y = \log_{10} x$ is usually expressed as $y =$ ________ and is called the ________ logarithmic function.

10. $y = \log_e x$ is usually expressed as $y =$ ________ and is called the ________ logarithmic function.

11. The function $y = ax^2 + bx + c$ or $f(x) = ax^2 + bx + c$, $a \neq 0$, is called a/an ________ function. The graph of this function is called a/an ________. The vertex, or turning point, occurs where $x =$ ________.

Exercise Set 7.6

Practice Exercises

In Exercises 1–6, use a table of coordinates to graph each exponential function. Begin by selecting −2, −1, 0, 1, and 2 for x.

1. $f(x) = 4^x$ **2.** $f(x) = 5^x$ **3.** $y = 2^{x+1}$

4. $y = 2^{x-1}$ **5.** $f(x) = 3^{x-1}$ **6.** $f(x) = 3^{x+1}$

In Exercises 7–8,

a. *Rewrite each equation in exponential form.*

b. *Use a table of coordinates and the exponential form from part (a) to graph each logarithmic function. Begin by selecting −2, −1, 0, 1, and 2 for y.*

7. $y = \log_4 x$ **8.** $y = \log_5 x$

In Exercises 9–14,

a. *Determine if the parabola whose equation is given opens upward or downward.*

b. *Find the vertex.*

c. *Find the x-intercepts.*

d. *Find the y-intercept.*

e. *Use (a)–(d) to graph the quadratic function.*

9. $y = x^2 + 8x + 7$ **10.** $y = x^2 + 10x + 9$

11. $f(x) = x^2 - 2x - 8$ **12.** $f(x) = x^2 + 4x - 5$

13. $y = -x^2 + 4x - 3$ **14.** $y = -x^2 + 2x + 3$

In Exercises 15–22,

a. *Create a scatter plot for the data in each table.*

b. *Use the shape of the scatter plot to determine if the data are best modeled by a linear function, an exponential function, a logarithmic function, or a quadratic function.*

15.

x	y
0	0
9	1
16	1.2
19	1.3
25	1.4

16.

x	y
0	0.3
8	1
15	1.2
18	1.3
24	1.4

17.

x	y
0	−3
1	2
2	7
3	12
4	17

18.

x	y
0	5
1	3
2	1
3	−1
4	−3

19.

x	y
0	4
1	1
2	0
3	1
4	4

20.

x	y
0	−4
1	−1
2	0
3	−1
4	−4

21.

x	y
0	−3
1	−2
2	0
3	4
4	12

22.

x	y
0	4
1	5
2	7
3	11
4	19

Practice Plus

In Exercises 23–24, use a table of coordinates to graph each exponential function. Begin by selecting −2, −1, 0, 1, and 2 for x. Based on your graph, describe the shape of a scatter plot that can be modeled by $f(x) = b^x, 0 < b < 1$.

23. $f(x) = (\frac{1}{2})^x$ (Equivalently, $y = 2^{-x}$)

24. $f(x) = (\frac{1}{3})^x$ (Equivalently, $y = 3^{-x}$)

In Exercises 25–26, use the directions for Exercises 7–8 to graph each logarithmic function. Based on your graph, describe the shape of a scatter plot that can be modeled by $f(x) = \log_b x, 0 < b < 1$.

25. $y = \log_{\frac{1}{2}} x$

26. $y = \log_{\frac{1}{3}} x$

In Exercises 27–28, use the directions for Exercises 9–14 to graph each quadratic function. Use the quadratic formula to find x-intercepts, rounded to the nearest tenth.

27. $f(x) = -2x^2 + 4x + 5$

28. $f(x) = -3x^2 + 6x - 2$

In Exercises 29–30, find the vertex for the parabola whose equation is given by writing the equation in the form $y = ax^2 + bx + c$.

29. $y = (x - 3)^2 + 2$

30. $y = (x - 4)^2 + 3$

Application Exercises

In 1900, Americans age 65 and over made up only 4.1% of the population. By 2010, that figure was 13%, or approximately 44.6 million people. Demographic projections indicate that by the year 2050, approximately 82 million Americans, or 20% of the U.S. population, will be at least 65. The bar graph shows the number of people in the United States age 65 and over, with projected figures for the year 2020 and beyond.

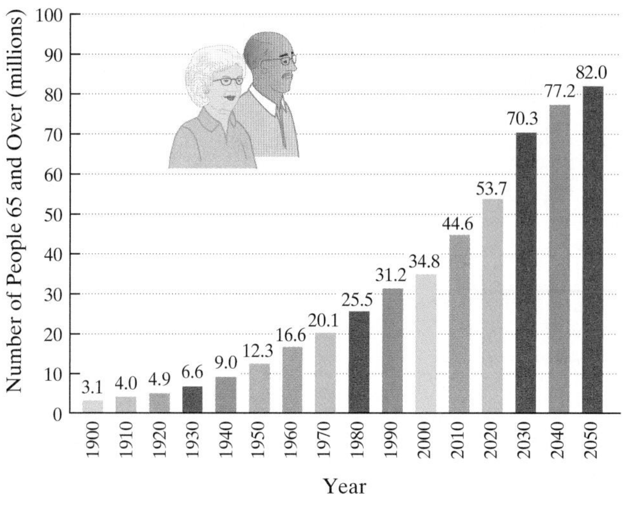

Source: U.S. Census Bureau

The graphing calculator screen displays an exponential function that models the U.S. population age 65 and over, y, in millions, x years after 1899. Use this information to solve Exercises 31–32.

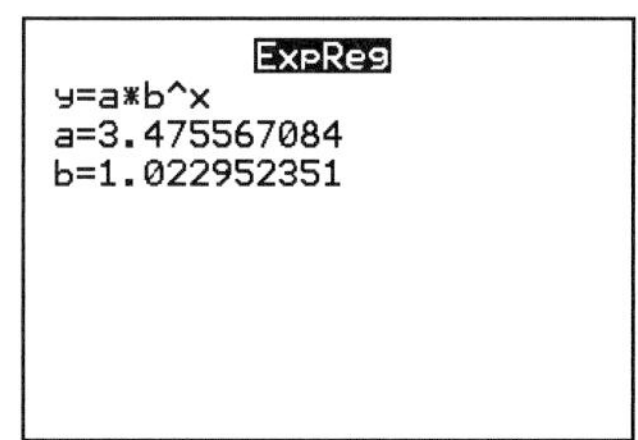

31. a. Explain why an exponential function was used to model the population data.

b. Use the graphing calculator screen to express the model in function notation, with numbers rounded to three decimal places.

c. According to the model in part (b), how many Americans age 65 and over were there in 2010? Use a calculator with a y^x key or a ∧ key, and round to one decimal place. Does this rounded number overestimate or underestimate the 2010 population displayed by the bar graph? By how much?

d. According to the model in part (b), how many Americans age 65 and over will there be in 2020? Round to one decimal place. Does this rounded number overestimate or underestimate the 2020 population projection displayed by the bar graph? By how much?

32. Refer to the graph showing the U.S. population age 65 and over and the graphing calculator screen on the previous page.
 a. Explain why an exponential function was used to model the population data.
 b. Use the graphing calculator screen to express the model in function notation, with numbers rounded to three decimal places.
 c. According to the model in part (b), how many Americans age 65 and over were there in 2000? Use a calculator with a $\boxed{y^x}$ key or a $\boxed{\wedge}$ key, and round to one decimal place. Does this rounded number overestimate or underestimate the 2000 population displayed by the bar graph? By how much?
 d. According to the model in part (b), how many Americans age 65 and over will there be in 2030? Round to one decimal place. Does this rounded number overestimate or underestimate the 2030 population projection displayed by the bar graph? By how much?

Use a calculator with an $\boxed{e^x}$ *key to solve Exercises 33–34.*

The bar graph shows the average cost of room and board at four-year public and private U.S. colleges for four selected years from 2011 through 2017.

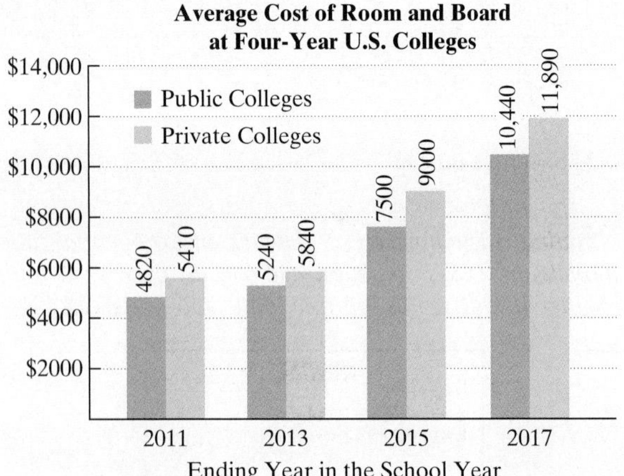

Source: U.S. Department of Education

The data can be modeled by

$$f(x) = 956x + 3176 \quad \text{and} \quad g(x) = 3904e^{0.134x},$$

in which $f(x)$ and $g(x)$ represent the average cost of room and board at public four-year colleges in the school year ending x years after 2010. Use these functions to solve Exercises 33–34. Where necessary, round answers to the nearest whole dollar.

33. a. According to the linear model, what was the average cost of room and board at public four-year colleges for the school year ending in 2017?
 b. According to the exponential model, what was the average cost of room and board at public four-year colleges for the school year ending in 2017?
 c. Which function is a better model for the data for the school year ending in 2017?

34. a. According to the linear model, what was the average cost of room and board at public four-year colleges for the school year ending in 2015?
 b. According to the exponential model, what was the average cost of room and board at public four-year colleges for the school year ending in 2015?
 c. Which function is a better model for the data for the school year ending in 2015?

The data in the following table indicate that between the ages of 1 and 11, the human brain does not grow linearly, or steadily. A scatter plot for the data is shown to the right of the table.

GROWTH OF THE HUMAN BRAIN

Age	Percentage of Adult Size Brain
1	30%
2	50%
4	78%
6	88%
8	92%
10	95%
11	99%

Source: Gerrig and Zimbardo, *Psychology and Life*, 18th Edition, Allyn and Bacon, 2008.

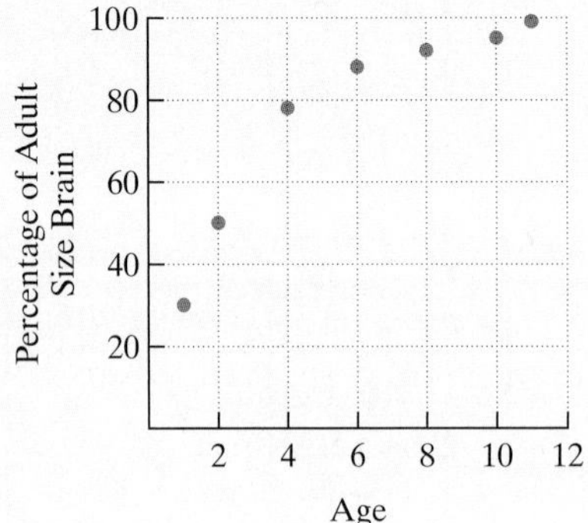

The graphing calculator screen displays the percentage of an adult size brain, y, for a child at age x, where $1 \le x \le 11$. Use this information to solve Exercises 35–36.

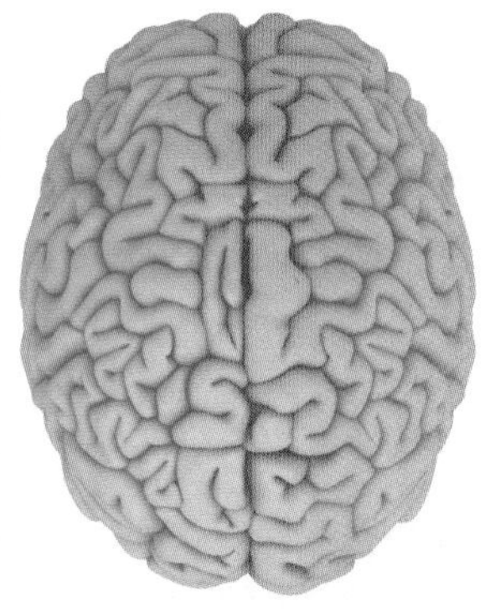

```
LnReg
y=a+blnx
a=31.95404756
b=28.94733911
```

35. a. Explain why a logarithmic function was used to model the data.
 b. Use the graphing calculator screen to express the model in function notation, with numbers rounded to the nearest whole number.
 c. According to the model in part (b), what percentage of an adult size brain does a child have at age 10? Use a calculator with an $\boxed{\text{LN}}$ key and round to the nearest whole percent. Does this overestimate or underestimate the percent displayed by the table? By how much?

36. a. Explain why a logarithmic function was used to model the data.
 b. Use the graphing calculator screen to express the model in function notation, with numbers rounded to the nearest whole number.
 c. According to the model in part (b), what percentage of an adult size brain does a child have at age 8? Use a calculator with an $\boxed{\text{LN}}$ key and round to the nearest whole percent. How does this compare with the percent displayed by the table?

The percentage of adult height attained by a girl who is x years old can be modeled by

$$f(x) = 62 + 35\log(x - 4),$$

where x represents the girl's age (from 5 to 15) and $f(x)$ represents the percentage of her adult height. Use the function to solve Exercises 37–38.

37. a. According to the model, what percentage of her adult height has a girl attained at age 13? Use a calculator with a LOG key and round to the nearest tenth of a percent.

b. Why was a logarithmic function used to model the percentage of adult height attained by a girl from ages 5 to 15, inclusive?

38. a. According to the model, what percentage of her adult height has a girl attained at age ten? Use a calculator with a LOG key and round to the nearest tenth of a percent.

b. Why was a logarithmic function used to model the percentage of adult height attained by a girl from ages 5 to 15, inclusive?

39. A ball is thrown upward and outward from a height of 6 feet. The table shows four measurements indicating the ball's height at various horizontal distances from where it was thrown. The graphing calculator screen displays a quadratic function that models the ball's height, y, in feet, in terms of its horizontal distance, x, in feet.

x, Ball's Horizontal Distance (feet)	y, Ball's Height (feet)
0	6
1	7.6
3	6
4	2.8

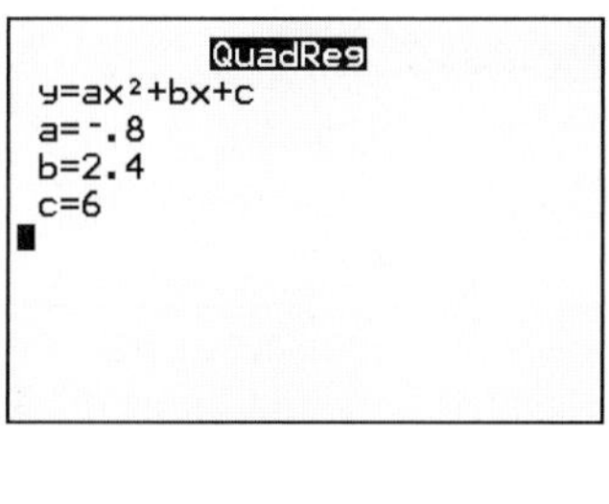

a. Explain why a quadratic function was used to model the data. Why is the value of a negative?

b. Use the graphing calculator screen to express the model in function notation.

c. Use the model from part (b) to determine the x-coordinate of the quadratic function's vertex. Then complete this statement: The maximum height of the ball occurs _________ feet from where it was thrown and the maximum height is _________ feet.

40. A ball is thrown upward and outward from a height of 6 feet. The table shows four measurements indicating the ball's height at various horizontal distances from where it was thrown. The graphing calculator screen displays a quadratic function that models the ball's height, y, in feet, in terms of its horizontal distance, x, in feet.

x, Ball's Horizontal Distance (feet)	y, Ball's Height (feet)
0	6
0.5	7.4
1.5	9
4	6

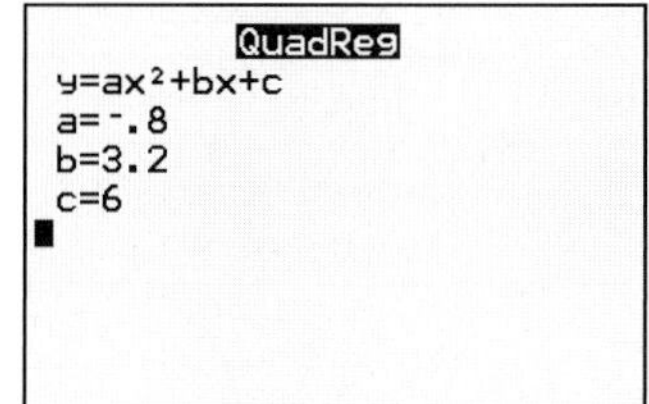

a. Explain why a quadratic function was used to model the data. Why is the value of a negative?

b. Use the graphing calculator screen to express the model in function notation.

c. Use the model from part (b) to determine the x-coordinate of the quadratic function's vertex. Then complete this statement: The maximum height of the ball occurs _________ feet from where it was thrown, and the maximum height is _________ feet.

Explaining the Concepts

41. What is a scatter plot?

42. What is an exponential function?

43. Describe the shape of a scatter plot that suggests modeling the data with an exponential function.

44. Describe the shape of a scatter plot that suggests modeling the data with a logarithmic function.

45. Would you prefer that your salary be modeled exponentially or logarithmically? Explain your answer.

46. Describe the shape of a scatter plot that suggests modeling the data with a quadratic function.

Critical Thinking Exercises

Make Sense? *In Exercises 47–50, determine whether each statement makes sense or does not make sense, and explain your reasoning.*

47. I'm looking at data that show the number of residential solar installations in the United States, and a linear function appears to be a better choice than an exponential function for modeling the number of new U.S. homes that generated solar power from 2010 through 2015.

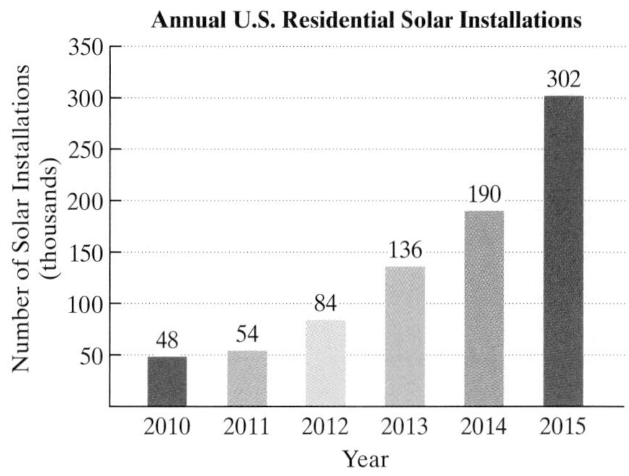

Source: Solar Energy Industries Association

48. I used two different functions to model the data in a scatter plot.

49. Drinking and driving is extremely dangerous because the risk of a car accident increases logarithmically as the concentration of alcohol in the blood increases.

50. This work by artist Scott Kim (1955–) has the same kind of symmetry as the graph of a quadratic function.

© 2011 Scott Kim, scottkim.com

In Exercises 51–53, the value of a in $y = ax^2 + bx + c$ and the vertex of the parabola are given. How many x-intercepts does the parabola have? Explain how you arrived at this number.

51. $a = -2$; vertex at $(4, 8)$

52. $a = 1$; vertex at $(2, 0)$

53. $a = 3$; vertex at $(3, 1)$

Technology Exercises

54. Use a graphing calculator to graph the exponential functions that you graphed by hand in Exercises 1–6. Describe similarities and differences between the graphs obtained by hand and those that appear in the calculator's viewing window.

55. Use a graphing calculator to graph the quadratic functions that you graphed by hand in Exercises 9–14.

Group Exercise

56. Each group member should consult an almanac, newspaper, magazine, or the Internet to find data that can be modeled by linear, exponential, logarithmic, or quadratic functions. Group members should select the two sets of data that are most interesting and relevant. Then consult a person who is familiar with graphing calculators to show you how to obtain a function that best fits each set of data. Once you have these functions, each group member should make one prediction based on one of the models, and then discuss a consequence of this prediction. What factors might change the accuracy of the prediction?

Chapter Summary, Review, and Test

SUMMARY – DEFINITIONS AND CONCEPTS	EXAMPLES
7.1 Graphing and Functions	
a. The rectangular coordinate system is formed using two number lines that intersect at right angles at their zero points. See Figure 7.1 on page 412. The horizontal line is the x-axis and the vertical line is the y-axis. Their point of intersection, $(0, 0)$, is the origin. Each point in the system corresponds to an ordered pair of real numbers, (x, y).	Ex. 1, p. 413
b. The graph of an equation in two variables is the set of all points whose coordinates satisfy the equation.	Ex. 2, p. 414; Ex. 3, p. 414
c. If an equation in x and y yields one value of y for each value of x, then y is a function of x, indicated by writing $f(x)$ for y.	Ex. 4, p. 416; Ex. 5, p. 417
d. The graph of a function is the graph of its ordered pairs.	Ex. 6, p. 418
e. The Vertical Line Test: If any vertical line intersects a graph in more than one point, the graph does not define y as a function of x.	Ex. 7, p. 419
7.2 Linear Functions and Their Graphs	
a. A function whose graph is a straight line is a linear function.	
b. The graph of $Ax + By = C$, a linear equation in two variables, is a straight line. The line can be graphed using intercepts and a checkpoint. To locate the x-intercept, set $y = 0$ and solve for x. To locate the y-intercept, set $x = 0$ and solve for y.	Ex. 1, p. 425
c. The slope of the line through (x_1, y_1) and (x_2, y_2) is $m = \frac{\text{Rise}}{\text{Run}} = \frac{y_2 - y_1}{x_2 - x_1} = \frac{y_1 - y_2}{x_1 - x_2}$.	Ex. 2, p. 426; Ex. 7, p. 432
d. The equation $y = mx + b$ is the slope-intercept form of the equation of a line, in which m is the slope and b is the y-intercept.	Ex. 3, p. 429; Ex. 4, p. 430; Ex. 8, p. 433
e. Horizontal and Vertical Lines **1.** The graph of $y = b$ is a horizontal line that intersects the y-axis at $(0, b)$. **2.** The graph of $x = a$ is a vertical line that intersects the x-axis at $(a, 0)$.	Ex. 5, p. 431; Ex. 6, p. 431

7.3 Systems of Linear Equations in Two Variables

a. Two equations in the form $Ax + By = C$ are called a system of linear equations. A solution of the system is an ordered pair that satisfies both equations in the system.	Ex. 1, p. 439
b. Linear systems with one solution can be solved by graphing. The coordinates of the point of intersection of the lines are the system's solution.	Ex. 2, p. 439
c. Systems of linear equations in two variables can be solved by eliminating a variable, using the substitution method (see the box on page 440) or the addition method (see the box on page 443).	Ex. 3, p. 440; Ex. 4, p. 441; Ex. 5, p. 443; Ex. 6, p. 444
d. When solving by substitution or addition, if the variable is eliminated and a false statement results, the linear system has no solution. If the variable is eliminated and a true statement results, the system has infinitely many solutions.	Ex. 7, p. 445; Ex. 8, p. 446
e. Functions of Business. A company produces and sells x units of a product. Revenue Function: $R(x) = (\text{price per unit sold})x$ Cost Function: $C(x) = \text{fixed cost} + (\text{cost per unit produced})x$ Profit Function: $P(x) = R(x) - C(x)$ The point of intersection of the graphs of R and C is the break-even point. The x-coordinate of the point reveals the number of units that a company must produce and sell so that the money coming in, the revenue, is equal to the money going out, the cost. The y-coordinate gives the amount of money coming in and going out.	Ex. 9, p. 447

7.4 Linear Inequalities in Two Variables

a. A linear inequality in two variables can be written in the form $Ax + By > C$, $Ax + By \geq C$, $Ax + By < C$, or $Ax + By \leq C$.	
b. The procedure for graphing a linear inequality in two variables is given in the box on page 454.	Ex. 1, p. 454; Ex. 2, p. 455
c. Some inequalities can be graphed without using test points, including $y > mx + b$ (the half-plane above $y = mx + b$), $y < mx + b$, $x > a$ (the half-plane to the right of $x = a$), $x < a$, $y > b$ (the half-plane above $y = b$), and $y < b$.	Ex. 3, p. 456
d. Graphing Systems of Linear Inequalities **1.** Graph each inequality in the system in the same rectangular coordinate system. **2.** Find the intersection of the individual graphs.	Ex. 5, p. 458; Ex. 6, p. 459

7.5 Linear Programming

a. An objective function is an algebraic expression in three variables describing a quantity that must be maximized or minimized.	Ex. 1, p. 462
b. Constraints are restrictions, expressed as linear inequalities.	Ex. 2, p. 463; Ex. 3, p. 464
c. Steps for solving a linear programming problem are given in the box on page 464.	Ex. 4, p. 465

7.6 Modeling Data: Exponential, Logarithmic, and Quadratic Functions

a. The exponential function with base b is defined by $y = b^x$ or $f(x) = b^x$, $b > 0$ and $b \neq 1$.	Ex. 1, p. 469
b. All exponential functions of the form $y = b^x$, where $b > 1$, have the shape of the graph in Figure 7.43 on page 469, making this function a good model for data points in a scatter plot that are increasing more and more rapidly.	Ex. 2, p. 471
c. The irrational number e, $e \approx 2.72$, appears in many applied exponential functions. The function $f(x) = e^x$ is called the natural exponential function.	Ex. 3, p. 472
d. The logarithmic function with base b is defined by $y = \log_b x$ or $f(x) = \log_b x$, $x > 0$, $b > 0$, and $b \neq 1$. $y = \log_b x$ means $b^y = x$, so a logarithm is an exponent.	Ex. 4, p. 474
e. $y = \log x$ means $y = \log_{10} x$, the common logarithmic function. $y = \ln x$ means $y = \log_e x$, the natural logarithmic function. Calculators contain LOG and LN keys for evaluating these functions.	

f. All logarithmic functions of the form $y = \log_b x$, where $b > 1$, have the shape of the graph in Figure 7.49 on page 474, making this function a good model for data points in a scatter plot that are increasing, but whose rate of increase is slowing down.	Ex. 5, p. 475
g. A quadratic function is any function of the form $y = ax^2 + bx + c$ or $f(x) = ax^2 + bx + c, a \neq 0$. The graph of a quadratic function is called a parabola.	Figure 7.53, p. 476
h. The x-coordinate of a parabola's vertex is $x = \frac{-b}{2a}$. The y-coordinate is found by substituting the x-coordinate into the parabola's equation. The vertex, or turning point, is the low point when the graph opens upward and the high point when the graph opens downward.	
i. The six steps for graphing a parabola are given in the box on page 477.	Ex. 6, p. 477
j. All quadratic functions have the shape of one of the graphs in Figure 7.53 on page 476, making this function a good model for data points in a scatter plot that are decreasing and then increasing, or vice versa.	Ex. 7, p. 478
k. Table 7.5 on page 479 summarizes the functions used to model various scatter plots.	

Review Exercises

7.1

In Exercises 1–4, plot the given point in a rectangular coordinate system.

1. $(2, 5)$
2. $(-4, 3)$
3. $(-5, -3)$
4. $(2, -5)$

Graph each equation in Exercises 5–7. Let $x = -3, -2, -1, 0, 1, 2$, and 3.

5. $y = 2x - 2$
6. $y = |x| + 2$
7. $y = x$

8. If $f(x) = 4x + 11$, find $f(-2)$.

9. If $f(x) = -7x + 5$, find $f(-3)$.

10. If $f(x) = 3x^2 - 5x + 2$, find $f(4)$.

11. If $f(x) = -3x^2 + 6x + 8$, find $f(-4)$.

In Exercises 12–13, evaluate $f(x)$ for the given values of x. Then use the ordered pairs $(x, f(x))$ from your table to graph the function.

12. $f(x) = \frac{1}{2}|x|$

x	$f(x) = \frac{1}{2}\lvert x\rvert$
−6	
−4	
−2	
0	
2	
4	
6	

13. $f(x) = x^2 - 2$

x	$f(x) = x^2 - 2$
−2	
−1	
0	
1	
2	

In Exercises 14–15, use the vertical line test to identify graphs in which y is a function of x.

14.

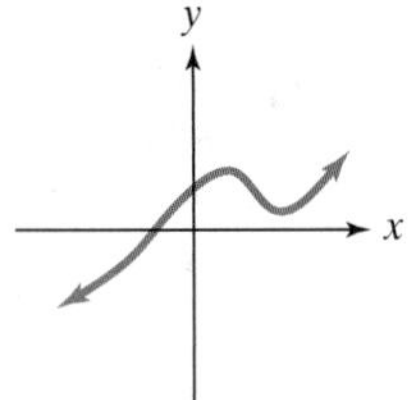

15.

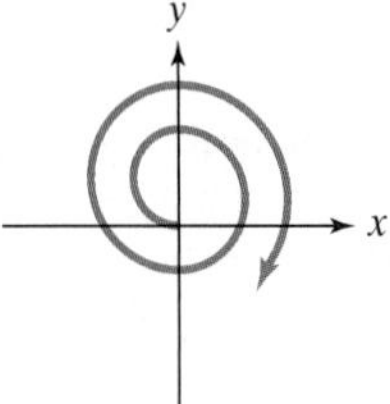

16. Whether on the slopes or at the shore, people are exposed to harmful amounts of the sun's skin-damaging ultraviolet (UV) rays. The quadratic function

$$D(x) = 0.8x^2 - 17x + 109$$

models the average time in which skin damage begins for burn-prone people, $D(x)$, in minutes, where x is the UV index, or measure of the sun's UV intensity. The graph of D is shown for a UV index from 1 (low) to 11 (high).

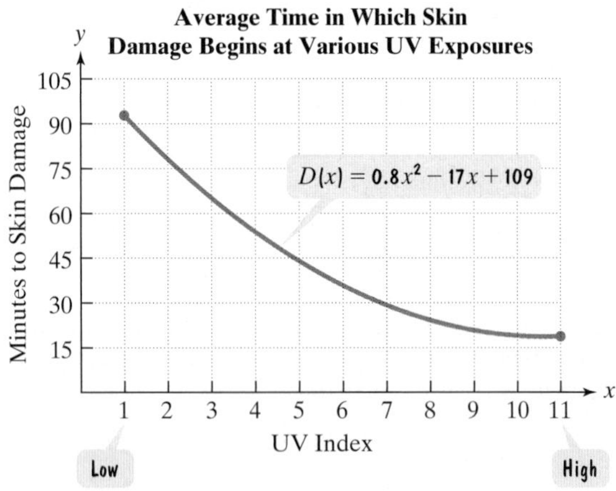

Source: National Oceanic and Atmospheric Administration

a. Find and interpret $D(1)$. How is this shown on the graph of D?

b. Find and interpret $D(10)$. How is this shown on the graph of D?

7.2

In Exercises 17–19, use the x- and y-intercepts to graph each linear equation.

17. $2x + y = 4$

18. $2x - 3y = 6$

19. $5x - 3y = 15$

In Exercises 20–23, calculate the slope of the line passing through the given points. If the slope is undefined, so state. Then indicate whether the line rises, falls, is horizontal, or is vertical.

20. $(3, 2)$ and $(5, 1)$

21. $(-1, 2)$ and $(-3, -4)$

22. $(-3, 4)$ and $(6, 4)$

23. $(5, 3)$ and $(5, -3)$

In Exercises 24–27, graph each linear function using the slope and y-intercept.

24. $y = 2x - 4$ **25.** $y = -\frac{2}{3}x + 5$

26. $f(x) = \frac{3}{4}x - 2$ **27.** $y = \frac{1}{2}x + 0$

In Exercises 28–30, **a.** *Write the equation in slope-intercept form;* **b.** *Identify the slope and the y-intercept;* **c.** *Use the slope and y-intercept to graph the line.*

28. $2x + y = 0$ **29.** $3y = 5x$ **30.** $3x + 2y = 4$

In Exercises 31–33, graph each horizontal or vertical line.

31. $x = 3$ **32.** $y = -4$ **33.** $x + 2 = 0$

34. Shown, again, is the scatter plot that indicates a relationship between the percentage of adult females in a country who are literate and the mortality of children under five. Also shown is a line that passes through or near the points.

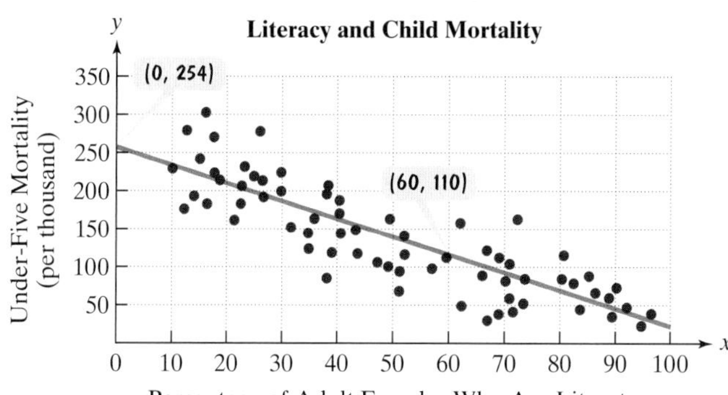

Source: United Nations

a. According to the graph, what is the y-intercept of the line? Describe what this represents in this situation.

b. Use the coordinates of the two points shown to compute the slope of the line. Describe what this means in terms of the rate of change.

c. Use the y-intercept from part (a) and the slope from part (b) to write a linear function that models child mortality, $f(x)$, per thousand, for children under five in a country where x% of adult women are literate.

d. Use the function from part (c) to predict the mortality rate of children under five in a country where 50% of adult females are literate.

7.3

In Exercises 35–37, solve each system by graphing. Check the coordinates of the intersection point in both equations.

35. $\begin{cases} x + y = 5 \\ 3x - y = 3 \end{cases}$ **36.** $\begin{cases} 2x - y = -1 \\ x + y = -5 \end{cases}$ **37.** $\begin{cases} y = -x + 5 \\ 2x - y = 4 \end{cases}$

In Exercises 38–40, solve each system by the substitution method.

38. $\begin{cases} 2x + 3y = 2 \\ x = 3y + 10 \end{cases}$ **39.** $\begin{cases} y = 4x + 1 \\ 3x + 2y = 13 \end{cases}$ **40.** $\begin{cases} x + 4y = 14 \\ 2x - y = 1 \end{cases}$

In Exercises 41–43, solve each system by the addition method.

41. $\begin{cases} x + 2y = -3 \\ x - y = -12 \end{cases}$ **42.** $\begin{cases} 2x - y = 2 \\ x + 2y = 11 \end{cases}$ **43.** $\begin{cases} 5x + 3y = 1 \\ 3x + 4y = -6 \end{cases}$

In Exercises 44–46, solve by the method of your choice. Identify systems with no solution and systems with infinitely many solutions, using set notation to express their solution sets.

44. $\begin{cases} y = -x + 4 \\ 3x + 3y = -6 \end{cases}$ **45.** $\begin{cases} 3x + y = 8 \\ 2x - 5y = 11 \end{cases}$

46. $\begin{cases} 3x - 2y = 6 \\ 6x - 4y = 12 \end{cases}$

47. A company is planning to manufacture computer desks. The fixed cost will be \$60,000 and it will cost \$200 to produce each desk. Each desk will be sold for \$450.

a. Write the cost function, C, of producing x desks.

b. Write the revenue function, R, from the sale of x desks.

c. Determine the break-even point. Describe what this means.

48. The graph shows the number of guns in private hands in the United States and the country's population, both expressed in millions, from 1995 through 2020, with projections from 2017 onward.

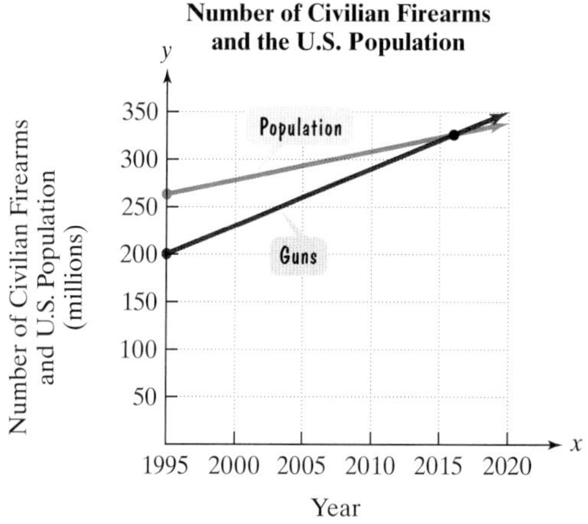

Source: Mother Jones

a. Use the graphs to estimate the point of intersection. In what year was there a gun for every man, woman, and child in the United States? What was the population and the number of firearms in that year?

b. In 1995, there were an estimated 200 million firearms in private hands. This has increased at an average rate of 6 million firearms per year. Write a function that models the number of civilian firearms in the United States, y, in millions, x years after 1995.

c. The function $y - 3x = 263$ models the U.S. population, y, in millions, x years after 1995. Use this model and the model you obtained in part (b) to determine the year in which there was a gun for every U.S. citizen. According to the models, what was the population and the number of firearms in that year?

d. How well do the models in parts (b) and (c) describe the point of intersection of the graphs that you estimated in part (a)?

7.4

In Exercises 49–55, graph each linear inequality.

49. $x - 3y \le 6$ **50.** $2x + 3y \ge 12$ **51.** $2x - 7y > 14$

52. $y > \frac{3}{5}x$ **53.** $y \le -\frac{1}{2}x + 2$ **54.** $x \le 2$

55. $y > -3$

In Exercises 56–61, graph the solution set of each system of linear inequalities.

56. $\begin{cases} 3x - y \le 6 \\ x + y \ge 2 \end{cases}$ **57.** $\begin{cases} x + y < 4 \\ x - y < 4 \end{cases}$ **58.** $\begin{cases} x \le 3 \\ y > -2 \end{cases}$

59. $\begin{cases} 4x + 6y \le 24 \\ y > 2 \end{cases}$ **60.** $\begin{cases} x + y \le 6 \\ y \ge 2x - 3 \end{cases}$ **61.** $\begin{cases} y < -x + 4 \\ y > x - 4 \end{cases}$

7.5

62. Find the value of the objective function $z = 2x + 3y$ at each corner of the graphed region shown. What is the maximum value of the objective function? What is the minimum value of the objective function?

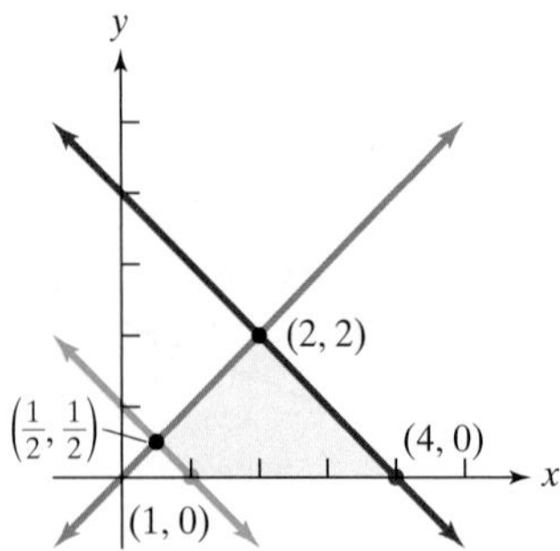

63. Consider the objective function $z = 2x + 3y$ and the following constraints:

$$x \le 6, y \le 5, x + y \ge 2, \underbrace{x \ge 0, y \ge 0}_{\text{Quadrant I and its boundary}}.$$

a. Graph the system of inequalities representing the constraints.

b. Find the value of the objective function at each corner of the graphed region.

c. Use the values in part (b) to determine the maximum and minimum values of the objective function and the values of x and y for which they occur.

64. A paper manufacturing company converts wood pulp to writing paper and newsprint. The profit on a unit of writing paper is \$500 and the profit on a unit of newsprint is \$350.

a. Let x represent the number of units of writing paper produced daily. Let y represent the number of units of newsprint produced daily. Write the objective function that models total daily profit.

b. The manufacturer is bound by the following constraints:
- Equipment in the factory allows for making at most 200 units of paper (writing paper and newsprint) in a day.
- Regular customers require at least 10 units of writing paper and at least 80 units of newsprint daily.

Write a system of inequalities that models these constraints.

c. Graph the inequalities in part (b). Use only the first quadrant and its boundary, because x and y must both be nonnegative. (*Suggestion:* Let each unit along the x- and y-axes represent 20.)

d. Evaluate the objective profit function at each of the three vertices of the graphed region.

e. Complete the missing portions of this statement: The company will make the greatest profit by producing ______ units of writing paper and ______ units of newsprint each day. The maximum daily profit is \$ _________.

7.6

In Exercises 65–66, use a table of coordinates to graph each exponential function. Begin by selecting −2, −1, 0, 1, and 2 for x.

65. $f(x) = 2^x$ **66.** $y = 2^{x+1}$

67. Graph $y = \log_2 x$ by rewriting the equation in exponential form. Use a table of coordinates and select −2, −1, 0, 1, and 2 for y.

In Exercises 68–69,

a. *Determine if the parabola whose equation is given opens upward or downward.*

b. *Find the vertex.*

c. *Find any x-intercepts.*

d. *Find the y-intercept.*

e. *Use (a)–(d) to graph the quadratic function.*

68. $y = x^2 - 6x - 7$ **69.** $f(x) = -x^2 - 2x + 3$

In Exercises 70–74,

a. *Create a scatter plot for the data in each table.*

b. *Use the shape of the scatter plot to determine if the data are best modeled by a linear function, an exponential function, a logarithmic function, or a quadratic function.*

70. AGE OF U.S. DRIVERS AND FATAL CRASHES

Age	Fatal Crashes per 100 Million Miles Driven
20	6.2
25	4.1
35	2.8
45	2.4
55	3.0
65	3.8
75	8.0

Source: Insurance Institute for Highway Safety

71. NUMBER OF JOBS IN THE U.S. SOLAR-ENERGY INDUSTRY

Year	Number of Jobs (thousands)
2010	94
2011	100
2012	119
2013	143
2014	174
2015	209
2016	240

Source: The Solar Foundation

72. INTENSITY AND LOUDNESS LEVEL OF VARIOUS SOUNDS

Intensity (watts per meter^2)	Loudness Level (decibels)
0.1 (loud thunder)	110
1 (rock concert, 2 yd from speakers)	120
10 (jackhammer)	130
100 (jet takeoff, 40 yd away)	140

73. NUMBER OF SERIOUSLY MENTALLY ILL ADULTS IN THE UNITED STATES

Year	Number of Seriously Mentally Ill Adults (millions)
2006	9.0
2008	9.2
2010	9.4
2012	9.6

Source: U.S. Census Bureau

74. HAMACHIPHOBIA

Generation	Percentage Who Won't Try Sushi	Percentage Who Don't Approve of Marriage Equality
Millennials	42	36
Gen X	52	49
Boomers	60	59
Silent/Greatest Generation	72	66

Source: Pew Research Center

75. Just browsing? Take your time. Researchers know, to the dollar, the average amount the typical consumer spends at the shopping mall. And the longer you stay, the more you spend.

Source: International Council of Shopping Centers Research

The data in the bar graph can be modeled by the functions

$$f(x) = 47x + 22 \quad \text{and} \quad g(x) = 42.2(1.56)^x,$$

where $f(x)$ and $g(x)$ model the average amount spent, in dollars, at a shopping mall after x hours.

a. What is the slope of the linear model? What does this mean in terms of the average amount spent at a shopping mall?

b. Which function, the linear or the exponential, is a better model for the average amount spent after 3.5 hours of browsing?

76. The bar graph shows that people with lower incomes are more likely to report that their health is fair or poor.

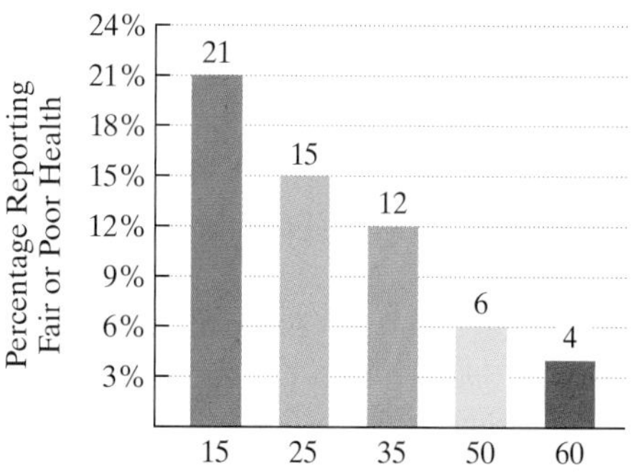

Source: William Kornblum and Joseph Julian, *Social Problems*, 12th Edition, Prentice Hall, 2007.

The data can be modeled by $f(x) = -0.4x + 25.4$ and $g(x) = 54.8 - 12.3 \ln x$, where $f(x)$ and $g(x)$ model the percentage of Americans reporting fair or poor health in terms of annual income, x, in thousands of dollars. Which function, the linear or the logarithmic, is a better model for an annual income of $60 thousand?

Chapter 7 Test

1. Graph $y = |x| - 2$. Let $x = -3, -2, -1, 0, 1, 2,$ and 3.
2. If $f(x) = 3x^2 - 7x - 5$, find $f(-2)$.

In Exercises 3–4, use the vertical line test to identify graphs in which y is a function of x.

3.

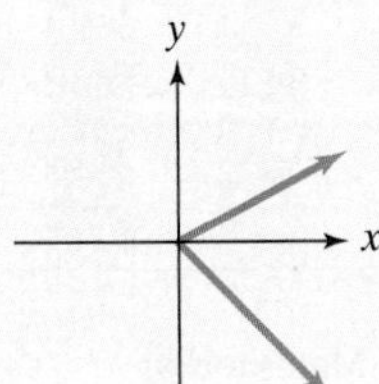

4.

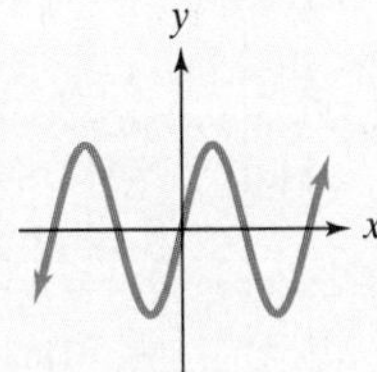

5. The graph shows the height, in meters, of an eagle in terms of its time, in seconds, in flight.

a. Is the eagle's height a function of time? Use the graph to explain why or why not.

b. Find $f(15)$. Describe what this means in practical terms.

c. What is a reasonable estimate of the eagle's maximum height?

d. During which period of time was the eagle descending?

6. Use the x- and y-intercepts to graph $4x - 2y = -8$.
7. Find the slope of the line passing through $(-3, 4)$ and $(-5, -2)$.

In Exercises 8–9, graph each linear function using the slope and y-intercept.

8. $y = \frac{2}{3}x - 1$
9. $f(x) = -2x + 3$

10. In a 2010 survey of more than 200,000 freshmen at 279 colleges, only 52% rated their emotional health high or above average, a drop from 64% in 1985.

Percentage of U.S. College Freshmen Rating Their Emotional Health High or Above Average

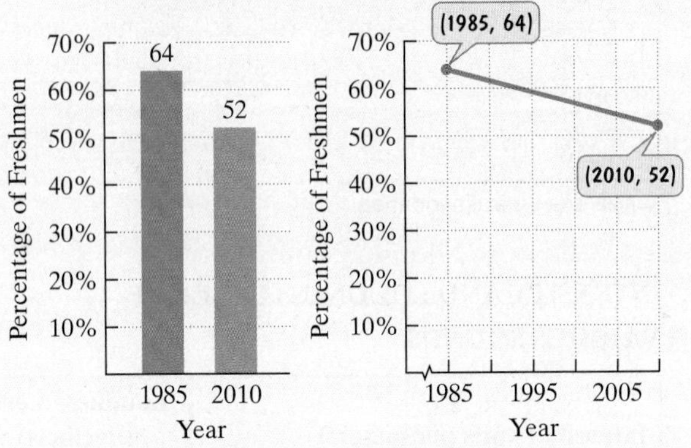

Source: UCLA Higher Education Research Institute

a. Find the slope of the line passing through the two points shown by the voice balloons.

b. Use your answer from part (a) to complete this statement:

For each year from 1985 through 2010, the percentage of U.S. college freshmen rating their emotional health high or above average decreased by ________. The rate of change was ________ per ________.

11. Studies show that texting while driving is as risky as driving with a 0.08 blood alcohol level, the standard for drunk driving. The bar graph shows the number of fatalities in the United States involving distracted driving from 2005 through 2008. Although the distracted category involves such activities as talking on cellphones, conversing with passengers, and eating, experts at the National Highway Traffic Safety Administration claim that texting while driving is the clearest menace because it requires looking away from the road. Shown to the right of the bar graph is a scatter plot with a line passing through two of the data points.

Number of Highway Fatalities in the United States Involving Distracted Driving

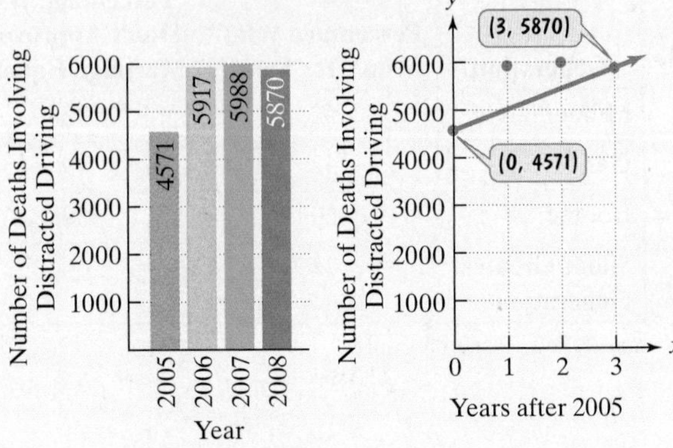

Source: National Highway Traffic Safety Administration

a. According to the scatter plot shown on the right, what is the y-intercept? Describe what this represents in this situation.

b. Use the coordinates of the two points shown in the scatter plot to compute the slope. What does this represent in terms of the rate of change in the number of highway fatalities involving distracted driving?

c. Use the y-intercept shown in the scatter plot and the slope from part (b) to write a linear function that models the number of highway fatalities involving distracted driving, $f(x)$, in the United States x years after 2005.

d. In 2010, surveys showed overwhelming public support to ban texting while driving, although at that time only 19 states and Washington, D.C., outlawed the practice. Without additional laws that penalize texting drivers, use the linear function you obtained from part (c) to project the number of fatalities in the United States in 2015 involving distracted driving.

12. Solve by graphing:

$$\begin{cases} x + y = 6 \\ 4x - y = 4. \end{cases}$$

13. Solve by substitution:

$$\begin{cases} x = y + 4 \\ 3x + 7y = -18. \end{cases}$$

14. Solve by addition:

$$\begin{cases} 5x + 4y = 10 \\ 3x + 5y = -7. \end{cases}$$

15. A company is planning to produce and sell a new line of computers. The fixed cost will be \$360,000 and it will cost \$850 to produce each computer. Each computer will be sold for \$1150.

a. Write the cost function, C, of producing x computers.

b. Write the revenue function, R, from the sale of x computers.

c. Determine the break-even point. Describe what this means.

Graph each linear inequality in Exercises 16–18.

16. $3x - 2y < 6$ **17.** $y \leq \frac{1}{2}x - 1$ **18.** $y > -1$

19. Graph the system of linear inequalities:

$$\begin{cases} 2x - y \leq 4 \\ 2x - y > -1. \end{cases}$$

20. Find the value of the objective function $z = 3x + 2y$ at each corner of the graphed region shown. What is the maximum value of the objective function? What is the minimum value of the objective function?

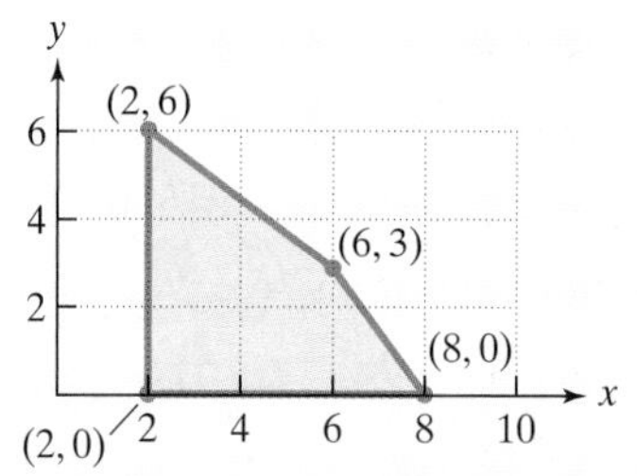

21. Find the maximum value of the objective function $z = 3x + 5y$ subject to the following constraints: $x \geq 0, y \geq 0, x + y \leq 6, x \geq 2$.

22. A manufacturer makes two types of jet skis, regular and deluxe. The profit on a regular jet ski is \$200 and the profit on the deluxe model is \$250. To meet customer demand, the company must manufacture at least 50 regular jet skis per week and at least 75 deluxe models. To maintain high quality, the total number of both models of jet skis manufactured by the company should not exceed 150 per week. How many jet skis of each type should be manufactured per week to obtain maximum profit? What is the maximum weekly profit?

23. Graph $f(x) = 3^x$. Use $-2, -1, 0, 1,$ and 2 for x and find the corresponding values for y.

24. Graph $y = \log_3 x$ by rewriting the equation in exponential form. Use a table of coordinates and select $-2, -1, 0, 1,$ and 2 for y.

25. Use the vertex and intercepts to graph the quadratic function $f(x) = x^2 - 2x - 8$.

In Exercises 26–29, determine whether the values in each table belong to an exponential function, a logarithmic function, a linear function, or a quadratic function.

26.

x	y
0	3
1	1
2	−1
3	−3
4	−5

27.

x	y
$\frac{1}{3}$	−1
1	0
3	1
9	2
27	3

28.

x	y
0	1
1	5
2	25
3	125
4	625

29.

x	y
0	12
1	3
2	0
3	3
4	12

30. The bar graph and the scatter plot show what it cost the United States Mint to make a penny for five selected years from 1982 through 2012. The data can be modeled by the functions

$$f(x) = 0.03x + 0.63 \text{ and } g(x) = 0.72(1.03)^x,$$

where $f(x)$ and $g(x)$ model what it cost to make a penny x years after 1982.

The Cost of Making a Penny

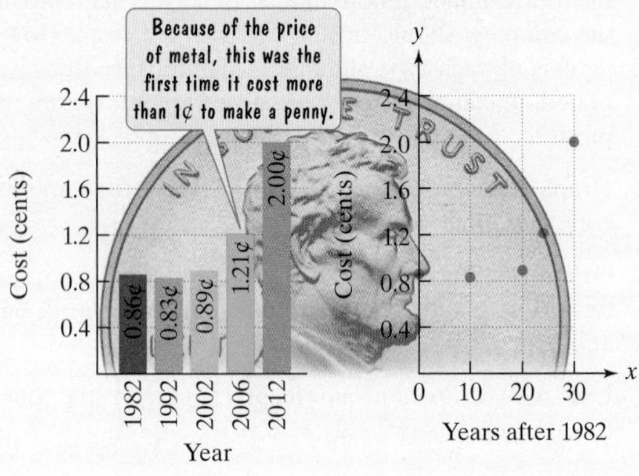

Source: U.S. Mint

a. What is the slope of the linear model? What does this mean in terms of the change in the cost of making a penny?

b. Based on the shape of the scatter plot, which function, the linear or the exponential, is a better model for the data? Explain your answer.

c. Use each function to find the cost of making a penny in 2012. Where necessary, round to two decimal places. Is either function a particularly good model for the 2012 data? Which function serves as a better model? Is this consistent with your answer in part (b)?

Personal Finance

8

"I realize, of course, that it's no shame to be poor, but it's no great honor either. So what would have been so terrible if I had a small fortune?"
—Tevye, a poor dairyman, in the musical *Fiddler on the Roof*

WE ALL WANT A WONDERFUL LIFE WITH FULFILLING WORK, GOOD HEALTH, AND LOVING RELATIONSHIPS. AND LET'S BE honest: Financial security, or even a small fortune, wouldn't hurt! Achieving this goal depends on understanding basic ideas about savings, loans, and investments. A solid understanding of the topics in this chapter can pay, literally, by making your financial goals a reality.

Here's where you'll find these applications:

A number of examples illustrate how to attain fortunes ranging from over a half-million dollars to $4 million through regular savings. See Example 3 in Section 8.5 and Exercises 33–36 in Exercise Set 8.5.

8.1 Percent, Sales Tax, and Discounts

WHAT AM I SUPPOSED TO LEARN?

After studying this section, you should be able to:

1 Express a fraction as a percent.
2 Express a decimal as a percent.
3 Express a percent as a decimal.
4 Solve applied problems involving sales tax and discounts.
5 Determine percent increase or decrease.
6 Investigate some of the ways percent can be abused.

"And if elected, it is my solemn pledge to cut your taxes by 10% for each of my first three years in office, for a total cut of 30%."

PERSONAL FINANCE INCLUDES EVERY area of your life that involves money. It's about what you do with your money and how financial management will affect your future. Because an understanding of *percent* plays an important role in personal finance, we open the chapter with a discussion on the meaning, uses, and abuses of percent.

Basics of Percent

Kinds of Textbooks College Students Prefer: Preferences per 100 Students

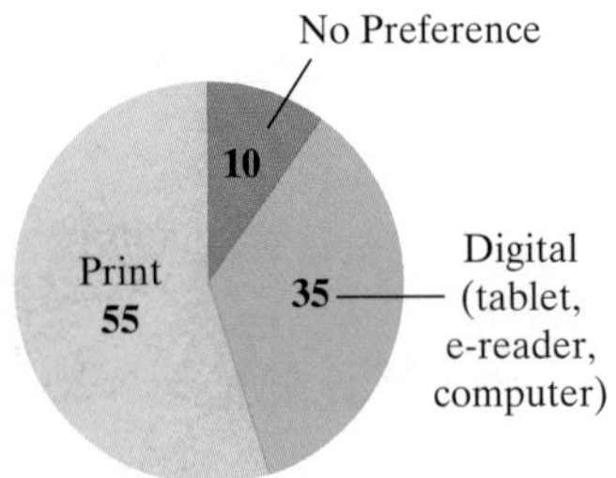

FIGURE 8.1
Source: Harris Interactive for Pearson Foundation

Percents are the result of expressing numbers as part of 100. The word *percent* means *per hundred*. For example, the circle graph in **Figure 8.1** shows that 55 out of every 100 college students prefer print textbooks. Thus, $\frac{55}{100} = 55\%$, indicating that 55% of college students prefer print textbooks. The percent sign, %, is used to indicate the number of parts out of 100 parts.

A fraction can be expressed as a percent using the following procedure:

EXPRESSING A FRACTION AS A PERCENT

1. Divide the numerator by the denominator.
2. Multiply the quotient by 100. This is done by moving the decimal point in the quotient two places to the right.
3. Add a percent sign.

1 *Express a fraction as a percent.*

EXAMPLE 1 Expressing a Fraction as a Percent

Express $\frac{5}{8}$ as a percent.

SOLUTION

Step 1 Divide the numerator by the denominator.

$$5 \div 8 = 0.625$$

Step 2 Multiply the quotient by 100.

$$0.625 \times 100 = 62.5$$

Step 3 Attach a percent sign.

$$62.5\%$$

Thus, $\frac{5}{8} = 62.5\%$.

CHECK POINT 1 Express $\frac{1}{8}$ as a percent.

2 *Express a decimal as a percent.*

Our work in Example 1 shows that $0.625 = 62.5\%$. This illustrates the procedure for expressing a decimal number as a percent.

EXPRESSING A DECIMAL NUMBER AS A PERCENT

1. Move the decimal point two places to the right.
2. Attach a percent sign.

GREAT QUESTION!

What's the difference between the word *percentage* and the word *percent*?

Dictionaries indicate that the word *percentage* has the same meaning as the word *percent*. Use the word that sounds better in the circumstance.

EXAMPLE 2 Expressing a Decimal as a Percent

Express 0.47 as a percent.

SOLUTION

Move decimal point two places to the right.

0.47 % Attach a percent sign.

Thus, 0.47 = 47%.

CHECK POINT 2 Express 0.023 as a percent.

3 *Express a percent as a decimal.*

We reverse the procedure of Example 2 to express a percent as a decimal number.

EXPRESSING A PERCENT AS A DECIMAL NUMBER

1. Move the decimal point two places to the left.
2. Remove the percent sign.

EXAMPLE 3 Expressing Percents as Decimals

Express each percent as a decimal:

a. 19% **b.** 180%.

SOLUTION

Use the two steps in the box.

a.

$$19\% = 19.\% = 0.19\%$$

The decimal point starts at the far right. The decimal point is moved two places to the left. The percent sign is removed.

Thus, 19% = 0.19.

b. 180% = 1.80% = 1.80 or 1.8

CHECK POINT 3 Express each percent as a decimal:

a. 67% **b.** 250%.

If a fraction is part of a percent, as in $\frac{1}{4}\%$, begin by expressing the fraction as a decimal, retaining the percent sign. Then, express the percent as a decimal number. For example,

$$\frac{1}{4}\% = 0.25\% = 00.25\% = 0.0025.$$

GREAT QUESTION!

Can I expect to have lots of zeros when expressing a small percent as a decimal?

Yes. Be careful with the zeros. For example,

$$\frac{1}{100}\% = 0.01\% = 00.01\% = 0.0001.$$

4 *Solve applied problems involving sales tax and discounts.*

Percent, Sales Tax, and Discounts

Many applications involving percent are based on the following formula:

A is P percent of B.

$$A = P \cdot B.$$

Note that the word *of* implies multiplication.

We can use this formula to determine the **sales tax** collected by states, counties, and cities on sales of items to customers. The sales tax is a percent of the cost of an item.

Sales tax amount = tax rate × item's cost

EXAMPLE 4 *Percent and Sales Tax*

Suppose that the local sales tax rate is 7.5% and you purchase a bicycle for $894.

a. How much tax is paid?

b. What is the bicycle's total cost?

SOLUTION

a. Sales tax amount = tax rate × item's cost

= 7.5% × $894 = 0.075 × $894 = $67.05

7.5% of the item's cost, or 7.5% of $894

The tax paid is $67.05.

b. The bicycle's total cost is the purchase price, $894, plus the sales tax, $67.05.

Total cost = $894.00 + $67.05 = $961.05

The bicycle's total cost is $961.05.

CHECK POINT 4 Suppose that the local sales tax rate is 6% and you purchase a computer for $1260.

a. How much tax is paid?

b. What is the computer's total cost?

None of us is thrilled about sales tax, but we do like buying things that are *on sale*. Businesses reduce prices, or **discount**, to attract customers and to reduce inventory. The discount rate is a percent of the original price.

Discount amount = discount rate × original price

EXAMPLE 5 *Percent and Sales Price*

A computer with an original price of $1460 is on sale at 15% off.

a. What is the discount amount?

b. What is the computer's sale price?

TECHNOLOGY

A calculator is useful, and sometimes essential, in this chapter. The keystroke sequence that gives the sale price in Example 5 is

1460 [−] .15 [×] 1460.

Press [=] or [ENTER] to display the answer, 1241.

SOLUTION

a. Discount amount = discount rate × original price

$= 15\% \times \$1460 = 0.15 \times \$1460 = \$219$

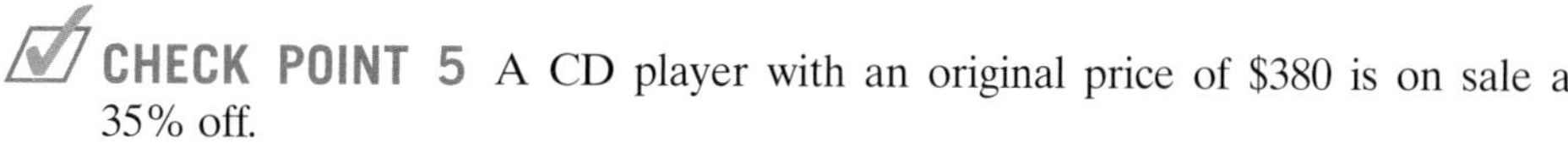

The discount amount is $219.

b. The computer's sale price is the original price, $1460, minus the discount amount, $219.

Sale price = $1460 − $219 = $1241

The computer's sale price is $1241.

CHECK POINT 5 A CD player with an original price of $380 is on sale at 35% off.

a. What is the discount amount?

b. What is the CD player's sale price?

GREAT QUESTION!

Do I have to determine the discount amount before finding the sale price?

No. For example, in Example 5 the computer is on sale at 15% off. This means that the sale price must be 100% − 15%, or 85%, of the original price.

Sale price = 85% × $1460 = 0.85 × $1460 = $1241

5 *Determine percent increase or decrease.*

Percent and Change

Percents are used for comparing changes, such as increases or decreases in sales, population, prices, and production. If a quantity changes, its **percent increase** or its **percent decrease** can be found as follows:

FINDING PERCENT INCREASE OR PERCENT DECREASE

1. Find the fraction for the percent increase or the percent decrease:

$$\frac{\text{amount of increase}}{\text{original amount}} \quad \text{or} \quad \frac{\text{amount of decrease}}{\text{original amount}}.$$

2. Find the percent increase or the percent decrease by expressing the fraction in step 1 as a percent.

EXAMPLE 6 *Finding Percent Increase and Decrease*

In 2000, world population was approximately 6 billion. **Figure 8.2** shows world population projections through the year 2150. The data are from the United Nations Family Planning Program and are based on optimistic or pessimistic expectations for successful control of human population growth.

a. Find the percent increase in world population from 2000 to 2150 using the high projection data.

b. Find the percent decrease in world population from 2000 to 2150 using the low projection data.

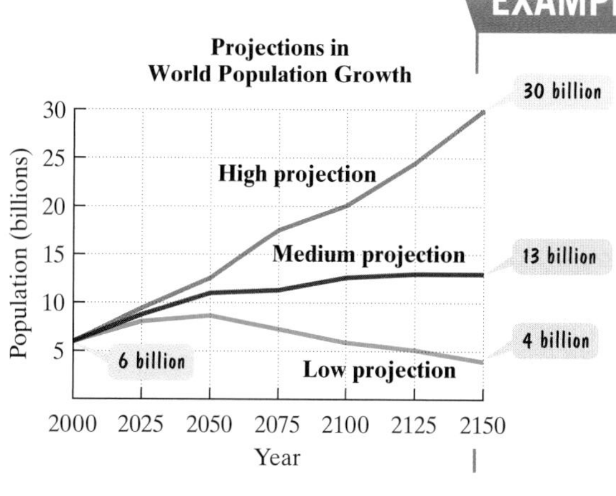

FIGURE 8.2
Source: United Nations

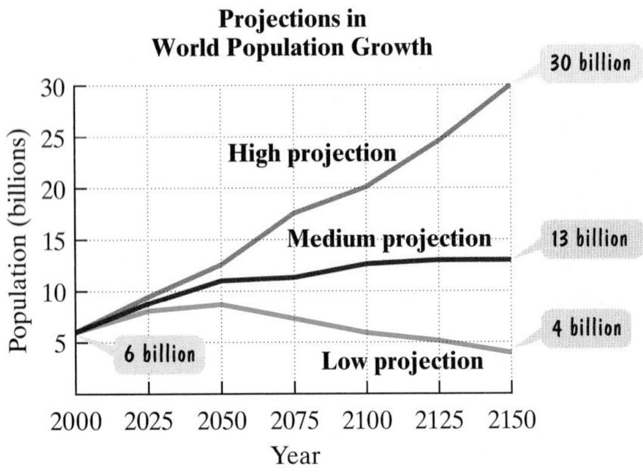

FIGURE 8.2 (repeated)

SOLUTION

a. Use the data shown on the blue, high-projection, graph.

$$\text{Percent increase} = \frac{\text{amount of increase}}{\text{original amount}}$$

$$= \frac{30 - 6}{6} = \frac{24}{6} = 4 = 400\%$$

The projected percent increase in world population is 400%.

b. Use the data shown on the green, low-projection, graph.

$$\text{Percent decrease} = \frac{\text{amount of decrease}}{\text{original amount}}$$

$$= \frac{6 - 4}{6} = \frac{2}{6} = \frac{1}{3} = 0.33\frac{1}{3} = 33\frac{1}{3}\%$$

The projected percent decrease in world population is $33\frac{1}{3}\%$.

In Example 6, we expressed the percent decrease as $33\frac{1}{3}\%$ because of the familiar conversion $\frac{1}{3} = 0.33\frac{1}{3}$. However, in many situations, rounding is needed. We suggest that you round to the nearest tenth of a percent. Carry the division in the fraction for percent increase or decrease to four places after the decimal point. Then round the decimal to three places, or to the nearest thousandth. Expressing this rounded decimal as a percent gives percent increase or decrease to the nearest tenth of a percent.

GREAT QUESTION!

I know that increasing 2 to 8 is a 300% increase. Does that mean decreasing 8 to 2 is a 300% decrease?

No. Notice the difference between the following examples:

- 2 is increased to 8.

$$\text{Percent increase} = \frac{\text{amount of increase}}{\text{original amount}} = \frac{6}{2} = 3 = 300\%$$

- 8 is decreased to 2.

$$\text{Percent decrease} = \frac{\text{amount of decrease}}{\text{original amount}} = \frac{6}{8} = \frac{3}{4} = 0.75 = 75\%$$

Although an increase from 2 to 8 is a 300% increase, a decrease from 8 to 2 is *not* a 300% decrease. **A percent decrease involving nonnegative quantities can never exceed 100%.** When a quantity is decreased by 100%, it is reduced to zero.

CHECK POINT 6

a. If 6 is increased to 10, find the percent increase.

b. If 10 is decreased to 6, find the percent decrease.

EXAMPLE 7 *Finding Percent Decrease*

A jacket regularly sells for \$135.00. The sale price is \$60.75. Find the percent decrease of the sale price from the regular price.

SOLUTION

$$\text{Percent decrease} = \frac{\text{amount of decrease}}{\text{original amount}}$$

$$= \frac{135.00 - 60.75}{135} = \frac{74.25}{135} = 0.55 = 55\%$$

The percent decrease of the sale price from the regular price is 55%. This means that the sale price of the jacket is 55% lower than the regular price.

CHECK POINT 7 A television regularly sells for \$940. The sale price is \$611. Find the percent decrease of the sale price from the regular price.

6 *Investigate some of the ways percent can be abused.*

Abuses of Percent

In our next examples, we look at a few of the many ways that percent can be used incorrectly. Confusion often arises when percent increase (or decrease) refers to a changing quantity that is itself a percent.

EXAMPLE 8 *Percents of Percents*

John Tesh, while he was still coanchoring *Entertainment Tonight*, reported that the PBS series *The Civil War* had an audience of 13% versus the usual 4% PBS audience, "an increase of more than 300%." Did Tesh report the percent increase correctly?

SOLUTION

We begin by finding the percent increase.

$$\text{Percent increase} = \frac{\text{amount of increase}}{\text{original amount}}$$

$$= \frac{13\% - 4\%}{4\%} = \frac{9\%}{4\%} = \frac{9}{4} = 2.25 = 225\%$$

The percent increase for PBS was 225%. This is not more than 300%, so Tesh did not report the percent increase correctly.

CHECK POINT 8 An episode of a television series had an audience of 12% versus its usual 10%. What was the percent increase for this episode?

EXAMPLE 9 *Promises of a Politician*

A politician states, "If you elect me to office, I promise to cut your taxes for each of my first three years in office by 10% each year, for a total reduction of 30%." Evaluate the accuracy of the politician's statement.

SOLUTION

To make things simple, let's assume that a taxpayer paid \$100 in taxes in the year previous to the politician's election. A 10% reduction during year 1 is 10% of \$100.

$$10\% \text{ of previous year tax} = 10\% \text{ of } \$100 = 0.10 \times \$100 = \$10$$

With a 10% reduction the first year, the taxpayer will pay only \$100 − \$10, or \$90, in taxes during the politician's first year in office.

The following table shows how we calculate the new, reduced tax for each of the first three years in office:

Year	Tax Paid the Year Before	10% Reduction	Taxes Paid This Year
1	\$100	$0.10 \times \$100 = \$10$	$\$100 - \$10 = \$90$
2	\$90	$0.10 \times \$90 = \$9$	$\$90 - \$9 = \$81$
3	\$81	$0.10 \times \$81 = \$8.10$	$\$81 - \$8.10 = \$72.90$

Now, we determine the percent decrease in taxes over the three years.

$$\text{Percent decrease} = \frac{\text{amount of decrease}}{\text{original amount}}$$

$$= \frac{\$100 - \$72.90}{\$100} = \frac{\$27.10}{\$100} = \frac{27.1}{100} = 0.271 = 27.1\%$$

The taxes decline by 27.1%, not by 30%. The politician is ill-informed in saying that three consecutive 10% cuts add up to a total tax cut of 30%. In our calculation, which serves as a counterexample to the promise, the total tax cut is only 27.1%.

CHECK POINT 9 Suppose you paid \$1200 in taxes. During year 1, taxes decrease by 20%. During year 2, taxes increase by 20%.

a. What do you pay in taxes for year 2?

b. How do your taxes for year 2 compare with what you originally paid, namely \$1200? If the taxes are not the same, find the percent increase or decrease.

Blitzer Bonus

Testing Your Financial Literacy

Scores have been falling on tests that measure financial literacy. Here are four items from a test given to high school seniors. Would you ace this one?

1. Which of the following is true about sales taxes?
- **A.** The national sales-tax percentage rate is 6%.
- **B.** The Federal Government will deduct it from your paycheck.
- **C.** You don't have to pay the tax if your income is very low.
- **D.** It makes things more expensive for you to buy.

58% of high school seniors answered incorrectly.

2. If you have caused an accident, which type of automobile insurance would cover damage to your own car?
- **A.** Comprehensive
- **B.** Liability
- **C.** Term
- **D.** Collision

63% of high school seniors answered incorrectly.

3. Which of the following types of investment would best protect the purchasing power of a family's savings in the event of a sudden increase in inflation?
- **A.** A 10-year bond issued by a corporation
- **B.** A certificate of deposit at a bank
- **C.** A 25-year corporate bond
- **D.** A house financed with a fixed-rate mortgage

64% of high school seniors answered incorrectly.

4. Sara and Joshua just had a baby. They received money as baby gifts and want to put it away for the baby's education. Which of the following tends to have the highest growth over periods of time as long as 18 years?
- **A.** A checking account
- **B.** Stocks
- **C.** A U.S. government savings bond
- **D.** A savings account

83% of high school seniors answered incorrectly.

Source: The Jump\$tart Coalition's 2008 Personal Financial Survey

Answers: 1. D; 2. D; 3. D; 4. B

Concept and Vocabulary Check

Fill in each blank so that the resulting statement is true.

1. Percents are the result of expressing numbers as part of ______.
2. To express $\frac{7}{8}$ as a percent, divide ____ by ____, multiply the quotient by ______, and attach ____________.
3. To express 0.1 as a percent, move the decimal point ______ places to the ______ and attach ____________.
4. To express 7.5% as a decimal, move the decimal point ______ places to the ______ and remove ______________.
5. To find the sales tax amount, multiply the ________ and the __________.
6. To find the discount amount, multiply the ____________ and the ____________.
7. The numerator of the fraction for percent increase is __________________ and the denominator of the fraction for percent increase is ________________.
8. The numerator of the fraction for percent decrease is __________________ and the denominator of the fraction for percent decrease is ________________.

Exercises 9–10 are based on items from a financial literacy survey from the Center for Economic and Entrepreneurial Literacy. Determine whether each statement is true or false. If the statement is false, make the necessary change(s) to produce a true statement.

9. Santa had to lay off 25% of his eight reindeer because of the bad economy, so only seven reindeer remained. (65% answered this question incorrectly. Santa might consider leaving *Thinking Mathematically* in stockings across the country.) ______
10. You spent 1% of your $50,000-per-year salary on gifts, so you spent $5000 on gifts for the year. ______

Exercise Set 8.1

Practice Exercises

In Exercises 1–10, express each fraction as a percent.

1. $\frac{2}{5}$ **2.** $\frac{3}{5}$ **3.** $\frac{1}{4}$ **4.** $\frac{3}{4}$
5. $\frac{3}{8}$ **6.** $\frac{7}{8}$ **7.** $\frac{1}{40}$ **8.** $\frac{3}{40}$
9. $\frac{9}{80}$ **10.** $\frac{13}{80}$

In Exercises 11–20, express each decimal as a percent.

11. 0.59 **12.** 0.96 **13.** 0.3844
14. 0.003 **15.** 2.87 **16.** 9.83
17. 14.87 **18.** 19.63
19. 100 **20.** 95

In Exercises 21–34, express each percent as a decimal.

21. 72% **22.** 38% **23.** 43.6%
24. 6.25% **25.** 130% **26.** 260%
27. 2% **28.** 6% **29.** $\frac{1}{2}$%
30. $\frac{3}{4}$% **31.** $\frac{5}{8}$% **32.** $\frac{1}{8}$%
33. $62\frac{1}{2}$% **34.** $87\frac{1}{2}$%

Use the percent formula, $A = PB$: A is P percent of B, to solve Exercises 35–38.

35. What is 3% of 200? **36.** What is 8% of 300?
37. What is 18% of 40? **38.** What is 16% of 90?

Practice Plus

Three basic types of percent problems can be solved using the percent formula $A = PB$.

Question	Given	Percent Formula
What is *P* percent of *B*?	*P* and *B*	Solve for *A*.
A is *P* percent of what?	*A* and *P*	Solve for *B*.
A is what percent of *B*?	*A* and *B*	Solve for *P*.

Exercises 35–38 involved using the formula to answer the first question. In Exercises 39–46, use the percent formula and the information in the previous column to answer the second or third question.

39. 3 is 60% of what?
40. 8 is 40% of what?
41. 24% of what number is 40.8?
42. 32% of what number is 51.2?
43. 3 is what percent of 15?
44. 18 is what percent of 90?
45. What percent of 2.5 is 0.3?
46. What percent of 7.5 is 0.6?

Application Exercises

47. Suppose that the local sales tax rate is 6% and you purchase a car for $32,800.
  a. How much tax is paid?
  b. What is the car's total cost?

48. Suppose that the local sales tax rate is 7% and you purchase a graphing calculator for $96.
  a. How much tax is paid?
  b. What is the calculator's total cost?

49. An exercise machine with an original price of $860 is on sale at 12% off.
  a. What is the discount amount?
  b. What is the exercise machine's sale price?

50. A dictionary that normally sells for $16.50 is on sale at 40% off.
  a. What is the discount amount?
  b. What is the dictionary's sale price?

The circle graph shows a breakdown of spending for the average U.S. household using 365 days worked as a basis of comparison. Use this information to solve Exercises 51–52. Round answers to the nearest tenth of a percent.

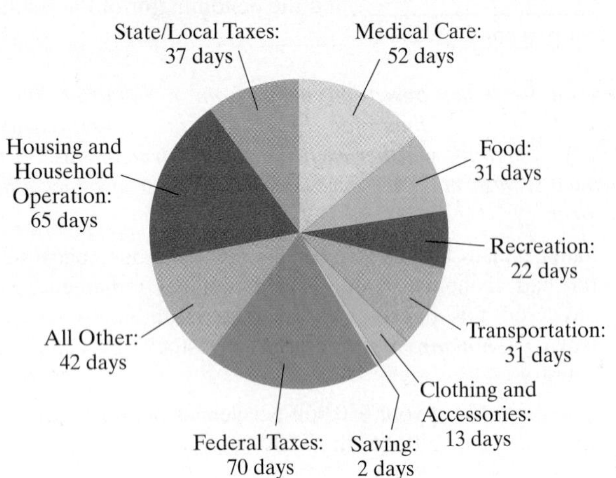

Source: The Tax Foundation

51. What percentage of work time does the average U.S. household spend paying for federal taxes?

52. What percentage of work time does the average U.S. household spend paying for state and local taxes?

The bar graph shows that life expectancy, the number of years newborns are expected to live, has increased dramatically since ancient times. Use this information to solve Exercises 53–54.

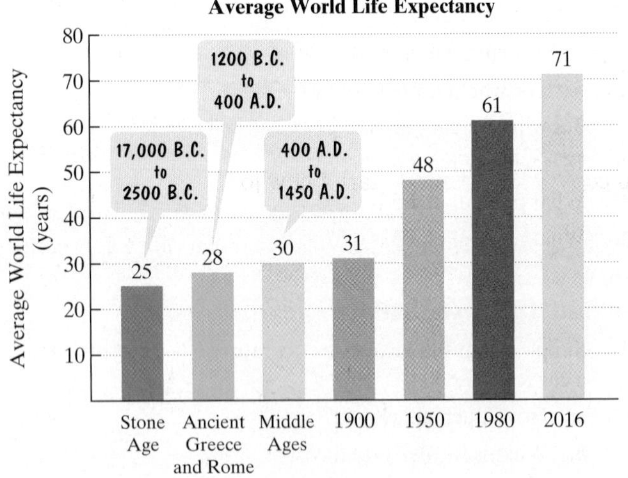

Source: The New York Times Upfront

53. Find the percent increase in average world life expectancy from the Stone Age to 2016.

54. Find the percent increase in average world life expectancy from the Stone Age to 1980.

55. A sofa regularly sells for \$840. The sale price is \$714. Find the percent decrease of the sale price from the regular price.

56. A FAX machine regularly sells for \$380. The sale price is \$266. Find the percent decrease of the sale price from the regular price.

57. Suppose that you have \$10,000 in a rather risky investment recommended by your financial advisor. During the first year, your investment decreases by 30% of its original value. During the second year, your investment increases by 40% of its first-year value. Your advisor tells you that there must have been a 10% overall increase of your original \$10,000 investment. Is your financial advisor using percentages properly? If not, what is your actual percent gain or loss of your original \$10,000 investment?

58. The price of a color printer is reduced by 30% of its original price. When it still does not sell, its price is reduced by 20% of the reduced price. The salesperson informs you that there has been a total reduction of 50%. Is the salesperson using percentages properly? If not, what is the actual percent reduction from the original price?

Explaining the Concepts

59. What is a percent?

60. Describe how to express a decimal number as a percent and give an example.

61. Describe how to express a percent as a decimal number and give an example.

62. Explain how to use the sales tax rate to determine an item's total cost.

63. Describe how to find percent increase and give an example.

64. Describe how to find percent decrease and give an example.

Critical Thinking Exercises

Make Sense? *In Exercises 65–68, determine whether each statement makes sense or does not make sense, and explain your reasoning.*

65. I have \$100 and my restaurant bill comes to \$80, which is not enough to leave a 20% tip.

66. I found the percent decrease in a jacket's price to be 120%.

67. My weight increased by 1% in January and 1% in February, so my increase in weight over the two months is 2%.

68. My rent increased from 20% to 30% of my income, so the percent increase is 10%.

69. What is the total cost of a \$720 iPad that is on sale at 15% off if the local sales tax rate is 6%?

70. A condominium is taxed based on its \$78,500 value. The tax rate is \$3.40 for every \$100 of value. If the tax is paid before March 1, 3% of the normal tax is given as a discount. How much tax is paid if the condominium owner takes advantage of the discount?

71. In January, each of 60 people purchased a \$500 washing machine. In February, 10% fewer customers purchased the same washing machine that had increased in price by 20%. What was the change in sales from January to February?

72. When you buy something, it actually costs more than you may think—at least in terms of how much money you must earn to buy it. For example, if you pay 28% of your income in taxes, how much money would you have to earn to buy a used car for \$7200?

8.2

WHAT AM I SUPPOSED TO LEARN?

After studying this section, you should be able to:

1 Determine gross income, adjustable gross income, and taxable income.
2 Calculate federal income tax.
3 Calculate FICA taxes.
4 Solve problems involving working students and taxes.

Income Tax

"THE TROUBLE WITH TRILLIONS" EPISODE of the *Simpsons* finds Homer frantically putting together his tax return two hours before the April 15th mailing deadline. In a frenzy, he shouts to his wife, "Marge, how many kids do we have, no time to count, I'll just estimate nine. If anyone asks, you need 24-hour nursing care, Lisa is a clergyman, Maggie is seven people, and Bart was wounded in Vietnam."

"Cool!" replies Bart.

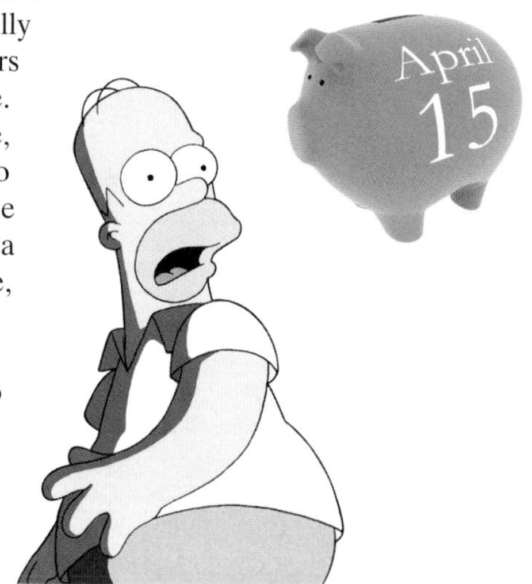

It isn't only cartoon characters who are driven into states of frantic agitation over taxes. The average American pays over $10,000 per year in income tax. Yes, it's important to pay Uncle Sam what you owe, but not a penny more. People who do not understand the federal tax system often pay more than they have to. In this section, you will learn how income taxes are determined and calculated, reinforcing the role of tax planning in personal finance.

Paying Income Tax

Income tax is a percentage of your income collected by the government to fund its services and programs. The federal government collects income tax, and most, but not all, state governments do, too. (Alaska, Florida, Nevada, South Dakota, Texas, Washington, and Wyoming have no state income tax.) Tax revenue pays for our national defense, fire and police protection, road construction, schools, libraries, and parks. Without taxes, the government would not be able to conduct medical research, provide medical care for the elderly, or send astronauts into space.

Income tax is automatically withheld from your paycheck by your employer. The precise amount withheld for federal income tax depends on how you fill out your W-4 form, which you complete when you start a new job.

Although the United States Congress determines federal tax laws, the Internal Revenue Service (IRS) is the government body that enforces the laws and collects taxes. The IRS is a branch of the Treasury Department.

1 *Determine gross income, adjustable gross income, and taxable income.*

Determining Taxable Income

Federal income taxes are a percentage of your *taxable income*, which is based on your earnings in the calendar year—January to December. When the year is over, you have until April 15th to file your tax return.

Calculating your federal income tax begins with **gross income**, or total income for the year. This includes income from wages, tips, interest or dividends from investments, unemployment compensation, profits from a business, rental income, and even game-show winnings. It does not matter whether these winnings are in cash or in the form of items such as cars or vacations.

The next step in calculating your federal income tax is to determine your **adjusted gross income**. Adjusted gross income is figured by taking gross income and subtracting certain allowable amounts, called **adjustments**. These untaxed portions of gross income include contributions to certain retirement accounts and tax-deferred savings plans, interest paid on student loans, and alimony payments. In a traditional tax-deferred retirement plan, you get to deduct the full amount of your contribution from your gross income. You pay taxes on the money later, when you withdraw it at retirement.

$$\text{Adjusted gross income} = \text{Gross income} - \text{Adjustments}$$

IRS rules detail exactly what can be subtracted from gross income to determine your adjusted gross income.

Blitzer Bonus

Willie Nelson's Adjustments

In 1990, the IRS sent singer Willie Nelson a bill for \$32 million. Egad! But as Willie said, "Thirty-two million ain't much if you say it fast." How did this happen? On bad advice, Willie got involved in a number of investments that he declared as adjustments. This reduced his adjusted gross income to a negligible amount that made paying taxes unnecessary. The IRS ruled these "adjustments" as blatant tax-avoidance schemes. Eventually, Willie and the IRS settled on a \$9 million payment.

Source: Arthur J. Keown, *Personal Finance*, Fourth Edition, Pearson, 2007.

You are entitled to certain *exemptions* and *deductions*, subtracted from your adjusted gross income, before calculating your taxes. An **exemption** is a fixed amount on your return for each person supported by your income. You are entitled to this fixed amount (\$4050 in 2016) for yourself and the same amount for each dependent.

A **standard deduction** is a lump-sum amount that you can subtract from your adjusted gross income. The IRS sets this amount. Most young people take the standard deduction because their financial situations are relatively simple and they are not eligible for numerous deductions associated with owning a home or making charitable contributions. **Itemized deductions** are deductions you list separately if you have incurred a large number of deductible expenses. Itemized deductions include interest on home mortgages, state income taxes, property taxes, charitable contributions, and medical expenses exceeding 7.5% of adjusted gross income. Taxpayers should choose the greater of a standard deduction or an itemized deduction.

Taxable income is figured by subtracting exemptions and deductions from adjusted gross income.

$$\text{Taxable income} = \text{Adjusted gross income} - (\text{Exemptions} + \text{Deductions})$$

EXAMPLE 1 *Gross Income, Adjusted Gross Income, and Taxable Income*

A single man earned wages of \$46,500, received \$1850 in interest from a savings account, received \$15,000 in winnings on a television game show, and contributed \$2300 to a tax-deferred savings plan. He is entitled to a personal exemption of \$4050 and a standard deduction of \$6300. The interest on his home mortgage was \$6500, he paid \$2100 in property taxes and \$1855 in state taxes, and he contributed \$3000 to charity.

a. Determine the man's gross income.

b. Determine the man's adjusted gross income.

c. Determine the man's taxable income.

SOLUTION

a. Gross income refers to this person's total income, which includes wages, interest from a savings account, and game-show winnings.

$$\text{Gross income} = \underbrace{\$46{,}500}_{\text{Wages}} + \underbrace{\$1850}_{\text{Earned interest}} + \underbrace{\$15{,}000}_{\text{Game-show winnings}} = \$63{,}350$$

The gross income is \$63,350.

b. Adjusted gross income is gross income minus adjustments. The adjustment in this case is the contribution of \$2300 to a tax-deferred savings plan. The full amount of this contribution is deducted from this year's gross income, although taxes will be paid on the money later when it is withdrawn, probably at retirement.

$$\text{Adjusted gross income} = \text{Gross income} - \text{Adjustments} = \$63{,}350 - \underbrace{\$2300}_{\text{Contribution to a tax-deferred savings plan}} = \$61{,}050$$

The adjusted gross income is \$61,050.

c. We need to subtract exemptions and deductions from the adjusted gross income to determine the man's taxable income. This taxpayer is entitled to a personal exemption of \$4050 and a standard deduction of \$6300. However, a deduction greater than \$6300 is obtained by itemizing deductions.

$$\text{Itemized deductions} = \underbrace{\$6500}_{\text{Interest on home mortgage}} + \underbrace{\$2100}_{\text{Property taxes}} + \underbrace{\$1855}_{\text{State taxes}} + \underbrace{\$3000}_{\text{Charity}} = \$13{,}455$$

We choose the itemized deductions of \$13,455 because they are greater than the standard deduction of \$6300.

$$\begin{aligned}\text{Taxable income} &= \text{Adjusted gross income} - (\text{Exemptions} + \text{Deductions})\\ &= \$61{,}050 - (\$4050 + \$13{,}455)\\ &= \$61{,}050 - \$17{,}505\\ &= \$43{,}545\end{aligned}$$

The taxable income is \$43,545.

SUMMARY OF KINDS OF INCOME ASSOCIATED WITH FEDERAL TAXES

Gross income is total income for the year.

Adjusted gross income = Gross income − Adjustments

Taxable income = Adjusted gross income − (Exemptions + Deductions)

 **CHECK POINT 1** A single woman earned wages of \$87,200, received \$2680 in interest from a savings account, and contributed \$3200 to a tax-deferred savings plan. She is entitled to a personal exemption of \$4050 and a standard deduction of \$6300. The interest on her home mortgage was \$11,700, she paid \$4300 in property taxes and \$5220 in state taxes, and she contributed \$15,000 to charity.

a. Determine the woman's gross income.

b. Determine the woman's adjusted gross income.

c. Determine the woman's taxable income.

2 *Calculate federal income tax.*

Calculating Federal Income Tax

A tax table is used to determine how much you owe based on your taxable income. However, you do not have to pay this much tax if you are entitled to any *tax credits.* **Tax credits** are sums of money that reduce the income tax owed by the full dollar-for-dollar amount of the credits.

Blitzer Bonus

Taking a Bite Out of Taxes

A tax credit is not the same thing as a tax deduction. A tax deduction reduces taxable income, saving only a percentage of the deduction in taxes. A tax credit reduces the income tax owed on the full dollar amount of the credit. There are credits available for everything from donating a kidney to buying a solar energy system. The American Opportunity Credit, included in the economic stimulus package of 2009, provides a tax credit of up to $2500 per student. The credit can be used to lower the costs of the first four years of college. You can claim the credit for up to 100% of the first $2000 in qualified college costs and 25% of the next $2000. Significantly, 40% of this credit is refundable. This means that even if you do not have any taxable income, you could receive a check from the government for up to $1000.

Credits are awarded for a variety of activities that the government wants to encourage. Because a tax credit represents a dollar-for-dollar reduction in your tax bill, it pays to know your tax credits. You can learn more about tax credits at www.irs.gov.

Most people pay part or all of their tax bill during the year. If you are employed, your employer deducts federal taxes through *withholdings* based on a percentage of your gross pay. If you are self-employed, you pay your tax bill through *quarterly estimated taxes.*

When you file your tax return, all you are doing is settling up with the IRS over the amount of taxes you paid during the year versus the federal income tax that you owe. Many people will have paid more during the year than they owe, in which case they receive a *tax refund.* Others will not have paid enough and need to send the rest to the IRS by the deadline.

CALCULATING FEDERAL INCOME TAX

Round all amounts to the nearest dollar.

1. Determine your adjusted gross income:

$$\text{Adjusted gross income} = \text{Gross income} - \text{Adjustments}.$$

Gross income: All income for the year, including wages, tips, earnings from investments, and unemployment compensation

Adjustments: Includes payments to tax-deferred savings plans and a percentage of college expenses

2. Determine your taxable income:

$$\text{Taxable income} = \text{Adjusted gross income} - (\text{Exemptions} + \text{Deductions}).$$

Exemptions: A fixed amount for yourself ($4050 in 2016) and the same amount for each dependent

Deductions: Choose the greater of a standard deduction or an itemized deduction, which includes interest on home mortgages, state income taxes, property taxes, charitable contributions, and medical expenses exceeding 7.5% of adjusted gross income.

3. Determine your income tax:

$$\text{Income tax} = \text{Tax computation} - \text{Tax credits.}$$

Use your taxable income and tax rates for your filing status (single, married, etc.) to determine this amount.

Lawmakers have enacted numerous tax credits to help defray college costs.

Table 8.1 shows 2016 tax rates, standard deductions, and exemptions for the four **filing status** categories described in the voice balloons. The tax rates in the left column, called **marginal tax rates**, are assigned to various income ranges, called margins. For example, suppose you are single and your taxable income is \$25,000. The singles column of the table shows that you must pay 10% tax on the first \$9275, which is

$$10\% \text{ of } \$9275 = 0.10 \times \$9275 = \$927.50.$$

You must also pay 15% tax on the remaining \$15,725 $(\$25{,}000 - \$9275 = \$15{,}725)$, which is

$$15\% \text{ of } \$15{,}725 = 0.15 \times \$15{,}725 = \$2358.75$$

Your total tax is $\$927.50 + \$2358.75 = \$3286.25$. In this scenario, your *marginal rate* is 15% and you are in the 15% *tax bracket*.

TABLE 8.1 2016 Marginal Tax Rates, Standard Deductions, and Exemptions

	Unmarried, divorced, or legally separated	Married and each partner files a separate tax return	Married and both partners file a single tax return	Unmarried and paying more than half the cost of supporting a child or parent
Tax Rate	**Single**	**Married Filing Separately**	**Married Filing Jointly**	**Head of Household**
10%	up to \$9275	up to \$9275	up to \$18,550	up to \$13,250
15%	\$9276 to \$37,650	\$9276 to \$37,650	\$18,551 to \$75,300	\$13,251 to \$50,400
25%	\$37,651 to \$91,150	\$37,651 to \$75,950	\$75,301 to \$151,900	\$50,401 to \$130,150
28%	\$91,151 to \$190,150	\$75,951 to \$115,725	\$151,901 to \$231,450	\$130,151 to \$210,800
33%	\$190,151 to \$413,350	\$115,726 to \$206,675	\$231,451 to \$413,350	\$210,801 to \$413,350
35%	\$413,351 to \$415,050	\$206,676 to \$233,475	\$413,351 to \$466,950	\$413,351 to \$441,000
39.6%	more than \$415,050	more than \$233,475	more than \$466,950	more than \$441,000
Standard Deduction	\$6300	\$6300	\$12,600	\$9300
Exemptions (per person)	\$4050	\$4050	\$4050	\$4050

SINGLE WOMAN WITH NO DEPENDENTS

Gross income: \$62,000

Adjustments: \$4000 paid to a tax-deferred IRA (Individual Retirement Account)

Deductions:
- \$7500: mortgage interest
- \$2200: property taxes
- \$2400: charitable contributions
- \$1500: medical expenses not covered by insurance

Tax credit: \$500

EXAMPLE 2 Computing Federal Income Tax

Calculate the federal income tax owed by a single woman with no dependents whose gross income, adjustments, deductions, and credits are given in the margin. Use the 2016 marginal tax rates in **Table 8.1**.

SOLUTION

Step 1 Determine the adjusted gross income.

$$\begin{aligned}\text{Adjusted gross income} &= \text{Gross income} - \text{Adjustments}\\ &= \$62{,}000 - \$4000\\ &= \$58{,}000\end{aligned}$$

SINGLE WOMAN WITH NO DEPENDENTS

Gross income: $62,000

Adjustments: $4000 paid to a tax-deferred IRA (Individual Retirement Account)

Deductions:

- $7500: mortgage interest
- $2200: property taxes
- $2400: charitable contributions
- $1500: medical expenses not covered by insurance

Tax credit: $500

Step 2 Determine the taxable income.

$$\begin{aligned}\text{Taxable income} &= \text{Adjusted gross income} - (\text{Exemptions} + \text{Deductions})\\ &= \$58{,}000 - (\$4050 + \text{Deductions})\end{aligned}$$

The singles column in **Table 8.1** shows a personal exemption of $4050.

The singles column in **Table 8.1** shows a $6300 standard deduction. A greater deduction can be obtained by itemizing.

Itemized Deductions

$7500 : mortgage interest
$2200 : property taxes
$2400 : charitable contributions
~~$1500 : medical expenses~~
$12,100 : total of deductible expenditures

Can only deduct amount in excess of 7.5% of adjusted gross income: $0.075 \times \$58{,}000 = \$4350$

We substitute $12,100 for deductions in the formula for taxable income.

$$\begin{aligned}\text{Taxable income} &= \text{Adjusted gross income} - (\text{Exemptions} + \text{Deductions})\\ &= \$58{,}000 - (\$4050 + \$12{,}100)\\ &= \$58{,}000 - \$16{,}150\\ &= \$41{,}850\end{aligned}$$

Tax Rate	Single
10%	up to $9275
15%	$9276 to $37,650
25%	$37,651 to $91,150

A portion of **Table 8.1** (repeated)

Step 3 Determine the income tax.

$$\begin{aligned}\text{Income tax} &= \text{Tax computation} - \text{Tax credits}\\ &= \text{Tax computation} - \$500\end{aligned}$$

We perform the tax computation using the singles rates in **Table 8.1**, partly repeated in the margin. Our taxpayer is in the 25% tax bracket because her taxable income, $41,850, is in the $37,651 to $91,150 income range. This means that she owes 10% on the first $9275 of her taxable income, 15% on her taxable income between $9276 and $37,650, inclusive, and 25% on her taxable income above $37,650.

10% marginal rate on first $9275 of taxable income

15% marginal rate on taxable income between $9276 and $37,650

25% marginal rate on taxable income above $37,650

$$\begin{aligned}\text{Tax computation} &= 0.10 \times \$9275 + 0.15 \times (\$37{,}650 - \$9275) + 0.25 \times (\$41{,}850 - \$37{,}650)\\ &= 0.10 \times \$9275 + 0.15 \times \$28{,}375 + 0.25 \times \$4200\\ &= \$927.50 + \$4256.25 + \$1050.00\\ &= \$6233.75\end{aligned}$$

We substitute $6233.75 for the tax computation in the formula for income tax.

$$\begin{aligned}\text{Income tax} &= \text{Tax computation} - \text{Tax credits}\\ &= \$6233.75 - \$500\\ &= \$5733.75\end{aligned}$$

The federal income tax owed is $5733.75.

CHECK POINT 2 Use the 2016 marginal tax rates in **Table 8.1** on page 507 to calculate the federal tax owed by a single man with no dependents whose gross income, adjustments, deductions, and credits are given as follows:

Gross income: \$40,000
Adjustments: \$1000
Deductions: \$3000: charitable contributions
\$1500: theft loss
\$300: cost of tax preparation

Tax credit: none.

3 *Calculate FICA taxes.*

Social Security and Medicare (FICA)

In addition to income tax, we are required to pay the federal government **FICA** (Federal Insurance Contributions Act) taxes that are used for Social Security and Medicare benefits. **Social Security** provides payments to eligible retirees, people with health problems, eligible dependents of deceased persons, and disabled citizens. **Medicare** provides health care coverage mostly to Americans 65 and older.

The 2016 FICA tax rates are given in **Table 8.2**.

TABLE 8.2 2016 FICA Tax Rates

Employee's Rates	Matching Rates Paid by the Employer	Self-Employed Rates
• 7.65% on first \$118,500 of income	• 7.65% on first \$118,500 paid in wages	• 15.3% on first \$118,500 of net profits
• 1.45% of income in excess of \$118,500	• 1.45% of wages paid in excess of \$118,500	• 2.9% of net profits in excess of \$118,500

Taxpayers are not permitted to subtract adjustments, exemptions, or deductions when determining FICA taxes.

Blitzer Bonus

Percents and Tax Rates

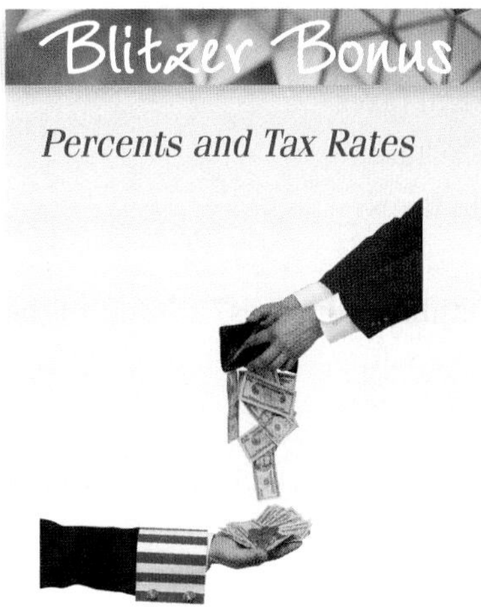

In 1944 and 1945, the highest marginal tax rate in the United States was a staggering 94%. The tax rate on the highest-income Americans remained at approximately 90% throughout the 1950s, decreased to 70% in the 1960s and 1970s, and to 50% in the early 1980s before reaching a post–World War II low of 28% in 1998. (*Source*: IRS)

EXAMPLE 3 Computing FICA Tax

If you are not self-employed and earn \$150,000, what are your FICA taxes?

SOLUTION

The tax rates are 7.65% on the first \$118,500 of income and 1.45% on income in excess of \$118,500.

(7.65% rate on the first \$118,500 of income; 1.45% rate on income in excess of \$118,500)

$$\begin{aligned}\text{FICA Tax} &= 0.0765 \times \$118{,}500 + 0.0145 \times (\$150{,}000 - \$118{,}500)\\ &= 0.0765 \times \$118{,}500 + 0.0145 \times \$31{,}500\\ &= \$9065.25 + \$456.75\\ &= \$9522.00\end{aligned}$$

The FICA taxes are \$9522.

CHECK POINT 3 If you are not self-employed and earn \$200,000, what are your FICA taxes?

4 *Solve problems involving working students and taxes.*

Blitzer Bonus

A Brief History of U.S. Income Tax

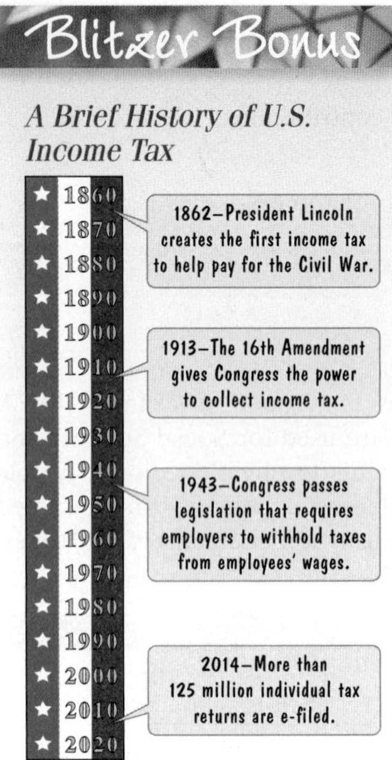

Working Students and Taxes

For those of you who work part-time, getting paid is great. However, because employers withhold federal and state taxes, as well as FICA, your paychecks probably contain less spending money than you had anticipated.

A pay stub attached to your paycheck provides a lot of information about the money you earned, including both your *gross pay* and your *net pay*. Gross pay, also known as **base pay**, is your salary prior to any withheld taxes for the pay period the check covers. Your gross pay is what you would receive if nothing were deducted. **Net pay** is the actual amount of your check after taxes have been withheld.

EXAMPLE 4 *Taxes for a Working Student*

You would like to have extra spending money, so you decide to work part-time at the local gym. The job pays \$15 per hour and you work 20 hours per week. Your employer withholds 10% of your gross pay for federal taxes, 7.65% for FICA taxes, and 3% for state taxes.

a. What is your weekly gross pay?

b. How much is withheld per week for federal taxes?

c. How much is withheld per week for FICA taxes?

d. How much is withheld per week for state taxes?

e. What is your weekly net pay?

f. What percentage of your gross pay is withheld for taxes? Round to the nearest tenth of a percent.

SOLUTION

a. Your weekly gross pay is the number of hours worked, 20, times your hourly wage, \$15 per hour.

$$\text{Gross pay} = 20\ \cancel{\text{hours}} \times \frac{\$15}{\cancel{\text{hour}}} = 20 \times \$15 = \$300$$

Your weekly gross pay, or what you would receive if nothing were deducted, is \$300.

b. Your employer withholds 10% of your gross pay for federal taxes.

$$\text{Federal taxes} = 10\%\ \text{of}\ \$300 = 0.10 \times \$300 = \$30$$

\$30 is withheld per week for federal taxes.

c. Your employer withholds 7.65% of your gross pay for FICA taxes.

$$\text{FICA taxes} = 7.65\%\ \text{of}\ \$300 = 0.0765 \times \$300 = \$22.95$$

\$22.95 is withheld per week for FICA taxes.

d. Your employer withholds 3% of your gross pay for state taxes.

$$\text{State taxes} = 3\%\ \text{of}\ \$300 = 0.03 \times \$300 = \$9$$

\$9 is withheld per week for state taxes.

e. Your weekly net pay is your gross pay minus the amounts withheld for federal, FICA, and state taxes.

Gross pay | Federal taxes | FICA taxes | State taxes

$$\begin{aligned}\text{Net pay} &= \$300 - (\$30 + \$22.95 + \$9)\\ &= \$300.00 - \$61.95\\ &= \$238.05\end{aligned}$$

Your weekly net pay is \$238.05. This is the actual amount of your paycheck.

f. Our work in part (e) shows that \$61.95 is withheld for taxes. The fractional part of your gross pay that is withheld for taxes is the amount that is withheld, \$61.95, divided by your gross pay, \$300.00. We then express this fraction as a percent.

Percent of gross pay withheld for taxes

$$= \frac{\$61.95}{\$300.00} = 0.2065 = 20.65\% \approx 20.7\%$$

Taxes (numerator); Gross pay (denominator)

Your employer takes \$61.95 from your weekly gross salary and sends the money to the government. This represents approximately 20.7% of your gross pay.

Figure 8.3 contains a sample pay stub for the working student in Example 4.

YOUR WORKPLACE 10 MAIN STREET ANY TOWN, STATE	YOUR NAME YOUR ADDRESS YOUR CITY, STATE, ZIP CODE SSN: 000-00-0000	Pay End Date: 0/00/16	
		Federal	State
		Single 1	Single 1

HOURS AND EARNINGS				
Current			Year to Date	
Base Rate (\$15 per hour)	Hours	Earnings	Hours	Earnings
	20	\$300	400	\$6000

TAXES		
Description:	Current	Year to Date
Federal	\$30.00	\$600.00
FICA	\$22.95	\$459.00
State	\$ 9.00	\$180.00
TOTAL	\$61.95	\$1239.00
	Current	Year to Date
TOTAL GROSS	\$300.00	\$6000.00
TOTAL DEDUCTIONS	\$ 61.95	\$1239.00
NET PAY	\$238.05	\$4761.00

FIGURE 8.3 A working student's sample pay stub

Pay stubs are attached to your paycheck and usually contain four sections:

- Personal information about the employee: This may include your name, your address, your social security number, and your marital status.
- Information about earnings: This includes your hourly wage, the number of hours worked in the current pay period, and the number of hours worked year-to-date, which is usually the total number of hours worked since January 1 of the current year (the first pay of the year may include the latter part of December), the amount earned during the current pay period, and the amount earned year-to-date.
- Information about tax deductions, summarizing withholdings for the current pay period and the withholdings year-to-date.
- Gross pay, total deductions, and net pay for the current period, and the year-to-date total for each.

CHECK POINT 4 You decide to work part-time at a local nursery. The job pays \$16 per hour and you work 15 hours per week. Your employer withholds 10% of your gross pay for federal taxes, 7.65% for FICA taxes, and 4% for state taxes.

a. What is your weekly gross pay?

b. How much is withheld per week for federal taxes?

c. How much is withheld per week for FICA taxes?

d. How much is withheld per week for state taxes?

e. What is your weekly net pay?

f. What percentage of your gross pay is withheld for taxes? Round to the nearest tenth of a percent.

Concept and Vocabulary Check

Fill in each blank so that the resulting statement is true.

1. Your _______ income is your total income for the year.
2. Subtracting certain allowable amounts from the income in Exercise 1 results in your _____________ income. These allowable amounts, or untaxed portions of income, are called ___________.
3. A fixed amount deducted on your tax return for each person supported by your income, including yourself, is called a/an ___________.
4. Your taxable income is your ____________ income minus the sum of your _____________ and ____________.
5. Sums of money that reduce federal income tax by the full dollar-for-dollar amount are called ________.
6. Taxes used for Social Security and Medicare benefits are called _______ taxes.
7. Your base pay, or ______ pay, is your salary prior to any withheld taxes.
8. The actual amount of your paycheck after taxes have been withheld is called your ______ pay.

In Exercises 9–12, determine whether each statement is true or false. If the statement is false, make the necessary change(s) to produce a true statement.

9. Federal income tax is a percentage of your gross income. ______
10. If tax credits are equal, federal tax tables show that the greater your taxable income, the more you pay. ______
11. People in some states are not required to pay state income taxes. ______
12. FICA tax is a percentage of your gross income. ______

Exercise Set 8.2

Practice and Application Exercises

In Exercises 1–2, find the gross income, the adjusted gross income, and the taxable income.

1. A taxpayer earned wages of $52,600, received $720 in interest from a savings account, and contributed $3200 to a tax-deferred retirement plan. He was entitled to a personal exemption of $4050 and had deductions totaling $7250.
2. A taxpayer earned wages of $23,500, received $495 in interest from a savings account, and contributed $1200 to a tax-deferred retirement plan. She was entitled to a personal exemption of $4050 and had deductions totaling $6450.

In Exercises 3–4, find the gross income, the adjusted gross income, and the taxable income. Base the taxable income on the greater of a standard deduction or an itemized deduction.

3. Suppose your neighbor earned wages of $86,250, received $1240 in interest from a savings account, and contributed $2200 to a tax-deferred retirement plan. She is entitled to a personal exemption of $4050 and a standard deduction of $6300. The interest on her home mortgage was $8900, she contributed $2400 to charity, and she paid $1725 in state taxes.
4. Suppose your neighbor earned wages of $319,150, received $1790 in interest from a savings account, and contributed $4100 to a tax-deferred retirement plan. He is entitled to a personal exemption of $4050 and the same exemption for each of his two children. He is also entitled to a standard deduction of $6300. The interest on his home mortgage was $51,235, he contributed $74,000 to charity, and he paid $12,760 in state taxes.

In Exercises 5–14, use the 2016 marginal tax rates in ***Table 8.1*** *on page 507 to compute the tax owed by each person or couple.*

5. a single man with a taxable income of $40,000
6. a single woman with a taxable income of $42,000
7. a married woman filing separately with a taxable income of $120,000
8. a married man filing separately with a taxable income of $130,000
9. a single man with a taxable income of $15,000 and a $2500 tax credit
10. a single woman with a taxable income of $18,000 and a $3500 tax credit
11. a married couple filing jointly with a taxable income of $250,000 and a $7500 tax credit
12. a married couple filing jointly with a taxable income of $400,000 and a $4500 tax credit
13. a head of household with a taxable income of $58,000 and a $6500 tax credit
14. a head of household with a taxable income of $46,000 and a $3000 tax credit

In Exercises 15–18, use the 2016 marginal tax rates in ***Table 8.1*** *on page 507 to calculate the income tax owed by each person.*

15. Single male, no dependents

    Gross income: $75,000
    Adjustments: $4000
    Deductions:
        $28,000 mortgage interest
        $4200 property taxes
        $3000 charitable contributions
    Tax credit: none

16. Single female, no dependents

    Gross income: $70,000
    Adjustments: $2000
    Deductions:
        $10,000 mortgage interest
        $2500 property taxes
        $1200 charitable contributions
    Tax credit: none

17. Unmarried head of household with two dependent children

Gross income: $50,000
Adjustments: none
Deductions:
$4500 state taxes
$2000 theft loss
Tax credit: $2000

18. Unmarried head of household with one dependent child

Gross income: $40,000
Adjustments: $1500
Deductions:
$3600 state taxes
$800 charitable contributions
Tax credit: $2500

In Exercises 19–24, use the 2016 FICA tax rates in ***Table 8.2*** *on page 509.*

19. If you are not self-employed and earn $120,000 what are your FICA taxes?

20. If you are not self-employed and earn $140,000 what are your FICA taxes?

21. If you are self-employed and earn $150,000, what are your FICA taxes?

22. If you are self-employed and earn $160,000, what are your FICA taxes?

23. To help pay for college, you worked part-time at a local restaurant, earning $20,000 in wages and tips.

a. Calculate your FICA taxes.

b. Use **Table 8.1** on page 507 to calculate your income tax. Assume you are single with no dependents, have no adjustments or tax credit, and you take the standard deduction.

c. Including both FICA and income tax, what percentage of your gross income are your federal taxes? Round to the nearest tenth of a percent.

24. To help pay for college, you worked part-time at a local restaurant, earning $18,000 in wages and tips.

a. Calculate your FICA taxes.

b. Use **Table 8.1** on page 507 to calculate your income tax. Assume you are single with no dependents, have no adjustments or tax credit, and take the standard deduction.

c. Including both FICA and income tax, what percentage of your gross income are your federal taxes? Round to the nearest tenth of a percent.

25. You decide to work part-time at a local supermarket. The job pays $16.50 per hour and you work 20 hours per week. Your employer withholds 10% of your gross pay for federal taxes, 7.65% for FICA taxes, and 3% for state taxes.

a. What is your weekly gross pay?

b. How much is withheld per week for federal taxes?

c. How much is withheld per week for FICA taxes?

d. How much is withheld per week for state taxes?

e. What is your weekly net pay?

f. What percentage of your gross pay is withheld for taxes? Round to the nearest tenth of a percent.

26. You decide to work part-time at a local veterinary hospital. The job pays $14.50 per hour and you work 20 hours per week. Your employer withholds 10% of your gross pay for federal taxes, 7.65% for FICA taxes, and 5% for state taxes.

a. What is your weekly gross pay?

b. How much is withheld per week for federal taxes?

c. How much is withheld per week for FICA taxes?

d. How much is withheld per week for state taxes?

e. What is your weekly net pay?

f. What percentage of your gross pay is withheld for taxes? Round to the nearest tenth of a percent

Explaining the Concepts

27. What is income tax?

28. What is gross income?

29. What is adjusted gross income?

30. What are exemptions?

31. What are deductions?

32. Under what circumstances should taxpayers itemize deductions?

33. How is taxable income determined?

34. What are tax credits?

35. What is the difference between a tax credit and a tax deduction?

36. What are FICA taxes?

37. How do you determine your net pay?

Critical Thinking Exercises

Make Sense? *In Exercises 38–42, determine whether each statement makes sense or does not make sense, and explain your reasoning.*

38. The only important thing to know about my taxes is whether I receive a refund or owe money on my return.

39. Because I am a student with a part-time job, federal tax law does not allow me to itemize deductions.

40. I'm paying less federal tax on my first dollars of earnings and more federal tax on my last dollars of earnings.

41. My employer withholds the same amount of federal tax on my first dollars of earnings and my last dollars of earnings.

42. Now that I'm a college student, I can choose a $4000 deduction or a $2500 credit to offset tuition and fees. I'll pay less federal taxes by selecting the $4000 deduction.

43. Suppose you are in the 10% tax bracket. As a college student, you can choose a $4000 deduction or a $2500 credit to offset tuition and fees. Which option will reduce your tax bill by the greater amount? What is the difference in your savings between the two options?

44. A common complaint about income tax is "I can't afford to work more because it will put me in a higher tax bracket." Is it possible that being in a higher bracket means you actually lose money? Explain your answer.

45. Because of the mortgage interest tax deduction, is it possible to save money buying a house rather than renting, even though rent payments are lower than mortgage payments? Explain your answer.

Group Exercises

The following topics are appropriate for either individual or group research projects. Use the Internet to investigate each topic.

46. Proposals to Simplify Federal Tax Laws and Filing Procedures

47. The Most Commonly Recommended Tax Saving Strategies

48. The Most Commonly Audited Tax Return Sections

49. Federal Tax Procedures Questioned over Issues of Fairness (Examples include the marriage penalty, the alternative minimum tax (AMT), and capital gains rates.)

8.3

WHAT AM I SUPPOSED TO LEARN?

After studying this section, you should be able to:

1 Calculate simple interest.

2 Use the future value formula.

Simple Interest

IN 1626, PETER MINUIT CONVINCED the Wappinger Indians to sell him Manhattan Island for $24. If the Native Americans had put the $24 into a bank account at a 5% interest rate compounded monthly, by the year 2020 there would have been well over $8 billion in the account!

Although you may not yet understand terms such as *interest rate* and *compounded monthly*, one thing seems clear: Money in certain savings accounts grows in remarkable ways. You, too, can take advantage of such accounts with astonishing results. In the next two sections, we will show you how.

1 *Calculate simple interest.*

Simple Interest

Interest is the amount of money that we get paid for lending or investing money, or that we pay for borrowing money. When we deposit money in a savings institution, the institution pays us interest for its use. When we borrow money, interest is the price we pay for the privilege of using the money until we repay it.

The amount of money that we deposit or borrow is called the **principal**. For example, if you deposit $2000 in a savings account, then $2000 is the principal. The amount of interest depends on the principal, the interest **rate**, which is given as a percent and varies from bank to bank, and the length of time for which the money is deposited. In this section, the rate is assumed to be annual (per year).

Simple interest involves interest calculated only on the principal. The following formula is used to find simple interest:

CALCULATING SIMPLE INTEREST

$$\text{Interest} = \text{principal} \times \text{rate} \times \text{time}$$

$$I = Prt$$

The rate, r, is expressed as a decimal when calculating simple interest.

Throughout this section and the chapter, keep in mind that all given rates are assumed to be *per year*, unless otherwise stated.

EXAMPLE 1 ***Calculating Simple Interest for a Year***

You deposit $2000 in a savings account at Hometown Bank, which has a rate of 6%. Find the interest at the end of the first year.

SOLUTION

To find the interest at the end of the first year, we use the simple interest formula.

$$I = Prt = (2000)(0.06)(1) = 120$$

Principal, or amount deposited, is $2000. Rate is 6% = 0.06. Time is 1 year.

At the end of the first year, the interest is $120. You can withdraw the $120 interest and you still have $2000 in the savings account.

CHECK POINT 1 You deposit \$3000 in a savings account at Yourtown Bank, which has a rate of 5%. Find the interest at the end of the first year.

EXAMPLE 2 Calculating Simple Interest for More Than a Year

A student took out a simple interest loan for \$1800 for two years at a rate of 8% to purchase a used car. What is the interest on the loan?

SOLUTION

To find the interest on the loan, we use the simple interest formula.

$$I = Prt = (1800)(0.08)(2) = 288$$

Principal, or amount borrowed, is \$1800. Rate is $8\% = 0.08$. Time is 2 years.

The interest on the loan is \$288.

CHECK POINT 2 A student took out a simple interest loan for \$2400 for two years at a rate of 7%. What is the interest on the loan?

Simple interest is used for many short-term loans, including automobile and consumer loans. Imagine that a short-term loan is taken for 125 days. The time of the loan is $\frac{125}{365}$ because there are 365 days in a year. However, before the modern use of calculators and computers, the **Banker's rule** allowed financial institutions to use 360 in the denominator of such a fraction because this simplified the interest calculation. Using the Banker's rule, the time, t, for a 125-day short-term loan is

$$\frac{125 \text{ days}}{360 \text{ days}} = \frac{125}{360}.$$

Compare the values for time, t, for a 125-day short-term loan using denominators of 360 and 365.

$$\frac{125}{360} \approx 0.347 \qquad \frac{125}{365} \approx 0.342$$

The denominator of 360 benefits the bank by resulting in a greater period of time for the loan, and consequently more interest.

With the widespread use of calculators and computers, government agencies and the Federal Reserve Bank calculate simple interest using 365 days in a year, as do many credit unions and banks. However, there are still some financial institutions that use the Banker's rule with 360 days in a year because it produces a greater amount of interest.

2 *Use the future value formula.*

Future Value: Principal Plus Interest

When a loan is repaid, the interest is added to the original principal to find the total amount due. In Example 2, at the end of two years, the student will have to repay

$$\text{principal} + \text{interest} = \$1800 + \$288 = \$2088.$$

In general, if a principal P is borrowed at a simple interest rate r, then after t years the amount due, A, can be determined as follows:

$$A = P + I = P + Prt = P(1 + rt).$$

The amount due, A, is called the **future value** of the loan. The principal borrowed now, P, is also known as the loan's **present value**.

CALCULATING FUTURE VALUE FOR SIMPLE INTEREST

The future value, A, of P dollars at simple interest rate r (as a decimal) for t years is given by

$$A = P(1 + rt).$$

EXAMPLE 3 Calculating Future Value

A loan of \$1060 has been made at 6.5% for three months. Find the loan's future value.

TECHNOLOGY

1060 [1 + (0.065)(0.25)]
On a Scientific Calculator:

1060 [×] [(] 1 [+] .065 [×] .25 [)] [=]

SOLUTION

The amount borrowed, or principal, P, is \$1060. The rate, r, is 6.5%, or 0.065. The time, t, is given as three months. We need to express the time in years because the rate is understood to be 6.5% per year. Because three months is $\frac{3}{12}$ of a year, $t = \frac{3}{12} = \frac{1}{4} = 0.25$.

The loan's future value, or the total amount due after three months, is

$$A = P(1 + rt) = 1060[1 + (0.065)(0.25)] \approx \$1077.23.$$

Rounded to the nearest cent, the loan's future value is \$1077.23.

CHECK POINT 3 A loan of \$2040 has been made at 7.5% for four months. Find the loan's future value.

EXAMPLE 4 Earning Money by Putting Your Wallet Away Today

Suppose you spend \$4 each day, five days per week, on gourmet coffee.

a. How much do you spend on this item in a year?

b. If you invested your yearly spending on gourmet coffee in a savings account with a rate of 5%, how much would you have after one year?

SOLUTION

a. Because you are spending \$4 each day, five days per week, you are spending

$$\frac{\$4}{\cancel{\text{day}}} \times \frac{5 \cancel{\text{days}}}{\text{week}} = \frac{\$4 \times 5}{\text{week}} = \frac{\$20}{\text{week}},$$

or \$20 each week on gourmet coffee. Assuming that this continues throughout the 52 weeks in the year, you are spending

$$\frac{\$20}{\cancel{\text{week}}} \times \frac{52 \cancel{\text{weeks}}}{\text{year}} = \frac{\$20 \times 52}{\text{year}} = \frac{\$1040}{\text{year}},$$

or \$1040 each year on gourmet coffee.

b. Now suppose you invest \$1040 in a savings account with a rate of 5%. To find your savings after one year, we use the future value formula for simple interest.

$$A = P(1 + rt) = 1040[1 + (0.05)(1)] = 1040(1.05) = 1092$$

Giving up the day-to-day expense of gourmet coffee can result in potential savings of \$1092.

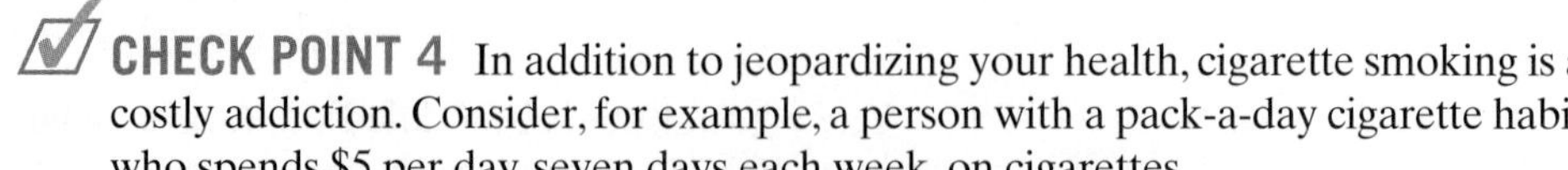

CHECK POINT 4 In addition to jeopardizing your health, cigarette smoking is a costly addiction. Consider, for example, a person with a pack-a-day cigarette habit who spends \$5 per day, seven days each week, on cigarettes.

a. How much is spent on this item in a year?

b. If this person invested the yearly spending on cigarettes in a savings account with a rate of 4%, how much would be saved after one year?

The formula for future value, $A = P(1 + rt)$, has four variables. If we are given values for any three of these variables, we can solve for the fourth.

EXAMPLE 5 *Determining a Simple Interest Rate*

You borrow \$2500 from a friend and promise to pay back \$2655 in six months. What simple interest rate will you pay?

SOLUTION

We use the formula for future value, $A = P(1 + rt)$. You borrow \$2500: $P = 2500$. You will pay back \$2655, so this is the future value: $A = 2655$. You will do this in six months, which must be expressed in years: $t = \frac{6}{12} = \frac{1}{2} = 0.5$. To determine the simple interest rate you will pay, we solve the future value formula for r.

$$A = P(1 + rt) \quad \text{This is the formula for future value.}$$

$$2655 = 2500[1 + r(0.5)] \quad \text{Substitute the given values.}$$

$$2655 = 2500 + 1250r \quad \text{Use the distributive property.}$$

$$155 = 1250r \quad \text{Subtract 2500 from both sides.}$$

$$\frac{155}{1250} = \frac{1250r}{1250} \quad \text{Divide both sides by 1250.}$$

$$r = 0.124 = 12.4\% \quad \text{Express } \frac{155}{1250} \text{ as a percent.}$$

You will pay a simple interest rate of 12.4%.

 **CHECK POINT 5** You borrow \$5000 from a friend and promise to pay back \$6800 in two years. What simple interest rate will you pay?

EXAMPLE 6 *Determining a Present Value*

You plan to save \$2000 for a trip to Europe in two years. You decide to purchase a certificate of deposit (CD) from your bank that pays a simple interest rate of 4%. How much must you put in this CD now in order to have the \$2000 in two years?

SOLUTION

We use the formula for future value, $A = P(1 + rt)$. We are interested in finding the principal, P, or the present value.

$$A = P(1 + rt) \quad \text{This is the formula for future value.}$$

$$2000 = P[1 + (0.04)(2)] \quad A\text{(future value)} = \$2000,\ r\text{(interest rate)} = 0.04, \text{ and } t = 2 \text{ (you want \$2000 in two years).}$$

$$2000 = 1.08P \quad \text{Simplify: } 1 + (0.04)(2) = 1.08.$$

$$\frac{2000}{1.08} = \frac{1.08P}{1.08} \quad \text{Divide both sides by 1.08.}$$

$$P \approx 1851.852 \quad \text{Simplify.}$$

To make sure you will have enough money for the vacation, let's round this principal *up* to \$1851.86. Thus, you should put \$1851.86 in the CD now to have \$2000 in two years.

GREAT QUESTION!

How should I round when finding present value?

When computing present value, round the principal, or present value, *up*. To round up to the nearest cent, add 1 to the hundredths digit, regardless of the digit to the right. In this way, you'll be sure to have enough money to meet future goals.

 **CHECK POINT 6** How much should you put in an investment paying a simple interest rate of 8% if you need \$4000 in six months?

Concept and Vocabulary Check

Fill in each blank so that the resulting statement is true.

1. The formula for calculating simple interest, I, is _________, where P is the _________, r is the _________, and t is the ______.
2. The future value, A, of P dollars at simple interest rate r for t years is given by the formula ______________.
3. The Banker's rule allows using _____ days in a year.

In Exercises 4–6, determine whether each statement is true or false. If the statement is false, make the necessary change(s) to produce a true statement.

4. Interest is the amount of money we get paid for borrowing money or that we pay for investing money. ______
5. In simple interest, only the original money invested or borrowed generates interest over time. ______
6. If \$4000 is borrowed at 7.6% for three months, the loan's future value is \$76. ______

Exercise Set 8.3

Practice Exercises

In Exercises 1–8, the principal P is borrowed at simple interest rate r for a period of time t. Find the simple interest owed for the use of the money. Assume 360 days in a year.

1. $P = \$4000, r = 6\%, t = 1$ year
2. $P = \$7000, r = 5\%, t = 1$ year
3. $P = \$180, r = 3\%, t = 2$ years
4. $P = \$260, r = 4\%, t = 3$ years
5. $P = \$5000, r = 8.5\%, t = 9$ months
6. $P = \$18{,}000, r = 7.5\%, t = 18$ months
7. $P = \$15{,}500, r = 11\%, t = 90$ days
8. $P = \$12{,}600, r = 9\%, t = 60$ days

In Exercises 9–14, the principal P is borrowed at simple interest rate r for a period of time t. Find the loan's future value, A, or the total amount due at time t.

9. $P = \$3000, r = 7\%, t = 2$ years
10. $P = \$2000, r = 6\%, t = 3$ years
11. $P = \$26{,}000, r = 9.5\%, t = 5$ years
12. $P = \$24{,}000, r = 8.5\%, t = 6$ years
13. $P = \$9000, r = 6.5\%, t = 8$ months
14. $P = \$6000, r = 4.5\%, t = 9$ months

In Exercises 15–20, the principal P is borrowed and the loan's future value, A, at time t is given. Determine the loan's simple interest rate, r, to the nearest tenth of a percent.

15. $P = \$2000, A = \$2150, t = 1$ year
16. $P = \$3000, A = \$3180, t = 1$ year
17. $P = \$5000, A = \$5900, t = 2$ years
18. $P = \$10{,}000, A = \$14{,}060, t = 2$ years
19. $P = \$2300, A = \$2840, t = 9$ months
20. $P = \$1700, A = \$1820, t = 6$ months

In Exercises 21–26, determine the present value, P, you must invest to have the future value, A, at simple interest rate r after time t. Round answers up to the nearest cent.

21. $A = \$6000, r = 8\%, t = 2$ years
22. $A = \$8500, r = 7\%, t = 3$ years
23. $A = \$14{,}000, r = 9.5\%, t = 6$ years
24. $A = \$16{,}000, r = 11.5\%, t = 5$ years
25. $A = \$5000, r = 14.5\%, t = 9$ months
26. $A = \$2000, r = 12.6\%, t = 8$ months

Practice Plus

27. Solve for r: $A = P(1 + rt)$.
28. Solve for t: $A = P(1 + rt)$.
29. Solve for P: $A = P(1 + rt)$.
30. Solve for P: $A = P(1 + \frac{r}{n})^{nt}$. (We will be using this formula in the next section.)

Application Exercises

31. In order to start a small business, a student takes out a simple interest loan for \$4000 for nine months at a rate of 8.25%.
    a. How much interest must the student pay?
    b. Find the future value of the loan.
32. In order to pay for baseball uniforms, a school takes out a simple interest loan for \$20,000 for seven months at a rate of 12%.
    a. How much interest must the school pay?
    b. Find the future value of the loan.
33. You borrow \$1400 from a friend and promise to pay back \$2000 in two years. What simple interest rate, to the nearest tenth of a percent, will you pay?

34. Treasury bills (T-bills) can be purchased from the U.S. Treasury Department. You buy a T-bill for \$981.60 that pays \$1000 in 13 weeks. What simple interest rate, to the nearest tenth of a percent, does this T-bill earn?

35. To borrow money, you pawn your guitar. Based on the value of the guitar, the pawnbroker loans you \$960. One month later, you get the guitar back by paying the pawnbroker \$1472. What annual interest rate did you pay?

36. To borrow money, you pawn your mountain bike. Based on the value of the bike, the pawnbroker loans you \$552. One month later, you get the bike back by paying the pawnbroker \$851. What annual interest rate did you pay?

37. A bank offers a CD that pays a simple interest rate of 6.5%. How much must you put in this CD now in order to have \$3000 for a home-entertainment center in two years?

38. A bank offers a CD that pays a simple interest rate of 5.5%. How much must you put in this CD now in order to have \$8000 for a kitchen remodeling project in two years?

Explaining the Concepts

39. Explain how to calculate simple interest.

40. What is the future value of a loan and how is it determined?

Critical Thinking Exercises

Make Sense? *In Exercises 41–43, determine whether each statement makes sense or does not make sense, and explain your reasoning.*

41. After depositing \$1500 in an account at a rate of 4%, my balance at the end of the first year was \$(1500)(0.04).

42. I saved money on my short-term loan for 90 days by finding a financial institution that used the Banker's rule rather than one that calculated interest using 365 days in a year.

43. I planned to save \$5000 in four years, computed the present value to be \$3846.153, so I rounded the principal to \$3846.15.

44. Use the future value formula to show that the time required for an amount of money P to double in value to $2P$ is given by

$$t = \frac{1}{r}.$$

45. You deposit \$5000 in an account that earns 5.5% simple interest.

a. Express the future value in the account as a linear function of time, t.

b. Determine the slope of the function in part (a) and describe what this means. Use the phrase "rate of change" in your description.

8.4 Compound Interest

WHAT AM I SUPPOSED TO LEARN?

After studying this section, you should be able to:

1 Use compound interest formulas.
2 Calculate present value.
3 Understand and compute effective annual yield.

SO, HOW DID THE PRESENT VALUE OF Manhattan in 1626—that is, the \$24 paid to the Native Americans—attain a future value of over \$8 billion in 2020, 394 years later, at a mere 5% interest rate? After all, the future value on \$24 for 394 years at 5% simple interest is

$$\begin{aligned} A &= P(1 + rt) \\ &= 24[1 + (0.05)(394)] = 496.8, \end{aligned}$$

or a paltry \$496.80, compared to over \$8 billion. To understand this dramatic difference in future value, we turn to the concept of *compound interest*.

1 *Use compound interest formulas.*

Compound Interest

Compound interest is interest computed on the original principal as well as on any accumulated interest. Many savings accounts pay compound interest. For example, suppose you deposit \$1000 in a savings account at a rate of 5%. **Table 8.3** on the next page shows how the investment grows if the interest earned is automatically added on to the principal.

TABLE 8.3 Calculating the Amount in an Account Subject to Compound Interest

		Use $A = P(1 + rt)$ with $r = 0.05$ and $t = 1$, or $A = P(1 + 0.05)$.
Year	**Starting Balance**	**Amount in the Account at Year's End**
1	\$1000	$A = \$1000(1 + 0.05) = \$1050$
2	\$1050 or $\$1000(1 + 0.05)$	$A = \$1050(1 + 0.05) = \$1102.50$ or $A = \$1000(1 + 0.05)(1 + 0.05) = \$1000(1 + 0.05)^2$
3	\$1102.50 or $\$1000(1 + 0.05)^2$	$A = \$1102.50(1 + 0.05) \approx \$1157.63$ or $A = \$1000(1 + 0.05)^2(1 + 0.05) = \$1000(1 + 0.05)^3$

Using inductive reasoning, the amount, A, in the account after t years is the original principal, \$1000, times $(1 + 0.05)^t$: $A = 1000(1 + 0.05)^t$.

If the original principal is P and the interest rate is r, we can use this same approach to determine the amount, A, in an account subject to compound interest.

CALCULATING THE AMOUNT IN AN ACCOUNT FOR COMPOUND INTEREST PAID ONCE A YEAR

If you deposit P dollars at rate r, in decimal form, subject to compound interest, then the amount, A, of money in the account after t years is given by

$$A = P(1 + r)^t.$$

The amount A is called the account's **future value** and the principal P is called its **present value**.

EXAMPLE 1 *Using the Compound Interest Formula*

You deposit \$2000 in a savings account at Hometown Bank, which has a rate of 6%.

a. Find the amount, A, of money in the account after three years subject to interest compounded once a year.

b. Find the interest.

SOLUTION

a. The amount deposited, or principal, P, is \$2000. The rate, r, is 6%, or 0.06. The time of the deposit, t, is three years. The amount in the account after three years is

$$A = P(1 + r)^t = 2000(1 + 0.06)^3 = 2000(1.06)^3 \approx 2382.03.$$

Rounded to the nearest cent, the amount in the savings account after three years is \$2382.03.

b. Because the amount in the account is \$2382.03 and the original principal is \$2000, the interest is \$2382.03 − \$2000, or \$382.03.

TECHNOLOGY

Here are the calculator keystrokes to compute $2000(1.06)^3$:

Many Scientific Calculators

2000 [×] 1.06 [y^x] 3 [=]

Many Graphing Calculators

2000 [×] 1.06 [∧] 3 [ENTER]

CHECK POINT 1 You deposit \$1000 in a savings account at a bank that has a rate of 4%.

a. Find the amount, A, of money in the account after five years subject to interest compounded once a year. Round to the nearest cent.

b. Find the interest.

Compound Interest Paid More Than Once a Year

The period of time between two interest payments is called the **compounding period**. When compound interest is paid once per year, the compounding period is one year. We say that the interest is **compounded annually**.

Most savings institutions have plans in which interest is paid more than once per year. If compound interest is paid twice per year, the compounding period is six months. We say that the interest is **compounded semiannually**. When compound interest is paid four times per year, the compounding period is three months and the interest is said to be **compounded quarterly**. Some plans allow for monthly compounding or daily compounding.

In general, when compound interest is paid n times a year, we say that there are **n compounding periods per year. Table 8.4** shows the three most frequently used plans in which interest is paid more than once a year.

TABLE 8.4 Interest Plans

Name	Number of Compounding Periods per Year	Length of Each Compounding Period
Semiannual Compounding	$n = 2$	6 months
Quarterly Compounding	$n = 4$	3 months
Monthly Compounding	$n = 12$	1 month

The following formula is used to calculate the amount in an account subject to compound interest with n compounding periods per year:

CALCULATING THE AMOUNT IN AN ACCOUNT FOR COMPOUND INTEREST PAID n TIMES A YEAR

If you deposit P dollars at rate r, in decimal form, subject to compound interest paid n times per year, then the amount, A, of money in the account after t years is given by

$$A = P\left(1 + \frac{r}{n}\right)^{nt}.$$

A is the account's **future value** and the principal P is its **present value**.

TECHNOLOGY

Here are the calculator keystrokes to compute

$$7500\left(1 + \frac{0.06}{12}\right)^{12\cdot 5}.$$

Many Scientific Calculators

7500 [×] [(] 1 [+] .06 [÷] 12 [)] [y^x] [(] 12 [×] 5 [)] [=]

Many Graphing Calculators

7500 [(] 1 [+] .06 [÷] 12 [)] [^] [(] 12 [×] 5 [)] [ENTER]

You may find it easier to compute the value of the exponent nt (12×5 or 60) before using these keystrokes. By entering 60 instead of 12×5, you're less likely to make a careless mistake by forgetting to enclose the exponent n times t in parentheses.

EXAMPLE 2 *Using the Compound Interest Formula*

You deposit \$7500 in a savings account that has a rate of 6%. The interest is compounded monthly.

a. How much money will you have after five years?

b. Find the interest after five years.

SOLUTION

a. The amount deposited, or principal, P, is \$7500. The rate, r, is 6%, or 0.06. Because interest is compounded monthly, there are 12 compounding periods per year, so $n = 12$. The time of the deposit, t, is five years. The amount in the account after five years is

$$A = P\left(1 + \frac{r}{n}\right)^{nt} = 7500\left(1 + \frac{0.06}{12}\right)^{12\cdot 5} = 7500(1.005)^{60} \approx 10{,}116.38.$$

Rounded to the nearest cent, you will have \$10,116.38 after five years.

b. Because the amount in the account is \$10,116.38 and the original principal is \$7500, the interest after five years is \$10,116.38 − \$7500, or \$2616.38.

GREAT QUESTION!

Can I use the formula for compound interest paid *n* times a year to calculate the amount in an account that pays compound interest only once a year?

Yes. If $n = 1$ (interest paid once a year), the formula

$$A = P\left(1 + \frac{r}{n}\right)^{nt}$$

becomes

$$A = P\left(1 + \frac{r}{1}\right)^{1t}, \text{ or}$$

$$A = P(1 + r)^t.$$

This shows that the amount in an account subject to annual compounding is just one application of the general formula for compound interest paid *n* times a year. With this general formula, you no longer need a separate formula for annual compounding.

CHECK POINT 2 You deposit $4200 in a savings account that has a rate of 4%. The interest is compounded quarterly.

a. How much money will you have after 10 years? Round to the nearest cent.

b. Find the interest after 10 years.

Continuous Compounding

Some banks use **continuous compounding**, where the compounding periods increase infinitely (compounding interest every trillionth of a second, every quadrillionth of a second, etc.). As n, the number of compounding periods in a year, increases without bound, the expression $\left(1 + \frac{1}{n}\right)^n$ approaches the irrational number e: $e \approx 2.71828$. This is illustrated in **Table 8.5**. As a result, the formula for the balance in an account with n compounding periods per year, $A = P(1 + \frac{r}{n})^{nt}$, becomes $A = Pe^{rt}$ with continuous compounding. Although continuous compounding sounds terrific, it yields only a fraction of a percent more interest over a year than daily compounding.

TABLE 8.5 As n Takes on Increasingly Large Values, the Expression $\left(1 + \frac{1}{n}\right)^n$ Approaches the Irrational Number e.

n	$\left(1 + \frac{1}{n}\right)^n$
1	2
2	2.25
5	2.48832
10	2.59374246
100	2.704813829
1000	2.716923932
10,000	2.718145927
100,000	2.718268237
1,000,000	2.718280469
1,000,000,000	2.718281827

FORMULAS FOR COMPOUND INTEREST

After t years, the balance, A, in an account with principal P and annual interest rate r (in decimal form) is given by the following formulas:

1. For n compounding periods per year: $A = P\left(1 + \frac{r}{n}\right)^{nt}$
2. For continuous compounding: $A = Pe^{rt}$.

TECHNOLOGY

You can compute e to a power using the $\boxed{e^x}$ key on your calculator. Use the key to enter e^1 and verify that e is approximately equal to 2.71828.

Scientific Calculators	Graphing Calculators
1 $\boxed{e^x}$	$\boxed{e^x}$ 1 $\boxed{\text{ENTER}}$

EXAMPLE 3 ***Choosing between Investments***

You decide to invest $8000 for six years and you have a choice between two accounts. The first pays 7% per year, compounded monthly. The second pays 6.85% per year, compounded continuously. Which is the better investment?

SOLUTION

The better investment is the one with the greater balance in the account after six years. Let's begin with the account with monthly compounding. We use the compound interest formula with $P = 8000$, $r = 7\% = 0.07$, $n = 12$ (monthly compounding means 12 compounding periods per year), and $t = 6$.

$$A = P\left(1 + \frac{r}{n}\right)^{nt} = 8000\left(1 + \frac{0.07}{12}\right)^{12 \cdot 6} \approx 12{,}160.84$$

The balance in this account after six years would be $12,160.84.

TECHNOLOGY

Here are the calculator keystrokes to compute

$8000e^{0.0685(6)}$:

Many Scientific Calculators

8000 [×] [(] .0685 [×] 6 [)] [e^x] [=]

Many Graphing Calculators

8000 [e^x] [(] .0685 [×] 6 [)] [ENTER]

For the second investment option, we use the formula for continuous compounding with $P = 8000$, $r = 6.85\% = 0.0685$, and $t = 6$.

$$A = Pe^{rt} = 8000e^{0.0685(6)} \approx 12{,}066.60$$

The balance in this account after six years would be \$12,066.60, slightly less than the previous amount. Thus, the better investment is the 7% monthly compounding option.

CHECK POINT 3 A sum of \$10,000 is invested at an annual rate of 8%. Find the balance in the account after five years subject to **a.** quarterly compounding and **b.** continuous compounding.

2 *Calculate present value.*

Planning for the Future with Compound Interest

Just as we did in Section 8.3, we can determine P, the principal or present value, that should be deposited now in order to have a certain amount, A, in the future. If an account earns compound interest, the amount of money that should be invested today to obtain a future value of A dollars can be determined by solving the compound interest formula for P:

CALCULATING PRESENT VALUE

If A dollars are to be accumulated in t years in an account that pays rate r compounded n times per year, then the present value, P, that needs to be invested now is given by

$$P = \frac{A}{\left(1 + \frac{r}{n}\right)^{nt}}.$$

Remember to round the principal *up* to the nearest cent when computing present value so there will be enough money to meet future goals.

EXAMPLE 4 *Calculating Present Value*

How much money should be deposited today in an account that earns 6% compounded monthly so that it will accumulate to \$20,000 in five years?

TECHNOLOGY

Here are the keystrokes for Example 4:

Many Scientific Calculators

20000 [÷] [(] 1 [+] .06 [÷] 12 [)]

[y^x] [(] 12 [×] 5 [)] [=]

Many Graphing Calculators

20000 [÷] [(] 1 [+] .06 [÷] 12 [)]

[∧] [(] 12 [×] 5 [)] [ENTER]

SOLUTION

The amount we need today, or the present value, is determined by the present value formula. Because the interest is compounded monthly, $n = 12$. Furthermore, A (the future value) $=$ \$20,000, r (the rate) $= 6\% = 0.06$, and t (time in years) $= 5$.

$$P = \frac{A}{\left(1 + \frac{r}{n}\right)^{nt}} = \frac{20{,}000}{\left(1 + \frac{0.06}{12}\right)^{12 \cdot 5}} \approx 14{,}827.4439$$

To make sure there will be enough money, we round the principal *up* to \$14,827.45. Approximately \$14,827.45 should be invested today in order to accumulate to \$20,000 in five years.

CHECK POINT 4 How much money should be deposited today in an account that earns 7% compounded weekly so that it will accumulate to \$10,000 in eight years?

Blitzer Bonus

The Time Value of Money

When you complete your education and begin earning money, it will be tempting to spend every penny earned. By doing this, you will fail to take advantage of the *time value of money*. The **time value of money** means that a dollar received today is worth more than a dollar received next year or the year after. This is because a sum of money invested today starts earning compound interest sooner than a sum of money invested some time in the future. **A significant way to increase your wealth is to spend less than you earn and invest the difference.** With time on your side, even a small amount of money can be turned into a substantial sum through the power of compounding. Make the time value of money work for you by postponing certain purchases now and investing the savings instead. Pay close attention to your spending habits as you study the time value of money.

3 *Understand and compute effective annual yield.*

Effective Annual Yield

As we've seen before, a common problem in financial planning is selecting the best investment from two or more investments. For example, is an investment that pays 8.25% interest compounded quarterly better than one that pays 8.3% interest compounded semiannually? Another way to answer the question is to compare the *effective rates* of the investments, also called their *effective annual yields*.

Blitzer Bonus

Doubling Your Money: The Rule of 72

Here's a shortcut for estimating the number of years it will take for your investment to double: Divide 72 by the effective annual yield without the percent sign. For example, if the effective annual yield is 6%, your money will double in approximately

$$\frac{72}{6}$$

years, or in 12 years.

EFFECTIVE ANNUAL YIELD

The **effective annual yield**, or the **effective rate**, is the simple interest rate that produces the same amount of money in an account at the end of one year as when the account is subject to compound interest at a stated rate.

EXAMPLE 5 *Understanding Effective Annual Yield*

You deposit $4000 in an account that pays 8% interest compounded monthly.

a. Find the future value after one year.

b. Use the future value formula for simple interest to determine the effective annual yield.

SOLUTION

a. We use the compound interest formula to find the account's future value after one year.

$$A = P\left(1+\frac{r}{n}\right)^{nt} = 4000\left(1+\frac{0.08}{12}\right)^{12\cdot 1} \approx \$4332.00$$

Principal is $4000.

Stated rate is 8% = 0.08.

Monthly compounding: $n = 12$
Time is one year: $t = 1$.

Rounded to the nearest cent, the future value after one year is $4332.00.

b. The effective annual yield, or effective rate, is a simple interest rate. We use the future value formula for simple interest to determine the simple interest rate that produces a future value of $4332 for a $4000 deposit after one year.

High-Yield Savings

Online banks can afford to pay the best rates on savings accounts by avoiding the expense of managing a brick-and-mortar bank. To search for the best deals, go to BankRate.com for bank accounts and iMoneyNet.com for money-market funds.

$$A = P(1 + rt)$$ This is the future value formula for simple interest.

$$4332 = 4000(1 + r \cdot 1)$$ Substitute the given values.

$$4332 = 4000 + 4000r$$ Use the distributive property.

$$332 = 4000r$$ Subtract 4000 from both sides.

$$\frac{332}{4000} = \frac{4000r}{4000}$$ Divide both sides by 4000.

$$r = \frac{332}{4000} = 0.083 = 8.3\%$$ Express r as a percent.

The effective annual yield, or effective rate, is 8.3%. This means that money invested at 8.3% simple interest earns the same amount in one year as money invested at 8% interest compounded monthly.

In Example 5, the stated 8% rate is called the **nominal rate**. The 8.3% rate is the effective rate and is a simple interest rate.

CHECK POINT 5 You deposit \$6000 in an account that pays 10% interest compounded monthly.

a. Find the future value after one year.

b. Determine the effective annual yield.

Generalizing the procedure of Example 5 and Check Point 5 gives a formula for effective annual yield:

CALCULATING EFFECTIVE ANNUAL YIELD

Suppose that an investment has a nominal interest rate, r, in decimal form, and pays compound interest n times per year. The investment's effective annual yield, Y, in decimal form, is given by

$$Y = \left(1 + \frac{r}{n}\right)^n - 1.$$

The decimal form of Y given by the formula should then be converted to a percent.

TECHNOLOGY

Here are the keystrokes for Example 6:

Many Scientific Calculators

[(] 1 [+] .05 [÷] 360 [)] [y^x] 360 [−] 1 [=]

Many Graphing Calculators

[(] 1 [+] .05 [÷] 360 [)] [∧] 360 [−] 1 [ENTER].

Some graphing calculators require that you press the right arrow key after entering the exponent 360.

Given the nominal rate and the number of compounding periods per year, some graphing calculators display the effective annual yield. The screen shows the calculation of the effective rate in Example 6 on the TI-84 Plus C.

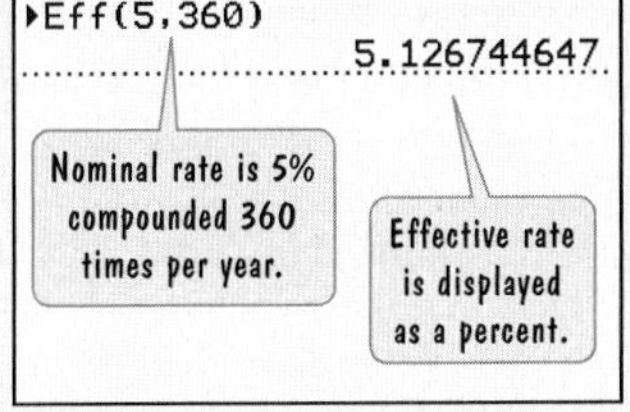

EXAMPLE 6 *Calculating Effective Annual Yield*

A passbook savings account has a nominal rate of 5%. The interest is compounded daily. Find the account's effective annual yield. (Assume 360 days in a year.)

SOLUTION

The rate, r, is 5%, or 0.05. Because interest is compounded daily and we assume 360 days in a year, $n = 360$. The account's effective annual yield is

$$Y = \left(1 + \frac{r}{n}\right)^n - 1 = \left(1 + \frac{0.05}{360}\right)^{360} - 1 \approx 0.0513 = 5.13\%.$$

The effective annual yield is 5.13%. Thus, money invested at 5.13% simple interest earns the same amount of interest in one year as money invested at 5% interest, the nominal rate, compounded daily.

CHECK POINT 6 What is the effective annual yield of an account paying 8% compounded quarterly?

The effective annual yield is often included in the information about investments or loans. Because it's the true interest rate you're earning or paying, it's the number you should pay attention to. **If you are selecting the best investment from two or more investments, the best choice is the account with the greatest effective annual yield.** However, there are differences in the types of accounts that you need to take into consideration. Some pay interest from the day of deposit to the day of withdrawal. Other accounts start paying interest the first day of the month that follows the day of deposit. Some savings institutions stop paying interest if the balance in the account falls below a certain amount.

When *borrowing money*, the effective rate or effective annual yield is usually called the **annual percentage rate**. If all other factors are equal and you are borrowing money, select the option with the least annual percentage rate.

Concept and Vocabulary Check

Fill in each blank so that the resulting statement is true.

1. Compound interest is interest computed on the original __________ as well as on any accumulated __________.

2. The formula $A = P\left(1 + \frac{r}{n}\right)^{nt}$ gives the amount of money, A, in an account after ____ years at rate ____ subject to compound interest paid ____ times per year.

3. If interest is compounded once a year, the formula in Exercise 2 becomes ______________.

4. If compound interest is paid twice per year, the compounding period is ____ months and the interest is compounded __________.

5. If compound interest is paid four times per year, the compounding period is ______ months and the interest is compounded __________.

6. When the number of compounding periods in a year increases without bound, this is known as __________ compounding.

7. In the formula

$$P = \frac{A}{\left(1 + \frac{r}{n}\right)^{nt}},$$

the variable ____ represents the amount that needs to be invested now in order to have ____ dollars accumulated in ____ years in an account that pays rate ____ compounded ____ times per year.

8. If you are selecting the best investment from two or more investments, the best choice is the account with the greatest ________________, which is the ________ interest rate that produces the same amount of money at the end of one year as when the account is subject to compound interest at a stated rate.

In Exercises 9–12, determine whether each statement is true or false. If the statement is false, make the necessary change(s) to produce a true statement.

9. Formulas for compound interest show that a dollar invested today is worth more than a dollar invested in the future.______

10. Formulas for compound interest show that if you make the decision to postpone certain purchases and save the money instead, small amounts of money can be turned into substantial sums over a period of years. ______

11. At a given annual interest rate, your money grows faster as the compounding period becomes shorter. ______

12. According to the Rule of 72 (see the Blitzer Bonus on page 524), an investment with an effective annual yield of 12% can double in six years. ______

Exercise Set 8.4

Here is a list of formulas needed to solve the exercises. Be sure you understand what each formula describes and the meaning of the variables in the formulas.

$$A = P\left(1 + \frac{r}{n}\right)^{nt} \qquad P = \frac{A}{\left(1 + \frac{r}{n}\right)^{nt}}$$

$$A = Pe^{rt} \qquad Y = \left(1 + \frac{r}{n}\right)^{n} - 1$$

Practice Exercises

In Exercises 1–12, the principal represents an amount of money deposited in a savings account subject to compound interest at the given rate.

a. *Find how much money there will be in the account after the given number of years. (Assume 360 days in a year.)*

b. *Find the interest earned.*

Round answers to the nearest cent.

Principal	Rate	Compounded	Time
1. $10,000	4%	annually	2 years
2. $8000	6%	annually	3 years
3. $3000	5%	semiannually	4 years
4. $4000	4%	semiannually	5 years
5. $9500	6%	quarterly	5 years
6. $2500	8%	quarterly	6 years
7. $4500	4.5%	monthly	3 years
8. $2500	6.5%	monthly	4 years
9. $1500	8.5%	daily	2.5 years
10. $1200	8.5%	daily	3.5 years
11. $20,000	4.5%	daily	20 years
12. $25,000	5.5%	daily	20 years

Solve Exercises 13–16 using appropriate compound interest formulas. Round answers to the nearest cent.

13. Find the accumulated value of an investment of $10,000 for five years at an interest rate of 5.5% if the money is **a.** compounded semiannually; **b.** compounded quarterly; **c.** compounded monthly; **d.** compounded continuously.

14. Find the accumulated value of an investment of $5000 for 10 years at an interest rate of 6.5% if the money is **a.** compounded semiannually; **b.** compounded quarterly; **c.** compounded monthly; **d.** compounded continuously.

15. Suppose that you have $12,000 to invest. Which investment yields the greater return over three years: 7% compounded monthly or 6.85% compounded continuously?

16. Suppose that you have $6000 to invest. Which investment yields the greater return over four years: 8.25% compounded quarterly or 8.3% compounded semiannually?

In Exercises 17–20, round answers up to the nearest cent.

17. How much money should be deposited today in an account that earns 6% compounded semiannually so that it will accumulate to $10,000 in three years?

18. How much money should be deposited today in an account that earns 7% compounded semiannually so that it will accumulate to $12,000 in four years?

19. How much money should be deposited today in an account that earns 9.5% compounded monthly so that it will accumulate to $10,000 in three years?

20. How much money should be deposited today in an account that earns 10.5% compounded monthly so that it will accumulate to $22,000 in four years?

21. You deposit $10,000 in an account that pays 4.5% interest compounded quarterly.

a. Find the future value after one year.

b. Use the future value formula for simple interest to determine the effective annual yield.

22. You deposit $12,000 in an account that pays 6.5% interest compounded quarterly.

a. Find the future value after one year.

b. Use the future value formula for simple interest to determine the effective annual yield.

In Exercises 23–28, a passbook savings account has a rate of 6%. Find the effective annual yield, rounded to the nearest tenth of a percent, if the interest is compounded

23. semiannually.

24. quarterly.

25. monthly.

26. daily. (Assume 360 days in a year.)

27. 1000 times per year.

28. 100,000 times per year.

In Exercises 29–32, determine the effective annual yield for each investment. Then select the better investment. Assume 360 days in a year. If rounding is required, round to the nearest tenth of a percent.

29. 8% compounded monthly; 8.25% compounded annually

30. 5% compounded monthly; 5.25% compounded quarterly

31. 5.5% compounded semiannually; 5.4% compounded daily

32. 7% compounded annually; 6.85% compounded daily

Practice Plus

In Exercises 33–36, how much more would you earn in the first investment than in the second investment? Round answers to the nearest dollar.

33.
- $25,000 invested for 40 years at 12% compounded annually
- $25,000 invested for 40 years at 6% compounded annually

34.
- $30,000 invested for 40 years at 10% compounded annually
- $30,000 invested for 40 years at 5% compounded annually

35.
- $50,000 invested for 30 years at 10% compounded annually
- $50,000 invested for 30 years at 5% compounded monthly

36.
- $20,000 invested for 30 years at 12% compounded annually
- $20,000 invested for 30 years at 6% compounded monthly

Application Exercises

Assume that the accounts described in the exercises have no other deposits or withdrawals except for what is stated. Round all answers to the nearest dollar, rounding up to the nearest dollar in present-value problems. Assume 360 days in a year.

37. At the time of a child's birth, $12,000 was deposited in an account paying 6% interest compounded semiannually. What will be the value of the account at the child's twenty-first birthday?

38. At the time of a child's birth, \$10,000 was deposited in an account paying 5% interest compounded semiannually. What will be the value of the account at the child's twenty-first birthday?

39. You deposit \$2600 in an account that pays 4% interest compounded once a year. Your friend deposits \$2200 in an account that pays 5% interest compounded monthly.

a. Who will have more money in their account after one year? How much more?

b. Who will have more money in their account after five years? How much more?

c. Who will have more money in their account after 20 years? How much more?

40. You deposit \$3000 in an account that pays 3.5% interest compounded once a year. Your friend deposits \$2500 in an account that pays 4.8% interest compounded monthly.

a. Who will have more money in their account after one year? How much more?

b. Who will have more money in their account after five years? How much more?

c. Who will have more money in their account after 20 years? How much more?

41. You deposit \$3000 in an account that pays 7% interest compounded semiannually. After 10 years, the interest rate is increased to 7.25% compounded quarterly. What will be the value of the account after 16 years?

42. You deposit \$6000 in an account that pays 5.25% interest compounded semiannually. After 10 years, the interest rate is increased to 5.4% compounded quarterly. What will be the value of the account after 18 years?

43. In 1626, Peter Minuit convinced the Wappinger Indians to sell him Manhattan Island for \$24. If the Native Americans had put the \$24 into a bank account paying compound interest at a 5% rate, how much would the investment have been worth in the year 2020 ($t = 394$ years) if interest were compounded **a.** monthly? **b.** 360 times per year?

44. In 1777, Jacob DeHaven loaned George Washington's army \$450,000 in gold and supplies. Due to a disagreement over the method of repayment (gold versus Continental money), DeHaven was never repaid, dying penniless. In 1989, his descendants sued the U.S. government over the 212-year-old debt. If the DeHavens used an interest rate of 6% and daily compounding (the rate offered by the Continental Congress in 1777), how much money did the DeHaven family demand in their suit? (*Hint:* Use the compound interest formula with $n = 360$ and $t = 212$ years.)

45. Will you earn more interest in one year by depositing \$2000 in a simple interest account that pays 6% or in an account that pays 5.9% interest compounded daily? How much more interest will you earn?

46. Will you earn more interest in one year by depositing \$1000 in a simple interest account that pays 7% or in an account that pays 6.9% interest compounded daily? How much more interest will you earn?

47. Two accounts each begin with a deposit of \$5000. Both accounts have rates of 5.5%, but one account compounds interest once a year while the other account compounds interest continuously. Make a table that shows the amount in each account and the interest earned after 1 year, 5 years, 10 years, and 20 years.

48. Two accounts each begin with a deposit of \$10,000. Both accounts have rates of 6.5%, but one account compounds interest once a year while the other account compounds interest continuously. Make a table that shows the amount in each account and the interest earned after 1 year, 5 years, 10 years, and 20 years.

49. Parents wish to have \$80,000 available for a child's education. If the child is now 5 years old, how much money must be set aside at 6% compounded semiannually to meet their financial goal when the child is 18?

50. A 30-year-old worker plans to retire at age 65. He believes that \$500,000 is needed to retire comfortably. How much should be deposited now at 7% compounded monthly to meet the \$500,000 retirement goal?

51. You would like to have \$75,000 available in 15 years. There are two options. Account A has a rate of 4.5% compounded once a year. Account B has a rate of 4% compounded daily. How much would you have to deposit in each account to reach your goal?

52. You would like to have \$150,000 available in 20 years. There are two options. Account A has a rate of 5.5% compounded once a year. Account B has a rate of 5% compounded daily. How much would you have to deposit in each account to reach your goal?

53. You invest \$1600 in an account paying 5.4% interest compounded daily. What is the account's effective annual yield? Round to the nearest hundredth of a percent.

54. You invest \$3700 in an account paying 3.75% interest compounded daily. What is the account's effective annual yield? Round to the nearest hundredth of a percent.

55. An account has a nominal rate of 4.2%. Find the effective annual yield, rounded to the nearest tenth of a percent, with quarterly compounding, monthly compounding, and daily compounding. How does changing the compounding period affect the effective annual yield?

56. An account has a nominal rate of 4.6%. Find the effective annual yield, rounded to the nearest tenth of a percent, with quarterly compounding, monthly compounding, and daily compounding. How does changing the compounding period affect the effective annual yield?

57. A bank offers a money market account paying 4.5% interest compounded semiannually. A competing bank offers a money market account paying 4.4% interest compounded daily. Which account is the better investment?

58. A bank offers a money market account paying 4.9% interest compounded semiannually. A competing bank offers a money market account paying 4.8% interest compounded daily. Which account is the better investment?

Explaining the Concepts

59. Describe the difference between simple and compound interest.

60. Give two examples that illustrate the difference between a compound interest problem involving future value and a compound interest problem involving present value.

61. What is effective annual yield?

62. Explain how to select the best investment from two or more investments.

Critical Thinking Exercises

Make Sense? *In Exercises 63–66, determine whether each statement makes sense or does not make sense, and explain your reasoning.*

63. My bank provides simple interest at 3.25% per year, but I can't determine if this is a better deal than a competing bank offering 3.25% compound interest without knowing the compounding period.

64. When choosing between two accounts, the one with the greater annual interest rate is always the better deal.

65. A bank can't increase compounding periods indefinitely without owing its customers an infinite amount of money.

66. My bank advertises a compound interest rate of 2.4%, although, without making deposits or withdrawals, the balance in my account increased by 2.43% in one year.

67. A depositor opens a new savings account with \$6000 at 5% compounded semiannually. At the beginning of year 3, an additional \$4000 is deposited. At the end of six years, what is the balance in the account?

68. A depositor opens a money market account with \$5000 at 8% compounded monthly. After two years, \$1500 is withdrawn from the account to buy a new computer. A year later, \$2000 is put in the account. What will be the ending balance if the money is kept in the account for another three years?

69. Use the future value formulas for simple and compound interest in one year to derive the formula for effective annual yield.

Group Exercise

70. This activity is a group research project intended for four or five people. Present your research in a seminar on the history of interest and banking. The seminar should last about 30 minutes. Address the following questions:

When was interest first charged on loans? How was lending money for a fee opposed historically? What is usury? What connection did banking and interest rates play in the historic European rivalries between Christians and Jews? When and where were some of the highest interest rates charged? What were the rates? Where does the word *interest* come from? What is the origin of the word *shylock*? What is the difference between usury and interest in modern times? What is the history of a national bank in the United States?

8.5 Annuities, Methods of Saving, and Investments

WHAT AM I SUPPOSED TO LEARN?

After studying this section, you should be able to:

1. Determine the value of an annuity.
2. Determine regular annuity payments needed to achieve a financial goal.
3. Understand stocks and bonds as investments.
4. Read stock tables.
5. Understand accounts designed for retirement savings.

ACCORDING TO THE *FORBES* *Billionaires List*, in 2016 the two richest Americans were Bill Gates (net worth: \$75 billion) and Warren Buffett (net worth: \$61 billion). In May 1965, Buffett's new company, Berkshire Hathaway, was selling one share of stock for \$18. By 2017, the price of a share had increased to \$260,000! If you had purchased one share in May 1965, your **return**, or percent increase, would be

Warren Buffett and Bill Gates

$$\frac{\text{amount of increase}}{\text{original amount}} = \frac{\$260{,}000 - \$18}{\$18} \approx 14{,}443.44 = 1{,}444{,}344\%.$$

What does a return of approximately 1,400,000% mean? If you had invested \$250 in Warren Buffett's company in May 1965, your shares would have been worth over \$3.5 million by December 2017.

Of course, investments that potentially offer outrageous returns come with great risk of losing part or all of the principal. The bottom line: Is there a safe way to save regularly and have an investment worth one million dollars or more? In this section, we consider such savings plans, some of which come with special tax treatment, as well as riskier investments in stocks and bonds.

1 *Determine the value of an annuity.*

Annuities

The compound interest formula

$$A = P(1 + r)^t$$

gives the future value, A, after t years, when a fixed amount of money, P, the principal, is deposited in an account that pays an annual interest rate r (in decimal form) compounded once a year. However, money is often invested in small amounts at periodic intervals. For example, to save for retirement, you might decide to place \$1000 into an Individual Retirement Account (IRA) at the end of each year until you retire. An **annuity** is a sequence of equal payments made at equal time periods. An IRA is an example of an annuity.

The **value of an annuity** is the sum of all deposits plus all interest paid. Our first example illustrates how to find this value.

EXAMPLE 1 *Determining the Value of an Annuity*

You deposit \$1000 into a savings plan at the end of each year for three years. The interest rate is 8% per year compounded annually.

a. Find the value of the annuity after three years.

b. Find the interest.

SOLUTION

a. The value of the annuity after three years is the sum of all deposits made plus all interest paid over three years.

This is the \$1000 deposit at year's end.

Value at end of year 1 = \$1000

This is the first-year deposit with interest earned for a year. This is the \$1000 deposit at year's end.

Value at end of year 2 = \$1000(1 + 0.08) + \$1000

= \$1080 + \$1000 = \$2080

Use $A = P(1 + r)^t$ with $r = 0.08$ and $t = 1$, or $A = P(1 + 0.08)$.

This is the second-year balance, \$2080, with interest earned for a year. This is the \$1000 deposit at year's end.

Value at end of year 3 = \$2080(1 + 0.08) + \$1000

= \$2246.40 + \$1000 = \$3246.40

The value of the annuity at the end of three years is \$3246.40.

b. You made three payments of \$1000 each, depositing a total of 3 × \$1000, or \$3000. Because the value of the annuity is \$3246.40, the interest is \$3246.40 − \$3000, or \$246.40.

CHECK POINT 1 You deposit \$2000 into a savings plan at the end of each year for three years. The interest rate is 10% per year compounded annually.

a. Find the value of the annuity after three years.

b. Find the interest.

Suppose that you deposit P dollars into an account at the end of each year. The account pays an annual interest rate, r, compounded annually. At the end of the first year, the account contains P dollars. At the end of the second year, P dollars is deposited again. At the time of this deposit, the first deposit has received interest earned during the second year. Thus, the value of the annuity after two years is

$$P + P(1 + r).$$

Deposit of P dollars at end of second year

First-year deposit of P dollars with interest earned for a year

The value of the annuity after three years is

$$P \quad + \quad P(1 + r) \quad + \quad P(1 + r)^2.$$

Deposit of P dollars at end of third year

Second-year deposit of P dollars with interest earned for a year

First-year deposit of P dollars with interest earned over two years

The value of the annuity after t years is

$$P + P(1 + r) + P(1 + r)^2 + P(1 + r)^3 + \cdots + P(1 + r)^{t-1}.$$

Deposit of P dollars at end of year t

First-year deposit of P dollars with interest earned over $t - 1$ years

Each term in this sum is obtained by multiplying the preceding term by $(1 + r)$. Thus, the terms form a geometric sequence. Using a formula for the sum of the terms of a geometric sequence, we can obtain the following formula that gives the value of this annuity:

VALUE OF AN ANNUITY: INTEREST COMPOUNDED ONCE A YEAR

If P is the deposit made at the end of each year for an annuity that pays an annual interest rate r (in decimal form) compounded once a year, the value, A, of the annuity after t years is

$$A = \frac{P[(1 + r)^t - 1]}{r}.$$

EXAMPLE 2 *Determining the Value of an Annuity*

Although you are a long way from retirement, the time to begin retirement savings is when you begin earning a paycheck and can take advantage of the time value of your money.

Suppose that when you are 35, you decide to save for retirement by depositing \$1000 into an IRA at the end of each year for 30 years. If you can count on an interest rate of 10% per year compounded annually,

a. How much will you have from the IRA after 30 years?

b. Find the interest.

Round answers to the nearest dollar.

SOLUTION

a. The amount that you will have from the IRA is its value after 30 years.

$$A = \frac{P[(1+r)^t - 1]}{r}$$ Use the formula for the value of an annuity.

$$A = \frac{1000[(1+0.10)^{30} - 1]}{0.10}$$ The annuity involves year-end deposits of \$1000: $P = 1000$. The interest rate is 10%: $r = 0.10$. The number of years is 30: $t = 30$. The Technology box shows how this computation can be done in a single step using parentheses keys.

$$= \frac{1000[(1.10)^{30} - 1]}{0.10}$$ Add inside parentheses: $1 + 0.10 = 1.10$.

$$\approx \frac{1000(17.4494 - 1)}{0.10}$$ Use a calculator to find $(1.10)^{30}$: 1.1 [y^x] 30 [=].

$$= \frac{1000(16.4494)}{0.10}$$ Simplify inside parentheses: $17.4494 - 1 = 16.4494$.

$$= 164{,}494$$ Use a calculator: 1000 [×] 16.4494 [÷] .10 [=].

After 30 years, you will have approximately \$164,494 from the IRA.

b. You made 30 payments of \$1000 each, depositing a total of 30 × \$1000, or \$30,000. Because the value of the annuity is approximately \$164,494, the interest is approximately

\$164,494 − \$30,000, or \$134,494.

The interest is nearly $4\frac{1}{2}$ times the amount of your payments, illustrating the power of compounding.

TECHNOLOGY

Here are the calculator keystrokes to compute

$$\frac{1000[(1+0.10)^{30} - 1]}{0.10}:$$

Many Scientific Calculators

1000 [×] [(] [(] 1 [+] .10 [)] [y^x] 30 [−] 1 [)] [÷] .10 [=]

Observe that the part of the numerator in brackets is entered as $((1 + 0.10)^{30} - 1)$ to maintain the order of operations.

Many Graphing Calculators

1000 [(] [(] 1 [+] .10 [)] [∧] 30 [−] 1 [)] [÷] .10 [ENTER]

(Some graphing calculators require that you press the right arrow key after entering the exponent, 30.)

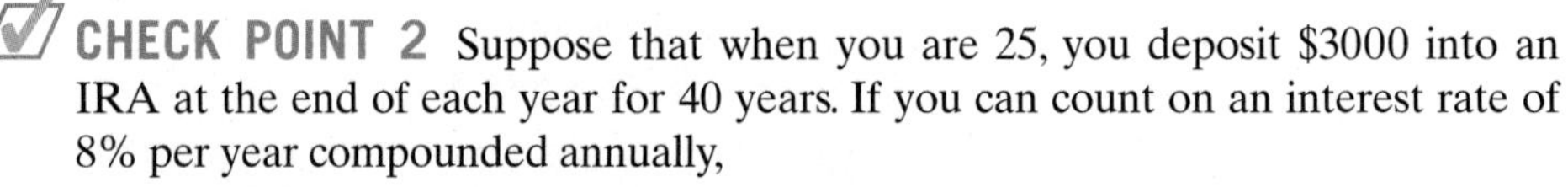

CHECK POINT 2 Suppose that when you are 25, you deposit \$3000 into an IRA at the end of each year for 40 years. If you can count on an interest rate of 8% per year compounded annually,

a. How much will you have from the IRA after 40 years?

b. Find the interest.

Round answers to the nearest dollar.

We can adjust the formula for the value of an annuity if equal payments are made at the end of each of n yearly compounding periods.

VALUE OF AN ANNUITY: INTEREST COMPOUNDED *n* TIMES PER YEAR

If P is the deposit made at the end of each compounding period for an annuity that pays an annual interest rate r (in decimal form) compounded n times per year, the value, A, of the annuity after t years is

$$A = \frac{P\left[\left(1+\frac{r}{n}\right)^{nt} - 1\right]}{\left(\frac{r}{n}\right)}.$$

GREAT QUESTION!

Can I use the formula for the value of an annuity with interest compounded *n* times per year to calculate the value of an annuity if interest is compounded only once a year?

Yes. If $n = 1$ (interest compounded once a year), the formula in the box at the right below becomes

$$A = \frac{P[(1+\frac{r}{1})^{1t} - 1]}{(\frac{r}{1})}, \text{ or}$$

$$A = \frac{P[(1+r)^t - 1]}{r}.$$

This was the boxed formula on page 531. This shows that the value of an annuity with interest compounded once a year is just one application of the general formula in the box on the right. With this general formula, you no longer need a separate formula for annual compounding.

EXAMPLE 3 *Determining the Value of an Annuity*

At age 25, to save for retirement, you decide to deposit \$200 at the end of each month into an IRA that pays 7.5% compounded monthly.

a. How much will you have from the IRA when you retire at age 65?

b. Find the interest.

Round answers to the nearest dollar.

SOLUTION

a. Because you are 25, the amount that you will have from the IRA when you retire at 65 is its value after 40 years.

$$A = \frac{P\left[\left(1+\frac{r}{n}\right)^{nt}-1\right]}{\left(\frac{r}{n}\right)}$$ Use the formula for the value of an annuity.

$$A = \frac{200\left[\left(1+\frac{0.075}{12}\right)^{12\cdot 40}-1\right]}{\left(\frac{0.075}{12}\right)}$$ The annuity involves month-end deposits of \$200: $P = 200$. The interest rate is 7.5%: $r = 0.075$. The interest is compounded monthly: $n = 12$. The number of years is 40: $t = 40$.

$$= \frac{200[(1+0.00625)^{480}-1]}{0.00625}$$ Using parentheses keys, these calculations can be performed in a single step on a calculator. Answers may vary slightly if you do the calculations in stages and round along the way.

$$= \frac{200[(1.00625)^{480}-1]}{0.00625}$$ Add inside parentheses: $1 + 0.00625 = 1.00625$.

$$\approx \frac{200(19.8989-1)}{0.00625}$$ Use a calculator to find $(1.00625)^{480}$: 1.00625 [y^x] 480 [=].

$$\approx 604{,}765$$

After 40 years, you will have approximately \$604,765 when retiring at age 65.

b. Interest = Value of the IRA − Total deposits

$\approx \$604{,}765 - \$200 \cdot 12 \cdot 40$

\$200 per month × 12 months per year × 40 years

$= \$604{,}765 - \$96{,}000 = \$508{,}765$

The interest is approximately \$508,765, more than five times the amount of your contributions to the IRA.

Annuities can be categorized by when payments are made. The formula used to solve Example 3 describes **ordinary annuities**, where payments are made at the end of each period. The formula assumes the same number of yearly payments and yearly compounding periods. An annuity plan in which payments are made at the beginning of each period is called an **annuity due**. The formula for the value of this type of annuity is slightly different than the one used in Example 3.

CHECK POINT 3 At age 30, to save for retirement, you decide to deposit \$100 at the end of each month into an IRA that pays 9.5% compounded monthly.

a. How much will you have from the IRA when you retire at age 65?

b. Find the interest.

Round answers to the nearest dollar.

2 *Determine regular annuity payments needed to achieve a financial goal.*

Planning for the Future with an Annuity

By solving the annuity formula for P, we can determine the amount of money that should be deposited at the end of each compounding period so that an annuity has a future value of A dollars. The following formula gives the regular payments, P, needed to reach a financial goal, A:

REGULAR PAYMENTS NEEDED TO ACHIEVE A FINANCIAL GOAL

The deposit, P, that must be made at the end of each compounding period into an annuity that pays an annual interest rate r (in decimal form) compounded n times per year in order to achieve a value of A dollars after t years is

$$P = \frac{A\left(\frac{r}{n}\right)}{\left[\left(1 + \frac{r}{n}\right)^{nt} - 1\right]}.$$

When computing regular payments needed to achieve a financial goal, round the deposit made at the end of each compounding period *up*. In this way, you won't fall slightly short of being able to meet future goals. In this section, we will round annuity payments up to the nearest dollar.

EXAMPLE 4 *Using Long-Term Planning to Achieve a Financial Goal*

Suppose that once you complete your college education and begin working, you would like to save \$20,000 over five years to use as a down payment for a home. You anticipate making regular, end-of-month deposits in an annuity that pays 6% compounded monthly.

a. How much should you deposit each month? Round up to the nearest dollar.

b. How much of the \$20,000 down payment comes from deposits and how much comes from interest?

SOLUTION

a. $$P = \frac{A\left(\frac{r}{n}\right)}{\left[\left(1 + \frac{r}{n}\right)^{nt} - 1\right]}$$

Use the formula for regular payments, P, needed to achieve a financial goal, A.

$$P = \frac{20{,}000\left(\frac{0.06}{12}\right)}{\left[\left(1 + \frac{0.06}{12}\right)^{12 \cdot 5} - 1\right]}$$

Your goal is to accumulate \$20,000 $(A = 20{,}000)$ over five years $(t = 5)$. The interest rate is 6% $(r = 0.06)$ compounded monthly $(n = 12)$.

$$\approx 287$$

Use a calculator and round up to the nearest dollar to be certain you do not fall short of your goal.

You should deposit \$287 each month to be certain of having \$20,000 for a down payment on a home.

GREAT QUESTION!

Which will earn me more money: a lump-sum deposit or an annuity?

In Example 4 of Section 8.4, we saw that a lump-sum deposit of approximately \$14,828 at 6% compounded monthly would accumulate to \$20,000 in five years. In Example 4 on the right, we see that total deposits of \$17,220 are required to reach the same goal. With the same interest rate, compounding period, and time period, a lump-sum deposit will generate more interest than an annuity. If you don't have a large sum of money to open an account, an annuity is a realistic, although more expensive, option to a lump-sum deposit for achieving the same financial goal.

b. Total deposits $= \$287 \cdot 12 \cdot 5 = \$17{,}220$

\$287 per month × 12 months per year × 5 years

$$\text{Interest} = \$20{,}000 - \$17{,}220 = \$2780$$

We see that \$17,220 of the \$20,000 comes from your deposits and the remainder, \$2780, comes from interest.

CHECK POINT 4 Parents of a baby girl are in a financial position to begin saving for her college education. They plan to have \$100,000 in a college fund in 18 years by making regular, end-of-month deposits in an annuity that pays 9% compounded monthly.

a. How much should they deposit each month? Round up to the nearest dollar.

b. How much of the \$100,000 college fund comes from deposits and how much comes from interest?

3 *Understand stocks and bonds as investments.*

Investments: Risk and Return

When you deposit money into a bank account, you are making a **cash investment**. Because bank accounts up to \$250,000 are insured by the federal government, there is no risk of losing the principal you've invested. The account's interest rate guarantees a certain percent increase in your investment, called its **return**.

All investments involve a trade-off between risk and return. The different types of bank accounts carry little or no risk, so investors must be willing to accept low returns. There are other kinds of investments that are riskier, meaning that it is possible to lose all or part of your principal. These investments, including *stocks* and *bonds*, give a reasonable expectation of higher returns to attract investors.

Stocks

Investors purchase **stock**, shares of ownership in a company. The shares indicate the percent of ownership. For example, if a company has issued a total of one million shares and an investor owns 20,000 of these shares, that investor owns

$$\frac{20{,}000 \text{ shares}}{1{,}000{,}000 \text{ shares}} = 0.02$$

or 2% of the company. Any investor who owns some percentage of the company is called a **shareholder**.

Buying or selling stock is referred to as **trading**. Shares of stock need both a seller and a buyer to be traded. Stocks are traded on a **stock exchange**. The price of a share of stock is determined by the law of supply and demand. If a company is prospering, investors will be willing to pay a good price for its stock, and so the stock price goes up. If the company does not do well, investors may decide to sell, and the stock price goes down. Stock prices indicate the performance of the companies they represent, as well as the state of the national and global economies.

There are two ways to make money by investing in stock:

- You sell the shares for more money than what you paid for them, in which case you have a **capital gain** on the sale of stock. (There can also be a capital loss by selling for less than what you paid, or if the company goes bankrupt.)
- While you own the stock, the company distributes all or part of its profits to shareholders as **dividends**. Each share is paid the same dividend, so the amount you receive depends on the number of shares owned. (Some companies reinvest all profits and do not distribute dividends.)

When more and more average Americans began investing and making money in stocks in the 1990s, the federal government cut the capital-gains tax rate. Long-term capital gains (profits on items held for more than a year before being sold) and dividends are taxed at lower rates than wages and interest earnings.

Bonds

People who buy stock become part owners in a company. In order to raise money and not dilute the ownership of current stockholders, companies sell **bonds**. People who buy a bond are **lending money** to the company from which they buy the bond. Bonds are a commitment from a company to pay the price an investor pays for the bond at the time it was purchased, called the **face value**, along with interest payments at a given rate.

There are many reasons for issuing bonds. A company might need to raise money for research on a drug that has the potential for curing AIDS, so it issues bonds. The U.S. Treasury Department issues 30-year bonds at a fixed 7% annual rate to borrow money to cover federal deficits. Local governments often issue bonds to borrow money to build schools, parks, and libraries.

Bonds are traded like stock, and their price is a function of supply and demand. If a company goes bankrupt, bondholders are the first to claim the company's assets. They make their claims before the stockholders, even though (unlike stockholders) they do not own a share of the company. Generally speaking, investing in bonds is less risky than investing in stocks, although the return is lower.

Mutual Funds

It is not an easy job to determine which stocks and bonds to buy or sell, or when to do so. Even IRAs can be funded by mixing stocks and bonds. Many small investors have decided that they do not have the time to stay informed about the progress of corporations, even with the help of online industry research. Instead, they invest in a **mutual fund**. A mutual fund is a group of stocks and/or bonds managed by a professional investor. When you purchase shares in a mutual fund, you give your money to the **fund manager**. Your money is combined with the money of other investors in the mutual fund. The fund manager invests this pool of money, buying and selling shares of stocks and bonds to obtain the maximum possible returns.

Investors in mutual funds own a small portion of many different companies, which may protect them against the poor performance of a single company. When comparing mutual funds, consider both the fees charged for investing and how well the fund manager is doing with the fund's money. Newspapers publish ratings from 1 (worst) to 5 (best) of mutual fund performance based on whether the manager is doing a good job with its investors' money. Two numbers are given. The first number compares the performance of the mutual fund to a large group of similar funds. The second number compares the performance to funds that are nearly identical. The best rating a fund manager can receive is 5/5; the worst is 1/1.

A listing of all the investments that a person holds is called a **financial portfolio**. Most financial advisors recommend a portfolio with a mixture of low-risk and high-risk investments, called a **diversified portfolio**.

4 *Read stock tables.*

Reading Stock Tables

Daily newspapers and online services give current stock prices and other information about stocks. We will use FedEx (Federal Express) stock to learn how to read these daily stock tables. Look at the following newspaper listing of FedEx stock.

52-Week High	52-Week Low	Stock	SYM	Div	Yld %	PE	Vol 100s	Hi	Lo	Close	Net Chg
99.46	34.02	FedEx	FDX	.44	1.0	19	37701	45	43.47	44.08	−1.60

52-Week High
99.46

The headings indicate the meanings of the numbers across the row.

The heading **52-Week High** refers to the *highest price* at which FedEx stock traded during the past 52 weeks. The highest price was $99.46 per share. This means that during the past 52 weeks at least one investor was willing to pay $99.46 for a share of FedEx stock. Notice that 99.46 represents a quantity in dollars, although the stock table does not show the dollar sign.

52-Week Low
34.02

The heading **52-Week Low** refers to the *lowest price* at which FedEx stock traded during the past 52 weeks. This price is $34.02.

Stock	SYM
FedEx	FDX

The heading **Stock** is the *company name*, FedEx. The heading **SYM** is the *symbol* the company uses for trading. FedEx uses the symbol FDX.

Div
.44

The heading **Div** refers to *dividends* paid per share to stockholders during the past year. FedEx paid a dividend of $0.44 per share. Once again, the dollar symbol does not appear in the table. Thus, if you owned 100 shares, you received a dividend of $\$0.44 \times 100$, or $44.00.

Yld %
1.0

The heading **Yld %** stands for *percent yield*. In this case, the percent yield is 1.0%. (The stock table does not show the percent sign.) This means that the dividends alone gave investors an annual return of 1.0%. This is much lower than the average inflation rate. However, this percent does not take into account the fact that FedEx stock prices might rise. If an investor sells shares for more than the purchase price, the gain will probably make FedEx stock a much better investment than a bank account.

In order to understand the meaning of the heading PE, we need to understand some of the other numbers in the table. We will return to this column.

Vol 100s
37701

The heading **Vol 100s** stands for *sales volume in hundreds*. This is the number of shares traded yesterday, in hundreds. The number in the table is 37,701. This means that yesterday, a total of $37{,}701 \times 100$, or 3,770,100 shares of FedEx were traded.

Hi
45

The heading **Hi** stands for the *highest price* at which FedEx stock traded *yesterday*. This number is 45. Yesterday, FedEx's highest trading price was $45 a share.

Lo
43.47

The heading **Lo** stands for the *lowest price* at which FedEx stock traded *yesterday*. This number is 43.47. Yesterday, FedEx's lowest trading price was $43.47 a share.

Close
44.08

The heading **Close** stands for the *price* at which shares last traded *when the stock exchange closed yesterday*. This number is 44.08. Thus, the price at which shares of FedEx traded when the stock exchange closed yesterday was $44.08 per share. This is called yesterday's **closing price**.

Net Chg
−1.60

The heading **Net Chg** stands for *net change*. This is the change in price from the market close two days ago to yesterday's market close. This number is −1.60. Thus, the price of a share of FedEx stock went down by $1.60. For some stock listings, the notation. . . appears under Net Chg. This means that there was *no change in price* for a share of stock from the market close two days ago to yesterday's market close.

PE
19

Now, we are ready to return to the heading **PE**, standing for the *price-to-earnings ratio*.

$$\text{PE ratio} = \frac{\text{Yesterday's closing price per share}}{\text{Annual earnings per share}}$$

This can also be expressed as

$$\text{Annual earnings per share} = \frac{\text{Yesterday's closing price per share}}{\text{PE ratio}}.$$

The PE ratio for FedEx is given to be 19. Yesterday's closing price per share was 44.08. We can substitute these numbers into the formula to find annual earnings per share:

Close	PE
44.08	19

$$\text{Annual earnings per share} = \frac{44.08}{19} = 2.32.$$

The annual earnings per share for FedEx were $2.32. The PE ratio, 19, tells us that yesterday's closing price per share, $44.08, is 19 times greater than the earnings per share, $2.32.

EXAMPLE 5 *Reading Stock Tables*

52-Week High	52-Week Low	Stock	SYM	Div	Yld %	PE	Vol 100s	Hi	Lo	Close	Net Chg
42.38	22.50	Disney	DIS	.21	.6	43	115900	32.50	31.25	32.50	...

Use the stock table for Disney to answer the following questions.

a. What were the high and low prices for the past 52 weeks?

b. If you owned 3000 shares of Disney stock last year, what dividend did you receive?

c. What is the annual return for dividends alone? How does this compare to a bank account offering a 3.5% interest rate?

d. How many shares of Disney were traded yesterday?

e. What were the high and low prices for Disney shares yesterday?

f. What was the price at which Disney shares last traded when the stock exchange closed yesterday?

g. What does the value or symbol in the net change column mean?

h. Compute Disney's annual earnings per share using

$$\text{Annual earnings per share} = \frac{\text{Yesterday's closing price per share}}{\text{PE ratio}}.$$

SOLUTION

a. We find the high price for the past 52 weeks by looking under the heading **High**. The price is listed in dollars, given as 42.38. Thus, the high price for a share of stock for the past 52 weeks was $42.38. We find the low price for the past 52 weeks by looking under the heading **Low**. This price is also listed in dollars, given as 22.50. Thus, the low price for a share of Disney stock for the past 52 weeks was $22.50.

b. We find the dividend paid for a share of Disney stock last year by looking under the heading **Div**. The price is listed in dollars, given as .21. Thus, Disney paid a dividend of $0.21 per share to stockholders last year. If you owned 3000 shares, you received a dividend of $0.21 × 3000, or $630.

c. We find the annual return for dividends alone by looking under the heading **Yld %**, standing for percent yield. The number in the table, .6, is a percent. This means that the dividends alone gave Disney investors an annual return of 0.6%. This is much lower than a bank account paying a 3.5% interest rate. However, if Disney shares increase in value, the gain might make Disney stock a better investment than the bank account.

d. We find the number of shares of Disney traded yesterday by looking under the heading **Vol 100s**, standing for sales volume in hundreds. The number in the table is 115,900. This means that yesterday, a total of 115,900 × 100, or 11,590,000 shares, were traded.

e. We find the high and low prices for Disney shares yesterday by looking under the headings **Hi** and **Lo**. Both prices are listed in dollars, given as 32.50 and 31.25. Thus, the high and low prices for Disney shares yesterday were $32.50 and $31.25, respectively.

f. We find the price at which Disney shares last traded when the stock exchange closed yesterday by looking under the heading **Close**. The price is listed in dollars, given as 32.50. Thus, when the stock exchange closed yesterday, the price of a share of Disney stock was $32.50.

g. The . . . under **Net Chg** means that there was no change in price in Disney stock from the market close two days ago to yesterday's market close. In part (f), we found that the price of a share of Disney stock at yesterday's close was \$32.50, so the price at the market close two days ago was also \$32.50.

h. We are now ready to use

$$\text{Annual earnings per share} = \frac{\text{Yesterday's closing price per share}}{\text{PE ratio}}$$

to compute Disney's annual earnings per share. We found that yesterday's closing price per share was \$32.50. We find the PE ratio under the heading **PE**. The given number is 43. Thus,

$$\text{Annual earnings per share} = \frac{\$32.50}{43} \approx \$0.76.$$

The annual earnings per share for Disney were \$0.76. The PE ratio, 43, tells us that yesterday's closing price per share, \$32.50, is 43 times greater than the earnings per share, approximately \$0.76.

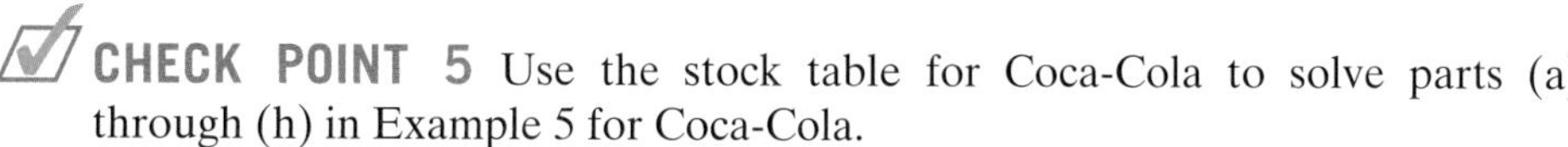

CHECK POINT 5 Use the stock table for Coca-Cola to solve parts (a) through (h) in Example 5 for Coca-Cola.

52-Week High	52-Week Low	Stock	SYM	Div	Yld %	PE	Vol 100s	Hi	Lo	Close	Net Chg
63.38	42.37	Coca-Cola	CocaCl	.72	1.5	37	72032	49.94	48.33	49.50	+0.03

Blitzer Bonus

The Bottom Line on Investments

Here are some investment suggestions from financial advisors:

- Do not invest money in the stock market that you will need within 10 years. Government bonds, CDs, and money-market accounts are more appropriate options for short-term goals.
- If you are a years old, approximately $(100 - a)\%$ of your investments should be in stocks. For example, at age 25, approximately $(100 - 25)\%$, or 75%, of your portfolio should be invested in stocks.
- Diversify your investments. Invest in a variety of different companies, as well as cushioning stock investments with cash investments and bonds. Diversification enables investors to take advantage of the stock market's superior returns while reducing risk to manageable levels.

Sources: Ralph Frasca, *Personal Finance*, Eighth Edition, Pearson, 2009; Eric Tyson, *Personal Finance for Dummies*, Sixth Edition, Wiley, 2010; Liz Pulliam Weston, *Easy Money*, Pearson, 2008.

5 *Understand accounts designed for retirement savings.*

Retirement Savings: Stashing Cash and Making Taxes Less Taxing

As you prepare for your future career, retirement probably seems very far away. However, we have seen that you can accumulate wealth much more easily if you have time to make your money work. As soon as you have a job and a paycheck, you should start putting some money away for retirement. Opening a retirement savings account early in your career is a smart way to gain more control over how you will spend a large part of your life.

You can use regular savings and investment accounts to save for retirement. There are also a variety of accounts designed specifically for retirement savings.

- A **traditional individual retirement account (IRA)** is a savings plan that allows you to set aside money for retirement, up to \$5500 per year for people under 50 and \$6500 per year for people 50 or older. You do not pay taxes on the money you deposit into the IRA. You can start withdrawing from your IRA when you are $59\frac{1}{2}$ years old. The withdrawals are taxed.
- A **Roth IRA** is a type of IRA with slightly different tax benefits. You pay taxes on the money you deposit into the IRA, but then you can withdraw your earnings tax-free when you are $59\frac{1}{2}$ years old. Although your contributions are not tax deductible, your earnings are never taxed, even after withdrawal.
- **Employer-sponsored retirement plans**, including 401(k) and 403(b) plans, are set up by the employer, who often makes some contribution to the plan on your behalf. These plans are not offered by all employers and are used to attract high-quality employees.

All accounts designed specifically for retirement savings have penalties for withdrawals before age $59\frac{1}{2}$.

EXAMPLE 6 *Dollars and Sense of Retirement Plans*

a. Suppose that between the ages of 25 and 35, you contribute \$4000 per year to a 401(k) and your employer matches this contribution dollar for dollar on your behalf. The interest rate is 8.5% compounded annually. What is the value of the 401(k) at the end of the 10 years?

b. After 10 years of working for this firm, you move on to a new job. However, you keep your accumulated retirement funds in the 401(k). How much money will you have in the plan when you reach age 65?

c. What is the difference between the amount of money you will have accumulated in the 401(k) at age 65 and the amount you contributed to the plan?

SOLUTION

a. We begin by finding the value of your 401(k) after 10 years.

$$A = \frac{P[(1+r)^t - 1]}{r}$$

Use the formula for the value of an annuity with interest compounded once a year.

$$A = \frac{8000[(1+0.085)^{10} - 1]}{0.085}$$

You contribute \$4000 and your employer matches this each year: $P = 4000 + 4000 = 8000$. The interest rate is 8.5%: $r = 0.085$. The time from age 25 to 35 is 10 years: $t = 10$.

$$\approx 118{,}681$$

Use a calculator.

The value of the 401(k) at the end of the 10 years is approximately \$118,681.

b. Now we find the value of your investment at age 65.

$$A = P(1+r)^t$$

Use the formula from Section 8.4 for future value with interest compounded once a year.

$$A = 118{,}681(1+0.085)^{30}$$

The value of the 401(k) is \$118,681: $P = 118{,}681$. The interest rate is 8.5%: $r = 0.085$. The time from age 35 to age 65 is 30 years: $t = 30$.

$$\approx 1{,}371{,}745$$

Use a calculator.

You will have approximately \$1,371,745 in the 401(k) when you reach age 65.

c. You contributed \$4000 per year to the 401(k) for 10 years, for a total of \$4000 × 10, or \$40,000. The difference between the amount you will have accumulated in the plan at age 65, \$1,371,745, and the amount you contributed, \$40,000, is

$$\$1{,}371{,}745 - \$40{,}000, \quad \text{or} \quad \$1{,}331{,}745.$$

Even when taxes are taken into consideration, we suspect you'll be quite pleased with your earnings from 10 years of savings.

CHECK POINT 6

a. Suppose that between the ages of 25 and 40, you contribute \$2000 per year to a 401(k) and your employer contributes \$1000 per year on your behalf. The interest rate is 8% compounded annually. What is the value of the 401(k), rounded to the nearest dollar, after 15 years?

b. After 15 years of working for this firm, you move on to a new job. However, you keep your accumulated retirement funds in the 401(k). How much money, to the nearest dollar, will you have in the plan when you reach age 65?

c. What is the difference between the amount of money you will have accumulated in the 401(k) and the amount you contributed to the plan?

Concept and Vocabulary Check

Fill in each blank so that the resulting statement is true.

1. A sequence of equal payments made at equal time periods is called a/an ________.

2. In the formula

$$A = \frac{P\left[\left(1 + \frac{r}{n}\right)^{nt} - 1\right]}{\left(\frac{r}{n}\right)},$$

____ is the deposit made at the end of each compounding period, ____ is the annual interest rate compounded ____ times per year, and A is the ______________ after ____ years.

3. Shares of ownership in a company are called ________. If you sell shares for more money than what you paid for them, you have a/an ________ gain on the sale. Some companies distribute all or part of their profits to shareholders as ________.

4. People who buy ________ are lending money to the company from which they buy them.

5. A listing of all the investments that a person holds is called a financial ________. To minimize risk, it should be ________, containing a mixture of low-risk and high-risk investments.

6. A group of investments managed by a professional investor is called a/an ____________.

7. With a/an ______ IRA, you pay taxes on the money you deposit, but you can withdraw your earnings tax-free beginning when you are _____ years old.

In Exercises 8–11, determine whether each statement is true or false. If the statement is false, make the necessary change(s) to produce a true statement.

8. With the same interest rate, compounding period, and time period, a lump-sum deposit will generate more interest than an annuity. ______

9. People who buy bonds are purchasing shares of ownership in a company. ______

10. Stocks are generally considered higher-risk investments than bonds. ______

11. A traditional IRA requires paying taxes when withdrawing money from the account at age $59\frac{1}{2}$ or older. ______

Exercise Set 8.5

Practice Exercises

Here are the formulas needed to solve the exercises. Be sure you understand what each formula describes and the meaning of the variables in the formulas.

$$A = \frac{P[(1+r)^t - 1]}{r} \qquad A = \frac{P\left[\left(1+\frac{r}{n}\right)^{nt} - 1\right]}{\left(\frac{r}{n}\right)} \qquad P = \frac{A\left(\frac{r}{n}\right)}{\left[\left(1+\frac{r}{n}\right)^{nt} - 1\right]}$$

In Exercises 1–10,

a. *Find the value of each annuity. Round to the nearest dollar.*

b. *Find the interest.*

	Periodic Deposit	Rate	Time
1.	\$2000 at the end of each year	5% compounded annually	20 years
2.	\$3000 at the end of each year	4% compounded annually	20 years
3.	\$4000 at the end of each year	6.5% compounded annually	40 years
4.	\$4000 at the end of each year	5.5% compounded annually	40 years
5.	\$50 at the end of each month	6% compounded monthly	30 years

	Periodic Deposit	Rate	Time
6.	\$60 at the end of each month	5% compounded monthly	30 years
7.	\$100 at the end of every six months	4.5% compounded semiannually	25 years
8.	\$150 at the end of every six months	6.5% compounded semiannually	25 years
9.	\$1000 at the end of every three months	6.25% compounded quarterly	6 years
10.	\$1200 at the end of every three months	3.25% compounded quarterly	6 years

In Exercises 11–18,

a. *Determine the periodic deposit. Round up to the nearest dollar.*

b. *How much of the financial goal comes from deposits and how much comes from interest?*

	Periodic Deposit	Rate	Time	Financial Goal
11.	\$? at the end of each year	6% compounded annually	18 years	\$140,000
12.	\$? at the end of each year	5% compounded annually	18 years	\$150,000
13.	\$? at the end of each month	4.5% compounded monthly	10 years	\$200,000
14.	\$? at the end of each month	7.5% compounded monthly	10 years	\$250,000
15.	\$? at the end of each month	7.25% compounded monthly	40 years	\$1,000,000
16.	\$? at the end of each month	8.25% compounded monthly	40 years	\$1,500,000
17.	\$? at the end of every three months	3.5% compounded quarterly	5 years	\$20,000
18.	\$? at the end of every three months	4.5% compounded quarterly	5 years	\$25,000

Exercises 19 and 20 refer to the stock tables for Goodyear (the tire company) and Dow Chemical given below. In each exercise, use the stock table to answer the following questions. Where necessary, round dollar amounts to the nearest cent.

a. *What were the high and low prices for a share for the past 52 weeks?*

b. *If you owned 700 shares of this stock last year, what dividend did you receive?*

c. *What is the annual return for the dividends alone? How does this compare to a bank offering a 3% interest rate?*

d. *How many shares of this company's stock were traded yesterday?*

e. *What were the high and low prices for a share yesterday?*

f. *What was the price at which a share last traded when the stock exchange closed yesterday?*

g. *What was the change in price for a share of stock from the market close two days ago to yesterday's market close?*

h. *Compute the company's annual earnings per share using*

$$\text{Annual earnings per share} = \frac{\text{Yesterday's closing price per share}}{\text{PE ratio}}.$$

19.

52-Week High	52-Week Low	Stock	SYM	Div	Yld %	PE	Vol 100s	Hi	Lo	Close	Net Chg
73.25	45.44	Goodyear	GT	1.20	2.2	17	5915	56.38	54.38	55.50	+1.25

20.

52-Week High	52-Week Low	Stock	SYM	Div	Yld %	PE	Vol 100s	Hi	Lo	Close	Net Chg
56.75	37.95	Dow Chemical	DOW	1.34	3.0	12	23997	44.75	44.35	44.69	+0.16

Practice Plus

Here are additional formulas that you will use to solve some of the remaining exercises. Be sure you understand what each formula describes and the meaning of the variables in the formulas.

$$A = P(1 + r)^t \qquad A = P\left(1 + \frac{r}{n}\right)^{nt}$$

In Exercises 21–22, round all answers to the nearest dollar.

21. Here are two ways of investing $30,000 for 20 years.

-

Lump-Sum Deposit	Rate	Time
$30,000	5% compounded annually	20 years

-

Periodic Deposit	Rate	Time
$1500 at the end of each year	5% compounded annually	20 years

a. After 20 years, how much more will you have from the lump-sum investment than from the annuity?

b. After 20 years, how much more interest will have been earned from the lump-sum investment than from the annuity?

22. Here are two ways of investing $40,000 for 25 years.

-

Lump-Sum Deposit	Rate	Time
$40,000	6.5% compounded annually	25 years

-

Periodic Deposit	Rate	Time
$1600 at the end of each year	6.5% compounded annually	25 years

a. After 25 years, how much more will you have from the lump-sum investment than from the annuity?

b. After 25 years, how much more interest will have been earned from the lump-sum investment than from the annuity?

23. Solve for P:

$$A = \frac{P[(1 + r)^t - 1]}{r}.$$

What does the resulting formula describe?

24. Solve for P:

$$A = \frac{P\left[\left(1 + \frac{r}{n}\right)^{nt} - 1\right]}{\left(\frac{r}{n}\right)}.$$

What does the resulting formula describe?

Application Exercises

In Exercises 25–30, round to the nearest dollar.

25. Suppose that you earned a bachelor's degree and now you're teaching high school. The school district offers teachers the opportunity to take a year off to earn a master's degree. To achieve this goal, you deposit $2000 at the end of each year in an annuity that pays 7.5% compounded annually.

a. How much will you have saved at the end of 5 years?

b. Find the interest.

26. Suppose that you earned a bachelor's degree and now you're teaching high school. The school district offers teachers the opportunity to take a year off to earn a master's degree. To achieve this goal, you deposit $2500 at the end of each year in an annuity that pays 6.25% compounded annually.

a. How much will you have saved at the end of 5 years?

b. Find the interest.

27. Suppose that at age 25, you decide to save for retirement by depositing $50 at the end of each month in an IRA that pays 5.5% compounded monthly.

a. How much will you have from the IRA when you retire at age 65?

b. Find the interest.

28. Suppose that at age 25, you decide to save for retirement by depositing $75 at the end of each month in an IRA that pays 6.5% compounded monthly.

a. How much will you have from the IRA when you retire at age 65?

b. Find the interest.

29. To offer scholarships to children of employees, a company invests $10,000 at the end of every three months in an annuity that pays 10.5% compounded quarterly.

a. How much will the company have in scholarship funds at the end of 10 years?

b. Find the interest.

30. To offer scholarships to children of employees, a company invests $15,000 at the end of every three months in an annuity that pays 9% compounded quarterly.

a. How much will the company have in scholarship funds at the end of 10 years?

b. Find the interest.

In Exercises 31–34, round up to the nearest dollar.

31. You would like to have $3500 in four years for a special vacation following college graduation by making deposits at the end of every six months in an annuity that pays 5% compounded semiannually.

a. How much should you deposit at the end of every six months?

b. How much of the $3500 comes from deposits and how much comes from interest?

32. You would like to have $4000 in four years for a special vacation following college graduation by making deposits at the end of every six months in an annuity that pays 7% compounded semiannually.

a. How much should you deposit at the end of every six months?

b. How much of the $4000 comes from deposits and how much comes from interest?

33. How much should you deposit at the end of each month into an IRA that pays 6.5% compounded monthly to have $2 million when you retire in 45 years? How much of the $2 million comes from interest?

34. How much should you deposit at the end of each month into an IRA that pays 8.5% compounded monthly to have $4 million when you retire in 45 years? How much of the $4 million comes from interest?

35. a. Suppose that between the ages of 22 and 40, you contribute \$3000 per year to a 401(k) and your employer contributes \$1500 per year on your behalf. The interest rate is 8.3% compounded annually. What is the value of the 401(k), rounded to the nearest dollar, after 18 years?

b. Suppose that after 18 years of working for this firm, you move on to a new job. However, you keep your accumulated retirement funds in the 401(k). How much money, to the nearest dollar, will you have in the plan when you reach age 65?

c. What is the difference between the amount of money you will have accumulated in the 401(k) and the amount you contributed to the plan?

36. a. Suppose that between the ages of 25 and 37, you contribute \$3500 per year to a 401(k) and your employer matches this contribution dollar for dollar on your behalf. The interest rate is 8.25% compounded annually. What is the value of the 401(k), rounded to the nearest dollar, after 12 years?

b. Suppose that after 12 years of working for this firm, you move on to a new job. However, you keep your accumulated retirement funds in the 401(k). How much money, to the nearest dollar, will you have in the plan when you reach age 65?

c. What is the difference between the amount of money you will have accumulated in the 401(k) and the amount you contributed to the plan?

Explaining the Concepts

37. What is an annuity?

38. What is meant by the value of an annuity?

39. What is stock?

40. Describe how to find the percent ownership that a shareholder has in a company.

41. Describe the two ways that investors make money with stock.

42. What is a bond? Describe the difference between a stock and a bond.

43. If an investor sees that the return from dividends for a stock is lower than the return for a no-risk bank account, should the stock be sold and the money placed in the bank account? Explain your answer.

44. What is a mutual fund?

45. What is the difference between a traditional IRA and a Roth IRA?

46. Write a problem involving the formula for regular payments needed to achieve a financial goal. The problem should be similar to Example 4 on page 534. However, the problem should be unique to your situation. Include something for which you would like to save, how much you need to save, and how long you have to achieve your goal. Then solve the problem.

Critical Thinking Exercises

Make Sense? *In Exercises 47–53, determine whether each statement makes sense or does not make sense, and explain your reasoning.*

47. By putting \$10 at the end of each month into an annuity that pays 3.5% compounded monthly, I'll be able to retire comfortably in just 30 years.

48. When I invest my money, I am making a trade-off between risk and return.

49. I have little tolerance for risk, so I must be willing to accept lower returns on my investments.

50. Diversification is like saying "don't put all your eggs in one basket."

51. Now that I've purchased bonds, I'm a shareholder in the company.

52. I've been promised a 20% return on an investment without any risk.

53. I appreciate my Roth IRA because I do not pay taxes on my deposits.

In Exercises 54–55,

a. *Determine the deposit at the end of each month. Round up to the nearest dollar.*

b. *Assume that the annuity in part (a) is a tax-deferred IRA belonging to a man whose gross income is \$50,000. Use **Table 8.1** on page 507 to calculate his taxes first with and then without the IRA. Assume the man is single with no dependents, has no tax credits, and takes the standard deduction.*

c. *What percent of his gross income are the man's federal taxes with and without the IRA? Round to the nearest tenth of a percent.*

54.

Periodic Deposit	Rate	Time	Financial Goal
\$? at the end of each month	8% compounded monthly	40 years	\$1,000,000

55.

Periodic Deposit	Rate	Time	Financial Goal
\$? at the end of each month	7% compounded monthly	40 years	\$650,000

56. How much should you deposit at the end of each month in an IRA that pays 8% compounded monthly to earn \$60,000 per year from interest alone, while leaving the principal untouched, when you retire in 30 years?

Group Exercises

57. Each group should have a newspaper with current stock quotations. Choose nine stocks that group members think would make good investments. Imagine that you invest \$10,000 in each of these nine investments. Check the value of your stock each day over the next five weeks and then sell the nine stocks after five weeks. What is the group's profit or loss over the five-week period? Compare this figure with the profit or loss of other groups in your class for this activity.

58. This activity is a group research project intended for four or five people. Use the research to present a seminar on investments. The seminar is intended to last about 30 minutes and should result in an interesting and informative presentation to the entire class. The seminar should include investment considerations, how to read the bond section of the newspaper, how to read the mutual fund section, and higher-risk investments.

59. Group members have inherited \$1 million. However, the group cannot spend any of the money for 10 years. As a group, determine how to invest this money in order to maximize the money you will make over 10 years. The money can be invested in as many ways as the group decides. Explain each investment decision. What are the risks involved in each investment plan?

8.6

WHAT AM I SUPPOSED TO LEARN?

After studying this section, you should be able to:

1 Compute the monthly payment and interest costs for a car loan.
2 Understand the types of leasing contracts.
3 Understand the pros and cons of leasing versus buying a car.
4 Understand the different kinds of car insurance.
5 Compare monthly payments on new and used cars.
6 Solve problems related to owning and operating a car.

Cars

TO THE GUYS AT RYDELL High in the musical *Grease!*, Kenickie's new car looks like a hunk of junk, but to him it's Greased Lightnin', a hot-rodding work of art on wheels. As with many teens, Kenickie's first car is a rite of passage—a symbol of emerging adulthood.

Our love affair with cars began in the early 1900s when Henry Ford cranked out the first Model T. Since then, we've admired cars to the point of identifying with the vehicles we drive. Cars can serve as status symbols, providing unique insights into a driver's personality.

In this section, we view cars from another vantage point—money. The money pit of owning a car ranges from financing the purchase to escalating costs of everything from fuel to tires to insurance. We open the section with the main reason people spend more money on a car than they can afford: financing.

The Mathematics of Financing a Car

A loan that you pay off with weekly or monthly payments, or payments in some other time period, is called an **installment loan**. The advantage of an installment loan is that the consumer gets to use a product immediately. The disadvantage is that the interest can add a substantial amount to the cost of a purchase.

Let's begin with car loans in which you make regular monthly payments, called **fixed installment loans**. Suppose that you borrow P dollars at interest rate r over t years.

The lender expects A dollars at the end of t years.

$$A = P\left(1 + \frac{r}{n}\right)^{nt}$$

You save the A dollars in an annuity by paying PMT dollars n times per year.

$$A = \frac{PMT\left[\left(1 + \frac{r}{n}\right)^{nt} - 1\right]}{\left(\frac{r}{n}\right)}$$

To find your regular payment amount, PMT, we set the amount the lender expects to receive equal to the amount you will save in the annuity:

$$P\left(1 + \frac{r}{n}\right)^{nt} = \frac{PMT\left[\left(1 + \frac{r}{n}\right)^{nt} - 1\right]}{\left(\frac{r}{n}\right)}.$$

Solving this equation for PMT, we obtain a formula for the loan payment for any installment loan, including payments on car loans.

1 *Compute the monthly payment and interest costs for a car loan.*

LOAN PAYMENT FORMULA FOR FIXED INSTALLMENT LOANS

The regular payment amount, PMT, required to repay a loan of P dollars paid n times per year over t years at an annual rate r is given by

$$PMT = \frac{P\left(\frac{r}{n}\right)}{\left[1 - \left(1 + \frac{r}{n}\right)^{-nt}\right]}.$$

EXAMPLE 1 Comparing Car Loans

Suppose that you decide to borrow \$20,000 for a new car. You can select one of the following loans, each requiring regular monthly payments:

Installment Loan A: three-year loan at 7%
Installment Loan B: five-year loan at 9%.

a. Find the monthly payments and the total interest for Loan A.

b. Find the monthly payments and the total interest for Loan B.

c. Compare the monthly payments and total interest for the two loans.

SOLUTION

For each loan, we use the loan payment formula to compute the monthly payments.

a. We first determine monthly payments and total interest for Loan A.

P, the loan amount, is \$20,000. Rate, r, is 7%. 12 payments per year. The loan is for 3 years.

$$PMT = \frac{P\left(\frac{r}{n}\right)}{\left[1 - \left(1 + \frac{r}{n}\right)^{-nt}\right]} = \frac{20{,}000\left(\frac{0.07}{12}\right)}{\left[1 - \left(1 + \frac{0.07}{12}\right)^{-12(3)}\right]} \approx 618$$

The monthly payments are approximately \$618.

Now we calculate the interest over three years, or 36 months.

Total interest over 3 years = Total of all monthly payments minus amount of the loan.

$$= \$618 \times 36 - \$20{,}000$$
$$= \$2248$$

The total interest paid over three years is approximately \$2248.

b. Next, we determine monthly payments and total interest for Loan B.

P, the loan amount, is \$20,000. Rate, r, is 9%. 12 payments per year. The loan is for 5 years.

$$PMT = \frac{P\left(\frac{r}{n}\right)}{\left[1 - \left(1 + \frac{r}{n}\right)^{-nt}\right]} = \frac{20{,}000\left(\frac{0.09}{12}\right)}{\left[1 - \left(1 + \frac{0.09}{12}\right)^{-12(5)}\right]} \approx 415$$

The monthly payments are approximately \$415.

TECHNOLOGY

Here are the calculator keystrokes to compute

$$\frac{20{,}000\,\left(\frac{0.07}{12}\right)}{\left[1 - \left(1 + \frac{0.07}{12}\right)^{-12(3)}\right]}.$$

Begin by simplifying the exponent, $-12(3)$, to -36 to avoid possible errors with parentheses:

$$\frac{20{,}000\,\left(\frac{0.07}{12}\right)}{\left[1 - \left(1 + \frac{0.07}{12}\right)^{-36}\right]}.$$

Scientific and graphing calculator keystrokes require placing parentheses around the expressions in both the numerator and the denominator.

Many Scientific Calculators

[(] 20000 [×] .07 [÷] 12 [)] [÷] [(] 1 [−] [(] 1 [+] .07 [÷] 12 [)] [y^x] 36 [+/−] [)] [=]

Many Graphing Calculators

[(] 20000 [×] .07 [÷] 12 [)] [÷] [(] 1 [−] [(] 1 [+] .07 [÷] 12 [)] [∧] [(−)] 36 [)] [ENTER]

Answers may vary if you do calculations in stages and round along the way.

Now we calculate the interest over five years, or 60 months.

Total interest over 5 years = Total of all monthly payments minus amount of the loan.

$$= \$415 \times 60 - \$20{,}000$$

$$= \$4900$$

The total interest paid over five years is approximately $4900.

c. **Table 8.6** compares the monthly payments and total interest for the two loans.

TABLE 8.6 Comparing Car Loans

$20,000 Loan	Monthly Payment	Total Interest
3-year loan at 7%	$618	$2248
5-year loan at 9%	$415	$4900

Monthly payments are less with the longer-term loan.

Interest is more with the longer-term loan.

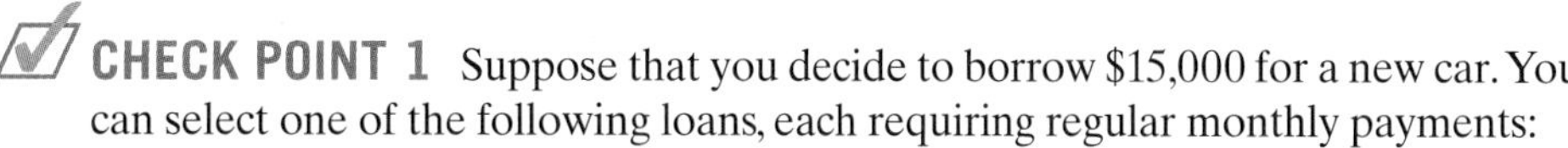

CHECK POINT 1 Suppose that you decide to borrow $15,000 for a new car. You can select one of the following loans, each requiring regular monthly payments:

Installment Loan A: four-year loan at 8%
Installment Loan B: six-year loan at 10%.

a. Find the monthly payments and the total interest for Loan A.

b. Find the monthly payments and the total interest for Loan B.

c. Compare the monthly payments and total interest for the two loans.

Blitzer Bonus

Financing Your Car

- Check out financing options. It's a good idea to get preapproved for a car loan through a bank or credit union before going to the dealer. You can then compare the loan offered by the dealer to your preapproved loan. Furthermore, with more money in hand, you'll have more negotiating power.
- Dealer financing often costs 1% or 2% more than a bank or credit union. Shop around for interest rates. Credit unions traditionally offer the best rates on car loans, more than 1.5% less on average than a bank loan.
- Put down as much money as you can. Interest rates generally decrease as the money you put down toward the car increases. Furthermore, you'll be borrowing less money, thereby paying less interest.
- A general rule is that you should spend no more than 20% of your net monthly income on a car payment.

2 *Understand the types of leasing contracts.*

The Leasing Alternative

Leasing is the practice of paying a specified amount of money over a specified time for the use of a product. Leasing is essentially a long-term rental agreement.

Leasing a car instead of buying one has become increasingly popular over the past several years. There are two types of leasing contracts:

- **A closed-end lease:** Each month, you make a fixed payment based on estimated usage. When the lease ends, you return the car and pay for mileage in excess of your estimate.

- **An open-end lease:** Each month you make a fixed payment based on the car's *residual value*. **Residual value** is the estimated resale value of the car at the end of the lease and is determined by the dealer. When the lease ends, you return the car and make a payment based on its appraised value at that time compared to its residual value. If the appraised value is less than the residual value stated in the lease, you pay all or a portion of the difference. If the appraised value is greater than or equal to the residual value, you owe nothing and you may receive a refund.

3 *Understand the pros and cons of leasing versus buying a car.*

Leasing a car offers both advantages and disadvantages over buying one.

Advantages of Leasing

- Leases require only a small down payment, or no down payment at all.
- Lease payments for a new car are lower than loan payments for the same car. Most people can lease a more expensive car than they would be able to buy.
- When the lease ends, you return the car to the dealer and do not have to be concerned about selling the car.

Disadvantages of Leasing

- When the lease ends, you do not own the car.
- Most lease agreements have mileage limits: 12,000 to 15,000 miles per year is common. If you exceed the number of miles allowed, there can be considerable charges.
- When mileage penalties and other costs at the end of the leasing period are taken into consideration, the total cost of leasing is almost always more expensive than financing a car.
- While leasing the car, you are responsible for keeping it in perfect condition. You are liable for any damage to the car.
- Leasing does not cover maintenance.
- There are penalties for ending the lease early.

Car leases tend to be extremely complicated. It can appear that there are as many lease deals as there are kinds of cars. A helpful pamphlet entitled "Keys to Vehicle Leasing" is published by the Federal Reserve Board. Copies are available on the Internet. Additional information can be found at online websites.

4 *Understand the different kinds of car insurance.*

The Importance of Auto Insurance

Who needs auto insurance? The simple answer is that if you own or lease a car, you do.

When you purchase **insurance**, you buy protection against loss associated with unexpected events. Different types of coverage are associated with auto insurance, but the one required by nearly every state is *liability*. There are two components of **liability coverage**:

- **Bodily injury liability** covers the costs of lawsuits if someone is injured or killed in an accident in which you are at fault.
- **Property damage liability** covers damage to other cars and property from negligent operation of your vehicle.

If you have a car loan or lease a car, you will also need *collision* and *comprehensive* coverage:

- **Collision coverage** pays for damage or loss of your car if you're in an accident.
- **Comprehensive coverage** protects your car from perils such as fire, theft, falling objects, acts of nature, and collision with an animal.

There is a big difference in auto insurance rates, so be sure to shop around. Insurance can be very expensive for younger drivers with limited driving experience. A poor driving record dramatically increases your insurance rates. Other factors that impact your insurance premium include where you live, the number of miles you drive each year, and the value of your car.

5 *Compare monthly payments on new and used cars.*

New or Used?

Who insists you need a new car? A new car loses an average of 12% of its value the moment it is driven off the dealer's lot. It's already a used car and you haven't even arrived home.

Used cars are a good option for many people. Your best buy is typically a two- to three-year-old car because the annual depreciation in price is greatest over the first few years. Furthermore, many sources of financing for used cars will loan money only on newer models that are less than five years old. Reputable car dealerships offer a good selection of used cars, with extended warranties and other perks.

The two most commonly used sources of pricing information for used cars are the *National Automobile Dealers Association Official Used Car Guide* (www.nada.com) and the *Kelley Blue Book Used Car Guide* (www.kbb.com). They contain the average retail price for many different makes of used cars.

EXAMPLE 2 *Saving Money with a Used Car*

Suppose that you are thinking about buying a car and have narrowed down your choices to two options:

- The new-car option: The new car costs \$25,000 and can be financed with a four-year loan at 7.9%.
- The used-car option: A three-year-old model of the same car costs \$14,000 and can be financed with a four-year loan at 8.45%.

What is the difference in monthly payments between financing the new car and financing the used car?

SOLUTION

We first determine the monthly payments for the new car that costs \$25,000, financed with a four-year loan at 7.9%.

P, the loan amount, is \$25,000.

Rate, *r*, is 7.9%.

12 payments per year

$$PMT = \frac{P\left(\frac{r}{n}\right)}{\left[1-\left(1+\frac{r}{n}\right)^{-nt}\right]} = \frac{25{,}000\left(\frac{0.079}{12}\right)}{\left[1-\left(1+\frac{0.079}{12}\right)^{-12(4)}\right]} \approx 609$$

The loan is for 4 years.

The monthly payments for the new car are approximately \$609. Now we determine the monthly payments for the used car that costs \$14,000, financed with a four-year loan at 8.45%.

P, the loan amount, is \$14,000.

Rate, *r*, is 8.45%.

12 payments per year

$$PMT = \frac{P\left(\frac{r}{n}\right)}{\left[1-\left(1+\frac{r}{n}\right)^{-nt}\right]} = \frac{14{,}000\left(\frac{0.0845}{12}\right)}{\left[1-\left(1+\frac{0.0845}{12}\right)^{-12(4)}\right]} \approx 345$$

The loan is for 4 years.

The monthly payments for the used car are approximately \$345. The difference in monthly payments between the new-car loan, \$609, and the used-car loan, \$345, is

\$609 − \$345, or \$264.

You save \$264 each month over a period of four years with the used-car option.

CHECK POINT 2 Suppose that you are thinking about buying a car and have narrowed down your choices to two options:

The new-car option: The new car costs \$19,000 and can be financed with a three-year loan at 6.18%.

The used-car option: A two-year-old model of the same car costs \$11,500 and can be financed with a three-year loan at 7.5%.

What is the difference in monthly payments between financing the new car and financing the used car?

6 *Solve problems related to owning and operating a car.*

The Money Pit of Car Ownership

Buying a car is a huge expense. To make matters worse, the car continues costing money after you purchase it. These costs include operating expenses such as fuel, maintenance, tires, tolls, parking, and cleaning. The costs also include ownership expenses such as insurance, license fees, registration fees, taxes, and interest on loans.

The significant expense of owning and operating a car is shown in **Table 8.7**. According to the American Automobile Association (AAA), the average yearly cost of owning and operating a car is just under \$9000.

TABLE 8.7 Annual Costs of Owning and Operating a Car in 2016*

Type of Car	Small Sedan	Medium Sedan	Minivan	SUV 4WD	Large Sedan
Cost per mile	44.9¢	58.1¢	62.5¢	70.8¢	71.0¢
Cost per year	\$6729	\$8716	\$9372	\$10,624	\$10,649

*Based on driving 15,000 miles per year.
Source: AAA

A large portion of a car's operating expenses involves the cost of gasoline. As the luster of big gas-guzzlers becomes less appealing, many people are turning to fuel-efficient hybrid cars that use a combination of gasoline and rechargeable batteries as power sources.

Our next example compares fuel expenses for a gas-guzzler and a hybrid. You can estimate the annual fuel expense for a vehicle if you know approximately how many miles the vehicle will be driven each year, how many miles the vehicle can be driven per gallon of gasoline, and how much a gallon of gasoline will cost.

THE COST OF GASOLINE

$$\text{Annual fuel expense} = \frac{\text{annual miles driven}}{\text{miles per gallon}} \times \text{price per gallon}$$

EXAMPLE 3 *Comparing Fuel Expenses*

Suppose that you drive 24,000 miles per year and gas averages \$4 per gallon.

a. What will you save in annual fuel expenses by owning a hybrid car averaging 50 miles per gallon rather than an SUV (sport utility vehicle) averaging 12 miles per gallon?

b. If you deposit your monthly fuel savings at the end of each month into an annuity that pays 7.3% compounded monthly, how much will you have saved at the end of six years?

SOLUTION

a. We use the formula for annual fuel expense.

$$\text{Annual fuel expense} = \frac{\text{annual miles driven}}{\text{miles per gallon}} \times \text{price per gallon}$$

$$\text{Annual fuel expense for the hybrid} = \frac{24{,}000}{50} \times \$4 = 480 \times \$4 = \$1920$$

The hybrid averages 50 miles per gallon.

$$\text{Annual fuel expense for the SUV} = \frac{24{,}000}{12} \times \$4 = 2000 \times \$4 = \$8000$$

The SUV averages 12 miles per gallon.

Your annual fuel expense is \$1920 for the hybrid and \$8000 for the SUV. By owning the hybrid rather than the SUV, you save

$$\$8000 - \$1920, \text{ or } \$6080$$

in annual fuel expenses.

b. Because you save \$6080 per year, you save

$$\frac{\$6080}{12} \approx \$507,$$

or approximately \$507 per month. Now you deposit \$507 at the end of each month into an annuity that pays 7.3% compounded monthly. We use the formula for the value of an annuity to determine your savings at the end of six years.

$$A = \frac{P\left[\left(1 + \frac{r}{n}\right)^{nt} - 1\right]}{\left(\frac{r}{n}\right)}$$

Use the formula for the value of an annuity.

$$A = \frac{507\left[\left(1 + \frac{0.073}{12}\right)^{12 \cdot 6} - 1\right]}{\left(\frac{0.073}{12}\right)}$$

The annuity involves month-end deposits of \$507: $P = 507$. The interest rate is 7.3%: $r = 0.073$. The interest is compounded monthly: $n = 12$. The number of years is 6: $t = 6$.

$$\approx 45{,}634$$

Use a calculator.

You will have saved approximately \$45,634 at the end of six years. This illustrates how driving a car that consumes less gas can yield significant savings for your future.

CHECK POINT 3 Suppose that you drive 36,000 miles per year and gas averages \$3.50 per gallon.

a. What will you save in annual fuel expenses by owning a hybrid car averaging 40 miles per gallon rather than an SUV averaging 15 miles per gallon?

b. If you deposit your monthly fuel savings at the end of each month into an annuity that pays 7.25% compounded monthly, how much will you have saved at the end of seven years? Round all computations to the nearest dollar.

Concept and Vocabulary Check

Fill in each blank so that the resulting statement is true.

1. In the formula

$$PMT = \frac{P\left(\frac{r}{n}\right)}{\left[1 - \left(1 + \frac{r}{n}\right)^{-nt}\right]},$$

_______ is the regular payment amount required to repay a loan of ____ dollars paid ____ times per year over ____ years at an annual interest rate ____.

2. The two types of contracts involved with leasing a car are called a/an __________ lease and a/an __________ lease.
3. The estimated resale value of a car at the end of its lease is called the car's __________.
4. There are two components of liability insurance. The component that covers costs if someone is injured or killed in an accident in which you are at fault is called __________ liability. The component that covers damage to other cars if you are at fault is called __________ liability.
5. The type of car insurance that pays for damage or loss of your car if you're in an accident is called ________ coverage.
6. The type of insurance that pays for damage to your car due to fire, theft, or falling objects is called __________ coverage.

In Exercises 7–12, determine whether each statement is true or false. If the statement is false, make the necessary changes(s) to produce a true statement.

7. The interest on a car loan can be determined by taking the difference between the total of all monthly payments and the amount of the loan. ______
8. When an open-end lease terminates and the car's appraised value is less than the residual value stated in the lease, you owe nothing. ______
9. One advantage to leasing a car is that you are not responsible for any damage to the car. ______
10. One disadvantage to leasing a car is that most lease agreements have mileage limits. ______
11. Collision coverage pays for damage to another car if you cause an accident. ______
12. Due to operating and ownership expenses, a car continues costing money after you buy it. ______

Exercise Set 8.6

Practice and Application Exercises

In Exercises 1–10, use

$$PMT = \frac{P\left(\frac{r}{n}\right)}{\left[1 - \left(1 + \frac{r}{n}\right)^{-nt}\right]}.$$

Round answers to the nearest dollar.

1. Suppose that you borrow \$10,000 for four years at 8% toward the purchase of a car. Find the monthly payments and the total interest for the loan.
2. Suppose that you borrow \$30,000 for four years at 8% for the purchase of a car. Find the monthly payments and the total interest for the loan.
3. Suppose that you decide to borrow \$15,000 for a new car. You can select one of the following loans, each requiring regular monthly payments:

   Installment Loan A: three-year loan at 5.1%
   Installment Loan B: five-year loan at 6.4%.

   a. Find the monthly payments and the total interest for Loan A.
   b. Find the monthly payments and the total interest for Loan B.
   c. Compare the monthly payments and the total interest for the two loans.
4. Suppose that you decide to borrow \$40,000 for a new car. You can select one of the following loans, each requiring regular monthly payments:

   Installment Loan A: three-year loan at 6.1%
   Installment Loan B: five-year loan at 7.2%.

   a. Find the monthly payments and the total interest for Loan A.
   b. Find the monthly payments and the total interest for Loan B.
   c. Compare the monthly payments and the total interest for the two loans.
5. Suppose that you are thinking about buying a car and have narrowed down your choices to two options:

   The new-car option: The new car costs \$28,000 and can be financed with a four-year loan at 6.12%.

   The used-car option: A three-year old model of the same car costs \$16,000 and can be financed with a four-year loan at 6.86%.

   What is the difference in monthly payments between financing the new car and financing the used car?
6. Suppose that you are thinking about buying a car and have narrowed down your choices to two options:

   The new-car option: The new car costs \$68,000 and can be financed with a four-year loan at 7.14%.

   The used-car option: A three-year old model of the same car costs \$28,000 and can be financed with a four-year loan at 7.92%.

   What is the difference in monthly payments between financing the new car and financing the used car?

7. Suppose that you decide to buy a car for $29,635, including taxes and license fees. You saved $9000 for a down payment and can get a five-year car loan at 6.62%. Find the monthly payment and the total interest for the loan.

8. Suppose that you decide to buy a car for $37,925, including taxes and license fees. You saved $12,000 for a down payment and can get a five-year loan at 6.58%. Find the monthly payment and the total interest for the loan.

9. Suppose that you are buying a car for $60,000, including taxes and license fees. You saved $10,000 for a down payment. The dealer is offering you two incentives:

Incentive A is $5000 off the price of the car, followed by a five-year loan at 7.34%.

Incentive B does not have a cash rebate, but provides free financing (no interest) over five years.

What is the difference in monthly payments between the two offers? Which incentive is the better deal?

10. Suppose that you are buying a car for $56,000, including taxes and license fees. You saved $8000 for a down payment. The dealer is offering you two incentives:

Incentive A is $10,000 off the price of the car, followed by a four-year loan at 12.5%.

Incentive B does not have a cash rebate, but provides free financing (no interest) over four years.

What is the difference in monthly payments between the two offers? Which incentive is the better deal?

In Exercises 11–14, use the formula

$$A = \frac{P\left[\left(1 + \frac{r}{n}\right)^{nt} - 1\right]}{\left(\frac{r}{n}\right)}.$$

Round all computations to the nearest dollar.

11. Suppose that you drive 40,000 miles per year and gas averages $4 per gallon.

a. What will you save in annual fuel expenses by owning a hybrid car averaging 40 miles per gallon rather than an SUV averaging 16 miles per gallon?

b. If you deposit your monthly fuel savings at the end of each month into an annuity that pays 5.2% compounded monthly, how much will you have saved at the end of six years?

12. Suppose that you drive 15,000 miles per year and gas averages $3.50 per gallon.

a. What will you save in annual fuel expenses by owning a hybrid car averaging 60 miles per gallon rather than an SUV averaging 15 miles per gallon?

b. If you deposit your monthly fuel savings at the end of each month into an annuity that pays 5.7% compounded monthly, how much will you have saved at the end of six years?

The table shows the expense of operating and owning three selected cars, by average costs per mile. Use the appropriate information in the table to solve Exercises 13–16.

AVERAGE ANNUAL COSTS OF OWNING AND OPERATING A CAR

	Average Costs per Mile		
Make and Model	**Operating**	**Ownership**	**Total**
Cadillac STS	$0.26	$0.72	$0.98
Honda Accord LX	$0.21	$0.34	$0.55
Toyota Corolla CE	$0.15	$0.25	$0.40

Source: Runzheimer International

13. a. If you drive 20,000 miles per year, what is the total annual expense for a Cadillac STS?

b. If the total annual expense for a Cadillac STS is deposited at the end of each year into an IRA paying 8.5% compounded yearly, how much will be saved at the end of six years?

14. a. If you drive 14,000 miles per year, what is the total annual expense for a Toyota Corolla CE?

b. If the total annual expense for a Toyota Corolla CE is deposited at the end of each year into an IRA paying 8.2% compounded yearly, how much will be saved at the end of six years?

15. If you drive 30,000 miles per year, by how much does the total annual expense for a Cadillac STS exceed that of a Toyota Corolla CE over six years?

16. If you drive 25,000 miles per year, by how much does the total annual expense for a Cadillac STS exceed that of a Honda Accord LX over six years?

Explaining the Concepts

17. If a three-year car loan has the same interest rate as a six-year car loan, how do the monthly payments and the total interest compare for the two loans?

18. What is the difference between a closed-end car lease and an open-end car lease?

19. Describe two advantages of leasing a car over buying one.

20. Describe two disadvantages of leasing a car over buying one.

21. What are the two components of liability coverage and what is covered by each component?

22. What does collision coverage pay for?

23. What does comprehensive coverage pay for?

24. How can you estimate a car's annual fuel expense?

Critical Thinking Exercises

Make Sense? *In Exercises 25–30, determine whether each statement makes sense or does not make sense, and explain your reasoning.*

25. If I purchase a car using money that I've saved, I can eliminate paying interest on a car loan, but then I have to give up the interest income I could have earned on my savings.

26. The problem with my car lease is that when it ends, I have to be concerned about selling the car.
27. Although lease payments for a new car are lower than loan payments for the same car, once I take mileage penalties and other costs into consideration, the total cost of leasing is more expensive than financing the car.
28. I've paid off my car loan, so I am not required to have liability coverage.
29. Buying a used car or a fuel-efficient car can yield significant savings for my future.
30. Because it is extremely expensive to own and operate a car, I plan to look closely at whether or not a car is essential and consider other modes of transportation.
31. Use the discussion at the bottom of page 545 to prove the loan payment formula shown in the box at the top of page 546. Work with the equation in which the amount the lender expects to receive is equal to the amount saved in the annuity. Multiply both sides of this equation by $\frac{r}{n}$ and then solve for PMT by dividing both sides by the appropriate expression. Finally, divide the numerator and the denominator of the resulting formula for PMT by $\left(1 + \frac{r}{n}\right)^{nt}$ to obtain the form of the loan payment formula shown in the box.
32. The unpaid balance of an installment loan is equal to the present value of the remaining payments. The unpaid balance, P, is given by

$$P = PMT\frac{\left[1 - \left(1 + \dfrac{r}{n}\right)^{-nt}\right]}{\left(\dfrac{r}{n}\right)},$$

where PMT is the regular payment amount, r is the annual interest rate, n is the number of payments per year, and t is the number of years remaining in the loan.

a. Use the loan payment formula to derive the unpaid balance formula.

b. The price of a car is \$24,000. You have saved 20% of the price as a down payment. After the down payment, the balance is financed with a 5-year loan at 9%. Determine the unpaid balance after three years. Round all calculations to the nearest dollar.

Group Exercises

33. Group members should go to the Internet and select a car that they might like to buy. Price the car and its options. Then find two loans with the best rates, but with different terms. For each loan, calculate the monthly payments and total interest.
34. **Student Loans**

Group members should present a report on federal loans to finance college costs, including Stafford loans, Perkins loans, and PLUS loans. Also include a discussion of grants that do not have to be repaid, such as Pell Grants and National Merit Scholarships. Refer to *Funding Your Education*, published by the Department of Education and available at studentaid.ed.gov. Use the loan repayment formula that we applied to car loans to determine regular payments and interest on some of the loan options presented in your report.

8.7 The Cost of Home Ownership

WHAT AM I SUPPOSED TO LEARN?

After studying this section, you should be able to:

1 Compute the monthly payment and interest costs for a mortgage.

2 Prepare a partial loan amortization schedule.

3 Solve problems involving what you can afford to spend for a mortgage.

4 Understand the pros and cons of renting versus buying.

THE BIGGEST SINGLE PURCHASE THAT MOST PEOPLE MAKE IN their lives is the purchase of a home. If you choose home ownership at some point in the future, it is likely that you will finance the purchase with an installment loan. Knowing the unique issues surrounding the purchase of a home, and whether or not this aspect of the American dream is right for you, can play a significant role in your financial future.

Mortgages

A **mortgage** is a long-term installment loan (perhaps up to 30, 40, or even 50 years) for the purpose of buying a home, and for which the property is pledged as security for payment. If payments are not made on the loan, the lender may take possession of the property. The **down payment** is the portion of the sale price of the home that the buyer initially pays to the seller. The minimum required down payment is computed as a percentage of the sale price. For example, suppose you decide to buy a \$220,000 home. The lender requires you to pay the seller 10% of the sale price. You must pay 10% of \$220,000, which is $0.10 \times 220{,}000$ or \$22,000, to the seller. Thus, \$22,000 is the down payment. The **amount of the mortgage** is the difference between the sale price and the down payment. For your \$220,000 home, the amount of the mortgage is \$220,000 − \$22,000, or \$198,000.

Monthly payments for a mortgage depend on the amount of the mortgage (the principal), the interest rate, and the duration of the mortgage. Mortgages can have a fixed interest rate or a variable interest rate. **Fixed-rate mortgages** have the same monthly principal and interest payment during the entire time of the loan. A loan like this that has a schedule for paying a fixed amount each period is called a **fixed installment loan**. **Variable-rate mortgages**, also known as **adjustable-rate mortgages** (ARMs), have payment amounts that change from time to time depending on changes in the interest rate. ARMs are less predictable than fixed-rate mortgages. They start out at lower rates than fixed-rate mortgages. Caps limit how high rates can go over the term of the loan.

1 *Compute the monthly payment and interest costs for a mortgage.*

Computations Involved with Buying a Home

Although monthly payments for a mortgage depend on the amount of the mortgage, the duration of the loan, and the interest rate, the interest is not the only cost of a mortgage. Most lending institutions require the buyer to pay one or more **points** at the time of closing—that is, the time at which the mortgage begins. A point is a one-time charge that equals 1% of the loan amount. For example, two points means that the buyer must pay 2% of the loan amount at closing. Often, a buyer can pay fewer points in exchange for a higher interest rate or more points for a lower rate. A document, called the **Truth-in-Lending Disclosure Statement**, shows the buyer the APR, or the annual percentage rate, for the mortgage. The APR takes into account the interest rate and points.

A monthly mortgage payment is used to repay the principal plus interest. In addition, lending institutions can require monthly deposits into an **escrow account**, an account used by the lender to pay real estate taxes and insurance. These deposits increase the amount of the monthly payment.

In the previous section, we used the loan payment formula for fixed installment loans to determine payments on car loans. Because a fixed-rate mortgage is a fixed installment loan, we use the same formula to compute the monthly payment for a mortgage.

LOAN PAYMENT FORMULA FOR FIXED INSTALLMENT LOANS

The regular payment amount, PMT, required to repay a loan of P dollars paid n times per year over t years at an annual rate r is given by

$$PMT = \frac{P\left(\frac{r}{n}\right)}{\left[1 - \left(1 + \frac{r}{n}\right)^{-nt}\right]}.$$

EXAMPLE 1 ***Computing the Monthly Payment and Interest Costs for a Mortgage***

The price of a home is \$195,000. The bank requires a 10% down payment and two points at the time of closing. The cost of the home is financed with a 30-year fixed-rate mortgage at 7.5%.

a. Find the required down payment.

b. Find the amount of the mortgage.

c. How much must be paid for the two points at closing?

d. Find the monthly payment (excluding escrowed taxes and insurance).

e. Find the total interest paid over 30 years.

SOLUTION

a. The required down payment is 10% of \$195,000 or

$$0.10 \times \$195{,}000 = \$19{,}500.$$

b. The amount of the mortgage is the difference between the price of the home and the down payment.

Amount of the mortgage = sale price − down payment

$$= \$195{,}000 - \$19{,}500$$
$$= \$175{,}500$$

c. To find the cost of two points on a mortgage of \$175,500, find 2% of \$175,500.

$$0.02 \times \$175{,}500 = \$3510$$

The down payment (\$19,500) is paid to the seller and the cost of two points (\$3510) is paid to the lending institution.

d. We are interested in finding the monthly payment for a \$175,500 mortgage at 7.5% for 30 years. We use the loan payment formula for installment loans.

P, the mortgage amount, is \$175,500. Fixed rate, *r*, is 7.5%. 12 payments per year. The mortgage time, *t*, is 30 years.

$$PMT = \frac{P\left(\frac{r}{n}\right)}{\left[1-\left(1+\frac{r}{n}\right)^{-nt}\right]} = \frac{175{,}500\left(\frac{0.075}{12}\right)}{\left[1-\left(1+\frac{0.075}{12}\right)^{-12(30)}\right]}$$

$$\approx 1227$$

The monthly mortgage payment for principal and interest is approximately \$1227. (Keep in mind that this payment does not include escrowed taxes and insurance.)

e. The total cost of interest over 30 years is equal to the difference between the total of all monthly payments and the amount of the mortgage. The total of all monthly payments is equal to the amount of the monthly payment multiplied by the number of payments. We found the amount of each monthly payment in (d): \$1227. The number of payments is equal to the number of months in a year, 12, multiplied by the number of years in the mortgage, 30: $12 \times 30 = 360$. Thus, the total of all monthly payments is $\$1227 \times 360$.

Now we can calculate the interest over 30 years.

Total interest paid = total of all monthly payments minus amount of the mortgage.

$$= \$1227 \times 360 - \$175{,}500$$
$$= \$441{,}720 - \$175{,}500 = \$266{,}220$$

The total interest paid over 30 years is approximately \$266,220.

TECHNOLOGY

Here are the calculator keystrokes to compute

$$\frac{175{,}500\left(\frac{0.075}{12}\right)}{\left[1-\left(1+\frac{0.075}{12}\right)^{-12(30)}\right]}.$$

Begin by simplifying the exponent, −12(30), to −360 to avoid possible errors with parentheses:

$$\frac{175{,}500\left(\frac{0.075}{12}\right)}{\left[1-\left(1+\frac{0.075}{12}\right)^{-360}\right]}.$$

Many Scientific Calculators

(175500 × .075 ÷ 12) ÷
(1 − (1 + .075 ÷ 12)
y^x 360 +/−) =

Many Graphing Calculators

(175500 × .075 ÷ 12) ÷
(1 − (1 + .075 ÷ 12)
∧ (−) 360) ENTER

CHECK POINT 1 In Example 1, the \$175,500 mortgage was financed with a 30-year fixed rate at 7.5%. The total interest paid over 30 years was approximately \$266,220.

a. Use the loan payment formula for installment loans to find the monthly payment if the time of the mortgage is reduced to 15 years. Round to the nearest dollar.

b. Find the total interest paid over 15 years.

c. How much interest is saved by reducing the mortgage from 30 to 15 years?

2 Prepare a partial loan amortization schedule.

Loan Amortization Schedules

When a mortgage loan is paid off through a series of regular payments, it is said to be **amortized**, which literally means "killed off." In working Check Point 1(c), were you surprised that nearly \$150,000 was saved when the mortgage was amortized over 15 years rather than over 30 years? What adds to the interest cost is the long period over which the loan is financed. **Although each payment is the same, with each successive payment the interest portion decreases and the portion applied toward paying off the principal increases.** The interest is computed using the simple interest formula $I = Prt$. The principal, P, is equal to the balance of the loan, which decreases each month. The rate, r, is the annual interest rate of the mortgage loan. Because a payment is made each month, the time, t, is

$$\frac{1 \text{ month}}{12 \text{ months}} = \frac{1 \cancel{\text{month}}}{12 \cancel{\text{months}}}$$

or $\frac{1}{12}$ of a year.

A document showing how the payment each month is split between interest and principal is called a **loan amortization schedule**. Typically, for each payment, this document includes the payment number, the interest for the payment, the amount of the payment applied to the principal, and the balance of the loan after the payment is applied.

EXAMPLE 2 *Preparing a Loan Amortization Schedule*

Prepare a loan amortization schedule for the first two months of the mortgage loan shown in the table below. Round entries to the nearest cent.

LOAN AMORTIZATION SCHEDULE

Annual % Rate: 9.5% **Amount of Mortgage: \$130,000** **Number of Monthly Payments: 180**		**Monthly Payment: \$1357.50** **Term: Years 15, Months 0**	
Payment Number	**Interest Payment**	**Principal Payment**	**Balance of Loan**
1			
2			

SOLUTION

We begin with payment number 1.

$$\begin{aligned}\text{Interest for the month} &= Prt = \$130{,}000 \times 0.095 \times \frac{1}{12} \approx \$1029.17\\ \text{Principal payment} &= \text{Monthly payment} - \text{Interest payment}\\ &= \$1357.50 - \$1029.17 = \$328.33\\ \text{Balance of loan} &= \text{Principal balance} - \text{Principal payment}\\ &= \$130{,}000 - \$328.33 = \$129{,}671.67\end{aligned}$$

Now, starting with a loan balance of \$129,671.67, we repeat these computations for the second payment.

$$\begin{aligned}\text{Interest for the month} &= Prt = \$129{,}671.67 \times 0.095 \times \frac{1}{12} \approx \$1026.57\\ \text{Principal payment} &= \text{Monthly payment} - \text{Interest payment}\\ &= \$1357.50 - \$1026.57 = \$330.93\\ \text{Balance of loan} &= \text{Principal balance} - \text{Principal payment}\\ &= \$129{,}671.67 - \$330.93 = \$129{,}340.74\end{aligned}$$

The results of these computations are included in **Table 8.8** on the next page, a partial loan amortization schedule. By using the simple interest formula month-to-month on the loan's balance, a complete loan amortization schedule for all 180 payments can be calculated.

Blitzer Bonus

The Mortgage Crisis

In 2006, the median U.S. home price jumped to \$206,000, up a stunning 15% in just one year and 55% over five years. This rise in home values made real estate an attractive investment to many people, including those with poor credit records and low incomes. Credit standards for mortgages were lowered and loans were made to high-risk borrowers. By 2008, America's raucous house party was over. A brief period of easy lending, especially lax mortgage practices from 2002 through 2006, exploded into the worst financial crisis since the Great Depression. The plunge in home prices wiped out trillions of dollars in home equity, setting off fears that foreclosures and tight credit could send home prices falling to the point that millions of families and thousands of banks might be thrust into insolvency.

TABLE 8.8 Loan Amortization Schedule

Annual % Rate: 9.5% Amount of Mortgage: $130,000 Number of Monthly Payments: 180		Monthly Payment: $1357.50 Term: Years 15, Months 0	
Payment Number	**Interest Payment**	**Principal Payment**	**Balance of Loan**
1	$1029.17	$ 328.33	$129,671.67
2	$1026.57	$ 330.93	$129,340.74
3	$1023.96	$ 333.54	$129,007.22
4	$1021.32	$ 336.18	$128,671.04
30	$ 944.82	$ 412.68	$118,931.35
31	$ 941.55	$ 415.95	$118,515.52
125	$ 484.62	$ 872.88	$ 60,340.84
126	$ 477.71	$ 879.79	$ 59,461.05
179	$ 21.26	$1336.24	$ 1347.74
180	$ 9.76	$ 1347.74	

Many lenders supply a loan amortization schedule like the one in Example 2 at the time of closing. Such a schedule shows how the buyer pays slightly less in interest and more in principal for each payment over the entire life of the loan.

CHECK POINT 2 Prepare a loan amortization schedule for the first two months of the mortgage loan shown in the following table. Round entries to the nearest cent.

Annual % Rate: 7.0% Amount of Mortgage: $200,000 Number of Monthly Payments: 240		Monthly Payment: $1550.00 Term: Years 20, Months 0	
Payment Number	**Interest Payment**	**Principal Payment**	**Balance of Loan**
1			
2			

Blitzer Bonus

Bittersweet Interest

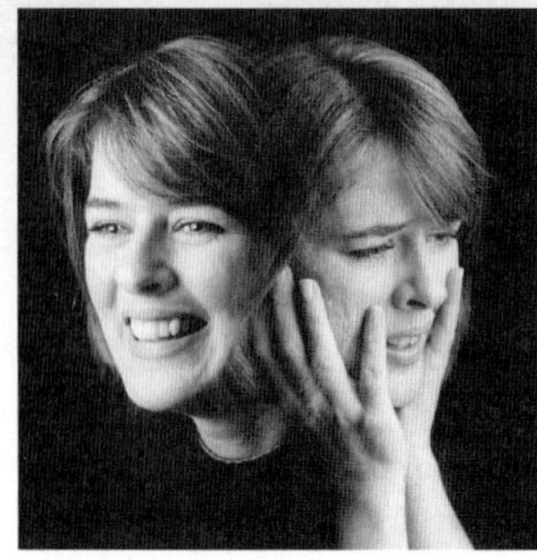

Looking at amortization tables, you could get discouraged by how much of your early mortgage payments goes toward interest and how little goes toward paying off the principal. Although you get socked with tons of interest in the early years of a loan, the one bright side to the staggering cost of a mortgage is the **mortgage interest tax deduction**. To make the cost of owning a home more affordable, the tax code permits deducting all the mortgage interest (but not the principal) that you pay per year on the loan. **Table 8.9** illustrates how this tax loophole reduces the cost of the mortgage.

TABLE 8.9 Tax Deductions for a $100,000 Mortgage at 7% for a Taxpayer in the 28% Tax Bracket

Year	Interest	Tax Savings	Net Cost of Mortgage
1	$6968	$1951	$5017
2	$6895	$1931	$4964
3	$6816	$1908	$4908
4	$6732	$1885	$4847
5	$6641	$1859	$4782

3 *Solve problems involving what you can afford to spend for a mortgage.*

Determining What You Can Afford

Here's the bottom line from most financial advisers:

- Spend no more than 28% of your gross monthly income for your mortgage payment.
- Spend no more than 36% of your gross monthly income for your total monthly debt, including mortgage payments, car payments, credit card bills, student loans, and medical debt.

Using these guidelines, **Table 8.10** shows the maximum monthly amount you could afford for mortgage payments and total credit obligations for a variety of income levels.

TABLE 8.10 Maximum Amount You Can Afford

Gross Annual Income	Monthly Mortgage Payment	Total Monthly Credit Obligations
\$20,000	\$467	\$600
\$30,000	\$700	\$900
\$40,000	\$933	\$1200
\$50,000	\$1167	\$1500
\$60,000	\$1400	\$1800
\$70,000	\$1633	\$2100
\$80,000	\$1867	\$2400
\$90,000	\$2100	\$2700
\$100,000	\$2333	\$3000

Source: Fannie Mae

Four Decades of Mortgages

Mortgage rates in 2017 were low relative to the yearly averages from the past three decades. The table shows the average rate for a fixed 30-year mortgage for selected years.

Year	Average Mortgage Rate
1981	18.63%
1985	12.43%
1988	10.34%
1991	9.25%
2000	8.05%
2002	5.83%
2006	6.41%
2011	4.45%
2017	4.46%

Source: Freddie Mac

EXAMPLE 3 *What Can You Afford?*

Suppose that your gross annual income is \$25,000.

a. What is the maximum amount you should spend each month on a mortgage payment?

b. What is the maximum amount you should spend each month for total credit obligations?

c. If your monthly mortgage payment is 80% of the maximum amount you can afford, what is the maximum amount you should spend each month for all other debt?

Round all computations to the nearest dollar.

SOLUTION

With a gross annual income of \$25,000, your gross monthly income is

$$\frac{\$25{,}000}{12},$$

or approximately \$2083.

a. You should spend no more than 28% of your gross monthly income, \$2083, on a mortgage payment.

$$28\% \text{ of } \$2083 = 0.28 \times \$2083 \approx \$583.$$

Your monthly mortgage payment should not exceed \$583.

b. You should spend no more than 36% of your gross monthly income, \$2083, for total monthly debt.

$$36\% \text{ of } \$2083 = 0.36 \times \$2083 \approx \$750.$$

Your total monthly credit obligations should not exceed \$750.

c. The problem's conditions state that your monthly mortgage payment is 80% of the maximum you can afford, which is \$583. This means that your monthly mortgage payment is 80% of \$583.

$$80\% \text{ of } \$583 = 0.8 \times \$583 \approx \$466.$$

In part (b), we saw that your total monthly debt should not exceed \$750. Because you are paying \$466 for your mortgage payment, this leaves \$750 − \$466, or \$284, for all other debt. Your monthly credit obligations, excluding mortgage payments, should not exceed \$284.

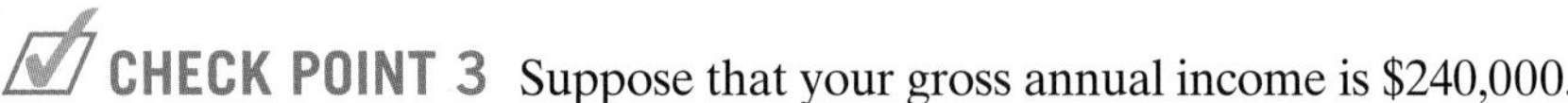

CHECK POINT 3 Suppose that your gross annual income is \$240,000.

a. What is the maximum amount you should spend each month on a mortgage payment?

b. What is the maximum amount you should spend each month for total credit obligations?

c. If your monthly mortgage payment is 90% of the maximum amount you can afford, what is the maximum amount you should spend each month for all other debt?

Round all computations to the nearest dollar.

4 *Understand the pros and cons of renting versus buying.*

Renting versus Buying

Nearly everyone is faced at some stage in life with the dilemma "should I rent or should I buy a home?" The rent-or-buy decision can be highly complex and is often based on lifestyle rather than finances. Aside from a changing economic climate, there are many factors to consider. Here are some advantages of both renting and buying to help smooth the way:

Benefits of Renting

- No down payment or points are required. You generally have a security deposit that is returned at the end of your lease.
- Very mobile: You can easily relocate, moving as often as you like and as your lease permits.
- Does not tie up hundreds of thousands of dollars that might be invested more safely and lucratively elsewhere. Most financial advisers agree that you should buy a home because you want to live in it, not because you want to fund your retirement.
- Does not clutter what you can afford for your total monthly debt with mortgage payments.
- May involve lower monthly expenses. You pay rent, whereas a homeowner pays the mortgage, taxes, insurance, and upkeep.
- Can provide amenities like swimming pools, tennis courts, and health clubs.
- Avoids the risk of falling housing prices.
- Does not require home repair, maintenance, and groundskeeping.
- There are no property taxes.
- Generally less costly than buying a home when staying in it for fewer than three years.

Benefits of Home Ownership

- Peace of mind and stability.
- Provides significant tax advantages, including deduction of mortgage interest and property taxes.
- There is no chance of rent increasing over time.
- Allows for freedom to remodel, landscape, and redecorate.
- You can build up **equity**, the difference between the home's value and what you owe on the mortgage, as the mortgage is paid off. The possibility of home appreciation is a potential source of cash in the form of home equity loans.
- When looking at seven-year time frames, the total cost of renting (monthly rent, renter's insurance, loss of potential interest on a security deposit) is more than twice the total cost of buying for home owners who itemize their tax deductions.

Source: Arthur J. Keown, *Personal Finance*, Fourth Edition, Pearson, 2007.

Blitzer Bonus

Reducing Rental Costs

Let's assume that one of your long-term financial goals involves home ownership. It's still likely that you'll be renting for a while once you complete your education and begin your first job. Other than living in a tent, here are some realistic suggestions for reducing costs on your first rental:

- Select a lower-cost rental. Who says you should begin your career in a large apartment with fancy amenities, private parking spots, and lakefront views? The less you spend renting, the more you can save for a down payment toward buying your own place. You will ultimately qualify for the most favorable mortgage terms by making a down payment of at least 20% of the purchase price of the property.
- Negotiate rental increases. Landlords do not want to lose good tenants who are respectful of their property and pay rent on time. Filling vacancies can be time consuming and costly.
- Rent a larger place with roommates. By sharing a rental, you will decrease rental costs and get more home for your rental dollars.

Concept and Vocabulary Check

Fill in each blank so that the resulting statement is true.

1. A long-term installment loan for the purpose of buying a home is called a/an __________. The portion of the sale price of the home that the buyer initially pays to the seller is called the ______________.

2. A document showing how each monthly installment payment is split between interest and principal is called a/an ______________________.

In Exercises 3–6, determine whether each statement is true or false. If the statement is false, make the necessary change(s) to produce a true statement.

3. Over the life of an installment loan, the interest portion increases and the portion applied to paying off the principal decreases with each successive payment. ______

4. Financial advisors suggest spending no more than 5% of your gross monthly income for your mortgage payment. ______

5. Renters are not required to pay property taxes. ______

6. Renters can build up equity as the rent is paid each month. ______

Exercise Set 8.7

Practice and Application Exercises

In Exercises 1–10, use

$$PMT = \frac{P\left(\frac{r}{n}\right)}{\left[1 - \left(1 + \frac{r}{n}\right)^{-nt}\right]}$$

to determine the regular payment amount, rounded to the nearest dollar.

1. The price of a home is $220,000. The bank requires a 20% down payment and three points at the time of closing. The cost of the home is financed with a 30-year fixed-rate mortgage at 7%.

a. Find the required down payment.

b. Find the amount of the mortgage.

c. How much must be paid for the three points at closing?

d. Find the monthly payment (excluding escrowed taxes and insurance).

e. Find the total cost of interest over 30 years.

2. The price of a condominium is $180,000. The bank requires a 5% down payment and one point at the time of closing. The cost of the condominium is financed with a 30-year fixed-rate mortgage at 8%.

a. Find the required down payment.

b. Find the amount of the mortgage.

c. How much must be paid for the one point at closing?

d. Find the monthly payment (excluding escrowed taxes and insurance).

e. Find the total cost of interest over 30 years.

3. The price of a small cabin is $100,000. The bank requires a 5% down payment. The buyer is offered two mortgage options: 20-year fixed at 8% or 30-year fixed at 8%. Calculate the amount of interest paid for each option. How much does the buyer save in interest with the 20-year option?

4. The price of a home is $160,000. The bank requires a 15% down payment. The buyer is offered two mortgage options: 15-year fixed at 8% or 30-year fixed at 8%. Calculate the amount of interest paid for each option. How much does the buyer save in interest with the 15-year option?

5. In terms of paying less in interest, which is more economical for a $150,000 mortgage: a 30-year fixed-rate at 8% or a 20-year fixed-rate at 7.5%? How much is saved in interest?

6. In terms of paying less in interest, which is more economical for a $90,000 mortgage: a 30-year fixed-rate at 8% or a 15-year fixed-rate at 7.5%? How much is saved in interest?

In Exercises 7–8, which mortgage loan has the greater total cost (closing costs + the amount paid for points + total cost of interest)? By how much?

7. A $120,000 mortgage with two loan options:

Mortgage A: 30-year fixed at 7% with closing costs of $2000 and one point

Mortgage B: 30-year fixed at 6.5% with closing costs of $1500 and four points

8. A $250,000 mortgage with two loan options:

Mortgage A: 30-year fixed at 7.25% with closing costs of $2000 and one point

Mortgage B: 30-year fixed at 6.25% with closing costs of $350 and four points

9. The cost of a home is financed with a $120,000 30-year fixed-rate mortgage at 4.5%.

a. Find the monthly payments and the total interest for the loan.

b. Prepare a loan amortization schedule for the first three months of the mortgage. Round entries to the nearest cent.

Payment Number	Interest	Principal	Loan Balance
1			
2			
3			

10. The cost of a home is financed with a $160,000 30-year fixed-rate mortgage at 4.2%.

a. Find the monthly payments and the total interest for the loan.

b. Prepare a loan amortization schedule for the first three months of the mortgage. Round entries to the nearest cent.

Payment Number	Interest	Principal	Loan Balance
1			
2			
3			

Use this advice from most financial advisers to solve Exercises 11–12.

- *Spend no more than 28% of your gross monthly income for your mortgage payment.*
- *Spend no more than 36% of your gross monthly income for your total monthly debt.*

Round all computations to the nearest dollar.

11. Suppose that your gross annual income is $36,000.

a. What is the maximum amount you should spend each month on a mortgage payment?

b. What is the maximum amount you should spend each month for total credit obligations?

c. If your monthly mortgage payment is 70% of the maximum you can afford, what is the maximum amount you should spend each month for all other debt?

12. Suppose that your gross annual income is $62,000.

a. What is the maximum amount you should spend each month on a mortgage payment?

b. What is the maximum amount you should spend each month for total credit obligations?

c. If your monthly mortgage payment is 90% of the maximum you can afford, what is the maximum amount you should spend each month for all other debt?

Explaining the Concepts

13. What is a mortgage?

14. What is a down payment?

15. How is the amount of a mortgage determined?

16. Describe why a buyer would select a 30-year fixed-rate mortgage instead of a 15-year fixed-rate mortage if interest rates are $\frac{1}{4}$% to $\frac{1}{2}$% lower on a 15-year mortgage.

17. Describe one advantage and one disadvantage of an adjustable-rate mortage over a fixed-rate mortgage.

18. What is a loan amortization schedule?

19. Describe what happens to the portions of payments going to principal and interest over the life of an installment loan.

20. Describe how to determine what you can afford for your monthly mortgage payment.

21. Describe two advantages of renting over home ownership.

22. Describe two advantages of home ownership over renting.

Critical Thinking Exercises

Make Sense? *In Exercises 23–26, determine whether each statement makes sense or does not make sense, and explain your reasoning.*

23. I use the same formula to determine mortgage payments and payments for car loans.

24. There must be an error in the loan amortization schedule for my mortgage because the annual interest rate is only 3.5%, yet the schedule shows that I'm paying more on interest than on the principal for many of my payments.

25. My landlord required me to pay 2 points when I signed my rental lease.

26. I include rental payments among my itemized tax deductions.

27. If your gross annual income is \$75,000, use appropriate computations to determine whether you could afford a \$200,000 30-year fixed-rate mortgage at 5.5%.

28. The partial loan amortization schedule shows payments 50–54. Although payment 50 is correct, there are errors in one or more of the payments from 51 through 54. Find the errors and correct them.

LOAN AMORTIZATION SCHEDULE

Annual % Rate: 6.0% **Amount of Mortgage: \$120,000** **Number of Monthly Payments: 180**		**Monthly Payment: \$1012.63** **Term: Years 15, Months 0**	
Payment Number	**Interest Payment**	**Principal Payment**	**Balance of Loan**
50	\$485.77	\$526.86	\$96,626.51
51	\$483.13	\$529.50	\$96,097.01
52	\$477.82	\$534.81	\$95,030.06
53	\$480.49	\$532.14	\$95,564.87
54	\$495.15	\$537.48	\$94,492.58

8.8

WHAT AM I SUPPOSED TO LEARN?

After studying this section, you should be able to:

1 Find the interest, the balance due, and the minimum monthly payment for credit card loans.
2 Understand the pros and cons of using credit cards.
3 Understand the difference between credit cards and debit cards.
4 Know what is contained in a credit report.
5 Understand credit scores as measures of creditworthiness.

Credit Cards

WOULD YOU LIKE TO BUY PRODUCTS WITH A CREDIT CARD? Although the card will let you use a product while paying for it, the costs associated with such cards, including their high interest rates, fees, and penalties, stack the odds in favor of your getting hurt by them. In 2016, the average credit-card debt per U.S. household was \$16,748, with \$1292 paid in interest per year. (*Source:* nerdwallet.com) One advantage of making a purchase with a credit card is that the consumer gets to use a product immediately. In this section, we will see that a significant disadvantage is that it can add a substantial amount to the cost of a purchase. When it comes to using a credit card, consumer beware!

Open-End Installment Loans

Using a credit card is an example of an open-end installment loan, commonly called **revolving credit**. Open-end loans differ from fixed installment loans such as car loans and mortgages in that there is no schedule for paying a fixed amount each period. Credit card loans require users to make only a minimum monthly payment that depends on the unpaid balance and the interest rate. Credit cards have high interest rates compared to other kinds of loans. The interest on credit cards is computed using the simple interest formula $I = Prt$. However, r represents the *monthly* interest rate and t is time in months rather than in years. A typical interest rate is 1.57% monthly. This is equivalent to a yearly rate of $12 \times 1.57\%$, or 18.84%. With such a high annual percentage rate, credit card balances should be paid off as quickly as possible.

Most credit card customers are billed every month. A typical billing period is May 1 through May 31, but it can also run from, say, May 5 through June 4. Customers receive a statement, called an **itemized billing**, that includes the unpaid balance on the first day of the billing period, the total balance owed on the last day of the billing period, a list of purchases and cash advances made during the billing period, any finance charges or other fees incurred, the date of the last day of the billing period, the payment due date, and the minimum payment required.

Customers who make a purchase during the billing period and pay the entire amount of the purchase by the payment due date are not charged interest. By contrast, customers who make cash advances using their credit cards must pay interest from the day the money is advanced until the day it is repaid.

1 *Find the interest, the balance due, and the minimum monthly payment for credit card loans.*

Interest on Credit Cards: The Average Daily Balance Method

Methods for calculating interest, or finance charges, on credit cards may vary and the interest can differ on credit cards that show the same annual percentage rate, or APR. The method used for calculating interest on most credit cards is called the *average daily balance method.*

Up to Our Ears in Debt

In 2016, total debt in the United States was about to surpass the amounts owed in 2007, the beginning of the Great Recession.

Type of Debt	Average Owed per U.S. Household
Credit Cards	\$16,748
Mortgages	\$176,222
Car Loans	\$28,948
Student Loans	\$49,905

Source: nerdwallet.com (2016 data)

THE AVERAGE DAILY BALANCE METHOD

Interest is calculated using $I = Prt$, where r is the monthly rate and t is one month. The principal, P, is the *average daily balance*. The **average daily balance** is the sum of the unpaid balances for each day in the billing period divided by the number of days in the billing period.

$$\text{Average daily balance} = \frac{\text{Sum of the unpaid balances for each day in the billing period}}{\text{Number of days in the billing period}}$$

In Example 1, we illustrate how to determine the average daily balance. At the conclusion of the example, we summarize the steps used in the computation.

EXAMPLE 1 *Balance Due on a Credit Card*

The issuer of a particular VISA card calculates interest using the average daily balance method. The monthly interest rate is 1.3% of the average daily balance. The following transactions occurred during the May 1–May 31 billing period.

Transaction Description	Transaction Amount
Previous balance, \$1350.00	
May 1 Billing date	
May 8 Payment	\$250.00 credit
May 10 Charge: Airline Tickets	\$375.00
May 20 Charge: Books	\$ 57.50
May 28 Charge: Restaurant	\$ 65.30
May 31 End of billing period	
Payment Due Date: June 9	

a. Find the average daily balance for the billing period. Round to the nearest cent.

b. Find the interest to be paid on June 1, the next billing date. Round to the nearest cent.

c. Find the balance due on June 1.

d. This credit card requires a \$10 minimum monthly payment if the balance due at the end of the billing period is less than \$360. Otherwise, the minimum monthly payment is $\frac{1}{36}$ of the balance due at the end of the billing period, rounded up to the nearest whole dollar. What is the minimum monthly payment due by June 9?

SOLUTION

a. We begin by finding the average daily balance for the billing period. First make a table that shows the beginning date of the billing period, each transaction date, and the unpaid balance for each date.

Date	Unpaid Balance	
May 1	\$1350.00	previous balance
May 8	\$1350.00 − \$250.00 = \$1100.00	\$250.00 payment
May 10	\$1100.00 + \$375.00 = \$1475.00	\$375.00 charge
May 20	\$1475.00 + \$57.50 = \$1532.50	\$57.50 charge
May 28	\$1532.50 + \$65.30 = \$1597.80	\$65.30 charge

We now extend our table by adding two columns. One column shows the number of days at each unpaid balance. The final column shows each unpaid balance multiplied by the number of days that the balance is outstanding.

Date	Unpaid Balance	Number of Days at Each Unpaid Balance	(Unpaid Balance) · (Number of Days)
May 1	\$1350.00	7	(\$1350.00)(7) = \$9450.00
May 8	\$1100.00	2	(\$1100.00)(2) = \$2200.00
May 10	\$1475.00	10	(\$1475.00)(10) = \$14,750.00
May 20	\$1532.50	8	(\$1532.50)(8) = \$12,260.00
May 28	\$1597.80	4	(\$1597.80)(4) = \$6391.20
		Total: 31	Total: \$45,051.20

There are 4 days at this unpaid balance, May 28, 29, 30, and 31, before the beginning of the next billing period, June 1.

This is the number of days in the billing period.

This is the sum of the unpaid balances for each day in the billing period.

Notice that we found the sum of the products in the final column of the table. This dollar amount, \$45,051.20, gives the sum of the unpaid balances for each day in the billing period.

Now we divide the sum of the unpaid balances for each day in the billing period, \$45,051.20, by the number of days in the billing period, 31. This gives the average daily balance.

Average daily balance

$$= \frac{\text{Sum of the unpaid balances for each day in the billing period}}{\text{Number of days in the billing period}}$$

$$= \frac{\$45{,}051.20}{31} \approx \$1453.26$$

The average daily balance is approximately \$1453.26.

b. Now we find the interest to be paid on June 1, the next billing date. The monthly interest rate is 1.3% of the average daily balance. The interest due is computed using $I = Prt$.

$$I = Prt = (\$1453.26)\,(0.013)\,(1) \approx \$18.89$$

The average daily balance serves as the principal.

Time, t, is measured in months, and $t = 1$ month.

The interest, or finance charge, for the June 1 billing will be \$18.89.

GREAT QUESTION!

How do credit card companies round when determining a minimum monthly payment?

They round *up* to the nearest dollar. For example, the quotient in part (d) is approximately 44.908, which rounds to 45. Because the minimum monthly payment is rounded up, \$45 would still be the payment if the approximate quotient had been 44.098.

c. The balance due on June 1, the next billing date, is the unpaid balance on May 31 plus the interest.

$$\text{Balance due} = \$1597.80 + \$18.89 = \$1616.69$$

Unpaid balance on May 31, obtained from the second table on the previous page

Interest, or finance charge, obtained from part (b)

The balance due on June 1 is \$1616.69.

d. Because the balance due, \$1616.69, exceeds \$360, the customer must pay a minimum of $\frac{1}{36}$ of the balance due.

$$\text{Minimum monthly payment} = \frac{\text{balance due}}{36} = \frac{\$1616.69}{36} \approx \$45$$

Rounded up to the nearest whole dollar, the minimum monthly payment due by June 9 is \$45.

The following box summarizes the steps used in Example 1 to determine the average daily balance. Calculating the average daily balance can be quite tedious when there are numerous transactions during a billing period.

DETERMINING THE AVERAGE DAILY BALANCE

Step 1 Make a table that shows the beginning date of the billing period, each transaction date, and the unpaid balance for each date.

Step 2 Add a column to the table that shows the number of days at each unpaid balance.

Step 3 Add a final column to the table that shows each unpaid balance multiplied by the number of days that the balance is outstanding.

Step 4 Find the sum of the products in the final column of the table. This dollar amount is the sum of the unpaid balances for each day in the billing period.

Step 5 Compute the average daily balance.

$$\text{Average daily balance} = \frac{\text{Sum of the unpaid balances for each day in the billing period}}{\text{Number of days in the billing period}}$$

CHECK POINT 1 A credit card company calculates interest using the average daily balance method. The monthly interest rate is 1.6% of the average daily balance. The following transactions occurred during the May 1–May 31 billing period.

Transaction Description	Transaction Amount
Previous balance, \$8240.00	
May 1 Billing date	
May 7 Payment	\$ 350.00 credit
May 15 Charge: Computer	\$ 1405.00
May 17 Charge: Restaurant	\$ 45.20
May 30 Charge: Clothing	\$ 180.72
May 31 End of billing period	
Payment Due Date: June 9	

Answer parts (a) through (d) in Example 1 on page 564 using this information.

2 *Understand the pros and cons of using credit cards.*

Credit Cards: Marvelous Tools or Snakes in Your Wallet?

Credit cards are convenient. Pay the entire balance by the due date for each monthly billing and you avoid interest charges. Carry over a balance and interest charges quickly add up. With this in mind, let's consider the positives and the negatives involved with credit card usage.

Advantages of Using Credit Cards

- Get to use a product before actually paying for it.
- No interest charges by paying the balance due at the end of each billing period.
- Responsible use is an effective way to build a good credit score. (See page 568 for a discussion of credit scores.)
- No need to carry around large amounts of cash.
- More convenient to use than checks.
- Offer consumer protections: If there is a disputed or fraudulent charge on your credit card statement, let the card issuer know and the amount is generally removed.
- Provide a source of temporary emergency funds.
- Extend shopping opportunities to purchases over the phone or the Internet.
- Simple tasks like renting a car or booking a hotel room can be difficult or impossible without a credit card.
- Monthly statements can help keep track of spending. Some card issuers provide an annual statement that aids in tax preparation.
- May provide amenities such as free miles toward air travel.
- Useful as identification when multiple pieces of identification are needed.

The Best Financial Advice for College Graduates

- Avoid credit-card debt.
- Create a budget.
- Start saving for retirement.
- Pay off student loans.
- Buy a used car.

Sources: ING Group and *USA Today*

Credit Card Woes

- High interest rates on unpaid balances. In 2009, interest rates were as high as 30%.
- No cap on interest rates. In 2009, the U.S. Senate defeated an amendment that would have imposed a 15% cap on credit card interest rates. (The Credit Card Act, passed by Congress in 2009, does restrict when issuers can raise rates on existing unpaid balances.) Your initial credit card interest rate is unlikely to go down, but it can sure go up.
- No cap on fees. *Consumer Reports* (October 2008) cited a credit card with an enticing 9.9% annual interest rate. But the fine print revealed a $29 account-setup fee, a $95 program fee, a $48 annual fee, and a $7 monthly servicing fee. Nearly 40% of the $40 billion in profits that U.S. card issuers earned in 2008 came from fees. Furthermore, issuers can hike fees at any time, for any reason. Read the fine print of a credit card agreement before you sign up.
- Easy to overspend. Purchases with credit cards can create the illusion that you are not actually spending money.
- Can serve as a tool for financial trouble. Using a credit card to buy more than you can afford and failing to pay the bill in full each month can result in serious debt. Fees and interest charges are added to the balance, which continues to grow, even if there are no new purchases.
- The minimum-payment trap: Credit-card debt is made worse by paying only the required minimum, a mistake made by 11% of credit-card debtors. Pay the minimum and most of it goes to interest charges.

Credit card statements now include a Minimum Payment Warning: "If you make only the minimum payment each period, you will pay more in interest and it will take you longer to pay off your balance."

Debit Cards

Credit cards have been around since the 1950s. Their plastic lookalikes, debit cards, were introduced in the mid-1970s. Although debit cards look like credit cards, the big difference is that debit cards are linked to your bank account. When you use a

3 *Understand the difference between credit cards and debit cards.*

debit card, the money you spend is deducted electronically from your bank account. It's similar to writing an electronic check, but there's no paper involved and the check gets "cashed" instantly.

Debit cards offer the convenience of making purchases with a piece of plastic without the temptation or ability to run up credit-card debt. You can't spend money you don't have because the card won't work if the money isn't in your checking or savings account.

Debit cards have drawbacks. They may not offer the protection a credit card does for disputed purchases. It's easy to rack up overdraft charges if your bank enrolls you in an "overdraft protection" program. That means your card won't be turned down if you do not have sufficient funds in your account to cover your purchase. You are spared the embarrassment of having your card rejected, but it will cost you fees of approximately $27 per overdraft.

Debit card purchases should be treated like those for which you use a check. Record all transactions and their amounts in your checkbook, including any cash received from an ATM. Always keep track of how much money is available in your account. Your balance can be checked at an ATM or online.

4 *Know what is contained in a credit report.*

Credit Reports and Credit Scores

As a college student, it is unlikely that you have a credit history. Once you apply for your first credit card, your personal *credit report* will begin. A **credit report** contains the following information:

- **Identifying Information**: This includes your name, social security number, current address, and previous addresses.
- **Record of Credit Accounts**: This includes details about all open or closed credit accounts, such as when each account was opened, the latest balance, and the payment history.
- **Public Record Information**: Any of your public records, such as bankruptcy information, appears in this section of the credit report.
- **Collection Agency Account Information:** Unpaid accounts are turned over to collection agencies. Information about such actions appear in this section of the credit report.
- **Inquiries**: Companies that have asked for your credit information because you applied for credit are listed here.

Organizations known as **credit bureaus** collect credit information on individual consumers and provide credit reports to potential lenders, employers, and others upon request. The three main credit bureaus are Equifax, Experian, and TransUnion. You can get your credit report from the three bureaus free at www.annualcreditreport.com.

5 *Understand credit scores as measures of creditworthiness.*

Credit bureaus use data from your credit report to create a *credit score*, which is used to measure your creditworthiness. **Credit scores**, or **FICO scores**, range from 300 to 850, with a higher score indicating better credit. **Table 8.11** contains ranges of credit scores and their measures of creditworthiness.

TABLE 8.11 Credit Scores and Their Significance

Scores	**Creditworthiness**
720–850	Very good to excellent; Best interest rates on loans
650–719	Good; Likely to get credit, but not the best interest rates on loans
630–649	Fair; May get credit, but only at higher rates
580–629	Poor; Likely to be denied credit by all but a high-interest lender
300–579	Bad; Likely to be denied credit

Your credit score will have an enormous effect on your financial life. Individuals with higher credit scores get better interest rates on loans because they are considered to have a lower risk of defaulting. A good credit score can save you thousands of dollars in interest charges over your lifetime.

Blitzer Bonus

College Students and Credit Cards

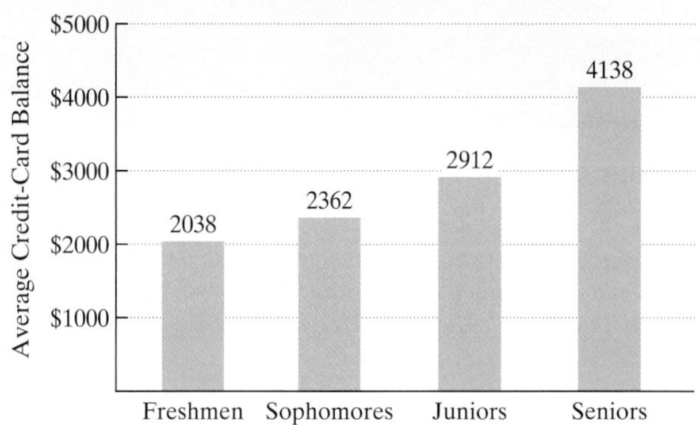

Source: Nellie Mae

If you have no credit history but are at least 18 years old with a job, you may be able to get a card with a limited amount of credit, usually \$500 to \$1000. Your chances of getting a credit card increase if you apply through a bank with which you have an account.

Prior to 2010, it was not necessary for college students to have a job to get a credit card. Credit card companies viewed college students as good and responsible customers who would continue to have a lifelong need for credit. The issuers anticipated retaining these students after graduation when their accounts would become more valuable. Many resorted to aggressive marketing tactics, offering everything from T-shirts to iPods to students who signed up.

Times have changed. In May 2009, President Barack Obama signed legislation that prohibits issuing credit cards to college students younger than 21 unless they can prove they are able to make payments or get a parent or guardian to co-sign. Because college students do not have much money, most won't be able to get a credit card without permission from their parents. The bill also requires lenders to get permission from the co-signer before increasing the card's credit limit.

Before credit card reform swiped easy plastic from college students, those who fell behind on their credit card bills often left college with blemished credit reports. This made it more difficult for them to rent an apartment, get a car loan, or even find a job. "A lot of kids got themselves in trouble," said Adam Levin, founder of Credit.com, a consumer Web site. "As much as college students are obsessed with GPAs, their credit score is the most important number they're going to have to deal with after graduation."

GREAT QUESTION!

What's the bottom line on responsible credit card use?

A critical component of financial success involves demonstrating that you can handle the responsibility of using a credit card. Responsible credit card use means:

- You pay the entire balance by the due date for each monthly billing.
- You only use the card to make purchases that you can afford.
- You save all receipts for credit card purchases and check each itemized billing carefully for any errors.
- You use your credit card sometimes, but also use cash, checks, or your debit card.

When you decide to apply for your first credit card, you should look for a card with no annual fee and a low interest rate. You can compare rates and fees at www.creditcards.com, www.bankrate.com, and www.indexcreditcards.com.

Concept and Vocabulary Check

Fill in each blank so that the resulting statement is true.

1. Using a credit card is an example of a/an __________ installment loan for which there is no schedule for paying a fixed amount each period.
2. The average daily balance for a credit card's billing period is ________ divided by ______________________________.
3. When you use a/an ______ card, the money you spend is deducted electronically from your bank account.
4. Details about when all your credit accounts were opened, their latest balance, and their payment history are included in a/an ____________.
5. Credit scores range from _____ to _____, with a higher score indicating ______________________.

In Exercises 6–9, determine whether each statement is true or false. If the statement is false, make the necessary change(s) to produce a true statement.

6. Interest on a credit card is calculated using $I = Prt$, where r is the monthly rate, t is one month, and P is the balance due. ______

7. When using a credit card, the money spent is deducted electronically from the user's bank account. ______

8. Credit reports contain bankruptcy information. ______

9. Higher credit scores indicate better credit. ______

Exercise Set 8.8

Practice and Application Exercises

Exercises 1–2 involve credit cards that calculate interest using the average daily balance method. The monthly interest rate is 1.5% of the average daily balance. Each exercise shows transactions that occurred during the March 1–March 31 billing period. In each exercise,

a. *Find the average daily balance for the billing period. Round to the nearest cent.*

b. *Find the interest to be paid on April 1, the next billing date. Round to the nearest cent.*

c. *Find the balance due on April 1.*

d. *This credit card requires a $10 minimum monthly payment if the balance due at the end of the billing period is less than $360. Otherwise, the minimum monthly payment is $\frac{1}{36}$ of the balance due at the end of the billing period, rounded up to the nearest whole dollar. What is the minimum monthly payment due by April 9?*

1.

Transaction Description	Transaction Amount
Previous balance, $6240.00	
March 1 Billing date	
March 5 Payment	$ 300 credit
March 7 Charge: Restaurant	$ 40
March 12 Charge: Groceries	$ 90
March 21 Charge: Car Repairs	$ 230
March 31 End of billing period	
Payment Due Date: April 9	

2.

Transaction Description	Transaction Amount
Previous balance, $7150.00	
March 1 Billing date	
March 4 Payment	$ 400 credit
March 6 Charge: Furniture	$ 1200
March 15 Charge: Gas	$ 40
March 30 Charge: Groceries	$ 50
March 31 End of billing period	
Payment Due Date: April 9	

Exercises 3–4 involve credit cards that calculate interest using the average daily balance method. The monthly interest rate is 1.2% of the average daily balance. Each exercise shows transactions that occurred during the June 1–June 30 billing period. In each exercise,

a. *Find the average daily balance for the billing period. Round to the nearest cent.*

b. *Find the interest to be paid on July 1, the next billing date. Round to the nearest cent.*

c. *Find the balance due on July 1.*

d. *This credit card requires a $30 minimum monthly payment if the balance due at the end of the billing period is less than $400. Otherwise, the minimum monthly payment is $\frac{1}{25}$ of the balance due at the end of the billing period, rounded up to the nearest whole dollar. What is the minimum monthly payment due by July 9?*

3.

Transaction Description	Transaction Amount
Previous balance, $2653.48	
June 1 Billing date	
June 6 Payment	$1000.00 credit
June 8 Charge: Gas	$ 36.25
June 9 Charge: Groceries	$ 138.43
June 17 Charge: Gas Charge: Groceries	$ 42.36 $ 127.19
June 27 Charge: Clothing	$ 214.83
June 30 End of billing period	
Payment Due Date: July 9	

4.

Transaction Description	Transaction Amount
Previous balance, $4037.93	
June 1 Billing date	
June 5 Payment	$ 350.00 credit
June 10 Charge: Gas	$ 31.17
June 15 Charge: Prescriptions	$ 42.50
June 22 Charge: Gas Charge: Groceries	$ 43.86 $ 112.91
June 29 Charge: Clothing	$ 96.73
June 30 End of billing period	
Payment Due Date: July 9	

In Exercises 5–10, use

$$PMT = \frac{P\left(\frac{r}{n}\right)}{\left[1 - \left(1 + \frac{r}{n}\right)^{-nt}\right]}$$

to determine the regular payment amount, rounded to the nearest dollar.

5. Suppose your credit card has a balance of $4200 and an annual interest rate of 18%. You decide to pay off the balance over two years. If there are no further purchases charged to the card,
   a. How much must you pay each month?
   b. How much total interest will you pay?
6. Suppose your credit card has a balance of $3600 and an annual interest rate of 16.5%. You decide to pay off the balance over two years. If there are no further purchases charged to the card,
   a. How much must you pay each month?
   b. How much total interest will you pay?
7. To pay off the $4200 credit card balance in Exercise 5, suppose that you can get a bank loan at 10.5% with a term of three years.
   a. How much will you pay each month? How does this compare with your credit card payment in Exercise 5?
   b. How much total interest will you pay? How does this compare with your total credit card interest in Exercise 5?
8. To pay off the $3600 credit card balance in Exercise 6, suppose that you can get a bank loan at 9.5% with a term of three years.
   a. How much will you pay each month? How does this compare with your credit card payment in Exercise 6?
   b. How much total interest will you pay? How does this compare with your total credit card interest in Exercise 6?
9. Rework Exercise 5 assuming you decide to pay off the balance over one year rather than two. How much more must you pay each month and how much less will you pay in total interest?
10. Rework Exercise 6 assuming you decide to pay off the balance over one year rather than two. How much more must you pay each month and how much less will you pay in total interest?

Explaining the Concepts

11. Describe the difference between a fixed installment loan and an open-end installment loan.
12. For a credit card billing period, describe how the average daily balance is determined. Why is this computation somewhat tedious when done by hand?
13. Describe two advantages of using credit cards.
14. Describe two disadvantages of using credit cards.
15. What is a debit card?
16. Describe what is contained in a credit report.
17. What are credit scores?
18. Describe two aspects of responsible credit card use.

Critical Thinking Exercises

Make Sense? *In Exercises 19–25, determine whether each statement makes sense or does not make sense, and explain your reasoning.*

19. I like to keep all my money, so I pay only the minimum required payment on my credit card.
20. One advantage of using credit cards is that there are caps on interest rates and fees.
21. The balance due on my credit card from last month to this month increased even though I made no new purchases.
22. My debit card offers the convenience of making purchases with a piece of plastic without the ability to run up credit-card debt.
23. My credit score is 630, so I anticipate an offer on a car loan at a very low interest rate.
24. In order to achieve financial success, I should consistently pay my entire credit card balance by the due date.
25. As a college student, I'm frequently enticed to sign up for credit cards with offers of free T-shirts or iPods.
26. A bank bills its credit card holders on the first of each month for each itemized billing. The card provides a 20-day period in which to pay the bill before charging interest. If the card holder wants to buy an expensive gift for a September 30 wedding but can't pay for it until November 5, explain how this can be done without adding an interest charge.

Group Exercises

27. **Cellphone Plans**

    If credit cards can cause financial woes, cellphone plans are not far behind. Group members should present a report on cellphone plans, addressing each of the following questions: What are the monthly fees for these plans and what features are included? What happens if you use the phone more than the plan allows? Are there higher rates for texting and Internet access? What additional charges are imposed by the carrier on top of the monthly fee? What are the termination fees if you default on the plan? What can happen to your credit report and your credit score in the event of early termination? Does the carrier use free T-shirts, phones, and other items to entice new subscribers into binding contracts? What suggestions can the group offer to avoid financial difficulties with these plans?
28. **Risky Credit Arrangements**

    Group members should present a report on the characteristics and financial risks associated with payday lending, tax refund loans, and pawn shops.

Chapter Summary, Review, and Test

SUMMARY – DEFINITIONS AND CONCEPTS	EXAMPLES
8.1 Percent, Sales Tax, and Discounts	
a. Percent means per hundred. Thus, $97\% = \frac{97}{100}$.	
b. To express a fraction as a percent, divide the numerator by the denominator, move the decimal point in the quotient two places to the right, and add a percent sign.	Ex. 1, p. 494
c. To express a decimal number as a percent, move the decimal point two places to the right and add a percent sign.	Ex. 2, p. 495
d. To express a percent as a decimal number, move the decimal point two places to the left and remove the percent sign.	Ex. 3, p. 495
e. The percent formula, $A = PB$, means A is P percent of B.	
f. Sales tax amount = tax rate × item's cost	Ex. 4, p. 496
g. Discount amount = discount rate × original price	Ex. 5, p. 496
h. The fraction for percent increase (or decrease) is $\frac{\text{amount of increase (or decrease)}}{\text{original amount}}$. Find the percent increase (or decrease) by expressing this fraction as a percent.	Ex. 6, p. 497; Ex. 7, p. 498; Ex. 8, p. 499; Ex. 9, p. 499
8.2 Income Tax	
a. Calculating Income Tax **1.** Determine adjusted gross income: Adjusted gross income = Gross income − Adjustments. **2.** Determine taxable income: Taxable income = Adjusted gross income − (Exemptions + Deductions). **3.** Determine the income tax: Income tax = Tax computation − Tax credits. See details in the box on pages 506–507.	Ex. 1, p. 504; Ex. 2, p. 507
b. FICA taxes are used for Social Security and Medicare benefits. FICA tax rates are given in Table 8.2 on page 509.	Ex. 3, p. 509
c. Gross pay is your salary prior to any withheld taxes for a pay period. Net pay is the actual amount of your check after taxes have been withheld.	Ex. 4, p. 510
8.3 Simple Interest	
a. Interest is the amount of money that we get paid for lending or investing money, or that we pay for borrowing money. The amount deposited or borrowed is the principal. The charge for interest, given as a percent, is the rate, assumed to be per year.	
b. Simple interest involves interest calculated only on the principal and is computed using $I = Prt$.	Ex. 1, p. 514; Ex. 2, p. 515
c. The future value, A, of P dollars at simple interest rate r for t years is $A = P(1 + rt)$.	Ex. 3, p. 516; Ex. 4, p. 516; Ex. 5, p. 517; Ex. 6, p. 517
8.4 Compound Interest	
a. Compound interest involves interest computed on the original principal as well as on any accumulated interest. The amount in an account for one compounding period per year is $A = P(1 + r)^t$. For n compounding periods per year, the amount is $A = P\left(1 + \frac{r}{n}\right)^{nt}$. For continuous compounding, the amount is $A = Pe^{rt}$, where $e \approx 2.72$.	Ex. 1, p. 520; Ex. 2, p. 521; Ex. 3, p. 522

b. Calculating Present Value Ex. 4, p. 523

If A dollars are to be accumulated in t years in an account that pays rate r compounded n times per year, then the present value, P, that needs to be invested now is given by

$$P = \frac{A}{\left(1 + \frac{r}{n}\right)^{nt}}.$$

c. Effective Annual Yield Ex. 5, p. 524; Ex. 6, p. 525

Effective annual yield is defined in the box on page 524. The effective annual yield, Y, for an account that pays rate r compounded n times per year is given by

$$Y = \left(1 + \frac{r}{n}\right)^{n} - 1.$$

8.5 Annuities, Methods of Saving, and Investments

a. An annuity is a sequence of equal payments made at equal time periods. The value of an annuity is the sum of all deposits plus all interest paid. Ex. 1, p. 530

b. The value of an annuity after t years is Ex. 2, p. 531; Ex. 3, p. 533

$$A = \frac{P\left[\left(1 + \frac{r}{n}\right)^{nt} - 1\right]}{\left(\frac{r}{n}\right)},$$

where interest is compounded n times per year. See the box on page 532.

c. The formula Ex. 4, p. 534

$$P = \frac{A\left(\frac{r}{n}\right)}{\left[\left(1 + \frac{r}{n}\right)^{nt} - 1\right]}$$

gives the deposit, P, into an annuity at the end of each compounding period needed to achieve a value of A dollars after t years. See the box on page 534.

d. The return on an investment is the percent increase in the investment.

e. Investors purchase stock, shares of ownership in a company. The shares indicate the percent of ownership. Trading refers to buying and selling stock. Investors make money by selling a stock for more money than they paid for it. They can also make money while they own stock if a company distributes all or part of its profits as dividends. Each share of stock is paid the same dividend.

f. Investors purchase a bond, lending money to the company from which they purchase the bond. The company commits itself to pay the price an investor pays for the bond at the time it was purchased, called its face value, along with interest payments at a given rate.

A listing of all the investments a person holds is called a financial portfolio. A portfolio with a mixture of low-risk and high-risk investments is called a diversified portfolio.

A mutual fund is a group of stocks and/or bonds managed by a professional investor, called the fund manager.

g. Reading stock tables is explained on pages 536–537. Ex. 5, p. 538

h. Accounts designed for retirement savings include traditional IRAs (requires paying taxes when withdrawing money at age $59\frac{1}{2}$ or older), Roth IRAs (requires paying taxes on deposits, but not withdrawals/earnings at age $59\frac{1}{2}$ or older), and employer-sponsored plans, such as 401(k) and 403(b) plans. Ex. 6, p. 540

8.6 Cars

a. A fixed installment loan is paid off with a series of equal periodic payments. A car loan is an example of a fixed installment loan. Ex. 1, p. 546; Ex. 2, p. 549

Loan Payment Formula for Fixed Installment Loans

$$PMT = \frac{P\left(\frac{r}{n}\right)}{\left[1 - \left(1 + \frac{r}{n}\right)^{-nt}\right]}$$

PMT is the regular payment amount required to repay a loan of P dollars paid n times per year over t years at an annual interest rate r.

b. Leasing is a long-term rental agreement. Leasing a car offers some advantages over buying one (small down payment; lower monthly payments; no concerns about selling the car when the lease ends). Leasing also offers some disadvantages over buying (mileage penalties and other costs at the end of the leasing period often make the total cost of leasing more expensive than buying; penalties for ending the lease early; not owning the car when the lease ends; liability for damage to the car). Ex. 3, p. 550

Auto insurance includes liability coverage: Bodily injury liability covers the costs of lawsuits if someone is injured or killed in an accident in which you are at fault. Property damage liability covers damage to other cars and property from negligent operation of your vehicle. If you have a car loan or lease a car, you also need collision coverage (pays for damage or loss of your car if you're in an accident) and comprehensive coverage (protects your car from fire, theft, and acts of nature).

The Cost of Gasoline

$$\text{Annual fuel expense} = \frac{\text{annual miles driven}}{\text{miles per gallon}} \times \text{price per gallon}$$

8.7 The Cost of Home Ownership

a. A mortgage is a long-term loan for the purpose of buying a home, and for which the property is pledged as security for payment. The term of the mortgage is the number of years until final payoff. The down payment is the portion of the sale price of the home that the buyer initially pays. The amount of the mortgage is the difference between the sale price and the down payment.

b. Fixed-rate mortgages have the same monthly payment during the entire time of the loan. Variable-rate mortgages, or adjustable-rate mortgages, have payment amounts that change from time to time depending on changes in the interest rate.

c. A point is a one-time charge that equals 1% of the amount of a mortgage loan.

d. The loan payment formula for fixed installment loans can be used to determine the monthly payment for a mortgage. Ex. 1, p. 555

e. Amortizing a mortgage loan is the process of making regular payments on the principal and interest until the loan is paid off. A loan amortization schedule is a document showing the following information for each mortgage payment: the payment number, the interest paid from the payment, the amount of the payment applied to the principal, and the balance of the loan after the payment. Such a schedule shows how the buyer pays slightly less in interest and more in principal for each payment over the entire life of the loan. Ex. 2, p. 557

f. Here are the guidelines for what you can spend for a mortgage: Ex. 3, p. 559

- Spend no more than 28% of your gross monthly income for your mortgage payment.
- Spend no more than 36% of your gross monthly income for your total monthly debt.

g. Home ownership provides significant tax advantages, including deduction of mortgage interest and property taxes. Renting does not provide tax benefits, although renters do not pay property taxes.

8.8 Credit Cards

a. A fixed installment loan is paid off with a series of equal periodic payments. An open-end installment loan is paid off with variable monthly payments. Credit card loans are open-end installment loans.

Most credit cards calculate interest using the average daily balance method. Interest is calculated using $I = Prt$, where P is the average daily balance, r is the monthly rate, and t is one month. Ex. 1, p. 564

$$\text{Average daily balance} = \frac{\text{Sum of the unpaid balances for each day in the billing period}}{\text{Number of days in the billing period}}$$

The steps needed to determine the average daily balance are given in the box on page 566.

b. One advantage of using a credit card is that there are no interest charges by paying the balance due at the end of each billing period. A disadvantage is the high interest rate on unpaid balances. Failing to pay the bill in full each month can result in serious debt.

c. The difference between debit cards and credit cards is that debit cards are linked to your bank account. When you use a debit card, the money you spend is deducted electronically from your account balance.

d. Credit reports include details about all open or closed credit accounts, such as when each account was opened, the latest balance, and the payment history. They also contain bankruptcy information and information about unpaid accounts that were turned over to collection agencies. Credit scores range from 300 to 850, with a higher score indicating better credit.

PERSONAL FINANCE FORMULAS

Simple Interest

$$I = Prt$$
$$A = P(1 + rt)$$

Compound Interest

$$A = P\left(1 + \frac{r}{n}\right)^{nt}$$

$$P = \frac{A}{\left(1 + \frac{r}{n}\right)^{nt}}$$

$$Y = \left(1 + \frac{r}{n}\right)^{n} - 1$$

Annuities

$$A = \frac{P\left[\left(1 + \frac{r}{n}\right)^{nt} - 1\right]}{\left(\frac{r}{n}\right)}$$

$$P = \frac{A\left(\frac{r}{n}\right)}{\left[\left(1 + \frac{r}{n}\right)^{nt} - 1\right]}$$

Amortization

$$PMT = \frac{P\left(\frac{r}{n}\right)}{\left[1 - \left(1 + \frac{r}{n}\right)^{-nt}\right]}$$

Be sure you understand what each formula in the box describes and the meaning of the variables in the formulas. Select the appropriate formula or formulas as you work the exercises in the Review Exercises and the Chapter 8 Test.

Review Exercises

8.1

In Exercises 1–3, express each fraction as a percent.

1. $\frac{4}{5}$ **2.** $\frac{1}{8}$ **3.** $\frac{3}{4}$

In Exercises 4–6, express each decimal as a percent.

4. 0.72 **5.** 0.0035 **6.** 4.756

In Exercises 7–12, express each percent as a decimal.

7. 65% **8.** 99.7% **9.** 150%

10. 3% **11.** 0.65% **12.** $\frac{1}{4}$%

13. What is 8% of 120?

14. Suppose that the local sales-tax rate is 6% and you purchase a backpack for $24.
- **a.** How much tax is paid?
- **b.** What is the backpack's total cost?

15. A television with an original price of $850 is on sale at 35% off.
- **a.** What is the discount amount?
- **b.** What is the television's sale price?

16. A college that had 40 students for each lecture course increased the number to 45 students. What is the percent increase in the number of students in a lecture course?

17. A dictionary regularly sells for $56.00. The sale price is $36.40. Find the percent decrease of the sale price from the regular price.

18. Consider the following statement:

> My investment portfolio fell 10% last year, but then it rose 10% this year, so at least I recouped my losses.

Is this statement true? In particular, suppose you invested $10,000 in the stock market last year. How much money would be left in your portfolio with a 10% fall and then a 10% rise? If there is a loss, what is the percent decrease, to the nearest tenth of a percent, in your portfolio?

8.2

In Exercises 19–20, find the gross income, the adjusted gross income, and the taxable income. In Exercise 20, base the taxable income on the greater of a standard deduction or an itemized deduction.

19. Your neighbor earned wages of $30,200, received $130 in interest from a savings account, and contributed $1100 to a tax-deferred retirement plan. He was entitled to a personal exemption of $4050 and had deductions totaling $8450.

20. Your neighbor earned wages of $86,400, won $350,000 on a television game show, and contributed $50,000 to a tax-deferred savings plan. She is entitled to a personal exemption of $4050 and a standard deduction of $6300. The interest on her home mortgage was $9200 and she contributed $95,000 to charity.

*In Exercises 21–22, use the 2016 marginal tax rates in **Table 8.1** on page 507 to compute the tax owed by each person or couple.*

21. A single woman with a taxable income of $600,000

22. A married couple filing jointly with a taxable income of $82,000 and a $7500 tax credit

23. Use the 2016 marginal tax rates in **Table 8.1** to calculate the income tax owed by the following person:
- Single, no dependents
- Gross income: $40,000
- $2500 paid to a tax-deferred IRA
- $6500 mortgage interest
- $1800 property taxes
- No tax credits

*Use the 2016 FICA tax rates in **Table 8.2** on page 509 to solve Exercises 24–25.*

24. If you are not self-employed and earn $86,000, what are your FICA taxes?

25. If you are self-employed and earn $260,000, what are your FICA taxes?

26. You decide to work part-time at a local clothing store. The job pays $14.00 per hour and you work 16 hours per week. Your employer withholds 10% of your gross pay for federal taxes, 7.65% for FICA taxes, and 4% for state taxes.
a. What is your weekly gross pay?
b. How much is withheld per week for federal taxes?
c. How much is withheld per week for FICA taxes?
d. How much is withheld per week for state taxes?
e. What is your weekly net pay?
f. What percentage of your gross pay is withheld for taxes? Round to the nearest tenth of a percent.

8.3

In Exercises 27–30, find the simple interest. (Assume 360 days in a year.)

	Principal	Rate	Time
27.	$6000	3%	1 year
28.	$8400	5%	6 years
29.	$20,000	8%	9 months
30.	$36,000	15%	60 days

31. In order to pay for tuition and books, a college student borrows $3500 for four months at 10.5% interest.
a. How much interest must the student pay?
b. Find the future value of the loan.

In Exercises 32–34, use the formula for future value with simple interest to find the missing quantity. Round dollar amounts to the nearest cent and rates to the nearest tenth of a percent.

32. $A = ?, P = \$12{,}000, r = 8.2\%, t = 9$ months

33. $A = \$5750, P = \$5000, r = ?, t = 2$ years

34. $A = \$16{,}000, P = ?, r = 6.5\%, t = 3$ years

35. You plan to buy a $12,000 sailboat in four years. How much should you invest now, at 7.3% simple interest, to have enough for the boat in four years? (Round up to the nearest cent.)

36. You borrow $1500 from a friend and promise to pay back $1800 in six months. What simple interest rate will you pay?

8.4

In Exercises 37–39, the principal represents an amount of money deposited in a savings account that provides the lender compound interest at the given rate.

a. *Find how much money, to the nearest cent, there will be in the account after the given number of years.*

b. *Find the interest earned.*

	Principal	Rate	Compounding Periods per Year	Time
37	$7000	3%	1	5 years
38	$30,000	2.5%	4	10 years
39	$2500	4%	12	20 years

40. Suppose that you have $14,000 to invest. Which investment yields the greater return over 10 years: 7% compounded monthly or 6.85% compounded continuously? How much more (to the nearest dollar) is yielded by the better investment?

In Exercises 41–42, round answers up to the nearest cent.

41. How much money should parents deposit today in an account that earns 7% compounded monthly so that it will accumulate to $100,000 in 18 years for their child's college education?

42. How much money should be deposited today in an account that earns 5% compounded quarterly so that it will accumulate to $75,000 in 35 years for retirement?

43. You deposit $2000 in an account that pays 6% interest compounded quarterly.
a. Find the future value, to the nearest cent, after one year.
b. Use the future value formula for simple interest to determine the effective annual yield. Round to the nearest tenth of a percent.

44. What is the effective annual yield, to the nearest hundredth of a percent, of an account paying 5.5% compounded quarterly? What does your answer mean?

45. Which investment is the better choice: 6.25% compounded monthly or 6.3% compounded annually?

8.5

In Exercises 46–48, round the value of each annuity to the nearest dollar.

46. A person who does not understand probability theory (see Chapter 11) wastes $10 per week on lottery tickets, averaging $520 per year. Instead of buying tickets, if this person deposits the $520 at the end of each year in an annuity paying 6% compounded annually,
a. How much would he or she have after 20 years?
b. Find the interest.

47. To save for retirement, you decide to deposit $100 at the end of each month in an IRA that pays 5.5% compounded monthly.
a. How much will you have from the IRA after 30 years?
b. Find the interest.

48. Suppose that you would like to have \$25,000 to use as a down payment for a home in five years by making regular deposits at the end of every three months in an annuity that pays 7.25% compounded quarterly.

a. Determine the amount of each deposit. Round up to the nearest dollar.

b. How much of the \$25,000 comes from deposits and how much comes from interest?

For Exercises 49–56, refer to the stock table for Harley Davidson (the motorcycle company). Where necessary, round dollar amounts to the nearest cent.

52-Week High	52-Week Low	Stock	SYM	Div	Yld %	PE
64.06	26.13	Harley Dav	HOG	.16	.3	41

Vol 100s	Hi	Lo	Close	Net Chg
5458	61.25	59.25	61	+1.75

49. What were the high and low prices for a share for the past 52 weeks?

50. If you owned 900 shares of this stock last year, what dividend did you receive?

51. What is the annual return for the dividends alone?

52. How many shares of this company's stock were traded yesterday?

53. What were the high and low prices for a share yesterday?

54. What was the price at which a share last traded when the stock exchange closed yesterday?

55. What was the change in price for a share of stock from the market close two days ago to yesterday's market close?

56. Compute the company's annual earnings per share using

$$\frac{\text{Yesterday's closing price per share}}{\text{PE ratio}}.$$

57. Explain the difference between investing in a stock and investing in a bond.

58. What is the difference between tax benefits for a traditional IRA and a Roth IRA?

8.6

59. Suppose that you decide to take a \$15,000 loan for a new car. You can select one of the following loans, each requiring regular monthly payments:

Loan A: three-year loan at 7.2%

Loan B: five-year loan at 8.1%.

a. Find the monthly payments and the total interest for Loan A.

b. Find the monthly payments and the total interest for Loan B.

c. Compare the monthly payments and interest for the longer-term loan to the monthly payments and interest for the shorter-term loan.

60. Describe two advantages of leasing a car.

61. Describe two disadvantages of leasing a car.

62. Two components of auto insurance are property damage liability and collision. What is the difference between these types of coverage?

63. Suppose that you drive 36,000 miles per year and gas averages \$3.60 per gallon.

a. What will you save in annual fuel expenses by owning a hybrid car averaging 40 miles per gallon rather than an SUV averaging 12 miles per gallon?

b. If you deposit your monthly fuel savings at the end of each month into an annuity that pays 5.2% compounded monthly, how much will you have saved at the end of six years? Round all computations to the nearest dollar.

8.7

In Exercises 64–66, round to the nearest dollar.

64. The price of a home is \$240,000. The bank requires a 20% down payment and two points at the time of closing. The cost of the home is financed with a 30-year fixed-rate mortgage at 7%.

a. Find the required down payment.

b. Find the amount of the mortgage.

c. How much must be paid for the two points at closing?

d. Find the monthly payment (excluding escrowed taxes and insurance).

e. Find the total cost of interest over 30 years.

65. In terms of paying less in interest, which is more economical for a \$70,000 mortgage: a 30-year fixed-rate at 8.5% or a 20-year fixed-rate at 8%? How much is saved in interest? Discuss one advantage and one disadvantage for each mortgage option.

66. Suppose that you need a loan of \$100,000 to buy a home. Here are your options:

Option A: 30-year fixed-rate at 8.5% with no closing costs and no points

Option B: 30-year fixed-rate at 7.5% with closing costs of \$1300 and three points.

a. Determine your monthly payments for each option and discuss how you would decide between the two options.

b. Which mortgage loan has the greater total cost (closing costs + the amount paid for points + total cost of interest)? By how much?

67. The cost of a home is financed with a \$300,000 30-year fixed rate mortgage at 6.5%.

a. Find the monthly payments, rounded to the nearest dollar, for the loan.

b. Prepare a loan amortization schedule for the first three months of the mortgage. Round entries to the nearest cent.

Payment Number	Interest	Principal	Loan Balance
1			
2			
3			

68. Use these guidelines to solve this exercise: Spend no more than 28% of your gross monthly income for your mortgage payment and no more than 36% for your total monthly debt. Round all computations to the nearest dollar. Suppose that your gross annual income is $54,000.

a. What is the maximum amount you should spend each month on a mortgage payment?

b. What is the maximum amount you should spend each month for total credit obligations?

c. If your monthly mortgage payment is 80% of the maximum amount you can afford, what is the maximum amount you should spend each month for all other debt?

69. Describe three benefits of renting over home ownership.

70. Describe three benefits of home ownership over renting.

8.8

71. A credit card issuer calculates interest using the average daily balance method. The monthly interest rate is 1.1% of the average daily balance. The following transactions occurred during the November 1–November 30 billing period.

Transaction Description		Transaction Amount
Previous balance, $4620.80		
November 1	Billing date	
November 7	Payment	$650.00 credit
November 11	Charge: Airline Tickets	$350.25
November 25	Charge: Groceries	$125.70
November 28	Charge: Gas	$ 38.25
November 30	End of billing period	
Payment Due Date: December 9		

a. Find the average daily balance for the billing period. Round to the nearest cent.

b. Find the interest to be paid on December 1, the next billing date. Round to the nearest cent.

c. Find the balance due on December 1.

d. This credit card requires a $10 minimum monthly payment if the balance due at the end of the billing period is less than $360. Otherwise, the minimum monthly payment is $\frac{1}{36}$ of the balance due at the end of the billing period, rounded up to the nearest whole dollar. What is the minimum monthly payment due by December 9?

72. In 2016, the average credit-card debt was $16,748. Suppose your card has this balance and an annual interest rate of 18%. You decide to pay off the balance over two years. If there are no further purchases charged to the card,

a. How much must you pay each month?

b. How much total interest will you pay?

Round answers to the nearest dollar.

73. Describe two advantages of using credit cards.

74. Describe two disadvantages of using credit cards.

75. How does a debit card differ from a credit card?

76. Is a credit report the same thing as a credit score? If not, what is the difference?

77. Describe two ways to demonstrate that you can handle the responsibility of using a credit card.

Chapter 8 Test

The box on page 575 summarizes the finance formulas you have worked with throughout the chapter. Where applicable, use the appropriate formula to solve an exercise in this test. Unless otherwise stated, round dollar amounts to the nearest cent and rates to the nearest tenth of a percent.

1. A CD player with an original price of $120 is on sale at 15% off.

a. What is the amount of the discount?

b. What is the sale price of the CD player?

2. You purchased shares of stock for $2000 and sold them for $3500. Find the percent increase, or your return, on this investment.

3. You earned wages of $46,500, received $790 in interest from a savings account, and contributed $1100 to a tax-deferred savings plan. You are entitled to a personal exemption of $4050 and a standard deduction of $6300. The interest on your home mortgage was $7300, you contributed $350 to charity, and you paid $1395 in state taxes.

a. Find your gross income.

b. Find your adjusted gross income.

c. Find your taxable income. Base your taxable income on the greater of the standard deduction or an itemized deduction.

4. Use the 2016 marginal tax rates in **Table 8.1** on page 507 to calculate the federal income tax owed by the following person:

- Single, no dependents
- Gross income: $36,500
- $2000 paid to a tax-deferred IRA
- $4700 mortgage interest
- $1300 property taxes
- No tax credits

5. Use FICA tax rates for people who are not self-employed, 7.65% on the first $118,500 of income and 1.45% on income in excess of $118,500, to answer this question: If a person is not self-employed and earns $150,000, what are that person's FICA taxes?

6. You decide to work part-time at a local stationery store. The job pays $10 per hour and you work 15 hours per week. Your employer withholds 10% of your gross pay for federal taxes, 7.65% for FICA taxes, and 3% for state taxes.
   a. What is your weekly gross pay?
   b. How much is withheld per week for federal taxes?
   c. How much is withheld per week for FICA taxes?
   d. How much is withheld per week for state taxes?
   e. What is your weekly net pay?
   f. What percentage of your gross pay is withheld for taxes? Round to the nearest tenth of a percent.
7. You borrow $2400 for three months at 12% simple interest. Find the amount of interest paid and the future value of the loan.
8. You borrow $2000 from a friend and promise to pay back $3000 in two years. What simple interest rate will you pay?
9. In six months, you want to have $7000 worth of remodeling done to your home. How much should you invest now, at 9% simple interest, to have enough money for the project? (Round up to the nearest cent.)
10. Find the effective annual yield, to the nearest hundredth of a percent, of an account paying 4.5% compounded quarterly. What does your answer mean?
11. You receive an inheritance of $20,000 and invest it in an account that pays 6.5% compounded monthly.
    a. How much, to the nearest dollar, will you have after 40 years?
    b. Find the interest.
12. You would like to have $3000 in four years for a special vacation by making a lump-sum investment in an account that pays 9.5% compounded semiannually. How much should you deposit now? Round up to the nearest dollar.
13. Suppose that you save money for a down payment to buy a home in five years by depositing $6000 in an account that pays 6.5% compounded monthly.
    a. How much, to the nearest dollar, will you have as a down payment after five years?
    b. Find the interest.
14. Instead of making the lump-sum deposit of $6000 described in Exercise 13, suppose that you decide to deposit $100 at the end of each month in an annuity that pays 6.5% compounded monthly.
    a. How much, to the nearest dollar, will you have as a down payment after five years?
    b. Find the interest.
    c. Why is less interest earned from this annuity than from the lump-sum deposit in Exercise 13? With less interest earned, why would one select the annuity rather than the lump-sum deposit?
15. Suppose that you want to retire in 40 years. How much should you deposit at the end of each month in an IRA that pays 6.25% compounded monthly to have $1,500,000 in 40 years? Round up to the nearest dollar. How much of the $1.5 million comes from interest?

Use the stock table for AT&T to solve Exercises 16–18.

52-Week High	52-Week Low	Stock	SYM	Div	Yld %	PE
26.50	24.25	AT&T	PNS	2.03	7.9	18

Vol 100s	Hi	Lo	Close	Net Chg
961	25.75	25.50	25.75	+0.13

16. What were the high and low prices for a share yesterday?
17. If you owned 1000 shares of this stock last year, what dividend did you receive?
18. Suppose that you bought 600 shares of AT&T, paying the price per share at which a share traded when the stock exchange closed yesterday. If the broker charges 2.5% of the price paid for all 600 shares, find the broker's commission.
19. Suppose that you drive 30,000 miles per year and gas averages $3.80 per gallon. What will you save in annual fuel expense by owning a hybrid car averaging 50 miles per gallon rather than a pickup truck averaging 15 miles per gallon?

Use this information to solve Exercises 20–25. The price of a home is $120,000. The bank requires a 10% down payment and two points at the time of closing. The cost of the home is financed with a 30-year fixed-rate mortgage at 8.5%.

20. Find the required down payment.
21. Find the amount of the mortgage.
22. How much must be paid for the two points at closing?
23. Find the monthly payment (excluding escrowed taxes and insurance). Round to the nearest dollar.
24. Find the total cost of interest over 30 years.
25. Prepare a loan amortization schedule for the first two months of the mortgage. Round entries to the nearest cent.

Payment Number	Interest	Principal	Loan Balance
1			
2			

26. Use these guidelines to solve this exercise. Spend no more than 28% of your gross monthly income for your mortgage payment and no more than 36% for your total monthly debt. Round all computations to the nearest dollar. Suppose that your gross annual income is $66,000.
    a. What is the maximum amount you should spend each month on a mortgage payment?
    b. What is the maximum amount you should spend each month for total credit obligations?
    c. If your monthly mortgage payment is 90% of the maximum amount you can afford, what is the maximum amount you should spend each month for all other debt?

27. A credit card issuer calculates interest using the average daily balance method. The monthly interest rate is 2% of the average daily balance. The following transactions occurred during the September 1–September 30 billing period.

Transaction Description		Transaction Amount
Previous balance, $3800.00		
September 1	Billing date	
September 5	Payment	$800.00 credit
September 9	Charge: Gas	$ 40.00
September 19	Charge: Clothing	$160.00
September 27	Charge: Airline Ticket	$200.00
September 30	End of billing period	
Payment Due Date: October 9		

a. Find the average daily balance for the billing period. Round to the nearest cent.

b. Find the interest to be paid on October 1, the next billing date. Round to the nearest cent.

c. Find the balance due on October 1.

d. Terms for the credit card require a $10 minimum monthly payment if the balance due is less than $360. Otherwise, the minimum monthly payment is $\frac{1}{36}$ of the balance due, rounded up to the nearest whole dollar. What is the minimum monthly payment due by October 9?

In Exercises 28–34, determine whether each statement is true or false. If the statement is false, make the necessary change(s) to produce a true statement.

28. By buying bonds, you purchase shares of ownership in a company.

29. A traditional IRA requires paying taxes when withdrawing money from the account at age $59\frac{1}{2}$ or older.

30. One advantage to leasing a car is that there are no penalties for ending the lease early.

31. If you cause an accident, collision coverage pays for damage to the other car.

32. Home ownership provides significant tax advantages, including deduction of mortgage interest.

33. Money spent using a credit card is deducted electronically from your bank account.

34. Credit scores range from 100 to 1000, with a higher score indicating better credit.

Measurement

YOU ARE FEELING CROWDED IN. PERHAPS IT WOULD BE A GOOD TIME TO MOVE TO A STATE WITH MORE ELBOW ROOM. BUT WHICH STATE? YOU CAN LOOK UP THE POPULATION OF EACH STATE, but that does not take into account the amount of land the population occupies. How is this land measured and how can you use this measure to select a place where wildlife outnumber humans?

In this chapter, we explore ways of measuring things in our English system, as well as in the metric system. Knowing how units of measure are used to describe your world can help you make decisions on issues ranging from where to live to alcohol consumption to determining proper dosages of medication.

Here's where you'll find these applications:

- Finding a state with lots of room to spread out is explored in Example 2 of Section 9.2.
- Measuring alcohol consumption is addressed in the Blitzer Bonus on page 606.
- Dosages of medication form the basis of Example 8 in Section 9.2 and Example 4 in Section 9.3.

9.1 Measuring Length; The Metric System

WHAT AM I SUPPOSED TO LEARN?

After studying this section, you should be able to:

1. Use dimensional analysis to change units of measurement.
2. Understand and use metric prefixes.
3. Convert units within the metric system.
4. Use dimensional analysis to change to and from the metric system.

HAVE YOU SEEN ANY OF THE *JURASSIC PARK* FILMS? The popularity of these movies reflects our fascination with dinosaurs and their incredible size. From end to end, the largest dinosaur from the Jurassic period, which lasted from 208 to 146 million years ago, was about 88 feet. To **measure** an object such as a dinosaur is to assign a number to its size. The number representing its measure from end to end is called its **length**. Measurements are used to describe properties of length, area, volume, weight, and temperature. Over the centuries, people have developed systems of measurement that are now accepted in most of the world.

Length

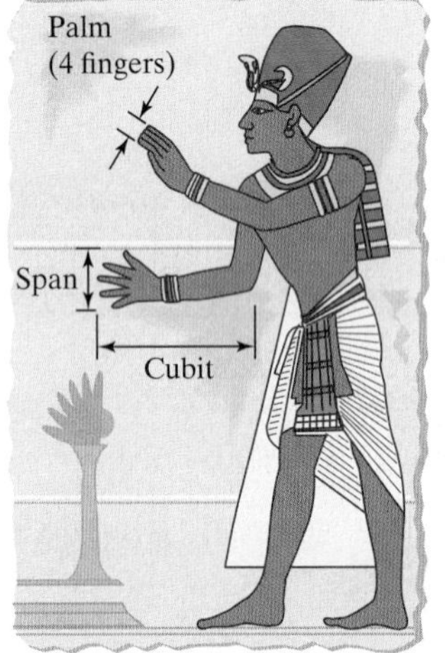

Linear units of measure were originally based on parts of the body. The Egyptians used the palm (equal to four fingers), the span (a handspan), and the cubit (length of forearm).

Every measurement consists of two parts: a number and a unit of measure. For example, if the length of a dinosaur is 88 feet, the number is 88 and the unit of measure is the foot. Many different units are commonly used in measuring length. The foot is from a system of measurement called the **English system**, which is generally used in the United States. In this system of measurement, length is expressed in such units as inches, feet, yards, and miles.

The result obtained from measuring length is called a **linear measurement** and is stated in **linear units**.

LINEAR UNITS OF MEASURE: THE ENGLISH SYSTEM

$$12 \text{ inches (in.)} = 1 \text{ foot (ft)}$$
$$3 \text{ feet} = 1 \text{ yard (yd)}$$
$$36 \text{ inches} = 1 \text{ yard}$$
$$5280 \text{ feet} = 1 \text{ mile (mi)}$$

Because many of us are familiar with the measures in the box, we find it simple to change from one measure to another, say from feet to inches. We know that there are 12 inches in a foot. To convert from 5 feet to a measure in inches, we multiply by 12. Thus, 5 feet $= 5 \times 12$ inches $= 60$ inches.

1 Use dimensional analysis to change units of measurement.

Another procedure used to convert from one unit of measurement to another is called **dimensional analysis**. Dimensional analysis uses *unit fractions*. A **unit fraction** has two properties: The numerator and denominator contain different units and the value of the unit fraction is 1. Here are some examples of unit fractions:

$$\frac{12 \text{ in.}}{1 \text{ ft}}; \quad \frac{1 \text{ ft}}{12 \text{ in.}}; \quad \frac{3 \text{ ft}}{1 \text{ yd}}; \quad \frac{1 \text{ yd}}{3 \text{ ft}}; \quad \frac{5280 \text{ ft}}{1 \text{ mi}}; \quad \frac{1 \text{ mi}}{5280 \text{ ft}}.$$

In each unit fraction, the numerator and denominator are equal measures, making the value of the fraction 1.

Let's see how to convert 5 feet to inches using dimensional analysis.

$$5 \text{ ft} = ? \text{ in.}$$

We need to eliminate feet and introduce inches. The unit we need to introduce, inches, must appear in the numerator of the fraction. The unit we need to eliminate, feet, must appear in the denominator. Therefore, we choose the unit fraction with inches in the numerator and feet in the denominator. The units divide out as follows:

$$5 \text{ ft} = \frac{5 \cancel{\text{ft}}}{1} \cdot \frac{12 \text{ in.}}{1 \cancel{\text{ft}}} = 5 \cdot 12 \text{ in.} = 60 \text{ in.}$$

unit fraction

DIMENSIONAL ANALYSIS

To convert a measurement to a different unit, multiply by a unit fraction (or by unit fractions). The given unit of measurement should appear in the denominator of the unit fraction so that this unit cancels upon multiplication. The unit of measurement that needs to be introduced should appear in the numerator of the fraction so that this unit will be retained upon multiplication.

EXAMPLE 1 *Using Dimensional Analysis to Change Units of Measurement*

Convert:

a. 40 inches to feet **b.** 13,200 feet to miles **c.** 9 inches to yards.

SOLUTION

a. Because we want to convert 40 inches to feet, feet should appear in the numerator and inches in the denominator. We use the unit fraction

$$\frac{1 \text{ ft}}{12 \text{ in.}}$$

and proceed as follows:

This period ends the sentence and is not part of the abbreviated unit.

$$40 \text{ in.} = \frac{40 \cancel{\text{in.}}}{1} \cdot \frac{1 \text{ ft}}{12 \cancel{\text{in.}}} = \frac{40}{12} \text{ ft} = 3\frac{1}{3} \text{ ft or } 3.\overline{3} \text{ ft.}$$

b. To convert 13,200 feet to miles, miles should appear in the numerator and feet in the denominator. We use the unit fraction

$$\frac{1 \text{ mi}}{5280 \text{ ft}}$$

and proceed as follows:

$$13{,}200 \text{ ft} = \frac{13{,}200 \cancel{\text{ft}}}{1} \cdot \frac{1 \text{ mi}}{5280 \cancel{\text{ft}}} = \frac{13{,}200}{5280} \text{mi} = 2\frac{1}{2} \text{ mi or } 2.5 \text{ mi.}$$

c. To convert 9 inches to yards, yards should appear in the numerator and inches in the denominator. We use the unit fraction

$$\frac{1 \text{ yd}}{36 \text{ in.}}$$

and proceed as follows:

$$9 \text{ in.} = \frac{9 \cancel{\text{in.}}}{1} \cdot \frac{1 \text{ yd}}{36 \cancel{\text{in.}}} = \frac{9}{36} \text{yd} = \frac{1}{4} \text{yd or } 0.25 \text{ yd.}$$

GREAT QUESTION!

When should I use a period for abbreviated units of measurement?

The abbreviations for units of measurement are written without a period, such as ft for feet, except for the abbreviation for inches (in.).

In each part of Example 1, we converted from a smaller unit to a larger unit. Did you notice that this results in a smaller number in the converted unit of measure? **Converting to a larger unit always produces a smaller number. Converting to a smaller unit always produces a larger number.**

CHECK POINT 1 Convert:

a. 78 inches to feet

b. 17,160 feet to miles

c. 3 inches to yards.

2 *Understand and use metric prefixes.*

Length and the Metric System

Although the English system of measurement is most commonly used in the United States, most industrialized countries use the metric system of measurement. One of the advantages of the metric system is that units are based on powers of ten, making it much easier than the English system to change from one unit of measure to another.

The basic unit for linear measure in the metric system is the meter (m). A meter is slightly longer than a yard, approximately 39 inches. Prefixes are used to denote a multiple or part of a meter. **Table 9.1** summarizes the more commonly used metric prefixes and their meanings.

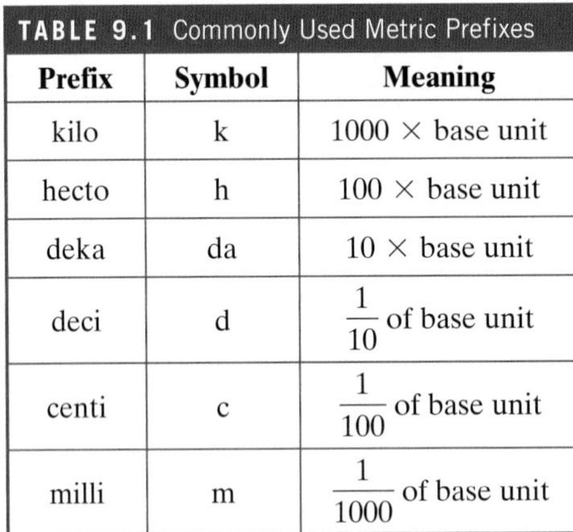

TABLE 9.1 Commonly Used Metric Prefixes

Prefix	Symbol	Meaning
kilo	k	$1000 \times$ base unit
hecto	h	$100 \times$ base unit
deka	da	$10 \times$ base unit
deci	d	$\frac{1}{10}$ of base unit
centi	c	$\frac{1}{100}$ of base unit
milli	m	$\frac{1}{1000}$ of base unit

GREAT QUESTION!

Does the "i" in the prefixes ending in "i" (deci, centi, milli) have any significance?

Prefixes ending in "i" are all fractional parts of one unit.

kilodollar

hectodollar

dekadollar

dollar

decidollar centidollar

Like our system of money, the metric system is based on powers of ten.

The prefixes kilo, centi, and milli are used more frequently than hecto, deka, and deci. **Table 9.2** applies all six prefixes to the meter. The first part of the symbol indicates the prefix and the second part (m) indicates meter.

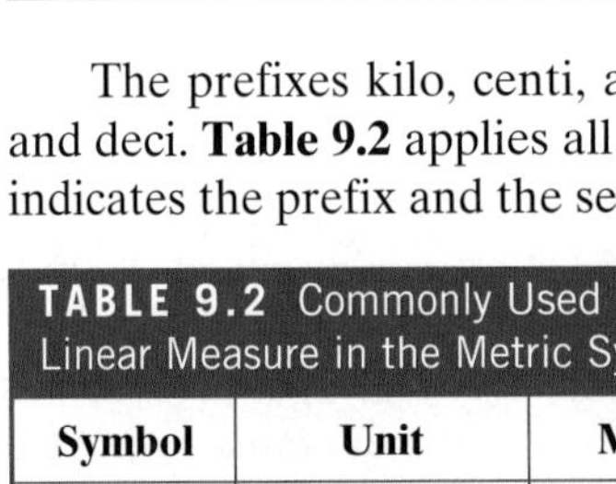

TABLE 9.2 Commonly Used Units of Linear Measure in the Metric System

Symbol	Unit	Meaning
km	kilometer	1000 meters
hm	hectometer	100 meters
dam	dekameter	10 meters
m	meter	1 meter
dm	decimeter	0.1 meter
cm	centimeter	0.01 meter
mm	millimeter	0.001 meter

Kilometer is pronounced kil'-oh-met-er with the accent on the FIRST syllable. If pronounced correctly, kilometers should sound something like "kill all meters."

In the metric system, the kilometer is used to measure distances comparable to those measured in miles in the English system. One kilometer is approximately 0.6 mile, and one mile is approximately 1.6 kilometers.

Metric units of centimeters and millimeters are used to measure what the English system measures in inches. **Figure 9.1** on the next page shows that a centimeter is less than half an inch; there are 2.54 centimeters in an inch. The smaller markings on the bottom scale are millimeters. A millimeter is approximately the thickness of a dime. The length of a bee or a fly may be measured in millimeters.

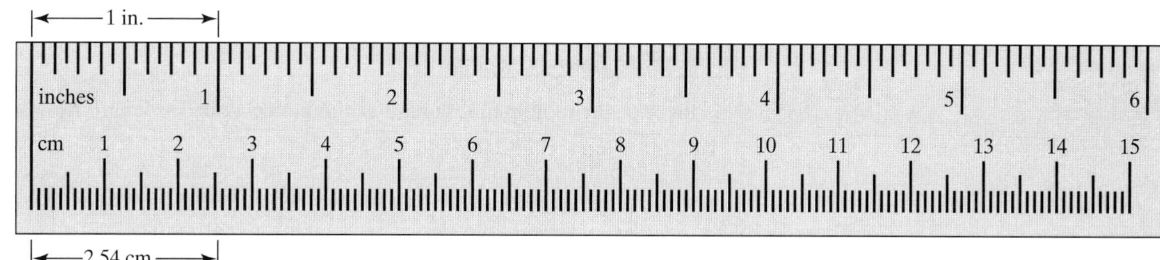

FIGURE 9.1

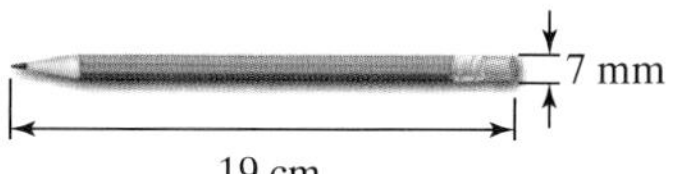

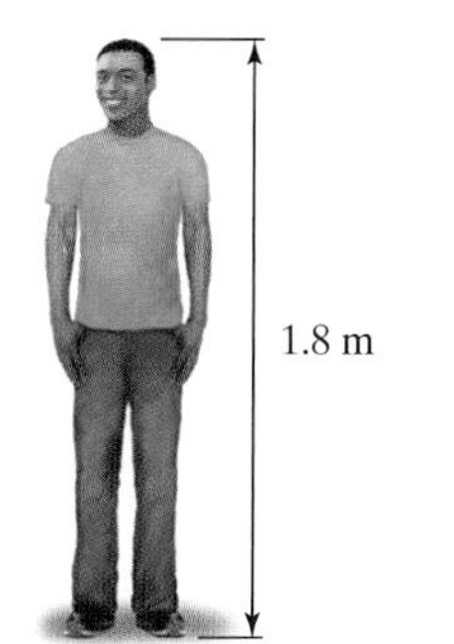

Those of us born in the United States have a good sense of what a length in the English system tells us about an object. An 88-foot dinosaur is huge, about 15 times the height of a 6-foot man. But what sense can we make of knowing that a whale is 25 meters long? The following lengths and the given approximations can help give you a feel for metric units of linear measure.

(1 meter $\approx$ 39 inches 1 kilometer $\approx$ 0.6 mile)

Item	**Approximate Length**
Width of lead in pencil	2 mm or 0.08 in.
Width of an adult's thumb	2 cm or 0.8 in.
Height of adult male	1.8 m or 6 ft
Typical room height	2.5 m or 8.3 ft
Length of medium-size car	5 m or 16.7 ft
Height of Empire State Building	381 m or 1270 ft
Average depth of ocean	4 km or 2.5 mi
Length of Manhattan Island	18 km or 11.25 mi
Distance from New York City to San Francisco	4800 km or 3000 mi
Radius of Earth	6378 km or 3986 mi
Distance from Earth to the Moon	384,401 km or 240,251 mi

3 *Convert units within the metric system.*

Although dimensional analysis can be used to convert from one unit to another within the metric system, there is an easier, faster way to accomplish this conversion. The procedure is based on the observation that successively smaller units involve division by 10 and successively larger units involve multiplication by 10.

CHANGING UNITS WITHIN THE METRIC SYSTEM

Use the following chart to find equivalent measures of length:

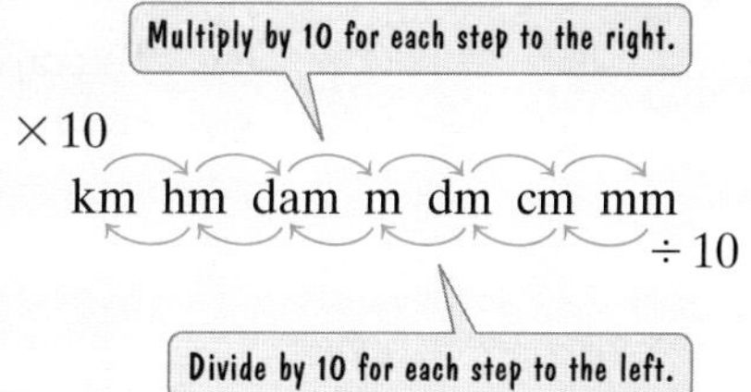

1. To change from a larger unit to a smaller unit (moving to the right in the diagram), multiply by 10 for each step to the right. Thus, move the decimal point in the given quantity one place to the right for each smaller unit until the desired unit is reached.
2. To change from a smaller unit to a larger unit (moving to the left in the diagram), divide by 10 for each step to the left. Thus, move the decimal point in the given quantity one place to the left for each larger unit until the desired unit is reached.

GREAT QUESTION!

Is there a way to help me remember the metric units for length from largest to smallest?

The following sentence should help:

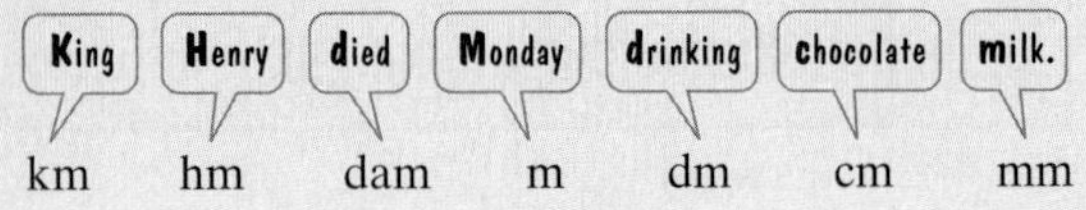

km hm dam m dm cm mm

EXAMPLE 2 Changing Units within the Metric System

a. Convert 504.7 meters to kilometers.

b. Convert 27 meters to centimeters.

c. Convert 704 mm to hm.

d. Convert 9.71 dam to dm.

SOLUTION

a. To convert from meters to kilometers, we start at meters and move three steps to the left to obtain kilometers:

km hm dam m dm cm mm.

Hence, we move the decimal point three places to the left:

$$504.7 \text{ m} = 0.5047 \text{ km}.$$

Thus, 504.7 meters converts to 0.5047 kilometer. Changing from a smaller unit of measurement (meter) to a larger unit of measurement (kilometer) results in an answer with a smaller number of units.

b. To convert from meters to centimeters, we start at meters and move two steps to the right to obtain centimeters:

km hm dam m dm cm mm.

Hence, we move the decimal point two places to the right:

$$27 \text{ m} = 2700 \text{ cm}.$$

Thus, 27 meters converts to 2700 centimeters. Changing from a larger unit of measurement (meter) to a smaller unit of measurement (centimeter) results in an answer with a larger number of units.

c. To convert from mm (millimeters) to hm (hectometers), we start at mm and move five steps to the left to obtain hm:

km hm dam m dm cm mm.

Hence, we move the decimal point five places to the left:

$$704 \text{ mm} = 0.00704 \text{ hm}.$$

d. To convert from dam (dekameters) to dm (decimeters), we start at dam and move two places to the right to obtain dm:

km hm dam m dm cm mm.

Hence, we move the decimal point two places to the right:

$$9.71 \text{ dam} = 971 \text{ dm}.$$

The First Meter

The French first defined the meter in 1791, calculating its length as a romantic one ten-millionth of a line running from the Equator through Paris to the North Pole. Today's meter, officially accepted in 1983, is equal to the length of the path traveled by light in a vacuum during the time interval of 1/299,794,458 of a second.

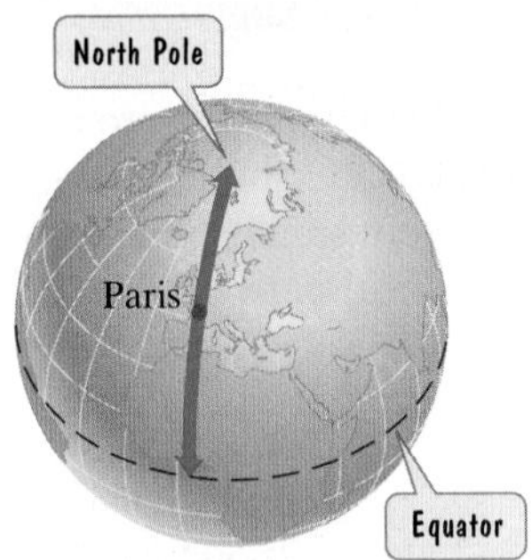

1 meter = 1/10,000,000 of the distance from the North Pole to the Equator on the meridian through Paris.

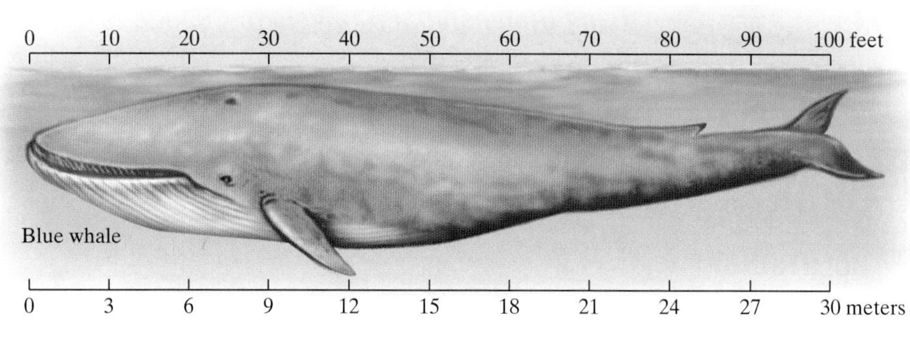

In Example 2(b), we showed that 27 meters converts to 2700 centimeters. This is the average length of the California blue whale, the longest of the great whales. Blue whales can have lengths that exceed 30 meters, making them over 100 feet long.

CHECK POINT 2

a. Convert 8000 meters to kilometers.
b. Convert 53 meters to millimeters.
c. Convert 604 cm to hm.
d. Convert 6.72 dam to cm.

Blitzer Bonus

Viruses and Metric Prefixes

Viruses are measured in attometers. An attometer is one quintillionth of a meter, or 10^{-18} meter, symbolized am. If a virus measures 1 am, you can place 10^{15} of them across a penciled 1 millimeter dot. If you were to enlarge each of these viruses to the size of the dot, they would stretch far into space, almost reaching Saturn.

Here is a list of all 20 metric prefixes. When applied to the meter, they range from the yottameter (10^{24} meters) to the yoctometer (10^{-24} meter).

LARGER THAN THE BASIC UNIT

Prefix	Symbol	Power of Ten	English Name
yotta-	Y	+24	septillion
zetta-	Z	+21	sextillion
exa-	E	+18	quintillion
peta-	P	+15	quadrillion
tera-	T	+12	trillion
giga-	G	+9	billion
mega-	M	+6	million
kilo-	k	+3	thousand
hecto-	h	+2	hundred
deca-	da	+1	ten

SMALLER THAN THE BASIC UNIT

Prefix	Symbol	Power of Ten	English Name
deci-	d	−1	tenth
centi-	c	−2	hundredth
milli-	m	−3	thousandth
micro-	μ	−6	millionth
nano-	n	−9	billionth
pico-	p	−12	trillionth
femto-	f	−15	quadrillionth
atto-	a	−18	quintillionth
zepto-	z	−21	sextillionth
yocto-	y	−24	septillionth

4 *Use dimensional analysis to change to and from the metric system.*

Although dimensional analysis is not necessary when changing units within the metric system, it is a useful tool when converting to and from the metric system. Some conversions are given in **Table 9.3**.

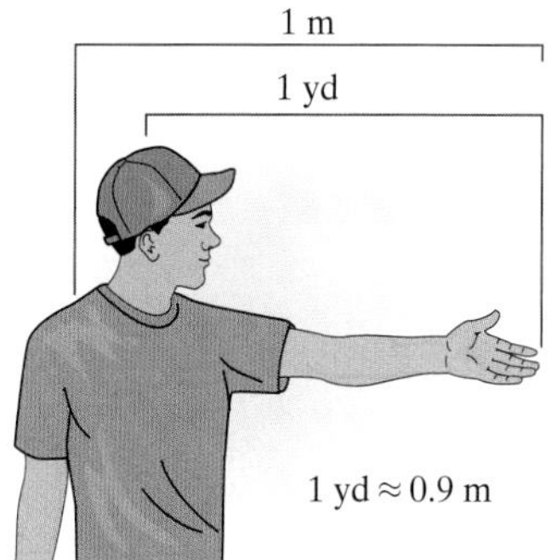

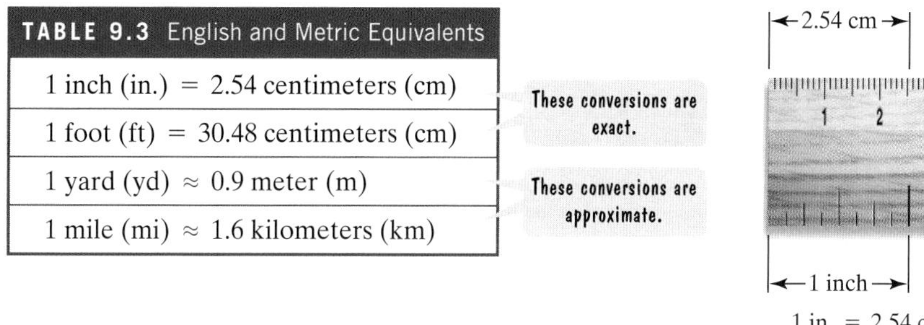

TABLE 9.3 English and Metric Equivalents

1 inch (in.) = 2.54 centimeters (cm)
1 foot (ft) = 30.48 centimeters (cm)
1 yard (yd) ≈ 0.9 meter (m)
1 mile (mi) ≈ 1.6 kilometers (km)

These conversions are exact.

These conversions are approximate.

2.54 cm

1 inch

1 in. = 2.54 cm

TABLE 9.3 English and Metric Equivalents (repeated)
1 inch (in.) = 2.54 centimeters (cm)
1 foot (ft) = 30.48 centimeters (cm)
1 yard (yd) ≈ 0.9 meter (m)
1 mile (mi) ≈ 1.6 kilometers (km)

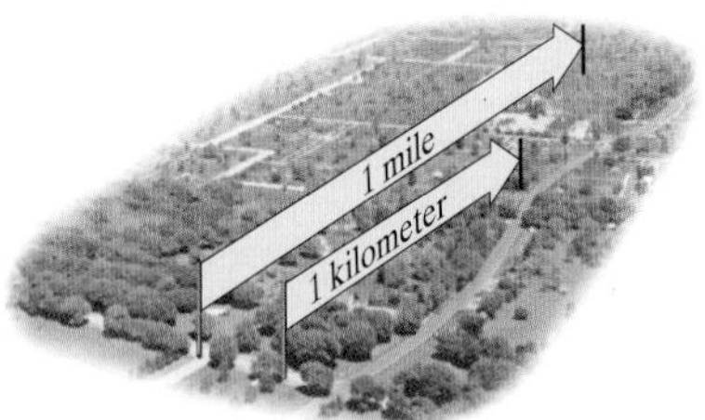

1 mi ≈ 1.6 km

GREAT QUESTION!

Is the conversion in part (b) approximate or exact?

It's approximate. More accurately,

$$1 \text{ mi} \approx 1.61 \text{ km}.$$

Thus,

$$\begin{aligned}125 \text{ mi} &\approx 125(1.61) \text{ km}\\ &\approx 201.25 \text{ km}.\end{aligned}$$

EXAMPLE 3 *Using Dimensional Analysis to Change to and from the Metric System*

a. Convert 8 inches to centimeters.

b. Convert 125 miles to kilometers.

c. Convert 26,800 millimeters to inches.

SOLUTION

a. To convert 8 inches to centimeters, we use a unit fraction with centimeters in the numerator and inches in the denominator:

$$\frac{2.54 \text{ cm}}{1 \text{ in.}}.$$ Table 9.3 shows that 1 in. = 2.54 cm.

We proceed as follows.

$$8 \text{ in.} = \frac{8 \text{ in.}}{1} \cdot \frac{2.54 \text{ cm}}{1 \text{ in.}} = 8(2.54) \text{ cm} = 20.32 \text{ cm}$$

b. To convert 125 miles to kilometers, we use a unit fraction with kilometers in the numerator and miles in the denominator:

$$\frac{1.6 \text{ km}}{1 \text{ mi}}.$$ Table 9.3 shows that 1 mi ≈ 1.6 km.

Thus,

$$125 \text{ mi} \approx \frac{125 \text{ mi}}{1} \cdot \frac{1.6 \text{ km}}{1 \text{ mi}} = 125(1.6) \text{ km} = 200 \text{ km}.$$

c. To convert 26,800 millimeters to inches, we observe that **Table 9.3** has only a conversion factor between inches and centimeters. We begin by changing millimeters to centimeters:

$$26{,}800 \text{ mm} = 2680.0 \text{ cm}$$

Now we need to convert 2680 centimeters to inches. We use a unit fraction with inches in the numerator and centimeters in the denominator:

$$\frac{1 \text{ in.}}{2.54 \text{ cm}}.$$

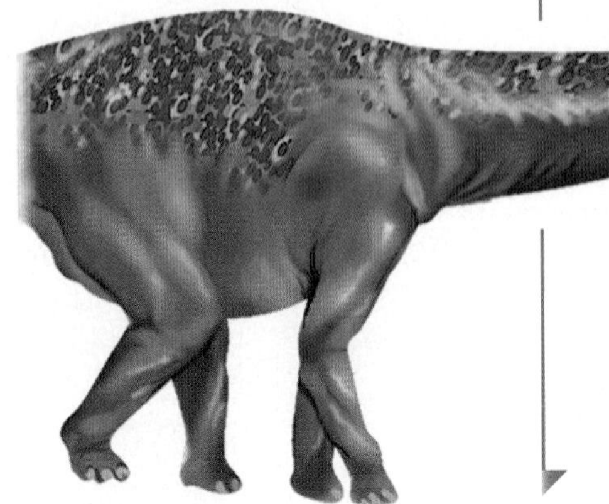

Thus,

$$26{,}800 \text{ mm} = 2680 \text{ cm} = \frac{2680 \text{ cm}}{1} \cdot \frac{1 \text{ in.}}{2.54 \text{ cm}} = \frac{2680}{2.54} \text{ in.} \approx 1055 \text{ in.}$$

This measure is equivalent to about 88 feet, the length of the largest dinosaur from the Jurassic period. The diplodocus, a plant eater, was 26.8 meters, approximately 88 feet, long.

CHECK POINT 3

a. Convert 8 feet to centimeters.

b. Convert 20 meters to yards.

c. Convert 30 meters to inches.

So far, we have used dimensional analysis to change units of length. Dimensional analysis may also be used to convert other kinds of measures, such as speed.

Blitzer Bonus

The Mars Climate Orbiter

If you think that using dimensional analysis to change to and from the metric system is no big deal, consider this: The Mars Climate Orbiter, launched on December 11, 1998, was lost when it crashed into Mars about a year later due to failure to properly convert English units into metric units. The price tag: $125 million!

EXAMPLE 4 *Using Dimensional Analysis*

a. The speed limit on many highways in the United States is 55 miles per hour (mi/hr). How many kilometers per hour (km/hr) is this?

b. If a high-speed train in Japan is capable of traveling at 200 kilometers per hour, how many miles per hour is this?

SOLUTION

a. To change miles per hour to kilometers per hour, we need to concentrate on changing miles to kilometers, so we need a unit fraction with kilometers in the numerator and miles in the denominator:

$$\frac{1.6 \text{ km}}{1 \text{ mi}}. \qquad \text{Table 9.3 shows that } 1 \text{ mi} \approx 1.6 \text{ km.}$$

Thus,

$$\frac{55 \text{ mi}}{\text{hr}} \approx \frac{55 \cancel{\text{mi}}}{\text{hr}} \cdot \frac{1.6 \text{ km}}{1 \cancel{\text{mi}}} = 55(1.6)\frac{\text{km}}{\text{hr}} = 88 \text{ km/hr}.$$

This shows that 55 miles per hour is approximately 88 kilometers per hour.

b. To change 200 kilometers per hour to miles per hour, we must convert kilometers to miles. We need a unit fraction with miles in the numerator and kilometers in the denominator:

$$\frac{1 \text{ mi}}{1.6 \text{ km}}. \qquad \text{Table 9.3 shows that } 1 \text{ mi} \approx 1.6 \text{ km.}$$

Thus,

$$\frac{200 \text{ km}}{\text{hr}} \approx \frac{200 \cancel{\text{km}}}{\text{hr}} \cdot \frac{1 \text{ mi}}{1.6 \cancel{\text{km}}} = \frac{200}{1.6}\frac{\text{mi}}{\text{hr}} = 125 \text{ mi/hr}.$$

A train capable of traveling at 200 kilometers per hour can therefore travel at about 125 miles per hour.

CHECK POINT 4 A road in Europe has a speed limit of 60 kilometers per hour. Approximately how many miles per hour is this?

Blitzer Bonus

Three Weird Units of Measure

- The Sheppey: 1 sheppey $\approx \frac{7}{8}$ mile

  A herd of sheep can look attractive from a distance, but as you get closer, the wool appears dirty and matted. Humorists Douglas Adams and John Lloyd (*The Meaning of Liff*) defined a sheppey as the minimum distance from a group of sheep so that they resemble cute balls of fluff, approximately $\frac{7}{8}$ of a mile or one sheppey.

- The Wheaton: 1 wheaton = 500,000 Twitter followers

  Named after actor Will Wheaton, one of the first celebrities to embrace Twitter and also one of the first to attract a massive number of followers. When half a million people subscribed to his Tweets, that number was dubbed a wheaton.

- The Smoot: 1 smoot = 5 ft 7 in.

  A smoot is exactly 5 feet 7 inches, the height of MIT freshman Oliver Smoot, used by his fraternity brothers to measure the length of Harvard Bridge. The brothers calculated the bridge to be 364.4 smoots plus the length of one of Oliver's ears.

  Source: Mental_Floss: The Book, First Edition, HarperCollins, 2011.

Concept and Vocabulary Check

Fill in each blank so that the resulting statement is true.

1. The result obtained from measuring length is called a/an _______ measurement and is stated in _______ units.
2. In the English system, _____ in. = 1 ft, _____ ft = 1 yd, _____ in. = 1 yd, and _______ ft = 1 mi.
3. Fractions such as $\frac{12 \text{ in.}}{1 \text{ ft}}$ and $\frac{1 \text{ yd}}{3 \text{ ft}}$ are called _______ fractions. The value of such fractions is _____.
4. In the metric system, 1 km = _______ m, 1 hm = _______ m, 1 dam = _____ m, 1 dm = _______ m, 1 cm = _______ m, and 1 mm = _______ m.

In Exercises 5–8, determine whether each statement is true or false. If the statement is false, make the necessary change(s) to produce a true statement.

5. Dimensional analysis uses powers of 10 to convert from one unit of measurement to another. _______
6. One of the advantages of the English system is that units are based on powers of 10. _______
7. There are 2.54 inches in a centimeter. _______
8. The height of an adult male is approximately 6 meters. _______

Exercise Set 9.1

Practice Exercises

In Exercises 1–16, use dimensional analysis to convert the quantity to the indicated unit. If necessary, round the answer to two decimal places.

1. 30 in. to ft
2. 100 in. to ft
3. 30 ft to in.
4. 100 ft to in.
5. 6 in. to yd
6. 21 in. to yd
7. 6 yd to in.
8. 21 yd to in.
9. 6 yd to ft
10. 12 yd to ft
11. 6 ft to yd
12. 12 ft to yd
13. 23,760 ft to mi
14. 19,800 ft to mi
15. 0.75 mi to ft
16. 0.25 mi to ft

In Exercises 17–26, use the diagram in the box on page 585 to convert the given measurement to the unit indicated.

17. 5 m to cm
18. 8 dam to m
19. 16.3 hm to m
20. 0.37 hm to m
21. 317.8 cm to hm
22. 8.64 hm to cm
23. 0.023 mm to m
24. 0.00037 km to cm
25. 2196 mm to dm
26. 71 dm to km

In Exercises 27–44, use the following English and metric equivalents, along with dimensional analysis, to convert the given measurement to the unit indicated.

English and Metric Equivalents

1 in. = 2.54 cm
1 ft = 30.48 cm
1 yd ≈ 0.9 m
1 mi ≈ 1.6 km

27. 14 in. to cm
28. 26 in. to cm
29. 14 cm to in.
30. 26 cm to in.
31. 265 mi to km
32. 776 mi to km
33. 265 km to mi
34. 776 km to mi
35. 12 m to yd
36. 20 m to yd
37. 14 dm to in.
38. 1.2 dam to in.
39. 160 in. to dam
40. 180 in. to hm
41. 5 ft to m
42. 8 ft to m
43. 5 m to ft
44. 8 m to ft

Use 1 mi ≈ 1.6 km *to solve Exercises 45–48.*

45. Express 96 kilometers per hour in miles per hour.
46. Express 104 kilometers per hour in miles per hour.
47. Express 45 miles per hour in kilometers per hour.
48. Express 50 miles per hour in kilometers per hour.

Practice Plus

In Exercises 49–52, use the unit fractions

$$\frac{36 \text{ in.}}{1 \text{ yd}} \text{ and } \frac{2.54 \text{ cm}}{1 \text{ in.}}.$$

49. Convert 5 yd to cm.
50. Convert 8 yd to cm.
51. Convert 762 cm to yd.
52. Convert 1016 cm to yd.

In Exercises 53–54, use the unit fractions

$$\frac{5280 \text{ ft}}{1 \text{ mi}}, \frac{12 \text{ in.}}{1 \text{ ft}}, \text{ and } \frac{2.54 \text{ cm}}{1 \text{ in.}}.$$

53. Convert 30 mi to km.
54. Convert 50 mi to km.
55. Use unit fractions to express 120 miles per hour in feet per second.
56. Use unit fractions to express 100 miles per hour in feet per second.

Application Exercises

In Exercises 57–66, selecting from millimeter, meter, and kilometer, determine the best unit of measure to express the given length.

57. A person's height

58. The length of a football field

59. The length of a bee

60. The distance from New York City to Washington, D.C.

61. The distance around a one-acre lot

62. The length of a car

63. The width of a book

64. The altitude of an airplane

65. The diameter of a screw

66. The width of a human foot

In Exercises 67–74, select the best estimate for the measure of the given item.

67. The length of a pen

a. 30 cm **b.** 19 cm **c.** 19 mm

68. The length of this page

a. 2.5 mm **b.** 25 mm **c.** 250 mm

69. The height of a skyscraper

a. 325 m **b.** 32.5 km **c.** 325 km **d.** 3250 km

70. The length of a pair of pants

a. 700 cm **b.** 70 cm **c.** 7 cm

71. The height of a room

a. 4 mm **b.** 4 cm **c.** 4 m **d.** 4 dm

72. The length of a rowboat

a. 4 cm **b.** 4 dm **c.** 4 m **d.** 4 dam

73. The width of an electric cord

a. 4 mm **b.** 4 cm **c.** 4 dm **d.** 4 m

74. The dimensions of a piece of printer paper

a. 22 mm by 28 mm **b.** 22 cm by 28 cm

c. 22 dm by 28 dm **d.** 22 m by 28 m

75. A baseball diamond measures 27 meters along each side. If a batter scored two home runs in a game, how many kilometers did the batter run?

76. If you jog six times around a track that is 700 meters long, how many kilometers have you covered?

77. The distance from the Earth to the Sun is about 93 million miles. What is this distance in kilometers?

78. The distance from New York City to Los Angeles is 4690 kilometers. What is the distance in miles?

Exercises 79–80 give the approximate length of some of the world's longest rivers. In each exercise, determine which is the longer river and by how many kilometers.

79. Nile: 4130 miles; Amazon: 6400 kilometers

80. Yangtze: 3940 miles; Mississippi: 6275 kilometers

Exercises 81–82 give the approximate height of some of the world's tallest mountains. In each exercise, determine which is the taller mountain and by how many meters. Round to the nearest meter.

81. K2: 8611 meters; Everest: 29,035 feet

82. Lhotse: 8516 meters; Kangchenjunga: 28,170 feet

Exercises 83–84 give the average rainfall of some of the world's wettest places. In each exercise, determine which location has the greater average rainfall and by how many inches. Round to the nearest inch.

83. Debundscha (Cameroon): 10,280 millimeters; Waialeale (Hawaii): 451 inches

84. Mawsynram (India): 11,870 millimeters; Cherrapunji (India): 498 inches

(Source for Exercises 79–84: Russell Ash, *The Top 10 of Everything 2009*)

Explaining the Concepts

85. Describe the two parts of a measurement.

86. Describe how to use dimensional analysis to convert 20 inches to feet.

87. Describe advantages of the metric system over the English system.

88. Explain how to change units within the metric system.

89. You jog 500 meters in a given period of time. The next day, you jog 500 yards over the same time period. On which day was your speed faster? Explain your answer.

90. What kind of difficulties might arise if the United States immediately eliminated all units of measure in the English system and replaced the system by the metric system?

91. The United States is the only Westernized country that does not use the metric system as its primary system of measurement. What reasons might be given for continuing to use the English system?

Critical Thinking Exercises

Make Sense? *In Exercises 92–95, determine whether each statement makes sense or does not make sense, and explain your reasoning.*

92. I can run 4000 meters in approximately one hour.

93. I ran 2000 meters and you ran 2000 yards in the same time, so I ran at a faster rate.

94. The most frequent use of dimensional analysis involves changing units within the metric system.

95. When multiplying by a unit fraction, I put the unit of measure that needs to be introduced in the denominator.

In Exercises 96–100, convert to an appropriate metric unit so that the numerical expression in the given measure does not contain any zeros.

96. 6000 cm

97. 900 m

98. 7000 dm

99. 11,000 mm

100. 0.0002 km

9.2 Measuring Area and Volume

WHAT AM I SUPPOSED TO LEARN?

After studying this section, you should be able to:

1. Use square units to measure area.
2. Use dimensional analysis to change units for area.
3. Use cubic units to measure volume.
4. Use English and metric units to measure capacity.

ARE YOU FEELING A BIT CROWDED IN? Although there are more people on the East Coast of the United States than there are bears, there are places in the Northwest where bears outnumber humans. The most densely populated state is New Jersey, averaging 1218.1 people per square mile. The least densely populated state is Alaska, averaging 1.3 persons per square mile. The U.S. average is 91.5 persons per square mile.

A square mile is one way of measuring the **area** of a state. A state's area is the region within its boundaries. Its **population density** is its population divided by its area. In this section, we discuss methods for measuring both area and volume.

1 *Use square units to measure area.*

Measuring Area

In order to measure a region that is enclosed by boundaries, we begin by selecting a *square unit*. A **square unit** is a square, each of whose sides is one unit in length, illustrated in **Figure 9.2**. The region in **Figure 9.2** is said to have an area of **one square unit**. The side of the square can be 1 inch, 1 centimeter, 1 meter, 1 foot, or one of any linear unit of measure. The corresponding units of area are the square inch (in.^{2}), the square centimeter (cm^2), the square meter (m^2), the square foot (ft^2), and so on. **Figure 9.3** illustrates 1 square inch and 1 square centimeter, drawn to actual size.

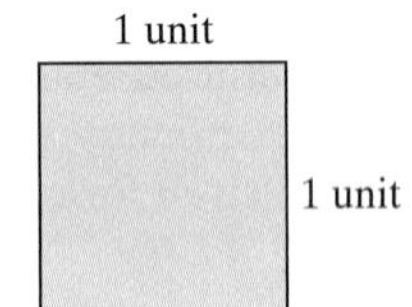

FIGURE 9.2 One square unit

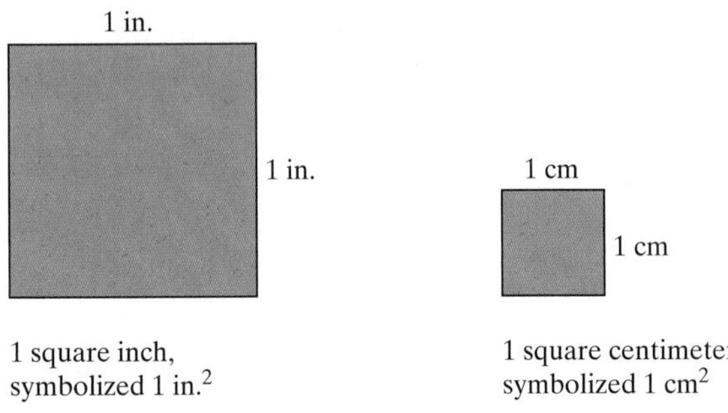

FIGURE 9.3 Common units of measurement for area, drawn to actual size

EXAMPLE 1 Measuring Area

What is the area of the region shown in **Figure 9.4**?

Square Unit

FIGURE 9.4

SOLUTION

We can determine the area of the region by counting the number of square units contained within the region. There are 12 such units. Therefore, the area of the region is 12 square units.

CHECK POINT 1 What is the area of the region represented by the first two rows in **Figure 9.4**?

Although there are 12 inches in one foot and 3 feet in one yard, these numerical relationships are not the same for square units.

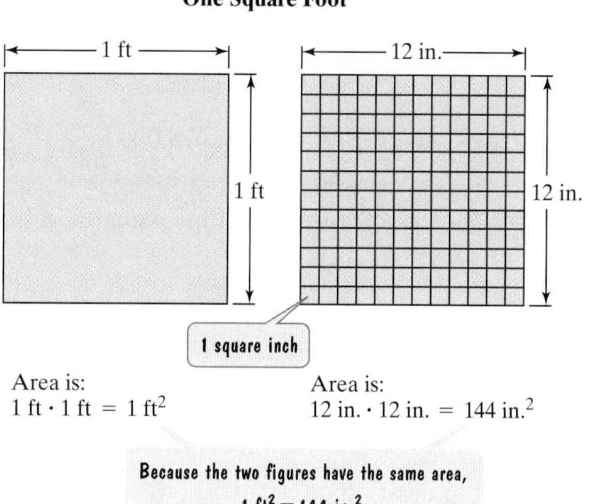

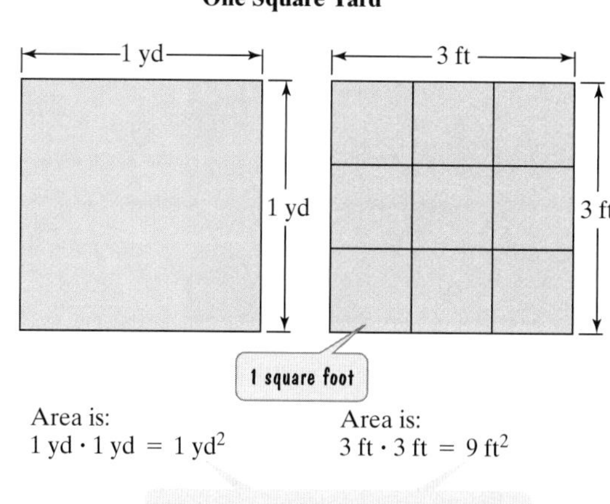

SQUARE UNITS OF MEASURE: THE ENGLISH SYSTEM

1 square foot (ft^2) = 144 square inches ($in.^2$)
1 square yard (yd^2) = 9 square feet (ft^2)
1 acre (a) = 43,560 ft^2 or 4840 yd^2
1 square mile (mi^2) = 640 acres

GREAT QUESTION!

Which square units of measure are frequently used in everyday situations?

A small plot of land is usually measured in square feet, rather than a small fraction of an acre. Curiously, square yards are rarely used. However, square yards are commonly used for carpet and flooring measures.

EXAMPLE 2 *Using Square Units to Compute Population Density*

After Alaska, Wyoming is the least densely populated state. The population of Wyoming is 568,158 and its area is 97,814 square miles. What is Wyoming's population density?

SOLUTION

We compute the population density by dividing Wyoming's population by its area.

$$\text{population density} = \frac{\text{population}}{\text{area}} = \frac{568{,}158 \text{ people}}{97{,}814 \text{ square miles}}$$

Using a calculator and rounding to the nearest tenth, we obtain a population density of 5.8 people per square mile. This means that there is an average of only 5.8 people for each square mile of area.

CHECK POINT 2 The population of California is 39,144,818 and its area is 163,695 square miles. What is California's population density? Round to the nearest tenth.

2 *Use dimensional analysis to change units for area.*

EXAMPLE 3 *Using Dimensional Analysis on Units of Area*

Your author wrote *Thinking Mathematically* in Point Reyes National Seashore, 40 miles north of San Francisco. The national park consists of 75,000 acres with miles of pristine surf-washed beaches, forested ridges, and bays bordered by white cliffs. How large is the national park in square miles?

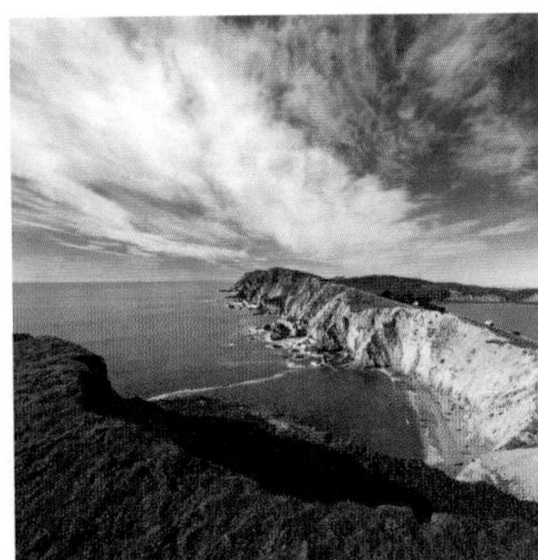

SOLUTION

We use the fact that 1 square mile = 640 acres to set up our unit fraction:

$$\frac{1 \text{ mi}^2}{640 \text{ acres}}.$$

Thus,

$$75{,}000 \text{ acres} = \frac{75{,}000 \text{ acres}}{1} \cdot \frac{1 \text{ mi}^2}{640 \text{ acres}} = \frac{75{,}000}{640} \text{ mi}^2 \approx 117 \text{ mi}^2.$$

The area of Point Reyes National Seashore is approximately 117 square miles.

CHECK POINT 3 The National Park Service administers approximately 84,000,000 acres of national parks. How large is this in square miles?

In Section 9.1, we saw that in most other countries, the system of measurement that is used is the metric system. In the metric system, the square centimeter is used instead of the square inch. The square meter replaces the square foot and the square yard.

The English system uses the acre and the square mile to measure large land areas, where one square mile = 640 acres. The metric system uses the hectare (symbolized ha and pronounced "hectair"). A hectare is about the area of two football fields placed side by side, approximately 2.5 acres. One square mile of land consists of approximately 260 hectares. Just as the hectare replaces the acre, the square kilometer is used instead of the square mile. One square kilometer is approximately 0.38 square mile.

Some basic approximate conversions for units of area are given in **Table 9.4**.

TABLE 9.4 English and Metric Equivalents for Area

1 square inch (in.^{2})	$\approx$ 6.5 square centimeters (cm^2)
1 square foot (ft^2)	$\approx$ 0.09 square meter (m^2)
1 square yard (yd^2)	$\approx$ 0.8 square meter (m^2)
1 square mile (mi^2)	$\approx$ 2.6 square kilometers (km^2)
1 acre	$\approx$ 0.4 hectare (ha)

EXAMPLE 4 *Using Dimensional Analysis on Units of Area*

A property in Italy is advertised at $545,000 for 6.8 hectares.

a. Find the area of the property in acres.

b. What is the price per acre?

SOLUTION

a. Using **Table 9.4**, we see that 1 acre $\approx$ 0.4 hectare. To convert 6.8 hectares to acres, we use a unit fraction with acres in the numerator and hectares in the denominator.

$$6.8 \text{ ha} \approx \frac{6.8 \text{ ha}}{1} \cdot \frac{1 \text{ acre}}{0.4 \text{ ha}} = \frac{6.8}{0.4} \text{ acres} = 17 \text{ acres}$$

The area of the property is approximately 17 acres.

b. The price per acre is the total price, \$545,000, divided by the number of acres, 17.

$$\text{price per acre} = \frac{\$545{,}000}{17 \text{ acres}} \approx \$32{,}059/\text{acre}$$

The price is approximately \$32,059 per acre.

CHECK POINT 4 A property in northern California is on the market at \$415,000 for 1.8 acres.

a. Find the area of the property in hectares.

b. What is price per hectare?

3 *Use cubic units to measure volume.*

Measuring Volume

A shoe box and a basketball are examples of three-dimensional figures. **Volume** refers to the amount of space occupied by such figures. In order to measure this space, we begin by selecting a *cubic unit*. Two such cubic units are shown in **Figure 9.5**.

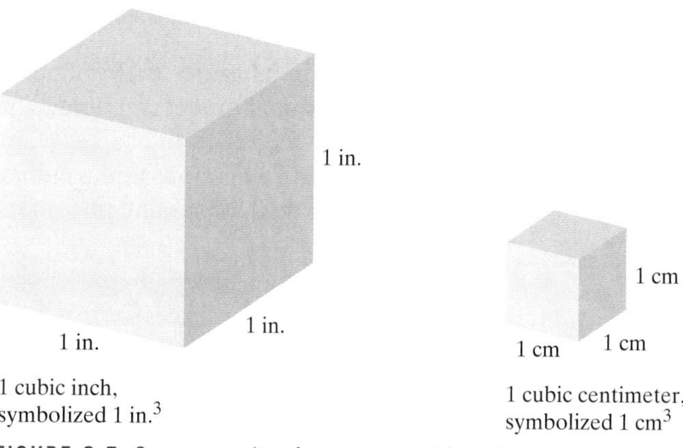

FIGURE 9.5 Common units of measurement for volume

The edges of a cube all have the same length. Other cubic units used to measure volume include 1 cubic foot (1 ft^3) and 1 cubic meter (1 m^3). One way to measure the volume of a solid is to calculate the number of cubic units contained in its interior.

EXAMPLE 5 Measuring Volume

What is the volume of the solid shown in **Figure 9.6**?

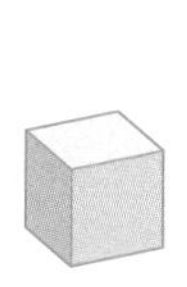

Cubic unit   Volume = ?

FIGURE 9.6

SOLUTION

We can determine the volume of the solid by counting the number of cubic units contained within the region. Because we have drawn a solid three-dimensional figure on a flat two-dimensional page, some of the small cubic units in the back right are hidden. The figures below show how the cubic units are used to fill the inside of the solid.

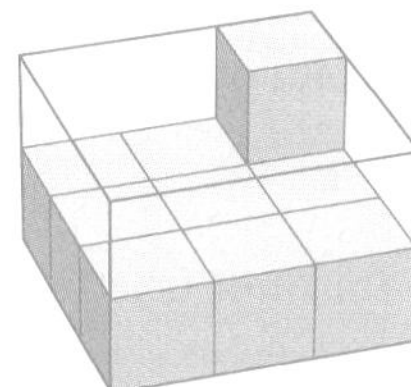

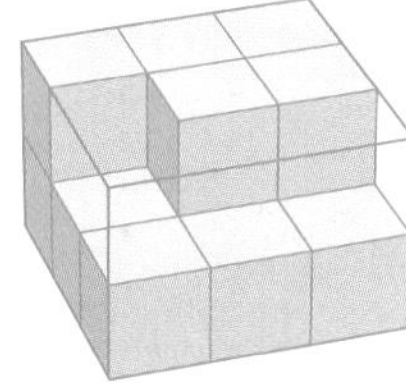

Do these figures help you to see that there are 18 cubic units inside the solid? The volume of the solid is 18 cubic units.

CHECK POINT 5 What is the volume of the region represented by the bottom row of blocks in **Figure 9.6**?

GREAT QUESTION!

I'm having difficulty seeing the detail in Figure 9.7. Can you help me out?

Cubing numbers is helpful:

$$3 \text{ ft} = 1 \text{ yd}$$
$$(3 \text{ ft})^3 = (1 \text{ yd})^3$$
$$3 \cdot 3 \cdot 3 \text{ ft}^3 = 1 \cdot 1 \cdot 1 \text{ yd}^3.$$

Conclusion: $27 \text{ ft}^3 = 1 \text{ yd}^3$

$$12 \text{ in.} = 1 \text{ ft}$$
$$(12 \text{ in.})^3 = (1 \text{ ft})^3$$
$$12 \cdot 12 \cdot 12 \text{ in.}^3 = 1 \cdot 1 \cdot 1 \text{ ft}^3$$

Conclusion: $1728 \text{ in.}^3 = 1 \text{ ft}^3$

We have seen that there are 3 feet in a yard, but 9 square feet in a square yard. Neither of these relationships holds for cubic units. **Figure 9.7** illustrates that there are 27 cubic feet in a cubic yard. Furthermore, there are 1728 cubic inches in a cubic foot.

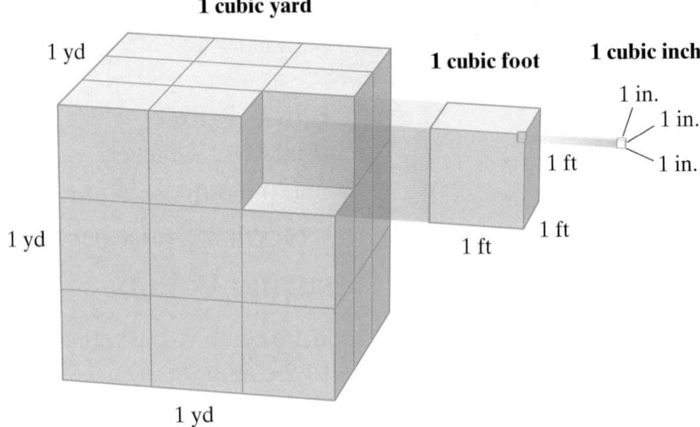

FIGURE 9.7 $27 \text{ ft}^3 = 1 \text{ yd}^3$ and $1728 \text{ in.}^3 = 1 \text{ ft}^3$

The measure of volume also includes the amount of fluid that a three-dimensional object can hold. This is often called the object's **capacity**. For example, we often refer to the capacity, in gallons, of a gas tank. A cubic yard has a capacity of about 200 gallons and a cubic foot has a capacity of about 7.48 gallons.

4 Use English and metric units to measure capacity.

Table 9.5 contains information about standard units of capacity in the English system.

TABLE 9.5 English Units for Capacity

2 pints (pt)	= 1 quart (qt)
4 quarts	= 1 gallon (gal)
1 gallon	= 128 fluid ounces (fl oz)
1 cup (c)	= 8 fluid ounces
Volume in Cubic Units	**Capacity**
1 cubic yard	about 200 gallons
1 cubic foot	about 7.48 gallons
231 cubic inches	about 1 gallon

EXAMPLE 6 *Volume and Capacity in the English System*

A swimming pool has a volume of 22,500 cubic feet. How many gallons of water does the pool hold?

SOLUTION

We use the fact that 1 cubic foot has a capacity of about 7.48 gallons to set up our unit fraction:

$$\frac{7.48 \text{ gal}}{1 \text{ ft}^3}.$$

We use this unit fraction to find the capacity of the 22,500 cubic feet.

$$22{,}500 \text{ ft}^3 \approx \frac{22{,}500 \cancel{\text{ft}^3}}{1} \cdot \frac{7.48 \text{ gal}}{1 \cancel{\text{ft}^3}} = 22{,}500(7.48) \text{ gal} = 168{,}300 \text{ gal}$$

The pool holds approximately 168,300 gallons of water.

CHECK POINT 6 A pool has a volume of 10,000 cubic feet. How many gallons of water does the pool hold?

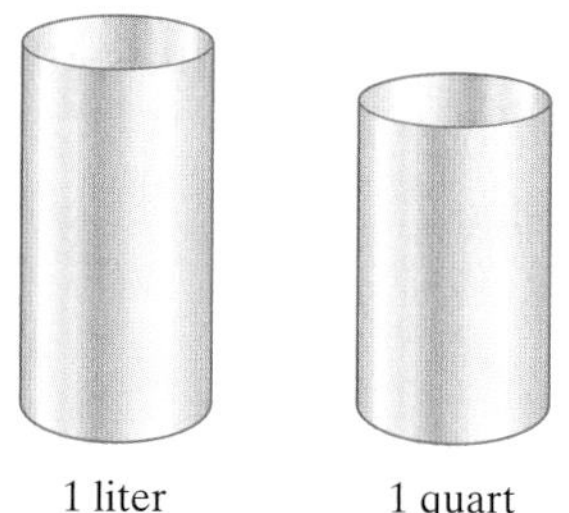

As we have come to expect, things are simpler when the metric system is used to measure capacity. The basic unit is the **liter**, symbolized by L. A liter is slightly larger than a quart.

$$1 \text{ liter} \approx 1.0567 \text{ quarts}$$

The standard metric prefixes are used to denote a multiple or part of a liter. **Table 9.6** applies these prefixes to the liter.

TABLE 9.6 Units of Capacity in the Metric System

Symbol	Unit	Meaning
kL	kiloliter	1000 liters
hL	hectoliter	100 liters
daL	dekaliter	10 liters
L	liter	1 liter ≈ 1.06 quarts
dL	deciliter	0.1 liter
cL	centiliter	0.01 liter
mL	milliliter	0.001 liter

The following list should help give you a feel for capacity in the metric system.

Item	Capacity
Average cup of coffee	250 mL
12-ounce can of soda	355 mL
One quart of fruit juice	0.95 L
One gallon of milk	3.78 L
Average gas capacity of a car (about 18.5 gallons)	70 L

10 cm = 1 dm
10 cm = 1 dm
10 cm = 1 dm

1 L = 1000 mL $1000 \text{ cm}^3 = 1 \text{ dm}^3$

FIGURE 9.8

Figure 9.8 shows a 1-liter container filled with water. The water in the liter container will fit exactly into the cube shown to its right. The volume of this cube is 1000 cubic centimeters, or equivalently, 1 cubic decimeter. Thus,

$$1000 \text{ cm}^3 = 1 \text{ dm}^3 = 1 \text{ L}.$$

Table 9.7 expands on this relationship between volume and capacity in the metric system.

TABLE 9.7 Volume and Capacity in the Metric System

Volume in Cubic Units		Capacity
1 cm^3	=	1 mL
$1 \text{ dm}^3 = 1000 \text{ cm}^3$	=	1 L
1 m^3	=	1 kL

A milliliter is the capacity of a cube measuring 1 centimeter on each side.

A liter is the capacity of a cube measuring 10 centimeters on each side.

EXAMPLE 7 Volume and Capacity in the Metric System

An aquarium has a volume of 36,000 cubic centimeters. How many liters of water does the aquarium hold?

SOLUTION

We use the fact that 1000 cubic centimeters corresponds to a capacity of 1 liter to set up our unit fraction:

$$\frac{1 \text{ L}}{1000 \text{ cm}^3}.$$

We use this unit fraction to find the capacity of the 36,000 cubic centimeters.

$$36{,}000\ \text{cm}^3 = \frac{36{,}000\ \cancel{\text{cm}^3}}{1} \cdot \frac{1\ \text{L}}{1000\ \cancel{\text{cm}^3}} = \frac{36{,}000}{1000}\ \text{L} = 36\ \text{L}$$

The aquarium holds 36 liters of water.

CHECK POINT 7 A fish pond has a volume of 220,000 cubic centimeters. How many liters of water does the pond hold?

Table 9.8 can be used to convert units of capacity to and from the metric system.

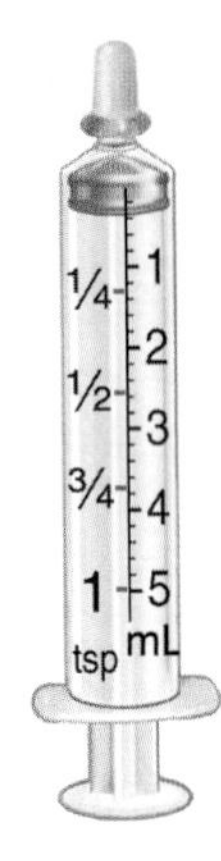

1 tsp ≈ 5 mL

TABLE 9.8 English and Metric Equivalents for Capacity

1 teaspoon (tsp)	≈ 5 milliliters (mL)
1 tablespoon (tbsp)	≈ 15 milliliters (mL)
1 fluid ounce (fl oz)	≈ 30 milliliters (mL)
1 cup (c)	≈ 0.24 liter (L)
1 pint (pt)	≈ 0.47 liter (L)
1 quart (qt)	≈ 0.95 liter (L)
1 gallon (gal)	≈ 3.8 liters (L)

Our next example involves measuring dosages of medicine in liquid form. We have seen that

$$1\ \text{cm}^3 = 1\ \text{mL}.$$

Dosages of liquid medication are measured using cubic centimeters, or milliliters as they are also called. In the United States, cc, rather than cm^3, denotes cubic centimeters.

EXAMPLE 8 *Measuring Dosages of Medicine in Liquid Form*

A physician orders 10 cc of the drug Lexapro (used to treat depression and anxiety) to be administered to a patient in liquid form.

a. How many milliliters of the drug should be administered?

b. How many fluid ounces of the drug should be administered?

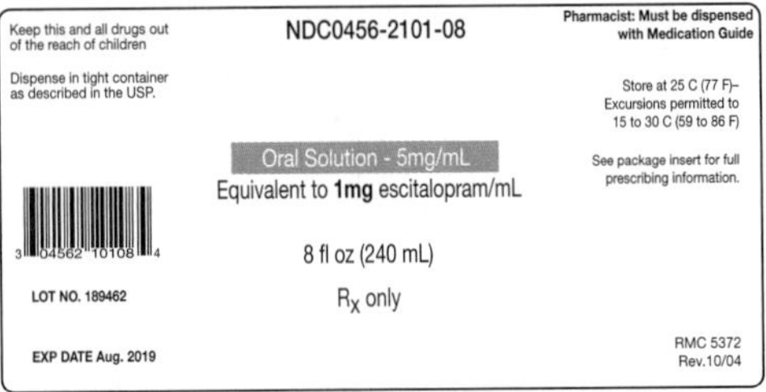

SOLUTION

a. We use a relationship between volume and capacity in the metric system: 1 cubic centimeter (cc) = 1 milliliter (mL). Because 10 cubic centimeters (cc) of the drug is to be administered, this is equivalent to 10 milliliters (mL) of Lexapro.

b. We now need to convert 10 milliliters to fluid ounces. **Table 9.8** shows that 1 fluid ounce (fl oz) ≈ 30 milliliters (mL). We use

$$\frac{1\ \text{fl oz}}{30\ \text{mL}}$$

as our unit fraction.

$$10 \text{ mL} \approx \frac{10 \cancel{\text{mL}}}{1} \cdot \frac{1 \text{ fl oz}}{30 \cancel{\text{mL}}} = \frac{10}{30} \text{ fl oz} \approx 0.33 \text{ fl oz}$$

Approximately 0.33 fluid ounce (fl oz) of Lexapro should be administered.

CHECK POINT 8 A physician orders 20 cc of the antibiotic Omnicef to be administered every 12 hours.

a. How many milliliters of the drug should be administered?

b. How many fluid ounces of the drug should be administered? Round to two decimal places.

Concept and Vocabulary Check

Fill in each blank so that the resulting statement is true.

1. Area is measured in ________ units and volume is measured in ________ units.
2. In the English system, 1 ft^2 = ________ in.^{2} and 1 yd^2 = ________ ft^2.
3. Because 1 mi^2 = 640 acres, the unit fraction needed to convert from acres to square miles is ________ and the unit fraction needed to convert from square miles to acres is ________.
4. In the English system, ____ pt = 1 qt and ____ qt = 1 gal.
5. The amount of fluid that a three-dimensional object can hold is called the object's ________, measured in the metric system using a basic unit called the ______.
6. A state's population density is its population divided by its ______.
7. 1 cm^3, or 1 cc, = ____ mL

In Exercises 8–10, determine whether each statement is true or false. If the statement is false, make the necessary change(s) to produce a true statement.

8. Because there are 3 feet in one yard, there are also 3 square feet in one square yard. ______
9. The English system uses in.^{2} to measure large land areas. ______
10. One quart is approximately 1.06 liters. ______

Exercise Set 9.2

Practice Exercises

In Exercises 1–4, use the given figure to find its area in square units.

1.

2.
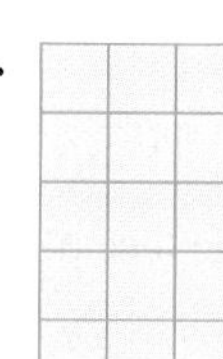

3.
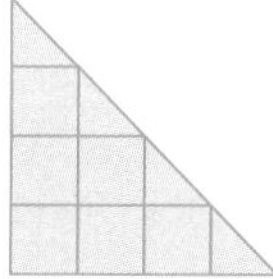

4.
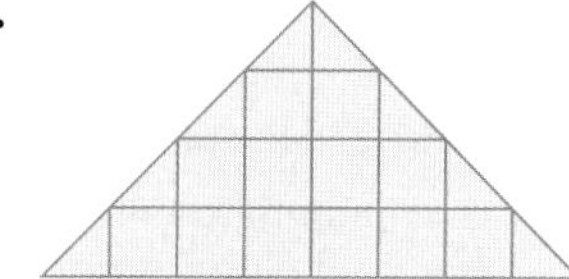

In Exercises 5–12, use ***Table 9.4*** *on page 594, along with dimensional analysis, to convert the given square unit to the square unit indicated. Where necessary, round answers to two decimal places.*

5. 14 cm^2 to in.^{2}

6. 20 m^2 to ft^2

7. 30 m^2 to yd^2

8. 14 mi^2 to km^2

9. 10.2 ha to acres

10. 20.6 ha to acres

11. 14 in.^{2} to cm^2

12. 20 in.^{2} to cm^2

In Exercises 13–14, use the given figure to find its volume in cubic units.

13.
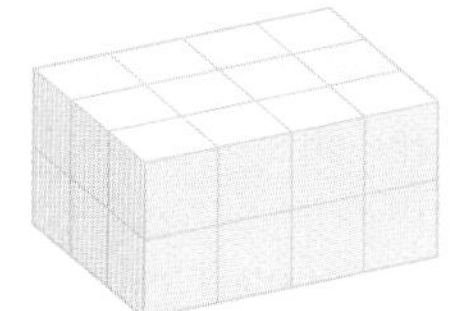

14.

In Exercises 15–22, use ***Table 9.5*** *on page 596, along with dimensional analysis, to convert the given unit to the unit indicated. Where necessary, round answers to two decimal places.*

15. 10,000 ft^3 to gal

16. 25,000 ft^3 to gal

17. 8 yd^3 to gal

18. 35 yd^3 to gal

19. 2079 in.^{3} to gal

20. 6237 in.^{3} to gal

21. 2700 gal to yd^3

22. 1496 gal to ft^3

In Exercises 23–32, use ***Table 9.7*** *on page 597, along with dimensional analysis, to convert the given unit to the unit indicated.*

23. 45,000 cm^3 to L
24. 75,000 cm^3 to L
25. 17 cm^3 to mL
26. 19 cm^3 to mL
27. 1.5 L to cm^3
28. 4.5 L to cm^3
29. 150 mL to cm^3
30. 250 mL to cm^3
31. 12 kL to dm^3
32. 16 kL to dm^3

In Exercises 33–48, use ***Table 9.8*** *on page 598, along with dimensional analysis, to convert the given unit to the unit indicated. Where necessary, round to two decimal places.*

33. 12 mL to tsp
34. 14 mL to tsp
35. 3 tbsp to mL
36. 4 tbsp to mL
37. 70 mL to fl oz
38. 80 mL to fl oz
39. 1.4 L to c
40. 2.6 L to c
41. 6 pt to L
42. 9 pt to L
43. 4 L to qt
44. 7 L to qt
45. 3 gal to L
46. 5 gal to L
47. 2000 mL to qt
48. 5000 mL to qt

Practice Plus

The bar graph shows the resident population and the land area of the United States for selected years from 1800 through 2010. Use the information shown by the graph to solve Exercises 49–52.

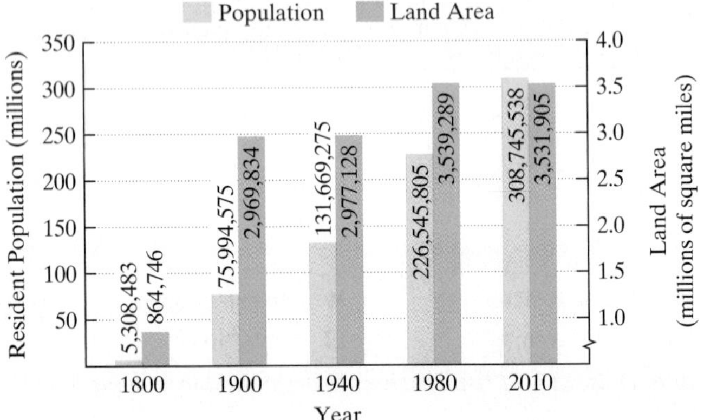

Source: U.S. Census Bureau

49. a. Find the population density of the United States, to the nearest tenth, in 1900 and in 2010.

b. Find the percent increase in population density, to the nearest tenth of a percent, from 1900 to 2010.

50. a. Find the population density of the United States, to the nearest tenth, in 1800 and in 2010.

b. Find the percent increase in population density, to the nearest tenth of a percent, from 1800 to 2010.

51. Find the population density of the United States, to the nearest tenth, expressed in people per square kilometer, in 1940.

52. Find the population density of the United States, to the nearest tenth, expressed in people per square kilometer, in 1980.

Application Exercises

In Exercises 53–54, find the population density, to the nearest tenth, for each state. Which state has the greater population density? How many more people per square mile inhabit the state with the greater density than inhabit the state with the lesser density?

53. Illinois population: 12,869,257 area: 57,914 mi^2
Ohio population: 11,544,951 area: 44,826 mi^2

54. New York population: 19,465,197 area: 54,555 mi^2
Rhode Island population: 1,051,302 area: 1545 mi^2

In Exercises 55–56, use the fact that 1 square mile = 640 acres to find the area of each national park to the nearest square mile.

55. Everglades National Park (Florida): 1,509,154 acres

56. Yosemite National Park (California): 761,268 acres

57. A property that measures 8 hectares is for sale.

a. How large is the property in acres?

b. If the property is selling for \$250,000, what is the price per acre?

58. A property that measures 100 hectares is for sale.

a. How large is the property in acres?

b. If the property is selling for \$350,000, what is the price per acre?

In Exercises 59–62, selecting from square centimeters, square meters, or square kilometers, determine the best unit of measure to express the area of the object described.

59. The top of a desk

60. A dollar bill

61. A national park

62. The wall of a room

In Exercises 63–66, select the best estimate for the measure of the area of the object described.

63. The area of the floor of a room
a. 25 cm^2 **b.** 25 m^2 **c.** 25 km^2

64. The area of a television screen
a. 2050 mm^2 **b.** 2050 cm^2 **c.** 2050 dm^2

65. The area of the face of a small coin
a. 6 mm^2 **b.** 6 cm^2 **c.** 6 dm^2

66. The area of a parcel of land in a large metropolitan area on which a house can be built
a. 900 cm^2 **b.** 900 m^2 **c.** 900 ha

67. A swimming pool has a volume of 45,000 cubic feet. How many gallons of water does the pool hold?

68. A swimming pool has a volume of 66,000 cubic feet. How many gallons of water does the pool hold?

69. A container of grapefruit juice has a volume of 4000 cubic centimeters. How many liters of juice does the container hold?

70. An aquarium has a volume of 17,500 cubic centimeters. How many liters of water does the aquarium hold?

Exercises 71–72 give the approximate area of some of the world's largest island countries. In each exercise, determine which country has the greater area and by how many square kilometers.

71. Philippines: 300,000 km^2; Japan: 145,900 mi^2

72. Iceland: 103,000 km^2; Cuba: 42,800 mi^2

Exercises 73–74 give the approximate area of some of the world's largest islands. In each exercise, determine which island has the greater area and by how many square miles. Round to the nearest square mile.

73. Baffin Island, Canada: 194,574 mi^2;

Sumatra, Indonesia: 443,070 km^2

74. Honshu, Japan: 87,805 mi^2;

Victoria Island, Canada: 217,300 km^2

(Source for Exercises 71–74: Russell Ash, *The Top 10 of Everything 2009*)

Exercises 75–76 involve dosages of the anti-inflammatory drug indomethacin, administered in liquid form. Refer to the appropriate entries in ***Table 9.8*** *on page 598 to convert given units to the unit indicated. Where necessary, round to two decimal places.*

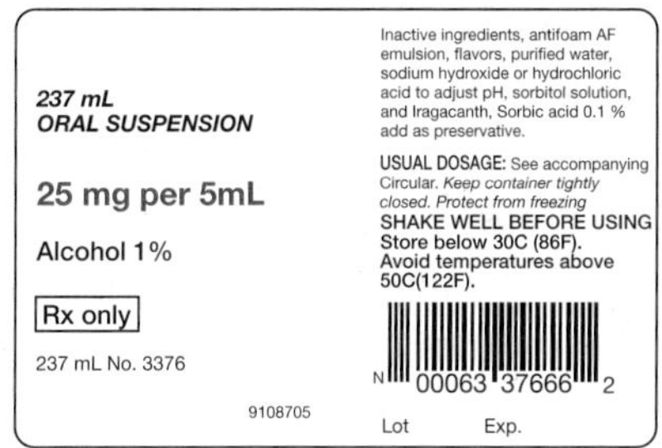

75. A physician orders 3 teaspoons daily of indomethacin in three equally divided doses.

a. How many milliliters of the drug should be administered daily?

b. How many cc of the drug should be administered daily?

c. How many fluid ounces of the drug should be administered daily?

d. How many fluid ounces of the drug should the patient receive in each divided dose?

76. A physician orders 4 teaspoons daily of indomethacin in four equally divided doses.

a. How many milliliters of the drug should be administered daily?

b. How many cc of the drug should be administered daily?

c. How many fluid ounces of the drug should be administered daily?

d. How many fluid ounces of the drug should the patient receive in each divided dose?

Explaining the Concepts

77. Describe how area is measured. Explain why linear units cannot be used.

78. New Mexico has a population density of 17 people per square mile. Describe what this means. Use an almanac or the Internet to compare this population density to that of the state in which you are now living.

79. Describe the difference between the following problems: How much fencing is needed to enclose a garden? How much fertilizer is needed for the garden?

80. Describe how volume is measured. Explain why linear or square units cannot be used.

81. For a swimming pool, what is the difference between the following units of measure: cubic feet and gallons? For each unit, write a sentence about the pool that makes the use of the unit appropriate.

82. If there are 10 decimeters in a meter, explain why there are not 10 cubic decimeters in a cubic meter.

Critical Thinking Exercises

Make Sense? *In Exercises 83–86, determine whether each statement makes sense or does not make sense, and explain your reasoning.*

83. The gas capacity of my car is approximately 40 meters.

84. The population density of Montana is approximately 6.5 people per mile.

85. Yesterday I drank 2000 cm^3 of water.

86. I read that one quart is approximately 0.94635295 liter.

87. Singapore has the highest population density of any country: 46,690 people per 1000 hectares. How many people are there per square mile?

88. Nebraska has a population density of 23.8 people per square mile and a population of 1,842,641. What is the area of Nebraska? Round to the nearest square mile.

89. A high population density is a condition common to extremely poor and extremely rich locales. Explain why this is so.

90. Although Alaska is the least densely populated state, over 90% of its land is protected federal property that is off-limits to settlement. A resident of Anchorage, Alaska, might feel hemmed in. In terms of "elbow room," what other important factor must be considered when calculating a state's population density?

91. Does an adult's body contain approximately 6.5 liters, kiloliters, or milliliters of blood? Explain your answer.

92. Is the volume of a coin approximately 1 cubic centimeter, 1 cubic millimeter, or 1 cubic decimeter? Explain your answer.

Group Exercise

93. If you could select any place in the world, where would you like to live? Look up the population and area of your ideal place, and compute its population density. Group members should share places to live and population densities. What trend, if any, does the group observe?

9.3 Measuring Weight and Temperature

WHAT AM I SUPPOSED TO LEARN?

After studying this section, you should be able to:

1. Apply metric prefixes to units of weight.
2. Convert units of weight within the metric system.
3. Use relationships between volume and weight within the metric system.
4. Use dimensional analysis to change units of weight to and from the metric system.
5. Understand temperature scales.

YOU ARE WATCHING CNN International on cable television. The temperature in Honolulu, Hawaii, is reported as 30°C. Are Honolulu's tourists running around in winter jackets? In this section, we will make sense of Celsius temperature readings, as we discuss methods for measuring temperature and weight.

Measuring Weight

You step on the scale at the doctor's office to check your weight, discovering that you are 150 pounds. Compare this to your weight on the Moon: 25 pounds. Why the difference? **Weight** is the measure of the gravitational pull on an object. The gravitational pull on the Moon is only about one-sixth the gravitational pull on Earth. Although your weight varies depending on the force of gravity, your mass is exactly the same in all locations. **Mass** is a measure of the quantity of matter in an object, determined by its molecular structure. On Earth, as your weight increases, so does your mass. In this section, measurements are assumed to involve everyday situations on the surface of Earth. Thus, we will treat weight and mass as equivalent, and refer strictly to weight.

UNITS OF WEIGHT: THE ENGLISH SYSTEM

16 ounces (oz) = 1 pound (lb)

2000 pounds (lb) = 1 ton (T)

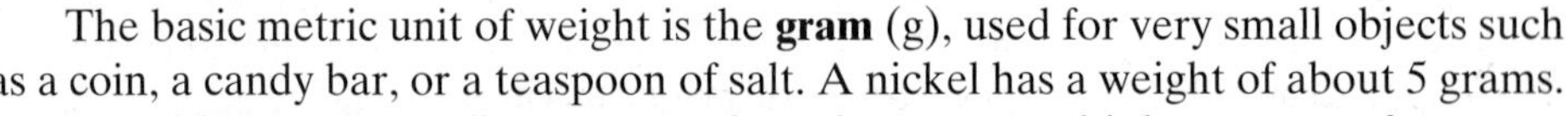

1 *Apply metric prefixes to units of weight.*

The basic metric unit of weight is the **gram** (g), used for very small objects such as a coin, a candy bar, or a teaspoon of salt. A nickel has a weight of about 5 grams.

As with meters, prefixes are used to denote a multiple or part of a gram. **Table 9.9** applies the common metric prefixes to the gram. The first part of the symbol indicates the prefix and the second part (g) indicates gram.

Weight of pineapple is 1 kg, or 1000 g.

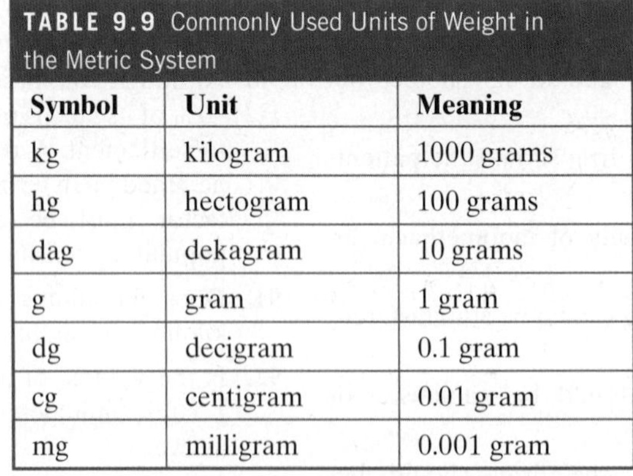

TABLE 9.9 Commonly Used Units of Weight in the Metric System

Symbol	Unit	Meaning
kg	kilogram	1000 grams
hg	hectogram	100 grams
dag	dekagram	10 grams
g	gram	1 gram
dg	decigram	0.1 gram
cg	centigram	0.01 gram
mg	milligram	0.001 gram

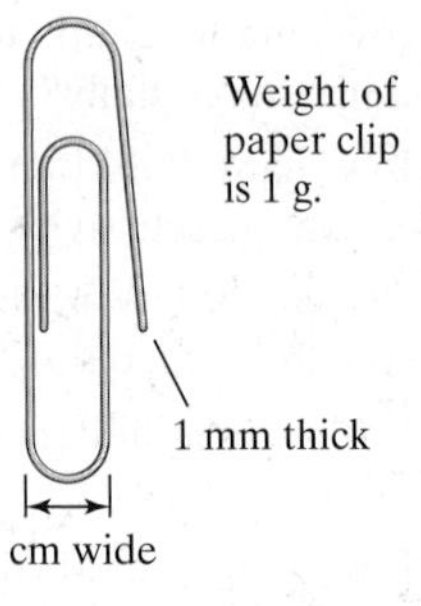

Weight of paper clip is 1 g.

In the metric system, the kilogram is the comparable unit to the pound in the English system. **A weight of 1 kilogram is approximately 2.2 pounds.** Thus, an average man has a weight of about 75 kilograms. Objects that we measure in pounds are measured in kilograms in most countries.

A milligram, equivalent to 0.001 gram, is an extremely small unit of weight and is used extensively in the pharmaceutical industry. If you look at the label on a bottle of tablets, you will see that the amounts of different substances in each tablet are expressed in milligrams.

The weight of a very heavy object is expressed in terms of the metric tonne (t), which is equivalent to 1000 kilograms, or about 2200 pounds. This is 10 percent more than the English ton (T) of 2000 pounds.

We change units of weight within the metric system exactly the same way that we changed units of length.

2 Convert units of weight within the metric system.

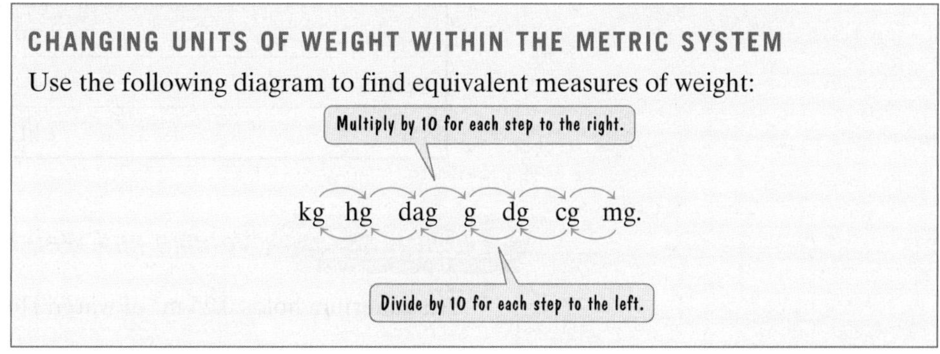

EXAMPLE 1 Changing Units within the Metric System

a. Convert 8.7 dg to mg.

b. Convert 950 mg to g.

SOLUTION

a. To convert from dg (decigrams) to mg (milligrams), we start at dg and move two steps to the right:

kg hg dag g dg cg mg.

Hence, we move the decimal point two places to the right:

$$8.7 \text{ dg} = 870 \text{ mg}.$$

b. To convert from mg (milligrams) to g (grams), we start at mg and move three steps to the left:

kg hg dag g dg cg mg.

Hence, we move the decimal point three places to the left:

$$950 \text{ mg} = 0.950 \text{ g}.$$

CHECK POINT 1

a. Convert 4.2 dg to mg.

b. Convert 620 cg to g.

---

3 Use relationships between volume and weight within the metric system.

We have seen a convenient relationship in the metric system between volume and capacity:

$$1000 \text{ cm}^3 = 1 \text{ dm}^3 = 1 \text{ L}.$$

This relationship can be extended to include weight based on the following:

One kilogram of water has a volume of 1 liter.

Thus,

$$1000 \text{ cm}^3 = 1 \text{ dm}^3 = 1 \text{ L} = 1 \text{ kg}.$$

Table 9.10 shows the relationships between volume and weight of water in the metric system.

TABLE 9.10 Volume and Weight of Water in the Metric System

Volume		Capacity		Weight
1 cm^3	=	1 mL	=	1 g
1 dm^3 = 1000 cm^3	=	1 *L*	=	1 kg
1 m^3	=	1 kL	=	1000 kg = 1 t

EXAMPLE 2 *Volume and Weight in the Metric System*

An aquarium holds 0.25 m^3 of water. How much does the water weigh?

SOLUTION

We use the fact that 1 m^3 of water = 1000 kg of water to set up our unit fraction:

$$\frac{1000 \text{ kg}}{1 \text{ m}^3}.$$

Thus,

$$0.25 \text{ m}^3 = \frac{0.25 \cancel{\text{m}^3}}{1} \cdot \frac{1000 \text{ kg}}{1 \cancel{\text{m}^3}} = 250 \text{ kg}.$$

The water weighs 250 kilograms.

CHECK POINT 2 An aquarium holds 0.145 m^3 of water. How much does the water weigh?

4 *Use dimensional analysis to change units of weight to and from the metric system.*

A problem like Example 2 involves more awkward computation in the English system. For example, if you know the aquarium's volume in cubic feet, you must also know that 1 cubic foot of water weighs about 62.5 pounds to determine the water's weight.

Dimensional analysis is a useful tool when converting units of weight between the English and metric systems. Some basic approximate conversions are given in **Table 9.11**.

1-ounce coin ≈ 28 grams

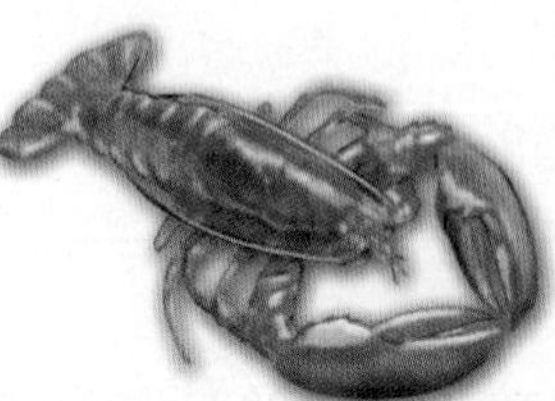

1-pound lobster ≈ 0.45 kilogram

TABLE 9.11 Weight: English and Metric Equivalents

1 ounce (oz)	≈ 28 grams (g)
1 pound (lb)	≈ 0.45 kilogram (kg)
1 ton (T)	≈ 0.9 tonne (t)

1-ton bison ≈ 0.9 tonne

EXAMPLE 3 *Using Dimensional Analysis*

a. Convert 160 pounds to kilograms.

b. Convert 300 grams to ounces.

Blitzer Bonus

The SI System

The metric system is officially called the Système International d'Unités, abbreviated the SI system. Burma, Liberia, and the United States are the only countries in the world not using the SI system.

In 1975, the United States prepared to adopt the SI system with the Metric Conversion Act. However, due to concerns in grassroots America, Congress made the conversion voluntary, effectively chasing it off. But it's not as bleak as you think; from metric tools to milligram measures of medication, we have been edging, inching, even centimetering toward the metric system.

SOLUTION

a. To convert 160 pounds to kilograms, we use a unit fraction with kilograms in the numerator and pounds in the denominator:

$$\frac{0.45 \text{ kg}}{1 \text{ lb}}. \quad \text{Table 9.11 shows that 1 lb} \approx 0.45 \text{ kg.}$$

Thus,

$$160 \text{ lb} \approx \frac{160 \cancel{\text{lb}}}{1} \cdot \frac{0.45 \text{ kg}}{1 \cancel{\text{lb}}} = 160(0.45) \text{ kg} = 72 \text{ kg}.$$

b. To convert 300 grams to ounces, we use a unit fraction with ounces in the numerator and grams in the denominator:

$$\frac{1 \text{ oz}}{28 \text{ g}}. \quad \text{Table 9.11 shows that 1 oz} \approx 28 \text{ g.}$$

Thus,

$$300 \text{ g} \approx \frac{300 \cancel{\text{g}}}{1} \cdot \frac{1 \text{ oz}}{28 \cancel{\text{g}}} = \frac{300}{28} \text{ oz} \approx 10.7 \text{ oz}.$$

CHECK POINT 3

a. Convert 120 pounds to kilograms.

b. Convert 500 grams to ounces.

EXAMPLE 4 Dosages of Medication and Weight in the Metric System

Drug dosage is frequently based on a patient's weight, measured in kilograms. For example, 20 milligrams of the drug Didronel (used to treat irregular bone formation) should be administered daily for each kilogram of a patient's weight. This can be expressed as 20 mg/kg. How many 400-milligram tablets should be given each day to a patient who weighs 180 pounds?

SOLUTION

- Convert 180 pounds to kilograms, using 1 lb ≈ 0.45 kg to set up the unit fraction.

$$180 \text{ lb} \approx \frac{180 \cancel{\text{lb}}}{1} \cdot \frac{0.45 \text{ kg}}{1 \cancel{\text{lb}}} = 180(0.45) \text{ kg} = 81 \text{ kg}$$

- Determine the dosage. We are given 20 mg/kg, meaning that 20 milligrams of the drug is to be given for each kilogram of a patient's weight. Multiply the patient's weight, 81 kilograms, by 20 to determine the dosage.

$$\text{Dosage} = \frac{81 \cancel{\text{kg}}}{1} \cdot \frac{20 \text{ mg}}{1 \cancel{\text{kg}}} = 81(20) \text{ mg} = 1620 \text{ mg}$$

- Determine the number of tablets that should be given each day. The patient should receive 1620 milligrams of Didronel daily. We are given that each tablet contains 400 milligrams.

$$\text{Number of tablets} = \frac{1620 \cancel{\text{mg}}}{400 \cancel{\text{mg}}} = 4.05$$

The patient should receive 4 tablets of Didronel daily.

CHECK POINT 4 The prescribed dosage of a drug is 6 mg/kg daily. How many 200-milligram tablets should be given each day to a patient who weighs 150 pounds?

Blitzer Bonus

Using the Metric System to Measure Blood-Alcohol Concentration

In Chapter 6, we presented a formula for determining blood-alcohol concentration. Blood-alcohol concentration (BAC) is measured in grams of alcohol per 100 milliliters of blood. To put this measurement into perspective, a 175-pound man has approximately 5 liters (5000 milliliters) of blood and a 12-ounce can of 6%-alcohol-by-volume beer contains about 15 grams of alcohol. Based on the time it takes for alcohol to be absorbed into the bloodstream, as well as its elimination at a rate of 10–15 grams per hour, **Figure 9.9** shows the effect on BAC of a number of drinks on individuals within weight classes. It is illegal to drive with a BAC at 0.08 g/100 mL, or 0.08 gram of alcohol per 100 milliliters of blood, or greater.

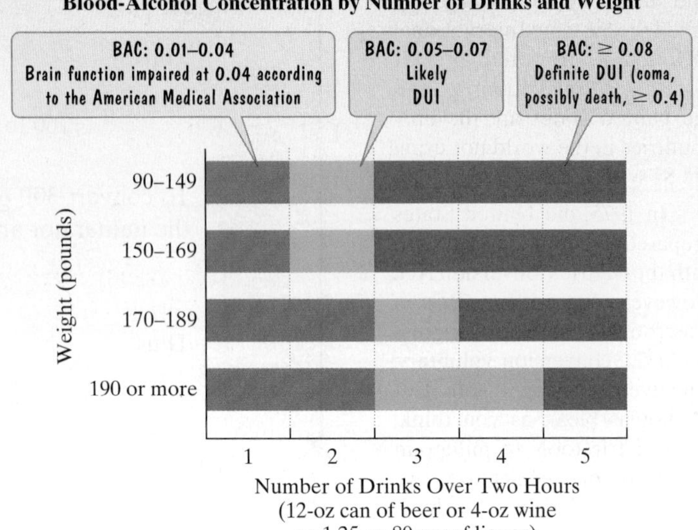

FIGURE 9.9
Source: Patrick McSharry, *Everyday Numbers*, Random House, 2002.

5 *Understand temperature scales.*

Measuring Temperature

You'll be leaving the cold of winter for a vacation to Hawaii. CNN International reports a temperature in Hawaii of 30°C. Should you pack a winter coat?

The idea of changing from Celsius readings—or is it Centigrade?—to familiar Fahrenheit readings can be disorienting. Reporting a temperature of 30°C doesn't have the same impact as the Fahrenheit equivalent of 86 degrees (don't pack the coat). Why these annoying temperature scales?

The **Fahrenheit temperature scale**, the one we are accustomed to, was established in 1714 by the German physicist Gabriel Daniel Fahrenheit. He took a mixture of salt and ice, then thought to be the coldest possible temperature, and called it 0 degrees. He called the temperature of the human body 96 degrees, dividing the space between 0 and 96 into 96 parts. Fahrenheit was wrong about body temperature. It was later found to be 98.6 degrees. On his scale, water froze (without salt) at 32 degrees and boiled at 212 degrees. The symbol ° was used to replace the word *degree.*

Twenty years later, the Swedish scientist Anders Celsius introduced another temperature scale. He set the freezing point of water at 0° and its boiling point at 100°, dividing the space into 100 parts. Degrees were called centigrade until 1948, when the name was officially changed to honor its inventor. However, *centigrade* is still commonly used in the United States.

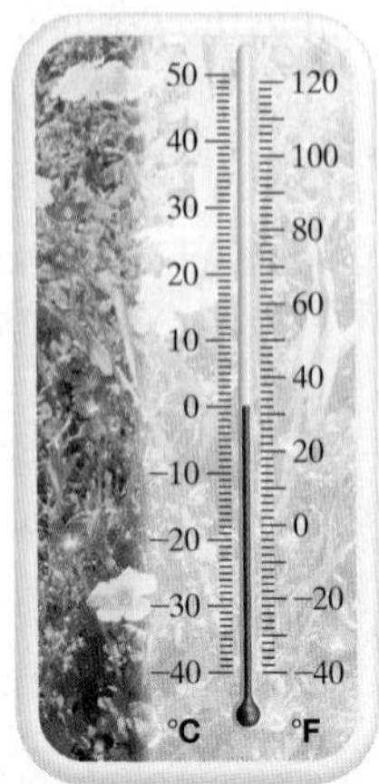

FIGURE 9.10 The Celsius scale is on the left and the Fahrenheit scale is on the right.

Figure 9.10 shows a thermometer that measures temperatures in both degrees Celsius (°C is the scale on the left) and degrees Fahrenheit (°F is the scale on the right). The thermometer should help orient you if you need to know what a temperature in °C means. For example, if it is 40°C, find the horizontal line representing this temperature on the left. Now read across to the °F scale on the right. The reading is above 100°, indicating heat wave conditions.

The following formulas can be used to convert from one temperature scale to the other:

FROM CELSIUS TO FAHRENHEIT

$$F = \frac{9}{5}C + 32$$

FROM FAHRENHEIT TO CELSIUS

$$C = \frac{5}{9}(F - 32)$$

GREAT QUESTION!

Because $\frac{9}{5} = 1.8$, can I use 1.8 instead of $\frac{9}{5}$ in the formula $F = \frac{9}{5}C + 32$?

Yes. The formula used to convert from Celsius to Fahrenheit can be expressed without the use of fractions:

$$F = 1.8C + 32.$$

Some students find this form of the formula easier to memorize.

EXAMPLE 5 *Converting from Celsius to Fahrenheit*

The bills from your European vacation have you feeling a bit feverish, so you decide to take your temperature. The thermometer reads 37°C. Should you panic?

SOLUTION

Use the formula

$$F = \frac{9}{5}C + 32$$

to convert 37°C from °C to °F. Substitute 37 for C in the formula and find the value of F.

$$F = \frac{9}{5}(37) + 32 = 66.6 + 32 = 98.6$$

No need to panic! Your temperature is 98.6°F, which is perfectly normal.

CHECK POINT 5 Convert 50°C from °C to °F.

EXAMPLE 6 *Converting from Fahrenheit to Celsius*

The temperature on a warm spring day is 77°F. Find the equivalent temperature on the Celsius scale.

SOLUTION

Use the formula

$$C = \frac{5}{9}(F - 32)$$

to convert 77°F from °F to °C. Substitute 77 for F in the formula and find the value of C.

$$C = \frac{5}{9}(77 - 32) = \frac{5}{9}(45) = 25$$

Thus, 77°F is equivalent to 25°C.

CHECK POINT 6 Convert 59°F from °F to °C.

Lake Baikal, Siberia, is one of the coldest places on Earth, reaching −76°F (−60°C) in winter. The lowest temperature possible is absolute zero. Scientists have cooled atoms to a few millionths of a degree above absolute zero.

Because temperature is a measure of heat, scientists do not find negative temperatures meaningful in their work. In 1948, the British physicist Lord Kelvin introduced a third temperature scale. He put 0 degrees at absolute zero, the coldest possible temperature, at which there is no heat and molecules stop moving. **Figure 9.11** on the next page illustrates the three temperature scales.

Kelvin	Celsius	Fahrenheit	
373.15 K	100° C	212° F	Water boils
273.15 K	0° C	32° F	Water freezes
0 K	−273.15° C	−459.67° F	Absolute zero

FIGURE 9.11 The three temperature scales

Figure 9.11 shows that water freezes at 273.15 K (read "K" or "Kelvins," not "degrees Kelvin") and boils at 373.15 K. The Kelvin scale is the same as the Celsius scale, except in its starting (zero) point. This makes it easy to go back and forth from Celsius to Kelvin.

FROM CELSIUS TO KELVIN

$$K = C + 273.15$$

FROM KELVIN TO CELSIUS

$$C = K - 273.15$$

Kelvin's scale was embraced by the scientific community. Today, it is the final authority, as scientists define Celsius and Fahrenheit in terms of Kelvins.

Blitzer Bonus

Running a 5 K Race?

A 5 K race means a race at 5 Kelvins. This is a race so cold that no one would be able to move because all the participants would be frozen solid! The proper symbol for a race five kilometers long is a 5 km race.

Concept and Vocabulary Check

Fill in each blank so that the resulting statement is true.

1. In the English system, ____ oz = 1 lb and ______ lb = 1 T.
2. The basic metric unit of weight is the ______, used for very small objects such as a coin, a candy bar, or a teaspoon of salt.
3. A weight of 1 kilogram is approximately _____ pounds.
4. One kilogram of water has a volume of ___ liter(s).
5. On the Fahrenheit temperature scale, water freezes at ____ degrees and boils at _____ degrees.
6. On the Celsius temperature scale, water freezes at ___ degrees and boils at _____ degrees.

In Exercises 7–10, determine whether each statement is true or false. If the statement is false, make the necessary change(s) to produce a true statement.

7. 1 gram = 1000 kilograms ______
8. 1 gram ≈ 28 ounces ______
9. The formula $F = \frac{9}{5}C + 32$ is used to convert from Fahrenheit to Celsius. ______
10. The formula $C = \frac{1}{9}(F - 32)$ is used to convert from Fahrenheit to Celsius. ______

Exercise Set 9.3

Practice Exercises

In Exercises 1–10, convert the given unit of weight to the unit indicated.

1. 7.4 dg to mg
2. 6.9 dg to mg
3. 870 mg to g
4. 640 mg to g
5. 8 g to cg
6. 7 g to cg
7. 18.6 kg to g
8. 0.37 kg to g
9. 0.018 mg to g
10. 0.029 mg to g

*In Exercises 11–18, use **Table 9.10** on page 604 to convert the given measurement of water to the unit indicated.*

11. $0.05\ m^3$ to kg
12. $0.02\ m^3$ to kg
13. 4.2 kg to cm^3
14. 5.8 kg to cm^3

15. 1100 m^3 to t
16. 1500 t to m^3
17. 0.04 kL to g
18. 0.03 kL to g

In Exercises 19–30, use the following equivalents, along with dimensional analysis, to convert the given measurement to the unit indicated. When necessary, round answers to two decimal places.

$$16 \text{ oz} = 1 \text{ lb}$$
$$2000 \text{ lb} = 1 \text{ T}$$
$$1 \text{ oz} \approx 28 \text{ g}$$
$$1 \text{ lb} \approx 0.45 \text{ kg}$$
$$1 \text{ T} \approx 0.9 \text{ t}$$

19. 36 oz to lb
20. 26 oz to lb
21. 36 oz to g
22. 26 oz to g
23. 540 lb to kg
24. 220 lb to kg
25. 80 lb to g
26. 150 lb to g
27. 540 kg to lb
28. 220 kg to lb
29. 200 t to T
30. 100 t to T

In Exercises 31–38, convert the given Celsius temperature to its equivalent temperature on the Fahrenheit scale. Where appropriate, round to the nearest tenth of a degree.

31. 10°C
32. 20°C
33. 35°C
34. 45°C
35. 57°C
36. 98°C
37. −5°C
38. −10°C

In Exercises 39–50, convert the given Fahrenheit temperature to its equivalent temperature on the Celsius scale. Where appropriate, round to the nearest tenth of a degree.

39. 68°F
40. 86°F
41. 41°F
42. 50°F
43. 72°F
44. 90°F
45. 23°F
46. 14°F
47. 350°F
48. 475°F
49. −22°F
50. −31°F

Practice Plus

51. The nine points shown below represent Celsius temperatures and their equivalent Fahrenheit temperatures. Also shown is a line that passes through the points.

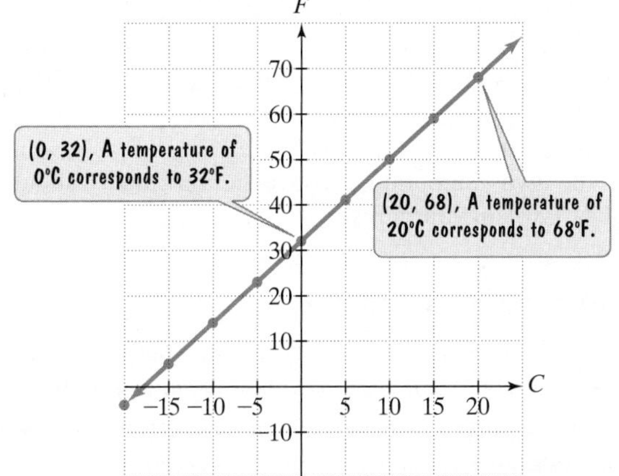

a. Use the coordinates of the two points identified by the voice balloons to compute the line's slope. Express the answer as a fraction reduced to lowest terms. What does this mean about the change in Fahrenheit temperature for each degree change in Celsius temperature?

b. Use the slope-intercept form of the equation of a line, $y = mx + b$, the slope from part (a), and the y-intercept shown by the graph to derive the formula used to convert from Celsius to Fahrenheit.

52. Solve the formula used to convert from Celsius to Fahrenheit for C and derive the other temperature conversion formula.

Application Exercises

In Exercises 53–59, selecting from milligram, gram, kilogram, and tonne, determine the best unit of measure to express the given item's weight.

53. A bee
54. This book
55. A tablespoon of salt
56. A Boeing 747
57. A stacked washer-dryer
58. A pen
59. An adult male

In Exercises 60–66, select the best estimate for the weight of the given item.

60. A newborn infant's weight
 a. 3000 kg b. 300 kg
 c. 30 kg d. 3 kg
61. The weight of a nickel
 a. 5 kg b. 5 g c. 5 mg
62. A person's weight
 a. 60 kg b. 60 g c. 60 dag
63. The weight of a box of cereal
 a. 0.5 kg b. 0.5 g c. 0.5 t
64. The weight of a glass of water
 a. 400 dg b. 400 g
 c. 400 dag d. 400 hg
65. The weight of a regular-size car
 a. 1500 dag b. 1500 hg c. 1500 kg d. 15,000 kg
66. The weight of a bicycle
 a. 140 kg b. 140 hg c. 140 dag d. 140 g
67. Six items purchased at a grocery store weigh 14 kilograms. One of the items is detergent weighing 720 grams. What is the total weight, in kilograms, of the other five items?
68. If a nickel weighs 5 grams, how many nickels are there in 4 kilograms of nickels?
69. If the cost to mail a letter is 47 cents for mail weighing up to one ounce and 21 cents for each additional ounce or fraction of an ounce, find the cost of mailing a letter that weighs 85 grams.

70. Using the information given below the pictured finback whale, estimate the weight, in tons and kilograms, of the killer whale.

71. Which is more economical: purchasing the economy size of a detergent at 3 kilograms for \$3.15 or purchasing the regular size at 720 grams for 60¢?

Exercises 72–73 ask you to determine drug dosage by a patient's weight. Use the fact that 1 lb ≈ 0.45 kg.

72. The prescribed dosage of a drug is 10 mg/kg daily, meaning that 10 milligrams of the drug should be administered daily for each kilogram of a patient's weight. How many 400-milligram tablets should be given each day to a patient who weighs 175 pounds?

73. The prescribed dosage of a drug is 15 mg/kg daily, meaning that 15 milligrams of the drug should be administered daily for each kilogram of a patient's weight. How many 200-milligram tablets should be given each day to a patient who weighs 120 pounds?

The label on a bottle of Emetrol ("for food or drink indiscretions") reads

Each 5 mL teaspoonful contains glucose, 1.87 g; levulose, 1.87 g; and phosphoric acid, 21.5 mg.

Use this information to solve Exercises 74–75.

74. a. Find the amount of glucose in the recommended dosage of two teaspoons.

b. If the bottle contains 4 ounces, find the quantity of glucose in the bottle. (1 oz ≈ 30 mL)

75. a. Find the amount of phosphoric acid in the recommended dosage of two teaspoons.

b. If the bottle contains 4 ounces, find the quantity of phosphoric acid in the bottle. (1 oz ≈ 30 mL)

In Exercises 76–79, select the best estimate of the Celsius temperature of

76. A very hot day.

a. 85°C **b.** 65°C **c.** 35°C **d.** 20°C

77. A warm winter day in Washington, D.C.

a. 10°C **b.** 30°C **c.** 50°C **d.** 70°C

78. A setting for a home thermostat.

a. 20°C **b.** 40°C **c.** 60°C **d.** 80°C

79. The oven temperature for cooking a roast.

a. 80°C **b.** 100°C **c.** 175°C **d.** 350°C

Exercises 80–81 give the average temperature of some of the world's hottest places. In each exercise, determine which location has the hotter average temperature and by how many degrees Fahrenheit. Round to the nearest tenth of a degree.

80. Assab, Eritrea: 86.8°F; Dalol, Ethiopia: 34.6°C

81. Berbera, Somalia: 86.2°F; Néma, Mauritania: 30.3°C

Exercises 82–83 give the average temperature of some of the world's coldest places. In each exercise, determine which location has the colder average temperature and by how many degrees Celsius. Round to the nearest tenth of a degree.

82. Plateau, Antarctica: −56.7°C;
Amundsen-Scott, Antarctica: −56.2°F

83. Eismitte, Greenland: −29.2°C;
Resolute, Canada: −11.6°F

(Source for Exercises 80–83: Russell Ash, *The Top 10 of Everything 2009*)

Explaining the Concepts

84. Describe the difference between weight and mass.

85. Explain how to use dimensional analysis to convert 200 pounds to kilograms.

86. Why do you think that countries using the metric system prefer the Celsius scale over the Fahrenheit scale?

87. Describe in words how to convert from Celsius to Fahrenheit.

88. Describe in words how to convert from Fahrenheit to Celsius.

89. If you decide to travel outside the United States, which one of the two temperature conversion formulas should you take? Explain your answer.

Critical Thinking Exercises

Make Sense? *In Exercises 90–93, determine whether each statement makes sense or does not make sense, and explain your reasoning.*

90. When I played high school basketball, I weighed 250 kilograms.

91. It's not realistic to convert 500 mg to g because measurements should not involve fractional units.

92. Both the English system and the metric system enable me to easily see relationships among volume, capacity, and weight.

93. A comfortable classroom temperature is 20°C.

In Exercises 94–101, determine whether each statement is true or false. If the statement is false, make the necessary change(s) to produce a true statement.

94. A 4-pound object weighs more than a 2000-gram object.

95. A 100-milligram object weighs more than a 2-ounce object.

96. A 50-gram object weighs more than a 2-ounce object.

97. A 10-pound object weighs more than a 4-kilogram object.

98. Flour selling at 3¢ per gram is a better buy than flour selling at 55¢ per pound.

99. The measures

32,600 g, 32.1 kg, 4 lb, 36 oz

are arranged in order, from greatest to least weight.

100. If you are taking aspirin to relieve cold symptoms, a reasonable dose is 2 kilograms four times a day.

101. A large dog weighs about 350 kilograms.

Group Exercise

102. Present a group report on the current status of the metric system in the United States. At present, does it appear that the United States will convert to the metric system? Who supports the conversion and who opposes it? Summarize each side's position. Give examples of how our current system of weights and measures is an economic liability. What are the current obstacles to metric conversion?

Chapter Summary, Review, and Test

SUMMARY – DEFINITIONS AND CONCEPTS	EXAMPLES
9.1 Measuring Length; The Metric System	
a. The result obtained from measuring length is called a linear measurement, stated in linear units.	
b. Linear Units: The English System 12 in. = 1 ft, 3 ft = 1 yd, 36 in. = 1 yd, 5280 ft = 1 mi	
c. Dimensional Analysis Multiply the given measurement by a unit fraction with the unit of measurement that needs to be introduced in the numerator and the unit of measurement that needs to be eliminated in the denominator.	Ex. 1, p. 583
d. Linear Units: The Metric System The basic unit is the meter (m), approximately 39 inches. 1 km = 1000 m, 1 hm = 100 m, 1 dam = 10 m, 1 dm = 0.1 m, 1 cm = 0.01 m, 1 mm = 0.001 m	
e. Changing Linear Units within the Metric System $\times 10$ km hm dam m dm cm mm $\div 10$	Ex. 2, p. 586
f. English and metric equivalents for length are given in **Table 9.3** on page 587.	Ex. 3, p. 588; Ex. 4, p. 589
9.2 Measuring Area and Volume	
a. The area measure of a plane region is the number of square units contained in the given region.	Ex. 1, p. 592
b. Square Units: The English System $1 \text{ ft}^2 = 144 \text{ in.}^2$, $1 \text{ yd}^2 = 9 \text{ ft}^2$, $1 \text{ a} = 43{,}560 \text{ ft}^2 = 4840 \text{ yd}^2$, $1 \text{ mi}^2 = 640 \text{ a}$	Ex. 2, p. 593; Ex. 3, p. 593
c. English and metric equivalents for area are given in **Table 9.4** on page 594.	Ex. 4, p. 594
d. The volume measure of a three-dimensional figure is the number of cubic units contained in its interior.	Ex. 5, p. 595
e. Capacity refers to the amount of fluid that a three-dimensional object can hold. English units for capacity include pints, quarts, and gallons: 2 pt = 1 qt; 4 qt = 1 gal; one cubic yard has a capacity of about 200 gallons. See **Table 9.5** on page 596.	Ex. 6, p. 596

f. The basic unit for capacity in the metric system is the liter (L). One liter is about 1.06 quarts. Prefixes for the liter are the same as throughout the metric system.	
g. **Table 9.7** on page 597 shows relationships between volume and capacity in the metric system. $1\text{ dm}^3 = 1000\text{ cm}^3 = 1\text{ L}$	Ex. 7, p. 597
h. English and metric equivalents for capacity are given in **Table 9.8** on page 598.	Ex. 8, p. 598

9.3 Measuring Weight and Temperature

a. Weight: The English System $16\text{ oz} = 1\text{ lb}, 2000\text{ lb} = 1\text{ T}$	
b. Units of Weight: The Metric System The basic unit is the gram (g). $1000\text{ grams }(1\text{ kg}) \approx 2.2\text{ lb},$ $1\text{ kg} = 1000\text{ g}, 1\text{ hg} = 100\text{ g}, 1\text{ dag} = 10\text{ g},$ $1\text{ dg} = 0.1\text{ g}, 1\text{ cg} = 0.01\text{ g}, 1\text{ mg} = 0.001\text{ g}$	
c. Changing Units of Weight within the Metric System 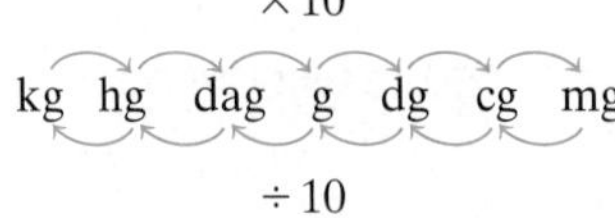	Ex. 1, p. 603
d. One kilogram of water has a volume of 1 liter. **Table 9.10** on page 604 shows relationships between volume and weight of water in the metric system.	Ex. 2, p. 604
e. English and metric equivalents for weight are given in **Table 9.11** on page 604.	Ex. 3, p. 604; Ex. 4, p. 605
f. Temperature Scales: Celsius to Fahrenheit: $F = \frac{9}{5}C + 32$	Ex. 5, p. 607
g. Temperature Scales: Fahrenheit to Celsius: $C = \frac{5}{9}(F - 32)$	Ex. 6, p. 607

Review Exercises

9.1

In Exercises 1–4, use dimensional analysis to convert the quantity to the indicated unit.

1. 69 in. to ft
2. 9 in. to yd
3. 21 ft to yd
4. 13,200 ft to mi

In Exercises 5–10, convert the given linear measurement to the metric unit indicated.

5. 22.8 m to cm
6. 7 dam to m
7. 19.2 hm to m
8. 144 cm to hm
9. 0.5 mm to m
10. 18 cm to mm

In Exercises 11–16, use the given English and metric equivalents, along with dimensional analysis, to convert the given measurement to the unit indicated. Where necessary, round answers to two decimal places.

11. 23 in. to cm
12. 19 cm to in.
13. 330 mi to km
14. 600 km to mi
15. 14 m to yd
16. 12 m to ft

$1\text{ in.} = 2.54\text{ cm}$
$1\text{ ft} = 30.48\text{ cm}$
$1\text{ yd} \approx 0.9\text{ m}$
$1\text{ mi} \approx 1.6\text{ km}$

17. Express 45 kilometers per hour in miles per hour.
18. Express 60 miles per hour in kilometers per hour.
19. Arrange from smallest to largest: 0.024 km, 2400 m, 24,000 cm.
20. If you jog six times around a track that is 800 meters long, how many kilometers have you covered?

9.2

21. Use the given figure to find its area in square units.

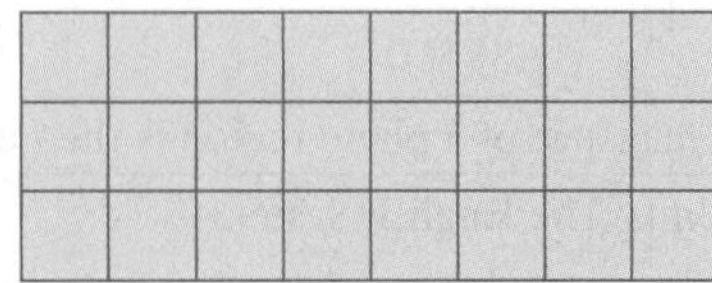

22. Singapore, with an area of 268 square miles and a population of 4,425,700, is one of the world's most densely populated countries. Find Singapore's population density, to the nearest tenth. Describe what this means.
23. Acadia National Park on the coast of Maine consists of 47,453 acres. How large is the national park in square miles? Round to the nearest square mile. $(1\text{ mi}^2 = 640\text{ a})$

24. Given 1 acre $\approx$ 0.4 hectare, use dimensional analysis to find the size of a property in acres measured at 7.2 hectares.

25. Using 1 $ft^2 \approx 0.09\ m^2$, convert 30 m^2 to ft^2.

26. Using 1 $mi^2 \approx 2.6\ km^2$, convert 12 mi^2 to km^2.

27. Which one of the following is a reasonable measure for the area of a flower garden in a person's yard?

a. 100 m^2

b. 0.4 ha

c. 0.01 km^2

28. Use the given figure to find its volume in cubic units.

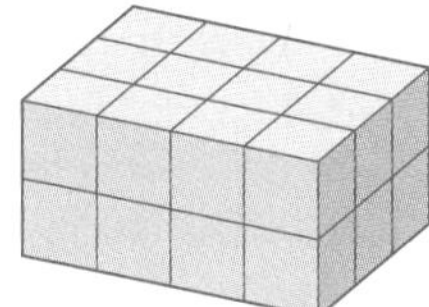

29. A swimming pool has a volume of 33,600 cubic feet. Given that 1 cubic foot has a capacity of about 7.48 gallons, how many gallons of water does the pool hold?

30. An aquarium has a volume of 76,000 cubic centimeters. How many liters of water does the aquarium hold?

In Exercises 31–35, use the given English and metric equivalents, along with dimensional analysis, to convert the given measurement to the unit indicated. Where necessary, round answers to two decimal places.

1 tsp	$\approx$ 5 mL
1 tbsp	$\approx$ 15 mL
1 fl oz	$\approx$ 30 mL
1 c	$\approx$ 0.24 L
1 pt	$\approx$ 0.47 L
1 qt	$\approx$ 0.95 L
1 gal	$\approx$ 3.8 L

31. 22 mL to tsp

32. 5.4 L to c

33. 8 L to qt

34. 6 gal to L

35. A physician orders 15 cc of medication.

a. How many milliliters of the drug should be administered?

b. How many fluid ounces of the drug should be administered?

36. The capacity of a one-quart container of juice is approximately

a. 0.1 kL **b.** 0.5 L **c.** 1 L **d.** 1 mL

37. There are 3 feet in a yard. Explain why there are not 3 square feet in a square yard. If helpful, illustrate your explanation with a diagram.

38. Explain why the area of Texas could not be measured in cubic miles.

9.3

In Exercises 39–42, convert the given unit of weight to the unit indicated.

39. 12.4 dg to mg

40. 12 g to cg

41. 0.012 mg to g

42. 450 mg to kg

In Exercises 43–44, use ***Table 9.10*** *on page 604 to convert the given measurement of water to the unit or units indicated.*

43. 50 kg to cm^3

44. 4 kL to dm^3 to g

45. Using 1 lb $\approx$ 0.45 kg, convert 210 pounds to kilograms.

46. Using 1 oz $\approx$ 28 g, convert 392 grams to ounces.

47. The prescribed dosage of a drug is 12 mg/kg daily. How many 400-milligram tablets should be given each day to a patient who weighs 220 pounds? (1 lb $\approx$ 0.45 kg)

48. If you are interested in your weight in the metric system, would it be best to report it in milligrams, grams, or kilograms? Explain why the unit you selected would be most appropriate. Explain why each of the other two units is not the best choice for reporting your weight.

49. Given 16 oz = 1 lb, use dimensional analysis to convert 36 ounces to pounds.

In Exercises 50–51, select the best estimate for the weight of the given item.

50. A dollar bill:

a. 1 g **b.** 10 g **c.** 1 kg **d.** 4 kg

51. A hamburger:

a. 3 kg **b.** 1 kg **c.** 200 g **d.** 5 g

In Exercises 52–56, convert the given Celsius temperature to its equivalent temperature on the Fahrenheit scale.

52. 15°C

53. 100°C

54. 5°C

55. 0°C

56. −25°C

In Exercises 57–62, convert the given Fahrenheit temperature to its equivalent temperature on the Celsius scale.

57. 59°F

58. 41°F

59. 212°F

60. 98.6°F

61. 0°F

62. 14°F

63. Is a decrease of 15° Celsius more or less than a decrease of 15° Fahrenheit? Explain your answer.

Chapter 9 Test

1. Change 807 mm to hm.
2. Given 1 inch = 2.54 centimeters, use dimensional analysis to change 635 centimeters to inches.
3. If you jog eight times around a track that is 600 meters long, how many kilometers have you covered?

In Exercises 4–6, write the most reasonable metric unit for length in each blank. Select from mm, cm, m, and km.

4. A human thumb is 20 ______ wide.
5. The height of the table is 45 ______.
6. The towns are 60 ______ apart.
7. If 1 mile ≈ 1.6 kilometers, express 80 miles per hour in kilometers per hour.
8. How many times greater is a square yard than a square foot?
9. Australia has a population of 22,992,654 and an area of 2,967,908 square miles. Find Australia's population density to the nearest tenth. Describe what this means.
10. Given 1 acre ≈ 0.4 hectare, use dimensional analysis to find the area of a property measured at 18 hectares.
11. The area of a dollar bill is approximately
    a. 10 cm^2
    b. 100 cm^2
    c. 1000 cm^2
    d. 1 m^2
12. There are 10 decimeters in a meter. Explain why there are not 10 cubic decimeters in a cubic meter. How many times greater is a cubic meter than a cubic decimeter?
13. The label on a bottle of Pepto Bismol indicates that the dosage should not exceed 16 tablespoons in 24 hours. How many fluid ounces is the maximum daily dose? (1 tbsp ≈ 15 mL; 1 fl oz ≈ 30 mL)
14. A swimming pool has a volume of 10,000 cubic feet. Given that 1 cubic foot has a capacity of about 7.48 gallons, how many gallons of water does the pool hold?
15. The capacity of a pail used to wash floors is approximately
    a. 3 L
    b. 12 L
    c. 80 L
    d. 2 kL
16. Change 137 g to kg.
17. The prescribed dosage of a drug is 10 mg/kg daily. How many 200-milligram tablets should be given each day to a patient who weighs 130 pounds? (1 lb ≈ 0.45 kg)

In Exercises 18–19, write the most reasonable metric unit for weight in each blank. Select from mg, g, and kg.

18. My suitcase weighs 20 ______.
19. I took a 350 ______ aspirin.
20. Convert 30°C to Fahrenheit.
21. Convert 176°F to Celsius.
22. Comfortable room temperature is approximately
    a. 70°C
    b. 50°C
    c. 30°C
    d. 20°C

Geometry

10

GEOMETRY IS THE STUDY OF THE SPACE YOU LIVE IN AND THE SHAPES THAT SURROUND YOU. YOU'RE EVEN MADE OF IT! THE HUMAN LUNG CONSISTS OF NEARLY 300 SPHERICAL AIR SACS, geometrically designed to provide the greatest surface area within the limited volume of our bodies. Viewed in this way, geometry becomes an intimate experience.

For thousands of years, people have studied geometry in some form to obtain a better understanding of the world in which they live. A study of the shape of your world will provide you with many practical applications and perhaps help to increase your appreciation of its beauty.

Here's where you'll find these applications:

- A relationship between geometry and the visual arts is developed in Section 10.3 (Tessellations: pages 640–642).
- Using geometry to describe nature's complexity is discussed in Section 10.7 (Fractals: page 682).

10.1 Points, Lines, Planes, and Angles

WHAT AM I SUPPOSED TO LEARN?

After studying this section, you should be able to:

1 Understand points, lines, and planes as the basis of geometry.

2 Solve problems involving angle measures.

3 Solve problems involving angles formed by parallel lines and transversals.

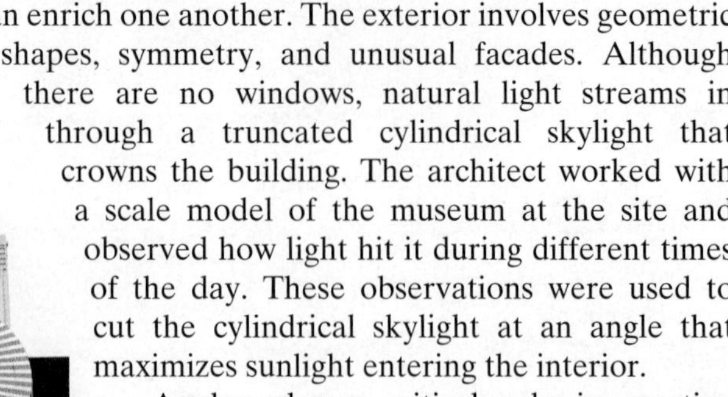

THE SAN FRANCISCO MUSEUM OF MODERN ART WAS constructed in 1995 to illustrate how art and architecture can enrich one another. The exterior involves geometric shapes, symmetry, and unusual facades. Although there are no windows, natural light streams in through a truncated cylindrical skylight that crowns the building. The architect worked with a scale model of the museum at the site and observed how light hit it during different times of the day. These observations were used to cut the cylindrical skylight at an angle that maximizes sunlight entering the interior.

Angles play a critical role in creating modern architecture. They are also fundamental in the study of geometry. The word "geometry" means "earth measure." Because it involves the mathematics of shapes, geometry connects mathematics to art and architecture. It also has many practical applications. You can use geometry at home when you buy carpet, build a fence, tile a floor, or determine whether a piece of furniture will fit through your doorway. In this chapter, we look at the shapes that surround us and their applications.

1 *Understand points, lines, and planes as the basis of geometry.*

Points, Lines, and Planes

Points, lines, and planes make up the basis of all geometry. Stars in the night sky look like points of light. Long stretches of overhead power lines that appear to extend endlessly look like lines. The top of a flat table resembles part of a plane. However, stars, power lines, and tabletops only approximate points, lines, and planes. Points, lines, and planes do not exist in the physical world. Representations of these forms are shown in **Figure 10.1**. A **point**, represented as a small dot, has no length, width, or thickness. No object in the real world has zero size. A **line**, connecting two points along the shortest possible path, has no thickness and extends infinitely in both directions. However, no familiar everyday object is infinite in length. A **plane** is a flat surface with no thickness and no boundaries. This page resembles a plane, although it does not extend indefinitely and it does have thickness.

• A

Point A

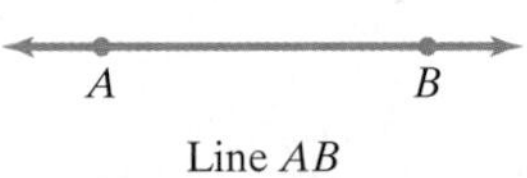

Line AB

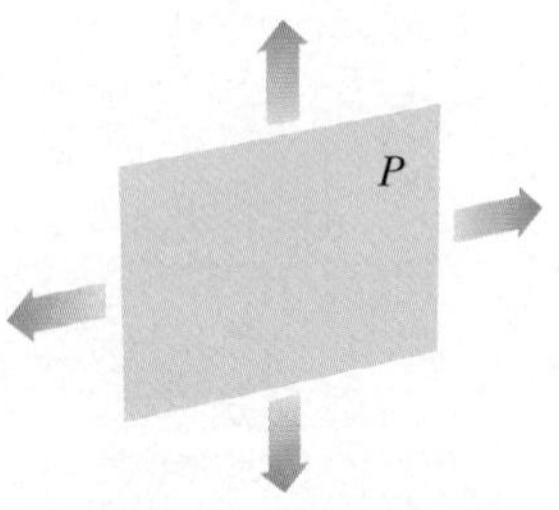

Plane P

FIGURE 10.1 Representing a point, a line, and a plane

A line may be named using any two of its points. In **Figure 10.2(a)**, line AB can be symbolized $\overleftrightarrow{AB}$ or $\overleftrightarrow{BA}$. Any point on the line divides the line into three parts—the point and two **half-lines**. **Figure 10.2(b)** illustrates half-line AB, symbolized $\overset{\circ}{\overrightarrow{AB}}$. The open circle above the A in the symbol and in the diagram indicates that point A is not included in the half-line. A **ray** is a half-line with its endpoint included. **Figure 10.2(c)** illustrates ray AB, symbolized $\overrightarrow{AB}$. The closed dot above the A in the diagram shows that point A is included in the ray. A portion of a line joining two points and including the endpoints is called a **line segment**. **Figure 10.2(d)** illustrates line segment AB, symbolized $\overline{AB}$ or $\overline{BA}$.

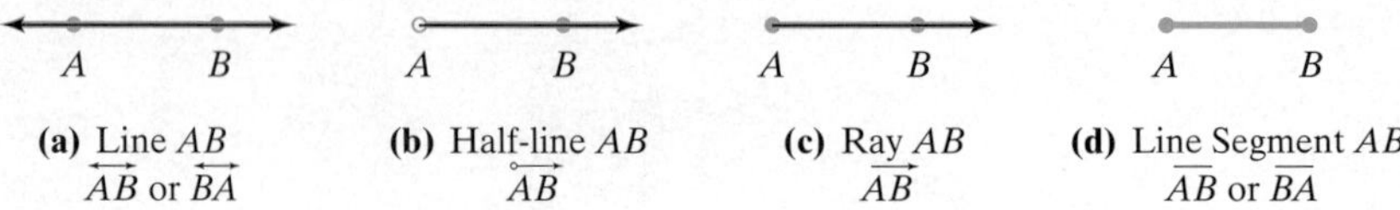

(a) Line AB $\overleftrightarrow{AB}$ or $\overleftrightarrow{BA}$ **(b)** Half-line AB $\overset{\circ}{\overrightarrow{AB}}$ **(c)** Ray AB $\overrightarrow{AB}$ **(d)** Line Segment AB $\overline{AB}$ or $\overline{BA}$

FIGURE 10.2 Lines, half-lines, rays, and line segments

Angles

An **angle**, symbolized $\measuredangle$, is formed by the union of two rays that have a common endpoint. One ray is called the **initial side** and the other the **terminal side**.

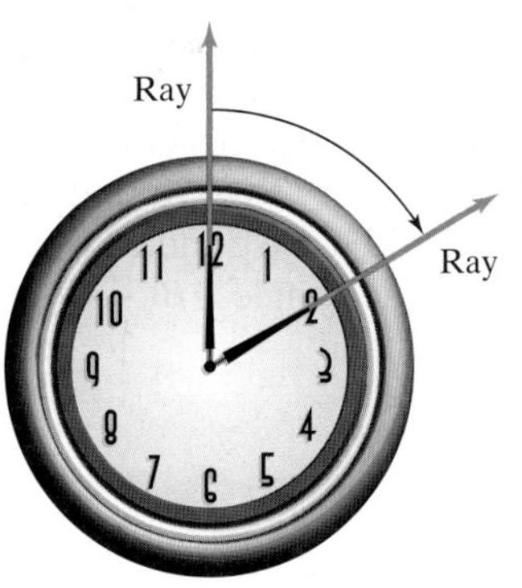

FIGURE 10.3 Clock with hands forming an angle

A rotating ray is often a useful way to think about angles. The ray in **Figure 10.3** rotates from 12 to 2. The ray pointing to 12 is the initial side and the ray pointing to 2 is the terminal side. The common endpoint of an angle's initial side and terminal side is the **vertex** of the angle.

Figure 10.4 shows an angle. The common endpoint of the two rays, B, is the vertex. The two rays that form the angle, $\overrightarrow{BA}$ and $\overrightarrow{BC}$, are the sides. The four ways of naming the angle are shown to the right of **Figure 10.4**.

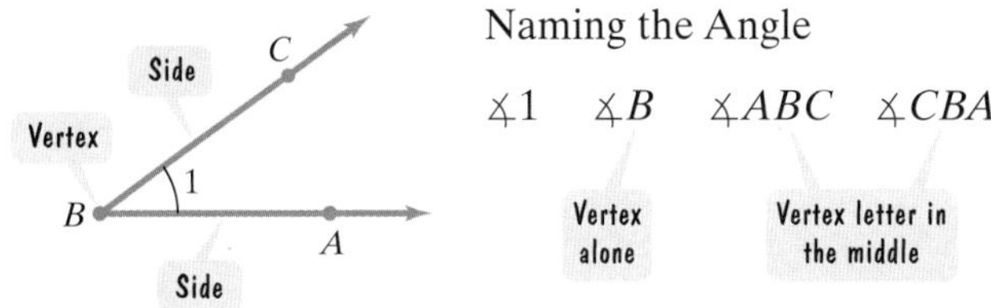

FIGURE 10.4 An angle: two rays with a common endpoint

2 *Solve problems involving angle measures.*

Measuring Angles Using Degrees

Angles are measured by determining the amount of rotation from the initial side to the terminal side. One way to measure angles is in **degrees**, symbolized by a small, raised circle °. Think of the hour hand of a clock. From 12 noon to 12 midnight, the hour hand moves around in a complete circle. By definition, the ray has rotated through 360 degrees, or 360°, shown in **Figure 10.5**. Using 360° as the amount of rotation of a ray back onto itself, **a degree, 1°, is $\frac{1}{360}$ of a complete rotation.**

FIGURE 10.5 A complete 360° rotation

EXAMPLE 1 *Using Degree Measure*

The hand of a clock moves from 12 to 2 o'clock, shown in **Figure 10.3**. Through how many degrees does it move?

SOLUTION

We know that one complete rotation is 360°. Moving from 12 to 2 o'clock is $\frac{2}{12}$, or $\frac{1}{6}$, of a complete revolution. Thus, the hour hand moves

$$\frac{1}{6} \times 360° = \frac{360°}{6} = 60°$$

in going from 12 to 2 o'clock.

CHECK POINT 1 The hand of a clock moves from 12 to 1 o'clock. Through how many degrees does it move?

Figure 10.6 shows angles classified by their degree measurement. An **acute angle** measures less than 90° [see **Figure 10.6(a)**]. A **right angle**, one-quarter of a complete rotation, measures 90° [**Figure 10.6(b)**]. Examine the right angle—do you see a small square at the vertex? This symbol is used to indicate a right angle. An **obtuse angle** measures more than 90°, but less than 180° [**Figure 10.6(c)**]. Finally, a **straight angle**, one-half a complete rotation, measures 180° [**Figure 10.6(d)**]. The two rays in a straight angle form a straight line.

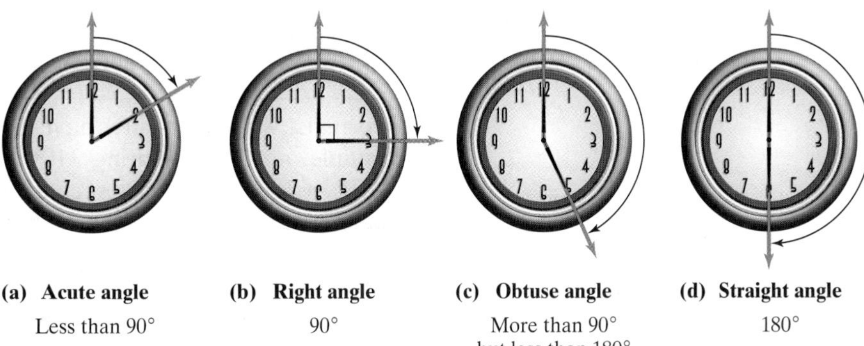

(a) Acute angle Less than 90° **(b) Right angle** 90° **(c) Obtuse angle** More than 90° but less than 180° **(d) Straight angle** 180°

FIGURE 10.6 Classifying angles by their degree measurements

Figure 10.7 illustrates a **protractor**, used for finding the degree measure of an angle. As shown in the figure, we measure an angle by placing the center point of the protractor on the vertex of the angle and the straight side of the protractor along one side of the angle. The measure of $\measuredangle ABC$ is then read as 50°. Observe that the measure is not 130° because the angle is obviously less than 90°. We indicate the angle's measure by writing $m\measuredangle ABC = 50°$, read "the measure of angle ABC is 50°."

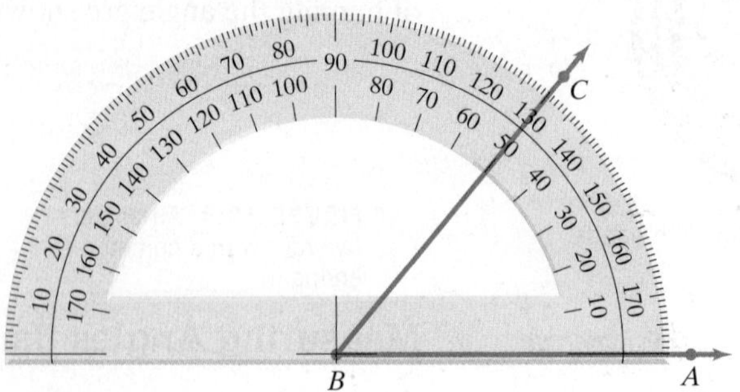

FIGURE 10.7 Using a protractor to measure an angle: $m\measuredangle ABC = 50°$

Two angles whose measures have a sum of 90° are called **complementary angles**. For example, angles measuring 70° and 20° are complementary angles because $70° + 20° = 90°$. For angles such as those measuring 70° and 20°, each angle is the **complement** of the other: The 70° angle is the complement of the 20° angle and the 20° angle is the complement of the 70° angle. **The measure of the complement can be found by subtracting the angle's measure from 90°.** For example, we can find the complement of a 25° angle by subtracting 25° from 90°: $90° - 25° = 65°$. Thus, an angle measuring 65° is the complement of one measuring 25°.

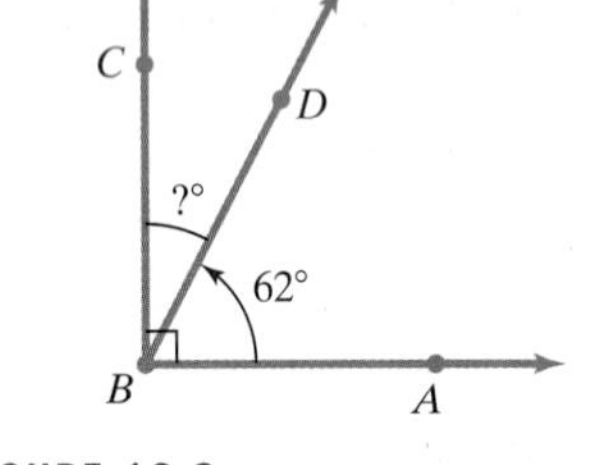

FIGURE 10.8

EXAMPLE 2 *Angle Measures and Complements*

Use **Figure 10.8** to find $m\measuredangle DBC$.

SOLUTION

The measure of $\measuredangle DBC$ is not yet known. It is shown as ?° in **Figure 10.8**. The acute angles $\measuredangle ABD$, which measures 62°, and $\measuredangle DBC$ form a right angle, indicated by the square at the vertex. This means that the measures of the acute angles add up to 90°. Thus, $\measuredangle DBC$ is the complement of the angle measuring 62°. The measure of $\measuredangle DBC$ is found by subtracting 62° from 90°:

$$m\measuredangle DBC = 90° - 62° = 28°.$$

The measure of $\measuredangle DBC$ can also be found using an algebraic approach.

$$m\measuredangle ABD + m\measuredangle DBC = 90°$$ The sum of the measures of complementary angles is 90°.

$$62° + m\measuredangle DBC = 90°$$ We are given $m\measuredangle ABD = 62°$.

$$m\measuredangle DBC = 90° - 62° = 28°$$ Subtract 62° from both sides of the equation.

CHECK POINT 2 In **Figure 10.8**, let $m\measuredangle DBC = 19°$. Find $m\measuredangle DBA$.

Two angles whose measures have a sum of 180° are called **supplementary angles**. For example, angles measuring 110° and 70° are supplementary angles because $110° + 70° = 180°$. For angles such as those measuring 110° and 70°, each angle is the **supplement** of the other: The 110° angle is the supplement of the 70° angle, and the 70° angle is the supplement of the 110° angle. **The measure of the supplement can be found by subtracting the angle's measure from 180°.** For example, we can find the supplement of a 25° angle by subtracting 25° from 180°: $180° - 25° = 155°$. Thus, an angle measuring 155° is the supplement of one measuring 25°.

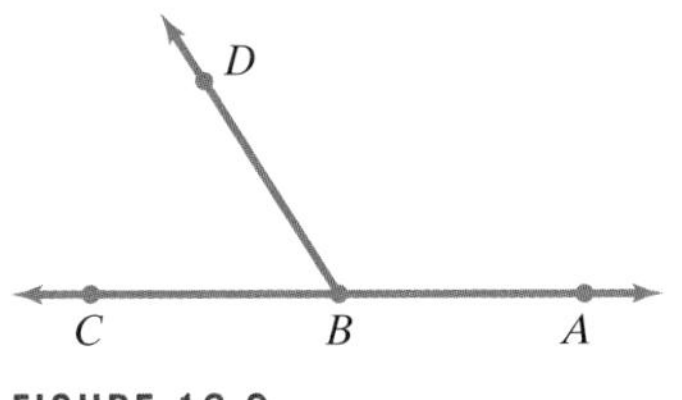

FIGURE 10.9

EXAMPLE 3 Angle Measures and Supplements

Figure 10.9 shows that $\measuredangle ABD$ and $\measuredangle DBC$ are supplementary angles. If $m\measuredangle ABD$ is 66° greater than $m\measuredangle DBC$, find the measure of each angle.

SOLUTION

Let $m\measuredangle DBC = x$. Because $m\measuredangle ABD$ is 66° greater than $m\measuredangle DBC$, then $m\measuredangle ABD = x + 66°$. We are given that these angles are supplementary.

$$m\measuredangle DBC + m\measuredangle ABD = 180°$$ The sum of the measures of supplementary angles is 180°.

$$x + (x + 66°) = 180°$$ Substitute the variable expressions for the measures.

$$2x + 66° = 180°$$ Combine like terms: $x + x = 2x$.

$$2x = 114°$$ Subtract 66° from both sides.

$$x = 57°$$ Divide both sides by 2.

Thus, $m\measuredangle DBC = 57°$ and $m\measuredangle ABD = 57° + 66° = 123°$.

CHECK POINT 3 In **Figure 10.9**, if $m\measuredangle ABD$ is 88° greater than $m\measuredangle DBC$, find the measure of each angle.

FIGURE 10.10

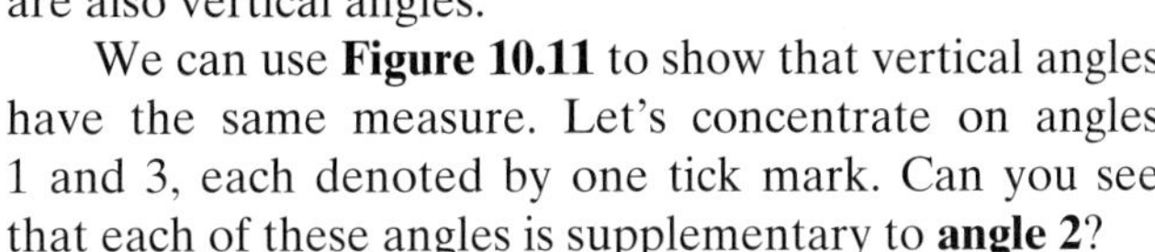

Figure 10.10 illustrates a highway sign that warns of a railroad crossing. When two lines intersect, the opposite angles formed are called **vertical angles**.

In **Figure 10.11**, there are two pairs of vertical angles. Angles 1 and 3 are vertical angles. Angles 2 and 4 are also vertical angles.

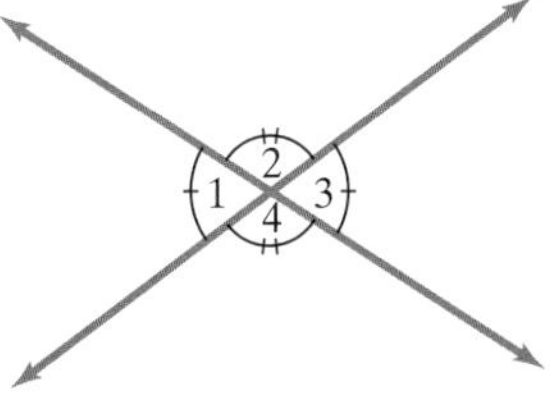

FIGURE 10.11

We can use **Figure 10.11** to show that vertical angles have the same measure. Let's concentrate on angles 1 and 3, each denoted by one tick mark. Can you see that each of these angles is supplementary to **angle 2**?

$$m\measuredangle 1 + m\measuredangle 2 = 180°$$ The sum of the measures of supplementary angles is 180°.

$$m\measuredangle 2 + m\measuredangle 3 = 180°$$

$$m\measuredangle 1 + m\measuredangle 2 = m\measuredangle 2 + m\measuredangle 3$$ Substitute $m\measuredangle 2 + m\measuredangle 3$ for 180° in the first equation.

$$m\measuredangle 1 = m\measuredangle 3$$ Subtract $m\measuredangle 2$ from both sides.

Using a similar approach, we can show that $m\measuredangle 2 = m\measuredangle 4$, each denoted by two tick marks in **Figure 10.11**.

> Vertical angles have the same measure.

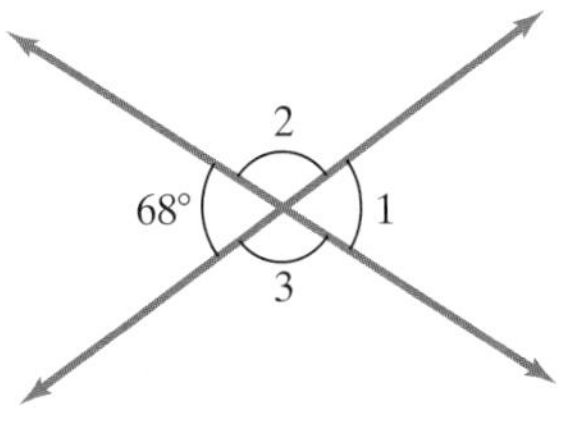

FIGURE 10.12

EXAMPLE 4 Using Vertical Angles

Figure 10.12 shows that the angle on the left measures 68°. Find the measures of the other three angles.

SOLUTION

Angle 1 and the angle measuring 68° are vertical angles. Because vertical angles have the same measure,

$$m\measuredangle 1 = 68°.$$

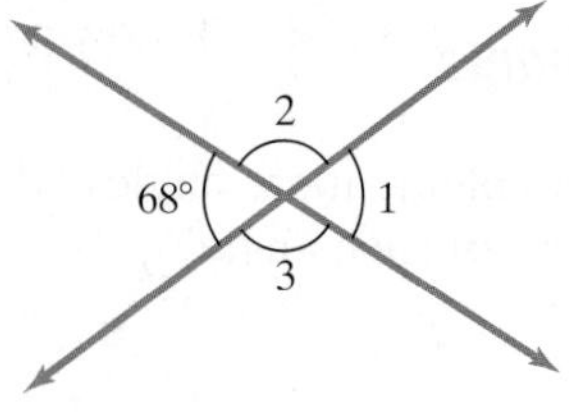

FIGURE 10.12 (repeated)

Angle 2 and the angle measuring 68° form a straight angle and are supplementary. Because their measures add up to 180°,

$$m\measuredangle 2 = 180° - 68° = 112°.$$

Angle 2 and angle 3 are also vertical angles, so they have the same measure. Because the measure of angle 2 is 112°,

$$m\measuredangle 3 = 112°.$$

CHECK POINT 4 In **Figure 10.12**, assume that the angle on the left measures 57°. Find the measures of the other three angles.

Parallel Lines

Parallel lines are lines that lie in the same plane and have no points in common. If two different lines in the same plane are not parallel, they have a single point in common and are called **intersecting lines.** If the lines intersect at an angle of 90°, they are called **perpendicular lines.**

If we intersect a pair of parallel lines with a third line, called a **transversal**, eight angles are formed, as shown in **Figure 10.13**. Certain pairs of these angles have special names, as well as special properties. These names and properties are summarized in **Table 10.1**.

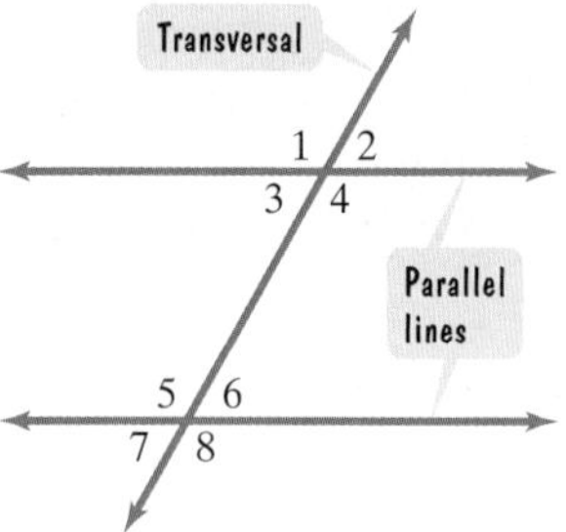

FIGURE 10.13

TABLE 10.1 Names of Angle Pairs Formed by a Transversal Intersecting Parallel Lines

Name	Description	Sketch	Angle Pairs Described	Property
Alternate interior angles	Interior angles that do not have a common vertex on alternate sides of the transversal		$\measuredangle 3$ and $\measuredangle 6$ $\measuredangle 4$ and $\measuredangle 5$	Alternate interior angles have the same measure. $m\measuredangle 3 = m\measuredangle 6$ $m\measuredangle 4 = m\measuredangle 5$
Alternate exterior angles	Exterior angles that do not have a common vertex on alternate sides of the transversal		$\measuredangle 1$ and $\measuredangle 8$ $\measuredangle 2$ and $\measuredangle 7$	Alternate exterior angles have the same measure. $m\measuredangle 1 = m\measuredangle 8$ $m\measuredangle 2 = m\measuredangle 7$
Corresponding angles	One interior and one exterior angle on the same side of the transversal		$\measuredangle 1$ and $\measuredangle 5$ $\measuredangle 2$ and $\measuredangle 6$ $\measuredangle 3$ and $\measuredangle 7$ $\measuredangle 4$ and $\measuredangle 8$	Corresponding angles have the same measure. $m\measuredangle 1 = m\measuredangle 5$ $m\measuredangle 2 = m\measuredangle 6$ $m\measuredangle 3 = m\measuredangle 7$ $m\measuredangle 4 = m\measuredangle 8$

3 *Solve problems involving angles formed by parallel lines and transversals.*

When two parallel lines are intersected by a transversal, the following relationships are true:

PARALLEL LINES AND ANGLE PAIRS

If parallel lines are intersected by a transversal,

- alternate interior angles have the same measure,
- alternate exterior angles have the same measure, and
- corresponding angles have the same measure.

Conversely, if two lines are intersected by a third line and a pair of alternate interior angles or a pair of alternate exterior angles or a pair of corresponding angles have the same measure, then the two lines are parallel.

EXAMPLE 5 *Finding Angle Measures When Parallel Lines Are Intersected by a Transversal*

In **Figure 10.14**, two parallel lines are intersected by a transversal. One of the angles ($\measuredangle 8$) has a measure of 35°. Find the measure of each of the other seven angles.

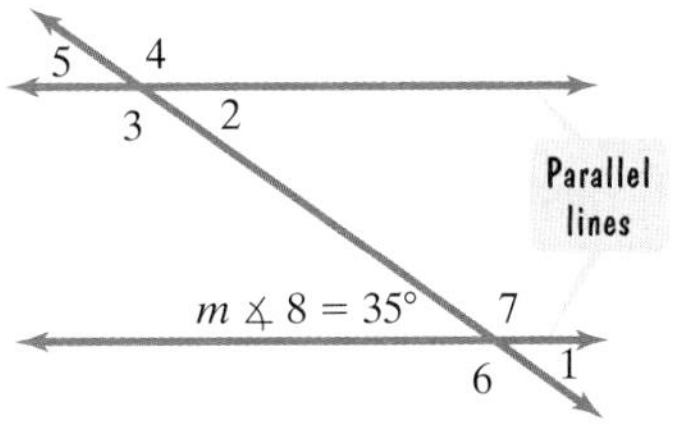

5 4 3 2 Parallel lines $m\measuredangle 8 = 35°$ 7 6 1

FIGURE 10.14

SOLUTION

Look carefully at **Figure 10.14** and fill in the angle measures as you read each line in this solution.

$m\measuredangle 1 = 35°$ — $\measuredangle 8$ and $\measuredangle 1$ are vertical angles and vertical angles have the same measure.

$m\measuredangle 6 = 180° - 35° = 145°$ — $\measuredangle 8$ and $\measuredangle 6$ are supplementary.

$m\measuredangle 7 = 145°$ — $\measuredangle 6$ and $\measuredangle 7$ are vertical angles, so they have the same measure.

$m\measuredangle 2 = 35°$ — $\measuredangle 8$ and $\measuredangle 2$ are alternate interior angles, so they have the same measure.

$m\measuredangle 3 = 145°$ — $\measuredangle 7$ and $\measuredangle 3$ are alternate interior angles. Thus, they have the same measure.

$m\measuredangle 5 = 35°$ — $\measuredangle 8$ and $\measuredangle 5$ are corresponding angles. Thus, they have the same measure.

$m\measuredangle 4 = 180° - 35° = 145°$ — $\measuredangle 4$ and $\measuredangle 5$ are supplementary.

GREAT QUESTION!

Is there more than one way to solve Example 5?

Yes. For example, once you know that $m\measuredangle 2 = 35°$, $m\measuredangle 5$ is also 35° because $\measuredangle 2$ and $\measuredangle 5$ are vertical angles.

CHECK POINT 5 In **Figure 10.14**, assume that $m\measuredangle 8 = 29°$. Find the measure of each of the other seven angles.

Concept and Vocabulary Check

Fill in each blank so that the resulting statement is true.

1. $\overleftrightarrow{AB}$ symbolizes _____ AB, $\overset{\circ\rightarrow}{AB}$ symbolizes ________ AB, $\overrightarrow{AB}$ symbolizes _____ AB, and $\overline{AB}$ symbolizes ____________ AB.
2. A/an __________ angle measures less than 90°, a/an __________ angle measures 90°, a/an __________ angle measures more than 90° and less than 180°, and a/an __________ angle measures 180°.
3. Two angles whose measures have a sum of 90° are called ______________ angles. Two angles whose measures have a sum of 180° are called ______________ angles.
4. When two lines intersect, the opposite angles are called ________ angles.
5. Lines that lie in the same plane and have no points in common are called ________ lines. If these lines are intersected by a third line, called a __________, eight angles are formed.
6. Lines that intersect at an angle of 90° are called ____________ lines.

In Exercises 7–12, determine whether each statement is true or false. If the statement is false, make the necessary change(s) to produce a true statement.

7. A ray extends infinitely in both directions. ______

8. An angle is formed by the union of two lines that have a common endpoint. ______

9. A degree, 1°, is $\frac{1}{90}$ of a complete rotation. ______

10. A ruler is used for finding the degree measure of an angle. ______

11. The measure of an angle's complement is found by subtracting the angle's measure from 90°. ______

12. Vertical angles have the same measure. ______

Exercise Set 10.1

Practice Exercises

1. The hour hand of a clock moves from 12 to 5 o'clock. Through how many degrees does it move?

2. The hour hand of a clock moves from 12 to 4 o'clock. Through how many degrees does it move?

3. The hour hand of a clock moves from 1 to 4 o'clock. Through how many degrees does it move?

4. The hour hand of a clock moves from 1 to 7 o'clock. Through how many degrees does it move?

In Exercises 5–10, use the protractor to find the measure of each angle. Then classify the angle as acute, right, straight, or obtuse.

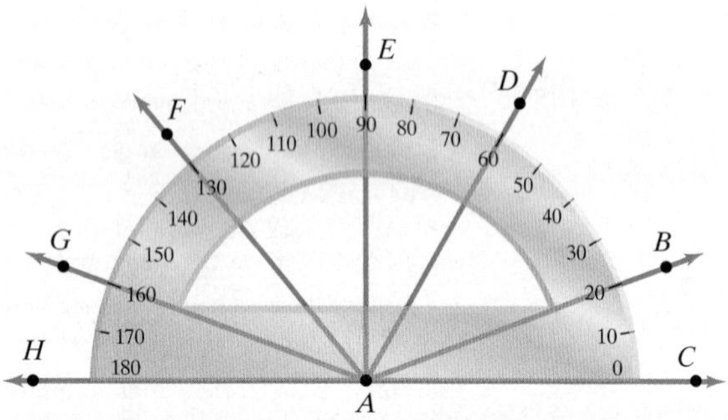

5. ∡ CAB **6.** ∡ CAF

7. ∡ HAB **8.** ∡ HAF

9. ∡ CAH **10.** ∡ HAE

In Exercises 11–14, find the measure of the angle in which a question mark with a degree symbol appears.

11.

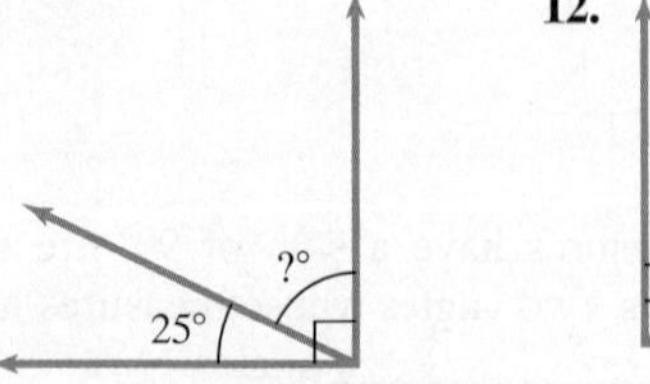

12.

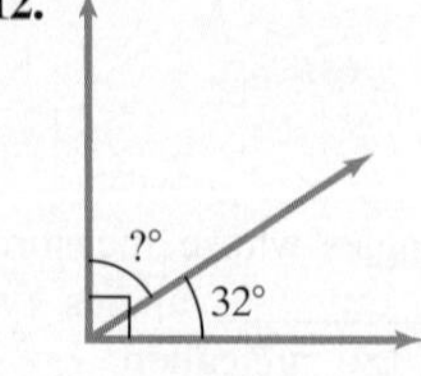

13.

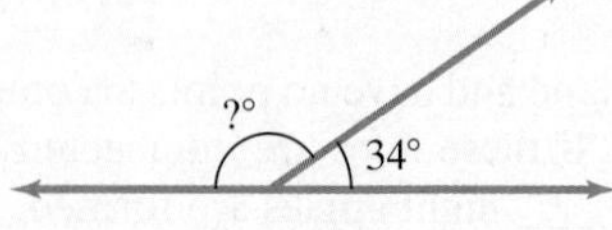

14.

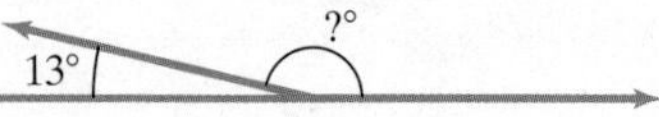

In Exercises 15–20, find the measure of the complement and the supplement of each angle.

15. 48° **16.** 52° **17.** 89°

18. 1° **19.** 37.4° **20.** $15\frac{1}{3}°$

In Exercises 21–24, use an algebraic equation to find the measures of the two angles described. Begin by letting x represent the degree measure of the angle's complement or its supplement.

21. The measure of the angle is 12° greater than its complement.

22. The measure of the angle is 56° greater than its complement.

23. The measure of the angle is three times greater than its supplement.

24. The measure of the angle is 81° more than twice that of its supplement.

In Exercises 25–28, find the measures of angles 1, 2, and 3.

25.

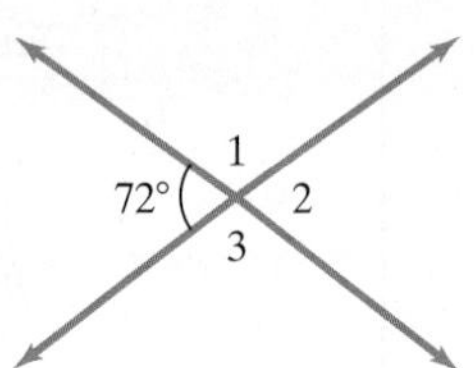

26.

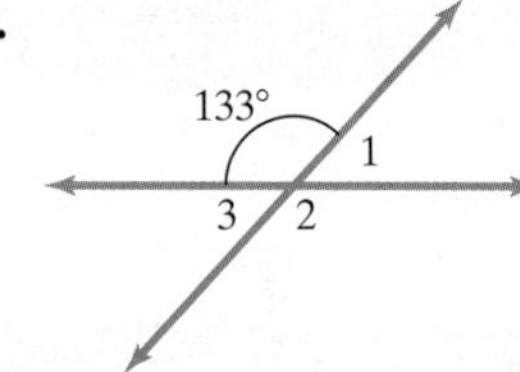

27.

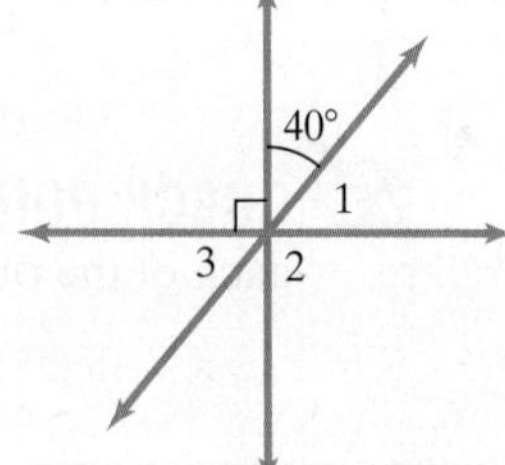

28.

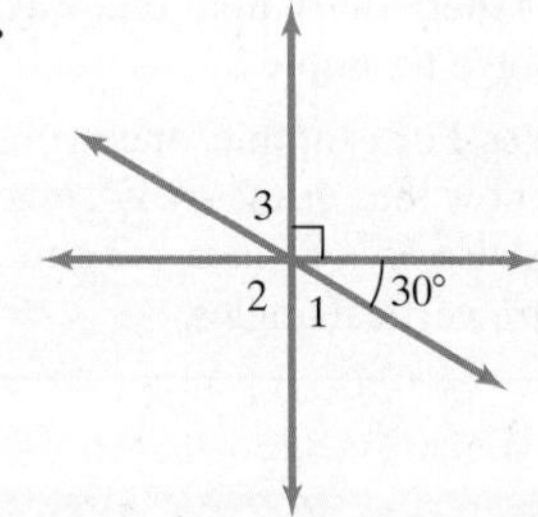

The figures for Exercises 29–30 show two parallel lines intersected by a transversal. One of the angle measures is given. Find the measure of each of the other seven angles.

29.

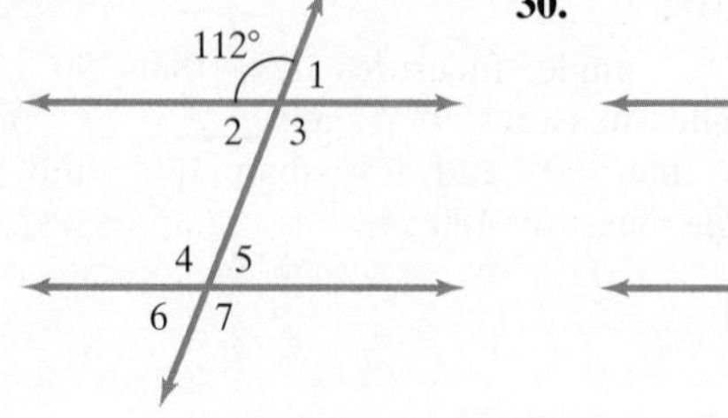

30.

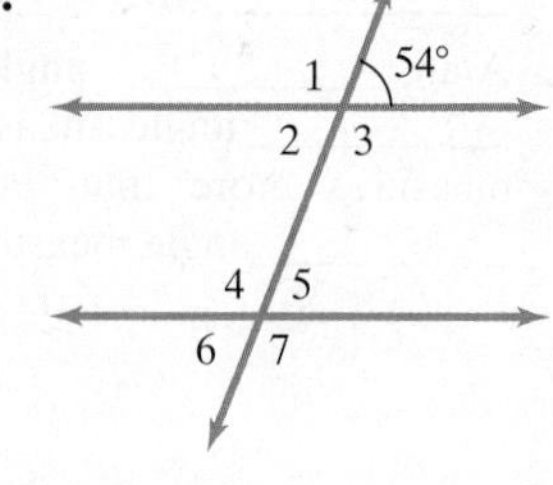

The figures for Exercises 31–34 show two parallel lines intersected by more than one transversal. Two of the angle measures are given. Find the measures of angles 1, 2, and 3.

31.

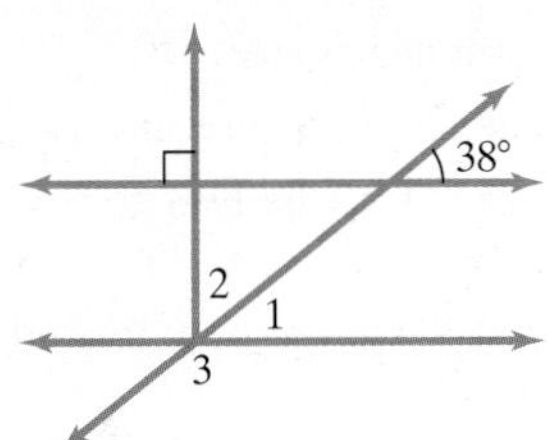

32.

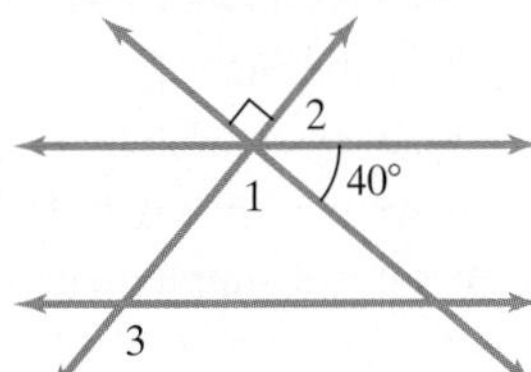

33.

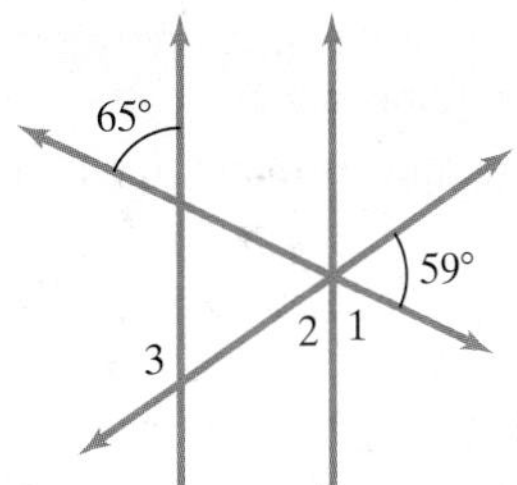

34.

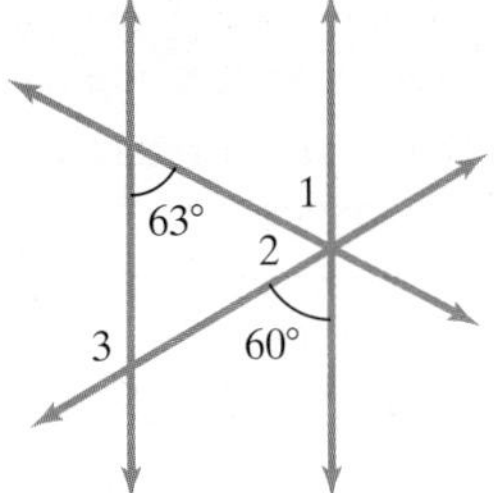

Use the following figure to determine whether each statement in Exercises 35–38 is true or false.

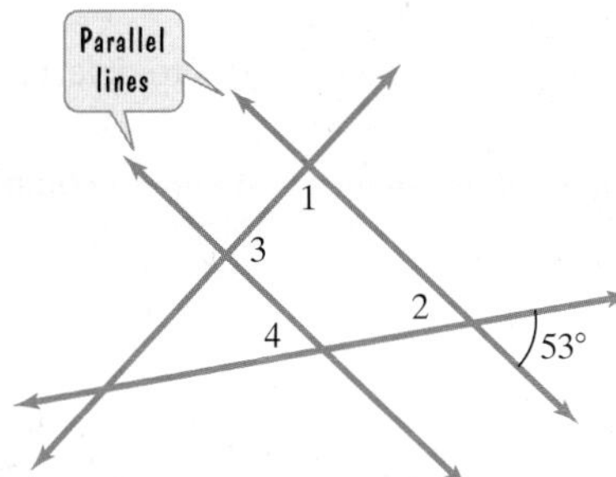

35. $m\measuredangle 2 = 37°$ **36.** $m\measuredangle 1 = m\measuredangle 2$

37. $m\measuredangle 4 = 53°$ **38.** $m\measuredangle 3 = m\measuredangle 4$

Use the following figure to determine whether each statement in Exercises 39–42 is true or false.

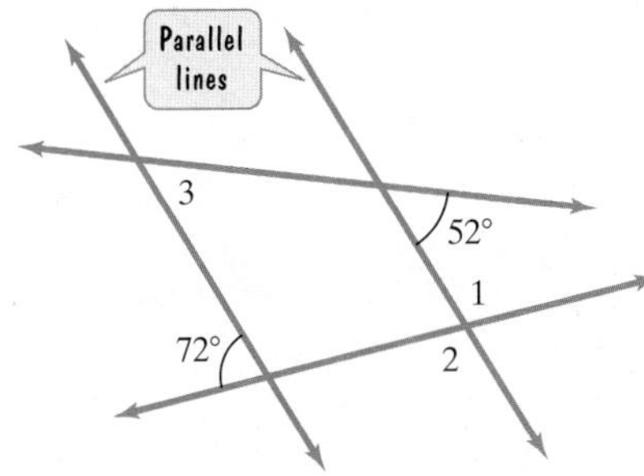

39. $m\measuredangle 1 = 38°$ **40.** $m\measuredangle 1 = 108°$

41. $m\measuredangle 2 = 52°$ **42.** $m\measuredangle 3 = 72°$

Practice Plus

In Exercises 43–46, use an algebraic equation to find the measure of each angle that is represented in terms of x.

43.

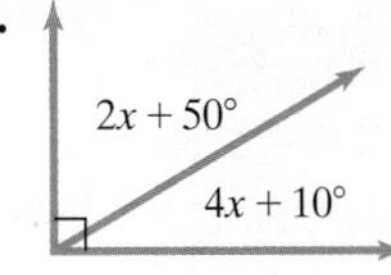

44.

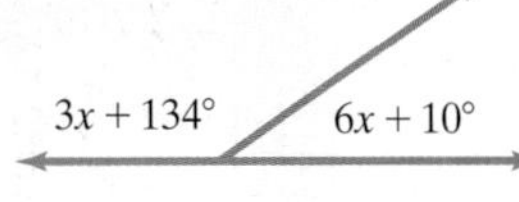

45.

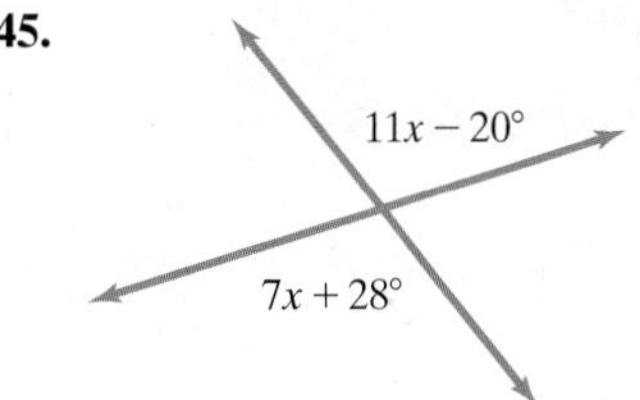

46.

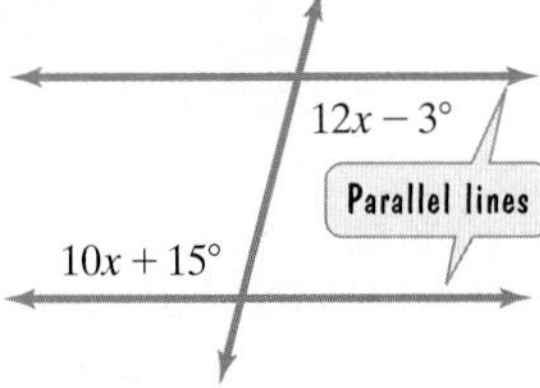

Because geometric figures consist of sets of points, we can apply set operations to obtain the union, ∪, or the intersection, ∩, of such figures. The union of two geometric figures is the set of points that belongs to either of the figures or to both figures. The intersection of two geometric figures is the set of points common to both figures. In Exercises 47–54, use the line shown to find each set of points.

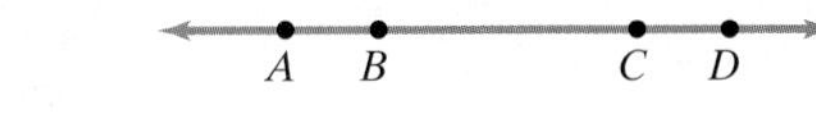

47. $\overline{AC} \cap \overline{BD}$ **48.** $\overline{AB} \cap \overline{BC}$

49. $\overline{AC} \cup \overline{BD}$ **50.** $\overline{AB} \cup \overline{BC}$

51. $\overrightarrow{BA} \cup \overrightarrow{BC}$ **52.** $\overrightarrow{CB} \cup \overrightarrow{CD}$

53. $\overrightarrow{AD} \cap \overrightarrow{DB}$ **54.** $\overrightarrow{AC} \cap \overrightarrow{CB}$

Application Exercises

55. The picture shows the top of an umbrella in which all the angles formed by the spokes have the same measure. Find the measure of each angle.

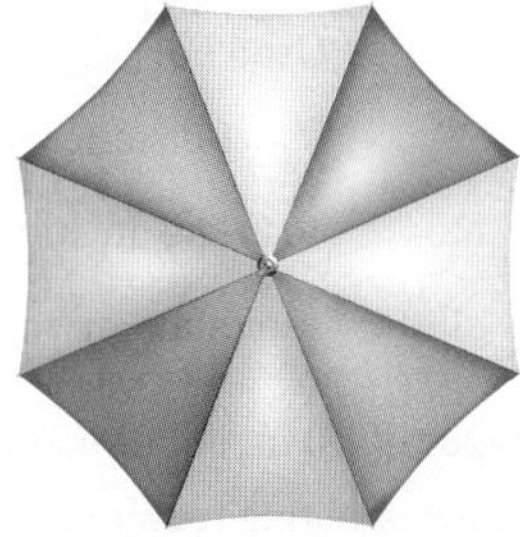

56. In the musical *Company*, composer Stephen Sondheim describes the marriage between two of the play's characters as "parallel lines who meet." What is the composer saying about this relationship?

57. The picture shows a window with parallel framing in which snow has collected in the corners. What property of parallel lines is illustrated by where the snow has collected?

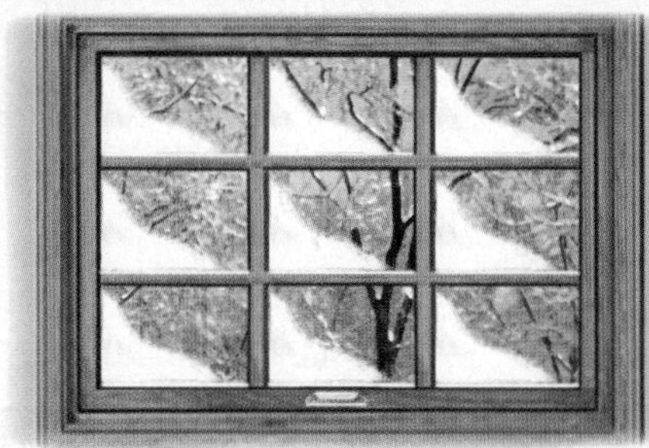

In Exercises 58–59, consider the following uppercase letters from the English alphabet:

A E F H N T X Z.

58. Which letters contain parallel line segments?

59. Which letters contain perpendicular line segments?

Angles play an important role in custom bikes that are properly fitted to the biking needs of cyclists. One of the angles to help find the perfect fit is called the hip angle. The figure indicates that the hip angle is created when you're sitting on the bike, gripping the handlebars, and your leg is fully extended. Your hip is the vertex, with one ray extending to your shoulder and the other ray extending to the front-bottom of your foot.

The table indicates hip angles for various biking needs. Use this information to pedal through Exercises 60–63.

Hip Angle	Used For
$85° \leq$ hip angle $\leq 89°$	short-distance aggressive racing
$91° \leq$ hip angle $\leq 115°$	long-distance riding
$116° \leq$ hip angle $\leq 130°$	mountain biking

60. Which type or types of biking require an acute hip angle?

61. Which type or types of biking require an obtuse hip angle?

62. A racer who had an 89° hip angle decides to switch to long-distance riding. What is the maximum difference in hip angle for the two types of biking?

63. A racer who had an 89° hip angle decides to switch to mountain biking. What is the minimum difference in hip angle for the two types of biking?

(Source for Exercises 60–63: *Scholastic Math*, January 11, 2010)

Explaining the Concepts

64. Describe the differences among lines, half-lines, rays, and line segments.

65. What is an angle and what determines its size?

66. Describe each type of angle: acute, right, obtuse, and straight.

67. What are complementary angles? Describe how to find the measure of an angle's complement.

68. What are supplementary angles? Describe how to find the measure of an angle's supplement.

69. Describe the difference between perpendicular and parallel lines.

70. If two parallel lines are intersected by a transversal, describe the location of the alternate interior angles, the alternate exterior angles, and the corresponding angles.

71. Describe everyday objects that approximate points, lines, and planes.

72. If a transversal is perpendicular to one of two parallel lines, must it be perpendicular to the other parallel line as well? Explain your answer.

Critical Thinking Exercises

Make Sense? *In Exercises 73–76, determine whether each statement makes sense or does not make sense, and explain your reasoning.*

73. I drew two lines that are not parallel and that intersect twice.

74. I used the length of an angle's sides to determine that it was obtuse.

75. I'm working with two angles that are not complementary, so I can conclude that they are supplementary.

76. The rungs of a ladder are perpendicular to each side, so the rungs are parallel to each other.

77. Use the figure to select a pair of complementary angles.

a. $\measuredangle 1$ and $\measuredangle 4$

b. $\measuredangle 3$ and $\measuredangle 6$

c. $\measuredangle 2$ and $\measuredangle 5$

d. $\measuredangle 1$ and $\measuredangle 5$

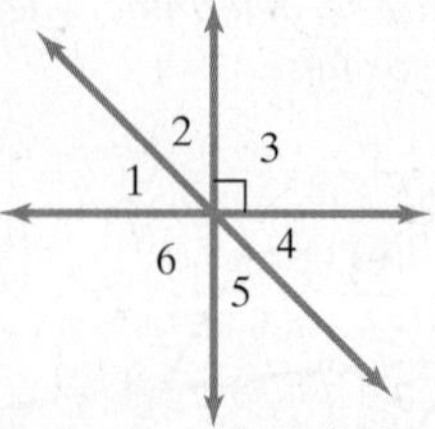

78. If $m\measuredangle AGB = m\measuredangle BGC$, and $m\measuredangle CGD = m\measuredangle DGE$, find $m\measuredangle BGD$.

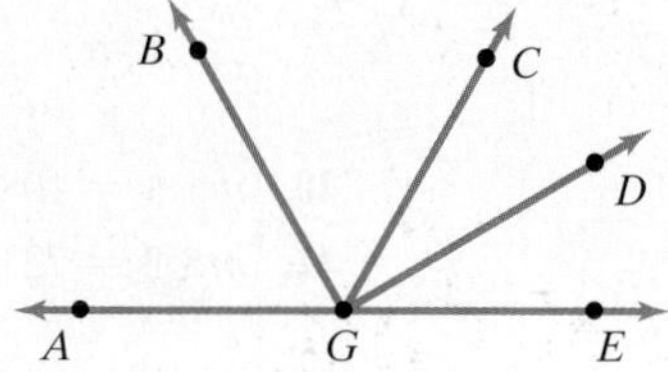

10.2

WHAT AM I SUPPOSED TO LEARN?

After studying this section, you should be able to:

1 Solve problems involving angle relationships in triangles.

2 Solve problems involving similar triangles.

3 Solve problems using the Pythagorean Theorem.

Triangles

The Walter Pyramid, California State University, Long Beach

IN CHAPTER 1, WE DEFINED *deductive reasoning* as the process of proving a specific conclusion from one or more general statements. A conclusion that is proved to be true through deductive reasoning is called a **theorem**. The Greek mathematician Euclid, who lived more than 2000 years ago, used deductive reasoning. In his 13-volume book, *Elements*, Euclid proved over 465 theorems about geometric figures. Euclid's work established deductive reasoning as a fundamental tool of mathematics. Here's looking at Euclid!

A **triangle** is a geometric figure that has three sides, all of which lie on a flat surface or plane. If you start at any point along the triangle and trace along the entire figure exactly once, you will end at the same point at which you started. Because the beginning point and ending point are the same, the triangle is called a **closed** geometric figure. Euclid used parallel lines to prove one of the most important properties of triangles: The sum of the measures of the three angles of any triangle is 180°. Here is how he did it. He began with the following general statement:

EUCLID'S ASSUMPTION ABOUT PARALLEL LINES

Given a line and a point not on the line, one and only one line can be drawn through the given point parallel to the given line.

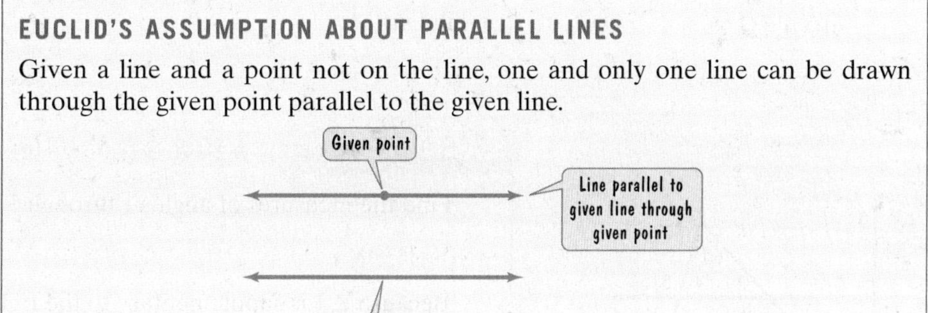

In **Figure 10.15**, triangle ABC represents any triangle. Using the general assumption given above, we draw a line through point B parallel to line AC.

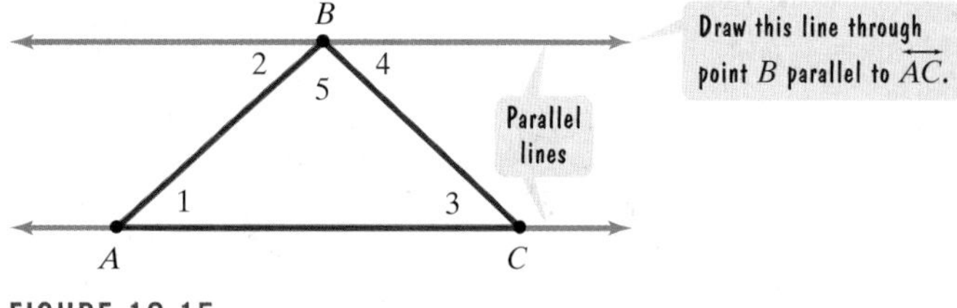

FIGURE 10.15

Because the lines are parallel, alternate interior angles have the same measure.

$$m\measuredangle 1 = m\measuredangle 2 \quad \text{and} \quad m\measuredangle 3 = m\measuredangle 4$$

Also observe that angles 2, 5, and 4 form a straight angle.

$$m\measuredangle 2 + m\measuredangle 5 + m\measuredangle 4 = 180°$$

Because $m\measuredangle 1 = m\measuredangle 2$, replace $m\measuredangle 2$ with $m\measuredangle 1$.

Because $m\measuredangle 3 = m\measuredangle 4$, replace $m\measuredangle 4$ with $m\measuredangle 3$.

$$m\measuredangle 1 + m\measuredangle 5 + m\measuredangle 3 = 180°$$

Because $\measuredangle 1$, $\measuredangle 5$, and $\measuredangle 3$ are the three angles of the triangle, this last equation shows that the measures of the triangle's three angles have a 180° sum.

THE ANGLES OF A TRIANGLE
The sum of the measures of the three angles of any triangle is 180°.

1 *Solve problems involving angle relationships in triangles.*

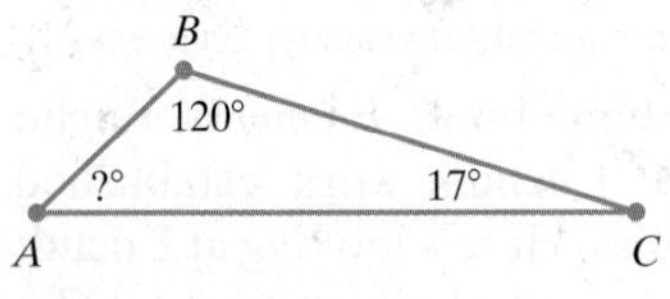

FIGURE 10.16

EXAMPLE 1 Using Angle Relationships in Triangles

Find the measure of angle A for triangle ABC in **Figure 10.16**.

SOLUTION

Because $m\angle A + m\angle B + m\angle C = 180°$, we obtain

$m\angle A + 120° + 17° = 180°$	The sum of the measures of a triangle's three angles is 180°.
$m\angle A + 137° = 180°$	Simplify: $120° + 17° = 137°$.
$m\angle A = 180° - 137°$	Find the measure of A by subtracting 137° from both sides of the equation.
$m\angle A = 43°$	Simplify.

CHECK POINT 1 In **Figure 10.16**, suppose that $m\angle B = 116°$ and $m\angle C = 15°$. Find $m\angle A$.

EXAMPLE 2 Using Angle Relationships in Triangles

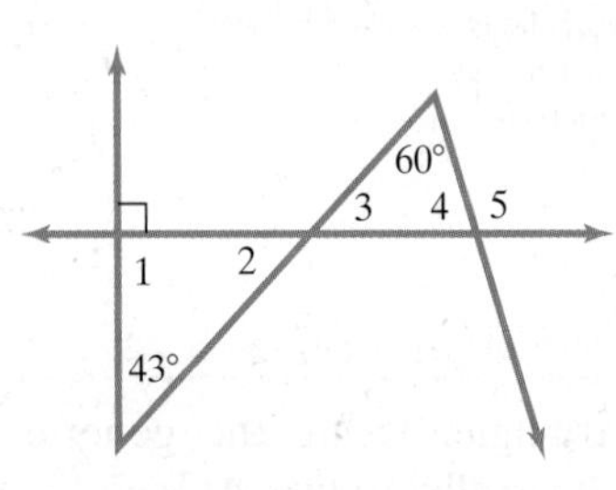

FIGURE 10.17

Find the measures of angles 1 through 5 in **Figure 10.17**.

SOLUTION

Because $\angle 1$ is supplementary to the right angle, $m\angle 1 = 90°$.

$m\angle 2$ can be found using the fact that the sum of the measures of the angles of a triangle is 180°.

$m\angle 1 + m\angle 2 + 43° = 180°$	The sum of the measures of a triangle's three angles is 180°.
$90° + m\angle 2 + 43° = 180°$	We previously found that $m\angle 1 = 90°$.
$m\angle 2 + 133° = 180°$	Simplify: $90° + 43° = 133°$.
$m\angle 2 = 180° - 133°$	Subtract 133° from both sides.
$m\angle 2 = 47°$	Simplify.

$m\angle 3$ can be found using the fact that vertical angles have equal measures: $m\angle 3 = m\angle 2$. Thus, $m\angle 3 = 47°$.

$m\angle 4$ can be found using the fact that the sum of the measures of the angles of a triangle is 180°. Refer to the triangle at the top of **Figure 10.17**.

$m\angle 3 + m\angle 4 + 60° = 180°$	The sum of the measures of a triangle's three angles is 180°.
$47° + m\angle 4 + 60° = 180°$	We previously found that $m\angle 3 = 47°$.
$m\angle 4 + 107° = 180°$	Simplify: $47° + 60° = 107°$.
$m\angle 4 = 180° - 107°$	Subtract 107° from both sides.
$m\angle 4 = 73°$	Simplify.

Finally, we can find $m\measuredangle 5$ by observing that angles 4 and 5 form a straight angle.

$$m\measuredangle 4 + m\measuredangle 5 = 180^\circ \quad \text{A straight angle measures } 180^\circ.$$
$$73^\circ + m\measuredangle 5 = 180^\circ \quad \text{We previously found that } m\measuredangle 4 = 73^\circ.$$
$$m\measuredangle 5 = 180^\circ - 73^\circ \quad \text{Subtract } 73^\circ \text{ from both sides.}$$
$$m\measuredangle 5 = 107^\circ \quad \text{Simplify.}$$

CHECK POINT 2 In **Figure 10.17**, suppose that the angle shown to measure 43° measures, instead, 36°. Further suppose that the angle shown to measure 60° measures, instead, 58°. Under these new conditions, find the measures of angles 1 through 5 in the figure.

GREAT QUESTION!

Does an isosceles triangle have exactly two equal sides or at least two equal sides?

We checked a variety of math textbooks and found two slightly different definitions for an isosceles triangle. One definition states that an isosceles triangle has *exactly* two sides of equal length. A second definition asserts that an isosceles triangle has *at least* two sides of equal length. (This makes an equilateral triangle, with three sides of the same length, an isosceles triangle.) In this book, we assume that an isosceles triangle has exactly two sides, but not three sides, of the same length.

Triangles can be described using characteristics of their angles or their sides.

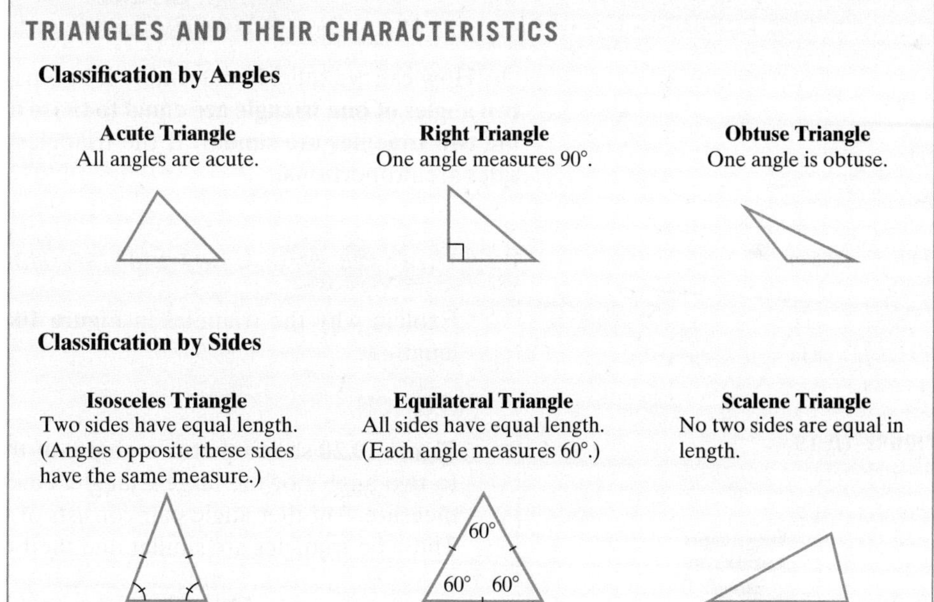

2 *Solve problems involving similar triangles.*

Pedestrian crossing

Similar Triangles

Shown in the margin is an international road sign. This sign is shaped just like the actual sign, although its size is smaller. Figures that have the same shape, but not the same size, are used in **scale drawings**. A scale drawing always pictures the exact shape of the object that the drawing represents. Architects, engineers, landscape gardeners, and interior decorators use scale drawings in planning their work.

Figures that have the same shape, but not necessarily the same size, are called **similar figures**. In **Figure 10.18**, triangles ABC and DEF are similar. Angles A and D measure the same number of degrees and are called **corresponding angles**. Angles C and F are corresponding angles, as are angles B and E. Angles with the same number of tick marks in **Figure 10.18** are the corresponding angles.

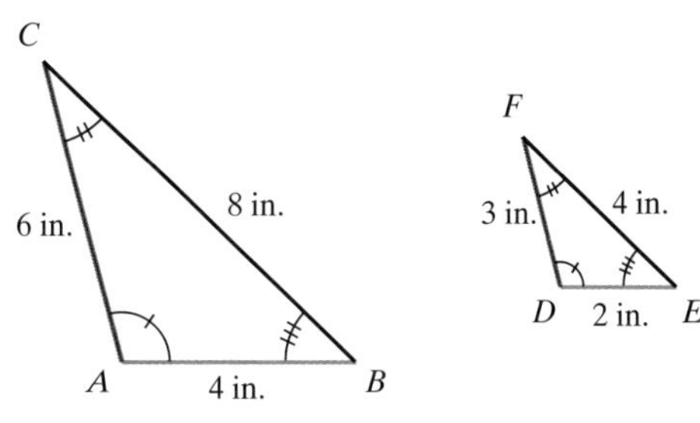

FIGURE 10.18

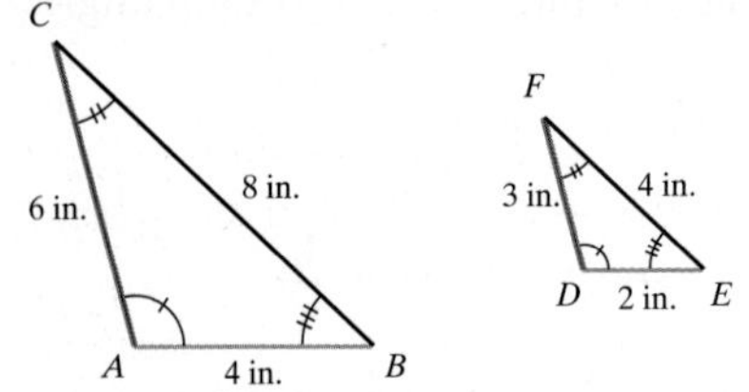

FIGURE 10.18 (repeated)

The sides opposite the corresponding angles are called **corresponding sides**. Thus, $\overline{CB}$ and $\overline{FE}$ are corresponding sides. $\overline{AB}$ and $\overline{DE}$ are also corresponding sides, as are $\overline{AC}$ and $\overline{DF}$. Corresponding angles measure the same number of degrees, but corresponding sides may or may not be the same length. For the triangles in **Figure 10.18**, each side in the smaller triangle is half the length of the corresponding side in the larger triangle.

The triangles in **Figure 10.18** illustrate what it means to be **similar triangles. Corresponding angles have the same measure and the ratios of the lengths of the corresponding sides are equal.**

$$\frac{\text{length of } \overline{AC}}{\text{length of } \overline{DF}} = \frac{6 \text{ in.}}{3 \text{ in.}} = \frac{2}{1}; \quad \frac{\text{length of } \overline{CB}}{\text{length of } \overline{FE}} = \frac{8 \text{ in.}}{4 \text{ in.}} = \frac{2}{1}; \quad \frac{\text{length of } \overline{AB}}{\text{length of } \overline{DE}} = \frac{4 \text{ in.}}{2 \text{ in.}} = \frac{2}{1}$$

In similar triangles, the lengths of the corresponding sides are proportional. Thus,

AC represents the length of $\overline{AC}$, DF the length of $\overline{DF}$, and so on.

$$\frac{AC}{DF} = \frac{CB}{FE} = \frac{AB}{DE}.$$

How can we quickly determine if two triangles are similar? **If the measures of two angles of one triangle are equal to those of two angles of a second triangle, then the two triangles are similar.** If the triangles are similar, then their corresponding sides are proportional.

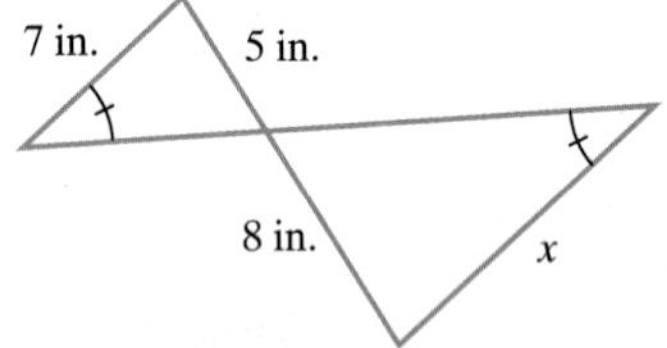

FIGURE 10.19

EXAMPLE 3 *Using Similar Triangles*

Explain why the triangles in **Figure 10.19** are similar. Then find the missing length, x.

SOLUTION

Vertical angles have the same measure.

7 in.

5 in.

8 in.

x

Given to have the same measure

FIGURE 10.20

Figure 10.20 shows that two angles of the small triangle are equal in measure to two angles of the large triangle. One angle pair is given to have the same measure. Another angle pair consists of vertical angles with the same measure. Thus, the triangles are similar and their corresponding sides are proportional.

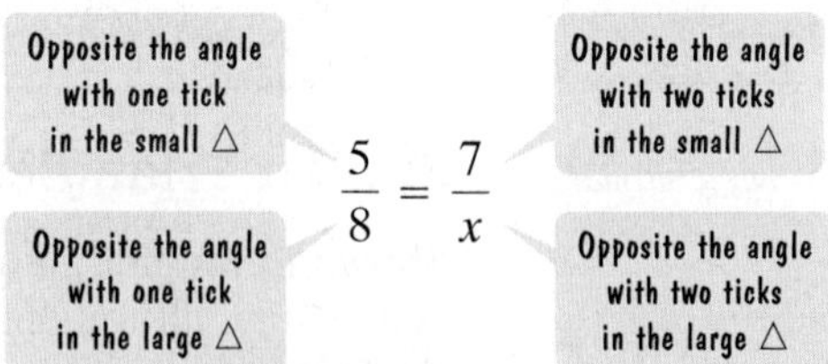

We solve $\frac{5}{8} = \frac{7}{x}$ for x by applying the cross-products principle for proportions that we discussed in Section 6.2: If $\frac{a}{b} = \frac{c}{d}$, then $ad = bc$.

$$5x = 8 \cdot 7 \quad \text{Apply the cross-products principle.}$$

$$5x = 56 \quad \text{Multiply: } 8 \cdot 7 = 56.$$

$$\frac{5x}{5} = \frac{56}{5} \quad \text{Divide both sides by 5.}$$

$$x = 11.2 \quad \text{Simplify.}$$

The missing length, x, is 11.2 inches.

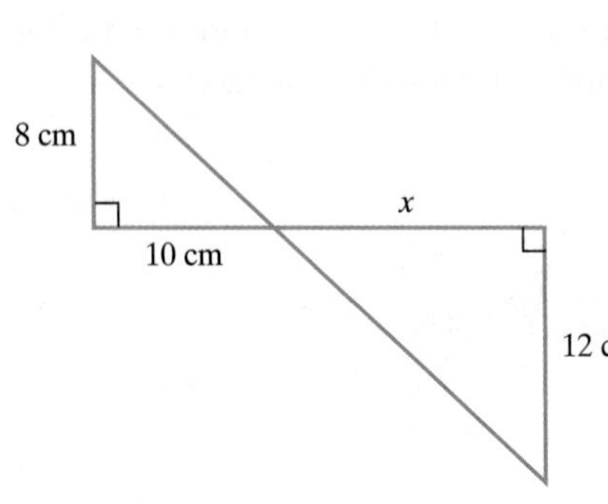

FIGURE 10.21

CHECK POINT 3 Explain why the triangles in **Figure 10.21** are similar. Then find the missing length, x.

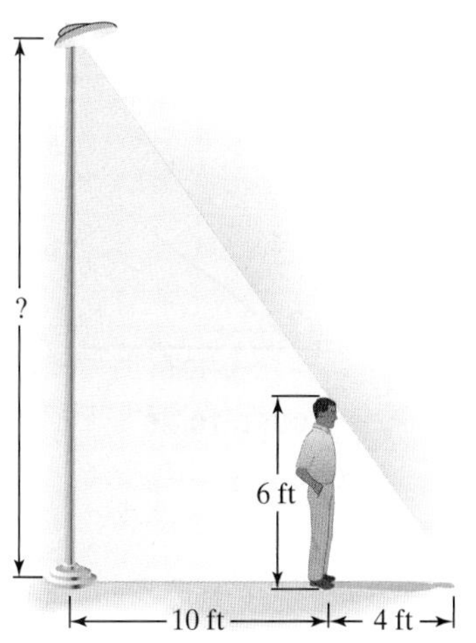

FIGURE 10.22

EXAMPLE 4 Problem Solving Using Similar Triangles

A man who is 6 feet tall is standing 10 feet from the base of a lamppost (see **Figure 10.22**). The man's shadow has a length of 4 feet. How tall is the lamppost?

SOLUTION

The drawing in **Figure 10.23** makes the similarity of the triangles easier to see. The large triangle with the lamppost on the left and the small triangle with the man on the left both contain 90° angles. They also share an angle. Thus, two angles of the large triangle are equal in measure to two angles of the small triangle. This means that the triangles are similar and their corresponding sides are proportional. We begin by letting x represent the height of the lamppost, in feet. Because corresponding sides of the two similar triangles are proportional,

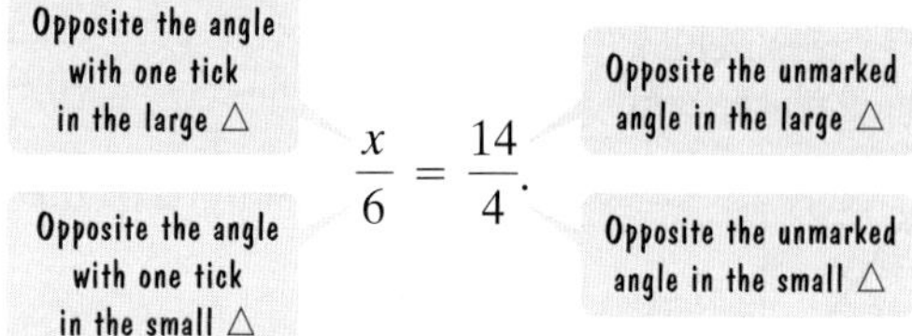

$$\frac{x}{6} = \frac{14}{4}.$$

We solve for x by applying the cross-products principle.

$$4x = 6 \cdot 14 \quad \text{Apply the cross-products principle.}$$
$$4x = 84 \quad \text{Multiply: } 6 \cdot 14 = 84.$$
$$\frac{4x}{4} = \frac{84}{4} \quad \text{Divide both sides by 4.}$$
$$x = 21 \quad \text{Simplify.}$$

The lamppost is 21 feet tall.

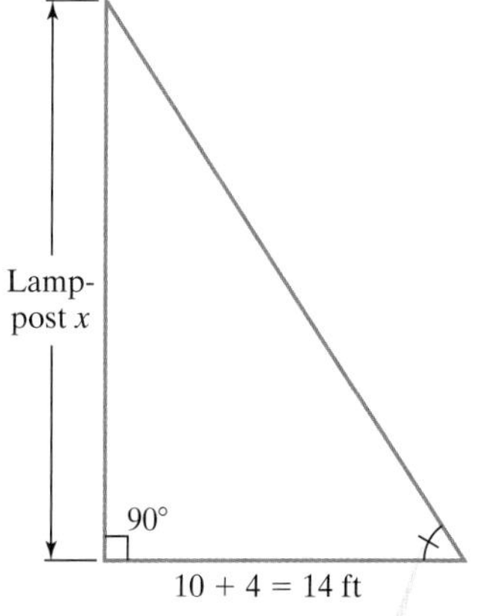

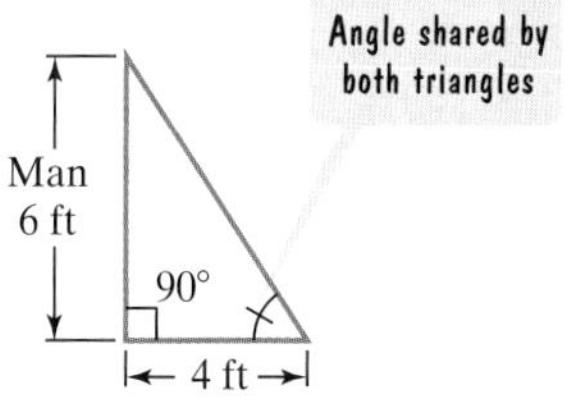

FIGURE 10.23

CHECK POINT 4 Find the height of the lookout tower shown in **Figure 10.24** using the figure that lines up the top of the tower with the top of a stick that is 2 yards long and 3.5 yards from the line to the top of the tower.

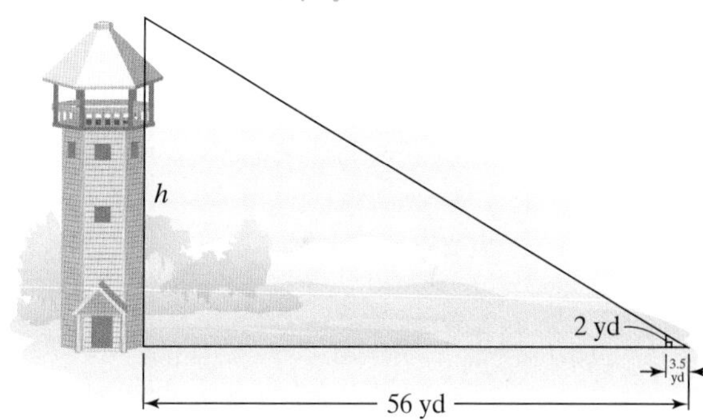

FIGURE 10.24

3 Solve problems using the Pythagorean Theorem.

The Pythagorean Theorem

The ancient Greek philosopher and mathematician Pythagoras (approximately 582–500 B.C.) founded a school whose motto was "All is number." Pythagoras is best remembered for his work with the **right triangle**, a triangle with one angle measuring 90°. The side opposite the 90° angle is called the **hypotenuse**. The other sides are called **legs**. Pythagoras found that if he constructed squares on each of the legs, as well as a larger square on the hypotenuse, the sum of the areas of the smaller squares is equal to the area of the larger square. This is illustrated in **Figure 10.25**.

This relationship is usually stated in terms of the lengths of the three sides of a right triangle and is called the **Pythagorean Theorem**.

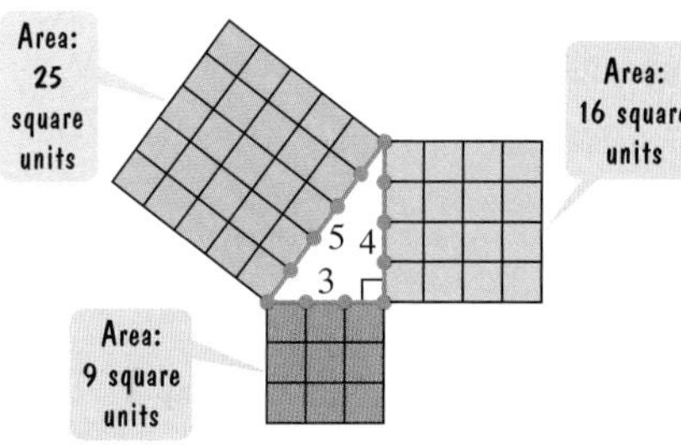

FIGURE 10.25 The area of the large square equals the sum of the areas of the smaller squares.

THE PYTHAGOREAN THEOREM

The sum of the squares of the lengths of the legs of a right triangle equals the square of the length of the hypotenuse.

If the legs have lengths a and b and the hypotenuse has length c, then

$$a^2 + b^2 = c^2.$$

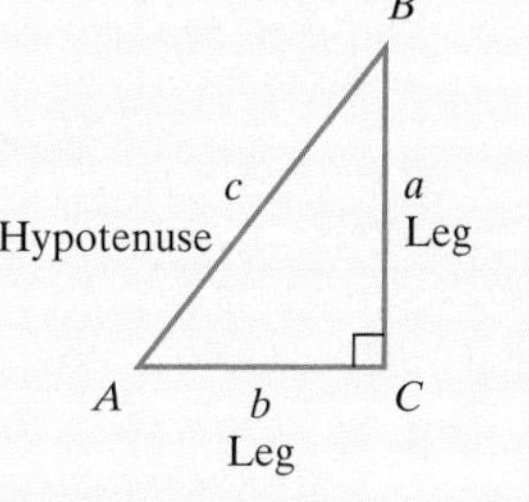

Blitzer Bonus

The Universality of Mathematics

Is mathematics discovered or invented? The "Pythagorean" Theorem, credited to Pythagoras, was known in China, from land surveying, and in Egypt, from pyramid building, centuries before Pythagoras was born. The same mathematics is often discovered/invented by independent researchers separated by time, place, and culture.

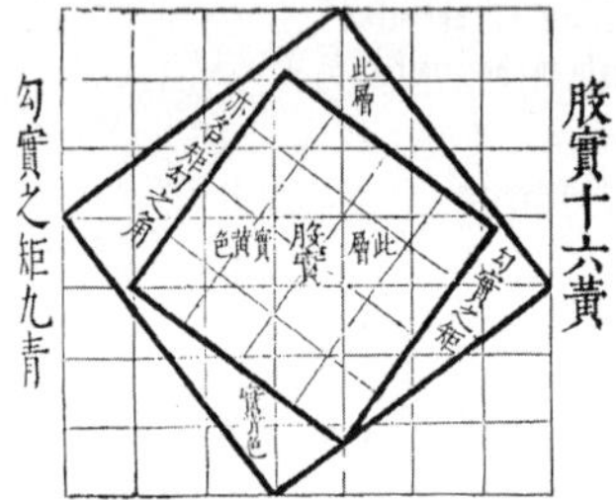

This diagram of the Pythagorean Theorem is from a Chinese manuscript dated as early as 1200 B.C.

EXAMPLE 5 Using the Pythagorean Theorem

Find the length of the hypotenuse c in the right triangle shown in **Figure 10.26**.

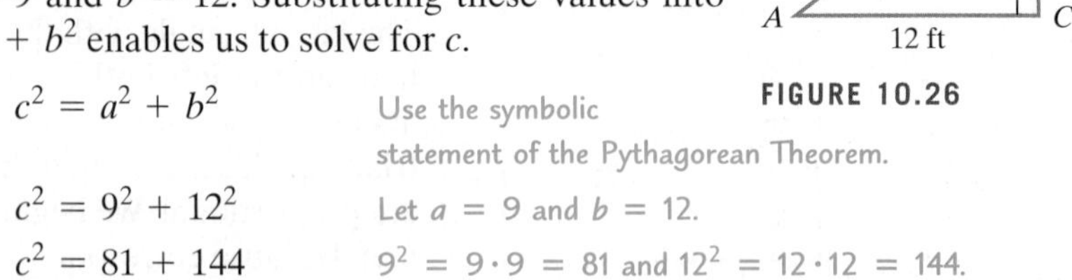

FIGURE 10.26

SOLUTION

Let $a = 9$ and $b = 12$. Substituting these values into $c^2 = a^2 + b^2$ enables us to solve for c.

$c^2 = a^2 + b^2$ Use the symbolic statement of the Pythagorean Theorem.

$c^2 = 9^2 + 12^2$ Let $a = 9$ and $b = 12$.

$c^2 = 81 + 144$ $9^2 = 9 \cdot 9 = 81$ and $12^2 = 12 \cdot 12 = 144$.

$c^2 = 225$ Add.

$c = \sqrt{225} = 15$ Solve for c by taking the positive square root of 225.

The length of the hypotenuse is 15 feet.

CHECK POINT 5 Find the length of the hypotenuse in a right triangle whose legs have lengths 7 feet and 24 feet.

EXAMPLE 6 Using the Pythagorean Theorem: Screen Math

Did you know that the size of a television screen refers to the length of its diagonal? If the length of the HDTV screen in **Figure 10.27** is 28 inches and its width is 15.7 inches, what is the size of the screen to the nearest inch?

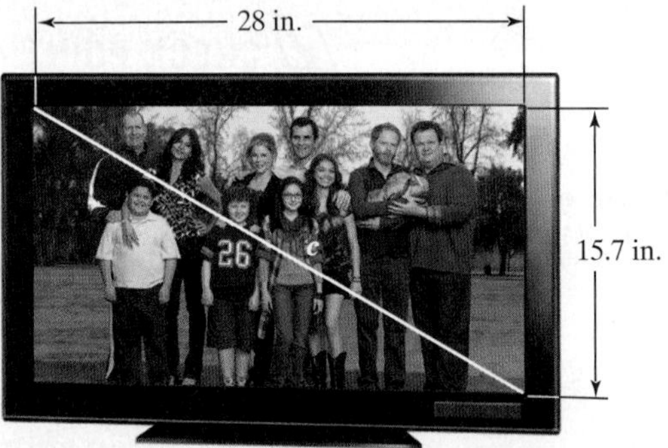

FIGURE 10.27

SOLUTION

Figure 10.27 shows that the length, width, and diagonal of the screen form a right triangle. The diagonal is the hypotenuse of the triangle. We use the Pythagorean Theorem with $a = 28$, $b = 15.7$, and solve for the screen size, c.

$c^2 = a^2 + b^2$ This is the symbolic statement of the Pythagorean Theorem.

$c^2 = 28^2 + 15.7^2$ Let $a = 28$ and $b = 15.7$.

$c^2 = 784 + 246.49$ $28^2 = 28 \cdot 28 = 784$ and $15.7^2 = 15.7 \cdot 15.7 = 246.49$

$c^2 = 1030.49$ Add.

$c = \sqrt{1030.49}$ Solve for c by taking the positive square root of 1030.49.

$c \approx 32$ Use a calculator and round to the nearest inch.
1030.49 [√] [=] or [√] 1030.49 [ENTER]

The screen size of the HDTV is 32 inches.

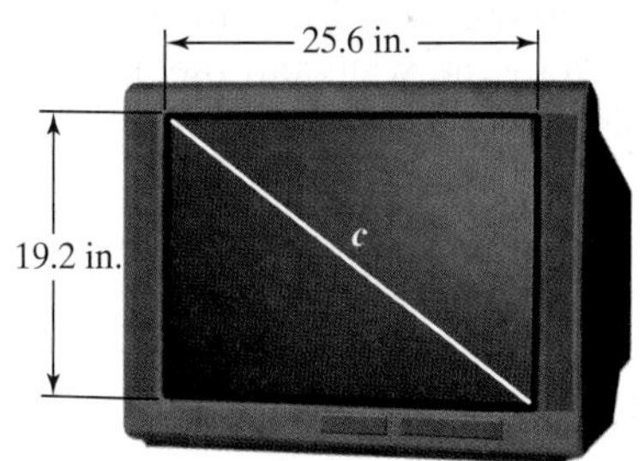

FIGURE 10.28

GREAT QUESTION!

Why did you include a decimal like 15.7 in the screen math example? Squaring 15.7 is awkward. Why not just use 16 inches for the width of the screen?

We wanted to use the actual dimensions for a 32-inch HDTV screen. In the Check Point that follows, we use the exact dimensions of an "old" TV screen (prior to HDTV). A calculator is helpful in squaring each of these dimensions.

CHECK POINT 6 **Figure 10.28** shows the dimensions of an old TV screen. What is the size of the screen?

Blitzer Bonus

Screen Math

That new 32-inch HDTV you want: How much larger than your old 32-incher is it? Actually, based on Example 6 and Check Point 6, it's smaller! **Figure 10.29** compares the screen area of the old 32-inch TV in Check Point 6 with the 32-inch HDTV in Example 6.

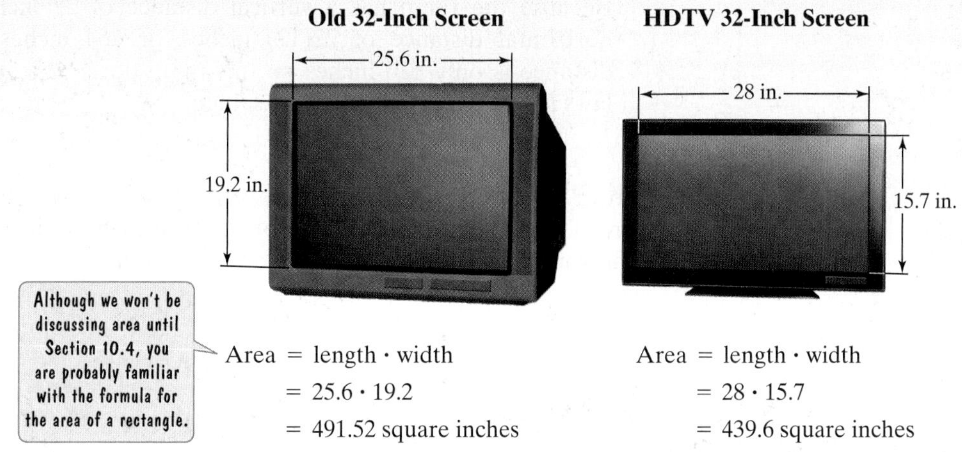

FIGURE 10.29

To make sure your HDTV has the same screen area as your old TV, it needs to have a diagonal measure, or screen size, that is 6% larger. Equivalently, take the screen size of the old TV and multiply by 1.06. If you have a 32-inch regular TV, this means the HDTV needs a 34-inch screen $(32 \times 1.06 = 33.92 \approx 34)$ if you don't want your new TV picture to be smaller than the old one.

EXAMPLE 7 *Using the Pythagorean Theorem*

a. A wheelchair ramp with a length of 122 inches has a horizontal distance of 120 inches. What is the ramp's vertical distance?

b. Construction laws are very specific when it comes to access ramps for the disabled. Every vertical rise of 1 inch requires a horizontal run of 12 inches. Does this ramp satisfy the requirement?

SOLUTION

a. The problem's conditions state that the wheelchair ramp has a length of 122 inches and a horizontal distance of 120 inches. **Figure 10.30** shows the right triangle that is formed by the ramp, the wall, and the ground. We can find x, the ramp's vertical distance, using the Pythagorean Theorem.

(leg)² plus (leg)² equals (hypotenuse)²

$$x^2 + 120^2 = 122^2$$

FIGURE 10.30

$x^2 + 120^2 = 122^2$ — This is the equation resulting from the Pythagorean Theorem.

$x^2 + 14{,}400 = 14{,}884$ — Square 120 and 122.

$x^2 = 484$ — Isolate x^2 by subtracting 14,400 from both sides.

$x = \sqrt{484} = 22$ — Solve for x by taking the positive square root of 484.

The ramp's vertical distance is 22 inches.

b. Every vertical rise of 1 inch requires a horizontal run of 12 inches. Because the ramp has a vertical distance of 22 inches, it requires a horizontal distance of 22(12) inches, or 264 inches. The horizontal distance is only 120 inches, so this ramp does not satisfy construction laws for access ramps for the disabled.

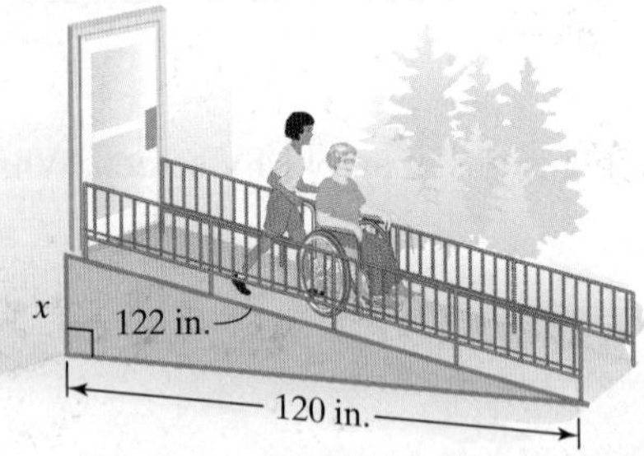

CHECK POINT 7 A radio tower is supported by two wires that are each 130 yards long and attached to the ground 50 yards from the base of the tower. How far from the ground are the wires attached to the tower?

Concept and Vocabulary Check

Fill in each blank so that the resulting statement is true.

1. The sum of the measures of the three angles of any triangle is ______.
2. A triangle in which each angle measures less than 90° is called a/an ______ triangle.
3. A triangle in which one angle measures more than 90° is called a/an ______ triangle.
4. A triangle with exactly two sides of the same length is called a/an ______ triangle.
5. A triangle whose sides are all the same length is called a/an ______ triangle.
6. A triangle that has no sides of the same length is called a/an ______ triangle.
7. Triangles that have the same shape, but not necessarily the same size, are called ______ triangles. For such triangles, corresponding angles have ______ and the lengths of the corresponding sides are ______.
8. The Pythagorean Theorem states that in any ______ triangle, the sum of the squares of the lengths of the ______ equals ______.

In Exercises 9–13, determine whether each statement is true or false. If the statement is false, make the necessary change(s) to produce a true statement.

9. Euclid's assumption about parallel lines states that given a line and a point not on the line, one and only one line can be drawn through the given point parallel to the given line. ______
10. A triangle cannot have both a right angle and an obtuse angle. ______
11. Each angle of an isosceles triangle measures 60°. ______
12. If the measures of two angles of one triangle are equal to those of two angles of a second triangle, then the two triangles have same size and shape. ______
13. In any triangle, the sum of the squares of the lengths of the two shorter sides equals the square of the length of the longest side. ______

Exercise Set 10.2

Practice Exercises

In Exercises 1–4, find the measure of angle A for the triangle shown.

1.

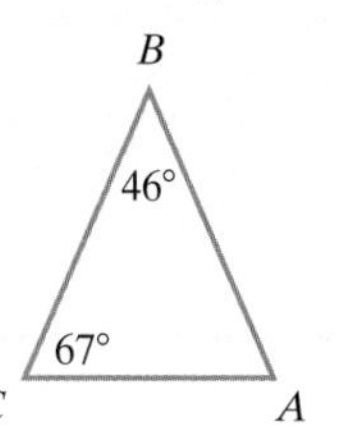

2.

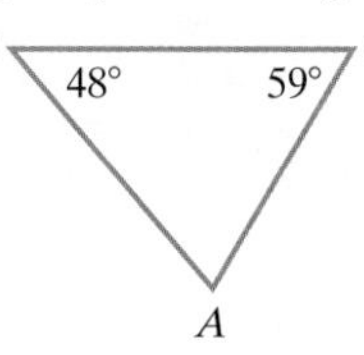

3.

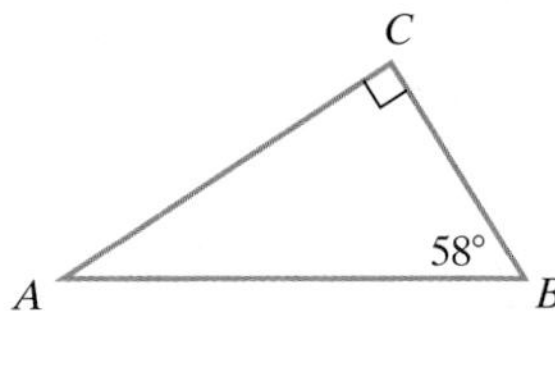

4.

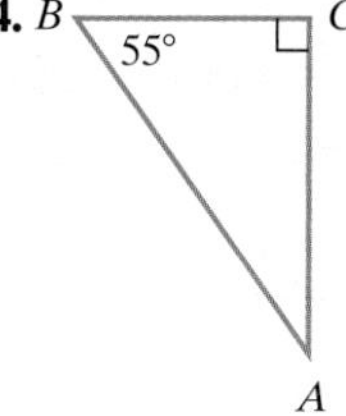

In Exercises 5–6, find the measures of angles 1 through 5 in the figure shown.

5.

6.

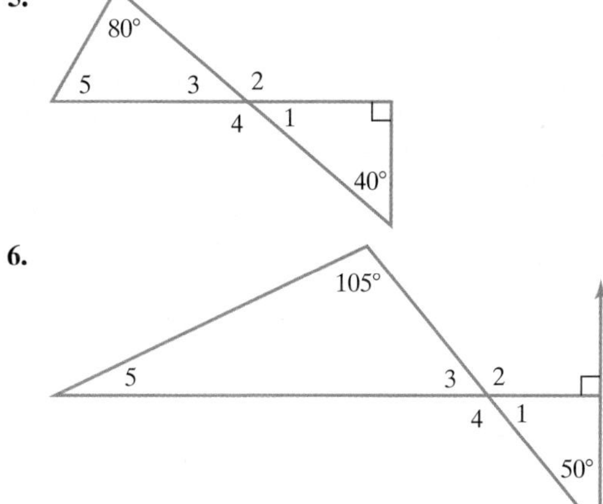

We have seen that isosceles triangles have two sides of equal length. The angles opposite these sides have the same measure. In Exercises 7–8, use this information to help find the measure of each numbered angle.

7.

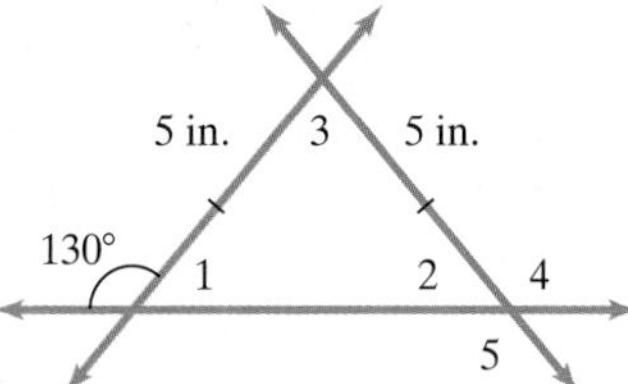

8.

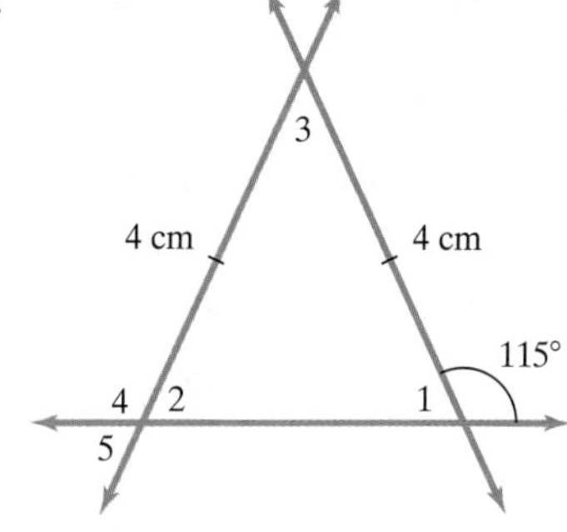

In Exercises 9–10, lines l and m are parallel. Find the measure of each numbered angle.

9.

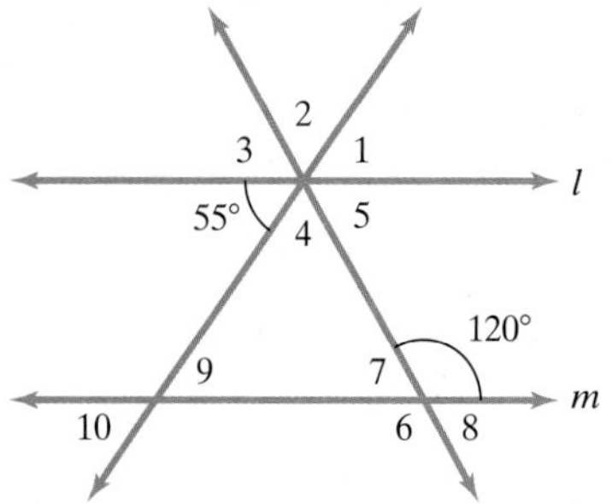

10.

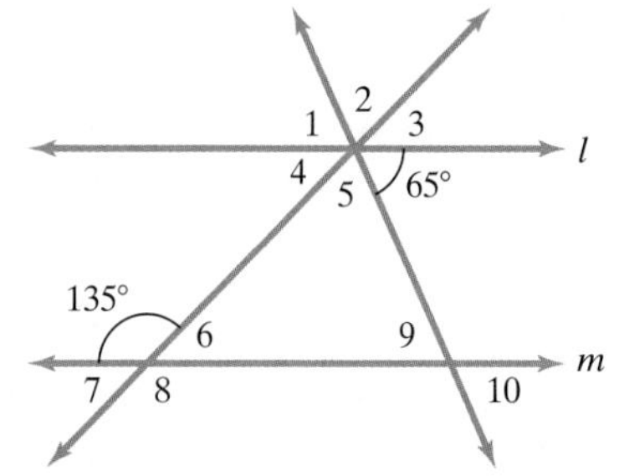

In Exercises 11–16, explain why the triangles are similar. Then find the missing length, x.

11.

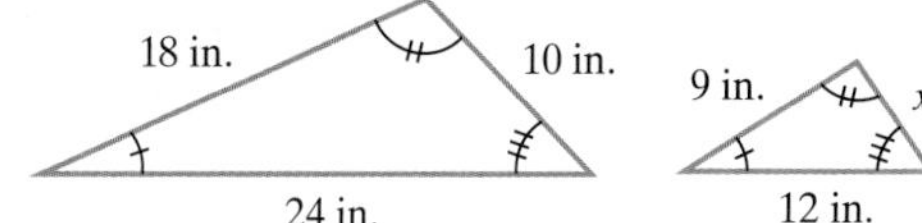

12.

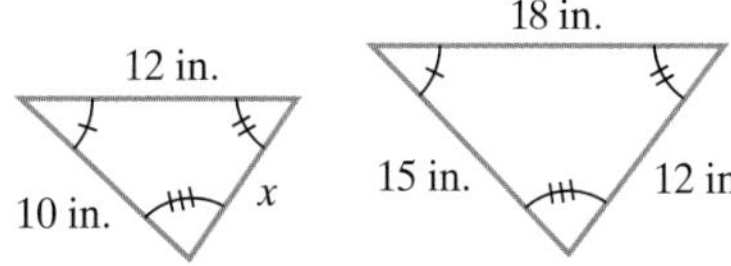

13.

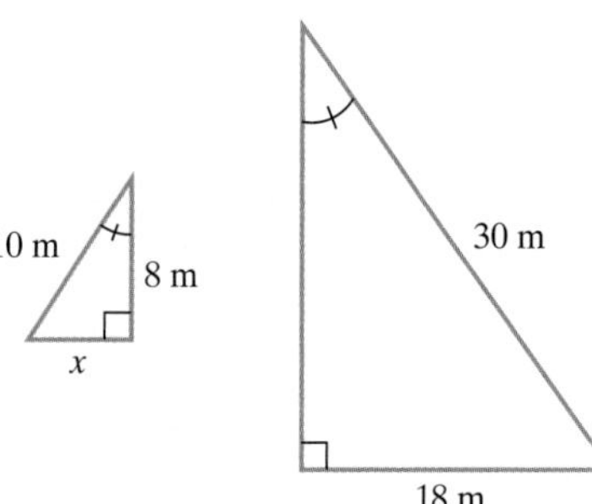

14.

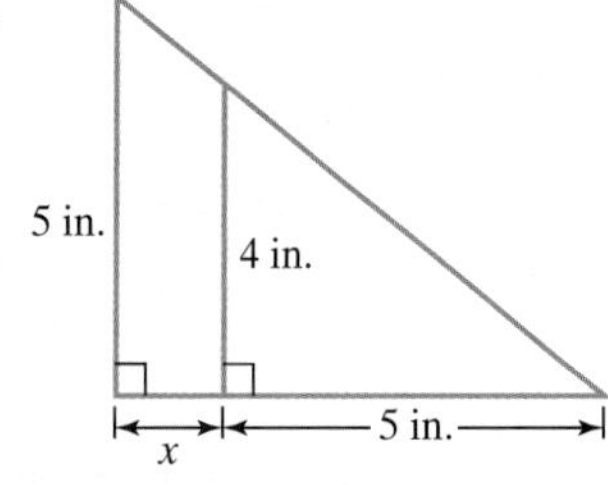

15.

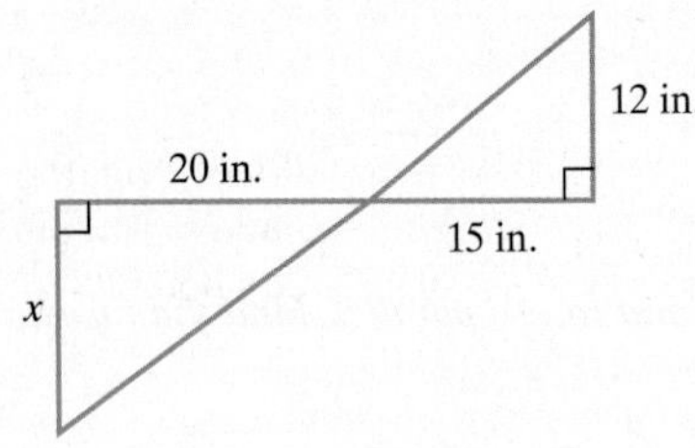

16.

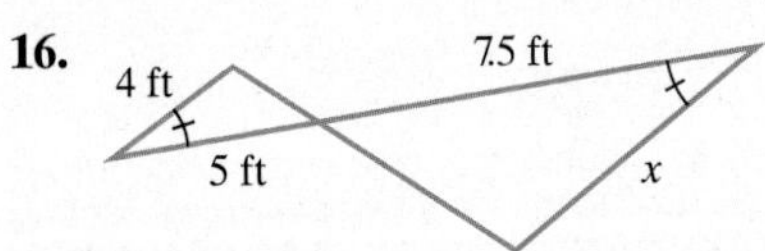

In Exercises 17–19, ΔABC and ΔADE are similar. Find the length of the indicated side.

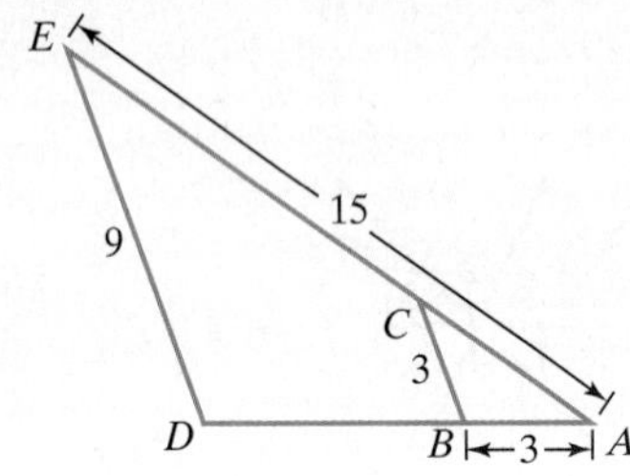

17. $\overline{CA}$ **18.** $\overline{DB}$ **19.** $\overline{DA}$

20. In the diagram for Exercises 17–19, suppose that you are not told that ΔABC and ΔADE are similar. Instead, you are given that $\overleftrightarrow{ED}$ and $\overleftrightarrow{CB}$ are parallel. Under these conditions, explain why the triangles must be similar.

In Exercises 21–26, use the Pythagorean Theorem to find the missing length in each right triangle. Use your calculator to find square roots, rounding, if necessary, to the nearest tenth.

21.

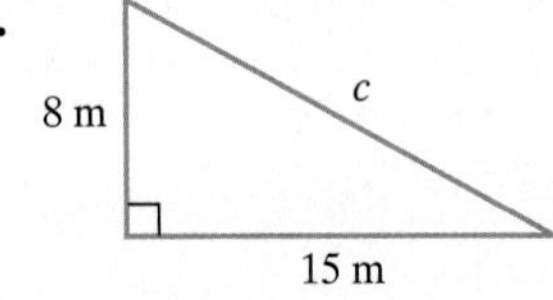

22.

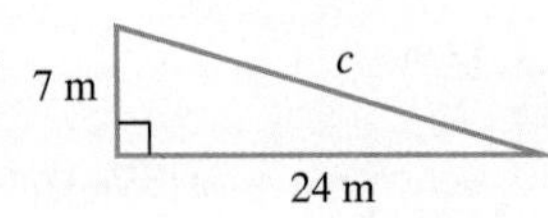

23.

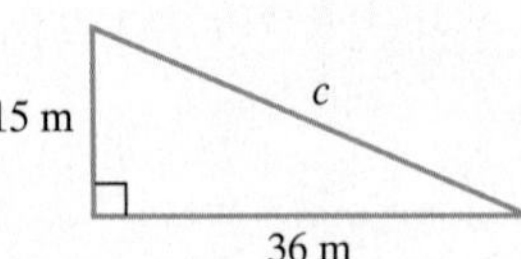

24.

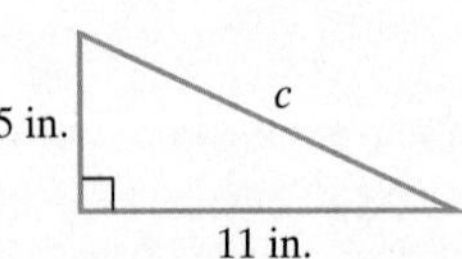

25.

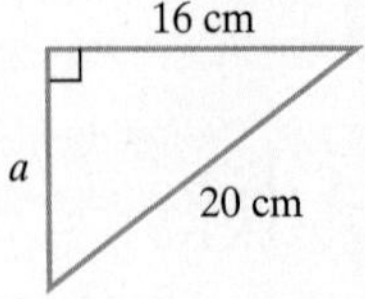

26.

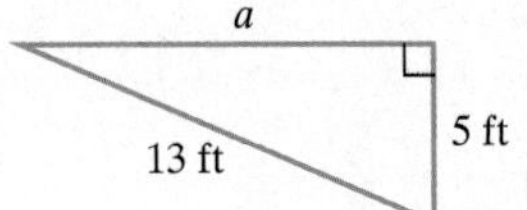

Practice Plus

Two triangles are **congruent** *if they have the same shape and the same size. In congruent triangles, the measures of corresponding angles are equal and the corresponding sides have the same length. The following triangles are congruent:*

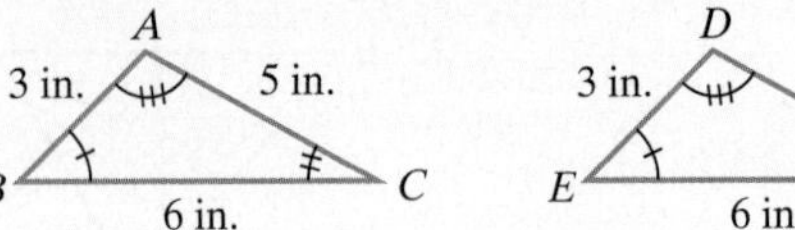

Any one of the following may be used to determine if two triangles are congruent.

Determining Congruent Triangles

1. **Side-Side-Side (SSS)**
   If the lengths of three sides of one triangle equal the lengths of the corresponding sides of a second triangle, then the two triangles are congruent.

SSS

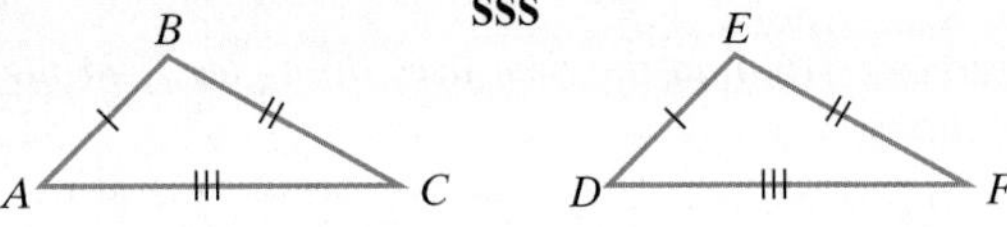

2. **Side-Angle-Side (SAS)**
   If the lengths of two sides of one triangle equal the lengths of the corresponding sides of a second triangle and the measures of the angles between each pair of sides are equal, then the two triangles are congruent.

SAS

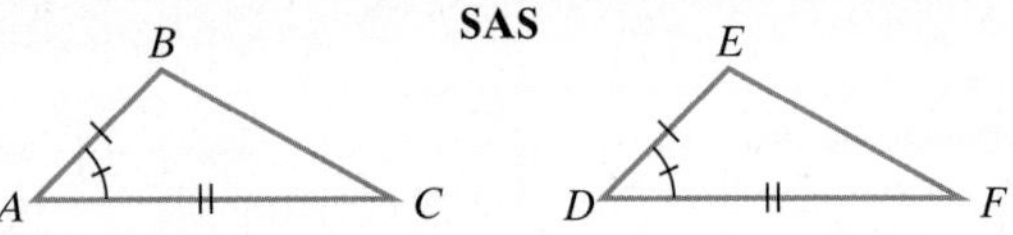

3. **Angle-Side-Angle (ASA)**
   If the measures of two angles of one triangle equal the measures of two angles of a second triangle and the lengths of the sides between each pair of angles are equal, then the two triangles are congruent.

ASA

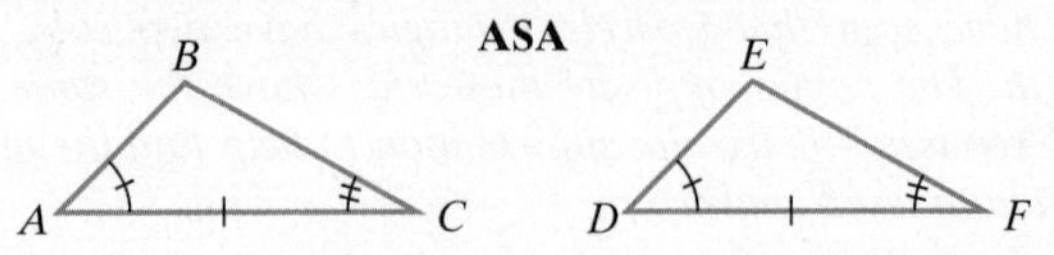

In Exercises 27–36, determine whether ΔI and ΔII are congruent. If the triangles are congruent, state the reason why, selecting from SSS, SAS, or ASA. (More than one reason may be possible.)

27.

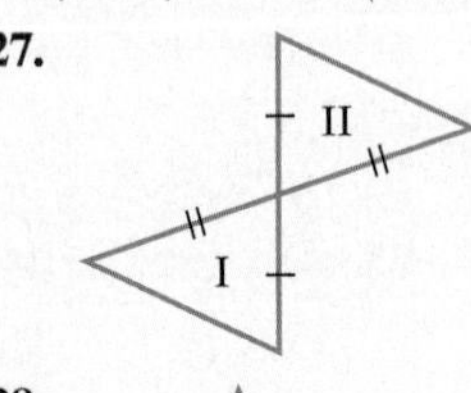

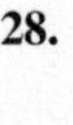

28.

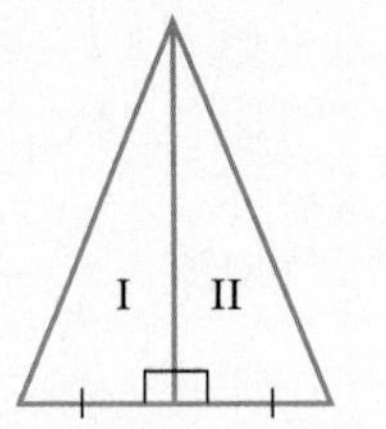

29.

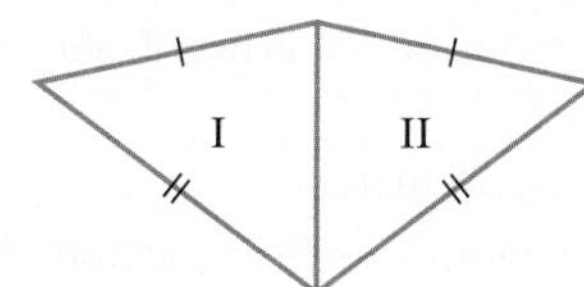

30.

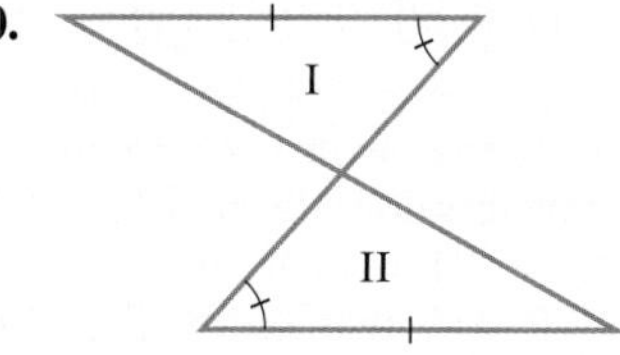

31.

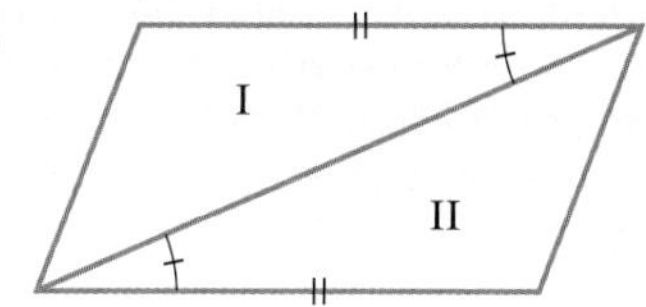

32.

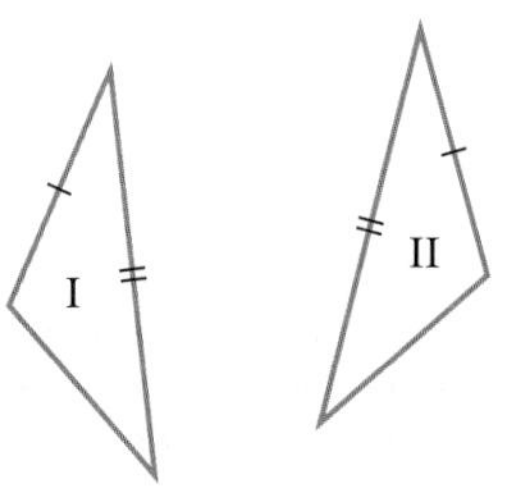

33.

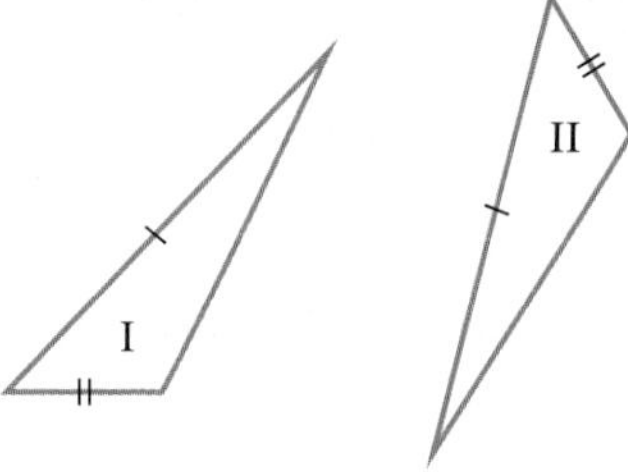

34.

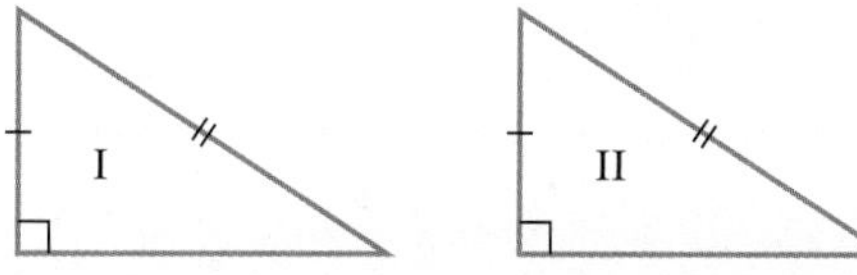

35.

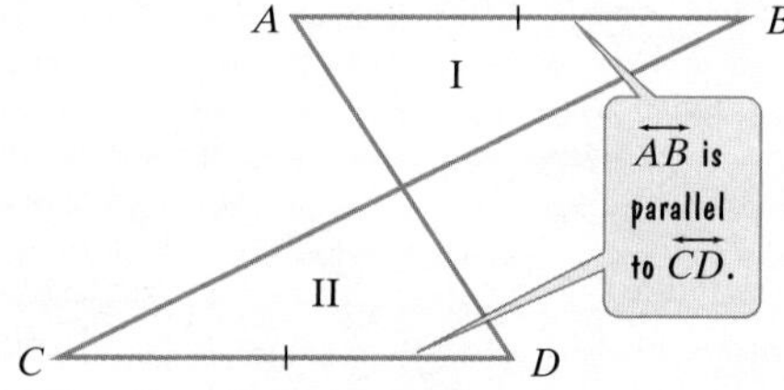

36.

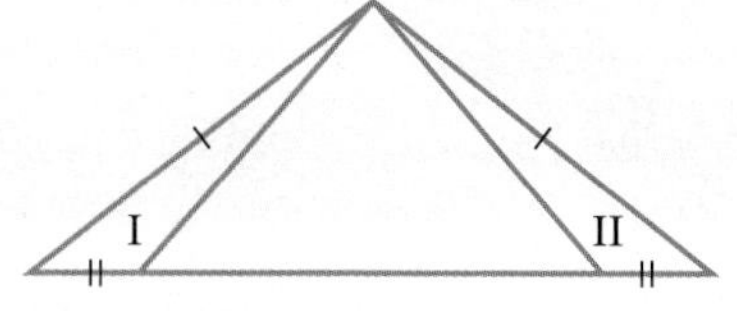

Application Exercises

Use similar triangles to solve Exercises 37–38.

37. A person who is 5 feet tall is standing 80 feet from the base of a tree and the tree casts an 86-foot shadow. The person's shadow is 6 feet in length. What is the tree's height?

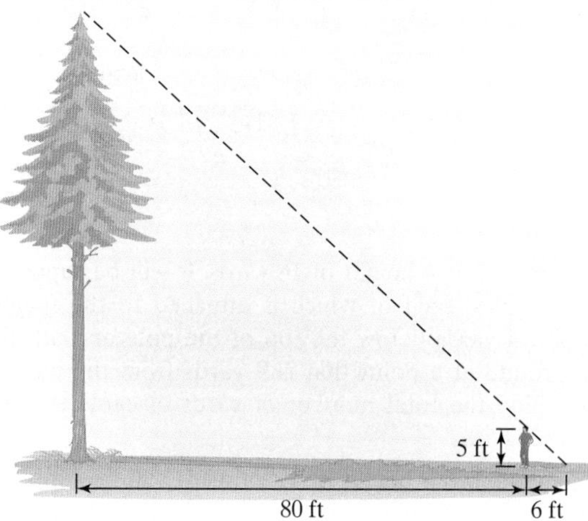

38. A tree casts a shadow 12 feet long. At the same time, a vertical rod 8 feet high casts a shadow that is 6 feet long. How tall is the tree?

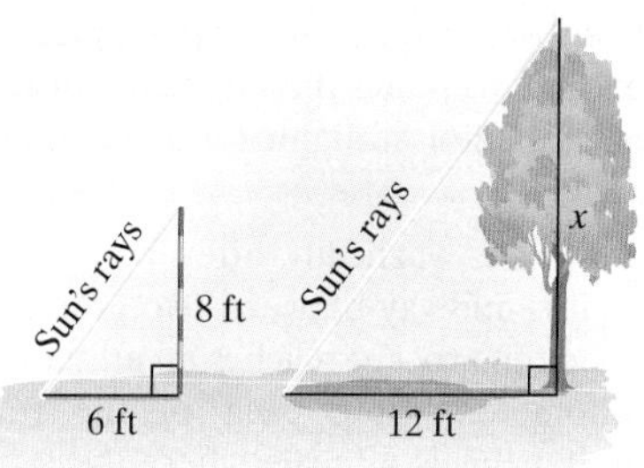

Use the Pythagorean Theorem to solve Exercises 39–46. Use your calculator to find square roots, rounding, if necessary, to the nearest tenth.

39. A baseball diamond is actually a square with 90-foot sides. What is the distance from home plate to second base?

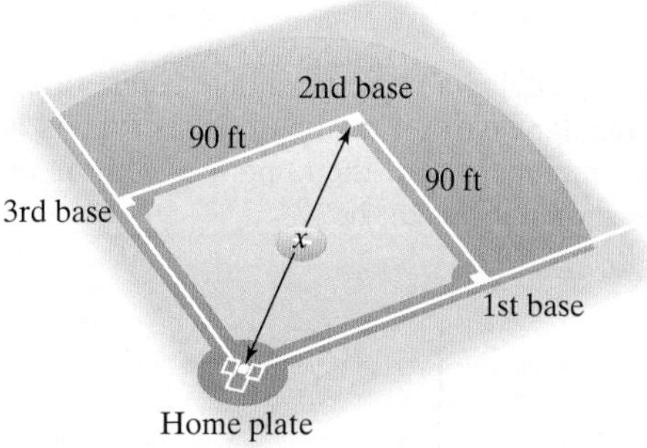

40. The base of a 20-foot ladder is 15 feet from the house. How far up the house does the ladder reach?

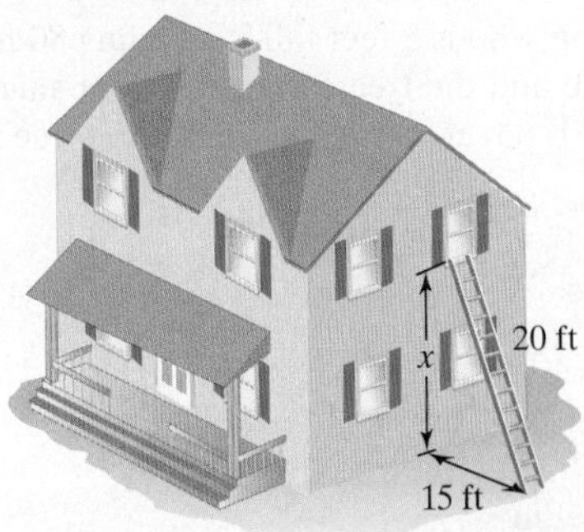

41. A flagpole has a height of 16 yards. It will be supported by three cables, each of which is attached to the flagpole at a point 4 yards below the top of the pole and attached to the ground at a point that is 9 yards from the base of the pole. Find the total number of yards of cable that will be required.

42. A flagpole has a height of 10 yards. It will be supported by three cables, each of which is attached to the flagpole at a point 4 yards below the top of the pole and attached to the ground at a point that is 8 yards from the base of the pole. Find the total number of yards of cable that will be required.

43. A rectangular garden bed measures 5 feet by 12 feet. A water faucet is located at one corner of the garden bed. A hose will be connected to the water faucet. The hose must be long enough to reach the opposite corner of the garden bed when stretched straight. Find the required length of hose.

44. A rocket ascends vertically after being launched from a location that is midway between two ground-based tracking stations. When the rocket reaches an altitude of 4 kilometers, it is 5 kilometers from each of the tracking stations. Assuming that this is a locale where the terrain is flat, how far apart are the two tracking stations?

45. If construction costs are \$150,000 per *kilometer*, find the cost of building the new road in the figure shown.

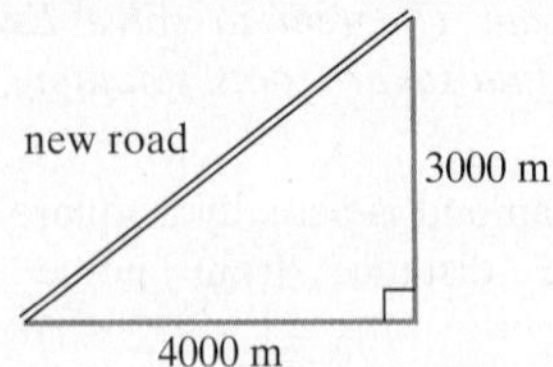

46. Picky, Picky, Picky This problem appeared on a high school exit exam:

Alex is building a ramp for a bike competition. He has two rectangular boards. One board is six meters and the other is five meters long. If the ramp has to form a right triangle, what should its height be?

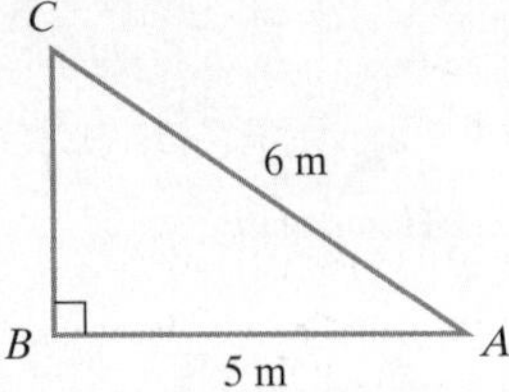

Students were asked to select the correct answer from the following options:

3 meters; 4 meters; 3.3 meters; 7.8 meters.

a. Among the available choices, which option best expresses the ramp's height? How many feet, to the nearest tenth of a foot, is this? Does a bike competition that requires riders to jump off these heights seem realistic? (ouch!)

b. Express the ramp's height to the nearest hundredth of a meter. By how many centimeters does this differ from the "correct" answer on the test? How many inches, to the nearest half inch, is this? Is it likely that a carpenter with a tape measure would make this error?

c. According to the problem, Alex has boards that measure 5 meters and 6 meters. A 6-meter board? How many feet, to the nearest tenth of a foot, is this? When was the last time you found a board of this length at Home Depot? (*Source: The New York Times*, April 24, 2005)

Explaining the Concepts

47. If the measures of two angles of a triangle are known, explain how to find the measure of the third angle.

48. Can a triangle contain two right angles? Explain your answer.

49. What general assumption did Euclid make about a point and a line in order to prove that the sum of the measures of the angles of a triangle is 180°?

50. What are similar triangles?

51. If the ratio of the corresponding sides of two similar triangles is 1 to 1 $\left(\frac{1}{1}\right)$, what must be true about the triangles?

52. What are corresponding angles in similar triangles?

53. Describe how to identify the corresponding sides in similar triangles.

54. In your own words, state the Pythagorean Theorem.

55. In the 1939 movie *The Wizard of Oz*, upon being presented with a Th.D. (Doctor of Thinkology), the Scarecrow proudly exclaims, "The sum of the square roots of any two sides of an isosceles triangle is equal to the square root of the remaining side." Did the Scarecrow get the Pythagorean Theorem right? In particular, describe four errors in the Scarecrow's statement.

Critical Thinking Exercises

Make Sense? *In Exercises 56–59, determine whether each statement makes sense or does not make sense, and explain your reasoning.*

56. I'm fencing off a triangular plot of land that has two right angles.

57. Triangle I is equilateral, as is triangle II, so the triangles are similar.

58. Triangle I is a right triangle, as is triangle II, so the triangles are similar.

59. If I am given the lengths of two sides of a triangle, I can use the Pythagorean Theorem to find the length of the third side.

60. Find the measure of angle R.

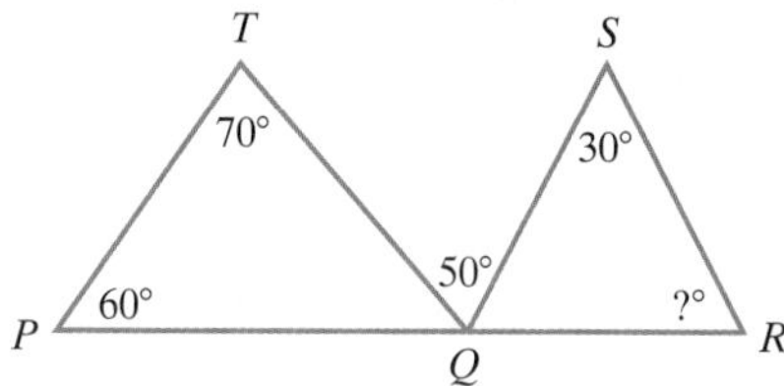

61. What is the length of $\overline{AB}$ in the accompanying figure?

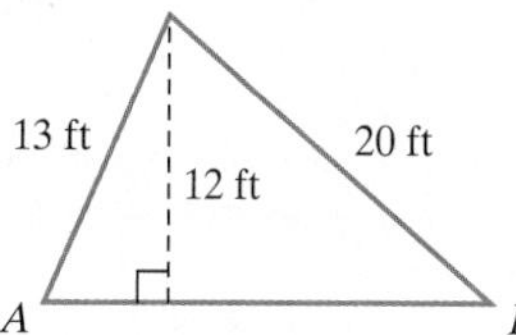

62. Use a quadratic equation to solve this problem. The width of a rectangular carpet is 7 meters shorter than the length, and the diagonal is 1 meter longer than the length. What are the carpet's dimensions?

10.3 Polygons, Perimeter, and Tessellations

WHAT AM I SUPPOSED TO LEARN?

After studying this section, you should be able to:

1. Name certain polygons according to the number of sides.
2. Recognize the characteristics of certain quadrilaterals.
3. Solve problems involving a polygon's perimeter.
4. Find the sum of the measures of a polygon's angles.
5. Understand tessellations and their angle requirements.

YOU HAVE JUST PURCHASED A BEAUTIFUL PLOT OF LAND IN THE COUNTRY, SHOWN IN Figure 10.31. In order to have more privacy, you decide to put fencing along each of its four sides. The cost of this project depends on the distance around the four outside edges of the plot, called its **perimeter**, as well as the cost for each foot of fencing.

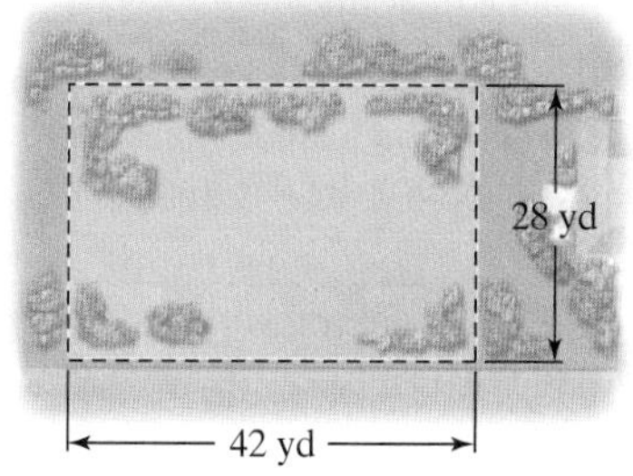

FIGURE 10.31

Your plot of land is a geometric figure: It has four straight sides that are line segments. The plot is on level ground, so that the four sides all lie on a flat surface, or plane. The plot is an example of a *polygon*. Any closed shape in the plane formed by three or more line segments that intersect only at their endpoints is a **polygon**.

1 *Name certain polygons according to the number of sides.*

A polygon is named according to the number of sides it has. We know that a three-sided polygon is called a **triangle**. A four-sided polygon is called a **quadrilateral**.

A polygon whose sides are all the same length and whose angles all have the same measure is called a **regular polygon**. **Table 10.2** provides the names of six polygons. Also shown are illustrations of regular polygons.

TABLE 10.2 Illustrations of Regular Polygons

Name	Picture	Name	Picture
Triangle 3 sides		Hexagon 6 sides	
Quadrilateral 4 sides		Heptagon 7 sides	
Pentagon 5 sides		Octagon 8 sides	

2 *Recognize the characteristics of certain quadrilaterals.*

Blitzer Bonus

Bees and Regular Hexagons

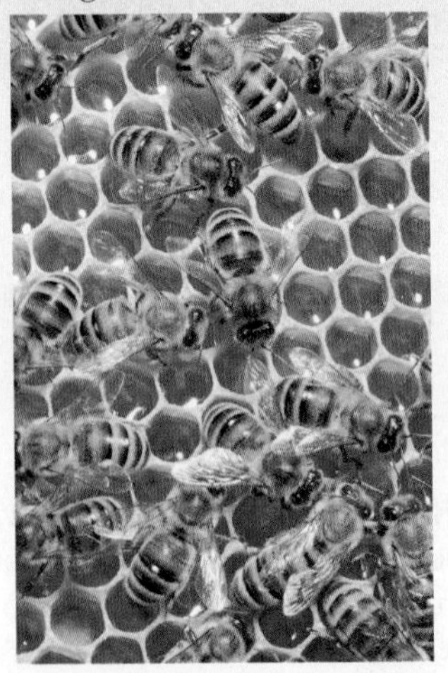

Bees use honeycombs to store honey and house larvae. They construct honey storage cells from wax. Each cell has the shape of a regular hexagon. The cells fit together perfectly, preventing dirt or predators from entering. Squares or equilateral triangles would fit equally well, but regular hexagons provide the largest storage space for the amount of wax used.

Quadrilaterals

The plot of land in **Figure 10.31** on the previous page is a four-sided polygon, or a quadrilateral. However, when you first looked at the figure, perhaps you thought of the plot as a rectangular field. A **rectangle** is a special kind of quadrilateral in which both pairs of opposite sides are parallel, have the same measure, and whose angles are right angles. **Table 10.3** presents some special quadrilaterals and their characteristics.

TABLE 10.3 Types of Quadrilaterals

Name	Characteristics	Representation
Parallelogram	Quadrilateral in which both pairs of opposite sides are parallel and have the same measure. Opposite angles have the same measure.	
Rhombus	Parallelogram with all sides having equal length.	
Rectangle	Parallelogram with four right angles. Because a rectangle is a parallelogram, opposite sides are parallel and have the same measure.	
Square	A rectangle with all sides having equal length. Each angle measures 90°, and the square is a regular quadrilateral.	
Trapezoid	A quadrilateral with exactly one pair of parallel sides.	

3 *Solve problems involving a polygon's perimeter.*

Perimeter

The **perimeter**, P, of a polygon is the sum of the lengths of its sides. Perimeter is measured in linear units, such as inches, feet, yards, meters, or kilometers.

Example 1 involves the perimeter of a rectangle. Because perimeter is the sum of the lengths of the sides, the perimeter of the rectangle shown in **Figure 10.32** is $l + w + l + w$. This can be expressed as

$$P = 2l + 2w.$$

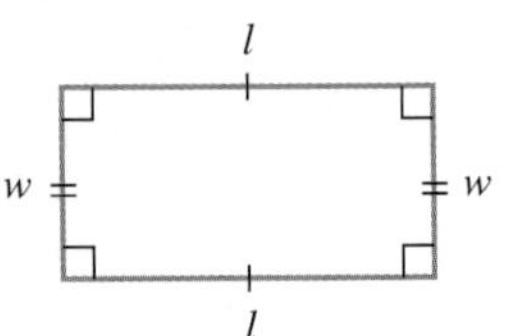

FIGURE 10.32 A rectangle with length l and width w

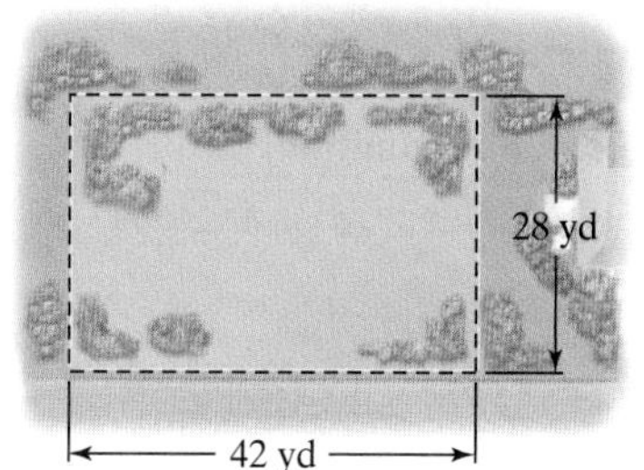

FIGURE 10.31 (repeated)

EXAMPLE 1 *An Application of Perimeter*

The rectangular field we discussed at the beginning of this section (see **Figure 10.31**) has a length of 42 yards and a width of 28 yards. If fencing costs \$5.25 per foot, find the cost to enclose the field with fencing.

SOLUTION

We begin by finding the perimeter of the rectangle in yards. Using 3 ft = 1 yd and dimensional analysis, we express the perimeter in feet. Finally, we multiply the perimeter, in feet, by \$5.25 because the fencing costs \$5.25 per foot.

The length, l, is 42 yards and the width, w, is 28 yards. The perimeter of the rectangle is determined using the formula $P = 2l + 2w$.

$$P = 2l + 2w = 2 \cdot 42 \text{ yd} + 2 \cdot 28 \text{ yd} = 84 \text{ yd} + 56 \text{ yd} = 140 \text{ yd}$$

Because 3 ft = 1 yd, we use the unit fraction $\frac{3 \text{ ft}}{1 \text{ yd}}$ to convert from yards to feet.

$$140 \text{ yd} = \frac{140 \cancel{\text{yd}}}{1} \cdot \frac{3 \text{ ft}}{1 \cancel{\text{yd}}} = 140 \cdot 3 \text{ ft} = 420 \text{ ft}$$

The perimeter of the rectangle is 420 feet. Now we are ready to find the cost of the fencing. We multiply 420 feet by \$5.25, the cost per foot.

$$\text{Cost} = \frac{420 \cancel{\text{feet}}}{1} \cdot \frac{\$5.25}{\cancel{\text{foot}}} = 420(\$5.25) = \$2205$$

The cost to enclose the field with fencing is \$2205.

CHECK POINT 1 A rectangular field has a length of 50 yards and a width of 30 yards. If fencing costs \$6.50 per foot, find the cost to enclose the field with fencing.

4 *Find the sum of the measures of a polygon's angles.*

The Sum of the Measures of a Polygon's Angles

We know that the sum of the measures of the three angles of any triangle is 180°. We can use inductive reasoning to find the sum of the measures of the angles of any polygon. Start by drawing line segments from a single point where two sides meet so that nonoverlapping triangles are formed. This is done in **Figure 10.33**.

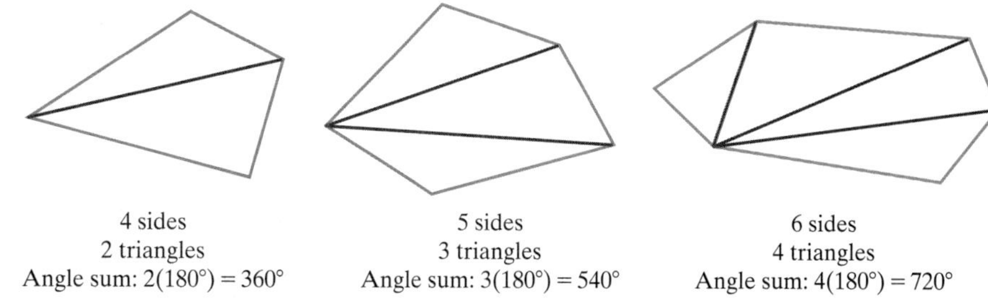

FIGURE 10.33

In each case, the number of triangles is two less than the number of sides of the polygon. Thus, for an n-sided polygon, there are $n - 2$ triangles. Because each triangle has an angle-measure sum of 180°, the sum of the measures for the angles in the $n - 2$ triangles is $(n - 2)180°$. Thus, the sum of the measures of the angles of an n-sided polygon is $(n - 2)180°$.

THE ANGLES OF A POLYGON

The sum of the measures of the angles of a polygon with n sides is

$$(n - 2)180°.$$

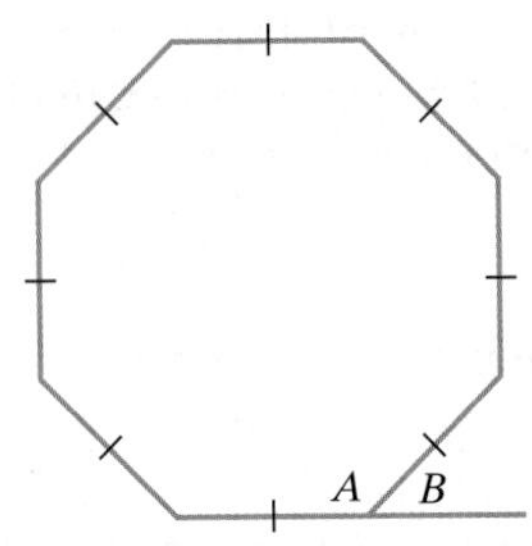

FIGURE 10.34 A regular octagon

EXAMPLE 2 Using the Formula for the Angles of a Polygon

a. Find the sum of the measures of the angles of an octagon.

b. **Figure 10.34** shows a regular octagon. Find the measure of angle A.

c. Find the measure of exterior angle B.

SOLUTION

a. An octagon has eight sides. Using the formula $(n - 2)180°$ with $n = 8$, we can find the sum of the measures of its eight angles.

The sum of the measures of an octagon's angles is

$$\begin{aligned}&(n - 2)180°\\ &= (8 - 2)180°\\ &= 6 \cdot 180°\\ &= 1080°.\end{aligned}$$

b. Examine the regular octagon in **Figure 10.34**. Note that all eight sides have the same length. Likewise, all eight angles have the same degree measure. Angle A is one of these eight angles. We find its measure by taking the sum of the measures of all eight angles, 1080°, and dividing by 8.

$$m\measuredangle A = \frac{1080°}{8} = 135°$$

c. Because $\measuredangle B$ is the supplement of $\measuredangle A$,

$$m\measuredangle B = 180° - 135° = 45°.$$

CHECK POINT 2

a. Find the sum of the measures of the angles of a 12-sided polygon.

b. Find the measure of an angle of a regular 12-sided polygon.

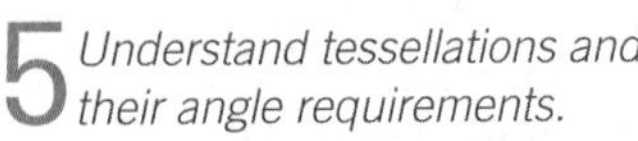

5 *Understand tessellations and their angle requirements.*

Tessellations

A relationship between geometry and the visual arts is found in an art form called *tessellations*. A **tessellation**, or **tiling**, is a pattern consisting of the repeated use of the same geometric figures to completely cover a plane, leaving no gaps and no overlaps. **Figure 10.35** shows eight tessellations, each consisting of the repeated use of two or more regular polygons.

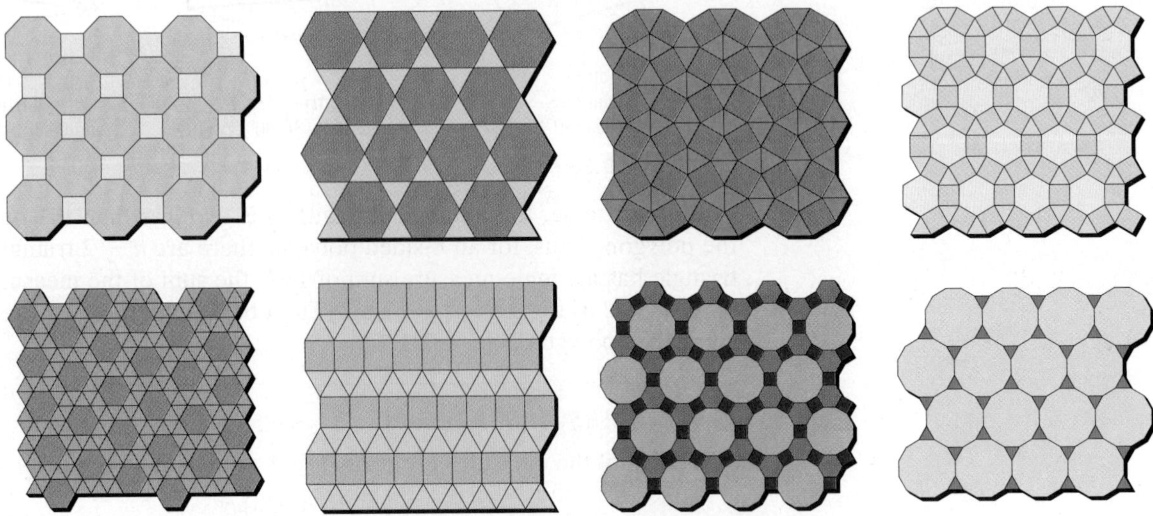

FIGURE 10.35 Eight tessellations formed by two or more regular polygons

In each tessellation in **Figure 10.35**, the same types of regular polygons surround every vertex (intersection point). Furthermore, **the sum of the measures of the angles that come together at each vertex is 360°, a requirement for the formation of a tessellation**. This is illustrated in the enlarged version of the tessellation in **Figure 10.36**. If you select any vertex and count the sides of the polygons that touch it, you'll see that vertices are surrounded by two regular octagons and a square. Can you see why the sum of the angle measures at every vertex is 360°?

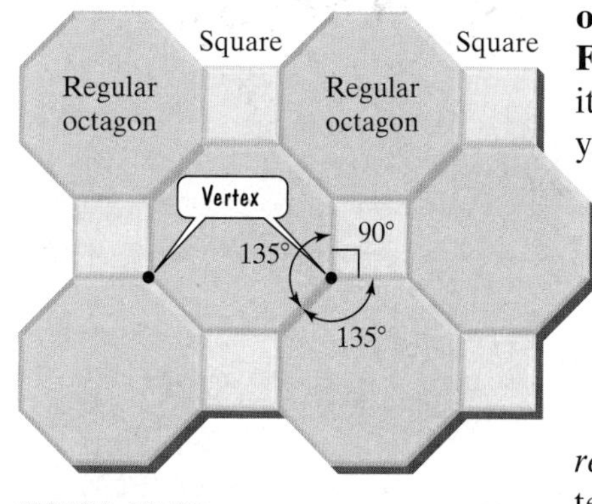

FIGURE 10.36

$$135° + 135° + 90° = 360°$$

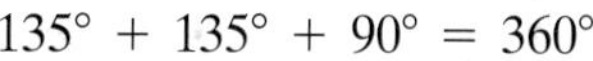

In Example 2, we found that each angle of a regular octagon measures 135°.

Each angle of a square measures 90°.

The most restrictive condition in creating tessellations is that just *one type of regular polygon* may be used. With this restriction, there are only three possible tessellations, made from equilateral triangles or squares or regular hexagons, as shown in **Figure 10.37**.

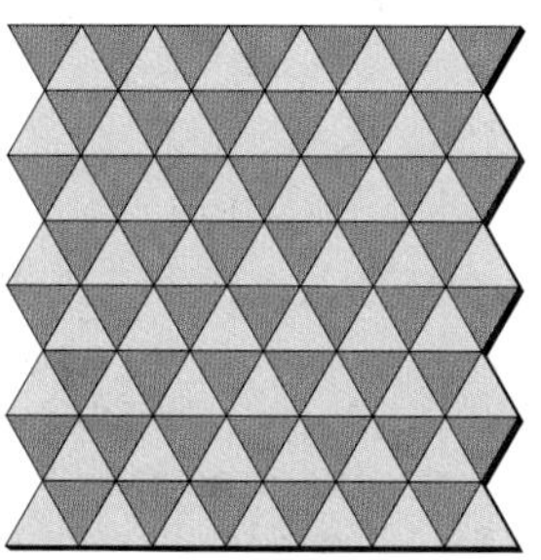

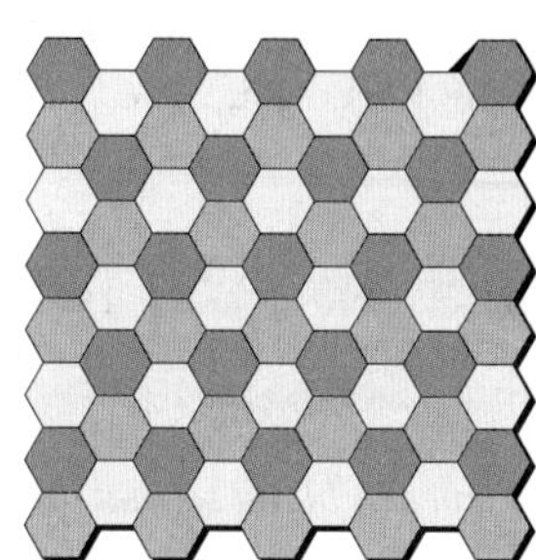

FIGURE 10.37 The three tessellations formed using one regular polygon

Each tessellation is possible because the angle sum at every vertex is 360°:

Six equilateral triangles at each vertex: $60° + 60° + 60° + 60° + 60° + 60° = 360°$

Four squares at each vertex: $90° + 90° + 90° + 90° = 360°$

Three regular hexagons at each vertex: $120° + 120° + 120° = 360°.$

Each angle measures $\frac{(n-2)180°}{n} = \frac{(6-2)180°}{6} = 120°.$

EXAMPLE 3 *Angle Requirements of Tessellations*

Explain why a tessellation cannot be created using only regular pentagons.

SOLUTION

Let's begin by applying $(n - 2)180°$ to find the measure of each angle of a regular pentagon. Each angle measures

$$\frac{(5-2)180°}{5} = \frac{3(180°)}{5} = 108°.$$

A requirement for the formation of a tessellation is that the measures of the angles that come together at each vertex is 360°. With each angle of a regular pentagon measuring 108°, **Figure 10.38** shows that three regular pentagons fill in $3 \cdot 108° = 324°$ and leave a $360° - 324°$, or a 36°, gap. Because the 360° angle requirement cannot be met, no tessellation by regular pentagons is possible.

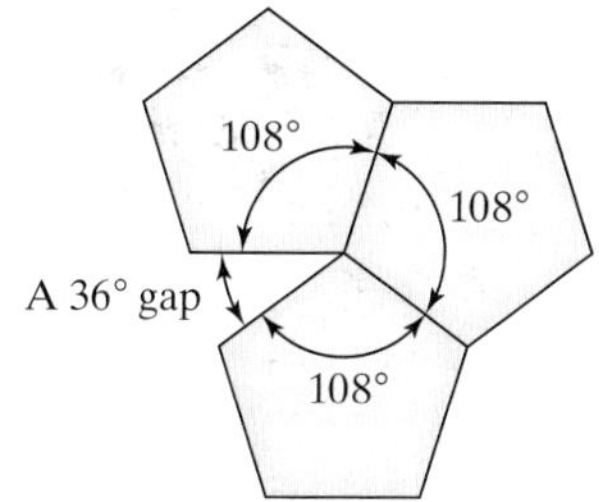

FIGURE 10.38

CHECK POINT 3 Explain why a tessellation cannot be created using only regular octagons.

"The regular division of the plane into congruent figures evoking an association in the observer with a familiar natural object is one of these hobbies or problems . . . I have embarked on this geometric problem again and again over the years, trying to throw light on different aspects each time. I cannot imagine what my life would be like if this problem had never occurred to me; one might say that I am head over heels in love with it, and I still don't know why."

—M. C. ESCHER

Tessellations that are not restricted to the repeated use of regular polygons are endless in number. They are prominent in Islamic art, Italian mosaics, quilts, and ceramics. The Dutch artist M. C. Escher (1898–1972) created a dazzling array of prints, drawings, and paintings using tessellations composed of stylized interlocking animals. Escher's art reflects the mathematics that underlies all things, while creating surreal manipulations of space and perspective that make gentle fun of consensus reality.

Symmetry Drawing E85 (1952). M.C. Escher's "Symmetry Drawing E85" www.mcescher.com

Concept and Vocabulary Check

Fill in each blank so that the resulting statement is true.

1. The distance around the sides of a polygon is called its ________.
2. A four-sided polygon is a/an ________, a five-sided polygon is a/an ________, a six-sided polygon is a/an ________, a seven-sided polygon is a/an ________, and an eight-sided polygon is a/an ________.
3. A polygon whose sides are all the same length and whose angles all have the same measure is called a/an ________ polygon.
4. Opposite sides of a parallelogram are ________ and ________.
5. A parallelogram with all sides of equal length without any right angles is called a/an ________.
6. A parallelogram with four right angles without all sides of equal length is called a/an ________.
7. A parallelogram with four right angles and all sides of equal length is called a/an ________.
8. A four-sided figure with exactly one pair of parallel sides is called a/an ________.
9. The perimeter, P, of a rectangle with length l and width w is given by the formula ________.
10. The sum of the measures of the angles of a polygon with n sides is ________.
11. A pattern consisting of the repeated use of the same geometric figures to completely cover a plane, leaving no gaps and no overlaps, is called a/an ________.

In Exercises 12–18, determine whether each statement is true or false. If the statement is false, make the necessary change(s) to produce a true statement.

12. Every parallelogram is a rhombus. ______
13. Every rhombus is a parallelogram. ______
14. All squares are rectangles. ______
15. Some rectangles are not squares. ______
16. No triangles are polygons. ______
17. Every rhombus is a regular polygon. ______
18. A requirement for the formation of a tessellation is that the sum of the measures of the angles that come together at each vertex is 360°. ______

Exercise Set 10.3

Practice Exercises

In Exercises 1–4, use the number of sides to name the polygon.

1.

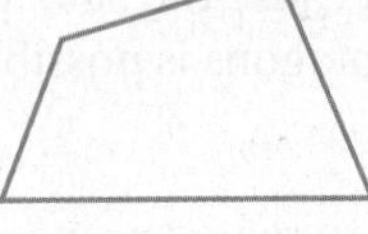

2.

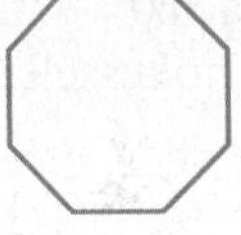

3.

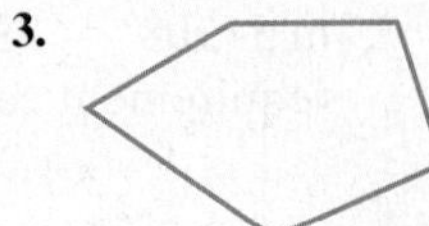

4.

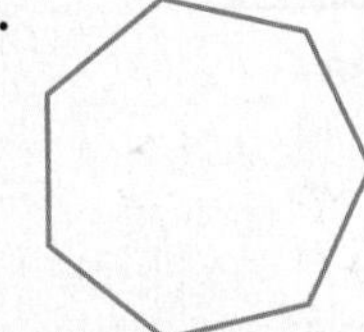

Use these quadrilaterals to solve Exercises 5–10.

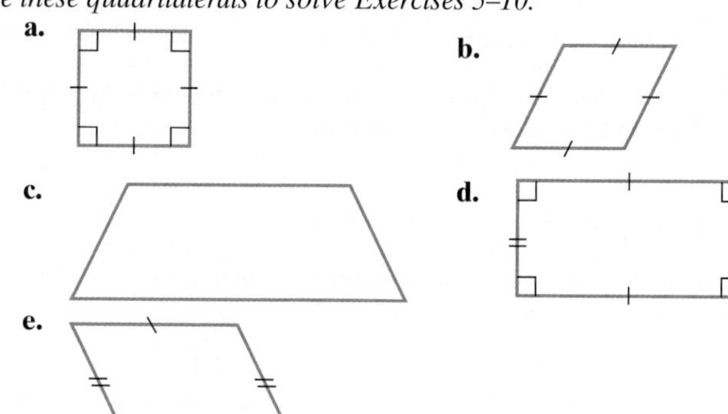

5. Which of these quadrilaterals have opposite sides that are parallel? Name these quadrilaterals.
6. Which of these quadrilaterals have sides of equal length that meet at a vertex? Name these quadrilaterals.
7. Which of these quadrilaterals have right angles? Name these quadrilaterals.
8. Which of these quadrilaterals do not have four sides of equal length? Name these quadrilaterals.
9. Which of these quadrilaterals is not a parallelogram? Name this quadrilateral.
10. Which of these quadrilaterals is/are a regular polygon? Name this/these quadrilateral(s).

In Exercises 11–20, find the perimeter of the figure named and shown. Express the perimeter using the same unit of measure that appears on the given side or sides.

11. Rectangle

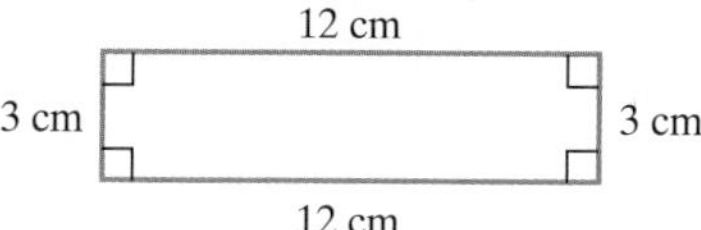

12. Parallelogram

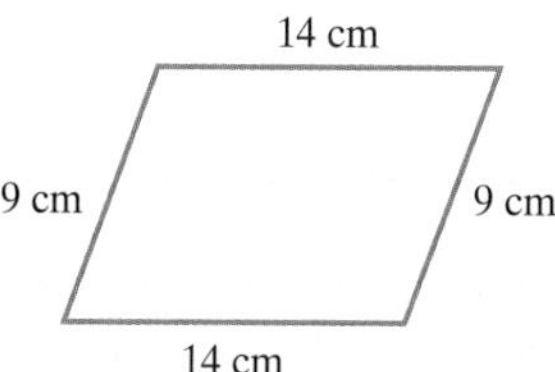

13. Rectangle

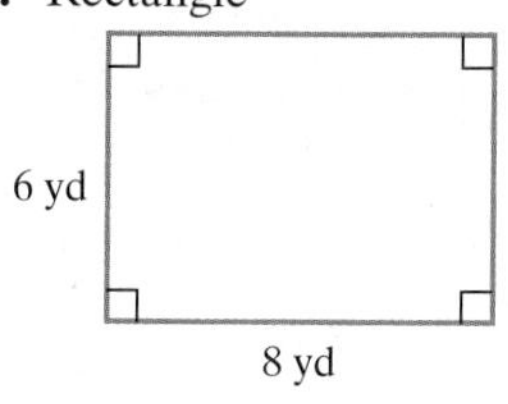

14. Rectangle

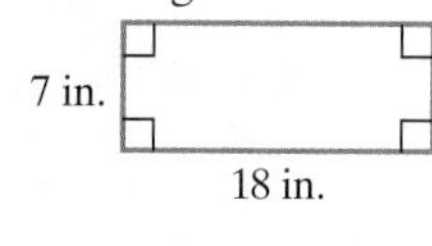

15. Square

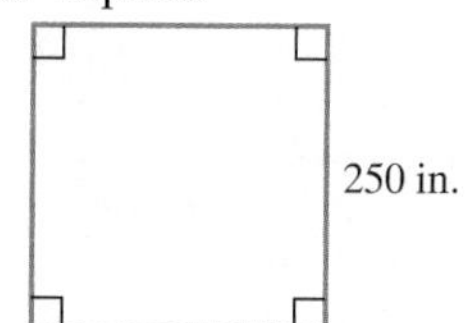

16. Square

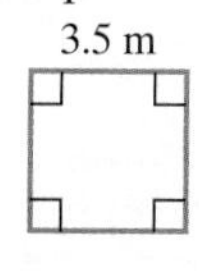

17. Triangle

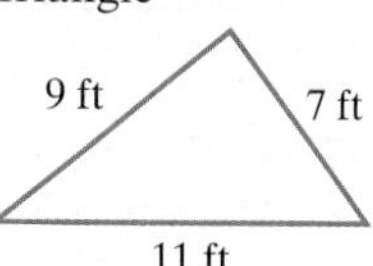

18. Triangle

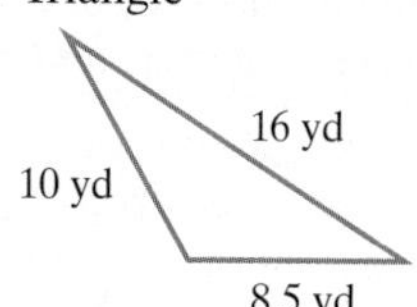

19. Equilateral triangle

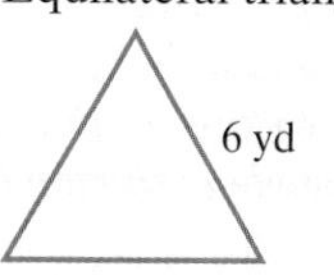

20. Regular hexagon

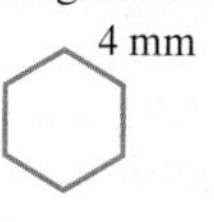

In Exercises 21–24, find the perimeter of the figure shown. Express the perimeter using the same unit of measure that appears on the given side or sides.

21.

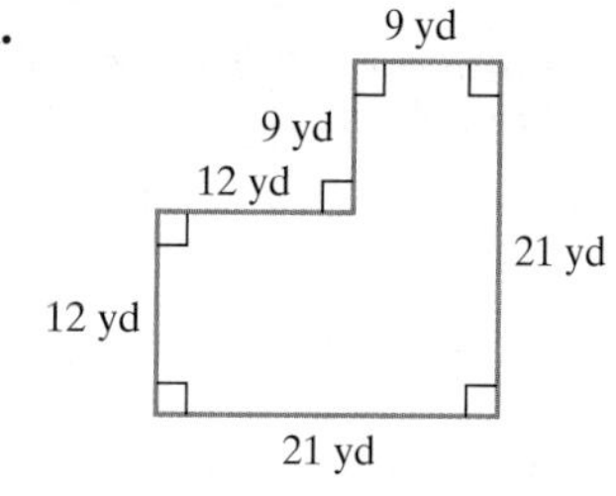

22.

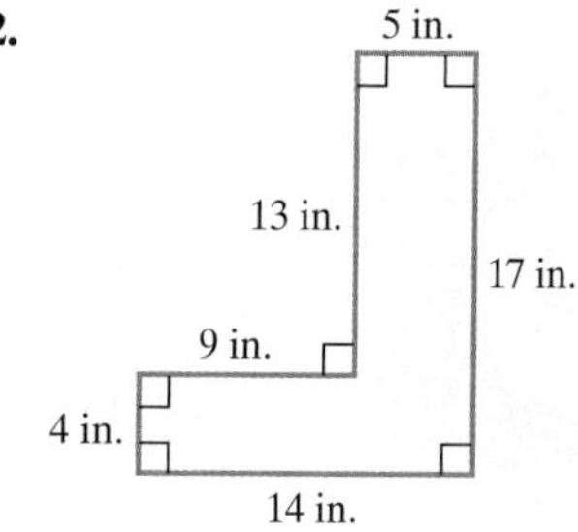

23.

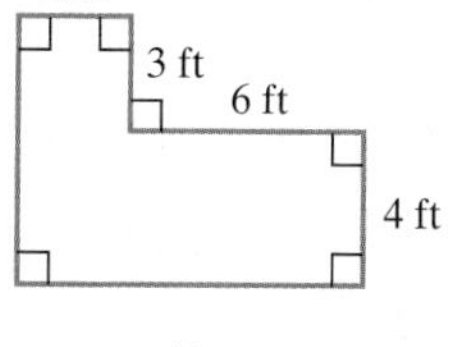

24.

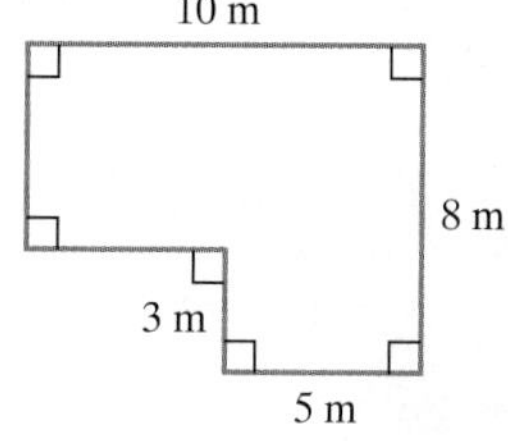

25. Find the sum of the measures of the angles of a five-sided polygon.
26. Find the sum of the measures of the angles of a six-sided polygon.
27. Find the sum of the measures of the angles of a quadrilateral.
28. Find the sum of the measures of the angles of a heptagon.

In Exercises 29–30, each figure shows a regular polygon. Find the measures of angle A and angle B.

29.

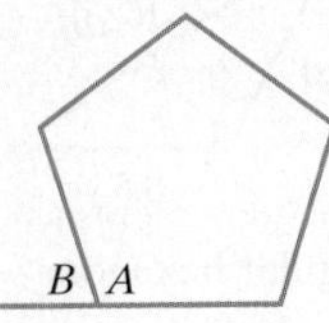

30.

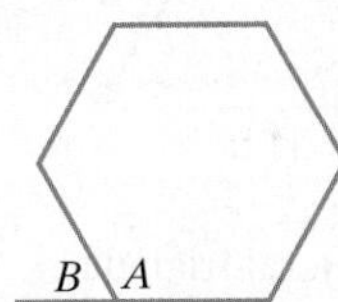

In Exercises 31–32, **a.** *Find the sum of the measures of the angles for the figure given;* **b.** *Find the measures of angle A and angle B.*

31.

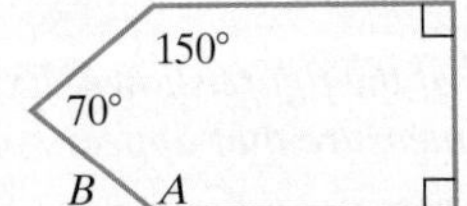

32.

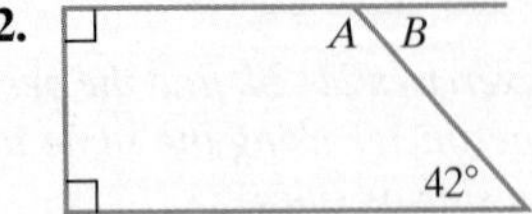

In Exercises 33–36, tessellations formed by two or more regular polygons are shown.

a. *Name the types of regular polygons that surround each vertex.*

b. *Determine the number of angles that come together at each vertex, as well as the measures of these angles.*

c. *Use the angle measures from part (b) to explain why the tessellation is possible.*

33.

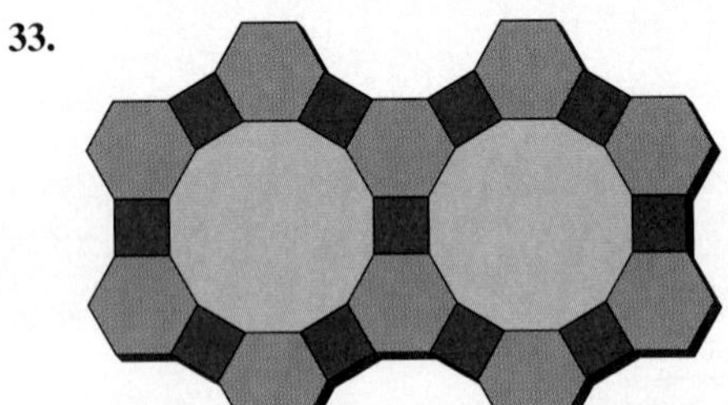

34.

35.

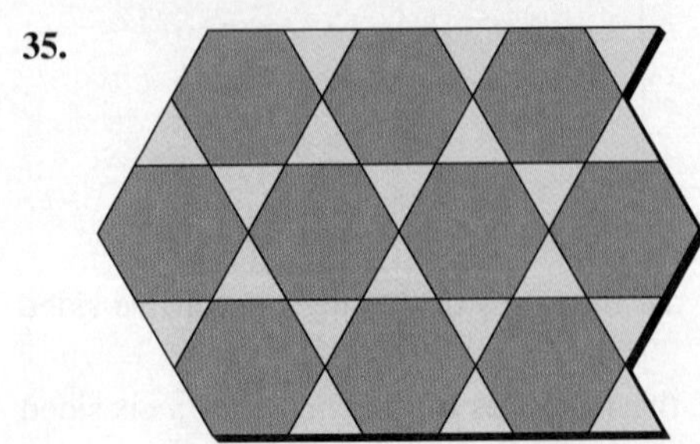

36.

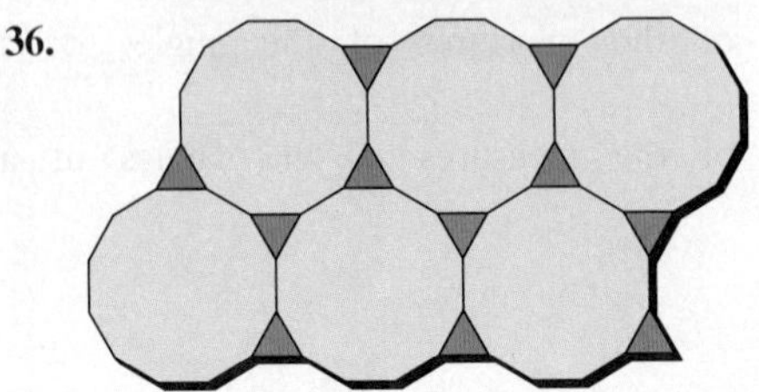

37. Can a tessellation be created using only regular nine-sided polygons? Explain your answer.

38. Can a tessellation be created using only regular ten-sided polygons? Explain your answer.

Practice Plus

In Exercises 39–42, use an algebraic equation to determine each rectangle's dimensions.

39. A rectangular field is four times as long as it is wide. If the perimeter of the field is 500 yards, what are the field's dimensions?

40. A rectangular field is five times as long as it is wide. If the perimeter of the field is 288 yards, what are the field's dimensions?

41. An American football field is a rectangle with a perimeter of 1040 feet. The length is 200 feet more than the width. Find the width and length of the rectangular field.

42. A basketball court is a rectangle with a perimeter of 86 meters. The length is 13 meters more than the width. Find the width and length of the basketball court.

In Exercises 43–44, use algebraic equations to find the measure of each angle that is represented in terms of x.

43.

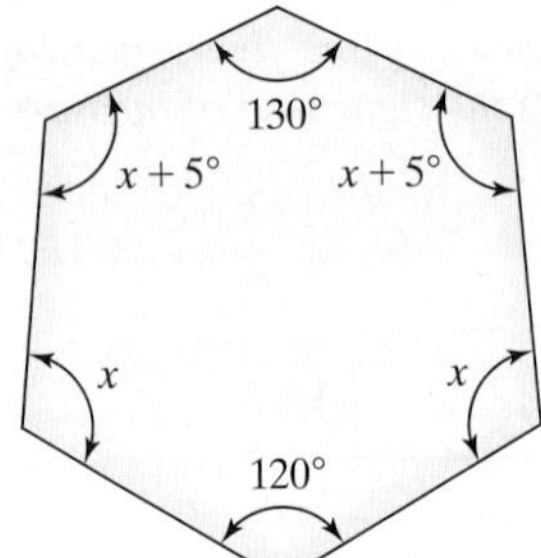

44.

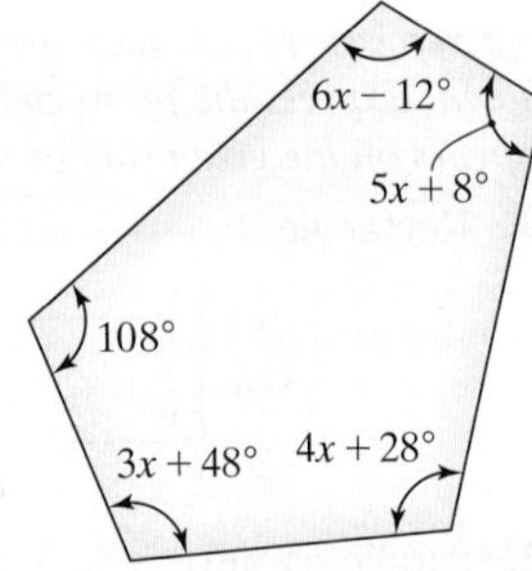

In the figure shown, the artist has cunningly distorted the "regular" polygons to create a fraudulent tessellation with discrepancies that are too subtle for the eye to notice. In Exercises 45–46, you will use mathematics, not your eyes, to observe the irregularities.

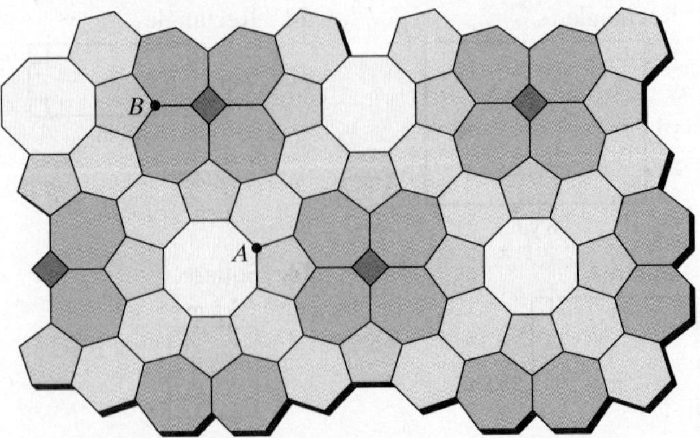

45. Find the sum of the angle measures at vertex A. Then explain why the tessellation is a fake.

46. Find the sum of the angle measures at vertex B. Then explain why the tessellation is a fake.

Application Exercises

47. A school playground is in the shape of a rectangle 400 feet long and 200 feet wide. If fencing costs $14 per yard, what will it cost to place fencing around the playground?

48. A rectangular field is 70 feet long and 30 feet wide. If fencing costs $8 per yard, how much will it cost to enclose the field?

In Exercises 49–50, refer to the appropriate English and metric equivalent in **Table 9.3** *on page 587.*

49. A protected wilderness area in the shape of a rectangle is 4 kilometers long and 3.2 kilometers wide. The forest is to be surrounded by a hiking trail that will cost $11,000 per mile to construct. What will it cost to install the trail?

50. A rectangular field is 27 meters long and 18 meters wide. If fencing costs $12 per yard, how much will it cost to enclose the field?

51. One side of a square flower bed is 8 feet long. How many plants are needed if they are to be spaced 8 inches apart around the outside of the bed?

52. What will it cost to place baseboard around the region shown if the baseboard costs $0.25 per foot? No baseboard is needed for the 2-foot doorway.

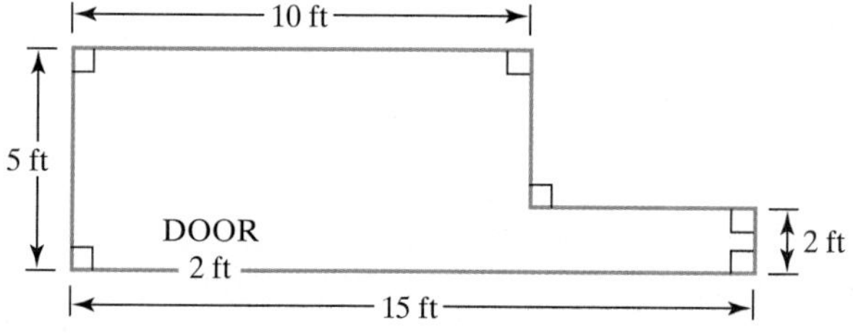

Explaining the Concepts

53. What is a polygon?

54. Explain why rectangles and rhombuses are also parallelograms.

55. Explain why every square is a rectangle, a rhombus, a parallelogram, a quadrilateral, and a polygon.

56. Explain why a square is a regular polygon, but a rhombus is not.

57. Using words only, describe how to find the perimeter of a rectangle.

58. Describe a practical situation in which you needed to apply the concept of a geometric figure's perimeter.

59. Describe how to find the measure of an angle of a regular pentagon.

Critical Thinking Exercises

Make Sense? *In Exercises 60–63, determine whether each statement makes sense or does not make sense, and explain your reasoning.*

60. I drew a polygon having two sides that intersect at their midpoints.

61. I find it helpful to think of a polygon's perimeter as the length of its boundary.

62. If a polygon is not regular, I can determine the sum of the measures of its angles, but not the measure of any one of its angles.

63. I used floor tiles in the shape of regular pentagons to completely cover my kitchen floor.

In Exercises 64–65, write an algebraic expression that represents the perimeter of the figure shown.

64.

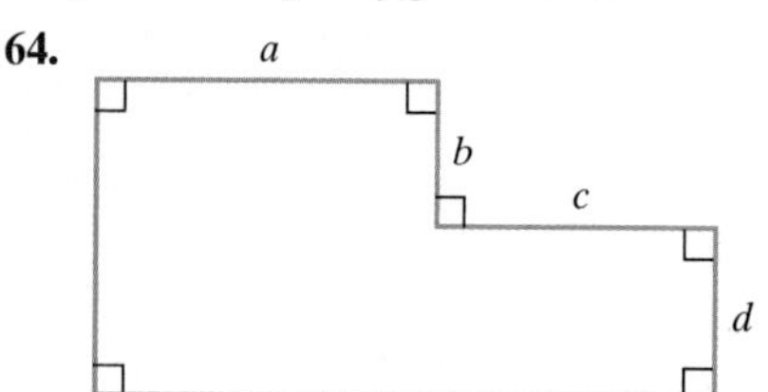

65.

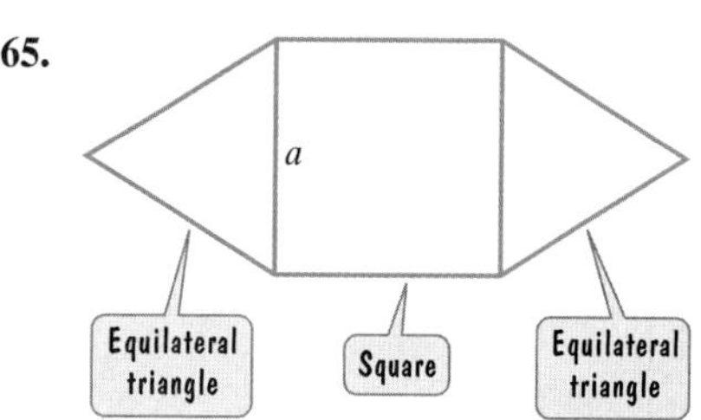

66. Find $m\measuredangle 1$ in the figure shown.

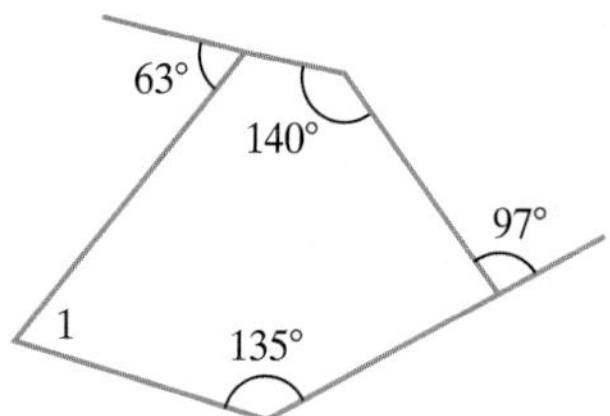

Group Exercise

67. Group members should consult sites on the Internet devoted to tessellations, or tilings, and present a report that expands upon the information in this section. Include a discussion of cultures that have used tessellations on fabrics, wall coverings, baskets, rugs, and pottery, with examples. Include the Alhambra, a fourteenth-century palace in Granada, Spain, in the presentation, as well as works by the artist M. C. Escher. Discuss the various symmetries (translations, rotations, reflections) associated with tessellations. Demonstrate how to create unique tessellations, including Escher-type patterns. Other than creating beautiful works of art, are there any practical applications of tessellations?

10.4

WHAT AM I SUPPOSED TO LEARN?

After studying this section, you should be able to:

1 Use area formulas to compute the areas of plane regions and solve applied problems.

2 Use formulas for a circle's circumference and area.

Area and Circumference

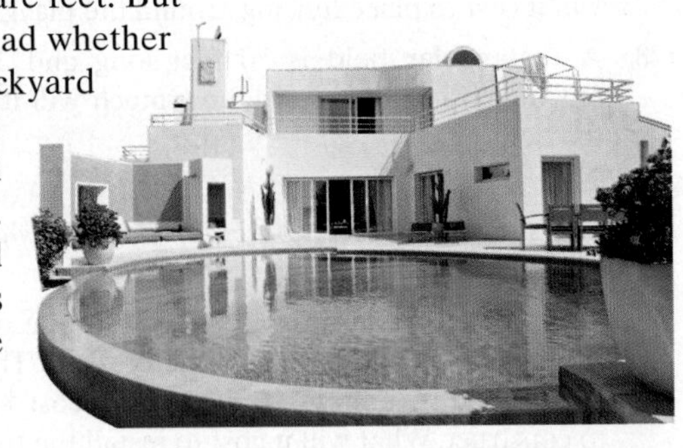

The size of a house is described in square feet. But how do you know from the real estate ad whether the 1200-square-foot home with the backyard pool is large enough to warrant a visit? Faced with hundreds of ads, you need some way to sort out the best bets. What does 1200 square feet mean and how is this area determined? In this section, we discuss how to compute the areas of plane regions.

1 *Use area formulas to compute the areas of plane regions and solve applied problems.*

Formulas for Area

In Section 9.2, we saw that the area of a two-dimensional figure is the number of square units, such as square inches or square miles, it takes to fill the interior of the figure. For example, **Figure 10.39** shows that there are 12 square units contained within the rectangular region. The area of the region is 12 square units. Notice that the area can be determined in the following manner:

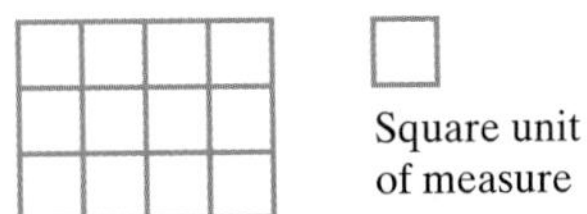

FIGURE 10.39 The area of the region on the left is 12 square units.

Distance across: 4 units; Distance down: 3 units

$$4 \text{ units} \times 3 \text{ units} = 4 \times 3 \times \text{units} \times \text{units}$$
$$= 12 \text{ square units.}$$

The area of a rectangular region, usually referred to as the area of a rectangle, is the product of the distance across (length) and the distance down (width).

AREA OF A RECTANGLE AND A SQUARE

The area, A, of a rectangle with length l and width w is given by the formula

$$A = lw.$$

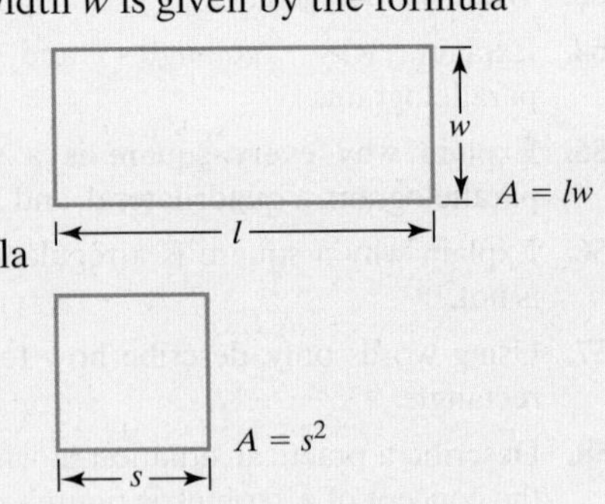

The area, A, of a square with one side measuring s linear units is given by the formula

$$A = s^2.$$

EXAMPLE 1 *Solving an Area Problem*

You decide to cover the path shown in **Figure 10.40** with bricks.

a. Find the area of the path.

b. If the path requires four bricks for every square foot, how many bricks are needed for the project?

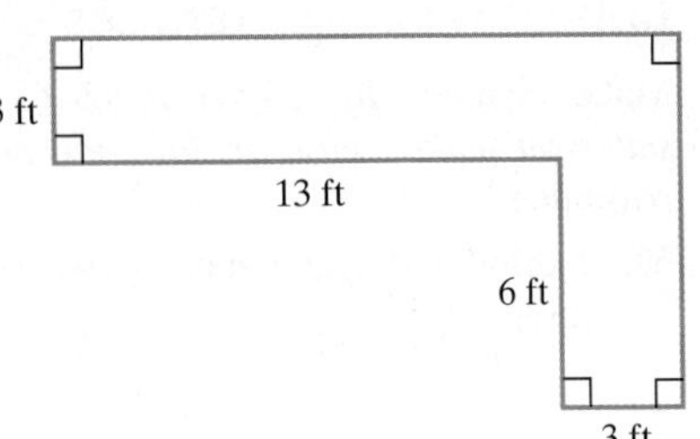

FIGURE 10.40

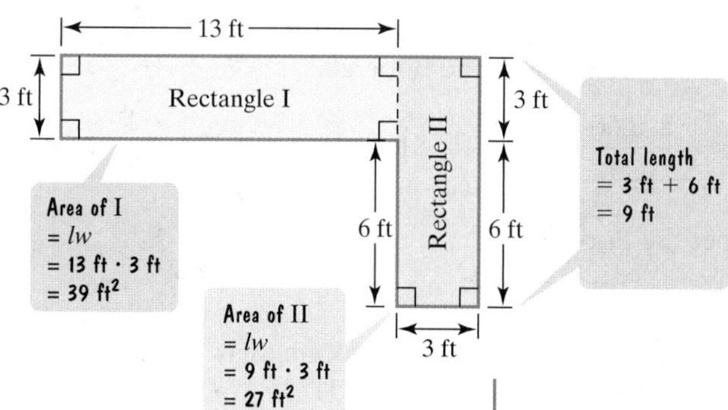

SOLUTION

a. Because we have a formula for the area of a rectangle, we begin by drawing a dashed line that divides the path into two rectangles. One way of doing this is shown at the left. We then use the length and width of each rectangle to find its area. The computations for area are shown in the green and blue voice balloons.

The area of the path is found by adding the areas of the two rectangles.

$$\text{Area of path} = 39 \text{ ft}^2 + 27 \text{ ft}^2 = 66 \text{ ft}^2$$

b. The path requires 4 bricks per square foot. The number of bricks needed for the project is the number of square feet in the path, its area, times 4.

$$\text{Number of bricks needed} = 66 \cancel{\text{ft}^2} \cdot \frac{4 \text{ bricks}}{\cancel{\text{ft}^2}} = 66 \cdot 4 \text{ bricks} = 264 \text{ bricks}$$

Thus, 264 bricks are needed for the project.

Appraising a House

A house is measured by an appraiser hired by a bank to help establish its value. The appraiser works from the outside, measuring off a rectangle. Then the appraiser adds the living spaces that lie outside the rectangle and subtracts the empty areas inside the rectangle. The final figure, in square feet, includes all the finished floor space in the house. Not included are the garage, outside porches, decks, or an unfinished basement.

A 1000-square-foot house is considered small, one with 2000 square feet average, and one with more than 2500 square feet pleasantly large. If a 1200-square-foot house has three bedrooms, the individual rooms might seem snug and cozy. With only one bedroom, the space may feel palatial!

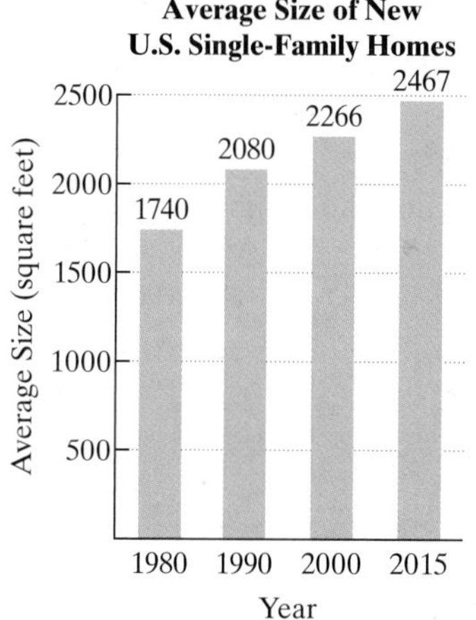

Source: U.S. Census Bureau

CHECK POINT 1 Find the area of the path described in Example 1, rendered on the right as a green region, by first measuring off a large rectangle as shown. The area of the path is the area of the large rectangle (the blue and green regions combined) minus the area of the blue rectangle. Do you get the same answer as we did in Example 1(a)?

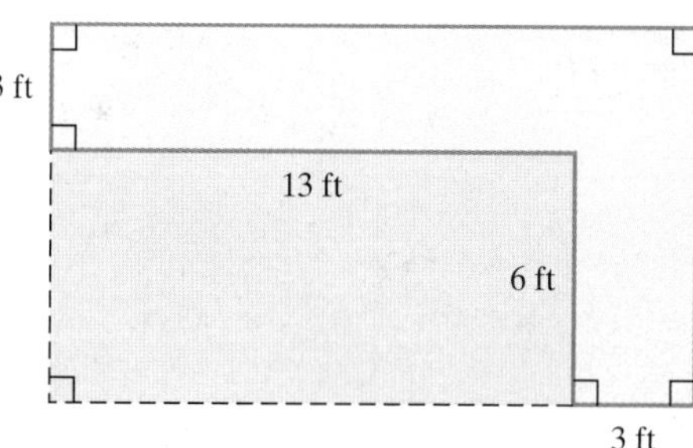

In Section 9.2, we saw that although there are 3 linear feet in a linear yard, there are 9 square feet in a square yard. If a problem requires measurement of area in square yards and the linear measures are given in feet, to avoid errors, first convert feet to yards. Then apply the area formula. This idea is illustrated in Example 2.

EXAMPLE 2 *Solving an Area Problem*

What will it cost to carpet a rectangular floor measuring 12 feet by 15 feet if the carpet costs $18.50 per square yard?

SOLUTION

We begin by converting the linear measures from feet to yards.

$$12 \text{ ft} = \frac{12 \cancel{\text{ft}}}{1} \cdot \frac{1 \text{ yd}}{3 \cancel{\text{ft}}} = \frac{12}{3} \text{ yd} = 4 \text{ yd}$$

$$15 \text{ ft} = \frac{15 \cancel{\text{ft}}}{1} \cdot \frac{1 \text{ yd}}{3 \cancel{\text{ft}}} = \frac{15}{3} \text{ yd} = 5 \text{ yd}$$

Next, we find the area of the rectangular floor in square yards.

$$A = lw = 5 \text{ yd} \cdot 4 \text{ yd} = 20 \text{ yd}^2$$

Finally, we find the cost of the carpet by multiplying the cost per square yard, \$18.50, by the number of square yards in the floor, 20.

$$\text{Cost of carpet} = \frac{\$18.50}{\cancel{yd^2}} \cdot \frac{20\ \cancel{yd^2}}{1} = \$18.50(20) = \$370$$

It will cost \$370 to carpet the floor.

CHECK POINT 2 What will it cost to carpet a rectangular floor measuring 18 feet by 21 feet if the carpet costs \$16 per square yard?

We can use the formula for the area of a rectangle to develop formulas for areas of other polygons. We begin with a parallelogram, a quadrilateral with opposite sides equal and parallel. The **height** of a parallelogram is the perpendicular distance between two of the parallel sides. Height is denoted by h in **Figure 10.41**. The **base**, denoted by b, is the length of either of these parallel sides.

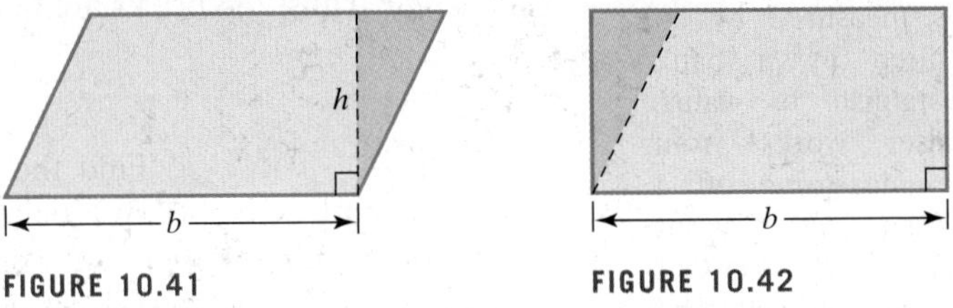

FIGURE 10.41

FIGURE 10.42

In **Figure 10.42**, the orange triangular region has been cut off from the right of the parallelogram and attached to the left. The resulting figure is a rectangle with length b and width h. Because bh is the area of the rectangle, it also represents the area of the parallelogram.

GREAT QUESTION!

Is the height of a parallelogram the length of one of its sides?

No. The height of a parallelogram is the perpendicular distance between two of the parallel sides. It is *not* the length of a side.

AREA OF A PARALLELOGRAM

The area, A, of a parallelogram with height h and base b is given by the formula

$$A = bh.$$

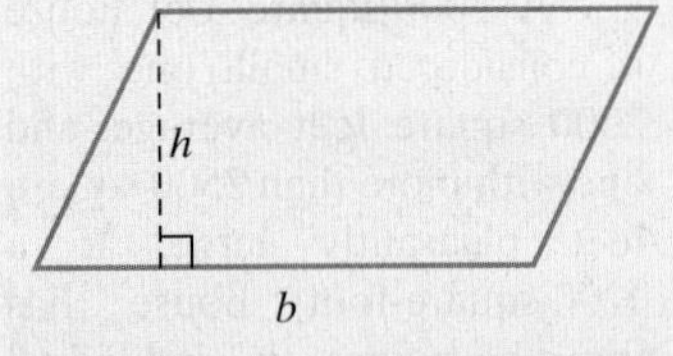

EXAMPLE 3 *Using the Formula for a Parallelogram's Area*

Find the area of the parallelogram in **Figure 10.43**.

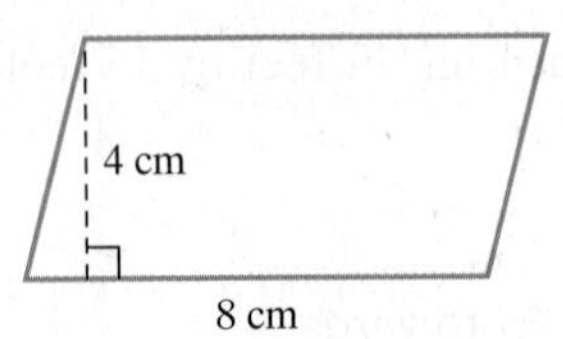

FIGURE 10.43

SOLUTION

As shown in the figure, the base is 8 centimeters and the height is 4 centimeters. Thus, $b = 8$ and $h = 4$.

$$A = bh$$
$$A = 8\text{ cm} \cdot 4\text{ cm} = 32\text{ cm}^2$$

The area is 32 square centimeters.

CHECK POINT 3 Find the area of a parallelogram with a base of 10 inches and a height of 6 inches.

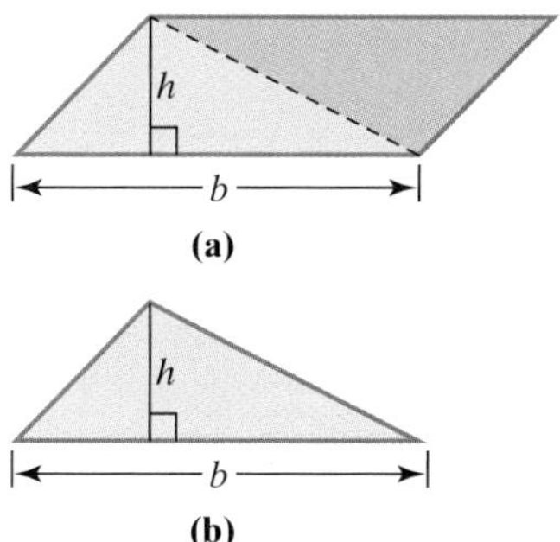

FIGURE 10.44

Figure 10.44 demonstrates how we can use the formula for the area of a parallelogram to obtain a formula for the area of a triangle. The area of the parallelogram in **Figure 10.44(a)** is given by $A = bh$. The diagonal shown in the parallelogram divides it into two triangles with the same size and shape. This means that the area of each triangle is one-half that of the parallelogram. Thus, the area of the triangle in **Figure 10.44(b)** is given by $A = \frac{1}{2}bh$.

AREA OF A TRIANGLE

The area, A, of a triangle with height h and base b is given by the formula

$$A = \frac{1}{2}bh.$$

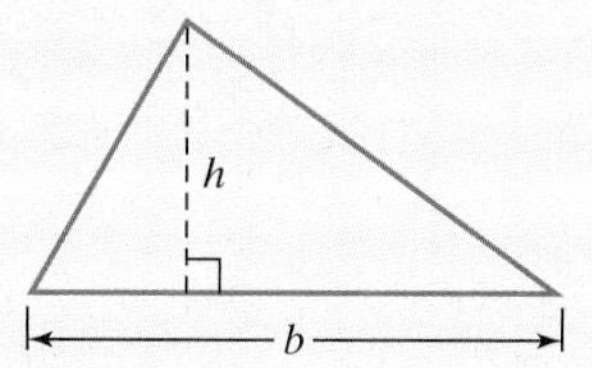

EXAMPLE 4 Using the Formula for a Triangle's Area

Find the area of each triangle in **Figure 10.45**.

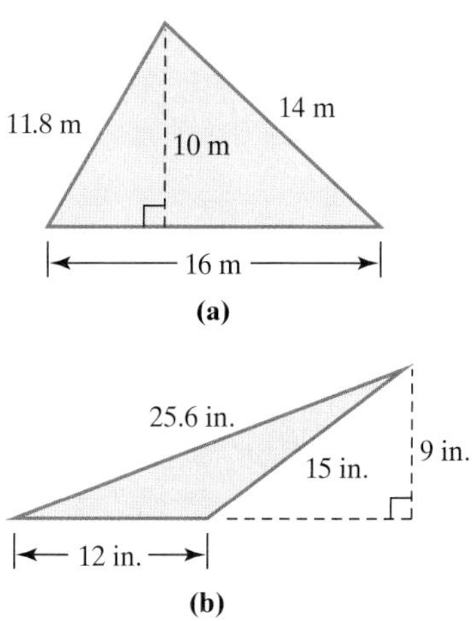

FIGURE 10.45

SOLUTION

a. In **Figure 10.45(a)**, the base is 16 meters and the height is 10 meters, so $b = 16$ and $h = 10$. We do not need the 11.8 meters or the 14 meters to find the area. The area of the triangle is

$$A = \frac{1}{2}bh = \frac{1}{2} \cdot 16 \text{ m} \cdot 10 \text{ m} = 80 \text{ m}^2.$$

The area is 80 square meters.

b. In **Figure 10.45(b)**, the base is 12 inches. The base line needs to be extended to draw the height. However, we still use 12 inches for b in the area formula. The height, h, is given to be 9 inches. The area of the triangle is

$$A = \frac{1}{2}bh = \frac{1}{2} \cdot 12 \text{ in.} \cdot 9 \text{ in.} = 54 \text{ in.}^2.$$

The area of the triangle is 54 square inches.

CHECK POINT 4 A sailboat has a triangular sail with a base of 12 feet and a height of 5 feet. Find the area of the sail.

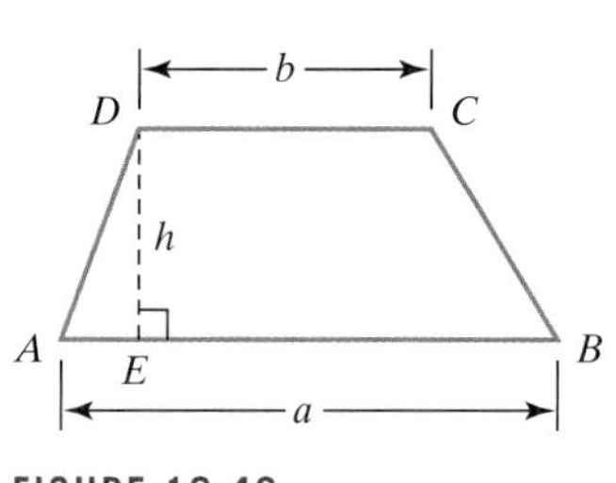

FIGURE 10.46

The formula for the area of a triangle can be used to obtain a formula for the area of a trapezoid. Consider the trapezoid shown in **Figure 10.46**. The lengths of the two parallel sides, called the **bases**, are represented by a (the lower base) and b (the upper base). The trapezoid's height, denoted by h, is the perpendicular distance between the two parallel sides.

In **Figure 10.47**, we have drawn line segment BD, dividing the trapezoid into two triangles, shown in yellow and orange. The area of the trapezoid is the sum of the areas of these triangles.

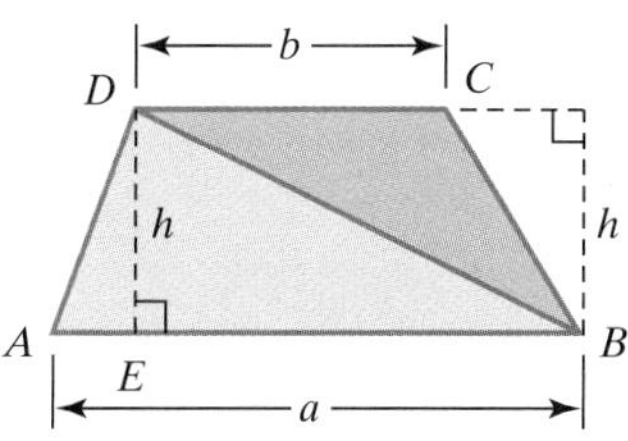

FIGURE 10.47

Area of trapezoid	=	Area of yellow △	plus	Area of orange △
A	$=$	$\frac{1}{2}ah$	$+$	$\frac{1}{2}bh$
	$=$	$\frac{1}{2}h(a+b)$		Factor out $\frac{1}{2}h$.

AREA OF A TRAPEZOID

The area, A, of a trapezoid with parallel bases a and b and height h is given by the formula

$$A = \frac{1}{2}h(a + b).$$

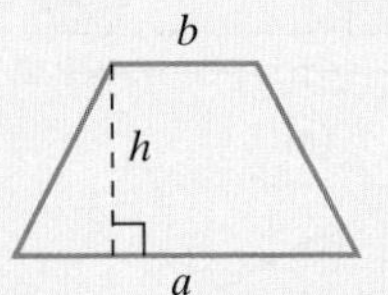

EXAMPLE 5 *Finding the Area of a Trapezoid*

Find the area of the trapezoid in **Figure 10.48**.

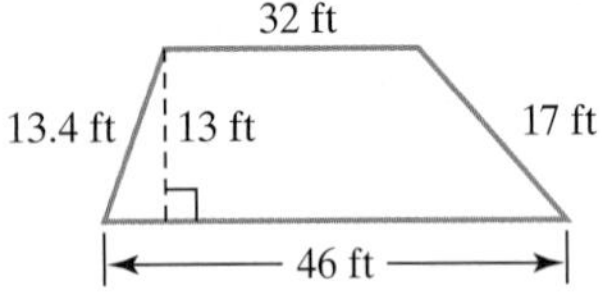

FIGURE 10.48

SOLUTION

The height, h, is 13 feet. The lower base, a, is 46 feet, and the upper base, b, is 32 feet. We do not use the 17-foot and 13.4-foot sides in finding the trapezoid's area.

$$A = \frac{1}{2}h(a + b) = \frac{1}{2} \cdot 13 \text{ ft} \cdot (46 \text{ ft} + 32 \text{ ft})$$
$$= \frac{1}{2} \cdot 13 \text{ ft} \cdot 78 \text{ ft} = 507 \text{ ft}^2$$

The area of the trapezoid is 507 square feet.

CHECK POINT 5 Find the area of a trapezoid with bases of length 20 feet and 10 feet and height 7 feet.

2 *Use formulas for a circle's circumference and area.*

It's a good idea to know your way around a circle. Clocks, angles, maps, and compasses are based on circles. Circles occur everywhere in nature: in ripples on water, patterns on a butterfly's wings, and cross sections of trees. Some consider the circle to be the most pleasing of all shapes.

A **circle** is a set of points in the plane equally distant from a given point, its **center**. **Figure 10.49** shows two circles. The **radius** (plural: radii), r, is a line segment from the center to any point on the circle. For a given circle, all radii have the same length. The **diameter**, d, is a line segment through the center whose endpoints both lie on the circle. For a given circle, all diameters have the same length. In any circle, the **length of the diameter is twice the length of the radius**.

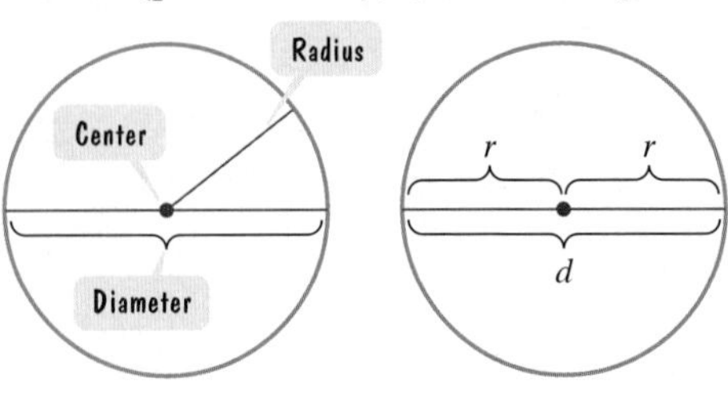

FIGURE 10.49

The point at which a pebble hits a flat surface of water becomes the center of a number of circular ripples.

The words *radius* and *diameter* refer to both the line segments in **Figure 10.49** as well as to their linear measures. The distance around a circle (its perimeter) is called its **circumference**, C. For all circles, if you divide the circumference by the diameter, or by twice the radius, you will get the same number. This ratio is the irrational number π and is approximately equal to 3.14:

$$\frac{C}{d} = \pi \quad \text{or} \quad \frac{C}{2r} = \pi.$$

Thus,

$$C = \pi d \quad \text{or} \quad C = 2\pi r.$$

FINDING THE DISTANCE AROUND A CIRCLE

The circumference, C, of a circle with diameter d and radius r is

$$C = \pi d \quad \text{or} \quad C = 2\pi r.$$

When computing a circle's circumference by hand, round π to 3.14. When using a calculator, use the $\boxed{\pi}$ key, which gives the value of π rounded to approximately 11 decimal places. In either case, calculations involving π give approximate answers. These answers can vary slightly depending on how π is rounded. The symbol $\approx$ (is approximately equal to) will be written in these calculations.

EXAMPLE 6 *Finding a Circle's Circumference*

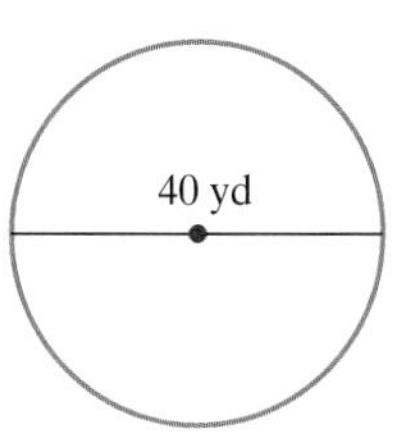

FIGURE 10.50

Find the circumference of the circle in **Figure 10.50**.

SOLUTION

The diameter is 40 yards, so we use the formula for circumference with d in it.

$$C = \pi d = \pi(40 \text{ yd}) = 40\pi \text{ yd} \approx 125.7 \text{ yd}$$

The distance around the circle is approximately 125.7 yards.

CHECK POINT 6 Find the circumference of a circle whose diameter measures 10 inches. Express the answer in terms of π and then round to the nearest tenth of an inch.

EXAMPLE 7 *Using the Circumference Formula*

FIGURE 10.51

How much trim, to the nearest tenth of a foot, is needed to go around the window shown in **Figure 10.51**?

SOLUTION

The trim covers the 6-foot bottom of the window, the two 8-foot sides, and the half-circle (called a semicircle) on top. The length needed is

$$6 \text{ ft} + 8 \text{ ft} + 8 \text{ ft} + \text{circumference of the semicircle.}$$

The circumference of the semicircle is half the circumference of a circle whose diameter is 6 feet.

Circumference of semicircle

$$\text{Circumference of semicircle} = \frac{1}{2}\pi d$$
$$= \frac{1}{2}\pi(6 \text{ ft}) = 3\pi \text{ ft} \approx 9.4 \text{ ft}$$

Rounding the circumference to the nearest tenth (9.4 feet), the length of trim that is needed is approximately

$$6 \text{ ft} + 8 \text{ ft} + 8 \text{ ft} + 9.4 \text{ ft},$$

or 31.4 feet.

CHECK POINT 7 In **Figure 10.51**, suppose that the dimensions are 10 feet and 12 feet for the window's bottom and side, respectively. How much trim, to the nearest tenth of a foot, is needed to go around the window?

The irrational number π is also used to find the area of a circle in square units. This is because the ratio of a circle's area to the square of its radius is π:

$$\frac{A}{r^2} = \pi.$$

Multiplying both sides of this equation by r^2 gives a formula for determining a circle's area.

FINDING THE AREA OF A CIRCLE

The area, A, of a circle with radius r is

$$A = \pi r^2.$$

EXAMPLE 8 *Problem Solving Using the Formula for a Circle's Area*

Which one of the following is the better buy: a large pizza with a 16-inch diameter for \$15.00 or a medium pizza with an 8-inch diameter for \$7.50?

SOLUTION

The better buy is the pizza with the lower price per square inch. The radius of the large pizza is $\frac{1}{2} \cdot 16$ inches, or 8 inches, and the radius of the medium pizza is $\frac{1}{2} \cdot 8$ inches, or 4 inches. The area of the surface of each circular pizza is determined using the formula for the area of a circle.

$$\text{Large pizza: } A = \pi r^2 = \pi(8 \text{ in.})^2 = 64\pi \text{ in.}^2 \approx 201 \text{ in.}^2$$

$$\text{Medium pizza: } A = \pi r^2 = \pi(4 \text{ in.})^2 = 16\pi \text{ in.}^2 \approx 50 \text{ in.}^2$$

For each pizza, the price per square inch is found by dividing the price by the area:

$$\text{Price per square inch for large pizza} = \frac{\$15.00}{64\pi \text{ in.}^2} \approx \frac{\$15.00}{201 \text{ in.}^2} \approx \frac{\$0.07}{\text{in.}^2}$$

$$\text{Price per square inch for medium pizza} = \frac{\$7.50}{16\pi \text{ in.}^2} \approx \frac{\$7.50}{50 \text{ in.}^2} = \frac{\$0.15}{\text{in.}^2}$$

The large pizza costs approximately \$0.07 per square inch and the medium pizza costs approximately \$0.15 per square inch. Thus, the large pizza is the better buy.

TECHNOLOGY

You can use your calculator to obtain the price per square inch for each pizza in Example 8. The price per square inch for the large pizza, $\frac{15}{64\pi}$, is approximated by one of the following sequences of keystrokes:

Many Scientific Calculators

15 [÷] [(] 64 [×] [π] [)] [=]

Many Graphing Calculators

15 [÷] [(] 64 [π] [)] [ENTER]

In Example 8, did you at first think that the price per square inch would be the same for the large and the medium pizzas? After all, the radius of the large pizza is twice that of the medium pizza, and the cost of the large is twice that of the medium. However, the large pizza's area, 64π square inches, is *four times the area* of the medium pizza's, 16π square inches. Doubling the radius of a circle increases its area by a factor of 2^2, or 4. In general, if the radius of a circle is increased by k times its original linear measure, the area is multiplied by k^2. The same principle is true for any two-dimensional figure: If the shape of the figure is kept the same while linear dimensions are increased k times, the area of the larger, similar, figure is k^2 times greater than the area of the original figure.

 **CHECK POINT 8** Which one of the following is the better buy: a large pizza with an 18-inch diameter for \$20 or a medium pizza with a 14-inch diameter for \$14?

Concept and Vocabulary Check

Fill in each blank so that the resulting statement is true.

1. The area, A, of a rectangle with length l and width w is given by the formula ________.
2. The area, A, of a square with one side measuring s linear units is given by the formula ________.
3. The area, A, of a parallelogram with height h and base b is given by the formula ________.
4. The area, A, of a triangle with height h and base b is given by the formula ________.
5. The area, A, of a trapezoid with parallel bases a and b and height h is given by the formula ____________.
6. The circumference, C, of a circle with diameter d is given by the formula ________.
7. The circumference, C, of a circle with radius r is given by the formula ________.
8. The area, A, of a circle with radius r is given by the formula ________.

In Exercises 9–13, determine whether each statement is true or false. If the statement is false, make the necessary change(s) to produce a true statement.

9. The area, A, of a rectangle with length l and width w is given by the formula $A = 2l + 2w$. ______
10. The height of a parallelogram is the perpendicular distance between two of the parallel sides. ______
11. The area of either triangle formed by drawing a diagonal in a parallelogram is one-half that of the parallelogram. ______
12. In any circle, the length of the radius is twice the length of the diameter. ______
13. The ratio of a circle's circumference to its diameter is the irrational number π. ______

Exercise Set 10.4

Practice Exercises

In Exercises 1–14, use the formulas developed in this section to find the area of each figure.

1.

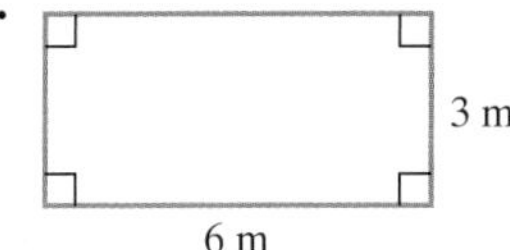

2.

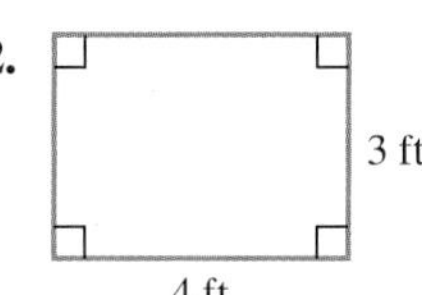

3.

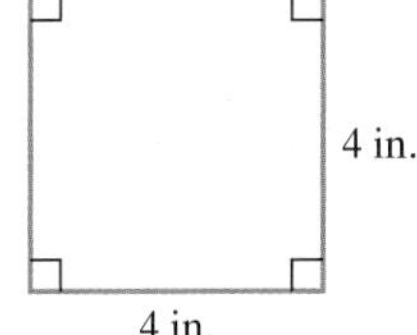

4.

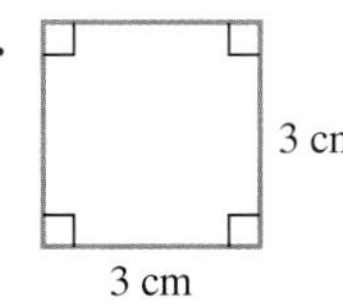

5.

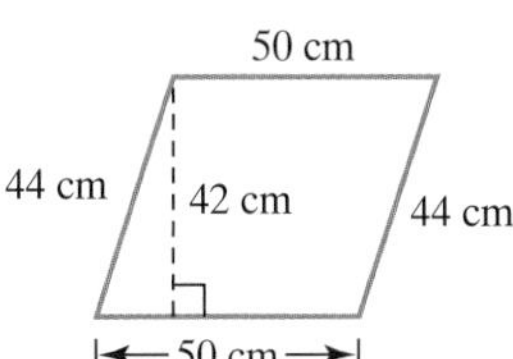

6.

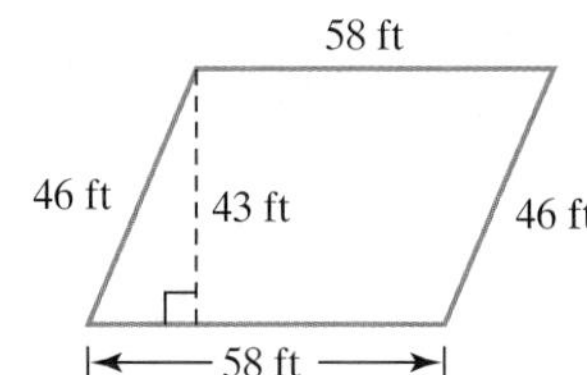

7.

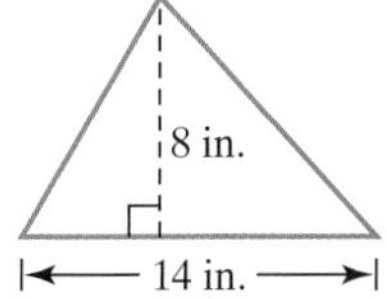

8.

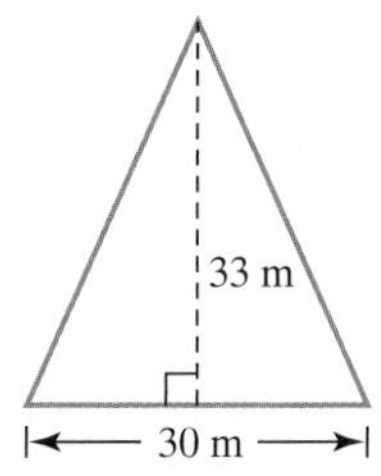

9.

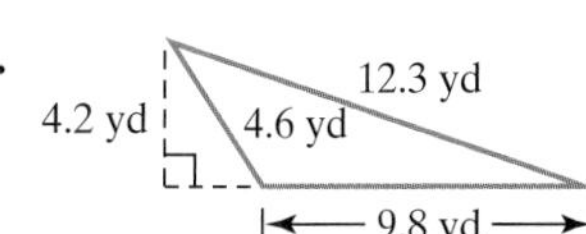

10.

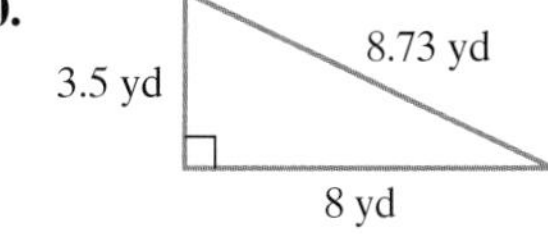

11.

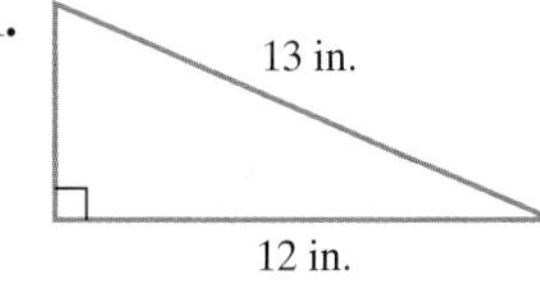

12.

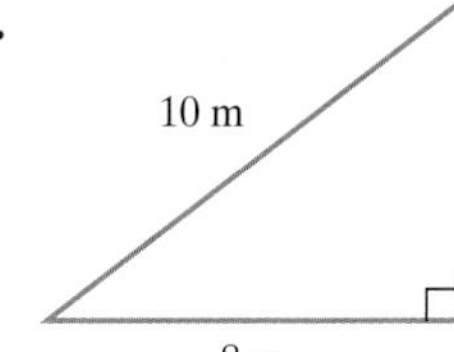

13.

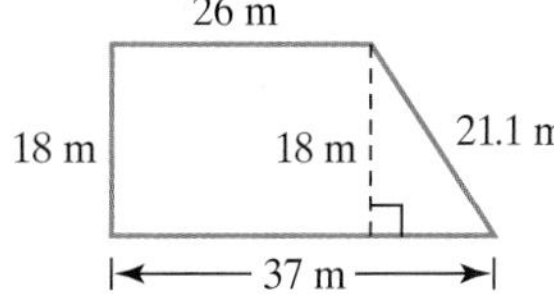

14.

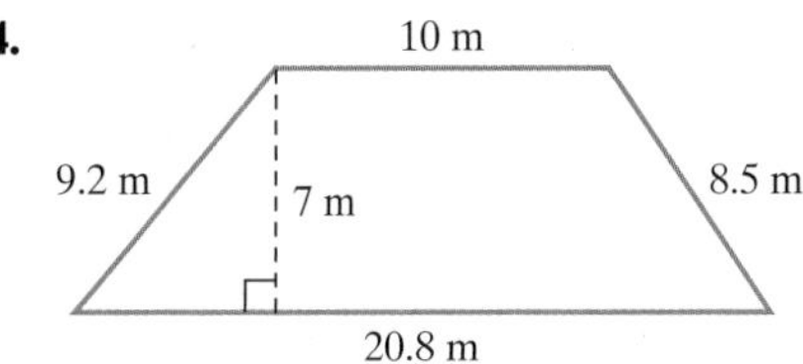

In Exercises 15–18, find the circumference and area of each circle. Express answers in terms of π and then round to the nearest tenth.

15.

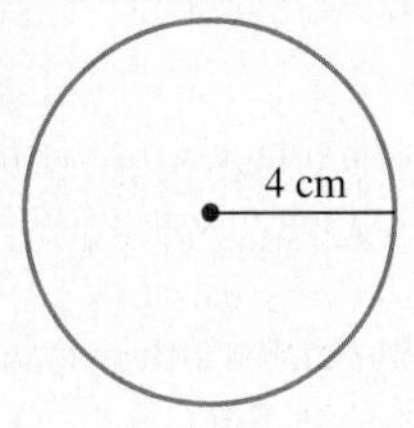

16.

17.

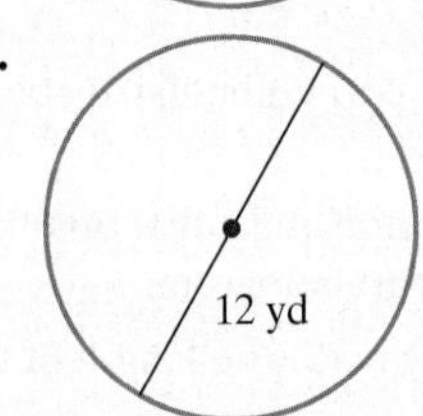

18.

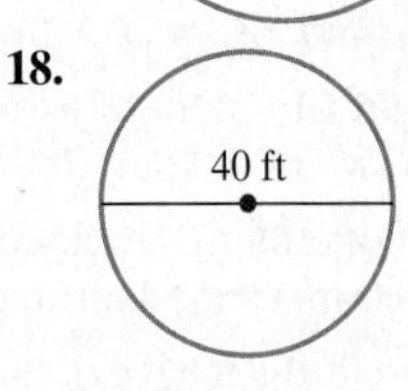

Find the area of each figure in Exercises 19–24. Where necessary, express answers in terms of π and then round to the nearest tenth.

19.

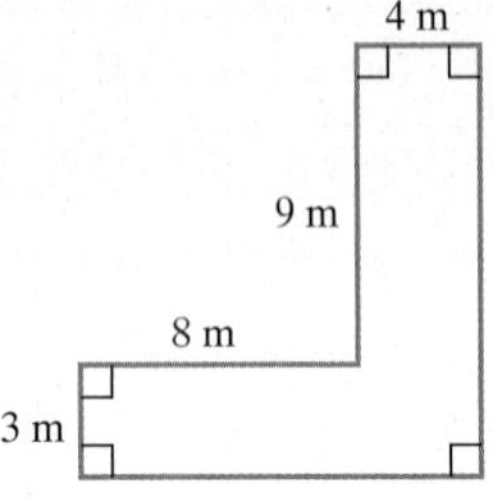

20.

21.

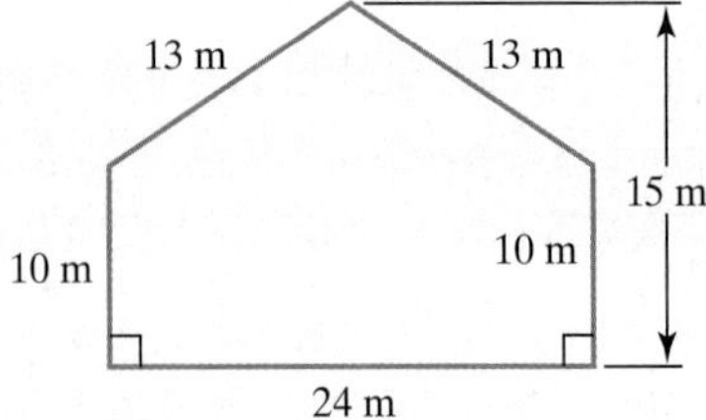

22.

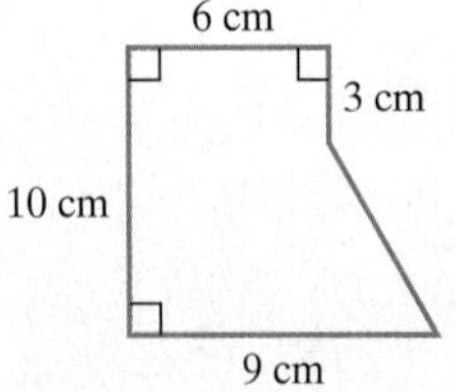

23.

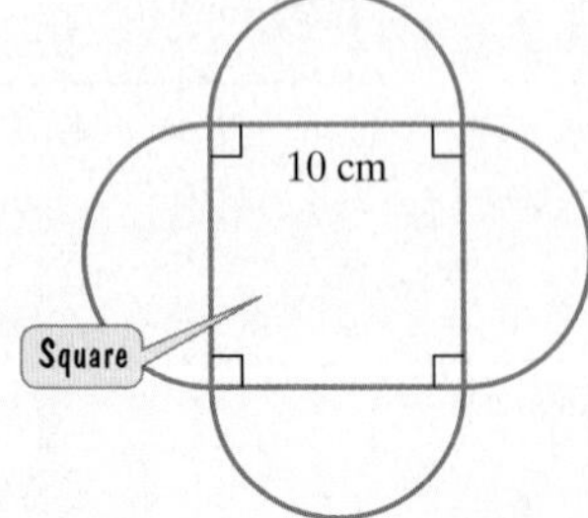

24.

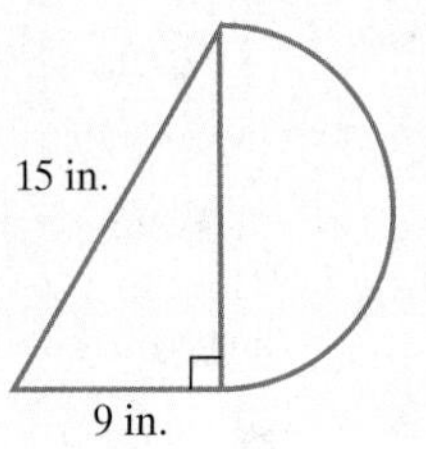

In Exercises 25–28, find a formula for the total area, A, of each figure in terms of the variable(s) shown. Where necessary, use π in the formula.

25.

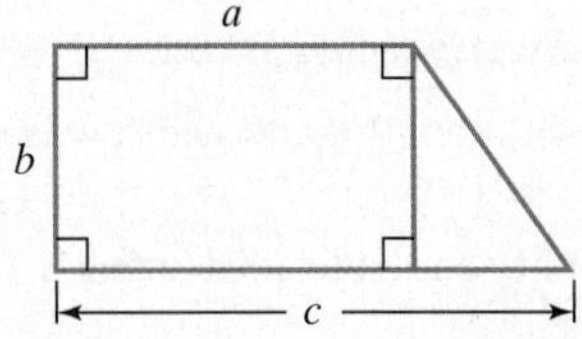

26.

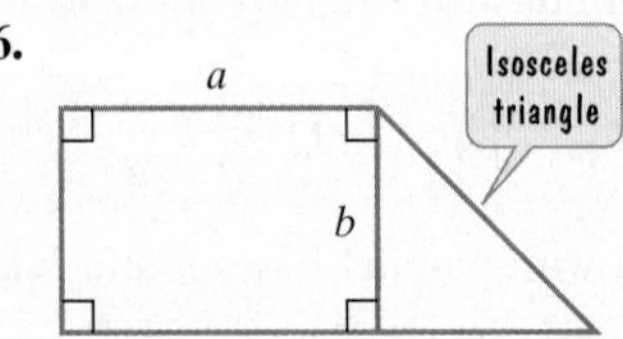

27.

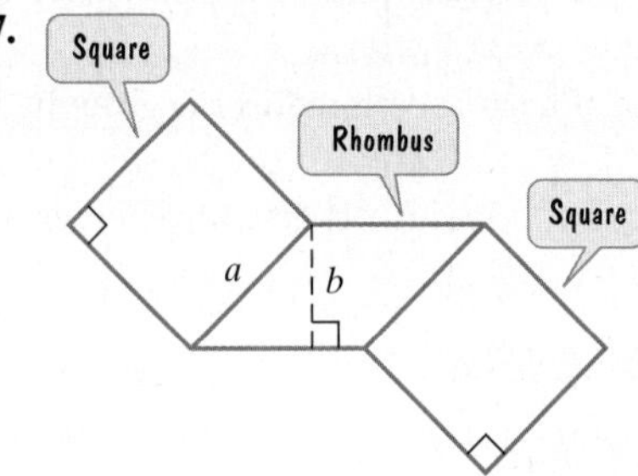

28.

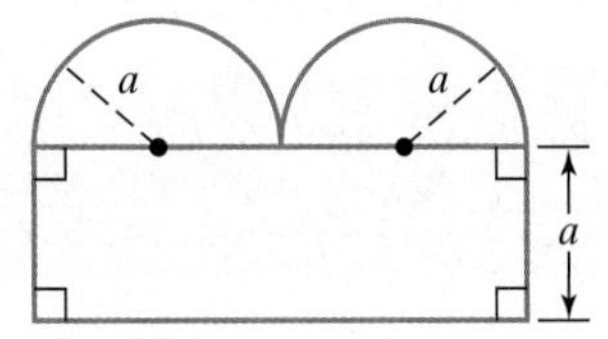

Practice Plus

In Exercises 29–30, find the area of each shaded region.

29.

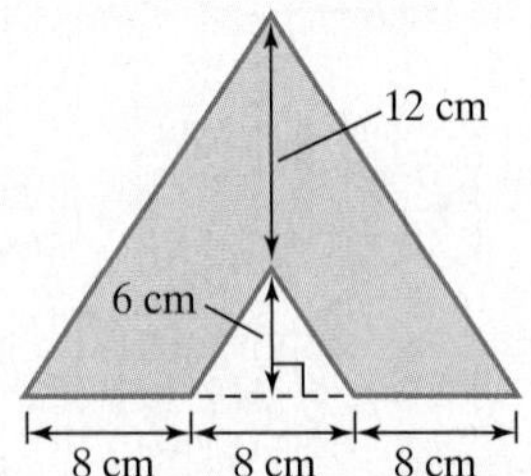

30.

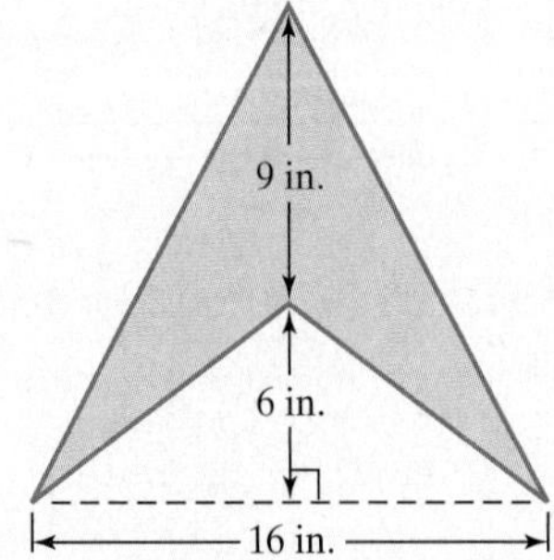

In Exercises 31–34, find the area of each shaded region in terms of π.

31.

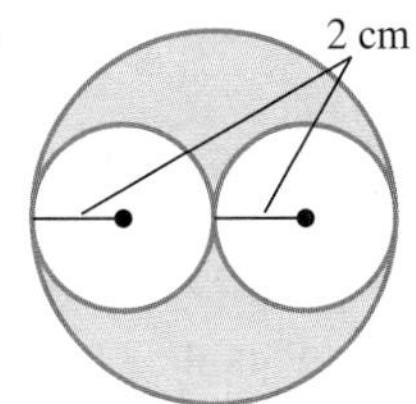

32.

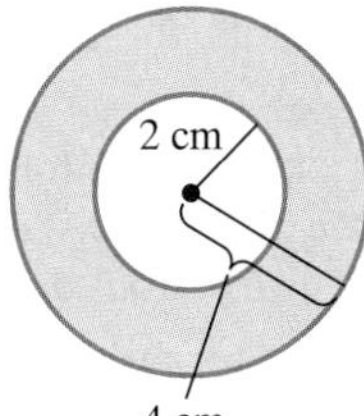

33.

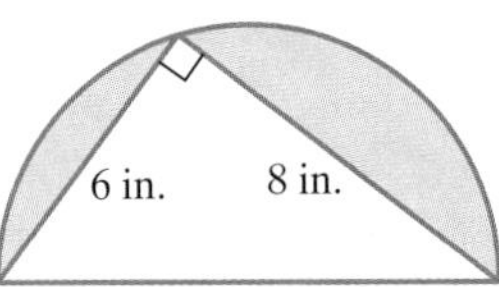

34.

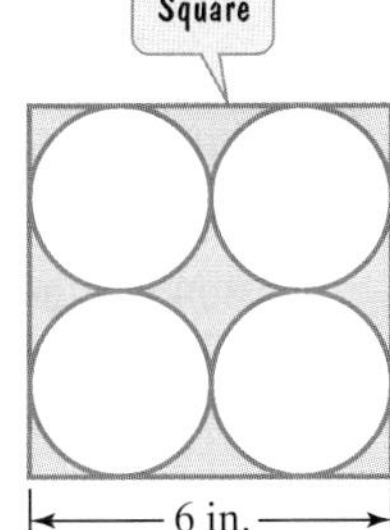

In Exercises 35–36, find the perimeter and the area of each figure. Where necessary, express answers in terms of π and round to the nearest tenth.

35.

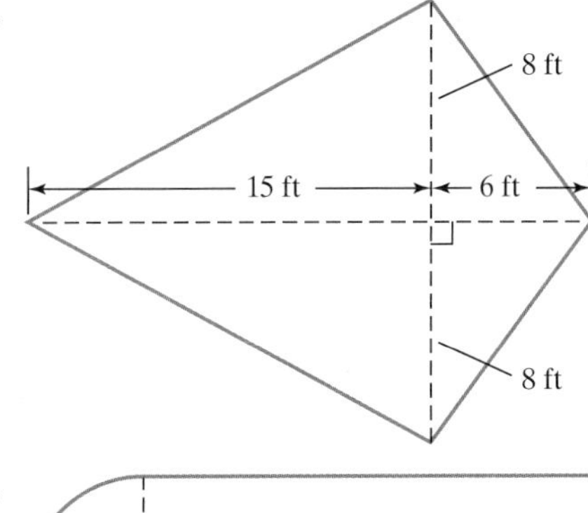

36.

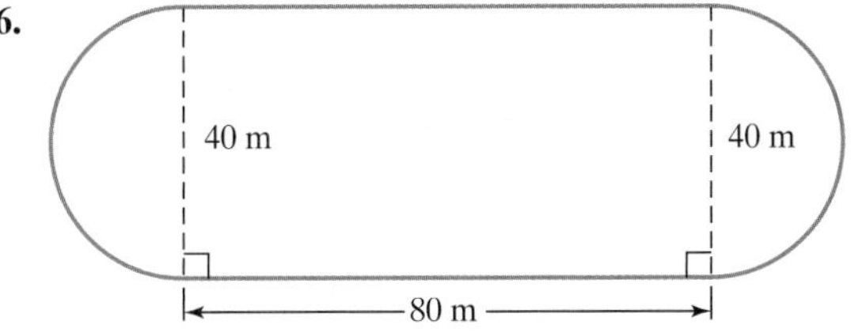

Application Exercises

37. What will it cost to carpet a rectangular floor measuring 9 feet by 21 feet if the carpet costs \$26.50 per square yard?

38. A plastering contractor charges \$18 per square yard. What is the cost of plastering 60 feet of wall in a house with a 9-foot ceiling?

39. A rectangular field measures 81 meters by 108 meters. What will it cost to cover the field with pallets of grass each covering 90 square yards and costing \$235 per pallet? (Hint: 1 yd $\approx$ 0.9 m)

40. A rectangular floor measures 20 feet by 25 feet. What will it cost to carpet the floor if the carpet costs \$28 per square meter? (Hint: 1 ft $\approx$ 0.3 m)

41. A rectangular kitchen floor measures 12 feet by 15 feet. A stove on the floor has a rectangular base measuring 3 feet by 4 feet, and a refrigerator covers a rectangular area of the floor measuring 4 feet by 5 feet. How many square feet of tile will be needed to cover the kitchen floor not counting the area used by the stove and the refrigerator?

42. A rectangular room measures 12 feet by 15 feet. The entire room is to be covered with rectangular tiles that measure 3 inches by 2 inches. If the tiles are sold at ten for \$1.20, what will it cost to tile the room?

43. The lot in the figure shown, except for the house, shed, and driveway, is lawn. One bag of lawn fertilizer costs \$25 and covers 4000 square feet.

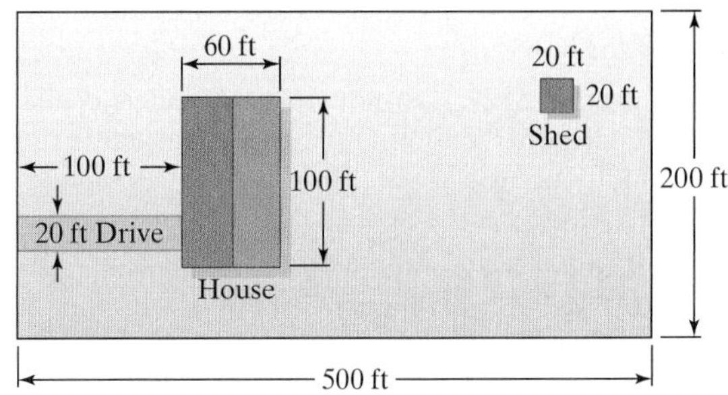

a. Determine the minimum number of bags of fertilizer needed for the lawn.

b. Find the total cost of the fertilizer.

44. Taxpayers with an office in their home may deduct a percentage of their home-related expenses. This percentage is based on the ratio of the office's area to the area of the home. A taxpayer with an office in a 2200-square-foot home maintains a 20 foot by 16 foot office. If the yearly utility bills for the home come to \$4800, how much of this is deductible?

45. You are planning to paint the house whose dimensions are shown in the figure.

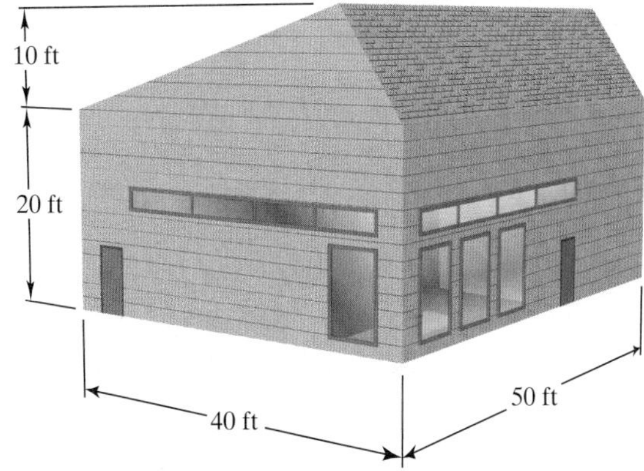

a. How many square feet will you need to paint? (There are four windows, each 8 feet by 5 feet; two windows, each 30 feet by 2 feet; and two doors, each 80 inches by 36 inches, that do not require paint.)

b. The paint that you have chosen is available in gallon cans only. Each can covers 500 square feet. If you want to use two coats of paint, how many cans will you need for the project?

c. If the paint you have chosen sells for \$26.95 per gallon, what will it cost to paint the house?

The diagram shows the floor plan for a one-story home. Use the given measurements to solve Exercises 46–48. (A calculator will be helpful in performing the necessary computations.)

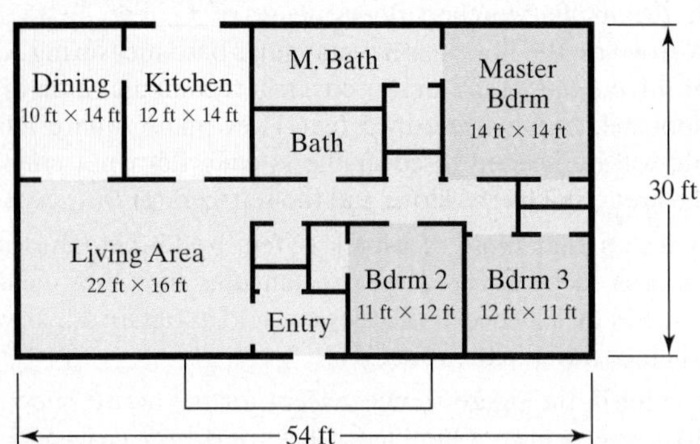

46. If construction costs \$95 per square foot, find the cost of building the home.

47. If carpet costs \$17.95 per square yard and is available in whole square yards only, find the cost of carpeting the three bedroom floors.

48. If ceramic tile costs \$26.95 per square yard and is available in whole square yards only, find the cost of installing ceramic tile on the kitchen and dining room floors.

In Exercises 49–50, express the required calculation in terms of π and then round to the nearest tenth.

49. How much fencing is required to enclose a circular garden whose radius is 20 meters?

50. A circular rug is 6 feet in diameter. How many feet of fringe is required to edge this rug?

51. How many plants spaced every 6 inches are needed to surround a circular garden with a 30-foot radius?

52. A stained glass window is to be placed in a house. The window consists of a rectangle, 6 feet high by 3 feet wide, with a semicircle at the top. Approximately how many feet of stripping, to the nearest tenth of a foot, will be needed to frame the window?

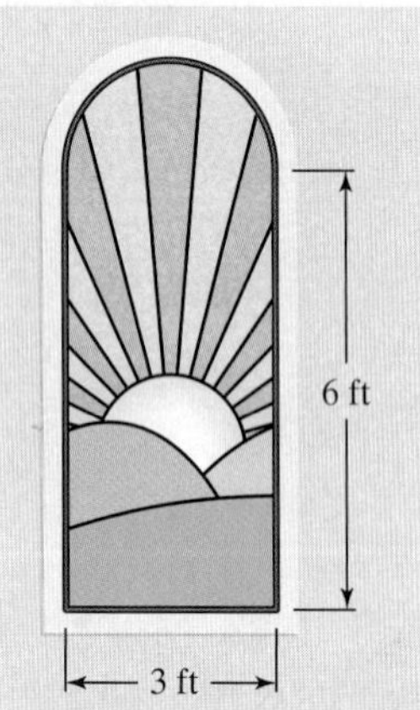

53. Which one of the following is a better buy: a large pizza with a 14-inch diameter for \$12 or a medium pizza with a 7-inch diameter for \$5?

54. Which one of the following is a better buy: a large pizza with a 16-inch diameter for \$12 or two small pizzas, each with a 10-inch diameter, for \$12?

Explaining the Concepts

55. Using the formula for the area of a rectangle, explain how the formula for the area of a parallelogram ($A = bh$) is obtained.

56. Using the formula for the area of a parallelogram ($A = bh$), explain how the formula for the area of a triangle $\left(A = \frac{1}{2}bh\right)$ is obtained.

57. Using the formula for the area of a triangle, explain how the formula for the area of a trapezoid is obtained.

58. Explain why a circle is not a polygon.

59. Describe the difference between the following problems: How much fencing is needed to enclose a circular garden? How much fertilizer is needed for a circular garden?

Critical Thinking Exercises

Make Sense? *In Exercises 60–63, determine whether each statement makes sense or does not make sense, and explain your reasoning.*

60. The house is a 1500-square-foot mansion with six bedrooms.

61. Because a parallelogram can be divided into two triangles with the same size and shape, the area of a triangle is one-half that of a parallelogram.

62. I used $A = \pi r^2$ to determine the amount of fencing needed to enclose my circular garden.

63. I paid \$10 for a pizza, so I would expect to pay approximately \$20 for the same kind of pizza with twice the radius.

64. You need to enclose a rectangular region with 200 feet of fencing. Experiment with different lengths and widths to determine the maximum area you can enclose. Which quadrilateral encloses the most area?

65. Suppose you know the cost for building a rectangular deck measuring 8 feet by 10 feet. If you decide to increase the dimensions to 12 feet by 15 feet, by how much will the cost increase?

66. A rectangular swimming pool measures 14 feet by 30 feet. The pool is surrounded on all four sides by a path that is 3 feet wide. If the cost to resurface the path is \$2 per square foot, what is the total cost of resurfacing the path?

67. A proposed oil pipeline will cross 16.8 miles of national forest. The width of the land needed for the pipeline is 200 feet. If the U.S. Forest Service charges the oil company \$32 per acre, calculate the total cost. (1 mile = 5280 feet and 1 acre = 43,560 square feet.)

10.5 Volume and Surface Area

WHAT AM I SUPPOSED TO LEARN?

After studying this section, you should be able to:

1 Use volume formulas to compute the volumes of three-dimensional figures and solve applied problems.

2 Compute the surface area of a three-dimensional figure.

BEFORE SHE RETIRED, YOU WERE CONSIDERING GOING TO Judge Judy's Web site and filling out a case submission form to appear on her TV show. The case involves your contractor, who promised to install a water tank that holds 500 gallons of water. Upon delivery, you noticed that capacity was not printed anywhere, so you decided to do some measuring. The tank is shaped like a giant tuna can, with a circular top and bottom. You measured the radius of each circle to be 3 feet and you measured the tank's height to be 2 feet 4 inches. You know that 500 gallons is the capacity of a solid figure with a volume of about 67 cubic feet. Now you need some sort of method to compute the volume of the water tank. In this section, we discuss how to compute the volumes of various solid, three-dimensional figures. Using a formula you will learn in the section, you can determine whether the evidence indicates you can win a case against the contractor if you appear on *Judge Judy*. Or do you risk joining a cast of bozos who entertain television viewers by being loudly castigated by the judge? (Before a possible ear-piercing "Baloney, sir, you're a geometric idiot!," we suggest working Exercise 45 in Exercise Set 10.5.)

1 *Use volume formulas to compute the volumes of three-dimensional figures and solve applied problems.*

Formulas for Volume

In Section 9.2, we saw that **volume** refers to the amount of space occupied by a solid object, determined by the number of cubic units it takes to fill the interior of that object. For example, **Figure 10.52** shows that there are 18 cubic units contained within the box. The volume of the box, called a **rectangular solid**, is 18 cubic units. The box has a length of 3 units, a width of 3 units, and a height of 2 units. The volume, 18 cubic units, may be determined by finding the product of the length, the width, and the height:

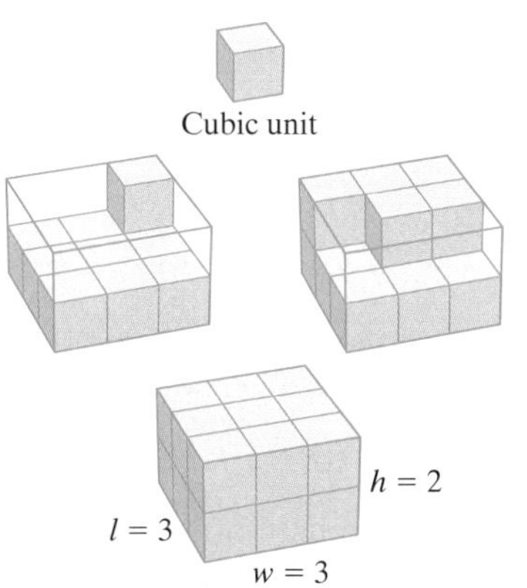

FIGURE 10.52 Volume = 18 cubic units

$$\text{Volume} = 3 \text{ units} \cdot 3 \text{ units} \cdot 2 \text{ units} = 18 \text{ units}^3.$$

In general, the volume, V, of a rectangular solid is the product of its length, l, its width, w, and its height, h:

$$V = lwh.$$

If the length, width, and height are the same, the rectangular solid is called a **cube**. Formulas for these boxlike shapes are given below.

VOLUMES OF BOXLIKE SHAPES

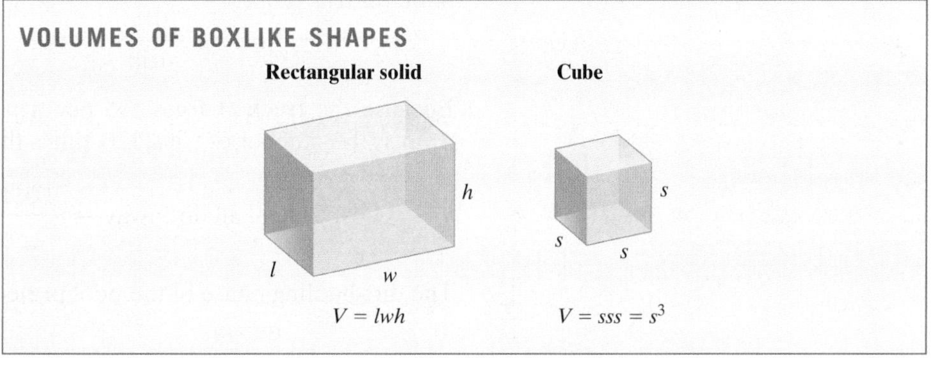

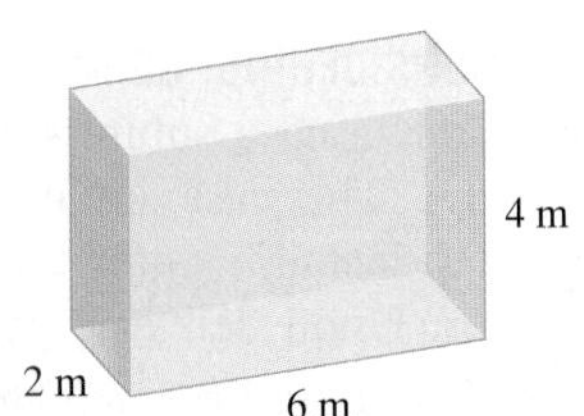

FIGURE 10.53

EXAMPLE 1 *Finding the Volume of a Rectangular Solid*

Find the volume of the rectangular solid in **Figure 10.53**.

SOLUTION

As shown in the figure, the length is 6 meters, the width is 2 meters, and the height is 4 meters. Thus, $l = 6$, $w = 2$, and $h = 4$.

$$V = lwh = 6\text{ m} \cdot 2\text{ m} \cdot 4\text{ m} = 48\text{ m}^3$$

The volume of the rectangular solid is 48 cubic meters.

CHECK POINT 1 Find the volume of a rectangular solid with length 5 feet, width 3 feet, and height 7 feet.

In Section 9.2, we saw that although there are 3 feet in a yard, there are 27 cubic feet in a cubic yard. If a problem requires measurement of volume in cubic yards and the linear measures are given in feet, to avoid errors, first convert feet to yards. Then apply the volume formula. This idea is illustrated in Example 2.

EXAMPLE 2 *Solving a Volume Problem*

You are about to begin work on a swimming pool in your yard. The first step is to have a hole dug that is 90 feet long, 60 feet wide, and 6 feet deep. You will use a truck that can carry 10 cubic yards of dirt and charges $35 per load. How much will it cost you to have all the dirt hauled away?

SOLUTION

We begin by converting feet to yards:

$$90\text{ ft} = \frac{90\text{ }\cancel{\text{ft}}}{1} \cdot \frac{1\text{ yd}}{3\text{ }\cancel{\text{ft}}} = \frac{90}{3}\text{ yd} = 30\text{ yd}.$$

Similarly, 60 ft = 20 yd and 6 ft = 2 yd. Next, we find the volume of dirt that needs to be dug out and hauled off.

$$V = lwh = 30\text{ yd} \cdot 20\text{ yd} \cdot 2\text{ yd} = 1200\text{ yd}^3$$

Now, we find the number of loads that the truck needs to haul off all the dirt. Because the truck carries 10 cubic yards, divide the number of cubic yards of dirt by 10.

$$\text{Number of truckloads} = \frac{1200\text{ yd}^3}{\frac{10\text{ yd}^3}{\text{trip}}} = \frac{1200\text{ }\cancel{\text{yd}^3}}{1} \cdot \frac{\text{trip}}{10\text{ }\cancel{\text{yd}^3}} = \frac{1200}{10}\text{ trips} = 120\text{ trips}$$

Because the truck charges $35 per trip, the cost to have all the dirt hauled away is the number of trips, 120, times the cost per trip, $35.

$$\text{Cost to haul all dirt away} = \frac{120\text{ }\cancel{\text{trips}}}{1} \cdot \frac{\$35}{\cancel{\text{trip}}} = 120(\$35) = \$4200$$

The dirt-hauling phase of the pool project will cost you $4200.

 **CHECK POINT 2** Find the volume, in cubic yards, of a cube whose edges each measure 6 feet.

FIGURE 10.54 The volume of a pyramid is $\frac{1}{3}$ the volume of a rectangular solid having the same base and the same height.

A rectangular solid is an example of a **polyhedron**, a solid figure bounded by polygons. A rectangular solid is bounded by six rectangles, called faces. By contrast, a **pyramid** is a polyhedron whose base is a polygon and whose sides are triangles. **Figure 10.54** shows a pyramid with a rectangular base drawn inside a rectangular solid. The contents of three pyramids with rectangular bases exactly fill a rectangular solid of the same base and height. Thus, the formula for the volume of the pyramid is $\frac{1}{3}$ that of the rectangular solid.

VOLUME OF A PYRAMID

The volume, V, of a pyramid is given by the formula

$$V = \frac{1}{3}Bh,$$

where B is the area of the base and h is the height (the perpendicular distance from the top to the base).

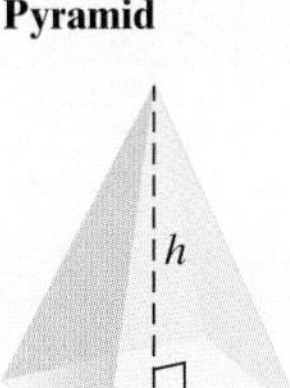

EXAMPLE 3 *Using the Formula for a Pyramid's Volume*

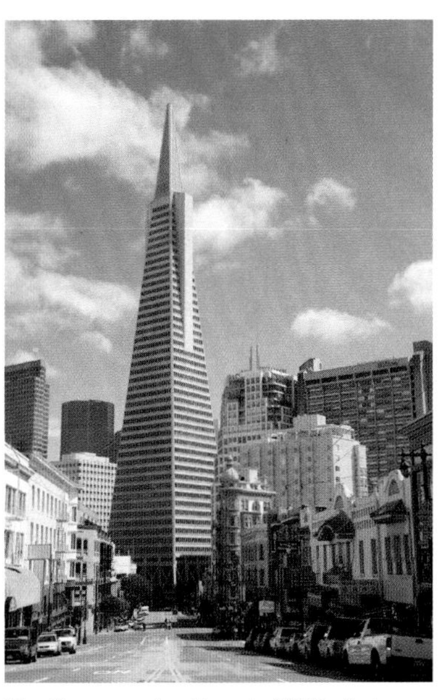

The Transamerica Tower's 3678 windows take cleaners one month to wash. Its foundation is sunk 15.5 m (52 ft) into the ground and is designed to move with earth tremors.

Capped with a pointed spire on top of its 48 stories, the Transamerica Tower in San Francisco is a pyramid with a square base. The pyramid is 256 meters (853 feet) tall. Each side of the square base has a length of 52 meters. Although San Franciscans disliked it when it opened in 1972, they have since accepted it as part of the skyline. Find the volume of the building.

SOLUTION

First find the area of the square base, represented as B in the volume formula. Because each side of the square base is 52 meters, the area of the square base is

$$B = 52 \text{ m} \cdot 52 \text{ m} = 2704 \text{ m}^2.$$

The area of the square base is 2704 square meters. Because the pyramid is 256 meters tall, its height, h, is 256 meters. Now we apply the formula for the volume of a pyramid:

$$V = \frac{1}{3}Bh = \frac{1}{3} \cdot \frac{2704 \text{ m}^2}{1} \cdot \frac{256 \text{ m}}{1} = \frac{2704 \cdot 256}{3} \text{ m}^3 \approx 230{,}741 \text{ m}^3.$$

The volume of the building is approximately 230,741 cubic meters.

The San Francisco pyramid is relatively small compared to the Great Pyramid outside Cairo, Egypt. Built in about 2550 B.C. by a labor force of 100,000, the Great Pyramid is approximately 11 times the volume of San Francisco's pyramid.

 **CHECK POINT 3** A pyramid is 4 feet tall. Each side of the square base has a length of 6 feet. Find the pyramid's volume.

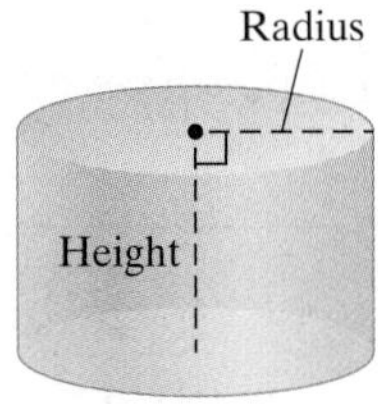

FIGURE 10.55

Not every three-dimensional figure is a polyhedron. Take, for example, the right circular cylinder shown in **Figure 10.55**. Its shape should remind you of a soup can or a stack of coins. The right circular cylinder is so named because the top and bottom are circles, and the side forms a right angle with the top and bottom. The formula for the volume of a right circular cylinder is given as follows:

VOLUME OF A RIGHT CIRCULAR CYLINDER

The volume, V, of a right circular cylinder is given by the formula

$$V = \pi r^2 h,$$

where r is the radius of the circle at either end and h is the height.

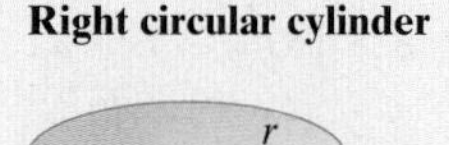

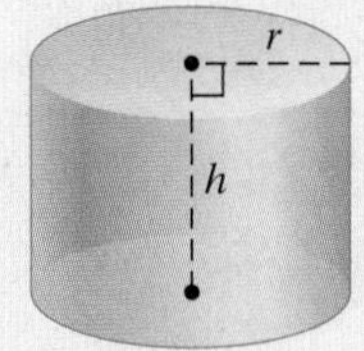

EXAMPLE 4 *Finding the Volume of a Cylinder*

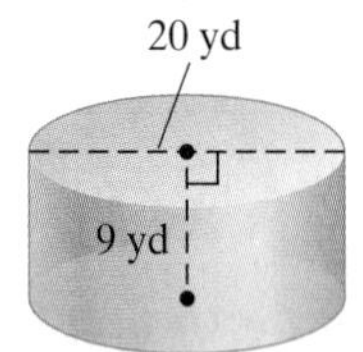

FIGURE 10.56

Find the volume of the cylinder in **Figure 10.56**.

SOLUTION

In order to find the cylinder's volume, we need both its radius and its height. Because the diameter is 20 yards, the radius is half this length, or 10 yards. The height of the cylinder is given to be 9 yards. Thus, $r = 10$ and $h = 9$. Now we apply the formula for the volume of a cylinder.

$$V = \pi r^2 h = \pi(10 \text{ yd})^2 \cdot 9 \text{ yd} = 900\pi \text{ yd}^3 \approx 2827 \text{ yd}^3$$

The volume of the cylinder is approximately 2827 cubic yards.

CHECK POINT 4 Find the volume, to the nearest cubic inch, of a cylinder with a diameter of 8 inches and a height of 6 inches.

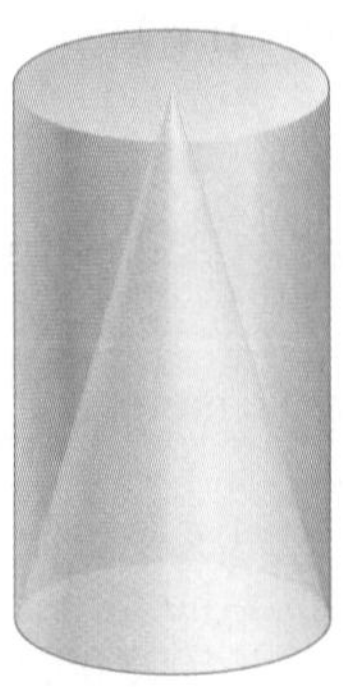

FIGURE 10.57

Figure 10.57 shows a **right circular cone** inside a cylinder, sharing the same circular base as the cylinder. The height of the cone, the perpendicular distance from the top to the circular base, is the same as that of the cylinder. Three such cones can occupy the same amount of space as the cylinder. Therefore, the formula for the volume of the cone is $\frac{1}{3}$ the volume of the cylinder.

VOLUME OF A CONE

The volume, V, of a right circular cone that has height h and radius r is given by the formula

$$V = \frac{1}{3}\pi r^2 h.$$

Cone

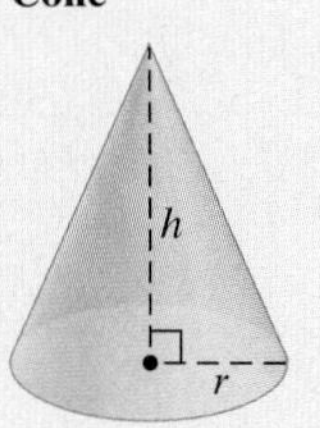

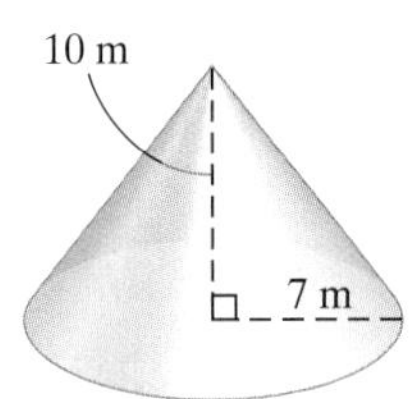

FIGURE 10.58

EXAMPLE 5 Finding the Volume of a Cone

Find the volume of the cone in **Figure 10.58**.

SOLUTION

The radius of the cone is 7 meters and the height is 10 meters. Thus, $r = 7$ and $h = 10$. Now we apply the formula for the volume of a cone.

$$V = \frac{1}{3}\pi r^2 h = \frac{1}{3}\pi(7 \text{ m})^2 \cdot 10 \text{ m} = \frac{490\pi}{3} \text{ m}^3 \approx 513 \text{ m}^3$$

The volume of the cone is approximately 513 cubic meters.

CHECK POINT 5 Find the volume, to the nearest cubic inch, of a cone with a radius of 4 inches and a height of 6 inches.

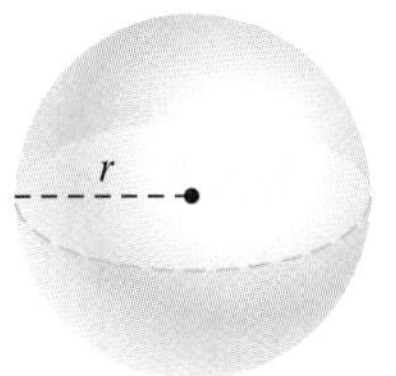

FIGURE 10.59

Figure 10.59 shows a *sphere*. Its shape may remind you of a basketball. The Earth is not a perfect sphere, but it's close. A **sphere** is the set of points in space equally distant from a given point, its **center**. Any line segment from the center to a point on the sphere is a **radius** of the sphere. The word *radius* is also used to refer to the length of this line segment. A sphere's volume can be found by using π and its radius.

VOLUME OF A SPHERE

The volume, V, of a sphere of radius r is given by the formula

$$V = \frac{4}{3}\pi r^3.$$

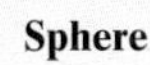

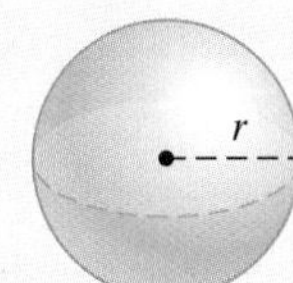

EXAMPLE 6 Applying Volume Formulas

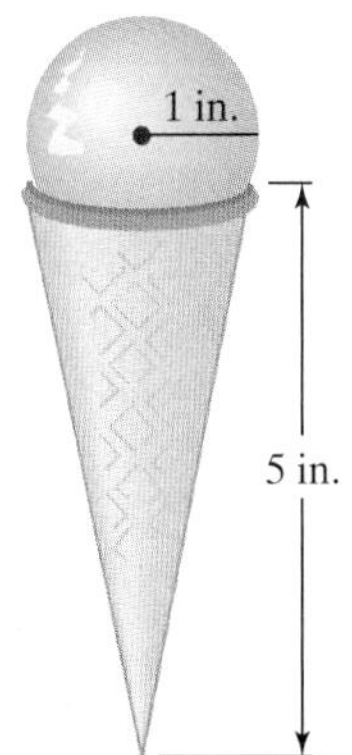

FIGURE 10.60

An ice cream cone is 5 inches deep and has a radius of 1 inch. A spherical scoop of ice cream also has a radius of 1 inch. (See **Figure 10.60**.) If the ice cream melts into the cone, will it overflow?

SOLUTION

The ice cream will overflow if the volume of the ice cream, a sphere, is greater than the volume of the cone. Find the volume of each.

$$V_{\text{cone}} = \frac{1}{3}\pi r^2 h = \frac{1}{3}\pi(1 \text{ in.})^2 \cdot 5 \text{ in.} = \frac{5\pi}{3} \text{ in.}^3 \approx 5 \text{ in.}^3$$

$$V_{\text{sphere}} = \frac{4}{3}\pi r^3 = \frac{4}{3}\pi(1 \text{ in.})^3 = \frac{4\pi}{3} \text{ in.}^3 \approx 4 \text{ in.}^3$$

The volume of the spherical scoop of ice cream is less than the volume of the cone, so there will be no overflow.

CHECK POINT 6 A basketball has a radius of 4.5 inches. If the ball is filled with 350 cubic inches of air, is this enough air to fill it completely?

2 *Compute the surface area of a three-dimensional figure.*

Surface Area

In addition to volume, we can also measure the area of the outer surface of a three-dimensional object, called its **surface area**. Like area, surface area is measured in square units. For example, the surface area of the rectangular solid in **Figure 10.61** is the sum of the areas of the six outside rectangles of the solid.

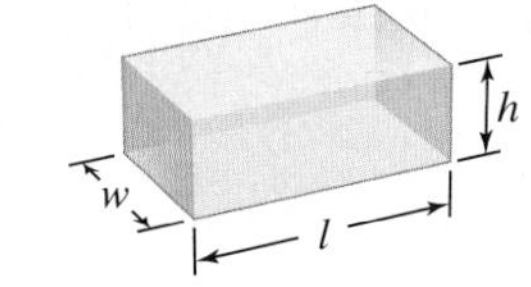

FIGURE 10.61

$$\text{Surface Area} = \underbrace{lw + lw}_{\text{Areas of top and bottom rectangles}} + \underbrace{lh + lh}_{\text{Areas of front and back rectangles}} + \underbrace{wh + wh}_{\text{Areas of rectangles on left and right sides}}$$

$$= 2lw + 2lh + 2wh$$

Formulas for the surface area, abbreviated SA, of three-dimensional figures are given in **Table 10.4**.

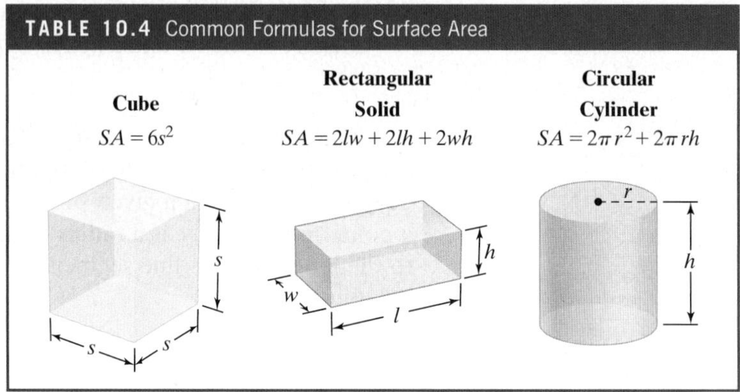

TABLE 10.4 Common Formulas for Surface Area

Cube	Rectangular Solid	Circular Cylinder
$SA = 6s^2$	$SA = 2lw + 2lh + 2wh$	$SA = 2\pi r^2 + 2\pi rh$

EXAMPLE 7 *Finding the Surface Area of a Solid*

Find the surface area of the rectangular solid in **Figure 10.62**.

3 yd
5 yd
8 yd

FIGURE 10.62

SOLUTION

As shown in the figure, the length is 8 yards, the width is 5 yards, and the height is 3 yards. Thus, $l = 8$, $w = 5$, and $h = 3$.

$$\begin{aligned} SA &= 2lw + 2lh + 2wh \\ &= 2 \cdot 8 \text{ yd} \cdot 5 \text{ yd} + 2 \cdot 8 \text{ yd} \cdot 3 \text{ yd} + 2 \cdot 5 \text{ yd} \cdot 3 \text{ yd} \\ &= 80 \text{ yd}^2 + 48 \text{ yd}^2 + 30 \text{ yd}^2 = 158 \text{ yd}^2 \end{aligned}$$

The surface area is 158 square yards.

CHECK POINT 7 If the length, width, and height shown in **Figure 10.62** are each doubled, find the surface area of the resulting rectangular solid.

Concept and Vocabulary Check

Fill in each blank so that the resulting statement is true.

1. The volume, V, of a rectangular solid with length l, width w, and height h is given by the formula ________.
2. The volume, V, of a cube with an edge that measures s linear units is given by the formula ________.
3. A solid figure bounded by polygons is called a/an ________.
4. The volume, V, of a pyramid with base area B and height h is given by the formula ________.
5. The volume, V, of a right circular cylinder with height h and radius r is given by the formula ________.
6. The volume, V, of a right circular cone with height h and radius r is given by the formula ________.
7. The volume, V, of a sphere of radius r is given by the formula ________.

In Exercises 8–14, determine whether each statement is true or false. If the statement is false, make the necessary change(s) to produce a true statement.

8. A cube is a rectangular solid with the same length, width, and height. ______
9. A cube is an example of a polyhedron. ______
10. The volume of a pyramid is $\frac{1}{2}$ the volume of a rectangular solid having the same base and the same height. ______
11. Some three-dimensional figures are not polyhedrons. ______
12. A sphere is the set of points in space equally distant from its center. ______
13. Surface area refers to the area of the outer surface of a three-dimensional object. ______
14. The surface area, SA, of a rectangular solid with length l, width w, and height h is given by the formula $SA = lw + lh + wh$. ______

Exercise Set 10.5

Practice Exercises

In Exercises 1–20, find the volume of each figure. If necessary, express answers in terms of π and then round to the nearest whole number.

1.

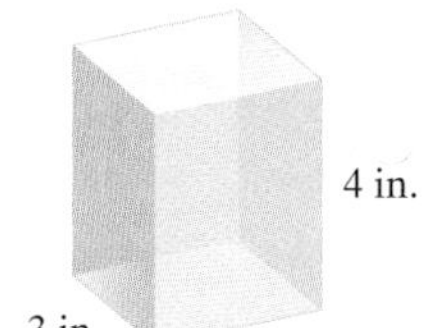

2.

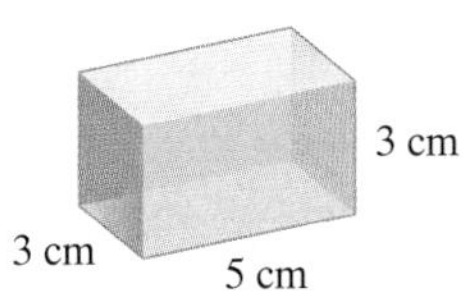

3.

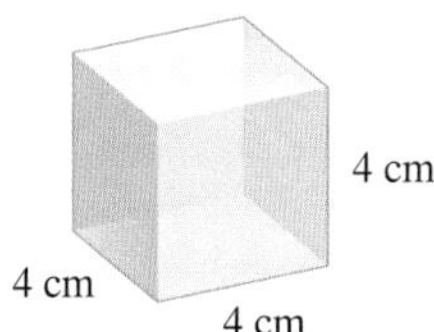

4.

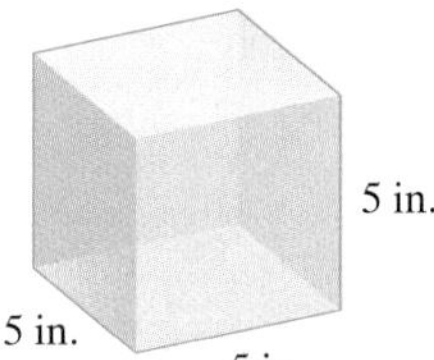

5.

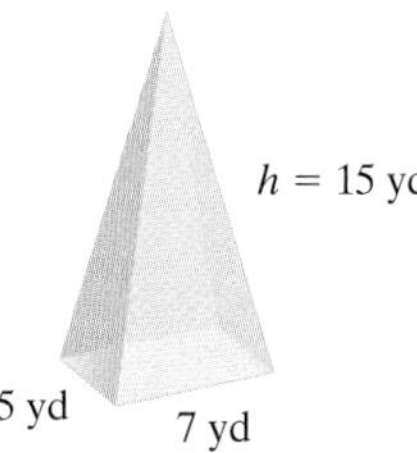

6.

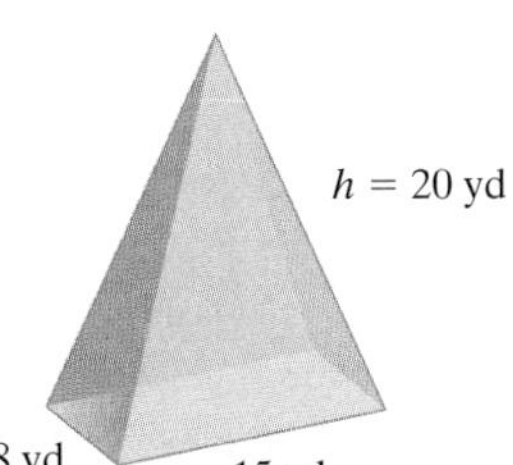

7.

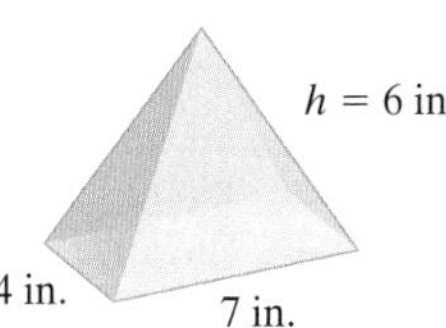

8.

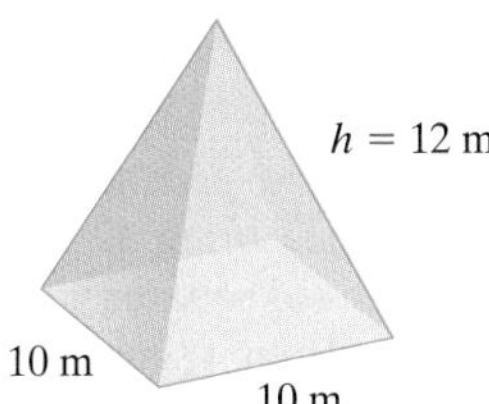

9.

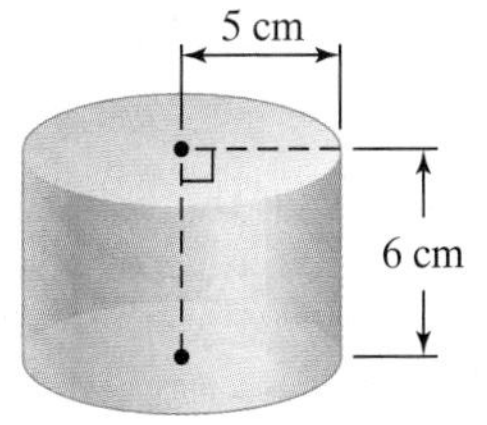

10.

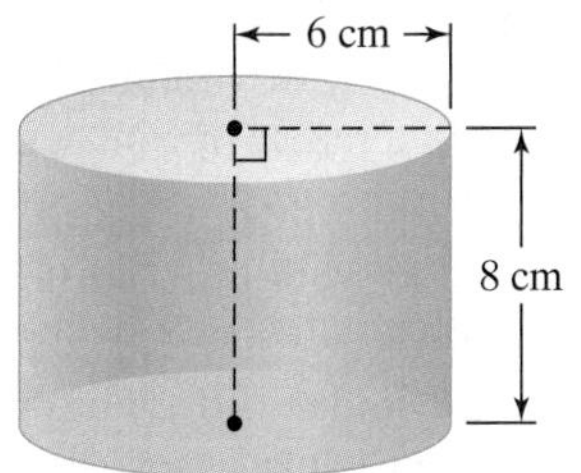

11.

24 in.
21 in.

12.

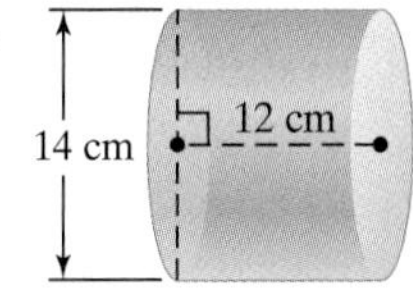

13.

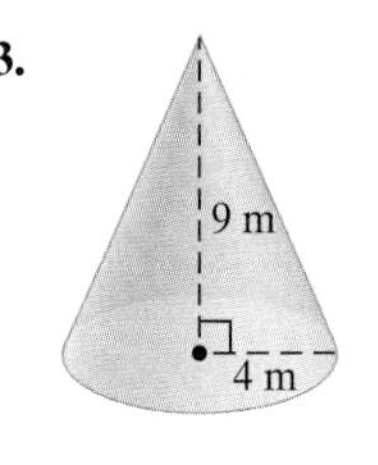

14.

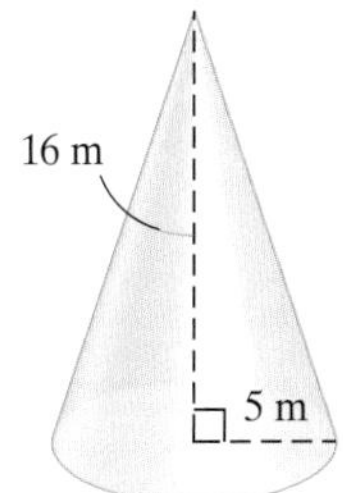

15.

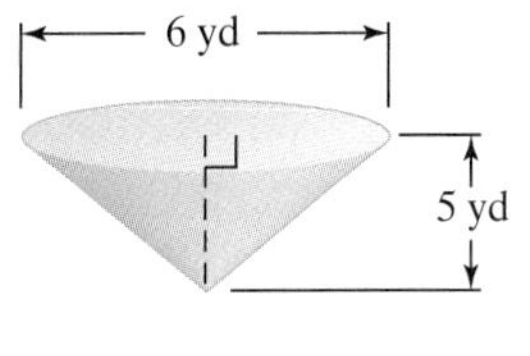

16.

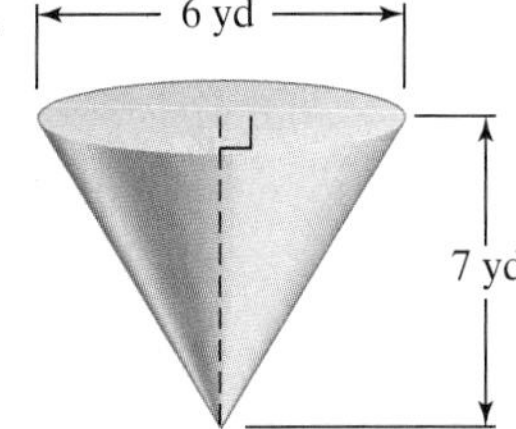

17.

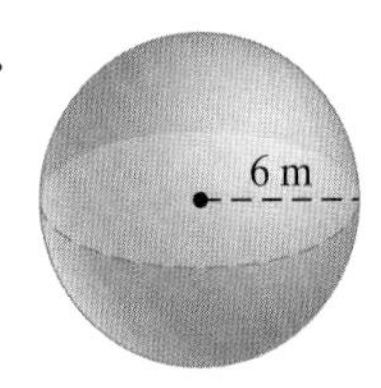

18.

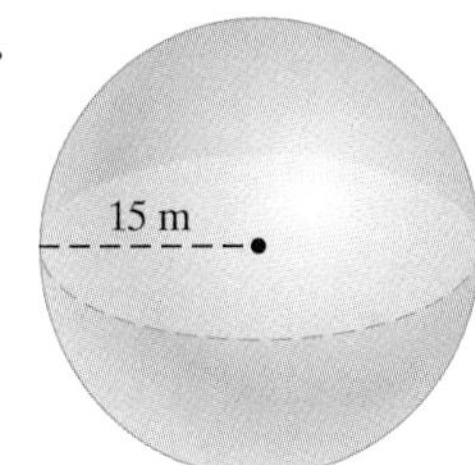

19.

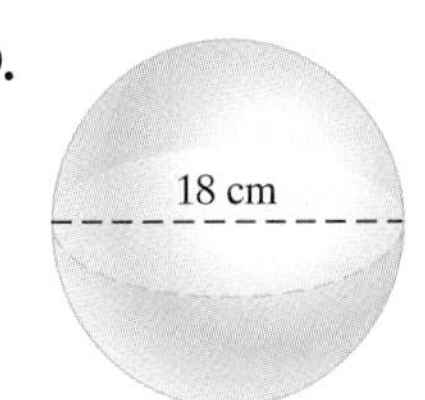

20.

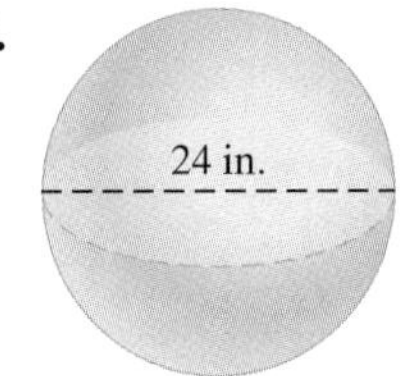

In Exercises 21–24, find the surface area of each figure.

21.

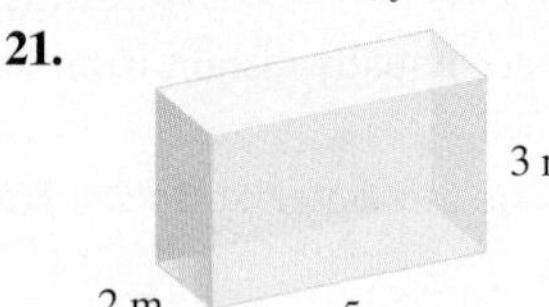

22.

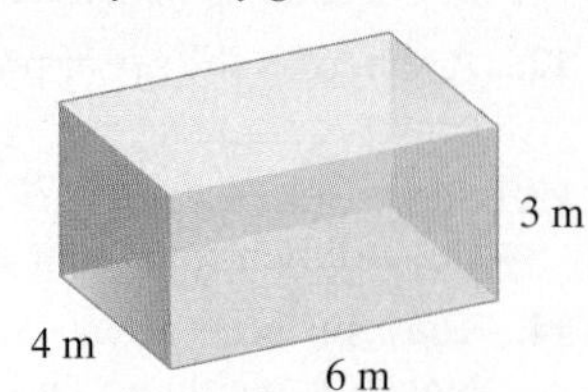

23.

24.

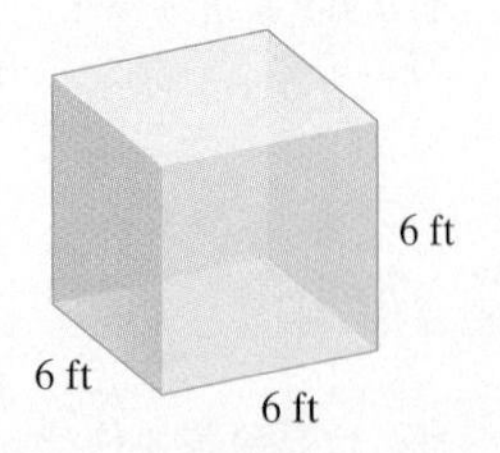

In Exercises 25–30, use two formulas for volume to find the volume of each figure. Express answers in terms of π and then round to the nearest whole number.

25.

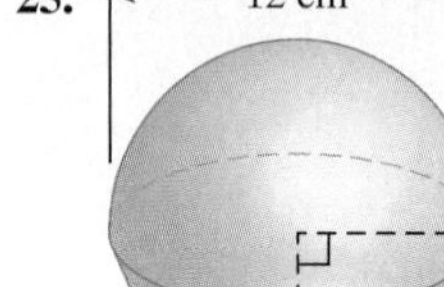

26.

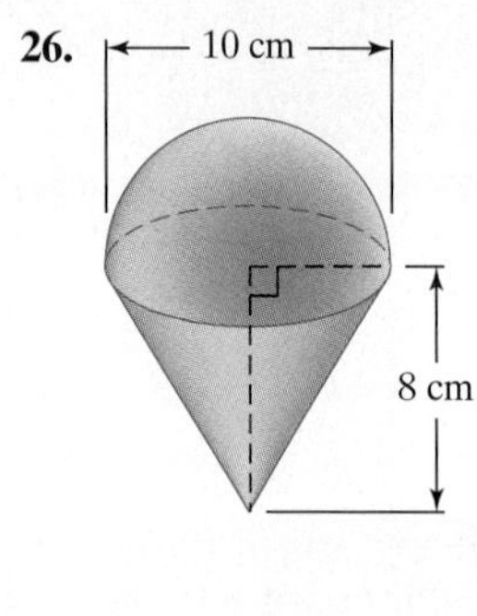

27.

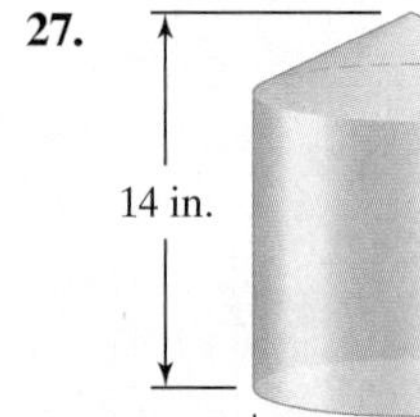

28.

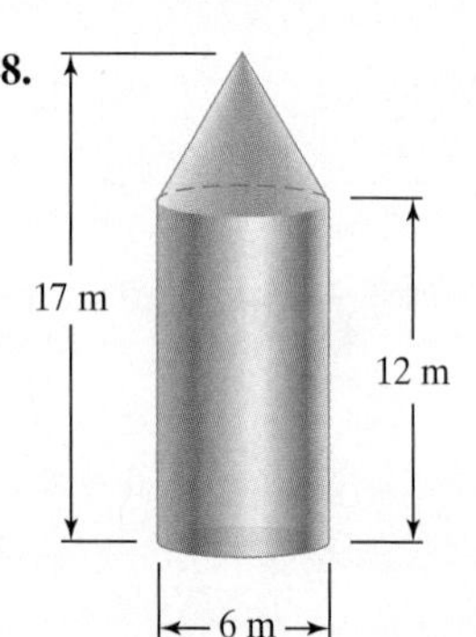

29.

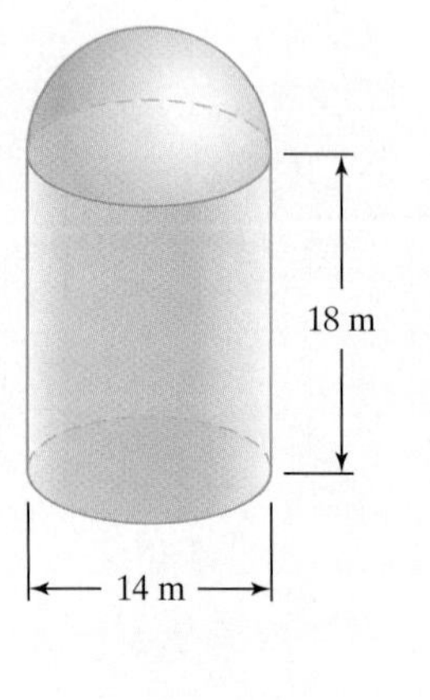

30.

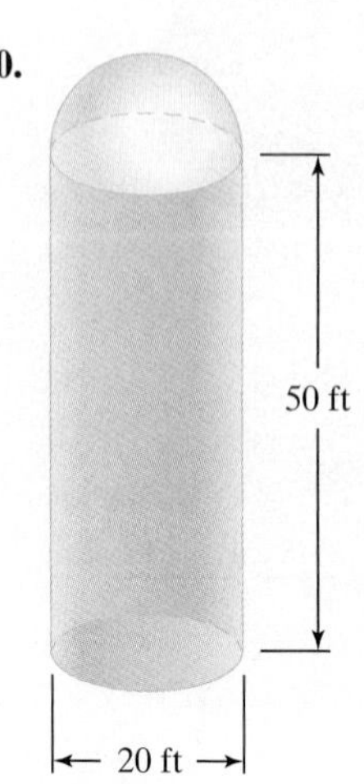

Practice Plus

31. Find the surface area and the volume of the figure shown.

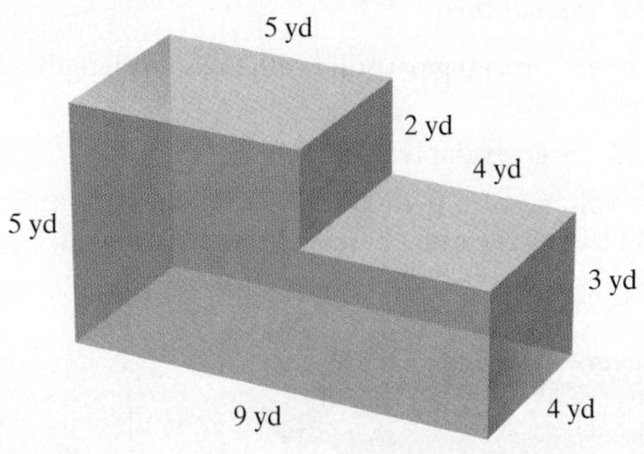

32. Find the surface area and the volume of the cement block in the figure shown.

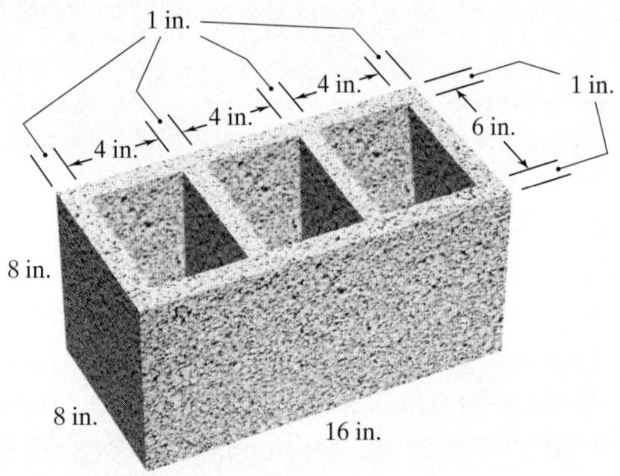

33. Find the surface area of the figure shown.

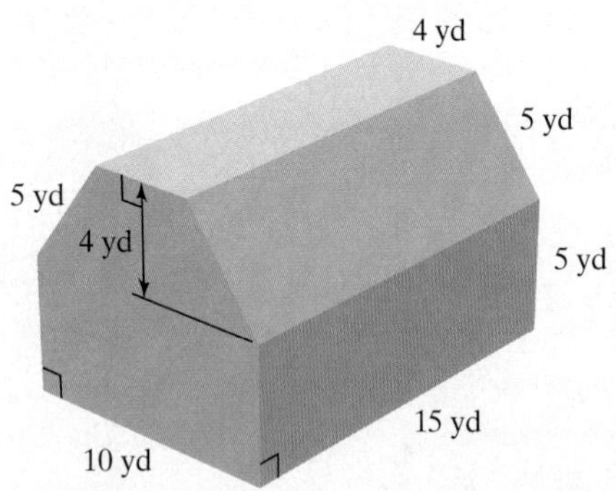

34. A machine produces open boxes using square sheets of metal measuring 12 inches on each side. The machine cuts equal-sized squares whose sides measure 2 inches from each corner. Then it shapes the metal into an open box by turning up the sides. Find the volume of the box.

35. Find the ratio, reduced to lowest terms, of the volume of a sphere with a radius of 3 inches to the volume of a sphere with a radius of 6 inches.

36. Find the ratio, reduced to lowest terms, of the volume of a sphere with a radius of 3 inches to the volume of a sphere with a radius of 9 inches.

37. A cylinder with radius 3 inches and height 4 inches has its radius tripled. How many times greater is the volume of the larger cylinder than the smaller cylinder?

38. A cylinder with radius 2 inches and height 3 inches has its radius quadrupled. How many times greater is the volume of the larger cylinder than the smaller cylinder?

Application Exercises

39. A building contractor is to dig a foundation 12 feet long, 9 feet wide, and 6 feet deep for a toll booth. The contractor pays $85 per load for trucks to remove the dirt. Each truck holds 6 cubic yards. What is the cost to the contractor to have all the dirt hauled away?

40. What is the cost of concrete for a walkway that is 15 feet long, 8 feet wide, and 9 inches deep if the concrete costs $30 per cubic yard?

41. A furnace is designed to heat 10,000 cubic feet. Will this furnace be adequate for a 1400-square-foot house with a 9-foot ceiling?

42. A water reservoir is shaped like a rectangular solid with a base that is 50 yards by 30 yards, and a vertical height of 20 yards. At the start of a three-month period of no rain, the reservoir was completely full. At the end of this period, the height of the water was down to 6 yards. How much water was used in the three-month period?

43. The Great Pyramid outside Cairo, Egypt, has a square base measuring 756 feet on a side and a height of 480 feet.

   a. What is the volume of the Great Pyramid, in cubic yards?

   b. The stones used to build the Great Pyramid were limestone blocks with an average volume of 1.5 cubic yards. How many of these blocks were needed to construct the Great Pyramid?

44. Although the Eiffel Tower in Paris is not a solid pyramid, its shape approximates that of a pyramid with a square base measuring 120 feet on a side and a height of 980 feet. If it were a solid pyramid, what would be the Eiffel Tower's volume, in cubic yards?

45. You are about to sue your contractor who promised to install a water tank that holds 500 gallons of water. You know that 500 gallons is the capacity of a tank that holds 67 cubic feet. The cylindrical tank has a radius of 3 feet and a height of 2 feet 4 inches. Does the evidence indicate you can win the case against the contractor if it goes to court?

46. Two cylindrical cans of soup sell for the same price. One can has a diameter of 6 inches and a height of 5 inches. The other has a diameter of 5 inches and a height of 6 inches. Which can contains more soup and, therefore, is the better buy?

47. A circular backyard pool has a diameter of 24 feet and is 4 feet deep. One cubic foot of water has a capacity of approximately 7.48 gallons. If water costs $2 per thousand gallons, how much, to the nearest dollar, will it cost to fill the pool?

48. The tunnel under the English Channel that connects England and France is the world's longest tunnel. There are actually three separate tunnels built side by side. Each is a half-cylinder that is 50,000 meters long and 4 meters high. How many cubic meters of dirt had to be removed to build the tunnel?

Explaining the Concepts

49. Explain the following analogy:

   In terms of formulas used to compute volume, a pyramid is to a rectangular solid just as a cone is to a cylinder.

50. Explain why a cylinder is not a polyhedron.

Critical Thinking Exercises

Make Sense? *In Exercises 51–54, determine whether each statement makes sense or does not make sense, and explain your reasoning.*

51. The physical education department ordered new basketballs in the shape of right circular cylinders.

52. When completely full, a cylindrical soup can with a diameter of 3 inches and a height of 4 inches holds more soup than a cylindrical can with a diameter of 4 inches and a height of 3 inches.

53. I found the volume of a rectangular solid in cubic inches and then divided by 12 to convert the volume to cubic feet.

54. Because a cylinder is a solid figure, I use cubic units to express its surface area.

55. What happens to the volume of a sphere if its radius is doubled?

56. A scale model of a car is constructed so that its length, width, and height are each $\frac{1}{10}$ the length, width, and height of the actual car. By how many times does the volume of the car exceed its scale model?

In Exercises 57–58, find the volume of the darkly shaded region. If necessary, round to the nearest whole number.

57. 58.

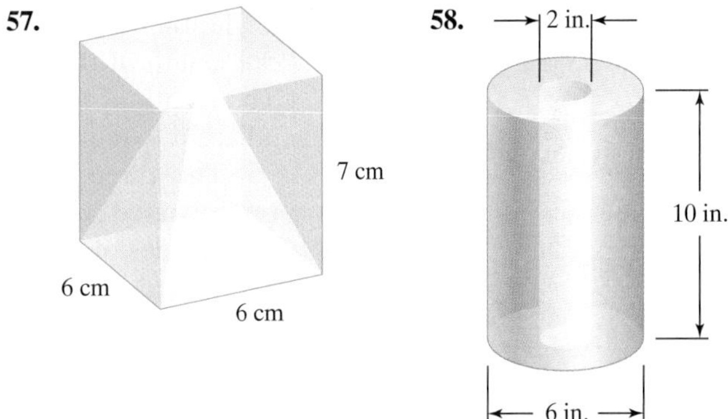

59. Find the surface area of the figure shown.

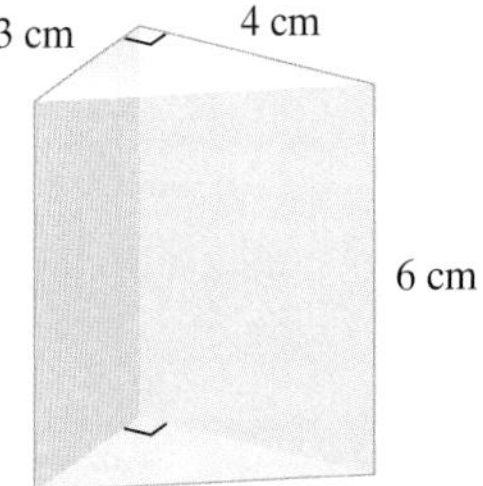

10.6 Right Triangle Trigonometry

WHAT AM I SUPPOSED TO LEARN?

After studying this section, you should be able to:

1. Use the lengths of the sides of a right triangle to find trigonometric ratios.
2. Use trigonometric ratios to find missing parts of right triangles.
3. Use trigonometric ratios to solve applied problems.

MOUNTAIN CLIMBERS HAVE FOREVER BEEN FASCINATED by reaching the top of Mount Everest, sometimes with tragic results. The mountain, on Asia's Tibet-Nepal border, is Earth's highest, peaking at an incredible 29,035 feet. The heights of mountains can be found using *trigonometry*. The word **trigonometry** means *measurement of triangles*. Trigonometry is used in navigation, building, and engineering. For centuries, Muslims used trigonometry and the stars to navigate across the Arabian desert to Mecca, the birthplace of the prophet Muhammad, the founder of Islam. The ancient Greeks used trigonometry to record the locations of thousands of stars and worked out the motion of the Moon relative to Earth. Today, trigonometry is used to study the structure of DNA, the master molecule that determines how we grow from a single cell to a complex, fully developed adult.

Ratios in Right Triangles

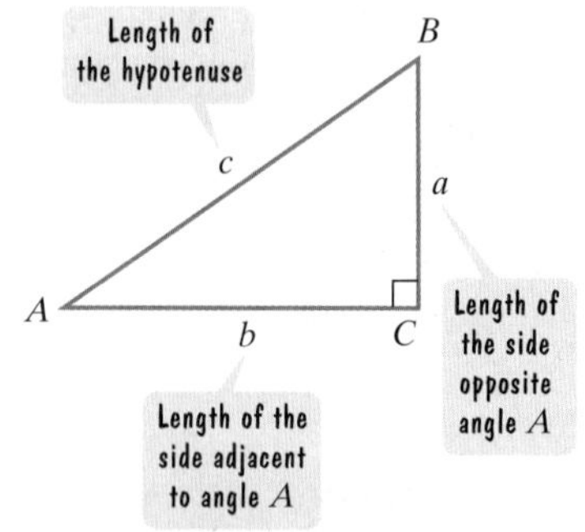

Length of the hypotenuse
B
c
a
A
b
C
Length of the side opposite angle A
Length of the side adjacent to angle A

FIGURE 10.63 Naming a right triangle's sides from the point of view of an acute angle

The right triangle forms the basis of trigonometry. If either acute angle of a right triangle stays the same size, the shape of the triangle does not change even if it is made larger or smaller. Because of properties of similar triangles, this means that the ratios of certain lengths stay the same regardless of the right triangle's size. These ratios have special names and are defined in terms of the **side opposite** an acute angle, the **side adjacent** to the acute angle, and the **hypotenuse**. In **Figure 10.63**, the length of the hypotenuse, the side opposite the 90° angle, is represented by c. The length of the side opposite angle A is represented by a. The length of the side adjacent to angle A is represented by b.

The three fundamental trigonometric ratios, **sine** (abbreviated sin), **cosine** (abbreviated cos), and **tangent** (abbreviated tan), are defined as ratios of the lengths of the sides of a right triangle. In the box that follows, when a side of a triangle is mentioned, we are referring to the *length* of that side.

1 *Use the lengths of the sides of a right triangle to find trigonometric ratios.*

TRIGONOMETRIC RATIOS

Let A represent an acute angle of a right triangle, with right angle C, shown in **Figure 10.63**. For angle A, the trigonometric ratios are defined as follows:

sine of A
$$\sin A = \frac{\text{side opposite angle } A}{\text{hypotenuse}} = \frac{a}{c}$$

cosine of A
$$\cos A = \frac{\text{side adjacent to angle } A}{\text{hypotenuse}} = \frac{b}{c}$$

tangent of A
$$\tan A = \frac{\text{side opposite angle } A}{\text{side adjacent to angle } A} = \frac{a}{b}.$$

GREAT QUESTION!

Is there a way to help me remember the definitions of the trigonometric ratios?

The word

SOHCAHTOA (pronounced: so-cah-tow-ah)

may be helpful in remembering the definitions of the three trigonometric ratios.

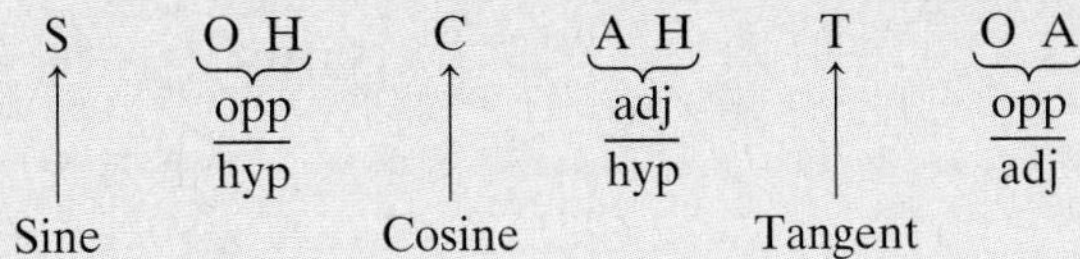

"Some Old Hog Came Around Here and Took Our Apples."

EXAMPLE 1 *Becoming Familiar with the Trigonometric Ratios*

Find the sine, cosine, and tangent of A in **Figure 10.64**.

FIGURE 10.64

SOLUTION

We begin by finding the measure of the hypotenuse c using the Pythagorean Theorem.

$$c^2 = a^2 + b^2 = 5^2 + 12^2 = 25 + 144 = 169$$

$$c = \sqrt{169} = 13$$

Now, we apply the definitions of the trigonometric ratios.

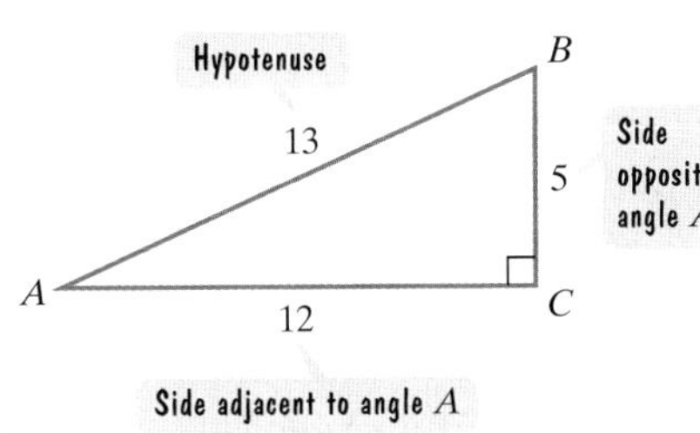

$$\sin A = \frac{\text{side opposite angle } A}{\text{hypotenuse}} = \frac{5}{13}$$

$$\cos A = \frac{\text{side adjacent to angle } A}{\text{hypotenuse}} = \frac{12}{13}$$

$$\tan A = \frac{\text{side opposite angle } A}{\text{side adjacent to angle } A} = \frac{5}{12}$$

GREAT QUESTION!

Can you clarify which is the opposite side and which is the adjacent side?

We know that the longest side of the right triangle is the hypotenuse. The two legs of the triangle are described by their relationship to each acute angle. The adjacent leg "touches" the acute angle by forming one of the angle's sides. The opposite leg is not a side of the acute angle because it lies directly opposite the angle.

CHECK POINT 1 Find the sine, cosine, and tangent of A in the figure shown.

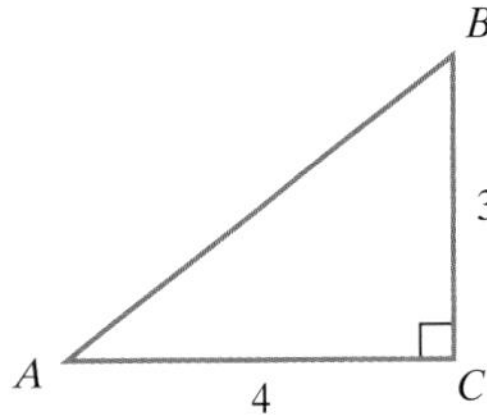

A BRIEF REVIEW *Equations with Fractions*

Finding missing parts of right triangles involves solving equations with fractions.

- The variable can appear in a fraction's numerator.

Example

Solve for a: $0.3 = \frac{a}{150}$

$$150(0.3) = 150\left(\frac{a}{150}\right)$$ Clear fractions by multiplying both sides by 150.

$$45 = a$$ Simplify: $150(0.3) = 45$ and $150\left(\frac{a}{150}\right) = \frac{\cancel{150}^{1}}{1} \cdot \frac{a}{\cancel{150}_{1}} = a.$

- The variable can appear in a fraction's denominator.

Example

Solve for c: $0.8 = \frac{150}{c}$

$$0.8c = \left(\frac{150}{c}\right)c$$ Clear fractions by multiplying both sides by c.

$$0.8c = 150$$ Simplify: $\left(\frac{150}{c}\right)c = \frac{150}{\cancel{c}_{1}} \cdot \frac{\cancel{c}^{1}}{1} = 150.$

We're still not done. We have not isolated c.

$$\frac{0.8c}{0.8} = \frac{150}{0.8}$$ Divide both sides by 0.8.

$$c = 187.5$$ Simplify.

2 *Use trigonometric ratios to find missing parts of right triangles.*

A scientific or graphing calculator in the degree mode will give you decimal approximations for the trigonometric ratios of any angle. For example, to find an approximation for tan 37°, the tangent of an angle measuring 37°, a keystroke sequence similar to one of the following can be used:

```
tan(37)
                .7535540501
```

Many Scientific Calculators: 37 [TAN]

Many Graphing Calculators: [TAN] 37 [ENTER].

The tangent of 37°, rounded to four decimal places, is 0.7536.

If we are given the length of one side and the measure of an acute angle of a right triangle, we can use trigonometry to solve for the length of either of the other two sides. Example 2 illustrates how this is done.

EXAMPLE 2 *Finding a Missing Leg of a Right Triangle*

Find a in the right triangle in **Figure 10.65**.

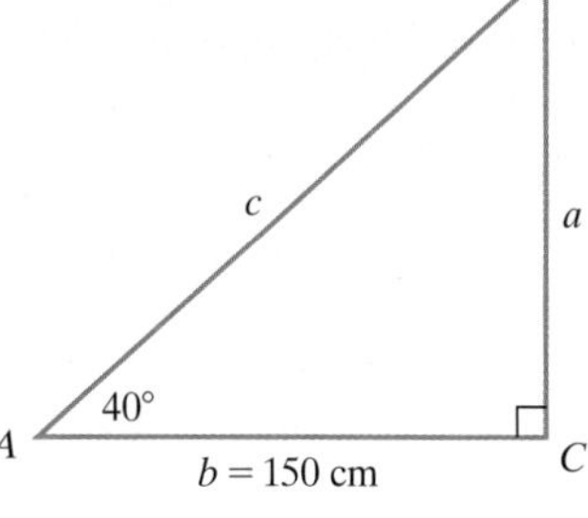

FIGURE 10.65

SOLUTION

We need to identify a trigonometric ratio that will make it possible to find a. Because we have a known angle, 40°, an unknown opposite side, a, and a known adjacent side, 150 cm, we use the tangent ratio.

$$\tan 40° = \frac{a}{150}$$

Side opposite the 40° angle

Side adjacent to the 40° angle

Now we solve for a by multiplying both sides by 150.

$$a = 150 \tan 40° \approx 126$$

The tangent ratio reveals that a is approximately 126 centimeters.

TECHNOLOGY

Here is the keystroke sequence for 150 tan 40°:

Many Scientific Calculators

150 [×] 40 [TAN] [=]

Many Graphing Calculators

150 [TAN] 40 [ENTER]

CHECK POINT 2 In **Figure 10.65**, let $m\measuredangle A = 62°$ and $b = 140$ cm. Find a to the nearest centimeter.

EXAMPLE 3 *Finding a Missing Hypotenuse of a Right Triangle*

Find c in the right triangle in **Figure 10.65**.

SOLUTION

In Example 2, we found a: $a \approx 126$. Because we are given that $b = 150$, it is possible to find c using the Pythagorean Theorem: $c^2 = a^2 + b^2$. However, if we made an error in computing a, we will perpetuate our mistake using this approach.

Instead, we will use the quantities given and identify a trigonometric ratio that will make it possible to find c. Refer to **Figure 10.65**. Because we have a known angle, 40°, a known adjacent side, 150 cm, and an unknown hypotenuse, c, we use the cosine ratio.

$$\cos 40° = \frac{150}{c}$$

Side adjacent to the 40° angle (150)

Hypotenuse (c)

$$c \cos 40° = 150 \quad \text{Multiply both sides by } c.$$

$$c = \frac{150}{\cos 40°} \quad \text{Divide both sides by } \cos 40°.$$

$$c \approx 196 \quad \text{Use a calculator.}$$

The cosine ratio reveals that the hypotenuse is approximately 196 centimeters.

TECHNOLOGY

Here is the keystroke sequence for $\frac{150}{\cos 40°}$:

Many Scientific Calculators

150 [÷] 40 [COS] [=]

Many Graphing Calculators

150 [÷] [COS] 40 [ENTER]

CHECK POINT 3 In **Figure 10.65**, let $m\measuredangle A = 62°$ and $b = 140$ cm. Find c to the nearest centimeter.

3 *Use trigonometric ratios to solve applied problems.*

Applications of the Trigonometric Ratios

Trigonometry was first developed to determine heights and distances that are inconvenient or impossible to measure. These applications often involve the angle made with an imaginary horizontal line. As shown in **Figure 10.66**, an angle formed by a horizontal line and the line of sight to an object that is above the horizontal line is called the **angle of elevation**. The angle formed by a horizontal line and the line of sight to an object that is below the horizontal line is called the **angle of depression**. Transits and sextants are instruments used to measure such angles.

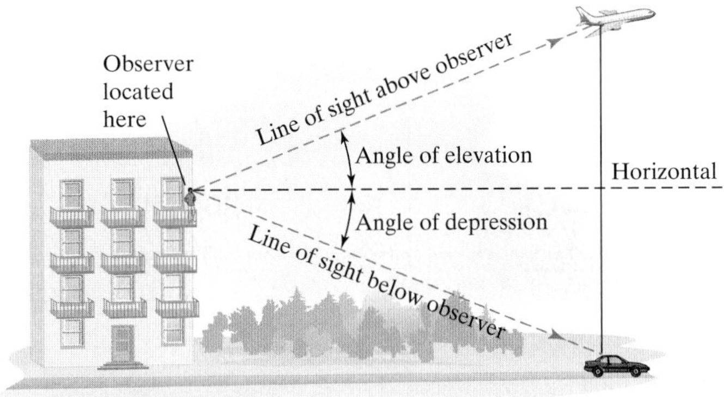

FIGURE 10.66

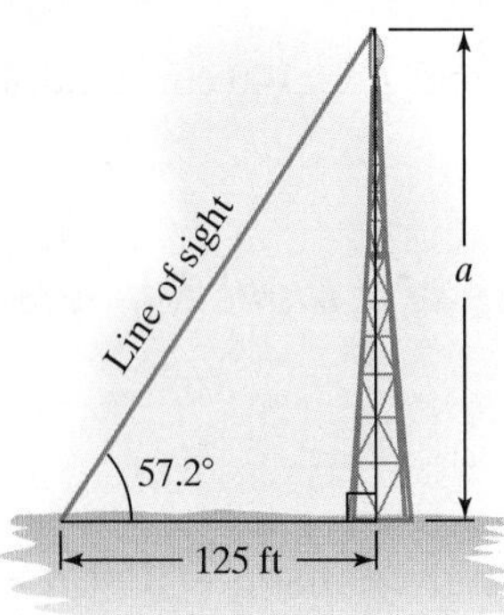

FIGURE 10.67 Determining height without using direct measurement

EXAMPLE 4 Problem Solving Using an Angle of Elevation

From a point on level ground 125 feet from the base of a tower, the angle of elevation to the top of the tower is 57.2°. Approximate the height of the tower to the nearest foot.

SOLUTION

A sketch is shown in **Figure 10.67**, where a represents the height of the tower. In the right triangle, we have a known angle, an unknown opposite side, and a known adjacent side. Therefore, we use the tangent ratio.

$$\tan 57.2^\circ = \frac{a}{125}$$

Side opposite the 57.2° angle

Side adjacent to the 57.2° angle

We solve for a by multiplying both sides of this equation by 125:

$$a = 125 \tan 57.2^\circ \approx 194.$$

The tower is approximately 194 feet high.

CHECK POINT 4 From a point on level ground 80 feet from the base of the Eiffel Tower, the angle of elevation is 85.4°. Approximate the height of the Eiffel Tower to the nearest foot.

If the measures of two sides of a right triangle are known, the measures of the two acute angles can be found using the **inverse trigonometric keys** on a calculator. For example, suppose that $\sin A = 0.866$. We can find the measure of angle A by using the *inverse sine* key, usually labeled [SIN⁻¹]. The key [SIN⁻¹] is not a button you will actually press; it is the secondary function for the button labeled [SIN].

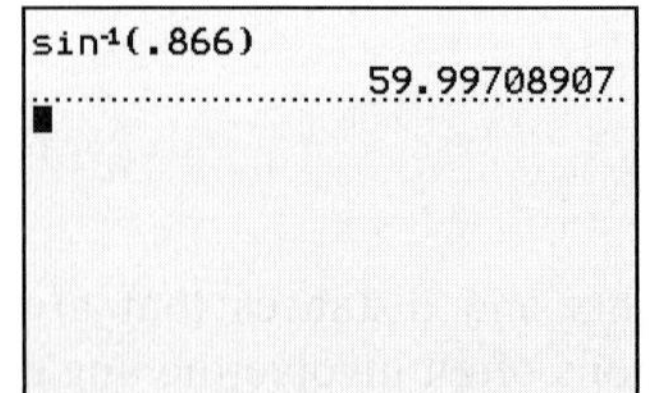

Many Scientific Calculators:

.866 [2nd] [SIN]

Pressing [2nd] [SIN] accesses the inverse sine key, [SIN⁻¹].

Many Graphing Calculators:

[2nd] [SIN] .866 [ENTER]

The display should show approximately 59.99°, which can be rounded to 60°. Thus, if $\sin A = 0.866$, then $m\measuredangle A \approx 60^\circ$.

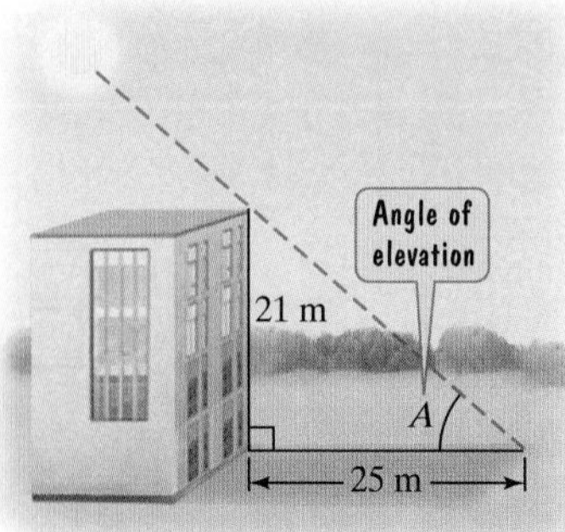

FIGURE 10.68

EXAMPLE 5 Determining the Angle of Elevation

A building that is 21 meters tall casts a shadow 25 meters long. Find the angle of elevation of the Sun to the nearest degree.

SOLUTION

The situation is illustrated in **Figure 10.68**. We are asked to find $m\measuredangle A$. We begin with the tangent ratio.

$$\tan A = \frac{\text{side opposite } A}{\text{side adjacent to } A} = \frac{21}{25}$$

```
tan⁻¹(21/25)
                40.03025927
```

With $\tan A = \frac{21}{25}$, we use the **inverse tangent** key, $\boxed{\text{TAN}^{-1}}$, to find A.

Many Scientific Calculators:

(21 ÷ 25) 2nd TAN

Pressing 2nd TAN accesses the inverse tangent key, TAN^{-1}.

Many Graphing Calculators:

2nd TAN (21 ÷ 25) ENTER

The display should show approximately 40. Thus, the angle of elevation of the Sun is approximately 40°.

CHECK POINT 5 A flagpole that is 14 meters tall casts a shadow 10 meters long. Find the angle of elevation of the Sun to the nearest degree.

Blitzer Bonus

The Mountain Man

In the 1930s, a *National Geographic* team headed by Brad Washburn used trigonometry to create a map of the 5000-square-mile region of the Yukon, near the Canadian border. The team started with aerial photography. By drawing a network of angles on the photographs, the approximate locations of the major mountains and their rough heights were determined. The expedition then spent three months on foot to find the exact heights. Team members established two base points a known distance apart, one directly under the mountain's peak. By measuring the angle of elevation from one of the base points to the peak, the tangent ratio was used to determine the peak's height. The Yukon expedition was a major advance in the way maps are made.

Concept and Vocabulary Check

Fill in each blank so that the resulting statement is true. Exercises 1–3 are based on the following right triangle:

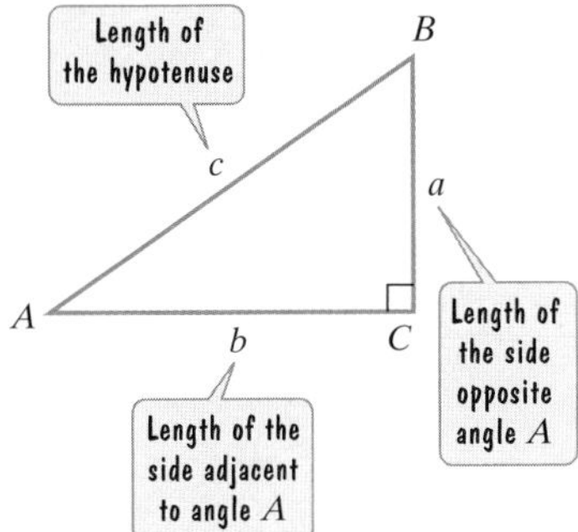

1. sin A, which represents the _____ of angle A, is defined by the length of the side ________ angle A divided by the length of the __________. This is represented by ___ in the figure shown.

2. cos A, which represents the _______ of angle A, is defined by the length of the side ___________ angle A divided by the length of the ___________. This is represented by ___ in the figure shown.

3. tan A, which represents the ________ of angle A, is defined by the length of the side ________ angle A divided by the length of the side ___________ angle A. This is represented by ___ in the figure shown.

4. An angle formed by a horizontal line and the line of sight to an object that is above the horizontal line is called the angle of _________.

5. An angle formed by a horizontal line and the line of sight to an object that is below the horizontal line is called the angle of _________.

In Exercises 6–9, determine whether each statement is true or false. If the statement is false, make the necessary change(s) to produce a true statement.

6. sin 30° increases as the size of a right triangle with an acute angle of 30° grows larger. ______

7. The side of a right triangle opposite an acute angle forms one of the acute angle's sides. ______

8. On a scientific calculator, the keystroke 47 TAN gives a decimal approximation for tan 47°. ______

9. The equation $\cos 40° = \frac{150}{c}$ is solved for c by multiplying both sides by 150. ______

Exercise Set 10.6

Practice Exercises

In Exercises 1–8, use the given right triangles to find ratios, in reduced form, for sin A, cos A, and tan A.

1.

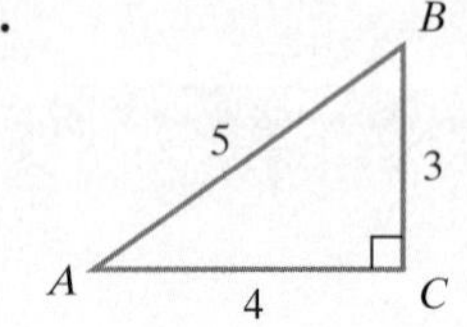

2.

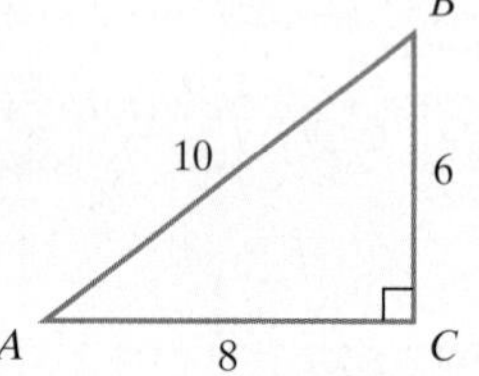

3.

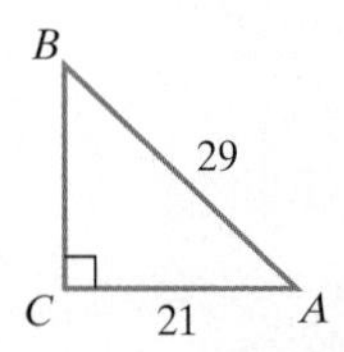

4.

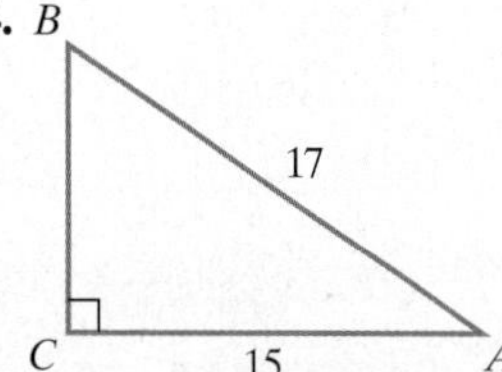

5.

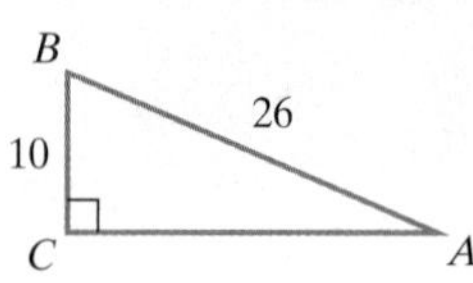

6.

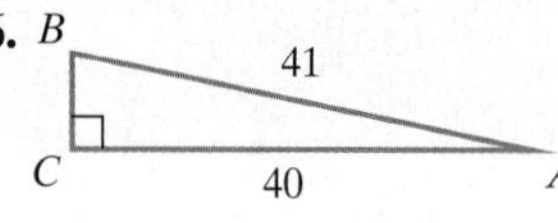

7.

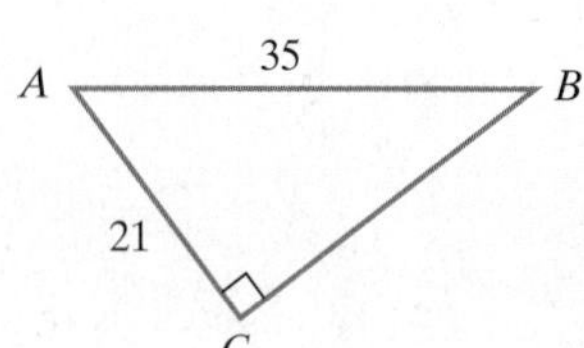

8.

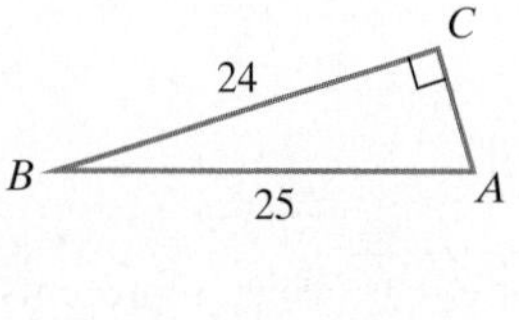

In Exercises 9–18, find the measure of the side of the right triangle whose length is designated by a lowercase letter. Round answers to the nearest whole number.

9.

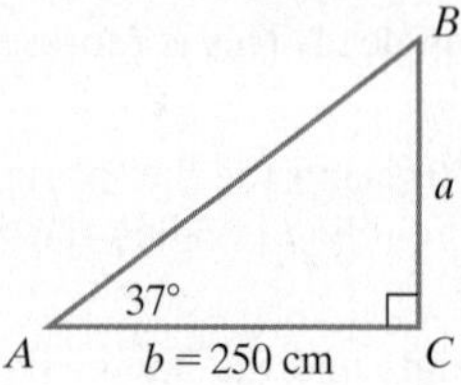

10.

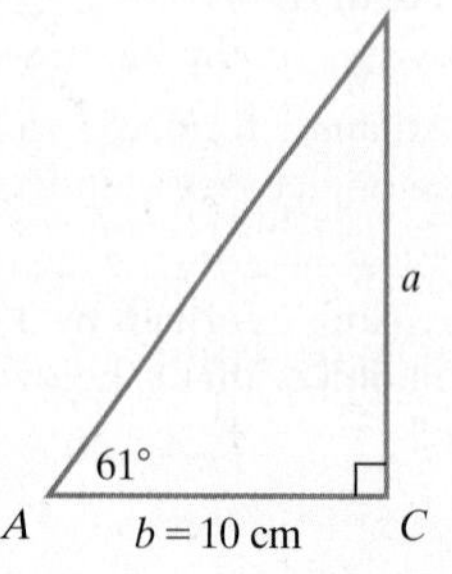

11.

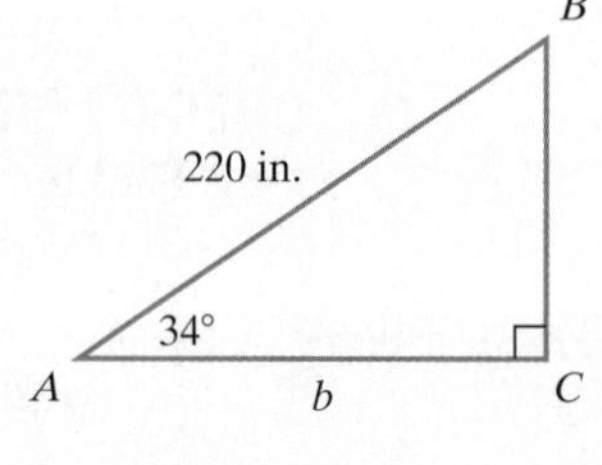

12.

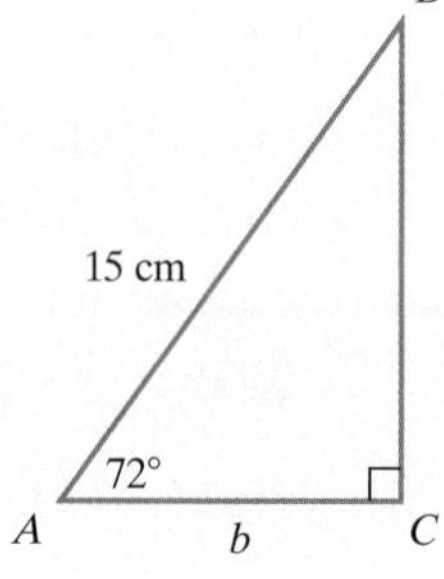

13.

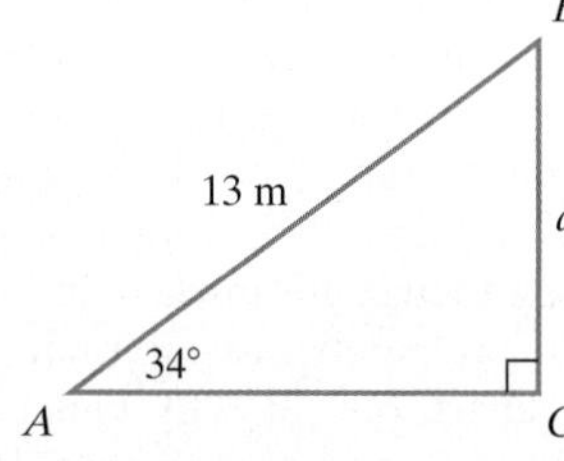

14.

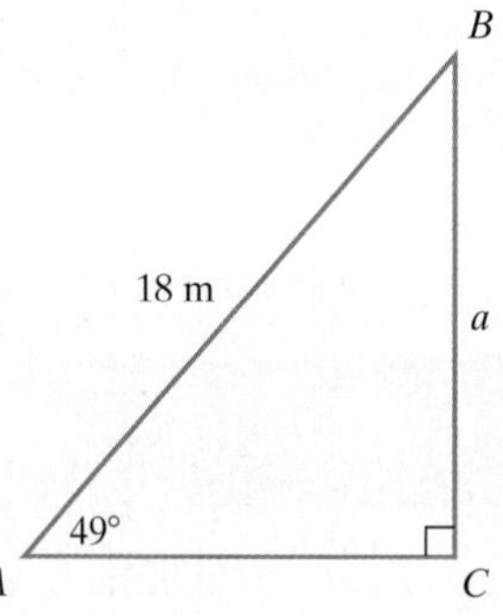

15.

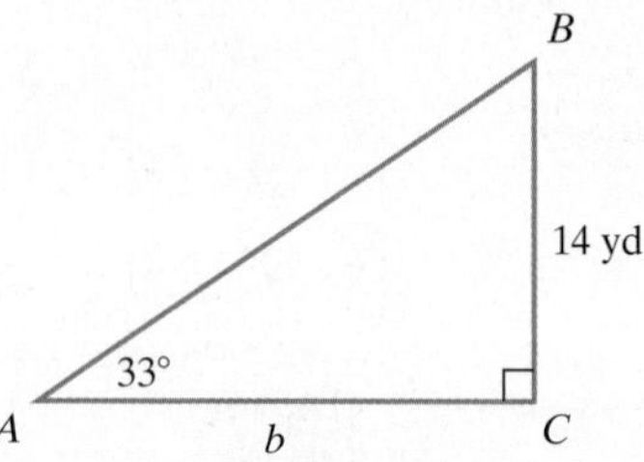

16.

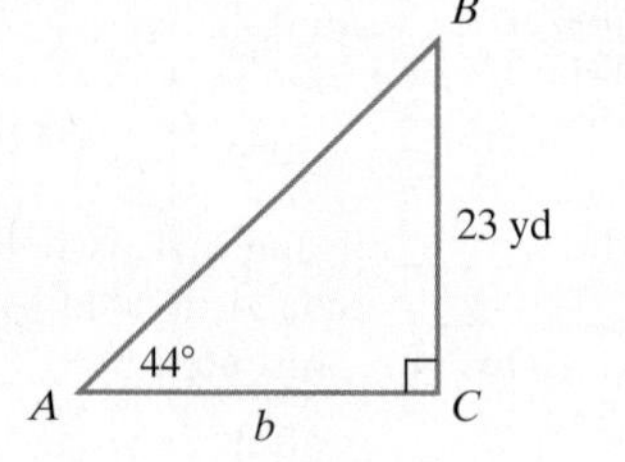

17.

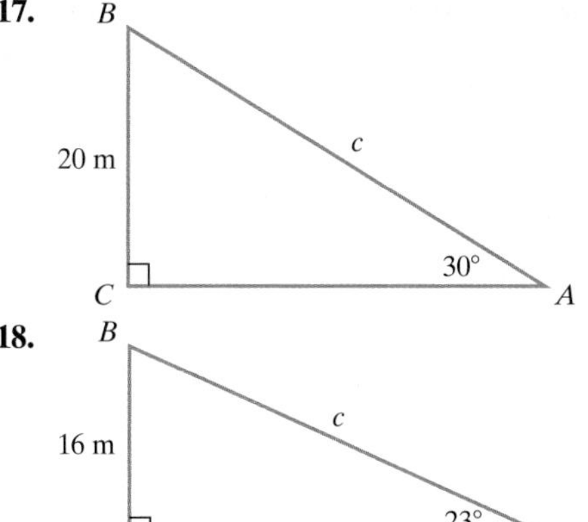

18.

In Exercises 19–22, find the measures of the parts of the right triangle that are not given. Round all answers to the nearest whole number.

19.

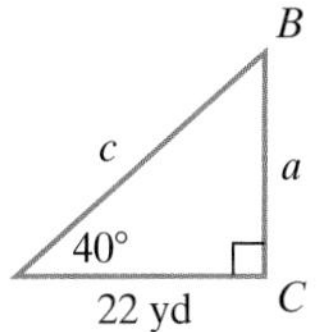

20.

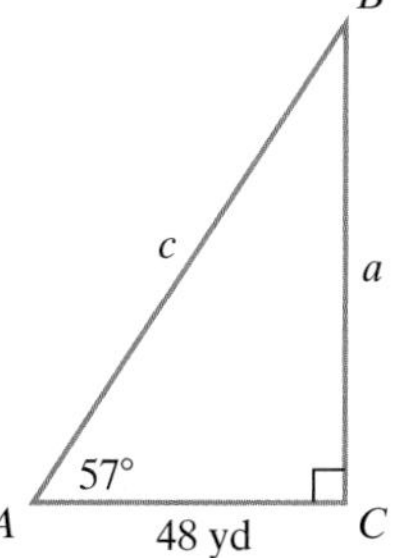

21.

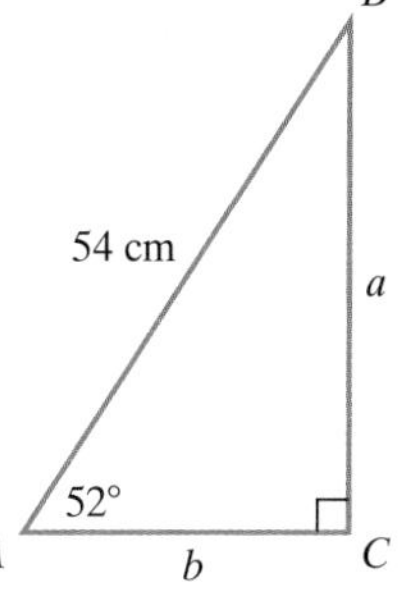

22.

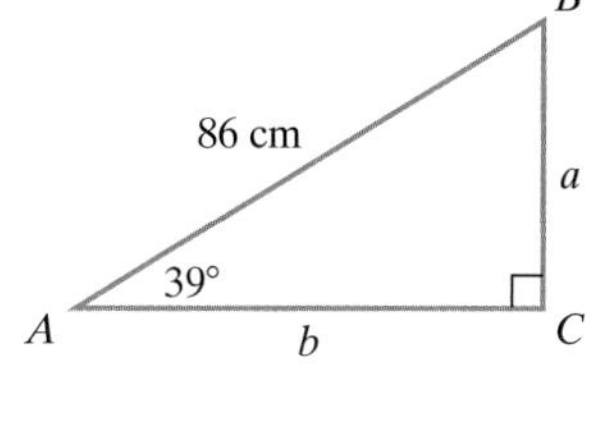

In Exercises 23–26, use the inverse trigonometric keys on a calculator to find the measure of angle A, rounded to the nearest whole degree.

23.

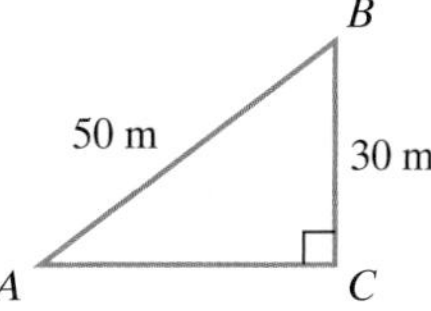

24.

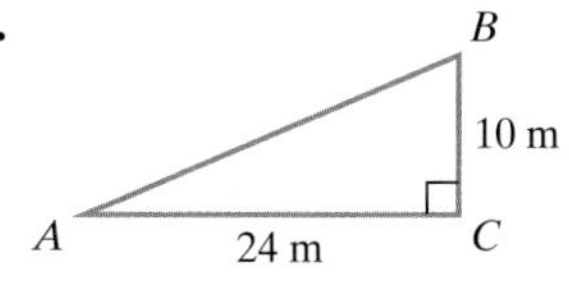

25.

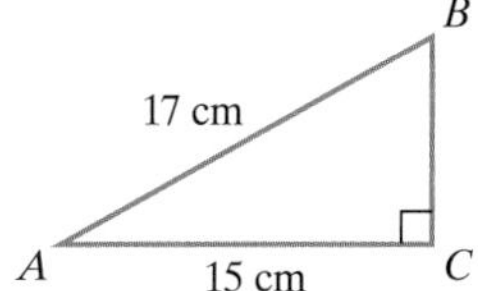

26.

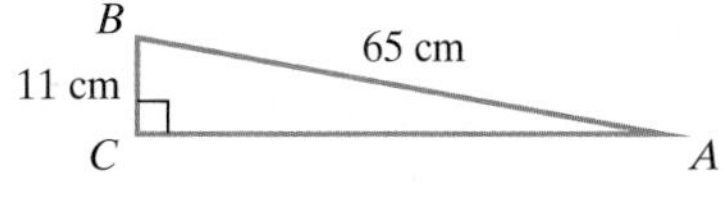

Practice Plus

In Exercises 27–34, find the length x to the nearest whole number.

27.

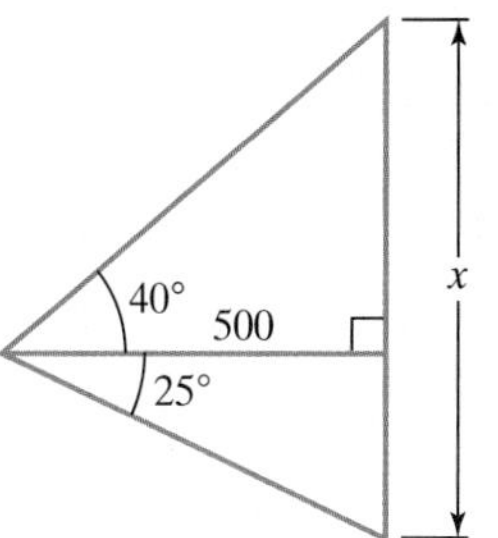

28.

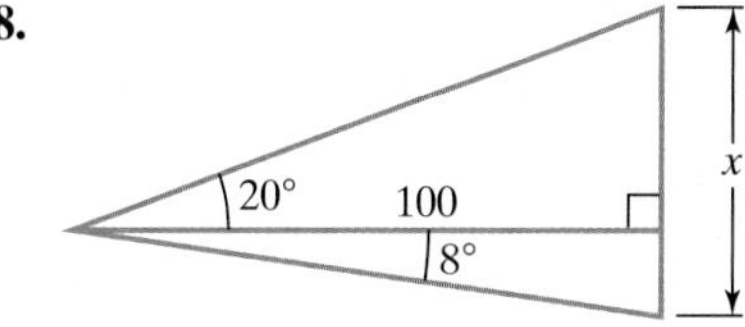

29.

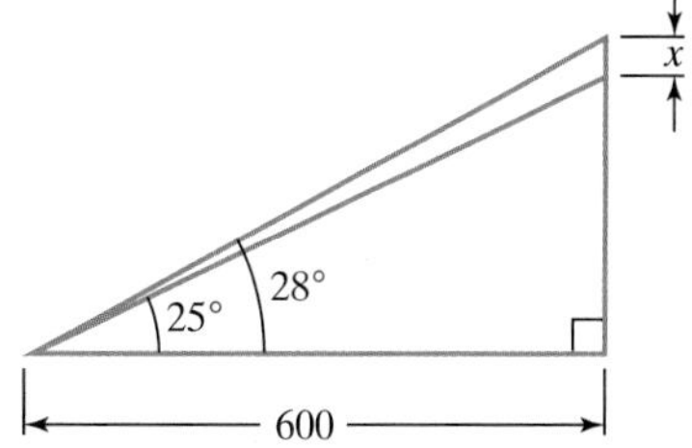

30.

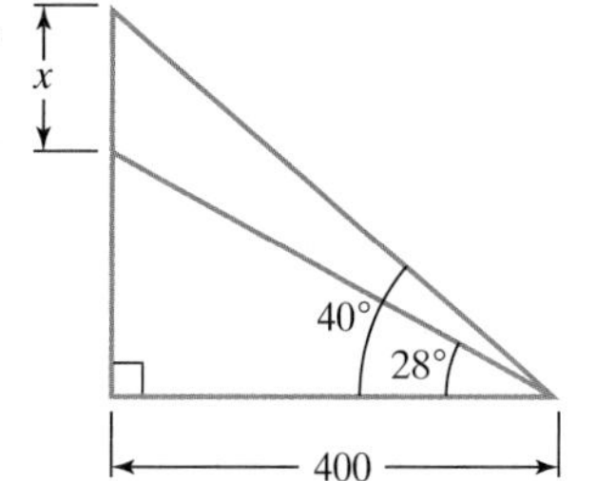

31.

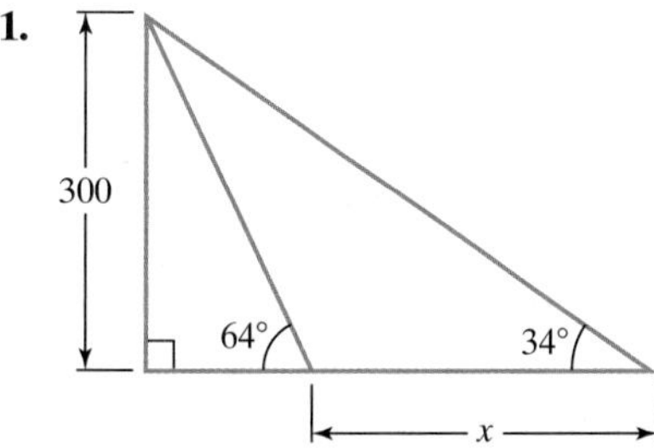

32.

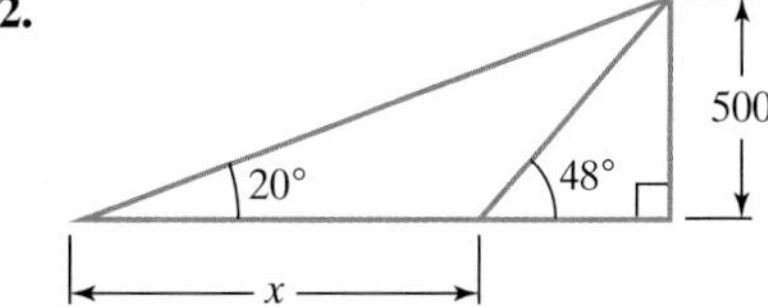

33.

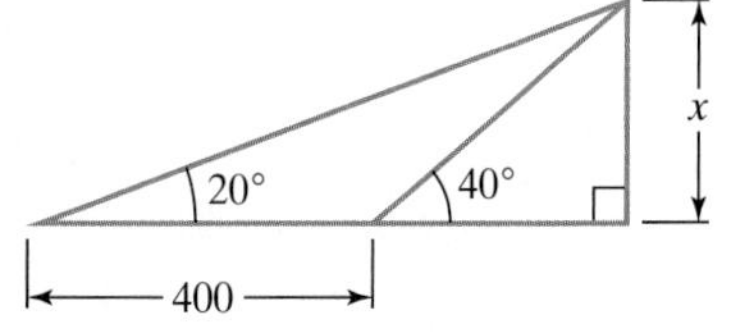

34.

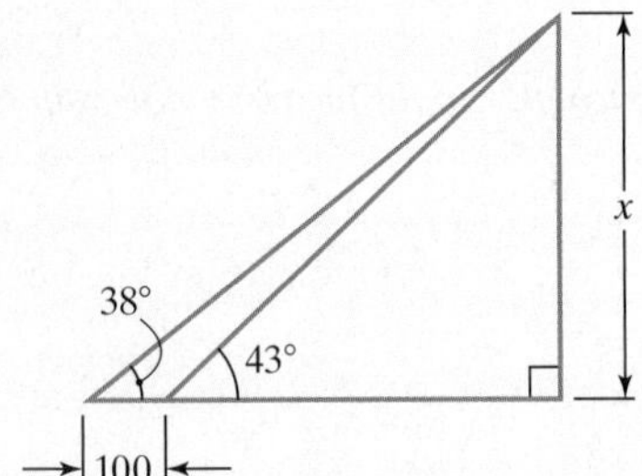

Application Exercises

35. To find the distance across a lake, a surveyor took the measurements shown in the figure. Use these measurements to determine how far it is across the lake. Round to the nearest yard.

36. At a certain time of day, the angle of elevation of the Sun is 40°. To the nearest foot, find the height of a tree whose shadow is 35 feet long.

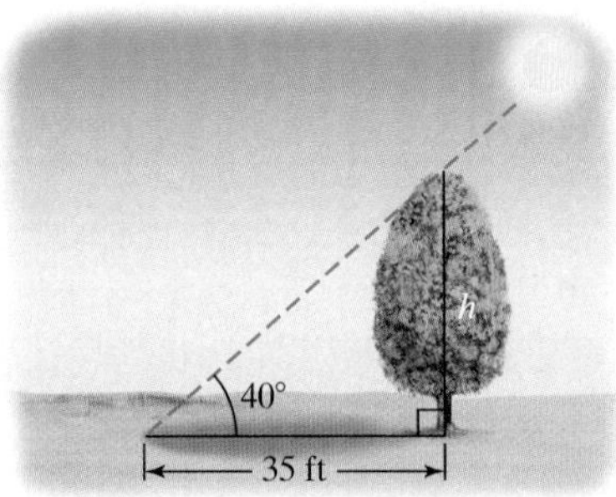

37. A plane rises from take-off and flies at an angle of 10° with the horizontal runway. When it has gained 500 feet in altitude, find the distance, to the nearest foot, the plane has flown.

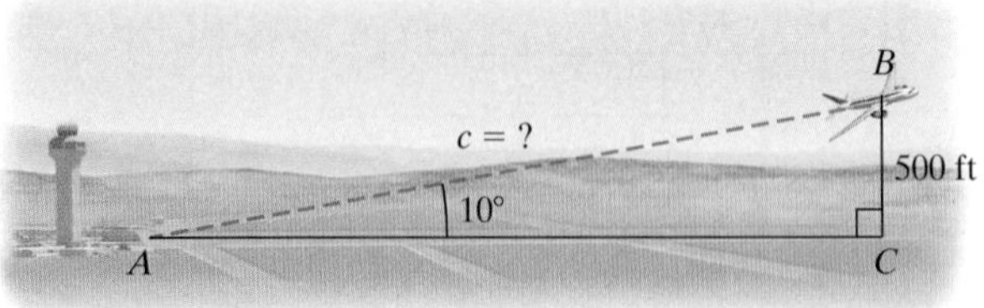

38. A road is inclined at an angle of 5°. After driving 5000 feet along this road, find the driver's increase in altitude. Round to the nearest foot.

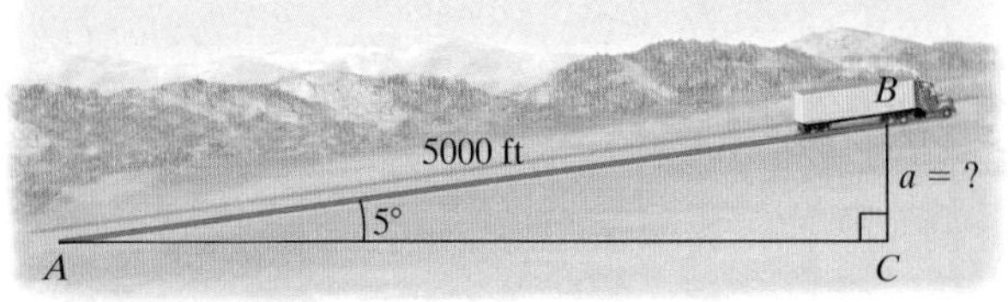

39. The tallest television transmitting tower in the world is in North Dakota. From a point on level ground 5280 feet (one mile) from the base of the tower, the angle of elevation to the top of the tower is 21.3°. Approximate the height of the tower to the nearest foot.

40. From a point on level ground 30 yards from the base of a building, the angle of elevation to the top of the building is 38.7°. Approximate the height of the building to the nearest foot.

41. The Statue of Liberty is approximately 305 feet tall. If the angle of elevation of a ship to the top of the statue is 23.7°, how far, to the nearest foot, is the ship from the statue's base?

42. A 200-foot cliff drops vertically into the ocean. If the angle of elevation of a ship to the top of the cliff is 22.3°, how far off shore, to the nearest foot, is the ship?

43. A tower that is 125 feet tall casts a shadow 172 feet long. Find the angle of elevation of the Sun to the nearest degree.

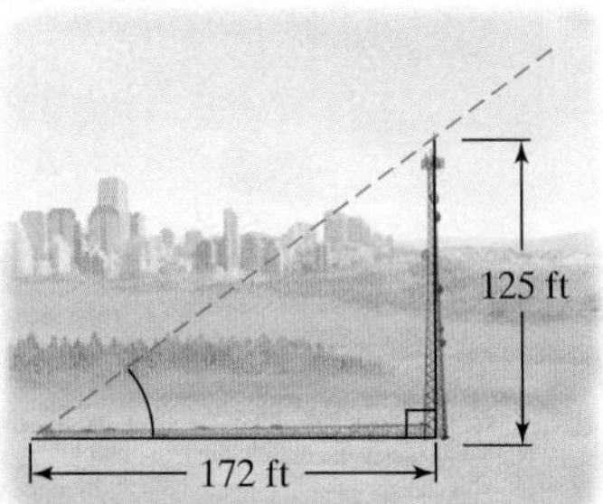

44. The Washington Monument is 555 feet high. If you stand one quarter of a mile, or 1320 feet, from the base of the monument and look to the top, find the angle of elevation to the nearest degree.

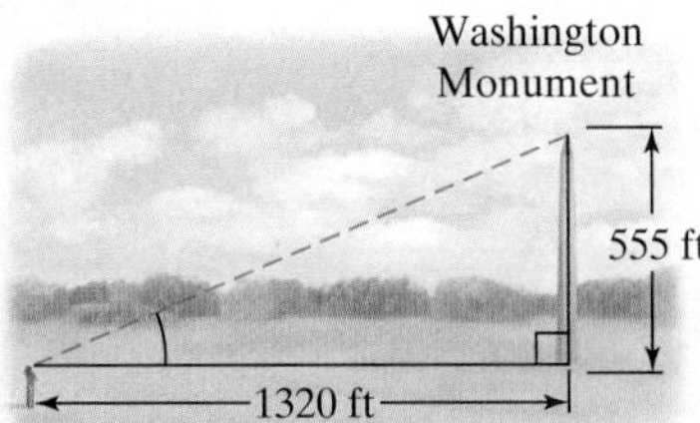

45. A helicopter hovers 1000 feet above a small island. The figure below shows that the angle of depression from the helicopter to point P is 36°. How far off the coast, to the nearest foot, is the island?

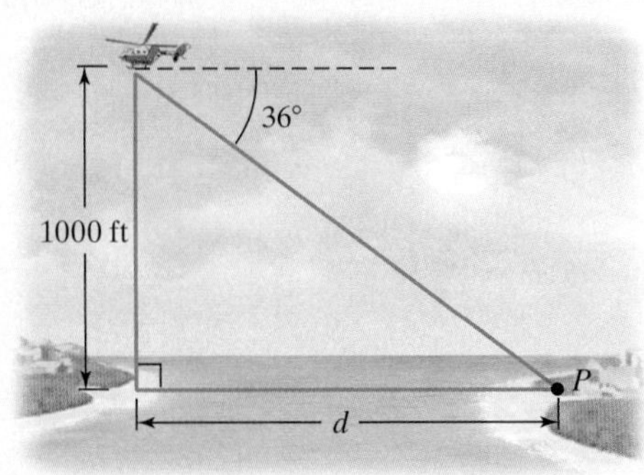

46. A police helicopter is flying at 800 feet. A stolen car is sighted at an angle of depression of 72°. Find the distance of the stolen car, to the nearest foot, from a point directly below the helicopter.

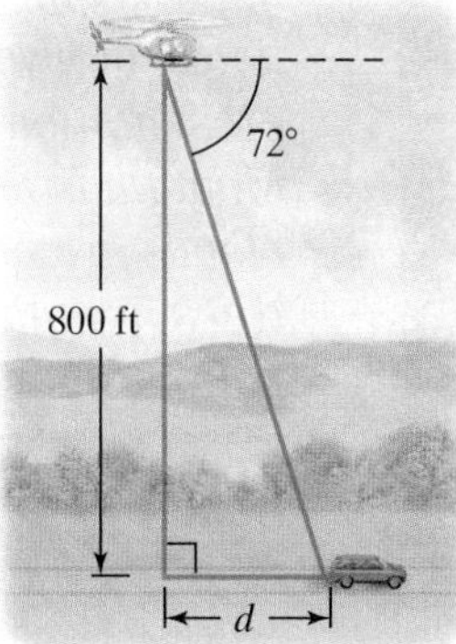

47. A wheelchair ramp is to be built beside the steps to the campus library. Find the angle of elevation of the 23-foot ramp, to the nearest tenth of a degree, if its final height is 6 feet.

48. A kite flies at a height of 30 feet when 65 feet of string is out. If the string is in a straight line, find the angle that it makes with the ground. Round to the nearest tenth of a degree.

Explaining the Concepts

49. If you are given the lengths of the sides of a right triangle, describe how to find the sine of either acute angle.

50. Describe one similarity and one difference between the sine ratio and the cosine ratio in terms of the sides of a right triangle.

51. If one of the acute angles of a right triangle is 37°, explain why the sine ratio does not increase as the size of the triangle increases.

52. If the measure of one of the acute angles and the hypotenuse of a right triangle are known, describe how to find the measure of the remaining parts of the triangle.

53. Describe what is meant by an angle of elevation and an angle of depression.

54. Give an example of an applied problem that can be solved using one or more trigonometric ratios. Be as specific as possible.

55. Use a calculator to find each of the following: sin 32° and cos 58°; sin 17° and cos 73°; sin 50° and cos 40°; sin 88° and cos 2°. Describe what you observe. Based on your observations, what do you think the *co* in *cosine* stands for?

56. Stonehenge, the famous "stone circle" in England, was built between 2750 B.C. and 1300 B.C. using solid stone blocks weighing over 99,000 pounds each. It required 550 people to pull a single stone up a ramp inclined at a 9° angle. Describe how right triangle trigonometry can be used to determine the distance the 550 workers had to drag a stone in order to raise it to a height of 30 feet.

Critical Thinking Exercises

Make Sense? *In Exercises 57–60, determine whether each statement makes sense or does not make sense, and explain your reasoning.*

57. For a given angle A, I found a slight increase in sin A as the size of the triangle increased.

58. I'm working with a right triangle in which the hypotenuse is the side adjacent to the acute angle.

59. Standing under this arch, I can determine its height by measuring the angle of elevation to the top of the arch and my distance to a point directly under the arch.

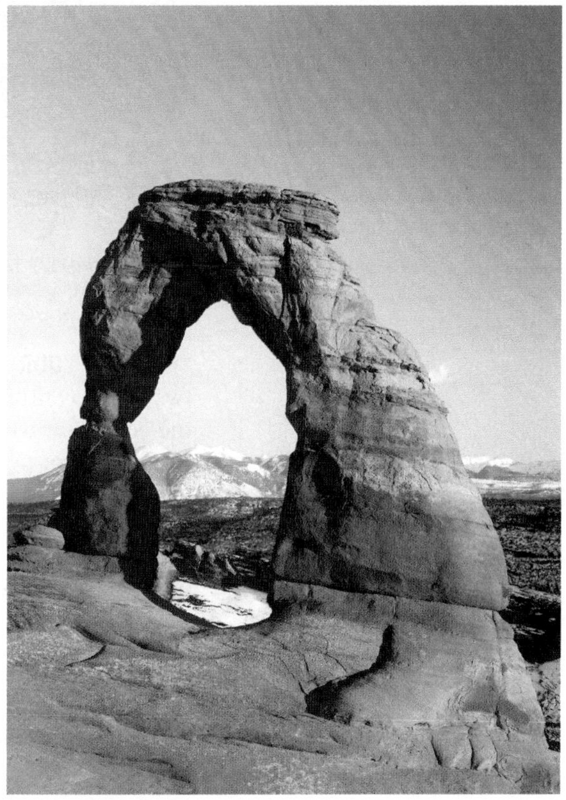

Delicate Arch in Arches National Park, Utah

60. A wheelchair ramp must be constructed so that the slope is not more than 1 inch of rise for every 1 foot of run, so I used the tangent ratio to determine the maximum angle that the ramp can make with the ground.

61. Explain why the sine or cosine of an acute angle cannot be greater than or equal to 1.

62. Describe what happens to the tangent of an angle as the measure of the angle gets close to 90°.

63. From the top of a 250-foot lighthouse, a plane is sighted overhead and a ship is observed directly below the plane. The angle of elevation of the plane is 22° and the angle of depression of the ship is 35°. Find **a.** the distance of the ship from the lighthouse; **b.** the plane's height above the water. Round to the nearest foot.

64. Sighting the top of a building, a surveyor measured the angle of elevation to be 22°. The transit is 5 feet above the ground and 300 feet from the building. Find the building's height. Round to the nearest foot.

10.7 Beyond Euclidean Geometry

WHAT AM I SUPPOSED TO LEARN?

After studying this section, you should be able to:

1 Gain an understanding of some of the general ideas of other kinds of geometries.

The Family (1962), Marisol Escobar. Digital Image © The Museum of Modern Art/Licensed by Scala/Art Resource, NY; Art © Estate of Marisol Escobar/Licensed by VAGA, New York, NY

"I love Euclidean geometry, but it is quite clear that it does not give a reasonable presentation of the world. Mountains are not cones, clouds are not spheres, trees are not cylinders, neither does lightning travel in a straight line. Almost everything around us is non-Euclidean."
—Benoit Mandelbrot

THINK OF YOUR FAVORITE NATURAL SETTING. DURING THE LAST QUARTER OF THE twentieth century, mathematicians developed a new kind of geometry that uses the computer to produce the diverse forms of nature that we see around us. These forms are not polygons, polyhedrons, circles, or cylinders. In this section, we move beyond Euclidean geometry to explore ideas which have extended geometry beyond the boundaries first laid down by the ancient Greek scholars.

1 *Gain an understanding of some of the general ideas of other kinds of geometries.*

The Geometry of Graphs

In the early 1700s, the city of Königsberg, Germany, was connected by seven bridges, shown in **Figure 10.69**. Many people in the city were interested in finding if it were possible to walk through the city so as to cross each bridge exactly once. After a few trials, you may be convinced that the answer is no. However, it is not easy to prove your answer by trial and error because there are a large number of ways of taking such a walk.

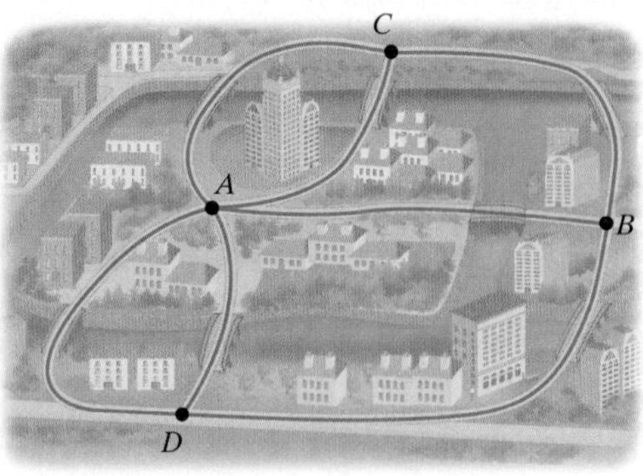

FIGURE 10.69

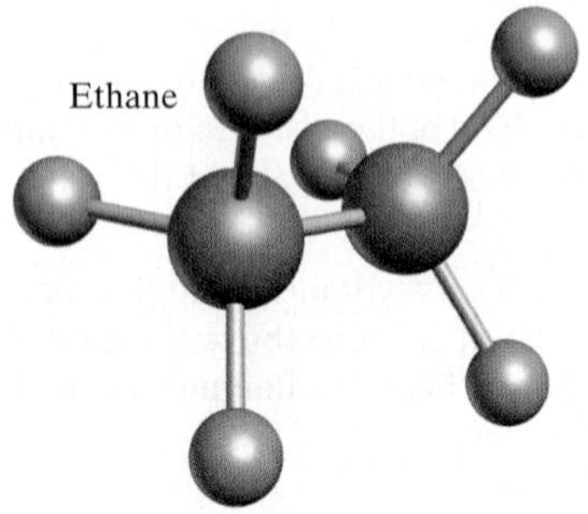

Graph theory is used to show how atoms are linked to form molecules.

The problem was taken to the Swiss mathematician Leonhard Euler (1707–1783). In the year 1736, Euler (pronounced "oil er") proved that it is not possible to stroll through the city and cross each bridge exactly once. His solution of the problem opened up a new kind of geometry called **graph theory**. Graph theory is now used to design city streets, analyze traffic patterns, and find the most efficient routes for public transportation.

Euler solved the problem by introducing the following definitions:

A **vertex** is a point. An **edge** is a line segment or curve that starts and ends at a vertex.

Vertices and edges form a **graph**. A vertex with an odd number of attached edges is an **odd vertex**. A vertex with an even number of attached edges is an **even vertex**.

Using these definitions, **Figure 10.70** shows that the graph in the Königsberg bridge problem (**Figure 10.69**) has four odd vertices.

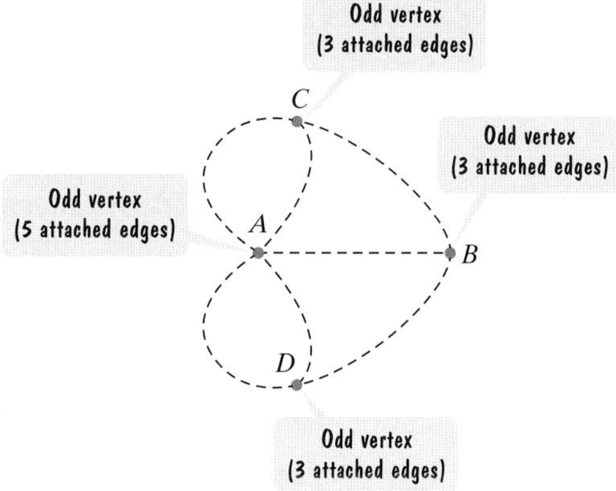

FIGURE 10.70

A graph is **traversable** if it can be traced without lifting the pencil from the paper and without tracing an edge more than once. Euler proved the following rules of traversability in solving the problem:

RULES OF TRAVERSABILITY

1. A graph with all even vertices is traversable. One can start at any vertex and end where one began.
2. A graph with two odd vertices is traversable. One must start at either of the odd vertices and finish at the other.
3. A graph with more than two odd vertices is not traversable.

Because the graph in the Königsberg bridge problem has four odd vertices, it cannot be traversed.

EXAMPLE 1 *To Traverse or Not to Traverse?*

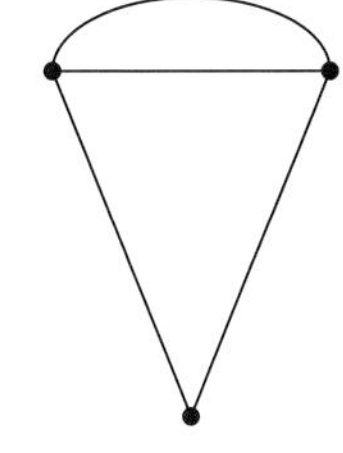

FIGURE 10.71

Consider the graph in **Figure 10.71**.

a. Is this graph traversable?

b. If it is, describe a path that will traverse it.

SOLUTION

a. Begin by determining whether each vertex is even or odd.

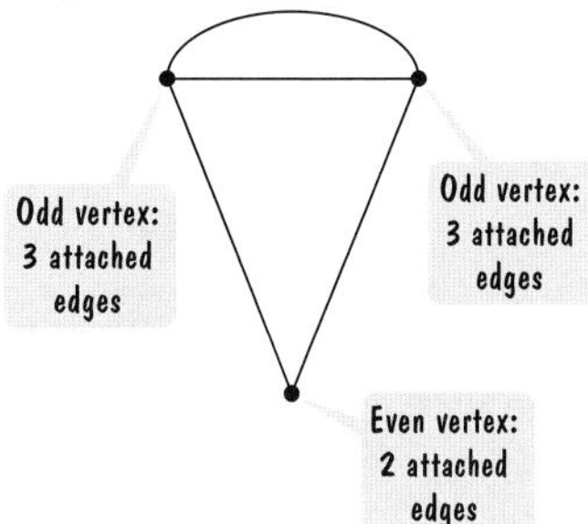

Because this graph has exactly two odd vertices, by Euler's second rule, it is traversable.

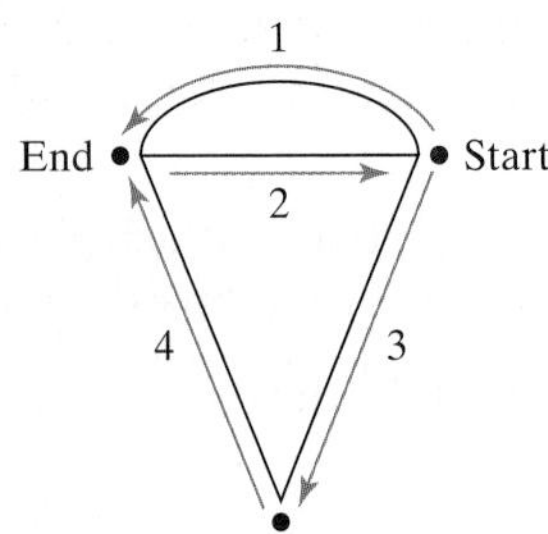

FIGURE 10.72

b. In order to find a path that traverses this graph, we again use Euler's second rule. It states that we must start at either of the odd vertices and finish at the other. One possible path is shown in **Figure 10.72**.

CHECK POINT 1 Create a graph with two even and two odd vertices. Then describe a path that will traverse it.

Graph theory is used to solve many kinds of practical problems. The first step is to create a graph that represents, or models, the problem. For example, consider a UPS driver trying to find the best way to deliver packages around town. By modeling the delivery locations as vertices and roads between locations as edges, graph theory reveals the driver's most efficient path. Graphs are powerful tools because they illustrate the important aspects of a problem and leave out unnecessary detail. A more detailed presentation on the geometry of graphs, called *graph theory*, can be found in Chapter 14.

Topology

A branch of modern geometry called **topology** looks at shapes in a completely new way. In Euclidean geometry, shapes are rigid and unchanging. In topology, shapes can be twisted, stretched, bent, and shrunk. Total flexibility is the rule.

A topologist does not know the difference between a doughnut and a coffee cup! This is because, topologically speaking, there is no difference. They both have one hole, and in this strange geometry of transformations, a doughnut can be flattened, pulled, and pushed to form a coffee cup.

In topology, objects are classified according to the number of holes in them, called their **genus**. The genus gives the largest number of complete cuts that can be made in the object without cutting the object into two pieces. Objects with the same genus are **topologically equivalent**.

GREAT QUESTION!

Can you remind me of the relationship between genus and holes?

The genus of an object is the same as the number of holes in the object.

The three shapes shown below (sphere, cube, irregular blob) have the same genus: 0. No complete cuts can be made without cutting these objects into two pieces.

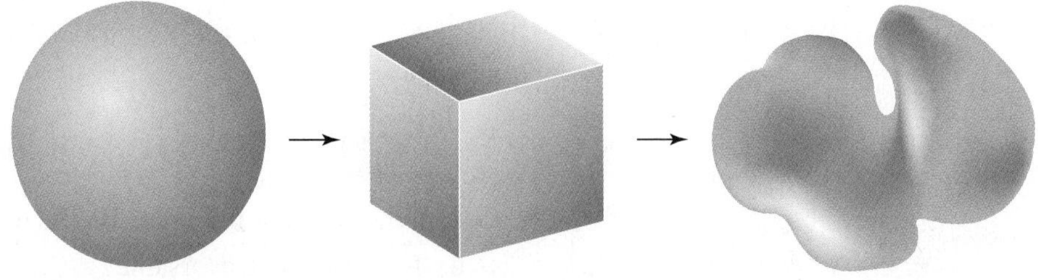

How can a doughnut be transformed into a coffee cup? It can be stretched and deformed to make a bowl part of the surface. Both the doughnut and the coffee cup have genus 1. One complete cut can be made without cutting these objects into two pieces.

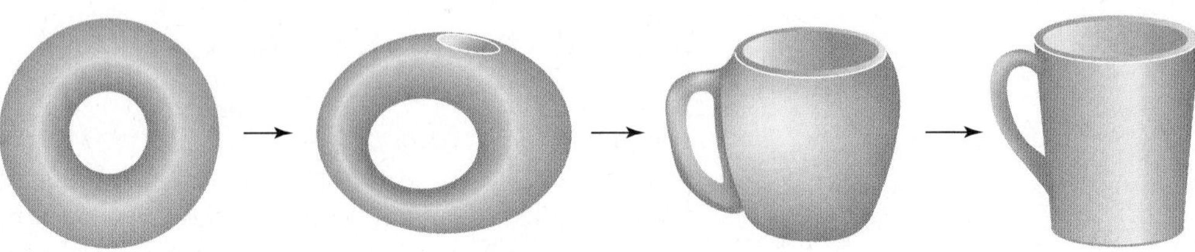

To a topologist, a lump with two holes in it is no different than a sugar bowl with two handles. Both have genus 2. Two complete cuts can be made without cutting these objects into two pieces.

Shown below is a transformation that results in a most unusual topological figure.

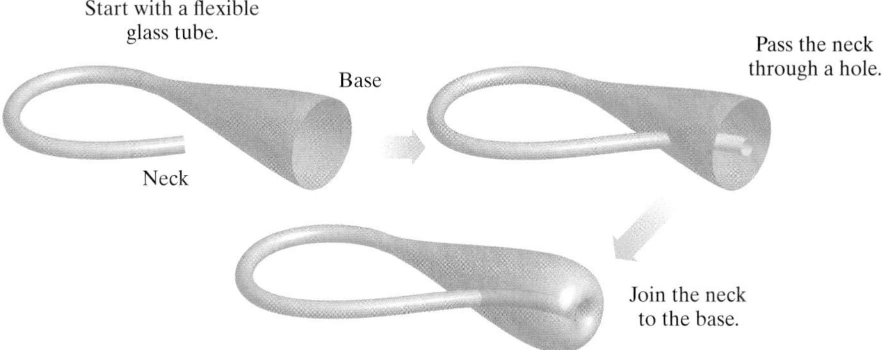

The figure that results is called a **Klein bottle**. You couldn't use one to carry water on a long hike. Because the inside surface loops back on itself to merge with the outside, it has neither an outside nor an inside and cannot hold water. A Klein bottle passes through itself without the existence of a hole, which is impossible in three-dimensional space. A true Klein bottle is visible only when generated on a computer.

Topology is more than an excursion in geometric fantasy. Topologists study knots and ways in which they are topologically equivalent. Knot theory is used to identify viruses and understand the ways in which they invade our cells. Such an understanding is a first step in finding vaccines for viruses ranging from the common cold to HIV.

Non-Euclidean Geometries

Imagine this: Earth is squeezed and compressed to the size of a golf ball. It becomes a **black hole**, a region in space where everything seems to disappear. Its pull of gravity is now so strong that the space near the golf ball can be thought of as a funnel with the black hole sitting at the bottom. Any object that comes near the funnel will be pulled into it by the immense gravity of the golf-ball-sized Earth. Then the object disappears! Although this sounds like science fiction, some physicists believe that there are billions of black holes in space.

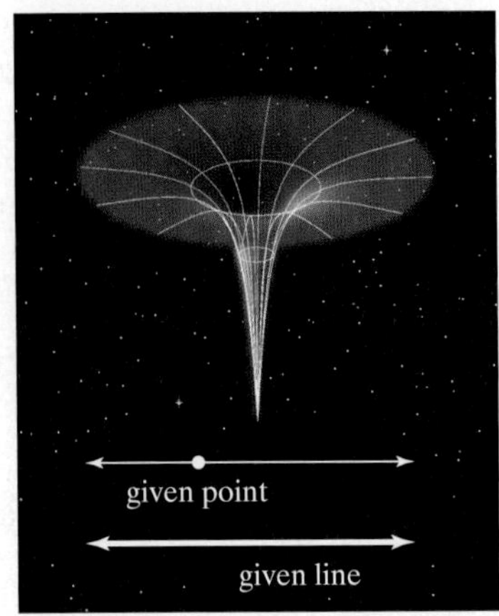

FIGURE 10.73

Is there a geometry that describes this unusual phenomenon? Yes, although it's not the geometry of Euclid. Euclidean geometry is based on the assumption that given a point not on a line, there is one line that can be drawn through the point parallel to the given line, shown as the thinner white line in **Figure 10.73**. We used this assumption to prove that the sum of the measures of the three angles of any triangle is 180°.

Hyperbolic geometry was developed independently by the Russian mathematician Nikolay Lobachevsky (1792–1856) and the Hungarian mathematician Janos Bolyai (1802–1860). It is based on the assumption that given a point not on a line, there are an infinite number of lines that can be drawn through the point parallel to the given line. The shapes in this non-Euclidean geometry are not represented on a plane. Instead, they are drawn on a funnel-like surface, called a *pseudosphere*, shown in **Figure 10.74**. On such a surface, the shortest distance between two points is a curved line. A triangle's sides are composed of arcs, and the sum of the measures of its angles is less than 180°. The shape of the pseudosphere looks like the distorted space near a black hole.

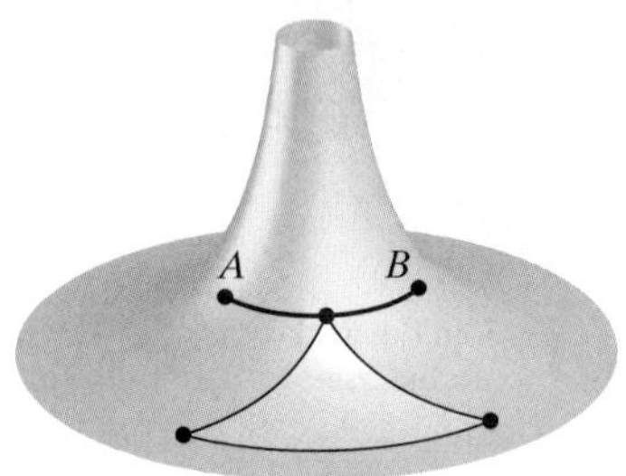

FIGURE 10.74

Once the ice had been broken by Lobachevsky and Bolyai, mathematicians were stimulated to set up other non-Euclidean geometries. The best known of these was **elliptic geometry**, proposed by the German mathematician Bernhard Riemann (1826–1866). Riemann began his geometry with the assumption that there are no parallel lines. As shown in **Figure 10.75**, elliptic geometry is on a sphere, and the sum of the measures of the angles of a triangle is greater than 180°.

Riemann's elliptic geometry was used by Albert Einstein in his theory of the universe. One aspect of the theory states that if you set off on a journey through space and keep going in the same direction, you will eventually come back to your starting point. The same thing happens on the surface of a sphere. This means that space itself is "curved," although the idea is difficult to visualize. It is interesting that Einstein used ideas of non-Euclidean geometry more than 50 years after the system was logically developed. Mathematics often moves ahead of our understanding of the physical world.

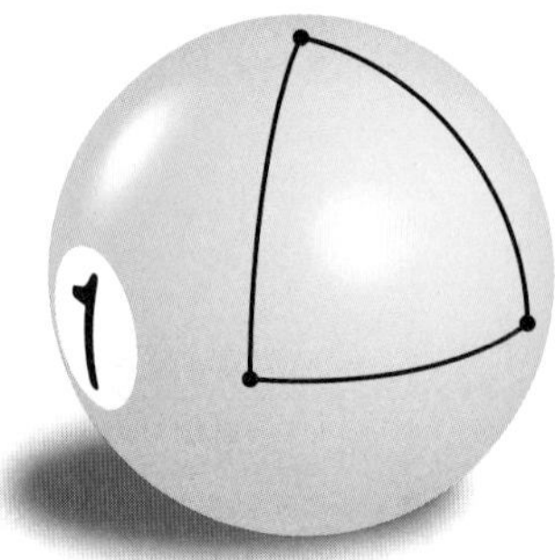

FIGURE 10.75

Blitzer Bonus

Colorful Puzzles

How many different colors are needed to color a plane map so that any two regions that share a common border are colored differently? Even though no map could be found that required more than four colors, a proof that four colors would work for every map remained elusive to mathematicians for over 100 years. The conjecture that at most four colors are needed, stated in 1852, was proved by American mathematicians Kenneth Appel and Wolfgang Haken of the University of Illinois in 1976. Their proof translated maps into graphs that represented each region as a vertex and each shared common border as an edge between corresponding vertices. The proof relied heavily on computer computations that checked approximately 1500 special cases, requiring 1200 hours of computer time. Although there are recent rumors about a flaw in the proof, no one has found a proof that does not enlist the use of computer calculations.

If you take a strip of paper, give it a single half-twist, and paste the ends together, the result is a one-sided surface called a Möbius strip. A map on a Möbius strip requires six colors to make sure that no adjacent areas are the same color.

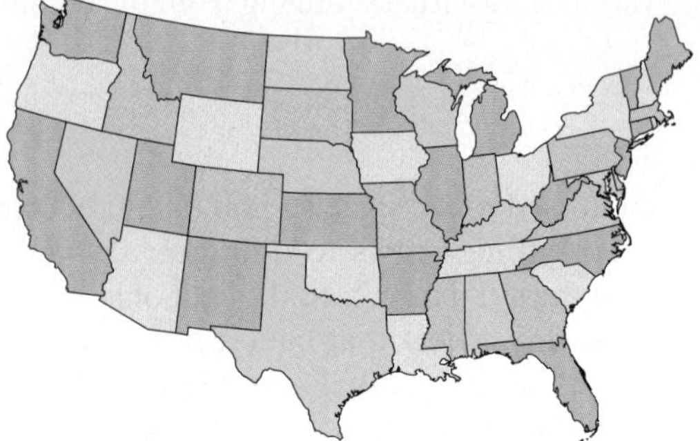

At most four colors are needed on a plane map.

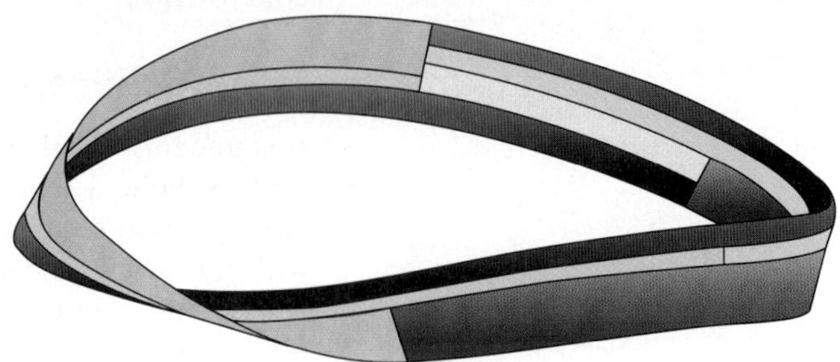

At most six colors are needed on a map drawn on a Möbius strip.

We have seen that Dutch artist M. C. Escher is known for combining art and geometry. Many of his images are based on non-Euclidean geometry. In the woodcut shown below, things enlarge as they approach the center and shrink proportionally as they approach the boundary. The shortest distance between two points is a curved line, and the triangles that are formed look like those from hyperbolic geometry.

Circle Limit III (1959). M.C. Escher's "Circle Limit III" © 2017 The M.C. Escher Company-The Netherlands. All rights reserved. www.mcescher.com

Table 10.5 compares Euclidean geometry with hyperbolic and elliptic geometry.

TABLE 10.5 Comparing Three Systems of Geometry

Euclidean Geometry **Euclid (300 B.C.)**	**Hyperbolic Geometry** **Lobachevsky, Bolyai (1830)**	**Elliptic Geometry** **Riemann (1850)**
Given a point not on a line, there is one and only one line through the point parallel to the given line.	Given a point not on a line, there are an infinite number of lines through the point that do not intersect the given line.	There are no parallel lines.
Geometry is on a plane:	Geometry is on a pseudosphere:	Geometry is on a sphere:
The sum of the measures of the angles of a triangle is 180°.	The sum of the measures of the angles of a triangle is less than 180°.	The sum of the measures of the angles of a triangle is greater than 180°.

Fractal Geometry

How can geometry describe nature's complexity in objects such as ferns, coastlines, or the human circulatory system? Any magnified portion of a fern repeats much of the pattern of the whole fern, as well as new and unexpected patterns. This property is called **self-similarity**.

Through the use of computers, a new geometry of natural shapes, called **fractal geometry**, has arrived. The word *fractal* is from the Latin word *fractus*, meaning "broken up," or "fragmented." The fractal shape shown in **Figure 10.76** is, indeed, broken up. This shape was obtained by repeatedly subtracting triangles from within triangles. From the midpoints of the sides of triangle ABC, remove triangle $A'B'C'$. Repeat the process for the three remaining triangles, then with each of the triangles left after that, and so on ad infinitum.

FIGURE 10.76

The process of repeating a rule again and again to create a self-similar fractal like the broken-up triangle is called **iteration**. The self-similar fractal shape in **Figure 10.77** was generated on a computer by iteration that involved subtracting pyramids within pyramids. Computers can generate fractal shapes by carrying out thousands or millions of iterations.

FIGURE 10.77 Daryl H. Hepting and Allan N. Snider, "Desktop Tetrahedron" 1990. Computer generated at the University of Regina in Canada.

In 1975, Benoit Mandelbrot (1924–2010), the mathematician who developed fractal geometry, published *The Fractal Geometry of Nature*. This book of beautiful computer graphics shows images that look like actual mountains, coastlines, underwater coral gardens, and flowers, all imitating nature's forms. Many of the realistic-looking landscapes that you see in movies are actually computer-generated fractal images.

These images are computer-generated fractals.

Concept and Vocabulary Check

Fill in each blank so that the resulting statement is true.

1. In the geometry of graphs, a point is called a/an ______. A line segment or curve that starts and ends at a point is called a/an ______. A set of points connected by line segments or curves is called a/an ______.

2. In the geometry of graphs, if you can trace a graph without lifting your pencil from the paper and without tracing an edge more than once, then the graph is called __________.

3. In topology, objects are classified according to the number of holes in them, called their ______.

4. In Euclidean geometry, a basic assumption states that given a line and a point not on the line, one and only one line may be drawn through the given point ________ to the given line.

5. By changing the basic assumption in Exercise 4, we obtain ____________ geometries. In one of these geometries, called elliptic geometry, there are no ________ lines.

6. In fractal geometry, magnified portions of an object repeat much of the pattern of the whole object, as well as new and unexpected patterns. This property is called ____________. The process of repeating a rule again and again to create such a fractal image is called ________.

In Exercises 7–10, determine whether each statement is true or false. If the statement is false, make the necessary change(s) to produce a true statement.

7. A graph that contains exactly three vertices, each of which is even, is traversable. ______

8. A graph that contains exactly three vertices, each of which is odd, is traversable. ______

9. A doughnut and a compact disc both have genus 1 and are topologically equivalent. ______

10. In non-Euclidean geometries, the sum of the measures of the angles of a triangle is not necessarily 180°. ______

Exercise Set 10.7

Practice and Application Exercises

For each graph in Exercises 1–6, **a.** *Is the graph traversable?* **b.** *If it is, describe a path that will traverse it.*

1.

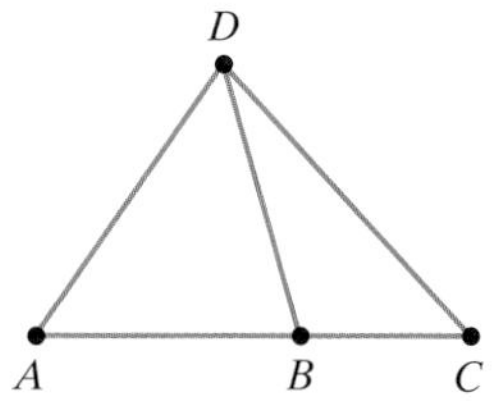

2.

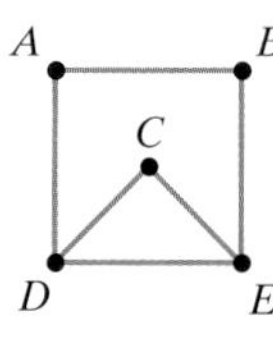

3.

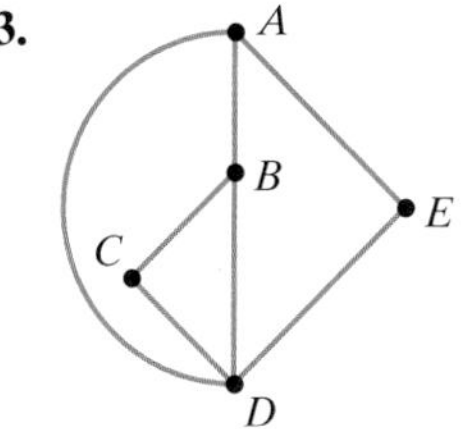

4.

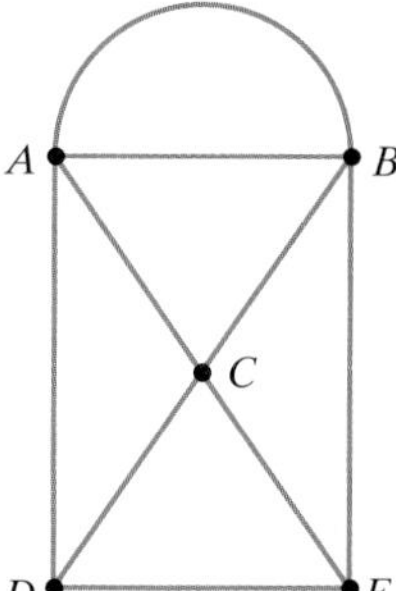

5.

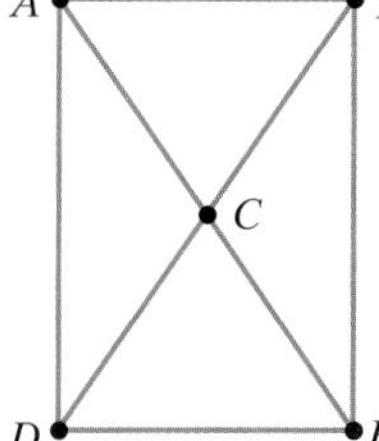

6.

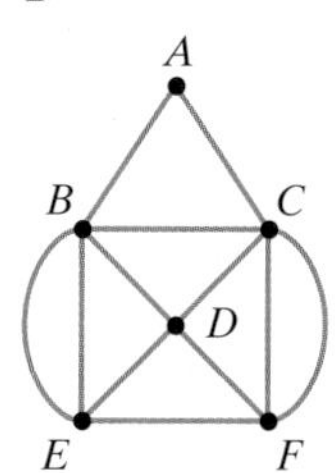

The model shows the way that carbon atoms (red) and hydrogen atoms (blue) link together to form a propane molecule. Use the model to answer Exercises 7–9.

Propane

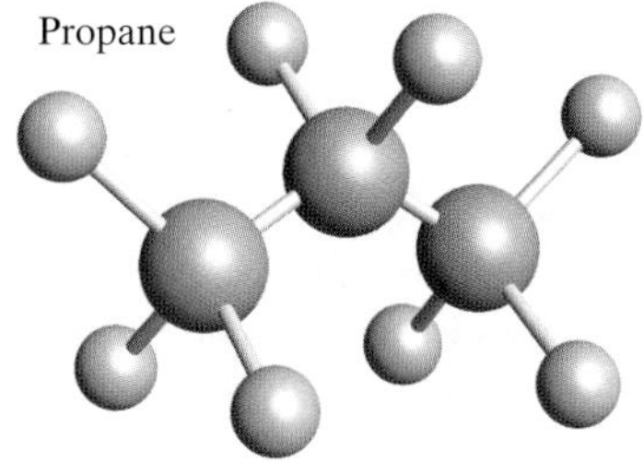

7. Using each of the three red carbon atoms as a vertex and each of the eight blue hydrogen atoms as a vertex, draw a graph for the propane molecule.

8. Determine if each vertex of the graph is even or odd.

9. Is the graph traversable?

The figure below on the left shows the floor plan of a four-room house. By representing rooms as vertices, the outside, E, as a vertex, and doors as edges, the figure below on the right is a graph that models the floor plan. Use these figures to solve Exercises 10–13.

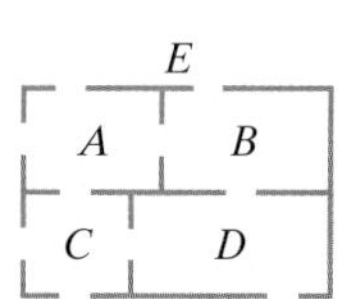

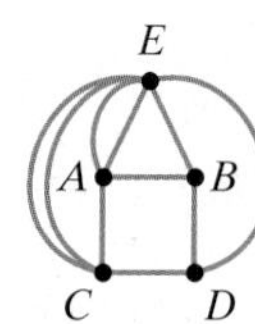

10. Use the floor plan to determine how many doors connect the outside, E, to room A. How is this shown in the graph?

11. Use the floor plan to determine how many doors connect the outside, E, to room C. How is this shown in the graph?

12. Is the graph traversable? Explain your answer.

13. If the graph is traversable, determine a path to walk through every room, using each door only once.

In Exercises 14–17, give the genus of each object.

14.

Pretzel

15.

Pitcher

16.

Wrench

17.

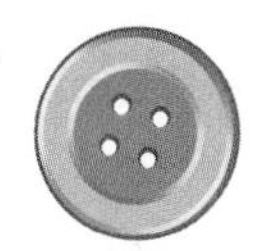

Button

18. In Exercises 14–17, which objects are topologically equivalent?

19. Draw (or find and describe) an object of genus 4 or more.

20. What does the figure shown illustrate about angles in a triangle when the triangle is drawn on a sphere?

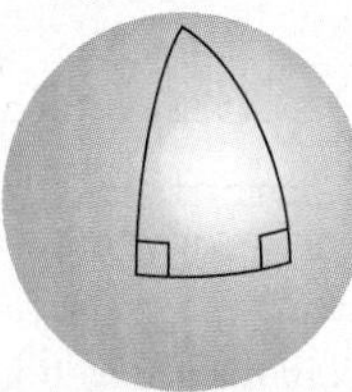

21. The figure shows a quadrilateral that was drawn on the surface of a sphere. Angles C and D are obtuse angles. What does the figure illustrate about the sum of the measures of the four angles in a quadrilateral in elliptic geometry?

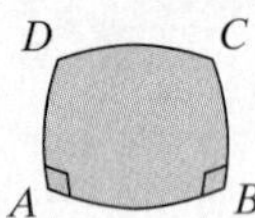

In Exercises 22–24, describe whether each figure shown exhibits self-similarity.

22.

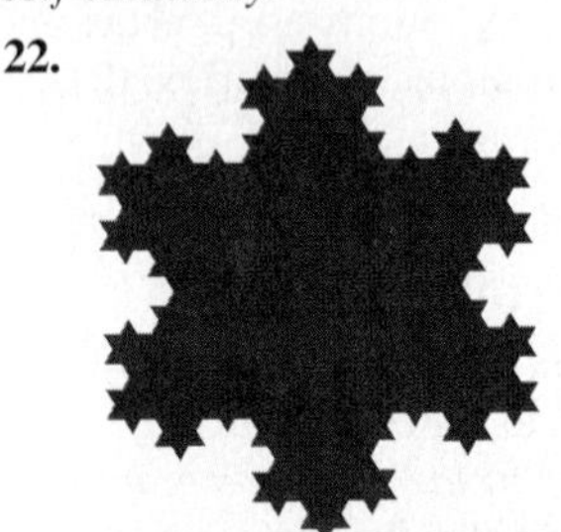

23.

24.

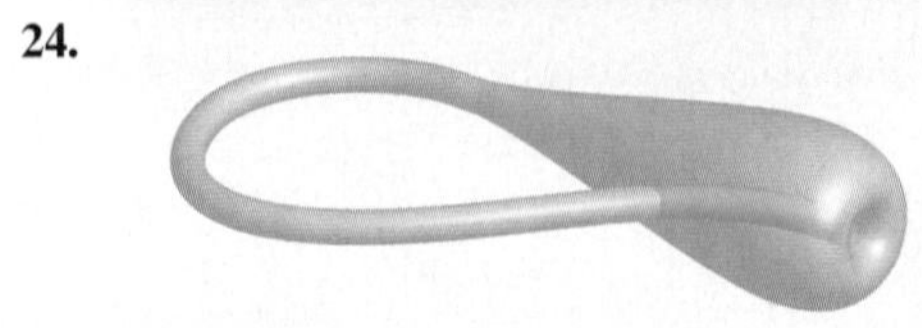

25. Consult an Internet site devoted to fractals and download two fractal images that imitate nature's forms. How is self-similarity shown in each image?

Explaining the Concepts

26. What is a graph?

27. What does it mean if a graph is traversable?

28. How do you determine whether or not a graph is traversable?

29. Describe one way in which topology is different than Euclidean geometry.

30. What is the genus of an object?

31. Describe how a rectangular solid and a sphere are topologically equivalent.

32. State the assumption that Euclid made about parallel lines that was altered in both hyperbolic and elliptic geometry.

33. How does hyperbolic geometry differ from Euclidean geometry?

34. How does elliptic geometry differ from Euclidean geometry?

35. What is self-similarity? Describe an object in nature that has this characteristic.

36. Some suggest that nature produces its many forms through a combination of both iteration and a touch of randomness. Describe what this means.

37. Find an Internet site devoted to fractals. Use the site to write a paper on a specific use of fractals.

38. Did you know that laid end to end, the veins, arteries, and capillaries of your body would reach over 40,000 miles? However, your vascular system occupies a very small fraction of your body's volume. Describe a self-similar object in nature, such as your vascular system, with enormous length but relatively small volume.

39. Describe a difference between the shapes of man-made objects and objects that occur in nature.

40. Explain what this short poem by Jonathan Swift has to do with fractal images:

So Nat'ralists observe, A Flea
Hath Smaller Fleas that on him prey
and these have smaller Fleas to bite'em
And so proceed, ad infinitum

41. In Tom Stoppard's play *Arcadia*, the characters talk about mathematics, including ideas from fractal geometry. Read the play and write a paper on one of the ideas that the characters discuss that is related to something presented in this section.

Critical Thinking Exercises

Make Sense? *In Exercises 42–45, determine whether each statement makes sense or does not make sense, and explain your reasoning.*

42. The most effective way to determine if a graph is traversable is by trial and error.

43. Objects that appear to be quite different can be topologically equivalent.

44. Non-Euclidean geometries can be used to prove that the sum of the measures of a triangle's angles is 180°.

45. Euclidean geometry serves as an excellent model for describing the diverse forms that arise in nature.

Group Exercises

46. This activity is suggested for two or three people. Research some of the practical applications of topology. Present the results of your research in a seminar to the entire class. You may also want to include a discussion of some of topology's more unusal figures, such as the Klein bottle and the Möbius strip.

47. Research non-Euclidean geometry and plan a seminar based on your group's research. Each group member should research one of the following five areas:

a. Present an overview of the history of the people who developed non-Euclidean geometry. Who first used the term and why did he never publish his work?

b. Present an overview of the connection between Saccheri quadrilaterals and non-Euclidean geometry. Describe the work of Girolamo Saccheri.

c. Describe how Albert Einstein applied the ideas of Gauss and Riemann. Discuss the notion of curved space and a fourth dimension.

d. Present examples of the work of M. C. Escher that provide ways of visualizing hyperbolic and elliptic geometry.

e. Describe how non-Euclidean geometry changed the direction of subsequent research in mathematics.

After all research has been completed, the group should plan the order in which each group member will speak. Each person should plan on taking about five minutes for his or her portion of the presentation.

48. Albert Einstein's theory of general relativity is concerned with the structure, or the geometry, of the universe. In order to describe the universe, Einstein discovered that he needed four variables: three variables to locate an object in space and a fourth variable describing time. This system is known as space-time.

Because we are three-dimensional beings, how can we imagine four dimensions? One interesting approach to visualizing four dimensions is to consider an analogy of a two-dimensional being struggling to understand three dimensions. This approach first appeared in a book called *Flatland* by Edwin Abbott, written around 1884.

Flatland describes an entire civilization of beings who are two dimensional, living on a flat plane, unaware of the existence of anything outside their universe. A house in Flatland would look like a blueprint or a line drawing to us. If we were to draw a closed circle around Flatlanders, they would be imprisoned in a cell with no way to see out or escape because there is no way to move up and over the circle. For a two-dimensional being moving only on a plane, the idea of up would be incomprehensible. We could explain that up means moving in a new direction, perpendicular to the two dimensions they know, but it would be similar to telling us we can move in the fourth dimension by traveling perpendicular to our three dimensions.

Group members should obtain copies of or excerpts from Edwin Abbott's *Flatland*. We especially recommend *The Annotated Flatland*, Perseus Publishing, 2002, with fascinating commentary by mathematician and author Ian Stewart. Once all group members have read the story, the following questions are offered for group discussion.

a. How does the sphere, the visitor from the third dimension, reflect the same narrow perspective as the Flatlanders?

b. What are some of the sociological problems raised in the story?

c. What happens when we have a certain way of seeing the world that is challenged by coming into contact with something quite different? Be as specific as possible, citing either personal examples or historical examples.

d. How are A. Square's difficulties in visualizing three dimensions similar to those of a three-dimensional dweller trying to visualize four dimensions?

e. How does the author reflect the overt sexism of his time?

f. What "upward not northward" ideas do you hold that, if shared, would result in criticism, rejection, or a fate similar to that of the narrator of *Flatland*?

Chapter Summary, Review, and Test

SUMMARY – DEFINITIONS AND CONCEPTS	EXAMPLES
10.1 Points, Lines, Planes, and Angles	
a. Line AB ($\overleftrightarrow{AB}$ or $\overleftrightarrow{BA}$), half-line AB ($\overset{\circ\rightarrow}{AB}$), ray AB ($\overrightarrow{AB}$), and line segment AB ($\overline{AB}$ or $\overline{BA}$) are represented in Figure 10.2 on page 616.	
b. Angles are measured in degrees. A degree, 1°, is $\frac{1}{360}$ of a complete rotation. Acute angles measure less than 90°, right angles 90°, obtuse angles more than 90° but less than 180°, and straight angles 180°.	Ex. 1, p. 617
c. Complementary angles are two angles whose measures have a sum of 90°. Supplementary angles are two angles whose measures have a sum of 180°.	Ex. 2, p. 618; Ex. 3, p. 619
d. Vertical angles have the same measure.	Ex. 4, p. 619
e. If parallel lines are intersected by a transversal, alternate interior angles, alternate exterior angles, and corresponding angles have the same measure.	Ex. 5, p. 621
10.2 Triangles	
a. The sum of the measures of the three angles of any triangle is 180°.	Ex. 1, p. 626; Ex. 2, p. 626
b. Triangles can be classified by angles (acute, right, obtuse) or by sides (isosceles, equilateral, scalene). See the box on page 627.	

c. Similar triangles have the same shape, but not necessarily the same size. Corresponding angles have the same measure and corresponding sides are proportional. If the measures of two angles of one triangle are equal to those of two angles of a second triangle, then the two triangles are similar.	Ex. 3, p. 628; Ex. 4, p. 629
d. The Pythagorean Theorem: The sum of the squares of the lengths of the legs of a right triangle equals the square of the length of the hypotenuse.	Ex. 5, p. 630; Ex. 6, p. 630; Ex. 7, p. 631

10.3 Polygons, Perimeter, and Tessellations

a. A polygon is a closed geometric figure in a plane formed by three or more line segments. Names of some polygons, given in Table 10.2 on page 638, include triangles (three sides), quadrilaterals (four sides), pentagons (five sides), hexagons (six sides), heptagons (seven sides), and octagons (eight sides). A regular polygon is one whose sides are all the same length and whose angles all have the same measure. The perimeter of a polygon is the sum of the lengths of its sides.	Ex. 1, p. 639
b. Types of quadrilaterals, including the parallelogram, rhombus, rectangle, square, and trapezoid, and their characteristics, are given in Table 10.3 on page 638.	
c. The sum of the measures of the angles of an n-sided polygon is $(n - 2)180°$. If the n-sided polygon is a regular polygon, then each angle measures $\frac{(n-2)180°}{n}$.	Ex. 2, p. 640
d. A tessellation is a pattern consisting of the repeated use of the same geometric figures to completely cover a plane, leaving no gaps and having no overlaps. The angle requirement for the formation of a tessellation is that the sum of the measures of the angles at each vertex must be 360°.	Ex. 3, p. 641

10.4 Area and Circumference

a. Formulas for Area Rectangle: $A = lw$; Square: $A = s^2$; Parallelogram: $A = bh$; Triangle: $A = \frac{1}{2}bh$; Trapezoid: $A = \frac{1}{2}h(a + b)$	Ex. 1, p. 646; Ex. 2, p. 647; Ex. 3, p. 648; Ex. 4, p. 649; Ex. 5, p. 650
b. Circles Circumference: $C = 2\pi r$ or $C = \pi d$ Area: $A = \pi r^2$	Ex. 6, p. 651; Ex. 7, p. 651; Ex. 8, p. 652;

10.5 Volume and Surface Area

a. Formulas for Volume Rectangular Solid: $V = lwh$; Cube: $V = s^3$; Pyramid: $V = \frac{1}{3}Bh$; Cylinder: $V = \pi r^2 h$; Cone: $V = \frac{1}{3}\pi r^2 h$ Sphere: $V = \frac{4}{3}\pi r^3$	Ex. 1, p. 658; Ex. 2, p. 658; Ex. 3, p. 659; Ex. 4, p. 660; Ex. 5, p. 661; Ex. 6, p. 661
b. Formulas for surface area are given in Table 10.4 on page 662.	Ex. 7, p. 662

10.6 Right Triangle Trigonometry

a. Trigonometric ratios, sin A, cos A, and tan A, are defined in the box on page 666.	Ex. 1, p. 667
b. Given one side and the measure of an acute angle of a right triangle, the trigonometric ratios can be used to find the measures of the other parts of the right triangle.	Ex. 2, p. 668; Ex. 3, p. 669; Ex. 4, p. 670
c. Given the measures of two sides of a right triangle, the measures of the acute angles can be found using the inverse trigonometric ratio keys $\boxed{\text{SIN}^{-1}}$, $\boxed{\text{COS}^{-1}}$, and $\boxed{\text{TAN}^{-1}}$, on a calculator.	Ex. 5, p. 670

10.7 Beyond Euclidean Geometry

a. A vertex is a point. An edge is a line segment or curve that starts and ends at a vertex. Vertices and edges form a graph. A vertex with an odd number of edges is odd; one with an even number of edges is even. A traversable graph is one that can be traced without removing the pencil from the paper and without tracing an edge more than once. A graph with all even vertices or one with exactly two odd vertices is traversable. With more than two odd vertices, the graph cannot be traversed.	Ex. 1, p. 677

b.	In topology, shapes are twisted, stretched, bent, and shrunk. The genus of an object refers to the number of holes in the object. Objects with the same genus are topologically equivalent.
c.	Non-Euclidean geometries, including hyperbolic geometry and elliptic geometry, are outlined in **Table 10.5** on page 681.
d.	Fractal geometry includes self-similar forms; magnified portions of such forms repeat much of the pattern of the whole form.

Review Exercises

10.1

In the figure shown, lines l and m are parallel. In Exercises 1–7, match each term with the numbered angle or angles in the figure.

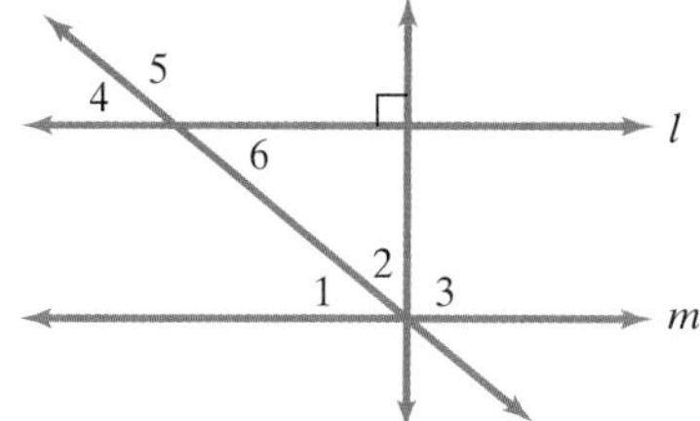

1. right angle
2. obtuse angle
3. vertical angles
4. alternate interior angles
5. corresponding angles
6. the complement of ∡1
7. the supplement of ∡6

In Exercises 8–9, find the measure of the angle in which a question mark with a degree symbol appears.

8.

115°
?°

9.

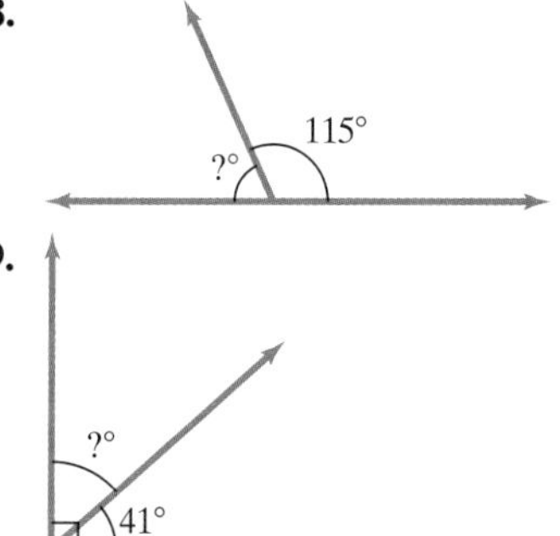

10. If an angle measures 73°, find the measure of its complement.

11. If an angle measures 46°, find the measure of its supplement.

12. In the figure shown, find the measures of angles 1, 2, and 3.

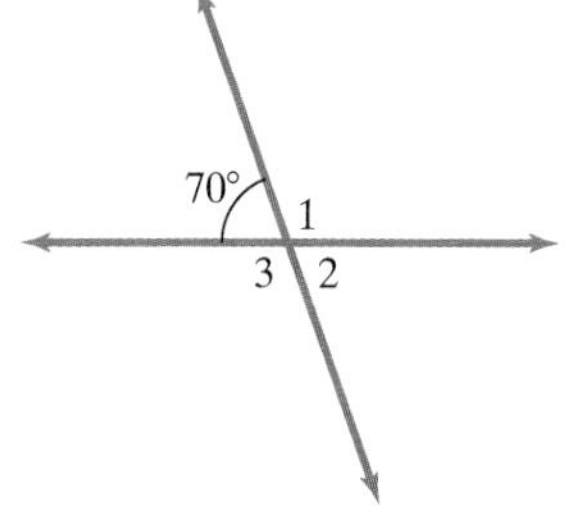

13. In the figure shown, two parallel lines are intersected by a transversal. One of the angle measures is given. Find the measure of each of the other seven angles.

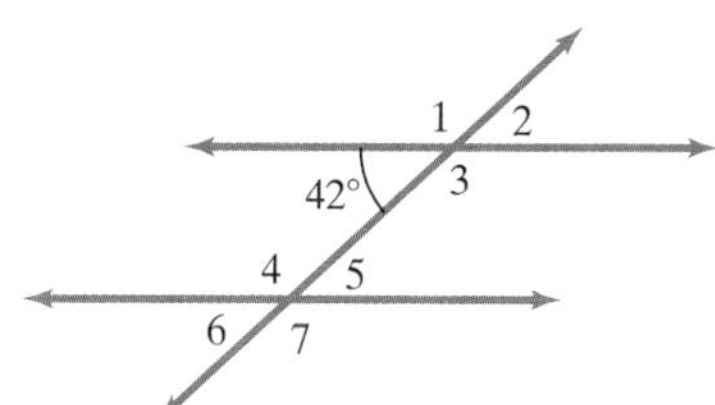

10.2

In Exercises 14–15, find the measure of angle A for the triangle shown.

14. B 60° A 48° C

15.

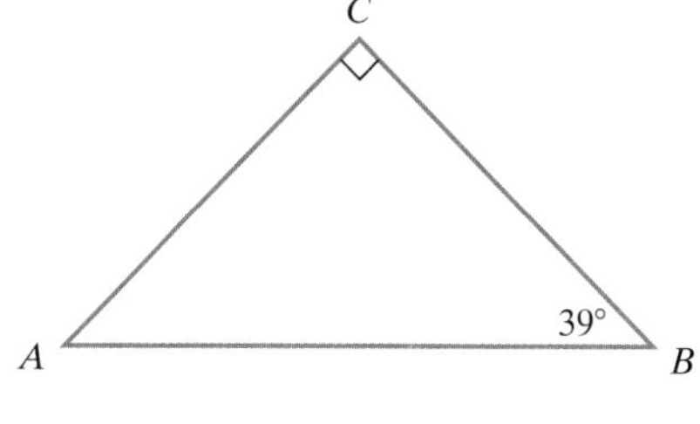

16. Find the measures of angles 1 through 5 in the figure shown.

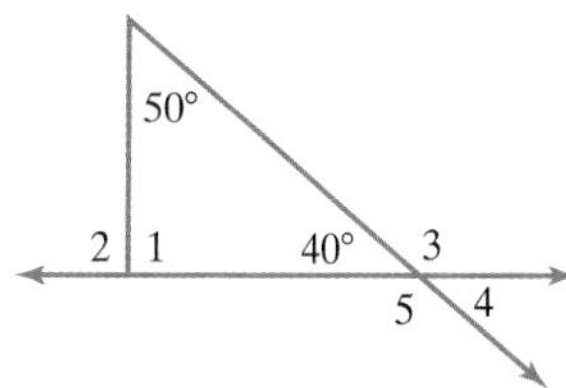

17. In the figure shown, lines l and m are parallel. Find the measure of each numbered angle.

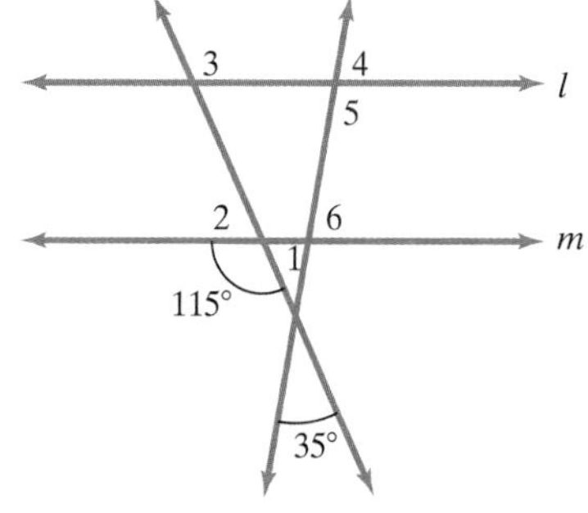

In Exercises 18–19, use similar triangles and the fact that corresponding sides are proportional to find the length of each side marked with an x.

18.

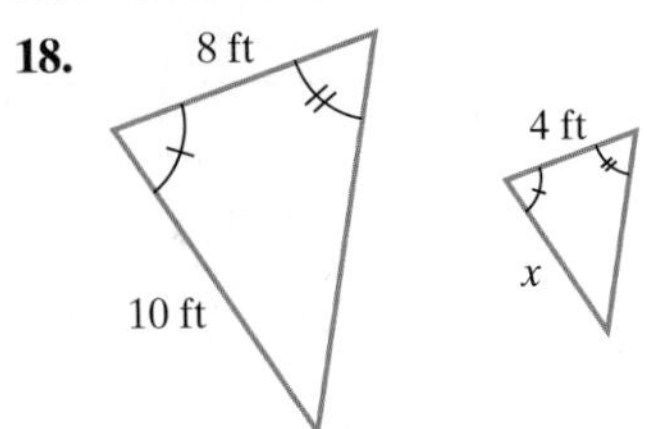

19.

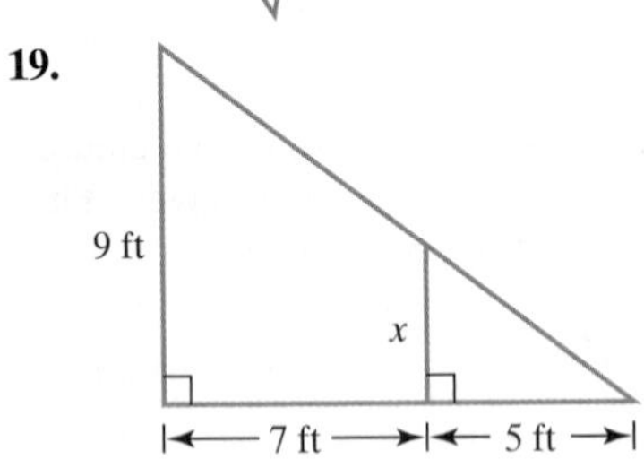

In Exercises 20–22, use the Pythagorean Theorem to find the missing length in each right triangle. Round, if necessary, to the nearest tenth.

20.

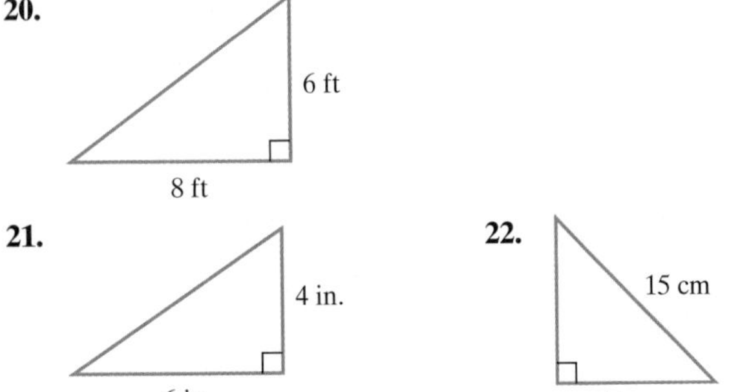

21. 4 in. 6 in.

22. 15 cm 11 cm

23. Find the height of the lamppost in the figure.

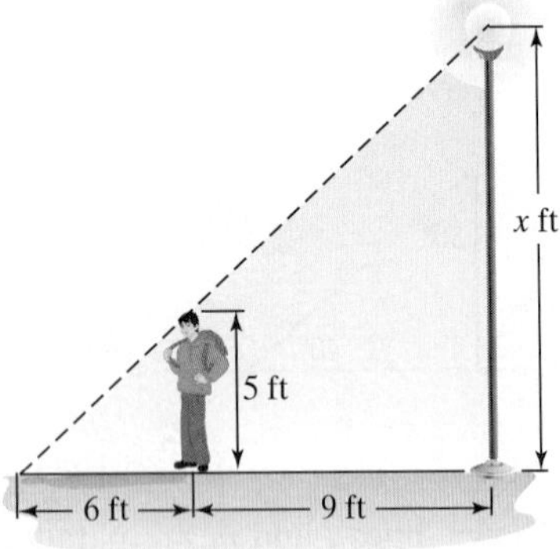

24. How far away from the building in the figure shown is the bottom of the ladder?

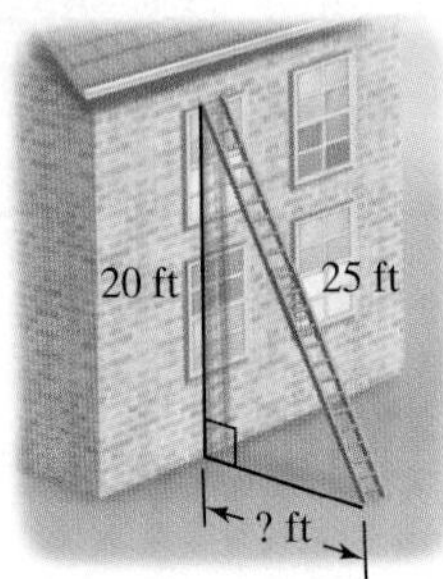

25. A vertical pole is to be supported by three wires. Each wire is 13 yards long and is anchored 5 yards from the base of the pole. How far up the pole will the wires be attached?

10.3

26. Write the names of all quadrilaterals that always have four right angles.

27. Write the names of all quadrilaterals with four sides always having the same measure.

28. Write the names of all quadrilaterals that do not always have four angles with the same measure.

In Exercises 29–31, find the perimeter of the figure shown. Express the perimeter using the same unit of measure that appears in the figure.

29.

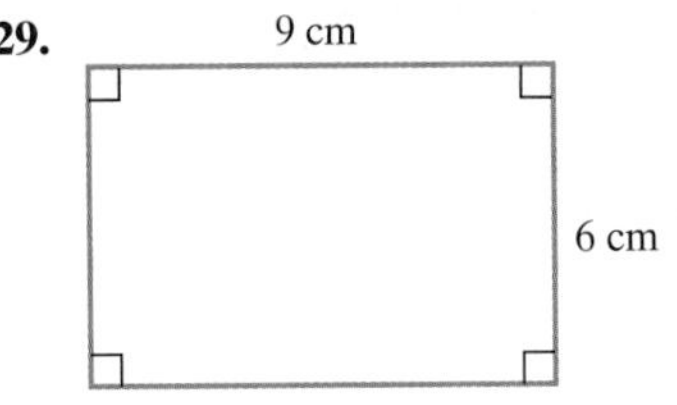

30.

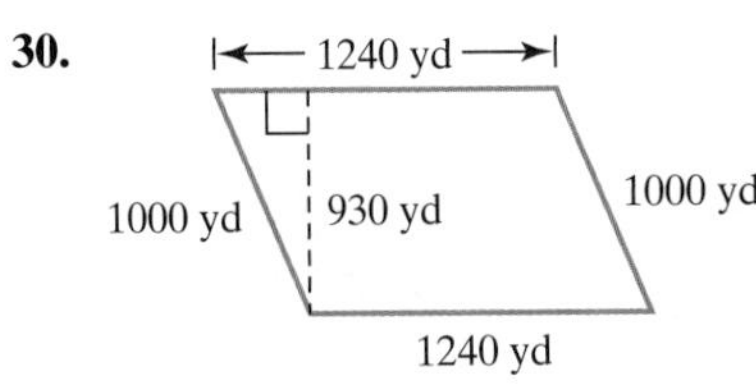

31.

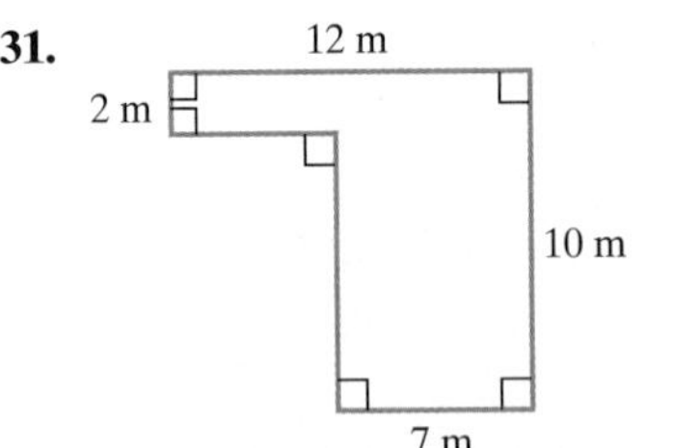

32. Find the sum of the measures of the angles of a 12-sided polygon.

33. Find the sum of the measures of the angles of an octagon.

34. The figure shown is a regular polygon. Find the measures of angle 1 and angle 2.

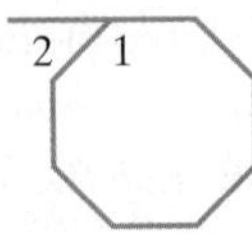

35. A carpenter is installing a baseboard around a room that has a length of 35 feet and a width of 15 feet. The room has four doorways and each doorway is 3 feet wide. If no baseboard is to be put across the doorways and the cost of the baseboard is $1.50 per foot, what is the cost of installing the baseboard around the room?

36. Use the following tessellation to solve this exercise.

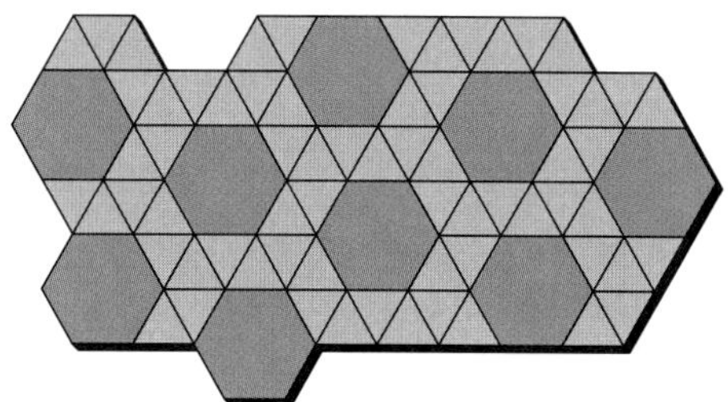

a. Name the types of regular polygons that surround each vertex.

b. Determine the number of angles that come together at each vertex, as well as the measures of these angles.

c. Use the angle measures from part (b) to explain why the tessellation is possible.

37. Can a tessellation be created using only regular hexagons? Explain your answer.

In Exercises 38–41, find the area of each figure.

38.

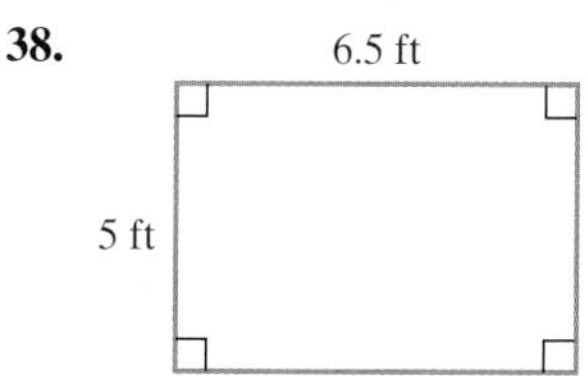

39.

5 m
6 m
4 m
6 m
5 m

40.

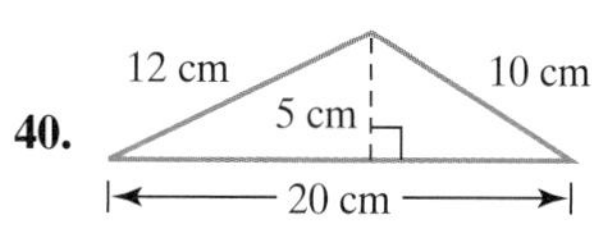

41.

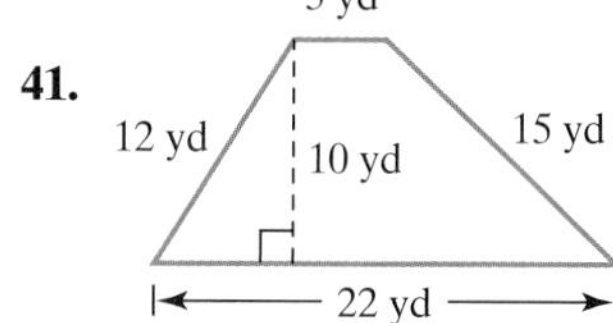

42. Find the circumference and the area of a circle with a diameter of 20 meters. Express answers in terms of π and then round to the nearest tenth.

In Exercises 43–44, find the area of each figure.

43.

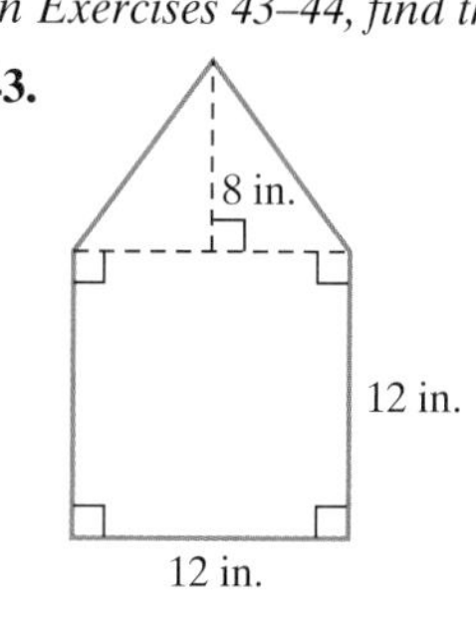

44.

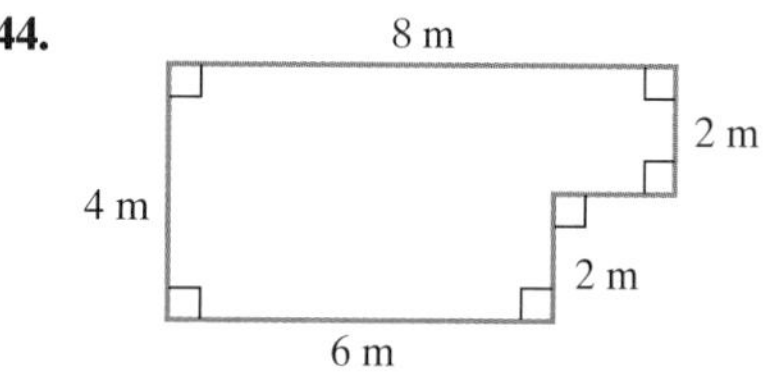

In Exercises 45–46, find the area of each shaded region. Where necessary, express answers in terms of π and then round to the nearest tenth.

45.

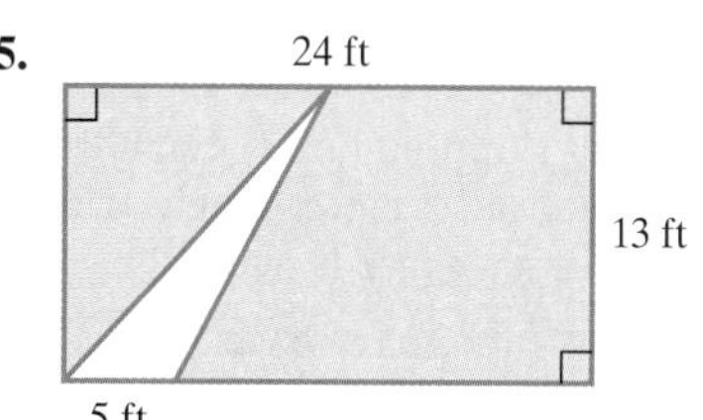

46.

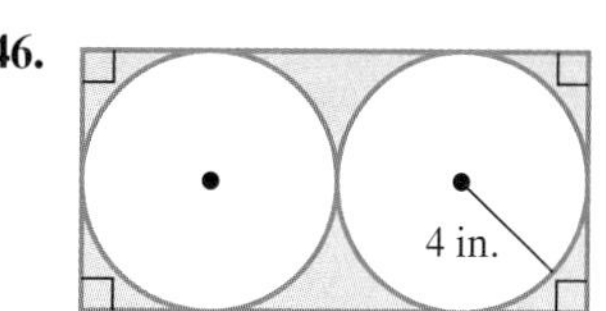

47. What will it cost to carpet a rectangular floor measuring 15 feet by 21 feet if the carpet costs $22.50 per square yard?

48. What will it cost to cover a rectangular floor measuring 40 feet by 50 feet with square tiles that measure 2 feet on each side if a package of 10 tiles costs $13?

49. How much fencing, to the nearest whole yard, is needed to enclose a circular garden that measures 10 yards across?

10.5

In Exercises 50–54, find the volume of each figure. Where necessary, express answers in terms of π and then round to the nearest whole number.

50.

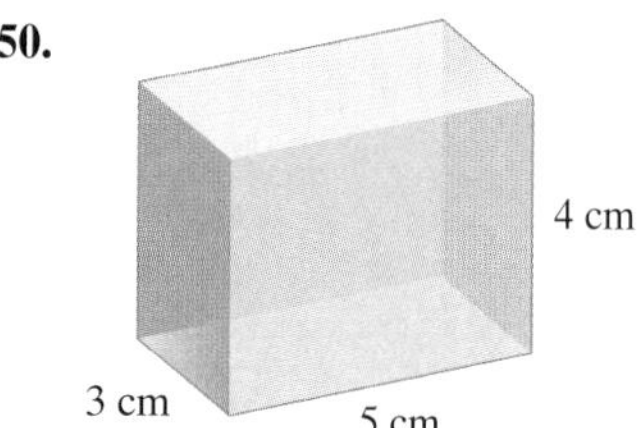

51. **52.**

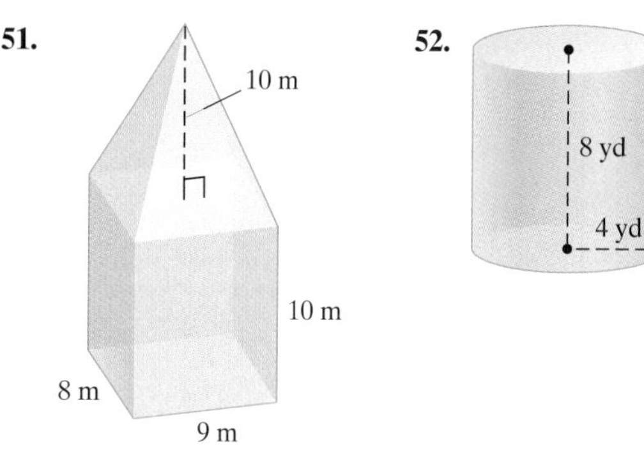

53. **54.**

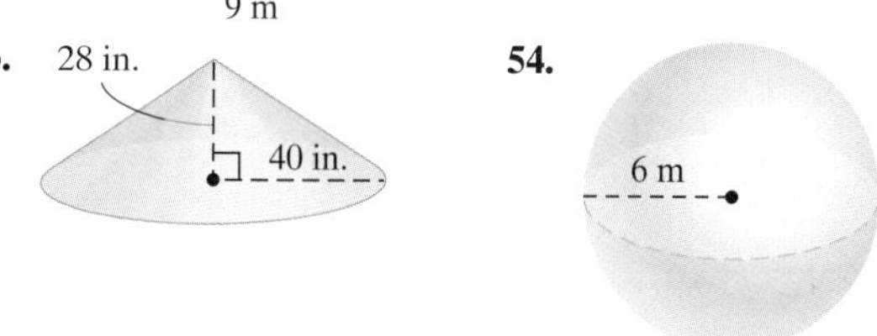

55. Find the surface area of the figure shown.

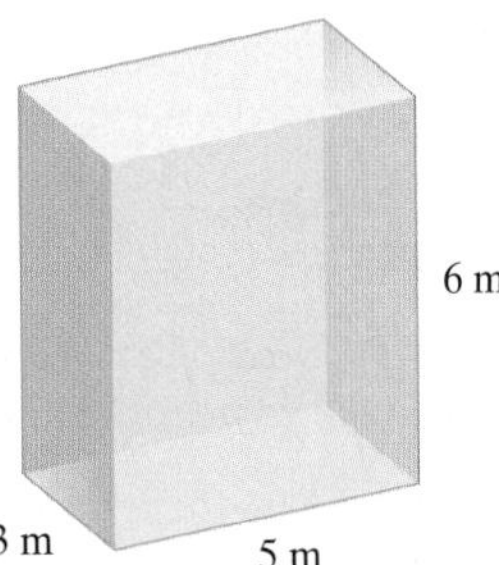

56. A train is being loaded with shipping boxes. Each box is 8 meters long, 4 meters wide, and 3 meters high. If there are 50 shipping boxes, how much space is needed?

57. An Egyptian pyramid has a square base measuring 145 meters on each side. If the height of the pyramid is 93 meters, find its volume.

58. What is the cost of concrete for a walkway that is 27 feet long, 4 feet wide, and 6 inches deep if the concrete is $40 per cubic yard?

10.6

59. Use the right triangle shown to find ratios, in reduced form, for sin A, cos A, and tan A.

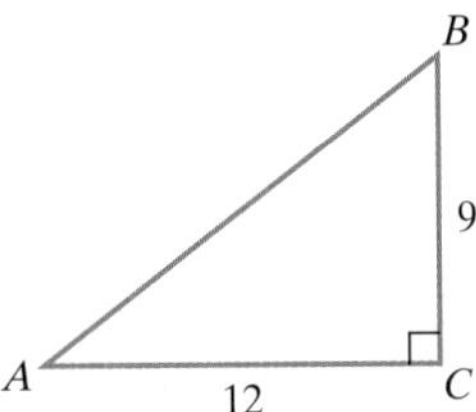

In Exercises 60–62, find the measure of the side of the right triangle whose length is designated by a lowercase letter. Round answers to the nearest whole number.

60.

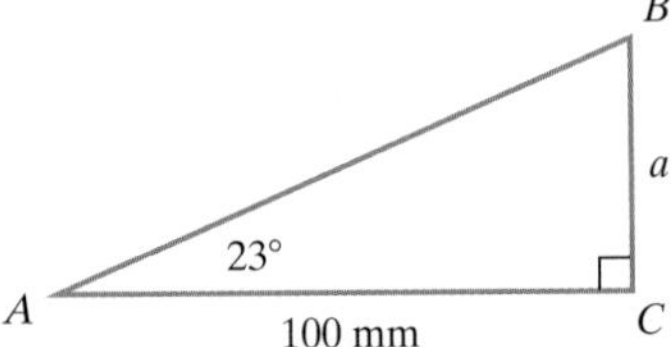

61.

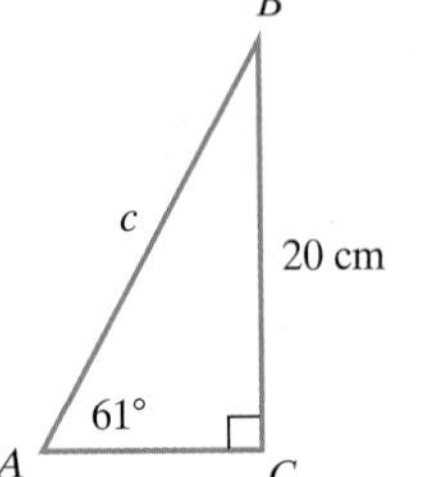

62.

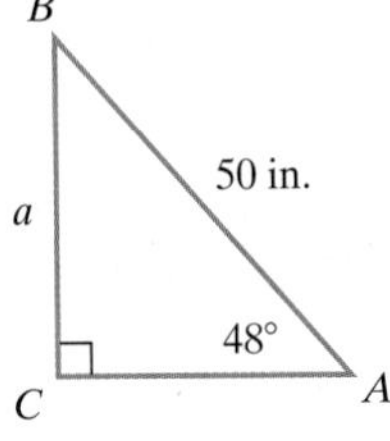

63. Find the measure of angle A in the right triangle shown. Round to the nearest whole degree.

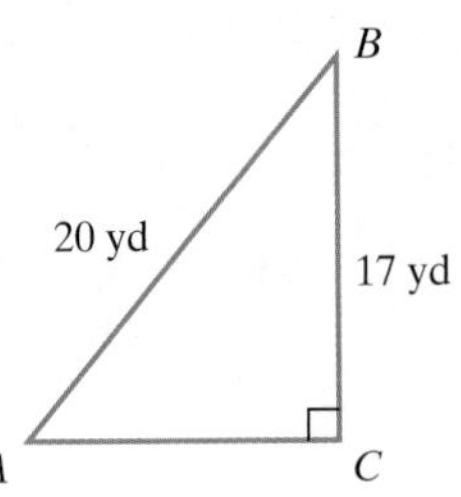

64. A hiker climbs for a half mile (2640 feet) up a slope whose inclination is 17°. How many feet of altitude, to the nearest foot, does the hiker gain?

65. To find the distance across a lake, a surveyor took the measurements in the figure shown. What is the distance across the lake? Round to the nearest meter.

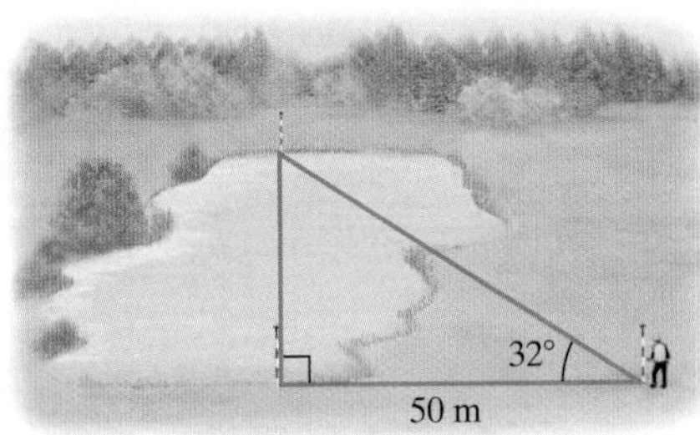

66. When a six-foot pole casts a four-foot shadow, what is the angle of elevation of the Sun? Round to the nearest whole degree.

10.7

For each graph in Exercises 67–68, determine whether it is traversable. If it is, describe a path that will traverse it.

67.

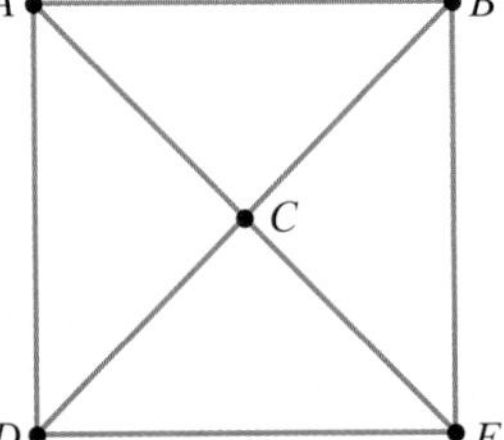

68. A, B, C, D

In Exercises 69–72, give the genus of each object. Which objects are topologically equivalent?

69.

70.

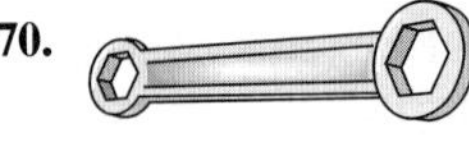

71.

72.

73. State Euclid's assumption about parallel lines that no longer applies in hyperbolic and elliptic geometry.

74. What is self-similarity? Describe an object in nature that has this characteristic.

Chapter 10 Test

1. If an angle measures 54°, find the measure of its complement and supplement.

In Exercises 2–4, use the figure shown to find the measure of angle 1.

2.

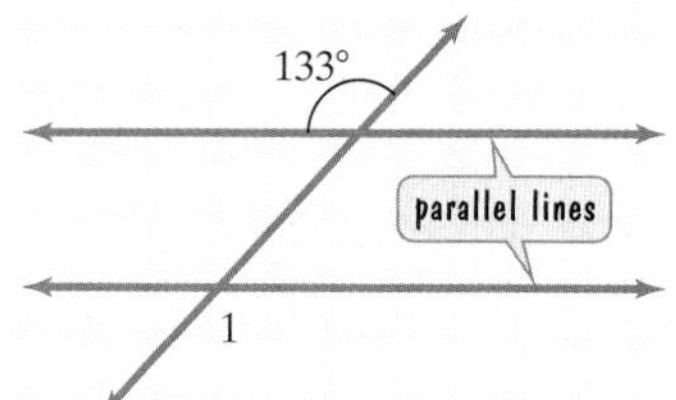

3.

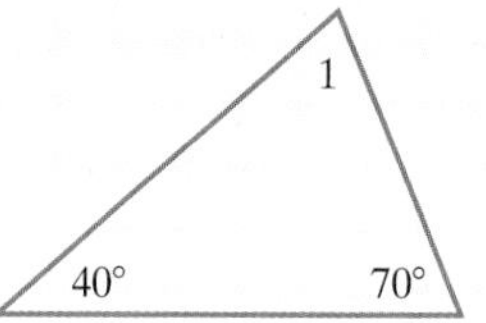

4.

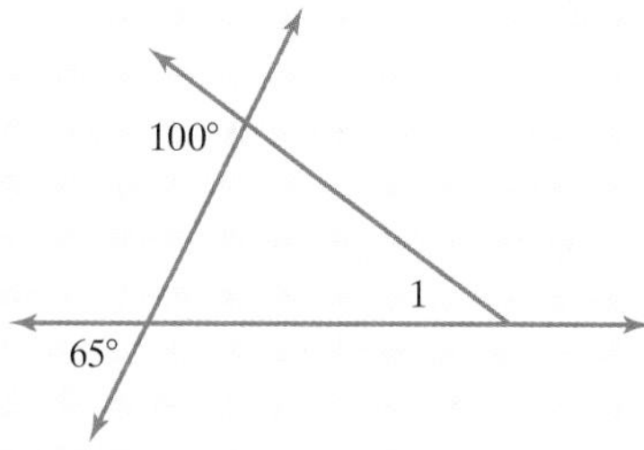

5. The triangles in the figure are similar. Find the length of the side marked with an x.

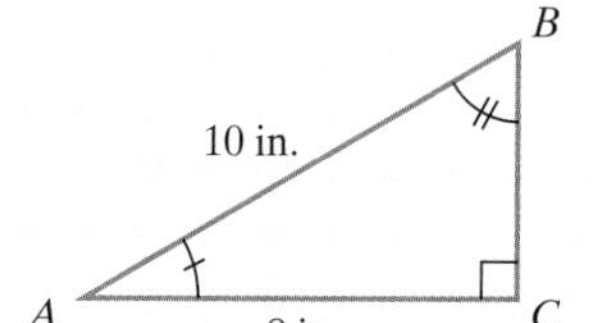

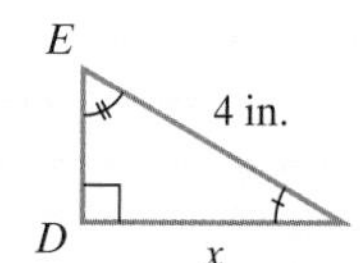

6. A vertical pole is to be supported by three wires. Each wire is 26 feet long and is anchored 24 feet from the base of the pole. How far up the pole should the wires be attached?

7. Find the sum of the measures of the angles of a ten-sided polygon.

8. Find the perimeter of the figure shown.

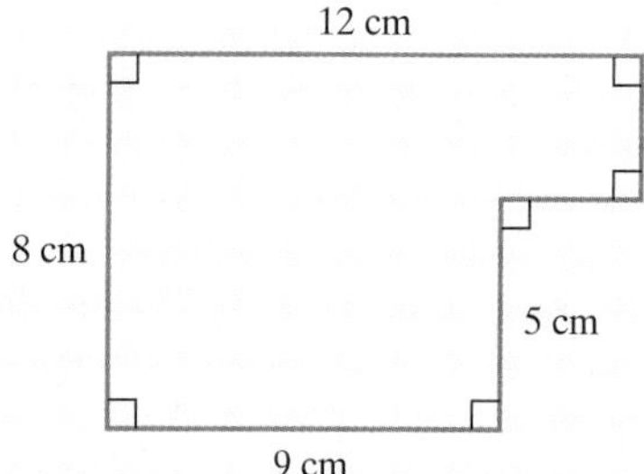

9. Which one of the following names a quadrilateral in which the sides that meet at each vertex have the same measure?

a. rectangle **b.** parallelogram
c. trapezoid **d.** rhombus

10. Use the following tessellation to solve this exercise.

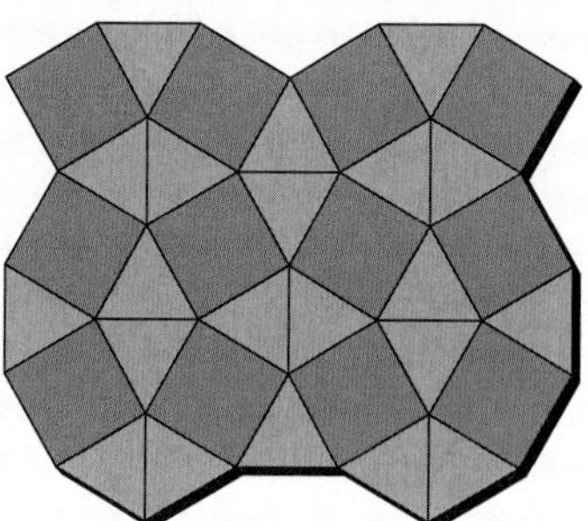

a. Name the types of regular polygons that surround each vertex.

b. Determine the number of angles that come together at each vertex, as well as the measures of these angles.

c. Use the angle measures from part (b) to explain why the tessellation is possible.

In Exercises 11–12, find the area of each figure.

11.

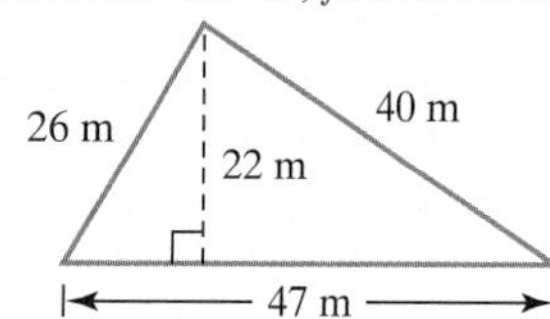

12.

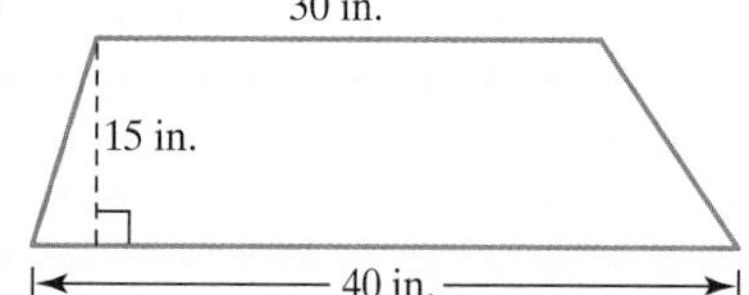

13. The right triangle shown has one leg of length 5 centimeters and a hypotenuse of length 13 centimeters.

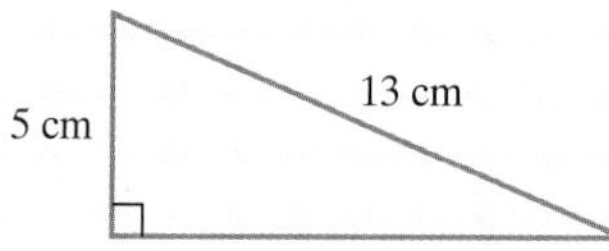

a. Find the length of the other leg.

b. What is the perimeter of the triangle?

c. What is the area of the triangle?

14. Find the circumference and area of a circle with a diameter of 40 meters. Express answers in terms of π and then round to the nearest tenth.

15. A rectangular floor measuring 8 feet by 6 feet is to be completely covered with square tiles measuring 8 inches on each side. How many tiles are needed to completely cover the floor?

In Exercises 16–18, find the volume of each figure. If necessary, express the answer in terms of π and then round to the nearest whole number.

16.

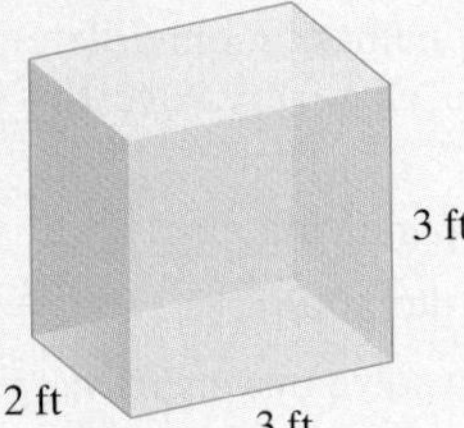

17.

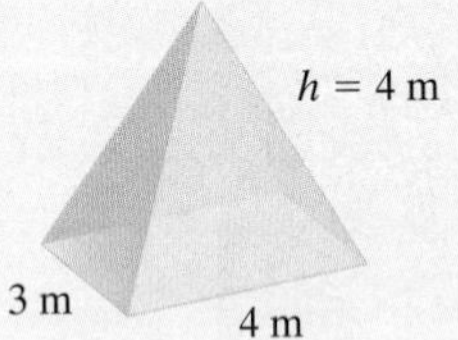

18.

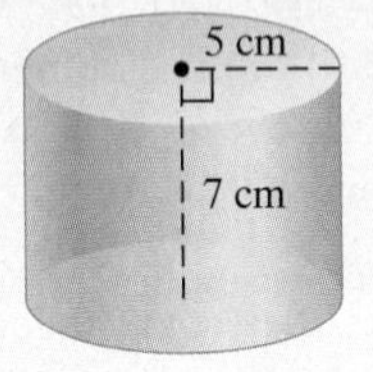

19. Find the measure, to the nearest whole number, of the side of the right triangle whose length is designated by c.

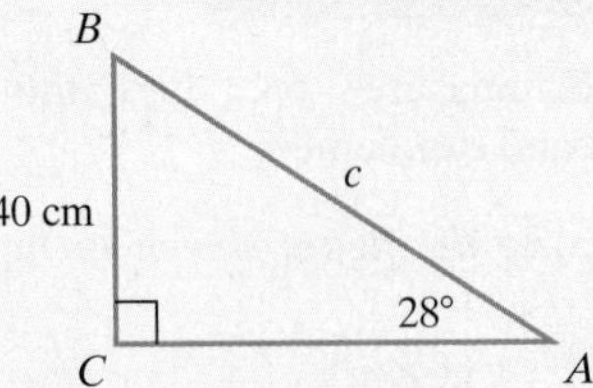

20. At a certain time of day, the angle of elevation of the Sun is 34°. If a building casts a shadow measuring 104 feet, find the height of the building to the nearest foot.

21. Determine if the graph shown is traversable. If it is, describe a path that will traverse it.

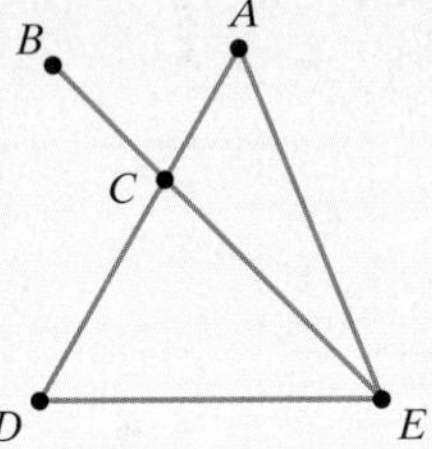

22. Describe a difference between the shapes of fractal geometry and those of Euclidean geometry.

Counting Methods and Probability Theory

11

TWO OF AMERICA'S BEST-LOVED PRESIDENTS, ABRAHAM LINCOLN AND JOHN F. KENNEDY, ARE LINKED BY A BIZARRE SERIES OF COINCIDENCES:

- Lincoln was elected president in 1860. Kennedy was elected president in 1960.
- Lincoln's assassin, John Wilkes Booth, was born in 1839. Kennedy's assassin, Lee Harvey Oswald, was born in 1939.
- Lincoln's secretary, named Kennedy, warned him not to go to the theater on the night he was shot. Kennedy's secretary, named Lincoln, warned him not to go to Dallas on the day he was shot.
- Booth shot Lincoln in a theater and ran into a warehouse. Oswald shot Kennedy from a warehouse and ran into a theater.
- Both Lincoln and Kennedy were shot from behind, with their wives present.
- Andrew Johnson, who succeeded Lincoln, was born in 1808. Lyndon Johnson, who succeeded Kennedy, was born in 1908.

Source: Edward Burger and Michael Starbird, *Coincidences, Chaos, and All That Math Jazz*, W.W. Norton and Company, 2005.

Amazing coincidences? A cosmic conspiracy? Not really. In this chapter, you will see how the mathematics of uncertainty and risk, called probability theory, numerically describes situations in which to expect the unexpected. By assigning numbers to things that are extraordinarily unlikely, we can logically analyze coincidences without erroneous beliefs that strange and mystical events are occurring. We'll even see how wildly inaccurate our intuition can be about the likelihood of an event by examining an "amazing" coincidence that is nearly certain.

Here's where you'll find these applications:

Coincidences are discussed in the Blitzer Bonus on page 750. Coincidences that are nearly certain are developed in Exercise 77 of Exercise Set 11.7.

11.1 The Fundamental Counting Principle

WHAT AM I SUPPOSED TO LEARN?

After studying this section, you should be able to:

1 Use the Fundamental Counting Principle to determine the number of possible outcomes in a given situation.

Have you ever imagined what your life would be like if you won the lottery? What changes would you make? Before you fantasize about becoming a person of leisure with a staff of obedient elves, think about this: The probability of winning top prize in the lottery is about the same as the probability of being struck by lightning. There are millions of possible number combinations in lottery games, but there is only one way of winning the grand prize. Determining the probability of winning involves calculating the chance of getting the winning combination from all possible outcomes. In this section, we begin preparing for the surprising world of probability by looking at methods for counting possible outcomes.

1 Use the Fundamental Counting Principle to determine the number of possible outcomes in a given situation.

The Fundamental Counting Principle with Two Groups of Items

It's early morning, you're groggy, and you have to select something to wear for your 8 A.M. class. (What *were* you thinking when you signed up for a class at that hour?!) Fortunately, your "lecture wardrobe" is rather limited—just two pairs of jeans to choose from (one blue, one black) and three T-shirts to choose from (one tan, one yellow, and one blue). Your early-morning dilemma is illustrated in **Figure 11.1**.

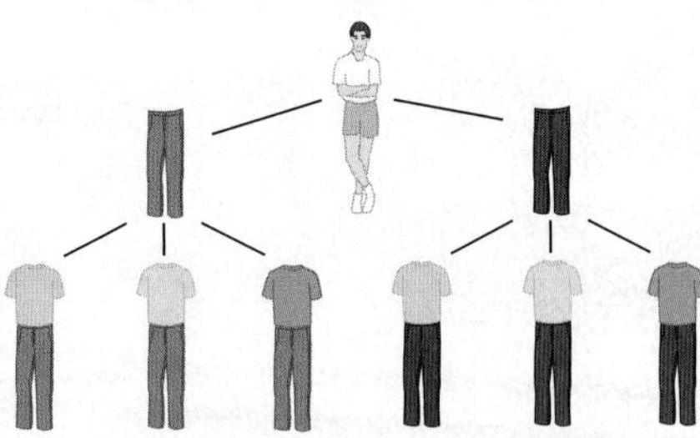

FIGURE 11.1 Selecting a wardrobe

The **tree diagram**, so named because of its branches, shows that you can form six different outfits from your two pairs of jeans and three T-shirts. Each pair of jeans can be combined with one of three T-shirts. Notice that the total number of outfits can be obtained by multiplying the number of choices for the jeans, 2, by the number of choices for the T-shirts, 3:

$$2 \cdot 3 = 6.$$

We can generalize this idea to any two groups of items—not just jeans and T-shirts—with the **Fundamental Counting Principle**.

THE FUNDAMENTAL COUNTING PRINCIPLE

If you can choose one item from a group of M items and a second item from a group of N items, then the total number of two-item choices is $M \cdot N$.

EXAMPLE 1 Applying the Fundamental Counting Principle

The Greasy Spoon Restaurant offers 6 appetizers and 14 main courses. In how many ways can a person order a two-course meal?

SOLUTION

Choosing from one of 6 appetizers and one of 14 main courses, the total number of two-course meals is

$$6 \cdot 14 = 84.$$

A person can order a two-course meal in 84 different ways.

CHECK POINT 1 A restaurant offers 10 appetizers and 15 main courses. In how many ways can you order a two-course meal?

EXAMPLE 2 Applying the Fundamental Counting Principle

This is the semester that you will take your required psychology and social science courses. Because you decide to register early, there are 15 sections of psychology from which you can choose. Furthermore, there are 9 sections of social science that are available at times that do not conflict with those for psychology. In how many ways can you create two-course schedules that satisfy the psychology–social science requirement?

SOLUTION

The number of ways that you can satisfy the requirement is found by multiplying the number of choices for each course. You can choose your psychology course from 15 sections and your social science course from 9 sections. For both courses you have

$$15 \cdot 9, \text{ or } 135$$

choices. Thus, you can satisfy the psychology–social science requirement in 135 ways.

CHECK POINT 2 Rework Example 2 given that the number of sections of psychology and nonconflicting sections of social science each decrease by 5.

The Fundamental Counting Principle with More Than Two Groups of Items

Whoops! You forgot something in choosing your lecture wardrobe—shoes! You have two pairs of sneakers to choose from—one black and one red, for that extra fashion flair! Your possible outfits including sneakers are shown in **Figure 11.2**.

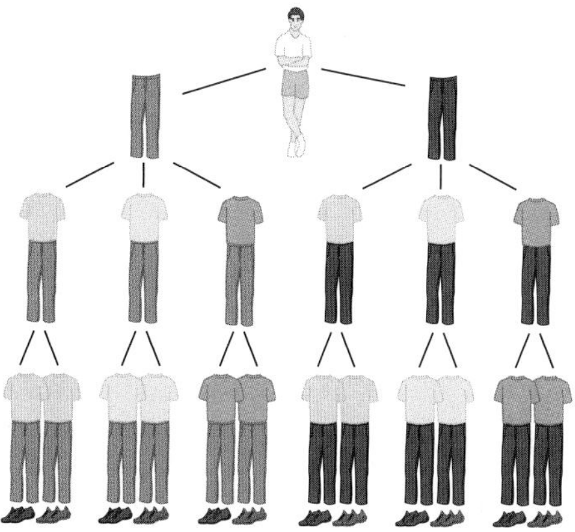

FIGURE 11.2 Increasing wardrobe selections

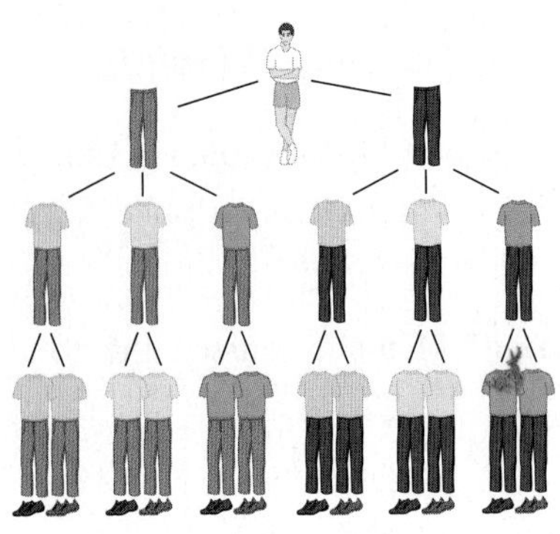

FIGURE 11.2 (repeated)

The tree diagram shows that you can form 12 outfits from your two pairs of jeans, three T-shirts, and two pairs of sneakers. Notice that the number of outfits can be obtained by multiplying the number of choices for jeans, 2, the number of choices for T-shirts, 3, and the number of choices for sneakers, 2:

$$2 \cdot 3 \cdot 2 = 12.$$

Unlike your earlier dilemma, you are now dealing with *three* groups of items. The Fundamental Counting Principle can be extended to determine the number of possible outcomes in situations in which there are three or more groups of items.

THE FUNDAMENTAL COUNTING PRINCIPLE

The number of ways in which a series of successive things can occur is found by multiplying the number of ways in which each thing can occur.

For example, if you own 30 pairs of jeans, 20 T-shirts, and 12 pairs of sneakers, you have

$$30 \cdot 20 \cdot 12 = 7200$$

choices for your wardrobe.

The number of possible ways of playing the first four moves on each side in a game of chess is 318,979,564,000.

EXAMPLE 3 Options in Planning a Course Schedule

Next semester you are planning to take three courses—math, English, and humanities. Based on time blocks and highly recommended professors, there are eight sections of math, five of English, and four of humanities that you find suitable. Assuming no scheduling conflicts, how many different three-course schedules are possible?

SOLUTION

This situation involves making choices with three groups of items.

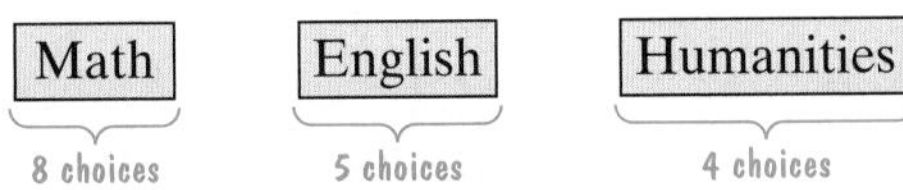

We use the Fundamental Counting Principle to find the number of three-course schedules. Multiply the number of choices for each of the three groups.

$$8 \cdot 5 \cdot 4 = 160$$

Thus, there are 160 different three-course schedules.

CHECK POINT 3 A pizza can be ordered with two choices of size (medium or large), three choices of crust (thin, thick, or regular), and five choices of toppings (ground beef, sausage, pepperoni, bacon, or mushrooms). How many different one-topping pizzas can be ordered?

EXAMPLE 4 Car of the Future

Car manufacturers are now experimenting with lightweight three-wheel cars, designed for one person and considered ideal for city driving. Intrigued? Suppose you could order such a car with a choice of nine possible colors, with or without air conditioning, electric or gas powered, and with or without an onboard computer. In how many ways can this car be ordered with regard to these options?

SOLUTION

This situation involves making choices with four groups of items.

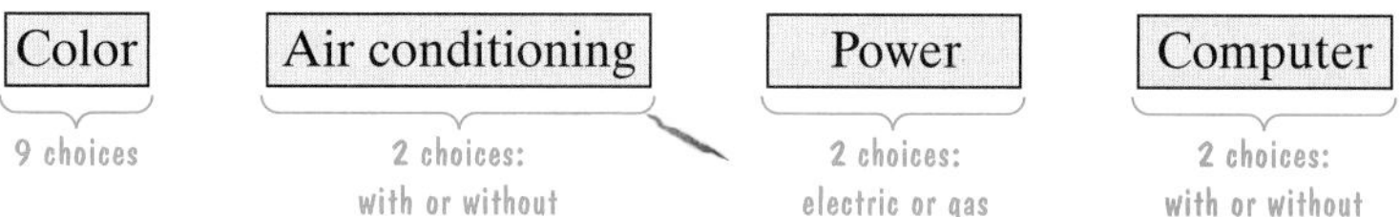

We use the Fundamental Counting Principle to find the number of ordering options. Multiply the number of choices for each of the four groups.

$$9 \cdot 2 \cdot 2 \cdot 2 = 72$$

Thus, the car can be ordered in 72 different ways.

 **CHECK POINT 4** The car in Example 4 is now available in ten possible colors. The options involving air conditioning, power, and an onboard computer still apply. Furthermore, the car is available with or without a global positioning system (for pinpointing your location at every moment). In how many ways can this car be ordered in terms of these options?

EXAMPLE 5 *A Multiple-Choice Test*

You are taking a multiple-choice test that has ten questions. Each of the questions has four answer choices, with one correct answer per question. If you select one of these four choices for each question and leave nothing blank, in how many ways can you answer the questions?

SOLUTION

This situation involves making choices with ten questions.

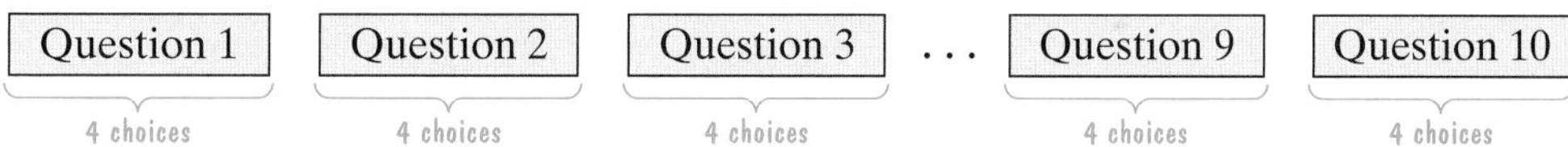

We use the Fundamental Counting Principle to determine the number of ways that you can answer the questions on the test. Multiply the number of choices, 4, for each of the ten questions.

$$4 \cdot 4 \cdot 4 \cdot 4 \cdot 4 \cdot 4 \cdot 4 \cdot 4 \cdot 4 \cdot 4 = 4^{10} = 1{,}048{,}576$$ Use a calculator: 4 [y^x] 10 [=].

Thus, you can answer the questions in 1,048,576 different ways.

Are you surprised that there are over one million ways of answering a ten-question multiple-choice test? Of course, there is only one way to answer the test and receive a perfect score. The probability of guessing your way into a perfect score involves calculating the chance of getting a perfect score, just one way, from all 1,048,576 possible outcomes. In short, prepare for the test and do not rely on guessing!

 **CHECK POINT 5** You are taking a multiple-choice test that has six questions. Each of the questions has three answer choices, with one correct answer per question. If you select one of these three choices for each question and leave nothing blank, in how many ways can you answer the questions?

Blitzer Bonus

Running Out of Telephone Numbers

By the year 2020, portable telephones used for business and pleasure will all be videophones. At that time, U.S. population is expected to be 323 million. Faxes, beepers, cellphones, computer phone lines, and business lines may result in certain areas running out of phone numbers. Solution: Add more digits!

With or without extra digits, we expect that the 2020 videophone greeting will still be "hello," a word created by Thomas Edison in 1877. Phone inventor Alexander Graham Bell preferred "ahoy," but "hello" won out, appearing in the *Oxford English Dictionary* in 1883.

Source: New York Times

EXAMPLE 6 ***Telephone Numbers in the United States***

Telephone numbers in the United States begin with three-digit area codes followed by seven-digit local telephone numbers. Area codes and local telephone numbers cannot begin with 0 or 1. How many different telephone numbers are possible?

SOLUTION

This situation involves making choices with ten groups of items.

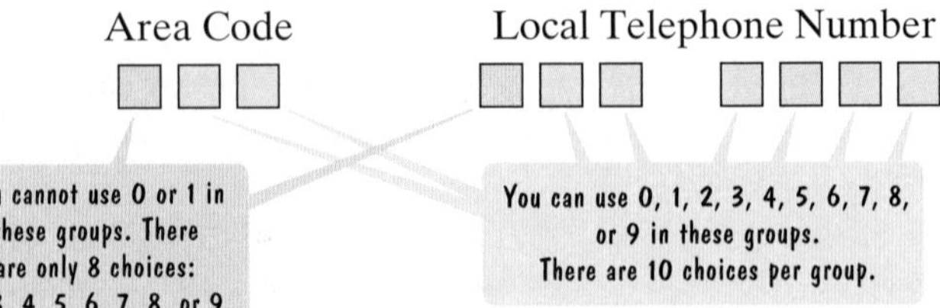

Here are the choices for each of the ten groups of items:

Area Code	Local Telephone Number
8 10 10	8 10 10 10 10 10 10.

We use the Fundamental Counting Principle to determine the number of different telephone numbers that are possible. The total number of telephone numbers possible is

$$8 \cdot 10 \cdot 10 \cdot 8 \cdot 10 \cdot 10 \cdot 10 \cdot 10 \cdot 10 \cdot 10 = 6{,}400{,}000{,}000.$$

There are six billion, four hundred million different telephone numbers that are possible.

CHECK POINT 6 An electronic gate can be opened by entering five digits on a keypad containing the digits 0, 1, 2, 3, . . . , 8, 9. How many different keypad sequences are possible if the digit 0 cannot be used as the first digit?

Concept and Vocabulary Check

Fill in each blank so that the resulting statement is true.

1. If you can choose one item from a group of M items and a second item from a group of N items, then the total number of two-item choices is _______.
2. The number of ways in which a series of successive things can occur is found by ___________ the number of ways in which each thing can occur. This is called the ___________________ Principle.

In Exercises 3–4, determine whether each statement is true or false. If the statement is false, make the necessary change(s) to produce a true statement.

3. If one item is chosen from M items, a second item is chosen from N items, and a third item is chosen from P items, the total number of three-item choices is $M + N + P$. ______
4. Regardless of the United States population, we will not run out of telephone numbers as long as we continue to add new digits. ______

Exercise Set 11.1

Practice and Application Exercises

1. A restaurant offers eight appetizers and ten main courses. In how many ways can a person order a two-course meal?
2. The model of the car you are thinking of buying is available in nine different colors and three different styles (hatchback, sedan, or station wagon). In how many ways can you order the car?
3. A popular brand of pen is available in three colors (red, green, or blue) and four writing tips (bold, medium, fine, or micro). How many different choices of pens do you have with this brand?
4. In how many ways can a casting director choose a female lead and a male lead from five female actors and six male actors?

5. A student is planning a two-part trip. The first leg of the trip is from San Francisco to New York, and the second leg is from New York to Paris. From San Francisco to New York, travel options include airplane, train, or bus. From New York to Paris, the options are limited to airplane or ship. In how many ways can the two-part trip be made?

6. For a temporary job between semesters, you are painting the parking spaces for a new shopping mall with a letter of the alphabet and a single digit from 1 to 9. The first parking space is A1 and the last parking space is Z9. How many parking spaces can you paint with distinct labels?

7. An ice cream store sells two drinks (sodas or milk shakes), in four sizes (small, medium, large, or jumbo), and five flavors (vanilla, strawberry, chocolate, coffee, or pistachio). In how many ways can a customer order a drink?

8. A pizza can be ordered with three choices of size (small, medium, or large), four choices of crust (thin, thick, crispy, or regular), and six choices of toppings (ground beef, sausage, pepperoni, bacon, mushrooms, or onions). How many one-topping pizzas can be ordered?

9. A restaurant offers the following limited lunch menu.

Main Course	Vegetables	Beverages	Desserts
Ham	Potatoes	Coffee	Cake
Chicken	Peas	Tea	Pie
Fish	Green beans	Milk	Ice cream
Beef		Soda	

If one item is selected from each of the four groups, in how many ways can a meal be ordered? Describe two such orders.

10. An apartment complex offers apartments with four different options, designated by A through D.

A	B	C	D
one bedroom	one bathroom	first floor	lake view
two bedrooms	two bathrooms	second floor	golf course view
three bedrooms			no special view

How many apartment options are available? Describe two such options.

11. Shoppers in a large shopping mall are categorized as male or female, over 30 or 30 and under, and cash or credit card shoppers. In how many ways can the shoppers be categorized?

12. There are three highways from city A to city B, two highways from city B to city C, and four highways from city C to city D. How many different highway routes are there from city A to city D?

13. A person can order a new car with a choice of six possible colors, with or without air conditioning, with or without automatic transmission, with or without power windows, and with or without a CD player. In how many different ways can a new car be ordered with regard to these options?

14. A car model comes in nine colors, with or without air conditioning, with or without a sun roof, with or without automatic transmission, and with or without antilock brakes. In how many ways can the car be ordered with regard to these options?

15. You are taking a multiple-choice test that has five questions. Each of the questions has three answer choices, with one correct answer per question. If you select one of these three choices for each question and leave nothing blank, in how many ways can you answer the questions?

16. You are taking a multiple-choice test that has eight questions. Each of the questions has three answer choices, with one correct answer per question. If you select one of these three choices for each question and leave nothing blank, in how many ways can you answer the questions?

17. In the original plan for area codes in 1945, the first digit could be any number from 2 through 9, the second digit was either 0 or 1, and the third digit could be any number except 0. With this plan, how many different area codes are possible?

18. The local seven-digit telephone numbers in Inverness, California, have 669 as the first three digits. How many different telephone numbers are possible in Inverness?

19. License plates in a particular state display two letters followed by three numbers, such as AT-887 or BB-013. How many different license plates can be manufactured for this state?

20. How many different four-letter radio station call letters can be formed if the first letter must be W or K?

21. A stock can go up, go down, or stay unchanged. How many possibilities are there if you own seven stocks?

22. A social security number contains nine digits, such as 074-66-7795. How many different social security numbers can be formed?

Explaining the Concepts

23. Explain the Fundamental Counting Principle.

24. **Figure 11.2** on page 695 shows that a tree diagram can be used to find the total number of outfits. Describe one advantage of using the Fundamental Counting Principle rather than a tree diagram.

25. Write an original problem that can be solved using the Fundamental Counting Principle. Then solve the problem.

Critical Thinking Exercises

Make Sense? *In Exercises 26–29, determine whether each statement makes sense or does not make sense, and explain your reasoning.*

26. I used the Fundamental Counting Principle to determine the number of five-digit ZIP codes that are available to the U.S. Postal Service.

27. The Fundamental Counting Principle can be used to determine the number of ways of arranging the numbers 1, 2, 3, 4, 5, . . . , 98, 99, 100.

28. I estimate there are approximately 10,000 ways to arrange the letters A, B, C, D, E, . . . , X, Y, Z.

29. There are more than 2000 possible sets of answers on an eleven-item true/false test.

30. How many four-digit odd numbers are there? Assume that the digit on the left cannot be 0.

31. In order to develop a more appealing hamburger, a franchise used taste tests with 12 different buns, 30 sauces, 4 types of lettuce, and 3 types of tomatoes. If the taste test was done at one restaurant by one tester who takes 10 minutes to eat each hamburger, approximately how long would it take the tester to eat all possible hamburgers?

11.2 Permutations

WHAT AM I SUPPOSED TO LEARN?

After studying this section, you should be able to:

1 Use the Fundamental Counting Principle to count permutations.
2 Evaluate factorial expressions.
3 Use the permutations formula.
4 Find the number of permutations of duplicate items.

WE OPEN THIS SECTION WITH SIX jokes about books. (Stay with us on this one.)

- *"Outside of a dog, a book is man's best friend. Inside of a dog, it's too dark to read."* *—Groucho Marx*
- *"I was in the campus bookstore the other day and there was half off all titles. I bought this liberal arts math book under the new title* Thinking.*"— Bob Blitzer*
- *"If a word in the dictionary was misspelled, how would we know?" —Steven Wright*
- *"I wrote a book under a pen name: Bic"—Henny Youngman*
- *"A bit of advice: Never read a pop-up book about giraffes." —Jerry Seinfeld*
- *"I honestly believe there is absolutely nothing like going to bed with a good book. Or a friend who's read one."—Phyllis Diller*

1 Use the Fundamental Counting Principle to count permutations.

We can use the Fundamental Counting Principle to determine the number of ways these jokes can be delivered. You can choose any one of the six jokes as the first one told. Once this joke is delivered, you'll have five jokes to choose from for the second joke told. You'll then have four jokes left to choose from for the third delivery. Continuing in this manner, the situation can be shown as follows:

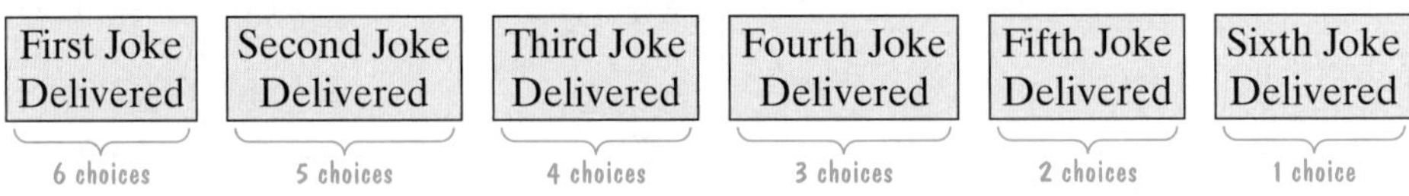

Using the Fundamental Counting Principle, we multiply the choices:

$$6 \cdot 5 \cdot 4 \cdot 3 \cdot 2 \cdot 1 = 720.$$

Thus, there are 720 different ways to deliver the six jokes about books. One of the 720 possible arrangements is

Seinfeld joke—Youngman joke—Blitzer joke—Marx joke—Wright joke—Diller joke.

Such an ordered arrangement is called a *permutation* of the six jokes.

A **permutation** is an ordered arrangement of items that occurs when

- No item is used more than once. (Each joke is told exactly once.)
- The order of arrangement makes a difference. (The order in which these jokes are told makes a difference in terms of how they are received.)

EXAMPLE 1 Counting Permutations

In how many ways can the six jokes about books be delivered if Bob Blitzer's joke is delivered first and Jerry Seinfeld's joke is told last?

SOLUTION

The conditions of Blitzer's joke first and Seinfeld's joke last can be shown as follows:

First Joke Delivered	Second Joke Delivered	Third Joke Delivered	Fourth Joke Delivered	Fifth Joke Delivered	Sixth Joke Delivered
1 choice: Blitzer					1 choice: Seinfeld

Now let's fill in the number of choices for positions two through five. You can choose any one of the four remaining jokes for the second delivery. Once you've chosen this joke, you'll have three jokes to choose from for the third delivery. This leaves only two jokes to choose from for the fourth delivery. Once this choice is made, there is just one joke left to choose for the fifth delivery.

First Joke Delivered	Second Joke Delivered	Third Joke Delivered	Fourth Joke Delivered	Fifth Joke Delivered	Sixth Joke Delivered
1 choice: Blitzer	4 choices	3 choices	2 choices	1 choice	1 choice: Seinfeld

We use the Fundamental Counting Principle to find the number of ways the six jokes can be delivered. Multiply the choices:

$$1 \cdot 4 \cdot 3 \cdot 2 \cdot 1 \cdot 1 = 24.$$

Thus, there are 24 different ways the jokes can be delivered if Blitzer's joke is told first and Seinfeld's joke is told last.

GREAT QUESTION!

I rolled my eyes at Bob Blitzer's bad joke. How come our author is doing shtick?

He had no choice. We originally had a joke by George Carlin, but we were not granted permission to use it.

CHECK POINT 1 How many ways can the six jokes about books be delivered if a man's joke is told first?

Blitzer Bonus

How to Pass the Time for $2\frac{1}{2}$ Million Years

If you were to arrange 15 different books on a shelf and it took you one minute for each permutation, the entire task would take 2,487,965 years.

Source: *Isaac Asimov's Book of Facts*

EXAMPLE 2 Counting Permutations

You need to arrange seven of your favorite books along a small shelf. How many different ways can you arrange the books, assuming that the order of the books makes a difference to you?

SOLUTION

You may choose any of the seven books for the first position on the shelf. This leaves six choices for second position. After the first two positions are filled, there are five books to choose from for third position, four choices left for the fourth position, three choices left for the fifth position, then two choices for the sixth position, and only one choice for the last position. This situation can be shown as follows:

First Shelf Position	Second Shelf Position	Third Shelf Position	Fourth Shelf Position	Fifth Shelf Position	Sixth Shelf Position	Seventh Shelf Position
7 choices	6 choices	5 choices	4 choices	3 choices	2 choices	1 choice

We use the Fundamental Counting Principle to find the number of ways you can arrange the seven books along the shelf. Multiply the choices:

$$7 \cdot 6 \cdot 5 \cdot 4 \cdot 3 \cdot 2 \cdot 1 = 5040.$$

Thus, you can arrange the books in 5040 ways. There are 5040 different possible permutations.

CHECK POINT 2 In how many ways can you arrange five books along a shelf, assuming that the order of the books makes a difference?

2 *Evaluate factorial expressions.*

Factorial Notation

The product in Example 2,

$$7 \cdot 6 \cdot 5 \cdot 4 \cdot 3 \cdot 2 \cdot 1,$$

is given a special name and symbol. It is called 7 **factorial**, and written 7!. Thus,

$$7! = 7 \cdot 6 \cdot 5 \cdot 4 \cdot 3 \cdot 2 \cdot 1.$$

In general, if n is a positive integer, then $n!$ (*n factorial*) is the product of all positive integers from n down through 1. For example,

$$1! = 1$$
$$2! = 2 \cdot 1 = 2$$
$$3! = 3 \cdot 2 \cdot 1 = 6$$
$$4! = 4 \cdot 3 \cdot 2 \cdot 1 = 24$$
$$5! = 5 \cdot 4 \cdot 3 \cdot 2 \cdot 1 = 120$$
$$6! = 6 \cdot 5 \cdot 4 \cdot 3 \cdot 2 \cdot 1 = 720.$$

FACTORIALS: 0 THROUGH 20

0!	1
1!	1
2!	2
3!	6
4!	24
5!	120
6!	720
7!	5040
8!	40,320
9!	362,880
10!	3,628,800
11!	39,916,800
12!	479,001,600
13!	6,227,020,800
14!	87,178,291,200
15!	1,307,674,368,000
16!	20,922,789,888,000
17!	355,687,428,096,000
18!	6,402,373,705,728,000
19!	121,645,100,408,832,000
20!	2,432,902,008,176,640,000

FACTORIAL NOTATION

If n is a positive integer, the notation $n!$ (read "n factorial") is the product of all positive integers from n down through 1.

$$n! = n(n-1)(n-2) \cdots (3)(2)(1)$$

0! (zero factorial), by definition, is 1.

$$0! = 1$$

EXAMPLE 3 *Using Factorial Notation*

Evaluate the following factorial expressions without using the factorial key on your calculator:

a. $\frac{8!}{5!}$ **b.** $\frac{26!}{21!}$ **c.** $\frac{500!}{499!}$.

SOLUTION

a. We can evaluate the numerator and the denominator of $\frac{8!}{5!}$. However, it is easier to use the following simplification:

$$\frac{8!}{5!} = \frac{8 \cdot 7 \cdot 6 \cdot \boxed{5 \cdot 4 \cdot 3 \cdot 2 \cdot 1}}{\boxed{5 \cdot 4 \cdot 3 \cdot 2 \cdot 1}} = \frac{8 \cdot 7 \cdot 6 \cdot \boxed{5!}}{\boxed{5!}} = \frac{8 \cdot 7 \cdot 6 \cdot \cancel{5!}}{\cancel{5!}} = 8 \cdot 7 \cdot 6 = 336.$$

TECHNOLOGY

Most calculators have a key or menu item for calculating factorials. Here are the keystrokes for finding 9!:

Many Scientific Calculators:

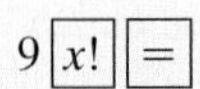

Many Graphing Calculators:

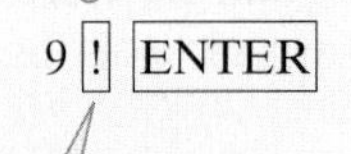

On TI graphing calculators, this is selected using the MATH PROB menu.

Because $n!$ becomes quite large as n increases, your calculator will display these larger values in scientific notation.

b. Rather than write out 26!, the numerator of $\frac{26!}{21!}$, as the product of all integers from 26 down to 1, we can express 26! as

$$26! = 26 \cdot 25 \cdot 24 \cdot 23 \cdot 22 \cdot 21!.$$

In this way, we can cancel 21! in the numerator and the denominator of the given expression.

$$\frac{26!}{21!} = \frac{26 \cdot 25 \cdot 24 \cdot 23 \cdot 22 \cdot 21!}{21!} = \frac{26 \cdot 25 \cdot 24 \cdot 23 \cdot 22 \cdot \cancel{21!}}{\cancel{21!}}$$

$$= 26 \cdot 25 \cdot 24 \cdot 23 \cdot 22 = 7{,}893{,}600$$

c. In order to cancel identical factorials in the numerator and the denominator of $\frac{500!}{499!}$, we can express 500! as $500 \cdot 499!$.

$$\frac{500!}{499!} = \frac{500 \cdot 499!}{499!} = \frac{500 \cdot \cancel{499!}}{\cancel{499!}} = 500$$

CHECK POINT 3 Evaluate without using a calculator's factorial key:

a. $\frac{9!}{6!}$ **b.** $\frac{16!}{11!}$ **c.** $\frac{100!}{99!}$

3 *Use the permutations formula.*

A Formula for Permutations

You are the coach of a little league baseball team. There are 13 players on the team (and lots of parents hovering in the background, dreaming of stardom for their little "Albert Pujols"). You need to choose a batting order having 9 players. The order makes a difference, because, for instance, if bases are loaded and "Little Albert" is fourth or fifth at bat, his possible home run will drive in three additional runs. How many batting orders can you form?

You can choose any of 13 players for the first person at bat. Then you will have 12 players from which to choose the second batter, then 11 from which to choose the third batter, and so on. The situation can be shown as follows:

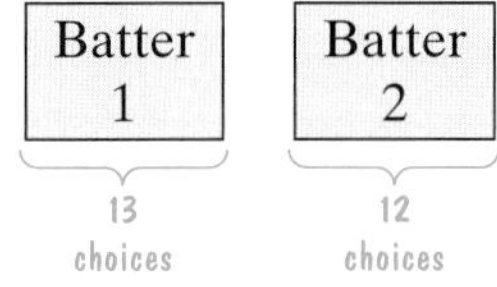

Batter 3 — 11 choices; Batter 4 — 10 choices; Batter 5 — 9 choices; Batter 6 — 8 choices; Batter 7 — 7 choices;

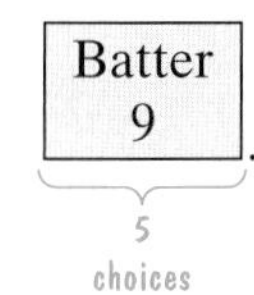

The total number of batting orders is

$$13 \cdot 12 \cdot 11 \cdot 10 \cdot 9 \cdot 8 \cdot 7 \cdot 6 \cdot 5 = 259{,}459{,}200.$$

Nearly 260 million batting orders are possible for your 13-player little league team. Each batting order is a permutation because the order of the batters makes a difference. The number of permutations of 13 players taken 9 at a time is 259,459,200.

We can obtain a formula for finding the number of permutations by rewriting our computation:

$$13 \cdot 12 \cdot 11 \cdot 10 \cdot 9 \cdot 8 \cdot 7 \cdot 6 \cdot 5$$

$$= \frac{13 \cdot 12 \cdot 11 \cdot 10 \cdot 9 \cdot 8 \cdot 7 \cdot 6 \cdot 5 \cdot \boxed{4 \cdot 3 \cdot 2 \cdot 1}}{\boxed{4 \cdot 3 \cdot 2 \cdot 1}} = \frac{13!}{4!} = \frac{13!}{(13-9)!}.$$

Thus, the number of permutations of 13 things taken 9 at a time is $\frac{13!}{(13-9)!}$. The special notation ${}_{13}P_9$ is used to replace the phrase "the number of permutations of 13 things taken 9 at a time." Using this new notation, we can write

$${}_{13}P_9 = \frac{13!}{(13-9)!}.$$

Let's take a moment to focus on the formula for $_{13}P_9$, the number of possible permutations of 13 things taken 9 at a time:

$$_{13}P_9 = \frac{13!}{(13-9)!}.$$

The numerator of this expression is the factorial of the number of items, 13 team members: 13!. The denominator is also a factorial. It is the factorial of the difference between the number of items, 13, and the number of items in each permutation, 9 batters: $(13-9)!$.

The notation $_nP_r$ means the **number of permutations of n things taken r at a time**. We can generalize from the situation in which 9 batters were taken from 13 players. By generalizing, we obtain the following formula for the number of permutations if r items are taken from n items:

GREAT QUESTION!

Do I have to use the formula for $_nP_r$ to solve permutation problems?

No. Because all permutation problems are also fundamental counting problems, they can be solved using the formula for $_nP_r$ or using the Fundamental Counting Principle.

PERMUTATIONS OF n THINGS TAKEN r AT A TIME

The number of possible permutations if r items are taken from n items is

$$_nP_r = \frac{n!}{(n-r)!}.$$

EXAMPLE 4 Using the Formula for Permutations

You and 19 of your friends have decided to form an Internet marketing consulting firm. The group needs to choose three officers—a CEO, an operating manager, and a treasurer. In how many ways can those offices be filled?

SOLUTION

Your group is choosing $r = 3$ officers from a group of $n = 20$ people (you and 19 friends). The order in which the officers are chosen matters because the CEO, the operating manager, and the treasurer each have different responsibilities. Thus, we are looking for the number of permutations of 20 things taken 3 at a time. We use the formula

$$_nP_r = \frac{n!}{(n-r)!}$$

with $n = 20$ and $r = 3$.

$$_{20}P_3 = \frac{20!}{(20-3)!} = \frac{20!}{17!} = \frac{20 \cdot 19 \cdot 18 \cdot 17!}{17!} = \frac{20 \cdot 19 \cdot 18 \cdot \cancel{17!}}{\cancel{17!}} = 20 \cdot 19 \cdot 18 = 6840$$

Thus, there are 6840 different ways of filling the three offices.

TECHNOLOGY

Graphing calculators have a menu item for calculating permutations, usually labeled $\boxed{_nP_r}$. For example, to find $_{20}P_3$, the keystrokes are

$20 \boxed{_nP_r} 3 \boxed{\text{ENTER}}$.

To access $\boxed{_nP_r}$ on a TI-84 Plus C, first press $\boxed{\text{MATH}}$ and then use the right or left arrow key to highlight $\boxed{\text{PROB}}$. Press $\boxed{2}$ for $\boxed{_nP_r}$.

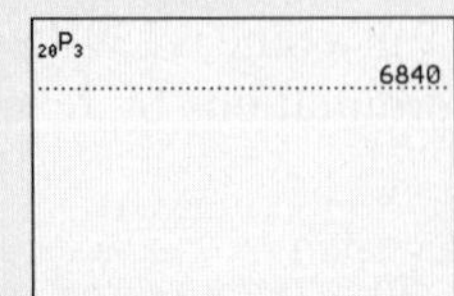

If you are using a scientific calculator, check your manual for the location of the menu item for calculating permutations and the required keystrokes.

CHECK POINT 4 A corporation has seven members on its board of directors. In how many different ways can it elect a president, vice-president, secretary, and treasurer?

EXAMPLE 5 Using the Formula for Permutations

You are working for the Sitcom Television Network. Your assignment is to help set up the television schedule for Monday evenings between 7 and 10 P.M. You need to schedule a show in each of six 30-minute time blocks, beginning with 7 to 7:30 and ending with 9:30 to 10:00. You can select from among the following situation comedies: *The Office, Seinfeld, That 70s Show, Cheers, The Big Bang Theory, Frasier, All in the Family, I Love Lucy, M*A*S*H, The Larry Sanders Show, Modern Family, Married ... With Children*, and *Curb Your Enthusiasm*. How many different programming schedules can be arranged?

SOLUTION

You are choosing $r = 6$ situation comedies from a collection of $n = 13$ classic sitcoms. The order in which the programs are aired matters. Family-oriented comedies have higher ratings when aired in earlier time blocks, such as 7 to 7:30. By contrast, comedies with adult themes do better in later time blocks. In short, we are looking for the number of permutations of 13 things taken 6 at a time. We use the formula

$$_nP_r = \frac{n!}{(n-r)!}$$

with $n = 13$ and $r = 6$.

$$_{13}P_6 = \frac{13!}{(13-6)!} = \frac{13!}{7!} = \frac{13 \cdot 12 \cdot 11 \cdot 10 \cdot 9 \cdot 8 \cdot \cancel{7!}}{\cancel{7!}} = 13 \cdot 12 \cdot 11 \cdot 10 \cdot 9 \cdot 8 = 1{,}235{,}520$$

There are 1,235,520 different programming schedules that can be arranged.

CHECK POINT 5 How many different programming schedules can be arranged by choosing five situation comedies from a collection of nine classic sitcoms?

4 *Find the number of permutations of duplicate items.*

Permutations of Duplicate Items

The number of permutations of the letters in the word SET is 3!, or 6. The six permutations are

SET, STE, EST, ETS, TES, TSE.

Are there also six permutations of the letters in the name ANA? The answer is no. Unlike SET, with three distinct letters, ANA contains three letters, of which the two As are duplicates. If we rearrange the letters just as we did with SET, we obtain

ANA, AAN, NAA, NAA, ANA, AAN.

Without the use of color to distinguish between the two As, there are only three distinct permutations: ANA, AAN, NAA.

There is a formula for finding the number of distinct permutations when duplicate items exist:

PERMUTATIONS OF DUPLICATE ITEMS

The number of permutations of n items, where p items are identical, q items are identical, r items are identical, and so on, is given by

$$\frac{n!}{p!\,q!\,r!\cdots}.$$

For example, ANA contains three letters ($n = 3$), where two of the letters are identical ($p = 2$). The number of distinct permutations is

$$\frac{n!}{p!} = \frac{3!}{2!} = \frac{3 \cdot \cancel{2!}}{\cancel{2!}} = 3.$$

We saw that the three distinct permutations are ANA, AAN, and NAA.

TECHNOLOGY

Parentheses are necessary to enclose the factorials in the denominator when using a calculator to find

$$\frac{11!}{4!\,4!\,2!}.$$

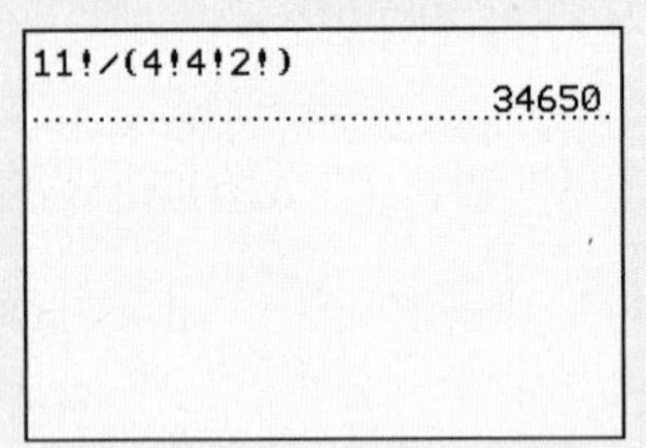

EXAMPLE 6 *Using the Formula for Permutations of Duplicate Items*

In how many distinct ways can the letters of the word MISSISSIPPI be arranged?

SOLUTION

The word contains 11 letters ($n = 11$), where four Is are identical ($p = 4$), four Ss are identical ($q = 4$), and 2 Ps are identical ($r = 2$). The number of distinct permutations is

$$\frac{n!}{p!\,q!\,r!} = \frac{11!}{4!\,4!\,2!} = \frac{11 \cdot 10 \cdot 9 \cdot 8 \cdot 7 \cdot 6 \cdot 5 \cdot \cancel{4!}}{\cancel{4!}\,4 \cdot 3 \cdot 2 \cdot 1 \cdot 2 \cdot 1} = 34{,}650$$

There are 34,650 distinct ways the letters in the word MISSISSIPPI can be arranged.

CHECK POINT 6 In how many ways can the letters of the word OSMOSIS be arranged?

Concept and Vocabulary Check

Fill in each blank so that the resulting statement is true.

1. 5!, called 5 ________, is the product of all positive integers from ____ down through ____. By definition, 0! = ____.
2. The number of possible permutations if r objects are taken from n items is ${}_nP_r$ = ________.
3. The number of permutations of n items, where p items are identical and q items are identical, is given by ______.

In Exercises 4–7, determine whether each statement is true or false. If the statement is false, make the necessary change(s) to produce a true statement.

4. A permutation occurs when the order of arrangement does not matter. ______
5. $8! = 8 + 7 + 6 + 5 + 4 + 3 + 2 + 1$ ______
6. Because all permutation problems are also Fundamental Counting problems, they can be solved using the formula for ${}_nP_r$ or using the Fundamental Counting Principle. ______
7. Because the words BET and BEE both contain three letters, the number of permutations of the letters in each word is 3!, or 6. ______

Exercise Set 11.2

Practice and Application Exercises

Use the Fundamental Counting Principle to solve Exercises 1–12.

1. Six performers are to present their comedy acts on a weekend evening at a comedy club. How many different ways are there to schedule their appearances?
2. Five singers are to perform on a weekend evening at a night club. How many different ways are there to schedule their appearances?
3. In the *Cambridge Encyclopedia of Language* (Cambridge University Press, 1987), author David Crystal presents five sentences that make a reasonable paragraph regardless of their order. The sentences are as follows:
   - Mark had told him about the foxes.
   - John looked out of the window.
   - Could it be a fox?
   - However, nobody had seen one for months.
   - He thought he saw a shape in the bushes.

   In how many different orders can the five sentences be arranged?
4. In how many different ways can a police department arrange eight suspects in a police lineup if each lineup contains all eight people?
5. As in Exercise 1, six performers are to present their comedy acts on a weekend evening at a comedy club. One of the performers insists on being the last stand-up comic of the evening. If this performer's request is granted, how many different ways are there to schedule the appearances?
6. As in Exercise 2, five singers are to perform at a night club. One of the singers insists on being the last performer of the evening. If this singer's request is granted, how many different ways are there to schedule the appearances?

7. You need to arrange nine of your favorite books along a small shelf. How many different ways can you arrange the books, assuming that the order of the books makes a difference to you?

8. You need to arrange ten of your favorite photographs on the mantel above a fireplace. How many ways can you arrange the photographs, assuming that the order of the pictures makes a difference to you?

In Exercises 9–10, use the five sentences that are given in Exercise 3.

9. How many different five-sentence paragraphs can be formed if the paragraph begins with "He thought he saw a shape in the bushes" and ends with "John looked out of the window"?

10. How many different five-sentence paragraphs can be formed if the paragraph begins with "He thought he saw a shape in the bushes" followed by "Mark had told him about the foxes"?

11. A television programmer is arranging the order in which five movies will be seen between the hours of 6 P.M. and 4 A.M. Two of the movies have a G rating, and they are to be shown in the first two time blocks. One of the movies is rated NC-17, and it is to be shown in the last of the time blocks, from 2 A.M. until 4 A.M. Given these restrictions, in how many ways can the five movies be arranged during the indicated time blocks?

12. A camp counselor and six campers are to be seated along a picnic bench. In how many ways can this be done if the counselor must be seated in the middle and a camper who has a tendency to engage in food fights must sit to the counselor's immediate left?

In Exercises 13–32, evaluate each factorial expression.

13. $\frac{9!}{6!}$ **14.** $\frac{12!}{10!}$ **15.** $\frac{29!}{25!}$

16. $\frac{31!}{28!}$ **17.** $\frac{19!}{11!}$ **18.** $\frac{17!}{9!}$

19. $\frac{600!}{599!}$ **20.** $\frac{700!}{699!}$ **21.** $\frac{104!}{102!}$

22. $\frac{106!}{104!}$ **23.** $7! - 3!$ **24.** $6! - 3!$

25. $(7 - 3)!$ **26.** $(6 - 3)!$ **27.** $\left(\frac{12}{4}\right)!$

28. $\left(\frac{45}{9}\right)!$ **29.** $\frac{7!}{(7-2)!}$ **30.** $\frac{8!}{(8-5)!}$

31. $\frac{13!}{(13-3)!}$ **32.** $\frac{17!}{(17-3)!}$

In Exercises 33–40, use the formula for ${}_nP_r$ to evaluate each expression.

33. ${}_9P_4$ **34.** ${}_7P_3$ **35.** ${}_8P_5$

36. ${}_{10}P_4$ **37.** ${}_6P_6$ **38.** ${}_9P_9$

39. ${}_8P_0$ **40.** ${}_6P_0$

Use an appropriate permutations formula to solve Exercises 41–56.

41. A club with ten members is to choose three officers—president, vice-president, and secretary-treasurer. If each office is to be held by one person and no person can hold more than one office, in how many ways can those offices be filled?

42. A corporation has seven members on its board of directors. In how many different ways can it elect a president, vice-president, secretary, and treasurer?

43. For a segment of a radio show, a disc jockey can play 7 songs. If there are 13 songs to select from, in how many ways can the program for this segment be arranged?

44. Suppose you are asked to list, in order of preference, the three best movies you have seen this year. If you saw 20 movies during the year, in how many ways can the three best be chosen and ranked?

45. In a race in which six automobiles are entered and there are no ties, in how many ways can the first three finishers come in?

46. In a production of *West Side Story*, eight actors are considered for the male roles of Tony, Riff, and Bernardo. In how many ways can the director cast the male roles?

47. Nine bands have volunteered to perform at a benefit concert, but there is only enough time for five of the bands to play. How many lineups are possible?

48. How many arrangements can be made using four of the letters of the word COMBINE if no letter is to be used more than once?

49. In how many distinct ways can the letters of the word DALLAS be arranged?

50. In how many distinct ways can the letters of the word SCIENCE be arranged?

51. How many distinct permutations can be formed using the letters of the word TALLAHASSEE?

52. How many distinct permutations can be formed using the letters of the word TENNESSEE?

53. In how many ways can the digits in the number 5,446,666 be arranged?

54. In how many ways can the digits in the number 5,432,435 be arranged?

In Exercises 55–56, a signal can be formed by running different colored flags up a pole, one above the other.

55. Find the number of different signals consisting of eight flags that can be made using three white flags, four red flags, and one blue flag.

56. Find the number of different signals consisting of nine flags that can be made using three white flags, five red flags, and one blue flag.

Explaining the Concepts

57. What is a permutation?

58. Explain how to find $n!$, where n is a positive integer.

59. Explain the best way to evaluate $\frac{900!}{899!}$ without a calculator.

60. Describe what ${}_nP_r$ represents.

61. Write a word problem that can be solved by evaluating 5!.

62. Write a word problem that can be solved by evaluating ${}_7P_3$.

63. If 24 permutations can be formed using the letters in the word BAKE, why can't 24 permutations also be formed using the letters in the word BABE? How is the number of permutations in BABE determined?

Critical Thinking Exercises

Make Sense? *In Exercises 64–67, determine whether each statement makes sense or does not make sense, and explain your reasoning.*

64. I used the formula for ${}_nP_r$ to determine the number of ways the manager of a baseball team can form a 9-player batting order from a team of 25 players.

65. I used the Fundamental Counting Principle to determine the number of permutations of the letters of the word ENGLISH.

66. I used the formula for ${}_nP_r$ to determine the number of ways people can select their 9 favorite baseball players from a team of 25 players.

67. I used the Fundamental Counting Principle to determine the number of permutations of the letters of the word SUCCESS.

68. Ten people board an airplane that has 12 aisle seats. In how many ways can they be seated if they all select aisle seats?

69. Six horses are entered in a race. If two horses are tied for first place, and there are no ties among the other four horses, in how many ways can the six horses cross the finish line?

70. Performing at a concert are eight rock bands and eight jazz groups. How many ways can the program be arranged if the first, third, and eighth performers are jazz groups?

71. Five men and five women line up at a checkout counter in a store. In how many ways can they line up if the first person in line is a woman, and the people in line alternate woman, man, woman, man, and so on?

72. How many four-digit odd numbers less than 6000 can be formed using the digits 2, 4, 6, 7, 8, and 9?

73. Express ${}_nP_{n-2}$ without using factorials.

11.3 Combinations

WHAT AM I SUPPOSED TO LEARN?

After studying this section, you should be able to:

1. Distinguish between permutation and combination problems.
2. Solve problems involving combinations using the combinations formula.

DISCUSSING THE TRAGIC DEATH OF actor Heath Ledger at age 28, *USA Today* (January 30, 2008) cited five people who had achieved cult-figure status after death. Made iconic by death were Marilyn Monroe (actress, 1927–1962), James Dean (actor, 1931–1955), Jim Morrison (musician and lead singer of The Doors, 1943–1971), Janis Joplin (blues/rock singer, 1943–1970), and Jimi Hendrix (guitar virtuoso, 1943–1970).

Imagine that you ask your friends the following question: "Of these five people, which three would you select to be included in a documentary featuring the best of their work?" You are not asking your friends to rank their three favorite artists in any kind of order—they should merely select the three to be included in the documentary.

One friend answers, "Jim Morrison, Janis Joplin, and Jimi Hendrix." Another responds, "Jimi Hendrix, Janis Joplin, and Jim Morrison." These two people have the same artists in their group of selections, even if they are named in a different order. We are interested *in which artists are named, not the order in which they are named*, for the documentary. Because the items are taken without regard to order, this is not a permutation problem. No ranking of any sort is involved.

Later on, you ask your roommate which three artists she would select for the documentary. She names Marilyn Monroe, James Dean, and Jimi Hendrix. Her selection is different from those of your two other friends because different entertainers are cited.

Mathematicians describe the group of artists given by your roommate as a *combination*. A **combination** of items occurs when

- The items are selected from the same group (the five stars who were made iconic by death).
- No item is used more than once. (You may view Jimi Hendrix as a guitar god, but your three selections cannot be Jimi Hendrix, Jimi Hendrix, and Jimi Hendrix.)
- The order of the items makes no difference. (Morrison, Joplin, Hendrix is the same group in the documentary as Hendrix, Joplin, Morrison.)

1 *Distinguish between permutation and combination problems.*

Do you see the difference between a permutation and a combination? A permutation is an ordered arrangement of a given group of items. A combination is a group of items taken without regard to their order. **Permutation** problems involve situations in which **order matters**. **Combination** problems involve situations in which the **order** of the items **makes no difference**.

EXAMPLE 1 Distinguishing between Permutations and Combinations

For each of the following problems, determine whether the problem is one involving permutations or combinations. (It is not necessary to solve the problem.)

a. Six students are running for student government president, vice-president, and treasurer. The student with the greatest number of votes becomes the president, the second highest vote-getter becomes vice-president, and the student who gets the third largest number of votes will be treasurer. How many different outcomes are possible for these three positions?

b. Six people are on the board of supervisors for your neighborhood park. A three-person committee is needed to study the possibility of expanding the park. How many different committees could be formed from the six people?

c. Baskin-Robbins offers 31 different flavors of ice cream. One of its items is a bowl consisting of three scoops of ice cream, each a different flavor. How many such bowls are possible?

SOLUTION

a. Students are choosing three student government officers from six candidates. The order in which the officers are chosen makes a difference because each of the offices (president, vice-president, treasurer) is different. Order matters. This is a problem involving permutations.

b. A three-person committee is to be formed from the six-person board of supervisors. The order in which the three people are selected does not matter because they are not filling different roles on the committee. Because order makes no difference, this is a problem involving combinations.

c. A three-scoop bowl of three different flavors is to be formed from Baskin-Robbins's 31 flavors. The order in which the three scoops of ice cream are put into the bowl is irrelevant. A bowl with chocolate, vanilla, and strawberry is exactly the same as a bowl with vanilla, strawberry, and chocolate. Different orderings do not change things, and so this is a problem involving combinations.

CHECK POINT 1 For each of the following problems, determine whether the problem is one involving permutations or combinations. (It is not necessary to solve the problem.)

a. How many ways can you select 6 free DVDs from a list of 200 DVDs?

b. In a race in which there are 50 runners and no ties, in how many ways can the first three finishers come in?

2 Solve problems involving combinations using the combinations formula.

A Formula for Combinations

We have seen that the notation ${}_nP_r$ means the number of permutations of n things taken r at a time. Similarly, the notation **${}_nC_r$ means the number of combinations of n things taken r at a time.**

We can develop a formula for ${}_nC_r$ by comparing permutations and combinations. Consider the letters A, B, C, and D. The number of permutations of these four letters taken three at a time is

$$ {}_4P_3 = \frac{4!}{(4-3)!} = \frac{4!}{1!} = \frac{4 \cdot 3 \cdot 2 \cdot 1}{1} = 24. $$

Here are the 24 permutations:

ABC,	ABD,	ACD,	BCD,
ACB,	ADB,	ADC,	BDC,
BAC,	BAD,	CAD,	CBD,
BCA,	BDA,	CDA,	CDB,
CAB,	DAB,	DAC,	DBC,
CBA,	DBA,	DCA,	DCB.
This column contains only one combination, ABC.	This column contains only one combination, ABD.	This column contains only one combination, ACD.	This column contains only one combination, BCD.

Because the order of items makes no difference in determining combinations, each column of six permutations represents one combination. There is a total of four combinations:

$$ \text{ABC}, \quad \text{ABD}, \quad \text{ACD}, \quad \text{BCD}. $$

Thus, ${}_4C_3 = 4$: The number of combinations of 4 things taken 3 at a time is 4. With 24 permutations and only four combinations, there are 6, or 3!, times as many permutations as there are combinations.

In general, there are $r!$ times as many permutations of n things taken r at a time as there are combinations of n things taken r at a time. Thus, we find the number of combinations of n things taken r at a time by dividing the number of permutations of n things taken r at a time by $r!$.

$$ {}_nC_r = \frac{{}_nP_r}{r!} = \frac{\dfrac{n!}{(n-r)!}}{r!} = \frac{n!}{(n-r)!\,r!} $$

GREAT QUESTION!

Do I have to use the formula for ${}_nC_r$ to solve combination problems?

Yes. The number of combinations if r items are taken from n items cannot be found using the Fundamental Counting Principle and requires the use of the formula shown in the yellow box.

TECHNOLOGY

Graphing calculators have a menu item for calculating combinations, usually labeled ${}_nC_r$. (On TI graphing calculators, ${}_nC_r$ is selected using the MATH PROB menu.) For example, to find ${}_8C_3$, the keystrokes on most graphing calculators are

8 [${}_nC_r$] 3 [ENTER].

To access [${}_nC_r$] on a TI-84 Plus C, first press [MATH] and then use the left or right arrow key to highlight [PROB]. Press [3] for [${}_nC_r$].

If you are using a scientific calculator, check your manual to see whether there is a menu item for calculating combinations.

If you use your calculator's factorial key to find $\frac{8!}{5!3!}$, be sure to enclose the factorials in the denominator with parentheses

8 [!] [÷] [(] 5 [!] [×] 3 [!] [)]

pressing [=] or [ENTER] to obtain the answer.

COMBINATIONS OF n THINGS TAKEN r AT A TIME

The number of possible combinations if r items are taken from n items is

$${}_nC_r = \frac{n!}{(n-r)!\,r!}.$$

EXAMPLE 2 Using the Formula for Combinations

A three-person committee is needed to study ways of improving public transportation. How many committees could be formed from the eight people on the board of supervisors?

SOLUTION

The order in which the three people are selected does not matter. This is a problem of selecting $r = 3$ people from a group of $n = 8$ people. We are looking for the number of combinations of eight things taken three at a time. We use the formula

$${}_nC_r = \frac{n!}{(n-r)!\,r!}$$

with $n = 8$ and $r = 3$.

$${}_8C_3 = \frac{8!}{(8-3)!\,3!} = \frac{8!}{5!\,3!} = \frac{8\cdot7\cdot6\cdot5!}{5!\cdot3\cdot2\cdot1} = \frac{8\cdot7\cdot6\cdot\cancel{5!}}{\cancel{5!}\cdot3\cdot2\cdot1} = 56$$

Thus, 56 committees of three people each can be formed from the eight people on the board of supervisors.

CHECK POINT 2 You volunteer to pet-sit for your friend who has seven different animals. How many different pet combinations are possible if you take three of the seven pets?

EXAMPLE 3 Using the Formula for Combinations

In poker, a person is dealt 5 cards from a standard 52-card deck. The order in which the 5 cards are received does not matter. How many different 5-card poker hands are possible?

SOLUTION

Because the order in which the 5 cards are dealt does not matter, this is a problem involving combinations. We are looking for the number of combinations of $n = 52$ cards drawn $r = 5$ at a time. We use the formula

$${}_nC_r = \frac{n!}{(n-r)!\,r!}$$

with $n = 52$ and $r = 5$.

$${}_{52}C_5 = \frac{52!}{(52-5)!\,5!} = \frac{52!}{47!\,5!} = \frac{52\cdot51\cdot50\cdot49\cdot48\cdot\cancel{47!}}{\cancel{47!}\cdot5\cdot4\cdot3\cdot2\cdot1} = 2{,}598{,}960$$

Thus, there are 2,598,960 different 5-card poker hands possible. It surprises many people that more than 2.5 million 5-card hands can be dealt from a mere 52 cards.

FIGURE 11.3 A royal flush

If you are a card player, it does not get any better than to be dealt the 5-card poker hand shown in **Figure 11.3**. This hand is called a *royal flush*. It consists of an ace, king, queen, jack, and 10, all of the same suit: all hearts, all diamonds, all clubs, or all spades. The probability of being dealt a royal flush involves calculating the number of ways of being dealt such a hand: just 4 of all 2,598,960 possible hands. In the next section, we move from counting possibilities to computing probabilities.

CHECK POINT 3 How many different 4-card hands can be dealt from a deck that has 16 different cards?

EXAMPLE 4 *Using the Formula for Combinations and the Fundamental Counting Principle*

In January 2017, the U.S. Senate consisted of 46 Democrats, 52 Republicans, and 2 Independents. How many distinct five-person committees can be formed if each committee must have 2 Democrats and 3 Republicans?

SOLUTION

The order in which the members are selected does not matter. Thus, this is a problem involving combinations.

We begin with the number of ways of selecting 2 Democrats out of 46 Democrats without regard to order. We are looking for the number of combinations of $n = 46$ people taken $r = 2$ people at a time. We use the formula

$$_nC_r = \frac{n!}{(n-r)!\,r!}$$

with $n = 46$ and $r = 2$.

We are picking 2 Democrats out of 46 Democrats.

$$_{46}C_2 = \frac{46!}{(46-2)!\,2!} = \frac{46!}{44!\,2!} = \frac{46 \cdot 45 \cdot \cancel{44!}}{\cancel{44!} \cdot 2 \cdot 1} = \frac{46 \cdot 45}{2 \cdot 1} = 1035$$

There are 1035 ways to choose two Democrats for a committee.

Next, we find the number of ways of selecting 3 Republicans out of 52 Republicans without regard to order. We are looking for the number of combinations of $n = 52$ people taken $r = 3$ people at a time. Once again, we use the formula

$$_nC_r = \frac{n!}{(n-r)!\,r!}.$$

This time, $n = 52$ and $r = 3$.

We are picking 3 Republicans out of 52 Republicans.

$$_{52}C_3 = \frac{52!}{(52-3)!\,3!} = \frac{52!}{49!\,3!} = \frac{52 \cdot 51 \cdot 50 \cdot \cancel{49!}}{\cancel{49!} \cdot 3 \cdot 2 \cdot 1} = \frac{52 \cdot 51 \cdot 50}{3 \cdot 2 \cdot 1} = 22{,}100$$

There are 22,100 ways to choose three Republicans for a committee.

We use the Fundamental Counting Principle to find the number of committees that can be formed.

$$_{46}C_2 \cdot {}_{52}C_3 = 1035 \cdot 22{,}100 = 22{,}873{,}500$$

Thus, 22,873,500 distinct committees can be formed.

CHECK POINT 4 A zoo has six male bears and seven female bears. Two male bears and three female bears will be selected for an animal exchange program with another zoo. How many five-bear collections are possible?

Concept and Vocabulary Check

Fill in each blank so that the resulting statement is true.

1. The number of possible combinations if r objects are taken from n items is ${}_nC_r =$ ________.
2. The formula for ${}_nC_r$ has the same numerator as the formula for ${}_nP_r$ but contains an extra factor of ____ in the denominator.

In Exercises 3–4, determine whether each statement is true or false. If the statement is false, make the necessary change(s) to produce a true statement.

3. Combination problems involve situations in which the order of the items makes a difference. ______
4. Permutation problems involve situations in which the order of the items does not matter. ______

Exercise Set 11.3

Practice Exercises

In Exercises 1–4, does the problem involve permutations or combinations? Explain your answer. (It is not necessary to solve the problem.)

1. A medical researcher needs 6 people to test the effectiveness of an experimental drug. If 13 people have volunteered for the test, in how many ways can 6 people be selected?
2. Fifty people purchase raffle tickets. Three winning tickets are selected at random. If first prize is \$1000, second prize is \$500, and third prize is \$100, in how many different ways can the prizes be awarded?
3. How many different four-letter passwords can be formed from the letters A, B, C, D, E, F, and G if no repetition of letters is allowed?
4. Fifty people purchase raffle tickets. Three winning tickets are selected at random. If each prize is \$500, in how many different ways can the prizes be awarded?

In Exercises 5–20, use the formula for ${}_nC_r$ to evaluate each expression.

5. ${}_6C_5$
6. ${}_8C_7$
7. ${}_9C_5$
8. ${}_{10}C_6$
9. ${}_{11}C_4$
10. ${}_{12}C_5$
11. ${}_8C_1$
12. ${}_7C_1$
13. ${}_7C_7$
14. ${}_4C_4$
15. ${}_{30}C_3$
16. ${}_{25}C_4$
17. ${}_5C_0$
18. ${}_6C_0$
19. $\dfrac{{}_7C_3}{{}_5C_4}$
20. $\dfrac{{}_{10}C_3}{{}_6C_4}$

Practice Plus

In Exercises 21–28, evaluate each expression.

21. $\dfrac{{}_7P_3}{3!} - {}_7C_3$
22. $\dfrac{{}_{20}P_2}{2!} - {}_{20}C_2$
23. $1 - \dfrac{{}_3P_2}{{}_4P_3}$
24. $1 - \dfrac{{}_5P_3}{{}_{10}P_4}$
25. $\dfrac{{}_7C_3}{{}_5C_4} - \dfrac{98!}{96!}$
26. $\dfrac{{}_{10}C_3}{{}_6C_4} - \dfrac{46!}{44!}$
27. $\dfrac{{}_4C_2 \cdot {}_6C_1}{{}_{18}C_3}$
28. $\dfrac{{}_5C_1 \cdot {}_7C_2}{{}_{12}C_3}$

Application Exercises

Use the formula for ${}_nC_r$ to solve Exercises 29–40.

29. An election ballot asks voters to select three city commissioners from a group of six candidates. In how many ways can this be done?
30. A four-person committee is to be elected from an organization's membership of 11 people. How many different committees are possible?
31. Of 12 possible books, you plan to take 4 with you on vacation. How many different collections of 4 books can you take?
32. There are 14 standbys who hope to get seats on a flight, but only 6 seats are available on the plane. How many different ways can the 6 people be selected?
33. You volunteer to help drive children at a charity event to the zoo, but you can fit only 8 of the 17 children present in your van. How many different groups of 8 children can you drive?
34. Of the 100 people in the U.S. Senate, 18 serve on the Foreign Relations Committee. How many ways are there to select Senate members for this committee (assuming party affiliation is not a factor in the selection)?
35. To win at LOTTO in the state of Florida, one must correctly select 6 numbers from a collection of 53 numbers (1 through 53). The order in which the selection is made does not matter. How many different selections are possible?
36. To win in the New York State lottery, one must correctly select 6 numbers from 59 numbers. The order in which the selection is made does not matter. How many different selections are possible?
37. In how many ways can a committee of four men and five women be formed from a group of seven men and seven women?
38. How many different committees can be formed from 5 professors and 15 students if each committee is made up of 2 professors and 10 students?
39. The U.S. Senate of the 109th Congress consisted of 55 Republicans, 44 Democrats, and 1 Independent. How many committees can be formed if each committee must have 4 Republicans and 3 Democrats?
40. A mathematics exam consists of 10 multiple-choice questions and 5 open-ended problems in which all work must be shown. If an examinee must answer 8 of the multiple-choice questions and 3 of the open-ended problems, in how many ways can the questions and problems be chosen?

In Exercises 41–60, solve by the method of your choice.

41. In a race in which six automobiles are entered and there are no ties, in how many ways can the first four finishers come in?
42. A book club offers a choice of 8 books from a list of 40. In how many ways can a member make a selection?

43. A medical researcher needs 6 people to test the effectiveness of an experimental drug. If 13 people have volunteered for the test, in how many ways can 6 people be selected?

44. Fifty people purchase raffle tickets. Three winning tickets are selected at random. If first prize is \$1000, second prize is \$500, and third prize is \$100, in how many different ways can the prizes be awarded?

45. From a club of 20 people, in how many ways can a group of three members be selected to attend a conference?

46. Fifty people purchase raffle tickets. Three winning tickets are selected at random. If each prize is \$500, in how many different ways can the prizes be awarded?

47. How many different four-letter passwords can be formed from the letters A, B, C, D, E, F, and G if no repetition of letters is allowed?

48. Nine comedy acts will perform over two evenings. Five of the acts will perform on the first evening. How many ways can the schedule for the first evening be made?

49. Using 15 flavors of ice cream, how many cones with three different flavors can you create if it is important to you which flavor goes on the top, middle, and bottom?

50. Baskin-Robbins offers 31 different flavors of ice cream. One of its items is a bowl consisting of three scoops of ice cream, each a different flavor. How many such bowls are possible?

51. A restaurant lunch special allows the customer to choose two vegetables from this list: okra, corn, peas, carrots, and squash. How many outcomes are possible if the customer chooses two different vegetables?

52. There are six employees in the stock room at an appliance retail store. The manager will choose three of them to deliver a refrigerator. How many three-person groups are possible?

53. You have three dress shirts, two ties, and two jackets. You need to select a dress shirt, a tie, and a jacket for work today. How many outcomes are possible?

54. You have four flannel shirts. You are going to choose two of them to take on a camping trip. How many outcomes are possible?

55. A chef has five brands of hot sauce. Three of the brands will be chosen to mix into gumbo. How many outcomes are possible?

56. In the Mathematics Department, there are four female professors and six male professors. Three female professors will be chosen to serve as mentors for a special program designed to encourage female students to pursue careers in mathematics. In how many ways can the professors be chosen?

57. Three are four Democrats and five Republicans on the county commission. From among their group they will choose a committee of two Democrats and two Republicans to examine a proposal to purchase land for a new county park. How many four-person groups are possible?

58. An office employs six customer service representatives. Each day, two of them are randomly selected and their customer interactions are monitored for the purposes of improving customer relations. In how many ways can the representatives be chosen?

59. A group of campers is going to occupy five campsites at a campground. There are 12 campsites from which to choose. In how many ways can the campsites be chosen?

60. Your mom and dad have driven to work in separate cars. When they arrive, there are seven empty spaces in the parking lot. They each choose a parking space. How many outcomes are possible?

(*Source for Exercises 51–60:* James Wooland, *CLAST Manual for Thinking Mathematically*)

Thousands of jokes have been told about marriage and divorce. Exercises 61–68 are based on the following observations:

- *"By all means, marry; if you get a good wife, you'll be happy. If you get a bad one, you'll become a philosopher." —Socrates*
- *"My wife and I were happy for 20 years. Then we met." —Rodney Dangerfield*
- *"Whatever you may look like, marry a man your own age. As your beauty fades, so will his eyesight." —Phyllis Diller*
- *"Why do Jewish divorces cost so much? Because they're worth it." —Henny Youngman*
- *"I think men who have a pierced ear are better prepared for marriage. They've experienced pain and bought jewelry." —Rita Rudner*
- *"For a while we pondered whether to take a vacation or get a divorce. We decided that a trip to Bermuda is over in two weeks, but a divorce is something you always have." —Woody Allen*

61. In how many ways can these six jokes be ranked from best to worst?

62. If Socrates's thoughts about marriage are excluded, in how many ways can the remaining five jokes be ranked from best to worst?

63. In how many ways can people select their three favorite jokes from these thoughts about marriage and divorce?

64. In how many ways can people select their two favorite jokes from these thoughts about marriage and divorce?

65. If the order in which these jokes are told makes a difference in terms of how they are received, how many ways can they be delivered if Socrates's comments are scheduled first and Dangerfield's joke is told last?

66. If the order in which these jokes are told makes a difference in terms of how they are received, how many ways can they be delivered if a joke by a woman (Rudner or Diller) is told first?

67. In how many ways can people select their favorite joke told by a woman (Rudner or Diller) and their two favorite jokes told by a man?

68. In how many ways can people select their favorite joke told by a woman (Rudner or Diller) and their three favorite jokes told by a man?

Explaining the Concepts

69. What is a combination?

70. Explain how to distinguish between permutation and combination problems.

71. Write a word problem that can be solved by evaluating ${}_7C_3$.

Critical Thinking Exercises

Make Sense? *In Exercises 72–75, determine whether each statement makes sense or does not make sense, and explain your reasoning.*

72. I used the formula for ${}_nC_r$ to determine the number of possible outcomes when ranking five politicians from most admired to least admired.

73. I used the formula for ${}_nC_r$ to determine how many four-letter passwords with no repeated letters can be formed using a, d, h, n, p, and w.

74. I solved a problem involving the number of possible outcomes when selecting from two groups, which required me to use both the formula for ${}_nC_r$ and the Fundamental Counting Principle.

75. I used the formula for ${}_nC_r$ to determine how many outcomes are possible when choosing four letters from a, d, h, n, p, and w.

76. Write a word problem that can be solved by evaluating ${}_{10}C_3 \cdot {}_7C_2$.

77. A 6/53 lottery involves choosing 6 of the numbers from 1 through 53 and a 5/36 lottery involves choosing 5 of the numbers from 1 through 36. The order in which the numbers are chosen does not matter. Which lottery is easier to win? Explain your answer.

78. If the number of permutations of n objects taken r at a time is six times the number of combinations of n objects taken r at a time, determine the value of r. Is there enough information to determine the value of n? Why or why not?

79. In a group of 20 people, how long will it take each person to shake hands with each of the other persons in the group, assuming that it takes three seconds for each shake and only 2 people can shake hands at a time? What if the group is increased to 40 people?

80. A sample of 4 telephones is selected from a shipment of 20 phones. There are 5 defective telephones in the shipment. How many of the samples of 4 phones do not include any of the defective ones?

11.4 Fundamentals of Probability

WHAT AM I SUPPOSED TO LEARN?

After studying this section, you should be able to:

1. Compute theoretical probability.
2. Compute empirical probability.

HOW MANY HOURS OF SLEEP DO YOU TYPICALLY GET each night? **Table 11.1** indicates that 75 million out of 300 million Americans are getting six hours of sleep on a typical night. The *probability* of an American getting six hours of sleep on a typical night is $\frac{75}{300}$. This fraction can be reduced to $\frac{1}{4}$, or expressed as 0.25, or 25%. Thus, 25% of Americans get six hours of sleep each night.

We find a probability by dividing one number by another. Probabilities are assigned to an *event*, such as getting six hours of sleep on a typical night. Events that are certain to occur are assigned probabilities of 1, or 100%. For example, the probability that a given individual will eventually die is 1. Although Woody Allen whined, "I don't want to achieve immortality through my work. I want to achieve it through not dying," death (and taxes) are always certain. By contrast, if an event cannot occur, its probability is 0. Regrettably, the probability that Elvis will return and serenade us with one final reprise of "Don't Be Cruel" (and we hope we're not) is 0.

Probabilities of events are expressed as numbers ranging from 0 to 1, or 0% to 100%. The closer the probability of a given event is to 1, the more likely it is that the event will occur. The closer the probability of a given event is to 0, the less likely it is that the event will occur.

100% or 1 — *Certain*
Likely
50% or $\frac{1}{2}$ — *50-50 Chance*
Unlikely
0% or 0 — *Impossible*

Possible Values for Probabilities

TABLE 11.1 The Hours of Sleep Americans Get on a Typical Night

Hours of Sleep	Number of Americans, in millions
4 or less	12
5	27
6	75
7	90
8	81
9	9
10 or more	6

Total: 300

Source: Discovery Health Media

1 *Compute theoretical probability.*

Theoretical Probability

You toss a coin. Although it is equally likely to land either heads up, denoted by H, or tails up, denoted by T, the actual outcome is uncertain. Any occurrence for which the outcome is uncertain is called an **experiment**. Thus, tossing a coin is an example of an experiment. The set of all possible outcomes of an experiment is the **sample space** of the experiment, denoted by S. The sample space for the coin-tossing experiment is

$$S = \{H, T\}.$$

Lands heads up — Lands tails up

An **event**, denoted by E, is any subset of a sample space. For example, the subset $E = \{T\}$ is the event of landing tails up when a coin is tossed.

Theoretical probability applies to situations like this, in which the sample space only contains equally likely outcomes, all of which are known. To calculate the theoretical probability of an event, we divide the number of outcomes resulting in the event by the total number of outcomes in the sample space.

COMPUTING THEORETICAL PROBABILITY

If an event E has $n(E)$ equally likely outcomes and its sample space S has $n(S)$ equally likely outcomes, the **theoretical probability** of event E, denoted by $P(E)$, is

$$P(E) = \frac{\text{number of outcomes in event } E}{\text{total number of possible outcomes}} = \frac{n(E)}{n(S)}.$$

How can we use this formula to compute the probability of a coin landing tails up? We use the following sets:

$$E = \{T\} \qquad S = \{H, T\}.$$

This is the event of landing tails up. — This is the sample space with all equally likely outcomes.

The probability of a coin landing tails up is

$$P(E) = \frac{\text{number of outcomes that result in tails up}}{\text{total number of possible outcomes}} = \frac{n(E)}{n(S)} = \frac{1}{2}.$$

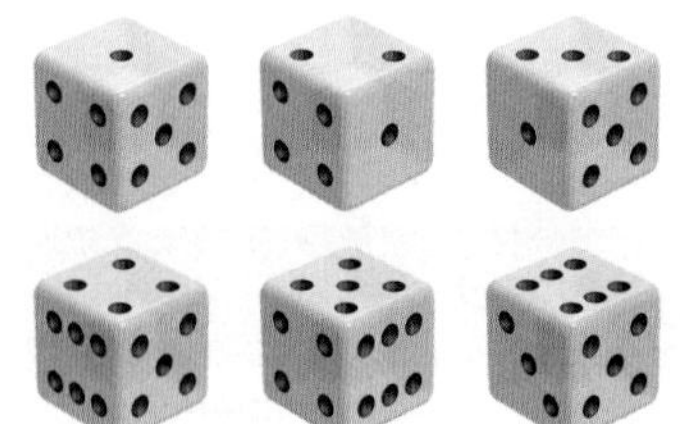

FIGURE 11.4 Outcomes when a die is rolled

Theoretical probability applies to many games of chance, including dice rolling, lotteries, card games, and roulette. We begin with rolling a die. **Figure 11.4** illustrates that when a die is rolled, there are six equally likely possible outcomes. The sample space can be shown as

$$S = \{1, 2, 3, 4, 5, 6\}.$$

EXAMPLE 1 *Computing Theoretical Probability*

A die is rolled once. Find the probability of rolling

a. a 3. **b.** an even number. **c.** a number less than 5.

d. a number less than 10. **e.** a number greater than 6.

SOLUTION

The sample space is $S = \{1, 2, 3, 4, 5, 6\}$ with $n(S) = 6$. We will use 6, the total number of possible outcomes, in the denominator of each probability fraction.

a. The phrase "rolling a 3" describes the event $E = \{3\}$. This event can occur in one way: $n(E) = 1$.

$$P(3) = \frac{\text{number of outcomes that result in 3}}{\text{total number of possible outcomes}} = \frac{n(E)}{n(S)} = \frac{1}{6}$$

The probability of rolling a 3 is $\frac{1}{6}$.

b. The phrase "rolling an even number" describes the event $E = \{2, 4, 6\}$. This event can occur in three ways: $n(E) = 3$.

$$P(\text{even number}) = \frac{\text{number of outcomes that result in an even number}}{\text{total number of possible outcomes}} = \frac{n(E)}{n(S)} = \frac{3}{6} = \frac{1}{2}$$

The probability of rolling an even number is $\frac{1}{2}$.

c. The phrase "rolling a number less than 5" describes the event $E = \{1, 2, 3, 4\}$. This event can occur in four ways: $n(E) = 4$.

$$P(\text{less than 5}) = \frac{\text{number of outcomes that are less than 5}}{\text{total number of possible outcomes}} = \frac{n(E)}{n(S)} = \frac{4}{6} = \frac{2}{3}$$

The probability of rolling a number less than 5 is $\frac{2}{3}$.

d. The phrase "rolling a number less than 10" describes the event $E = \{1, 2, 3, 4, 5, 6\}$. This event can occur in six ways: $n(E) = 6$. Can you see that all of the possible outcomes are less than 10? This event is certain to occur.

$$P(\text{less than 10}) = \frac{\text{number of outcomes that are less than 10}}{\text{total number of possible outcomes}} = \frac{n(E)}{n(S)} = \frac{6}{6} = 1$$

The probability of any certain event is 1.

e. The phrase "rolling a number greater than 6" describes an event that cannot occur, or the empty set. Thus, $E = \varnothing$ and $n(E) = 0$.

$$P(\text{greater than 6}) = \frac{\text{number of outcomes that are greater than 6}}{\text{total number of possible outcomes}} = \frac{n(E)}{n(S)} = \frac{0}{6} = 0$$

The probability of an event that cannot occur is 0.

In Example 1, there are six possible outcomes, each with a probability of $\frac{1}{6}$:

$$P(1) = \frac{1}{6} \quad P(2) = \frac{1}{6} \quad P(3) = \frac{1}{6} \quad P(4) = \frac{1}{6} \quad P(5) = \frac{1}{6} \quad P(6) = \frac{1}{6}.$$

The sum of these probabilities is 1: $\frac{1}{6} + \frac{1}{6} + \frac{1}{6} + \frac{1}{6} + \frac{1}{6} + \frac{1}{6} = 1$. In general, **the sum of the theoretical probabilities of all possible outcomes in the sample space is 1**.

CHECK POINT 1 A die is rolled once. Find the probability of rolling

a. a 2.

b. a number less than 4.

c. a number greater than 7.

d. a number less than 7.

Our next example involves a standard 52-card deck, illustrated in **Figure 11.5**. The deck has four suits: Hearts and diamonds are red, and clubs and spades are black. Each suit has 13 different face values—A(ace), 2, 3, 4, 5, 6, 7, 8, 9, 10, J(jack), Q(queen), and K(king). Jacks, queens, and kings are called **picture cards** or **face cards**.

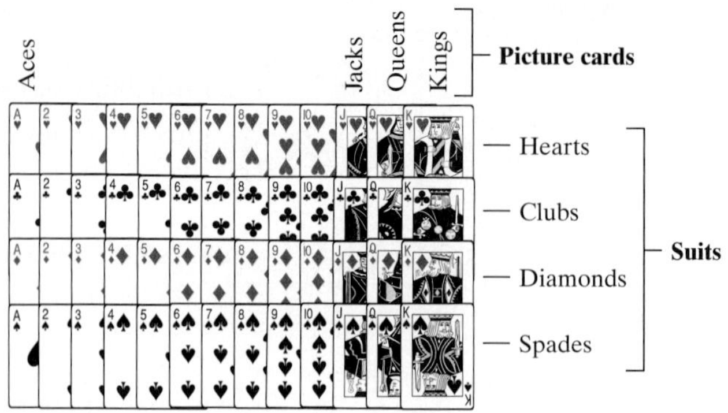

FIGURE 11.5 A standard 52-card deck

EXAMPLE 2 *Probability and a Deck of 52 Cards*

You are dealt one card from a standard 52-card deck. Find the probability of being dealt

a. a king. **b.** a heart. **c.** the king of hearts.

SOLUTION

Because there are 52 cards in the deck, the total number of possible ways of being dealt a single card is 52. The number of outcomes in the sample space is 52: $n(S) = 52$. We use 52 as the denominator of each probability fraction.

a. Let E be the event of being dealt a king. Because there are four kings in the deck, this event can occur in four ways: $n(E) = 4$.

$$P(\text{king}) = \frac{\text{number of outcomes that result in a king}}{\text{total number of possible outcomes}} = \frac{n(E)}{n(S)} = \frac{4}{52} = \frac{1}{13}$$

The probability of being dealt a king is $\frac{1}{13}$.

b. Let E be the event of being dealt a heart. Because there are 13 hearts in the deck, this event can occur in 13 ways: $n(E) = 13$.

$$P(\text{heart}) = \frac{\text{number of outcomes that result in a heart}}{\text{total number of possible outcomes}} = \frac{n(E)}{n(S)} = \frac{13}{52} = \frac{1}{4}$$

The probability of being dealt a heart is $\frac{1}{4}$.

c. Let E be the event of being dealt the king of hearts. Because there is only one card in the deck that is the king of hearts, this event can occur in just one way: $n(E) = 1$.

$$P(\text{king of hearts}) = \frac{\text{number of outcomes that result in the king of hearts}}{\text{total number of possible outcomes}} = \frac{n(E)}{n(S)} = \frac{1}{52}$$

The probability of being dealt the king of hearts is $\frac{1}{52}$.

CHECK POINT 2 You are dealt one card from a standard 52-card deck. Find the probability of being dealt

a. an ace. **b.** a red card. **c.** a red king.

Blitzer Bonus

Try to Bench Press This at the Gym

You have a deck of cards for every permutation of the 52 cards. If each deck weighed only as much as a single hydrogen atom, all the decks together would weigh a billion times as much as the Sun.

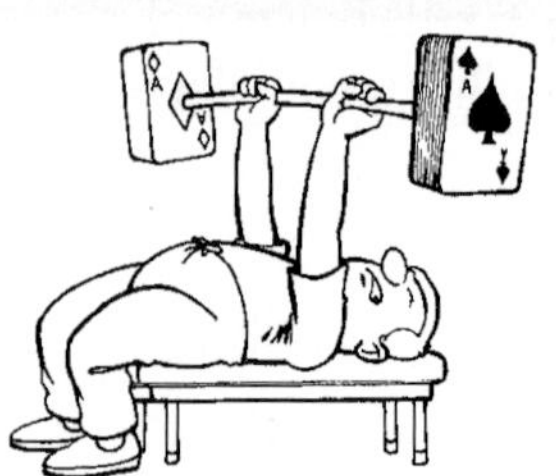

Source: Isaac Asimov's Book of Facts

Probabilities play a valuable role in the science of genetics. Example 3 deals with cystic fibrosis, an inherited lung disease occurring in about 1 out of every 2000 births among Caucasians and in about 1 out of every 250,000 births among non-Caucasians.

EXAMPLE 3 *Probabilities in Genetics*

Each person carries two genes that are related to the absence or presence of the disease cystic fibrosis. Most Americans have two normal genes for this trait and are unaffected by cystic fibrosis. However, 1 in 25 Americans carries one normal gene and one defective gene. If we use c to represent a defective gene and C a normal gene, such a carrier can be designated as Cc. Thus, CC is a person who neither carries nor has cystic fibrosis, Cc is a carrier who is not actually sick, and cc is a person sick with the disease. **Table 11.2** shows the four equally likely outcomes for a child's genetic inheritance from two parents who are both carrying one cystic fibrosis gene. One of the genes is passed on to the child from each parent.

TABLE 11.2 Cystic Fibrosis and Genetic Inheritance

		Second Parent	
		C	c
First Parent	C	CC	Cc
	c	cC	cc

Shown in the table are the four possibilities for a child whose parents each carry one cystic fibrosis gene.

If each parent carries one cystic fibrosis gene, what is the probability that their child will have cystic fibrosis?

SOLUTION

Table 11.2 shows that there are four equally likely outcomes. The sample space is $S = \{CC, Cc, cC, cc\}$ and $n(S) = 4$. The phrase "will have cystic fibrosis" describes only the cc child. Thus, $E = \{cc\}$ and $n(E) = 1$.

$$P(\text{cystic fibrosis}) = \frac{\text{number of outcomes that result in cystic fibrosis}}{\text{total number of possible outcomes}} = \frac{n(E)}{n(S)} = \frac{1}{4}$$

If each parent carries one cystic fibrosis gene, the probability that their child will have cystic fibrosis is $\frac{1}{4}$.

CHECK POINT 3 Use **Table 11.2** in Example 3 to solve this exercise. If each parent carries one cystic fibrosis gene, find the probability that their child will be a carrier of the disease who is not actually sick.

2 *Compute empirical probability.*

Empirical Probability

Theoretical probability is based on a set of equally likely outcomes and the number of elements in the set. By contrast, *empirical probability* applies to situations in which we observe how frequently an event occurs. We use the following formula to compute the empirical probability of an event:

COMPUTING EMPIRICAL PROBABILITY

The empirical probability of event E is

$$P(E) = \frac{\text{observed number of times } E \text{ occurs}}{\text{total number of observed occurrences}}.$$

EXAMPLE 4 Computing Empirical Probability

In 2015, there were approximately 254 million Americans ages 15 or older. **Table 11.3** shows the distribution, by marital status and gender, of this population. Numbers in the table are expressed in millions.

TABLE 11.3 Marital Status of the U.S. Population, Ages 15 or Older, 2015, in Millions

	Married	Never Married	Divorced	Widowed	Total
Male	66	43	11	3	123
Female	67	38	15	11	131
Total	133	81	26	14	254

Total male: $66 + 43 + 11 + 3 = 123$

Total female: $67 + 38 + 15 + 11 = 131$

Total married: $66 + 67 = 133$

Total never married: $43 + 38 = 81$

Total divorced: $11 + 15 = 26$

Total widowed: $3 + 11 = 14$

Total adult population: $123 + 131 = 254$

Source: U.S. Census Bureau

GREAT QUESTION!

What do you mean by saying that one person is *randomly selected* from the population?

This means that every person in the population has an equal chance of being chosen. We'll have much more to say about random selections in Chapter 12, Statistics.

If one person is randomly selected from the population described in **Table 11.3**, find the probability, to the nearest hundredth, that the person

a. is divorced. **b.** is female.

SOLUTION

a. The probability of selecting a divorced person is the observed number of divorced people, 26 (million), divided by the total number in the population, 254 (million).

$$P(\text{selecting a divorced person from the population})$$
$$= \frac{\text{number of divorced people}}{\text{total number in the population}} = \frac{26}{254} \approx 0.10$$

The empirical probability of selecting a divorced person from the population in **Table 11.3** is approximately 0.10.

b. The probability of selecting a female is the observed number of females, 131 (million), divided by the total number in the population, 254 (million).

$$P(\text{selecting a female from the population})$$
$$= \frac{\text{number of females}}{\text{total number in the population}} = \frac{131}{254} \approx 0.52$$

The empirical probability of selecting a female from the population in **Table 11.3** is approximately 0.52.

CHECK POINT 4 If one person is randomly selected from the population described in **Table 11.3**, find the probability, expressed as a decimal rounded to the nearest hundredth, that the person

a. has never been married.

b. is male.

In certain situations, we can establish a relationship between the two kinds of probability. Consider, for example, a coin that is equally likely to land heads or tails. Such a coin is called a **fair coin**. Empirical probability can be used to determine whether a coin is fair. Suppose we toss a coin 10, 50, 100, 1000, 10,000, and 100,000 times. We record the number of heads observed, shown in **Table 11.4** at the top of the next page. For each of the six cases in the table, the empirical probability of heads is determined by dividing the number of heads observed by the number of tosses.

TABLE 11.4 Empirical Probabilities of Heads as the Number of Tosses Increases

Number of Tosses	Number of Heads Observed	Empirical Probability of Heads, or $P(H)$
10	4	$P(H) = \frac{4}{10} = 0.4$
50	27	$P(H) = \frac{27}{50} = 0.54$
100	44	$P(H) = \frac{44}{100} = 0.44$
1000	530	$P(H) = \frac{530}{1000} = 0.53$
10,000	4851	$P(H) = \frac{4851}{10,000} = 0.4851$
100,000	49,880	$P(H) = \frac{49,880}{100,000} = 0.4988$

A pattern is exhibited by the empirical probabilities in the right-hand column of **Table 11.4**. As the number of tosses increases, the empirical probabilities tend to get closer to 0.5, the theoretical probability. These results give us no reason to suspect that the coin is not fair.

Table 11.4 illustrates an important principle when observing uncertain outcomes such as the event of a coin landing on heads. As an experiment is repeated more and more times, the empirical probability of an event tends to get closer to the theoretical probability of that event. This principle is known as the **law of large numbers**.

Concept and Vocabulary Check

Fill in each blank so that the resulting statement is true.

1. The set of all possible outcomes of an experiment is called the ____________ of the experiment.
2. The theoretical probability of event E, denoted by _______, is the ______________________ divided by the ______________________.
3. A standard bridge deck has _____ cards with four suits: ________ and __________ are red, and ______ and _________ are black.
4. Probability that is based on situations in which we observe how frequently an event occurs is called _________ probability.

In Exercises 5–8, determine whether each statement is true or false. If the statement is false, make the necessary change(s) to produce a true statement.

5. If an event is certain to occur, its probability is 1. ______
6. If an event cannot occur, its probability is –1. ______
7. The sum of the probabilities of all possible outcomes in an experiment is 1. ______
8. If an experiment is repeated more and more times, the theoretical probability of an event tends to get closer to the empirical probability of that event. ______

Exercise Set 11.4

Practice and Application Exercises

In Exercises 1–54, express each probability as a fraction reduced to lowest terms.

In Exercises 1–10, a die is rolled. The set of equally likely outcomes is {1, 2, 3, 4, 5, 6}. *Find the probability of rolling*

1. a 4.
2. a 5.
3. an odd number.
4. a number greater than 3.
5. a number less than 3.
6. a number greater than 4.
7. a number less than 20.
8. a number less than 8.
9. a number greater than 20.
10. a number greater than 8.

In Exercises 11–20, you are dealt one card from a standard 52-card deck. Find the probability of being dealt

11. a queen.
12. a jack.
13. a club.
14. a diamond.
15. a picture card.
16. a card greater than 3 and less than 7.
17. the queen of spades.
18. the ace of clubs.
19. a diamond and a spade.
20. a card with a green heart.

In Exercises 21–26, a fair coin is tossed two times in succession. The set of equally likely outcomes is {HH, HT, TH, TT}. Find the probability of getting

21. two heads.
22. two tails.
23. the same outcome on each toss.

24. different outcomes on each toss.

25. a head on the second toss.

26. at least one head.

In Exercises 27–34, you select a family with three children. If M represents a male child and F a female child, the set of equally likely outcomes for the children's genders is {MMM, MMF, MFM, MFF, FMM, FMF, FFM, FFF}. Find the probability of selecting a family with

27. exactly one female child.

28. exactly one male child.

29. exactly two male children.

30. exactly two female children.

31. at least one male child.

32. at least two female children.

33. four male children.

34. fewer than four female children.

In Exercises 35–40, a single die is rolled twice. The 36 equally likely outcomes are shown as follows:

	Second Roll					
First Roll	⚀	⚁	⚂	⚃	⚄	⚅
⚀	(1, 1)	(1, 2)	(1, 3)	(1, 4)	(1, 5)	(1, 6)
⚁	(2, 1)	(2, 2)	(2, 3)	(2, 4)	(2, 5)	(2, 6)
⚂	(3, 1)	(3, 2)	(3, 3)	(3, 4)	(3, 5)	(3, 6)
⚃	(4, 1)	(4, 2)	(4, 3)	(4, 4)	(4, 5)	(4, 6)
⚄	(5, 1)	(5, 2)	(5, 3)	(5, 4)	(5, 5)	(5, 6)
⚅	(6, 1)	(6, 2)	(6, 3)	(6, 4)	(6, 5)	(6, 6)

Find the probability of getting

35. two even numbers.

36. two odd numbers.

37. two numbers whose sum is 5.

38. two numbers whose sum is 6.

39. two numbers whose sum exceeds 12.

40. two numbers whose sum is less than 13.

Use the spinner shown to answer Exercises 41–48. Assume that it is equally probable that the pointer will land on any one of the ten colored regions. If the pointer lands on a borderline, spin again.

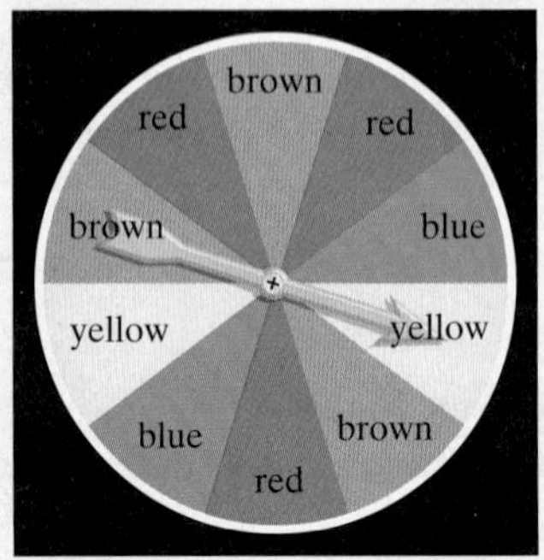

Find the probability that the spinner lands in

41. a red region.

42. a yellow region.

43. a blue region.

44. a brown region.

45. a region that is red or blue.

46. a region that is yellow or brown.

47. a region that is red and blue.

48. a region that is yellow and brown.

Exercises 49–54 deal with sickle cell anemia, an inherited disease in which red blood cells become distorted and deprived of oxygen. Approximately 1 in every 500 African-American infants is born with the disease; only 1 in 160,000 white infants has the disease. A person with two sickle cell genes will have the disease, but a person with only one sickle cell gene will have a mild, nonfatal anemia called sickle cell trait. (Approximately 8%–10% of the African-American population has this trait.)

		Second Parent	
		S	*s*
First	*S*	*SS*	*Ss*
Parent	*s*	*sS*	*ss*

If we use s to represent a sickle cell gene and S a healthy gene, the above table shows the four possibilities for the children of two Ss parents. Each parent has only one sickle cell gene, so each has the relatively mild sickle cell trait. Find the probability that these parents give birth to a child who

49. has sickle cell anemia.

50. has sickle cell trait.

51. is healthy.

In Exercises 52–54, use the following table that shows the four possibilities for the children of one healthy, SS, parent, and one parent with sickle cell trait, Ss.

		Second Parent (with Sickle Cell Trait)	
		S	*s*
Healthy	*S*	*SS*	*Ss*
First Parent	*S*	*SS*	*Ss*

Find the probability that these parents give birth to a child who

52. has sickle cell anemia.

53. has sickle cell trait.

54. is healthy.

The table shows the distribution, by age and gender, of the 29.3 million Americans who live alone. Use the data in the table to solve Exercises 55–60.

NUMBER OF PEOPLE IN THE UNITED STATES LIVING ALONE, IN MILLIONS

	Ages 15–24	Ages 25–34	Ages 35–44	Ages 45–64	Ages 65–74	Ages ≥ 75	Total
Male	0.7	2.2	2.6	4.3	1.3	1.4	12.5
Female	0.8	1.6	1.6	5.0	2.9	4.9	16.8
Total	1.5	3.8	4.2	9.3	4.2	6.3	29.3

Source: U.S. Census Bureau

Find the probability, expressed as a decimal rounded to the nearest hundredth, that a randomly selected American living alone is

55. male.

56. female.

57. in the 25–34 age range.

58. in the 35–44 age range.

59. a woman in the 15–24 age range.

60. a man in the 45–64 age range.

The table shows the number of Americans who moved in a recent year, categorized by where they moved and whether they were an owner or a renter. Use the data in the table, expressed in millions, to solve Exercises 61–66.

NUMBER OF PEOPLE IN THE UNITED STATES WHO MOVED, IN MILLIONS

	Moved to Same State	Moved to Different State	Moved to Different Country
Owner	11.7	2.8	0.3
Renter	18.7	4.5	1.0

Source: U.S. Census Bureau

Use the above table to find the probability, expressed as a decimal rounded to the nearest hundredth, that a randomly selected American who moved was

61. an owner.

62. a renter.

63. a person who moved within the same state.

64. a person who moved to a different country.

65. a renter who moved to a different state.

66. an owner who moved to a different state.

The table shows the educational attainment of the U.S. population, ages 25 and over. Use the data in the table, expressed in millions, to solve Exercises 67–70.

EDUCATIONAL ATTAINMENT, IN MILLIONS, OF THE UNITED STATES POPULATION, AGES 25 AND OVER

	Less Than 4 Years High School	4 Years High School Only	Some College (Less Than 4 Years)	4 Years College (or More)	Total
Male	14	25	20	23	82
Female	15	31	24	22	92
Total	29	56	44	45	174

Source: U.S. Census Bureau

Find the probability, expressed as a simplified fraction, that a randomly selected American, age 25 or over,

67. had less than four years of high school.

68. had four years of high school only.

69. was a woman with four years of college or more.

70. was a man with four years of college or more.

Explaining the Concepts

71. What is the sample space of an experiment? What is an event?

72. How is the theoretical probability of an event computed?

73. Describe the difference between theoretical probability and empirical probability.

74. Give an example of an event whose probability must be determined empirically rather than theoretically.

75. Use the definition of theoretical probability to explain why the probability of an event that cannot occur is 0.

76. Use the definition of theoretical probability to explain why the probability of an event that is certain to occur is 1.

77. Write a probability word problem whose answer is one of the following fractions: $\frac{1}{6}$ or $\frac{1}{4}$ or $\frac{1}{3}$.

78. The president of a large company with 10,000 employees is considering mandatory cocaine testing for every employee. The test that would be used is 90% accurate, meaning that it will detect 90% of the cocaine users who are tested, and that 90% of the nonusers will test negative. This also means that the test gives 10% false positive. Suppose that 1% of the employees actually use cocaine. Find the probability that someone who tests positive for cocaine use is, indeed, a user.

Hint: Find the following probability fraction:

$$\frac{\text{the number of employees who test positive and are cocaine users}}{\text{the number of employees who test positive}}.$$

This fraction is given by

$$\frac{90\% \text{ of } 1\% \text{ of } 10{,}000}{\text{the number who test positive who actually use cocaine plus the number who test positive who do not use cocaine}}.$$

What does this probability indicate in terms of the percentage of employees who test positive who are not actually users? Discuss these numbers in terms of the issue of mandatory drug testing. Write a paper either in favor of or against mandatory drug testing, incorporating the actual percentage accuracy for such tests.

Critical Thinking Exercises

Make Sense? *In Exercises 79–82, determine whether each statement makes sense or does not make sense, and explain your reasoning.*

79. Assuming the next U.S. president will be a Democrat or a Republican, the probability of a Republican president is 0.5.

80. The probability that I will go to graduate school is 1.5.

81. When I toss a coin, the probability of getting heads *or* tails is 1, but the probability of getting heads *and* tails is 0.

82. When I am dealt one card from a standard 52-card deck, the probability of getting a red card *or* a black card is 1, but the probability of getting a red card *and* a black card is 0.

83. The target in the figure shown contains four squares. If a dart thrown at random hits the target, find the probability that it will land in an orange region.

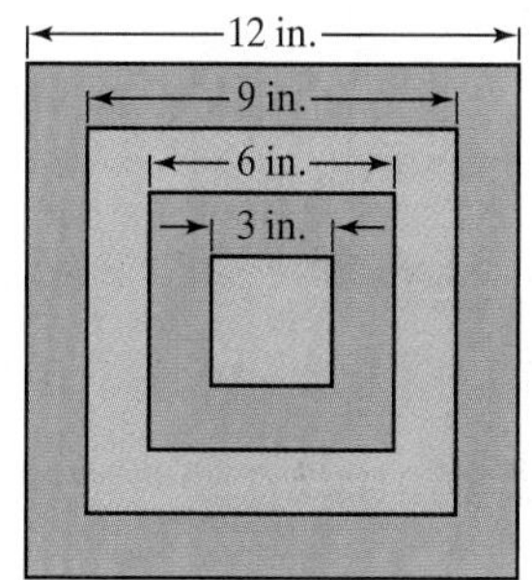

84. Some three-digit numbers, such as 101 and 313, read the same forward and backward. If you select a number from all three-digit numbers, find the probability that it will read the same forward and backward.

11.5 Probability with the Fundamental Counting Principle, Permutations, and Combinations

WHAT AM I SUPPOSED TO LEARN?

After studying this section, you should be able to:

1. Compute probabilities with permutations.
2. Compute probabilities with combinations.

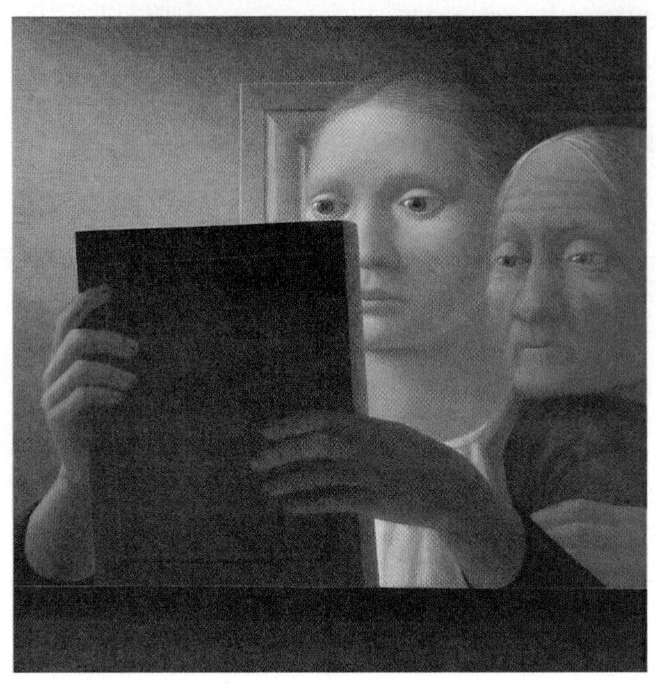

George Tooker (1920–2011) ***"Mirror II"***
Addison Gallery of American Art, Phillips Academy, Andover, MA/Art Resource, NY; © Estate of George Tooker. Courtesy of DC Moore Gallery, New York

PROBABILITY OF DYING AT ANY GIVEN AGE

Age	Probability of Male Death	Probability of Female Death
10	0.00013	0.00010
20	0.00140	0.00050
30	0.00153	0.00050
40	0.00193	0.00095
50	0.00567	0.00305
60	0.01299	0.00792
70	0.03473	0.01764
80	0.07644	0.03966
90	0.15787	0.11250
100	0.26876	0.23969
110	0.39770	0.39043

Source: George Shaffner, *The Arithmetic of Life and Death*

ACCORDING TO ACTUARIAL TABLES, THERE IS NO YEAR IN WHICH DEATH IS as likely as continued life, at least until the age of 115. Until that age, the probability of dying in any one year ranges from a low of 0.00009 for a girl at age 11 to a high of 0.465 for either gender at age 114. For a healthy 30-year-old, how does the probability of dying this year compare to the probability of winning the jackpot in a lottery game? In this section, we provide the surprising answer to this question, as we study probability with the Fundamental Counting Principle, permutations, and combinations.

1 *Compute probabilities with permutations.*

Probability with Permutations

EXAMPLE 1 *Probability and Permutations*

We return to the six jokes about books by Groucho Marx, Bob Blitzer, Steven Wright, Henny Youngman, Jerry Seinfeld, and Phyllis Diller that opened Section 11.2. Suppose that each joke is written on one of six cards. The cards are placed in a hat and then six cards are drawn, one at a time. The order in which the cards are drawn determines the order in which the jokes are delivered. What is the probability that a man's joke will be delivered first and a man's joke will be delivered last?

Groucho Marx

SOLUTION

We begin by applying the definition of probability to this situation.

$$P(\text{man's joke first, man's joke last}) = \frac{\text{number of permutations with man's joke first, man's joke last}}{\text{total number of possible permutations}}$$

Bob Blitzer

Steven Wright

Henny Youngman

Jerry Seinfeld

Phyllis Diller

We can use the Fundamental Counting Principle to find the total number of possible permutations. This represents the number of ways the six jokes can be delivered.

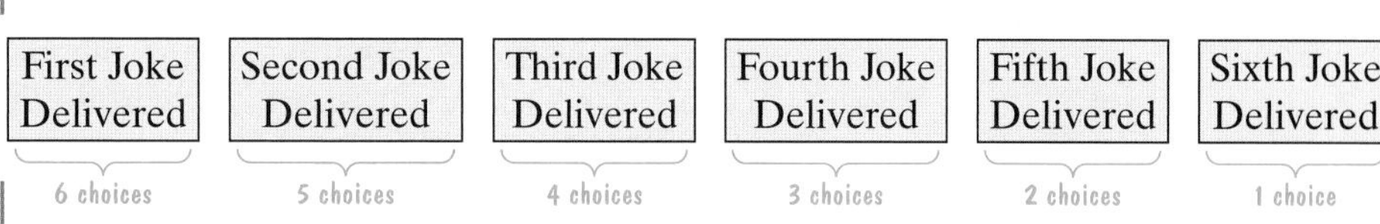

There are $6 \cdot 5 \cdot 4 \cdot 3 \cdot 2 \cdot 1$, or 720 possible permutations. Equivalently, there are 720 different ways to deliver the six jokes about books.

We can also use the Fundamental Counting Principle to find the number of permutations with a man's joke delivered first and a man's joke delivered last. These conditions can be shown as follows:

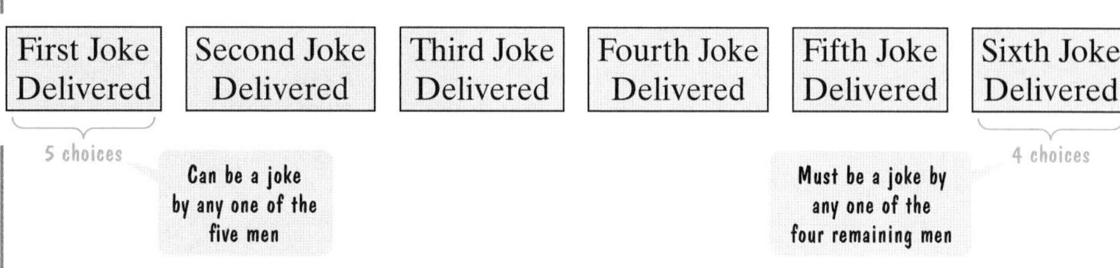

Now let's fill in the number of choices for positions two through five.

Thus, there are $5 \cdot 4 \cdot 3 \cdot 2 \cdot 1 \cdot 4$, or 480 possible permutations. Equivalently, there are 480 ways to deliver the jokes with a man's joke told first and a man's joke delivered last.

Now we can return to our probability fraction.

$$P(\text{man's joke first, man's joke last}) = \frac{\text{number of permutations with man's joke first, man's joke last}}{\text{total number of possible permutations}} = \frac{480}{720} = \frac{2}{3}$$

The probability of a man's joke delivered first and a man's joke told last is $\frac{2}{3}$.

CHECK POINT 1 Consider the six jokes about books by Groucho Marx, Bob Blitzer, Steven Wright, Henny Youngman, Jerry Seinfeld, and Phyllis Diller. As in Example 1, each joke is written on one of six cards which are randomly drawn one card at a time. The order in which the cards are drawn determines the order in which the jokes are delivered. What is the probability that a joke by a comic whose first name begins with G is told first and a man's joke is delivered last?

2 *Compute probabilities with combinations.*

Probability with Combinations

In 2015, Americans spent $70.15 billion on state and multi-state lotteries, more than on books, music, video games, movies, and sporting events combined. With each lottery drawing, the probability that someone will win the jackpot is relatively high. If there is no winner, it is virtually certain that eventually someone will be graced with millions of dollars. So, why are you so unlucky compared to this undisclosed someone? In Example 2, we provide an answer to this question.

EXAMPLE 2 Probability and Combinations: Powerball

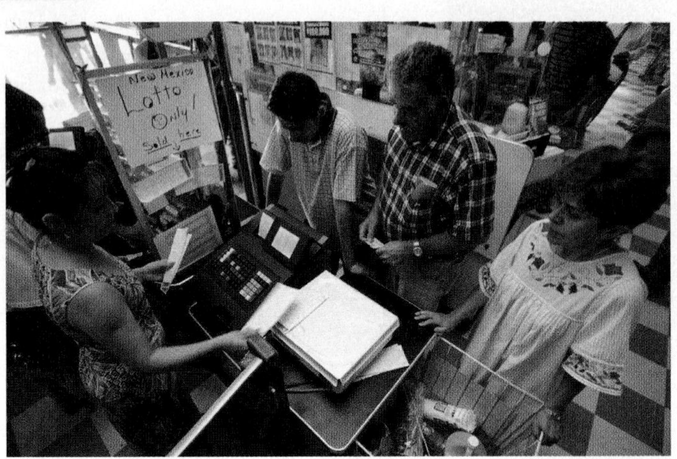

Powerball is a multi-state lottery played in most U.S. states. It is the first lottery game to randomly draw numbers from two drums. The game is set up so that each player chooses five different numbers from 1 to 69 and one Powerball number from 1 to 26. Twice per week 5 white balls are drawn randomly from a drum with 69 white balls, numbered 1 to 69, and then one red Powerball is drawn randomly from a drum with 26 red balls, numbered 1 to 26. A player wins the jackpot by matching all five numbers drawn from the white balls in any order and matching the number on the red Powerball. With one $2 Powerball ticket, what is the probability of winning the jackpot?

SOLUTION

Because the order of the five numbers shown on the white balls does not matter, this is a situation involving combinations. We begin with the formula for probability.

$$P(\text{winning the jackpot}) = \frac{\text{number of ways of winning the jackpot}}{\text{total number of possible combinations}}$$

We can use the combinations formula

$${}_nC_r = \frac{n!}{(n-r)!\,r!}$$

to find the total number of possible combinations in the first part of the Powerball lottery. We are selecting $r = 5$ numbers from a collection of $n = 69$ numbers from the drum of white balls.

$${}_{69}C_5 = \frac{69!}{(69-5)!\,5!} = \frac{69!}{64!\,5!} = \frac{69 \cdot 68 \cdot 67 \cdot 66 \cdot 65 \cdot \cancel{64!}}{\cancel{64!} \cdot 5 \cdot 4 \cdot 3 \cdot 2 \cdot 1} = 11{,}238{,}513$$

There are 11,238,513 number combinations in the first part of Powerball.

Next, we must determine the number of ways of selecting the red Powerball. Because there are 26 red Powerballs in the second drum, there are 26 possible combinations of numbers.

Comparing the Probability of Dying to the Probability of Winning Powerball's Jackpot

As a healthy nonsmoking 30-year-old, your probability of dying this year is approximately 0.001. Divide this probability by the probability of winning the Powerball jackpot with one ticket

$$\left(\frac{1}{292{,}201{,}338} \approx 0.000000003422\right):$$

$$\frac{0.001}{0.000000003422} \approx 292{,}227.$$

A healthy 30-year-old is nearly 292,000 times more likely to die this year than to win the Powerball jackpot. It's no wonder that people who calculate probabilities often refer to lotteries as a "tax on stupidity."

We can use the Fundamental Counting Principle to find the total number of possible number combinations in Powerball.

$${}_{69}C_5 \cdot 26 = 11{,}238{,}513 \cdot 26 = 292{,}201{,}338$$

Combinations of white balls from the first drum

Combinations of red Powerballs from the second drum

There are 292,201,338 number combinations in Powerball. If a person buys one $2 ticket, that person has selected only one combination of the numbers. With one Powerball ticket, there is only one way of winning the jackpot.

Now we can return to our probability fraction.

$$P(\text{winning the jackpot}) = \frac{\text{number of ways of winning the jackpot}}{\text{total number of possible combinations}} = \frac{1}{292{,}201{,}338}$$

The probability of winning the jackpot with one Powerball ticket is $\frac{1}{292{,}201{,}338}$ or about 1 in 292 million.

Suppose that a person buys 5000 different Powerball tickets. Because that person has selected 5000 different combinations of the Powerball numbers, the probability of winning the jackpot is

$$\frac{5000}{292{,}201{,}338} \approx 1.71 \times 10^{-5} = 0.0000171.$$

The chances of winning the jackpot are about 171 in ten million. At $2 per Powerball ticket, it is highly probable that our Powerball player will be $10,000 poorer. Knowing a little probability helps a lotto.

CHECK POINT 2 Hitting the jackpot in Powerball is not the only way to win a monetary prize. For example, a minimum award of $50,000 is given to a player who correctly matches four of the five numbers drawn from the 69 white balls and the one number drawn from the 26 red Powerballs. Find the probability of winning this consolation prize. Express the answer as a fraction.

EXAMPLE 3 Probability and Combinations

A club consists of five men and seven women. Three members are selected at random to attend a conference. Find the probability that the selected group consists of

a. three men.

b. one man and two women.

SOLUTION

12 Club Members

5 Men 7 Women

Select 3

The order in which the three people are selected does not matter, so this is a problem involving combinations.

a. We begin with the probability of selecting three men.

$$P(3\text{ men}) = \frac{\text{number of ways of selecting 3 men}}{\text{total number of possible combinations}}$$

12 Club Members

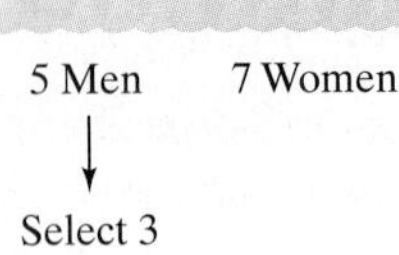

First, we consider the denominator of the probability fraction. We are selecting $r = 3$ people from a total group of $n = 12$ people (five men and seven women). The total number of possible combinations is

$$_{12}C_3 = \frac{12!}{(12-3)!\,3!} = \frac{12!}{9!\,3!} = \frac{12 \cdot 11 \cdot 10 \cdot \cancel{9!}}{\cancel{9!} \cdot 3 \cdot 2 \cdot 1} = 220.$$

Thus, there are 220 possible three-person selections.

Next, we consider the numerator of the probability fraction. We are interested in the number of ways of selecting three men from five men. We are selecting $r = 3$ men from a total group of $n = 5$ men. The number of possible combinations of three men is

$$_5C_3 = \frac{5!}{(5-3)!\,3!} = \frac{5!}{2!\,3!} = \frac{5 \cdot 4 \cdot \cancel{3!}}{2 \cdot 1 \cdot \cancel{3!}} = 10.$$

Thus, there are ten ways of selecting three men from five men. Now we can fill in the numbers in the numerator and the denominator of our probability fraction.

$$P(3 \text{ men}) = \frac{\text{number of ways of selecting 3 men}}{\text{total number of possible combinations}} = \frac{10}{220} = \frac{1}{22}$$

The probability that the group selected to attend the conference consists of three men is $\frac{1}{22}$.

12 Club Members

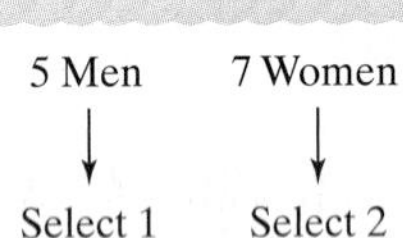

b. We set up the fraction for the probability that the selected group consists of one man and two women.

$$P(1 \text{ man}, 2 \text{ women}) = \frac{\text{number of ways of selecting 1 man and 2 women}}{\text{total number of possible combinations}}$$

The denominator of this fraction is the same as the denominator in part (a). The total number of possible combinations is found by selecting $r = 3$ people from $n = 12$ people: $_{12}C_3 = 220$.

Next, we move to the numerator of the probability fraction. The number of ways of selecting $r = 1$ man from $n = 5$ men is

$$_5C_1 = \frac{5!}{(5-1)!\,1!} = \frac{5!}{4!\,1!} = \frac{5 \cdot \cancel{4!}}{\cancel{4!} \cdot 1} = 5.$$

The number of ways of selecting $r = 2$ women from $n = 7$ women is

$$_7C_2 = \frac{7!}{(7-2)!\,2!} = \frac{7!}{5!\,2!} = \frac{7 \cdot 6 \cdot \cancel{5!}}{\cancel{5!} \cdot 2 \cdot 1} = 21.$$

By the Fundamental Counting Principle, the number of ways of selecting 1 man and 2 women is

$$_5C_1 \cdot {_7C_2} = 5 \cdot 21 = 105.$$

Now we can fill in the numbers in the numerator and the denominator of our probability fraction.

$$P(1 \text{ man}, 2 \text{ women}) = \frac{\text{number of ways of selecting 1 man and 2 women}}{\text{total number of possible combinations}} = \frac{_5C_1 \cdot {_7C_2}}{_{12}C_3} = \frac{105}{220} = \frac{21}{44}$$

The probability that the group selected to attend the conference consists of one man and two women is $\frac{21}{44}$.

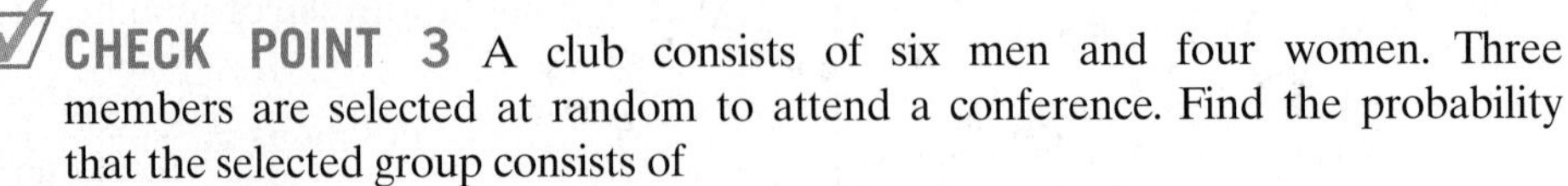

CHECK POINT 3 A club consists of six men and four women. Three members are selected at random to attend a conference. Find the probability that the selected group consists of

a. three men.

b. two men and one woman.

Concept and Vocabulary Check

Fill in each blank so that the resulting statement is true.

1. Six stand-up comics, A, B, C, D, E, and F, are to perform on a single evening at a comedy club. The order of performance is determined by random selection. The probability that comic E will perform first is the number of ____________ with comic E performing first divided by ____________________________.
2. The probability of winning a lottery with one lottery ticket is the number of ways of winning, which is precisely ___, divided by the total number of possible ____________.

In Exercises 3–4, determine whether each statement is true or false. If the statement is false, make the necessary change(s) to produce a true statement.

3. When working problems involving probability with permutations, the denominators of the probability fractions consist of the total number of possible permutations. ______
4. When working problems involving probability with combinations, the numerators of the probability fractions consist of the total number of possible combinations. ______

Exercise Set 11.5

Practice and Application Exercises

1. Martha, Lee, Nancy, Paul, and Armando have all been invited to a dinner party. They arrive randomly, and each person arrives at a different time.
   a. In how many ways can they arrive?
   b. In how many ways can Martha arrive first and Armando last?
   c. Find the probability that Martha will arrive first and Armando last.
2. Three men and three women line up at a checkout counter in a store.
   a. In how many ways can they line up?
   b. In how many ways can they line up if the first person in line is a woman, and then the line alternates by gender—that is a woman, a man, a woman, a man, and so on?
   c. Find the probability that the first person in line is a woman and the line alternates by gender.
3. Six stand-up comics, A, B, C, D, E, and F, are to perform on a single evening at a comedy club. The order of performance is determined by random selection. Find the probability that
   a. Comic E will perform first.
   b. Comic C will perform fifth and comic B will perform last.
   c. The comedians will perform in the following order: D, E, C, A, B, F.
   d. Comic A or comic B will perform first.
4. Seven performers, A, B, C, D, E, F, and G, are to appear in a fund raiser. The order of performance is determined by random selection. Find the probability that
   a. D will perform first.
   b. E will perform sixth and B will perform last.
   c. They will perform in the following order: C, D, B, A, G, F, E.
   d. F or G will perform first.
5. A group consists of four men and five women. Three people are selected to attend a conference.
   a. In how many ways can three people be selected from this group of nine?
   b. In how many ways can three women be selected from the five women?
   c. Find the probability that the selected group will consist of all women.
6. A political discussion group consists of five Democrats and six Republicans. Four people are selected to attend a conference.
   a. In how many ways can four people be selected from this group of eleven?
   b. In how many ways can four Republicans be selected from the six Republicans?
   c. Find the probability that the selected group will consist of all Republicans.

Mega Millions is a multi-state lottery played in most U.S. states. As of this writing, the top cash prize was $656 million, going to three lucky winners in three states. Players pick five different numbers from 1 to 56 and one number from 1 to 46. Use this information to solve Exercises 7–10. Express all probabilities as fractions.

7. A player wins the jackpot by matching all five numbers drawn from the white balls (1 through 56) and matching the number on the gold Mega Ball® (1 through 46). What is the probability of winning the jackpot?
8. A player wins a minimum award of $10,000 by correctly matching four numbers drawn from the white balls (1 through 56) and matching the number on the gold Mega Ball® (1 through 46). What is the probability of winning this consolation prize?
9. A player wins a minimum award of $150 by correctly matching three numbers drawn from the white balls (1 through 56) and matching the number on the gold Mega Ball® (1 through 46). What is the probability of winning this consolation prize?
10. A player wins a minimum award of $10 by correctly matching two numbers drawn from the white balls (1 through 56) and matching the number on the gold Mega Ball® (1 through 46). What is the probability of winning this consolation prize?
11. A box contains 25 transistors, 6 of which are defective. If 6 are selected at random, find the probability that
    a. all are defective.
    b. none are defective.

12. A committee of five people is to be formed from six lawyers and seven teachers. Find the probability that
   a. all are lawyers.
   b. none are lawyers.
13. A city council consists of six Democrats and four Republicans. If a committee of three people is selected, find the probability of selecting one Democrat and two Republicans.
14. A parent-teacher committee consisting of four people is to be selected from fifteen parents and five teachers. Find the probability of selecting two parents and two teachers.

Exercises 15–20 involve a deck of 52 cards. If necessary, refer to the picture of a deck of cards, as shown in **Figure 11.5** *on page 718.*

15. A poker hand consists of five cards.
   a. Find the total number of possible five-card poker hands.
   b. A diamond flush is a five-card hand consisting of all diamonds. Find the number of possible diamond flushes.
   c. Find the probability of being dealt a diamond flush.
16. A poker hand consists of five cards.
   a. Find the total number of possible five-card poker hands.
   b. Find the number of ways in which four aces can be selected.
   c. Find the number of ways in which one king can be selected.
   d. Use the Fundamental Counting Principle and your answers from parts (b) and (c) to find the number of ways of getting four aces and one king.
   e. Find the probability of getting a poker hand consisting of four aces and one king.
17. If you are dealt 3 cards from a shuffled deck of 52 cards, find the probability that all 3 cards are picture cards.
18. If you are dealt 4 cards from a shuffled deck of 52 cards, find the probability that all 4 are hearts.
19. If you are dealt 4 cards from a shuffled deck of 52 cards, find the probability of getting two queens and two kings.
20. If you are dealt 4 cards from a shuffled deck of 52 cards, find the probability of getting three jacks and one queen.

Explaining the Concepts

21. If people understood the mathematics involving probabilities and lotteries, as you now do, do you think they would continue to spend hundreds of dollars per year on lottery tickets? Explain your answer.
22. Write and solve an original problem involving probability and permutations.
23. Write and solve an original problem involving probability and combinations whose solution requires $\frac{_{14}C_{10}}{_{20}C_{10}}$.

Critical Thinking Exercises

Make Sense? *In Exercises 24–27, determine whether each statement makes sense or does not make sense, and explain your reasoning.*

24. When solving probability problems using the Fundamental Counting Principle, I find it easier to reduce the probability fraction if I leave the numerator and denominator as products of numbers.
25. I would never choose the lottery numbers 1, 2, 3, 4, 5, 6 because the probability of winning with six numbers in a row is less than winning with six random numbers.
26. From an investment point of view, a state lottery is a very poor place to put my money.
27. When finding the probability of randomly selecting two men and one woman from a group of ten men and ten women, I used the formula for $_nC_r$ three times.
28. An apartment complex offers apartments with four different options, designated by A through D. There are an equal number of apartments with each combination of options.

A	B	C	D
one bedroom two bedrooms three bedrooms	one bathroom two bathrooms	first floor second floor	lake view golf course view no special view

If there is only one apartment left, what is the probability that it is precisely what a person is looking for, namely two bedrooms, two bathrooms, first floor, and a lake or golf course view?

29. Suppose that it is a drawing in which the Powerball jackpot is promised to exceed \$700 million. If a person purchases 292,201,338 tickets at \$2 per ticket (all possible combinations), isn't this a guarantee of winning the jackpot? Because the probability in this situation is 1, what's wrong with doing this?
30. The digits 1, 2, 3, 4, and 5 are randomly arranged to form a three-digit number. (Digits are not repeated.) Find the probability that the number is even and greater than 500.
31. In a five-card poker hand, what is the probability of being dealt exactly one ace and no picture cards?

Group Exercise

32. Research and present a group report on state and multi-state lotteries. Include answers to some or all of the following questions. As always, make the report interesting and informative. Which states do not have lotteries? Why not? How much is spent per capita on lotteries? What are some of the lottery games? What is the probability of winning the jackpot and various consolation prizes in these games? What income groups spend the greatest amount of money on lotteries? If your state has one or more lotteries, what does it do with the money it makes? Is the way the money is spent what was promised when the lotteries first began?

11.6 Events Involving *Not* and *Or*; Odds

WHAT AM I SUPPOSED TO LEARN?

After studying this section, you should be able to:

1. Find the probability that an event will not occur.
2. Find the probability of one event or a second event occurring.
3. Understand and use odds.

WHAT ARE YOU MOST AFRAID OF? A SHARK ATTACK? AN AIRPLANE CRASH? THE Harvard Center for Risk Analysis helps to put these fears in perspective. According to the Harvard Center, the odds in favor of a fatal shark attack are 1 to 280 million and the odds against a fatal airplane accident are 3 million to 1.

There are several ways to express the likelihood of an event. For example, we can determine the probability of a fatal shark attack or a fatal airplane accident. We can also determine the *odds in favor* and the *odds against* these events. In this section, we expand our knowledge of probability and explain the meaning of odds.

1 *Find the probability that an event will not occur.*

Probability of an Event Not Occurring

If we know $P(E)$, the probability of an event E, we can determine the probability that the event will not occur, denoted by $P(\text{not } E)$. The event *not* E is the **complement** of E because it is the set of all outcomes in the sample space S that are not outcomes in the event E. In any experiment, an event must occur or its complement must occur. Thus, the sum of the probability that an event will occur and the probability that it will not occur is 1:

$$P(E) + P(\text{not } E) = 1.$$

Solving for $P(E)$ or for $P(\text{not } E)$, we obtain the following formulas:

COMPLEMENT RULES OF PROBABILITY

- The probability that an event E will not occur is equal to 1 minus the probability that it will occur.

$$P(\text{not } E) = 1 - P(E)$$

- The probability that an event E will occur is equal to 1 minus the probability that it will not occur.

$$P(E) = 1 - P(\text{not } E)$$

Using set notation, if E' is the complement of E, then $P(E') = 1 - P(E)$ and $P(E) = 1 - P(E')$.

GREAT QUESTION!

Do I have to use the formula for *P*(not *E*) to solve Example 1?

No. Here's how to work the example without using the formula:

$$P(\text{not a queen}) = \frac{\text{number of ways a non-queen can occur}}{\text{total number of outcomes}} = \frac{48}{52} = \frac{4 \cdot 12}{4 \cdot 13} = \frac{12}{13}.$$

With 4 queens, 52 − 4 = 48 cards are not queens.

EXAMPLE 1 *The Probability of an Event Not Occurring*

If you are dealt one card from a standard 52-card deck, find the probability that you are not dealt a queen.

SOLUTION

Because

$$P(\text{not } E) = 1 - P(E)$$

then

$$P(\text{not a queen}) = 1 - P(\text{queen}).$$

There are four queens in a deck of 52 cards. The probability of being dealt a queen is $\frac{4}{52} = \frac{1}{13}$. Thus,

$$P(\text{not a queen}) = 1 - P(\text{queen}) = 1 - \frac{1}{13} = \frac{13}{13} - \frac{1}{13} = \frac{12}{13}.$$

The probability that you are not dealt a queen is $\frac{12}{13}$.

CHECK POINT 1 If you are dealt one card from a standard 52-card deck, find the probability that you are not dealt a diamond.

EXAMPLE 2 *The Probability of an Event Not Occurring*

It may surprise you to see the data showing how little actual football there is in a televised National Football League (NFL) game. The circle graph in **Figure 11.6** shows the time breakdown, in minutes, for various aspects of an average 190-minute NFL TV broadcast. What is the probability that a minute of the broadcast is not devoted to game action, or actual football? Express the probability as a simplified fraction.

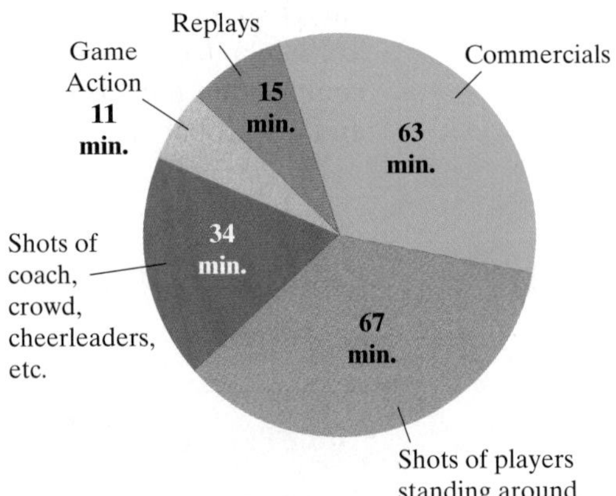

FIGURE 11.6
Source: *Wall Street Journal*

SOLUTION

We could compute the probability that a minute is not devoted to game action by adding the numbers in each of the four sectors representing times not devoted to game action and dividing this sum by 190 (minutes). However, it is easier to use complements.

We use the probability that a minute of the broadcast is devoted to game action to find the probability that a minute of the broadcast is not devoted to actual football.

P(not devoted to game action)

$$= 1 - P(\text{devoted to game action})$$

The graph shows 11 minutes devoted to game action.

$$= 1 - \frac{11}{190}$$

This number, 190 minutes, was given, but can be obtained by adding the numbers in the five sectors.

$$= \frac{190}{190} - \frac{11}{190} = \frac{179}{190}$$

The probability that a minute of an NFL broadcast is not devoted to game action is $\frac{179}{190}$.

CHECK POINT 2 Use the data in **Figure 11.6** to find the probability that a minute of an NFL broadcast is not devoted to commercials.

EXAMPLE 3 *Using the Complement Rules*

The circle graph in **Figure 11.7** shows the distribution, by age group, of the 214 million car drivers in the United States, with all numbers rounded to the nearest million. If one driver is randomly selected from this population, find the probability that the person is less than 80 years old.

Express probabilities as simplified fractions.

Number of U.S. Car Drivers, by Age Group

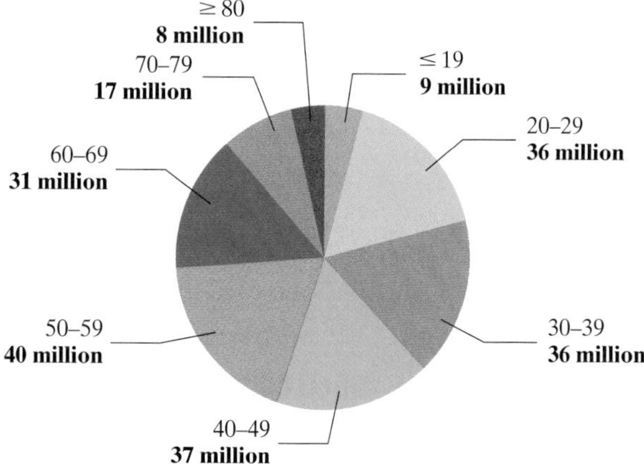

FIGURE 11.7
Source: U.S. Census Bureau

SOLUTION

As in Example 2, we could compute the probability that a randomly selected driver is less than 80 years old by adding the numbers in each of the seven sectors representing drivers less than 80 and dividing this sum by 214 million. However, it is easier to use complements. The complement of selecting a driver less than 80 years old is selecting a driver 80 or older.

$$P(\text{less than 80 years old}) = 1 - P(\text{80 or older})$$

$$= 1 - \frac{8}{214}$$

The graph shows 8 million drivers 80 or older.

$$= \frac{214}{214} - \frac{8}{214} = \frac{206}{214} = \frac{103}{107}$$

The probability that a randomly selected driver is less than 80 years old is $\frac{103}{107}$.

CHECK POINT 3 If one driver is randomly selected from the population represented in **Figure 11.7**, find the probability, expressed as a simplified fraction, that the person is at least 20 years old.

2 *Find the probability of one event or a second event occurring.*

Or Probabilities with Mutually Exclusive Events

Suppose that you randomly select one card from a deck of 52 cards. Let A be the event of selecting a king and B be the event of selecting a queen. Only one card is selected, so it is impossible to get both a king and a queen. The events of selecting a king and a queen cannot occur simultaneously. They are called *mutually exclusive events*.

MUTUALLY EXCLUSIVE EVENTS

If it is impossible for events A and B to occur simultaneously, the events are said to be **mutually exclusive**.

In general, if A and B are mutually exclusive events, the probability that either A or B will occur is determined by adding their individual probabilities.

***OR* PROBABILITIES WITH MUTUALLY EXCLUSIVE EVENTS**

If A and B are mutually exclusive events, then

$$P(A \text{ or } B) = P(A) + P(B).$$

Using set notation, $P(A \cup B) = P(A) + P(B)$.

EXAMPLE 4 *The Probability of Either of Two Mutually Exclusive Events Occurring*

If one card is randomly selected from a deck of cards, what is the probability of selecting a king or a queen?

SOLUTION

We find the probability that either of these mutually exclusive events will occur by adding their individual probabilities.

$$P(\text{king or queen}) = P(\text{king}) + P(\text{queen}) = \frac{4}{52} + \frac{4}{52} = \frac{8}{52} = \frac{2}{13}$$

The probability of selecting a king or a queen is $\frac{2}{13}$.

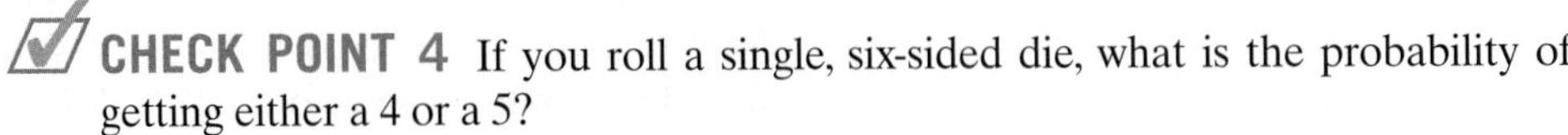

CHECK POINT 4 If you roll a single, six-sided die, what is the probability of getting either a 4 or a 5?

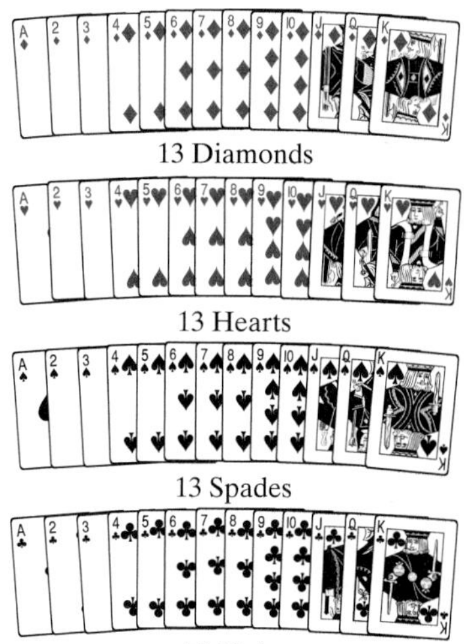

FIGURE 11.8 A deck of 52 cards

Or Probabilities with Events That Are Not Mutually Exclusive

Consider the deck of 52 cards shown in **Figure 11.8**. Suppose that these cards are shuffled and you randomly select one card from the deck. What is the probability of selecting a diamond or a picture card (jack, queen, king)? Begin by adding their individual probabilities.

$$P(\text{diamond}) + P(\text{picture card}) = \frac{13}{52} + \frac{12}{52}$$

There are 13 diamonds in the deck of 52 cards.

There are 12 picture cards in the deck of 52 cards.

However, this sum is not the probability of selecting a diamond or a picture card. The problem is that there are three cards that are *simultaneously* diamonds and picture cards, shown in **Figure 11.9**. The events of selecting a diamond and selecting a picture card are not mutually exclusive. It is possible to select a card that is both a diamond and a picture card.

FIGURE 11.9 Three diamonds are picture cards.

The situation is illustrated in the Venn diagram in **Figure 11.10**. Why can't we find the probability of selecting a diamond or a picture card by adding their individual probabilities? The Venn diagram shows that three of the cards, the three diamonds that are picture cards, get counted twice when we add the individual probabilities. First the three cards get counted as diamonds and then they get counted as picture cards. In order to avoid the error of counting the three cards twice, we need to subtract the probability of getting a diamond and a picture card, $\frac{3}{52}$, as follows:

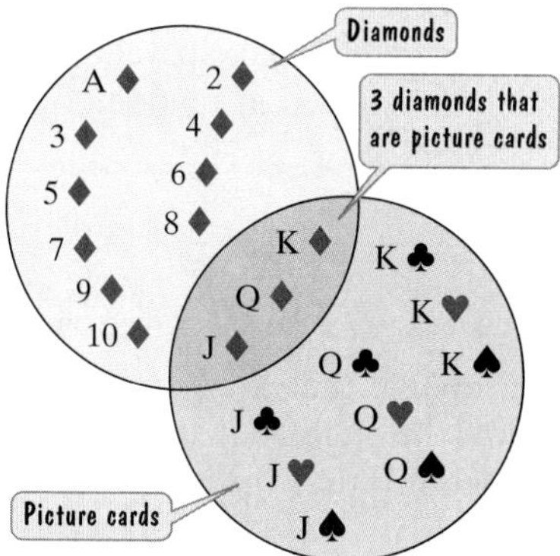

FIGURE 11.10

$P(\text{diamond or picture card})$

$$= P(\text{diamond}) + P(\text{picture card}) - P(\text{diamond and picture card})$$

$$= \frac{13}{52} + \frac{12}{52} - \frac{3}{52} = \frac{13 + 12 - 3}{52} = \frac{22}{52} = \frac{11}{26}.$$

Thus, the probability of selecting a diamond or a picture card is $\frac{11}{26}$.

In general, if A and B are events that are not mutually exclusive, the probability that A or B will occur is determined by adding their individual probabilities and then subtracting the probability that A and B occur simultaneously.

OR PROBABILITIES WITH EVENTS THAT ARE NOT MUTUALLY EXCLUSIVE

If A and B are not mutually exclusive events, then

$$P(A \text{ or } B) = P(A) + P(B) - P(A \text{ and } B).$$

Using set notation,

$$P(A \cup B) = P(A) + P(B) - P(A \cap B).$$

EXAMPLE 5 An *Or* Probability with Events That Are Not Mutually Exclusive

In a group of 25 baboons, 18 enjoy grooming their neighbors, 16 enjoy screeching wildly, while 10 enjoy grooming their neighbors and screeching wildly. If one baboon is selected at random from the group, find the probability that it enjoys grooming its neighbors or screeching wildly.

SOLUTION

It is possible for a baboon in the group to enjoy both grooming its neighbors and screeching wildly. Ten of the brutes are given to engage in both activities. These events are not mutually exclusive.

$$P\binom{\text{grooming}}{\text{or screeching}} = P(\text{grooming}) + P(\text{screeching}) - P\binom{\text{grooming}}{\text{and screeching}}$$

$$= \frac{18}{25} + \frac{16}{25} - \frac{10}{25}$$

18 of the 25 baboons enjoy grooming.

16 of the 25 baboons enjoy screeching.

10 of the 25 baboons enjoy both.

$$= \frac{18 + 16 - 10}{25} = \frac{24}{25}$$

The probability that a baboon in the group enjoys grooming its neighbors or screeching wildly is $\frac{24}{25}$.

 **CHECK POINT 5** In a group of 50 students, 23 take math, 11 take psychology, and 7 take both math and psychology. If one student is selected at random, find the probability that the student takes math or psychology.

EXAMPLE 6 *An* **Or** *Probability with Events That Are Not Mutually Exclusive*

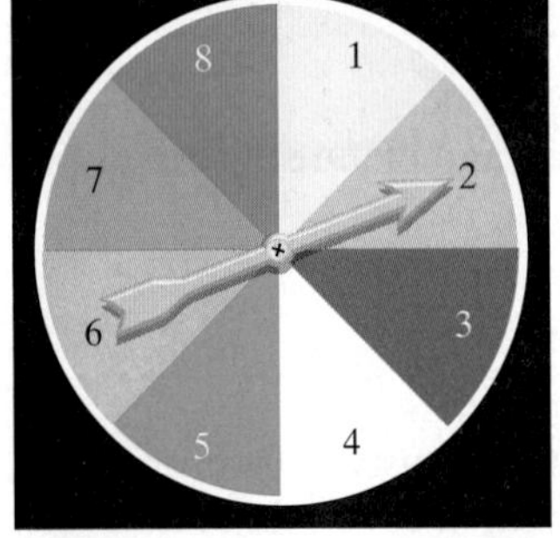

FIGURE 11.11 It is equally probable that the pointer will land on any one of the eight regions.

Figure 11.11 illustrates a spinner. It is equally probable that the pointer will land on any one of the eight regions, numbered 1 through 8. If the pointer lands on a borderline, spin again. Find the probability that the pointer will stop on an even number or on a number greater than 5.

SOLUTION

It is possible for the pointer to land on a number that is both even and greater than 5. Two of the numbers, 6 and 8, are even and greater than 5. These events are not mutually exclusive. The probability of landing on a number that is even or greater than 5 is calculated as follows:

$$P\binom{\text{even or}}{\text{greater than 5}} = P(\text{even}) + P(\text{greater than 5}) - P\binom{\text{even and}}{\text{greater than 5}}$$

$$= \frac{4}{8} + \frac{3}{8} - \frac{2}{8}$$

Four of the eight numbers, 2, 4, 6, and 8, are even.

Three of the eight numbers, 6, 7, and 8, are greater than 5.

Two of the eight numbers, 6 and 8, are even and greater than 5.

$$= \frac{4 + 3 - 2}{8} = \frac{5}{8}.$$

The probability that the pointer will stop on an even number or a number greater than 5 is $\frac{5}{8}$.

 **CHECK POINT 6** Use **Figure 11.11** to find the probability that the pointer will stop on an odd number or a number less than 5.

EXAMPLE 7 **Or** *Probabilities with Real-World Data*

Table 11.5 at the top of the next page, first presented in Section 11.4, shows the marital status of the U.S. population in 2015. Numbers in the table are expressed in millions.

TABLE 11.5 Marital Status of the U.S. Population, Ages 15 or Older, 2015, in Millions

	Married	Never Married	Divorced	Widowed	Total
Male	66	43	11	3	123
Female	67	38	15	11	131
Total	133	81	26	14	254

Source: U.S. Census Bureau

If one person is randomly selected from the population represented in **Table 11.5**, find the probability that

a. the person is divorced or male.

b. the person is married or divorced.

Express probabilities as simplified fractions and as decimals rounded to the nearest hundredth.

SOLUTION

a. It is possible to select a person who is both divorced and male. Thus, these events are not mutually exclusive.

$P(\text{divorced or male})$

$$= P(\text{divorced}) + P(\text{male}) - P(\text{divorced and male})$$

$$= \frac{26}{254} + \frac{123}{254} - \frac{11}{254}$$

Of the 254 million Americans, 26 million are divorced.

Of the 254 million Americans, 123 million are male.

Of the 254 million Americans, 11 million are divorced and male.

$$= \frac{26 + 123 - 11}{254} = \frac{138}{254} = \frac{2 \cdot 69}{2 \cdot 127} = \frac{69}{127} \approx 0.54$$

The probability of selecting a person who is divorced or male is $\frac{69}{127}$, or approximately 0.54.

b. It is impossible to select a person who is both married and divorced. These events are mutually exclusive.

$P(\text{married or divorced})$

$$= P(\text{married}) + P(\text{divorced})$$

$$= \frac{133}{254} + \frac{26}{254}$$

Of the 254 million Americans, 133 million are married.

Of the 254 million Americans, 26 million are divorced.

$$= \frac{133 + 26}{254} = \frac{159}{254} \approx 0.63$$

The probability of selecting a person who is married or divorced is $\frac{159}{254}$, or approximately 0.63.

CHECK POINT 7 If one person is randomly selected from the population represented in **Table 11.5**, find the probability that

a. the person is married or female.

b. the person is divorced or widowed.

Express probabilities as simplified fractions and as decimals rounded to the nearest hundredth.

3 Understand and use odds.

Odds

If we know the probability of an event E, we can also speak of the *odds in favor*, or the *odds against*, the event. The following definitions link together the concepts of odds and probabilities:

PROBABILITY TO ODDS

If $P(E)$ is the probability of an event E occurring, then

1. The **odds in favor of E** are found by taking the probability that E will occur and dividing by the probability that E will not occur.

$$\text{Odds in favor of } E = \frac{P(E)}{P(\text{not } E)}$$

2. The **odds against E** are found by taking the probability that E will not occur and dividing by the probability that E will occur.

$$\text{Odds against } E = \frac{P(\text{not } E)}{P(E)}$$

The odds against E can also be found by reversing the ratio representing the odds in favor of E.

EXAMPLE 8 From Probability to Odds

You roll a single, six-sided die.

a. Find the odds in favor of rolling a 2.

b. Find the odds against rolling a 2.

SOLUTION

Let E represent the event of rolling a 2. In order to determine odds, we must first find the probability of E occurring and the probability of E not occurring. With $S = \{1, 2, 3, 4, 5, 6\}$ and $E = \{2\}$, we see that

$$P(E) = \frac{1}{6}$$

$$\text{and } P(\text{not } E) = 1 - \frac{1}{6} = \frac{6}{6} - \frac{1}{6} = \frac{5}{6}.$$

Now we are ready to construct the ratios for the odds in favor of E and the odds against E.

a. $$\text{Odds in favor of } E \text{ (rolling a 2)} = \frac{P(E)}{P(\text{not } E)} = \frac{\frac{1}{6}}{\frac{5}{6}} = \frac{1}{6} \cdot \frac{6}{5} = \frac{1}{5}$$

The odds in favor of rolling a 2 are $\frac{1}{5}$. The ratio $\frac{1}{5}$ is usually written 1:5 and is read "1 to 5." Thus, the odds in favor of rolling a 2 are 1 to 5.

b. Now that we have the odds in favor of rolling a 2, namely, $\frac{1}{5}$ or 1:5, we can find the odds against rolling a 2 by reversing this ratio. Thus,

$$\text{Odds against } E\text{(rolling a 2)} = \frac{5}{1} \text{ or } 5{:}1.$$

The odds against rolling a 2 are 5 to 1.

GREAT QUESTION!

When you computed the odds in Example 8(a), the denominators of the two probabilities divided out. Will this always occur?

Yes.

CHECK POINT 8 You are dealt one card from a 52-card deck.

a. Find the odds in favor of getting a red queen.

b. Find the odds against getting a red queen.

EXAMPLE 9 From Probability to Odds

The winner of a raffle will receive a new sports utility vehicle. If 500 raffle tickets were sold and you purchased ten tickets, what are the odds against your winning the car?

SOLUTION

Let E represent the event of winning the SUV. Because you purchased ten tickets and 500 tickets were sold,

$$P(E) = \frac{10}{500} = \frac{1}{50} \text{ and } P(\text{not } E) = 1 - \frac{1}{50} = \frac{50}{50} - \frac{1}{50} = \frac{49}{50}.$$

Now we are ready to construct the ratio for the odds against E (winning the SUV).

$$\text{Odds against } E = \frac{P(\text{not } E)}{P(E)} = \frac{\frac{49}{50}}{\frac{1}{50}} = \frac{49}{50} \cdot \frac{50}{1} = \frac{49}{1}$$

The odds against winning the SUV are 49 to 1, written 49:1.

CHECK POINT 9 The winner of a raffle will receive a two-year scholarship to the college of his or her choice. If 1000 raffle tickets were sold and you purchased five tickets, what are the odds against your winning the scholarship?

House Odds

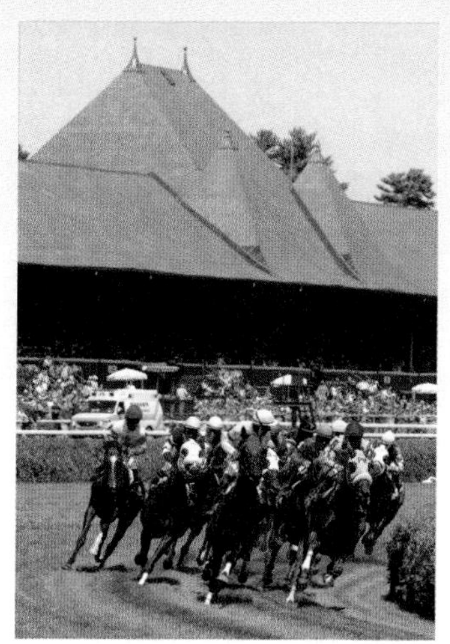

The odds given at horse races and at all games of chance are usually *odds against*. If the odds in favor of a particular horse winning a race are 2 to 5, then the odds against winning are 5 to 2. At a horse race, the odds on this horse are given simply as 5 to 2. These **house odds** tell a gambler what the payoff is on a bet. For every \$2 bet on the horse, the gambler would win \$5 if the horse won, in addition to having the \$2 bet returned.

Odds enable us to play and bet fairly on games. For example, we have seen that the odds in favor of getting 2 when you roll a die are 1 to 5. Suppose this is a gaming situation and you bet \$1 on a 2 turning up. In terms of your bet, there is one favorable outcome, rolling 2, and five unfavorable outcomes, rolling 1, 3, 4, 5, or 6. The odds in favor of getting 2, 1 to 5, compares the number of favorable outcomes, one, to the number of unfavorable outcomes, five.

Using odds in a gaming situation where money is waged, we can determine if the game is *fair*. If the odds in favor of an event E are a to b, the **game is fair** if a bet of \$$a$ is lost if event E does not occur, but a win of \$$b$ (as well as returning the bet of \$$a$) is realized if event E does occur. For example, the odds in favor of getting 2 on a die roll are 1 to 5. If you bet \$1 on a 2 turning up and the game is fair, you should win \$5 (and have your bet of \$1 returned) if a 2 turns up.

Now that we know how to convert from probability to odds, let's see how to convert from odds to probability. Suppose that the odds in favor of event E occurring are a to b. This means that

$$\frac{P(E)}{P(\text{not } E)} = \frac{a}{b} \quad \text{or} \quad \frac{P(E)}{1 - P(E)} = \frac{a}{b}.$$

By solving the equation on the right for $P(E)$, we obtain the formula for converting from odds to probability.

ODDS TO PROBABILITY

If the odds in favor of event E are a to b, then the probability of the event is given by

$$P(E) = \frac{a}{a + b}.$$

EXAMPLE 10 From Odds to Probability

The odds in favor of a particular horse winning a race are 2 to 5. What is the probability that this horse will win the race?

SOLUTION

Because odds in favor, a to b, means a probability of $\frac{a}{a+b}$, then odds in favor, 2 to 5, means a probability of

$$\frac{2}{2+5} = \frac{2}{7}.$$

The probability that this horse will win the race is $\frac{2}{7}$.

CHECK POINT 10 The odds against a particular horse winning a race are 15 to 1. Find the odds in favor of the horse winning the race and the probability of the horse winning the race.

Blitzer Bonus

Big Fears and Their Odds

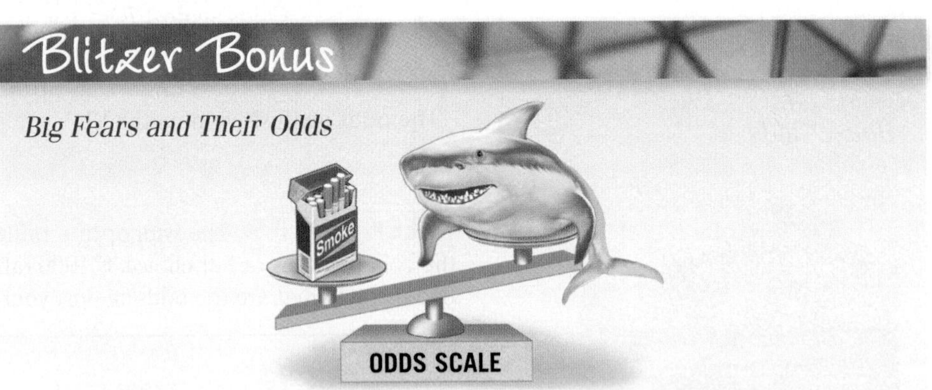

The Fear	Odds in Favor
Fatal shark attack	1 to 280,000,000
Fatal airplane accident	1 to 3,000,000
Losing your job	1 to 252
Home burglary at night	1 to 181
Developing cancer	1 to 7
Catching a sexually transmitted disease	1 to 4
Developing heart disease	1 to 4
Dying from tobacco-related illnesses (smokers)	1 to 2

Source: David Ropeik, Harvard Center for Risk Analysis

Concept and Vocabulary Check

Fill in each blank so that the resulting statement is true.

1. Because $P(E) + P(\text{not } E) = 1$, then $P(\text{not } E) =$ ______ and $P(E) =$ ______.
2. If it is impossible for events A and B to occur simultaneously, the events are said to be ______. For such events, $P(A \text{ or } B) =$ ______.
3. If it is possible for events A and B to occur simultaneously, then $P(A \text{ or } B) =$ ______.
4. The odds in favor of E can be found by taking the probability that ______ and dividing by the probability that ______.
5. If the odds in favor of event E are a to b, then the probability of the event, $P(E)$, is given by the formula ______.

In Exercises 6–9, determine whether each statement is true or false. If the statement is false, make the necessary change(s) to produce a true statement.

6. The probability that an event will not occur is equal to the probability that it will occur minus 1. ______
7. The probability of A or B can always be found by adding the probability of A and the probability of B. ______
8. The odds against E can always be found by reversing the ratio representing the odds in favor of E. ______
9. According to the National Center for Health Statistics, the lifetime odds in favor of dying from heart disease are 1 to 5, so the probability of dying from heart disease is $\frac{1}{5}$. ______

Exercise Set 11.6

Practice and Application Exercises

In Exercises 1–6, you are dealt one card from a 52-card deck. Find the probability that you are not dealt

1. an ace. **2.** a 3. **3.** a heart. **4.** a club. **5.** a picture card. **6.** a red picture card.

In 5-card poker, played with a standard 52-card deck, ${}_{52}C_5$, or 2,598,960, different hands are possible. The probability of being dealt various hands is the number of different ways they can occur divided by 2,598,960. Shown in Exercises 7–10 are various types of poker hands and their probabilities. In each exercise, find the probability of not being dealt this type of hand.

Type of Hand	Illustration	Number of Ways the Hand Can Occur	Probability
7. Straight flush: 5 cards with consecutive numbers, all in the same suit (excluding royal flush)		36	$\frac{36}{2,598,960}$
8. Four of a kind: 4 cards with the same number, plus 1 additional card		624	$\frac{624}{2,598,960}$
9. Full house: 3 cards of one number and 2 cards of a second number		3744	$\frac{3744}{2,598,960}$
10. Flush: 5 cards of the same suit (excluding royal flush and straight flush)		5108	$\frac{5108}{2,598,960}$

The graph shows the probability of cardiovascular disease, by age and gender. Use the information in the graph to solve Exercises 11–12. Express all probabilities as decimals, rounded to two decimal places.

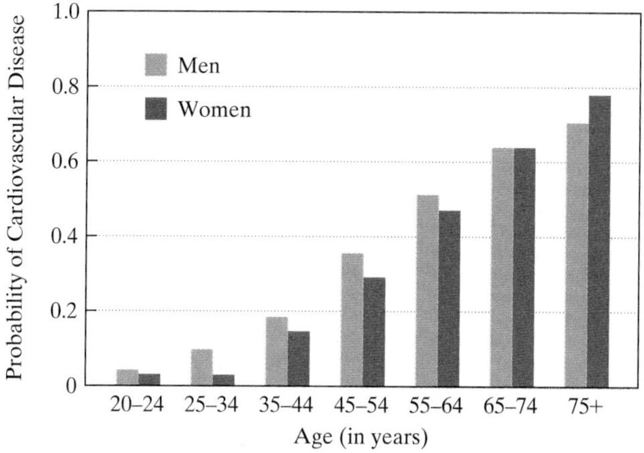

Source: American Heart Association

11. a. What is the probability that a randomly selected man between the ages of 25 and 34 has cardiovascular disease?

b. What is the probability that a randomly selected man between the ages of 25 and 34 does not have cardiovascular disease?

12. a. What is the probability that a randomly selected woman, 75 or older, has cardiovascular disease?

b. What is the probability that a randomly selected woman, 75 or older, does not have cardiovascular disease?

The table shows the distribution, by age, of a random sample of 3000 American moviegoers ages 12 through 74. Use this distribution to solve Exercises 13–16.

AGE DISTRIBUTION OF THE U.S. MOVIEGOER AUDIENCE

Ages	Number
12–24	900
25–44	1080
45–64	840
65–74	180

Source: Nielsen survey of 3000 American moviegoers ages 12–74

If one moviegoer is randomly selected from this population, find the probability, expressed as a simplified fraction, that

13. the moviegoer is not in the 25–44 age range.

14. the moviegoer is not in the 45–64 age range.

15. the moviegoer's age is less than 65.

16. the moviegoer's age is at least 25.

In Exercises 17–22, you randomly select one card from a 52-card deck. Find the probability of selecting

17. a 2 or a 3.

18. a 7 or an 8.

19. a red 2 or a black 3.

20. a red 7 or a black 8.

21. the 2 of hearts or the 3 of spades.

22. the 7 of hearts or the 8 of spades.

23. The mathematics faculty at a college consists of 8 professors, 12 associate professors, 14 assistant professors, and 10 instructors. If one faculty member is randomly selected, find the probability of choosing a professor or an instructor.

24. A political discussion group consists of 30 Republicans, 25 Democrats, 8 Independents, and 4 members of the Green party. If one person is randomly selected from the group, find the probability of choosing an Independent or a Green.

In Exercises 25–26, a single die is rolled. Find the probability of rolling

25. an even number or a number less than 5.

26. an odd number or a number less than 4.

In Exercises 27–30, you are dealt one card from a 52-card deck. Find the probability that you are dealt

27. a 7 or a red card.

28. a 5 or a black card.

29. a heart or a picture card.

30. a card greater than 2 and less than 7, or a diamond.

In Exercises 31–34, it is equally probable that the pointer on the spinner shown will land on any one of the eight regions, numbered 1 through 8. If the pointer lands on a borderline, spin again.

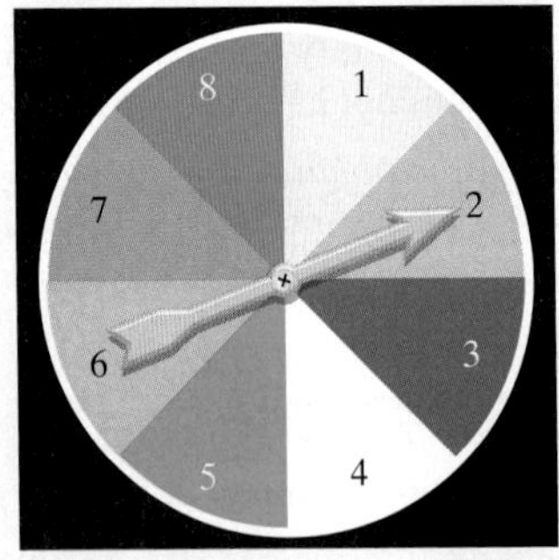

Find the probability that the pointer will stop on

31. an odd number or a number less than 6.

32. an odd number or a number greater than 3.

33. an even number or a number greater than 5.

34. an even number or a number less than 4.

Use this information to solve Exercises 35–38. The mathematics department of a college has 8 male professors, 11 female professors, 14 male teaching assistants, and 7 female teaching assistants. If a person is selected at random from the group, find the probability that the selected person is

35. a professor or a male.

36. a professor or a female.

37. a teaching assistant or a female.

38. a teaching assistant or a male.

39. In a class of 50 students, 29 are Democrats, 11 are business majors, and 5 of the business majors are Democrats. If one student is randomly selected from the class, find the probability of choosing a Democrat or a business major.

40. A student is selected at random from a group of 200 students in which 135 take math, 85 take English, and 65 take both math and English. Find the probability that the selected student takes math or English.

The table shows the educational attainment of the U.S. population, ages 25 and over. Use the data in the table, expressed in millions, to solve Exercises 41–48.

EDUCATIONAL ATTAINMENT OF THE U.S. POPULATION, AGES 25 AND OVER, IN MILLIONS

	Less Than 4 Years High School	4 Years High School Only	Some College (Less than 4 years)	4 Years College (or More)	Total
Male	14	25	20	23	82
Female	15	31	24	22	92
Total	29	56	44	45	174

Source: U.S. Census Bureau

Find the probability, expressed as a simplified fraction, that a randomly selected American, age 25 or over,

41. has not completed four years (or more) of college.

42. has not completed four years of high school.

43. has completed four years of high school only or less than four years of college.

44. has completed less than four years of high school or four years of high school only.

45. has completed four years of high school only or is a man.

46. has completed four years of high school only or is a woman.

Find the odds in favor and the odds against a randomly selected American, age 25 and over, with

47. four years (or more) of college.

48. less than four years of high school.

Scrabble Tiles. *The game of Scrabble has 100 tiles. The diagram shows the number of tiles for each letter and the letter's point value.*

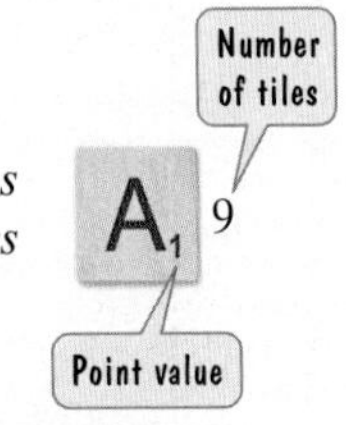

A_1 9	F_4 2	K_5 1	P_3 2	U_1 4	Z_{10} 1
B_3 2	G_2 3	L_1 4	Q_{10} 1	V_4 2	2
C_3 2	H_4 2	M_3 2	R_1 6	W_4 2	
D_2 4	I_1 9	N_1 6	S_1 4	X_8 1	
E_1 12	J_8 1	O_1 8	T_1 6	Y_4 2	

In Exercises 49–58, one tile is drawn from Scrabble's 100 tiles, pictured at the bottom of the previous page. Find each probability, expressed as a simplified fraction.

49. a. Find the probability of selecting an A.

b. Find the probability of not selecting an A.

50. a. Find the probability of selecting an E.

b. Find the probability of not selecting an E.

51. Find the probability of selecting a letter worth 4 points.

52. Find the probability of selecting a letter worth 3 points.

53. Find the probability of selecting an N or a D.

54. Find the probability of selecting an R or a C.

55. Find the probability of selecting a letter that precedes E or a letter worth 3 points.

56. Find the probability of selecting a letter that follows V or a letter worth 8 points.

57. Find the probability of selecting one of the letters needed to spell the word AND.

58. Find the probability of selecting one of the letters needed to spell the word QUIT.

In Exercises 59–62, refer to the diagram showing the Scrabble tiles. If one tile is drawn from the 100 tiles, find the odds in favor and the odds against

59. selecting a Z.

60. selecting a G.

61. selecting a letter worth one point.

62. selecting a letter worth two points.

In Exercises 63–64, a single die is rolled. Find the odds

63. in favor of rolling a number greater than 2.

64. in favor of rolling a number less than 5.

65. against rolling a number greater than 2.

66. against rolling a number less than 5.

The circle graphs show the percentage of millennial men and millennial women who consider marriage an important part of adult life. Use the information shown to solve Exercises 67–68.

Marriage Is an Important Part of Being an Adult

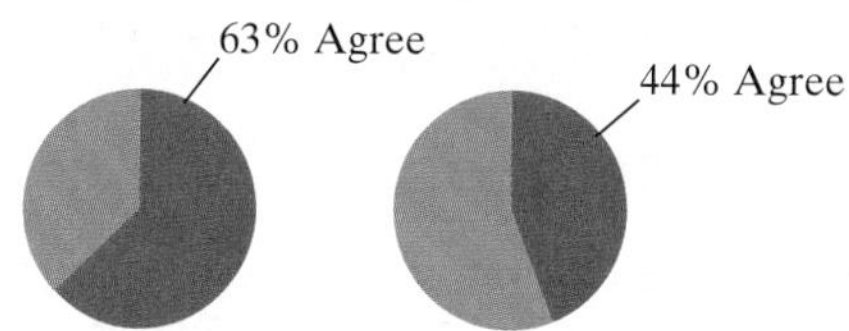

Source: *Rolling Stone*

67. a. What are the odds in favor of a millennial man agreeing that marriage is an important part of adult life?

b. What are the odds against a millennial man agreeing that marriage is an important part of adult life?

68. a. What are the odds in favor of a millennial woman agreeing that marriage is an important part of adult life?

b. What are the odds against a millennial woman agreeing that marriage is an important part of adult life?

In Exercises 69–78, one card is randomly selected from a deck of cards. Find the odds

69. in favor of drawing a heart.

70. in favor of drawing a picture card.

71. in favor of drawing a red card.

72. in favor of drawing a black card.

73. against drawing a 9.

74. against drawing a 5.

75. against drawing a black king.

76. against drawing a red jack.

77. against drawing a spade greater than 3 and less than 9.

78. against drawing a club greater than 4 and less than 10.

79. The winner of a raffle will receive a 21-foot outboard boat. If 1000 raffle tickets were sold and you purchased 20 tickets, what are the odds against your winning the boat?

80. The winner of a raffle will receive a 30-day all-expense-paid trip throughout Europe. If 5000 raffle tickets were sold and you purchased 30 tickets, what are the odds against your winning the trip?

Of the 38 plays attributed to Shakespeare, 18 are comedies, 10 are tragedies, and 10 are histories. In Exercises 81–88, one play is randomly selected from Shakespeare's 38 plays. Find the odds

81. in favor of selecting a comedy.

82. in favor of selecting a tragedy.

83. against selecting a history.

84. against selecting a comedy.

85. in favor of selecting a comedy or a tragedy.

86. in favor of selecting a tragedy or a history.

87. against selecting a tragedy or a history.

88. against selecting a comedy or a history.

89. If you are given odds of 3 to 4 in favor of winning a bet, what is the probability of winning the bet?

90. If you are given odds of 3 to 7 in favor of winning a bet, what is the probability of winning the bet?

91. Based on his skills in basketball, it was computed that when Michael Jordan shot a free throw, the odds in favor of his making it were 21 to 4. Find the probability that when Michael Jordan shot a free throw, he missed it. Out of every 100 free throws he attempted, on the average how many did he make?

92. The odds in favor of a person who is alive at age 20 still being alive at age 70 are 193 to 270. Find the probability that a person who is alive at age 20 will still be alive at age 70.

Exercises 93–94 give the odds against various flight risks. Use these odds to determine the probability of the underlined event for those in flight. (Source: Men's Health)

93. odds against contracting an airborne disease: 999 to 1

94. odds against deep-vein thrombosis (blood clot in the leg): 28 to 1

Explaining the Concepts

95. Explain how to find the probability of an event not occurring. Give an example.

96. What are mutually exclusive events? Give an example of two events that are mutually exclusive.

97. Explain how to find *or* probabilities with mutually exclusive events. Give an example.

98. Give an example of two events that are not mutually exclusive.

99. Explain how to find *or* probabilities with events that are not mutually exclusive. Give an example.

100. Explain how to find the odds in favor of an event if you know the probability that the event will occur.

101. Explain how to find the probability of an event if you know the odds in favor of that event.

Critical Thinking Exercises

Make Sense? *In Exercises 102–105, determine whether each statement makes sense or does not make sense, and explain your reasoning.*

102. The probability that Jill will win the election is 0.7 and the probability that she will not win is 0.4.

103. The probability of selecting a king from a deck of 52 cards is $\frac{4}{52}$ and the probability of selecting a heart is $\frac{13}{52}$, so the probability of selecting a king or a heart is $\frac{4}{52} + \frac{13}{52}$.

104. The probability of selecting a king or a heart from a deck of 52 cards is the same as the probability of selecting the king of hearts.

105. I estimate that the odds in favor of most students getting married before receiving an undergraduate degree are 9:1.

106. In Exercise 39, find the probability of choosing **a.** a Democrat who is not a business major; **b.** a student who is neither a Democrat nor a business major.

107. On New Year's Eve, the probability of a person driving while intoxicated or having a driving accident is 0.35. If the probability of driving while intoxicated is 0.32 and the probability of having a driving accident is 0.09, find the probability of a person having a driving accident while intoxicated.

108. The formula for converting from odds to probability is given in the box on page 739. Read the paragraph that precedes this box and derive the formula.

11.7 Events Involving *And*; Conditional Probability

WHAT AM I SUPPOSED TO LEARN?

After studying this section, you should be able to:

1 Find the probability of one event and a second event occurring.

2 Compute conditional probabilities.

YOU ARE CONSIDERING a job offer in South Florida. The job offer is just what you wanted and you are excited about living in the midst of Miami's tropical diversity. However, there is just one thing: the risk of hurricanes. You expect to stay in Miami ten years and buy a home. What is the probability that South Florida will be hit by a hurricane at least once in the next ten years?

In this section, we look at the probability that an event occurs at least once by expanding our discussion of probability to events involving *and*. (We'll return to South Florida and hurricanes in Example 3 and Exercise 25 in the Exercise Set.)

1 Find the probability of one event and a second event occurring.

And Probabilities with Independent Events

Consider tossing a fair coin two times in succession. The outcome of the first toss, heads or tails, does not affect what happens when you toss the coin a second time. For example, the occurrence of tails on the first toss does not make tails more likely or less likely to occur on the second toss. The repeated toss of a coin produces *independent events* because the outcome of one toss does not affect the outcome of others.

GREAT QUESTION!

What's the difference between *independent events* and *mutually exclusive events*?

Mutually exclusive events cannot occur at the same time. Independent events occur at different times, although they have no effect on each other.

INDEPENDENT EVENTS

Two events are **independent events** if the occurrence of either of them has no effect on the probability of the other.

When a fair coin is tossed two times in succession, the set of equally likely outcomes is

{heads heads, heads tails, tails heads, tails tails}.

We can use this set, {heads heads, heads tails, tails heads, tails tails}, to find the probability of getting heads on the first toss and heads on the second toss:

$$P(\text{heads and heads}) = \frac{\text{number of ways two heads can occur}}{\text{total number of possible outcomes}} = \frac{1}{4}.$$

We can also determine the probability of two heads, $\frac{1}{4}$, without having to list all the equally likely outcomes. The probability of heads on the first toss is $\frac{1}{2}$. The probability of heads on the second toss is also $\frac{1}{2}$. The product of these probabilities, $\frac{1}{2} \cdot \frac{1}{2}$, results in the probability of two heads, namely $\frac{1}{4}$. Thus,

$$P(\text{heads and heads}) = P(\text{heads}) \cdot P(\text{heads}).$$

In general, if two events are independent, we can calculate the probability of the first occurring and the second occurring by multiplying their probabilities.

***And* PROBABILITIES WITH INDEPENDENT EVENTS**

If A and B are independent events, then

$$P(A \text{ and } B) = P(A) \cdot P(B).$$

EXAMPLE 1 *Independent Events on a Roulette Wheel*

FIGURE 11.12 A U.S. roulette wheel

Figure 11.12 shows a U.S. roulette wheel that has 38 numbered slots (1 through 36, 0, and 00). Of the 38 compartments, 18 are black, 18 are red, and 2 are green. A play has the dealer (called the "croupier") spin the wheel and a small ball in opposite directions. As the ball slows to a stop, it can land with equal probability on any one of the 38 numbered slots. Find the probability of red occurring on two consecutive plays.

SOLUTION

The wheel has 38 equally likely outcomes and 18 are red. Thus, the probability of red occurring on a play is $\frac{18}{38}$, or $\frac{9}{19}$. The result that occurs on each play is independent of all previous results. Thus,

$$P(\text{red and red}) = P(\text{red}) \cdot P(\text{red}) = \frac{9}{19} \cdot \frac{9}{19} = \frac{81}{361} \approx 0.224.$$

The probability of red occurring on two consecutive plays is $\frac{81}{361}$.

Some roulette players incorrectly believe that if red occurs on two consecutive plays, then another color is "due." Because the events are independent, the outcomes of previous spins have no effect on any other spins.

 **CHECK POINT 1** Find the probability of green occurring on two consecutive plays on a roulette wheel.

The *and* rule for independent events can be extended to cover three or more independent events. Thus, if A, B, and C are independent events, then

$$P(A \text{ and } B \text{ and } C) = P(A) \cdot P(B) \cdot P(C).$$

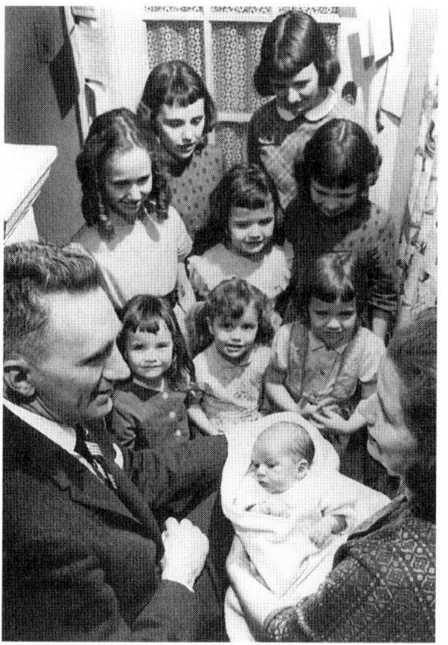

EXAMPLE 2 *Independent Events in a Family*

The picture in the margin shows a family that had nine girls in a row. Find the probability of this occurrence.

SOLUTION

If two or more events are independent, we can find the probability of them all occurring by multiplying their probabilities. The probability of a baby girl is $\frac{1}{2}$, so the probability of nine girls in a row is $\frac{1}{2}$ used as a factor nine times.

$$P(\text{nine girls in a row}) = \frac{1}{2}\cdot\frac{1}{2}\cdot\frac{1}{2}\cdot\frac{1}{2}\cdot\frac{1}{2}\cdot\frac{1}{2}\cdot\frac{1}{2}\cdot\frac{1}{2}\cdot\frac{1}{2}$$
$$= \left(\frac{1}{2}\right)^9 = \frac{1}{512}$$

The probability of a run of nine girls in a row is $\frac{1}{512}$. (If another child is born into the family, this event is independent of the other nine and the probability of a girl is still $\frac{1}{2}$.)

CHECK POINT 2 Find the probability of a family having four boys in a row.

TABLE 11.6 The Saffir/Simpson Hurricane Scale

Category	Winds (Miles per Hour)
1	74–95
2	96–110
3	111–130
4	131–155
5	> 155

Now let us return to the hurricane problem that opened this section. The Saffir/Simpson scale assigns numbers 1 through 5 to measure the disaster potential of a hurricane's winds. **Table 11.6** describes the scale. According to the National Hurricane Center, the probability that South Florida will be hit by a category 1 hurricane or higher in any single year is $\frac{5}{19}$, or approximately 0.263. In Example 3, we explore the risks of living in "Hurricane Alley."

EXAMPLE 3 *Hurricanes and Probabilities*

If the probability that South Florida will be hit by a hurricane in any single year is $\frac{5}{19}$,

a. What is the probability that South Florida will be hit by a hurricane in three consecutive years?

b. What is the probability that South Florida will not be hit by a hurricane in the next ten years?

SOLUTION

a. The probability that South Florida will be hit by a hurricane in three consecutive years is

$$P(\text{hurricane and hurricane and hurricane})$$
$$= P(\text{hurricane})\cdot P(\text{hurricane})\cdot P(\text{hurricane}) = \frac{5}{19}\cdot\frac{5}{19}\cdot\frac{5}{19} = \frac{125}{6859} \approx 0.018.$$

b. We will first find the probability that South Florida will not be hit by a hurricane in any single year.

$$P(\text{no hurricane}) = 1 - P(\text{hurricane}) = 1 - \frac{5}{19} = \frac{14}{19} \approx 0.737$$

The probability of not being hit by a hurricane in a single year is $\frac{14}{19}$. Therefore, the probability of not being hit by a hurricane ten years in a row is $\frac{14}{19}$ used as a factor ten times.

$$P(\text{no hurricanes for ten years})$$
$$= P\begin{pmatrix}\text{no hurricane}\\\text{for year 1}\end{pmatrix}\cdot P\begin{pmatrix}\text{no hurricane}\\\text{for year 2}\end{pmatrix}\cdot P\begin{pmatrix}\text{no hurricane}\\\text{for year 3}\end{pmatrix}\cdot\ldots\cdot P\begin{pmatrix}\text{no hurricane}\\\text{for year 10}\end{pmatrix}$$
$$= \frac{14}{19}\cdot\frac{14}{19}\cdot\frac{14}{19}\cdot\ldots\cdot\frac{14}{19}$$
$$= \left(\frac{14}{19}\right)^{10} \approx (0.737)^{10} \approx 0.047$$

The probability that South Florida will not be hit by a hurricane in the next ten years is approximately 0.047.

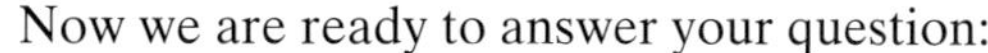

GREAT QUESTION!

When solving probability problems, how do I decide whether to use the *or* formulas or the *and* formulas?

- *Or* problems usually have the word *or* in the statement of the problem. These problems involve only one selection.

Example:

If one person is selected, find the probability of selecting a man or a Canadian.

- *And* problems often do not have the word *and* in the statement of the problem. These problems involve more than one selection.

Example:

If two people are selected, find the probability that both are men.

Now we are ready to answer your question:

What is the probability that South Florida will be hit by a hurricane at least once in the next ten years?

Because $P(\text{not } E) = 1 - P(E)$,

$$P(\text{no hurricane for ten years}) = 1 - P(\text{at least one hurricane in ten years}).$$

In our logic chapter, we saw that the negation of "at least one" is "no."

Equivalently,

$$\begin{aligned} P(\text{at least one hurricane in ten years}) &= 1 - P(\text{no hurricane for ten years}) \\ &= 1 - 0.047 = 0.953. \end{aligned}$$

With a probability of 0.953, it is nearly certain that South Florida will be hit by a hurricane at least once in the next ten years.

THE PROBABILITY OF AN EVENT HAPPENING AT LEAST ONCE

$$P(\text{event happening at least once}) = 1 - P(\text{event does not happen})$$

CHECK POINT 3 If the probability that South Florida will be hit by a hurricane in any single year is $\frac{5}{19}$,

a. What is the probability that South Florida will be hit by a hurricane in four consecutive years?

b. What is the probability that South Florida will not be hit by a hurricane in the next four years?

c. What is the probability that South Florida will be hit by a hurricane at least once in the next four years?

Express all probabilities as fractions and as decimals rounded to three places.

And Probabilities with Dependent Events

5 chocolate-covered cherries lie within the 20 pieces.

Chocolate lovers, please help yourselves! There are 20 mouth-watering tidbits to select from. What's that? You want 2? And you prefer chocolate-covered cherries? The problem is that there are only 5 chocolate-covered cherries, and it's impossible to tell what is inside each piece. They're all shaped exactly alike. At any rate, reach in, select a piece, enjoy, choose another piece, eat, and be well. There is nothing like savoring a good piece of chocolate in the midst of all this chit-chat about probability and hurricanes.

Another question? You want to know what your chances are of selecting 2 chocolate-covered cherries? Well, let's see. Five of the 20 pieces are chocolate-covered cherries, so the probability of getting one of them on your first selection is $\frac{5}{20}$, or $\frac{1}{4}$. Now, suppose that you did choose a chocolate-covered cherry on your first pick. Eat it slowly; there's no guarantee that you'll select your favorite on the second selection. There are now only 19 pieces of chocolate left. Only 4 are chocolate-covered cherries. The probability of getting a chocolate-covered cherry on your second try is 4 out of 19, or $\frac{4}{19}$. This is a different probability than the $\frac{1}{4}$ probability on your first selection. Selecting a chocolate-covered cherry the first time changes what is in the candy box. The probability of what you select the second time *is* affected by the outcome of the first event. For this reason, we say that these are *dependent events*.

Once a chocolate-covered cherry is selected, only 4 chocolate-covered cherries lie within the remaining 19 pieces.

DEPENDENT EVENTS

Two events are **dependent events** if the occurrence of one of them has an effect on the probability of the other.

The probability of selecting two chocolate-covered cherries in a row can be found by multiplying the $\frac{1}{4}$ probability on the first selection by the $\frac{4}{19}$ probability on the second selection:

P(chocolate-covered cherry and chocolate-covered cherry)

$$= P(\text{chocolate-covered cherry}) \cdot P\begin{pmatrix}\text{chocolate-covered cherry} \\ \text{given that one was selected}\end{pmatrix}$$

$$= \frac{1}{4} \cdot \frac{4}{19} = \frac{1}{19}.$$

The probability of selecting two chocolate-covered cherries in a row is $\frac{1}{19}$. This is a special case of finding the probability that each of two dependent events occurs.

***And* PROBABILITIES WITH DEPENDENT EVENTS**

If A and B are dependent events, then

$$P(A \text{ and } B) = P(A) \cdot P(B \text{ given that } A \text{ has occurred}).$$

EXAMPLE 4 *An* And *Probability with Dependent Events*

Good news: You won a free trip to Madrid and can take two people with you, all expenses paid. Bad news: Ten of your cousins have appeared out of nowhere and are begging you to take them. You write each cousin's name on a card, place the cards in a hat, and select one name. Then you select a second name without replacing the first card. If three of your ten cousins speak Spanish, find the probability of selecting two Spanish-speaking cousins.

SOLUTION

Because $P(A \text{ and } B) = P(A) \cdot P(B \text{ given that } A \text{ has occurred})$, then

P(two Spanish-speaking cousins)

$$= P(\text{speaks Spanish and speaks Spanish})$$

$$= P(\text{speaks Spanish}) \cdot P\begin{pmatrix}\text{speaks Spanish given that a Spanish-speaking} \\ \text{cousin was selected first}\end{pmatrix}$$

$$= \frac{3}{10} \cdot \frac{2}{9}$$

There are ten cousins, three of whom speak Spanish.

After picking a Spanish-speaking cousin, there are nine cousins left, two of whom speak Spanish.

$$= \frac{6}{90} = \frac{1}{15} \approx 0.067.$$

The probability of selecting two Spanish-speaking cousins is $\frac{1}{15}$.

GREAT QUESTION!

You solved Example 4 using the *and* probability formula with dependent events. Because cousins are being *selected*, can I also solve the problem using the combinations formula?

Yes. Here's how it works:

P(two Spanish speakers)

$$= \frac{\text{number of ways of selecting 2 Spanish-speaking cousins}}{\text{number of ways of selecting 2 cousins}}$$

$$= \frac{{}_3C_2}{{}_{10}C_2}$$

2 Spanish speakers selected from 3 Spanish-speaking cousins

2 cousins selected from 10 cousins

$$= \frac{3}{45} = \frac{1}{15}$$

CHECK POINT 4 You are dealt two cards from a 52-card deck. Find the probability of getting two kings.

The multiplication rule for dependent events can be extended to cover three or more dependent events. For example, in the case of three such events,

$P(A \text{ and } B \text{ and } C)$

$$= P(A) \cdot P(B \text{ given that } A \text{ occurred}) \cdot P(C \text{ given that } A \text{ and } B \text{ occurred}).$$

EXAMPLE 5 ***An* And *Probability with Three Dependent Events***

Three people are randomly selected, one person at a time, from five freshmen, two sophomores, and four juniors. Find the probability that the first two people selected are freshmen and the third is a junior.

SOLUTION

$P(\text{first two are freshmen and the third is a junior})$

$$= P(\text{freshman}) \cdot P\left(\begin{matrix}\text{freshman given that a}\\ \text{freshman was selected first}\end{matrix}\right) \cdot P\left(\begin{matrix}\text{junior given that a freshman was}\\ \text{selected first and a freshman was}\\ \text{selected second}\end{matrix}\right)$$

$$= \frac{5}{11} \cdot \frac{4}{10} \cdot \frac{4}{9}$$

There are 11 people, five of whom are freshmen.

After picking a freshman, there are 10 people left, four of whom are freshmen.

After the first two selections, 9 people are left, four of whom are juniors.

$$= \frac{8}{99}$$

The probability that the first two people selected are freshmen and the third is a junior is $\frac{8}{99}$.

CHECK POINT 5 You are dealt three cards from a 52-card deck. Find the probability of getting three hearts.

2 *Compute conditional probabilities.*

Conditional Probability

We have seen that for any two dependent events A and B,

$$P(A \text{ and } B) = P(A) \cdot P(B \text{ given that } A \text{ occurs}).$$

The probability of B given that A occurs is called *conditional probability*, denoted by $P(B|A)$.

CONDITIONAL PROBABILITY

The probability of event B, assuming that the event A has already occurred, is called the **conditional probability** of B, given A. This probability is denoted by $P(B|A)$.

It is helpful to think of the conditional probability $P(B|A)$ as the **probability that event *B* occurs if the sample space is restricted to the outcomes associated with event *A*.**

Blitzer Bonus

Coincidences

The phone rings and it's the friend you were just thinking of. You're driving down the road and a song you were humming in your head comes on the radio. Although these coincidences seem strange, perhaps even mystical, they're not. Coincidences are bound to happen. Ours is a world in which there are a great many potential coincidences, each with a low probability of occurring. When these surprising coincidences happen, we are amazed and remember them. However, we pay little attention to the countless number of non-coincidences: How often do you think of your friend and she doesn't call, or how often does she call when you're not thinking about her? By noticing the hits and ignoring the misses, we incorrectly perceive that there is a relationship between the occurrence of two independent events.

Another problem is that we often underestimate the probabilities of coincidences in certain situations, acting with more surprise than we should when they occur. For example, in a group of only 23 people, the probability that two individuals share a birthday (same month and day) is greater than $\frac{1}{2}$. Above 50 people, the probability of any two people sharing a birthday approaches certainty. You can verify the probabilities behind the coincidence of shared birthdays in relatively small groups by working Exercise 77 in Exercise Set 11.7.

EXAMPLE 6 Finding Conditional Probability

A letter is randomly selected from the letters of the English alphabet. Find the probability of selecting a vowel, given that the outcome is a letter that precedes h.

SOLUTION

We are looking for

$$P(\text{vowel}|\text{letter precedes h}).$$

This is the probability of a vowel if the sample space is restricted to the set of letters that precede h. Thus, the sample space is given by

$$S = \{a, b, c, d, e, f, g\}.$$

There are seven possible outcomes in the sample space. We can select a vowel from this set in one of two ways: a or e. Therefore, the probability of selecting a vowel, given that the outcome is a letter that precedes h, is $\frac{2}{7}$.

$$P(\text{vowel}|\text{letter precedes h}) = \frac{2}{7}$$

CHECK POINT 6 A letter is randomly selected from the letters of the English alphabet. Find the probability of selecting a letter that precedes h, given that the outcome is a vowel. (Do not include the letter y among the vowels.)

EXAMPLE 7 Finding Conditional Probability

You are dealt one card from a 52-card deck.

a. Find the probability of getting a heart, given that the card you were dealt is a red card.

b. Find the probability of getting a red card, given that the card you were dealt is a heart.

SOLUTION

a. We begin with

$$P(\text{heart}\,|\,\text{red card}).$$

Probability of getting a heart if the sample space is restricted to the set of red cards

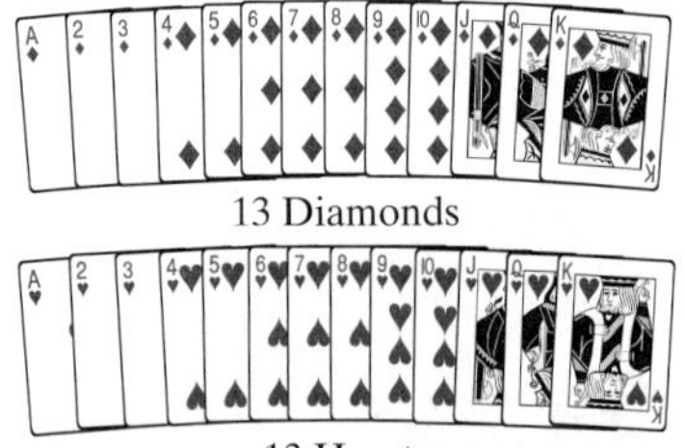

FIGURE 11.13

The sample space is shown in **Figure 11.13**. There are 26 outcomes in the sample space. We can get a heart from this set in 13 ways. Thus,

$$P(\text{heart}|\text{red card}) = \frac{13}{26} = \frac{1}{2}.$$

b. We now find

$$P(\text{red card}\,|\,\text{heart}).$$

Probability of getting a red card if the sample space is restricted to the set of hearts

FIGURE 11.14

The sample space is shown in **Figure 11.14**. There are 13 outcomes in the sample space. All of the outcomes are red. We can get a red card from this set in 13 ways. Thus,

$$P(\text{red card}|\text{heart}) = \frac{13}{13} = 1.$$

Example 7 illustrates that $P(\text{heart} \mid \text{red card})$ is not equal to $P(\text{red card} \mid \text{heart})$. In general, $P(B \mid A) \neq P(A \mid B)$.

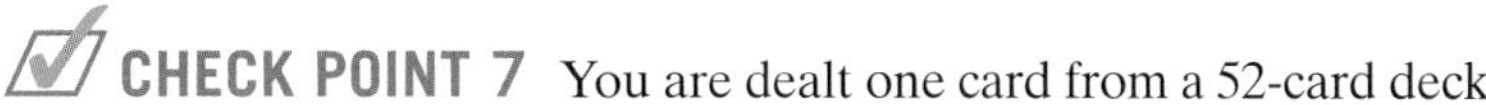

CHECK POINT 7 You are dealt one card from a 52-card deck.

a. Find the probability of getting a black card, given the card you were dealt is a spade.

b. Find the probability of getting a spade, given the card you were dealt is a black card.

EXAMPLE 8 *Conditional Probabilities with Real-World Data*

When women turn 40, their gynecologists typically remind them that it is time to undergo mammography screening for breast cancer. The data in **Table 11.7** are based on 100,000 U.S. women, ages 40 to 49, who participated in mammography screening.

TABLE 11.7 Mammography Screening on 100,000 U.S. Women, Ages 40 to 49

	Breast Cancer	No Breast Cancer	Total
Positive Mammogram	720	6944	7664
Negative Mammogram	80	92,256	92,336
Total	800	99,200	100,000

Source: Gerd Gigerenzer, *Calculated Risks.* Simon and Schuster, 2002.

Assuming that these numbers are representative of all U.S. women ages 40 to 49, find the probability that a woman in this age range

a. has a positive mammogram, given that she does not have breast cancer.

b. does not have breast cancer, given that she has a positive mammogram.

SOLUTION

a. We begin with the probability that a U.S. woman aged 40 to 49 has a positive mammogram, given that she does not have breast cancer:

$$P(\text{positive mammogram} \mid \text{no breast cancer}).$$

This is the probability of a positive mammogram if the data are restricted to women without breast cancer:

	No Breast Cancer
Positive Mammogram	6944
Negative Mammogram	92,256
Total	99,200

Within the restricted data, there are 6944 women with positive mammograms and 6944 + 92,256, or 99,200 women without breast cancer. Thus,

$$P(\text{positive mammogram} \mid \text{no breast cancer}) = \frac{6944}{99{,}200} = 0.07.$$

Among women without breast cancer, the probability of a positive mammogram is 0.07.

GREAT QUESTION!

What's the difference between the two conditional probabilities that you determined in Example 8?

Example 8 shows that the probability of a positive mammogram among women without breast cancer, 0.07, is not the same as the probability of not having breast cancer among women with positive mammograms, approximately 0.9. We have seen that the conditional probability of B, given A, is not the same as the conditional probability of A, given B:

$$P(B|A) \neq P(A|B).$$

b. Now, we find the probability that a U.S. woman aged 40 to 49 does not have breast cancer, given that she has a positive mammogram:

$$P(\text{no breast cancer} \,|\, \text{positive mammogram}).$$

This is the probability of not having breast cancer if the data are restricted to women with positive mammograms:

	Breast Cancer	No Breast Cancer	Total
Positive Mammogram	720	6944	7664

Within the restricted data, there are 6944 women without breast cancer and 720 + 6944, or 7664 women with positive mammograms. Thus,

$$P(\text{no breast cancer}|\text{positive mammogram}) = \frac{6944}{7664} \approx 0.906.$$

Among women with positive mammograms, the probability of not having breast cancer is $\frac{6944}{7664}$, or approximately 0.906.

The conditional probability in Example 8(b) indicates a probability of approximately 0.9 that a woman aged 40 to 49 who has a positive mammogram is actually cancer-free. The likely probability of this false positive changed the age at which the Amercian Cancer Society recommends women start getting regular mammograms. The American Cancer Society now advises that women at an average risk for breast cancer begin screening at age 45 (it had previously recommended starting at 40) and to transition at age 55 to a schedule of every other year.

CHECK POINT 8 Use the data in **Table 11.7** on the previous page to find the probability that a U.S. woman aged 40 to 49

a. has a positive mammogram, given that she has breast cancer.

b. has breast cancer, given that she has a positive mammogram.

Express probabilities as decimals and, if necessary, round to three decimal places.

We have seen that $P(B|A)$ is the probability that event B occurs if the sample space is restricted to event A. Thus,

$$P(B|A) = \frac{\text{number of outcomes of } B \text{ that are in the restricted sample space } A}{\text{number of outcomes in the restricted sample space } A}.$$

This can be stated in terms of the following formula:

A FORMULA FOR CONDITIONAL PROBABILITY

$$P(B|A) = \frac{n(B \cap A)}{n(A)} = \frac{\text{number of outcomes common to } B \text{ and } A}{\text{number of outcomes in } A}$$

Concept and Vocabulary Check

Fill in each blank so that the resulting statement is true.

1. If the occurrence of one event has no effect on the probability of another event, the events are said to be ______________. For such events, $P(A \text{ and } B) =$ ___________.
2. The probability of an event occurring at least once is equal to 1 minus the probability that ___________________.
3. If the occurrence of one event has an effect on the probability of another event, the events are said to be ____________. For such events, $P(A \text{ and } B) =$ ________________________.
4. The probability of event B, assuming that event A has already occurred, is called the ___________ probability of B, given A. This probability is denoted by _________.

In Exercises 5–8, determine whether each statement is true or false. If the statement is false, make the necessary change(s) to produce a true statement.

5. *And* probabilities can always be determined using the formula $P(A \text{ and } B) = P(A) \cdot P(B)$. ______
6. Probability problems with the word *or* involve more than one selection. ______
7. The probability that an event happens at least once can be found by subtracting the probability that the event does not happen from 1. ______
8. $P(B|A)$ is the probability that event B occurs if the sample space is restricted to the outcomes associated with event A. ______

Exercise Set 11.7

Practice and Application Exercises

Exercises 1–26 involve probabilities with independent events.

Use the spinner shown to solve Exercises 1–10. It is equally probable that the pointer will land on any one of the six regions. If the pointer lands on a borderline, spin again. If the pointer is spun twice, find the probability it will land on

1. green and then red.
2. yellow and then green.
3. yellow and then yellow.
4. red and then red.
5. a color other than red each time.
6. a color other than green each time.

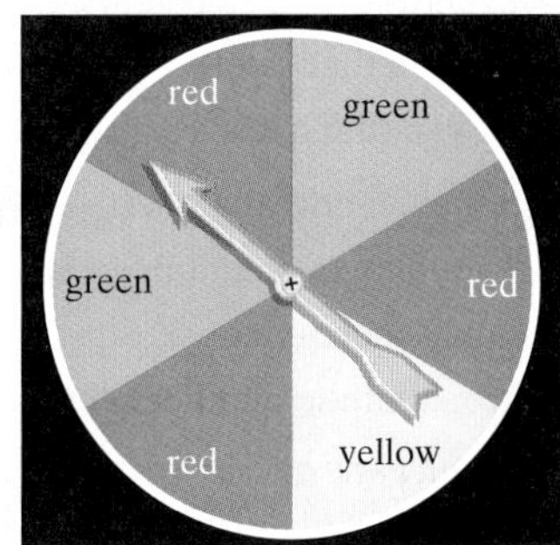

If the pointer is spun three times, find the probability it will land on

7. green and then red and then yellow.
8. red and then red and then green.
9. red every time.
10. green every time.

In Exercises 11–14, a single die is rolled twice. Find the probability of rolling

11. a 2 the first time and a 3 the second time.
12. a 5 the first time and a 1 the second time.
13. an even number the first time and a number greater than 2 the second time.
14. an odd number the first time and a number less than 3 the second time.

In Exercises 15–20, you draw one card from a 52-card deck. Then the card is replaced in the deck, the deck is shuffled, and you draw again. Find the probability of drawing

15. a picture card the first time and a heart the second time.
16. a jack the first time and a club the second time.
17. a king each time.
18. a 3 each time.
19. a red card each time.
20. a black card each time.
21. If you toss a fair coin six times, what is the probability of getting all heads?
22. If you toss a fair coin seven times, what is the probability of getting all tails?

In Exercises 23–24, a coin is tossed and a die is rolled. Find the probability of getting

23. a head and a number greater than 4.
24. a tail and a number less than 5.
25. The probability that South Florida will be hit by a major hurricane (category 4 or 5) in any single year is $\frac{1}{16}$. (*Source:* National Hurricane Center)
    a. What is the probability that South Florida will be hit by a major hurricane two years in a row?
    b. What is the probability that South Florida will be hit by a major hurricane in three consecutive years?
    c. What is the probability that South Florida will not be hit by a major hurricane in the next ten years?
    d. What is the probability that South Florida will be hit by a major hurricane at least once in the next ten years?
26. The probability that a region prone to flooding will flood in any single year is $\frac{1}{10}$.
    a. What is the probability of a flood two years in a row?
    b. What is the probability of flooding in three consecutive years?
    c. What is the probability of no flooding for ten consecutive years?
    d. What is the probability of flooding at least once in the next ten years?

Exercises 27–52 involve probabilities with dependent events.

In Exercises 27–30, we return to our box of chocolates. There are 30 chocolates in the box, all identically shaped. Five are filled with coconut, 10 with caramel, and 15 are solid chocolate. You randomly select one piece, eat it, and then select a second piece. Find the probability of selecting

27. two solid chocolates in a row.

28. two caramel-filled chocolates in a row.

29. a coconut-filled chocolate followed by a caramel-filled chocolate.

30. a coconut-filled chocolate followed by a solid chocolate.

In Exercises 31–36, consider a political discussion group consisting of 5 Democrats, 6 Republicans, and 4 Independents. Suppose that two group members are randomly selected, in succession, to attend a political convention. Find the probability of selecting

31. two Democrats.

32. two Republicans.

33. an Independent and then a Republican.

34. an Independent and then a Democrat.

35. no Independents.

36. no Democrats.

In Exercises 37–42, an ice chest contains six cans of apple juice, eight cans of grape juice, four cans of orange juice, and two cans of mango juice. Suppose that you reach into the container and randomly select three cans in succession. Find the probability of selecting

37. three cans of apple juice.

38. three cans of grape juice.

39. a can of grape juice, then a can of orange juice, then a can of mango juice.

40. a can of apple juice, then a can of grape juice, then a can of orange juice.

41. no grape juice.

42. no apple juice.

In Exercises 43–52, we return to the game of Scrabble that has 100 tiles. Shown in blue are the number of tiles for each letter. The letter's point value is printed on the tile.

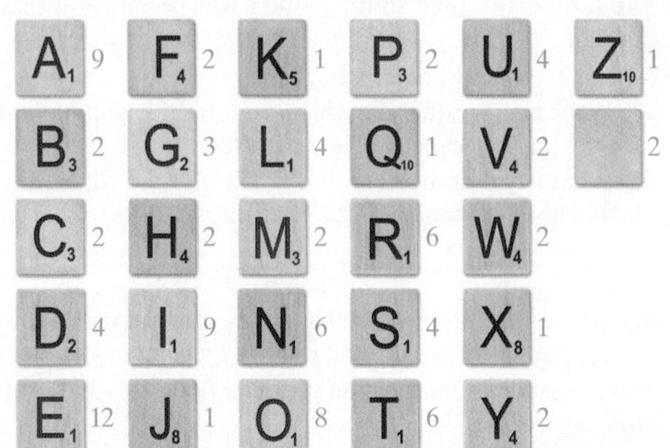

In Exercises 43–48, two tiles are drawn in succession from Scrabble's 100 tiles. Find each probability, expressed as a simplified fraction.

43. Find the probability of selecting an A on the first draw and an F on the second draw.

44. Find the probability of selecting an E on the first draw and a J on the second draw.

45. Find the probability of selecting a tile worth 4 points on the first draw and a tile worth 3 points on the second draw.

46. Find the probability of selecting a tile worth 3 points on the first draw and a tile worth 4 points on the second draw.

47. Find the probability of selecting a C or a D on the second draw, given that a B was selected on the first draw.

48. Find the probability of selecting an L or an M on the second draw, given that a K was selected on the first draw.

In Exercises 49–52, three tiles are drawn in succession from Scrabble's tiles. Find each probability, expressed as a simplified fraction.

49. Find the probability of selecting three Ls in a row.

50. Find the probability of selecting three Rs in a row.

51. Find the probability of selecting a letter worth 10 points on the first draw, 5 points on the second draw, and 1 point on the third draw.

52. Find the probability of selecting a letter worth 1 point on the first draw, 10 points on the second draw, and 5 points on the third draw.

In Exercises 53–60, the numbered disks shown are placed in a box and one disk is selected at random.

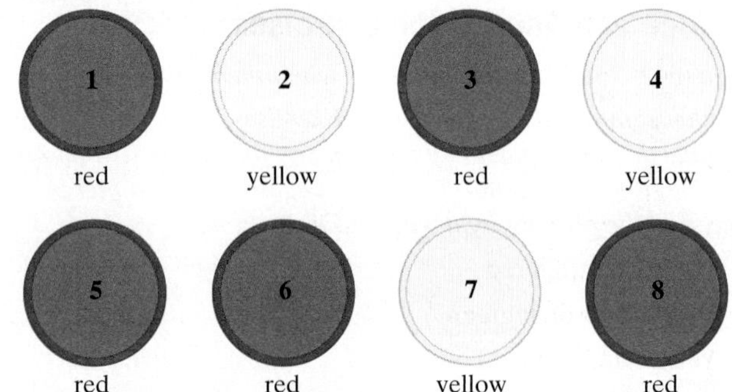

Find the probability of selecting

53. a 3, given that a red disk is selected.

54. a 7, given that a yellow disk is selected.

55. an even number, given that a yellow disk is selected.

56. an odd number, given that a red disk is selected.

57. a red disk, given that an odd number is selected.

58. a yellow disk, given that an odd number is selected.

59. a red disk, given that the number selected is at least 5.

60. a yellow disk, given that the number selected is at most 3.

The table shows the outcome of car accidents in Florida for a recent year by whether or not the driver wore a seat belt. Use the data to solve Exercises 61–64. Express probabilities as fractions and as decimals rounded to three places.

CAR ACCIDENTS IN FLORIDA

	Wore Seat Belt	No Seat Belt	Total
Driver Survived	412,368	162,527	574,895
Driver Died	510	1601	2111
Total	412,878	164,128	577,006

Source: Alan Agresti and Christine Franklin, *Statistics*, Prentice Hall, 2007

61. Find the probability of surviving a car accident, given that the driver wore a seat belt.

62. Find the probability of not surviving a car accident, given that the driver did not wear a seat belt.

63. Find the probability of wearing a seat belt, given that a driver survived a car accident.

64. Find the probability of not wearing a seat belt, given that a driver did not survive a car accident.

Shown again is the table indicating the marital status of the U.S. population in 2015. Numbers in the table are expressed in millions. Use the data in the table to solve Exercises 65–76. Express probabilities as simplified fractions and as decimals rounded to the nearest hundredth.

MARITAL STATUS OF THE U.S. POPULATION, AGES 15 OR OLDER, 2015, IN MILLIONS

	Married	Never Married	Divorced	Widowed	Total
Male	66	43	11	3	123
Female	67	38	15	11	131
Total	133	81	26	14	254

If one person is selected from the population described in the table, find the probability that the person

65. is not divorced.

66. is not widowed.

67. is widowed or divorced.

68. has never been married or is divorced.

69. is male or divorced.

70. is female or divorced.

71. is male, given that this person is divorced.

72. is female, given that this person is divorced.

73. is widowed, given that this person is a woman.

74. is divorced, given that this person is a man.

75. has never been married or is married, given that this person is a man.

76. has never been married or is married, given that this person is a woman.

77. Probabilities and Coincidence of Shared Birthdays

Use a calculator to solve this exercise. Round probabilities to three decimal places.

a. If two people are selected at random, the probability that they do not have the same birthday (day and month) is $\frac{365}{365} \cdot \frac{364}{365}$. Explain why this is so. (Ignore leap years and assume 365 days in a year.)

b. If three people are selected at random, find the probability that they all have different birthdays.

c. If three people are selected at random, find the probability that at least two of them have the same birthday.

d. If 20 people are selected at random, find the probability that at least 2 of them have the same birthday.

e. Show that if 23 people are selected at random, the probability that at least 2 of them have the same birthday is greater than $\frac{1}{2}$.

Blitzer Bonus

More Probabilities of Shared Birthdays

- If 253 people are selected at random, the probability that at least 2 of them have the same birthday (day, month, *and year*) is approximately $\frac{1}{2}$.
- If 18 people are selected at random, the probability that at least 3 of them have the same birthday (day and month) is approximately $\frac{1}{2}$.
- If 14 people are selected at random, the probability that at least 2 of them are born within a day of each other is approximately $\frac{1}{2}$.
- If 7 people are selected at random, the probability that at least 2 of them are born within a week of each other is approximately $\frac{1}{2}$.

Source: Gina Kolata, *The New York Times Book of Mathematics*, Sterling Publishing, 2013

Explaining the Concepts

78. Explain how to find *and* probabilities with independent events. Give an example.

79. Explain how to find *and* probabilities with dependent events. Give an example.

80. What does $P(B|A)$ mean? Give an example.

In Exercises 81–85, write a probability problem involving the word "and" whose solution results in the probability fractions shown.

81. $\frac{1}{2} \cdot \frac{1}{2}$ **82.** $\frac{1}{6} \cdot \frac{1}{6} \cdot \frac{1}{6}$ **83.** $\frac{1}{2} \cdot \frac{1}{6}$ **84.** $\frac{13}{52} \cdot \frac{12}{51}$ **85.** $\frac{1}{4} \cdot \frac{3}{5}$

Critical Thinking Exercises

Make Sense? *In Exercises 86–89, determine whether each statement makes sense or does not make sense, and explain your reasoning.*

86. If a fourth child is born into a family with three boys, the odds in favor of a girl are better than 1:1.

87. In a group of five men and five women, the probability of randomly selecting a man is $\frac{1}{2}$, so if I select two people from the group, the probability that both are men is $\frac{1}{2} \cdot \frac{1}{2}$.

88. I found the probability of getting rain at least once in ten days by calculating the probability that none of the days have rain and subtracting this probability from 1.

89. I must have made an error calculating probabilities because $P(A|B)$ is not the same as $P(B|A)$.

90. If the probability of being hospitalized during a year is 0.1, find the probability that no one in a family of five will be hospitalized in a year.

91. If a single die is rolled five times, what is the probability it lands on 2 on the first, third, and fourth rolls, but not on either of the other rolls?

92. Nine cards numbered from 1 through 9 are placed into a box and two cards are selected without replacement. Find the probability that both numbers selected are odd, given that their sum is even.

93. If a single die is rolled twice, find the probability of rolling an odd number and a number greater than 4 in either order.

Group Exercises

94. Do you live in an area prone to catastrophes, such as earthquakes, fires, tornados, hurricanes, or floods? If so, research the probability of this catastrophe occurring in a single year. Group members should then use this probability to write and solve a problem similar to Exercise 25 in this Exercise Set.

95. Group members should use the table for Exercises 65–76 to write and solve four probability problems different than those in the exercises. Two should involve *or* (one with events that are mutually exclusive and one with events that are not), one should involve *and*—that is, events in succession—and one should involve conditional probability.

11.8 Expected Value

WHAT AM I SUPPOSED TO LEARN?

After studying this section, you should be able to:

1. Compute expected value.
2. Use expected value to solve applied problems.
3. Use expected value to determine the average payoff or loss in a game of chance.

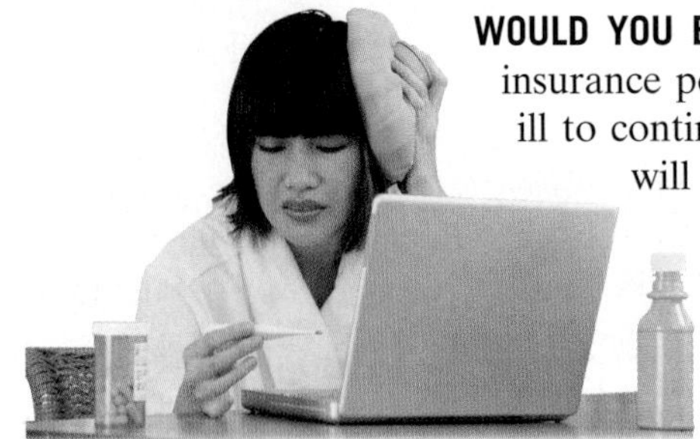

WOULD YOU BE WILLING TO SPEND $50 A YEAR FOR AN insurance policy that pays $200,000 if you become too ill to continue your education? It is unlikely that this will occur. Insurance companies make money by compensating us for events that have a low probability. If one in every 5000 students needs to quit college due to serious illness, the probability of this event is $\frac{1}{5000}$. Multiplying the amount of the claim, $200,000, by its probability, $\frac{1}{5000}$, tells the insurance company what to expect to pay out on average for each policy:

$$\underbrace{\$200{,}000}_{\text{Amount of the claim}} \times \underbrace{\frac{1}{5000}}_{\text{Probability of paying the claim}} = \$40.$$

Over the long run, the insurance company can expect to pay $40 for each policy it sells. By selling the policy for $50, the expected profit is $10 per policy. If 400,000 students choose to take out this insurance, the company can expect to make 400,000 × $10, or $4,000,000.

1 *Compute expected value.*

Expected value is a mathematical way to use probabilities to determine what to expect in various situations over the long run. Expected value is used to determine premiums on insurance policies, weigh the risks versus the benefits of alternatives in business ventures, and indicate to a player of any game of chance what will happen if the game is played repeatedly.

The standard way to find expected value is to multiply each possible outcome by its probability, and then add these products. We use the letter E to represent expected value.

EXAMPLE 1 *Computing Expected Value*

Find the expected value for the outcome of the roll of a fair die.

SOLUTION

The outcomes are 1, 2, 3, 4, 5, and 6, each with a probability of $\frac{1}{6}$. The expected value, E, is computed by multiplying each outcome by its probability and then adding these products.

$$E = 1\cdot\frac{1}{6} + 2\cdot\frac{1}{6} + 3\cdot\frac{1}{6} + 4\cdot\frac{1}{6} + 5\cdot\frac{1}{6} + 6\cdot\frac{1}{6}$$

$$= \frac{1+2+3+4+5+6}{6} = \frac{21}{6} = 3.5$$

The expected value of the roll of a fair die is 3.5. This means that if the die is rolled repeatedly, there is an average of 3.5 dots per roll over the long run. This expected value cannot occur on a single roll of the die. However, it is a long-run average of the various outcomes that can occur when a fair die is rolled.

 **CHECK POINT 1** It is equally probable that a pointer will land on any one of four regions, numbered 1 through 4. Find the expected value for where the pointer will stop.

EXAMPLE 2 *Computing Expected Value*

Find the expected value for the number of girls for a family with three children.

SOLUTION

A family with three children can have 0, 1, 2, or 3 girls. There are eight ways these outcomes can occur.

No girls : Boy Boy Boy — One way
One girl : Girl Boy Boy, Boy Girl Boy, Boy Boy Girl — Three ways
Two girls : Girl Girl Boy, Girl Boy Girl, Boy Girl Girl — Three ways
Three girls : Girl Girl Girl — One way

TABLE 11.8 Outcomes and Probabilities for the Number of Girls in a Three-Child Family

Outcome: Number of Girls	Probability
0	$\frac{1}{8}$
1	$\frac{3}{8}$
2	$\frac{3}{8}$
3	$\frac{1}{8}$

Table 11.8 shows the probabilities for 0, 1, 2, and 3 girls.

The expected value, E, is computed by multiplying each outcome by its probability and then adding these products.

$$E = 0 \cdot \frac{1}{8} + 1 \cdot \frac{3}{8} + 2 \cdot \frac{3}{8} + 3 \cdot \frac{1}{8} = \frac{0 + 3 + 6 + 3}{8} = \frac{12}{8} = \frac{3}{2} = 1.5$$

The expected value is 1.5. This means that if we record the number of girls in many different three-child families, the average number of girls for all these families will be 1.5. In a three-child family, half the children are expected to be girls, so the expected value of 1.5 is consistent with this observation.

CHECK POINT 2 A fair coin is tossed four times in succession. **Table 11.9** shows the probabilities for the different number of heads that can arise. Find the expected value for the number of heads.

TABLE 11.9

Number of Heads	Probability
0	$\frac{1}{16}$
1	$\frac{4}{16}$
2	$\frac{6}{16}$
3	$\frac{4}{16}$
4	$\frac{1}{16}$

2 *Use expected value to solve applied problems.*

Applications of Expected Value

Empirical probabilities can be determined in many situations by examining what has occurred in the past. For example, an insurance company can tally various claim amounts over many years. If 15% of these amounts are for a \$2000 claim, then the probability of this claim amount is 0.15. By studying sales of similar houses in a particular area, a realtor can determine the probability that he or she will sell a listed house, another agent will sell the house, or the listed house will remain unsold. Once probabilities have been assigned to all possible outcomes, expected value can indicate what is expected to happen in the long run. These ideas are illustrated in Examples 3 and 4.

EXAMPLE 3 *Determining an Insurance Premium*

An automobile insurance company has determined the probabilities for various claim amounts for drivers ages 16 through 21, shown in **Table 11.10**.

a. Calculate the expected value and describe what this means in practical terms.

b. How much should the company charge as an average premium so that it does not lose or gain money on its claim costs?

TABLE 11.10 Probabilities for Auto Claims

Amount of Claim (to the nearest \$2000)	Probability
\$0	0.70
\$2000	0.15
\$4000	0.08
\$6000	0.05
\$8000	0.01
\$10,000	0.01

SOLUTION

a. The expected value, E, is computed by multiplying each outcome by its probability, and then adding these products.

$$\begin{aligned} E &= \$0(0.70) + \$2000(0.15) + \$4000(0.08) + \$6000(0.05) \\ &\quad + \$8000(0.01) + \$10{,}000(0.01) \\ &= \$0 + \$300 + \$320 + \$300 + \$80 + \$100 \\ &= \$1100 \end{aligned}$$

The expected value is \$1100. This means that in the long run the average cost of a claim is \$1100. The insurance company should expect to pay \$1100 per car insured to people in the 16–21 age group.

b. At the very least, the amount that the company should charge as an average premium for each person in the 16–21 age group is \$1100. In this way, it will not lose or gain money on its claims costs. It's quite probable that the company will charge more, moving from break-even to profit.

CHECK POINT 3 Work Example 3 again if the probabilities for claims of \$0 and \$10,000 are reversed. Thus, the probability of a \$0 claim is 0.01 and the probability of a \$10,000 claim is 0.70.

Business decisions are interpreted in terms of dollars and cents. In these situations, **expected value is calculated by multiplying the gain or loss for each possible outcome by its probability. The sum of these products is the expected value.**

EXAMPLE 4 *Expectation in a Business Decision*

You are a realtor considering listing a \$500,000 house. The cost of advertising and providing food for other realtors during open showings is anticipated to cost you \$5000. The house is quite unusual, and you are given a four-month listing. If the house is unsold after four months, you lose the listing and receive nothing. You anticipate that the probability you sell your own listed house is 0.3, the probability that another agent sells your listing is 0.2, and the probability that the house is unsold after 4 months is 0.5. If you sell your own listed house, the commission is a hefty \$30,000. If another realtor sells your listing, the commission is \$15,000. The bottom line: You will not take the listing unless you anticipate earning at least \$6000. Should you list the house?

THE REALTOR'S SUMMARY SHEET

My Cost:	**\$5000**
My Possible Income:	
I sell house:	\$30,000
Another agent sells house:	\$15,000
House unsold after 4 months:	\$0
The Probabilities:	
I sell house:	0.3
Another agent sells house:	0.2
House unsold after 4 months:	0.5
My Bottom Line: I take the listing only if I anticipate earning at least \$6000.	

SOLUTION

Shown in the margin is a summary of the amounts of money and probabilities that will determine your decision. The expected value in this situation is the sum of each income possibility times its probability. The expected value represents the amount you can anticipate earning if you take the listing. If the expected value is not at least \$6000, you should not list the house.

The possible incomes listed in the margin, \$30,000, \$15,000, and \$0, do not take into account your \$5000 costs. Because of these costs, each amount needs to be reduced by \$5000. Thus, you can gain \$30,000 − \$5000, or \$25,000, or you can gain \$15,000 − \$5000, or \$10,000. Because \$0 − \$5000 = −\$5000, you can also lose \$5000. **Table 11.11** summarizes possible outcomes if you take the listing, and their respective probabilities.

TABLE 11.11 Gains, Losses, and Probabilities for Listing a \$500,000 House

Outcome	Gain or Loss	Probability
Sells house	\$25,000	0.3
Another agent sells house	\$10,000	0.2
House doesn't sell	−\$5000	0.5

The expected value, E, is computed by multiplying each gain or loss in **Table 11.11** by its probability, and then adding these results.

$$E = \$25{,}000(0.3) + \$10{,}000(0.2) + (-\$5000)(0.5)$$
$$= \$7500 + \$2000 + (-\$2500) = \$7000$$

You can expect to earn \$7000 by listing the house. Because the expected value exceeds \$6000, you should list the house.

CHECK POINT 4 The SAT is a multiple-choice test. Each question has five possible answers. The test taker must select one answer for each question or not answer the question. One point is awarded for each correct response and $\frac{1}{4}$ point is subtracted for each wrong answer. No points are added or subtracted for answers left blank. **Table 11.12** summarizes the information for the outcomes of a random guess on an SAT question. Find the expected point value of a random guess. Is there anything to gain or lose on average by guessing? Explain your answer.

TABLE 11.12 Gains and Losses for Guessing on the SAT

Outcome	Gain or Loss	Probability
Guess correctly	1	$\frac{1}{5}$
Guess incorrectly	$-\frac{1}{4}$	$\frac{4}{5}$

3 Use expected value to determine the average payoff or loss in a game of chance.

Expected Value and Games of Chance

Expected value can be interpreted as the average payoff in a contest or game when either is played a large number of times. **To find the expected value of a game, multiply the gain or loss for each possible outcome by its probability. Then add the products.**

EXAMPLE 5 Expected Value as Average Payoff

A game is played using one die. If the die is rolled and shows 1, 2, or 3, the player wins nothing. If the die shows 4 or 5, the player wins \$3. If the die shows 6, the player wins \$9. If there is a charge of \$1 to play the game, what is the game's expected value? Describe what this means in practical terms.

SOLUTION

Because there is a charge of \$1 to play the game, a player who wins \$9 gains \$9 − \$1, or \$8. A player who wins \$3 gains \$3 − \$1, or \$2. If the player gets \$0, there is a loss of \$1 because \$0 − \$1 = −\$1. The outcomes for the die, with their respective gains, losses, and probabilities, are summarized in **Table 11.13**.

TABLE 11.13 Gains, Losses, and Probabilities in a Game of Chance

Outcome	Gain or Loss	Probability
1, 2, or 3	−\$1	$\frac{3}{6}$
4 or 5	\$2	$\frac{2}{6}$
6	\$8	$\frac{1}{6}$

Expected value, E, is computed by multiplying each gain or loss in **Table 11.13** by its probability and then adding these results.

$$E = (-\$1)\left(\frac{3}{6}\right) + \$2\left(\frac{2}{6}\right) + \$8\left(\frac{1}{6}\right)$$

$$= \frac{-\$3 + \$4 + \$8}{6} = \frac{\$9}{6} = \$1.50$$

The expected value is \$1.50. This means that in the long run, a player can expect to win an average of \$1.50 for each game played. However, this does not mean that the player will win \$1.50 on any single game. It does mean that if the game is played repeatedly, then, in the long run, the player should expect to win about \$1.50 per play on the average. If 1000 games are played, one could expect to win \$1500. However, if only three games are played, one's net winnings can range between −\$3 and \$24, even though the expected winnings are \$1.50(3), or \$4.50.

Gambling It Away

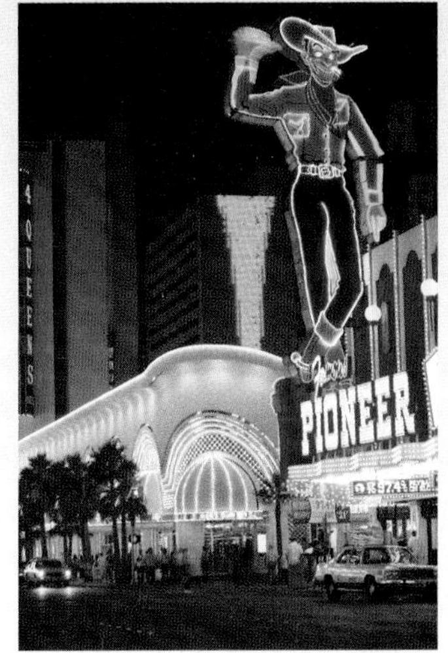

Games in gambling casinos have probabilities and payoffs that result in negative expected values. This means that players will lose money in the long run. With no clocks on the walls or windows to look out of, casinos bet that players will forget about time and play longer. This brings casino owners closer to what they can expect to earn, while negative expected values sneak up on losing patrons. (No wonder they provide free beverages!)

A Potpourri of Gambling-Related Data

- Population of Las Vegas: 535,000
- Slot Machines in Las Vegas: 150,000
- Average Earnings per Slot Machine per Year: \$100,000
- Average Amount Lost per Hour in Las Vegas Casinos: \$696,000
- Average Amount an American Adult Loses Gambling per Year: \$350
- Average Amount a Nevada Resident Loses Gambling per Year: \$1000

Source: Paul Grobman, *Vital Statistics*, Plume, 2005.

CHECK POINT 5 A charity is holding a raffle and sells 1000 raffle tickets for \$2 each. One of the tickets will be selected to win a grand prize of \$1000. Two other tickets will be selected to win consolation prizes of \$50 each. Fill in the missing column in **Table 11.14**. Then find the expected value if you buy one raffle ticket. Describe what this means in practical terms. What can you expect to happen if you purchase five tickets?

TABLE 11.14 Gains, Losses, and Probabilities in a Raffle

Outcome	Gain or Loss	Probability
Win Grand Prize		$\frac{1}{1000}$
Win Consolation Prize		$\frac{2}{1000}$
Win Nothing		$\frac{997}{1000}$

FIGURE 11.15 A U.S. roulette wheel

Unlike the game in Example 5, games in gambling casinos are set up so that players will lose in the long run. These games have negative expected values. Such a game is roulette, French for "little wheel." We first saw the roulette wheel in Section 11.7. It is shown again in **Figure 11.15**. Recall that the wheel has 38 numbered slots (1 through 36, 0, and 00). In each play of the game, the dealer spins the wheel and a small ball in opposite directions. The ball is equally likely to come to rest in any one of the slots, which are colored black, red, or green. Gamblers can place a number of different bets in roulette. Example 6 illustrates one gambling option.

EXAMPLE 6 *Expected Value and Roulette*

One way to bet in roulette is to place \$1 on a single number. If the ball lands on that number, you are awarded \$35 and get to keep the \$1 that you paid to play the game. If the ball lands on any one of the other 37 slots, you are awarded nothing and the \$1 that you bet is collected. Find the expected value for playing roulette if you bet \$1 on number 20. Describe what this means.

SOLUTION

Table 11.15 contains the two outcomes of interest: the ball landing on your number, 20, and the ball landing elsewhere (in any one of the other 37 slots). The outcomes, their respective gains, losses, and probabilities, are summarized in the table.

TABLE 11.15 Playing One Number with a 35 to 1 Payoff in Roulette

Outcome	Gain or Loss	Probability
Ball lands on 20	\$35	$\frac{1}{38}$
Ball does not land on 20	−\$1	$\frac{37}{38}$

Expected value, E, is computed by multiplying each gain or loss in **Table 11.15** by its probability and then adding these results.

$$E = \$35\left(\frac{1}{38}\right) + (-\$1)\left(\frac{37}{38}\right) = \frac{\$35 - \$37}{38} = \frac{-\$2}{38} \approx -\$0.05$$

The expected value is approximately −\$0.05. This means that in the long run, a player can expect to lose about 5¢ for each game played. If 2000 games are played, one could expect to lose \$100.

Blitzer Bonus

How to Win at Roulette

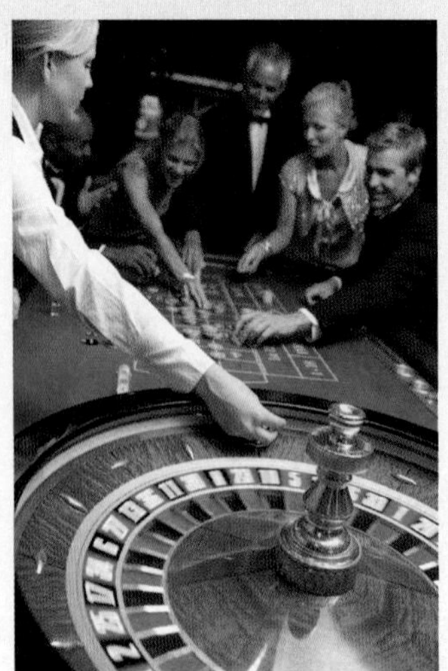

A player may win or lose at roulette, but in the long run the casino always wins. Casinos make an average of three cents on every dollar spent by gamblers. There *is* a way to win at roulette in the long run: Own the casino.

CHECK POINT 6 In the game of one-spot keno, a card is purchased for \$1. It allows a player to choose one number from 1 to 80. A dealer then chooses twenty numbers at random. If the player's number is among those chosen, the player is paid \$3.20, but does not get to keep the \$1 paid to play the game. Find the expected value of a \$1 bet. Describe what this means.

Concept and Vocabulary Check

Fill in each blank so that the resulting statement is true.

1. A mathematical way to use probabilities to determine what to expect in various situations over the long run is called __________ value. The standard way to find this value is to multiply each possible outcome by its __________, and then ______ these products.

2. To find the expected value of a game of chance, multiply the monetary gain or ______ for each possible outcome by its __________, and then ______ these products.

In Exercises 3–4, determine whether each statement is true or false. If the statement is false, make the necessary change(s) to produce a true statement.

3. Business decisions are made based on expected values of zero. ______

4. Games in gambling casinos are set up so that expected values are negative. ______

Exercise Set 11.8

Practice and Application Exercises

In Exercises 1–2, the numbers that each pointer can land on and their respective probabilities are shown. Compute the expected value for the number on which each pointer lands.

1.

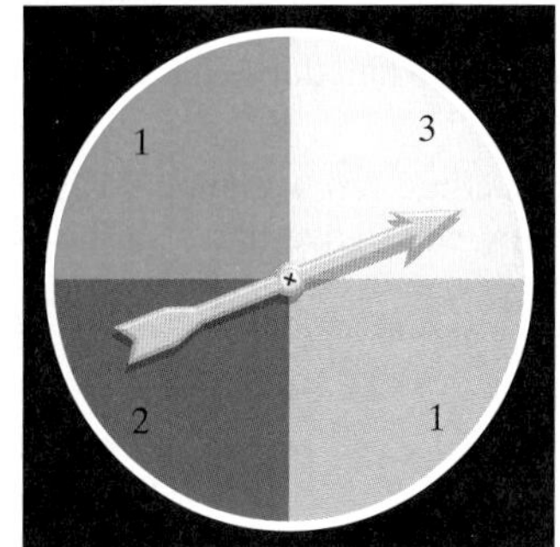

Outcome	Probability
1	$\frac{1}{2}$
2	$\frac{1}{4}$
3	$\frac{1}{4}$

2.

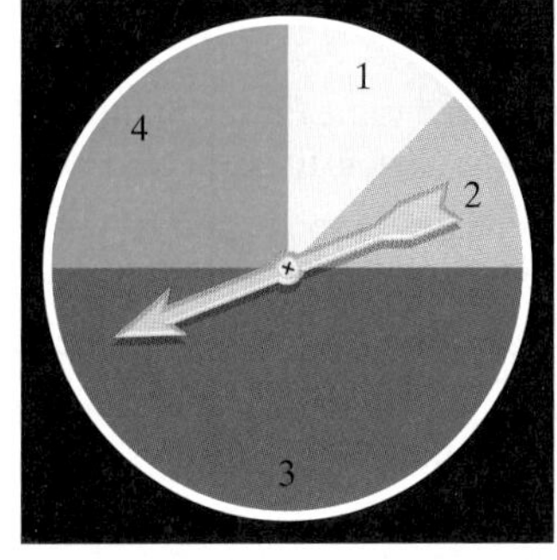

Outcome	Probability
1	$\frac{1}{8}$
2	$\frac{1}{8}$
3	$\frac{1}{2}$
4	$\frac{1}{4}$

The tables in Exercises 3–4 show claims and their probabilities for an insurance company.

a. *Calculate the expected value and describe what this means in practical terms.*

b. *How much should the company charge as an average premium so that it breaks even on its claim costs?*

c. *How much should the company charge to make a profit of $50 per policy?*

3. **PROBABILITIES FOR HOMEOWNERS' INSURANCE CLAIMS**

Amount of Claim (to the nearest $50,000)	Probability
$0	0.65
$50,000	0.20
$100,000	0.10
$150,000	0.03
$200,000	0.01
$250,000	0.01

4. **PROBABILITIES FOR MEDICAL INSURANCE CLAIMS**

Amount of Claim (to the nearest $20,000)	Probability
$0	0.70
$20,000	0.20
$40,000	0.06
$60,000	0.02
$80,000	0.01
$100,000	0.01

5. An architect is considering bidding for the design of a new museum. The cost of drawing plans and submitting a model is $10,000. The probability of being awarded the bid is 0.1, and anticipated profits are $100,000, resulting in a possible gain of this amount minus the $10,000 cost for plans and a model. What is the expected value in this situation? Describe what this value means.

6. A construction company is planning to bid on a building contract. The bid costs the company $1500. The probability that the bid is accepted is $\frac{1}{5}$. If the bid is accepted, the company will make $40,000 minus the cost of the bid. Find the expected value in this situation. Describe what this value means.

7. It is estimated that there are 27 deaths for every 10 million people who use airplanes. A company that sells flight insurance provides $100,000 in case of death in a plane crash. A policy can be purchased for $1. Calculate the expected value and thereby determine how much the insurance company can make over the long run for each policy that it sells.

8. A 25-year-old can purchase a one-year life insurance policy for $10,000 at a cost of $100. Past history indicates that the probability of a person dying at age 25 is 0.002. Determine the company's expected gain per policy.

Exercises 9–10 are related to the SAT, described in Check Point 4 on page 759.

9. Suppose that you can eliminate one of the possible five answers. Modify the two probabilities shown in the final column in **Table 11.12** by finding the probabilities of guessing correctly and guessing incorrectly under these circumstances. What is the expected point value of a random guess? Is it advantageous to guess under these circumstances?

10. Suppose that you can eliminate two of the possible five answers. Modify the two probabilities shown in the final column in **Table 11.12** by finding the probabilities of guessing correctly and guessing incorrectly under these circumstances. What is the expected point value of a random guess? Is it advantageous to guess under these circumstances?

11. A store specializing in mountain bikes is to open in one of two malls. If the first mall is selected, the store anticipates a yearly profit of $300,000 if successful and a yearly loss of $100,000 otherwise. The probability of success is $\frac{1}{2}$. If the second mall is selected, it is estimated that the yearly profit will be $200,000 if successful; otherwise, the annual loss will be $60,000. The probability of success at the second mall is $\frac{3}{4}$. Which mall should be chosen in order to maximize the expected profit?

12. An oil company is considering two sites on which to drill, described as follows:

Site A:	Profit if oil is found: $80 million
	Loss if no oil is found: $10 million
	Probability of finding oil: 0.2
Site B:	Profit if oil is found: $120 million
	Loss if no oil is found: $18 million
	Probability of finding oil: 0.1

Which site has the larger expected profit? By how much?

13. In a product liability case, a company can settle out of court for a loss of $350,000, or go to trial, losing $700,000 if found guilty and nothing if found not guilty. Lawyers for the company estimate the probability of a not-guilty verdict to be 0.8.
    a. Find the expected value of the amount the company can lose by taking the case to court.
    b. Should the company settle out of court?

14. A service that repairs air conditioners sells maintenance agreements for $80 a year. The average cost for repairing an air conditioner is $350 and 1 in every 100 people who purchase maintenance agreements have air conditioners that require repair. Find the service's expected profit per maintenance agreement.

Exercises 15–19 involve computing expected values in games of chance.

15. A game is played using one die. If the die is rolled and shows 1, the player wins $5. If the die shows any number other than 1, the player wins nothing. If there is a charge of $1 to play the game, what is the game's expected value? What does this value mean?

16. A game is played using one die. If the die is rolled and shows 1, the player wins $1; if 2, the player wins $2; if 3, the player wins $3. If the die shows 4, 5, or 6, the player wins nothing. If there is a charge of $1.25 to play the game, what is the game's expected value? What does this value mean?

17. Another option in a roulette game (see Example 6 on page 760) is to bet $1 on red. (There are 18 red compartments, 18 black compartments, and 2 compartments that are neither red nor black.) If the ball lands on red, you get to keep the $1 that you paid to play the game and you are awarded $1. If the ball lands elsewhere, you are awarded nothing and the $1 that you bet is collected. Find the expected value for playing roulette if you bet $1 on red. Describe what this number means.

18. The spinner on a wheel of fortune can land with an equal chance on any one of ten regions. Three regions are red, four are blue, two are yellow, and one is green. A player wins $4 if the spinner stops on red and $2 if it stops on green. The player loses $2 if it stops on blue and $3 if it stops on yellow. What is the expected value? What does this mean if the game is played ten times?

19. For many years, organized crime ran a numbers game that is now run legally by many state governments. The player selects a three-digit number from 000 to 999. There are 1000 such numbers. A bet of $1 is placed on a number, say number 115. If the number is selected, the player wins $500. If any other number is selected, the player wins nothing. Find the expected value for this game and describe what this means.

Explaining the Concepts

20. What does the expected value for the outcome of the roll of a fair die represent?

21. Explain how to find the expected value for the number of girls for a family with two children. What is the expected value?

22. How do insurance companies use expected value to determine what to charge for a policy?

23. Describe a situation in which a business can use expected value.

24. If the expected value of a game is negative, what does this mean? Also describe the meaning of a positive and a zero expected value.

25. The expected value for purchasing a ticket in a raffle is −$0.75. Describe what this means. Will a person who purchases a ticket lose $0.75?

Critical Thinking Exercises

Make Sense? *In Exercises 26–29, determine whether each statement makes sense or does not make sense, and explain your reasoning.*

26. I found the expected value for the number of boys for a family with five children to be 2.5. I must have made an error because a family with 2.5 boys cannot occur.

27. Here's my dilemma: I can accept a $1000 bill or play a dice game ten times. For each roll of the single die,
   - I win $500 for rolling 1 or 2.
   - I win $200 for rolling 3.
   - I lose $300 for rolling 4, 5, or 6.

   Based on expected value, I should accept the $1000 bill.

28. I've lost a fortune playing roulette, so I'm bound to reduce my losses if I play the game a little longer.

29. My expected value in a state lottery game is $7.50.

30. A popular state lottery is the 5/35 lottery, played in Arizona, Connecticut, Illinois, Iowa, Kentucky, Maine, Massachusetts, New Hampshire, South Dakota, and Vermont. In Arizona's version of the game, prizes are set: First prize is $50,000, second prize is $500, and third prize is $5. To win first prize, you must select all five of the winning numbers, numbered from 1 to 35. Second prize is awarded to players who select any four of the five winning numbers, and third prize is awarded to players who select any three of the winning numbers. The cost to purchase a lottery ticket is $1. Find the expected value of Arizona's "Fantasy Five" game, and describe what this means in terms of buying a lottery ticket over the long run.

31. Refer to the probabilities of dying at any given age in the table presented on page 724 to solve this exercise. A 20-year-old woman wants to purchase a $200,000 one-year life insurance policy. What should the insurance company charge the woman for the policy if it wants an expected profit of $60?

Group Exercise

32. This activity is a group research project intended for people interested in games of chance at casinos. The research should culminate in a seminar on games of chance and their expected values. The seminar is intended to last about 30 minutes and should result in an interesting and informative presentation made to the entire class.

    Each member of the group should research a game available at a typical casino. Describe the game to the class and compute its expected value. After each member has done this, so that class members now have an idea of those games with the greatest and smallest house advantages, a final group member might want to research and present ways for currently treating people whose addiction to these games has caused their lives to swirl out of control.

Chapter Summary, Review, and Test

SUMMARY – DEFINITIONS AND CONCEPTS	EXAMPLES
11.1 The Fundamental Counting Principle	
The number of ways in which a series of successive things can occur is found by multiplying the number of ways in which each thing can occur.	Ex. 1, p. 695; Ex. 2, p. 695; Ex. 3, p. 696; Ex. 4, p. 696; Ex. 5, p. 697; Ex. 6, p. 698
11.2 Permutations	
a. A permutation from a group of items occurs when no item is used more than once and the order of arrangement makes a difference. The Fundamental Counting Principle can be used to determine the number of permutations possible.	Ex. 1, p. 701; Ex. 2, p. 701
b. Factorial Notation $n! = n(n-1)(n-2)\cdots(3)(2)(1)$ and $0! = 1$	Ex. 3, p. 702
c. Permutations Formula: The number of permutations possible if r items are taken from n items is ${}_nP_r = \frac{n!}{(n-r)!}$.	Ex. 4, p. 704; Ex. 5, p. 704
d. Permutations of Duplicate Items: The number of permutations of n items, where p items are identical, q items are identical, r items are identical, and so on, is $\frac{n!}{p!\,q!\,r!\,\ldots}$.	Ex. 6, p. 706
11.3 Combinations	
a. A combination from a group of items occurs when no item is used more than once and the order of items makes no difference.	Ex. 1, p. 709
b. Combinations Formula: The number of combinations possible if r items are taken from n items is ${}_nC_r = \frac{n!}{(n-r)!\,r!}$.	Ex. 2, p. 711; Ex. 3, p. 711; Ex. 4, p. 712

11.4 Fundamentals of Probability

a. Theoretical probability applies to experiments in which the set of all equally likely outcomes, called the sample space, is known. An event is any subset of the sample space.

b. The theoretical probability of event E with sample space S is

$$P(E) = \frac{\text{number of outcomes in } E}{\text{total number of possible outcomes}} = \frac{n(E)}{n(S)}.$$

Ex. 1, p. 716; Ex. 2, p. 718; Ex. 3, p. 719

c. Empirical probability applies to situations in which we observe the frequency of the occurrence of an event.

d. The empirical probability of event E is

$$P(E) = \frac{\text{observed number of times } E \text{ occurs}}{\text{total number of observed occurrences}}.$$

Ex. 4, p. 720

11.5 Probability with the Fundamental Counting Principle, Permutations, and Combinations

a. Probability of a permutation

$$= \frac{\text{the number of ways the permutation can occur}}{\text{total number of possible permutations}}.$$

Ex. 1, p. 724

b. Probability of a combination

$$= \frac{\text{the number of ways the combination can occur}}{\text{total number of possible combinations}}.$$

Ex. 2, p. 726; Ex. 3, p. 727

11.6 Events Involving *Not* and *Or*; Odds

a. Complement Rules of Probability

$$P(\text{not } E) = 1 - P(E) \text{ and } P(E) = 1 - P(\text{not } E)$$

Ex. 1, p. 732; Ex. 2, p. 732; Ex. 3, p. 733

b. If it is impossible for events A and B to occur simultaneously, the events are mutually exclusive.

c. If A and B are mutually exclusive events, then $P(A \text{ or } B) = P(A) + P(B)$.

Ex. 4, p. 734; Ex. 7(b), p. 736

d. If A and B are not mutually exclusive events, then

$$P(A \text{ or } B) = P(A) + P(B) - P(A \text{ and } B).$$

Ex. 5, p. 735; Ex. 6, p. 736; Ex. 7 (a), p. 736

e. Probability to Odds

1. Odds in favor of $E = \dfrac{P(E)}{P(\text{not } E)}$ **2.** Odds against $E = \dfrac{P(\text{not } E)}{P(E)}$

Ex. 8, p. 738; Ex. 9, p. 739

f. Odds to Probability

If odds in favor of E are a to b, then $P(E) = \dfrac{a}{a+b}$.

Ex. 10, p. 740

11.7 Events Involving *And*; Conditional Probability

a. Two events are independent if the occurrence of either of them has no effect on the probability of the other.

b. If A and B are independent events,

$$P(A \text{ and } B) = P(A) \cdot P(B).$$

Ex. 1, p. 745

c. The probability of a succession of independent events is the product of each of their probabilities.

Ex. 2, p. 745; Ex. 3, p. 746

d. Two events are dependent if the occurrence of one of them has an effect on the probability of the other.

e. If A and B are dependent events,

$$P(A \text{ and } B) = P(A) \cdot P(B \text{ given that } A \text{ has occurred}).$$

Ex. 4, p. 748

f. The multiplication rule for dependent events can be extended to cover three or more dependent events. In the case of three such events,

$$P(A \text{ and } B \text{ and } C) = P(A) \cdot P(B \text{ given } A \text{ occurred}) \cdot P(C \text{ given } A \text{ and } B \text{ occurred}).$$

Ex. 5, p. 749

g. The conditional probability of B, given A, written $P(B\|A)$, is the probability of B if the sample space is restricted to A. $$P(B\|A) = \frac{n(B \cap A)}{n(A)} = \frac{\text{number of outcomes common to } B \text{ and } A}{\text{number of outcomes in } A}$$	Ex. 6, p. 750; Ex. 7, p. 750; Ex. 8, p. 751

11.8 Expected Value

a. Expected value, E, is found by multiplying every possible outcome by its probability and then adding these products.	Ex. 1, p. 756; Ex. 2, p. 757; Ex. 3, p. 757
b. In situations involving business decisions, expected value is calculated by multiplying the gain or loss for each possible outcome by its probability. The sum of these products is the expected value.	Ex. 4, p. 758
c. In a game of chance, expected value is the average payoff when the game is played a large number of times. To find the expected value of a game, multiply the gain or loss for each possible outcome by its probability. Then add the products.	Ex. 5, p. 759; Ex. 6, p. 760

Review Exercises

11.1

1. A restaurant offers 20 appetizers and 40 main courses. In how many ways can a person order a two-course meal?

2. A popular brand of pen comes in red, green, blue, or black ink. The writing tip can be chosen from extra bold, bold, regular, fine, or micro. How many different choices of pens do you have with this brand?

3. In how many ways can first and second prize be awarded in a contest with 100 people, assuming that each prize is awarded to a different person?

4. You are answering three multiple-choice questions. Each question has five answer choices, with one correct answer per question. If you select one of these five choices for each question and leave nothing blank, in how many ways can you answer the questions?

5. A stock can go up, go down, or stay unchanged. How many possibilities are there if you own five stocks?

6. A person can purchase a condominium with a choice of five kinds of carpeting, with or without a pool, with or without a porch, and with one, two, or three bedrooms. How many different options are there for the condominium?

11.2 and 11.3

In Exercises 7–10, evaluate each factorial expression.

7. $\frac{16!}{14!}$

8. $\frac{800!}{799!}$

9. $5! - 3!$

10. $\frac{11!}{(11-3)!}$

In Exercises 11–12, use the formula for $_nP_r$ to evaluate each expression.

11. $_{10}P_6$

12. $_{100}P_2$

In Exercises 13–14, use the formula for $_nC_r$ to evaluate each expression.

13. $_{11}C_7$

14. $_{14}C_5$

In Exercises 15–17, does the problem involve permutations or combinations? Explain your answer. (It is not necessary to solve the problem.)

15. How many different 4-card hands can be dealt from a 52-card deck?

16. How many different ways can a director select from 20 male actors to cast the roles of Mark, Roger, Angel, and Collins in the musical *Rent?*

17. How many different ways can a director select 4 actors from a group of 20 actors to attend a workshop on performing in rock musicals?

In Exercises 18–28, solve each problem using an appropriate method.

18. Six acts are scheduled to perform in a variety show. How many different ways are there to schedule their appearances?

19. A club with 15 members is to choose four officers—president, vice-president, secretary, and treasurer. In how many ways can these offices be filled?

20. An election ballot asks voters to select four city commissioners from a group of ten candidates. In how many ways can this be done?

21. In how many distinct ways can the letters of the word TORONTO be arranged?

22. From the 20 CDs that you've bought during the past year, you plan to take 3 with you on vacation. How many different sets of three CDs can you take?

23. You need to arrange seven of your favorite books along a small shelf. Although you are not arranging the books by height, the tallest of the books is to be placed at the left end and the shortest of the books at the right end. How many different ways can you arrange the books?

24. Suppose you are asked to list, in order of preference, the five favorite CDs you purchased in the past 12 months. If you bought 20 CDs over this time period, in how many ways can the five favorite be ranked?

25. In how many ways can five airplanes line up for departure on a runway?

26. How many different 5-card hands can be dealt from a deck that has only hearts (13 different cards)?

27. A political discussion group consists of 12 Republicans and 8 Democrats. In how many ways can 5 Republicans and 4 Democrats be selected to attend a conference on politics and social issues?

28. In how many ways can the digits in the number 335,557 be arranged?

11.4

In Exercises 29–32, a die is rolled. Find the probability of rolling

29. a 6.

30. a number less than 5.

31. a number less than 7.

32. a number greater than 6.

In Exercises 33–37, you are dealt one card from a 52-card deck. Find the probability of being dealt

33. a 5.

34. a picture card.

35. a card greater than 4 and less than 8.

36. a 4 of diamonds.

37. a red ace.

In Exercises 38–40, suppose that you reach into a bag and randomly select one piece of candy from 15 chocolates, 10 caramels, and 5 peppermints. Find the probability of selecting

38. a chocolate.

39. a caramel.

40. a peppermint.

41. Tay-Sachs disease occurs in 1 of every 3600 births among Jews from central and eastern Europe, and in 1 in 600,000 births in other populations. The disease causes abnormal accumulation of certain fat compounds in the spinal cord and brain, resulting in paralysis, blindness, and mental impairment. Death generally occurs before the age of five. If we use t to represent a Tay-Sachs gene and T a healthy gene, the table below shows the four possibilities for the children of one healthy, TT, parent, and one parent who carries the disease, Tt, but is not sick.

		Second Parent	
		T	t
First	T	TT	Tt
Parent	T	TT	Tt

a. Find the probability that a child of these parents will be a carrier without the disease.

b. Find the probability that a child of these parents will have the disease.

The table shows the employment status of the U.S. civilian labor force, ages 16 and over, by gender, for a recent year. Use the data in the table, expressed in millions, to solve Exercises 42–44.

EMPLOYMENT STATUS OF THE U.S. LABOR FORCE, AGES 16 OR OLDER, IN MILLIONS

	Employed	Unemployed	Not in Labor Force	Total
Male	74	8	34	116
Female	66	6	52	124
Total	140	14	86	240

Source: U.S. Bureau of Labor Statistics

Find the probability, expressed as a simplified fraction, that a randomly selected person from the civilian labor force represented in the table

42. is employed.

43. is female.

44. is an unemployed male.

11.5

45. If cities A, B, C, and D are visited in random order, each city visited once, find the probability that city D will be visited first, city B second, city A third, and city C last.

In Exercises 46–49, suppose that six singers are being lined up to perform at a charity. Call the singers A, B, C, D, E, and F. The order of performance is determined by writing each singer's name on one of six cards, placing the cards in a hat, and then drawing one card at a time. The order in which the cards are drawn determines the order in which the singers perform. Find the probability that

46. singer C will perform last.

47. singer B will perform first and singer A will perform last.

48. the singers will perform in the following order: F, E, A, D, C, B.

49. the performance will begin with singer A or C.

50. A lottery game is set up so that each player chooses five different numbers from 1 to 20. If the five numbers match the five numbers drawn in the lottery, the player wins (or shares) the top cash prize. What is the probability of winning the prize

a. with one lottery ticket?

b. with 100 different lottery tickets?

51. A committee of four people is to be selected from six Democrats and four Republicans. Find the probability that

a. all are Democrats.

b. two are Democrats and two are Republicans.

52. If you are dealt 3 cards from a shuffled deck of red cards (26 different cards), find the probability of getting exactly 2 picture cards.

11.6

In Exercises 53–57, a die is rolled. Find the probability of

53. not rolling a 5.

54. not rolling a number less than 4.

55. rolling a 3 or a 5.

56. rolling a number less than 3 or greater than 4.

57. rolling a number less than 5 or greater than 2.

In Exercises 58–63, you draw one card from a 52-card deck. Find the probability of

58. not drawing a picture card.

59. not drawing a diamond.

60. drawing an ace or a king.

61. drawing a black 6 or a red 7.

62. drawing a queen or a red card.

63. drawing a club or a picture card.

In Exercises 64–69, it is equally probable that the pointer on the spinner shown will land on any one of the six regions, numbered 1 through 6, and colored as shown. If the pointer lands on a borderline, spin again. Find the probability of

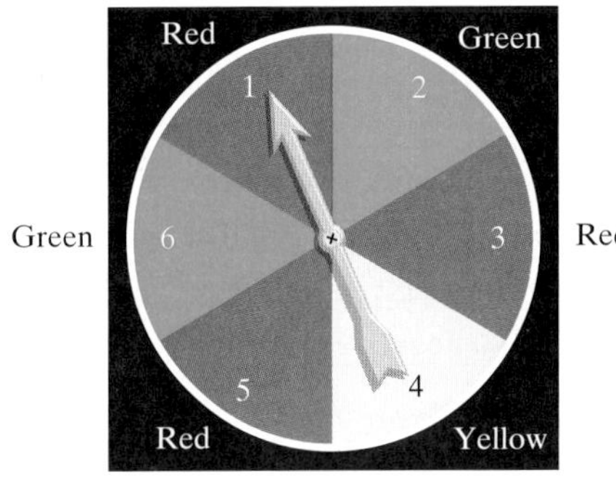

64. not stopping on 4.

65. not stopping on yellow.

66. not stopping on red.

67. stopping on red or yellow.

68. stopping on red or an even number.

69. stopping on red or a number greater than 3.

Use this information to solve Exercises 70–71. At a workshop on police work and the African-American community, there are 50 African-American male police officers, 20 African-American female police officers, 90 white male police officers, and 40 white female police officers. If one police officer is selected at random from the people at the workshop, find the probability that the selected person is

70. African American or male.

71. female or white.

Suppose that a survey of 350 college students is taken. Each student is asked the type of college attended (public or private) and the family's income level (low, middle, high). Use the data in the table to solve Exercises 72–75. Express probabilities as simplified fractions.

	Public	Private	Total
Low	120	20	140
Middle	110	50	160
High	22	28	50
Total	252	98	350

Find the probability that a randomly selected student in the survey

72. attends a public college.

73. is not from a high-income family.

74. is from a middle-income or a high-income family.

75. attends a private college or is from a high-income family.

76. One card is randomly selected from a deck of 52 cards. Find the odds in favor and the odds against getting a queen.

77. The winner of a raffle will receive a two-year scholarship to any college of the winner's choice. If 2000 raffle tickets were sold and you purchased 20 tickets, what are the odds against your winning the scholarship?

78. The odds in favor of a candidate winning an election are given at 3 to 1. What is the probability that this candidate will win the election?

11.7

Use the spinner shown to solve Exercises 79–83. It is equally likely that the pointer will land on any one of the six regions, numbered 1 through 6, and colored as shown. If the pointer lands on a borderline, spin again. If the pointer is spun twice, find the probability it will land on

79. yellow and then red. **80.** 1 and then 3.

81. yellow both times.

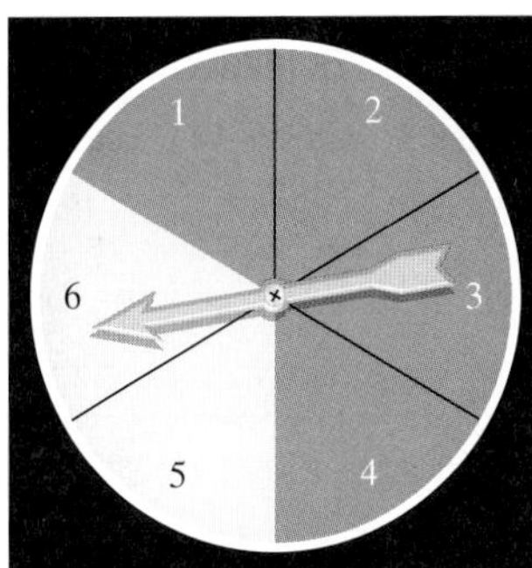

If the pointer is spun three times, find the probability it will land on

82. yellow and then 4 and then an odd number.

83. red every time.

84. What is the probability of a family having five boys born in a row?

85. The probability of a flood in any given year in a region prone to flooding is 0.2.

a. What is the probability of a flood two years in a row?

b. What is the probability of a flood for three consecutive years?

c. What is the probability of no flooding for four consecutive years?

d. What is the probability of a flood at least once in the next four years?

In Exercises 86–87, two students are selected from a group of four psychology majors, three business majors, and two music majors. The two students are to meet with the campus cafeteria manager to voice the group's concerns about food prices and quality. One student is randomly selected and leaves for the cafeteria manager's office. Then, a second student is selected. Find the probability of selecting

86. a music major and then a psychology major.

87. two business majors.

88. A final visit to the box of chocolates: It's now grown to a box of 50, of which 30 are solid chocolate, 15 are filled with jelly, and 5 are filled with cherries. The story is still the same: They all look alike. You select a piece, eat it, select a second piece, eat it, and help yourself to a final sugar rush. Find the probability of selecting a solid chocolate followed by two cherry-filled chocolates.

89. A single die is tossed. Find the probability that the tossed die shows 5, given that the outcome is an odd number.

90. A letter is randomly selected from the letters of the English alphabet. Find the probability of selecting a vowel, given that the outcome is a letter that precedes k.

91. The following numbers are each written on a colored chip. The chips are placed into a bag and one chip is selected at random. Find the probability of selecting

a. an odd number, given that a red chip is selected.

b. a yellow chip, given that the number selected is at least 3.

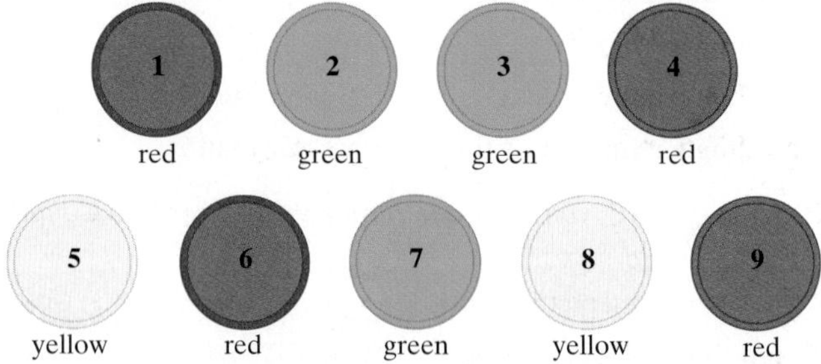

The data in the table are based on 145 Americans tested for tuberculosis. Use the data to solve Exercises 92–99. Express probabilities as simplified fractions.

	TB	No TB
Positive Screening Test	9	11
Negative Screening Test	1	124

Source: Deborah J. Bennett, *Randomness*, Harvard University Press, 1998.

Find the probability that a randomly selected person from this group

92. does not have TB.

93. tests positive.

94. does not have TB or tests positive.

95. does not have TB, given a positive test.

96. has a positive test, given no TB.

97. has TB, given a negative test.

Suppose that two people are randomly selected, in succession, from this group. Find the probability of selecting

98. two people with TB.

99. two people with positive screening tests.

11.4 and 11.6–11.7

The table shows the distribution, by age and gender, of the 33,632 deaths in the United States involving firearms in 2013.

DEATHS IN THE UNITED STATES INVOLVING FIREARMS, 2013

	Under Age 5	Ages 5–14	Ages 15–19	Ages 20–24	Ages 25–44	Ages 45–64	Ages 65–74	Age ≥ 75	Total
Male	57	246	1847	3651	9849	8161	2461	2514	28,786
Female	25	81	210	479	1669	1766	379	237	4846
Total	82	327	2057	4130	11,518	9927	2840	2751	33,632

Source: National Safety Council

In Exercises 100–106, use the data in the table to find the probability, expressed as a fraction and as a decimal rounded to three places, that a firearm death in the United States

100. involved a male.

101. involved a person in the 25–44 age range.

102. involved a person less than 75 years old.

103. involved a person in the 20–24 age range or in the 25–44 age range.

104. involved a female or a person younger than 5.

105. involved a person in the 20–24 age range, given that this person was a male.

106. involved a male, given that this person was at least 75.

11.8

107. The numbers that the pointer can land on and their respective probabilities are shown below.

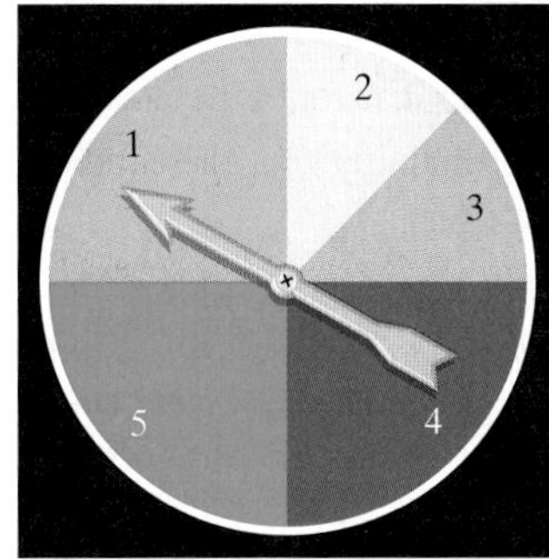

Outcome	Probability
1	$\frac{1}{4}$
2	$\frac{1}{8}$
3	$\frac{1}{8}$
4	$\frac{1}{4}$
5	$\frac{1}{4}$

Compute the expected value for the number on which the pointer lands.

108. The table shows claims and their probabilities for an insurance company.

LIFE INSURANCE FOR AN AIRLINE FLIGHT

Amount of Claim	Probability
$0	0.9999995
$1,000,000	0.0000005

a. Calculate the expected value and describe what this value means.

b. How much should the company charge to make a profit of $9.50 per policy?

109. A construction company is planning to bid on a building contract. The bid costs the company $3000. The probability that the bid is accepted is $\frac{1}{4}$. If the bid is accepted, the company will make $30,000 minus the cost of the bid. Find the expected value in this situation. Describe what this value means.

110. A game is played using a fair coin that is tossed twice. The sample space is {*HH*, *HT*, *TH*, *TT*}. If exactly one head occurs, the player wins $5, and if exactly two tails occur, the player also wins $5. For any other outcome, the player receives nothing. There is a $4 charge to play the game. What is the expected value? What does this value mean?

Chapter 11 Test

1. A person can purchase a particular model of a new car with a choice of ten colors, with or without automatic transmission, with or without four-wheel drive, with or without air conditioning, and with two, three, or four radio-CD speakers. How many different options are there for this model of the car?

2. Four acts are scheduled to perform in a variety show. How many different ways are there to schedule their appearances?

3. In how many ways can seven airplanes line up for a departure on a runway if the plane with the greatest number of passengers must depart first?

4. A human resource manager has 11 applicants to fill three different positions. Assuming that all applicants are equally qualified for any of the three positions, in how many ways can this be done?

5. From the ten books that you've recently bought but not read, you plan to take four with you on vacation. How many different sets of four books can you take?

6. In how many distinct ways can the letters of the word ATLANTA be arranged?

In Exercises 7–9, one student is selected at random from a group of 12 freshmen, 16 sophomores, 20 juniors, and 2 seniors. Find the probability that the person selected is

7. a freshman.

8. not a sophomore.

9. a junior or a senior.

10. If you are dealt one card from a 52-card deck, find the probability of being dealt a card greater than 4 and less than 10.

11. Seven movies (A, B, C, D, E, F, and G) are being scheduled for showing. The order of showing is determined by random selection. Find the probability that film C will be shown first, film A next-to-last, and film E last.

12. A lottery game is set up so that each player chooses six different numbers from 1 to 15. If the six numbers match the six numbers drawn in the lottery, the player wins (or shares) the top cash prize. What is the probability of winning the prize with 50 different lottery tickets?

In Exercises 13–14, it is equally probable that the pointer on the spinner shown will land on any one of the eight colored regions. If the pointer lands on a borderline, spin again.

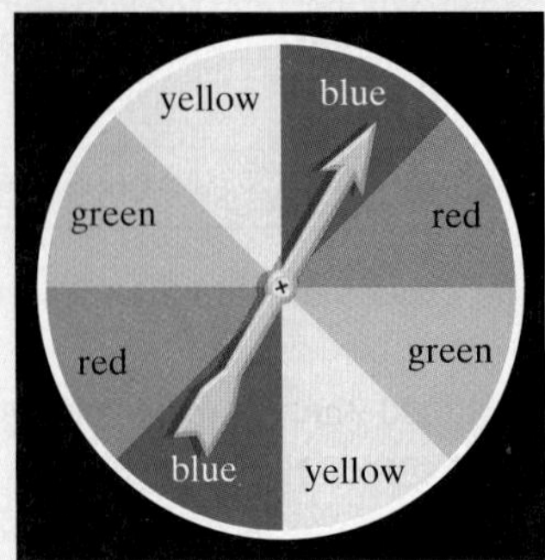

13. If the spinner is spun once, find the probability that the pointer will land on red or blue.

14. If the spinner is spun twice, find the probability that the pointer lands on red on the first spin and blue on the second spin.

15. A region is prone to flooding once every 20 years. The probability of flooding in any one year is $\frac{1}{20}$. What is the probability of flooding for three consecutive years?

16. One card is randomly selected from a deck of 52 cards. Find the probability of selecting a black card or a picture card.

17. A group of students consists of 10 male freshmen, 15 female freshmen, 20 male sophomores, and 5 female sophomores. If one person is randomly selected from the group, find the probability of selecting a freshman or a female.

18. A box contains five red balls, six green balls, and nine yellow balls. Suppose you select one ball at random from the box and do not replace it. Then you randomly select a second ball. Find the probability that both balls selected are red.

19. A quiz consisting of four multiple-choice questions has four available options (a, b, c, or d) for each question. If a person guesses at every question, what is the probability of answering *all* questions correctly?

20. A group is comprised of 20 men and 15 women. If one person is randomly selected from the group, find the odds against the person being a man.

21. The odds against a candidate winning an election are given at 1 to 4.

a. What are the odds in favor of the candidate winning?

b. What is the probability that the candidate will win the election?

A class is collecting data on eye color and gender. They organize the data they collected into the table shown. Numbers in the table represent the number of students from the class that belong to each of the categories. Use the data to solve Exercises 22–26. Express probabilities as simplified fractions.

	Brown	**Blue**	**Green**
Male	22	18	10
Female	18	20	12

Find the probability that a randomly selected student from this class

22. does not have brown eyes.

23. has brown eyes or blue eyes.

24. is female or has green eyes.

25. is male, given the student has blue eyes.

26. If two people are randomly selected, in succession, from the students in this class, find the probability that they both have green eyes.

27. An architect is considering bidding for the design of a new theater. The cost of drawing plans and submitting a model is \$15,000. The probability of being awarded the bid is 0.2, and anticipated profits are \$80,000, resulting in a possible gain of this amount minus the \$15,000 cost for plans and models. What is the expected value if the architect decides to bid for the design? Describe what this value means.

28. A game is played by selecting one bill at random from a bag that contains ten \$1 bills, five \$2 bills, three \$5 bills, one \$10 bill, and one \$100 bill. The player gets to keep the selected bill. There is a \$20 charge to play the game. What is the expected value? What does this value mean?

12 Statistics

SOME RANDOM STATISTICAL FACTOIDS:

- 28% of liberals have insomnia, compared with 16% of conservatives. (*Mother Jones*)
- 17% of American workers would reveal company secrets for money, and 8% have done it already. (Monster.com)
- 31% of American adults find giving up their smartphone for a day more difficult than giving up their significant other. (Microsoft)
- Between the ages of 18 and 24, 73% of Americans have used text messages to send suggestive pictures, compared with 55% ages 25–29, 52% ages 30–34, 42% ages 35–44, 26% ages 45–54, 10% ages 55–64, and 7% ages 65 and older. (*Time*)
- 49% of Americans cite a "lot" of stress at age 22, compared with 45% at 42, 35% at 58, 29% at 62, and 20% by 70. (*Proceedings of the National Academy of Sciences*)
- 34% of American adults believe in ghosts. (AP/Ipsos)
- Unwillingness to eat sushi correlates nearly perfectly with disapproval of marriage equality. (Pew Research Center)

Statisticians collect numerical data from subgroups of populations to find out everything imaginable about the population as a whole, including whom they favor in an election, what they watch on TV, how much money they make, or what worries them. Comedians and statisticians joke that 62.38% of all statistics are made up on the spot. Because statisticians both record and influence our behavior, it is important to distinguish between good and bad methods for collecting, presenting, and interpreting data.

Throughout this chapter, you will gain an understanding of where data come from and how these numbers are used to make decisions. We'll return to the bizarre sushi/marriage equality correlation in Exercises 5 and 35 of Exercise Set 12.6.

12.1 Sampling, Frequency Distributions, and Graphs

WHAT AM I SUPPOSED TO LEARN?

After studying this section, you should be able to:

1 Describe the population whose properties are to be analyzed.
2 Select an appropriate sampling technique.
3 Organize and present data.
4 Identify deceptions in visual displays of data.

*M*A*S*H* took place in the early 1950s, during the Korean War. By the final episode, the show had lasted four times as long as the Korean War.

AT THE END OF THE TWENTIETH century, there were 94 million households in the United States with television sets. The television program viewed by the greatest percentage of such households in that century was the final episode of *M*A*S*H*. Over 50 million American households watched this program.

Numerical information, such as the information about the top three TV shows of the twentieth century, shown in **Table 12.1**, is called **data**. The word **statistics** is often used when referring to data. However, statistics has a second meaning: Statistics is also a method for collecting, organizing, analyzing, and interpreting data, as well as drawing conclusions based on the data. This methodology divides statistics into two main areas. **Descriptive statistics** is concerned with collecting, organizing, summarizing, and presenting data. **Inferential statistics** has to do with making generalizations about and drawing conclusions from the data collected.

TABLE 12.1 TV Programs with the Greatest U.S. Audience Viewing Percentage of the Twentieth Century

Program	Total Households	Viewing Percentage
1. *M*A*S*H* Feb. 28, 1983	50,150,000	60.2%
2. *Dallas* Nov. 21, 1980	41,470,000	53.3%
3. *Roots Part 8* Jan. 30, 1977	36,380,000	51.1%

Source: Nielsen Media Research

1 *Describe the population whose properties are to be analyzed.*

Populations and Samples

Consider the set of all American TV households. Such a set is called the *population*. In general, a **population** is the set containing all the people or objects whose properties are to be described and analyzed by the data collector.

The population of American TV households is huge. At the time of the *M*A*S*H* conclusion, there were nearly 84 million such households. Did over 50 million American TV households really watch the final episode of *M*A*S*H*? A friendly phone call to each household ("So, how are you? What's new? Watch any good television last night? If so, what?") is, of course, absurd. A **sample**, which is a subset or subgroup of the population, is needed. In this case, it would be appropriate to have a sample of a few thousand TV households to draw conclusions about the population of all TV households.

Blitzer Bonus

A Sampling Fiasco

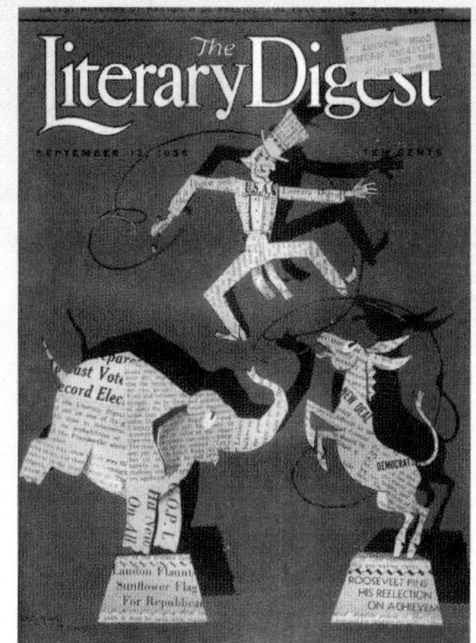

In 1936, the *Literary Digest* mailed out over ten million ballots to voters throughout the country. The results poured in, and the magazine predicted a landslide victory for Republican Alf Landon over Democrat Franklin Roosevelt. However, the prediction of the *Literary Digest* was wrong. Why? The mailing lists the editors used included people from their own subscriber list, directories of automobile owners, and telephone books. As a result, its sample was anything but random. It excluded most of the poor, who were unlikely to subscribe to the *Literary Digest*, or to own a car or telephone in the heart of the Depression. Prosperous people in 1936 were more likely to be Republican than the poor. Thus, although the sample was massive, it included a higher percentage of affluent individuals than the population as a whole did. A victim of both the Depression and the 1936 sampling fiasco, the *Literary Digest* folded in 1937.

EXAMPLE 1 *Populations and Samples*

A group of hotel owners in a large city decide to conduct a survey among citizens of the city to discover their opinions about casino gambling.

a. Describe the population.

b. One of the hotel owners suggests obtaining a sample by surveying all the people at six of the largest nightclubs in the city on a Saturday night. Each person will be asked to express his or her opinion on casino gambling. Does this seem like a good idea?

SOLUTION

a. The population is the set containing all the citizens of the city.

b. Questioning people at six of the city's largest nightclubs is a terrible idea. The nightclub subset is probably more likely to have a positive attitude toward casino gambling than the population of all the city's citizens.

CHECK POINT 1 A city government wants to conduct a survey among the city's homeless to discover their opinions about required residence in city shelters from midnight until 6 A.M.

a. Describe the population.

b. A city commissioner suggests obtaining a sample by surveying all the homeless people at the city's largest shelter on a Sunday night. Does this seem like a good idea? Explain your answer.

Random Sampling

There is a way to use a small sample to make generalizations about a large population: Guarantee that every member of the population has an equal chance to be selected for the sample. Surveying people at six of the city's largest nightclubs does not provide this guarantee. Unless it can be established that all citizens of the city frequent these clubs, which seems unlikely, this sampling scheme does not permit each citizen an equal chance of selection.

> **RANDOM SAMPLES**
>
> A **random sample** is a sample obtained in such a way that every element in the population has an equal chance of being selected for the sample.

Suppose that you are elated with the quality of one of your courses. Although it's an auditorium section with 120 students, you feel that the professor is lecturing right to you. During a wonderful lecture, you look around the auditorium to see if any of the other students are sharing your enthusiasm. Based on body language, it's hard to tell. You really want to know the opinion of the population of 120 students taking this course. You think about asking students to grade the course on an A to F scale, anticipating a unanimous A. You cannot survey everyone. Eureka! Suddenly you have an idea on how to take a sample. Place cards numbered from 1 through 120, one number per card, in a box. Because the course has assigned seating by number, each numbered card corresponds to a student in the class. Reach in and randomly select six cards. Each card, and therefore each student, has an equal chance of being selected. Then use the opinions about the course from the six randomly selected students to generalize about the course opinion for the entire 120-student population.

2 *Select an appropriate sampling technique.*

Your idea is precisely how random samples are obtained. In random sampling, each element in the population must be identified and assigned a number. The numbers are generally assigned in order. The way to sample from the larger numbered population is to generate random numbers using a computer or calculator. Each numbered element from the population that corresponds to one of the generated random numbers is selected for the sample.

Call-in polls on radio and television are not reliable because those polled do not represent the larger population. A person who calls in is likely to have feelings about an issue that are consistent with the politics of the show's host. For a poll to be accurate, the sample must be chosen randomly from the larger population. The A. C. Nielsen Company uses a random sample of approximately 5000 TV households to measure the percentage of households tuned in to a television program.

EXAMPLE 2 *Selecting an Appropriate Sampling Technique*

We return to the hotel owners in the large city who are interested in how the city's citizens feel about casino gambling. Which of the following would be the most appropriate way to select a random sample?

a. Randomly survey people who live in the oceanfront condominiums in the city.

b. Survey the first 200 people whose names appear in the city's telephone directory.

c. Randomly select neighborhoods of the city and then randomly survey people within the selected neighborhoods.

SOLUTION

Keep in mind that the population is the set containing all the city's citizens. A random sample must give each citizen an equal chance of being selected.

a. Randomly selecting people who live in the city's oceanfront condominiums is not a good idea. Many hotels lie along the oceanfront, and the oceanfront property owners might object to the traffic and noise as a result of casino gambling. Furthermore, this sample does not give each citizen of the city an equal chance of being selected.

b. If the hotel owners survey the first 200 names in the city's telephone directory, all citizens do not have an equal chance of selection. For example, individuals whose last name begins with a letter toward the end of the alphabet have no chance of being selected.

c. Randomly selecting neighborhoods of the city and then randomly surveying people within the selected neighborhoods is an appropriate technique. Using this method, each citizen has an equal chance of being selected.

In summary, given the three options, the sampling technique in part (c) is the most appropriate.

Surveys and polls involve data from a sample of some population. Regardless of the sampling technique used, the sample should exhibit characteristics typical of those possessed by the target population. This type of sample is called a **representative sample**.

CHECK POINT 2 Explain why the sampling technique described in Check Point 1(b) on page 773 is not a random sample. Then describe an appropriate way to select a random sample of the city's homeless.

Blitzer Bonus

The United States Census

A census is a survey that attempts to include the entire population. The U.S. Constitution requires a census of the American population every ten years. When the Founding Fathers invented American democracy, they realized that if you are going to have government by the people, you need to know who and where they are. Nowadays about $400 billion per year in federal aid is distributed based on the Census numbers, for everything from jobs to bridges to schools. For every 100 people not counted, states and communities could lose as much as $130,000 annually, or $1300 per person each year, so this really matters.

Although the Census generates volumes of statistics, its main purpose is to give the government block-by-block population figures. The U.S. Census is not foolproof. The 1990 Census missed 1.6% of the American population, including an estimated 4.4% of the African-American population, largely in inner cities. Only 67% of households responded to the 2000 Census, even after door-to-door canvassing. About 6.4 million people were missed and 3.1 million were counted twice. Although the 2010 Census was one of the shortest forms in history, counting each person was not an easy task, particularly with concerns about immigration status and privacy of data.

Of course, there would be more than $400 billion to spread around if it didn't cost so much to count us in the first place: about $15 billion for the 2010 Census. That included $338 million for ads in 28 languages, a Census-sponsored NASCAR entry, and $2.5 million for a Super Bowl ad. The ads were meant to boost the response rate, since any household that did not mail back its form got visited by a Census worker, another pricey item. In all, the cost of the 2010 Census worked out to appoximately $49 per person.

3 *Organize and present data.*

Frequency Distributions

After data have been collected from a sample of the population, the next task facing the statistician is to present the data in a condensed and manageable form. In this way, the data can be more easily interpreted.

Suppose, for example, that researchers are interested in determining the age at which adolescent males show the greatest rate of physical growth. A random sample of 35 ten-year-old boys is measured for height and then remeasured each year until they reach 18. The age of maximum yearly growth for each subject is as follows:

12, 14, 13, 14, 16, 14, 14, 17, 13, 10, 13, 18, 12, 15, 14, 15, 15, 14, 14, 13, 15, 16, 15, 12, 13, 16, 11, 15, 12, 13, 12, 11, 13, 14, 14.

A piece of data is called a **data item**. This list of data has 35 data items. Some of the data items are identical. Two of the data items are 11 and 11. Thus, we can say that the **data value** 11 occurs twice. Similarly, because five of the data items are 12, 12, 12, 12, and 12, the data value 12 occurs five times.

Collected data can be presented using a **frequency distribution**. Such a distribution consists of two columns. The data values are listed in one column. Numerical data are generally listed from smallest to largest. The adjacent column is labeled **frequency** and indicates the number of times each value occurs.

EXAMPLE 3 Constructing a Frequency Distribution

Construct a frequency distribution for the data of the age of maximum yearly growth for 35 boys:

12, 14, 13, 14, 16, 14, 14, 17, 13, 10, 13, 18, 12, 15, 14, 15, 15, 14, 14, 13, 15, 16, 15, 12, 13, 16, 11, 15, 12, 13, 12, 11, 13, 14, 14.

SOLUTION

It is difficult to determine trends in the data above in their current format. Perhaps we can make sense of the data by organizing them into a frequency distribution. Let us create two columns. One lists all possible data values, from smallest (10) to largest (18). The other column indicates the number of times the value occurs in the sample. The frequency distribution is shown in **Table 12.2**.

The frequency distribution indicates that one subject had maximum growth at age 10, two at age 11, five at age 12, seven at age 13, and so on. The maximum growth for most of the subjects occurred between the ages of 12 and 15. Nine boys experienced maximum growth at age 14, more than at any other age within the sample. The sum of the frequencies, 35, is equal to the original number of data items.

The trend shown by the frequency distribution in **Table 12.2** indicates that the number of boys who attain their maximum yearly growth at a given age increases until age 14 and decreases after that. This trend is not evident in the data in their original format.

TABLE 12.2 A Frequency Distribution for a Boy's Age of Maximum Yearly Growth

Age of Maximum Growth	Number of Boys (Frequency)
10	1
11	2
12	5
13	7
14	9
15	6
16	3
17	1
18	1
Total:	$n = 35$

35 is the sum of the frequencies.

CHECK POINT 3 Construct a frequency distribution for the data showing final course grades for students in a precalculus course, listed alphabetically by student name in a grade book:

F, A, B, B, C, C, B, C, A, A, C, C, D, C, B, D, C, C, B, C.

A frequency distribution that lists all possible data items can be quite cumbersome when there are many such items. For example, consider the following data items. These are statistics test scores for a class of 40 students.

82	47	75	64	57	82	63	93
76	68	84	54	88	77	79	80
94	92	94	80	94	66	81	67
75	73	66	87	76	45	43	56
57	74	50	78	71	84	59	76

It's difficult to determine how well the group did when the grades are displayed like this. Because there are so many data items, one way to organize these data so that the results are more meaningful is to arrange the grades into groups, or **classes**, based on something that interests us. Many grading systems assign an A to grades in the 90–100 class, B to grades in the 80–89 class, C to grades in the 70–79 class, and so on. These classes provide one way to organize the data.

Looking at the 40 statistics test scores, we see that they range from a low of 43 to a high of 94. We can use classes that run from 40 through 49, 50 through 59, 60 through 69, and so on up to 90 through 99, to organize the scores. In Example 4, we go through the data and tally each item into the appropriate class. This method for organizing data is called a **grouped frequency distribution**.

EXAMPLE 4 Constructing a Grouped Frequency Distribution

Use the classes 40–49, 50–59, 60–69, 70–79, 80–89, and 90–99 to construct a grouped frequency distribution for the 40 test scores on the previous page.

SOLUTION

We use the 40 given scores and tally the number of scores in each class.

Tallying Statistics Test Scores

Test Scores (Class)	Tally	Number of Students (Frequency)
40–49	|||	3
50–59	~~||||~~		6
60–69	~~||||~~		6
70–79	~~||||~~ ~~||||~~		11
80–89	~~||||~~ ||||	9
90–99	~~||||~~	5

The second score in the list, 47, is shown as the first tally in this row.

The first score in the list, 82, is shown as the first tally in this row.

TABLE 12.3 A Grouped Frequency Distribution for Statistics Test Scores

Class	Frequency
40–49	3
50–59	6
60–69	6
70–79	11
80–89	9
90–99	5
Total:	$n = 40$

40, the sum of the frequencies, is the number of data items.

Omitting the tally column results in the grouped frequency distribution in **Table 12.3**. The distribution shows that the greatest frequency of students scored in the 70–79 class. The number of students decreases in classes that contain successively lower and higher scores. The sum of the frequencies, 40, is equal to the original number of data items.

The leftmost number in each class of a grouped frequency distribution is called the **lower class limit**. For example, in **Table 12.3**, the lower limit of the first class is 40 and the lower limit of the third class is 60. The rightmost number in each class is called the **upper class limit**. In **Table 12.3**, 49 and 69 are the upper limits for the first and third classes, respectively. Notice that if we take the difference between any two consecutive lower class limits, we get the same number:

$$50 - 40 = 10,\ 60 - 50 = 10,\ 70 - 60 = 10,\ 80 - 70 = 10,\ 90 - 80 = 10.$$

The number 10 is called the **class width**.

When setting up class limits, each class, with the possible exception of the first or last, should have the same width. Because each data item must fall into exactly one class, it is sometimes helpful to vary the width of the first or last class to allow for items that fall far above or below most of the data.

CHECK POINT 4 Use the classes in **Table 12.3** to construct a grouped frequency distribution for the following 37 exam scores:

73	58	68	75	94	79	96	79
87	83	89	52	99	97	89	58
95	77	75	81	75	73	73	62
69	76	77	71	50	57	41	98
77	71	69	90	75.			

TABLE 12.2 A Frequency Distribution for a Boy's Age of Maximum Yearly Growth (repeated)

Age of Maximum Growth	Number of Boys (Frequency)
10	1
11	2
12	5
13	7
14	9
15	6
16	3
17	1
18	1
Total:	$n = 35$

35 is the sum of the frequencies.

Histograms and Frequency Polygons

Take a second look at the frequency distribution for the age of a boy's maximum yearly growth in **Table 12.2**. A bar graph with bars that touch can be used to visually display the data. Such a graph is called a **histogram**. **Figure 12.1** illustrates a histogram that was constructed using the frequency distribution in **Table 12.2**. A series of rectangles whose heights represent the frequencies are placed next to each other. For example, the height of the bar for the data value 10, shown in **Figure 12.1**, is 1. This corresponds to the frequency for 10 given in **Table 12.2**. The higher the bar, the more frequent the age. The break along the horizontal axis, symbolized by ⺊, eliminates listing the ages 1 through 9.

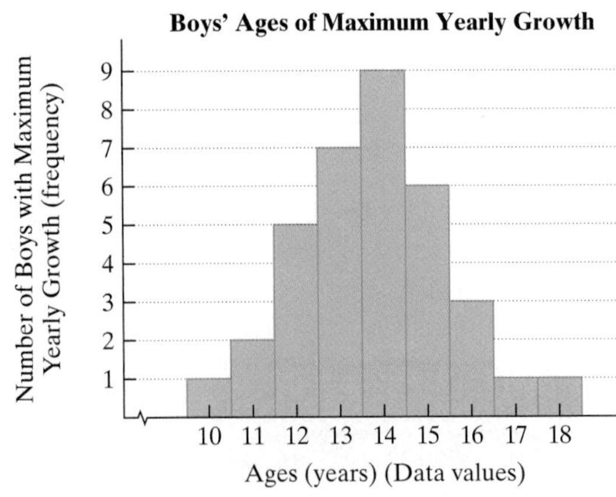

FIGURE 12.1 A histogram for a boy's age of maximum yearly growth.

A line graph called a **frequency polygon** can also be used to visually convey the information shown in **Figure 12.1**. The axes are labeled just like those in a histogram. Thus, the horizontal axis shows data values and the vertical axis shows frequencies. Once a histogram has been constructed, it's fairly easy to draw a frequency polygon. **Figure 12.2** shows a histogram with a dot at the top of each rectangle at its midpoint. Connect each of these midpoints with a straight line. To complete the frequency polygon at both ends, the lines should be drawn down to touch the horizontal axis. The completed frequency polygon is shown in **Figure 12.3**.

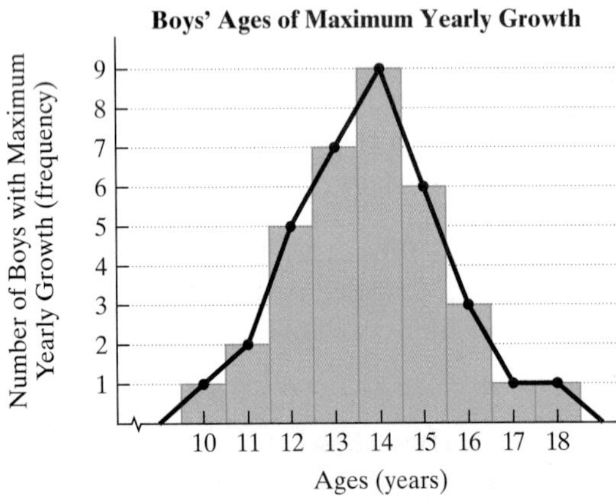

FIGURE 12.2 A histogram with a superimposed frequency polygon.

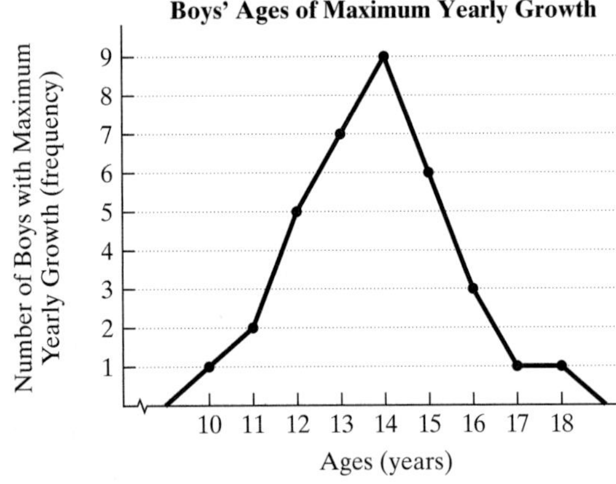

FIGURE 12.3 A frequency polygon

Stem-and-Leaf Plots

A unique way of displaying data uses a tool called a **stem-and-leaf plot**. Example 5 illustrates how we sort the data, revealing the same visual impression created by a histogram.

EXAMPLE 5 *Constructing a Stem-and-Leaf Plot*

Use the data showing statistics test scores for 40 students to construct a stem-and-leaf plot:

82	47	75	64	57	82	63	93
76	68	84	54	88	77	79	80
94	92	94	80	94	66	81	67
75	73	66	87	76	45	43	56
57	74	50	78	71	84	59	76.

SOLUTION

The plot is constructed by separating each data item into two parts. The first part is the *stem*. The **stem** consists of the tens digit. For example, the stem for the score of 82 is 8. The second part is the *leaf*. The **leaf** consists of the units digit for a given value. For the score of 82, the leaf is 2. The possible stems for the 40 scores are 4, 5, 6, 7, 8, and 9, entered in the left column of the plot.

Begin by entering each data item in the first row:

82 47 75 64 57 82 63 93.

Entering 82: Stems	Leaves	**Adding 47:** Stems	Leaves	**Adding 75:** Stems	Leaves	**Adding 64:** Stems	Leaves
4		4	7	4	7	4	7
5		5		5		5	
6		6		6		6	4
7		7		7	5	7	5
8	2	8	2	8	2	8	2
9		9		9		9	

Adding 57: Stems	Leaves	**Adding 82:** Stems	Leaves	**Adding 63:** Stems	Leaves	**Adding 93:** Stems	Leaves
4	7	4	7	4	7	4	7
5	7	5	7	5	7	5	7
6	4	6	4	6	43	6	43
7	5	7	5	7	5	7	5
8	2	8	22	8	22	8	22
9		9		9		9	3

We continue in this manner and enter all the data items. **Figure 12.4** shows the completed stem-and-leaf plot. If you turn the page so that the left margin is on the bottom and facing you, the visual impression created by the enclosed leaves is the same as that created by a histogram. An advantage over the histogram is that the stem-and-leaf plot preserves exact data items. The enclosed leaves extend farthest to the right when the stem is 7. This shows that the greatest frequency of students scored in the 70s.

A Stem-and-Leaf Plot for 40 Test Scores

Tens digit   Units digit

Stems	Leaves
4	7 5 3
5	7 4 6 7 0 9
6	4 3 8 6 7 6
7	5 6 7 9 5 3 6 4 8 1 6
8	2 2 4 8 0 0 1 7 4
9	3 4 2 4 4

FIGURE 12.4 A stem-and-leaf plot displaying 40 test scores

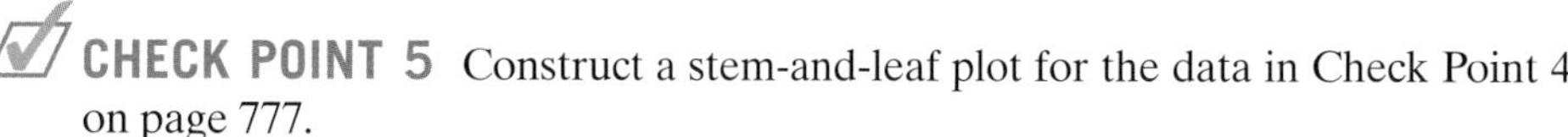

CHECK POINT 5 Construct a stem-and-leaf plot for the data in Check Point 4 on page 777.

4 Identify deceptions in visual displays of data.

A Statistical Deception

CLAIM

In 2016, 94 million Americans were out of the labor force.

REALITY

More than 88 million Americans who did not have a job in 2016 didn't want one. This included large numbers of retirees, stay-at-home parents, students, and people who were disabled.

Deceptions in Visual Displays of Data

Benjamin Disraeli, Queen Victoria's prime minister, stated that there are "lies, damned lies, and statistics." The problem is not that statistics lie, but rather that liars use statistics. Graphs can be used to distort the underlying data, making it difficult for the viewer to learn the truth. One potential source of misunderstanding is the scale on the vertical axis used to draw the graph. This scale is important because it lets a researcher "inflate" or "deflate" a trend. For example, both graphs in **Figure 12.5** present identical data for the percentage of people in the United States living below the poverty level from 2001 through 2005. The graph on the left stretches the scale on the vertical axis to create an overall impression of a poverty rate increasing rapidly over time. The graph on the right compresses the scale on the vertical axis to create an impression of a poverty rate that is slowly increasing, and beginning to level off, over time.

Percentage of People in the United States Living below the Poverty Level, 2001–2005

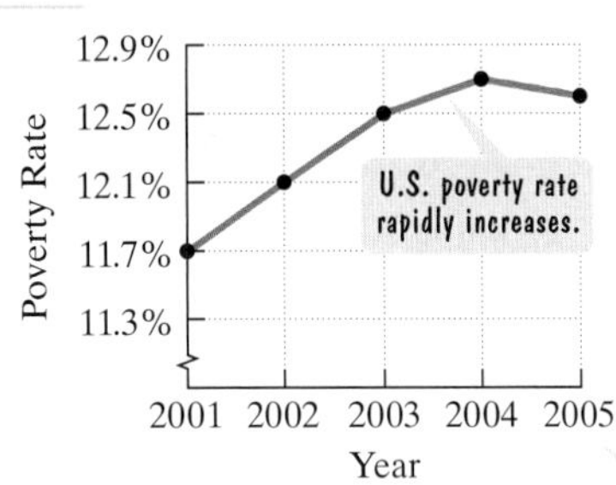

Year	Poverty Rate
2001	11.7%
2002	12.1%
2003	12.5%
2004	12.7%
2005	12.6%

The graph in both figures present these data.

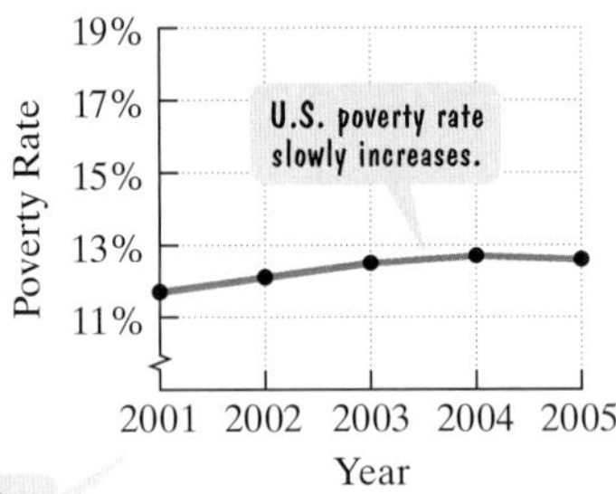

FIGURE 12.5
Source: U.S. Census Bureau

There is another problem with the data in **Figure 12.5**. Look at **Table 12.4** that shows the poverty rate from 2001 through 2015. Depending on the time frame chosen, the data can be interpreted in various ways. Carefully choosing a time frame can help represent data trends in the most positive or negative light.

TABLE 12.4 U.S. Poverty Rate from 2001 to 2015

Year	Poverty Rate
2001	11.7%
2002	12.1%
2003	12.5%
2004	12.7%
2005	12.6%
2006	12.3%
2007	12.5%
2008	13.2%
2009	14.3%
2010	15.1%
2011	15.0%
2012	15.0%
2013	14.5%
2014	14.8%
2015	13.5%

THINGS TO WATCH FOR IN VISUAL DISPLAYS OF DATA

1. Is there a title that explains what is being displayed?
2. Are numbers lined up with tick marks on the vertical axis that clearly indicate the scale? Has the scale been varied to create a more or less dramatic impression than shown by the actual data?
3. Do too many design and cosmetic effects draw attention from or distort the data?
4. Has the wrong impression been created about how the data are changing because equally spaced time intervals are not used on the horizontal axis? Furthermore, has a time interval been chosen that allows the data to be interpreted in various ways?
5. Are bar sizes scaled proportionately in terms of the data they represent?
6. Is there a source that indicates where the data in the display came from? Do the data come from an entire population or a sample? Was a random sample used and, if so, are there possible differences between what is displayed in the graph and what is occurring in the entire population? (We'll discuss these *margins of error* in Section 12.4.) Who is presenting the visual display, and does that person have a special case to make for or against the trend shown by the graph?

Table 12.5 contains two examples of misleading visual displays.

TABLE 12.5 Examples of Misleading Visual Displays

Graphic Display	Presentation Problems
Purchasing Power of the Diminishing Dollar $1.00 63¢ 54¢ 48¢ 43¢ $1.00 $0.50 0 1980 1990 1995 2000 2005 *Source:* U.S. Bureau of Labor Statistics	Although the length of each dollar bill is proportional to its spending power, the visual display varies both the length *and width* of the bills to show the diminishing power of the dollar over time. Because our eyes focus on the *areas* of the dollar-shaped bars, this creates the impression that the purchasing power of the dollar diminished even more than it really did. If the area of the dollar were drawn to reflect its purchasing power, the 2005 dollar would be approximately twice as large as the one shown in the graphic display.
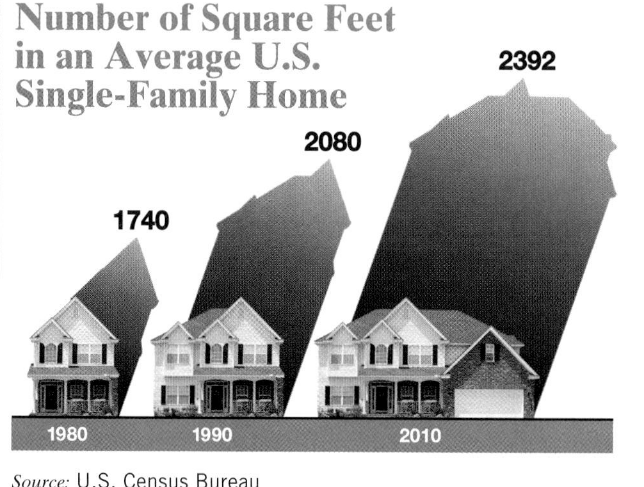 *Source:* U.S. Census Bureau	Cosmetic effects of homes with equal heights, but different frontal additions and shadow lengths, make it impossible to tell if they proportionately depict the given areas. Time intervals on the horizontal axis are not uniform in size, making it appear that dwelling swelling has been linear from 1980 through 2010. The data indicate that this is not the case. There was a greater increase in area from 1980 through 1990, averaging 34 square feet per year, than from 1990 through 2010, averaging 15.6 square feet per year.

Concept and Vocabulary Check

Fill in each blank so that the resulting statement is true.

1. A sample obtained in such a way that every member of the population has an equal chance of being selected is called a/an ________ sample.
2. If data values are listed in one column and the adjacent column indicates the number of times each value occurs, the data presentation is called a/an ________________.
3. If the data presentation in Exercise 2 is varied by organizing the data into classes, the data presentation is called a/an ________________________. If one class in such a distribution is 80–89, the lower class limit is _____ and the upper class limit is _____.
4. Data can be displayed using a bar graph with bars that touch each other. This visual presentation of the data is called a/an __________. The heights of the bars represent the ___________ of the data values.
5. If the midpoints of the tops of the bars for the data presentation in Exercise 4 are connected with straight lines, the resulting line graph is a data presentation called a/an ________________. To complete such a graph at both ends, the lines are drawn down to touch the ______________.
6. A data presentation that separates each data item into two parts is called a/an ________________.

In Exercises 7–10, determine whether each statement is true or false. If the statement is false, make the necessary change(s) to produce a true statement.

7. A sample is the set of all the people or objects whose properties are to be described and analyzed by the data collector. ______

8. A call-in poll on radio or television is not reliable because the sample is not chosen randomly from a larger population. ______

9. One disadvantage of a stem-and-leaf plot is that it does not display the data items. ______

10. A deception in the visual display of data can result by stretching or compressing the scale on a graph's vertical axis. ______

Exercise Set 12.1

Practice and Application Exercises

1. The government of a large city needs to determine whether the city's residents will support the construction of a new jail. The government decides to conduct a survey of a sample of the city's residents. Which one of the following procedures would be most appropriate for obtaining a sample of the city's residents?

a. Survey a random sample of the employees and inmates at the old jail.

b. Survey every fifth person who walks into City Hall on a given day.

c. Survey a random sample of persons within each geographic region of the city.

d. Survey the first 200 people listed in the city's telephone directory.

2. The city council of a large city needs to know whether its residents will support the building of three new schools. The council decides to conduct a survey of a sample of the city's residents. Which procedure would be most appropriate for obtaining a sample of the city's residents?

a. Survey a random sample of teachers who live in the city.

b. Survey 100 individuals who are randomly selected from a list of all people living in the state in which the city in question is located.

c. Survey a random sample of persons within each neighborhood of the city.

d. Survey every tenth person who enters City Hall on a randomly selected day.

A questionnaire was given to students in an introductory statistics class during the first week of the course. One question asked, "How stressed have you been in the last $2\frac{1}{2}$ weeks, on a scale of 0 to 10, with 0 being not at all stressed and 10 being as stressed as possible?" The students' responses are shown in the frequency distribution. Use this frequency distribution to solve Exercises 3–6.

Stress Rating	Frequency	Stress Rating	Frequency
0	2	6	13
1	1	7	31
2	3	8	26
3	12	9	15
4	16	10	14
5	18		

Source: Journal of Personality and Social Psychology, 69, 1102–1112

3. Which stress rating describes the greatest number of students? How many students responded with this rating?

4. Which stress rating describes the least number of students? How many responded with this rating?

5. How many students were involved in this study?

6. How many students had a stress rating of 8 or more?

7. A random sample of 30 college students is selected. Each student is asked how much time he or she spent on homework during the previous week. The following times (in hours) are obtained:

16, 24, 18, 21, 18, 16, 18, 17, 15, 21, 19, 17, 17, 16, 19, 18, 15, 15, 20, 17, 15, 17, 24, 19, 16, 20, 16, 19, 18, 17.

Construct a frequency distribution for the data.

8. A random sample of 30 male college students is selected. Each student is asked his height (to the nearest inch). The heights are as follows:

72, 70, 68, 72, 71, 71, 71, 69, 73, 71, 73, 75, 66, 67, 75, 74, 73, 71, 72, 67, 72, 68, 67, 71, 73, 71, 72, 70, 73, 70.

Construct a frequency distribution for the data.

A college professor had students keep a diary of their social interactions for a week. Excluding family and work situations, the number of social interactions of ten minutes or longer over the week is shown in the following grouped frequency distribution. Use this information to solve Exercises 9–16.

Number of Social Interactions	Frequency
0–4	12
5–9	16
10–14	16
15–19	16
20–24	10
25–29	11
30–34	4
35–39	3
40–44	3
45–49	3

Source: Society for Personality and Social Psychology

9. Identify the lower class limit for each class.

10. Identify the upper class limit for each class.

11. What is the class width?

12. How many students were involved in this study?

13. How many students had at least 30 social interactions for the week?

14. How many students had at most 14 social interactions for the week?

15. Among the classes with the greatest frequency, which class has the least number of social interactions?

16. Among the classes with the smallest frequency, which class has the least number of social interactions?

17. As of 2017, the following are the ages, in chronological order, at which U.S. presidents were inaugurated:

57, 61, 57, 57, 58, 57, 61, 54, 68, 51, 49, 64, 50, 48, 65, 52, 56, 46, 54, 49, 50, 47, 55, 55, 54, 42, 51, 56, 55, 51, 54, 51, 60, 62, 43, 55, 56, 61, 52, 69, 64, 46, 54, 47, 70.
Source: Time Almanac

Construct a grouped frequency distribution for the data. Use 41–45 for the first class and use the same width for each subsequent class.

18. The IQ scores of 70 students enrolled in a liberal arts course at a college are as follows:

102, 100, 103, 86, 120, 117, 111, 101, 93, 97, 99, 95, 95, 104, 104, 105, 106, 109, 109, 89, 94, 95, 99, 99, 103, 104, 105, 109, 110, 114, 124, 123, 118, 117, 116, 110, 114, 114, 96, 99, 103, 103, 104, 107, 107, 110, 111, 112, 113, 117, 115, 116, 100, 104, 102, 94, 93, 93, 96, 96, 111, 116, 107, 109, 105, 106, 97, 106, 107, 108.

Construct a grouped frequency distribution for the data. Use 85–89 for the first class and use the same width for each subsequent class.

19. Construct a histogram and a frequency polygon for the data involving stress ratings in Exercises 3–6.

20. Construct a histogram and a frequency polygon for the data in Exercise 7.

21. Construct a histogram and a frequency polygon for the data in Exercise 8.

The histogram shows the distribution of starting salaries (rounded to the nearest thousand dollars) for college graduates based on a random sample of recent graduates.

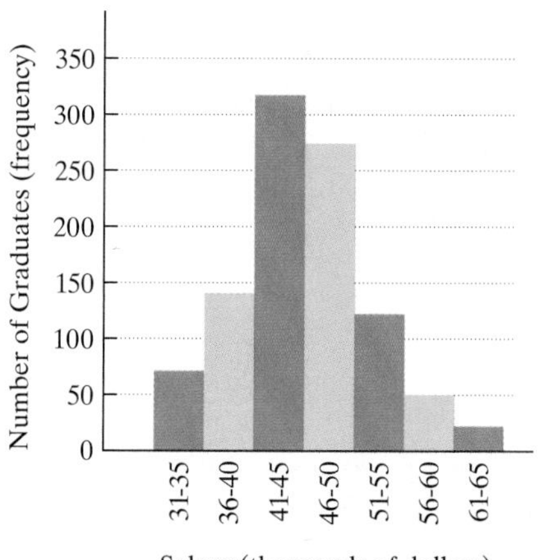

In Exercises 22–25, determine whether each statement is true or false according to the graph at the bottom of the previous column.

22. The graph is based on a sample of approximately 500 recent college graduates.

23. More college graduates had starting salaries in the \$51,000–\$55,000 range than in the \$36,000–\$40,000 range.

24. If the sample is truly representative, then for a group of 400 college graduates, we can expect about 28 of them to have starting salaries in the \$31,000–\$35,000 range.

25. The percentage of starting salaries falling above those shown by any rectangular bar is equal to the percentage of starting salaries falling below that bar.

The frequency polygon shows a distribution of IQ scores.

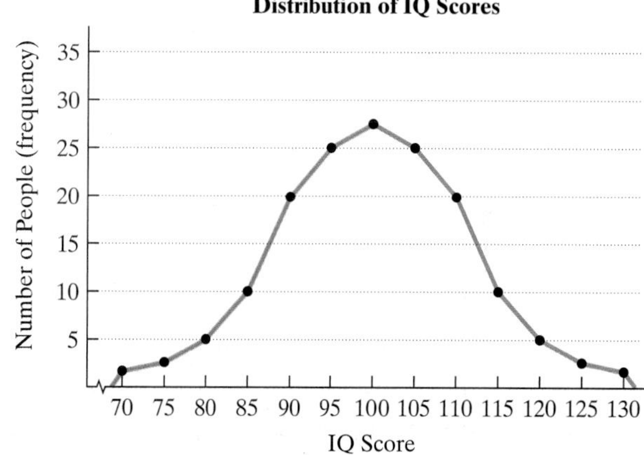

In Exercises 26–29, determine whether each statement is true or false according to the graph.

26. The graph is based on a sample of approximately 50 people.

27. More people had an IQ score of 100 than any other IQ score, and as the deviation from 100 increases or decreases, the scores fall off in a symmetrical manner.

28. More people had an IQ score of 110 than a score of 90.

29. The percentage of scores above any IQ score is equal to the percentage of scores below that score.

30. Construct a stem-and-leaf plot for the data in Exercise 17 showing the ages at which U.S. presidents were inaugurated.

31. A random sample of 40 college professors is selected from all professors at a university. The following list gives their ages:

63, 48, 42, 42, 38, 59, 41, 44, 45, 28, 54, 62, 51, 44, 63, 66, 59, 46, 51, 28, 37, 66, 42, 40, 30, 31, 48, 32, 29, 42, 63, 37, 36, 47, 25, 34, 49, 30, 35, 50.

Construct a stem-and-leaf plot for the data. What does the shape of the display reveal about the ages of the professors?

32. In "Ages of Oscar-Winning Best Actors and Actresses" (*Mathematics Teacher* magazine) by Richard Brown and Gretchen Davis, the stem-and-leaf plots shown on the right compare the ages of 30 actors and 30 actresses at the time they won the award.

a. What is the age of the youngest actor to win an Oscar?

b. What is the age difference between the oldest and the youngest actress to win an Oscar?

c. What is the oldest age shared by two actors to win an Oscar?

d. What differences do you observe between the two stem-and-leaf plots? What explanations can you offer for these differences?

Actors	Stems	Actresses
	2	146667
98753221	3	00113344455778
88776543322100	4	11129
6651	5	
210	6	011
6	7	4
	8	0

In Exercises 33–37, describe what is misleading in each visual display of data.

33. **World Population, in Billions**

Source: U.S. Census Bureau

34. **Book Title Output in the United States**

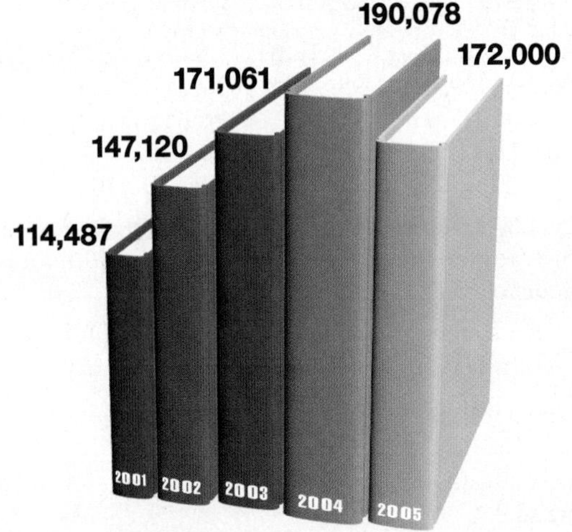

Source: R. R. Bowker

35. **Percentage of the World's Computers in Use, by Country**

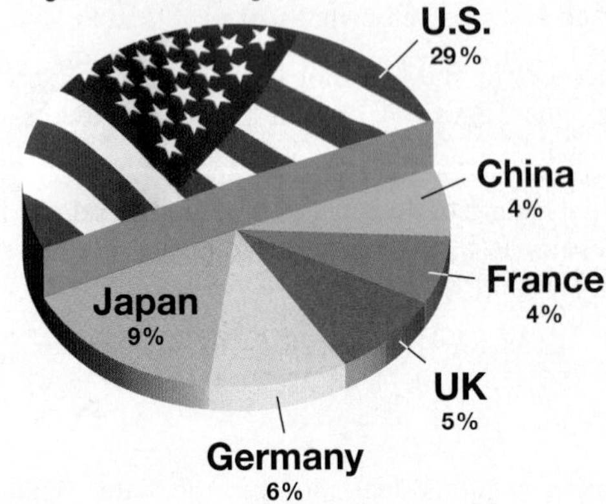

Source: Computer Industry Almanac

36. **Percentage of U.S. Households Watching ABC, CBS, and NBC in Prime Time**

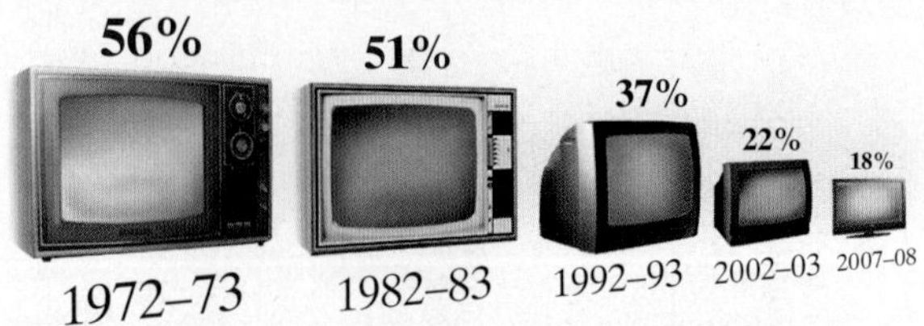

Source: Nielsen Media Research

37.

Source: *Entertainment Weekly*

Explaining the Concepts

38. What is a population? What is a sample?

39. Describe what is meant by a random sample.

40. Suppose you are interested in whether or not the students at your college would favor a grading system in which students may receive final grades of A+, A, A−, B+, B, B−, C+, C, C−, and so on. Describe how you might obtain a random sample of 100 students from the entire student population.

41. For Exercise 40, would questioning every fifth student as he or she is leaving the campus library until 100 students are interviewed be a good way to obtain a random sample? Explain your answer.

42. What is a frequency distribution?

43. What is a histogram?

44. What is a frequency polygon?

45. Describe how to construct a frequency polygon from a histogram.

46. Describe how to construct a stem-and-leaf plot from a set of data.

47. Describe two ways that graphs can be misleading.

Critical Thinking Exercises

Make Sense? *In Exercises 48–51, determine whether each statement makes sense or does not make sense, and explain your reasoning.*

48. The death rate from this new strain of flu is catastrophic because 25% of the people hospitalized with the disease have died.

49. The following graph indicates that for the period from 2000 through 2010, the percentage of female college freshmen describing their health as "above average" has rapidly decreased.

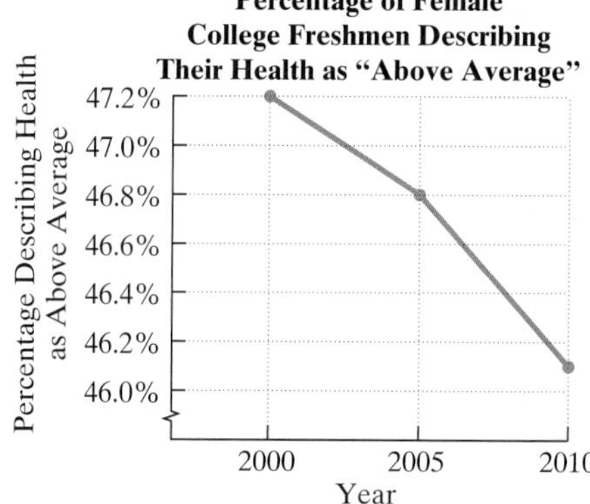

Source: John Macionis, *Sociology, Fourteenth Edition*, Pearson, 2012.

50. A public radio station needs to survey its contributors to determine their programming interests, so they should select a random sample of 100 of their largest contributors.

51. Improperly worded questions can steer respondents toward answers that are not their own.

52. Construct a grouped frequency distribution for the following data, showing the length, in miles, of the 25 longest rivers in the United States. Use five classes that have the same width.

2540	2340	1980	1900	1900
1460	1450	1420	1310	1290
1280	1240	1040	990	926
906	886	862	800	774
743	724	692	659	649

Source: U.S. Department of the Interior

Group Exercises

53. The classic book on distortion using statistics is *How to Lie with Statistics* by Darrell Huff. This activity is designed for five people. Each person should select two chapters from Huff's book and then present to the class the common methods of statistical manipulation and distortion that Huff discusses.

54. Each group member should find one example of a graph that presents data with integrity and one example of a graph that is misleading. Use newspapers, magazines, the Internet, books, and so forth. Once graphs have been collected, each member should share his or her graphs with the entire group. Be sure to explain why one graph depicts data in a forthright manner and how the other graph misleads the viewer.

12.2 Measures of Central Tendency

WHAT AM I SUPPOSED TO LEARN?

After studying this section, you should be able to:

1 Determine the mean for a data set.
2 Determine the median for a data set.
3 Determine the mode for a data set.
4 Determine the midrange for a data set.

DURING A LIFETIME, AMERICANS AVERAGE TWO WEEKS KISSING.

But wait, there's more:

- 130: The average number of "Friends" for a Facebook user
- 12: The average number of cars an American owns during a lifetime
- 300: The average number of times a 6-year-old child laughs each day
- 550: The average number of hairs in the human eyebrow
- 28: The average number of years in the lifespan of a citizen during the Roman Empire
- 6,000,000: The average number of dust mites living in a U.S. bed.

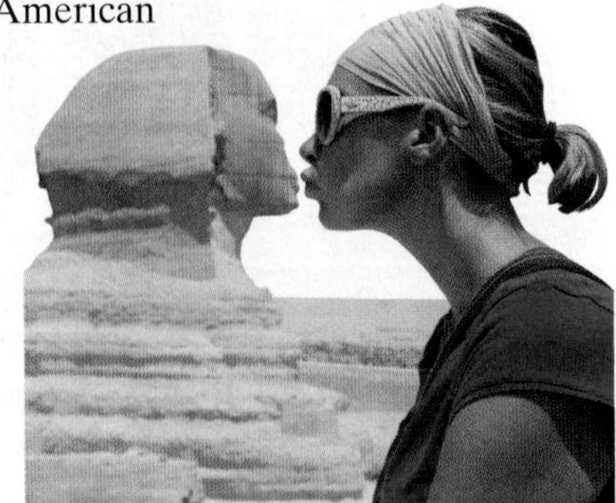

Source: *Listomania*, Harper Design

These numbers represent what is "average" or "typical" in a variety of situations. In statistics, such values are known as **measures of central tendency** because they are generally located toward the center of a distribution. Four such measures are discussed in this section: the mean, the median, the mode, and the midrange. Each measure of central tendency is calculated in a different way. Thus, it is better to use a specific term (mean, median, mode, or midrange) than to use the generic descriptive term "average."

1 *Determine the mean for a data set.*

The Mean

By far the most commonly used measure of central tendency is the *mean*. The **mean** is obtained by adding all the data items and then dividing the sum by the number of items. The Greek letter sigma, Σ, called a **symbol of summation**, is used to indicate the sum of data items. The notation Σx, read "the sum of x," means to add all the data items in a given data set. We can use this symbol to give a formula for calculating the mean.

THE MEAN

The **mean** is the sum of the data items divided by the number of items.

$$\text{Mean} = \frac{\Sigma x}{n},$$

where Σx represents the sum of all the data items and n represents the number of items.

The mean of a sample is symbolized by $\bar{x}$ (read "x bar"), while the mean of an entire population is symbolized by μ (the lowercase Greek letter *mu*). Unless otherwise indicated, the data sets throughout this chapter represent samples, so we will use $\bar{x}$ for the mean: $\bar{x} = \frac{\Sigma x}{n}$.

EXAMPLE 1 ***Calculating the Mean***

Are inventors born or made? It would be nice to think we could all be great inventors, but history says otherwise. **Figure 12.6**, based on a sample of adults in ten selected countries, shows wide agreement that inventiveness is a quality that can be learned.

Using Means to Compare How the U.S. Measures Up

- Mean Life Expectancy

78	82
U.S.	ITALY

- Mean Cost of an Angiogram

\$798	\$35
U.S.	CANADA

- Mean Number of Minutes Spent Online or Watching TV Daily

430	410
U.S.	ENGLAND

- Mean Number of Working Hours per Week

35.1	27.8
U.S.	GERMANY

- Mean Size of a Steak Served at Restaurants

13 ounces	8 ounces
U.S.	ENGLAND

Source: Time, USA Today

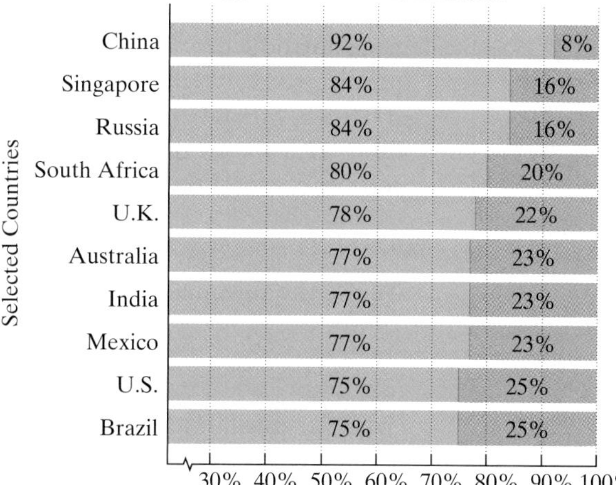

FIGURE 12.6 *Source: Time*

Find the mean percentage of adults in the ten countries who agree that inventiveness can be learned.

SOLUTION

We find the mean, $\bar{x}$, by adding the percents for the countries and dividing this sum by 10, the number of data items.

$$\bar{x} = \frac{\Sigma x}{n} = \frac{92 + 84 + 84 + 80 + 78 + 77 + 77 + 77 + 75 + 75}{10} = \frac{799}{10} = 79.9$$

The mean percentage of adults in the ten countries who agree that inventiveness can be learned is 79.9%.

One and only one mean can be calculated for any group of numerical data. The mean may or may not be one of the actual data items. In Example 1, the mean is 79.9%, although no data item is 79.9%.

CHECK POINT 1 Use **Figure 12.6** to find the mean percentage of adults in the ten countries who agree that inventiveness is inherited.

In Example 1, some of the data items were identical. We can use multiplication when computing the sum for these identical items.

$$\bar{x} = \frac{92 + 84 + 84 + 80 + 78 + 77 + 77 + 77 + 75 + 75}{10}$$

$$= \frac{92 \cdot 1 + 84 \cdot 2 + 80 \cdot 1 + 78 \cdot 1 + 77 \cdot 3 + 75 \cdot 2}{10}$$

The data value 84 has a frequency of 2.
The data value 77 has a frequency of 3.
The data value 75 has a frequency of 2.

When many data values occur more than once and a frequency distribution is used to organize the data, we can use the following formula to calculate the mean:

CALCULATING THE MEAN FOR A FREQUENCY DISTRIBUTION

$$\text{Mean} = \bar{x} = \frac{\Sigma xf}{n},$$

where

x represents a data value.
f represents the frequency of that data value.
Σxf represents the sum of all the products obtained by multiplying each data value by its frequency.
n represents the *total frequency* of the distribution.

TABLE 12.6 Students' Stress-Level Ratings

Stress Rating x	Frequency f
0	2
1	1
2	3
3	12
4	16
5	18
6	13
7	31
8	26
9	15
10	14

Source: Journal of Personality and Social Psychology, 69, 1102–1112

EXAMPLE 2 *Calculating the Mean for a Frequency Distribution*

In the previous Exercise Set, we mentioned a questionnaire given to students in an introductory statistics class during the first week of the course. One question asked, "How stressed have you been in the last $2\frac{1}{2}$ weeks, on a scale of 0 to 10, with 0 being not at all stressed and 10 being as stressed as possible?" **Table 12.6** shows the students' responses. Use this frequency distribution to find the mean of the stress-level ratings.

SOLUTION

We use the formula for the mean, $\bar{x}$:

$$\bar{x} = \frac{\Sigma xf}{n}.$$

First, we must find xf, obtained by multiplying each data value, x, by its frequency, f. Then, we need to find the sum of these products, Σxf. We can use the frequency distribution to organize these computations. Add a third column in which each data value is multiplied by its frequency. This column, shown on the right, is headed xf. Then, find the sum of the values, Σxf, in this column.

x	f	xf
0	2	$0 \cdot 2 = 0$
1	1	$1 \cdot 1 = 1$
2	3	$2 \cdot 3 = 6$
3	12	$3 \cdot 12 = 36$
4	16	$4 \cdot 16 = 64$
5	18	$5 \cdot 18 = 90$
6	13	$6 \cdot 13 = 78$
7	31	$7 \cdot 31 = 217$
8	26	$8 \cdot 26 = 208$
9	15	$9 \cdot 15 = 135$
10	14	$10 \cdot 14 = 140$
Totals:	$n = 151$	$\Sigma xf = 975$

Σxf is the sum of the numbers in the third column.

This value, the sum of the numbers in the second column, is the total frequency of the distribution.

Now, substitute these values into the formula for the mean, $\bar{x}$. Remember that n is the *total frequency* of the distribution, or 151.

$$\bar{x} = \frac{\Sigma xf}{n} = \frac{975}{151} \approx 6.46$$

The mean of the 0 to 10 stress-level ratings is approximately 6.46. Notice that the mean is greater than 5, the middle of the 0 to 10 scale.

CHECK POINT 2 Find the mean, $\bar{x}$, for the data items in the frequency distribution. (In order to save space, we've written the frequency distribution horizontally.)

Score, x	30	33	40	50
Frequency, f	3	4	4	1

2 *Determine the median for a data set.*

The Median

The *median* age in the United States is 37.8. The oldest state by median age is Maine (44.6) and the youngest state is Utah (30.6). To find these values, researchers begin with appropriate random samples. The data items—that is, the ages—are arranged from youngest to oldest. The median age is the data item in the middle of each set of ranked, or ordered, data.

THE MEDIAN

To find the **median** of a group of data items,

1. Arrange the data items in order, from smallest to largest.
2. If the number of data items is odd, the median is the data item in the middle of the list.
3. If the number of data items is even, the median is the mean of the two middle data items.

GREAT QUESTION!

What exactly does the median do with the data?

The median splits the data items down the middle, like the median strip in a road.

EXAMPLE 3 *Finding the Median*

Find the median for each of the following groups of data:

a. 84, 90, 98, 95, 88

b. 68, 74, 7, 13, 15, 25, 28, 59, 34, 47.

SOLUTION

a. Arrange the data items in order, from smallest to largest. The number of data items in the list, five, is odd. Thus, the median is the middle number.

84, 88, 90, 95, 98

Middle data item

The median is 90. Notice that two data items lie above 90 and two data items lie below 90.

b. Arrange the data items in order, from smallest to largest. The number of data items in the list, ten, is even. Thus, the median is the mean of the two middle data items.

7, 13, 15, 25, 28, 34, 47, 59, 68, 74

Middle data items are 28 and 34.

$$\text{Median} = \frac{28 + 34}{2} = \frac{62}{2} = 31$$

The median is 31. Five data items lie above 31 and five data items lie below 31.

7 13 15 25 28 | 34 47 59 68 74

Five data items lie below 31. Five data items lie above 31.

Median is 31.

CHECK POINT 3 Find the median for each of the following groups of data:

a. 28, 42, 40, 25, 35

b. 72, 61, 85, 93, 79, 87.

If a relatively long list of data items is arranged in order, it may be difficult to identify the item or items in the middle. In cases like this, the median can be found by determining its position in the list of items.

GREAT QUESTION!

Does the formula $\frac{n+1}{2}$ give the value of the median?

No. The formula gives the *position* of the median, and not the actual value of the median. When finding the median, be sure to first arrange the data items in order from smallest to largest.

POSITION OF THE MEDIAN

If n data items are arranged in order, from smallest to largest, the median is the value in the

$$\frac{n+1}{2}$$

position.

EXAMPLE 4 *Finding the Median Using the Position Formula*

Table 12.7 gives the nine longest words in the English language. Find the median number of letters for the nine longest words.

TABLE 12.7 The Nine Longest Words in the English Language

Word	Number of Letters
Pneumonoultramicroscopicsilicovolcanoconiosis A lung disease caused by breathing in volcanic dust	45
Supercalifragilisticexpialidocious Meaning "wonderful", from song of this title in the movie *Mary Poppins*	34
Floccinaucinihilipilification Meaning "the action or habit of estimating as worthless"	29
Trinitrophenylmethylnitramine A chemical compound used as a detonator in shells	29
Antidisestablishmentarianism Meaning "opposition to the disestablishment of the Church of England"	28
Electroencephalographically Relating to brain waves	27
Microspectrophotometrically Relating to the measurement of light waves	27
Immunoelectrophoretically Relating to measurement of immunoglobulin	25
Spectroheliokinematograph A 1930s' device for monitoring and filming solar activity	25

Source: Chris Cole, rec.puzzles archive

SOLUTION

We begin by listing the data items from smallest to largest.

25, 25, 27, 27, 28, 29, 29, 34, 45

There are nine data items, so $n = 9$. The median is the value in the

$$\frac{n+1}{2}\text{ position} = \frac{9+1}{2}\text{ position} = \frac{10}{2}\text{ position} = \text{fifth position.}$$

We find the median by selecting the data item in the fifth position.

Position 3 Position 4

25, 25, 27, 27, 28, 29, 29, 34, 45

Position 1 Position 2 Position 5

The median is 28. Notice that four data items lie above 28 and four data items lie below it. The median number of letters for the nine longest words in the English language is 28.

CHECK POINT 4 Find the median for the following group of data items:

$$1, 2, 2, 2, 3, 3, 3, 3, 3, 5, 6, 7, 7, 10, 11, 13, 19, 24, 26.$$

TABLE 12.8 Hours and Minutes per Day Spent Sleeping and Eating in Selected Countries

Country	Sleeping	Eating
France	8:50	2:15
U.S.	8:38	1:14
Spain	8:34	1:46
New Zealand	8:33	2:10
Australia	8:32	1:29
Turkey	8:32	1:29
Canada	8:29	1:09
Poland	8:28	1:34
Finland	8:27	1:21
Belgium	8:25	1:49
United Kingdom	8:23	1:25
Mexico	8:21	1:06
Italy	8:18	1:54
Germany	8:12	1:45
Sweden	8:06	1:34
Norway	8:03	1:22
Japan	7:50	1:57
S. Korea	7:49	1:36

Source: Organization for Economic Cooperation and Development

EXAMPLE 5 *Finding the Median Using the Position Formula*

Table 12.8 gives the mean number of hours and minutes per day spent sleeping and eating in 18 selected countries. Find the median number of hours and minutes per day spent sleeping for these countries.

SOLUTION

Reading from the bottom to the top of **Table 12.8**, the data items for sleeping appear from smallest to largest. There are 18 data items, so $n = 18$. The median is the value in the

$$\frac{n+1}{2}\text{ position} = \frac{18+1}{2}\text{ position} = \frac{19}{2}\text{ position} = 9.5\text{ position}.$$

This means that the median is the mean of the data items in positions 9 and 10.

Position 3 Position 4 Position 7 Position 8

7:49, 7:50, 8:03, 8:06, 8:12, 8:18, 8:21, 8:23, **8:25, 8:27,** 8:28, 8:29, 8:32, 8:32, 8:33, 8:34, 8:38, 8:50

Position 1 Position 2 Position 5 Position 6 Position 9 Position 10

$$\text{Median} = \frac{8{:}25 + 8{:}27}{2} = \frac{16{:}52}{2} = 8{:}26$$

The median number of hours per day spent sleeping for the 18 countries is 8 hours, 26 minutes.

CHECK POINT 5 Arrange the data items for eating in **Table 12.8** from smallest to largest. Then find the median number of hours and minutes per day spent eating for the 18 countries.

When individual data items are listed from smallest to largest, you can find the median by identifying the item or items in the middle or by using the $\frac{n+1}{2}$ formula for its position. However, the formula for the position of the median is more useful when data items are organized in a frequency distribution.

EXAMPLE 6 Finding the Median for a Frequency Distribution

The frequency distribution for the stress-level ratings of 151 students is repeated below using a horizontal format. Find the median stress-level rating.

Stress rating / Number of college students

x	0	1	2	3	4	5	6	7	8	9	10
f	2	1	3	12	16	18	13	31	26	15	14

Total: $n = 151$

SOLUTION

There are 151 data items, so $n = 151$. The median is the value in the

$$\frac{n+1}{2}\text{ position} = \frac{151+1}{2}\text{ position} = \frac{152}{2}\text{ position} = 76\text{th position.}$$

We find the median by selecting the data item in the 76th position. The frequency distribution indicates that the data items begin with

$$0, 0, 1, 2, 2, 2, \ldots.$$

We can write the data items all out and then select the median, the 76th data item. A more efficient way to proceed is to count down the frequency column in the distribution until we identify the 76th data item:

x	f	We count down the frequency column.
0	2	1, 2
1	1	3
2	3	4, 5, 6
3	12	7, 8, 9, 10, 11, 12, 13, 14, 15, 16, 17, 18
4	16	19, 20, 21, 22, 23, 24, 25, 26, 27, 28, 29, 30, 31, 32, 33, 34
5	18	35, 36, 37, 38, 39, 40, 41, 42, 43, 44, 45, 46, 47, 48, 49, 50, 51, 52
6	13	53, 54, 55, 56, 57, 58, 59, 60, 61, 62, 63, 64, 65
7	31	66, 67, 68, 69, 70, 71, 72, 73, 74, 75, 76
8	26	
9	15	
10	14	

Stop counting. We've reached the 76th data item.

The 76th data item is 7. The median stress-level rating is 7.

CHECK POINT 6 Find the median for the following frequency distribution.

Age at presidential inauguration / Number of U.S. presidents assuming office in the 20th century with the given age

x	42	43	46	51	52	54	55	56	60	61	64	69
f	1	1	1	3	1	2	2	2	1	2	1	1

Statisticians generally use the median, rather than the mean, when reporting income. Why? Our next example will help to answer this question.

EXAMPLE 7 Comparing the Median and the Mean

Five employees in the assembly section of a television manufacturing company earn salaries of \$19,700, \$20,400, \$21,500, \$22,600, and \$23,000 annually. The section manager has an annual salary of \$95,000.

a. Find the median annual salary for the six people.

b. Find the mean annual salary for the six people.

SOLUTION

a. To compute the median, first arrange the salaries in order:

$$\$19{,}700,\quad \$20{,}400,\quad \$21{,}500,\quad \$22{,}600,\quad \$23{,}000,\quad \$95{,}000.$$

Because the list contains an even number of data items, six, the median is the mean of the two middle items.

$$\text{Median} = \frac{\$21{,}500 + \$22{,}600}{2} = \frac{\$44{,}100}{2} = \$22{,}050$$

The median annual salary is \$22,050.

b. We find the mean annual salary by adding the six annual salaries and dividing by 6.

$$\text{Mean} = \frac{\$19{,}700 + \$20{,}400 + \$21{,}500 + \$22{,}600 + \$23{,}000 + \$95{,}000}{6}$$

$$= \frac{\$202{,}200}{6} = \$33{,}700$$

The mean annual salary is \$33,700.

In Example 7, the median annual salary is \$22,050 and the mean annual salary is \$33,700. Why such a big difference between these two measures of central tendency? The relatively high annual salary of the section manager, \$95,000, pulls the mean salary to a value considerably higher than the median salary. When one or more data items are much greater than the other items, these extreme values can greatly influence the mean. In cases like this, the median is often more representative of the data.

This is why the median, rather than the mean, is used to summarize the incomes, by gender and race, shown in **Figure 12.7**. Because no one can earn less than \$0, the distribution of income must come to an end at \$0 for each of these eight groups. By contrast, there is no upper limit on income on the high side. In the United States, the wealthiest 20% of the population earn about 50% of the total income. The relatively few people with very high annual incomes tend to pull the mean income to a value considerably greater than the median income. Reporting mean incomes in **Figure 12.7** would inflate the numbers shown, making them nonrepresentative of the millions of workers in each of the eight groups.

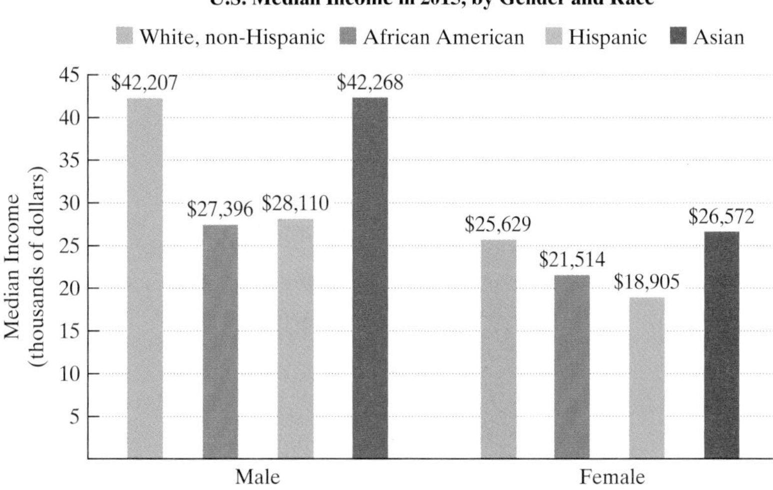

FIGURE 12.7
Source: U.S. Census Bureau

CHECK POINT 7 **Table 12.9** shows the net worth, in millions of 2016 dollars, for ten U.S. presidents from Johnson through Trump.

TABLE 12.9 Net Worth for Ten U.S. Presidents

President	Net Worth (millions of dollars)
Johnson	\$98
Nixon	\$15
Ford	\$7
Carter	\$7
Reagan	\$13
Bush	\$23
Clinton	\$38
Bush	\$20
Obama	\$5
Trump	\$3500 (i.e. \$3.5 billion)

Source: Time

a. Find the mean net worth, in millions of dollars, for the ten presidents.

b. Find the median net worth, in millions of dollars, for the ten presidents.

c. Describe why one of the measures of central tendency is greater than the other.

3 *Determine the mode for a data set.*

The Mode

Let's take one final look at the frequency distribution for the stress-level ratings of 151 college students.

Stress rating

Number of college students

x	0	1	2	3	4	5	6	7	8	9	10
f	2	1	3	12	16	18	13	31	26	15	14

7 is the stress rating with the greatest frequency.

The data value that occurs most often in this distribution is 7, the stress rating for 31 of the 151 students. We call 7 the *mode* of this distribution.

THE MODE

The **mode** is the data value that occurs most often in a data set. If more than one data value has the highest frequency, then each of these data values is a mode. If there is no data value that occurs most often, then the data set has no mode.

EXAMPLE 8 *Finding the Mode*

Find the mode for each of the following groups of data:

a. 7, 2, 4, 7, 8, 10 **b.** 2, 1, 4, 5, 3 **c.** 3, 3, 4, 5, 6, 6.

SOLUTION

a. 7, 2, 4, 7, 8, 10

7 occurs most often.

The mode is 7.

b. 2, 1, 4, 5, 3

Each data item occurs the same number of times.

There is no mode.

c. 3, 3, 4, 5, 6, 6

Both 3 and 6 occur most often.

The modes are 3 and 6. The data set is said to be **bimodal**.

CHECK POINT 8 Find the mode for each of the following groups of data:

a. 3, 8, 5, 8, 9, 10

b. 3, 8, 5, 8, 9, 3

c. 3, 8, 5, 6, 9, 10.

4 *Determine the midrange for a data set.*

TABLE 12.10 Ten Hottest U.S. Cities

City	Mean Temperature
Key West, FL	77.8°
Miami, FL	75.9°
West Palm Beach, FL	74.7°
Fort Myers, FL	74.4°
Yuma, AZ	74.2°
Brownsville, TX	73.8°
Phoenix, AZ	72.6°
Vero Beach, FL	72.4°
Orlando, FL	72.3°
Tampa, FL	72.3°

Source: National Oceanic and Atmospheric Administration

The Midrange

Table 12.10 shows the ten hottest cities in the United States. Because temperature is constantly changing, you might wonder how the mean temperatures shown in the table are obtained.

First, we need to find a representative daily temperature. This is obtained by adding the lowest and highest temperatures for the day and then dividing this sum by 2. Next, we take the representative daily temperatures for all 365 days, add them, and divide the sum by 365. These are the mean temperatures that appear in **Table 12.10**.

Representative daily temperature,

$$\frac{\text{lowest daily temperature} + \text{highest daily temperature}}{2},$$

is an example of a measure of central tendency called the *midrange*.

THE MIDRANGE

The **midrange** is found by adding the lowest and highest data values and dividing the sum by 2.

$$\text{Midrange} = \frac{\text{lowest data value} + \text{highest data value}}{2}$$

EXAMPLE 9 *Finding the Midrange*

Newsweek magazine examined factors that affect women's lives, including justice, health, education, economics, and politics. Using these five factors, the magazine graded each of 165 countries on a scale from 0 to 100. The 12 best places to be a woman and the 12 worst places to be a woman are shown in **Table 12.11**.

TABLE 12.11 Women in the World

Best Places to Be a Woman		Worst Places to Be a Woman	
Country	**Score**	**Country**	**Score**
Iceland	100.0	Chad	0.0
Canada	99.6	Afghanistan	2.0
Sweden	99.2	Yemen	12.1
Denmark	95.3	Democratic Republic of the Congo	13.6
Finland	92.8	Mali	17.6
Switzerland	91.9	Solomon Islands	20.8
Norway	91.3	Niger	21.2
United States	89.8	Pakistan	21.4
Australia	88.2	Ethiopia	23.7
Netherlands	87.7	Sudan	26.1
New Zealand	87.2	Guinea	28.5
France	87.2	Sierra Leone	29.0

Source: Newsweek

Find the midrange score among the 12 best countries to be a woman.

SOLUTION

Refer to **Table 12.11** on the previous page.

$$\text{Midrange} = \frac{\text{best place with the lowest score} + \text{best place with the highest score}}{2}$$

$$= \frac{87.2 + 100.0}{2} = \frac{187.2}{2} = 93.6$$

The midrange score among the 12 best countries to be a woman is 93.6.

We can find the mean score among the 12 best countries to be a woman by adding up the 12 scores and then dividing the sum by 12. By doing so, we can determine that the mean score is approximately 92.5. It is much faster to calculate the midrange, which is often used as an estimate for the mean.

CHECK POINT 9 Use **Table 12.11** on the previous page to find the midrange score among the 12 worst countries to be a woman.

EXAMPLE 10 *Finding the Four Measures of Central Tendency*

Suppose your six exam grades in a course are

52, 69, 75, 86, 86, and 92.

Compute your final course grade (90–100 = A, 80–89 = B, 70–79 = C, 60–69 = D, below 60 = F) using the

a. mean. **b.** median. **c.** mode. **d.** midrange.

SOLUTION

a. The mean is the sum of the data items divided by the number of items, 6.

$$\text{Mean} = \frac{52 + 69 + 75 + 86 + 86 + 92}{6} = \frac{460}{6} \approx 76.67$$

Using the mean, your final course grade is C.

b. The six data items, 52, 69, 75, 86, 86, and 92, are arranged in order. Because the number of data items is even, the median is the mean of the two middle items.

$$\text{Median} = \frac{75 + 86}{2} = \frac{161}{2} = 80.5$$

Using the median, your final course grade is B.

c. The mode is the data value that occurs most frequently. Because 86 occurs most often, the mode is 86. Using the mode, your final course grade is B.

d. The midrange is the mean of the lowest and highest data values.

$$\text{Midrange} = \frac{52 + 92}{2} = \frac{144}{2} = 72$$

Using the midrange, your final course grade is C.

CHECK POINT 10 *Consumer Reports* magazine gave the following data for the number of calories in a meat hot dog for each of 17 brands:

173, 191, 182, 190, 172, 147, 146, 138, 175, 136, 179, 153, 107, 195, 135, 140, 138.

Find the mean, median, mode, and midrange for the number of calories in a meat hot dog for the 17 brands. If necessary, round answers to the nearest tenth of a calorie.

Concept and Vocabulary Check

Fill in each blank so that the resulting statement is true.

1. $\frac{\Sigma x}{n}$, the sum of all the data items divided by the number of data items, is the measure of central tendency called the _______.
2. The measure of central tendency that is the data item in the middle of ranked, or ordered, data is called the _______.
3. If n data items are arranged in order, from smallest to largest, the data item in the middle is the value in ______ position.
4. A data value that occurs most often in a data set is the measure of central tendency called the _______.
5. The measure of central tendency that is found by adding the lowest and highest data values and dividing the sum by 2 is called the _________.

In Exercises 6–9, determine whether each statement is true or false. If the statement is false, make the necessary change(s) to produce a true statement.

6. Numbers representing what is average or typical about a data set are called measures of central tendency. ______
7. When finding the mean, it is necessary to arrange the data items in order. ______
8. If one or more data items are much greater than the other items, the mean, rather than the median, is more representative of the data. ______
9. A data set can contain more than one median, or no median at all. ______

Exercise Set 12.2

Practice Exercises

In Exercises 1–8, find the mean for each group of data items.

1. 7, 4, 3, 2, 8, 5, 1, 3
2. 11, 6, 4, 0, 2, 1, 12, 0, 0
3. 91, 95, 99, 97, 93, 95
4. 100, 100, 90, 30, 70, 100
5. 100, 40, 70, 40, 60
6. 1, 3, 5, 10, 8, 5, 6, 8
7. 1.6, 3.8, 5.0, 2.7, 4.2, 4.2, 3.2, 4.7, 3.6, 2.5, 2.5
8. 1.4, 2.1, 1.6, 3.0, 1.4, 2.2, 1.4, 9.0, 9.0, 1.8

In Exercises 9–12, find the mean for the data items in the given frequency distribution.

9.

Score x	Frequency f
1	1
2	3
3	4
4	4
5	6
6	5
7	3
8	2

10.

Score x	Frequency f
1	2
2	4
3	5
4	7
5	6
6	4
7	3

11.

Score x	Frequency f
1	1
2	1
3	2
4	5
5	7
6	9
7	8
8	6
9	4
10	3

12.

Score x	Frequency f
1	3
2	4
3	6
4	8
5	9
6	7
7	5
8	2
9	1
10	1

In Exercises 13–20, find the median for each group of data items.

13. 7, 4, 3, 2, 8, 5, 1, 3
14. 11, 6, 4, 0, 2, 1, 12, 0, 0
15. 91, 95, 99, 97, 93, 95
16. 100, 100, 90, 30, 70, 100
17. 100, 40, 70, 40, 60
18. 1, 3, 5, 10, 8, 5, 6, 8
19. 1.6, 3.8, 5.0, 2.7, 4.2, 4.2, 3.2, 4.7, 3.6, 2.5, 2.5
20. 1.4, 2.1, 1.6, 3.0, 1.4, 2.2, 1.4, 9.0, 9.0, 1.8

Find the median for the data items in the frequency distribution in

21. Exercise 9.
22. Exercise 10.
23. Exercise 11.
24. Exercise 12.

In Exercises 25–32, find the mode for each group of data items. If there is no mode, so state.

25. 7, 4, 3, 2, 8, 5, 1, 3
26. 11, 6, 4, 0, 2, 1, 12, 0, 0
27. 91, 95, 99, 97, 93, 95
28. 100, 100, 90, 30, 70, 100
29. 100, 40, 70, 40, 60
30. 1, 3, 5, 10, 8, 5, 6, 8
31. 1.6, 3.8, 5.0, 2.7, 4.2, 4.2, 3.2, 4.7, 3.6, 2.5, 2.5
32. 1.4, 2.1, 1.6, 3.0, 1.4, 2.2, 1.4, 9.0, 9.0, 1.8

Find the mode for the data items in the frequency distribution in

33. Exercise 9.
34. Exercise 10.
35. Exercise 11.
36. Exercise 12.

In Exercises 37–44, find the midrange for each group of data items.

37. 7, 4, 3, 2, 8, 5, 1, 3
38. 11, 6, 4, 0, 2, 1, 12, 0, 0
39. 91, 95, 99, 97, 93, 95
40. 100, 100, 90, 30, 70, 100
41. 100, 40, 70, 40, 60
42. 1, 3, 5, 10, 8, 5, 6, 8
43. 1.6, 3.8, 5.0, 2.7, 4.2, 4.2, 3.2, 4.7, 3.6, 2.5, 2.5
44. 1.4, 2.1, 1.6, 3.0, 1.4, 2.2, 1.4, 9.0, 9.0, 1.8

Find the midrange for the data items in the frequency distribution in

45. Exercise 9.
46. Exercise 10.
47. Exercise 11.
48. Exercise 12.

Practice Plus

In Exercises 49–54, use each display of data items to find the mean, median, mode, and midrange.

49.

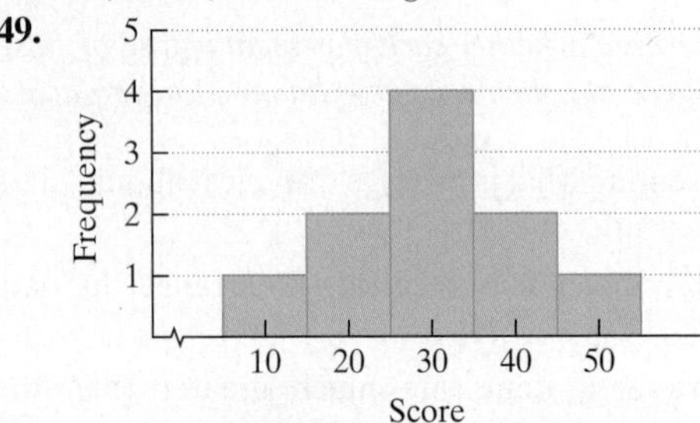

50.

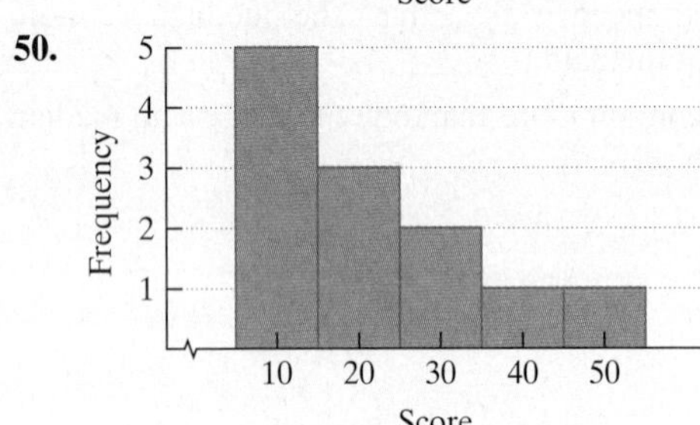

51.

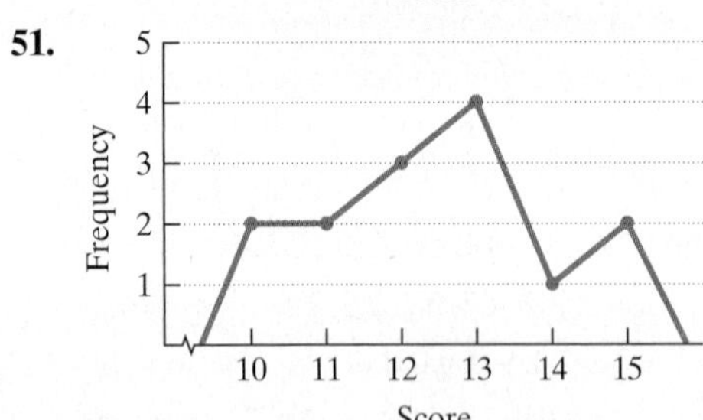

52.

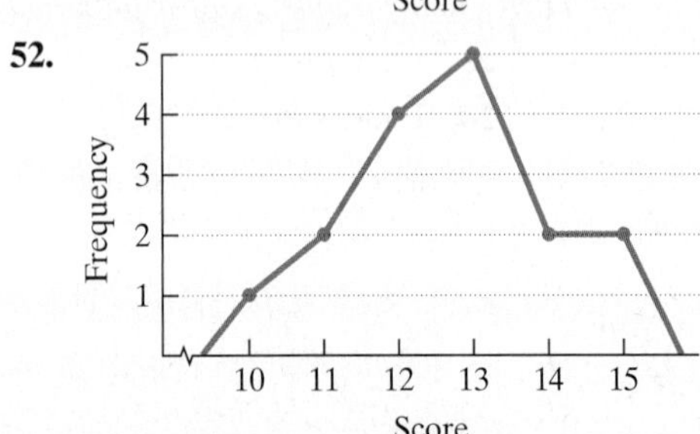

53.

Stems	Leaves
2	1 4 5
3	0 1 1 3
4	2 5

54.

Stems	Leaves
2	8
3	2 4 4 9
4	0 1 5 7

Application Exercises

Exercises 55–57 present data on a variety of topics. For each data set described in boldface, find the

a. *mean.* **b.** *median.*

c. *mode (or state that there is no mode).*

d. *midrange.*

55. Net Worth of the Richest Under 35

Billionaire	Net Worth (billions of dollars)
John Collison, 26 (Stripe)	\$1.1
Evan Spiegel, 26 (Snap)	\$4.0
Patrick Collison, 28 (Stripe)	\$1.1
Bobby Murphy, 28 (Snap)	\$4.0
Dustin Moskovitz, 32 (Facebook)	\$10.7
Mark Zuckerberg, 32 (Facebook)	\$56.0
Nathan Blecharczyk, 33 (Airbnb)	\$3.8
Ryan Graves, 33 (Uber)	\$1.6
Kevin Systrom, 33 (Instagram)	\$1.2
Drew Houston, 34 (Dropbox)	\$1.0
Liu Ruopeng, 34 (Technology)	\$1.3
Eduardo Saverin, 34 (Facebook)	\$7.9
Kirill Shamalov, 34 (Petrochemicals)	\$1.3
Evan Sharp, 34 (Pinterest)	\$1.0
Ben Silbermann, 34 (Pinterest)	\$1.6

Source: Forbes March 28, 2017

56. Net Worth for the First 13 U.S. Presidents

President	Net Worth (millions of 2016 dollars)
Washington	\$580
Adams	\$21
Jefferson	\$234
Madison	\$112
Monroe	\$30
Adams	\$21
Jackson	\$131
Van Buren	\$29
Harrison	\$6
Tyler	\$57
Polk	\$11
Taylor	\$7
Fillmore	\$4

Source: Time

57. Number of Social Interactions of College Students In Exercise Set 12.1, we presented a grouped frequency distribution showing the number of social interactions of ten minutes or longer over a one-week period for a group of college students. (These interactions excluded family and work situations.) Use the frequency distribution shown to solve this exercise. (This distribution was obtained by replacing the classes in the grouped frequency distribution previously shown with the midpoints of the classes.)

Social interactions in a week

Number of college students

x	2	7	12	17	22	27	32	37	42	47
f	12	16	16	16	10	11	4	3	3	3

The weights (to the nearest five pounds) of 40 randomly selected male college students are organized in a histogram with a superimposed frequency polygon. Use the graph to answer Exercises 58–61.

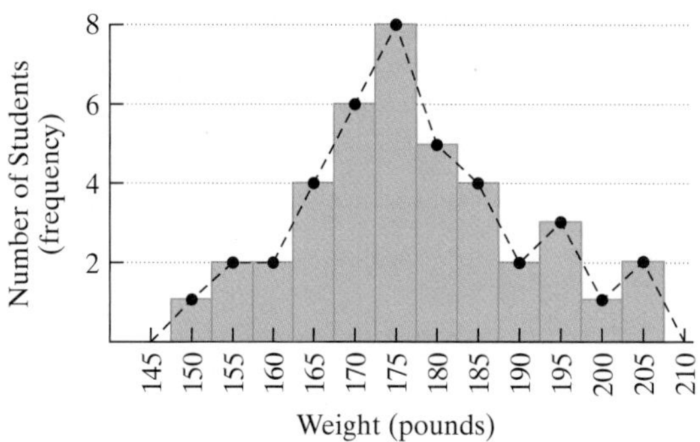

58. Find the mean weight.

59. Find the median weight.

60. Find the modal weight.

61. Find the midrange weight.

62. An advertisement for a speed-reading course claimed that the "average" reading speed for people completing the course was 1000 words per minute. Shown below are the actual data for the reading speeds per minute for a sample of 24 people who completed the course.

1000	900	800	1000	900	850
650	1000	1050	800	1000	850
700	750	800	850	900	950
600	1100	950	700	750	650

a. Find the mean, median, mode, and midrange. (If you prefer, first organize the data in a frequency distribution.)

b. Which measure of central tendency was given in the advertisement?

c. Which measure of central tendency is the best indicator of the "average" reading speed in this situation? Explain your answer.

63. In one common system for finding a grade-point average, or GPA,

$$A = 4, B = 3, C = 2, D = 1, F = 0.$$

The GPA is calculated by multiplying the number of credit hours for a course and the number assigned to each grade, and then adding these products. Then divide this sum by the total number of credit hours. Because each course grade is weighted according to the number of credits of the course, GPA is called a *weighted mean*. Calculate the GPA for this transcript:

Sociology: 3 cr. A; Biology: 3.5 cr. C; Music: 1 cr. B; Math: 4 cr. B; English: 3 cr. C.

Explaining the Concepts

64. What is the mean and how is it obtained?

65. What is the median and how is it obtained?

66. What is the mode and how is it obtained?

67. What is the midrange and how is it obtained?

68. The "average" income in the United States can be given by the mean or the median.

a. Which measure would be used in anti-U.S. propaganda? Explain your answer.

b. Which measure would be used in pro-U.S. propaganda? Explain your answer.

69. In a class of 40 students, 21 have examination scores of 77%. Which measure or measures of central tendency can you immediately determine? Explain your answer.

70. You read an article that states, "Of the 411 players in the National Basketball Association, only 138 make more than the average salary of \$3.12 million." Is \$3.12 million the mean or the median salary? Explain your answer.

71. A student's parents promise to pay for next semester's tuition if an A average is earned in chemistry. With examination grades of 97%, 97%, 75%, 70%, and 55%, the student reports that an A average has been earned. Which measure of central tendency is the student reporting as the average? How is this student misrepresenting the course performance with statistics?

72. According to the National Oceanic and Atmospheric Administration, the coldest city in the United States is International Falls, Minnesota, with a mean Fahrenheit temperature of 36.8°. Explain how this mean is obtained.

Critical Thinking Exercises

Make Sense? *In Exercises 73–76, determine whether each statement makes sense or does not make sense, and explain your reasoning.*

73. I'm working with a data set for which neither the mean nor the median is one of the data items.

74. I made a distribution of the heights of the 12 players on our basketball team. Because one player is much taller than the others, the team's median height is greater than its mean height.

75. Although the data set 1, 1, 2, 3, 3, 3, 4, 4 has a number of repeated items, there is only one mode.

76. If professors use the same test scores for a particular student and calculate measures of central tendency correctly, they will always agree on the student's final course grade.

77. Give an example of a set of six examination grades (from 0 to 100) with each of the following characteristics:

a. The mean and the median have the same value, but the mode has a different value.

b. The mean and the mode have the same value, but the median has a different value.

c. The mean is greater than the median.

d. The mode is greater than the mean.

e. The mean, median, and mode have the same value.

f. The mean and mode have values of 72.

78. On an examination given to 30 students, no student scored below the mean. Describe how this occurred.

Group Exercises

79. Select a characteristic, such as shoe size or height, for which each member of the group can provide a number. Choose a characteristic of genuine interest to the group. For this characteristic, organize the data collected into a frequency distribution and a graph. Compute the mean, median, mode, and midrange. Discuss any differences among these values. What happens if the group is divided (men and women, or people under a certain age and people over a certain age) and these measures of central tendency are computed for each of the subgroups? Attempt to use measures of central tendency to discover something interesting about the entire group or the subgroups.

80. A book on spotting bad statistics and learning to think critically about these influential numbers is *Damn Lies and Statistics* by Joel Best (University of California Press, 2001). This activity is designed for six people. Each person should select one chapter from Best's book. The group report should include examples of the use, misuse, and abuse of statistical information. Explain exactly how and why bad statistics emerge, spread, and come to shape policy debates. What specific ways does Best recommend to detect bad statistics?

12.3 Measures of Dispersion

WHAT AM I SUPPOSED TO LEARN?

After studying this section, you should be able to:

1 Determine the range for a data set.

2 Determine the standard deviation for a data set.

WHEN YOU THINK OF HOUSTON, TEXAS, AND Honolulu, Hawaii, do balmy temperatures come to mind? Both cities have a mean temperature of 75°. However, the mean temperature does not tell the whole story. The temperature in Houston differs seasonally from a low of about 40° in January to a high of close to 100° in July and August. By contrast, Honolulu's temperature varies less throughout the year, usually ranging between 60° and 90°.

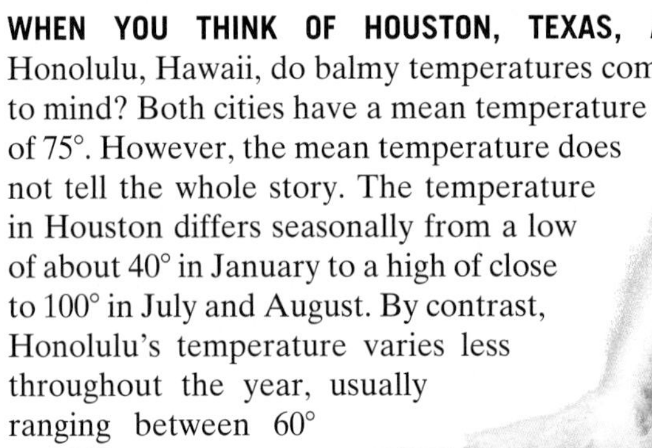

Measures of dispersion are used to describe the spread of data items in a data set. Two of the most common measures of dispersion, the *range* and the *standard deviation*, are discussed in this section.

The Range

1 *Determine the range for a data set.*

A quick but rough measure of dispersion is the **range**, the difference between the highest and lowest data values in a data set. For example, if Houston's hottest annual temperature is 103° and its coldest annual temperature is 33°, the range in temperature is

$$103° - 33°, \quad \text{or} \quad 70°.$$

If Honolulu's hottest day is 89° and its coldest day 61°, the range in temperature is

$$89° - 61°, \quad \text{or} \quad 28°.$$

THE RANGE

The **range**, the difference between the highest and lowest data values in a data set, indicates the total spread of the data.

$$\text{Range} = \text{highest data value} - \text{lowest data value}$$

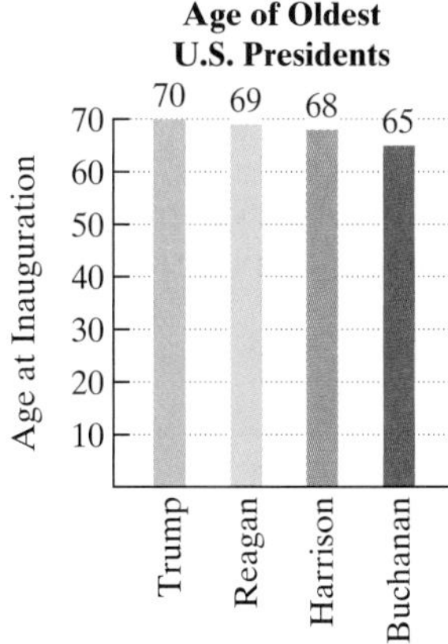

FIGURE 12.8
Source: Internet Public Library

EXAMPLE 1 Computing the Range

Figure 12.8 shows the age of the four oldest U.S presidents at the start of their first term. Find the age range for the four oldest presidents.

SOLUTION

$$\text{Range} = \text{highest data value} - \text{lowest data value}$$
$$= 70 - 65 = 5$$

The range is 5 years.

CHECK POINT 1 Find the range for the following group of data items:

4, 2, 11, 7.

The Standard Deviation

A second measure of dispersion, and one that is dependent on *all* of the data items, is called the **standard deviation**. The standard deviation is found by determining how much each data item differs from the mean.

In order to compute the standard deviation, it is necessary to find by how much each data item deviates from the mean. First compute the mean, $\bar{x}$. Then subtract the mean from each data item, $x - \bar{x}$. Example 2 shows how this is done. In Example 3, we will use this skill to actually find the standard deviation.

EXAMPLE 2 Preparing to Find the Standard Deviation; Finding Deviations from the Mean

Find the deviations from the mean for the four data items 70, 69, 68, and 65, shown in **Figure 12.8**.

SOLUTION

First, calculate the mean, $\bar{x}$.

$$\bar{x} = \frac{\Sigma x}{n} = \frac{70 + 69 + 68 + 65}{4} = \frac{272}{4} = 68$$

The mean age for the four oldest U.S. presidents is 68 years. Now, let's find by how much each of the four data items in **Figure 12.8** differs from 68, the mean. For Trump, who was 70 at his inauguration, the computation is shown as follows:

$$\text{Deviation from mean} = \text{data item} - \text{mean}$$
$$= x - \bar{x}$$
$$= 70 - 68 = 2.$$

This indicates that Trump's inaugural age exceeds the mean by two years.

The computation for Buchanan, who was 65 at the start of his first term, is given by

$$\text{Deviation from mean} = \text{data item} - \text{mean}$$
$$= x - \bar{x}$$
$$= 65 - 68 = -3.$$

This indicates that Buchanan's inaugural age is three years below the mean.

The deviations from the mean for each of the four given data items are shown in **Table 12.12**.

TABLE 12.12 Deviations from the Mean

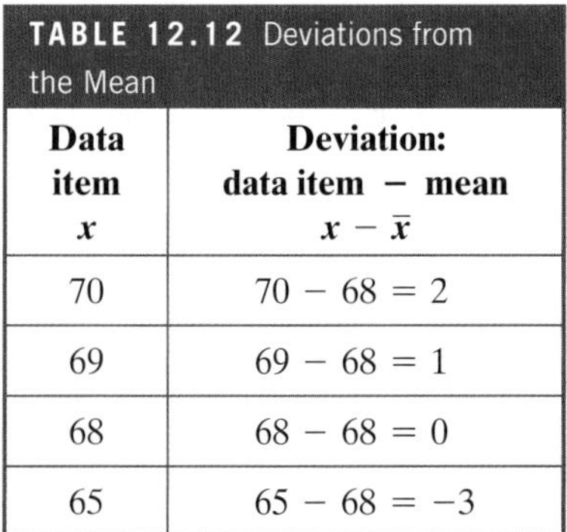

Data item x	Deviation: data item − mean $x - \bar{x}$
70	$70 - 68 = 2$
69	$69 - 68 = 1$
68	$68 - 68 = 0$
65	$65 - 68 = -3$

CHECK POINT 2 Compute the mean for the following group of data items:

$$2, 4, 7, 11.$$

Then find the deviations from the mean for the four data items. Organize your work in table form just like **Table 12.12**. Keep track of these computations. You will be using them in Check Point 3.

TABLE 12.12 Deviations from the Mean (repeated)

Data item x	Deviation: data item − mean $x - \bar{x}$
70	$70 - 68 = 2$
69	$69 - 68 = 1$
68	$68 - 68 = 0$
65	$65 - 68 = -3$

The sum of the deviations from the mean for a set of data is always zero: $\Sigma(x - \bar{x}) = 0$. For the deviations from the mean shown in **Table 12.12**,

$$2 + 1 + 0 + (-3) = 3 + (-3) = 0.$$

This shows that we cannot find a measure of dispersion by finding the mean of the deviations, because this value is always zero. However, a kind of average of the deviations from the mean, called the **standard deviation**, can be computed. We do so by squaring each deviation and later introducing a square root in the computation. Here are the details on how to find the standard deviation for a set of data:

2 Determine the standard deviation for a data set.

COMPUTING THE STANDARD DEVIATION FOR A DATA SET

1. Find the mean of the data items.
2. Find the deviation of each data item from the mean:
$$\text{data item} - \text{mean}.$$
3. Square each deviation:
$$(\text{data item} - \text{mean})^2.$$
4. Sum the squared deviations:
$$\Sigma(\text{data item} - \text{mean})^2.$$
5. Divide the sum in step 4 by $n - 1$, where n represents the number of data items:
$$\frac{\Sigma(\text{data item} - \text{mean})^2}{n - 1}.$$
6. Take the square root of the quotient in step 5. This value is the standard deviation for the data set.
$$\text{Standard deviation} = \sqrt{\frac{\Sigma(\text{data item} - \text{mean})^2}{n - 1}}$$

The standard deviation of a sample is symbolized by s, while the standard deviation of an entire population is symbolized by σ (the lowercase Greek letter *sigma*). Unless otherwise indicated, data sets represent samples, so we will use s for the standard deviation:

$$s = \sqrt{\frac{\Sigma(x - \bar{x})^2}{n - 1}}.$$

The computation of the standard deviation can be organized using a table with three columns:

Data item x	Deviation: $x - \bar{x}$ Data item − mean	(Deviation)^2: $(x - \bar{x})^2$ (Data item − mean)^2

In Example 2, we worked out the first two columns of such a table. Let's continue working with the data for the ages of the four oldest U.S. presidents and compute the standard deviation.

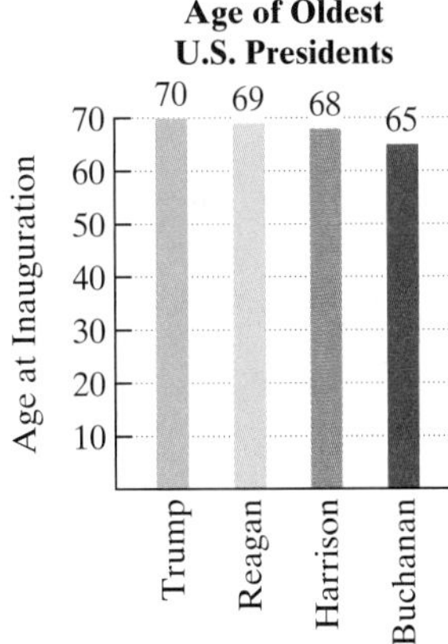

FIGURE 12.8 (repeated)

EXAMPLE 3 Computing the Standard Deviation

Figure 12.8, showing the age of the four oldest U.S. presidents at the start of their first term, is repeated in the margin. Find the standard deviation for the ages of the four presidents.

SOLUTION

Step 1 Find the mean. From our work in Example 2, the mean is 68: $\bar{x} = 68$.

Step 2 Find the deviation of each data item from the mean: data item − mean or $x - \bar{x}$. This, too, was done in Example 2 for each of the four data items.

Step 3 Square each deviation: (data item − mean)^2 or $(x - \bar{x})^2$. We square each of the numbers in the (data item − mean) column, shown in **Table 12.13**. Notice that squaring the difference always results in a nonnegative number.

TABLE 12.13 Computing the Standard Deviation

Data item x	Deviation: data item − mean $x - \bar{x}$	(Deviation)^2: (data item − mean)^2 $(x - \bar{x})^2$
70	$70 - 68 = 2$	$2^2 = 2 \cdot 2 = 4$
69	$69 - 68 = 1$	$1^2 = 1 \cdot 1 = 1$
68	$68 - 68 = 0$	$(0)^2 = 0 \cdot 0 = 0$
65	$65 - 68 = -3$	$(-3)^2 = (-3)(-3) = 9$
Totals:	$\Sigma(x - \bar{x}) = 0$	$\Sigma(x - \bar{x})^2 = 14$

The sum of the deviations for a set of data is always zero.

Adding the four numbers in the third column gives the sum of the squared deviations: Σ(data item − mean)^2.

TECHNOLOGY

Almost all scientific and graphing calculators compute the standard deviation of a set of data. Using the data items in Example 3,

70, 69, 68, 65,

the keystrokes for obtaining the standard deviation on many scientific calculators are as follows:

70 [Σ+] 69 [Σ+] 68 [Σ+]

65 [Σ+] [2nd] [$\sigma n - 1$].

Graphing calculators require that you specify if data items are from an entire population or a sample of the population.

Step 4 Sum the squared deviations: Σ(data item − mean)^2. This step is shown in **Table 12.13**. The squares in the third column were added, resulting in a sum of 14: $\Sigma(x - \bar{x})^2 = 14$.

Step 5 Divide the sum in step 4 by $n - 1$, where n represents the number of data items. The number of data items is 4 so we divide by 3.

$$\frac{\Sigma(x - \bar{x})^2}{n - 1} = \frac{\Sigma(\text{data item} - \text{mean})^2}{n - 1} = \frac{14}{4 - 1} = \frac{14}{3} \approx 4.67$$

Step 6 The standard deviation, s, is the square root of the quotient in step 5.

$$s = \sqrt{\frac{\Sigma(x - \bar{x})^2}{n - 1}} = \sqrt{\frac{\Sigma(\text{data item} - \text{mean})^2}{n - 1}} \approx \sqrt{4.67} \approx 2.16$$

The standard deviation for the four oldest U.S. presidents is approximately 2.16 years.

 **CHECK POINT 3** Find the standard deviation for the group of data items in Check Point 2 on page 802. Round to two decimal places.

GREAT QUESTION!

Computing standard deviation involves lots of steps. Does my calculator have the capability of doing all the work for me?

Yes. Most scientific and graphing calculators allow you to enter a data set and then compute statistics, such as the mean and standard deviation, by choosing the appropriate commands. In practice, data sets are large and a calculator or other software is crucial for computing standard deviation efficiently. However, in this text, the data sets are small and we encourage you to compute standard deviation by hand. You will achieve a greater understanding of standard deviation by working through the steps and recalling the reasons behind them.

Example 4 illustrates that as the spread of data items increases, the standard deviation gets larger.

EXAMPLE 4 *Computing the Standard Deviation*

Find the standard deviation of the data items in each of the samples shown below.

Sample A	**Sample B**
17, 18, 19, 20, 21, 22, 23	5, 10, 15, 20, 25, 30, 35

SOLUTION

Begin by finding the mean for each sample.

Sample A:

$$\text{Mean} = \frac{17 + 18 + 19 + 20 + 21 + 22 + 23}{7} = \frac{140}{7} = 20$$

Sample B:

$$\text{Mean} = \frac{5 + 10 + 15 + 20 + 25 + 30 + 35}{7} = \frac{140}{7} = 20$$

Although both samples have the same mean, the data items in sample B are more spread out. Thus, we would expect sample B to have the greater standard deviation. The computation of the standard deviation requires that we find $\Sigma(\text{data item} - \text{mean})^2$, shown in **Table 12.14**.

TABLE 12.14 Computing Standard Deviations for Two Samples

Sample A			Sample B		
Data item x	**Deviation: data item − mean $x - \bar{x}$**	**(Deviation)²: (data item − mean)² $(x - \bar{x})^2$**	**Data item x**	**Deviation: data item − mean $x - \bar{x}$**	**(Deviation)²: (data item − mean)² $(x - \bar{x})$**
17	$17 - 20 = -3$	$(-3)^2 = 9$	5	$5 - 20 = -15$	$(-15)^2 = 225$
18	$18 - 20 = -2$	$(-2)^2 = 4$	10	$10 - 20 = -10$	$(-10)^2 = 100$
19	$19 - 20 = -1$	$(-1)^2 = 1$	15	$15 - 20 = -5$	$(-5)^2 = 25$
20	$20 - 20 = 0$	$0^2 = 0$	20	$20 - 20 = 0$	$0^2 = 0$
21	$21 - 20 = 1$	$1^2 = 1$	25	$25 - 20 = 5$	$5^2 = 25$
22	$22 - 20 = 2$	$2^2 = 4$	30	$30 - 20 = 10$	$10^2 = 100$
23	$23 - 20 = 3$	$3^2 = 9$	35	$35 - 20 = 15$	$15^2 = 225$
Totals:		$\Sigma(x - \bar{x})^2 = 28$			$\Sigma(x - \bar{x})^2 = 700$

Each sample contains seven data items, so we compute the standard deviation by dividing the sums in **Table 12.14**, 28 and 700, by 7 − 1, or 6. Then we take the square root of each quotient.

$$\text{Standard deviation} = \sqrt{\frac{\Sigma(x - \bar{x})^2}{n - 1}} = \sqrt{\frac{\Sigma(\text{data item} - \text{mean})^2}{n - 1}}$$

Sample A:

$$s = \sqrt{\frac{28}{6}} \approx 2.16$$

Sample B:

$$s = \sqrt{\frac{700}{6}} \approx 10.80$$

Sample A has a standard deviation of approximately 2.16 and sample B has a standard deviation of approximately 10.80. The data in sample B are more spread out than those in sample A.

CHECK POINT 4 Find the standard deviation of the data items in each of the samples shown below. Round to two decimal places.

Sample A: 73, 75, 77, 79, 81, 83

Sample B: 40, 44, 92, 94, 98, 100

Figure 12.9 illustrates four sets of data items organized in histograms. From left to right, the data items are

Figure 12.9(a): 4, 4, 4, 4, 4, 4, 4
Figure 12.9(b): 3, 3, 4, 4, 4, 5, 5
Figure 12.9(c): 3, 3, 3, 4, 5, 5, 5
Figure 12.9(d): 1, 1, 1, 4, 7, 7, 7.

Each data set has a mean of 4. However, as the spread of the data items increases, the standard deviation gets larger. Observe that when all the data items are the same, the standard deviation is 0.

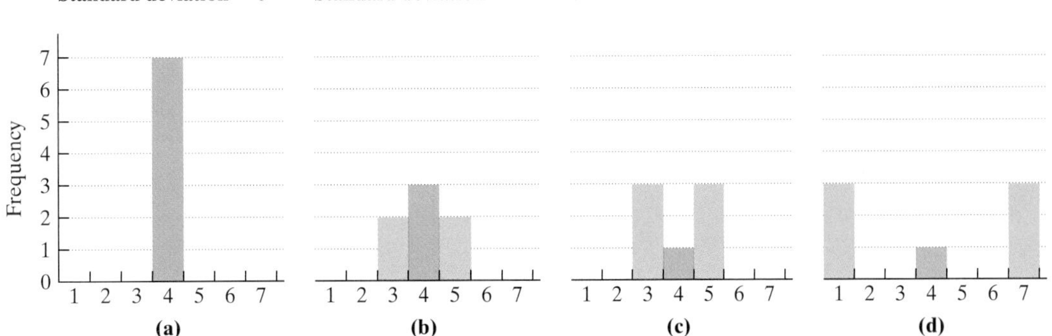

FIGURE 12.9 The standard deviation gets larger with increased dispersion among data items. In each case, the mean is 4.

EXAMPLE 5 Interpreting Standard Deviation

Two fifth-grade classes have nearly identical mean scores on an aptitude test, but one class has a standard deviation three times that of the other. All other factors being equal, which class is easier to teach, and why?

SOLUTION

The class with the smaller standard deviation is easier to teach because there is less variation among student aptitudes. Course work can be aimed at the average student without too much concern that the work will be too easy for some or too difficult for others. By contrast, the class with greater dispersion poses a greater challenge. By teaching to the average student, the students whose scores are significantly above the mean will be bored; students whose scores are significantly below the mean will be confused.

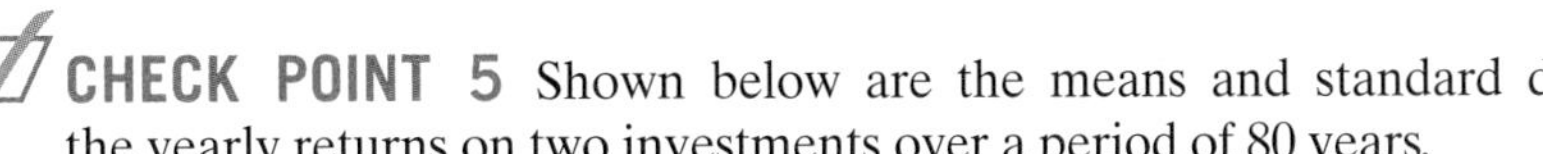

CHECK POINT 5 Shown below are the means and standard deviations of the yearly returns on two investments over a period of 80 years.

Investment	Mean Yearly Return	Standard Deviation
Small-Company Stocks	17.5%	33.3%
Large-Company Stocks	12.4%	20.4%

a. Use the means to determine which investment provided the greater yearly return.

b. Use the standard deviations to determine which investment had the greater risk. Explain your answer.

Concept and Vocabulary Check

Fill in each blank so that the resulting statement is true.

1. The difference between the highest and lowest data values in a data set is called the ______.
2. The formula

$$\sqrt{\frac{\Sigma(\text{data item} - \text{mean})^2}{n-1}}$$

gives the value of the ______________ for a data set.

In Exercises 3–5, determine whether each statement is true or false. If the statement is false, make the necessary change(s) to produce a true statement.

3. Measures of dispersion are used to describe the spread of data items in a data set. ______
4. The sum of the deviations from the mean for a data set is always zero. ______
5. Measures of dispersion get smaller as the spread of data items increases. ______

Exercise Set 12.3

Practice Exercises

In Exercises 1–6, find the range for each group of data items.

1. 1, 2, 3, 4, 5 **2.** 16, 17, 18, 19, 20

3. 7, 9, 9, 15 **4.** 11, 13, 14, 15, 17

5. 3, 3, 4, 4, 5, 5 **6.** 3, 3, 3, 4, 5, 5, 5

In Exercises 7–10, a group of data items and their mean are given.

a. *Find the deviation from the mean for each of the data items.*

b. *Find the sum of the deviations in part (a).*

7. 3, 5, 7, 12, 18, 27; Mean = 12

8. 84, 88, 90, 95, 98; Mean = 91

9. 29, 38, 48, 49, 53, 77; Mean = 49

10. 60, 60, 62, 65, 65, 65, 66, 67, 70, 70; Mean = 65

In Exercises 11–16, find a. the mean; b. the deviation from the mean for each data item; and c. the sum of the deviations in part (b).

11. 85, 95, 90, 85, 100

12. 94, 62, 88, 85, 91

13. 146, 153, 155, 160, 161

14. 150, 132, 144, 122

15. 2.25, 3.50, 2.75, 3.10, 1.90

16. 0.35, 0.37, 0.41, 0.39, 0.43

In Exercises 17–26, find the standard deviation for each group of data items. Round answers to two decimal places.

17. 1, 2, 3, 4, 5 **18.** 16, 17, 18, 19, 20

19. 7, 9, 9, 15 **20.** 11, 13, 14, 15, 17

21. 3, 3, 4, 4, 5, 5 **22.** 3, 3, 3, 4, 5, 5, 5

23. 1, 1, 1, 4, 7, 7, 7 **24.** 6, 6, 6, 6, 7, 7, 7, 4, 8, 3

25. 9, 5, 9, 5, 9, 5, 9, 5 **26.** 6, 10, 6, 10, 6, 10, 6, 10

In Exercises 27–28, compute the mean, range, and standard deviation for the data items in each of the three samples. Then describe one way in which the samples are alike and one way in which they are different.

27. Sample A: 6, 8, 10, 12, 14, 16, 18
Sample B: 6, 7, 8, 12, 16, 17, 18
Sample C: 6, 6, 6, 12, 18, 18, 18

28. Sample A: 8, 10, 12, 14, 16, 18, 20
Sample B: 8, 9, 10, 14, 18, 19, 20
Sample C: 8, 8, 8, 14, 20, 20, 20

Practice Plus

In Exercises 29–36, use each display of data items to find the standard deviation. Where necessary, round answers to two decimal places.

29.

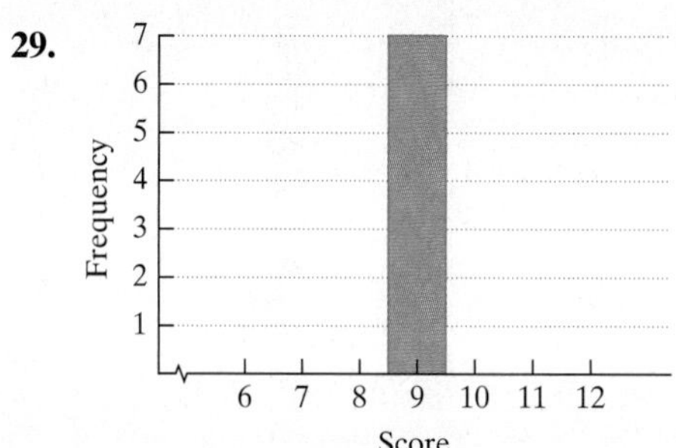

30.

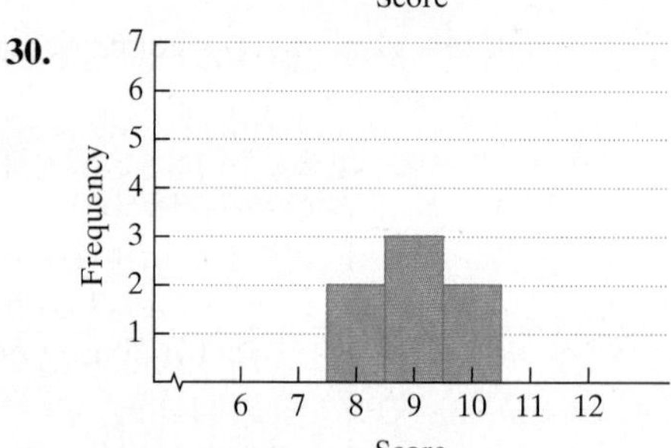

31.

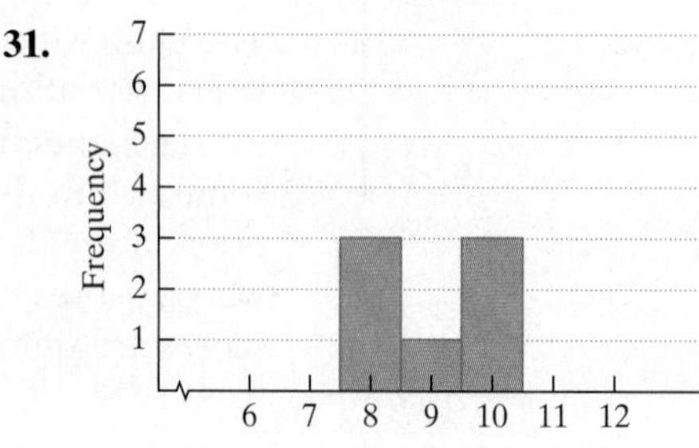

32.

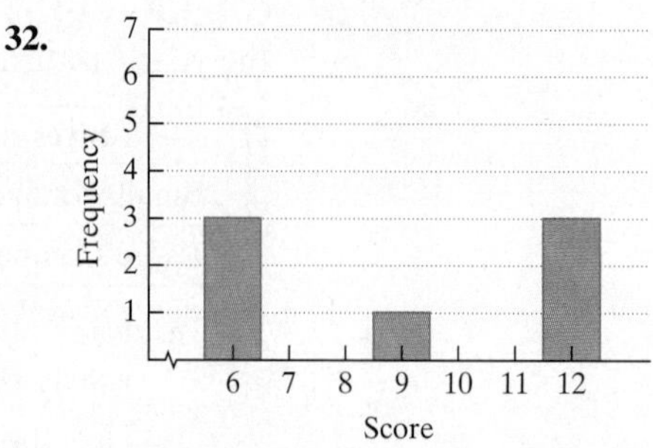

33.

Stems	Leaves
0	5
1	0 5
2	0 5

34.

Stems	Leaves
0	4 8
1	2 6
2	0

35.

Stems	Leaves
1	8 9 9 8 7 8
2	0 1 0 2

36.

Stems	Leaves
1	3 5 3 8 3 4
2	3 0 0 4

Application Exercises

37. The data sets give the number of platinum albums for the five male artists and the five female artists in the United States with the most platinum albums through a recent year. (Platinum albums sell one million units or more.)

MALE ARTISTS WITH THE MOST PLATINUM ALBUMS

Artist	Platinum Albums
Garth Brooks	145
Elvis Presley	104
Billy Joel	80
Michael Jackson	71
Elton John	65

FEMALE ARTISTS WITH THE MOST PLATINUM ALBUMS

Artist	Platinum Albums
Mariah Carey	64
Madonna	63
Barbra Streisand	61
Whitney Houston	54
Celine Dion	48

Source: RIAA

a. Without calculating, which data set has the greater mean number of platinum albums? Explain your answer.

b. Verify your conjecture from part (a) by calculating the mean number of platinum albums for each data set.

c. Without calculating, which data set has the greater standard deviation? Explain your answer.

d. Verify your conjecture from part (c) by calculating the standard deviation for each data set. Round answers to two decimal places.

38. The data sets give the ages of the first six U.S. presidents and the last six U.S. presidents (through Donald Trump).

AGE OF FIRST SIX U.S. PRESIDENTS AT INAUGURATION

President	Age
Washington	57
J. Adams	61
Jefferson	57
Madison	57
Monroe	58
J. Q. Adams	57

AGE OF LAST SIX U.S. PRESIDENTS AT INAUGURATION

President	Age
Reagan	69
G. H. W. Bush	64
Clinton	46
G. W. Bush	54
Obama	47
Trump	70

Source: Time Almanac

a. Without calculating, which set has the greater standard deviation? Explain your answer.

b. Verify your conjecture from part (b) by calculating the standard deviation for each data set. Round answers to two decimal places.

Explaining the Concepts

39. Describe how to find the range of a data set.

40. Describe why the range might not be the best measure of dispersion.

41. Describe how the standard deviation is computed.

42. Describe what the standard deviation reveals about a data set.

43. If a set of test scores has a standard deviation of zero, what does this mean about the scores?

44. Two classes took a statistics test. Both classes had a mean score of 73. The scores of class A had a standard deviation of 5 and those of class B had a standard deviation of 10. Discuss the difference between the two classes' performance on the test.

45. A sample of cereals indicates a mean potassium content per serving of 93 milligrams and a standard deviation of 2 milligrams. Write a description of what this means for a person who knows nothing about statistics.

46. Over a one-month period, stock A had a mean daily closing price of 124.7 and a standard deviation of 12.5. By contrast, stock B had a mean daily closing price of 78.2 and a standard deviation of 6.1. Which stock was more volatile? Explain your answer.

Critical Thinking Exercises

Make Sense? *In Exercises 47–50, determine whether each statement makes sense or does not make sense, and explain your reasoning.*

47. The mean can be misleading if you don't know the spread of data items.

48. The standard deviation for the weights of college students is greater than the standard deviation for the weights of 3-year-old children.

49. I'm working with data sets with different means and the same standard deviation.

50. I'm working with data sets with the same mean and different standard deviations.

51. Describe a situation in which a relatively large standard deviation is desirable.

52. If a set of test scores has a large range but a small standard deviation, describe what this means about students' performance on the test.

53. Use the data 1, 2, 3, 5, 6, 7. Without actually computing the standard deviation, which of the following best approximates the standard deviation?

a. 2 **b.** 6 **c.** 10 **d.** 20

54. Use the data 0, 1, 3, 4, 4, 6. Add 2 to each of the numbers. How does this affect the mean? How does this affect the standard deviation?

Group Exercises

55. As a follow-up to Group Exercise 79 on page 800, the group should reassemble and compute the standard deviation for each data set whose mean you previously determined. Does the standard deviation tell you anything new or interesting about the entire group or subgroups that you did not discover during the previous group activity?

56. Group members should consult a current almanac or the Internet and select intriguing data. The group's function is to use statistics to tell a story. Once "intriguing" data are identified, as a group

a. Summarize the data. Use words, frequency distributions, and graphic displays.

b. Compute measures of central tendency and dispersion, using these statistics to discuss the data.

12.4 The Normal Distribution

WHAT AM I SUPPOSED TO LEARN?

After studying this section, you should be able to:

1. Recognize characteristics of normal distributions.
2. Understand the 68–95–99.7 Rule.
3. Find scores at a specified standard deviation from the mean.
4. Use the 68–95–99.7 Rule.
5. Convert a data item to a z-score.
6. Understand percentiles and quartiles.
7. Use and interpret margins of error.
8. Recognize distributions that are not normal.

OUR HEIGHTS ARE ON THE RISE! IN ONE million B.C., the mean height for men was 4 feet 6 inches. The mean height for women was 4 feet 2 inches. Because of improved diets and medical care, the mean height for men is now 5 feet 10 inches and for women it is 5 feet 5 inches. Mean adult heights are expected to plateau by 2050.

Mean Adult Heights

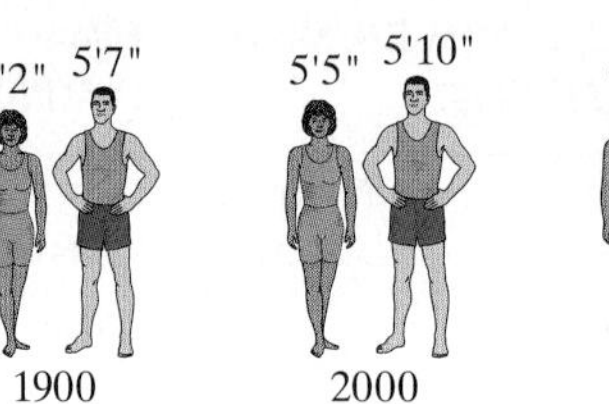

Source: National Center for Health Statistics

1 Recognize characteristics of normal distributions.

Suppose that a researcher selects a random sample of 100 adult men, measures their heights, and constructs a histogram for the data. The graph is shown in **Figure 12.10(a)** below. **Figure 12.10(b)** and **(c)** illustrate what happens as the sample size increases. In **Figure 12.10(c)**, if you were to fold the graph down the middle, the left side would fit the right side. As we move out from the middle, the heights of the bars are the same to the left and right. Such a histogram is called **symmetric**. As the sample size increases, so does the graph's symmetry. If it were possible to measure the heights of all adult males, the entire population, the histogram would approach what is called the **normal distribution**, shown in **Figure 12.10(d)**. This distribution is also called the **bell curve** or the **Gaussian distribution**, named for the German mathematician Carl Friedrich Gauss (1777–1855).

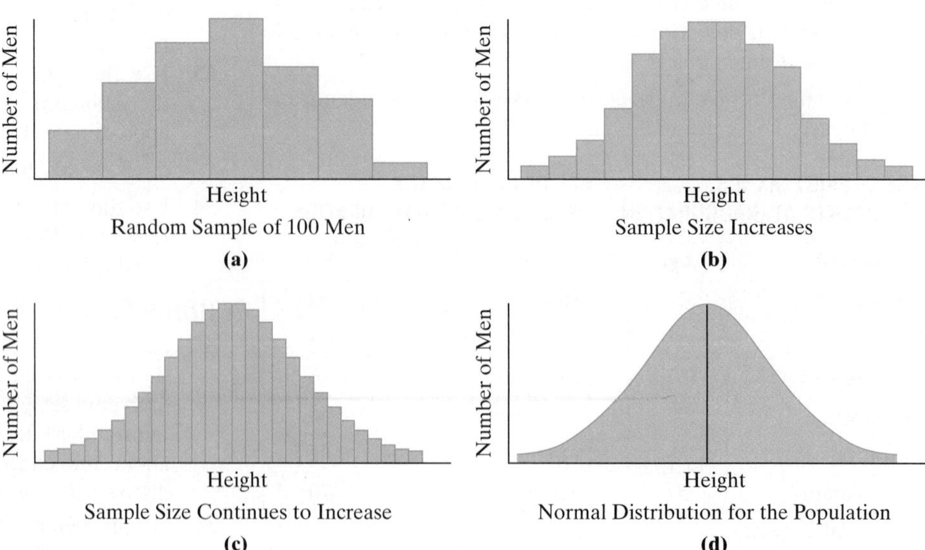

FIGURE 12.10 Heights of adult males

Figure 12.10(d) illustrates that the normal distribution is bell shaped and symmetric about a vertical line through its center. Furthermore, **the mean, median, and mode of a normal distribution are all equal** and located at the center of the distribution.

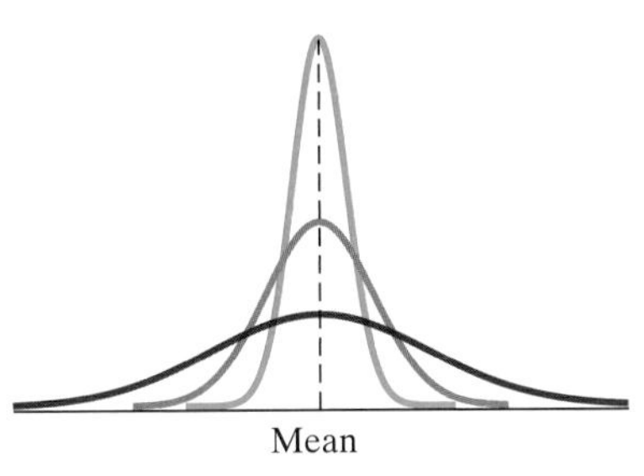

FIGURE 12.11

The shape of the normal distribution depends on the mean and the standard deviation. **Figure 12.11** illustrates three normal distributions with the same mean, but different standard deviations. As the standard deviation increases, the distribution becomes more dispersed, or spread out, but retains its symmetric bell shape.

The normal distribution provides a wonderful model for all kinds of phenomena because many sets of data items closely resemble this population distribution. Examples include heights and weights of adult males, intelligence quotients, SAT scores, prices paid for a new car model, and life spans of light bulbs. In these distributions, the data items tend to cluster around the mean. The more an item differs from the mean, the less likely it is to occur.

The normal distribution is used to make predictions about an entire population using data from a sample. In this section, we focus on the characteristics and applications of the normal distribution.

2 *Understand the 68–95–99.7 Rule.*

The Standard Deviation and *z*-Scores in Normal Distributions

The standard deviation plays a crucial role in the normal distribution, summarized by the **68–95–99.7 Rule**. This rule is illustrated in **Figure 12.12**.

THE 68–95–99.7 RULE FOR THE NORMAL DISTRIBUTION

1. Approximately 68% of the data items fall within 1 standard deviation of the mean (in both directions).
2. Approximately 95% of the data items fall within 2 standard deviations of the mean.
3. Approximately 99.7% of the data items fall within 3 standard deviations of the mean.

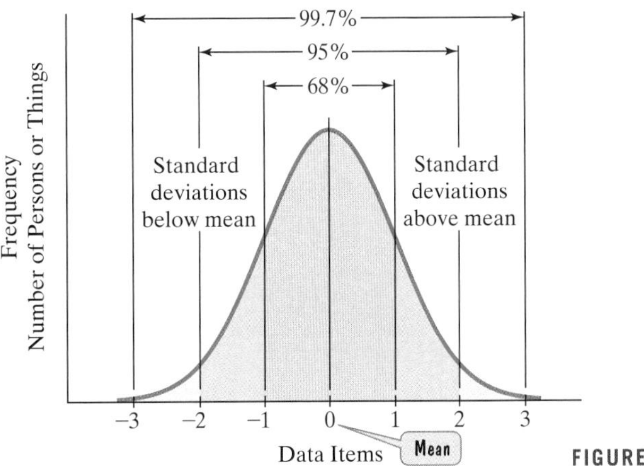

FIGURE 12.12

Well-Worn Steps and the Normal Distribution

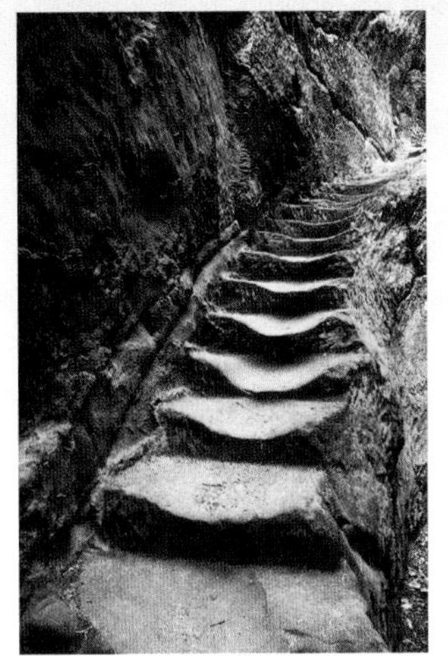

These ancient steps each take on the shape of a normal distribution when the picture is viewed upside down. The center of each step is more worn than the outer edges. The greatest number of people have walked in the center, making this the mean, median, and mode for where people have walked.

Figure 12.12 illustrates that a very small percentage of the data in a normal distribution lies more than 3 standard deviations above or below the mean. As we move from the mean, the curve falls rapidly, and then more and more gradually, toward the horizontal axis. The tails of the curve approach, but never touch, the horizontal axis, although they are quite close to the axis at 3 standard deviations from the mean. The range of the normal distribution is infinite. No matter how far out from the mean we move, there is always the probability (although very small) of a data item occurring even farther out.

3 *Find scores at a specified standard deviation from the mean.*

EXAMPLE 1 Finding Scores at a Specified Standard Deviation from the Mean

Male adult heights in North America are approximately normally distributed with a mean of 70 inches and a standard deviation of 4 inches. Find the height that is

a. 2 standard deviations above the mean.

b. 3 standard deviations below the mean.

SOLUTION

a. First, let us find the height that is 2 standard deviations above the mean.

$$\text{Height} = \text{mean} + 2 \cdot \text{standard deviation}$$
$$= 70 + 2 \cdot 4 = 70 + 8 = 78$$

A height of 78 inches is 2 standard deviations above the mean.

b. Next, let us find the height that is 3 standard deviations below the mean.

$$\text{Height} = \text{mean} - 3 \cdot \text{standard deviation}$$
$$= 70 - 3 \cdot 4 = 70 - 12 = 58$$

A height of 58 inches is 3 standard deviations below the mean.

The distribution of male adult heights in North America is illustrated as a normal distribution in **Figure 12.13**.

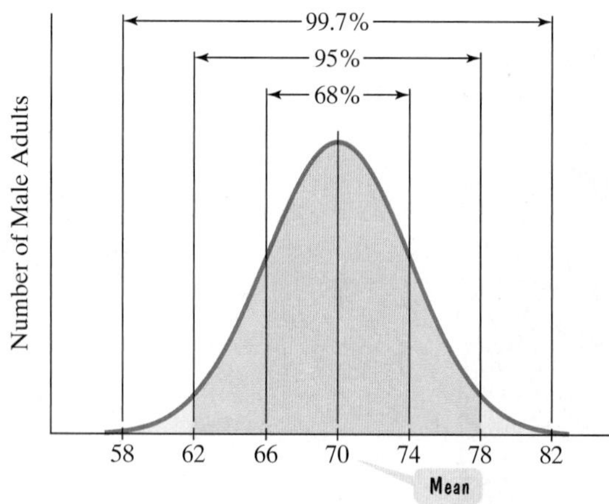

FIGURE 12.13

CHECK POINT 1 Female adult heights in North America are approximately normally distributed with a mean of 65 inches and a standard deviation of 3.5 inches. Find the height that is

a. 3 standard deviations above the mean.

b. 2 standard deviations below the mean.

4 *Use the 68–95–99.7 Rule.*

EXAMPLE 2 Using the 68–95–99.7 Rule

Use the distribution of male adult heights in **Figure 12.13** on the previous page to find the percentage of men in North America with heights

a. between 66 inches and 74 inches. **b.** between 70 inches and 74 inches.

c. above 78 inches.

SOLUTION

a. The 68–95–99.7 Rule states that approximately 68% of the data items fall within 1 standard deviation, 4, of the mean, 70.

$$\text{mean} - 1 \cdot \text{standard deviation} = 70 - 1 \cdot 4 = 70 - 4 = 66$$

$$\text{mean} + 1 \cdot \text{standard deviation} = 70 + 1 \cdot 4 = 70 + 4 = 74$$

Figure 12.13 shows that 68% of male adults have heights between 66 inches and 74 inches.

b. The percentage of men with heights between 70 inches and 74 inches is not given directly in **Figure 12.13**. Because of the distribution's symmetry, the percentage with heights between 66 inches and 70 inches is the same as the percentage with heights between 70 and 74 inches. **Figure 12.14** indicates that 68% have heights between 66 inches and 74 inches. Thus, half of 68%, or 34%, of men have heights between 70 inches and 74 inches.

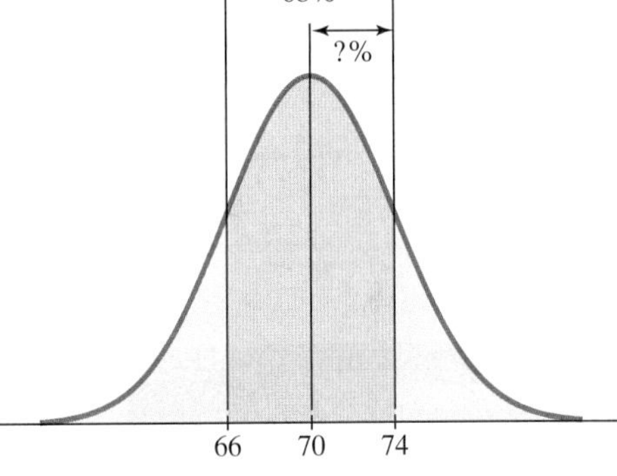

FIGURE 12.14 What percentage have heights between 70 inches and 74 inches?

c. The percentage of men with heights above 78 inches is not given directly in **Figure 12.13**. A height of 78 inches is 2 standard deviations, $2 \cdot 4$, or 8 inches, above the mean, 70 inches. The 68–95–99.7 Rule states that approximately 95% of the data items fall within 2 standard deviations of the mean. Thus, approximately 100% − 95%, or 5%, of the data items are farther than 2 standard deviations from the mean. The 5% of the data items are represented by the two shaded green regions in **Figure 12.15**. Because of the distribution's symmetry, half of 5%, or 2.5%, of the data items are more than 2 standard deviations above the mean. This means that 2.5% of men have heights above 78 inches.

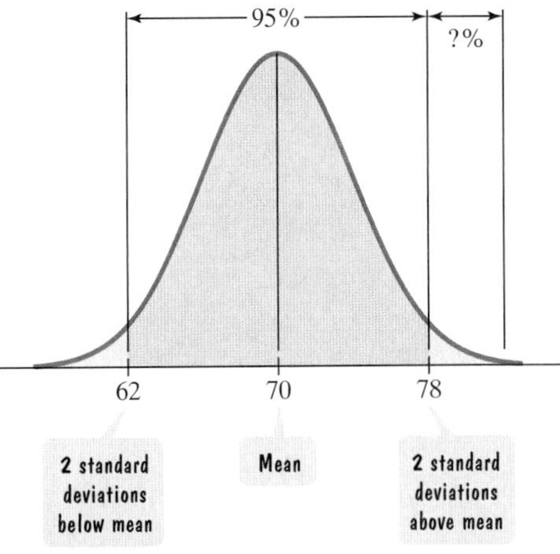

FIGURE 12.15 What percentage have heights above 78 inches?

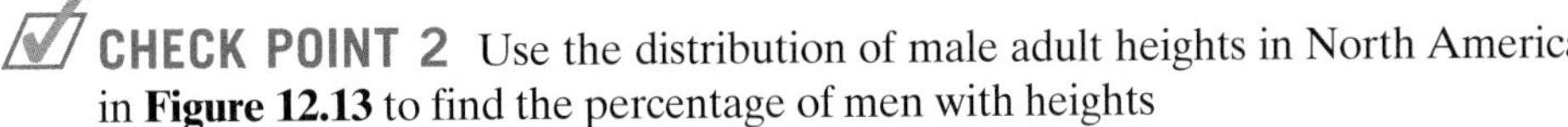

CHECK POINT 2 Use the distribution of male adult heights in North America in **Figure 12.13** to find the percentage of men with heights

a. between 62 inches and 78 inches.

b. between 70 inches and 78 inches. **c.** above 74 inches.

Because the normal distribution of male adult heights in North America has a mean of 70 inches and a standard deviation of 4 inches, a height of 78 inches lies 2 standard deviations above the mean. In a normal distribution, a ***z*-score** describes how many standard deviations a particular data item lies above or below the mean. Thus, the *z*-score for the data item 78 is 2.

The following formula can be used to express a data item in a normal distribution as a *z*-score:

5 *Convert a data item to a z-score.*

> **COMPUTING *z*-SCORES**
>
> A *z*-score describes how many standard deviations a data item in a normal distribution lies above or below the mean. The *z*-score can be obtained using
>
> $$z\text{-score} = \frac{\text{data item} - \text{mean}}{\text{standard deviation}}.$$
>
> Data items above the mean have positive *z*-scores. Data items below the mean have negative *z*-scores. The *z*-score for the mean is 0.

EXAMPLE 3 *Computing z-Scores*

The mean weight of newborn infants is 7 pounds, and the standard deviation is 0.8 pound. The weights of newborn infants are normally distributed. Find the *z*-score for a weight of

a. 9 pounds. **b.** 7 pounds. **c.** 6 pounds.

SOLUTION

We compute the *z*-score for each weight by using the *z*-score formula. The mean is 7 and the standard deviation is 0.8.

a. The *z*-score for a weight of 9 pounds, written z_9, is

$$z_9 = \frac{\text{data item} - \text{mean}}{\text{standard deviation}} = \frac{9 - 7}{0.8} = \frac{2}{0.8} = 2.5.$$

The *z*-score of a data item greater than the mean is always positive. A 9-pound infant is a chubby little tyke, with a weight that is 2.5 standard deviations above the mean.

b. The *z*-score for a weight of 7 pounds is

$$z_7 = \frac{\text{data item} - \text{mean}}{\text{standard deviation}} = \frac{7 - 7}{0.8} = \frac{0}{0.8} = 0.$$

The *z*-score for the mean is always 0. A 7-pound infant is right at the mean, deviating 0 pounds above or below it.

c. The *z*-score for a weight of 6 pounds is

$$z_6 = \frac{\text{data item} - \text{mean}}{\text{standard deviation}} = \frac{6 - 7}{0.8} = \frac{-1}{0.8} = -1.25.$$

The *z*-score of a data item less than the mean is always negative. A 6-pound infant's weight is 1.25 standard deviations below the mean.

Figure 12.16 shows the normal distribution of weights of newborn infants. The horizontal axis is labeled in terms of weights and z-scores.

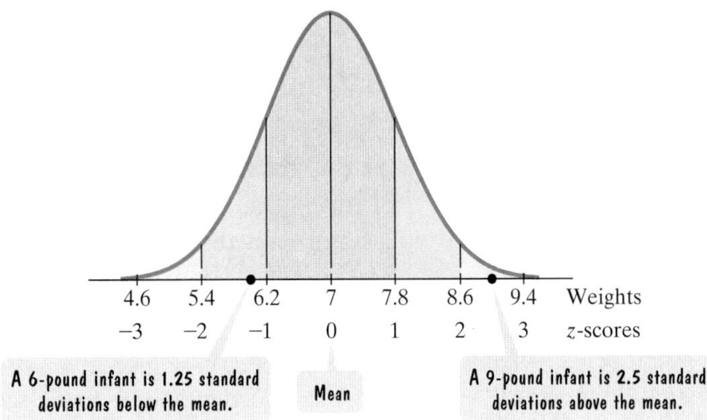

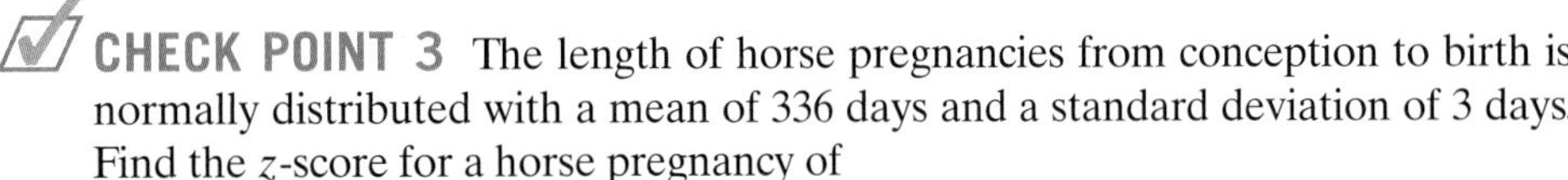

FIGURE 12.16 Infants' weights are normally distributed

CHECK POINT 3 The length of horse pregnancies from conception to birth is normally distributed with a mean of 336 days and a standard deviation of 3 days. Find the z-score for a horse pregnancy of

a. 342 days. **b.** 336 days. **c.** 333 days.

In Example 4, we consider two normally distributed sets of test scores, in which a higher score generally indicates a better result. To compare scores on two different tests in relation to the mean on each test, we can use z-scores. The better score is the item with the greater z-score.

EXAMPLE 4 *Using and Interpreting z-Scores*

A student scores 70 on an arithmetic test and 66 on a vocabulary test. The scores for both tests are normally distributed. The arithmetic test has a mean of 60 and a standard deviation of 20. The vocabulary test has a mean of 60 and a standard deviation of 2. On which test did the student have the better score?

SOLUTION

To answer the question, we need to find the student's z-score on each test, using

$$z = \frac{\text{data item} - \text{mean}}{\text{standard deviation}}.$$

The arithmetic test has a mean of 60 and a standard deviation of 20.

$$z\text{-score for } 70 = z_{70} = \frac{70 - 60}{20} = \frac{10}{20} = 0.5$$

The vocabulary test has a mean of 60 and a standard deviation of 2.

$$z\text{-score for } 66 = z_{66} = \frac{66 - 60}{2} = \frac{6}{2} = 3$$

The arithmetic score, 70, is half a standard deviation above the mean, whereas the vocabulary score, 66, is 3 standard deviations above the mean. The student did much better than the mean on the vocabulary test.

CHECK POINT 4 The SAT (Scholastic Aptitude Test) has a mean of 500 and a standard deviation of 100. The ACT (American College Test) has a mean of 18 and a standard deviation of 6. Both tests measure the same kind of ability, with scores that are normally distributed. Suppose that you score 550 on the SAT and 24 on the ACT. On which test did you have the better score?

Blitzer Bonus

The IQ Controversy

Is intelligence something we are born with or is it a quality that can be manipulated through education? Can it be measured accurately and is IQ the way to measure it? There are no clear answers to these questions.

In a study by Carolyn Bird (*Pygmalion in the Classroom*), a group of third-grade teachers was told that they had classes of students with IQs well above the mean. These classes made incredible progress throughout the year. In reality, these were not gifted kids, but, rather, a random sample of all third-graders. It was the teachers' expectations, and not the IQs of the students, that resulted in increased performance.

EXAMPLE 5 *Understanding z-Scores*

Intelligence quotients (IQs) on the Stanford-Binet intelligence test are normally distributed with a mean of 100 and a standard deviation of 16.

a. What is the IQ corresponding to a z-score of -1.5?

b. Mensa is a group of people with high IQs whose members have z-scores of 2.05 or greater on the Stanford-Binet intelligence test. What is the IQ corresponding to a z-score of 2.05?

SOLUTION

a. We begin with the IQ corresponding to a z-score of -1.5. The negative sign in -1.5 tells us that the IQ is $1\frac{1}{2}$ standard deviations below the mean.

$$\begin{aligned}\text{IQ} &= \text{mean} - 1.5 \cdot \text{standard deviation}\\ &= 100 - 1.5(16) = 100 - 24 = 76\end{aligned}$$

The IQ corresponding to a z-score of -1.5 is 76.

b. Next, we find the IQ corresponding to a z-score of 2.05. The positive sign implied in 2.05 tells us that the IQ is 2.05 standard deviations above the mean.

$$\begin{aligned}\text{IQ} &= \text{mean} + 2.05 \cdot \text{standard deviation}\\ &= 100 + 2.05(16) = 100 + 32.8 = 132.8\end{aligned}$$

The IQ corresponding to a z-score of 2.05 is 132.8. (An IQ score of at least 133 is required to join Mensa.)

CHECK POINT 5 Use the information in Example 5 to find the IQ corresponding to a z-score of

a. -2.25.

b. 1.75.

6 *Understand percentiles and quartiles.*

Percentiles and Quartiles

A z-score measures a data item's position in a normal distribution. Another measure of a data item's position is its **percentile**. Percentiles are often associated with scores on standardized tests. If a score is in the 45th percentile, this means that 45% of the scores are less than this score. If a score is in the 95th percentile, this indicates that 95% of the scores are less than this score.

PERCENTILES

If n% of the items in a distribution are less than a particular data item, we say that the data item is in the ***n*th percentile** of the distribution.

EXAMPLE 6 Interpreting Percentile

The cutoff IQ score for Mensa membership, 132.8, is in the 98th percentile. What does this mean?

SOLUTION

Because 132.8 is in the 98th percentile, this means that 98% of IQ scores fall below 132.8. Caution: A score in the 98th percentile does *not* mean that 98% of the answers are correct. Nor does it mean that the score was 98%.

CHECK POINT 6 A student scored in the 75th percentile on the SAT. What does this mean?

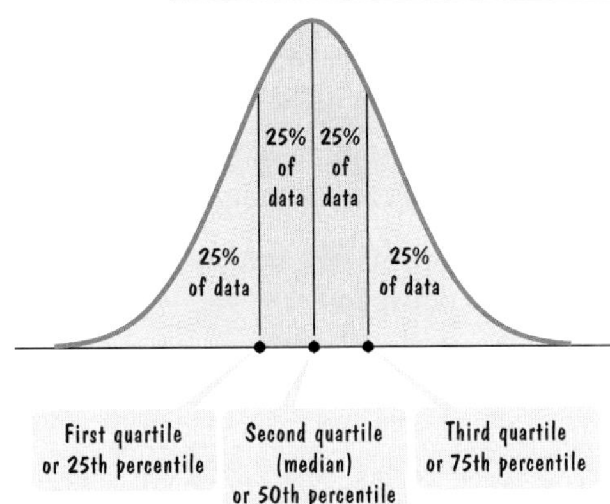

FIGURE 12.17 Quartiles

Three commonly encountered percentiles are the *quartiles.* **Quartiles** divide data sets into four equal parts. The 25th percentile is the **first quartile**: 25% of the data fall below the first quartile. The 50th percentile is the **second quartile**: 50% of the data fall below the second quartile, so the second quartile is equivalent to the median. The 75th percentile is the **third quartile**: 75% of the data fall below the third quartile. **Figure 12.17** illustrates the concept of quartiles for the normal distribution.

7 *Use and interpret margins of error.*

Polls and Margins of Error

What activities do you dread? Reading math textbooks with an infinity of framed cows on the cover? (Be kind!) No, that's not America's most-dreaded activity. In a random sample of 1000 U.S. adults, 46% of those questioned responded, "Public speaking." The problem is that this is a single random sample. Do 46% of adults in the entire U.S. population dread public speaking?

Statisticians use properties of the normal distribution to estimate the probability that a result obtained from a single sample reflects what is truly happening in the population. If you look at the results of a poll like the one shown in **Figure 12.18**, you will observe that a *margin of error* is reported. Surveys and opinion polls often give a margin of error. Let's use our understanding of the normal distribution to see how to calculate and interpret margins of error.

Activities U.S. Adults Say They Dread

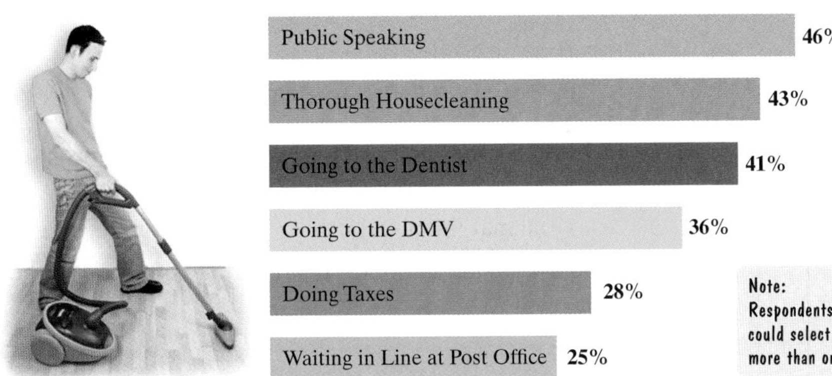

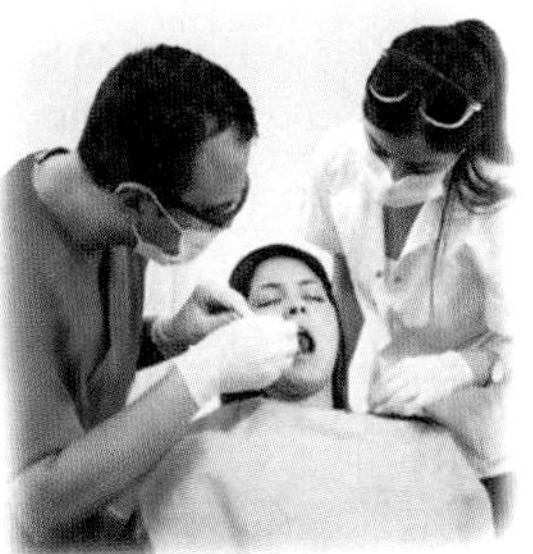

FIGURE 12.18 *Source:* TNS survey of 1000 adults, March 2010

Note the margin of error.

Suppose that $p\%$ of the population of U.S. adults dread public speaking. Instead of taking only one random sample of 1000 adults, we repeat the process of selecting a random sample of 1000 adults hundreds of times. Then, we calculate the percentage of adults for each sample who dread public speaking. With random sampling, we expect to find the percentage in many of the samples close to $p\%$, with relatively few samples having percentages far from $p\%$. **Figure 12.19** shows that the percentages of U.S adults from the hundreds of samples can be modeled by a normal distribution. The mean of this distribution is the actual population percent, $p\%$, and is the most frequent result from the samples.

Mathematicians have shown that the standard deviation of a normal distribution of samples like the one in **Figure 12.19** is approximately $\frac{1}{2\sqrt{n}} \times 100\%$, where n is the sample size. Using the 68–95–99.7 Rule, approximately 95% of the samples have a percentage within 2 standard deviations of the true population percentage, $p\%$:

$$2 \text{ standard deviations} = 2 \cdot \frac{1}{2\sqrt{n}} \times 100\% = \frac{1}{\sqrt{n}} \times 100\%.$$

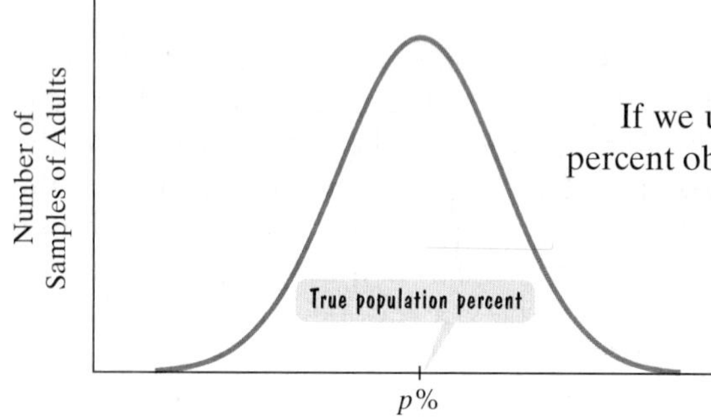

FIGURE 12.19 Percentage of U.S. adults who dread public speaking.

If we use a single random sample of size n, there is a 95% probability that the percent obtained will lie within two standard deviations, or $\frac{1}{\sqrt{n}} \times 100\%$, of the true population percent. We can be 95% confident that the true population percent lies between

$$\text{the sample percent} - \frac{1}{\sqrt{n}} \times 100\%$$

and

$$\text{the sample percent} + \frac{1}{\sqrt{n}} \times 100\%.$$

We call $\pm \frac{1}{\sqrt{n}} \times 100\%$ the **margin of error**.

MARGIN OF ERROR IN A SURVEY

If a statistic is obtained from a random sample of size n, there is a 95% probability that it lies within $\frac{1}{\sqrt{n}} \times 100\%$ of the true population percent, where $\pm\frac{1}{\sqrt{n}} \times 100\%$ is called the **margin of error**.

TABLE 12.15 Activities U.S. Adults Dread

Activity	Percentage Who Dread the Activity
Public speaking	46%
Thorough house-cleaning	43%
Going to the dentist	41%
Going to the DMV	36%
Doing taxes	28%
Waiting in line at the post office	25%

Source: TNS survey of 1000 adults, March 2010

EXAMPLE 7 *Using and Interpreting Margin of Error*

Table 12.15 shows that in a random sample of 1000 U.S adults, 46% of those questioned said that they dread public speaking.

a. Verify the margin of error that was given for this survey.

b. Write a statement about the percentage of adults in the U.S. population who dread public speaking.

SOLUTION

a. The sample size is $n = 1000$. The margin of error is

$$\pm\frac{1}{\sqrt{n}} \times 100\% = \pm\frac{1}{\sqrt{1000}} \times 100\% \approx \pm 0.032 \times 100\% = \pm 3.2\%.$$

b. There is a 95% probability that the true population percentage lies between

$$\text{the sample percent} - \frac{1}{\sqrt{n}} \times 100\% = 46\% - 3.2\% = 42.8\%$$

and

$$\text{the sample percent} + \frac{1}{\sqrt{n}} \times 100\% = 46\% + 3.2\% = 49.2\%.$$

We can be 95% confident that between 42.8% and 49.2% of all U.S adults dread public speaking.

Blitzer Bonus

A Caveat Giving a True Picture of a Poll's Accuracy

Unlike the precise calculation of a poll's margin of error, certain polling imperfections cannot be determined exactly. One problem is that people do not always respond to polls honestly and accurately. Some people are embarrassed to say "undecided," so they make up an answer. Other people may try to respond to questions in the way they think will make the pollster happy, just to be "nice." Perhaps the following caveat, applied to the poll in Example 7, would give the public a truer picture of its accuracy:

> The poll results are 42.8% to 49.2% at the 95% confidence level, but it's only under ideal conditions that we can be 95% confident that the true numbers are within 3.2% of the poll's results. The true error span is probably greater than 3.2% due to limitations that are inherent in this and every poll, but, unfortunately, this additional error amount cannot be calculated precisely. Warning: Five percent of the time—that's one time out of 20—the error will be greater than 3.2%. We remind readers of the poll that things occurring "only" 5% of the time do, indeed, happen.

We suspect that the public would tire of hearing this.

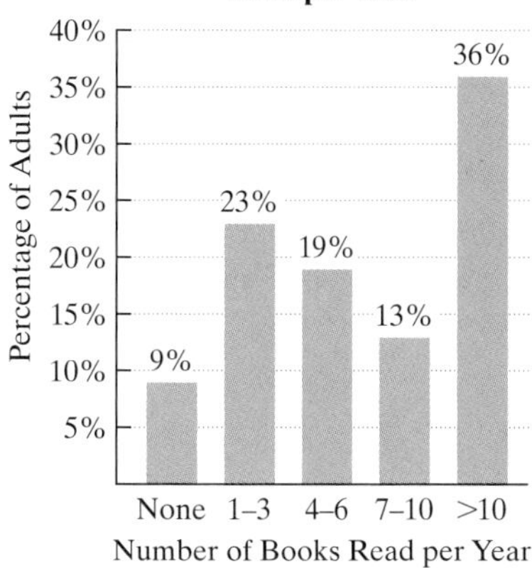

FIGURE 12.20
Source: Harris Poll of 2513 U.S. adults ages 18 and older

CHECK POINT 7 A Harris Poll of 2513 U.S. adults ages 18 and older asked the question

How many books do you typically read in a year?

The results of the poll are shown in **Figure 12.20**.

a. Find the margin of error for this survey. Round to the nearest tenth of a percent.

b. Write a statement about the percentage of U.S. adults who read more than ten books per year.

c. Why might some people not respond honestly and accurately to the question in this poll?

8 *Recognize distributions that are not normal.*

Other Kinds of Distributions

Although the normal distribution is the most important of all distributions in terms of analyzing data, not all data can be approximated by this symmetric distribution with its mean, median, and mode all having the same value.

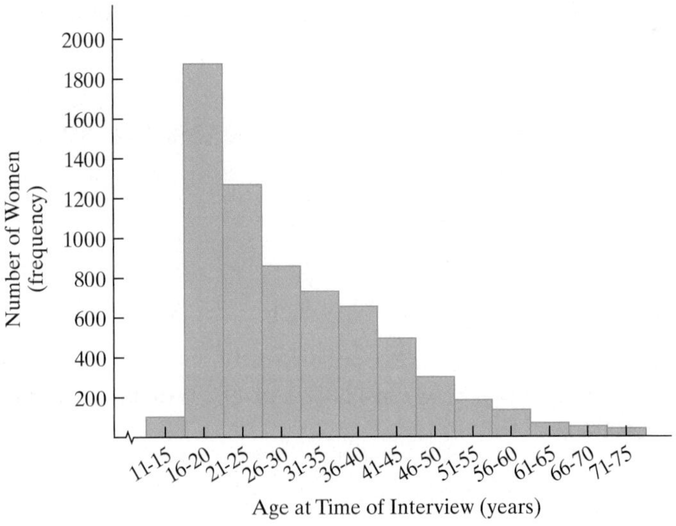

FIGURE 12.21 Histogram of the ages of females interviewed by Kinsey and his associates.

The histogram in **Figure 12.21** represents the frequencies of the ages of women interviewed by Kinsey and his colleagues in their study of female sexual behavior. This distribution is not symmetric. The greatest frequency of women interviewed was in the 16–20 age range. The bars get shorter and shorter after this. The shorter bars fall on the right, indicating that relatively few older women were included in Kinsey's interviews.

In our discussion of measures of central tendency, we mentioned that the median, rather than the mean, is used to summarize income. **Figure 12.22** illustrates the population distribution of weekly earnings in the United States. There is no upper limit on weekly earnings. The relatively few people with very high weekly incomes tend to pull the mean income to a value greater than the median. The most frequent income, the mode, occurs toward the low end of the data items. The mean, median, and mode do not have the same value, and a normal distribution is not an appropriate model for describing weekly earnings in the United States.

The distribution in **Figure 12.22** is called a *skewed distribution.* A distribution of data is **skewed** if a large number of data items are piled up at one end or the other, with a "tail" at the opposite end. In the distribution of weekly earnings in **Figure 12.22**, the tail is to the right. Such a distribution is said to be **skewed to the right**.

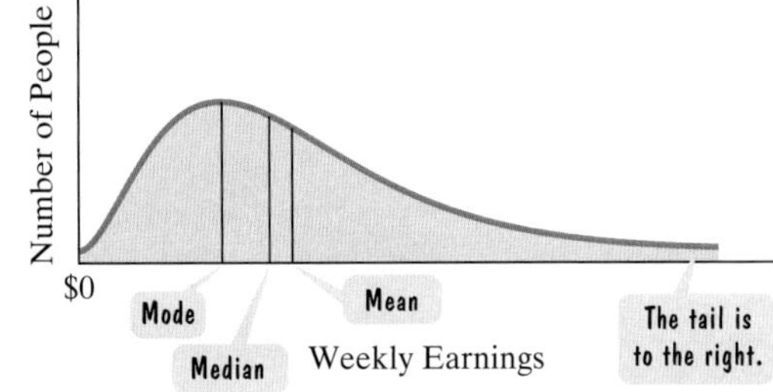

FIGURE 12.22 Skewed to the right

By contrast to the distribution of weekly earnings, the distribution in **Figure 12.23** has more data items at the high end of the scale than at the low end. The tail of this distribution is to the left. The distribution is said to be **skewed to the left**. In many colleges, an example of a distribution skewed to the left is based on the student ratings of faculty teaching performance. Most professors are given rather high ratings, while only a few are rated as terrible. These low ratings pull the value of the mean lower than the median.

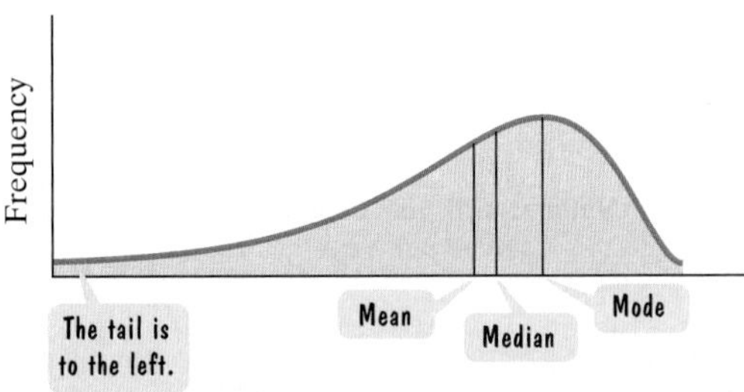

FIGURE 12.23 Skewed to the left

GREAT QUESTION!

What's the bottom line on the relationship between the mean and the median for skewed distributions?

If the data are skewed to the right, the mean is greater than the median. If the data are skewed to the left, the mean is less than the median.

Concept and Vocabulary Check

Fill in each blank so that the resulting statement is true.

1. In a normal distribution, approximately ____% of the data items fall within 1 standard deviation of the mean, approximately ____% of the data items fall within 2 standard deviations of the mean, and approximately _____% of the data items fall within 3 standard deviations of the mean.
2. A z-score describes how many standard deviations a data item in a normal distribution lies above or below the ______.
3. If $n\%$ of the items in a distribution are less than a particular data item, we say that the data item is in the nth _________ of the distribution.
4. If a statistic is obtained from a random sample of size n, there is a 95% probability that it lies within $\frac{1}{\sqrt{n}} \times 100\%$ of the true population percent, where $\pm \frac{1}{\sqrt{n}} \times 100\%$ is called the ____________.

In Exercises 5–8, determine whether each statement is true or false. If the statement is false, make the necessary change(s) to produce a true statement.

5. The mean, median, and mode of a normal distribution are all equal. ______

6. In a normal distribution, the z-score for the mean is 0. ______

7. The z-score for a data item in a normal distribution is obtained using

$$z\text{-score} = \frac{\text{data item} - \text{standard deviation}}{\text{mean}}.$$ ______

8. A score in the 50th percentile on a standardized test is the median. ______

Exercise Set 12.4

Practice and Application Exercises

The scores on a test are normally distributed with a mean of 100 and a standard deviation of 20. In Exercises 1–10, find the score that is

1. 1 standard deviation above the mean.

2. 2 standard deviations above the mean.

3. 3 standard deviations above the mean.

4. $1\frac{1}{2}$ standard deviations above the mean.

5. $2\frac{1}{2}$ standard deviations above the mean.

6. 1 standard deviation below the mean.

7. 2 standard deviations below the mean.

8. 3 standard deviations below the mean.

9. one-half a standard deviation below the mean.

10. $2\frac{1}{2}$ standard deviations below the mean.

Not everyone pays the same price for the same model of a car. The figure illustrates a normal distribution for the prices paid for a particular model of a new car. The mean is \$17,000 and the standard deviation is \$500.

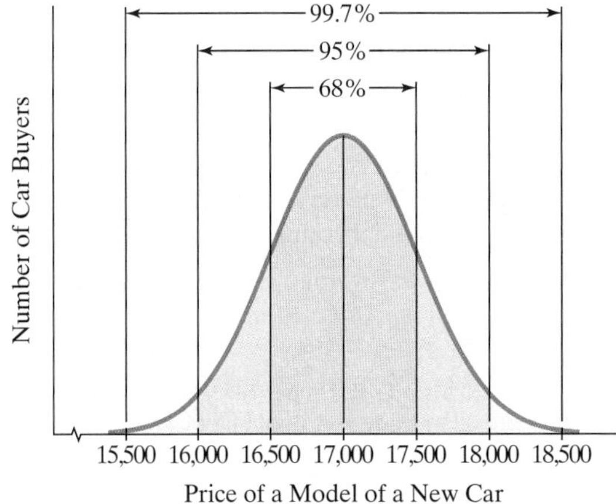

In Exercises 11–22, use the 68–95–99.7 Rule, illustrated in the figure, to find the percentage of buyers who paid

11. between \$16,500 and \$17,500.

12. between \$16,000 and \$18,000.

13. between \$17,000 and \$17,500.

14. between \$17,000 and \$18,000.

15. between \$16,000 and \$17,000.

16. between \$16,500 and \$17,000.

17. between \$15,500 and \$17,000.

18. between \$17,000 and \$18,500.

19. more than \$17,500.

20. more than \$18,000.

21. less than \$16,000.

22. less than \$16,500.

Intelligence quotients (IQs) on the Stanford-Binet intelligence test are normally distributed with a mean of 100 and a standard deviation of 16. In Exercises 23–32, use the 68–95–99.7 Rule to find the percentage of people with IQs

23. between 68 and 132.

24. between 84 and 116.

25. between 68 and 100.

26. between 84 and 100.

27. above 116.

28. above 132.

29. below 68.

30. below 84.

31. above 148.

32. below 52.

A set of data items is normally distributed with a mean of 60 and a standard deviation of 8. In Exercises 33–48, convert each data item to a z-score.

33. 68 **34.** 76 **35.** 84

36. 92 **37.** 64 **38.** 72

39. 74 **40.** 78 **41.** 60

42. 100 **43.** 52 **44.** 44

45. 48 **46.** 40 **47.** 34

48. 30

Scores on a dental anxiety scale range from 0 (no anxiety) to 20 (extreme anxiety). The scores are normally distributed with a mean of 11 and a standard deviation of 4. In Exercises 49–56, find the z-score for the given score on this dental anxiety scale.

49. 17 **50.** 18

51. 20 **52.** 12

53. 6 **54.** 8

55. 5 **56.** 1

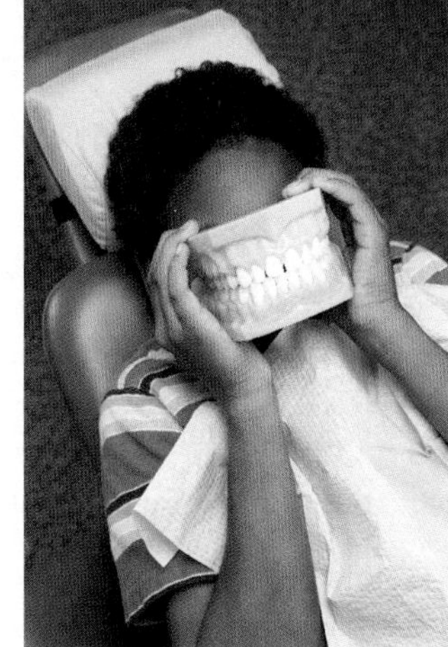

Intelligence quotients on the Stanford-Binet intelligence test are normally distributed with a mean of 100 and a standard deviation of 16. Intelligence quotients on the Wechsler intelligence test are normally distributed with a mean of 100 and a standard deviation of 15. Use this information to solve Exercises 57–58.

57. Use z-scores to determine which person has the higher IQ: an individual who scores 128 on the Stanford-Binet or an individual who scores 127 on the Wechsler.

58. Use z-scores to determine which person has the higher IQ: an individual who scores 150 on the Stanford-Binet or an individual who scores 148 on the Wechsler.

A set of data items is normally distributed with a mean of 400 and a standard deviation of 50. In Exercises 59–66, find the data item in this distribution that corresponds to the given z-score.

59. $z = 2$

60. $z = 3$

61. $z = 1.5$

62. $z = 2.5$

63. $z = -3$

64. $z = -2$

65. $z = -2.5$

66. $z = -1.5$

67. Reducing Gun Violence The data in the bar graph are from a random sample of 814 American adults. The graph shows four proposals to reduce gun violence in the United States and the percentage of surveyed adults who favored each of these proposals.

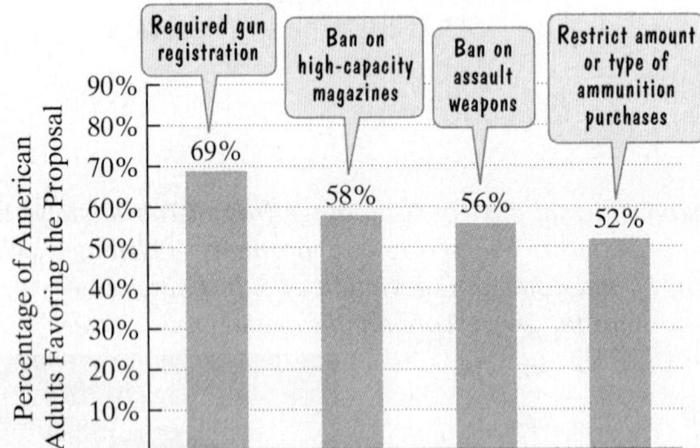

Source: Time/CNN poll using a sample of 814 American adults, January 14–15, 2013

a. Find the margin of error, to the nearest tenth of a percent, for this survey.

b. Write a statement about the percentage of adults in the U.S population who favor required gun registration to reduce gun violence.

68. How to Blow Your Job Interview The data in the bar graph at the top of the next column are from a random sample of 1910 job interviewers. The graph shows the top interviewer turnoffs and the percentage of surveyed interviewers who were offended by each of these behaviors.

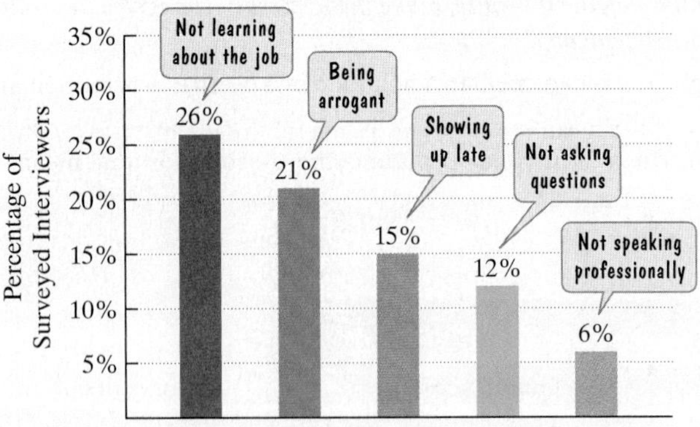

Source: Scott Erker, PhD., and Kelli Buczynski, "Are You Failing the Interview? 2009 Survey of Global Interviewing Practices and Perceptions." Development Dimensions International.

a. Find the margin of error, to the nearest tenth of a percent, for this survey.

b. Write a statement about the percentage of interviewers in the population who are turned off by a job applicant being arrogant.

69. Using a random sample of 4000 TV households, Nielsen Media Research found that 60.2% watched the final episode of *M*A*S*H*.

a. Find the margin of error in this percent.

b. Write a statement about the percentage of TV households in the population that tuned into the final episode of *M*A*S*H*.

70. Using a random sample of 4000 TV households, Nielsen Media Research found that 51.1% watched *Roots, Part 8*.

a. Find the margin of error in this percent.

b. Write a statement about the percentage of TV households in the population that tuned into *Roots, Part 8*.

71. In 1997, Nielsen Media Research increased its random sample to 5000 TV households. By how much, to the nearest tenth of a percent, did this improve the margin of error over that in Exercises 69 and 70?

72. If Nielsen Media Research were to increase its random sample from 5000 to 10,000 TV households, by how much, to the nearest tenth of a percent, would this improve the margin of error?

The histogram shows murder rates per 100,000 residents and the number of U.S. states that had these rates for a recent year. Use this histogram to solve Exercises 73–74.

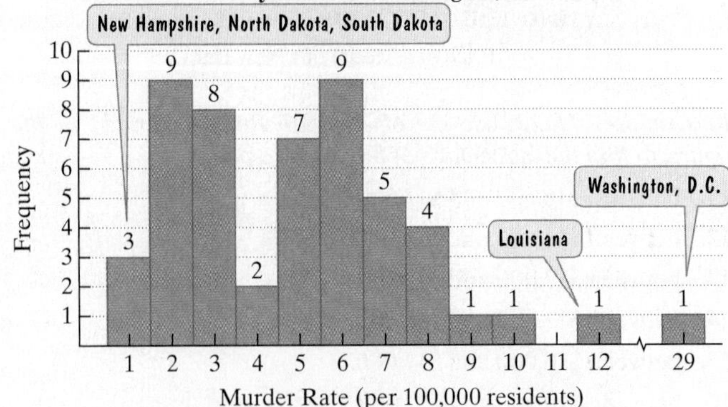

Source: FBI, *Crime in the United States*

73. a. Is the shape of this distribution best classified as normal, skewed to the right, or skewed to the left?

b. Calculate the mean murder rate per 100,000 residents for the 50 states and Washington, D.C.

c. Find the median murder rate per 100,000 residents for the 50 states and Washington, D.C.

d. Are the mean and median murder rates consistent with the shape of the distribution that you described in part (a)? Explain your answer.

e. The standard deviation for the data is approximately 4.2. If the distribution were roughly normal, what would be the z-score, rounded to one decimal place, for Washington, D.C.? Does this seem unusually high? Explain your answer.

74. a. Find the median murder rate per 100,000 residents for the 50 states and Washington, D.C.

b. Find the first quartile by determining the median of the lower half of the data. (This is the median of the items that lie below the median that you found in part (a).)

c. Find the third quartile by determining the median of the upper half of the data. (This is the median of the items that lie above the median that you found in part (a).)

d. Use the following numerical scale:

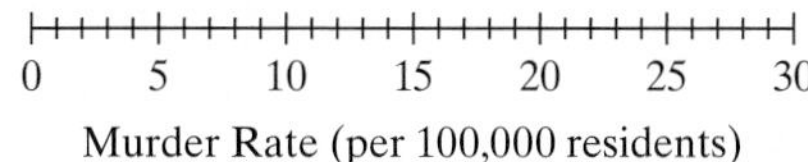

Above this scale, show five points, each at the same height. (The height is arbitrary.) Each point should represent one of the following numbers:

lowest data value, first quartile, median, third quartile, highest data value.

e. A **box-and-whisker plot** consists of a rectangular box extending from the first quartile to the third quartile, with a dashed line representing the median, and line segments (or whiskers) extending outward from the box to the lowest and highest data values:

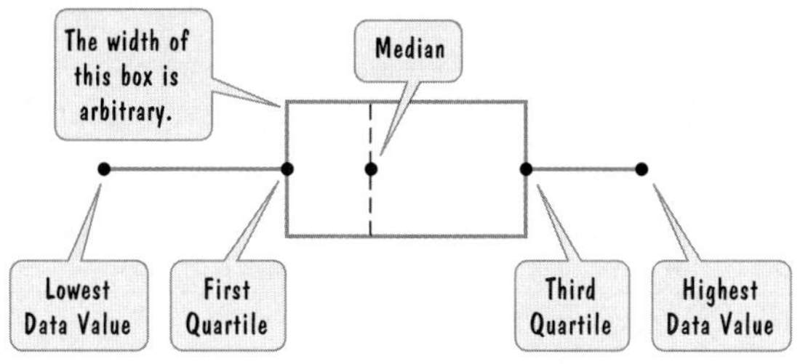

Use your graph from part (d) to create a box-and-whisker plot for U.S. murder rates per 100,000 residents.

f. If one of the whiskers in a box-and-whisker plot is clearly longer, the distribution is usually skewed in the direction of the longer whisker. Based on this observation, does your box-and-whisker plot in part (e) indicate that the distribution is skewed to the right or skewed to the left?

g. Is the shape of the distribution of scores shown by the given histogram consistent with your observation in part (f)?

Explaining the Concepts

75. What is a symmetric histogram?

76. Describe the normal distribution and discuss some of its properties.

77. Describe the 68–95–99.7 Rule.

78. Describe how to determine the z-score for a data item in a normal distribution.

79. What does a z-score measure?

80. Give an example of both a commonly occurring and an infrequently occurring z-score. Explain how you arrived at these examples.

81. Describe when a z-score is negative.

82. If you score in the 83rd percentile, what does this mean?

83. If your weight is in the third quartile, what does this mean?

84. Two students have scores with the same percentile, but for different administrations of the SAT. Does this mean that the students have the same score on the SAT? Explain your answer.

85. Give an example of a phenomenon that is normally distributed. Explain why. (Try to be creative and not use one of the distributions discussed in this section.) Estimate what the mean and the standard deviation might be and describe how you determined these estimates.

86. Give an example of a phenomenon that is not normally distributed and explain why.

Critical Thinking Exercises

Make Sense? *In Exercises 87–90, determine whether each statement makes sense or does not make sense, and explain your reasoning.*

87. The heights of the men on our college basketball team are normally distributed with a mean of 6 feet 3 inches and a standard deviation of 1 foot 2 inches.

88. I scored in the 50th percentile on a standardized test, so my score is the median.

89. A poll administered to a random sample of 1150 voters shows 51% in favor of candidate A, so I'm 95% confident that candidate A will win the election.

90. My math teacher gave a very difficult exam for which the distribution of scores was skewed to the right.

Group Exercise

91. For this activity, group members will conduct interviews with a random sample of students on campus. Each student is to be asked, "What is the worst thing about being a student?" One response should be recorded for each student.

a. Each member should interview enough students so that there are at least 50 randomly selected students in the sample.

b. After all responses have been recorded, the group should organize the four most common answers. For each answer, compute the percentage of students in the sample who felt that this is the worst thing about being a student.

c. Find the margin of error for your survey.

d. For each of the four most common answers, write a statement about the percentage of all students on your campus who feel that this is the worst thing about being a student.

12.5 Problem Solving with the Normal Distribution

WHAT AM I SUPPOSED TO LEARN?

After studying this section, you should be able to:

1 Solve applied problems involving normal distributions.

WE HAVE SEEN THAT MALE HEIGHTS IN NORTH AMERICA ARE approximately normally distributed with a mean of 70 inches and a standard deviation of 4 inches. Suppose we are interested in the percentage of men with heights below 80 inches:

$$z_{80} = \frac{\text{data item} - \text{mean}}{\text{standard deviation}} = \frac{80 - 70}{4} = \frac{10}{4} = 2.5.$$

Because this z-score is not an integer, the 68–95–99.7 Rule is not helpful in finding the percentage of data items that fall below 2.5 standard deviations of the mean. In this section, we will use a table that contains numerous z-scores and their percentiles to solve a variety of problems involving the normal distribution.

1 *Solve applied problems involving normal distributions.*

Problem Solving Using z-Scores and Percentiles

Table 12.16 gives a percentile interpretation for z-scores.

TABLE 12.16 z-Scores and Percentiles

z-Score	Percentile	z-Score	Percentile	z-Score	Percentile	z-Score	Percentile
−4.0	0.003	−1.0	15.87	0.0	50.00	1.1	86.43
−3.5	0.02	−0.95	17.11	0.05	51.99	1.2	88.49
−3.0	0.13	−0.90	18.41	0.10	53.98	1.3	90.32
−2.9	0.19	−0.85	19.77	0.15	55.96	1.4	91.92
−2.8	0.26	−0.80	21.19	0.20	57.93	1.5	93.32
−2.7	0.35	−0.75	22.66	0.25	59.87	1.6	94.52
−2.6	0.47	−0.70	24.20	0.30	61.79	1.7	95.54
−2.5	0.62	−0.65	25.78	0.35	63.68	1.8	96.41
−2.4	0.82	−0.60	27.43	0.40	65.54	1.9	97.13
−2.3	1.07	−0.55	29.12	0.45	67.36	2.0	97.72
−2.2	1.39	−0.50	30.85	0.50	69.15	2.1	98.21
−2.1	1.79	−0.45	32.64	0.55	70.88	2.2	98.61
−2.0	2.28	−0.40	34.46	0.60	72.57	2.3	98.93
−1.9	2.87	−0.35	36.32	0.65	74.22	2.4	99.18
−1.8	3.59	−0.30	38.21	0.70	75.80	2.5	99.38
−1.7	4.46	−0.25	40.13	0.75	77.34	2.6	99.53
−1.6	5.48	−0.20	42.07	0.80	78.81	2.7	99.65
−1.5	6.68	−0.15	44.04	0.85	80.23	2.8	99.74
−1.4	8.08	−0.10	46.02	0.90	81.59	2.9	99.81
−1.3	9.68	−0.05	48.01	0.95	82.89	3.0	99.87
−1.2	11.51	0.0	50.00	1.0	84.13	3.5	99.98
−1.1	13.57					4.0	99.997

TWO ENTRIES FROM TABLE 12.16

z-Score	Percentile
2.5	99.38
0.0	50.00

The portion of the table in the margin indicates that the corresponding percentile for a z-score of 2.5 is 99.38. This tells us that 99.38% of North American men have heights that are less than 80 inches, or $z = 2.5$.

In a normal distribution, the mean, median, and mode all have a corresponding z-score of 0. **Table 12.16** shows that the percentile for a z-score of 0 is 50.00. Thus,

50% of the data items in a normal distribution are less than the mean, median, and mode. Consequently, 50% of the data items are greater than or equal to the mean, median, and mode.

Table 12.16 can be used to find the percentage of data items that are less than any data item in a normal distribution. Begin by converting the data item to a z-score. Then, use the table to find the percentile for this z-score. This percentile is the percentage of data items that are less than the data item in question.

EXAMPLE 1 *Finding the Percentage of Data Items Less Than a Given Data Item*

According to the Department of Health and Education, cholesterol levels are normally distributed. For men between 18 and 24 years, the mean is 178.1 (measured in milligrams per 100 milliliters) and the standard deviation is 40.7. What percentage of men in this age range have a cholesterol level less than 239.15?

SOLUTION

If you are familiar with your own cholesterol level, you probably recognize that a level of 239.15 is fairly high for a young man. Because of this, we would expect most young men to have a level less than 239.15. Let's see if this is so. **Table 12.16** requires that we use z-scores. We compute the z-score for a 239.15 cholesterol level by using the z-score formula.

$$z_{239.15} = \frac{\text{data item} - \text{mean}}{\text{standard deviation}} = \frac{239.15 - 178.1}{40.7} = \frac{61.05}{40.7} = 1.5$$

A PORTION OF TABLE 12.16

z-Score	Percentile
1.4	91.92
1.5	93.32
1.6	94.52

A man between 18 and 24 with a 239.15 cholesterol level is 1.5 standard deviations above the mean, illustrated in **Figure 12.24(a)**. The question mark indicates that we must find the percentage of men with a cholesterol level less than $z = 1.5$, the z-score for a 239.15 cholesterol level. **Table 12.16** gives this percentage as a percentile. Find 1.5 in the z-score column in the right portion of the table. The percentile given to the right of 1.5 is 93.32. Thus, 93.32% of men between 18 and 24 have a cholesterol level less than 239.15, shown in **Figure 12.24(b)**.

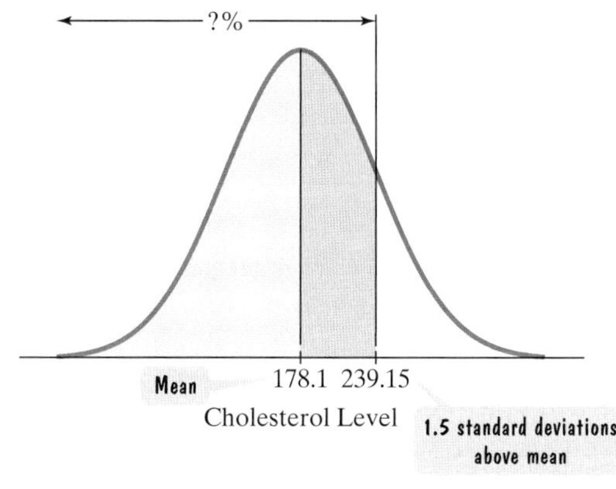

FIGURE 12.24(a)

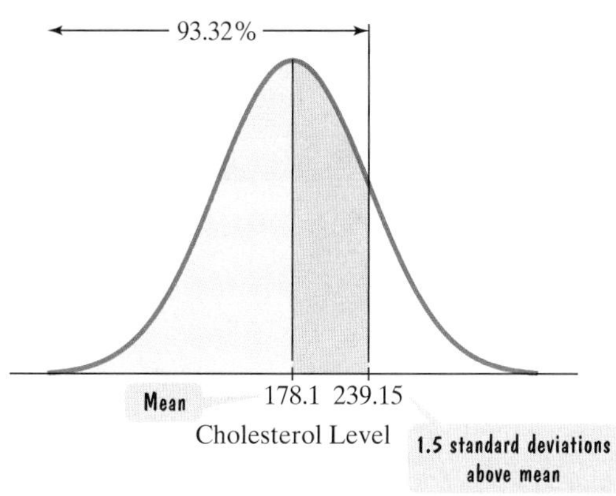

FIGURE 12.24(b)

CHECK POINT 1 The distribution of monthly charges for cellphone plans in the United States is approximately normal with a mean of \$62 and a standard deviation of \$18. What percentage of plans have charges that are less than \$83.60?

The normal distribution accounts for all data items, meaning 100% of the scores. This means that **Table 12.16** can also be used to find the percentage of data items that are greater than any data item in a normal distribution. Use the percentile in the table to determine the percentage of data items less than the data item in question. Then subtract this percentage from 100% to find the percentage of data items greater than the item in question. In using this technique, we will treat the phrases "greater than" and "greater than or equal to" as equivalent.

EXAMPLE 2 ***Finding the Percentage of Data Items Greater Than a Given Data Item***

Lengths of pregnancies of women are normally distributed with a mean of 266 days and a standard deviation of 16 days. What percentage of children are born from pregnancies lasting more than 274 days?

SOLUTION

A PORTION OF TABLE 12.16

z-Score	Percentile
0.45	67.36
0.50	69.15
0.55	70.88

Table 12.16 requires that we use z-scores. We compute the z-score for a 274-day pregnancy by using the z-score formula.

$$z_{274} = \frac{\text{data item} - \text{mean}}{\text{standard deviation}} = \frac{274 - 266}{16} = \frac{8}{16} = 0.5$$

A 274-day pregnancy is 0.5 standard deviation above the mean. **Table 12.16** gives the percentile corresponding to 0.50 as 69.15. This means that 69.15% of pregnancies last less than 274 days, illustrated in **Figure 12.25**. We must find the percentage of pregnancies lasting more than 274 days by subtracting 69.15% from 100%.

$$100\% - 69.15\% = 30.85\%$$

Thus, 30.85% of children are born from pregnancies lasting more than 274 days.

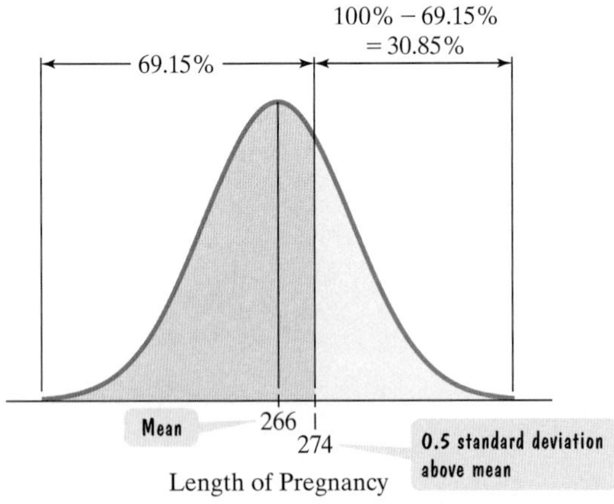

FIGURE 12.25

CHECK POINT 2 Female adult heights in North America are approximately normally distributed with a mean of 65 inches and a standard deviation of 3.5 inches. What percentage of North American women have heights that exceed 69.9 inches?

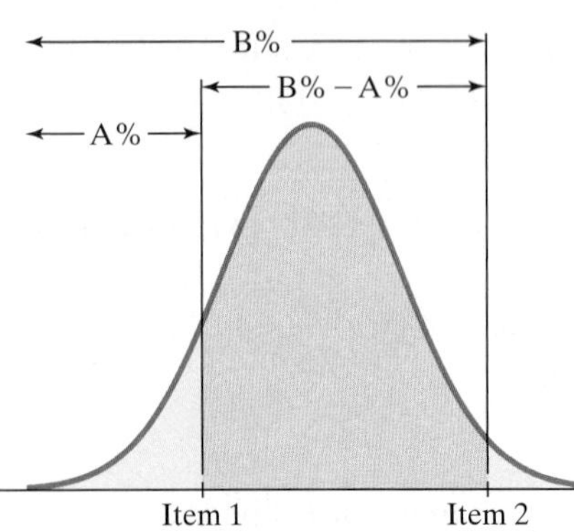

FIGURE 12.26 The percentile for data item 1 is A. The percentile for data item 2 is B. The percentage of data items between item 1 and item 2 is B% − A%.

We have seen how **Table 12.16** is used to find the percentage of data items that are less than or greater than any given item. The table can also be used to find the percentage of data items *between* two given items. Because the percentile for each item is the percentage of data items less than the given item, the percentage of data between the two given items is found by subtracting the lesser percent from the greater percent. This is illustrated in **Figure 12.26**.

FINDING THE PERCENTAGE OF DATA ITEMS BETWEEN TWO GIVEN ITEMS IN A NORMAL DISTRIBUTION

1. Convert each given data item to a z-score:
$$z = \frac{\text{data item} - \text{mean}}{\text{standard deviation}}.$$
2. Use **Table 12.16** to find the percentile corresponding to each z-score in step 1.
3. Subtract the lesser percentile from the greater percentile and attach a % sign.

EXAMPLE 3 Finding the Percentage of Data Items between Two Given Data Items

The amount of time that self-employed Americans work each week is normally distributed with a mean of 44.6 hours and a standard deviation of 14.4 hours. What percentage of self-employed individuals in the United States work between 37.4 and 80.6 hours per week?

A PORTION OF TABLE 12.16

z-Score	Percentile
−0.55	29.12
−0.50	30.85
−0.45	32.64

A PORTION OF TABLE 12.16

z-Score	Percentile
2.4	99.18
2.5	99.38
2.6	99.53

SOLUTION

Step 1 Convert each given data item to a z-score.

$$z_{37.4} = \frac{\text{data item} - \text{mean}}{\text{standard deviation}} = \frac{37.4 - 44.6}{14.4} = \frac{-7.2}{14.4} = -0.5$$

$$z_{80.6} = \frac{\text{data item} - \text{mean}}{\text{standard deviation}} = \frac{80.6 - 44.6}{14.4} = \frac{36}{14.4} = 2.5$$

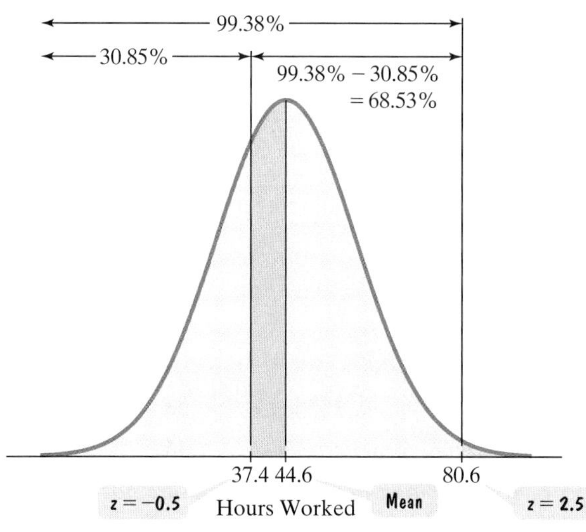

FIGURE 12.27

Step 2 Use Table 12.16 to find the percentile corresponding to these z-scores. The percentile given to the right of −0.50 is 30.85. This means that 30.85% of self-employed Americans work less than 37.4 hours per week.

Table 12.16 also gives the percentile corresponding to $z = 2.5$. Find 2.5 in the z-score column in the far-right portion of the table. The percentile given to the right of 2.5 is 99.38. This means that 99.38% of self-employed Americans work less than 80.6 hours per week.

Step 3 Subtract the lesser percentile from the greater percentile and attach a % sign. Subtracting percentiles, we obtain

$$99.38 - 30.85 = 68.53.$$

Thus, 68.53% of self-employed Americans work between 37.4 and 80.6 hours per week. The solution is illustrated in **Figure 12.27**.

CHECK POINT 3 The distribution for the life of refrigerators is approximately normal with a mean of 14 years and a standard deviation of 2.5 years. What percentage of refrigerators have lives between 11 years and 18 years?

Our work in Examples 1 through 3 is summarized as follows:

COMPUTING PERCENTAGE OF DATA ITEMS FOR NORMAL DISTRIBUTIONS

Description of Percentage	Graph	Computation of Percentage
Percentage of data items less than a given data item with $z = b$	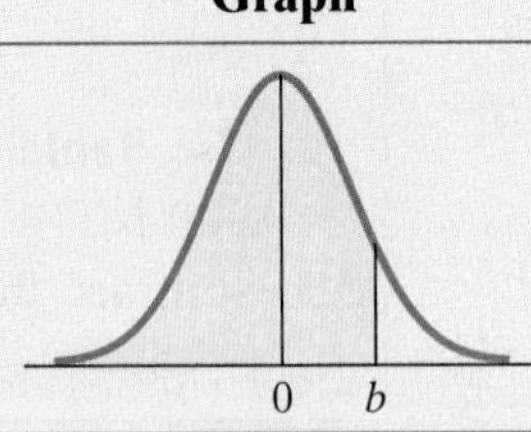	Use the table percentile for $z = b$ and add a % sign.
Percentage of data items greater than a given data item with $z = a$	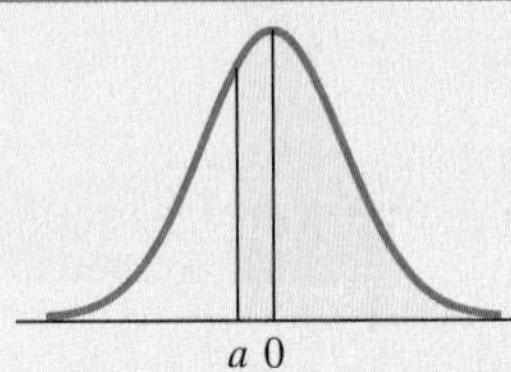	Subtract the table percentile for $z = a$ from 100 and add a % sign.
Percentage of data items between two given data items with $z = a$ and $z = b$	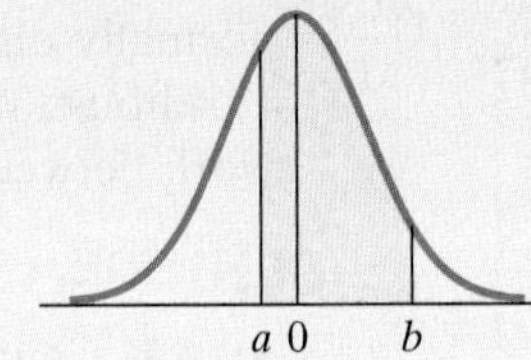	Subtract the table percentile for $z = a$ from the table percentile for $z = b$ and add a % sign.

Concept and Vocabulary Check

Use the information shown below to fill in each blank so that the resulting statement is true.

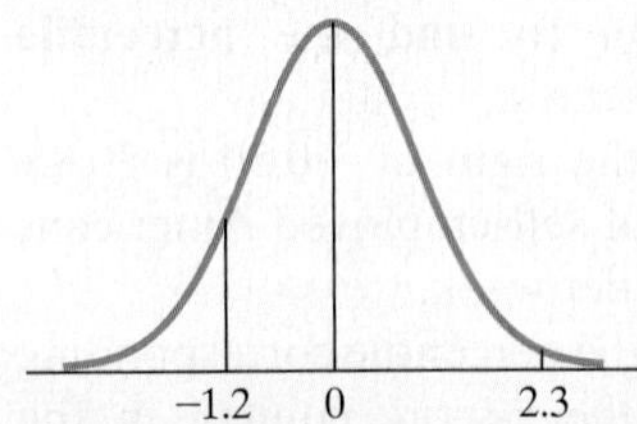

z-Score	Percentile
−1.2	11.51
2.3	98.93

1. The percentage of scores less than $z = 2.3$ is ________.
2. The percentage of scores greater than $z = 2.3$ is ________.
3. The percentage of scores greater than $z = -1.2$ is ________.
4. The percentage of scores between $z = -1.2$ and $z = 2.3$ is ________.
5. True or False: The 68–95–99.7 Rule cannot be used if z-scores are not integers. ________

Exercise Set 12.5

Practice and Application Exercises

Use **Table 12.16** *on page 822 to solve Exercises 1–16.*

In Exercises 1–8, find the percentage of data items in a normal distribution that lie **a.** *below and* **b.** *above the given z-score.*

1. $z = 0.6$ **2.** $z = 0.8$

3. $z = 1.2$ **4.** $z = 1.4$

5. $z = -0.7$ **6.** $z = -0.4$

7. $z = -1.2$ **8.** $z = -1.8$

In Exercises 9–16, find the percentage of data items in a normal distribution that lie between

9. $z = 0.2$ and $z = 1.4$. **10.** $z = 0.3$ and $z = 2.1$.

11. $z = 1$ and $z = 3$. **12.** $z = 2$ and $z = 3$.

13. $z = -1.5$ and $z = 1.5$. **14.** $z = -1.2$ and $z = 1.2$.

15. $z = -2$ and $z = -0.5$. **16.** $z = -2.2$ and $z = -0.3$.

Systolic blood pressure readings are normally distributed with a mean of 121 and a standard deviation of 15. (A reading above 140 is considered to be high blood pressure.) In Exercises 17–26, begin by converting any given blood pressure reading or readings into z-scores. Then use **Table 12.16** *on page 822 to find the percentage of people with blood pressure readings*

17. below 142. **18.** below 148.

19. above 130. **20.** above 133.

21. above 103. **22.** above 100.

23. between 142 and 154. **24.** between 145 and 157.

25. between 112 and 130. **26.** between 109 and 133.

The weights for 12-month-old baby boys are normally distributed with a mean of 22.5 pounds and a standard deviation of 2.2 pounds. In Exercises 27–30, use **Table 12.16** *on page 822 to find the percentage of 12-month-old baby boys who weigh*

27. more than 25.8 pounds.

28. more than 23.6 pounds.

29. between 19.2 and 21.4 pounds.

30. between 18.1 and 19.2 pounds.

Practice Plus

The table shows selected ages of licensed drivers in the United States and the corresponding percentiles.

AGES OF U.S. DRIVERS

Age	Percentile
75	98
65	88
55	77
45	60
35	37
25	14
20	5

Source: Department of Transportation

In Exercises 31–36, use the information given by the table to find the percentage of U.S. drivers who are

31. younger than 55.

32. younger than 45.

33. at least 25.

34. at least 35.

35. at least 65 and younger than 75.

36. at least 20 and younger than 65.

Explaining the Concepts

37. Explain when it is necessary to use a table showing z-scores and percentiles rather than the 68–95–99.7 Rule to determine the percentage of data items less than a given data item.

38. Explain how to use a table showing z-scores and percentiles to determine the percentage of data items between two z-scores.

Critical Thinking Exercises

Make Sense? *In Exercises 39–42, determine whether each statement makes sense or does not make sense, and explain your reasoning.*

39. I'm using a table showing z-scores and percentiles that has positive percentiles corresponding to positive z-scores and negative percentiles corresponding to negative z-scores.

40. My table showing z-scores and percentiles displays the percentage of data items less than a given value of z.

41. My table showing z-scores and percentiles does not display the percentage of data items greater than a given value of z.

42. I can use a table showing z-scores and percentiles to verify the three approximate numbers given by the 68–95–99.7 Rule.

43. Find two z-scores so that 40% of the data in the distribution lies between them. (More than one answer is possible.)

44. A woman insists that she will never marry a man as short or shorter than she, knowing that only one man in 400 falls into this category. Assuming a mean height of 69 inches for men with a standard deviation of 2.5 inches (and a normal distribution), approximately how tall is the woman?

45. The placement test for a college has scores that are normally distributed with a mean of 500 and a standard deviation of 100. If the college accepts only the top 10% of examinees, what is the cutoff score on the test for admission?

12.6 Scatter Plots, Correlation, and Regression Lines

WHAT AM I SUPPOSED TO LEARN?

After studying this section, you should be able to:

1. Make a scatter plot for a table of data items.
2. Interpret information given in a scatter plot.
3. Compute the correlation coefficient.
4. Write the equation of the regression line.
5. Use a sample's correlation coefficient to determine whether there is a correlation in the population.

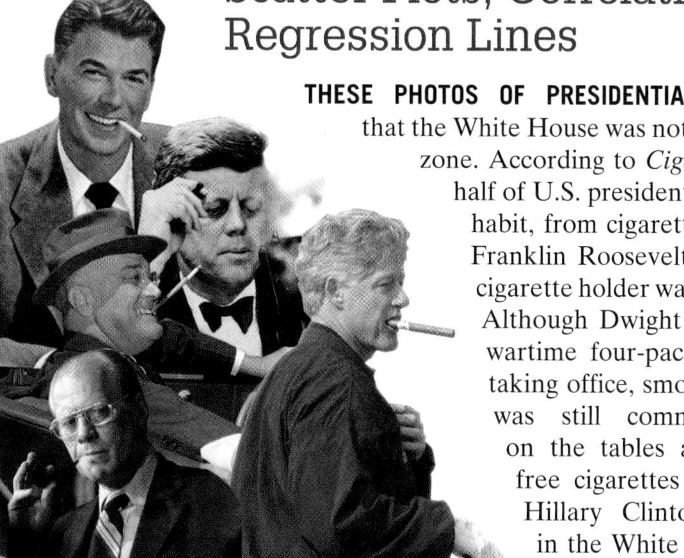

THESE PHOTOS OF PRESIDENTIAL PUFFING INDICATE that the White House was not always a no-smoking zone. According to *Cigar Aficionado*, nearly half of U.S. presidents have had a nicotine habit, from cigarettes to pipes to cigars. Franklin Roosevelt's stylish way with a cigarette holder was part of his mystique. Although Dwight Eisenhower quit his wartime four-pack-a-day habit before taking office, smoking in the residence was still common, with ashtrays on the tables at state dinners and free cigarettes for guests. In 1993, Hillary Clinton banned smoking in the White House, although Bill Clinton's cigars later made a sordid cameo in the Lewinsky scandal. Barack Obama quit smoking before entering the White House, but had "fallen off the wagon occasionally" as he admitted in a *Meet the Press* interview.

Changing attitudes toward smoking, both inside and outside the White House, date back to 1964 and an equation in two variables. To understand the mathematics behind this turning point in public health, we need to explore situations involving data collected on two variables.

Up to this point in the chapter, we have studied situations in which data sets involve a single variable, such as height, weight, cholesterol level, and length of pregnancies. By contrast, the 1964 study involved data collected on two variables from 11 countries—annual cigarette consumption for each adult male and deaths per million males from lung cancer. In this section, we consider situations in which there are two data items for each randomly selected person or thing. Our interest is in determining whether or not there is a relationship between the two variables and, if so, the strength of that relationship.

1 *Make a scatter plot for a table of data items.*

Scatter Plots and Correlation

Is there a relationship between education and prejudice? With increased education, does a person's level of prejudice tend to decrease? Notice that we are interested in two quantities—years of education and level of prejudice. For each person in our sample, we will record the number of years of school completed and the score on a test measuring prejudice. Higher scores on this 1-to-10 test indicate greater prejudice. Using x to represent years of education and y to represent scores on a test measuring prejudice, **Table 12.17** shows these two quantities for a random sample of ten people.

TABLE 12.17 Recording Two Quantities in a Sample of Ten People

Respondent	A	B	C	D	E	F	G	H	I	J
Years of education (x)	12	5	14	13	8	10	16	11	12	4
Score on prejudice test (y)	1	7	2	3	5	4	1	2	3	10

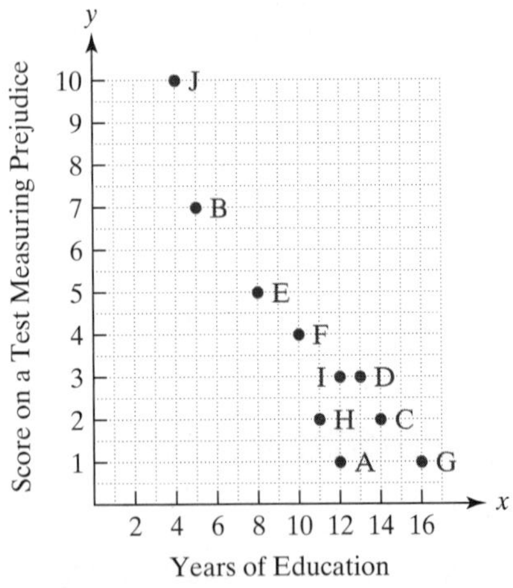

FIGURE 12.28 A scatter plot for education-prejudice data.

When two data items are collected for every person or object in a sample, the data items can be visually displayed using a *scatter plot*. A **scatter plot** is a collection of data points, one data point per person or object. We can make a scatter plot of the data in **Table 12.17** by drawing a horizontal axis to represent years of education and a vertical axis to represent scores on a test measuring prejudice. We then represent each of the ten respondents with a single point on the graph. For example, the dot for respondent A is located to represent 12 years of education on the horizontal axis and 1 on the prejudice test on the vertical axis. Plotting each of the ten pieces of data in a rectangular coordinate system results in the scatter plot shown in **Figure 12.28**.

A scatter plot like the one in **Figure 12.28** can be used to determine whether two quantities are related. If there is a clear relationship, the quantities are said to be **correlated**. The scatter plot shows a downward trend among the data points, although there are a few exceptions. People with increased education tend to have a lower score on the test measuring prejudice. **Correlation** is used to determine if there is a relationship between two variables and, if so, the strength and direction of that relationship.

Correlation and Causal Connections

Correlations can often be seen when data items are displayed on a scatter plot. Although the scatter plot in **Figure 12.28** indicates a correlation between education and prejudice, we cannot conclude that increased education causes a person's level of prejudice to decrease. There are at least three possible explanations:

1. The correlation between increased education and decreased prejudice is simply a coincidence.
2. Education usually involves classrooms with a variety of different kinds of people. Increased exposure to diversity in the classroom setting, which accompanies increased levels of education, might be an underlying cause for decreased prejudice.
3. Education, the process of acquiring knowledge, requires people to look at new ideas and see things in different ways. Thus, education causes one to be more tolerant and less prejudiced.

This list represents three possibilities. Perhaps you can provide a better explanation about decreasing prejudice with increased education.

Establishing that one thing causes another is extremely difficult, even if there is a strong correlation between these things. For example, as the air temperature increases, there is an increase in the number of people stung by jellyfish at the beach. This does not mean that an increase in air temperature causes more people to be stung. It might mean that because it is hotter, more people go into the water. With an increased number of swimmers, more people are likely to be stung. In short, correlation is not necessarily causation.

2 Interpret information given in a scatter plot.

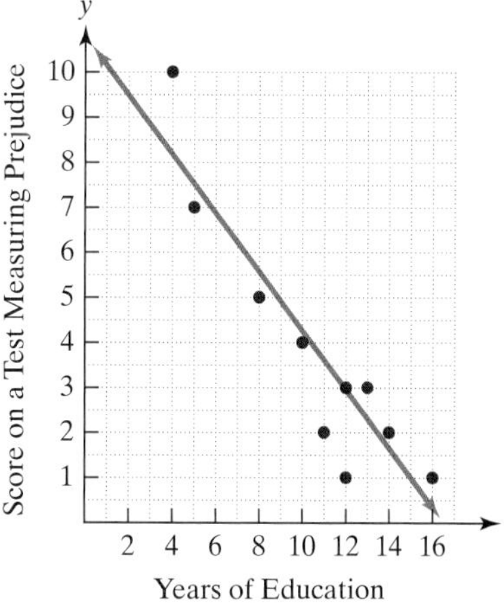

FIGURE 12.29 A scatter plot with a regression line.

Regression Lines and Correlation Coefficients

Figure 12.29 shows the scatter plot for the education-prejudice data. Also shown is a straight line that seems to approximately "fit" the data points. Most of the data points lie either near or on this line. A line that best fits the data points in a scatter plot is called a **regression line**. The regression line is the particular line in which the spread of the data points around it is as small as possible.

A measure that is used to describe the strength and direction of a relationship between variables whose data points lie on or near a line is called the **correlation coefficient**, designated by r. **Figure 12.30** shows scatter plots and correlation coefficients. Variables are **positively correlated** if they tend to increase or decrease together, as in **Figure 12.30(a)**, **(b)**, and **(c)**. By contrast, variables are **negatively correlated** if one variable tends to decrease while the other increases, as in **Figure 12.30(e)**, **(f)**, and **(g)**. **Figure 12.30** illustrates that a correlation coefficient, r, is a number between -1 and 1, inclusive. **Figure 12.30(a)** shows a value of 1. This indicates a **perfect positive correlation** in which all points in the scatter plot lie precisely on the regression line that rises from left to right. **Figure 12.30(g)** shows a value of -1. This indicates a **perfect negative correlation** in which all points in the scatter plot lie precisely on the regression line that falls from left to right.

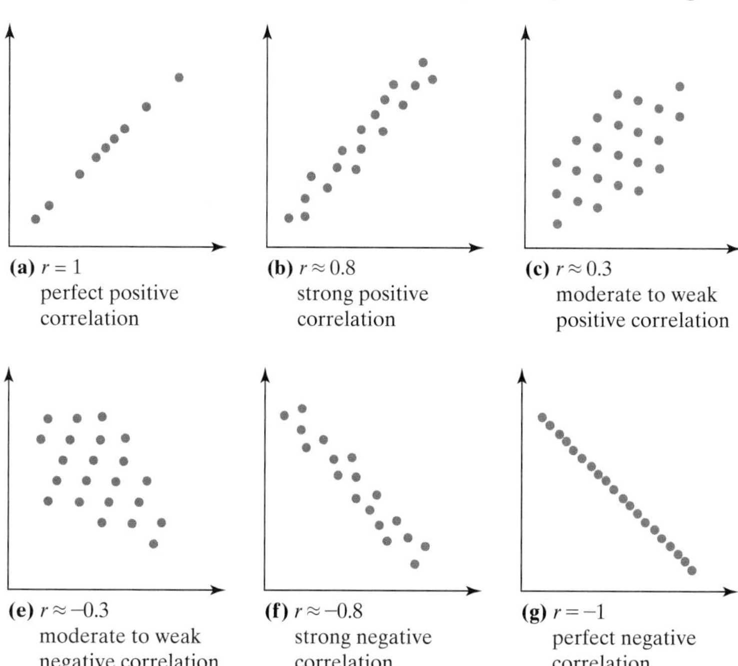
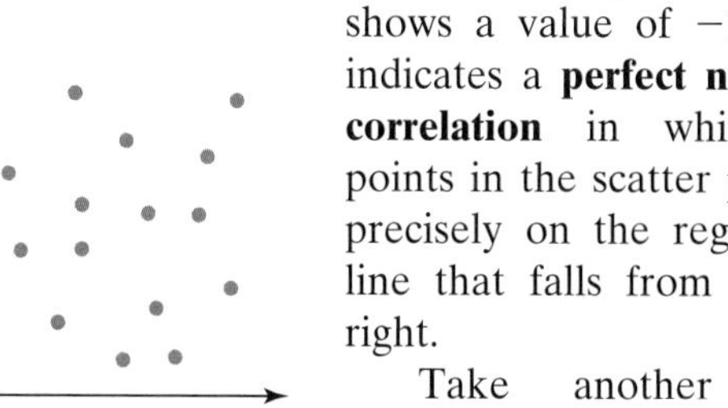

FIGURE 12.30 Scatter plots and correlation coefficients

Take another look at **Figure 12.30**. If r is between 0 and 1, as in **(b)** and **(c)**, the two variables are positively correlated, but not perfectly. Although all the data points will not lie on the regression line, as in **(a)**, an increase in one variable tends to be accompanied by an increase in the other. Negative correlations are also illustrated in **Figure 12.30**. If r is between 0 and -1, as in **(e)** and **(f)**, the two variables are negatively correlated, but not perfectly. Although all the data points will not lie on the regression line, as in **(g)**, an increase in one variable tends to be accompanied by a decrease in the other.

Blitzer Bonus

Beneficial Uses of Correlation Coefficients

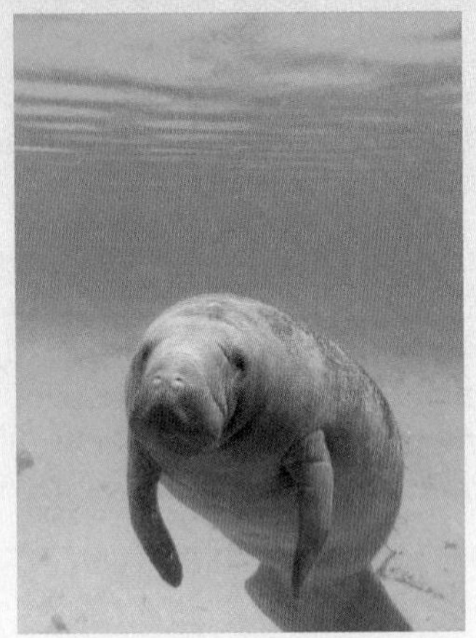

- A Florida study showed a high positive correlation between the number of powerboats and the number of manatee deaths. Many of these deaths were seen to be caused by boats' propellers gashing into the manatees' bodies. Based on this study, Florida set up coastal sanctuaries where powerboats are prohibited so that these large gentle mammals that float just below the water's surface could thrive.
- Researchers studied how psychiatric patients readjusted to their community after their release from a mental hospital. A moderate positive correlation ($r = 0.38$) was found between patients' attractiveness and their post-discharge social adjustment. The better-looking patients were better off. The researchers suggested that physical attractiveness plays a role in patients' readjustment to community living because good-looking people tend to be treated better by others than homely people are.

EXAMPLE 1 *Interpreting a Correlation Coefficient*

In a 1971 study involving 232 subjects, researchers found a relationship between the subjects' level of stress and how often they became ill. The correlation coefficient in this study was 0.32. Does this indicate a strong relationship between stress and illness?

SOLUTION

The correlation coefficient $r = 0.32$ means that as stress increases, frequency of illness also tends to increase. However, 0.32 is only a moderate correlation, illustrated in **Figure 12.30(c)** on the previous page. There is not, based on this study, a strong relationship between stress and illness. In this study, the relationship is somewhat weak.

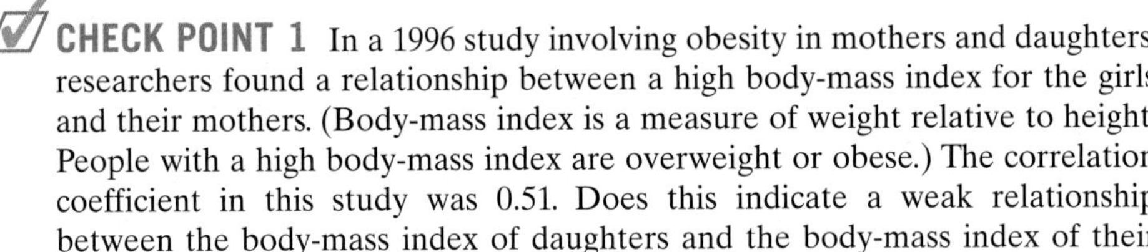

CHECK POINT 1 In a 1996 study involving obesity in mothers and daughters, researchers found a relationship between a high body-mass index for the girls and their mothers. (Body-mass index is a measure of weight relative to height. People with a high body-mass index are overweight or obese.) The correlation coefficient in this study was 0.51. Does this indicate a weak relationship between the body-mass index of daughters and the body-mass index of their mothers?

How to Obtain the Correlation Coefficient and the Equation of the Regression Line

The easiest way to find the correlation coefficient and the equation of the regression line is to use a graphing or statistical calculator. Graphing calculators have statistical menus that enable you to enter the x and y data items for the variables. Based on this information, you can instruct the calculator to display a scatter plot, the equation of the regression line, and the correlation coefficient.

We can also compute the correlation coefficient and the equation of the regression line by hand using formulas. First, we compute the correlation coefficient.

COMPUTING THE CORRELATION COEFFICIENT BY HAND

The following formula is used to calculate the correlation coefficient, r:

$$r = \frac{n(\Sigma xy) - (\Sigma x)(\Sigma y)}{\sqrt{n(\Sigma x^2) - (\Sigma x)^2}\sqrt{n(\Sigma y^2) - (\Sigma y)^2}}.$$

In the formula,

n = the number of data points, (x, y)

Σx = the sum of the x-values

Σy = the sum of the y-values

Σxy = the sum of the product of x and y in each pair

Σx^2 = the sum of the squares of the x-values

Σy^2 = the sum of the squares of the y-values

$(\Sigma x)^2$ = the square of the sum of the x-values

$(\Sigma y)^2$ = the square of the sum of the y-values

When computing the correlation coefficient by hand, organize your work in five columns:

x	y	xy	x^2	y^2

Find the sum of the numbers in each column. Then, substitute these values into the formula for r. Example 2 illustrates computing the correlation coefficient for the education-prejudice test data.

3 *Compute the correlation coefficient.*

EXAMPLE 2 Computing the Correlation Coefficient

Shown below are the data involving the number of years of school, x, completed by ten randomly selected people and their scores on a test measuring prejudice, y. Recall that higher scores on the measure of prejudice (1 to 10) indicate greater levels of prejudice. Determine the correlation coefficient between years of education and scores on a prejudice test.

Respondent	A	B	C	D	E	F	G	H	I	J
Years of education (x)	12	5	14	13	8	10	16	11	12	4
Score on prejudice test (y)	1	7	2	3	5	4	1	2	3	10

SOLUTION

As suggested, organize the work in five columns.

x	y	xy	x^2	y^2
12	1	12	144	1
5	7	35	25	49
14	2	28	196	4
13	3	39	169	9
8	5	40	64	25
10	4	40	100	16
16	1	16	256	1
11	2	22	121	4
12	3	36	144	9
4	10	40	16	100
$\Sigma x = 105$	$\Sigma y = 38$	$\Sigma xy = 308$	$\Sigma x^2 = 1235$	$\Sigma y^2 = 218$
Add all values in the x-column.	Add all values in the y-column.	Add all values in the xy-column.	Add all values in the x^2-column.	Add all values in the y^2-column.

We use these five sums to calculate the correlation coefficient.

Another value in the formula for r that we have not yet determined is n, the number of data points (x, y). Because there are ten items in the x-column and ten items in the y-column, the number of data points (x, y) is ten. Thus, $n = 10$.

In order to calculate r, we also need to find the square of the sum of the x-values and the y-values:

$$(\Sigma x)^2 = (105)^2 = 11{,}025 \quad \text{and} \quad (\Sigma y)^2 = (38)^2 = 1444.$$

TECHNOLOGY

Graphing Calculators, Scatter Plots, and Regression Lines

You can use a graphing calculator to display a scatter plot and the regression line. After entering the x and y data items for years of education and scores on a prejudice test, the calculator shows the scatter plot of the data and the regression line.

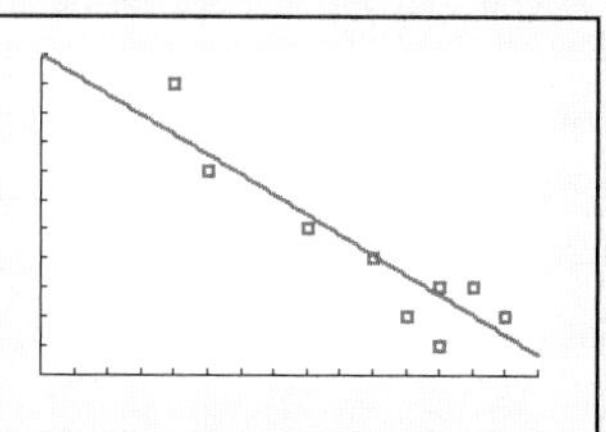

Also displayed below is the regression line's equation and the correlation coefficient, r. The slope shown below is approximately -0.69. The negative slope reinforces the fact that there is a negative correlation between the variables in Example 2.

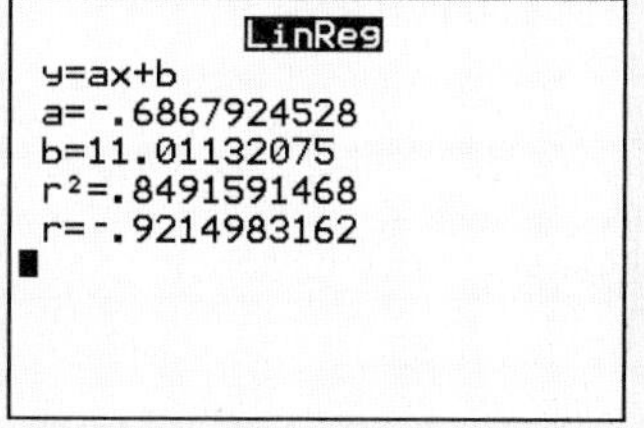

We are ready to determine the value for r. We use the sums obtained on the previous page, with $n = 10$.

$$r = \frac{n(\Sigma xy) - (\Sigma x)(\Sigma y)}{\sqrt{n(\Sigma x^2) - (\Sigma x)^2}\sqrt{n(\Sigma y^2) - (\Sigma y)^2}}$$

$$= \frac{10(308) - 105(38)}{\sqrt{10(1235) - 11{,}025}\sqrt{10(218) - 1444}}$$

$$= \frac{-910}{\sqrt{1325}\sqrt{736}}$$

$$\approx -0.92$$

The value for r, approximately -0.92, is fairly close to -1 and indicates a strong negative correlation. This means that the more education a person has, the less prejudiced that person is (based on scores on the test measuring levels of prejudice).

CHECK POINT 2 The points in the scatter plot in **Figure 12.31** show the number of firearms per 100 persons and the number of deaths per 100,000 persons for the ten industrialized countries with the highest death rates. Use the data displayed by the voice balloons to determine the correlation coefficient between these variables. Round to two decimal places. What does the correlation coefficient indicate about the strength and direction of the relationship between firearms per 100 persons and deaths per 100,000 persons?

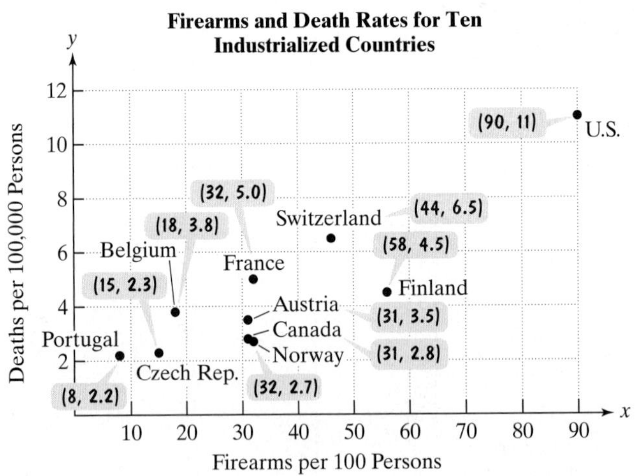

FIGURE 12.31
Source: International Action Network on Small Arms

---

4 *Write the equation of the regression line.*

Once we have determined that two variables are related, we can use the equation of the regression line to determine the exact relationship. Here is the formula for writing the equation of the line that best fits the data:

WRITING THE EQUATION OF THE REGRESSION LINE BY HAND

The equation of the regression line is

$$y = mx + b,$$

where

$$m = \frac{n(\Sigma xy) - (\Sigma x)(\Sigma y)}{n(\Sigma x^2) - (\Sigma x)^2} \quad \text{and} \quad b = \frac{\Sigma y - m(\Sigma x)}{n}.$$

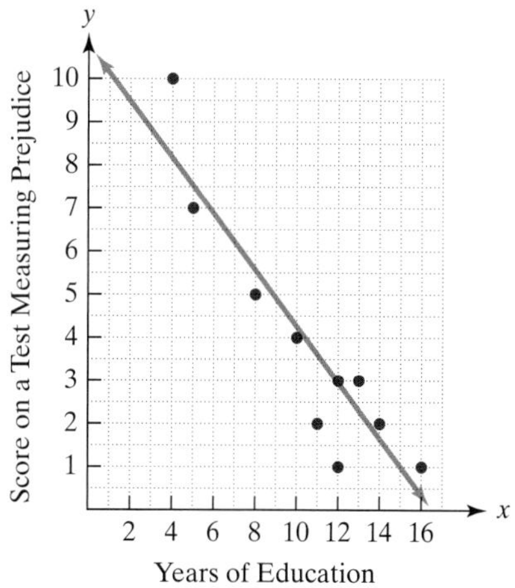

FIGURE 12.29 (repeated)

EXAMPLE 3 *Writing the Equation of the Regression Line*

a. Shown, again, in **Figure 12.29** is the scatter plot and the regression line for the data in Example 2. Use the data to find the equation of the regression line that relates years of education and scores on a prejudice test.

b. Approximately what score on the test can be anticipated by a person with nine years of education?

SOLUTION

a. We use the sums obtained in Example 2. We begin by computing m.

$$m = \frac{n(\Sigma xy) - (\Sigma x)(\Sigma y)}{n(\Sigma x^2) - (\Sigma x)^2} = \frac{10(308) - 105(38)}{10(1235) - (105)^2} = \frac{-910}{1325} \approx -0.69$$

With a negative correlation coefficient, it makes sense that the slope of the regression line is negative. This line falls from left to right, indicating a negative correlation.

Now, we find the y-intercept, b.

$$b = \frac{\Sigma y - m(\Sigma x)}{n} = \frac{38 - (-0.69)(105)}{10} = \frac{110.45}{10} \approx 11.05$$

Using $m \approx -0.69$ and $b \approx 11.05$, the equation of the regression line, $y = mx + b$, is

$$y = -0.69x + 11.05,$$

where x represents the number of years of education and y represents the score on the prejudice test.

b. To anticipate the score on the prejudice test for a person with nine years of education, substitute 9 for x in the regression line's equation.

$$y = -0.69x + 11.05$$
$$y = -0.69(9) + 11.05 = 4.84$$

A person with nine years of education is anticipated to have a score close to 5 on the prejudice test.

GREAT QUESTION!

Why is $b \approx 11.05$ in Example 3, but $b \approx 11.01$ in the Technology box on page 831?

In Example 3, we rounded the value of m when we calculated b. The value of b on the calculator screen on page 831 is more accurate.

CHECK POINT 3 Use the data in **Figure 12.31** of Check Point 2 on page 832 to find the equation of the regression line. Round m and b to one decimal place. Then use the equation to project the number of deaths per 100,000 persons in a country with 80 firearms per 100 persons.

5 *Use a sample's correlation coefficient to determine whether there is a correlation in the population.*

The Level of Significance of r

In Example 2, we found a strong negative correlation between education and prejudice, computing the correlation coefficient, r, to be -0.92. However, the sample size ($n = 10$) was relatively small. With such a small sample, can we truly conclude that a correlation exists in the population? Or could it be that education and prejudice are not related? Perhaps the results we obtained were simply due to sampling error and chance.

Mathematicians have identified values to determine whether r, the correlation coefficient for a sample, can be attributed to a relationship between variables in the population. These values are shown in the second and third columns of

TABLE 12.18 Values for Determining Correlations in a Population

n	$\alpha = 0.05$	$\alpha = 0.01$
4	0.950	0.990
5	0.878	0.959
6	0.811	0.917
7	0.754	0.875
8	0.707	0.834
9	0.666	0.798
10	0.632	0.765
11	0.602	0.735
12	0.576	0.708
13	0.553	0.684
14	0.532	0.661
15	0.514	0.641
16	0.497	0.623
17	0.482	0.606
18	0.468	0.590
19	0.456	0.575
20	0.444	0.561
22	0.423	0.537
27	0.381	0.487
32	0.349	0.449
37	0.325	0.418
42	0.304	0.393
47	0.288	0.372
52	0.273	0.354
62	0.250	0.325
72	0.232	0.302
82	0.217	0.283
92	0.205	0.267
102	0.195	0.254

The larger the sample size, n, the smaller is the value of r needed for a correlation in the population.

Table 12.18. They depend on the sample size, n, listed in the left column. If $|r|$, the absolute value of the correlation coefficient computed for the sample, is greater than the value given in the table, a correlation exists between the variables in the population. The column headed $\alpha = 0.05$ denotes a **significance level of 5%**, meaning that there is a 0.05 probability that, when the statistician says the variables are correlated, they are actually not related in the population. The column on the right, headed $\alpha = 0.01$, denotes a **significance level of 1%**, meaning that there is a 0.01 probability that, when the statistician says the variables are correlated, they are actually not related in the population. Values in the $\alpha = 0.01$ column are greater than those in the $\alpha = 0.05$ column. Because of the possibility of sampling error, there is always a probability that when we say the variables are related, there is actually not a correlation in the population from which the sample was randomly selected.

EXAMPLE 4 *Determining a Correlation in the Population*

In Example 2, we computed $r = -0.92$ for $n = 10$. Can we conclude that there is a negative correlation between education and prejudice in the population?

SOLUTION

Begin by taking the absolute value of the calculated correlation coefficient.

$$|r| = |-0.92| = 0.92$$

Now, look to the right of $n = 10$ in **Table 12.18**. Because 0.92 is greater than both of these values (0.632 and 0.765), we may conclude that a correlation does exist between education and prejudice in the population. (There is a probability of at most 0.01 that the variables are not really correlated in the population and our results could be attributed to chance.)

Blitzer Bonus

Cigarettes and Lung Cancer

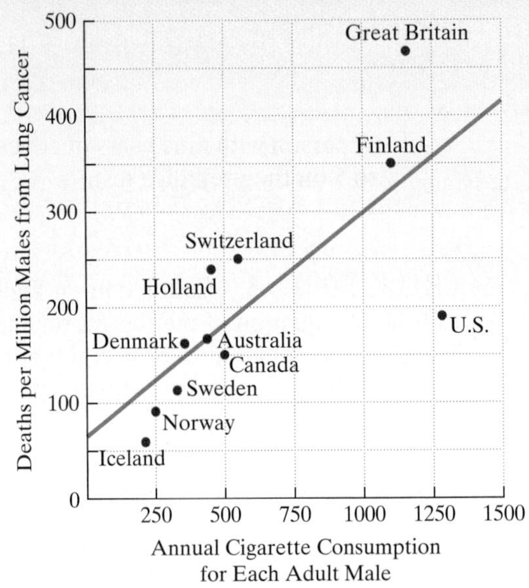

Source: Smoking and Health, Washington, D.C., 1964

This scatter plot shows a relationship between cigarette consumption among males and deaths due to lung cancer per million males. The data are from 11 countries and date back to a 1964 report by the U.S. Surgeon General. The scatter plot can be modeled by a line whose slope indicates an increasing death rate from lung cancer with increased cigarette consumption. At that time, the tobacco industry argued that in spite of this regression line, tobacco use is not the cause of cancer. Recent data do, indeed, show a causal effect between tobacco use and numerous diseases.

CHECK POINT 4 If you worked Check Point 2 correctly, you should have found that $r \approx 0.89$ for $n = 10$. Can you conclude that there is a positive correlation for all industrialized countries between firearms per 100 persons and deaths per 100,000 persons?

Concept and Vocabulary Check

Fill in each blank so that the resulting statement is true.

1. A set of points representing data is called a/an ________.
2. The line that best fits a set of points is called a/an ________.
3. A measure that is used to describe the strength and direction of a relationship between variables whose data points lie on or near a line is called the ________, ranging from $r =$ ____ to $r =$ ____.

In Exercises 4–7, determine whether each statement is true or false. If the statement is false, make the necessary change(s) to produce a true statement.

4. If $r = 0$, there is no correlation between two variables. ______
5. If $r = 1$, changes in one variable cause changes in the other variable. ______
6. If $r = -0.1$, there is a strong negative correlation between two variables. ______
7. A significance level of 5% means that there is a 0.05 probability that when a statistician says that variables are correlated, they are actually not related in the population. ______

Exercise Set 12.6

Practice and Application Exercises

In Exercises 1–8, make a scatter plot for the given data. Use the scatter plot to describe whether or not the variables appear to be related.

1.

x	1	6	4	3	7	2
y	2	5	3	3	4	1

2.

x	2	1	6	3	4
y	4	5	10	8	9

3.

x	8	6	1	5	4	10	3
y	2	4	10	5	6	2	9

4.

x	4	5	2	1
y	1	3	5	4

5. **HAMACHIPHOBIA**

	Percentage Who	
Generation	Won't Try Sushi x	Don't Approve of Marriage Equality y
Millennials	42	36
Gen X	52	49
Boomers	60	59
Silent/Greatest Generation	72	66

Source: Pew Research Center

6. **TREASURED CHEST: FILMS OF MATTHEW MCCONAUGHEY**

Film	Minutes Shirtless x	Opening Weekend Gross (millions of dollars) y
We Are Marshall	0	6.1
EDtv	0.8	8.3
Reign of Fire	1.6	15.6
Sahara	1.8	18.1
Fool's Gold	14.6	21.6

Source: Entertainment Weekly

7. **TEENAGE DRUG USE**

	Percentage Who Have Used	
Country	Marijuana x	Other Illegal Drugs y
Czech Republic	22	4
Denmark	17	3
England	40	21
Finland	5	1
Ireland	37	16
Italy	19	8
Northern Ireland	23	14
Norway	6	3
Portugal	7	3
Scotland	53	31
United States	34	24

Source: De Veaux et al., *Intro Stats*, Pearson, 2009.

8. LITERACY AND HUNGER

	Percentage Who Are	
	Literate	Undernourished
Country	x	y
Cuba	100	2
Egypt	71	4
Ethiopia	36	46
Grenada	96	7
Italy	98	2
Jamaica	80	9
Jordan	91	6
Pakistan	50	24
Russia	99	3
Togo	53	24
Uganda	67	19

Source: The Penguin State of the World Atlas

The scatter plot in the figure shows the relationship between the percentage of married women of child-bearing age using contraceptives and births per woman in selected countries. Use the scatter plot to determine whether each of the statements in Exercises 9–18 is true or false.

Contraceptive Prevalence and Average Number of Births per Woman, Selected Countries

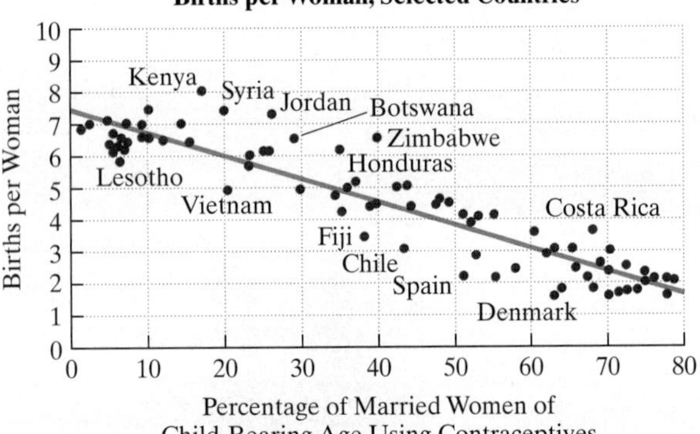

Source: Population Reference Bureau

9. There is a strong positive correlation between contraceptive use and births per woman.

10. There is no correlation between contraceptive use and births per woman.

11. There is a strong negative correlation between contraceptive use and births per woman.

12. There is a causal relationship between contraceptive use and births per woman.

13. With approximately 43% of women of child-bearing age using contraceptives, there are three births per woman in Chile.

14. With 20% of women of child-bearing age using contraceptives, there are six births per woman in Vietnam.

15. No two countries have a different number of births per woman with the same percentage of married women using contraceptives.

16. The country with the greatest number of births per woman also has the smallest percentage of women using contraceptives.

17. Most of the data points do not lie on the regression line.

18. The number of selected countries shown in the scatter plot is approximately 20.

Just as money doesn't buy happiness for individuals, the two don't necessarily go together for countries either. However, the scatter plot does show a relationship between a country's annual per capita income and the percentage of people in that country who call themselves "happy." Use the scatter plot to determine whether each of the statements in Exercises 19–26 is true or false.

Per Capita Income and National Happiness

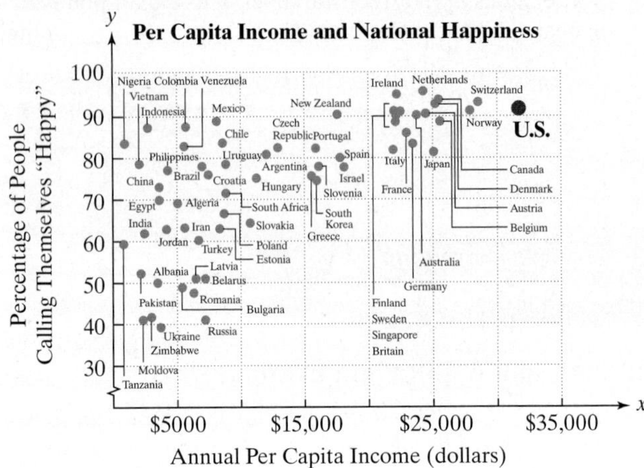

Source: Richard Layard, *Happiness: Lessons from a New Science*, Penguin

19. There is no correlation between per capita income and the percentage of people who call themselves "happy."

20. There is an almost-perfect positive correlation between per capita income and the percentage of people who call themselves "happy."

21. There is a positive correlation between per capita income and the percentage of people who call themselves "happy."

22. As per capita income decreases, the percentage of people who call themselves "happy" also tends to decrease.

23. The country with the lowest per capita income has the least percentage of people who call themselves "happy."

24. The country with the highest per capita income has the greatest percentage of people who call themselves "happy."

25. A reasonable estimate of the correlation coefficient for the data is 0.8.

26. A reasonable estimate of the correlation coefficient for the data is −0.3.

Use the scatter plots shown, labeled (a)–(f), to solve Exercises 27–30.

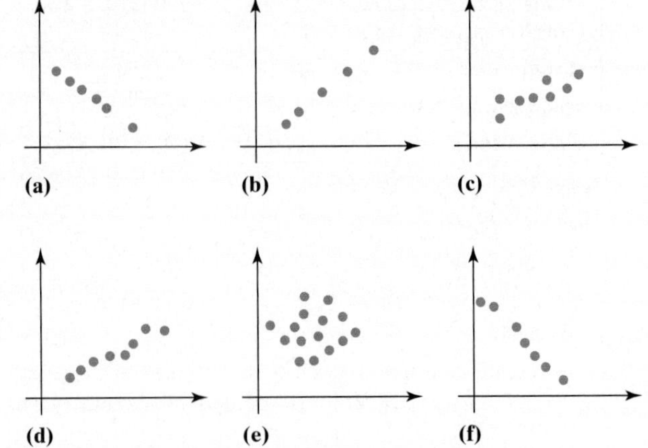

27. Which scatter plot indicates a perfect negative correlation?

28. Which scatter plot indicates a perfect positive correlation?

29. In which scatter plot is $r = 0.9$?

30. In which scatter plot is $r = 0.01$?

Compute r, the correlation coefficient, rounded to two decimal places, for the data in

31. Exercise 1.

32. Exercise 2.

33. Exercise 3.

34. Exercise 4.

35. Use the data in Exercise 5 to solve this exercise.

a. Determine the correlation coefficient, rounded to two decimal places, between the percentage of people who won't try sushi and the percentage who don't approve of marriage equality.

b. Find the equation of the regression line for the percentage who won't try sushi and the percentage who don't approve of marriage equality. Round m and b to two decimal places.

c. What percentage of people, to the nearest percent, can we anticipate do not approve of marriage equality in a generation where 30% won't try sushi?

36. Use the data in Exercise 6 to solve this exercise.

a. Determine the correlation coefficient, rounded to two decimal places, between the minutes Matthew McConaughey appeared shirtless in a film and the film's opening weekend gross.

b. Find the equation of the regression line for the minutes McConaughey appeared shirtless in a film and the film's opening weekend gross. Round m and b to two decimal places.

c. What opening weekend gross, to the nearest tenth of a million dollars, can we anticipate in a McConaughey film in which he appears shirtless for 20 minutes?

37. Use the data in Exercise 7 to solve this exercise.

a. Determine the correlation coefficient, rounded to two decimal places, between the percentage of teenagers who have used marijuana and the percentage who have used other drugs.

b. Find the equation of the regression line for the percentage of teenagers who have used marijuana and the percentage who have used other drugs. Round m and b to two decimal places.

c. What percentage of teenagers, to the nearest percent, can we anticipate using illegal drugs other than marijuana in a country where 10% of teenagers have used marijuana?

38. Use the data in Exercise 8 to solve this exercise.

a. Determine the correlation coefficient, rounded to two decimal places, between the percentage of people in a country who are literate and the percentage who are undernourished.

b. Find the equation of the regression line for the percentage who are literate and the percentage who are undernourished. Round m and b to two decimal places.

c. What percentage of people, to the nearest percent, can we anticipate are undernourished in a country where 60% of the people are literate?

In Exercises 39–45, the correlation coefficient, r, is given for a sample of n data points. Use the $\alpha = 0.05$ column in **Table 12.18** *on page 834 to determine whether or not we may conclude that a correlation does exist in the population. (Using the $\alpha = 0.05$ column, there is a probability of 0.05 that the variables are not really correlated in the population and our results could be attributed to chance. Ignore this possibility when concluding whether or not there is a correlation in the population.)*

39. $n = 20, r = 0.5$

40. $n = 27, r = 0.4$

41. $n = 12, r = 0.5$

42. $n = 22, r = 0.04$

43. $n = 72, r = -0.351$

44. $n = 37, r = -0.37$

45. $n = 20, r = -0.37$

46. In the 1964 study on cigarette consumption and deaths due to lung cancer (see the Blitzer Bonus on page 834), $n = 11$ and $r = 0.73$. What can you conclude using the $\alpha = 0.05$ column in **Table 12.18** on page 834?

Explaining the Concepts

47. What is a scatter plot?

48. How does a scatter plot indicate that two variables are correlated?

49. Give an example of two variables with a strong positive correlation and explain why this is so.

50. Give an example of two variables with a strong negative correlation and explain why this is so.

51. What is meant by a regression line?

52. When all points in a scatter plot fall on the regression line, what is the value of the correlation coefficient? Describe what this means.

For the pairs of quantities in Exercises 53–56, describe whether a scatter plot will show a positive correlation, a negative correlation, or no correlation. If there is a correlation, is it strong, moderate, or weak? Explain your answers.

53. Height and weight

54. Number of days absent and grade in a course

55. Height and grade in a course

56. Hours of television watched and grade in a course

57. Explain how to use the correlation coefficient for a sample to determine if there is a correlation in the population.

Critical Thinking Exercises

Make Sense? *In Exercises 58–61, determine whether each statement makes sense or does not make sense, and explain your reasoning.*

58. I found a strong positive correlation for the data in Exercise 7 relating the percentage of teenagers in various countries who have used marijuana and the percentage who have used other drugs. I concluded that using marijuana leads to the use of other drugs.

59. I found a strong negative correlation for the data in Exercise 8 relating the percentage of people in various countries who are literate and the percentage who are undernourished. I concluded that an increase in literacy causes a decrease in undernourishment.

60. I'm working with a data set for which the correlation coefficient and the slope of the regression line have opposite signs.

61. I read that there is a correlation of 0.72 between IQ scores of identical twins reared apart, so I would expect a significantly lower correlation, approximately 0.52, between IQ scores of identical twins reared together.

62. Give an example of two variables with a strong correlation, where each variable is not the cause of the other.

Technology Exercise

63. Use the linear regression feature of a graphing calculator to verify your work in any two exercises from Exercises 35–38, parts (a) and (b).

What explanations can you offer for the correlation coefficient in part (a)?

Group Exercises

64. The group should select two variables related to people on your campus that it believes have a strong positive or negative correlation. Once these variables have been determined,

a. Collect at least 30 ordered pairs of data (x, y) from a sample of people on your campus.

b. Draw a scatter plot for the data collected.

c. Does the scatter plot indicate a positive correlation, a negative correlation, or no relationship between the variables?

d. Calculate r. Does the value of r reinforce the impression conveyed by the scatter plot?

e. Find the equation of the regression line.

f. Use the regression line's equation to make a prediction about a y-value given an x-value.

g. Are the results of this project consistent with the group's original belief about the correlation between the variables, or are there some surprises in the data collected?

65. What is the opinion of students on your campus about . . .? Group members should begin by deciding on some aspect of college life around which student opinion can be polled. The poll should consist of the question, "What is your opinion of . . .?" Be sure to provide options such as excellent, good, average, poor, horrible, or a 1-to-10 scale, or possibly grades of A, B, C, D, F. Use a random sample of students on your campus and conduct the opinion survey. After collecting the data, present and interpret the data using as many of the skills and techniques learned in this chapter as possible.

Chapter Summary, Review, and Test

SUMMARY – DEFINITIONS AND CONCEPTS	EXAMPLES
12.1 Sampling, Frequency Distributions, and Graphs	
a. A population is the set containing all objects whose properties are to be described and analyzed. A sample is a subset of the population.	Ex. 1, p. 773
b. Random samples are obtained in such a way that each member of the population has an equal chance of being selected.	Ex. 2, p. 774
c. Data can be organized and presented in frequency distributions, grouped frequency distributions, histograms, frequency polygons, and stem-and-leaf plots.	Ex. 3, p. 776; Ex. 4, p. 777; Figures 12.2 and 12.3, p. 778; Ex. 5, p. 778
d. The box on page 780 lists some things to watch for in visual displays of data.	Table 12.5, p. 781
12.2 Measures of Central Tendency	
a. The mean, $\bar{x}$, is the sum of the data items divided by the number of items: $\bar{x} = \frac{\Sigma x}{n}$.	Ex. 1, p. 786
b. The mean, $\bar{x}$, of a frequency distribution is computed using $\bar{x} = \frac{\Sigma xf}{n}$, where x is a data value, f is its frequency, and n is the total frequency of the distribution.	Ex. 2, p. 788
c. The median of ranked data is the item in the middle or the mean of the two middlemost items. The median is the value in the $\frac{n+1}{2}$ position in the list of ranked data.	Ex. 3, p. 789; Ex. 4, p. 790; Ex. 5, p. 791; Ex. 6, p. 792

d. When one or more data items are much greater than or much less than the other items, these extreme values greatly influence the mean, often making the median more representative of the data.	Ex. 7, p. 792
e. The mode of a data set is the value that occurs most often. If there is no such value, there is no mode. If more than one data value has the highest frequency, then each of these data values is a mode.	Ex. 8, p. 794
f. The midrange is computed using $\frac{\text{lowest data value} + \text{highest data value}}{2}$.	Ex. 9, p. 795; Ex. 10, p. 796

12.3 Measures of Dispersion

a. Range = highest data value − lowest data value	Ex. 1, p. 801
b. Standard deviation $= \sqrt{\frac{\Sigma(\text{data item} - \text{mean})^2}{n-1}}$ This is symbolized by $s = \sqrt{\frac{\Sigma(x-\bar{x})^2}{n-1}}$.	Ex. 2, p. 801; Ex. 3, p. 803; Ex. 4, p. 804
c. As the spread of data items increases, the standard deviation gets larger.	Ex. 5, p. 805

12.4 The Normal Distribution

a. The normal distribution is a theoretical distribution for the entire population. The distribution is bell shaped and symmetric about a vertical line through its center, where the mean, median, and mode are located.	
b. The 68–95–99.7 Rule Approximately 68% of the data items fall within 1 standard deviation of the mean. Approximately 95% of the data items fall within 2 standard deviations of the mean. Approximately 99.7% of the data items fall within 3 standard deviations of the mean.	Ex. 1, p. 810; Ex. 2, p. 811
c. A z-score describes how many standard deviations a data item in a normal distribution lies above or below the mean. $z\text{-score} = \frac{\text{data item} - \text{mean}}{\text{standard deviation}}$	Ex. 3, p. 812; Ex. 4, p. 813; Ex. 5, p. 814
d. If n% of the items in a distribution are less than a particular data item, that data item is in the nth percentile of the distribution. The 25th percentile is the first quartile, the 50th percentile, or the median, is the second quartile, and the 75th percentile is the third quartile.	Ex. 6, p. 815; Figure 12.17, p. 815
e. If a statistic is obtained from a random sample of size n, there is a 95% probability that it lies within $\frac{1}{\sqrt{n}} \times 100\%$ of the true population statistic. $\pm\frac{1}{\sqrt{n}} \times 100\%$ is called the margin of error.	Ex. 7, p. 816
f. A distribution of data is skewed if a large number of data items are piled up at one end or the other, with a "tail" at the opposite end.	Figure 12.22, p. 818; Figure 12.23, p. 818

12.5 Problem Solving with the Normal Distribution

a. A table showing z-scores and their percentiles can be used to find the percentage of data items less than or greater than a given data item in a normal distribution, as well as the percentage of data items between two given items. See the boxed summary on computing percentage of data items on page 826.	Ex. 1, p. 823; Ex. 2, p. 824; Ex. 3, p. 825

12.6 Scatter Plots, Correlation, and Regression Lines

a. A plot of data points is called a scatter plot. If the points lie approximately along a line, the line that best fits the data is called a regression line.	
b. A correlation coefficient, r, measures the strength and direction of a possible relationship between variables. If $r = 1$, there is a perfect positive correlation, and if $r = -1$, there is a perfect negative correlation. If $r = 0$, there is no relationship between the variables. **Table 12.18** on page 834 indicates whether r denotes a correlation in the population.	Ex. 1, p. 830; Ex. 4, p. 834
c. The formula for computing the correlation coefficient, r, is given in the box on page 830. The equation of the regression line is given in the box on page 832.	Ex. 2, p. 831; Ex. 3, p. 833

Review Exercises

12.1

1. The government of a large city wants to know if its citizens will support a three-year tax increase to provide additional support to the city's community college system. The government decides to conduct a survey of the city's residents before placing a tax increase initiative on the ballot. Which one of the following is most appropriate for obtaining a sample of the city's residents?
   - **a.** Survey a random sample of persons within each geographic region of the city.
   - **b.** Survey a random sample of community college professors living in the city.
   - **c.** Survey every tenth person who walks into the city's government center on two randomly selected days of the week.
   - **d.** Survey a random sample of persons within each geographic region of the state in which the city is located.

A random sample of ten college students is selected and each student is asked how much time he or she spent on homework during the previous weekend. The following times, in hours, are obtained:

$$8, 10, 9, 7, 9, 8, 7, 6, 8, 7.$$

Use these data items to solve Exercises 2–4.

2. Construct a frequency distribution for the data.
3. Construct a histogram for the data.
4. Construct a frequency polygon for the data.

The 50 grades on a physiology test are shown. Use the data to solve Exercises 5–6.

44	24	54	81	18
34	39	63	67	60
72	36	91	47	75
57	74	87	49	86
59	14	26	41	90
13	29	13	31	68
63	35	29	70	22
95	17	50	42	27
73	11	42	31	69
56	40	31	45	51

5. Construct a grouped frequency distribution for the data. Use 0–39 for the first class, 40–49 for the second class, and make each subsequent class width the same as the second class.
6. Construct a stem-and-leaf plot for the data.

7. Describe what is misleading about the size of the barrels in the following visual display.

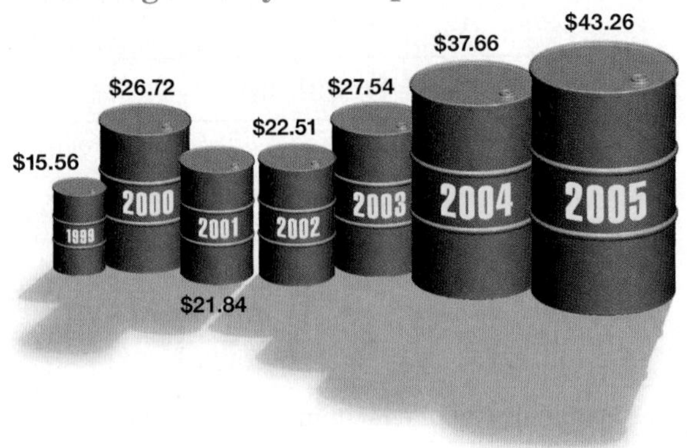

Source: U.S. Department of Energy

12.2

In Exercises 8–9, find the mean for each group of data items.

8. 84, 90, 95, 89, 98
9. 33, 27, 9, 10, 6, 7, 11, 23, 27
10. Find the mean for the data items in the given frequency distribution.

Score x	Frequency f
1	2
2	4
3	3
4	1

In Exercises 11–12, find the median for each group of data items.

11. 33, 27, 9, 10, 6, 7, 11, 23, 27
12. 28, 16, 22, 28, 34
13. Find the median for the data items in the frequency distribution in Exercise 10.

In Exercises 14–15, find the mode for each group of data items. If there is no mode, so state.

14. 33, 27, 9, 10, 6, 7, 11, 23, 27
15. 582, 585, 583, 585, 587, 587, 589
16. Find the mode for the data items in the frequency distribution in Exercise 10.

In Exercises 17–18, find the midrange for each group of data items.

17. 84, 90, 95, 88, 98
18. 33, 27, 9, 10, 6, 7, 11, 23, 27
19. Find the midrange for the data items in the frequency distribution in Exercise 10.

20. A student took seven tests in a course, scoring between 90% and 95% on three of the tests, between 80% and 89% on three of the tests, and below 40% on one of the tests. In this distribution, is the mean or the median more representative of the student's overall performance in the course? Explain your answer.

21. The data items below are the ages of U.S. presidents at the time of their first inauguration.

57 61 57 57 58 57 61 54 68 51 49 64 50 48 65

52 56 46 54 49 51 47 55 55 54 42 51 56 55 51

54 51 60 62 43 55 56 61 52 69 64 46 54 47 70

a. Organize the data in a frequency distribution.

b. Use the frequency distribution to find the mean age, median age, modal age, and midrange age of the presidents when they were inaugurated.

12.3

In Exercises 22–23, find the range for each group of data items.

22. 28, 34, 16, 22, 28

23. 312, 783, 219, 312, 426, 219

24. The mean for the data items 29, 9, 8, 22, 46, 51, 48, 42, 53, 42 is 35. Find **a.** the deviation from the mean for each data item and **b.** the sum of the deviations in part (a).

25. Use the data items 36, 26, 24, 90, and 74 to find **a.** the mean, **b.** the deviation from the mean for each data item, and **c.** the sum of the deviations in part (b).

In Exercises 26–27, find the standard deviation for each group of data items.

26. 3, 3, 5, 8, 10, 13

27. 20, 27, 23, 26, 28, 32, 33, 35

28. A test measuring anxiety levels is administered to a sample of ten college students with the following results. (High scores indicate high anxiety.)

10, 30, 37, 40, 43, 44, 45, 69, 86, 86

Find the mean, range, and standard deviation for the data.

29. Compute the mean and the standard deviation for each of the following data sets. Then, write a brief description of similarities and differences between the two sets based on each of your computations.

Set A: 80, 80, 80, 80 Set B: 70, 70, 90, 90

30. Describe how you would determine

a. which of the two groups, men or women, at your college has a higher mean grade point average.

b. which of the groups is more consistently close to its mean grade point average.

12.4

The scores on a test are normally distributed with a mean of 70 and a standard deviation of 8. In Exercises 31–33, find the score that is

31. 2 standard deviations above the mean.

32. $3\frac{1}{2}$ standard deviations above the mean.

33. $1\frac{1}{4}$ standard deviations below the mean.

The ages of people living in a retirement community are normally distributed with a mean age of 68 years and a standard deviation of 4 years. In Exercises 34–40, use the 68–95–99.7 Rule to find the percentage of people in the community whose ages

34. are between 64 and 72.

35. are between 60 and 76.

36. are between 68 and 72.

37. are between 56 and 80.

38. exceed 72.

39. are less than 72.

40. exceed 76.

A set of data items is normally distributed with a mean of 50 and a standard deviation of 5. In Exercises 41–45, convert each data item to a z-score.

41. 50 **42.** 60

43. 58 **44.** 35

45. 44

46. A student scores 60 on a vocabulary test and 80 on a grammar test. The data items for both tests are normally distributed. The vocabulary test has a mean of 50 and a standard deviation of 5. The grammar test has a mean of 72 and a standard deviation of 6. On which test did the student have the better score? Explain why this is so.

The number of miles that a particular brand of car tires lasts is normally distributed with a mean of 32,000 miles and a standard deviation of 4000 miles. In Exercises 47–49, find the data item in this distribution that corresponds to the given z-score.

47. $z = 1.5$

48. $z = 2.25$

49. $z = -2.5$

50. Using a random sample of 2281 American adults ages 18 and older, an Adecco survey asked respondents if they would be willing to sacrifice a percentage of their salary in order to work for an environmentally friendly company. The poll indicated that 31% of the respondents said "yes," 39% said "no," and 30% declined to answer.

a. Find the margin of error, to the nearest tenth of a percent, for this survey.

b. Write a statement about the percentage of American adults who would be willing to sacrifice a percentage of their salary in order to work for an environmentally friendly company.

51. The histogram indicates the frequencies of the number of syllables per word for 100 randomly selected words in Japanese.

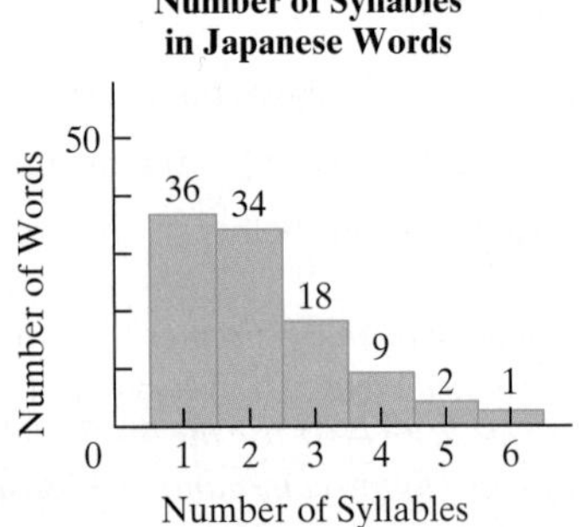

a. Is the shape of this distribution best classified as normal, skewed to the right, or skewed to the left?

b. Find the mean, median, and mode for the number of syllables in the sample of Japanese words.

c. Are the measures of central tendency from part (b) consistent with the shape of the distribution that you described in part (a)? Explain your answer.

12.5

The mean cholesterol level for all men in the United States is 200 and the standard deviation is 15. In Exercises 52–55, use **Table 12.16** *on page 822 to find the percentage of U.S. men whose cholesterol level*

52. is less than 221.

53. is greater than 173.

54. is between 173 and 221.

55. is between 164 and 182.

Use the percentiles for the weights of adult men over 40 to solve Exercises 56–58.

Weight	Percentile
235	86
227	third quartile
180	second quartile
173	first quartile

Find the percentage of men over 40 who weigh

56. less than 227 pounds.

57. more than 235 pounds.

58. between 227 and 235 pounds.

12.6

In Exercises 59–60, make a scatter plot for the given data. Use the scatter plot to describe whether or not the variables appear to be related.

59.

x	1	3	4	6	8	9
y	1	2	3	3	5	5

60.

Country	Canada	U.S.	Mexico	Brazil	Costa Rica
Life expectancy in years, x	81	78	76	72	77
Infant deaths per 1000 births, y	5.1	6.3	19.0	23.3	9.0

Denmark	China	Egypt	Pakistan	Bangla-desh	Australia	Japan	Russia
78	73	72	64	63	82	82	66
4.4	21.2	28.4	66.9	57.5	4.8	2.8	10.8

Source: U.S. Bureau of the Census International Database

The scatter plot shows the relationship between the percentage of adult females in a country who are literate and the mortality of children under five. Also shown is the regression line. Use this information to determine whether each of the statements in Exercises 61–67 is true or false.

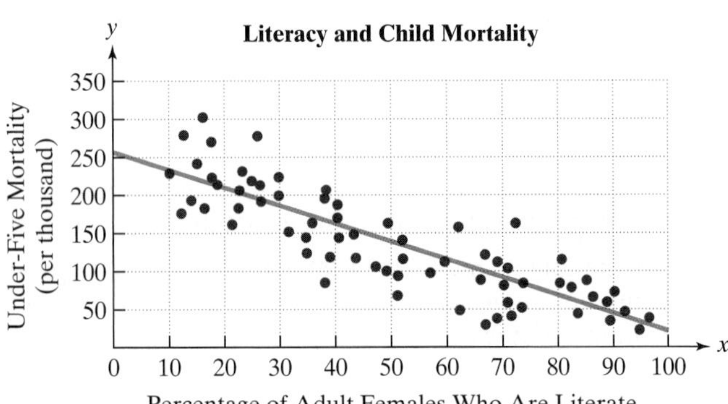

Source: United Nations

61. There is a perfect negative correlation between the percentage of adult females who are literate and under-five mortality.

62. As the percentage of adult females who are literate increases, under-five mortality tends to decrease.

63. The country with the least percentage of adult females who are literate has the greatest under-five mortality.

64. No two countries have the same percentage of adult females who are literate but different under-five mortalities.

65. There are more than 20 countries in this sample.

66. There is no correlation between the percentage of adult females who are literate and under-five mortality.

67. The country with the greatest percentage of adult females who are literate has an under-five mortality rate that is less than 50 children per thousand.

68. Which one of the following scatter plots indicates a correlation coefficient of approximately −0.9?

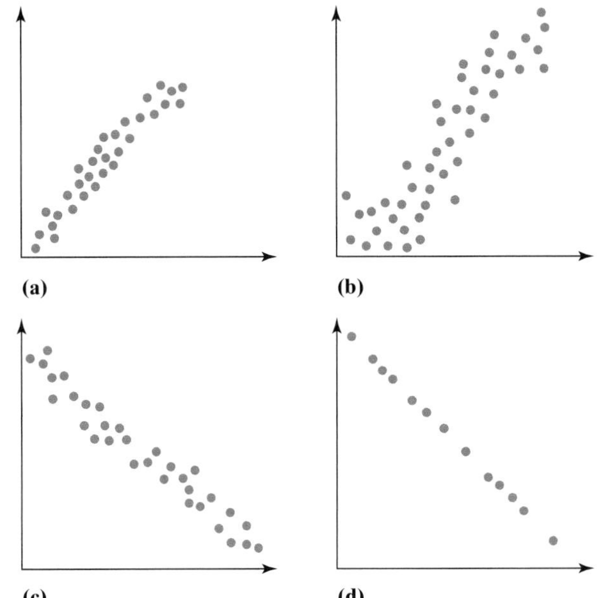

69. Use the data in Exercise 59 to solve this exercise.

a. Compute r, the correlation coefficient, rounded to the nearest thousandth.

b. Find the equation of the regression line.

70. The graph, based on Nielsen Media Research data taken from random samples of Americans at various ages, indicates that as we get older, we watch more television. (The graph includes "traditional TV" viewing on set-up boxes and does include viewing via TV-connected devices.)

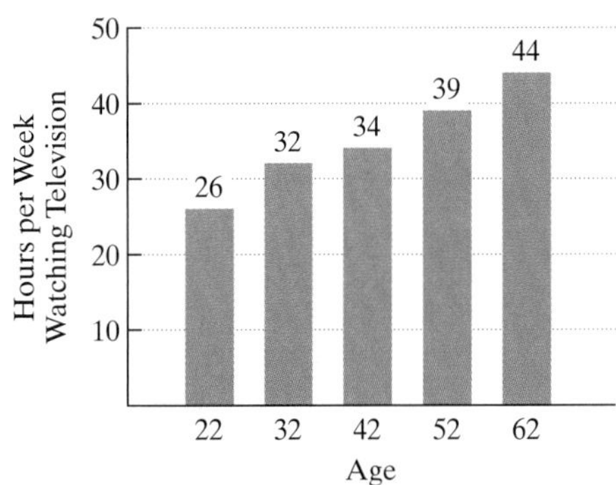

Source: Nielsen Media Research

a. Let x represent one's age and let y represent hours per week watching television. Calculate the correlation coefficient.

b. Using **Table 12.18** on page 834 and the $\alpha = 0.05$ column, determine whether there is a correlation between age and time spent watching television in the American population.

Chapter 12 Test

1. Politicians in the Florida Keys need to know if the residents of Key Largo think the amount of money charged for water is reasonable. The politicians decide to conduct a survey of a sample of Key Largo's residents. Which procedure would be most appropriate for a sample of Key Largo's residents?

a. Survey all water customers who pay their water bills at Key Largo City Hall on the third day of the month.

b. Survey a random sample of executives who work for the water company in Key Largo.

c. Survey 5000 individuals who are randomly selected from a list of all people living in Georgia and Florida.

d. Survey a random sample of persons within each neighborhood of Key Largo.

Use these scores on a ten-point quiz to solve Exercises 2–4.

8, 5, 3, 6, 5, 10, 6, 9, 4, 5, 7, 9, 7, 4, 8, 8

2. Construct a frequency distribution for the data.

3. Construct a histogram for the data.

4. Construct a frequency polygon for the data.

Use the 30 test scores listed below to solve Exercises 5–6.

79	51	67	50	78
62	89	83	73	80
88	48	60	71	79
89	63	55	93	71
41	81	46	50	61
59	50	90	75	61

5. Construct a grouped frequency distribution for the data. Use 40–49 for the first class and use the same width for each subsequent class.

6. Construct a stem-and-leaf display for the data.

7. The graph shows the percentage of students in the United States through grade 12 who were home-schooled in 1999 and 2007. What impression does the roofline in the visual display imply about what occurred in 2000 through 2006? How might this be misleading?

Source: National Center for Education Statistics

Use the six data items listed below to solve Exercises 8–11.

$$3, 6, 2, 1, 7, 3$$

8. Find the mean.
9. Find the median.
10. Find the midrange.
11. Find the standard deviation.

Use the frequency distribution shown to solve Exercises 12–14.

Score x	Frequency f
1	3
2	5
3	2
4	2

12. Find the mean.
13. Find the median.
14. Find the mode.
15. The annual salaries of four salespeople and the owner of a bookstore are

$$\$17{,}500, \$19{,}000, \$22{,}000, \$27{,}500, \$98{,}500.$$

Is the mean or the median more representative of the five annual salaries? Briefly explain your answer.

According to the American Freshman, *the number of hours that college freshmen spend studying each week is normally distributed with a mean of 7 hours and a standard deviation of 5.3 hours. In Exercises 16–17, use the 68–95–99.7 Rule to find the percentage of college freshmen who study*

16. between 7 and 12.3 hours each week.
17. more than 17.6 hours each week.
18. IQ scores are normally distributed in the population. Who has a higher IQ: a student with a 120 IQ on a scale where 100 is the mean and 10 is the standard deviation, or a professor with a 128 IQ on a scale where 100 is the mean and 15 is the standard deviation? Briefly explain your answer.
19. Use the z-scores and the corresponding percentiles shown at the top of the next column to solve this exercise. Test scores are normally distributed with a mean of 74 and a standard deviation of 10. What percentage of the scores are above 88?

z-Score	Percentile
1.1	86.43
1.2	88.49
1.3	90.32
1.4	91.92
1.5	93.32

20. Use the percentiles in the table shown below to find the percentage of scores between 630 and 690.

Score	Percentile
780	99
750	87
720	72
690	49
660	26
630	8
600	1

21. Using a random sample of 100 students from a campus of approximately 12,000 students, 60% of the students in the sample said they were very satisfied with their professors.
    a. Find the margin of error in this percent.
    b. Write a statement about the percentage of the entire population of students from this campus who are very satisfied with their professors.
22. Make a scatter plot for the given data. Use the scatter plot to describe whether or not the variables appear to be related.

x	1	4	3	5	2
y	5	2	2	1	4

The scatter plot shows the number of minutes each of 16 people exercise per week and the number of headaches per month each person experiences. Use the scatter plot to determine whether each of the statements in Exercises 23–25 is true or false.

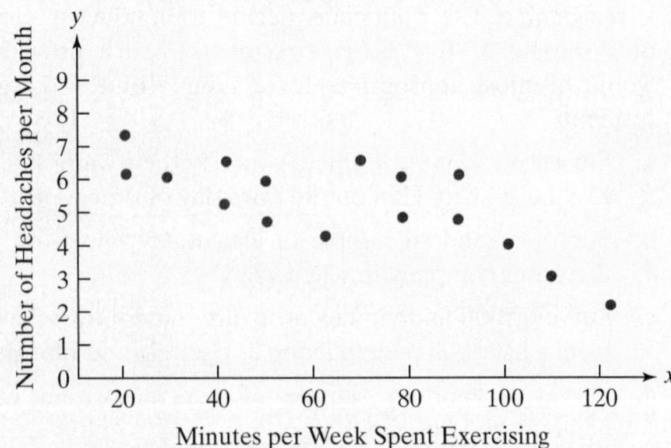

23. An increase in the number of minutes devoted to exercise causes a decrease in headaches.
24. There is a perfect negative correlation between time spent exercising and number of headaches.
25. The person who exercised most per week had the least number of headaches per month.
26. Is the relationship between the price of gas and the number of people visiting our national parks a positive correlation, a negative correlation, or is there no correlation? Explain your answer.

Voting and Apportionment

13

AS CITIZENS OF A FREE SOCIETY, WE HAVE BOTH A RIGHT AND A DUTY TO VOTE. THE POLITICAL, SOCIAL, FINANCIAL, AND ENVIRONMENTAL CHOICES WE make in elections can affect every aspect of our lives. However, voting is only part of the story. The second part—methods for counting our votes—lies at the heart of the democratic process.

In this chapter, you will learn the role that mathematics plays in finding our collective voice when faced with more than two options in an election. You will see that determining the winning option may depend on the method used for counting votes, as well as on the issues and the preferences of the voters. Is there a method for counting votes that is always fair? In exploring voting theory, as well as the ideal of "one person, one vote," you will begin to understand some of the mathematical paradoxes of our democracy. This will increase your ability to function as a more fully aware citizen.

Here's where you'll find these applications:

Voting methods and their mathematical paradoxes are developed in the first half of the chapter. We conclude the chapter with a discussion of apportionment methods, such as how to determine the number of representatives each state has in the U.S. House of Representatives and the flaws that can occur with these methods.

13.1 Voting Methods

WHAT AM I SUPPOSED TO LEARN?

After studying this section, you should be able to:

1. Understand and use preference tables.
2. Use the plurality method to determine an election's winner.
3. Use the Borda count method to determine an election's winner.
4. Use the plurality-with-elimination method to determine an election's winner.
5. Use the pairwise comparison method to determine an election's winner.

Orson Welles (center), star and director of the highly innovative *Citizen Kane*

IN THE DRAMATIC ARTS, OURS IS THE ERA OF THE MOVIES. AS INDIVIDUALS and as a nation, we've grown up with them. Our images of love, war, family, country—even of things that terrify us—owe much to what we've seen on screen.

To celebrate one hundred years of movie making (1896–1996), more than 1500 ballots were sent to leaders in the industry to rank the top American films. The winner? *Citizen Kane* (1941). Controversial outcome? You bet. In this section, you will see how an outcome can be greatly influenced by the voting method used. We begin our discussion of voting theory with a group of students who love movies and are attempting to elect a president of their club. We will use this example throughout the section. Its importance is not in the example itself, but in its implications for conducting fair elections in a democracy.

"What is a vote if it isn't an expression of hope?"
—Lawrence O'Donnell, television journalist

Preference Tables

1 Understand and use preference tables.

Four candidates are running for president of the Student Film Institute: Paul (P), Rita (R), Sarah (S), and Tim (T). Each of the club's 37 members submits a secret ballot indicating his or her first, second, third, and fourth choice for president. The 37 ballots are shown in **Figure 13.1**.

Ballot 1	Ballot 2	Ballot 3	Ballot 4	Ballot 5	Ballot 6	Ballot 7	Ballot 8	Ballot 9	Ballot 10	Ballot 11		
1st P	1st S	1st R	1st P	1st R	1st S	1st R	1st S	1st P	1st S	1st P		
2nd R	2nd R	2nd T	2nd R	2nd T	2nd R	2nd T	2nd R	2nd R	2nd R	2nd R		
3rd S	3rd T	3rd S	3rd S	3rd S	3rd T	3rd S	3rd T	3rd S	3rd T	3rd S		
4th T	4th P	4th P	4th T	4th P	4th P	4th P	4th P	4th T	4th P	4th T		
Ballot 12	**Ballot 13**	**Ballot 14**	**Ballot 15**	**Ballot 16**	**Ballot 17**	**Ballot 18**	**Ballot 19**	**Ballot 20**	**Ballot 21**	**Ballot 22**	**Ballot 23**	**Ballot 24**
1st T	1st P	1st T	1st S	1st P	1st T	1st R	1st P	1st S	1st P	1st P	1st T	1st P
2nd S	2nd R	2nd S	2nd R	2nd R	2nd S	2nd T	2nd R	2nd T	2nd R	2nd R	2nd S	2nd R
3rd R	3rd S	3rd R	3rd T	3rd S	3rd R	3rd S	3rd S	3rd R	3rd S	3rd S	3rd R	3rd S
4th P	4th T	4th P	4th P	4th T	4th P	4th P	4th T	4th P	4th T	4th T	4th P	4th T
Ballot 25	**Ballot 26**	**Ballot 27**	**Ballot 28**	**Ballot 29**	**Ballot 30**	**Ballot 31**	**Ballot 32**	**Ballot 33**	**Ballot 34**	**Ballot 35**	**Ballot 36**	**Ballot 37**
1st T	1st P	1st P	1st S	1st P	1st S	1st T	1st S	1st P	1st T	1st T	1st S	1st S
2nd S	2nd R	2nd R	2nd R	2nd R	2nd R	2nd S	2nd R	2nd R	2nd S	2nd S	2nd R	2nd R
3rd R	3rd S	3rd S	3rd T	3rd S	3rd T	3rd R	3rd T	3rd S	3rd R	3rd R	3rd T	3rd T
4th P	4th T	4th T	4th P	4th T	4th P	4th P	4th P	4th T	4th P	4th P	4th P	4th P

FIGURE 13.1 The 37 ballots for the Student Film Institute election

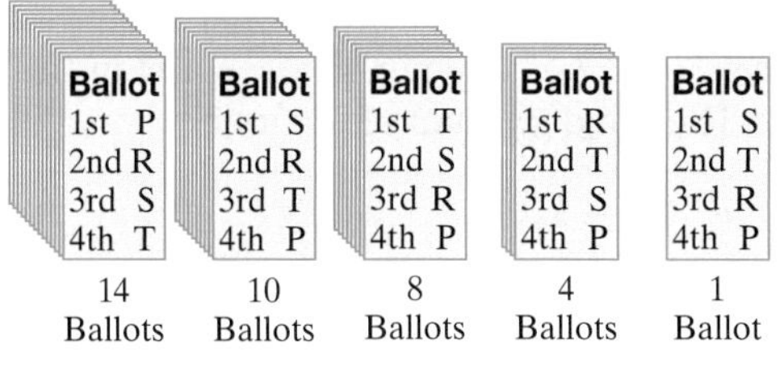

FIGURE 13.2 The 37 election ballots placed into identical stacks

Do you see that several club members had identical ballots? For example, ballots 3, 5, 7, and 18 ranked the candidates exactly the same way, namely RTSP. We can group together these identical ballots. **Figure 13.2** shows the 37 ballots placed in identical stacks.

Preference ballots are ballots in which a voter is asked to rank all the candidates in order of preference. Rather than placing the preference ballots in identical stacks, we can use a *preference table* to summarize the results of the election. A **preference table** shows how often each particular outcome occurred. **Table 13.1** shows a preference table for the 37 ballots cast for president of the Student Film Institute.

There were 14 + 10 + 8 + 4 + 1, or 37, voters in the election.

TABLE 13.1 Preference Table for the Student Film Institute Election

Number of Votes	14	10	8	4	1
First Choice	P	S	T	R	S
Second Choice	R	R	S	T	T
Third Choice	S	T	R	S	R
Fourth Choice	T	P	P	P	P

14 ballots rank the candidates P, R, S, T.

10 ballots rank the candidates S, R, T, P.

8 ballots rank the candidates T, S, R, P.

4 ballots rank the candidates R, T, S, P.

1 ballot ranks the candidates S, T, R, P.

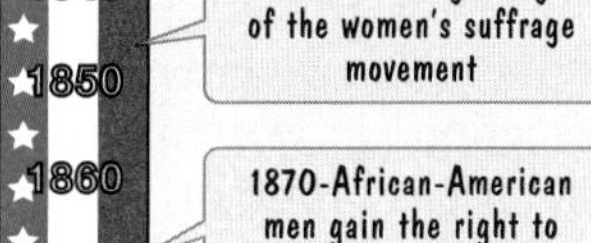

A Brief History of the Right to Vote

1840 1850 1860 1870 1880 1890 1900 1910 1920 1930 1940 1950 1960 1970 1980 1990 2000 2010 2020

1848-The beginning of the women's suffrage movement

1870-African-American men gain the right to vote with the ratification of the 15th Amendment.

1894-Congress repeals laws enforcing the 15th Amendment.

1920-Women gain the right to vote with the ratification of the 19th Amendment.

1965-The Voting Rights Act prohibits the South's practice of excluding African Americans from voting.

1971-The voting age is lowered to 18 with the ratification of the 26th Amendment.

1993-The National Voter Registration Act simplifies the registration process.

2014-By a 5 to 4 decision, the U.S. Supreme Court strikes down key passages of the 1965 Voting Rights Act.

EXAMPLE 1 *Understanding a Preference Table*

Four candidates are running for mayor of Smallville: Antonio (A), Bob (B), Carmen (C), and Donna (D). The voters were asked to rank all the candidates in order of preference. **Table 13.2** shows the preference table for this election.

TABLE 13.2 Preference Table for the Smallville Mayoral Election

Number of Votes	130	120	100	150
First Choice	A	D	D	C
Second Choice	B	B	B	B
Third Choice	C	C	A	A
Fourth Choice	D	A	C	D

a. How many people voted in the election?

b. How many people selected the candidates in this order: D, B, A, C?

c. How many people selected Donna (D) as their first choice for mayor?

SOLUTION

a. We find the number of people who voted in the election by adding the numbers in the row labeled *Number of Votes*:

$$130 + 120 + 100 + 150 = 500.$$

Thus, 500 people voted in the election.

b. We find the number of people selecting the candidates in the order D, B, A, C by referring to the third column of letters in the preference table. Above this column is the number 100. Thus, 100 people voted for the candidates in the order D, B, A, C.

TABLE 13.2 Preference Table for the Smallville Mayoral Election (repeated)

Number of Votes	130	120	100	150
First Choice	A	D	D	C
Second Choice	B	B	B	B
Third Choice	C	C	A	A
Fourth Choice	D	A	C	D

c. We find the number of people who voted for D as their first choice by reading across the row that says *First Choice*. When you see a D in this row, write the number above it. Then find the sum of the numbers:

$$120 + 100 = 220.$$

Thus, 220 voters selected Donna as their first choice for mayor.

CHECK POINT 1 Four candidates are running for student body president: Alan (A), Bonnie (B), Carlos (C), and Samir (S). The students were asked to rank all the candidates in order of preference. **Table 13.3** shows the preference table for this election.

TABLE 13.3 Preference Table for the Election of Student Body President

Number of Votes	2100	1305	765	40
First Choice	S	A	S	B
Second Choice	A	S	A	S
Third Choice	B	B	C	A
Fourth Choice	C	C	B	C

a. How many students voted in the election?

b. How many students selected the candidates in this order: B, S, A, C?

c. How many students selected Samir (S) as their first choice for student body president?

Now that you know how to summarize the results of an election using a preference table, it's time to decide the outcome of elections in general and the Student Film Institute election in particular. We will discuss four of the most popular voting methods:

1. the plurality method
2. the Borda count method
3. the plurality-with-elimination method
4. the pairwise comparison method.

2 *Use the plurality method to determine an election's winner.*

The Plurality Method

If there are three or more candidates in an election, it often happens that no single candidate receives a majority of first-place votes—that is, more than 50% of the votes. Using the **plurality method**, the candidate with the most first-place votes wins. The only information that we use from the preference ballots are the votes for first place.

THE PLURALITY METHOD

The candidate (or candidates, if there is more than one) with the most first-place votes is the winner.

EXAMPLE 2 *Using the Plurality Method*

Table 13.1, repeated at the top of the next page, shows the preference table for the four candidates running for president of the Student Film Institute: Paul (P), Rita (R), Sarah (S), and Tim (T). Who is declared the winner using the plurality method?

SOLUTION

The candidate with the most first-place votes is the winner. When using **Table 13.1**, we only need to look at the row indicating the number of first-choice votes.

P received 14 first-place votes.
S received 10 + 1 = 11 first-place votes.
T received 8 first-place votes.
R received 4 first-place votes.

TABLE 13.1 Preference Table for the Student Film Institute Election

Number of Votes	**14**	**10**	**8**	**4**	**1**
First Choice	P	S	T	R	S
Second Choice	R	R	S	T	T
Third Choice	S	T	R	S	R
Fourth Choice	T	P	P	P	P

The voice balloon to the left of the preference table indicates that P (Paul) is the winner and the new president of the Student Film Institute. Although Paul received the most first-place votes, he received only $\frac{14}{37}$, or approximately 38%, of the first-place votes, which is less than a majority.

CHECK POINT 2 **Table 13.2** on page 847 shows the preference table for the four candidates running for mayor of Smallville: Antonio (A), Bob (B), Carmen (C), and Donna (D). Who is declared the winner using the plurality method?

3 *Use the Borda count method to determine an election's winner.*

The Borda Count Method

The **Borda count method**, developed by the French mathematician and naval captain Jean-Charles de Borda (1733–1799), assigns points to each candidate based on how they are ranked by the voters.

THE BORDA COUNT METHOD

Each voter ranks the candidates from the most favorable to the least favorable. Each last-place vote is given 1 point, each next-to-last-place vote is given 2 points, each third-from-last-place vote is given 3 points, and so on. The points are totaled for each candidate separately. The candidate with the most points is the winner.

When using the plurality method, all the information in the preference table not related to first place is ignored. Example 3 illustrates that this is not the case when using the Borda count method. The Borda count method takes into account all the information provided by the voters' preferences.

EXAMPLE 3 *Using the Borda Count Method*

Table 13.1, repeated below, shows the preference table for the four candidates running for president of the Student Film Institute: Paul (P), Rita (R), Sarah (S), and Tim (T). Who is declared the winner using the Borda count method?

TABLE 13.1 Preference Table for the Student Film Institute Election

Number of Votes	**14**	**10**	**8**	**4**	**1**
First Choice	P	S	T	R	S
Second Choice	R	R	S	T	T
Third Choice	S	T	R	S	R
Fourth Choice	T	P	P	P	P

SOLUTION

Because there are four candidates, a first-place vote is worth 4 points, a second-place vote is worth 3 points, a third-place vote is worth 2 points, and a fourth-place vote is worth 1 point. **Table 13.4** shows the points produced by the votes in the preference table.

TABLE 13.4 Points for the Student Film Institute Election

Number of Votes	**14**	**10**	**8**	**4**	**1**
First Choice: 4 points	P: $14 \times 4 = 56$ pts	S: $10 \times 4 = 40$ pts	T: $8 \times 4 = 32$ pts	R: $4 \times 4 = 16$ pts	S: $1 \times 4 = 4$ pts
Second Choice: 3 points	R: $14 \times 3 = 42$ pts	R: $10 \times 3 = 30$ pts	S: $8 \times 3 = 24$ pts	T: $4 \times 3 = 12$ pts	T: $1 \times 3 = 3$ pts
Third Choice: 2 points	S: $14 \times 2 = 28$ pts	T: $10 \times 2 = 20$ pts	R: $8 \times 2 = 16$ pts	S: $4 \times 2 = 8$ pts	R: $1 \times 2 = 2$ pts
Fourth Choice: 1 point	T: $14 \times 1 = 14$ pts	P: $10 \times 1 = 10$ pts	P: $8 \times 1 = 8$ pts	P: $4 \times 1 = 4$ pts	P: $1 \times 1 = 1$ pt

Now we read down each column and total the points for each candidate separately.

$$\begin{aligned} &\text{P gets } 56 + 10 + 8 + 4 + 1 = 79 \text{ points.}\\ &\text{R gets } 42 + 30 + 16 + 16 + 2 = 106 \text{ points.}\\ &\text{S gets } 28 + 40 + 24 + 8 + 4 = 104 \text{ points.}\\ &\text{T gets } 14 + 20 + 32 + 12 + 3 = 81 \text{ points.} \end{aligned}$$

Because R (Rita) has received the most points, she is the winner and the new president of the Student Film Institute.

In Examples 2 and 3, the plurality method and the Borda count method produced different election winners. This illustrates the importance of choosing a voting system prior to an election.

CHECK POINT 3 **Table 13.2** on page 847 shows the preference table for the four candidates running for mayor of Smallville: Antonio (A), Bob (B), Carmen (C), and Donna (D). Who is declared the winner using the Borda count method?

4 *Use the plurality-with-elimination method to determine an election's winner.*

The Plurality-with-Elimination Method

The **plurality-with-elimination method** is based on "survival of the fittest" and may involve a series of elections. If a candidate receives a majority of votes, that candidate is the "fittest" and is declared the winner. If no candidate receives a majority of votes, the candidate with the least number of votes is eliminated and a second election is held. The process is repeated until a candidate receives a majority.

Multiple elections are not necessary when using preference ballots. We remove the eliminated candidate from the preference table and assume that the relative preferences of a voter are not affected by the elimination of this candidate.

THE PLURALITY-WITH-ELIMINATION METHOD USING PREFERENCE BALLOTS

The candidate with the majority of first-place votes is the winner. If no candidate receives a majority of first-place votes, eliminate the candidate (or candidates if there is a tie) with the fewest first-place votes from the preference table. Move the candidates in each column below the eliminated candidate up one place. The candidate with the majority of first-place votes in the new preference table is the winner. If no candidate receives a majority of first-place votes, repeat this process until a candidate receives a majority.

Blitzer Bonus

The Academy Awards

Since 1927, the Academy of Motion Pictures of Arts and Sciences has given its annual Academy Awards ("Oscars") for various achievements in connection with motion pictures. In its early years, the studio bosses controlled the nominations and virtually handpicked the winners. It took years for Oscar to clean house; the Academy did not even start using sealed envelopes until 1941. Since then, the Academy, which has grown from a scant 36 members in 1927 to between 6000 and 6500 voting members of today, has tried to discourage ad campaigns to influence voting. It hasn't succeeded, of course.

The election process for best picture takes place in two stages. First comes the nomination stage, in which the top five pictures are nominated. This stage involves a variation of the plurality-with-elimination method, where the picture with the fewest first-place votes is eliminated. Once the top pictures are nominated, each Academy member votes for one candidate, and the winner is chosen by the plurality method. Oscar's imprimatur can add $25 million or more to a winning film's gross.

EXAMPLE 4 Using the Plurality-with-Elimination Method

Table 13.1, repeated below, shows the preference table for the four candidates running for president of the Student Film Institute: Paul (P), Rita (R), Sarah (S), and Tim (T). Who is declared the winner using the plurality-with-elimination method?

TABLE 13.1 Preference Table for the Student Film Institute Election

Number of Votes	**14**	**10**	**8**	**4**	**1**
First Choice	P	S	T	R	S
Second Choice	R	R	S	T	T
Third Choice	S	T	R	S	R
Fourth Choice	T	P	P	P	P

SOLUTION

Recall that there are 14 + 10 + 8 + 4 + 1, or 37, people voting. In order to receive a majority, a candidate must receive more than 50% of the first-place votes, meaning 19 or more votes. The number of first-place votes for each candidate is

P(Paul) = 14   S(Sarah) = 10 + 1 = 11   T(Tim) = 8   R(Rita) = 4.

We see that no candidate received a majority of first-place votes. Because Rita received the fewest first-place votes, she is eliminated in the first round.

The new preference table with R eliminated is shown in **Table 13.5**. Compare this table with **Table 13.1**. For each column, each candidate below R moves up one place while the positions of candidates above R remain unchanged. Thus, once Rita is eliminated, the order of preference is preserved.

TABLE 13.5 Preference Table for the Student Film Institute Election with R Eliminated

Number of Votes	**14**	**10**	**8**	**4**	**1**
First Choice	P	S	T	T	S
Second Choice	S	T	S	S	T
Third Choice	T	P	P	P	P

Recall that to receive a majority, a candidate must get 19 or more first-place votes. The number of first-place votes for each candidate is now

P(Paul) = 14   S(Sarah) = 10 + 1 = 11   T(Tim) = 8 + 4 = 12.

Once again, no candidate receives a majority of first-place votes. Because Sarah received the fewest first-place votes, she is eliminated in the second round.

The new preference table with S eliminated is shown in **Table 13.6**. Compare this table with **Table 13.5**. For each column, each candidate below S moves up one place while the positions of candidates above S remain unchanged.

TABLE 13.6 Preference Table for the Student Film Institute Election with R and S Eliminated

Number of Votes	**14**	**10**	**8**	**4**	**1**
First Choice	P	T	T	T	T
Second Choice	T	P	P	P	P

TABLE 13.6 (repeated)

Number of Votes	14	10	8	4	1
First Choice	P	T	T	T	T
Second Choice	T	P	P	P	P

Keep in mind that a candidate must get 19 or more first-place votes to receive a majority. The number of first-place votes for each candidate is now

$$\text{P(Paul)} = 14 \qquad \text{T(Tim)} = 10 + 8 + 4 + 1 = 23.$$

Because T (Tim) has received the majority of first-place votes, he is the winner and the new president of the Student Film Institute.

We have now used three different voting methods in the same election. Each method resulted in a different winner. Can you now see how important it is to choose a voting system before an election takes place?

CHECK POINT 4 **Table 13.2** on page 847 shows the preference table for the four candidates running for mayor of Smallville: Antonio (A), Bob (B), Carmen (C), and Donna (D). Who is declared the winner using the plurality-with-elimination method?

Blitzer Bonus

Choosing a New Voting System in San Francisco

A little-noticed proposition approved in 2003 by San Francisco voters offers a glimpse of how we might vote in the future. Instead of casting their ballots for just one candidate, San Franciscans rank the candidates in most local races according to their first, second, and third choices. The election's winner is declared using the plurality-with-elimination method, also known as instant-runoff voting. San Francisco approved this new voting system after a runoff election for city attorney drew an abysmally low 16.6% of registered voters.

By 2012, San Francisco's ranked-choice voting system using the plurality-with-elimination method had fulfilled its promise, avoiding 15 runoff elections that would have cost taxpayers millions to administer. Elsewhere, in 2011, St. Paul, Minnesota, allowed six rankings with voting machines similar to San Francisco's. Portland, Maine, allowed 15 rankings. By 2013, 20 states had considered ranked-choice voting in various elections. The biggest hurdle involved electronic voting machines that lacked tabulating software.

5 *Use the pairwise comparison method to determine an election's winner.*

The Pairwise Comparison Method

Using the **pairwise comparison method**, every candidate is compared one-on-one with every other candidate. For example, the candidates running for president of the Student Film Institute are Paul, Rita, Sarah, and Tim. Thus, we make the following comparisons:

Paul vs. Rita Paul vs. Sarah Paul vs. Tim
Rita vs. Sarah Rita vs. Tim Sarah vs. Tim.

We must make six comparisons.

Is there a formula for how many comparisons are needed when there are n candidates? The answer is yes. We are basically counting the number of combinations possible if two candidates are taken from n candidates. In Section 11.3, we saw that the number of possible combinations if r items are taken from n items is

$${}_nC_r = \frac{n!}{(n-r)!\,r!}.$$

The number of comparisons needed if two candidates are taken from n candidates is

$${}_nC_2 = \frac{n!}{(n-2)!\,2!} = \frac{n(n-1)\cancel{(n-2)!}}{\cancel{(n-2)!}\,2\cdot 1} = \frac{n(n-1)}{2}.$$

THE NUMBER OF COMPARISONS USING THE PAIRWISE COMPARISON METHOD

In an election with n candidates, the number of comparisons, C, that must be made is

$$C = \frac{n(n-1)}{2}.$$

Blitzer Bonus

The Heisman Trophy

2010 Heisman Trophy winner, Cam Newton, from Auburn University

The Heisman Trophy, given each year since 1935 to the best college football player in the country, is chosen using a variation of the Borda count method. Voting is done by approximately 950 sports journalists and former Heisman winners. Each preference ballot allows for three names (out of thousands of college football players). A player receives three points for each first-place vote, two points for each second-place vote, and one point for each third-place vote.

Heisman Winners with the Most Points

1968—O. J. Simpson, USC, 2853
2005—Reggie Bush, USC, 2541
2006—Troy Smith, Ohio, 2540
1976—Tony Dorsett, Pittsburgh, 2357
1998—Ricky Williams, Texas, 2355
1993—Charlie Ward, FSU, 2310
2010—Cam Newton, Auburn, 2263
1984—Doug Flutie, Boston College, 2240

Source: ESPN Sports Almanac

How do we make each of the $\frac{n(n-1)}{2}$ comparisons using the pairwise comparison method? We use the preference table and award points, as described in the following box:

THE PAIRWISE COMPARISON METHOD

Voters rank all the candidates and the results are summarized in a preference table. The table is used to make a series of comparisons in which each candidate is compared to each of the other candidates. For each pair of candidates, X and Y, use the table to determine how many voters prefer X to Y and vice versa. If a majority prefer X to Y, then X receives 1 point. If a majority prefer Y to X, then Y receives 1 point. If the candidates tie, then each receives $\frac{1}{2}$ point. After all comparisons have been made, the candidate receiving the most points is the winner.

EXAMPLE 5 *Using the Pairwise Comparison Method*

Table 13.1, repeated below, shows the preference table for the four candidates running for president of the Student Film Institute: Paul (P), Rita (R), Sarah (S), and Tim (T). Who is declared the winner using the pairwise comparison method?

TABLE 13.1 Preference Table for the Student Film Institute Election

Number of Votes	**14**	**10**	**8**	**4**	**1**
First Choice	P	S	T	R	S
Second Choice	R	R	S	T	T
Third Choice	S	T	R	S	R
Fourth Choice	T	P	P	P	P

GREAT QUESTION!

When using the pairwise comparison method, how do I list all the comparisons?

Start by listing all the candidates. Then pair each candidate with every other person whose name follows in the list.

SOLUTION

Because there are four candidates, $n = 4$ and the number of comparisons we must make is

$$C = \frac{n(n-1)}{2} = \frac{4(4-1)}{2} = \frac{4 \cdot 3}{2} = \frac{12}{2} = 6.$$

Although you can make each of the six comparisons using **Table 13.1** on the previous page, we will show the table six times to make certain you can follow the details behind each comparison. With P, R, S, and T, the comparisons are P versus R, P versus S, P versus T, R versus S, R versus T, and S versus T.

P vs. R

14	10	8	4	1
[P]	S	T	[R]	S
(R)	[R]	S	T	T
S	T	[R]	S	[R]
T	(P)	(P)	(P)	(P)

14 voters prefer P to R. 10 + 8 + 4 + 1, or 23, voters prefer R to P.

Conclusion: R wins this comparison and gets 1 point.

P vs. S

14	10	8	4	1
[P]	[S]	T	R	[S]
R	R	[S]	T	T
(S)	T	R	[S]	R
T	(P)	(P)	(P)	(P)

14 voters prefer P to S. 10 + 8 + 4 + 1, or 23, voters prefer S to P.

Conclusion: S wins this comparison and gets 1 point.

P vs. T

14	10	8	4	1
[P]	S	[T]	R	S
R	R	S	[T]	[T]
S	[T]	R	S	R
(T)	(P)	(P)	(P)	(P)

14 voters prefer P to T. 10 + 8 + 4 + 1, or 23, voters prefer T to P.

Conclusion: T wins this comparison and gets 1 point.

R vs. S

14	10	8	4	1
P	[S]	T	[R]	[S]
[R]	(R)	[S]	T	T
(S)	T	(R)	(S)	(R)
T	P	P	P	P

14 + 4, or 18, voters prefer R to S. 10 + 8 + 1, or 19, voters prefer S to R.

Conclusion: S wins this comparison and gets 1 point.

R vs. T

14	10	8	4	1
P	S	[T]	[R]	S
[R]	[R]	S	(T)	[T]
S	(T)	(R)	S	(R)
(T)	P	P	P	P

14 + 10 + 4, or 28, voters prefer R to T. 8 + 1, or 9, voters prefer T to R.

Conclusion: R wins this comparison and gets 1 point.

S vs. T

14	10	8	4	1
P	[S]	[T]	R	[S]
R	R	(S)	[T]	(T)
[S]	(T)	R	(S)	R
(T)	P	P	P	P

14 + 10 + 1, or 25, voters prefer S to T. 8 + 4, or 12, voters prefer T to S.

Conclusion: S wins this comparison and gets 1 point.

Now let's use each of the six conclusions and add points for the six comparisons.

R gets $1 + 1 = 2$ points.

S gets $1 + 1 + 1 = 3$ points.

T gets 1 point.

After all comparisons have been made, the candidate receiving the most points is S (Sarah). Sarah is the winner and the new president of the Student Film Institute.

Blitzer Bonus

Using Democracy to Ensure a Candidate's Election

Donald Saari, author of *Chaotic Elections! A Mathematician Looks at Voting,* often begins his lectures by joking with the audience, while simultaneously telling them that a voting method can affect an election's outcome:

> For a price, I will serve as a consultant for your group for your next election. Tell me who you want to win and I will design a "democratic procedure" which ensures the election of your candidate.

CHECK POINT 5 **Table 13.2** on page 847 shows the preference table for the four candidates running for mayor of Smallville: Antonio (A), Bob (B), Carmen (C), and Donna (D). Make 6 comparisons (A vs. B, A vs. C, A vs. D, B vs. C, B vs. D, and C vs. D) and determine the winner using the pairwise comparison method.

A tie-breaking method should be determined before an election takes place. For example, prior to an election it can be announced that the plurality method will be used to determine the winner and the pairwise comparison method will decide the winner in the event that the plurality method leads to a tie.

GREAT QUESTION!

The examples in this section involved ballots where voters choose more than one candidate, ranking their choices in order of preference. Which voting methods require candidates to be ranked?

All candidates must be ranked by voters when using the Borda count and pairwise comparison methods. Ranking is not necessary with the plurality method, in which voters need only select one candidate. Ranking is optional in the plurality-with-elimination method. If voters do not rank all the candidates and simply vote for one candidate, multiple elections can result. If they do rank the candidates, multiple elections can be avoided.

In this section we have seen that the chosen voting method can affect the election results. **Table 13.7** summarizes the four voting methods we discussed. Which of these methods is the best? Unfortunately, each method has serious flaws, as you will see in the next section.

TABLE 13.7 Summary of Voting Methods

Voting Method	How the Winning Candidate Is Determined
Plurality Method	The candidate with the most first-place votes is the winner.
Borda Count Method	Voters rank all candidates from the most favorable to the least favorable. Each last-place vote receives 1 point, each next-to-last-place vote 2 points, and so on. The candidate with the most points is the winner.
Plurality-with-Elimination Method	The candidate with the majority (over 50%) of first-place votes is the winner. If no candidate receives a majority, eliminate the candidate with the fewest first-place votes. Either hold another election or adjust the preference table. Continue this process until a candidate receives a majority of first-place votes. That candidate is the winner.
Pairwise Comparison Method	Voters rank all the candidates. A series of comparisons is made in which each candidate is compared to each of the other candidates. The preferred candidate in each comparison receives 1 point; in case of a tie, each receives $\frac{1}{2}$ point. The candidate with the most points is the winner.

Concept and Vocabulary Check

Fill in each blank so that the resulting statement is true.

1. Ballots in which voters are asked to rank all the candidates are called __________ ballots. A table that shows how often each particular outcome occurred and that summarizes an election's results is called a/an __________ table.
2. If there are three or more candidates in an election, it often happens that no single candidate receives more than 50% of the first-place votes, or a/an ________ of votes.
3. The voting method in which each candidate is compared with each of the other candidates is called the ______________ method. The preferred candidate in each comparison receives ___ point(s). If the candidates tie, then each receives ___ point(s). The candidate receiving the __________ is the winner.
4. In an election with n candidates, the number of comparisons, C, that must be made is determined by the formula $C =$ ________.
5. The voting method in which the candidate receiving the most first-place votes is declared the winner is called the ________ method.
6. The voting method that may involve a series of elections or eliminations using a preference table until a candidate receives a majority of the votes is called the ________-with-elimination method.
7. One of the voting methods requires voters to rank all candidates from the most favorable to the least favorable. Each last-place vote receives 1 point, each next-to-last-place vote 2 points, and so on. This voting method is called the ___________ method. The candidate with the __________ is the winner.
8. True or False: The choice of voting method can affect an election's outcome. ______

Exercise Set 13.1

Practice and Application Exercises

In Exercises 1–2, the preference ballots for three candidates (A, B, and C) are shown. Fill in the number of votes in the first row of the given preference table.

1. ABC BCA BCA CBA
CBA ABC ABC BCA
BCA CBA ABC ABC
BCA ABC ABC CBA

Number of Votes			
First Choice	A	B	C
Second Choice	B	C	B
Third Choice	C	A	A

2. ABC CBA BCA BCA
ABC BCA BCA ABC
ABC ABC BCA CBA
BCA ABC ABC ABC

Number of Votes			
First Choice	A	B	C
Second Choice	B	C	B
Third Choice	C	A	A

In Exercises 3–4, four students are running for president of the Let's-Talk-Politics Club: Ann (A), Barbara (B), Charlie (C), and Devon (D). The preference ballots for the four candidates are shown. Construct a preference table to illustrate the results of the election.

3. ABCD BDCA CBDA ABCD CBDA CBDA
ABCD ABCD CBAD CBAD ABCD CBDA

4. ABCD CBAD CBAD ABCD BDCA CBAD
CBAD ABCD CBDA CBAD ABCD CBDA

5. Your class is given the option of choosing a day for the final exam. The students in the class are asked to rank the three available days, Monday (M), Wednesday (W), and Friday (F). The results of the election are shown in the following preference table.

Number of Votes	14	8	3	1
First Choice	F	F	W	M
Second Choice	W	M	F	W
Third Choice	M	W	M	F

a. How many students voted in the election?

b. How many students selected the days in this order: F, M, W?

c. How many students selected Friday as their first choice for the final?

d. How many students selected Wednesday as their first choice for the final?

6. Students at your college are given the option of choosing a topic for which a speaker will be selected. Students are asked to rank three topics: Technology (T), Environmental Issues (E), and Terrorism in the Name of Religion (R). The results of the election are shown in the following preference table.

Number of Votes	70	30	10	5
First Choice	R	T	T	E
Second Choice	E	R	E	T
Third Choice	T	E	R	R

a. How many students voted?

b. How many students selected the topics in this order: T, E, R?

c. How many students selected technology as their first choice for a speaker's topic?

d. How many students selected environmental issues as their second choice for a speaker's topic?

7. The theater society members are voting for the kind of play they will perform next semester: a comedy (C), a drama (D), or a musical (M). Their votes are summarized in the following preference table.

Number of Votes	10	6	6	4	2	2
First Choice	M	C	D	C	D	M
Second Choice	C	M	C	D	M	D
Third Choice	D	D	M	M	C	C

Which type of play is selected using the plurality method?

8. The travel club members are voting for the American city they will visit next semester: New York (N), San Francisco (S), or Chicago (C). Their votes are summarized in the following preference table.

Number of Votes	16	8	6	4
First Choice	S	N	N	C
Second Choice	N	S	C	N
Third Choice	C	C	S	S

Which city is selected using the plurality method?

9. Four professors are running for chair of the Natural Science Division: Professors Darwin (D), Einstein (E), Freud (F), and Hawking (H). The votes of the professors in the natural science division are summarized in the following preference table.

Number of Votes	30	22	20	12	2
First Choice	D	E	F	H	H
Second Choice	H	F	E	E	F
Third Choice	F	H	H	F	D
Fourth Choice	E	D	D	D	E

Who is declared the new division chair using the plurality method?

10. Four professors are running for president of the League of Innovation: Professors Disney (D), Ford (F), Gates (G), and Sarnoff (S). The votes of the members in the League of Innovation are summarized in the following preference table.

Number of Votes	30	22	18	10	2
First Choice	F	G	S	D	G
Second Choice	D	D	G	S	S
Third Choice	G	S	D	G	D
Fourth Choice	S	F	F	F	F

Who is declared the new president using the plurality method?

11. Use the preference table shown in Exercise 7. Which type of play is selected using the Borda count method?

12. Use the preference table shown in Exercise 8. Which city is selected using the Borda count method?

13. Use the preference table shown in Exercise 9. Who is declared the new division chair using the Borda count method?

14. Use the preference table shown in Exercise 10. Who is declared the new president using the Borda count method?

15. Use the preference table shown in Exercise 7. Which type of play is selected using the plurality-with-elimination method?

16. Use the preference table shown in Exercise 8. Which city is selected using the plurality-with-elimination method?

17. Use the preference table shown in Exercise 9. Who is declared the new division chair using the plurality-with-elimination method?

18. Use the preference table shown in Exercise 10. Who is declared the new president using the plurality-with-elimination method?

In Exercises 19–22, suppose that the pairwise comparison method is used to determine the winner in an election.

19. If there are five candidates, how many comparisons must be made?

20. If there are six candidates, how many comparisons must be made?

21. If there are eight candidates, how many comparisons must be made?

22. If there are nine candidates, how many comparisons must be made?

23. Use the preference table shown in Exercise 7. Which type of play is selected using the pairwise comparison method?

24. Use the preference table shown in Exercise 8. Which city is selected using the pairwise comparison method?

25. Use the preference table shown in Exercise 9. Who is declared the new division chair using the pairwise comparison method?

26. Use the preference table shown in Exercise 10. Who is declared the new president using the pairwise comparison method?

In Exercises 27–30, 72 voters are asked to rank four brands of soup: A, B, C, and D. The votes are summarized in the following preference table.

Number of Votes	34	30	6	2
First Choice	A	B	C	D
Second Choice	B	C	D	B
Third Choice	C	D	B	C
Fourth Choice	D	A	A	A

27. Determine the winner using the plurality method.

28. Determine the winner using the Borda count method.

29. Determine the winner using the plurality-with-elimination method.

30. Determine the winner using the pairwise comparison method.

The programmers at the Theater Channel need to select a live musical to introduce their new network. The five choices are Cabaret (C), The Producers (P), Rent (R), Sweeney Todd (S), or West Side Story (W). The 22 programmers rank their choices, summarized in the following preference table. Use the table to solve Exercises 31–34.

Number of Votes	5	5	4	3	3	2
First Choice	C	S	C	W	W	P
Second Choice	R	R	P	P	R	S
Third Choice	P	W	R	R	S	C
Fourth Choice	W	P	S	S	C	R
Fifth Choice	S	C	W	C	P	W

31. Determine which musical is selected using the Borda count method.

32. Determine which musical is selected using the plurality method.

33. Determine which musical is selected using the pairwise comparison method.

34. Determine which musical is selected using the plurality-with-elimination method.

35. Five candidates, A, B, C, D, and E, are running for chair of the Psychology Department. The votes of the department members are summarized in the following preference table.

Number of Votes	5	5	3	3	3	2
First Choice	A	C	D	A	B	D
Second Choice	B	E	C	D	E	C
Third Choice	C	D	B	B	A	B
Fourth Choice	D	A	E	C	C	A
Fifth Choice	E	B	A	E	D	E

a. Who is declared the new department chair using the Borda count method?

b. Suppose that Professor E withdraws from the running before the votes are counted. Write the new preference table. Now who is declared the new department chair using the Borda count method?

36. Rework parts (a) and (b) of Exercise 35 using the pairwise comparison method.

37. Three candidates, A, B, and C, are running for mayor. Election rules stipulate that the plurality method will determine the winner. In the event that the plurality method leads to a tie, the Borda count method will decide the winner. The election results are summarized in the following preference table. Under these rules, which candidate becomes the new mayor?

Number of Votes	12,000	7500	4500
First Choice	C	A	A
Second Choice	B	B	C
Third Choice	A	C	B

38. Three candidates, A, B, and C, are running for mayor. Election rules stipulate that the pairwise comparison method will determine the winner. In the event that the pairwise comparison method leads to a tie, the Borda count method will decide the winner. The election results are summarized in the following preference table. Under these rules, which candidate becomes the new mayor?

Number of Votes	60,000	40,000	40,000	20,000	20,000
First Choice	A	C	B	A	C
Second Choice	B	A	C	C	B
Third Choice	C	B	A	B	A

Explaining the Concepts

39. What is a preference ballot?

40. Describe what is contained in a preference table. What does the table show?

41. Describe the plurality method. Why is ranking not necessary when using this method?

42. Describe the Borda count method. Is it possible to use this method without ranking the candidates? Explain.

43. What is the plurality-with-elimination method? Why is it advantageous to rank the candidates when using this method?

44. What is the pairwise comparison method? Is it possible to use this method without ranking the candidates? Explain.

45. Describe the process used to determine how many comparisons must be made with the pairwise comparison method.

46. Why is it important to choose a voting system before an election takes place?

47. Playwright Tom Stoppard wrote, "It's not the voting that's democracy; it's the counting." Explain what he meant by this.

48. Based on the answers to Exercises 27–30, if you were at the market, which brand of soup, A, B, C, or D, would you select? Explain your choice.

Critical Thinking Exercises

Make Sense? *In Exercises 49–52, determine whether each statement makes sense or does not make sense, and explain your reasoning.*

49. A candidate has a majority of the vote, yet lost the election using the plurality method.

50. A candidate has a majority of the vote, yet lost the election using the plurality-with-elimination method.

51. A candidate has a plurality of the vote, yet lost the election using the Borda count method.

52. A candidate won the election using the plurality-with-elimination method, yet lost the election when the votes were counted by the pairwise comparison method.

In Exercises 53–56, construct a preference table for an election with candidates A, B, and C satisfying the given condition.

53. A wins using the Borda count method.

54. B wins using the plurality-with-elimination method.

55. C wins using the pairwise comparison method.

56. B wins using all four methods discussed in this section.

Group Exercises

57. Research and present a group report on how voting is conducted for the Academy Awards. Describe the single transferable voting method, a variation of the plurality-with-elimination method, in the nomination stage for best picture. (Members of the Irish Senate are also elected by this method.) Be sure to describe some of the more bizarre occurrences at the Oscar ceremonies.

58. Research and present a group report on how voting is conducted for one or more of the following awards: the Heisman Trophy, the Nobel Prize, the Grammy, the Tony, the Emmy, the Pulitzer Prize, or any event or award that the group finds intriguing. Be sure to discuss how the nominees are selected, who participates in the voting, the voting system or systems used, and who counts the results.

13.2 Flaws of Voting Methods

WHAT AM I SUPPOSED TO LEARN?

After studying this section, you should be able to:

1 Use the majority criterion to determine a voting system's fairness.

2 Use the head-to-head criterion to determine a voting system's fairness.

3 Use the monotonicity criterion to determine a voting system's fairness.

4 Use the irrelevant alternatives criterion to determine a voting system's fairness.

5 Understand Arrow's Impossibility Theorem.

IT'S NOT FAIR! THE OLYMPIC GAMES CAN GENERATE \$10 billion in spending for the host city. With all that money at stake, no wonder the selection process has a questionable past. In 1998, evidence surfaced to indicate what many Olympic insiders had been whispering for years: that bid cities had showered members of the International Olympic Committee (IOC) with gifts—fully paid shopping trips for spouses, college scholarships for children, and even cash in envelopes for members. A 2004 BBC documentary, *Buying the Games,* investigated the taking of bribes by IOC members in the bidding process for the 2012 Summer Olympics.

The voting method used to select the winning Olympic host city is a variation of the plurality-with-elimination method. If the International Olympic Committee can eliminate the abuse and bribery that have plagued its selection process, can a winning city be fairly elected? Not necessarily. In the early 1950s, the mathematical economist Kenneth Arrow (1921–2017) proved that **a method for determining election results that is democratic and always fair is a mathematical impossibility**.

What exactly is meant by "democratic" and "fair"? In this section, we will look at four criteria that mathematicians and political scientists have agreed a fair voting system should meet. These four **fairness criteria** are the *majority criterion,* the *head-to-head criterion,* the *monotonicity criterion,* and the *irrelevant alternatives criterion.*

1 *Use the majority criterion to determine a voting system's fairness.*

The Majority Criterion

Most people would agree that if a single candidate receives more than half of the first-place votes in an election, then that candidate should be declared the winner. This requirement for a fair and democratic election is called the **majority criterion**.

GREAT QUESTION!

Can you remind me of the difference between a *majority* and a *plurality*?

A *majority* refers to receiving more than half the first-place votes. A *plurality* refers to receiving the most first-place votes.

THE MAJORITY CRITERION

If a candidate receives a majority of first-place votes in an election, then that candidate should win the election.

Example 1 shows that the Borda count method may not satisfy the majority criterion.

TABLE 13.8 Preference Table for Selecting a New College President

Number of Votes	6	3	2
First Choice	E	G	F
Second Choice	F	H	G
Third Choice	G	F	H
Fourth Choice	H	E	E

EXAMPLE 1 *The Borda Count Method Violates the Majority Criterion*

The 11 members of the Board of Trustees of your college must hire a new college president. The four finalists for the job, E, F, G, and H, are ranked by the 11 members. The preference table is shown in **Table 13.8**. The board members agree to use the Borda count method to determine the winner.

a. Which candidate has a majority of first-place votes?

b. Which candidate is declared the new college president using the Borda count method?

SOLUTION

a. There are 11 first-place votes. Because a majority involves more than half of these votes, any candidate with six or more first-place votes receives a majority. The first-choice row in **Table 13.8** shows that candidate E received six first-place votes. Thus, candidate E has a majority of first-place votes. By the majority criterion, candidate E should be the new college president.

b. Using the Borda count method with four candidates, a first-place vote is worth 4 points, a second-place vote 3 points, a third-place vote 2 points, and a fourth-place vote 1 point. **Table 13.9** shows the points produced by the votes in the preference table.

TABLE 13.9 Points for the College President Election

Number of Votes	6	3	2
First Choice: 4 points	E: $6 \times 4 = 24$ pts	G: $3 \times 4 = 12$ pts	F: $2 \times 4 = 8$ pts
Second Choice: 3 points	F: $6 \times 3 = 18$ pts	H: $3 \times 3 = 9$ pts	G: $2 \times 3 = 6$ pts
Third Choice: 2 points	G: $6 \times 2 = 12$ pts	F: $3 \times 2 = 6$ pts	H: $2 \times 2 = 4$ pts
Fourth Choice: 1 point	H: $6 \times 1 = 6$ pts	E: $3 \times 1 = 3$ pts	E: $2 \times 1 = 2$ pts

Now we read down the columns and total the points for each candidate.

E gets $24 + 3 + 2 = 29$ points. F gets $18 + 6 + 8 = 32$ points.

G gets $12 + 12 + 6 = 30$ points. H gets $6 + 9 + 4 = 19$ points.

Because candidate F has received the most points, candidate F is declared the new college president using the Borda count method.

In Example 1, although a majority of members of the Board of Trustees preferred candidate E, the Borda count method yielded a different choice, candidate F. In a situation like this, the Borda count method is said to *violate* the majority criterion.

CHECK POINT 1 The 14 members of the school board must hire a new principal. The four finalists for the job, A, B, C, and D, are ranked by the 14 members. The preference table is shown in **Table 13.10**. The board members agree to use the Borda count method to determine the winner.

TABLE 13.10 Preference Table for Selecting a New Principal

Number of Votes	**6**	**4**	**2**	**2**
First Choice	A	B	B	A
Second Choice	B	C	D	B
Third Choice	C	D	C	D
Fourth Choice	D	A	A	C

a. Which candidate has a majority of first-place votes?

b. Which candidate is declared the new principal using the Borda count method?

We have seen that an advantage of the Borda count method is that it takes into account all the information about the voters' preferences in the preference table. However, a candidate who has a majority of first-place votes can lose an election with the Borda count method. The other three voting methods that we have discussed—the plurality method, the plurality-with-elimination method, and the pairwise comparison method—never violate the majority criterion. Unfortunately, there are other fairness criteria that are violated by each of these three systems.

2 *Use the head-to-head criterion to determine a voting system's fairness.*

The Head-to-Head Criterion

Most people would agree that if there is one candidate who is favored by the voters when compared, in turn, to each of the other candidates, then that candidate should win the election. This requirement of a fair and democratic election is called the **head-to-head criterion.**

THE HEAD-TO-HEAD CRITERION

If a candidate is favored when compared separately—that is, head-to-head—with every other candidate, then that candidate should win the election.

Example 2 shows that the plurality method may violate the head-to-head criterion.

EXAMPLE 2 *The Plurality Method Violates the Head-to-Head Criterion*

Twenty-two people are asked to taste-test and rank three different brands, A, B, and C, of tuna fish. The results are summarized in the preference table in **Table 13.11**.

TABLE 13.11 Preference Table for Three Brands of Tuna Fish

Number of Votes	**8**	**6**	**4**	**4**
First Choice	A	C	C	B
Second Choice	B	B	A	A
Third Choice	C	A	B	C

a. Which brand is favored over the other two using a head-to-head comparison?

b. Which brand wins the taste test using the plurality method?

The Marquis de Condorcet (1743–1794)

"Either no individual of the human species has any true rights, or all have the same. And he who votes against the rights of another, of whatever religion, color, or sex, has thereby abjured his own."
—Marquis de Condorcet

The head-to-head criterion is also known as the **Condorcet criterion**, named after the Marquis de Condorcet, one of the leading mathematicians, sociologists, economists, and political thinkers of France at the time of the American and French revolutions. Condorcet believed that mathematics could be used for the benefit of the people and that principles of fair government could be discovered mathematically. He believed in abolishing slavery, advocated free speech, and supported early feminists lobbying for equal rights.

Condorcet analyzed voting methods and discovered that sometimes there is no clear way to choose a winner of an election. He showed that it is possible to have three candidates, A, B, and C, for whom voters prefer A to B and B to C, but then prefer C to A.

Condorcet was arrested for his liberal and humanitarian views and died in prison by suicide or possibly murder.

SOLUTION

a. We begin by comparing brands A and B using **Table 13.11**. A is favored over B in columns 1 and 3, giving A 8 + 4, or 12, votes. B is favored over A in columns 2 and 4, giving B 6 + 4, or 10, votes. Because A has 12 votes and B has 10 votes, A is favored when compared to B.

Now let's use **Table 13.11** to compare brands A and C. A is favored over C in columns 1 and 4, giving A 8 + 4, or 12, votes. C is favored over A in columns 2 and 3, giving C 6 + 4, or 10, votes. Because A has 12 votes and C has 10 votes, A is favored when compared to C.

We see that A is favored over the other two brands using a head-to-head comparison.

b. Using the plurality method, the brand with the most first-place votes is the winner. Look at the row indicating the first choice. A received 8 votes, B received 4 votes, and C received 6 + 4, or 10, votes. Brand C wins the taste test using the plurality method.

In Example 2, although brand A was favored over the other two brands using a head-to-head comparison, the plurality method yielded a different winner, brand C. Thus, in this situation, the plurality method violates the head-to-head criterion.

Is the plurality method the only voting method with the potential to violate the head-to-head criterion? The answer is no. The Borda count method and the plurality-with-elimination method both have the potential to violate the head-to-head criterion. Can this occur with the pairwise comparison method? No. A candidate who defeats each of the other candidates in a head-to-head vote wins every pairwise comparison and thus gets the greatest number of points under this method.

 **CHECK POINT 2** Seven people are asked to listen to and rate three different pairs of stereo speakers, A, B, and C. The results are summarized in **Table 13.12**.

TABLE 13.12 Preference Table for Three Pairs of Stereo Speakers

Number of Votes	3	2	2
First Choice	A	B	C
Second Choice	B	A	B
Third Choice	C	C	A

a. Which brand is favored over all others using a head-to-head comparison?

b. Which brand wins the listening test using the plurality method?

The Monotonicity Criterion

Elections are generally preceded by a campaign and discussions of the candidates' strengths and weaknesses. In some situations, a preliminary election in which votes do not count, called a **straw vote**, is taken as a measure of voters' intentions. Most people would agree that if a candidate wins the straw poll, which we can consider the first election, and then gains additional support without losing any of the original support, then that candidate should win the second election. This requirement for a fair and democratic election is called the **monotonicity criterion**.

THE MONOTONICITY CRITERION

If a candidate wins an election and, in a reelection, the only changes are changes that favor the candidate, then that candidate should win the reelection.

Example 3 shows that the plurality-with-elimination method may not satisfy the monotonicity criterion.

3 *Use the monotonicity criterion to determine a voting system's fairness.*

Choosing the Olympic Host City

Like cardinals sequestered to vote for a new pope, the members of the International Olympic Committee lock themselves into a conference room and do not leave until they can proclaim a winner from among the candidate cities. Think of it as a sort of papal consistory on steroids. The voting method used to select the winner is the plurality-with-elimination method with a slight difference: Instead of indicating their preferences all at once, the 100-plus IOC members let their preferences be known one round at a time. Here is the voting that took place in the selection of the site for the 2012 Summer Olympics. In each round, the delegates voted for just one city, and the city with the fewest votes was eliminated. (In all but the final round, some members abstained from voting.)

First Round		Second Round	
London	22	London	27
Paris	21	Paris	25
Madrid	20	Madrid	32
New York	19	New York	16
Moscow	15		

Moscow eliminated

New York eliminated

Third Round		Fourth Round	
London	39	London	54
Paris	33	Paris	50
Madrid	31		

Madrid eliminated

Paris eliminated

London was chosen as the host city for the 2012 Summer Olympics.

EXAMPLE 3 *The Plurality-with-Elimination Method Violates the Monotonicity Criterion*

The 58 members of the Student Activity Council are meeting to elect a keynote speaker to launch student involvement week. The choices are Bill Gates (G), Howard Stern (S), or Oprah Winfrey (W). A straw vote is taken, and the results are given in **Table 13.13**. After a lengthy discussion, all but eight students vote in exactly the same way. The eight voters, shown in the last column of **Table 13.13**, all changed their ballots to make Oprah Winfrey (W) their first choice. The results of the second election are given in **Table 13.14**. Notice that the first column has increased by eight votes.

TABLE 13.13 Preference Table for the Straw Vote

Number of Votes	20	16	14	8
First Choice	W	S	G	G
Second Choice	G	W	S	W
Third Choice	S	G	W	S

These 8 voters changed their ballots to make Oprah Winfrey (W) their first choice.

TABLE 13.14 Preference Table for the Second Election

Number of Votes	28	16	14
First Choice	W	S	G
Second Choice	G	W	S
Third Choice	S	G	W

a. Using the plurality-with-elimination method, which speaker wins the first election?

b. Using the plurality-with-elimination method, which speaker wins the second election?

c. Does this violate the monotonicity criterion?

SOLUTION

a. There are 58 people voting. No speaker receives a majority of votes (30 or more votes) in **Table 13.13**, the first election (the straw vote). Using the plurality-with-elimination method, because S (Howard Stern) receives the fewest first-place votes, he is eliminated in the first round. The new preference table with S eliminated is shown in **Table 13.15**. Because W (Oprah Winfrey) has received a majority of first-place votes, she is the winner of the straw poll.

b. Now we focus on **Table 13.14**, the preference table for the second election. No speaker receives a majority of votes (30 or more votes) in this election. Using the plurality-with-elimination method, because G (Bill Gates) receives the fewest first-place votes, he is eliminated in the first round. The new preference table with G eliminated is shown in **Table 13.16**. Because S (Howard Stern) has received a majority of first-place votes, he is the winner of the second election.

TABLE 13.15 Preference Table for the Straw Vote with S Eliminated

Number of Votes	20	16	14	8
First Choice	W	W	G	G
Second Choice	G	G	W	W

TABLE 13.16 Preference Table for the Second Election with G Eliminated

Number of Votes	28	16	14
First Choice	W	S	S
Second Choice	S	W	W

c. Oprah Winfrey (W) won the first election. She then gained additional support with the eight voters who changed their ballots to make her their first choice. However, she lost the second election. This violates the monotonicity criterion.

Example 3 illustrates the bizarre possibility that you can actually do worse by doing better! The example illustrates that the plurality-with-elimination method has the potential to violate the monotonicity criterion. Of the four voting methods we have studied, the only one that cannot violate the monotonicity criterion is the plurality method.

CHECK POINT 3 An election with 120 voters and three candidates, A, B, and C, is to be decided using the plurality-with-elimination method. **Table 13.17** shows the results of a straw poll. After the straw poll, 12 voters all changed their ballots to make candidate A their first choice. The results of the second election are given in **Table 13.18**.

TABLE 13.17 Preference Table for the Straw Vote

Number of Votes	42	34	28	16
First Choice	A	C	B	B
Second Choice	B	A	C	A
Third Choice	C	B	A	C

TABLE 13.18 Preference Table for the Second Election

Number of Votes	54	34	28	4
First Choice	A	C	B	B
Second Choice	B	A	C	A
Third Choice	C	B	A	C

a. Using the plurality-with-elimination method, which candidate wins the first election?

b. Using the plurality-with-elimination method, which candidate wins the second election?

c. Does this violate the monotonicity criterion? Explain your answer.

4 *Use the irrelevant alternatives criterion to determine a voting system's fairness.*

The Irrelevant Alternatives Criterion

A candidate wins an election. Then one or more of the other candidates are removed from the ballot and a recount is done. Most people would agree that the previous winner should still be declared the winner. This requirement for a fair and democratic election is called the **irrelevant alternatives criterion**.

THE IRRELEVANT ALTERNATIVES CRITERION

If a candidate wins an election and, in a recount, the only changes are that one or more of the other candidates are removed from the ballot, then that candidate should still win the election.

Each of the four voting methods we have discussed may violate the irrelevant alternatives criterion. Example 4 shows that the pairwise comparison method may not satisfy the criterion.

EXAMPLE 4 *The Pairwise Comparison Method Violates the Irrelevant Alternatives Criterion*

Four candidates, E, F, G, and H, are running for mayor of Bolinas. The election results are shown in **Table 13.19**.

TABLE 13.19 Preference Table for the Mayor of Bolinas

Number of Votes	160	100	80	20
First Choice	E	G	H	H
Second Choice	F	F	E	E
Third Choice	G	H	G	F
Fourth Choice	H	E	F	G

TABLE 13.19 Preference Table for the Mayor of Bolinas (repeated)

Number of Votes	160	100	80	20
First Choice	E	G	H	H
Second Choice	F	F	E	E
Third Choice	G	H	G	F
Fourth Choice	H	E	F	G

a. Using the pairwise comparison method, who wins this election?

b. Prior to the announcement of the election results, candidates F and G both withdraw from the running. Using the pairwise comparison method, which candidate is declared mayor of Bolinas with F and G eliminated from the preference table?

c. Does this violate the irrelevant alternatives criterion?

SOLUTION

a. Because there are four candidates, $n = 4$, and the number of comparisons we must make is

$$C = \frac{n(n-1)}{2} = \frac{4(4-1)}{2} = \frac{4 \cdot 3}{2} = \frac{12}{2} = 6.$$

The following table shows the results of these six comparisons. Use **Table 13.19** to verify each of these results.

Comparison	Vote Results	Conclusion
E vs. F	260 voters prefer E to F. 100 voters prefer F to E.	E wins and gets 1 point.
E vs. G	260 voters prefer E to G. 100 voters prefer G to E.	E wins and gets 1 point.
E vs. H	160 voters prefer E to H. 200 voters prefer H to E.	H wins and gets 1 point.
F vs. G	180 voters prefer F to G. 180 voters prefer G to F.	It's a tie. F gets $\frac{1}{2}$ point. G gets $\frac{1}{2}$ point.
F vs. H	260 voters prefer F to H. 100 voters prefer H to F.	F wins and gets 1 point.
G vs. H	260 voters prefer G to H. 100 voters prefer H to G.	G wins and gets 1 point.

Thus, E gets 2 points, F and G each get $1\frac{1}{2}$ points, and H gets 1 point. Therefore, E is the winner when candidates F and G are included.

b. Once F and G both withdraw from the running, only two candidates, E and H, remain. Keep in mind that E was the winning candidate prior to this withdrawal. The new preference table is shown in **Table 13.20**. Using the pairwise comparison test, there are two candidates and the number of comparisons we must make is

$$C = \frac{n(n-1)}{2} = \frac{2(2-1)}{2} = \frac{2 \cdot 1}{2} = 1.$$

The one comparison we must make is E to H. The voice balloons below **Table 13.20** show that H wins, defeating E by 200 votes to 160 votes. H gets 1 point, E gets 0 points, and candidate H is the new mayor of Bolinas.

TABLE 13.20 Preference Table for the Mayor of Bolinas with F and G Removed

Number of Votes	160	100	80	20
First choice	E	H	H	H
Second choice	H	E	E	E

160 voters prefer E to H.

200 voters prefer H to E.

c. The first election count produced E as the winner. Then, even though F and G were not the winning candidates, their removal from the ballots produced H, not E, as the winner. This violates the irrelevant alternatives criterion.

CHECK POINT 4 Four candidates, A, B, C, and D, are running for mayor. The election results are shown in **Table 13.21**.

a. Using the pairwise comparison method, who wins this election?

b. Prior to the announcement of the election results, candidates B and C both withdraw from the running. Using the pairwise comparison method, which candidate is declared mayor with B and C eliminated from the preference table?

c. Does this violate the irrelevant alternatives criterion? Explain your answer.

TABLE 13.21 Preference Table for Mayor

Number of Votes	150	90	90	30
First Choice	A	C	D	D
Second Choice	B	B	A	A
Third Choice	C	D	C	B
Fourth Choice	D	A	B	C

Table 13.22 summarizes the four fairness criteria we have discussed. **Table 13.23** shows which voting methods satisfy the criteria.

TABLE 13.22 Summary of Fairness Criteria

Criterion	Description
Majority Criterion	If a candidate receives a majority of first-place votes in an election, then that candidate should win the election.
Head-to-Head Criterion	If a candidate is favored when compared head-to-head with every other candidate, then that candidate should win the election.
Monotonicity Criterion	If a candidate wins an election and, in a reelection, the only changes are changes that favor the candidate, then that candidate should win the reelection.
Irrelevant Alternatives Criterion	If a candidate wins an election and, in a recount, the only changes are that one or more of the other candidates are removed from the ballot, then that candidate should still win the election.

TABLE 13.23 Voting Methods and Whether They Satisfy the Fairness Criteria

	Voting Method			
Fairness Criteria	**Plurality Method**	**Borda Count Method**	**Plurality-with-Elimination Method**	**Pairwise Comparison Method**
Majority Criterion	Always satisfies	May not satisfy	Always satisfies	Always satisfies
Head-to-Head Criterion	May not satisfy	May not satisfy	May not satisfy	Always satisfies
Monotonicity Criterion	Always satisfies	May not satisfy	May not satisfy	May not satisfy
Irrelevant Alternatives Criterion	May not satisfy	May not satisfy	May not satisfy	May not satisfy

5 *Understand Arrow's Impossibility Theorem.*

The Search for a Fair Voting System

We have seen that none of the voting methods discussed here always satisfies each of the fairness criteria. Is there another democratic voting method that does satisfy all four criteria—a perfectly fair voting system? For elections involving more than two candidates, the answer is no. No perfectly fair voting method exists. How do we know? In 1951, economist Kenneth Arrow proved the now famous **Arrow's Impossibility Theorem**: There does not exist, and will never exist, any democratic voting system that satisfies all of the fairness criteria.

Kenneth Arrow (1921–2017). Awarded 1972 Nobel Prize in Economics

ARROW'S IMPOSSIBILITY THEOREM

It is mathematically impossible for any democratic voting system to satisfy each of the four fairness criteria.

In 1972, Arrow was awarded the Nobel Prize in Economics for his pioneering work. Arrow's discipline is now known as *social-choice theory,* a field that combines mathematics, economics, and political science.

Concept and Vocabulary Check

Fill in each blank so that the resulting statement is true.

1. If a candidate receives more than half the first-place votes in an election, then that candidate should be declared the winner. This criterion is called the ________ criterion.
2. If a candidate is favored when compared separately with every other candidate in an election, then that candidate should be declared the winner. This criterion is called the __________ criterion.
3. If a candidate wins an election and, in a reelection, the only changes are changes that favor the candidate, then that candidate should win the reelection. This criterion is called the ____________ criterion.
4. If a candidate wins an election and, in a recount, the only changes are that one or more of the other candidates are removed from the ballot, then that candidate should still win the election. This criterion is called the ______________ criterion.
5. A voting method that may not satisfy any of the fairness criteria is the ___________ method.
6. True or False: It is impossible for any democratic voting system to satisfy each of the fairness criteria. ______

Exercise Set 13.2

Practice and Application Exercises

1. Voters in a small town are considering four proposals, A, B, C, and D, for the design of affordable housing. The winning design is to be determined by the Borda count method. The preference table for the election is shown.

Number of Votes	300	120	90	60
First Choice	D	C	C	A
Second Choice	A	A	A	D
Third Choice	B	B	D	B
Fourth Choice	C	D	B	C

a. Which design has a majority of first-place votes?

b. Using the Borda count method, which design will be used for the affordable housing?

c. Is the majority criterion satisfied? Explain your answer.

2. Fifty-three people are asked to taste-test and rank three different brands of yogurt, A, B, and C. The preference table shows the rankings of the 53 voters.

Number of Votes	27	24	2
First Choice	A	B	C
Second Choice	C	C	B
Third Choice	B	A	A

a. Which brand has a majority of first-place votes?

b. Suppose that the Borda count method is used to determine the winner. Which brand wins the taste test?

c. Is the majority criterion satisfied? Explain your answer.

3. MTV's *Real World* is considering three cities for its new season: Amsterdam (A), Rio de Janeiro (R), or Vancouver (V). Programming executives and the show's production team vote to decide where the new season will be taped. The winning city is to be determined by the plurality method. The preference table for the election is shown at the top of the next column.

Number of Votes	12	9	4	4
First Choice	A	V	V	R
Second Choice	R	R	A	A
Third Choice	V	A	R	V

a. Which city is favored over all others using a head-to-head comparison?

b. Which city wins the vote using the plurality method?

c. Is the head-to-head criterion satisfied? Explain your answer.

4. A computer company is considering opening a new branch in Atlanta (A), Boston (B), or Chicago (C). Senior managers vote to decide where the new branch will be located. The winning city is to be determined by the plurality method. The preference table for the election is shown.

Number of Votes	20	19	5
First Choice	A	B	C
Second Choice	B	C	B
Third Choice	C	A	A

a. Which city is favored over all others using a head-to-head comparison?

b. Which city wins the vote using the plurality method?

c. Is the head-to-head criterion satisfied? Explain your answer.

5. A town is voting on an ordinance dealing with smoking in public spaces. The options are (A) permit unrestricted smoking; (B) permit smoking in designated areas only; and (C) ban all smoking in public places. The winner is to be determined by the Borda count method. The preference table for the election is shown.

Number of Votes	120	60	30	30	30
First Choice	A	C	B	C	B
Second Choice	C	B	A	A	C
Third Choice	B	A	C	B	A

a. Which option is favored over all others using a head-to-head comparison?

b. Which option wins the vote using the Borda count method?

c. Is the head-to-head criterion satisfied? Explain your answer.

6. A town is voting on an ordinance dealing with nudity at its public beaches. The options are (A) make clothing optional at all beaches; (B) permit nudity at designated beaches only; and (C) permit no nudity at public beaches. The winner is to be determined by the Borda count method. The preference table for the election is shown.

Number of Votes	200	80	80
First Choice	C	B	A
Second Choice	A	A	B
Third Choice	B	C	C

a. Which option is favored over all others using a head-to-head comparison?

b. Which option wins the vote using the Borda count method?

c. Is the head-to-head criterion satisfied? Explain your answer.

7. The following preference table gives the results of a straw vote among three candidates, A, B, and C.

Number of Votes	10	8	7	4
First Choice	C	B	A	A
Second Choice	A	C	B	C
Third Choice	B	A	C	B

a. Using the plurality-with-elimination method, which candidate wins the straw vote?

b. In the actual election, the four voters in the last column who voted A, C, B, in that order, change their votes to C, A, B. Using the plurality-with-elimination method, which candidate wins the actual election?

c. Is the monotonicity criterion satisfied? Explain your answer.

8. The preference table gives the results of a straw vote among three candidates, A, B, and C.

Number of Votes	14	12	10	6
First Choice	C	B	A	A
Second Choice	A	C	B	C
Third Choice	B	A	C	B

a. Using the plurality-with-elimination method, which candidate wins the straw vote?

b. In the actual election, the six voters in the last column who voted A, C, B, in that order, change their votes to C, A, B. Using the plurality-with-elimination method, which candidate wins the actual election?

c. Is the monotonicity criterion satisfied? Explain your answer.

9. Members of the Student Activity Committee at a college are considering three film directors to speak at a campus arts festival: Ron Howard (H), Spike Lee (L), and Steven Spielberg (S). Committee members vote for their preferred speaker. The winner is to be selected by the pairwise comparison method. The preference table for the election is shown.

Number of Votes	10	8	5
First Choice	H	L	S
Second Choice	S	S	L
Third Choice	L	H	H

a. Using the pairwise comparison method, who is selected as the speaker?

b. Prior to the announcement of the speaker, Ron Howard informs the committee that he will not be able to participate due to other commitments. Construct a new preference table for the election with H eliminated. Using the new table and the pairwise comparison method, who is selected as the speaker?

c. Is the irrelevant alternatives criterion satisfied? Explain your answer.

10. Members of the Student Activity Committee at a college are considering three actors to speak at a campus festival on women in the arts: Whoopi Goldberg (G), Julia Roberts (R), and Meryl Streep (S). Committee members vote for their preferred speaker. The winner is to be selected by the pairwise comparison method. The preference table for the election is shown.

Number of Votes	12	8	6
First Choice	S	R	G
Second Choice	G	G	R
Third Choice	R	S	S

a. Using the pairwise comparison method, who is selected as the speaker?

b. Prior to the announcement of the speaker, Meryl Streep informs the committee that she will not be able to participate due to other commitments. Construct a new preference table for the election with S eliminated. Using the new table and the pairwise comparison method, who is selected as the speaker?

c. Is the irrelevant alternatives criterion satisfied? Explain your answer.

In Exercises 11–18, the preference table for an election is given. Use the table to answer the questions that follow it.

11.

Number of Votes	20	16	10	4
First Choice	D	C	C	A
Second Choice	A	A	B	B
Third Choice	B	B	D	D
Fourth Choice	C	D	A	C

a. Using the Borda count method, who is the winner?

b. Is the majority criterion satisfied? Explain your answer.

12. Number of Votes	20	15	3	1
First Choice	A	B	C	D
Second Choice	B	C	D	B
Third Choice	C	D	B	C
Fourth Choice	D	A	A	A

a. Using the Borda count method, who is the winner?

b. Is the majority criterion satisfied? Explain your answer.

13. Number of Votes	24	18	10	8	8	2
First Choice	D	C	A	A	C	B
Second Choice	A	A	D	B	B	C
Third Choice	B	B	B	C	D	A
Fourth Choice	C	D	C	D	A	D

a. Using the plurality-with-elimination method, who is the winner?

b. Is the head-to-head criterion satisfied? Explain your answer.

14. Number of Votes	18	10	9	7	4	3
First Choice	A	C	C	B	D	D
Second Choice	D	B	D	D	B	C
Third Choice	C	D	B	C	A	A
Fourth Choice	B	A	A	A	C	B

a. Using the plurality-with-elimination method, who is the winner?

b. Is the head-to-head criterion satisfied? Explain your answer.

15. Number of Votes	14	8	4
First Choice	A	B	D
Second Choice	B	D	A
Third Choice	C	C	C
Fourth Choice	D	A	B

a. Using the Borda count method, who is the winner?

b. Is the majority criterion satisfied? Explain your answer.

c. Is the head-to-head criterion satisfied? Explain your answer.

d. Suppose that candidate C drops out of the race. Using the Borda count method, who among the remaining candidates wins the election? Is the irrelevant alternatives criterion satisfied? Explain your answer.

16. Number of Votes	14	12	10	6
First Choice	A	B	C	D
Second Choice	B	A	B	C
Third Choice	C	C	A	B
Fourth Choice	D	D	D	A

a. Using the plurality-with-elimination method, who is the winner?

b. The six voters on the right all move candidate A from last place on their preference lists to first place on their preference lists. Construct a new preference table for the election. Using this table and the plurality-with-elimination method, who is the winner? Is the monotonicity criterion satisfied? Explain your answer.

17. Number of Votes	16	14	12	4	2
First Choice	A	D	D	C	E
Second Choice	B	B	B	A	A
Third Choice	C	A	E	B	D
Fourth Choice	D	C	C	D	B
Fifth Choice	E	E	A	E	C

a. Using the Borda count method, who is the winner?

b. Is the majority criterion satisfied? Explain your answer.

c. Is the head-to-head criterion satisfied? Explain your answer.

18. Number of Votes	6	6	6	6	2	2	2	2
First Choice	A	B	A	B	C	C	E	B
Second Choice	D	A	C	A	E	B	D	A
Third Choice	E	D	D	C	B	A	A	E
Fourth Choice	C	E	E	D	D	D	C	C
Fifth Choice	B	C	B	E	A	E	B	D

a. Using the pairwise comparison method, who is the winner?

b. Suppose that candidate C drops out, but the winner is still chosen using the pairwise comparison method. Is the irrelevant alternatives criterion satisfied? Explain your answer.

19. The preference table shows the results of an election among three candidates, A, B, and C.

Number of Votes	7	3	2
First Choice	A	B	C
Second Choice	B	C	B
Third Choice	C	A	A

a. Using the plurality method, who is the winner?

b. Is the majority criterion satisfied? Explain your answer.

c. Is the head-to-head criterion satisfied? Explain your answer.

d. The two voters on the right both move candidate A from last place on their preference lists to first place on their preference lists. Construct a new preference table for the election. Using the table and the plurality method, who is the winner?

e. Suppose that candidate C drops out, but the winner is still chosen by the plurality method. Is the irrelevant alternatives criterion satisfied? Explain your answer.

f. Do your results from parts (b) through (e) contradict Arrow's Impossibility Theorem? Explain your answer.

Explaining the Concepts

20. Describe the majority criterion.

21. Describe the head-to-head criterion.

22. Describe the monotonicity criterion.

23. Describe the irrelevant alternatives criterion.

24. In your own words, state Arrow's Impossibility Theorem.

25. Write a realistic voting problem similar to Exercise 5 or 6 that would be relevant to your college or the area of the country in which you live. Describe the voting method that you will use to determine the winning option. What problems might occur due to flaws in this method?

26. Is it possible to have election results using a particular voting method that satisfy all four fairness criteria? If so, does this contradict Arrow's Impossibility Theorem?

27. Kenneth Arrow wrote, "Well, okay, since we can't find perfection, let's at least try to find a method that works well most of the time." Explain what he meant by this.

Critical Thinking Exercises

Make Sense? *In Exercises 28–31, determine whether each statement makes sense or does not make sense, and explain your reasoning.*

28. My candidate received a majority of first-place votes and lost the election.

29. My candidate was favored when compared head-to-head with every other candidate and lost the election.

30. A candidate won an election and, in a reelection, the only changes were changes that favored the candidate, so I'm certain that this candidate won the reelection.

31. There's a new voting method, called *approval voting*, where voters approve or disapprove of each candidate. Because the candidate with the most approval votes wins, this method satisfies each of the four fairness criteria.

In Exercises 32–35, construct a preference table for an election among three candidates, A, B, and C, with the given characteristics. Do not use any of the tables from this section.

32. The Borda count winner violates the majority criterion.

33. The winner by the plurality method violates the head-to-head criterion.

34. The winner by the plurality method violates the irrelevant alternatives criterion.

35. The winner by the plurality-with-elimination method violates the monotonicity criterion.

Group Exercise

36. Citizen-initiated ballot measures often present voters with controversial issues over which they do not think alike. Here's one your author would like to initiate:

Please rank each of the following options regarding permitting dogs on national park trails.

i. Unleashed dogs accompanied by their caregivers should be permitted on designated national park trails.

ii. Leashed dogs accompanied by their caregivers should be permitted on designated national park trails.

iii. No dogs should be permitted on any national park trails.

Your author was not happy with the fact that he could not take his dog running with him on the park trails at Point Reyes National Seashore. Of course, that is his issue. For this project, group members should write a ballot measure, perhaps controversial, like the sample above, but dealing with an issue of relevance to your campus and community. Rather than holding an election, use a random sample of students on your campus, administer the ballot, and have them rank their choices.

a. Use each of the four voting methods to determine the winning option for your ballot measure.

b. Check to see if any of the four fairness criteria are violated.

13.3 Apportionment Methods

WHAT AM I SUPPOSED TO LEARN?

After studying this section, you should be able to:

1 Find standard divisors and standard quotas.

2 Understand the apportionment problem.

3 Use Hamilton's method.

4 Understand the quota rule.

5 Use Jefferson's method.

6 Use Adams's method.

7 Use Webster's method.

The Signing of the Constitution

The Senate of the United States shall be composed of two Senators from each state . . .

—Article I. Section 3. Constitution of the United States

Representatives . . . shall be apportioned among the several states . . . according to their respective numbers . . .

—Article I. Section 2. Constitution of the United States

It is the summer of 1787. Delegates from the 13 states are meeting in Philadelphia to draft a constitution for a new nation. The most heated debate concerns the makeup of the new legislature.

The smaller states, led by New Jersey, insist that all states have the same number of representatives. The larger states, led by Virginia, want some form of proportional representation based on population. The delegates compromise on this issue by creating a Senate, in which each state has two senators, and a House of Representatives, in which each state has a number of representatives based on its population.

According to the Constitution of the United States, "Representatives ... shall be apportioned among the several states ... according to their respective numbers ...." But how are these calculations to be done, and what sorts of problems can arise? In this section, you will learn to divide, on an equitable basis, those things that are not individually divisible. By doing this, you will get to look at a unique role that mathematics played in United States history.

1 *Find standard divisors and standard quotas.*

Standard Divisors and Standard Quotas

Our discussion in the first part of this section deals with the following simple but important example: The Republic of Margaritaville is composed of four states, A, B, C, and D. According to the country's constitution, the congress will have 30 seats, divided among the four states according to their respective populations. **Table 13.24** shows each state's population.

TABLE 13.24 Population of Margaritaville by State

State	A	B	C	D	Total
Population (in thousands)	275	383	465	767	1890

Before determining a method to allocate the 30 seats in a fair manner, we introduce two important definitions. The first quantity we define is called the *standard divisor*.

THE STANDARD DIVISOR

The **standard divisor** is found by dividing the total population under consideration by the number of items to be allocated.

$$\text{Standard divisor} = \frac{\text{total population}}{\text{number of allocated items}}$$

For the population of Margaritaville in **Table 13.24**, the standard divisor is found as follows:

$$\text{Standard divisor} = \frac{\text{total population}}{\text{number of allocated items}} = \frac{1890}{30} = 63.$$

The congress will have 30 seats.

In situations dealing with allocating congressional seats to states based on their populations, the standard divisor gives the number of people per seat in congress on a national basis. Thus, in Margaritaville, there are 63 thousand people for each seat in congress.

The second quantity we define is called the *standard quota*.

THE STANDARD QUOTA

The **standard quota** for a particular group is found by dividing that group's population by the standard divisor.

$$\text{Standard quota} = \frac{\text{population of a particular group}}{\text{standard divisor}}$$

Senate Voting Power

"The Senate of the United States shall be composed of two Senators from each state."

Most Senate Voting Power

1. WYOMING
Population: **576,000**
People per Senator: **288,000**

2. VERMONT
Population: **626,000**
People per Senator: **312,000**

3. NORTH DAKOTA
Population: **700,000**
People per Senator: **350,000**

4. ALASKA
Population: **731,000**
People per Senator: **365,500**

Least Senate Voting Power

1. CALIFORNIA
Population: **38 million**
People per Senator: **19 million**

2. TEXAS
Population: **26 million**
People per Senator: **13 million**

3. NEW YORK
Population: **20 million**
People per Senator: **10 million**

4. FLORIDA
Population: **19 million**
People per Senator: **9.5 million**

Source: U.S. Census Bureau

In computing standard divisors and standard quotas, we will round to the nearest hundredth—that is, to two decimal places.

EXAMPLE 1 *Finding Standard Quotas*

Find the standard quotas for states A, B, C, and D in Margaritaville and complete **Table 13.25**.

TABLE 13.25 Population of Margaritaville by State

State	A	B	C	D	Total
Population	275	383	465	767	1890
Standard Quota					

SOLUTION

The standard quotas are obtained by dividing each state's population by the standard divisor. We previously computed the standard divisor and found it to be 63. Thus, we use 63 in the denominator for each of the four computations for the standard quota.

$$\text{Standard quota for state A} = \frac{\text{population of state A}}{\text{standard divisor}} = \frac{275}{63} \approx 4.37$$

$$\text{Standard quota for state B} = \frac{\text{population of state B}}{\text{standard divisor}} = \frac{383}{63} \approx 6.08$$

$$\text{Standard quota for state C} = \frac{\text{population of state C}}{\text{standard divisor}} = \frac{465}{63} \approx 7.38$$

$$\text{Standard quota for state D} = \frac{\text{population of state D}}{\text{standard divisor}} = \frac{767}{63} \approx 12.17$$

Table 13.26 shows the standard quotas for each state in the Republic of Margaritaville.

TABLE 13.26 Standard Quotas for Each State in Margaritaville

State	A	B	C	D	Total
Population	275	383	465	767	1890
Standard Quota	4.37	6.08	7.38	12.17	30

Notice that the sum of all the standard quotas is 30, the total number of seats in the congress.

GREAT QUESTION!

Does the sum of the standard quotas always give the exact number of allocated items?

No. Due to rounding, the sum of the standard quotas can be slightly above or slightly below the total number of allocated items.

CHECK POINT 1 The Republic of Amador is composed of five states, A, B, C, D, and E. According to the country's constitution, the congress will have 200 seats, divided among the five states according to their respective populations. **Table 13.27** shows each state's population.

TABLE 13.27 Population of Amador by State

State	A	B	C	D	E	Total
Population (in thousands)	1112	1118	1320	1515	4935	10,000
Standard Quota						

a. Find the standard divisor.

b. Find the standard quota for each state and complete **Table 13.27**.

2 *Understand the apportionment problem.*

The Apportionment Problem

The standard quotas in **Table 13.26** on the previous page represent each state's exact fair share of the 30 seats for the congress of Margaritaville. However, what do we mean by saying that state A has 4.37 seats in congress? Seats have to be given out in whole numbers. The **apportionment problem** is to determine a method for rounding standard quotas into whole numbers so that the sum of the numbers is the total number of allocated items.

Can we possibly round standard quotas down to the nearest whole number or up to the nearest whole number and solve the apportionment problem? The **lower quota** is the standard quota rounded down to the nearest whole number. The **upper quota** is the standard quota rounded up to the nearest whole number. **Table 13.28** shows the lower and upper quotas.

GREAT QUESTION!

How do I find lower and upper quotients if a standard quotient is already a whole number?

In the unusual case that a standard quota is a whole number, the lower and upper quotas are both equal to the standard quota.

TABLE 13.28 Standard Quotas, Lower Quotas, and Upper Quotas for Each State in Margaritaville

State	A	B	C	D	Total
Population	275	383	465	767	1890
Standard quota	4.37	6.08	7.38	12.17	30
Lower quota	4	6	7	12	29
Upper quota	5	7	8	13	33

These totals should be 30 because Margaritaville has 30 seats in congress.

GREAT QUESTION!

What's the meaning of the word *apportion* in everyday English?

According to one dictionary, to *apportion* means "to divide and assign in due and proper proportion or according to some plan."

The voice balloon shows that by rounding to lower or upper quotas, we have not solved the apportionment problem. We are giving out either 29 or 33 seats in congress. With 29 seats, where will the extra seat go? With 33 seats, where are those three extra seats going to come from?

We now discuss four different **apportionment methods**—that is, methods for solving the apportionment problem. The methods are called *Hamilton's method, Jefferson's method, Adams's method,* and *Webster's method.*

3 *Use Hamilton's method.*

Hamilton's Method

Hamilton's method of apportionment proceeds in three steps.

HAMILTON'S METHOD

1. Calculate each group's standard quota.
2. Round each standard quota down to the nearest whole number, thereby finding the lower quota. Initially, give to each group its lower quota.
3. Give the surplus items, one at a time, to the groups with the largest decimal parts in their standard quotas until there are no more surplus items.

Alexander Hamilton (1757–1804)

Alexander Hamilton was the nation's first secretary of the treasury in George Washington's administration. Hamilton and Thomas Jefferson were bitter political adversaries. In 1804, Hamilton was killed in a duel with Aaron Burr, who served as vice president under Jefferson.

For example, consider the lower quotas in the Margaritaville example. We initially give each state its lower quota, shown in **Table 13.29**. Because there are 30 seats in congress, can you see that there is one surplus seat?

TABLE 13.29 Standard Quotas and Lower Quotas for Each State in Margaritaville

State	A	B	C	D	Total
Population	275	383	465	767	1890
Standard quota	4.37	6.08	7.38	12.17	30
Lower quota	4	6	7	12	29

There are 30 seats in congress. Who will get the extra seat?

Hamilton Mania

Creator Lin-Manuel Miranda's hip-hop musical *Hamilton* uses a young, multiracial cast to liberate the Founding Fathers from their own legend. The show, almost entirely sung-through, transforms esoteric Cabinet debates between rivals Hamilton and Jefferson into riveting rap battles. Although there are no songs about apportionment methods (standard quotas are not very musical), the show is expansive enough to include torchy show tunes, high-camp pop, and nods to hip-hop classics.

The surplus seat goes to the state with the greatest decimal part in its standard quota. The greatest decimal part is 0.38, corresponding to state C. Thus, state C receives the additional seat, and is assigned 8 seats in congress.

Table 13.30 summarizes the congressional seat assignments for the states.

TABLE 13.30 Solving Margaritaville's Apportionment Problem: Hamilton's Method

State	A	B	C	D	Total
Lower Quota	4	6	7	12	29
Hamilton's Apportionment	4	6	8	12	30

EXAMPLE 2 Using Hamilton's Method

A rapid transit service operates 130 buses along six routes, A, B, C, D, E, and F. The number of buses assigned to each route is based on the average number of daily passengers per route, given in **Table 13.31**. Use Hamilton's method to apportion the buses.

TABLE 13.31

Route	A	B	C	D	E	F	Total
Average Number of Passengers	4360	5130	7080	10,245	15,535	22,650	65,000

SOLUTION

Before applying Hamilton's method, we must compute the standard divisor.

$$\text{Standard divisor} = \frac{\text{total population}}{\text{number of allocated items}}$$

$$= \frac{\text{total number of passengers}}{\text{number of buses}} = \frac{65{,}000}{130} = 500$$

The standard divisor, 500, represents the average number of passengers per bus per day. Using this value, **Table 13.32** shows the three steps in Hamilton's method.

TABLE 13.32 Apportionment of Buses Using Hamilton's Method

Route	Passengers	Standard Quota	Lower Quota	Decimal Part	Surplus Buses	Final Apportionment
A	4360	8.72	8	0.72 (largest)	1	9
B	5130	10.26	10	0.26		10
C	7080	14.16	14	0.16		14
D	10,245	20.49	20	0.49 (next largest)	1	21
E	15,535	31.07	31	0.07		31
F	22,650	45.30	45	0.30		45
Total	65,000	130	128			130

Step 1 Calculate each standard quota:
$$\text{standard quota} = \frac{\text{population of a group}}{\text{standard divisor}} = \frac{\text{number of passengers}}{500}$$

Step 2 Round down and find each lower quota. The sum, 128, indicates we must assign two extra buses.

Step 3 Give the two surplus buses, one at a time, to the routes with the largest decimal parts.

The apportionment of buses, shown in the final column of **Table 13.32**, is computed by giving each route its lower quota of buses plus any surplus buses.

CHECK POINT 2 Refer to Check Point 1 on page 871. Use Hamilton's method to apportion the 200 congressional seats.

---

4 *Understand the quota rule.*

Have you noticed that every apportionment in Hamilton's method is either the lower quota or the upper quota? This is an important criterion for fairness, called the **quota rule**.

> **THE QUOTA RULE**
>
> A group's apportionment should be either its upper quota or its lower quota. An apportionment method that guarantees that this will always occur is said to **satisfy the quota rule.**

5 *Use Jefferson's method.*

Jefferson's Method

Congress approved Hamilton's method as the apportionment method to be used following the 1790 census. However, the method was vetoed by President Washington. The veto was sustained and a method proposed by Thomas Jefferson was adopted. (In the 1850s, Congress resurrected Hamilton's method; it was used until 1900.)

What could possibly concern President Washington about Hamilton's method? Perhaps it was the fact that in allocating congressional seats, some states would be chosen over others for preferential treatment. These would be the states with the greatest fractional parts in their standard quotas. Ideally, we should be able to replace the standard divisor by some other divisor, called the *modified divisor*. After dividing every state's population by this divisor and then rounding the modified quotas down, we should be left with no surplus items. Thus, no group (or state) could receive preferential treatment.

Thomas Jefferson (1743–1826)

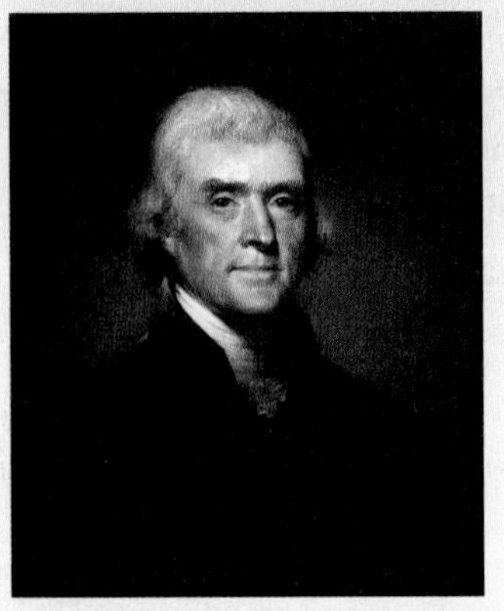

"Science is my passion, politics my duty."

Thomas Jefferson, third president of the United States (1801–1809), was the author of the Declaration of Independence at age 33 and secretary of state under Washington. Although he suffered most of his life from migraine headaches, Jefferson was an outstanding politician, writer, philosopher, and inventor. Known as "long Tom," the $6'2\frac{1}{2}''$ Jefferson often scandalized visitors with his informal attire, a worn brown coat and carpet slippers. Under his presidency, the area of the country doubled through the Louisiana Purchase in 1803.

> **JEFFERSON'S METHOD**
>
> 1. Find a **modified divisor**, d, such that when each group's **modified quota** (group's population divided by d) is rounded down to the nearest whole number, the sum of the whole numbers for all the groups is the number of items to be apportioned. The modified quotients that are rounded down are called **modified lower quotas.**
> 2. Apportion to each group its modified lower quota.

How do we find the modified divisor, d, that makes Jefferson's method work? Before turning to this issue, let's illustrate the method using calculations based on a given value of the modified divisor.

EXAMPLE 3 *Using Jefferson's Method*

As in Example 2, a rapid transit service operates 130 buses along six routes, A, B, C, D, E, and F. The number of buses assigned to each route is based on the average number of daily passengers per route, repeated in **Table 13.31**. Use Jefferson's method with $d = 486$ to apportion the buses.

TABLE 13.31 (repeated)

Route	A	B	C	D	E	F	Total
Average Number of Passengers	4360	5130	7080	10,245	15,535	22,650	65,000

SOLUTION

Using $d = 486$, **Table 13.33** illustrates Jefferson's method.

TABLE 13.33 Apportionment of Buses Using Jefferson's Method with $d = 486$

Route	Passengers	Modified Quota	Modified Lower Quota	Final Apportionment
A	4360	8.97	8	8
B	5130	10.56	10	10
C	7080	14.57	14	14
D	10,245	21.08	21	21
E	15,535	31.97	31	31
F	22,650	46.60	46	46
Total	65,000		130	130

Modified quota $= \frac{\text{number of passengers}}{486}$

Round each modified quota down to the nearest whole number. The sum, 130, indicates there are no surplus buses to assign.

The apportionment of buses, shown in the final column of **Table 13.33**, is determined by giving each route a number of buses equal to its modified lower quota.

CHECK POINT 3 Refer to Check Point 1 on page 871. Use Jefferson's method with $d = 49.3$ to apportion the 200 congressional seats.

Unfortunately, there is no formula for the modified divisor, d. Furthermore, there is usually more than one modified divisor that will make Jefferson's method work. You can find one of these values using trial and error. Begin with the fact that **in Jefferson's method, a modified divisor is slightly less than the standard divisor**. Pick a number d that you think might work. Carry out the calculations required by Jefferson's method: Divide each group's population by d, round down to the nearest whole number, and find the sum of the whole numbers. With luck, the sum is the number of items to be apportioned. Otherwise, change the value of d (make it greater if the sum is too high and less if the sum is too low) and try again. In most cases, with two or three guesses, you will find a modified divisor, d, that works.

Jefferson's method was adopted in 1791. In the apportionment of 1832, New York, with a standard quota of 38.59, received 40 seats in the House of Representatives. Because 40 is neither the upper quota nor the lower quota of 38.59, it was discovered that Jefferson's method violated the quota rule. The apportionment of 1832 was the last time the House of Representatives was apportioned using Jefferson's method.

6 *Use Adams's method.*

Adams's Method

At the same time that Jefferson's method was found to be problematic because of its violation of the quota rule, John Quincy Adams proposed a method that is the mirror image of it. Rather than using modified lower quotas as in Jefferson's method, Adams suggested using modified upper quotas.

In Example 4, we will use trial and error to find a modified divisor, d, that satisfies Adams's method. **When using Adams's method, begin with a modified divisor that is slightly greater than the standard divisor.**

ADAMS'S METHOD

1. Find a **modified divisor**, *d*, such that when each group's **modified quota** (group's population divided by *d*) is rounded up to the nearest whole number, the sum of the whole numbers for all the groups is the number of items to be apportioned. The modified quotas that are rounded up are called **modified upper quotas**.
2. Apportion to each group its modified upper quota.

Blitzer Bonus

John Quincy Adams (1767–1848)

John Quincy Adams, the son of John Adams, second president of the United States, served as the one-term sixth U.S. president from 1825 through 1829. In the electoral free-for-all of 1824, Adams finished second, behind General Andrew Jackson, but because none of the four candidates had a majority, the choice was thrown to the House of Representatives. There, speaker Henry Clay swung the election to Adams. When Adams appointed Clay as secretary of state, the Jackson forces cried "corrupt bargain"—a charge that doomed Adams's bid for reelection in 1828. After leaving the White House, Adams served in the House of Representatives for 18 years and was at his desk when he suffered a fatal stroke.

EXAMPLE 4 Using Adams's Method

Consider, again, the rapid transit service that operates 130 buses along six routes, A, B, C, D, E, and F. The number of buses assigned to each route is based on the average number of daily passengers per route, repeated in **Table 13.31**. Use Adams's method to apportion the buses.

TABLE 13.31 (repeated)

Route	A	B	C	D	E	F	Total
Average Number of Passengers	4360	5130	7080	10,245	15,535	22,650	65,000

SOLUTION

In Example 2, we found that the standard divisor is 500. We begin by guessing at a possible modified divisor, *d*, that we hope will work. This value of *d* must be greater than 500 so that the modified quotas will be less than the standard quotas. When rounded up, the sum of these whole numbers should be 130, the number of buses to be apportioned. Perhaps a good guess might be $d = 512$. **Table 13.34** shows the calculations using Adams's method with $d = 512$.

TABLE 13.34 Calculations Using Adams's Method with $d = 512$

Route	Passengers	Modified Quota	Modified Upper Quota
A	4360	8.52	9
B	5130	10.02	11
C	7080	13.83	14
D	10,245	20.01	21
E	15,535	30.34	31
F	22,650	44.24	45
Total	65,000		131

Modified quota $= \frac{\text{number of passengers}}{512}$

The sum should be 130, not 131.

Because the sum of the modified upper quotas is too high, we need to lower the modified quotas a bit. We tried a higher divisor, $d = 513$, and obtained 129 for the sum of modified upper quotas. Because this sum is too low, we increase the modified divisor from $d = 512$ a bit and try $d = 512.7$. **Table 13.35** at the top of the next page shows the calculations required by Adams's method using this value.

TABLE 13.35 Calculations Using Adams's Method with $d = 512.7$

Route	Passengers	Modified Quota	Modified Upper Quota	Final Apportionment
A	4360	8.50	9	9
B	5130	10.01	11	11
C	7080	13.81	14	14
D	10,245	19.98	20	20
E	15,535	30.30	31	31
F	22,650	44.18	45	45
Total	65,000		130	130

Modified quota $= \frac{\text{number of passengers}}{512.7}$

This sum is precisely the number of buses to be apportioned.

The apportionment of buses, shown in the final column of **Table 13.35**, is determined by giving each route a number of buses equal to its modified upper quota.

CHECK POINT 4 Refer to Check Point 1 on page 871. Use Adams's method to apportion the 200 congressional seats. In guessing at a value for d, begin with $d = 50.5$. If necessary, modify this value as we did in Example 4.

Recall that by Jefferson's method, a delighted New York, with a standard quota of 38.59, received 40 seats. This exceeds 39, the upper quota, and is called an **upper-quota violation**. Adams thought he could avoid this violation, which he did. Unfortunately, his method can result in an assignment of seats in the House of Representatives that is below a state's lower quota. This flaw in Adams's method is called a **lower-quota violation**. With Adams's method, all violations of the quota rule are of this kind.

Webster's Method

In 1832, Daniel Webster suggested an apportionment method that sounds like a compromise between Jefferson's method, where modified quotas are rounded down, and Adams's method, where modified quotas are rounded up. Let's round the modified quotas in the usual way that we round decimals: If the fractional part is less than 0.5, round down to the nearest whole number. If the fractional part is greater than or equal to 0.5, round up to the nearest whole number. Webster believed that this was the only fair way to round numbers.

WEBSTER'S METHOD

1. Find a **modified divisor**, d, such that when each group's **modified quota** (group's population divided by d) is rounded to the nearest whole number, the sum of the whole numbers for all the groups is the number of items to be apportioned. The modified quotas that are rounded are called **modified rounded quotas**.
2. Apportion to each group its modified rounded quota.

In many ways, Webster's method works similarly to Jefferson's and Adams's methods. However, **when using Webster's method, the modified divisor, d, can be less than, greater than, or equal to the standard divisor**. Thus, it may take a bit longer to find a modified divisor that satisfies Webster's method.

Blitzer Bonus

Daniel Webster (1782–1852)

Daniel Webster served in the U.S. House of Representatives (1813–1817; 1823–1827), where he made his name as one of the nation's best-known orators. In the Senate (1827–1841; 1845–1850) and as secretary of state (1841–1843; 1850–1852), he was influential in maintaining the Union. His great disappointment was never becoming president.

7 *Use Webster's method.*

Dirty Presidential Elections

1796: John Adams vs. Thomas Jefferson
Jefferson supporters accuse Adams of having "sesquipedality of belly," meaning it was 18 inches long.

1800: John Adams vs. Thomas Jefferson
Jefferson describes Adams as "a hideous hermaphrodite with neither the force and firmness of a man, nor the gentleness and sensibility of a woman."

1828: Andrew Jackson vs. John Quincy Adams
A pamphlet published by Adams's supporters calls Jackson "a cockfighter, drunkard, thief, liar, and husband of a very fat wife."

2016: Hillary Clinton vs. Donald Trump
Clinton describes Trump's foreign policy experience as "running the Miss Universe pageant in Russia."

Source: Time, November 21, 2016

EXAMPLE 5 *Using Webster's Method*

We return, for the last time, to the rapid transit service that operates 130 buses along six routes, A, B, C, D, E, and F. The number of buses assigned to each route is based on the average number of daily passengers per route, repeated in **Table 13.31**. Use Webster's method to apportion the buses.

TABLE 13.31 (repeated)

Route	A	B	C	D	E	F	Total
Average Number of Passengers	4360	5130	7080	10,245	15,535	22,650	65,000

SOLUTION

From Example 2, we know that the standard divisor is 500. We begin by trying $d = 502$, using a modified divisor greater than the standard divisor. After rounding the modified quotients to the nearest integer and taking the resulting sum, we obtain 129, rather than the desired 130. Because 129 is too low, this suggests that the modified divisor is too large, so we should guess at a modified divisor less than 500, the standard divisor. We try $d = 498$. **Table 13.36** shows the calculations required by Webster's method using this value.

TABLE 13.36 Calculations Using Webster's Method with $d = 498$

Route	Passengers	Modified Quota	Modified Rounded Quotas	Final Apportionment
A	4360	8.76	9	9
B	5130	10.30	10	10
C	7080	14.22	14	14
D	10,245	20.57	21	21
E	15,535	31.19	31	31
F	22,650	45.48	45	45
Total	65,000		130	130

Modified quota $= \dfrac{\text{number of passengers}}{498}$

This sum is precisely the number of buses to be apportioned.

The apportionment of buses, shown in the final column of **Table 13.36**, is determined by giving each route a number of buses equal to its modified rounded quota.

CHECK POINT 5 Refer to Check Point 1 on page 871. Use Webster's method with $d = 49.8$ to apportion the 200 congressional seats.

In Examples 2–5, we apportioned the buses using four different methods. Although the final apportionment using Hamilton's method is the same as the one obtained using Webster's method, this is not always the case. Like voting methods, different apportionment methods applied to the same situation can produce different results.

Although Webster's method can violate the quota rule, such violations are rare. Many experts consider Webster's method the best overall apportionment method available. The method currently used to apportion the U.S. House of Representatives is known as the *Huntington-Hill method*. (See Exercise 50.) However, some experts believe that in our lifetimes, Webster's method will make a comeback and replace Huntington-Hill.

Table 13.37 summarizes the four apportionment methods discussed in this section.

TABLE 13.37 Summary of Apportionment Methods

Method	Divisor	Apportionment
Hamilton's	Standard divisor $= \dfrac{\text{total population}}{\text{number of allocated items}}$	Round each standard quota down to the nearest whole number. Initially give each group its lower quota. Give surplus items, one at a time, to the groups with the largest decimal parts.
Jefferson's	The modified divisor is less than the standard divisor.	Round each group's modified quota down to the nearest whole number. Apportion to each group its modified lower quota.
Adams's	The modified divisor is greater than the standard divisor.	Round each group's modified quota up to the nearest whole number. Apportion to each group its modified upper quota.
Webster's	The modified divisor may be less than, greater than, or equal to the standard divisor.	Round each group's modified quota to the nearest whole number. Apportion to each group its modified rounded quota.

Blitzer Bonus

The Electoral College

The framers of the Constitution believed that the opinion of the majority sometimes had to be tempered by the wisdom of elected representatives. Consequently, the president and the vice president are the only elected U.S. officials who are not chosen by a direct vote of the people, but rather by the Electoral College.

Here is an outline of how the Electoral College works.

- There is a total of 538 electoral votes.
- The votes are divided by the states and among the District of Columbia. The number of electoral votes for each state is equal to its total number of congressional senators and representatives.
- The candidate who wins a majority of popular votes in a given state earns all the votes of that state's Electoral College members.
- The candidate who wins the majority, not merely a plurality, of the electoral votes is declared the winner. A presidential candidate needs at least 270 of the 538 electoral votes to win. If no candidate wins, then the election is turned over to the House of Representatives, who vote for one of the top three contenders.

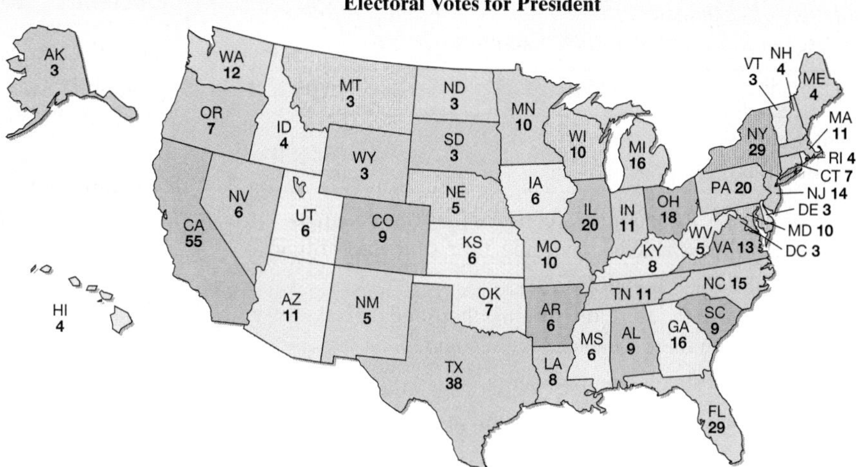

These figures, based on the 2010 Census, were used in the 2016 U.S. presidential election.
Source: Federal Election Commission

As we know from the 2016 election, it is possible for a presidential candidate who has not won the popular vote to win the election. If a candidate wins the popular vote in large states (ones with lots of electoral votes) by only a slim margin and loses the popular vote in smaller states by a wide margin, it is possible that the popular vote winner will actually lose the election. Is Donald Trump the only U.S. president who lost the popular vote but won the election? The answer is no. It happened in 2000, when Democrat Al Gore won the popular vote, but Republican George W. Bush narrowly won the electoral vote and the White House. It happened three other times: in 1824, 1876, and 1888.

Concept and Vocabulary Check

Fill in each blank so that the resulting statement is true.

1. The total population under consideration divided by the number of items to be allocated is called the standard ________. When each group's population is divided by this divisor, a standard ________ for that group is obtained.
2. The apportionment problem is to determine a method for rounding standard ________ into whole numbers so that the sum of the numbers is the total number of allocated items.
3. A standard quota rounded down to the nearest whole number is called a/an ________ quota. A standard quota rounded up to the nearest whole number is called a/an ________ quota. An important criterion for fairness, called the quota rule, states that a group's apportionment should be either its ________ or its ________.
4. The apportionment method that involves rounding each standard quota down to the nearest whole number is called ________ method. Surplus items are given, one at a time, to the groups with the largest ________ parts in their standard quotas until there are no more surplus items.
5. Jefferson's method, Adams's method, and Webster's method require the use of a ________ divisor. The apportionment method in which this divisor is less than the standard divisor is ________ method. The apportionment method in which this divisor is greater than the standard divisor is ________ method. The apportionment method in which this divisor may be less than, greater than, or equal to the standard divisor is ________ method.
6. When each group's population is divided by the modified divisor, a modified ________ for that group is obtained. These numbers are rounded down to the nearest whole number using ________ method, rounded up to the nearest whole number using ________ method, and rounded to the nearest whole number using ________ method.
7. True or False: Different apportionment methods applied to the same situation can produce different results. ______

Exercise Set 13.3

Practice and Application Exercises

Throughout this Exercise Set, in computing standard divisors, standard quotas, and modified quotas, round to the nearest hundredth when necessary.

A small country is comprised of four states, A, B, C, and D. The population of each state, in thousands, is given in the following table. Use this information to solve Exercises 1–4.

State	A	B	C	D	Total
Population (in thousands)	138	266	534	662	1600

1. According to the country's constitution, the congress will have 80 seats, divided among the four states according to their respective populations.
   a. Find the standard divisor, in thousands. How many people are there for each seat in congress?
   b. Find each state's standard quota.
   c. Find each state's lower quota and upper quota.
2. According to the country's constitution, the congress will have 200 seats, divided among the four states according to their respective populations.
   a. Find the standard divisor, in thousands. How many people are there for each seat in congress?
   b. Find each state's standard quota.
   c. Find each state's lower quota and upper quota.
3. Use Hamilton's method to find each state's apportionment of congressional seats in Exercise 1.
4. Use Hamilton's method to find each state's apportionment of congressional seats in Exercise 2.
5. A university is composed of five schools. The enrollment in each school is given in the following table.

School	Humanities	Social Science	Engineering	Business	Education
Enrollment	1050	1410	1830	2540	3580

There are 300 new computers to be apportioned among the five schools according to their respective enrollments. Use Hamilton's method to find each school's apportionment of computers.

6. A university is composed of five schools. The enrollment in each school is given in the following table.

School	Liberal Arts	Education	Business	Engineering	Sciences
Enrollment	1180	1290	2140	2930	3320

There are 300 new computers to be apportioned among the five schools according to their respective enrollments. Use Hamilton's method to find each school's apportionment of computers.

7. A small country is composed of five states, A, B, C, D, and E. The population of each state is given in the following table. Congress will have 57 seats, divided among the five states according to their respective populations. Use Jefferson's method with $d = 32{,}920$ to apportion the 57 congressional seats.

State	A	B	C	D	E
Population	126,316	196,492	425,264	526,664	725,264

8. A small country is comprised of four states, A, B, C, and D. The population of each state, in thousands, is given in the following table. Congress will have 400 seats, divided among the four states according to their respective populations. Use Jefferson's method with $d = 7.82$ to apportion the 400 congressional seats.

State	A	B	C	D
Population (in thousands)	424	664	892	1162

9. An HMO has 150 doctors to be apportioned among four clinics. The HMO decides to apportion the doctors based on the average weekly patient load for each clinic, given in the following table. Use Jefferson's method to apportion the 150 doctors. (*Hint*: Find the standard divisor. A modified divisor that is less than this standard divisor will work.)

Clinic	A	B	C	D
Average Weekly Patient Load	1714	5460	2440	5386

10. An HMO has 70 doctors to be apportioned among six clinics. The HMO decides to apportion the doctors based on the average weekly patient load for each clinic, given in the following table. Use Jefferson's method to apportion the 70 doctors. (*Hint:* A modified divisor between 39 and 40 will work.)

Clinic	A	B	C	D	E	F
Average Weekly Patient Load	316	598	396	692	426	486

11. The police department in a large city has 180 new officers to be apportioned among six high-crime precincts. Crimes by precinct are shown in the following table. Use Adams's method with $d = 16$ to apportion the new officers among the precincts.

Precinct	A	B	C	D	E	F
Crimes	446	526	835	227	338	456

12. Four people pool their money to buy 60 shares of stock. The amount that each person contributes is shown in the following table. Use Adams's method with $d = 108$ to apportion the shares of stock.

Person	A	B	C	D
Contribution	$2013	$187	$290	$3862

13. Three people pool their money to buy 30 shares of stock. The amount that each person contributes is shown in the following table. Use Adams's method to apportion the shares of stock. (*Hint:* Find the standard divisor. A modified divisor that is greater than this standard divisor will work.)

Person	A	B	C
Amount	$795	$705	$525

14. Refer to Exercise 9. Use Adams's method to apportion the 150 doctors. (*Hint:* Find the standard divisor. A modified divisor that is greater than this standard divisor will work.)

15. Twenty sections of bilingual math courses, taught in both English and Spanish, are to be offered in introductory algebra, intermediate algebra, and liberal arts math. The preregistration figures for the number of students planning to enroll in these bilingual sections are given in the following table. Use Webster's method with $d = 29.6$ to determine how many bilingual sections of each course should be offered.

Course	Introductory Algebra	Intermediate Algebra	Liberal Arts Math
Enrollment	130	282	188

16. Refer to the populations of the four states in Exercise 8. Unlike Exercise 8, Congress now has 314 seats to divide among the states according to their respective populations. Use Webster's method with $d = 9.98$ to apportion the 314 congressional seats.

17. A rapid transit service operates 200 buses along five routes, A, B, C, D, and E. The number of buses assigned to each route is based on the average number of daily passengers per route, given in the following table. Use Webster's method to apportion the buses. (*Hint*: A modified divisor between 55 and 56 will work.)

Route	A	B	C	D	E
Average Number of Passengers	1087	1323	1592	1596	5462

18. Refer to Exercise 11. Use Webster's method to apportion the 180 new officers among the precincts. (*Hint:* A modified divisor between 15 and 16 will work.)

A hospital has a nursing staff of 250 nurses working in four shifts: A (7:00 A.M. to 1:00 P.M.), B (1:00 P.M. to 7:00 P.M.), C (7:00 P.M. to 1:00 A.M.), and D (1:00 A.M. to 7:00 A.M.). The number of nurses apportioned to each shift is based on the average number of patients per shift, given in the following table. Use this information to solve Exercises 19–22.

Shift	A	B	C	D
Average Number of Patients	453	650	547	350

19. Use Hamilton's method to apportion the 250 nurses among the shifts at the hospital.

20. Use Jefferson's method to apportion the 250 nurses among the shifts in the hospital. (*Hint:* A modified divisor that works should be less than the standard divisor in Exercise 19.)

21. Use Adams's method to apportion the 250 nurses among the shifts in the hospital. (*Hint:* A modified divisor that works should be greater than the standard divisor in Exercise 19.)

22. Use Webster's method to apportion the 250 nurses among the shifts in the hospital. (*Hint:* A modified divisor that works is equal to the standard divisor that you found in Exercise 19.)

The table shows the 1790 United States census. In 1793, at the direction of President George Washington, 105 seats in the House of Representatives were to be divided among the 15 states according to their 1790 populations. Use this information to solve Exercises 23–26.

1790 UNITED STATES CENSUS

Connecticut	236,841	New York	331,589
Delaware	55,540	North Carolina	353,523
Georgia	70,835	Pennsylvania	432,879
Kentucky	68,705	Rhode Island	68,446
Maryland	278,514	South Carolina	206,236
Massachusetts	475,327	Vermont	85,533
New Hampshire	141,822	Virginia	630,560
New Jersey	179,570		

23. Use Hamilton's method to find each state's apportionment of congressional seats.

24. Use Jefferson's method with $d = 33{,}000$ to find each state's apportionment of congressional seats.

25. Use Adams's method with $d = 36{,}100$ to find each state's apportionment of congressional seats.

26. Use Webster's method with $d = 34{,}500$ to find each state's apportionment of congressional seats.

Explaining the Concepts

27. Describe how to find a standard divisor.

28. Describe how to determine a standard quota for a particular group.

29. How is the lower quota found from a standard quota?

30. How is the upper quota found from a standard quota?

31. Describe the apportionment problem.

32. In your own words, describe Hamilton's method of apportionment.

33. What is the quota rule?

34. Explain why Hamilton's method satisfies the quota rule.

35. Describe the difference between how modified quotas are rounded using Jefferson's method and Adams's method.

36. Suppose that you guess at a modified divisor, d, using Jefferson's method. How do you determine if your guess satisfies the method?

37. Describe the difference between the modified divisor, d, in terms of the standard divisor using Jefferson's method and Adams's method.

38. In allocating congressional seats, how does Hamilton's method choose some states over others for preferential treatment? Explain how this is avoided in Jefferson's and Adams's methods.

39. How are modified quotas rounded using Webster's method?

40. Why might it take longer to guess at a modified divisor that works using Webster's method than using Jefferson's method or Adams's method?

41. In this Exercise Set, we have used apportionment methods to divide congressional seats, assign computers to schools, assign doctors to clinics, divide police officers among precincts, divide shares of stock, assign sections of bilingual math, assign buses to city routes, and assign nurses to hospital shifts. Describe another situation that requires the use of apportionment methods.

Critical Thinking Exercises

Make Sense? *In Exercises 42–45, determine whether each statement makes sense or does not make sense, and explain your reasoning.*

42. An apportionment method is used to determine each state's representation in the U.S. House of Representatives.

43. An apportionment method is used to determine each state's representation in the U.S. Senate.

44. The mathematics required by Adams's method is more complicated than the mathematics required by Jefferson's method.

45. A modified quota of 46.01 and a final apportionment of 47 is a violation of the quota rule.

46. In 1880, there were 300 congressional seats in the U.S. House of Representatives. The population of Alabama was 1,262,505, and its standard quota was 7.671. Find, to the nearest whole number, the U.S. population in 1880.

A small country is composed of three states, A, B, and C. The country's constitution specifies that congressional seats will be divided among the three states according to their respective populations. In Exercises 47–49, write an apportionment problem satisfying the given criterion.

47. Hamilton's method and Jefferson's method result in the same apportionment.

48. Hamilton's method and Adams's method result in the same apportionment.

49. Hamilton's method and Webster's method result in the same apportionment.

Group Exercises

50. The method currently used to apportion the U.S. House of Representatives is known as the **Huntington-Hill method**, and more commonly as the **method of equal proportions**. Research and present a group report on this method. Include the history of how the method came into use and describe how the method works.

51. Research and present a group report on a brief history of apportionment in the United States.

13.4 Flaws of Apportionment Methods

WHAT AM I SUPPOSED TO LEARN?

After studying this section, you should be able to:

1. Understand and illustrate the Alabama paradox.
2. Understand and illustrate the population paradox.
3. Understand and illustrate the new-states paradox.
4. Understand Balinski and Young's Impossibility Theorem.

Posters from the controversial election of 1876. The candidates were Rutherford B. Hayes (Republican) and Samuel J. Tilden (Democrat).

THE VERY MENTION OF FLORIDA OUTRAGED THE DEMOCRATS. FLORIDA'S CONTESTED electoral votes helped elect a Republican president who had lost the popular vote.

No, we're not referring to the controversial election of George W. Bush in 2000, but rather the 1876 election of Republican Rutherford B. Hayes over Democrat Samuel J. Tilden. Tilden won the popular vote with 4,300,590 votes, whereas Hayes received 4,036,298 votes. In 1872, a power grab among the states resulted in an illegal apportionment of the House of Representatives that was not based on Hamilton's method. Two seats that would have gone to New York and Illinois under Hamilton's method instead went to Florida and New Hampshire. Four years later, Rutherford B. Hayes became president of the United States based on this unconstitutional apportionment, winning the electoral vote by a margin of one vote, 185 to 184. If Hamilton's method had been used in the congressional apportionment, New York would have had one more vote, taking the Hayes votes away from Florida or New Hampshire. In short, if the 1872 apportionment of the House of Representatives had been carried out according to the legal Hamilton's method, Tilden would have won the electoral vote and, therefore, the presidency.

Hamilton's method may appear to be a fair and reasonable apportionment method. It is the only method that satisfies the quota rule—that is, a group's apportionment is always its upper quota or its lower quota. Unfortunately, it is also the only method that can produce some serious problems, called the *Alabama paradox,* the *population paradox,* and the *new-states paradox*. In this section, we discuss each of these flaws. We conclude by seeing if there is an ideal apportionment method that can guarantee the states their fair share of seats in the House of Representatives. At stake is equal representation, avoiding election debacles such as the one that occurred in 1876, and the reality that many government policies are based on the number of representatives for each state.

The Alabama Paradox

1 *Understand and illustrate the Alabama paradox.*

After the 1880 census, two possible sizes for the House of Representatives were under consideration: 299 members or 300 members. The chief clerk of the U.S. Census Office used Hamilton's method to compute apportionments for both House sizes. He found that adding one more seat to the House in order to have 300 seats would actually decrease the number of seats for Alabama, from 8 seats to 7 seats. This was the first time this paradoxical behavior was observed in Congressional apportionment; as a result, it is called the **Alabama paradox**.

"No invasions of the Constitution are fundamentally so dangerous as the tricks played on their own numbers."
—Thomas Jefferson

THE ALABAMA PARADOX

An increase in the total number of items to be apportioned results in the loss of an item for a group.

EXAMPLE 1 *Illustrating the Alabama Paradox*

A small country with a population of 10,000 is composed of three states. According to the country's constitution, the congress will have 200 seats, divided among the three states according to their respective populations. **Table 13.38** shows each state's population. Use Hamilton's method to show that the Alabama paradox occurs if the number of seats is increased to 201.

TABLE 13.38

State	A	B	C	Total
Population	5015	4515	470	10,000

SOLUTION

We begin with 200 seats in the congress. First we compute the standard divisor.

$$\text{Standard divisor} = \frac{\text{total population}}{\text{number of allocated items}} = \frac{10{,}000}{200} = 50$$

Using this value, **Table 13.39** shows the apportionment for each state using Hamilton's method.

TABLE 13.39 Apportionment of 200 Congressional Seats Using Hamilton's Method

State	Population	Standard Quota	Lower Quota	Decimal Part	Surplus Seats	Final Apportionment
A	5015	100.3	100	0.3		100
B	4515	90.3	90	0.3		90
C	470	9.4	9	0.4	1	10
Total	10,000	200	199	largest		200

Standard quota $= \frac{\text{population}}{50}$

This sum indicates we must assign one extra seat.

Now let's see what happens with 201 seats in congress. First we compute the standard divisor.

$$\text{Standard divisor} = \frac{\text{total population}}{\text{number of allocated items}} = \frac{10{,}000}{201} \approx 49.75$$

Using 49.75 as the standard divisor, **Table 13.40** shows the apportionment for each state using Hamilton's method.

TABLE 13.40 Apportionment of 201 Congressional Seats Using Hamilton's Method

State	Population	Standard Quota	Lower Quota	Decimal Part	Surplus Seats	Final Apportionment
A	5015	100.80	100	largest 0.80	1	101
B	4515	90.75	90	next largest 0.75	1	91
C	470	9.45	9	0.45		9
Total	10,000	201	199			201

Standard quota $= \frac{\text{population}}{49.75}$

This sum indicates we must assign two extra seats.

The final apportionments are summarized in **Table 13.41**. When the number of seats increased from 200 to 201, state C's apportionment actually decreased, from 10 to 9. This is an example of the Alabama paradox. In this situation, the larger states, A and B, benefit at the expense of the smaller state, C.

TABLE 13.41 Illustrating the Alabama Paradox

State	Apportionment with 200 Seats	Apportionment with 201 Seats
A	100	101
B	90	91
C	10	9

State C's apportionment decreased from 10 to 9.

CHECK POINT 1 **Table 13.42** shows the populations of the four states in a country with a population of 20,000. Use Hamilton's method to show that the Alabama paradox occurs if the number of seats in congress is increased from 99 to 100.

TABLE 13.42

State	A	B	C	D	Total
Population	2060	2080	7730	8130	20,000

2 *Understand and illustrate the population paradox.*

The Population Paradox

The issue of power and representation among the states has been a serious concern since the drafting of the Constitution. In the early 1900s, Virginia was growing much faster than Maine—about 60% faster—yet Virginia lost a seat to Maine in the House of Representatives. This paradox, called the **population paradox**, illustrates another serious flaw of Hamilton's method.

THE POPULATION PARADOX

Group A loses items to group B, even though the population of group A grew at a faster rate than that of group B.

EXAMPLE 2 *Illustrating the Population Paradox*

A small country with a population of 10,000 is composed of three states. There are 11 seats in the congress, divided among the three states according to their respective populations. Using Hamilton's method, **Table 13.43** shows the apportionment of congressional seats for each state.

TABLE 13.43 Apportionment of 11 Congressional Seats Using Hamilton's Method

State	Population	Standard Quota	Lower Quota	Hamilton's Apportionment
A	540	0.59	0	0
B	2430	2.67	2	3
C	7030	7.73	7	8
Total	10,000	10.99	9	11

Blitzer Bonus

Hamilton's Method in Other Countries

Parliament in session in Brussels, Belgium

The discovery of the Alabama paradox in 1880 resulted in the termination of Hamilton's method in the U.S. House of Representatives. However, Hamilton's method, also known as the method of largest remainders, is still used to apportion at least part of the legislatures in Austria, Belgium, El Salvador, Indonesia, Liechtenstein, and Madagascar.

The population of the country increases, shown in **Table 13.44**.

TABLE 13.44

State	A	B	C	Total
Original Population	540	2430	7030	10,000
New Population	560	2550	7890	11,000

a. Find the percent increase in the population of each state.

b. Use Hamilton's method to show that the population paradox occurs.

SOLUTION

a. Recall that the fraction for percent increase is the amount of increase divided by the original amount. The percent increase in the population of each state is determined as follows:

$$\text{State A: } \frac{560 - 540}{540} = \frac{20}{540} \approx 0.037 = 3.7\%$$

(20: increase in population; 540: original population)

$$\text{State B: } \frac{2550 - 2430}{2430} = \frac{120}{2430} \approx 0.049 = 4.9\%$$

$$\text{State C: } \frac{7890 - 7030}{7030} = \frac{860}{7030} \approx 0.122 = 12.2\%$$

All three states had an increase in their populations. State C is increasing at a faster rate than state B, while state B is increasing at a faster rate than state A.

b. We need to use Hamilton's method to find the apportionment for each state with its new population. First we compute the standard divisor.

$$\text{Standard divisor} = \frac{\text{total population}}{\text{number of allocated items}} = \frac{11{,}000}{11} = 1000$$

Using this value, **Table 13.45** shows the apportionment for each state using Hamilton's method.

TABLE 13.45 Apportionment of 11 Congressional Seats for the New Populations Using Hamilton's Method

State	Population	Standard Quota	Lower Quota	Decimal Part	Surplus Seats	Final Apportionment
A	560	0.56	0	0.56	1	1
B	2550	2.55	2	0.55		2
C	7890	7.89	7	0.89	1	8
Total	11,000	11	9			11

Standard quota $= \frac{\text{population}}{1000}$

This sum indicates we must assign two extra seats.

The final apportionments are summarized in **Table 13.46** at the top of the next page. The good news is that state A now has a congressional seat. The bad news is that state B has lost a seat to state A, even though state B's population grew at a faster rate than that of state A. This is an example of the population paradox.

U.S. Presidents Elected without a Majority of the Total Votes Cast

President Woodrow Wilson, 1913–1921

Year	President	Electoral Percent	Popular Percent
1824	John Q. Adams*	31.8%	29.8%
1844	James K. Polk (D)	61.8	49.3
1848	Zachary Taylor (W)	56.2	47.3
1856	James Buchanan (D)	58.7	45.3
1860	Abraham Lincoln (R)	59.4	39.9
1876	Rutherford B. Hayes (R)*	50.1	47.9
1880	James A. Garfield (R)	57.9	48.3
1884	Grover Cleveland (D)	54.6	48.8
1888	Benjamin Harrison (R)*	58.1	47.8
1892	Grover Cleveland (D)	62.4	46.0
1912	Woodrow Wilson (D)	81.9	41.8
1916	Woodrow Wilson (D)	52.1	49.3
1948	Harry S. Truman (D)	57.1	49.5
1960	John F. Kennedy (D)	56.4	49.7
1968	Richard M. Nixon (R)	56.1	43.4
1992	William J. Clinton (D)	68.8	43.0
1996	William J. Clinton (D)	70.4	49.0
2000	George W. Bush (R)*	50.3	47.9
2016	Donald J. Trump (R)*	56.9	46.1

* elected despite losing the popular vote

Source: Federal Election Commission

TABLE 13.46 Illustrating the Population Paradox

State	Growth Rate	Original Apportionment	New Apportionment
A	3.7%	0	1
B	4.9%	3	2
C	12.2%	8	8

State B lost a seat to state A.

CHECK POINT 2 A small country has 100 seats in the congress, divided among the three states according to their respective populations. **Table 13.47** shows each state's population before and after the country's population increase.

TABLE 13.47

State	A	B	C	Total
Original Population	19,110	39,090	141,800	200,000
New Population	19,302	39,480	141,800	200,582

a. Use Hamilton's method to apportion the 100 congressional seats using the original population.

b. Find the percent increase in the populations of states A and B. (State C did not have any change in population.)

c. Use Hamilton's method to apportion the 100 congressional seats using the new population. Show that the population paradox occurs.

3 *Understand and illustrate the new-states paradox.*

The New-States Paradox

Another flaw in Hamilton's method was discovered in 1907 when Oklahoma became a state. Previously, the House of Representatives had 386 seats. Based on its population, it was determined that Oklahoma should have 5 seats. Thus, the House size was changed from 386 to 391. The intent was to leave the number of seats unchanged for the other states. However, when the apportionment was recalculated using Hamilton's method, Maine gained a seat (4 instead of 3) and New York lost a seat (from 38 to 37). The addition of Oklahoma, with its fair share of seats, forced New York to give a seat to Maine. The **new-states paradox** occurs when the addition of a new state affects the apportionments of other states.

THE NEW-STATES PARADOX

The addition of a new group changes the apportionments of other groups.

EXAMPLE 3 *Illustrating the New-States Paradox*

A school district has two high schools: East High, with an enrollment of 1688 students, and West High, with an enrollment of 7912 students. The school district has a counseling staff of 48 counselors, who are apportioned to the schools using Hamilton's method. This apportionment results in 8 counselors assigned to East High and 40 counselors assigned to West High, shown in **Table 13.48** at the top of the next page. The standard divisor is

$$\frac{\text{total population}}{\text{number of allocated items}} = \frac{9600}{48} = 200.$$

There are 200 students per counselor.

TABLE 13.48 Apportionment of 48 Counselors Using Hamilton's Method

School	Population	Standard Quota	Lower Quota	Hamilton's Apportionment
East High	1688	8.44	8	8
West High	7912	39.56	39	40
Total	9600	48	47	48

Suppose that a new high school, North High, with a population of 1448 students, is added to the district. Using the standard divisor of 200 students per counselor, the school district decides to hire 7 new counselors for North High. Show that the new-states paradox occurs when the counselors are reapportioned.

SOLUTION

The new total population is the previous student population, 9600, plus the student population at North High, 1448: $9600 + 1448 = 11{,}048$. The new number of counselors is the previous number, 48, plus the new counselors hired at North High, 7: $48 + 7 = 55$. Thus, when North High is added, the new standard divisor is

$$\frac{\text{total population}}{\text{number of allocated items}} = \frac{11{,}048}{55} \approx 200.87.$$

The new standard quotas and Hamilton's apportionment are shown in **Table 13.49**.

TABLE 13.49 Apportionment of 55 Counselors Using Hamilton's Method

School	Population	Standard Quota	Lower Quota	Decimal Part	Surplus Counselors	Final Apportionment
East High	1688	8.40	8	0.40 (largest)	1	9
West High	7912	39.39	39	0.39		39
North High	1448	7.21	7	0.21		7
Total	11,048	55	54			55

standard quota $= \frac{\text{school population}}{200.87}$

This sum indicates we must assign one extra counselor.

Before North High was added to the district, East High was assigned 8 counselors and West High was assigned 40 counselors. After the addition of a new school and an increase in the total number of counselors to be apportioned, East High was assigned 9 counselors and West High was assigned 39 counselors. Thus, West High ended up losing a counselor to East High. The addition of a new high school changed the apportionments of the other two high schools. This is an example of the new-states paradox.

CHECK POINT 3

a. A school district has two high schools, East High, with an enrollment of 2574 students, and West High, with an enrollment of 9426 students. The school district has a counseling staff of 100 counselors. Use Hamilton's method to apportion the counselors to the two schools.

b. Suppose that a new high school, North High, with a population of 750 students, is added to the district. The school district decides to hire 6 new counselors for North High. Use Hamilton's method to show that the new-states paradox occurs when the counselors are reapportioned.

4 *Understand Balinski and Young's Impossibility Theorem.*

The Search for an Ideal Apportionment Method

We have seen that although Hamilton's method satisfies the quota rule, it can produce paradoxes. By contrast, Jefferson's, Adams's, and Webster's methods can all violate the quota rule, but do not produce paradoxes. Furthermore, Hamilton's and Jefferson's methods can favor large states, while Adams's and Webster's methods can favor small states. For many years, scholars inside and outside Congress hoped mathematicians would eventually devise an ideal apportionment method—one that satisfies the quota rule, does not produce any paradoxes, and treats large and small states without favoritism.

Is there an ideal apportionment method? The answer, unfortunately, is no. In 1980, mathematicians Michel L. Balinski and H. Peyton Young proved that there is no apportionment method that avoids all paradoxes and at the same time satisfies the quota rule. Their theorem is known as **Balinski and Young's Impossibility Theorem**.

BALINSKI AND YOUNG'S IMPOSSIBILITY THEOREM

There is no perfect apportionment method. Any apportionment method that does not violate the quota rule must produce paradoxes, and any apportionment method that does not produce paradoxes must violate the quota rule.

Because any apportionment method must be flawed, the politics of representation can play as large a role as mathematics when Congress discusses an apportionment method. The United States House of Representatives has been apportioned approximately 20 times and the choice of an apportionment method can ultimately be a political decision.

Table 13.50 compares the four apportionment methods we discussed and serves as an example of Balinski and Young's Impossibility Theorem.

TABLE 13.50 Flaws of Apportionment Methods

	Apportionment Method: Is It Flawed?			
Flaw	**Hamilton**	**Jefferson**	**Adams**	**Webster**
May not satisfy the quota rule (apportionment should always be the upper or lower quota)	No	Yes	Yes	Yes
May produce the Alabama paradox (an increase in total apportioned items results in the loss of an item for a group)	Yes	No	No	No
May produce the population paradox (group A loses items to group B, even though group A's population grew at a faster rate than group B's)	Yes	No	No	No
May produce the new-states paradox (the addition of a new group changes the apportionments of other groups)	Yes	No	No	No
In the House of Representatives, apportionment method favors large states	Yes	Yes	No	No
In the House of Representatives, apportionment method favors small states	No	No	Yes	Yes

Blitzer Bonus

Campaign Posters as Art

Blitzer Bonus

The 2016 Presidential Election

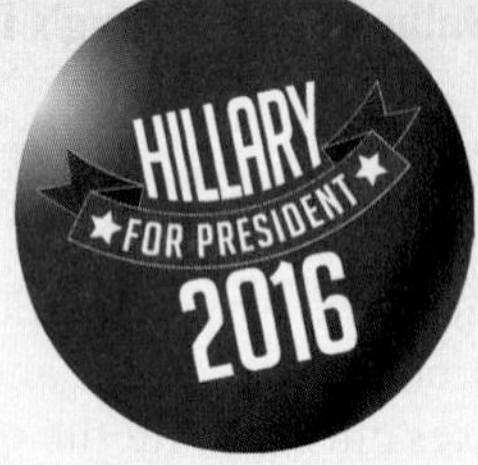

On November 8, 2016, real estate billionaire Donald Trump (R) defeated former first lady, U.S. senator, and Secretary of State Hillary Clinton (D) to win the White House. Pollsters correctly predicted Hillary Clinton's popular-vote victory, but failed to envision Donald Trump's rise to the presidency through a triumph in the Electoral College, capturing close battleground states won by Barack Obama in 2012, including Florida, Ohio, Pennsylvania, and Iowa. Donald Trump became the first U.S. president who had no experience serving in either elected office or the military. The results of the 2016 presidential election are shown in **Table 13.51**.

TABLE 13.51 Results of the 2016 Presidential Election

Party	Candidates	Popular Vote	Percent	Electoral Vote
Democrat	Hillary Clinton and Tim Kaine	65,853,625	48.2%	232
Republican	Donald Trump and Michael Pence	62,985,106	46.1%	306

Source: Federal Election Commission

In the 2016 election, there were numerous statewide ballot measures that presented voters with politically diverse, emotional, and controversial issues. Here is a sample of key ballot initiatives:

Alabama

Right to Work
✔ Yes 70%
No 30%
Membership in a labor union cannot be used as a requirement for employment.

California

Adult Film Health
Yes 46%
✔ No 54%
Would have required use of condoms in adult films made in California, and film producers to pay for vaccinations and test for sexually transmitted infections.

Colorado

Medical Aid in Dying
✔ Yes 65%
No 35%
Allows patient with diagnosis of death within 6 months to receive prescription for fatal doses of medication.

Maine

Rank-Choice Voting
✔ Yes 52%
No 48%
Creates new voting system for all but presidential race. Voters rank candidates; 2nd place vote will be counted if voter's 1st place choice is eliminated but no other candidate has majority.

Massachusetts

Legalize Marijuana
✔ Yes 54%
No 46%
Creates a Cannabis Control Commission to oversee licensing and regulation of recreational marijuana.

Missouri

Voter ID
✔ Yes 63%
No 37%
Amends state constitution to add that voters may be required to produce a photo ID to verify residence.

Nebraska

Reinstate the Death Penalty
Retain 39%
✔ Repeal 61%
Legislature banned death penalty in 2015; initiative would overturn the ban and reinstate the death penalty.

Oklahoma

Execution Methods
✔ Yes 66%
No 34%
Allows death penalty by any method not prohibited by U.S. Constitution; if method is deemed invalid, death penalty remains in force until a valid execution method is found.

Source: USA Today

Concept and Vocabulary Check

Fill in each blank so that the resulting statement is true.

1. If an increase in the total number of items to be apportioned results in the loss of an item for a group, the __________ paradox occurs.
2. If group A loses items to group B, even though the population of group A grew at a faster rate than that of group B, the __________ paradox occurs.
3. If the addition of a new group changes the apportionment of other groups, the __________ paradox occurs.
4. True or False: There is no perfect apportionment method that satisfies the quota rule and avoids any paradoxes. _______

Exercise Set 13.4

Practice and Application Exercises

1. The mathematics department has 30 teaching assistants to be divided among three courses, according to their respective enrollments. The table shows the courses and the number of students enrolled in each course.

Course	College Algebra	Statistics	Liberal Arts Math	Total
Enrollment	978	500	322	1800

   a. Apportion the teaching assistants using Hamilton's method.
   b. Use Hamilton's method to determine if the Alabama paradox occurs if the number of teaching assistants is increased from 30 to 31. Explain your answer.

2. A school district has 57 new laptop computers to be divided among four schools, according to their respective enrollments. The table shows the number of students enrolled in each school.

School	A	B	C	D	Total
Enrollment	5040	4560	4040	610	14,250

   a. Apportion the laptop computers using Hamilton's method.
   b. Use Hamilton's method to determine if the Alabama paradox occurs if the number of laptop computers is increased from 57 to 58. Explain your answer.

3. The table shows the populations of three states in a country with a population of 20,000. Use Hamilton's method to show that the Alabama paradox occurs if the number of seats in congress is increased from 40 to 41.

State	A	B	C	Total
Population	680	9150	10,170	20,000

4. The table at the top of the next column shows the populations, in thousands, of the three states in a country with a population of 3760 thousand. Use Hamilton's method to show that the Alabama paradox occurs if the number of seats in congress is increased from 24 to 25.

State	A	B	C	Total
Population (in thousands)	530	990	2240	3760

5. A small country has 24 seats in the congress, divided among the three states according to their respective populations. The table shows each state's population, in thousands, before and after the country's population increase.

State	A	B	C	Total
Original Population (in thousands)	530	990	2240	3760
New Population (in thousands)	680	1250	2570	4500

   a. Use Hamilton's method to apportion the 24 congressional seats using the original population.
   b. Find the percent increase, to the nearest tenth of a percent, in the population of each state.
   c. Use Hamilton's method to apportion the 24 congressional seats using the new population. Does the population paradox occur? Explain your answer.

6. A country has 200 seats in the congress, divided among the five states according to their respective populations. The table shows each state's population, in thousands, before and after the country's population increase.

State	A	B	C	D	E	Total
Original Population (in thousands)	2224	2236	2640	3030	9870	20,000
New Population (in thousands)	2424	2436	2740	3130	10,070	20,800

   a. Use Hamilton's method to apportion the 200 congressional seats using the original population.
   b. Find the percent increase, to the nearest tenth of a percent, in the population of each state.
   c. Use Hamilton's method to apportion the 200 congressional seats using the new population. What do you observe about the percent increases for states A and B and their respective changes in apportioned seats? Is this the population paradox?

7. A town has 40 mail trucks and four districts in which mail is distributed. The trucks are to be apportioned according to each district's population. The table shows these populations before and after the town's population increase. Use Hamilton's method to show that the population paradox occurs.

District	A	B	C	D	Total
Original Population	1188	1424	2538	3730	8880
New Population	1188	1420	2544	3848	9000

8. A town has five districts in which mail is distributed and 50 mail trucks. The trucks are to be apportioned according to each district's population. The table shows these populations before and after the town's population increase. Use Hamilton's method to show that the population paradox occurs.

District	A	B	C	D	E	Total
Original Population	780	1500	1730	2040	2950	9000
New Population	780	1500	1810	2040	2960	9090

9. A corporation has two branches, A and B. Each year the company awards 100 promotions within its branches. The table shows the number of employees in each branch.

Branch	A	B	Total
Employees	1045	8955	10,000

a. Use Hamilton's method to apportion the promotions.

b. Suppose that a third branch, C, with the number of employees shown in the table below, is added to the corporation. The company adds five new yearly promotions for branch C. Use Hamilton's method to determine if the new-states paradox occurs when the promotions are reapportioned.

Branch	A	B	C	Total
Employees	1045	8955	525	10,525

10. A corporation has three branches, A, B, and C. Each year the company awards 60 promotions within its branches. The table shows the number of employees in each branch.

Branch	A	B	C	Total
Employees	209	769	2022	3000

a. Use Hamilton's method to apportion the promotions.

b. Suppose that a fourth branch, D, with the number of employees shown in the table below, is added to the corporation. The company adds five new yearly promotions for branch D. Use Hamilton's method to determine if the new-states paradox occurs when the promotions are reapportioned.

Branch	A	B	C	D	Total
Employees	209	769	2022	260	3260

11. a. A country has two states, state A, with a population of 9450, and state B, with a population of 90,550. The congress has 100 seats, divided between the two states according to their respective populations. Use Hamilton's method to apportion the congressional seats to the states.

b. Suppose that a third state, state C, with a population of 10,400, is added to the country. The country adds 10 new congressional seats for state C. Use Hamilton's method to show that the new-states paradox occurs when the congressional seats are reapportioned.

12. a. A country has three states, state A, with a population of 99,000, state B, with a population of 214,000, and state C, with a population of 487,000. The congress has 50 seats, divided among the three states according to their respective populations. Use Hamilton's method to apportion the congressional seats to the states.

b. Suppose that a fourth state, state D, with a population of 116,000, is added to the country. The country adds seven new congressional seats for state D. Use Hamilton's method to show that the new-states paradox occurs when the congressional seats are reapportioned.

13. In Exercise 12, use Jefferson's method with $d = 15{,}500$ to solve parts (a) and (b). Does the new-states paradox occur? Explain your answer.

Explaining the Concepts

14. What is the Alabama paradox?

15. What is the population paradox?

16. What is the new-states paradox?

17. According to Balinski and Young's Impossibility Theorem, can the democratic ideal of "one person, one vote" ever be perfectly achieved? Explain your answer.

Critical Thinking Exercises

Make Sense? *In Exercises 18–21, determine whether each statement makes sense or does not make sense, and explain your reasoning.*

18. The county hired seven new doctors to apportion among its three clinics. Although our local clinic has the same proportion of the county's patients as it did before the doctors were hired, it now has one fewer doctor.

19. The population of state A grew at a faster rate than that of state B, yet state A lost an apportioned seat in the legislature to state B.

20. Balinski and Young's Theorem shows that "one person, one vote" is mathematically possible.

21. It's just a matter of time until mathematicians devise an ideal apportionment method that satisfies the quota rule and does not produce any paradoxes.

22. Give an example of a country with three states in which the Alabama paradox occurs using Hamilton's method when the number of seats in congress is increased from 3 to 4.

Chapter Summary, Review, and Test

SUMMARY – DEFINITIONS AND CONCEPTS | EXAMPLES

13.1 Voting Methods

a. Preference ballots are ballots in which a voter is asked to rank all the candidates in order of preference. Preference tables show how often each particular outcome occurred. Ex. 1, p. 847

b. The four voting methods—the plurality method, the Borda count method, the plurality-with-elimination method, and the pairwise comparison method—are summarized in **Table 13.7** on page 855. Ex. 2, p. 848; Ex. 3, p. 849; Ex. 4, p. 851; Ex. 5, p. 853

13.2 Flaws of Voting Methods

a. The four fairness criteria for voting methods—the majority criterion, the head-to-head criterion, the monotonicity criterion, and the irrelevant alternatives criterion—are summarized in **Table 13.22** on page 865. **Table 13.23** on page 865 shows which voting methods satisfy which criteria. Ex. 1, p. 859; Ex. 2, p. 860; Ex. 3, p. 862; Ex. 4, p. 863

b. Arrow's Impossibility Theorem states that it is mathematically impossible for any democratic voting system to satisfy each of the four fairness criteria.

13.3 Apportionment Methods

a. Standard divisors and standard quotas are defined as follows: Ex. 1, p. 871

$$\text{Standard divisor} = \frac{\text{total population}}{\text{number of allocated items}}$$

$$\text{Standard quota} = \frac{\text{population of a particular group}}{\text{standard divisor}}.$$

b. The lower quota is the standard quota rounded down to the nearest whole number. The upper quota is the standard quota rounded up to the nearest whole number.

c. The apportionment problem is to determine a method for rounding standard or modified quotas into whole numbers so that the sum of the numbers is the total number of allocated items.

d. The four apportionment methods discussed in the text—Hamilton's method, Jefferson's method, Adams's method, and Webster's method—are summarized in **Table 13.37** on page 879. The quota rule states that a group's apportionment should be either its upper quota or its lower quota. Ex. 2, p. 873; Ex. 3, p. 874; Ex. 4, p. 876; Ex. 5, p. 878

13.4 Flaws of Apportionment Methods

a. Although Hamilton's method satisfies the quota rule, it may produce paradoxes, including the Alabama paradox, the population paradox, and the new-states paradox. These paradoxes are summarized in **Table 13.50** on page 889. Ex. 1, p. 884; Ex. 2, p. 885; Ex. 3, p. 887

b. Although Jefferson's, Adams's, and Webster's methods do not produce paradoxes, they may not satisfy the quota rule.

c. Balinski and Young's Impossibility Theorem states that there is no perfect apportionment method. Any method that does not violate the quota rule must produce paradoxes, and any method that does not produce paradoxes must violate the quota rule.

Review Exercises

13.1

1. The 12 preference ballots for four candidates (A, B, C, and D) are shown. Construct a preference table to illustrate the results of the election.

ABCD BDCA CBDA ABCD CBDA ABCD
BDCA BDCA CBAD CBAD ABCD CBDA

In Exercises 2–5, your class is given the option of choosing a day for the final exam. Students are asked to rank the four available days, Monday (M), Tuesday (T), Wednesday (W), and Friday (F). The results of the election are shown in the preference table.

Number of Votes	9	5	4	2	2	1
First Choice	T	T	M	T	W	W
Second Choice	W	W	T	M	F	M
Third Choice	F	M	F	W	T	F
Fourth Choice	M	F	W	F	M	T

(In Exercises 2–5, be sure to refer to the preference table at the bottom of the previous page.)

2. How many students voted in the election?

3. How many students selected the days in this order: M, T, F, W?

4. How many students selected Tuesday as their first choice for the final?

5. How many students selected Wednesday as their second choice for the final?

In Exercises 6–9, the Theater Society members are voting for the kind of play they will perform next semester: a comedy (C), a drama (D), or a musical (M). Their votes are summarized in the following preference table.

Number of Votes	10	8	4	2
First Choice	C	M	M	D
Second Choice	D	C	D	M
Third Choice	M	D	C	C

6. Which type of play is selected using the plurality method?

7. Which type of play is selected using the Borda count method?

8. Which type of play is selected using the plurality-with-elimination method?

9. Which type of play is selected using the pairwise comparison method?

In Exercises 10–13, four candidates, A, B, C, and D, are running for chair of the Natural Science Division. The votes of the division members are summarized in the following preference table.

Number of Votes	40	30	6	2
First Choice	A	B	C	D
Second Choice	B	C	D	B
Third Choice	C	D	B	C
Fourth Choice	D	A	A	A

10. Determine the winner using the plurality method.

11. Determine the winner using the Borda count method.

12. Determine the winner using the plurality-with-elimination method.

13. Determine the winner using the pairwise comparison method.

13.2

In Exercises 14–16, voters in a small town are considering four proposals, A, B, C, and D, for the design of affordable housing. Their votes are summarized in the following preference table.

Number of Votes	1500	600	300
First Choice	A	B	C
Second Choice	B	D	B
Third Choice	C	C	D
Fourth Choice	D	A	A

14. Using the Borda count method, which design will be used for the affordable housing?

15. Which design has a majority of first-place votes? Based on your answer to Exercise 14, is the majority criterion satisfied?

16. Which design is favored over all others using a head-to-head comparison? Based on your answer to Exercise 14, is the head-to-head criterion satisfied?

Use the following preference table to solve Exercises 17–18.

Number of Votes	1500	700	300
First Choice	B	A	C
Second Choice	C	B	A
Third Choice	A	C	B

17. Using the plurality-with-elimination method, which candidate wins the election?

18. Which candidate is favored over all others using a head-to-head comparison? Based on your answer to Exercise 17, is the head-to-head criterion satisfied?

Use the following preference table to solve Exercises 19–23.

Number of Votes	180	100	40	30
First Choice	A	B	D	C
Second Choice	B	D	B	B
Third Choice	C	A	C	A
Fourth Choice	D	C	A	D

19. Using the plurality method, which candidate wins the election?

20. Using the Borda count method, which candidate wins the election?

21. Using the plurality-with-elimination method, which candidate wins the election?

22. Using the pairwise comparison method, which candidate wins the election?

23. Which candidate has a majority of first-place votes? Based on your answers to Exercises 19–22, which voting system(s) violate the majority criterion in this election?

Use the following preference table, which shows the results of a straw vote, to solve Exercises 24–25.

Number of Votes	500	400	350	200
First Choice	B	A	C	C
Second Choice	C	B	A	B
Third Choice	A	C	B	A

24. Using the plurality-with-elimination method, which candidate wins the straw vote?

25. In the actual election, the 200 voters in the last column who voted C, B, A, in that order, change their votes to B, C, A. Using the plurality-with-elimination method, which candidate wins the actual election? Based on your answer to Exercise 24, is the monotonicity criterion satisfied?

Use the following preference table to solve Exercises 26–29.

Number of Votes	400	250	200
First Choice	A	C	B
Second Choice	B	B	C
Third Choice	C	A	A

26. Using the plurality method, which candidate wins the election?
27. Suppose that candidate B drops out. Construct a new preference table for the election with B eliminated. Using the new table and the plurality method, which candidate wins the election? Based on your answer to Exercise 26, is the irrelevant alternatives criterion satisfied by the plurality method?
28. Using the Borda count method, which candidate wins the election?
29. If candidate C drops out, which candidate wins the election using the Borda count method? Based on your answer to Exercise 28, is the irrelevant alternatives criterion satisfied by the Borda count method?

13.3

In Exercises 30–36, an HMO has 40 doctors to be apportioned among four clinics. The HMO decides to apportion the doctors based on the average weekly patient load for each clinic, given in the following table.

Clinic	A	B	C	D
Average Weekly Patient Load	275	392	611	724

30. Find the standard divisor.
31. Find each clinic's standard quota.
32. Find each clinic's lower quota and upper quota.
33. Use Hamilton's method to apportion the doctors to the clinics.
34. Use Jefferson's method with $d = 48$ to apportion the doctors to the clinics.
35. Use Adams's method with $d = 52$ to apportion the doctors to the clinics.
36. Use Webster's method with $d = 49.95$ to apportion the doctors to the clinics.

In Exercises 37–40, a country is composed of four states, A, B, C, and D. The population of each state is given in the following table. Congress will have 200 seats, divided among the four states according to their respective populations.

State	A	B	C	D
Population	3320	10,060	15,020	19,600

37. Use Hamilton's method to apportion the congressional seats.
38. Use Jefferson's method to apportion the congressional seats. (*Hint*: A modified divisor that works should be less than the standard divisor in Exercise 37.)
39. Use Adams's method to apportion the congressional seats. (*Hint*: A modified divisor that works should be greater than the standard divisor in Exercise 37.)
40. Use Webster's method to apportion the congressional seats. (*Hint:* A modified divisor that works should lie between 240 and 241.)

13.4

41. A school district has 150 new laptop computers to be divided among three schools, according to their respective enrollments. The table shows the number of students enrolled in each school.

School	A	B	C	Total
Enrollment	370	3365	3765	7500

    a. Apportion the laptop computers using Hamilton's method.
    b. Use Hamilton's method to determine if the Alabama paradox occurs if the number of laptop computers is increased from 150 to 151.

42. A country has 100 seats in the congress, divided among the three states according to their respective populations. The table shows each state's population before and after the country's population increase.

State	A	B	C	Total
Original Population	143,796	41,090	15,114	200,000
New Population	143,796	41,420	15,304	200,520

    a. Use Hamilton's method to apportion the 100 congressional seats using the original population.
    b. Find the percent increase, to the nearest tenth of a percent, in the population of state B and state C.
    c. Use Hamilton's method to apportion the 100 congressional seats using the new population. Does the population paradox occur?

43. A corporation has two branches, A and B. Each year the company awards 33 promotions within its branches. The table shows the number of employees in each branch.

Branch	A	B	Total
Employees	372	1278	1650

    a. Use Hamilton's method to apportion the promotions.
    b. Suppose that a third branch, C, with the number of employees shown in the table, is added to the corporation. The company adds seven new yearly promotions for branch C. Use Hamilton's method to determine if the new-states paradox occurs when the promotions are reapportioned.

Branch	A	B	C	Total
Employees	372	1278	355	2005

44. Is the following statement true or false?

    There are perfect voting methods, as well as perfect apportionment methods.

    Explain your answer.

Chapter 13 Test

In Exercises 1–8, three candidates, A, B, and C, are running for mayor of a small town. The results of the election are shown in the following preference table.

Number of Votes	1200	900	900	600
First Choice	A	C	B	B
Second Choice	B	A	C	A
Third Choice	C	B	A	C

1. How many people voted in the election?
2. How many people selected the candidates in this order: B, A, C?
3. How many people selected candidate B as their first choice?
4. How many people selected candidate A as their second choice?
5. Determine the winner using the plurality method.
6. Determine the winner using the Borda count method.
7. Determine the winner using the plurality-with-elimination method.
8. Determine the winner using the pairwise comparison method.

Use the following preference table to solve Exercises 9–10.

Number of Votes	240	160	60
First Choice	A	C	D
Second Choice	B	B	A
Third Choice	C	D	C
Fourth Choice	D	A	B

9. Determine the winner using the Borda count method.
10. Which candidate has a majority of first-place votes? Based on your answer to Exercise 9, is the majority criterion satisfied?

Use the following preference table to solve Exercises 11–12.

Number of Votes	1500	1000	1000
First Choice	A	B	C
Second Choice	B	A	B
Third Choice	C	C	A

11. Determine the winner using the plurality method.
12. Which candidate is favored over all others using a head-to-head comparison? Based on your answer to Exercise 11, is the head-to-head criterion satisfied?

Use the following preference table, which shows the results of a straw vote, to solve Exercises 13–14.

Number of Votes	70	60	50	30
First Choice	C	B	A	A
Second Choice	A	C	B	C
Third Choice	B	A	C	B

13. Determine the winner using the plurality-with-elimination method.
14. In the actual election, the 30 voters in the last column who voted A, C, B, in that order, change their votes to C, A, B. Using the plurality-with-elimination method, which candidate wins the election? Based on your answer to Exercise 13, is the monotonicity criterion satisfied?
15. Suppose that the plurality method is used on the following preference table. If candidate C drops out, is the irrelevant alternatives criterion satisfied? Explain your answer.

Number of Votes	90	75	45
First Choice	B	C	A
Second Choice	C	A	C
Third Choice	A	B	B

In Exercises 16–24, an HMO has 10 doctors to be apportioned among three clinics. The HMO decides to apportion the doctors based on the average weekly patient load for each clinic, given in the following table.

Clinic	A	B	C
Average Weekly Patient Load	119	165	216

16. Find the standard divisor.
17. Find each clinic's standard quota.
18. Find each clinic's lower quota and upper quota.
19. Use Hamilton's method to apportion the doctors to the clinics.
20. Use Jefferson's method to apportion the doctors to the clinics.
21. Use Adams's method to apportion the doctors to the clinics.
22. Use Webster's method with $d = 47.7$ to apportion the doctors to the clinics.
23. Suppose that the HMO hires one new doctor. Using Hamilton's method, does the Alabama paradox occur when the number of doctors is increased from 10 to 11?
24. As given in the original description, assume that the HMO has 10 doctors. Suppose that a fourth clinic, clinic D, with the average weekly patient load shown in the table, is added to the HMO. The administration hires two new doctors for clinic D. Use Hamilton's method to determine if the new-states paradox occurs when the doctors are reapportioned.

Clinic	A	B	C	D
Average Weekly Patient Load	119	165	216	110

25. Write one sentence for a person not familiar with the mathematics of voting and apportionment that summarizes the two impossibility theorems discussed in this chapter.

Graph Theory

14

- A more user-friendly map for a city's subway system is needed.
- Public works administrators must find the most efficient routes for snowplows, garbage trucks, and street sweepers.
- A sales director traveling to branch offices needs to minimize the cost of the trip.
- You need to run a series of errands and return home using the shortest route.
- After a heavy snow, campus services must clear a minimum number of sidewalks and still ensure that students walking from building to building will be able to do so along cleared paths.
- You need to represent the parent-child relationships for five generations of your family.

These unrelated problems can be solved using special diagrams, called *graphs*. Graph theory was launched by the Swiss mathematician Leonhard Euler (1707–1783). Euler (pronounced "oil er") analyzed a puzzle involving a city with seven bridges. Would it be possible to walk through the city and cross each of its bridges exactly once? Although graph theory developed from this puzzle, you will see how graphs are used to solve a diverse series of realistic problems faced by businesses and individuals.

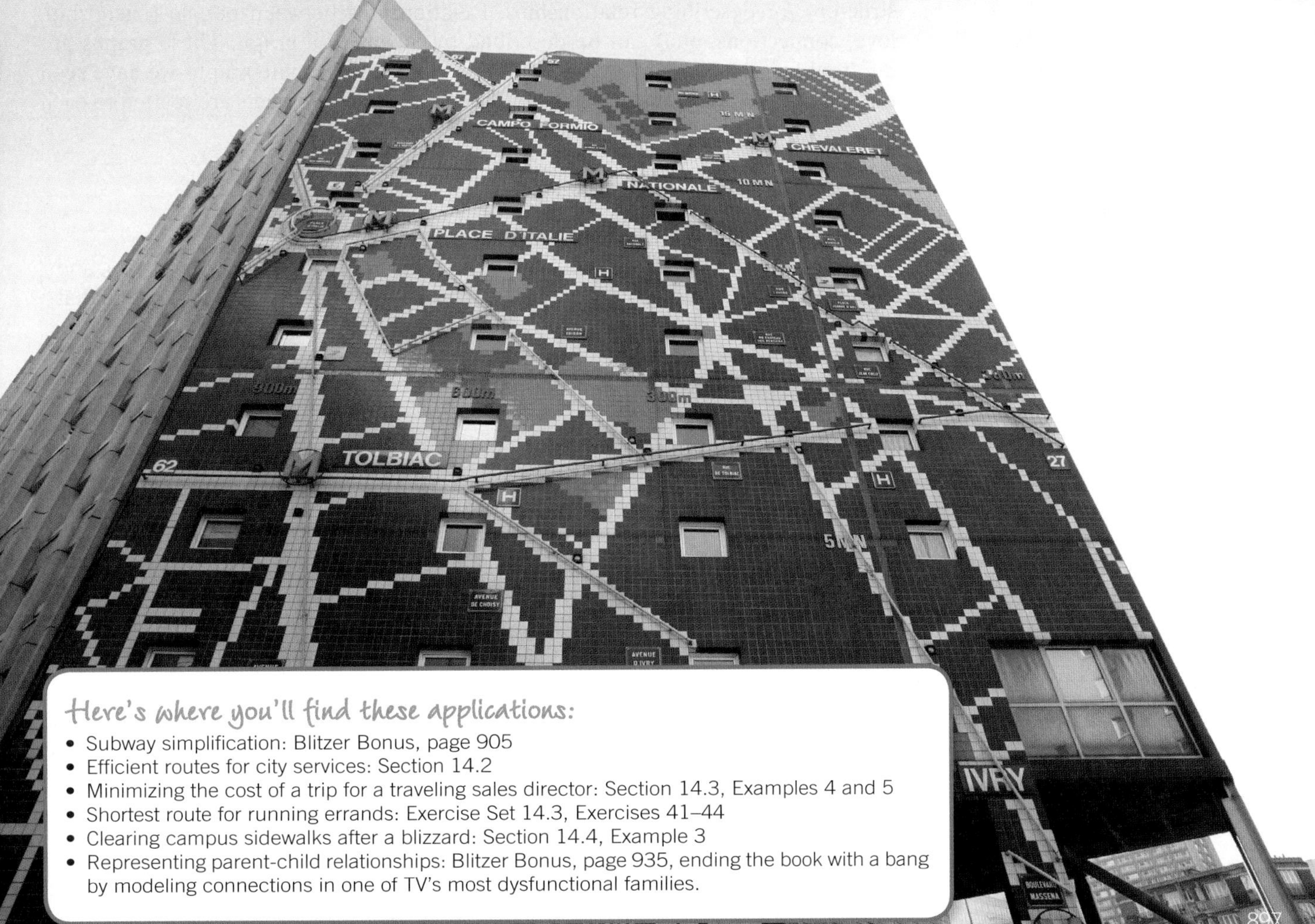

Here's where you'll find these applications:

- Subway simplification: Blitzer Bonus, page 905
- Efficient routes for city services: Section 14.2
- Minimizing the cost of a trip for a traveling sales director: Section 14.3, Examples 4 and 5
- Shortest route for running errands: Exercise Set 14.3, Exercises 41–44
- Clearing campus sidewalks after a blizzard: Section 14.4, Example 3
- Representing parent-child relationships: Blitzer Bonus, page 935, ending the book with a bang by modeling connections in one of TV's most dysfunctional families.

14.1 Graphs, Paths, and Circuits

WHAT AM I SUPPOSED TO LEARN?

After studying this section, you should be able to:

1 Understand relationships in a graph.
2 Model relationships using graphs.
3 Understand and use the vocabulary of graph theory.

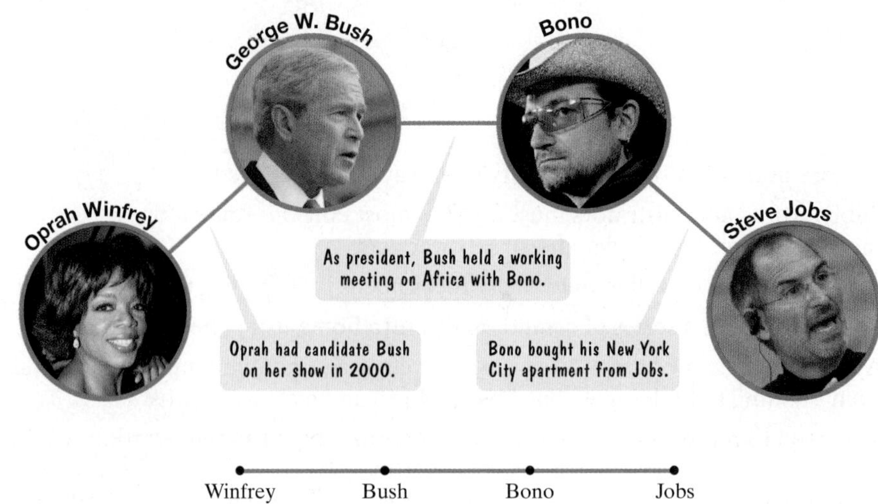

FIGURE 14.1
Source: Time, May 8, 2006

GREAT QUESTION!

Graphs were briefly introduced in Section 10.7. Do I have to read that section before covering this chapter?

No. We open this chapter by revisiting the definitions of graph theory.

THE "SIX DEGREES OF SEPARATION" THEORY STATES THAT ANYONE IN THE WORLD CAN be joined to any other person through a chain of no more than four intermediate connections. **Figure 14.1** indicates that only two intermediate connections link Oprah Winfrey to the late Steve Jobs.

The diagram in **Figure 14.1** is an example of a *graph*. A graph provides a structure for describing relationships. Relationships between people (friendship, love, connections, etc.) can be described by means of a graph. These graphs are different from the line, bar, circle, and rectangular coordinate graphs we have seen throughout the book. Instead they tell us how a group of objects are related to each other. Let's begin with the definition of a graph.

1 *Understand relationships in a graph.*

DEFINITION OF A GRAPH

A **graph** consists of a finite set of points, called **vertices** (singular is *vertex*), and line segments or curves, called **edges,** that start and end at vertices. An edge that starts and ends at the same vertex is called a **loop**.

Figure 14.2 shows an example of a graph. This graph has four vertices, labeled A, B, C, and D. (We will often use uppercase letters to refer to vertices.) If there is not more than one edge between two vertices, we can use the two vertices to refer to that edge. For example, the edge that connects vertex A to vertex D can be called edge AD or edge DA. The loop on the lower right can be referred to as edge CC or loop CC.

We will always indicate vertices by black dots. In **Figure 14.2,** the point where edges AC and DB cross has no black dot. Thus, this point is not a vertex. **Not every point where two edges cross is a vertex.** You might find it useful to think of one of the edges as passing above the other.

In a graph, the important information is which vertices are connected by edges. Two graphs are **equivalent** if they have the same number of vertices connected to each other in the same way. The placement of the vertices and the shapes of the edges are unimportant.

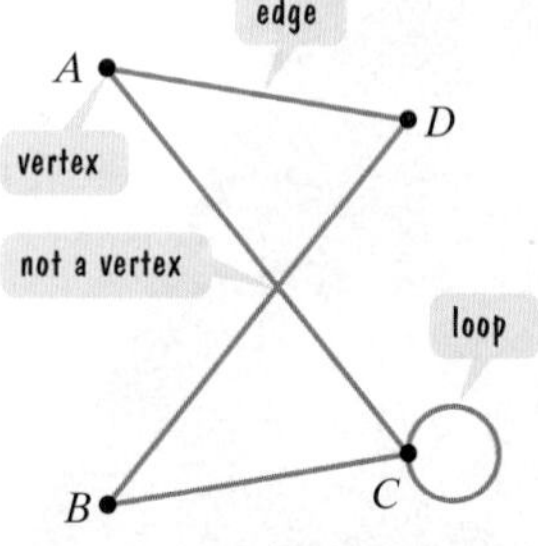

FIGURE 14.2 An example of a graph.

EXAMPLE 1 Understanding Relationships in Graphs

Explain why **Figures 14.3(a)** and **(b)** show equivalent graphs.

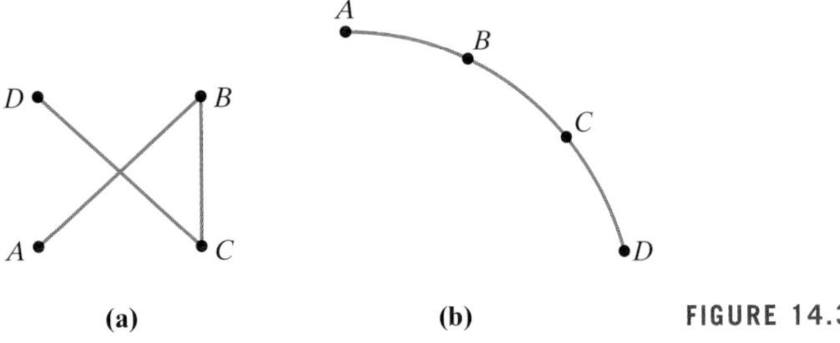

FIGURE 14.3

SOLUTION

In both figures, the vertices are A, B, C, and D. Both graphs have an edge that connects vertex A to vertex B (edge AB or BA), an edge that connects vertex B to vertex C (edge BC or CB), and an edge that connects vertex C to vertex D (edge CD or DC). Because the two graphs have the same number of vertices connected to each other in the same way, they are equivalent.

CHECK POINT 1 Explain why **Figures 14.4(a)** and **(b)** show equivalent graphs.

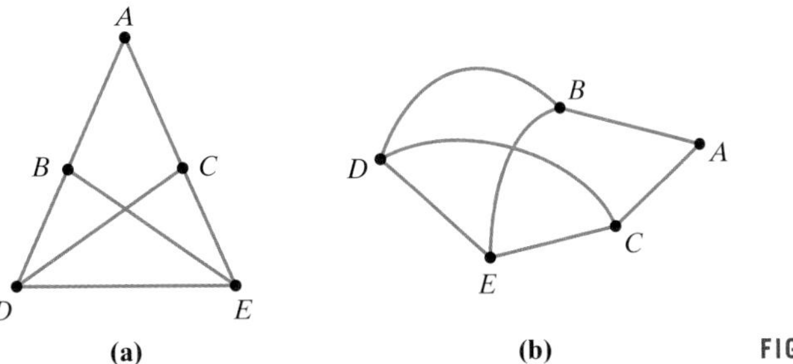

FIGURE 14.4

2 *Model relationships using graphs.*

Modeling Relationships Using Graphs

Examples 2 through 5 illustrate a number of different situations that we can represent, or model, using graphs. We begin with a situation that was briefly discussed in Chapter 10. Our graph will tell how a group of land masses are related to each other.

EXAMPLE 2 Modeling Königsberg with a Graph

In the early 1700s, the city of Königsberg, Germany, was located on both banks and two islands of the Pregel River. **Figure 14.5** shows that the town's sections were connected by seven bridges. Draw a graph that models the layout of Königsberg. Use vertices to represent the land masses and edges to represent the bridges.

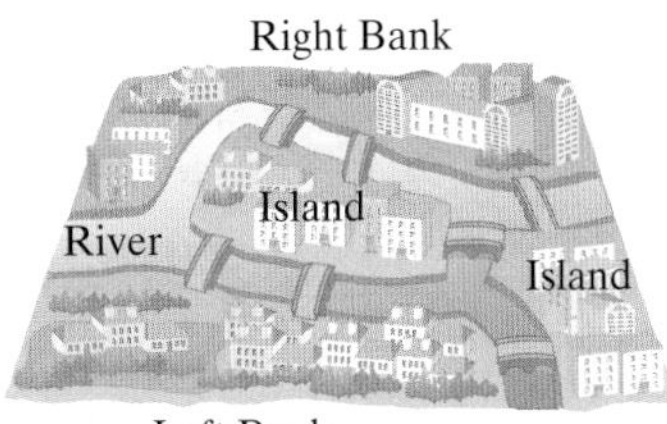

FIGURE 14.5 The layout of Königsberg.

SOLUTION

In drawing the graph, the only thing that matters is the relationship between land masses and bridges: which land masses are connected to each other and by how many bridges. We label each of the four land masses with an uppercase letter, shown in **Figure 14.6(a)**.

Next, we use points to represent the land masses. **Figure 14.6(b)** shows vertices A, B, L, and R. The precise placement of these vertices is not important.

Now we are ready to draw the edges that represent the bridges. There are two bridges that connect island A to right bank R. Therefore, we draw two edges connecting vertex A to vertex R, shown in the graph in **Figure 14.6(b)**. Can you see that there is only one bridge that connects island A to island B? Thus, we draw one edge connecting vertex A to vertex B in our graph. There are also two bridges connecting island A to left bank L, so we draw two edges connecting vertex A to vertex L. Continuing in this manner, we obtain the graph shown in **Figure 14.6(c)**.

GREAT QUESTION!

When modeling relationships using graphs, is the shape of the graph important?

No. The important part is how the vertices are connected to each other. Remember that a graph can be drawn in many equivalent ways.

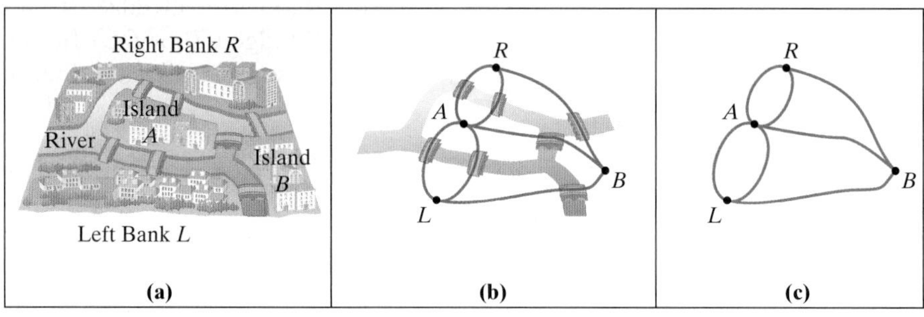

FIGURE 14.6 Creating a graph that models the layout of Königsberg

CHECK POINT 2 The city of Metroville is located on both banks and three islands of the Metro River. **Figure 14.7** shows that the town's sections are connected by five bridges. Draw a graph that models the layout of Metroville.

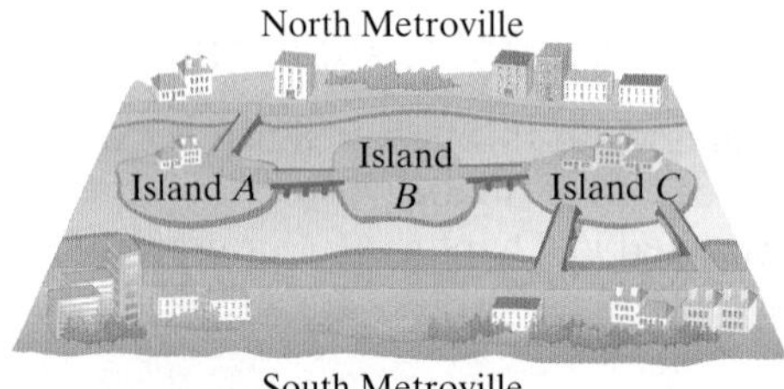

FIGURE 14.7

In the next example, our graph will tell how a group of states are related to each other.

EXAMPLE 3 *Modeling Bordering Relationships for the New England States*

The map in **Figure 14.8** shows the New England states. Draw a graph that models which New England states share a common border. Use vertices to represent the states and edges to represent common borders.

FIGURE 14.8 The New England states.

SOLUTION

Because the states are labeled with their abbreviations, we can use the abbreviations in **Figure 14.9(a)** to label each vertex. We use points to represent these vertices, shown in **Figure 14.9(b)**. The precise placement of these vertices is not important.

Now we are ready to draw the edges that represent common borders. Whenever two states share a common border, we connect the respective vertices with an edge. For example, Rhode Island shares a common border with Connecticut and with Massachusetts. Therefore, we draw an edge that connects vertex RI to vertex CT and one that connects vertex RI to vertex MA, shown in **Figure 14.9(c)**. Can you see that Massachusetts shares a common border with four states, resulting in edges connecting MA to CT, MA to RI, MA to VT, and MA to NH? Continuing in this manner, we obtain the graph shown in **Figure 14.9(c)**.

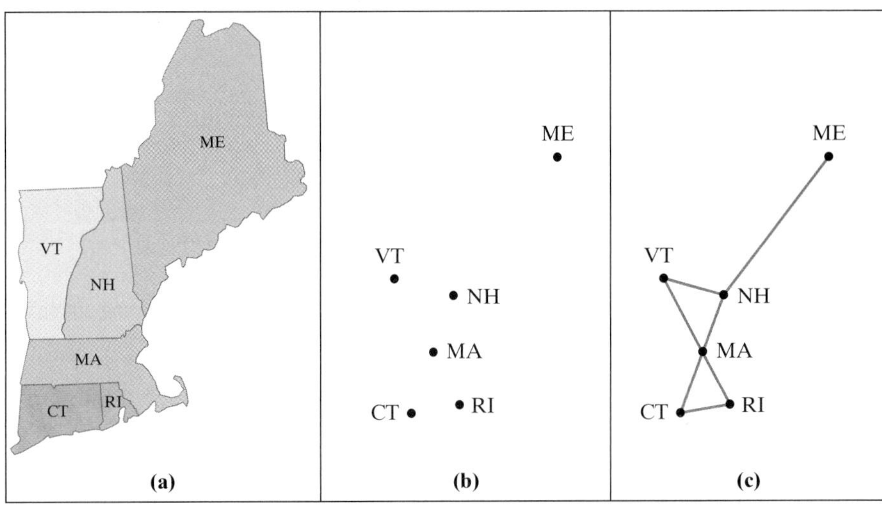

FIGURE 14.9 Creating a graph that models bordering relationships among New England states

CHECK POINT 3 Create a graph that models the bordering relationships among the five states shown in **Figure 14.10**.

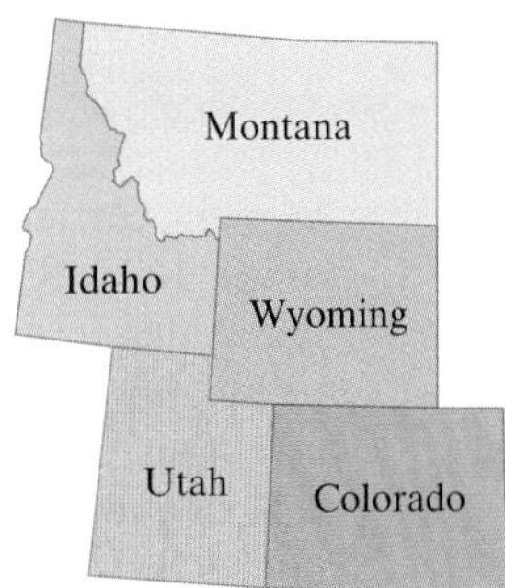

FIGURE 14.10

In the next example, our graph will tell how rooms in a floor plan are related to each other.

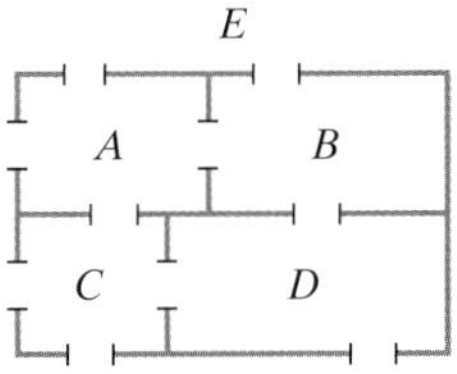

EXAMPLE 4 *Modeling Connecting Relationships in a Floor Plan*

The floor plan of a four-room house is shown in **Figure 14.11**. The rooms are labeled *A*, *B*, *C*, and *D*. The outside of the house is labeled *E*. The openings represent doors. Draw a graph that models the connecting relationships in the floor plan. Use vertices to represent the rooms and the outside, and edges to represent the connecting doors.

E *A* *B* *C* *D*

FIGURE 14.11

SOLUTION

Because the rooms and outside are labeled, we can use the letters in **Figure 14.12(a)** to label each vertex. We use points to represent these vertices, shown in **Figure 14.12(b)**.

Now we are ready to draw the edges that represent the doors. Two doors connect the outside, *E*, with room *A*, so we draw two edges that connect vertex *E* to vertex *A*, shown in **Figure 14.12(c)**. Only one door connects the outside, *E*, with room *B*, so we draw one edge from vertex *E* to vertex *B*. Two doors connect the outside, *E*, to room *C*; we draw two edges from *E* to *C*. Counting doors from the outside to each room and doors between the rooms, the graph is completed as shown in **Figure 14.12(c)**.

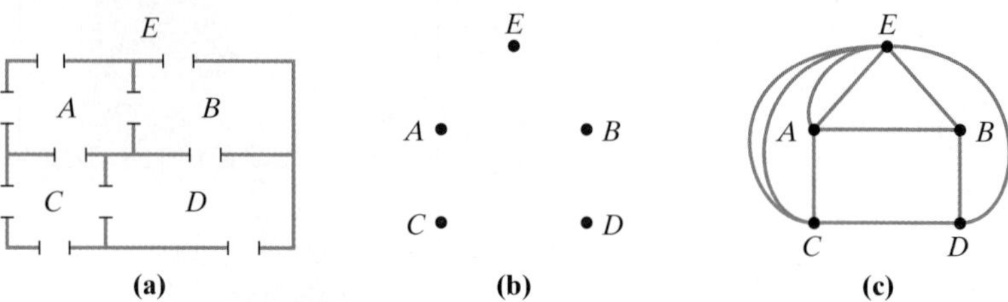

FIGURE 14.12 Creating a graph that models connecting relationships between rooms

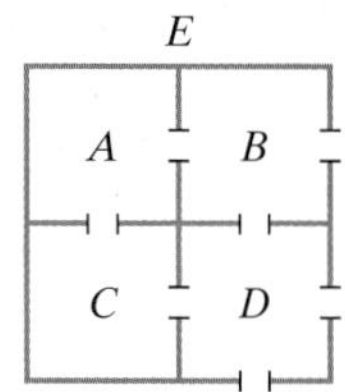
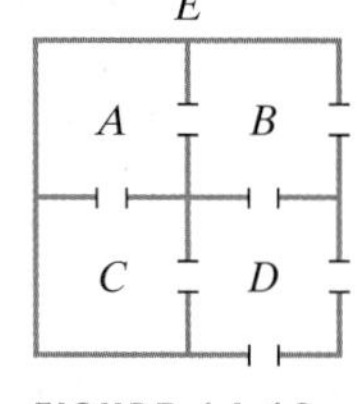

FIGURE 14.13

CHECK POINT 4 The floor plan of a four-room house is shown in **Figure 14.13**. The rooms are labeled *A*, *B*, *C*, and *D*. The outside of the house is labeled *E*. Draw a graph that models the connecting relationships in the floor plan.

In setting up routes for mail carriers, the post office is interested in delivering mail along the route that involves the least amount of walking or driving. Graphs play an important role in problems concerned with finding efficient ways to route services such as mail delivery, garbage collection, and police protection. The first step in the planning and design of delivery routes is creating graphs that model neighborhoods.

EXAMPLE 5 Modeling Walking Relationships for a Neighborhood's Streets

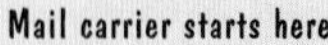

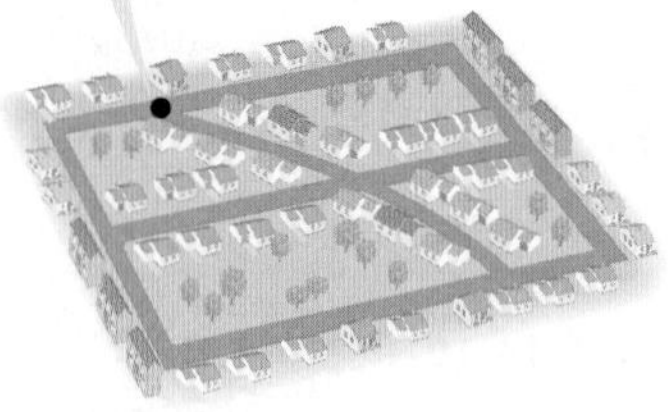

FIGURE 14.14

A mail carrier delivers mail to the four-block neighborhood shown in **Figure 14.14**. She parks her truck at the intersection shown in the figure and then walks to deliver mail to each of the houses. The streets on the outside of the neighborhood have houses on one side only. By contrast, the interior streets have houses on both sides of the street. On these streets, the mail carrier must walk down the street twice, covering each side separately. Draw a graph that models the streets of the neighborhood walked by the mail carrier. Use vertices to represent the street intersections and corners. Use one edge if streets must be covered only once and two edges for streets that must be covered twice.

SOLUTION

We begin by labeling each of the corners and street intersections with an uppercase letter, shown in **Figure 14.15(a)** at the top of the next page.

Next, we use points to represent the corners and street intersections. **Figure 14.15(b)** shows vertices *A* through *I*.

Now we are ready to draw the edges that represent walking relationships between corners and street intersections. The street from *B* to *A* must be covered only once, so we draw an edge that connects vertex *A* to vertex *B*, shown in **Figure 14.15(c)**. The same is true for the street from *A* to *D*. By contrast, the street from *D* to *E* has houses on both sides and must be walked twice. Therefore, we draw two edges that connect vertex *D* to vertex *E*. Continuing in this manner, the graph is completed as shown in **Figure 14.15(c)**.

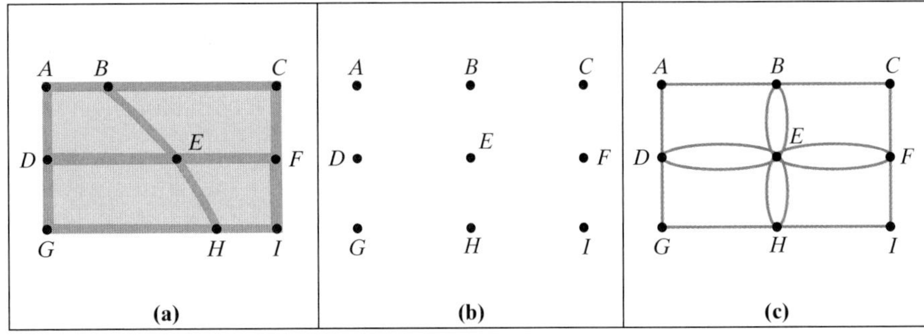

FIGURE 14.15

CHECK POINT 5 A security guard needs to walk the streets of the neighborhood shown in **Figure 14.14**. Unlike the postal worker, the guard would like to walk down each street once, whether or not the street has houses on both sides. Draw a graph that models the streets of the neighborhood walked by the security guard.

3 *Understand and use the vocabulary of graph theory.*

The Vocabulary of Graph Theory

Every branch of mathematics has its own unique vocabulary. We conclude this section by introducing some of the terms used in graph theory.

The **degree of a vertex** is the number of edges at that vertex. If a loop connects a vertex to itself, that loop contributes 2 to the degree of the vertex. On a graph, the degree of each vertex is found by counting the number of line segments or curves attached to the vertex. **Figure 14.16** illustrates a graph and the degree of each vertex.

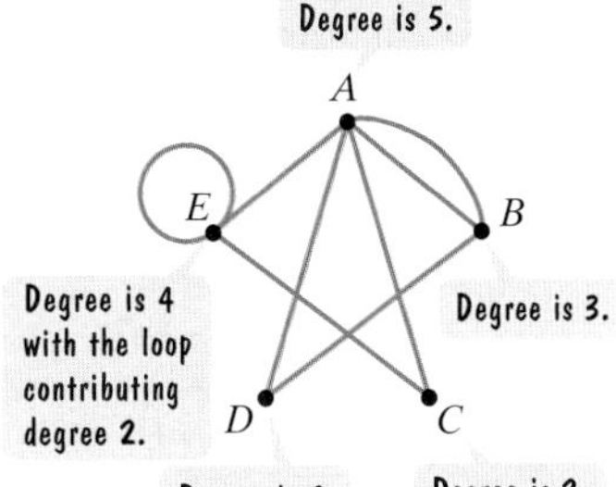

FIGURE 14.16 The degrees of a graph's vertices.

A vertex with an even number of edges attached to it is an **even vertex**. For example, in **Figure 14.16**, vertices *E*, *D*, and *C* are even. A vertex with an odd number of edges attached to it is an **odd vertex**. In **Figure 14.16**, vertices *A* and *B* are odd.

Two vertices in a graph are said to be **adjacent vertices** if there is at least one edge connecting them. It is helpful to think of adjacent vertices as *connected* vertices.

GREAT QUESTION!

If vertices are near each other, must they be adjacent?

Not necessarily. In **Figure 14.16**, vertices *C* and *D* are not adjacent because there is no edge connecting them.

EXAMPLE 6 *Identifying Adjacent Vertices*

List the pairs of adjacent vertices for the graph in **Figure 14.16**.

SOLUTION

A systematic way to approach this problem is to first list all the pairs of adjacent vertices involving vertex *A*, then all those involving *B* but not *A*, then all those involving *C* but neither *A* nor *B*, then those involving *D* but not *A*, *B*, or *C*. Finally, check to see if *E* is adjacent to itself. (It is.) Thus, the adjacent vertices are *A* and *B*, *A* and *C*, *A* and *D*, *A* and *E*, *B* and *D*, *C* and *E*, and *E* and *E*.

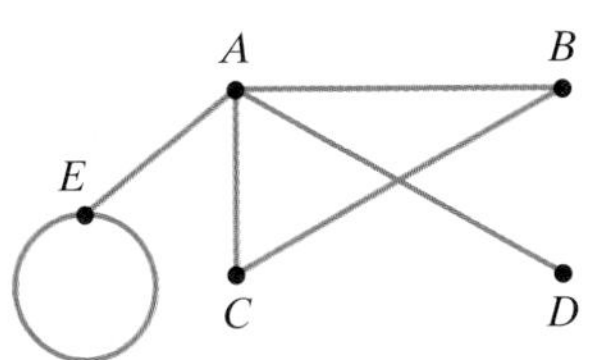

FIGURE 14.17

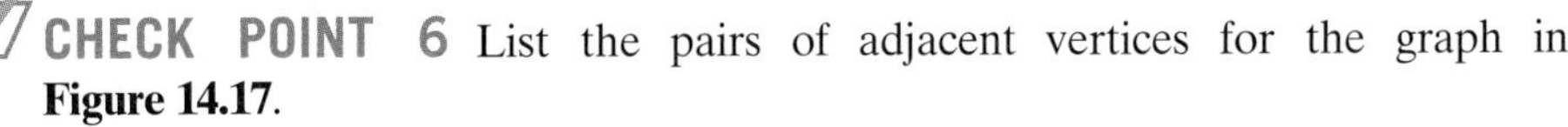

CHECK POINT 6 List the pairs of adjacent vertices for the graph in **Figure 14.17**.

We can use adjacent vertices to describe movement along a graph. A **path** in a graph is a sequence of adjacent vertices and the edges connecting them. Although a vertex can appear on the path more than once, **an edge can be part of a path only once**. For example, a path along a graph, shown in red, is illustrated in **Figure 14.18**.

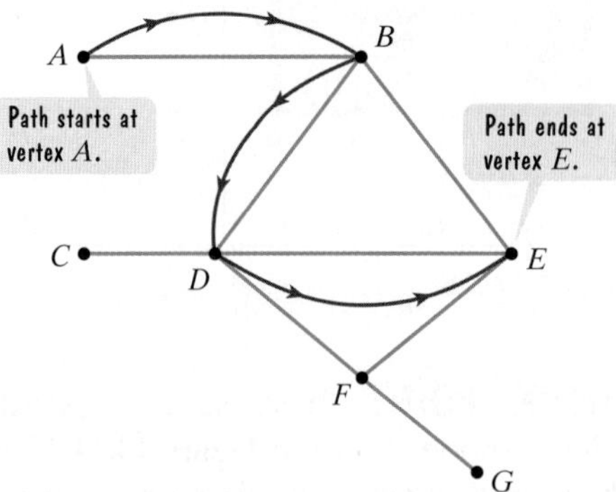

FIGURE 14.18 A path along a graph

You can think of this path as movement from vertex *A* to vertex *B* to vertex *D* to vertex *E*. We can refer to this path using a sequence of vertices separated by commas. Thus, the path shown in **Figure 14.18** is described by *A, B, D, E*.

The graph in **Figure 14.18** has many paths. Can you see that a path does not have to include every vertex and every edge of a graph?

A **circuit** is a path that begins and ends at the same vertex. In **Figure 14.19**, the path given by *B, D, F, E, B* is a circuit. Observe that every circuit is a path. However, because not every path ends at the same vertex where it starts, not every path is a circuit.

FIGURE 14.19 A circuit along a graph

The words *connected* and *disconnected* are used to describe graphs. A graph is **connected** if for any two of its vertices there is at least one path connecting them. Thus, a graph is connected if it consists of one piece. If a graph is not connected, it is said to be **disconnected**. A disconnected graph is made up of pieces that are by themselves connected. Such pieces are called the **components** of the graph. **Figure 14.20** illustrates connected and disconnected graphs.

Connected Graphs

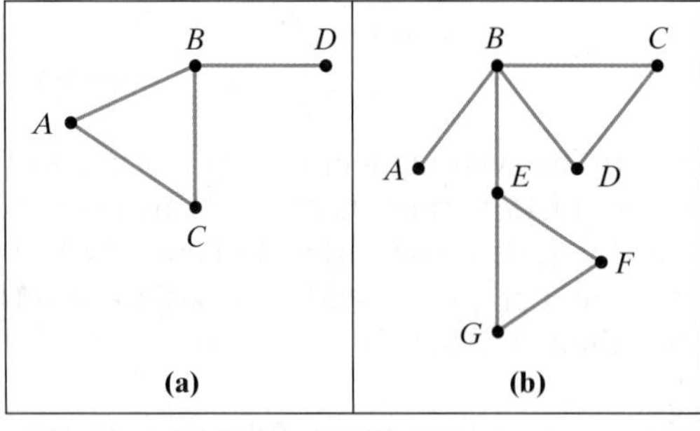

Disconnected Graphs

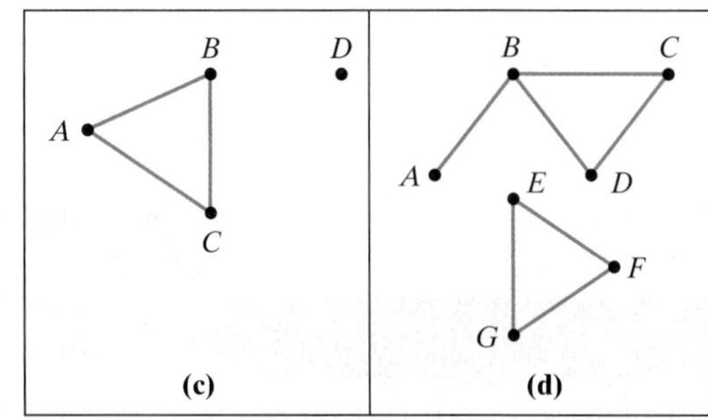

FIGURE 14.20 Examples of two connected graphs and two disconnected graphs

GREAT QUESTION!

Is the definition of *bridge* in graph theory the same as *bridges* connecting sections of a city such as Königsberg?

No. In our graph for the layout of Königsberg [**Figure 14.6(c)** on page 900], none of the edges that model the city's bridges are bridges as defined in graph theory.

A cliché says that if you burn a bridge behind you, you'll never get back to where you were. This cliché tells us something about the word *bridge* in graph theory. A **bridge** is an edge that if removed from a connected graph would leave behind a disconnected graph. For example, compare **Figures 14.20(a)** and **(c)**. If edge *BD* were removed from **Figure 14.20(a),** vertex *D* would be isolated from the rest of the graph, leaving behind the disconnected graph in **Figure 14.20(c)**. Thus, *BD* is a bridge for the graph in **Figure 14.20(a)**.

Next, compare **Figures 14.20(b)** and **(d)**. If edge *BE* were removed from **Figure 14.20(b),** the graph would become disconnected, shown by the two components in **Figure 14.20(d)**. Thus, *BE* is a bridge for the graph in **Figure 14.20(b)**.

Blitzer Bonus

Subway Simplification

London's subway system is as sprawling as the city it serves. Drawing the system on a map made it difficult for users to follow. Planning a trip with it was about as easy as finding one's way into and out of a maze—and more difficult if the trip involved changing trains.

In 1931, draftsman Henry C. Beck found a solution, which led to the London underground map so widely used today. Beck's proposal for a user-friendly subway map was to abandon the idea that it should be a literal representation of how the lines ran underground. Instead, it should show which vertices (subway stations) are connected by edges (underground routes). Beck knew that two graphs are equivalent if they have the same number of vertices connected to each other in the same way. The important information was the relationship between the stations. The shapes of the edges were unimportant. For simplicity, Beck used horizontal, vertical, and diagonal lines to represent edges. He also enlarged the central, most complex, part of the system in relation to the simple parts at the outskirts. In 1933, a few experimental copies of the simplified graph that modeled the underground system were printed. They were an immediate success.

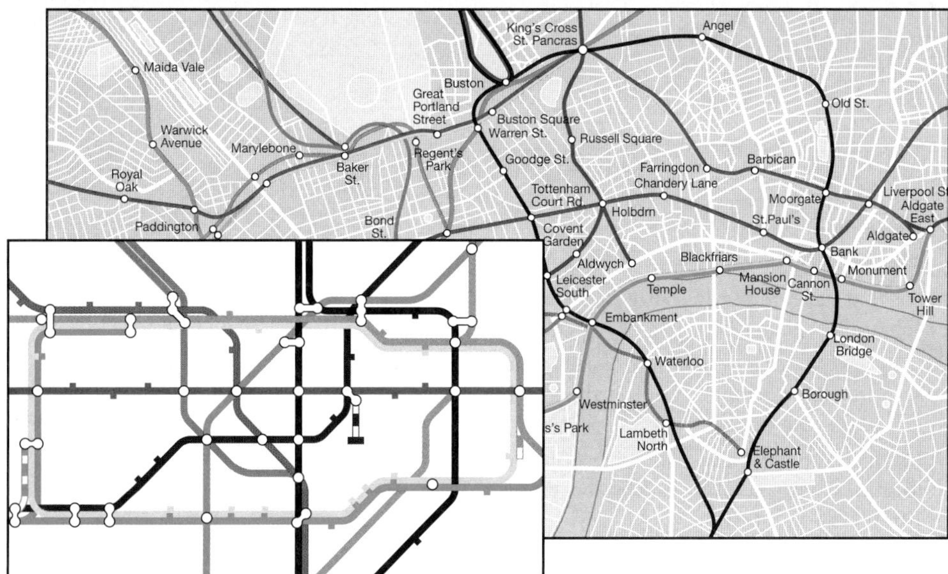

Tangled Lines London's subway system is difficult to use on a conventional map. However, a simple graph that models the system (inset) helps travelers find their way.

Concept and Vocabulary Check

Fill in each blank so that the resulting statement is true.

1. A finite set of points connected by line segments or curves is called a/an _______. The points are called _________. The line segments or curves are called _______. Such a line segment or curve that starts and ends at the same point is called a/an ______.

2. Two graphs that have the same number of vertices connected to each other in the same way are called __________.

3. The number of edges at a vertex is called the ________ of the vertex.

4. If there is at least one edge connecting two vertices in a graph, the vertices are called _________. A sequence of such vertices and the edges connecting them is called a/an ______. If this sequence of vertices and connecting edges begins and ends at the same vertex, it is called a/an _______.

5. If an edge is removed from a connected graph and leaves behind a disconnected graph, such an edge is called a/an ________.

6. True or False: A graph can be drawn in many equivalent ways. ______

7. True or False: An edge can be a part of a path only once. ______

8. True or False: Every circuit is a path. ______

9. True or False: Every path is a circuit. ______

Exercise Set 14.1

Practice and Application Exercises

The graph models the baseball schedule for a week. The vertices represent the teams. Each game played during that week is represented as an edge between two teams. Use the information in the graph to solve Exercises 1–4.

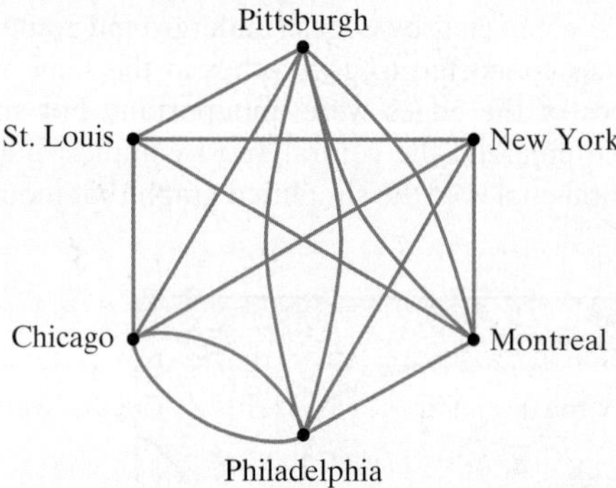

1. How many games are scheduled for Pittsburgh during the week? List the teams that they are playing. How many times are they playing each of these teams?
2. How many games are scheduled for Montreal during the week? List the teams that they are playing. How many times are they playing each of these teams?
3. Do the positions of New York and Montreal correspond to their geographic locations on a map? If not, is the graph drawn incorrectly? Explain your answer.
4. Do the positions of New York and Chicago correspond to their geographic locations on a map? If not, is the graph drawn incorrectly? Explain your answer.

In Exercises 5–6, draw two equivalent graphs for each description.

5. The vertices are *A*, *B*, *C*, and *D*. The edges are *AB*, *BC*, *BD*, *CD*, and *CC*.
6. The vertices are *A*, *B*, *C*, and *D*. The edges are *AD*, *BC*, *DC*, *BB*, and *DB*.

In Exercises 7–8, explain why the two figures show equivalent graphs. Then draw a third equivalent graph.

7.

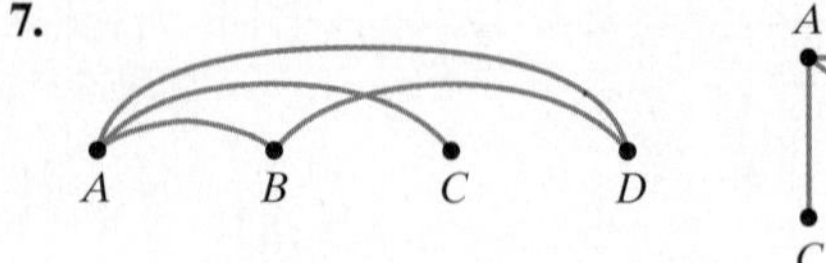

8.

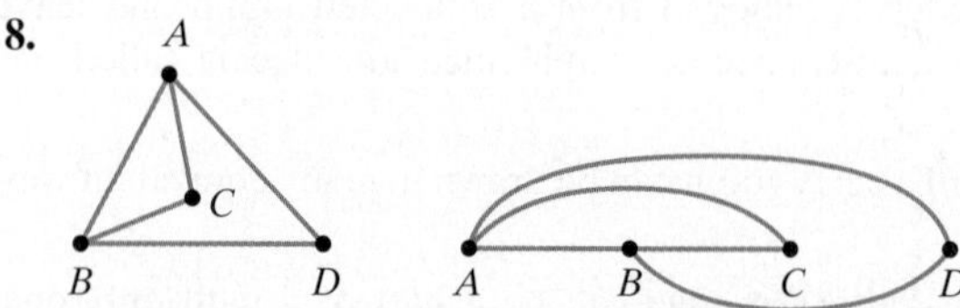

9. Eight students form a math homework group. The students in the group are Zeb, Stryder, Amy, Jed, Evito, Moray, Carrie, and Oryan. Prior to forming the group, Stryder was friends with everyone but Moray. Moray was friends with Zeb, Amy, Carrie, and Evito. Jed was friends with Stryder, Evito, Oryan, and Zeb. Draw a graph that models pairs of friendships among the eight students prior to forming the math homework group.

10. An environmental action group has six members, A, B, C, D, E, and F. The group has three committees: The Preserving Open Space Committee (B, D, and F), the Fund Raising Committee (B, C, and D), and the Wetlands Protection Committee (A, C, D, and E). Draw a graph that models the common members among committees. Use vertices to represent committees and edges to represent common members.

In Exercises 11–12, draw a graph that models the layout of the city shown in each map. Use vertices to represent the land masses and edges to represent the bridges.

11. The City of Gothamville

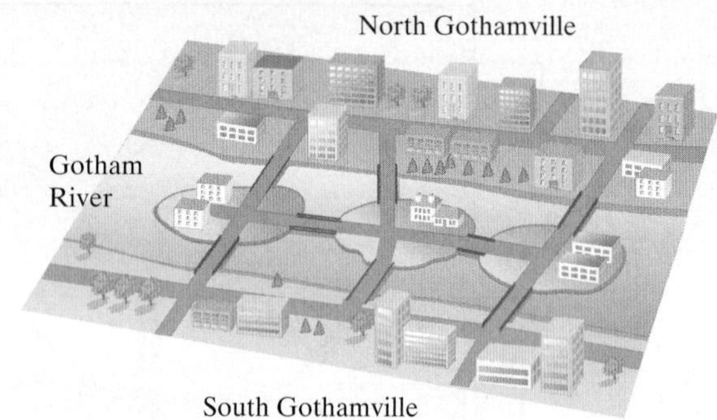

12. The City of Wisdomville

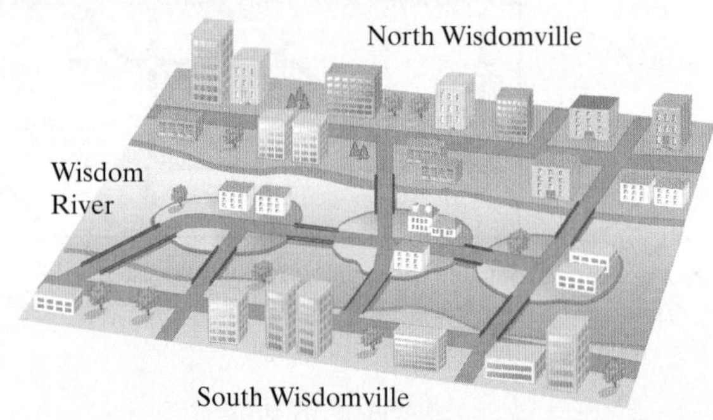

In Exercises 13–14, create a graph that models the bordering relationships among the states shown in each map. Use vertices to represent the states and edges to represent common borders.

13.

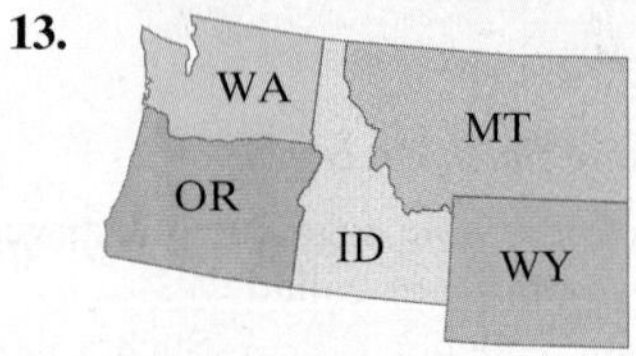

14.

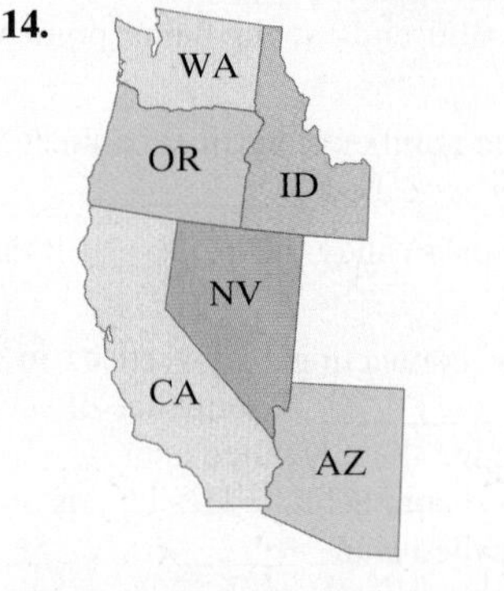

In Exercises 15–18, draw a graph that models the connecting relationships in each floor plan. Use vertices to represent the rooms and the outside, and edges to represent the connecting doors.

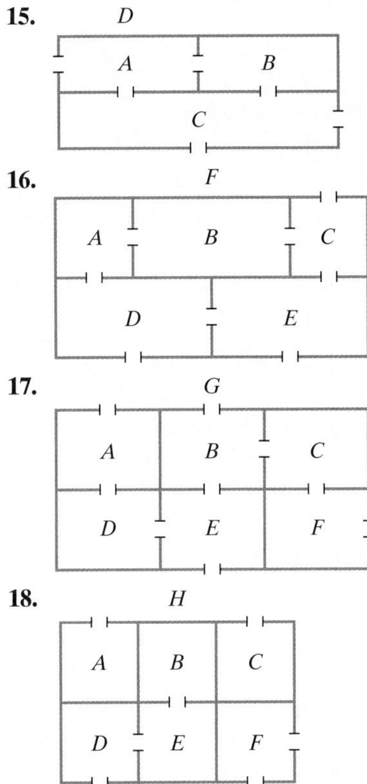

In Exercises 19–20, a security guard needs to walk the streets of the neighborhood shown in each figure. The guard is to walk down each street once, whether or not the street has houses on both sides. Draw a graph that models the neighborhood. Use vertices to represent the street intersections and corners. Use edges to represent the streets the security guard needs to walk.

19.

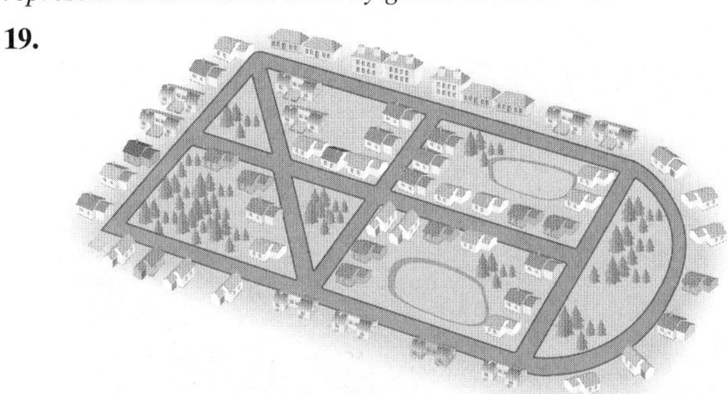

20.

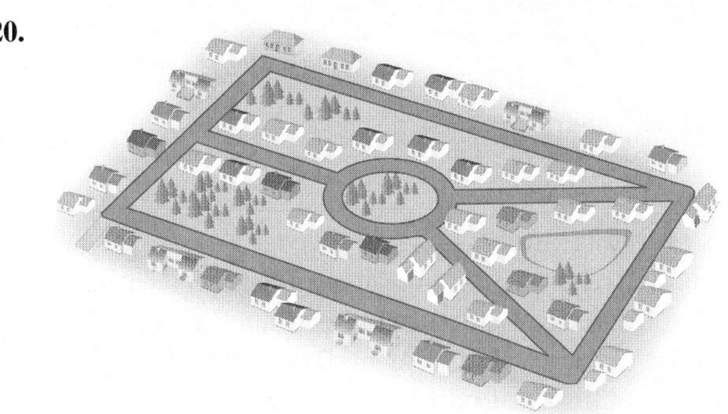

In Exercises 21–22, a mail carrier is to walk the streets of the neighborhood shown in Exercises 19 and 20, respectively. Unlike the security guard, the mail carrier must walk down each street with houses on both sides twice, covering each side separately. Draw a graph that models the neighborhood for the mail carrier. Use vertices to represent the street intersections and corners. Use edges to represent the streets the mail carrier needs to walk, with one edge for streets covered once and two edges for streets covered twice.

21. See the figure for Exercise 19.

22. See the figure for Exercise 20.

In Exercises 23–33, use the following graph.

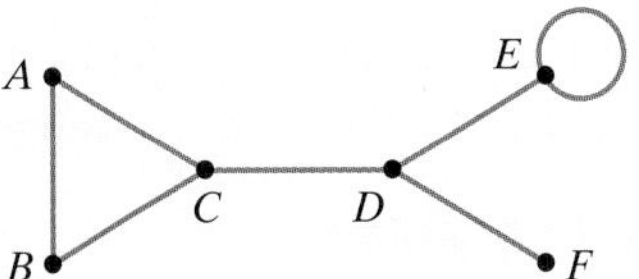

23. Find the degree of each vertex in the graph.

24. Identify the even vertices and identify the odd vertices.

25. Which vertices are adjacent to vertex A?

26. Which vertices are adjacent to vertex D?

27. Use vertices to describe two paths that start at vertex A and end at vertex D.

28. Use vertices to describe two paths that start at vertex B and end at vertex D.

29. Which edges shown on the graph are not included in the following path: E, E, D, C, B, A?

30. Which edges shown on the graph are not included in the following path: E, E, D, C, A, B?

31. Explain why edge CD is a bridge.

32. Explain why edge DE is a bridge.

33. Identify an edge on the graph other than those in Exercises 31 and 32 that is a bridge.

In Exercises 34–48, use the following graph.

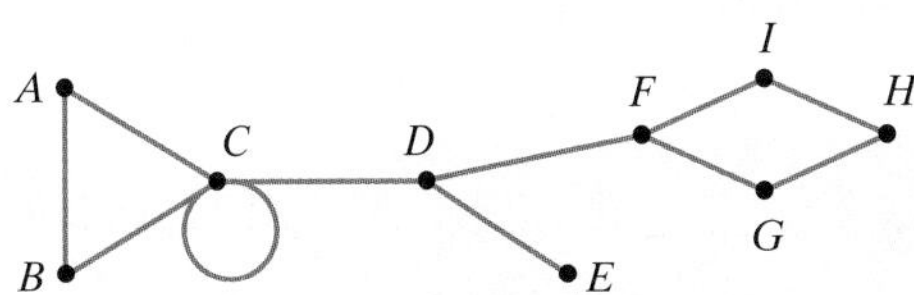

34. Find the degree of each vertex in the graph.

35. Identify the even vertices and identify the odd vertices.

36. Which vertices (or vertex) are adjacent to vertex E?

37. Which vertices are adjacent to vertex F?

38. Use vertices to describe two paths that start at vertex A and end at vertex F.

39. Use vertices to describe two paths that start at vertex B and end at vertex F.

40. Use vertices to describe a circuit that begins and ends at vertex F.

41. Use vertices to describe a circuit that begins and ends at vertex G.

42. Use vertices to describe a path from vertex H to vertex E, passing through vertex I, but not through vertex G.

43. Use vertices to describe a path from vertex A to vertex I, passing through vertices B and G, as well as through the loop.

(In Exercises 44–48, continue to refer to the graph toward the bottom of the previous page.)

44. Explain why *A, C, D, E, D* is not a path.

45. Explain why *G, F, D, E, D* is not a path.

46. Explain why *A, C, D, G* is not a path.

47. Explain why *H, I, F, E* is not a path.

48. Identify three edges that are bridges. Then show the components of the resulting graph once each bridge is removed.

In Exercises 49–52, draw a graph with the given characteristics.

49. The graph has four even vertices.

50. The graph has six odd vertices.

51. The graph has four odd vertices and at least one loop.

52. The graph has eight vertices and exactly one bridge.

Explaining the Concepts

53. What is a graph? Define vertices and edges as part of your description.

54. Describe how to determine whether or not a point where two of a graph's edges cross is a vertex.

55. What are equivalent graphs?

56. Because a graph can be drawn in many equivalent ways, describe the important part in drawing a graph.

57. What is meant by the degree of a graph's vertex and how is it determined?

58. Describe how to determine if a vertex is even or odd.

59. What are adjacent vertices? If two vertices are near each other in a graph, are they necessarily adjacent? Explain your answer.

60. What is a path in a graph?

61. What is a circuit? Describe the difference between a path and a circuit.

62. What is a connected graph?

63. What is a bridge?

64. Describe a situation involving relationships that can be modeled with a graph.

65. Describe one advantage and one disadvantage of using the graph for the baseball schedule in Exercises 1–4 rather than consulting a newspaper or television guide.

Critical Thinking Exercises

Make Sense? *In Exercises 66–69, determine whether each statement makes sense or does not make sense, and explain your reasoning.*

66. These graphs cannot be equivalent because they do not look alike.

67. A graph can be used to model diplomatic relations among the countries of the Americas.

68. The path along this graph is not a circuit.

69. The circuit along this graph is not a path.

70. Draw a graph with six vertices and two bridges.

71. Use the information in Exercise 10 to draw a graph that models which pairs of members are on the same committee. (The number of edges should match the number of committees on which two members serve together.)

72. Use inductive reasoning to make a conjecture that compares the sum of the degrees of the vertices of a graph and the number of edges in that graph.

Group Exercises

73. Group members should determine a relationship that exists among some, but not all, members. Did some of you know one another before the course began? Do some of you have the same academic major? Be as creative as possible in determining this relationship. Then create a graph that serves as a model for describing this relationship.

74. Each group member should select a favorite television series, movie, or novel with an extensive and interesting cast of characters. Determine a relationship that exists among some, but not all, of the characters. Represent the characters as vertices and use edges to show these relationships. Each person should share his or her model with the group, including aspects of the model that give rise to good comedy or drama.

14.2 Euler Paths and Euler Circuits

WHAT AM I SUPPOSED TO LEARN?

After studying this section, you should be able to:

1 Understand the definition of an Euler path.

2 Understand the definition of an Euler circuit.

3 Use Euler's Theorem.

4 Solve problems using Euler's Theorem.

5 Use Fleury's Algorithm to find possible Euler paths and Euler circuits.

SINCE THE 1970S, MUNICIPAL SERVICES IN NEW YORK City, including garbage collection, curb sweeping, and snow removal, have been scheduled and organized using graph models. By determining efficient routes for snowplows, garbage trucks, and street sweepers, New York is able to save tens of millions of dollars per year. Such routes must ensure that each street is serviced exactly once without having service vehicles retrace portions of their route.

In the previous section, you learned how to model physical settings with graphs. Now you will see how paths and circuits can solve problems such as constructing efficient routes along city streets for garbage pickup or mail delivery. To solve such problems, we begin by defining special kinds of paths and circuits.

1 *Understand the definition of an Euler path.*

Euler Paths and Euler Circuits

We have seen that a path in a graph is a sequence of adjacent vertices and the edges connecting them. Recall that an edge can be part of a path only once. If a path passes through each edge of a graph exactly one time, it is called an *Euler path.*

> **DEFINITION OF AN EULER PATH**
>
> An **Euler path** is a path that travels through *every edge* of a graph once and only once. Each edge must be traveled and no edge can be retraced.

We know that names of vertices separated by commas can be used to designate paths. As we discuss Euler paths, you can use a pencil to trace these paths. We can also indicate a starting vertex and number the edges to illustrate Euler paths. For example, examine the graph in **Figure 14.21**.

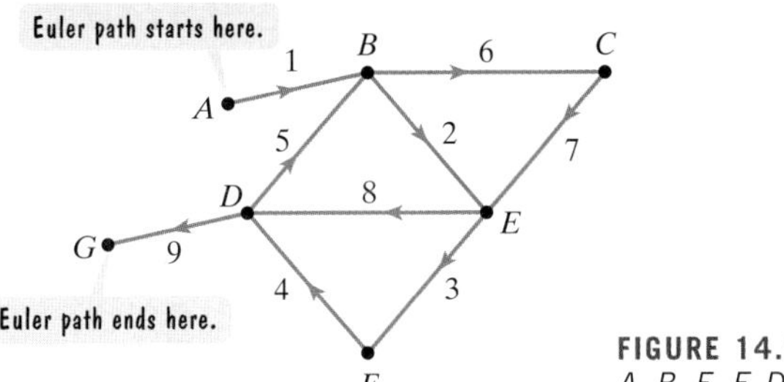

FIGURE 14.21 An Euler path: *A, B, E, F, D, B, C, E, D, G.*

The path *A, B, E, F, D, B, C, E, D, G* is an Euler path because each edge is traveled once. Trace this path with your pencil. Now try using the numbers along the edges. The voice balloon indicates a starting vertex, *A,* and the arrow shows which way to trace first. When you arrive at the next vertex, *B,* take the next numbered edge, 2. When you arrive at the next vertex, *E,* take the next numbered edge, 3. Continue in this manner until numbered edge 9 ends the path at vertex *G.*

We will use vertices as well as numbered edges to designate Euler paths. Do you see why the Euler path in **Figure 14.21** is not a circuit? A circuit must begin and end at the same vertex. The Euler path in **Figure 14.21** begins at vertex *A* and ends at vertex *G.*

If an Euler path begins and ends at the same vertex, it is called an *Euler circuit.*

2 *Understand the definition of an Euler circuit.*

> **DEFINITION OF AN EULER CIRCUIT**
>
> An **Euler circuit** is a circuit that travels through every edge of a graph once and only once. Like all circuits, an Euler circuit must begin and end at the same vertex.

The graph in **Figure 14.22** shows an Euler circuit.

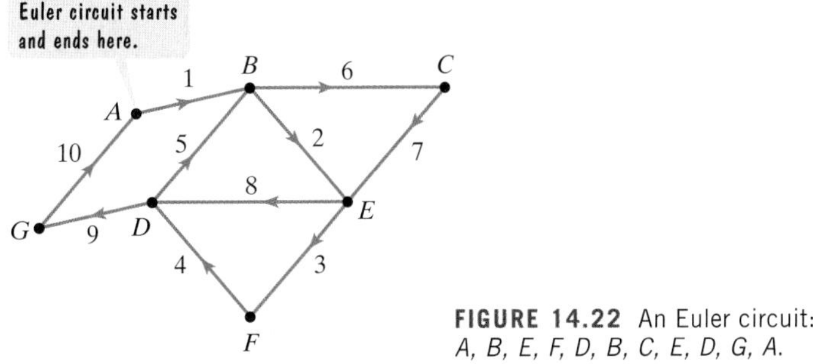

FIGURE 14.22 An Euler circuit: *A, B, E, F, D, B, C, E, D, G, A.*

The path *A, B, E, F, D, B, C, E, D, G, A,* shown with numbered edges 1 through 10, is an Euler circuit. Do you see why? Each edge is traced only once. Furthermore, the path begins and ends at the same vertex, *A*. Notice that **every Euler circuit is an Euler path**. However, **not every Euler path is an Euler circuit**.

3 Use Euler's Theorem.

Some graphs have no Euler paths. Other graphs have several Euler paths. Furthermore, some graphs with Euler paths have no Euler circuits. **Euler's Theorem** is used to determine if a graph contains Euler paths or Euler circuits.

GREAT QUESTION!

What's the bottom line on Euler's Theorem?

In all three parts of Euler's Theorem, it is the *number of odd vertices* that determines if a graph has Euler paths and Euler circuits.

EULER'S THEOREM

The following statements are true for connected graphs:

1. If a graph has exactly two odd vertices, then it has at least one Euler path, but no Euler circuit. Each Euler path must start at one of the odd vertices and end at the other one.
2. If a graph has no odd vertices (all even vertices), it has at least one Euler circuit (which, by definition, is also an Euler path). An Euler circuit can start and end at any vertex.
3. If a graph has more than two odd vertices, then it has no Euler paths and no Euler circuits.

EXAMPLE 1 Using Euler's Theorem

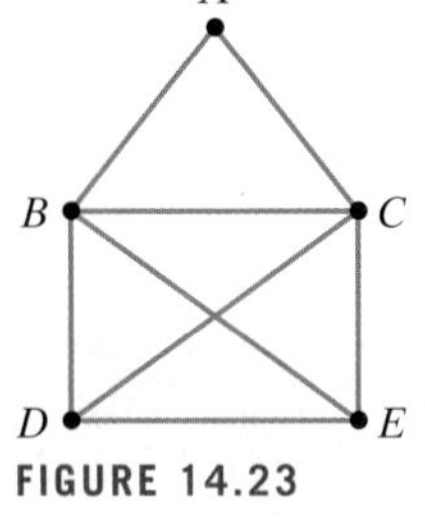

FIGURE 14.23

a. Explain why the graph in **Figure 14.23** has at least one Euler path.

b. Use trial and error to find one such path.

SOLUTION

a. In **Figure 14.24**, we count the number of edges at each vertex to determine if the vertex is odd or even. We see that there are exactly two odd vertices, namely D and E. By the first statement in Euler's Theorem, the graph has at least one Euler path, but no Euler circuit.

b. Euler's Theorem tells us that a possible Euler path must start at one of the odd vertices and end at the other one. We will use trial and error to determine an Euler path, starting at vertex D and ending at vertex E. **Figure 14.25** shows an Euler path: D, C, B, E, C, A, B, D, E. Trace this path and verify the numbers along the edges.

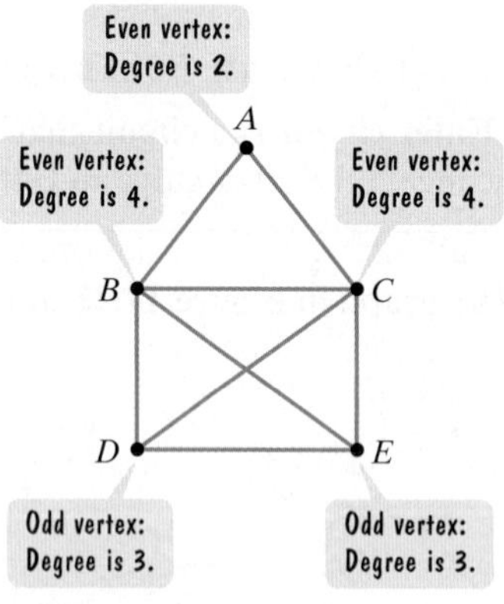

FIGURE 14.24 The graph has two odd vertices.

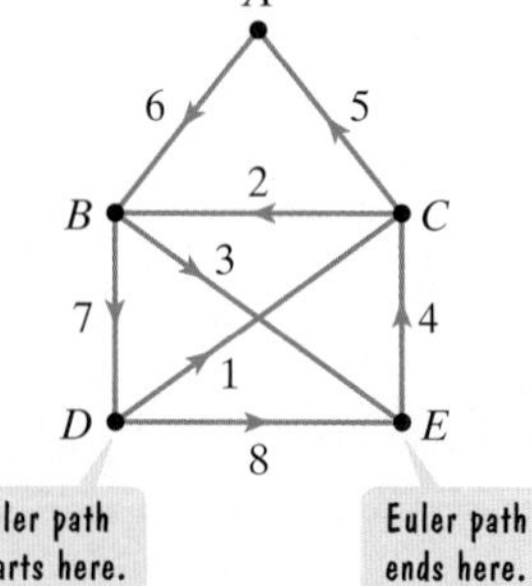

FIGURE 14.25 An Euler path: D, C, B, E, C, A, B, D, E.

CHECK POINT 1 Refer to the graph in **Figure 14.23**. Use trial and error to find an Euler path that starts at E and ends at D. Number the edges of the graph to indicate the path. Then use vertex letters separated by commas to name the path.

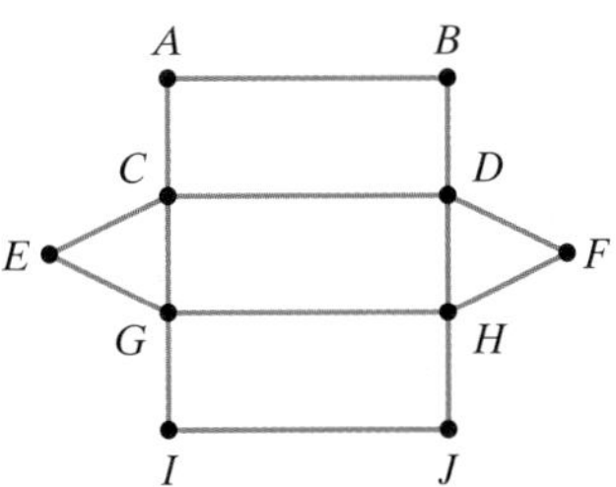

FIGURE 14.26

EXAMPLE 2 Using Euler's Theorem

a. Explain why the graph in **Figure 14.26** has at least one Euler circuit.

b. Use trial and error to find one such circuit.

SOLUTION

a. In **Figure 14.27**, we count the number of edges at each vertex to determine if the vertex is odd or even. We see that the graph has no odd vertices. By the second statement in Euler's Theorem, the graph has at least one Euler circuit.

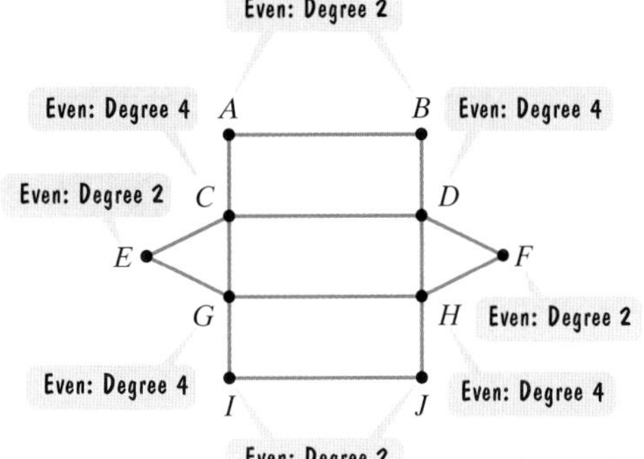

FIGURE 14.27 The graph has no odd vertices.

b. An Euler circuit can start and end at any vertex. We will use trial and error to determine an Euler circuit, starting and ending at vertex *H*. Remember that you must trace every edge exactly once and start and end at *H*. **Figure 14.28** shows an Euler circuit. Trace this circuit using the vertices in the figure's caption and verify the numbers along the edges.

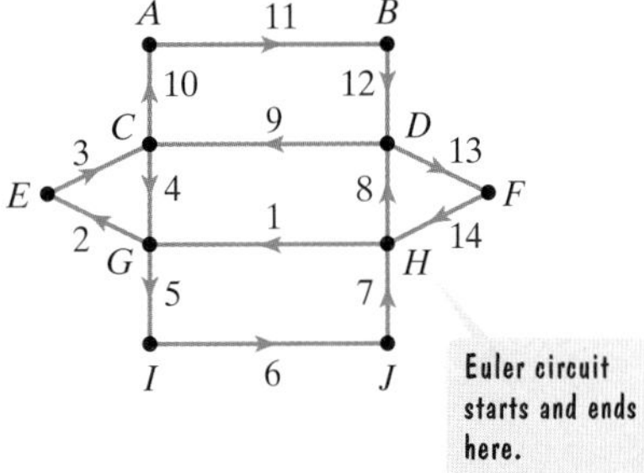

FIGURE 14.28 An Euler circuit: *H, G, E, C, G, I, J, H, D, C, A, B, D, F, H*.

CHECK POINT 2 Refer to the graph in **Figure 14.26**. Use trial and error to find an Euler circuit that starts and ends at *G*. Number the edges of the graph to indicate the circuit. Then use vertex letters separated by commas to name the circuit.

Figure 14.29 shows a graph with one even vertex and four odd vertices. Because the graph has more than two odd vertices, by the third statement in Euler's Theorem, it has no Euler paths and no Euler circuits.

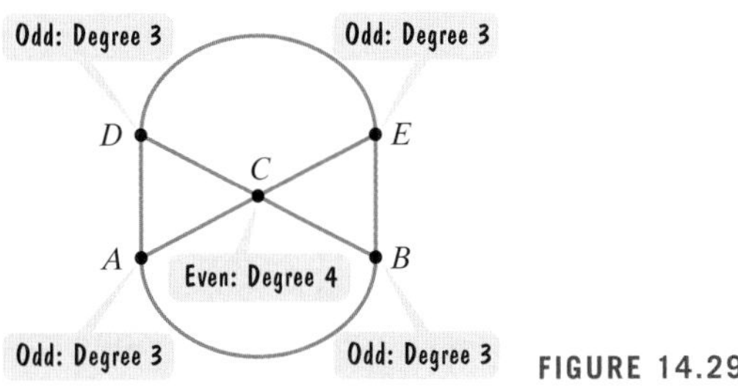

FIGURE 14.29

4 *Solve problems using Euler's Theorem.*

Applications of Euler's Theorem

We can use Euler's Theorem to solve problems involving situations modeled by graphs. For example, we return to the graph that models the layout of Königsberg, shown in **Figure 14.30(c)**. People in the city were interested in finding if it were possible to walk through the city so as to cross each bridge exactly once. Translated into the language of graph theory, we are interested in whether the graph in **Figure 14.30(c)** has an Euler path. Count the number of odd vertices. The graph has four odd vertices. Because it has more than two odd vertices, it has no Euler paths. Thus, there is no possible way anyone in Königsberg can walk across all of the bridges without having to recross some of them.

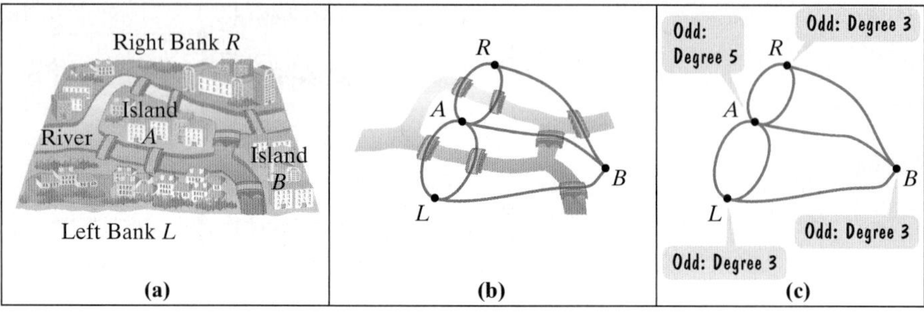

FIGURE 14.30 Revisiting the graph that models the layout of Königsberg

EXAMPLE 3 *Applying Euler's Theorem*

Figure 14.31(c) shows the graph we developed to model the connecting relationships in a floor plan. Recall from our work in Section 14.1 that *A*, *B*, *C*, and *D* represent the rooms and *E* represents the outside of the house. The edges represent the connecting doors.

a. Is it possible to find a path that uses each door exactly once?

b. If it is possible, use trial and error to show such a path on the graph in **Figure 14.31(c)** and on the floor plan in **Figure 14.31(a)**.

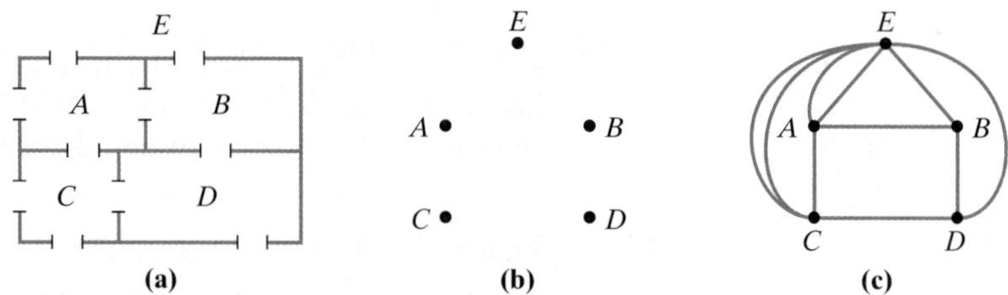

FIGURE 14.31 Revisiting the graph that models the layout of a floor plan

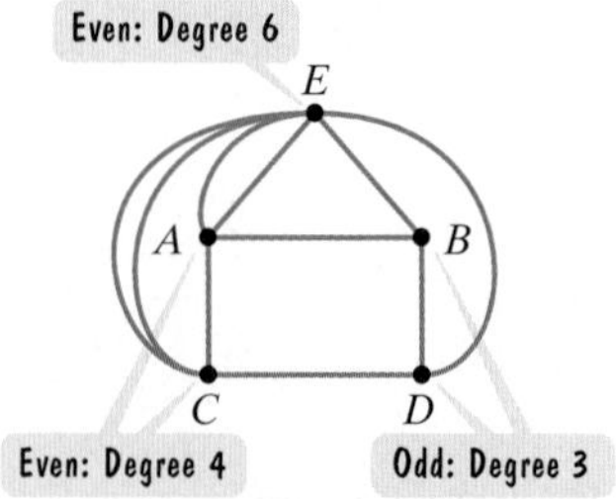

FIGURE 14.32 The graph has two odd vertices.

SOLUTION

a. A path that uses each door (or edge) exactly once means that we are looking for an Euler path or an Euler circuit on the graph in **Figure 14.31(c)**. **Figure 14.32** indicates that there are exactly two odd vertices, namely *B* and *D*. By Euler's Theorem, the graph has at least one Euler path, but no Euler circuit. It is possible to find a path that uses each door exactly once. It is not possible to begin and end the walk in the same place.

b. Euler's Theorem tells us that a possible Euler path must start at one of the odd vertices and end at the other one. We will use trial and error to determine an Euler path, starting at vertex B (room B in the floor plan) and ending at vertex D (room D in the floor plan). **Figure 14.33(a)** shows an Euler path on the graph. **Figure 14.33(b)** translates the path into a walk through the rooms.

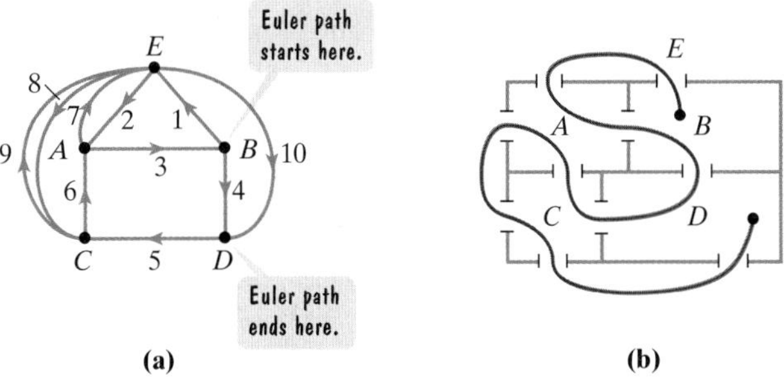

FIGURE 14.33 An Euler path: $B, E, A, B, D, C, A, E, C, E, D$ shown on a graph and on a floor plan.

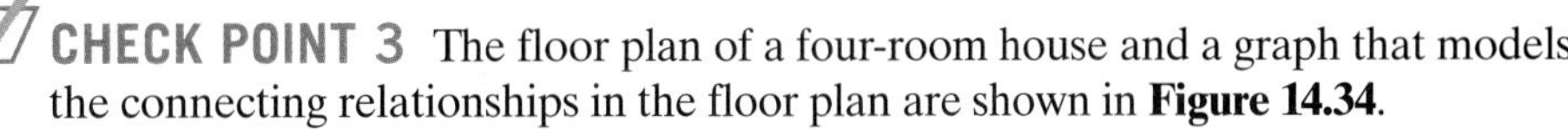

CHECK POINT 3 The floor plan of a four-room house and a graph that models the connecting relationships in the floor plan are shown in **Figure 14.34**.

a. Is it possible to find a path that uses each door exactly once?

b. If it is possible, use trial and error to show such a path on the graph in **Figure 14.34(b)** and the floor plan in **Figure 14.34(a)**.

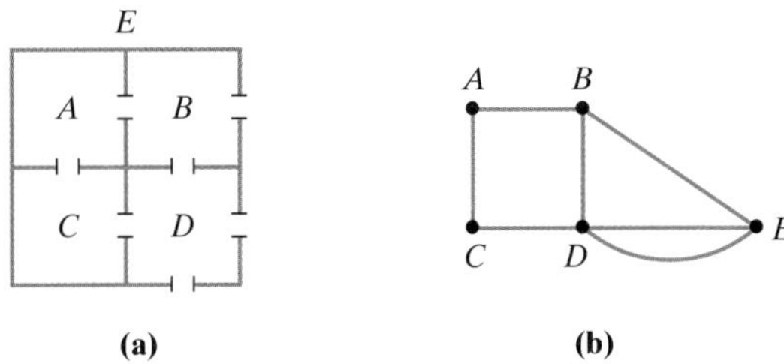

FIGURE 14.34 A floor plan and a graph that models its connecting relationships.

5 Use Fleury's Algorithm to find possible Euler paths and Euler circuits.

Finding Possible Euler Paths and Euler Circuits without Using Trial and Error

Euler's Theorem enables us to count a graph's odd vertices and determine if it has an Euler path or an Euler circuit. Unfortunately, the theorem provides almost no assistance in finding an actual Euler path or circuit. There is, however, an algorithm, or procedure, for finding such paths and circuits, called **Fleury's Algorithm**.

FLEURY'S ALGORITHM

If Euler's Theorem indicates the existence of an Euler path or Euler circuit, one can be found using the following procedure:

1. If the graph has exactly two odd vertices (and therefore an Euler path), choose one of the two odd vertices as the starting point. If the graph has no odd vertices (and therefore an Euler circuit), choose any vertex as the starting point.
2. Number edges as you trace through the graph according to the following rules:
   - After you have traveled over an edge, erase it. (This is because you must travel each edge exactly once.) Show the erased edges as dashed lines.
   - When faced with a choice of edges to trace, choose an edge that is not a bridge. Travel over an edge that is a bridge only if there is no alternative.

GREAT QUESTION!

What's the bottom line on Fleury's Algorithm?

Don't cross a bridge unless you have to!

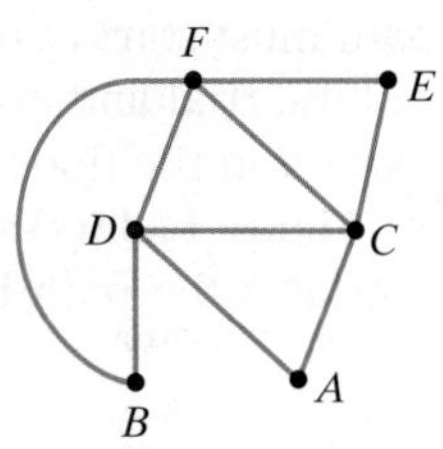

FIGURE 14.35

EXAMPLE 4 Using Fleury's Algorithm

The graph in **Figure 14.35** has at least one Euler circuit. Find one by Fleury's Algorithm.

SOLUTION

The graph has no odd vertices, so we can begin at any vertex. We choose vertex A as the starting point and proceed as follows.

Step 1

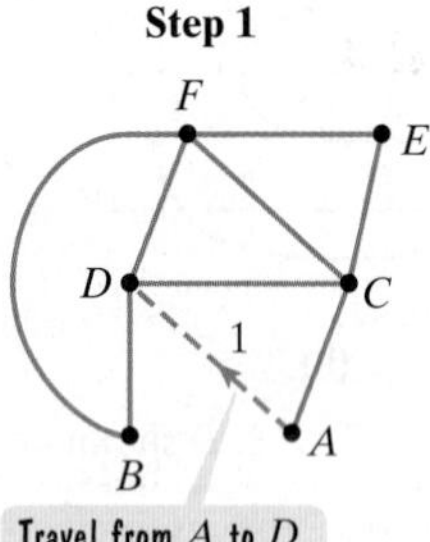

We could also travel from A to C.

Step 2

Travel from D to C and erase edge DC.

CA is now a bridge. If it were removed, vertex A would be isolated from the rest of the graph.

We could also travel from D to F or B.

Step 3

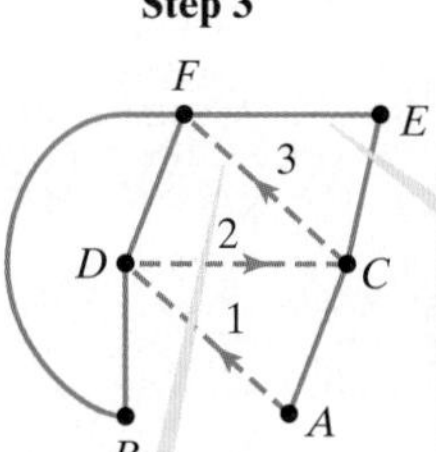

We could also travel from C to E, but not from C to A. Don't cross the bridge.

Step 4

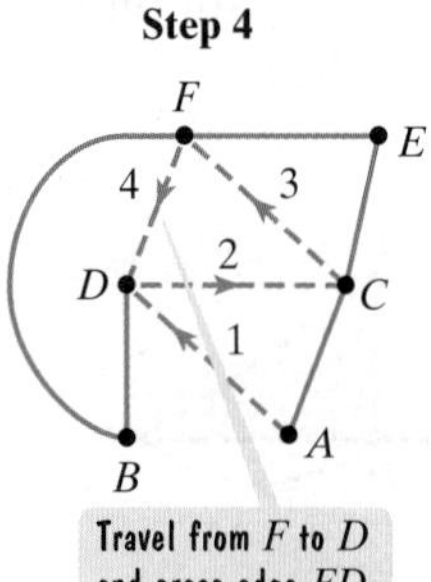

We could also travel from F to B, but not from F to E. Don't cross the bridge.

Step 5

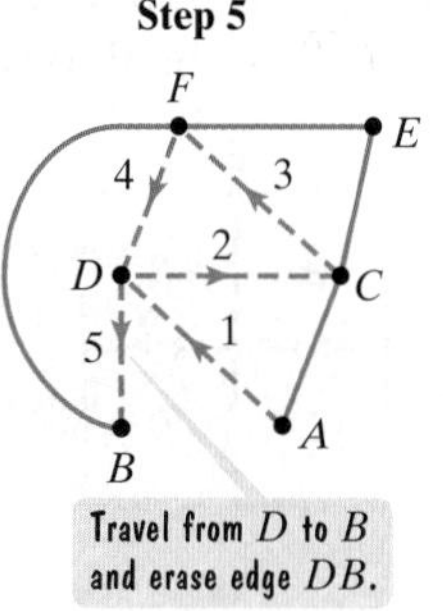

There are no other choices.

Steps 6, 7, 8, 9

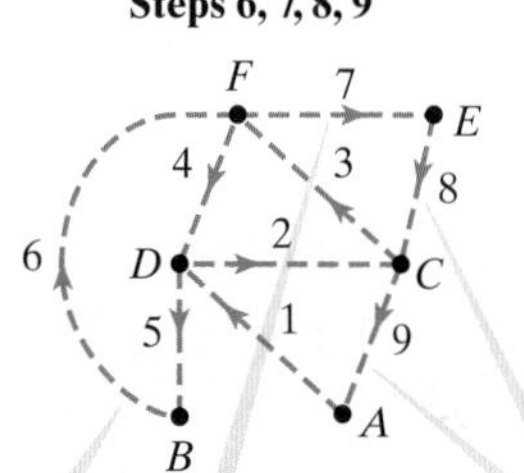

There are no other choices at each step.

The completed Euler circuit is shown in **Figure 14.36**.

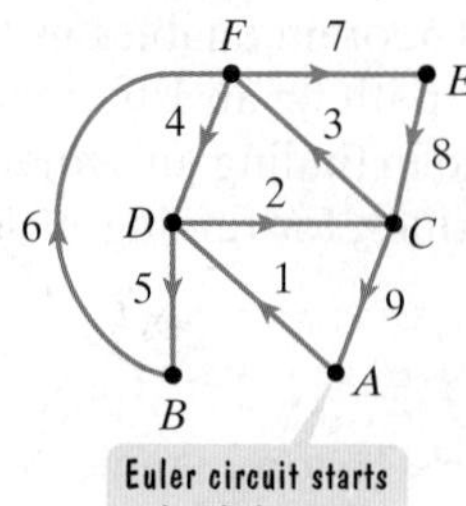

FIGURE 14.36 An Euler circuit: $A, D, C, F, D, B, F, E, C, A$.

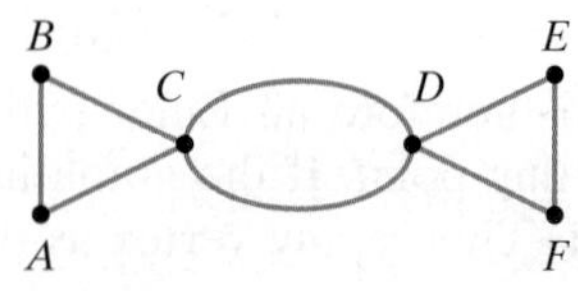

FIGURE 14.37

CHECK POINT 4 The graph in **Figure 14.37** has at least one Euler circuit. Find one by Fleury's Algorithm.

In the Exercise Set, we will revisit many of the graph models from the previous section. In solving problems with these models, you can use trial and error or Fleury's Algorithm to find possible Euler paths and Euler circuits. The procedure of Fleury's Algorithm is useful as graphs become more complicated.

Here is a summary, in chart form, of Euler's Theorem that you might find helpful in the Exercise Set that follows the Concept and Vocabulary Check.

SUMMARY OF EULER'S THEOREM

Number of Odd Vertices	Euler Paths	Euler Circuits
none (all even)	at least one	at least one
exactly two	at least one	none
more than two	none	none

Concept and Vocabulary Check

Fill in each blank so that the resulting statement is true.

1. A path that travels through every edge of a graph exactly once is called a/an _______ path.
2. A circuit that travels through every edge of a graph exactly once is called a/an _______ circuit.
3. A connected graph has at least one Euler path, but no Euler circuit, if the graph has exactly _____ odd vertices.
4. A connected graph has at least one Euler circuit, which, by definition, is also an Euler path, if the graph has ____ odd vertices. An Euler circuit can start and end at _____ vertex.
5. A connected graph has no Euler paths and no Euler circuits if the graph has more than two _____ vertices.
6. A connected graph has even vertices A, B, and C, and odd vertices, D and E. Each Euler path must begin at vertex D and end at vertex ____, or begin at vertex ____ and end at vertex _____.
7. Euler's Theorem enables us to count a graph's odd vertices and determine if it has an Euler path or an Euler circuit. A procedure for finding such paths and circuits is called ________ Algorithm. When using this algorithm and faced with a choice of edges to trace, choose an edge that is not a/an _______. Travel over such an edge only if there is no alternative.
8. True or False: Every Euler circuit is an Euler path ______
9. True or False: Every Euler path is an Euler circuit. ______
10. True or False: Euler's Theorem provides a procedure for finding Euler paths and Euler circuits. ______

Exercise Set 14.2

Practice Exercises

In Exercises 1–6, use the graph shown. In each exercise, a path along the graph is described. Trace this path with a pencil and number the edges. Then determine if the path is an Euler path, an Euler circuit, or neither. Explain your answer.

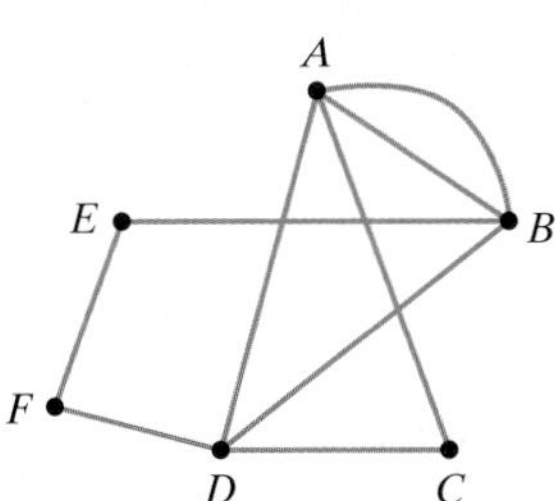

1. *F, E, B, A, C, D, A, B, D*
2. *F, E, B, A, B, D, C, A, D*
3. *F, E, B, A, C, D, A, B, D, F*
4. *F, E, B, A, B, D, C, A, D, F*
5. *A, B, A, C, D, B, E, F, D*
6. *A, B, A, C, D, F, E, B, D*

In Exercises 7–8, a graph is given.

a. *Explain why the graph has at least one Euler path.*

b. *Use trial and error or Fleury's Algorithm to find one such path.*

7.
A B C D E

8.
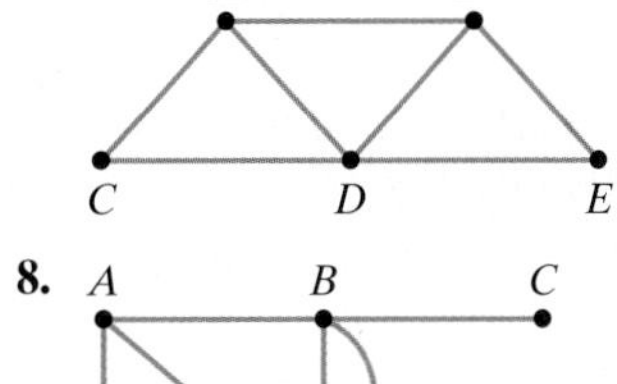

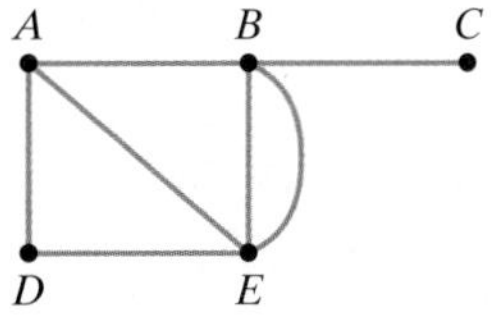

In Exercises 9–10, a graph is given.

a. *Explain why the graph has at least one Euler circuit.*

b. *Use trial and error or Fleury's Algorithm to find one such circuit.*

9.

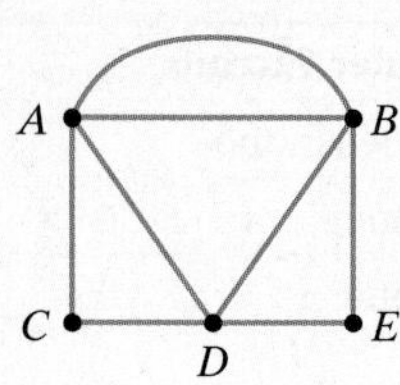

10.

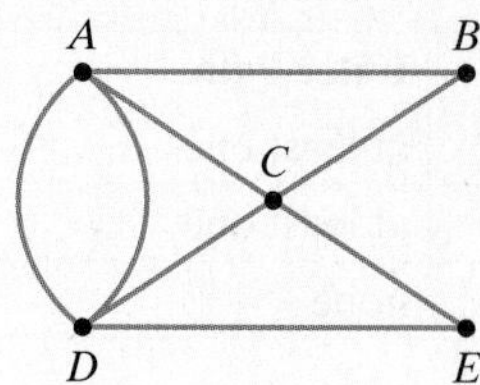

In Exercises 11–12, a graph is given. Explain why each graph has no Euler paths and no Euler circuits.

11.

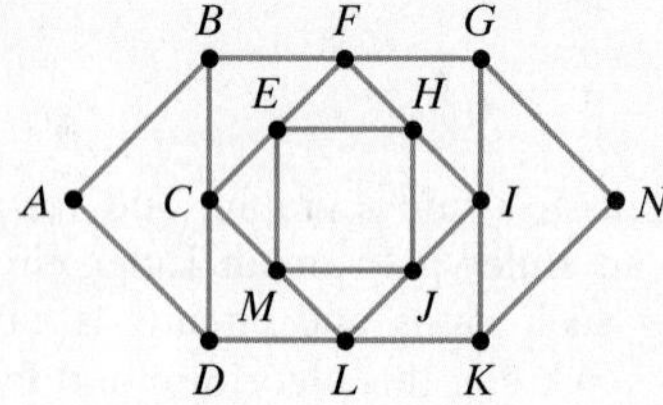

12.

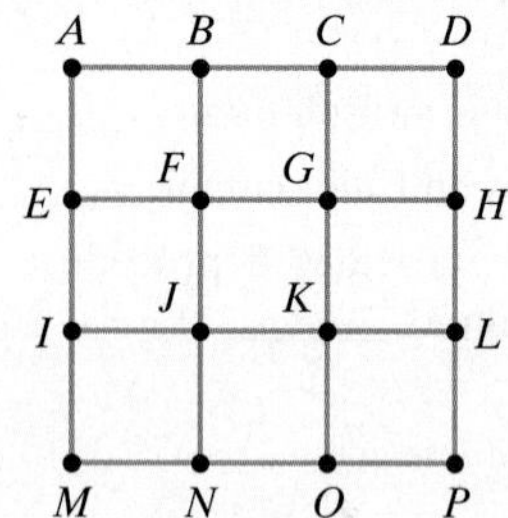

In Exercises 13–18, a connected graph is described. Determine whether the graph has an Euler path (but not an Euler circuit), an Euler circuit, or neither an Euler path nor an Euler circuit. Explain your answer.

13. The graph has 60 even vertices and no odd vertices.

14. The graph has 80 even vertices and no odd vertices.

15. The graph has 58 even vertices and two odd vertices.

16. The graph has 78 even vertices and two odd vertices.

17. The graph has 57 even vertices and four odd vertices.

18. The graph has 77 even vertices and four odd vertices.

In Exercises 19–32, a graph is given.

a. *Determine whether the graph has an Euler path, an Euler circuit, or neither.*

b. *If the graph has an Euler path or circuit, use trial and error or Fleury's Algorithm to find one.*

19.

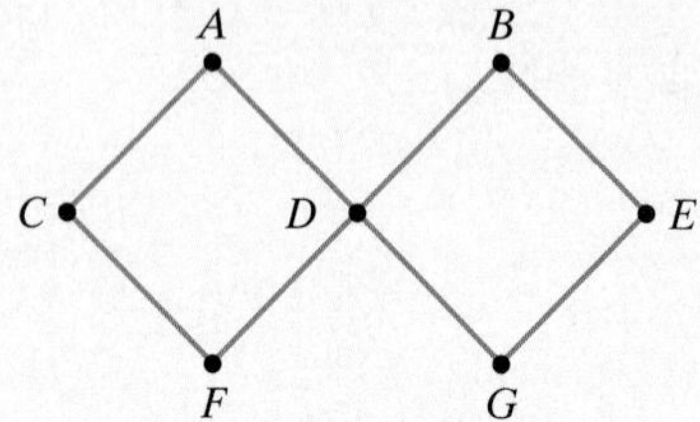

20.

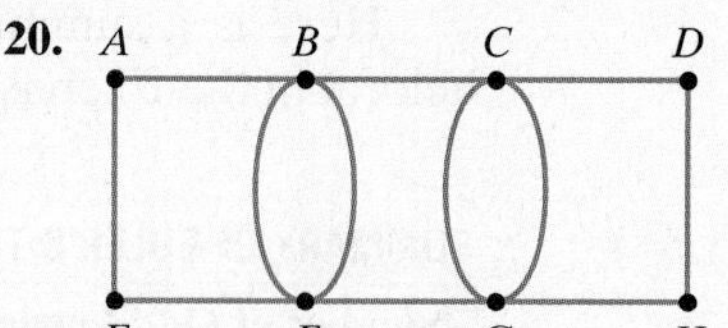

21.

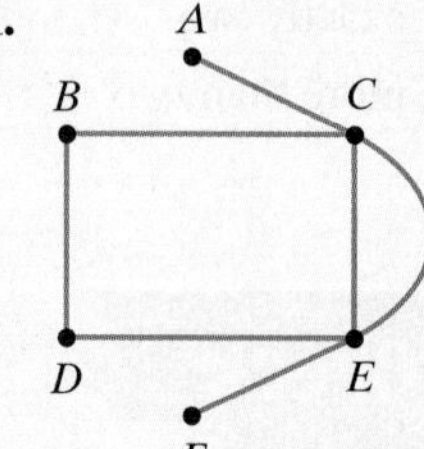

22.

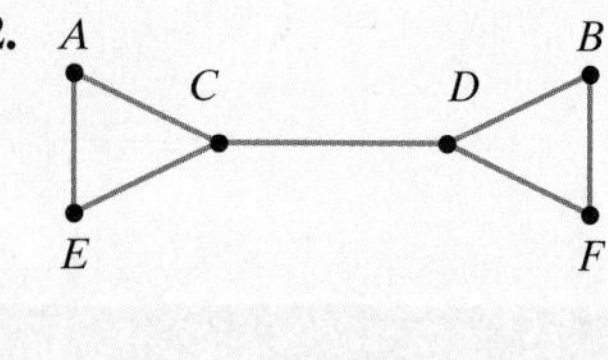

23.

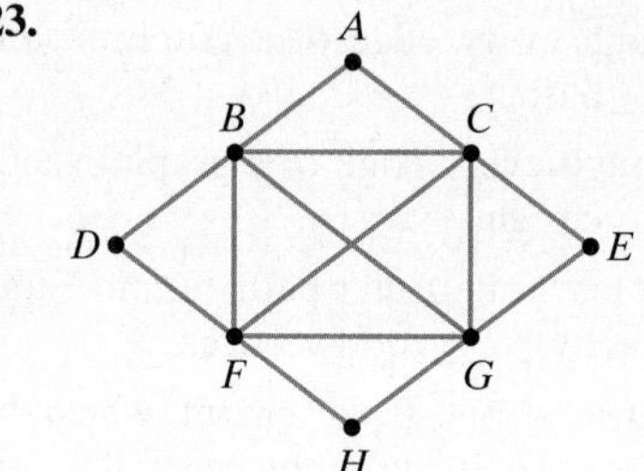

24.

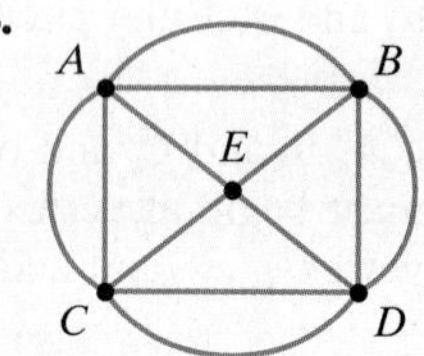

25.

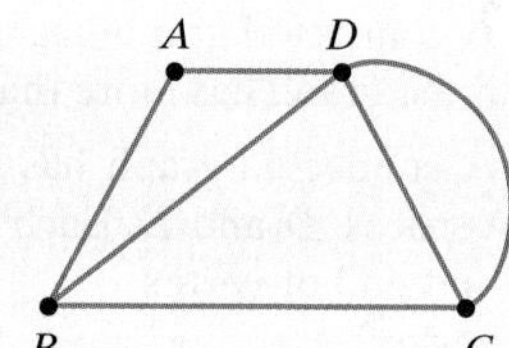

26.

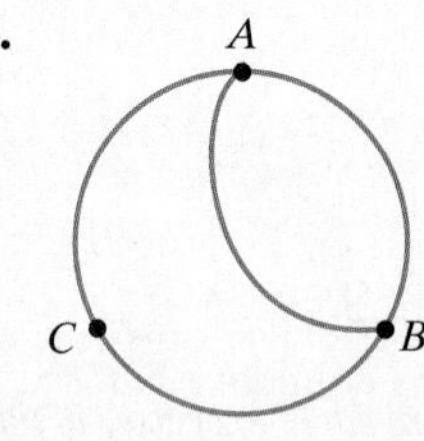

27.

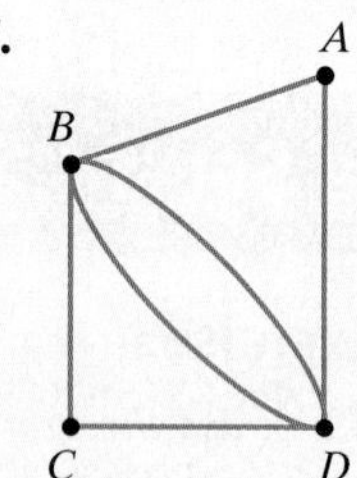

28.

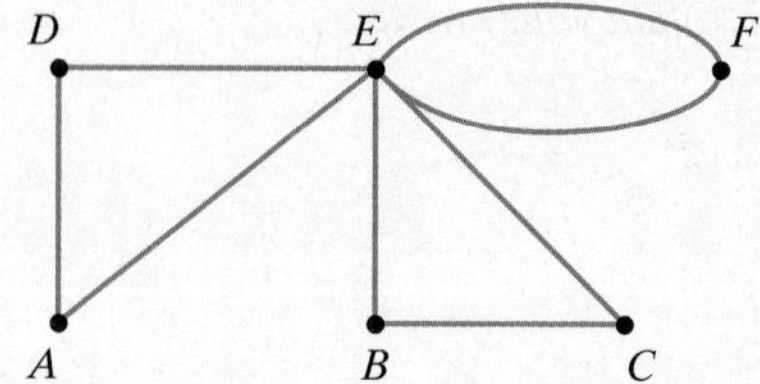

29.

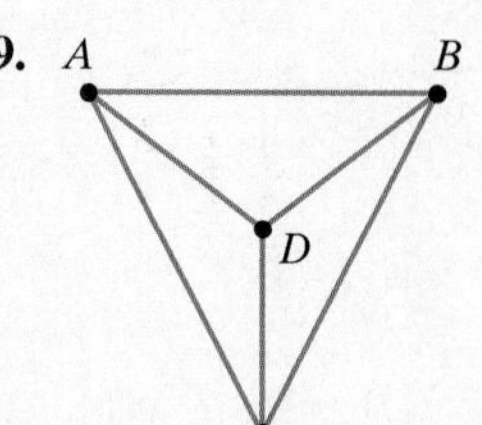

30.

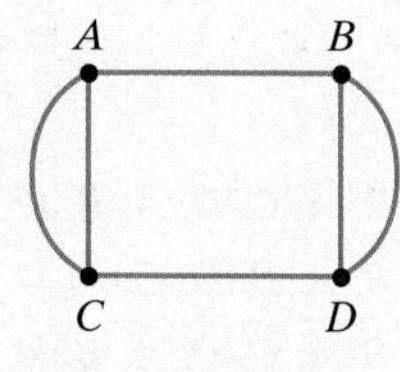

31.

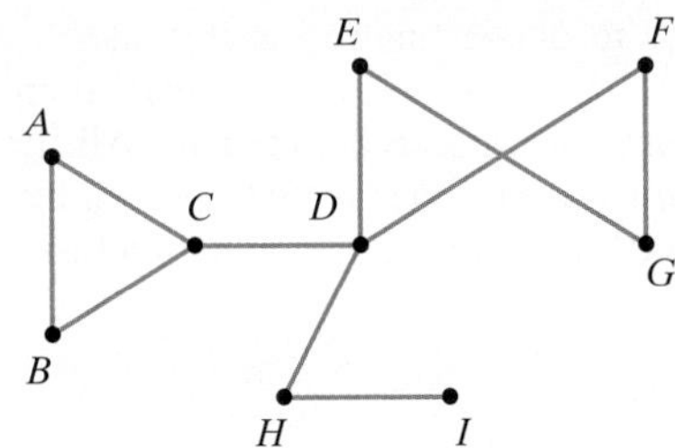

32.

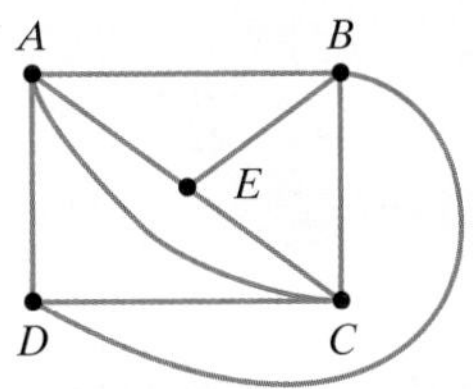

In Exercises 33–36, use Fleury's Algorithm to find an Euler path.

33.

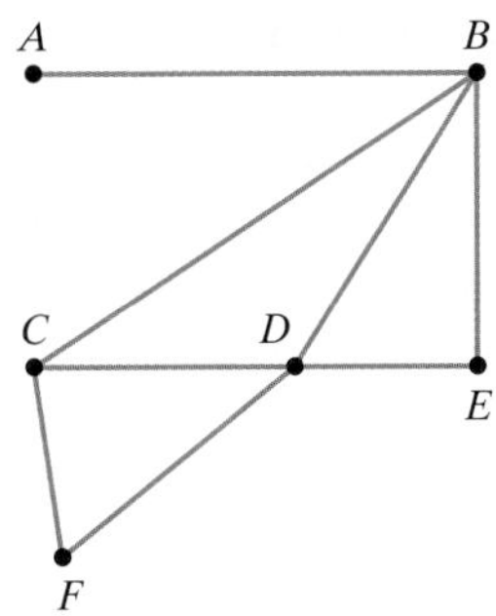

34.

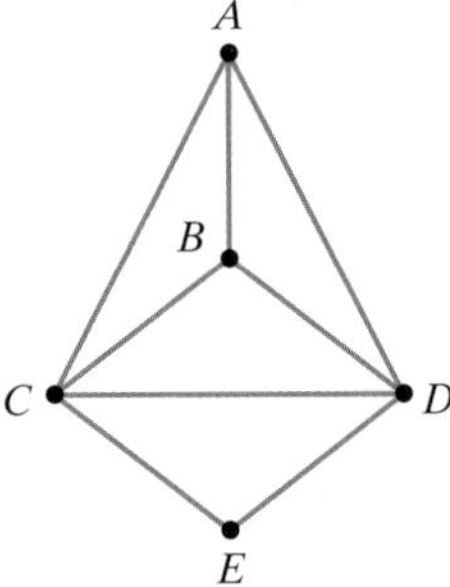

35.

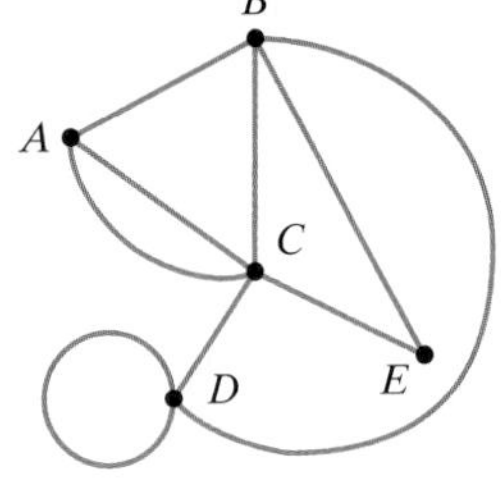

36.

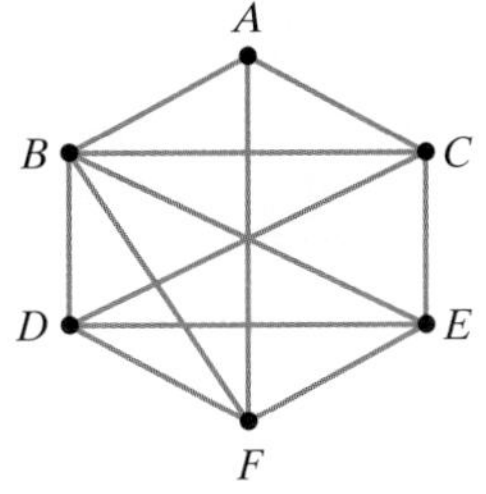

In Exercises 37–40, use Fleury's Algorithm to find an Euler circuit.

37.

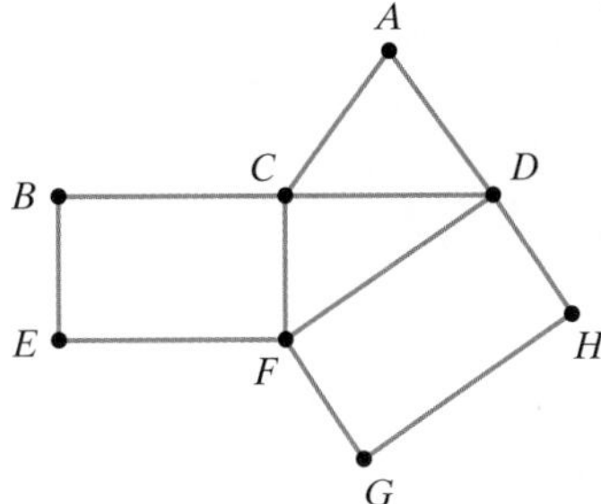

38.

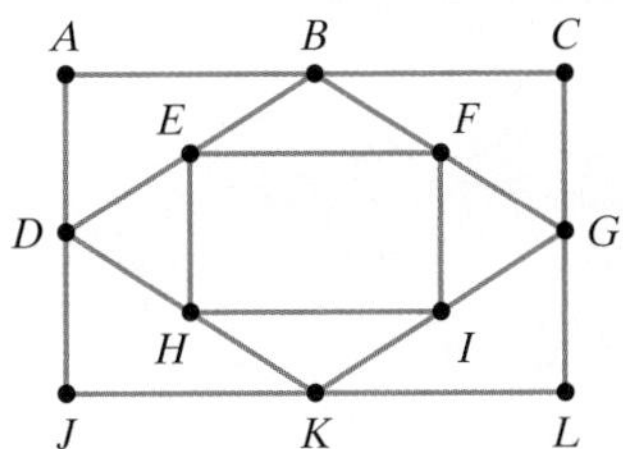

39.

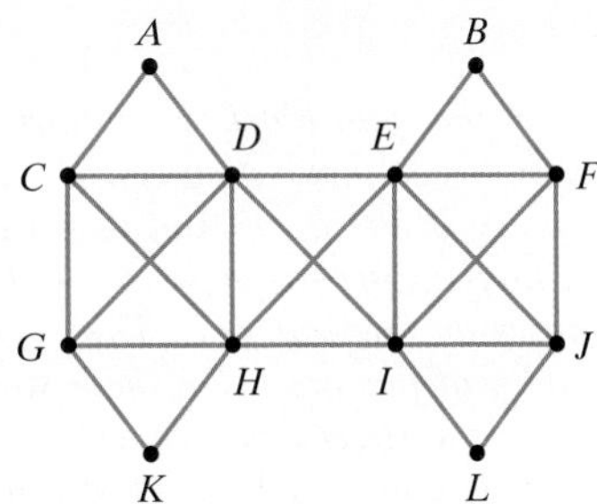

40.

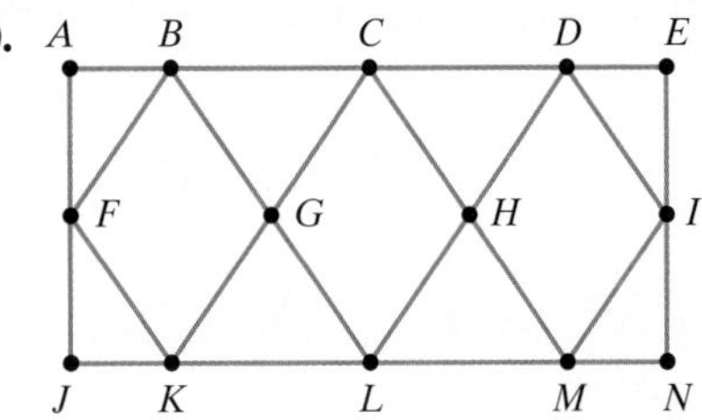

Practice Plus

In Exercises 41–44, a graph is given.

a. *Modify the graph by removing the least number of edges so that the resulting graph has an Euler circuit.*

b. *Find an Euler circuit for the modified graph.*

41.

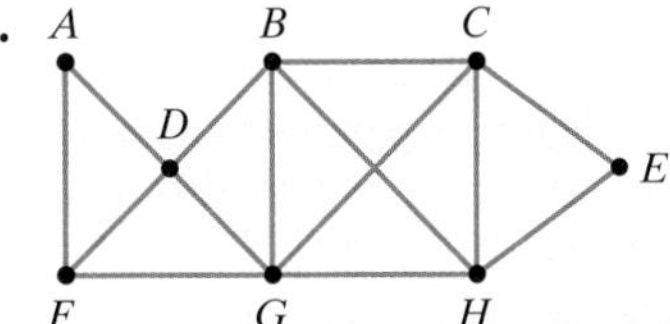

42.

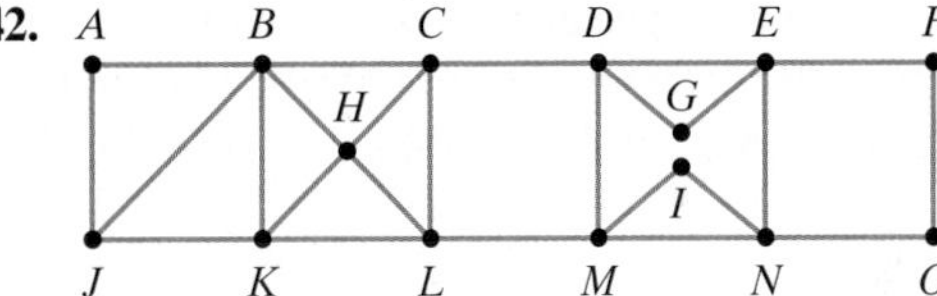

43.

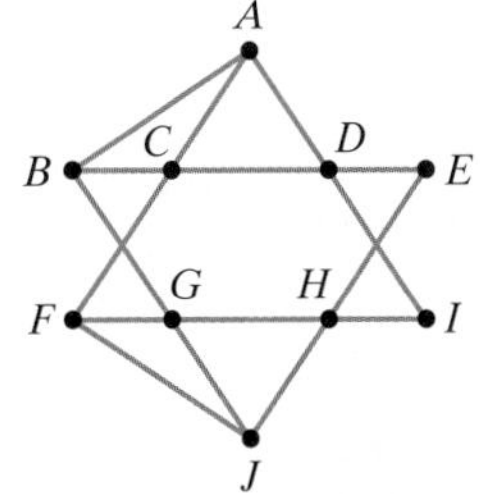

44.

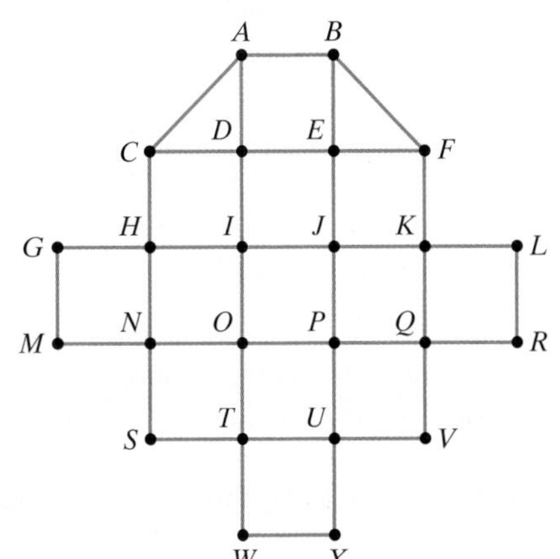

Application Exercises

In Exercises 45–48, we revisit the four-block neighborhood discussed in the previous section. Recall that a mail carrier parks her truck at the intersection shown in the figure and then walks to deliver mail to each of the houses. The streets on the outside of the neighborhood have houses on one side only. The interior streets have houses on both sides of the street. On these streets, the mail carrier must walk down the street twice, covering each side of the street separately. A graph that models the streets of the neighborhood walked by the mail carrier is shown.

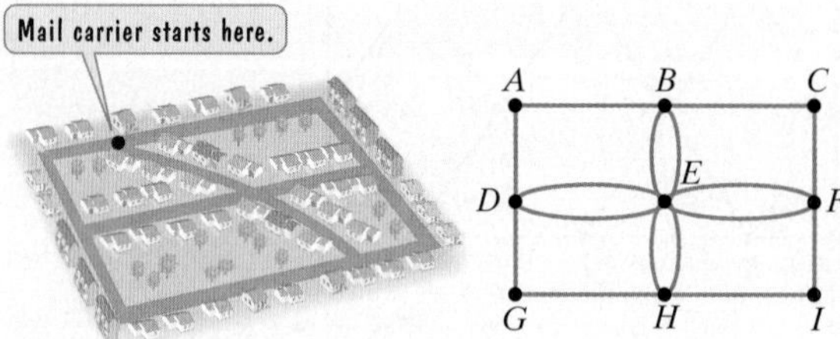

45. Use Euler's Theorem to explain why it is possible for the mail carrier to park at B, deliver mail to each house without retracing her route, and then return to B.

46. Use trial and error or Fleury's Algorithm to find an Euler circuit that starts and ends at vertex B on the graph that models the neighborhood.

47. Use your Euler circuit in Exercise 46 to show the mail carrier the route she should follow on the map of the neighborhood. Use the method of your choice, such as red lines with arrows, but be sure that the route is clearly designated.

48. A security guard needs to walk the streets of the neighborhood. Unlike the postal worker, the guard is to walk down each street once, whether or not the street has houses on both sides. Draw a graph that models the streets of the neighborhood walked by the security guard. Then determine whether the residents in the neighborhood will be able to establish a route for the security guard so that each street is walked exactly once. If this is possible, use your map to show where the guard should begin the walk.

49. A security guard needs to walk the streets of the neighborhood shown, walking each street once.

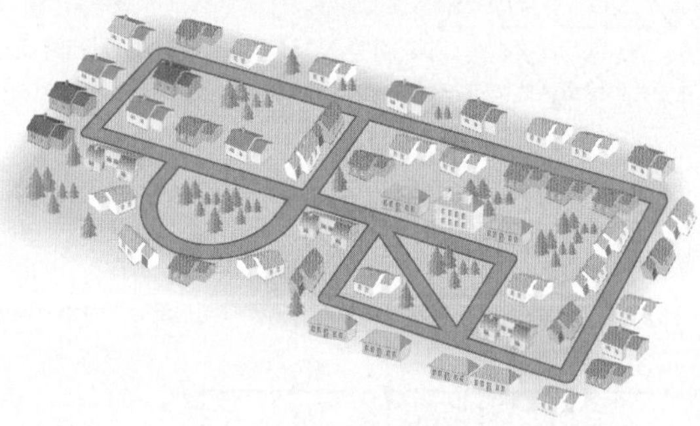

a. Draw a graph that models the streets of the neighborhood walked by the security guard.

b. Determine whether the residents in the neighborhood will be able to establish a route for the security guard so that each street is walked exactly once. If this is possible, use your map to show where the guard should begin the walk.

50. A mail carrier needs to deliver mail to each house in the three-block neighborhood shown. He plans to park at one of the street intersections and walk to deliver mail. All streets have houses on both sides. This means that the mail carrier must walk down every street twice, delivering mail to each side separately.

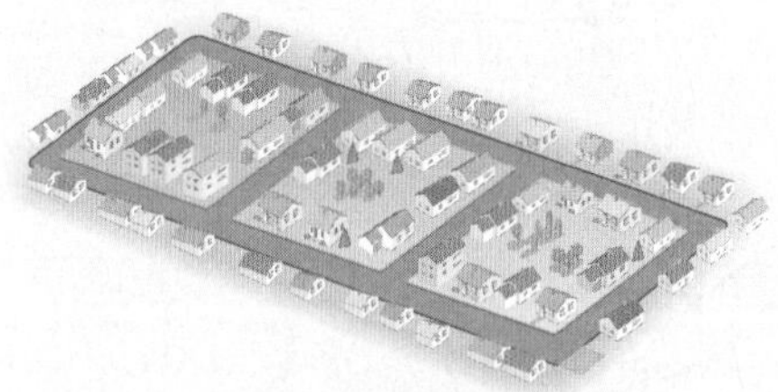

a. Draw a graph that models the streets of the neighborhood walked by the mail carrier.

b. Use the graph to determine if the carrier can park at an intersection, deliver mail to each house without retracing the side of any street, and return to the parked truck.

c. If such a walk is possible, show the path on your graph in part (a). Then trace this route on the neighborhood map in a manner that is clear to the mail carrier.

In Exercises 51–52, the layout of a city with land masses and bridges is shown.

a. *Draw a graph that models the layout of the city. Use vertices to represent the land masses and edges to represent the bridges.*

b. *Use the graph to determine if the city residents would be able to walk across all of the bridges without crossing the same bridge twice.*

c. *If such a walk is possible, show the path on your graph in part (a). Then trace this route on the city map in a manner that is clear to the city's residents.*

51.

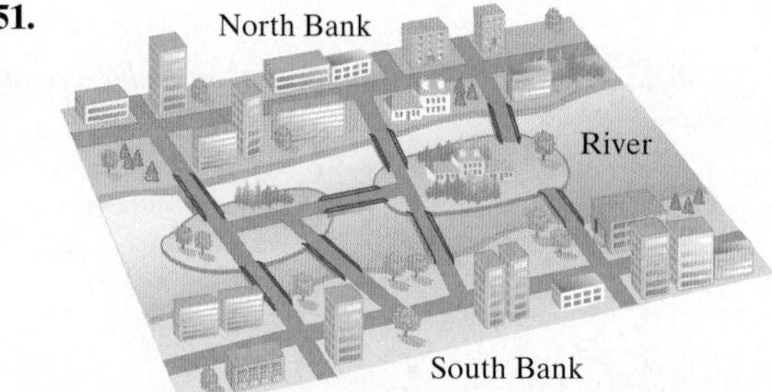

52.

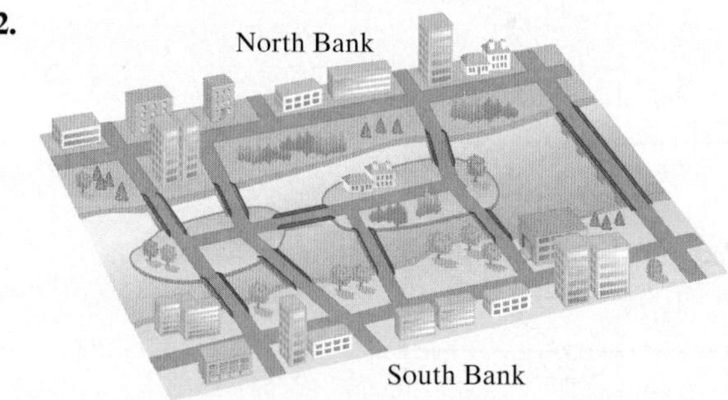

53. The figure shows a map of a portion of New York City with the bridge and tunnel connections. Use a graph to determine if it is possible to visit Manhattan, Long Island, Staten Island, and New Jersey using each bridge or tunnel only once.

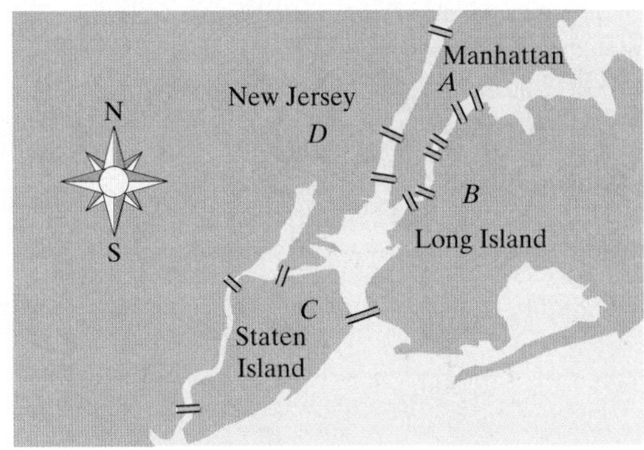

In Exercises 54–55, a floor plan is shown.

a. *Draw a graph that models the connecting relationships in each floor plan. Use vertices to represent the rooms and the outside, and edges to represent the connecting doors.*

b. *Use your graph to determine if it is possible to find a path that uses each door only once.*

c. *If such a path is possible, show it on your graph in part (a). Then trace this route on the floor plan in a manner that is clear to a person strolling through the house.*

54.

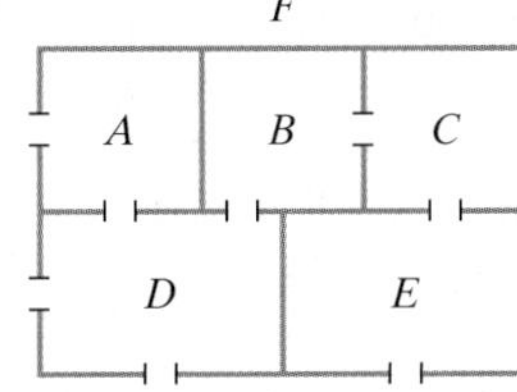

55.

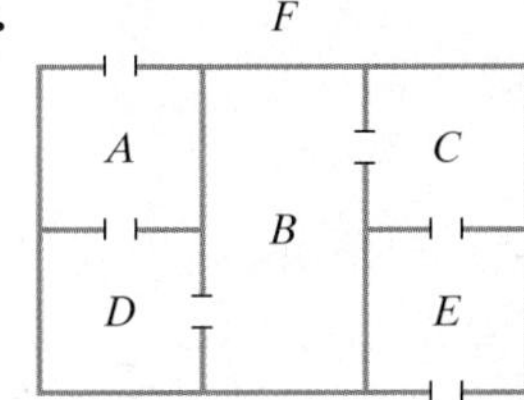

In Exercises 56–58, we revisit the map of the New England states discussed in the previous section. Shown in the figure are the map and the graph that models the bordering relationships among the six states.

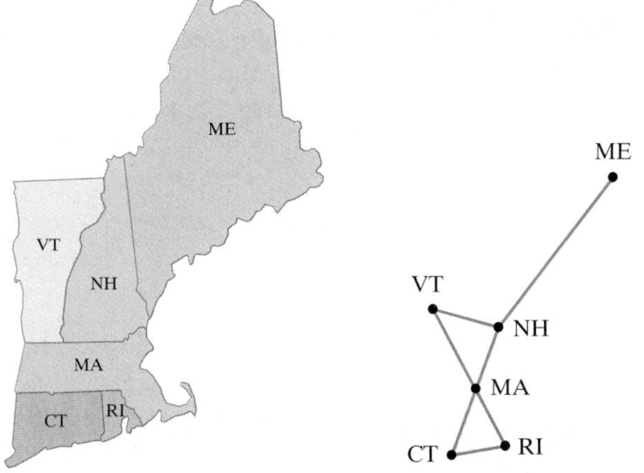

56. Use Euler's Theorem to explain why it is possible to find a path that crosses each common state border exactly once.

57. Use trial and error or Fleury's Algorithm to find an Euler path on the graph that models the map.

58. Use your Euler path in Exercise 57 to show the route on the map of the New England states.

In Exercises 59–60, a map is shown.

a. *Draw a graph that models each map. Use vertices to represent the states and edges to represent common borders.*

b. *Use your graph to determine if it is possible to find a path that crosses each common state border exactly once.*

c. *If such a trip is possible, show the path on your graph in part (a). Then trace this route on the map in a manner that is clear to people driving through these states.*

d. *Determine if it is possible to find a path that crosses each common state border exactly once, starting and ending the trip in the same state.*

59.

60.

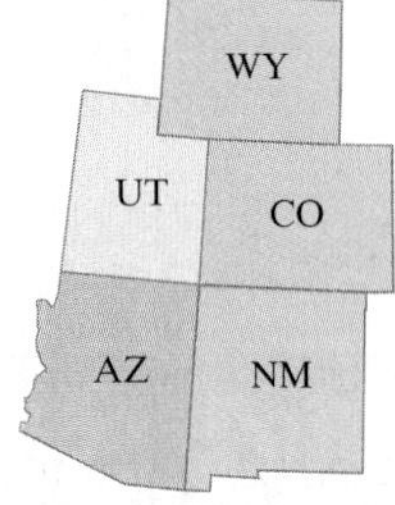

Explaining the Concepts

61. What is an Euler path?

62. What is an Euler circuit?

63. How do you determine if a graph has at least one Euler path, but no Euler circuit?

64. How do you determine if a graph has at least one Euler circuit?

65. How do you determine if a graph has no Euler paths and no Euler circuits?

66. What is the purpose of Fleury's Algorithm?

67. If a graph has at least one Euler path, but no Euler circuit, which vertex should be chosen as the starting point for a path?

68. Explain why it is important that the director of municipal services (police patrols, garbage collection, curb sweeping, snow removal) of a large city have a knowledge of graph theory.

Critical Thinking Exercises

Make Sense? *In Exercises 69–72, determine whether each statement makes sense or does not make sense, and explain your reasoning.*

69. I'm working with a graph whose vertices are all even, so an Euler circuit must exist.

70. Euler's Theorem is useful in finding Euler paths or Euler circuits.

71. I use Fleury's Algorithm to determine if a graph contains Euler paths or Euler circuits.

72. A UPS driver trying to find the best way to deliver packages around town can model the delivery locations as the vertices on a graph and the roads between the locations as edges. Now the driver needs to find an Euler path.

73. Refer to the layout of Königsberg in **Figure 14.30** on page 912. Suppose that the citizens of the city decide to build a new bridge. Where should it be located to make it possible to walk through the city and cross each bridge exactly once? If such a location is not possible, so state and explain why.

74. A police car would like to patrol each street in the neighborhood shown on the map exactly once. Use a numbering system from 1 through 31, one number per street, to show the police how this can be done.

75. Use Fleury's Algorithm to find an Euler circuit.

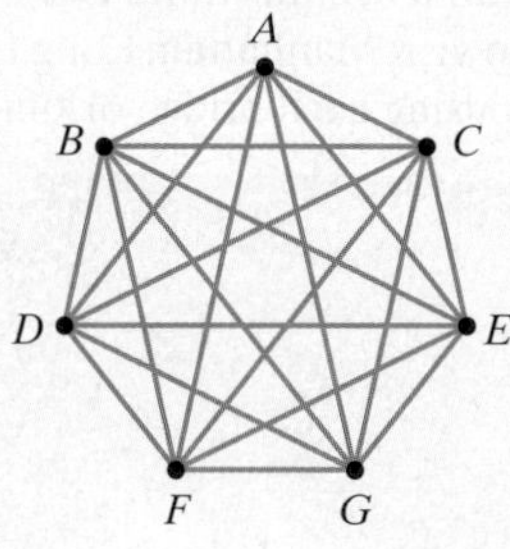

14.3 Hamilton Paths and Hamilton Circuits

WHAT AM I SUPPOSED TO LEARN?

After studying this section, you should be able to:

1. Understand the definitions of Hamilton paths and Hamilton circuits.
2. Find the number of Hamilton circuits in a complete graph.
3. Understand and use weighted graphs.
4. Use the Brute Force Method to solve traveling salesperson problems.
5. Use the Nearest Neighbor Method to approximate solutions to traveling salesperson problems.

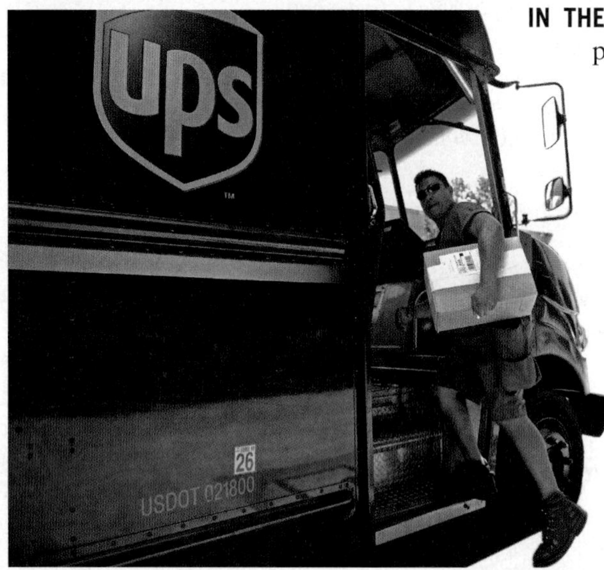

IN THE LAST SECTION, WE STUDIED paths and circuits that covered every edge of a graph. Using every edge of a graph exactly once is helpful for garbage collection, curb sweeping, or snow removal. But what about a United Parcel Service (UPS) driver trying to find the best way to deliver packages around town? We can model the delivery locations as the vertices on a graph and the roads between the locations as the edges. The driver is only concerned with finding a path that passes through each vertex of the graph once. It is not important to use every road, or edge, between delivery locations. In this section, we focus on paths and circuits that contain each vertex of a graph exactly once.

1 Understand the definitions of Hamilton paths and Hamilton circuits.

Hamilton Paths and Hamilton Circuits

Trying to find the best way to deliver packages around town or planning the best route by which to run a series of errands can be modeled and solved using paths and circuits that pass through each vertex of a graph exactly once. These kinds of paths and circuits are named after the Irish mathematician William Rowan Hamilton (1805–1865).

GREAT QUESTION!

What's the relationship between an Euler circuit and a Hamilton circuit?

If you take the word *edge* in the definition of an Euler circuit and replace it by the word *vertex,* you obtain the definition of a Hamilton circuit. However, the two kinds of circuits are otherwise unrelated.

HAMILTON PATHS AND CIRCUITS

A path that passes through each vertex of a graph exactly once is called a **Hamilton path**. If a Hamilton path begins and ends at the same vertex and passes through all other vertices exactly once, it is called a **Hamilton circuit**.

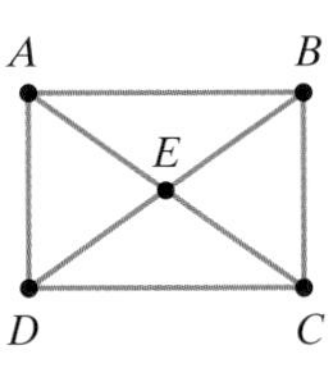

FIGURE 14.38

EXAMPLE 1 *Examples of Hamilton Paths and Hamilton Circuits*

a. Find a Hamilton path for the graph in **Figure 14.38**.

b. Find a Hamilton circuit for the graph in **Figure 14.38**.

SOLUTION

a. A Hamilton path must pass through each vertex exactly once. The graph has many Hamilton paths. An example of such a path is

$$A, B, C, D, E.$$

b. A Hamilton circuit must pass through every vertex exactly once and begin and end at the same vertex. The graph has many Hamilton circuits. An example of such a circuit is

$$A, B, C, D, E, A.$$

The graph in **Figure 14.38** has many Hamilton paths and circuits. However, because it has four vertices of odd degree, it has no Euler paths and no Euler circuits. When it comes to Hamilton circuits and Euler circuits, a graph can have one or the other, or both, or neither.

E
A C
F
B D
G
FIGURE 14.39

CHECK POINT 1

a. Find a Hamilton path that begins at vertex E for the graph in **Figure 14.39**.

b. Find a Hamilton circuit that begins at vertex E for the graph in **Figure 14.39**.

Consider the graph in **Figure 14.40**. It has many Hamilton paths, such as

$$A, B, E, C, D \quad \text{or} \quad C, D, E, A, B.$$

However, it has no Hamilton circuits. Regardless of the starting point, it is necessary to pass through vertex E more than once to return to the starting point.

A **complete graph** is a graph that has an edge between each pair of its vertices. Can you see that the graph in **Figure 14.40** is not complete? There is no edge between vertex A and vertex D. Nor is there an edge between vertex B and vertex C. Complete graphs are significant because **every complete graph with three or more vertices has a Hamilton circuit**. The graph in **Figure 14.40** is not complete and has no Hamilton circuits.

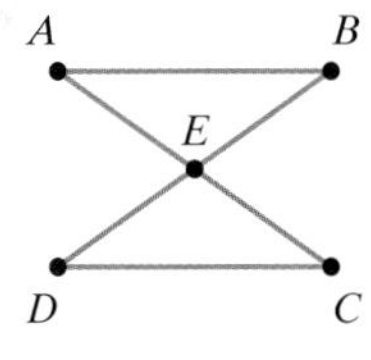

FIGURE 14.40

2 *Find the number of Hamilton circuits in a complete graph.*

The Number of Hamilton Circuits in a Complete Graph

The graph in **Figure 14.41** has an edge between each pair of its four vertices. Thus, the graph is complete and has a Hamilton circuit. Actually, it has a complete repertoire of Hamilton circuits. For example, one Hamilton circuit is

$$A, B, C, D, A.$$

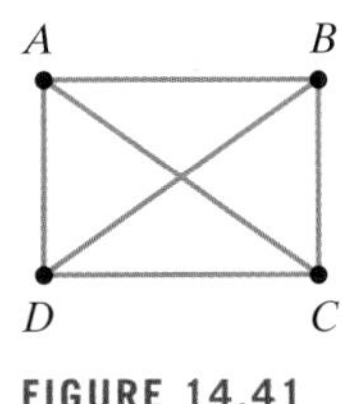

FIGURE 14.41

Any two circuits that pass through the same vertices in the same order will be considered to be the same. For example, here are four different sequences of letters that produce the same Hamilton circuit on the graph in **Figure 14.41**:

$$A, B, C, D, A \text{ and } B, C, D, A, B \text{ and } C, D, A, B, C \text{ and } D, A, B, C, D.$$

The Hamilton circuit passing through A, B, C, and D clockwise along the four outside edges can be written in four ways.

In order to avoid this duplication, in forming a Hamilton circuit, we can always assume that it begins at A.

A B D C

FIGURE 14.41 (repeated)

There are six different, or unique, Hamilton circuits for the graph with four vertices in **Figure 14.41**. They are given as follows:

To build these six Hamilton circuits, start with A, form all ordered arrangements (permutations) of the remaining three letters, and then return to A.

A, B, C, D, A
A, B, D, C, A
A, C, B, D, A
A, C, D, B, A
A, D, B, C, A
A, D, C, B, A

Each of these six different Hamilton circuits can be written in four ways.

How many different Hamilton circuits are there in a complete graph? For a complete graph with four vertices, the preceding list indicates that the number of circuits depends upon the number of permutations of the three letters B, C, and D. If a graph has n vertices, once we list vertex A, there are $n - 1$ remaining letters. The number of Hamilton circuits depends upon the number of permutations of the $n - 1$ letters.

Recall from our work in Chapter 11 that factorial notation is useful in determining the number of ordered arrangements, or permutations. If n is a positive integer, then $n!$ (n factorial) is the product of all positive integers from n down to 1. For example, $4! = 4 \cdot 3 \cdot 2 \cdot 1 = 24$ and $6! = 6 \cdot 5 \cdot 4 \cdot 3 \cdot 2 \cdot 1 = 720$.

We can find the number of ordered arrangements of $n - 1$ letters using the Fundamental Counting Principle, discussed in Section 11.1. With $n - 1$ letters to arrange, there are $n - 1$ choices for first position. Once you've chosen the first letter, you'll have $n - 2$ choices for second position. Continuing in this manner, there are $n - 3$ choices for third position, $n - 4$ choices for fourth position, down to only one choice (the one remaining letter) for last position. We multiply the number of choices to find the total number of ordered arrangements:

$$(n - 1)(n - 2)(n - 3) \cdot \cdots \cdot 1.$$

This product, by definition, is $(n - 1)!$. This expression therefore describes the number of Hamilton circuits in a complete graph with n vertices.

THE NUMBER OF HAMILTON CIRCUITS IN A COMPLETE GRAPH

The number of Hamilton circuits in a complete graph with n vertices is

$$(n - 1)!.$$

EXAMPLE 2 *Determining the Number of Hamilton Circuits*

Determine the number of Hamilton circuits in a complete graph with

a. four vertices. **b.** five vertices. **c.** eight vertices.

SOLUTION

In each case, we use the expression $(n - 1)!$. For four vertices, substitute 4 for n in the expression. For five and eight vertices, substitute 5 and 8, respectively, for n.

a. A complete graph with four vertices has $(4 - 1)! = 3! = 3 \cdot 2 \cdot 1 = 6$ Hamilton circuits. These are the six circuits that we listed at the top of the page.

b. A complete graph with five vertices has

$$(5 - 1)! = 4! = 4 \cdot 3 \cdot 2 \cdot 1 = 24 \text{ Hamilton circuits.}$$

c. A complete graph with eight vertices has

$$(8 - 1)! = 7! = 7 \cdot 6 \cdot 5 \cdot 4 \cdot 3 \cdot 2 \cdot 1 = 5040 \text{ Hamilton circuits.}$$

As the number of vertices in a complete graph increases, notice how quickly the number of Hamilton circuits increases.

 **CHECK POINT 2** Determine the number of Hamilton circuits in a complete graph with

a. three vertices. **b.** six vertices.

c. ten vertices.

3 *Understand and use weighted graphs.*

Weighted Graphs and the Traveling Salesperson Problem

Sales directors for large companies are often required to visit regional offices in a number of different cities. How can these visits be scheduled in the cheapest possible way?

For example, a sales director who lives in city A is required to fly to regional offices in cities B, C, and D. Other than starting and ending the trip in city A, there are no restrictions as to the order in which the other three cities are visited. The one-way fares between each of the four cities are given in **Table 14.1**. A graph that models this information is shown in **Figure 14.42**. The vertices represent the cities. The airfare between each pair of cities is shown as a number on the respective edge.

TABLE 14.1 One-Way Airfares

	A	*B*	*C*	*D*
A	*	$190	$124	$157
B	$190	*	$126	$155
C	$124	$126	*	$179
D	$157	$155	$179	*

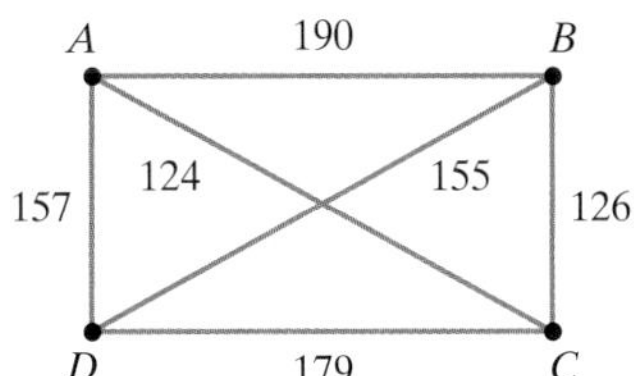

FIGURE 14.42 Modeling **Table 14.1** with a graph.

GREAT QUESTION!

When modeling problems with weighted graphs, how long should I make the edges?

It doesn't matter. In a weighted graph, the lengths of the edges do not have to be proportional to the weights.

A graph whose edges have numbers attached to them is called a **weighted graph**. The numbers shown along the edges of a weighted graph are called the **weights** of the edges. **Figure 14.42** is an example of a complete, weighted graph. The weight of edge AB is 190, modeling a $190 airfare from city A to city B. The sales director needs to find the least expensive way to visit cities B, C, and D once, and return home to A. Our goal is to find the Hamilton circuit with the lowest associated cost.

EXAMPLE 3 *Understanding the Information in a Weighted Graph*

Use the weighted graph in **Figure 14.42** to find the cost of the trip for the Hamilton circuit A, B, D, C, A.

SOLUTION

The trip described by the Hamilton circuit A, B, D, C, A involves the sum of four costs:

$$\$190 + \$155 + \$179 + \$124 = \$648.$$

- $190: This is the cost from A to B in A, B, D, C, A.
- $155: This is the cost from B to D in A, B, D, C, A.
- $179: This is the cost from D to C in A, B, D, C, A.
- $124: This is the cost from C to A in A, B, D, C, A.

The cost of the trip is $648.

 **CHECK POINT 3** Use the weighted graph in **Figure 14.42** to find the cost of the trip for the Hamilton circuit A, C, B, D, A.

4 Use the Brute Force Method to solve traveling salesperson problems.

The traveling salesperson problem can be stated as follows:

> **THE TRAVELING SALESPERSON PROBLEM**
>
> The **traveling salesperson problem** is the problem of finding a Hamilton circuit in a complete, weighted graph for which the sum of the weights of the edges is a minimum. Such a Hamilton circuit is called the **optimal Hamilton circuit** or the **optimal solution**.

One method for finding an optimal Hamilton circuit is called the **Brute Force Method**.

> **THE BRUTE FORCE METHOD OF SOLVING TRAVELING SALESPERSON PROBLEMS**
>
> The optimal solution is found using the following steps:
>
> 1. Model the problem with a complete, weighted graph.
> 2. Make a list of all possible Hamilton circuits.
> 3. Determine the sum of the weights of the edges for each of these Hamilton circuits.
> 4. The Hamilton circuit with the minimum sum of weights is the optimal solution.

EXAMPLE 4 *Using the Brute Force Method*

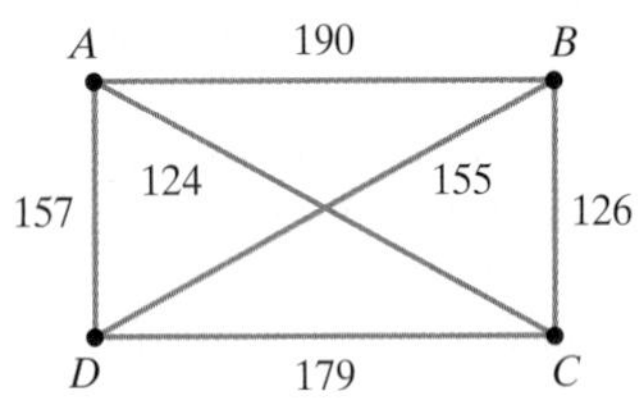

FIGURE 14.42 (repeated)

Use the complete, weighted graph in **Figure 14.42**, repeated in the margin, to find the optimal solution. Describe what this means for the sales director who starts at A, flies once to each of B, C, and D, and returns home to A.

SOLUTION

The graph has four vertices. Thus, using $(n - 1)!$, there are $(4 - 1)! = 3! = 6$ possible Hamilton circuits. The six possible Hamilton circuits and their total costs are shown in **Table 14.2**.

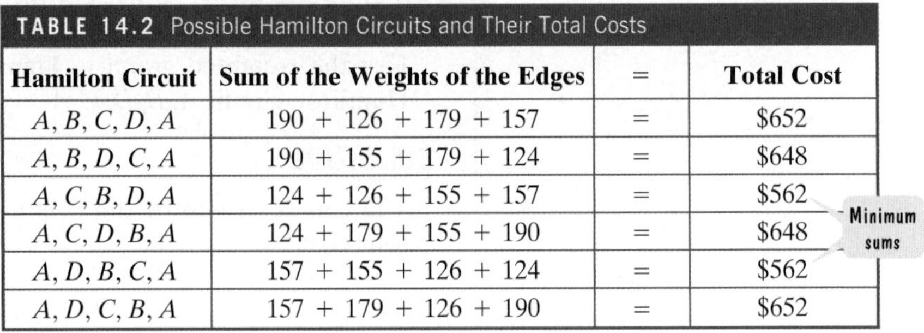

TABLE 14.2 Possible Hamilton Circuits and Their Total Costs

Hamilton Circuit	Sum of the Weights of the Edges	=	Total Cost
A, B, C, D, A	190 + 126 + 179 + 157	=	\$652
A, B, D, C, A	190 + 155 + 179 + 124	=	\$648
A, C, B, D, A	124 + 126 + 155 + 157	=	\$562
A, C, D, B, A	124 + 179 + 155 + 190	=	\$648
A, D, B, C, A	157 + 155 + 126 + 124	=	\$562
A, D, C, B, A	157 + 179 + 126 + 190	=	\$652

The voice balloon in the last column indicates that two Hamilton circuits have the minimum cost of \$562. The optimal solution is either

$$A, C, B, D, A \quad \text{or} \quad A, D, B, C, A.$$

For the sales director, this means that either route shown in **Figure 14.43** at the top of the next page is the least expensive way to visit the regional offices in cities B, C, and D. Notice that the route in **Figure 14.43(b)** involves visiting the cities in the reverse order of the route in **Figure 14.43(a)**. Although these are different Hamilton circuits, because the one-way airfares are the same in either direction, the cost is the same regardless of the direction flown.

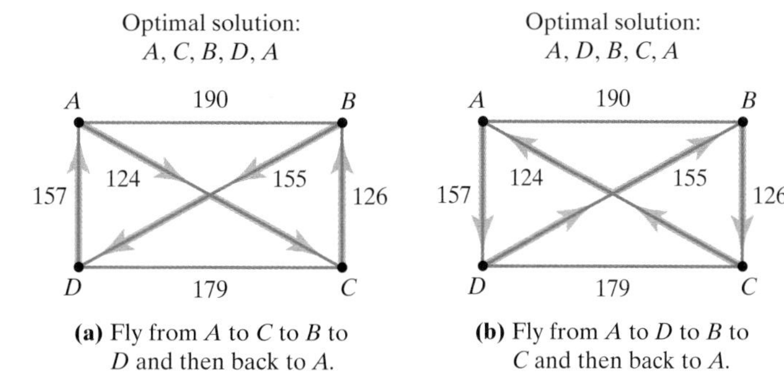

(a) Fly from A to C to B to D and then back to A.

(b) Fly from A to D to B to C and then back to A.

FIGURE 14.43 Options for an optimal solution

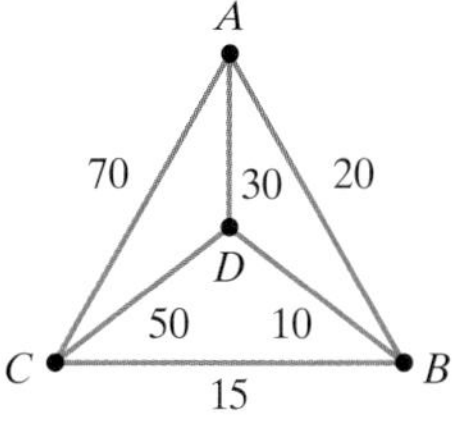

FIGURE 14.44

CHECK POINT 4 Use the Brute Force Method to find the optimal solution for the complete, weighted graph in **Figure 14.44**. List Hamilton circuits as in **Table 14.2**.

As the number of vertices increases, the rapidly increasing number of possible Hamilton circuits makes the Brute Force Method impractical. Unfortunately, mathematicians have not been able to establish another method for solving the traveling salesperson problem and they do not know whether it is even possible to find such a method. There are, however, a number of methods for finding approximate solutions.

Blitzer Bonus

The Brute Force Method and Supercomputers

Suppose that a supercomputer can find the sum of the weights of one billion, or 10^9, Hamilton circuits per second. Because there are 31,536,000 seconds in a year, the computer can calculate the sums for approximately 3.2×10^{16} Hamilton circuits in one year. The table shows that as the number of vertices increases, the Brute Force Method is useless even with a powerful computer.

COMPUTER TIME NEEDED TO SOLVE THE TRAVELING SALESPERSON PROBLEM

Number of Vertices	Number of Hamilton Circuits	Time Needed by a Supercomputer to Find Sums of All Hamilton Circuits
18	$17! \approx 3.6 \times 10^{14}$	$\approx$0.01 year $\approx$ 3.7 days
19	$18! \approx 6.4 \times 10^{15}$	$\approx$0.2 year $\approx$ 73 days
20	$19! \approx 1.2 \times 10^{17}$	$\approx$3.8 years
21	$20! \approx 2.4 \times 10^{18}$	$\approx$76 years
22	$21! \approx 5.1 \times 10^{19}$	$\approx$1597 years
23	$22! \approx 1.1 \times 10^{21}$	$\approx$35,125 years

5 *Use the Nearest Neighbor Method to approximate solutions to traveling salesperson problems.*

Suppose a sales director who lives in city A is required to fly to regional offices in ten other cities and then return home to city A. With $(11 - 1)!$, or 3,628,800, possible Hamilton circuits, a list is out of the question. What do you think of this option? Start at city A. From there, fly to the city to which the airfare is cheapest. Then from there fly to the next city to which the airfare is cheapest, and so on. From the last of the ten cities, fly home to city A.

By continually taking an edge with the smallest weight, we can find approximate solutions to traveling salesperson problems. This method is called the **Nearest Neighbor Method**.

THE NEAREST NEIGHBOR METHOD OF FINDING APPROXIMATE SOLUTIONS TO TRAVELING SALESPERSON PROBLEMS

The optimal solution can be approximated using the following steps:

1. Model the problem with a complete, weighted graph.
2. Identify the vertex that serves as the starting point.
3. From the starting point, choose the edge with the smallest weight. Move along this edge to the second vertex. (If there is more than one edge with the smallest weight, choose one.)
4. From the second vertex, choose the edge with the smallest weight that does not lead to a vertex already visited. Move along this edge to the third vertex.
5. Continue building the circuit, one vertex at a time, by moving along the edge with the smallest weight until all vertices are visited.
6. From the last vertex, return to the starting point.

EXAMPLE 5 *Using the Nearest Neighbor Method*

A sales director who lives in city A is required to fly to regional offices in cities B, C, D, and E. The complete, weighted graph showing the one-way airfares is given in **Figure 14.45**. Use the Nearest Neighbor Method to find an approximate solution. What is the total cost?

FIGURE 14.45

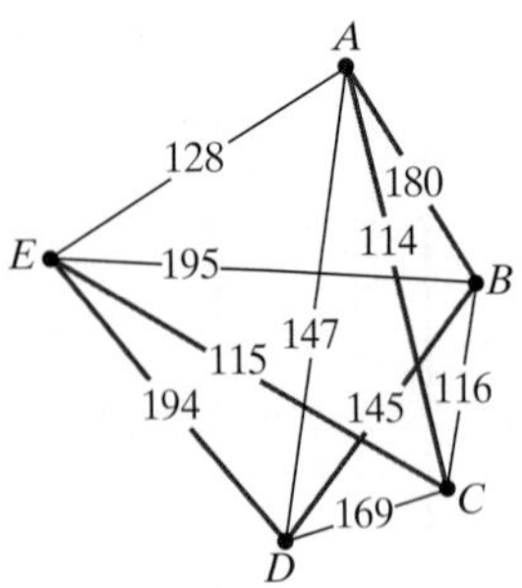

FIGURE 14.46 An approximate solution: A, C, E, D, B, A.

SOLUTION

The Nearest Neighbor Method, illustrated in **Figure 14.46**, is carried out as follows:

- Start at A.
- Choose the edge with the smallest weight: 114. Move along this edge to C. (cost: \$114)
- From C, choose the edge with the smallest weight that does not lead to A: 115. Move along this edge to E. (cost: \$115)
- From E, choose the edge with the smallest weight that does not lead to a city already visited: 194. Move along this edge to D. (cost: \$194)
- From D, there is little choice but to fly to B, the only city not yet visited. (cost: \$145)
- From B, close the circuit and return home to A. (cost: \$180)

An approximate solution is the Hamilton circuit

$$A, C, E, D, B, A.$$

The circuit is shown in **Figure 14.46**. The total cost is

$$\$114 + \$115 + \$194 + \$145 + \$180 = \$748.$$

The Brute Force Method can be applied to the 24 possible Hamilton circuits in Example 5. How does the actual solution compare with the \$748 total cost obtained by the Nearest Neighbor Method? The actual solution is A, D, B, C, E, A or the reverse order A, E, C, B, D, A, and the cost is \$651. This shows that the Nearest Neighbor Method does not always give the optimal solution. Of the 24 Hamilton circuits, eight result in a total cost that is less than the \$748 we obtained in Example 5.

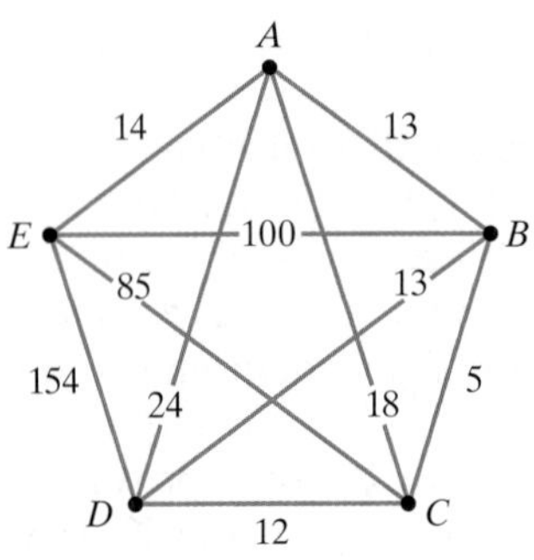

FIGURE 14.47

CHECK POINT 5 Use the Nearest Neighbor Method to approximate the optimal solution for the complete, weighted graph in **Figure 14.47**. Begin the circuit at vertex A. What is the total weight of the resulting Hamilton circuit?

Concept and Vocabulary Check

Fill in each blank so that the resulting statement is true.

1. A path that passes through each vertex of a graph exactly once is called a/an _________ path. Such a path that begins and ends at the same vertex and passes through all other vertices exactly once is called a/an _________ circuit.
2. A graph that has an edge between each pair of its vertices is called a/an _________ graph. If such a graph has n vertices, the number of Hamilton circuits in the graph is given by the factorial expression _________.
3. A graph whose edges have numbers attached to them is called a/an _________ graph. The numbers shown along the edges of such a graph are called the _________ of the edges. The problem of finding a Hamilton circuit for which the sum of these numbers is a minimum is called the _________ salesperson problem. Such a Hamilton circuit is called the _________ solution.
4. A method that determines the solution to the salesperson problem in Exercise 3 involves listing all Hamilton circuits and selecting the circuit with the minimum sum of weights. This method is called the _________ Method.
5. A method that approximates the solution to the salesperson problem in Exercise 3 is called the _________ Method. This method involves continually choosing an edge with the smallest _________.
6. True or False: Every complete graph that has a Hamilton circuit has at least one Euler circuit. ______
7. True or False: In a weighted graph, the lengths of the edges are proportional to their weights. ______
8. True or False: The Nearest Neighbor Method provides exact solutions to traveling salesperson problems. ______

Exercise Set 14.3

Practice Exercises

In Exercises 1–4, use the graph shown.

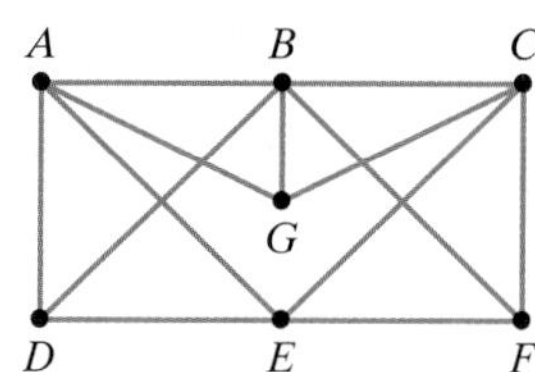

1. Find a Hamilton path that begins at A and ends at B.
2. Find a Hamilton path that begins at G and ends at E.
3. Find a Hamilton circuit that begins as $A, B, \ldots$.
4. Find a Hamilton circuit that begins as $A, G, \ldots$.

In Exercises 5–8, use the graph shown.

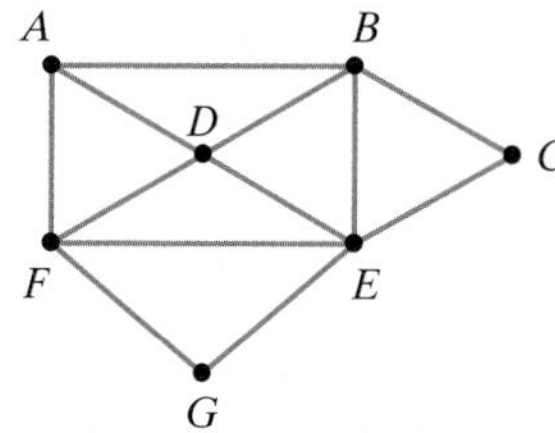

5. Find a Hamilton path that begins at A and ends at D.
6. Find a Hamilton path that begins at A and ends at G.
7. Find a Hamilton circuit that begins at A and ends with the pair of vertices D, A.
8. Find a Hamilton circuit that begins at F and ends with the pair of vertices D, F.

For each graph in Exercises 9–14,

a. *Determine if the graph must have Hamilton circuits. Explain your answer.*

b. *If the graph must have Hamilton circuits, determine the number of such circuits.*

9.
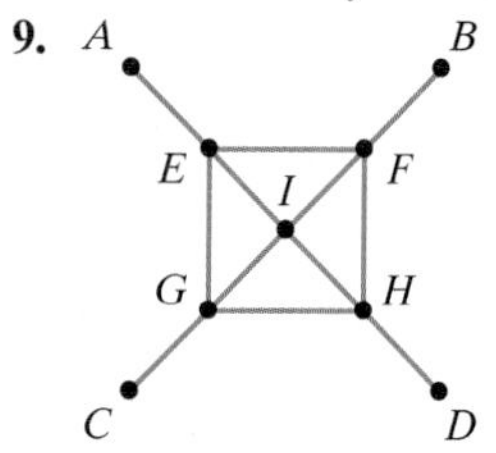

10.
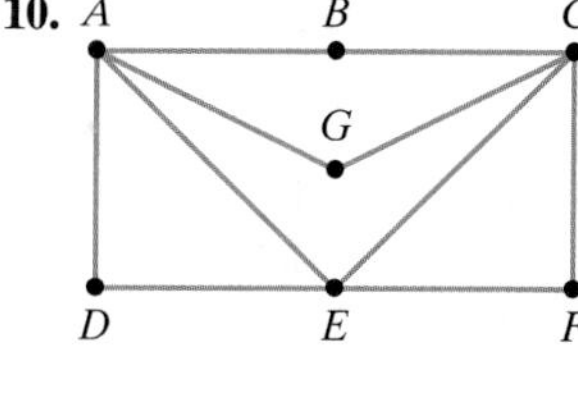

11.
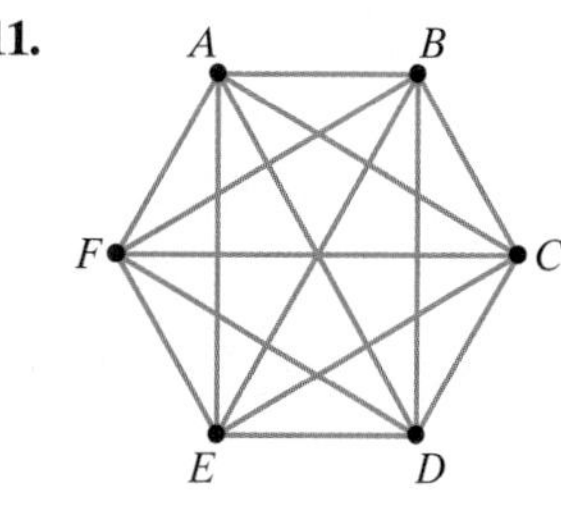

12.
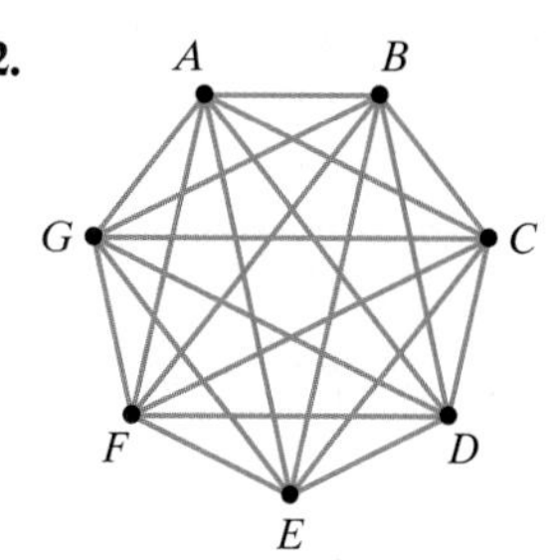

13.
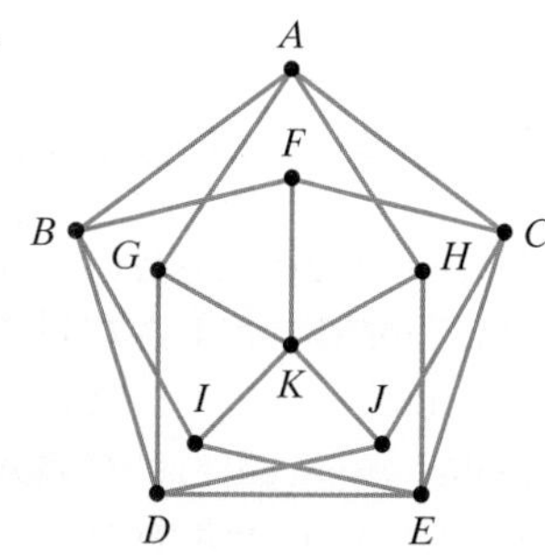

14.

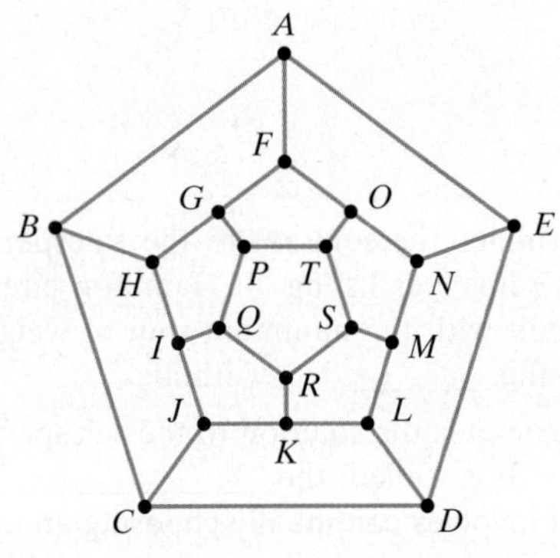

In Exercises 15–18, determine the number of Hamilton circuits in a complete graph with the given number of vertices.

15. 3 **16.** 4 **17.** 12 **18.** 13

In Exercises 19–24, use the complete, weighted graph shown.

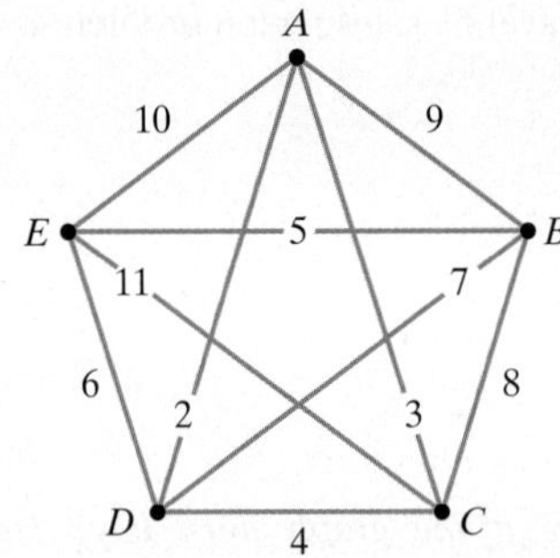

19. Find the weight of edge CE.

20. Find the weight of edge BD.

21. Find the total weight of the Hamilton circuit

A, B, C, E, D, A.

22. Find the total weight of the Hamilton circuit

A, B, D, C, E, A.

23. Find the total weight of the Hamilton circuit

A, B, D, E, C, A.

24. Find the total weight of the Hamilton circuit

A, B, E, C, D, A.

In Exercises 25–34, use the complete, weighted graph shown.

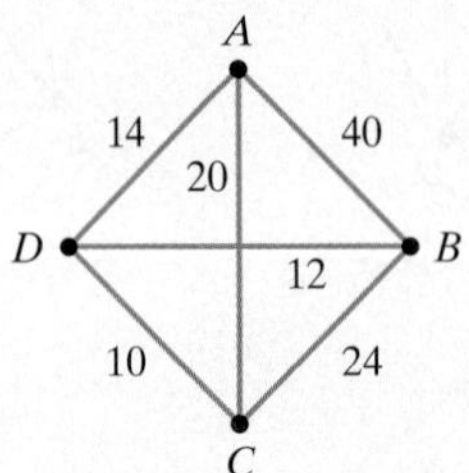

25. Find the total weight of the Hamilton circuit A, B, C, D, A.

26. Find the total weight of the Hamilton circuit A, B, D, C, A.

27. Find the total weight of the Hamilton circuit A, C, B, D, A.

28. Find the total weight of the Hamilton circuit A, C, D, B, A.

29. Find the total weight of the Hamilton circuit A, D, B, C, A.

30. Find the total weight of the Hamilton circuit A, D, C, B, A.

31. Use your answers from Exercises 25–30 and the Brute Force Method to find the optimal solution.

32. Use the Nearest Neighbor Method, with starting vertex A, to find an approximate solution. What is the total weight of the Hamilton circuit?

33. Use the Nearest Neighbor Method, with starting vertex B, to find an approximate solution. What is the total weight of the Hamilton circuit?

34. Use the Nearest Neighbor Method, with starting vertex C, to find an approximate solution. What is the total weight of the Hamilton circuit?

Practice Plus

In Exercises 35–38, a graph is given.

a. *Modify the graph by adding the least number of edges so that the resulting graph is complete. Determine the number of Hamilton circuits for the modified graph.*

b. *Give two Hamilton circuits for the modified graph in part (a).*

c. *Modify the given graph by removing the least number of edges so that the resulting graph has an Euler circuit.*

d. *Find an Euler circuit for the modified graph in part (c).*

35. **36.**

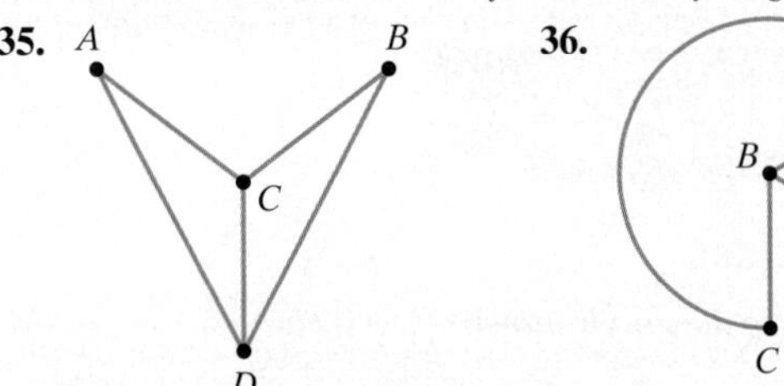

37.

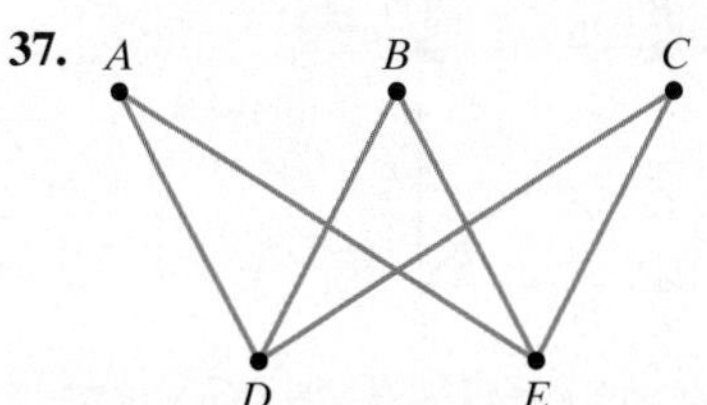

38.

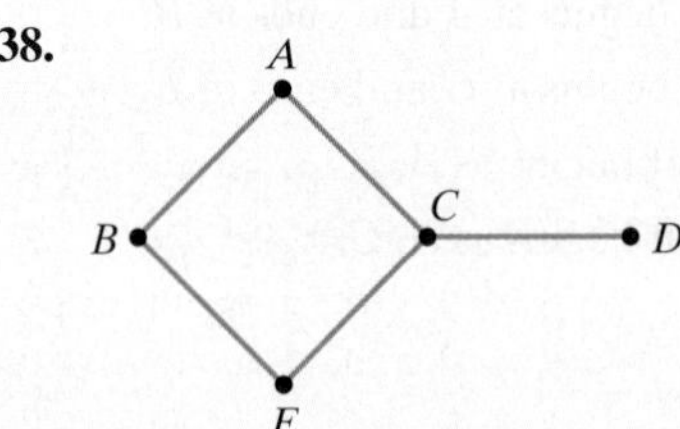

Application Exercises

In Exercises 39–40, a sales director who lives in city A is required to fly to regional offices in cities B, C, D, and E. The weighted graph shows the one-way airfares between any two cities.

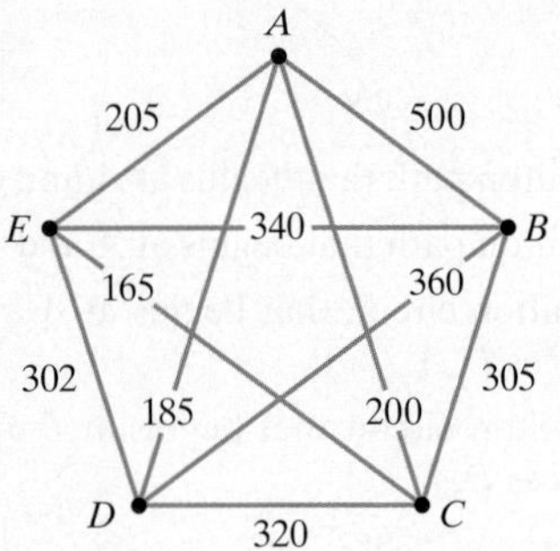

39. Use the Brute Force Method to find the optimal solution. Describe what this means for the sales director. (*Hint*: Because the airfares are the same in either direction, you need only compute the total cost for 12 Hamilton circuits:

A, B, C, D, E, A; A, B, C, E, D, A;
A, B, D, C, E, A; A, B, D, E, C, A;
A, B, E, C, D, A; A, B, E, D, C, A;
A, C, B, D, E, A; A, C, B, E, D, A;
A, C, D, B, E, A; A, C, E, B, D, A;
A, D, B, C, E, A; A, D, C, B, E, A.)

40. Use the Nearest Neighbor Method, with starting vertex A, to find an approximate solution. What is the total cost for this Hamilton circuit?

You have five errands to run around town, in no particular order. You plan to start and end at home. You must go to the post office, deposit a check at the bank, drop off dry cleaning, visit a friend at the hospital, and get a flu shot. The map shows your home and the locations of your five errands. Each block represents one mile. Also shown is a weighted graph with distances on the appropriate edges. In Exercises 41–44, your goal is to run the errands and return home using the shortest route.

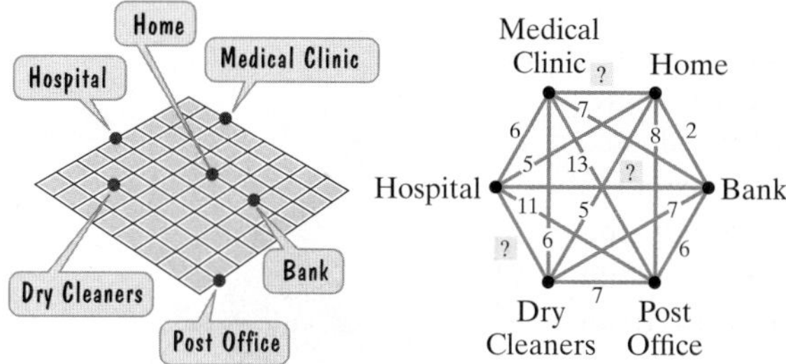

41. Use the map to fill in the three missing weights in the graph.

42. If each Hamilton circuit represents a route to run your errands, how many different routes are possible?

43. Using the Brute Force Method, the optimal solution is Home, Bank, Post Office, Dry Cleaners, Hospital, Medical Clinic, Home. What is the total length of the shortest route?

44. Use the Nearest Neighbor Method to find an approximate solution. What is the total length of the shortest route using this solution? How does this compare with your answer to Exercise 43?

In Exercises 45–47, you have three errands to run around town, although in no particular order. You plan to start and end at home. You must go to the bank, the post office, and the market. Distances, in miles, between any two of these locations are given in the table.

DISTANCES (IN MILES) BETWEEN LOCATIONS

	Home	Bank	Post Office	Market
Home	*	3	5.5	3.5
Bank	3	*	4	5
Post Office	5.5	4	*	4.5
Market	3.5	5	4.5	*

45. Create a complete, weighted graph that models the information in the table.

46. Use the Brute Force Method to find the shortest route to run your errands and return home. What is the minimum distance you can travel?

47. Use the Nearest Neighbor Method to approximate the shortest route to run your errands and return home. What is the minimum distance given by the Hamilton circuit? How does this compare with your answer to Exercise 46?

Explaining the Concepts

48. What is a Hamilton path? How does this differ from an Euler path?

49. What is a Hamilton circuit? How does this differ from an Euler circuit?

50. What is a complete graph?

51. How can you look at a graph and determine if it has a Hamilton circuit?

52. Describe how to determine the number of Hamilton circuits in a complete graph.

53. What is a weighted graph and what are the weights?

54. What is the traveling salesperson problem? What is the optimal solution?

55. Describe the Brute Force Method of solving traveling salesperson problems.

56. Why is the Brute Force Method impractical for large numbers of vertices?

57. Describe the Nearest Neighbor Method for approximating the optimal solution to traveling salesperson problems.

58. Describe a practical example of a traveling salesperson problem other than those involving one-way airfares or distances between errand locations.

59. An efficient solution for solving traveling salesperson problems has eluded mathematicians for more than 50 years. What explanations can you offer for this?

Critical Thinking Exercises

Make Sense? *In Exercises 60–63, determine whether each statement makes sense or does not make sense, and explain your reasoning.*

60. I'm working with a complete graph that has 25 Hamilton circuits.

61. I'm amazed by the power of my new computer, so this evening I plan to use it to find an optimal Hamilton circuit for a complete, weighted graph with 20 vertices.

62. City planners need to solve the traveling salesperson problem to determine efficient routes along city streets for garbage pickup.

63. Exercises 45–47 illustrate that solving the traveling salesperson problem can be relevant to my life even if I'm not a salesperson.

64. A complete graph has 120 distinct Hamilton circuits. How many vertices does the graph have?

65. Ambassadors from countries *A, B, C, D, E,* and *F* are to be seated around a circular conference table. Friendly relations among the various countries are described as follows: *A* has friendly relations with *B* and *F*. *B* has friendly relations with *A, C,* and *E*. *C* has friendly relations with *B, D, E,* and *F*. *E* has friendly relations with *B, C, D,* and *F*. All friendly relations are mutual. Using vertices to represent countries and edges to represent friendly relations, draw a graph that models the information given. Then use a Hamilton circuit to devise a seating arrangement around the table so that the ambassadors from *B* and *E* are seated next to each other, and each ambassador represents a country that has friendly relations with the countries represented by the ambassadors next to him or her.

Group Exercise

66. In this group exercise, you will create and solve a traveling salesperson problem similar to Exercises 39 and 40. Consult the graph given for these exercises as you work on this activity.

a. Group members should agree on four cities to be visited.

b. As shown in the graph for Exercises 39 and 40, assume that you are located at *A*. Let *B, C, D,* and *E* represent each of the four cities you have agreed upon. Consult the Internet and use one-way airfares between cities to create a weighted graph.

c. As you did in Exercise 39, use the Brute Force Method to find the optimal solution to visiting each of your chosen cities and returning home.

d. As you did in Exercise 40, use the Nearest Neighbor Method to approximate the optimal solution. How much money does the group save using the optimal solution instead of the approximation?

14.4 Trees

WHAT AM I SUPPOSED TO LEARN?

After studying this section, you should be able to:

1. Understand the definition and properties of a tree.
2. Find a spanning tree for a connected graph.
3. Find the minimum spanning tree for a weighted graph.

I need to call the Weather Channel to let them know I just shoveled three feet of "partly cloudy" off my driveway. This late spring storm caught all of us by surprise. I was amazed that classes were not canceled. Somehow the campus groundskeeper managed to shovel the minimum total length of walkways while ensuring that students could walk from building to building.

The general theme of this section is finding an efficient network linking a set of points. Think about the cleared campus sidewalks. Because of the unexpected storm, there is not enough time to shovel every sidewalk. Instead, campus services must clear a select number of sidewalks so that by moving from building to building, students can reach any location without having to trudge through mountains of snow. Finding efficient networks is accomplished using special kinds of graphs, called *trees*.

1 Understand the definition and properties of a tree.

Trees

The campus groundskeeper is interested in a graph that passes through each vertex (campus building) exactly once with the smallest possible number of edges (cleared sidewalks). A graph with the smallest number of edges that allows all vertices to be reached from all other vertices is called a *tree*.

Figure 14.48 shows examples of graphs that are trees.

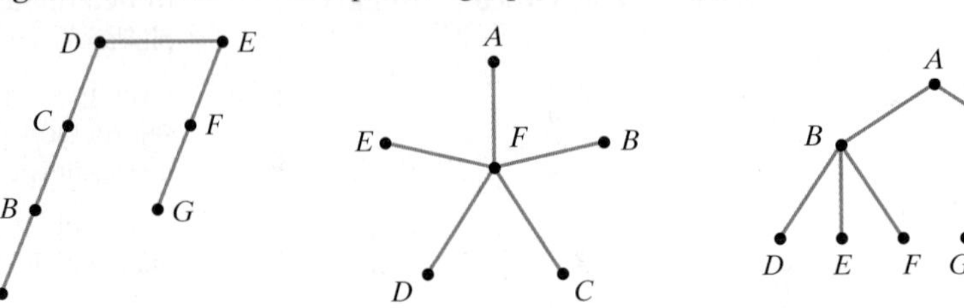

FIGURE 14.48 Examples of graphs that are trees.

Notice that each graph is connected. This is a requirement because we must be able to reach any vertex from any other vertex. Furthermore, no graph contains any circuits. This is because vertices must be reached with the smallest number of edges. Thus, the graphs in **Figure 14.49** are not trees. The circuits create redundant connections—take away each redundant connection and all vertices can still be reached from all other vertices.

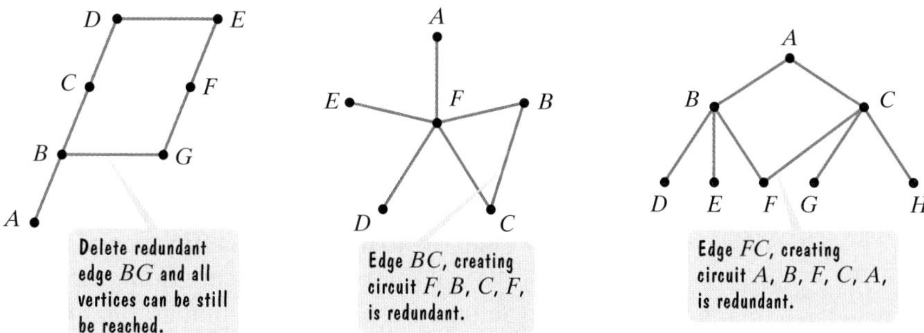

FIGURE 14.49 Examples of graphs that are not trees.

Using these ideas, we are now ready to define a tree and list some of its properties.

GREAT QUESTION!

What's the difference between a *connected* graph and a *complete* graph?

A connected graph has at least one path between each pair of its vertices. Because a tree has exactly one path joining any two vertices, it is connected. By contrast, a complete graph has an edge between each pair of its vertices. A tree is not a complete graph. In clearing campus sidewalks, there is not enough time to create an edge (a cleared path) between each pair of buildings (vertices).

DEFINITION AND PROPERTIES OF A TREE

A **tree** is a graph that is connected and has no circuits. All trees have the following properties:

1. There is one and only one path joining any two vertices.
2. Every edge is a bridge.
3. A tree with n vertices must have $n - 1$ edges.

Property 3 is a numerical property of trees that relates the number of vertices and the number of edges. The total number of edges is always one less than the number of vertices. For example, a tree with 5 vertices must have $5 - 1$, or 4, edges.

EXAMPLE 1 Identifying Trees

Which graph in **Figure 14.50** is a tree? Explain why the other two graphs shown are not trees.

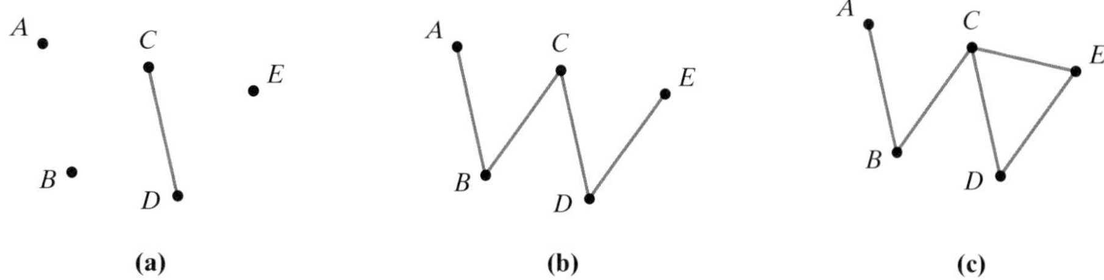

FIGURE 14.50 Identifying a tree

SOLUTION

The graph in **Figure 14.50(b)** is a tree. It is connected and has no circuits. There is only one path joining any two vertices. Every edge is a bridge; if removed, each edge would create a disconnected graph. Finally, the graph has 5 vertices and $5 - 1$, or 4, edges.

The graph in **Figure 14.50(a)** is not a tree because it is disconnected. There are five vertices and only one edge; a tree with five vertices must have four edges.

The graph in **Figure 14.50(c)** is not a tree because it has a circuit, namely *C, D, E, C*. There are five vertices and five edges; a tree with five vertices must have exactly four edges.

CHECK POINT 1 Which graph in **Figure 14.51** is a tree? Explain why the other two graphs shown are not trees.

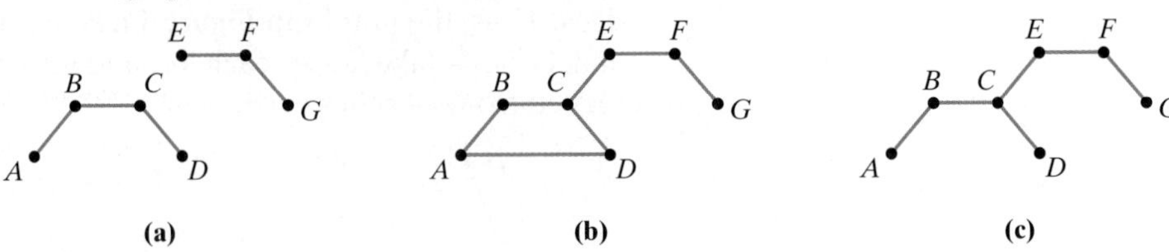

FIGURE 14.51

2 *Find a spanning tree for a connected graph.*

Spanning Trees

One way to increase the efficiency of a network is to remove redundant connections. We are interested in a **subgraph**, meaning a set of vertices and edges chosen from among those of the original graph. **Figure 14.52** illustrates a graph and two possible subgraphs.

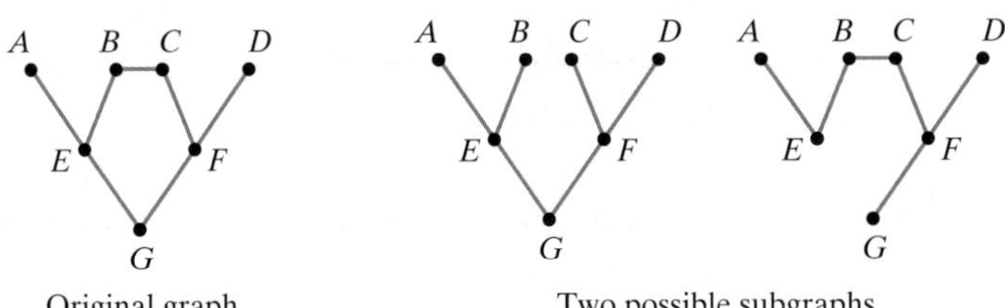

FIGURE 14.52 A graph and two possible subgraphs

The original graph is a connected graph with seven vertices and seven edges. Each subgraph has all seven of the vertices of the connected graph and six of its edges. Although the original graph is not a tree, each subgraph is a tree—that is, each is connected and contains no circuits.

A subgraph that contains all of a connected graph's vertices, is connected, and contains no circuits is called a **spanning tree**. The two subgraphs in **Figure 14.52** are spanning trees for the original graph. By removing redundant connections, the spanning trees increase the efficiency of the network modeled by the original graph.

It is always possible to start with a connected graph, retain all of its vertices, and remove edges until a spanning tree remains. Being a tree, the spanning tree must have one less edge than it has vertices.

EXAMPLE 2 *Finding a Spanning Tree*

Find a spanning tree for the graph in **Figure 14.53**.

SOLUTION

A possible spanning tree must contain all eight vertices shown in the connected graph in **Figure 14.53**. The spanning tree must have one less edge than it has vertices, so it must have seven edges. The graph in **Figure 14.53** has 12 edges, so we need to remove five edges. We break the inner rectangular circuit by removing edge FG. We break the outer rectangular circuit by removing all four of its edges while retaining the edges leading to vertices A, B, C, and D. This leaves us the spanning tree in **Figure 14.54**. Notice that each edge is a bridge and no circuits are present.

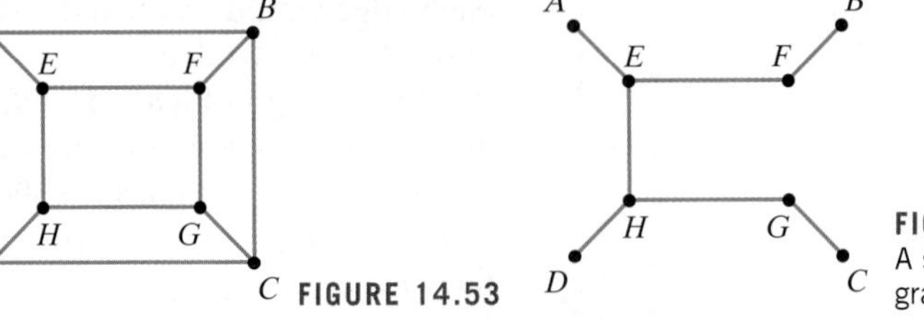

FIGURE 14.53

FIGURE 14.54 A spanning tree for the graph in **Figure 14.53**.

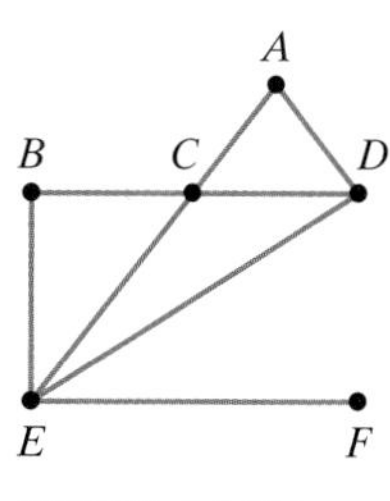

FIGURE 14.55

CHECK POINT 2 Find a spanning tree for the graph in **Figure 14.55**.

GREAT QUESTION!

Can a connected graph have more than one spanning tree?

Yes. Most connected graphs have many possible spanning trees. You can think of each spanning tree as a bare skeleton holding together the connected graph. There are many such skeletons.

3 *Find the minimum spanning tree for a weighted graph.*

Minimum Spanning Trees and Kruskal's Algorithm

Many applied problems involve creating the most efficient network for a weighted graph. The weights often model distances, costs, or time, which we want to minimize. We do this by finding a *minimum spanning tree.*

MINIMUM SPANNING TREES

The **minimum spanning tree** for a weighted graph is a spanning tree with the smallest possible total weight.

Figures 14.56(b) and **(c)** show two spanning trees for the weighted graph in **Figure 14.56(a)**. The total weight for the spanning tree in **Figure 14.56(c)**, 107, is less than that in **Figure 14.56(b)**, 119. Is this the minimum spanning tree, or should we continue to explore other possible spanning trees whose total weight might be less than 107?

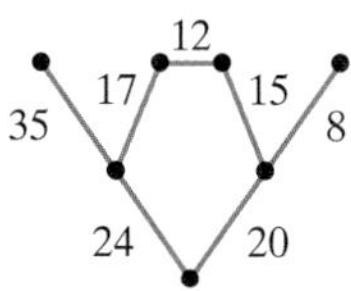

(a) Original weighted graph

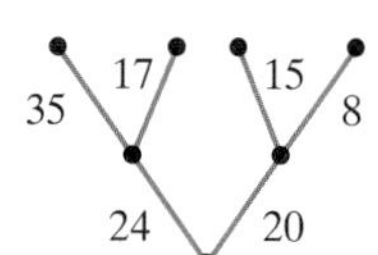

(b) A spanning tree with weight $35 + 24 + 20 + 8 + 17 + 15 = 119$

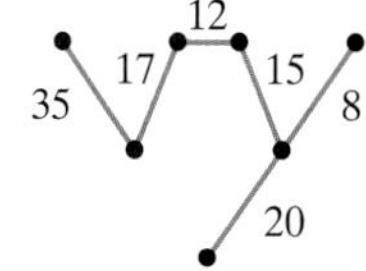

(c) A spanning tree with weight $35 + 17 + 12 + 15 + 20 + 8 = 107$

FIGURE 14.56 A weighted graph and two spanning trees with their weights.

A very simple graph can have many spanning trees. Finding the minimum spanning tree by finding all possible spanning trees and comparing their weights would be too time-consuming. In 1956, the American mathematician Joseph Kruskal discovered a procedure that will always yield a minimum spanning tree for a weighted graph. The basic idea in **Kruskal's Algorithm** is to always pick the edge with the smallest available weight, but avoid creating any circuits.

KRUSKAL'S ALGORITHM

Here is a procedure for finding the minimum spanning tree from a weighted graph:

1. Find the edge with the smallest weight in the graph. If there is more than one, pick one at random. Mark it in red (or using any other designation).
2. Find the next-smallest edge in the graph. If there is more than one, pick one at random. Mark it in red.
3. Find the next-smallest unmarked edge in the graph that does not create a red circuit. If there is more than one, pick one at random. Mark it in red.
4. Repeat step 3 until all vertices have been included. The red edges are the desired minimum spanning tree.

EXAMPLE 3 Using Kruskal's Algorithm

Seven buildings on a college campus are connected by the sidewalks shown in **Figure 14.57**. The weighted graph in **Figure 14.58** represents buildings as vertices, sidewalks as edges, and sidewalk lengths as weights.

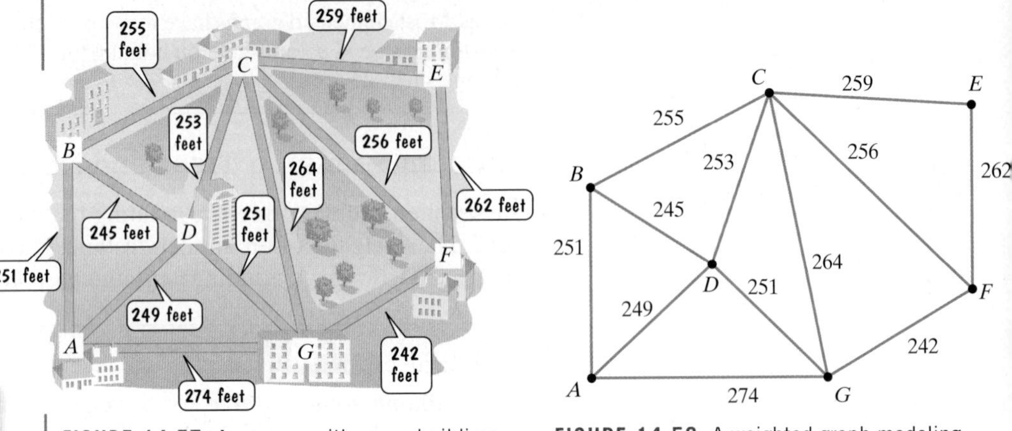

FIGURE 14.57 A campus with seven buildings and connecting sidewalks.

FIGURE 14.58 A weighted graph modeling the campus in **Figure 14.57**.

A heavy snow has fallen and the sidewalks need to be cleared quickly. Campus services decides to clear as little as possible and still ensure that students walking from building to building will be able to do so along cleared paths. Determine the shortest series of sidewalks to clear. What is the total length of the sidewalks that need to be cleared?

SOLUTION

Campus services wants to keep the total length of cleared sidewalks to a minimum and still have a cleared path connecting any two buildings. Thus, they are seeking a minimum spanning tree for the weighted graph in **Figure 14.58**. We find this minimum spanning tree using Kruskal's Algorithm. Refer to **Figure 14.59** as you read the steps in the algorithm.

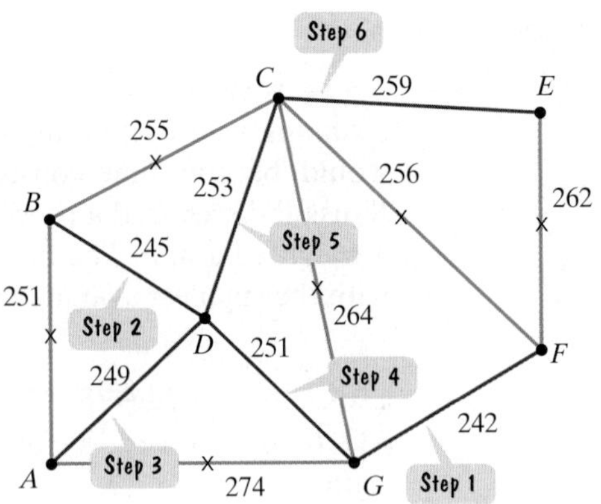

FIGURE 14.59 Finding a minimum spanning tree.

Step 1 Find the edge with the smallest weight. Select edge GF (length: 242 feet) by marking it in red.

Step 2 Find the next-smallest edge in the graph. Select edge BD (length: 245 feet) by marking it in red.

Step 3 Find the next-smallest edge in the graph. Select edge AD (length: 249 feet) by marking it in red.

Step 4 Find the next-smallest edge in the graph that does not create a circuit. The next-smallest edges are AB and DG (length of each: 251 feet). Do not select AB—it creates a circuit. Select edge DG by marking it in red.

Blitzer Bonus

Beyond Following Roads and Sidewalks

This stamp, issued by the Japanese post office, celebrates the construction of a trans-Pacific fiber-optic cable line linking Japan, Guam, and Hawaii. Can you see how finding the shortest distance connecting the three islands is different than using minimum spanning trees to provide optimal routes? Connections along minimum spanning trees must lie along prescribed roads or sidewalks. In building many fiber-optic cable systems, pipelines, and high-speed rail systems, as well as in the design of computer chips, there are no connecting roads. Graph theory can be used to solve problems involving the shortest network connecting three or more points when there are no roads or sidewalks joining these points. The graph shown on the stamp indicates the network the telephone companies used to link the three islands with the least amount of cable.

Step 5 Find the next-smallest edge in the graph that does not create a circuit. Select edge CD (length: 253 feet) by marking it in red. Notice that this does not create a circuit.

Step 6 Find the next-smallest edge in the graph that does not create a circuit. The next-smallest edge is BC (length: 255 feet), but this creates a circuit. Discard BC. The next-smallest edge is CF (length: 256 feet), but this also creates a circuit. Discard CF. The next-smallest edge is CE (length: 259 feet). This does not create a circuit, so select edge CE by marking it in red.

Can you see that the minimum spanning tree in **Figure 14.59** is completed? The red subgraph contains all of the graph's seven vertices, is connected, contains no circuits, and has $7 - 1$, or 6, edges. Therefore, the red subgraph in **Figure 14.59** shows the shortest series of sidewalks to clear. From the figure, we see that there are

$$242 + 245 + 249 + 251 + 253 + 259,$$

or 1499 feet of sidewalks that need to be cleared.

CHECK POINT 3 Use Kruskal's Algorithm to find the minimum spanning tree for the graph in **Figure 14.60**. Give the total weight of the minimum spanning tree.

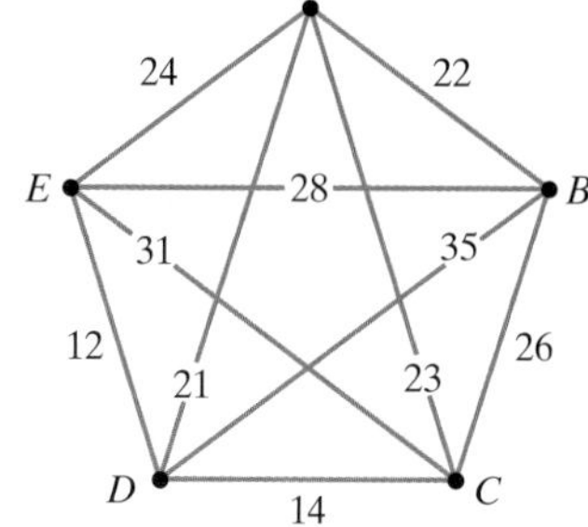

FIGURE 14.60

Blitzer Bonus

A Family Tree: The Sopranos

The Sopranos: Only on cable could they have gotten away with this violent and satiric profile of an angst-ridden mafia family man and his dysfunctional domestic clan. **Figure 14.61** shows three generations of HBO's Soprano family. **Figure 14.62** models this information with a tree. The vertices represent some of the pictured people. The edges represent parent-child relationships. Genealogists use trees like the one in **Figure 14.62** to show the relationship between members of different generations of a family.

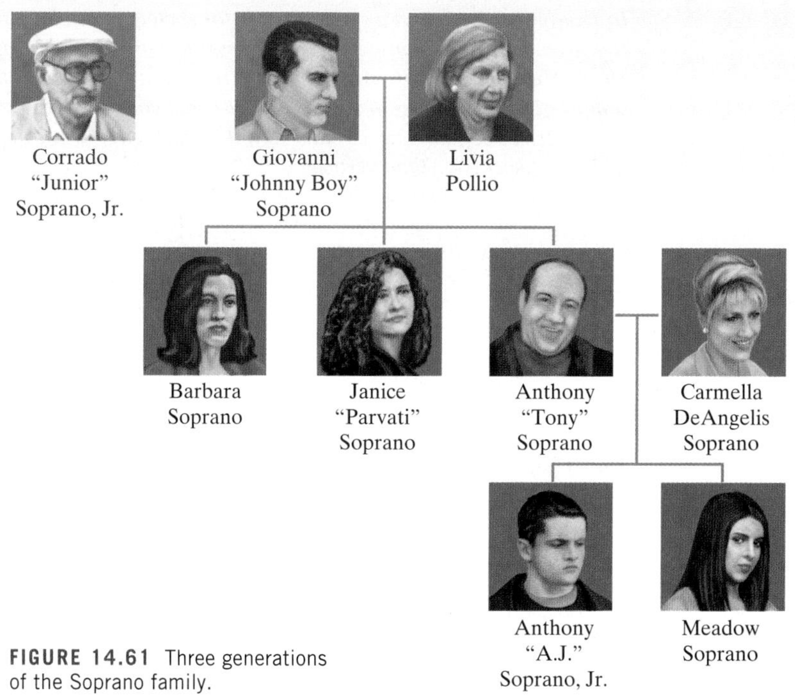

FIGURE 14.61 Three generations of the Soprano family.

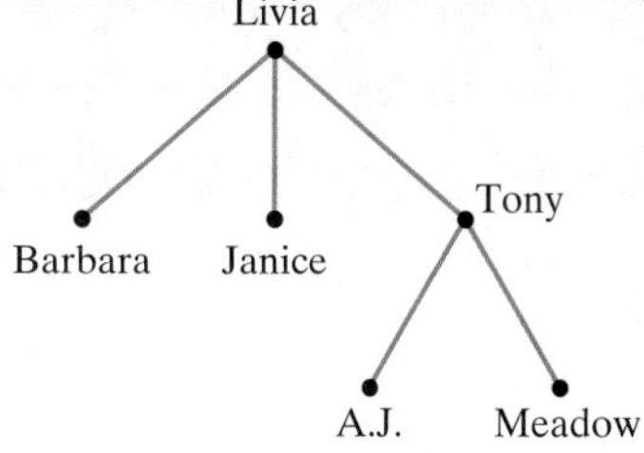

FIGURE 14.62 A tree modeling Soprano parent-child relationships.

Concept and Vocabulary Check

Fill in each blank so that the resulting statement is true.

1. A graph that is connected and has no circuits is called a/an ______. For such a graph,
   - every edge is a/an ________.
   - if there are n vertices, there must be ________ edges.
2. A tree that is created from another connected graph and that contains all of the connected graph's vertices, is connected, and contains no circuits is called a/an ________ tree.
3. A tree that is created from a weighted graph and that has the smallest possible weight is called the ________________ tree.
4. A procedure that yields the tree in Exercise 3 is called ____________ Algorithm. The idea of the algorithm is to always pick the edge with the smallest available ________, but avoid creating any ________.
5. True or False: A tree is a complete graph. ______
6. True or False: Most connected graphs have many possible spanning trees. ______

Exercise Set 14.4

Practice Exercises

In Exercises 1–10, determine whether each graph is a tree. If the graph is not a tree, give a reason why.

1. A B C D

2. A B C D E F

3.

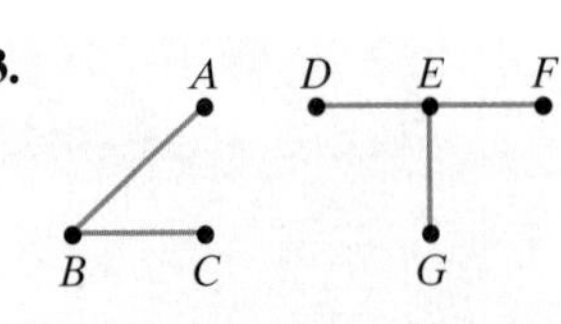

4. A B C D E F

5.

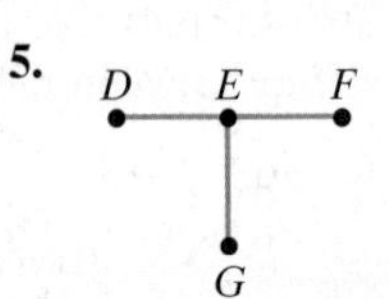

6. A B C D

7.

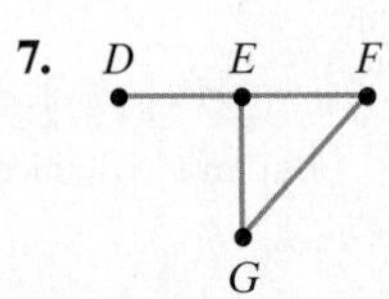

8. A B C D

9.

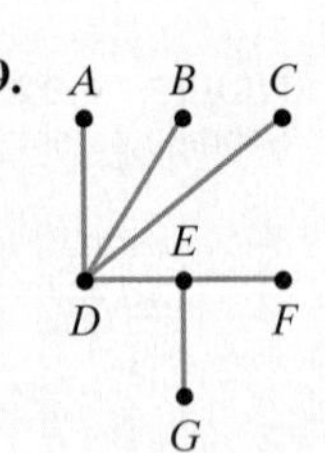

10.

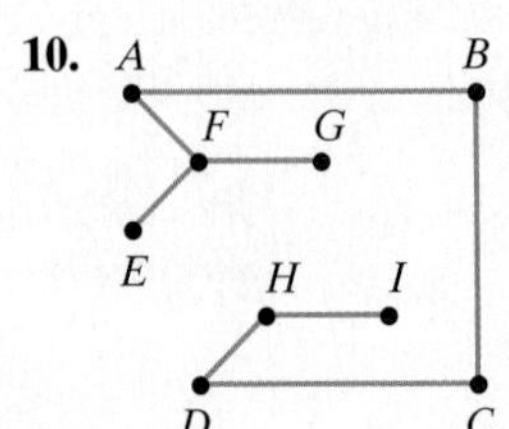

In Exercises 11–16, a graph with no loops or more than one edge between any two vertices is described. Which one of the following applies to the description?

i. *The described graph is a tree.*
ii. *The described graph is not a tree.*
iii. *The described graph may or may not be a tree.*

11. The graph has five vertices, and there is exactly one path from any vertex to any other vertex.
12. The graph has five vertices, is connected, and every edge is a bridge.
13. The graph has eight vertices and five edges.
14. The graph has nine vertices and six edges.
15. The graph has five vertices and four edges.
16. The graph has four vertices and three edges.

In Exercises 17–22, find a spanning tree for each connected graph. Because many spanning trees are possible, answers will vary. Do not be concerned if your spanning tree is not the same as the sample given in the answer section. Be sure that your spanning tree is connected, contains no circuits, and that each edge is a bridge, with one fewer edge than vertices.

17. A B C D E F

18.

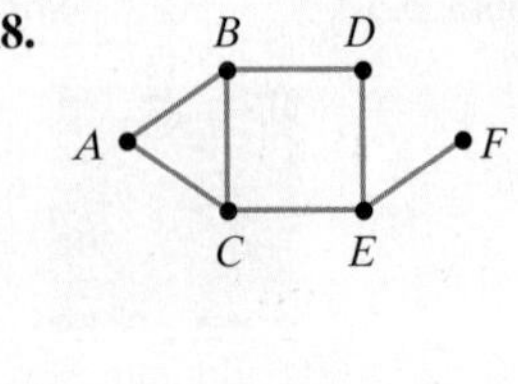

19. A E C D B F

20.

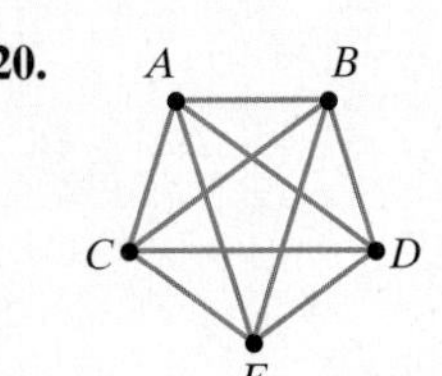

21.

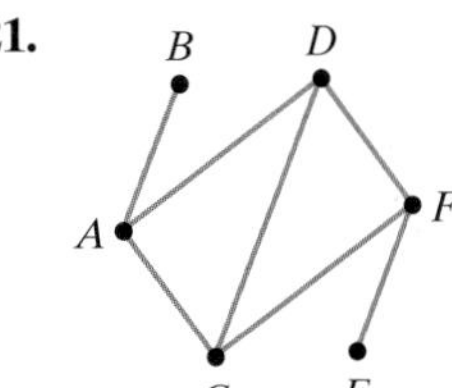

22.

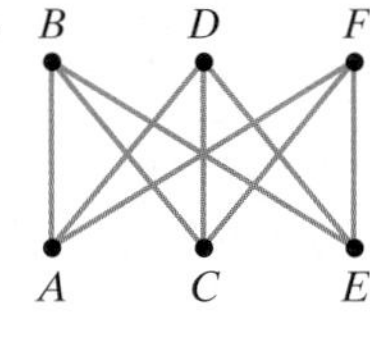

In Exercises 23–30, use Kruskal's Algorithm to find the minimum spanning tree for each weighted graph. Give the total weight of the minimum spanning tree.

23.

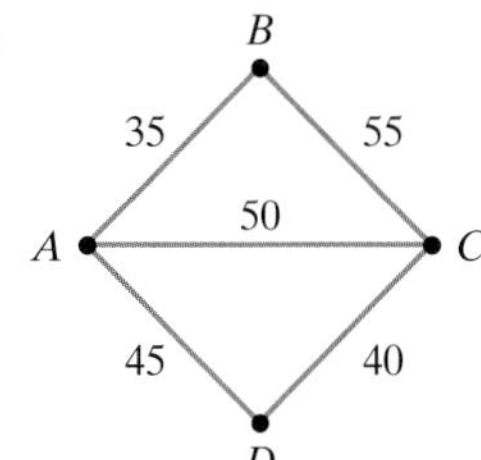

24.

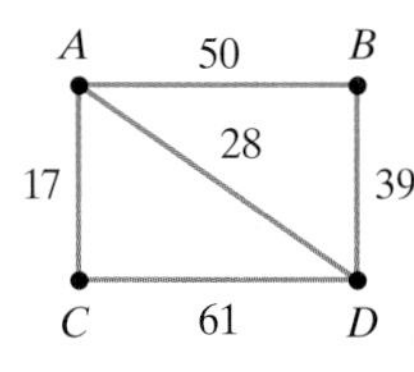

25.

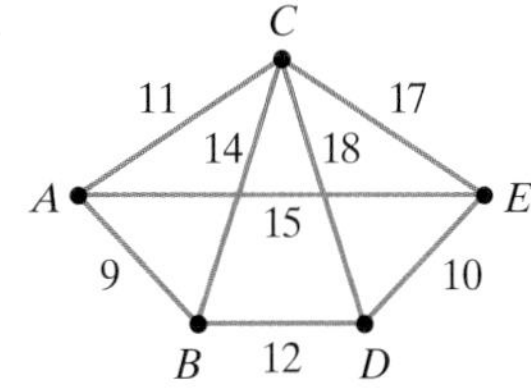

26.

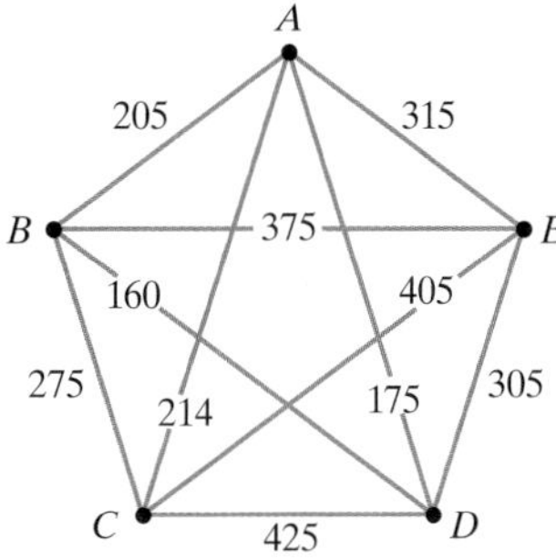

27.

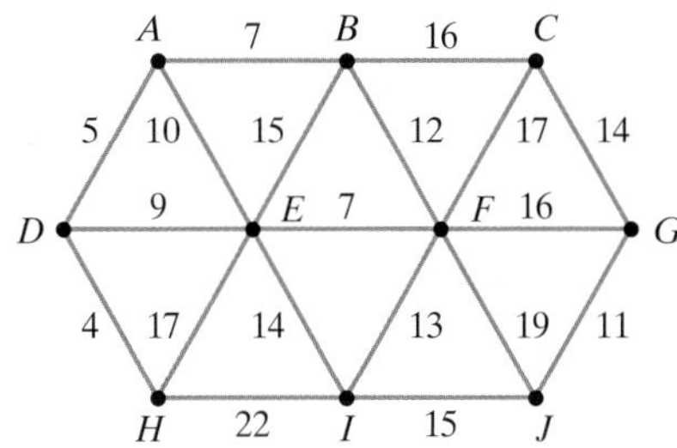

28.

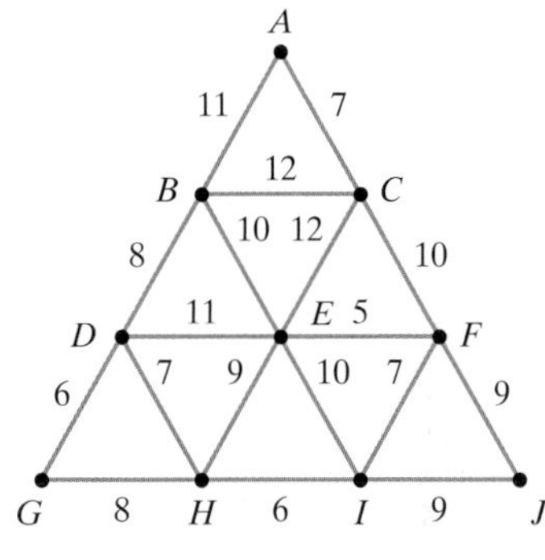

29.

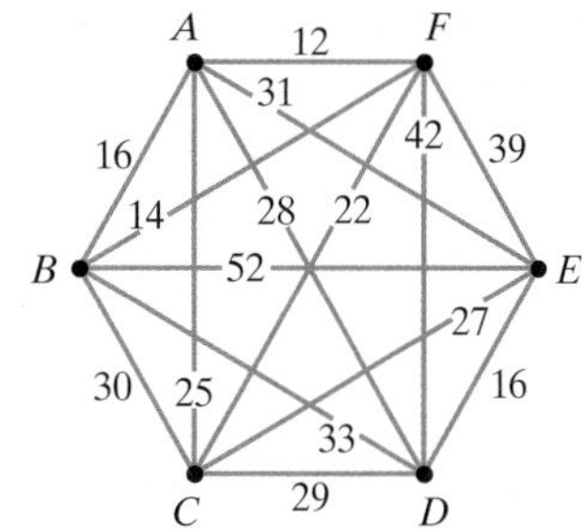

30.

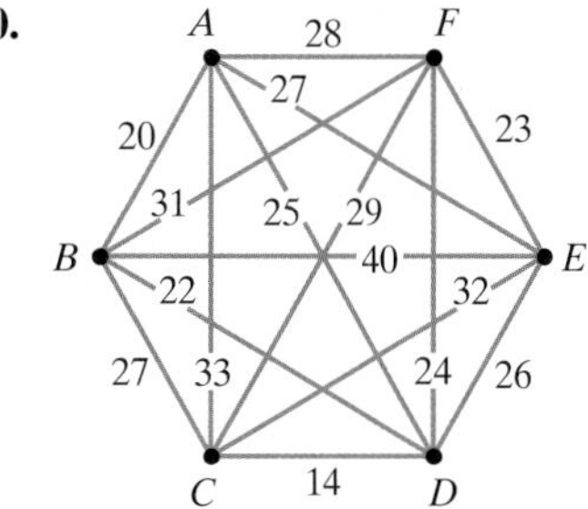

Practice Plus

In Exercises 31–32, find four spanning trees for each connected graph.

31.

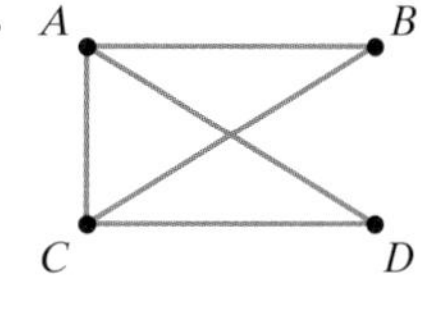

32.

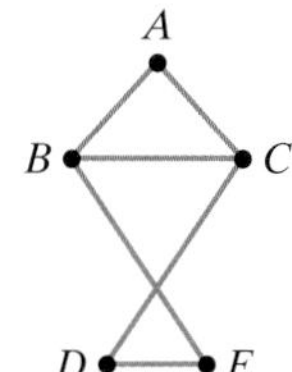

In Exercises 33–36, use a modification of Kruskal's Algorithm to find the maximum spanning tree for each weighted graph. Give the total weight of the maximum spanning tree.

33. Use the graph in Exercise 25.

34. Use the graph in Exercise 26.

35. Use the graph in Exercise 27.

36. Use the graph in Exercise 28.

Application Exercises

37. Use a tree to model the parent-child relationships in the following family:

> Peter has three children: Zoila, Keanu, and Sandra. Zoila has two children: Sean and Helen. Keanu has no children. Sandra has one child: Martin.

Use vertices to model the people and edges to represent the parent-child relationships.

38. Use a tree to model the employee relationships among the chief administrators of a large community college system:

> Three campus vice presidents report directly to the college president. On two campuses, the academic dean, the dean for administration, and the dean of student services report directly to the vice president. On the third campus, only the academic dean and the dean for administration report directly to the vice president.

39. A college campus plans to provide awnings above its sidewalks to shelter students from the rain as they walk from the parking lot and between buildings. To save money, awnings will not be placed over all of the sidewalks shown in the figure. Just enough awnings will be placed over a select number of sidewalks to ensure that students walking from building to building will be able to do so without getting wet.

a. Use a weighted graph to model the given map. Represent buildings as vertices, sidewalks as edges, and sidewalk lengths as weights.

b. Use Kruskal's Algorithm to find a minimum spanning tree that allows students to move between the parking lot and any buildings shown without getting wet. What is the total length of the sidewalks that need to be sheltered by awnings?

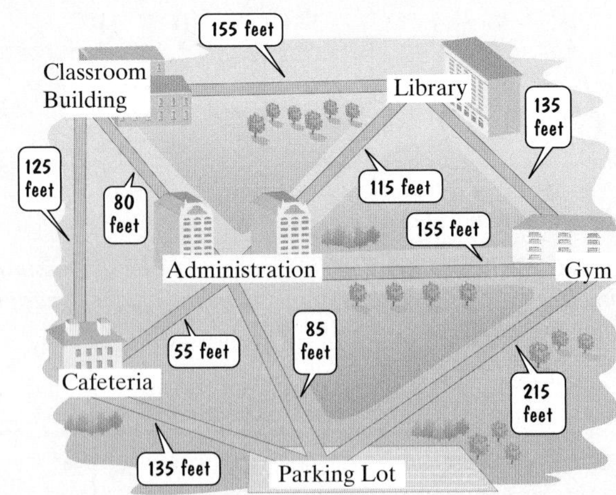

40. A fiber-optic cable system is to be installed along highways connecting six cities. The cities include Boston, Cincinnati, Cleveland, Detroit, New York, and Philadelphia. The highway distances, in miles, between the cities are given in the following table.

a. Model this information with a weighted, complete graph.

b. Use Kruskal's Algorithm to find a minimum spanning tree that would connect each city using the smallest amount of cable. Determine the total length of cable needed.

	Boston	Cincinnati	Cleveland	Detroit	New York	Philadelphia
Boston	*	840	628	734	222	296
Cincinnati	840	*	244	269	647	567
Cleveland	628	244	*	170	473	413
Detroit	734	269	170	*	641	573
New York	222	647	473	641	*	101
Philadelphia	296	567	413	573	101	*

41. The graph shows a proposed layout of an irrigation system. The main water source is located at vertex C. The other vertices represent the proposed sprinkler head locations. The edges indicate all possible choices for installing underground pipes. The weights show the distances, in feet, between proposed sprinkler head locations. The sprinkler system will work if each sprinkler head is connected to the main water source through one or more underground pipes. Use Kruskal's Algorithm to determine the layout of the irrigation system and the smallest number of feet of underground pipes.

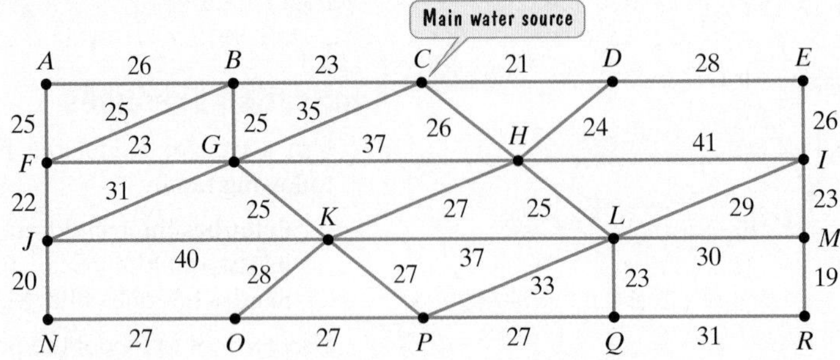

42. The graph shows a proposed layout of a bike trail system to be installed between the towns shown. The vertices represent the ten towns, designated A through J. The edges indicate all possible choices for building the trail. The weights show the distances, in miles, between bike trailheads connecting the towns. Use Kruskal's Algorithm to determine the minimum mileage for the bike trail and the layout of the trail system.

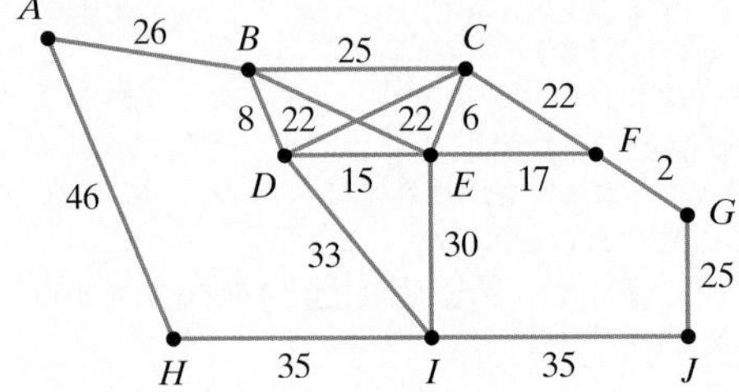

Explaining the Concepts

43. What is a tree?
44. If a graph is given, how do you determine whether or not it is a tree?
45. Describe the relationship between the number of vertices and the number of edges in a tree.
46. What is a subgraph?
47. What is a spanning tree?
48. Describe how to obtain a spanning tree for a connected graph.
49. What is the minimum spanning tree for a weighted graph?
50. In your own words, briefly describe how to find the minimum spanning tree using Kruskal's Algorithm.
51. Describe a practical problem that can be solved using Kruskal's Algorithm.

Critical Thinking Exercises

Make Sense? *In Exercises 52–55, determine whether each statement makes sense or does not make sense, and explain your reasoning.*

52. The tree that I've drawn has the same number of vertices and edges.
53. Starting from my home, I need to go to the post office, the bank, the medical clinic, and the dry cleaners. Because my goal is to run errands and return home using the shortest route, I need to determine the minimum spanning tree.
54. Using the shortest possible line length to connect ten towns with telephone lines involves finding the minimum spanning tree.
55. Although the Nearest Neighbor Method does not always give the Hamilton circuit for which the sum of the weights is a minimum, Kruskal's Algorithm always gives the spanning tree with the smallest possible total weight.
56. Find all the possible spanning trees for the given graph.

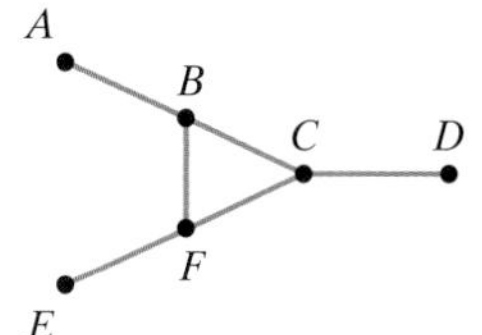

57. Give an example of a tree with six vertices whose degrees are 1, 1, 2, 2, 2, and 2.

Group Exercise

58. Group members should determine a project whose installation would enhance the quality of life on campus or in your community. For example, the project might involve installing awnings over campus sidewalks, building a community bike path, creating a community hiking trail, or installing a metrorail system providing easy access to your community's most desirable locations. The project that you determine should be one that can be carried out most efficiently using a minimum spanning tree. Begin by defining the project and its locations (vertices). Group members should then research the distances between the various locations in the project. Once these distances have been determined, the group should reassemble and create a graph that models the project. Then find a minimum spanning tree that serves as the most efficient way to carry out your project.

Chapter Summary, Review, and Test

SUMMARY – DEFINITIONS AND CONCEPTS	EXAMPLES
14.1 Graphs, Paths, and Circuits	
a. A graph consists of a finite set of points, called vertices, and line segments or curves, called edges, that start and end at vertices. A loop is an edge that starts and ends at the same vertex. Equivalent graphs have the same number of vertices connected to each other in the same way.	Ex. 1, p. 899; Ex. 2, p. 899; Ex. 3, p. 900; Ex. 4, p. 901; Ex. 5, p. 902
b. The Vocabulary of Graph Theory	
1. The degree of a vertex is the number of edges at that vertex. A loop connecting a vertex to itself contributes 2 to the degree of that vertex.	
2. A vertex with an even number of edges attached is an even vertex; one with an odd number of edges attached is an odd vertex.	
3. Two vertices are adjacent if there is at least one edge connecting them.	Ex. 6, p. 903
4. A path in a graph is a sequence of adjacent vertices and the edges connecting them. An edge can be part of a path only once.	Figure 14.18, p. 904
5. A circuit is a path that begins and ends at the same vertex.	Figure 14.19, p. 904
6. A graph is connected if for any two of its vertices there is at least one path connecting them. A connected graph consists of one piece. A graph that is not connected, made up of components, is called disconnected.	Figure 14.20, p. 904
7. A bridge is an edge that if removed from a connected graph would leave behind a disconnected graph.	

14.2 Euler Paths and Euler Circuits

a. An Euler path is a path that travels through every edge of a graph once and only once.	Figure 14.21, p. 909
b. An Euler circuit is a circuit that travels through every edge of a graph once and only once.	Figure 14.22, p. 909
c. Euler's Theorem determines if a graph contains Euler paths or Euler circuits. • A graph with exactly two odd vertices has at least one Euler path, but no Euler circuit. Each Euler path starts at one odd vertex and ends at the other odd vertex. • A graph with no odd vertices has at least one Euler circuit that can start and end at any vertex. • A graph with more than two odd vertices has no Euler paths and no Euler circuits.	Ex. 1, p. 910; Ex. 2, p. 911; Ex. 3, p. 912
d. If Euler's Theorem indicates the existence of Euler paths or Euler circuits, trial and error or Fleury's Algorithm can be used to find such paths or circuits. Fleury's Algorithm is explained in the box on page 913.	Ex. 4, p. 914

14.3 Hamilton Paths and Hamilton Circuits

a. A path that passes through each vertex of a graph exactly once is called a Hamilton path. If a Hamilton path begins and ends at the same vertex and passes through all other vertices exactly once, it is called a Hamilton circuit.	Ex. 1, p. 921
b. A complete graph is a graph that has an edge between each pair of its vertices. A complete graph with n vertices has $(n-1)!$ Hamilton circuits.	Ex. 2, p. 922
c. A graph whose edges have numbers, called weights, attached to them is called a weighted graph.	Ex. 3, p. 923
d. The traveling salesperson problem is the problem of finding a Hamilton circuit in a complete, weighted graph for which the sum of the weights of the edges is a minimum. Such a Hamilton circuit is called the optimal Hamilton circuit or the optimal solution.	
e. The Brute Force Method can be used to solve the traveling salesperson problem. The method, given in the box on page 924, involves determining the sum of the weights of the edges for all possible Hamilton circuits.	Ex. 4, p. 924
f. The Nearest Neighbor Method can be used to approximate the optimal solution to traveling salesperson problems. The method, given in the box on page 926, involves continually taking an edge with the smallest weight.	Ex. 5, p. 926

14.4 Trees

a. A tree is a graph that is connected and has no circuits.	Ex. 1, p. 931
b. Properties of Trees **1.** There is one and only one path joining any two vertices. **2.** Every edge is a bridge. **3.** A tree with n vertices has $n-1$ edges.	
c. A subgraph is a set of vertices and edges chosen from among those of the original graph.	
d. A spanning tree is a subgraph that contains all of a connected graph's vertices, is connected, and contains no circuits.	Ex. 2, p. 932
e. The minimum spanning tree for a weighted graph is a spanning tree with the smallest possible total weight.	
f. Kruskal's Algorithm, described in the box at the bottom of page 933, gives a procedure for finding the minimum spanning tree from a weighted graph. The basic idea is to always pick the smallest available edge, but avoid creating any circuits.	Ex. 3, p. 934

Review Exercises

14.1

1. Explain why the two figures show equivalent graphs. Then draw a third equivalent graph.

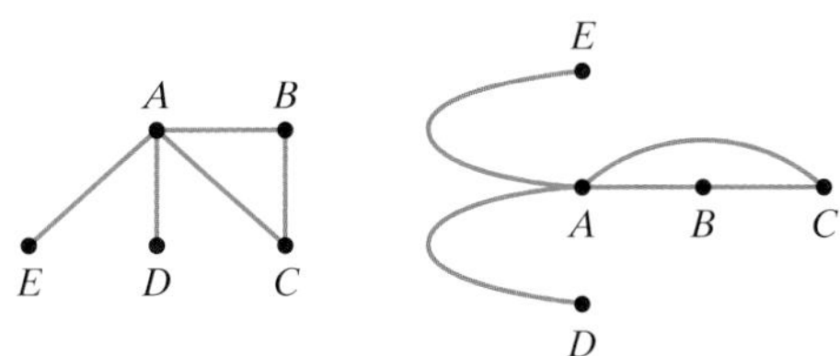

In Exercises 2–8, use the following graph.

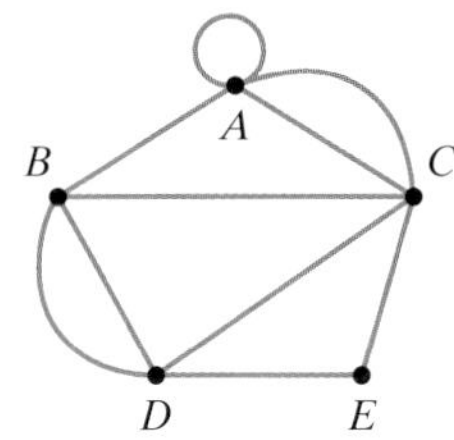

2. Find the degree of each vertex in the graph.
3. Identify the even vertices and identify the odd vertices.
4. Which vertices are adjacent to vertex D?
5. Use vertices to describe two paths that start at vertex E and end at vertex A.
6. Use vertices to describe a circuit that begins and ends at vertex E.
7. Is the graph connected? Explain your answer.
8. Is any edge in the graph a bridge? Explain your answer.
9. List all edges that are bridges in the graph shown.

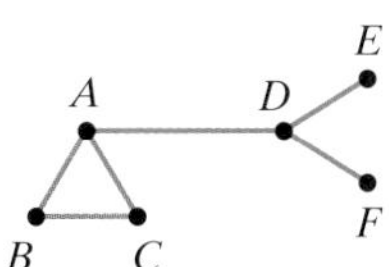

10. Draw a graph that models the layout of the city shown in the map. Use vertices to represent the land masses and edges to represent the bridges.

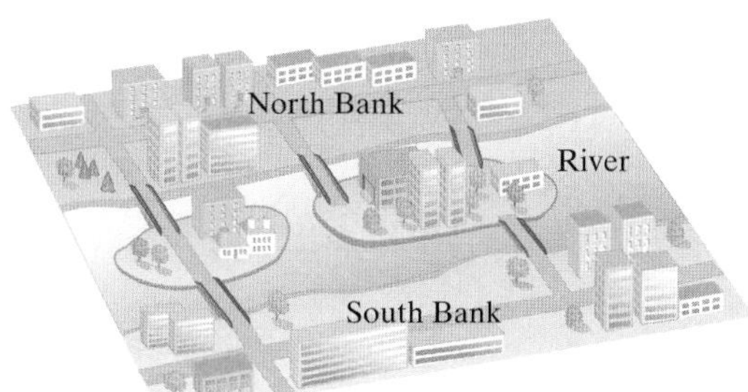

11. Draw a graph that models the bordering relationships among the states shown in the map. Use vertices to represent the states and edges to represent common borders.

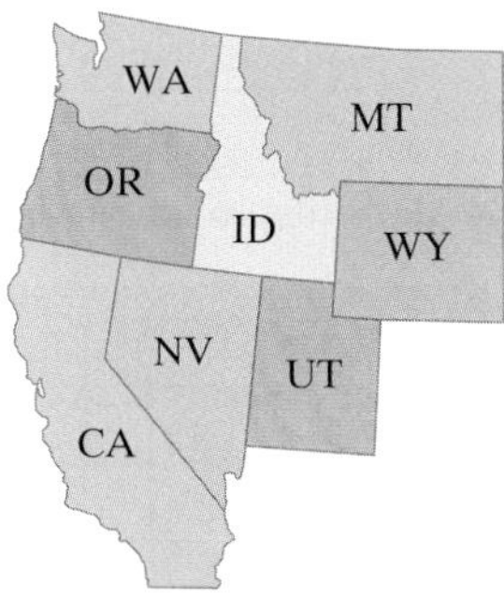

12. Draw a graph that models the connecting relationships in the floor plan. Use vertices to represent the rooms and the outside, and edges to represent the connecting doors.

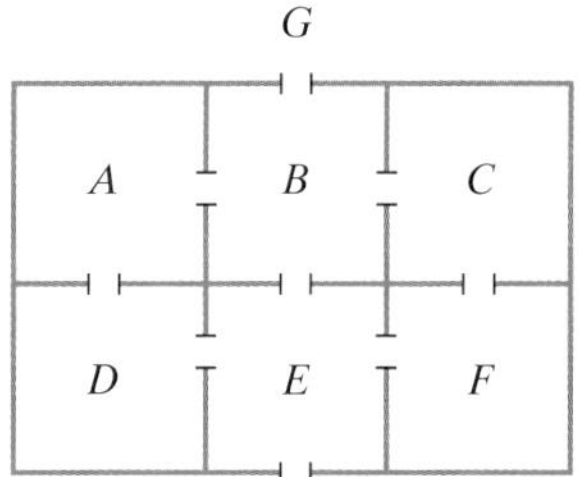

14.2

In Exercises 13–15, a graph is given.

a. *Determine whether the graph has an Euler path, an Euler circuit, or neither.*

b. *If the graph has an Euler path or circuit, use trial and error or Fleury's Algorithm to find one.*

13.

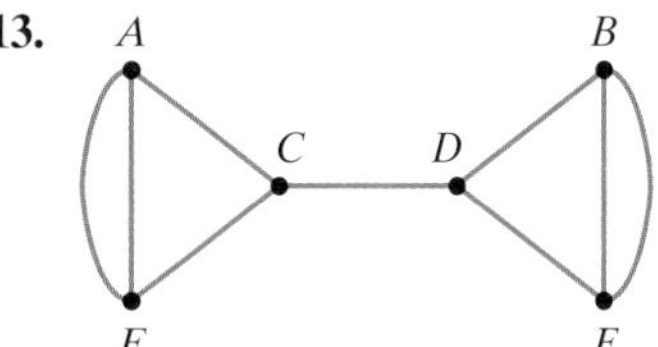

14.

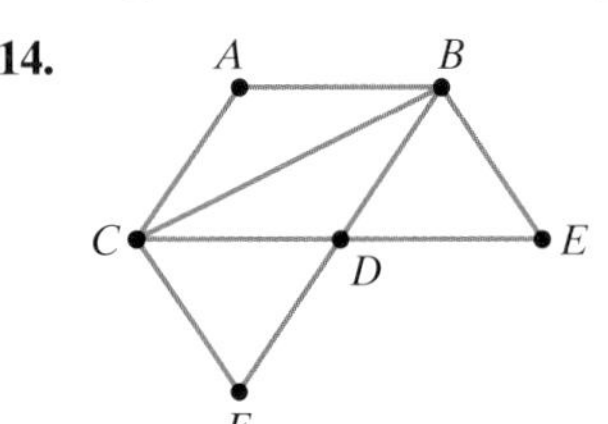

15.

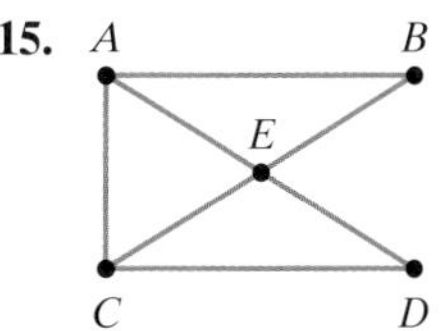

16. Use Fleury's Algorithm to find an Euler path.

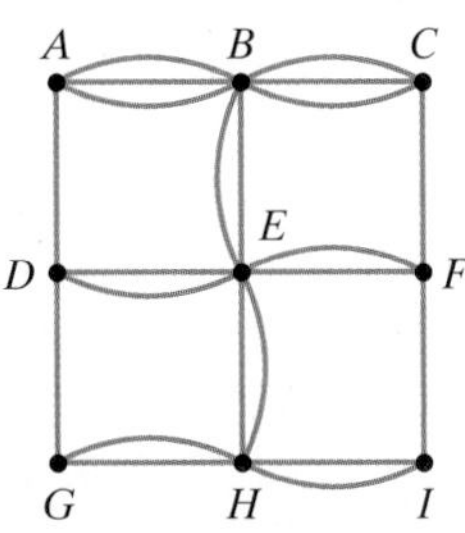

17. Use Fleury's Algorithm to find an Euler circuit.

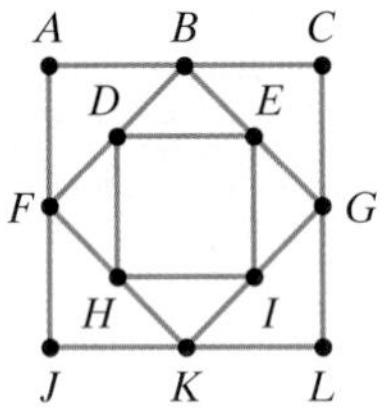

18. Refer to Exercise 10.

a. Use your graph to determine if the city residents would be able to walk across all of the bridges without crossing the same bridge twice.

b. If such a walk is possible, show the path on your graph. Then trace this route on the city map in a manner that is clear to the city's residents.

c. Use your graph to determine if there is a path that crosses each bridge exactly once and begins and ends on the same island. Explain your answer.

19. Refer to Exercise 11. Use your graph to determine if it is possible to find a path that crosses each common state border exactly once. Explain your answer.

20. Refer to Exercise 12.

a. Use your graph to determine if it is possible to find a path that uses each door exactly once.

b. If such a path is possible, show it on your graph. Then trace this route on the floor plan in a manner that is clear to a person strolling through the house.

21. A security guard needs to walk the streets of the neighborhood in the figure shown. The guard is to walk down each street once, whether or not the street has houses on both sides.

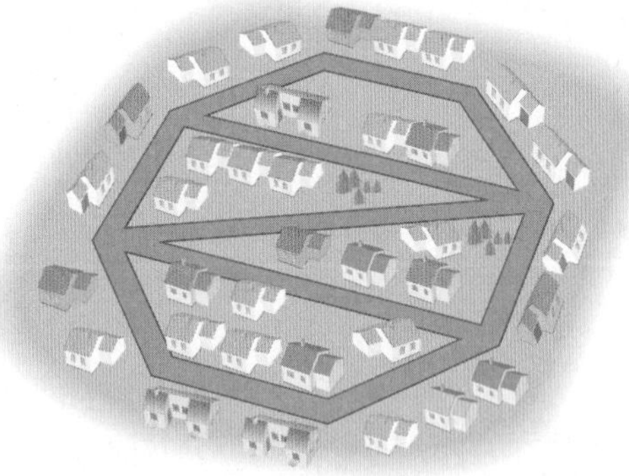

a. Draw a graph that models the streets of the neighborhood walked by the security guard.

b. Use your graph to determine if the security guard can walk each street exactly once. Explain your answer.

c. If so, use the map to show where the guard should begin and where the guard should end the walk.

14.3

In Exercises 22–23, use the graph shown.

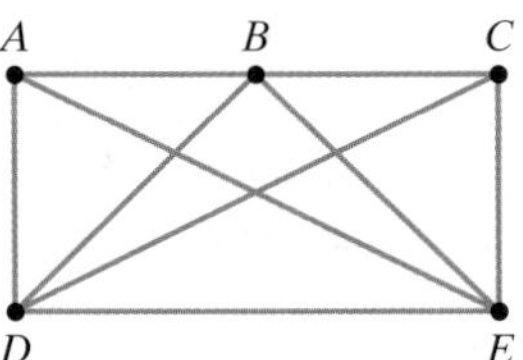

22. Find a Hamilton circuit that begins as $A, E, \ldots$.

23. Find a Hamilton circuit that begins as $D, B, \ldots$.

For each graph in Exercises 24–27,

a. *Determine if the graph must have Hamilton circuits. Explain your answer.*

b. *If the graph must have Hamilton circuits, determine the number of such circuits.*

24.

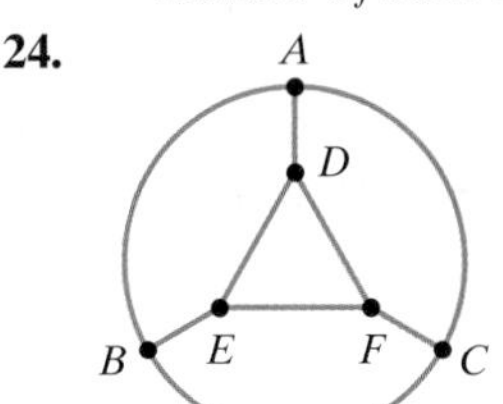

25.

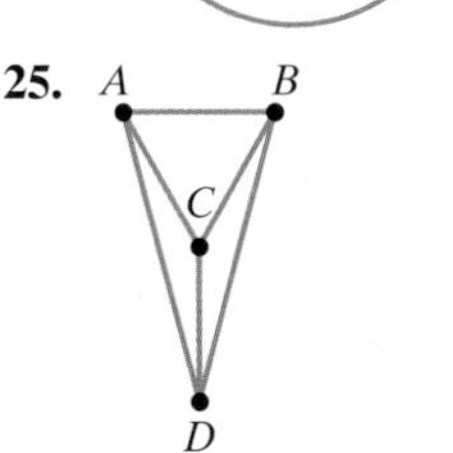

26.

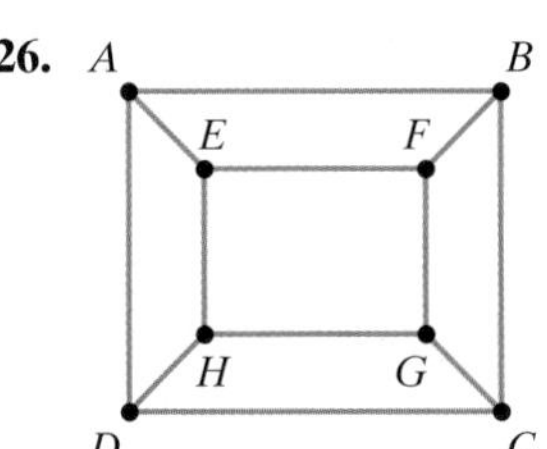

27.

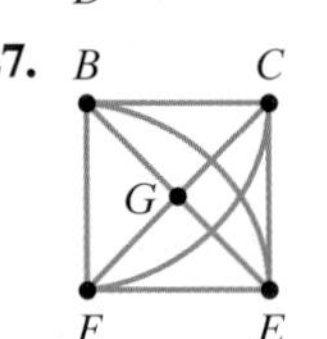

In Exercises 28–30, use the complete, weighted graph shown.

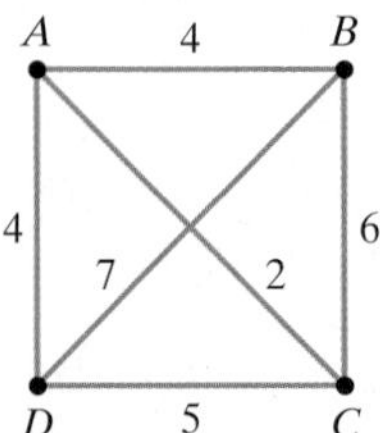

28. Find the total weight of each of the six possible Hamilton circuits:

A, B, C, D, A A, B, D, C, A A, C, B, D, A

A, C, D, B, A A, D, B, C, A A, D, C, B, A.

29. Use your answers from Exercise 28 and the Brute Force Method to find the optimal solution.

30. Use the Nearest Neighbor Method, with starting vertex A, to find an approximate solution. What is the total weight of the Hamilton circuit?

31. Use the Nearest Neighbor Method to find a Hamilton circuit that begins at vertex A in the given graph. What is the total weight of the Hamilton circuit?

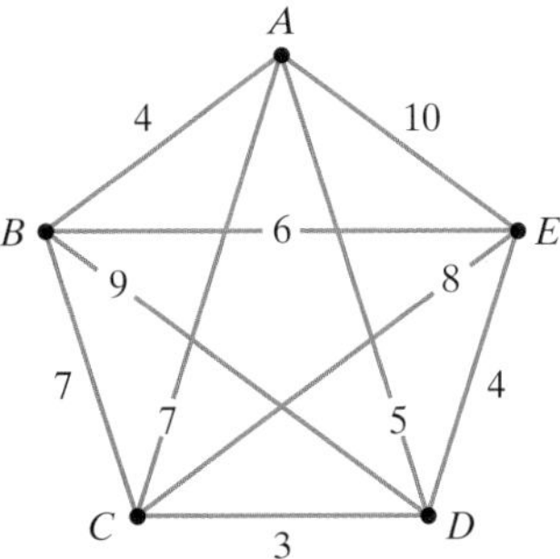

The table shows the one-way airfares between cities. Use this information to solve Exercises 32–33.

	A	**B**	**C**	**D**	**E**
A	*	$220	$240	$290	$430
B	$220	*	$260	$320	$360
C	$240	$260	*	$180	$300
D	$290	$320	$180	*	$250
E	$430	$360	$300	$250	*

32. Draw a graph that models the information in the table. Use vertices to represent the cities and weights on appropriate edges to show airfares.

33. A sales director who lives in city A is required to fly to regional offices in cities B, C, D, and E, and then return to city A. Use the Nearest Neighbor Method to approximate the optimal route. What is the total cost for this Hamilton circuit?

In Exercises 34–36, determine whether each graph is a tree. If the graph is not a tree, give a reason why.

34.

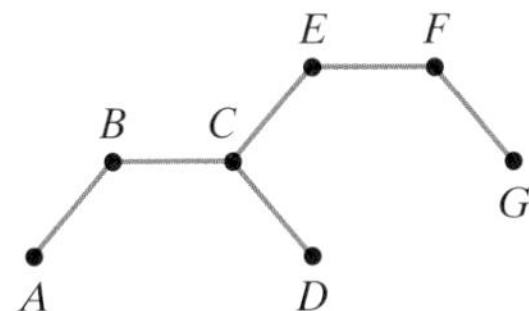

35.

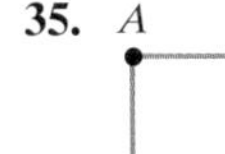

36.

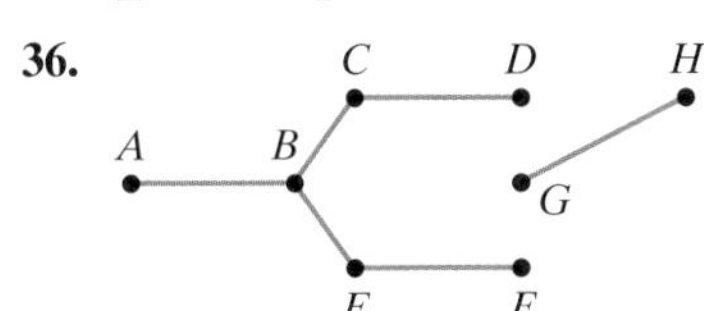

In Exercises 37–38, find a spanning tree for each connected graph.

37.

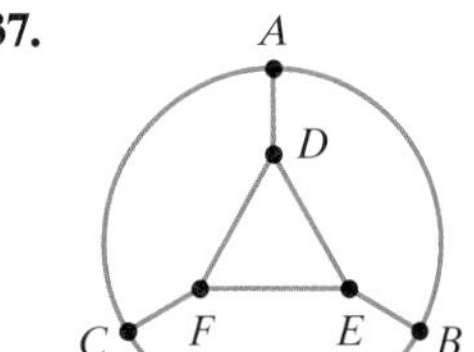

38.

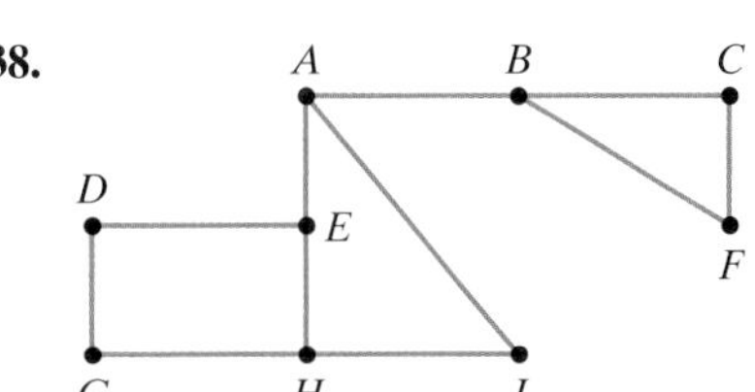

In Exercises 39–40, use Kruskal's Algorithm to find the minimum spanning tree for each weighted graph. Give the total weight of the minimum spanning tree.

39.

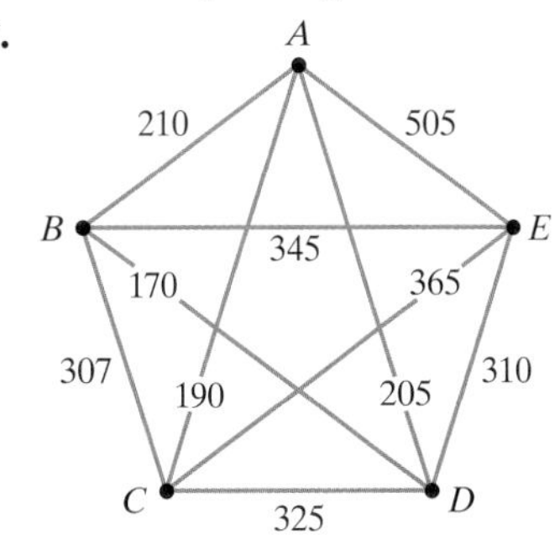

40.

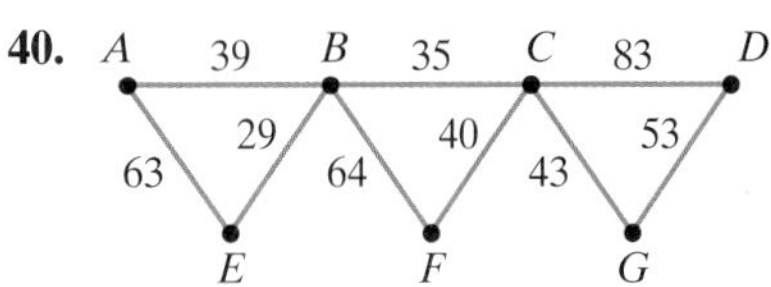

41. A fiber-optic cable system is to be installed along highways connecting nine cities. The weighted graph shows the highway distances, in miles, between the cities, designated A through I. Use Kruskal's Algorithm to determine the layout of the cable system and the smallest length of cable needed.

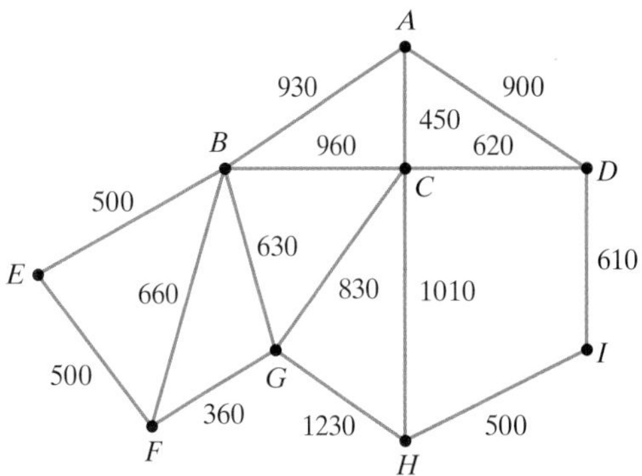

Chapter 14 Test

In Exercises 1–4, use the following graph.

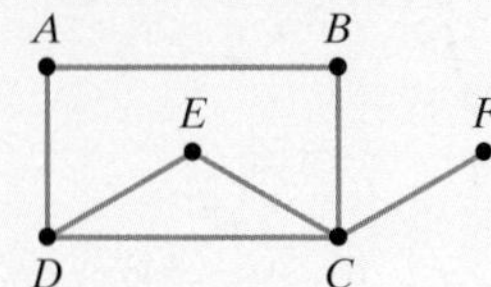

1. Find the degree of each vertex.
2. Use vertices to describe two paths that start at vertex A and end at vertex E.
3. Use vertices to describe a circuit that starts at vertex B and passes through vertex E.
4. List all edges that are bridges.
5. Draw a graph that models the bordering relationships among the countries shown in the map. Use vertices to represent the countries and edges to represent common borders.

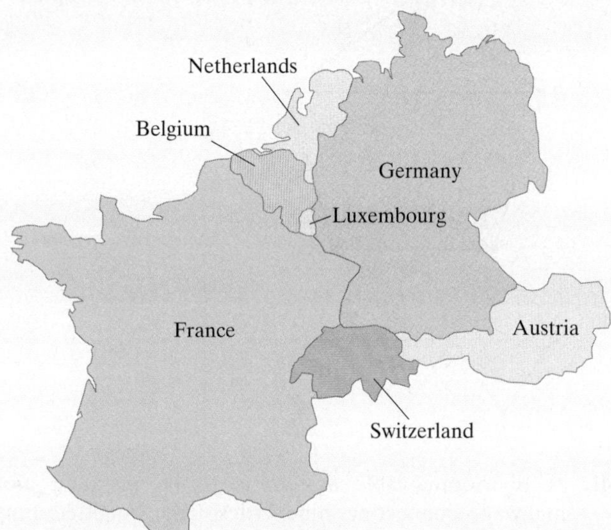

In Exercises 6–8, a graph is given.

a. *Determine whether the graph has an Euler path, an Euler circuit, or neither.*

b. *If the graph has an Euler path or circuit, use trial and error or Fleury's Algorithm to find one.*

6.

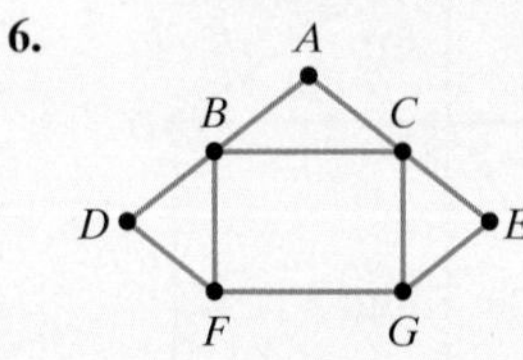

7.

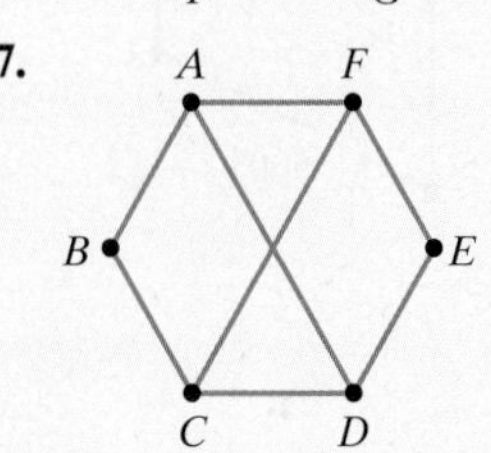

8.

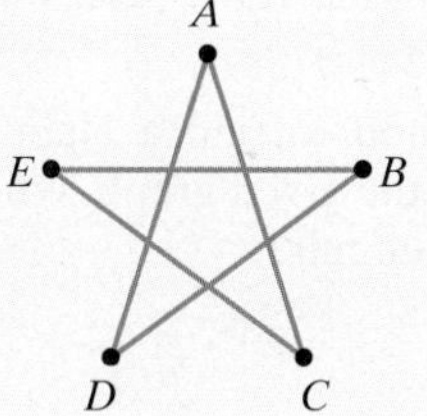

9. Use Fleury's Algorithm to find an Euler circuit.

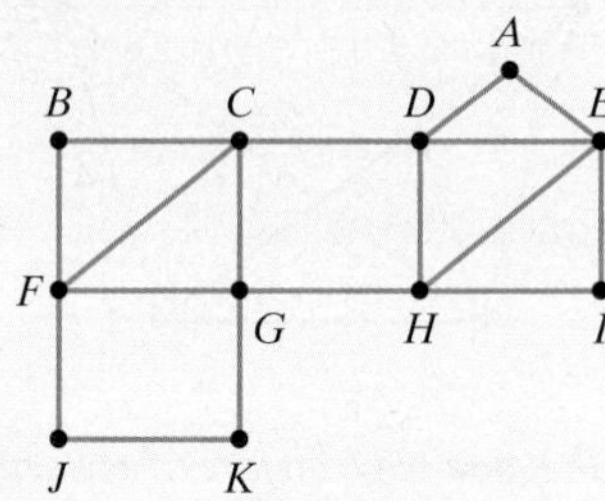

10. **a.** Draw a graph that models the layout of the city shown in the map.

b. Use your graph to determine if the city residents would be able to walk across all of the bridges without crossing the same bridge twice. Explain your answer.

c. If such a walk is possible, where should it begin?

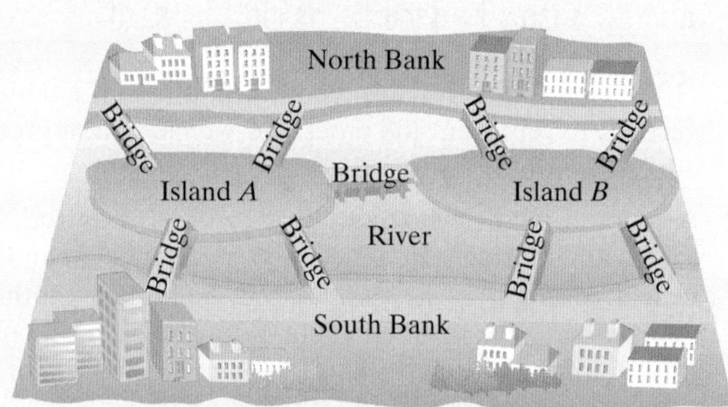

11. **a.** Draw a graph that models the connecting relationships in the floor plan shown below.

b. Use your graph to determine if it is possible to find a path that uses each door exactly once. Explain your answer.

c. If such a path is possible, where should it begin?

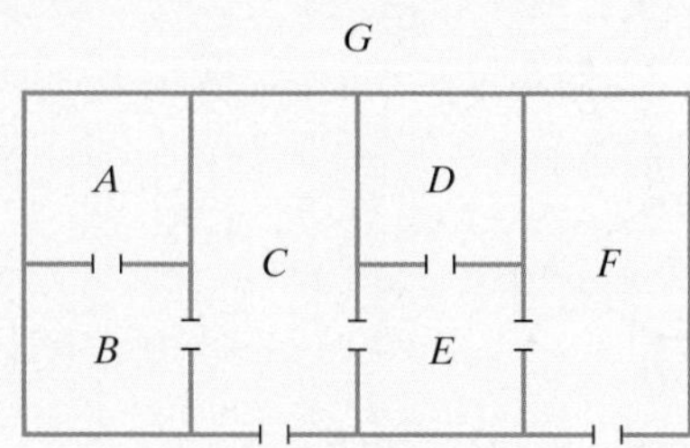

12. a. Draw a graph that models the streets of the neighborhood for a police officer who is to walk each street once.

b. Use your map to determine if the officer can walk each street exactly once. Explain your answer.

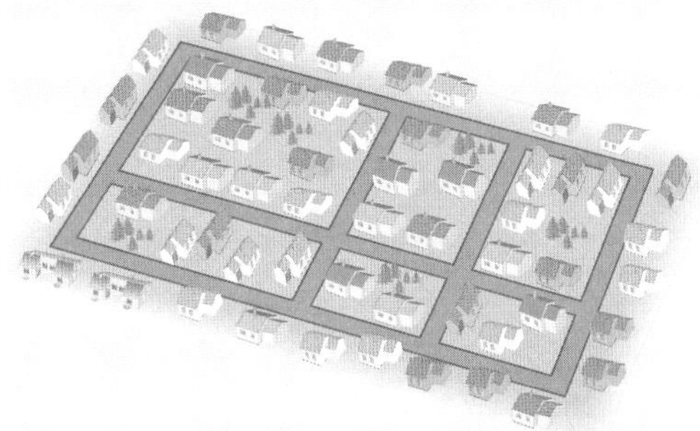

13. Find two Hamilton circuits in the graph shown. One circuit should begin as $A, B, \ldots$. The second circuit should begin as $A, F, \ldots$.

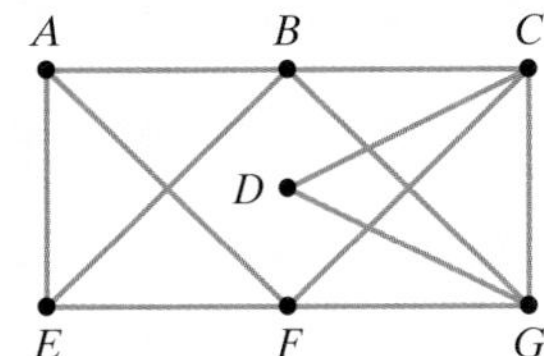

14. How many Hamilton circuits are there in a complete graph with five vertices?

The table shows the one-way airfares between cities. Use this information to solve Exercises 15–16.

	A	*B*	*C*	*D*
A	*	\$460	\$200	\$210
B	\$460	*	\$720	\$680
C	\$200	\$720	*	\$105
D	\$210	\$680	\$105	*

15. Draw a graph that models the information in the table. Use vertices to represent the cities and weights on appropriate edges to show airfares.

16. A sales director who lives in city *A* is required to fly to regional offices in cities *B*, *C*, and *D*, and then return to city *A*. Use the Brute Force Method to find the optimal route. What is the total cost for this Hamilton circuit? (*Hint*: The six possible Hamilton circuits are

A, B, C, D, A; *A, B, D, C, A*;
A, C, B, D, A; *A, C, D, B, A*;
A, D, B, C, A; *A, D, C, B, A*.)

17. Use the Nearest Neighbor Method to find a Hamilton circuit that begins at vertex *A* in the given graph. What is the total weight of the Hamilton circuit?

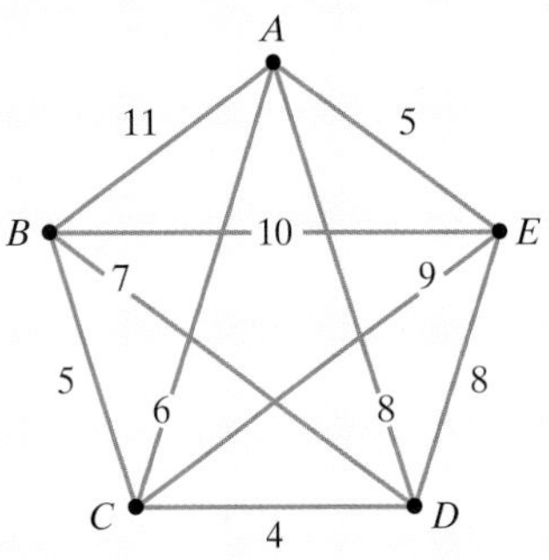

18. Is the graph shown a tree? Explain your answer.

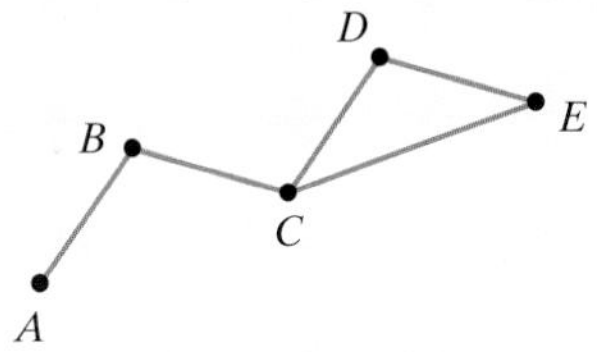

19. Find a spanning tree for the graph shown.

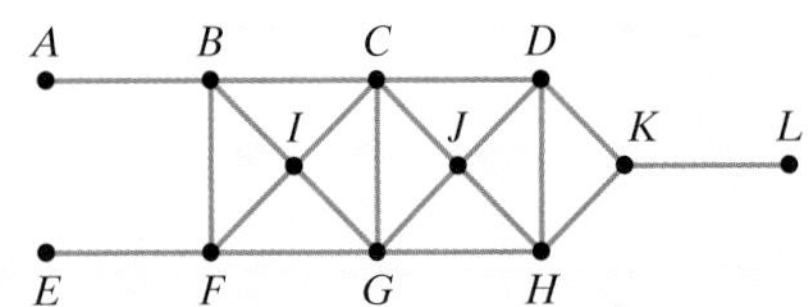

20. Use Kruskal's Algorithm to find the minimum spanning tree for the weighted graph shown. Give the total weight of the minimum spanning tree.

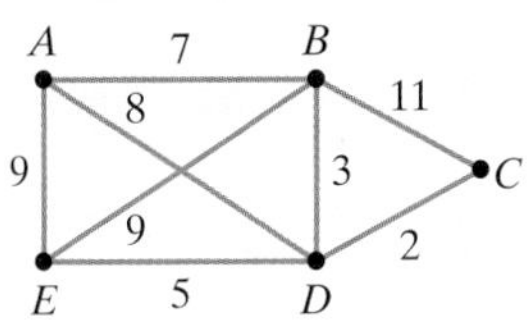

Answers to Selected Exercises

CHAPTER 1

Section 1.1

Check Point Exercises

1. Answers will vary; an example is $40 \times 40 = 1600$. **2. a.** Each number in the list is obtained by adding 6 to the previous number.; 33 **b.** Each number in the list is obtained by multiplying the previous number by 5.; 1250 **c.** To get the second number, multiply the previous number by 2. Then multiply by 3 and then by 4. Then multiply by 2, then by 3, and then by 4, repeatedly.; 3456 **d.** To get the second number, add 8 to the previous number. Then add 8 and then subtract 14. Then add 8, then add 8, and then subtract 14, repeatedly.; 7 **3. a.** Starting with the third number, each number is the sum of the previous two numbers.; 76 **b.** Starting with the second number, each number is one less than twice the previous number.; 257

4. The figures alternate between rectangles and triangles, and the number of appendages follows the pattern: one, two, three, one, two, three, etc.;

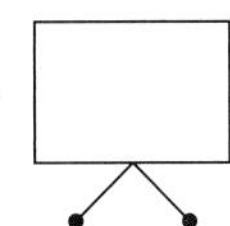

5. a. The result of the process is two times the original number selected.
b. Using n to represent the original number, we have

Select a number:	n
Multiply the number by 4:	$4n$
Add 6 to the product:	$4n + 6$
Divide this sum by 2:	$\frac{4n+6}{2} = 2n + 3$
Subtract 3 from the quotient:	$2n + 3 - 3 = 2n$.

Blitzer Bonus: Are You Smart Enough to Work at Google?

1. SSSS **2.** 3 1 2 2 1 1

Concept and Vocabulary Check

1. counterexample **2.** deductive **3.** inductive **4.** true

Exercise Set 1.1

1. Answers will vary; an example is: Barack Obama was younger than 65 at the time of his inauguration. **3.** Answers will vary; an example is: 3 multiplied by itself is 9, which is not even. **5.** Answers will vary; an example is: Adding 1 to the numerator and denominator of $\frac{1}{2}$ results in $\frac{2}{3}$, which is not equal to $\frac{1}{2}$. **7.** Answers will vary; an example is: When -1 is added to itself, the result is -2, which is less than -1. **9.** Each number in the list is obtained by adding 4 to the previous number.; 28 **11.** Each number in the list is obtained by subtracting 5 from the previous number.; 12 **13.** Each number in the list is obtained by multiplying the previous number by 3.; 729 **15.** Each number in the list is obtained by multiplying the previous number by 2.; 32 **17.** The numbers in the list alternate between 1 and numbers obtained by multiplying the number prior to the previous number by 2.; 32 **19.** Each number in the list is obtained by subtracting 2 from the previous number.; -6 **21.** Each number in the list is obtained by adding 4 to the denominator of the previous fraction.; $\frac{1}{22}$ **23.** Each number in the list is obtained by multiplying the previous number by $\frac{1}{3}$.; $\frac{1}{81}$ **25.** The second number is obtained by adding 4 to the first number. The third number is obtained by adding 5 to the second number. The number being added to the previous number increases by 1 each time.; 42 **27.** The second number is obtained by adding 3 to the first number. The third number is obtained by adding 5 to the second number. The number being added to the previous number increases by 2 each time.; 51 **29.** Starting with the third number, each number is the sum of the previous two numbers.; 71 **31.** To get the second number, add 5 to the previous number. Then add 5 and then subtract 7. Then add 5, then add 5, and then subtract 7, repeatedly.; 18 **33.** The second number is obtained by multiplying the first number by 2. The third number is obtained by subtracting 1 from the second number. Then multiply by 2 and then subtract 1, repeatedly.; 33 **35.** Each number in the list is obtained by multiplying the previous number by $-\frac{1}{4}$.; $\frac{1}{4}$ **37.** For each pair in the list, the second number is obtained by subtracting 4 from the first number.; -1

39. The pattern is: square, triangle, circle, square, triangle, circle, etc.;

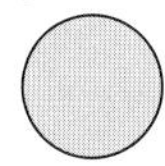

41. Each figure contains the letter of the alphabet following the letter in the previous figure with one more occurrence than in the previous figure.;

d	d	d
d	d	

43. a. The result of the process is two times the original number selected.
b. Using n to represent the original number, we have

Select a number:	n
Multiply the number by 4:	$4n$
Add 8 to the product:	$4n + 8$
Divide this sum by 2:	$\frac{4n+8}{2} = 2n + 4$
Subtract 4 from the quotient:	$2n + 4 - 4 = 2n$.

45. a. The result of the process is 3.
b. Using n to represent the original number, we have

Select a number:	n
Add 5:	$n + 5$
Double the result:	$2(n + 5) = 2n + 10$
Subtract 4:	$2n + 10 - 4 = 2n + 6$
Divide by 2:	$\frac{2n+6}{2} = n + 3$
Subtract n:	$n + 3 - n = 3$.

47. $1 + 2 + 3 + 4 + 5 + 6 = \frac{6 \times 7}{2}; 21 = 21$ **49.** $1 + 3 + 5 + 7 + 9 + 11 = 6 \times 6; 36 = 36$ **51.** $98{,}765 \times 9 + 3 = 888{,}888$; correct **53.** $165 \times 3367 = 555{,}555$; correct **55.** b **57.** deductive reasoning: Answers will vary. **59.** inductive reasoning; Answers will vary. **61. a.** 28, 36, 45, 55, 66 **b.** 36, 49, 64, 81, 100 **c.** 35, 51, 70, 92, 117 **d.** square **67.** makes sense **69.** makes sense

71. a.

16	3	11
5	10	15
9	17	4

The sums are all 30. **b.**

17	5	14
9	12	15
10	19	7

The sums are all 36. **c.** For any values of a, b, and c, the sums of all rows, all columns, and both diagonals are the same. **d.** The sum of the expressions in each row, each column, and each diagonal is $3a$. **e.** Add the variable expressions in a, b, and c, in each row, each column, and each diagonal. The sum is always $3a$.

73. a. The result is a three or four-digit number in which the thousands and hundreds places represent the month of the birthday and the tens and ones places represent the day of the birthday. **b.** $5[4(5M + 6) + 9] + D - 165 = 100M + D$ **75. a.** 10,101; 20,202; 30,303; 40,404 **b.** In the multiplications, the first factor is always 3367, and the second factors are consecutive multiples of 3, beginning with $3 \times 1 = 3$.; The second and fourth digits of the products are always 0; the first, third, and last digits are the same within each product; this digit is 1 in the first product and increases by 1 in each subsequent product. **c.** $3367 \times 15 = 50{,}505$; $3367 \times 18 = 60{,}606$ **d.** inductive reasoning; Answers will vary.

Section 1.2

Check Point Exercises

1. a. 7,000,000,000 **b.** 7,480,000,000 **2. a.** 3.1 **b.** 3.1416 **3. a.** \$3, \$2, \$6, \$5, \$3, \$3, and \$4; ≈ \$26 **b.** no **4. a.** ≈ \$2000 per wk **b.** ≈ \$100,000 per yr **5. a.** 0.48×2148.72 **b.** $0.5 \times 2100 = 1050$; Your family spent approximately \$1050 on heating and cooling last year. **6. a.** ≈ 0.15 year for each subsequent birth year **b.** ≈ 86.1 years **7. a.** 22% **b.** 1994 through 1998 **c.** 1982 and 1994 **d.** 2014; 13% **8. a.** \$1123 **b.** $T = 15{,}518 + 1123x$ **c.** \$37,978

Concept and Vocabulary Check

1. estimation **2.** circle graph **3.** mathematical model **4.** true **5.** true **6.** false

Exercise Set 1.2

1. a. 39,144,800 **b.** 39,145,000 **c.** 39,140,000 **d.** 39,100,000 **e.** 39,000,000 **f.** 40,000,000 **3.** 2.718 **5.** 2.71828 **7.** 2.718281828 **9.** $360 + 600 = 960$; 955; reasonably well **11.** $9 + 1 + 19 = 29$; 29.23; quite well **13.** $32 - 11 = 21$; 20.911; quite well **15.** $40 \times 6 = 240$; 218.185; not so well **17.** $0.8 \times 400 = 320$; 327.06; reasonably well **19.** $48 \div 3 = 16$; ≈ 16.49; quite well **21.** 30% of 200,000 is 60,000.; 59,920.96; quite well **23.** ≈ \$43 **25.** ≈ \$40,000 per yr **27.** ≈ \$24,000 **29.** ≈ \$1000 **31.** ≈ \$30 per hr **33.** ≈ 700,800 hr **35.** ≈ 40; ≈ 42.03; quite reasonable **37.** b **39.** c **41.** ≈ 3 hr **43.** $\approx 0.10 \times 16{,}000{,}000 = 1{,}600{,}000$ high school teenagers **45. a.** ≈ 85 people per 100 **b.** ≈ 5400 people **47. a.** ≈ 0.5% per year **b.** 29.7% **49. a.** ≈ 37% **b.** 55; ≈38% **c.** 25 **51. a.** 1.4 ppm per year **b.** $C = 310 + 1.4x$ **c.** 450 ppm **67.** does not make sense **69.** does not make sense **71.** a **73.** b **75.** ≈ 667 days; ≈ 1.8 yr

Section 1.3

Check Point Exercises

1. the amount of money given to the cashier **2.** The 128-ounce bottle at approximately 4¢ an ounce is the better value. **3.** 14 months **4.** 5 combinations **5.** 6 outfits **6.** A route that will cost less than \$1460 is A, D, E, C, B, A.

Trick Questions

1. 12 **2.** 12 **3.** sister and brother **4.** match

Concept and Vocabulary Check

1. understand **2.** devise a plan **3.** false **4.** false

Exercise Set 1.3

1. the price of the computer **3.** the number of words on the page **5.** unnecessary information: weekly salary of \$350; extra pay: \$180 **7.** unnecessary information: \$20 given to the parking attendant; charge: \$4.50 **9. a.** 24-ounce box for \$4.59 **b.** 22¢ per ounce for the 15.3-ounce box and \$3.06 per pound for the 24-ounce box **c.** no; Answers will vary. **11.** \$3000 **13.** \$2.01 less **15. a.** 6 **b.** 5 **17.** \$50 **19.** \$90 **21.** \$4525 **23.** \$565 **25.** 4 mi **27.** \$104 **29.** \$14,300 **31.** 5 ways **33.** 6 ways **35.** 9 ways **37.** 10 different total scores **39.** B owes \$18 and C owes \$2.; A is owed \$14, D is owed \$4, and E is owed \$2.; B should give A \$14 and D \$4, while C should give E \$2. **41.** 4 ways **43.** Andy, Darnell, Caleb, Beth, Ella **45.** Home, Bank, Post Office, Dry Cleaners, Home

47. Sample answer: CO, WY, UT, AZ, NM, CO, UT

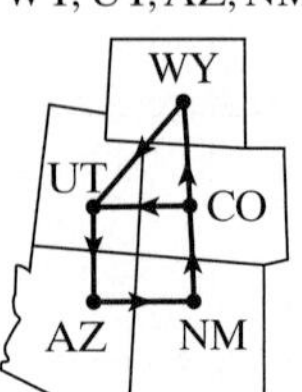

49. Bob's major is psychology.

51. a.

5	22	18
28	15	2
12	8	25

b.

4	9	8
11	7	3
6	5	10

53.

9	6	7
8	1	4
3	2	5

55.

```
     156
 28)4368
    28
    156
    140
     168
     168
       0
```

61. makes sense **63.** does not make sense **65.** the dentist with poor dental work **67.** Friday

69. Sample answer:

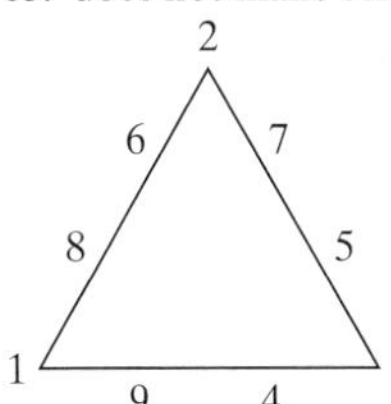

71. There is no missing dollar; in the end, the customers paid a total of \$27 of which \$25 went to the restaurant and \$2 was stolen by the waiter.

73. Answers will vary; an example is

State	A	B	C	D
Congressional Seats	4	6	8	12

Chapter 1 Review Exercises

1. deductive reasoning: Answers will vary. **2.** inductive reasoning; Answers will vary. **3.** Each number in the list is obtained by adding 5 to the previous number.; 24 **4.** Each number in the list is obtained by multiplying the previous number by 2.; 112 **5.** The successive differences are consecutive counting numbers beginning with 2.; 21 **6.** Each number in the list is obtained by writing a fraction with a denominator that is one more than the denominator of the previous fraction before it is reduced to lowest terms.; $\frac{3}{8}$ **7.** Each number in the list is obtained by multiplying the previous number by $-\frac{1}{2}$.; $\frac{5}{2}$ **8.** Each number in the list is obtained by subtracting 60 from the previous number; −200. **9.** Each number beginning with the third number is the sum of the two previous numbers.; 42 **10.** To get the second number, multiply the first number by 3. Then multiply the second number by 2 to get the third number. Then multiply by 3 and then by 2, repeatedly.; 432 **11.** The figures alternate between squares and circles, and in each figure the tick mark has been rotated 90° clockwise from its position in the previous figure.;
12. $2 + 4 + 8 + 16 + 32 = 64 - 2$; correct **13.** $444 \div 12 = 37$; correct

14. a. The result is the original number.

b. Using n to represent the original number, we have

Select a number:	n
Double the number:	$2n$
Add 4 to the product:	$2n + 4$
Divide the sum by 2:	$\frac{2n+4}{2} = n + 2$

Subtract 2 from the quotient: $n + 2 - 2 = n$.

15. a. 923,187,500 **b.** 923,187,000 **c.** 923,200,000 **d.** 923,000,000 **e.** 900,000,000 **16. a.** 1.5 **b.** 1.51 **c.** 1.507 **d.** 1.5065917 **17.** ≈ 16; 15.71; quite reasonable **18.** ≈ 450; 432.67; somewhat reasonable **19.** ≈ 5; ≈ 4.76; quite reasonable **20.** ≈ 2400; 2397.0548; quite reasonable **21.** ≈ \$18.00 **22.** ≈ \$800 **23.** ≈ \$60 **24.** ≈ 6,000,000 students **25.** b **26.** c **27. a.** Asian; ≈ 122 **b.** ≈ 990 million **28. a.** 0.4% per yr **b.** 34% **29. a.** 115 beats per min; 10 min **b.** 64 beats per min; 8 min **c.** between 9 and 10 minutes **d.** 9 min **30. a.** 2.65 million **b.** $p = 203.3 + 2.65x$ **c.** 335.8 million **31.** the weight of the child **32.** unnecessary information: \$20 given to driver; cost of trip: \$8.00 **33.** You will save \$13.50 with the 200-message package. **34.** 8 lb **35.** \$885 **36.** solar; \$5000 **37.** 5.75 hr or 5 hr 45 min **38.** \$15,500 **39.** 6 combinations

Chapter 1 Test

1. deductive **2.** inductive **3.** Each number in the list is obtained by adding 5 to the previous number.; 20
4. Each number in the list is obtained by multiplying the previous number by $\frac{1}{2}$.; $\frac{1}{96}$ **5.** $3367 \times 15 = 50{,}505$
6. The outer figure is always a square; the inner figure follows the pattern: triangle, circle, square, triangle, circle, square, etc.; the number of appendages on the outer square alternates between 1 and 2.

7. a. The original number is doubled.

b. Using n to represent the original number, we have

Select a number:	n
Multiply the number by 4:	$4n$
Add 8 to the product:	$4n + 8$
Divide the sum by 2:	$\frac{4n+8}{2} = 2n + 4$

Subtract 4 from the quotient: $2n + 4 - 4 = 2n$.

8. 3,300,000 **9.** 706.38 **10.** ≈ \$90 **11.** ≈ \$25,000 per person **12.** ≈ 60 **13.** ≈ 50 billion pounds **14.** a
15. a. 2001; ≈ 1275 discharges **b.** 2010; 275 discharges **c.** between 2001 and 2002 **d.** 1997
16. a. 0.8% per year **b.** $p = 27 + 0.8x$ **c.** 59% **17.** Estes Rental; \$12 **18.** \$14,080 **19.** 26 weeks
20. Belgium; 160,000 people

CHAPTER 2

Section 2.1

Check Point Exercises

1. L is the set of the first six lowercase letters of the alphabet. **2.** $M = \{\text{April, August}\}$ **3.** $O = \{1, 3, 5, 7, 9\}$ **4. a.** not the empty set **b.** empty set **c.** not the empty set **d.** not the empty set **5. a.** true **b.** true **c.** false **6. a.** $A = \{1, 2, 3\}$ **b.** $B = \{15, 16, 17, 18, \ldots\}$ **c.** $O = \{1, 3, 5, 7, \ldots\}$ **7. a.** $\{1, 2, 3, 4, \ldots, 199\}$ **b.** $\{51, 52, 53, 54, \ldots, 200\}$
8. a. $n(A) = 5$ **b.** $n(B) = 1$ **c.** $n(C) = 8$ **d.** $n(D) = 0$ **9.** No; the sets do not contain the same number of distinct elements.
10. a. true **b.** false

Concept and Vocabulary Check

1. roster; set-builder **2.** empty; ∅ **3.** is an element **4.** natural numbers **5.** cardinal; $n(A)$ **6.** equivalent **7.** equal

Exercise Set 2.1

1. well defined; set **3.** not well defined; not a set **5.** well defined; set **7.** the set of planets in our solar system **9.** the set of months that begin with J **11.** the set of natural numbers greater than 5 **13.** the set of natural numbers between 6 and 20, inclusive **15.** {winter, spring, summer, fall} **17.** {September, October, November, December} **19.** {1, 2, 3} **21.** {1, 3, 5, 7, 9, 11} **23.** {1, 2, 3, 4, 5} **25.** {6, 7, 8, 9, . . . } **27.** {7, 8, 9, 10} **29.** {10, 11, 12, 13, . . . , 79} **31.** {2} **33.** not the empty set **35.** empty set **37.** not the empty set **39.** empty set **41.** empty set **43.** not the empty set **45.** not the empty set **47.** true **49.** true **51.** false **53.** true **55.** false **57.** false **59.** true **61.** false **63.** false **65.** true **67.** 5 **69.** 15 **71.** 0 **73.** 1 **75.** 4 **77.** 5 **79.** 0 **81. a.** not equivalent; Answers will vary. **b.** not equal; Answers will vary. **83. a.** equivalent; Answers will vary. **b.** not equal; Answers will vary. **85. a.** equivalent; Answers will vary. **b.** equal; Answers will vary. **87. a.** equivalent; Answers will vary. **b.** not equal; Answers will vary. **89. a.** equivalent; Answers will vary. **b.** equal; Answers will vary. **91.** infinite **93.** finite **95.** finite **97.** $\{x \mid x \in \mathbf{N} \text{ and } x \geq 61\}$ **99.** $\{x \mid x \in \mathbf{N} \text{ and } 61 \leq x \leq 89\}$ **101.** Answers will vary; an example is: {0, 1, 2, 3} and {1, 2, 3, 4}. **103.** impossible; Answers will vary. **105.** {Engineering, Accounting} **107.** {Philosophy, Nursing, Journalism} **109.** ∅ or { } **111.** {Group 7} **113.** {12, 19} **115.** {20, 21} **117.** no one-to-one correspondence; not equivalent **125.** does not make sense **127.** makes sense **129.** false **131.** true **133.** false **135.** false

Section 2.2

Check Point Exercises

1. a. $\not\subseteq$ **b.** $\subseteq$ **c.** $\subseteq$ **2. a.** $\subseteq, \subset$ **b.** $\subseteq, \subset$ **3.** yes **4. a.** 16; 15 **b.** 64; 63

Concept and Vocabulary Check

1. $A \subseteq B$; every element in set A is also an element in set B **2.** $A \subset B$; sets A and B are not equal **3.** the empty set; subset **4.** 2^n **5.** 2^{n-1}

Exercise Set 2.2

1. $\subseteq$ **3.** $\not\subseteq$ **5.** $\not\subseteq$ **7.** $\not\subseteq$ **9.** $\subseteq$ **11.** $\not\subseteq$ **13.** $\subseteq$ **15.** $\not\subseteq$ **17.** $\subseteq$ **19.** both **21.** $\subseteq$ **23.** neither **25.** both **27.** $\subseteq$ **29.** $\subseteq$ **31.** both **33.** both **35.** both **37.** neither **39.** $\subseteq$ **41.** true **43.** false; Explanations will vary. **45.** true **47.** false; Explanations will vary. **49.** true **51.** false; Explanations will vary. **53.** true **55.** ∅, {border collie}, {poodle}, {border collie, poodle} **57.** ∅, {t}, {a}, {b}, {t, a}, {t, b}, {a, b}, {t, a, b} **59.** ∅ and {0} **61.** 16; 15 **63.** 64; 63 **65.** 128; 127 **67.** 8; 7 **69.** false; The set {1, 2, 3, . . . , 1000} has $2^{1000} - 1$ proper subsets. **71.** true **73.** false; $\varnothing \subseteq \{\varnothing, \{\varnothing\}\}$ **75.** true **77.** true **79.** true **81.** false; The set of subsets of {a, e, i, o, u} contains 32 elements. **83.** false; $D \subseteq T$ **85.** true **87.** false; If $x \in W$, then $x \in D$. **89.** true **91.** true **93.** 32 **95.** 64 **97.** 256 **105.** does not make sense **107.** does not make sense **109.** false **111.** false **113.** \$0.00, \$0.05, \$0.10, \$0.15, \$0.25, \$0.30, \$0.35, and \$0.40

Section 2.3

Check Point Exercises

1. a. {1, 5, 6, 7, 9} **b.** {1, 5, 6} **c.** {7, 9} **2. a.** {a, b, c, d} **b.** {e} **c.** {e, f, g} **d.** {f, g} **3.** {b, c, e} **4. a.** {7, 10} **b.** ∅ **c.** ∅ **5. a.** {1, 3, 5, 6, 7, 10, 11} **b.** {1, 2, 3, 4, 5, 6, 7} **c.** {1, 2, 3} **6. a.** {a, d} **b.** {a, d} **7. a.** {5} **b.** {2, 3, 7, 11, 13, 17, 19} **c.** {2, 3, 5, 7, 11, 13} **d.** {17, 19} **e.** {5, 7, 11, 13, 17, 19} **f.** {2, 3} **8.** 28

Concept and Vocabulary Check

1. Venn diagrams **2.** complement; A' **3.** intersection; $A \cap B$ **4.** union; $A \cup B$ **5.** $n(A) + n(B) - n(A \cap B)$ **6.** true **7.** false **8.** true **9.** false

Exercise Set 2.3

1. the set of all composers **3.** the set of all brands of soft drinks **5.** {c, d, e} **7.** {b, c, d, e, f} **9.** {6, 7, 8, 9, . . . , 20} **11.** {2, 4, 6, 8, . . . , 20} **13.** {21, 22, 23, 24, . . . } **15.** {1, 3, 5, 7, . . . } **17.** {1, 3} **19.** {1, 2, 3, 5, 7} **21.** {2, 4, 6} **23.** {4, 6} **25.** {1, 3, 5, 7} or A **27.** {1, 2, 4, 6, 7} **29.** {1, 2, 4, 6, 7} **31.** {4, 6} **33.** {1, 3, 5, 7} or A **35.** ∅ **37.** {1, 2, 3, 4, 5, 6, 7} or U **39.** {1, 3, 5, 7} or A **41.** {g, h} **43.** {a, b, g, h} **45.** {b, c, d, e, f} or C **47.** {c, d, e, f} **49.** {a, g, h} or A **51.** {a, b, c, d, e, f, g, h} or U **53.** {a, b, c, d, e, f, g, h} or U **55.** {c, d, e, f} **57.** {a, g, h} or A **59.** ∅ **61.** {a, b, c, d, e, f, g, h} or U **63.** {a, g, h} or A **65.** {a, c, d, e, f, g, h} **67.** {1, 3, 4, 7} **69.** {1, 2, 3, 4, 5, 6, 7, 8, 9} **71.** {3, 7} **73.** {1, 4, 8, 9} **75.** {8, 9} **77.** {1, 4} **79.** {Δ, two, four, six} **81.** {Δ, #, \$, two, four, six} **83.** 6 **85.** 5 **87.** {#, \$, two, four, six, 10, 01} **89.** {two, four, six} **91.** 4 **93.** 31 **95.** 27 **97.** {1, 2, 3, 4, 5, 6, 7, 8} or U **99.** {1, 3, 5, 7} or A **101.** {1, 7} **103.** {1, 3, 5, 6, 7, 8} **105.** {23, 29, 31, 37, 41, 43, 53, 59, 61, 67, 71} **107.** 60 **109.** {Ashley, Mike, Josh} **111.** {Ashley, Mike, Josh, Emily, Hanna, Ethan} **113.** {Ashley} **115.** {Jacob} **117.** III **119.** I **121.** II **123.** I **125.** IV **127.** II **129.** III **131.** I **133.** {2001} **135.** {1969, 1985, 2001, 2015} **137.** ∅ **139.** 283 people **153.** makes sense **155.** makes sense **157.** true **159.** false **161.** false **163.** true

165.

U
A
B

167.

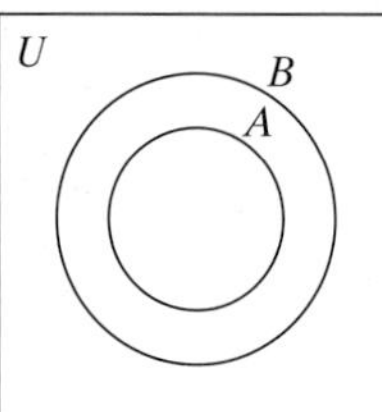

Section 2.4

Check Point Exercises

1. a. {a, b, c, d, f} **b.** {a, b, c, d, f} **c.** {a, b, d}
2. a. {5, 6, 7, 8, 9} **b.** {1, 2, 5, 6, 7, 8, 9, 10, 12} **c.** {5, 6, 7} **d.** {3, 4, 6, 8, 11} **e.** {1, 2, 3, 5, 6, 7, 8, 9, 10, 11, 12}
3.

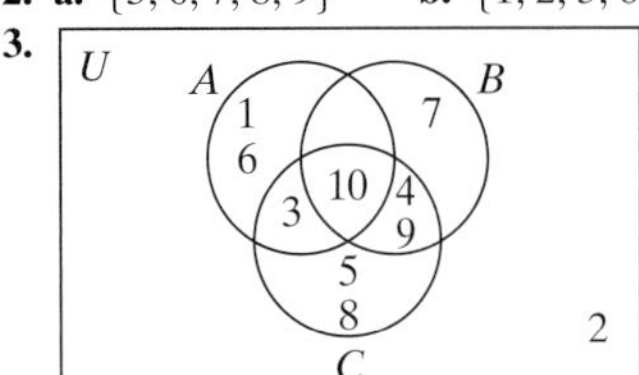

4. a. IV **b.** IV **c.** $(A \cup B)' = A' \cap B'$ **5. a.** II, IV, and V **b.** II, IV, and V **c.** $A \cap (B \cup C) = (A \cap B) \cup (A \cap C)$

Concept and Vocabulary Check

1. inside parentheses **2.** eight **3.** false **4.** true

Exercise Set 2.4

1. {1, 2, 3, 5, 7} **3.** {1, 2, 3, 5, 7} **5.** {2} **7.** {2} **9.** ∅ **11.** {4, 6} **13.** {a, b, g, h} **15.** {a, b, g, h} **17.** {b}
19. {b} **21.** ∅ **23.** {c, d, e, f} **25.** II, III, V, and VI **27.** I, II, IV, V, VI, and VII **29.** II and V **31.** I, IV, VII, and VIII
33. {1, 2, 3, 4, 5, 6, 7, 8} **35.** {1, 2, 3, 4, 5, 6, 7, 8, 9, 10, 11} **37.** {12, 13} **39.** {4, 5, 6} **41.** {6} **43.** {1, 2, 3, 4, 5, 7, 8, 9, 10, 11, 12, 13}
45.

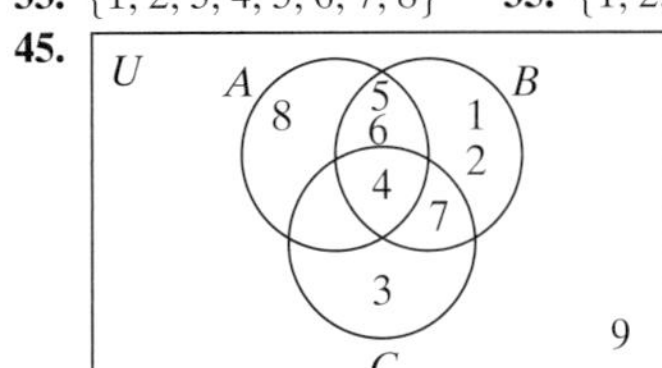

47.

49. a. II **b.** II **c.** $A \cap B = B \cap A$ **51. a.** I, III, and IV **b.** IV **c.** no; Answers will vary. **53.** not equal **55.** not equal
57. equal **59. a.** II, IV, V, VI, and VII **b.** II, IV, V, VI, and VII **c.** $(A \cap B) \cup C = (A \cup C) \cap (B \cup C)$ **61. a.** II, IV, and V
b. I, II, IV, V, and VI **c.** no; Answers will vary. **63.** not true **65.** true; theorem **67.** true; theorem **69. a.** {c, e, f}; {c, e, f}
b. {1, 3, 5, 7, 8}; {1, 3, 5, 7, 8} **c.** $A \cup (B' \cap C') = (A \cup B') \cap (A \cup C')$ **d.** theorem **71.** $(A \cap B)' \cap (A \cup B)$ **73.** $A' \cup B$
75. $(A \cap B) \cup C$ **77.** $A' \cap (B \cup C)$ **79.** {Ann, Jose, Al, Gavin, Amy, Ron, Grace} **81.** {Jose} **83.** {Lily, Emma}
85. {Lily, Emma, Ann, Jose, Lee, Maria, Fred, Ben, Sheila, Ellen, Gary} **87.** {Lily, Emma, Al, Gavin, Amy, Lee, Maria}
89. {Al, Gavin, Amy} **91.** The set of students who scored 90% or above on exam 1 and exam 3 but not on exam 2 **93.** I **95.** V
97. VI **99.** III **101.** IV **103.** VI
105.

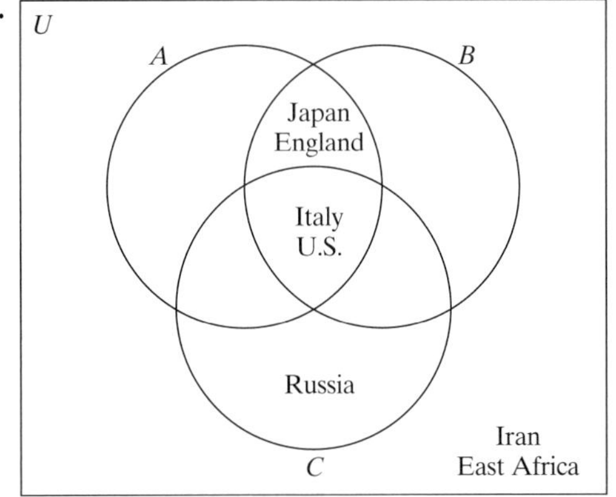

109. does not make sense **111.** makes sense **113.** AB^+ **115.** no

Section 2.5

Check Point Exercises

1. a. 75 **b.** 90 **c.** 20 **d.** 145 **e.** 55 **f.** 70 **g.** 30 **h.** 175 **2. a.** 750 **b.** 140
3.

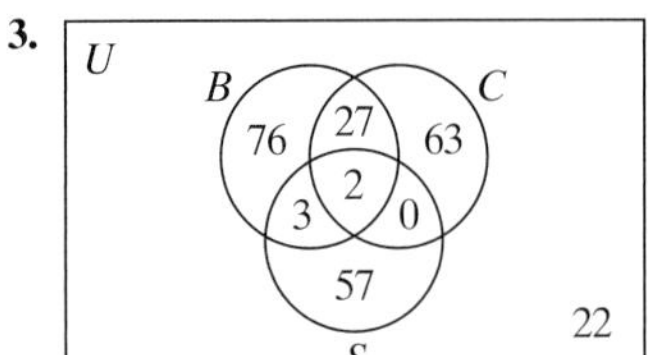

4. a. 63 **b.** 3 **c.** 136 **d.** 30 **e.** 228 **f.** 22

Concept and Vocabulary Check

1. *and/but* **2.** *or* **3.** *not* **4.** innermost; subtraction **5.** true **6.** true **7.** false **8.** true

Exercise Set 2.5

1. 26 **3.** 17 **5.** 37 **7.** 7 **9.** I: 14; III: 22; IV: 5 **11.** 17 **13.** 6 **15.** 28 **17.** 9 **19.** 3 **21.** 19 **23.** 21 **25.** 34 **27.** I: 5; II: 1; III: 4; IV: 3; VI: 1; VII: 8; VIII: 6 **29.** I: 5; II: 10; III: 3; IV: 4; V: 7; VI: 1; VII: 6; VIII: 2 **31.** impossible; There are only 10 elements in set *A* but there are 13 elements in set *A* that are also in sets *B* or *C*. A similar problem exists for set *C*. **33.** 18 **35.** 9 **37.** 9 **39. a.** 100 **b.** 740

41. a–d.

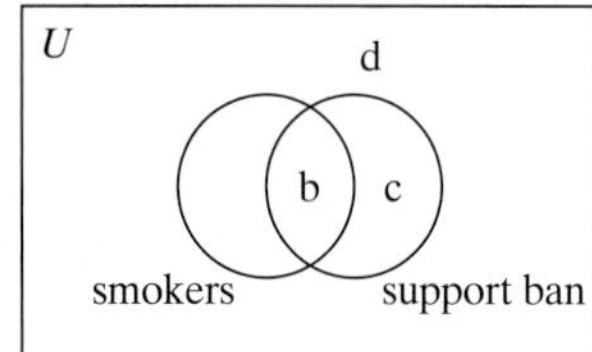

43. a–d.

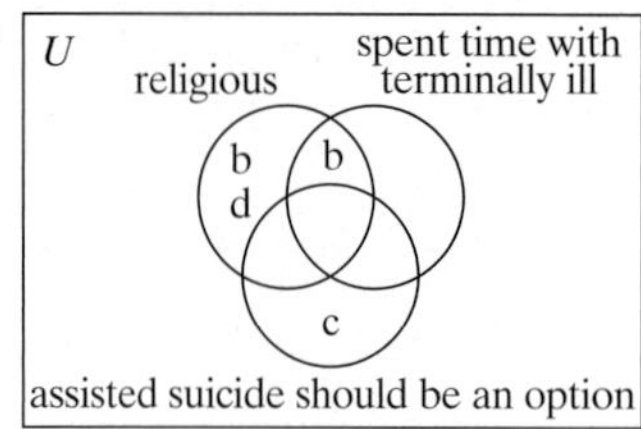

e. Answers will vary.

45.

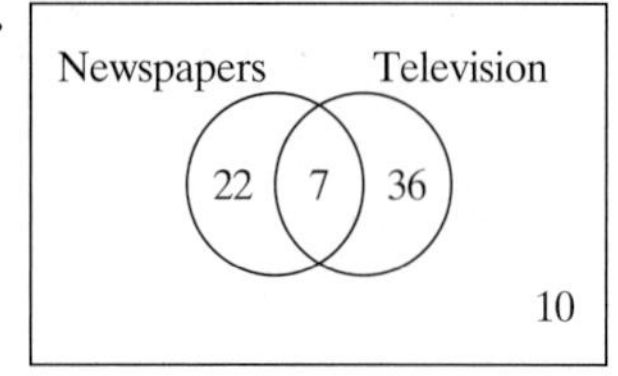

a. 22 **b.** 36 **c.** 65 **d.** 10

47.

U Rock Classical 23 7 17 7 5 3 12 Jazz 6

a. 23 **b.** 3 **c.** 32 **d.** 52 **e.** 22 **f.** 6

49. a. 1500 **b.** 1135 **c.** 56 **d.** 327 **e.** 526 **f.** 366 **g.** 1191 **53.** does not make sense **55.** does not make sense **57.** false **59.** false **61. a.** 0 **b.** 30 **c.** 60

Chapter 2 Review Exercises

1. the set of days of the week beginning with the letter T **2.** the set of natural numbers between 1 and 10, inclusive **3.** $\{m, i, s\}$ **4.** $\{8, 9, 10, 11, 12\}$ **5.** $\{1, 2, 3, \ldots, 30\}$ **6.** not empty **7.** empty set **8.** $\in$ **9.** $\notin$ **10.** 12 **11.** 15 **12.** $\neq$ **13.** $\neq$ **14.** equivalent **15.** both **16.** finite **17.** infinite **18.** $\subseteq$ **19.** $\not\subseteq$ **20.** $\subseteq$ **21.** $\subseteq$ **22.** both **23.** false; Answers will vary. **24.** false; Answers will vary. **25.** true **26.** false; Answers will vary. **27.** true **28.** false; It has $2^1 = 2$ subsets. **29.** true **30.** $\varnothing, \{1\}, \{5\}, \{1, 5\}; \{1, 5\}$ **31.** 32; 31 **32.** 8; 7 **33.** $\{1, 2, 4\}$ **34.** $\{1, 2, 3, 4, 6, 7, 8\}$ **35.** $\{5\}$ **36.** $\{6, 7, 8\}$ **37.** $\{6, 7, 8\}$ **38.** $\{4, 5, 6\}$ **39.** $\{2, 3, 6, 7\}$ **40.** $\{1, 4, 5, 6, 8, 9\}$ **41.** $\{4, 5\}$ **42.** $\{1, 2, 3, 6, 7, 8, 9\}$ **43.** $\{2, 3, 7\}$ **44.** $\{6\}$ **45.** $\{1, 2, 3, 4, 5, 6, 7, 8, 9\}$ **46.** 33 **47.** $\{1, 2, 3, 4, 5\}$ **48.** $\{1, 2, 3, 4, 5, 6, 7, 8\}$ or U **49.** $\{c, d, e, f, k, p, r\}$ **50.** $\{f, p\}$ **51.** $\{c, d, f, k, p, r\}$ **52.** $\{c, d, e\}$ **53.** $\{a, b, c, d, e, g, h, p, r\}$ **54.** $\{f\}$

55.

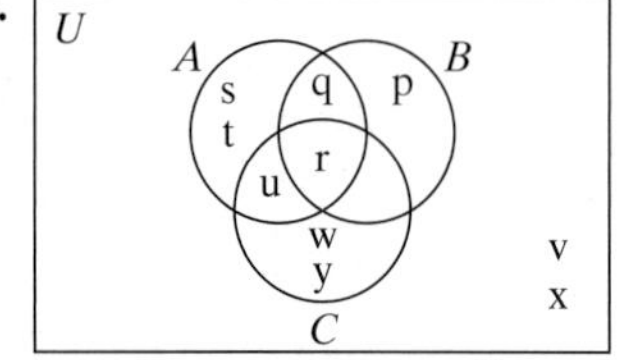

56. Use Figure 2.23.

Set	Regions in the Venn Diagram
A	I, II
B	II, III
$A \cup B$	I, II, III
$(A \cup B)'$	IV

Set	Regions in the Venn Diagram
A'	III, IV
B'	I, IV
$A' \cap B'$	IV

Since $(A \cup B)'$ and $A' \cap B'$ are represented by the same region, $(A \cup B)' = A' \cap B'$.

57. Use Figure 2.24.

Set	Regions in the Venn Diagram
A	I, II, IV, V
B	II, III, V, VI
C	IV, V, VI, VII
$B \cup C$	II, III, IV, V, VI, VII
$A \cap (B \cup C)$	II, IV, V

Set	Regions in the Venn Diagram
A	I, II, IV, V
B	II, III, V, VI
C	IV, V, VI, VII
$B \cap C$	V, VI
$A \cup (B \cap C)$	I, II, IV, V, VI

Since $A \cap (B \cup C)$ and $A \cup (B \cap C)$ are not represented by the same regions, $A \cap (B \cup C) = A \cup (B \cap C)$ is not a theorem.

58. United States: V; Italy: IV; Turkey: VIII; Norway: V; Pakistan: VIII; Iceland: V; Mexico: I
59. a. 490 men **b.** 510 men **60. a.** 250 **b.** 800 **c.** 200 **61. a.** 50 **b.** 26 **c.** 130 **d.** 46 **e.** 0

Chapter 2 Test

1. {18, 19, 20, 21, 22, 23, 24} **2.** false; Answers will vary. **3.** true **4.** true **5.** false; Answers will vary. **6.** true **7.** false; Answers will vary. **8.** false; Answers will vary. **9.** false; Answers will vary. **10.** ∅, {6}, {9}, {6, 9}; {6, 9} **11.** {a, b, c, d, e, f} **12.** {a, b, c, d, f, g} **13.** {b, c, d} **14.** {a, e} **15.** 5 **16.** {b, c, d, i, j, k} **17.** {a} **18.** {a, f, h}

19.

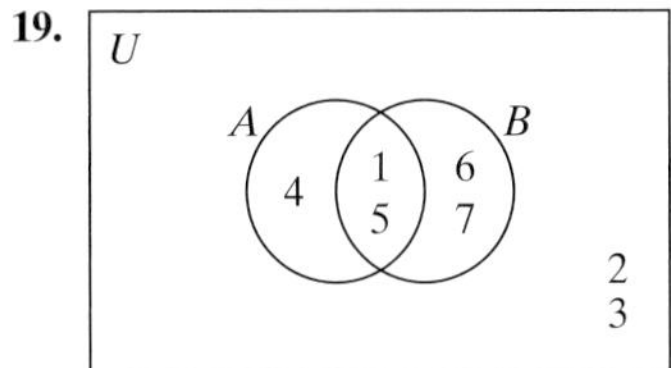

20. theorem **21. a.** V **b.** VII **c.** IV **d.** I **e.** VI

22. a.

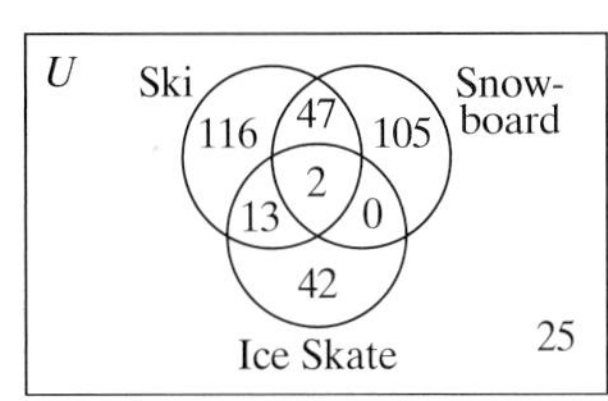

b. 263 **c.** 25 **d.** 62 **e.** 0 **f.** 147 **g.** 116

CHAPTER 3

Section 3.1

Check Point Exercises

1. a. Paris is not the capital of Spain. **b.** July is a month. **2. a.** $\sim p$ **b.** $\sim q$ **3.** Chicago O'Hare is not the world's busiest airport. **4.** Some new tax dollars will not be used to improve education.; At least one new tax dollar will not be used to improve education.

Concept and Vocabulary Check

1. true; false **2.** false; true **3.** $\sim p$; not p **4.** quantified **5.** There are no *A* that are not *B*. **6.** There exists at least one *A* that is a *B*. **7.** All *A* are not *B*. **8.** Not all *A* are *B*. **9.** Some *A* are not *B*. **10.** No *A* are *B*.

Exercise Set 3.1

1. statement **3.** statement **5.** not a statement **7.** statement **9.** not a statement **11.** statement **13.** statement **15.** It is not raining. **17.** Facts cease to exist when they are ignored. **19.** Chocolate in moderation is good for the heart. **21.** $\sim p$ **23.** $\sim r$ **25.** Listening to classical music does not make infants smarter. **27.** Sigmund Freud's father was 20 years older than his mother. **29. a.** There are no whales that are not mammals. **b.** Some whales are not mammals. **31. a.** At least one student is a business major. **b.** No students are business majors. **33. a.** At least one thief is not a criminal. **b.** All thieves are criminals. **35. a.** All Democratic presidents have not been impeached. **b.** Some Democratic presidents have been impeached. **37. a.** All seniors graduated. **b.** Some seniors did not graduate. **39. a.** Some parrots are not pets. **b.** All parrots are pets. **41. a.** No atheist is a churchgoer. **b.** Some atheists are churchgoers. **43.** Some Africans have Jewish ancestry. **45.** Some rap is not hip-hop. **47. a.** All birds are parrots. **b.** false; Some birds are not parrots. **49. a.** No college students are business majors. **b.** false; Some college students are business majors. **51. a.** All people like Sara Lee. **b.** Some people don't like Sara Lee. **53. a.** No safe thing is exciting. **b.** Some safe things are exciting. **55. a.** Some great actors are not a Tom Hanks. **b.** All great actors are a Tom Hanks. **57.** b **59.** true **61.** false; Some college students in the United States are willing to marry without romantic love. **63.** true **65.** false; The sentence "5% of college students in Australia are willing to marry without romantic love" is a statement. **75.** does not make sense **77.** does not make sense

Section 3.2

Check Point Exercises

1. a. $q \wedge p$ **b.** $\sim p \wedge q$ **2. a.** $p \vee q$ **b.** $q \vee \sim p$ **3. a.** $\sim p \rightarrow \sim q$ **b.** $q \rightarrow \sim p$ **4.** $\sim p \rightarrow q$ **5. a.** $q \leftrightarrow p$ **b.** $\sim p \leftrightarrow \sim q$ **6. a.** It is not true that he earns $105,000 yearly and that he is often happy. **b.** He is not often happy and he earns $105,000 yearly. **c.** It is not true that if he is often happy then he earns $105,000 yearly. **7. a.** If the plant is fertilized and watered, then it will not wilt. **b.** The plant is fertilized, and if it is watered then it will not wilt. **8.** p: There is too much homework.; q: A teacher is boring.; r: I take the class. **a.** $(p \vee q) \rightarrow \sim r$ **b.** $p \vee (q \rightarrow \sim r)$

Concept and Vocabulary Check

1. $p \wedge q$; conjunction **2.** $p \vee q$; disjunction **3.** $p \rightarrow q$; conditional **4.** $p \leftrightarrow q$; biconditional **5.** true **6.** false **7.** true **8.** false **9.** true **10.** true **11.** true **12.** true **13.** false

Exercise Set 3.2

1. $p \wedge q$ **3.** $q \wedge \sim p$ **5.** $\sim q \wedge p$ **7.** $p \vee q$ **9.** $p \vee \sim q$ **11.** $p \rightarrow q$ **13.** $\sim p \rightarrow \sim q$ **15.** $\sim q \rightarrow \sim p$ **17.** $p \rightarrow q$ **19.** $p \rightarrow \sim q$ **21.** $q \rightarrow \sim p$ **23.** $p \rightarrow \sim q$ **25.** $q \rightarrow \sim p$ **27.** $p \leftrightarrow q$ **29.** $\sim q \leftrightarrow \sim p$ **31.** $q \leftrightarrow p$ **33.** The heater is not working and the house is cold. **35.** The heater is working or the house is not cold. **37.** If the heater is working, then the house is not cold. **39.** The heater is working if and only if the house is not cold. **41.** It is July 4th and we are not having a barbecue. **43.** It is not July 4th or we are having a barbecue. **45.** If we are having a barbecue, then it is not July 4th. **47.** It is not July 4th if and only if we are having a barbecue. **49.** It is not true that Romeo loves Juliet and Juliet loves Romeo. **51.** Romeo does not love Juliet but Juliet loves Romeo. **53.** It is not true that Juliet loves Romeo or Romeo loves Juliet; Neither Juliet loves Romeo nor Romeo loves Juliet. **55.** Juliet does not love Romeo or Romeo loves Juliet.

57. Romeo does not love Juliet and Juliet does not love Romeo. **59.** $(p \wedge q) \vee r$ **61.** $(p \vee \sim q) \rightarrow r$ **63.** $r \leftrightarrow (p \wedge \sim q)$ **65.** $(p \wedge \sim q) \rightarrow r$ **67.** If the temperature is above 85° and we have finished studying, then we will go to the beach. **69.** The temperature is above 85°, and if we have finished studying then we will go to the beach. **71.** $\sim r \rightarrow (\sim p \vee \sim q)$; If we do not go to the beach, then the temperature is not above 85° or we have not finished studying. **73.** If we do not go to the beach then we have not finished studying, or the temperature is above 85°. **75.** $r \leftrightarrow (p \wedge q)$; We will go to the beach if and only if the temperature is above 85° and we have finished studying. **77.** The temperature is above 85° if and only if we have finished studying, and we go to the beach. **79.** If we do not go to the beach, then it is not true that both the temperature is above 85° and we have finished studying. **81.** p: I like the teacher.; q: The course is interesting.; r: I miss class.; $(p \vee q) \rightarrow \sim r$ **83.** p: I like the teacher.; q: The course is interesting.; r: I miss class.; $p \vee (q \rightarrow \sim r)$ **85.** p: I like the teacher.; q: The course is interesting.; r: I miss class.; $r \leftrightarrow \sim (p \wedge q)$ **87.** p: I like the teacher.; q: The course is interesting.; r: I miss class.; $(p \rightarrow \sim r) \leftrightarrow q$ **89.** p: I like the teacher.; q: The course is interesting.; r: I miss class.; s: I spend extra time reading the book.; $(\sim p \wedge r) \rightarrow (\sim q \vee s)$ **91.** p: Being French is necessary for being a Parisian.; q: Being German is necessary for being a Berliner.; $\sim p \rightarrow \sim q$ **93.** p: You file an income tax report.; q: You file a complete statement of earnings.; r: You are a taxpayer.; s: You are an authorized tax preparer.; $(r \vee s) \rightarrow (p \wedge q)$ **95.** p: You are wealthy.; q: You are happy.; r: You live contentedly.; $\sim (p \rightarrow (q \wedge r))$ **97.** $[p \rightarrow (q \vee r)] \leftrightarrow (p \wedge r)$ **99.** $(p \rightarrow p) \leftrightarrow [(p \wedge p) \rightarrow \sim p]$ **101.** $\sim p \rightarrow q$ **103.** $(p \rightarrow q) \wedge \sim q$ **105.** $((p \rightarrow q) \wedge \sim p) \rightarrow \sim q$ **115.** makes sense **117.** does not make sense **119.** p: Shooting unarmed civilians is morally justifiable.; q: Bombing unarmed civilians is morally justifiable.; $((p \leftrightarrow q) \wedge \sim p) \rightarrow \sim q$

Section 3.3

Check Point Exercises

1. a. false **b.** true **c.** false **d.** true

2.

p	q	$p \vee q$	$\sim(p \vee q)$
T	T	T	F
T	F	T	F
F	T	T	F
F	F	F	T

$\sim(p \vee q)$ is true when both p and q are false; otherwise, $\sim(p \vee q)$ is false.

3.

p	q	$\sim p$	$\sim q$	$\sim p \wedge \sim q$
T	T	F	F	F
T	F	F	T	F
F	T	T	F	F
F	F	T	T	T

$\sim p \wedge \sim q$ is true when both p and q are false; otherwise, $\sim p \wedge \sim q$ is false.

4.

p	q	$\sim q$	$p \wedge \sim q$	$\sim p$	$(p \wedge \sim q) \vee \sim p$
T	T	F	F	F	F
T	F	T	T	F	T
F	T	F	F	T	T
F	F	T	F	T	T

$(p \wedge \sim q) \vee \sim p$ is false when both p and q are true; otherwise, $(p \wedge \sim q) \vee \sim p$ is true.

5.

p	$\sim p$	$p \wedge \sim p$
T	F	F
F	T	F

$p \wedge \sim p$ is false in all cases.

6. p: I study hard.; q: I ace the final.; r: I fail the course.

a.

p	q	r	$q \vee r$	$p \wedge (q \vee r)$
T	T	T	T	T
T	T	F	T	T
T	F	T	T	T
T	F	F	F	F
F	T	T	T	F
F	T	F	T	F
F	F	T	T	F
F	F	F	F	F

b. false **7.** true

Concept and Vocabulary Check

1. opposite **2.** both p and q are true **3.** p and q are false **4.** true **5.** true **6.** true **7.** false **8.** true

Exercise Set 3.3

1. true **3.** false **5.** false **7.** false **9.** true **11.** true **13.** true **15.** true

17.

p	$\sim p$	$\sim p \wedge p$
T	F	F
F	T	F

19.

p	q	$\sim p$	$\sim p \wedge q$
T	T	F	F
T	F	F	F
F	T	T	T
F	F	T	F

21.

p	q	$p \vee q$	$\sim(p \vee q)$
T	T	T	F
T	F	T	F
F	T	T	F
F	F	F	T

23.

p	q	$\sim p$	$\sim q$	$\sim p \wedge \sim q$
T	T	F	F	F
T	F	F	T	F
F	T	T	F	F
F	F	T	T	T

25.

p	q	$\sim q$	$p \vee \sim q$
T	T	F	T
T	F	T	T
F	T	F	F
F	F	T	T

27.

p	q	$\sim p$	$\sim p \vee q$	$\sim(\sim p \vee q)$
T	T	F	T	F
T	F	F	F	T
F	T	T	T	F
F	F	T	T	F

29.

p	q	$\sim p$	$p \vee q$	$(p \vee q) \wedge \sim p$
T	T	F	T	F
T	F	F	T	F
F	T	T	T	T
F	F	T	F	F

31.

p	q	$\sim p$	$\sim q$	$p \wedge \sim q$	$\sim p \vee (p \wedge \sim q)$
T	T	F	F	F	F
T	F	F	T	T	T
F	T	T	F	F	T
F	F	T	T	F	T

33.

p	q	$\sim p$	$\sim q$	$p \vee q$	$\sim p \vee \sim q$	$(p \vee q) \wedge (\sim p \vee \sim q)$
T	T	F	F	T	F	F
T	F	F	T	T	T	T
F	T	T	F	T	T	T
F	F	T	T	F	T	F

35.

p	q	$\sim q$	$p \wedge \sim q$	$p \wedge q$	$(p \wedge \sim q) \vee (p \wedge q)$
T	T	F	F	T	T
T	F	T	T	F	T
F	T	F	F	F	F
F	F	T	F	F	F

37.

p	q	r	$\sim q$	$\sim q \vee r$	$p \wedge (\sim q \vee r)$
T	T	T	F	T	T
T	T	F	F	F	F
T	F	T	T	T	T
T	F	F	T	T	T
F	T	T	F	T	F
F	T	F	F	F	F
F	F	T	T	T	F
F	F	F	T	T	F

39.

p	q	r	$\sim p$	$\sim q$	$r \wedge \sim p$	$(r \wedge \sim p) \vee \sim q$
T	T	T	F	F	F	F
T	T	F	F	F	F	F
T	F	T	F	T	F	T
T	F	F	F	T	F	T
F	T	T	T	F	T	T
F	T	F	T	F	F	F
F	F	T	T	T	T	T
F	F	F	T	T	F	T

41.

p	q	r	$p \vee q$	$\sim(p \vee q)$	$\sim r$	$\sim(p \vee q) \wedge \sim r$
T	T	T	T	F	F	F
T	T	F	T	F	T	F
T	F	T	T	F	F	F
T	F	F	T	F	T	F
F	T	T	T	F	F	F
F	T	F	T	F	T	F
F	F	T	F	T	F	F
F	F	F	F	T	T	T

43. a. p: You did the dishes.; q: You left the room a mess.; $\sim p \wedge q$ **b.** See truth table for Exercise 19. **c.** The statement is true when p is false and q is true. **45. a.** p: I bought a meal ticket.; q: I used it.; $\sim(p \wedge \sim q)$

b.

p	q	$\sim q$	$p \wedge \sim q$	$\sim(p \wedge \sim q)$
T	T	F	F	T
T	F	T	T	F
F	T	F	F	T
F	F	T	F	T

c. Answers will vary; an example is: The statement is true when p and q are true.

47. a. p: The student is intelligent.; q: The student is an overachiever.; $(p \vee q) \wedge \sim q$

b.

p	q	$\sim q$	$p \vee q$	$(p \vee q) \wedge \sim q$
T	T	F	T	F
T	F	T	T	T
F	T	F	T	F
F	F	T	F	F

c. The statement is true when p is true and q is false.

49. a. p: Married people are healthier than single people.; q: Married people are more economically stable than single people.; r: Children of married people do better on a variety of indicators.; $(p \wedge q) \wedge r$

b.

p	q	r	$p \wedge q$	$(p \wedge q) \wedge r$
T	T	T	T	T
T	T	F	T	F
T	F	T	F	F
T	F	F	F	F
F	T	T	F	F
F	T	F	F	F
F	F	T	F	F
F	F	F	F	F

c. The statement is true when p, q, and r are all true.

51. a. p: I go to office hours.; q: I ask questions.; r: My professor remembers me.; $(p \wedge q) \vee \sim r$

b.

p	q	r	$\sim r$	$p \wedge q$	$(p \wedge q) \vee \sim r$
T	T	T	F	T	T
T	T	F	T	T	T
T	F	T	F	F	F
T	F	F	T	F	T
F	T	T	F	F	F
F	T	F	T	F	T
F	F	T	F	F	F
F	F	F	T	F	T

c. Answers will vary; an example is: The statement is true when p, q, and r are all true.

53. false **55.** true **57.** true **59.** false **61.** true

63.

p	q	$\sim[\sim(p \wedge \sim q) \vee \sim(\sim p \vee q)]$
T	T	F
T	F	F
F	T	F
F	F	F

65.

p	q	r	$[(p \wedge \sim r) \vee (q \wedge \sim r)] \wedge \sim(\sim p \vee r)$
T	T	T	F
T	T	F	T
T	F	T	F
T	F	F	T
F	T	T	F
F	T	F	F
F	F	T	F
F	F	F	F

67. p: You notice this notice.; q: You notice this notice is not worth noticing.; $(p \vee \sim p) \wedge q$

p	q	$\sim p$	$p \vee \sim p$	$(p \vee \sim p) \wedge q$
T	T	F	T	T
T	F	F	T	F
F	T	T	T	T
F	F	T	T	F

The statement is true when q is true.

69. p: $x \leq 3$; q: $x \geq 7$; $\sim(p \vee q) \wedge (\sim p \wedge \sim q)$

p	q	$\sim p$	$\sim q$	$p \vee q$	$\sim(p \vee q)$	$\sim p \wedge \sim q$	$\sim(p \vee q) \wedge (\sim p \wedge \sim q)$
T	T	F	F	T	F	F	F
T	F	F	T	T	F	F	F
F	T	T	F	T	F	F	F
F	F	T	T	F	T	T	T

The statement is true when both p and q are false.

71. Prices peaked in 2010 and prices did not remain steady from 1995 through 2000.; false **73.** Prices did not peak in 2010 and prices were more than $2.00 per gallon in 1990.; false **75.** Prices peaked in 2010 or prices did not remain steady from 1995 through 2000.; true **77.** Prices did not peak in 2010 or prices were more than $2.00 per gallon in 1990.; false **79.** Prices peaked in 2010 and prices remained steady from 1995 through 2000, or prices were more than $2.00 per gallon in 1990.; true **81.** p: More than 12% named business.; q: 9% named engineering.; $p \vee \sim q$; true **83.** p: 5.9% named teaching.; q: 8.9% named nursing.; r: 12% named business.; $(p \vee q) \wedge \sim r$; true **85. a.** Hora Gershwin **b.** Bolera Mozart does not have a master's degree in music. Cha-Cha Bach does not either play three instruments or have five years experience playing with a symphony orchestra. **95.** does not make sense **97.** does not make sense

101.

p	q	$p \veebar q$
T	T	F
T	F	T
F	T	T
F	F	F

Section 3.4

Check Point Exercises

1.

p	q	$\sim p$	$\sim q$	$\sim p \rightarrow \sim q$
T	T	F	F	T
T	F	F	T	T
F	T	T	F	F
F	F	T	T	T

2. The table shows that $[(p \rightarrow q) \wedge \sim q] \rightarrow \sim p$ is always true; therefore, it is a tautology.

p	q	$\sim p$	$\sim q$	$p \rightarrow q$	$(p \rightarrow q) \wedge \sim q$	$[(p \rightarrow q) \wedge \sim q] \rightarrow \sim p$
T	T	F	F	T	F	T
T	F	F	T	F	F	T
F	T	T	F	T	F	T
F	F	T	T	T	T	T

3. a. p: You use Hair Grow.; q: You apply it daily.; r: You go bald.

p	q	r	$\sim r$	$p \wedge q$	$(p \wedge q) \to \sim r$
T	T	T	F	T	F
T	T	F	T	T	T
T	F	T	F	F	T
T	F	F	T	F	T
F	T	T	F	F	T
F	T	F	T	F	T
F	F	T	F	F	T
F	F	F	T	F	T

b. no

4.

p	q	$p \vee q$	$\sim p$	$\sim p \to q$	$(p \vee q) \leftrightarrow (\sim p \to q)$
T	T	T	F	T	T
T	F	T	F	T	T
F	T	T	T	T	T
F	F	F	T	F	T

Because all cases are true, the statement is a tautology.

5. true

Concept and Vocabulary Check

1. p is true and q is false **2.** tautology; implications; self-contradiction **3.** p and q have the same truth value
4. false **5.** false **6.** true **7.** true

Exercise Set 3.4

1.

p	q	$\sim q$	$p \to \sim q$
T	T	F	F
T	F	T	T
F	T	F	T
F	F	T	T

3.

p	q	$q \to p$	$\sim(q \to p)$
T	T	T	F
T	F	T	F
F	T	F	T
F	F	T	F

5.

p	q	$p \wedge q$	$p \vee q$	$(p \wedge q) \to (p \vee q)$
T	T	T	T	T
T	F	F	T	T
F	T	F	T	T
F	F	F	F	T

7.

p	q	$p \to q$	$\sim q$	$(p \to q) \wedge \sim q$
T	T	T	F	F
T	F	F	T	F
F	T	T	F	F
F	F	T	T	T

9.

p	q	r	$p \vee q$	$(p \vee q) \to r$
T	T	T	T	T
T	T	F	T	F
T	F	T	T	T
T	F	F	T	F
F	T	T	T	T
F	T	F	T	F
F	F	T	F	T
F	F	F	F	T

11.

p	q	r	$p \wedge q$	$r \to (p \wedge q)$
T	T	T	T	T
T	T	F	T	T
T	F	T	F	F
T	F	F	F	T
F	T	T	F	F
F	T	F	F	T
F	F	T	F	F
F	F	F	F	T

13.

p	q	r	$\sim q$	$\sim r$	$\sim q \to p$	$\sim r \wedge (\sim q \to p)$
T	T	T	F	F	T	F
T	T	F	F	T	T	T
T	F	T	T	F	T	F
T	F	F	T	T	T	T
F	T	T	F	F	T	F
F	T	F	F	T	T	T
F	F	T	T	F	F	F
F	F	F	T	T	F	F

15.

p	q	r	$\sim q$	$p \wedge r$	$\sim(p \wedge r)$	$\sim q \vee r$	$\sim(p \wedge r) \rightarrow (\sim q \vee r)$
T	T	T	F	T	F	T	T
T	T	F	F	F	T	F	F
T	F	T	T	T	F	T	T
T	F	F	T	F	T	T	T
F	T	T	F	F	T	T	T
F	T	F	F	F	T	F	F
F	F	T	T	F	T	T	T
F	F	F	T	F	T	T	T

17.

p	q	$\sim q$	$p \leftrightarrow \sim q$
T	T	F	F
T	F	T	T
F	T	F	T
F	F	T	F

19.

p	q	$p \leftrightarrow q$	$\sim(p \leftrightarrow q)$
T	T	T	F
T	F	F	T
F	T	F	T
F	F	T	F

21.

p	q	$p \leftrightarrow q$	$(p \leftrightarrow q) \rightarrow p$
T	T	T	T
T	F	F	T
F	T	F	T
F	F	T	F

23.

p	q	$\sim p$	$\sim p \leftrightarrow q$	$\sim p \rightarrow q$	$(\sim p \leftrightarrow q) \rightarrow (\sim p \rightarrow q)$
T	T	F	F	T	T
T	F	F	T	T	T
F	T	T	T	T	T
F	F	T	F	F	T

25.

p	q	$p \wedge q$	$q \rightarrow p$	$(p \wedge q) \wedge (q \rightarrow p)$	$[(p \wedge q) \wedge (q \rightarrow p)] \leftrightarrow (p \wedge q)$
T	T	T	T	T	T
T	F	F	T	F	T
F	T	F	F	F	T
F	F	F	T	F	T

27.

p	q	r	$\sim r$	$p \leftrightarrow q$	$(p \leftrightarrow q) \rightarrow \sim r$
T	T	T	F	T	F
T	T	F	T	T	T
T	F	T	F	F	T
T	F	F	T	F	T
F	T	T	F	F	T
F	T	F	T	F	T
F	F	T	F	T	F
F	F	F	T	T	T

29.

p	q	r	$p \wedge r$	$q \vee r$	$\sim(q \vee r)$	$(p \wedge r) \leftrightarrow \sim(q \vee r)$
T	T	T	T	T	F	F
T	T	F	F	T	F	T
T	F	T	T	T	F	F
T	F	F	F	F	T	F
F	T	T	F	T	F	T
F	T	F	F	T	F	T
F	F	T	F	T	F	T
F	F	F	F	F	T	F

31.

p	q	r	$\sim q$	$\sim q \wedge p$	$r \vee (\sim q \wedge p)$	$\sim p$	$[r \vee (\sim q \wedge p)] \leftrightarrow \sim p$
T	T	T	F	F	T	F	F
T	T	F	F	F	F	F	T
T	F	T	T	T	T	F	F
T	F	F	T	T	T	F	F
F	T	T	F	F	T	T	T
F	T	F	F	F	F	T	F
F	F	T	T	F	T	T	T
F	F	F	T	F	F	T	F

33. neither

p	q	$p \rightarrow q$	$(p \rightarrow q) \wedge q$	$[(p \rightarrow q) \wedge q] \rightarrow p$
T	T	T	T	T
T	F	F	F	T
F	T	T	T	F
F	F	T	F	T

35. tautology

p	q	$\sim p$	$\sim q$	$p \rightarrow q$	$(p \rightarrow q) \wedge \sim q$	$[(p \rightarrow q) \wedge \sim q] \rightarrow \sim p$
T	T	F	F	T	F	T
T	F	F	T	F	F	T
F	T	T	F	T	F	T
F	F	T	T	T	T	T

37. neither

p	q	$\sim q$	$p \vee q$	$(p \vee q) \wedge p$	$[(p \vee q) \wedge p] \rightarrow \sim q$
T	T	F	T	T	F
T	F	T	T	T	T
F	T	F	T	F	T
F	F	T	F	F	T

39. tautology

p	q	$\sim p$	$p \rightarrow q$	$\sim p \vee q$	$(p \rightarrow q) \rightarrow (\sim p \vee q)$
T	T	F	T	T	T
T	F	F	F	F	T
F	T	T	T	T	T
F	F	T	T	T	T

41. self-contradiction

p	q	$\sim p$	$\sim q$	$p \wedge q$	$\sim p \vee \sim q$	$(p \wedge q) \wedge (\sim p \vee \sim q)$
T	T	F	F	T	F	F
T	F	F	T	F	T	F
F	T	T	F	F	T	F
F	F	T	T	F	T	F

43. neither

p	q	$\sim p$	$\sim q$	$p \wedge q$	$\sim(p \wedge q)$	$\sim p \wedge \sim q$	$\sim(p \wedge q) \leftrightarrow (\sim p \wedge \sim q)$
T	T	F	F	T	F	F	T
T	F	F	T	F	T	F	F
F	T	T	F	F	T	F	F
F	F	T	T	F	T	T	T

45. neither

p	q	$p \to q$	$q \to p$	$(p \to q) \leftrightarrow (q \to p)$
T	T	T	T	T
T	F	F	T	F
F	T	T	F	F
F	F	T	T	T

47. tautology

p	q	$\sim p$	$p \to q$	$\sim p \vee q$	$(p \to q) \leftrightarrow (\sim p \vee q)$
T	T	F	T	T	T
T	F	F	F	F	T
F	T	T	T	T	T
F	F	T	T	T	T

49. tautology

p	q	$p \leftrightarrow q$	$p \to q$	$q \to p$	$(q \to p) \wedge (p \to q)$	$(p \leftrightarrow q) \leftrightarrow [(q \to p) \wedge (p \to q)]$
T	T	T	T	T	T	T
T	F	F	F	T	F	T
F	T	F	T	F	F	T
F	F	T	T	T	T	T

51. neither

p	q	r	$\sim p$	$p \wedge q$	$\sim p \vee r$	$(p \wedge q) \leftrightarrow (\sim p \vee r)$
T	T	T	F	T	T	T
T	T	F	F	T	F	F
T	F	T	F	F	T	F
T	F	F	F	F	F	T
F	T	T	T	F	T	F
F	T	F	T	F	T	F
F	F	T	T	F	T	F
F	F	F	T	F	T	F

53. tautology

p	q	r	$p \to q$	$q \to r$	$(p \to q) \wedge (q \to r)$	$p \to r$	$[(p \to q) \wedge (q \to r)] \to (p \to r)$
T	T	T	T	T	T	T	T
T	T	F	T	F	F	F	T
T	F	T	F	T	F	T	T
T	F	F	F	T	F	F	T
F	T	T	T	T	T	T	T
F	T	F	T	F	F	T	T
F	F	T	T	T	T	T	T
F	F	F	T	T	T	T	T

55. neither

p	q	r	$[(q \to r) \wedge (r \to \sim p)] \leftrightarrow (q \wedge p)$
T	T	T	F
T	T	F	F
T	F	T	T
T	F	F	F
F	T	T	F
F	T	F	T
F	F	T	F
F	F	F	F

57. a. p: You do homework right after class.; q: You fall behind.; $(p \rightarrow \sim q) \wedge (\sim p \rightarrow q)$

b.

p	q	$\sim p$	$\sim q$	$p \rightarrow \sim q$	$\sim p \rightarrow q$	$(p \rightarrow \sim q) \wedge (\sim p \rightarrow q)$
T	T	F	F	F	T	F
T	F	F	T	T	T	T
F	T	T	F	T	T	T
F	F	T	T	T	F	F

c. Answers will vary; an example is: The statement is false when p and q are both true.

59. a. p: You cut and paste from the Internet.; q: You cite the source.; r: You are charged with plagiarism.; $(p \wedge \sim q) \rightarrow r$

b.

p	q	r	$\sim q$	$p \wedge \sim q$	$(p \wedge \sim q) \rightarrow r$
T	T	T	F	F	T
T	T	F	F	F	T
T	F	T	T	T	T
T	F	F	T	T	F
F	T	T	F	F	T
F	T	F	F	F	T
F	F	T	T	F	T
F	F	F	T	F	T

c. The statement is false when p is true and q and r are false.

61. a. p: You are comfortable in your room.; q: You are honest with your roommate.; r: You enjoy the college experience.; $(p \leftrightarrow q) \vee \sim r$

b.

p	q	r	$\sim r$	$p \leftrightarrow q$	$(p \leftrightarrow q) \vee \sim r$
T	T	T	F	T	T
T	T	F	T	T	T
T	F	T	F	F	F
T	F	F	T	F	T
F	T	T	F	F	F
F	T	F	T	F	T
F	F	T	F	T	T
F	F	F	T	T	T

c. Answers will vary; an example is: The statement is false when p and r are true and q is false.

63. a. p: I enjoy the course.; q: I choose the class based on the professor.; r: I choose the class based on the course description.; $p \leftrightarrow (q \wedge \sim r)$

b.

p	q	r	$\sim r$	$q \wedge \sim r$	$p \leftrightarrow (q \wedge \sim r)$
T	T	T	F	F	F
T	T	F	T	T	T
T	F	T	F	F	F
T	F	F	T	F	F
F	T	T	F	F	T
F	T	F	T	T	F
F	F	T	F	F	T
F	F	F	T	F	T

c. Answers will vary; an example is: The statement is false when p, q, and r are all true.

65. false **67.** true **69.** false **71.** true **73.** true

75. $(p \rightarrow q) \leftrightarrow [(p \wedge q) \rightarrow \sim p]$

p	q	$(p \rightarrow q) \leftrightarrow [(p \wedge q) \rightarrow \sim p]$
T	T	F
T	F	F
F	T	T
F	F	T

77. $[p \to (\sim q \vee r)] \leftrightarrow (p \wedge r)$

p	q	r	$[p \to (\sim q \vee r)] \leftrightarrow (p \wedge r)$
T	T	T	T
T	T	F	T
T	F	T	T
T	F	F	F
F	T	T	F
F	T	F	F
F	F	T	F
F	F	F	F

79. p: You love a person.; q: You marry that person.; $(q \to p) \wedge (\sim p \to \sim q)$

p	q	$\sim p$	$\sim q$	$q \to p$	$\sim p \to \sim q$	$(q \to p) \wedge (\sim p \to \sim q)$
T	T	F	F	T	T	T
T	F	F	T	T	T	T
F	T	T	F	F	F	F
F	F	T	T	T	T	T

The statement is false when p is false and q is true.

81. p: You are happy.; q: You live contentedly.; r: You are wealthy.; $\sim[r \to (p \wedge q)]$

p	q	r	$p \wedge q$	$r \to (p \wedge q)$	$\sim[r \to (p \wedge q)]$
T	T	T	T	T	F
T	T	F	T	T	F
T	F	T	F	F	T
T	F	F	F	T	F
F	T	T	F	F	T
F	T	F	F	T	F
F	F	T	F	F	T
F	F	F	F	T	F

Answers will vary; an example is: The statement is false when p, q, and r are all true.

83. p: There was an increase in the percentage who believed in God.; q: There was a decrease in the percentage who believed in Heaven.; r: There was an increase in the percentage who believed in the devil.; $(p \wedge q) \to r$; true **85.** p: There was a decrease in the percentage who believed in God.; q: There was an increase in the percentage who believed in Heaven.; r: The percentage believing in the devil decreased.; $(p \leftrightarrow q) \vee r$; true
87. p: Fifteen percent are capitalists.; q: Thirty-four percent are members of the upper-middle class.; r: The number of working poor exceeds the number belonging to the working class.; $(p \vee \sim q) \leftrightarrow r$; false **89.** p: There are more people in the lower-middle class than in the capitalist and upper-middle classes combined.; q: One percent are capitalists.; r: Thirty-four percent belong to the upper-middle class.; $p \to (q \wedge r)$; false **97.** makes sense **99.** does not make sense **101.** Answers will vary; examples are $p \to q$, $\sim p$, $(p \to q) \vee \sim p$, and $\sim p \to [(p \to q) \vee \sim p]$.

Section 3.5

Check Point Exercises

1. a.

p	q	$\sim q$	$p \vee q$	$\sim q \to p$
T	T	F	T	T
T	F	T	T	T
F	T	F	T	T
F	F	T	F	F

The statements are equivalent since their truth values are the same.

b. If I don't lose my scholarship, then I attend classes.

2.

p	$\sim p$	$\sim(\sim p)$	$\sim[\sim(\sim p)]$
T	F	T	F
F	T	F	T

Since the truth values are the same, $\sim[\sim(\sim p)] \equiv \sim p$.

3. c **4. a.** If you're not driving too closely, then you can't read this. **b.** If it's not time to do the laundry, then you have clean underwear. **c.** If supervision during exams is required, then some students are not honest. **d.** $q \to (p \vee r)$ **5.** Converse: If you don't see a Club Med, then you are in Iran.; Inverse: If you are not in Iran, then you see a Club Med.; Contrapositive: If you see a Club Med, then you are not in Iran.

Concept and Vocabulary Check

1. equivalent; $\equiv$ **2.** $\sim q \to \sim p$ **3.** $q \to p$ **4.** $\sim p \to \sim q$ **5.** equivalent; converse; inverse **6.** false **7.** false

Exercise Set 3.5

1. a.

p	q	$\sim p$	$p \vee q$	$\sim p \rightarrow q$
T	T	F	T	T
T	F	F	T	T
F	T	T	T	T
F	F	T	F	F

b. The United States supports the development of solar-powered cars or it will suffer increasing atmospheric pollution.

3. not equivalent **5.** equivalent **7.** equivalent **9.** not equivalent **11.** not equivalent **13.** equivalent **15.** a **17.** c **19.** Converse: If I am in Illinois, then I am in Chicago.; Inverse: If I am not in Chicago, then I am not in Illinois.; Contrapositive: If I am not in Illinois, then I am not in Chicago. **21.** Converse: If I cannot hear you, then the stereo is playing.; Inverse: If the stereo is not playing, then I can hear you.; Contrapositive: If I can hear you, then the stereo is not playing. **23.** Converse: If you die, you don't laugh.; Inverse: If you laugh, you don't die.; Contrapositive: If you don't die, you laugh. **25.** Converse: If all troops were withdrawn, then the president is telling the truth.; Inverse: If the president is not telling the truth, then some troops were not withdrawn.; Contrapositive: If some troops were not withdrawn, then the president was not telling the truth. **27.** Converse: If some people suffer, then all institutions place profit above human need.; Inverse: If some institutions do not place profit above human need, then no people suffer.; Contrapositive: If no people suffer, then some institutions do not place profit above human need. **29.** Converse: $\sim r \rightarrow \sim q$; Inverse: $q \rightarrow r$; Contrapositive: $r \rightarrow q$ **31.** If a person is a scientist, then he or she knows some math.; Converse: If a person knows some math, then he or she is a scientist.; Inverse: If a person is not a scientist, then he or she doesn't know any math.; Contrapositive: If a person doesn't know any math, then he or she is not a scientist. **33.** If a person is an atheist, then he or she is not in a foxhole.; Converse: If a person is not in a foxhole, then he or she is an atheist.; Inverse: If a person is not an atheist, then he or she is in a foxhole.; Contrapositive: If a person is in a foxhole, then he or she is not an atheist. **35.** If a person is an attorney, then he or she has passed the bar exam.; Converse: If a person has passed the bar exam, then he or she is an attorney.; Inverse: If a person is not an attorney, then he or she has not passed the bar exam.; Contrapositive: If a person has not passed the bar exam, then he or she is not an attorney. **37.** If a person is a pacifist, then he or she is not a warmonger.; Converse: If a person is not a warmonger, then he or she is a pacifist.; Inverse: If a person is not a pacifist, then he or she is a warmonger.; Contrapositive: If a person is a warmonger, then he or she is not a pacifist. **39. a.** true **b.** Converse: If the corruption rating is 9.6, then the country is Finland.; Inverse: If the country is not Finland, then the corruption rating is not 9.6.; Contrapositive: If the corruption rating is not 9.6, then the country is not Finland.; The converse and inverse are not necessarily true, and the contrapositive is true. **47.** does not make sense **49.** makes sense

Section 3.6

Check Point Exercises

1. You do not have a fever and you have the flu. **2.** Bart Simpson is not a cartoon character or Tony Soprano is not a cartoon character. **3.** You do not leave by 5 P.M. and you arrive home on time. **4. a.** Some horror movies are not scary or none are funny. **b.** Your workouts are not strenuous and you get stronger. **5.** If we cannot swim or we can sail, it is windy.

Concept and Vocabulary Check

1. $p \wedge \sim q$; antecedent; and; consequent **2.** $\sim p \vee \sim q$; $\sim p \wedge \sim q$ **3.** or **4.** and **5.** false

Exercise Set 3.6

1. I am in Los Angeles and not in California. **3.** It is purple and it is a carrot. **5.** He doesn't and I won't. **7.** There is a blizzard and some schools are not closed. **9.** $\sim q \wedge r$ **11.** Australia is not an island or China is not an island. **13.** My high school did not encourage creativity or it did not encourage diversity. **15.** Jewish scripture does not give a clear indication of a heaven and it does not give a clear indication of an afterlife. **17.** The United States has eradicated neither poverty nor racism. **19.** $p \vee \sim q$ **21.** If you do not succeed, you did not attend lecture or did not study. **23.** If his wife does not cook and his child does not cook, then he does. **25.** $(\sim q \wedge r) \rightarrow \sim p$ **27.** I'm going to neither Seattle nor San Francisco. **29.** I do not study and I pass. **31.** I am going or he is not going. **33.** A bill does not become law or it receives majority approval. **35.** $\sim p \wedge q$ **37.** $\sim p \vee (\sim q \wedge \sim r)$ **39.** none **41.** none **43.** a and b **45.** a and b **47.** If there is no pain, there is no gain.; Converse: If there is no gain, then there is no pain.; Inverse: If there is pain, then there is gain.; Contrapositive: If there is gain, then there is pain.; Negation: There is no pain and there is gain. **49.** If you follow Buddha's "Middle Way," then you are neither hedonistic nor ascetic.; Converse: If you are neither hedonistic nor ascetic, then you follow Buddha's "Middle Way."; Inverse: If you do not follow Buddha's "Middle Way," then you are either hedonistic or ascetic.; Contrapositive: If you are either hedonistic or ascetic, then you do not follow Buddha's "Middle Way."; Negation: You follow Buddha's "Middle Way" and you are either hedonistic or ascetic. **51.** $p \wedge (\sim r \vee s)$ **53.** $\sim p \vee (r \wedge s)$ **55. a.** false **b.** Smoking does not reduce life expectancy by 2370 days or heart disease does not reduce life expectancy by 1247 days. **c.** true **57. a.** true **b.** Homicide does not reduce life expectancy by 74 days and fire reduces life expectancy by 25 days. **c.** false **59. a.** true **b.** Drowning reduces life expectancy by 10 times the number of days as airplane accidents and drowning reduces life expectancy by 24 days. **c.** false **65.** makes sense **67.** makes sense **69.** Contrapositive: If no one is eating turkey, then it is not Thanksgiving.; Negation: It is Thanksgiving and no one is eating turkey.

Section 3.7

Check Point Exercises

1. valid **2.** valid **3.** invalid **4. a.** valid **b.** invalid **c.** valid **5.** valid **6.** Some people do not lead.

Concept and Vocabulary Check

1. valid **2.** q; valid; $[(p \rightarrow q) \wedge p] \rightarrow q$ **3.** $\sim p$; valid; $[(p \rightarrow q) \wedge \sim q] \rightarrow \sim p$ **4.** $p \rightarrow r$; valid; $[(p \rightarrow q) \wedge (q \rightarrow r)] \rightarrow (p \rightarrow r)$ **5.** q; valid; $[(p \vee q) \wedge \sim p] \rightarrow q$ **6.** p **7.** $\sim q$ **8.** false **9.** true **10.** false

Exercise Set 3.7

1. invalid **3.** valid **5.** valid **7.** invalid **9.** invalid **11.** valid **13.** valid

15. $\begin{array}{r} p \to \sim q \\ \underline{q} \\ \therefore \sim p \end{array}$ valid **17.** $\begin{array}{r} p \vee q \\ \underline{q} \\ \therefore \sim p \end{array}$ invalid **19.** $\begin{array}{r} p \to q \\ \underline{q} \\ \therefore p \end{array}$ invalid **21.** $\begin{array}{r} p \to q \\ \underline{\sim p \to q} \\ \therefore q \end{array}$ valid **23.** $\begin{array}{r} p \vee q \\ \underline{\sim q} \\ \therefore p \end{array}$ valid **25.** $\begin{array}{r} p \to q \\ \underline{\sim p} \\ \therefore \sim q \end{array}$ invalid **27.** $\begin{array}{r} p \to q \\ \underline{q \to r} \\ \therefore p \to r \end{array}$ valid **29.** $\begin{array}{r} p \to q \\ \underline{q \to r} \\ \therefore r \to p \end{array}$ invalid

31. $\begin{array}{r} (p \wedge q) \to r \\ \underline{p \wedge \sim r} \\ \therefore \sim q \end{array}$ valid **33.** $\begin{array}{r} (p \vee q) \to r \\ \underline{\sim r} \\ \therefore \sim p \wedge \sim q \end{array}$ valid **35.** $\begin{array}{r} (p \vee q) \to r \\ \underline{r} \\ \therefore p \vee q \end{array}$ invalid **37.** $\begin{array}{r} (p \wedge q) \to r \\ \underline{\sim p \vee \sim q} \\ \therefore \sim r \end{array}$ invalid **39.** $\begin{array}{r} p \to q \\ \underline{\sim p \to r} \\ \therefore q \vee r \end{array}$ valid **41.** $\begin{array}{r} p \to q \\ q \to \sim r \\ \underline{r} \\ \therefore \sim p \end{array}$ valid

43. My best friend is not a chemist. **45.** They were dropped from prime time. **47.** Some electricity is not off. **49.** If I vacation in Paris, I gain weight.

51. $\begin{array}{r} p \to q \\ \underline{\sim p} \\ \therefore \sim q \end{array}$ invalid **53.** $\begin{array}{r} p \vee q \\ \underline{\sim p} \\ \therefore q \end{array}$ valid **55.** $\begin{array}{r} p \to q \\ \underline{q} \\ \therefore p \end{array}$ invalid **57.** $\begin{array}{r} p \to q \\ \underline{\sim q} \\ \therefore \sim p \end{array}$ valid **59.** $\begin{array}{r} p \to q \\ \underline{\sim p} \\ \therefore \sim q \end{array}$ invalid **61.** $\begin{array}{r} p \to r \\ q \to r \\ \underline{p \vee q} \\ \therefore r \end{array}$ valid **63.** h **65.** i **67.** c **69.** a **71.** j

73. d **83.** does not make sense **85.** does not make sense **87.** People sometimes speak without being spoken to. **89.** The doctor either destroys the base on which the placebo rests or jeopardizes a relationship built on trust.

Section 3.8

Check Point Exercises

1. valid **2.** invalid **3.** valid **4.** invalid **5.** invalid **6.** invalid

Concept and Vocabulary Check

1. All *A* are *B*. **2.** No *A* are *B*. **3.** Some *A* are *B*. **4.** Some *A* are not *B*. **5.** false **6.** false

Exercise Set 3.8

1. valid **3.** invalid **5.** valid **7.** invalid **9.** invalid **11.** valid **13.** valid **15.** invalid **17.** invalid **19.** valid **21.** valid **23.** valid **25.** invalid **27.** invalid **29.** valid **31.** invalid **33.** invalid **35.** invalid **37.** valid **39.** valid **43.** makes sense **45.** does not make sense **47.** b **49.** Some teachers are amusing people.

Chapter 3 Review Exercises

1. $(p \wedge q) \to r$; If the temperature is below 32° and we have finished studying, we will go to the movies. **2.** $\sim r \to (\sim p \vee \sim q)$; If we do not go to the movies, then the temperature is not below 32° or we have not finished studying. **3.** The temperature is below 32°, and if we finish studying, we will go to the movies. **4.** We will go to the movies if and only if the temperature is below 32° and we have finished studying. **5.** It is not true that both the temperature is below 32° and we have finished studying. **6.** We will not go to the movies if and only if the temperature is not below 32° or we have not finished studying. **7.** $(p \wedge q) \vee r$ **8.** $(p \vee \sim q) \to r$ **9.** $q \to (p \leftrightarrow r)$ **10.** $r \leftrightarrow (p \wedge \sim q)$ **11.** $p \to r$ **12.** $q \to \sim r$ **13.** Some houses are not made with wood. **14.** Some students major in business. **15.** No crimes are motivated by passion. **16.** All Democrats are registered voters. **17.** Some new taxes will not be used for education.

18. neither

p	q	$\sim p$	$\sim p \wedge q$	$p \wedge (\sim p \wedge q)$
T	T	F	F	T
T	F	F	F	T
F	T	T	T	T
F	F	T	F	F

19. neither

p	q	$\sim p$	$\sim q$	$\sim p \vee \sim q$
T	T	F	F	F
T	F	F	T	T
F	T	T	F	T
F	F	T	T	T

20. neither

p	q	$\sim p$	$\sim p \vee q$	$p \to (\sim p \vee q)$
T	T	F	T	T
T	F	F	F	F
F	T	T	T	T
F	F	T	T	T

21. neither

p	q	$\sim q$	$p \leftrightarrow \sim q$
T	T	F	F
T	F	T	T
F	T	F	T
F	F	T	F

22. tautology

p	q	$\sim p$	$\sim q$	$p \vee q$	$\sim(p \vee q)$	$\sim p \wedge \sim q$	$\sim(p \vee q) \rightarrow (\sim p \wedge \sim q)$
T	T	F	F	T	F	F	T
T	F	F	T	T	F	F	T
F	T	T	F	T	F	F	T
F	F	T	T	F	T	T	T

23. neither

p	q	r	$\sim r$	$p \vee q$	$(p \vee q) \rightarrow \sim r$
T	T	T	F	T	F
T	T	F	T	T	T
T	F	T	F	T	F
T	F	F	T	T	T
F	T	T	F	T	F
F	T	F	T	T	T
F	F	T	F	F	T
F	F	F	T	F	T

24. neither

p	q	r	$p \wedge q$	$p \wedge r$	$(p \wedge q) \leftrightarrow (p \wedge r)$
T	T	T	T	T	T
T	T	F	T	F	F
T	F	T	F	T	F
T	F	F	F	F	T
F	T	T	F	F	T
F	T	F	F	F	T
F	F	T	F	F	T
F	F	F	F	F	T

25. neither

p	q	r	$r \rightarrow p$	$q \vee (r \rightarrow p)$	$p \wedge [q \vee (r \rightarrow p)]$
T	T	T	T	T	T
T	T	F	T	T	T
T	F	T	T	T	T
T	F	F	T	T	T
F	T	T	F	T	F
F	T	F	T	T	F
F	F	T	F	F	F
F	F	F	T	T	F

26. a. p: I'm in class.; q: I'm studying.; $(p \vee q) \wedge \sim p$

b.

p	q	$\sim p$	$p \vee q$	$(p \vee q) \wedge \sim p$
T	T	F	T	F
T	F	F	T	F
F	T	T	T	T
F	F	T	F	F

c. The statement is true when p is false and q is true.

27. a. p: You spit from a truck.; q: It's legal.; r: You spit from a car.; $(p \rightarrow q) \wedge (r \rightarrow \sim q)$

b.

p	q	r	$\sim q$	$p \rightarrow q$	$r \rightarrow \sim q$	$(p \rightarrow q) \wedge (r \rightarrow \sim q)$
T	T	T	F	T	F	F
T	T	F	F	T	T	T
T	F	T	T	F	T	F
T	F	F	T	F	T	F
F	T	T	F	T	F	F
F	T	F	F	T	T	T
F	F	T	T	T	T	T
F	F	F	T	T	T	T

c. The statement is true when p and q are both false.

28. false **29.** true **30.** true **31.** false **32.** p: The 2000 diversity index was 47.; q: The index increased from 2000 to 2010.; $p \wedge \sim q$; false
33. p: The diversity index decreased from 1980 through 2010.; q: The index was 55 in 1980.; r: The index was 34 in 2010.; $p \rightarrow (q \wedge r)$; true
34. p: The diversity increased by 6 from 1980 to 1990.; q: The diversity index increased by 7 from 1990 to 2000.; r: The index was at a maximum in 2010.; $(p \leftrightarrow q) \vee \sim r$; true

35. a.

p	q	$\sim p$	$\sim p \vee q$	$p \rightarrow q$
T	T	F	T	T
T	F	F	F	F
F	T	T	T	T
F	F	T	T	T

b. If the triangle is isosceles, then it has two equal sides.

36. c **37.** not equivalent **38.** equivalent **39.** Converse: If I am in the South, then I am in Atlanta.; Inverse: If I am not in Atlanta, then I am not in the South.; Contrapositive: If I am not in the South, then I am not in Atlanta. **40.** Converse: If today is not a holiday, then I am in class.; Inverse: If I am not in class, then today is a holiday.; Contrapositive: If today is a holiday, then I am not in class. **41.** Converse: If I pass all courses, then I worked hard.; Inverse: If I don't work hard, then I don't pass some courses.; Contrapositive: If I do not pass some course, then I did not work hard. **42.** Converse: $\sim q \rightarrow \sim p$; Inverse: $p \rightarrow q$.; Contrapositive: $q \rightarrow p$ **43.** An argument is sound and it is not valid. **44.** I do not work hard and I succeed. **45.** $\sim r \wedge \sim p$ **46.** Chicago is not a city or Maine is not a city. **47.** Ernest Hemingway was neither a musician nor an actor. **48.** If a number is not 0, the number is positive or negative. **49.** I do not work hard and I succeed. **50.** She is using her car or she is not taking a bus. **51.** $p \wedge \sim q$ **52.** a and c **53.** a and b **54.** a and c **55.** none **56.** invalid **57.** valid

58. $\begin{array}{l} p \rightarrow q \\ \underline{\quad q} \\ \therefore p \end{array}$ invalid **59.** $\begin{array}{l} p \vee q \\ \underline{\quad q} \\ \therefore \sim p \end{array}$ invalid **60.** $\begin{array}{l} p \vee q \\ \underline{\sim p} \\ \therefore q \end{array}$ valid **61.** $\begin{array}{l} p \rightarrow q \\ \underline{\sim q} \\ \therefore \sim p \end{array}$ valid **62.** $\begin{array}{l} p \rightarrow \sim q \\ \underline{\sim p \rightarrow q} \\ \therefore p \leftrightarrow \sim q \end{array}$ valid **63.** $\begin{array}{l} p \rightarrow \sim q \\ \underline{r \rightarrow q} \\ \therefore \sim r \rightarrow p \end{array}$ invalid

64. invalid **65.** valid **66.** valid **67.** invalid **68.** invalid **69.** valid

Chapter 3 Test

1. If I'm registered and I'm a citizen, then I vote. **2.** I don't vote if and only if I'm not registered or I'm not a citizen. **3.** I'm neither registered nor a citizen. **4.** $(p \wedge q) \vee \sim r$ **5.** $(\sim p \vee \sim q) \rightarrow \sim r$ **6.** $r \rightarrow q$ **7.** Some numbers are not divisible by 5. **8.** No people wear glasses.

9.

p	q	$\sim p$	$\sim p \vee q$	$p \wedge (\sim p \vee q)$
T	T	F	T	T
T	F	F	F	F
F	T	T	T	F
F	F	T	T	F

10.

p	q	$\sim p$	$\sim q$	$p \wedge q$	$\sim(p \wedge q)$	$(\sim p \vee \sim q)$	$\sim(p \wedge q) \leftrightarrow (\sim p \vee \sim q)$
T	T	F	F	T	F	F	T
T	F	F	T	F	T	T	T
F	T	T	F	F	T	T	T
F	F	T	T	F	T	T	T

11.

p	q	r	$q \vee r$	$p \leftrightarrow (q \vee r)$
T	T	T	T	T
T	T	F	T	T
T	F	T	T	T
T	F	F	F	F
F	T	T	T	F
F	T	F	T	F
F	F	T	T	F
F	F	F	F	T

12. p: You break the law.; q: You change the law.; $(p \wedge q) \rightarrow \sim p$

p	q	$\sim p$	$p \wedge q$	$(p \wedge q) \rightarrow \sim p$
T	T	F	T	F
T	F	F	F	T
F	T	T	F	T
F	F	T	F	T

The statement is false when p and q are both true. **13.** true

14. true **15.** p: There was an increase in the percentage spent on food.; q: There was an increase in the percentage spent on health care.; r: By 2010, the percentage spent on health care was more than triple the percentage spent on food.; $\sim p \vee (q \wedge r)$; true **16.** b **17.** If it snows, then it is not August. **18.** Converse: If I cannot concentrate, then the radio is playing.; Inverse: If the radio is not playing, then I can concentrate. **19.** It is cold and we use the pool. **20.** The test is not today and the party is not tonight. **21.** The banana is not green or it is ready to eat. **22.** a and b **23.** a and c **24.** invalid **25.** valid **26.** invalid **27.** invalid **28.** valid **29.** invalid

CHAPTER 4

Section 4.1

Check Point Exercises

1. a. $(4 \times 10^3) + (0 \times 10^2) + (2 \times 10^1) + (6 \times 1)$ or $(4 \times 1000) + (0 \times 100) + (2 \times 10) + (6 \times 1)$
b. $(2 \times 10^4) + (4 \times 10^3) + (2 \times 10^2) + (3 \times 10^1) + (2 \times 1)$ or $(2 \times 10{,}000) + (4 \times 1000) + (2 \times 100) + (3 \times 10) + (2 \times 1)$
2. a. 6073 **b.** 80,900 **3. a.** 12,031 **b.** 468,721 **4. a.** 80,293 **b.** 290,490

Concept and Vocabulary Check

1. numeral **2.** Hindu-Arabic **3.** 7; 10,000,000 **4.** expanded; positional **5.** $10^4; 10^3; 10^2; 10^1; 1$ **6.** 10; 60
7. $(10 + 1) \times 60^2 + (1 + 1) \times 60^1 + (10 + 10 + 1 + 1) \times 1$ **8.** $(10 \times 60^3) + (1 \times 60^2) + (2 \times 60^1) + (11 \times 1)$
9. 18×20^4 **10.** $2 \times 18 \times 20 = 720; 3 \times 20 = 60; 9 \times 1 = 9$; 789; 789

Exercise Set 4.1

1. 25 **3.** 8 **5.** 81 **7.** 100,000 **9.** $(3 \times 10^1) + (6 \times 1)$ **11.** $(2 \times 10^2) + (4 \times 10^1) + (9 \times 1)$
13. $(7 \times 10^2) + (0 \times 10^1) + (3 \times 1)$ **15.** $(4 \times 10^3) + (8 \times 10^2) + (5 \times 10^1) + (6 \times 1)$
17. $(3 \times 10^3) + (0 \times 10^2) + (7 \times 10^1) + (0 \times 1)$ **19.** $(3 \times 10^4) + (4 \times 10^3) + (5 \times 10^2) + (6 \times 10^1) + (9 \times 1)$
21. $(2 \times 10^8) + (3 \times 10^7) + (0 \times 10^6) + (0 \times 10^5) + (0 \times 10^4) + (7 \times 10^3) + (0 \times 10^2) + (0 \times 10^1) + (4 \times 1)$ **23.** 73
25. 385 **27.** 528,743 **29.** 7002 **31.** 600,002,007 **33.** 23 **35.** 1262 **37.** 1833 **39.** 11,523 **41.** 75,851 **43.** 2,416,271
45. 655,261 **47.** 14 **49.** 6846 **51.** 3048 **53.** 14,411 **55.** 75,610 **57.** 15,842,203 **59.** 29,520,224
61. $(4 \times 10^4) + (6 \times 10^3) + (2 \times 10^2) + (2 \times 10^1) + (5 \times 1)$ **63.** $(2 \times 10^3) + (2 \times 10^2) + (9 \times 10^1) + (9 \times 1)$
65. 0.4759 **67.** 0.700203 **69.** 5000.03 **71.** 30,700.05809 **73.** 9734 **75.** 8097 **77.** 365 is the number of days in a non–leap year.
87. does not make sense **89.** makes sense **91.** <<<<VVVV <<VVVVVVV

Section 4.2

Check Point Exercises

1. 487 **2.** 51 **3.** 2772 **4.** 11_{five} **5.** 1031_{seven} **6.** 110011_{two} **7.** 42023_{five}

Concept and Vocabulary Check

1. 5; 0, 1, 2, 3, and 4 **2.** $5^2; 5^1; 1$ **3.** 2; 0 and 1 **4.** $2^3; 2^2; 2^1; 1$ **5.** 13 **6.** 731 **7.** two; eight; sixteen

Exercise Set 4.2

1. 23 **3.** 42 **5.** 30 **7.** 11 **9.** 455 **11.** 28,909 **13.** 8342 **15.** 53 **17.** 44,261 **19.** 12_{five} **21.** 14_{seven} **23.** 10_{two}
25. 101_{two} **27.** 1000_{two} **29.** 31_{four} **31.** 101_{six} **33.** 322_{five} **35.** 1230_{four} **37.** 10011_{two} **39.** 111001_{two} **41.** 1011010_{two}
43. 12010_{three} **45.** 1442_{six} **47.** 4443_{seven} **49.** <<<<< <<<<<VV **51.** VVVVVV <<<VV <<VVVVVV

53. •
———
≡
••
———

55. • • •
• • • •
═
• • •
═
• • • •

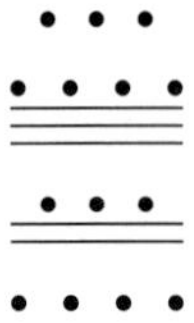

57. 25_{seven} **59.** 623_{eight} **61.** 1000110 **63.** 1101101 **65.** PAL **67.** 10011011101111101101

73. does not make sense **75.** makes sense **77.** 887_{nine}; 1000_{nine} **79.** 11111011_{two}; 673_{eight}; $3A6_{\text{twelve}}$

Section 4.3

Check Point Exercises

1. 131_{five} **2.** 1110_{two} **3.** 13_{five} **4.** 1605_{seven} **5.** 201_{seven} **6.** 23_{four}

Concept and Vocabulary Check

1. 1; 1; 11 **2.** 1; 1; 11 **3.** 1; 5; 15; 1; 5 **4.** 7; 7; 10; 5; 10; 5; 3; 1; 15 **5.** 2; 1; 21; 1; 2; 2; 2; 0; 20; 201 **6.** 2; 20 **7.** true

Exercise Set 4.3

1. 102_{four} **3.** 110_{two} **5.** 1310_{five} **7.** 1302_{seven} **9.** 15760_{nine} **11.** 23536_{seven} **13.** 13_{four} **15.** 4_{five} **17.** 206_{eight}
19. 366_{seven} **21.** 10_{two} **23.** 111_{three} **25.** 152_{six} **27.** 11_{two} **29.** 4011_{seven} **31.** 3114_{eight} **33.** 312_{four} **35.** 20_{four}
37. 41_{five} remainder of 1_{five} **39.** 1000110_{two} **41.** 110000_{two} **43.** 110111_{two} **45.** $71BE_{\text{sixteen}}$ **47.** 01100; 11101
49. 10001; 11001 **51.** 01100; 00110; 00100; 11011 **53.** The circuit in Exercise 47 **57.** makes sense **59.** makes sense
61. 8 hours, 13 minutes, 36 seconds **63.** 56_{seven}

Section 4.4

Check Point Exercises

1. 300,222 **2.** 𓆼𓆼𓍢𓍢𓍢𓍢𓍢𓎆𓎆𓎆𓎆𓎆𓎆𓏺𓏺𓏺 **3.** 1361 **4.** 1447 **5.** CCCXCIX **6.** 二
千
六
百
九
十
三 **7.** 885

Concept and Vocabulary Check

1. 1000 + 100 + 100 + 10 + 1 + 1 = 1212 **2.** 10 + 10 + 1 + 1 + 1; dccccbbaaa **3.** true **4.** add; +; 110 **5.** subtract; −; 40
6. 1000; 50 × 1000 = 50,000 **7.** true **8.** H
Y
D
X
F **9.** true **10.** VPC **11.** 6000; 500; 30; 5; 36, 535 **12.** true

Exercise Set 4.4

1. 322 **3.** 300,423 **5.** 132 **7.** 𓍢𓍢𓍢𓍢𓎆𓎆𓏺𓏺𓏺 **9.** 𓆼𓍢𓍢𓍢𓍢𓍢𓍢𓍢𓎆𓎆𓎆𓎆𓏺𓏺𓏺𓏺𓏺𓏺

11. 𓂭𓂭𓆼𓆼𓆼𓍢𓍢𓍢𓍢𓍢𓎆𓎆𓎆𓎆𓏺𓏺𓏺𓏺𓏺𓏺𓏺𓏺 **13.** 11 **15.** 16 **17.** 40 **19.** 59 **21.** 146 **23.** 1621 **25.** 2677

27. 9466 **29.** XLIII **31.** CXXIX **33.** MDCCCXCVI **35.** $\overline{\text{VI}}$DCCCXCII **37.** 88 **39.** 527 **41.** 2776

43. 四
十
三 **45.** 五
百
八
十
三 **47.** 四
千
八
百
七
十 **49.** 12 **51.** 234 **53.** $\mu\gamma$ **55.** $\upsilon\pi\gamma$ **57.** MMCCCXXIV; 二
千
三
百
二
十
四

59. a. 𓆼𓍢𓍢𓍢𓍢𓍢𓍢𓍢𓎆𓎆𓎆𓎆𓏺 **b.** 一
千
七
百
四
十
一 **61.** 3104_{five} **63.** 1232_{five} **65.** 𓎆𓎆𓎆𓎆𓎆𓎆𓎆𓎆𓎆𓏺𓏺𓏺𓏺𓏺𓏺𓏺𓏺

67. 1776; Declaration of Independence **69.** 4,640,224 **77.** does not make sense **79.** makes sense

81. Preceding: 𓍢𓍢𓎆𓎆𓎆𓎆𓎆𓎆𓎆𓎆𓎆𓏺𓏺𓏺𓏺𓏺𓏺𓏺𓏺𓏺

Following: 𓍢𓍢𓍢

Chapter 4 Review Exercises

1. 121 **2.** 343 **3.** $(4 \times 10^2) + (7 \times 10^1) + (2 \times 1)$ **4.** $(8 \times 10^3) + (0 \times 10^2) + (7 \times 10^1) + (6 \times 1)$
5. $(7 \times 10^4) + (0 \times 10^3) + (3 \times 10^2) + (2 \times 10^1) + (9 \times 1)$ **6.** 706,953 **7.** 740,000,306 **8.** 673 **9.** 8430 **10.** 2331
11. 65,536 **12.** Each position represents a particular value. The symbol in each position tells how many of that value are represented.
13. 19 **14.** 6 **15.** 325 **16.** 805 **17.** 4051 **18.** 560 **19.** 324_{five} **20.** 10101_{two} **21.** 122112_{three} **22.** 26452_{seven}
23. 111140_{six} **24.** $3A2_{\text{twelve}}$ **25.** 132_{seven} **26.** 1401_{eight} **27.** 110000_{two} **28.** $AD0_{\text{sixteen}}$ **29.** 5_{six} **30.** 345_{seven} **31.** 11_{two}
32. 2304_{five} **33.** 222_{four} **34.** 354_{seven} **35.** 1102_{five} **36.** 133_{four} **37.** 12_{five} **38.** 1246 **39.** 12,432
40. 𓆼𓆼𓍢𓍢𓍢𓍢𓎆𓎆𓎆𓎆𓎆𓎆𓎆𓎆𓏺𓏺𓏺𓏺𓏺𓏺 **41.** 𓂭𓂭𓂭𓆼𓆼𓆼𓆼𓍢𓍢𓍢𓍢𓍢𓎆𓎆𓎆𓎆𓎆𓎆𓎆𓏺𓏺𓏺 **42.** 2314
43. DDDDDCCCCBBBBBBBBBAA **44.** Answers will vary. **45.** 163 **46.** 1034 **47.** 1990 **48.** XLIX **49.** MMCMLXV
50. If symbols increase in value from left to right, subtract the value of the symbol on the left from the symbol on the right. Answers will vary.

51. 554 **52.** 8253 **53.** 二 百 七 十 四 **54.** 三 千 五 百 八 十 七 **55.** 365 **56.** 4520 **57.** G Y I X C **58.** F Z H Y E X D **59.** Answers will vary. **60.** 653

61. 678 **62.** $\upsilon\nu\gamma$ **63.** $\pi\beta$ **64.** 357 **65.** 37,894 **66.** 80,618 **67.** WRG **68.** lfVQC **69.** Answers will vary.

Chapter 4 Test

1. 729 **2.** $(5 \times 10^2) + (6 \times 10^1) + (7 \times 1)$ **3.** $(6 \times 10^4) + (3 \times 10^3) + (0 \times 10^2) + (2 \times 10^1) + (8 \times 1)$ **4.** 7493 **5.** 400,206
6. A number represents "How many?" whereas a numeral is a symbol used to write a number.
7. A symbol for zero is needed for a place holder when there are no values for a position.
8. 72,731 **9.** 1560 **10.** 113 **11.** 223 **12.** 53 **13.** 2212_{three} **14.** 111000_{two} **15.** 24334_{five} **16.** 1212_{five} **17.** 414_{seven}
18. 250_{six} **19.** 221_{five} **20.** 20,303 **21.**  **22.** 1994 **23.** CDLIX **24.** Answers will vary.

CHAPTER 5

Section 5.1

Check Point Exercises

1. b **2.** $2^3 \cdot 3 \cdot 5$ **3.** 75 **4.** 96 **5.** 90 **6.** 5:00 P.M.

Concept and Vocabulary Check

1. prime **2.** composite **3.** greatest common divisor **4.** least common multiple **5.** false **6.** true **7.** false **8.** false

Exercise Set 5.1

1. a. yes **b.** no **c.** yes **d.** no **e.** no **f.** yes **g.** no **h.** no **i.** no **3. a.** yes **b.** yes **c.** yes **d.** no **e.** yes **f.** yes **g.** no **h.** no **i.** yes **5. a.** yes **b.** no **c.** yes **d.** no **e.** no **f.** no **g.** no **h.** no **i.** no **7. a.** yes **b.** yes **c.** yes **d.** no **e.** yes **f.** yes **g.** yes **h.** no **i.** yes **9. a.** yes **b.** yes **c.** yes **d.** yes **e.** yes **f.** yes **g.** yes **h.** yes **i.** yes **11.** true; $5 + 9 + 5 + 8 = 27$ which is divisible by 3. **13.** true; the last two digits of 10,612 form the number 12 and 12 is divisible by 4. **15.** false **17.** true; 104,538 is an even number so it is divisible by 2. $1 + 0 + 4 + 5 + 3 + 8 = 21$ which is divisible by 3 so it is divisible by 3. Any number divisible by 2 and 3 is divisible by 6. **19.** true; the last three digits of 20,104 form the number 104 and 104 is divisible by 8. **21.** false **23.** true; $5 + 1 + 7 + 8 + 7 + 2 = 30$ which is divisible by 3 so it is divisible by 3. The last two digits of 517,872 form the number 72 and 72 is divisible by 4 so it is divisible by 4. Any number divisible by 3 and 4 is divisible by 12. **25.** $3 \cdot 5^2$ **27.** $2^3 \cdot 7$ **29.** $3 \cdot 5 \cdot 7$ **31.** $2^2 \cdot 5^3$ **33.** $3 \cdot 13 \cdot 17$ **35.** $3 \cdot 5 \cdot 59$ **37.** $2^5 \cdot 3^2 \cdot 5$ **39.** $2^2 \cdot 499$ **41.** $3 \cdot 5^2 \cdot 7^2$ **43.** $2^3 \cdot 3 \cdot 5^2 \cdot 11 \cdot 13$ **45.** 14 **47.** 2 **49.** 12 **51.** 24 **53.** 38 **55.** 15 **57.** 168 **59.** 336 **61.** 540 **63.** 360 **65.** 3420 **67.** 4560 **69.** 8 **71.** 6 **73.** 2, 6 **75.** perfect **77.** not perfect **79.** not an emirp **81.** emirp **83.** not a Germain prime **85.** not a Germain prime **87.** 648; 648; Answers will vary. An example is: The product of the greatest common divisor and least common multiple of two numbers equals the product of the two numbers. **89.** The numbers are the prime numbers between 2 and 97, inclusive. **91.** 12 **93.** 10 **95.** July 1 **97.** 90 min or $1\frac{1}{2}$ hr **111.** does not make sense **113.** does not make sense **115.** GCD: $2^{14} \cdot 3^{25} \cdot 5^{30}$; LCM: $2^{17} \cdot 3^{37} \cdot 5^{31}$ **117.** 2:20 A.M. on the third day **119.** c **121.** no

Section 5.2

Check Point Exercises

1. 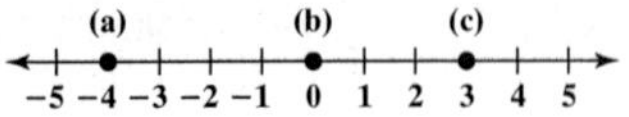

2. a. > **b.** < **c.** < **d.** < **3. a.** 8 **b.** 6 **c.** −8 **4. a.** 37 **b.** −4 **c.** −24 **5. a.** 8 years **b.** 14 years **6. a.** 25 **b.** −25 **c.** −64 **d.** 81 **7.** 36 **8.** 82

Concept and Vocabulary Check

1. $\{\ldots, -3, -2, -1, 0, 1, 2, 3, \ldots\}$ **2.** left **3.** the distance from 0 to a **4.** additive inverses **5.** false **6.** true **7.** true **8.** false

Exercise Set 5.2

1. 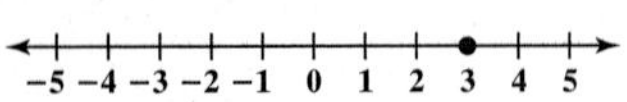

3. 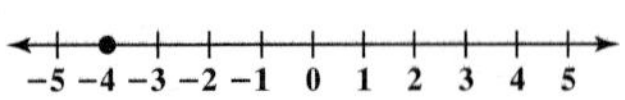

5. < **7.** < **9.** > **11.** < **13.** 14 **15.** 14 **17.** 300,000 **19.** −12 **21.** 4 **23.** −3 **25.** −5 **27.** −18 **29.** 0 **31.** 5 **33.** −7 **35.** 14 **37.** 11 **39.** −9 **41.** −28 **43.** −54 **45.** 21 **47.** −12 **49.** 13 **51.** 0 **53.** 25 **55.** 25 **57.** 64 **59.** −125 **61.** 625 **63.** −81 **65.** 81 **67.** −3 **69.** −7 **71.** 30 **73.** 0 **75.** undefined **77.** −20 **79.** −31 **81.** 25 **83.** 13 **85.** 13 **87.** 33 **89.** 32 **91.** −24 **93.** 0

95. −32 **97.** 45 **99.** 14 **101.** 14 **103.** −36 **105.** 88 **107.** −29 **109.** −10 **111.** $-10-(-2)^3$; −2 **113.** $[2(7-10)]^2$; 36 **115.** 20,602 ft **117.** shrink by 5 years **119.** shrink by 11 years **121.** no change **123.** 25 years **125.** 1 year **127. a.** $128 billion; surplus **b.** − $615 billion; deficit **c.** $743 billion **129.** $65 billion **131.** 7°F **133.** −2°F **145.** makes sense **147.** makes sense **149.** $(8-2)\cdot 3-4=14$ **151.** −36

Section 5.3

Check Point Exercises

1. $\frac{4}{5}$ **2.** $\frac{21}{8}$ **3.** $1\frac{2}{3}$ **4. a.** 0.375 **b.** $0.\overline{45}$ **5. a.** $\frac{9}{10}$ **b.** $\frac{43}{50}$ **c.** $\frac{53}{1000}$ **6.** $\frac{2}{9}$ **7.** $\frac{79}{99}$ **8. a.** $\frac{8}{33}$ **b.** $\frac{3}{2}$ or $1\frac{1}{2}$ **c.** $\frac{51}{10}$ or $5\frac{1}{10}$ **9. a.** $\frac{36}{55}$ **b.** $-\frac{4}{3}$ or $-1\frac{1}{3}$ **c.** $\frac{3}{2}$ or $1\frac{1}{2}$ **10. a.** $\frac{2}{3}$ **b.** $\frac{3}{2}$ or $1\frac{1}{2}$ **c.** $-\frac{9}{4}$ or $-2\frac{1}{4}$ **11.** $\frac{19}{20}$ **12.** $-\frac{17}{60}$ **13.** $\frac{3}{4}$ **14.** $\frac{5}{12}$ **15.** $2\frac{4}{5}$ eggs; 3 eggs

Concept and Vocabulary Check

1. rational; integers; zero **2.** improper; the numerator is greater than the denominator **3.** terminate/stop; have repeating digits **4.** reciprocal/multiplicative inverse **5.** false **6.** false **7.** true **8.** false

Exercise Set 5.3

1. $\frac{2}{3}$ **3.** $\frac{5}{6}$ **5.** $\frac{4}{7}$ **7.** $\frac{5}{9}$ **9.** $\frac{9}{10}$ **11.** $\frac{14}{19}$ **13.** $\frac{19}{8}$ **15.** $-\frac{38}{5}$ **17.** $\frac{199}{16}$ **19.** $4\frac{3}{5}$ **21.** $-8\frac{4}{9}$ **23.** $35\frac{11}{20}$ **25.** 0.75 **27.** 0.35 **29.** 0.875 **31.** $0.\overline{81}$ **33.** $3.\overline{142857}$ **35.** $0.\overline{285714}$ **37.** $\frac{3}{10}$ **39.** $\frac{2}{5}$ **41.** $\frac{39}{100}$ **43.** $\frac{41}{50}$ **45.** $\frac{29}{40}$ **47.** $\frac{5399}{10{,}000}$ **49.** $\frac{7}{9}$ **51.** 1 **53.** $\frac{4}{11}$ **55.** $\frac{257}{999}$ **57.** $\frac{21}{88}$ **59.** $-\frac{7}{120}$ **61.** $\frac{3}{2}$ **63.** 6 **65.** $\frac{10}{3}$ **67.** $-\frac{14}{15}$ **69.** 6 **71.** $\frac{5}{11}$ **73.** $\frac{2}{3}$ **75.** $\frac{2}{3}$ **77.** $\frac{7}{10}$ **79.** $\frac{9}{10}$ **81.** $\frac{53}{120}$ **83.** $\frac{1}{2}$ **85.** $\frac{7}{12}$ **87.** $-\frac{71}{150}$ **89.** $\frac{53}{12}$ or $4\frac{5}{12}$ **91.** $\frac{7}{6}$ or $1\frac{1}{6}$ **93.** $-\frac{5}{2}$ or $-2\frac{1}{2}$ **95.** $\frac{11}{14}$ **97.** $\frac{4}{15}$ **99.** $-\frac{9}{40}$ **101.** $-1\frac{1}{36}$ **103.** $-19\frac{3}{4}$ **105.** $\frac{7}{24}$ **107.** $\frac{7}{12}$ **109.** $-\frac{3}{4}$ **111.** Both are equal to $\frac{169}{36}$. **113.** $\frac{1}{2^2\cdot 3}$ **115.** $-\frac{289}{2^4\cdot 5^4\cdot 7}$ **117.** $0.\overline{54}$; $0.58\overline{3}$; < **119.** $-0.8\overline{3}$; $-0.\overline{8}$; > **121.** $\frac{4}{25}$ **123. a.** $\frac{13}{20}$ **b.** $\frac{11}{100}$ **125.** $\frac{1}{3}$ cup butter, $\frac{5}{2}=2.5$ ounces unsweetened chocolate, $\frac{3}{4}$ cup sugar, 1 teaspoon vanilla, 1 egg, $\frac{1}{2}$ cup flour **127.** $\frac{5}{6}$ cup butter, $\frac{25}{4}=6.25$ ounces unsweetened chocolate, $\frac{15}{8}=1\frac{7}{8}$ cups sugar, $\frac{5}{2}=2\frac{1}{2}$ teaspoons vanilla, $\frac{5}{2}=2\frac{1}{2}$ eggs, $\frac{5}{4}=1\frac{1}{4}$ cups flour **129.** 24 brownies **131.** $3\frac{2}{3}$ c **133. a.** D, E, G, A, B **b.** There are black keys to the left of the keys for the notes D, E, G, A, and B. **135.** $16\frac{7}{16}$ in. **137.** $\frac{1}{3}$ of the business **139.** $1\frac{3}{20}$ mi; $\frac{7}{20}$ mi **141.** $\frac{1}{10}$ **155.** makes sense **157.** makes sense **159.** $\frac{1}{1\cdot 2}+\frac{1}{2\cdot 3}+\frac{1}{3\cdot 4}+\frac{1}{4\cdot 5}+\frac{1}{5\cdot 6}=\frac{5}{6}$

Section 5.4

Check Point Exercises

1. a. $2\sqrt{3}$ **b.** $2\sqrt{15}$ **c.** cannot be simplified **2. a.** $\sqrt{30}$ **b.** 10 **c.** $2\sqrt{3}$ **3. a.** 4 **b.** $2\sqrt{2}$ **4. a.** $18\sqrt{3}$ **b.** $-5\sqrt{13}$ **c.** $8\sqrt{10}$ **5. a.** $3\sqrt{3}$ **b.** $-13\sqrt{2}$ **6. a.** $\frac{5\sqrt{10}}{2}$ **b.** $\frac{\sqrt{14}}{7}$ **c.** $\frac{5\sqrt{2}}{6}$

Concept and Vocabulary Check

1. terminating; repeating **2.** π **3.** $\sqrt{n}$; n **4.** $\sqrt{49}\cdot\sqrt{6}=7\sqrt{6}$ **5.** coefficient **6.** $(8+10)\sqrt{3}=18\sqrt{3}$ **7.** $5\sqrt{2}+4\sqrt{2}=9\sqrt{2}$ **8.** rationalizing the denominator **9.** $\sqrt{7}$ **10.** $\sqrt{3}$

Exercise Set 5.4

1. 3 **3.** 5 **5.** 8 **7.** 11 **9.** 13 **11. a.** 13.2 **b.** 13.15 **c.** 13.153 **13. a.** 133.3 **b.** 133.27 **c.** 133.270 **15. a.** 1.8 **b.** 1.77 **c.** 1.772 **17.** $2\sqrt{5}$ **19.** $4\sqrt{5}$ **21.** $5\sqrt{10}$ **23.** $14\sqrt{7}$ **25.** $\sqrt{42}$ **27.** 6 **29.** $3\sqrt{2}$ **31.** $2\sqrt{13}$ **33.** 3 **35.** $3\sqrt{5}$ **37.** $-4\sqrt{3}$ **39.** $13\sqrt{3}$ **41.** $-2\sqrt{13}$ **43.** $2\sqrt{5}$ **45.** $7\sqrt{2}$ **47.** $3\sqrt{5}$ **49.** $2\sqrt{2}$ **51.** $34\sqrt{2}$ **53.** $-\frac{3}{2}\sqrt{3}$ **55.** $11\sqrt{3}$ **57.** $\frac{5\sqrt{3}}{3}$ **59.** $3\sqrt{7}$ **61.** $\frac{2\sqrt{30}}{5}$ **63.** $\frac{5\sqrt{3}}{2}$ **65.** $\frac{\sqrt{10}}{5}$ **67.** $20\sqrt{2}-5\sqrt{3}$ **69.** $-7\sqrt{7}$ **71.** $\frac{43\sqrt{2}}{35}$

73. $\frac{5\sqrt{6}}{6}$ **75.** $6\sqrt{3}$ miles; 10.4 miles **77.** 70 mph; He was speeding. **79. a.** 41 in. **b.** 40.6 in.; quite well **81. a.** $5.8\sqrt{10}$ **b.** 18.3; underestimates by 5.8 million **83.** $0.6R_f$; 60 weeks **93.** does not make sense **95.** does not make sense **97.** false; Changes will vary. **99.** true **101.** < **103.** > **105.** −7 and −6 **107.** Answers will vary; an example is $(3+\sqrt{2})-(1+\sqrt{2})=2$

Section 5.5

Check Point Exercises

1. a. $\sqrt{9}$ **b.** $0, \sqrt{9}$ **c.** $-9, 0, \sqrt{9}$ **d.** $-9, -1.3, 0, 0.\overline{3}, \sqrt{9}$ **e.** $\frac{\pi}{2}, \sqrt{10}$ **f.** $-9, -1.3, 0, 0.\overline{3}, \frac{\pi}{2}, \sqrt{9}, \sqrt{10}$ **2. a.** associative property of multiplication **b.** commutative property of addition **c.** distributive property of multiplication over addition **d.** commutative property of multiplication **e.** identity property of addition **f.** inverse property of multiplication **3. a.** yes **b.** no **4. a.** The entries in the body of the table are all elements of the set. **b.** $(2\oplus 2)\oplus 3 = 2\oplus(2\oplus 3)$
$0\oplus 3 = 2\oplus 1$
$3 = 3$

c. 0 **d.** The inverse of 0 is 0, the inverse of 1 is 3, the inverse of 2 is 2, and the inverse of 3 is 1.
e. $1\oplus 3 = 3\oplus 1$; $3\oplus 2 = 2\oplus 3$
$0 = 0$ $\quad 1 = 1$

Concept and Vocabulary Check

1. rational; irrational **2.** closure **3.** $ab = ba$ **4.** $(a+b)+c = a+(b+c)$ **5.** $a(b+c) = ab+ac$ **6.** identity **7.** identity; 1 **8.** multiplicative inverse or reciprocal; multiplicative identity **9.** $1\oplus 4 = 0; 3\oplus 3 = 1; 4\oplus 2 = 1$ **10.** true

Exercise Set 5.5

1. a. $\sqrt{100}$ **b.** $0, \sqrt{100}$ **c.** $-9, 0, \sqrt{100}$ **d.** $-9, -\frac{4}{5}, 0, 0.25, 9.2, \sqrt{100}$ **e.** $\sqrt{3}$ **f.** $-9, -\frac{4}{5}, 0, 0.25, \sqrt{3}, 9.2, \sqrt{100}$ **3. a.** $\sqrt{64}$ **b.** $0, \sqrt{64}$ **c.** $-11, 0, \sqrt{64}$ **d.** $-11, -\frac{5}{6}, 0, 0.75, \sqrt{64}$ **e.** $\sqrt{5}, \pi$ **f.** $-11, -\frac{5}{6}, 0, 0.75, \sqrt{5}, \pi, \sqrt{64}$ **5.** 0 **7.** Answers will vary; an example is: $\frac{1}{2}$. **9.** Answers will vary; an example is: 1. **11.** Answers will vary; an example is: $\sqrt{2}$. **13.** 4 **15.** 6 **17.** 4; 3 **19.** 7 **21.** $30+5\sqrt{2}$ **23.** $3\sqrt{7}+\sqrt{14}$ **25.** $5\sqrt{3}+3$ **27.** $2\sqrt{3}+6$ **29.** commutative property of addition **31.** associative property of addition **33.** commutative property of addition **35.** distributive property of multiplication over addition **37.** associative property of multiplication **39.** identity property of multiplication **41.** inverse property of addition **43.** inverse property of multiplication **45.** Answers will vary; an example is: $1-2=-1$. **47.** Answers will vary; an example is: $4\div 8 = \frac{1}{2}$. **49.** Answers will vary; an example is: $\sqrt{2}\cdot\sqrt{2}=2$. **51. a.** The entries in the body of the table are all elements of the set. **b.** $(4\oplus 6)\oplus 7 = 4\oplus(6\oplus 7)$
$2\oplus 7 = 4\oplus 5$
$1 = 1$

c. 0 **d.** The inverse of 0 is 0, the inverse of 1 is 7, the inverse of 2 is 6, the inverse of 3 is 5, the inverse of 4 is 4, the inverse of 5 is 3, the inverse of 6 is 2, and the inverse of 7 is 1. **e.** $5\oplus 6 = 6\oplus 5$; $4\oplus 7 = 7\oplus 4$
$3 = 3$ $\quad 3 = 3$

53. true **55.** false **57.** distributive property; commutative property of addition; associative property of addition; distributive property; commutative property of addition **59. a.** $c\triangle(d\square e) = c\triangle c = e; (c\triangle d)\square(c\triangle e) = b\square d = e$ **b.** distributive property **61.** b **63.** c **65.** d **67.** vampire **69.** not a vampire **71.** narcissistic **73.** not narcissistic **75. a.** distributive property **b.** approximately 108 mg; Answers will vary. **77.** sevenfold rotational symmetry **89.** makes sense **91.** makes sense **93.** false **95.** false **97.** false **99.** false

Section 5.6

Check Point Exercises

1. a. 1 **b.** 1 **c.** 1 **d.** −1 **2. a.** $\frac{1}{81}$ **b.** $\frac{1}{216}$ **c.** $\frac{1}{12}$ **3. a.** 7,400,000,000 **b.** 0.000003017 **4. a.** 7.41×10^9 **b.** 9.2×10^{-8} **5. a.** 4.1×10^9 **6.** 520,000 **7.** 0.0000023 **8. a.** 1.872×10^4 **b.** 18,720 **9.** $60,000

Concept and Vocabulary Check

1. add **2.** multiply **3.** subtract **4.** one **5.** a number greater than or equal to 1 and less than 10; 10 to an integer power **6.** false **7.** false **8.** false **9.** true **10.** false

Exercise Set 5.6

1. $2^5 = 32$ **3.** $4^3 = 64$ **5.** $2^6 = 64$ **7.** $1^{20} = 1$ **9.** $4^2 = 16$ **11.** $2^4 = 16$ **13.** 1 **15.** 1 **17.** −1 **19.** $\frac{1}{4}$ **21.** $\frac{1}{64}$ **23.** $\frac{1}{32}$ **25.** $3^2 = 9$ **27.** $3^{-2} = \frac{1}{9}$ **29.** $2^{-4} = \frac{1}{16}$ **31.** $\frac{1}{x^{16}}$ **33.** $\frac{1}{x^2}$ **35.** $\frac{1}{x^{12}}$ **37.** $\frac{2}{5}$ **39.** $-\frac{6x^2}{y^3}$ **41.** $-\frac{5y^8}{x^6}$ **43.** 270 **45.** 912,000 **47.** 80,000,000 **49.** 100,000 **51.** 0.79 **53.** 0.0215 **55.** 0.000786 **57.** 0.00000318 **59.** 3.7×10^2 **61.** 3.6×10^3 **63.** 3.2×10^4 **65.** 2.2×10^8 **67.** 2.7×10^{-2} **69.** 3.7×10^{-3} **71.** 2.93×10^{-6} **73.** 8.2×10^7 **75.** 4.1×10^5 **77.** 2.1×10^{-6} **79.** 600,000 **81.** 60,000 **83.** 0.123 **85.** 30,000 **87.** 3,000,000 **89.** 0.03 **91.** 0.0021

93. $(8.2 \times 10^7)(3.0 \times 10^9) = 2.46 \times 10^{17}$ **95.** $(5.0 \times 10^{-4})(6.0 \times 10^6) = 3 \times 10^3$ **97.** $\frac{9.5 \times 10^6}{5 \times 10^2} = 1.9 \times 10^4$ **99.** $\frac{8 \times 10^{-5}}{2 \times 10^2} = 4 \times 10^{-7}$ **101.** $\frac{4.8 \times 10^{11}}{1.2 \times 10^{-4}} = 4 \times 10^{15}$ **103.** $\frac{11}{18}$ **105.** $\frac{99}{25} = 3\frac{24}{25}$ **107.** 2.5×10^{-3} **109.** 8×10^{-5} **111. a.** 3.18×10^{12} **b.** 3.20×10^8 **c.** \$9938 **113. a.** 1.89×10^{13} **b.** 6×10^4 **c.** 3.15×10^8 or 315,000,000 **115.** 1.06×10^{-18} g **117.** 4.064×10^9 **129.** makes sense **131.** does not make sense **133.** false **135.** false **137.** false **139.** false

Section 5.7

Check Point Exercises

1. 100, 120, 140, 160, 180, and 200 **2.** 8, 5, 2, −1, −4, and −7 **3.** −34 **4. a.** $a_n = 0.35n + 15.65$ **b.** 23% **5.** 12, −6, 3, $-\frac{3}{2}$, $\frac{3}{4}$, and $-\frac{3}{8}$ **6.** 3645 **7.** $a_n = 3(2)^{n-1}$; 384

Concept and Vocabulary Check

1. arithmetic; common difference **2.** $a_n = a_1 + (n-1)d$; the first term; the common difference **3.** geometric; common ratio **4.** $a_n = a_1r^{n-1}$; first term; the common ratio

Exercise Set 5.7

1. 8, 10, 12, 14, 16, and 18 **3.** 200, 220, 240, 260, 280, and 300 **5.** −7, −3, 1, 5, 9, and 13 **7.** −400, −100, 200, 500, 800, and 1100 **9.** 7, 4, 1, −2, −5, and −8 **11.** 200, 140, 80, 20, −40, and −100 **13.** $\frac{5}{2}$, 3, $\frac{7}{2}$, 4, $\frac{9}{2}$, and 5 **15.** $\frac{3}{2}$, $\frac{7}{4}$, 2, $\frac{9}{4}$, $\frac{5}{2}$, and $\frac{11}{4}$ **17.** 4.25, 4.55, 4.85, 5.15, 5.45, and 5.75 **19.** 4.5, 3.75, 3, 2.25, 1.5, and 0.75 **21.** 33 **23.** 252 **25.** 67 **27.** 955 **29.** −82 **31.** −142 **33.** −43 **35.** −248 **37.** $\frac{23}{2}$ **39.** 1.75 **41.** $a_n = 1 + (n-1)4$; 77 **43.** $a_n = 7 + (n-1)(-4)$; −69 **45.** $a_n = 9 + (n-1)2$; 47 **47.** $a_n = -20 + (n-1)(-4)$; −96 **49.** 4, 8, 16, 32, 64, and 128 **51.** 1000, 1000, 1000, 1000, 1000, and 1000 **53.** 3, −6, 12, −24, 48, and −96 **55.** 10, −40, 160, −640, 2560, and −10,240 **57.** 2000, −2000, 2000, −2000, 2000, and −2000 **59.** −2, 6, −18, 54, −162, and 486 **61.** −6, 30, −150, 750, −3750, and 18,750 **63.** $\frac{1}{4}$, $\frac{1}{2}$, 1, 2, 4, and 8 **65.** $\frac{1}{4}$, $\frac{1}{8}$, $\frac{1}{16}$, $\frac{1}{32}$, $\frac{1}{64}$, and $\frac{1}{128}$ **67.** $-\frac{1}{16}$, $\frac{1}{4}$, −1, 4, −16, and 64 **69.** 2, 0.2, 0.02, 0.002, 0.0002, and 0.00002 **71.** 256 **73.** $2{,}324{,}522{,}934 \approx 2.32 \times 10^9$ **75.** 50 **77.** 320 **79.** −2 **81.** 486 **83.** $\frac{3}{64}$ **85.** $-\frac{2}{27}$ **87.** $\approx -1.82 \times 10^{-9}$ **89.** 0.1 **91.** $a_n = 3(4)^{n-1}$; 12,288 **93.** $a_n = 18\left(\frac{1}{3}\right)^{n-1}$; $\frac{2}{81}$ **95.** $a_n = 1.5(-2)^{n-1}$; 96 **97.** $a_n = 0.0004(-10)^{n-1}$; 400 **99.** arithmetic; 18 and 22 **101.** geometric; 405 and 1215 **103.** arithmetic; 13 and 18 **105.** geometric; $\frac{3}{16}$ and $\frac{3}{32}$ **107.** arithmetic; $\frac{5}{2}$ and 3 **109.** geometric; 7 and −7 **111.** arithmetic; −49 and −63 **113.** geometric; $25\sqrt{5}$ and 125 **115.** arithmetic; 310 **117.** geometric; 59,048 **119.** geometric; −1023 **121.** arithmetic; 80 **123.** 5050 **125. a.** $a_n = 0.6n + 17.8$ **b.** 41.8% **127.** Company A; \$1400 **129.** \$16,384 **131.** \$3,795,957 **133. a.** approximately 1.01 for all but one division **b.** $a_n = 33.87(1.01)^{n-1}$ **c.** 41.33 million **143.** makes sense **145.** makes sense **147.** false **149.** true **151.** false **153.** true

Chapter 5 Review Exercises

1. 2: yes; 3: yes; 4: yes; 5: no; 6: yes; 8: yes; 9: no; 10: no; 12: yes **2.** 2: yes; 3: yes; 4: no; 5: yes; 6: yes; 8: no; 9: yes; 10: yes; 12: no **3.** $3 \cdot 5 \cdot 47$ **4.** $2^6 \cdot 3 \cdot 5$ **5.** $3 \cdot 5^2 \cdot 7 \cdot 13$ **6.** GCD: 6; LCM: 240 **7.** GCD: 6; LCM: 900 **8.** GCD: 2; LCM: 27,432 **9.** 12 **10.** 11:48 A.M. **11.** $<$ **12.** $>$ **13.** 860 **14.** 53 **15.** 0 **16.** −3 **17.** −11 **18.** −15 **19.** 1 **20.** 99 **21.** −15 **22.** 9 **23.** −4 **24.** −16 **25.** −16 **26.** 10 **27.** −2 **28.** 17 **29.** \$658 billion **30.** $\frac{8}{15}$ **31.** $\frac{6}{25}$ **32.** $\frac{11}{12}$ **33.** $\frac{64}{11}$ **34.** $-\frac{23}{7}$ **35.** $5\frac{2}{5}$ **36.** $-1\frac{8}{9}$ **37.** 0.8 **38.** $0.\overline{428571}$ **39.** 0.625 **40.** 0.5625 **41.** $\frac{3}{5}$ **42.** $\frac{17}{25}$ **43.** $\frac{147}{250}$ **44.** $\frac{21}{2500}$ **45.** $\frac{5}{9}$ **46.** $\frac{34}{99}$ **47.** $\frac{113}{999}$ **48.** $\frac{21}{50}$ **49.** $\frac{35}{6}$ **50.** $\frac{8}{3}$ **51.** $-\frac{1}{4}$ **52.** $\frac{2}{3}$ **53.** $\frac{43}{36}$ **54.** $\frac{37}{60}$ **55.** $\frac{11}{15}$ **56.** $\frac{5}{16}$ **57.** $-\frac{2}{5}$ **58.** $-\frac{20}{3}$ or $-6\frac{2}{3}$ **59.** $\frac{15}{112}$ **60.** $\frac{27}{40}$ **61.** $11\frac{1}{4}$ or about 11 pounds **62.** $\frac{5}{12}$ of the tank **63.** $2\sqrt{7}$ **64.** $6\sqrt{2}$ **65.** $5\sqrt{6}$ **66.** $10\sqrt{3}$ **67.** $4\sqrt{3}$ **68.** $5\sqrt{2}$ **69.** $2\sqrt{3}$ **70.** 3 **71.** $5\sqrt{5}$ **72.** $-6\sqrt{11}$ **73.** $7\sqrt{2}$ **74.** $-17\sqrt{3}$ **75.** $12\sqrt{2}$ **76.** $6\sqrt{5}$ **77.** $\frac{\sqrt{6}}{3}$ **78.** $8\sqrt{3} \approx 13.9$ feet per second **79. a.** $\sqrt{81}$ **b.** 0, $\sqrt{81}$ **c.** −17, 0, $\sqrt{81}$ **d.** −17, $-\frac{9}{13}$, 0, 0.75, $\sqrt{81}$ **e.** $\sqrt{2}$, π **f.** −17, $-\frac{9}{13}$, 0, 0.75, $\sqrt{2}$, π, $\sqrt{81}$ **80.** Answers will vary; an example is: 0. **81.** Answers will vary; an example is: $\frac{1}{2}$. **82.** Answers will vary; an example is: $\sqrt{2}$. **83.** commutative property of addition **84.** associative property of multiplication **85.** distributive property of multiplication over addition **86.** commutative property of multiplication **87.** commutative property of multiplication **88.** commutative property of addition **89.** inverse property of multiplication **90.** identity property of multiplication **91.** Answers will vary; an example is: $2 \div 6 = \frac{1}{3}$. **92.** Answers will vary; an example is: $0 - 2 = -2$. **93. a.** The entries in the body of the table are all elements of the set.

b. $(4 \oplus 2) \oplus 3 = 4 \oplus (2 \oplus 3)$
$1 \oplus 3 = 4 \oplus 0$
$4 = 4$

c. 0 **d.** The inverse of 0 is 0, the inverse of 1 is 4, the inverse of 2 is 3, the inverse of 3 is 2, and the inverse of 4 is 1.
e. $3 \oplus 4 = 4 \oplus 3$; $3 \oplus 2 = 2 \oplus 3$
$2 = 2$ $\quad 0 = 0$

94. 216 **95.** 64 **96.** 16 **97.** 729 **98.** 25 **99.** 1 **100.** 1 **101.** $\frac{1}{216}$ **102.** $\frac{1}{16}$ **103.** $\frac{1}{49}$ **104.** 27
105. 460 **106.** 37,400 **107.** 0.00255 **108.** 0.0000745 **109.** 7.52×10^3 **110.** 3.59×10^6 **111.** 7.25×10^{-3}
112. 4.09×10^{-7} **113.** 4.2×10^{13} **114.** 9.7×10^{-5} **115.** 390 **116.** 1,150,000 **117.** 0.023 **118.** 40
119. $(6.0 \times 10^4)(5.4 \times 10^5) = 3.24 \times 10^{10}$ **120.** $(9.1 \times 10^4)(4 \times 10^{-4}) = 3.64 \times 10^1$ **121.** $\frac{8.4 \times 10^6}{4 \times 10^3} = 2.1 \times 10^3$
122. $\frac{3 \times 10^{-6}}{6 \times 10^{-8}} = 5 \times 10^1$ **123.** 1.3×10^{12} **124.** 3.2×10^7 **125.** 40,625 years **126.** 2.88×10^{13} **127.** 7, 11, 15, 19, 23, and 27
128. $-4, -9, -14, -19, -24$, and -29 **129.** $\frac{3}{2}, 1, \frac{1}{2}, 0, -\frac{1}{2}$, and -1 **130.** 20 **131.** -30 **132.** -38 **133.** $a_n = -7 + (n-1)4$; 69
134. $a_n = 200 + (n-1)(-20)$; -180 **135.** 3, 6, 12, 24, 48, and 96 **136.** $\frac{1}{2}, \frac{1}{4}, \frac{1}{8}, \frac{1}{16}, \frac{1}{32}$, and $\frac{1}{64}$ **137.** 16, -8, 4, -2, 1, and $-\frac{1}{2}$ **138.** 54
139. $\frac{1}{2}$ **140.** -48 **141.** $a_n = 1(2)^{n-1}$; 128 **142.** $a_n = 100\left(\frac{1}{10}\right)^{n-1}$; $\frac{1}{100{,}000}$ **143.** arithmetic; 24 and 29 **144.** geometric; 162 and 486
145. geometric; $\frac{1}{256}$ and $\frac{1}{1024}$ **146.** arithmetic; -28 and -35 **147. a.** $a_n = 1.63n + 38.73$ **b.** \$81.11 **148. a.** approximately 1.02 for each division **b.** $a_n = 15.98(1.02)^{n-1}$ **c.** 28.95 million

Chapter 5 Test

1. 2, 3, 4, 6, 8, 9, and 12 **2.** $2^2 \cdot 3^2 \cdot 7$ **3.** GCD: 24; LCM: 144 **4.** 1 **5.** -4 **6.** -32 **7.** $0.58\overline{3}$ **8.** $\frac{64}{99}$ **9.** $\frac{1}{5}$ **10.** $\frac{37}{60}$
11. $-\frac{19}{2}$ **12.** $\frac{7}{12}$ **13.** $5\sqrt{2}$ **14.** $9\sqrt{2}$ **15.** $3\sqrt{2}$ **16.** $-7, -\frac{4}{5}, 0, 0.25, \sqrt{4}$, and $\frac{22}{7}$ **17.** commutative property of addition
18. distributive property of multiplication over addition **19.** 243 **20.** 64 **21.** $\frac{1}{64}$ **22.** 7500 **23.** $\frac{4.9 \times 10^4}{7 \times 10^{-3}} = 7 \times 10^6$
24. $\$5.36 \times 10^{10}$ **25.** 3.07×10^8 **26.** $\approx \$175$ **27.** 1, -4, -9, -14, -19, and -24 **28.** 22 **29.** 16, 8, 4, 2, 1, and $\frac{1}{2}$ **30.** 320

CHAPTER 6

Section 6.1

Check Point Exercises

1. 608 **2.** -2 **3.** -94 **4.** 2662 calories; underestimates by 38 calories **5.** $3x - 21$ **6.** $38x^2 + 23x$ **7.** $2x + 36$

Concept and Vocabulary Check

1. evaluating **2.** equation **3.** terms **4.** coefficient **5.** factors **6.** like terms

Exercise Set 6.1

1. 27 **3.** 23 **5.** 29 **7.** -2 **9.** -21 **11.** -10 **13.** 140 **15.** 217 **17.** 176 **19.** 30 **21.** 44 **23.** 22 **25.** 27 **27.** -12 **29.** 69 **31.** -33 **33.** -8 **35.** 10°C **37.** 60 ft **39.** distributive property; commutative property of addition; associative property of addition; commutative property of addition **41.** $17x$ **43.** $-3x^2$ **45.** $3x + 15$ **47.** $8x - 12$ **49.** $15x + 16$ **51.** $27x - 10$ **53.** $29y - 29$ **55.** $8y - 12$ **57.** $-8x^2 + 10x$ **59.** $16y - 25$ **61.** $-10x + 6$ **63.** $12x^2 + 11$ **65.** $-2x^2 - 9$ **67. a.** 140 beats per minute **b.** 160 beats per minute **69.** 1924; underestimates by 76 calories **71. a.** 22% **b.** 18.7%; less than the estimate **81.** makes sense **83.** makes sense **85.** false **87.** false **89.** false **91.** true **93. a.** \$50.50, \$5.50, and \$1.00 per clock, respectively **b.** no; Answers will vary.

Section 6.2

Check Point Exercises

1. {6} **2.** {−2} **3.** {2} **4.** {5} **5.** {6} **6.** $\frac{37}{10}$ or 3.7; If a horizontal line is drawn from 10 on the scale for level of depression until it touches the blue line graph for the low-humor group and then a vertical line is drawn from that point on the blue line graph to the scale for the intensity of the negative life event, the vertical line will touch the scale at 3.7. **7. a.** {15} **b.** {5} **8.** \$5880 **9.** 720 deer **10.** $\varnothing$ **11.** $\{x \mid x \text{ is a real number}\}$

Concept and Vocabulary Check

1. linear **2.** equivalent **3.** $b + c$ **4.** bc **5.** apply the distributive property **6.** simplified; solved **7.** least common denominator; 12 **8.** proportion **9.** $ad = bc$ **10.** no; $\varnothing$ **11.** true; $\{x \mid x \text{ is a real number}\}$ **12.** false **13.** false **14.** false **15.** false

Exercise Set 6.2

1. {10} **3.** {−17} **5.** {12} **7.** {9} **9.** {−3} **11.** $\left\{-\frac{1}{4}\right\}$ **13.** {3} **15.** {11} **17.** {−17} **19.** {11} **21.** $\left\{\frac{7}{5}\right\}$ **23.** $\left\{\frac{25}{3}\right\}$ **25.** {8} **27.** {−3} **29.** {2} **31.** {−4} **33.** $\left\{-\frac{1}{5}\right\}$ **35.** {−4} **37.** {5} **39.** {6} **41.** {1} **43.** {−57} **45.** {18} **47.** {1} **49.** {24} **51.** {−6} **53.** {20} **55.** {5} **57.** {−7} **59.** {14} **61.** {27} **63.** {−15} **65.** {−9} **67.** {34} **69.** {−49} **71.** {10} **73.** ∅ **75.** $\{x|x$ is a real number$\}$ **77.** $\left\{\frac{2}{3}\right\}$ **79.** $\{x|x$ is a real number$\}$ **81.** ∅ **83.** {0} **85.** ∅ **87.** {0} **89.** $\left\{\frac{7}{2}\right\}$ **91.** {0} **93.** 2 **95.** 161 **97.** {−2} **99.** ∅ **101.** {10} **103.** {−2} **105.** 142 lb; 13 lb **107. a.** 49%; overestimates by 1% **b.** 2020 **109.** 6.25 qt **111.** 5.4 ft **113.** 20,489 fur seal pups **125.** makes sense **127.** does not make sense **131.** yes; Answers will vary.

Section 6.3

Check Point Exercises

1. associate's degree: \$33 thousand; bachelor's degree: \$47 thousand; master's degree: \$59 thousand **2.** by 67 years after 1969, or in 2036 **3.** 20 bridge crossings per month **4.** \$1200 **5.** $w = \frac{P - 2l}{2}$ **6.** $m = \frac{T - D}{p}$

Concept and Vocabulary Check

1. $x + 658.6$ **2.** $31 + 2.4x$ **3.** $4 + 0.15x$ **4.** $x - 0.15x$ or $0.85x$ **5.** isolated on one side **6.** subtract b; divide by m

Exercise Set 6.3

1. 6 **3.** 25 **5.** 120 **7.** 320 **9.** 19 and 45 **11.** $x - (x + 4); -4$ **13.** $6(-5x); -30x$ **15.** $5x - 2x; 3x$ **17.** $8x - (3x + 6); 5x - 6$ **19.** TV: 9 years; sleeping: 28 years **21.** lawyer: \$132 thousand; architect: \$78 thousand **23.** 7 years after 2014; 2021 **25. a.** births: 384,000; deaths: 156,000 **b.** 83 million **c.** approximately 4 years **27.** after 5 years **29.** after 5 months; \$165 **31.** 30 times **33.** \$600 of merchandise; \$580 **35.** 2019; 22,300 students **37.** \$420 **39.** \$150 **41.** \$467.20 **43.** $L = \frac{A}{W}$ **45.** $b = \frac{2A}{h}$ **47.** $P = \frac{I}{rt}$ **49.** $m = \frac{E}{c^2}$ **51.** $m = \frac{y - b}{x}$ **53.** $a = \frac{2A}{h} - b$ **55.** $r = \frac{S - P}{Pt}$ **57.** $x = \frac{C - By}{A}$ **59.** $n = \frac{a_n - a_1}{d} + 1$ or $n = \frac{a_n - a_1 + d}{d}$ **65.** does not make sense **67.** does not make sense **69.** \$200 **71.** 10 problems **73.** 36 plants **75.** $x = \frac{bc}{ad}$

Section 6.4

Check Point Exercises

1. a. **b.** **c.**

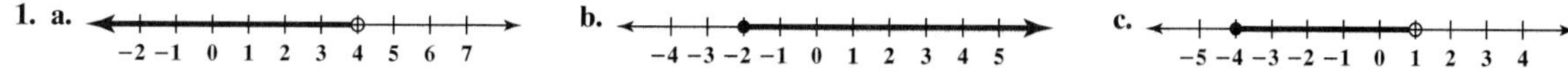

2. $\{x|x \le 4\}$;

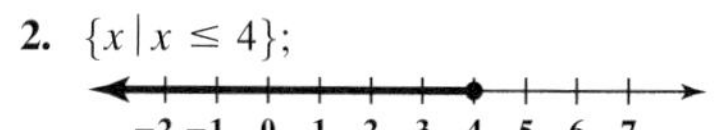

3. a. $\{x|x < 8\}$;

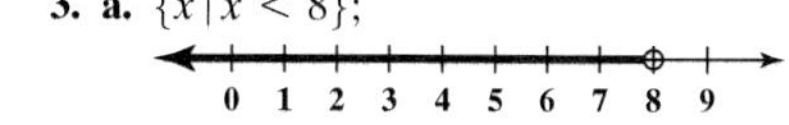

b. $\{x|x > -3\}$;

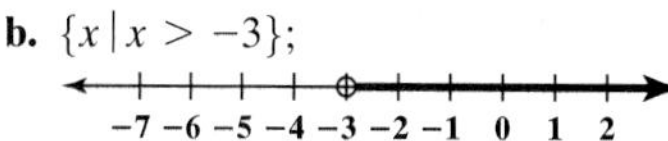

4. $\{x|x < -6\}$;

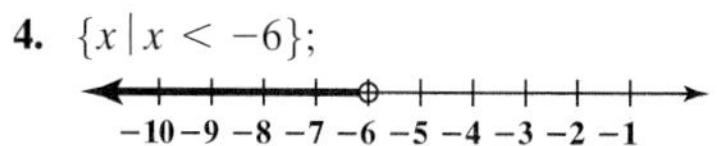

5. $\{x|x \ge 1\}$; **6.** $\{x|-1 \le x < 4\}$;

7. at least 83%

Concept and Vocabulary Check

1. is not included; is included **2.** $\{x|x < a\}$ **3.** $\{x|x < a \le b\}$ **4.** negative **5.** false **6.** false **7.** false **8.** true

Exercise Set 6.4

1.

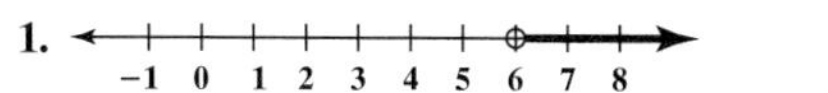

3.

5.

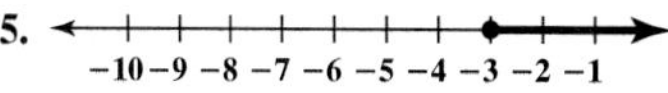

7.

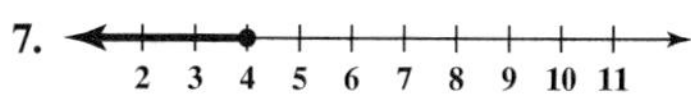

9.

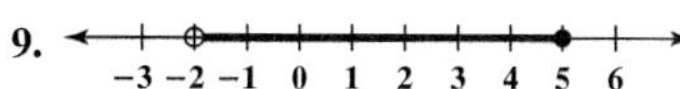

11.

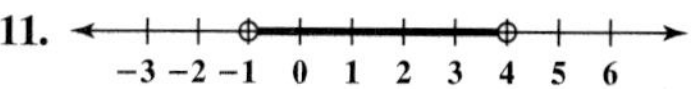

13. $\{x|x > 5\}$;

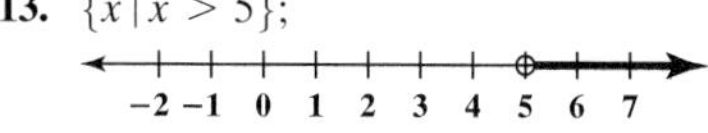

15. $\{x|x \le 5\}$;

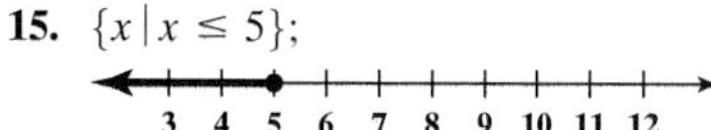

17. $\{x|x < 3\}$;

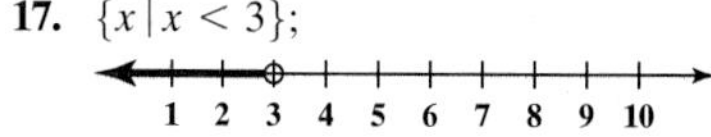

19. $\{x|x < 5\}$;

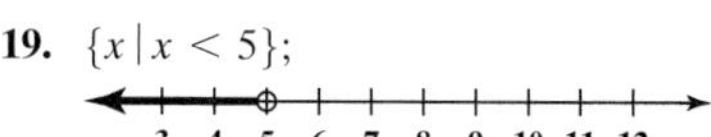

21. $\{x|x \ge -5\}$; **23.** $\{x|x > 5\}$;

25. $\{x \mid x < 5\}$;

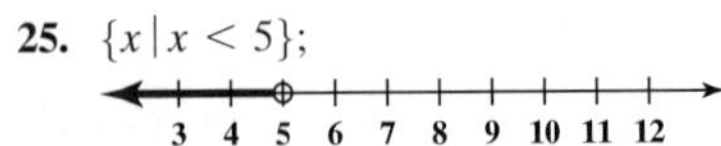

27. $\{x \mid x < 8\}$;
29. $\{x \mid x > -6\}$;
31. $\{x \mid x > -5\}$;
33. $\{x \mid x \le 5\}$;
35. $\{x \mid x \le 3\}$;
37. $\{x \mid x < 16\}$;
39. $\{x \mid x > -3\}$;
41. $\{x \mid x \ge -2\}$;
43. $\{x \mid x > -4\}$;
45. $\{x \mid x \ge 4\}$;
47. $\left\{x \,\middle|\, x > \frac{11}{3}\right\}$;
49. $\{x \mid x > 2\}$;
51. $\{x \mid x < 3\}$;
53. $\left\{x \,\middle|\, x > \frac{5}{3}\right\}$;
55. $\{x \mid x \ge -10\}$;
57. $\{x \mid x < -6\}$;
59. $\{x \mid 3 < x < 5\}$;
61. $\{x \mid -1 \le x < 3\}$;
63. $\{x \mid -5 < x \le -2\}$;
65. $\{x \mid 3 \le x < 6\}$;

67. $x > \frac{C - By}{A}$ **69.** $x < \frac{C - By}{A}$ **71.** $\{x \mid x \le 5\}$ **73.** $\{x \mid x \le 14\}$ **75.** $\{x \mid x \le 50\}$ **77.** $\{x \mid 0 < x < 4\}$
79. intimacy $\ge$ passion or passion $\le$ intimacy **81.** commitment $>$ passion or passion $<$ commitment **83.** 9; after 3 years
85. a. years after 2016 **b.** years after 2020 **c.** years after 2020 **87. a.** at least 96 **b.** less than 66 **89.** at most 1280 mi
91. at most 29 bags **93.** between 80 and 110 minutes, inclusive **99.** makes sense **101.** makes sense **103.** more than 720 miles

Section 6.5

Check Point Exercises

1. $x^2 + 11x + 30$ **2.** $28x^2 - x - 15$ **3.** $(x + 2)(x + 3)$ **4.** $(x + 5)(x - 2)$ **5.** $(5x - 4)(x - 2)$ **6.** $(3y - 1)(2y + 7)$
7. $\{-6, 3\}$ **8.** $\{-2, 8\}$ **9.** $\left\{-4, \frac{1}{2}\right\}$ **10.** $\left\{-\frac{1}{2}, \frac{1}{4}\right\}$ **11.** $\left\{\frac{3 + \sqrt{7}}{2}, \frac{3 - \sqrt{7}}{2}\right\}$ **12.** approximately 26 years old

Concept and Vocabulary Check

1. $2x^2$; $3x$; $10x$; 15 **2.** $(x + 3)(x + 10)$ **3.** $(x - 3)(x - 6)$ **4.** $(x - 6)(x + 5)$ **5.** $(x + 2)(x - 7)$ **6.** $(4x + 1)(2x - 3)$
7. $(4x + 5)(3x - 4)$ **8.** $(x - 1)(2x - 3)$ **9.** $(2x + 3)(3x + 4)$ **10.** quadratic **11.** $A = 0$ or $B = 0$ **12.** subtracting 18
13. $x = \frac{-b \pm \sqrt{b^2 - 4ac}}{2a}$; quadratic formula **14.** false **15.** false **16.** false **17.** false **18.** false

Exercise Set 6.5

1. $x^2 + 8x + 15$ **3.** $x^2 - 2x - 15$ **5.** $2x^2 + 3x - 2$ **7.** $12x^2 - 43x + 35$ **9.** $(x + 2)(x + 3)$ **11.** $(x - 5)(x + 3)$
13. $(x - 5)(x - 3)$ **15.** $(x - 12)(x + 3)$ **17.** prime **19.** $(x + 1)(x + 16)$ **21.** $(2x + 1)(x + 3)$ **23.** $(2x - 5)(x - 6)$
25. $(3x + 2)(x - 1)$ **27.** $(3x - 28)(x + 1)$ **29.** $(3x - 4)(2x - 1)$ **31.** $(2x + 3)(2x + 5)$ **33.** $\{-3, 8\}$ **35.** $\left\{-\frac{5}{4}, 2\right\}$
37. $\{-5, -3\}$ **39.** $\{-3, 5\}$ **41.** $\{-3, 7\}$ **43.** $\{-8, -1\}$ **45.** $\{6\}$ **47.** $\left\{-\frac{1}{2}, 4\right\}$ **49.** $\left\{-2, \frac{9}{5}\right\}$ **51.** $\left\{-\frac{7}{2}, -\frac{1}{3}\right\}$ **53.** $\{-5, -3\}$
55. $\left\{\frac{-5 + \sqrt{13}}{2}, \frac{-5 - \sqrt{13}}{2}\right\}$ **57.** $\{-2 + \sqrt{10}, -2 - \sqrt{10}\}$ **59.** $\{-2 + \sqrt{11}, -2 - \sqrt{11}\}$ **61.** $\{-3, 6\}$ **63.** $\left\{-\frac{2}{3}, \frac{3}{2}\right\}$
65. $\{1 + \sqrt{11}, 1 - \sqrt{11}\}$ **67.** $\left\{\frac{1 + \sqrt{57}}{2}, \frac{1 - \sqrt{57}}{2}\right\}$ **69.** $\left\{\frac{-3 + \sqrt{3}}{6}, \frac{-3 - \sqrt{3}}{6}\right\}$ **71.** $\left\{\frac{3}{2}\right\}$ **73.** $\left\{-\frac{2}{3}, 4\right\}$
75. $\{-3, 1\}$ **77.** $\left\{1, \frac{5}{2}\right\}$ **79.** $\{\pm\sqrt{6}\}$ **81.** $1 + \sqrt{7}$ **83.** 9 teams **85. a.** 11.5%; overestimates by 1.1 **b.** 2020 **87. a.** $\frac{1}{\Phi - 1}$
b. $\Phi = \frac{1 + \sqrt{5}}{2}$ **c.** $\frac{1 + \sqrt{5}}{2}$ **93.** does not make sense **95.** does not make sense **99.** 3 and 4 **101.** $\{-3\sqrt{3}, \sqrt{3}\}$

Chapter 6 Review Exercises

1. 33 **2.** 15 **3.** −48 **4.** 55; It's the same. **5.** $17x - 15$ **6.** $5y - 13$ **7.** $14x^2 + x$ **8.** {6} **9.** {2} **10.** {12} **11.** $\left\{-\frac{13}{3}\right\}$ **12.** {2} **13.** ∅ **14.** $\{x \mid x \text{ is a real number}\}$ **15.** {5} **16.** {−65} **17.** $\left\{\frac{2}{5}\right\}$ **18.** {3} **19.** 324 teachers **20.** 287 trout **21. a.** $22,000 **b.** $21,809; reasonably well **c.** $21,726; reasonably well **d.** 2020 **22.** engineering: $76 thousand; accounting: $63 thousand; marketing: $57 thousand **23.** by 40 years after 1980, or in 2020 **24.** 5 GB **25.** $60 **26.** $10,000 in sales **27.** $x = \frac{By + C}{A}$ **28.** $h = \frac{2A}{b}$ **29.** $B = 2A - C$ **30.** $g = \frac{s - vt}{t^2}$ **31.** $\{x \mid x < 4\}$; **32.** $\{x \mid x > -8\}$; **33.** $\{x \mid x \geq -3\}$; **34.** $\{x \mid x > 6\}$; **35.** $\{x \mid x \geq 4\}$; **36.** $\{x \mid x \leq 2\}$; **37.** $\left\{x \middle| -\frac{3}{4} < x \leq 1\right\}$; **38.** at least 64 **39.** $x^2 + 4x - 45$ **40.** $12x^2 - 13x - 14$ **41.** $(x - 4)(x + 3)$ **42.** $(x - 5)(x - 3)$ **43.** prime **44.** $(3x - 2)(x - 5)$ **45.** $(2x - 5)(3x + 2)$ **46.** prime **47.** {−7, 2} **48.** {−4, 8} **49.** $\left\{-8, \frac{1}{2}\right\}$ **50.** {−5, −2} **51.** {1, 3} **52.** $\left\{\frac{5 + \sqrt{41}}{2}, \frac{5 - \sqrt{41}}{2}\right\}$ **53.** $\left\{-3, \frac{1}{2}\right\}$ **54.** $\left\{\frac{3 + 2\sqrt{6}}{3}, \frac{3 - 2\sqrt{6}}{3}\right\}$ **55. a.** 277; underestimates by 14 **b.** 2023

Chapter 6 Test

1. −44 **2.** $14x - 4$ **3.** {−5} **4.** {2} **5.** ∅ **6.** {−15} **7.** $y = \frac{Ax + A}{B}$ **8. a.** 21%; overestimates by 0.2 **b.** 20.7%; underestimates by 0.1 **c.** 2021 **9.** $\left\{\frac{15}{2}\right\}$ **10.** {3} **11.** 6000 tule elk **12.** love: 630; thanks: 243; sorry: 211 **13.** after 7 years **14.** 200 text messages **15.** $50 **16.** $\{x \mid x \leq -3\}$; **17.** $\{x \mid x > 7\}$; **18.** $\left\{x \middle| -2 \leq x < \frac{5}{2}\right\}$; **19.** at least 92 **20.** $6x^2 - 7x - 20$ **21.** $(2x - 5)(x - 2)$ **22.** {−9, 4} **23.** $\left\{\frac{-2 + \sqrt{2}}{2}, \frac{-2 - \sqrt{2}}{2}\right\}$ **24.** 2018 **25.** 2018 **26.** quite well

CHAPTER 7

Section 7.1

Check Point Exercises

1. A(−2, 4), C(−3, 0), B(4, −2), D(0, −3)

2.

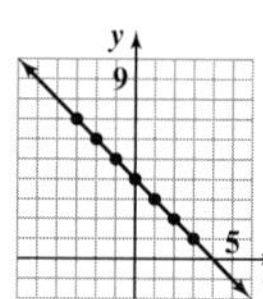

3. a. $y = 2x$

x	(x, y)
0	(0, 0)
2	(2, 4)
4	(4, 8)
6	(6, 12)
8	(8, 16)
10	(10, 20)
12	(12, 24)

$y = 10 + x$

x	(x, y)
0	(0, 10)
2	(2, 12)
4	(4, 14)
6	(6, 16)
8	(8, 18)
10	(10, 20)
12	(12, 22)

b.

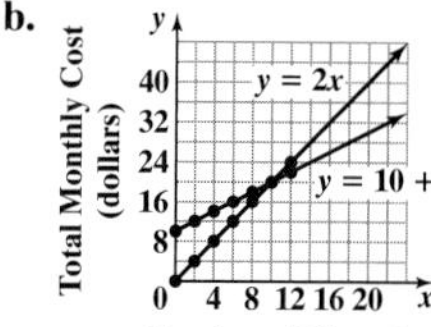

c. (10, 20); When the bridge is used 10 times during a month, the cost is $20 with or without the discount pass.

4. a. 29 **b.** 65 **c.** 46 **5. a.** 190 ft **b.** 191 ft **6.** ; The graph of g is the graph of f shifted down 3 units.

7. a. function **b.** function **c.** not a function **8. a.** 0 to 3 hours **b.** 3 to 13 hours **c.** 0.05 mg per 100 ml; after 3 hours **d.** None of the drug is left in the body. **e.** No vertical line intersects the graph in more than one point.

Concept and Vocabulary Check

1. x-axis **2.** y-axis **3.** origin **4.** quadrants; four **5.** x-coordinate; y-coordinate **6.** solution; satisfies **7.** y; x; function **8.** x; 6 **9.** more than once; function

Exercise Set 7.1

1.
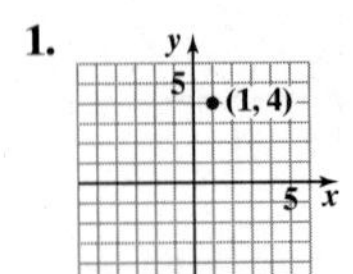

3.
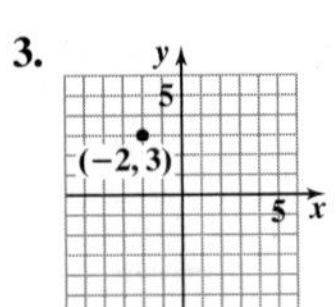

5.
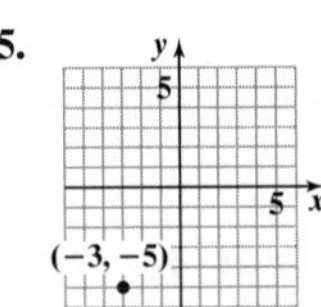

7.
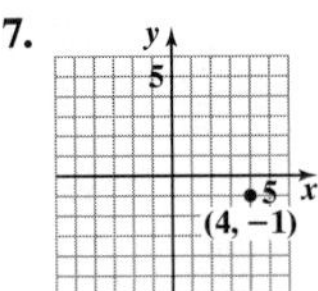

9.
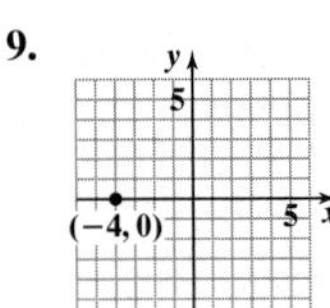

11.
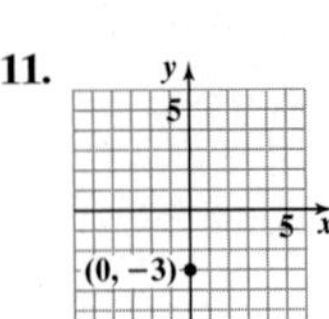

13.
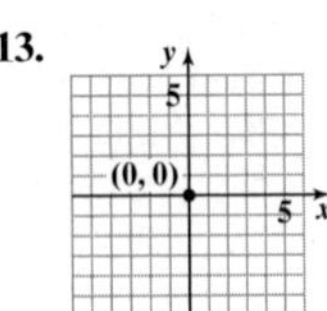

15.
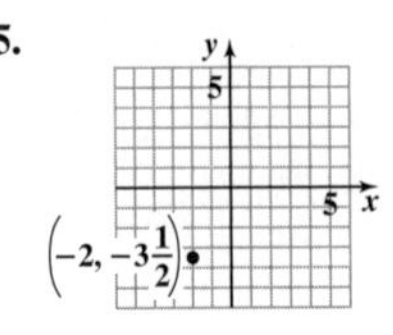

17.
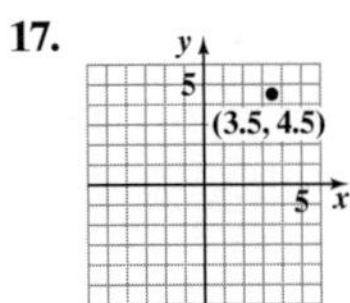

19.
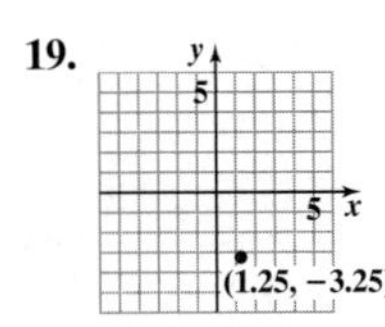

21.
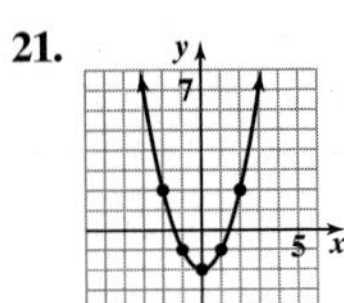

23.
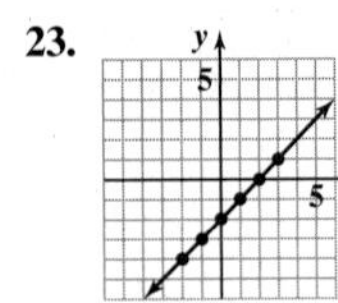

25.
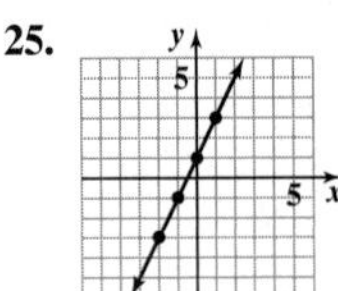

27.
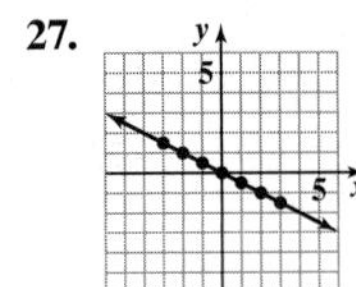

29.
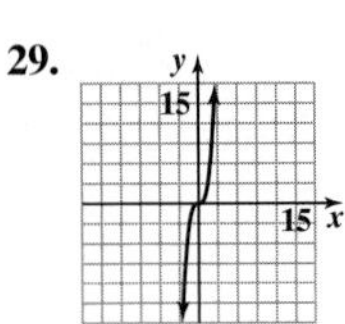

31.
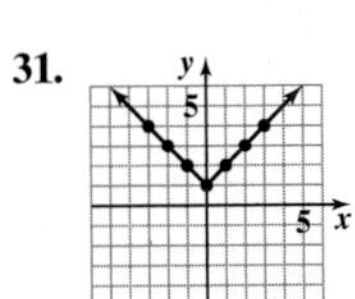

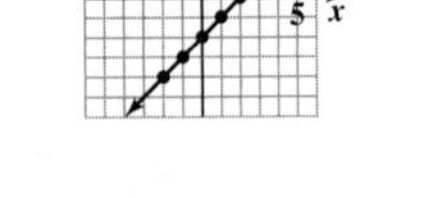
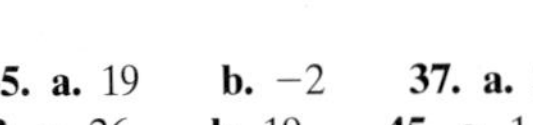

33. a. 4 **b.** −3 **35. a.** 19 **b.** −2 **37. a.** 5 **b.** 5 **39. a.** −14 **b.** −7 **41. a.** 53 **b.** 8 **43. a.** 26 **b.** 19 **45. a.** 1 **b.** −1

47.
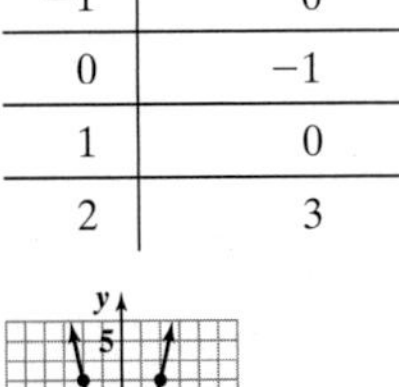

x	$f(x) = x^2 - 1$
−2	3
−1	0
0	−1
1	0
2	3

49.

x	$f(x) = x - 1$
−2	−3
−1	−2
0	−1
1	0
2	1

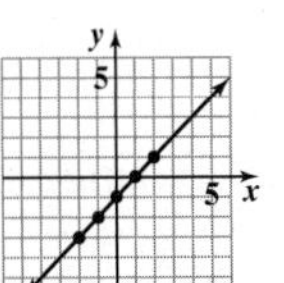

51.

x	$f(x) = (x - 2)^2$
0	4
1	1
2	0
3	1
4	4

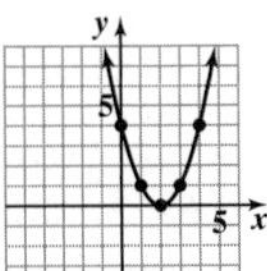

53.

x	$f(x) = x^3 + 1$
−3	−26
−2	−7
−1	0
0	1
1	2

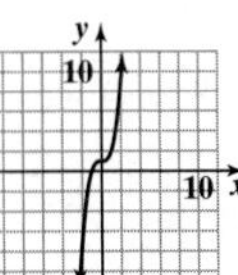

55. function **57.** function **59.** not a function **61.** function **63.** −2; 10 **65.** −38

67. $y = 2x + 4$
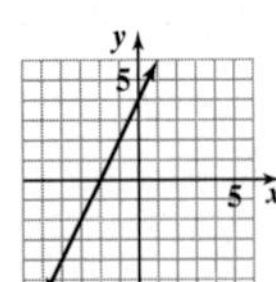

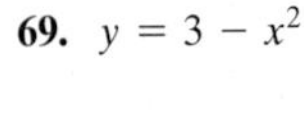
69. $y = 3 - x^2$
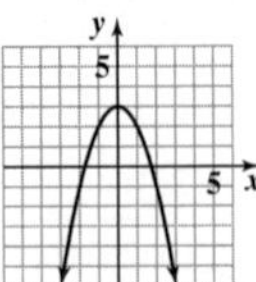

71. (2, 7); The football is 7 feet above the ground when it is 2 yards from the quarterback. **73.** (6, 9.25) **75.** 12 feet; 15 yards **77. a.** 81; According to the function, women's earnings were 81% of men's earning 30 years after 1980, or in 2010.; (30, 81) **b.** underestimates by 2% **79.** 440; For 20-year-old drivers, there are 440 accidents per 50 million miles driven.; (20, 440) **81.** $x = 45$; $y = 190$; The minimum number of accidents is 190 per 50 million miles driven and is attributed to 45-year-old drivers. **89.** makes sense **91.** makes sense **93.** −2 **95.** 1

Section 7.2

Check Point Exercises

1. 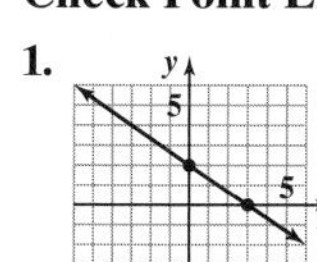 **2. a.** $m = 6$ **b.** $m = -\frac{7}{5}$ **3.** 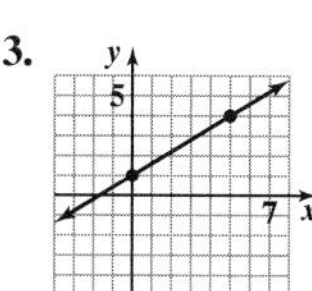 **4.** **5.** 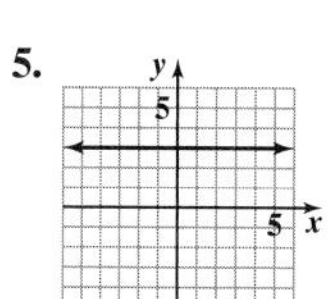 **6.** 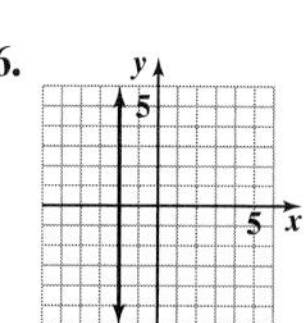

7. slope: -0.85; For the period from 1970 through 2010, the percentage of married men ages 20 to 24 decreased by 0.85 per year. The rate of change is -0.85% per year. **8. a.** $M(x) = 0.13x + 23.2$ **b.** 31

Concept and Vocabulary Check

1. x-intercept **2.** y-intercept **3.** $\frac{y_2 - y_1}{x_2 - x_1}$ **4.** $y = mx + b$; slope; y-intercept **5.** -4; 3 **6.** $(0, 3)$; 2; 5 **7.** horizontal **8.** vertical

Exercise Set 7.2

1. 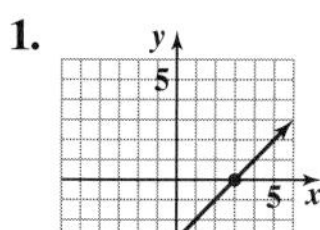 **3.** 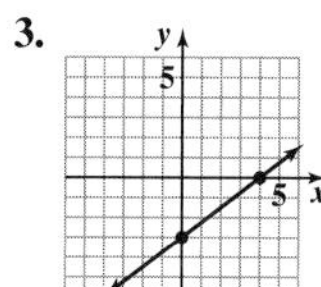 **5.** 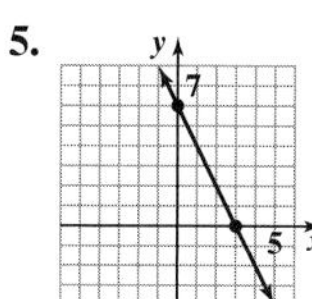 **7.** 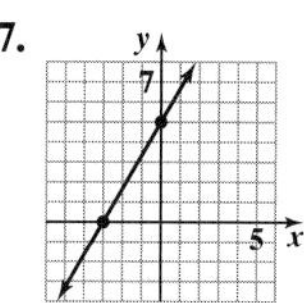

 **9.** -1; falls  **11.** $\frac{1}{4}$; rises  **13.** -5; falls **15.** undefined; vertical **17.** -4; falls **19.** 0; horizontal

21. 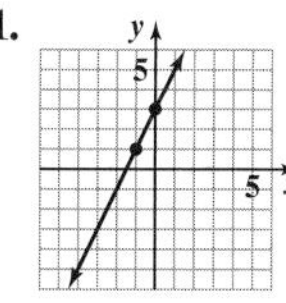 **23.** 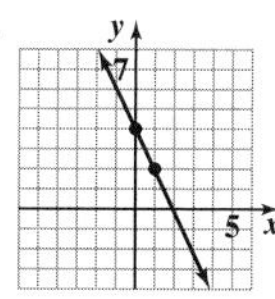 **25.** 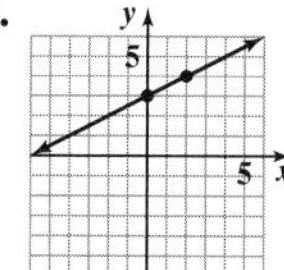 **27.** 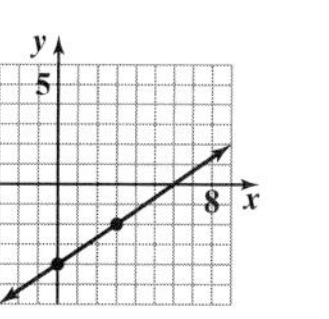 **29.** 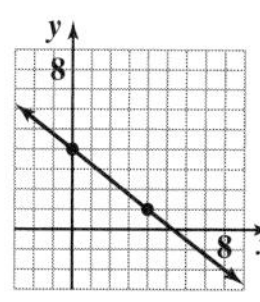 **31.** 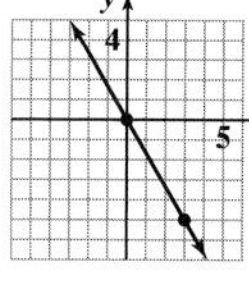

33. a. $y = -3x$ or $y = -3x + 0$ **b.** slope $= -3$; y-intercept $= 0$ **c.** 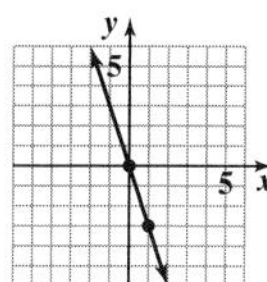

35. a. $y = \frac{4}{3}x$ or $y = \frac{4}{3}x + 0$ **b.** slope $= \frac{4}{3}$; y-intercept $= 0$ **c.** 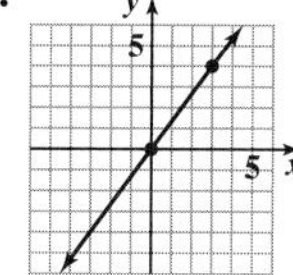

37. a. $y = -2x + 3$ **b.** slope $= -2$; y-intercept $= 3$ **c.** 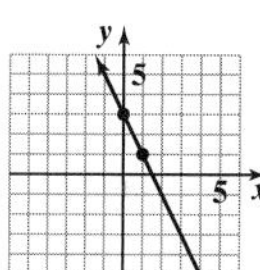

39. a. $y = -\frac{7}{2}x + 7$ **b.** slope $= -\frac{7}{2}$; y-intercept $= 7$  **c.** 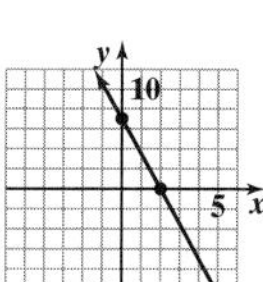 **41.** 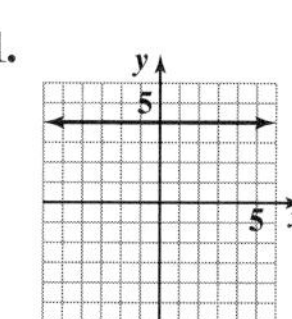 **43.** 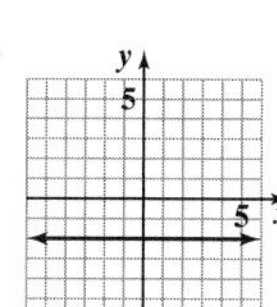 **45.** 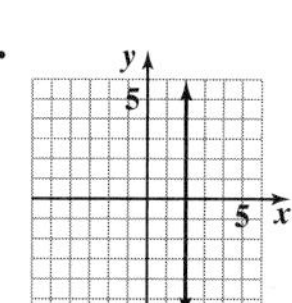 **47.** 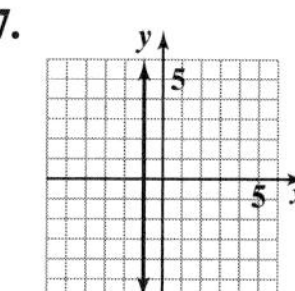

49. $m = -\frac{a}{b}$; falls **51.** undefined slope; vertical **53.** $m = -\frac{A}{B}$; $b = \frac{C}{B}$ **55.** -2 **57.** m_1, m_3, m_2, m_4 **59. a.** -0.5 **b.** 0.5; -0.5; year of aging **61.** $p(x) = -0.25x + 22$ **63. a.** $M(x) = 1.47x + 22$ **b.** $W(x) = 0.73x + 19$ **c.** men: \$36.7 thousand; women: \$26.3 thousand; by the points (10, 36.7) and (10, 26.3), respectively; \$10.4 thousand **75.** does not make sense **77.** makes sense **79.** false **81.** false

Section 7.3

Check Point Exercises

1. not a solution **2.** 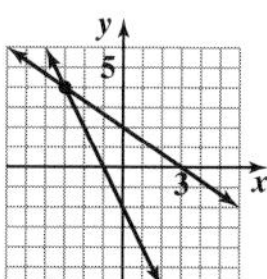 ; $\{(-3, 4)\}$ **3.** $\{(6, 11)\}$ **4.** $\{(1, -2)\}$ **5.** $\{(2, -1)\}$ **6.** $\{(2, -1)\}$ **7.** $\varnothing$

8. $\{(x, y) \mid y = 4x - 4\}$ or $\{(x, y) \mid 8x - 2y = 8\}$ **9. a.** $C(x) = 300{,}000 + 30x$ **b.** $R(x) = 80x$ **c.** (6000, 480,000); The company will break even if it produces and sells 6000 pairs of shoes.

Concept and Vocabulary Check

1. satisfies both equations in the system **2.** the intersection point **3.** $\left\{\left(\frac{1}{3}, -2\right)\right\}$ **4.** -2 **5.** -3 **6.** $\varnothing$; parallel **7.** $\{(x, y) \mid x = 3y + 2\}$ or $\{(x, y) \mid 5x - 15y = 10\}$; are identical or coincide **8.** revenue; profit **9.** break-even point

Exercise Set 7.3

1. solution **3.** not a solution

5.
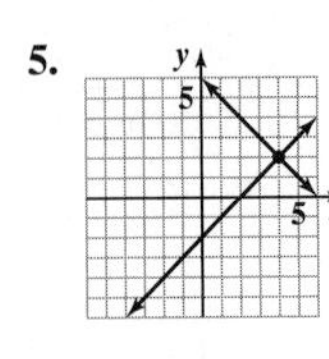
$\{(4, 2)\}$

7.
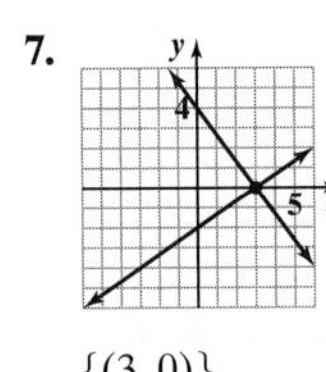
$\{(3, 0)\}$

9.
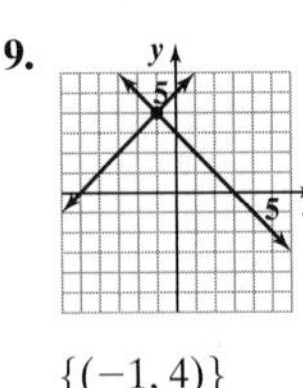
$\{(-1, 4)\}$

11.
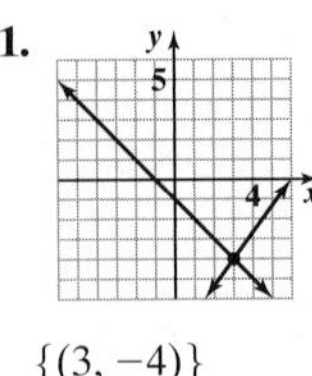
$\{(3, -4)\}$

13. $\{(1, 3)\}$ **15.** $\{(5, 1)\}$ **17.** $\{(2, 1)\}$ **19.** $\{(-1, 3)\}$ **21.** $\{(4, 5)\}$ **23.** $\left\{\left(-4, \frac{5}{4}\right)\right\}$ **25.** $\{(2, -1)\}$ **27.** $\{(3, 0)\}$ **29.** $\{(-4, 3)\}$ **31.** $\{(3, 1)\}$ **33.** $\{(1, -2)\}$ **35.** $\{(-5, -2)\}$ **37.** $\varnothing$ **39.** $\{(x, y) \mid y = 3x - 5\}$ or $\{(x, y) \mid 21x - 35 = 7y\}$ **41.** $\{(1, 4)\}$ **43.** $\{(x, y) \mid x + 3y = 2\}$ or $\{(x, y) \mid 3x + 9y = 6\}$ **45.** $\begin{cases} y = x - 4 \\ y = -\frac{1}{3}x + 4 \end{cases}$ **47.** $\left\{\left(\frac{1}{a}, 3\right)\right\}$ **49.** $m = -4, b = 3$ **51.** 500 radios **53.** -6000; When the company produces and sells 200 radios, the loss is \$6000. **55. a.** $P(x) = 20x - 10{,}000$ **b.** \$190,000 **57. a.** $C(x) = 18{,}000 + 20x$ **b.** $R(x) = 80x$ **c.** (300, 24,000); When 300 canoes are produced and sold, both revenue and cost are \$24,000. **59. a.** $C(x) = 30{,}000 + 2500x$ **b.** $R(x) = 3125x$ **c.** (48, 150,000); For 48 sold-out performances, both cost and revenue are \$150,000. **61. a.** 1000 gallons; \$4 **b.** \$4; 1000; 1000 **63. a.** $y = 0.45x + 0.8$ **b.** $y = 0.15x + 2.6$ **c.** week 6; 3.5 symptoms; by the intersection point (6, 3.5) **75.** makes sense **77.** does not make sense **81.** the twin who always lies

Section 7.4

Check Point Exercises

1.
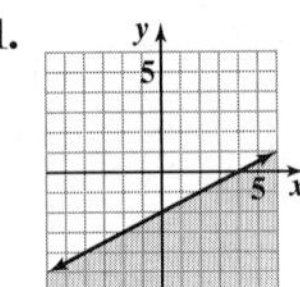

2.
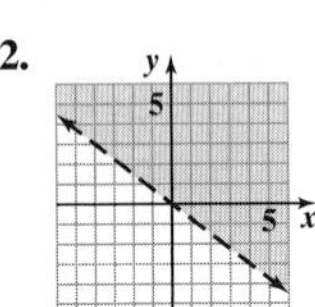

3. a.
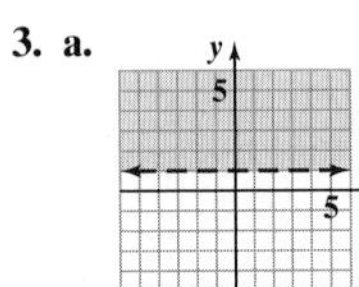

b.
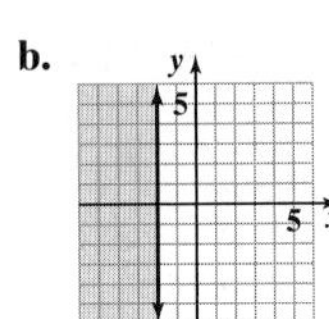

4. Point $B = (66, 130)$; $4.9(66) - 130 \geq 165$, or $193.4 \geq 165$, is true; $3.7(66) - 130 \leq 125$, or $114.2 \leq 125$, is true.

5.
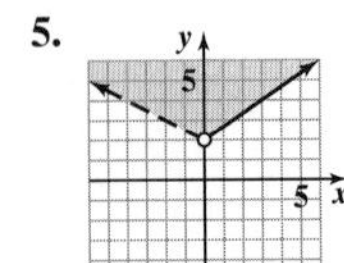

6.
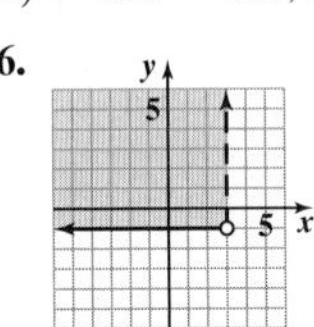

Concept and Vocabulary Check

1. solution; x; y; $5 > 1$ **2.** graph **3.** half-plane **4.** false **5.** true **6.** false **7.** $x - y < 1$; $2x + 3y \geq 12$ **8.** false

Exercise Set 7.4

1.
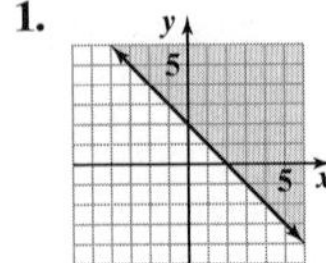

3.
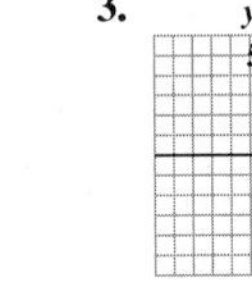

5.
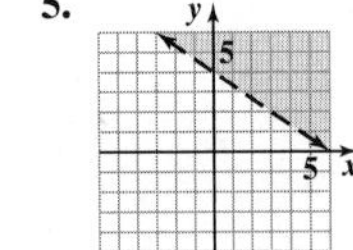

7.
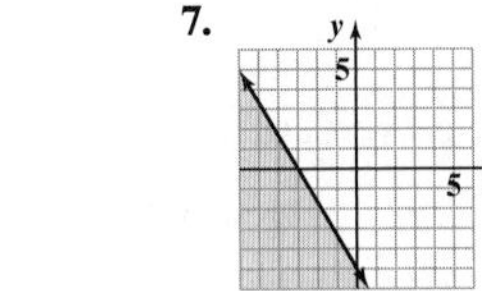

9.
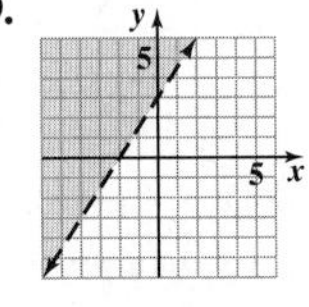

11.

13.
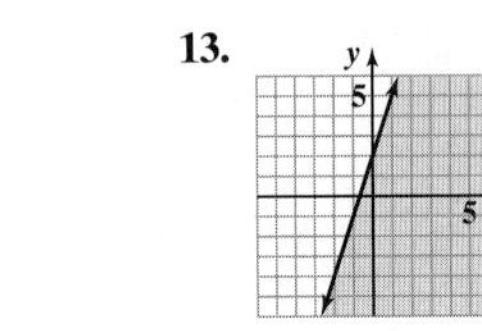

15.
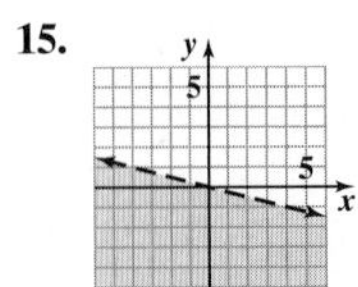

17.
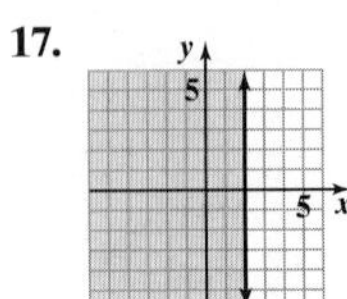

19.
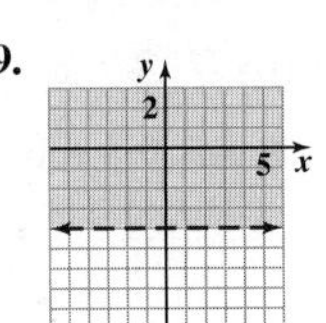

21.

23.
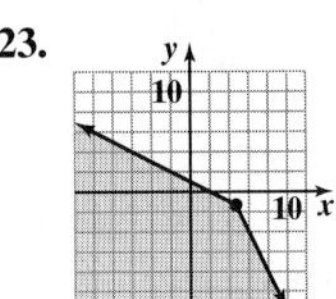

25.
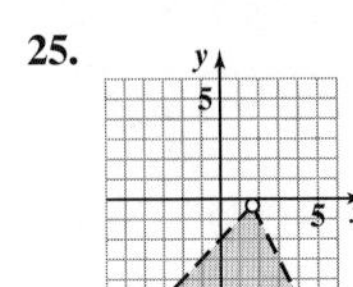

27.
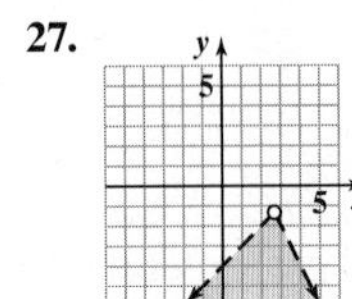

29.
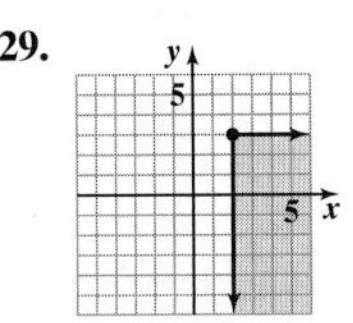

31. 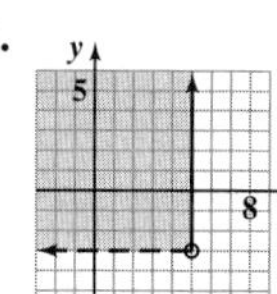
33. 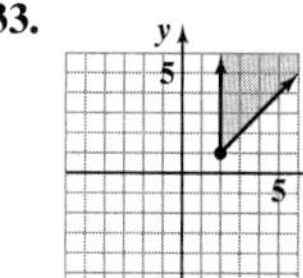
35. 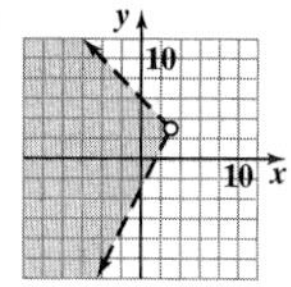
37. 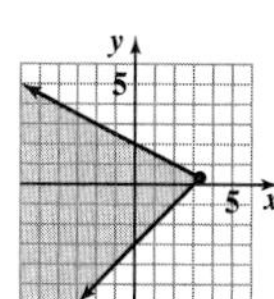
39. $y \geq -2x + 4$ 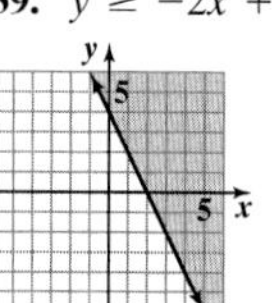

41. $\begin{cases} x + y \leq 4 \\ 3x + y \leq 6 \end{cases}$

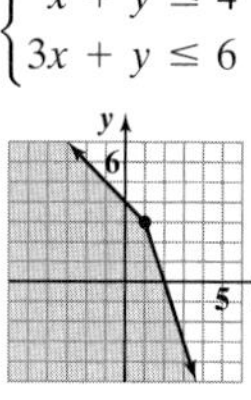

43. 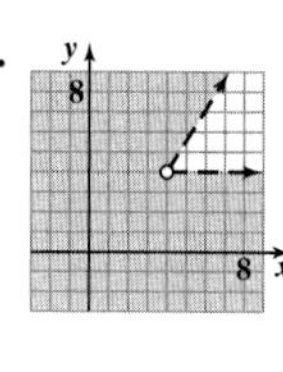

45. Point $A = (66, 160)$; $5.3(66) - 160 \geq 180$, or $189.8 \geq 180$, is true; $4.1(66) - 160 \leq 140$, or $110.6 \leq 140$, is true. **47.** no

49. a. $50x + 150y > 2000$

b. 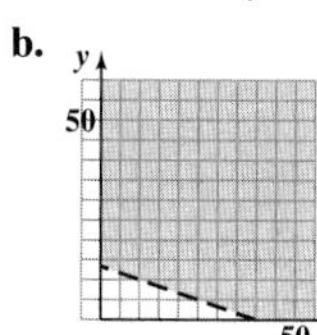

51. a. 27.1 **b.** overweight **59.** does not make sense **61.** makes sense **63.** $\begin{cases} y > x - 3 \\ y \leq x \end{cases}$ **65.** no solution
67. infinitely many solutions

Section 7.5

Check Point Exercises

1. $z = 25x + 55y$ **2.** $x + y \leq 80$ **3.** $30 \leq x \leq 80$; $10 \leq y \leq 30$; objective function: $z = 25x + 55y$; constraints: $\begin{cases} x + y \leq 80 \\ 30 \leq x \leq 80 \\ 10 \leq y \leq 30 \end{cases}$
4. 50 bookshelves and 30 desks; \$2900

Concept and Vocabulary Check

1. linear programming **2.** objective **3.** constraints; corner

Exercise Set 7.5

1. (1, 2): 17; (2, 10): 70; (7, 5): 65; (8, 3): 58; maximum: $z = 70$; minimum: $z = 17$
3. (0, 0): 0; (0, 8): 400; (4, 9): 610; (8, 0): 320; maximum: $z = 610$; minimum: $z = 0$

5. a. 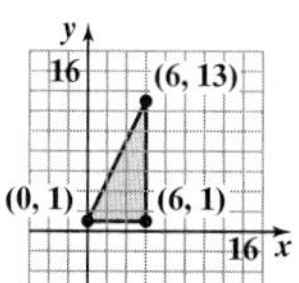

b. (0, 1): 1; (6, 13): 19; (6, 1): 7
c. maximum of 19 at $x = 6$ and $y = 13$

7. a. 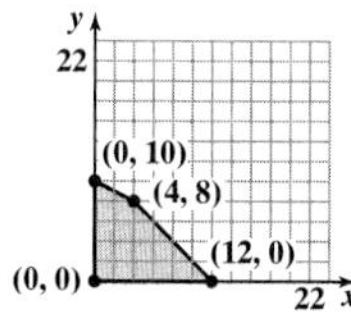

b. (0, 10): 100; (4, 8): 104; (12, 0): 72; (0, 0): 0
c. maximum of 104 at $x = 4$ and $y = 8$

9. a. 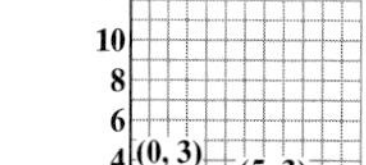
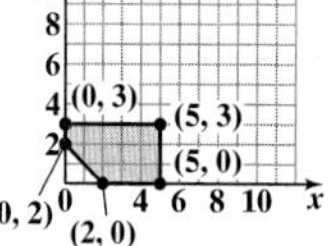

b. (0, 3): −6; (0, 2): −4; (2, 0): 10; (5, 0): 25; (5, 3): 19
c. maximum value: 25 at $x = 5$ and $y = 0$

11. a. 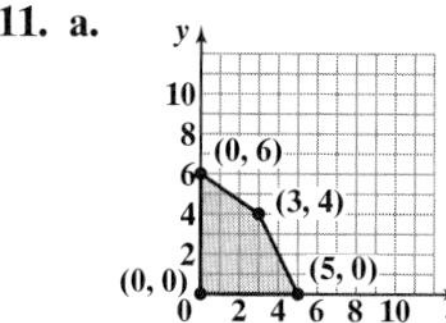

b. (0, 6): 72; (0, 0): 0; (5, 0): 50; (3, 4): 78
c. maximum value: 78 at $x = 3$ and $y = 4$

13. a. $z = 15x + 10y$

b. $\begin{cases} x + y \leq 20 \\ x \geq 3 \\ x \leq 8 \end{cases}$

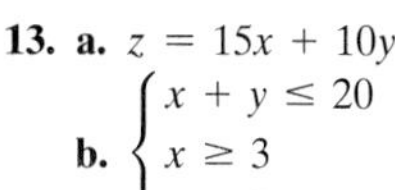

c.

d. (3, 0): 45; (8, 0): 120; (3, 17): 215; (8, 12): 240
e. 8; 12; \$240

15. 300 cartons of food and 200 cartons of clothing **17.** 50 students and 100 parents **25.** makes sense **27.** makes sense

Section 7.6

Check Point Exercises

1.

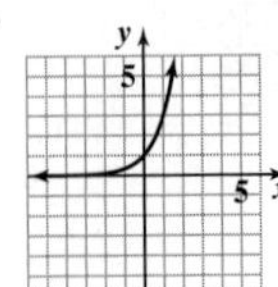

2. a. the exponential function g **b.** the linear function f **3.** 6.8%

4. $x = 3^y$;

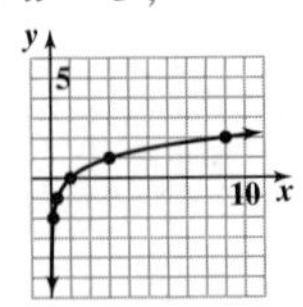

5. 34°; extremely well

6.

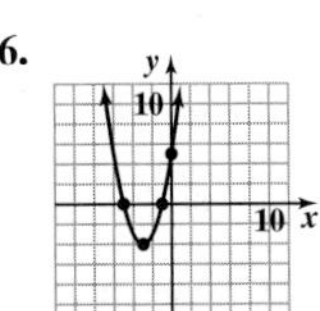

7. 10.8 ft; Answers will vary.; (8, 10.8)

Concept and Vocabulary Check

1. scatter plot; regression **2.** logarithmic **3.** exponential **4.** quadratic **5.** linear **6.** b^x **7.** 2.72; natural **8.** $b^y = x$ **9.** $\log x$; common **10.** $\ln x$; natural **11.** quadratic; parabola; $\frac{-b}{2a}$

Exercise Set 7.6

1.

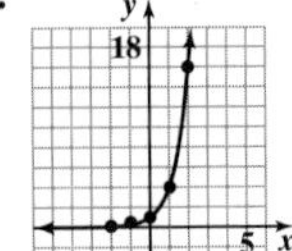

3.

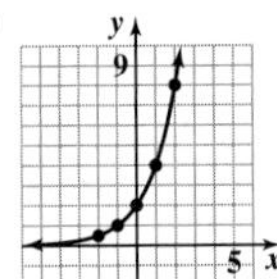

5.

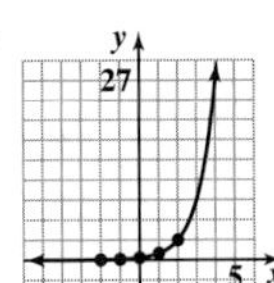

7. a. $x = 4^y$ **b.**

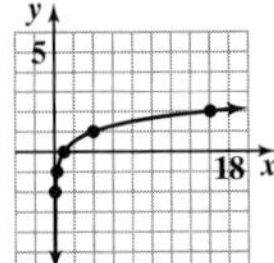

9. a. upward **b.** $(-4, -9)$ **c.** $(-7, 0)$ and $(-1, 0)$ **d.** $(0, 7)$ **e.**

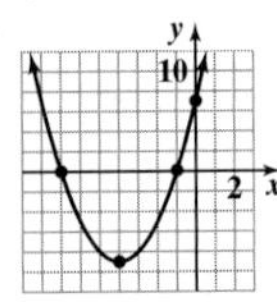

11. a. upward **b.** $(1, -9)$ **c.** $(-2, 0)$ and $(4, 0)$ **d.** $(0, -8)$ **e.**

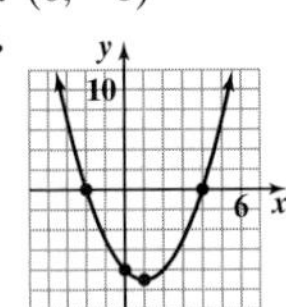

13. a. downward **b.** $(2, 1)$ **c.** $(1, 0)$ and $(3, 0)$ **d.** $(0, -3)$ **e.**

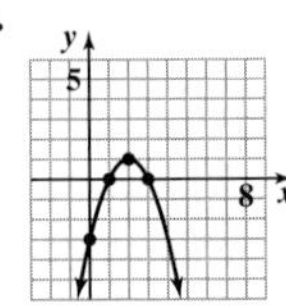

15. a.

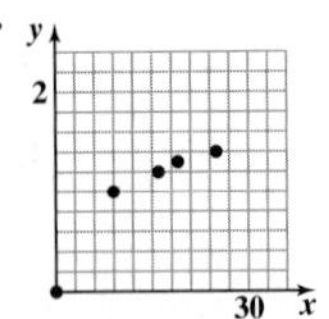

b. logarithmic

17. a.

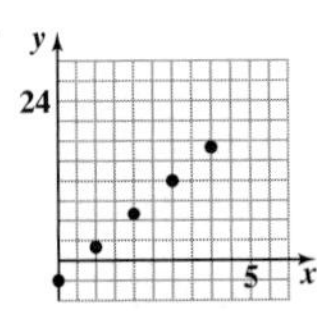

b. linear

19. a.

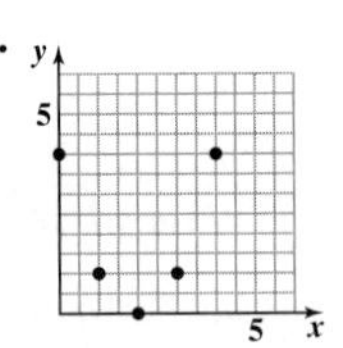

b. quadratic

21. a.

b. exponential

23.

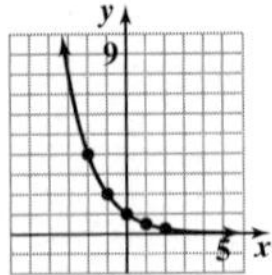

decreasing, although rate of decrease is slowing down

25.

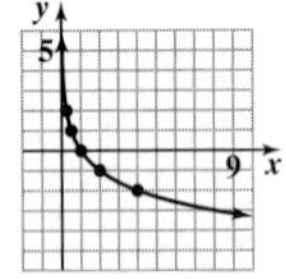

decreasing, although rate of decrease is slowing down

27. a. downward **b.** $(1, 7)$ **c.** $(-0.9, 0)$ and $(2.9, 0)$ **d.** $(0, 5)$ **e.**

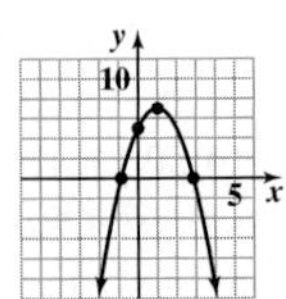

29. $(3, 2)$ **31. a.** The data are increasing more and more rapidly. **b.** $f(x) = 3.476(1.023)^x$ **c.** 43.4 million; underestimates by 1.2 million **d.** 54.5 million; overestimates by 0.8 million **33. a.** \$9868 **b.** \$9974 **c.** the exponential model **35. a.** The data increase rapidly and then begin to level off. **b.** $f(x) = 32 + 29 \ln x$ **c.** 99%; overestimates by 4% **37. a.** 95.4% **b.** Height increases rapidly at first and then more slowly. **39. a.** The data increase and then decrease.; The graph of the quadratic function modeling the data opens down. **b.** $f(x) = -0.8x^2 + 2.4x + 6$ **c.** 1.5; 7.8 **47.** does not make sense **49.** does not make sense **51.** 2 **53.** 0

Chapter 7 Review Exercises

1. 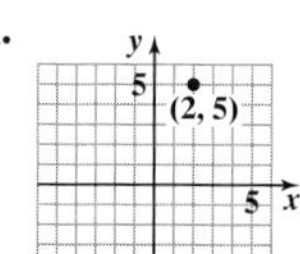

2. 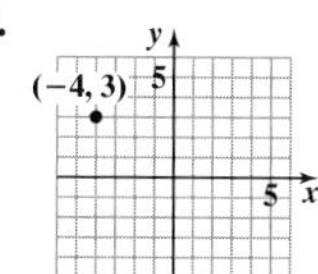

3. 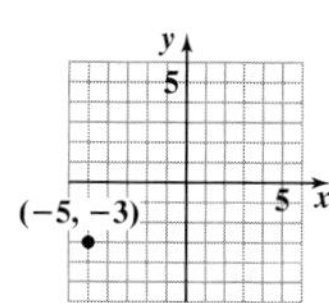

4. 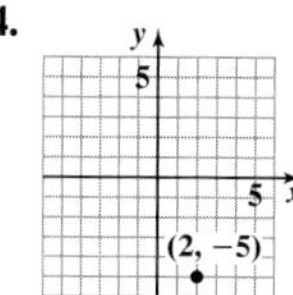

5. 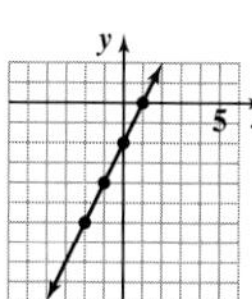

6. 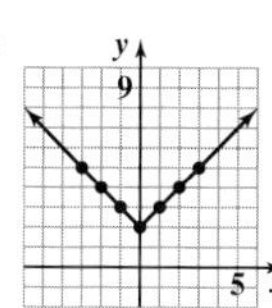

7. 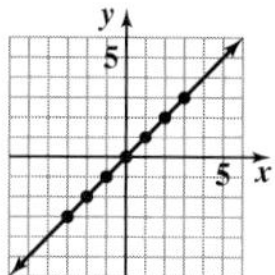

8. 3 **9.** 26 **10.** 30 **11.** −64

12.

x	$f(x) = \frac{1}{2}\lvert x \rvert$
−6	3
−4	2
−2	1
0	0
2	1
4	2
6	3

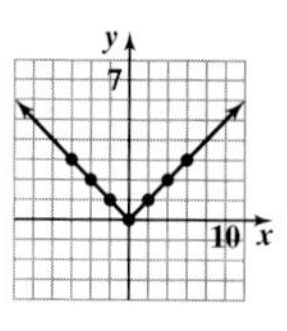

13.

x	$f(x) = x^2 - 2$
−2	2
−1	−1
0	−2
1	−1
2	2

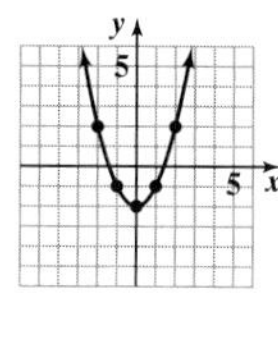

14. y is a function of x. **15.** y is not a function of x.

16. a. $D(1) = 92.8$; Skin damage begins for burn-prone people after 92.8 minutes when the sun's UV index is 1.; by the point (1, 92.8)
b. $D(10) = 19$; Skin damage begins for burn-prone people after 19 minutes when the sun's UV index is 10.; by the point (10, 19)

17. 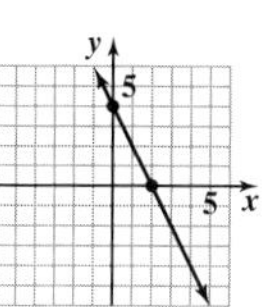

18. 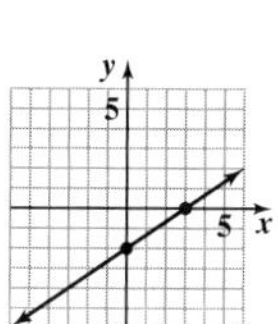

19. 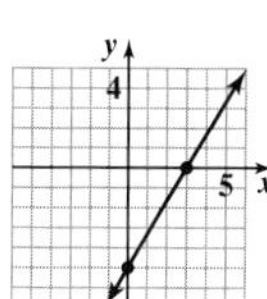

20. $-\frac{1}{2}$; falls **21.** 3; rises **22.** 0; horizontal **23.** undefined; vertical

24. 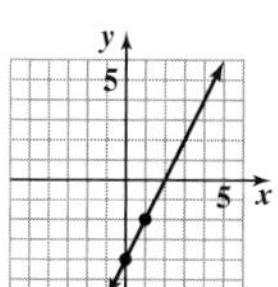

25. 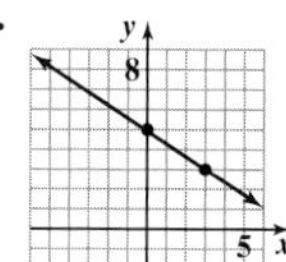

26. 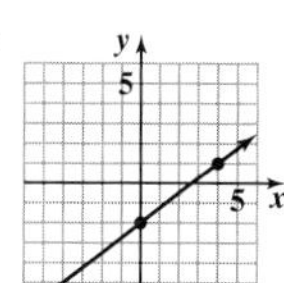

27. 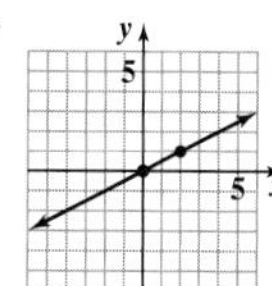

28. a. $y = -2x$
b. slope: −2; y-intercept: 0
c. 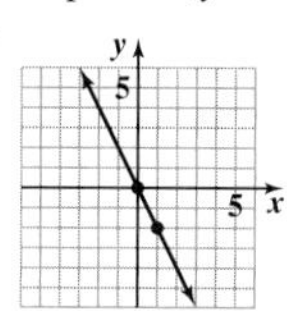

29. a. $y = \frac{5}{3}x$
b. slope: $\frac{5}{3}$; y-intercept: 0
c. 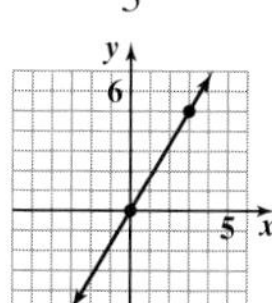

30. a. $y = -\frac{3}{2}x + 2$
b. slope: $-\frac{3}{2}$; y-intercept: 2
c. 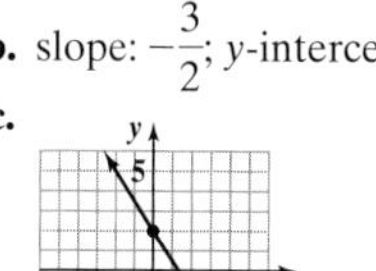

31. 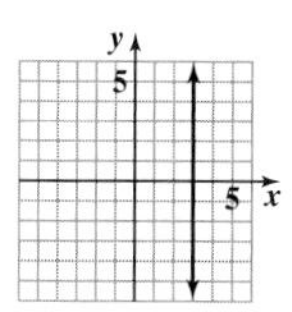

32. 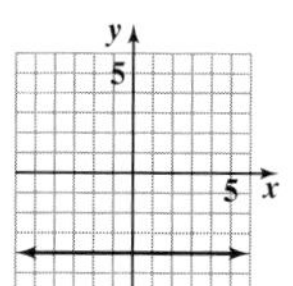

33. 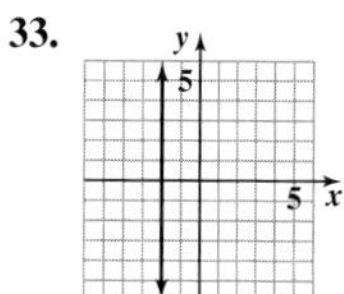

34. a. 254; If no women in a country are literate, the mortality rate of children under 5 is 254 per thousand. **b.** -2.4; For each 1% of adult females who are literate, the mortality rate of children under 5 decreases by 2.4 per thousand. **c.** $f(x) = -2.4x + 254$ **d.** 134 per thousand **35.** $\{(2, 3)\}$ **36.** $\{(-2, -3)\}$ **37.** $\{(3, 2)\}$ **38.** $\{(4, -2)\}$ **39.** $\{(1, 5)\}$ **40.** $\{(2, 3)\}$ **41.** $\{(-9, 3)\}$ **42.** $\{(3, 4)\}$ **43.** $\{(2, -3)\}$ **44.** $\varnothing$ **45.** $\{(3, -1)\}$ **46.** $\{(x, y) \mid 3x - 2y = 6\}$ or $\{(x, y) \mid 6x - 4y = 12\}$ **47. a.** $C(x) = 60{,}000 + 200x$ **b.** $R(x) = 450x$ **c.** (240, 108,000); This means the company will break even if it produces and sells 240 desks. **48. a.** Answers will vary.; approximately (2016, 325); 2016; 325 million **b.** $y = 6x + 200$ **c.** 2016; 326 million **d.** quite well, although answers may vary

49. 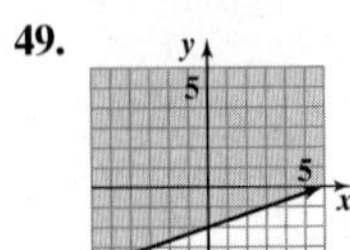

50. 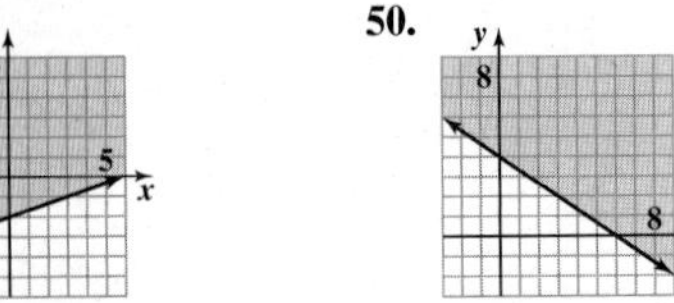

51. 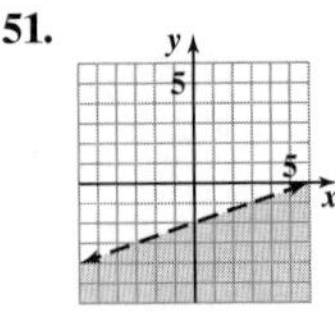

52. 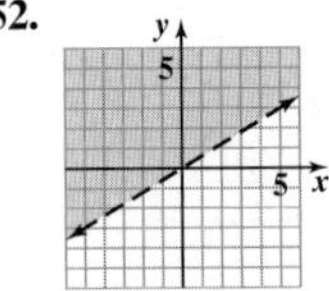

53. 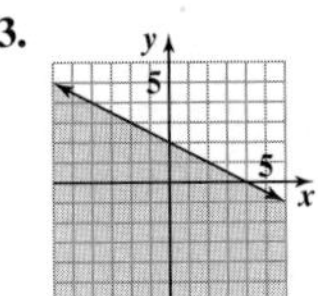

54. 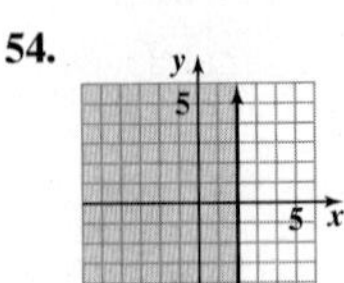

55. 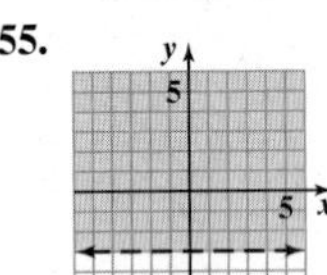

56. 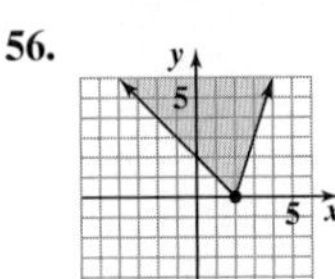

57. 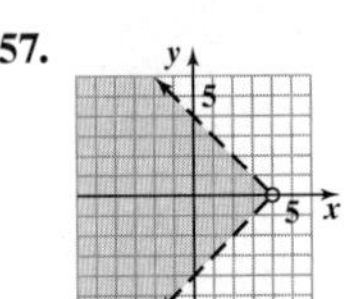

58. 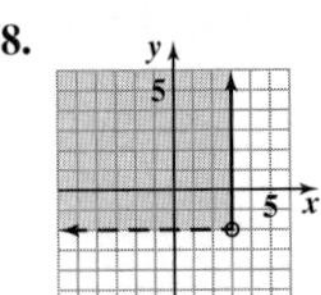

59. 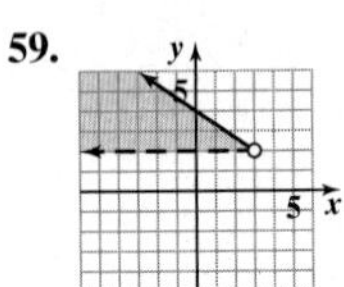

60. 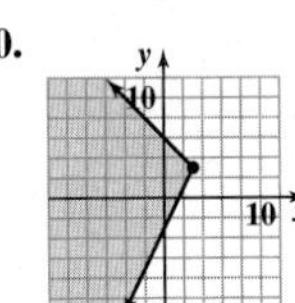

61. 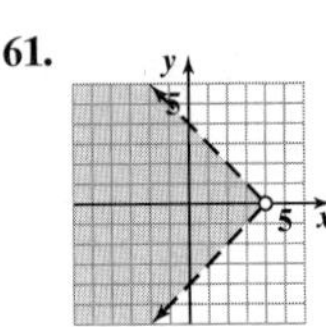

62. $(1, 0)$: 2; $\left(\frac{1}{2}, \frac{1}{2}\right)$: $\frac{5}{2}$; $(2, 2)$: 10; $(4, 0)$: 8; maximum: 10; minimum: 2

63. a. 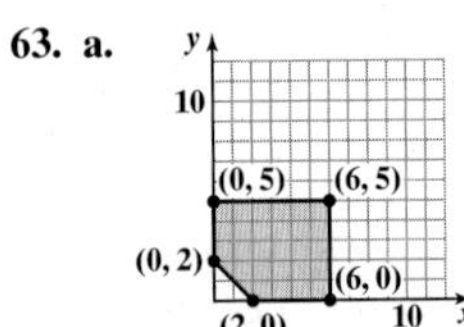

b. (0, 2): 6; (0, 5): 15; (6, 5): 27; (6, 0): 12; (2, 0): 4
c. maximum of 27 at $x = 6$ and $y = 5$; minimum of 4 at $x = 2$ and $y = 0$

64. a. $z = 500x + 350y$
b. 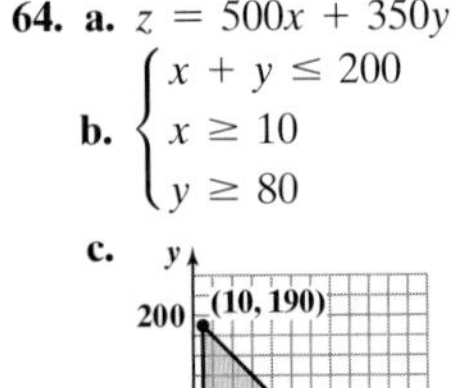
$$\begin{cases} x + y \le 200 \\ x \ge 10 \\ y \ge 80 \end{cases}$$
c. 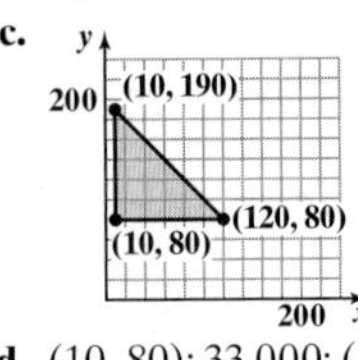

d. (10, 80): 33,000; (10, 190): 71,500; (120, 80): 88,000
e. 120; 80; 88,000

65. 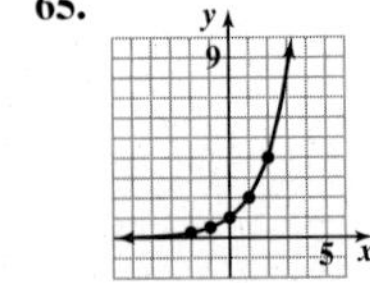

66. 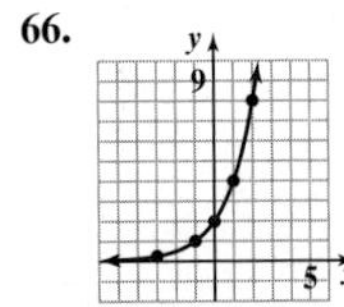

67. $x = 2^y$; 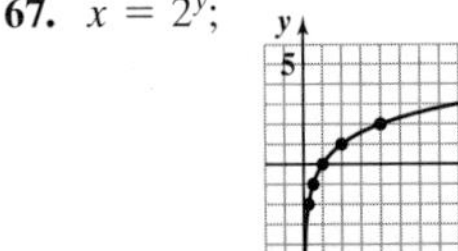

68. a. upward
b. $(3, -16)$
c. $(-1, 0)$ and $(7, 0)$
d. $(0, -7)$
e. 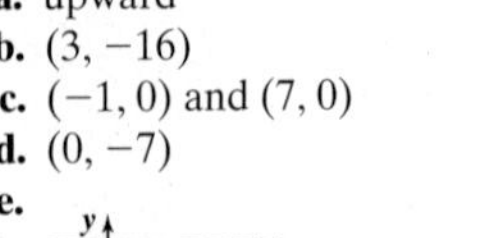 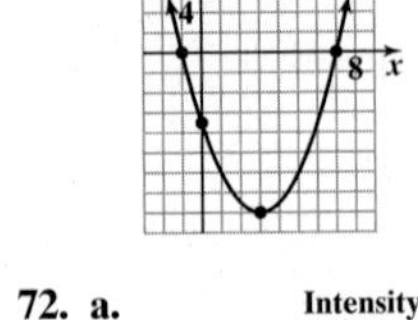

69. a. downward
b. $(-1, 4)$
c. $(-3, 0)$ and $(1, 0)$
d. $(0, 3)$
e. 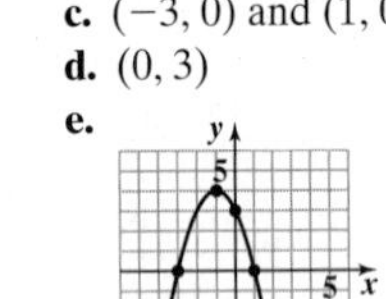

70. a. 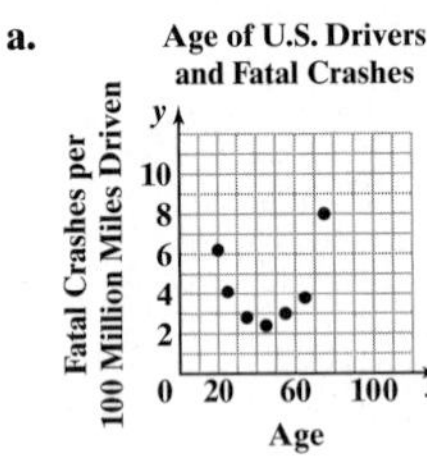

b. quadratic

71. a. 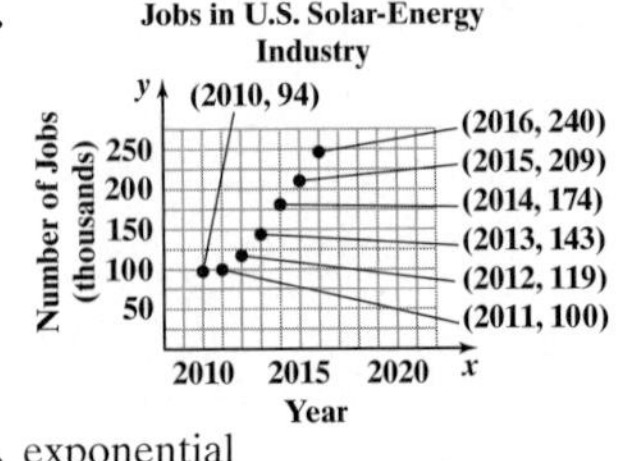

b. exponential

72. a. 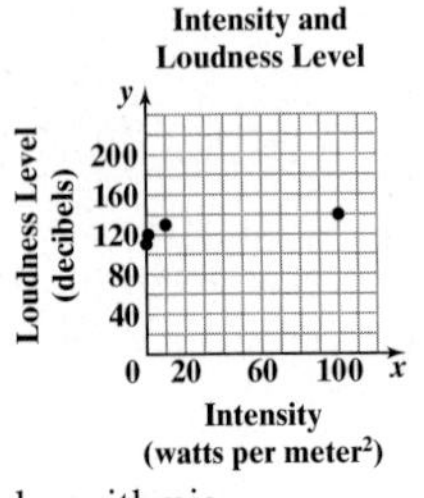

b. logarithmic

73. a. 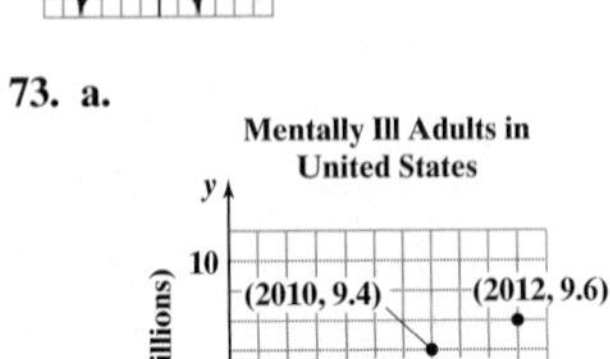 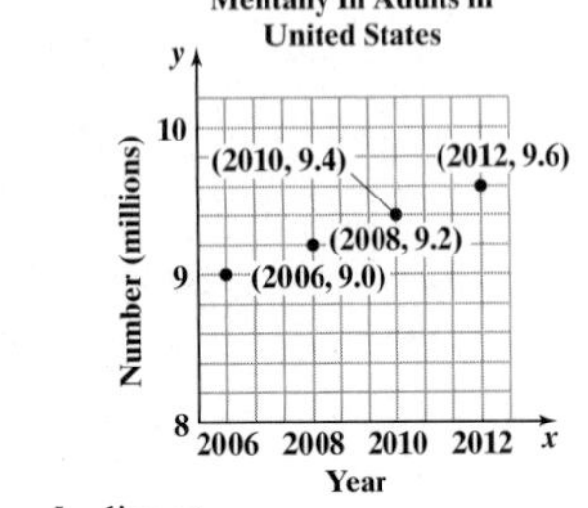

b. linear

74. a. 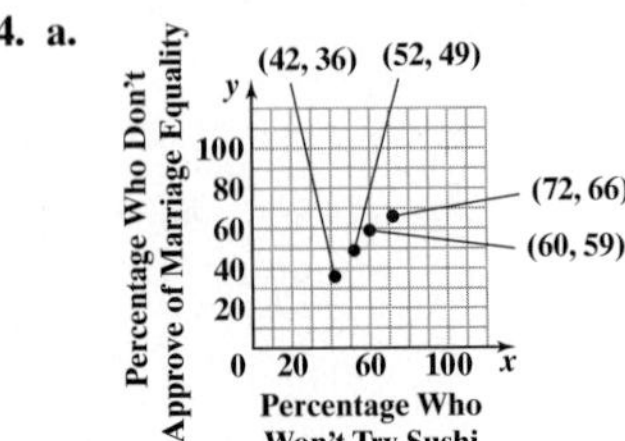

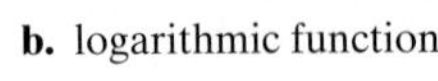
b. logarithmic function

75. a. 47; For each additional hour spent at a shopping mall, the average amount spent increases by \$47. The rate of change is \$47 per hour. **b.** the exponential function **76.** the logarithmic function

Chapter 7 Test

1. **2.** 21 **3.** y is not a function of x. **4.** y is a function of x. **5. a.** Yes; No vertical line intersects the graph in more than one point. **b.** 0; The eagle was on the ground after 15 seconds. **c.** 45 meters **d.** between 3 and 12 seconds

6. 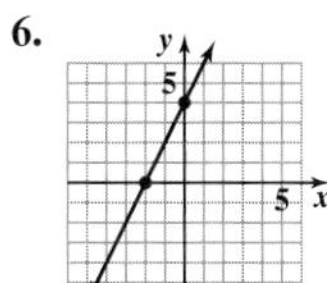

7. $m = 3$

8. 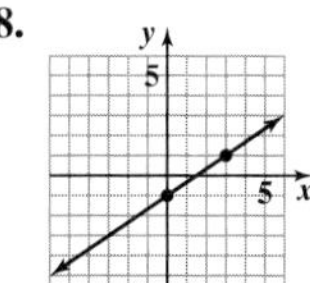

9. 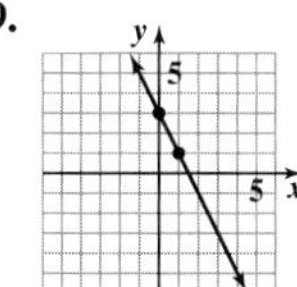

10. a. -0.48 **b.** 0.48; -0.48%; year **11. a.** 4571; There were 4571 deaths involving distracted driving 0 years after 2005, or in 2005. **b.** 433; The rate of change in the number of deaths involving distracted driving is 433 deaths per year. **c.** $f(x) = 433x + 4571$ **d.** 8901 deaths

12. $\{(2, 4)\}$ **13.** $\{(1, -3)\}$ **14.** $\{(6, -5)\}$ **15. a.** $C(x) = 360{,}000 + 850x$ **b.** $R(x) = 1150x$ **c.** (1200, 1,380,000); The company will break even if it produces and sells 1200 computers.

16. 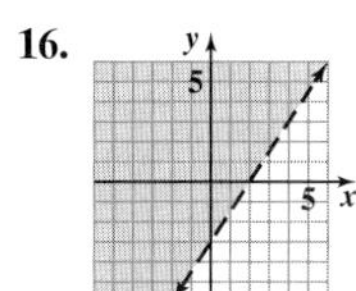

17. 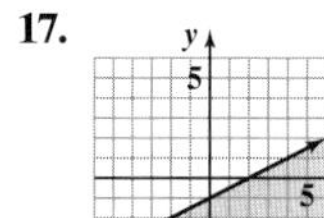

18. 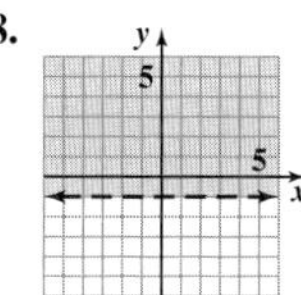

19. 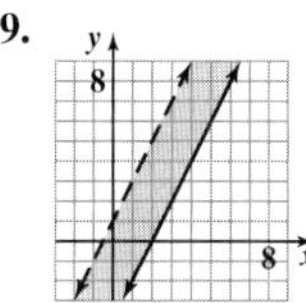

20. (2, 0): 6; (2, 6): 18; (6, 3): 24; (8, 0): 24; maximum: 24; minimum: 6 **21.** 26 **22.** 50 regular jet skis, 100 deluxe jet skis; \$35,000

23. 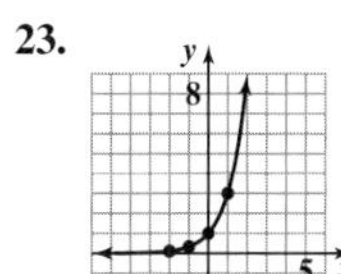

24. $x = 3^y$; 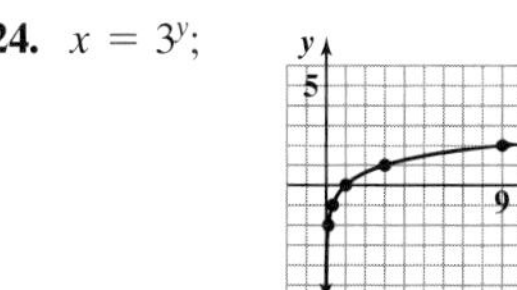

25. 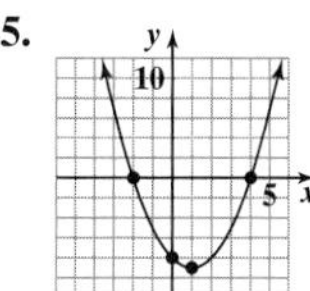

26. linear **27.** logarithmic **28.** exponential **29.** quadratic **30. a.** 0.03; The cost of making a penny is increasing an average of 0.03 cent per year. **b.** exponential; The cost is rising more and more rapidly. **c.** linear: 1.53¢; exponential: 1.75¢; not really; exponential; yes

CHAPTER 8

Section 8.1

Check Point Exercises

1. 12.5% **2.** 2.3% **3. a.** 0.67 **b.** 2.5 **4. a.** \$75.60 **b.** \$1335.60 **5. a.** \$133 **b.** \$247 **6. a.** $66\frac{2}{3}\%$ **b.** 40% **7.** 35% **8.** 20% **9. a.** \$1152 **b.** 4% decrease

Concept and Vocabulary Check

1. 100 **2.** 7; 8; 100; a percent sign **3.** two; right; a percent sign **4.** two; left; the percent sign **5.** tax rate; item's cost **6.** discount rate; original price **7.** the amount of increase; the original amount **8.** the amount of decrease; the original amount **9.** false; Changes will vary. **10.** false; Changes will vary.

Exercise Set 8.1

1. 40% **3.** 25% **5.** 37.5% **7.** 2.5% **9.** 11.25% **11.** 59% **13.** 38.44% **15.** 287% **17.** 1487% **19.** 10,000% **21.** 0.72 **23.** 0.436 **25.** 1.3 **27.** 0.02 **29.** 0.005 **31.** 0.00625 **33.** 0.625 **35.** 6 **37.** 7.2 **39.** 5 **41.** 170 **43.** 20% **45.** 12% **47. a.** \$1968 **b.** \$34,768 **49. a.** \$103.20 **b.** \$756.80 **51.** 19.2% **53.** 184% **55.** 15% **57.** no; 2% loss **65.** does not make sense **67.** does not make sense **69.** \$648.72 **71.** \$2400 increase

Section 8.2

Check Point Exercises

1. a. \$89,880 **b.** \$86,680 **c.** \$46,410 **2.** \$3833.75 **3.** \$10,247 **4. a.** \$240 **b.** \$24 **c.** \$18.36 **d.** \$9.60 **e.** \$188.04 **f.** 21.7%

Concept and Vocabulary Check

1. gross **2.** adjusted gross; adjustments **3.** exemption **4.** adjusted gross; exemptions; deductions **5.** credits **6.** FICA **7.** gross **8.** net **9.** false; Changes will vary. **10.** true **11.** true **12.** true

Exercise Set 8.2

1. gross income: \$53,320; adjusted gross income: \$50,120; taxable income: \$38,820 **3.** gross income: \$87,490; adjusted gross income: \$85,290; taxable income: \$68,215 **5.** \$5771.25 **7.** \$27,306.50 **9.** −\$713.75, or a \$713.75 refund **11.** \$50,413 **13.** \$2297.50 **15.** \$4298.75 **17.** \$1620 **19.** \$9087 **21.** \$19,044 **23. a.** \$1530 **b.** \$983.75 **c.** 12.6% **25. a.** \$330 **b.** \$33 **c.** \$25.25 **d.** \$9.90 **e.** \$261.85 **f.** 20.7% **39.** does not make sense **41.** makes sense **43.** The \$2500 credit reduces taxes by the greater amount; the difference is \$2100. **45.** yes; Answers will vary.

Section 8.3

Check Point Exercises

1. \$150 **2.** \$336 **3.** \$2091 **4. a.** \$1820 or \$1825 **b.** \$1892.80 or \$1898 **5.** 18% **6.** \$3846.16

Concept and Vocabulary Check

1. $I = Prt$; principal; rate; time **2.** $A = P(1 + rt)$ **3.** 360 **4.** false; Changes will vary. **5.** true **6.** false; Changes will vary.

Exercise Set 8.3

1. \$240 **3.** \$10.80 **5.** \$318.75 **7.** \$426.25 **9.** \$3420 **11.** \$38,350 **13.** \$9390 **15.** 7.5% **17.** 9% **19.** 31.3% **21.** \$5172.42 **23.** \$8917.20 **25.** \$4509.59 **27.** $r = \frac{A - P}{Pt}$ **29.** $P = \frac{A}{1 + rt}$ **31. a.** \$247.50 **b.** \$4247.50 **33.** 21.4% **35.** 640% **37.** \$2654.87 **41.** does not make sense **43.** does not make sense **45. a.** $A = 275t + 5000$ **b.** slope: 275; Answers will vary; an example is: The rate of change of the future value per year is \$275.

Section 8.4

Check Point Exercises

1. a. \$1216.65 **b.** \$216.65 **2. a.** \$6253.23 **b.** \$2053.23 **3. a.** \$14,859.47 **b.** \$14,918.25 **4.** \$5714.25 **5. a.** \$6628.28 **b.** 10.5% **6.** ≈ 8.24%

Concept and Vocabulary Check

1. principal; interest **2.** $t; r; n$ **3.** $A = P(1 + r)^t$ **4.** six; semiannually **5.** three; quarterly **6.** continuous **7.** $P; A; t; r; n$ **8.** effective annual yield; simple **9.** true **10.** true **11.** true **12.** true

Exercise Set 8.4

1. a. \$10,816 **b.** \$816 **3. a.** \$3655.21 **b.** \$655.21 **5. a.** \$12,795.12 **b.** \$3295.12 **7. a.** \$5149.12 **b.** \$649.12 **9. a.** \$1855.10 **b.** \$355.10 **11. a.** \$49,189.30 **b.** \$29,189.30 **13. a.** \$13,116.51 **b.** \$13,140.67 **c.** \$13,157.04 **d.** \$13,165.31 **15.** 7% compounded monthly **17.** \$8374.85 **19.** \$7528.59 **21. a.** \$10,457.65 **b.** 4.6% **23.** 6.1% **25.** 6.2% **27.** 6.2% **29.** 8.3%; 8.25%; 8% compounded monthly **31.** 5.6%; 5.5%; 5.5% compounded semiannually **33.** \$2,069,131 **35.** \$649,083 **37.** \$41,528 **39. a.** you; \$391 **b.** you; \$340 **c.** your friend; \$271 **41.** \$9186 or \$9187 **43. a.** ≈ \$8,280,100,000 **b.** ≈\$8,614,300,000 **45.** 5.9% compounded daily; \$2

47.

	Once a Year		Continuous	
Years	**Amount**	**Interest**	**Amount**	**Interest**
1	\$5275	\$275	\$5283	\$283
5	\$6535	\$1535	\$6583	\$1583
10	\$8541	\$3541	\$8666	\$3666
20	\$14,589	\$9589	\$15,021	\$10,021

49. \$37,096 **51. a.** account A: \$38,755 **b.** account B: \$41,163 **53.** 5.55% **55.** 4.3%; 4.3%; 4.3%; As the number of compounding periods increases, the effective annual yield increases slightly. However, with the rates rounded to the nearest tenth of a percent, this increase is not evident. **57.** 4.5% compounded semiannually **63.** does not make sense **65.** does not make sense **67.** \$12,942.94 or \$12,942.95

69. $A = P\left(1 + \frac{r}{n}\right)^{nt}$

Substitute this value of A into the future value formula for simple interest. Since the interest rates are two different values, use Y, the effective yield, for the interest rate on the right side of the equation.

$$P\left(1 + \frac{r}{n}\right)^{nt} = P\,(1 + Yt)$$

$$\left(1 + \frac{r}{n}\right)^{nt} = (1 + Yt) \qquad \text{Divide each side by } P.$$

$$\left(1 + \frac{r}{n}\right)^{n} = 1 + Y \qquad \text{Let } t = 1.$$

$$Y = \left(1 + \frac{r}{n}\right)^{n} - 1 \qquad \text{Subtract 1 from each side and interchange the sides.}$$

Section 8.5

Check Point Exercises

1. a. \$6620 **b.** \$620 **2. a.** \$777,169 or \$777,170 **b.** \$657,169 or \$657,170 **3. a.** \$333,946 or \$334,288 **b.** \$291,946 or \$292,288 **4. a.** \$187 **b.** deposits: \$40,392; interest: \$59,608 **5. a.** high: \$63.38; low: \$42.37 **b.** \$2160 **c.** 1.5%; This is much lower than a bank account paying 3.5%. **d.** 7,203,200 shares **e.** high: \$49.94; low: \$48.33 **f.** \$49.50 **g.** The closing price is up \$0.03 from the previous day's closing price. **h.** ≈ \$1.34 **6. a.** \$81,456 **b.** \$557,849 **c.** \$527,849

Concept and Vocabulary Check

1. annuity **2.** *P; r; n;* value of the annuity; *t* **3.** stock; capital; dividends **4.** bonds **5.** portfolio; diversified **6.** mutual fund **7.** Roth; $59\frac{1}{2}$ **8.** true **9.** false; Changes will vary. **10.** true **11.** true

Exercise Set 8.5

1. a. \$66,132 **b.** \$26,132 **3. a.** \$702,528 **b.** \$542,528 **5. a.** \$50,226 **b.** \$32,226 **7. a.** \$9076 **b.** \$4076 **9. a.** \$28,850 **b.** \$4850 **11. a.** \$4530 **b.** deposits: \$81,540; interest: \$58,460 **13. a.** \$1323 **b.** deposits: \$158,760; interest: \$41,240 **15. a.** \$356 **b.** deposits: \$170,880; interest: \$829,120 **17. a.** \$920 **b.** deposits: \$18,400; interest: \$1600 **19. a.** high: \$73.25; low: \$45.44 **b.** \$840 **c.** 2.2%; Answers will vary. **d.** 591,500 shares **e.** high: \$56.38; low: \$54.38 **f.** \$55.50 **g.** \$1.25 increase **h.** ≈ \$3.26 **21. a.** \$30,000 **b.** \$30,000 **23.** $P = \frac{Ar}{[(1+r)^t - 1]}$; the deposit at the end of each year that yields *A* dollars after *t* years with interest rate *r* compounded annually **25. a.** \$11,617 **b.** \$1617 **27. a.** \$87,052 **b.** \$63,052 **29. a.** \$693,031 **b.** \$293,031 **31. a.** \$401 **b.** deposits: \$3208; interest: \$292 **33.** \$620; \$1,665,200 **35. a.** \$173,527 **b.** \$1,273,733 **c.** \$1,219,733 **47.** does not make sense **49.** makes sense **51.** does not make sense **53.** does not make sense **55. a.** \$248 **b.** with: \$5037.35; without: \$5683.75 **c.** with: 10.1%; without: 11.4%

Section 8.6

Check Point Exercises

1. a. monthly payment: \$366; interest: \$2568 **b.** monthly payment: \$278; interest: \$5016 **c.** Monthly payments are lower with the longer-term loan, but there is more interest with this loan. **2.** \$222 **3. a.** \$5250 **b.** \$47,746

Concept and Vocabulary Check

1. *PMT; P; n; t; r* **2.** closed-end; open-end **3.** residual value **4.** bodily injury; property damage **5.** collision **6.** comprehensive **7.** true **8.** false; Changes will vary. **9.** false; Changes will vary. **10.** true **11.** false; Changes will vary. **12.** true

Exercise Set 8.6

1. monthly payment: \$244; interest: \$1712 **3. a.** monthly payment: \$450; interest: \$1200 **b.** monthly payment: \$293; interest: \$2580 **c.** Monthly payments are lower with the longer-term loan, but there is more interest with this loan. **5.** \$277 **7.** monthly payment: \$405; interest: \$3665 **9.** \$65; Incentive B is the better deal. **11. a.** \$6000 **b.** \$42,142 **13. a.** \$19,600 **b.** \$145,609 **15.** \$104,400 **25.** makes sense **27.** makes sense **29.** makes sense

31.

$$P\left(1+\frac{r}{n}\right)^{nt} = \frac{PMT\left[\left(1+\frac{r}{n}\right)^{nt}-1\right]}{\left(\frac{r}{n}\right)}$$

$$P\left(\frac{r}{n}\right)\left(1+\frac{r}{n}\right)^{nt} = PMT\left[\left(1+\frac{r}{n}\right)^{nt}-1\right]$$

$$\frac{P\left(\frac{r}{n}\right)\left(1+\frac{r}{n}\right)^{nt}}{\left[\left(1+\frac{r}{n}\right)^{nt}-1\right]} = PMT$$

$$\frac{P\left(\frac{r}{n}\right)}{\left[1-\left(1+\frac{r}{n}\right)^{-nt}\right]} = PMT$$

Section 8.7

Check Point Exercises

1. a. \$1627 **b.** \$117,360 **c.** \$148,860

2.

Payment Number	Interest Payment	Principal Payment	Balance of Loan
1	\$1166.67	\$383.33	\$199,616.67
2	\$1164.43	\$385.57	\$199,231.10

3. a. \$5600 **b.** \$7200 **c.** \$2160

Concept and Vocabulary Check

1. mortgage; down payment **2.** loan amortization schedule **3.** false; Changes will vary. **4.** false; Changes will vary. **5.** true **6.** false; Changes will vary.

Exercise Set 8.7

1. a. $44,000 **b.** $176,000 **c.** $5280 **d.** $1171 **e.** $245,560 **3.** $60,120 **5.** 20-year at 7.5%; $106,440 **7.** Mortgage A; $11,300 **9. a.** monthly payment: $608; total interest: $98,880

b.

Payment Number	Interest	Principal	Loan Balance
1	$450.00	$158.00	$119,842.00
2	$449.41	$158.59	$119,683.41
3	$448.81	$159.19	$119,524.22

11. a. $840 **b.** $1080 **c.** $492 **23.** makes sense **25.** does not make sense **27.** Yes

Section 8.8

Check Point Exercises

1. a. $8761.76 **b.** $140.19 **c.** $9661.11 **d.** $269

Concept and Vocabulary Check

1. open-end **2.** the sum of the unpaid balances for each day in the billing period; the number of days in the billing period **3.** debit **4.** credit report **5.** 300; 850; better credit(worthiness) **6.** false; Changes will vary. **7.** false; Changes will vary **8.** true **9.** true

Exercise Set 8.8

1. a. $6150.65 **b.** $92.26 **c.** $6392.26 **d.** $178 **3. a.** $2057.22 **b.** $24.69 **c.** $2237.23 **d.** $90 **5. a.** $210 **b.** $840 **7. a.** $137; lower monthly payment **b.** $732; less interest **9.** monthly payment: $385; total interest: $420; $175 more each month; $420 less interest **19.** does not make sense **21.** makes sense **23.** does not make sense **25.** does not make sense

Chapter 8 Review Exercises

1. 80% **2.** 12.5% **3.** 75% **4.** 72% **5.** 0.35% **6.** 475.6% **7.** 0.65 **8.** 0.997 **9.** 1.50 **10.** 0.03 **11.** 0.0065 **12.** 0.0025 **13.** 9.6 **14. a.** $1.44 **b.** $25.44 **15. a.** $297.50 **b.** $552.50 **16.** 12.5% increase **17.** 35% decrease **18.** no; $9900; 1% decrease **19.** gross income: $30,330; adjusted gross income: $29,230; taxable income: $16,730 **20.** gross income: $436,400; adjusted gross income: $386,400; taxable income: $278,150 **21.** $193,769.95 **22.** $4542.50 **23.** $3308.75 **24.** $6579 **25.** $22,234 **26. a.** $224 **b.** $22.40 **c.** $17.14 **d.** $8.96 **e.** $175.50 **f.** 21.7% **27.** $180 **28.** $2520 **29.** $1200 **30.** $900 **31. a.** $122.50 **b.** $3622.50 **32.** $12,738 **33.** 7.5% **34.** $13,389.12 **35.** $9287.93 **36.** 40% **37. a.** $8114.92 **b.** $1114.92 **38. a.** $38,490.80 **b.** $8490.80 **39. a.** $5556.46 **b.** $3056.46 **40.** 7% compounded monthly; $362 **41.** $28,469,44 **42.** $13,175.19 **43. a.** $2122.73 **b.** 6.1% **44.** 5.61%; Answers will vary; an example is: The same amount of money would earn 5.61% in a simple interest account for a year. **45.** 6.25% compounded monthly **46. a.** $19,129 **b.** $8729 **47. a.** $91,361 **b.** $55,361 **48. a.** $1049 **b.** $20,980; $4020 **49. a.** high: $64.06; low: $26.13 **50.** $144 **51.** 0.3% **52.** 545,800 shares **53.** high: $61.25; low: $59.25 **54.** $61.00 **55.** $1.75 increase **56.** $1.49 **59. a.** monthly payment: $465; total interest: $1740 **b.** monthly payment: $305; total interest: $3300 **c.** Longer term has lower monthly payment but greater total interest. **63. a.** $7560 **b.** $53,099 **64. a.** $48,000 **b.** $192,000 **c.** $3840 **d.** $1277 **e.** $267,720 **65. a.** 20-year at 8%; $53,040; Answers will vary. **66. a.** option A: $769: option B: $699: Answers will vary. **b.** Mortgage A: $20,900 **67. a.** $1896

b.

Payment Number	Interest	Principal	Loan Balance
1	$1625.00	$271.00	$299,729.00
2	$1623.53	$272.47	$299,456.53
3	$1622.06	$273.94	$299,182.59

68. a. $1260 **b.** $1620 **c.** $612 **71. a.** $4363.27 **b.** $48.00 **c.** $4533.00 **d.** $126 **72. a.** $836 **b.** $3316

Chapter 8 Test

1. a. $18 **b.** $102 **2.** 75% **3. a.** $47,290 **b.** $46,190 **c.** $33,095 **4.** $3158.75 **5.** $9522 **6. a.** $150 **b.** $15 **c.** $11.48 **d.** $4.50 **e.** $119.02 **f.** 20.7% **7.** $72; $2472 **8.** 25% **9.** $6698.57 **10.** 4.58%; Answers will vary; an example is: The same amount of money would earn 4.58% in a simple interest account for a year. **11. a.** $267,392 **b.** $247,392 **12.** $2070 **13. a.** $8297 **b.** $2297 **14. a.** $7067 **b.** $1067 **c.** Only part of the $6000 is invested for the entire five years.; Answers will vary. **15.** $704 per month; $1,162,080 interest **16.** high: $25.75; low: $25.50 **17.** $2030 **18.** $386.25 **19.** $5320 **20.** $12,000 **21.** $108,000 **22.** $2160 **23.** $830 **24.** $190,800

25.

Payment Number	Interest	Principal	Loan Balance
1	$765.00	$65.00	$107,935.00
2	$764.54	$65.46	$107,869.54

26. a. $1540 **b.** $1980 **c.** $594 **27. a.** $3226.67 **b.** $64.53 **c.** $3464.53 **d.** $97 **28.** false; Changes will vary. **29.** true **30.** false; Changes will vary. **31.** false; Changes will vary. **32.** true **33.** false; Changes will vary. **34.** false; Changes will vary.

CHAPTER 9

Section 9.1

Check Point Exercises

1. a. 6.5 ft **b.** 3.25 mi **c.** $\frac{1}{12}$ yd **2. a.** 8 km **b.** 53,000 mm **c.** 0.0604 hm **d.** 6720 cm **3. a.** 243.84 cm **b.** ≈ 22.22 yd **c.** ≈ 1181.1 in. **4.** ≈ 37.5 mi/hr

Concept and Vocabulary Check

1. linear; linear **2.** 12; 3; 36; 5280 **3.** unit; 1 **4.** 1000; 100; 10; 0.1; 0.01; 0.001 **5.** false; Changes will vary. **6.** false; Changes will vary. **7.** false; Changes will vary. **8.** false; Changes will vary.

Exercise Set 9.1

1. 2.5 ft **3.** 360 in. **5.** $\frac{1}{6}$ yd ≈ 0.17 yd **7.** 216 in. **9.** 18 ft **11.** 2 yd **13.** 4.5 mi **15.** 3960 ft **17.** 500 cm **19.** 1630 m **21.** 0.03178 hm **23.** 0.000023 m **25.** 21.96 dm **27.** 35.56 cm **29.** ≈ 5.51 in. **31.** ≈ 424 km **33.** ≈ 165.625 mi **35.** ≈ 13.33 yd **37.** ≈ 55.12 in. **39.** 0.4064 dam **41.** 1.524 m **43.** ≈ 16.40 ft **45.** ≈ 60 mi/hr **47.** ≈ 72 km/hr **49.** 457.2 cm **51.** $8\frac{1}{3}$ yd **53.** 48.28032 km **55.** 176 ft/sec **57.** meter **59.** millimeter **61.** meter **63.** millimeter **65.** millimeter **67.** b **69.** a **71.** c **73.** a **75.** 0.216 km **77.** ≈ 148.8 million km **79.** Nile; ≈ 208 km **81.** Everest; ≈ 239 m **83.** Waialeale; ≈ 46 in. **93.** makes sense **95.** does not make sense **97.** 9 hm **99.** 11 m or 1.1 dam

Section 9.2

Check Point Exercises

1. 8 square units **2.** 239.1 people per square mile **3.** 131,250 mi^2 **4. a.** ≈ 0.72 ha **b.** ≈ $576,389 per hectare **5.** 9 cubic units **6.** ≈ 74,800 gal **7.** 220 L **8. a.** 20 mL **b.** ≈ 0.67 fl oz

Concept and Vocabulary Check

1. square; cubic **2.** 144; 9 **3.** $\frac{1 \text{ mi}^2}{640 \text{ acres}}; \frac{640 \text{ acres}}{1 \text{ mi}^2}$ **4.** 2; 4 **5.** capacity; liter **6.** area **7.** 1 **8.** false; Changes will vary. **9.** false; Changes will vary. **10.** false; Changes will vary.

Exercise Set 9.2

1. 16 square units **3.** 8 square units **5.** ≈ 2.15 in.^{2} **7.** ≈ 37.5 yd^2 **9.** ≈ 25.5 acres **11.** ≈ 91 cm^2 **13.** 24 cubic units **15.** ≈ 74,800 gal **17.** ≈ 1600 gal **19.** ≈ 9 gal **21.** ≈ 13.5 yd^3 **23.** 45 L **25.** 17 mL **27.** 1500 cm^3 **29.** 150 cm^3 **31.** 12,000 dm^3 **33.** ≈ 2.4 tsp **35.** ≈ 45 mL **37.** ≈ 2.33 fl oz **39.** ≈ 5.83 c **41.** ≈ 2.82 L **43.** ≈ 4.21 qt **45.** ≈ 11.4 L **47.** ≈ 2.11 qt **49. a.** 1900: 25.6 people per square mile; 2010: 87.4 people per square mile **b.** 241.4% **51.** 17.0 people per square kilometer **53.** Illinois: 222.2 people per square mile; Ohio: 257.6 people per square mile; Ohio has the greater population density by 35.4 people per square mile. **55.** ≈ 2358 mi^2 **57. a.** ≈ 20 acres **b.** ≈ $12,500 per acre **59.** square centimeters **61.** square kilometers **63.** b **65.** b **67.** ≈ 336,600 gal **69.** 4 L **71.** Japan; ≈ 79,000 km^2 **73.** Baffin Island; ≈ 24,162 mi^2 **75. a.** ≈ 15 mL **b.** ≈ 15 cc **c.** 0.5 fl oz **d.** ≈ 0.17 fl oz **83.** does not make sense **85.** makes sense **87.** ≈ 11,952.64 people per square mile **89.** Answers will vary. **91.** 6.5 liters; Answers will vary.

Section 9.3

Check Point Exercises

1. a. 420 mg **b.** 6.2 g **2.** 145 kg **3. a.** ≈ 54 kg **b.** ≈ 17.9 oz **4.** 2 tablets **5.** 122°F **6.** 15°C

Concept and Vocabulary Check

1. 16; 2000 **2.** gram **3.** 2.2 **4.** 1 **5.** 32; 212 **6.** 0; 100 **7.** false; Changes will vary. **8.** false; Changes will vary. **9.** false; Changes will vary. **10.** false; Changes will vary.

Exercise Set 9.3

1. 740 mg **3.** 0.87 g **5.** 800 cg **7.** 18,600 g **9.** 0.000018 g **11.** 50 kg **13.** 4200 cm^3 **15.** 1100 t **17.** 40,000 g **19.** 2.25 lb **21.** ≈ 1008 g **23.** ≈ 243 kg **25.** ≈ 36,000 g **27.** ≈ 1200 lb **29.** ≈ 222.22 T **31.** 50°F **33.** 95°F **35.** 134.6°F **37.** 23°F **39.** 20°C **41.** 5°C **43.** 22.2°C **45.** − 5°C **47.** 176.7°C **49.** − 30°C **51. a.** $\frac{9}{5}$; Fahrenheit temperature increases by $\frac{9}{5}$° for each 1° change in Celsius temperature. **b.** $F = \frac{9}{5}C + 32$ **53.** milligram **55.** gram **57.** kilogram **59.** kilogram **61.** b **63.** a **65.** c **67.** 13.28 kg **69.** $1.10 **71.** Purchasing the regular size is more economical. **73.** 4 tablets **75. a.** 43 mg **b.** ≈ 516 mg **77.** a **79.** c **81.** Néma; ≈ 0.3 °F **83.** Eismitte; ≈ 5°C **91.** does not make sense **93.** makes sense **95.** false **97.** true **99.** true **101.** false

Chapter 9 Review Exercises

1. 5.75 ft **2.** 0.25 yd **3.** 7 yd **4.** 2.5 mi **5.** 2280 cm **6.** 70 m **7.** 1920 m **8.** 0.0144 hm **9.** 0.0005 m **10.** 180 mm **11.** 58.42 cm **12.** ≈ 7.48 in. **13.** ≈ 528 km **14.** ≈ 375 mi **15.** ≈ 15.56 yd **16.** ≈ 39.37 ft **17.** ≈ 28.13 mi/hr **18.** ≈ 96 km/hr **19.** 0.024 km, 24,000 cm, 2400 m **20.** 4.8 km **21.** 24 square units **22.** 16,513.8 people per square mile; In Singapore, there is an average of 16,513.8 people for each square mile. **23.** 74 mi^2 **24.** ≈ 18 acres **25.** ≈ 333.33 ft^2 **26.** ≈ 31.2 km^2 **27.** a **28.** 24 cubic units **29.** ≈ 251,328 gal **30.** 76 L **31.** ≈ 4.4 tsp **32.** ≈ 22.5 c **33.** ≈ 8.42 qt **34.** ≈ 22.8 L **35. a.** 15 mL **b.** ≈ 0.5 fl oz **36.** c **39.** 1240 mg **40.** 1200 cg **41.** 0.000012 g **42.** 0.00045 kg **43.** 50,000 cm^3 **44.** 4000 dm^3, 4,000,000 g **45.** 94.5 kg **46.** 14 oz **47.** 3 tablets **48.** kilograms; Answers will vary. **49.** 2.25 lb **50.** a **51.** c **52.** 59 °F **53.** 212°F **54.** 41°F **55.** 32°F **56.** − 13°F **57.** 15°C **58.** 5°C **59.** 100°C **60.** 37°C **61.** ≈ − 17.8°C **62.** − 10°C **63.** more; Answers will vary.

Chapter 9 Test

1. 0.00807 hm **2.** 250 in. **3.** 4.8 km **4.** mm **5.** cm **6.** km **7.** ≈ 128 km/hr **8.** 9 times **9.** 7.7 people per square mile; In Australia, there is an average of 7.7 people for each square mile. **10.** ≈ 45 acres **11.** b **12.** Answers will vary.; 1000 times **13.** 8 fl oz **14.** ≈ 74,800 gal **15.** b **16.** 0.137 kg **17.** 3 tablets **18.** kg **19.** mg **20.** 86°F **21.** 80°C **22.** d

CHAPTER 10

Section 10.1

Check Point Exercises

1. 30° **2.** 71° **3.** 46°, 134° **4.** $m\measuredangle 1 = 57°, m\measuredangle 2 = 123°$, and $m\measuredangle 3 = 123°$ **5.** $m\measuredangle 1 = 29°, m\measuredangle 2 = 29°, m\measuredangle 3 = 151°, m\measuredangle 4 = 151°, m\measuredangle 5 = 29°, m\measuredangle 6 = 151°$, and $m\measuredangle 7 = 151°$

Concept and Vocabulary Check

1. line; half-line; ray; line segment **2.** acute; right; obtuse; straight **3.** complementary; supplementary **4.** vertical **5.** parallel; transversal **6.** perpendicular **7.** false; Changes will vary. **8.** false; Changes will vary. **9.** false; Changes will vary. **10.** false; Changes will vary. **11.** true **12.** true

Exercise Set 10.1

1. 150° **3.** 90° **5.** 20°; acute **7.** 160°; obtuse **9.** 180°; straight **11.** 65° **13.** 146° **15.** 42°; 132° **17.** 1°; 91° **19.** 52.6°; 142.6° **21.** $x + x + 12° = 90°$; 39°, 51° **23.** $x + 3x = 180°$; 45°, 135° **25.** $m\measuredangle 1 = 108°; m\measuredangle 2 = 72°; m\measuredangle 3 = 108°$ **27.** $m\measuredangle 1 = 50°; m\measuredangle 2 = 90°; m\measuredangle 3 = 50°$ **29.** $m\measuredangle 1 = 68°; m\measuredangle 2 = 68°; m\measuredangle 3 = 112°; m\measuredangle 4 = 112°; m\measuredangle 5 = 68°; m\measuredangle 6 = 68°; m\measuredangle 7 = 112°$ **31.** $m\measuredangle 1 = 38°; m\measuredangle 2 = 52°; m\measuredangle 3 = 142°$ **33.** $m\measuredangle 1 = 65°; m\measuredangle 2 = 56°; m\measuredangle 3 = 124°$ **35.** false **37.** true **39.** false **41.** false **43.** 60°, 30° **45.** 112°, 112° **47.** $\overline{BC}$ **49.** $\overrightarrow{AD}$ **51.** $\overleftrightarrow{AD}$ **53.** $\overline{AD}$ **55.** 45° **57.** When two parallel lines are intersected by a transversal, corresponding angles have the same measure. **59.** E, F, H, and T **61.** long-distance riding and mountain biking **63.** 27° **73.** does not make sense **75.** does not make sense **77.** d

Section 10.2

Check Point Exercises

1. 49° **2.** $m\measuredangle 1 = 90°; m\measuredangle 2 = 54°; m\measuredangle 3 = 54°; m\measuredangle 4 = 68°; m\measuredangle 5 = 112°$ **3.** Two angles of the small triangle are equal to two angles of the large triangle. One angle pair is given to have the same measure (right angles). Another angle pair consists of vertical angles with the same measure.; 15 cm **4.** 32 yd **5.** 25 ft **6.** 32 in. **7.** 120 yd

Concept and Vocabulary Check

1. 180° **2.** acute **3.** obtuse **4.** isosceles **5.** equilateral **6.** scalene **7.** similar; the same measure; proportional **8.** right; legs; the square of the length of the hypotenuse **9.** true **10.** true **11.** false; Changes will vary. **12.** false; Changes will vary. **13.** false; Changes will vary.

Exercise Set 10.2

1. 67° **3.** 32° **5.** $m\measuredangle 1 = 50°; m\measuredangle 2 = 130°; m\measuredangle 3 = 50°; m\measuredangle 4 = 130°; m\measuredangle 5 = 50°$ **7.** $m\measuredangle 1 = 50°; m\measuredangle 2 = 50°; m\measuredangle 3 = 80°; m\measuredangle 4 = 130°; m\measuredangle 5 = 130°$ **9.** $m\measuredangle 1 = 55°; m\measuredangle 2 = 65°; m\measuredangle 3 = 60°; m\measuredangle 4 = 65°; m\measuredangle 5 = 60°; m\measuredangle 6 = 120°; m\measuredangle 7 = 60°; m\measuredangle 8 = 60°; m\measuredangle 9 = 55°; m\measuredangle 10 = 55°$ **11.** The three angles of the large triangle are given to have the same measures as the three angles of the small triangle.; 5 in. **13.** Two angles of the large triangle are given to have the same measures as two angles of the small triangle.; 6 m **15.** One angle pair is given to have the same measure (right angles). Another angle pair consists of vertical angles with the same measure.; 16 in. **17.** 5 **19.** 9 **21.** 17 m **23.** 39 m **25.** 12 cm **27.** congruent; SAS **29.** congruent; SSS **31.** congruent; SAS **33.** not necessarily congruent **35.** congruent; ASA **37.** 71.7 ft **39.** $90\sqrt{2}$ ft ≈ 127.3 ft **41.** 45 yd **43.** 13 ft **45.** $750,000 **57.** makes sense **59.** does not make sense **61.** 21 ft

Section 10.3

Check Point Exercises

1. $3120 **2. a.** 1800° **b.** 150° **3.** Each angle of a regular octagon measures 135°. 360° is not a multiple of 135°.

Concept and Vocabulary Check

1. perimeter **2.** quadrilateral; pentagon; hexagon; heptagon; octagon **3.** regular **4.** equal in measure; parallel **5.** rhombus **6.** rectangle **7.** square **8.** trapezoid **9.** $P = 2l + 2w$ **10.** $(n - 2)180°$ **11.** tessellation **12.** false; Changes will vary. **13.** true **14.** true **15.** true **16.** false; Changes will vary. **17.** false; Changes will vary. **18.** true

Exercise Set 10.3

1. quadrilateral **3.** pentagon **5.** a: square; b: rhombus; d: rectangle; e: parallelogram **7.** a: square; d: rectangle **9.** c: trapezoid **11.** 30 cm **13.** 28 yd **15.** 1000 in. **17.** 27 ft **19.** 18 yd **21.** 84 yd **23.** 32 ft **25.** 540° **27.** 360° **29.** 108°; 72° **31. a.** 540° **b.** $m\measuredangle A = 140°$; $m\measuredangle B = 40°$ **33. a.** squares, hexagons, dodecagons (12-sided polygons) **b.** 3 angles; 90°, 120°, 150° **c.** The sum of the measures is 360°. **35. a.** triangles, hexagons **b.** 4 angles; 60°, 60°, 120°, 120° **c.** The sum of the measures is 360°. **37.** no; Each angle of the polygon measures 140°. 360° is not a multiple of 140°. **39.** 50 yd by 200 yd **41.** 160 ft by 360 ft **43.** 115°, 115°, 120°, 120° **45.** If the polygons were all regular polygons, the sum would be 363°. The sum is not 360°. **47.** \$5600 **49.** \$99,000 **51.** 48 **61.** makes sense **63.** does not make sense **65.** $6a$

Section 10.4

Check Point Exercises

1. 66 ft^2; yes **2.** \$672 **3.** 60 in.^{2} **4.** 30 ft^2 **5.** 105 ft^2 **6.** 10π in.; 31.4 in. **7.** 49.7 ft **8.** large pizza

Concept and Vocabulary Check

1. $A = lw$ **2.** $A = s^2$ **3.** $A = bh$ **4.** $A = \frac{1}{2}bh$ **5.** $A = \frac{1}{2}h(a + b)$ **6.** $C = \pi d$ **7.** $C = 2\pi r$ **8.** $A = \pi r^2$ **9.** false; Changes will vary. **10.** true **11.** true **12.** false; Changes will vary. **13.** true

Exercise Set 10.4

1. 18 m^2 **3.** 16 in.^{2} **5.** 2100 cm^2 **7.** 56 in.^{2} **9.** 20.58 yd^2 **11.** 30 in.^{2} **13.** 567 m^2 **15.** C: 8π cm, $\approx$ 25.1 cm; A: 16π cm^2, $\approx$ 50.3 cm^2 **17.** C: 12π yd, $\approx$ 37.7 yd; A: 36π yd^2, $\approx$ 113.1 yd^2 **19.** 72 m^2 **21.** 300 m^2 **23.** $100 + 50\pi$; 257.1 cm^2 **25.** $A = ab + \frac{1}{2}(c - a)b$ or $A = \frac{1}{2}(a + c)b$ **27.** $A = 2a^2 + ab$ **29.** 192 cm^2 **31.** 8π cm^2 **33.** $(12.5\pi - 24)$ in.^{2} **35.** perimeter: 54 ft; area: 168 ft^2 **37.** \$556.50 **39.** \$28,200 **41.** 148 ft^2 **43. a.** 23 bags **b.** \$575 **45. a.** 3680 ft^2 **b.** 15 cans of paint **c.** \$404.25 **47.** \$933.40 **49.** 40π m; 125.7 m **51.** 377 plants **53.** large pizza **61.** makes sense **63.** does not make sense **65.** by a factor of $\frac{9}{4}$ **67.** \$13,032.73

Section 10.5

Check Point Exercises

1. 105 ft^3 **2.** 8 yd^3 **3.** 48 ft^3 **4.** 302 in.^{3} **5.** 101 in.^{3} **6.** no **7.** 632 yd^2

Concept and Vocabulary Check

1. $V = lwh$ **2.** $V = s^3$ **3.** polyhedron **4.** $V = \frac{1}{3}Bh$ **5.** $V = \pi r^2 h$ **6.** $V = \frac{1}{3}\pi r^2 h$ **7.** $V = \frac{4}{3}\pi r^3$ **8.** true **9.** true **10.** false; Changes will vary. **11.** true **12.** true **13.** true **14.** false; Changes will vary.

Exercise Set 10.5

1. 36 in.^{3} **3.** 64 cm^3 **5.** 175 yd^3 **7.** 56 in.^{3} **9.** 150π cm^3, 471 cm^3 **11.** 3024π in.^{3}, 9500 in.^{3} **13.** 48π m^3, 151 m^3 **15.** 15π yd^3, 47 yd^3 **17.** 288π m^3, 905 m^3 **19.** 972π cm^3, 3054 cm^3 **21.** 62 m^2 **23.** 96 ft^2 **25.** 324π cm^3, 1018 cm^3 **27.** 432π in.^{3}, 1357 in.^{3} **29.** $\frac{3332}{3}\pi$ m^3, 3489 m^3 **31.** surface area: 186 yd^2; volume: 148 yd^3 **33.** 666 yd^2 **35.** $\frac{1}{8}$ **37.** 9 times **39.** \$340 **41.** no **43. a.** 3,386,880 yd^3 **b.** 2,257,920 blocks **45.** yes **47.** \$27 **51.** does not make sense **53.** does not make sense **55.** The volume is multiplied by 8. **57.** 168 cm^3 **59.** 84 cm^2

Section 10.6

Check Point Exercises

1. $\sin A = \frac{3}{5}$; $\cos A = \frac{4}{5}$; $\tan A = \frac{3}{4}$ **2.** 263 cm **3.** 298 cm **4.** 994 ft **5.** 54°

Concept and Vocabulary Check

1. sine; opposite; hypotenuse; $\frac{a}{c}$ **2.** cosine; adjacent to; hypotenuse; $\frac{b}{c}$ **3.** tangent; opposite; adjacent to; $\frac{a}{b}$ **4.** elevation **5.** depression **6.** false; Changes will vary. **7.** false; Changes will vary. **8.** true **9.** false; Changes will vary.

Exercise Set 10.6

1. $\sin A = \frac{3}{5}$; $\cos A = \frac{4}{5}$; $\tan A = \frac{3}{4}$ **3.** $\sin A = \frac{20}{29}$; $\cos A = \frac{21}{29}$; $\tan A = \frac{20}{21}$ **5.** $\sin A = \frac{5}{13}$; $\cos A = \frac{12}{13}$; $\tan A = \frac{5}{12}$ **7.** $\sin A = \frac{4}{5}$; $\cos A = \frac{3}{5}$; $\tan A = \frac{4}{3}$ **9.** 188 cm **11.** 182 in. **13.** 7 m **15.** 22 yd **17.** 40 m **19.** $m\measuredangle B = 50°$, $a = 18$ yd, $c = 29$ yd **21.** $m\measuredangle B = 38°$, $a = 43$ cm, $b = 33$ cm **23.** 37° **25.** 28° **27.** 653 units **29.** 39 units **31.** 298 units **33.** 257 units **35.** 529 yd **37.** 2879 ft **39.** 2059 ft **41.** 695 ft **43.** 36° **45.** 1376 ft **47.** 15.1° **57.** does not make sense **59.** makes sense **63. a.** 357 ft **b.** 394 ft

Section 10.7

Check Point Exercises

1. Answers will vary.

Concept and Vocabulary Check

1. vertex; edge; graph **2.** traversable **3.** genus **4.** parallel **5.** non-Euclidean; parallel **6.** self-similarity; iteration **7.** true **8.** false; Changes will vary. **9.** true **10.** true

Exercise Set 10.7

1. a. traversable **b.** sample path: *D, A, B, D, C, B* **3. a.** traversable **b.** sample path: *A, D, C, B, D, E, A, B* **5.** not traversable **7.** **9.** no

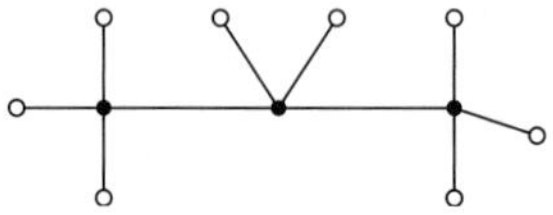

11. 2 **13.** sample path: *B, E, A, B, D, C, A, E, C, E, D* **15.** 2 **17.** 4 **19.** Answers will vary. **21.** The sum of the angles of such a quadrilateral is greater than 360°. **23.** yes **43.** makes sense **45.** does not make sense

Chapter 10 Review Exercises

1. $\measuredangle 3$ **2.** $\measuredangle 5$ **3.** $\measuredangle 4$ and $\measuredangle 6$ **4.** $\measuredangle 1$ and $\measuredangle 6$ **5.** $\measuredangle 4$ and $\measuredangle 1$ **6.** $\measuredangle 2$ **7.** $\measuredangle 5$ **8.** 65° **9.** 49° **10.** 17° **11.** 134° **12.** $m\measuredangle 1 = 110°$; $m\measuredangle 2 = 70°$; $m\measuredangle 3 = 110°$ **13.** $m\measuredangle 1 = 138°$; $m\measuredangle 2 = 42°$; $m\measuredangle 3 = 138°$; $m\measuredangle 4 = 138°$; $m\measuredangle 5 = 42°$; $m\measuredangle 6 = 42°$; $m\measuredangle 7 = 138°$ **14.** 72° **15.** 51° **16.** $m\measuredangle 1 = 90°$; $m\measuredangle 2 = 90°$; $m\measuredangle 3 = 140°$; $m\measuredangle 4 = 40°$; $m\measuredangle 5 = 140°$ **17.** $m\measuredangle 1 = 80°$; $m\measuredangle 2 = 65°$; $m\measuredangle 3 = 115°$; $m\measuredangle 4 = 80°$; $m\measuredangle 5 = 100°$; $m\measuredangle 6 = 80°$ **18.** 5 ft **19.** 3.75 ft **20.** 10 ft **21.** 7.2 in. **22.** 10.2 cm **23.** 12.5 ft **24.** 15 ft **25.** 12 yd **26.** rectangle, square **27.** rhombus, square **28.** parallelogram, rhombus, trapezoid **29.** 30 cm **30.** 4480 yd **31.** 44 m **32.** 1800° **33.** 1080° **34.** $m\measuredangle 1 = 135°$; $m\measuredangle 2 = 45°$ **35.** $132 **36. a.** triangles, hexagons **b.** 5 angles; 60°, 60°, 60°, 60°, 120° **c.** The sum of the measures is 360°. **37.** yes; Each angle of a regular hexagon measures 120°, and 360° is a multiple of 120°. **38.** 32.5 ft^2 **39.** 20 m^2 **40.** 50 cm^2 **41.** 135 yd^2 **42.** $C = 20\pi$ m ≈ 62.8 m; $A = 100\pi$ m^2 ≈ 314.2 m^2 **43.** 192 in.^{2} **44.** 28 m^2 **45.** 279.5 ft^3 **46.** $(128 - 32\pi)$ in.^{2}; 27.5 in.^{2} **47.** $787.50 **48.** $650 **49.** 31 yd **50.** 60 cm^3 **51.** 960 m^3 **52.** 128π yd^3, 402 yd^3 **53.** $\frac{44{,}800\pi}{3}$ in.^{3}, 46,914 in.^{3} **54.** 288π m^3, 905 m^3 **55.** 126 m^2 **56.** 4800 m^3 **57.** 651,775 m^3 **58.** $80 **59.** $\sin A = \frac{3}{5}$; $\cos A = \frac{4}{5}$; $\tan A = \frac{3}{4}$ **60.** 42 mm **61.** 23 cm **62.** 37 in. **63.** 58° **64.** 772 ft **65.** 31 m **66.** 56° **67.** not traversable **68.** traversable; sample path: *A, B, C, D, A, B, C, D, A* **69.** 0 **70.** 2 **71.** 1 **72.** 2 **73.** Answers will vary. **74.** Answers will vary.

Chapter 10 Test

1. 36°;126° **2.** 133° **3.** 70° **4.** 35° **5.** 3.2 in. **6.** 10 ft **7.** 1440° **8.** 40 cm **9.** d **10. a.** triangles, squares **b.** 5 angles; 60°, 60°, 60°, 90°, 90° **c.** The sum of the measures is 360°. **11.** 517 m^2 **12.** 525 in.^{2} **13. a.** 12 cm **b.** 30 cm **c.** 30 cm^2 **14.** C: 40π m, 125.7 m; A: 400π m^2, 1256.6 m^2 **15.** 108 tiles **16.** 18 ft^3 **17.** 16 m^3 **18.** 175π cm^3, 550 cm^3 **19.** 85 cm **20.** 70 ft **21.** traversable; sample path: *B, C, A, E, C, D, E* **22.** Answers will vary.

CHAPTER 11

Section 11.1

Check Point Exercises

1. 150 **2.** 40 **3.** 30 **4.** 160 **5.** 729 **6.** 90,000

Concept and Vocabulary Check

1. $M \cdot N$ **2.** multiplying; Fundamental Counting **3.** false; Changes will vary. **4.** true

Exercise Set 11.1

1. 80 **3.** 12 **5.** 6 **7.** 40 **9.** 144; Answers will vary. **11.** 8 **13.** 96 **15.** 243 **17.** 144 **19.** 676,000 **21.** 2187 **27.** makes sense **29.** makes sense **31.** 720 hr

Section 11.2

Check Point Exercises

1. 600 **2.** 120 **3. a.** 504 **b.** 524,160 **c.** 100 **4.** 840 **5.** 15,120 **6.** 420

Concept and Vocabulary Check

1. factorial; 5; 1; 1 **2.** $\frac{n!}{(n-r)!}$ **3.** $\frac{n!}{p!q!}$ **4.** false; Changes will vary. **5.** false; Changes will vary. **6.** true **7.** false; Changes will vary.

Exercise Set 11.2

1. 720 **3.** 120 **5.** 120 **7.** 362,880 **9.** 6 **11.** 4 **13.** 504 **15.** 570,024 **17.** 3,047,466,240 **19.** 600 **21.** 10,712 **23.** 5034 **25.** 24 **27.** 6 **29.** 42 **31.** 1716 **33.** 3024 **35.** 6720 **37.** 720 **39.** 1 **41.** 720 **43.** 8,648,640 **45.** 120 **47.** 15,120 **49.** 180 **51.** 831,600 **53.** 105 **55.** 280 **65.** makes sense **67.** does not make sense **69.** 360 **71.** 14,400 **73.** $\frac{n(n-1)\cdots 3\cdot 2\cdot 1}{2} = n(n-1)\cdots 3$

Section 11.3

Check Point Exercises

1. a. combinations **b.** permutations **2.** 35 **3.** 1820 **4.** 525

Concept and Vocabulary Check

1. $\frac{n!}{(n-r)!r!}$ **2.** $r!$ **3.** false; Changes will vary. **4.** false; Changes will vary.

Exercise Set 11.3

1. combinations **3.** permutations **5.** 6 **7.** 126 **9.** 330 **11.** 8 **13.** 1 **15.** 4060 **17.** 1 **19.** 7 **21.** 0 **23.** $\frac{3}{4}$ **25.** −9499 **27.** $\frac{3}{68}$ **29.** 20 **31.** 495 **33.** 24,310 **35.** 22,957,480 **37.** 735 **39.** 4,516,932,420 **41.** 360 **43.** 1716 **45.** 1140 **47.** 840 **49.** 2730 **51.** 10 **53.** 12 **55.** 10 **57.** 60 **59.** 792 **61.** 720 **63.** 20 **65.** 24 **67.** 12 **73.** does not make sense **75.** makes sense **77.** The 5/36 lottery is easier to win.; Answers will vary. **79.** 570 sec or 9.5 min; 2340 sec or 39 min

Section 11.4

Check Point Exercises

1. a. $\frac{1}{6}$ **b.** $\frac{1}{2}$ **c.** 0 **d.** 1 **2. a.** $\frac{1}{13}$ **b.** $\frac{1}{2}$ **c.** $\frac{1}{26}$ **3.** $\frac{1}{2}$ **4. a.** 0.32 **b.** 0.48

Concept and Vocabulary Check

1. sample space **2.** $P(E)$; number of outcomes in E; total number of possible outcomes **3.** 52; hearts; diamonds; clubs; spades **4.** empirical **5.** true **6.** false; Changes will vary. **7.** true **8.** false; Changes will vary.

Exercise Set 11.4

1. $\frac{1}{6}$ **3.** $\frac{1}{2}$ **5.** $\frac{1}{3}$ **7.** 1 **9.** 0 **11.** $\frac{1}{13}$ **13.** $\frac{1}{4}$ **15.** $\frac{3}{13}$ **17.** $\frac{1}{52}$ **19.** 0 **21.** $\frac{1}{4}$ **23.** $\frac{1}{2}$ **25.** $\frac{1}{2}$ **27.** $\frac{3}{8}$ **29.** $\frac{3}{8}$ **31.** $\frac{7}{8}$ **33.** 0 **35.** $\frac{1}{4}$ **37.** $\frac{1}{9}$ **39.** 0 **41.** $\frac{3}{10}$ **43.** $\frac{1}{5}$ **45.** $\frac{1}{2}$ **47.** 0 **49.** $\frac{1}{4}$ **51.** $\frac{1}{4}$ **53.** $\frac{1}{2}$ **55.** 0.43 **57.** 0.13 **59.** 0.03 **61.** 0.38 **63.** 0.78 **65.** 0.12 **67.** $\frac{1}{6}$ **69.** $\frac{11}{87}$ **79.** does not make sense **81.** makes sense **83.** $\frac{3}{8} = 0.375$

Section 11.5

Check Point Exercises

1. $\frac{2}{15}$ **2.** $\frac{320}{229{,}201{,}338} = \frac{160}{114{,}600{,}669}$ **3. a.** $\frac{1}{6}$ **b.** $\frac{1}{2}$

Concept and Vocabulary Check

1. permutations; the total number of possible permutations **2.** 1; combinations **3.** true **4.** false; Changes will vary.

Exercise Set 11.5

1. a. 120 **b.** 6 **c.** $\frac{1}{20}$ **3. a.** $\frac{1}{6}$ **b.** $\frac{1}{30}$ **c.** $\frac{1}{720}$ **d.** $\frac{1}{3}$ **5. a.** 84 **b.** 10 **c.** $\frac{5}{42}$ **7.** $\frac{1}{175{,}711{,}536}$ **9.** $\frac{2125}{29{,}285{,}256}$ **11. a.** $\frac{1}{177{,}100} \approx 0.00000565$ **b.** $\frac{27{,}132}{177{,}100} \approx 0.153$ **13.** $\frac{3}{10} = 0.3$ **15. a.** 2,598,960 **b.** 1287 **c.** $\frac{1287}{2{,}598{,}960} \approx 0.000495$ **17.** $\frac{11}{1105} \approx 0.00995$ **19.** $\frac{36}{270{,}725} \approx 0.000133$ **25.** does not make sense **27.** makes sense **29.** The prize is shared among all winners. You are guaranteed to win but not to win $700 million. **31.** $\frac{235{,}620}{2{,}598{,}960} \approx 0.0907$

Section 11.6

Check Point Exercises

1. $\frac{3}{4}$ **2.** $\frac{127}{190}$ **3.** $\frac{205}{214}$ **4.** $\frac{1}{3}$ **5.** $\frac{27}{50}$ **6.** $\frac{3}{4}$ **7. a.** $\frac{197}{254} \approx 0.78$ **b.** $\frac{20}{127} \approx 0.16$ **8. a.** 2:50 or 1:25 **b.** 50:2 or 25:1 **9.** 199:1 **10.** 1:15; $\frac{1}{16}$

Concept and Vocabulary Check

1. $1 - P(E)$; $1 - P(\text{not } E)$ **2.** mutually exclusive; $P(A) + P(B)$ **3.** $P(A) + P(B) - P(A \text{ and } B)$ **4.** E will occur; E will not occur **5.** $P(E) = \frac{a}{a+b}$ **6.** false; Changes will vary. **7.** false; Changes will vary. **8.** true **9.** false; Changes will vary.

Exercise Set 11.6

1. $\frac{12}{13}$ **3.** $\frac{3}{4}$ **5.** $\frac{10}{13}$ **7.** $\frac{2{,}598{,}924}{2{,}598{,}960} \approx 0.999986$ **9.** $\frac{2{,}595{,}216}{2{,}598{,}960} \approx 0.998559$ **11. a.** 0.10 **b.** 0.90 **13.** $\frac{16}{25}$ **15.** $\frac{47}{50}$ **17.** $\frac{2}{13}$ **19.** $\frac{1}{13}$ **21.** $\frac{1}{26}$ **23.** $\frac{9}{22}$ **25.** $\frac{5}{6}$ **27.** $\frac{7}{13}$ **29.** $\frac{11}{26}$ **31.** $\frac{3}{4}$ **33.** $\frac{5}{8}$ **35.** $\frac{33}{40}$ **37.** $\frac{4}{5}$ **39.** $\frac{7}{10}$ **41.** $\frac{43}{58}$ **43.** $\frac{50}{87}$ **45.** $\frac{113}{174}$ **47.** 15:43; 43:15 **49. a.** $\frac{9}{100}$ **b.** $\frac{91}{100}$ **51.** $\frac{1}{10}$ **53.** $\frac{1}{10}$ **55.** $\frac{21}{100}$ **57.** $\frac{19}{100}$ **59.** 1:99; 99:1 **61.** 17:8; 8:17 **63.** 2:1 **65.** 1:2 **67. a.** 63:37 **b.** 37:63 **69.** 1:3 **71.** 1:1 **73.** 12:1 **75.** 25:1 **77.** 47:5 **79.** 49:1 **81.** 9:10 **83.** 14:5 **85.** 14:5 **87.** 9:10 **89.** $\frac{3}{7}$ **91.** $\frac{4}{25}$; 84 **93.** $\frac{1}{1000}$ **103.** does not make sense **105.** does not make sense **107.** 0.06

Section 11.7

Check Point Exercises

1. $\frac{1}{361} \approx 0.00277$ **2.** $\frac{1}{16}$ **3. a.** $\frac{625}{130{,}321} \approx 0.005$ **b.** $\frac{38{,}416}{130{,}321} \approx 0.295$ **c.** $\frac{91{,}905}{130{,}321} \approx 0.705$ **4.** $\frac{1}{221} \approx 0.00452$ **5.** $\frac{11}{850} \approx 0.0129$ **6.** $\frac{2}{5}$ **7. a.** 1 **b.** $\frac{1}{2}$ **8. a.** $\frac{9}{10} = 0.9$ **b.** $\frac{45}{479} \approx 0.094$

Concept and Vocabulary Check

1. independent; $P(A) \cdot P(B)$ **2.** the event does not occur **3.** dependent; $P(A) \cdot P(B \text{ given that } A \text{ occurred})$ **4.** conditional; $P(B|A)$ **5.** false; Changes will vary. **6.** false; Changes will vary. **7.** true **8.** true

Exercise Set 11.7

1. $\frac{1}{6}$ **3.** $\frac{1}{36}$ **5.** $\frac{1}{4}$ **7.** $\frac{1}{36}$ **9.** $\frac{1}{8}$ **11.** $\frac{1}{36}$ **13.** $\frac{1}{3}$ **15.** $\frac{3}{52}$ **17.** $\frac{1}{169}$ **19.** $\frac{1}{4}$ **21.** $\frac{1}{64}$ **23.** $\frac{1}{6}$ **25. a.** $\frac{1}{256} \approx 0.00391$ **b.** $\frac{1}{4096} \approx 0.000244$ **c.** ≈ 0.524 **d.** ≈ 0.476 **27.** $\frac{7}{29}$ **29.** $\frac{5}{87}$ **31.** $\frac{2}{21}$ **33.** $\frac{4}{35}$ **35.** $\frac{11}{21}$ **37.** $\frac{1}{57}$ **39.** $\frac{8}{855}$ **41.** $\frac{11}{57}$ **43.** $\frac{1}{550}$ **45.** $\frac{4}{495}$ **47.** $\frac{2}{33}$ **49.** $\frac{1}{40{,}425}$ **51.** $\frac{17}{121{,}275}$ **53.** $\frac{1}{5}$ **55.** $\frac{2}{3}$ **57.** $\frac{3}{4}$ **59.** $\frac{3}{4}$ **61.** $\frac{412{,}368}{412{,}878} \approx 0.999$ **63.** $\frac{412{,}368}{574{,}895} \approx 0.717$ **65.** $\frac{114}{127} \approx 0.90$ **67.** $\frac{20}{127} \approx 0.16$ **69.** $\frac{69}{127} \approx 0.54$ **71.** $\frac{11}{26} \approx 0.42$ **73.** $\frac{11}{131} \approx 0.08$ **75.** $\frac{109}{123} \approx 0.89$ **77. a.** Answers will vary. **b.** $\frac{365}{365} \cdot \frac{364}{365} \cdot \frac{363}{365} \approx 0.992$ **c.** ≈ 0.008 **d.** 0.411 **e.** 23 people, $1 - \frac{365}{365} \cdot \frac{364}{365} \cdot \frac{363}{365} \cdot \cdots \cdot \frac{343}{365} \approx 0.507$ **87.** does not make sense **89.** does not make sense **91.** $\frac{25}{7776} \approx 0.00322$ **93.** $\frac{11}{36}$

Section 11.8

Check Point Exercises

1. 2.5 **2.** 2 **3. a.** \$8000; In the long run, the average cost of a claim is \$8000. **b.** \$8000 **4.** 0; no; Answers will vary. **5.** table entries: \$998, \$48, and −\$2; expected value: −\$0.90; In the long run, a person can expect to lose an average of \$0.90 for each ticket purchased.; Answers will vary. **6.** −\$0.20; In the long run, a person can expect to lose an average of \$0.20 for each card purchased.

Concept and Vocabulary Check

1. expected; probability; add **2.** loss; probability; add **3.** false; Changes will vary. **4.** true

Exercise Set 11.8

1. 1.75 **3. a.** \$29,000; In the long run, the average cost of a claim is \$29,000. **b.** \$29,000 **c.** \$29,050 **5.** \$0; Answers will vary. **7.** \$0.73 **9.** $\frac{1}{16} = 0.0625$; yes **11.** the second mall **13. a.** \$140,000 **b.** no **15.** ≈ −\$0.17; In the long run, a person can expect to lose an average of about \$0.17 for each game played. **17.** ≈ −\$0.05; In the long run, a person can expect to lose an average of about \$0.05 for each game played. **19.** −\$0.50; In the long run, a person can expect to lose an average of \$0.50 for each game played. **27.** makes sense **29.** does not make sense **31.** \$160

Chapter 11 Review Exercises

1. 800 **2.** 20 **3.** 9900 **4.** 125 **5.** 243 **6.** 60 **7.** 240 **8.** 800 **9.** 114 **10.** 990 **11.** 151,200 **12.** 9900 **13.** 330 **14.** 2002 **15.** combinations **16.** permutations **17.** combinations **18.** 720 **19.** 32,760 **20.** 210 **21.** 420 **22.** 1140 **23.** 120 **24.** 1,860,480 **25.** 120 **26.** 1287 **27.** 55,440 **28.** 60 **29.** $\frac{1}{6}$ **30.** $\frac{2}{3}$ **31.** 1 **32.** 0 **33.** $\frac{1}{13}$ **34.** $\frac{3}{13}$ **35.** $\frac{3}{13}$ **36.** $\frac{1}{52}$ **37.** $\frac{1}{26}$ **38.** $\frac{1}{2}$ **39.** $\frac{1}{3}$ **40.** $\frac{1}{6}$ **41. a.** $\frac{1}{2}$ **b.** 0 **42.** $\frac{7}{12}$ **43.** $\frac{31}{60}$ **44.** $\frac{1}{30}$ **45.** $\frac{1}{24}$ **46.** $\frac{1}{6}$ **47.** $\frac{1}{30}$ **48.** $\frac{1}{720}$ **49.** $\frac{1}{3}$ **50. a.** $\frac{1}{15{,}504} \approx 0.0000645$ **b.** $\frac{100}{15{,}504} \approx 0.00645$ **51. a.** $\frac{1}{14}$ **b.** $\frac{3}{7}$ **52.** $\frac{3}{26}$ **53.** $\frac{5}{6}$ **54.** $\frac{1}{2}$ **55.** $\frac{1}{3}$ **56.** $\frac{2}{3}$ **57.** 1 **58.** $\frac{10}{13}$ **59.** $\frac{3}{4}$ **60.** $\frac{2}{13}$ **61.** $\frac{1}{13}$ **62.** $\frac{7}{13}$ **63.** $\frac{11}{26}$ **64.** $\frac{5}{6}$ **65.** $\frac{5}{6}$ **66.** $\frac{1}{2}$ **67.** $\frac{2}{3}$ **68.** 1 **69.** $\frac{5}{6}$ **70.** $\frac{4}{5}$ **71.** $\frac{3}{4}$ **72.** $\frac{18}{25}$ **73.** $\frac{6}{7}$ **74.** $\frac{3}{5}$ **75.** $\frac{12}{35}$ **76.** in favor: 1:12; against: 12:1 **77.** 99:1 **78.** $\frac{3}{4}$ **79.** $\frac{2}{9}$ **80.** $\frac{1}{36}$ **81.** $\frac{1}{9}$ **82.** $\frac{1}{36}$ **83.** $\frac{8}{27}$ **84.** $\frac{1}{32}$ **85. a.** 0.04 **b.** 0.008 **c.** 0.4096 **d.** 0.5904 **86.** $\frac{1}{9}$ **87.** $\frac{1}{12}$ **88.** $\frac{1}{196}$ **89.** $\frac{1}{3}$ **90.** $\frac{3}{10}$ **91. a.** $\frac{1}{2}$ **b.** $\frac{2}{7}$ **92.** $\frac{27}{29}$ **93.** $\frac{4}{29}$ **94.** $\frac{144}{145}$ **95.** $\frac{11}{20}$ **96.** $\frac{11}{135}$ **97.** $\frac{1}{125}$ **98.** $\frac{1}{232}$ **99.** $\frac{19}{1044}$ **100.** $\frac{28{,}786}{33{,}632} \approx 0.856$ **101.** $\frac{11{,}518}{33{,}632} \approx 0.342$ **102.** $\frac{30{,}881}{33{,}632} \approx 0.918$ **103.** $\frac{15{,}648}{33{,}632} \approx 0.465$ **104.** $\frac{4903}{33{,}632} \approx 0.146$ **105.** $\frac{3651}{28{,}786} \approx 0.127$ **106.** $\frac{2514}{2751} \approx 0.914$ **107.** 3.125 **108. a.** \$0.50; In the long run, the average cost of a claim is \$0.50. **b.** \$10.00 **109.** \$4500; Answers will vary. **110.** −\$0.25; In the long run, a person can expect to lose an average of \$0.25 for each game played.

Chapter 11 Test

1. 240 **2.** 24 **3.** 720 **4.** 990 **5.** 210 **6.** 420 **7.** $\frac{6}{25}$ **8.** $\frac{17}{25}$ **9.** $\frac{11}{25}$ **10.** $\frac{5}{13}$ **11.** $\frac{1}{210}$ **12.** $\frac{10}{1001} \approx 0.00999$ **13.** $\frac{1}{2}$ **14.** $\frac{1}{16}$ **15.** $\frac{1}{8000} = 0.000125$ **16.** $\frac{8}{13}$ **17.** $\frac{3}{5}$ **18.** $\frac{1}{19}$ **19.** $\frac{1}{256}$ **20.** 3:4 **21. a.** 4:1 **b.** $\frac{4}{5}$ **22.** $\frac{3}{5}$ **23.** $\frac{39}{50}$ **24.** $\frac{3}{5}$ **25.** $\frac{9}{19}$ **26.** $\frac{7}{150}$ **27.** \$1000; Answers will vary. **28.** −\$12.75; In the long run, a person can expect to lose an average of \$12.75 for each game played.

CHAPTER 12

Section 12.1

Check Point Exercises

1. a. the set containing all the city's homeless **b.** no: People already in the shelters are probably less likely to be against mandatory residence in the shelters. **2.** By selecting people from a shelter, homeless people who do not go to the shelters have no chance of being selected. An appropriate method would be to randomly select neighborhoods of the city and then randomly survey homeless people within the selected neighborhood.

3.

Grade	Frequency
A	3
B	5
C	9
D	2
F	1
	20

4.

Class	Frequency
40–49	1
50–59	5
60–69	4
70–79	15
80–89	5
90–99	7
	37

5.

Stem	Leaves
4	1
5	8 2 8 0 7
6	8 2 9 9
7	3 5 9 9 7 5 5 3 3 6 7 1 7 1 5
8	7 3 9 9 1
9	4 6 9 7 5 8 0

Concept and Vocabulary Check

1. random **2.** frequency distribution **3.** grouped frequency distribution; 80; 89 **4.** histogram; frequencies **5.** frequency polygon; horizontal axis **6.** stem-and-leaf plot **7.** false; Changes will vary. **8.** true **9.** false; Changes will vary. **10.** true

Exercise Set 12.1

1. c **3.** 7; 31 **5.** 151

7.

Time Spent on Homework (in hours)	Number of Students
15	4
16	5
17	6
18	5
19	4
20	2
21	2
22	0
23	0
24	2
	30

9. $0, 5, 10, \ldots, 40, 45$ **11.** 5 **13.** 13 **15.** the 5–9 class

17.

Age at Inauguration	Number of Presidents
41–45	2
46–50	9
51–55	15
56–60	9
61–65	7
66–70	3
	45

19. a.

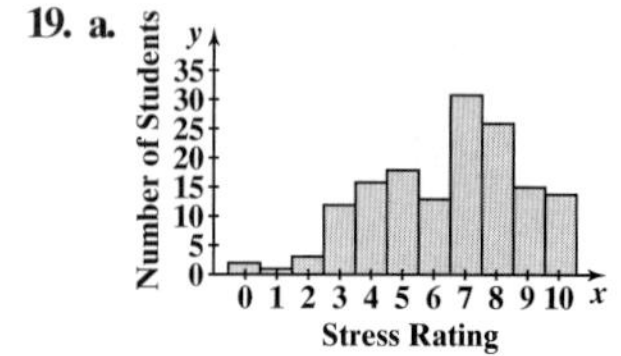

b.

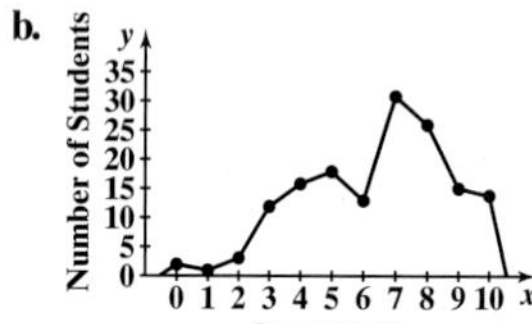

21.

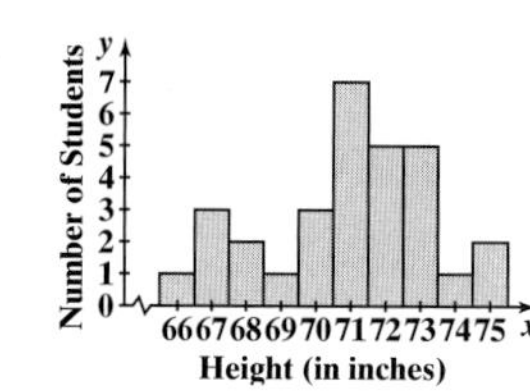

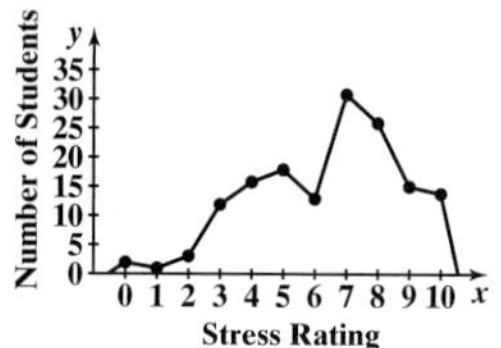

23. false **25.** false **27.** true **29.** false

31.

Stem	Leaves
2	8 8 9 5
3	8 7 0 1 2 7 6 4 0 5
4	8 2 2 1 4 5 4 6 2 0 8 2 7 9
5	9 4 1 9 1 0
6	3 2 3 6 6 3

The greatest number of college professors are in their 40s.

33. Time intervals on the horizontal axis do not represent equal amounts of time. **35.** Percentages do not add up to 100%. **37.** It is not clear whether the bars or the actors represent the box-office receipts. **49.** does not make sense **51.** makes sense

Section 12.2

Check Point Exercises

1. 20.1% **2.** 36 **3. a.** 35 **b.** 82 **4.** 5 **5.** 1:06, 1:09, 1:14, 1:21, 1:22, 1:25, 1:29, 1:29, 1:34, 1:34, 1:36, 1:45, 1:46, 1:49, 1:54, 1:57, 2:10, 2:15; median: 1 hour, 34 minutes **6.** 54.5 **7. a.** \$372.6 million **b.** \$17.5 million **c.** Trump's net worth was much greater than the other presidents'. **8. a.** 8 **b.** 3 and 8 **c.** no mode **9.** 14.5 **10.** mean: 158.6 cal; median: 153 cal; mode: 138 cal; midrange: 151 cal

Concept and Vocabulary Check

1. mean **2.** median **3.** $\frac{n+1}{2}$ **4.** mode **5.** midrange **6.** true **7.** false; Changes will vary. **8.** false; Changes will vary. **9.** false; Changes will vary.

Exercise Set 12.2

1. 4.125 **3.** 95 **5.** 62 **7.** ≈ 3.45 **9.** ≈ 4.71 **11.** ≈ 6.26 **13.** 3.5 **15.** 95 **17.** 60 **19.** 3.6 **21.** 5 **23.** 6 **25.** 3 **27.** 95 **29.** 40 **31.** 2.5, 4.2 (bimodal) **33.** 5 **35.** 6 **37.** 4.5 **39.** 95 **41.** 70 **43.** 3.3 **45.** 4.5 **47.** 5.5 **49.** mean: 30; median: 30; mode: 30; midrange: 30 **51.** mean: approximately 12.4; median: 12.5; mode: 13; midrange: 12.5 **53.** mean: approximately 31.3; median: 31; mode: 31; midrange: 33 **55. a.** $\approx$ \$6.5 billion **b.** \$1.6 billion **c.** \$1.0, \$1.1, \$1.3, \$1.6, and \$4.0 billion all occur twice. **d.** \$28.5 billion **57. a.** ≈ 17.27 **b.** 17 **c.** 7, 12, 17 **d.** 24.5 **59.** 175 lb **61.** 177.5 lb **63.** ≈ 2.76 **73.** makes sense **75.** makes sense **77.** Sample answers: **a.** 75, 80, 80, 90, 91, 94 **b.** 50, 80, 80, 85, 90, 95 **c.** 70, 75, 80, 85, 90, 100 **d.** 75, 80, 85, 90, 95, 95 **e.** 75, 80, 85, 85, 90, 95 **f.** 68, 70, 72, 72, 74, 76

Section 12.3

Check Point Exercises

1. 9 **2.** mean: 6;

Data item	Deviation
2	−4
4	−2
7	1
11	5

3. ≈3.92 **4.** sample A: 3.74; sample B: 28.06 **5. a.** small-company stocks **b.** small-company stocks; Answers will vary.

Concept and Vocabulary Check

1. range **2.** standard deviation **3.** true **4.** true **5.** false; Changes will vary.

Exercise Set 12.3

1. 4 **3.** 8 **5.** 2

7. a.

Data item	Deviation
3	−9
5	−7
7	−5
12	0
18	6
27	15

b. 0

9. a.

Data item	Deviation
29	−20
38	−11
48	−1
49	0
53	4
77	28

b. 0

11. a. 91

b.

Data item	Deviation
85	−6
95	4
90	−1
85	−6
100	9

c. 0

13. a. 155

b.

Data item	
146	−9
153	−2
155	0
160	5
161	6

c. 0

15. a. 2.70

b.

Data item	Deviation
2.25	−0.45
3.50	0.80
2.75	0.05
3.10	0.40
1.90	−0.80

c. 0

17. ≈1.58 **19.** ≈3.46 **21.** ≈0.89 **23.** 3 **25.** ≈2.14
27. *Sample A*: mean: 12; range: 12; standard deviation: ≈4.32
Sample B: mean: 12; range: 12; standard deviation: ≈5.07
Sample C: mean: 12; range: 12; standard deviation: 6
The samples have the same mean and range, but different standard deviations.
29. 0 **31.** 1 **33.** 7.91 **35.** 1.55 **37. a.** male artists; All of the data items for the men are greater than the greatest data item for the women. **b.** male artists: 93; female artists: 58 **c.** male artists; There is greater spread in the data for the men. **d.** male artists: 32.64; female artists: 6.82 **47.** makes sense **49.** makes sense **53.** a

Section 12.4

Check Point Exercises

1. a. 75.5 in. **b.** 58 in. **2. a.** 95% **b.** 47.5% **c.** 16% **3. a.** 2 **b.** 0 **c.** −1 **4.** ACT **5. a.** 64 **b.** 128 **6.** 75% of the scores on the SAT are less than this student's score. **7. a.** ± 2.0% **b.** We can be 95% confident that between 34% and 38% of Americans read more than ten books per year. **c.** Sample answer: Some people may be embarrassed to admit that they read few or no books per year.

Concept and Vocabulary Check

1. 68; 95; 99.7 **2.** mean **3.** percentile **4.** margin of error **5.** true **6.** true **7.** false; Changes will vary. **8.** true

Exercise Set 12.4

1. 120 **3.** 160 **5.** 150 **7.** 60 **9.** 90 **11.** 68% **13.** 34% **15.** 47.5% **17.** 49.85% **19.** 16% **21.** 2.5% **23.** 95% **25.** 47.5% **27.** 16% **29.** 2.5% **31.** 0.15% **33.** 1 **35.** 3 **37.** 0.5 **39.** 1.75 **41.** 0 **43.** −1 **45.** −1.5 **47.** −3.25 **49.** 1.5 **51.** 2.25 **53.** −1.25 **55.** −1.5 **57.** The person who scores 127 on the Wechsler has the higher IQ. **59.** 500 **61.** 475 **63.** 250 **65.** 275 **67. a.** ±3.5% **b.** We can be 95% confident that between 65.5% and 72.5% of the population favor required gun registration as a means to reduce gun violence. **69. a.** ±1.6% **b.** We can be 95% confident that between 58.6% and 61.8% of all TV households watched the final episode of *M*A*S*H*. **71.** 0.2% **73. a.** skewed to the right **b.** 5.3 murders per 100,000 residents **c.** 5 murders per 100,000 residents **d.** yes; The mean is greater than the median, which is consistent with a distribution skewed to the right. **e.** 5.6; yes; For a normal distribution, almost 100% of the z-scores are between −3 and 3. **87.** does not make sense **89.** does not make sense

Section 12.5

Check Point Exercises

1. 88.49% **2.** 8.08% **3.** 83.01%

Concept and Vocabulary Check

1. 98.93% **2.** 1.07% **3.** 88.49% **4.** 87.42% **5.** true

Exercise Set 12.5

1. a. 72.57% **b.** 27.43% **3. a.** 88.49% **b.** 11.51% **5. a.** 24.2% **b.** 75.8% **7. a.** 11.51% **b.** 88.49% **9.** 33.99% **11.** 15.74% **13.** 86.64% **15.** 28.57% **17.** 91.92% **19.** 27.43% **21.** 88.49% **23.** 6.69% **25.** 45.14% **27.** 6.68% **29.** 24.17% **31.** 77% **33.** 86% **35.** 10% **39.** does not make sense **41.** makes sense **45.** 630

Section 12.6

Check Point Exercises

1. This indicates a moderate relationship. **2.** 0.89; There is a moderately strong positive relationship between the two quantities. **3.** $y = 0.1x + 0.8$; 8.8 deaths per 100,000 people **4.** yes

Concept and Vocabulary Check

1. scatter plot **2.** regression line **3.** correlation coefficient; -1; 1 **4.** true **5.** false; Changes will vary. **6.** false; Changes will vary. **7.** true

Exercise Set 12.6

1.

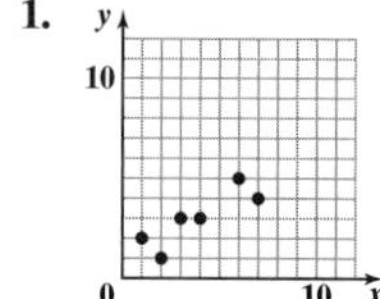

There appears to be a positive correlation.

3.

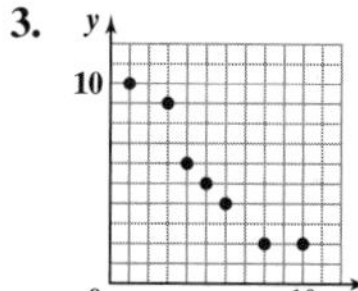

There appears to be a negative correlation.

5.

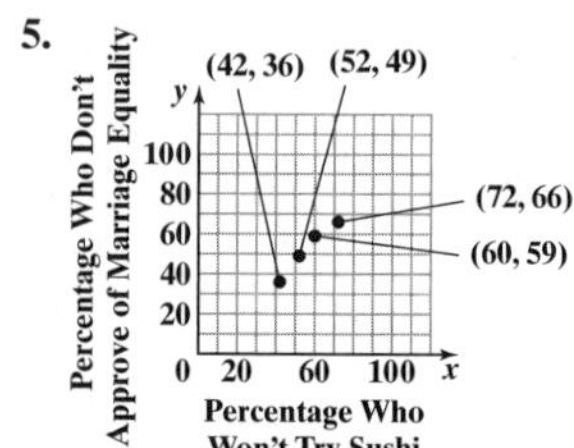

There appears to be a positive correlation.

7.

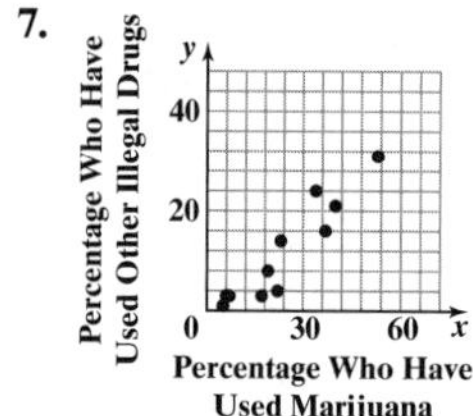

There appears to be a positive correlation.

9. false **11.** true **13.** true **15.** false **17.** true **19.** false **21.** true **23.** false **25.** false **27.** a **29.** d **31.** 0.85 **33.** -0.95 **35. a.** 0.98 **b.** Answers will vary. **c.** $y = 1.01x - 4.47$ **d.** 26% **37. a.** 0.93 **b.** $y = 0.62x - 3.07$ **c.** 3% **39.** A correlation does exist. **41.** A correlation does not exist. **43.** A correlation does exist. **45.** A correlation does not exist. **59.** does not make sense **61.** does not make sense

Chapter 12 Review Exercises

1. a

2.

Time Spent on Homework (in hours)	Number of Students
6	1
7	3
8	3
9	2
10	1
	10

3.

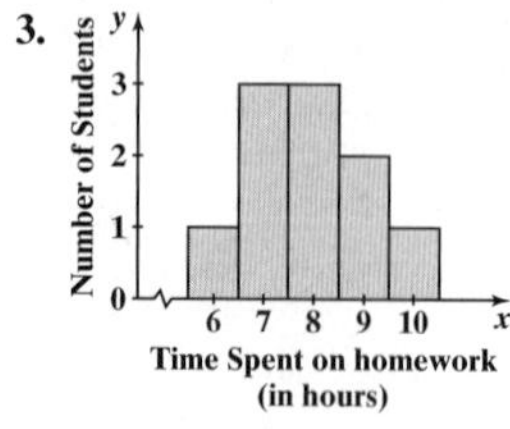

4.

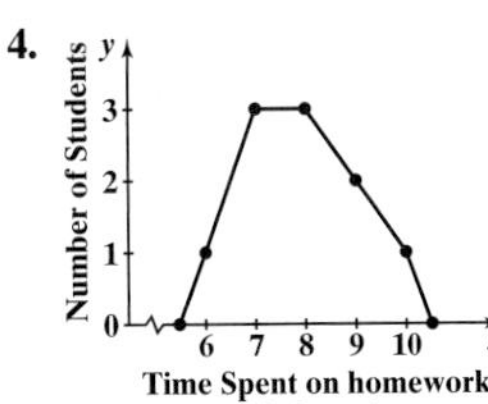

5.

Grades	Number of Students
0–39	19
40–49	8
50–59	6
60–69	6
70–79	5
80–89	3
90–99	3
	50

6.

Stem	Leaves
1	8 4 3 3 7 1
2	4 6 9 9 2 7
3	4 9 6 1 5 1 1
4	4 7 9 1 2 2 0 5
5	4 7 9 0 6 1
6	3 7 0 8 3 9
7	2 5 4 0 3
8	1 7 6
9	1 0 5

7. Sizes of barrels are not scaled proportionally in terms of the data they represent. **8.** 91.2 **9.** 17 **10.** 2.3 **11.** 11 **12.** 28 **13.** 2 **14.** 27 **15.** 585, 587 (bimodal) **16.** 2 **17.** 91 **18.** 19.5 **19.** 2.5

21. a.

Age at First Inauguration	Number of Presidents
42	1
43	1
44	0
45	0
46	2
47	2
48	1
49	2
50	1
51	5
52	2
53	0
54	5
55	4
56	3
57	4
58	1
59	0
60	1
61	3
62	1
63	0
64	2
65	1
66	0
67	0
68	1
69	1
70	1
	45

b. mean: 55 yr; median: 55 yr; mode: 51 yr, 54 yr (bimodal); midrange: 56 yr **22.** 18 **23.** 564

24. a.

Data item	Deviation
29	−6
9	−26
8	−27
22	−13
46	11
51	16
48	13
42	7
53	18
42	7

b. 0

25. a. 50

b.

Data item	Deviation
36	−14
26	−24
24	−26
90	40
74	24

c. 0

26. ≈ 4.05 **27.** ≈ 5.13 **28.** mean: 49; range: 76; standard deviation: ≈ 24.32 **29.** Set A: mean: 80; standard deviation: 0; Set B: mean: 80; standard deviation: ≈ 11.55; Answers will vary. **30.** Answers will vary. **31.** 86 **32.** 98 **33.** 60 **34.** 68% **35.** 95% **36.** 34% **37.** 99.7% **38.** 16% **39.** 84% **40.** 2.5% **41.** 0 **42.** 2 **43.** 1.6 **44.** -3 **45.** -1.2 **46.** vocabulary test **47.** 38,000 miles **48.** 41,000 miles **49.** 22,000 miles **50. a.** $\pm 2.1\%$ **b.** We can be 95% confident that between 28.9% and 33.1% of American adults would be willing to sacrifice a percentage of their salary to work for an environmentally friendly company. **51. a.** skewed to the right **b.** mean: 2.1 syllables; median: 2 syllables; mode: 1 syllable **c.** yes; The mean is greater than the median, which is consistent with a distribution skewed to the right. **52.** 91.92% **53.** 96.41% **54.** 88.33% **55.** 10.69% **56.** 75% **57.** 14% **58.** 11%

59.

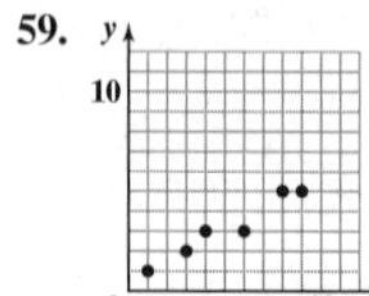

There appears to be a positive correlation.

60.

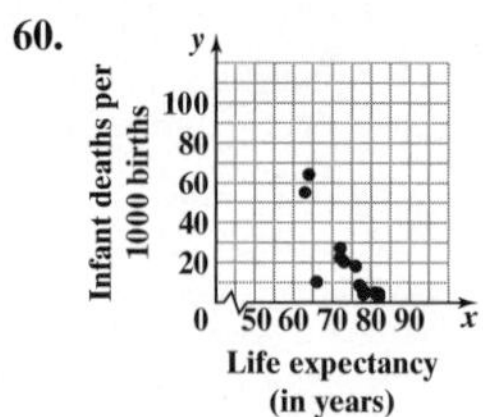

There appears to be a negative correlation.

61. false **62.** true **63.** false **64.** false **65.** true **66.** false **67.** true **68.** c **69. a.** 0.972 **b.** $y = 0.509x + 0.537$ **70. a.** 0.99 **b.** There is a correlation.

Chapter 12 Test

1. d

2.

Score	Frequency
3	1
4	2
5	3
6	2
7	2
8	3
9	2
10	1
	16

3.

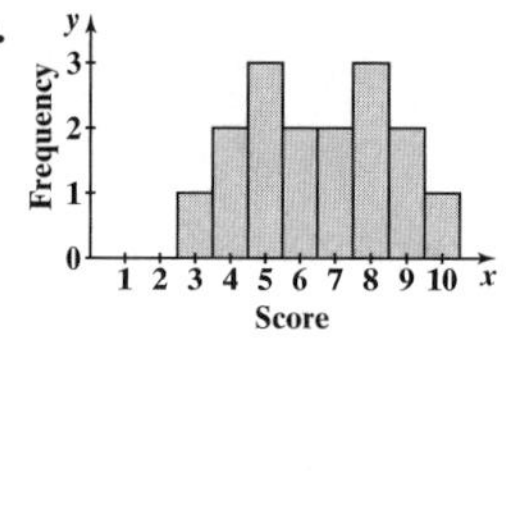

4.

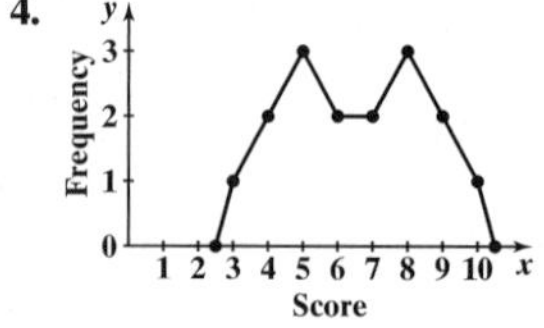

5.

Class	Frequency
40–49	3
50–59	6
60–69	6
70–79	7
80–89	6
90–99	2
	30

6.

Stem	Leaves
4	8 1 6
5	1 0 5 0 9 0
6	7 2 0 3 1 1
7	9 8 3 1 9 1 5
8	9 3 0 8 9 1
9	3 0

7. The roofline gives the impression that the percentage of home-schooled students grew at the same rate each year between the years shown, which might not have happened. **8.** ≈ 3.67 **9.** 3 **10.** 4 **11.** ≈ 2.34 **12.** 2.25 **13.** 2 **14.** 2 **15.** Answers will vary. **16.** 34% **17.** 2.5% **18.** student **19.** 8.08% **20.** 41% **21. a.** $\pm 10\%$ **b.** We can be 95% confident that between 50% and 70% of all students are very satisfied with their professors.

22.

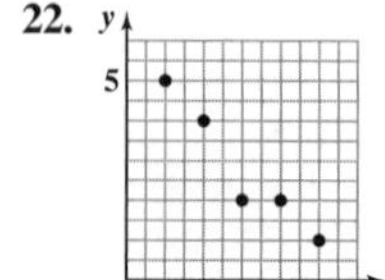

; There appears to be a strong negative correlation.

23. false **24.** false **25.** true **26.** Answers will vary.

CHAPTER 13

Section 13.1

Check Point Exercises

1. a. 4210 **b.** 40 **c.** 2865 **2.** Donna **3.** Bob **4.** Carmen **5.** Bob

Concept and Vocabulary Check

1. preference; preference **2.** majority **3.** pairwise comparison; 1; $\frac{1}{2}$; most points **4.** $\frac{n(n-1)}{2}$ **5.** plurality **6.** plurality **7.** Borda count; most points **8.** True

Exercise Set 13.1

1. 7; 5; 4 **3.**

Number of Votes	5	1	4	2
First Choice	A	B	C	C
Second Choice	B	D	B	B
Third Choice	C	C	D	A
Fourth Choice	D	A	A	D

5. a. 26 **b.** 8 **c.** 22 **d.** 3 **7.** musical **9.** Darwin **11.** Comedy **13.** Freud **15.** Comedy **17.** Einstein **19.** 10 **21.** 28 **23.** Comedy **25.** Hawking **27.** A **29.** B **31.** *Rent* **33.** *Rent* **35. a.** C **b.** A **37.** C **49.** does not make sense **51.** makes sense

Section 13.2

Check Point Exercises

1. a. A **b.** B **2. a.** B **b.** A **3. a.** A **b.** C **c.** yes **4. a.** A **b.** D **c.** yes

Concept and Vocabulary Check

1. majority **2.** head-to-head **3.** monotonicity **4.** irrelevant alternatives **5.** Borda count **6.** True

Exercise Set 13.2

1. a. D **b.** A **c.** no **3. a.** A **b.** V **c.** no **5. a.** A **b.** C **c.** no **7. a.** C **b.** B **c.** no

9. a. S **b.**

Number of Votes	15	8
First Choice	S	L
Second Choice	L	S

; S **c.** yes

11. a. A **b.** no **13. a.** C **b.** no **15. a.** B **b.** no **c.** no **d.** A; no **17. a.** B **b.** no **c.** no

19. a. A **b.** yes **c.** yes **d.**

Number of Votes	7	3	2
First Choice	A	B	A
Second Choice	B	C	C
Third Choice	C	A	B

; A **e.** yes **f.** no

29. makes sense **31.** does not make sense

Section 13.3

Check Point Exercises

1. a. 50 **b.** 22.24; 22.36; 26.4; 30.3; 98.7; 200 **2.** 22; 22; 27; 30; 99 **3.** 22; 22; 26; 30; 100 **4.** 22; 23; 27; 30; 98 **5.** 22; 22; 27; 30; 99

Concept and Vocabulary Check

1. divisor; quota **2.** quotas **3.** lower; upper; lower quota; upper quota **4.** Hamilton's; decimal **5.** modified; Jefferson's; Adams's; Webster's **6.** quota; Jefferson's; Adams's; Webster's **7.** True

Exercise Set 13.3

1. a. 20; 20,000 **b.** 6.9; 13.3; 26.7; 33.1 **c.**

State	Lower Quota	Upper Quota
A	6	7
B	13	14
C	26	27
D	33	34

3. 7; 13; 27; 33 **5.** 30; 41; 53; 73; 103 **7.** 3; 5; 12; 15; 22 **9.** 17; 55; 24; 54 **11.** 28; 33; 53; 15; 22; 29 **13.** 12; 10; 8 **15.** 4; 10; 6 **17.** 20; 24; 29; 29; 98 **19.** 57; 81; 68; 44 **21.** 57; 81; 68; 44 **23.** 7; 2; 2; 2; 8; 14; 4; 5; 10; 10; 13; 2; 6; 2; 18 **25.** 7; 2; 2; 2; 8; 14; 4; 5; 10; 10; 12; 2; 6; 3; 18 **43.** does not make sense **45.** does not make sense

Section 13.4

Check Point Exercises

1. B's apportionment decreases from 11 to 10. **2. a.** 10; 19; 71 **b.** State A: 1.005%; State B: 0.998% **c.** State A loses a seat to state B, even though the population of state A is increasing at a faster rate. **3. a.** 21; 79 **b.** With North High added to the district, West High loses a counselor to East High.

Concept and Vocabulary Check

1. Alabama **2.** population **3.** new-states **4.** True

Exercise Set 13.4

1. a. 16; 8; 6 **b.** 17; 9; 5; Liberal Arts Math loses a teaching assistant when the total number of teaching assistants is raised from 30 to 31. This is an example of the Alabama paradox. **3.** State A loses a seat when the total number of seats increases from 40 to 41. **5. a.** 4; 6; 14 **b.** 28.3%; 26.3%; 14.7% **c.** 3; 7; 14; A loses a seat while B gains, even though A has a faster increasing population. The population paradox does occur. **7.** C loses a truck to B even though C increased in population faster than B. **9. a.** 10; 90 **b.** Branch B loses a promotion when branch C is added. This means the new-states paradox has occurred. **11. a.** 9; 91 **b.** State B loses a seat when state C is added. **13. a.** 6; 13; 31 **b.** The new-states paradox does not occur. As long as the modified divisor, *d*, remains the same, adding a new state cannot change the number of seats held by existing states. **19.** makes sense **21.** does not make sense

Chapter 13 Review Exercises

1.

Number of Votes	4	3	3	2
First Choice	A	B	C	C
Second Choice	B	D	B	B
Third Choice	C	C	D	A
Fourth Choice	D	A	A	D

2. 23 **3.** 4 **4.** 16 **5.** 14 **6.** Musical **7.** Comedy **8.** Musical **9.** Musical **10.** A **11.** B **12.** A **13.** A **14.** B **15.** A; no **16.** A; no **17.** B **18.** B; yes **19.** A **20.** B **21.** A **22.** A **23.** A; Borda count method **24.** B **25.** A; no **26.** A

27.

Number of Votes	400	450
First Choice	A	C
Second Choice	C	A

C; no **28.** B **29.** B still wins. The irrelevant alternatives criterion is satisfied. **30.** 50.05 **31.** 5.49; 7.83; 12.21; 14.47 **32.** A: 5, 6; B: 7, 8; C: 12, 13; D: 14, 15 **33.** 6; 8; 12; 14 **34.** 5; 8; 12; 15 **35.** 6; 8; 12; 14 **36.** 6; 8; 12; 14 **37.** 14; 42; 62; 82 **38.** 13; 42; 63; 82 **39.** 14; 42; 63; 81 **40.** 14; 42; 62; 82 **41. a.** 8; 67; 75 **b.** The Alabama paradox occurs. **42. a.** 72; 20; 8 **b.** 0.8%; 1.3% **c.** 72; 21; 7; The population paradox occurs. **43. a.** 7; 26 **b.** The new-states paradox does not occur. **44.** False. Answers will vary.

Chapter 13 Test

1. 3600 **2.** 600 **3.** 1500 **4.** 1500 **5.** B **6.** B **7.** A **8.** A **9.** A **10.** A; yes **11.** A **12.** B; no **13.** C **14.** B; no **15.** no **16.** 50 **17.** 2.38; 3.3; 4.32 **18.** A: 2, 3; B: 3, 4; C; 4, 5 **19.** 3; 3; 4 **20.** 2; 3; 5 **21.** 3; 3; 4 **22.** 2; 3; 5 **23.** yes **24.** The new-states paradox does not occur. **25.** Answers will vary.

CHAPTER 14

Section 14.1

Check Point Exercises

1. The two graphs have the same number of vertices connected to each other in the same way.

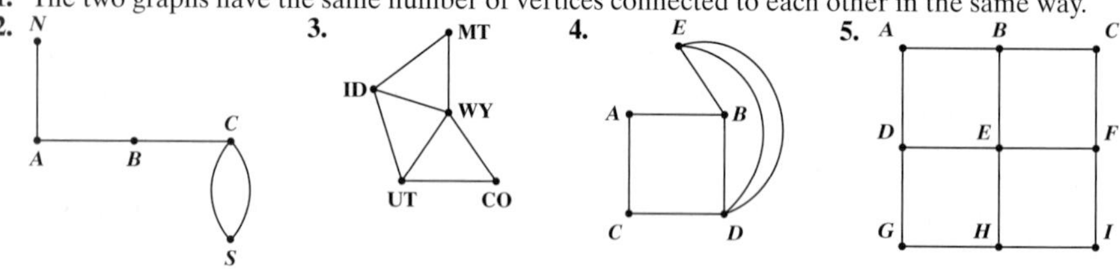

6. *A* and *B*, *A* and *C*, *A* and *D*, *A* and *E*, *B* and *C*, and *E* and *E*

Concept and Vocabulary Check

1. graph; vertices; edges; loop **2.** equivalent **3.** degree **4.** adjacent; path; circuit **5.** bridge **6.** true **7.** true **8.** true **9.** false

Exercise Set 14.1

1. 6; St. Louis, Chicago, Philadelphia, Montreal; 1, 1, 2, 2 **3.** No; no; Answers will vary.

5. Possible answers:

7. The two graphs have the same number of vertices connected in the same way, so they are the equivalent. Possible answer:

9.

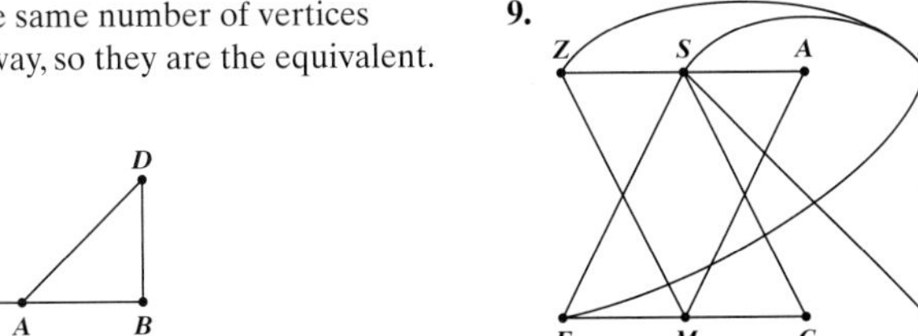

11.

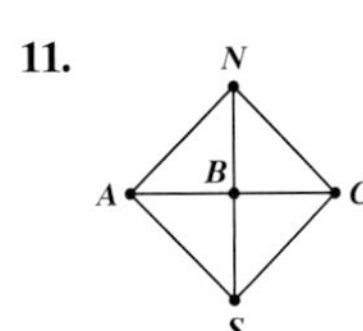

13.

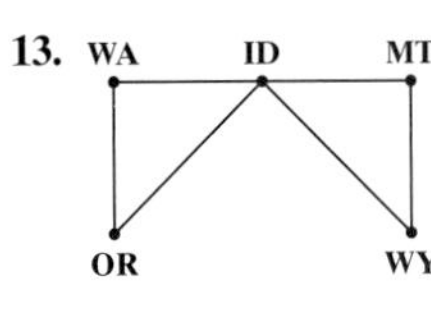

15.

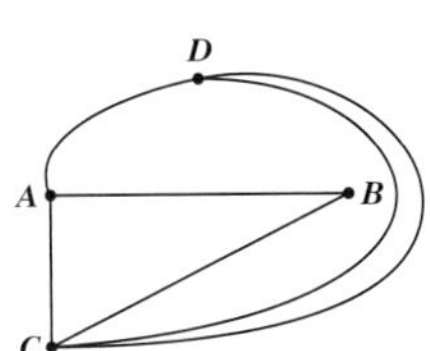

17.

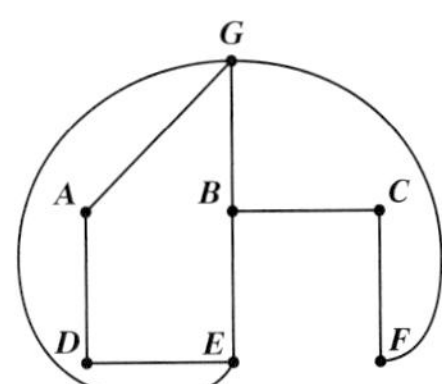

19.

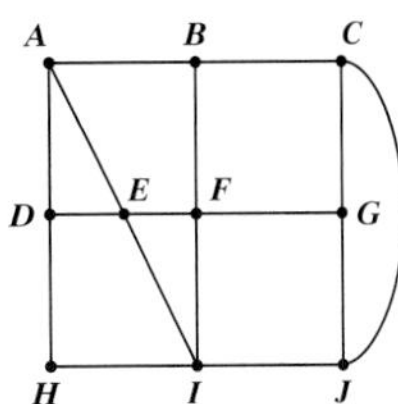

21.

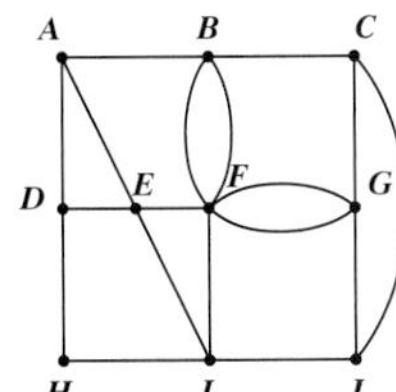

23. A: 2; B: 2; C: 3; D: 3; E: 3; F: 1 **25.** B and C **27.** $A, C, D; A, B, C, D$ **29.** AC and DF **31.** While edge CD is included, the graph is connected. If we remove CD, the graph will be disconnected. **33.** DF **35.** Even: A, B, G, H, I; odd: C, D, E, F **37.** D, G, and I **39.** Possible answers: B, C, D, F; B, A, C, D, F **41.** G, F, I, H, G **43.** $A, B, C, C, D, F, G, H, I$ **45.** G, F, D, E, D requires that edge DE be traversed twice. This is not allowed within a path. **47.** No edge connects F and E.

49. Possible answer:

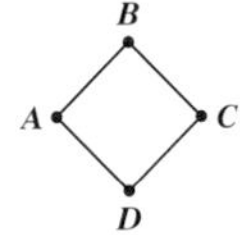

51. Possible answer:

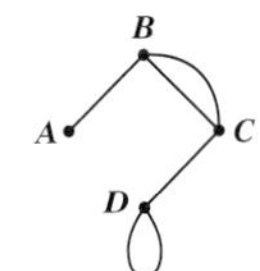

67. makes sense **69.** does not make sense **71.**

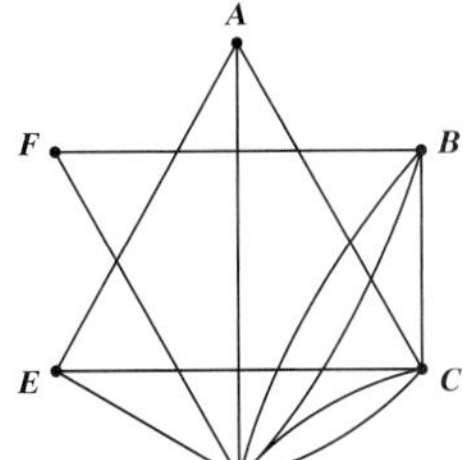

Section 14.2

Check Point Exercises

1. ; $E, C, D, E, B, C, A, B, D$ **2.**

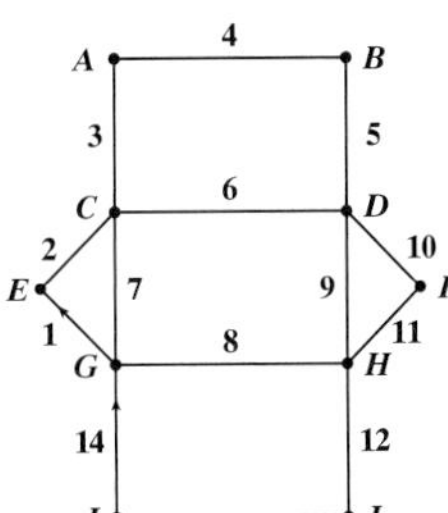

; $G, E, C, A, B, D, C, G, H, D, F, H, J, I, G$

3. a. yes **b.**

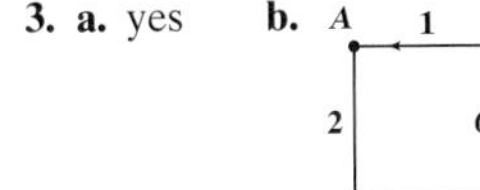

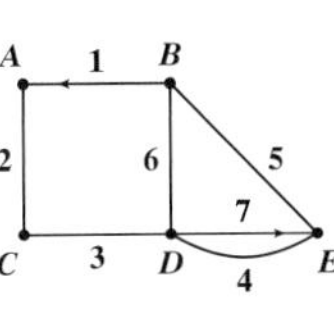

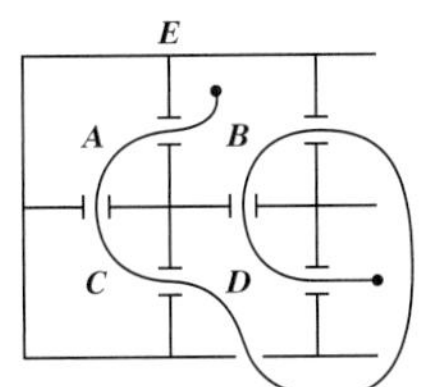

4.

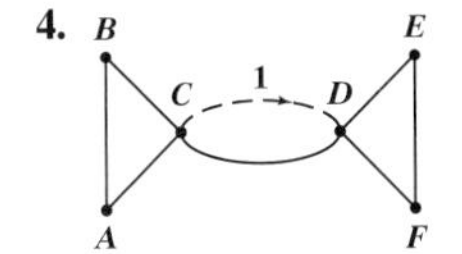

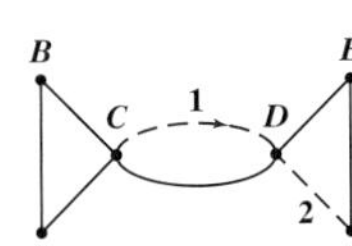

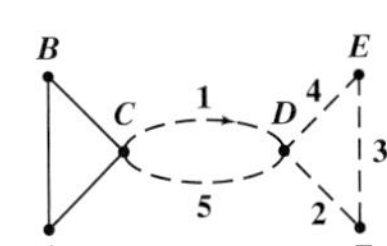

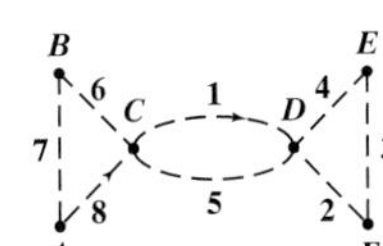

Concept and Vocabulary Check

1. Euler **2.** Euler **3.** two **4.** no; any **5.** odd **6.** *E*; *E*; *D* **7.** Fleury's; bridge **8.** true **9.** false **10.** false

Exercise Set 14.2

1.

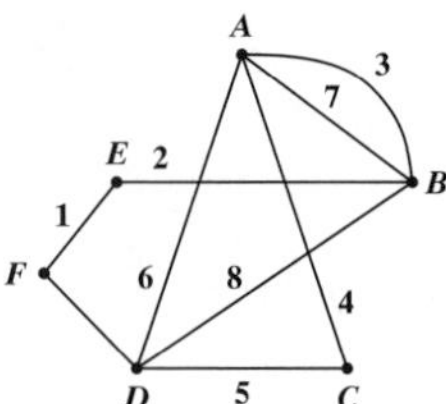

neither

3.

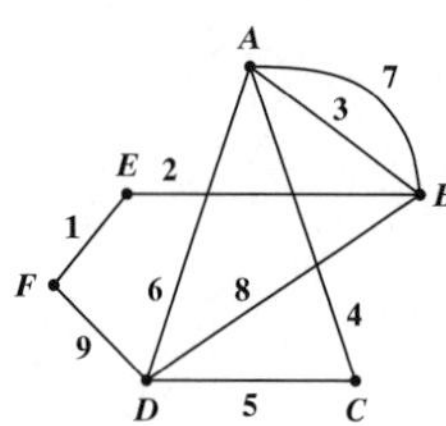

Euler circuit

5.

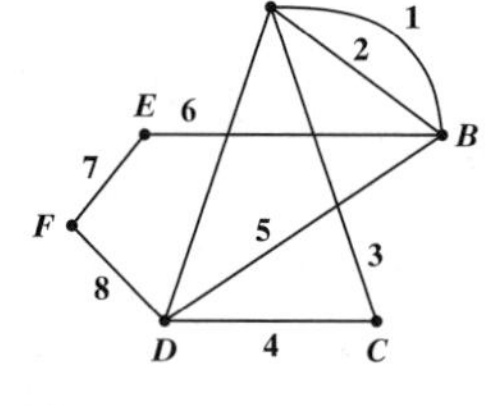

neither

7. a. There are two odd vertices.
b. Possible answer:

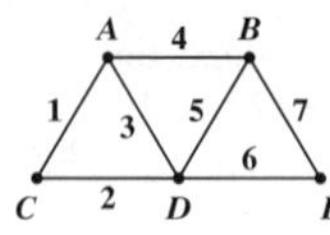

9. a. There are no odd vertices.
b. Possible answer:

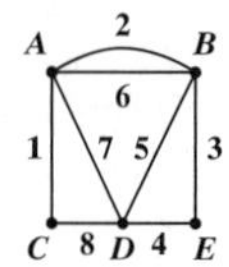

11. There are more than two odd vertices. **13.** Euler circuit **15.** Euler path, but no Euler circuit **17.** neither an Euler path nor an Euler circuit

19. a. Euler circuit
b. Possible answer:

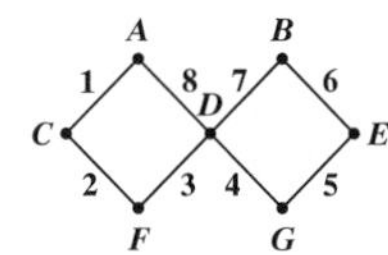

21. a. Euler path
b. Possible answer:

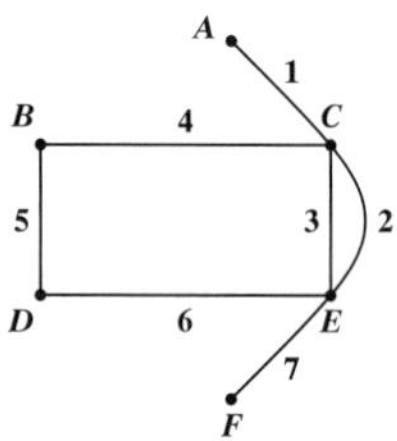

23. a. neither

25. a. Euler path
b. Possible answer:

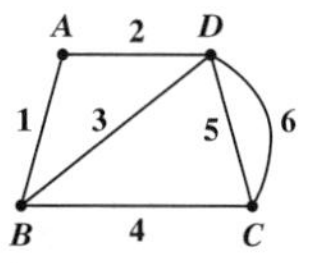

27. a. Euler circuit
b. Possible answer:

29. a. neither

31. a. Euler path
b. Possible answer:

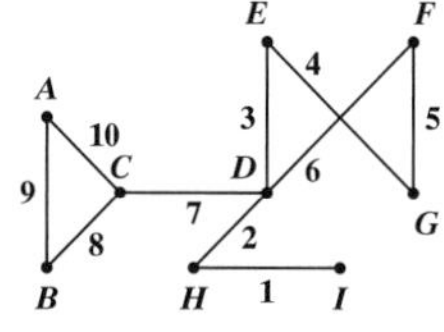

33. Possible answer:

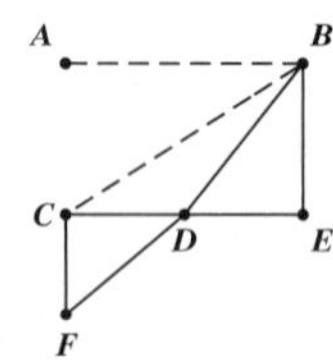

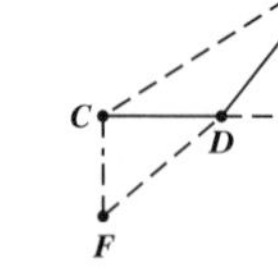

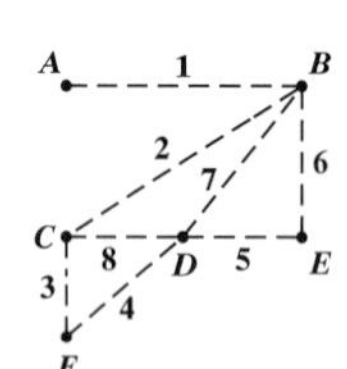

35. Possible answer:

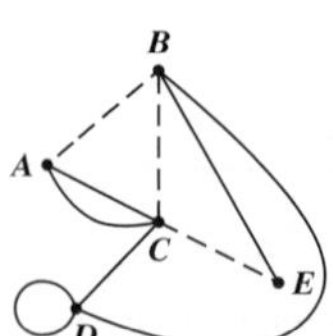

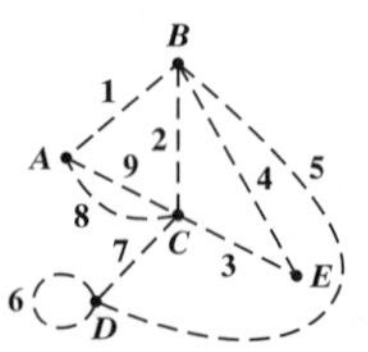

37. Possible answer:

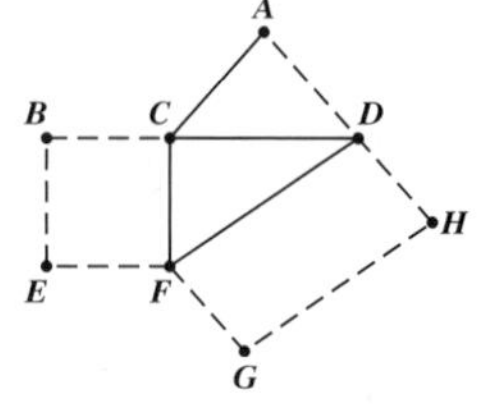

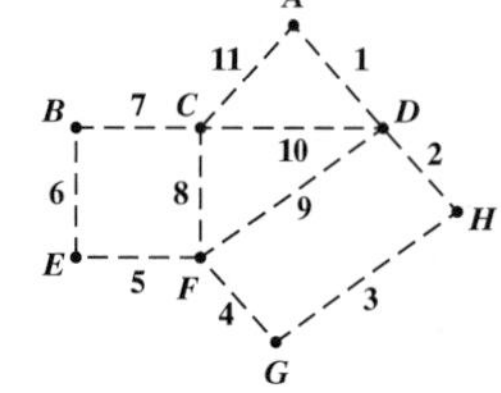

39. Possible answer:

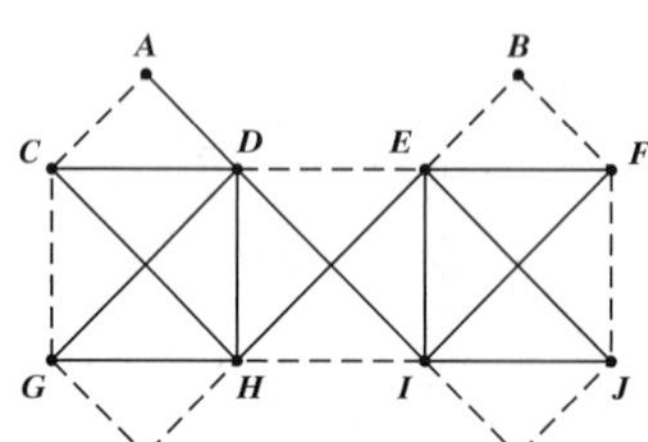

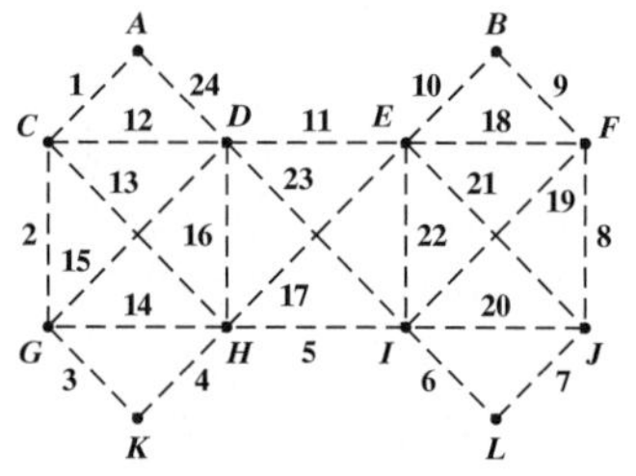

41. a. Remove *FG*. **b.** Sample answer: *EC, CB, BD, DF, FA, AD, DG, GH, HC, CG, GB, BH, HE*
43. a. Remove *BA* and *FJ*. **b.** Sample answer: *CA, AD, DI, IH, HG, GF, FC, CD, DE, EH, HJ, JG, GB, BC*
45. The graph has an Euler circuit. **47.**

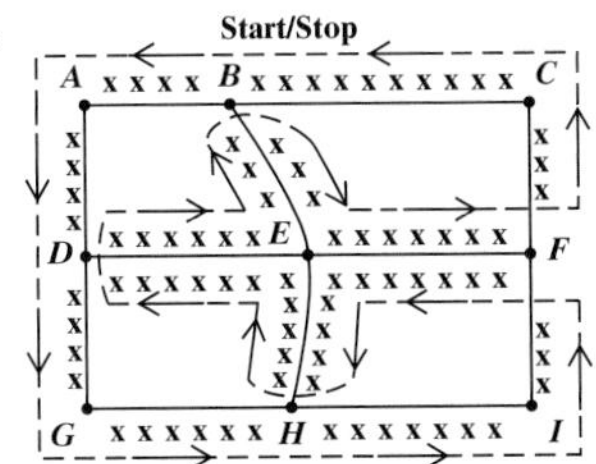

49. a. (graph with vertices A, B, C, D, E, F, G, H, I, J, K) **b.** Yes; begin at *B* or *E*.

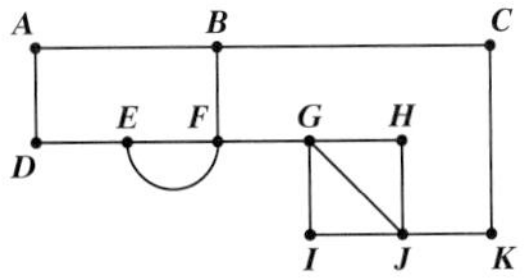

51. a.

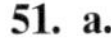

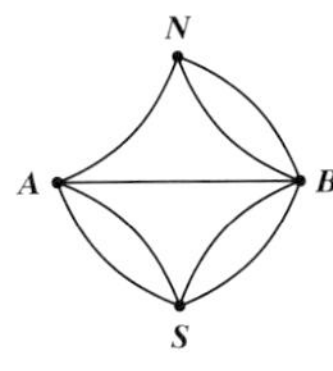

b. Yes, they can.
c. Possible answer:

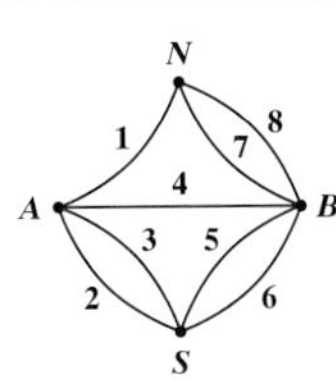

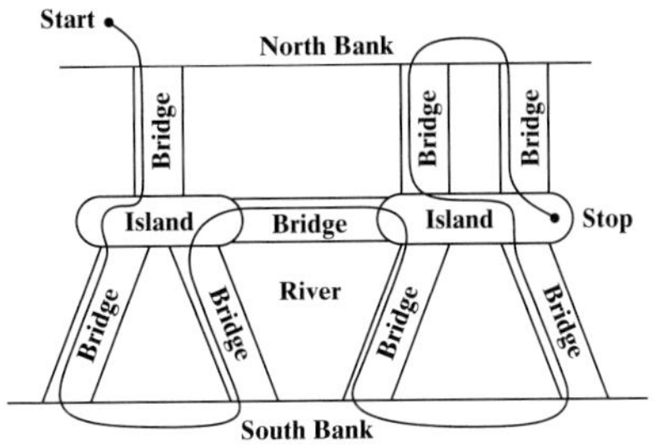

53.

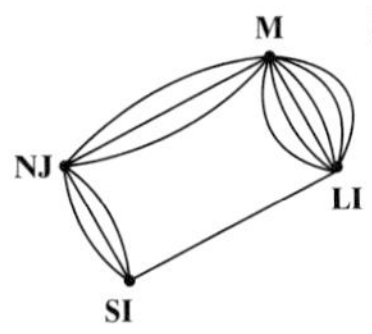

; It is possible.

55. a.

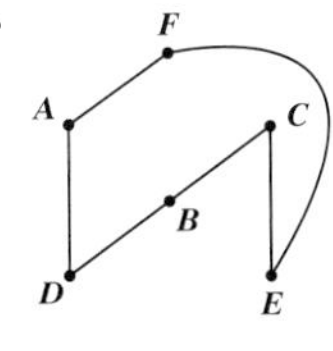

b. It is possible.
c. Possible answer:

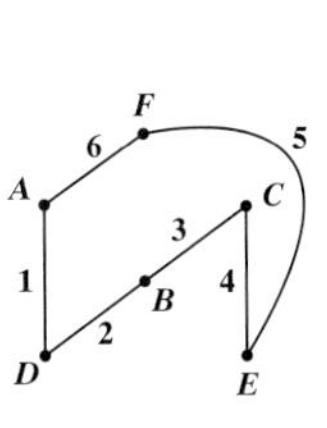

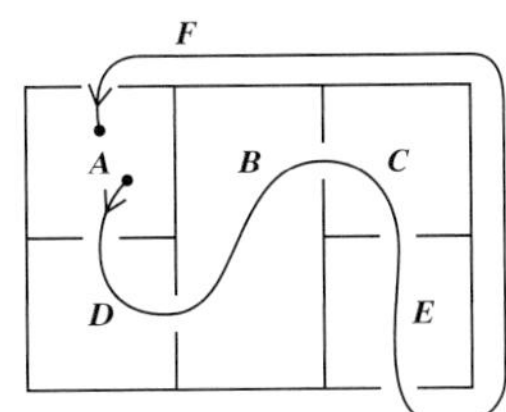

57. Possible answer:

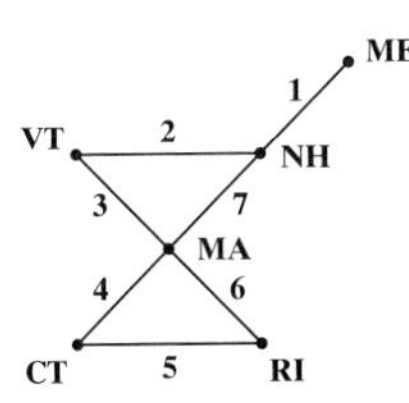

59. a.

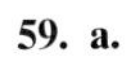

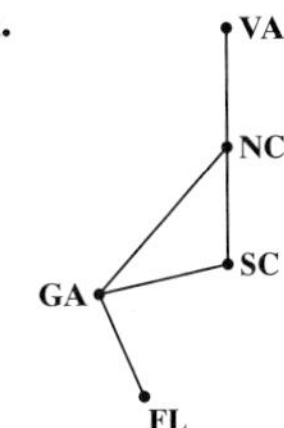

b. not possible
d. not possible

69. makes sense **71.** does not make sense

Section 14.3

Check Point Exercises

1. a. Possible answer: *E, C, D, G, B, A, F* **b.** Possible answer: *E, C, D, G, B, F, A, E* **2. a.** 2 **b.** 120 **c.** 362,880 **3.** $562

4.

Hamilton Circuit	Sum of the Weights of the Edges	=	Total Cost
A, B, C, D, A	20 + 15 + 50 + 30	=	$115
A, B, D, C, A	20 + 10 + 50 + 70	=	$150
A, C, B, D, A	70 + 15 + 10 + 30	=	$125
A, C, D, B, A	70 + 50 + 10 + 20	=	$150
A, D, B, C, A	30 + 10 + 15 + 70	=	$125
A, D, C, B, A	30 + 50 + 15 + 20	=	$115

A, B, C, D, A; *A, D, C, B, A*

5. *A, B, C, D, E, A*; 198

Concept and Vocabulary Check

1. Hamilton; Hamilton **2.** complete; $(n-1)!$ **3.** weighted; weights; traveling; optimal **4.** Brute Force **5.** Nearest Neighbor; weight **6.** false **7.** false **8.** false

Exercise Set 14.3

1. Possible answer: *A, G, C, F, E, D, B* **3.** Possible answer: *A, B, G, C, F, E, D, A* **5.** Possible answer: *A, F, G, E, C, B, D* **7.** Possible answer: *A, B, C, E, G, F, D, A* **9. a.** no **11. a.** yes **b.** 120 **13. a.** no **15.** 2 **17.** 39,916,800 **19.** 11 **21.** 36 **23.** 36 **25.** 88 **27.** 70 **29.** 70 **31.** *A, C, B, D, A* and *A, D, B, C, A* **33.** *B, D, C, A, B*; 82 **35. a.** Add *AB*.; 6 **b.** Sample answers: *AD, DB, BC, CA; CA, AB, BD, DC* **c.** Remove *CD*. **d.** Sample answer: *AC, CB, BD, DA* **37. a.** Add *AB, AC, BC*, and *DE*.; 24 **b.** Sample answers: *AB, BC, CE, ED, DA; AB, BC, CD, DE, EA* **c.** Remove *BD* and *BE*. **d.** Sample answer: *AE, EC, CD, DA* **39.** *A, D, B, C, E, A*

41.

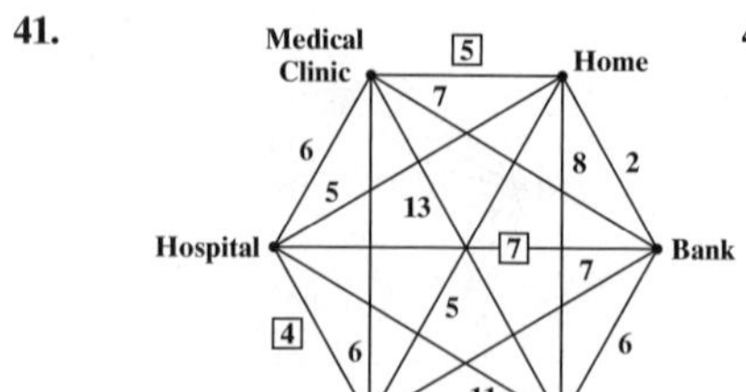

43. 30 **45.**

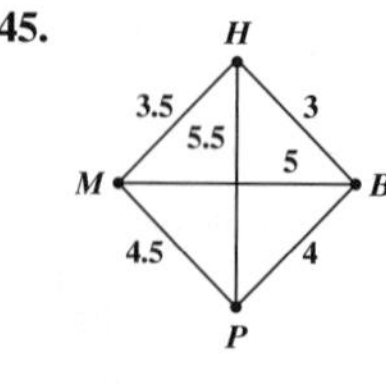

47. *H, B, P, M, H*; 15 mi; same

61. does not make sense **63.** makes sense

Section 14.4

Check Point Exercises

1. Figure 14.51 (c) **2.** Possible answer:

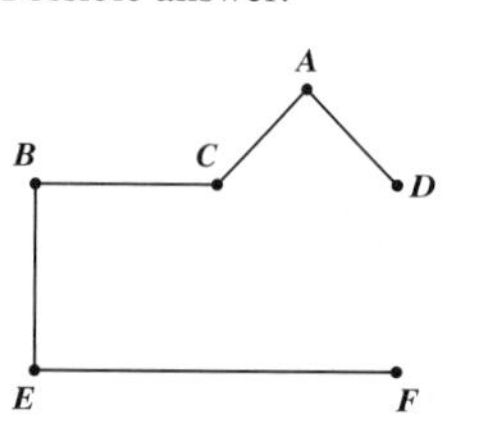

3.

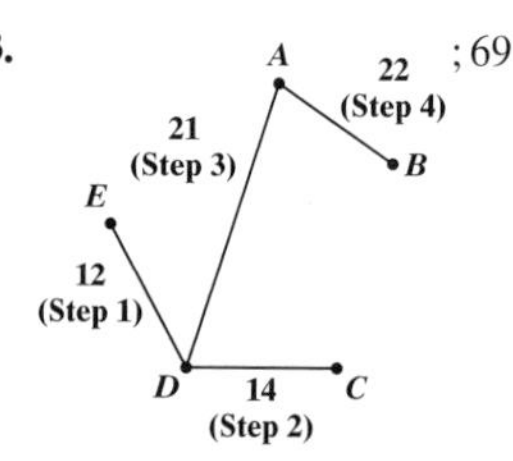

; 69

Concept and Vocabulary Check

1. tree; bridge; $n-1$ **2.** spanning **3.** minimum spanning **4.** Kruskal's; weight; circuits **5.** false **6.** true

Exercise Set 14.4

1. yes **3.** no **5.** yes **7.** no **9.** yes **11.** i **13.** ii **15.** iii

17.

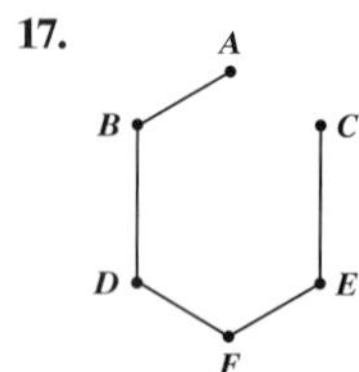

19.

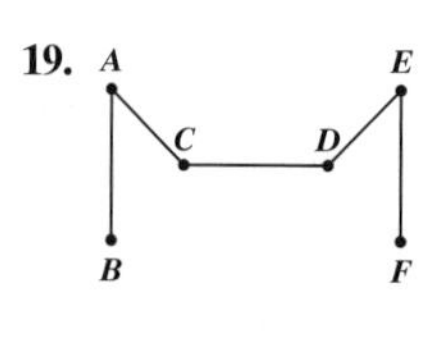

21.

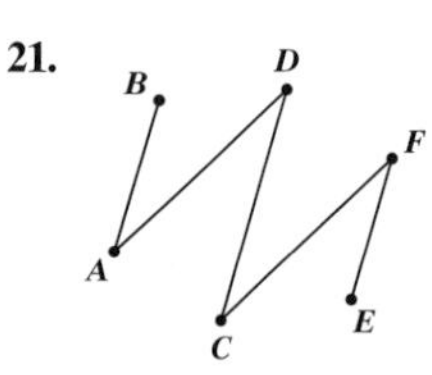

23.

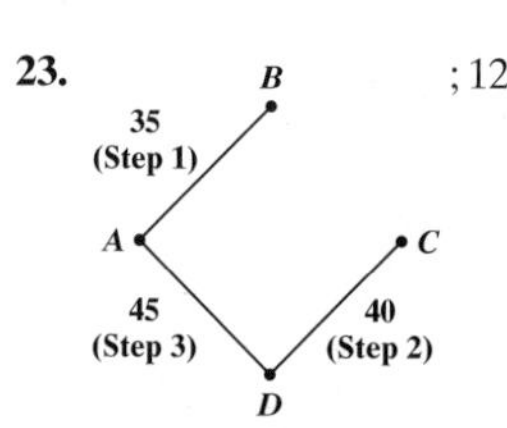

; 120

25.

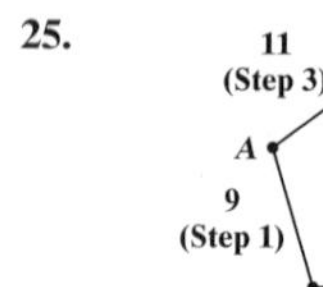

; 42 **27.**

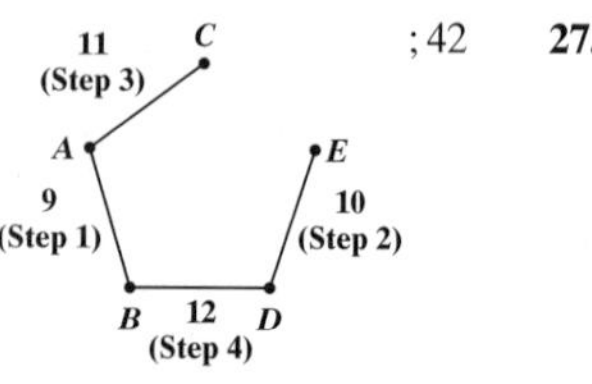

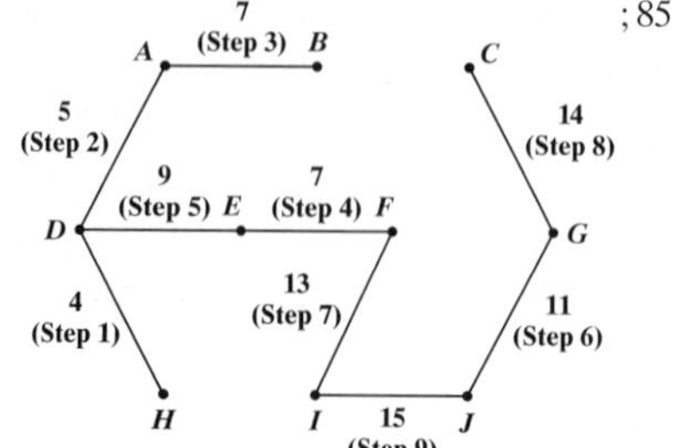

; 85 **29.**

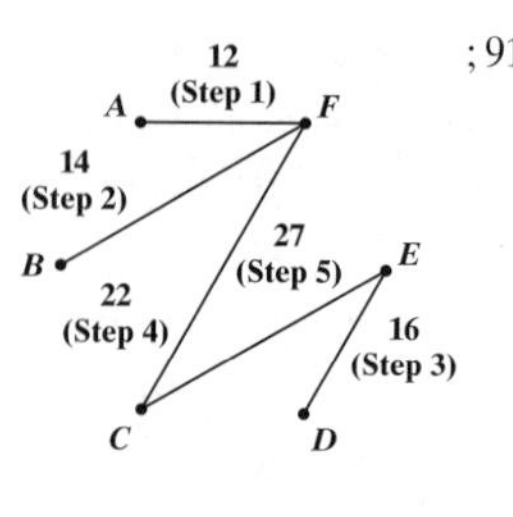

; 91

31. Sample answers: *AB, AC, CD; AB, BC, CD; AB, AD, CD; AB, AD, BC* **33.** *AE, BC, CD, CE*; 64 **35.** Sample answer: *AE, BC, BE, CF, DE, EH, FG, FJ, HI*; 141

37.

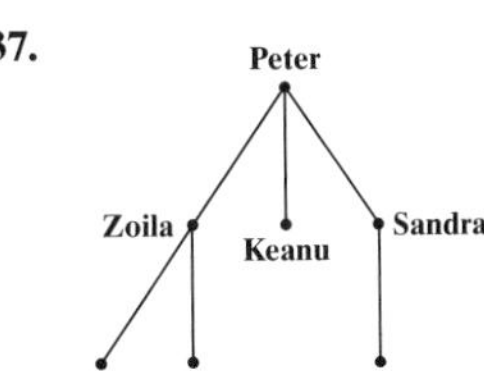

39. a.

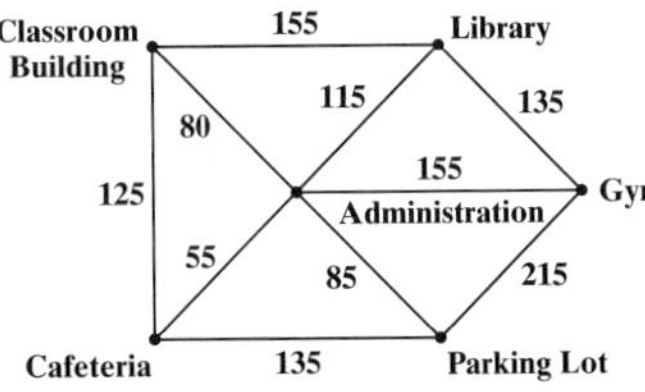

b.

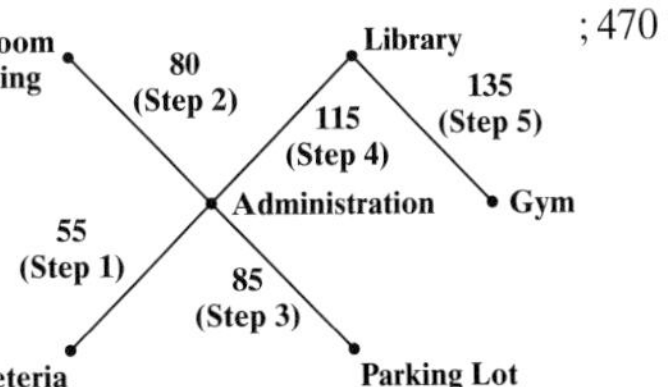

; 470 ft

41.

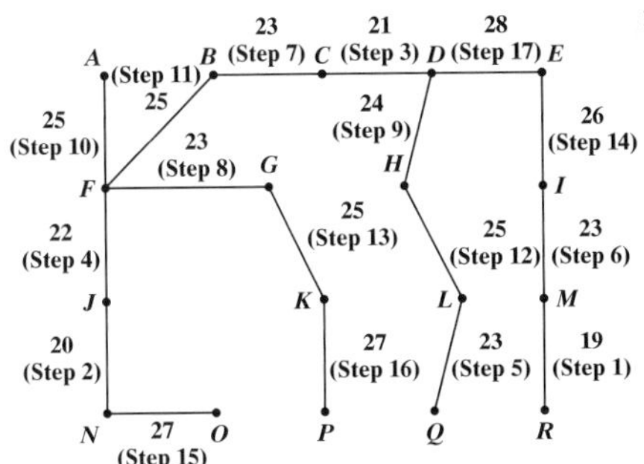

; 406 ft

53. does not make sense **55.** makes sense

Chapter 14 Review Exercises

1. Both graphs have the same number of vertices, and these vertices are connected in the same ways.
Possible answer:

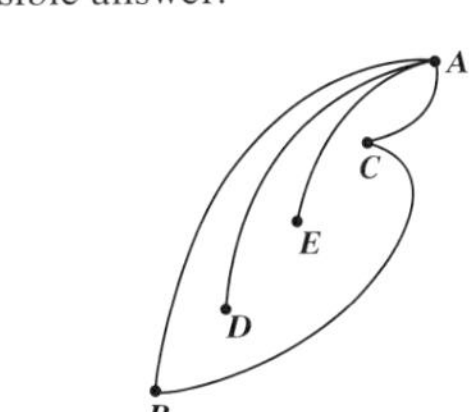

2. *A*: 5; *B*: 4; *C*: 5; *D*: 4; *E*: 2 **3.** Even: *B*, *D*, *E*; odd: *A*, *C* **4.** *B*, *C*, and *E* **5.** Possible answer: *E*, *D*, *B*, *A* and *E*, *C*, *A* **6.** Possible answer: *E*, *D*, *C*, *E* **7.** yes **8.** no **9.** *AD*, *DE*, and *DF*

10.

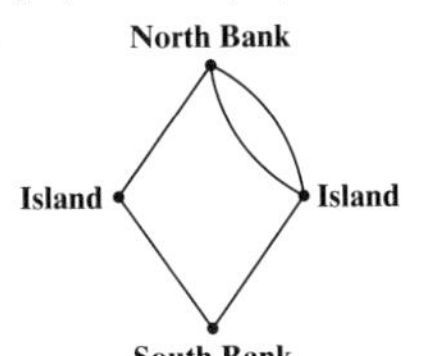

11.

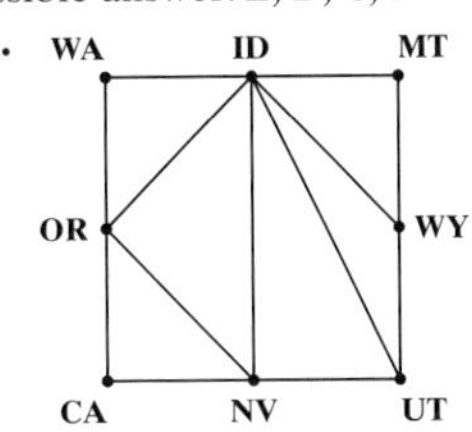

12.

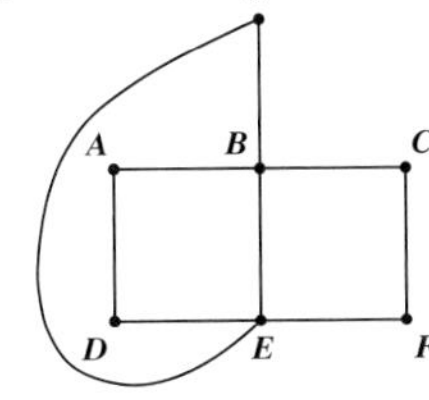

13. a. neither
14. a. Euler circuit
b. Possible answer:

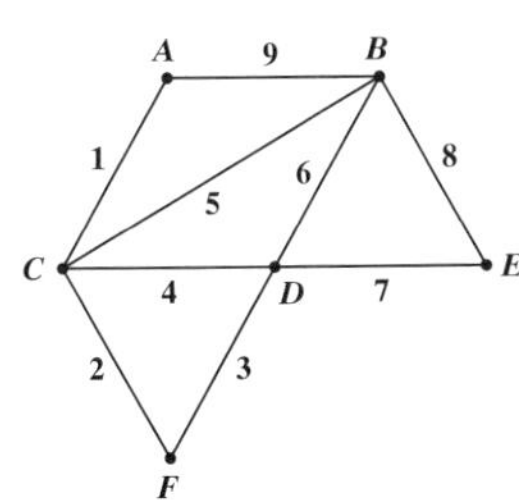

15. a. Euler path
b. Possible answer:

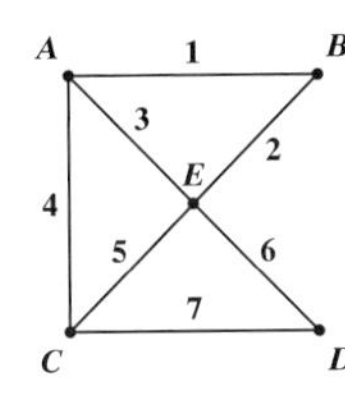

16.

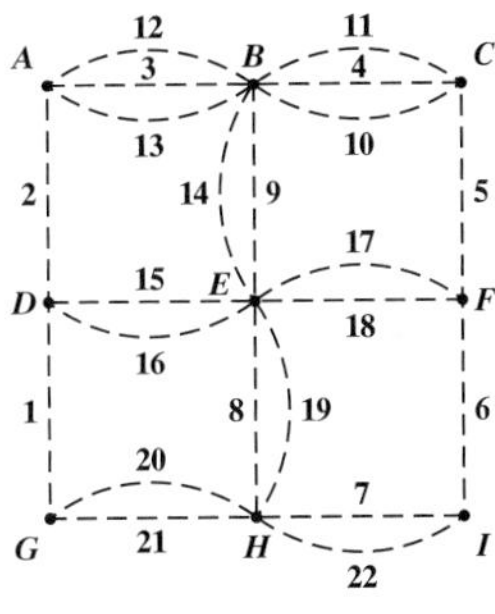

17.

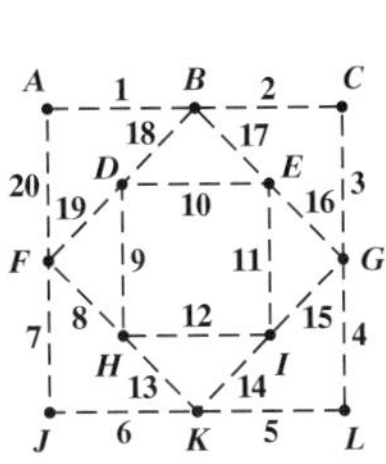

18. a. yes **b.**

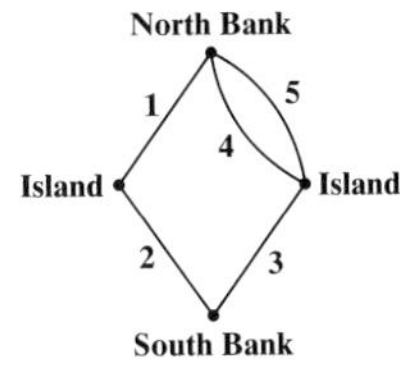

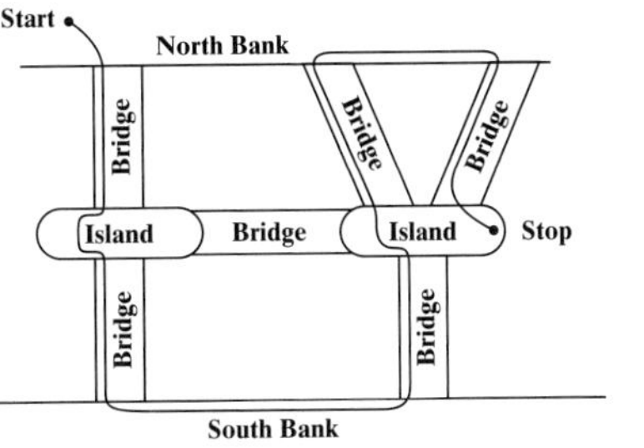

c. no

19. It is possible. **20. a.** yes **b.** Possible answer:

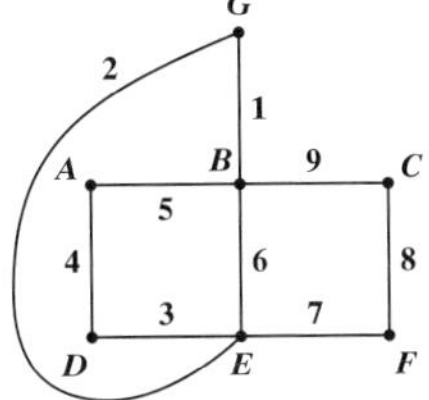

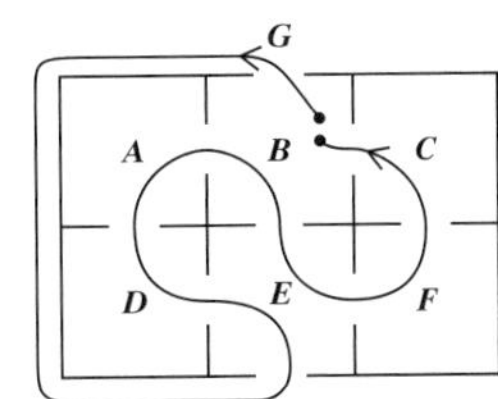

21. a.

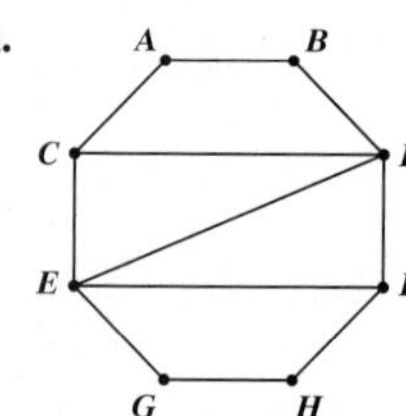

b. yes
c. The guard should begin at *C* and end at *F*, or vice versa.

22. Possible answer: *A, E, C, B, D, A*
23. Possible answer: *D, B, A, E, C, D*
24. a. no
25. a. yes **b.** 6
26. a. no
27. a. yes **b.** 24

28. *A, B, C, D, A*: 19; *A, B, D, C, A*: 18; *A, C, B, D, A*: 19; *A, C, D, B, A*: 18; *A, D, B, C, A*: 19; *A, D, C, B, A*: 19
29. *A, B, D, C, A* and *A, C, D, B, A* **30.** *A, C, D, B, A*; 18 **31.** *A, B, E, D, C, A*; 24

32.

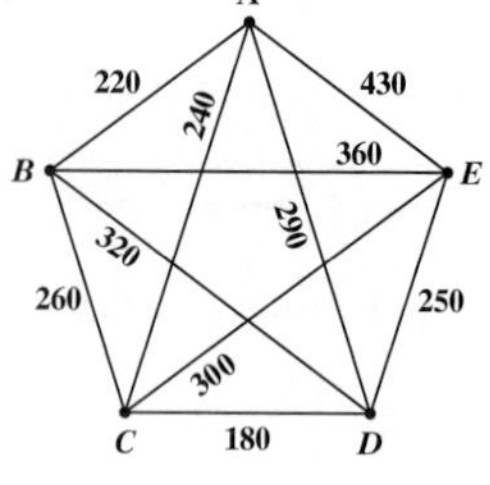

33. *A, B, C, D, E, A*; $1340 **34.** yes **35.** No. It has a circuit. **36.** No. It is disconnected.
37. Possible answer:

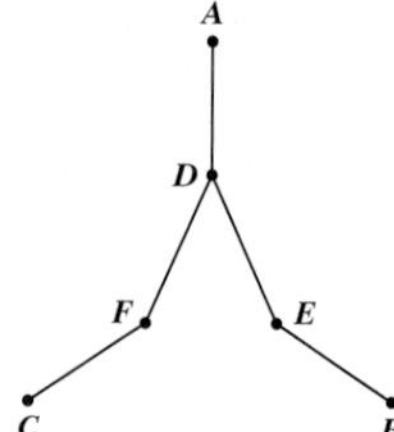

38. Possible answer:

A B C
D E F
G H I

39.

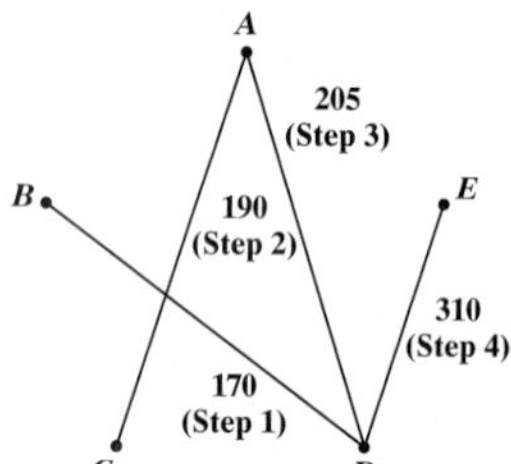

; 875

40.

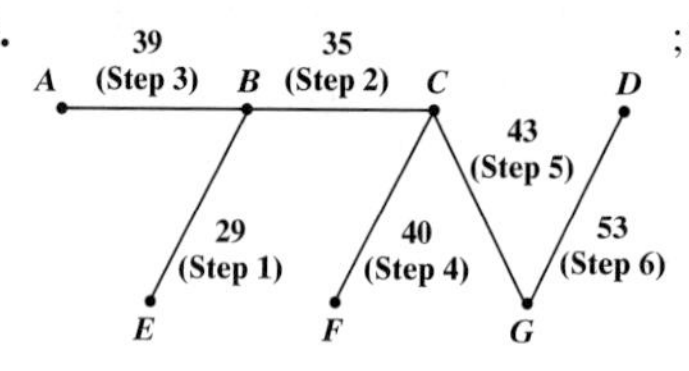

; 239

41.

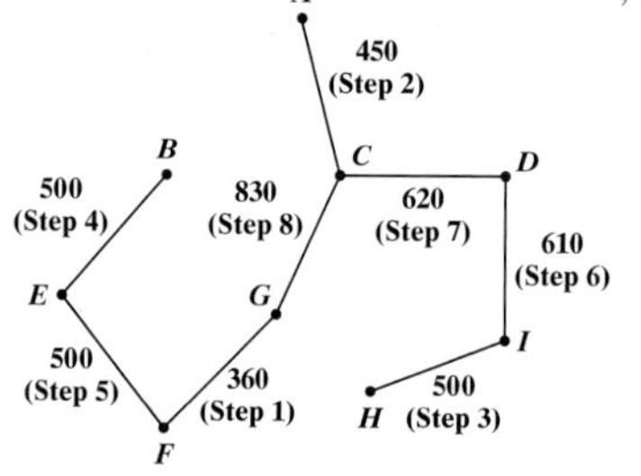

; 4370 mi

Chapter 14 Test

1. *A*: 2; *B*: 2; *C*: 4; *D*: 3; *E*: 2; *F*: 1 **2.** Possible answer: *A, D, E* and *A, B, C, E* **3.** *B, A, D, E, C, B* **4.** *CF*

5.

Netherlands
Belgium
Luxembourg
France
Germany
Switzerland
Austria

6. a. Euler path **b.** Possible answer:

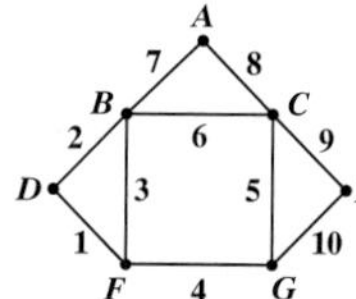

7. a. neither

8. a. Euler circuit **b.** Possible answer:

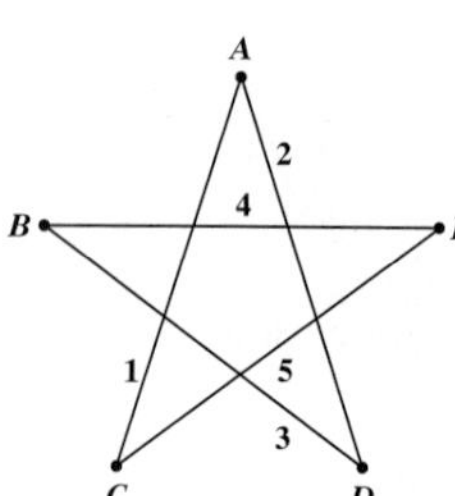

9.

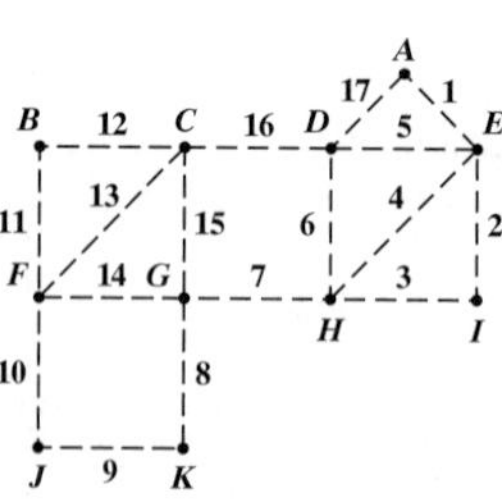

10. a.

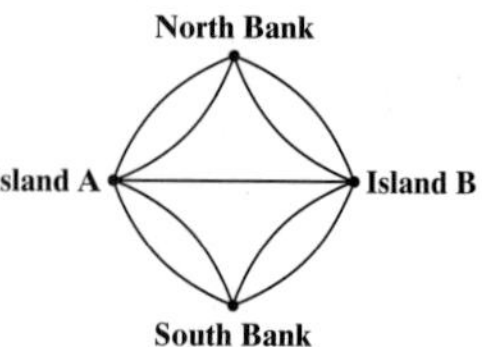

b. yes **c.** on one of the islands

11. a. 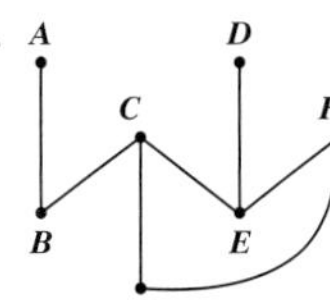

b. no

12. a. 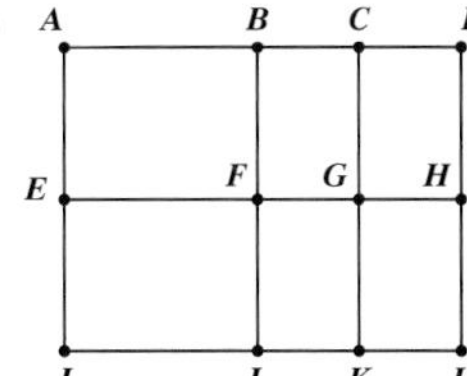

b. no

13. *A*, *B*, *C*, *D*, *G*, *F*, *E*, *A* and *A*, *F*, *G*, *D*, *C*, *B*, *E*, *A*

14. 24

15. 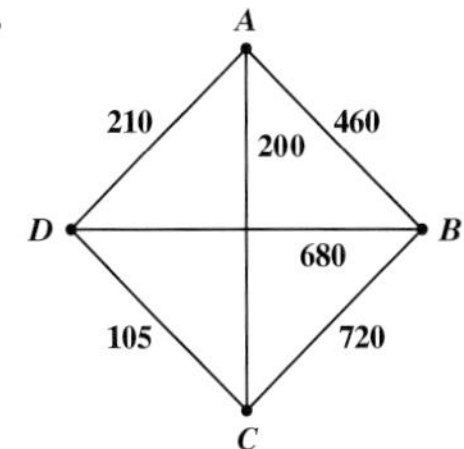

16. *A*, *B*, *D*, *C*, *A* or *A*, *C*, *D*, *B*, *A*; $1445

17. *A*, *E*, *D*, *C*, *B*, *A*; 33

18. no

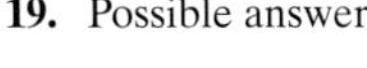
19. Possible answer:

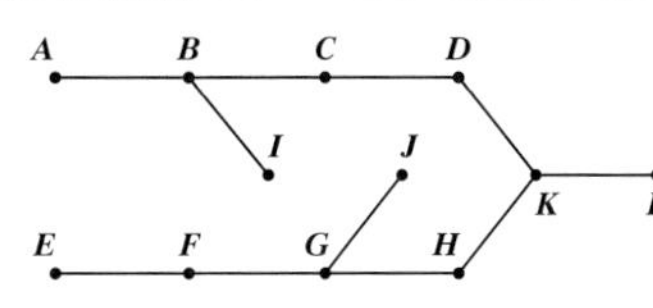

20. 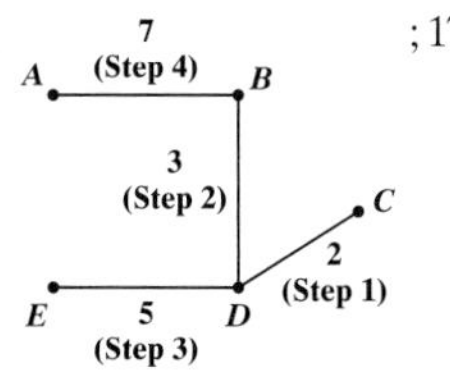

; 17

Subject Index

T

U

V

Credits

FM **p.** vi Robert Blitzer

CHAPTER 1 **p. 1** (Cap) Focal Point/Shutterstock **p. 1** (Child) Jose Luis Pelaez Inc/Blend Images/Alamy Stock Photo **p. 2** Richard F. Voss **p. 7** Science History Images/Alamy Stock Photo **p. 8** AF Archive/Alamy Stock Photo **p. 8** (BB Box) Reprinted by permission of ICM Partners Copyright (2012) by William Poundstone. **p. 14** Mauricio Graiki/Shutterstock **p. 17** Ken Fisher/The Image Bank/Getty Images **p. 23** (Cap) Stephen Coburn/Shutterstock **p. 23** (Child) Iofoto/Shutterstock **p. 26** Martin Gardner. Reproduced with permission from James Gardner, Martin Gardner Literary interest. **p. 30** (Bottom) Ristoviita/123RF **p. 30** (Top) Martin Bond/Alamy Stock Photo **p. 35** Mike Hewitt/Getty Images Sport/Getty Images **p. 44** Andreas Nilsson/Fotolia

CHAPTER 2 **p. 49** Erics Mandes/Shutterstock **p. 50** Moodboard/SuperStock **p. 51** Aroon Thaewchatturat/Alamy Stock Photo **p. 52** Pearson Education, Inc., **p. 64** Nicole Ackerman **p. 67** Dave Stroup **p. 70** Time and Time Again (1981), P.J. Crook. Acrylic on canvas and wood (91.4X122). Private collection/Bridgeman Art Library **p. 73** AdMedia/Splash News/Newscom **p. 78** (Left) Zuma Press, Inc./Alamy Stock Photo **p. 78** (Middle Left) Peter Brooker/Alamy Stock Photo **p. 78** (Middle Right) Everett Collection Inc/Alamy Stock Photo **p. 78** (Middle) Creative Collection Tolbert Photo/Alamy Stock Photo **p. 78** (Right) Hyperstar/Alamy Stock Photo **p. 87** Tatniz/Shutterstock **p. 99** Cesar Okada/ E+/Getty images **p. 105** Ermolaev Alexandr/Fotolia

CHAPTER 3 **p. 117** Alistair Scott/Alamy Stock Photo **p. 118** (Left) Granamour Weems Collection/Alamy Stock Photo **p. 118** (Right) 20th Century Fox Television/Photo 12/Alamy Stock Photo **p. 122** AF Archive/Alamy Stock Photo **p. 126** Glowimages RM/Alamy Stock Photo **p. 139** (Bottom Left) William D. Bird/Everett Collection Inc/Alamy Stock Photo **p. 139** (Bottom Middle) Gregorio T. Binuya//Everett Collection Inc/Alamy Stock Photo **p. 139** (Bottom Right) Michael Germana/Everett Collection Inc/Alamy Stock Photo **p. 139** (Top Left) Dee Cercone/Everett Collection Inc/Alamy Stock Photo **p. 139** (Top Middle) Dee Cercone/Everett Collection Inc/Alamy Stock Photo **p. 139** (Top Right) Desiree Navarro/Everett Collection Inc/Alamy Stock Photo **p. 140** (Bottom) Martine Mouchy/Stone/Getty Images **p. 140** (Top) Sborisov/Fotolia **p. 150** Photo 12/Alamy Stock Photo **p. 154** Fotolia **p. 159** Pearson Education, Inc., **p. 166** (Left Middle) AF Archive/Alamy Stock Photo **p. 166** (Left) Allstar Picture Library/Alamy Stock Photo **p. 166** (Right Middle) Moviestore Collection Ltd/Alamy Stock Photo **p. 166** (Right) Pictorial Press Ltd/Alamy Stock Photo **p. 166** (Top) Douglas Pizac/AP Images **p. 176** Sean Locke Photography/Shutterstock **p. 178** CBS Photo Archive/Getty images **p. 181** Banque d'Images, ADAGP/Art Resource, NY; © 2017 C. Herscovici/Artists Rights Society (ARS), NY **p. 184** Nick Ut/AP Images **p. 186** Paramount/Everett Collection **p. 192** Bettmann/Getty Images **p. 199** (Left) Interfoto/Alamy Stock Photo **p. 199** (Right) North Wind Picture Archives/Alamy Stock Photo **p. 206** Scala/Art Resource, NY

CHAPTER 4 **p. 215** Shizuo Kambayashi/AP Images **p. 216** Dinosmichail/Fotolia **p. 218** Martine Beck-Coppola/RMN-Grand Palais/Art Resource, NY **p. 219** Frank Boston/Fotolia **p. 220** Deniza 40x/Shutterstock **p. 224** Petr Vaclavek/Fotolia **p. 226** Mary Evans Picture Library/The Image Works **p. 231** Compuinfoto/Fotolia **p. 234** David R. Frazier/Photolibrary, Inc./Alamy Stock Photo **p. 237** Pasieka/Science Photo Library/Alamy Stock Photo **p. 240** Craig Fujii/AP Images **p. 242** Lee Snider/The Image Works

CHAPTER 5 **p. 251** Domen Colja/Alamy Stock Photo **p. 252** Chase Jarvis/DigitalVision/Getty Images **p. 262** Addison Gallery of American Art, Phillips Academy, Andover, MA/Art Resource, NY; © Estate of George Tooker. Courtesy of DC Moore Gallery, New York. **p. 266** (Left) Alon Brik/Shutterstock **p. 266** (Right) Kelly Richardson/Shutterstock **p. 276** Ariel Skelley/Blend Images/Getty Images **p. 288** Robert Voets/CBS Television/Album/Newscom **p. 291** Marc Rasmus/Image Broker/Alamy Stock Photo **p. 293** (Bottom) Joshua Haviv/Fotolia **p. 293** (Top) Chitose Suzuki/AP Images **p. 298** John G. Ross/Science Source **p. 299** Digital Image © The Museum of Modern Art/Licensed by Scala/Art Resource, NY; © 2017 Salvador Dalí, Fundació Gala-Salvador Dalí, Artists Rights Society **p. 304** Ronald Grant/HBO/Your Face Goes Here Entertainment/Everett Collection **p. 305** Blitzer, Robert F **p. 308** © 1985 by Roz Chast/The New Yorker Collection/The Cartoon Bank **p. 309** (Bottom) Dennisvdw/iStock/Getty Images Plus/Getty Images **p. 309** (Top) Kenneth Libbrecht/Science Source **p. 312** Lian Deng/Shutterstock **p. 314** © 2007 Gordon Caulkins **p. 317** John Trotter/Zuma Press/Newscom **p. 323** Photobank/Fotolia **p. 326** Galdzer/iStock/Getty Images **p. 329** U.S. Census Bureau **p. 332** Jokerpro/Shutterstock **p. 335** Pearson Education, Inc., **p. 341** Anna Om/Shutterstock

CHAPTER 6 **p. 343** John McCoy/Los Angeles Daily News/Zuma Press Inc/Alamy Stock Photo **p. 344** Flying Colours Ltd/DigitalVision/Getty Images **p. 346** (Left) Image Source/AGE Fotostock **p. 346** (Right) Erik Isakson/Tetra Images/Alamy Stock Photo **p. 350** Ikon Images/SuperStock **p. 354** AF Archive/Alamy Stock Photo **p. 364** Collie/iStock/Getty Images **p. 369** (Bottom Left Middle) Red Chopsticks/Getty Images **p. 369** (Bottom Left) ER Productions/Corbis Premium/Alamy Stock Photo **p. 369** (Bottom Right Middle) West Coast Surfer/Moodboard/AGE Fotostock **p. 369** (Bottom Right) Creatas/Getty Images **p. 369** (Top Left) Stuart Ramson/AP Images **p. 369** (Top Middle) S_Bukley/Newscom **p. 369** (Top Right) Icon Sports Media 181/Newscom **p. 372** Blitzer, Robert F. **p. 380** Steve Hamblin/Alamy Stock Photo **p. 386** Pearson

Education, Inc., **p. 390** Brian Goodman/Shutterstock **p. 398** Brian Goodman/Shutterstock **p. 402** (Bottom) Christian Delbert/Fotolia **p. 402** (Middle) Peter Barritt/SuperStock **p. 408** (Left) Paramount Pictures/Everett Collection **p. 408** (Right) Regency Enterprises/Photos 12/Alamy Stock Photo

CHAPTER 7 **p. 411** (Bottom Left) AF Archive/Alamy Stock Photo **p. 411** (Bottom Right) Nancy Kaszerman/Zuma Press, Inc./Alamy Stock Photo **p. 411** (Top Left) Carolina Hidalgo/Tampa Bay Times/ZUMA Press, Inc./Alamy Stock Photo **p. 411** (Top Right) L.McAfee/Lebrecht Music and Arts Photo Library/Alamy Stock Photo **p. 412** Iofoto/Shutterstock **p. 417** © Warren Miller/The New Yorker Collection/The Cartoon Bank **p. 424** Matthew Ward/Dorling Kindersley, Ltd. **p. 434** Soo Hee Kim/Shutterstock **p. 438** Photographee.eu/Fotolia **p. 447** Sun/Newscom **p. 451** Blitzer, Robert F. **p. 453** TerryJ/E+/Getty Images **p. 462** AP Images **p. 463** PJF Military Collection/Alamy Stock Photo **p. 468** Sharie Kennedy/LWA/Blend Images/Alamy Stock Photo **p. 471** Ladi Kirn/Alamy Stock Photo **p. 470** (Left) Bob Goldberg Feature Photo Service/Newscom **p. 470** (Right) Bruce Rolff/Shutterstock **p. 476** (Left) Pascal Rondeau/Getty Images **p. 476** (Right) Copyright © Scott Kim, scottkim.com **p. 482** Science Photo Library/SuperStock **p. 484** Copyright © Scott Kim, scottkim.com

CHAPTER 8 **p. 493** Brian Hagiwara/Exactostock-1555/SuperStock **p. 494** MariusFM77/E+/Getty Images **p. 504** Pictorial Press Ltd/Alamy Stock Photo **p. 503** (Left) Everett Collection **p. 503** (Right) Alexsl/iStock/Getty Images **p. 506** (Bottom) Hurst Photo/Shutterstock **p. 506** (Left) Thomas J. Peterson/Alamy Stock Photo **p. 506** (Right) TinCup/Alamy Stock Photo **p. 509** Jgroup/iStock/Getty Images **p. 514** North Wind Picture Archives/Alamy Stock Photo **p. 519** Dononeg/iStock/Getty Images **p. 529** Jeff Christensen/Reuters/Alamy Stock Photo **p. 545** Moviestore Collection Ltd/Alamy Stock Photo **p. 554** Grafissimo/E+/Getty Images **p. 558** Mel Yates/Taxi/Getty Images **p. 563** PhotoMediaGroup/Shutterstock **p. 567** (Left) IB Photography/Shutterstock **p. 567** (Right) Martin Harvey/Alamy Stock Photo

CHAPTER 9 **p. 581** EcoPrint/Shutterstock **p. 584** (Bottom) U.S. Department of the Treasury **p. 584** (Top) U.S. Department of the Treasury **p. 582** Dave King/Dorling Kindersley, Ltd. **p. 592** Radius Images/Alamy Stock Photo **p. 584** (Bottom Left) Pearson Education, Inc., **p. 584** (Bottom Middle) U.S. Department of the Treasury **p. 584** (Bottom Right) Pearson Education, Inc., **p. 584** (Top Middle) U.S. Department of the Treasury **p. 594** Mtilghma/Fotolia **p. 602** (Girl) LivingImages/iStock/Getty Images **p. 602** (Beach) Anna Jedynak/123RF **p. 607** Fotolia

CHAPTER 10 **p. 615** Shawn Hempel/Shutterstock **p. 616** Mattes René/Hemis/Alamy Stock Photo **p. 625** Everett Collection/Alamy Stock Photo **p. 630** Fox/Photo 12/Alamy Stock Photo **p. 636** Pictorial Press Ltd/Alamy Stock Photo **p. 637** WDG Photo/Shutterstock **p. 638** Lotus Studio/Shutterstock **p. 642** M.C. Escher's "Symmetry Drawing E85" © 2017 The M.C. Escher Company-The Netherlands. All rights reserved. www.mcescher.com **p. 646** Red Cover/Alamy Stock Photo **p. 650** Benjamin Shearn/The Image Bank/Getty Images **p. 657** Newscom **p. 659** Uyen Le/Photodisc/Getty Images **p. 666** Koonyongyut/iStock/Getty Images **p. 671** Senior Airman Christophe/AGE Fotostock/Superstock **p. 675** (Bottom) Paulbriden/Fotolia **p. 675** (Top) Pecold/Shutterstock **p. 676** The Family (1962), Marisol Escobar. Digital Image © The Museum of Modern Art/Licensed by Scala/Art Resource, NY; Art © Estate of Marisol Escobar/Licensed by VAGA, New York, NY **p. 679** SSPL/The Image Works **p. 681** M.C. Escher's "Circle Limit III" © 2017 The M.C. Escher Company-The Netherlands. All rights reserved. www.mcescher.com **p. 682** (Left) Daryl H. Hepting, Ph.D. **p. 682** (Middle) Gregory Sams/Science Source **p. 682** (Right) KeilaNeokow EliVokounova/Shutterstock **p. 684** Iren Udvarhazi/Fotolia

CHAPTER 11 **p. 693** (Left) Painting/Alamy Stock Photo **p. 693** (Background) Illusionstudio/Shutterstock **p. 693** (Middle) John Parrot/Stocktrek Images, Inc./Alamy Stock Photo **p. 693** (Right Middle) John Parrot/Stocktrek Images, Inc./Alamy Stock Photo **p. 693** (Right) John Parrot/Stocktrek Images, Inc./Alamy Stock Photo **p. 693** (Left Middle) Painting/Alamy Stock Photo **p. 694** Ingemar Edfalk/Blend Images/Super Stock **p. 696** Monkey Business/Fotolia **p. 700** Craftvision/E+/Getty Images **p. 708** (Left) AF Archive/Alamy Stock Photo **p. 708** (Right) Marka/Press/Alamy Stock Photo **p. 715** Jupiter Images/Stockbyte/Getty Images **p. 721** Ronstik/iStock/Getty Images **p. 724** (Bottom) W. Breeze/Hulton Archive/Getty Images **p. 724** (Top) Addison Gallery of American Art, Phillips Academy, Andover, MA/Art Resource, NY; © Estate of George Tooker. Courtesy of DC Moore Gallery, New York. **p. 725** (Bottom Middle) Bryan Bedder/Getty Images **p. 725** (Bottom) Hulton Archive/Getty Images **p. 725** (Middle) CBS Photo Archive/Getty Images **p. 725** (Top Middle) Wenn Ltd/Alamy Stock Photo **p. 725** (Top) Robert Blitzer **p. 726** Jack Kurtz/UPI Photo Service/Newscom **p. 731** F. Bettex/Lookandprint.com/Alamy Stock Photo **p. 736** Manoj Shah/The Image Bank/Getty Images **p. 739** Scott Serio/Icon SMI 137/Newscom **p. 744** Tom Salyer/Alamy Stock Photo **p. 745** Bettmann/Getty Images **p. 756** Hill Street Studios/Blend Images/Getty Images **p. 759** Car Culture/Getty Images **p. 760** Monkey Business Images/Getty Images

CHAPTER 12 **p. 771** Mstay/Getty Images **p. 772** 20th Century Fox Film Corp./Everett Collection **p. 773** Literary Digest Magazine **p. 785** Pearson Education, Inc., **p. 786** Alex Bramwell/Moment Select/Getty Images **p. 800** EpicStockMedia/Shutterstock **p. 808** Marmaduke St. John/Alamy Stock Photo **p. 809** Peter Barritt/Alamy Stock Photo **p. 812** Janice Richard/E+/Getty Images **p. 815** Kemalbas/iStock/Getty Images **p. 819** XiXinXing/iStock/Getty Images **p. 822** Boris Spremo/Toronto Star/ZUMApress/Newscom **p. 824** Erik Isakson/Blend Images/Getty Images **p. 827** (Bottom Left) Dennis Brack/DanitaDelimont.com "Danita Delimont Photography"/Newscom **p. 827** (Bottom Right) Mark Wilson/AP Images **p. 827** (Middle Left) Everett Collection Inc/Alamy Stock Photo **p. 827** (Middle Right) Library of Congress Prints and Photographs Division [LC-USZ62-133118] **p. 827** (Top) Newscom **p. 830** Kipling Brock/Shutterstock **p. 835** Shirley Buchan/Pacificcoastnews/Newscom

CHAPTER 13 **p. 845** Jgroup/iStock/Getty Images **p. 846** Everett Sloane/Snap/Entertainment Pictures/Alamy Stock Photo **p. 851** Tinseltown/Shutterstock **p. 853** Will Schneekloth/ZUMApress/Newscom **p. 858** Wolfgang Kumm/Dpa Picture

Alliance/Alamy Stock Photo **p. 861** Josse Christophel/Alamy Stock Photo **p. 865** Wang Zhou Bj/Imaginechina/AP Images **p. 869** Charles Phelps Cushing/ClassicStock/Alamy Stock Photo **p. 872** John Parrot/Stocktrek Images, Inc/Alamy Stock Photo **p. 873** Joseph Marzullo/JM11/Wenn/Newscom **p. 874** GL Archive/Alamy Stock Photo **p. 876** World History Archive/Alamy Stock Photo **p. 877** Everett Collection Inc/Alamy Stock Photo **p. 883** (Left) Paris Pierce/Alamy Stock Photo **p. 883** (Right) Paris Pierce/Alamy Stock Photo **p. 886** Nicolas Maeterlinck/AFP/Getty Images **p. 887** Everett Collection Inc/Alamy Stock Photo **p. 889** (Bottom Middle) Paris Pierce/Alamy Stock Photo **p. 889** (Bottom) Flab/Alamy Stock Photo **p. 889** (Top Middle) Niday Picture Library/Alamy Stock Photo **p. 889** (Top) Science History Images/Alamy Stock Photo **p. 890** (Bottom Left) Michele Paccione/Shutterstock **p. 890** (Bottom Right) Michele Paccione/Shutterstock **p. 890** (Top Left) Todd Strand/Independent Picture Service/Alamy Stock Photo **p. 890** (Top Middle Left) Carsten Reisinger/Shutterstock **p. 890** (Top Middle Right) Carsten Reisinger/Shutterstock **p. 890** (Top Right) PremiumArt/Shutterstock

CHAPTER 14

p. 897 Gilles Targat/Photo 12/Alamy Stock Photo **p. 898** (Bottom Left) Stephane Cardinale/Corbis Entertainment/Getty Images **p. 898** (Bottom Right) Peer Grimm/Dpa Picture Alliance Archive/Alamy Stock Photo **p. 898** (Top Left) Cherie Cullen/615 Collection/Alamy Stock Photo **p. 898** (Top Right) Chris Wattie/Reuters **p. 908** Peter Foley/Epa/Newscom **p. 920** Kevork Djansezian/Getty Images **p. 930** David Epperson/Photodisc/Getty Images